THE MICROPALAEONTOLOGY OF OCEANS

THE MICROPALAEONTOLOGY OF OCEANS

PROCEEDINGS OF THE SYMPOSIUM HELD IN CAMBRIDGE FROM 10 TO 17 SEPTEMBER 1967 UNDER THE TITLE 'MICROPALAEONTOLOGY OF MARINE BOTTOM SEDIMENTS'

EDITED BY

B. M. FUNNELL & W. R. RIEDEL

ON BEHALF OF

SCIENTIFIC COMMITTEE ON OCEANIC RESEARCH

WORKING GROUP 19

CAMBRIDGE

AT THE UNIVERSITY PRESS

1971

CAMBRIDGE UNIVERSITY PRESS
Cambridge, New York, Melbourne, Madrid, Cape Town, Singapore,
São Paulo, Delhi, Dubai, Tokyo, Mexico City

Cambridge University Press
The Edinburgh Building, Cambridge CB2 8RU, UK

Published in the United States of America by Cambridge University Press, New York

www.cambridge.org
Information on this title: www.cambridge.org/9780521187480

First published 1971
First paperback edition 2010

A catalogue record for this publication is available from the British Library

Library of Congress Catalogue Card Number: 71-96089

ISBN 978-0-521-07642-5 Hardback
ISBN 978-0-521-18748-0 Paperback

CONTRIBUTORS

M. S. Barash	Institute of Oceanology, U.S.S.R. Academy of Sciences, Moscow.
A. W. H. Bé	Lamont-Doherty Geological Observatory, Palisades, New York.
C. W. Beklemishev	Institute of Oceanology, U.S.S.R. Academy of Sciences, Moscow.
W. A. Berggren	Woods Hole Ocenographic Institution, Woods Hole, Massachusetts, U.S.A.
P. L. Bezrukov	Institute of Oceanology, U.S.S.R. Academy of Sciences, Moscow.
M. Black	Department of Geology, University of Cambridge.
H. M. Bolli	Geologisches Institut, Eidgenössische Technische Hochschule, Zürich, Switzerland.
E. Boltovskoy	Museo Argentino de Ciencias Naturales, Buenos Aires, Argentina.
M. N. Bramlette	Scripps Institution of Oceanography, La Jolla, California, U.S.A.
R. E. Casey	Department of Geology, San Fernando Valley Stage College, Northridge, California, U.S.A.
C. Chen	Lamont-Doherty Geological Observatory, Palisades, New York.
D. Curry	Eastbury Grange, Northwood, Middlesex, U.K.
B. M. Funnell	School of Environmental Sciences, University of East Anglia, Norwich, U.K.
K. R. Gaarder	Institutt for Marin Biologi, Universitetet i Oslo, Norway.
J. J. and C. R. Groot	Department of Geology, University of Delaware, U.S.A.
J. D. Hays	Lamont-Doherty Geological Observatory, Palisades, New York.
N. I. Hendey	'Pyxilla', St Agnes, Cornwall, U.K.
Y. Herman	Department of Geology, Washington State University, U.S.A.
A. P. Jousé	Institute of Oceanology, U.S.S.R. Academy of Sciences, Moscow.
T. Kanaya	Institute of Geology and Paleontology, Tohoku University, Japan.
E. V. Koreneva	Institute of Geology, U.S.S.R. Academy of Sciences, Moscow.
O. G. Kozlova†	Institute of Oceanology, U.S.S.R. Academy of Sciences, Moscow.
A. P. Lisitzin	Institute of Oceanology, U.S.S.R. Academy of Sciences, Moscow.
A. Longinelli	Laboratoria Geologia Nucleare, Pisa, Italy.
J. A. McGowan	Scripps Institution of Oceanography, California, U.S.A.
A. and R. McIntyre	Lamont-Doherty Geological Observatory, Palisades, New York
E. Martini	Geologisch - Paläontologisches Institut der Universität, Frankfurt, Germany.
V. V. Muhina	Institute of Oceanology, U.S.S.R. Academy of Sciences, Moscow.
C. A. Nigrini	P.O. Box 779, Kanata, Ontario, Canada.
E. Olausson	Oceanografiska Institutionen Göteborgs Universitet, Göteborg, Sweden.
F. L. Parker	Scripps Institution of Oceanography, La Jolla, California, U.S.A.
M. G. Petrushevskaya	Zoological Institute, U.S.S.R. Academy of Sciences, Leningrad, U.S.S.R.
H. S. Puri	Division of Geology, Florida Board of Conservation, Tallahassee, Florida, U.S.A.
V. V. Reshetnjak	Zoological Institute, U.S.S.R. Academy of Sciences, Leningrad, U.S.S.R.
Z. Reiss	Department of Geology, The Hebrew University, Jerusalem, Israel.
W. R. Riedel	Scripps Institution of Oceanography, La Jolla, California, U.S.A.

CONTRIBUTORS

H. J. Semina	Institute of Oceanology, U.S.S.R. Academy of Sciences, Moscow.
A. Soutar	Scripps Institution of Oceanography, La Jolla, California, U.S.A.
F. M. Swain	Department of Geology and Geophysics, University of Minnesota, Minneapolis, Minnesota, U.S.A.
D. Tolderlund	Lamont-Doherty Geological Observatory, Palisades, New York.
M. G. Uschakova	Institute of Oceanology, U.S.S.R. Academy of Sciences, Moscow.
D. Wall	Woods Hole Oceanographic Institution, Woods Hole, Massachusetts, U.S.A.
D. B. Williams	British Museum (Natural History), London.

† deceased

CONTENTS

CONTENTS

CONTENTS

PREFACE

This volume has been compiled from contributions presented at an international symposium held in Cambridge, England, 10-17 September 1967, under the title 'Micropalaeontology of Marine Bottom Sediments'. The symposium was organized by Working Group 19 of the Scientific Committee on Oceanic Research (S.C.O.R.), of the International Council of Scientific Unions, supported by funds from the Scientific Committee on Oceanic Research, United Nations Economic Scientific and Cultural Organisation (Intergovernmental Oceanographic Commission), British Petroleum and the Royal Society.

The members of Working Group 19 were: E. Seibold (chairman; F. R. Germany), G. Deflandre (France), B. M. Funnell (U.K.), A. P. Jousé (U.S.S.R.), T. Kanaya (Japan) and W. R. Riedel (U.S.A.). Seventy-three persons from 17 different countries attended the symposium, and a total of 45 authors have contributed to the 52 papers in this book.

The volume has been edited on behalf of Working Group 19 by B. M. Funnell and W. R. Riedel. As a matter of policy all papers have been retained as near as possible in the form in which they were submitted. This has resulted in some inconsistencies between papers, but this was thought preferable to editorial intervention, as such discrepancies often reflect real differences in approach and level of investigation in different fields. In the circumstances it is emphasized that it is the authors who are responsible for statements contained in their papers. In spite of such inconsistencies and unevenness of treatment as may be apparent, we believe that the systematic way in which papers have been assembled for this volume will nevertheless ensure its usefulness in advancing the study of micropalaeontology of oceans, and as a work of reference for many years to come.

We wish to thank the responsible officers of Scientific Committee on Oceanic Research for their initiation of plans for this symposium and their assistance in the arrangements which ensured its success, the authorities of Gonville and Caius College, and the Department of Geology and Mineralogy and Petrology of the University of Cambridge for the excellent facilities provided during the symposium, and the Cambridge University Press for its care in preparing this complex volume for publication.

M. Black, M. J. Fisher, Mrs J. K. Friend, R. B. D. Leidy and A. T. S. Ramsay assisted materially with arrangements for the symposium, and also gave help in the early stages of preparing this volume for publication. Mrs J. K. Friend and Mrs L. Harvey assisted with the Index.

We would also like to thank the authors for their generous initial response and their patience in the long process of preparing this book for press. A few papers presented at the symposium, but not included here, have been published elsewhere. They are as follows:

Bilal Ul Haq, U.Z. 1969. The structure of Eocene coccoliths and discoasters from a Tertiary deep-sea core in the central Pacific. *Stockholm Contrib. Geol.* 21 (1), 1-20.

Chen, C. 1968. Pleistocene pteropods in pelagic sediments. *Nature, Lond.* **219**, 1145-9.

Hasle, G. 1968. The valve processes of the centric diatom genus *Thalassiosira*. *Nytt Mag. Bot.* **15** (3).

SECTION A

DISTRIBUTION OF LIVING ORGANISMS IN THE OCEANS

1. OCEANIC BIOGEOGRAPHY OF THE PACIFIC[1]

J. A. McGOWAN

Summary: Temporal, non-seasonal variations in the abundance of species of both terrestrial and marine organisms have been observed. Many contemporary ecologists are studying the causes of such variability. While unusual climatic extremes can sometimes be associated with fluctuations in population sizes, very often no such association is apparent. Since there are theoretical, experimental, and field studies which indicate that such variability can be due to biological causes alone, it is pertinent to determine the extent and type of species interaction occurring. But the most common approach to the study of population variability is the examination of single species populations, and a few of the immediate factors that appear to affect their dynamics most. Although many such studies have been made, there is still no acceptable general theory that can be utilized to explain such variability in natural populations. This fact has led many ecologists to believe that a better understanding of the causes underlying variations in natural populations may be obtained by studying more complete systems of interacting species. These are known as ecosystems or communities, and they may be studied from both the structural (that is, the proportion of various species present), and the energetic viewpoint (that is, the rates at which energy is fixed, transformed, and transferred through the food chain). However, the 'system' or community must first be identified and its physical dimensions determined. Studies of structure and dynamics done in areas of transition between communities, or in ecotones, are very likely to yield results that differ qualitatively and quantitatively from studies done in areas where stable, climax communities occur, or where assemblages of species are progressing, by succession, towards the climax situation. Since we cannot see into the ocean, it is difficult to identify these areas in the pelagic realm. The approach of ecological biogeography must be used as the preliminary step to the determination of structure, dynamics, and the relative stability of both. The objectives of many oceanic biogeographers, and the questions they are attempting to answer are as follows:

1. What species are present?
2. What are the main patterns of species distribution and abundance?
3. What maintains the shape of these patterns?
4. How and why did the patterns develop?
5. What are the communities?

Studies pertinent to the above five questions have been done and are now being done. However, it is becoming apparent that the field and laboratory methods used to obtain estimates of abundance are lacking in both precision and accuracy. Improvements in these methods are clearly necessary.

INTRODUCTION

Temporal and areal variations in the abundance of plants and animals are demonstrable facts. Although even the most casual observer can see these variations, achieving a knowledge of the causes of the variability has frequently been very difficult. While in some cases population fluctuations of single species, particularly insects, certain game animals, and commercially important fish are thought to

[1] Contribution from the Scripps Institution of Oceanography, University of California, San Diego, California.

be fairly well documented, there is still no acceptable, general theory that can be utilized to explain all such variability in nature. Many contemporary ecologists are studying this variability and its causes.

There are two general approaches to the problem. Perhaps the oldest and certainly the most popular, is the study of single species populations, their immediate physical environment and occasionally their main competitor, grazer or predator. In many cases the impetus for such studies has been economic necessity. For instance, insects and other pests do tremendous damage to food crops, game animals form the basis of sporting and fur industries and, of course, commercial fish are of great importance. While such studies may have originated out of need, they were given a very firm basis by the theoretical studies of Verhulst (1838) and Pearl (1930) who examined, mathematically, the population consequences of various birth, growth, and death rates, and by Lotka (1956) and Volterra (1928) who developed mathematical treatments which predicted the outcome of competition between two species and of predator-prey interactions. Later, experimental studies of single species population growth and of the role of competition and predator-prey interactions were carried out by Gause (1934), Park (1962), Frank (1957), Slobodkin (1962) and others. From the viewpoint of the zoogeographer or palaeontologist perhaps the most interesting results of these studies were that it was shown that species populations can fluctuate widely due to biotic factors alone. Further, some of the treatments of predator-prey interactions in which spatial heterogenity or immigration was introduced, showed that persistent oscillations in the relative abundance of species were possible while the physical conditions were held constant or altered only slightly. Evidence from nature of such temporal fluctuations (i.e. non-seasonal) has been somewhat more difficult to obtain, partly because of the difficulties inherent in monitoring entire populations, partly because of the necessity for long, time-series observations and partly due to the fact that populations in nature are open systems in which immigration and emigration are common. However, some such data are available.

Elton (1942) has reviewed the records of population fluctuations of field mice (voles) in Europe and North America, predatory mammals in Canada and lemmings in the arctic. While many of the records of mice and lemming population changes are not quantitative, those of the foxes and lynx (Elton and Nicholson 1942) are based on the numbers of pelts sold to Hudson Bay Company and Moravian Mission trading posts. Many of the records cover time spans of 90 to 100 years and do not appear to be unduly influenced by variations in hunting pressure. The amplitude of the variation of population sizes, as indicated by these records, is a factor of one hundred in some cases.

These studies and others concerning fluctuations in wildlife populations in North America have been reviewed by Keith (1963). He points out that what appear to be cyclic oscillations in abundance are common; that these may be synchronous over very large areas; that the amplitudes are frequently greater than a factor of ten and sometimes as much as one hundred; and that the period of some approximates ten years. Keith also discussed the various hypotheses that have been presented as to the 'cause' of the cycle. These are: 'random fluctuation theory', 'meteorological (climatic) theories', 'overpopulation theories', and 'other theories'. It is his opinion that none of these is wholly satisfactory and that all are open to criticism.

However, in a somewhat more extensive discussion of the same subject Lack (1954, p. 276) concludes that 'the main density dependent control of numbers probably comes through variations in the death rate', and while 'climatic factors can produce or modify density-dependent mortality in various ways' (p. 203), 'the critical mortality factors are food shortage, predation and disease, one of which may be paramount, though they often act together'. It is of particular interest here to emphasize the point that these oscillations in abundance show no obvious correlation with climatic factors, particularly temperature. In almost direct opposition to the views summarized above, are those of Andrewartha and Birch (1954). They challenge the notion that density-dependent factors (i.e., competition, predation, disease, etc.) regulate the variations in the size of populations and present the arguments for climatic control. However, Hairston, Smith and Slobodkin (1960) have countered these by presenting a series of general, widely accepted observations which show a pattern of population control.

Ehrlich and Birch (1967) have challenged the contentions of Hairston *et al.*, that the observations reported are 'general, widely accepted' and show a 'pattern of population control'. Slobodkin, Smith and Hairston (1967) have clarified their original (1960) paper in a reply to Ehrlich and Birch (1967). For a complete version of this debate, the reader must study the original papers, but in summary, it appears that much of the argument is semantic, and further, Ehrlich and Birch do not really deny that the sizes of natural populations are frequently controlled and regulated by biotic mechanisms. They merely object to the original conclusions of Hairston *et al.*, as being too sweeping. For example, they do not agree that 'the decomposers as a group must be food-limited' and present alternative models nor do they admit that producers are 'limited by their own exhaustion of a resource,' since cases exist where herbivores are known to have reduced plant populations and phytogeographic observations imply (but do not prove) a climatic limitation of range. When Ehrlich and Birch deny 'that there is a basis for inferring that, in general, plants are limited by their own exhaustion of a resource', they are apparently unaware of the numerous, intensive studies of marine phytoplankton populations that show just that (Redfield, Ketchum and Richards 1963; Ryther 1963; Riley 1963). Further arguments turn on whether herbivores and carnivores are limited only by their food supply or not. In general it seems that these discussions are more concerned with the type of biological control mechanisms rather than with climatic versus biotic control, and I can only agree with the statement made by Ehrlich and Birch (1967, p. 103), 'If we can draw any general conclusion from the work which has been done on natural populations, it is that single neat "control" mechanisms are unlikely to explain fluctuations in the size of single populations, let alone all organisms of a trophic level'.

Long-term records of fluctuations in abundance of certain marine organisms exist. Uda (1957, 1961) has graphed abundance estimates of the Japanese sardine as far back as the fifteenth-century, and other fishes to the seventeenth-century. These estimates are, of course, based on catch records, and as an index of population size are subject to error because of possible variations in fishing intensity. Further there appear to be no accompanying records of climatic or hydrographic events reaching this far back. However, Uda's graph does show very large amplitude changes in the size of the catch with some peak periods lasting as long as eighty years. Other time series from the ocean are more amenable to analysis. Allen (1941) presented the results of a twenty year study of dinoflagellate abundance in the waters off La Jolla, as sampled at the end of the pier at Scripps Institution of Oceanography, and at Point Hueneme, one hundred miles to the north. A standardized method of sampling was used on a daily basis, and daily surface temperatures were recorded. Although Allen was able to show that within any one year the peak abundance of dinoflagellates (due primarily to one species, *Prorocentrum micans*) tended to be in the spring of the year, there were great fluctuations in the magnitude of the peak from year to year. The range between years at La Jolla was from 2.47×10^4 to 2.08×10^6 cells per litre while that at Point Hueneme was 1.20×10^4 to 4.78×10^5 cells per litre. A linear regression analysis of Allen's data (Figs. 1.1, 1.2) shows that at La Jolla there is no consistent relationship between the average yearly variations in dinoflagellate abundance and variations in the average temperature for the first quarter of the years (i.e. slope of line not significantly different from 0; $p > .05$). At Point Hueneme, however, there was a negative correlation of dinoflagellate abundance with temperature. This implies that during the spring of years of unusually cool waters there was a tendency for increased population sizes. This is the season of upwelling in the California current and it seems highly probable that the population of dinoglagellates was responding to an increased supply of nutrients due to this factor. Plant nutrients are generally considered to be non-conservative factors in the sea since their concentrations can be radically altered by phytoplankton. In the parlance of ecologists, nutrients would be a density-dependent factor. That surface temperature changes, *per se,* are not a good index of the influence of this factor on dinoflagellates is shown by Allen's La Jolla data. Other data on fluctuations in abundance of marine species do not cover this long a time span. However, Murphy (1965) has analysed the pronounced shift in relative abundance between the sardine, *Sardinops caerulea,* and anchovy, *Engraulis mordax*, in the California current over a ten-year period and Schaefer (1967) has reviewed the work on the variations of abun-

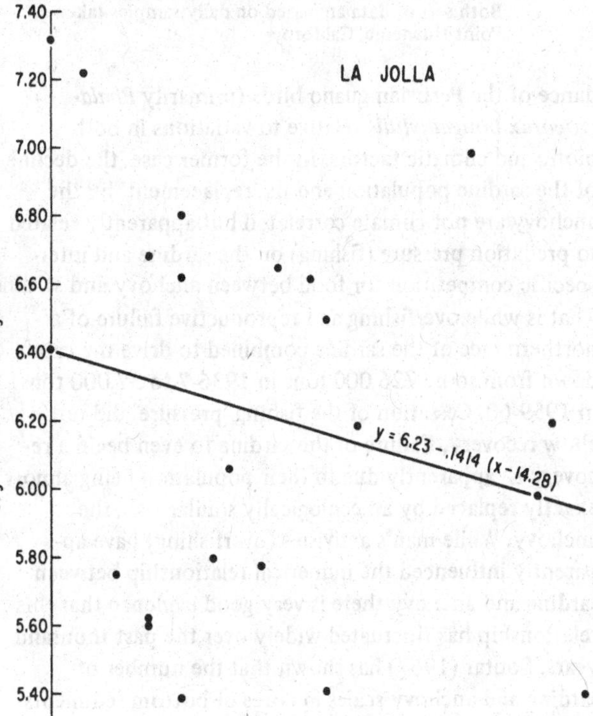
LA JOLLA
7.40
7.20
7.00
6.80
6.60
6.40
6.20
6.00
5.80
5.60
5.40
Log Number Dinoflagellates
y = 6.23 - .1414 (x - 14.28)
13
14
15
16
Temperature °C

Fig. 1.1. **A linear regression of the mean temperature of the surface waters for the first quarter of the years 1920 to 1940 versus the log numbers of dinoflagellates present. Both sets of data are based on daily samples taken at La Jolla, California.**

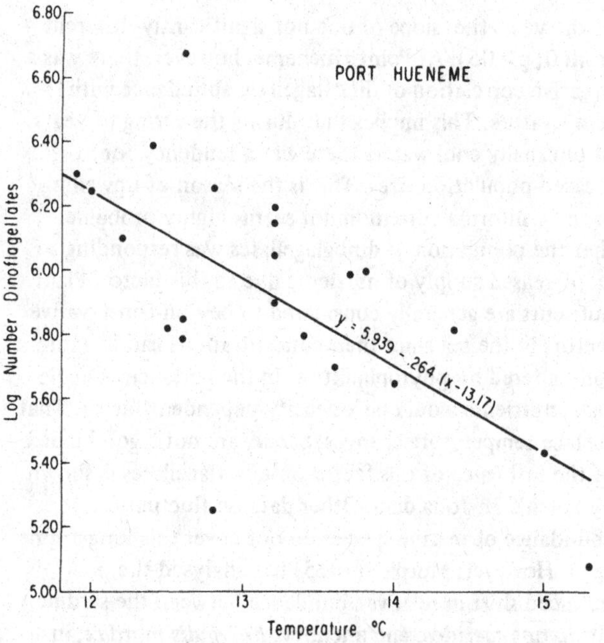
PORT HUENEME
Log Number Dinoflagellates
Temperature °C
y = 5.939 - .264 (x-13.17)
6.80
6.60
6.40
6.20
6.00
5.80
5.60
5.40
5.20
5.00
12
13
14
15

Fig. 1.2. A linear regression of the mean temperature of the surface waters for the first quarter of the years 1920 to 1940 versus the log numbers of dinoflagellates present. Both sets of data are based on daily samples taken at Point Hueneme, California

dance of the Peruvian guano birds (primarily *Phalacrocorax bougainvillii*) relative to variations in both biotic and climatic factors. In the former case, the decline of the sardine population and its 'replacement' by the anchovy are not climate correlated but apparently related to predation pressure (fishing) on the sardine and interspecific competition for food between anchovy and sardine. That is while overfishing and reproductive failure of a northern race of the sardine combined to drive the catch down from some 726,000 tons in 1936-7 to 37,000 tons in 1959-60. Cessation of the fishing 'pressure' did not allow recovery. Failure of the sardine to even begin a recovery is apparently due to their population being almost exactly replaced by an ecologically similar fish, the anchovy. While man's activities (overfishing) have apparently influenced the numerical relationship between sardine and anchovy there is very good evidence that this relationship has fluctuated widely over the past thousand years. Soutar (1967) has shown that the number of sardine and anchovy scales in cores of bottom sediments is highly variable, is not random with time and that the anchovy was almost always more abundant than the sardine. The Peruvian guano bird populations seem to be limited by the availability of suitable nesting sites and degree of protection of young birds from predators. Declines in the bird populations have been associated with changes in the availability of their food, the anchoveta, brought about by El Niño, a climatic event.

In general the evidence from long-term field studies of plant and animal single species populations show that large variations in abundance are common. Very often there appear to be cyclic oscillations. In most cases there are no very clear cut relationships to climatic events. That biotic factors can be responsible for such oscillations has been shown both theoretically and experimentally. The field evidence for biotic (density-dependent) control is obscure in some cases, but very clear in others.

The difficulty in interpreting this type of information and the lack of development of any general theory resulting from single species studies has led many ecologists to believe that a better understanding of natural populations may be achieved by studying more complete systems. This second approach, known variously as ecosystem studies, tropho-dynamics, or the community approach, has had its theorists (Lotka 1956; Elton 1927; Lindeman 1942; von Bertalanffy 1950; Bray 1958, and others) as well. Their works and their pertinence to marine studies have been well reviewed by Hedgpeth (1957) and Fager (1963). Essentially this approach states that organisms, in nature, do not live in a biological vacuum, that they are, as a matter of fact, continuously exposed to other organisms, and that these other organisms are just as much a part of their environment as are physical-chemical properties. Further concepts are that the community-ecosystem is organized, has structure, order, and displays pattern in the sense that Hutchinson has defined the word (1953). These 'recurrent organized systems' (Fager 1963) are the product of evolutionary adaptation. The structure is considered both in the sense of energy relations and in the sense of species composition; the species tend to be specialists, that is, they perform specific roles. The energy 'flows' through trophic levels from producers to herbivores, to several levels of carnivores, and to degraders, with a certain amount dissipated at each level. The measurement of the efficiency of transfer and transformation of this energy and the study of factors influencing efficiencies is a major field of modern ecology. It is felt that an understanding of these processes will lead to a general understanding of the variations in abundance of organisms both on the land and in the sea.

Because communities are obviously complex (i.e. there are many interacting species) some ecologists have chosen to make simplifying assumptions. Commonly the size categories of animals from the sampled populations are assigned to trophic levels and time series measures of changes in their biomass or calorific values are made. However, as Fager (1963, p. 417) has pointed out, 'the thermodynamic approach involves the danger that guesses at functions may seem obvious and yet be quite incorrect'. A further sophistication is to classify the organisms in the samples as far as possible (usually to Order or Family) and then assign a number or letter designation to individuals that appear to be of the same species. But 'it is, however, an almost universal experience that such a procedure results in failure to separate morphologically similar species which may have quite different functions' (Fager 1963, p. 418). There are further problems with such attempts to simplify the system in the case of zooplankton. For instance, many larval and juvenile forms do not process energy and matter through the same pathways as do adults, food habits may shift seasonally or by area and there seem to be many omnivores. This multiplicity of possible pathways of the flow of energy through the system may be a complicating feature in the study of communities but MacArthur (1955, 1965) has shown that this very diversity of pathways may, as a matter of fact, confer a great deal of stability on the community in general. That is to say, that in communities where the diversity of pathways is great, fluctuations in population sizes due to biotic interactions and climatic instability tend to be damped out. The study of community structure in terms of diversity (Fisher, Corbet and Williams 1943; Preston 1948; Williams 1964) is then, an important aspect of the general problem. The term 'diversity' as used in this context refers not merely to the length of the species list but rather to the relative abundances of the species *and* to the total number of species present. A practical method of expressing this has been suggested by Simpson (1949) who points out that when two individuals have been drawn independently and randomly from a (mixed) population, the probability that they will be the same species is related to numbers of species present and to their relative proportions. That is:

$$\ell = \frac{\Sigma n\,(n\text{-}1)}{N\,(N\text{-}1)},$$

where N is the total number of individuals and $n, n_2 \ldots n_x$ are the members of individuals of species $1, 2, \ldots x$. Another measure of diversity (the Shannon-Wiener measure) is based on the 'degree of uncertainty' associated with the identity of an individual specimen drawn at random from a sample of a mixed population. That is

$$H = -\sum_{i=1}^{s} P_i \log P_i,$$

where P_i is the proportion of the total individuals belonging to the *i*-th species and S is the total number of species. There have been rather few community structure studies in terms of diversity in the case of marine plankton. Margalef (1958) and Patten (1959, 1964, 1966) using the concepts of information theory and data derived from phytoplankton samples taken in estuaries have examined temporal and spatial variations in the diversity of populations. Margalef has shown that diversity in the estuary of the Vigo in Spain, is high in the early spring of the year, tends to decrease in the late spring, and then increases once again during the late summer. This increase does not quite reach the high levels of early spring, but is, however, maintained during the late summer, fall and early winter. Margalef interprets these as stages of succession which approach and reach a climax (i.e. stable) situation. Accompanying these seasonal changes in diversity of the populations is a pronounced change in the spatial heterogeneity of diversity within the entire estuary. In June the heterogeneity was much less than in September. A further interesting aspect of this study is that in August, during the climax period, Margalef was able to show a diversity gradient running along the major east-west axis of the estuary. Diversity was low at the head, increased to a maximum in the north-central part and decreased generally at its opening to the sea. Margalef interprets the zone of highest diversity as being an area of contact between 'a community belonging to the interior of the river's mouth and another coming in from the ocean. The peculiar form of the mixed zone results from the fact that waters from the Atlantic penetrate principally along the northern shores of the mouth of the river'. This latter observation is of particular importance to the study of the zoogeography of communities of the open ocean, for as we shall see later, a similar phenomenon occurs on a much larger scale there. Patten (1963, 1966) in an extensive study of the phytoplankton from a single station in the estuary of the York River, Virginia, also uses the concepts of information theory in the measurement of diversity, its relation to a variety of environmental factors and to the energetics of the community. The mathematics in his presentations are involved (and apparently, on occasion, incorrect; see Patten 1964) and the manner of notation sufficiently obscure so that his results are not entirely clear (at least to the present author). However, he seems to conclude that a high productive capacity is associated with high diversity and that certain advantages accrue to a diverse community. These are the achievement of stability through optimal use of the energy 'investment', and the ability to maintain optimal diversity structure in an environment which is 'comparatively unstructured'. There is, further, the strong implication that communities tend to move by succession in this direction. Although Margalef and Patten have shown that community structure and aspects of their dynamics *can* be described in terms of information theory, it seems doubtful that it is essential to do so. They have revealed a number of interesting aspects of communities and have probably increased our insight into the problem of their organization. There are, however, some fairly serious criticisms that may be made of their choice of 'communities' and places in which to sample them. Assemblages of phytoplankton species by themselves do not constitute an organized community in an energetic or evolutionary sense. Hutchinson (1967, pp. 357-60) has discussed the possible role of selective feeding as an influence on niche diversification. Riley (1963) and Steel (1961) have shown that grazing by herbivores can very strongly influence the rate of change of phytoplankton populations and there is reason to believe (Mullin 1966) that zooplankton herbivores may exercise a considerable degree of selectivity of type of plant they feed upon. Therefore, two components of diversity, numbers of species, and abundance of species may be markedly affected by biotic factors which were not considered by either Margalef or Patten. Furthermore, it is reasonable to expect that species tend to adapt more fully (i.e. become more efficient at particular roles) in the presence of other species which are a *frequent* part of their environment rather than species which are only very occasionally their associates. Fager (1957, 1963) has defined the word 'frequent', used in this sense, and has emphasized that the degree of fidelity of a species to its community is important in any consideration of community structure. We have been given no information by either Margalef or Patten of the degree of fidelity of the phytoplankton species to the assemblages studied in these two estuaries. Therefore, because the complexity of energy pathways has not been well evaluated, and because the species sampled in these studies may or may not be frequently associated over their entire distributional ranges, there is reasonable doubt that distinct, functional communities have, as a matter of fact, been studied. Furthermore, estuaries are so complex hydrographically that one might expect advection of water and populations into and out of the area of study to be great. This combined with the fact that samples from estuaries are very likely to contain combinations of species that have evolved separately in fresh water, brackish water, neritic, and oceanic environments, makes it seem even more unlikely that functionally organized communities were studied.

Hutchinson (1967) has compiled and reviewed a tremendous amount of information pertinent to the study of community-ecosystems from the limnological literature. As he points out (p. 232), lakes may be 'ordinarily considered an ecosystem, in a single biotope, supporting a single biocoenosis'. Because they are such self-contained and relatively simple units faunistically, they represent, to some degree, special cases of the general problem. Generalizations about community diversity and stability based on information derived from the study of lakes may not have wide applicability in the terrestrial or marine environments.

An entirely different approach to the description of community structure and function has been taken by Elton (1966) for a small woodland (Wytham Woods) near Oxford, England. In a twenty-year study, characterized by great attention to climate and details of the life histories, behaviour, population densities, food habits, spatial relationships, mortality factors, competition, etc., of both plants (where appropriate) and animals, he has achieved a real understanding of the pattern of the structure of this community. Further the biogeography and natural history of many of the species dealt with were well known; quantitative sampling methods were used frequently and the role of dead and decaying organic material thoroughly investigated. This approach has been called by Elton the 'ecological survey'. His conclusions are summarized in a series of fifteen propositions and there can be little doubt that his study has produced some new concepts of the structure of communities.

The energetics of the Wytham Woods community has not been well studied but Elton makes no claim of having a complete picture (p. 380) of community organization and function. However, he points out that 'without such a spatially and functionally defined plan of communities any study of "productivity" in terrestrial communities is bound to be shaky, to say the least', and further that 'an ecological survey of this kind can put in the many interrelationships that are not described by the foodchains alone'. It is this author's opinion that Elton's statements hold for studies of marine plankton as well!

There are certain inherent difficulties in this approach for the ocean. The terrestrial ecologist, whether he realizes it or not, has a very good intuitive impression of what communities are or should be, where their boundaries lie, what sort of plants and animals are in them and who does what to whom. It is then a fairly 'simple' matter to test and quantify his intuitive impressions by planning appropriate sampling programmes and by using appropriate sampling methods. He is in this fortunate position because he can *see* the system he is dealing with. Also, in Europe and North America at least, there is a backlog of taxonomic, physiological, and life history information on terrestrial species that is extremely useful to the student of communities. Further, it is a rather simple matter to walk into the woods or grasslands to observe, enumerate, and sample. Lastly, many terrestrial or fresh water species may be cultured in the laboratory and are amenable to experimental procedures.

Almost none of these is true for the open ocean and an attempt to apply the ecological survey approach must, of necessity, start from a considerably lower base line if it is to progress in any truly meaningful way. But since the open oceans are so different in so many ways from the terrestrial, lake, or even littoral environments there is no question that they must be studied from this (or a similar) viewpoint. This is, after all, the world's largest habitat; it is relatively undisturbed by man, many physical variables are quite stable, and if climax communities really exist they should be extremely well developed, especially, in the great, stable, Central water masses of the major oceans.

There is, however, the problem of establishing the base line from which to proceed. Although some pertinent information has been in existence for many years, much of it is of a qualitative nature and large parts of the world's oceans are essentially unknown biologically. Much of the available data have been reviewed by Raymont (1963), but it is obvious that they cannot be synthesized into anything remotely resembling Elton's 'Pattern of Animal Communities'.

It is clear that some sort of a systematic attack on the problem must be made and it is the purpose of the following sections of this discussion to indicate that such an attempt is being made. It is still in its early phases but there has been a reasonable amount of progress. The sequence of steps in this attack are essentially those of biogeography as outlined by Darlington (1957, pp. vii, preface) and somewhat modified for the ocean. These steps are in the form of five questions that need to be answered. They are as follows:

1. What species are present?
2. What are the main patterns of species distribution and abundance?
3. What maintains the shape of these patterns?
4. How and why did the patterns develop?
5. What are the communities?

This outline of a procedure differs very little from a similar discussion of the problem presented by Russell (1935) over thirty years ago. In it he points out the necessity for exact taxonomic work, for determinations of the broad patterns of distribution and abundance, and for an understanding of the 'interactions between species'. Quoting from Hardy (1924) Russell says, 'It is only by having some concrete representation of the normal marine community that one can attempt to predict the effect which an abnormal increase of one or other particular form may have in the balance of the whole'.

WHAT SPECIES ARE PRESENT?

There can be little doubt that most species, as defined morphologically by competent taxonomists, represent populations that are, to a marked degree, behaviourally and physiologically distinct. It is primarily at this level that specialized adaptations to interactions are most evident. Further it is at this level that the concept of niche is meaningful to the studies of communities. Species are the primary units of ecology in the same sense that elements are of chemistry. It is, therefore, of extreme importance that we have a firm grasp of the faunistics and floristics of the ocean. The degree that we know these units of our system will influence the validity of generalizations about patterns and communities. There are probably no practical ways in which one can demonstrate, with certainty, that all the species of a particular taxon are known. However, there are some ways of looking at this problem which will at least give us an indication of how well we have answered the question.

For the case of holoplanktonic animals Russell (1935) presented a list of the numbers of species in seven important major taxa. Species of these taxa would probably make up most of the biomass of zooplankton anywhere in the world ocean. It is instructive to attempt to make a similar list today, thirty-two years later. There has been a vast increase in the numbers of samples taken in all of the oceans since 1935. For instance, Scripps Institution alone has accumulated over 30,000 large zooplankton samples from a variety of depths at several thousand different localities in the Pacific and Indian oceans. Other institutions from many countries have carried out similar intensive sampling programmes in all of the oceans of the world. Although there is no good estimate of the number of plankton samples taken and examined by 1935, Russell points out that the majority of them were from coastal regions. It, therefore, seems to be a fairly safe guess that there has been at least an order of magnitude increase in the number of samples from oceanic regions. It is, of course, true that by no means all of these have been examined by taxonomists, but if the efforts at the author's own institution may be used as an index, an estimate of from 25 to 40 per cent is reasonable. Added to this must be the fact that collecting techniques have improved and taxonomy itself is a somewhat more developed discipline. Thus it seems that the effort devoted to answering the question 'What species are present?' has increased enormously in the past thirty years.

An examination of Table 1.1 indicates that there has been a very small increment in the estimate of number of species present compared to the increment of effort made to discover them. As a matter of fact, in some groups there are fewer species (this is presumably due to the fact that with a better knowledge of the group certain synonymies were discovered). The information in this table is open to several interpretations but it is my opinion that it constitutes very strong evidence that holoplanktonic zooplankton are taxonomically well known at the species

Table 1.1. Estimated number of species of holoplanktonic animals and meroplanktonic coelenterates

	(1935)	(1967)	
Coelenterata			
Antho-, Lepto- and Scyphomedusae . . .			
Meroplankton	535	656	Russell, 1935
Trachy- and Narcomedusae	134	110	Kramp, 1961
Siphonophora			
Calycophorae	72	96	Alvariño after Totton, 1965
Physophorae	34	37	Sears, 1953
Ctenophora	80	?	
Nemertea	34	?	
Polychaeta			
Tomopteridae	44	10	Tebble, 1962
Chaetognatha	30	55-57	Alvariño, 1967; David, 1963
Crustacea			
Cladocera	7	?	
Copepoda	754	1,200	Fleminger (Pers. Comm.)
Amphipoda			
Hyperiidea	292	?	
Gammaridea	44	?	
Euphausiacea	85	83	Brinton, 1967
Mollusca			
Pteropoda			
Thecosomata	51	35	Tesch, 1946, 1948; van der Spoel, 1966; McGowan, 1967
Gymnosomata	41	31	Provot-Fol, 1954
Heteropoda	90	24	Tesch, 1949; McGowan, 1967
Tunicata			
Appendicularia	61	68	*Zoo. Record*
Thaliacea			
Doliolidae	12	8-12	Thomson, 1948; Berner, 1967
Salpidae	25	22-25	Yount, Foxton
Pyrosomidae	8	2-10	Thomson, 1948; *Zoo. Record*
Total	2433		
(Excluding meroplanktonic Medusae)	1898		

level and that there are probably few new species to be discovered. These latter will certainly not be numerically dominant members of any large, important food chain.

If the above conclusion can be accepted as being substantially true, there is a further interesting comparison to be made. Mayr, Linsley and Usinger (1953) have presented a table of the estimated number of known species of all recent animals. The total, 1,120,310, minus the estimated number of zooplankton species, is still very much larger than the number for the open ocean habitat. Thus in spite of the fact that this habitat constitutes some 70 percent of the earth's surface and an even larger proportion of the volume of living space, it has relatively few species. It is, of course, not entirely fair to compare only the zooplankton of the oceanic regions to a total that includes vertebrates as well. However, Marshall (1963) has pointed out that the total number of species of meso- and bathypelagic fishes of the world's oceans is no more than the number of species in the Amazon River drainage system. Hubbs (personal communication) estimates that there are more than 1,000, but probably considerably less than 10,000 species of pelagic oceanic fishes (i.e. both epi- and bathypelagic). Adding to these the species of marine mammals, reptiles, and birds does not really alter the picture very much.

The implications of this observation are of importance to the study of evolution and community dynamics. For instance, whatever the mechanism(s) of speciation are, this process must be considerably more difficult or perhaps slower in the open ocean. On the other hand if planktonic animals have speciated as freely as the terrestrial, there must have been a great many more extinctions per unit time. Another implied situation that arises from this observation is that if the Gausian hypothesis of competitive exclusion holds and there is only 'one species to a niche' then there must be many fewer niches per unit area than on the land. Either that is true or almost all oceanic species live everywhere in the oceans and the number of species per unit area (or volume) is similar to that of the land. This could be the case if there were many relatively small community ecosystems each with its own complement of species on the land, but only a very few, very large ones in the open ocean. In other words, endemicity on the land must be on an entirely different scale than in the sea. While it is true, as we shall see later, that most oceanic species have enormous populations distributed over huge areas, it does not seem likely that this can account for the disparity. While there probably are many communities of fairly small areal extent on the land, few of these contain species strictly endemic to only that community. As Elton points out, terrestrial communities such as Wytham Woods are common throughout England and Western Europe and, therefore, the species combination found there repeats itself over and over in a sort of mosaic covering a large extent of land.

It would seem that the most probable situation is that there really are fewer niches per unit of space in the sea than on the land. The physical structure of the open ocean habitat is not directly comparable to the land in very many ways, for example, there is no problem of humidity in the sea. However, in the few ways that the two may be compared, the ocean is a much less complex place than the land. The amplitude of temperature variations on a unit volume basis is much smaller in the sea; further there are no leaves or bark to hide under, ground to burrow in or rocks to dodge behind in the open ocean. This type of 'cover' is of great importance to terrestrial animals. Other physical features of the open ocean are similarly monotonous. This lack of heterogenity of the physical structure of the ocean has probably had a profound influence on the type of species interaction that has led to niche diversification. Because of this implied, relative simplicity of oceanic communities, and in spite of the problems involved in studying them, they should be far more amenable to being ultimately understood than those of any other habitat.

There are two additional important considerations before we can consider oceanic communities to be relatively simple in the sense of there being fewer links in them. These are: the degree of diversification of the phytoplankton, and the degree to which intraspecific varieties of zooplankton species can occupy different niches. Apparently diatoms and dinoflagellates are rather well known taxonomically, but the small naked flagellates and coccolithophores are not. It is suspected that these latter two groups may be important contributors to the primary productivity of most tropical and sub-tropical oceanic regions. It is clear that, until they have been more thoroughly studied, no conclusions may be drawn about the degree of simplicity of the planktonic flora. However, it seems rather unlikely that when this work has been done the total number of phytoplanktonic species will equal or even approach the estimated 200,000 species of Spermatophyta (Spector 1956), most of which are terrestrial or freshwater. Studies of intraspecific variability of oceanic zooplankton species are not numerous, but there is reason to believe that such variability may be fairly common. If this proves to be a dominant feature of species populations, it could modify our concepts of diversity in the ocean. Since, by definition, intra-specific variants (sub-species, varieties, forma, etc.) tend to intergrade we might expect their roles in food chains to be quite similar. While this might allow the species that does vary to occupy more than one community, it would seem that it would perform a very similar function in each of these. This could tend to increase complexity by adding more links per community, but would not increase a total, cumulative species list for *all* such communities of the open ocean. The inference drawn from the comparison of the terrestrial and oceanic species lists could, therefore, be quite incorrect. If this sort of situation prevails, then the factors influencing population dynamics in one part of the species range would not be the same as in another part! Thus, it is of great importance that taxonomic and biogeographic studies at the species *and* sub-specific level be continued and intensified for all oceanic organisms.

WHAT ARE THE MAIN PATTERNS OF SPECIES DISTRIBUTION AND ABUNDANCE?

Historical biogeography is concerned primarily with determining the geographic areas of the earth in which endemicity of the higher taxonomic categories is evident. It is based primarily on Family and Order (Darlington 1957) and a similar situation prevails in the study of the biogeography of the littoral and shelf fauna of the oceans. However, Ekman (1953, p. 319) points out 'while the main regions of the (shelf fauna) are in the first instance characterized by special genera and families, pelagic fauna are characterized by species whose genera also occur in other main regions'. In other words: the main faunistic regions of the high oceanic pelagic fauna are more weakly characterized taxonomically than the main regions of the shelf. In this sense, the zooplankton (and probably the phytoplankton) of the world's oceans have no biogeography. For instance: species of the genera *Sapphirina, Copilia, Pleuromamma, Euphausia, Salpa, Doliolum, Pyrosoma,* and *Limacina* are found in almost all of the major bodies of water (Ekman 1953). While some genera are endemic to certain oceans or even parts of oceans there are very few families and almost no orders that show such a pattern. Therefore, biogeographic studies of plankton must be done at the species (or sub-specific) level and cannot give us much insight into the problem of determining the centers of origin and routes of dispersal of major taxonomic groups. The purposes in studying the distributions of these species is primarily ecological. Because of this outlook, it is of fundamental importance that biogeographic studies of zoo- and phytoplankton be quantitative. That is to say, that patterns of distribution of the abundance of species must be determined. Further, because most genera are so widespread, the *entire* pattern of abundance to the limits of the ranges should be mapped if broad ecological inferences are to be drawn. It would at first seem that this is an 'impossible' task, but substantial progress in accomplishing this task has been made.

One of the earliest attempts to establish the pattern of species distribution of a group of organisms was that of Giesbrecht (1892) who, on the basis of studies on copepods, divided the pelagic fauna into three 'main regions': a warm-water region and a northern and southern cold-water region. Ekman (1953, p. 325) citing Chun (1897), Moser (1915), Lohmann (1933), Meisenheimer (1906), Apstein (1894), Piroznikov (1937), Kofoid (1934), and Haecker (1908), says that 'most plankton geographers concurred with this view'. But Steuer (1933), who also studied copepods, came to somewhat different conclusions. In a summary of the results of his own studies plus those of others, he proposed twelve distributional regions and subregions (*Verbreitungsgebiete*) for the high sea copepods and pointed out that other groups fit this scheme as well (Fig. 1.3). Russell (1935) and Hedgpeth (1957) have followed Steuer's proposal.

None of these presentations indicate the relative abundances of individuals within a species, nor does the purpose in determining the regions seem very clearly stated. It is, therefore, appropriate to review the status of our knowledge of the pattern of abundances as it stands today. Ekman (1953, p. 312) estimates that between 1,500 and 2,000 plankton samples, taken on the great oceanographic expeditions of the late nineteenth and early twentieth century, had been examined by the late 1930s. As pointed out above the number of samples examined is now at least ten times that. Although these new

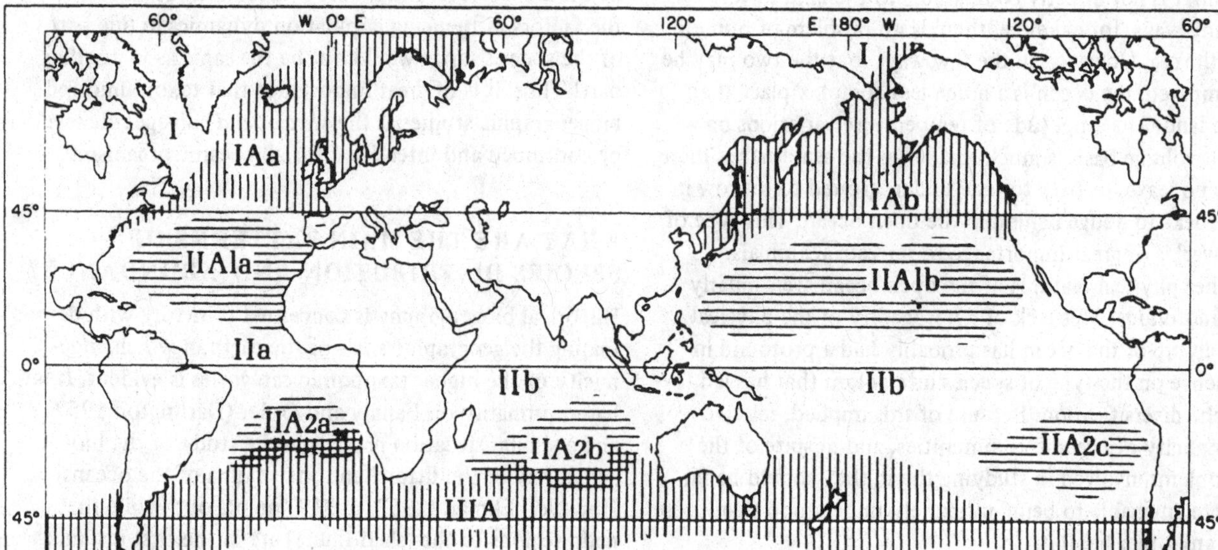
60°
W 0° E
60°
120°
E 180° W
120°
60°
45°
0°
I
IAa
IAb
IIAla
IIAlb
IIa
IIb
IIb
IIA2a
IIA2b
IIA2c
IIIA
III

Fig. 1.3. The distributional regions and subregions of the world ocean according to Steuer (1933):

I	Circumpolar Arctic region	II A 1b	Northern subtropical Pacific subregion
I A a	Subarctic Atlantic subregion	II A 2a	Southern subtropical Atlantic subregion
I A b	Subarctic Pacific subregion	II A 2b	Southern subtropical Indian subregion
II a	Tropical Atlantic region	II A 2c	Southern subtropical Pacific subregion
II b	Tropical Indopacific region	III	Circumpolar Antarctic region
II A 1a	Northern subtropical Atlantic subregion	III A	Circumpolar subantarctic subregion

samples are not particularly well distributed over the world's oceans, much new information has been developed, particularly for the Pacific and Antarctic. Further, these more recent studies are ecologically oriented and in many of them, estimates of abundance are an integral part of the presentations. Because of this, it is possible to attempt to define the major habitats of some oceans and to examine the properties of these in terms of variations of abundance of the inhabitants. Thus, we have an opportunity to take Macfadyen's advice (1957, p. 13) and attempt to 'detect the relevant factors in (an animal's) surroundings from the animal's point of view'. The assumption is that organisms tend to be most abundant in those places best suited for their existence. It is clear that we must have a better understanding of the features that characterize an animal's habitat for 'much earlier work in animal ecology belongs to a phase of what might be called blissful ignorance, in which it was generally believed that the occurrence of species could be readily explained by the combination of a few physical factors. While work correlating the distributions of species with physical measurements is not without its use in defining the extreme range of species, it is now generally recognized that such work has a limited application' (Macfadyen 1957, p. 14).

Johnson and Brinton (1963) have reviewed some of the information on the patterns of distribution of holoplanktonic animals in the Pacific. While they did emphasize the fact that species of other groups showed similar patterns, most of the evidence was derived from data on the Euphausiacea. It is, therefore, important to point out that many other groups of pelagic organisms, from protozoa to vertebrates, show very similar patterns. Although recent studies of the distribution and abundance of phytoplankton species in the oceanic regions are not nearly so numerous there is good evidence that they too follow the same scheme.

Phytoplankton

Graham and Bronikowsky (1944) showed distributional charts for 49 species of the genus *Ceratium* in the Atlantic and Pacific oceans. The number of localities sampled in the Atlantic was small (46) and little can be said about distributional patterns there. However, in the Pacific over 120 localities covering a broad range of latitude were sampled. These authors divided the areas sampled into five '*Ceratium* Life Zones'; two were in the Atlantic: a 'cold north Atlantic' zone and a 'warm Atlantic' zone. The Pacific was divided into three zones or regions; a 'cold north Pacific', a 'warm Pacific' and a 'southeast Pacific' region. These regions were based primarily on the qualitative species presence or absence data, but to some degree, estimates of relative abundance were also used, in that locality records were scored as abundant, common, occasional, or rare. These estimates, however, were not included on the charts of individual species distributions. An examination of their 49 charts shows that while these 'life zones' certainly can be used to describe the results there is considerably more information in them than this simple scheme would imply. Thus, it is obvious that in the Pacific *Ceratium cephalotum, C. pentagonum tenerum,* and *C. falcatum,* are limited to the mid-latitudes of both the North and South Pacific (Fig. 1.14) and were not found in either the equatorial or subarctic zones. On the other hand *Ceratium lunula* was found only in the equatorial waters (Fig. 1.16) and *C. deflexum* primarily there. *Ceratium arcticum* was limited to subarctic Pacific waters with the exception of a 'rare' record at 100 metres depth at about 32° N. *Ceratium tripos atlanticum* and *C. filicorne* show peculiar distributions, the former being quite frequent in an area of the North Pacific called the Transition Zone by Brinton (1962), Johnson and Brinton (1963) and myself (McGowan 1963) and in the Peru Current (Fig. 1.9). The latter species *Ceratium filicorne* was restricted to the higher latitudes sampled of the South Pacific. The remaining species dealt with by Graham and Bronikowsky show a high frequency of occurrence in all of the warmer waters of the Pacific, that is, from 40° N to as far south as the sampling went (approximately 40° S). This is not to say, however, that variations in their abundance did not occur.

Smayda (1958) has discussed some of the general features of phytoplankton biogeography. He has mapped the occurrences (but not the relative abundances) of three species he considers to be representative of arctic, antarctic, tropical and cosmopolitan distributions and lists a number of others for each category. In an examination of the 'factors affecting the distribution of phytoplankton' Smayda states 'the distribution of marine phytoplankton is governed by essentially the same variables that affect their abundance'. He lists these as 'light, temperature, salinity, pollution, fertility and currents'. If it is so, that distribution and abundance are but two sides of the same coin, it would seem that another important factor has been left out of this list, namely, grazing by zooplankton. Riley (1963) and Steel (1961) have clearly shown that along with light and nutrients this is of importance!

Kanaya and Koizumi (1966) have done an extensive study of the species distribution of diatom frustules recovered from the surface layers of deep-sea cores taken in the Pacific. In a review of their own data and of the information presented by other authors, they point out that there is very good reason to believe 'the species composition of a diatom assemblage on the sea bottom depends to a large extent on the species composition of the plankton in the sea water above'. If this can be accepted as being true, at least qualitatively, then their subsequent determinations of the distribution of diatom assemblages (thanatocoenoses) of the North Pacific is of great interest. Using objective criteria developed by Fager (1957) to determine which species occur together more 'frequently' than with other species, they have derived five recurrent groups. These groups included fifty five of the seventy five species and varieties available for analysis. On plotting the areal distribution of samples containing $n, n-1, n-2$, or $n-3$ members of the group (n=number of species in a group) it was discovered that definite patterns were formed. Group I, containing twenty-two taxa was distributed mainly in the equatorial zone of the eastern Pacific. However, plotting $n-1, n-2$, and $n-3$ members of this group successively increased its latitudinal boundaries to approximately 40° N. Group II consisted of known subarctic species. Group III, not shown on their chart, was limited to the Antarctic and Subantarctic. Group IV consisted of seven species common in the sea of Okhotsk and the Bering Sea, and group V with five species was purely equatorial. In addition to these it was possible to recognize a Central Assemblage consisting of group I, $n-3$ only; an assemblage of both Subarctic and Central groups; and a transitional assemblage occupying an area between the Subarctic and Central distributions but with no identifiable group. These results are very similar to the species patterns discussed by Johnson and Brinton (1963) and the recurrent groups of zooplankton species that have been detected in the North Pacific (Fager and McGowan 1963). As Kanaya and Koizumi point out, the patterns formed by these assemblages look very much like the water mass patterns of the Pacific as presented by Sverdrup, Johnson and Fleming (1942).

Zooplankton

Oceanwide, quantitative, species distribution patterns in the Pacific have been determined for at least eight important taxa of zooplankton. The taxa studied are Euphausiacea, Chaetognatha, Foraminifera, Heteropoda, Thecosomata, and the copepod genera *Eucalanus, Rhincalanus* and *Clausocalanus.* The total number of species in these groups in the Pacific is 194 and all species in each taxa have been mapped. In addition to these the taxa, Polychaeta, Tripylea (Radiolaria), and Thaliacea have been studied for large sectors of the Pacific which contain several of the major water masses. Further the distribution of certain species of additional taxa of zooplankton (Amphipoda, other Copepoda) have been determined in the eastern North Pacific sector.

On the basis of these studies there are a number of generalizations that may be made about the patterns of zooplankton species distributions in the Pacific. These are:

1. There is a remarkable amount of agreement between most of the species of all of the taxa as to the position and shape of their distributional boundaries, (Figs. 1.30-1.32).
2. A large number of species in each group show patterns of distribution whose boundaries are almost identical with the boundaries of physical water masses (Table 1.2 and Figs. 1.4-1.18).
3. A number of species have areas of highest levels of abundance within a water mass but whose boundaries extend somewhat beyond the boundary of the water mass (Table 1.2 *a* and *b* and Figs. 1.21-1.23).
4. Many species are found throughout several water masses (Tables 1.2 *a* and *b*, Figs. 1.24-1.28).
5. Some species are limited to certain large parts of some water masses (Tables 1.2 *a* and *b*, Fig. 1.29).

Table 1.2*a*. The areas of the North Pacific in which the listed species have been shown to occur

E.N.P. = Eastern North Pacific; W.T.P. = Western Tropical Pacific; E.T.P. = Eastern Tropical Pacific; T zone = Transitional zone

Organism	Subarctic	Transitional	Central	Equatorial	Eastern Tropic Pacific	Warm water Cosmopolites	Comments	References
PROTOZOA								
Foraminifera								
Globigerina quinqueloba	+							Bradshaw 1959
Globigerinoides minuta	+							Bradshaw 1959
Globigerina pachyderma	+							Bradshaw 1959
Globorotalia truncatulinoides			+					Bradshaw 1959

Table 1.2*a (continued)*

Organisms	Subarctic	Transitional	Central	Equatorial	Eastern Tropic Pacific	Warm water Cosmopolites	Comments	References
Pulleniatina obliquiloculata				+				Bradshaw 1959
Sphaeroidinella dehiscens				+				Bradshaw 1959
Globigerina conglomerata				+				Bradshaw 1959
Globorotalia tumida				+				Bradshaw 1959
Globorotalia hirsuta				+				Bradshaw 1959
Globigerinella aequilateralis						+	'Pure'	Bradshaw 1959
Globigerinoides conglobata						+	'Pure'	Bradshaw 1959
Globigerinoides rubra						+	'Pure'	Bradshaw 1959
Orbulina universa						+	'Pure'	Bradshaw 1959
Globigerinoides sacculifera						+	Peak at equator	Bradshaw 1959
Globorotalia menardii						+	Peak at equator	Bradshaw 1959
Globigerina eggeri						+	Edge effect	Bradshaw 1959
Hastigerina pelagica						+	Edge effect	Bradshaw 1959
Radiolaria								
Castanidium apsteini	+							Kling 1966 (E.N.P. only studied)
Castanidium variabile	+		•				Doubtful may be deep central too	Kling 1966
Haeckeliana porcellana	+						Doubtful may be deep central too	Kling 1966
Castanea amphora			+					Kling 1966
Castanissa brevidentata			+					Kling 1966
Castanella thomsoni			+				T. zone w/upwelled water?	Kling 1966
Castanea henseni			+				T. zone w/upwelled water?	Kling 1966
Castanea globosa			+				T. zone w/upwelled water?	Kling 1966
Castanidium longispinum				+				Kling 1966
Castanella aculeata				+				Kling 1966
CHAETOGNATHA								
Sagitta elegans	+							Bieri 1959
Eukrohnia hamata	+							Alvarino 1962
Sagitta scrippsae		+						Alvarino 1962
Sagitta pseudoserratodentata			+					Bieri 1959
Sagitta californica			+				Crossing W.T.P.	Bieri 1959
Sagitta ferox				+				Bieri 1959; Alvarino 1962
Sagitta robusta				+			Patchy	Bieri 1959; Alvarino 1962
Sagitta regularis				+				Bieri 1959
Sagitta hexaptera						+		Bieri 1959
Sagitta enflata						+	Peak at equator	Bieri 1959
Pterosagitta draco						+	Peak at equator	Bieri 1959
Sagitta pacifica						+	Peak at equator	Bieri 1959
Sagitta minima						+	Edge effect	Bieri 1959
ANNELIDA								
Tomopteris septentrionalis						+		Tebble 1962
Tomopteris pacifica	+							Tebble 1962
Poeobius meseres	+							McGowan 1960

Table 1.2*b*. The areas of the North and South Pacific in which the listed species have been shown to occur

Organism	Subarctic	Transitional	Central	Equatorial	Eastern Tropical	Warm water Cosmopolites	Subantarctic	Antarctic	Comments	References
ARTHROPODA										
Copepoda										
Calanus pacificus	+								May be T. zone	Brodsky 1965
Calanus plumchrus	+									Brodsky 1960
Calanus tonsus	+									Johnson and Brinton 1963
Calanus cristatus	+									Johnson and Brinton 1963
Eucalanus bungii bungii	+									Johnson and Brinton 1963
Eucalanus elongatus hyalinus		+							South Pacific also	Lang 1965
Eucalanus bungii californicus		+								Lang 1965
Clausocalanus pergens		+								Frost and Fleminger, in press
Clausocalanus lividus			+							Frost and Fleminger, in press
Eucalanus subcrassus				+						Lang 1964
Rhincalanus cornutus				+						Lang 1964
Eucalanus inermis					+					Lang 1964
Eucalanus crassus						+			Patchy, 'pure'	Lang 1964
Rhincalanus nasutus						+			Very patchy, almost pure equatorial	Lang 1964
Eucalanus attenuatus						+			Peak at equator; some edge effect	Lang 1964
Eucalanus subtenuis						+			Patchy, peak at equator; some edge effect	Lang 1964
Clausocalanus arcuicornis						+				Frost and Fleminger, in press
Eucalanus longiceps							+			Lang 1964
Rhincalanus gigas								+		Lang 1964
Clausocalanus laticeps								+		Frost and Fleminger, in press
Euphausiacea										
Thysanoessa longipes	+									Brinton 1962
Euphausia pacifica	+									Brinton 1962
Thysanopoda acutifrons		+								Brinton 1962
Thysanoessa gregaria		+								Brinton 1962
Euphausia gibboides		+								Brinton 1962
Nematoscelis difficilismegalops		+								Brinton 1962
Nematoscelis atlantica			+							Brinton 1962
Euphausia brevis			+							Brinton 1962
Euphausia hemigibba			+							Brinton 1962
Euphausia gibba			+							Brinton 1962
Euphausia mutica			+							Brinton 1962
Stylocheiron suhmii			+						Crossing in W.T.P.	Brinton 1962
Euphausia diomediae				+						Brinton 1962
Euphausia distinguenda				+						Brinton 1962
Nematoscelis gracilis				+						Brinton 1962
Euphausia distinguenda					+					Brinton 1962
Euphausia eximia					+					Brinton 1962
Euphausia lamelligera					+					Brinton 1962
Euphausia tenera						+			Peak at equator	Brinton 1962
Stylocheiron abbreviatum						+			avoids E.T.P.	Brinton 1962
Euphausia superba								+		Marr 1962
Amphipoda										
Parathimisto pacifica	+									Bowman 1960

Table 1.2*b (continued)*

Organisms	Subarctic	Transitional	Central	Equatorial	Eastern Tropical	Warm water Cosmopolites	Subantarctic	Antarctic	Comments	References
MOLLUSCA										
Pteropoda										
Limacina helicina	+									McGowan 1963
Clio polita	+									
Corolla pacifica		+								McGowan, unpub.; Beklemishev 1961
Clio balantium		+								
Cavolinia inflexa			+							McGowan 1960
Clio pyramidata			+						Crossing in W.T.P.	McGowan 1960
Styliola subula			+						Crossing in W.T.P.	McGowan 1960
Limacina lesueuri			+						Crossing in W.T.P.	McGowan 1960
Clio n.sp.				+						McGowan, in press
Cavolinia uncinata				+					Very patchy	McGowan 1960
Limacina trochiformis					+					McGowan 1960
Limacina inflata						+				McGowan 1960
Cavolinia longirostris						+			Very patchy; almost pure equatorial	McGowan 1960
Cavolinia gibbosa						+			Very patchy avoids E.T.P.	McGowan 1960
Hyalocylix striata						+				McGowan 1960
Creseis virgula						+			Edge effect	McGowan 1960
Creseis acicula						+				McGowan 1960
Cavolinia tridentata						+			Peak at equator	McGowan 1960
Diacria trispinosa						+			Peak at equator	McGowan 1960
Limacina bulimoides						+			Avoids E.T.P.	McGowan 1960
Clio antarctica						+				McGowan, unpub.
Heteropoda										
Caranaria japonica		+								McGowan, unpub.
Gymnosomata										
Clione limacina	+									McGowan, unpub.

The water masses of the Pacific, as defined by Sverdrup, Johnson, and Fleming (1942, pp. 698-733) are: Subarctic, North Pacific Central (East and West), Equatorial, South Pacific Central (East and West) and Subantarctic. In addition to these, there are two broad 'Transition Zones' between the Subarctic and Central water masses in the north and between Subantarctic and Central water masses of the south.

Tables 1.2 *a* and *b* indicate that a large proportion of the species of the taxa studied are essentially limited to water masses. Another group of species occupy both the Equatorial and the Central water masses but some of these tend to be most abundant in the equatorial portion only, or near the 'edges' of their range. There are two distinct patterns that do not seem to conform to the classical water mass distributions; these have been called the Transition Zone and the Eastern Tropical Pacific by Johnson and Brinton (1963), Beklemishev and Burkov (1958), Beklemishev (1961a) and McGowan (1963). However, there are reasons to doubt that the classical water mass patterns as described by Sverdrup, Johnson, and Fleming are complete. As these authors point out, the distribution of water mass patterns and the definitions on which these are based came from relatively few data. There are today, reasons to believe that both the Transition Zone and the Eastern Tropical Pacific are, in many ways, unique bodies of water with meteorological and hydrological forces influencing them in a manner that tends to maintain this uniqueness (Dodimead, Favorite and Hirano, 1963; Wooster and Cromwell 1958; McGowan 1963, 1967). That is to say, in some ways they are distinct, large, masses of water that may be identified.

It appears certain that a number of species are endemic to an area of the North Pacific that Johnson and Brinton (1963) have called the 'Transition Zone'. In addition to these endemics other species, somewhat more broadly distributed, are most abundant here. This is an apparent exception to the general rule that species tend to occupy

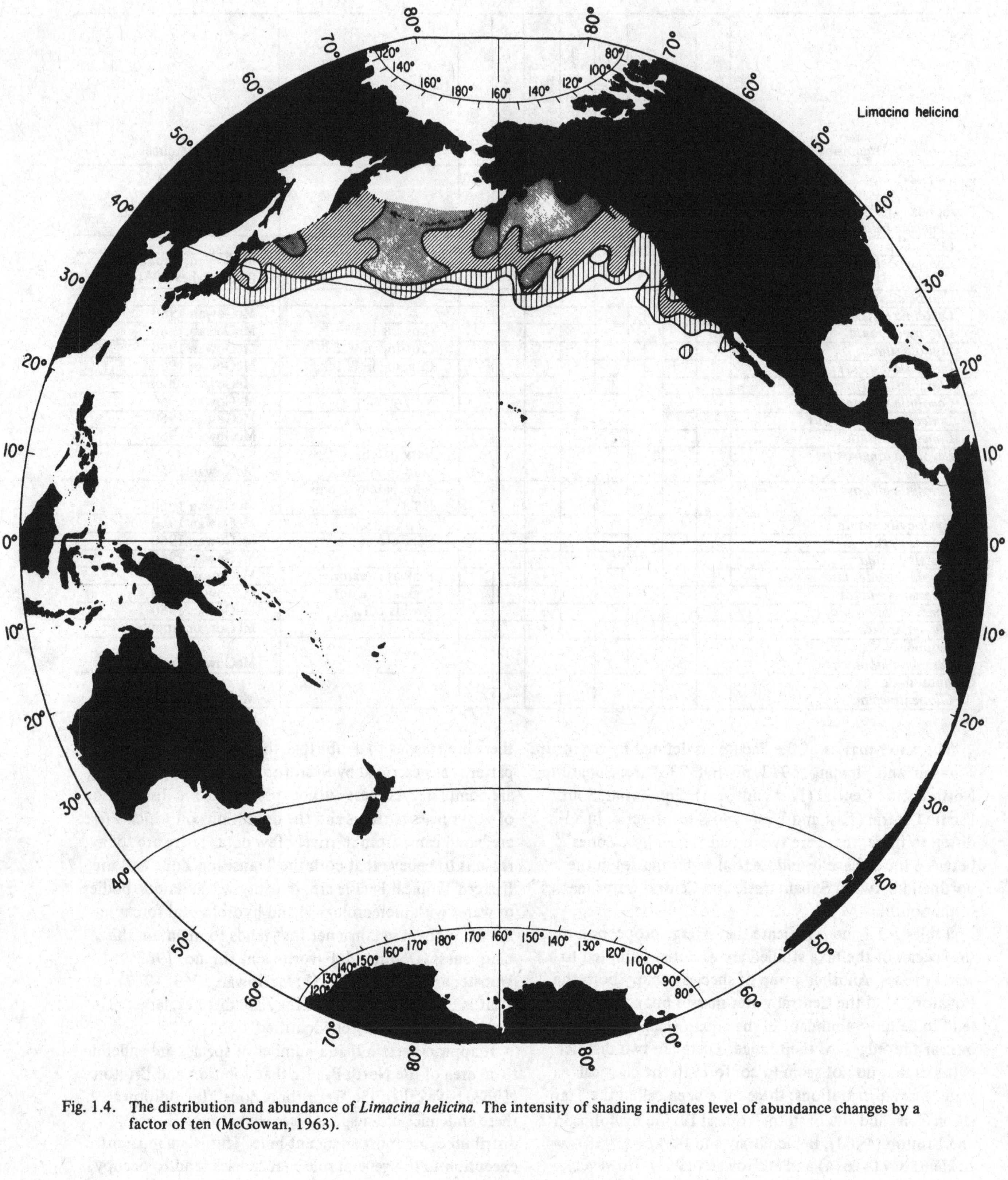
Limacina helicina
80°
70°
60°
50°
40°
30°
20°
10°
0°
10°
20°
30°
40°
50°
60°
70°
80°
120°
140°
160°
180°
160°
140°
120°
100°
80°
150°
170°
180°
170°
160°
150°
140°
130°
120°
110°
100°
90°
80°

Fig. 1.4. The distribution and abundance of *Limacina helicina*. The intensity of shading indicates level of abundance changes by a factor of ten (McGowan, 1963).

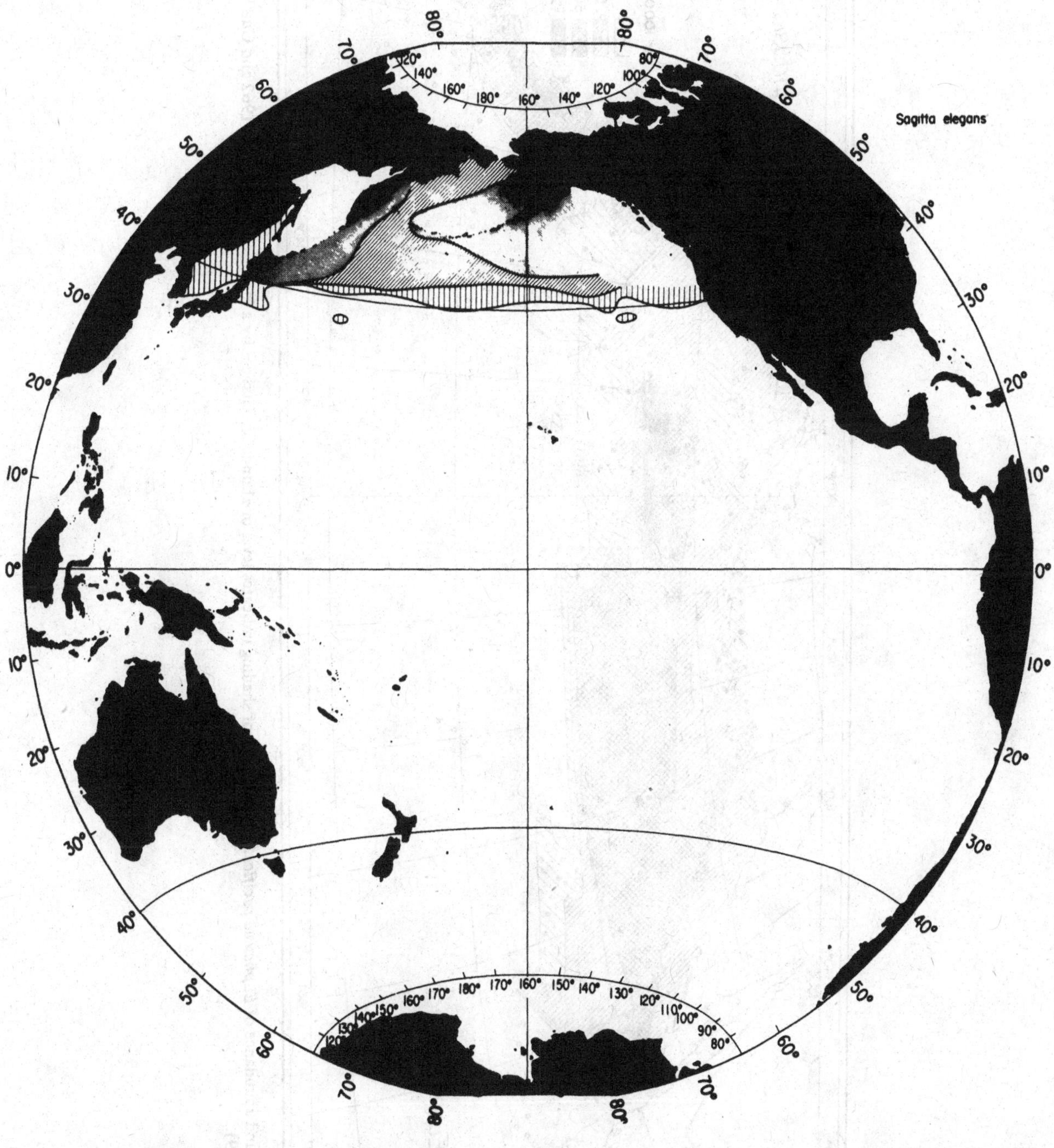
Sagitta elegans

Fig. 1.5. The distribution and abundance of *Sagitta elegans.* The intensity of shading indicates level of abundance changes by a factor of ten (Bieri 1959).

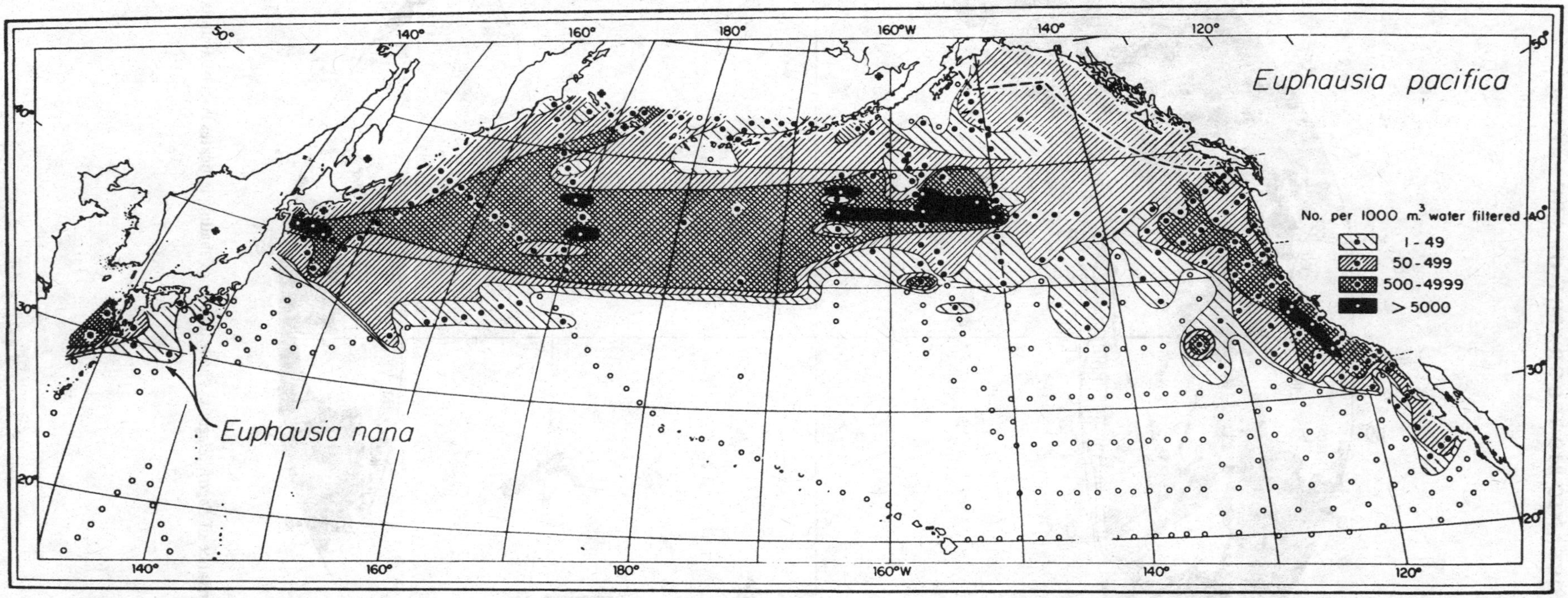
Euphausia pacifica
No. per 1000 m.³ water filtered
1 - 49
50 - 499
500 - 4999
> 5000
Euphausia nana
140°
160°
180°
160°W
140°
120°
50°
40°
30°
20°

Fig. 1.6. The distribution and abundance of *Euphausia pacifica*. The intensity of shading indicates level of abundance changes by a factor of ten (Brinton 1962 and Gulf of Alaska from Banner 1949).

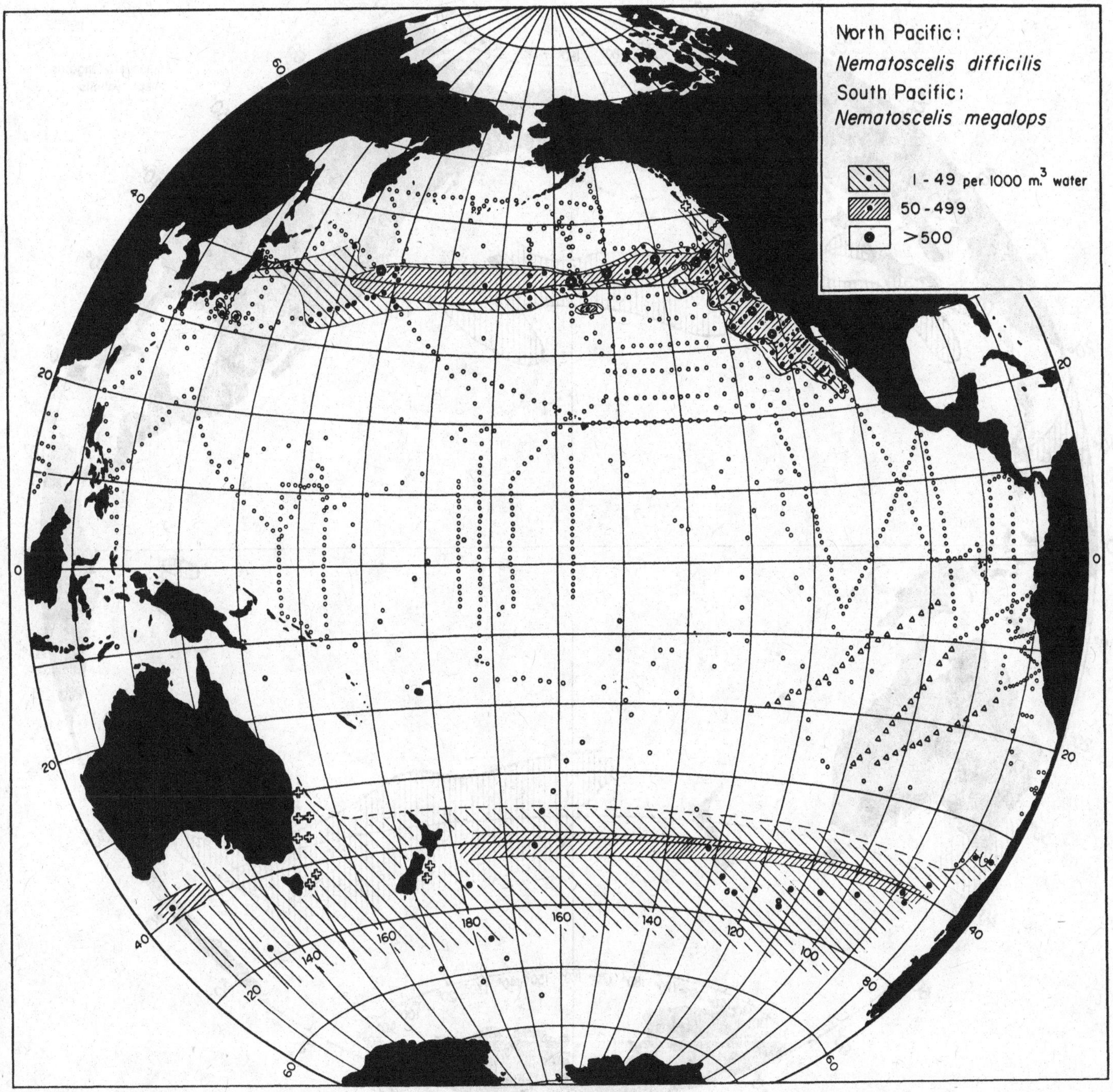
North Pacific:
Nematoscelis difficilis
South Pacific:
Nematoscelis megalops
1 - 49 per 1000 m.³ water
50 - 499
>500
60
40
20
0
20
40
60
180
160
140
120
100
80
60
160
140
120

Fig. 1.7. The distribution and abundance of *Nematoscelis difficilis* and *N. megalops.* The intensity of shading indicates level of abundance changes by a factor of ten (Brinton 1962).

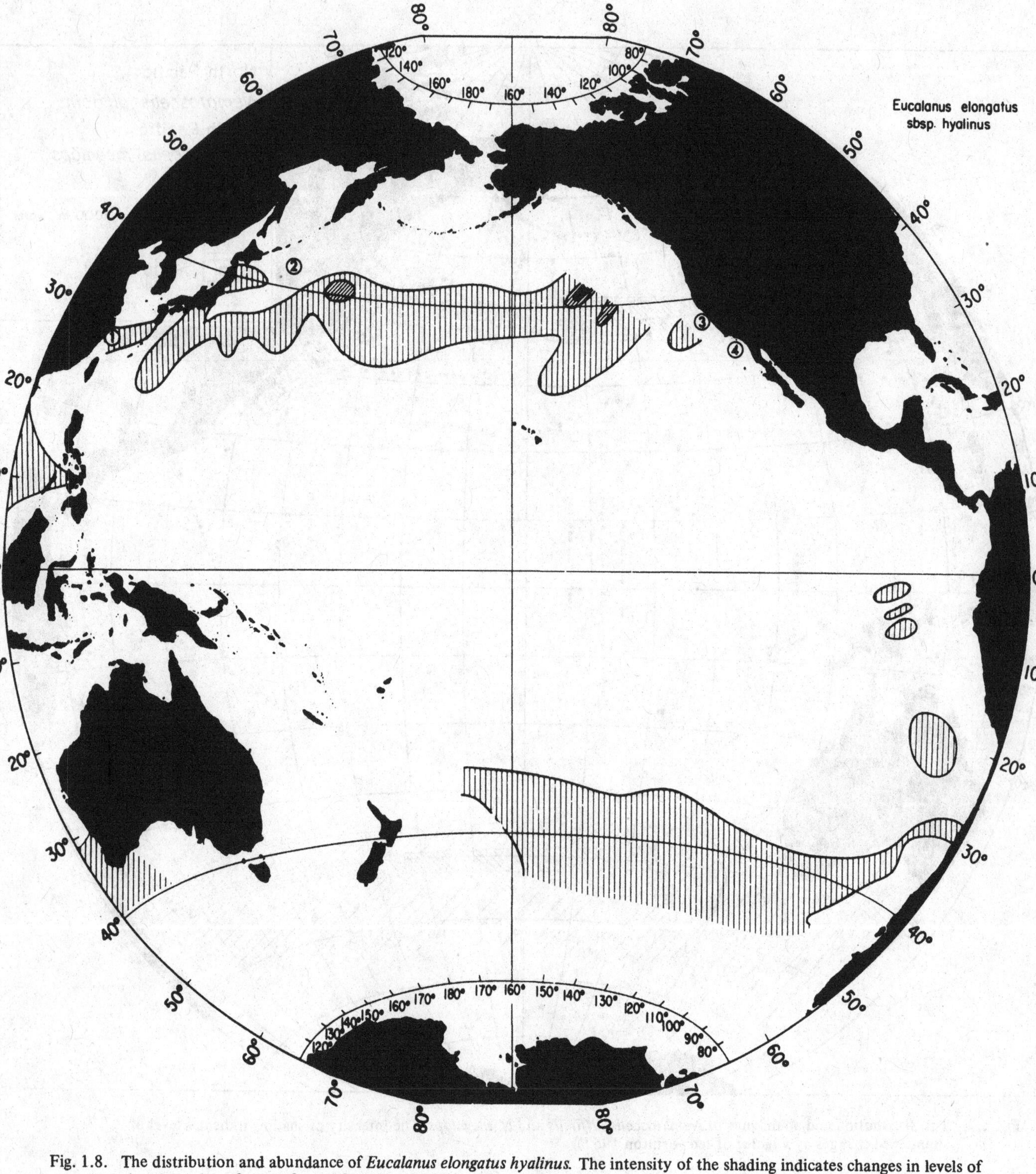
Eucalanus elongatus
sbsp. hyalinus

Fig. 1.8. The distribution and abundance of *Eucalanus elongatus hyalinus*. The intensity of the shading indicates changes in levels of abundance by a factor of ten (after Lang 1965).

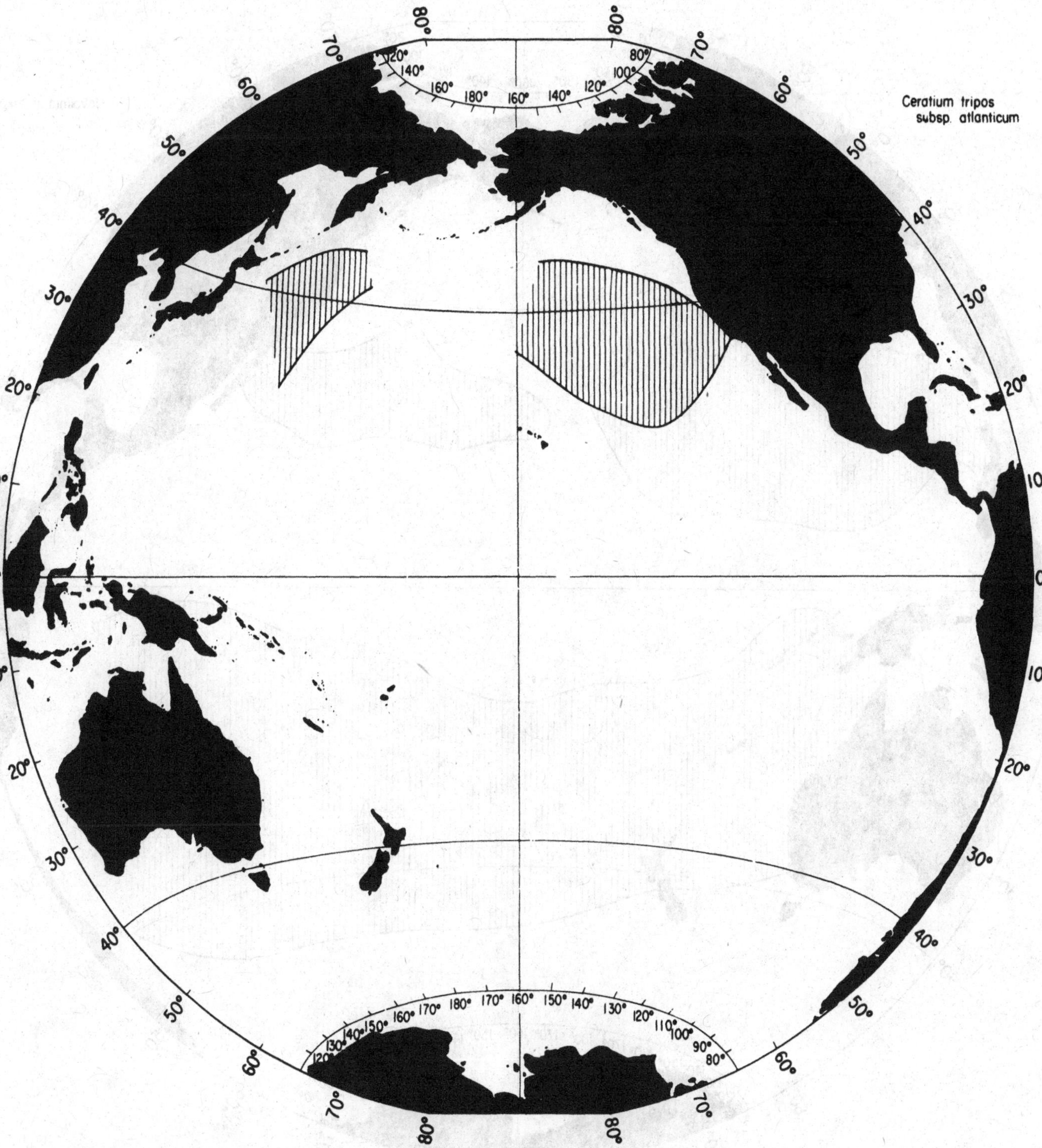
Ceratium tripos
subsp. atlanticum

Fig. 1.9. The distribution of *Ceratium tripos atlanticum* (after Graham and Bronikowsky 1944).

3

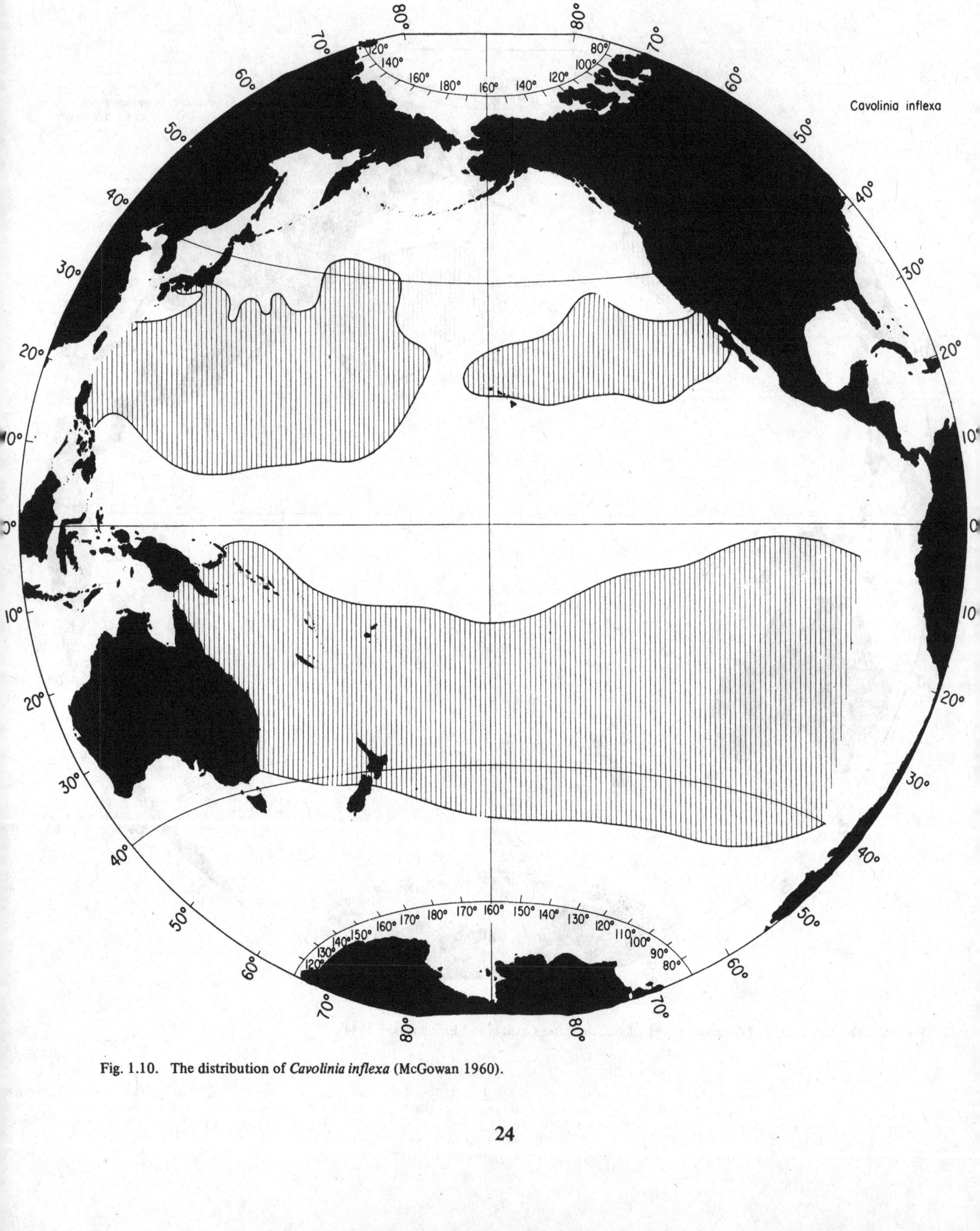
Cavolinia inflexa

Fig. 1.10. The distribution of *Cavolinia inflexa* (McGowan 1960).

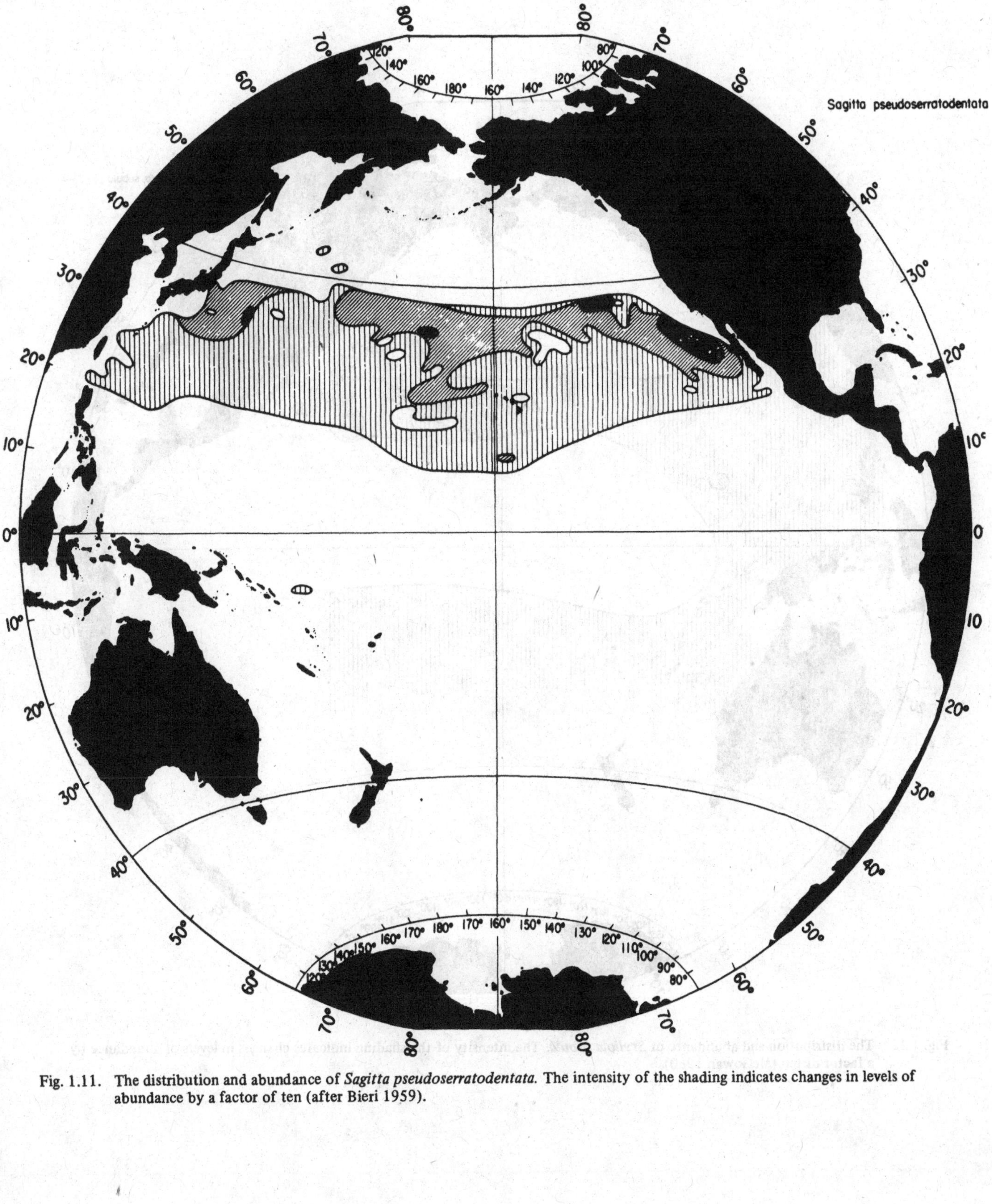
Sagitta pseudoserratodentata

Fig. 1.11. The distribution and abundance of *Sagitta pseudoserratodentata.* The intensity of the shading indicates changes in levels of abundance by a factor of ten (after Bieri 1959).

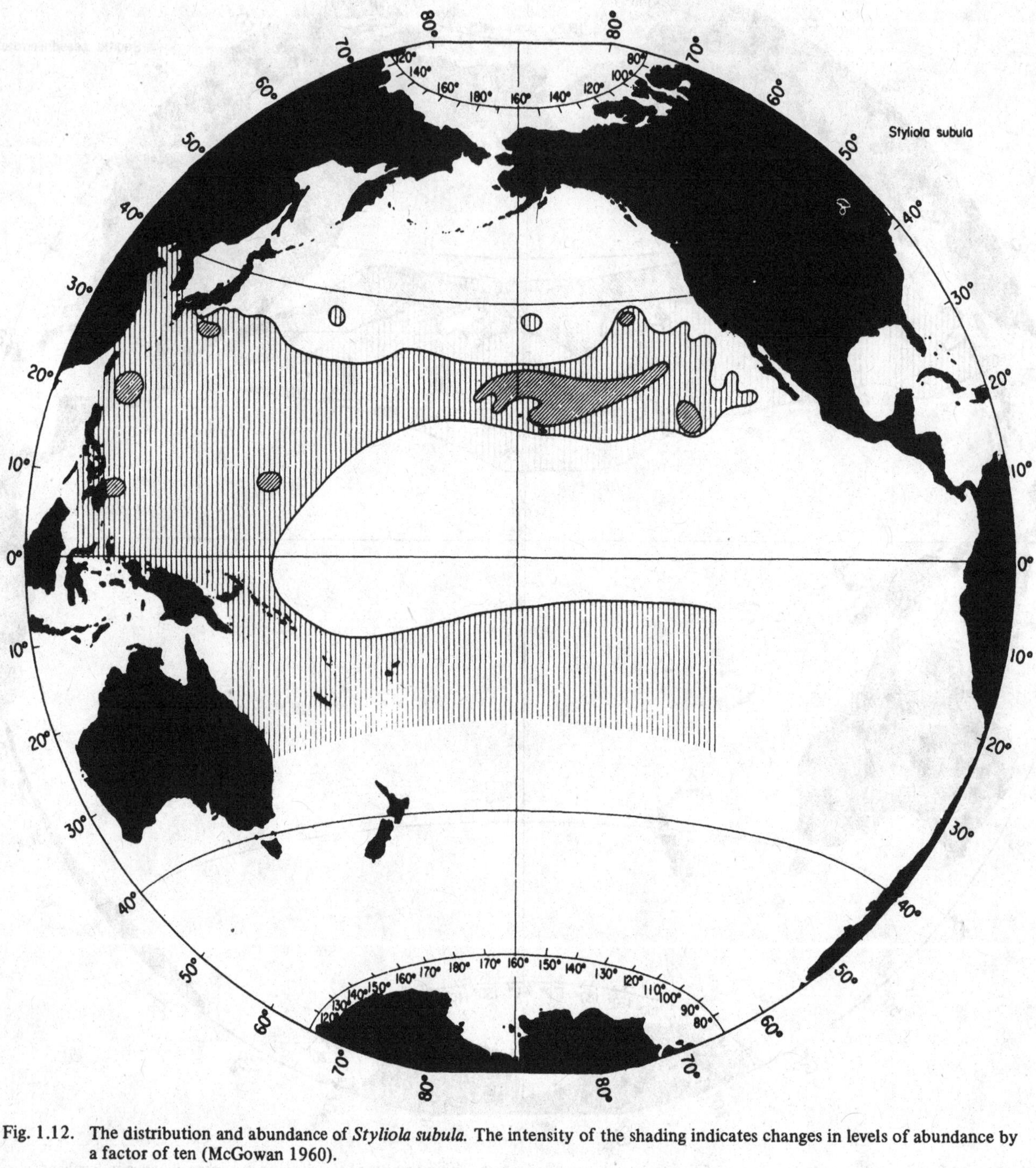
Styliola subula

Fig. 1.12. The distribution and abundance of *Styliola subula.* The intensity of the shading indicates changes in levels of abundance by a factor of ten (McGowan 1960).

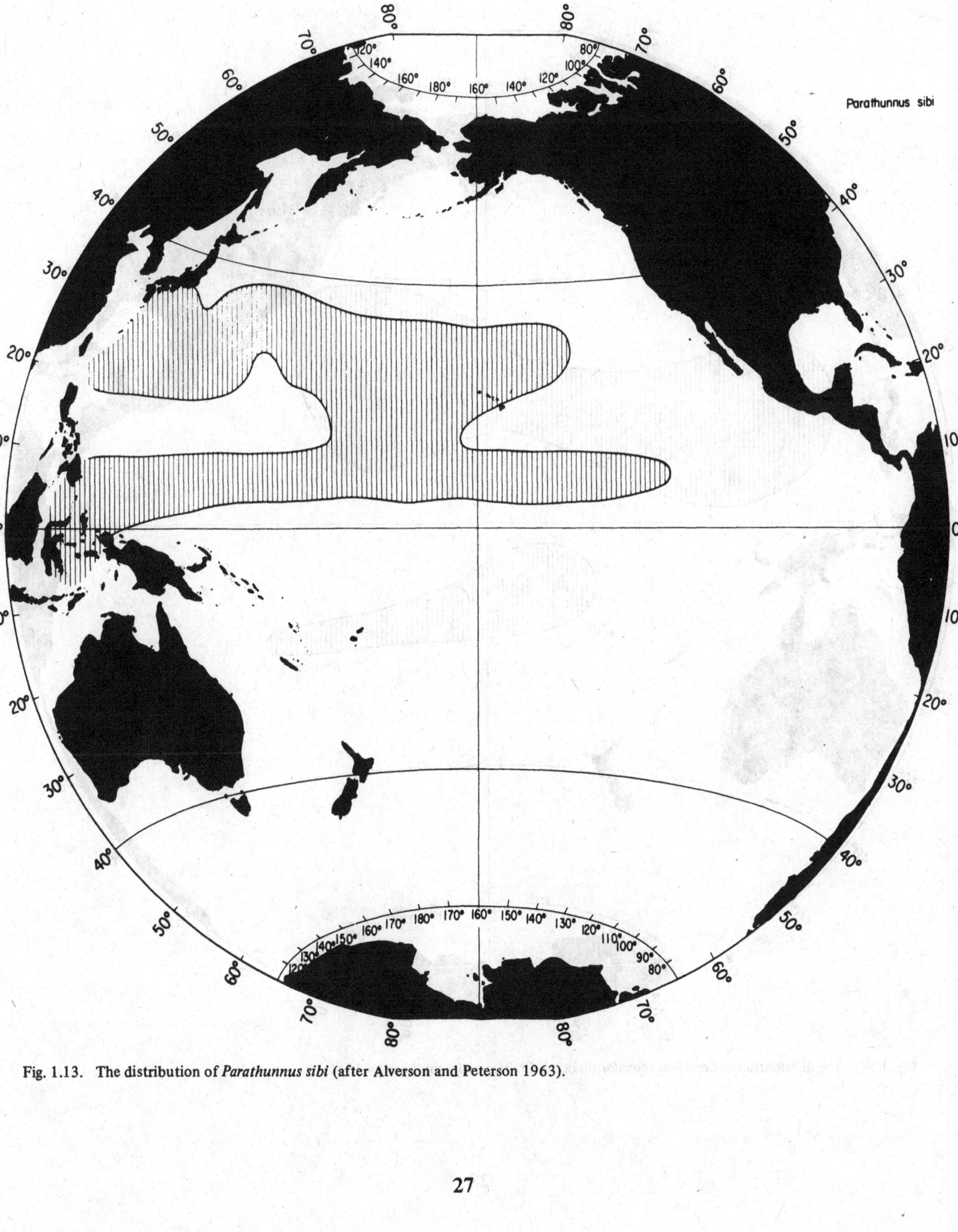
Parathunnus sibi
80°
70°
60°
50°
40°
30°
20°
10°
0°
10°
20°
30°
40°
50°
60°
70°
80°
120°
140°
160°
180°
160°
140°
120°
100°
80°
130°
140°
150°
160°
170°
180°
170°
160°
150°
140°
130°
120°
110°
100°
90°
80°

Fig. 1.13. The distribution of *Parathunnus sibi* (after Alverson and Peterson 1963).

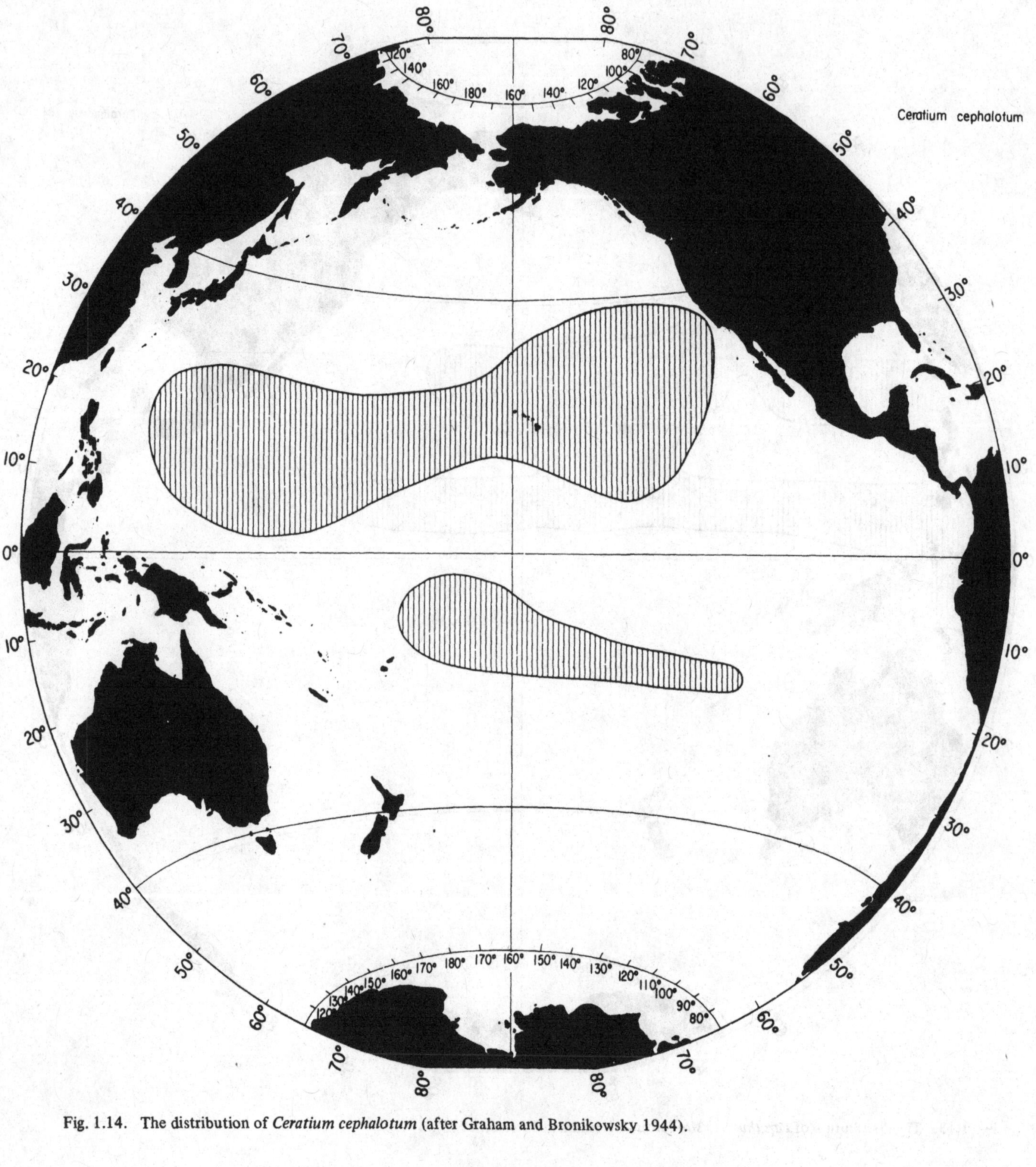
Ceratium cephalotum

Fig. 1.14. The distribution of *Ceratium cephalotum* (after Graham and Bronikowsky 1944).

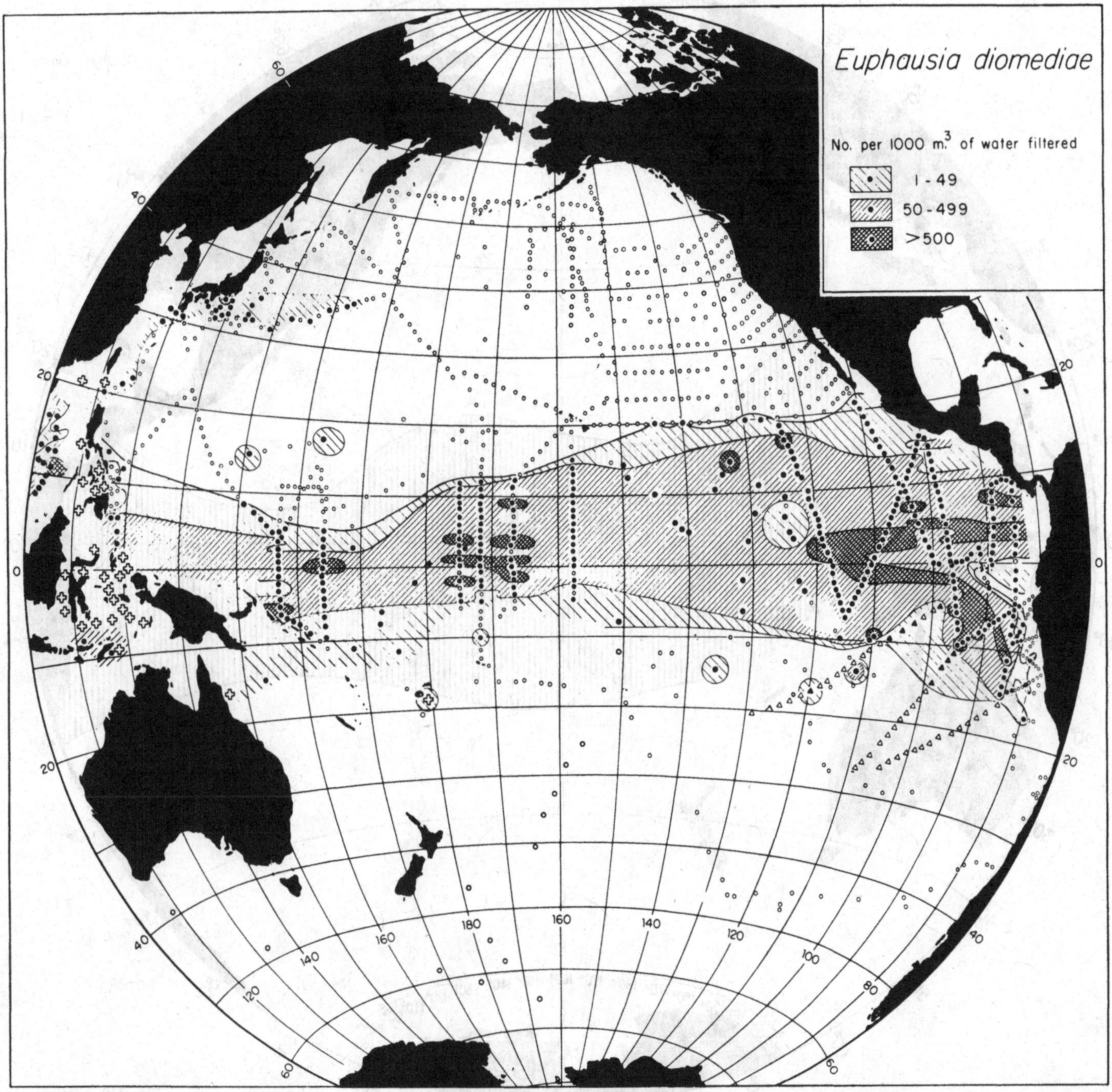
Euphausia diomediae
No. per 1000 m.^{3} of water filtered
1 - 49
50 - 499
>500

Fig. 1.15 The distribution and abundance of *Euphausia diomediae.* The intensity of the shading indicates changes in levels of abundance by a factor of ten (Brinton 1962).

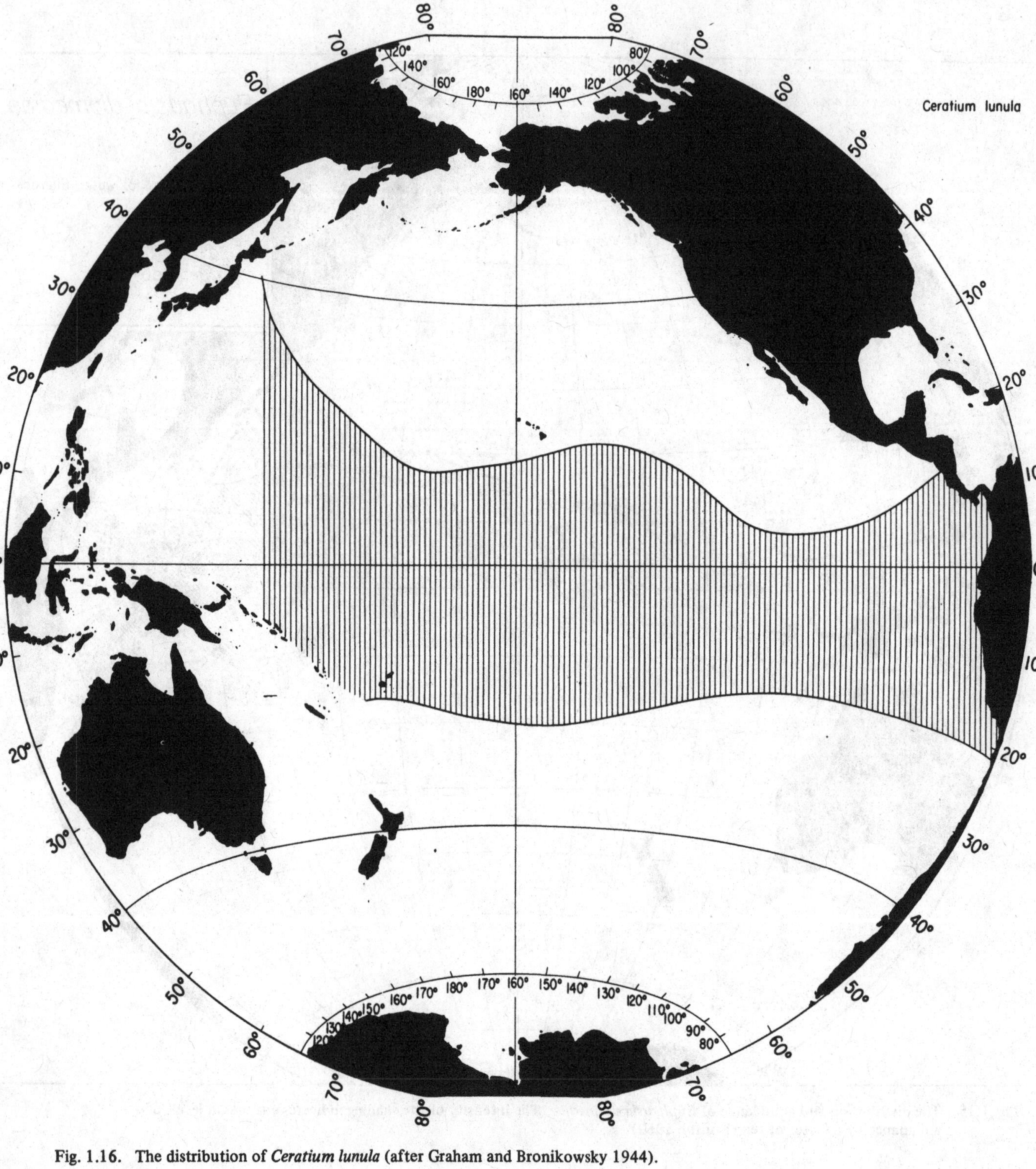
Ceratium lunula
80°
70°
60°
50°
40°
30°
20°
10°
0°
10°
20°
30°
40°
50°
60°
70°
80°
120°
140°
160°
180°
160°
140°
120°
100°
80°
120°
130°
140°
150°
160°
170°
180°
170°
160°
150°
140°
130°
120°
110°
100°
90°
80°

Fig. 1.16. The distribution of *Ceratium lunula* (after Graham and Bronikowsky 1944).

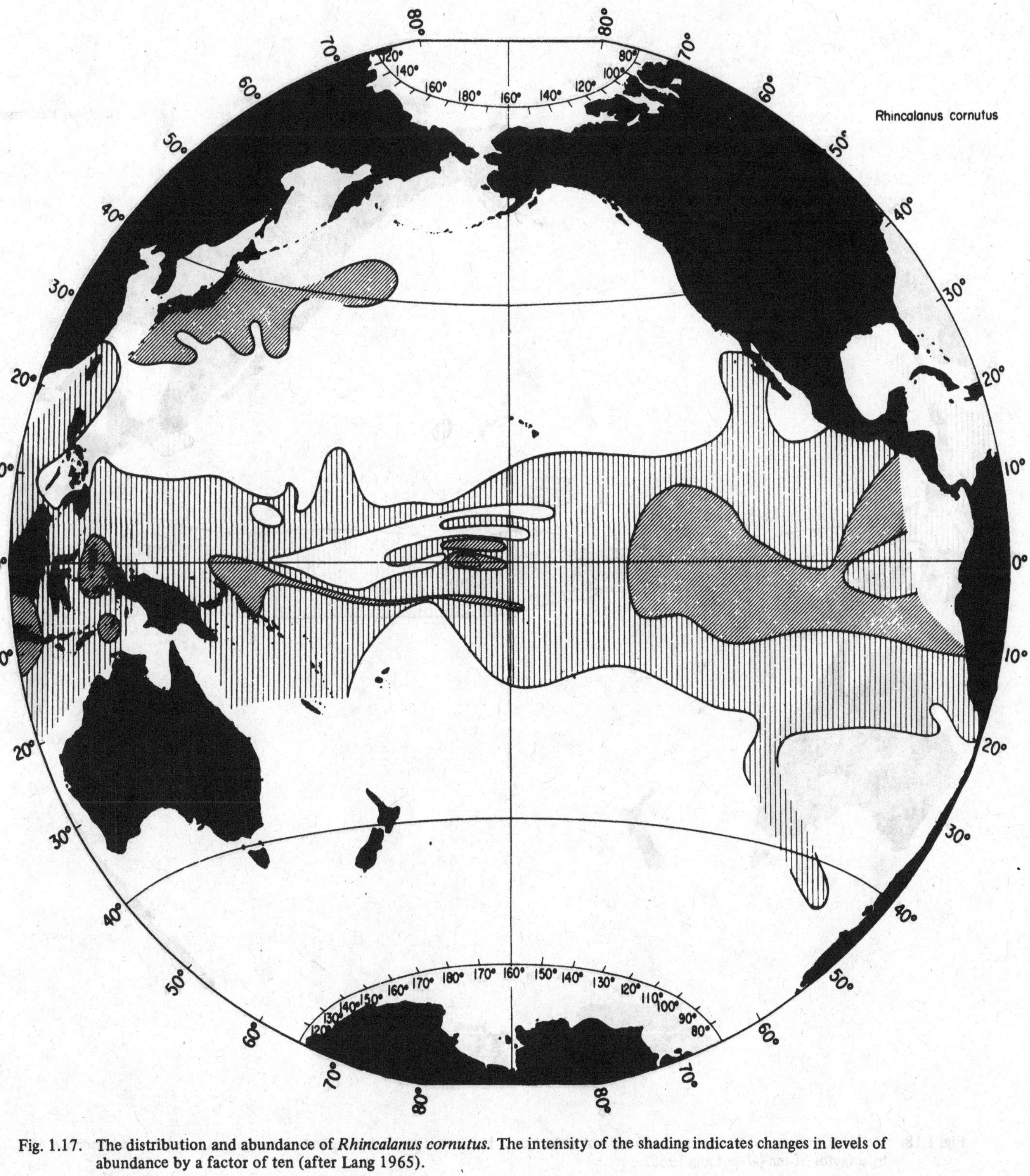
Rhincalanus cornutus
80°
70°
60°
50°
40°
30°
20°
10°
0°
120°
140°
160°
180°
160°
140°
120°
100°
80°
170°
150°
130°
110°
90°

Fig. 1.17. The distribution and abundance of *Rhincalanus cornutus.* The intensity of the shading indicates changes in levels of abundance by a factor of ten (after Lang 1965).

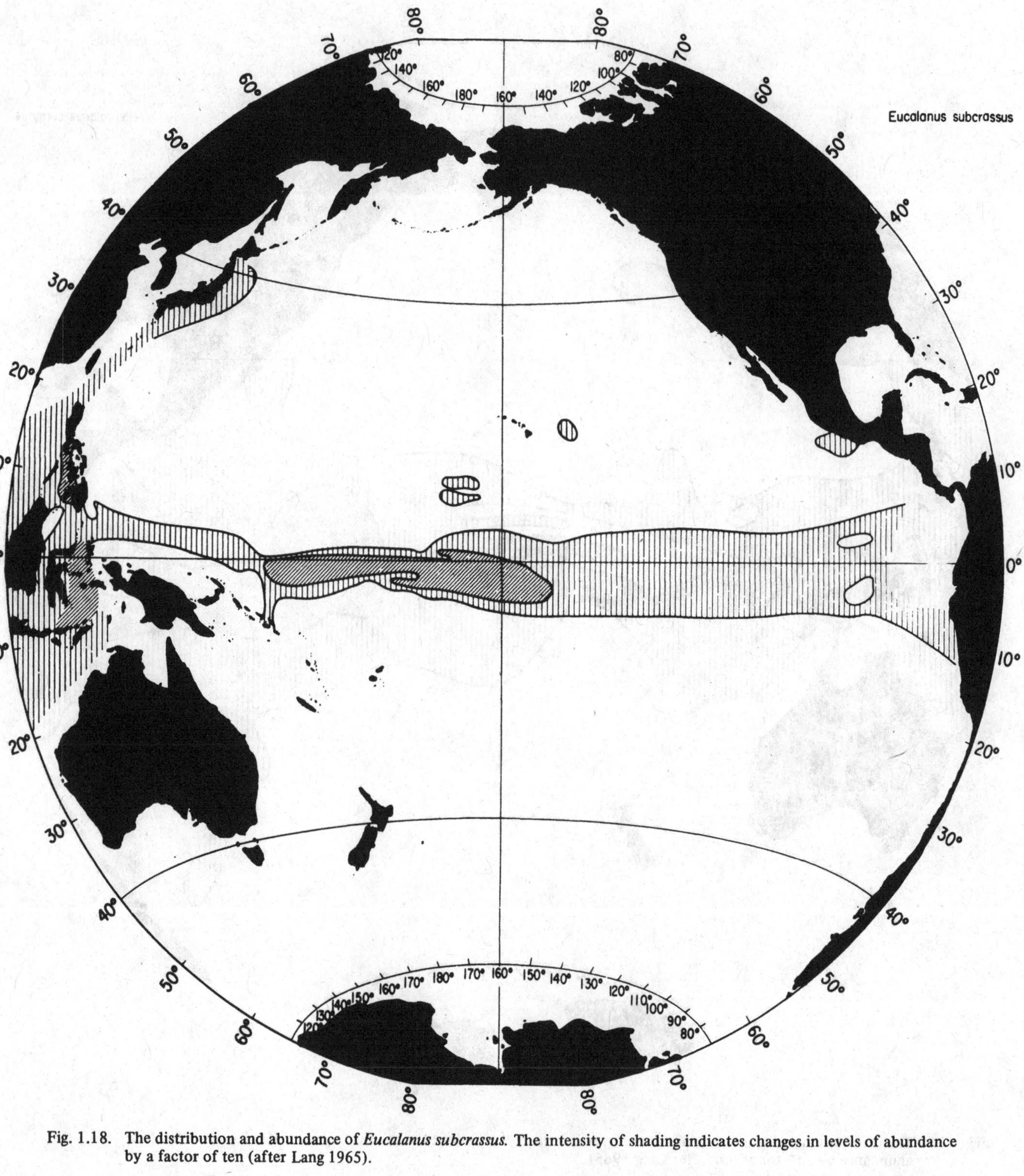
Eucalanus subcrassus
80°
70°
60°
50°
40°
30°
20°
10°
0°
120°
140°
160°
180°
100°

Fig. 1.18. The distribution and abundance of *Eucalanus subcrassus.* The intensity of shading indicates changes in levels of abundance by a factor of ten (after Lang 1965).

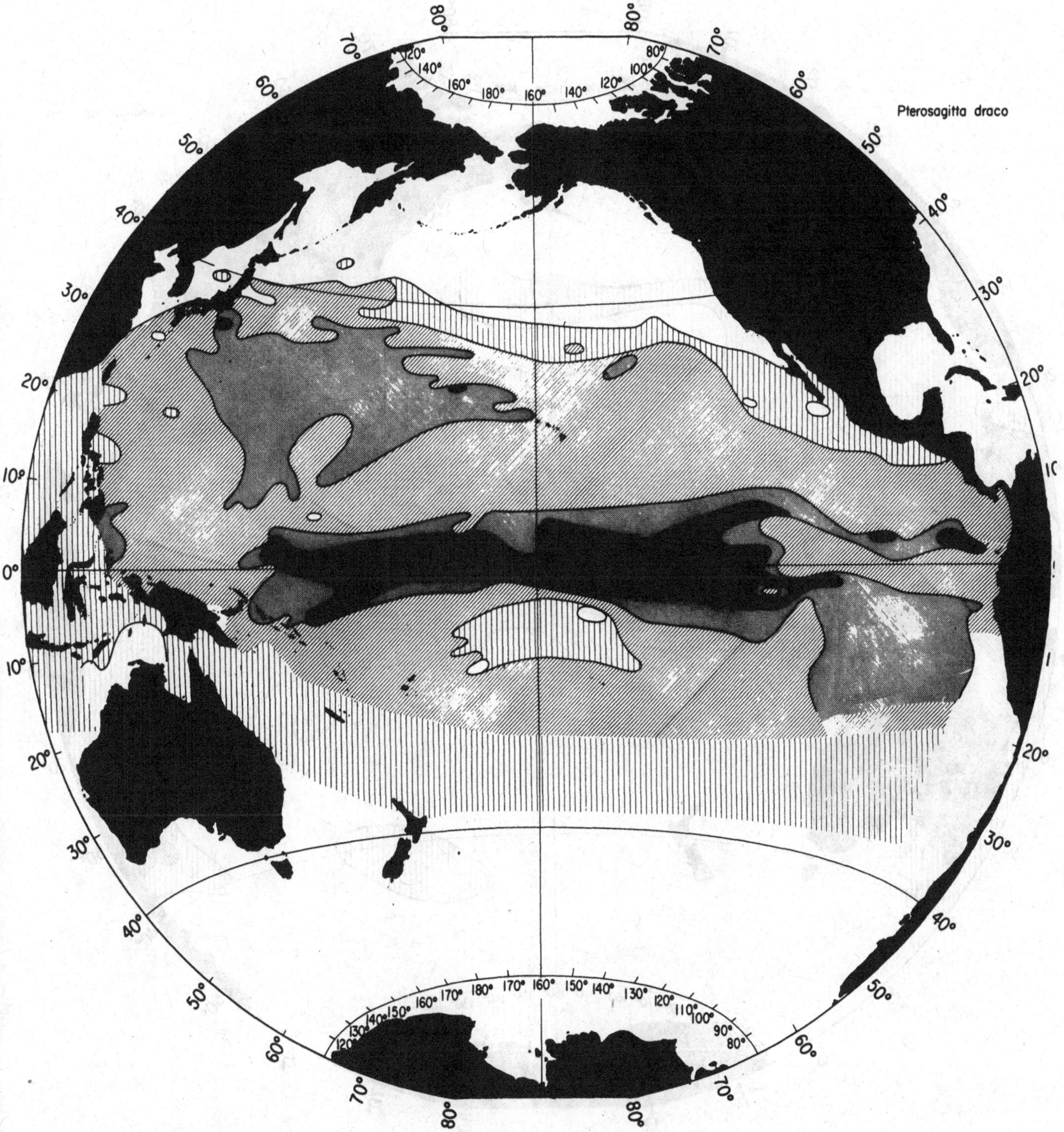
Pterosagitta draco

Fig. 1.19. The distribution and abundance of *Pterosagitta draco*. The intensity of the shading indicates changes in levels of abundance by a factor of ten (after Bieri 1959).

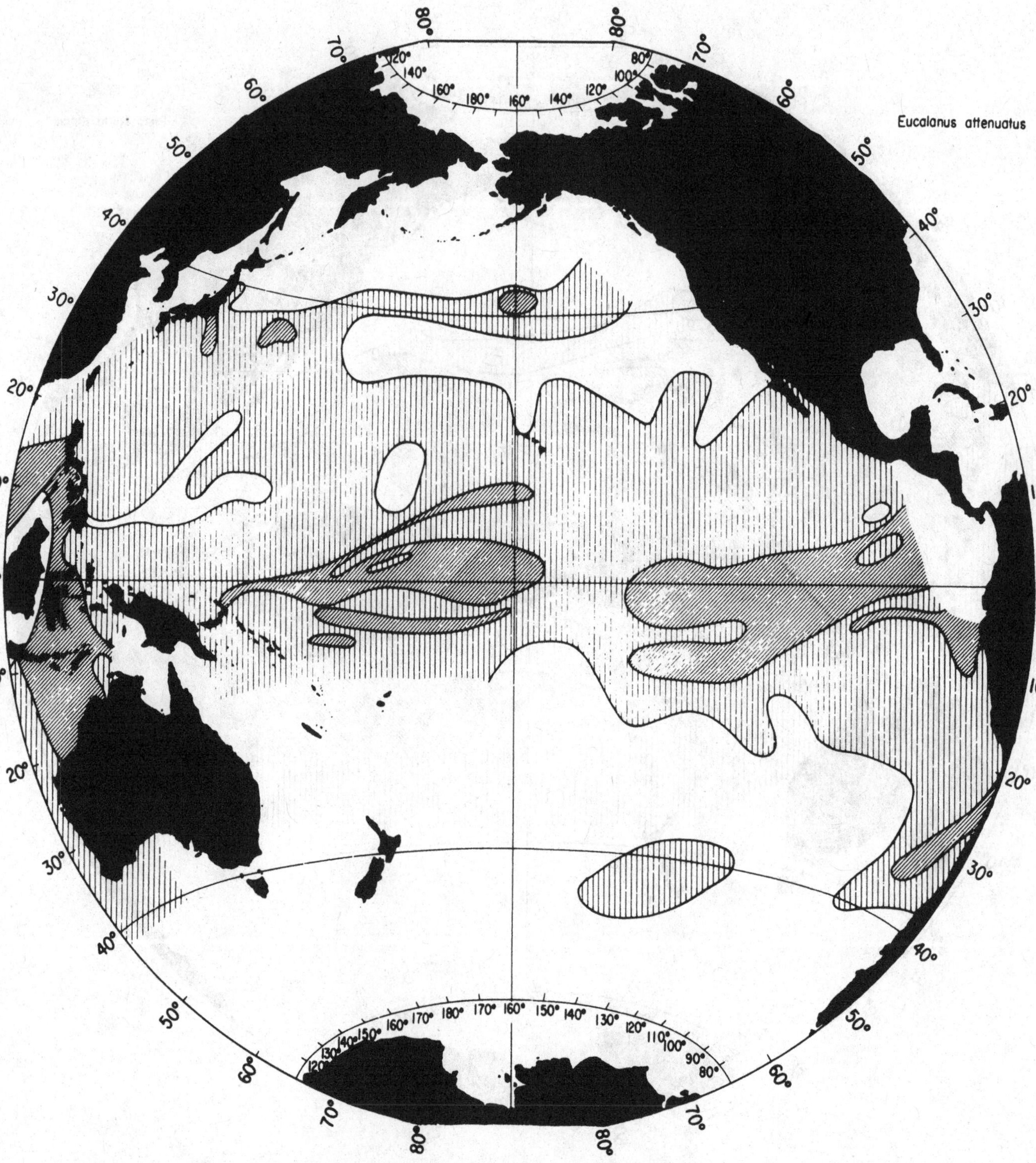
Eucalanus attenuatus

Fig. 1.20. The distribution and abundance of *Eucalanus attenuatus.* The intensity of the shading indicates changes in levels of abundance by a factor of ten (after Lang 1965).

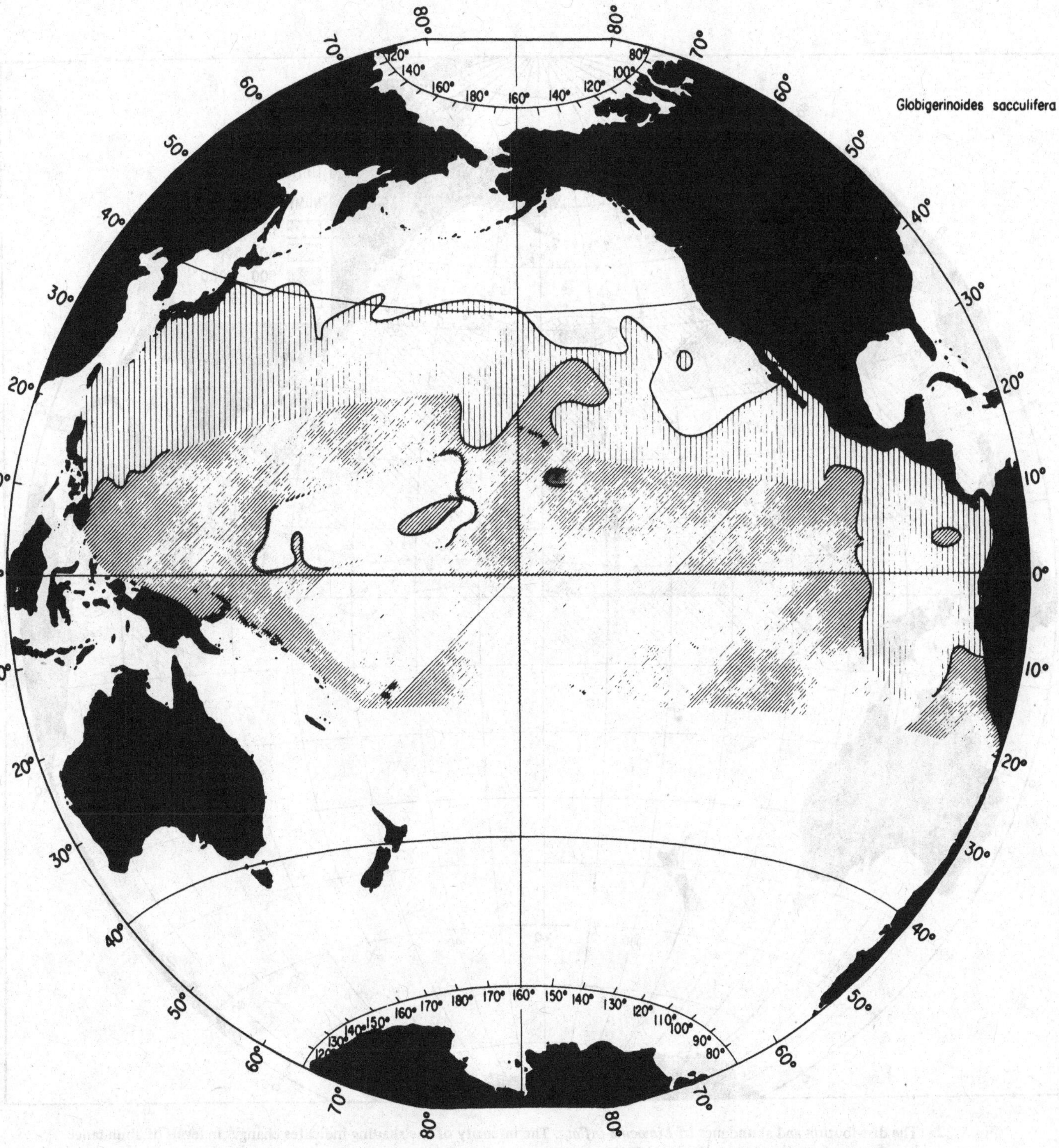
Globigerinoides sacculifera

Fig. 1.21. The distribution and abundance of *Globigerinoides sacculifera.* The intensity of the shading indicates changes in levels of abundance by a factor of ten (after Bradshaw 1959).

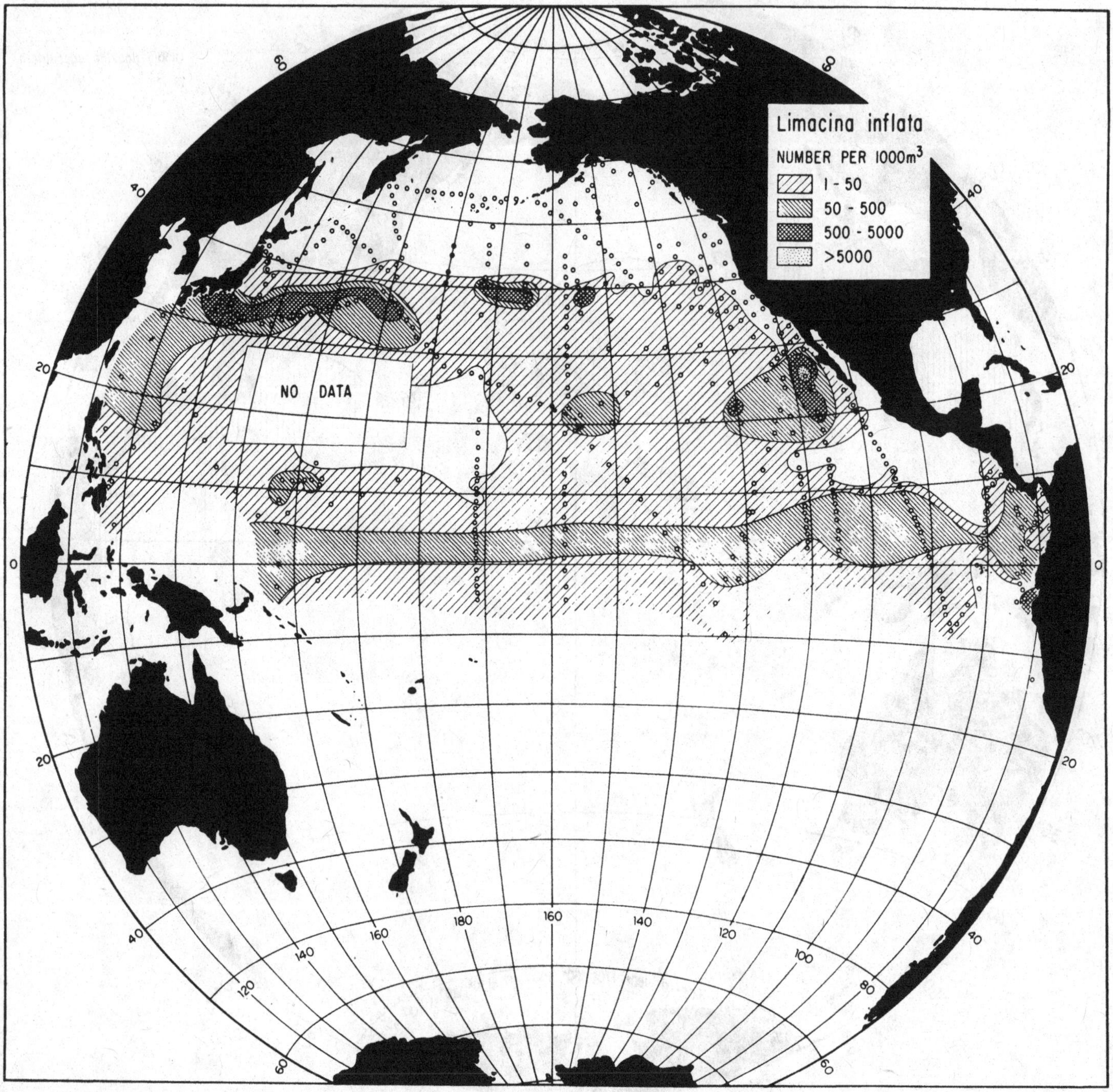

Limacina inflata
NUMBER PER 1000m^3
1 - 50
50 - 500
500 - 5000
>5000
NO DATA

Fig. 1.22. The distribution and abundance of *Limacina inflata.* The intensity of the shading indicates changes in levels of abundance by a factor of ten (McGowan 1960).

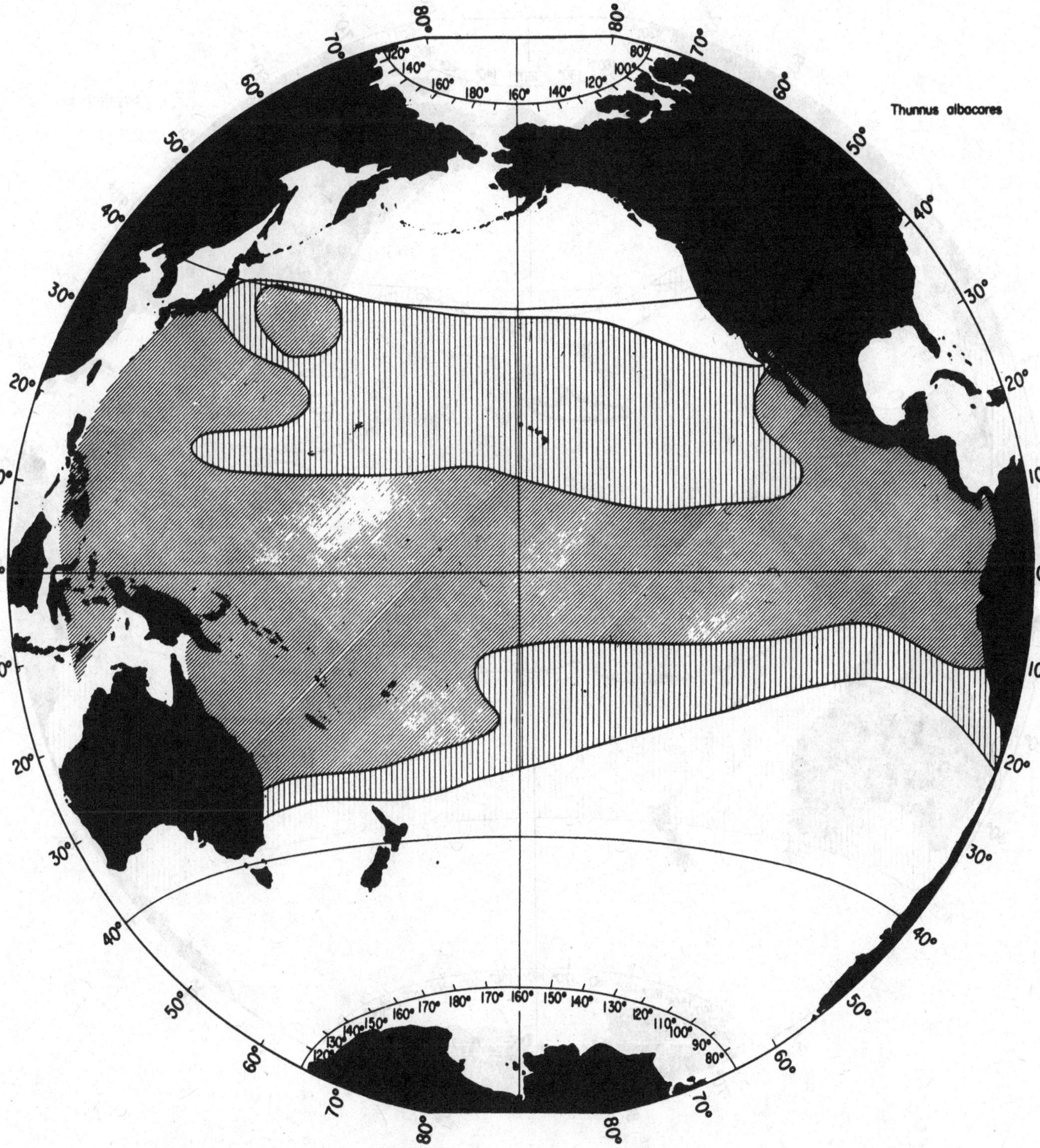
Thunnus albacares

Fig. 1.23. The distribution of *Thunnus albacares* (after Schaefer *et al.* 1963).

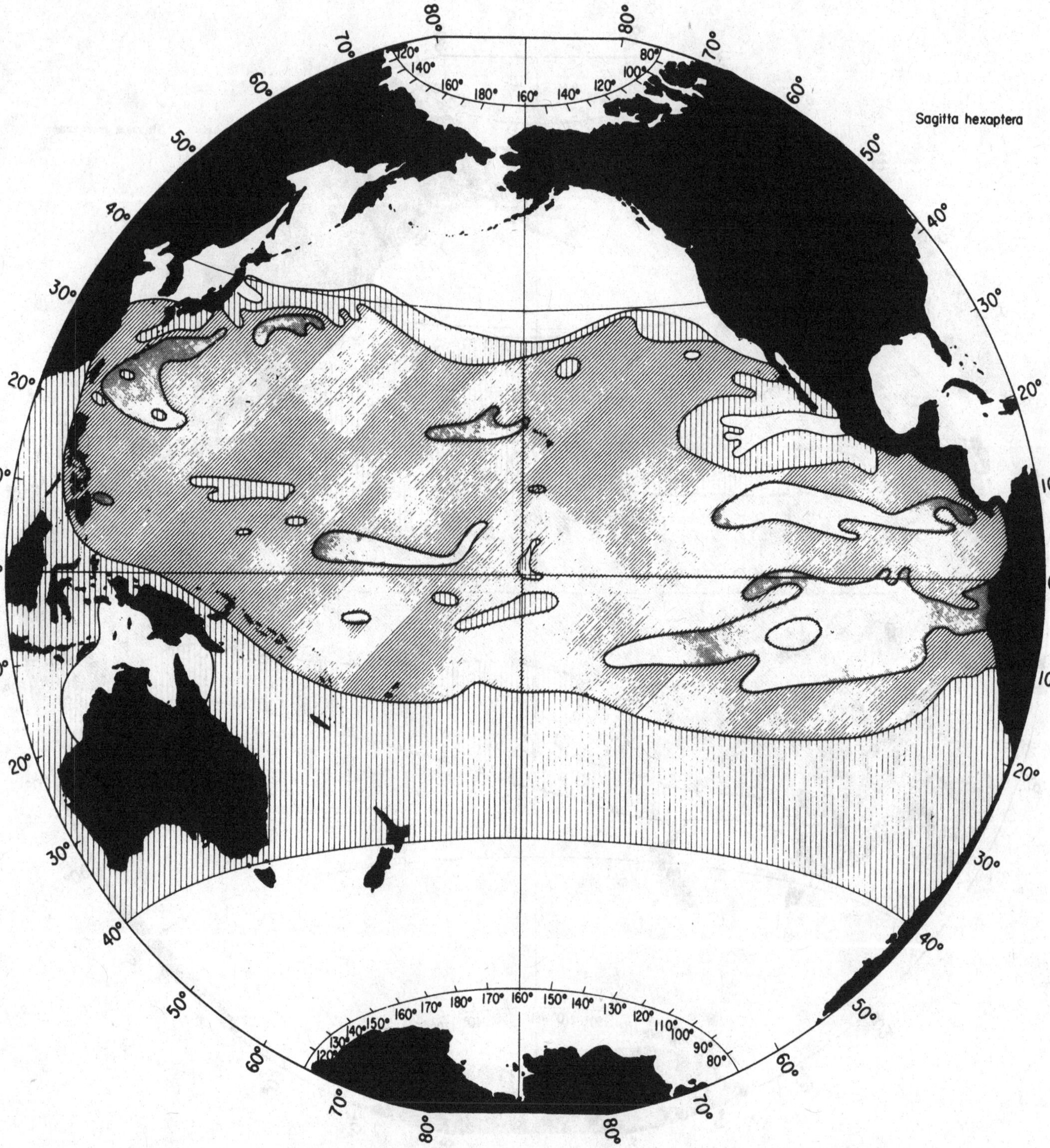
Sagitta hexaptera

Fig. 1.24. The distribution and abundance of *Sagitta hexaptera.* The intensity of the shading indicates changes in levels of abundance by a factor of ten (after Bieri 1959).

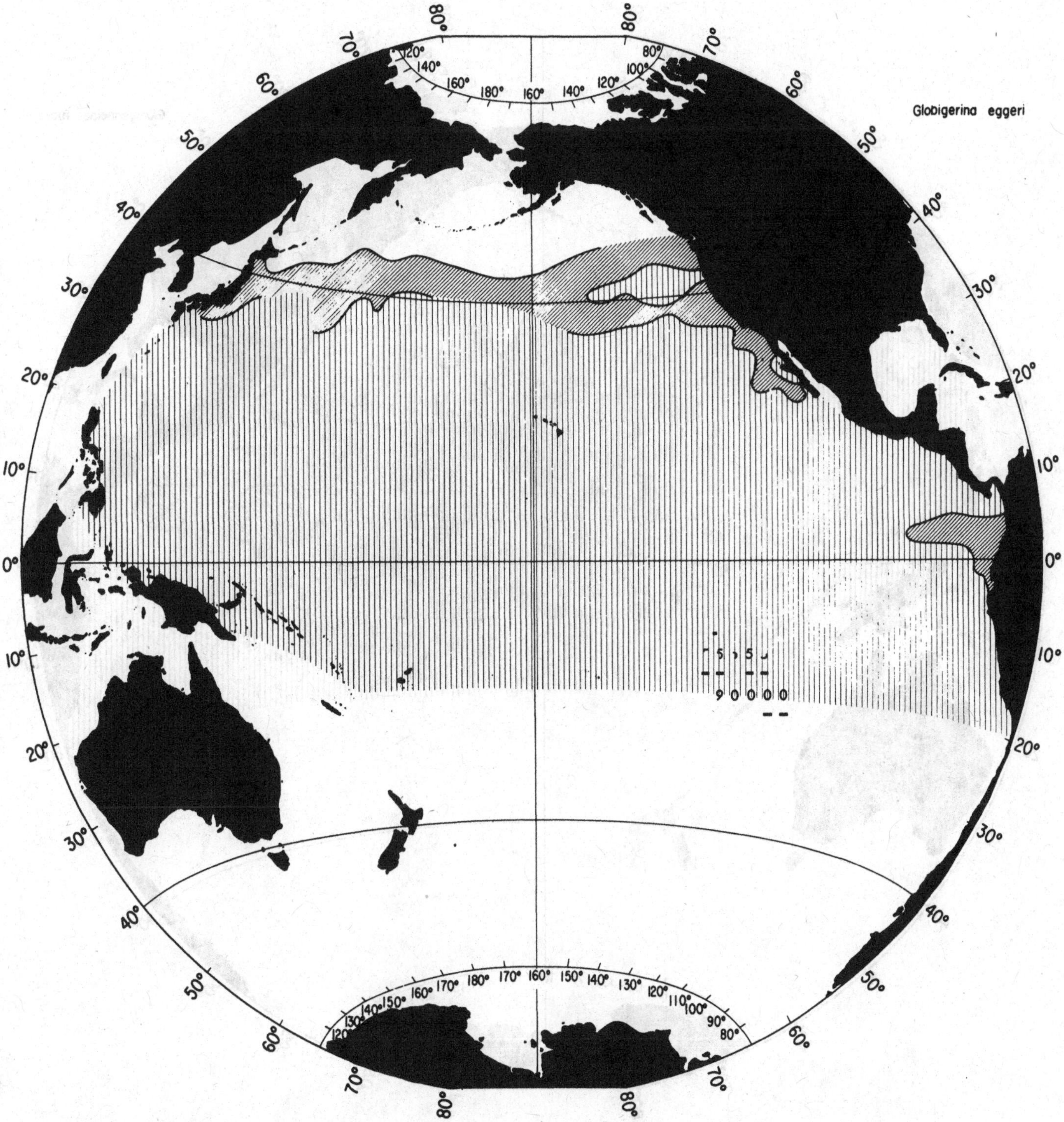
Globigerina eggeri
80°
70°
60°
50°
40°
30°
20°
10°
0°
10°
20°
30°
40°
50°
60°
70°
80°
120°
140°
160°
180°
160°
140°
120°
100°
80°
150°
170°
180°
170°
160°
150°
140°
130°
120°
110°
100°
90°
80°

Fig. 1.25. The distribution and abundance of *Globigerina eggeri*. The intensity of the shading indicates changes in levels of abundance by a factor of ten (after Bradshaw 1959).

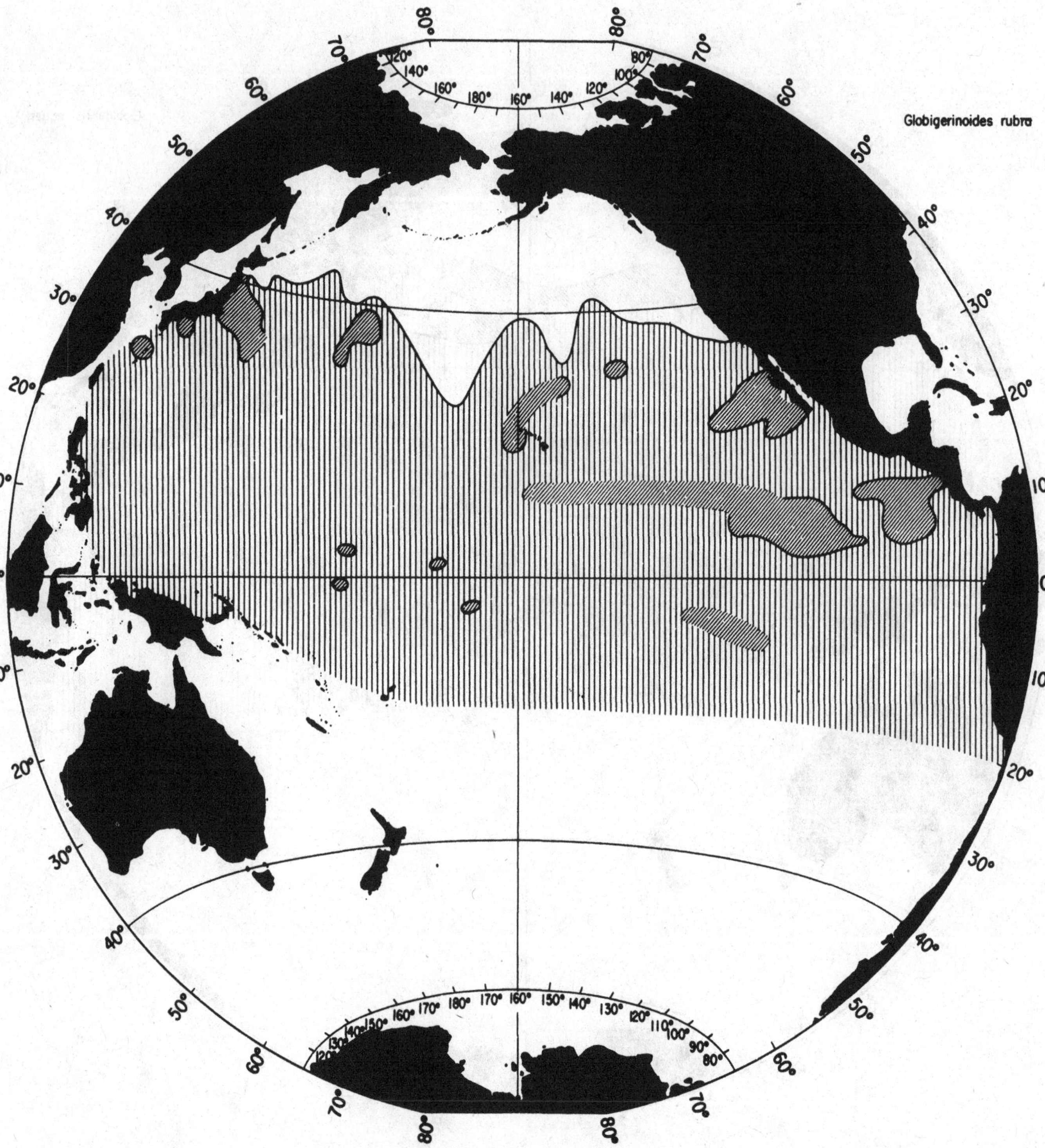
Globigerinoides rubra
80°
70°
60°
50°
40°
30°
20°
10°
0°
120°
140°
160°
180°
160°
140°
120°
100°
80°
170°
150°
130°
110°
90°

Fig. 1.26. The distribution and abundance of *Globigerinoides rubra.* The intensity of the shading indicates changes in levels of abundance by a factor of ten (after Bradshaw 1959).

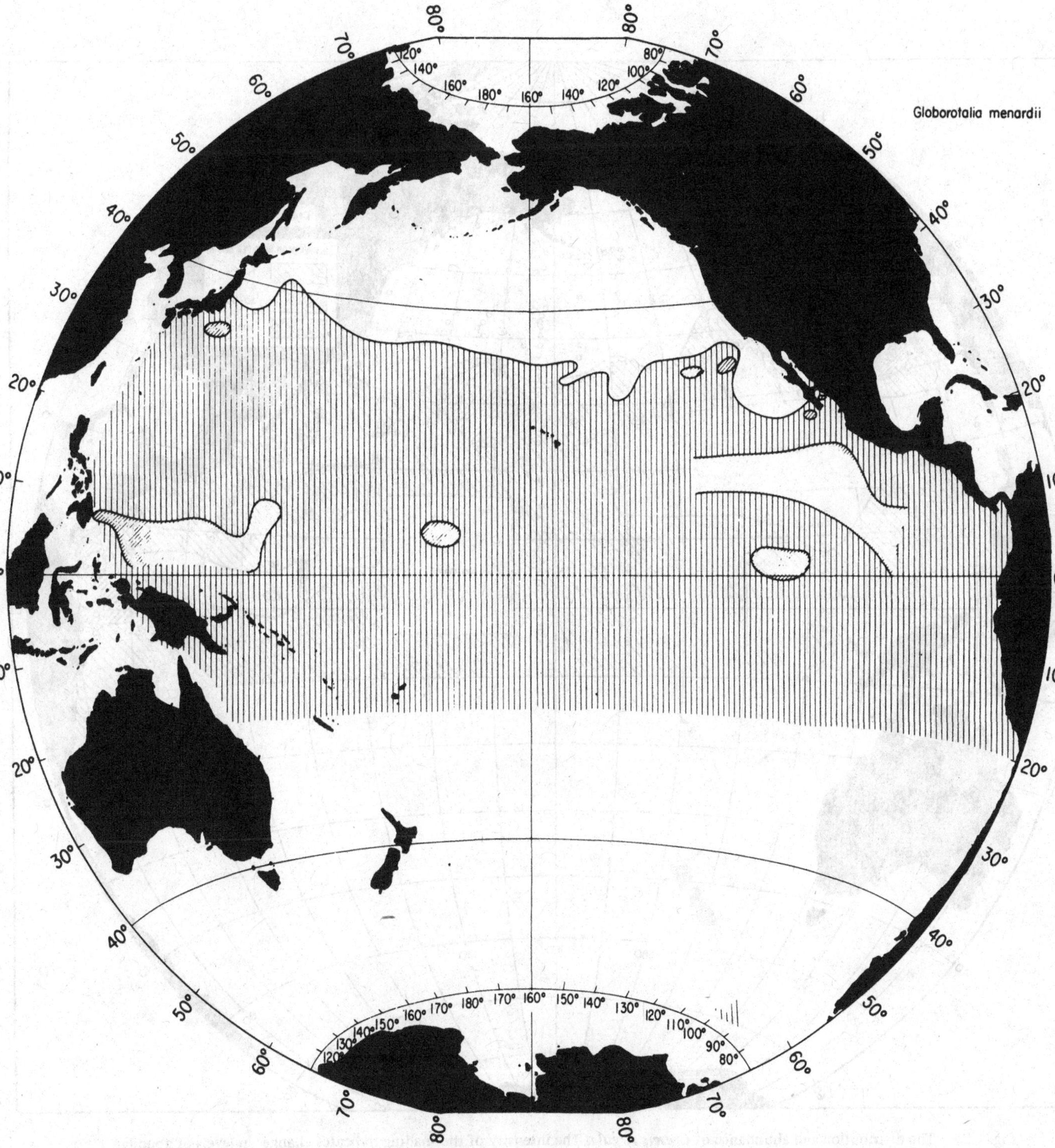
Globorotalia menardii
80°
70°
60°
50°
40°
30°
20°
10°
0°
10°
20°
30°
40°
50°
60°
70°
80°
120°
140°
160°
180°
160°
140°
120°
100°
80°
120°
130°
140°
150°
160°
170°
180°
170°
160°
150°
140°
130°
120°
110°
100°
90°
80°

Fig. 1.27. The distribution and abundance of *Globorotalia menardii.* The intensity of the shading indicates changes in levels of abundance by a factor of ten (after Bradshaw 1959).

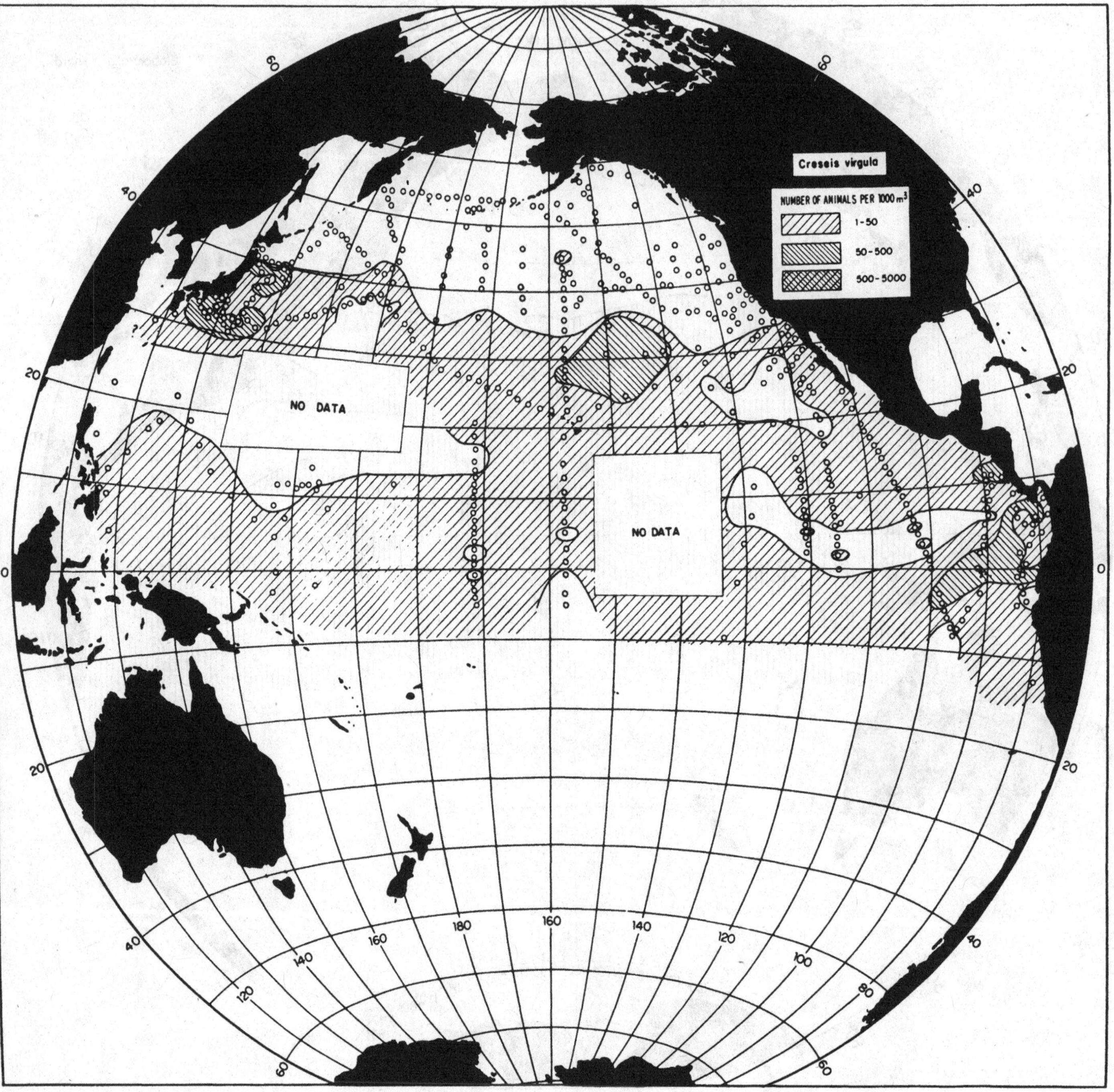

Creseis virgula
NUMBER OF ANIMALS PER 1000 m³
1-50
50-500
500-5000
NO DATA
NO DATA
60
40
20
0
20
40
60
60
40
20
0
20
40
60
120
140
160
180
160
140
120
100
80

Fig. 1.28. The distribution and abundance of *Creseis virgula*. The intensity of the shading indicates changes in levels of abundance by a factor of ten (McGowan 1960).

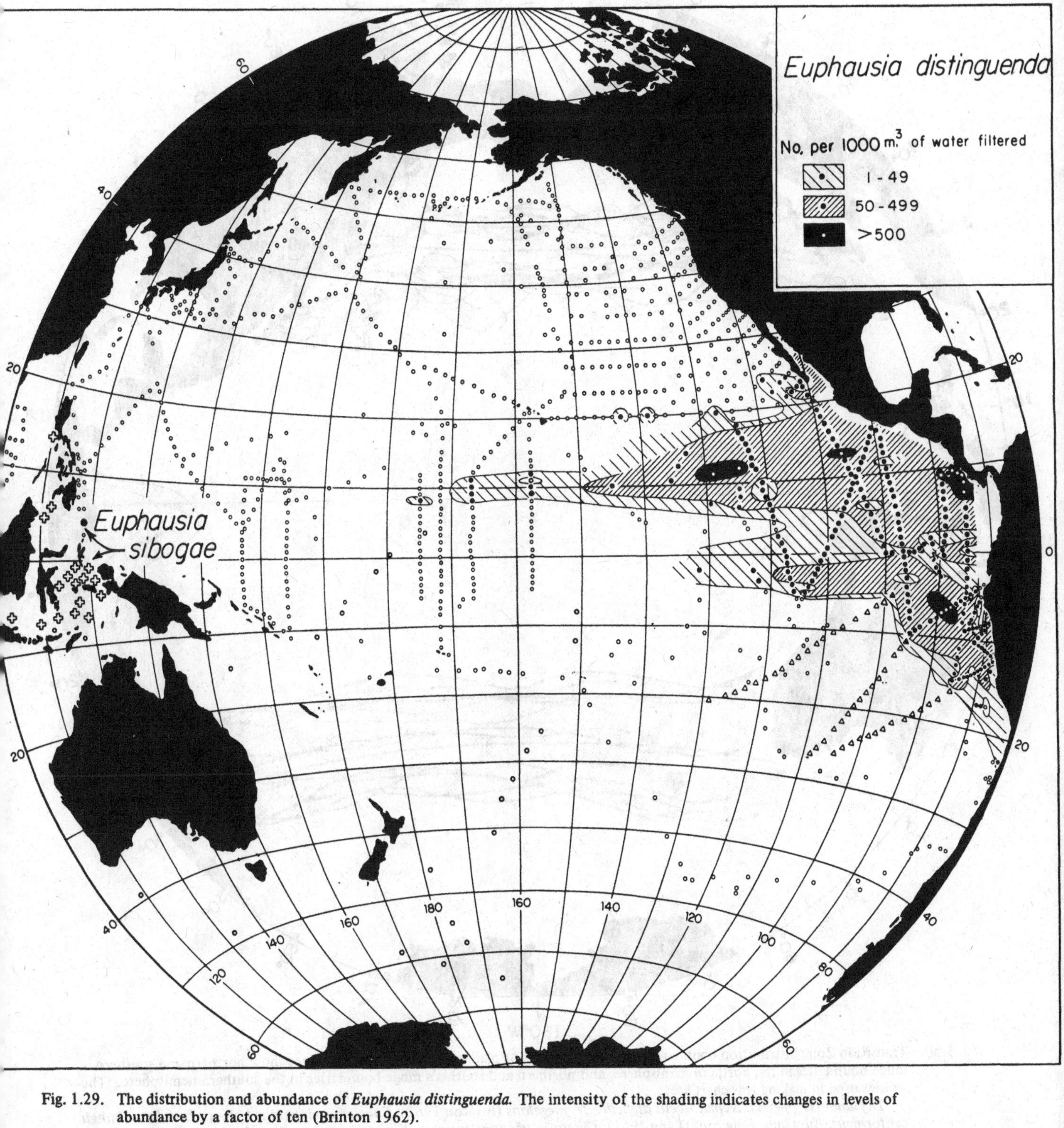

Euphausia distinguenda
No. per 1000 m.³ of water filtered
1 - 49
50 - 499
>500
Euphausia sibogae
60
40
20
0
20
40
60
180
160
140
120
100
80

Fig. 1.29. The distribution and abundance of *Euphausia distinguenda.* The intensity of the shading indicates changes in levels of abundance by a factor of ten (Brinton 1962).

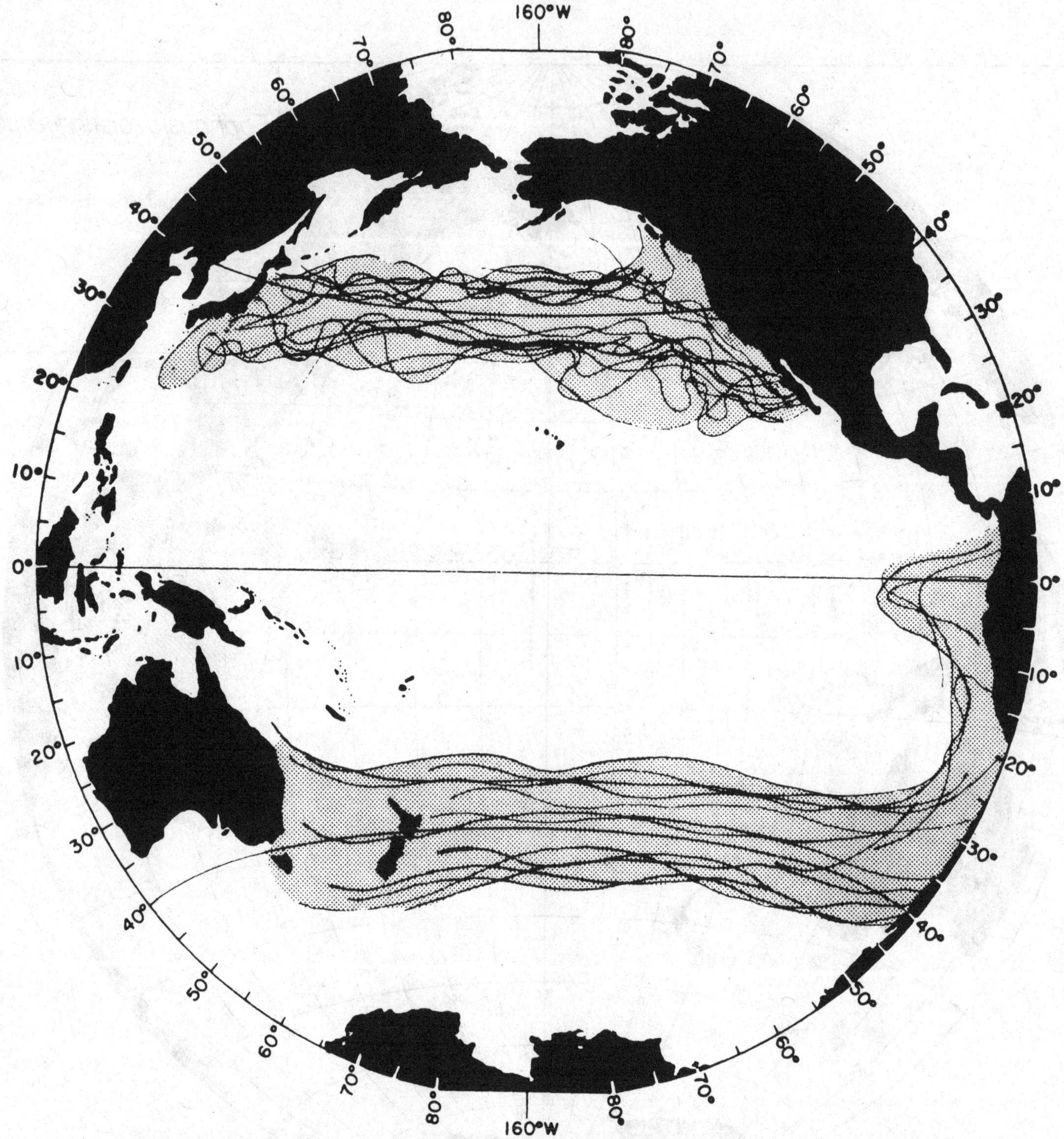
160°W
80°
70°
60°
50°
40°
30°
20°
10°
0°
10°
20°
30°
40°
50°
60°
70°
80°
160°W

Fig. 1.30. Transition Zone distribution boundaries for twelve species of planktonic organisms. The lines show northern and southern range boundaries in the northern hemisphere, and northern and southern range boundaries in the southern hemisphere. The species used in making this chart are:

Thysanoessa gregaria, Nematoscelis difficilis, N. megalops (Brinton 1962); *Eucalanus elongatus hylinus, Eucalanus bungii californicus, Eucalanus longiceps* (Lang 1965); *Ceratium tripos atlanticum* (Graham and Bronikowsky 1944); *Sagitta scrippsae* (Alvarino 1962); *Clausocalanus pergens* (Frost and Fleminger, in press); *Corolla spectabilis, Caranaria japonica* (McGowan, unpublished) *Limacina helicina* type B (McGowan 1963). Note that the width of the southern hemisphere zone is greater than that of the northern and that part of it is continuous with the Humbolt-Peru Currents.

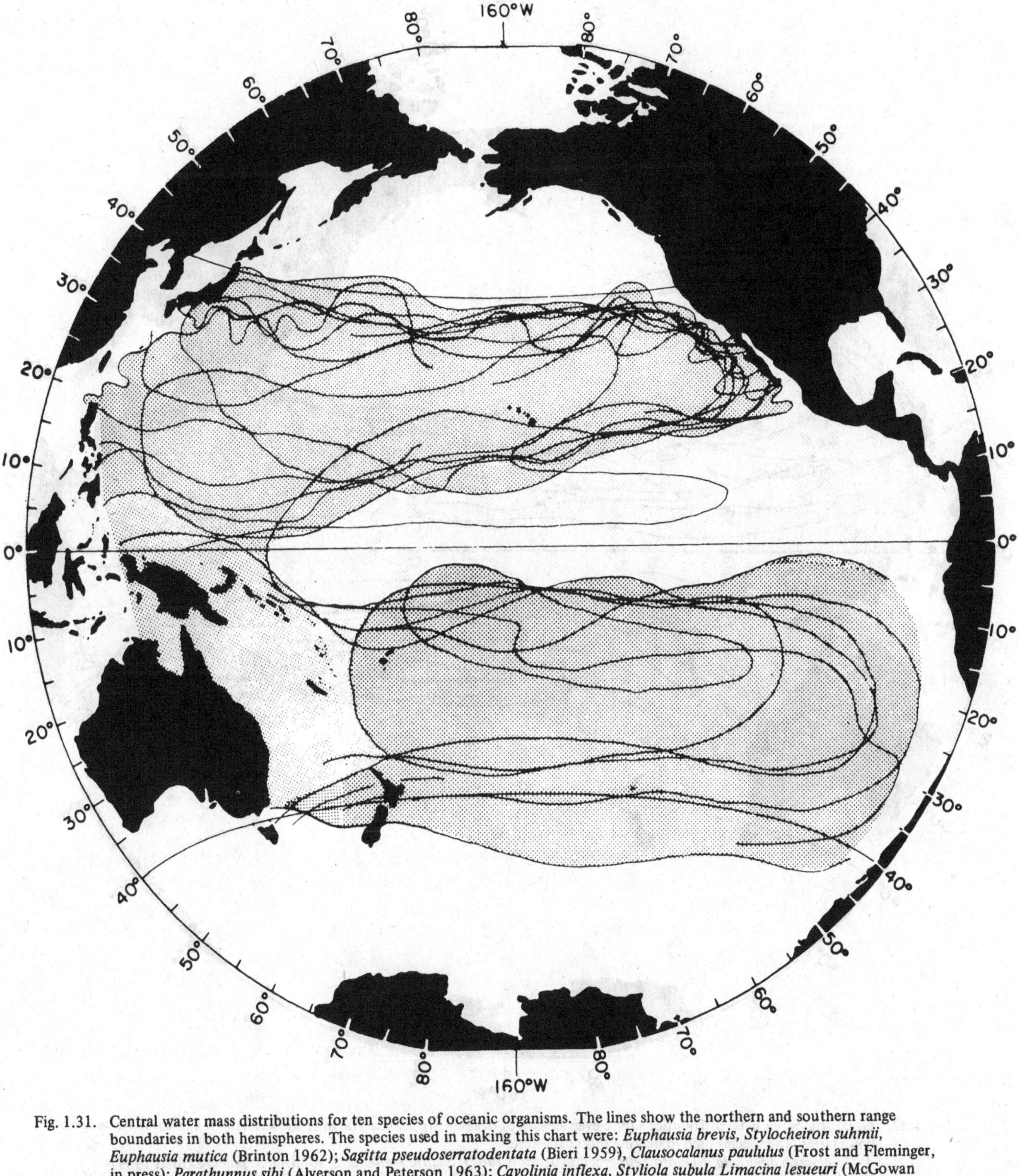
160°W
80°
70°
60°
50°
40°
30°
20°
10°
0°
10°
20°
30°
40°
50°
60°
70°
80°
160°W

Fig. 1.31. Central water mass distributions for ten species of oceanic organisms. The lines show the northern and southern range boundaries in both hemispheres. The species used in making this chart were: *Euphausia brevis, Stylocheiron suhmii, Euphausia mutica* (Brinton 1962); *Sagitta pseudoserratodentata* (Bieri 1959), *Clausocalanus paululus* (Frost and Fleminger, in press); *Parathunnus sibi* (Alverson and Peterson 1963); *Cavolinia inflexa, Styliola subula Limacina lesueuri* (McGowan 1960); *Ceratium cephalotum* (Graham and Bronikowsky 1944). Note that while the northern central patterns impinge on the coastal regions of North America, the southern central patterns do not impinge on the coastal regions of South America. Further, some central species have disjunct north-south ranges (i.e. they are amphitropical) others have continuous distributions in the equatorial far western Pacific.

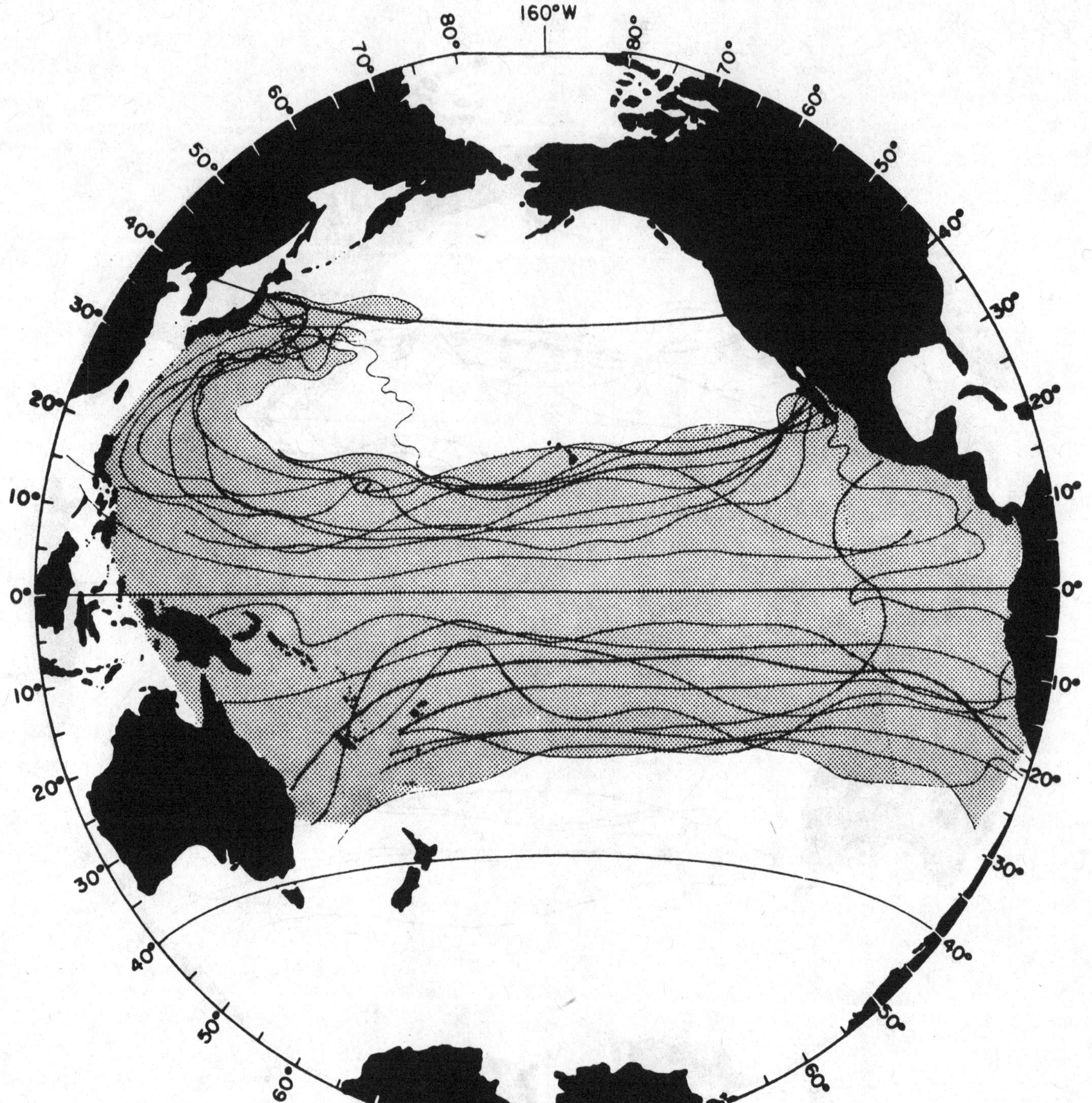
160°W
80°
70°
60°
50°
40°
30°
20°
10°
0°
10°
20°
30°
40°
50°
60°
70°
80°
160°W

Fig. 1.32. Equatorial water mass distributions for eleven species of oceanic organisms. The lines show northern and southern range boundaries. The species used in making this chart were:

Euphausia diomedia, Nemotoscelis gracilis (Brinton 1962); *Sagitta regularis, S. robusta, S. pulchra* (Bieri 1959), *Eucalanus subcrassus, Rhincalanus cornutus* (Lang 1965); *Clausocalanus ferrani* (Frost and Fleminger in press); *Ceratium lunula* (Graham and Bronikowsky 1944); *Clio* n. sp., *Cavolinia uncinata* (McGowan 1960). Note that while some equatorial species have 10° ranges symmetrical about the equator, others have approximately 46° ranges symmetrical about the equator. Many species ranges extend to the north with the western boundary current, the Kuroshio and its extension.

large areas called water masses, and have their populations conserved by the semi-enclosed gyre systems of these water masses. It is, therefore, important to consider this area in some detail. Sverdrup, Johnson and Fleming (1942, pp. 713-18, esp. p. 715 and top of p. 716), state that 'the term Subarctic water should be applied strictly only to the water to the north of 45° N' (p. 713). A transition takes place south of here to about 34° N (on a transect from the Aleutians to Hawaii). In this area of transition, some Central Water is present but the T-S curves do not generally overlap those of the Central Water. This summary and Fig. 209A in Sverdrup *et al.* (1942) appear to be the first indication of the existence of a transition zone in the North Pacific. However, the biogeographic studies done subsequently have shown that between 35° N and 45° N (Figs. 1.30 and 1.35) a distinctive fauna exists. The problem arises, why should a group of species from several of the major taxa be adapted to live in an area which cannot be characterized in the same way as the major habitats of the other species within these taxa? In other words most species live in water masses that may be well defined by T-S envelopes, but the transition zone has no unique envelope of such curves. Although it has never been claimed that the shape of T-S curves is the factor of these major habitats to which species have adapted the curves do seem to be a reliable index of these habitats. There is nothing very obvious about the Transition Zone species that would lead one to expect that their ecological requirements differ, in kind, from those of their relatives, and we are forced to the conclusion that T-S curves as an index of habitats are imperfect. However, in spite of our inability to define it by means of T-S curves, this zone does have some unique characteristics. Dodimead *et al.* (1963) have carefully reviewed the hydrography of the northern North Pacific and they point out that while definitions of various zones based on T-S curves are 'inconclusive' there are *other* features of the hydrography that may be used to define 'domains' in this area. 'The concept of a domain contains the ideas of consistent properties, structure, behaviour (flow, heating and cooling, etc.), climatic locality and continuity' (Dodimead *et al.* 1963, p. 24). Thus, the domains are considered to be unique, identifiable bodies of water whose areal extent can be mapped and whose spatial variability can be studied. Several specific domains have been identified, the properties of which in their upper zones (*ca.* 75-150 metres) are primarily influenced by seasonal heating and cooling, precipitation, evaporation and wind mixing; and in their lower zones by internal processes such as advection, diffusion, etc. These authors specifically define a Transitional Domain by a number of properties, and state that it 'lies just north of the Subarctic boundary' (Fig. 1.33). They

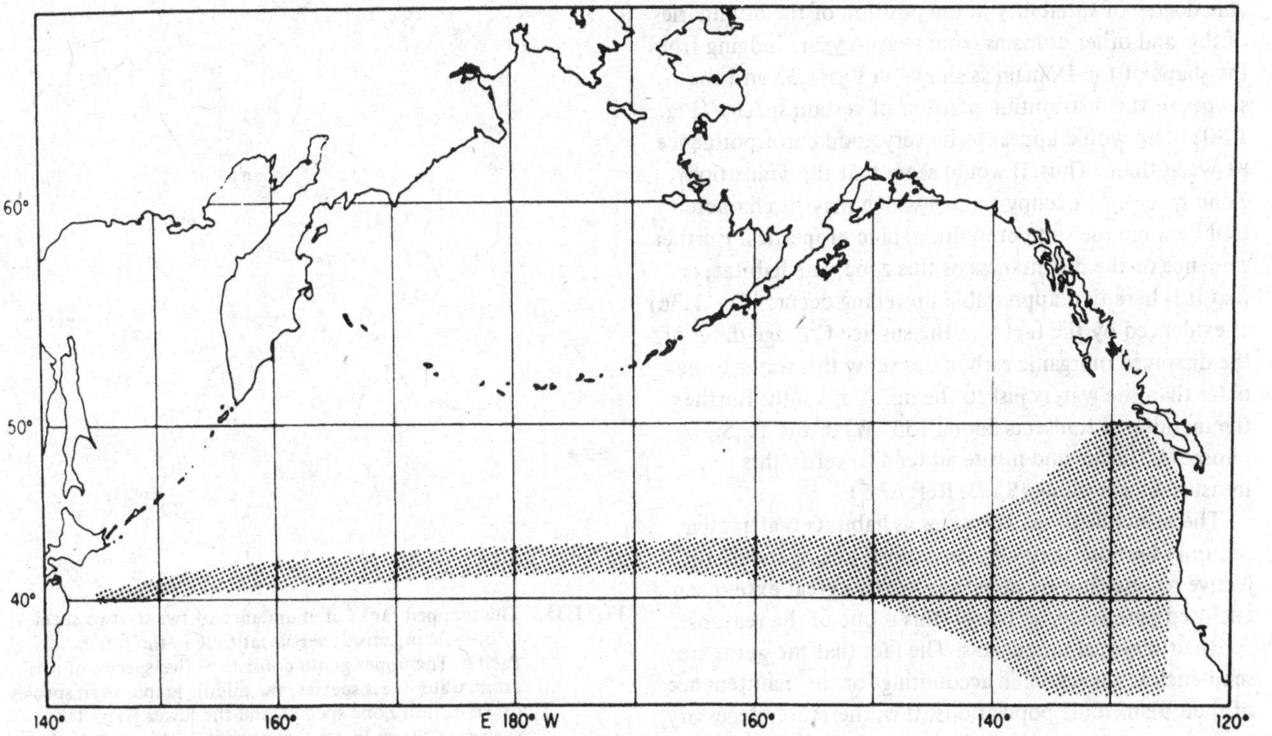
60°
50°
40°
140°
160°
E 180° W
160°
140°
120°

Fig. 1.33. A schematic diagram of the Transitional Domain of Dodimead, Favorite and Hirano (1963).

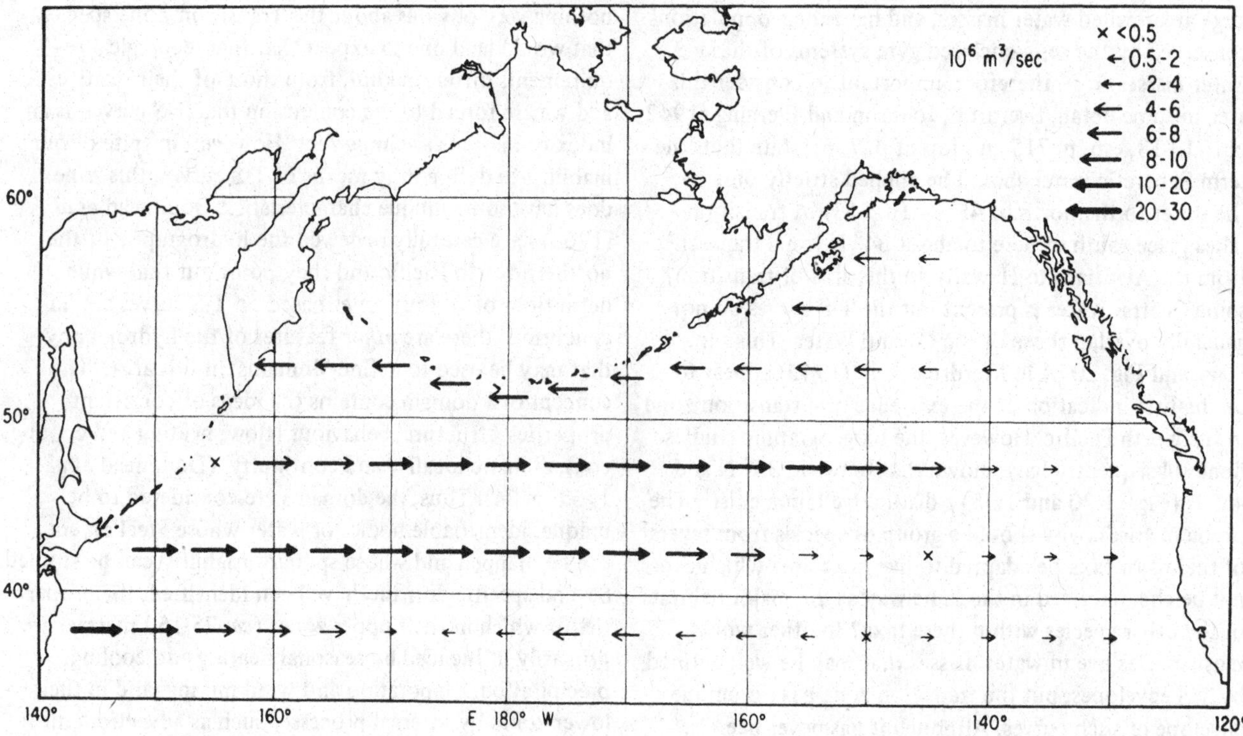

× <0.5
← 0.5-2
← 2-4
← 4-6
← 6-8
← 8-10
← 10-20
← 20-30
10⁶ m³/sec
60°
50°
40°
140°
160°
E 180° W
160°
140°
120°

Fig. 1.34. An estimate of the zonal component of the wind driven transport for July - August 1958 from Dodimead *et al.* (1963) after Fofonoff (1960).

are, however, careful to point out that there is a considerable degree of variability in the position of the boundaries of this and other domains from year to year. Judging from the shape of the Domain as shown in Fig. 1.33 and the shapes of the distribution patterns of certain species (Fig. 1.30) there would appear to be very good correspondence between them. Thus, it would seem that the Transition Zone species do occupy an area which may be characterized by a unique suite of hydrographic properties. Further evidence of the distinctness of this zone, as a habitat, is that it is here that appreciable upwelling occurs (Fig. 1.36) as evidenced by the fact that the surface C^{14} age dates of the dissolved inorganic carbonates show this water to be older than the waters just to the north or south. Further the meridional transects (along 155° W) of the T., S., phospate, silicate and nitrite all tend to verify this measure of upwelling (S.I.O. Ref. 67-5).

The principle of water masses as habitats is attractive not only because they may be defined in a relatively objective manner but because they are to a great extent semi-enclosed gyre systems. Indeed this is one of the reasons for their physical uniqueness. The fact that the gyres are semi-enclosed is useful in accounting for the maintenance of their planktonic populations. It is, therefore, necessary to demonstrate some sort of circulation mechanism within the Transitional Domain for the maintenance of its

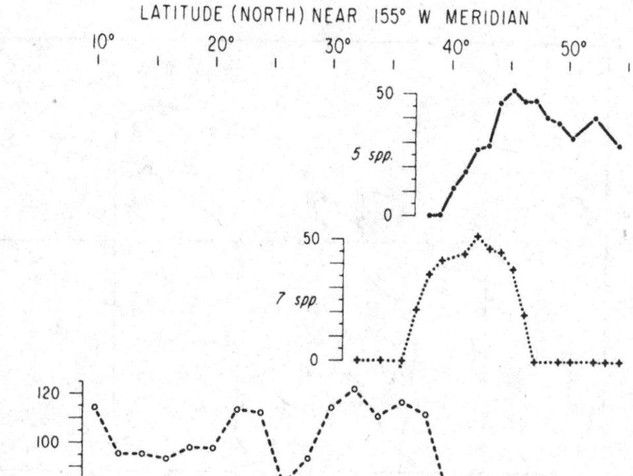

LATITUDE (NORTH) NEAR 155° W MERIDIAN
10°
20°
30°
40°
50°
50
5 spp.
0
50
7 spp.
0
120
100
10 spp.
50
10°
20°
30°
40°
50°

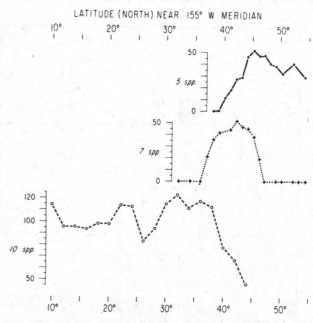

Fig. 1.35. The summed ranks of abundance of twenty-two species of oceanic organisms versus latitude in the North Pacific. The upper graph consists of five species of subarctic water-mass species, the middle graph seven species of Transition Zone species, and the lower graph ten species of Warm Water Cosmopolites. It is evident that while many of the species ranges overlap, the peaks of the summed ranks are well separated.

endemic plankton species. Previous considerations of this problem (McGowan 1963; Beklemishev 1961*b*) have used the circulation as deduced by the set and drift of vessels (Defant 1961, p. 343) or the geostrophic currents as deduced from charts of the geopotential topography (Reid 1962). There are, however, other methods such as direct current measurements, or theoretical considerations of the effect of wind stress on the sea surface. Fofonoff (1960) has presented a series of transport computations for the North Pacific based on wind stress and Dodimead *et al.* (1963, Fig. 142) have shown a chart of the zonal component of the wind-driven transport based on these. This chart is extremely interesting in terms of the present discussion because it shows in the eastern-most sector of the Transitional Domain (Fig. 1.34) very little transport ($< 0.5 \times 10^6$ cubic metres per sec) in an easterly direction, and further west at latitudes of about 38° N, there is actually *westward* transport (that is, a countercurrent!) as large as $4\text{-}6 \times 10^5$ cubic metres per sec. If this is generally the case in this area, it is relatively easy to account for the persistence of the Transition Zone (that is, Domain) fauna.

Nekton

There are relatively few accounts of the zoogeography of fish, squid, or pelagic mammals, based on quantitative sampling techniques, which cover very large portions of the oceans. However, in the Pacific, there is some reasonably good qualitative information on the species distributions of these forms. This information suggests strongly that many of the oceanic nektonic species have patterns of distribution which are similar to those of the phytoplankton and zooplankton. For example, experimental fishing in the oceanic North Pacific has shown that salmon (five species) are primarily Subarctic water-mass fish (Havanan and Tanonaka 1959). Extensive dip netting and mid-water trawling on various expeditions of the Scripps Institution of Oceanography indicate that *Gonatus fabricii*

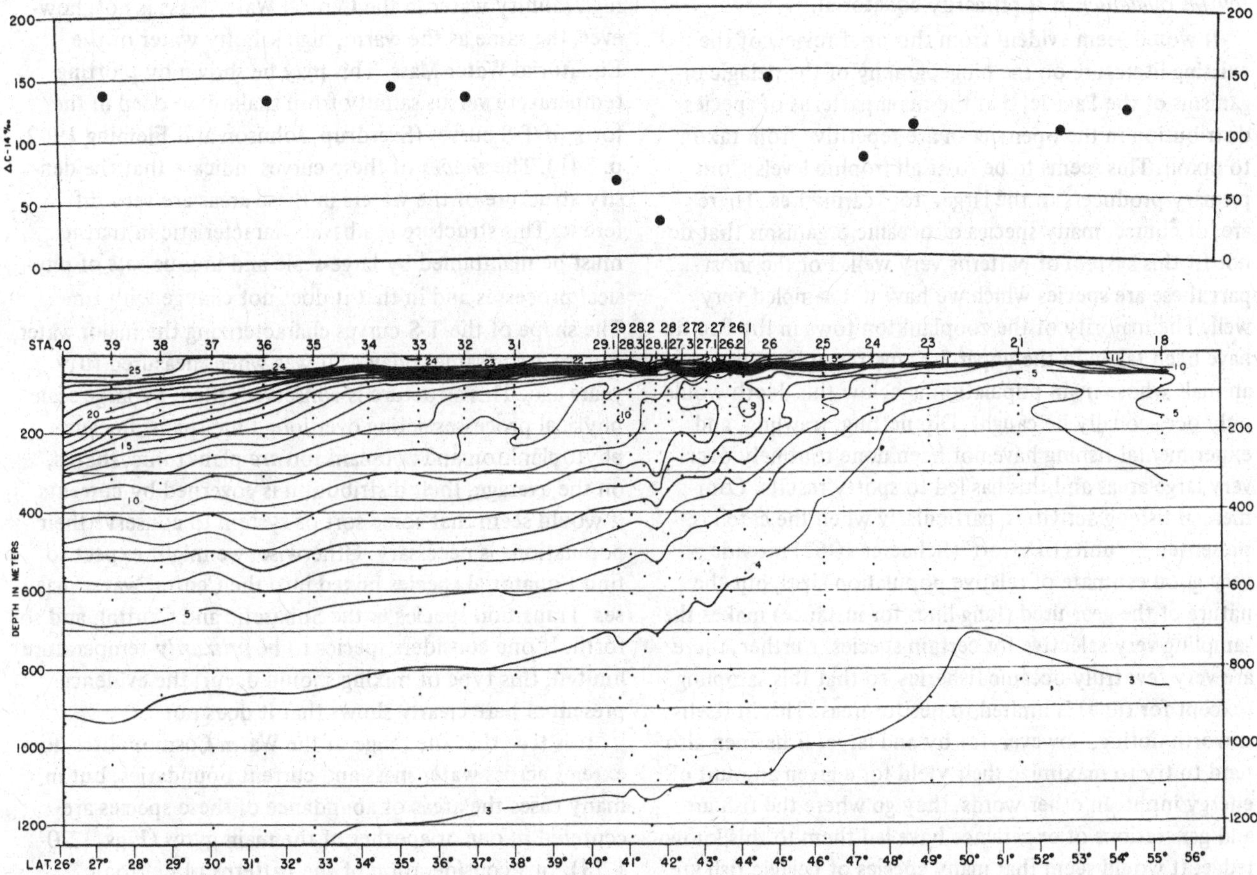
ΔC-14 ‰
DEPTH IN METERS
STA 40
LAT. 26° 27° 28° 29° 30° 31° 32° 33° 34° 35° 36° 37° 38° 39° 40° 41° 42° 43° 44° 45° 46° 47° 48° 49° 50° 51° 52° 53° 54° 55° 56°

Fig. 1.36. The meridional temperature section at 155° W longitude and the ΔC^{14} (in percent of the standard) of the surface water along the same section. Both sets of data were taken simultaneously during August-September 1964. (Anonymous, 1967, SIO Ref. 67-5).

(*Cephalopoda*) is subarctic. Data from commercial and experimental fisheries show that while the albacore, *Thunnus germo* is broadly distributed in warmer waters, it is most frequent in the Transition Zone of the open Pacific (Brock 1959). The yellow fin tuna *Thunnus albacares* occurs throughout the temperate and tropical Pacific but becomes abruptly absent at approximately 40° N. Further, the catch rate shows it to be more frequent in the Equatorial water mass (Fig. 1.23). On the other hand, the big eye tuna (*Parathunnes sibi*) which ranges over much the same area seems to be more frequent in the North Pacific Central water (Fig. 1.13). Thus, there is a sort of species replacement evident here. Skipjack tuna (*Katsuwonas pelamis*) seem to be primarily equatorial, avoiding the Central water but becoming abundant again in the Transition Zone (Brock 1959). In this respect, it is similar to *Limacina inflata* and several other zooplankton species. Unpublished studies of the distribution of ommastrephid squid indicate that they too have a similar pattern, in that *Symplectoteuthis luminosa* has been caught most frequently in the Transition Zone and *Symplectoteuthis oualaniensis* is primarily equatorial.

It would seem evident from this brief review of the existing literature on the biogeography of the pelagic organisms of the Pacific, that the main patterns of species distribution in the open ocean are repetitive from taxon to taxon. This seems to be so at all trophic levels from primary producers to the large, 'top' carnivores. There are, of course, many species of oceanic organisms that do not fit this system of patterns very well. For the most part these are species which we have not sampled very well. The majority of the zooplankton tows in the Pacific have been taken in the upper 300 metres only; thus, animals whose main population is below this depth would only occasionally be caught. Dip netting, trawling, and experimental fishing have not been done routinely over very large areas and this has led to spotty results. Commercial fishing activities, particularly when the catch is presented in 'units of effort' (Schaefer 1967) provide a very good estimate of relative population sizes, but the nature of the gear used (long-lines for instance) makes the sampling very selective for certain species. Further, there are very few truly oceanic fisheries, so that this sampling (except for tuna) is limited to neritic areas. This in itself is worth noting, however, for by and large, fishermen also tend to try to maximize their yield for a given amount of energy input. In other words, they go where the fish are, and generations of experience have led them to this knowledge. It would seem that many species of pelagic fish are neritic, as are many species of phytoplankton and zooplankton. While these too have 'patterns' which are not inconsistent with the water mass patterns (Brinton 1962) they do differ in detail and as yet we do not understand these details.

WHAT MAINTAINS THE PATTERNS?

It is obvious that no very substantial answer to this question can be had until we have much more knowledge of the ecology of the species involved. There are, however, some statements that can be made which are helpful. Essentially the processes that maintain the shape of the patterns of species distributions must be related to those that maintain the shape of the water mass patterns. The shape of these patterns is due to the major circulation of waters and the fact that certain physical processes act on the water masses. For example, it is evident that the Subarctic Water Mass is a semi-enclosed gyre where precipitation and run-off exceed evaporation and where the insolation is low so that this water is always cool. Thus, it has low salinity, cool water as opposed to the Central Water Mass, also enclosed by a large gyre system, which has warm, high salinity water. The vertical distribution of warm, high salinity water in the Central Water Mass is not, however, the same as the warm, high salinity water of the Equatorial Water Mass. This may be shown by plotting temperature versus salinity from shallow to deep in the form of T-S curves (Sverdrup, Johnson and Fleming 1942, p. 741). The *shapes* of these curves indicate that the density structure of the waters in these areas are very different. This structure is a basic characteristic in that it must be maintained by large-scale and unique sets of physical processes and in that it does not change with time. The shape of the T-S curves characterizing the major water masses is the same today as it was when measured fifty years ago. This structure is conserved, then, by large scale physical processes acting over long time spans. Because phytoplankton and zooplankton *are* planktonic, that is, on the average, their distribution is governed by currents, it would seem that some sort of system to conserve their populations is necessary. Otherwise, we might expect to find Equatorial species mixed into the Central water masses, Transition species in the Subarctic and Central, and so forth. If one considers species to be *primarily* temperature limited, this type of mixing should occur; the evidence presented here clearly shows that it does not.

It is true that the range of the Warm Cosmopolites does extend across water mass and current boundaries, but in many cases the areas of abundance of these species are centered in one or another of the main gyres (Figs. 1.20-1.23). In a consideration of the patterns of nektonic species it was indicated that many of these are broadly ranging but seem to have their centres of abundance in particular areas, and that these areas agree with the water mass-gyre concept. It cannot really be argued that the populations of these large, powerful animals are conserved by the comparatively sluggish currents of the gyres, there must be ecological reasons why this is so. It is probable that these species are interacting, through intermediate species, with those lower in the food chain whose populations *are* conserved in this manner. This would imply a degree of behavioural specialization, perhaps in the selection of food. For instance, a *Thunnus albacares-Symplectoteuthis oualanensis-Lampanyctus omostigma-L. parvicauda*-zooplankton linkage is possible in the Equatorial Water Mass. Similar interaction linkages between species adapted to one another's presence may well go far to explain the numerous other cases of warm cosmopolites whose population centres are more circumscribed than their overall ranges.

While the water mass-gyre system probably acts as a conserver of populations there is a large amount of 'leakage' along the edges of water masses and particularly in the eastern (i.e. California Current) and western (i.e.

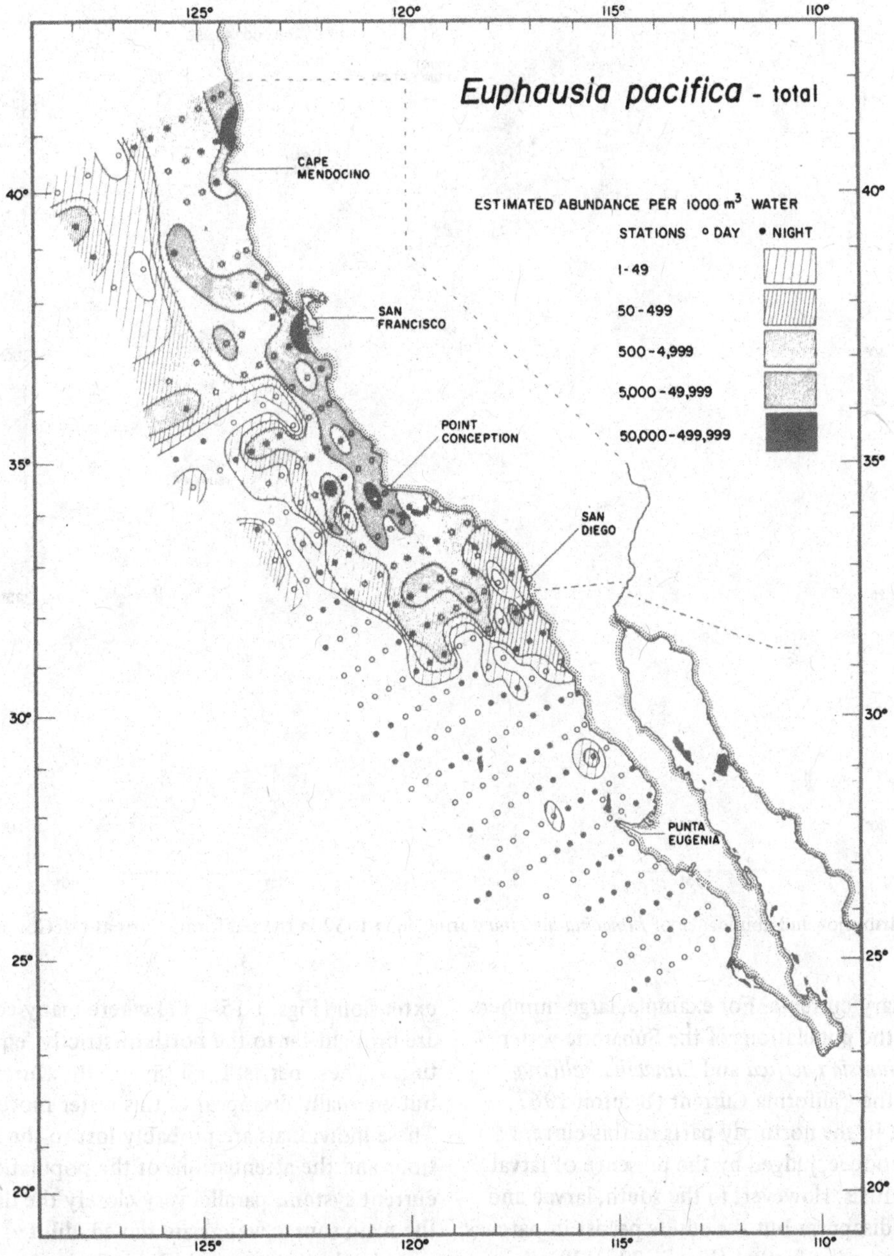
125°
120°
115°
110°
Euphausia pacifica - total
CAPE MENDOCINO
SAN FRANCISCO
POINT CONCEPTION
SAN DIEGO
PUNTA EUGENIA
ESTIMATED ABUNDANCE PER 1000 m³ WATER
STATIONS ○ DAY ● NIGHT
1-49
50-499
500-4,999
5,000-49,999
50,000-499,999
40°
35°
30°
25°
20°

Fig. 1.37. The distribution and abundance of *Euphausia pacifica* during June-July 1958 in the California Current (Brinton 1967).

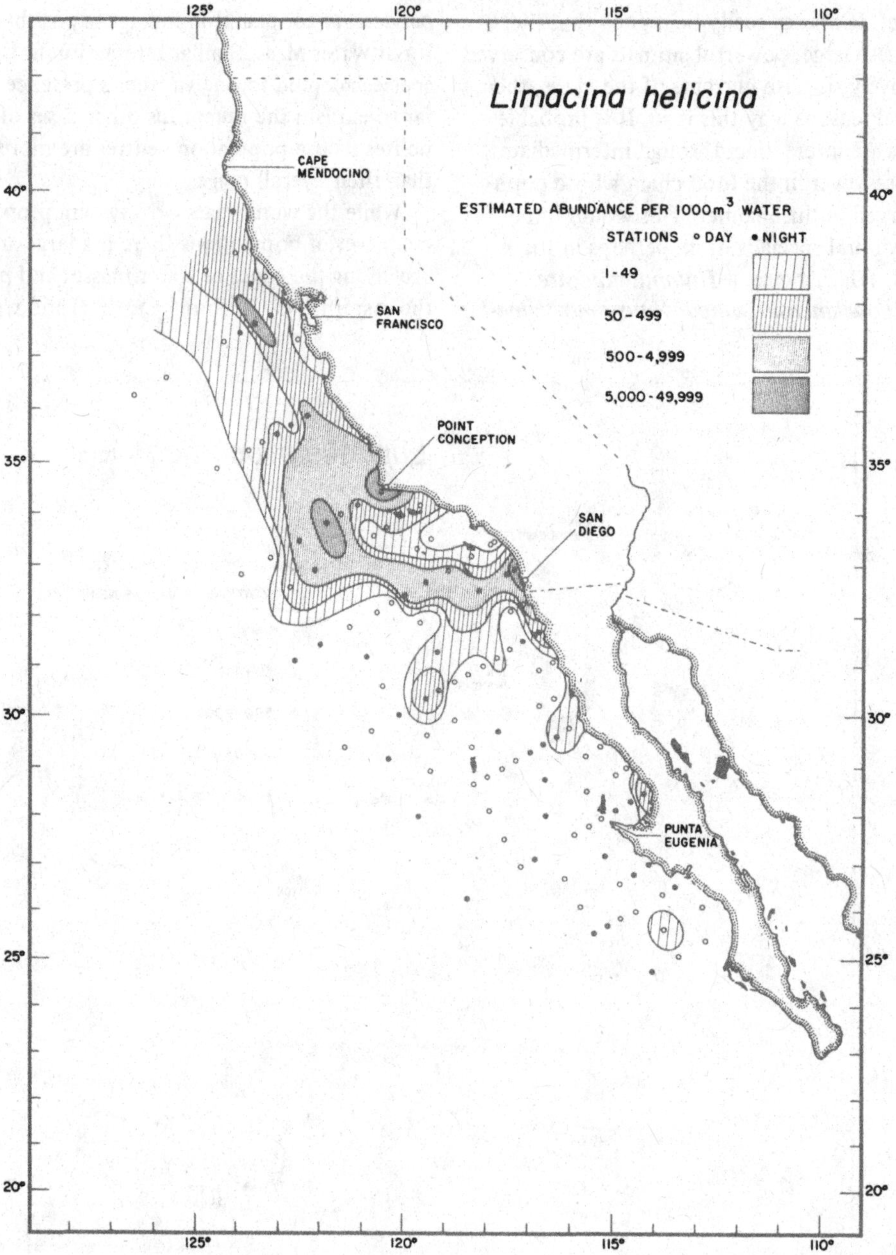
Limacina helicina
ESTIMATED ABUNDANCE PER 1000m³ WATER
STATIONS ○ DAY ● NIGHT
1-49
50-499
500-4,999
5,000-49,999
CAPE MENDOCINO
SAN FRANCISCO
POINT CONCEPTION
SAN DIEGO
PUNTA EUGENIA
125°
120°
115°
110°
40°
35°
30°
25°
20°

Fig. 1.38. The distribution and abundance of *Limacina helicina* during June 1952 in the California Current (McGowan 1967).

Kuroshio) boundary currents. For example, large numbers of individuals of the populations of the Subarctic water mass species *Euphausia pacifica* and *Limacina helicina* extend well into the California Current (Brinton 1967; McGowan 1967). In the northerly parts of this current these species reproduce, judged by the presence of larval and young individuals. However, to the south, larvae and juveniles tend to disappear but the adults persist in patches for several hundred miles further (Figs. 1.37, 1.38). A similar situation appears to occur in the Kuroshio and its extension (Figs. 1.15-1.17) where many equatorial species are brought far to the north of strictly 'equatorial' latitudes. These persist for a time in the Kuroshio extension. but gradually disappear as this water moves eastward. These individuals are probably lost to the main populations and the attenuations of the populations in these two current systems parallel very closely the degree to which the main source waters are mixed, diluted and gradually completely transformed (Reid, Roden and Wyllie 1958; Sverdrup, Johnson and Fleming 1942). Thus, it would

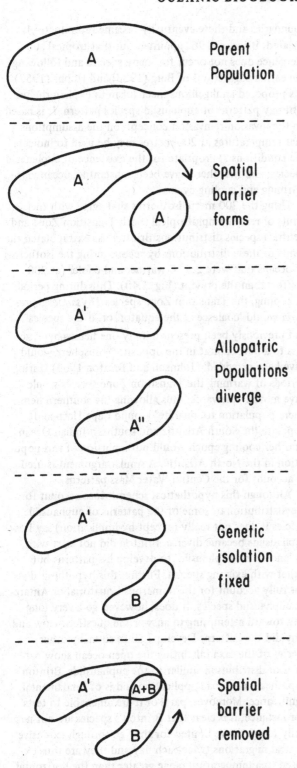

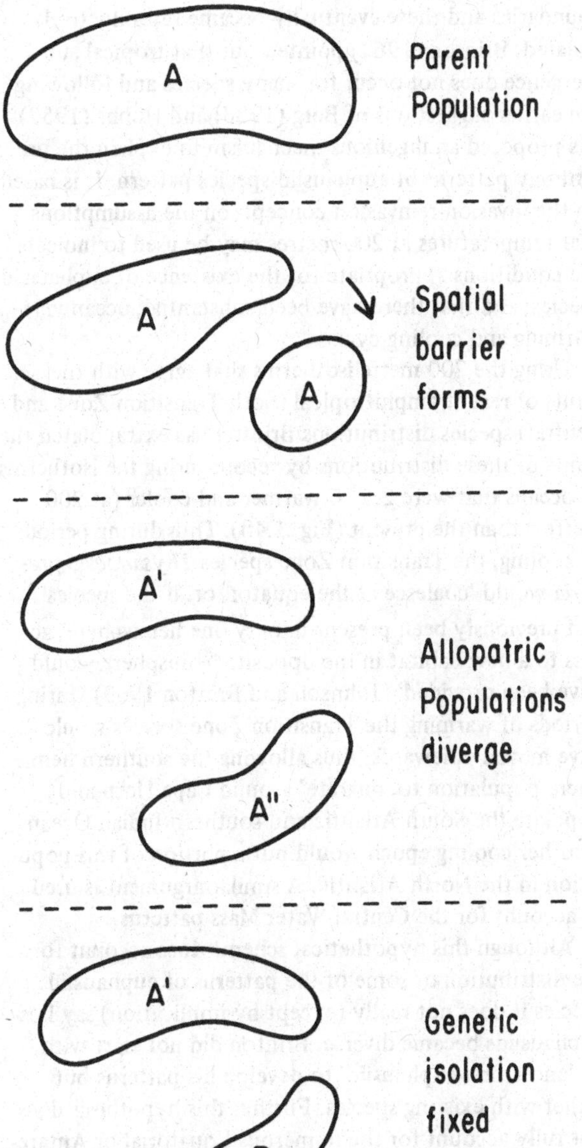
A
Parent
Population
A
Spatial
barrier
forms
A
A'
Allopatric
Populations
diverge
A''
A'
Genetic
isolation
fixed
B
A'
A+B
B
Spatial
barrier
removed

Fig. 1.39. An example of how allopatric speciation might occur. A portion of a parent population A, splits off and becomes isolated spatially. The two isolated populations diverge genetically, at first to the subspecies level A'; A'' and eventually to separate species A'; B. Since these may now be reproductively isolated, subsequent mixing of the two populations when the spatial barrier is removed results in sympatric distributions of closely related species.

appear that opportunities exist for species from one water mass to 'invade' another, for instance, the equatorial species should be able to colonize the Central Water Mass since there would appear to be no real physical or temperature barriers to prevent it. This has not happened and perhaps the reason may be that suggested by Elton (1958): stable, climax communities are difficult to invade.

HOW AND WHY DID THE PATTERNS DEVELOP?

Like many other questions about Nature, this one cannot be answered by direct observation or experimentation. The best that can be done, in cases like these, is to assemble what seem to be the most pertinent facts, suggest hypotheses based on these, test the validity of each, discarding some and accepting others. To the extent that some types of hypotheses *are* testable they are useful; non-testable hypotheses are no help at all.

It is obvious that there are two important principles involved in the approach to this problem. The first involves our concepts of the mechanisms of speciation, and we are not concerned here with how one particular line has changed through time, but rather how two, or more, new species are formed from a single parent population. Because of our concepts of how this may have happened, the second principle involves our ideas of what the ancient circulation systems and water-mass structures were like. While these two phenomena are tied together and cannot have operated independently, it is convenient to discuss them separately.

Due primarily to the works of Mayr (1942, 1963) the orthodox concept of speciation states that most, if not all, speciation of sexually reproducing animals occurred allopatrically. This means that in order for a new, genetically distinct, reproductively isolated, population (a *species*) to arise, there must have been a period when segments of the parent population were spatially isolated from one another. This isolation must have been complete enough to severely reduce, or completely eliminate any gene flow between the two populations. Further, the isolation must have been maintained long enough for a large number of genetic changes to have built up in each population so that later (in geologic time), should they again become contiguous or overlapping, there would be no possibility of exchange of genetic material (Fig. 1.39). Mayr cites many cases that seem to support this viewpoint. A less popular concept is that of sympatric speciation. This states, merely, that a new species can arise, under certain conditions, from a parent population without geographical, physical isolation or even very much small-scale spatial isolation. I have cited (McGowan 1963) some of the evidence for this, and Ehrlich and Holm (1963), Ford (1964, pp. 68-83) and Maynard Smith (1966) have discussed further possibilities. It would seem that this is among the class of natural phenomena that are simply not open to direct measurement. Although the weight of the evidence seems to be in favour of allopatric speciation, it is difficult to visualize how this 'works' in the case of oceanic zooplankton. What is required here, is that a portion of a water mass becomes 'detached' somehow and with its contained population drifts away from the parent water mass. It must, however, maintain its properties (i.e. not mix) for a long enough period of time so that the populations in it and adapted to it, may evolve toward adaptation to new conditions brought about by eventual mixing. This body of water must be large enough to contain a sizeable population of zooplankton, for the probabilities of mutations occurring which are adaptive are much greater in large populations than in small ones. Although eddies do break away from water masses along their periphery, particularly in western and eastern boundary currents (Wooster and Reid 1963; Takenouti 1960) these are not particularly large nor do they persist unmixed for very long. As I have already pointed out in the case of the Kuroshio extension and the California Current, the contained populations are rather quickly extinguished.

There are, however, certain spatial barriers extant today, as Brinton (1962) and Johnson and Brinton (1963) have pointed out, and these do act as isolating mechanisms. The continents separate equatorial Indo-Pacific species populations from those of the Atlantic. Also the northern Transition Zone and Central species populations of the Pacific are isolated from those of the Atlantic. Within the Pacific the amphitropical North and South Central species and the North and South Transition Zone species are geographically isolated, as are the true Bipolar species. Thus large, allopatric populations of the *same* species exist in the same ocean. The question is: how did these come about?

Hedgpeth (1957, pp. 377-80) has discussed the several hypotheses that have been suggested to account for bipolarity and these would apply to amphitropicality as well. These are: 1. The relict theory, in which the existing, separate populations are considered remnants of a former continuous distribution; 2. The theory of tropical submergence, which states that the disjunct populations are not really separate at all but continuous in deeper water, following temperature; and 3. The migration theory, in which equatorial species gave rise to acclimatized and eventually adapted subpopulations at their high latitude boundaries and these eventually became reproductively isolated. Brinton (1962) pointed out that tropical submergence does not occur for many species and following the earlier suggestions of Berg (1933) and Hubbs (1952) has proposed an ingenious mechanism to explain the present day patterns of euphausiid species pattern. It is based on the invasion-reinvasion concept; on the assumptions that temperatures at 200 metres may be used to indicate the conditions appropriate for the existence of euphausiid species; and that there have been substantial, oceanwide warming and cooling cycles.

Using the 200 metre isotherms that agree with the limits of recent amphitropical (both Transition Zone and Central) species distributions Brinton has extrapolated the limits of these distributions by recontouring the isotherms of oceans that were 2.5° C warmer and cooler (at 200 metres) than the present (Fig. 1.40). Thus during periods of cooling, the Transition Zone species *Thysanoessa gregaria* would 'coalesce at the equator, or, if the species had previously been present in only one hemisphere, access to a new habitat in the opposite hemisphere would have been provided' (Johnson and Brinton 1963) During periods of warming the Transition Zone species would have moved polewards, thus allowing the southern hemisphere population to 'migrate' around Cape Horn and populate the South Atlantic and southern Indian Ocean. Another cooling epoch would put a portion of this population in the North Atlantic. A similar argument is used to account for the Central Water Mass patterns.

Although this hypothetical scheme does account for the distribution of some of the patterns of euphausiid species it does not really (except by implication) say how euphausiids became diverse. Brinton did not start with an 'ancestoral euphausiid' to develop his patterns but rather with existing species. Further this hypothesis does not fully account for the numerous Equatorial or Antarctic euphausiid species. It does, however, go a very long way toward attempting to answer the question: How and Why did the Patterns Develop? Since so many other species of the taxa inhabiting the open ocean show patterns of distribution similar to the euphausiids, Brinton's hypothesis has broad application and is of fundamental significance. Moreover, parts of it are amenable to tests. For instance, it is clear that Brinton's species are not necessarily temperature limited for they go through extensive vertical migrations twice each day and they are thus exposed to a temperature range greater than the horizontal range of temperature either at the surface or at 200 metres in the area where they live. The horizontal range in temperature at 200 metres circumscribing the range of *Thysanoessa gregaria* is 7° to 11° C. While it could be argued that 7° C is the minimum *limiting* the population,

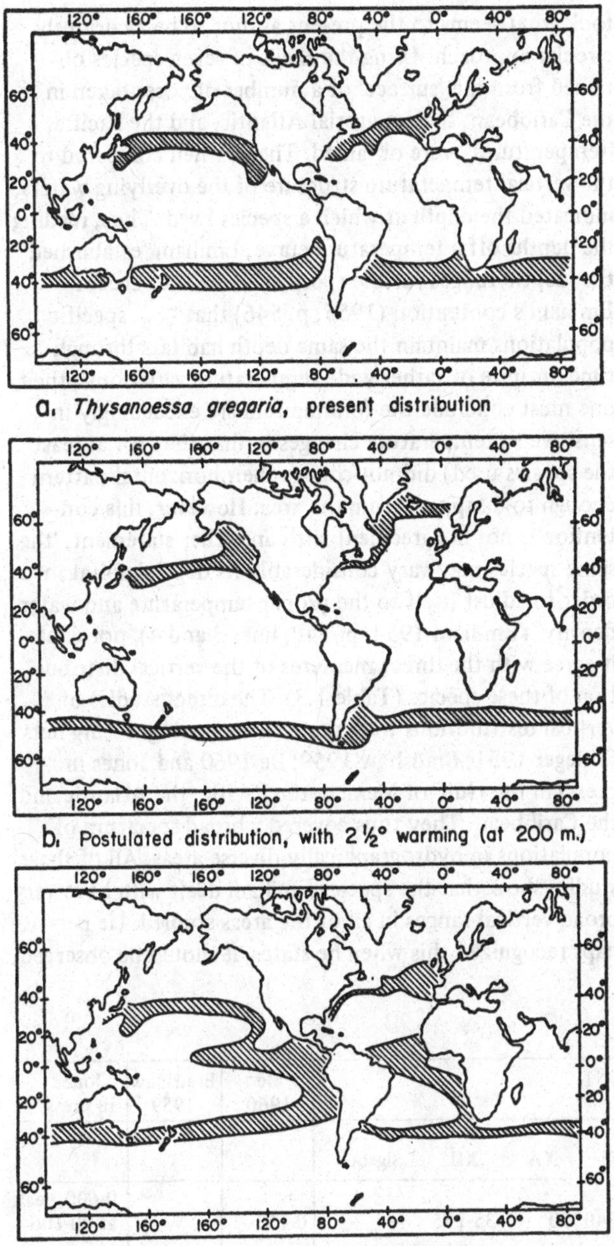

a, *Thysanoessa gregaria*, present distribution

b, Postulated distribution, with 2 ½° warming (at 200 m.)

c, Postulated distribution, with 2 ½° cooling (at 200 m.)

Fig. 1.40. Brinton's postulated distributions of *Thysanoessa gregaria* with warming and cooling of the world ocean. For details see text (Brinton 1962).

it cannot be argued that 11° C is the limiting maximum, for it is clear that these euphausiids spend a large fraction of every twenty-four hours at temperatures higher than this (Brinton 1962, pp. 147-8). But Brinton does not claim that these isotherms are limiting (Brinton 1962, pp. 244-5) the species but merely that they can characterize the range or potential range of this species and, by inference, other species. While there is obviously no possibility for experimental verification of this, certain observations may be pertinent. There are other zones in the Pacific contiguous with this one, with the temperature range 7-11° C and this species does not live there. There must then, be some other factors of importance and should these have changed, independently of temperature, they might have had an influence on the patterns. That such changes are of importance to animal populations has been discussed. During the years 1957 to 1958 the mixed layer of the California Current averaged 1-3° C warmer than the preceding seven year period (Reid 1960). The zooplankton populations had been rather well sampled during this entire period. Brinton (1967) has studied the distribution and abundance of euphausiid species before, during, and after the warm period. While there is good evidence that some southern species (e.g. *Euphausia eximia*), extended their range to the north this was only in a narrow band near the coast and amounted to 2° 30′ of latitude. A Subarctic species (*Euphausia pacifica*) did not retract its range consistently during this period. Thus, a general faunistic change did not occur in the California Current. Of course, an average 2° C warming of the mixed layer of the California Current is a much more trivial event than a general ocean-wide cooling to depths of 200 metres. Just how such a cooling came about is not mentioned by Brinton. For instance, it could have happened if the water, sinking at higher latitudes and spreading equatorially at 200 metres were chilled. But chilling by 2.5° C at the point of sinking would mean that the water was more dense and would merely sink deeper. It would take, probably, a very long time for this to eventually affect the 200 metre level. The consequences would be a much more abrupt thermocline in the area and a tendency for increased stratification and *less* mixing through the pycnocline, and a greater temperature differential for the euphausiids to migrate through. Exactly how this might affect the populations is not known.

If, on the other hand, the cooling was general over the entire surface of the oceans, the water sinking at the high latitude source of the 200 metre Transition Zone water would not go deeper. As a matter of fact, with uniform 2.5° C cooling, it seems unlikely that there would be any change in the general circulation scheme of any part of the oceans and, therefore, unlikely that the two Transition Zones would 'coalesce' at the equator as Brinton has suggested. Other factors such as an increase in the wind stress and changes in the evaporation-precipitation ratios and perhaps even a sea-level change would be necessary. Thus, the process would seem to be somewhat more complex than a simple warming-cooling cycle could account for. In spite of these criticisms, this hypothesis remains one that should be investigated. The most promising place to look for pertinent evidence is in the fossil record.

There have been a number of attempts to examine the fossil record of the floor of the deep sea in terms of palaeoclimates and palaeo-oceanography. Since it is apparently difficult to obtain a long enough sequence of samples (deep-sea cores) going much farther back in time than the the Pleistocene epoch, one must assume that the shapes of the present patterns were formed during the late Tertiary and Quaternary periods. Studies pertaining directly to the present discussion are those of Emiliani and his colleagues. It had been shown that the ratio O^{18}/O^{16} in calcium carbonate laid down by some organisms in their shells were dependent *inter alia* on the temperature at the time of shell secretion. On the assumption that this also held true for planktonic foraminifera, Emiliani (1955) measured these ratios from a large number of samples of the shells of species of foraminifera taken from different layers found in deep-sea cores. These data showed a variation in isotopic ratios which when converted to temperatures, showed an amplitude of slightly over 9° C.

However, in order to interpret this information in terms of palaeotemperatures of the ocean, it is necessary to estimate the age of the samples and since ocean temperatures vary with depth, to establish the depth at which the living species exist and lay down their shells. Having established the latter it is necessary to assume that the species always existed at this depth in spite of any temperature changes that might have taken place at that depth (Emiliani 1955, p. 546). Radio-carbon dates were obtained for various parts of the cores. In order to establish the depths at which the various species lived, Emiliani (1954) took what seems to the present author to be a curiously circular approach. Using the shells of seven species obtained from the 'surface' of a number of cores taken in the Caribbean, the equatorial Atlantic, and the Pacific, 'temperatures' were obtained. These, when compared to the vertical temperature structure of the overlying waters, indicated the depth at which a species lived. Thus, reading the depths off a temperature curve, Emiliani 'established' the 'depth' ranges for the species. If one is to accept Emiliani's contention (1955, p. 546) that '. . . specific populations maintain the same depth habitats through time, in spite of rather wide temperature variations', then one must conclude the foraminifera are exceedingly insensitive to temperature changes in that they (or at least the species used) did not change their horizontal pattern enough to disappear from the area. However, this contention is not in agreement with an earlier statement, 'the same species may vary considerably its depth habitat in order to adjust itself to the proper temperature and water density' (Emiliani 1954, p. 149, lines 3 and 4), nor does it agree with the direct measures of the vertical distribution of these species (Table 1.3). The direct studies of vertical distributions were done with opening-closing nets (Phleger 1951; Bradshaw 1959; Bé 1960 and Jones in press) in the Gulf of Mexico, the Pacific, the Atlantic and the Caribbean. They thus covered a broad spectrum of populations in hydrographically diverse areas. All of these studies show that the species Emiliani dealt with have very broad vertical ranges in all of the areas studied. He perhaps recognized this when he states 'it should be observed

Table 1.3

	Emiliani 1954, 1955	Phleger 1951								Bé 1960	Bradshaw 1959	Jones in press
		I	III	VI	VII-VIII	IX	XA	XII	Sigsbee			
Globigerinoides sacculifer	0-25	-	-	-	175	27-340	50	35-175	-	0-150		0-600, peak at 80-100
Globigerinoides ruber	0-45	23-200	25-770	30-462	20-350	25-340	50-425	35-675	40-1500	10-1000	0-400	0-600, peak at 40-90
Globorotalia menardii	43, 130	-	-	40	350	25-340	425	70	40-1500	0-1000	10-200	-
Globorotalia tumida	55, 62 130, 140	-	200	75-145	30-350	30-100	175	-	1500-2000	-	-	-
Globigerina inflata	40-140	-	-	-	-	-	-	-	-	75-200	25-400	-
Globigerina dubia	33-100	-	-	-	-	-	-	-	-	-	-	-
Pulleniatina obliquiloculata	15-35, 110	40-45	25-770	30-145	30-350	27-115	-	70	27-115	0-300	-	0-300
	Caribbean	Gulf of Mexico								Bermuda	Pacific	Caribbean

The depth range, in metres, of the species listed, according to various authors

that the depth values thus obtained and shown . . . merely indicate the depth at which the maximum density of a given foraminiferal population occurred' (Emiliani 1954, p. 152). However, Phleger (1951, pp. 27, 28, 31, 33, 34) has shown that this is frequently not the case. His data indicate that the peak population densities for *Globigerinoides ruber* and *Globorotalia menardii* are often at depths greater than 200 metres. Furthermore, all of these latter authors have shown that if one considers the entire water column, a very large proportion of the populations of most of Emiliani's species occur at depths greater than the population peaks. In some cases this may be as great as 70 per cent. It has also been shown (Bé and Ericson 1963, Bé 1960, 1965) that several species in different genera continue to secrete calcium carbonate at depths greater than their population peaks and that the size of the individuals at these depths is greater than those from shallower depths. Whether Emiliani's species do this is not known. It is the present author's opinion that it is very likely that they do. It is also a relatively safe assumption that mortalities of foraminifera occur at all stages of the life history. Although the exact proportions of various size categories reaching the sediment is probably altered by the types of mortality factors, there can be little doubt that some of these individuals are from the older, deeper living fractions of the populations. Since Emiliani (1955) used only the large size fraction for his analyses, it seems evident that his isotopic ratios are some sort of a skewed distribution of the temperatures from the entire water column through which the species lived. This bias may be toward the shallower depths if most of the calcium carbonate of each shell is secreted there, on the other hand, it may not. We are left then with the problem of determining just what it was that was measured. The only thing that seems clear is that a variation in the isotopic ratios of sea water may have taken place thoughout the Pleistocene. Shackleton (1967) has calculated the effect that the waxing and waning of the continental ice sheets would have on the isotopic composition of the oceans. Since 'the isotopic compositions of atmospheric precipitation, and hence that of a continental ice sheet, is different from that of the ocean' (Shackleton 1967, p. 15) one would expect this effect to have an influence. He shows that even without assuming any decrease in air temperature during the glacial stages, his calculation can account for the record of isotopic change as shown by fossil, planktonic foraminifera. It is thus extremely doubtful that Emiliani's results can be used to support Brinton's hypothesis.

Other studies on possible temperature and/or climatic changes in the Atlantic and Pacific have been done. Phleger, Parker and Peirson (1953) have examined the frequency of occurrence of the shells of species of planktonic foraminifera in a series of cores taken at mid- and low-latitudes in the Atlantic. They obtained an indication of the present relationship between species frequency and latitude by an examination of the top 1/2 or 1 centimetre of these cores. They based their frequency distributions on counts of not less than 300 individuals. They then obtained estimates of the proportion of each species in progressively deeper sections of the core. On the basis of changes in the relative proportions of some species through the core length they deduced that 'Faunal variations in the mid-latitude cores are shifts to faunas which appear to be characteristic of somewhat higher latitudes' (Phleger, Parker and Peirson 1953, p. 108). They based their recognition of this higher latitude assemblage largely on 'the decrease in percentage *Globorotalia truncatulinoides* and an increase in percentage of *Globigerina pachyderma*'. Other criteria used are: 'the disappearance of *Globorotalia menardii-tumida* group and *Pulleniatina obliquiloculata,* slight decrease in percentage of *Globigerina inflata* and *Globigerinoides sacculifera* and frequency of *Globigerina bulloides'.* They further state that 'these latter criteria do not seem to apply to many cores'.

I have tabulated the rank order of abundance of the ten most abundant species, using their data, from a series of five cores taken very close to one another at mid-latitudes. Only the surface samples of the cores were used in this tabulation (Table 1.4). It may be seen in this table that *G. truncatulinoides* had ranks that varied between 3 and 6.5, that is it 'decreased' in some cores relative to others. The rank of *G. pachyderma* varied between 9 to not present among the first 10, that is 'decreased', *G. menardii-tumida* 'disappeared' twice and *G. sacculifera* shifted its relative position rather markedly. Thus within one area in which the hydrographic conditions are comparatively uniform, there are apparent changes in the relative frequencies of the 'characteristic' species. Phleger (1951) has pointed out that in the plankton, foraminiferal species populations seem to be very patchy. It does not seem, however, that this patchiness would be reflected in a sample of the top centimeter, or so, of a core. This represents a time span of the order of 1000 years and patchiness should be integrated. There are good reasons to believe (see section on Sampling) that a count of 300 individuals is simply not large enough to discriminate the rank order (and therefore the proportions) of any but the first two or three species in the order of dominance.

In a series of papers Ericson and colleagues (Ericson, Wollin and Wollin 1954; Ericson and Wollin 1956a, 1956b; Ericson, Ewing and Wollin 1963, 1964) have reported on the results of similar faunal analyses and have reached similar conclusions, namely that there have been large latitudinal shifts through the Pleistocene of species and

Table 1.4. Rank order of the ten most abundant species from the surfaces of five closely spaced cores from 'Mid-latitudes' (Phleger, Parker and Peirson 1953)

Species	Core 288	Core 289	Core 290	Core 292	Core 293
Globigerina bulloides	1	1	1	1	1
Globigerina inflata	2	2	2	2	2
Globigerina eggeri	3.5	3	3	4	4
Globigerinoides ruber	3.5	7	5	3	6
Globorotalia scitula	5	4	4	5.5	5
Globorotalia truncatulinoides	6	6	6.5	5.5	3
Globorotalia hirsuta	7	5	8	9	8
Globigerina pachyderma	PNR*	PNR	9.5	9	PNR
Globigerina quinqueloba	8.5	10.5	PNR	7	8
Globigerinita glutinata	8.5	10.5	PNR	9	PNR
Globorotalia menardii-tumida	PNR	PNR	PNR	Absent	Absent
Globigerinella aequilateralis	10.5	8.5	PNR	PNR	PNR
Globigerinoides sacculifer	10.5	8.5	PNR	Absent	8
Orbulina universa	PNR	PNR	6.5	PNR	PNR
Globorotalia punctulata	PNR	PNR	9.5	PNR	Absent

*PNR = species present but did not rank among first 10.

subspecies assemblages. They interpret these to mean that the climate of the ocean has varied considerably, especially in its average surface temperature. Since Ericson *et al.* do not count species abundances or use known sample sizes, it is impossible to interpret their data. However, in a tabulation of their subjective estimates of whether species and subspecies were *V*ery abundant, *A*bundant, *C*ommon, *F*requent, *R*are or absent (*X*), from the tops of three of their cores from tropical latitudes, it may be seen that the relative proportions of their 'climate indicators' vary. For example, *G. p. punctulata* a cold 'indicator' varies from absent to common, and *G. scitula* another cold 'indicator' from rare to abundant (Table 1.5).

Blackman and Somayajulu (1966) estimated the abundance of nineteen species from counts of from 300 to 500 individuals sampled from the tops of 56 cores taken in the South Pacific. Faunal groups were determined on the basis of a vector analysis in which the proportional contribution of compositionally extreme samples were used. Composition in this sense refers to both qualitative *and* quantitative aspects. While there are no reasons to question the qualitative aspects of this study the determinations of the proportions in which various species occur is variable in the tops of cores taken quite close to one another (i.e. DWBG 98C and DWBG 114, Blackman 1966). The degree to which this affects the selection of end members and their proportional contribution to the other samples is not clear. It should however be clarified, for Blackman and Somayajulu used the faunal groupings derived in this manner to show displacements of a high latitude fauna toward the equator at various times during the Pleistocene.

Although there are some questionable aspects of the faunistic papers reviewed above, I do not share the view expressed by Emiliani and Milliman (1966, p. 126) that the works of Ericson *et al.* (and, by implication, all the other authors using similar techniques) are 'altogether meaningless'. There are some species abundance changes in some cores that cannot be readily explained by inadequate counting procedures. Further, the changes in coiling direction through time of certain species seem to be statistically significant. The question here is, should these changes be interpreted in terms of climatic (hydrographic) changes in the ocean? I have pointed out earlier that very wide, non-seasonal, fluctuations in terrestrial, planktonic and nektonic species populations are known and that these cannot be correlated with major climatic events. The periodicity of these 'cycles' varies, some at about four years, others at about ten and still others in between. I know of no ecological reason why 100; 1,000; or 10,000-year cycles are not equally possible. In view of the fact that foraminiferal Pleistocene 'faunal shifts' are based entirely on extant species, and that in any one area the species at the tops of cores are generally present throughout the cores, the evidence for warming and cooling or water-mass migration must be regarded as tenuous.

Changes in coiling direction of gastropod molluscs as well as foraminifera has been noted (Ford 1964, p. 170). *Partula suturalis*, a land snail on parts of the island of Moorea was wholly dextral in the last century, but it has colonized areas subsequently and sinistrals are common in these. There have been no accompanying large climatic changes.

It would appear that species interaction of some sort could just as well 'explain' the observed real changes in the microfossils of oceanic cores, and we are not therefore in a particularly good position to support Brinton's hypothesis. It seems evident that an examination of the fossil record on the floor of the deep sea is our only hope of acquiring the type of information necessary for a consideration of his or several other important hypotheses about the oceans. However, if the present is truly the key to the past, it may be that we must know a great deal about the present in order to get the necessary information about the past. We do *not*, at present, know much about what factors influence: the population sizes, proportion of species present, birth rates, death rates, niches occupied, or spatial distributions of planktonic foraminifera or much of anything else that lives in the open ocean. Approaches to the determination of these factors can and should be made.

WHAT ARE THE COMMUNITIES?

If one is to take the approach that biotic interactions between species are at least as important as the action of the physical environment in influencing temporal and areal variations in abundance and diversity, then one must have a very firm grasp of which species frequently live together. In other words, do functional communities exist and if so where? The history of attempts to define communities is extensive and has been reviewed by Fager (1957, 1963). Essentially the problem has been to reduce the number of subjective decisions necessary to define a community. While the subjective impresssions of a highly skilled naturalist as to what constitutes a community are

Table 1.5. Estimates of abundance of 18 species and varieties from the tops of three cores from equatorial latitudes (Ericson and Wollin 1956)

	14° 59′ N A-172-6 top	16° 36′ N A-179-4 top	00° 10′ N A-180-73 top
G. m. menardii	A	V	V
G. m. tumida	C	A	A
G. m. flexuosa	X	X	X
G. p. hirsuta	X	X	X
G. p. punctulata	X	F	C
G. truncatulinoides R	1	58	21
G. truncatulinoides L	1	21	0
G. scitula	R	C	A
G. inflata	X	X	X
G. bulloides	X	X	X
G. pachyderma	X	X	X
G. eggeri	C	V	C
G. ruber	A	V	V
G. sacculifer	A	V	V
G. conglobatus	F	F	X
O. universa	A	A	R
Pul. obliquiloculata	C	C	V
Sph. dehiscens	C	C	R

often very useful, they are not very reproducible. It is also true that highly skilled naturalists often disagree on what are or are not communities. It seems essential, therefore, that some sort of operational definition is provided. With this, even though it is imperfect, the broad outlines of food webs may be indicated and their temporal and areal patterns determined. Once this is done, further details may be added; for instance, a more exact determination of the extent and nature of the interactions or an examination of the diversity-stability relationships. While the naturalistic approach has been taken in studies of marine plankton, there have been rather few objective studies.

Williamson (1961) has developed a non-parametric, statistical method of grouping together 'entities'. He calls this method Principal Component Analysis. He uses it to examine the proposition that groups of species can be distinguished in the plankton and that the members of each group show similar fluctuations in abundance from year to year. Data on the relative abundance of twenty-three entities (some species, some genera, total chaetognaths, total dinoflagellates, etc.) as derived from samples taken with the Small Plankton Indicator in the North Sea over a period of eleven years were used in his analysis. The data were tabulated in the form of a 23 x 11 matrix, that is, an entity-year matrix. The rank order of each entity was entered for each year and the rows (entities) ranked and the sum of the ranks of the columns (years) calculated. A Spearman rank correlation coefficient was calculated for between years and a secondary, 11 x 11, year-year correlation matrix constructed. On the basis of this secondary correlation matrix Williamson showed that some years were more alike, in that certain kinds of organisms were abundant, than other years. This grouping of years is as follows:

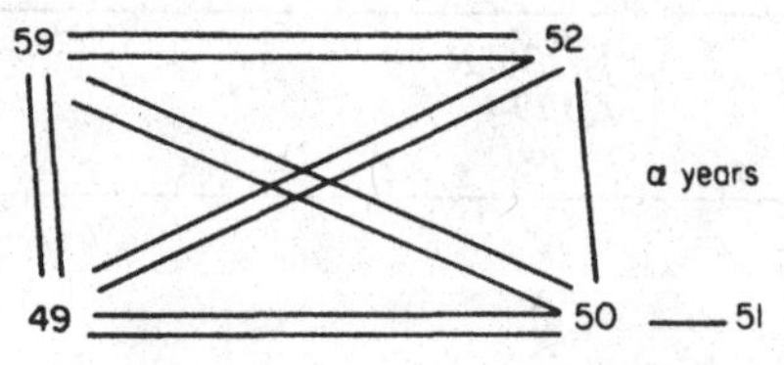
59
52
49
50
51
α years

and

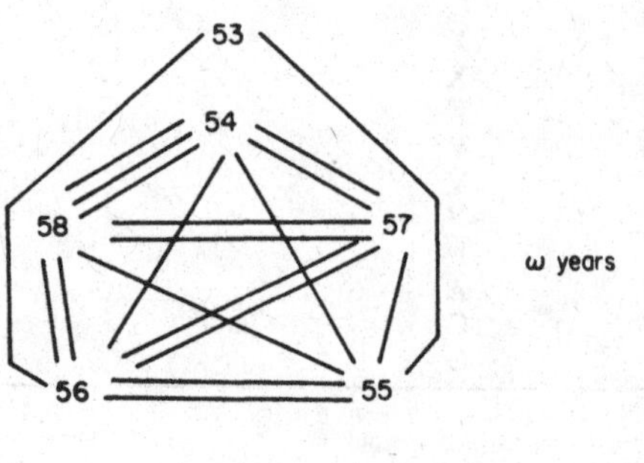
53
54
58
57
56
55
ω years

The number of lines connecting a pair of years is based on a statistical measure of their similarity. Using a very similar method he did the same thing for his entities. In this case a two-dimensional arrangement of the grouped entities was possible so that each was near those with which it has positive correlations and far from those with which it has negative correlations. They fell naturally into an elliptical area with each group in a distinct part of the ellipse. There are four groupings in this ellipse (A, B, M, Z) and they are ordered A-Z on the major axis. This means, according to Williamson, that there has been one factor with much more effect on the abundances of the entities than any other. The α years are the years when A group dominated and in the ω years, Z group was abundant. Group A tends to be made up of neritic forms in the North Sea and Z group more oceanic. B and M are not, at least in the North Sea, 'geographical groups'. In a further analysis, it was found that variations along the axes of the ellipse were associated with: a measure of vertical mixing in spring (48 per cent), temperature anomaly in spring (15 per cent), a component of oceanic circulation (9 per cent) and another, unidentified property.

Williamson's technique seems to be particularly useful in analysing sets of samples where all entities are almost always present. It cannot be used where entities often tend to be either present or absent. Since plankton data are almost never normally distributed, this technique is also appropriate because, being non-parametric, no clumsy transformations of the data are required to 'normalize' them for more orthodox statistical treatments. As applied to the study of communities, there are some obvious faults with this study. For instance, many of the entities are large taxa, the area studied is one of mixing of oceanic, North Sea and neritic waters and the samples taken were very small. In the latter case, one needs some assurance that the samples are 'representative' of the area they are intended to represent; in other words, what is the relationship of the samples to the universe being sampled? However, Williamson states quite clearly that his intention was 'to explore, so that the data are summarized in the simplest and most straightforward way . . . based on the fewest possible assumptions' (1961, pp. 207-8).

Colebrook (1965) has applied Williamson's technique to a series of Continuous Plankton Recorder samples taken from a larger area of the North Sea and from the northeastern Atlantic. He developed five groupings of species: a Northern oceanic, a Southern oceanic, a Neritic and a Northern and a Southern Intermediate, which correspond with Williamson's groups. Colebrook developed four components of the physical environment that accounted for a total of 71 per cent of the variance of the distributions of the species. The first C_1 was identified with surface salinity (33 per cent), the second C_2 (21 per cent) an index of vertical stability, C_3 (11 per cent) mixed oceanic and coastal waters and the fourth unidentified (6 per cent). He does not state that there is a cause and effect relationship between groups and components but merely that they may be related. It is clear from these studies that groups of entities, and species, *do* tend to recur with one another in this area and one can now examine the effects that they may have on one another and the role of the physical environment. It may be possible to separate these two factors.

Fager (1957) has described a somewhat different approach to the same problem. He points out that the typical ecological survey produces a mass of data involving many species and many localities and is frequently difficult to interpret in terms of deciding which species make up a nearly constant part of each other's biological environment. Often many of the species are present in some samples but not others and, therefore, procedures for grouping similar to Williamson's cannot be used. Fager (1963) has reviewed some of the procedures that are useful where there are a number of non-occurrences. Many of these involve, as the primary step, the use of correlation coefficients based on the variations in abundance of pairs of species. However, he points out that since raw abundance data are seldom normally distributed, some transformation of these data is required. Also with many zero values, the correlation coefficients of different pairs of species will be based on different numbers of observations. The significance of such coefficients changes with a change in the number of observations on which they are based, thus a further transformation is required. All of this results in some very highly 'derived' data with which to work. A further and very serious criticism of the use of correlation coefficients is that pairs of species which are always together but whose abundances vary differently will be either uncorrelated or negatively correlated. Thus, a predator whose population is growing rather slowly (due to a low birth rate or a rather low ecological efficiency) at the expense of a prey whose population is declining because of the predation will not be correlated with its prey! It, therefore, 'seems better . . . to use methods which put together species which are frequently part of each other's environment and *then* to look at abundance relations within these groups' (Fager 1963, p. 420). This may be done by using only presence and absence data. With this information all possible pairs of species, from a sampling survey, which have a high frequency of co-

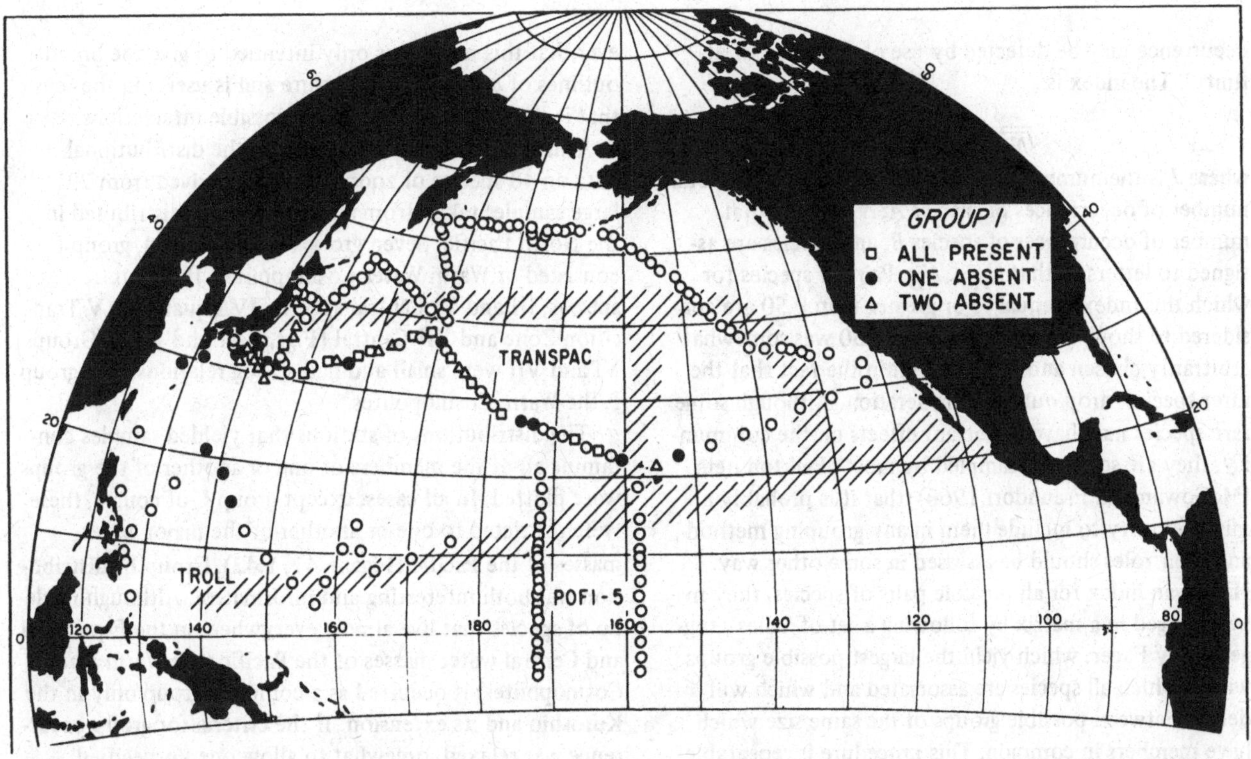
GROUP I
ALL PRESENT
ONE ABSENT
TWO ABSENT
TRANSPAC
TROLL
POFI-5
60
40
20
0
120
140
160
180
160
140
120
100
80
40
20

Fig. 1.41. The distribution of stations from which plankton samples contained n, $n-1$ and $n-2$ members of group 1 (Fager and McGowan 1963).

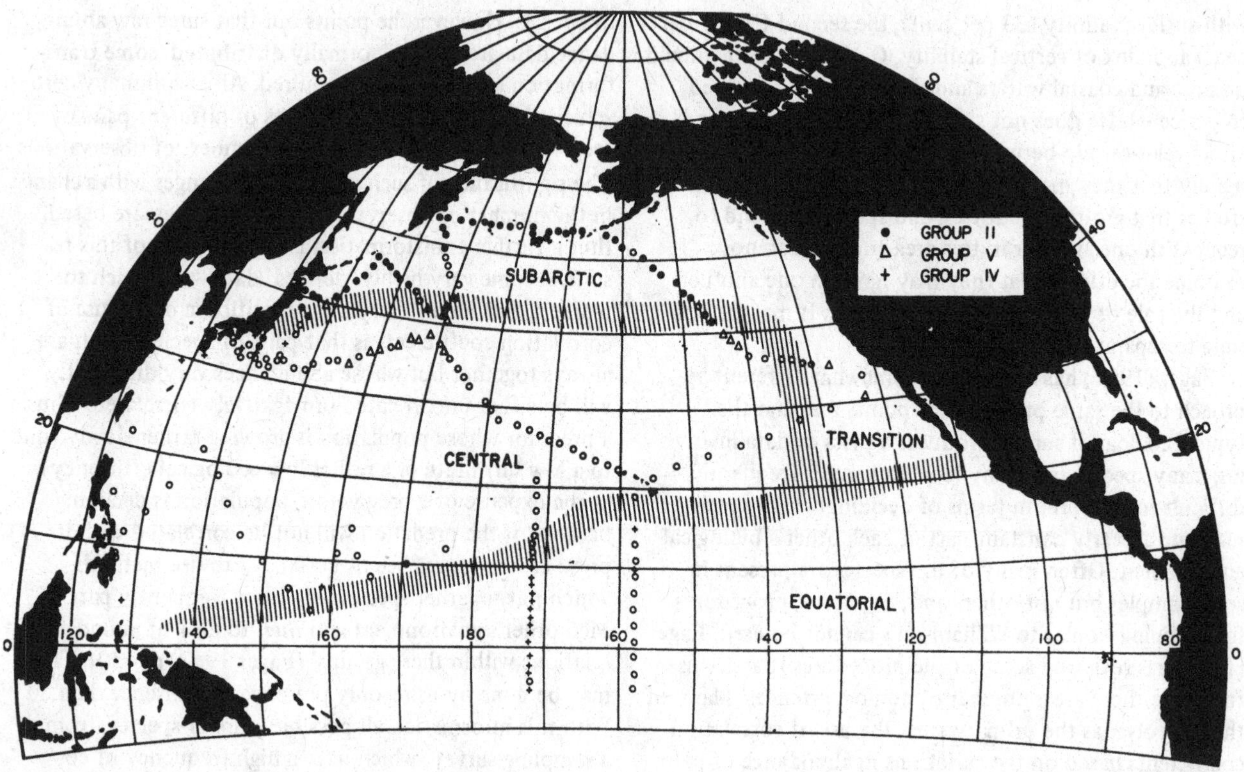
GROUP II
GROUP V
GROUP IV
SUBARCTIC
CENTRAL
TRANSITION
EQUATORIAL

Fig. 1.42. The distribution of stations from which plankton samples yielded *n* members of groups II, IV, and V (Fager and McGowan 1963).

occurrence may be detected by use of an 'index of affinity'. The index is

$$\frac{J}{\sqrt{N_A N_B}}\ 2\ \frac{1}{\sqrt{N_B}},$$

where J is the number of joint occurrences, N_A the total number of occurrences of species A; N_B is the total number of occurrences of species B, and species are assigned to letters to that $N_A < N_B$. Pairs of species for which this index is equal to or greater than 0.50 are considered to show 'affinity'. The level 0.50 was somewhat arbitrarily chosen and it does result in the fact that the rarer species 'drop out' of consideration. Although some rare species may have important effects on the community, they are so poorly sampled by most plankton nets (McGowan and Fraundorf 1966) that it is probably a mistake to try to include them in any grouping method and their roles should be assessed in some other way. Having an index for all possible pairs of species, they may be arranged in a matrix by following a set of 'rules', suggested by Fager, which yield the largest possible groups within which all species are associated and which will decide between possible groups of the same size which have members in common. This procedure is repeatable by anyone using the same set of data and the results are, therefore, reproducible. It should be emphasized, however, that this method is only intended to give the broad outlines of community structure and is useful in the sense that it shows where to look for possible interactions.

This method has been applied to the distributional data on 48 species of zooplankton as derived from 291 large samples taken from stations broadly distributed in the North Pacific. Seven groups were detected, group I consisted of Warm Water Cosmopolites, II Subarctic species, III and VI Central species, IV Equatorial, V Transition Zone and VII Central (Figs. 1.41 and 1.42). Groups VI and VII were small and had strong relations with group I, the Warm Cosmopolites.

The distributions of stations that yielded samples containing all of the members of one or another of the groups were plotted. In all cases, except group I, of course, these were restricted to one or another of the major water masses of the Pacific (Figs. 1.41, 1.42). Group I's distribution was both interesting and informative. Although made up of species that live almost everywhere in the Equatorial and Central water masses of the Pacific (the Warm Water Cosmopolites) it occurred as a complete group only in the Kuroshio and its extension. If the criteria for group occurrence was relaxed somewhat to allow one unspecified species to drop out ($n-1$, with n=number of species in the group) or two unspecified ($n-2$) species to drop out, the pattern did not change. Thus, the distribution of this group does not generally overlap the distributions of Equatorial or Central groups and, of course, does not occupy any of the classical water masses. A tentative explanation of this was suggested by Fager and the present author. It is of considerable significance that Sheard (1965), using the recurrent grouping method of Fager (1957) plus an independent method, found a grouping of species in the eastern Australian slope region almost identical with group I. That is, group I occurs not only in the Kuroshio and its extension but in the slope current off northeastern Australia.

A further important aspect of this study of recurrent groups was that a clear-cut group (group V, Fig. 1.42) appeared and was present as an intact entity only in the Transition Zone.

In an attempt to use this information to gain more insight into community ecology, an examination was made to determine the degree of agreement among the species of a group as to which were the 'best places to live' within their range and which the 'worst'. This was based on the assumption that species will tend to be most abundant in areas where they find the habitat most suitable and least abundant in those areas that are least suitable. For this purpose, the relative abundances were used in a Concordance test which is a simple, non-parametric procedure. The abundances of species are ranked in rows, the columns are stations where they all occurred, for example:

	Stations							
Species	26	27	28	29	30	31	32	
A	7	6	4	5	2	3	1	
B	7	6	5	4	3	1	2	
C	6	7	4	5	3	1	2	Ranks
D	7	6	5	3	4	2	1	
E	7	5	4	6	1	3	2	
Σ_R	34	30	22	23	13	10	8	

Species A to E may be ranked for stations 26 to 32 where they occurred 'together' in zooplankton samples. The ranks summed Σ_R, and the concordance W, may be calculated and its significance tested (Kendall 1955). In the above example $W = 0.860, F = 24.57, p < 0.0005$. In other words, there are real differences in the order of Σ_R and, therefore, there is a substantial amount of agreement between species A to E that stations 32 and 31 are 'best' places to live while stations 26 and 27 are the 'worst'. In concordance tests performed on the North Pacific data, group I showed very significant concordance ($p<0.005$), and the best places to live were clustered together in the core of the Kuroshio while the worst were on the periphery of the group's range. This is just about what one would expect if the members of a community have similar reactions to environmental conditions. Group II also showed concordance ($p < 0.005$) as did groups IV ($p < 0.10$) and V ($p < 0.15$). Groups III and VI did not show agreement.

In a further attempt to determine which physical parameters were associated with best and worst places to live, the Σ_R of the species of group II were used as the dependent variable in a multiple regression analysis. Hydrographic and other data were treated as the ten independent variables, these included shape of *T-S* curve, depth of thermocline, time of day, oxygen concentration, temperatures at two depths, salinities at two depths, latitude and longitude. The overall regression was significant ($p < 0.05$) and accounted for about 56 per cent of the variability, but only two of the independent variables had significant regression coefficients. These were shape of T-S curve ($p < 0.01$) and lower salinities at 10 metres ($p < 0.05$). Neither of these can be the actual 'cause' of the changes in relative abundance but both are associated with the history and 'quality' of the water. Although the other regression coefficients were not significant, they are interesting as suggesting possible trends. Many of these trends suggest that more thorough vertical mixing creates more favourable conditions for these organisms.

It is evident that the properties of the physical environment that Fager and the present author found to be associated with best place to live are very similar to the *factors* Williamson (1961) found to vary along the axes of his A-Z ellipse and that Colebrook (1965) called *components.* Williamson's results were unknown to us at the time we performed our analysis and our results were not available to Colebrook at the time he did his study.

A third attempt at a recurrent group type of community description has been made using samples from within the California Current only. The range of sampling was from about 40° 15′ N to 23° 50′ N and reached approximately 430 miles seaward. Species of the same taxa used in 1963 by Fager and the present author, plus ninety-eight additional species of Copepoda, Thaliacea, Siphonophora and Hydrozoan medusae, were included. The grouping procedure yielded fifteen groups, ten of which included only two or three species per group. This unexpected result was further complicated by the fact that when the positions of the samples containing n members of any group were plotted on a chart of the California Current, no discernible patterns were evident. There were, however, two probable exceptions, group 14 occurred in a coherent pattern of nine northern, near-shore stations which resembled the Subarctic species patterns in this area. It was also made up of Subarctic species. Group 13 occurred at 9 stations on the outer fringes of the central sector of the area, i.e. a typical Central water distribution and contained known Central Water Mass species.

These results were in a sense disappointing, for if one cannot 'depend', in this area, on finding patterns of structurally organized communities, there would seem to be little hope of determining energetic relationships or for understanding the biology of population interactions. The probable reason for this apparent chaos is that this area of the California Current is a vast zone of mixing of many pelagic faunal elements. For instance, it is evident from Figs. 1.30, 1.31 and 1.32 that here is where Transition Zone, Central, and Equatorial faunal boundaries overlap strongly, and added to these are the Subarctic and Warm Water Cosmopolite elements. Further, there are *endemic* neritic planktonic and nektonic species present (i.e. *Sagitta euneritica, Nyctiphanes simplex, Engraulus mordax, Loligo opalescens*, and others) which mix with the oceanic groups. Thus, it would seem reasonable to assume that the species combinations found in this very large area (1.3 x 10^6 km^2) are primarily a product of advective physical processes, rather than having evolved through biological interaction, the combinations are therefore far less consistent. Because of this, studies in this area, such as those of Patten or Margalef mentioned earlier, on the relation of stability to diversity are unlikely to be particularly meaningful.

However, studies of the structure of communities done in more hydrographically stable areas are informative and have led to consistent and reasonable results. Although the actual causes of the variations in abundance have not really been determined in any of these studies, they seem to involve, somehow, the history of the water and the populations. It is clear that no one simple physical parameter, temperature for instance, is by itself a causal factor. It is also clear that more kinds of organisms need to be included in these analyses (phytoplankton, fish and squid for example), and that we need more purely biological information on life histories, feeding, population dynamics, physiology and taxonomy. These latter studies, interesting for their own sake, have the added advantage of direction and purpose. They can contribute *directly* to the question, what causes the temporal and areal variations in abundance of oceanic organisms?

SAMPLING THE PLANKTON

Most surveys of large areas of the ocean which are intended to reveal the distribution and abundance of plankton species utilize a grid pattern of sampling. At each point in the grid a single phyto- or zooplankton sample is taken. By use of vertical or oblique net tows, the populations at a series of depths are 'integrated' or, in some cases, discrete samples are taken at a series of depths with either nets or water bottles. Whichever method is used, a single sample is meant to represent a very much larger population. If we are to study spatial variations in abundance, population growth rates and declines, age structure, diversity, the affects of herbivore species on producer species, etc., we must have some measure of the degree of accuracy of the sample. Samples are, by definition, a statistical entity and in order for them to be useful, the relationship of the sample to the universe being sampled must be known. In the case of plankton studies, measures of precision have sometimes been used as an indication of accuracy. But these are not necessarily the same thing. A set of replicate samples may yield data in which the variance about the mean is quite small and hence the precision good; however all of these samples may be biased in some systematic way so that the accuracy of measurement may be very low. This is a particularly serious problem if the bias is different for different species, and this certainly could be the case. Since estimates of population sizes and changes in sizes are essential for a further understanding of the ecology and biogeography of oceanic organisms, we must be able to place statistical confidence limits, in terms of accuracy, on such estimates. At present there are very few cases where this can be done.

There are two major sources of error in estimating the abundance of plankton; one, in the field, is due to inadequacies in the sampling device and/or programme; the other, in the laboratory, is due to inadequacies in the methods used to count the sample. It is convenient to discuss these separately; it is not intended to review the principles of sampling, but merely to point out some of the problems that seem to affect plankton studies most.

One can visualize plankton species as being spatially distributed in three ways; even, random, or clumped (Hutchinson 1953). As a matter of fact, there have been many studies that have shown both phyto- and zooplankton to be highly clumped (i.e. patchy) (Hardy and Gunther 1936; Windsor and Clarke 1940; Tonolli 1949; Barnes and Marshall 1951; Cushing 1953; Holmes and Widrig 1956; Barnes and Hasle 1957; Cassie 1959*a*, *b*, etc.) This sort of a situation makes it very difficult to detect real spatial or temporal variations in the mean population sizes of the area under study or even at a single 'time-series' station. This is especially true if the patchiness is on a scale similar to the length of the tow or smaller than the distance between grid points. Unfortunately, we have very little information on the scale(s) of patchiness for oceanic species, and it does not seem to be a good assumption that all zooplankton species will have a similar size of patch or mean distance between patches. In an attempt to gain some idea of the seriousness of this problem, Wiebe (Wiebe and Holland 1968) has devised a computer model of plankton patchiness and 'sampled' it with various sizes of nets. In separate runs, he sampled patch sizes of 7.5, 10, 12.5, 15, and 25 metres on one side with nets of 25, 50, 100, and 200 centimetres in mouth diameters. The patch centres were distributed at random (RD_1) in the area (a 'box') being sampled. The 'tows' were horizontal and 500 metres in length. The results (Fig. 1.43) show that very serious under *and* over estimates of the population were made by all net sizes. The small nets were worse in this regard than the larger, in that the 95 per cent confidence limits of their estimates were distributed more widely about the mean of the estimates. If these results represent what might be happening in the field, one could simply take the error into consideration when interpreting the results of a sampling programme. However, when Wiebe used other random distributions (RD_2) the results changed. Thus, it is evident that in order to design sampling programme for accuracy, we must have some information on the size of patches, the distance between patches, the density of organisms within and without the patches and on the temporal longevity of patches. Wiebe has attempted to measure real patchiness, in two dimensions, in the ocean and has preliminary evidence, for a number of species, that extreme clumping occurs on scales similar to those used in his experiment.

There are additional sources of error in field sampling. The mesh aperture of nets must of course be small enough to retain the size of organism being sampled. However, Smith, Counts and Clutter (in press) have shown that with nets having a low ratio of total mesh aperture area to mouth area (4.8:1 with 333μ mesh), clogging can very quickly reduce the filtration efficiency of the net (40 per cent in five minutes in some cases). Further, flow metres as presently rigged in most nets do not give an accurate measure of the amount of water filtered. Another source of error results from the fact that some relatively small zooplankters can apparently avoid being caught by the net. Barkley (1964) has shown, theoretically, how this can happen and concludes that the best way to diminish the effects of avoidance is to increase the size of the net, not increase the speed of tow. Fleminger and Clutter (1965) have presented the results of an experimental study done in a large pool, which they interpreted in terms of avoidance, and Fraundorf and the present author have presented the results (McGowan and Fraundorf 1966) of a field study which we interpreted as showing that certain species avoid nets. It would seem that there is such a phenomenon as 'avoidance' and this is greater with small nets than with large ones. However, Wiebe's model can account for *all* of the variance between replicates of a series of nets of various sizes or nets of the same size, as deduced from previously published studies. It is possible that avoidance, mesh selection, clogging and inaccurate flow meters are trivial sources of error compared

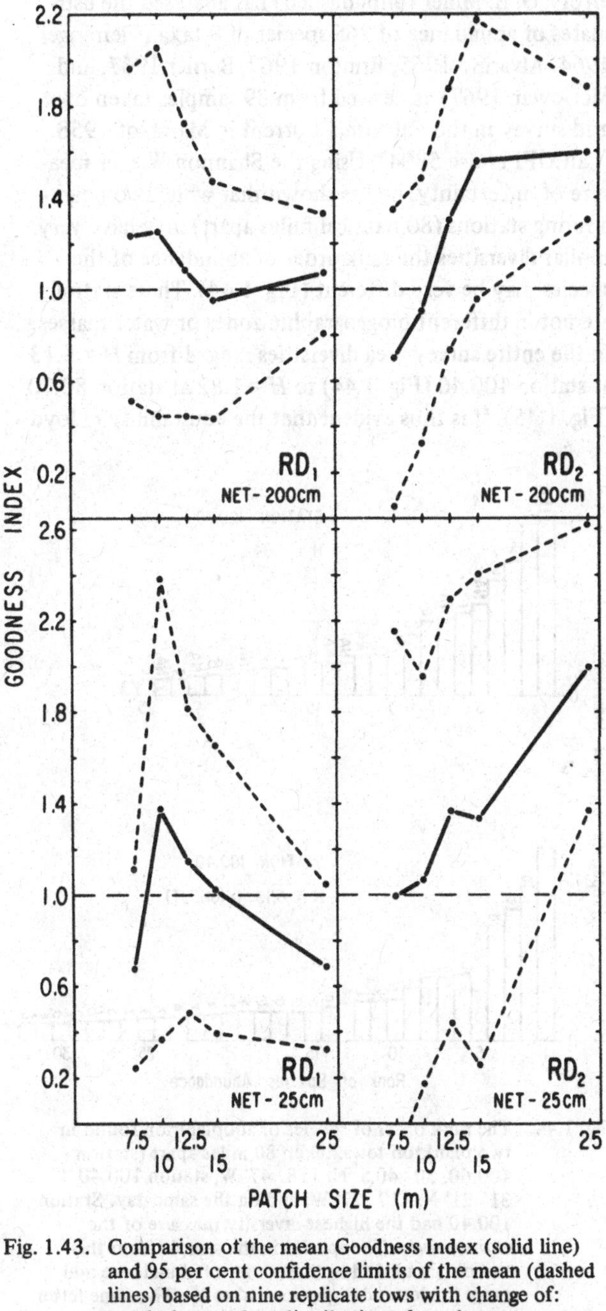
GOODNESS INDEX
2.2
1.8
1.4
1.0
0.6
0.2
2.6
RD1
NET-200cm
RD2
NET-200cm
RD1
NET-25cm
RD2
NET-25cm
7.5
10
12.5
15
25
PATCH SIZE (m)

Fig. 1.43. Comparison of the mean Goodness Index (solid line) and 95 per cent confidence limits of the mean (dashed lines) based on nine replicate tows with change of: patch size, random distribution of patch centres (RD_1, RD_2) and net size. Goodness Index is the estimated numbers of individuals in the model based on 'tows', divided by the known number in the model (Wiebe and Holland 1968).

to patchiness. Until we know more about the dimensions of plankton species patches, I strongly recommend that replicate samples be taken at every sampling point.

Compared to the sources of error from field sampling, the error in enumeration of species in the sample are probably small. But there are certain aspects of enumeration and identification that can lead to quite serious errors. C. B. Miller (unpublished) has analysed the estimates of abundance of 265 species of 9 taxa (Fleminger 1964, Alvariño 1965, Brinton 1967, Berner 1967, and McGowan 1967) as derived from 39 samples taken on a grid survey in the California Current in March of 1958 (CalCOFI cruise 5804). Using the Shannon-Wiener measure of uncertainty, he has shown that while two neighbouring stations (80 nautical miles apart) may have very similar diversities the rank order of abundance of the species may be very different (Fig. 1.44). These stations are not in different biogeographic zones or water masses. In the entire survey area diversities ranged from $H = 5.13$ at station 100.40 (Fig. 1.44) to $H = 1.82$ at station 80.70 (Fig. 1.45). It is thus evident that the equitability (Lloyd

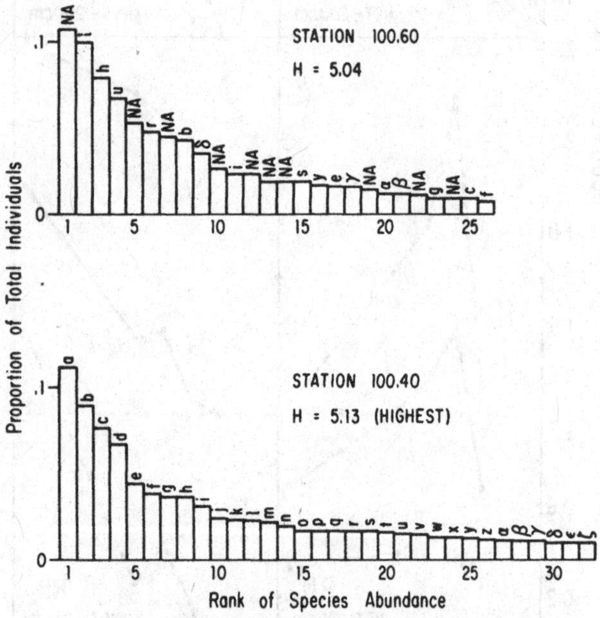
STATION 100.60
H = 5.04
STATION 100.40
H = 5.13 (HIGHEST)
Proportion of Total Individuals
Rank of Species Abundance

Fig. 1.44. The rank order of species of zooplankton found in two plankton tows, taken 80 miles apart (station 100.60, 30° 40.5′ N, 118° 47′ W; station 100.40, 31° 21′ N; 117° 27′ W) and on the same day. Station 100.40 had the highest diversity measure of the entire survey. The species in the sample from this station are ranked in their order of abundance and carry the letter designations 'a' to 'ξ'. The same letter designations for the same species are used in station 100.60. It may be seen that the rank order of the species are very different in the two samples. NA = species did not rank among the first 32 species at station 100.40.

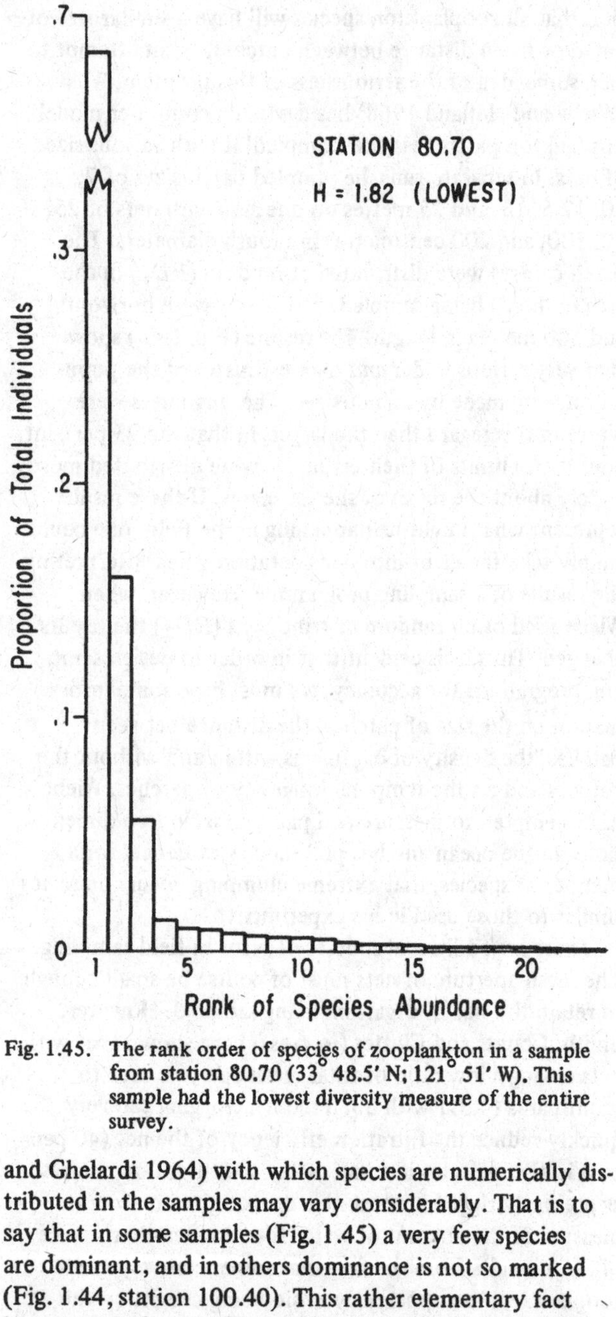
STATION 80.70
H = 1.82 (LOWEST)
Proportion of Total Individuals
Rank of Species Abundance

Fig. 1.45. The rank order of species of zooplankton in a sample from station 80.70 (33° 48.5′ N; 121° 51′ W). This sample had the lowest diversity measure of the entire survey.

and Ghelardi 1964) with which species are numerically distributed in the samples may vary considerably. That is to say that in some samples (Fig. 1.45) a very few species are dominant, and in others dominance is not so marked (Fig. 1.44, station 100.40). This rather elementary fact has some profound effects on our abilities to detect the 'correct' rank order of the species present in the sample. Miller and Wiebe (unpublished) programmed a computer to sample, at random, populations of both high and low equitabilities in an attempt to find out how many individuals must be identified and enumerated in order to obtain the 'correct' rank order (i.e. the species proportions) in the sample. Table 1.6 presents some of the results of this study for populations containing ten species. The table shows the results of taking five replicate subsamples of 50, 500, and 5,000 from populations of both low and

Table 1.6. Examples of rankings based on samples of 50, 500, and 5,000 from two distributions of numbers of individuals/species – one of high and one of low equitability. N.S. = no sample

Population rank	Low equitability, Sample number 1	2	3	4	5	High equitability 1	2	3	4	5
Sample size = 50										
1	1	1	1	1	1	1	1.5	1	3	2.5
2	7.5	12	4	3	3.5	2.5	1.5	2	7	2.5
3	N.S.	12	2	3	N.S.	2.5	3	4.5	1	1
4	2.5	2.5	4	6	N.S.	21	9.5	20.5	7	4.5
5	7.5	12	N.S.	N.S.	3.5	8.5	19.5	9.5	7	N.S.
6	N.S.	2.5	4	3	7.5	21	5.5	4.5	N.S.	N.S.
7	2.5	N.S.	N.S.	N.S.	N.S.	21	19.5	9.5	3	4.5
8	N.S.	12	N.S.	6	N.S.	21	5.5	9.5	11	6
9	7.5	N.S.	N.S.	N.S.	7.5	21	19.5	4.5	18.5	N.S.
10	7.5	12	N.S.	N.S.	N.S.	8.5	N.S.	9.5	18.5	N.S.
Sample size = 500										
1	1	1	1	1	1	1	1	1	1	1
2	2.5	2	4	2	2	2	2	3	3	2
3	4	4	2	3	3	3	3	4	4	3
4	2.5	3	3	4	4	4	4	2	2	4
5	5.5	5	5	5	6	7	6	7	9	5
6	7	6	6.5	6	5	5	7	6	8	6
7	5.5	9.5	6.5	7	7.5	6	5	9.5	6.5	7
8	8	7	14.5	28.5	13	8.5	10.5	5	5	8
9	10	8	9	10	10.5	8.5	12	8	6.5	11
10	13	14	18.5	8.5	10.5	9	8.5	9.5	10.5	14.5
Sample size = 5,000										
1	1	1	1	1	1	1	1	1	1	1
2	2	3	3	2	3	2	2	2	2	2
3	3	2	3	3.5	2	3	3	3	3	3
4	4	4	4	3.5	4	4	4	4	4	4
5	5	5	5	6	6	5	5	5	5	5
6	6	6	6	5	5	7	8	6	8	7
7	7	7	7	7	7	6	6	8	7	6
8	8	9	8	8	8	8	7	7	6	8
9	9	10	9	9.5	9	9	9	9	9	9
10	11	8	11	11	10	15	10	10	12	11

high equitability. The true ranks are shown in the column on the left and the body of the table shows the ranks as developed by replicate subsampling at the three levels. It is obvious that, at sample sizes of 50, one cannot accurately estimate the 'correct' proportions of even the second dominant species. At sample sizes of 500 the second dominant is often incorrect especially in the high equitability population, but the low equitability case is not much better. At sample sizes of 5,000 the sixth dominant is often incorrectly placed at high equitabilities and the ninth at low equitabilities. Miller and Wiebe expanded this study to include samples sizes of 50, 250, 500, 1,000, 2,000, 3,000, 4,000, and 5,000 individuals in replicates of 50 (Fig. 1.46). This figure shows that at sample sizes of less than 500, the proportions of species present in the sample cannot be correctly estimated for any but the first three most abundant species. If one is concerned with a determination of, say, the relative proportions of species that tend to rank sixth or seventh in the order of dominance in a set of samples, one must count and enumerate no less than 4,000 individuals in each sample. It is of interest to compare these theoretical results with those of Phleger, Parker and Peirson (1953) or Ericson and Wollin (1956). The former authors were concerned with determinations of the relative proportions of members of populations containing from fifteen to eighteen entities (foraminiferal species and subspecies) and used sample sizes of 300 individuals or more. It is apparent that (Table 1.4) only the two most abundant species had ranks that agreed consistently in a set of five cores that may be considered replicates. By the time the fourth ranking species is reached rather large inconsistencies in the rank order

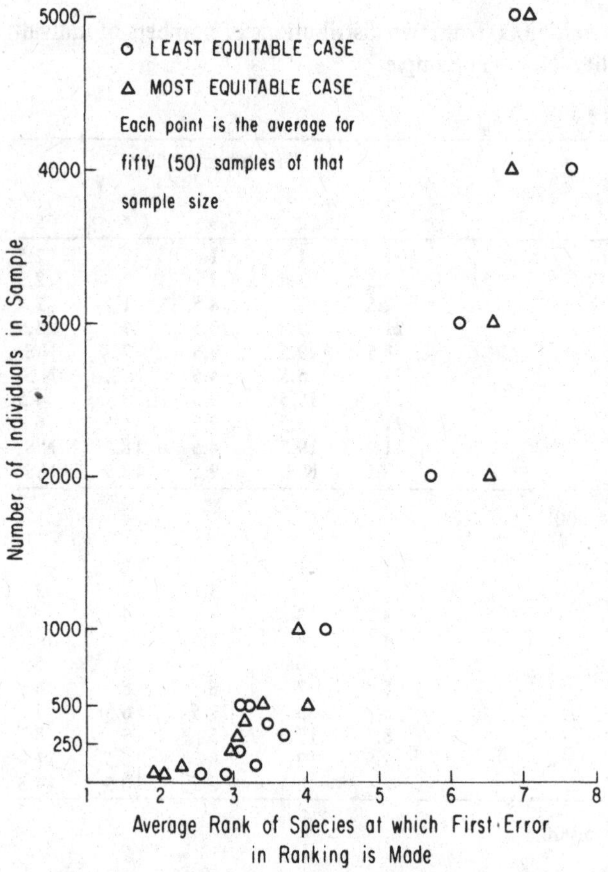

○ LEAST EQUITABLE CASE
△ MOST EQUITABLE CASE
Each point is the average for fifty (50) samples of that sample size
Number of Individuals in Sample
5000
4000
3000
2000
1000
500
250
1
2
3
4
5
6
7
8
Average Rank of Species at which First Error in Ranking is Made

Fig. 1.46. The average rank at which the first error in ranking species is made versus size of sample used for the 'count'. This is a computer model study in which a highly equitable population (H = 5.04, Fig. 1.44) and and inequitable population (H = 1.82, Fig. 1.45) were used and 50 replicate 'counts' were made at each sample size (Miller and Wiebe, in prep.).

appear. This is just about what would be predicted from Miller and Wiebe's study.

A further aspect of the problem of sampling plankton is the question of the size of sample needed (or alternatively numbers of samples needed) to define the fauna of a particular area. This is of particular interest to biogeographers, evolutionists, and palaeontologists for since most of the species present in any area are not very abundant (see Figs. 1.44 and 1.45 for example), it is necessary to have some assurance that sampling schemes are able to detect the presence of organisms whose population densities are quite low. Biogeographers need such information for the definition of species boundaries and the estimation of diversity gradients, evolutionists for the measurement of the degree of allopatry and palaeontologists for the calibration of species used as climatic indicators. This latter case is of particular importance, for a small protozoan may have a very low population density in a particular area, but its turnover rate may be high enough to provide a substantial number of fossils to the underlying sediment over the course of hundreds or thousands of years. For instance, Berger and Soutar (1967) have shown that turnover times are relatively short for certain species (twenty seven to seventy three days) and may differ among species by a factor of 2.7. Thus a species could be about one-third as abundant as some other species but could be contributing an equal number of shells per unit time to the sediment. Rather few studies of oceanic organisms have been done from this viewpoint. In a study of the effects of size of sampling device (McGowan and Fraundorf 1966) on numbers of species caught in a particular parcel of water, it was shown that nets smaller than 140 centimetres in diameter underestimate the length of the species list and that nets smaller than 40 centimetres in diameter very seriously underestimate this list. However, even the 140 centimetre net underestimated, for four replicates with this net yielded a cumulative species list of 99 while the cumulative list for all 24 tows, taken during the course of the experiment, showed a total of 140 species present in the water parcel. An additional study designed to determine the effects of size of sample (i.e. volume of water filtered) utilized a single size of net (100 cm) which was towed for various lengths of time so that the volumes of water filtered increased in approximately 100 cubic metre increments from 128 cubic metres to 1,510 cubic metres. All individuals and all species in three taxa, pelagic mollusca, Euphausiacia and larval fishes, were identified and enumerated. The information on numbers of species per sample was plotted on a graph of species versus sample size, and a regression line was fitted to the scatter diagram (Fig. 1.47). It is evident that in the case of molluscs and fish larvae, the number of species discovered to be present is dependent on the amount of water filtered. However, it is important to note that here too the largest sample seriously underestimated the number of species present as determined from a cumulative species list. The small arrows in Fig. 1.47 show the points on the regression lines where the cumulative number of species of larval fish and molluscs would fall. This may be interpreted to mean that sample sizes of from 7,000 to 10,000 cubic metres are necessary to define the fauna of even a particular parcel of water let alone an area of several thousand square kilometres. In the above studies there were a certain number of species common to all samples and these, predictably, were the most abundant species present. Thus small nets taking small samples can probably frequently detect the presence of abundant species, however, at certain times even abundant species become relatively rare. Near the periphery of their ranges abundant species almost always become rare. Since we frequently wish to compare varia-

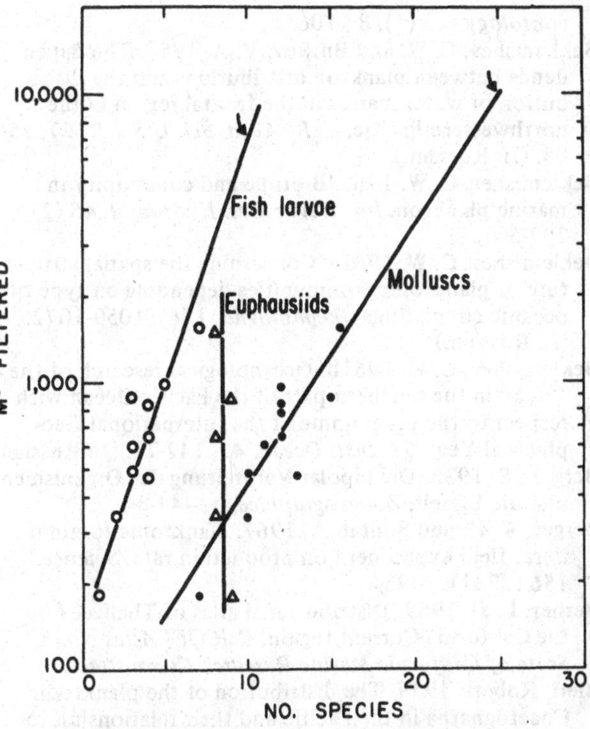
10,000
1,000
100
0
10
20
30
M³ FILTERED
NO. SPECIES
Fish larvae
Molluscs
Euphausiids

Fig. 1.47. The number of species of pelagic molluscs, Euphausiids and larval fishes found in samples of various sizes (i.e. cubic metres water filtered) taken while following a drogue. The number of species of larval fish and pelagic molluscs found are dependent on sample size, while the number of euphausiid species found is not. The arrows indicate the cumulative number of species of molluscs and fish and imply that sample sizes of from 7,500 to 10,000 cubic metres are necessary for a 'complete' species list.

tions in abundance with biological or physical gradients in the environment, it is of some importance to be able to estimate species abundance at low population densities, and as pointed out above, biogeographers need to have some knowledge of the validity of their negative data. In other words, it is important to be able to decide when a species can be considered to be 'biologically' absent. We, therefore, need to understand the sampling statistics of rare events.

Fisher and Yates (1963, pp. 1-3) have discussed the uses of χ^2 in determining the mean of a Poisson series in which the first 1/2 n terms (*n* the number of degrees of freedom equals twice the number of terms) constitute a given fraction (usually 5 per cent) of the whole. Their example (Example 1, p. 2) has to do with estimating the mean number of events (floods) occurring over a long time (twenty centuries) from the observed number (4). They wish to know the mean number for which only one century in twenty would have four serious floods or less, and the mean number for which only one century in twenty would have four floods or more. The assumption being that these events are distributed as a Poisson series over the time period. This method can be adapted for use in zoogeographic studies (McGowan 1960). For instance, if out of a series of twenty-one samples in an area none were positive for some species, what is the mean abundance of the species in the area, for which those results could be expected? There are several assumptions implicit here, one mathematical assumption is that the species is randomly distributed at low densities, other assumptions are that the sampling gear used is appropriate and that there are reasons to believe that the species might be present. An example of the latter case would be the area just to the south of the subarctic distributions in the California Current where we believe subarctic species are being swept out of their 'normal' range by this eastern boundary current but where twenty-one rather widely spaced net tows failed to catch any. We have the first term of the Poisson series so $n = 2$ and at the 5 per cent level, the χ^2 value is 5.99: since the mean of the whole Poisson series is $1/2\,\chi^2$, the expected mean density of the species is approximately three individuals per twenty-one tows. If these tows sampled, say, 500 cubic metres of water each, the expected upper limit of the average density of the species in the area would be three individuals per 10,500 cubic metres of water. If there were only five such samples, the density would be three individuals per 2,500 cubic metres and if there were 210 samples, the expected upper limit would be three individuals per 105,000 cubic metres. It is clear that in the latter case, the species if it were a small zooplankter, say *Limacina helicina,* could be considered biologically absent. That is three individuals in this volume of habitat could hardly be considered able to replace themselves or to serve as 'seed stock' for a new population, nor could they be an important competitor or prey for other species. On the other hand, a large carnivore, say yellow fin tuna, could at this density be important ecologically, and could, because of their ability to see and recognize one another, come together in a school and become the nucleus of a breeding population. Therefore, the decision as to what is the appropriate size of sample (number of negative samples) to imply 'biological absence' depends to a great degree on the characteristics of the species involved. In any event this is a useful device to apply near suspected species boundaries, and might be helpful in deciding how distinct the boundaries may be considered to be, given the amount of sampling done on either side.

Acknowledgements

The information in this paper has come from many sources and I have tried to cite these whenever possible. However, I have also derived much information from discus-

sions with colleagues; particularly E. Brinton, E. W. Fager, P. H. Wiebe, C. B. Miller and B. W. Frost. The latter three have allowed me to use some of their unpublished studies. This work was supported by the Scripps Institution's Marine Life Research Programme and the National Science Foundation, Grant GB-2861.
Received April 1968.

BIBLIOGRAPHY

Allen, W. E. 1941. Twenty years statistical study of marine plankton dinoflagellates of southern California. *The American Midland Naturalist* **26** (3), 603-35

Alvariño, A. 1961. Taxonomic revision of *Sagitta robusta* and *Sagitta ferox* Doncaster, and notes on their distribution in the Pacific. *Pacific Science,* **16** (2).

Alvariño, A. 1962. Two new Pacific chaetognaths. *Bull. Scripps. Inst. Oceanography, Univ. of Calif.* **8** (1), 1-50.

Alvariño, A. 1965. Distributional atlas of chaetognatha in the California Current region. *CalCOFI Atlas No. 2. State of California Marine Research Committee.*

Alverson, F. G. and Peterson, C. L. 1963. Synopsis of biological data on bigeye tuna *Parathunnus sibi* (Temminck and Schlegel 1844). *Proc. World Sci. Meet. Biol. Tunas and Rel. Species FAO Fish Rpts.* 2 (6), 482-514.

Andrewartha, H. G. and Birch, L. C. 1954. *The distribution and abundance of animals.* Univ. Chicago Press. 782 pp.

Anonymous. 1967. Data report, physical, chemical, biological data. Ursa Major Expedition. *Univ. Calif. Scripps Inst. of Oceanography, SIO Ref.* **67-5**.

Apstein, C. 1894. Verteilung der Salpen. *Ergebn. d. Plankton Exp. d. Humbolt-Stiftg.* **2**.

Barkley, R. A. 1964. The theoretical effectiveness of towed net samplers as related to sampler size and to swimming speed of organisms. *Jour. Conseil. Int. Explor. Mer.* **29** (2), 146-65.

Barnes, H. and Marshall, S. M. 1951. On the variability of replicate plankton samples and some applications of 'contagious' series to the statistical distribution of catches over restricted periods. *J. Mar. Biol. Assn. U.K.* **30**, 233-305.

Barnes, H. and Hasle, G. R. 1957. A statistical study of the distribution of some species of dinoflagellates in the polluted inner Oslo fjord. *Nytt. Mag. Bot.* **5**, 113-24.

Bé, A. W. H. 1959. Ecology of recent planktonic foraminifera: I areal distribution in the western North Atlantic. *Micropaleontology,* **5** (1), 77-100.

Bé, A. W. H. 1960. Ecology of recent planktonic foraminifera: II Bathymetric and seasonal distribution in the Sargasso Sea off Bermuda. *Micropaleontology,* **6** (4), 373-92.

Bé, A. W. H. and Ericson, D. B. 1963. Aspects of calcification in planktonic foraminifera (Sarcodina). *Annals of the New York Academy of Sciences,* **109** (1), 65-81.

Bé, A. W. H. 1965. The influence of depth on shell growth in *Globigerinoides sacculifer* (Brady). *Micropaleontology,* **11** (1), 81-97.

Bé, A. W. H. and Hamlin, H. 1967. Ecology of recent planktonic foraminifera. Part 3 Distribution in the north Atlantic during the summer of 1962. *Micropaleontology,* **13** (1), 87-106.

Beklemishev, C. W. and Burkov, V. A. 1958. The dependence between plankton distributions and the distribution of water masses in the frontal region in the northwestern Pacific. *C. R. Acad. Sci. U.S.S.R.* **27**, 55-65. (In Russian.)

Beklemishev, C. W. 1960. Biotope and community in marine plankton. *Int. Revue Ges. Hydrobiol.* **45** (2), 297-301.

Beklemishev, C. W. 1961a. Concerning the spatial structure of planktonic communities depending on type of oceanic circulation. *Okeanologiia,* **1** (6), 1059-1072. (In Russian.)

Beklemishev, C. W. 1961b. Oceanological research of the *Vityaz* in the northern part of the Pacific Ocean with respect to the programme of the International Geophysical Year. *Tr. Inst. Ocean.* **45**, 142-71. (In Russian.)

Berg, L. S. 1933. Die bipolar Verbreitung der Organismen und die Eiszeit. *Zoogeographica,* **1**, 444-84.

Berger, W. H. and Soutar, A. 1967. Planktonic foraminifera: field experiment on production rate. *Science,* **156** (3781), 1495-7.

Berner, L. D. 1967. Distributional atlas of Thaliacea in the California Current region. *CalCOFI Atlas No. 8. State of California Marine Research Committee.*

Bieri, Robert. 1959. The distribution of the planktonic Chaetognatha in the Pacific and their relationship to the water masses. *Limnol. and Oceanog.* **4** (1), 1-28.

Blackman, Abner. 1966. Pleistocene stratigraphy of cores from the southeast Pacific Ocean. Ph.D. Thesis, Univ. of California, San Diego, 200 pp.

Blackman, Abner and Somayajulu, B. L. K. 1966. Pacific pleistocene: faunal analyses and geochronology. *Science,* **154** (3751), 886-9.

Bowman, T. E. 1960. The pelagic amphipod genus *Parathemisto* (Hyperiidea: Hyperiidae) in the North Pacific and adjacent Arctic Ocean *Proc. U.S. Nat. Mus., Smithsonian Inst.* **112** (3439), 343-92.

Bradshaw, J. S. 1959. Ecology of living planktonic foraminifera in the north and equatorial Pacific Ocean. *Contrib. Cush. Fd. Foram. Res.* **10** (2), 25-64.

Bray, J. R. 1958. Notes towards an ecologic theory. *Ecology,* **39**, 770-6.

Brinton, E. 1962. The distribution of Pacific euphausiids. *Bull. Scripps Inst. Oceanogr. of the Univ. of Calif.* **8** (2), 51-270.

Brinton, E. 1967. Distributional atlas of Euphausiacea (Crustacea) in the California Current region, Part I. *CalCOFI Atlas No. 5 State of California Marine Research Committee.*

Brock, V. E. 1959. The tuna resource in relation to oceanographic features. *Circ. U.S. Fish and Wildlife Serv.* **65**, 1-11.

Brodsky, K.A. 1960. Zoogeographical zones of the south part of the Pacific Ocean and bipolar distributions of certain Calanoida. *Trud. Okean. Kom.* **10** (4), 8-13. (In Russian.)

Brodsky, K. A. 1965. Variability and systematics of the species of the genus Calanus (Copepoda). 1. *Calanus pacificus* Brodsky, 1948, and *Calanus sinicus* Brodsky. sp.n. *Akad. Nauk U.S.S.R. Zool. Inst.* **3** (11), 22-71 (In Russian.)

Cassie, R. M. 1959a. Microdistribution of plankton. *New Zealand Jour. Sci.* 2 (3), 398-409.

Cassie, R. M. 1959b. Some correlations in replicate plankton samples. *New Zealand Jour. Sci.* 2 (4), 473-84.

Cassie, R. M. 1963. Microdistribution of plankton. *Oceanogr. Mar. Biol. Ann. Rev.* (H. Barnes ed.), 1 223-52.

Chun, C. 1897. *Die Beziehungen zwischen dem Arktischen und Antarktischen Plankton.* Stuttgart.

Colebrook, J. M. 1965. Continuous plankton records: a principle component analysis of the geographical distribution of zooplankton. *Bull. Marine Ecol.* **6** (3), 78-100.

Cushing, D. H. 1953. Studies on plankton populations. *J. Conseil Perm. Intern. Expl. Mer.* **19**, 3-22.

Cushing, D. H. 1962. Patchiness. *Rappt. Proces-Verbaux Reunions, Conseil Perm. Intern. Expl. Mer.* **153**, 152-63.

Darlington, P. J. 1957. *Zoogeography: the geographical distribution of animals.* John Wiley & Sons, Inc., New York.

Defant, A. 1961. *Physical Oceanography* I, Pergamon Press. 729 pp.

Dodimead, A. J., Favorite, F. and Hirano, T. 1963. Review of the oceanography of the subarctic Pacific region in: Salmon of the North Pacific Ocean Part II. *International North Pacific Fisheries Comm. Bull.* **13**, 1-195.

Ehrlich, P. R. and Holm, R. W. 1963. *The process of evolution.* McGraw-Hill, New York. 347 pp.

Ehrlich, P. R. and Birch, L. C. 1967. The 'balance of nature' and 'population control'. *Amer. Nat.* **101**, 97-107.

Ekman, S. 1953. *Zoogeography of the sea.* Sidgwick and Jackson, London. 417 pp.

Elton, C. 1927. *Animal ecology.* London.

Elton, C. 1942. *Voles, mice, and lemmings: problems in population dynamics.* Oxford.

Elton, C. and Nicholson, M. 1942. The ten year cycle in numbers of the lynx in Canada. *Jour. An. Ecol.* **11** (2), 215-44.

Elton, C. S. 1958. *The ecology of invasions by animals and plants.* London.

Elton, C. S. 1966. *The pattern of animal communities.* Methuen and Co., London.

Emiliani, C. 1954. Depth habits of some species of pelagic foraminifera as indicated by oxygen isotope ratios. *Amer. Jour. Sci.* **252**, 149-58.

Emiliani, C. 1955. Pleistocene temperatures. *Jour. Geol.* **63** (6), 538-78.

Emiliani, C. 1966. Palaeotemperature analysis of Caribbean cores P6304-8 and P6304-9 and a generalized temperature curve for the past 425,000 years. *Jour. Geol.* **74** (2), 109-62.

Emiliani, C. and Milliman, J. D. 1966. Deep-sea sediments and their geological record. *Earth Science Review,* **1**, 105-32.

Ericson, D. B., Wollin, G. and Wollin, J. 1954. Coiling direction of *Globorotalia truncatulinoides* in deep-sea cores. *Deep-Sea Research,* 2, 152-8.

Ericson, D. B. and Wollin, G. 1956a. Correlation of six cores from the equatorial Atlantic and the Caribbean. *Deep-Sea Research,* 2, 104-25.

Ericson, D. B. and Wollin, G. 1956b. Micropalaentological and isotopic determinations of pleistocene climates. *Micropaleontology,* 2 (3), 257-70.

Ericson, D. B. 1959. Coiling direction, of *Globigerina pachyderma* as a climatic index. *Science,* **130** (3369), 219-20.

Ericson, D. B., Ewing, M. and Wollin, G. 1963. Pliocene-Pleistocene boundary in deep-sea sediments. *Science,* **139** (3556), 727-37.

Ericson, D. B., Ewing, M. and Wollin, G. 1964. The pleistocene epoch in deep-sea sediments. *Science,* **146** (3645), 723-32.

Fager, E. W. 1957. Determination and analysis of recurrent groups. *Ecology,* **38** (4), 586-95.

Fager, E. W. 1963. Communities of organisms. *The Sea,* 2, (M. N. Hill *ed.*). 415-37. Interscience Pub., New York.

Fager, E. W. and McGowan, J. A. 1963. Zooplankton species groups in the North Pacific. *Science,* **140** (3566), 453-60.

Fisher, R. A., Corbet, A. S. and Williams, C. B. 1943. The relation between the number of species and the number of individuals in a random sample of an animal population. *Jour. Animal. Ecol.* **12**, 42-58.

Fisher, R. A. and Yates, F. 1963. *Statistical tables for biological, agricultural, and medical research.* Sixth ed. Hafner Pub. Co., New York.

Fleminger, A. 1964. Distributional atlas of calanoid copepods in the California Current. *CalCOFI Atlas No. 2., State of California Marine Research Committee.*

Fleminger, A. and Clutter, R. I. 1965. Avoidance of towed nets by zooplankton. *Limnol. and Oceanog.* **10**, 96-104.

Fofonoff, N. P. 1960. Transport computations for the North Pacific Ocean 1958. *Fish. Res. Bd. Canada, Ms Rept. Oceanogr. and Limnol. Series* **80**, 11 pp.

Ford, E. B. 1964. *Ecological genetics.* Methuen Co., London.

Frank, P. 1957. Coaction in laboratory populations of two species of Daphnia. *Ecology,* 38 (3), 510-19.

Frost, B. W. and Fleminger, A. (in press). A revision of the genus *Clausocalanus* (Copepoda: Calanoida) with remarks on distributional patterns in diagnostic characters. *Bull. Scripps Inst. Oceanography, Univ. of California.*

Gause, G. F. 1934. *The struggle for existence.* Williams & Williams, Baltimore. 163 pp.

Giesbrecht, W. 1892. Systematik und Faunistik der pelagischen Copepoden des Golfes von Neapel und der angrenzenden Meersabschnitte. *Fauna und Flora Golf. Neapel,* **19**.

Graham, H. W. and Bronikowsky, N. 1944. The genus *Ceratium* in the Pacific and north Atlantic Oceans. *Sci. Res. Cruise VII Carnegie, Biology V, Carnegie Inst. Wash., Pub.* **565**.

Haecker, V. 1908. Tiefsee-Radiolarian. *Wiss. Ergebn. Deutsch. Tiefsee Exp.* **14**.

Hairston, N. G., Smith, F. E. and Slobodkin, L. B. 1960. Community structure, population control and competition. *The Amer. Natur.* **94**, 421-5.

Hardy, A. C. 1924. The herring in relation to its animate environment. Part I The food and feeding habits of the herring. *Fish. Invest. London. Ser. II,* **7** (3), 1-53.

Hardy, A. C. and Gunther, E. R. 1936. Plankton of the

south Georgia whaling grounds and adjacent waters, 1926-1927. *Discovery Rpts.* **11**, 1-456.

Havanan, M. G. and Tanonaka, G. K. 1959. Experimental fishing to determine distributions of salmon in the North Pacific Ocean and Bering Sea, 1956. *U.S. Fish and Wildlife Serv., Spec. Sci. Rpt. Fisheries No.* **302**, 1-22.

Hedgpeth, J. W. 1957. Concepts of marine ecology. *Treatise on marine ecology and palaeocology*, **I**, 29-52.

Hedgpeth, J. W. 1964. Evolution of community structure. In *Approaches to Paleoecology*. (J. Imbrie and N. Newell, *eds.*), 11-18. John Wiley & Sons, Inc.

Holmes, R. W. and Widrig, T. M. 1956. The enumeration and collection of marine phytoplankton. *J. Conseil Perm. Inter. Expl. Mer.* **22**, 19-32.

Hubbs, C. L. 1952. Antitropical distribution of fishes and other organisms. *Seventh Pac. Sci. Cong.* **3**, 324-9.

Hubbs, C. L., Mead, G. W. and Wilimovsky, N. J. 1953. The widespread and probably antitropical distribution and the relationship of the bathypelagic Iniomous fish *Anotopterus pharao. Bull. Scripps Inst. Oceanog. Univ. of California*, **6** (5), 173-98.

Hutchinson, G. E. 1953. The concept of pattern in ecology. *Proc. Acad. Nat. Sci. Phil.* **105**, 1-12.

Hutchinson, G. E. 1957. Concluding remarks. *Cold Spring Harbor Symposia on Quantitative Biology* **22**, 415-27.

Hutchinson, G. E. 1967. *A treatise on limnology. Introduction to lake biology and the limnoplankton*, **2**. John Wiley & Sons, Inc., New York.

Johnson, M. W. and Brinton, E. 1963. Biological species, water masses and currents. In *The Sea, vol. 2*, (M. N. Hill *ed.*), 381-414. Intersci. Pub., New York.

Jones, J. I. (in press). The relationship of planktonic foraminiferal populations to water masses in the western Caribbean and lower Gulf of Mexico. *Proc. II Mexican Oceanographic Cong., 1967.*

Kanaya, Taro and Koizumi, Itaru. 1966. Interpretation of diatom thanatocoenoses from the north Pacific applied to a study of core V20-130 (Studies of a deep-sea core V20-130 Part IV), *Science Rpts. of the Tohoku University, Sendai, Second Series (Geology)*, **37** (2), 89-130.

Keith, Lloyd B. 1963. *Wildlife's ten-year cycle.* The University of Wisconsin Press, Madison.

Kendall, M. G. 1955. *Rank correlation methods* (2nd ed.). Chas. Griffen, London.

Kling, S. A. 1966. Castanellid and Circoporid radiolarians: Systematics and zoogeography in the eastern north Pacific. *Ph.D. dissertation, Univ. Calif. San Diego, La Jolla, California.*

Kofoid, C. A. 1934. The distribution of pelagic ciliates in the eastern tropical Pacific. *Fifth Pacific Sci. Congress,* **3**.

Lack, D. L. 1954. *The natural regulation of animal numbers.* Clarendon Press, Oxford.

Lang, B. T. 1965. Taxonomic review and geographical survey of the copepod genera *Eucalanus* and *Rhincalanus* in the Pacific Ocean. Ph.D. dissertation, Univ. Calif., San Diego, La Jolla, California.

Lindeman, R. L. 1942. The tropho-dynamic aspect of ecology. *Ecology* **23** (4), 399-418.

Lloyd, M. and Ghelardi, R. J. 1964. A table for calculating the 'equitability' component of species diversity. *Jour. Anim. Ecol.* **33**, 217-25.

Lohmann, H. 1933. Appendiculariae, Kükenthal and Krumbach. *Handbuch d. Zool.* **5** (2).

Lotka, Alfred J. 1956. *Elements of mathematical biology.* Dover Pub., Inc., New York.

MacArthur, R. H. 1955. Fluctuations of animal populations and a measure of community stability. *Ecology* **36** (3), 533-36.

MacArthur, R. 1965. Patterns of species diversity. *Biol. Rev.* **40**, 510-33.

Macfadyen, A. 1957. *Animal ecology aims and methods.* Sir Isaac Pitman & Sons, London.

McGowan, J. A. 1960. The systematics, distribution, and abundance of the Euthecosomata of the north Pacific. Ph.D. dissertation, Univ. of Calif., San Diego, La Jolla, California.

McGowan, J. A. 1963. Geographical variation in *Linacina helicina* in the North Pacific. Speciation in the sea. *Systematics Assn Pub. No. 5, Brit. Mus. Nat. Hist., London,* 109-28.

McGowan, J. A. and Fraundorf, V. J. 1966. The relationship between size of net used and estimates of zooplankton diversity. *Limnol. and Oceanog.* **11** (4), 456-69.

McGowan, J. A. 1967. Distributional atlas of pelagic molluscs in the California Current region. *CalCOFI Atlas No. 6. State of California Marine Research Committee.*

Margalef, D. R. 1958. Information theory in ecology. *General systems,* **3**, 36-71.

Marr, J. W. S. 1962. The natural history and geography of the Antarctic krill (*Euphausia superba* Dana) *Discovery Rpts.* **32**, 33-464.

Marshall, N. B. 1963. Diversity, distribution, and speciation of deep-sea fishes. Speciation in the sea. *Systematics Assn Pub. No. 5, Brit. Mus. Nat. Hist., London,* 181-95.

Maynard Smith, J. 1966. Sympatric speciation. *The Amer. Naturalist.* **100** (916), 637-50.

Mayr, E. 1942. *Systematics and the origin of species.* Colum. Univ. Press.

Mayr, E., Linsley, E. G. and Usinger, R. L. 1953. *Methods and principles of systematic zoology.* McGraw-Hill, Inc.

Mayr, E. 1963. *Animal species and evolution.* Harvard Univ. Press, Cambridge, Mass.

Meisenheimer, J. 1906. Die Pteropoden der Deutschen Sudpolar-Expedition 1901-03. *Deuttsche Südp. Exp.* **9** (*Zool.* **1**).

Moser, F. 1915. Die geographische Verbreitung und das Entwicklungszentrum der Rohrenquallen. *S. B. Ges. Naturf.* Freunde Berlin.

Mullin, M. 1966. Selective feeding by calanoid copepods from the Indian Ocean. *Some Contemporary Studies in Marine Science.* (Harold Barnes *ed.*), George Allen & Unwin, London: 545-54.

Murphy, G. I. 1965. Population dynamics of the Pacific sardine (*Sardinops caerulea*). Ph.D. dissertation, Univ. of Calif, San Diego, La Jolla, California.

Parin, N. V. 1960. The range of the saury (Cololabis Saira Brev. Scombresocidae, Pisces) and effects of oceanographic features on its distribution. *Doklady Akademii*

Nauk U.S.S.R. **130** (3), 649-52. (In Russian.)

Park, T. 1962. Beetles, competition and populations. *Science,* **138**, 1369-75.

Patten, B. C. 1959. An introduction to the cybernetics of the ecosystem trophodynamic aspect. *Ecology,* **40** (2), 221-31.

Patten, B. C. 1963. Plankton: optimum diversity structure of a summer community. *Science,* **140**, 894-8.

Patten, B. C. 1964. The plankton community. *Science,* **144**, 557-8.

Patten, B. C. 1966. The biocoenetic process in an estuarine phytoplankton community. *Oak Ridge National Lab., ORNL-3946, UC-48, Biology and Medicine.*

Pearl, R. 1930. *The biology of population growth.* Knopf, New York.

Phleger, F. B. 1951. Ecology of foraminifera, northwest Gulf of Mexico. Pt. 1, Foraminifera distribution. *Geol. Soc. Amer. Mem.* **46**, 1-88.

Phleger, F. B. and Parker, F. L. 1951. Ecology of foraminifera, northwest Gulf of Mexico. Part I Foraminifera distribution, Part II Foraminifera species. *Geol. Soc. Amer. Mem.* **46**, 1-64.

Phleger, F. B., Parker, F. L. and Peirson, J. F. 1953. North Atlantic foraminifera. *Repts. Swedish Deep-Sea Exped.* **7**, 1947-1948. **7** (1), 122 pp.

Piroznikov, P. L. 1937. A contribution to the study of the origin of the northern elements in the fauna of the Caspian Sea. *C.R. Acad. Sci. U.S.S.R.* **15**.

Preston, F. W. 1948. The commoness and rarity of species. *Ecology,* **29**, 254-283.

Raymont, J. E. G. 1963. *Plankton and productivity in the oceans.* MacMillan Co. New York.

Redfield, A. C., Ketchum, B. H. and Richards, F. A. 1963. The influence of organisms on the composition of sea water. *The Sea,* (M. N. Hill *ed.*), 2, 26-77. Intersci. Pub. New York.

Reid, J. L., Roden, G. I. and Wyllie, J. G. 1958. Studies of the California Current system. Prog. *Rep. Calif. Coop. Ocean. Fish. Invest., 1 July 1956 to 1 January 1958*, 27-57.

Reid, J. L. 1960. Oceanography of the northeastern Pacific during the last ten years. *Calif. Coop. Ocean. Fish. Invest. Rpts,* **7**, 77-90.

Reid, J. L. 1962. On circulation, phosphate-phosphorus content and zooplankton *Volumes* in the upper part of the Pacific Ocean *Limnol. and Oceanog.* **1** (2), 287-306.

Riley, G. A. 1963. Theory of food chain relations in the ocean. *The Sea,* **2**, (M. N. Hill *ed.*), 438-63. Intersci. Pub. New York.

Russell, F. S. 1935. A review of some aspects of zooplankton research. Rappt. Proces-. Verbaux. Reunions Conseil. Intern. Expl. Perm. Mer. **95**, 5-30.

Ryther, J. H. 1963. Geographic variations in productivity. *The Sea,* **2**, (M. H. Hill *ed.*), 347-80. Intersci. Pub. New York.

Schaefer, M. B., Broadhead, G. C. and Orange, C. J. 1963. Synopsis on the biology of yellow fin tuna. *Thunnus albacares* (Bonnaterre 1788) (Pacific Ocean). *Proc. World Sci. Meet. on the Biol. Tunas and Rel. sp. FAO Fish. Rpts.* **2** (6), 538-61.

Schaefer, M. B. 1967. Dynamics of the fishery for the anchoveta *Engraulis ringens* off Peru. *Instituto del Mar del Peru,* **1** (5), 190-303.

Shackleton, N. 1967. Oxygen isotope analyses and Pleistocene temperatures re-assessed, *Nature, Lond.* **215**, 15-17.

Sheard, K. 1965. Species groups in the zooplankton of eastern Australian slope waters, 1938-41. *Aust. Jour. Mar. Freshwater. Res.* **16** (2), 219-254.

Simpson, E. H. 1949. Measurement of diversity. *Nature, Lond.* **163**, 688 pp.

Slobodkin, L. B. 1962. *Growth and regulation of animal populations.* Holt, Richart & Winston, New York.

Slobodkin, L. B., Smith, F. E. and Hairston, N. G. 1967. Regulation in terrestrial ecosystems and the implied balance of Nature. *The Amer. Naturalist,* **101**, 109-24.

Smayda, T. J. 1958. Biogeographical studies of marine phytoplankton. *Oikos,* **9** (2), 158-91.

Smith, P. E., Counts, R. C. and Clutter, R. I. (in press). Changes in filtering efficiency of plankton nets due to clogging under tow. *J. Conseil Perm. Intern. Expl. Mer.* **32** (7).

Soutar, A. 1967. The accumulation of fish debris in certain California coastal sediments. *Calif. Coop. Ocean. Fish. Invest. Rpts,* **11**, 136-39.

Spector, W. S. 1956. *Handbook of biological data.* W. B. Saunders Co., Phil. 533 pp.

Steel, J. H. 1961. Primary production in: Oceanography. *A.A.A.S. Pub.* **67**, 519-38.

Steuer, Adolf. 1933. Zur planmässigen Erforschung der geographischen Verbreitung des Haliplanktons, besonders der Copepoden. Zoogeographica; *International Review for Comp. and Causal Animal Geog.* **1** (3), 269-302.

Sverdrup, H. V., Johnson, M. W. and Fleming, R. H. 1942. *The oceans their physics, chemistry, and general biology.* Prentice-Hall Inc., New York.

Takenouti, Y. 1960. The 1957-1958 oceanographic changes in the western Pacific. *Calif. Coop. Ocean. Fish. Invest. Rpts,* **7**, 67-76.

Tebble, Norman. 1960. The distribution of pelagic polychaetes in the south Atlantic Ocean. *Discovery Rpts.* **30**, 161-300.

Tebble, Norman, 1962. The distribution of pelagic polychaetes across the North Pacific Ocean. *Bull. British Museum (Natural History) Zoology,* **7** (9), London.

Tonolli, V. 1949. Stuttura spaziale del popolamento mesoplanctico, etergeneita della densita dei popolamenti orizzontale e sua variazione in funzione della quota. *Mem. Inst. ital. Idrobiol.* **5**, 189-208.

Uda, M. 1957. A consideration on the long years trend of the fisheries fluctuation in relation to sea conditions. *Bull. Japanese Soc. Sci. Fish.* **23**, 368-72.

Uda, M. 1961. Fisheries oceanography in Japan, especially on the principles of fish distribution, concentration, dispersal and fluctuation. *Calif. Coop. Ocean. Fish. Invest. Rpts,* **8**, 25-31.

Verhulst, P. F. 1838. Notice sur la loi que la population suit dans son accroissement. *Corr. Math. et Phys.* **10**, 113 pp.

Volterra, V. 1928. Variations and fluctuations of the number of individuals in animal species living together.

J. Conseil Perm. Intern. Expl. Mer, 3, 1-51.
von Bertallanfy, L. 1950. Theory of open systems in biology. *Science,* 3, 23-9.
Wiebe, P. H. and Holland, W. R. 1968. Plankton patchiness: effects on repeated net tow. *Limnol. and Oceanog.* 13 (2), 315-21.
Williams, C. B. 1964. *Patterns in the balance of nature.* Academic Press, London.
Williamson, M. H. 1961. An ecological survey of a Scottish herring fishery. Part IV: changes in the plankton during the period 1949 to 1959. Appendix: a method for studying the relations of plankton variations to hydrography. *Bull. Mar. Ecol.* 5, 207-29.
Windsor, C. P. and Clarke, G. L. 1940. A statistical study of variation in the catch of plankton nets. *J. Mar. Res.* 3, 1-34.
Wisner, R. L. 1959. Distribution and differentiation of the North Pacific Myctophid fish, *Tarletonbeania taylori. Copeia,* 1-7.
Wooster, W. S. and Cromwell, T. 1958. An oceanographic description of the eastern tropical Pacific. *Bull. Scripps Inst. Oceanog. Univ. Calif., San Diego, La Jolla, California,* 7 (3), 169-282.
Wooster, W. S. and Reid, J. L. 1963. Eastern boundary currents. *The Sea,* 2, (M. N. Hill *ed.*), 253-80. Intersci. Pub. New York.

2. DISTRIBUTION OF PLANKTON AS RELATED TO MICROPALAEONTOLOGY

C. W. BEKLEMISHEV

Summary: Pelagic communities affect the underlying thanatocoenoses in two ways: (1) dead plankters descend to the bottom, and (2) benthonic communities depend very much, through the type of sediment, on pelagic productivity. Pelagic communities of the oceans have many depth zones and the difference between adjacent communities can be most easily seen in the most variable zone. This proves to be the zone of sub-surface water, as the isolation of large-scale gyres is most complete there. The rule 'one gyre – one species' almost never holds. Consequently, complex communities exist which are made up of terminal, terminal-cyclical and cyclical bioceonoses. These are basic features of the biological structure of the oceans. Eutrophic and oligotrophic areas also fit this pattern.

Palaeontologists study thanatocoenoses. Thanatocoenoses are made up of portions of both pelagic and bottom communities. It is the communities rather than individual species that should be investigated in order to understand thanatocoenoses. This is especially clear in the case of productivity, which is a quality of communities, not of species. The type of productivity cycle of a community may affect the underlying thanatocoenosis.

Productive cycles of pelagic communities may or may not be well balanced. If the productive cycle is *not* well balanced there is an excess of production which is not utilized in mid-water and which can therefore reach the bottom. In the *well*-balanced oligotrophic communities little organic matter reaches the bottom. In eutrophic communities more organic matter reaches the bottom, even if their productive cycles are fairly well balanced, since the balance between production and grazing is never complete. Thus the efficiency of productive cycles affects the rate of sedimentation of organic matter.

In bottom areas which differ in the rate of sedimentation of organic matter, benthonic communities differ too. If the rate is high, eutrophic communities develop, which are dominated by deposit feeders, if it is low, oligotrophic communities develop, which are dominated by suspension feeders (Sokolova and Neyman 1966) (see Fig. 2.1). Pelagic productivity can therefore affect sedimentation rate, bottom life and, eventually, the benthonic portion of thanatocoenoses.

Pelagic communities are also reflected on the bottom, and thus these also influence the thanatocoenoses. To quote an example: in the tropical parts of the Indian and Pacific oceans squid beaks are numerous in the bottom sediment exactly underneath areas rich in plankton. Sometimes this resemblance is very close and can even be seen in minor trends of distribution (Belyaev 1962, 1968) (see Fig. 2.2).

Oceanic pelagic communities are made up of many depth zones. One might expect that the deeper zones would contribute less to the thanatocoenoses, since the concentration of plankton decreases with depth. The layer 0-500 metres contains 65 per cent of the plankton biomass found in the 0-4000 metre layer (Vinogradov 1965) (see Fig. 2.3). On the other hand, in the boreal Pacific Reshetnjak (1955) found five zones differing in the specific composition of Phaeodaria: 0-50, 50-200, 200-1,000, 1,000-2,000, and 2,000-8,000 metres. The Phaeodaria are most abundant both in numbers and in species in the 200-1,000 metre zone (i.e. in the warm intermediate layer). For crustaceans and arrow-worms the layers are different: 0-200, 200-500, 500-6,000, and 6,000-8,000 metres (Bierstein, Vinogradov and Tchindonova 1954), but these animals are not so well preserved in the sediment. It is the quantity of durable organisms, that are not necessarily surface living, that counts.

Polar and sub-polar waters are much more isothermal than tropical waters. This difference results in differences between the vertical structure of temperate and tropical communities. In the tropics, the main zones of pelagic communities are chiefly limited by thermal boundaries (Beklemishev 1966), while in the temperate waters they

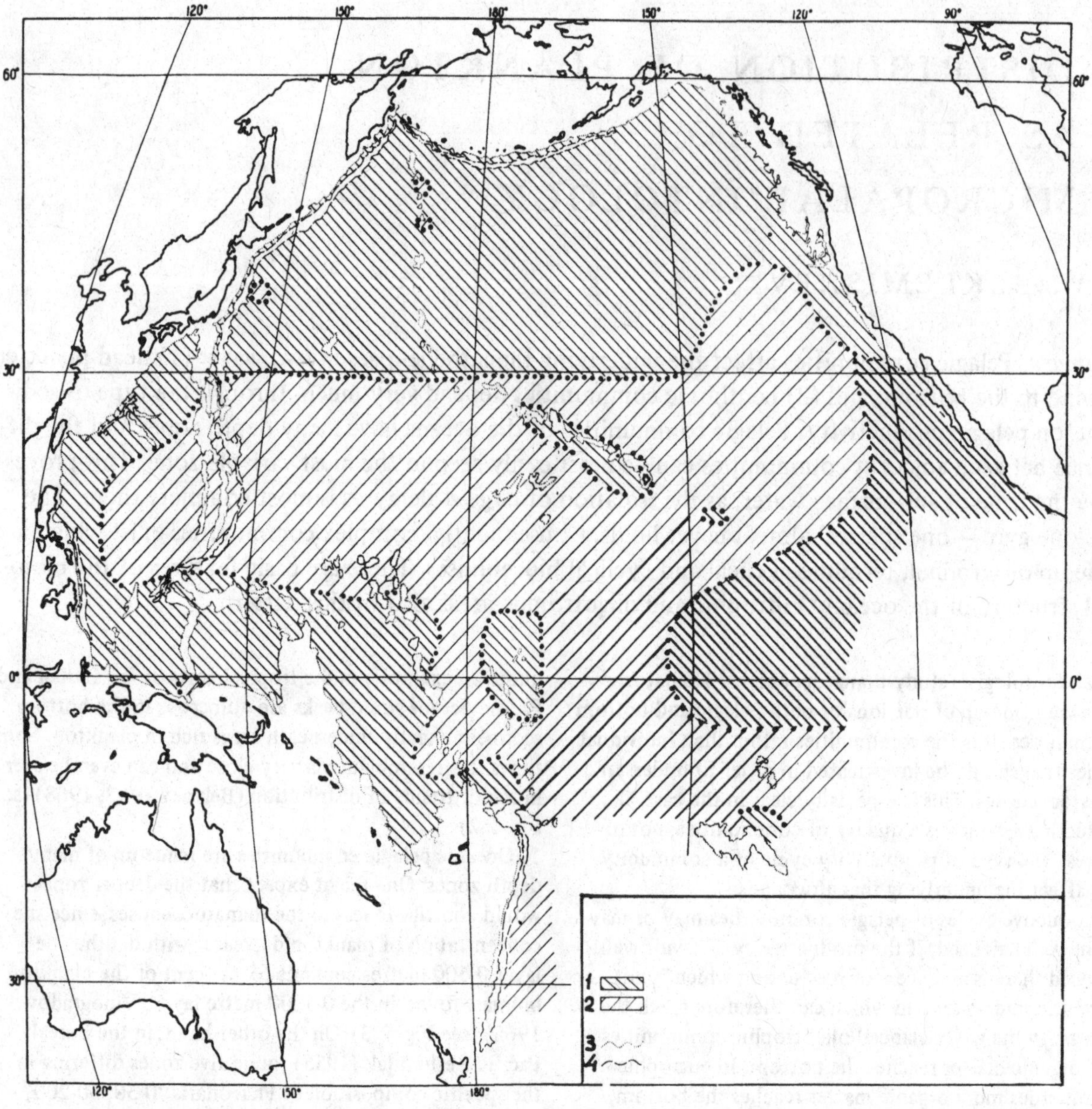
120°
150°
180°
150°
120°
90°
60°
30°
0°
1
2
3
4

Fig. 2.1. Eutrophic and oligotrophic areas of the North Pacific bottom communities (dominated by deposit feeders and suspension feeders respectively) between 3,000 and 6,000 metres. 1 = eutrophic areas; 2 = oligotrophic areas; 3 = 3,000 metre depth contour; 4 = 6,000 metre depth contour (after Sokolova and Neyman 1966).

are limited by the boundaries of water masses (Vinogradov 1965) (see Fig. 2.4).

Now consider the whole community, i.e. sequence of zones from the surface to the bottom (Fig. 2.5). Both in temperate and tropical waters their lateral boundaries coincide with the boundaries of water masses in the large-scale gyres. These gyres are best isolated from each other in the intermediate waters (Burkov 1966). Consequently, there are as many communities in the ocean as large-scale gyres in this layer. Perhaps this is why ranges of phytoplankton are more extensive than those of zooplankton that migrate between zones. Deep-sea plankton species have more extensive ranges again.

The boundaries of gyres fluctuate, and so do the boundaries of ranges. Consequently, maps of the average distribution of species based on all occurrences, are not ranges in the exact biogeographical meaning of the word, but are fully comparable with the boundaries of thanatocoenoses.

Large-scale communities are not quite homogeneous. Temporary small-scale variations always occur, and segments of communities may have peculiarities due to their

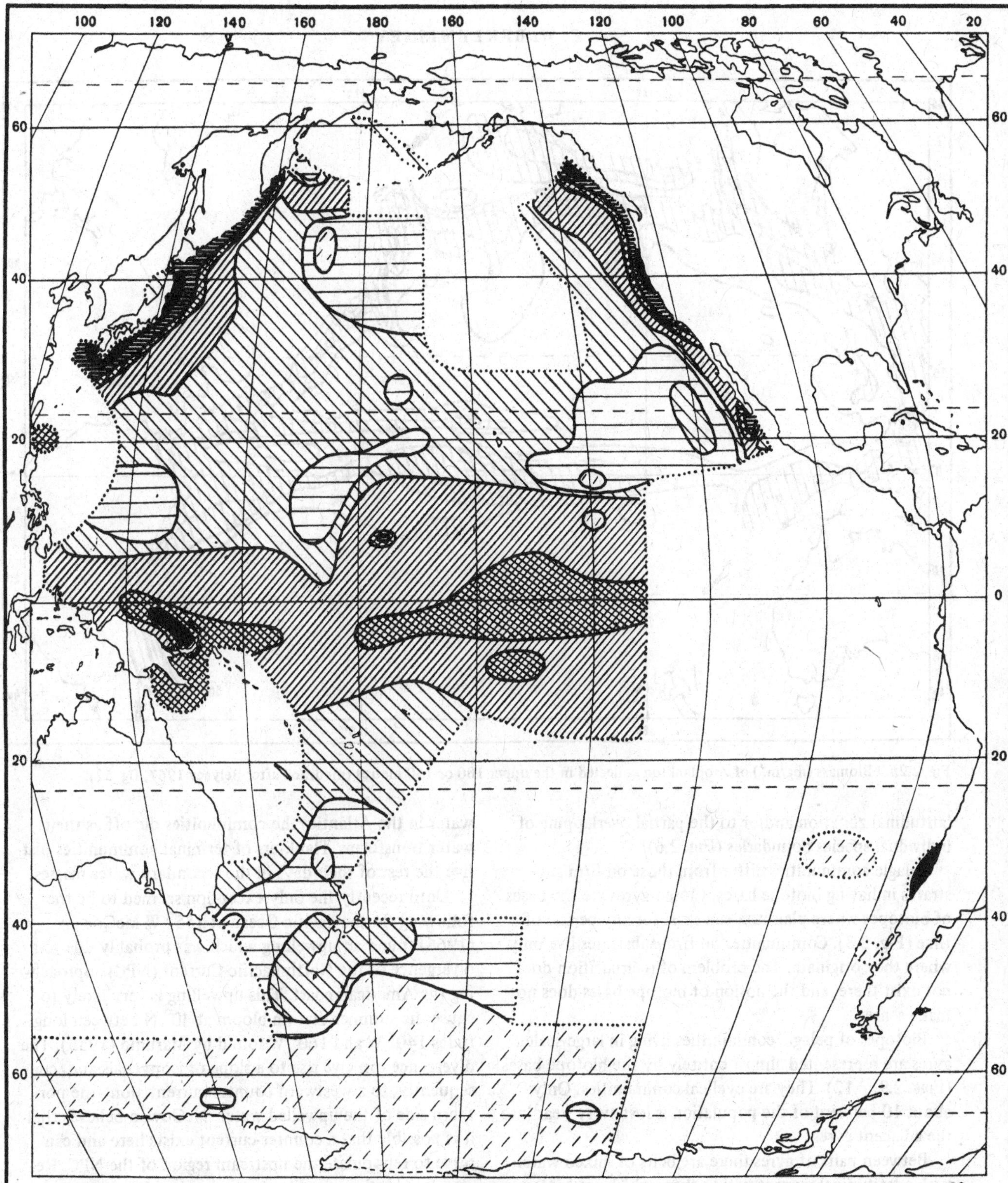
100 120 140 160 180 160 140 120 100 80 60 40 20
60
40
20
0
20
40
60
1
2
3
4
5
6
7
8

Fig. 2.2*a*. Number of squid beaks per square metre in bottom sediment of the Pacific Ocean beyond the 300 metre depth contour. 1 = more than 1,000; 2 = 500-1,000; 3 = 300-500; 4 = 100-300; 5 = 25-100; 6 = less than 25; 7 = no beaks; 8 = boundaries of surveyed areas (after Belyaev 1968).

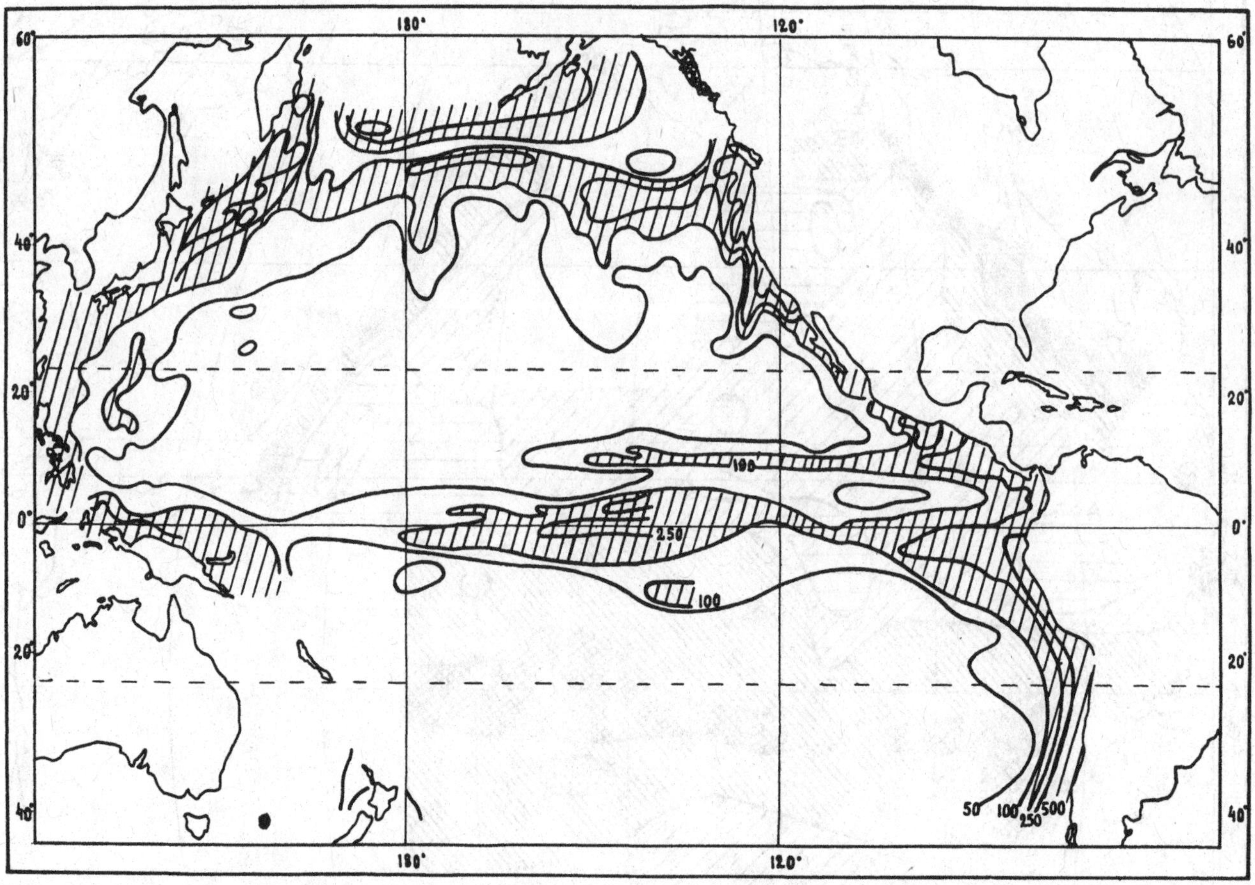
180°
120°
60°
40°
20°
0°
20°
40°
190
250
100
50
100
250
500
180°
120°

Fig. 2.2*b*. Biomass (mg/m^3) of zooplankton collected in the upper 100 or 150 metres (modified after Belyaev 1967, fig. 52).

latitudinal zonation and/or to the partial overlapping of individual species boundaries (Fig. 2.6).

Pelagic communities differ from those on firm substrates in having biotope bases. Closed gyres are the bases of biotopes where plankton can exist for any period of time (Fig. 2.8). Communities on firm substrates live only where they originate. The problem of recirculation does not exist there, and the notion of biotope bases does not make sense.

Biotopes of pelagic communities living in large-scale gyres are represented almost entirely by the biotope bases (Figs. 2.9, 2.12). They are cyclical communities. Only about 10 per cent of the population is lost every year into the adjacent gyres.

Between pairs of gyres there are belts of mixed water with a latitudinal current within them which are biotopes of mixed, secondary, or transition zone communities (Figs. 2.9, 2.12). These are terminal communities as their biotopes have no bases at all. No recirculation is possible here. Terminal communities disintegrate when the water which carried them diverges approaching the coast. In submerged secondary water masses (like the Mediterranean water in the Atlantic) the communities die off as their water transforms. Plankters of terminal communities just live the rest of their days in the secondary water masses.

Until recently the only exception seemed to be the following. In the Pacific Ocean at 155° W McGowan (1966) found an upwelling which was probably due to a divergence of the North Pacific Current (NPC) approaching the American coast. This upwelling is very likely to cause the summer diatom bloom at 40° N between longitudes 140° W and 160° W found by Marumo (1955). The divergence can give rise to a slope of isopycnals and, consequently, to a westward counter-current along the northern side of the upwelled water. McGowan believes that it is possible that a counter-current exists here and can serve to repopulate the upstream region of the NPC. Recirculation of species like *Lampanyctus ritteri* can seemingly be achieved by this counter-current. It is, nevertheless, hardly possible that the counter-current exists in the NPC proper since there is neither noticeable divergence, nor any upwelling in it. Consequently, no return mechanism is apparent in the NPC. If so, the plankton of the NPC forms a terminal community.

Nearer to the coast, especially in neutral cols between the large-scale gyres (Fig. 2.7), medium-scale, stable eddies can be populated by terminal-cyclical communities. Much water is lost here through the secondary water masses (about 50 per cent per year). These are distant-neritic communities. The myctophid *Lampanyctus ritteri* mentioned above is a distant-neritic species (Figs. 2.9, 2.10, 2.11).

There are very few species which have the bases of their ranges in one gyre only (Fig. 2.12). Most of them inhabit several gyres. Thus complex communities exist with several bases of biotopes. The simplest of them comprise: (1) a primary cyclical community in a large-scale gyre, (2) two terminal communities, and (3) four terminal-cyclical communities, which this primary community shares with two other adjacent primary communities (Fig. 2.13). In the ocean there are typically five primary communities: two temperate, two central and one equatorial (living in two adjacent gyres) (Fig. 2.14). Central communities are the most oligotrophic among them. Their productive cycle is balanced better than any others.

The entire complex community of the whole ocean looks rather like a huge figure 8 (Fig. 2.15). The empty places of it are halostases of central waters with oligotrophic communities, and the peripheries of halostases are made up of less well-balanced and/or more productive equatorial, distant-neritic and transition communities. Temperate communities are even more productive (Fig. 2.16).

On the bottom, under the eutrophic pelagic communities, deposit feeders live on a plethora of descending organic matter. Under the oligotrophic communities live benthonic suspension feeders. The specific composition of eutrophic pelagic communities differs from that of oligotrophic communities. Since the endemic plankters of both are deposited, and since the bottom communities differ in feeding habits too, the spatial structure of the whole thanatocoenosis is the same as that of the pelagic communities. Some additional variation is produced by the modifying effect of bottom relief.

Received September 1967.

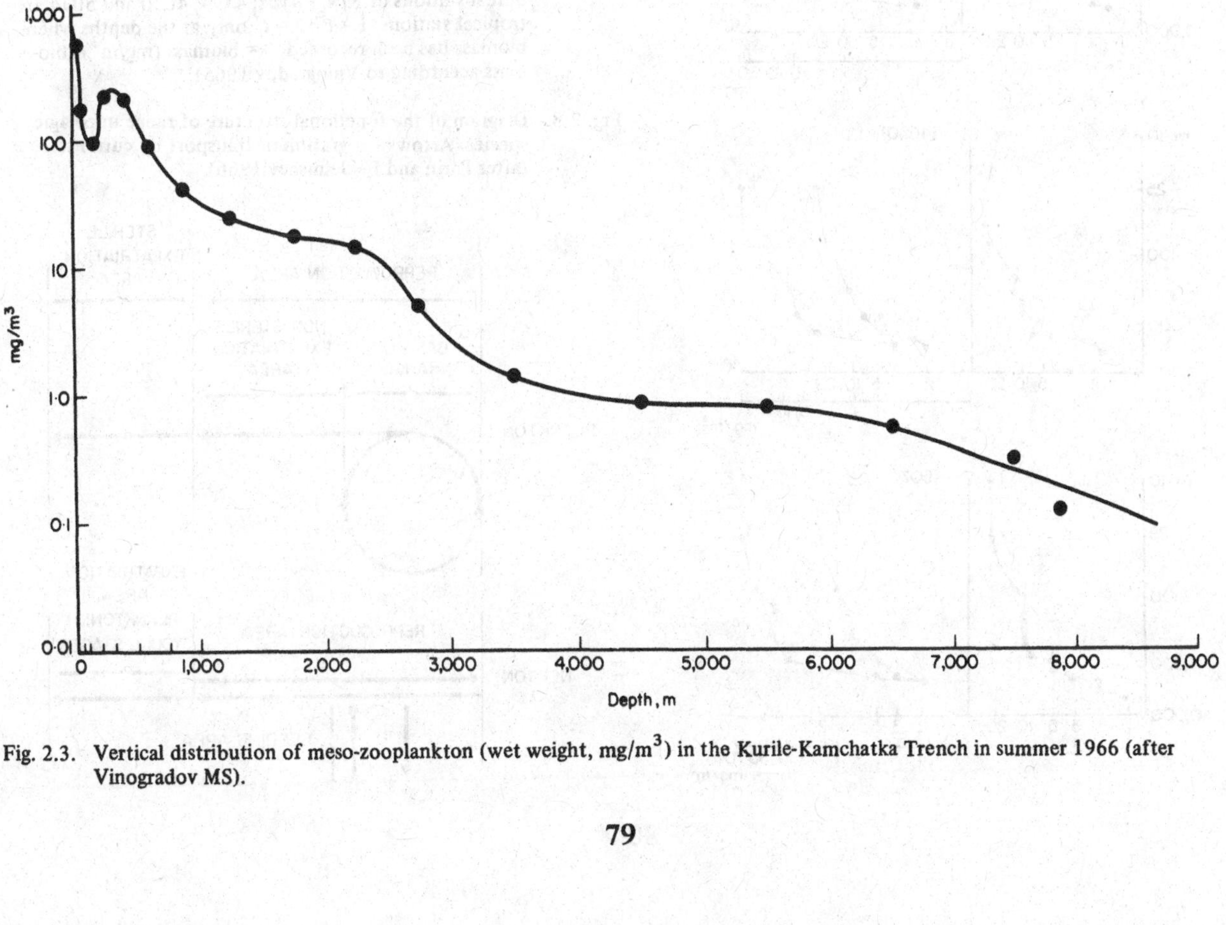
1,000
100
10
1·0
0·1
0·01
mg/m³
0
1,000
2,000
3,000
4,000
5,000
6,000
7,000
8,000
9,000
Depth, m

Fig. 2.3. Vertical distribution of meso-zooplankton (wet weight, mg/m^3) in the Kurile-Kamchatka Trench in summer 1966 (after Vinogradov MS).

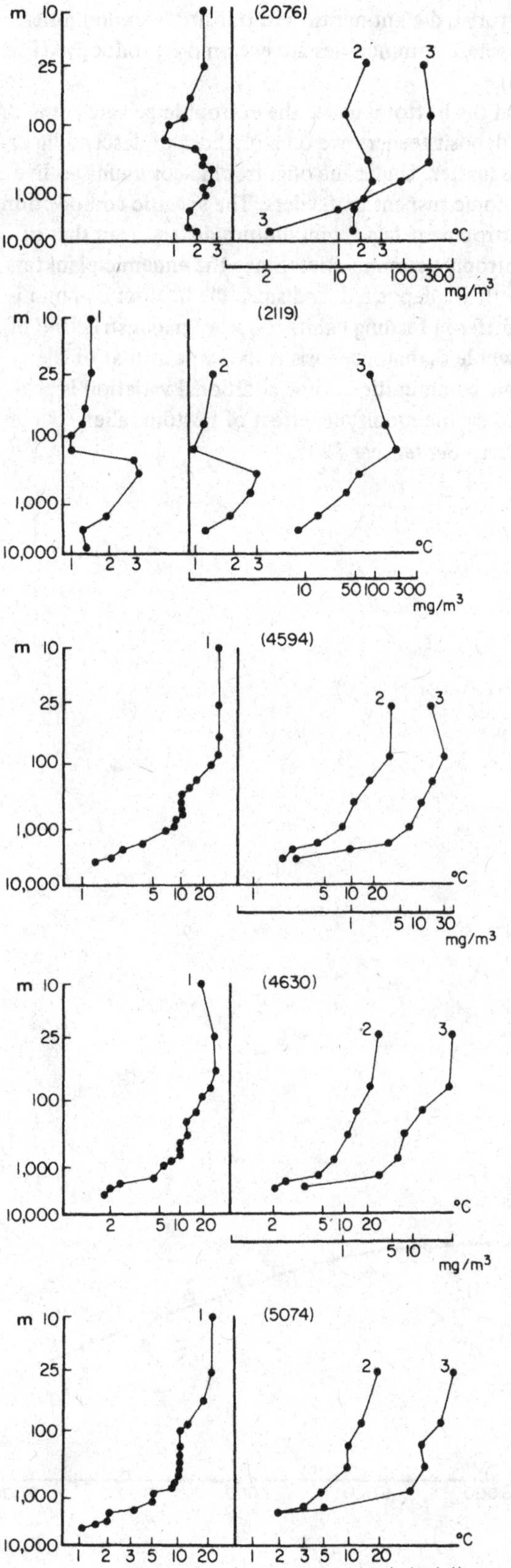

m
10
25
100
1,000
10,000
(2076)
(2119)
(4594)
(4630)
(5074)
°C
mg/m³

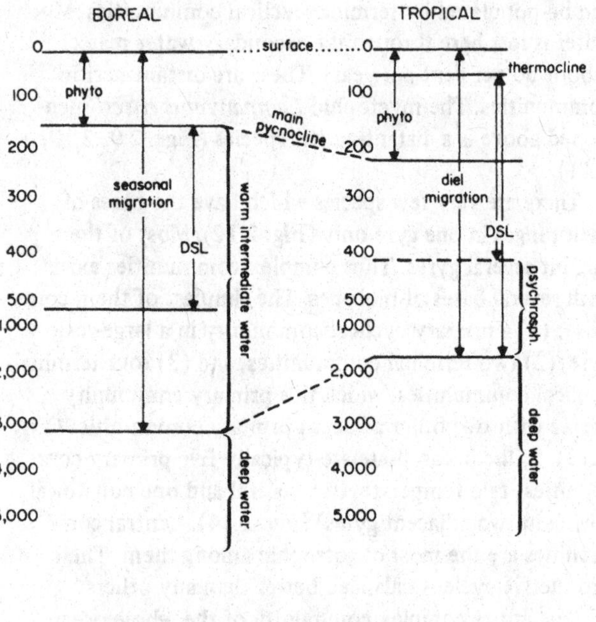

BOREAL
TROPICAL
surface
thermocline
phyto
main
pycnocline
seasonal
migration
diel
migration
DSL
warm intermediate water
psychrosph.
deep water
0
100
200
300
400
500
1,000
2,000
3,000
4,000
5,000

Fig. 2.5. **Main zones of the pelagic environment in the North Pacific. Horizontal lines represent faunal boundaries; some additional biological and hydrographical features are also shown. DSL - deep scattering layer.**

Fig. 2.4. **Dependence of plankton biomass and temperature on depth (after Beklemishev 1966). 2076 and 2119 are boreal stations of R/V *Vityaz*; 4594, 4630 and 5074 are tropical stations; 1 = t°; 2 = t° only at the depths where biomass has been recorded; 3 = biomass (mg/m³) (biomass according to Vinogradov 1965).**

Fig. 2.8. **Diagram of the functional structure of range in pelagic species. Arrows - migration or transport by currents (after Parin and Beklemishev 1966).**

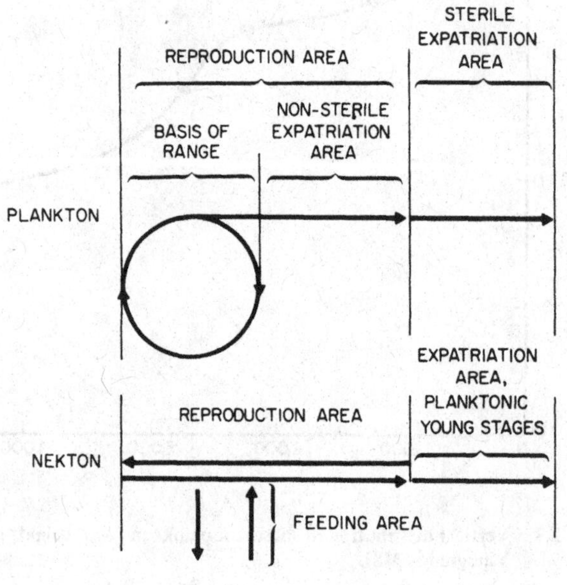

STERILE
EXPATRIATION
AREA
REPRODUCTION AREA
NON-STERILE
EXPATRIATION
AREA
BASIS OF
RANGE
PLANKTON
EXPATRIATION
AREA,
PLANKTONIC
YOUNG STAGES
REPRODUCTION AREA
NEKTON
FEEDING AREA

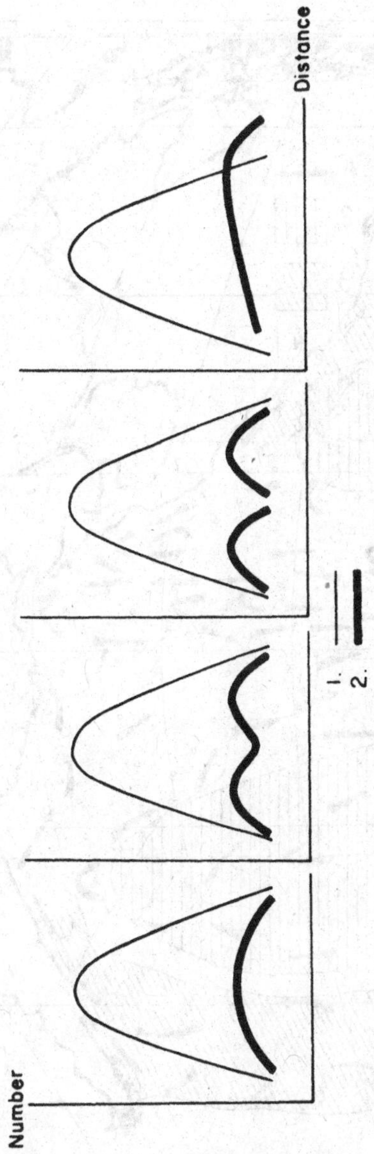
Distance
Number
1.
2.

Fig. 2.6 Influence of competition on the distribution of species (after Rabotnov 1963). 1 = quantitative distribution of species over their ranges without competition; 2 = the same with competition.

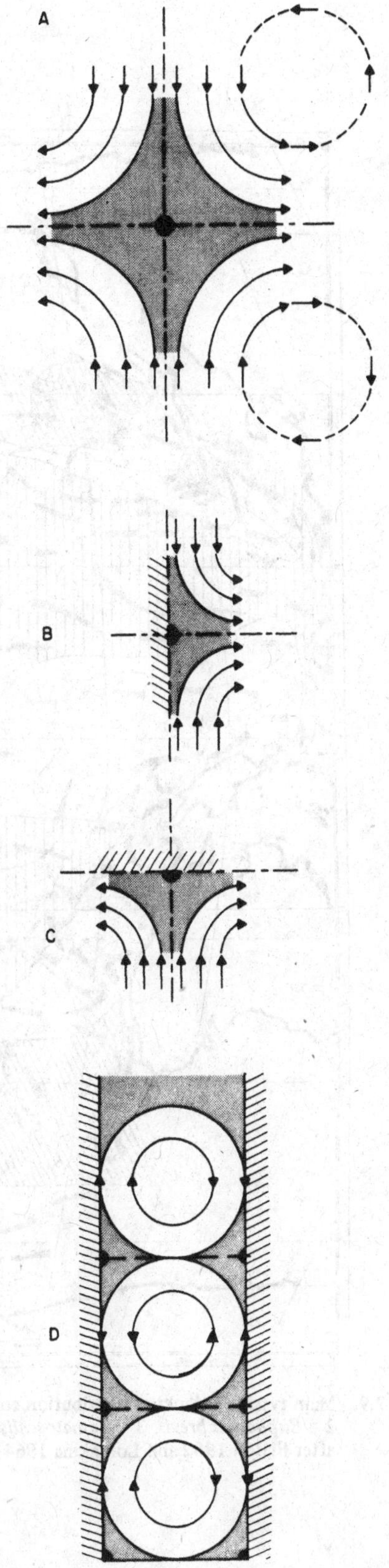
A
B
C
D

Fig. 2.7. Deformation field (A, B, C- adapted from Defant, 1961; D - original). Neutral cols are dotted.

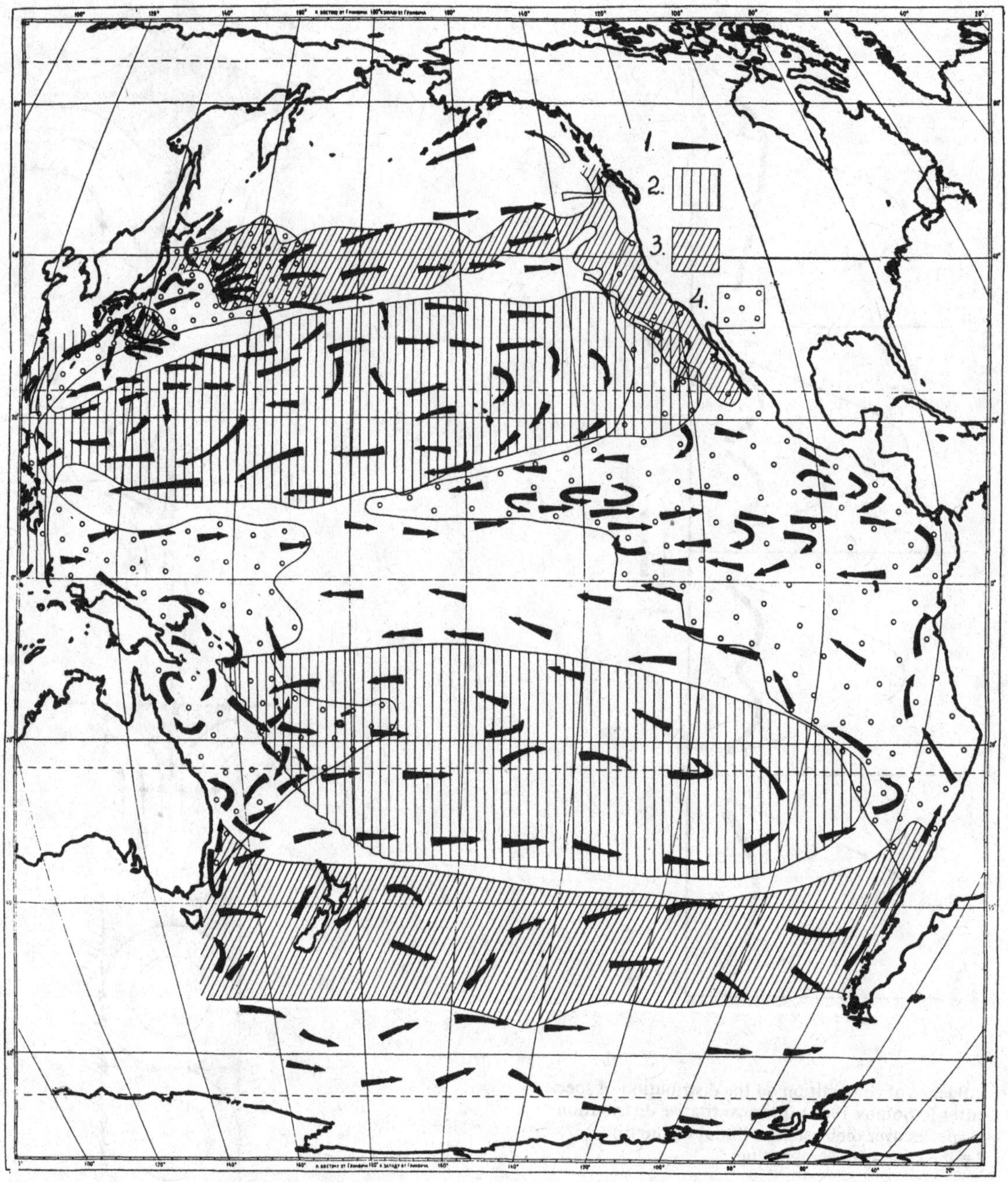
1.
2.
3.
4.

Fig. 2.9. Main types of plankton distribution compared with surface currents (from Beklemishev 1966). 1 = currents (diagrammatic); 2 = *Euphausia brevis;* 3 = *Nematoscelis megalops* and *N. difficilis*; 4 = distant neritic areas (currents after Burkov 1966; species after Briton 1962 and Lomakina 1964).

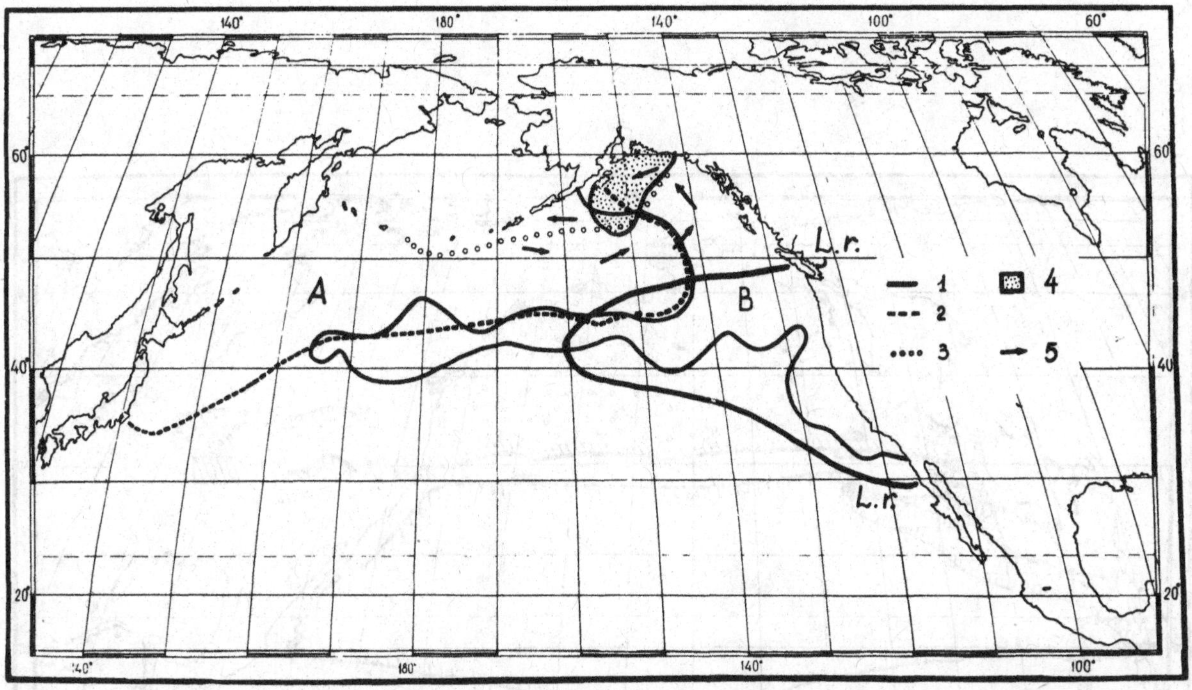
140°
180°
140°
100°
60°
60
40
20
A
B
L.r.
1
2
3
4
5

Fig. 2.10. Distribution of *Limacina helicina* and *Lampanyctus ritteri* in the North Pacific. 1 = *L. helicina,* form B; 2 = *L. helicina,* form A; 3 = average position of sub-polar divergence at 25 metres for August; 4 = northern part of the range of *L. helicina*, form AB; 5 = surface currents; *L.r.* = *Lampanyctus ritteri* (hydrographic data after Burkov 1963; species after McGowan 1963 and Parin 1961).

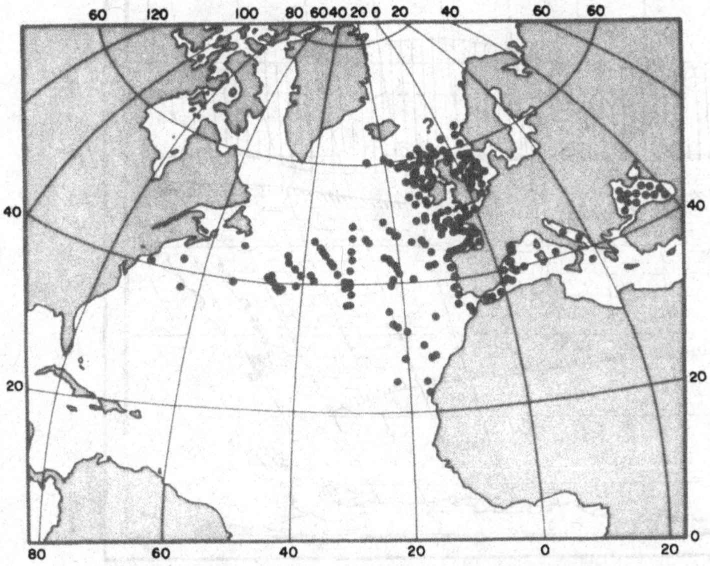
60
120
100
80 60 40 20 0 20
40
60
60
40
20
0
80
60
40
20
0
20
?

Fig. 2.11. Distribution of a distant neritic copepod, *Calanus helgolandicus* (compounded from the data of Jaschnov 1961; Semenova 1962; Grice and Hart 1962; Grice 1963 and Petipa, MS, for the Black and Mediterranean Seas). Question mark, expatriation area, (after Jaschnow 1961).

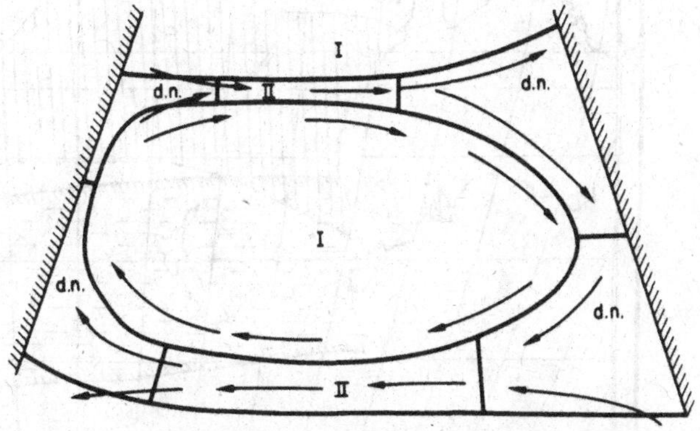
I
d.n.
II
d.n.
I
d.n.
d.n.
II

Fig. 2.13. Diagram of an architectonic complex of pelagic communities. I = primary, oceanic, cyclical communities; II = secondary, oceanic, terminal communities; d.n. = secondary, distant-neritic, terminal-cyclical communities; arrows show currents (after Beklemishev 1966).

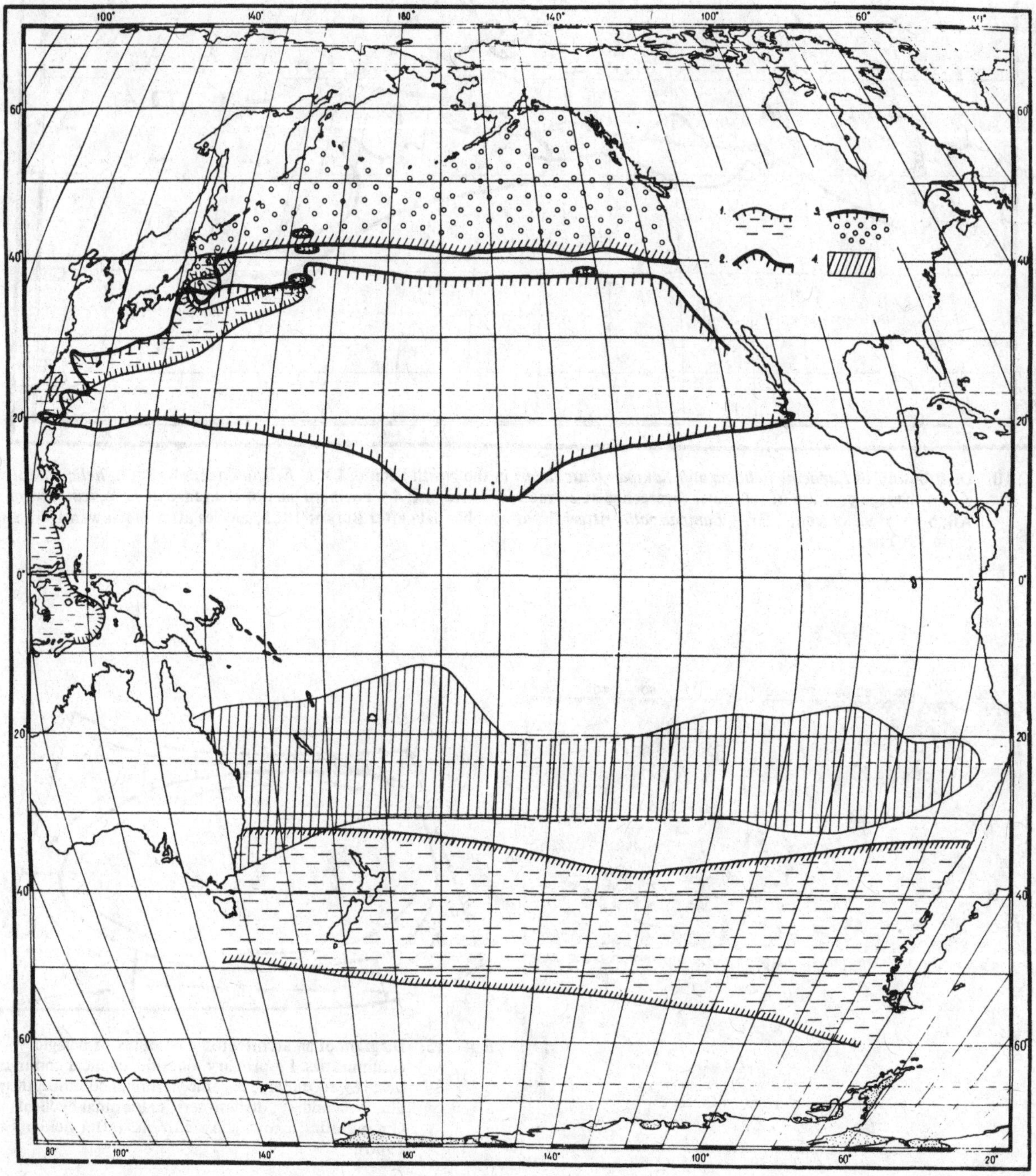

Fig. 2.12. Oceanic species with restricted ranges (modified after Brinton 1962 and Bieri 1959). 1 = *Euphausia similis*; 2 = *Sagitta pseudoserratodentata*; 3 = *S. elegans*; 4 = *Euphausia gibba*.

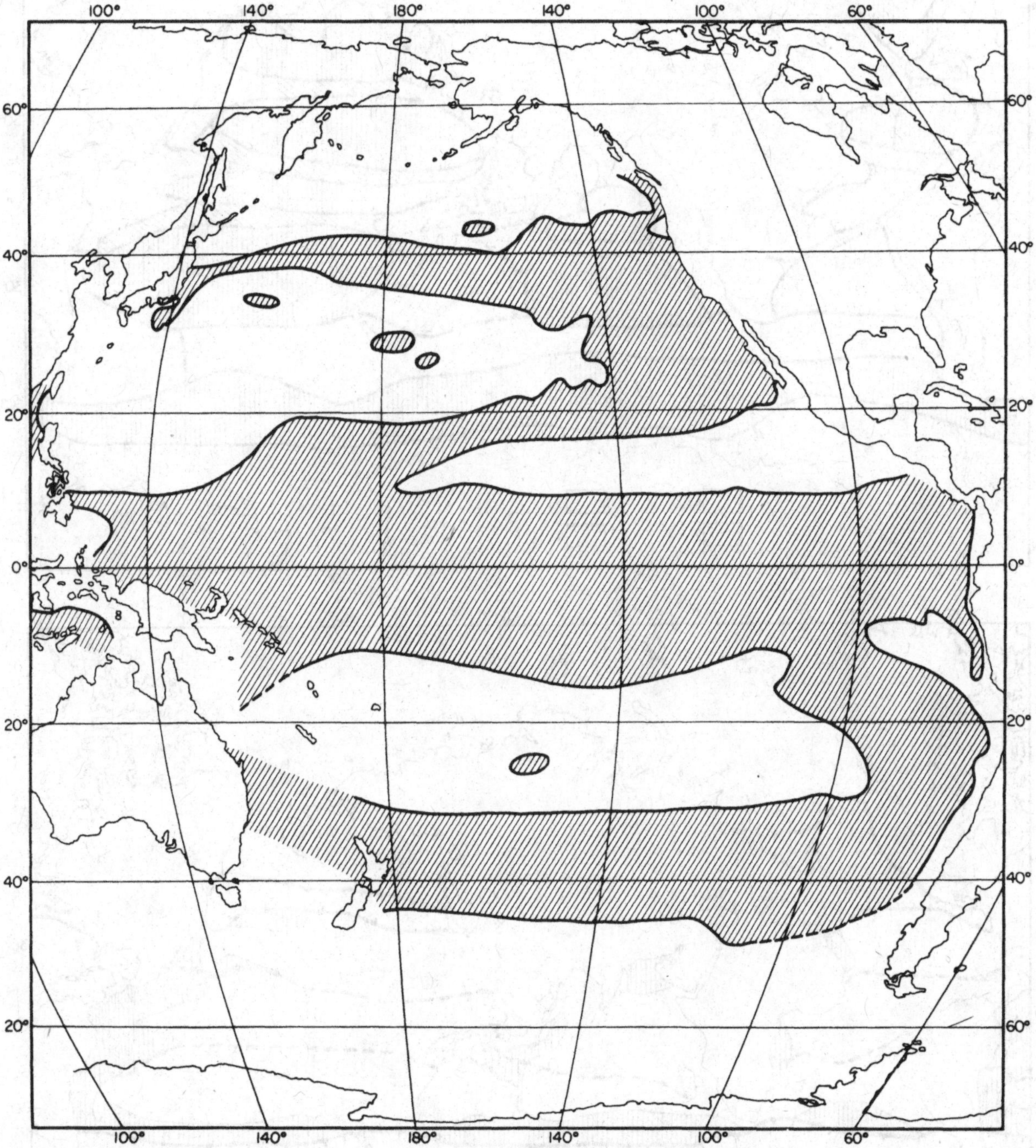
100°
140°
180°
140°
100°
60°
60°
40°
20°
0°
20°
40°
20°
60°
40°
20°
0°
20°
40°
60°
100°
140°
180°
140°
100°
60°

Fig. 2.15. *Stylocheiron longicorne* long form (after Brinton 1962): an index of eutrophic conditions.

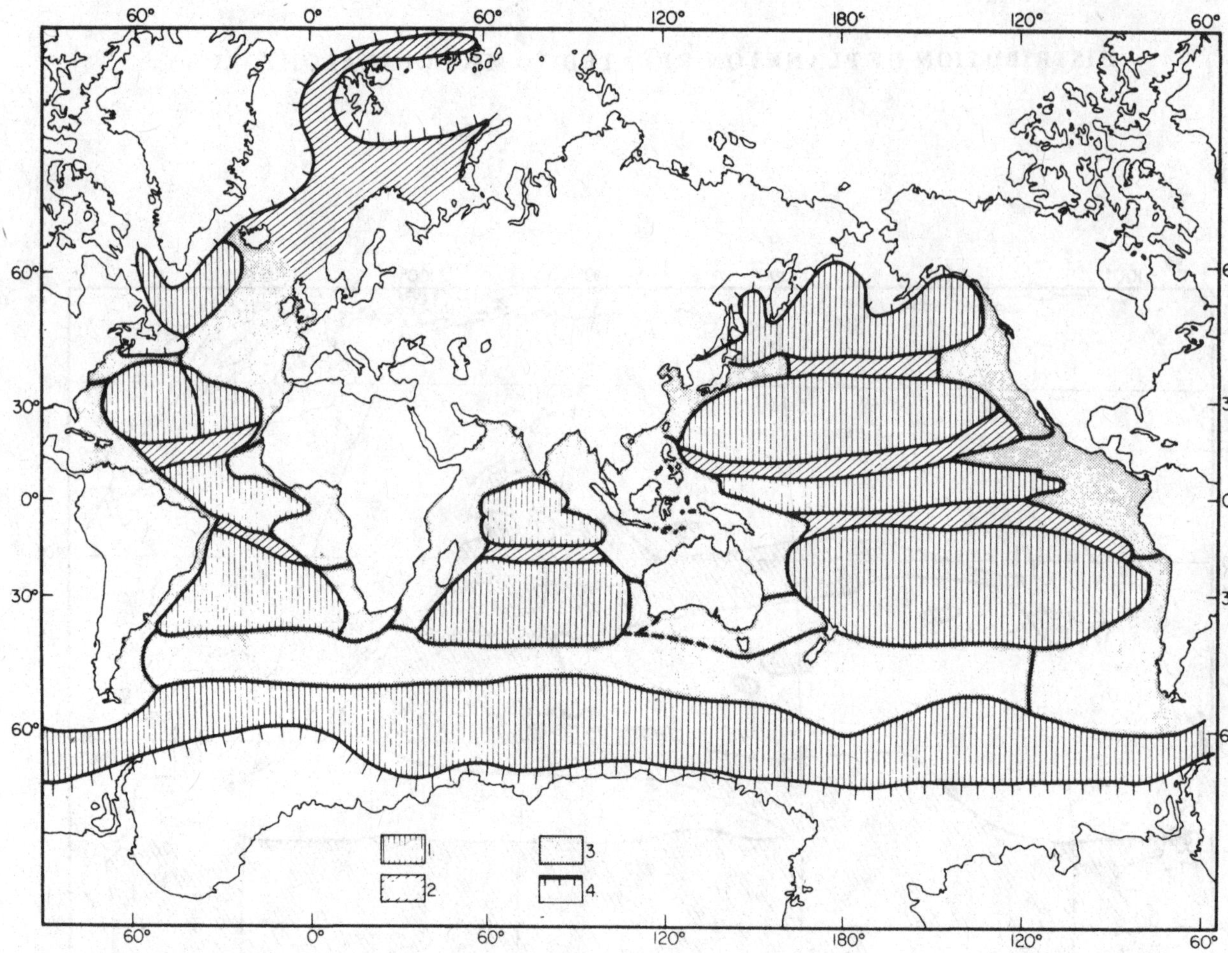
60°
0°
60°
120°
180°
120°
60°
30°
0°
30°
60°
1.
2.
3.
4.

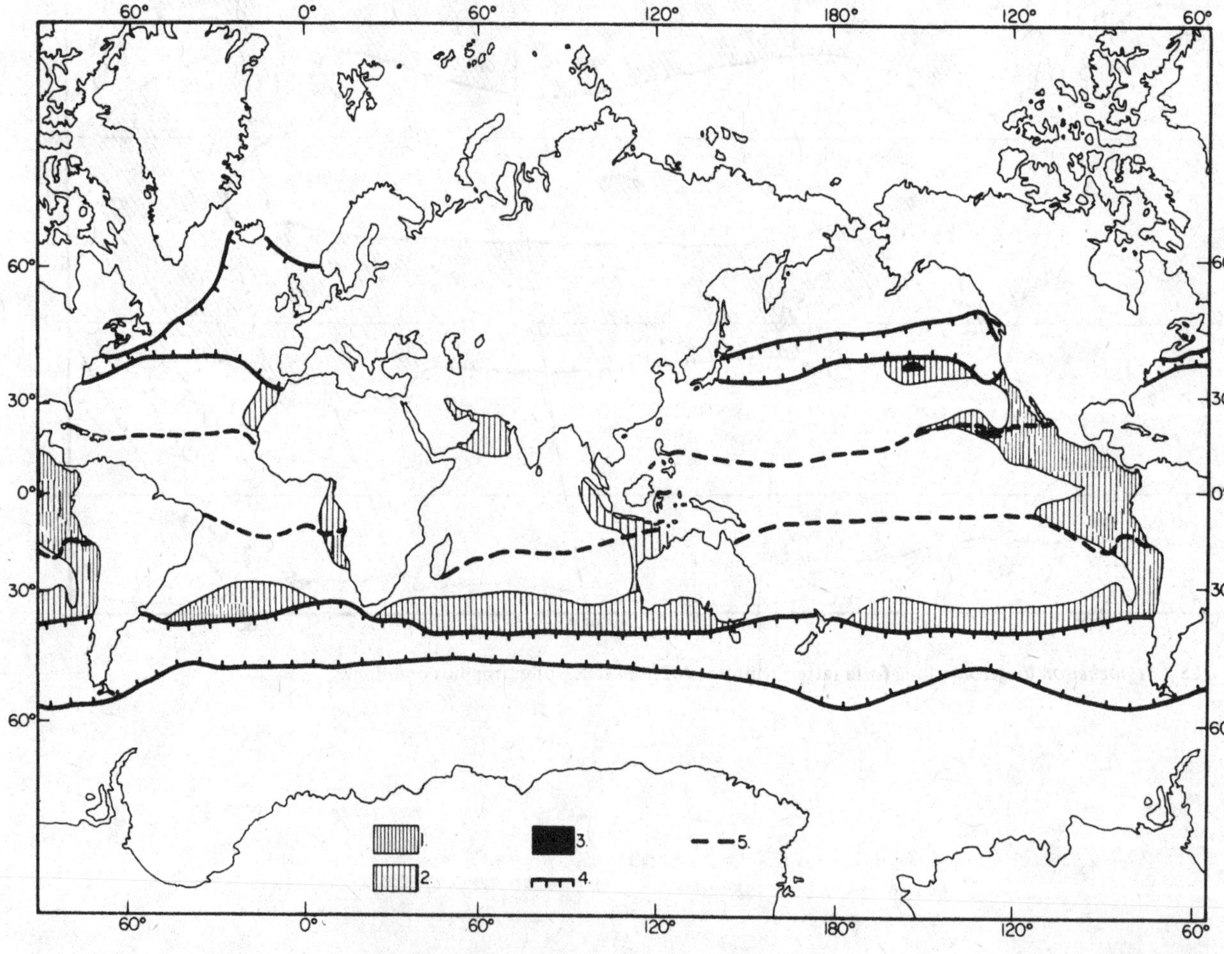
60°
0°
60°
120°
180°
120°
60°
30°
0°
30°
60°
1.
2.
3.
4.
5.

BIBLIOGRAPHY

Beklemishev, C. W. 1966. *Ecologia i biogeographia pelagiali v sviazi s izucheniem glubinnykh zvukorasseivayushchikh sloyev.* Institut Okeanologii Moscow (dissertation, MS).

Belyaev, G. M. 1962. Klyuvy golovonigikh molluscov v okeanicheskikh donnykh osadkakh. *Okeanologia* 2 (2), 311-26.

Belyaev, G.M. In Kort, V.G. (ed.) 1967. *The Pacific Ocean, Biology of the Pacific Ocean, Book 1, Plankton.* 1-266. Nauka: Moscow.

Belyaev, G.M. 1970. Klyuvy kalmarov v donnykh osadkakh Tikhogo okeana. *Tr. Inst. Oceanol.* **89.**

Bieri, R. 1959. The distribution of the planktonic Chaetognatha in the Pacific and their relationship to the water masses. *Limn. and Oceanol.* 4 (1), 1-28.

Bierstein, J. A., Vinogradov, M. E. and Tchindonova, J. G. 1954. Verticalnaya zonalnost planktona Kurilo-Kamchatskoy vpadiny. *Dokl. AN SSSR.* **95** (2), 389-92.

Brinton, E. 1962. The distribution of Pacific euphausiids. *Bull. Scripps Inst. Oceanogr.* **8** (2), 51-269.

Burkov, V. A. 1963. Circulatsia vod severnoy chasti Tikhogo okeana. *Okeanologia,* 3 (5), 761-76.

Burkov, V. A. 1966. Struktura techeniy Tikhogo okeana i ikh nomenklatura. *Okeanologia,* 6 (I), 3-14.

Defant, A. 1961. *Physical oceanography,* 1, XVI+729; 2, VIII+598. Pergamon Press; Oxford.

Grice, G. D. 1963. Deep water copepods from the western North Atlantic with notes on five species. *Bull. Mar. Sci. Gulf Carribb.* **13** (4), 493-501.

Grice, G. D. and Hart, A. D. 1962. The abundance, seasonal occurrence and distribution of the epizooplankton between New York and Bermuda. *Ecol. Monogr.* **32** (4), 287-309.

Jaschnov, V. A. 1961. Vodnye massy i plankton. I. Vidy *Calanus finmarchicus* s.1. kak indikatory vodnykh mass. *Zool. Journ.* **40** (9), 1314-34.

Lomakina, N. B. 1964. Fauna evfausiid (Euphausiacea) antarcticheskoy i notalnoy oblastey. *Issl. fauny morey, II (X), Res. biol. issl. Sov. Ant. Exp. (1955-1958),* 2, 254-334.

Marumo, R. 1955. Analysis of water masses by distributions of the microplankton. 1. Distribution of the microplankton and their relation to water masses in the North Pacific Ocean in the summer of 1954. *Jour. Ocean. Soc. Japan,* **11** (3), 133-7.

McGowan, J. A. 1963. Geographical variation in *Limacina helicina* in the North Pacific. *Speciation in the Sea, Syst. Assoc. Publ.* **5**, 109-28.

McGowan, J. A. 1966. The biological and physical structure of the Pacific Sub-Arctic, oceanic faunal boundary. *Second Int. Ocean. Congr., 30 May-9 June 1966, Moscow, Abstr. of papers,* 249.

Parin, N. V. 1961. Raspredeleniye glubokovodnykh ryb verkhnego sloya bathypelagiali v subarcticheskikh vodakh severnoy chasti Tikhogo okeana. *Tr. Inst. Okeanol.* **45**, 259-78.

Parin, N. V. and Beklemishev, C. W. 1966. Znacheniye mnogoletnikh izmeneniy circulatsii vod Tikhogo okeana dla rasprostraneniya pelagicheskikh zhivotnykh. *Hydrobiol. Jour.* 2 (1), 3-10.

Rabotnov, T. A. 1963. Opyt ispolzovaniya principa niepreryvnosti rastitelnogo pokrova pri izuchenii rastitelnosti shtata Wisconsin (U.S.A.). *Bull. Soc. Natur. Moscou, ser. Biol.* **68** (4), 147-51.

Reshetnjak, V. V. 1955. Verticalnoye raspredeleniye radiolariy Kurilo-Kamchatskoy vpadiny. *Tr. Zool. Inst. AN S.S.S.R.* 21, 94-101.

Semenova, T. N. 1962. Zooplankton rayona Newfoundlandskikh banok vesnoy 1960. In *Sov. rybokhoz. issl. v s.-z. chasti Atlanticheskogo okeana, VNIRO-PINRO, Moscow,* 201-10.

Sokolova, M. N. and Neyman, A.A. 1966. Trophicheskiye grouppirovki donnoy fauny i zakonomernosti ikh raspredelenia v okeane. In *Ecologia vodnykh organismov,* 42-50. Nauka: Moscow.

Vinogradov, M. E. 1965. *Verticalnoye raspredeleniye okeanicheskogo zooplanktona.* Institut Okeanologii, Moscow (dissertation MS).

Fig. 2.14. Pelagic communities of the world's oceans. 1 = cyclical; 2 = terminal; 3 = terminal-cyclical; 4 = ice-neritic (after Beklemishev 1966).

Fig. 2.16. Major biogeographical boundaries and impoverished faunal areas of the world's oceans 1-3 = impoverished faunal areas (the lighter the hatching the more species); 4 = boundaries of Transition Zones; 5 = boundaries of provinces in the Tropical Region (after Beklemishev 1966). Impoverished faunal areas tend to be more productive.

3. OCEANOGRAPHIC CONDITIONS AFFECTING THE CELL SIZE OF PHYTOPLANKTON

H. J. SEMINA

A first attempt at answering the questions, 'How does phytoplankton cell size differ throughout the oceans?' and 'If so, to what are such differences due?' has been made using data collected by R. V. *Vityaz* along a section at 174° W in the Pacific Ocean.

The section along 174° can be divided into five parts differing in average cell size of total phytoplankton. The southern-most and the northern-most parts, 40° S-23° S and 10° N-30° N respectively, have small cells, whereas 23° S-13° S and 3° N-10° N have cells which are comparatively large. Between 13° S and 3° N cell size varies much at individual stations, but the average size is similar to that at the ends of the section. This geographical distribution of cell size is characteristic of the diatoms. In peridinians and coccolithophorids it is less noticeable. The size distribution in net hauls is similar to that in bottle samples.

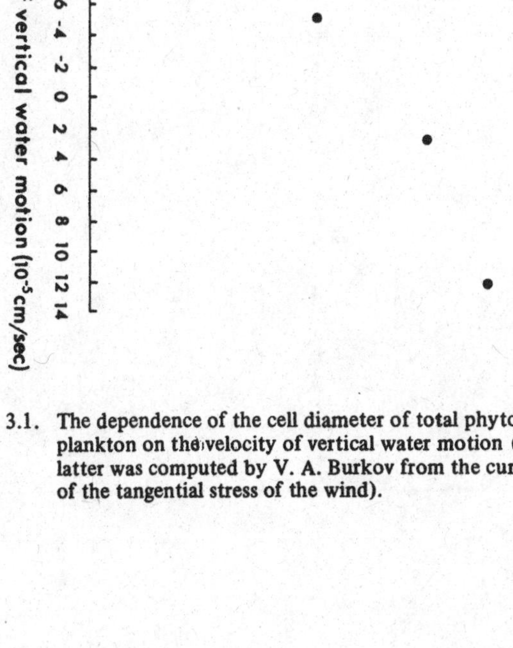
average cell diameter (μ)
0 2 4 6 8 10 12 14 16 18 20 22 24 26
velocity of vertical water motion (10⁻⁵cm/sec)
-14 -12 -10 -8 -6 -4 -2 0 2 4 6 8 10 12 14

Fig. 3.1. The dependence of the cell diameter of total phytoplankton on the velocity of vertical water motion (the latter was computed by V. A. Burkov from the curve of the tangential stress of the wind).

The increased cell size between 23° S and 13° S is due to the genus *Rhizosolenia*, between 3° N and 10° N, mostly to large peridinians. In other places Pennales and coccolithophorids predominate.

We could not detect any correlation between the distribution of cell size in the 174° W section and phosphate concentrations.

A comparison of the geographical distribution of cell size with vertical water motion shows that small cells live in areas with descending water motion, and larger cells in areas with ascending water motion. The dependence of cell size upon the velocity of ascending and descending water motion is shown in Fig. 3.1 ($r = 0.90; P = 0.01$). It can be seen that the smallest cells live in areas with the strongest descending motion. As the downward velocity decreases and the upward velocity increases the size of phytoplankton cells increases. The largest cells live at maximum upward velocity.

Thus the size of phytoplankton cells appears to depend directly on the upward component of water motion. The size of phytoplankton cells may therefore prove to be of interest as an index of their environment.

Papers containing an account of the methods and materials on which this note is based will be published elsewhere [1,2].

Received September 1967.

[1] *Sarsia*, 34, 267-272 (1968).
[2] *Okeanologia*, 9, 479-487 (1969).

4. THE DISTRIBUTION OF MARINE DINOFLAGELLATES IN RELATION TO PHYSICAL AND CHEMICAL CONDITIONS

D. B. WILLIAMS

Summary: The factors influencing productivity and species distribution in dinoflagellate phytoplankton are briefly discussed. Distribution of the genus *Ceratium* is used to illustrate the effects of these factors.

INTRODUCTION

The purpose of this paper is to consider the ways in which the physical and chemical factors of the marine environment control the distribution of dinoflagellates in the oceans, and the relevance of these factors to the dinoflagellate micropalaeontology of marine sediments. As will appear from the information set out below, reliable and useful data are sparse, but in spite of this, certain statements and predictions can be made.

METHODS OF STUDY

Dinoflagellates are usually caught, with diatoms and other phytoplankton, either by very fine mesh phytoplankton nets (200 mesh bolting silk), or by centrifuging a water sample. The first records of dinoflagellates were made by early microscopists who were attempting to satisfy a natural scientific curiosity as to what organisms existed in the world. Later, when the estimation of fisheries potential became important, studies were initiated on the numbers of phytoplankton species in various regions. The tedium of this method of estimating productivity has later been overcome by means of other techniques (which often give more accurate results) such as measurement of the amount of chlorophyll in a water supply, or measurement of the uptake of radioactive-labelled carbon by the organisms in a water sample.

After the initial outburst of investigation, when many of the known species were named, there was a lull in the collection of data, except in shelf regions where it had applications to commercial fisheries (e.g. Paulsen 1913). Since then, a few major world expeditions (Peters 1932; Nielsen 1934; Graham and Bronikowsky 1944) have attempted to determine the distribution of these organisms in relation to the physical and chemical properties of the waters in which they are found.

MORPHOLOGY AND TAXONOMY

The dinoflagellates are biflagellate members of the Protista being placed in the order Dinoflagellida of the class Mastigophora, or the class Dinophyta in the division Pyrrophyta of the Algae - depending on whether one's point of view on these organisms is zoological or botanical. Since most of the members of the group possess chloroplasts and are thus capable of photosynthesis, and many of them also have a cellulose cell wall, many workers, the present author included, prefer to consider them as plants.

Some are not enclosed in a cellulose wall, having an elastic pellicle instead, but all conform to a basic morphological pattern. The two flagella arise close to each other on the ventral side of the cell, and, unlike those of other biflagellate algae, operate in different directions. One encircles the cell, lying in a girdle (cingulum or transverse furrow) and causes the organism to rotate about its long axis as the cell is driven forward by the second flagellum. The latter lies in a sulcus (ventral groove, longitudinal furrow) and extends in a posterior direction beyond the cell. The girdle divides the cell into two parts, the epicone (or epitheca) anterior to the girdle and the hypocone (or hypotheca) posterior to it.

There are three major orders, the Gymnodiniales with a 'naked' protoplast (elastic pellicle), where the epicone and hypocone are approximately equal size; the Peridiniales with an 'armoured' protoplast (cellulose cell wall), also with approximately equal hypotheca and epitheca; and the Dinophysidiales also armoured, but with the epitheca very much reduced and with the cell somewhat compressed laterally. Of these three orders, only members of the Peridiniales appear in significant proportions as fossils, though possible members of the Gymnodiniales and Dinophysidiales are known. The cellulose walls of the armoured groups are made up of plates which conform to a consistent pattern within a species or genus.

Reproduction is normally by binary fission, but encystment has also been proved recently by morphological (Evitt 1961) and developmental (Wall 1966) studies. Since the cellulose of the armoured dinoflagellates is easily decomposed by bacteria, it is only the cysts, whose walls are of a more resistant substance, which are found in sediments as fossils.

FACTORS LIKELY TO AFFECT THE GROWTH OF DINOFLAGELLATES

The same factors which affect the rates of growth of other phytoplankton also influence those of dinoflagellates. This section is therefore a discussion of agencies affecting phytoplankton productivity in general.

Incident radiation

Being photosynthetic organisms, it is obvious that the potential numbers of phytoplankton in any given part of the sea will be strongly affected by the available incident radiation. Such radiation will be a function of both the latitude and the time of year. Given this function, and the relationship for energy transformations involved in photosynthesis, it is possible to predict potential productivity at any selected latitude. This has been illustrated by Ryther (1963, fig. 1). Although the idealistic situation is so strongly influenced by other factors that it is hardly ever found in nature, it does throw light on the strong seasonal changes in productivity found in high latitudes.

Effective radiation

Even the clearest sea-water is not completely transparent, and there is always a depth at which the phytoplankton cannot produce enough oxygen in photosynthesis to compensate for that used by normal respiration. This is known as the compensation depth and is the lower limit of the photic zone. It varies for different species, is deepest at midday, and cannot be considered to exist after nightfall. The compensation depth is also affected by the presence of particulate matter, dissolved salts, and the organisms themselves. The latter seems to be the most important factor of all in affecting the compensation depth, since Nielsen and Jensen (1957) found that the thickness of the photic zone was least where productivity was highest. Variations in the effective radiation are therefore less important than might at first be suspected.

Temperature

Temperature of the photic zone is related to incident radiation, although the pattern is disturbed by transport of warm water towards the poles and cold water towards the equator by surface currents. Since the regions of highest productivity are found in the high latitudes, although extreme cold can inhibit growth, temperature can only be considered to have a small effect on productivity, but there is little doubt that it can and does influence the distribution of individual species.

Salinity

Although one of the most frequently measured attributes of the oceans, there is no evidence that salinity itself affects productivity.

Major nutrients

The major nutrients are oxygen, carbon dioxide, carbon, nitrogen, phosphorus and silicon. The last of these is of no importance to dinoflagellates though it is essential for diatoms and silicoflagellates, nitrogen is required for protein synthesis, and phosphorus is an essential element for energy transformations within the cell. Oxygen is also required for these energy transformations, and carbon dioxide (and water) are necessary for carbohydrate synthesis.

The cycle of nutrients in the sea has been extensively discussed by Redfield, Ketchum and Richards (1963), and only a few of the more important points will be repeated here. Oxygen is always present in sufficient quantities in the photic zone for the respiration of animals and plants. Similarly there is usually enough carbon dioxide relative to the phosphorus and nitrogen for photosynthesis to take place. The latter two elements may therefore have a limiting effect on productivity in the photic zone. It has recently been shown (see Redfield *et al.* 1963) that productivity is ultimately limited by the nitrogen concentration, since while the atomic ratio of nitrogen to phosphorus in sea water is more or less constant at 15 : 1, the atomic ratio in phytoplankton is approximately 16 : 1. Growth of phytoplankton will use up the available nutrients, and unless these are replaced, growth will cease. Some of the phytoplankton will be eaten by zooplankton and some will die. A small quantity of the nutrients will therefore be returned almost immediately to the photic zone, by excretion and bacterial decay, but the majority will be carried below, to be taken up by deeper water. Agencies which bring these deep waters to the surface thus constitute the ultimate and dominant influence on productivity.

Hydrodynamic and other physical controls

The mixing of deep waters with the surface water, and the consequent nutrient enrichment of the latter is largely controlled by hydrodynamic effects of the circulation of oceanic waters. One of the areas where this is well marked is at the lines of discontinuity between surface water masses, particularly in regions of divergence. The withdrawal of surface water in opposite directions as a result of wind-driven currents requires replacement of water from below to maintain the hydrostatic equilibrium. Regions of divergence are therefore often regions of enhanced productivity. The turbulence occuring where a shear is produced between two subparallel opposed surface currents can also bring nutrients to the surface, with a corresponding increase in productivity.

The eastern margins of the major oceanic basins frequently show enhanced productivity. This arises as a combination of internal circulation in the eastern boundary current systems – a direct dynamic effect – and a wind stress parallel to the coast that causes an offshore transport of water which is replaced by upwelling of deep water (Wooster and Reid 1963).

The high productivity, admittedly seasonal, of sub-polar regions is ascribed to different factors for the sub-Arctic and sub-Antarctic (Ryther 1963). In the North, enrichment is due to the following sequence of events. Winter cooling causes sinking of denser water, bringing deep water to the surface by means of convection. The well-known spring bloom of phytoplankton in the sub-Arctic results from the stratification which develops from warming of the surface water – this prevents phytoplankton from being carried below the compensation depth which would prevent efficient growth. In the South polar regions, the enrichment comes from the upwelling to the surface of Atlantic Intermediate Water in response to the sinking of cold waters both at continental margin and at the subtropical convergence. The upwelled Intermediate Water is less dense than the waters beneath it, and is therefore very stable. It is so rich in nutrients that even in full summer they are not exhausted by plant growth (Ryther 1963).

Minor nutrients

Recently it has been realised that many regions of the seas are not supporting the crop of phytoplankton which nutrient measurements suggest they could. For example, the productivity round California is less than round the British Isles, though there are more nutrients in the former (Provasoli 1963). It has been suggested that these anomalies may be due to shortages of trace elements and/or vitamins, or presence of antibiotic substances. Vitamin requirements are markedly different for different species of phytoplankton and therefore, besides controlling the general fertility of the waters, they also affect the distribution of species between different regions (Provasoli 1963).

FACTORS AFFECTING THE DISTRIBUTION OF PHYTOPLANKTON SPECIES

When attention is turned to individual species, their distribution may be seen to be controlled by factors which had only minor importance in the control of productivity.

Perhaps the most obvious of these is temperature. It may operate both directly and indirectly on the organisms. By their response to temperature species may be divided into temperature tolerant species which are of cosmopolitan distribution, and temperature intolerant species which have a much more restricted distribution, usually to warm water. The real reason for the temperature intolerance of some species may often be open to doubt, since besides the direct effect of temperature, other features of the marine environment, such as viscosity and density, will also be affected by temperature. The difference between the relative importance of dinoflagellates and diatoms in the tropics and sub-polar regions may well be due to the difference in viscosity in the two regions; diatoms are capable of very little active movement, so the increased viscosity of cold and temperate regions is of undoubted advantage in retaining them in the photic zone, while the dinoflagellates being equipped with flagella are much better suited for life in a less viscous environment. Many of the tropical species of *Ceratium* were found by Peters (1932, fig. 3) to have longer horns and thinner thecae than the cold water species, thus showing adaptations to both reduced viscosities and densities.

At the individual species level, salinity is also of undoubted importance, though genera such as *Gymnodinium* and *Peridinium* may be found in both fresh water and the open ocean. Wood (1954) recognizes brackish and estuarine species of *Ceratium,* but points out that this tolerance of reduced salinity by a given species is only applicable to restricted regions. For example, in Australian waters *C. furca* fills the brackish water niche occupied by *C. hirundinella* in European waters, though both are cosmopolitan in waters of normal salinity.

In studies of zooplankton it is customary to relate the distribution of the organisms to the temperature and salinity of the localities at which they were found (see Johnson and Brinton 1963) by means of T-S diagrams. This implies that the zooplankton species are related in their distributions to the distribution of the water masses. A similar view is held on the distribution of dinoflagellates; Wood (1954) uses them to identify water masses, but to the present author's knowledge, the technique of plotting species of dinoflagellates on T-S diagrams has not been employed. This is probably due to the lack of such thorough research on their distribution as has been carried out on zooplankton. The apparent restriction of certain assemblages of dinoflagellates to water masses is however probably not directly due to the salinity-temperature combination but to some trace element property of the water mass.

THE DISTRIBUTION OF 'CERATIUM' SPECIES

Because of their constant participation in the phytoplankton, the species of the genus *Ceratium* have long attracted the attention of marine biologists. Their large size (relative to other dinoflagellates) and their wide variability has made them particularly attractive for study. In addition they may be caught by coarse mesh nets such as are used in the continuous plankton recorder (*Bull. mar. Ecol.* 1953, *et seq.*). Data on *Ceratium* species with data on the stations at which they are found are given by Karsten (1905, 1906, 1907), Schroder (1906), Böhm (1931), Peters (1932), Nielsen (1934), Graham and Bronikowsky (1944), and Wood (1954).

Of these papers, the most important, because of the area which they cover, are those by Peters (1932) and Graham and Bronikowsky (1944). Peters considers the distribution of *Ceratium* as found on the cruise of the German research ship *Meteor* – a cruise in the South Atlantic on which much modern oceanographic knowledge is based. Using the species, subspecies and varities of *Ceratium* he defined six natural regions. These regions have a close correspondence to the natural regions of the ocean recognized on a hydrographic basis by both Schott (1936) and Hela and Laevastu (1962). Peters concluded that temperatures between 15° C and 27° C had little influence on the distribution of species, though he recognized warm-, temperate- and cold-water forms. Only one form, *C. pentagonum robustum* was found in the cold 'southern boundary region'. In contrast, there were thirty-three species found in warm water, and twenty-one from warm to cool waters. He could distinguish no effects which were definitely due to salinity though he suggested that there were more species of *Ceratium* represented in high salinity waters than elsewhere. The high numbers of species in areas deficient in phosphate, and vice versa, made him suggest that nutrients might suppress certain species. On the other hand, he also found that phosphate-rich waters had far more individuals of the species which were represented. This illustrates a general feature which has been noted both in other groups of organisms, and in phytoplankton but related to other causes (e.g. Hulburt 1963). Where there is a high diversity the absolute numbers are usually small, where there is a low diversity the numbers are usually large.

Graham and Bronikowsky (1944) worked on material from the North Atlantic and Pacific Oceans, collected on the last cruise of the *Carnegie*. In general their observations support those of Peters (1932), especially with regard to the relationship between diversity and absolute frequency and its relationship to temperature. They define five natural zones of *Ceratium*, two in the North Atlantic and three in the Pacific. The data they give on environmental details (salinity, temperature and phosphate) are excellent, being arranged in tables for each species for all the stations at which it was recorded. Graham and Bronikowsky's (1944) conclusions on the role of phosphate are different from those of Peters, though they observed the same relationship between phosphate and number of species. Their studies 'tend to indicate that the phosphate content of the water has no direct effect on the horizontal distribution of *Ceratium* species, at least not as regards absolute values. There are indications however that the relative values in a given region bear some relation to the *Ceratium* flora. Perhaps some factor associated with an *increase* of phosphate is significant in the distribution of *Ceratium* species' (Graham and Bronikowsky 1944, p. 7). Study of both their charts and Peters' (1932) suggests certain species of *Ceratium* require a high initial level of nutrients for growth, but that when these nutrients are available they grow at such a rate as to completely dominate the assemblage.

Both the publications consider currents to be important in the distribution of *Ceratium* species, but their arguments seem to suggest that it is the current systems and their associated water masses which are of dominant importance. Graham and Bronikowsky do however cite the transportation of species away from their source areas by currents, mentioning two tolerant tropical species *C. hexacanthum* and *C. extensum* which had been carried across the Atlantic to Iceland and Britain, a considerable distance and drop in temperature from their optimum habitat.

THE RELEVANCE OF THE DISTRIBUTION OF DINOFLAGELLATES TO THE MICROPALAEONTOLOGY OF MARINE SEDIMENTS

Only the cysts of dinoflagellates are found as fossils. This means that unlike foraminifera, the forms found in the sediments cannot be correlated with those in the water without a detailed knowledge of the life-histories of the species. This we do not have, though work currently in progress at Woods Hole Oceanographic Institution has succeeded in relating some forms of cysts to their motile stages (Wall 1966; Wall and Dale 1966; Wall, Guillard and Dale 1967).

Secondly, the group of dinoflagellates for which the best distribution data are available, the genus *Ceratium*, has not yet had any Recent fossil cysts ascribed to it.

It would however be highly surprising if other genera and species of dinoflagellates did not behave in a similar way to *Ceratium*, and therefore we may expect to find similar distribution patterns for the cysts in the sediments. This latter aspect is considered in the paper on distribution of dinoflagellate cysts in sediments.

Received March 1968.

BIBLIOGRAPHY

(This bibliography includes only a selection of the many publications on dinoflagellates, most of those which are omitted are taxonomic. The most useful for identification of species are marked with an asterisk (*).)

Böhm, A. 1931. Distribution and variability of *Ceratium* in the northern and western Pacific. *Bull. Bernice P. Bishop Mus.* **87**, 46 pp.

Evitt, W. R. 1961. Observations on the morphology of fossil dinoflagellates. *Micropaleontology*, **7**, 385-420.

Graham, H. W. 1942. Studies in the morphology, taxonomy, and ecology of the Peridiniales. *Publs Carnegie Instn.* **542**, 129 pp.

Graham, H. W. and Bronikowsky, N. 1944. The genus *Ceratium* in the Pacific and North Atlantic Oceans. *Publs Carnegie Instn.* **565**, 209 pp.

*Hela, I. and Laevastu, T. 1962. *Fisheries Hydrography*. London, 137 pp.

Hulburt, E. M. 1963. The diversity of phytoplanktonic populations in oceanic, coastal and estuarine regions. *J. marine Res.* **21**, 81-93.

Johnson, M. W. and Brinton, E. 1963. Biological species, water masses and currents. *The Sea* (M. N. Hill *ed.*), **2** 381-414. John Wiley and Sons: New York.

Karsten, G. 1905. Das Phytoplankton des Antarktischen Meeres nach dem Material der deutschen Tiefsee-Expedition 1898-1899. *Wiss. Ergebn. dt. Tiefsee-Exped. 'Valdivia'*, **2** (2), 1-136.

Karsten, G. 1906. Das Phytoplankton des Atlantischen Oceans nach dem Material der deutschen Tiefsee-Expedition 1898-1899. *Wiss. Ergebn. dt. Tiefsee-Exped. 'Valdivia'*, **2** (2), 1-136.

Karsten, G. 1907. Das Indische Phytoplankton nach dem Material der deutschen Tiefsee-Expedition 1898-1899. *Wiss. Ergebn. dt. Tiefsee-Exped. 'Valdivia'*, **2** (2), 223-548.

*Kofoid, C. A. and Skogsberg, T. 1928. The Dinoflagellata: The Dinophysidae. *Mem. Mus. comp. Zool. Harvard*, **51**, 766 pp.

*Lebour, M. V. 1925. *The Dinoflagellates of the northern seas.* Plymouth, 172 pp.

Nielsen, E. S. 1934. Untersuchungen über die Verbreitung, Biologie und Variation der Ceratien im Südlichen Stillen Ozean. *Dana Rep.* **1** (4), 67 pp.

Nielsen, E. S. and Jensen, E. S. 1957. Primary organic production, the autotrophic production of organic matter in the oceans. *Galathea Rep.* **1**, 49-136.

Paulsen, O. 1908. Peridiniales. In Brandt, K. and Apstein, C. (eds.), *Nordisches Plankton.* Kiel, **18**, 124 pp.

Paulsen, O. 1913. Peridiniales ceterae. *Bull. plankt.* **3**, 251-90.

Peters, N. 1932. Die Bevölkerung des Südatlantischen Ozeans mit Ceratien. *Wiss. Ergebn. dt. atlant. Exped. 'Meteor'*, **11**, 1-69.

Provasoli, L. 1963. Organic regulation of phytoplankton fertility. *The Sea* (M. N. Hill *ed.*), **2**, 165-219. John Wiley and Sons: New York.

Redfield, A. C., Ketchum, B. H. and Richards, F. A. 1963. The influence of organisms on the composition of sea water. *The Sea* (M. H. Hill *ed.*) **2**, 26-77. John Wiley and Sons: New York.

Ryther, J. H. 1963. Geographic variations in productivity. In *The Sea* (M. H. Hill *ed.*), **2**, 347-80. John Wiley and Sons: New York.

*Schiller, J. 1933, 1937. Dinoflagellatae. In *Kryptogamen-Flora von Deutschland, Oesterreich, und der Schweiz,* (Rabenhorst *ed.*), **10** (3,1), 617 pp; **10** (3,2), 589 pp. Leipzig.

Schott, G. 1936. Die Aufteilung der Drei Ozeane im natürliche Regionen. *Petermanns geogr. Mitt.* (6), 165-70, 218-22.

Schroder, B. 1906. Beiträge zur Kenntnis der Phytoplanktons warmer Meere. *Vjschr. naturf. Ges. Zürich,* **51**, 319-77.

Wall, D. 1966. Modern hystrichospheres and dinoflagellate cysts from the Woods Hole region. *Grana palynol,* **6**, 297-314.

Wall, D. and Dale, B. 1966. 'Living fossils' in western Atlantic plankton. *Nature, Lond.* **211**, 1025-6.

Wall, D., Guillard, R. R. L. and Dale, B. 1967. Marine dinoflagellate cultures from resting spores. *Phycologia,* **6**, 83-6.

Wood, E. J. F. 1954. Dinoflagellates in the Australian region. *Aust. J. mar. Freshwat. Res.* **5**, 171-351.

Wooster, W. S. and Reid, J. L. 1963. Eastern boundary currents. In *The Sea* (M. N. Hill *ed.*), **2**, 253-80. John Wiley and Sons: New York.

5. COMMENTS ON THE DISTRIBUTION OF COCCOLITHOPHORIDS IN THE OCEANS

K. R. GAARDER

Summary: The general characteristics of the coccolithophorid cell are reviewed. Literature on coccolithophorid distribution in the Atlantic, Pacific and Indian Oceans, as well as in the Mediterranean area, is considered. It reveals the inadequacy of our knowledge of coccolithophorid vegetation in major parts of the oceans. Differences in methods as well as numerous cases of unsatisfactory taxonomic treatment are serious obstacles to biogeographic studies on species. Preliminary distribution charts are presented for *Calyptrosphaera oblonga* Lohm., *Discosphaera tubifera* (Murr. et Blackm.) Lohm., *Rhabdosphaera clavigera* Murr. et Blackm., *Syracosphaera pulchra* Lohm. and *Umbellosphaera irregularis* Paasche.

GENERAL CHARACTERISTICS

Today coccolithophorids are generally accepted as plants and referred to the Haptophyceae, introduced by Christensen (1962) and adopted by Parke and Dixon (1964). They are unicellular planktonic organisms, mostly with a diameter $< 20\ \mu$, many $< 10\ \mu$. Usually they have two golden-brown chromatophores, first reported by Weber-van Bosse (1901), and a nucleus in between (Ostenfeld 1900), two acronematic flagella of about the same length and a haptonema (Parke and Adams 1960), which has given the class its name. The plasmalemma can be covered by a layer of isomorphic or heteromorphic, finely sculptured, overlapping plate-scales (Parke and Adams 1960), presumably of an organic nature (Parke 1961, Green and Jennings 1967).

At least in one phase of the life cycle the cell bears a more or less rigid outer skeleton, consisting of another type of 'plate' – coccoliths, encrusted with calcite, in some cases with aragonite and or vaterite (Wilbur and Watabe 1963). Outside, a 'skin' of unidentified nature has been observed (Manton and Leedale 1963). Plate-scales and coccoliths are formed within internal vesicles, the coccolith formation being a light-dependent process (Paasche 1964).

Propagation may take place either by fission, mostly longitudinal, of the mother cell and subsequent regeneration of coccoliths, or, by repeated divisions of the mother cell inside the crust of coccoliths and release of motile or non-motile daughter cells. These may or may not regenerate coccoliths of the original or a different type (Parke and Adams 1960). Rayns (1962) observed that in *Cricosphaera carterae* a motile diploid phase alternates with a haploid non-motile phase. Resting spores were observed by Kamptner (1937*b*).

Classification of coccolithophorids is based on (1) coccolith morphology, (2) the mode of their arrangement on the cell, and (3) the shape of the cell. The cell is most frequently spherical or oval, sometimes sub-cylindrical, tapering towards one or both ends (*Calciopappus, Anoplosolenia*), with or without appendices presumably serving as a flotation apparatus.

The deposition of calcite in the coccoliths may take place in two ways, (1) as simple calcite microcrystals, mostly as rhombohedrons, in holococcoliths, (2) as a diversity of morphologically different elements, with the properties of single crystals, arranged together to form the various types of heterococcoliths (see Braarud *et al.* 1955).

The heterococcoliths are more rigid in their construction than are the holococcoliths. The latter tend to disintegrate more or less completely when coccolithophorids are digested by herbivores, or, simply during the post-mortem sinking of the cell to the sea floor. Holococcoliths are, therefore, seldom identified in bottom sediments.

Most coccolithophorids are photosynthetic and have their main occurrence within the euphotic layer. It has, however, been reported that some may exist heterotrophically, at least in one phase of their life-cycle (Bernard 1948). Some may live saprophytically, as they have been found attached to copepods in decomposition. Phagotrophy has been demonstrated experimentally by Parke and Adams (1960); epiphytism was observed by Dangeard (1934). Symbiosis with diatoms in *Brenneckella* was assumed by Lohmann (1912) but questioned by Gaarder and Hasle (1962).

Only few coccolithophorids live in fresh water while some seem to prefer brackish water. The ecological features of these groups have been discussed by Kamptner (1930). Most of the marine coccolithophorids are strictly oceanic, a few are restricted to inshore waters *(Cricosphaera* spp.) and some are indifferent in this respect (see Smayda 1958*a*). Some have a worldwide distribution, others are known only from very restricted areas.

Compared with the other phytoplankton groups, especially diatoms and dinoflagellates, the coccolithophorids have received little attention by marine biologists. It is a small group as regards number of taxa (so far about 200) but, quantitatively, it has proved to be of very great importance and appears to be a major food source for lower animals.

In surveys coccolithophorids have very often been found to come next to diatoms in number and in oceanic areas they may even outnumber this group (Hasle 1960; Semina pers. comm.), a situation which also may occur in inshore waters.

DISTRIBUTION

Owing to their characteristic outer skeleton, composed of coccoliths of diverse morphology, the coccolithophorids may fairly easily be recognized as a group. Identification to genus and species is, however, in most cases hampered by the minuteness of the cells and the coccoliths. The morphological features of the coccoliths, on which the taxonomy at present is mainly based, cannot always be discerned in a light microscope, and electron microscopy is required in order to bring out essential details. For this reason it is advisable to refrain from classifying specimens to genus or species if control by electron microscopy cannot be obtained. Even the very common *Coccolithus* (*Pontosphaera*) *huxleyi* may easily be confused with other minute species, e.g. *Gephyrocapsa oceanica* or the small form of *Cyclococcolithus fragilis* (Hasle 1960). A distinction between these species is of biogeographical significance as it has been found that in some areas *Gephyrocapsa oceanica* can be equal to *Coccolithus huxleyi* in abundance and, at least at some seasons, outnumber it, e.g. in the Gulf of Panama (Smayda 1963, 1966).

The Pacific Ocean

In the papers mentioned above, Smayda gave the first extensive report on the seasonal variation in the coccolithophorid population within an area of the Pacific Ocean. The coccolithophorids showed little regional variation as to quantity and composition but varied considerably from the upwelling to the rainy season. His list of twenty-three species contrasts with that of five species in Marshall's (1933) area of Australian inshore waters. This may partly be attributed to the different methods used. (For methods, see Braarud 1958.) From inshore waters in the western part of the ocean there is a single report (Smirnova 1959). She recorded *Acanthoica* sp. and *Syracosphaera* sp. besides the cosmopolitan *Coccolithus* (*Pontosphaera*) *huxleyi* from the Okhotsk Sea and the area around the Kurile Islands. Norris (1961) recorded thirty-one different species during qualitative studies of surface waters between New Zealand and Tonga Islands. The most extensive list was presented by Hasle (1960) from Equatorial oceanic waters. Several were also checked by electron microscope. She recorded thirty-three species. One of the species occurs in the list under the names *Cyclococcolithus mirabilis* and *C. sibogae*. According to our present knowledge it should be named *Umbilicosphaera sibogae* nov. comb. Gaarder, with basionym *Coccosphaera sibogae* (Weber-van Bosse 1901, p. 140, fig. 17).

The Indian Ocean

Few observations are available from this area. Weber-van Bosse (1901) observed the first living coccolithophorid with chromatophores and she also depicted division stages. Weber (1902) recorded a few others from the same Indonesian waters.

Silva (1960) made a quantitative study of phytoplankton from surface waters off the coast of Mozambique. She treated the coccolithophorids as a group and found that it came closest to diatoms in numbers. Lecal (in Bernard and Lecal 1960) described a number of new species and forms from the western part of the ocean, and Taylor (1966) presented a list of seven species, based upon net samples from off the South African coast. Travers and Travers (1965) reported on the group as a whole and a floristic study from coastal waters west of Madagascar has been announced. Norris (1965) described live cells of *Ceratolithus* Kpt., widely distributed in the Arabian Sea.

The Atlantic Ocean

The Atlantic is generally regarded as the ocean with the richest coccolithophorid flora, and next to the Mediterranean the most thoroughly investigated area. Lohmann (1920), on the *Deutschland* Expedition 1911, made an intensive study in the central part of the North Atlantic and along the eastern coast of South America and described a long series of new taxa. Hentschel (1932, 1936), who participated in the *Meteor* Expedition with fourteen crossings of the South Atlantic, found good agreement with Lohmann's observation. In most cases he recorded higher numbers than recorded by Lohmann but, even these, are small compared with those of Hasle and Smayda, a fact which may be attributed to methodological causes.

In the northern Atlantic Gran (1912) had already reported on the rich coccolithophorid flora, especially in the Sargasso Sea. He also named a number of new taxa, regrettably without giving full descriptions. Some others were reported by Gaarder (1954) from samples collected by Gran. During an ecological study of the Sargasso Sea, further information on variations in the coccolithophorid flora has been obtained during the last decade by Hulburt (1962, 1963*a*, *b*, 1966), Hulburt and collaborators (Hulburt, Ryther and Guillard 1960; Hulburt and Rodman 1963) and Marshall (1966). McIntyre and Bé (1967) have described a new species, *Coccolithus neohelis* from surface waters off Bermuda with coccoliths of a type thought to be extinct.

From the northern-most areas, the North Sea, Norwegian coastal waters, and the Norwegian Sea, papers by Braarud and collaborators (e.g. Braarud, Gaarder and Grøntved 1953; Braarud, Gaarder and Nordli 1958; Braarud, Føyn, Hjelmfoss and Øverland 1967; Halldal 1953; Paasche 1960; Ramsfjell 1960; Smayda 1958*b*) demonstrate the differences between inshore waters with seasonal mass occurrences of a few species, and offshore waters which have a more diverse flora during summer, caused by the influence of more saline and warmer Atlantic water brought in by currents. The seasonal variations in the population have been given much attention.

The Mediterranean

This is a classical area for coccolithophorid studies. Lohmann (1902) published the first monograph on the basis of observations off Syracuse (Sicily). This excellent paper has been the base for all later work on coccolithophorids. Schiller (1930), after several years' study of the coccolithophorids in the Adriatic Sea, published an extensive monograph, reporting on all taxa described until then. The Adriatic Sea studies were continued by Kamptner (1941), who also recorded coccolithophorids from other parts of the Mediterranean (1937*a*, 1944). Recently Saugestad (1967) has undertaken a phytoplankton study in the area around Sardinia and Corsica. The coccolithophorids were absolutely predominant and forty species were recorded. Bernhard and Rampi (1965) have studied the microdistribution of coccolithophorids in the Ligurian Sea.

From the Mediterranean and adjacent areas of the Atlantic, Bernard and Lecal have published papers recording a series of coccolithophorid taxa described as new; as yet they have not been recorded by other authors. Some of them can probably be included in earlier known species after checking by the electron microscope. (For literature see Bernard 1967; Lecal 1967; Loeblich and Tappan 1966.)

During studies in the Black Sea, several Russian biologists have recorded coccolithophorids as a group, dominated by *Coccolithus* (*Pontosphaera*) *huxleyi*. Morozova-Vodyanitskaya and Belogorskaya (1957) have published a list of eighteen species at a salinity of 17-18‰. Later, Mikhaĭlova (1965) recorded *Calciosolenia granii* var. *cylindrothecaeformis* Schiller, which we now would call *Anoplosolenia brasiliensis* (Lohm.) Defl. from the same area. Valkanov (1962) reported the occurrence of a littoral coccolithophorid, *Hymenomonas coccolithophora* Conrad, which according to electron microscope observations should most probably be classified as a *Cricosphaera* (see Braarud 1960).

This brief review of contributions to our knowledge of the distribution of coccolithophorids in the sea should suffice to demonstrate how great the obstacles are for presenting even the coarser features of the biogeography of the modern representatives of the group. Observations are extremely scarce from such vast areas as the Indian Ocean and the Pacific Ocean, and only in a few restricted regions have all-year investigations been undertaken. Still more serious are the inadequacies in the taxonomic treatment of the group apparent in many of the available publications.

In Figs. 5.1-5.5 distribution charts are presented for a few selected species. Even if the species concerned are assumed to be so easily recognizable that no serious errors have been introduced due to misidentification, they have to be regarded as highly preliminary.

The charts are mainly based on the *Meteor* observations by Hentschel (1936, Figs. 21-6). The *Meteor* route and stations in 1925-27 are marked, and the isolines for one cell/litre have been reproduced to show the areas where the species in question have been recorded. In Fig. 5.3, the one cell isolines from Hentschel (1936, Figs. 24 and 25), representing *Rhabdosphaera stylifera* Lohm. and *R. clavigera* Murr. et Blackm., have been combined. Additional records have been obtained from Lohmann (1920), presenting the observations from the *Deutschland* Expedition in 1911, as well as from the following publications: Bernhard and Rampi 1965; Braarud, Gaarder and Grøntved 1953; Ercegović 1936; Gran 1910 (unpublished journal from the *Michael Sars* Expedition); Gaarder 1954; Gran and Braarud 1935; Halldal 1953; Halldal and Markali 1955; Hulburt 1962, 1963*a, b*; Hulburt and Rodman 1963; Hulburt, Ryther and Guillard 1960; Kamptner 1937*a, b,* 1941, 1944; Lecal 1954, 1955, 1965, 1967; Lecal-Schlauder 1951; Lohmann 1902; Markali and Paasche 1955; Marshall 1966; Morozova-Vodyanitskaya and Belogorskaya 1957; Ostenfeld and Paulsen 1904;

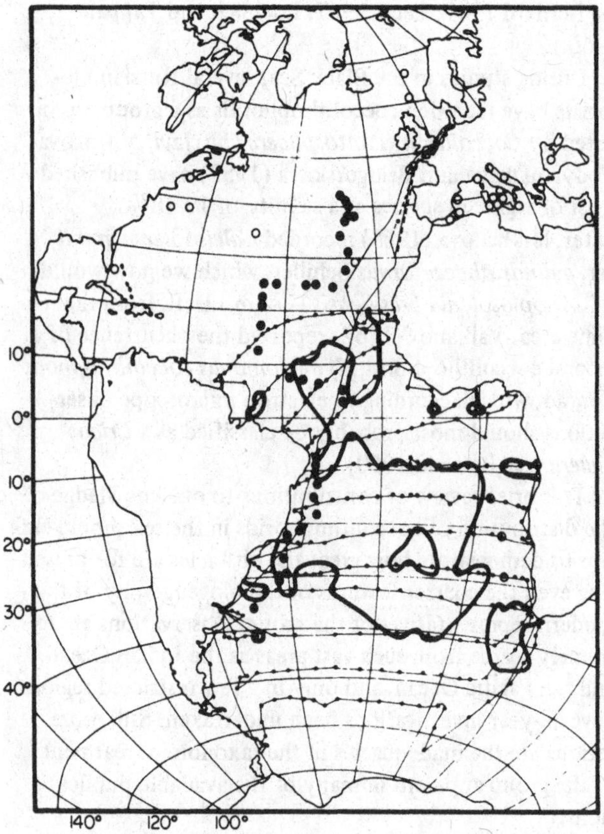
10°
0°
10°
20°
30°
40°
140°
120°
100°

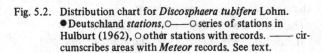

Fig. 5.1. Distribution chart for *Calyptrosphaera oblonga* Lohm. ● *Deutschland* stations, ○ other stations with records. —— circumscribes areas with *Meteor* records. See text.

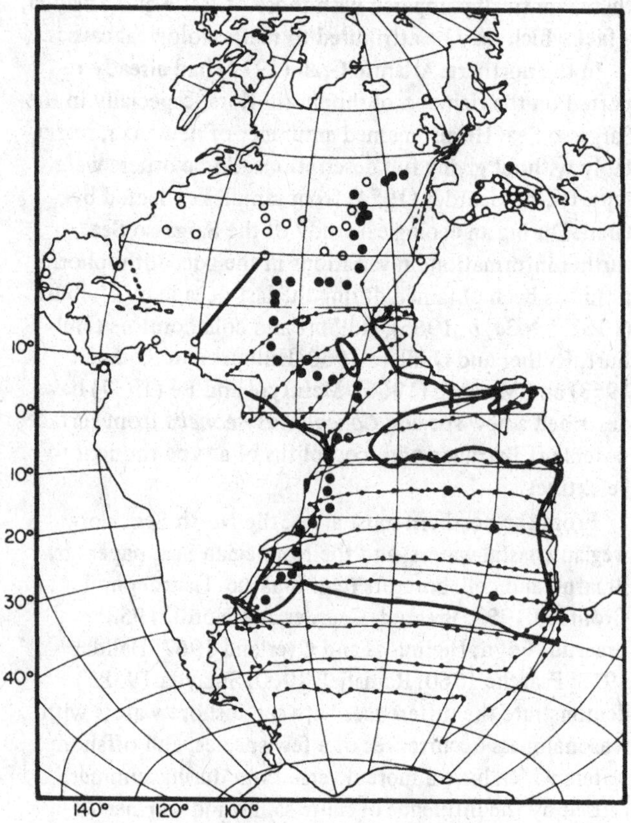
10°
0°
10°
20°
30°
40°
140°
120°
100°

Fig. 5.2. Distribution chart for *Discosphaera tubifera* Lohm. ● Deutschland *stations,* ○——○ series of stations in Hulburt (1962), ○ other stations with records. —— circumscribes areas with *Meteor* records. See text.

Saugestad 1967; Schiller 1930; Smayda 1966; Taylor 1966; Travers 1965. The author's unpublished records from the Faroe-Shetland Channel, Gulf of Mexico, Marseille, Spanish Bay and Split are also included.

A similar distribution chart for *Ophiaster* Gran has been published by the author (Gaarder 1967).

In compilation of the literature the bibliography collected by Loeblich and Tappan (1966) has been very useful. The author is indebted to cand. real. Berit Riddervold Heimdal for assistance in preparation of the charts and to Professor Trygve Braarud for helpful criticism of the manuscript.
Received September 1967.

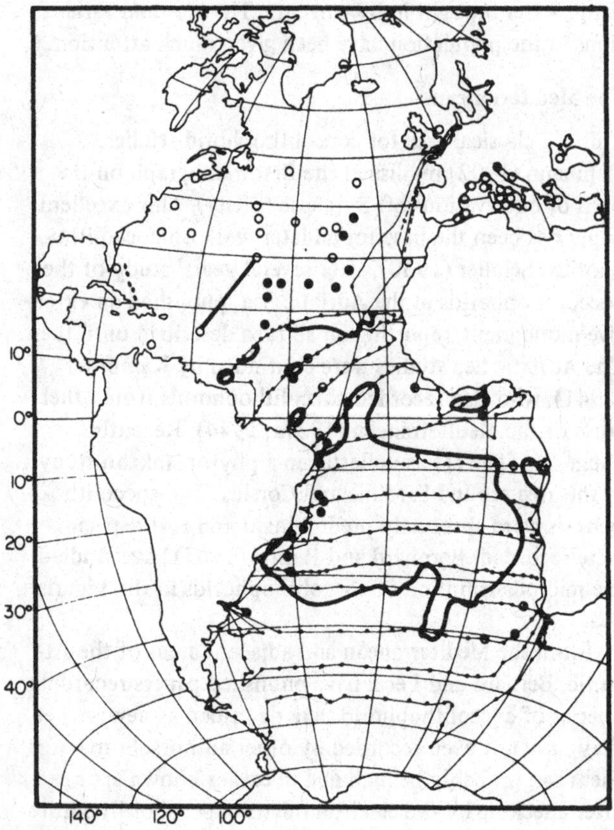
10°
0°
10°
20°
30°
40°
140°
120°
100°

Fig. 5.3. Distribution chart for *Rhabdosphaera clavigera* Murr. et Blackm. = *R. stylifera* Lohm. ● *Deutschland* stations, ○——○ series of stations in Hulburt (1962), ○ other stations with records. —— circumscribes areas with *Meteor* records. See text.

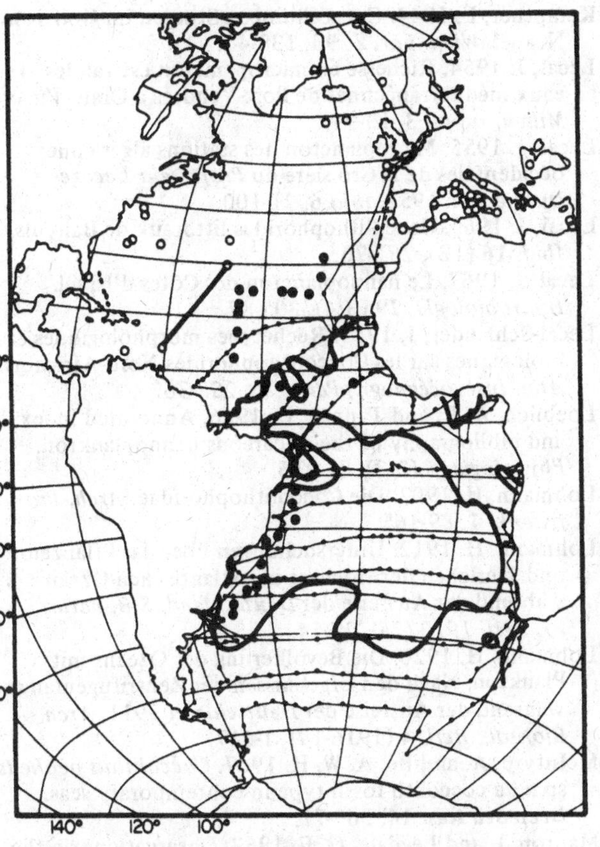
10°
0°
10°
20°
30°
40°
140°
120°
100°

Fig. 5.4. Distribution chart for *Syracosphaera pulchra* Lohm. ● *Deutschland* stations, ○—○ series of stations in Hulburt (1962), ○ other stations with records. —— circumscribes areas with *Meteor* records. See text.

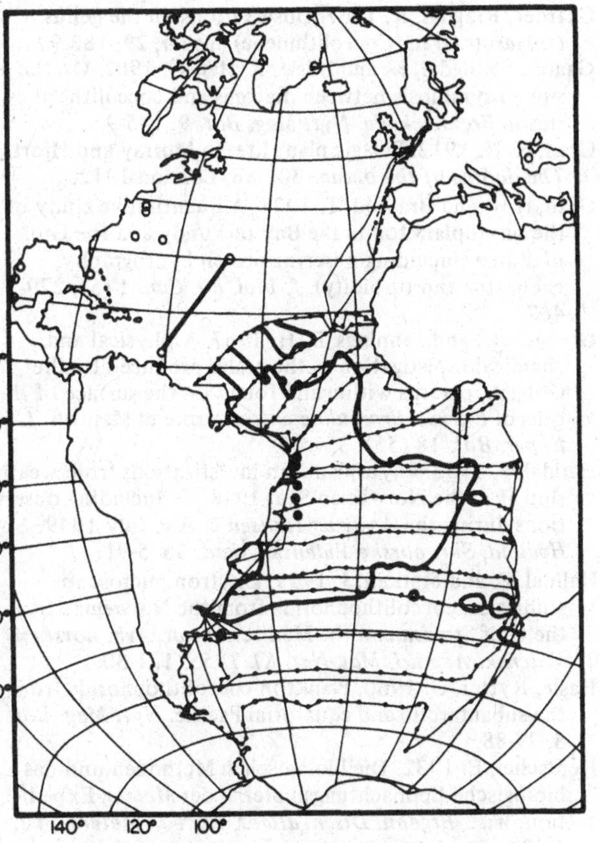
10°
0°
10°
20°
30°
40°
140°
120°
100°

Fig. 5.5. Distribution chart for *Umbellosphaera irregularis* Paasche = *Heyneckia barkowii* Lohm. ● *Deutschland* stations, ○—○ series of stations in Hulburt (1962), ○ other stations with records. —— circumscribes areas with *Meteor* records. See text.

BIBLIOGRAPHY

Bernard, F. 1948. Recherches sur le cycle du *Coccolithus fragilis* Lohm., Flagellé dominant des mers chaudes. *J. Cons.* 15, 177-88.

Bernard, F. 1967. Contribution à l'étude du nannoplancton, de 0 à 3000 m, dans les zones atlantiques lusitanienne et mauritanienne (Campagnes de la *Calypso*, 1960, et du *Coriolis*, 1964). *Pelagos*, 7, 1-81.

Bernard, F. and Lecal, J. 1960. Plancton unicellulaire récolté dans l'océan Indien par le *Charcot* (1950) et le *Norsel* (1955-56). *Bull. Inst. océanogr. (Monaco)*, **1166**, 1-59.

Bernhard, M. and Rampi, L. 1965. Horizontal microdistribution of marine phytoplankton in the Ligurian Sea. *Botanica Gothoburgensia III. Proc. Fifth Mar. Biol. Symp. 1965*, 13-24.

Braarud, T. 1958. Counting methods for determination of the standing crop of phytoplankton. *Rapp. Cons. Explor. Mer*, **144**, 17-19.

Braarud, T. 1960. On the coccolithophorid genus *Cricosphaera* n.gen. *Nytt Mag. Bot.* 8, 211-12.

Braarud, T., Deflandre, G., Halldal, P. and Kamptner, E. 1955. Terminology, nomenclature, and systematics of the Coccolithophoridae. *Micropaleontology*, 1 (2), 157-9.

Braarud, T., Føyn, B., Hjelmfoss, P. and Øverland, Aa. 1967. The natural history of the Hardangerfjord. 5. The phytoplankton in 1955-56. *Sarsia* (in press).

Braarud, T., Gaarder, Ringdal, K. and Grøntved, J. 1953. The phytoplankton of the North Sea and adjacent waters in May 1948. *Rapp. Cons. Explor. Mer*, 133, 1-87.

Braarud, T., Gaarder, Ringdal, K. and Nordli, O. 1958. Seasonal changes in the phytoplankton at various points off the Norwegian West Coast. *Fiskeridir. Skr. Havundersøk.* 12 (3), 1-77.

Christensen, T. 1962. Alger. In *Botanik*, 2 (2), 1-178. København.

Dangeard, P. 1934. Sur l'épiphytisme d'une Coccolithinée, rencontrée à Roscoff. *Act. Soc. linn. Bordeaux Proc.-Verb. 1934*, 45-7.

Ercegović, A. 1936. Études qualitative et quantitative du phytoplancton dans les eaux cotières de l'Adriatique Oriental moyen au cours de l'année 1934. *Acta adriat.* 1 (9), 1-126.

Gaarder, Ringdal, K. 1954. Coccolithineae, Silicoflagellatae, Pterospermataceae and other forms from the *Michael Sars* North Atlantic Deep-Sea Expedition 1910. *Rep. Sars. N. Atl. Deep Sea Exped.* 2 (4), 1-20.

Gaarder, Ringdal, K. 1967. Observations on the genus *Ophiaster* Gran (Coccolithineae). *Sarsia*, **29**, 183-92.
Gaarder, Ringdal, K. and Hasle, Rytter, G. 1962. On the assumed symbiosis between diatoms and coccolithophorids in *Brenneckella. Nytt Mag. Bot.* **9**, 145-9.
Gran, H. H. 1912. Pelagic plant life. In Murray and Hjort: *The depths of the ocean.* 307-86. London 1912.
Gran, H.H. and Braarud, T. 1935. A quantitative study of the phytoplankton in the Bay of Fundy and the Gulf of Maine (including observations on hydrography, chemistry and turbidity). *J. biol. Bd. Can.* **1** (5), 279-467.
Green, J. C. and Jennings, D. H. 1967. A physical and chemical investigation of the scales produced by the Golgi Apparatus within and found on the surface of the cells of *Chrysochromulina chiton* Parke et Manton. *J. Exper. Bot.* **18** (55), 359-70.
Halldal, P. 1953. Phytoplankton investigations from weather ship *M* in the Norwegian Sea, 1948-49. Including observations during the *Armauer Hansen* cruise, July 1949. *Hvalråd, Skr. norske Vidensk. Akad.* **38**, 5-91.
Halldal, P. and Markali, J. 1955. Electron microscope studies on coccolithophorids from the Norwegian Sea, the Gulf Stream and the Mediterranean. *Avh. norske Vidensk. Akad. I. Mat.-Nat. Kl. 1955*, **1**, 1-30.
Hasle, Rytter, G. 1960. Plankton coccolithophorids from the subantarctic and equatorial Pacific. *Nytt Mag. Bot.* **8**, 77-88.
Hentschel, E. 1932. Die biologischen Methoden und das biologische Beobachtungsmaterial der *Meteor*-Expedition. *Wiss. Ergebn. Dtsch. atlant. Exped. 'Meteor'*, **10**, 1-274.
Hentschel, E. 1936. Allgemeine Biologie des Südatlantischen Ozeans. *Ibid.* **11**, 1-344.
Hulburt, E. M. 1962. Phytoplankton in the southwestern Sargasso Sea and North Equatorial Current, February 1961. *Limnol. & Oceanogr.* **7** (3), 307-15.
Hulburt, E. M. 1963*a*. Distribution of phytoplankton in coastal waters of Venezuela. *Ecology*, **44** (1), 169-71.
Hulburt, E. M. 1963*b*. The diversity of phytoplanktonic populations in oceanic, coastal, and estuarine regions. *J. Mar. Res.* **21** (2), 81-93.
Hulburt, E. M. 1966. The distribution of phytoplankton, and its relationship to hydrography, between southern New England and Venezuela. *Ibid.* **24** (1), 67-81.
Hulburt, E. M. and Rodman, J. 1963. Distribution of phytoplankton species with respect to salinity between the coast of southern New England and Bermuda. *Limnol. & Oceanogr.* **8** (2), 263-9.
Hulburt, E. M., Ryther, J. H. and Guillard, R. R. L. 1960. The phytoplankton of the Sargasso Sea off Bermuda. *J. Cons.* **25** (2), 115-28.
Kamptner, E. 1930. Die Kalkflagellaten des Süsswassers und ihre Beziehungen zu jenen des Brackwassers und des Meeres. *Int. Rev. Hydrobiol.* **24**, 147-63.
Kamptner, E. 1937a. Neue und bemerkenswerte Coccolithineen aus dem Mittelmeer. *Arch. Protistenk.* **89** (3), 279-316.
Kamptner, E. 1937b. Über Dauersporen bei marinen Coccolithineen. *S.B. Akad. Wiss. Wien*, **146**, 67-76.
Kamptner, E. 1941. Die Coccolithineen der Südwestküste von Istrien. *Ann. naturh. Mus. Wien*, **51**, 54-149.
Kamptner, E. 1944. Coccolithineen-Studien im Golf von Neapel. *Wien. bot. Z.* **93**, 138-47.
Lecal, J. 1954. Richesse en microplancton estival des eaux méditerranéennes de Port-Vendres à Oran. *Vie et Milieu, suppl.* **3**, 13-95.
Lecal, J. 1955. Microplancton des stations algériennes occidentales de la Croisière du *Professeur Lacaze-Duthiers* en 1952. *Ibid.* **6**, 21-100.
Lecal, J. 1965. Coccolithophorides littoraux de Banyuls. *Ibid.* **16** (1B), 251-70.
Lecal, J. 1967. Le nannoplancton des Côtes d'Israël. *Hydrobiologia*, **29** (3/4), 305-87.
Lecal-Schlauder, J. 1951. Recherches morphologiques et biologiques sur les Coccolithophorides Nord-Africains. *Ann. Inst. océanogr., Paris*, **26**, 255-362.
Loeblich, A. R. and Tappan, H. 1966. Annotated index and bibliography of the calcareous nannoplankton. *Phycologia*, **5** (2/3), 81-216.
Lohmann, H. 1902. Die Coccolithophoridae. *Arch. Protistenk.* **1**, 89-165.
Lohmann, H. 1912. Untersuchungen über das Pflanzen- und Tierleben der Hochsee im Atlantischen Ozean während der Ausreise der *Deutschland. S.B. naturf. Fr. Berl. 1912* (2a), 23-54.
Lohmann, H. 1920. Die Bevölkerung des Ozeans mit Plankton. Nach den Ergebnissen der Zentrifugenfänge während der Ausreise der *Deutschland* 1911. *Arch. Biontol., Berl.* **4** (1916-19), 1-617.
McIntyre, A. and Bé, A. W. H. 1967. *Coccolithus neohelis* sp.n., a coccolith fossil type in contemporary seas. *Deep-Sea Res.* **14**, 369-71.
Manton, I. and Leedale, G. F. 1963. Observations on the micro-anatomy of *Crystallolithus hyalinus* Gaarder and Markali. *Arch. Mikrobiol.* **47**, 115-36.
Markali, J. and Paasche, E. 1955. On two species of *Umbellosphaera*, a new marine coccolithophorid genus. *Nytt Mag. Bot.* **4**, 95-100.
Marshall, H. G. 1966. Observations on the vertical distribution of coccolithophores in the northwestern Sargasso Sea. *Limnol. & Oceanogr.* **11** (3), 432-5.
Marshall, S. M. 1933. The production of microplankton in the Great Barrier Reef region. *Sci. Rep. Gr. Barrier Reef Exped.* **2** (5), 111-57.
Mikhaĭlova, N.F. 1965. Novyĭ dlya Chernogo morya vid kokkolitoforid *Calciosolenia granii* var. *cylindrothecaeformis* Schiller (A species of coccolithophorid new for the Black Sea). *Trud. sevastopol. biol. Stants.* **15**, 50-2.
Morozova-Vodyanitskaya, N. V. and Belogorskaya, E. V. 1957. O znachenii kokkolitoforid i osobenno pontosfery v planktone Chernogo morya (On the importance of the coccolithophorids and particularly Pontosphaera in the plankton of the Black Sea). *Ibid.* **9**, 14-21.
Norris, R. E. 1961. Observations on phytoplankton organisms collected on the N.Z.O.I. Pacific cruise, September 1958. *N.Z.J. Sci.* **4** (1), 162-88.
Norris, R. E. 1965. Living cells of *Ceratolithus cristatus* (Coccolithophorineae). *Arch. Protistenk.* **108**, 19-24.
Ostenfeld, C. 1900. Über Coccosphaera. *Zool. Anz.* **23**, 198-200.
Ostenfeld, C. H. and Paulsen, O. 1904. Planktonprøver fra Nord-Atlanterhavet (c. 58°-60° N. Br.), samlede i 1899 af Dr. K. J. V. Steenstrup. *Medd. Grønland*, **26**, 143-210.

Paasche, E. 1960. Phytoplankton distribution in the Norwegian Sea in June, 1954, related to hydrography and compared with primary production data. *Fiskeridir. Skr. Havundersøk.* **12** (11), 1-77.

Paasche, E. 1964. A tracer study of the inorganic carbon uptake during coccolith formation and photosynthesis in the coccolithophorid *Coccolithus huxleyi. Physiol. Plant. Suppl.* **3**, 1-82.

Parke, M. 1961. Some remarks concerning the class Chrysophyceae. *Brit. phyc. Bull.* **2** (2), 47-55.

Parke, M. and Adams, I. 1960. The motile (*Crystallolithus hyalinus* Gaarder and Markali) and non-motile phases in the life history of *Coccolithus pelagicus* (Wallich) Schiller. *J. Mar. biol. Ass. U.K.* **39**, 263-74.

Parke, M. and Dixon, P. S. 1964. A revised check-list of British marine algae. *Ibid.* **44**, 499-542.

Ramsfjell, E. 1960. Phytoplankton distribution in the Norwegian Sea in June, 1952 and 1953. *Fiskeridir. Skr. Havundersøk.* **12** (10), 1-112.

Rayns, D. G. 1962. Alternation of generations in a coccolithophorid, *Cricosphaera carterae* (Braarud & Fagerl.) Braarud. *J. Mar. biol. Ass. U.K.* **42**, 481-4.

Saugestad, A. 1967. Planteplankton i vestlige Middelhav mars-april 1961. Manuscript, University of Oslo.

Schiller, J. 1930. Coccolithineae. In *Rabenorst's Kryptgamenflora von Deutschland, Österreich und der Schweiz.* **10**, 89-267. Leipzig.

Silva, E. S. 1960. O microplâncton de superfície nos meses de setembro e outubro na estação de Inhaca (Moçambique). *Mem. Junta Invest. Ultram. 2. Sér.*, **18**, 1-50.

Smayda, T. J. 1958a. Biogeographical studies of marine phytoplankton. *Oikos* **9** (2), 158-91.

Smayda, T. J. 1958b. Phytoplankton studies around Jan Mayen Island March-April, 1955. *Nytt Mag. Bot.* **6**, 75-96.

Smayda, T. J. 1963. A quantitative analysis of the phytoplankton of the Gulf of Panama. I. Results of the regional phytoplankton surveys during July and November, 1957 and March, 1958. *Bull. Inter-Amer. trop. Tuna Comm.* **7** (3), 191-253.

Smayda, T. J. 1966. A quantitative analysis of the phytoplankton of the Gulf of Panama. III. General ecological conditions, and the phytoplankton dynamics at 80° 45' N, 79° 23' W from November 1954 to May 1957. *Ibid.* **11** (5), 354-612.

Smirnova, L. I. 1959. Fitoplankton Okhotskogo morya i Prikuril'-skogo raĭona (Phytoplankton of the Okhotsk Sea and region of the Kurile Islands). *Trud. Inst. Okeanol.* **30**, 3-51.

Taylor, F. J. R. 1966. Phytoplankton of the south western Indian Ocean. *Nova Hedwigia*, **12**, 433-76.

Travers, A. 1965. Microplancton récolté en un point fixe de la Mer Ligure (Bouée-Laboratoire du Comexo) pendant l'année 1964. *Rec. Trav. St. Mar. End. Bull.* **39** (55), 11-50.

Travers, A. and Travers, M. 1965. Introduction a l'étude du phytoplancton et des Tintinnides de la région de Tuléar (Madagascar). *Ibid.* (1965) suppl. **4**, 125-62.

Valkanov, A. 1962. Ueber die Entwicklung von *Hymenomonas coccolithophora* Conrad. *Rev. algol. N.S.* **6**, 220-6.

Weber, M. 1902. Introduction et description de l'expédition. In *Siboga-Expeditie. (Monographie.)* **1**, 1-159. Leiden: Brill.

Weber-van Bosse, A. 1901. Études sur les algues de l'archipel Malaisien. III. Note préliminaire sur les résultats algologiques de l'expédition du Siboga. *Ann. Jard. bot. Buitenz.* **17**, *Ser. 2*, **2**, 126-41.

Wilbur, K. M. and Watabe, N. 1963. Experimental studies on calcification in molluscs and the alga *Coccolithus huxleyi. Ann. N.Y. Acad. Sci.* **109**, 82-112.

6. DISTRIBUTION AND ECOLOGY OF LIVING PLANKTONIC FORAMINIFERA IN SURFACE WATERS OF THE ATLANTIC AND INDIAN OCEANS[1]

A. W. H. BÉ AND D. S. TOLDERLUND

Summary: The distributional patterns and relative abundance of twenty seven species of living planktonic Foraminifera are described, based on 703 surface plankton tows in the North and South Atlantic and Indian Oceans.

The species are grouped into five faunal zones, i.e. arctic-antarctic, subarctic-subantarctic, transitional, subtropical, and tropical. The bipolar and anti-tropical nature of the species distributions are evident from the striking similarity of the foraminiferal faunas in reciprocal latitudinal zones between the northern and southern hemispheres.

By comparing samples from surface waters with those from a greater depth range in our plankton repository, we are able to distinguish between surface and subsurface dwellers. Correlations between surface temperatures and species abundances were made to define the maximum and optimum temperature ranges for the individual species.

INTRODUCTION

Planktonic Foraminifera are one of the most common groups of pelagic organisms in the open ocean. Their global distribution through passive transport by ocean currents, coupled with their prolific productivity and sensitivity to environmental variations, has led to their utilization for interpreting ancient marine conditions. The calcite skeletons, which are abundantly preserved in oceanic sediments over approximately 47 per cent of the total sea floor (Sverdrup, Johnson and Fleming 1942), constitute one of our best means for deducing Cenozoic climatic histories and sedimentary successions. This is made possible primarily due to the fact that the majority of the modern species have rather specific environmental requirements. Latitudinal or geographic restrictions of species groups today are largely imposed by their dependence on particular water masses.

This study concerns the distributional patterns of planktonic Foraminifera in the surface waters of the North and South Atlantic and Indian Oceans. Such knowledge should contribute toward an understanding of (1) their broad zoogeographic zonations, and the smaller scale fluctuations resulting from the effects of seasonal and vertical distributions; (2) their life-cycles including rates of reproduction and mortality; (3) their relationship to water masses; (4) their trophic relationship to other groups of pelagic organisms; and (5) their relationship to fossil assemblages in bottom sediments.

[1] Lamont-Doherty Geological Observatory Contribution No.1526.

PREVIOUS INVESTIGATIONS

The pioneering investigations of Brady (1884), Murray (1897) and Rhumbler (1901, 1911) have established the importance of planktonic Foraminifera in the pelagic biotope. They made fundamental observations on the distributions of these organisms in oceanic water and deep-sea sediments, their food requirements and life processes.

Quantitative studies on the regional distribution of living species were initiated by Schott (1935) in the equatorial Atlantic, based on samples collected by the *Meteor* Expedition. This was followed by similar investigations in the Gulf of Mexico (Phleger 1951), in the North Atlantic (Bé 1959, 1960*b*; Bé and Hamlin 1967; Cifelli 1962, 1965, 1967), in the equatorial Atlantic (Boltovskoy 1964; Jones 1967) and in the Caribbean Sea (Jones 1966).

Distributional and ecological studies have also been

made on a broad scale in the North and equatorial Pacific by Bradshaw (1959), in the Indian Ocean by Belyaeva (1964) and in the Antarctic Ocean by Boltovskoy (1966*a*) and Bé (in press). Data of more limited coverage were obtained for the northeastern Pacific by Smith (1963, 1964), in the southeastern Pacific by Parker (1960) and the southwestern Atlantic Ocean by Boltovskoy (1959, 1962, 1966*b*).

A large amount of information is available from the many investigations on the distribution of planktonic foraminifera in bottom sediments, and such studies will be quoted in the following sections whenever they are pertinent to our discussion of the distribution of living species.

METHODS OF COLLECTION AND ANALYSIS

The present study is based on 703 surface plankton samples collected in the upper 10 metres of water of the Indian and Atlantic Oceans (Table 6.1 and Fig. 6.1). Although they were obtained over a twelve-year period between 1955 and 1966, we have striven as much as possible to standardize sample collecting. All tows were obtained with nylon plankton nets having a 200μ mesh aperture (NITEX 202, 86 meshes per inch), 50 centimetres square mouth opening and a length of 3 metres. A Tsurumi-Seiki-Kosakusho flowmeter, suspended within the mouth opening of the net, was used to provide quantitative measurement of the volume of water strained for each tow.

The net was attached either to a bathythermograph winch or a rope line and sampling duration averaged about thirty minutes. In plankton-rich waters the duration was shortened to prevent net clogging. The samples were preserved in 5 per cent sea-water formalin buffered with hexamethylene-tetramine. When plankton density was unusually high, the sample required 10 per cent or more formalin with additional buffering to prevent dissolution of the calcareous shells.

The majority of the plankton tows were collected on cruises of R. V. *Vema* and R. V. *Robert Conrad* of the Lamont Geological Observatory. Both ships made extensive cruises in the Indian Ocean, while participating in the International Indian Ocean Expedition. With the cooperation of the U.S. Coast Guard we were able to obtain 160 samples in the North Atlantic. About half of them were taken during the 1962 cadet training cruise of USCGC *Yakutat* and on USCGC *Casco* during its 1963 participation in the Equalant Expedition in the tropical Atlantic. The results from the *Yakutat* cruise have already been published by Bé and Hamlin (1967) and the data are incorporated in the present study. We have also used many surface tows that were collected by U.S. Coast Guard vessels *en route* to and from the ocean weather stations Bravo, Charlie, Delta and Echo.

The aforementioned samples supply the broad, geographic coverage needed for a study of the horizontal distributional patterns of planktonic Foraminifera. They complement our seasonal distribution study at stations Bravo (56° 30′ N, 51° 00′ W,) Charlie (52° 45′ N, 35° 30′ W), Delta (44° N, 41° W), Echo (35° N, 48° W), and off Bermuda, which will be the basis of a separate publication (Tolderlund and Bé, in preparation). Future work will also include the analysis of deep plankton tows collected in the upper 300 metres of water.

The same laboratory procedures and species synonomies used by Bé and Hamlin (1967) were followed. Most species are common and well described; the species names used in this paper are those included in a taxonomic key by Bé (1967) and the reader is referred to that key for species illustrations.

Space limitation and the general similarity of the distributional patterns between *relative* and *absolute* abundance prompted us to present only species maps of *relative* abundance (i.e. per cent of total planktonic foraminiferal population). This measure was especially useful in characterizing the faunal composition of particular oceanic regions, and in tracing the course of transported species groups. It is also important as the only direct method of comparing species composition of living populations with their fossil assemblages in bottom sediments, since the fossil concentrations cannot be expressed in terms of absolute abundance (number of specimens per unit volume of water).

An IBM (Model 1802) Data Acquisition and Control System was used to process the 703 samples by 27 species data matrix used in this study. The raw species counts for a given sample were first transferred to punch cards along with associated ecological and station data. Relative abundance values were then generated from this raw data matrix. These, in turn, were plotted on an IBM 1627 (Model 2) X-Y plotter on a species by species basis using a Mercator projection plotting programme. These maps were then contoured and the contours transferred to the species maps shown in Figs. 6.5-6.16.

FAUNAL PROVINCES

The twenty seven or more living species of planktonic foraminifera are distributed in rather distinct latitudinal belts and they can thus be grouped into several zoogeographic provinces which are primarily influenced by ecology and climate (Table 6.2 and Figs. 6.2 and 6.3.). Many plankton investigators, such as Meisenheimer (1905), Brandt and Apstein (1901-28), Steuer (1910, 1933), Russell (1939), Bogorov (1955), Brodsky (1957), Johnson and Brinton (1963) and others, have noted a close relation-

Table 6.1. Surface plankton samples collected in 0-10 m depth range in the North and South Atlantic and Indian Oceans. Cruise tracks and plankton stations are shown in Fig. 6.1.

Cruise		No. Samples	Inclusive dates
Vema-14	N. Atlantic Ocean	10	16-11-57 to 26-11-57
	S. Atlantic Ocean	42	20-12-57 to 6-4-58
	Indian Ocean	14	15-4-58 to 30-5-58
Vema-15	N. Atlantic Ocean	32	18-10-58 to 11-7-59
	S. Atlantic Ocean	11	16-2-59 to 8-5-59
Vema-16	N. Atlantic Ocean	34	8-10-59 to 19-9-60
	S. Atlantic Ocean	25	23-11-59 to 20-12-59
	Indian Ocean	15	30-12-59 to 6-3-60
Vema-17	N. Atlantic Ocean	23	20-12-60 to 26-9-61
Vema-18	N. Atlantic Ocean	5	9-1-62 to 14-1-62
	S. Atlantic Ocean	61	16-1-62 to 30-5-62
	Indian Ocean	21	4-6-62 to 7-7-62
Vema-19	N. Atlantic Ocean	12	16-10-63 to 21-11-63
	S. Atlantic Ocean	36	21-9-63 to 15-10-63
	Indian Ocean	34	17-7-63 to 11-9-63
Vema -20	N. Atlantic Ocean	1	17-11-64
	S. Atlantic Ocean	26	17-10-64 to 15-11-64
	Indian Ocean	11	25-8-64 to 17-9-64
Vema-22	N. Atlantic Ocean	13	17-5-66 to 11-6-66
Vema-23	N. Atlantic Ocean	74	30-7-66 to 13-12-66
Misc. *Vema and Conrad*	N. Atlantic Ocean	14	
	Indian Ocean	7	1955 to 1965
USCGC *Yakutat*	N. Atlantic Ocean	59	10-6-62 to 9-8-62
USCGC *Casco*	N. Atlantic Ocean	23	24-7-63 to 18-8-63
	S. Atlantic Ocean	22	
Misc. Coast Guard		78	1959 to 1963
Subtotal N. Atlantic Ocean		378	
Subtotal S. Atlantic Ocean		223	
Subtotal Indian Ocean		102	
Grand total of samples		703	

ship between the faunistic and hydrographic features of the major oceanic water masses and that their geographical variations are particularly strong in the epipelagic rather than in the meso- or bathypelagic zones. In the broadest sense three major biogeographical provinces can be distinguished: a warm-water region, located approximately between 40° N and 40° S latitude, lies between the northern and southern cold-water regions. All holoplanktonic groups including the Foraminifera have distinct species assemblages that occur in either the warm- or cold-water regions but very rarely in both. Our delineations of faunal provinces shown in Fig. 6.2 correspond in general to the major hydrographic regions of the World Ocean and the reader is referred to G. Schott (1935, 1942), Sverdrup, Johnson and Fleming (1942), Deacon (1945), Defant (1961), Dietrich (1963, 1964) and Fairbridge (1966), for a comparison between the faunal and hydrographic regional subdivisions.

Many cold-water species have a bipolar distribution, inhabiting both the northern and southern cold-water provinces. Their present disjunct distribution has been variously explained. One of the most plausible explanations is that from a former continuous distribution, the recent isolation of bipolar species is caused by warming of the ocean waters since the Würm glacial epoch. This has consequently resulted in the reduction of their geographical range. Stenothermal cold-water species, such as the chaetognath *Eukrohnia hamata* (Alvariño 1965), that have a

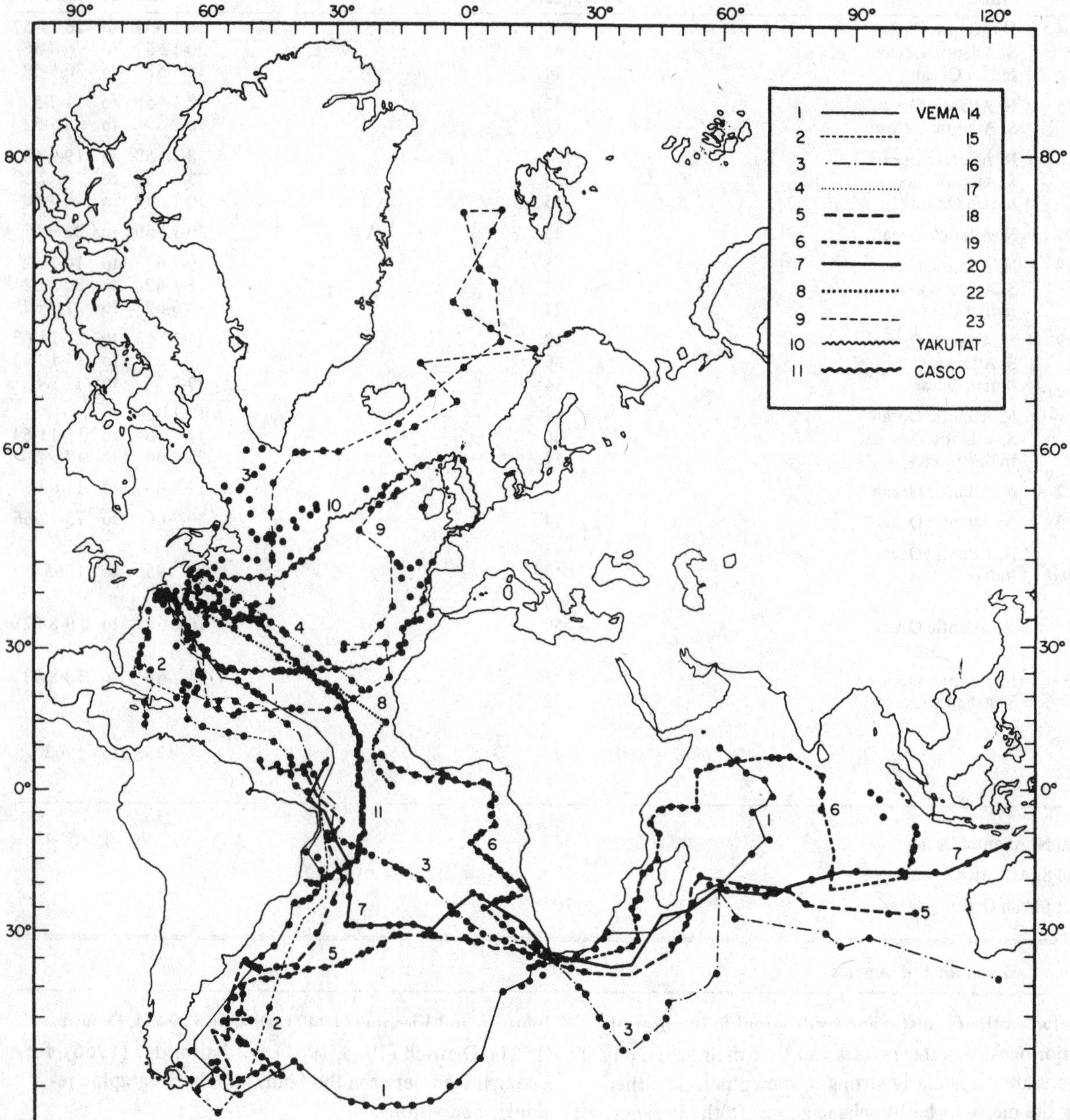
90°
60°
30°
0°
30°
60°
90°
120°
80°
60°
30°
0°
30°
60°
70°
1 VEMA 14
2 15
3 16
4 17
5 18
6 19
7 20
8 22
9 23
10 YAKUTAT
11 CASCO

Fig. 6.1. Cruise tracks and stations where plankton samples were collected for this study. (See also Table 6.1.)

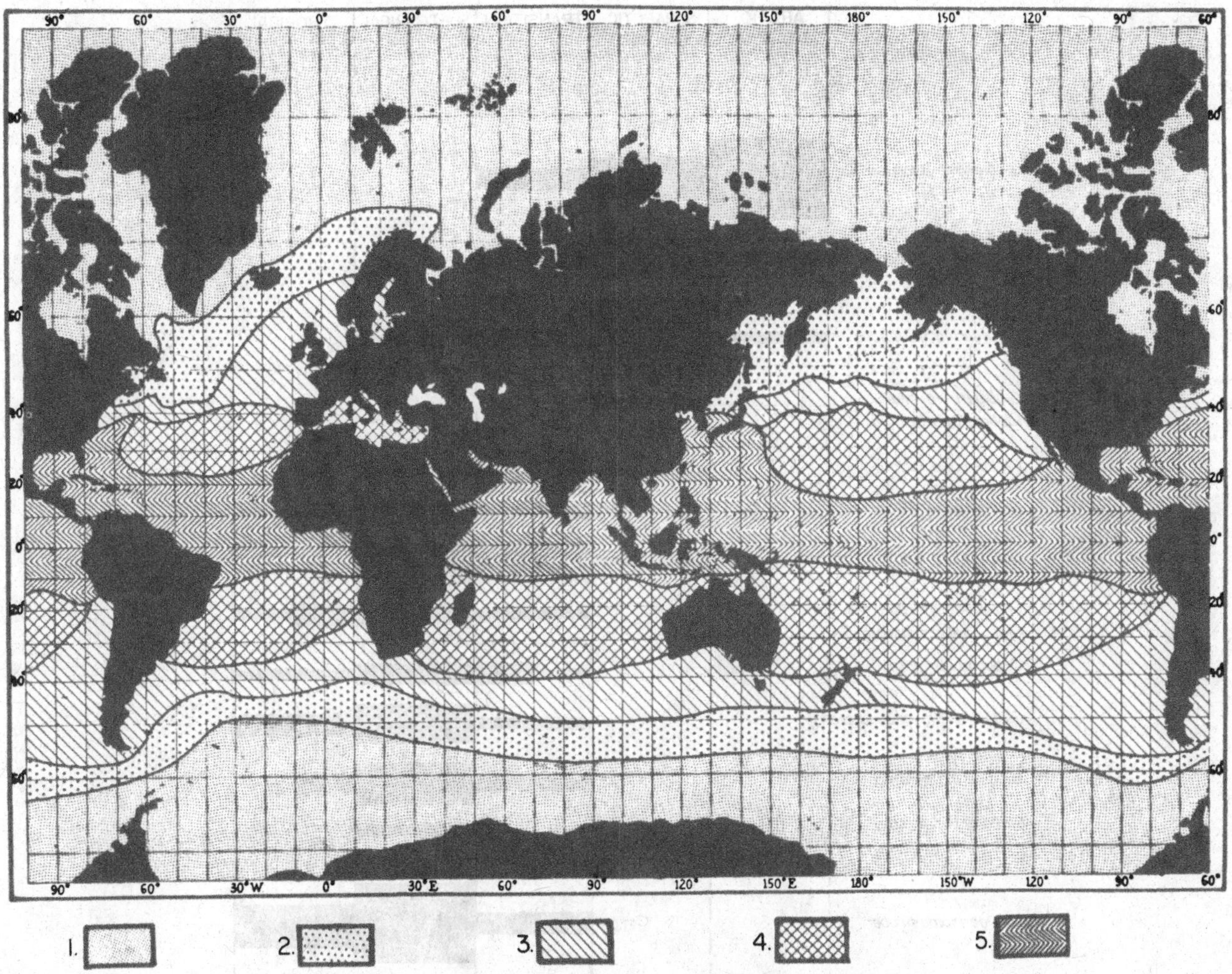
1.
2.
3.
4.
5.
WORLD DISTRIBUTIONAL ZONES OF PLANKTONIC FORAMINIFERA

Fig. 6.2. Five major world distributional zones of planktonic foraminifera. North Pacific zonations according to Bradshaw (1959). 1. Arctic and antarctic. 2. Subarctic and subantarctic. 3. Transition Zone. 4. Subtropical. 5. Tropical.

continuous distribution by 'submerging' below the tropical-subtropical waters are very rare. None of the planktonic foraminiferal species is known to exhibit deep tropical submergence. Even rarer are the eurythermal, cosmopolitan species that have a world-wide distribution in all water types. *Globigerinita glutinata* is a rare example in this category.

The majority of the planktonic foraminiferal species (22) live in tropical and subtropical waters. The tropical species (10) are *Globigerinoides sacculifer, Globorotalia menardii, G. tumida, Pulleniatina obliquiloculata, Candeina nitida, Hastigerinella digitata, 'Sphaeroidinella dehiscens', Globoquadrina conglomerata, Globigerinella adamsi* and *Globoquadrina hexagona.* The latter three species inhabit the Indian and Pacific Oceans only. *G. hexagona* and *G. conglomerata* are known from Pleistocene deep-sea sediments in the Atlantic Ocean, but they have apparently disappeared since from that ocean. The tropical species are transported to mid-latitudes via the warm currents along the eastern margins of the continents (Gulf Stream, Brazil and Agulhas Current, etc.).

The subtropical species (12) that live in the oligotrophic central areas of the oceans are *Globorotalia hirsuta, G. truncatulinoides, Globigerinoides ruber, Globigerina rubescens, G. falconensis, Globigerinoides conglobatus, Hastigerina pelagica, Globigerinita glutinata* and *Globorotalia crassaformis. Globoquadrina dutertrei, Globigerinella aequilateralis* and *Orbulina universa* are subtropical species commonly found at the edges of the central water masses, near continental margins and areas of upwelling, and they are transported into transitional waters.

Many of the aforementioned tropical and subtropical species do occur commonly in subtropical and tropical waters, respectively, indicating that the faunistic as well as

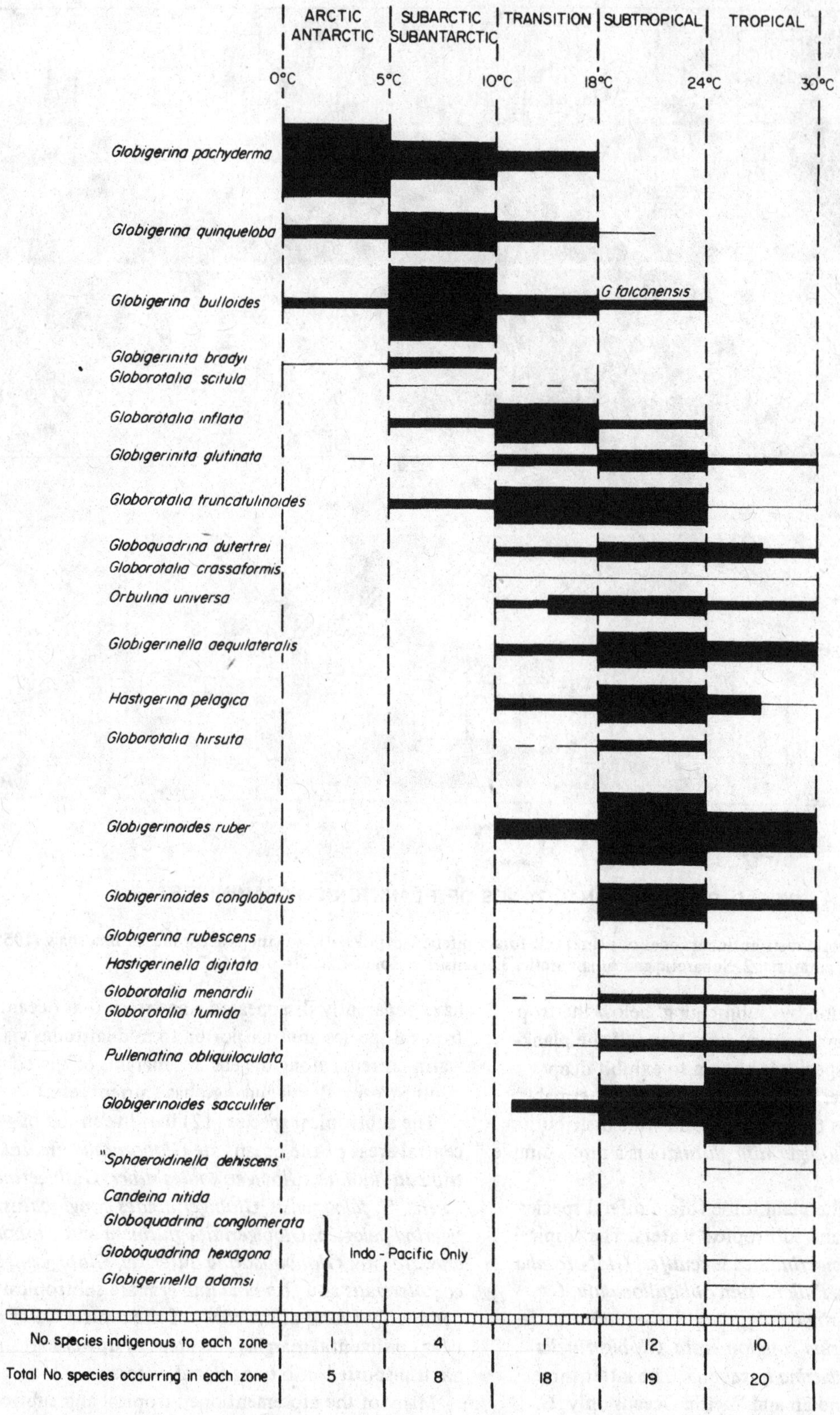
ARCTIC ANTARCTIC
SUBARCTIC SUBANTARCTIC
TRANSITION
SUBTROPICAL
TROPICAL
0°C
5°C
10°C
18°C
24°C
30°C
Globigerina pachyderma
Globigerina quinqueloba
Globigerina bulloides
G falconensis
Globigerinita bradyi
Globorotalia scitula
Globorotalia inflata
Globigerinita glutinata
Globorotalia truncatulinoides
Globoquadrina dutertrei
Globorotalia crassaformis
Orbulina universa
Globigerinella aequilateralis
Hastigerina pelagica
Globorotalia hirsuta
Globigerinoides ruber
Globigerinoides conglobatus
Globigerina rubescens
Hastigerinella digitata
Globorotalia menardii
Globorotalia tumida
Pulleniatina obliquiloculata
Globigerinoides sacculifer
"Sphaeroidinella dehiscens"
Candeina nitida
Globoquadrina conglomerata
Globoquadrina hexagona
Globigerinella adamsi
Indo - Pacific Only
No. species indigenous to each zone
1
4
1
12
10
Total No. species occurring in each zone
5
8
18
19
20

Fig. 6.3. Species assemblages and ranges in the five world distributional zones. Varying thickness represents relative abundance within each zone.

hydrographic differences between these regions are comparatively small.

The regions in the northern and southern hemisphere where cold-water and warm-water species overlap in distribution and where the greatest faunistic contrasts occur are designated 'Transition Zones' (Fig. 6.2). Hereafter, we shall use the term 'transitional' to refer to the waters of the Transition Zones. Sharp faunal boundaries between subpolar and subtropical species are encountered here among many zooplankton and phytoplankton groups as well as among the Foraminifera (Hensen *et al.* 1892-1926). The positions of these faunal boundaries may be sharply defined at any one time, but temporal changes result in a belt of varying width.

The Transition Zones are narrow and comparatively well-defined in the western part of the oceans, but they broaden into more diffuse regions in an eastward direction. Completely different foraminiferal species assemblages may be encountered at stations within a few miles of each other, as for example at the boundaries of the Gulf Stream - Labrador Currents (including slope waters) (Bé 1959; Bé and Hamlin 1967); Cifelli 1962, 1967), the Kuroshio-Oyashio Currents (Bradshaw 1959), and the Falkland-Brazil Currents (Boltovskoy 1966*b*).

While subtropical tropical and subpolar species do occur together in transitional waters, only one species *Globorotalia inflata* appears to be indigenous to and a good indicator species of the Transition Zone. In the Southern Transition Zone, our boundaries correspond very closely to Boltovskoy's (1966*b*) Subtropical-Subantarctic Convergence Zone. He too found that *Globorotalia inflata* is the most abundant species in this zone; *Globigerina pachyderma* and *G. bulloides* are common within and south of it, while *Globigerinoides trilobus* (= *G. sacculifer*) is dominant to the north.

The cold-water fauna can be divided into the subpolar species (*Globigerina quinqueloba*, right-coiling *Globigerina pachyderma, Globigerina bulloides* sensu stricto, *Globorotalia scitula* and *Globigerinita bradyi*) and a single polar species (left-coiling *G. pachyderma*). Some species appear to be more cold-tolerant in the South Pacific (Boltovskoy 1966*a*; Bé (in press) than in the northern oceans. For example, *Globorotalia truncatulinoides, G. inflata* and *G. glutinata* are commonly encountered in subantarctic waters, whereas they inhabit subtropical or transitional waters in the northern hemisphere. *Globigerina quinqueloba* and *Globigerinita bradyi* are also frequent south of the Antarctic Polar Front (i.e. Antarctic Convergence), but they are predominantly found in subarctic waters in the North Pacific and North Atlantic. *Globorotalia cavernula* is a rare subantarctic species, which apparently occurs only in the southern hemisphere.

Figure 6.3 summarizes the distributional ranges of twenty-seven species of planktonic foraminifera and indicate the zoogeographic province where each species has its maximum concentration. It should be noted that most species appear in two or more zones and that the overlapping distributions are particularly frequent between polar and subpolar species on the one hand and between subtropical and tropical species on the other, while in the Transition Zones a high degree of mixing takes place between subpolar and tropical-subtropical species. Thus the ratio of indigenous species to total species present is 1 : 5 in arctic-antarctic waters, 4 : 8 in subarctic-subantarctic waters, 1 : 18 in transitional waters, 12 : 19 in subtropical waters, and 10 : 20 in tropical waters. The high faunal diversity in the warm-water regions is strongly evident and suggests that here evolution has proceeded more rapidly than in the colder waters.

The bipolarity and anti-tropicality of the foraminiferal species is also quite obvious and is even reflected in the reciprocal distribution patterns of the coiling directions of *G. pachyderma* and, to a lesser degree, *G. truncatulinoides* in the North and South Atlantic (Figs. 6.6 and 6.12). We attribute the larger number of species occurring in the cold waters of the southern hemisphere to the more continuous ocean circulation in contrast to the land-locked and restricted Arctic Ocean where only a single species lives. The greater species diversity in the Indian and Pacific Oceans may also be partly due to the relative lack of a land barrier between these oceans. At present tropical-subtropical species cannot so easily circumvent Cape Horn and Cape of Good Hope or pass through the Mediterranean Sea in order to establish a population interchange between the Atlantic and Indo-Pacific Oceans.

BATHYMETRIC DISTRIBUTION

From quantitative comparisons of foraminiferal abundance between approximately 700 surface (0-10 metre) and 500 (0-300 metres) plankton tows from the Atlantic and Indian Oceans we can make several general remarks about the vertical distribution of planktonic foraminifera (Table 6.3). Firstly, the majority of the spinose species are surface dwellers, or at least prefer to live in the upper part of the euphotic zone. In this category belong all the species of *Globigerinoides* (*G. sacculifer, G. ruber* and *G. conglobatus*) and some *Globigerina* species (*G. quinqueloba* and *G. rubescens*), all of whose maximum concentrations usually occur in the upper 10 metres of water. However, adults are not precluded from deeper habitats. It is our general impression that adult tests of *Globigerinoides* and *Globigerina* species from sub-surface depths have fewer spines or lack them completely, but we are not certain whether this is a function of greater fragility of adult spines and/or destruction by net abrasion during the sampling operation.

Secondly, most of the non-spinose species live preferentially at sub-surface depths, i.e. below 50 metres, and in a number of instances well below 500 metres. All species of *Globorotalia* (*G. menardii, G. tumida, G. inflata, G. truncatulinoides, G. hirsuta* and *G. scitula*) exhibit a considerable range in vertical distribution, spending their early stages in the euphotic zone and descending to meso- or bathypelagic depths in late ontogeny. In subantarctic waters we have collected considerable numbers of *G. scitula gigantea* from a depth range between 500 and 1,000 metres. *G. crassaformis* is a particularly consistent inhabitant of deeper subtropical waters.

A very large number of planktonic foraminifera of all species probably die in the epipelagic zone (0-300 metres) and only a relatively small proportion of them survive to live at great depths and continue secreting thick calcitic tests. The shell thickening process as an adaptation to increasingly deeper habitats was described for *G. pachyderma, G. truncatulinoides, G. menardii,* and *G. sacculifer* in a series of earlier investigations (Bé 1960*a*; Bé and Ericson 1963; Bé 1965; Bé, McIntyre and Breger 1966).

The *Globoquadrina species* (*G. dutertrei, G. conglomerata, G. hexagona*) and *G. pachyderma* (which because of its lack of spines belongs more properly under *Globoquadrina* than *Globigerina*) are also more abundant in sub-

Table 6.2. Species composition of the five world distributional zones of planktonic foraminifera shown in Fig. 6.2 and 6.3. The species are listed under the zone where their highest concentrations are observed, but they are not necessarily limited to these areas.

NORTHERN AND SOUTHERN COLD-WATER REGIONS

1. Arctic and Antarctic Zones:
   - *Globigerina pachyderma* (Ehrenberg): left-coiling variety; right-coiling in subarctic and subantarctic zones.
2. Subarctic and Subantarctic Zones:
   - *Globigerina quinqueloba* Natland
   - *Globigerina bulloides* d'Orbigny
   - *Globigerinita bradyi* Wiesner
   - *Globorotalia scitula* (Brady)

TRANSITION ZONES

3. Northern and Southern Transition Zones between Cold-water and Warm-water Regions:
   - *Globorotalia inflata* (d'Orbigny), with mixed occurrences of subpolar and tropical-subtropical species

WARM-WATER REGION

4. Northern and Southern Subtropical Zones (usually located in central water masses between 20° N and 40° N or between 20° S and 40° S lat.):
   - *Globigerinoides ruber* (d'Orbigny): pink variety in Atlantic Ocean only.
   - *Globigerinoides conglobatus* (Brady): autumn species
   - *Hastigerina pelagica* (d'Orbigny)
   - *Globigerinita glutinata* (Egger)
   - *Globorotalia truncatulinoides* (d'Orbigny) } winter species
   - *Globorotalia hirsuta* (d'Orbigny) } winter species
   - *Globigerina rubescens* Hofker
   - *Globigerinella aequilateralis* (Brady) } prefer outer margins of subtropical central water masses and into Transitional Zone
   - *Orbulina universa* d'Orbigny } prefer outer margins of subtropical central water masses and into Transitional Zone
   - *Globoquadrina dutertrei* (d'Orbigny) } prefer outer margins of subtropical central water masses and into Transitional Zone
   - *Globigerina falconensis* Blow
   - *Globorotalia crassaformis* (Galloway and Wissler)
5. Tropical Zone:
   - *Globigerinoides sacculifer* (Brady) incl. *'Sphaeroidinella dehiscens'* (Parker and Jones)
   - *Globorotalia menardii* (d'Orbigny)
   - *Globorotalia tumida* (Brady)
   - *Pulleniatina obliquiloculata* (Parker and Jones)
   - *Candeina nitida* d'Orbigny
   - *Hastigerinella digitata* (Rhumbler)
   - *Globoquadrina conglomerata* (Schwager) } restricted to Indo-Pacific
   - *Globigerinella adamsi* (Banner and Blow) } restricted to Indo-Pacific
   - *Globoquadrina hexagona* (Natland) } restricted to Indo-Pacific

Most species listed under the Subtropical Zones are also common in the tropical waters.

Table 6.3. Depth habitats based on comparison of relative abundance between surface and deep plankton tows

Surface species (max. abundance in 0-10 metre tows)	Equally abundant in 0-10 metre and 0-300 metre tows
Globigerinoides ruber	*Globigerinella aequilateralis*
Globigerinoides sacculifer	*Globigerina bulloides*
Globigerinoides conglobatus	*Globigerinita glutinata*
Globigerina quinqueloba	*Pulleniatina obliquiloculata*
Globigerina rubescens	*Orbulina universa*
	Candeina nitida
Sub-surface species (more abundant in 0-300 metre than in 0-10 metre tows)	
Globorotalia truncatulinoides	*Globorotalia scitula*
Globorotalia inflata	*Globoquadrina dutertrei*
Globorotalia hirsuta	*'Globigerina' pachyderma* (adult)
Globorotalia menardii	*Hastigerina pelagica*
Globorotalia tumida	*Hastigerinella digitata*
Globorotalia crassaformis	*'Sphaeroidinella' dehiscens*

surface waters.

The only spinose species that prefer deeper waters are *H. pelagica* and *H. digitata.* We consider the latter as a bathypelagic isomorph, or late ontogenetic stage of the former, while *'S. dehiscens'* is an aberrant terminal stage of *G. sacculifer* (Bé 1965).

Thirdly, six species (*G. aequilateralis, G. bulloides, G. glutinata, P. obliquiloculata, O. universa* and *C. nitida*) occur in approximately equivalent concentrations in surface as in deep tows and therefore are not known to have any preferred depth habitats.

TEMPERATURE RANGES OF INDIVIDUAL SPECIES

Temperature ranges have been worked out for twenty species (Fig. 6.4). In order to establish these ranges, a surface temperature map was constructed using the actual temperature data gathered at every station. Each species distribution map was then superimposed over this map and the maximum and minimum temperatures determined for each species in the North Atlantic, South Atlantic and Indian Oceans. In addition, the temperatures included by the area or areas of maximum relative abundance for each species were also noted for each of these regions. If such an area of maximum abundance did not occur in a particular region, or if the data were insufficient, then only the temperature range is shown.

It should be kept in mind that the temperature ranges shown in Fig. 6.4 are subject to several artifacts most of which are related to the problems of sampling density and seasonal bias. For example, the fact that *G. conglobatus* does not have an area of maximum abundance in the Indian Ocean equivalent to those of the North and South Atlantic is probably an artifact resulting from a lack of austral autumn samples.

There is also the artifact created by the fact that the total temperature ranges over our sampling period in the three oceans do not coincide. The range observed in the North Atlantic was 1-29° C; in the South Atlantic it was 0-27° C; and in the Indian Ocean it was 3-30° C.

The fragmentation of the maximum abundance ranges in a given ocean in Fig. 6.4 can also be attributed to two other causes. The first being the preference of many species for sub-surface depths. A comparison with Table 6.3 reveals that there is distinctly less fragmentation in the maximum abundance ranges of the 'surface' species. Yet, the splitting of the South Atlantic graph of surface-dwelling *G. sacculifer* points up the other cause, i.e. the effect of particular ocean currents. In this case, the cooler range of maximum abundance represents the samples occurring in the area of the Benguela Current.

The overall temperature range and the range of peak abundance of individual species have been included under their descriptive sections.

DISTRIBUTIONAL PATTERNS OF INDIVIDUAL SPECIES

Globigerina pachyderma (Ehrenberg)

Globigerina pachyderma is the best indicator species of polar waters. It is the only species in the Arctic Ocean and the dominant form in the arctic waters north of the Arctic Circle, the Labrador Sea and Antarctic waters. High concentrations are also found in sub-polar and transitional (i.e. cold-temperate) waters. The relatively low frequencies between Newfoundland and Iceland are due to the dominance of *Globigerina bulloides* and *G. quinqueloba.*

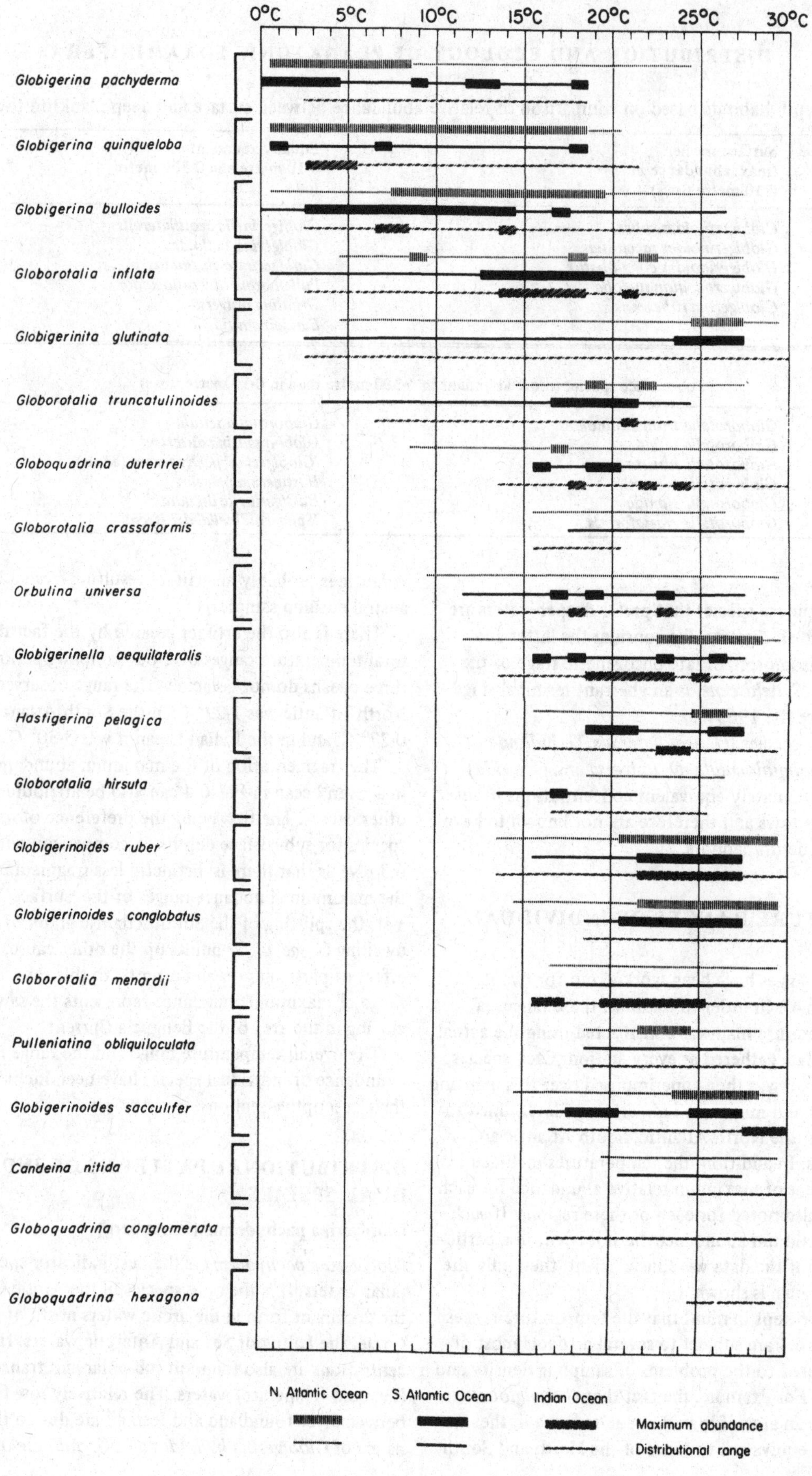
0°C
5°C
10°C
15°C
20°C
25°C
30°C
Globigerina pachyderma
Globigerina quinqueloba
Globigerina bulloides
Globorotalia inflata
Globigerinita glutinata
Globorotalia truncatulinoides
Globoquadrina dutertrei
Globorotalia crassaformis
Orbulina universa
Globigerinella aequilateralis
Hastigerina pelagica
Globorotalia hirsuta
Globigerinoides ruber
Globigerinoides conglobatus
Globorotalia menardii
Pulleniatina obliquiloculata
Globigerinoides sacculifer
Candeina nitida
Globoquadrina conglomerata
Globoquadrina hexagona
N. Atlantic Ocean
S. Atlantic Ocean
Indian Ocean
Maximum abundance
Distributional range

There is a problem in recognizing juveniles of *G. pachyderma*, which do not resemble the typical, thickly encrusted adults. Juveniles have four and a half to five hemispherical chambers per whorl, a large aperture, open umbilicus, and thin-walled test. We believe that the compact, thick-walled adult test with four coalescing chambers per whorl, reduced final chamber and constricted aperture is a product of shell thickening as the organism descends to deeper habitats (Bé 1960).

It is not certain whether *G. pachyderma* has true elongate spines. As far as we know it probably does not possess spines, and thus would have greater affinity to the nonspinose *Globoquadrina* than the spinose *Globigerina.*

Interesting distribution patterns have been found for the percentage ratios of left- versus right-coiling *G. pachyderma* tests, which apparently are temperature-dependent. Ericson (1959) observed a right- and a left-dominant coiling province in North Atlantic fossil assemblages, separated by the present 7.2° C surface isotherm (April). This line runs from south of Newfoundland, south of Iceland to Scotland. In our study of living populations the boundary between right- and left-coiling provinces (Fig. 6.6) agrees with Ericson's data, except in the Norwegian Sea where our results show predominantly right-coiling populations. Ninety per cent or more left-coiling individuals are present in high-Arctic waters and the northern Labrador Sea.

Right-coiling *G. pachyderma* is abundant in subarctic and transitional waters from the Norwegian Sea to New York and Cape Hatteras. In the Transitional Zone, as in the region of the slope waters and Gulf Stream off New York, the right-coiling variety grades morphologically into *Globoquadrina dutertrei* (= *Globigerina eggeri*) and this prompted Cifelli (1961) to name the intergrade *Globigerina incompta*, which he later (1965) considered as *Globigerina pachyderma incompta.* We believe that the morphological gradations from left-coiling *G. pachyderma* in polar waters, to the right-coiling variety in subpolar regions, to right-coiling *G. dutertrei* in transitional and subtropical waters are evidence that the three forms are genetically linked as a cline.

It is interesting to note that similar distributional relations in coiling preference exist in the southern hemisphere, where left-coiling populations occur mostly in antarctic and subantarctic waters. Ninety per cent or more of *G. pachyderma* shells are sinistral south of the Antarctic Polar Front and, in fact, define the area of maximum abundance shown in Fig. 6.5. Left-coiling populations occur mostly between 0° and 9° C, being most abundant below 4° C.

Dextral populations in the southern hemisphere, are likewise found in subantarctic and transitional waters with surface temperatures between 10° and 18° C. They define the area of maximum concentration over the Argentine continental shelf north of the Falkland Islands and the region of occurrence in the Benguela Current. The boundary between left- and right-coiling varieties is well defined north of the Falkland Islands, as evidenced in our plankton collections as well as those examined by Boltovskoy (1966*b*).

G. pachyderma occurs over a range of surface temperatures from 0° to 24° C, but its peak abundance is primarily found between 0° and 9° C.

There is no evidence for a present-day interchange between the northern and southern populations of the Atlantic Ocean, although a connection between them existed during the glacial epochs of the Pleistocene as shown by the abundant occurrence of *G. pachyderma* in equatorial Atlantic cores (Ericson and Wollin 1956).

Globigerina quinqueloba (Natland)

Globigerina quinqueloba occurs predominantly in subarctic and subantarctic waters, but is also frequently found in arctic, antarctic and transitional waters.

This small, epipelagic species may be confused with juvenile *G. pachyderma.* We consider *G. quinqueloba's* possession of spines, at least in early ontogeny, and the absence thereof in *G. pachyderma* as the major distinguishing criterion. However, when the former's spines are broken off during sampling or in preservation our problem is compounded. *G. quinqueloba* does not always have the diagnostic flap-like final chamber that extends over the umbilicus and constricts the aperture and, unfortunately, this prominent feature is also lacking in the earlier stages. *G. quinqueloba* has smaller pores and a higher pore concentration than *G. pachyderma*, but these features are not easily distinguishable.

In the North Atlantic *G. quinqueloba* is very abundant in subarctic waters from south of Spitsbergen to Iceland to the slope waters off the northeastern United States. It is commonly associated with *Globigerina bulliodes. G. quinqueloba* is not so frequently encountered in the Labrador Sea and rapidly disappears from the warmer waters within and south of the Transition Zone.

In the South Atlantic and Indian Oceans, the areas of maximum abundance are much more localized. Although this is partially an artifact of sampling density, the distinction appears to be real. The area of the Falkland Current,

Fig. 6.4. Total and optimum surface temperature ranges of individual species of planktonic foraminifera, based on correlations of surface temperatures and species relative abundance data.

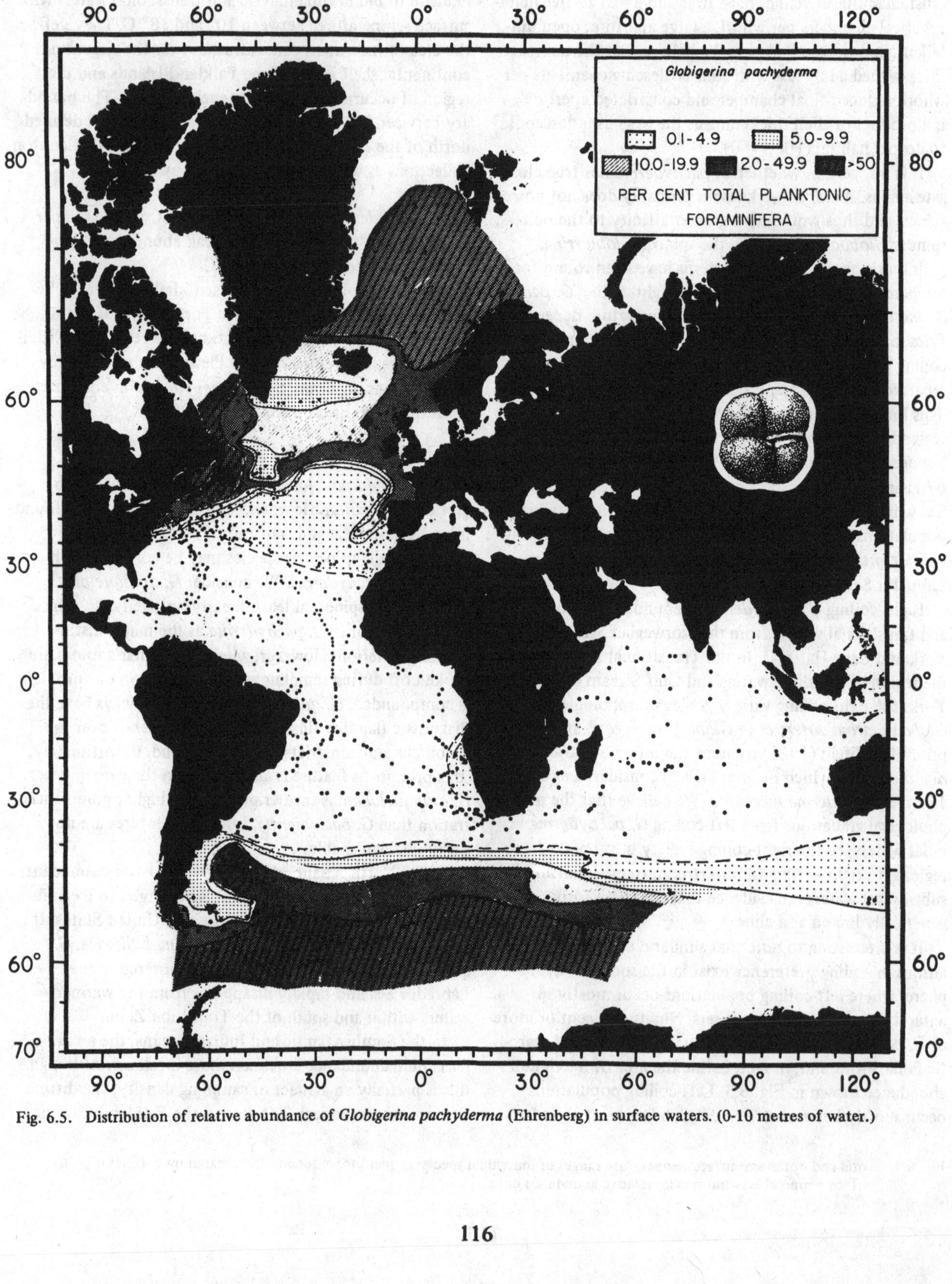
Globigerina pachyderma
0.1-4.9
5.0-9.9
10.0-19.9
20-49.9
>50
PER CENT TOTAL PLANKTONIC
FORAMINIFERA
90°
60°
30°
0°
30°
60°
90°
120°
80°
60°
30°
0°
30°
60°
70°

Fig. 6.5. Distribution of relative abundance of *Globigerina pachyderma* (Ehrenberg) in surface waters. (0-10 metres of water.)

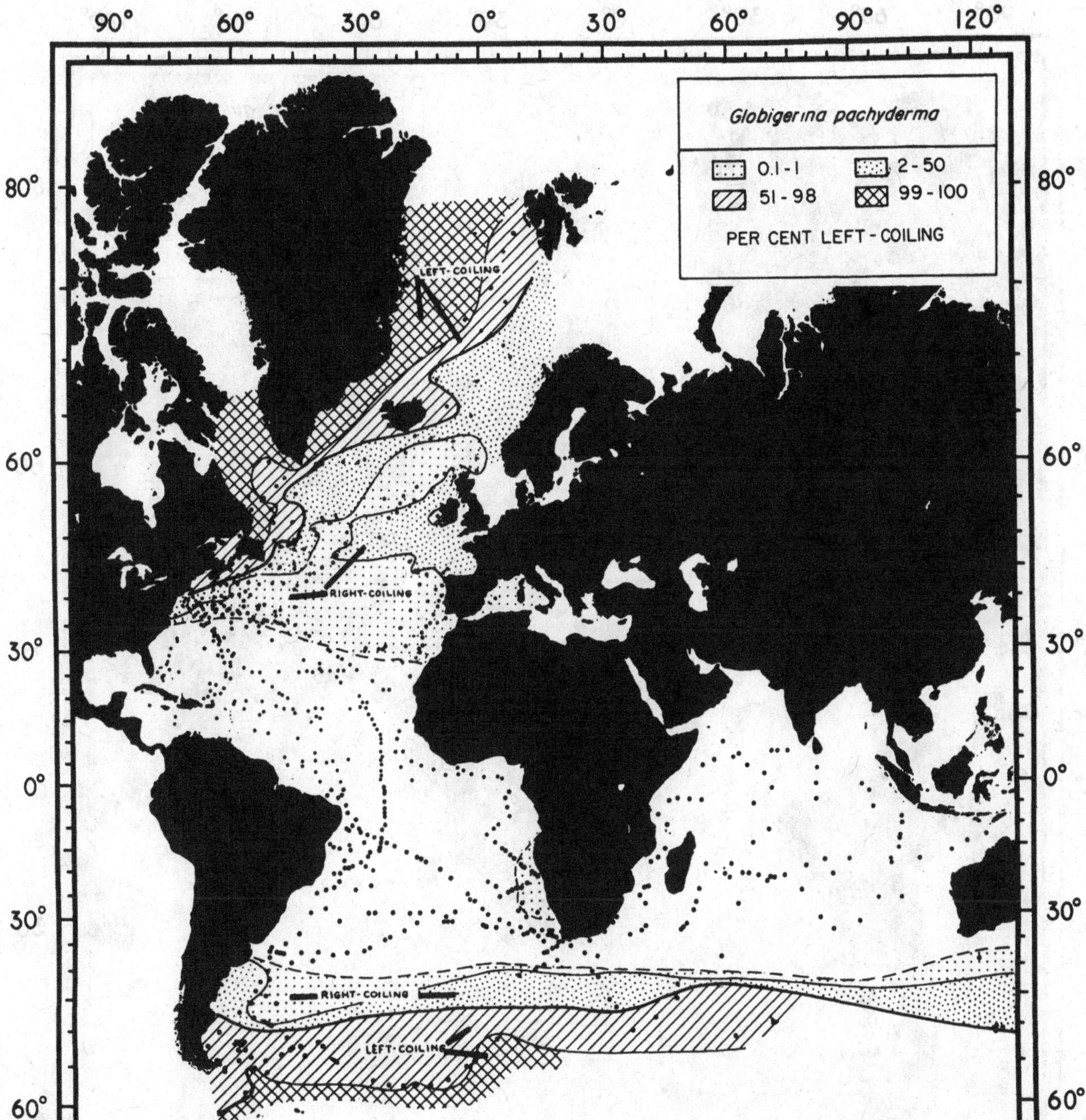
90°
60°
30°
0°
30°
60°
90°
120°
80°
60°
30°
0°
30°
60°
70°
Globigerina pachyderma
0.1-1
2-50
51-98
99-100
PER CENT LEFT-COILING
LEFT-COILING
RIGHT-COILING
RIGHT-COILING
LEFT-COILING

Fig. 6.6. Distribution of percentage ratios of left- versus right-coiling *Globigerina pachyderma* (Ehrenberg) in surface waters. (0-10 metres of water.)

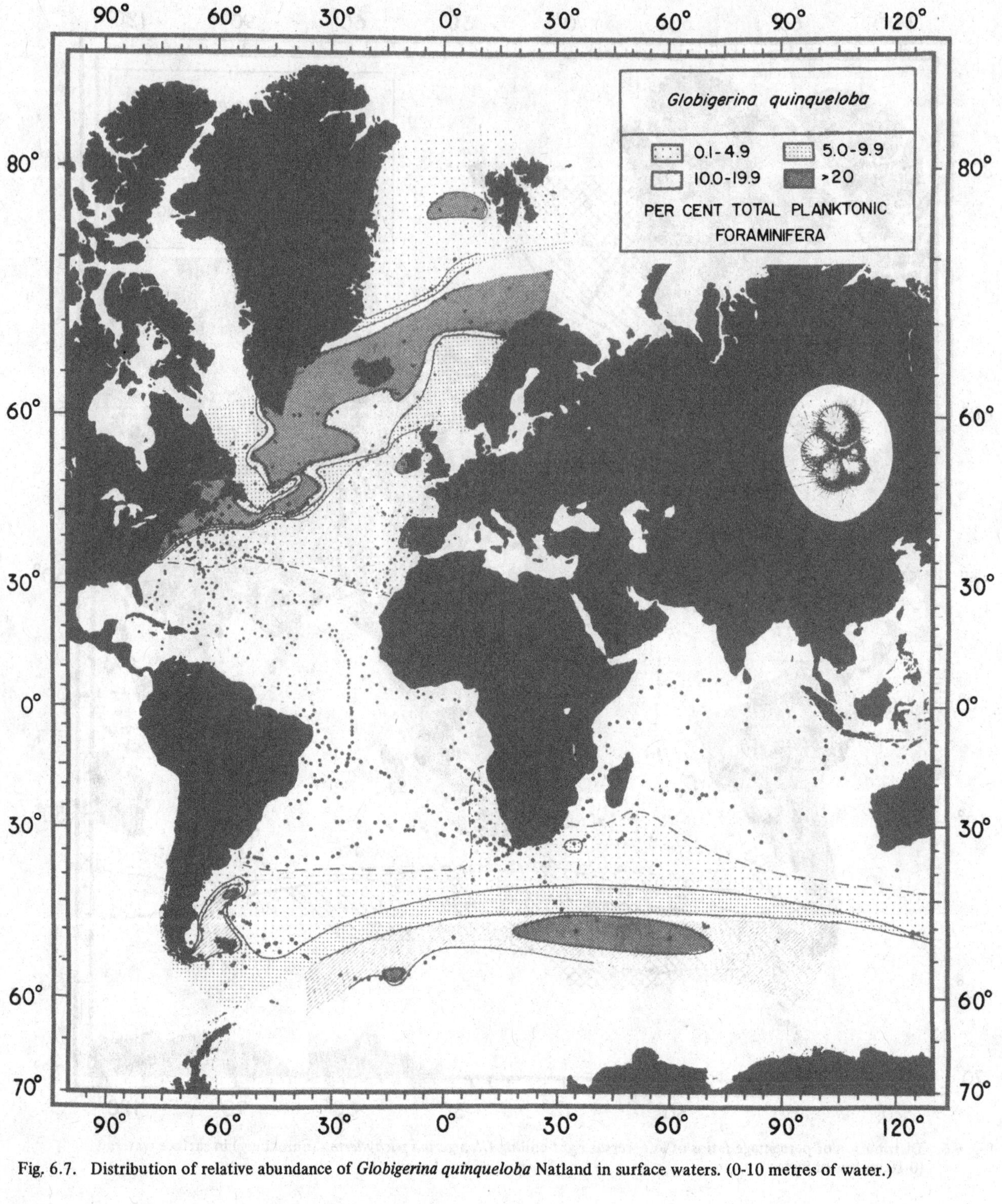
90°
60°
30°
0°
30°
60°
90°
120°
80°
60°
30°
0°
30°
60°
70°
Globigerina quinqueloba
0.1-4.9
5.0-9.9
10.0-19.9
>20
PER CENT TOTAL PLANKTONIC FORAMINIFERA

Fig. 6.7. Distribution of relative abundance of *Globigerina quinqueloba* Natland in surface waters. (0-10 metres of water.)

in particular, corresponds almost exactly to that of maximum abundance in the North Atlantic in terms of season and temperature, and yet only moderate abundance is found. *G. quinqueloba* is at times very common south of the Antarctic Polar Front. Frequencies up to 69 per cent are found south of Africa at approximately 50° S latitude. Its northern distributional limit is the Transition Zone, and it occurs in small percentages in the Benguela and Agulhas Currents.

There does not appear to be any continuous distribution between the populations of the North and South Atlantic.

G. quinqueloba has a surface temperature range of 1° to 21° C, but occurs predominantly in waters colder than 12° C. In the Antarctic region its greatest frequency is encountered between 1° and 5° C.

Globigerina bulloides d'Orbigny

Globigerina bulloides is a dominant species in subarctic and subantarctic waters, where it frequently exceeds 50 per cent of the foraminiferal populations. It has a widespread distribution and appears commonly in transitional and antarctic waters, and, to a lesser extent, in arctic waters between Spitsbergen and Iceland.

Its areal distribution is in part dependent on the taxonomist's concept of the morphological variability of this species. In our opinion, *Globigerina bulloids* (sensu stricto) of subpolar waters grades imperceptibly into *Globigerina falconensis* Blow of subtropical waters. *Globigerina calida* Parker may be an intermediate form between *G. bulloides* and *Globigerinella aequilateralis.* For these reasons, we have lumped *calida*-like forms under the latter species and *falconensis*-like forms under *G. bulloides*, thus restricting those spinose forms with four spherical chambers per whorl and a wide, umbilical aperture to *G. bulloides.*

We believe that there is a distinct possibility that the relatively high concentrations and frequent occurrence of *'G. bulloides'* reported from subtropical-tropical waters of the Indian Ocean (Belyaeva 1964), Pacific Ocean (Bradshaw 1959) and Atlantic Ocean (W. Schott 1935; Cifelli 1967) may be *G. falconensis* or *G. calida* or some other morphological variant of *G. bulloides* (sensu stricto) of high latitudes.

In the North Atlantic, *G. bulloides* is a prolific species in transitional and subarctic waters from New York to Spain and from the Labrador Sea to the Norwegian Sea. It grades southwards into *G. falconensis* in the northern Sargasso Sea.

In the South Atlantic and Indian Ocean, the species occurs abundantly (up to 80-90 per cent) in a wide subantarctic belt from Cape Horn to south of Australia, approximately between 40° S and 55° S latitude. Highest concentrations are generally observed during spring and summer (October to March), when *G. bulloides* extends south of the Antarctic Polar Front (Bé, in press). *G. bulloides* is carried northward along the Benguela Current. Occurrences in isolated areas of upwelling as off the coasts of Somalia and Java are puzzling, and as yet unexplained.

The northern and southern distributional limits of *G. bulloides* could not be determined from our present samples, but their decrease in arctic and antarctic waters suggests that they may not extend much farther north and south of the area under study.

It is not certain whether an interchange exists between the northern and southern populations in the Atlantic Ocean. If there is such a connection, it would most likely be found along the African side.

G. bulloides occurs over a surface temperature range of 0° to 27° C, with its peak abundance between 3° and 19° C.

Globorotalia inflata (d'Orbigny)

Globoratalia inflata is an excellent indicator and the only indigenous species of the Northern and Southern Transition Zones that separate the subpolar and subtropical waters.

In the North Atlantic the highest frequencies (up to 59 per cent) were encountered in the slope waters off the northeastern U.S. and the boundary region between the Labrador and the Gulf Stream-North Atlantic Currents. It is common northeastward to the British Isles and is transported southward by the Canaries Current. W. Schott (1935) reported a frequency higher than 10 per cent off Mauritania, N.W. Africa, but this was not observed by us. *G. inflata* rapidly diminishes south and north of the Transition Zone.

In the southern hemisphere, *G. inflata* occurs in a distinct, linear belt (roughly between 35° S and 45° S latitude) which nicely defines the zone of transitional waters. A prominent tongue extends along the region of the Benguela Current. The samples defining this belt were taken during the austral summer and autumn months. Although the continuity of this belt depends somewhat on our methods of extrapolation in lieu of sampling density, there does appear to be a real and major difference between the northern and southern hemispheres in terms of the areal extent of the zones of maximum abundance. We can find no seasonal or temperature bias to explain this difference, but there may be a difference in depth habitat between the southern and northern populations.

The sparsity of *G. inflata* in the North Pacific is anomalous by comparison with its common occurrence in the North and South Atlantic, and Indian Ocean. Bradshaw (1959) reported it predominantly from the subtropical,

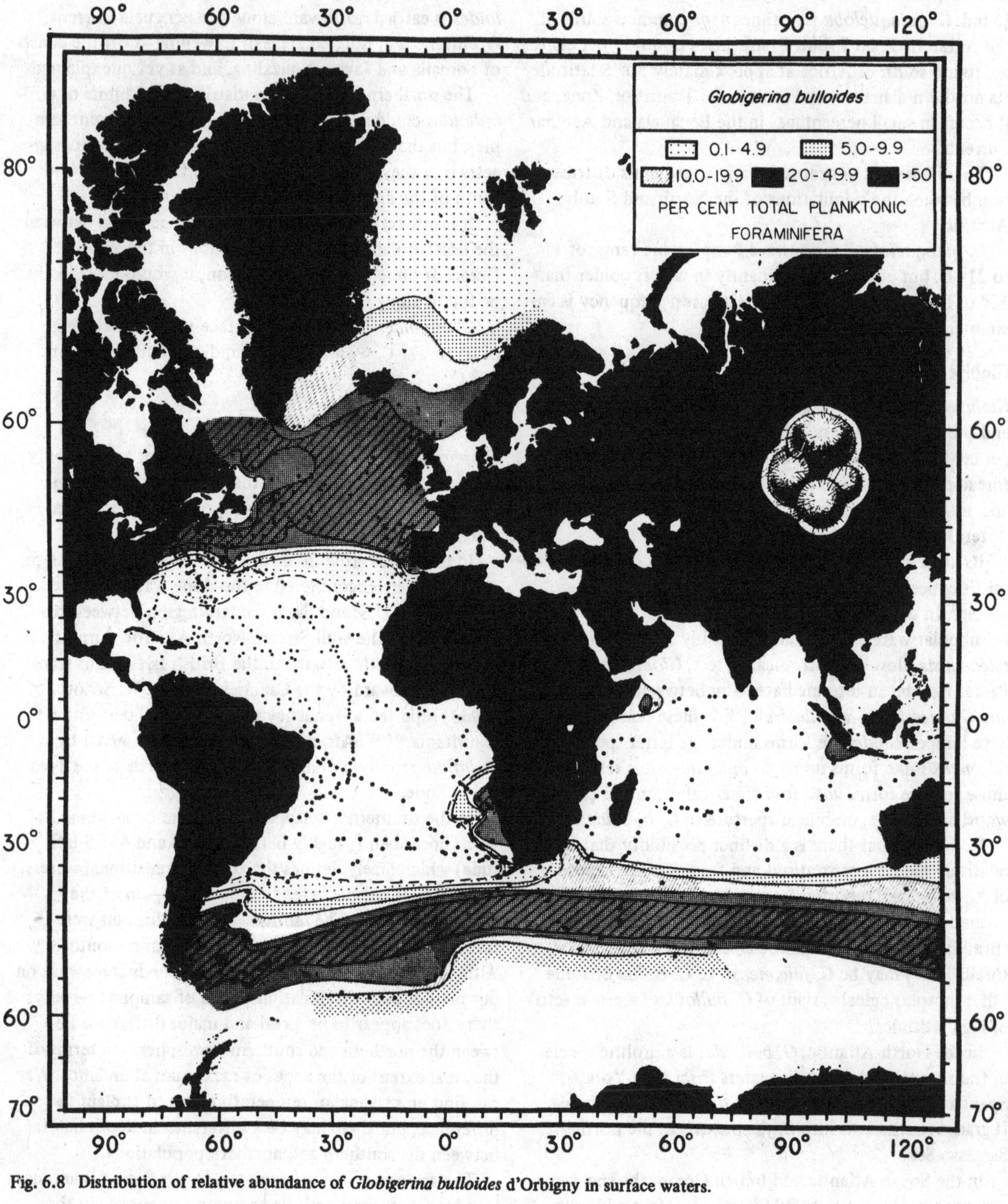
Globigerina bulloides
0.1-4.9
5.0-9.9
10.0-19.9
20-49.9
>50
PER CENT TOTAL PLANKTONIC FORAMINIFERA
90°
60°
30°
0°
30°
60°
90°
120°
80°
60°
30°
0°
30°
60°
70°

Fig. 6.8 Distribution of relative abundance of *Globigerina bulloides* d'Orbigny in surface waters.

central waters of the North Pacific. Its unusually low abundance there may be partly due to the preponderant sampling during the summer season. We conjecture that if winter and spring samples north of 20° N were to be examined, this species might well appear in dense concentrations.

The 10 per cent isopleth in the bottom sediments of the North Atlantic (Schott 1966, Fig. 4) is situated considerably farther south (Cape Hatteras to Dakar) than the equivalent isopleth based on our surface plankton data. The species is apparently much more widely distributed in the sediments than in the overlying waters. We interpret this as due to a northward shift of the population, especially in the eastern Atlantic, caused by a warming trend during the recent past.

It is not certain whether a connection exists between the northern and southern populations in the Atlantic Ocean.

G. inflata occurs over a surface temperature range of 1° to 27° C, with its peak abundance primarily between 13° and 19° C. In the South Pacific (Bé, in press) it proliferates in waters ranging between 2° and 6° C north of the Antarctic Polar Front, but it rapidly diminishes in waters below 2° C.

Globigerinita glutinata (Egger)

Globigerinita glutinata is one of the most ubiquitous species, whose distribution ranges from antarctic through tropical to subarctic waters. Although seldom a dominant species, it is usually present in low frequencies (less than 5 per cent) throughout the area under investigation. Its greatest abundance is found in subtropical waters.

In the North Atlantic, the highest relative abundance (up to 27 per cent) was encountered in the southwestern Sargasso Sea. Other rich but localized concentrations were found in slope and subarctic waters, Antilles Current and equatorial Atlantic.

In the South Atlantic and Indian Oceans, *G. glutinata* occurs with relatively high frequencies (5-10 per cent) in a continuous belt from Brazil to Cape of Good Hope to West Australia. This region includes central and transitional waters of both oceans. It is also abundant between Somalia and the Maldive Islands.

G. glutinata is present in low frequencies in subantarctic waters, but it is known to cross the Antarctic Polar Front in reduced numbers (Bé, in press). Its cosmopolitan distribution from the Antarctic to Subarctic Zones makes *G. glutinata* an enigmatic species. Judging from its areas of highest abundance and most common occurrence, one might consider it a 'subtropical' species with a cosmopolitan distribution.

G. glutinata occurs over a surface temperature range of 3° to 30° C, with its peak abundance between 24° and 27° C.

Globorotalia truncatulinoides (d'Orbigny)

Globorotalia truncatulinoides is predominantly a subtropical species inhabiting the central water-masses of the oceans. It has a wide distributional range, from the Northern Transition Zone through tropical waters to the Antarctic Polar Front.

In the central North Atlantic, this species inhabits the epipelagic zone, especially in sub-surface levels, between November and May. It reaches maximum abundance during the winter (January and February) in the northwestern Sargasso Sea (Bé and Ericson 1963). For the remaining months of the year it is very rare in the upper 300 metres of water, and we believe that the surviving individuals mature at bathypelagic depths (Bé and Lott 1964). Thus the seasonal factor and sub-surface occurrence are responsible for a much more limited distribution of *G. truncatulinoides* than would appear if our study were based on deeper hauls obtained during the winter season.

In the South Atlantic this species lives in transitional waters, whereas it occurs mainly in the subtropical region of the central Indian Ocean. The highest frequencies in these areas are observed during late autumn and early winter (May-June).

G. truncatulinoides is common in subantarctic waters of the South Pacific sector (Boltovskoy 1966*a*; Bé, in press), and therefore it seems to be more cold-tolerant here than in the North Atlantic and North Pacific. High frequencies of *G. truncatulinoides* occur between 4° and 18° C in the subantarctic Pacific waters, but in the North Atlantic the maximum concentrations are encountered between 19° and 22° C. Bradshaw's (1959) study also indicates its affinity to the subtropical waters of the North Pacific and corroborates our view that the northern hemisphere populations are on the average living in warmer waters than their South Pacific counterparts.

Kennett (1967) observed that the forms near the Antarctic Polar Front possess lower spires than those in the warmer subtropical waters, and Boltovskoy (1959, plate 3) also illustrated similar subantarctic forms in the South Atlantic. We agree with their observations.

The coiling ratios of *G. truncatulinoides* (Fig. 6.12) show interesting distributional patterns. Ericson, Wollin and Wollin (1954) reported a province of dominantly sinistral tests in surface sediments of the central North Atlantic from New England and Nova Scotia to northwest Africa, bordered by a region with mostly dextral tests in the northeastern Atlantic, and another dextral province in the equatorial Atlantic and Caribbean area. Our study of living populations agrees in general with Ericson *et al.*'s results. Living *G. truncatulinoides* in the tropical Atlantic are rare but predominantly right-coiling.

In the South Atlantic the sinistral variety is dominant in transitional and subantarctic waters, whereas in the Indian Ocean it occurs in the subtropical region as well. The apparent temperature dependence of coiling direction is much more evident in the southern hemisphere than in the North Atlantic. It should be noted that the transition in the South Atlantic and Indian Ocean from right- to left-coiling *G. truncatulinoides* takes place more abruptly than that shown by *G. pachyderma.* The majority of samples in these regions have either exclusively right- or left-coiling shells, whereas *G. pachyderma* exhibits a more gradational mixture between sinistral and dextral populations.

Schott (1966) observed high percentages (over 10 per cent) of empty *G. truncatulinoides* tests in surface sediments in the eastern and central North Atlantic, and low frequencies (*ca.* 1 per cent) in the equatorial Atlantic, Caribbean Sea, Gulf Stream and North Atlantic Currents. But living *G. truncatulinoides* are most frequent in the western North Atlantic.

Belyaeva (1964) found this species abundant (100-500 specimens per gramme) in a NW-SE zone in the central Indian Ocean from 20° S to 50° S lat. The belt swings south of Australia and this agrees very well with our observations of the living populations.

G. truncatulinoides occurs over a surface temperature range of 4° to 27° C, with its peak occurrence between 17° and 22° C.

Globoquadrina dutertrei (d'Orbigny)

Globoquadrina dutertrei is a subtropical-tropical species with its distributional limits located within the Transition Zones. It is especially abundant in major current systems near continental margins, and is rare in the central waters of the Atlantic and Indian Oceans. It is not a dominant species, and seldom exceeds frequencies higher than 20 per cent. The species grades into *G. pachyderma*, as discussed in more detail under the latter species.

In the North Atlantic, *G. dutertrei* is found most commonly (over 5 per cent) in the Gulf Stream and slope waters off the maritime provinces of Canada and the northeastern U.S.A., and in the North Equatorial and Antilles Currents. W. Schott (1935) noted a tongue of high frequency (over 10 per cent) extending from northwest Africa and the Cape Verde Islands westward along the North Equatorial Current, but we have not observed such density off Africa.

Jones (1967) found it most abundant (18 per cent) between 50 and 100 metres within the Equatorial Atlantic Undercurrent. *G. dutertrei* was fourth in order of abundance of the six species studied by him. Cifelli (1967) reported low percentages, and it occurred in only four out of forty-one North Atlantic stations.

In the South Atlantic, *G. dutertrei's* highest relative abundance is in the waters off Angola and off the northeast coast of Brazil.

In the Indian Ocean, the highest frequencies are again encountered near the margins of land masses, such as the region off eastern Africa from Somalia to Madagascar to Cape of Good Hope, as well as the waters south of Java. In these areas *G. dutertrei* does occasionally constitute greater than 20 per cent of the foraminiferal population. Elsewhere in the Indian Ocean, it is ubiquitous but sparse (below 5 per cent).

In the North Pacific, Bradshaw (1959) observed highest frequencies in transitional waters, the California Current, and along the boundary of the Kuroshio-Oyashio Currents. His observations are in agreement with our findings that this species thrives near continental margins and in strong currents.

G. dutertrei occurs over a surface temperature range of 9° to 30° C, with its peak abundance between 16° and 24° C.

Globorotalia crassaformis (Galloway and Wissler)

Globorotalia crassaformis has a marked preference for water depths greater than 100 metres and is considered a mesopelagic species in subtropical-tropical waters. We have seldom observed it in surface tows, but it is more frequent in 0-300 metres or deeper tows.

Jones (1967) found the highest abundance of *G. crassaformis* below the Equatorial Undercurrent in the oxygen-minimum waters, while low percentages were encountered in the upper 100 metres. Cifelli (1967) observed its greatest frequency (up to 13 per cent) in the South Equatorial Current. In the Indian Ocean, near Madagascar, an intergrade between *G. crassaformis* and *G. hirsuta* has been encountered.

G. crassaformis occurs over a surface temperature range of 16° to 27° C.

Orbulina universa d'Orbigny

O. universa is an ubiquitous species in subtropical, tropical and transitional waters, whose highest frequencies occur in the surface layers of strong current systems and upwelling regions near continental margins. Peak abundances are found on the eastern sides of the oceans, as for example in the Portugal Current, North Equatorial Current, Equatorial Counter Current, Benguela Current and eastern half of the Indian Central Water. Bradshaw (1959) also reported very high percentages (over 50 per cent) in the eastern North Pacific in the California Current and neighbouring region.

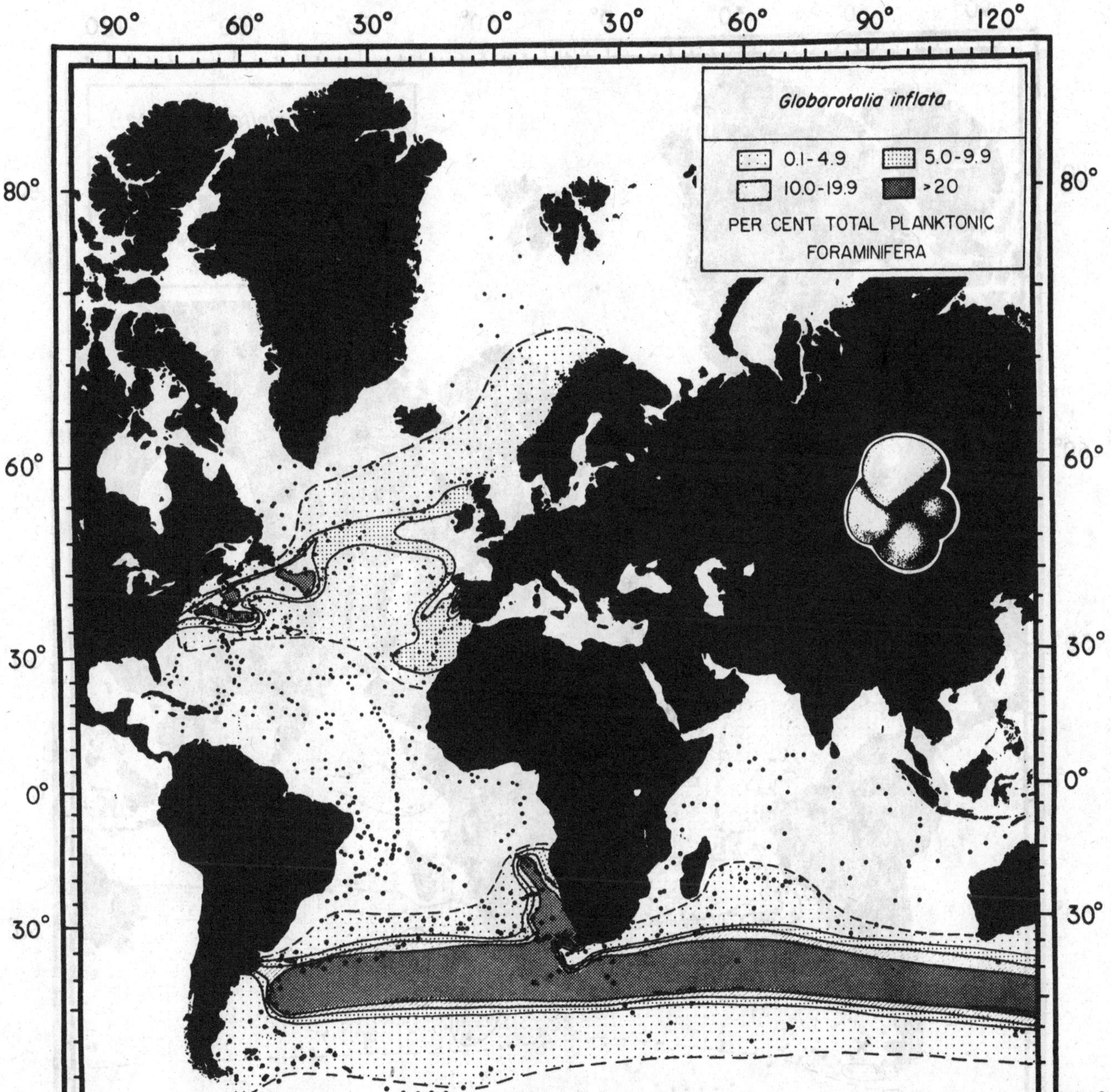
Globorotalia inflata
0.1-4.9
5.0-9.9
10.0-19.9
>20
PER CENT TOTAL PLANKTONIC FORAMINIFERA
90°
60°
30°
0°
30°
60°
90°
120°
80°
60°
30°
0°
30°
60°
70°

Fig. 6.9. Distribution of relative abundance of *Globorotalia inflata* (d'Orbigny) in surface waters. (0-10 metres of water.)

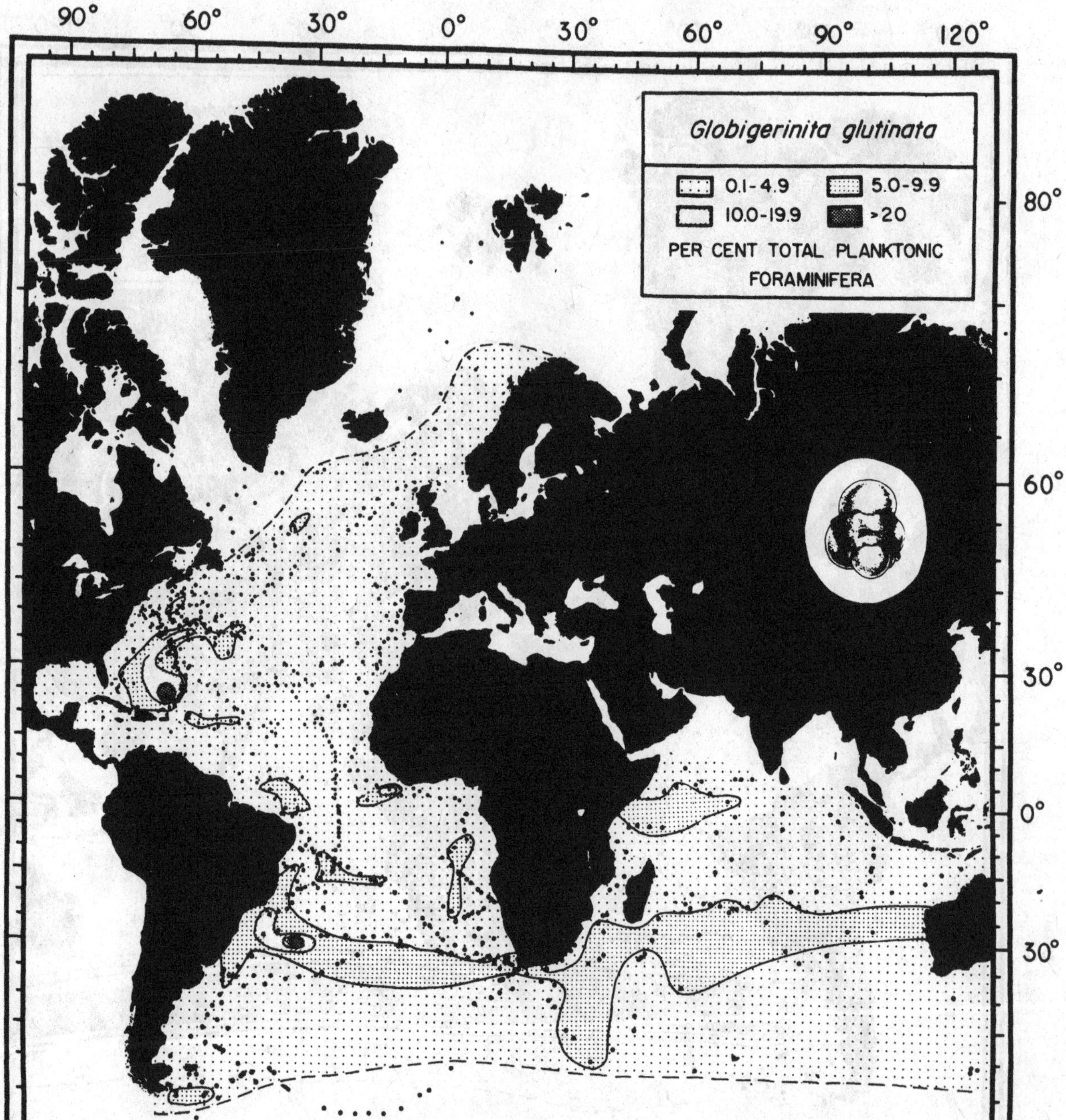
Globigerinita glutinata
0.1-4.9
5.0-9.9
10.0-19.9
>20
PER CENT TOTAL PLANKTONIC
FORAMINIFERA
90°
60°
30°
0°
30°
60°
90°
120°
80°
60°
30°
0°
30°
60°
70°

Fig. 6.10. Distribution of relative abundance of *Globigerinita glutinata* (Egger) in surface waters. (0-10 metres of water.)

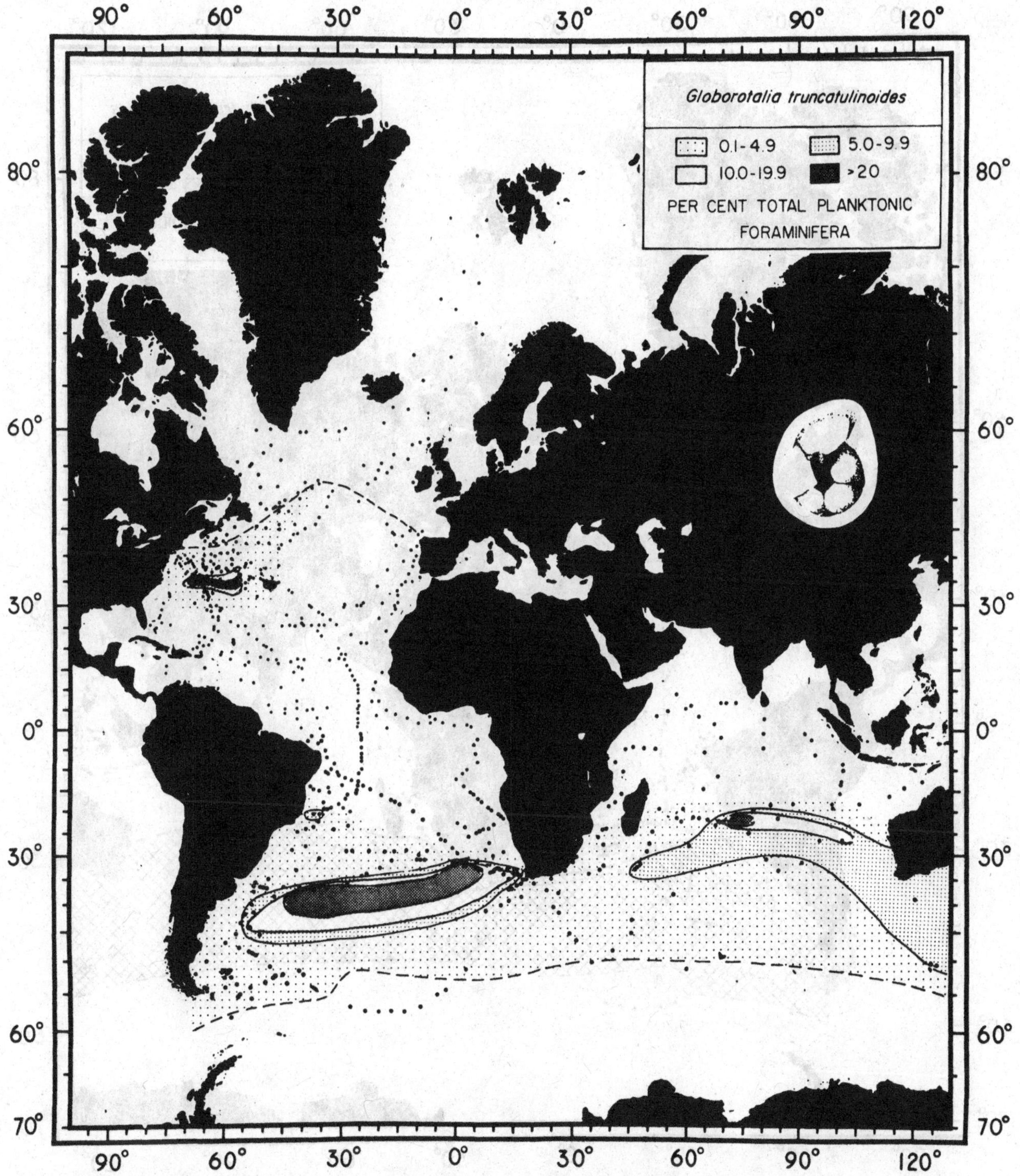
90°
60°
30°
0°
30°
60°
90°
120°
80°
60°
30°
0°
30°
60°
70°
Globorotalia truncatulinoides
0.1-4.9
5.0-9.9
10.0-19.9
>20
PER CENT TOTAL PLANKTONIC FORAMINIFERA

Fig. 6.11. Distribution of relative abundance of *Globorotalia truncatulinoides* (d'Orbigny) in surface waters. (0-10 metres of water.)

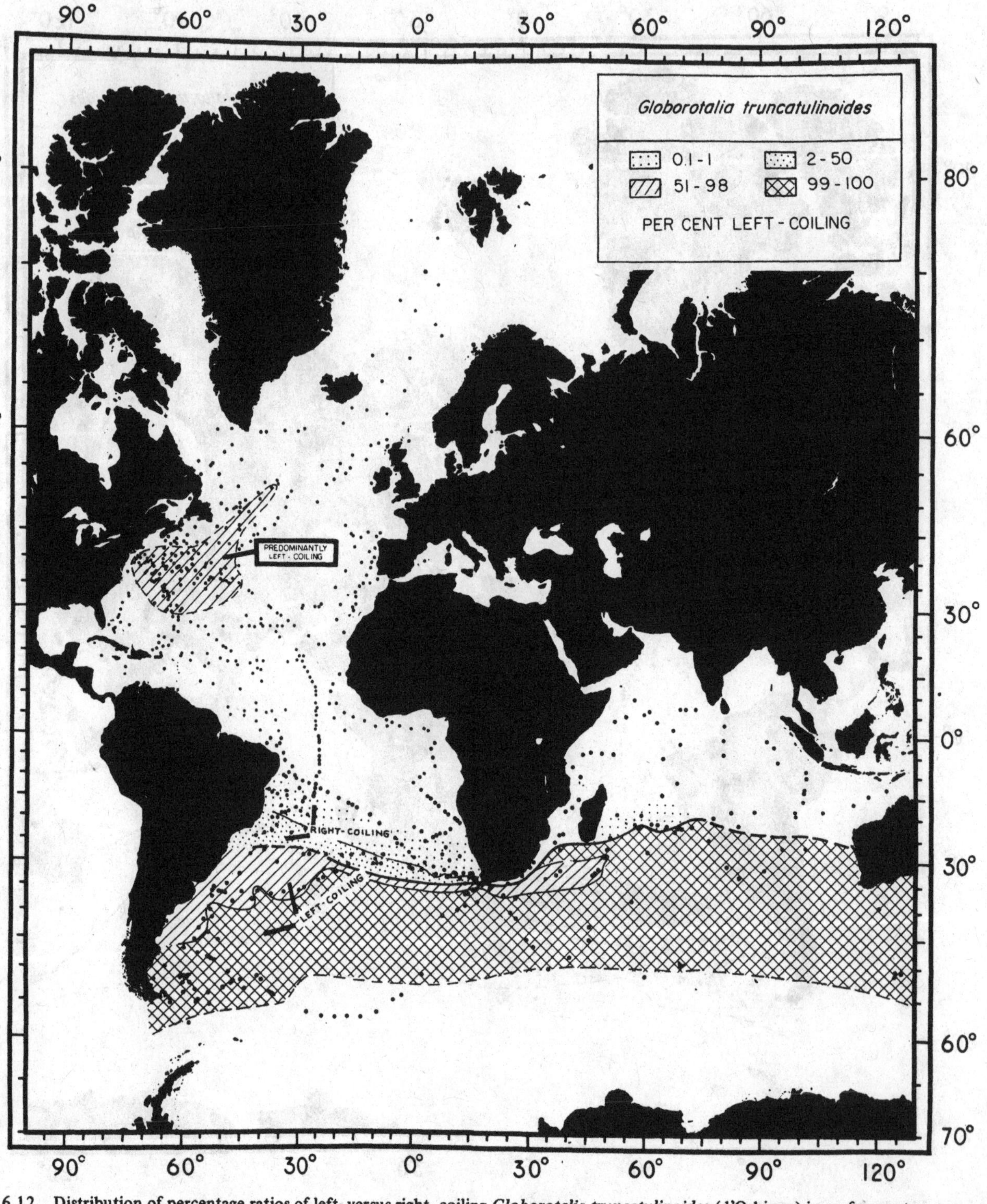

90° 60° 30° 0° 30° 60° 90° 120°
Globorotalia truncatulinoides
0.1-1
2-50
51-98
99-100
PER CENT LEFT-COILING
80°
60°
30°
0°
30°
60°
70°
PREDOMINANTLY LEFT-COILING
RIGHT-COILING
LEFT-COILING

Fig. 6.12. Distribution of percentage ratios of left- versus right- coiling *Globorotalia truncatulinoides* (d'Orbigny) in surface waters. (0-10 metres of water.)

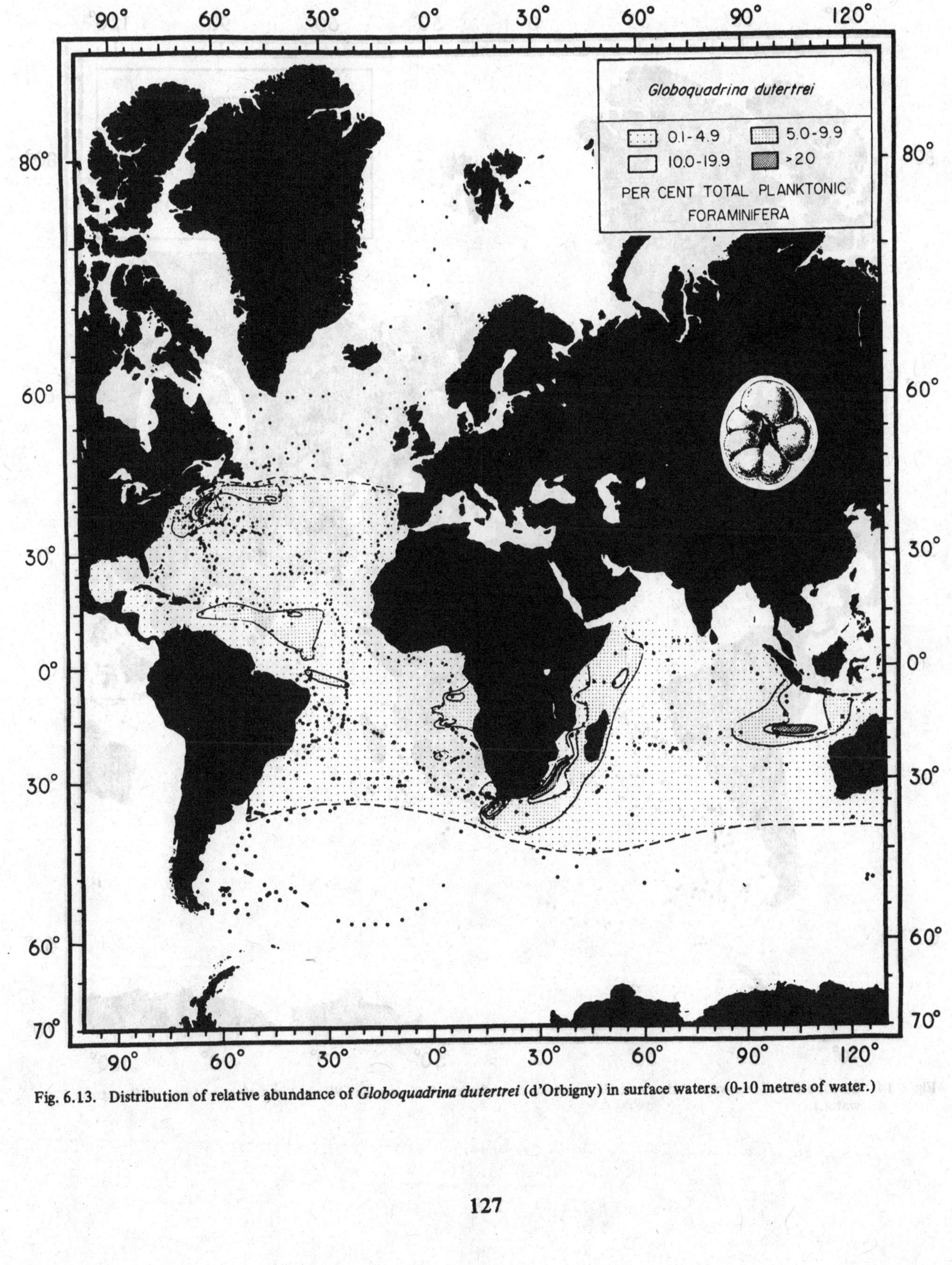
Globoquadrina dutertrei
0.1-4.9
5.0-9.9
10.0-19.9
>20
PER CENT TOTAL PLANKTONIC FORAMINIFERA
90°
60°
30°
0°
30°
60°
90°
120°
80°
60°
30°
0°
30°
60°
70°

Fig. 6.13. Distribution of relative abundance of *Globoquadrina dutertrei* (d'Orbigny) in surface waters. (0-10 metres of water.)

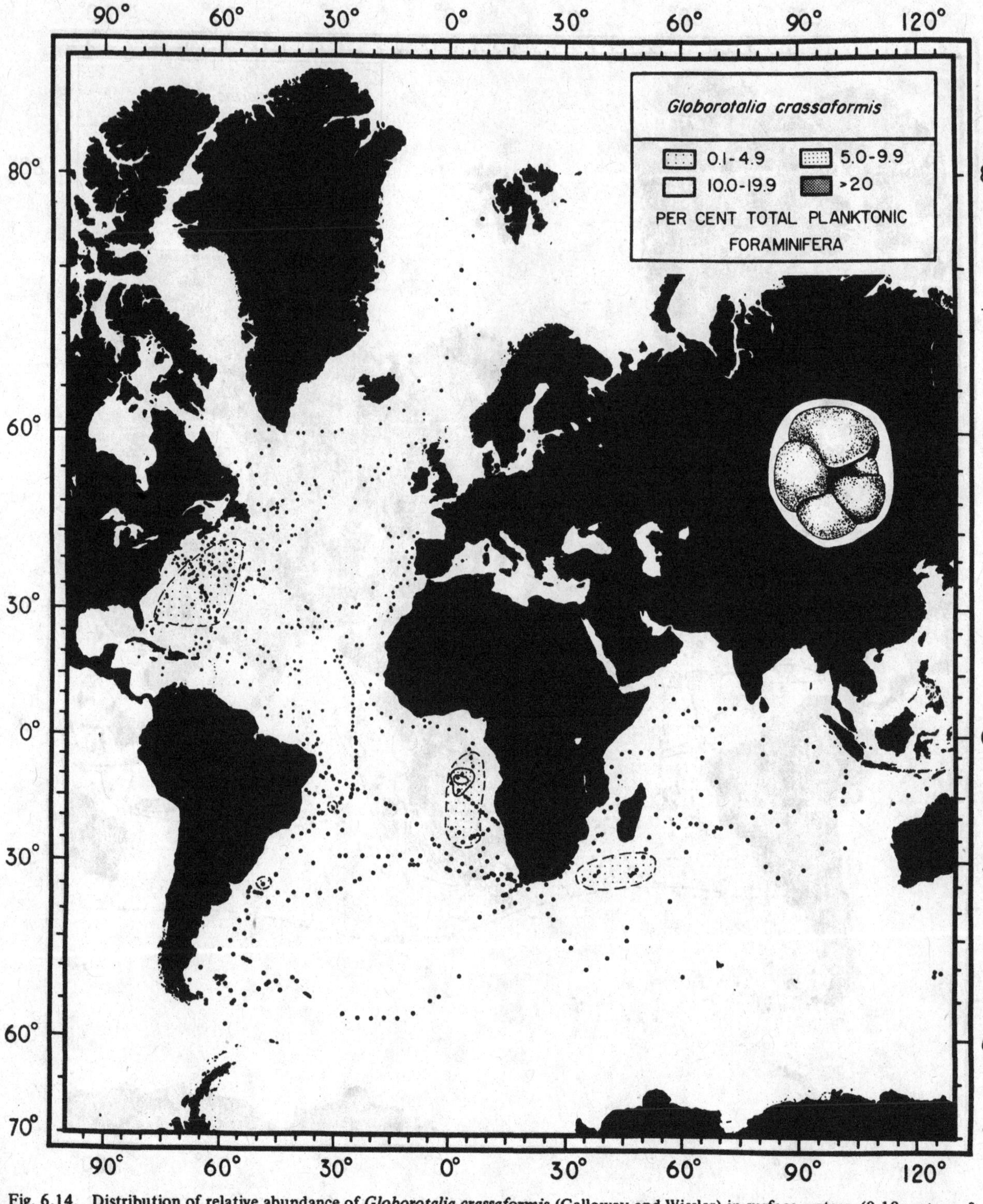
Globorotalia crassaformis
0.1-4.9
5.0-9.9
10.0-19.9
>20
PER CENT TOTAL PLANKTONIC FORAMINIFERA
90°
60°
30°
0°
30°
60°
90°
120°
80°
60°
30°
0°
30°
60°
70°

Fig. 6.14. Distribution of relative abundance of *Globorotalia crassaformis* (Galloway and Wissler) in surface waters. (0-10 metres of water.)

We noted high concentrations in the Gulf Stream and western Sargasso Sea. Its presence in decreasing frequencies along stations in the North Atlantic Current indicates its diminution during northeastward transport from its subtropical origin.

In the South Atlantic, exceptionally high frequencies (up to 41 per cent) are observed in the Benguela Current region and the eastern half of the South Atlantic central waters.

In the Indian Ocean, *O. universa* is abundant and widespread over a large region of subtropical waters between Mauritius and Australia. Our observations agree with those of Belyaeva (1964), who also found *O. universa* to be most abundant in central waters. Frequencies of 5 per cent or higher are encountered in the Mozambique-Agulhas Current.

The distributional limits of *O. universa* are probably located on the poleward boundaries of the Northern and Southern Transition Zones. The specimens in transitional waters appear to be smaller in size than those in lower latitudes, but more detailed study is necessary to document this.

O. universa occurs over a surface temperature range of 10° to 30° C, with peak abundance between 17° and 23° C.

Globigerinella aequilateralis (Brady)

Globigerinella aequilateralis is a common subtropical-tropical species, which occurs abundantly in strong current systems, transitional waters and areas of upwelling. It is, however, rare in the central waters of the North and South Atlantic, but is somewhat more common, in the central Indian Ocean. The species is equally plentiful in surface and 0-300 metre tows. We have included under *G. aequilateralis* forms that resemble *Globigerina calida* Parker, which appear occasionally in our samples.

In the North Atlantic, *G. aequilateralis* is most prolific in the Gulf Stream and western Sargasso Sea, especially from November to May, and it is transported northeastward as far as the Norwegian Sea. Whereas most subtropical species cannot survive in the cooler transitional waters, *G. aequilateralis* and *O. universa* are exceptional in their tolerance to a wide temperature range. Rather high frequencies (5-20 per cent) of *G. aequilateralis* occur in the North Equatorial Current, Antilles Current, Equatorial Counter Current and Guinea Current.

In the South Atlantic, unusually dense concentrations are encountered in the Benguela Current and the neighbouring waters to the north.

In the Indian Ocean, *G. aequilateralis* is most plentiful in the Agulhas-Somali Current and the region south and southeast of Madagascar, and is also common throughout the central waters and the equatorial region. Belyaeva (1964) found highest concentrations off southwestern India and in the central Indian Ocean.

Bradshaw's (1959) Pacific data support our belief that *G. aequilateralis* preferred habitat is in current systems or along continental margins. He found it to be widespread in equatorial and subtropical waters (1-20 per cent), but the highest frequencies (over 20 per cent) occur in the western equatorial area and the southern California Current.

The distributional limits of *G. aequilateralis* are quite similar to those of *O. universa* and fall generally on the poleward boundaries of the Northern and Southern Transition Zones.

G. aequilateralis occurs over a surface temperature range of 12° to 30° C, with peak abundance primarily between 19° and 28° C.

Hastigerina pelagica (d'Orbigny)

Hastigerina pelagica is a subtropical species whose areas of maximum concentration are located close to the margins of its distributional boundaries. In the North Atlantic it is most prolific from April to August in the northwestern Sargasso Sea. Its highest abundances in central waters of the South Atlantic and the Agulhas Current in the Indian Ocean occur near its southern limit of distribution.

In the Pacific Bradshaw (1959) likewise observed that the areas of greatest concentrations in the east-central and western subtropical regions are close to the northern distributional limit of this species.

Besides the aforementioned regions, *H. pelagica* is a rare (less than 5 per cent) but ubiquitous species in tropical and subtropical waters. Our observations agree with those of Jones (1967), who regards *H pelagica*, as well as *Globorotalia crassaformis* and *G. truncatulinoides*, as having marked preferences for water depths greater than 100 metres in the equatorial Atlantic. We consider *Hastigerinella digitata* (Rhumbler) a bathypelagic isomorph of *H. pelagica* that lives below 1,000 metres. We have received many specimens of *H. digitata* of extraordinarily large size (over 3 millimetres long) that were collected by the Dana-II Expedition in open-net plankton tows from 0-3,000 metres depth range off the coast of Senegal (13° 31′ N, 18° 03′ W).

H. pelagica occurs over a surface temperature range of 16° to 29° C, with peak abundance primarily between 20° and 26° C.

Globorotalia hirsuta (d'Orbigny)

Globorotalia hirsuta is a subtropical species which in many respects has both distribution and ecological requirements similar to *G. truncatulinoides.* Both occur most abundantly in sub-surface depths of the central water-masses during

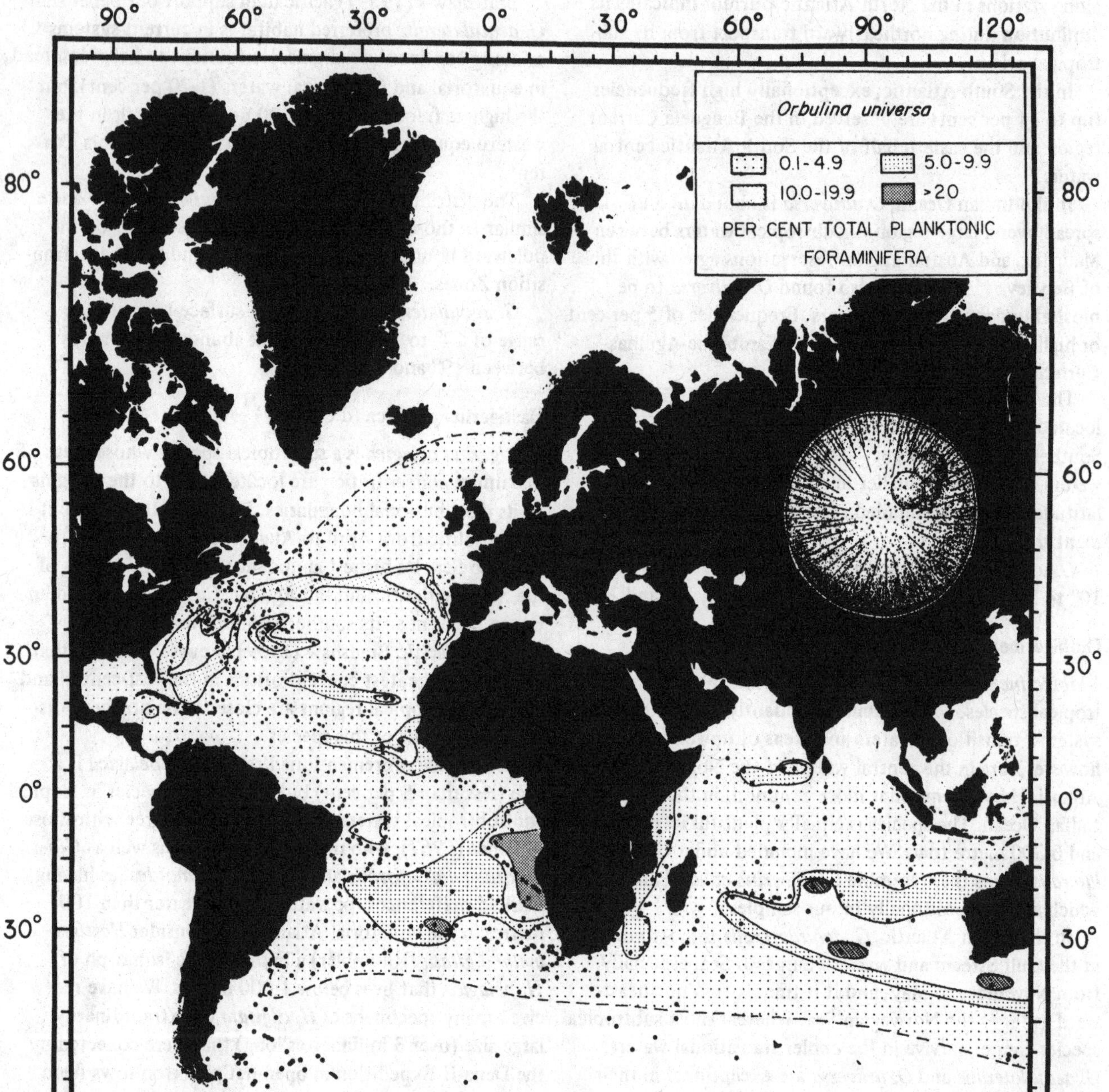
Orbulina universa
0.1-4.9
5.0-9.9
10.0-19.9
>20
PER CENT TOTAL PLANKTONIC FORAMINIFERA
90°
60°
30°
0°
30°
60°
90°
120°
80°
60°
30°
0°
30°
60°
70°

Fig. 6.15. Distribution of relative abundance of *Orbulina universa* d'Orbigny in surface waters. (0-10 metres of water.)

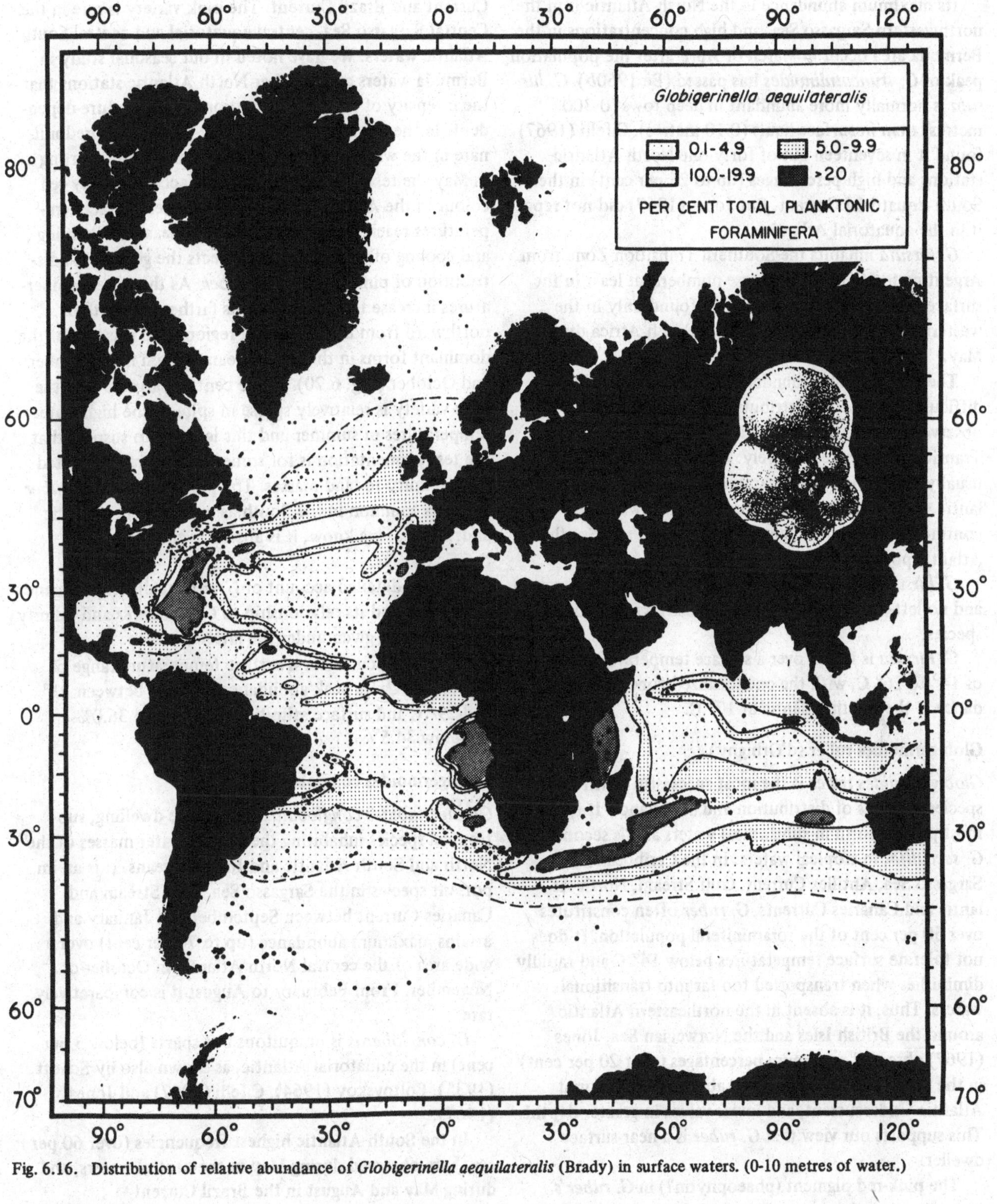
90°
60°
30°
0°
30°
60°
90°
120°
80°
60°
30°
0°
30°
60°
70°
Globigerinella aequilateralis
0.1-4.9
5.0-9.9
10.0-19.9
>20
PER CENT TOTAL PLANKTONIC FORAMINIFERA

Fig. 6.16. Distribution of relative abundance of *Globigerinella aequilateralis* (Brady) in surface waters. (0-10 metres of water.)

the winter season, so that our maps are to some extent artifacts of sampling depth and season.

Its maximum abundance in the North Atlantic is in the northwestern Sargasso Sea, and high concentrations in the Bermuda area occur in March or April after the population peak of *G. truncatulinoides* has passed (Bé 1960*b*). *G. hirsuta* is normally more abundant in deep tows (0-300 metres) than in surface hauls (0-10 metres). Cifelli (1967) found it in seventeen out of forty-one North Atlantic stations and high percentages (up to 25 per cent) in the South Equatorial Current. But Jones (1967) did not report it in the equatorial Atlantic.

G. hirsuta inhabits the Southern Transition Zone from Argentina to Mauritius in sparse numbers, at least in the surface layers. High frequencies are found only in the western South Atlantic and west of South Africa during May.

The generally low abundance of *G. hirsuta* makes it difficult to define its distributional limits. We have not observed it north or south of the Northern and Southern Transition Zones, respectively. And, although it is not usually present in surface samples in the equatorial Atlantic, its occasional appearance in deep tows points to a continuous distribution between the North and South Atlantic populations.

G. hirsuta has predominantly a dextral coiling direction and no left- or right-coiling provinces are known for this species.

G. hirsuta is found over a surface temperature range of 14° to 26° C, with the one localized zone of peak abundance in the South Atlantic at 17° C.

Globigerinoides ruber (d'Orbigny)

Globigerinoides ruber is the most successful warm-water species in terms of distribution and abundance. It is the most prolific form in subtropical waters and is second to *G. sacculifer* in tropical waters. In the Caribbean Sea, Sargasso Sea, Antilles Current, Gulf Stream, North Atlantic and Canaries Currents, *G. ruber* often constitutes over 50 per cent of the foraminiferal population. It does not tolerate surface temperatures below 14° C and rapidly diminishes when transported too far into transitional waters. Thus, it is absent in the northeastern Atlantic around the British Isles and the Norwegian Sea. Jones (1967) observed maximum percentages (over 20 per cent) in the upper 50 metres of water above the Equatorial Atlantic Undercurrent, and lower values in greater depths. This supports our view that *G. ruber* is a near-surface dweller.

The pink-red pigment (phaeophytin?) in *G. ruber's* tests appears to be restricted to populations in the North and South Atlantic and is associated predominantly with populations near land masses, such as the ones in the Caribbean Sea, Antilles Current, Gulf Stream, Canaries Current and Brazil Current. The pink variety is rare in the Central Sargasso Sea, central equatorial and central South Atlantic waters. We have noted in our seasonal study in Bermuda waters and at other North Atlantic stations that the intensity of pigment coloration is temperature-dependent. In the western North Atlantic white tests predominate in the winter and early spring months, but starting in May the tests become orange, and acquire a ruby-red colour in the August-October period when surface temperatures reach maximum values. The seasonal warming and cooling of the waters also affects the geographic distribution of pink-coloured *G. ruber.* As the water temperatures increase this form extends farther and farther northward from the Caribbean region until it is one of the dominant forms in the Gulf Stream system in September and October (Fig. 6.20). In the central Sargasso Sea, the pink variety is relatively sparse in spite of the high water temperatures in summer and this leads us to suspect that the test pigmentation is for some reason directly related to proximity to land masses. The pink variety of *G. ruber* is not as common in the South as in the North Atlantic and, as far as we know, it is absent in the Indian and Pacific Oceans.

A more detailed account of *G. ruber's* distribution and abundance and its relationship to *G. sacculifer* and salinity distribution is given under the latter species.

G. ruber occurs over a surface temperature range of 14° to 30° C. Its peak abundance is found between 21° and 29° C and surface salinities either above 36.0‰ or below 34.5‰.

Globigerinoides conglobatus (Brady)

Globigerinoides conglobatus is a surface-dwelling, subtropical species inhabiting the central water masses of the North and South Atlantic and Indian Oceans. It is an important species in the Sargasso Sea, Gulf Stream and Canaries Current between September and January and attains maximum abundance (up to 74 per cent) over a wide area of the central North Atlantic in October or November. From February to August it is comparatively rare.

G. conglobatus is ubiquitous but sparse (below 5 per cent) in the equatorial Atlantic, as shown also by Schott (1935), Boltovskoy (1964), Cifelli (1967) and Jones (1967).

In the South Atlantic highest frequencies (over 60 per cent) are observed during May in the central waters, and during May and August in the Brazil Current.

In the Indian Ocean rich concentrations occur in February, May and September in the central waters and South

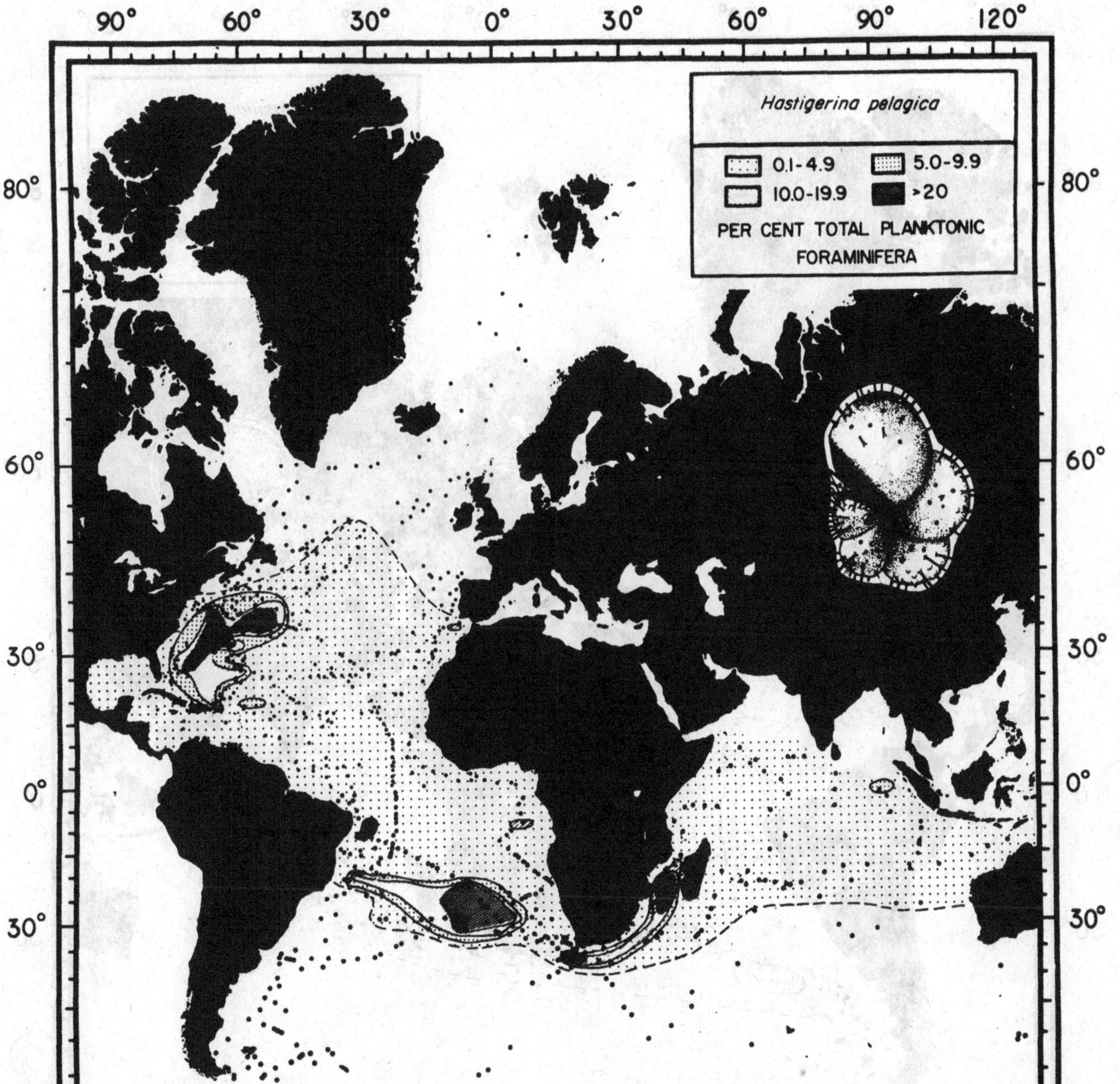
Hastigerina pelagica
0.1-4.9
5.0-9.9
10.0-19.9
>20
PER CENT TOTAL PLANKTONIC FORAMINIFERA
90°
60°
30°
0°
30°
60°
90°
120°
80°
60°
30°
0°
30°
60°
70°

Fig. 6.17. Distribution of relative abundance of *Hastigerina pelagica* (d'Orbigny) in surface waters. (0-10 metres of water.)

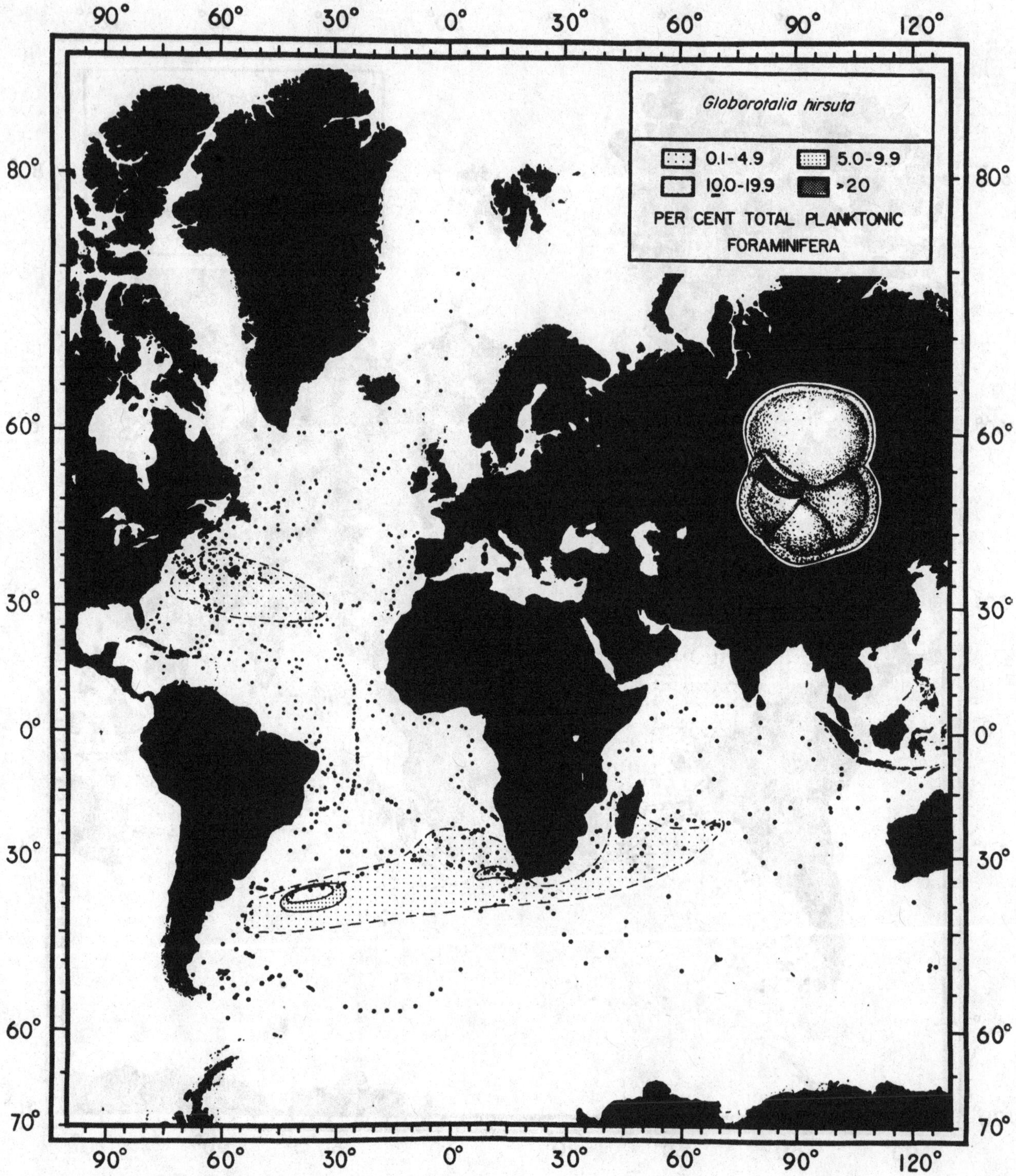
90°
60°
30°
0°
30°
60°
90°
120°
Globorotalia hirsuta
0.1-4.9
5.0-9.9
10.0-19.9
>20
PER CENT TOTAL PLANKTONIC
FORAMINIFERA
80°
60°
30°
0°
30°
60°
70°

Fig. 6.18. Distribution of relative abundance of *Globorotalia hirsuta* (d'Orbigny) in surface waters. (0-10 metres of water.)

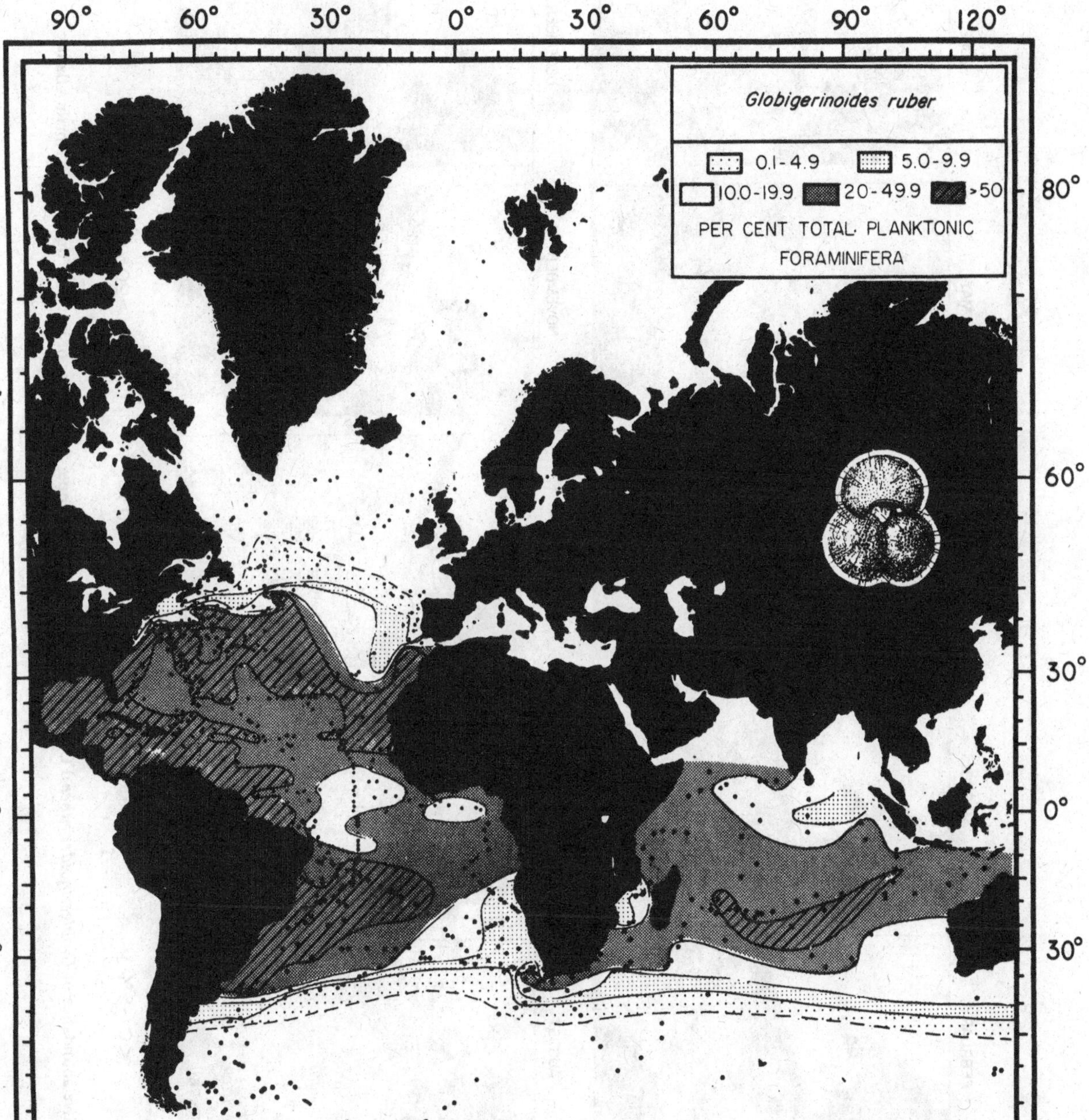
Globigerinoides ruber
0.1-4.9
5.0-9.9
10.0-19.9
20-49.9
>50
PER CENT TOTAL PLANKTONIC FORAMINIFERA
90°
60°
30°
0°
30°
60°
90°
120°
80°
60°
30°
0°
30°
60°
70°

Fig. 6.19. Distribution of relative abundance of *Globigerinoides ruber* (d'Orbigny) in surface waters. (0-10 metres of water.)

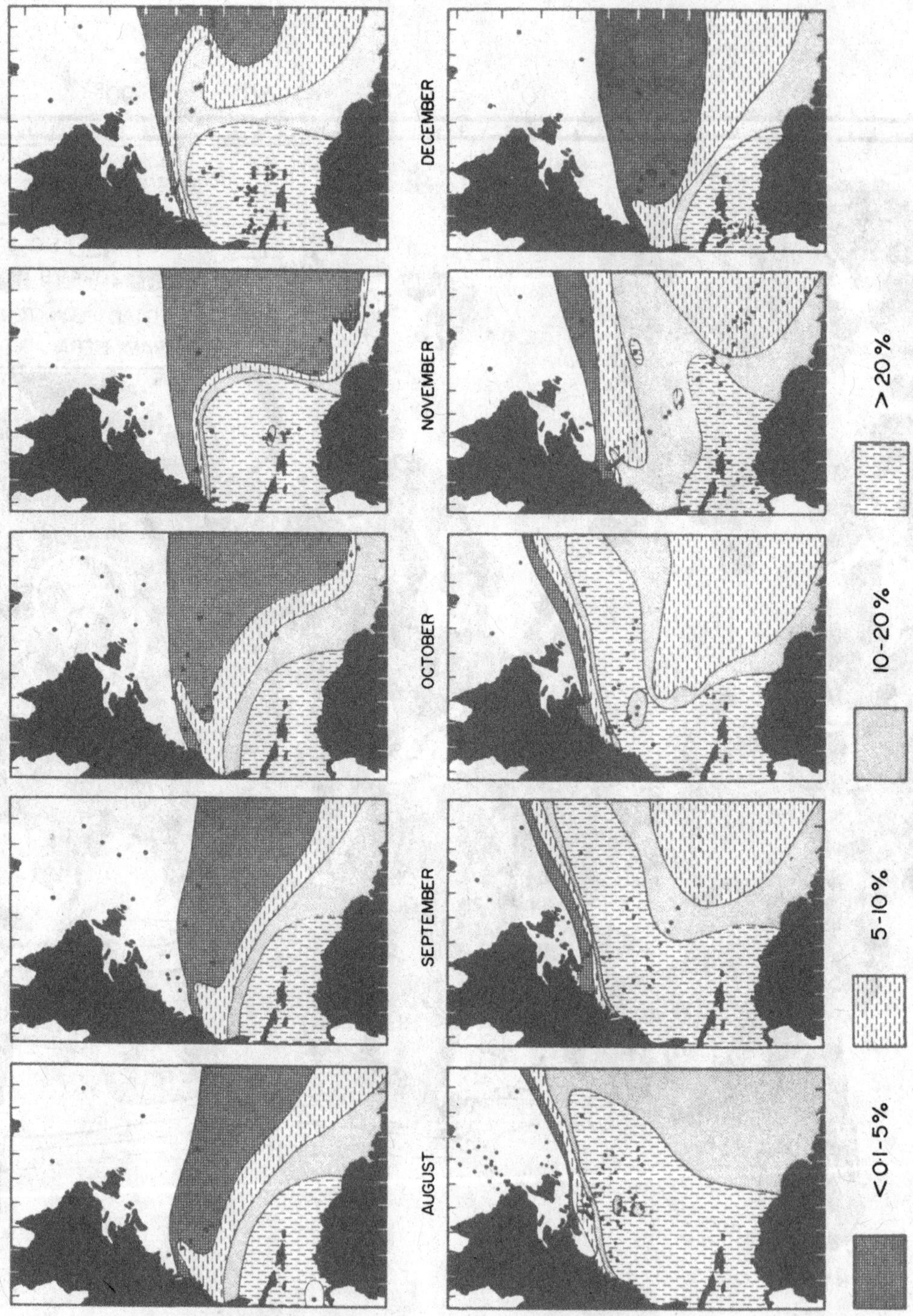
JANUARY
FEBRUARY
MARCH
APRIL
MAY
JUNE
JULY
AUGUST
SEPTEMBER
OCTOBER
NOVEMBER
DECEMBER
<0·1-5%
5-10%
10-20%
>20%

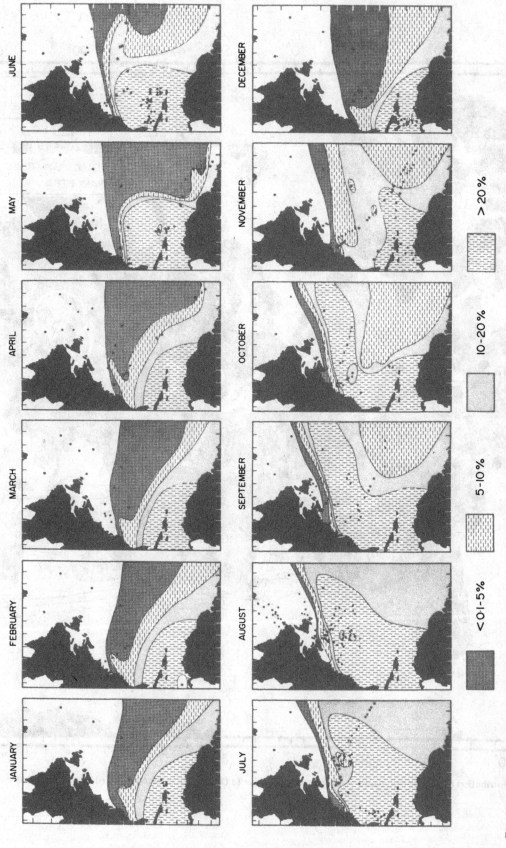

Fig. 6.20. Seasonal variations in relative abundance of *Globigerinoides ruber* d'Orbigny (pink variety) in the upper 300 metres of water of the western North Atlantic.

Equatorial Current. Elsewhere in tropical and subtropical waters it is generally present in small numbers.

The northern and southern distributional limits lie within the Transition Zones, but *G. conglobatus'* seasonal occurrence makes its distribution appear more restricted. *G. conglobatus* occurs over a surface temperature range of 15° to 30° C, with peak abundance between 21° and 29°C.

Globorotalia menardii (d'Orbigny)

Globorotalia menardii is a tropical-subtropical species that lives preferentially in sub-surface waters below 50 metres and probably descends to bathypelagic depths in late ontogeny (Bé, McIntyre and Breger 1967). It is a good warm-water indicator species and easily recognizable, but it is not as abundant as *G. sacculifer* or *G. ruber. G. menardii* cannot be readily differentiated from *Globorotalia tumida* in plankton tows, which frequently contains juvenile, thin-shelled forms that have not yet developed the thickly encrusted test. Sometimes it is possible to recognize juvenile *G. tumida* by the highly angular-conical chambers and elongate tests, but the wide morphological overlap between these forms and those with flat shells prompted us to include *'tumida'* varieties under *G. menardii.*

This species seldom reaches proportions exceeding 5 per cent in surface waters of the North and South Atlantic. The higher concentrations are usually associated with strong current systems, such as the Gulf Stream, North Equatorial and Equatorial Counter Current and Brazil Current. Jones (1967) found the maximum frequencies (over 25 per cent) beneath the Equatorial Atlantic Undercurrent or below a depth of 50 metres. This sub-surface distributional preference would account for the generally low percentages in our surface tows. It is particularly rare in the central water masses of the North and South Atlantic.

G. menardii appears to be much more abundant in the surface waters of the Indian Ocean than in the Atlantic Ocean. High frequencies occur (up to 54 per cent) southeast of Madagascar where they belong to a wide *'menardii'*-rich belt of the South Equatorial Current. The species is also common between Indonesia and Australia.

The distributional limits of *G. menardii* lie within the Northern and Southern Transition Zones. In the western North Atlantic it coincides with the boundary between the Gulf Stream - Labrador Currents and continues to southern Portugal. The species is absent in the Mediterranean bottom sediments (Parker 1958; Todd 1958) and presumably also in the overlying waters.

G. menardii occurs over a surface temperature range of 16° to 30° C, with peak abundance primarily between 20° and 25° C.

Pulleniatina obliquiloculata (Parker and Jones)

Pulleniatina obliquiloculata is another tropical-subtropical species whose highly seasonal occurrence makes its apparent geographic distribution more limited than it actually is during its season of peak abundance.

In the North Atlantic very high concentrations are found in the Gulf Stream System and northwestern Sargasso Sea from November to February and especially during November and December, becoming rare for the remainder of the year. Jones (1967) reported that, although it is comparatively rare in the Caribbean region and tropical Atlantic, *P. obliquiloculata* is a good indicator species of the Equatorial Atlantic Undercurrent and constitutes 4 per cent of the planktonic foraminifera. Cifelli (1967) found higher proportions (up to 8 per cent) in the Equatorial Counter Current than in the Sargasso Sea or other tropical Atlantic areas.

In the South Atlantic, *P. obliquiloculata* is generally quite rare and makes up 5 per cent or more of the foraminiferal population at only three stations. No seasonal preferences are obvious in the South Atlantic, but this may be due to our inadequate sample coverage during winter months.

P. obliquiloculata occurs over a surface temperature range of 19° to 30° C, with peak abundance between 22° and 24° C.

Globigerinoides sacculifer (Brady)

Globigerinoides sacculifer is, next to *Globigerinoides ruber*, the most prolific and widespread species among the warm-water planktonic foraminifera. It attains its highest frequencies (over 50 per cent) in tropical waters, especially in the central equatorial regions of the Atlantic, Indian and Pacific Oceans. *G. sacculifer* makes up 20 per cent or more of the planktonic foraminifera in all tropical seas and in wide areas of subtropical waters. Being a surface dweller the species shows a higher relative abundance on our map than would be the case if deeper tows were examined. *G. sacculifer* occurs over a surface temperature range of 15° to 30° C, with peak abundance primarily between 24° and 30° C. We have included the forms with and without the sac-like final chamber under *G. sacculifer. 'Sphaeroidinella dehiscens'* is considered an aberrant, terminal stage of *G. sacculifer* (Bé 1965), but the former's general absence in surface waters makes it an insignificant variety in our present study.

Salinity as an ecologic factor influencing the relative abundance of G. sacculifer *and* G. ruber. *G. sacculifer* and *G. ruber* are the two most successful warm-water species and may be considered as competitive species vying to dominate the same ecological niche. Both occur together in approximately the same geographic zones and water depths, and they differ mainly in the distributional patterns of their peak abundance areas.

We believe that salinity may be one of the most important factors in explaining the inverse distributional relationship in peak abundance between *G. sacculifer* and *G. ruber*. *G. sacculifer* is most suited by the intermediate surface salinities in all three oceans, i.e. those ranging from 34.5 to 36.0‰. On the other hand, *G. ruber* is a euryhaline species whose maximum frequencies occur at the more extreme salinity values, i.e. either 36.0‰ or below 34.5‰. For example, *G. sacculifer's* relative sparsity (less than 20 per cent) in the central and western South Atlantic off Brazil coincides with comparatively high surface salinities ranging from 36.5‰ to 37.0‰. Apparently, the rich populations of *G. sacculifer* in the South Equatorial Current that are transported southward via the Brazil Current cannot tolerate the high salinities they encounter at about 10° S lat. off Brazil. It is a curious fact that *G. ruber* replaces *G. sacculifer* as the dominant species in this high-salinity region.

In the Indian Ocean our data show *G. sacculifer* and *G. ruber* occurring over the same areas, but the former is more abundant in the northeastern quadrant, while *G. ruber* is more abundant in the higher salinities of the central waters. Belyaeva (1964) noted that *G. ruber* is generally a more abundant species than *G. sacculifer* in plankton tows, but that they appear in equal quantities in bottom sediments.

If *G. ruber* is the most abundant and widespread warm-water species in the North and South Atlantic, *G. sacculifer* overshadows it as the dominant species in the warm-water region of the Pacific Ocean. Bradshaw (1959) and Parker (1960) showed that *G. sacculifer* occurs in frequencies of more than 30 per cent over almost the entire tropical Pacific, between 20° N and 20° S lat., except in the eastern portion off Central America. The area of highest relative abundance (over 80 per cent) was observed in the equatorial Pacific in the Micronesia region.

Considering 34.5‰ to 36.0‰ as the optimum surface salinity range for *G. sacculifer*, we can now possibly explain why its area of maximum relative abundance (over 30 per cent) is very extensive in the Pacific Ocean, intermediate in the Indian Ocean and comparatively limited in the North and South Atlantic and Mediterranean Sea. Apparently, the higher average salinity in the mid-latitudes of the Atlantic and the Mediterranean Sea limits the 'maximum' distribution of *G. sacculifer*, while the 34.5-36.0‰ range in the Pacific and Indian Oceans is favourable for *G. sacculifer's* extensive distribution and high abundance.

In contrast, *G. ruber's* areas of maximum relative abundance (over 50 per cent) are found in high-salinity regions (over 36.0‰), such as the Caribbean Sea, the Gulf Stream system, central and northern Sargasso Sea, Canaries Currents, the Brazil Current, and central waters of the South Atlantic and Indian Ocean. Likewise, the percentage of *G. ruber* tests is high in Mediterranean bottom sediments (Parker 1958; Todd 1958) and increases eastwards following a similar trend of increase in salinity. In the eastern Mediterranean the surface salinities exceed 39‰. *G. sacculifer* is comparatively sparse in Mediterranean sediments and shows an inverse relationship to *G. ruber*.

On the other hand, *G. ruber* also attains frequencies over 20 per cent in low-salinity waters (less than 34.5‰), such as in the eastern Pacific off Central America, the Kuroshio Current, and the Guinea Current off Nigeria (Bradshaw 1959; Schott 1935).

In summary, the difference in average salinities of the subtropical-tropical Pacific and Atlantic Oceans may explain the generally inverse distributional relationship between *G. sacculifer* and *G. ruber*, but it does not explain why the latter species prefers the low-salinity area in the Pacific while it is generally a 'high-salinity' species in the Atlantic Ocean.

Candeina nitida d'Orbigny

Candeina nitida is one of the rarest species, seldom exceeding 5 per cent of the foraminiferal population. It is more abundant in tropical than subtropical waters. We encountered it in small numbers in the North Equatorial Current and the Gulf Stream System as far north as 48° N and 40° W.

In the South Atlantic extremely high frequencies (up to 57 per cent) were found in May at five stations in the Brazil Current. The tests were wrapped in peculiar, 'cocoon-like' envelopes. In the Indian Ocean *C. nitida* occurs sporadically in tropical waters. Belyaeva (1964) encountered it at only two stations.

C. nitida occurs over a surface temperature range of 20° to 30° C.

Globoquadrina conglomerata (Schwager)

Globoquadrina conglomerata is sparsely distributed in surface waters of the tropical and subtropical Indian Ocean. The highest frequency (22 per cent) is found at a tropical station west of the Maldive Islands, but the percentages range only from 0.1 to 3 per cent at eighteen other stations.

The species is transported around the Cape of Good Hope by the Agulhas Current and enters the South Atlantic for a limited distance. *G. conglomerata* is otherwise absent in the South and North Atlantic.

Bradshaw (1959) noted that this species is confined to

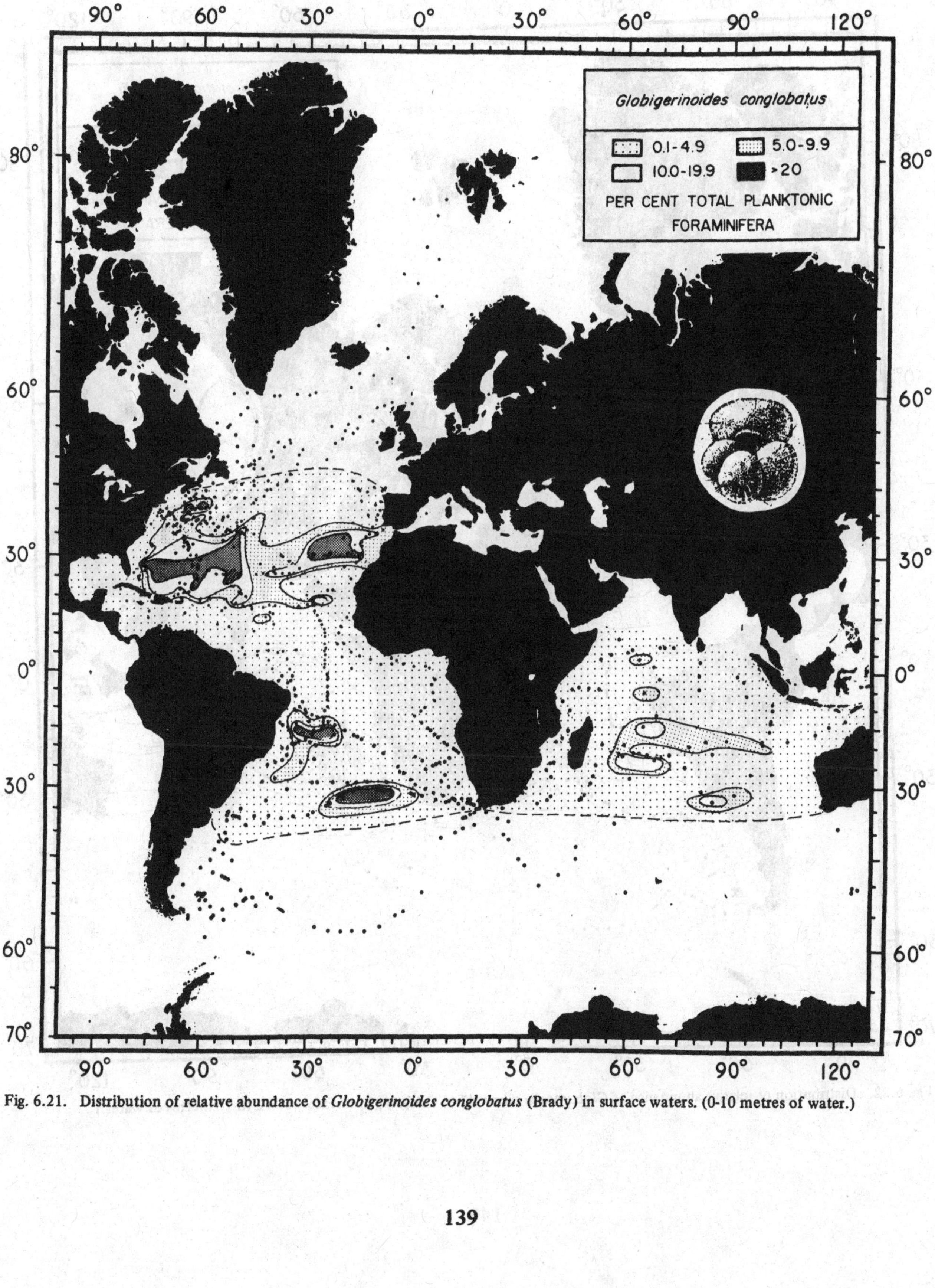
90°
60°
30°
0°
30°
60°
90°
120°
Globigerinoides conglobatus
0.1-4.9
5.0-9.9
10.0-19.9
>20
PER CENT TOTAL PLANKTONIC FORAMINIFERA
80°
60°
30°
0°
30°
60°
70°

Fig. 6.21. Distribution of relative abundance of *Globigerinoides conglobatus* (Brady) in surface waters. (0-10 metres of water.)

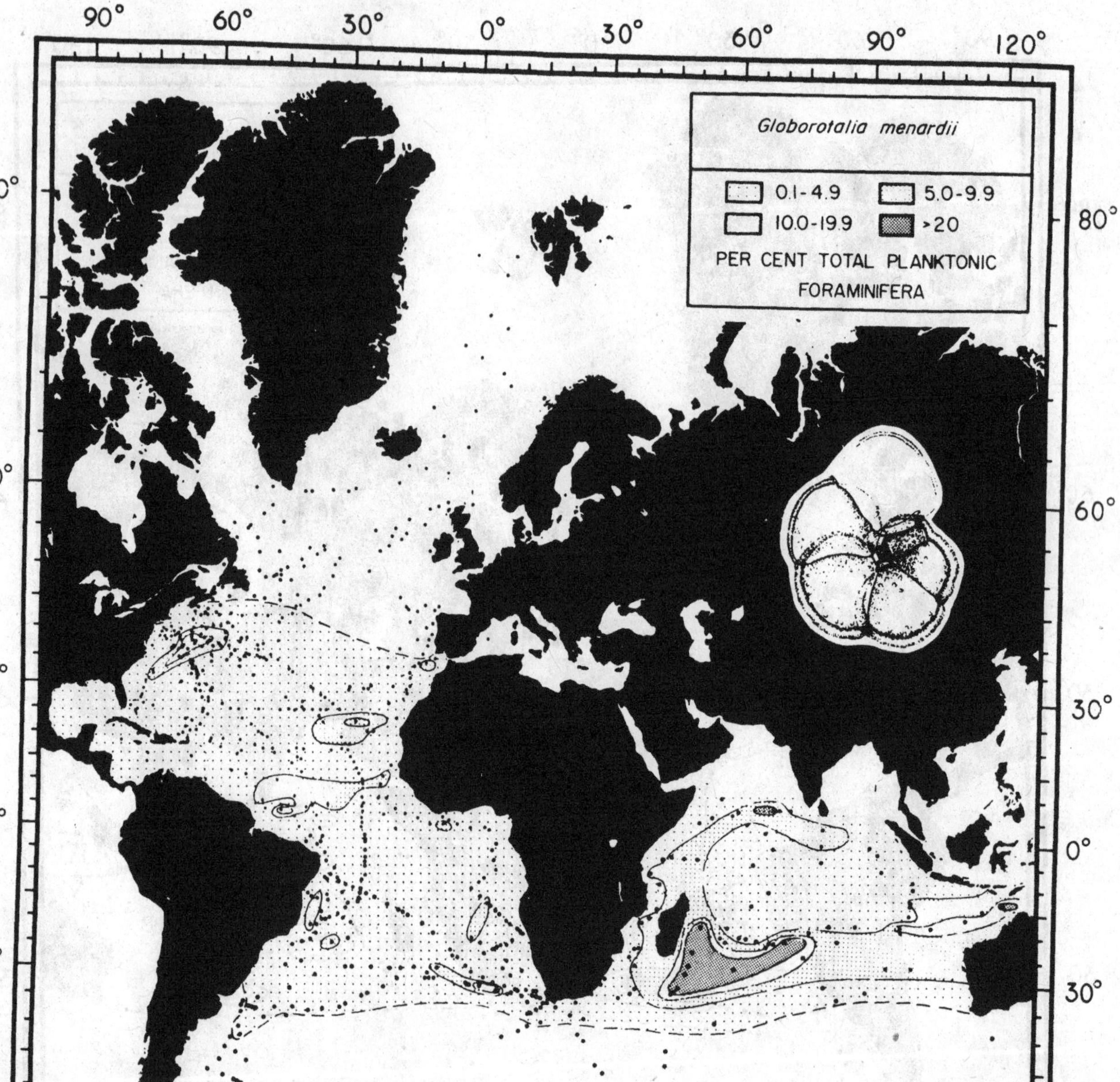
90°
60°
30°
0°
30°
60°
90°
120°
80°
60°
30°
0°
30°
60°
70°
Globorotalia menardii
0.1-4.9
5.0-9.9
10.0-19.9
>20
PER CENT TOTAL PLANKTONIC FORAMINIFERA

Fig. 6.22. Distribution of relative abundance of *Globorotalia menardii* (d'Orbigny) in surface waters. (0-10 metres of water.)

90°
60°
30°
0°
30°
60°
90°
120°
80°
60°
30°
0°
30°
60°
70°
Pulleniatina obliquiloculata
0.1-4.9
5.0-9.9
10.0-19.9
>20
PER CENT TOTAL PLANKTONIC FORAMINIFERA

Fig. 6.23. Distribution of relative abundance of *Pulleniatina obliquiloculata* (Parker and Jones) in surface waters. (0-10 metres of water.)

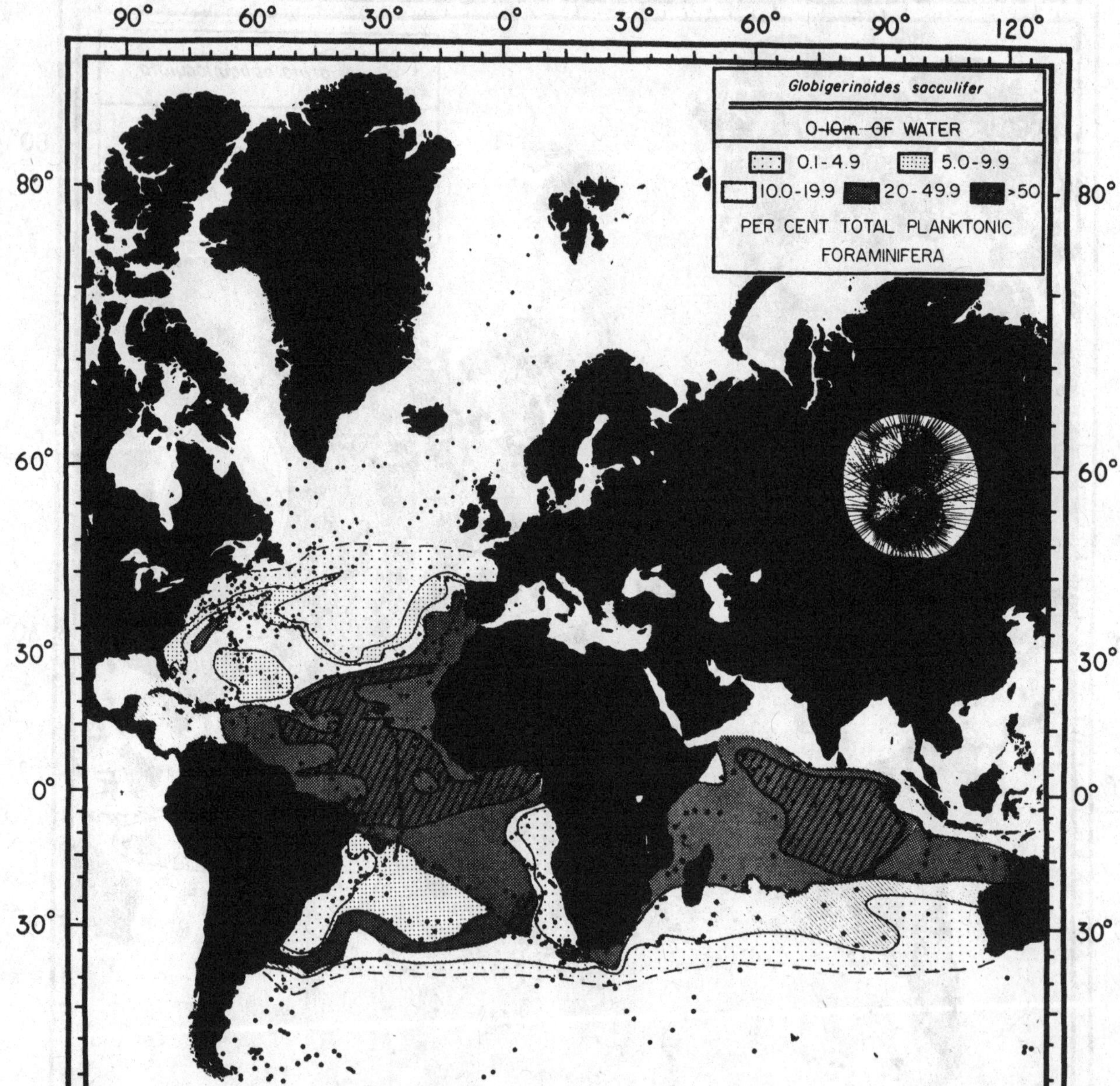

90°
60°
30°
0°
30°
60°
90°
120°
Globigerinoides sacculifer
0-10m OF WATER
0.1-4.9
5.0-9.9
10.0-19.9
20-49.9
>50
PER CENT TOTAL PLANKTONIC
FORAMINIFERA
80°
60°
30°
0°
30°
60°
70°

Fig. 6.24. Distribution of relative abundance of *Globigerinoides sacculifer* (Brady) in surface waters. (0-10 metres of water.)

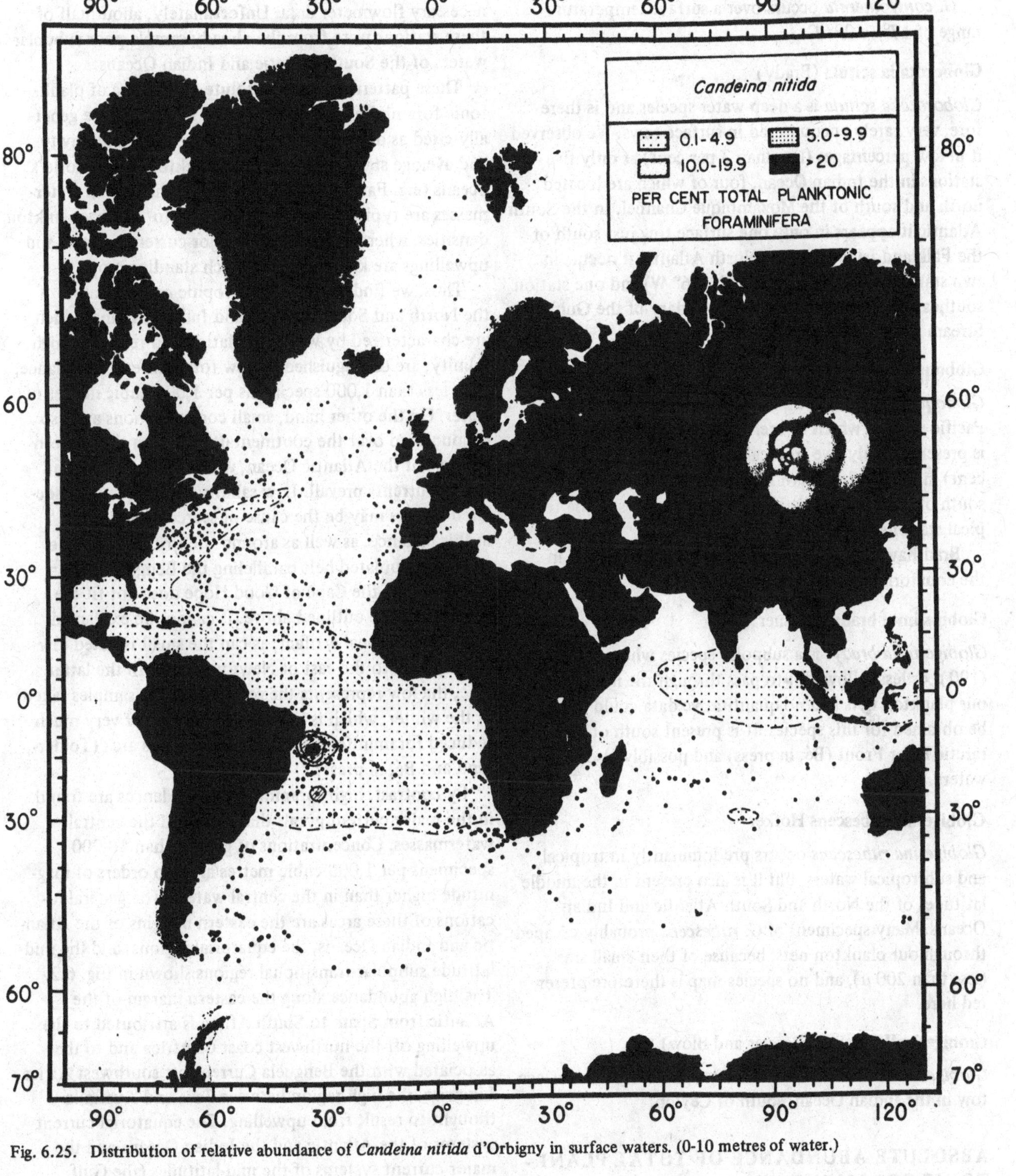
90°
60°
30°
0°
30°
60°
90°
120°
80°
60°
30°
0°
30°
60°
70°
Candeina nitida
0.1-4.9
5.0-9.9
10.0-19.9
>20
PER CENT TOTAL PLANKTONIC FORAMINIFERA

Fig. 6.25. Distribution of relative abundance of *Candeina nitida* d'Orbigny in surface waters. (0-10 metres of water.)

the equatorial Pacific. Thus, *G. conglomerata*, together with *G. adamsi* and *G. hexagona*, should be considered Indo-Pacific species.

G. conglomerata occurs over a surface temperature range of 17° to 29° C.

Globorotalia scitula (Brady)

Globorotalia scitula is a deep-water species and is therefore, very rarely encountered in surface tows. We observed it in low percentages (less than 3 per cent) at only five stations in the Indian Ocean, four of which are located north and south of the Mozambique Channel. In the South Atlantic it appears in only one surface tow just south of the Falkland Islands. In the North Atlantic it occurs in two stations west of Spain (44° N, 16° W) and one station southeast of Nova Scotia at the boundary of the Gulf Stream and Labrador Current.

Globoquadrina hexagona (Natland)

Globoquadrina hexagona is a tropical-subtropical Indo-Pacific species, which is rarely found in surface waters and is present in only five tropical stations (less than 3 per cent) in the northwestern Indian Ocean, and one station south of Capetown. The surface temperatures at the tropical stations ranged from 25° to 29° C.

Bradshaw (1959) observed the highest frequency in the equatorial west-central Pacific.

Globigerinita bradyi Wiesner

Globigerinita bradyi is a subpolar species whose small size (190 μ or less) allowed it to pass through the meshes of our plankton nets and no quantitative data could therefore be obtained for this species. It is present south of the Antarctic Polar Front (Bé, in press) and possibly also in arctic waters.

Globigerina rubescens Hofker

Globigerina rubescens occurs predominantly in tropical and subtropical waters, but it is also present in the middle latitudes of the North and South Atlantic and Indian Oceans. Many specimens of *G. rubescens* probably escaped through our plankton nets, because of their small size (less than 200 μ), and no species map is therefore presented here.

Globigerinella adamsi (Banner and Blow)

Globigerinella adamsi was encountered in only one surface tow in the Indian Ocean, south of Ceylon.

ABSOLUTE ABUNDANCE OF TOTAL PLANKTONIC FORAMINIFERA

The geographical variation in absolute abundance of the total planktonic foraminiferal population in surface waters is shown in Fig. 6.27. It is worth noticing that about fifty stations were deleted from this map as they lacked the necessary flowmeter data. Unfortunately, about half of these stations were from the already sample-poor subpolar waters of the South Atlantic and Indian Oceans.

These patterns of total absolute abundance of planktonic foraminifera are basically the same as those generally cited as depicting the primary organic productivity and average standing crop of zooplankton in the world's oceans (e.g. Fairbridge 1966, p. 272). The central watermasses are typically regions of low phyto- and zooplankton densities, whereas the areas of major current systems and upwellings are known for their rich standing crops.

Thus, we find that the oligotrophic central waters of the North and South Atlantic and Indian Oceans, which are characterized by weak circulation and relatively high salinity, are distinguished by low foraminiferal abundance, with less than 1,000 specimens per 1,000 cubic metres of water. On the other hand, small concentrations are also encountered over the continental shelves on the western margins of the Atlantic Ocean, where low salinities and strong currents prevail. Low salinity or other shelf-associated factors may be the cause of scarce foraminifera in this instance, as well as around the British Isles. The sparsely populated belt paralleling the Benguela region and rounding the Cape of Good Hope does not fit the general scheme outlined above. It cannot be explained away as a seasonal artifact, as can the small isolated low density area to the east of Newfoundland. In the latter case, the low represents the results of a few samples taken in the winter, which is known to be a time of very much reduced foraminiferal concentration in this area (Tolderlund and Bé, in press).

By contrast, high foraminiferal abundances are found in the major current gyrals that surround the central watermasses. Concentrations of greater than 10,000 specimens per 1,000 cubic metres are two orders of magnitude higher than in the central waters. The general locations of these areas are the eastern margins of the Atlantic and Indian Oceans, the equatorial regions, and the mid-latitude subpolar-transitional regions shown in Fig. 6.2. The high abundance along the eastern margin of the Atlantic from Spain to South Africa is attributed to the upwelling off the northwest coast of Africa and to that associated with the Benguela Current off southwest Africa. Likewise, the high region between Java and Australia is thought to result from upwelling. The equatorial current systems of the Atlantic and the Indian Oceans and the major current systems of the mid-latitudes (the Gulf Stream-North Atlantic Current and the West Wind Drift) are also very rich regions. Although the subpolar-transit-

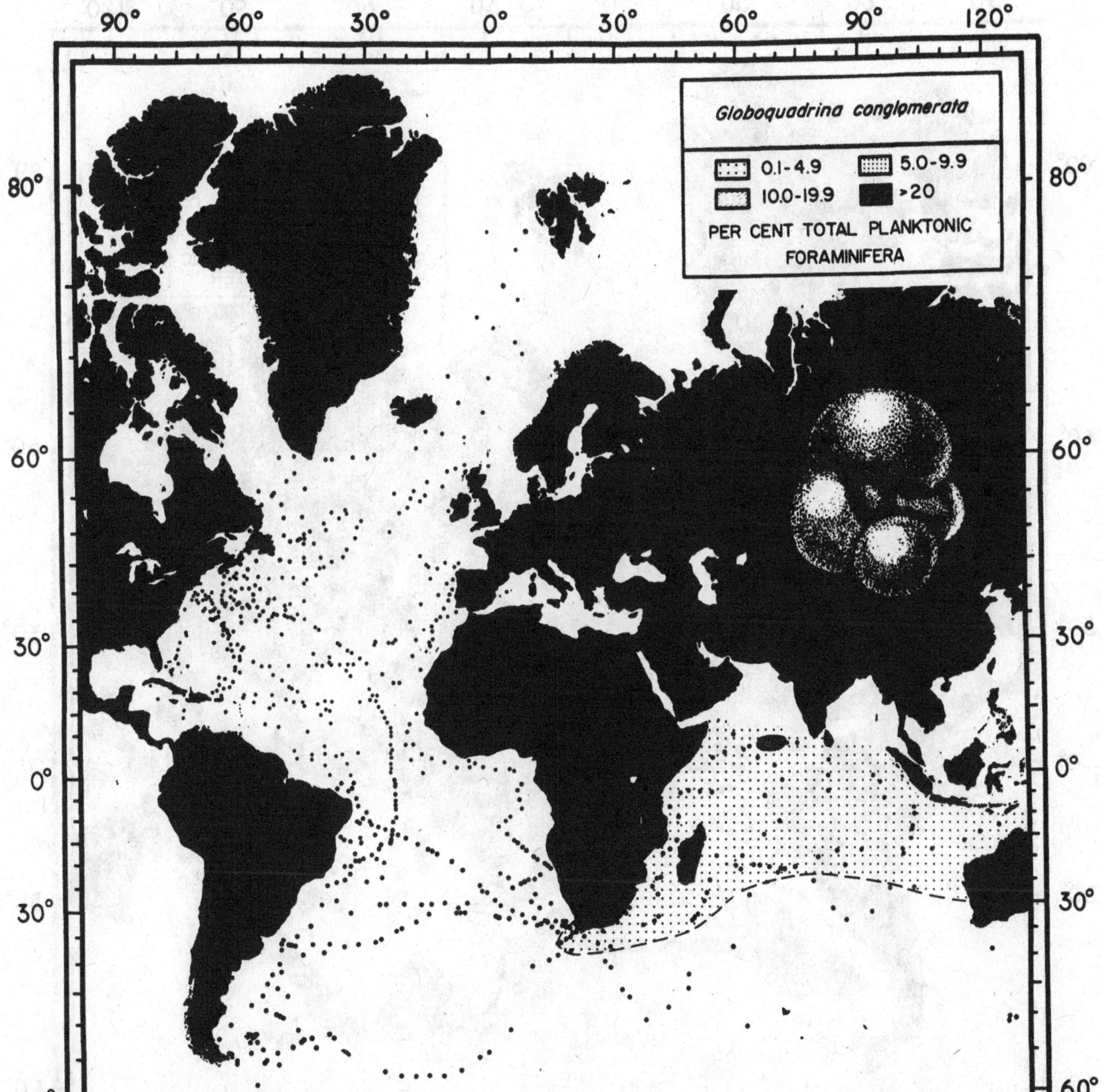
90°
60°
30°
0°
30°
60°
90°
120°
Globoquadrina conglomerata
0.1-4.9
5.0-9.9
10.0-19.9
>20
PER CENT TOTAL PLANKTONIC FORAMINIFERA
80°
60°
30°
0°
30°
60°
70°

Fig. 6.26. Distribution of relative abundance of *Globoquadrina conglomerata* (Natland) in surface waters. (0-10 metres of water.)

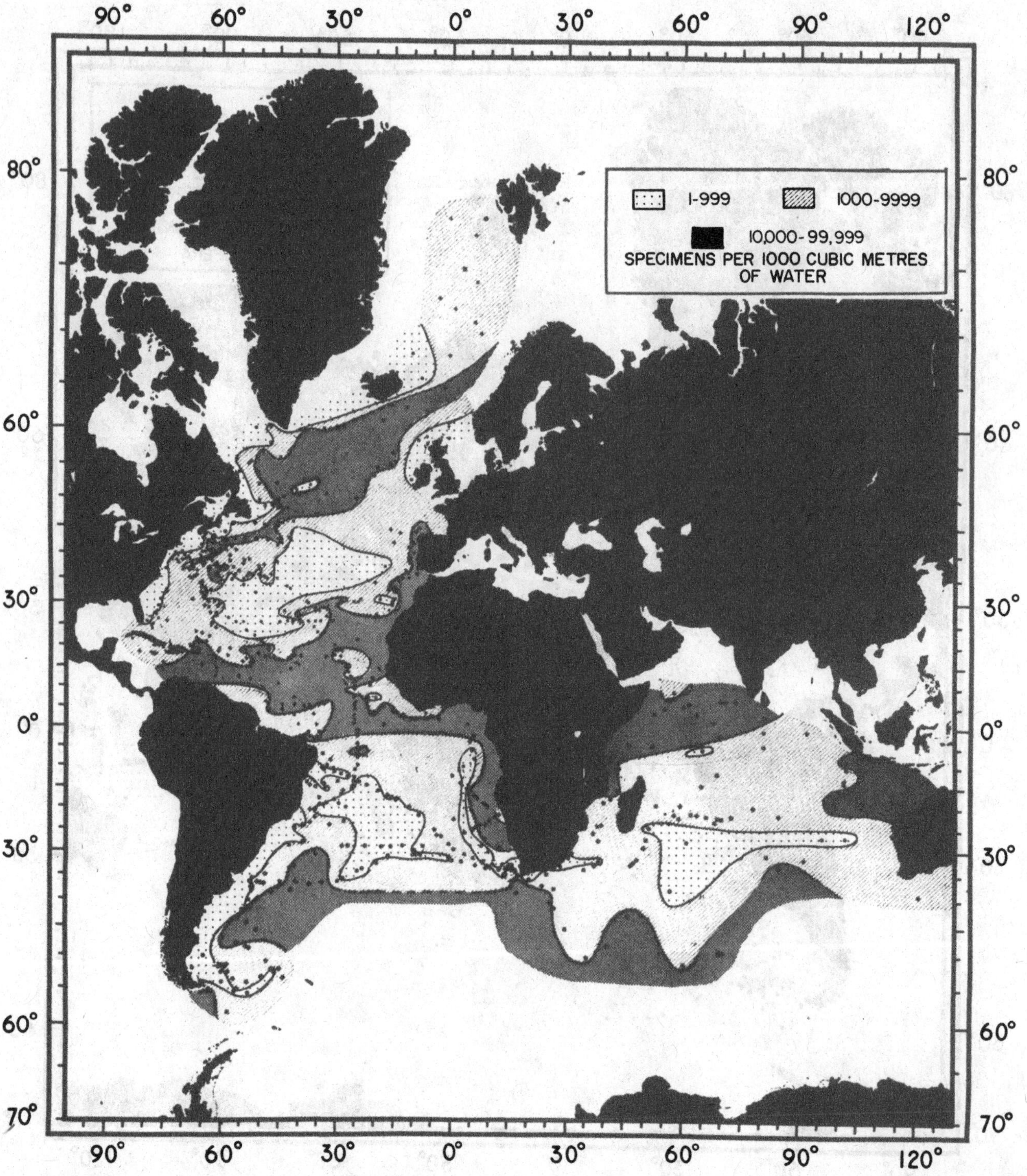
90°
60°
30°
0°
30°
60°
90°
120°
80°
60°
30°
0°
30°
60°
70°
1-999
1000-9999
10,000-99,999
SPECIMENS PER 1000 CUBIC METRES
OF WATER

Fig. 6.27. Absolute abundance of total planktonic foraminifera in surface waters. (0-10 metres of water.)

ional waters have a slightly higher average abundance (30,000 specimens per 1,000 cubic metres) than the equatorial waters (25,000 per 1,000 cubic metres), the seasonal factor must be taken into account. Again, as mentioned above, the winter months in the region of the North Atlantic Current are known to be times of foraminiferal sparsity, as least in the surface waters. Moreover, the majority of the samples which delineate the high in this area are from the months of June-October. Thus, since foraminiferal productivity in tropical waters is not subject to the seasonal fluctuations that affect high-latitude populations, and tropical species probably have shorter generation times and possess larger shells, our findings substantiate the generally recognized observation that a greater thickness of calcareous sediments accumulate in equatorial regions (Sverdrup *et al.* 1942, p. 1007). It is worth emphasizing that the values in Fig. 6.27 represent 'standing crop' or population abundance at a particular time and place. The ultimate contribution of foraminiferal shells to the ocean floor depends on their productivity, modified by destructive agents such as scavengers and inorganic solution.

Bradshaw (1959) found the highest foraminiferal concentrations in the equatorial and boreal North Pacific waters. The important effect of mesh size is well illustrated by Bradshaw when he compared the total absolute abundance from two sets of samples taken with plankton nets having 310 μ and 70 μ mesh apertures. The upper limit of abundance for the latter net was two orders of magnitude higher than of the coarser net. This may explain why our data collected with nets having a 200 μ mesh aperture are essentially intermediate in magnitude to those of Bradshaw.

Thus, our results show basic agreement with the investigations of Bradshaw (1959) and Belyaeva (1964), indicating that low concentrations are to be found in the central watermasses and that high population densities characterize the major current systems of the world's oceans.

ACKNOWLEDGEMENTS

We wish to thank Professor Maurice Ewing for his continuous encouragement and support that has made the present study possible. We are grateful to all our Lamont colleagues, too numerous to mention by name, who helped directly and indirectly in collecting the hundreds of plankton samples on numerous cruises of R. V. *Vema* and R. V. *Robert Conrad.*

Special thanks are due to the officers and men of the U.S. Coast Guard of the Eastern Area who have participated in our plankton sampling programme at Ocean Stations *Bravo, Charlie, Delta, Echo* and *Bermuda.*

We are grateful for the support that this study received from the National Science Foundation (grants GB-4219, GB-6991, GA-668 and GA-1205). Field collecting was also made possible by funds from the Office of Naval Research (Contract N000 14-67-A-0108-0004).

Received February 1968

BIBLIOGRAPHY

Alvariño, A. 1965. Chaetognaths. In *Annual review of oceanography and marine biology,* vol. 3 (H. Barnes *ed.)* 115-94. George Allen and Unwin: London.

Bé, A. W. H. 1959. Ecology of recent planktonic foraminifera. Part 1. Areal distribution in the western North Atlantic. *Micropaleont.* 5 (1), 77-100.

Bé, A. W. H. 1960a. Some observations on Arctic planktonic foraminifera. *Cushman Found. Foram. Res. Contr.* **11** (2), 64-8.

Bé, A. W. H. 1960b. Ecology of recent planktonic foraminifera. Part 2. Bathymetric and seasonal distributions in the Sargasso Sea off Bermuda. *Micropaleont.* **6** (4), 373-92.

Bé, A. W. H. and Ericson, D. B. 1963. Aspects of calcification in planktonic foraminifera. In *Comparative Biology of Calcified Tissues, N.Y. Acad. Sci., Trans.* **109** (1), 65-81.

Bé, A. W. H. and Lott, L. 1964. Shell growth and structure of planktonic foraminifera. *Science,* **145**, (3634), 823-4.

Bé, A. W. H. 1965. The influence of depth on shell growth in *Globigerinoides sacculifer* (Brady). *Micropaleont.* **11**, (1), 81-97.

Bé, A. W. H. and McIntyre, A. and Breger, D. L. 1966. The shell microstructure of a planktonic foraminifer, *Globorotalia menardii* (d' Orbigny). *Eclogae Geol. Helv.* **59**, (2), 885-96.

Bé, A. W. H. 1967. Foraminifera, Families: *Globigerinidae* and *Globorotaliidae.* Fiche no. **108**. In *Fiches d'Identification du Zooplancton,* Redigé par J. H. Fraser. *Cons. Internat. Explor. Mer,* Charlottenlund, Denmark.

Bé, A. W. H and Hamlin, W. H. 1967. Ecology of recent planktonic foraminifera. Part 3. Distribution in the North Atlantic during the summer of 1962. *Micropaleont.* 13, (1), 87-106.

Bé, A. W. H. (in press) Planktonic Foraminifera. In *Marine Invertebrates South of 35° S Latitude.* Antarct. Map Folio Ser., Am. Geogr. Soc.; New York.

Belyaeva, N. V. 1964. Distribution of planktonic foraminifera in the water and on the floor of the Indian Ocean. *Trudy Inst. Okean., Acad. Sci. U.S.S.R.* **68**, 12-83. (In Russian).

Boltovskoy, E. 1959. Foraminifera as biological indicators in the study of ocean currents. *Micropaleont.* **5**, (4), 473-81.

Boltovskoy, E. 1962. Planktonic foraminifera as indicators of different water masses in the South Atlantic. *Micropaleont.* **8**, (3), 403-8.

Boltovskoy, E. 1964. Distribución de los foraminiferos planctonicos vivos en el Atlántico Ecuatorial, parte oeste (Expedición *Equalant*). Argent., Serv., Hidrogr. Naval, Publ. **H. 639**, 1-54.

Boltovskoy, E. 1966a. Zonación en las latitudes altas del Pacifico sur según los foraminiferos planctónicos vivos. *Rev. Mus. Argent Cien. Nat. Bernardino Rivadavia, Hidrobiologia*, 2 (1), 1-56.

Boltovskoy, E. 1966b. La zona de convergencia subtropical subantárctica en el océano Atlántico (parte occidental). (Un estudio en base a la investigación de Foraminiferos-indicadores). *Argent., Serv. Hidrogr. Naval, Publ.* **H. 640**, 1-69.

Bogorov, B. G. 1955 Regularities of plankton distribution in Northwest Pacific. *UNESCO Symp. Phys. Oceanog., Proc., Tokyo,* 260-76.

Bradshaw, J. S. 1959. Ecology of living planktonic foraminifera of the North and Equatorial Pacific Ocean. *Cushman Found. Foram. Res., Contr.* **10** (2), 25-64.

Brady, H. B. 1884. Report on the foraminifera dredged by H.M.S. *Challenger* during the years 1873-1876. *Rept. Voy. 'Challenger', Zool.* **9**, 1-814.

Brodsky, K. A. 1957. The Copepoda (Calanoida) fauna and zoogeographic division into districts in the northern part of the Pacific Ocean and of the adjacent waters. *Zool. Inst., Acad. Sci., U.S.S.R., Moscow* (in Russian).

Brandt, K. and Apstein, C. (*Eds.*) 1901-28. Nordisches Plankton (many authors). Lipsius and Tischer: Kiel and Leipzig.

Cifelli, R. 1961. *Globigerina incompta,* a new species of pelagic foraminifera from the North Atlantic. *Cushman Found. Foram. Res., Contr.* **12** (3), 83-6.

Cifelli, R. 1962. Some dynamic aspects of the distribution of planktonic foraminifera in the western North Atlantic. *J. Mar. Res.* **20** (3), 201-13.

Cifelli, R. 1965. Planktonic foraminifera from the western North Atlantic. *Smithonian Misc. Coll.* **148** (4), publ. 4599, 1-35.

Cifelli, R. 1967. Distributional analysis of North Atlantic foraminifera collected in 1961 during Cruises 17 and 21 of the R. V. *Chain. Cushman Found. Foram. Res., Contr.* **18** (3), 118-27.

Deacon, G. E. R. 1945. Water circulation and surface boundaries in the oceans. *Quart. Jour. Roy. Meteor. Soc.* **71** (307-8), 11-25.

Defant, A. 1961. *Physical Oceanography. Vols. 1 and 2.* Pergamon Press: New York.

Dietrich, G. 1963. *General oceanography.* Interscience Publishers: New York.

Dietrich, G. 1964. Ocean Polar Front Survey in the North Atlantic. In, *Research in geophysics*, vol. 2, (Odishaw, Hugh *ed.*), pp. 291-309. The M.I.T. Press: Cambridge, Mass.

Ericson, D. B. 1959. Coiling direction of *Globigerina pachyderma* as a climatic index. *Science,* **130** (3369), 219-20.

Ericson, D. B. and Wollin, G. and Wollin, J. 1954. Coiling direction of *Globorotalia truncatulinoides* in deep-sea cores. *Deep-sea Res.* **2** (2). 152-8.

Ericson, D. B. and Wollin, G. 1956. Correlation of six cores from the equatorial Atlantic and the Caribbean. *Deep-sea Res.* **3** (2), 104-25.

Fairbridge, R. W. (*ed.*). 1966. *The encyclopedia of oceanography.* Reinhold Publ. Corp.: New York.

Hensen, V. (*ed.*) 1892-1926. *Ergebnisse der Plankton-Expedition der Humboldtstiftung.* (Many volumes and parts by many authors) Lipsius and Tischer: Kiel and Leipzig.

Johnson, M. W. and Brinton, E. 1963. Biological species, water-masses and currents. In *The Sea,* vol. 2 (M. N. Hill (*ed.*).) pp. 381-414. John Wiley and Sons: New York.

Jones, J. I. 1966. The distribution and variation of living pelagic foraminifera in the Caribbean Sea. *Carib. Geol. Conf., 3rd (1962), Proc.* 178-83.

Jones, J. I. 1967. Significance of distribution of planktonic foraminifera in the Equatorial Atlantic Undercurrent. *Micropaleont.* **13** (4), 489-501.

Kennett, J. 1967. Recent latitudinal variation in the planktonic foraminifera *Globigerina pachyderma* (Ehrenberg) and *Globorotalia truncatulinoides* (d'Orbigny). *Geol. Soc. Am., 1967 Annual Mtg., Abstracts,* 117-18.

Mackintosh, N. A. 1934. Distribution of the macroplankton in the Atlantic Sector of the Antarctic. *Discovery Repts.* **9**, 65-160.

Meisenheimer, J. 1905. Die tiergeographischen Regionen des Pelagials, auf Grund der Verbreitung der Pteropoden. *Zool. Anz.* **29**, 155-63.

Murray, J. 1897. On the distribution of the pelagic Foraminifera at the surface and on the floor of the ocean. *Natural Science/(ecology),* **11**, 17-27.

Parker, F. L. 1958. Eastern Mediterranean Foraminifera. *Swedish Deep-Sea Exped., Repts.* **8** (4), 219-83.

Parker, F. L. 1960. Living planktonic Foraminifera from the equatorial and southeast Pacific. *Tohoku Univ., Sci. Rept., Spec. (Hanzawa),* **4**, 71-82.

Phleger, F. B. 1951. Ecology of foraminifera, northwest Gulf of Mexico; Part I - Foraminifera distribution. *Geol. Soc. Amer., Mem.* **46** (1), 1-88.

Rhumbler, L. 1901. Nordische Plankton-Foraminiferen. In Brandt, K., *Nordisches Plankton.* Lipsius and Tischer: Kiel and Leipzig. **1** (14), 32 pp.

Rhumbler, L. 1911. Die Foraminiferen (Thalamorphoren) der Plankton Expedition; Teil 1 - Die allgemeinen Organisationsverhältnisse der Foraminiferen. *Plankton-Exped. Humboldt-Stiftung, Ergebn.* **3** (1), 1-331.

Russell, F. A. 1939. Hydrographical and biological conditions in the North Sea as indicated by plankton organisms. *J. Cons. Explor. Mer,* **14**, 171-92.

Schott, G. 1935. *Geographie des Indischen und Stillen Ozeans.* 413 pp. Boysen: Hamburg.

Schott, G. 1942. *Geographie des Atlantischen Ozeans.* 438 pp. Third ed. Boysen: Hamburg.

Schott, W. 1935. Die Foraminiferen in dem äquatorialen Teil des Atlantischen Ozeans. *Deutsch. Atlant. Exped. Meteor 1925-1927, Wiss., Ergebn.* **3** (3), 43-134.

Schott, W. 1966. Foraminiferenfauna und Stratigraphie der Tiefsee-sedimente im Nordatlantischen Ozean. *Swedish Deep-Sea Exped., Repts.* **7**, Sed. Cores from N. Atlantic Oc. **8** (7), 357-469.

Smith, A. B. 1963. Distribution of living planktonic foraminifera in the northeastern Pacific. *Cushman Found. Foram. Res., Contr.* **14** (1), 1-15.

Smith, A. B. 1964. Living planktonic foraminifera collected along an east-west traverse in the North Pacific. *Cushman Found. Foram. Res., Contr.* **15** (4), 131-4.

Steuer, A. 1910. *Planktonkunde.* B. G. Teubner: Leipzig and Berlin.

Steuer, A. 1933. Zur planmässigen Erforschung der geographischen Verbreitung des Haliplanktons, besonders der Copepoden. *Zoogeographica,* 1, 269-302.

Sverdrup, H. U., Johnson, M. W. and Fleming, R. H. 1942. *The Oceans.* 1087 pp. Prentice Hall: Englewood Cliffs, N.J.

Todd, R. 1958. Foraminifera from Western Mediterranean deep-sea cores. *Swedish Deep-Sea Exped., Repts.* 8 (3), 169-215.

Tolderlund, D.S. and Bé, A.W.H. (in preparation). Seasonal distributions of planktonic foraminifera in the surface waters of five stations in the western North Atlantic.

7. DISTRIBUTION OF POLYCYSTINE RADIOLARIA IN THE OCEANS IN RELATION TO PHYSICAL AND CHEMICAL CONDITIONS

R. E. CASEY

Summary: Radiolarian biogeographical studies are at a stage at which some synthesis is necessary. In an attempt to do this selected papers on radiolarian distribution are briefly reviewed and commented on. A scheme of geographical zones for living polycystine radiolarians related to physical and chemical conditions is then proposed for the Pacific. A general scheme is also proposed which may be applicable to other oceans and their adjacent seas.

INTRODUCTION

Radiolaria are marine planktonic protozoans. The polycystine radiolarians possess skeletons of almost pure opaline silica, and comprise the suborders Spumellaria and Nassellaria (Ehrenberg 1838).

The first notable work on the distribution of living polycystins was that by E. Haeckel on material from the voyage of H.M.S. *Challenger*. In the ensuing publication (1887) he states that radiolarians occur in all seas, at all climatic zones, and from surface layers to great depths.

In expanding these general comments by Haeckel, the present author deals with polycystine radiolarian distribution in the following order: daily fluctuations in distribution, seasonal distribution, vertical distribution, horizontal distribution, and other distributional phenomena. The present author then proposes a scheme of biogeographical zones for polycystins for the Pacific and a general one that may be suitable for the world ocean. The author acknowledges the invaluable assistance of the works of Yu. A. Orlov (1962) and W. R. Riedel (1967) in preparing sections dealing with some aspects of radiolarian distribution.

DAILY FLUCTUATIONS IN DISTRIBUTION

Few records of daily fluctuations in the distribution of radiolarians exist. One such record comes from Wolfenden (1905), who recorded no vertical response of certain radiolarians to rain. (N.B. Acantharians supposedly descend from the surface during rains or storms, but these radiolarians possess a hydrostatic adaptation not found in polycystins.) Haeckel (1887) states that radiolarians in his Pelagic Faunal zone swim at the surface but may descend to a depth of 20 to 30 fathoms.

Off Southern California, U.S.A., a shallow daily fluctuation in polycystins has been noted (Casey 1966). During late afternoon radiolarian density in the surface waters increased, and at first this appeared to be a vertical migration. However, the timing corresponds with the afternoon on-shore winds that are common in the area. The increase in density is therefore probably caused by a stirring of the mixed layer bringing radiolarians from a deeper preferred depth. This shallow mixing is confirmed by bathythermograph records taken simultaneously with the collections.

Although the evidence is inconclusive for active vertical migration or other daily movements of polycystins, a movement of some type may occur. Many polycystins possess a locomotory sarcoflagellum.

SEASONAL DISTRIBUTION

Cleve (1900) discusses the seasonal distribution of Atlantic plankton organisms including polycystins. He shows that forms are brought into regions seasonally during fluctuations of current systems, and may even dive below other waters of less density and still be maintained within their 'endemic' or characteristic waters.

Popofsky (1908), working in the Antarctic, observed a number of radiolarians that exhibit seasonal fluctuations in their distribution, inasmuch as these radiolarians were present only during summer, apparently as a result of a southward displacement of warm currents.

Mielck (1913) notes that in the North Sea and north-east Atlantic, polycystins are ordinarily restricted to depths greater than 100 metres in spring and summer, but appear in autumn and winter to rise above 100 metres. He finds these distributions to be related to the influx of waters from surrounding areas.

Schewiakoff (1926) notes that Mediterranean polycystins rise from deep water to the surface in colder seasons. During this period they gather in masses in the 0-50 metre layer (whereas Acantharians, on the contrary, move from surface layers to the 50-200 metre zone, possibly because of their possessing a hydrostatic apparatus permitting better depth control).

These last two papers by Mielck and Schewiakoff may illustrate a winter rise to the surface, and this rise may be explained by a stirring of the mixed layer.

Bernstein (1934) finds that in the north Kara Sea some species are endemic, whereas others are brought in seasonally from the Barents Sea by the inflow of Atlantic water.

In the upper 200 metres of the water column off the coast of Southern California, Casey (1966) finds a distinct seasonal change in radiolarian assemblages which is apparently dependent upon local circulation patterns that bring in radiolarians from different surrounding water masses during different seasons. Few major fluctuations are seen in deeper zones (below 200 metres). One fluctuation occurs during the oceanographic winter when radiolarians from depths as great as 400 metres are brought to the surface, probably associated with the local winter upwelling.

Seasonal changes in radiolarian populations obviously do exist, and are dramatic in regions where oceanographic conditions undergo seasonal fluctuations. Winter upwelling or stirring of the mixed layer are probably illustrated in the works of Mielck (1913), Schewiakoff (1926), and Casey (1966); the radiolarians may be passively brought to the surface with the upwelled water. Horizontal seasonal movements were observed by Cleve (1900), Popofsky (1908), Mielck (1913), Schewiakoff (1926), and Casey (1966); these movements are all related to the physical oceanography of the areas studied and are in areas where one would expect physical changes during the various seasons.

VERTICAL DISTRIBUTION

Haeckel (1887) states that certain species of radiolarians are limited to particular bathymetric faunal zones. Haeckel recognizes three zones: Pelagic Faunal, Zonarial Faunal, and Abyssal Faunal. Radiolarians of the Pelagic Faunal zones occurred from the surface to about 46 metres. This fauna consists primarily of the 'Order' Porulosida (suborder Spumellaria, plus the Acantharia[1]) and includes a few members of the 'Order' Oculosida (suborder Nassellaria, plus the Phaeodaria[1]). Skeletons of these forms have thinner skeletal structures than abyssal forms. Haeckel states that many of these radiolarians in the Pelagic Fauna possess either incomplete skeletons or none at all, and within the same family the dimensions of the pelagic species seem to be smaller than those of related abyssal forms.

The Zonarial Fauna inhabits various bathymetric zones between the Pelagic Fauna and the Abyssal Fauna. The 'Order' Porulosida predominates in the upper portions of the Zonarial Faunal zone from 46 metres to 3,656 metres. At greater depths this group is replaced gradually by the 'Order' Osculosida, which predominates from about 3,656 metres to just above the ocean floor. The Abyssal Fauna, comprising forms which float only a short distance above the bottom of the deep sea, is composed mainly of the 'Order' Osculosida.

Lo-Bianco (1903) distinguishes four depth zones in the Mediterranean Sea which he believes may be correlated with degrees of illumination in these zones. These zones are as follows:

1. 0-50 metres - Phaeoplankton (illuminated zone).
2. 50-400 metres - Knephoplankton (partially illuminated zone).
3. 400-1,500 metres - Skotoplankton (barely illuminated zone).
4. 1,500-5,000 metres - Nyctoplankton (non-illuminated zone).

Haecker (1907) indicates that some Sphaerellaria and Cyrtellaria are restricted to 400-5,000 metres. He finds these deep-water forms to have massive skeletons, reduced radial spines, and a laterally compressed form. Haecker (1908*a*) distinguishes a Colloid zone (0-50 metres), a Challengerid zone (50-350 or 400 metres), a Tuscarorid zone (350 or 400-1,000 or 1,500 metres), and a Pharyngellid zone (1,000 or 1,500-5,000 metres). Haecker correlates his zones with those of Lo-Bianco. Although these are basically phaeodarian zones, the polycystins also show a similar zonation. Haecker (1908*b*) states that species are apparently related to environmental conditions. He surmises that surface-living forms are in general small and spherical and possess delicate skeletons; whereas, deep-living forms are generally bilaterally symmetrical with larger and heavier skeletons. (Haecker describes some Antarctic phaeodarians that live in surface waters of the Antarctic and dive beneath temperate and tropical waters to the north.)

Popofsky (1913), working with warm water nassel-

[1] These groups are not polycystine radiolarians.

larians, finds that almost all nassellarians obtained in the Atlantic and Southwest Indian Oceans occur at depths from 0 to 400 metres.

Reshetnjak's study (1955) in the Kuril-Kamchatka Deep resulted in the designation of stenobathic and eurybathic forms. Stenobathic forms occur strictly in one or several horizons, and eurybathic forms are distributed throughout a range of depths but usually show a depth of greatest abundance.

Reshetnjak discusses over 100 species, 45 per cent of which she refers to as stenobathic forms. Although these are mainly phaeodarian zones, some polycystins also occur in these zones. Her zonation is as follows:

1. Surface Radiolaria, 0-50 metres.
2. Subsurface forms, 50-200 metres.
3. Moderately deep forms, 200-1,000 metres.
4. Bathypelagic forms, 1,000-2,000 metres.
5. Abyssal forms, 4,000-8,000 metres.
6. A transitional fauna, from 50 to 1,000 metres.

Stenobathic forms, according to Reshetnjak, are related to each of these zones. Eurybathic forms are either distributed throughout the entire water depth from 50 to 8,000 metres or exhibit a more limited distribution from 50 to 2,000 metres.

Reshetnjak finds that radiolarians display greatest density and diversity between 200 and 2,000 metres. She also notes that at great depths the protozoan plankton is composed almost exclusively of Radiolaria.

Reshetnjak correlates her depth zones with water masses that have previously been established for the region. Many of Reshetnjak's shallow radiolarians occur in Haecker's (1908*a*) deeper tows and show a tropical submergence, probably due to the diving of water masses.

Hulsemann (1963) finds two distinct depth zones in the Arctic – a shallow zone of endemic forms, and a deeper zone of cosmopolitan forms.

Kling (1966) discusses the distribution of living Castanellid and Circoporid radiolarians in the eastern North Pacific Ocean. He finds forms zoned with depth, water masses, and hydrographic conditions.

Petrushevskaya (1966), working on polycystins in plankton and bottom sediments, contributed a series of significant findings; those which pertain to depth are as follows:

1. In the central Pacific Ocean the species composition is nearly homogeneous from the surface to a depth of 100-300 metres. However, at greater depths in the Pacific a general impoverishment of the fauna is observed and new species appear which are not encountered at the surface.
2. Polycystins are found in all layers from 0 to 5,000 metres. Their concentrations in the upper water layers reach 16,000 specimens per cubic metre of water.
3. Almost no radiolarians occur in the surface layers of the Antarctic waters south of the Antarctic Convergence.
4. Below the surface layer in the Antarctic (0-30 metres) Nassellaria and Spumellaria are encountered in greatest abundance between 200 and 400 metres, and below this their density decreases.

Petrushevskaya (1967), working on the distribution of radiolarians in the Antarctic, states that radiolarians are rarely found at the surface (0-30 metres). This is contrary to what she finds in the tropical Pacific (Petrushevskaya 1966) where the surface zone abounds with nassellarian and spumellarian species. The greatest number of radiolarians and the greatest variety of species are found at the 150-400 metre horizons. Down to 800 metres no increase in number of specimens and varieties of species could be noticed.

Casey (1966), investigating waters off Southern California, notices a distinct vertical zonation of polycystins into three major depth zones from the surface to 1,000 metres. (Within the major depth zones inhabited by a given species, juvenile forms occupy the shallowest regions of that depth zone.) From 0 to 200 metres there is a distinct zone, the fauna of which fluctuates throughout the year. From 200 to 400 or 500 metres is another zone, and from 400 or 500-1,000 metres (1,000 metres is the deepest depth sampled) is a third major zone. During the summer, in addition to different assemblages of radiolarians, smaller zones also occur within the 0-200 metre zone (from 0-25 metres, 25-50 metres, 50-125 metres, and 125-200 metres). The 25- and 50-metre breaks correlate with thermoclines at these depths, the 125-metre break with a weak pycnocline, and the 200-metre break correlates with surface water overlying Pacific Central water. The physical and chemical barriers mentioned may also act as shear zones, so that one 'micro-water mass' may be pushed over or under another. These shear zones may help to account for the distinct faunal breaks observed.

It is evident that a pronounced vertical zonation of polycystins occurs in thinner zones at the surface and thicker zones at depth. Also certain species are eurybathic, whereas others are stenobathic; and within the depth range inhabited by a certain species juvenile forms are usually found at the shallowest depth.

If the results of all these works are combined, divisions in depth distributions seem to occur at approximately 50, 200, 400, and 1,000 metres in temperate and tropical waters. These breaks may be delineated by physical and chemical conditions in the following ways:

1. The 50-metre and perhaps the 200-metre faunal

breaks may correlate with either thermoclines or pycnoclines as is seen in waters off Southern California (Casey 1966). These breaks may be also influenced by light penetration, if these forms are truly dependent upon associations with symbiotic zooxanthellae.

2. The 200-metre division may correspond with the bottom of the mixed layer, the bottom of the surface water masses (water masses in the sense of Muromtsev, 1963), or the beginning of the permanent thermocline.
3. The break at about 400 metres may correspond with a water mass boundary such as Central Water on top of Intermediate Water (as occurs off Southern California), or surface water on top of Central Water, as might occur in equatorial regions.
4. The breaks deeper than the above may also correlate with water mass boundaries, even though at depth all waters approximate a single water type. Therefore, deeper boundaries may in some way be correlated to pressure, biological associations, the biologic history of the water, microconstituents of the waters, oxygen minimum zones, deep water circulation, or many other parameters, some of which may never be measured by our present sampling techniques.

Although the present author separates vertical and horizontal distributions, these are closely related, and many forms which exist in shallow waters in high latitudes submerge to occupy deeper zones at lower latitudes. However, Reshetnyak (1955) observed faunal breaks as deep as 4,000 metres in high latitude waters, and others, such as Riedel (1958) and Petrushevskaya (1966), have suggested that certain polycystins are deep dwellers. Because of these observations, the author believes there is good evidence to suggest the controls mentioned under 4 above.

HORIZONTAL DISTRIBUTION

Haeckel (1887) has determined that the richest development of forms and the greatest number of species occur in the tropics. The 'frigid zones' possess few genera and species. From the tropics the abundance of species seems to diminish regularly towards the poles, and more rapidly in the northern than in the southern hemisphere. Also, the southern hemisphere seems to possess more species than the northern. The Pacific Ocean radiolarian fauna seem to be quantitatively and qualitatively richer than that of any other ocean.

Cleve (1900) gives examples of many planktonic forms that are neritic, but none of these are polycystins. The present author can find no account of endemic, neritic polycystins.

Popofsky published a series of studies on radiolarians (1907, 1908, 1912, and 1913), in which he distinguished warm-water and cold-water species of polycystins. He further states that cold- and warm-water forms of some species are morphologically distinct. He finds several species to have a bipolar distribution and some to have different open-sea and coastal forms.

Riedel (1958) compares polycystins from Antarctic sediments with those from sediments of other oceans and finds that:

1. Certain species are endemic to the Antarctic.
2. Eight species are bipolar, cold-water types.
3. Thirteen species are apparently cosmopolitan in distribution.
4. The majority of species in shallower samples (to about 200 metres) are restricted either to Antarctic waters or have a bipolar, cold-water pattern of distribution.
5. Many of the apparently cosmopolitan species are found only in deeper samples (below 200 metres).

Hulsemann (1963) states that the shallow zone she found in the Arctic is a zone of endemic forms and the deeper zone is one of cosmopolitan forms.

Hays (1965) notes a break in the distribution of different polycystine radiolarian faunas in the Antarctic that seems to correlate with the Antarctic Convergence zone.

Clarke (1965) and Nigrini (née Clarke) (1967), investigating radiolarians in Recent pelagic sediments from the Indian and Atlantic Oceans, distinguishes warm, cool, and cold-water assemblages and finds a pronounced faunal break at the Antarctic Convergence zone.

Kling (1966) analyses the distribution of living Castanellid and Circoporid radiolarians in the eastern North Pacific Ocean, and finds species that may be correlated with Sub-Arctic, Transition, Central and Equatorial-Central zones. He relates the distributions to water masses of the region, and also notes that some forms dive with the Central and Intermediate Water Masses of the North Pacific Ocean.

Casey (1966), working on the distribution of polycystins in waters off the Southern California coast, relates local species distribution to the distribution in the north Pacific by comparing his samples with sediment samples from the north Pacific. Off the Southern California coast he finds a seasonal change in radiolarian assemblages in the upper 200 metres. During summer radiolarians from equatorial regions are dominant. During winter radiolarians from Sub-Arctic regions are dominant. At the end of the summer invasion of equatorial radiolarians, and before the winter invasion of Sub-Arctic radiolarians, a short (one month) invasion of radiolarians from the central regions of the Pacific occurs. All these invasions correlate with the local and general pattern of water mass and circulation patterns in the region. Two deeper zones are noted, one from 200 to 400 or 500 metres and another from 400 or 500 to 1,000 metres; both have characteristic polycystine assemblages. Due to the characteristic species and the hydrographic conditions observed during collection periods the 200-400 or 500-metre fauna is considered to represent forms 'endemic' to the Central Water Mass of the North Pacific Ocean that dives at the Subtropical Convergence. The deeper zone (400 or 500-1,000 metres) contains forms 'endemic' to the North Pacific Intermediate Water Mass that dives at the Polar Convergence. These findings have been substantiated by additional work on both plankton and sediment samples from the North and South Pacific, which are discussed in greater detail in the paper on 'Radiolarians as indicators of past and present water masses'. (Chapter 23, this volume, p. 331.)

Petrushevskaya (1966) states that many species are believed to be correlated with more or less definite hydrological conditions. Also some suborders, families, genera and, of course, species are correlated with tropical, temperate, and polar regions. Petrushevskaya states that tropical and temperate radiolarian faunas do not transgress the zone of the Antarctic Convergence. The Antarctic Divergence may also play a substantial role in the total quantitative distribution of radiolarians and may affect the quantitative relation between individual species. However, the Antarctic Divergence is not a boundary in their distribution. The species characteristic of the surface layer of Antarctic waters are restricted by the zone of the Antarctic Convergence.

The current systems set up perimeters for radiolarian distribution. Not only do currents separate biogeographical regions, such as a central from an equatorial and polar region, but the currents themselves contain characteristic assemblages. The author has noticed one example in the California Current system, which shows a succession of species from north to south. This succession may be due simply to the mixing of water masses and therefore of the faunas themselves, or may have some subtle basis such as the biological history of the water of the California Current itself. The obvious changes in the current system from north to south are the decrease in Transition and Subarctic forms, an increase in Equatorial forms, and individual, subspecific morphological changes (Casey 1967).

Many polycystine species are limited in a horizontal sense to the same basic biogeographical zones as are most other marine organisms; such as, polar, temperate, and tropical. However, the correlation becomes meaningful when radiolarian distributions are correlated to hydrographic conditions as Hulsemann (1963), Kling (1966), Casey (1966, 1967, and 1970), Petrushevskaya (1966, and 1967), and others have done.

OTHER DISTRIBUTIONAL PHENOMENA

Patchiness

The author knows of no study devoted to determining whether polycystins exhibit 'patchiness'. On a summer cruise in Southern California waters, the author found small, elongated, cylindrical colonies of collosphaerid radiolarians floating in animal streams on the water. This patchiness is easy to explain as the colonies were subjected to the dynamics of an air-water interface and were undoubtedly passively collected into patches much in the same manner as some water slicks are.

Density

Haeckel (1887), Popofsky (1908, 1912, 1913), and Petrushevskaya (1966) find species of polycystins to be very dense in plankton of equatorial waters. In the central and polar areas, Petrushevskaya (1966) notes a decrease in densities as compared to her samples from the tropical Indian Ocean. She records densities of 15,000 radiolarians per cubic metre at 100 metres, and above and below this depth the density diminishes rapidly. About 10° north of this station (in the Pacific) she finds densities of 6,000 per cubic metre at 100 metres, 2,000-3,000 per cubic metre down to about 500 metres, and about 1,000 per cubic metre below 500 metres. In the Antarctic, Petrushevskaya states that radiolarians probably occur in densities of tens or hundreds per cubic metre.

Casey (1966) calculated the following densities off Southern California in April 1963: 280 per cubic metre from 0 to 100 metres, 190 per cubic metre from 100 to 500 metres, and 33 per cubic metre from 500 to 1,000 metres. In August 1963, the following densities were found: 74 per cubic metre for 0-25 metres, 120 per cubic metre from 25 to 50 metres, 380 per cubic metre from 50 to 123 metres, 250 per cubic metre from 123 to 194 metres and 29 per cubic metre from 480 to 960 metres. These values are much lower than those of Petrushevskaya, probably because the author's data are from net tows while Petrushevskaya's are taken by filtering a known amount of sea water and therefore may be closer to actual densities in the plankton.

Evidence from previous work suggests that the greatest densities of polycystine radiolarians occur at about 100 metres in temperate and tropical waters and that densities drop rapidly below 500 metres. For Antarctic waters, Petrushevskaya (1967) states that the greatest densities are between 150 and 400 metres.

Following this brief review of the distribution of polycystins, the author proposes biogeographic zones applicable to these organisms.

PROPOSED BIOGEOGRAPHIC ZONES SUITABLE FOR USE WITH POLYCYSTINS

The author (Casey 1970) examined sediment and plankton samples from the North and South Pacific. From the sediment studies the following major zones are found from north to south: Subarctic, Transition, North Central, Equatorial, South Central, Subantarctic, and Antarctic. Using both plankton and sediment samples the following faunal zones may be distinguished: Subarctic shallow (probably a Subarctic-Arctic shallow), Transition shallow, North Central shallow, Equatorial South Central shallow, Subantarctic shallow, and Antarctic shallow. All shallow zones extend from the surface to 100 or 200 metres, and perhaps to as much as 400 metres in the lower latitudes. These zones correlate fairly well with Muromtsev's surface water masses (Muromtsev 1963). The shallow radiolarian faunal zones also seem to correlate well with the following hydrographic conditions:

1. Subarctic faunal zone: waters north of the North Pacific Drift and the Subarctic Convergence (also sometimes termed Arctic or Polar Convergence).
2. Transition faunal zone: the North Pacific Drift waters bounded on the north by the Subarctic Convergence and on the south by the Subtropical Convergence.
3. North Central 'shallow' faunal zone: waters within the large anticyclonic circulation pattern of the North Pacific, which could be divided easily into two parts (east and west) as the circulation and Central Water Masses are.
4. Equatorial faunal zone: the regions occupied by the North and South Equatorial Current systems.
5. South Central 'shallow' faunal zone: waters within the large anticyclonic circulation pattern of the South Pacific, which could be divided easily into two parts (east and west) as the circulation and Central Water Masses are.
6. Subantarctic faunal zone: waters bounded to the north by the Subtropical Convergence (the Subantarctic Convergence) and to the south by the Polar Convergence (the Antarctic Convergence).
7. Antarctic faunal zone: waters bounded by the Polar Convergence on the north and the Antarctic Continent on the south.

Within these broad zones mentioned are subzones, which are shown as large dots on Fig. 7.1, and which usually occur in regions of relative oceanographic stability. Contained in these subzones are not only faunas characteristic of the major zone, but also different densities of species, and species that are absent or rare in the balance of the major zone (Casey 1970).

Below the layers of shallow zones (down to 200 or 400 metres in some regions) other zones exist, which probably correlate well with the distribution of water masses. Although many species are characteristic, indeed endemic to certain water masses, many others are not; for example, forms which seem to be eurybathic or deep stenobathic. However, those forms which are endemic to water masses below the shallow zones may be classified as follows: the North Transition-Central fauna, which is endemic to waters of the North Pacific Central Water Mass (or masses, east and west), is shallow at the area of formation and dives with the water mass at the North Pacific Subtropical Convergence; the Subarctic-Intermediate fauna endemic to waters of the North Pacific Intermediate Water Mass, which is shallow north of the Subarctic Convergence and dives with the waters at the Convergence (Casey 1970). There may also be faunas associated with Common Water and Antarctic Bottom water, although the author has no conclusive evidence for this. This scheme of radiolarian distribution for the Pacific is illustrated in Fig. 7.1.

There appears to be sufficient evidence to propose a distributional scheme for polycystins that may be applicable to the World Ocean and the adjacent seas, as follows:

I. Shallow Water Faunal Zones (approximately 0-200 metres)
  1. Polar
     (1) Subarctic
     (2) Antarctic
  2. Subpolar
     (1) Transition (North Pacific and North Atlantic)
     (2) Subantarctic
  3. Central
     (1) North (may be divided into east and west where appropriate)
     (2) South (may be divided into east and west where appropriate)
  4. Equatorial
  5. Special (adjacent seas such as Arctic and Mediterranean shallow water faunas)

II. Deep Water Faunal Zones (diving below or existing below Shallow Water Faunas)
  1. Central
     (1) Transition-Central
     (2) Subantarctic-Central
  2. Intermediate
     (1) Subarctic-Intermediate
     (2) Antarctic-Intermediate

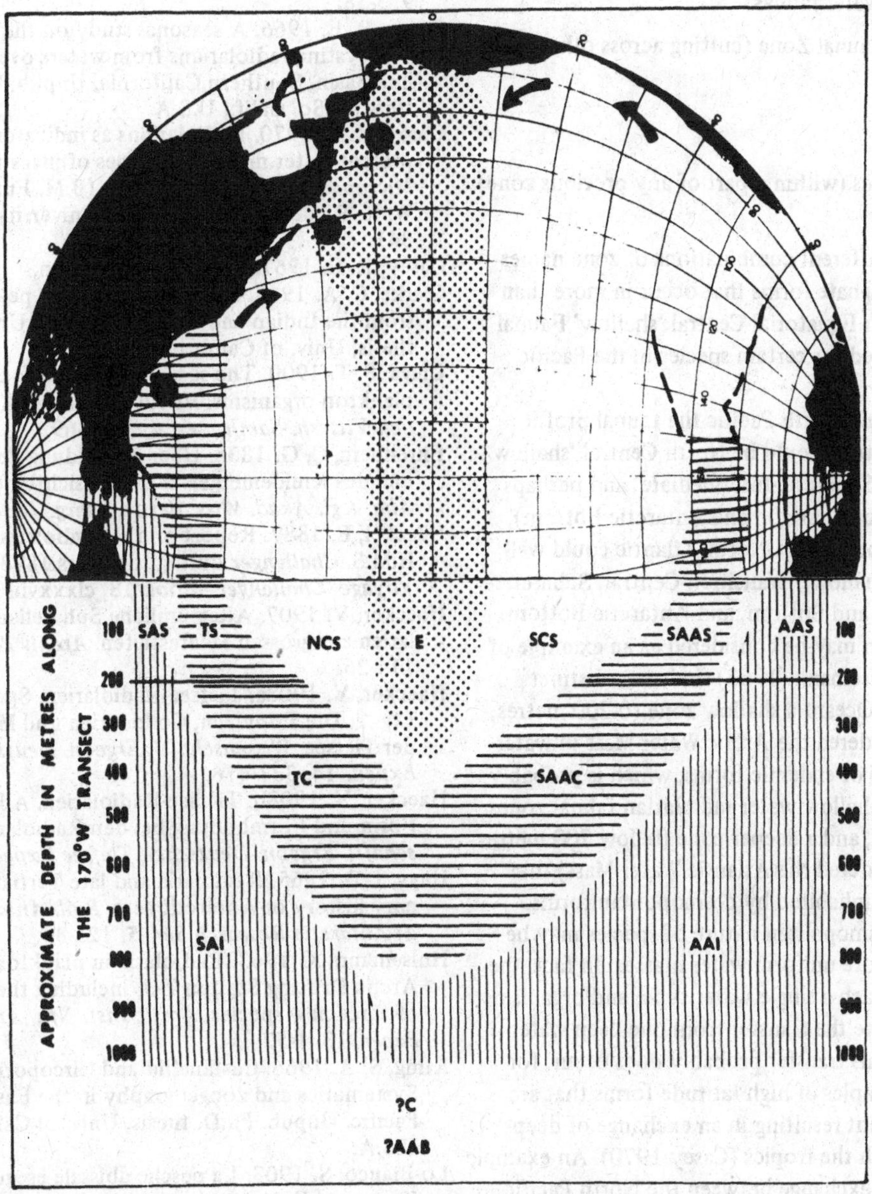
APPROXIMATE DEPTH IN METRES ALONG THE 170°W TRANSECT
100
200
300
400
500
600
700
800
900
1000
SAS
TS
NCS
E
SCS
SAAS
AAS
TC
SAAC
SAI
AAI
?C
?AAB

Fig. 7.1. Proposed biogeographic zones for polycystine radiolarians in the Pacific Ocean

AAS – Antarctic Shallow Faunal Zone
AAI – Antarctic-Intermediate Faunal Zone
SAAS – Subantarctic Shallow Faunal Zone
SAAC – Subantarctic-Central Faunal Zone
SCS – South Central Shallow Faunal Zone
E – Equatorial Faunal Zone
NCS – North Central Shallow Faunal Zone
TC – Transition-Central Faunal Zone
TS – Transition Shallow Faunal Zone
SAI – Subarctic-Intermediate Faunal Zone
SAS – Subarctic Shallow Faunal Zone

(Perhaps a C – Common Faunal Zone may exist below the Intermediate Zones, and an AAB – Antarctic Bottom Faunal Zone below the Common Zone. The large dots represent subzones mentioned in the text.

3. Also Common, Deep and Bottom, and Bottom Faunal Zones (in oceans where appropriate)
4. Special (adjacent seas, such as Arctic and Mediterranean deep-water faunas)

III. Cosmopolitan Faunal Zone (cutting across other faunal zones)
1. Shallow
2. Deep

IV. Narrow Endemics (within a part of any previous zone)

In this scheme different combinations of zone names may be used to designate forms that occur in more than one zone, such as an Equatorial-Central 'shallow' Faunal Zone, previously used for certain species in the Pacific (Casey 1970).

Whereas in the mid-North Pacific the faunal profile from surface to bottom would be North Central 'shallow', Transition-Central, Subarctic-Intermediate, and perhaps Antarctic Bottom (or Common and Antarctic Bottom), the faunal profile for the mid-North Atlantic could well be North Central 'shallow', Transition-Central, Subarctic-Intermediate, Deep and Bottom, and Antarctic Bottom.

The Arctic Ocean may be considered as an example of an adjacent sea. Hulsemann (1963) finds two distinct zones in the Arctic Ocean: a shallow zone (0-200 metres, which may be considered the Arctic Water Mass in water-mass classifications) of endemic forms, which is probably similar to endemic shallow water radiolarian faunal zones of the major oceans, and a deeper zone (below 200 metres, which may be considered the Atlantic Water Mass) that she considers to be inhabited by cosmopolitan forms.

At depth the cosmopolitan nature of species may be related to fewer, more uniform water masses. In fact, the water masses approach a single water type, and organisms should easily tolerate the transfer from one deep water mass to another. This transfer probably does occur, for there are good examples of high latitude forms that are tropically submergent resulting in an exchange of deep-water forms beneath the tropics (Casey 1970). An example of this might be an exchange between the North Pacific Intermediate Water and the Antarctic Intermediate Water faunas.

In conclusion, the author believes that radiolarians are sensitive to minute physical and chemical conditions in the ocean and may be used as indicators of many such conditions. Their potential as indicators is enhanced by the large number of species, a tendency toward endemism, and subspecific morphological changes that may be correlated with changes in oceanographic conditions.

Received January 1968.

BIBLIOGRAPHY

Bernstein, T. 1934. Zooplankton des nordlichen Teiles des Kareschen Meeres. *Trudy Arktichesk. Inst., Leningrad,* **9**, 3-58.

Casey, R. E. 1966. A seasonal study on the distribution of polycystine radiolarians from waters overlying the Catalina Basin, Southern California. Unpub. Ph.D. thesis, Univ. of So. Calif., U.S.A.

Casey, R. E. 1970. Radiolarians as indicators of past and present water masses – a series of investigations. In *The Micropalaeontology of Oceans* (B.M. Funnell and W. R. Riedel *eds.*), Cambridge University Press. pp.331-341.

Casey, R. E. 1967. Paper in preparation.

Clarke, C. A. 1965. Radiolaria in recent pelagic sediments from the Indian and Atlantic Oceans. Unpub. Ph.D. thesis, Univ. of Cambridge. England.

Cleve, P. T. 1900. The seasonal distribution of Atlantic plankton organisms. *Handl. Göteborgs Kgl. Vettensk-och Vitterh.-Samh., ser. 4,* 3, 1-369.

Ehrenberg, C. G. 1838. Über die Bildung der Kreidefelsen und des Kreidemergels durch unsichtbare Organismen. *Abh. Kgl. Akad. Wiss. Berlin, Jahrg. 1838,* 59-147.

Haeckel, E. 1887. Report on the Radiolaria collected by H.M.S. *Challenger* during the years 1873-76. *Rept. Voyage 'Challenger', Zool.* **18**, clxxxviii + 1803 pp.

Haecker, V. 1907. Altertümliche Spharellarien und Crytellarien aus grossen Meerestiefen. *Archiv Protistenk.* **10**, 114-26.

Haecker, V. 1908a. Tiefsee-Radiolarien. Spezieller Teil, Lfg. 2. Die Tripyleen, Collodarien und Mikroradiolarien der Tiefsee. *Wissenschaftl. Ergebn. Deutschen Tiefsee Exped.* **14**, 337-476.

Haecker, V. 1908b. Tiefsee-Radiolarien. Allgemeiner Teil. Form und Formbildung bei den Radiolarien. *Wissenschaftl. Ergebn. Deutschen Tiefsee Exped.* **14**, 477-706.

Hays, J. D. 1965. Radiolaria and late Tertiary and Quaternary history of Antarctic seas. *Biol. Antarctic Seas* (vol. 2) *Antarctic Research Ser.* **5**, 125-84.

Hulsemann, K. 1963. Radiolaria in plankton from the Arctic Drifting Station *T-3,* including the description of three new species. *Arctic Inst. Nth. Amer., Tech. Paper,* **13**, 1-52.

Kling, S. A. 1966. Castanellid and Circoporid Radiolarians: Systematics and zoogeography in the Eastern North Pacific. Unpub. Ph.D. thesis, Univ. of Calif., San Diego, U.S.A.

Lo-Bianco, S. 1903. La pesche abissale eseguite da F. A. Krupp col Yacht *Puritan* nelle adiacenze di Capri ed in altre localita del Mediterraneo. *Mitt. Zool. Sta. Neapol,* **16** (1-2), 109-279.

Mielck, W. 1913. Résumé des observations sur le plankton des mers explorées par le *Conseil* pendant les années 1902-1908: Radiolaria. *Conseil Permanent Internat. pur l'Exploration de la Mer, Bull. Trimest.* **3**, 303-402.

Muromtsev, A. M. 1963. *The principal hydrological features of the Pacific Ocean.* Israel Prog. for Sci. Transl. 417 pp.

Nigrini, C. 1967. Radiolaria in pelagic sediments from the Indian and Atlantic Oceans. *Bull. Scripps Instn. Oceanogr.* **11**, 1-106.

Orlov, Yu. A. 1962. *Fundamentals of palaeontology, a*

manual for palaeontologists and geologists of the U.S.S.R., gen. pt. Protozoa. Israel Prog. for Sci. Transl., 728 pp.

Petrushevskaya, M. G. 1966. Radiolyarii v planktone i v donnykh osadkakh. In *Geokhimiya Kremnesema,* 219-45. Nauka: Moscow.

Petrushevskaya, M. G. 1967. Nassellaria antarkticheskoe oblasti (Antarctic spumelline and nasseline radiolarians). *Issled. Fauny Morei* **4** (12) (*Rez. biol. Issled. sov. antarkt. Eksped. 1955-58*), 3, 5-186.

Popofsky, A. 1907. Neue Radiolarien der deutschen Sudpolar-Expedition. *Zool. Anz.* **31** (23), 697-705.

Popofsky, A. 1908. Die Radiolarien der Antarktis. *Dt. Sudpol. Exped. 1901-1903,* **10** (3), 183-305.

Popofsky, A. 1912. Die Sphaerellarien des Warmwassergebietes. *Dt. Sudpol. Exped. 1901-1903,* **13**, (2), 73-159.

Popofsky, A. 1913. Die Nassellarien des Warmwassergebietes. *Dt. Sudpol. Exped. 1901-1903,* **14**, 217-416.

Reshetnjak, V. V. 1955. Vertikalnoe raspredelenie radiolyarii Kurilo-Kamchatskoi vpadiny. *Trudy Zool. Inst.,* Akad. Nauk U.S.S.R. 21, 94-101.

Riedel, W. R. 1958. Radiolaria in Antarctic sediments. *Repts. B.A.N.Z. Antarctic Research Exped., ser. B,* **6** (10), 217-55.

Riedel, W. R. 1967. *An annotated and indexed bibliography of polycystine radiolaria.* Privately printed by W. R. Riedel, Scripp's Institution of Oceanography, La Jolla, California, U.S.A. 220 pp.

Schewiakoff, W. 1926. Acantharia: *Fauna e Flora del Golfo di Napoli. Monogr.* **37**, 755 pp.

Wolfenden, R. N. 1905. Biscayan plankton collected during a cruise of H.M.S. *Research,* 1900. Part VI, the colloid Radiolaria. *Trans. Linnean Soc., London,* **10**, 131-5.

8. DISTRIBUTION OF SHELL-BEARING PTEROPODS IN THE OCEANS

CHIN CHEN

ABSTRACT

Shell-bearing pteropods are a major group of holoplanktonic gastropods. They are widely distributed in the open ocean, from polar to tropical regions. Most species inhabit the upper 500 metres while a few live in deeper waters.

Several pteropod species are characteristic of the following regions or currents: *Limacina helicina* occurs mainly in the polar regions, and *Clio sulcata* is found in the south polar area. *Limacina retroversa* is mainly found in the Labrador Current and Subarctic waters, whereas *Clio antarctica* lives in the south subpolar area. *Clio pyramidata*, with several subspecies, has been reported as a cosmopolitan species, but its highest population density is found in the North Atlantic temperate region. *Limacina inflata* is characteristic of the Sargasso Sea, while *L. bulimoides* is abundant in the Benguela Current. All species of *Cavolina* have their main distribution in the equatorial region. *Creseis acicula* ranks first in the shell-bearing pteropod population in the Gulf of Mexico. *Limacina trochiformis* and *Creseis conica* are major pteropod species in the Gulf Stream.

There are seasonal fluctuations of pteropod species in Bermuda waters. The tropical species, *Limacina trochiformis, Creseis acicula,* and *C. conica,* are abundant in the upper 50 metres during the summer months, whereas the Sargasso species, *Limacina inflata,* is present at a depth of 50-500 metres. During the rest of the year, *L. inflata* is dominant in the upper 500 metres.

There are two currents within a radius of 100 nautical miles off Cape Hatteras. The southward movement of the Labrador Current along the continental slope meets the northeastward movement of the Gulf Stream at about 35°-36° N. In summer, a thin, warm, surface layer (bearing very few specimens of Gulf Stream pteropod species) overlies thick, cold, low-salinity slope water (carrying very abundant Labrador pteropod species) in the area of 35.5° to 37° N. In winter, the cold Labrador Current surfaces at about 36° N. The vertical slope of the isotherms indicates the area separating warm and cold pteropod species.

There are four common shell-bearing pteropod species in the Antarctic. *Limacina helicina* and *Clio sulcata* are mainly distributed south of the Antarctic Convergence, whereas *Limacina retroversa* and *Clio antarctica* are mainly found north of it. These two groups of species are mixed in the area of the Convergence, indicating a transition zone between the two faunal groups. The degree of faunal mixing is similar to the mixture of the water masses.

Received September 1967.

BIBLIOGRAPHY

Chen, C. 1966. Calcareous zooplankton in the Scotia Sea and Drake Passage. *Nature, Lond.* **212**, 678-81.

Chen, C. and Bé, A. 1964. Seasonal distributions of euthecosomatous pteropods in the surface waters of five stations in the western North Atlantic. *Bull. of Mar. Sci. of Gulf and Car.* **14**, 185-220.

Leal-Rodriguez, D. G. 1965. Distribución de pteropodes en Veracruz. *Ver. Anales del Instituto de Biologia,* **36**, 249-51.

Massey, A. L. 1909. The Pteropoda and Heteropoda of the coasts of Ireland. *Fisheries Ireland, Sci. Invest. 1907,* **2**, 1-52.

Tesch, J. J. 1946. The thecosomatous pteropods. I. Atlantic. *Dana Rep.* **28**, 1-82.

9. DISTRIBUTION OF OSTRACODES IN THE OCEANS

H. S. PURI

Summary: Ostracodes live in an environment which is controlled by temperature, bottom topography, transparency, salinity, hydrogen ion concentration, alkalinity, dissolved oxygen and depth. Distinct depth assemblages, which are modified by salinity, food supply, bottom and turbidity currents, can be universally recognized. These assemblages are limnetic, (mixo-) oligohaline, (mixo-) mesohaline, (mixo-) polyhaline, euhaline, inner neritic, outer neritic, bathyal and abyssal.

There is a striking relationship between faunal distribution patterns and mean grain size and organic carbon content of the sediments. Both surface and bottom temperature affect distribution patterns of the various assemblages.

INTRODUCTION

Ostracodes, as mentioned above, live in an environment in which the controlling factors are temperature, bottom topography, depth transparency and bottom currents, salinity, hydrogen ion concentration (pH), alkalinity, total phosphate and dissolved oxygen (O_2). Food supply and gross relationships with other life also affect total populations. Striking relationships have been observed between faunal distribution patterns and mean grain size, organic carbon content of sediments and temperature. Salinity and density are the most important factors which affect ostracodes in the oceans. Of the biological factors, food supply and nature of the substrate (algae, sea grasses, sponges, etc.) seem to have a limiting effect on the distribution of various ostracode communities. Other factors that play an important role in zoogeography are palaeogeography, dispersal ability, reproduction, evolution, morphological adaptation to suitable habitats, population structure of the communities and effectiveness of barriers.

ECOLOGICAL FACTORS

Salinity

Salinity has a profound effect on the distribution of ostracodes. This effect is most significant in Mixohaline (salinity 0.5 to ± 30‰) waters as is amply shown in classic studies by Elofson (1941) in the Baltic Sea and the Skagerrak, by Wagner (1957) in the North Sea, and by Swain (1955) in the San Antonia Bay area in the Gulf of Mexico. In the North Sea, five distinct assemblages can be recognized from the work of Wagner (1957): i.e. Limnetic (fresh water, salinity range 0.2-0.5‰), (Mixo-) oligohaline (salinity range 0.2-0.5‰, 2-3‰), (Mixo-) mesohaline A (salinity range 2-10‰), (Mixo-) mesohaline B (salinity range > 10‰), (Mixo-) polyhaline (salinity range > 17‰), Euhaline (salinity range > 32‰).

Salinity has a direct effect on the number of species in marine and marginal environments. Generally, the number of species increases with increasing salinity as is shown by Elofson (1941). The greatest increase in the Baltic Sea and Skagerrak occurs at salinity range between 2 and 10‰ (18 species) to 10-17‰ (42 species), and the number of species gradually rises to 78 with the increase in salinity to euhaline at 33‰. With the decrease in the number of species from euhaline to mixohaline waters, the number of individuals progressively increases. This phenomenon is world wide.

Temperature

Water temperature varies with latitude and climate. A 'sensitive phase' of temperature requirement during reproduction of ostracodes may affect sex ratios. The occurrence of Boreal species in the Arctic region could be attributed to the warmer temperature of the Gulf Stream. Studies of the temperature tolerances of individual species under controlled conditions are needed.

Substrate

The nature of bottom sediments, grain size, sorting coefficient, presence or absence of bottom vegetation, has a considerable effect on the distribution of ostracodes.

Depth

There seems to be a definite relationship between depth and ostracode distribution although many factors, such as temperature, salinity, density and substrate, which are closely related to depth, may influence the distribution of ostracodes. In the marine environment progressive increase in depth is accompanied by an increase in salinity, density, fine-grained sediments, and a decrease in bottom temperature. Thus, in an ideal situation, a shallow depth (0-30 metres) assemblage will coincide with a (mixo-) mesohaline (salinity range > 10‰) eurythermal assemblage. This situation is reported from the Skagerrak (see Wagner 1957, fig. 3) where a 10-125 metre assemblage is restricted to a substrate which consists of a mixture of sand-mud and coarse sand in (mixo-) polyhaline (salinity range > 17‰) water, which is also eurythermal. Another depth assemblage (125-700 metres) lives on a substrate of mud (and mixture of mud and slight sand) in stenohaline (salinity > 32‰) water which is also stenothermal.

Since increase in depth of water also corresponds to an increase in pressure, pressure itself could be the limiting ecological factor. There are however certain species which are eurybathic.

Organic carbon

Sand-mud mixtures usually contain a higher percentage of organic carbon than clean sands. Off the west coast of Florida offshore areas composed of a clean sand substrate support few ostracodes and are generally regions of low productivity. Sand-mud mixtures, however, support a greater number of benthonic ostracodes; this coincides with an increased amount of organic carbon and is considered to be a primary controlling factor (Hulings and Puri 1964). This also holds true for the Gulf of Naples.

Palaeogeography and evolution

In order to have a clear concept of the distribution of modern ostracodes, it is necessary to have some understanding of palaeogeography. The nature and location of ancient seas have affected the distribution of the modern fauna inasmuch as all modern species are descendants of the Miocene, Pliocene and Pleistocene faunas. Relict faunas have been reported from the Caspian Sea, the Sea of Azov and the Black Sea, where ancient euhaline species became acclimatized to living in either ultrahaline or mixohaline waters. In the Mediterranean, colder Boreal species have adapted to the waters of the Gulf of Naples.

In the Gulf of Mexico over 50 per cent of the Miocene, Pliocene and Pleistocene species persist into the Recent. Such a close relationship with ancient faunas makes it necessary to understand the evolution, palaeoecology and distribution of these long-ranging species. For example, the Tamiami (upper Miocene) and Caloosahatchee (Pleistocene) faunas south of Lake Okeechobee in Florida are more closely related to the modern Caribbean fauna, while the fauna from sediments of similar ages in western Florida and the western Gulf are closely related to the Gulf of Mexico. Such a relationship becomes clearer when it is realized that during the Miocene the Gulf of Mexico waters communicated with the North Atlantic through the Suwannee Straits. This fact also explains the occurrence of the cosmopolitan deep water North Atlantic fauna in the Gulf of Mexico.

Dispersal and migration

The distribution of marine ostracodes is accomplished by two means of dispersal. Active dispersal is carried out by planktonic forms which are mostly either swimmers or floaters. Since benthonic ostracodes either creep or crawl and are not equipped with active means of locomotion, they are distributed by passive dispersal. In oligohaline (and fresh water) ostracodes passive distribution of eggs and larvae can be carried out by wind, drifting vegetation, birds, tides, storms, etc. The wide distribution of marginal marine ostracodes can be attributed to tides, strong winds, storms and currents. Tides have a significant effect in estuaries, lagoons which have an open connection with the sea, and tidal streams. High salinity wedges carry with them euhaline species which may adapt themselves to a mixohaline environment. The boundaries of such a mixohaline environment are in a constant state of flux and adjustment, and the faunas which live under such rigorous circumstances are morphologically better equipped to expand to nearby estuaries, lagoons, tidal rivers and bays. This perhaps accounts for the very wide distribution of relatively uniform mixohaline and oligohaline assemblages.

Birds are mentioned by Klie (1926) as agents of passive dispersal. The very wide distribution of certain species of *Cyprideis* in the Americas coincides with the paths of migratory waterfowl, and shore and other birds are considered to be significant agents of passive distribution. (Sandberg 1964).

There is correlation between the mode of reproduction (parthenogenetic or amphigonic) and the distribution of ostracodes. Marine species are sometimes parthenogenetic although amphigonic populations are prevalent. Hartmann (personal communication 1965) found several thousand females of *Callistocythe fischeri* and only one male during an entire year in Chile.

Bottom and ocean currents play important roles in the distribution of ostracodes. In the Gulf of Naples, nearshore shallow water ostracodes are believed to have drifted by currents to the deeper offshore environment. In the North

Atlantic, the distribution of *Krithe bartonensis* auctt. coincides with the course of the Gulf Stream and the North Atlantic Drift (Neale 1964). Similarly the occurrence of a boreal fauna of *Philomedes brenda* (Baird), *Heterocyprideis sorbyana* (Jones), *Eucytheridea papillosa* (Bosquet), *Echinocythereis ? septentrionalis* (Brady), and *'Cythereis' dunelmensis* (Norman) in the Arctic Bay of Onega on the White Sea (Akatova 1957) could be attributed to the effect of the eastern branch of the Gulf Stream. Similarly both the Jutland and Baltic currents affect the coastal fauna of the Skagerrak region, where a branch of the high salinity (30-34‰) Jutland Current enters the Baltic and meets the low salinity (10-15‰) Baltic Current. This causes extensive layering and considerable fluctuations in the water. In the Baltic Current, the fauna is mixohaline and eurythermal, and in the deeper parts of the Skagerrak region it is euhaline and stenothermal (Ekman 1953, p. 106; Elofson 1941).

Reproduction and sex ratios

Most marine and marginal marine species are amphigonic, the males are, however, much scarcer than the females. There is no published record of a truly marine parthenogenetic population in modern ostracodes, although it is known that a large number of species have been described only from the female populations. Consequently computation of male-female ratios in modern marine ostracodes is meaningless. (However, in recent studies by Pokorny (1961, 1964), evidence has been presented which favours this in the Trachyleberididae during the Cretaceous.) This poses some very difficult problems in population analysis in modern ostracodes.

The factors that control sex ratios and sex determination in ostracodes are unknown. Studies on other Crustaceans have established that sex and sex ratios are largely a function of temperature and to a lesser extent of salinity. In the case of *Gammarus duebani* (Kinne 1961) other environmental factors may be responsible. There is a 'sensitive phase' before oviposition during which temperature seems to control sex. In the case of *G. duebeni*, 'females kept at a constant salinity of 10 ± 0.4‰ and under normal annual temperature fluctuation . . . produced males if the temperature during "sensitive phase" happened to be below 5° C; they produced females if the temperature happened to be above 6° C and mixed broods if the temperature happened to be between 5° and 6° C' (Kinne 1961). Unless experimental work is done under controlled conditions to determine the factors that control sex in ostracodes, the sex ratios cannot be applied to modern marine populations with any degree of accuracy.

Adaptive morphology

There is a relationship between particle size of the substrate and the ornamentation of the ostracode carapace. Coarsely ornamented forms usually live among large sand grains. Thick, heavy, highly ornamented, large carapaces and genera like *Carinocythereis, Costa, Mutilus, Quadracythere, Cytheridea* and *Urocythereis* characterize nearshore, coarse-grained sediments. Ventral flattening of the carapace in *Microcythere* is attributed to ease of climbing on to sand grains. In areas affected by strong currents, the animal has a tendency to develop a stronger and more reticulate carapace. Several genera such as *Bosquetina, Pterygocythereis, Cytherura* and *Cytheropteron*, have lateral spines which aid in supporting the carapace on soft, fine-grained sediments. Ostracodes with smooth or subdued ornamentation are generally found living on fine sediments. Most of the species which live in mud have feather-like bristles on the limbs as well as the antennae which assist the animal's forward motion on the mud surface.

Genera like *Polycope* and *Parvocythere* live in the interstitial water of small cavities and are adapted morphologically to this habitat by reduction of both internal and external structures. Such adaptation in *Parvocythere* seems to be independent of salinity, since two species of this genus live in two different hemispheres, one in a salinity of 1-20‰ and the other (*P. hartmanni*) in 17-18‰ (Marinov 1964, p. 86). Genera like *Paradoxostoma* and *Paracytherois* have developed lips with a sucking disk in order to feed on vegetation. Planktonic forms, such as species of *Conchoecia* and *Archiconchoecia*, have thin carapaces and some of the appendages are furnished with long natatory bristles. Other adaptations to this mode of life include long processes on the carapace (e.g. *Thaumatocypris* and several species of *Conchoecia*). The development of the rostral incisure in some forms is also an adaptation to planktonic life.

Population structure

Ostracode population densities at the sediment-water interface have been studied in detail in some areas. They generally show some correlation with the relative size of the individuals. The lowest densities (1-50), in the Gulf of Naples, are composed of individuals with large carapaces which inhabit a coarse-grained substrate. The largest densities (500-1,000) are generally composed of individuals of much smaller size, and their distribution pattern coincides with areas of either fine-grained substrate or increased vegetation; this effect may well be due to an increased food supply.

Size of the standing crop of ostracodes is also influenced by the biotope. A marine biotope, with many ecological niches, supports relatively small populations consisting of many species. Brackish-water biotopes, which are subject to daily and/or seasonal fluctuations in salinity, support dense populations of the small number of species which are able to survive this rigorous environment. Ostracode individuals which live under such conditions are very large (e.g. *Cyprideis*), compared with individuals of *Cytherura*, living in a marine biotope such as *Amphioxus.*

Food supply

Food supply is one of the most important factors in the distribution of ostracodes, but very little is known about feeding habits in marine benthonic ostracodes. Genera like *Paradoxostoma* feed on *Posidonia* and algae and have sucking discs in their lips; species of *Cytherella* are filter feeders and some species of *Cypridina* and *Entocythere* are parasitic and live on the gills of fishes. Elofson (1941) mentioned that algae and diatoms were common in the intestines of forty-five species; while mud-burrowing forms contained some Foraminifera and bristles of polychaetes. The effect of the availability, nature and amount of food at the sediment-water interface, and its relationship to the productivity of benthonic ostracodes is essential to a clearer understanding of their distribution patterns. Swain (1955) thought that variations in ostracode populations in San Antonio Bay were affected by food supply. This is also true in the Gulf of Naples, where the largest total populations are associated with calcareous algae, *Posidonia* and other forms of vegetation.

DEPTH ZONATION

Well-defined distribution patterns are noticed when ostracode assemblages are plotted against depth. This is most evident from the available data from the Gulf of Mexico and the Mediterranean – two classic areas which have been studied in much detail. Information on the zonation of ostracodes in the eastern Gulf of Mexico, from fresh water to a depth of 240 feet can be found in Benda and Puri (1962) and Benson and Coleman (1963). In the Mediterranean Sea, the marine and marginal marine environments support remarkably uniform ostracode faunas in belts parallel to the coast (Puri, Bonaduce and Gervasio 1968). In shallow water (average depth 2 metres), littoral lagoons, like Lago di Patria, where salinity seldom exceeds 13-14‰, typical brackish assemblages occur. In the Gulf of Naples itself two shallow shelf (depth up to 100 metres and three inner neritic assemblages are recognized (Puri, Bonaduce and Malloy 1964); this data is summarized on Fig. 9.1.

A distinct *bathyal* fauna occurs in the Mediterranean. The following is a list of the most common forms present between depths of 250-800 metres: *Polycope fragilis, P. rostrata* (150-255 m), *P. maculata* (860 m), *P.* n.sp. *A.* (265-1,226 m), *Argilloecia* ? *clavata* (495-818 m), *A.* ? *caudata* (1266 m), *A. bulbifera* (265 m), *A.* ? *minor* (265-1,266 m), sp.m. (818 m), *Bairdia* ? *minor* (911 m), *B. reticulata* (1,226 m), *Sclerochilus contortus* (495-1,330 m), *S. aeguus* (281 m), *Pseudocythere* aff. *P. caudata* (598 m), *Cytherura dispar* (422-2,159 m), *C. rara* (495-1,033 m), *C.* n.sp. *E.* (281 m), *Semicytherura muelleri* (281 m), *S. paradoxa* (64-1,226 m), *S.* n.sp. *B.* (281-764 m), *Hemicytherura videns* (281-2,443 m), *Cytherois* n.sp. *A.* (911 m), *Paracytherois* n.sp. *X.* (281 m), *Paradoxostoma cylindricum* (265 m), *Microcythere gibba* (265 m), *M. hians* (265 - 1,226 m), *M.* cf. *M. nana* (265 m), *Xestoleberis* ? *rara* (133-422 m), *X. communis* (1,226 m), *X.* ? *plana* (495 m), *X.* n.sp. *A.* (265-860 m), *Loxoconcha ovulata* (495-1,225 m), *Leptocythere* n.sp. *L.* (422-764 m), *Mutilus speyeri* (281-911 m), *Carinocythereis rubra* (495 m), *'Cythereis' polygonata* (911-1,226 m), ? *Costa* n.sp. 1 (495 m), *Cytherella* n.sp. 2 (281 m), *Bythocythere* n.sp. *B.* (801 m).

Two shallow-water (less than 100 metres) forms, *Polycopsis compressa* and *Occultocythereis dohrni,* were found in association with a typical bathyal assemblage.

The following species occur in the Mediterranean 'abyssal' fauna between depths of 800-4,285 metres: ? ?*Conchoecia* sp. (2,049-2,499 m), *Polycope* n.sp. 3 (217-2,488 m), *P.* n.sp. 1 (265-2,723 m), *P.* cf. *P. dentata* (150-4,285 m), *Polycopsis* n.sp. 1 (228-2,384 m), *Erythrocypris* (228-2,975 m), *Bairdia formosa* (2,250 m), *Pseudocythere caudata* (150-2,574 m), *Hemicytherura videns gracilicosta* (1,226 m), *Paradoxostoma versicolor* (1,130-1,226 m), *P. maculatum* (2,049 m), *Loxoconcha* n.sp. *M.* (1,225 m), *Krithe* n.sp. 1 (1,226-2,550 m), *K.* (?) *similis* (885 m), *Caudites* n.sp. *A.* (1,387 m), *Bythocypris bosquetiana* (150-3,881 m), *B. obtusata* (150-2,905 m), *Argilloecia* n.sp. 1 (192-1,884 m), *Cytheropteron rotundatum* (69-3,427 m), *latum* (69-2,382), *Cytherois* n.sp. *A.* (911 m), *Paracytherois striata* (150-1,225 m), *Xestoleberis dispar* (69-4,285 m).

There is no *true* abyssal fauna in the Mediterranean inasmuch as the ornate hemicytherids and trachyleberids, which are so typical of abyssal assemblages elsewhere, are absent. Some of the above species are doubtless Archibenthic forms (200-1,000 m depth) which have become adapted to living in deeper water as the temperature of the water below 1,000 metres is about 13° C, which is about 1° lower than that at a depth of 300 metres.

Nine samples of sediment cores from depths over 3,000 metres were examined from the Mediterranean Sea. Two of the cores, V10-15 (depth 3434.9 m) from the Tyrrhenian Sea, and V10-61 (depth 4,216.3 m) from a small abyssal plain southeast of Rhodes, did not contain ostracodes, although Foraminifera were noticed in both of these cores. The substrate at V10-61 consists of a clay, while V10-61 consists of *Globigerina* ooze.

Generally, the limit between bathyal and abyssal faunas

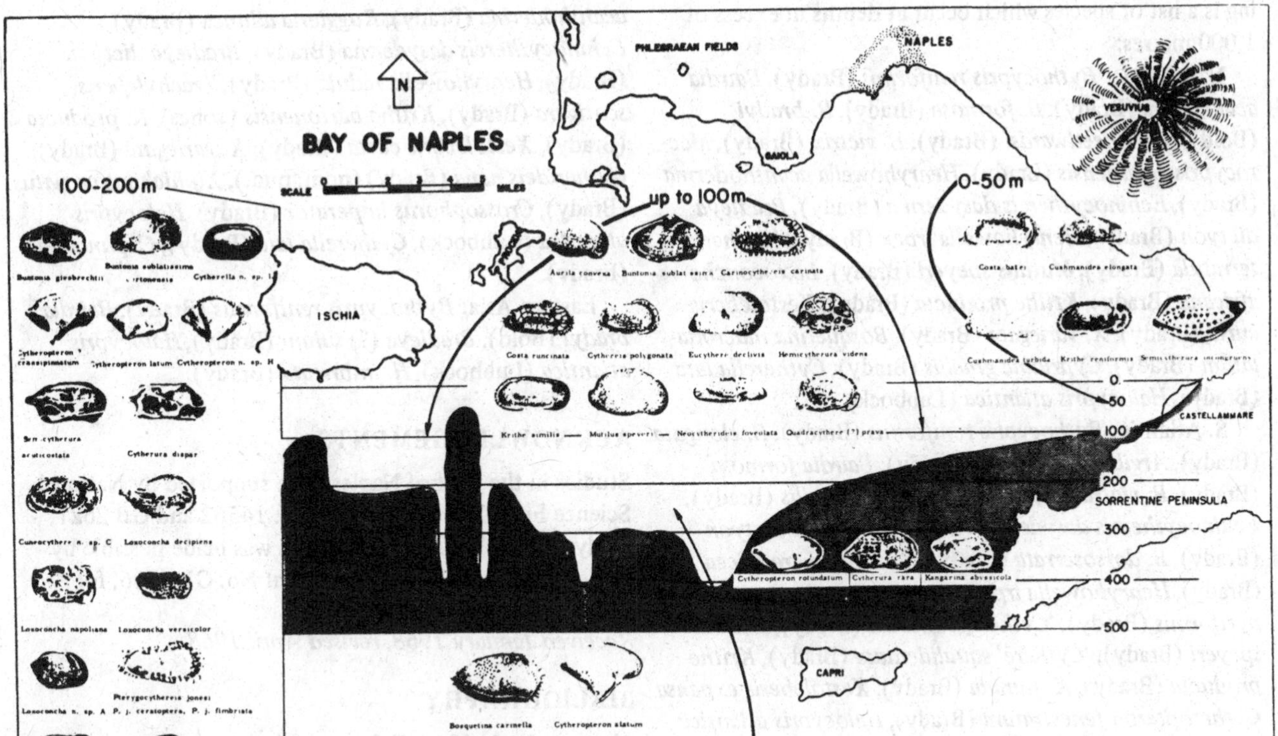

N
BAY OF NAPLES
SCALE
MILES
PHLEGRAEAN FIELDS
NAPLES
VESUVIUS
GAIOLA
ISCHIA
PROCIDA
CAPRI
CASTELLAMMARE
SORRENTINE PENINSULA
100-200m
up to 100m
0-50m
200-300m
0
50
100
200
300
400
500
Buntonia giesbrechtii
Buntonia sublatissima
Cytherella n. sp. 1
Cytheropteron latum
Cytheropteron n. sp. H
Cytherura dispar
Cuneocythere n. sp. C
Loxoconcha decipiens
Loxoconcha napoliana
Loxoconcha versicolor
Pterygocythereis jonesi
Loxoconcha n. sp. A
P. j. ceratoptera, P. j. fimbriata
Cytherura acuminata
Tetracytherura angulosa
Aurila convexa
Carinocythereis carinata
Costa runcinata
Cythereis polygonata
Eucythere declivis
Hemicytherura videns
Krithe similis
Neocytherideis foveolata
Quadracythere (?) prava
Costa batei
Cushmanidea elongata
Cushmanidea turbida
Krithe reniformis
Cytheropteron rotundatum
Cytherura rara
Kangarina abyssicola
Bosquetina carinella
Cytheropteron alatum
Henryhowella sarsi
Krithe n. sp. B
Semicytherura paradoxa

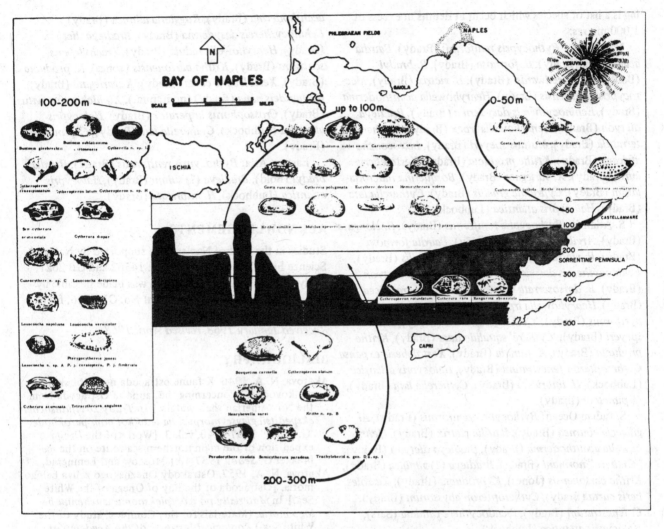

Fig. 9.1. Distribution of characteristic species with depth in Gulf of Naples (data from Puri, Bonaduce and Malloy 1964).

occurs at a depth of about 3,000 metres and a temperature of 4° C, coinciding with a great change in the benthonic fauna. However, this is not reflected in the ostracode fauna. The ostracode abyssal fauna does not appear to be of any great antiquity since it consists mainly of species of *Polycope* and *Bythocypris bosquetiana*, which also are common constituents of the bathyal fauna. Nature of the substrate, rather than depth, may be controlling the vertical distribution of the ostracode fauna since the presence of *Globigerina* ooze coincides with the occurrence of a stable abyssal fauna. Deep-sea abyssal faunas seem to be controlled by the availability of nutrients and consequently abyssal plains closer to land masses have a higher proportion of ostracode species. However, pelagic life seems to control the ostracode assemblages in the Mediterranean since a *Globigerina* ooze substrate, even at depths less than 1,000 metres (e.g. V10-6, depth 818 m), has a similar fauna to ooze at much greater depths, except that the number of species decreases progressively from ten species distributed over six genera in V10-6, 818 metres, to four species distributed over three genera in deeper waters.

The voyage of H.M.S. *Challenger* (December 1872 to May 1876), first discovered the abyssal ostracode fauna of the oceans (Brady 1880). The abyssal fauna, as described by Brady, is cosmopolitan in nature, and several species (e.g. *Bythocypris reniformis* Brady), were reported from the north and south Atlantic, South Indian Ocean, Austral-Asia and Eastern Asia. The modern concept of ostracode species has been very much refined since Brady's time and some of the forms included by him in a single species represent several species. We are at the present designating lectotypes of the *Challenger* species in the hope that it will not only clarify several taxonomic problems but also stabilize the species from the *Challenger* collection. The following is a list of species which occur at depths in excess of 1,000 metres:

N. Atlantic: *Bythocypris reniformis* (Brady), *Bairdia acanthigera* (Brady), *B. formosa* (Brady), *B. bradyi* (Bold), *B. milneedwardsi* (Brady), *B. victrix* (Brady), *Macrocypris canariensis* (Brady), *Henryhowella acanthoderma* (Brady), *Echinocythereis dasyderma* (Brady), *Bradleya dictyon* (Brady), *Henryhowella irpex* (Brady), *'Cythere' serratula* (Brady), *Mutilus speyeri* (Brady), *Loxoconcha africana* (Brady), *Krithe producta* (Brady), *Xestoleberis curta* (Brady), *X. variegata* (Brady), *Bosquetina mucronalatum* (Brady), *Cypridina gracilis* (Brady), *Cytherella lata* (Brady), *Halocypris atlantica* (Lubbock).

S. Atlantic: *Bythocypris reniformis* (Brady), *B. elongata* (Brady), *Argillaecia eburnea* (Brady), *Bairdia formosa* (Brady), *B. victrix* (Brady), *Macrocypris similis* (Brady), *Echinocythereis dasyderma* (Brady), *Bradleya dictyon* (Brady), *B. dorsoserrata* (Brady), *Schizocythere ericea* (Brady), *Henryhowella irpex* (Brady), *Quasibuntonia pyriformis* (Brady), *'Cythere' serratula* (Brady), *Mutilus speyeri* (Brady), *'Cythere' squalidentata* (Brady), *Krithe producta* (Brady), *K. tumida* (Brady), *Xestoleberis expansa* *Cytheropteron fenestratum* (Brady), *Halocypris atlantica* (Lubbock), *H. imbricata* (Brady), *Cytherella lata* (Brady), *C. punctata* (Brady).

S. Indian Ocean: *Bythocypris reniformis* (Brady), *Argillaecia eburnea* (Brady), *Bairdia victrix* (Brady), *Henryhowella acanthoderma* (Brady), *Bradleya dictyon* (Brady), *"Cythere" normani* (Brady), *Bradleya* (?) *viminea* (Brady), *Krithe bartonensis* (Jones), *K. producta* (Brady), *Xestoleberis curta* (Brady), *Cytheropteron abyssorum* (Brady), *C. fenestratum* (Brady), *Pseudocythere caudata* (Sars), *Halocypris atlantica* (Lubbock).

N. Pacific: *Bairdia abyssicola* (Brady), *B. minima* (Brady), *Henryhowella acanthoderma* (Brady), *Trachyleberis circumdentata* (Brady), *Echinocythereis dasyderma* (Brady), *Bradleya dictyon* (Brady), *B.* (?) *suhmi* (Brady), *Xestoleberis curta* (Brady), *Bosquetina mucronalatum* (Brady), *Halocypris atlantica* (Lubbock), *H. imbricata* (Brady).

S. Pacific: *Bairdia bradyi* (Bold), *B. hirsuta* (Brady), *Henryhowella acanthoderma* (Brady), *Trachyleberis circumdentata* (Brady), *Echinocythereis dasyderma* (Brady), *Bradleya dictyon* (Brady), *'Cythere' normani* (Brady), *Trachyleberis scutigera* (Brady), *Ambocythere stolonifera* (Brady), *Henryhowella sulcatoperforata* (Brady), *Krithe producta* (Brady), *Xestoleberis curta* (Brady), *Bosquetina mucronalatum* (Brady), *Halocypris atlantica* (Lubbock), *Cytherella punctata* (Brady).

Austral-Asia: *Bythocypris reniformis* (Brady), *Bairdia exaltata* (Brady), *B. formosa* (Brady), *B. bradyi* (Bold), *B. minima* (Brady), *B. victrix* (Brady), *Henryhowella acanthoderma* (Brady), *Ruggieria adunca* (Brady), *Echinocythereis dasyderma* (Brady), *Bradleya dictyon* (Brady), *Henryhowella radula* (Brady), *Trachyleberis scutigera* (Brady), *Krithe bartonensis* (Jones), *K. producta* (Brady), *Xestoleberis curta* (Brady), *X. variegata* (Brady), *Cytherideis nana* (Brady) (nom. nud.), *Xiphichilus arcuatus* (Brady), *Crossophorus imperator* (Brady), *Halocypris atlantica* (Lubbock), *Cytherella lata* (Brady), *C. punctata* (Brady).

Eastern Asia: *Bythocypris reniformis* (Brady), *Bairdia bradyi* (Bold), *Bradleya* (?) *suhmi* (Brady), *Halocypris atlantica* (Lubbock), *H. imbricata* (Brady).

ACKNOWLEDGEMENTS

Studies in the Gulf of Naples were supported by National Science Foundation Grant Nos. G. 14562 and GB 2621; study of the *Challenger* Ostracoda was made possible by National Science Foundation Grant No. GB 6706, for which gratitude is expressed.
Received January 1968, revised April 1968.

BIBLIOGRAPHY

Akatova, N. A. 1946. K faune ostracoda novosibirskogo melkovod'ia. [Concerning the fauna of Ostracoda from the Novosiberian shelf waters.] In *Trudy Dreifuinshchei ekspeditsii glavsevmorputi na ledo kol'nom parokhode* 'G. *Sedov*' 1937-1940, vol. 3. [Works of the *Dreifa* expedition of the main northern sea route on the ice-breaker *G. Sedov* 1937-40.] Moscow and Leningrad.

Akatova, N. A. 1957. Ostrakody Onezhskogo zaliva belogo moria. [Ostracoda of the Bay of Onega of the White Sea.] In *Materially po Kompleksnomu izucheniiu belogo moria.* (Materials for the 'Complex' study of the White Sea.) *Zoological Institute of the Academy of Sciences of the U.S.S.R.* Issue 1.

Benda, W. K. and Puri, H. S. 1962. The distribution of Foraminifera and Ostracoda off the Gulf Coast of the Cape Romano area, Florida. *Trans. Gulf. Coast Assoc. Geol. Soc.* 12, 303-41.

Benson, R. H. and Coleman, J. L. 1963. Recent marine ostracods from the Eastern Gulf of Mexico. *Univ. of Kansas Palaeontological Contr. Arthropoda,* Art. 2, 1-52.

Brady, G. S. 1880. Report on the Ostracoda dredged by H.M.S. *Challenger* during the years 1873-6. *Rept. Sci. Results, Voyage H.M.S. 'Challenger'. Zool.* 1 (3), 1-184, pls. 1-44.

Brady, G. S. and Norman, A. M. 1889. A monograph of the marine and freshwater Ostracoda of the North Atlantic and of North-Western Europe. Section 1, Podocopa. *Sci. Trans. Roy. Dublin Soc.* 4 (Series II), 63-270, pls. 8-23.

Brady, G. S. and Norman, A. M. 1896. A monography of the marine and freshwater Ostracoda of the North Atlantic and of North-Western Europe. Part II, Sec. II-IV, Myodocopa, Cladocopa and Platycopa. *Ibid.* 5 (Series II), 621-746, pls. 50-68.

Ekman, S. 1953. *Zoogeography of the Sea.* Sidgwick and Jackson: London. 417 pp.

Elofson, O. 1941. Zur Kenntnis der marinen Ostracoden Schwedens mit besonderer Berücksichtigung des Skageraks. *Zool. Bidr. Uppsala,* **19**, 215-534.

Emery, K. O., Heezen, B. C. and Allan, T. D. 1966. Bathymetery of the Eastern Mediterranean Sea. *Deep-Sea Research,* **13**, 173-92.

Hulings, N. C. and Puri, H. S. 1964. The ecology of shallow water ostracods of the West Coast of Florida. *Pubbl. Staz. Zool. Napoli,* **33** Suppl., 308-44.

Kinne, O. 1961. Geschlechtsbestimmung des Flohkrebses *Gammarus duebeni* Lillj. (Amphipoda) ist temperaturabhängig–eine Entgegnung. *Crustaceana,* **3**, 56-69.

Klie, W. 1926. Ostracoda. *Biologie der Tiere Deutschlands,* **22** 1-56.

Marinov, T. 1964a. Untersuchungen über die Ostracodenfauna des Schwarzen Meeres. *Kieler Meeresfors.* **20** (1), 82-91.

Marinov, T. 1964b. Beitrag zur Ostracodenfauna des Schwarzen Meeres. *Bult. Inst. cent. Rech. Sci. pisccul. et de pêcherie, Acad. Sci., Bulg.* **4**. 29-60.

McKenzie, K. G. 1963. A brackish-water ostracod fauna from Lago di Patria, near Napoli. *Ann. Inst. Mus. Zool., Univ. Napoli,* **15** (1), 1-14, 2 pls.

Muller, G. W. 1894. Die Ostracoden des Golfes von Neapel und der angrenzeden Meeres-Abschnitte. *Fauna und Flora des Golfes von Neapel.* **21**, Monographie, i-viii, 1-404, pls. 1-40.

Neale, J. W. 1964. Some factors influencing the distribution of Recent British Ostracoda. *Pubbl. Staz. Zool. Napoli,* **33**, Suppl., 247-307.

Pokorny, V. 1961. Sex ratio in fossil ostracods, a new stratigraphical tool. *Casopis Mineral. Geol.* **6**, 185-6.

Pokorny, V. 1964. Some palaeoecological problems in marine ostracod faunas, demonstrated on the upper cretaceous ostracodes of Bohemia, Czechoslovakia. *Pubbl. Staz. Zool. Napoli,* **33**, Suppl., 462-79.

Price, W. A. 1954. Shorelines and coasts of the Gulf of Mexico. *Fishery Bull.* **89**, 39-65.

Purasjoki, K. J. 1948. *Cyprilla humilus* G. O. Sars, an interesting ostracod discovery from Finland. *Comm. Biol. Sci. Fenn.* 1-7.

Puri, H. S. 1960. Recent Ostracoda from the west coast of Florida. *Gulf Coast Assoc. Geol. Soc. Trans.* **10**, 107-49.

Puri, H. S. and Benda, W. K. 1970. Distribution of Ostracoda in the Lower Boca Ciega Bay, Florida. *Fla. Geol. Survey Bull.* (in press).

Puri, H. S., Bonaduce, G. and Malloy, J. 1964. Ecology of the Gulf of Naples. *Pubbl. Staz. Zool. Napoli,* **33**, Suppl., 87-199.

Puri, H. S., Bonaduce, G. and Gervasio, A. M. 1968. Distribution of Ostracoda in the Mediterranean: *Hull Ostracoda Symposium* (in press).

Puri, H. S. and Hulings, N. C. 1957. Recent ostracod facies from Panama City to Florida Bay Area. *Trans. Gulf. Coast Assoc. Geol. Soc.* **7**, 167-90.

Ryan, W. B. Workum, F. and Hershey, J. B. 1965. Sediments on the Tyrrhenian Abyssal Plain: *Geol. Soc. American, Bull.* **76**, 1261-82.

Ryan, W. B. F. and Heezen, B. C. 1965. Ionian Sea Submarine canyons and the 1908 Messina Turbidity Current. *Bull. geol. Soc. Am.* **76**, 915-32.

Sandberg, P. 1964. Notes on some Tertiary and Recent brackish-water Ostracoda, *Publ. Stat. Zool. Napoli,* **33**, Supl., 496-514.

Swain, F. M. 1955. Ostracoda of San Antonio Bay, Texas. *Jour. Palaeontology,* **29**, 561-646.

Wagner, C. W. 1957. *Sur les ostracodes du quaternaire récent des pays-bas et leur utilisation dans l'étude géologique des dépots holocènes.* Mouton and Co. The Hague.

SECTION B

ACCUMULATION AND DISTRIBUTION OF MICROFOSSIL REMAINS IN OCEAN BOTTOM SEDIMENTS

10. DISTRIBUTION OF SILICEOUS MICROFOSSILS IN SUSPENSION AND IN BOTTOM SEDIMENTS

A. P. LISITZIN

Summary: The analysis of over 20,000 samples of suspended matter and of more than 2,000 ocean sediment samples was used as a basis for studying the quantitative distribution and qualitative composition of amorphous silica (opal). Apart from chemical analyses, thorough microscopic studies of biogenous material were made; along with the analysis of fine fractions by X-ray diffraction and the electron microscope.

Opal distribution in the surface water (0-200 metres) is controlled by plankton productivity. Three major belts of high productivity are distinguished: southern, which encompasses the globe, equatorial, which is especially clearly defined in the Pacific and Atlantic oceans, and northern, which is usually discontinuous and poorly developed. Waters with insignificant amounts of suspended silica lie between the above-mentioned belts. Practically all silica in suspension is biogenous (the remains of diatoms, radiolarians and silicoflagellates). More than 70 per cent of silica in suspension consists of diatoms; in the equatorial belt radiolarians increase in importance.

Study of opal in suspension from the ocean depths demonstrates that only 1/10-1/100 of the initial number of diatom frustules reaches the ocean bottom. The variety of species decreases with depth: finely silicified planktonic forms do not reach the bottom in a number of places. Radiolarians and silicoflagellates are much better preserved. Despite considerable loss during settling, the silica belts established for the surface suspension are reflected on the ocean bottom, giving rise to global belts of silica accumulation in sediments.

Unlike calcium carbonate, no critical depth exists for amorphous silica. Siliceous sediments are encountered at all depths down to the greatest.

The highest rates of recent silica accumulation were recorded in the Gulf of California where they constitute about 50 grammes per square centimetre per 1,000 years. The maximum rate for sea basins was observed in the Bering and Okhotsk seas (1.5-3 grammes per square centimetre per 1,000 years). Within the southern belt of silica accumulation it does not usually exceed 1-1.9 grammes per square centimetre per 1,000 years, in the tropical (arid) zones of the oceans the values drop to 0.05 and lower. It is shown that the response of silica accumulation to climatic changes may be variable. In the northern belt glaciation resulted in a sharp decrease in silica accumulation rates, in the equatorial belt it increased rates of accumulation, whereas in the southern belt rates changed very little.

The annual silica supply from land is 0.3 milliard tons. Proceeding from the primary production of organic carbon and the ratio of amorphous silica to organic carbon in suspension, the silica amounts utilized annually by organisms to construct their tests were determined to be 80-160 milliard tons per year. The total amount of a compound supplied to the ocean is approximately equal to the quantity settling to the ocean bottom per year. Therefore the amount of silica accumulating in sediments is about 0.3 milliard tons per year. Thus 1/20-1/50 of silica used by plankton for their tests near the ocean surface is buried in the bottom sediments. The major part of the silica is dissolved and enters the chemical cycle again. These data show the large scale on which recent silica accumulation proceeds in the world's oceans.

INTRODUCTION

Comprehensive studies of the quantitative distribution and the qualitative composition of silica in suspension and in bottom sediments of the world's oceans, conducted over the last few years reveal its biogenic nature. Hypotheses of chemical precipitation of silica, as well as volcanogenic or sorptive ones, which were once very popular among petrologists, have not been corroborated by recent data.

Knowing the amount of amorphous silica (opal) in bottom sediments or in suspended matter, one may judge of the role in sedimentation of different silica-accumulating organisms. Two basic methods have been suggested for opal content determinations – chemical and X-ray diffraction; both methods usually give similar results. However, they do not enable one to determine the particular rôle of various organisms in sedimentation. For studying the qualitative composition of siliceous microfossils, microscopic analysis is necessary. During the last few years much data has been obtained on the quantitative distribution and specific composition of diatoms, silicoflagellates and radiolarians in suspension from the ocean surface, and from its water column, as well as from the surface layer of bottom sediments and from sediment cores. This makes it possible not only to define the rôle of various organisms in recent silica accumulation, but also to distinguish in different parts of the ocean the most abundant and important families and species, and to identify their complexes (associations). Yet in such studies an essential gap is due to the variable resistance of skeletons of different organisms to solution in sea water and in bottom sediments. Studies of suspension from the ocean surface show, in particular, that the greater part of biogenous silica falls within the smaller than 0.01 mm fraction, which makes up usually over 70 per cent, and sometimes over 90 per cent of the total suspension. The size of diatom frustules ranges between 0.2 mm and 0.02 mm, that of radiolarians between 0.25 mm and 0.05 mm and of silicoflagellates between 0.1 mm and 0.02 mm (Lisitzin 1964). In bottom sediments the greatest part of the biogenous silica is concentrated in the pelitic ($<$ 0.01 mm) fraction. This is confirmed both by chemical analysis of separate fractions and by their electron microscopic examination. Thus a considerable, usually the major part of siliceous skeletons in suspension and in bottom sediments is in a fine-grained state, and escapes the attention of biologists and palaeontologists. Therefore, it is essential for a correct judgement of the actual role of silica-accumulating organisms that account should be taken both of the total silica content (by the application of chemical and X-ray methods) and of its microscopic identification.

METHODS AND DATA

The distribution of silica in suspension and in bottom sediments has been studied in all the main regions of the world's oceans, by the application of unified methods for data collecting and processing. In addition to the author's own materials collected mainly on cruises of the research vessels *Vityaz* and *Ob* in 1950-65, all available data from other research workers have been used.

Suspension, as we understand it, is composed of particles occurring in the water column in a suspended state and having sizes from 1 mm to 0.01μ. To study the composition of the suspension quantitatively and qualitatively, a method of membrane ultrafiltration and separation of large volumes of water was used, both from the surface and from depth. In this case the finest material contained in the water (up to 0.05-1 gramme per cubic metre) is extracted quantitatively from many tons of water. When obtaining water samples for membrane filtration 5-10 litre plastic bathometers were used, along with 200 litre bathometers and submerged pumps (Lisitzin and Glazunov 1960; Lisitzin 1960*a*). The filtration devices retain the particles larger than 0.7μ – and, as a result of clogging of the filter, also practically all the particles of the colloidal fraction. This permits the trapping of the total sea water suspension including the finest fractions of siliceous organisms (Lisitzin 1955*b*, 1956, 1959*a*, 1960*a*, 1961*a*, *b*, 1962*a*, *b*, 1964, 1966*a*). The ultrafiltration method enables one to trap quantitatively tens of milligrammes, and (using the most powerful installations) grammes of suspension. The membrane filters, after clarification with Canada balsam, can be used for microscopic counts. A number of microchemical methods has been developed for the analysis of filters; in particular, a method for opal content determination. About 20,000 suspension samples have been studied from vertical sections from more than 1,000 stations.

To obtain large amounts of suspension a separation method was used employing powerful devices processing up to 100-200 tons of water per day both *en route* and at stations. By applying this method, suspension was collected from the surface almost uninterruptedly along expedition tracks. Besides, large quantities of suspension were sampled at a number of stations with the aid of submerged pumps from depths down to 200 metres (Lisitzin and

Zhivago 1958*b*; Lisitzin 1960*b*).

Sediment samples were taken by an 'Ocean-50' bottom grab (Lisitzin and Udintsev 1952), as well as by gravity and piston corers, and a large-diameter corer (200 millimetres inner diameter). Sediment samples have been studied from more than 2,000 stations.

The principal method for determination of the amorphous silica content of suspension and sediments was a chemical method using a double extraction technique with 5 per cent sodium carbonate solution. This method is widely used in the U.S.S.R. for the analysis of ancient rocks (Ponomarev 1961). An X-ray method of opal determination was also used on a small scale (Goldberg 1958; Calvert 1966). Suspension samples were subjected to thorough microscopic examinations by micropalaeontologists to determine diatoms (Kozlova 1964; Kozlova and Muhina 1966), silicoflagellates (Kozlova and Muhina 1966), and radiolarians (Petrushevskaya 1966; Kruglikova 1966). For the Pacific bottom sediments such estimations were made by Riedel (Riedel 1959). As a result of many years' study data on suspension and sediments have been collected covering all the world's oceans from the Arctic to Antarctica, and from the ocean surface down to depths of 8,000-10,000 metres.

QUANTITATIVE DISTRIBUTION AND PRODUCTION OF SILICA IN SUSPENSION AND IN THE SURFACE WATERS OF THE WORLD'S OCEANS

Most of the organisms forming siliceous skeletons inhabit the surface (0-200 metre) layer of the ocean. Deeper, only dead frustules of these organisms and rare living radiolarians are encountered. To understand the laws governing the distribution of siliceous microfossils it is necessary to study their distributions in the surface layers of the ocean and along vertical sections down to the bottom.

The content of suspended amorphous silica (opal) in the surface (0-5 metres) water of the world's oceans, according to data from over 300 analyses, is shown in Figs. 10.1 and 10.2. Each determination, on suspension obtained by the separator, refers not to a single point station, but is an integrated sample for a section of the ship's route 200-1,000 kilometres long. Suspension collections fall mainly within the spring-summer season of the Southern Hemisphere.

When silica concentration is calculated in per cent of the dry suspension its content ranges within wide limits – from 0 to 30-35 per cent, i.e., low-siliceous (10-30 per cent amorphous silica) and siliceous (over 30 per cent) suspension is found in the surface water (Bezrukov and Lisitzin 1960).

Amorphous silica concentrations are very regularly distributed. Highly distinctive is the southern belt of siliceous suspension encompassing the entire globe, much more feebly developed are the equatorial belt and the northern belt traced by separate determinations in the Northern Pacific. The vast space between these belts is occupied by water containing less than 0.5 per cent opal in suspension.

When calculating silica concentration per volume of water from which suspension was obtained (in microgrammes per litre) some minute differences are observed due to changes in suspension concentration, yet the basic laws of enrichment of the surface water in biogenous opal appear to be the same (Fig. 10.2). Opal content of the surface water of the world's oceans, according to our determinations, range between 0.13 microgrammes per litre and 1,086 microgrammes per litre.

As a rule the amount of suspended (biogenous) silica is five to fifty times less than that of dissolved silica, and only in the Antarctic is this difference as little as three or four times. Thus the organisms succeed in converting to suspension only an insignificant part of the silicic acid (only H_4SiO_4 is assimilated).

As microscopic studies show, the most abundant siliceous organisms in suspension are diatom algae, which are at the same time the main producers of organic matter in the ocean. The pattern of quantitative distribution of diatoms in suspension, calculated in million cells per 1 gramme of suspension (Fig. 10.3), may be correlated with Fig. 10.1 where silica content is expressed in per cent of dry suspension. In the same way, the pattern in Fig. 10.4 where the amount of diatoms is expressed as the number of cells per cubic metre of water, may be correlated with the pattern in Fig. 10.2. These correlations indicate that, in spite of minute differences, a general relationship exists: the siliceous suspension belts coincide with the diatom suspension belts, which points to the dominant role of diatoms.

Microscopic examination of suspension also shows that the basic role in recent silica accumulation belongs to diatom algae; over 70 per cent, and sometimes more than 90 per cent of the silica in suspension is composed of them.

The second major producers of amorphous silica in suspension are radiolarians. Notwithstanding widespread opinion to the contrary, they are found from the Arctic to the Antarctic, being most abundant in the equatorial zone. In the equatorial Pacific their content in suspension is about 16,000 specimens per cubic metre, while in Antarctic waters it is much less – tens and hundreds of specimens per cubic metre. A distinct boundary in both quantitative distribution and qualitative composition of radiolarians is formed by the Antarctic Convergence.

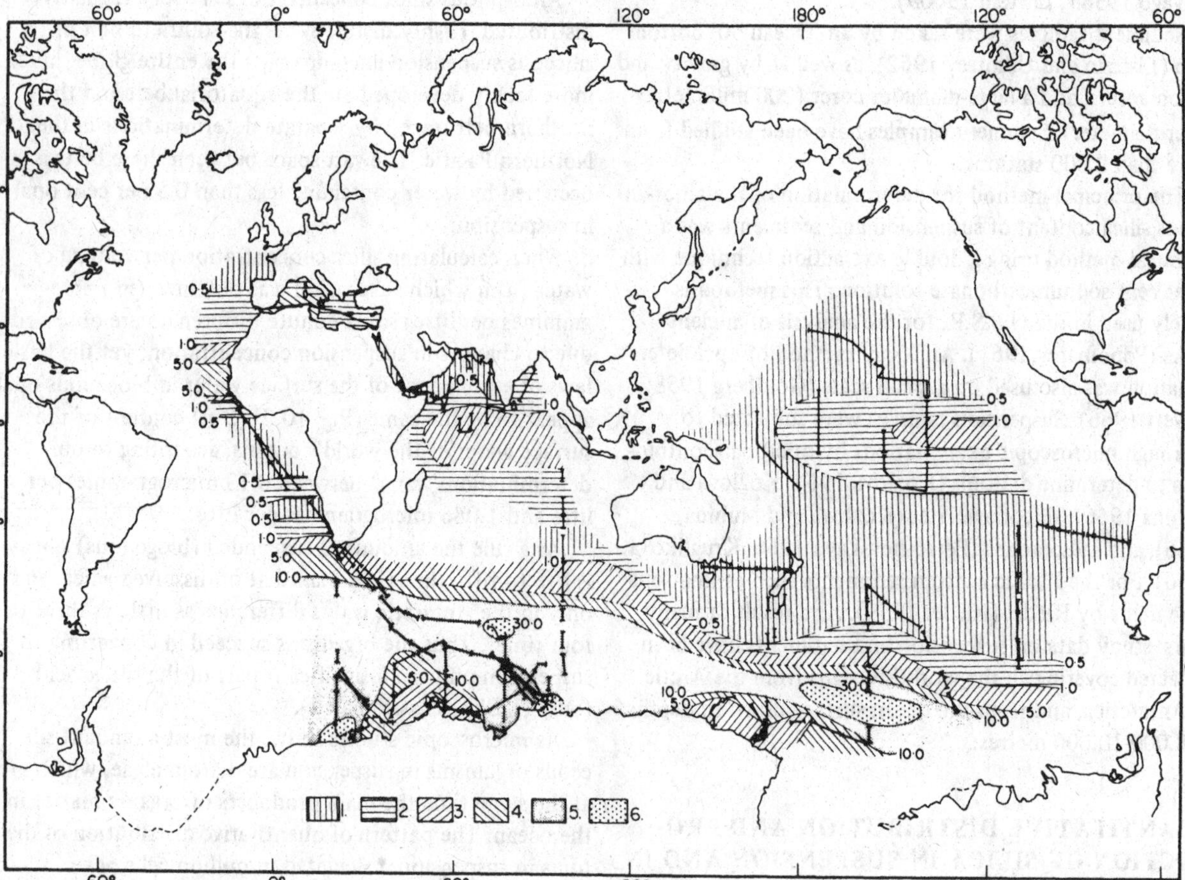
1.
2.
3.
4.
5.
6.

Fig. 10.1. Amorphous silica distribution in the surface suspended matter of the ocean. Samples were obtained from the 0-5 metre layer with the aid of the separator. Measured in per cent of the dry material. 1. less than 0.5%; 2. from 0.5 - 1%; 3. from 1 to 5%; 4. from 5 to 10%; 5. from 10 to 30%; 6. more than 30%.

The third place in importance is occupied by silicoflagellates (they are found in amounts of 0 to 111,000 cells per cubic metre in temperate and cold waters). The northern and southern belts of siliceous suspension are chiefly diatomaceous with an admixture of radiolarians, and within the equatorial belt the role of radiolarians increases sharply. All three groups of siliceous organisms are encountered in minimum amounts in the arid zones of the ocean.

To estimate the contribution of each of the three major groups of silica-concentrating planktonic organisms to the formation of siliceous suspension, results were plotted for the two main regions of silica accumulation in the Pacific Ocean: Antarctic and tropical. The above mentioned regions are characterized by the following values (Table 10.1).

To turn from the numbers to the more significant weight relationships we have made use of Petrushevskaya's (1966) data on weight determinations of tropical radiolarians. According to her data, 1 milligramme of silica corresponds to 1,000-3,000 radiolarian skeletons. Our determinations of diatom weight characteristics show that 1 milligramme of silica corresponds to 3,300-260,000 cells. Thus on an average, one radiolarian skeleton is (to a first approximation) fifty times as heavy as one diatom cell. Direct determinations of the skeleton weights for silicoflagellates are not available. Yet, from an analysis of their structure and sizes the conclusion can be drawn that they are at least five times as light as diatoms. Hence, the silica-concentrating marine planktonic organisms may be arranged by their weight in the following order: silicoflagellates:diatoms:radiolarians = 1:5:250.

Using these approximate data one may estimate rough weight relationships between silica-concentrators in the surface suspension of the ocean (Table 10.2).

The actual silica concentration in suspension depends not only on the initial amount of siliceous organisms but also on dilution by terrigenous (in the northern and southern zones) or by carbonate (in the equatorial zone) material. When calculating in units of an open system (milligrammes per litre) the effect of dilution is eliminated.

Since diatoms are the basic organic carbon producers in the oceans, their zones of maximum concentration coincide with zones of highest primary production. Therefore, the causes of the latitudinal zonation in silica distribution in suspended matter are the same as for organic carbon (supply of essential elements from depths to the photosynthetic zone, and solar radiation). The input into the surface waters is determined by global and local divergences. The largest of the global divergences is the Antarctic Divergence and it is here that the basic present-day silica accumulation proceeds. Local divergences are found off the western coasts of the continents, and are particularly clearly defined near South-West Africa and North America.

The pattern of primary plankton production for the world's oceans was drawn up by Gessner (1959) and a more precise variant of the pattern for the Pacific Ocean was given by Koblentz-Mishke (1965) (Fig. 10.5). Despite an inevitable discrepancy due to different methods of analysis, both patterns show essentially the same zonal picture of organic carbon distribution, which is close to that of opal distribution and the quantity of diatoms in the surface suspension.

The author has shown that a rather constant relation

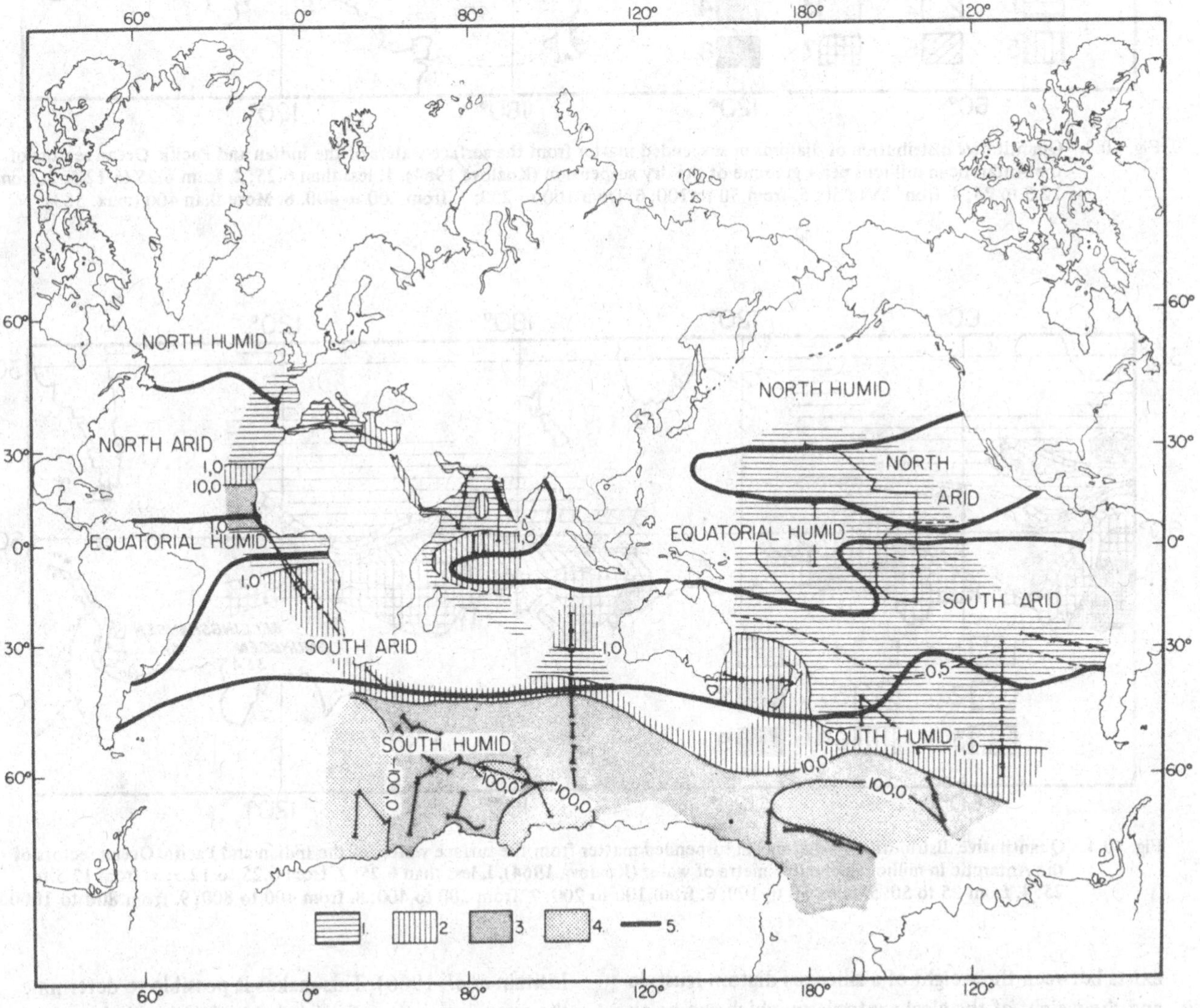
60°
0°
80°
120°
180°
120°
60°
30°
0°
30°
60°
NORTH HUMID
NORTH ARID
EQUATORIAL HUMID
SOUTH ARID
SOUTH HUMID
NORTH HUMID
NORTH
ARID
EQUATORIAL HUMID
SOUTH ARID
SOUTH HUMID
1,0
10,0
100,0
0,5
1.
2.
3.
4.
5.

Fig. 10.2. Amorphous silica distribution in suspended matter from the surface waters of the oceans and the location of the basic climatic zones. Samples were obtained from the 0-5 metre layer with the aid of the separator. Calculated in microgrammes per litre. The zones are marked on the basis of the relationship between the annual precipitation and evaporation as well as by temperature (Lisitzin 1966, 1967). 1. less than 1; 2. from 1 to 10; 3. from 10 to 100; 4. more than 100; 5. boundaries of basic climatic zones.

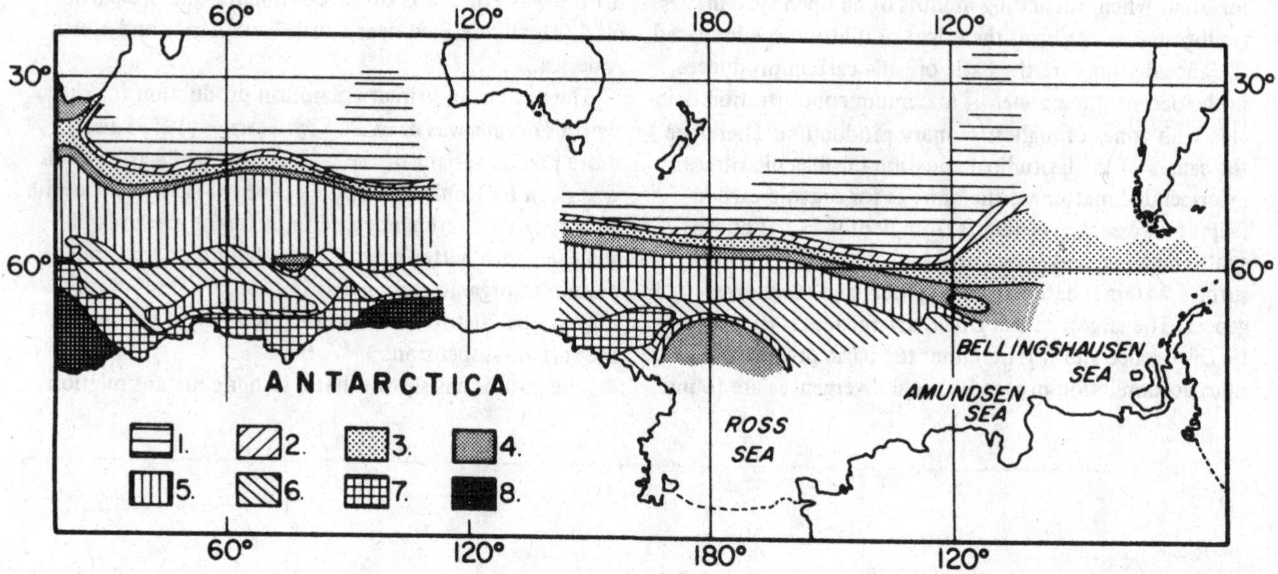
60°
120°
180
120°
30°
30°
60°
60°
ANTARCTICA
BELLINGSHAUSEN SEA
AMUNDSEN SEA
ROSS SEA
1.
2.
3.
4.
5.
6.
7.
8.
60°
120°
180°
120°

Fig. 10.3. Quantitative distribution of diatoms in suspended matter from the surface waters of the Indian and Pacific Ocean sectors of the Antarctic in millions per 1 gramme of the dry suspension (Kozlova 1964). 1. less than 6.25; 2. from 6.25 to 12.5; 3. from 12.5 to 25; 4. from 25 to 50; 5. from 50 to 100; 6. from 100 to 200; 7. from 200 to 400. 8. More than 400 (max. 583).

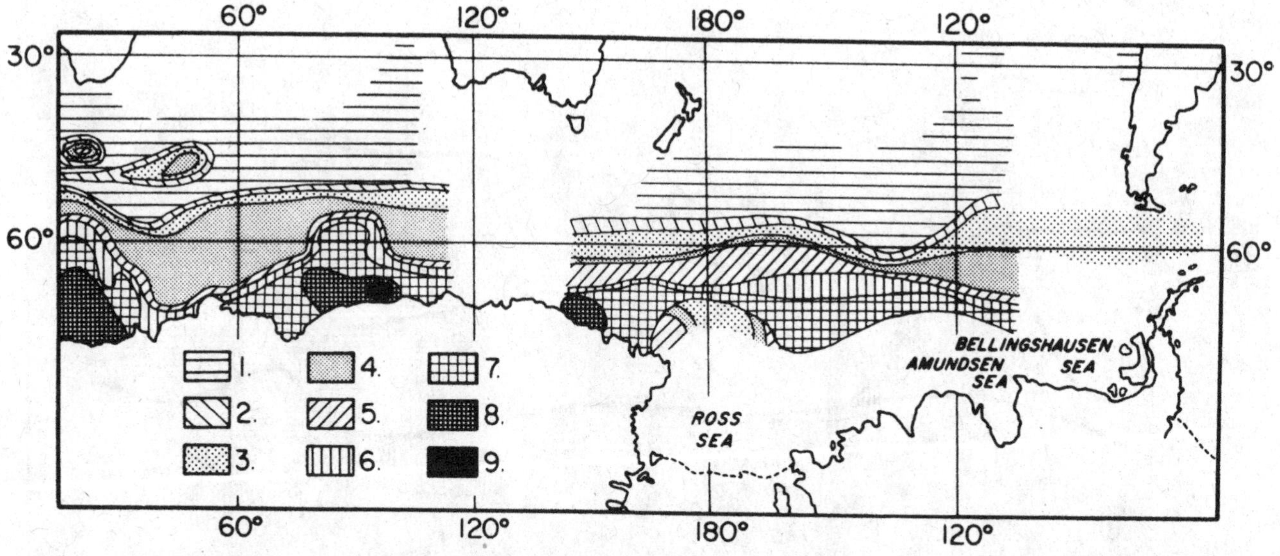
60°
120°
180°
120°
30°
30°
60°
60°
BELLINGSHAUSEN SEA
AMUNDSEN SEA
ROSS SEA
1.
2.
3.
4.
5.
6.
7.
8.
9.
60°
120°
180°
120°

Fig. 10.4. Quantitative distribution of diatoms in suspended matter from the surface waters of the Indian and Pacific Ocean sectors of the Antarctic in millions per cubic metre of water (Kozlova 1964). 1. less than 6.25; 2. from 6.25 to 12.5; 3. from 12.5 to 25; 4. from 25 to 50; 5. from 50 to 100; 6. from 100 to 200; 7. from 200 to 400; 8. from 400 to 800; 9. from 800 to 1000.

exists between the weight of a siliceous diatom frustule and the weight of the algal protoplasm which can be expressed as amorphous silica/organic carbon (Lisitzin 1964). The mean value of this ratio for diatom algae is 2.3, and for diatomaceous suspension from the Antarctic (based on the analysis of fifty samples) 1.85 (Lisitzin 1964; Lisitzin *et al.* 1966). This makes it possible to determine the annual quantity of amorphous silica bonded by diatoms into opal frustules in different parts of the ocean (Fig. 10.6). Three pronounced basic zones of silica accumulation can be traced when studying suspension – southern, northern and equatorial. In the equatorial zone

Table 10.1. Amounts of silica-concentrating organisms in suspension*

	Antarctic region		Tropical region	
	Amount	% of the total amount	Amount	% of the total amount
Diatoms	6,767,256	99.44	8,625	31.86-75.84
Radiolarians	tens-hundreds		several hundreds -16,000	2.64-59.10
Silicoflagellates	38,100	0.56	2,448	9.04-21.52

*To compile this table data obtained by Petrushevskaya (1966) and by Kozlova and Muhina (1966) have been used.

Table 10.2. Weight-relationships between silica-concentrators in the surface suspension from the Pacific Ocean (in weight per cent)

Siliceous organisms	Antarctic region %	Tropical region %
Diatoms	11.8-99.8	1.1-35.8
Radiolarians	0.01- 0.1	62.2-98.8
Silicoflagellates	0.1	0.1- 2.0

Table 10.3. Settling velocities of diatom frustules

Diatom species	Frustule size μ	Temperature in C°	Settling velocity	Source
Freshwater *Melosira baikalensis*	-	-	3.4 metres per day	Votintsev 1953
Marine diatoms	20-50	-	0.2-1.2 metres per day	Skopintsev 1949
Resting spores	-	-	7 metres per day	Gessner 1948
Medium-sized diatoms	10-20	20	2.8×10^{-3} cm/sec	Lohman 1942
Medium-sized diatoms	-	-	$1.7\text{-}6 \times 10^{-3}$ cm/sec	Margelef 1961
Diatoms from the Gulf of California	-	-	1.3×10^{-3} cm/sec.	Calvert 1966

diatoms convert to suspended opal between five and ten times less dissolved silica per year than in the southern belt. Thus photosynthesis is an energetic spring of recent silica accumulation processes.

QUANTITATIVE DISTRIBUTION AND PRODUCTION OF SILICA IN SUSPENSION IN THE OCEAN DEPTHS

On the death of planktonic organisms their skeletons descend to the bottom. The settling velocity of the dead siliceous organisms is determined mainly by their size and, to a lesser extent, by water density. In the ocean, currents and density boundaries (thermocline) are also of great importance.

Data on the settling velocities of diatom frustules of different species and sizes are contained in Table 10.3. The settling velocity determines not only the residence time of a particle in the water column and, thus, the possibility of its solution, but also the possibility of its deposition in the bottom sediments of any granulometric composition. Usually diatom frustules settle at a rate of $1\text{-}6 \times 10^{-3}$ centimetres per second. From observations carried out in the Gulf of California, Calvert (1966) infers that

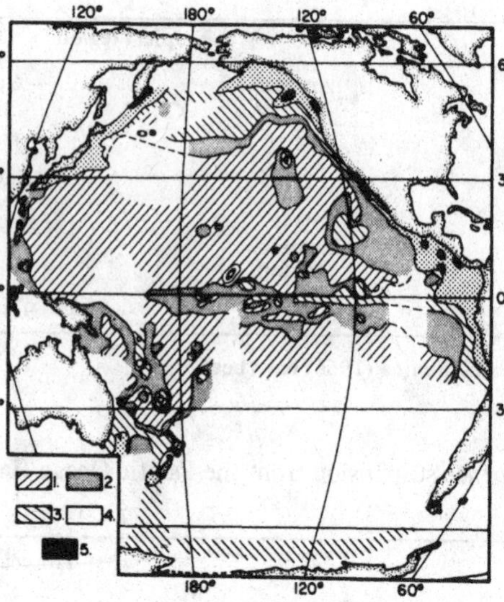
120°
180°
120°
60°
60°
30°
0°
30°
1.
2.
3.
4.
5.
180°
120°
60°

Fig. 10.5. Primary production in the Pacific Ocean (Koblentz-Mishke 1965*a*), and the distribution of phosphates in μg-atom/L at the surface.
Primary production:
At the surface
(Milligrammes of carbon per cubic metre per day)
1. less than 2
2. 2-5
3. 5-10
4. 10-100
5. more than 100

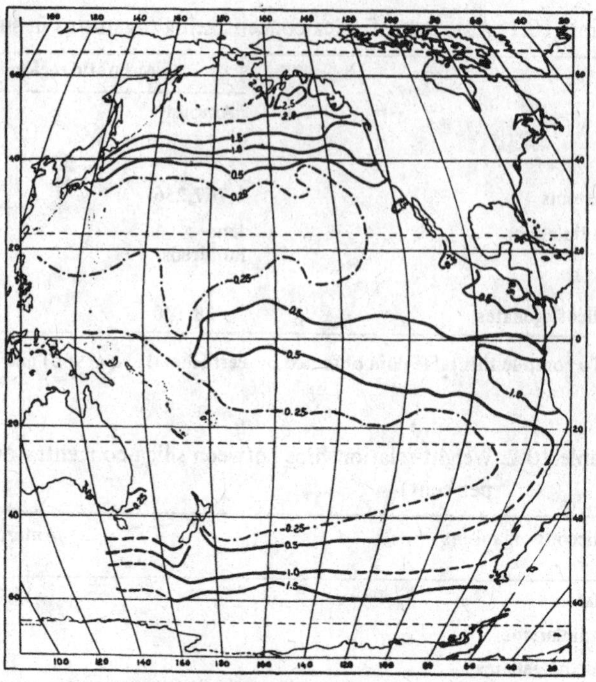
0.25
0.5
1.0
1.5

In the water column
(Milligrammes of carbon per square metre per day)
1. less than 100
2. 100-150
3. 150-250
4. 250-650
5. more than 650

about 59-71 per cent of the total opal has a settling velocity of 1.3×10^{-3} centimetres per second, equivalent to the settling velocity of 4μ quartz spheres. Our observations of the settling velocities of opal remains in suspended matter, and in bottom sediments, made in granulometric analyses of thousands of samples, show that this velocity ranges within rather wide limits and corresponds to equivalent quartz spheres of 1 to 50μ (0.053×10^{-3} to 133×10^{-3} centimetres per sec, at 5° C). Most of the siliceous material in suspension and in bottom sediments settles at a rate equivalent to that of $1\text{-}5\mu$ quartz spheres. This accounts for the fact that the coarsest and heaviest frustules reach a depth of 5,000 metres in 30-100 days, and the finest fraction in many tens of years.

Suspension studies also show that only an insignificant proportion of the frustules descend freely to the ocean floor as mineral particles do. The majority enter food chains and are utilized by zooplankton for food. Copepods break coarse and medium-sized diatom frustules into 5μ fragments and bond them into lumps which, in their turn, may be utilized by deep-water plankton and benthos for food. Thus the ultimate fate of diatom frustules in the water column appears to be closely related to food chains (Moore 1931; Tchindonova 1959; Sokolova 1959).

Suspension studies on vertical sections indicate that the diatom frustules are destroyed most rapidly in the upper 100 metre water layer. At greater depths the rate of the destruction (solution and breaking down by zooplankton) decreases rapidly. Species with thin frustules (*Porosira dichotomica, Chaetoceros, Thalassiosira tcherniai, Fragilariopsis cylindrus,* etc.) are mainly destroyed at 200-500 metre depths, separate fragments being a result. Owing to the solution of thin-frustule forms, the deep-water suspension and bottom sediments appear to be enriched in forms with coarse frustules. Preservation of diatom frustules in the water column and under laboratory conditions has been dealt with in a number of papers (Lewin 1961; Lisitzin 1964, 1966*a*).

Silicoflagellate skeletons are usually well preserved in the water column, yet they occur rarely in surface waters, playing a minor rôle in deep-water suspension. Living radiolarians, unlike diatoms or silicoflagellates, are encountered in suspension from the surface down to a depth of 5 kilometres, their maximum development being confined to the upper kilometre. In Antarctic regions radiolarians are not found in the 0-25 metre layer, whereas they are common at 200-400 metre depths. The distribution of radiolarians in cold waters and the underlying

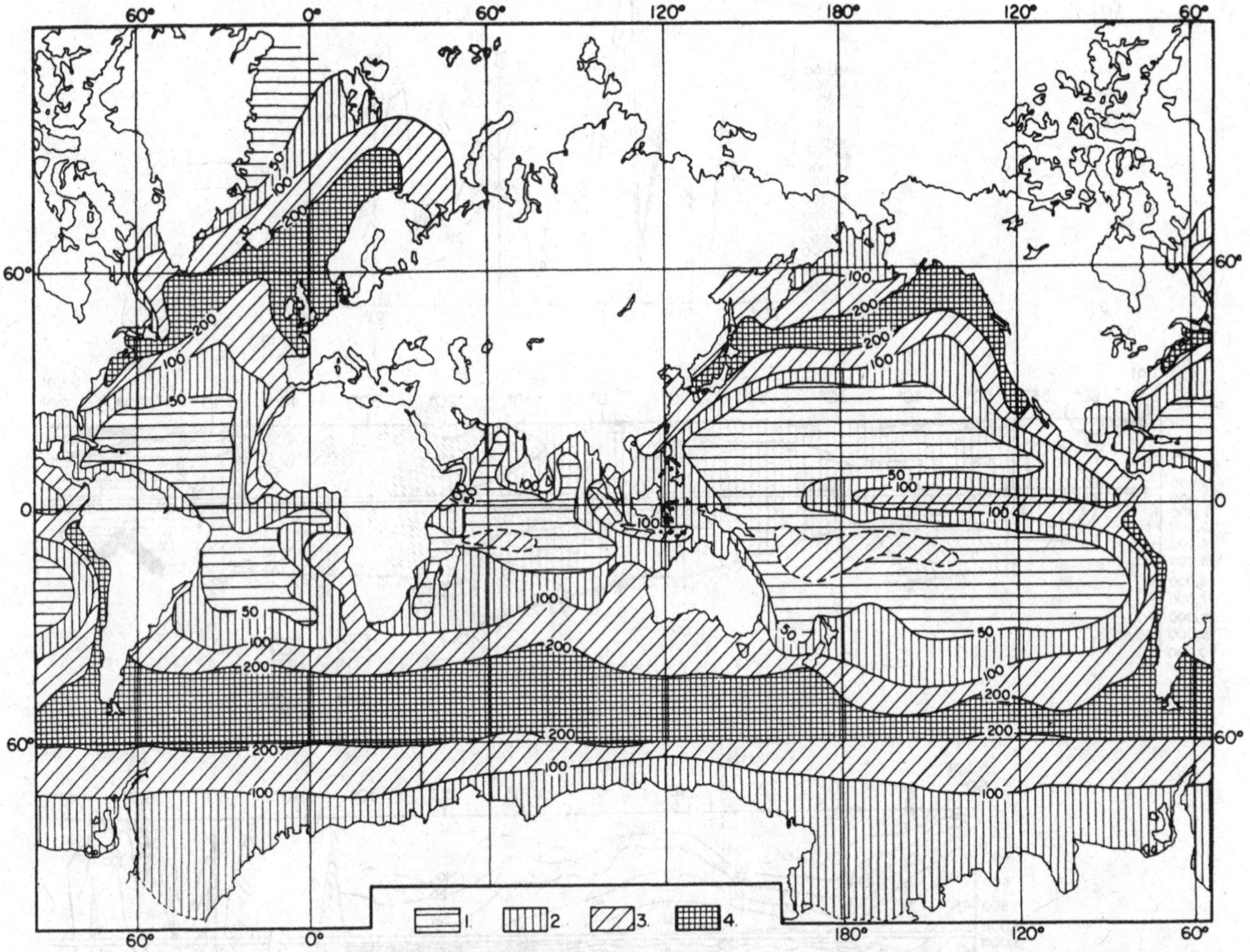
60°
0°
60°
120°
180°
120°
60°
60°
0
60°
50
100
200
1.
2
3.
4.

Fig. 10.6. The pattern of absolute masses, or the annual production of silica. Based on data on primary production by carbon-14 and oxygen as well as on the relationship between amorphous silica and organic carbon in suspended matter from different climatic zones (Lisitzin, *et al.* 1966) in grammes of amorphous silica per square metre per year. 1. less than 100; 2. from 100 to 250; 3. from 250 to 500; 4. more than 500.

sediments is inconsistent with a widespread idea about the warm-water character of radiolarian faunas.

Empty skeletons of dead radiolarians in the surface waters amount to approximately 1/10 of their total quantity, and in near-bottom layers up to 1/3, i.e. even the near-bottom layers of the ocean are inhabited by living radiolarians, which ensures their good preservation. Radiolarians are especially well preserved in Antarctic waters where empty skeletons constitute 5-20 per cent of the total number, while in the tropical zone they constitute 10-30 per cent. Thus radiolarians, and to a smaller extent silicoflagellates, are very well preserved in the deep-water suspension, and reach the bottom without any appreciable loss.

The most complicated fate appears to be that of the diatom frustules, which are most important for silica accumulation. Fig. 10.7 shows the quantitative distribution of unbroken diatom frustules along a meridional section through the Pacific Ocean, as compared with dissolved silica distribution and the number of living cells in the 0-100 metre layer. One can see from the section two well-defined regions of enrichment of the deep-sea suspension in diatom frustules, corresponding to areas of greatest amorphous silica content in the surface suspension – the northern and southern belts of silica concentration. The equatorial belt is not pronounced in this section; it is marked on the sections running farther east where the equatorial divergence is more distinct. Within these belts, waters at a depth of 4 kilometres contain 0.5-0.05 million frustules per cubic metre. The arid zones, mentioned above when considering amorphous silica distribution at the surface are also marked; in the deeper waters of the area neither diatom frustules, nor silicoflagellates are generally found.

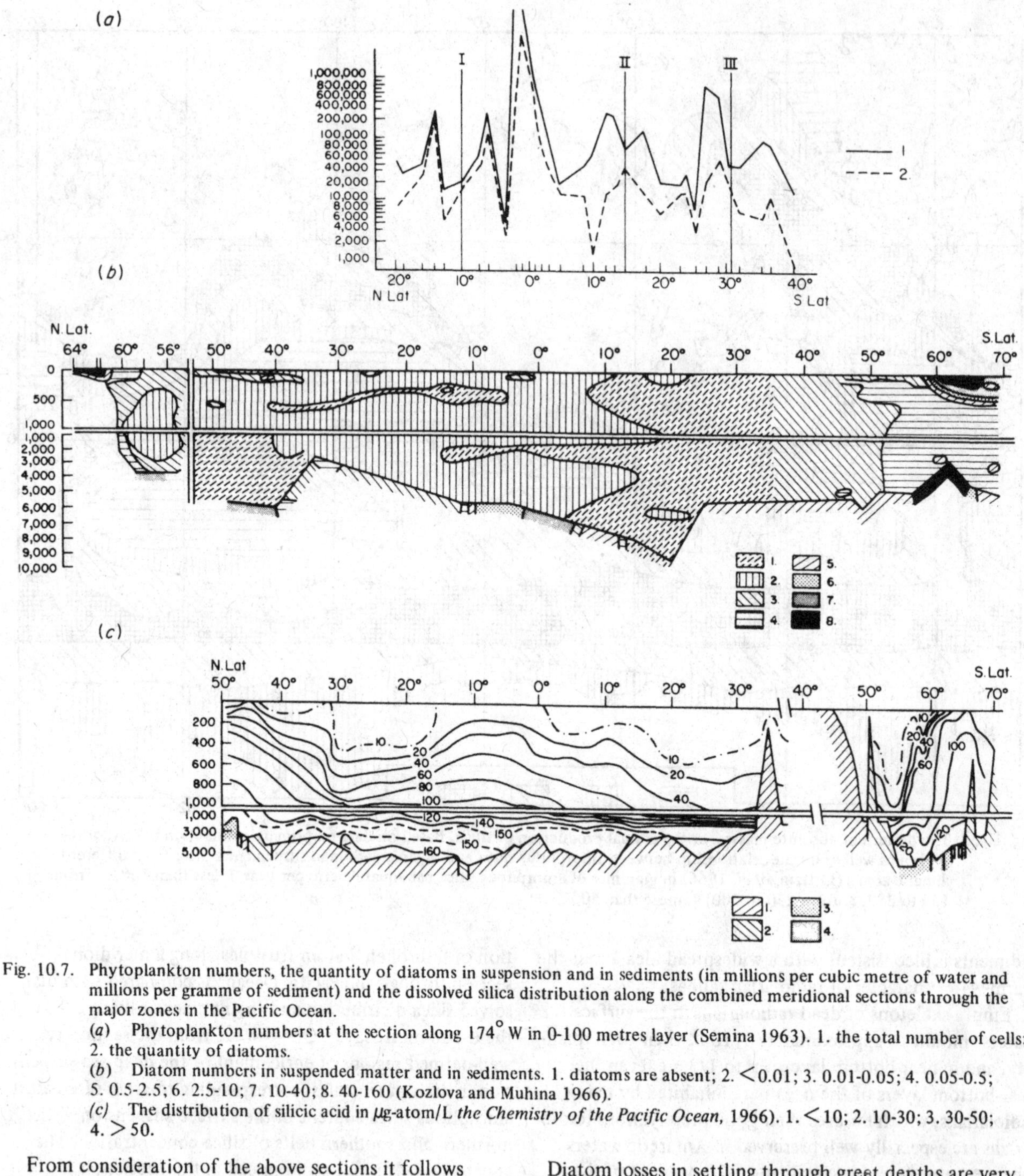
(a)
1,000,000
800,000
600,000
400,000
200,000
100,000
80,000
60,000
40,000
20,000
10,000
8,000
6,000
4,000
2,000
1,000
I
II
III
1
2
20° N Lat
10°
0°
10°
20°
30°
40° S Lat
(b)
N. Lat.
64°
60°
56°
50°
40°
30°
20°
10°
0°
10°
20°
30°
40°
50°
60°
70°
S. Lat.
0
500
1,000
1,000
2,000
3,000
4,000
5,000
6,000
7,000
8,000
9,000
10,000
1.
2.
3.
4.
5.
6.
7.
8.
(c)
N. Lat
50°
40°
30°
20°
10°
0°
10°
20°
30°
40°
50°
60°
70°
S. Lat.
200
400
600
800
1,000
1,000
3,000
5,000
10
20
40
60
80
100
120
140
150
160
100
1.
2.
3.
4.

Fig. 10.7. Phytoplankton numbers, the quantity of diatoms in suspension and in sediments (in millions per cubic metre of water and millions per gramme of sediment) and the dissolved silica distribution along the combined meridional sections through the major zones in the Pacific Ocean.
(*a*) Phytoplankton numbers at the section along 174° W in 0-100 metres layer (Semina 1963). 1. the total number of cells; 2. the quantity of diatoms.
(*b*) Diatom numbers in suspended matter and in sediments. 1. diatoms are absent; 2. <0.01; 3. 0.01-0.05; 4. 0.05-0.5; 5. 0.5-2.5; 6. 2.5-10; 7. 10-40; 8. 40-160 (Kozlova and Muhina 1966).
(*c*) The distribution of silicic acid in μg-atom/L *the Chemistry of the Pacific Ocean*, 1966). 1. <10; 2. 10-30; 3. 30-50; 4. >50.

From consideration of the above sections it follows that the quantity of frustules decreases with depth, yet the regions of maximum development of diatoms in the photosynthetic layer are as it were projected on to the bottom. Thus, despite the considerable destruction of frustules occurring in the water column, under the influence of biotic and abiotic factors, silica belts in suspension, though somewhat distorted, may also be traced in the near-bottom water and in the bottom sediments.

Diatom losses in settling through great depths are very uneven (Table 10.4), and in a number of cases tens and hundreds of times fewer frustules reach the near-bottom layers compared to their amounts in the surface suspension. As stated above, silicoflagellate and radiolarian skeletons reach the bottom without any significant loss. These conclusions, based on microscopic studies of suspension, are also confirmed by microchemical studies of membrane filters (Chumakov 1961; Bogoyavlensky

Table 10.4 Amount, specific composition and preservation of diatoms in suspension and in bottom sediments from the different climatic zones of the oceans

	Southern belt of accumulation		Southern arid zone	Equatorial belt	Northern arid zone	Northern belt of silica accumulation (the northern Pacific, the Bering Sea)	
	Neritic complex	Oceanic complex				Neritic complex	Oceanic complex
I. *Suspension*							
(a) *Amount*							
Number of frustules in the surface water (million per cubic metre)	40-69	0.05-22	0.001-0.06	0.007-20	0.001-0.05	52-167 mid. 73	0.06-
Number of frustules in the 4-5,000 metre layer	0.06-5.6	0.006-10	0.006-0.08	—	—	—	—
Percentage of the surface-layer numbers		25-40	5-50	15-64	0.4-30	1-5	5-27-75
Number of frustules in 1 gramme of suspension (million per gramme)	100-380	15-220	0.01-0.8	7.8-20	0-0.03	1.7-6	
(b) *Specific composition*							
Number of species at the surface	34	41	54	23	19-34	13	34
Number of species in the 4-5,000 metre layer	10-12	20	22-27	5-12	9	6	
Percentage of the surface-layer numbers	30	50	40-50	20-50	12-50	30-50	70
II. *Sediment*							
(a) *Amount*							
Amount in the surface sediment layer (million per gramme)	0.05-2.2	40-220	0-0.015	7.8-20	0-0.4	2-6	4.5-26 mid. 11
Percentage of the amount in the surface suspension	0.06-5.4	50-100	2.5			1-5	
Percentage of the amount in suspension from a 4-5,000 metre depth							
(b) *Specific composition*							
Number of species in the surface layer				5-12		6	
Percentage of the species number in the surface suspension	30-40	60-70	40	20-50		50	75
Percentage of the species number in the near bottom layer suspension				100		100	

1966). During the settling of diatom frustules to the ocean bottom their specific composition changes in the direction of a general decrease of the number of species. Table 10.4 shows that only some of the species from the active layer reach the bottom sediments, usually from 20 to 70 per cent. The best preserved appear to be coarsely silicified forms of the oceanic complex of the Northern and Southern Hemispheres (up to 70 per cent and sometimes even up to 100 per cent of the initial number of species). A decrease in the numbers of diatom species in the near-bottom suspension, as well as a decrease in the frustule numbers, is attributable to solution processes affecting, first of all, finely silicified forms (Fig. 10.8).

All this shows that both the quantitative distribution of frustules and their taxonomic composition changes markedly compared to those of the surface layers. For radiolarians such changes are practically non-existent, whereas for silicoflagellates a two fold and even greater decrease in the number of skeletons is observed.

QUANTITATIVE DISTRIBUTION AND PRODUCTION OF SILICEOUS MICROFOSSILS IN THE SURFACE LAYER OF BOTTOM SEDIMENTS

In order to compile a map of amorphous silica distribution in the bottom sediments of the world's oceans (Fig. 10.9) new bathymetric maps of the seas and oceans, as well as new maps of sediment types, drawn at the Institute of Oceanology of the U.S.S.R. Academy of Sciences, were used.

On the basis of over 2,000 analyses the basic zones of recent silica accumulation, and of regions low in silica, can be shown quite reliably.

Amorphous silica content of recent sediments ranges from fractions of one per cent to the maximum value of

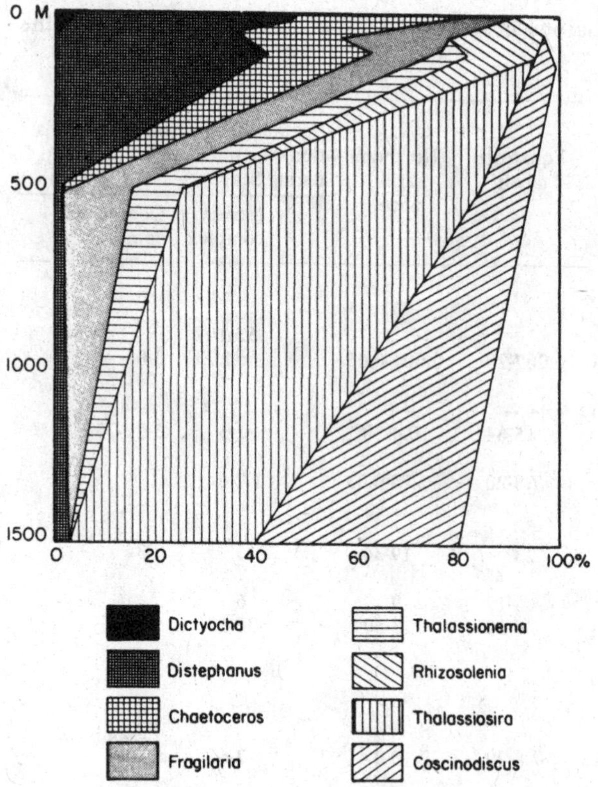

0 M
500
1000
1500
0
20
40
60
80
100%
Dictyocha
Distephanus
Chaetoceros
Fragilaria
Thalassionema
Rhizosolenia
Thalassiosira
Coscinodiscus

Fig. 10.8. Changes of the quantity and specific composition of diatoms and silicoflagellates in suspended matter from 0 to 1,500 metres in the vicinity of Japan (Oshite).

72 per cent (station 275 of the *Ob* cruise – 52° 45′ S, 62° 19′ E; depth 4,746 metres).

Three major belts of recent silica accumulation can be distinguished:

1. a southern belt encompassing the globe in an almost uninterrupted band in the southern Hemisphere;
2. a northern belt in the Pacific Ocean, Sea of Okhotsk, Bering and Japan seas. In the Atlantic Ocean it is not well developed;
3. an equatorial (more precisely, near-equatorial) belt which is well defined in the Pacific and Indian Oceans and more feebly developed in the Atlantic Ocean.

The southern belt is characterized by its large width and by the highest silica content. More than three-quarters of the silica of the world's oceans accumulates here (Lisitzin 1966*a*). The siliceous sediment belt (within the 10 per cent amorphous silica isolines) is 900-2000 kilometres wide, its northern boundary coinciding with the Convergence and with the middle boundary of iceberg distribution. Diatom oozes penetrate south beyond the south polar circle and in a number of places they even appear on the Antarctic shelf (Lisitzin 1963). Within this belt amorphous silica concentrations are far from evenly distributed; siliceous sediments are never found on steep scarps or on underwater elevations. Crests of underwater elevations are here usually capped with foraminiferal sediments. Siliceous sediments of the southern belt have been thoroughly studied and mapped (Lisitzin and Zhivago 1958*a*; Lisitzin 1960*a*; 1961*c*, 1966*a, c; Atlas of the Antarctic*, 1966).

The northern belt differs from the southern one in its lower silica concentrations and discontinuity. It embraces the northern part of the Pacific including the Far Eastern seas – Bering, Okhotsk and Japan. More than 500 analyses are available for the northern Pacific (Bezrukov *et al.* 1961; Gershanovich 1964). This belt was discovered long ago (Bailey 1856; Belknap 1874; Murray and Irvine 1890-1; Murray and Renard 1891), though the diatoms in the sediments have been determined only by microscopic counts.

Silica concentrations in the sediments of the northern Pacific, as the analyses show, do not usually exceed 10-20 per cent, and only seldom do they reach 30 per cent. The contours of the siliceous sediment belt differ from those shown on earlier maps (Murray and Renard 1891; Schott 1935; Sverdrup, Johnson and Fleming 1942; Revelle 1944; Revelle *et al.* 1955; Arrhènius 1963).

Silica content is substantially higher in the sediments of the Bering Sea (up to 33-37 per cent) (Lisitzin 1955*c*, 1958, 1959*b*, 1961*d*, 1966*a, b*; Gershanovich 1962) and particularly so in the Sea of Okhotsk (up to 56 per cent) (Bezrukov 1955, 1960; Lisitzin 1966*a*). In the Japan Sea siliceous sediments are found only in the northern portion, where their maximum values slightly exceed 20 per cent (Strakhov *et al.* 1954; Solovyev 1960; Lisitzin 1966*a*). Silica content of the bottom sediments in the Yellow, East-China and South-China seas is no more than 2-3 per cent. In the northern part of the Atlantic Ocean, owing to the influence of the Gulf Stream, the silica belt is interrupted – bottom sediments here usually contain less than 3 per cent amorphous silica and only at a few stations up to 5-7 per cent (Shurko 1966). The Mediterranean Sea sediments contain a maximum of about 3 per cent, and often below 1 per cent amorphous silica (Emelyanov 1966). Bottom sediments of the Arctic Ocean and of the Arctic seas usually contain no more than 0.5 per cent (Belov and Lapina 1961), and only in the marginal parts up to 4-7 per cent. This is attributable to the ice cover of the ocean which hampers photosynthesis, as well as to the abundant terrigenous supply from the Siberian rivers.

The equatorial belt is made up of separate patches of various sizes. Correlations between silica distribution maps on the one hand, and the maps of sediment types and of bathymetry, on the other, indicate that (unlike the northern and southern belts) the distribution of silica sediments here is closely related to depth. They are found only at depths exceeding the critical one at which calcium car-

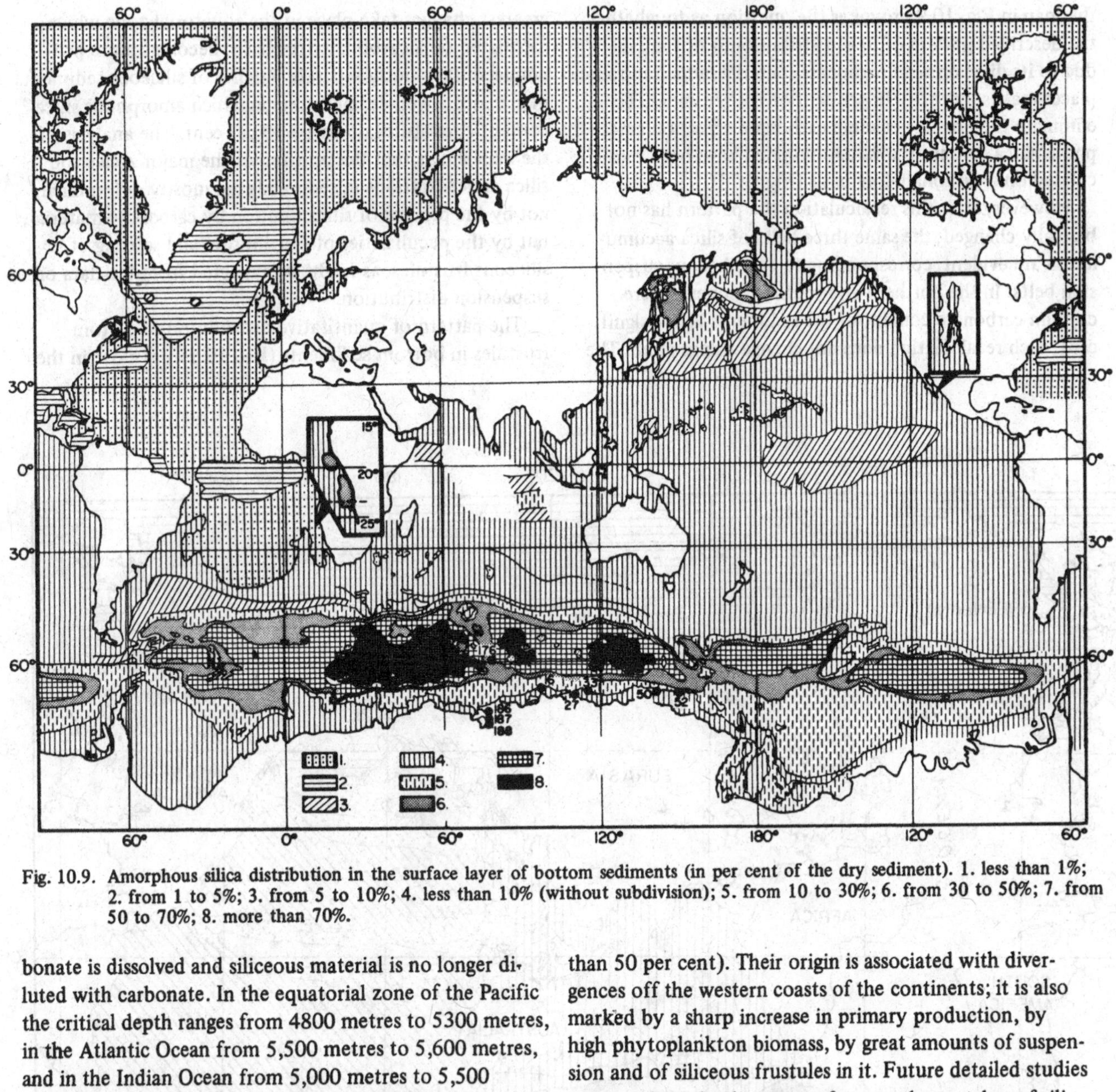

60°
0°
60°
120°
180°
120°
60°
60°
30°
0°
30°
60°
15°
20°
25°
1.
2.
3.
4.
5.
6.
7.
8.

Fig. 10.9. Amorphous silica distribution in the surface layer of bottom sediments (in per cent of the dry sediment). 1. less than 1%; 2. from 1 to 5%; 3. from 5 to 10%; 4. less than 10% (without subdivision); 5. from 10 to 30%; 6. from 30 to 50%; 7. from 50 to 70%; 8. more than 70%.

bonate is dissolved and siliceous material is no longer diluted with carbonate. In the equatorial zone of the Pacific the critical depth ranges from 4800 metres to 5300 metres, in the Atlantic Ocean from 5,500 metres to 5,600 metres, and in the Indian Ocean from 5,000 metres to 5,500 metres. Delineation of the extreme boundaries of separate patches of siliceous sediment yields a zone stretching from 20° N to 20° S and concentrated on the Equator. This zone is best developed in the Indian and Pacific oceans. No analytical confirmation of the existence of such a belt in the Atlantic Ocean has been so far obtained.

Two more patches of siliceous sediments should be noted, which are not included in the belts of silica accumulation already distinguished. One of these patches has been found in the Gulf of California (maximum amorphous silica content up to 65 per cent; Calvert 1966) and the second, off Southwest Africa, near the mouth of the Orange River (maximum amorphous silica content is more than 50 per cent). Their origin is associated with divergences off the western coasts of the continents; it is also marked by a sharp increase in primary production, by high phytoplankton biomass, by great amounts of suspension and of siliceous frustules in it. Future detailed studies may result in the discovery of some other patches of siliceous sediments of minor extent associated with local divergences.

When comparing the map in Fig. 10.9 with those in Figs. 10.1 and 10.2 one notices that the position of the major zones of silica concentration in suspension is the same for bottom sediments.

As in suspension, wide zones of non-siliceous sediments, corresponding to the northern and southern arid zones of the ocean, are encountered in the bottom sediments. Siliceous pelagic sediments are not found between 20° S and 45° S, and between 20° S and 40° S; arid zones are associated with non-siliceous sediments. Examination of the map in Fig. 10.9 provokes the question as to whether the described pattern of opal distribution is not in fact due to its dilution with carbonate material, which in some places makes up to 90-95 per cent of the sediment. To eliminate the possible calcium carbonate effect on amorphous silica content the values obtained have been recalculated on a carbonate-free basis (Fig. 10.10).

However, after this recalculation the pattern has not basically changed; the same three belts of silica accumulation are evident, corresponding to the siliceous suspension belts. In the northern and southern zones, where calcium carbonate content of the sediments is not significant, such recalculation does not change things at all. The greatest changes take place in the equatorial zone which, calculated on a carbonate-free basis, becomes more pronounced. Previously separate patches of siliceous sediments merge into a well-defined zone in which amorphous silica content constitutes up to 10-30 per cent. The analysis of the map shows that the position of the major zones and silica concentrations in sediments are mostly determined, not by the process of silica dilution by carbonate material, but by the peculiarities of distribution and settling of siliceous frustules, as might be expected from the data on suspension distribution.

The pattern of quantitative distribution of diatom frustules in bottom sediments (Fig. 10.11), is close in the

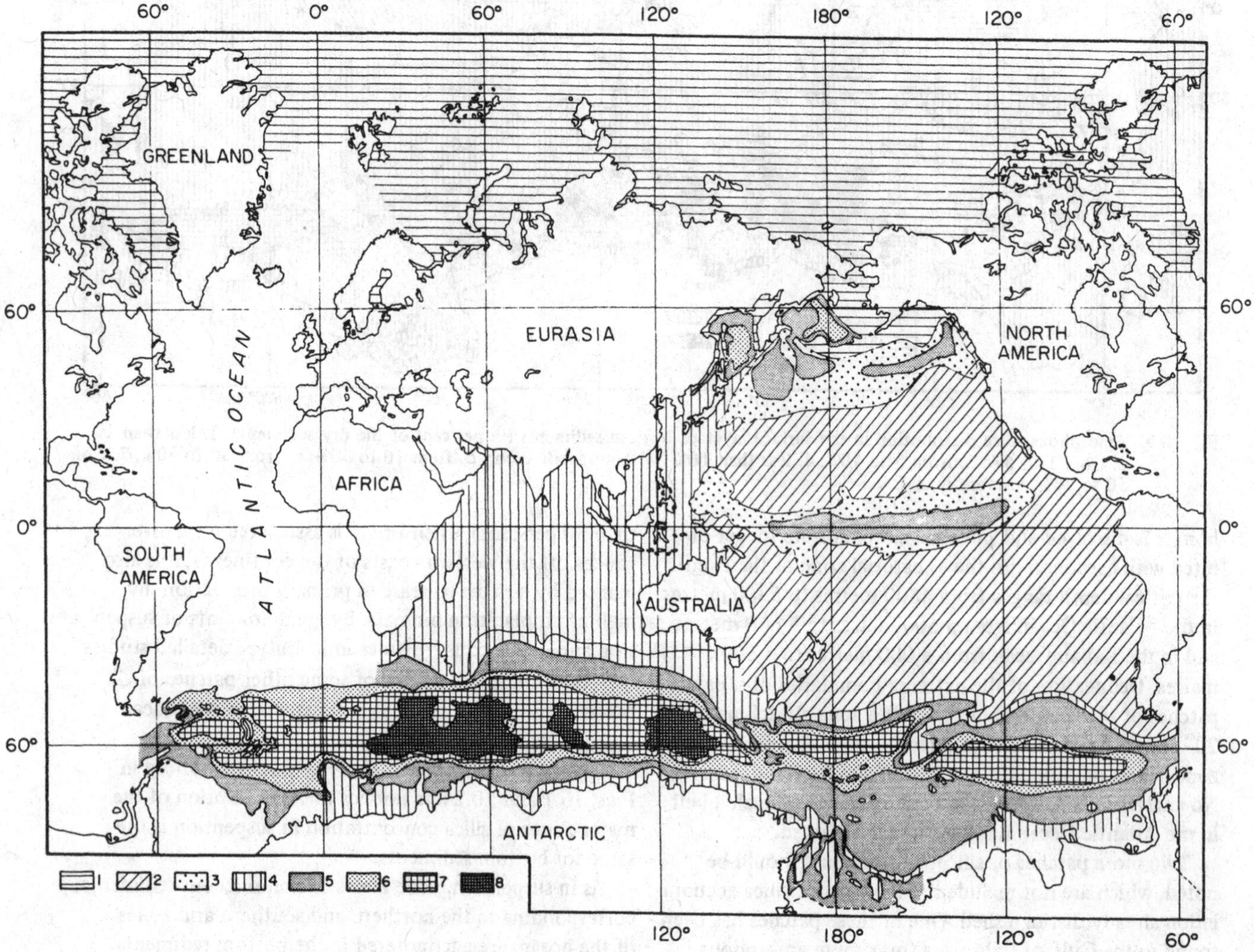
60°
0°
60°
120°
180°
120°
60°
GREENLAND
EURASIA
NORTH AMERICA
ATLANTIC OCEAN
AFRICA
SOUTH AMERICA
AUSTRALIA
ANTARCTIC
1
2
3
4
5
6
7
8

Fig. 10.10. Amorphous silica distribution in the surface layer of bottom sediments expressed on a carbonate-free basis (in per cent of dry sediment). 1. less than 1%; 2. from 1 to 5%; 3. from 5 to 10%; 4. less than 10% (without subdivision); 5. from 10 to 30%; 6. from 30 to 50%; 7. from 50 to 70%; 8. more than 70%.

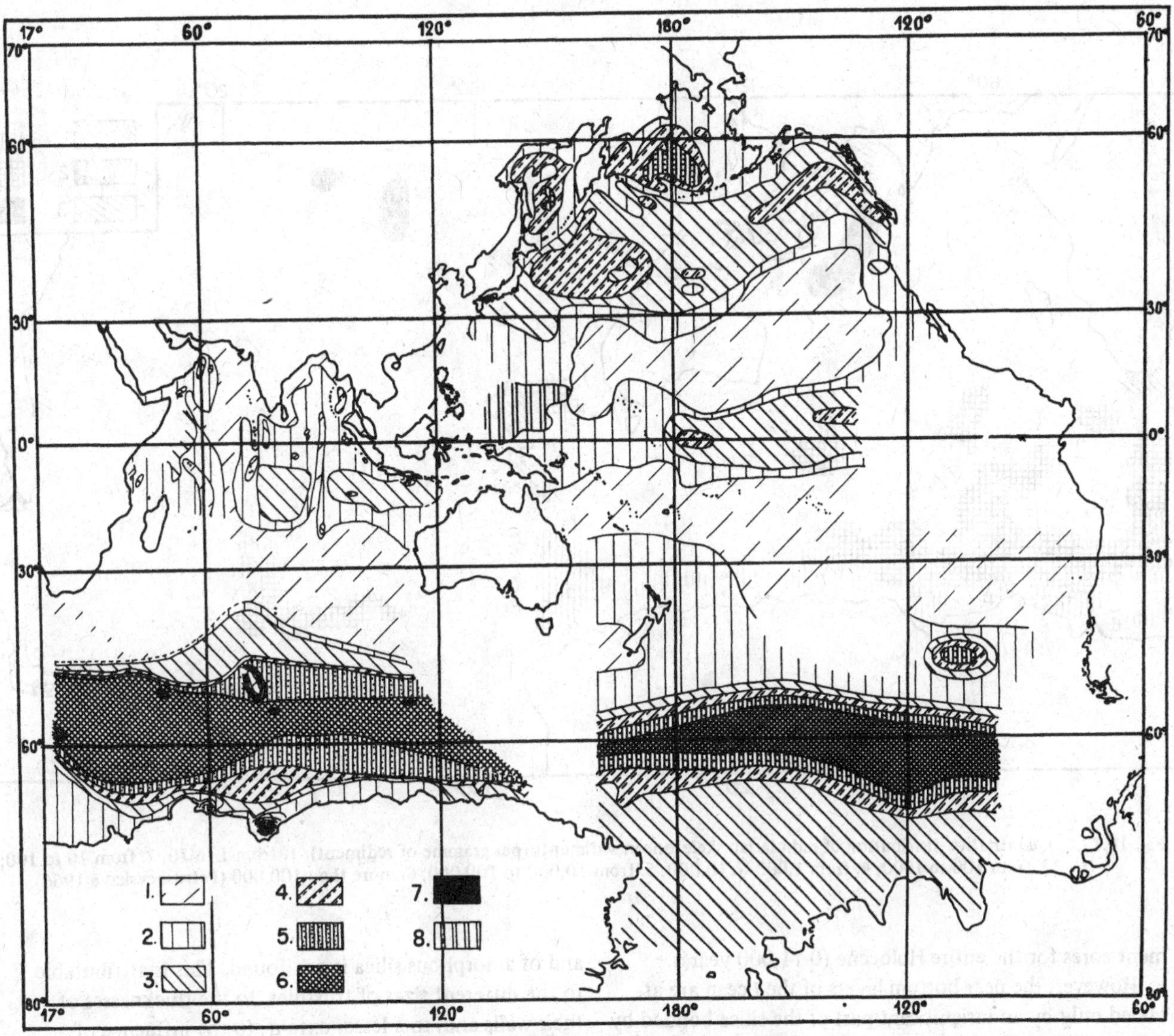
17°
60°
120°
180°
120°
60°
70°
60°
30°
0°
30°
60°
80°
1.
2.
3.
4.
5.
6.
7.
8.

Fig. 10.11. Quantitative distribution of diatom frustules in the surface layer of bottom sediments (in millions per gramme of sediment). 1. no diatoms; 2. 0.1-1; 3. 1-5; 4. 5-15; 5. 15-45; 6. 45-100; 7. > 100; 8. *Ethmodiscus* oozes.

main to the maps of amorphous silica distribution in Figs. 10.1, 10.2, 10.10. The same zones are distinguished here as for amorphous silica in suspension and in sediments. There are some differences which will be discussed below.

The pattern of quantitative distribution of radiolarians in sediments (Fig. 10.12) shows that radiolarians are of greatest importance in recent sedimentation in the equatorial zone; in the temperate humid zones their content drops by tens of times compared to that in the equatorial zone. Yet radiolarians are found down to the Antarctic coasts.

The maximum numbers of silicoflagellates in bottom sediments have been recorded in the equatorial zone (0.6-1.3 million per gramme). In the arid zones their quantity ranges from 0 to 0.4 million per gramme; in the northern and southern belts of silica accumulation (humid zones) it increases again up to 0.1-5.5 million per gramme with a decrease on the shelf to 0.06-0.19 million per gramme.

Thus within all three belts of recent silica accumulation an increase in the content of all major siliceous organisms is observed both in suspension and in bottom sediments.

In the equatorial zone the rôle of radiolarians is particularly great; in a number of places they are more significant than diatoms, which dominate in all other zones.

Thus the siliceous belts of the Earth, which were shown to characterize the suspension in the phytosynthetic zone, are also found throughout the whole water column and in bottom sediments. They are also seen when studying sedi-

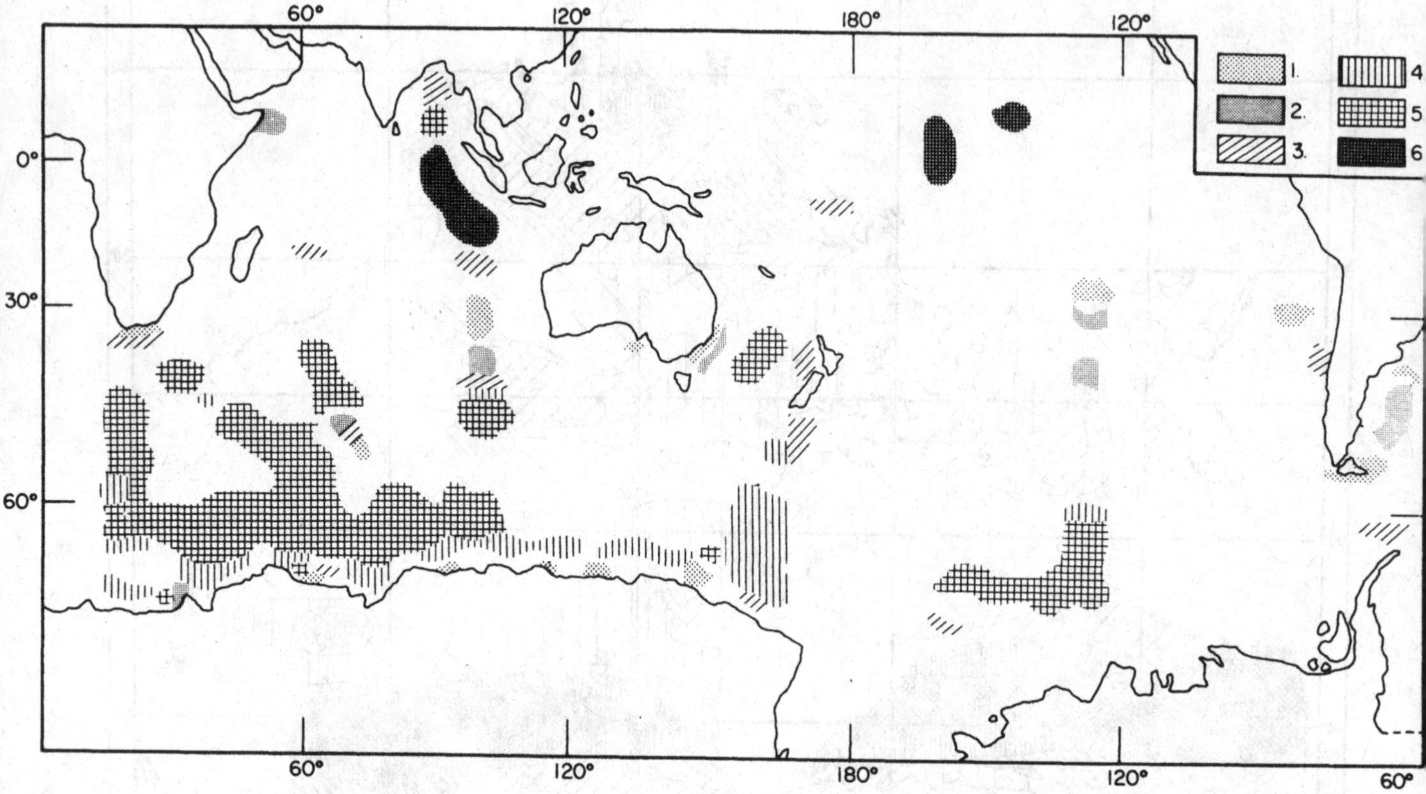
60°
120°
180°
120°
0°
30°
60°
1.
2.
3.
4
5
6

Fig. 10.12. Quantitative distribution of radiolarian skeletons in sediments (per gramme of sediment). 1. from 1 to 10; 2. from 10 to 100; 3. from 100 to 1000; 4. from 1,000 to 10,000; 5. from 10,000 to 100,000; 6. more than 100,000 (Petrushevskaya 1966).

ment cores for the entire Holocene (0-11,000 years).

However, the near-bottom layers of the ocean are attained only by an insignificant part of the silica bonded by organisms into frustules at the surface. In the oceans this usually amounts to 1/10-1/100.

The remaining part of silica is dissolved, and enters the chemical cycle again. Thus both the quantitative distribution and specific composition of diatoms in bottom sediments differs essentially from the biocoenoses in surface waters. This, however, does not rule out the possibility of reconstructing biocoenoses on the basis of studies of thanatocoenoses, since the most resistant and representative forms (indicators) are preserved in the sediments.

When the amounts of diatom frustules, radiolarian and silicoflagellate skeletons are correlated with amorphous silica contents of suspension and bottom sediments, numerous discrepancies are revealed. In particular, they can be seen clearly from the curves (Fig. 10.13).

The scatter in values here is very great, and a direct quantitative relation between the amounts of frustules and of amorphous silica is not found. This is attributable to the different sizes of frustules, to the thicknesses of their walls and, to a lesser extent, to the influence of radiolarians and silicoflagellates and to fine division of the frustules in suspension and in bottom sediments.

One can see from Figs. 10.14 and 10.15 that in all the belts of silica accumulation most of the silica appears to be concentrated in the fraction smaller than 0.01 millimetres, into which fine diatom detritus falls. The numbers of frustules in suspension and in bottom sediments which are estimated by palaeontologists represent only an insignificant part of their quantity in the form of fragments.

With increasing depth the amount of preserved frustules generally decreases as can be seen from Fig. 10.7, if the bottom plane is presumed shifted from 500 metres to 5,000 metres depths. Unlike calcium carbonate, no critical depth for silica is found to exist below which it does not penetrate; siliceous remains are encountered in bottom sediments down to maximum depths. An increase in depth, all other things being equal, results in a decrease of

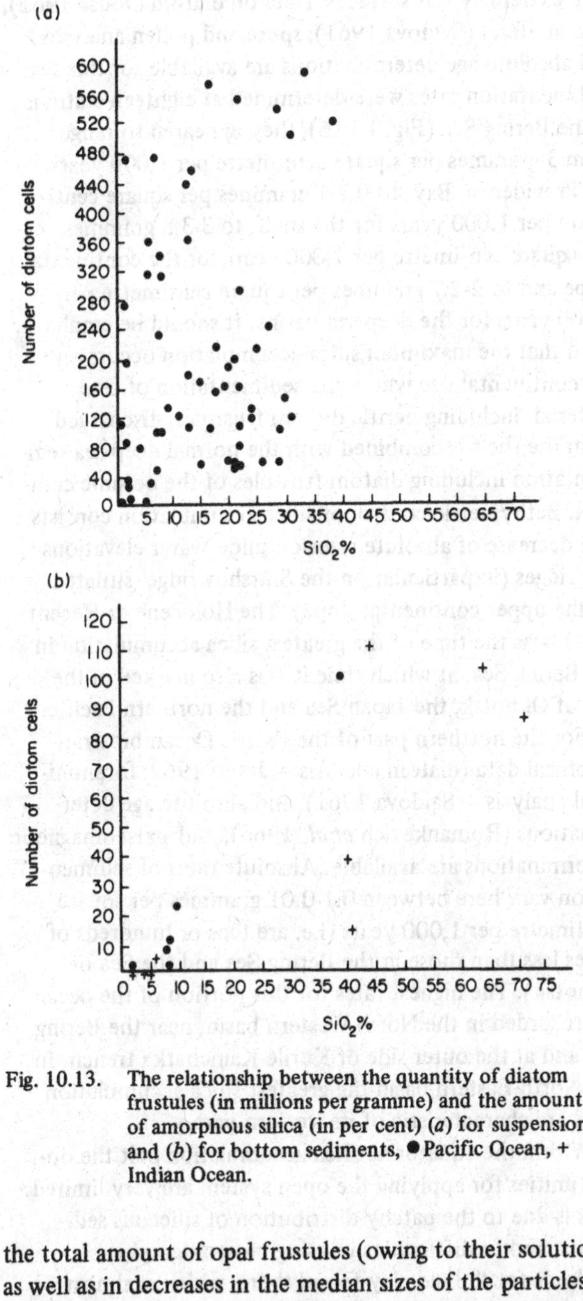
(a)
Number of diatom cells
SiO_2%
(b)
Number of diatom cells
SiO_2%

Fig. 10.13. **The relationship between the quantity of diatom frustules (in millions per gramme) and the amount of amorphous silica (in per cent) (*a*) for suspension, and (*b*) for bottom sediments, ● Pacific Ocean, + Indian Ocean.**

the total amount of opal frustules (owing to their solution), as well as in decreases in the median sizes of the particles and in numbers of species. The influence of depth is especially marked for neritic assemblages, where finely silicified forms prevail, and is more feebly marked for oceanic complexes where the preservation of frustules is higher.

ABSOLUTE MASSES OF SILICEOUS MICROFOSSIL ACCUMULATIONS

As noted above, the actual intensity of accumulation of opal remains seldom coincides with the percentage of amorphous silica, and even less so with the amount of frustules in sediments. This is accounted for by carbonate, terrigenous, and sometimes volcanogenic material, acting as dilutants settling to the bottom simultaneously with the silica. After recalculating the sediment on a non-carbonate basis, silica content depends on the intensity of terrigenous accumulation, to which it is inversely proportional.

To determine the real intensity of silica accumulation it is necessary to present the data in an open system using the method of absolute masses. In so doing, the rate of accumulation of any component is determined independently of other sediment components. Estimates are made of silica amounts accumulating on one square centimetre of the ocean bottom during a certain period of time (Holocene, per 1,000 years, or per year).

To solve this problem reliable determinations of ages, or rates of sedimentation, are necessary. These can be obtained either by biostratigraphic methods (diatom, fora-

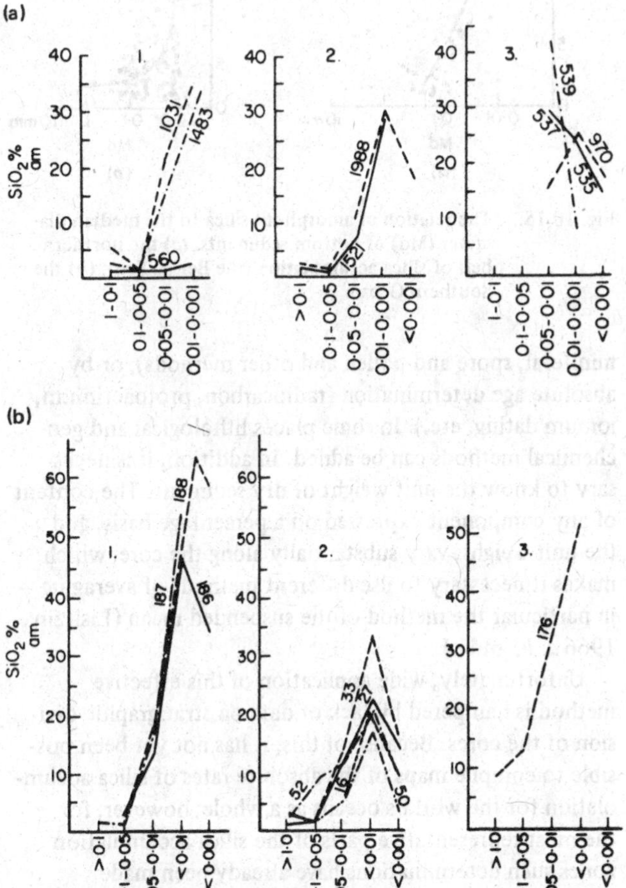
(a)
SiO_2% am
(b)
SiO_2% am

Fig. 10.14. **Silica distribution by different fractions of bottom sediments in the northern (*a*) and southern (*b*) silica accumulation belts. 1. neritic diatom complexes; 2. mixed complexes; 3. oceanic complex. The figures correspond to the numbers of the oceanographic stations of the research vessels *Vityaz* and *Ob*.**

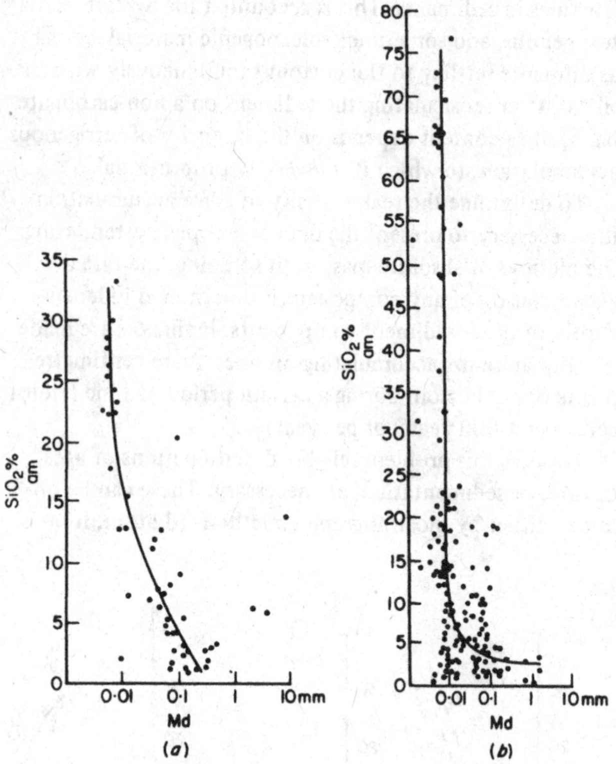
SiO_2%
am
Md
(a)
(b)
0·01
0·1
1
10mm

Fig. 10.15. The relation of amorphous silica to the median diameter (Md) of bottom sediments. (*a*) the northern belt of silica accumulation (the Bering Sea); (*b*) the Southern Ocean.

miniferal, spore and pollen and other methods), or by absolute age determination (radiocarbon, protoactinium, ionium dating, etc.). In some places lithological and geochemical methods can be added. In addition, it is necessary to know the unit weight of dry sediment. The content of any component expressed on a percentage basis, and the unit weight, vary substantially along the core, which makes it necessary to use different methods of averaging – in particular the method of the suspended mean (Lisitzin 1966 *a, b,* and *c*).

Unfortunately, wide application of this effective method is hampered by lack of data on stratigraphic division of the cores. Because of this, it has not yet been possible to compile maps of the absolute rates of silica accumulation for the world's oceans as a whole; however, for the most representative parts of the silica accumulation zones such determinations have already been made (Lisitzin 1966 *a, b* and *c*). In particular, reliable data have been obtained for the northern and southern belts of silica accumulation; less reliable ones for the equatorial belt of the Pacific Ocean.

The Bering Sea is especially favourable for studying the absolute rates of silica accumulation. Its deep-water basins have extremely flat surfaces. Data on diatom (Jousé 1962), foraminiferal (Saidova 1961), spore and pollen analyses and absolute age determinations are available for this sea. Sedimentation rates were determined at eighteen stations in the Bering Sea (Fig. 10.16); they appeared to range from 5 grammes per square centimetre per 1,000 years for Providence Bay, to 0.2-1 grammes per square centimetre per 1,000 years for the shelf, to 3-3.8 grammes per square centimetre per 1,000 years for the continental slope and to 2-2.7 grammes per square centimetre per 1,000 years for the deep-sea basins. It should be emphasized that the maximum silica accumulation occurs on the continental rise where the sedimentation of fine material, including neritic diatom frustules, discharged from the shelf is combined with the normal deep-sea sedimentation including diatom frustules of the oceanic complex. Bathymetric control of silica accumulation consists of a decrease of absolute rates on underwater elevations and ridges (in particular on the Shirshov ridge, situated on the upper continental slope). The Holocene or Recent epoch was the time of the greatest silica accumulation in the Bering Sea, at which time it was also marked in the Sea of Okhotsk, the Japan Sea and the northern Pacific.

For the northern part of the Pacific Ocean biostratigraphical data (diatom analysis – Jousé 1962; foraminiferal analysis – Saidova 1961), and absolute age determinations (Romankevich *et al.* 1966), and palaeomagnetic determinations are available. Absolute rates of sedimentation vary here between 0.1-0.01 grammes per square centimetre per 1,000 years (i.e. are tens or hundreds of times less than those in the Bering Sea and the Sea of Okhotsk). The highest rates for this portion of the ocean are recorded in the North-Western basin, near the Bering Sea and at the outer side of Kurile-Kamchatka trench. In the North-Eastern basin the greatest silica accumulation rates are characteristic of its western part.

Within the equatorial silica accumulation belt the opportunities for applying the open system are very limited. This is due to the patchy distribution of siliceous sediments, and to difficulties in using biostratigraphic methods for the red clay found there, which is almost completely devoid of microfossils. The equatorial zone of the Pacific is divided, by island ranges and underwater highs, into a number of basins, which often have no deep-water relationships. Therefore sharp contrasts in sedimentation rates are natural in this zone. Values usually range between 0.01 and 0.001 grammes per square centimetre per 1,000 years, i.e. are ten times lower than those in the northern part of the ocean. In some small basins, where *Ethmodiscus* oozes are accumulating, the absolute rates may show a sharp increase – up to several grammes, even to 10 grammes per square centimetre per 1,000 years,

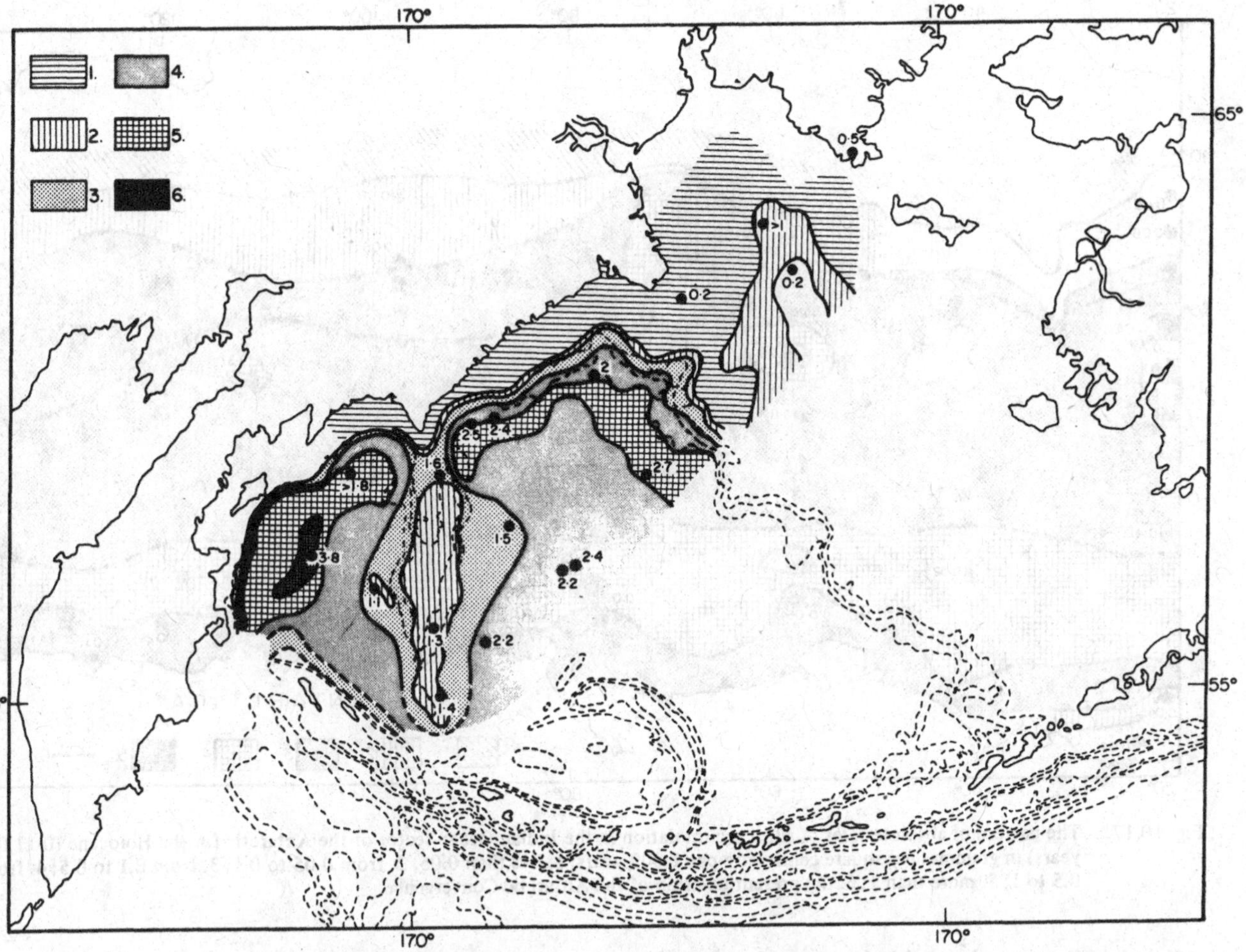

170°
170°
1.
2.
3.
4.
5.
6.
65°
55°
55°
0·5
>1
0·2
0·2
2
2·5
2·4
1·6
2·7
>1·8
3·8
1·5
2·4
2·2
1·1
1·3
2·2
1·4
170°
170°

Fig. 10.16. The pattern of absolute rates of silica accumulation in the Bering Sea for the Holocene (0-11,000 years) in grammes per square centimetre per 1,000 years. 1. less than 1; 2. from 1 to 1.5; 3. from 1.5 to 2; 4. from 2 to 2.5; 5. from 2.5 to 3; 6. more than 3.

which is attributable to the flow of these sediments down into separate bottom depressions (Jousé *et al.* 1959; Hanzawa 1935).

The relative importance of radiolarians in the total amorphous silica supply to the sediments increases considerably here; in the Indian Ocean and in the eastern Pacific they become dominant, but the sedimentation rates are very low.

Within the southern silica accumulation belt of the Indian and Pacific oceans the accumulation rate ranges from 0.1-0.5 grammes per square centimetre per 1,000 years to 1-2.9 grammes per square centimetre per 1,000 years on the lower part of the continental slope, and around 0.1-0.5 grammes per square centimetre per 1,000 years in the vast areas of the basins within the zone of diatom oozes (Fig. 10.17). At the northern margin of the belt absolute masses decrease rapidly to 0.05-0.1 grammes per square centimetre per 1,000 years, and in the arid zone to less than 0.05 grammes per square centimetre per 1,000 years. Thus, as in the Bering Sea, not only studies of diatoms in suspension and in bottom sediments, but also determinations of absolute rates of silica accumulation show here a discharge of neritic diatom detritus from the shelf to the peripheral parts of the deep-sea basins, and a sharp increase of silica accumulation rates in such places. Also marked is an increase in silica accumulation rates in the peripheral parts of the underwater elevations (for instance near the Kerguelen ridge). This is associated with the fact that the hydrodynamic conditions above the ridge cause the diatom suspension to move towards the edges of the ridge, where it settles along with the local material. Thus the influence of the underwater ridge here is similar to that of the shelf.

It is interesting to correlate the characteristic rates of amorphous silica accumulation in all the major siliceous belts of the Earth with the distribution of diatoms and

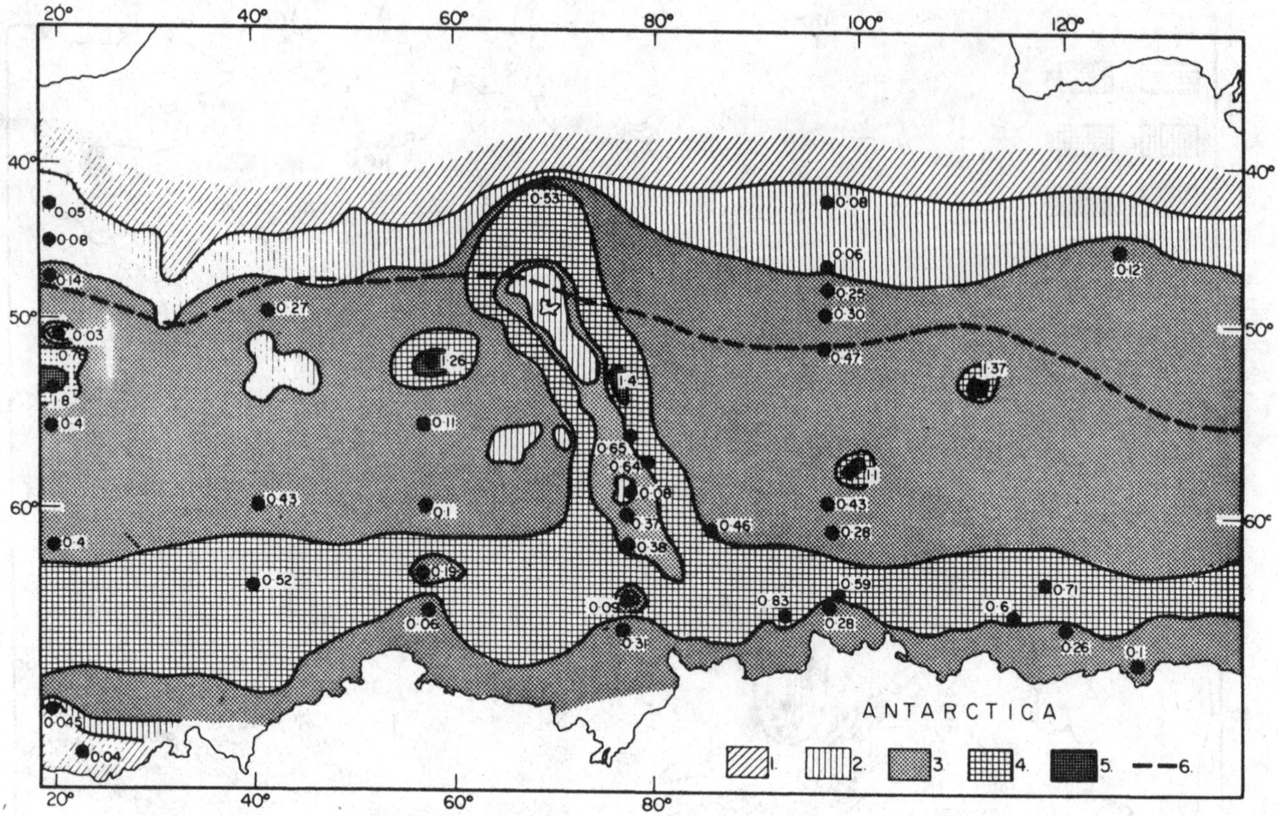
ANTARCTICA

Fig. 10.17. The pattern of absolute rates of silica accumulation in the Indian Ocean sector of the Antarctic for the Holocene (0-11,000 years) in grammes per square centimetre per 1,000 years. 1. less than 0.05; 2. from 0.05 to 0.1; 3. from 0.1 to 0.5; 4. from 0.5 to 1; 5. more than 1; 6. the present position of the Antarctic Convergence.

radiolarians in suspension and in bottom sediments (Fig. 10.18). It can be seen from the sections that the maximum silica accumulation rates for the Holocene are observed in the Bering Sea (1.5-3 grammes per square centimetre per 1,000 years). Within the arid zones they drop to values below 0.05 grammes per square centimetre per 1,000 years, and there is a considerable increase within the equatorial belt. In the southern belt the absolute rates do not usually exceed 1-1.9 grammes per square centimetre per 1,000 years. The highest silica accumulation rates are recorded in the Gulf of California where they vary between 0.6 and 174 grammes per square centimetre per 1,000 years with an average of about 50 grammes per square centimetre per 1,000 years (Calvert 1966).

During the Wurm glaciation Far-Eastern seas were covered with ice and conditions there remind one of those in the Arctic basin today. Beyond the ice edge in the northern Pacific during the glaciation a slight increase in absolute rates is found. Silica accumulation increases considerably during the glaciation in the equatorial zone. Within the southern silica accumulation belt at that time the maximum was displaced 500-800 kilometres northward compared to its present position, which is also a result of the severe ice conditions, but the silica accumulation rate here during the glaciation did not exceed the recent one.

Thus the response of silica accumulation to climatic changes may be different. Within the northern belt the glaciation has resulted in a sharp decrease of silica accumulation rates, within the equatorial belt it caused a sharp increase, and within the southern, although the maximum appears to have been displaced, the intensity remained the same.

Similar results have been obtained on a section through the southern and equatorial zones of the Indian Ocean. In the Atlantic Ocean the northern zone of silica accumulation is absent and the equatorial one is weaker, apparently owing to feeble divergence.

ELEMENTS OF THE SILICA BUDGET

On the basis of a primary phytoplankton production of 35-70 milliard tons organic carbon per year (Winberg 1960) and on a amorphous silica/organic carbon ratio equal to 2.3 (Lisitizin 1964), one can determine the amount

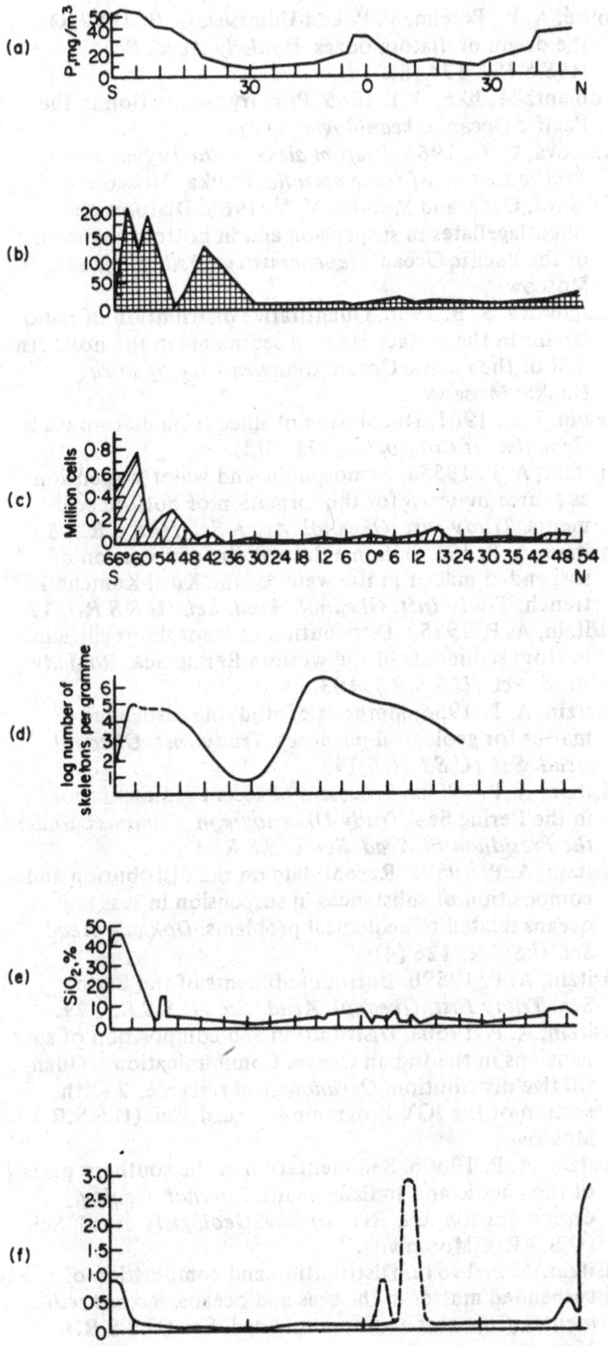
(a)
P, mg/m³
50
40
20
0
S
30
0
30
N
(b)
200
150
100
50
0
(c)
Million cells
0·8
0·6
0·4
0·2
0
66°60 54 48 42 36 30 24 18 12 6 0° 6 12 13 24 30 35 42 48 54
S
N
(d)
log number of skeletons per gramme
6
5
4
3
2
1
(e)
SiO₂, %
50
40
30
20
10
0
(f)
3·0
2·5
2·0
1·5
1·0
0·5
0
(g)
3·0
2·5
2·0
1·5
1·0
0·5
0
66°60 54 48 42 36 30 24 18 12 6 0° 6 12 18 24 30 36 42 48 54°
S
N

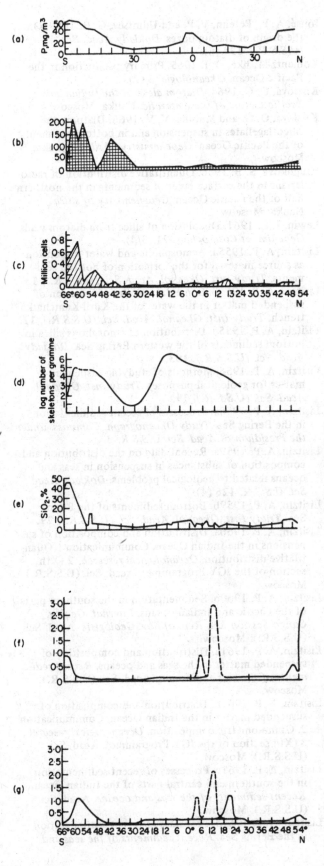

Fig. 10.18. Zonation of recent silica accumulation. (*a*) Phosphorus distribution along a meridional sector through the Pacific Ocean. (*b*) The amount of diatom frustules in suspended matter in the 0-25 metre layer (in millions cells/m^3) along a section through the Pacific Ocean. (*c*) The amount of diatom frustules in the surface layer of sediments along a section through the Pacific Ocean. (*d*) The amount of radiolarians in suspended matter along a meridional section through the Indian Ocean. (*e*) The amount of amorphous silica in the surface layer of sediments (in per cent of the dry sediment). (*f*) Absolute rates of silica accumulation in grammes per square centimetre per 1,000 years for the Holocene (0-11,000 years) along a meridional section through the Pacific Ocean. (*g*) The same for the last glaciation (Wurm 11-80,000 years).

of silica which is utilized by planktonic organisms in constructing their skeletons. This amount comprises up to 80-160 milliard tons amorphous silica per year. The annual silica supply from land makes up about 0.3 milliard tons (Brujevic 1953), and its total reserves in the world's oceans are 5,480 milliard tons (Strakhov 1963).

In calculating geochemical budgets one usually proceeds from the standpoint that the total amount of an element supplied to the ocean is equal to the quantity settling annually to the bottom. Thus the silica amount deposited annually cannot exceed 0.3 milliard tons.

Data on losses in the silica content of siliceous planktonic organisms during settling from the surface to the bottom, and at the bottom proper, show that these budget estimations are valid. Only 1/10 and in some places 1/50-1/100 of the silica utilized annually by plankton near the ocean surface is buried in bottom sediments. Therefore, it is not necessary to include volcanic sources to make up for the seeming deficiency of silica entering the oceans from the land.

These data can also be used to draw conclusions about the present scale on which biogenic processes occur in the oceans. Their influence is such that they involve not only all silica supplied from land, but also a tenfold greater amount of amorphous silica most of which is subsequently returned to the exchangeable part of the budget. This conversion mechanism for great masses of silica also involves some other elements and compounds (both those composing the frustules and protoplasm, and sorbed ones).

Silica acts in this case as a transport medium, delivering large amounts of elements and compounds to the depths. After the frustules (the carriers) are dissolved, elements and compounds associated with them pass into deep waters and into sediments. The action of this mechanism has been overlooked by petrologists, since it cannot be observed when studying sediments and rocks. Thus, recent silica accumulation is a most important process operating on a planetary scale.

Received November 1967.

BIBLIOGRAPHY

Atlas of the Antarctic, Vol. I, 1966 (maps of suspended matter and of bottom sediments), Vol. II, 1969. Leningrad.

Arrhenius, G. 1963. Pelagic Sediments. In *The Sea,* 3, (M. N. Hill *ed.*), 655-727. London.

Bailey, J. W. 1851. Microscopical examination of soundings made off the coast of the United States by the Coast Survey. *Smithsonian Contr. Knowlege,* 2 (2).

Bailey, J. W. 1856. Notice of microscopic forms found in the soundings of the Sea of Kamchatka. *Amer. Journ. Sci. Arts, Ser. 2,* 22, 1-6.

Belknap, C. E. 1874. Deep sea soundings in the north Pacific ocean by U.S.S. *Tuscarora.* Washington.

Belov, N. A. and Lapina, N. N. 1961. *Bottom sediments of the Arctic basin.* Marine Transport Publishers: Leningrad.

Bezrukov, P. L. 1955. On the distribution and rate of sedimentation of silica sediments in the Sea of Okhotsk. *Doklady Acad. Sci. (U.S.S.R.),* **103** (3).

Bezrukov, P. L. 1960. Bottom sediments of the Sea of Okhotsk. *Trudy Inst. Okeanol. Acad. Sci. (U.S.S.R.),* 32.

Bezrukov, P. L. and Lisitzin, A. P. 1960. Classification of bottom sediments in recent marine basins. *Trudy Inst. Okeanol. Acad. Sci. (U.S.S.R.),* 32.

Bezrukov, P. L., Lisitzin, A. P,, Romankevich, E. A. and Skornyakova, N. S. 1961. Recent sedimentation in the northern part of the Pacific Ocean. *Recent sediments of the seas and oceans,* Acad. Sci. (U.S.S.R.) Moscow.

Bogoyavlensky, A. N. 1966. Distribution and migration of dissolved silica in the oceans. *Geochemistry of silica.* Nauka: Moscow.

Brujevic, S. W. 1953. On the geochemistry of silica in the sea. *Izv. Acad. Sci. (U.S.S.R.), ser. Geol.* **4**.

Calvert, S. E. 1966. Origin of diatom-rich varved sediments from the Gulf of California. *Journ. of Geol.* **74** (5).

Chumakov, V. D. 1961. On the suspended silica content of the sea water. *Trudy of the Soviet Antarctic expedition,* 19.

Emelyanov, E. M. 1966. Distribution of authigenous silica in suspension and in recent sediments of the Mediterranean Sea.*Geochemistry of silica.* Nauka: Moscow.

Gershanovich, D. E. 1962. New data on recent sediments of the Bering Sea. *Trudy VNIRO,* **46**.

Gershanovich, D. E. 1964. Recent sediments of the Bay of Alaska. *Internat. Geol. Congress, session XXII. Reports of Soviet Geologists.*

Gessner, F. 1959. *Hydrobotanic.* Berlin.

Goldberg, E. D. 1958. Determination of opal in marine sediments. *Journ. Mar. Res.* **17**, 178-82.

Hanzawa, S. 1935. Diatom (*Ethmodiscus*) ooze obtained from the tropical Southwestern North Pacific Ocean. *Records Oceanogr. work in Japan,* **7** (1).

Jousé, A. P. 1962. *Stratigraphic and palaeogeographic studies in the Northwestern Pacific.* Acad. Sci. (U.S.S.R.): Moscow.

Jousé, A. P., Koroleva, G. S. and Nagayeva, G. A. 1962. Diatom algae in the surface layer of the bottom sediments from the Indian sector of the Antarctic. *Trudy Inst. Okeanol. Acad. Sci. (U.S.S.R.),* **61**.

Jousé, A. P., Petelin, V. P. and Udintsev, G. B. 1959. On the origin of diatom oozes. *Doklady Acad. Sci. (U.S.S.R.),* **124** (6).

Koblentz-Mishke, O. I. 1965. Primary production in the Pacific Ocean. *Okeanologia,* **5** (2).

Kozlova, O. G. 1964. *Diatom algae of the Indian and Pacific sectors of the Antarctic.* Nauka: Moscow.

Kozlova, O. G. and Muhina, V. V. 1966. Diatoms and silicoflagellates in suspension and in bottom sediments of the Pacific Ocean. *Geochemistry of silica,* Nauka: Moscow.

Kruglikova, S. B. 1966. Quantitative distribution of radiolarians in the surface layer of sediments in the northern half of the Pacific Ocean. *Geochemistry of silica.* Nauka: Moscow.

Lewin, J. C. 1961. Dissolution of silica from diatom walls. *Geochim. et Cosmochim.* **21** (3/4).

Lisitzin, A. P. 1955a. Atmospheric and water suspension as source material for the formation of bottom sediments. *Trudy Inst. Okeanol. Acad. Sci. (U.S.S.R.),* 13.

Lisitzin, A. P. 1955b. Some data on the distribution of suspended matter in the water of the Kuril-Kamchatka trench. *Trudy Inst. Okeanol. Acad. Sci. (U.S.S.R.),* 12.

Lisitzin, A. P. 1955c. Distribution of amorphous silica in bottom sediments of the western Bering Sea. *Doklady Acad. Sci. (U.S.S.R.),* **103**.

Lisitzin, A. P. 1956. Methods of studying suspended matter for geological purposes. *Trudy Inst. Okeanol. Acad. Sci. (U.S.S.R.),* **19**.

Lisitzin, A. P. 1958a. Processes of recent sedimentation in the Bering Sea. *Trudy Okeanograph. Commiss. under the Presidium of Acad. Sci. U.S.S.R.* 3.

Lisitzin, A. P. 1959a. Recent data on the distribution and composition of substances in suspension in seas and oceans related to geological problems. *Doklady Acad. Sci. U.S.S.R.* **126** (4).

Lisitzin, A. P. 1959b. Bottom sediments of the Bering Sea. *Trudy Inst. Okeanol. Acad. Sci. (U.S.S.R.),* 29.

Lisitzin, A. P. 1960a. Distribution and composition of suspensions in the Indian Ocean. Communication 1.Quantitative distribution. *Oceanological research,* 2 (Xth section of the IGY Programme). Acad. Sci. (U.S.S.R.) Moscow.

Lisitzin, A. P. 1960b. Sedimentation in the southern parts of the Pacific and Indian oceans. *Internat. Geolog. Congr. Session XII. Rep. of Sov. Geologists.* Acad. Sci. (U.S.S.R.): Moscow.

Lisitzin, A. P. 1961a. Distribution and composition of suspended matter in the seas and oceans. *Recent sediments of the seas and oceans.* Acad. Sci. (U.S.S.R.): Moscow.

Lisitzin, A. P. 1961b. Distribution and composition of suspended matter in the Indian Ocean. Communication 2. Granulometric composition. *Oceanological research,* 3 (Xth section of the IGY Programme). Acad. Sci. (U.S.S.R.): Moscow.

Lisitzin, A. P. 1961c. Processes of recent sedimentation in the southern and central parts of the Indian Ocean. *Recent sediments of the seas and oceans.* Acad. Sci. (U.S.S.R.): Moscow.

Lisitzin, A. P. 1961d. Processes of recent sedimentation in the Bering Sea. *Recent sediments of the seas and oceans.* Acad. Sci. (U.S.S.R.): Moscow.

Lisitzin, A. P. 1962a. Distribution and composition of suspended matter in the Indian Ocean. Communication 3. Correlation between the granulometric compositions of suspensions and of bottom sediments. *Oceanological research,* **5** (Xth section of the IGY Programme). Acad. Sci. (U.S.S.R.): Moscow.

Lisitzin, A. P. 1962b. Experience of the application of the submerged pumps for large volume deep-water sampling. *Trudy Inst. Oceanol. Acad. Sci. (U.S.S.R.),* **55**.

Lisitzin, A. P. 1963. Bottom sediments of the Antarctic shelf. *Delta and shallow-water marine bottom sediments.* Acad. Sci. (U.S.S.R.): Moscow.

Lisitzin, A. P. 1964. Distribution and chemical composition of suspended matter in the Indian Ocean. *Oceanological research,* **10** (Xth Section of the IGY Programme). Acad. Sci. (U.S.S.R.): Moscow.

Lisitzin, A. P. 1966a. Main regularities in the distribution of recent siliceous sediments and their relation to the climatic zonality. *Geochemistry of silica.* Nauka: Moscow.

Lisitzin, A. P. 1966b. *Processes of recent sedimentation in the Bering Sea.* Nauka: Moscow.

Lisitzin, A. P. 1966c. Distribution of silica in Quaternary sediments as related to the climatic zonality of the geological past. *Geochemistry of silica.* Nauka: Moscow.

Lisitzin, A. P., Belyaev, Yu. I., Bogdanov, Yu. A. and Bogoyavlensky, A. N. 1966. Regularities in the distribution and forms of silica suspended in the waters of the world's Oceans. *Geochemistry of silica.* Nauka: Moscow.

Lisitzin, A. P. and Glazunov, V. A. 1960. Design and experience of operating of 200 litre sampling bottles. *Trudy Inst. Okeanol. Acad. Sci. (U.S.S.R.),* **44**.

Lisitzin, A. P. and Udintsev, G. B. 1952. New grab sampler 'Ocean-50'. *Meteorology and Hydrology*, **8**.

Lisitzin, A. P. and Zhivago, A. V. 1958a. Bottom relief and sediments in the southern part of the Pacific Ocean. Communication 1, *Izv. an S.S.S.R. ser. Geograph.*, **2**. Communication 2. *Izv. Acad. Sci. (U.S.S.R.), ser. Geograph.* **3**.

Listizin, A. P. and Zhivago, A. V. 1958b. Modern methods of studying bottom geomorphology and sediments of the Antarctic Seas. *Izv. Acad. Sci. (U.S.S.R.), ser. geograph.* **6**.

Moore, H. B. 1931. Muds of the Clyde Sea area. *Journ. Mar. Biol. Ass. U.K.* **17** (2).

Murray, I. and Irvine, R. 1890-1. On silica and siliceous remains of organisms in modern seas. *Proc. Roy. Soc. Edinb.* **18**, 229.

Murray, D. and Renard, A. F. 1891. Deep-sea deposits *Sci. Rep. 'Challenger' Exped. 1873-76.* London.

Oshite, K. *Suspended Matter in the Sea Water of South-Eastern Part of Hokkaido* (In Japanese).

Petrushevskaya, M. G. 1966. Radiolarians in plankton and in bottom sediments *Geochemistry of sicila.* Nauka: Moscow.

Ponomarev, A. I. 1961. *Methods of chemical analysis of siliceous and carbonate rocks.* Acad. Sci. (U.S.S.R.) Moscow.

Revelle, R. R. 1944. Marine bottom samples collected in the Pacific Ocean by the *Carnegie* on its seventh cruise. 1928-1929. *Oceanography,* **2**.

Revelle, R., Bramlette, M., Arrhenius, G. and Goldberg, E. D. 1955. Pelagic sediments of the Pacific. *Geol. Soc. Amer. Spec. Paper,* **62**.

Riedel, W. R. 1959. Siliceous organic remains in pelagic sediments. *Silica in Sediments.* Soc. Econ. Paleont. Mineral., Spec. Publ. **7**: Tulsa.

Romankevich, E. A., Bezrukov, P. L., Baranov, E. I. and Khristianova, L. A. 1966. Stratigraphy and absolute age of the deep-sea sediments in the Western Pacific. *Okeanologia,* **14** (Results of researches in the International Geophysical Year projects). Nauka: Moscow.

Saidova, Kh. M. 1961. *Foraminiferal ecology and palaeogeography of the far-eastern seas of the U.S.S.R. and of the north-western Pacific.* Acad. Sci. (U.S.S.R.) Moscow.

Schott, G. 1935. *Geographic des Indischen und Stillen Ozeans.* Hamburg.

Semina, H. J. 1963. Phytoplankton of the central Pacific on the section along the 174° W meridian. Part II. The number of phytoplankton cells. *Trudy Inst. Okeanol. Acad. Sci. (U.S.S.R.),* **71**.

Shurko, I. I. 1966. Amorphous silica in the bottom sediments of the North Atlantic. *Geochemistry of silica.* Nauka: Moscow.

Skornyakova, N. S. 1961. Sediments in the North-eastern Pacific. *Trudy Inst. Okeanol. Acad. Sci. (U.S.S.R.),* **45**.

Sokolova, M. N. 1959. Some ecological peculiarities of the deep-water bottom invertebrates. *Progress in Oceanology,* Acad. Sci. (U.S.S.R.). Moscow.

Solovyev, A. V. 1960. Typical features of sedimentation process in the Sea of Japan. *Internat. Geol. Review,* **4**, 17-23.

Strakhov, N. M. (*ed.*). 1957. Methods for studying sedimentary rocks, 1, 2. *Gosgeoltehizdat.*

Strakhov, N. M. 1963. Types of lithogenesis and their evolution during the Earth's history. *Gosgeoltehizdat.*

Strakhov, N. M., Brodskaya, N. G., Knyazeva, L. M., Razzhivina, A. N., Rateev, M. A., Sapozhnikov, D. G. and Shishova, E. S. 1954. *Sedimentation in recent basins.* Acad. Sci. (U.S.S.R.): Moscow.

Sverdrup, H. U., Johnson, M. W., Fleming, R. H. 1942. *The oceans.* New Jersey.

Tchindonova, Yu, G. 1959. Feeding of some groups of the deep-water macroplankton in the North-Western Pacific. *Trudy Inst. Okeanol. Acad. Sci. (U.S.S.R.),* **30**.

Winberg, G. G. 1960. *Primary production of reservoirs.* Publish. Acad. Sci. Belorussian SSR, Minsk.

11. DISTRIBUTION OF CARBONATE MICROFOSSILS IN SUSPENSION AND IN BOTTOM SEDIMENTS

A. P. LISITZIN

Summary: About half the area of the sediments of the world's oceans consists of recent carbonate sediments. To study the quantitative distribution of calcium carbonate in suspension some 300 samples were analysed. Over ten thousand analyses were used to map calcium carbonate distribution in ocean sediments. At the same time microscopic studies of carbonate remains in suspension and in sediments were made; along with X-ray diffraction and electron microscopic analyses of the fine fractions.

Maximum amounts of carbonate plankton are found below the surface layer at 10-200 metre depths. Suspension analysis shows that the maximum concentrations of calcium carbonate (in milligrammes per litre) are recorded near the Convergence zone in the Southern Hemisphere and in the vicinity of the polar front in the Northern Hemisphere. High values are also observed in the equatorial zone. In the tropical (arid) zones, where the purest carbonate sediments accumulate, calcium carbonate content of suspension is extremely low. This is confirmed not only by the results of chemical analyses, but also by counts of Foraminifera and coccolithophorids in suspension. If the quantity of Foraminifera in the arid zones is taken to be equal to unity, it shows two to ten times increase in the neighbouring humid zones (temperate and equatorial) and decreases by tens of times in the ice zones. Thus three belts of carbonate in suspension can be distinguished.

Studies of Foraminifera from the ocean depths show that they do not dissolve during settling, therefore the major belts of carbonate in suspension are reflected on the ocean bottom in the form of belts of high rates of carbonate accumulation. Study of the laws of calcium carbonate distribution in sediments indicates that at great depths calcium carbonate solution proceeds in the surface layer of the sediment. The carbonate sediment belts prove to be broken into separate patches by regions where depths exceed the critical depth. Along with this, carbonate material in the vicinity of the continents is diluted by terrigenous material, and within the silica accumulation belts by silica. All this complicates the pattern of calcium carbonate distribution in sediments.

The critical (compensation) depth is that at which calcium carbonate decreases to 10 per cent and corresponds to the replacement of carbonate sediments by red clay with increasing depth.

A detailed analysis of critical depth positions in the main climatic zones of the oceans shows that they change systematically. The most important factors are:

1. phytoplankton productivity (determined by climatic zonation);
2. calcium carbonate dilution with carbonate-free (terrigenous and siliceous) material;
3. the solution of tests at depths below 3,700-4,000 metres;
4. mechanical differentiation.

If the influence of diluting material is eliminated the climatic zonation of calcium carbonate distribution in sediments becomes distinct, with a vertical zonation superimposed on it. The apparent disorderly pattern of calcium carbonate distribution in sediments turns out to be regular.

Calcium carbonate distribution studies by absolute masses show that three latitudinal belts of high carbonate accumulation rates (which are broken into separate patches by the regions of great depths) can be traced in the bottom sediments. In the Atlantic and Indian oceans calcium carbonate accumulation rates are 10-30 grammes per square centimetre per 1,000 years, i.e. they considerably exceed the rates of recent silica accumulation. These belts are separated by zones with low absolute masses of calcium carbonate (0.1-1 gramme per square centimetre per 1,000 years and lower) which correspond to the arid zones. Absolute masses in the ice zones are also insignificant. The rate of recent biogenous carbonate accumulation in the ocean is, on the average, 4.5 times higher than that of biogenous silica accumulation. The annual deposition of terrigenous (clastic and clayey) material over the ocean bottom is four to seven times in excess of biogenous (carbonate and siliceous) sedimentation.

INTRODUCTION

Calcium carbonate is the most important biogenous component in recent sedimentation. According to various authors, from 48 to 55 per cent of the total area of Recent bottom sediments in the oceans is covered by carbonate sediments (Pacific Ocean – 36.2 per cent, Indian Ocean – 54.3 per cent, Atlantic Ocean – 67.5 per cent; Bezrukov, Lisitzin, Petelin and Skornyakova 1961). Carbonates are also of very great significance in the accumulations of older sedimentary strata.

A number of studies has shown that in the seas and oceans the warmer waters are saturated, and in some places oversaturated, with calcium carbonate; hence, chemogenic carbonate accumulation may be expected there. The colder waters are undersaturated with calcium carbonate. Chemogenic deposition is completely precluded for magnesium carbonate, since sea water is undersaturated with it everywhere. Despite the high calcium carbonate content of the warmer waters, its chemogenic deposition is not observed (except for one or two localities). Many papers have shown that Recent carbonate accumulation in the pelagic parts of seas and oceans is biogenous. Only in lagoons and bays, as well as in the vicinity of the Great Bahama Bank, do conditions exist for the chemical-biological accumulation of aragonite crystals.

In the near-shore areas of seas and oceans and on shelves, benthonic organisms are of major importance for calcium carbonate accumulation: e.g. molluscs, bryozoans, coral reefs, etc. In the pelagic parts of the oceans the situation is reversed; the dominant role in carbonate accumulation there belongs to planktonic organisms: foraminifers, coccolithophorids, and more rarely pteropods. The tests of these organisms are composed of calcium carbonate while the magnesium carbonate content is usually insignificant. Thus the calcium carbonate content of the pelagic bottom sediments may be indicative of the amount of foraminiferal material in the sediment. That is why, at shallow oceanic depths, where foraminiferal tests are well preserved, a close relation is observed between calcium carbonate content and the number of tests in one gramme of the sediment – the so-called 'Schott foraminiferal number'. Yet, it is shown below that the foraminiferal number may vary over wide limits depending on the conditions of test preservation. In some places with a low foraminiferal number the sediments are entirely composed of finely-dispersed foraminiferal material.

Data obtained in recent years on calcium carbonate distribution and composition in the ocean water column (in suspension and in plankton), as well as in bottom sediments, make it possible to draw a number of conclusions that are essential for micropalaeontology.

METHODS AND DATA

Suspended matter was collected by the same methods of separation and membrane ultrafiltration as for amorphous silica studies (see Lisitzin, chapter 10). Carbonate material was determined in the same samples as the silica, which enables one to judge the role of the biogenous factor in sedimentation in different parts of the ocean.

Since the preservation of different micro-organisms with carbonate skeletons in suspension varies greatly, data from chemical and X-ray diffraction analyses were used for quantitative estimations. The genetic composition of the carbonate remains was determined by microscopic studies of the suspension. X-ray diffraction methods, immersion and staining techniques were used to study the mineralogy of the carbonates.

Bottom sediments and suspended matter were studied by the same methods (except for the microchemical analyses of the membrane filters) which ensures comparability of the data obtained.

For quantitative determinations of the calcium carbonate both in the sediments and in the suspensions obtained by the separator the classical method of Knopp-Frezenius was used (Strakhov 1957; Ponomarev 1961). The modern variation of the method provides ±0.1-0.2 per cent accuracy in the determinations. Knopp's method was employed, combined with analysis of hydrochloride extracts, for calcium carbonate and magnesium carbonate determinations. Microchemical and X-ray diffraction methods were used for the analyses of calcium carbonate in suspension made on the membrane filters.

Calcium carbonate was determined in more than 300 samples obtained with the industrial separators in different parts of the ocean, as well as on a great number of the membrane filters from the surface and along vertical sections extending to the bottom. All this resulted in studies covering all the major areas of the oceans.

Calcium carbonate studies of the surface layer of bottom sediments are based on more than 10,000 reliable determinations, about one half of which were conducted in the U.S.S.R. by applying standardized methods. The map of calcium carbonate in sediments is based on about 3,000 analyses for the Atlantic Ocean, on more than 1,500 analyses for the Pacific Ocean, and on more than 800 analyses for the Indian Ocean. In addition, a great number of analyses were made for individual seas.

The application of standardized methods for different parts of the oceans, often very distant from one another, made it possible to evaluate calcium carbonate determinations previously made by other research workers. For this purpose graphs of dependence were constructed using the data obtained from the same station or from the adjacent region. In so doing a considerable amount of data had, unfortunately, to be rejected. In particular, the semi-quantitative determinations of Trask (Trask 1932, about 1,500 samples) appeared to be highly inaccurate. A

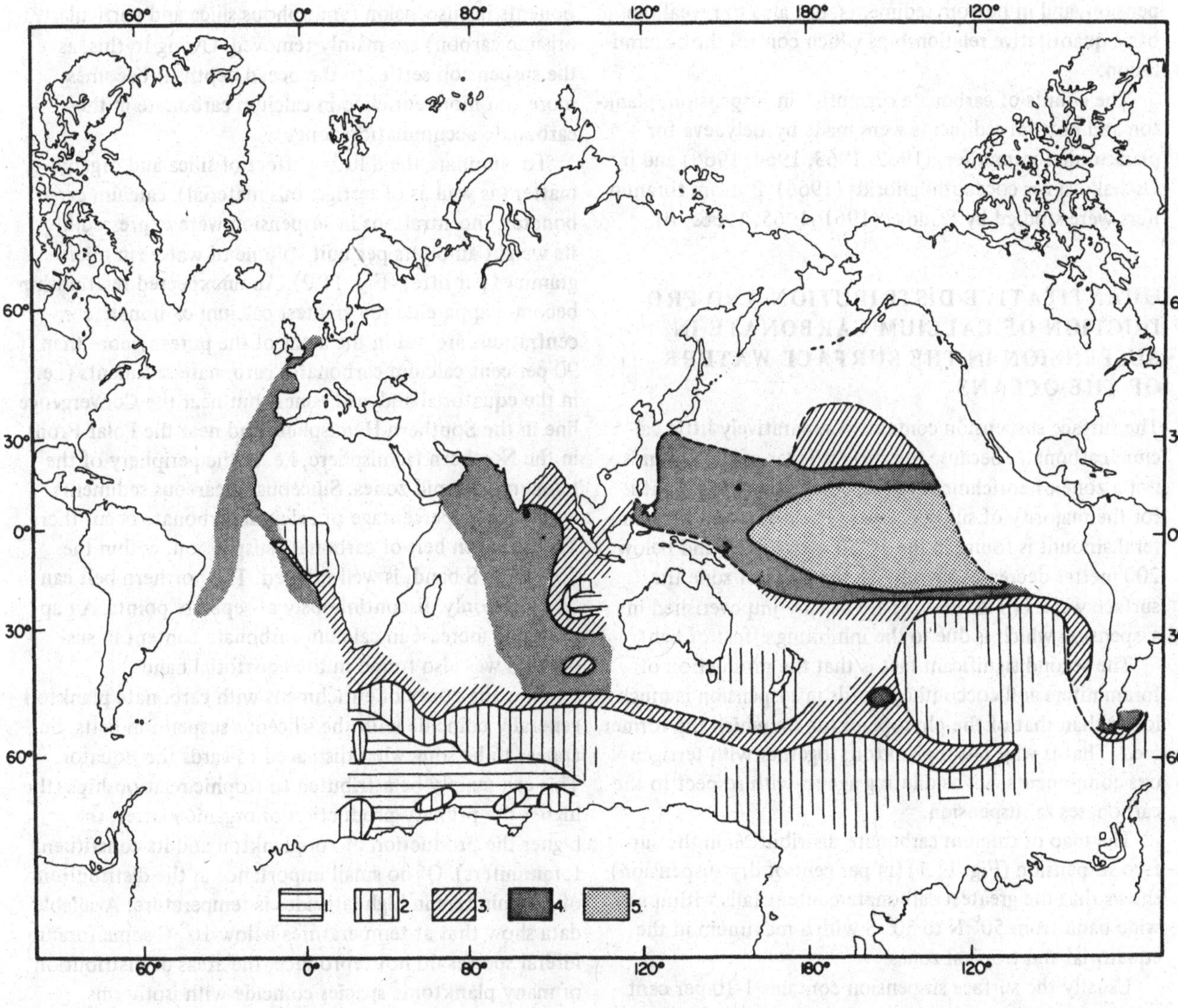
60°
0°
80°
120°
180°
120°
60°
30°
0°
30°
60°
1.
2.
3.
4.
5.

Fig. 11.1. Calcium carbonate distribution in the surface suspended matter (0-5 metre layer) measured in per cent of the dry suspension 1. less than 1%; 2. from 1% to 3%; 3. from 3% to 5%; 4. from 5% to 10%; 5. more than 10%.

significant discrepancy was found in a number of cases in the analyses of Revelle (1944) for the Pacific Ocean. The accuracy of his determinations varies between ±1 and ±5 per cent, sometimes exceeding ±10 per cent. The accuracy of calcium carbonate determinations in the samples of the Swedish Deep-Sea Expedition 1947-8 is from ±0.4 to 4 per cent (Arrhenius 1952).

Over 5,000 determinations of calcium carbonate in sediment cores were used to find the rate of carbonate deposition. The counts of foraminifers and coccolithophorids in the cores were made by the same palaeontologists that studied the distribution of microfossils in suspension and in the surface layer of sediments. The application of standardized accurate methods of analysis to a considerable number of samples made it possible not only to compile reliable maps of calcium carbonate distribution in suspension, and in bottom sediments, but also to reveal the basic quantitative relationships which control the accumulation.

The counts of carbonate organisms in suspension, plankton and bottom sediments were made by Belyaeva for planktonic foraminifers (1962, 1963, 1964, 1966) and by Uschakova for coccolithophorids (1966). Bottom foraminifers were studied by Saidova (1961, 1965, 1966*a*, *b*).

QUANTITATIVE DISTRIBUTION AND PRODUCTION OF CALCIUM CARBONATE IN SUSPENSION IN THE SURFACE WATERS OF THE OCEANS

The surface suspension contains comparatively little calcium carbonate, because the surface layer of the ocean is not a zone of enrichment for carbonate organisms as it is for the majority of siliceous ones. The maximum foraminiferal amount is found in the 10-50 metre layer and below 200 metres decreases sharply. In the tropical zone the surface waters (0-20 metres) are usually impoverished in suspension which is due to the inhibiting effect of light.

The second significant fact is that the production of foraminifers and coccolithophorids in suspension is much lower than that of the phytoplankton on which the former feed. That is why phytoplankton, together with terrigenous components, act as diluting agents with respect to the carbonates in suspension.

The map of calcium carbonate distribution in the surface suspension (Fig. 11.1) (in per cent of dry suspension) shows that the greatest carbonate content falls within a wide band from 50° N to 50° S with a maximum in the equatorial and tropical zones.

Usually the surface suspension contains 1-10 per cent calcium carbonate, which is attributable, as the microscopic and chemical analyses show, to the dilution of carbonates with amorphous silica and with organic carbon. The suspension in cold Antarctic waters contains, as a rule, less than 1 per cent calcium carbonate. In the Pacific Ocean the maximum values (up to 3 per cent calcium carbonate) are found near the Equator. The surface suspension contains only 5-10 per cent calcium carbonate, even in areas of Recent carbonate accumulation such as the Mediterranean and the Red Sea.

Thus the amount of the diluting biogenous material, amorphous silica and organic carbon in suspension is usually much greater than that of calcium carbonate, this being true both of regions with intensive carbonate sedimentation and (particularly) outside such regions. It is shown below that in settling to the ocean bottom not only are new contributions of foraminifers from those inhabiting the 0-200 metre layer added, but the unstable components in suspension (amorphous silica and particularly organic carbon) are mainly removed. Owing to this, as the suspension settles to the ocean depths it becomes more and more enriched in calcium carbonate in the carbonate accumulation zones.

To eliminate the diluting effect of silica and organic matter (as well as of terrigenous material), calcium carbonate concentrations in suspension were expressed in its weight amounts per unit volume of water (in microgrammes per litre) (Fig. 11.2). An unexpected relationship becomes apparent: the greatest calcium carbonate concentrations are not in the areas of the purest (more than 90 per cent calcium carbonate) carbonate sediments (i.e. in the equatorial and arid zones) but near the Convergence line in the Southern Hemisphere and near the Polar Front in the Northern Hemisphere, i.e. at the periphery of the temperate humid zones. Siliceous-calcareous sediments with a lower percentage of calcium carbonate occur there. The southern belt of carbonate suspension, within the 40° S-60° S band, is well marked. The northern belt can be traced only discontinuously at separate points. An appreciable increase in calcium carbonate content in suspension was also found in the equatorial band.

Thus the zones of enrichment with carbonate plankton generally coincide with the siliceous suspension belts, but appear to be somewhat displaced towards the Equator. This can mainly be attributed to trophic relationships (the higher the primary production of organic matter, the higher the production of zooplankton and its constituent foraminifers). Of no small importance in the distribution of foraminifers in high latitudes is temperature. Available data show that at temperatures below 10° C some foraminiferal species do not reproduce; the areas of distribution of many planktonic species coincide with isotherms (Bradshaw 1957, 1959). With increasing temperature the number of species increases too: at temperatures of 8°-

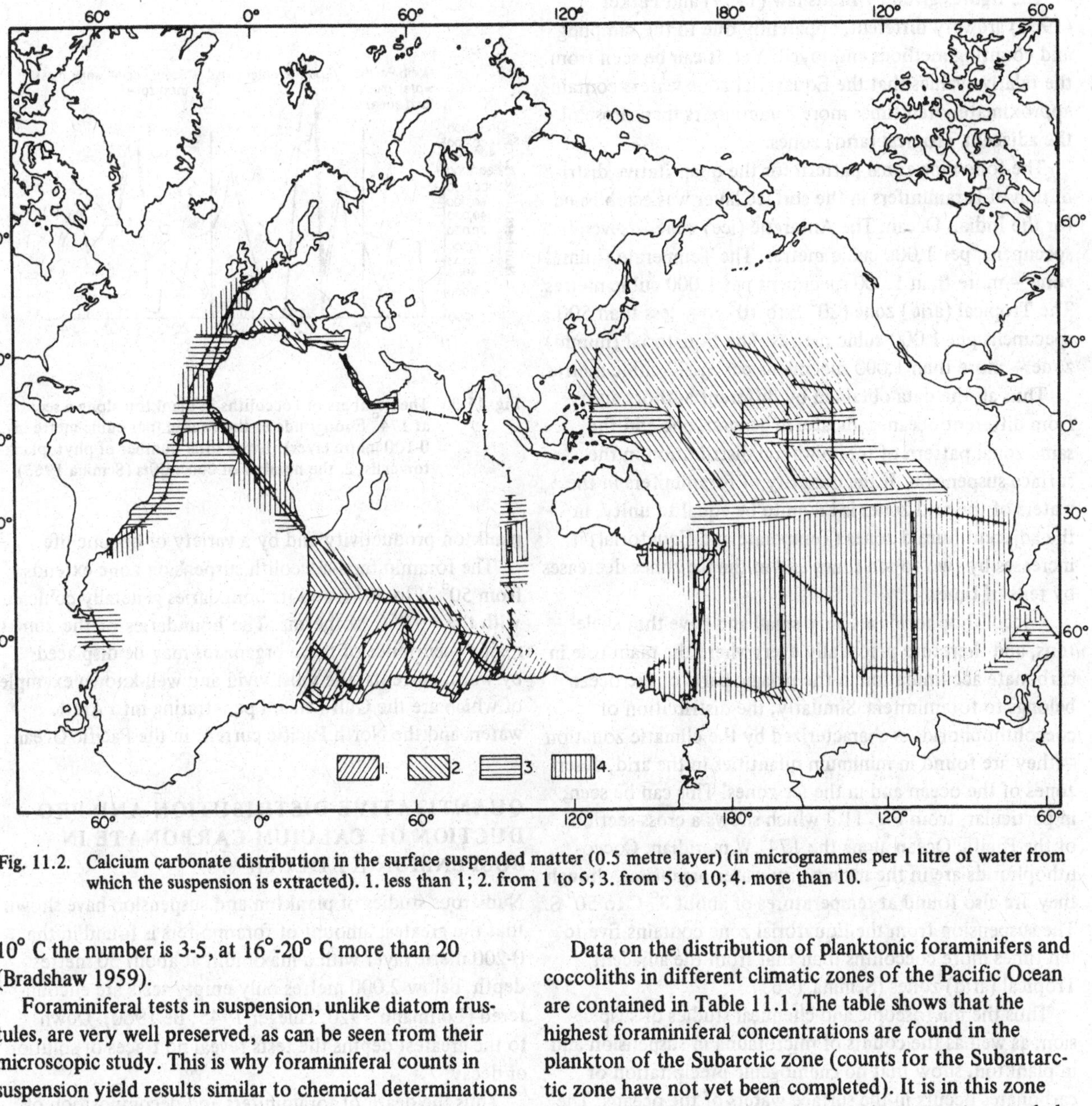
60°
0°
60°
120°
180°
120°
60°
60°
30°
0°
30°
60°
1.
2.
3.
4.

Fig. 11.2. Calcium carbonate distribution in the surface suspended matter (0.5 metre layer) (in microgrammes per 1 litre of water from which the suspension is extracted). 1. less than 1; 2. from 1 to 5; 3. from 5 to 10; 4. more than 10.

10° C the number is 3-5, at 16°-20° C more than 20 (Bradshaw 1959).

Foraminiferal tests in suspension, unlike diatom frustules, are very well preserved, as can be seen from their microscopic study. This is why foraminiferal counts in suspension yield results similar to chemical determinations of calcium carbonate.

Data on the distribution of planktonic foraminifers and coccoliths in different climatic zones of the Pacific Ocean are contained in Table 11.1. The table shows that the highest foraminiferal concentrations are found in the plankton of the Subarctic zone (counts for the Subarctic zone have not yet been completed). It is in this zone that a value of 100 specimens per cubic metre is recorded.

Table 11.1 The distribution of planktonic carbonate-concentrating organisms in the Pacific Ocean (cells per cubic metre).

Zones	Coccoliths (after Semina 1963)	Foraminifers	
		After Bradshaw (1959)	After Parker (1960)
Subarctic	—	After 100	—
Mixed zone		1-100	
Northern tropical	62,170	10-100	Less than 0.1
Equatorial	293,340	—	0.1-1.0
Southern tropical	15,530	—	Less than 0.1

The figures given by Bradshaw (1959) and Parker (1960) are very different, apparently due to the sampling and counting methods employed. Yet, it can be seen from the relative values that the Equatorial zone waters contain approximately ten times more foraminifers than those of the adjacent Tropical (arid) zones.

The following zonal pattern for the quantitative distribution of foraminifers in the surface layer was established for the Indian Ocean. The Antarctic (ice) zone – several specimens per 1,000 cubic metres. The Temperate (humid) zone – more than 1,000 specimens per 1,000 cubic metres. The Tropical (arid) zone (20° S to 40° S) – less than 500 specimens per 1,000 cubic metres. The Equatorial (humid) zone – more than 1,000 specimens per 1,000 cubic metres.

Thus all the data obtained by different authors and from different oceans indicate, in general, one and the same zonal pattern of foraminiferal distribution in the surface suspension. If the quantity of foraminifers in the waters of the arid zones is taken to be equal to unity, in the adjacent humid zones (Temperate and Equatorial) it increases by two to ten times and in the ice zones decreases by tens of times.

Coccolithophorids are very small and have thin skeletons, and therefore practically everywhere the main role in carbonate accumulation in the pelagic parts of the ocean belongs to foraminifers. Similarly, the distribution of coccolithophorids is characterized by the climatic zonation – they are found in minimum quantities in the arid, desert zones of the ocean and in the ice zones. This can be seen, in particular, from Fig. 11.3 which shows a cross-section of the Pacific Ocean along the 174° W meridian. Coccolithophorids are in the main warm-water organisms, though they are also found at temperatures of about 3° C to 50° S. The suspension from the Equatorial zone contains five to ten times more coccoliths than that from the adjacent Tropical (arid) zones (Semina 1963).

Thus the microscopic and chemical studies of suspension, as well as the counts of microfauna in suspension and in plankton, show that no chemogenic precipitation of carbonates occurs in the surface waters of the oceans. The main role in calcium carbonate accumulation belongs to foraminifers, followed by coccolithophorids. Pteropods are of distinctly subordinate significance. Despite the favourable conditions for calcium carbonate concentration existing in the arid zones of the ocean (waters are two and a half to three times oversaturated with calcium carbonate), the highest carbonate planktonic concentrations and the largest calcium carbonate amounts in suspension (in microgrammes per litre) are found not in these regions, but in the three humid zones. Most of all in the Equatorial zone, and somewhat less in the Temperate zones (nearest the Equator), i.e. the places distinguished by high phytoplankton productivity and by a variety of organic life.

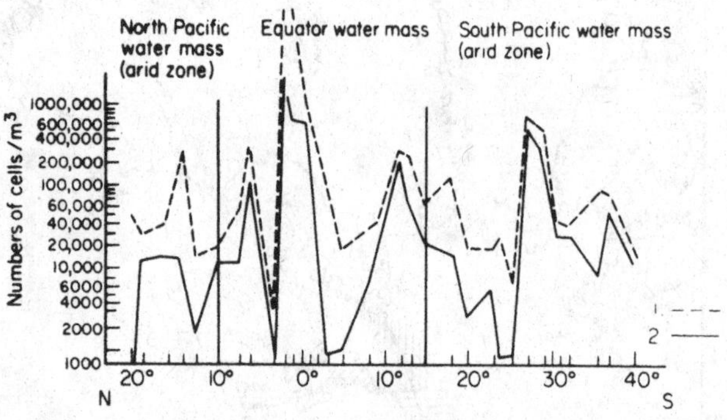
North Pacific water mass (arid zone)
Equator water mass
South Pacific water mass (arid zone)
Numbers of cells/m³
1000,000
600,000
400,000
200,000
100,000
60,000
40,000
20,000
10,000
6000
4000
2000
1000
20°
10°
0°
10°
20°
30°
40°
N
S
1
2

Fig. 11.3. The numbers of coccoliths in plankton along a section at 174° E longitude in the Pacific (per cubic metre in 0-100 metre layer). 1. the total number of phytoplankton cells; 2. the number of coccoliths (Semina 1963).

The foraminiferal-coccolith suspension zone extends from 50° N to 50° S and its boundaries generally coincide with the + 10° C isotherms. The boundaries of the zone of major occurrence of these organisms may be displaced by warm currents, the most vivid and well-known examples of which are the Gulf Stream penetrating into Arctic waters and the North Pacific current in the Pacific Ocean.

QUANTITATIVE DISTRIBUTION AND PRODUCTION OF CALCIUM CARBONATE IN SUSPENSION AT DEPTH

Numerous studies of plankton and suspension have shown that the greatest amount of foraminifers is found in the 0-200 metre layer with a maximum at about 50 metres depth; below 2,000 metres only empty tests are encountered (Lohmann 1920; Phleger 1945; Bé 1960). Down to the greatest depths the tests reveal no traces of solution or decay.

Thus the death of foraminifers and decomposition of protoplasm in them proceeds within the 0-2,000 metre depth range. There is evidence that some planktonic foraminiferal species can inhabit greater depths but this, apparently, may take place only in the areas of a considerable enrichment of suspension in organic matter, i.e. mainly in the continental slope areas. As compared to the surface layer, the absolute quantities of tests in the deep waters are not great – 124 specimens per 1,000 cubic metres at most.

The time of settling of foraminiferal tests, coccoliths and pteropods depends on their sizes and hydraulic properties. The most widely distributed sizes of the planktonic foraminiferal tests are 0.05-0.25 millimetres. The settling velocity through the water column at temperature 5° C ranges from 0.15 centimetres per second to 1.7-2 centimetres per second, i.e. they reach 5 kilometres depth in a few days, while the settling of coccoliths takes many years.

The size of coccolithophorids ranges between 5 and 100μ with an average size of 20-40μ. Coccolithophorids are heterotrophic organisms. Some cases have been observed where organic detritus became covered with coccolithophorids at depth, which increased the settling velocity of such a colony to the ocean bottom. Fig. 11.4 shows

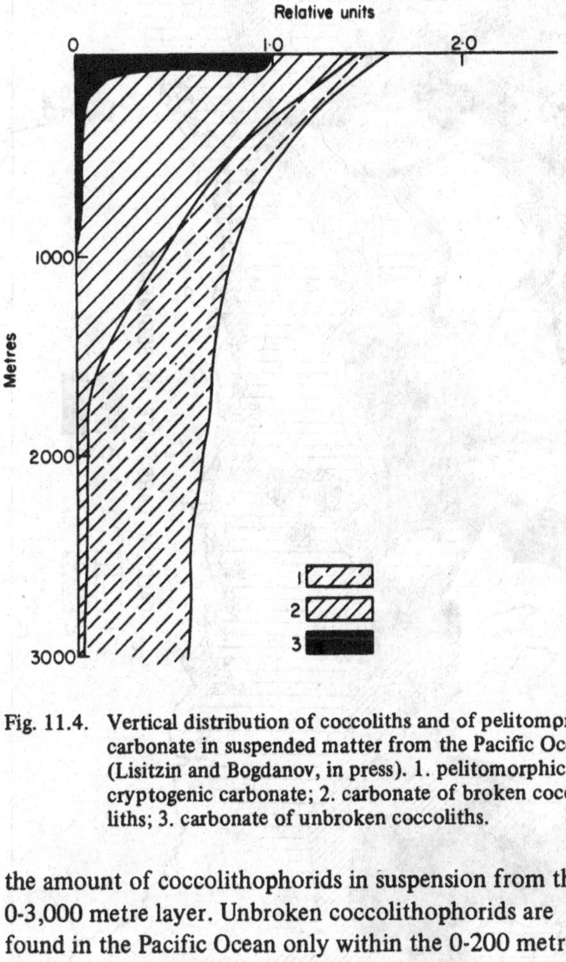
Relative units
0
1·0
2·0
1000
2000
3000
Metres
1
2
3

Fig. 11.4. Vertical distribution of coccoliths and of pelitomorphic carbonate in suspended matter from the Pacific Ocean (Lisitzin and Bogdanov, in press). 1. pelitomorphic cryptogenic carbonate; 2. carbonate of broken coccoliths; 3. carbonate of unbroken coccoliths.

the amount of coccolithophorids in suspension from the 0-3,000 metre layer. Unbroken coccolithophorids are found in the Pacific Ocean only within the 0-200 metre layer, below which they are rare. Only fragments of these algae are encountered at depth – coccoliths and rhabdoliths as well as fine carbonate, the content of which increases at depth simultaneously with a decrease in the amount of skeletons. Thus, unlike foraminifers, the coccolithophorid material is characterized by a general decrease in content in the vertical direction due to its gradual conversion into finely-dispersed carbonate (like diatom silica). The ocean bottom is reached by only part of the initial coccolith material. According to Vikhrenko and Nikolaeva (1962), the change in the coccolith amount in the vertical direction in the Atlantic Ocean differs from that in the Pacific; in the former ocean coccoliths are preserved to the greatest depths.

The size of pteropod shells varies, usually from 0.5 millimetres to 3-4 millimetres; their maximum amounts are found in the 0-200 metre layer. Experiments made by Vinogradov (1961) show that they settle in oceanic water at a rate of one metre per 40-70 seconds (water temperature is about 30° C). It might be assumed that such molluscs can reach the ocean bottom in two to three days. Yet, due to the existence of the picnocline and water turbulence the actual settling velocity of pteropods appears to be considerably lower. This was confirmed when studying the preservation of mollusc bodies. Below 2,000 metres practically all the shells are empty which indicates a residence of many days in the 0-2,000 metre water layer.

Microchemical determinations of calcium carbonate in suspension from the ocean depths (G. I. Tikhonov under the guidance of A. N. Bogoyavlensky) also show that the highest calcium carbonate concentrations are found in the 0-200 metre layer in the Equatorial zone and at the periphery of the Temperate humid zones. A decrease in the calcium carbonate content of suspension is noted in the vertical direction which is due, first of all, to the solution of the finest fractions of coccolithophorids (Fig. 11.4).

Of basic significance is the fact that the high calcium carbonate concentrations mentioned for the carbonate belts at the surface, despite the solution of part of the coccoliths, are preserved down to the great depths. Therefore, it can be expected that the carbonate accumulation belts formed in suspension at the ocean surface should be reflected in the bottom sediments. Since the distribution of calcium carbonate in suspension, like that of silica, is determined mainly by the climatic zonation, the distribution of carbonates in the bottom sediments should be subject to the same zonation.

QUANTITATIVE DISTRIBUTION AND GENESIS OF CALCIUM CARBONATE IN THE SURFACE LAYER OF THE BOTTOM SEDIMENTS

The processing of extensive material on calcium carbonate content of the bottom sediments of the world's oceans has resulted in a set of maps compiled for the Pacific (Lisitzin and Petelin 1967), Atlantic (Lisitzin *et al.* 1967) and Indian (Bezrukov and Lisitzin 1966) oceans as well as for the Antarctic (the Atlas of the Antarctic, v.I, 1966; v.II, 1969; Lisitzin 1969). The maps are based on the collection

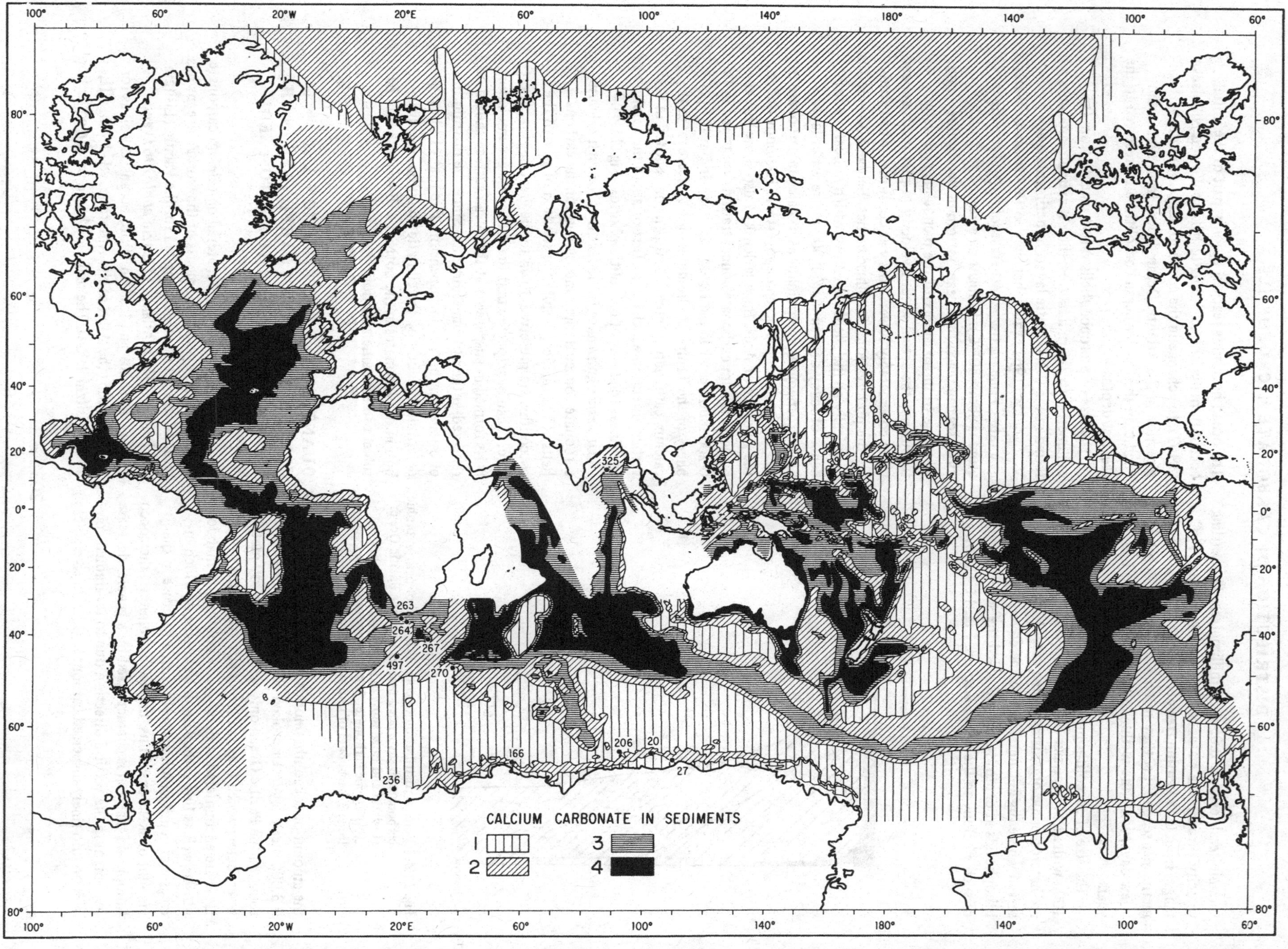
CALCIUM CARBONATE IN SEDIMENTS
1
2
3
4
325
263
264
267
497
270
166
206
20
27
236

Fig. 11.5. Map of calcium carbonate distribution in bottom sediments of the seas and oceans (in per cent of dry sediment). 1, less than 1%; 2, from 1% to 30%; 3, from 30% to 70%;

of new bathymetric maps of the world's oceans drawn at the Institute of Oceanology, Academy of Sciences of the U.S.S.R. The new maps of the types of sediments of the Pacific (Bezrukov, Lisitzin, Petelin and Skornyakova 1966), Indian (Bezrukov and Lisitzin 1966) and the Atlantic (Emelyanov *et al.* 1966) oceans and the Antarctic (Lisitzin 1960*a* and *b*, 1961 *a, b* and *c*, 1969) have also been incorporated in the above-mentioned maps.

Use was also made of all the available material on calcium carbonate distribution in the sediments taken from foreign papers. Earlier the same methods had been used to compile maps of calcium carbonate distribution in the bottom sediments of the Bering Sea (Lisitzin 1959*b*, 1961*d*, 1966; Lisitzin and Petelin 1953), the Sea of Okhotsk (Bezrukov 1960), the Japan Sea (Strakhov *et al.* 1954; Gershanovich 1956), the northwestern Pacific (Bezrukov, Lisitzin, Romankevich and Skornyakova 1961), the Mediterranean Sea (Emelyanov 1965), the Black Sea (Strakhov *et al.* 1954), the Arctic Ocean (Belov and Lapina 1961), etc.

Examination of these maps indicates many differences from those previously compiled.

When generalizing the above material a number of regularities were revealed which had been disregarded before. First of all, while examining the summary map (Fig. 11.5) attention is drawn to the fact that the carbonate sediment belts which were noted in suspension from the surface and from depths are not marked in the bottom sediments. The carbonate accumulation zones (from 50° N to 50° S) in suspension and in sediments coincide.

The pattern of calcium carbonate distribution in bottom sediments appears to be more complicated than that in suspension. In a number of places the non-carbonate sediments correspond to the zones of carbonate deep-sea suspension and vice versa. These facts are extremely important for lithological studies, especially for micropalaeontology. They show that in far from all cases are the carbonate planktonic biocoenoses reflected in the bottom sediments. They are often distorted or the planktonic remains disappear from the sediments altogether.

Calcium carbonate concentrations in the marine bottom sediments range over very wide limits; i.e. from 0-90/98 per cent. Within small areas, mainly, on seamounts and ridges and on the scarps of the continental slope, separate patches of the low-carbonate sediments are traced from the Arctic to the Antarctic. They are also found in some places on the Antarctic shelf, near the ice shores (Lisitzin 1961*c*, 1963, 1969). Foraminiferal sediments are also encountered on the bottom rises of the Arctic under drift ice.

The following three regularities of calcium carbonate distribution in bottom sediments have been established (Lisitzin 1966):

1. The supply of the greatest amounts of carbonate material with planktonic organisms proceeds in the pelagic parts of the ocean only in strictly defined zones. Their position is controlled by the climatic zonation (the zones of high productivity coincide with global divergences).
2. Calcium carbonate content depends on depth, i.e. it is under strict bathymetric control. In the oceans there exist critical depths, below which carbonates do not penetrate, since the rate of calcium carbonate solution increases sharply there.
3. Carbonate material in the bottom sediments is diluted with terrigenous and siliceous material. The actual calcium carbonate content depends on the relationship between the rates of supply of carbonate and diluting material. That this is so is confirmed by the studies of calcium carbonate distribution by the method of absolute masses (open system).

The location of zones of greatest carbonate supply by planktonic organisms was considered above. To this it should be added that in calcium carbonate accumulation in the bottom sediments, along with plankton, the benthonic organisms take part – molluscs, bryozoans, corals, benthonic foraminifers, etc. The first three groups of organisms are found in maximum amounts in the 0-200 metre depth range and are of no great importance in the pelagic sediment area. At depths down to 500 metres benthonic foraminifers prevail everywhere, at 500-1,000 metre depths the relation between them varies from place to place, and from 1,000 metres depth down to the critical depths planktonic foraminifers are dominant (Belyaeva and Saidova 1965). Below the critical depths the role of the benthonic agglutinating forms, which are of no importance for calcium carbonate accumulation, rises again.

The relation of calcium carbonate distribution to depth was noted as far back as the voyage of H.M.S. *Challenger* (Murray and Renard 1891). Attempts to find a direct relationship between depth and calcium carbonate content were made by Pia (1933), though before him André had pointed out that if such a relationship did exist for the Atlantic sediments it was to a very small extent and Pratje (1932) denied the existence of such a relationship altogether. Later the dependence of calcium carbonate on depth was discussed by many authors (Fig. 11.6). The curves of different authors for various basins indicate a sharp decrease in calcium carbonate amounts at depths exceeding 4,000-5,000 metres, yet they are very different. The causes for the relation between calcium carbonate and depth were interpreted in different ways. Use was made of assumptions about the influence of water temperature,

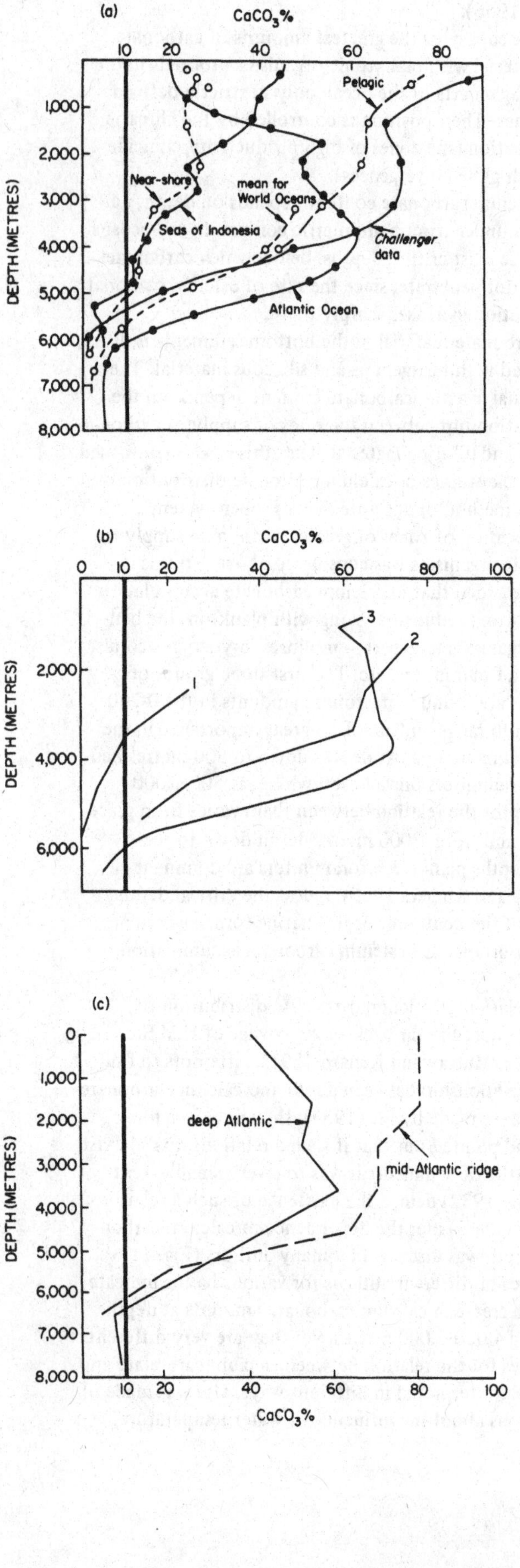
(a)
CaCO3%
10 20 40 60 80
Pelagic
1,000
2,000
Near-shore
mean for
World Oceans
3,000
Seas of Indonesia
4,000
Challenger
data
5,000
Atlantic Ocean
6,000
7,000
8,000
DEPTH (METRES)
(b)
CaCO3%
0 10 20 40 60 80 100
3
2
1
2,000
4,000
6,000
DEPTH (METRES)
(c)
0
1,000
2,000
deep Atlantic
3,000
mid-Atlantic ridge
4,000
5,000
6,000
7,000
8,000
DEPTH (METRES)
10 20 40 60 80 100
CaCO3%

Fig. 11.6. Types of vertical distribution of calcium carbonate in sediments, based on the data of various authors. (*a*) Neeb's data (1943). The seas of Indonesia based on data from the *Snellius* voyage. For the Atlantic Ocean from the data of Pia (1933), the mean for the World Oceans (Trask 1937), separately for the near-shore (closer than 800 kilometres from the shores) and for the pelagic (farther than 800 kilometres from the shores) parts. (*b*). 1. for the Pacific Ocean from 10° to 50° N latitude (Revelle 1944); 2. for the Pacific Ocean from 10° N to 50° S latitudes (Revelle (1944); 3. for the Atlantic Ocean (Pia 1933). (*c*) For the Atlantic Ocean as a whole, and separately for the mid-Atlantic ridge (Turekian 1964).

carbon dioxide content of the water, pH, organic carbon content of the sediments, the destructive action of mud-eaters, the effects of volcanism, etc.

When comparing the maps of calcium carbonate distribution in the bottom sediments with bathymetric maps one notices that the relation between depth and calcium carbonate, though undoubtedly existing, shows up differently in different climatic zones. That is why the analysis of this relation should be made, not for the oceans as a whole, but separately for their major climatic zones.

The depth at which the amount of calcium carbonate decreases to 10 per cent is called the critical (or compensation) depth and corresponds to the replacement of low-carbonate sediments with increasing depth by non-calcareous red clays.

For graphic analysis we have used about 5,200 reliable determinations made for the pelagic sediments and approximately as many determinations made for the seas.

All available reliable analytical data on the calcium carbonate content of the sediments was processed by the graphic method for the following latitudinal zones: 70°-40° N, 40°-20° N, 20° N -20° S, 20°-40° S, 40°-70° S. The most complicated proved to be the Equatorial zone (20° N -20° S) where the need arose for a more detailed analysis distinguishing 10 narrow latitudinal zones. The change of critical depth in a longitudinal direction (within four belts in each of the oceans) was also studied in the Equatorial zone. Within each latitudinal zone at a given depth or over a range of depths the scatter in calcium carbonate amounts may be very significant. However, the maximum concentration of calcium carbonate and the scatter of points are noticed to decrease everywhere at depths exceeding 4,500-5,000 metres. From many graphs one may also see the decrease in the maximum concentration in the upper portion corresponding to the shelves. In other zones this is not pronounced. If in each latitudinal zone a curve is drawn connecting the highest values, i.e. the most pure high-carbonate sediments, one may gain an idea of the relation between calcium carbonate and depth.

The examination of a series of the connecting curves

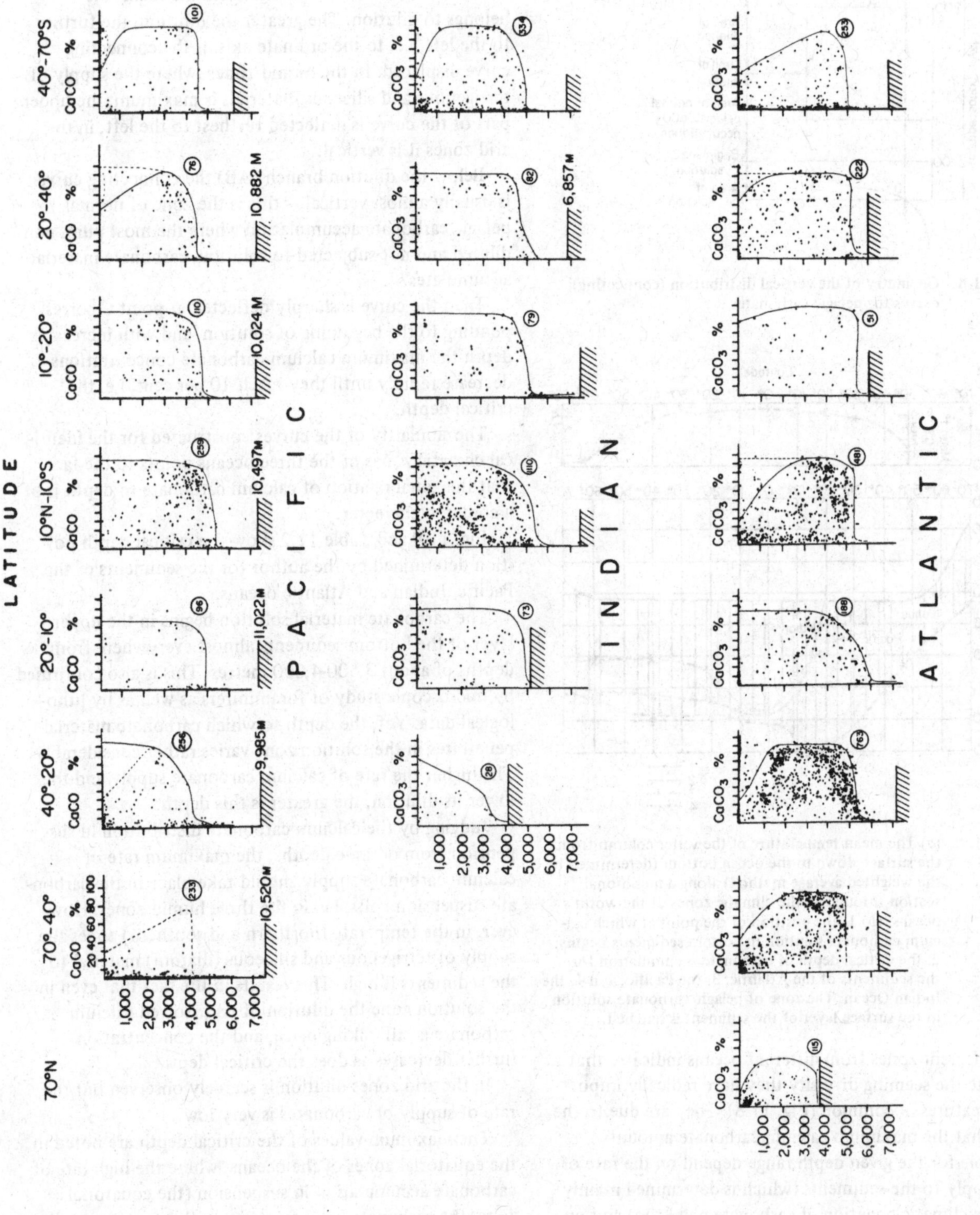
LATITUDE
70°N
70°-40°
40°-20°
20°-10°
10°N-10°S
10°-20°
20°-40°
40°-70°S
PACIFIC
INDIAN
ATLANTIC
CaCO₃ %
1,000
2,000
3,000
4,000
5,000
6,000
7,000
10,520 M
9,985 M
11,022 M
10,497 M
10,024 M
10,882 M
6,857 M

Fig. 11.7. Graphs of vertical distribution of calcium carbonate for different latitudinal zones of the world's oceans.

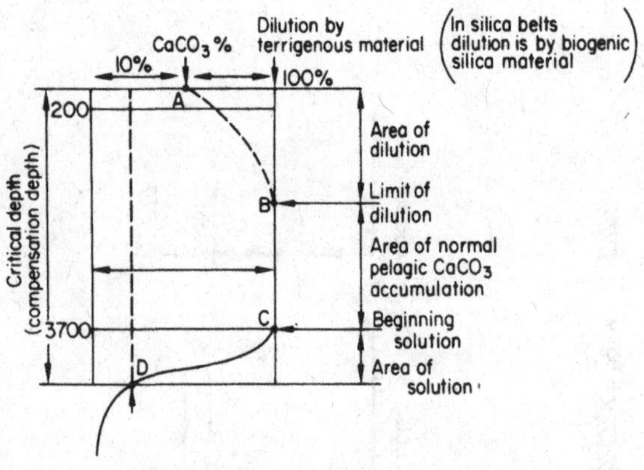
Dilution by terrigenous material
(In silica belts dilution is by biogenic silica material)
CaCO3 %
10%
100%
200
3700
Critical depth (compensation depth)
A
B
C
D
Area of dilution
Limit of dilution
Area of normal pelagic CaCO3 accumulation
Beginning solution
Area of solution

Fig. 11.8. Geometry of the vertical distribution (connecting) curves for pelagic carbonates.

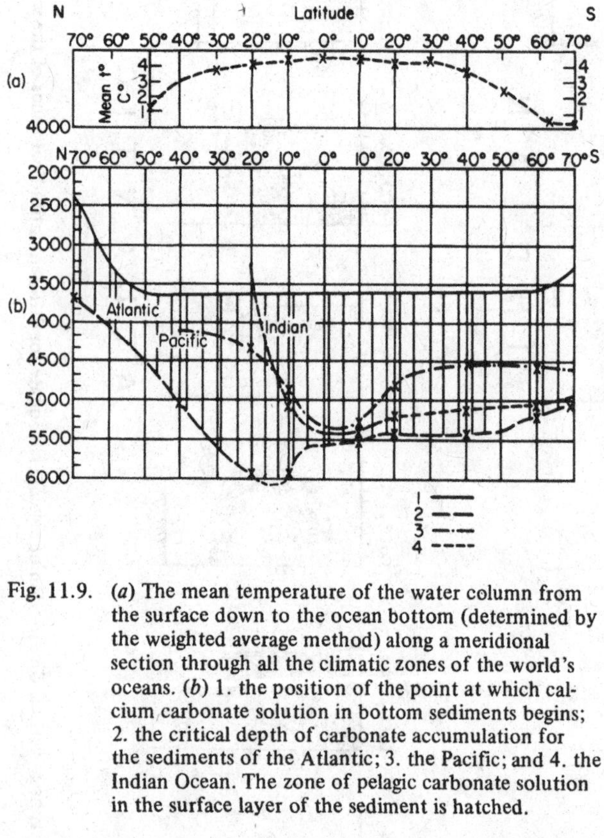
N
Latitude
S
70° 60° 50° 40° 30° 20° 10° 0° 10° 20° 30° 40° 50° 60° 70°
(a)
Mean t° C°
4000
N 70° 60° 50° 40° 30° 20° 10° 0° 10° 20° 30° 40° 50° 60° 70° S
2000
2500
3000
3500
(b)
4000
4500
5000
5500
6000
Atlantic
Pacific
Indian
1
2
3
4

Fig. 11.9. (*a*) The mean temperature of the water column from the surface down to the ocean bottom (determined by the weighted average method) along a meridional section through all the climatic zones of the world's oceans. (*b*) 1. the position of the point at which calcium carbonate solution in bottom sediments begins; 2. the critical depth of carbonate accumulation for the sediments of the Atlantic; 3. the Pacific; and 4. the Indian Ocean. The zone of pelagic carbonate solution in the surface layer of the sediment is hatched.

for different zones from different oceans indicates that despite the seeming diversity they have radically important features in common (Fig. 11.8). They are due to the fact that the maximum calcium carbonate amounts possible for the given depth range depend on the rate of its supply to the sediments (which is determined mainly by the climatic zonation of carbonate plankton) and on the factors decreasing the concentration – on the dilution with terrigenous and siliceous material, and at great depths primarily on solution.

For the upper portion of the curves, at least to 3,700 metres, solution is not essential, and the main rôle belongs to dilution. The greater the dilution, the further to the left, i.e. to the ordinate axis, is the connecting curve displaced. In the humid zones, where the supply of terrigenous and siliceous materials is maximum, the upper part of the curve is deflected furthest to the left, in the arid zones it is vertical.

Below the dilution branch (AB) the connecting curve is usually almost vertical – this is the zone of normal pelagic carbonate accumulation where the most pure, non-diluted and not-subjected-to-solution, carbonate material accumulates.

Then the curve is sharply deflected at point C corresponding to the beginning of solution, and with increasing depth the maximum calcium carbonate concentrations decrease rapidly until they reach 10 per cent, i.e. the critical depth.

The similarity of the curves constructed for the identical climatic zones of the three oceans points to the fact that the zonal relation of calcium carbonate to depth is of a common character.

Fig. 11.9 and Table 11.2 show the critical depth position determined by the author for the sediments of the Pacific, Indian and Atlantic oceans.

The carbonate material solution begins in the upper layer of the bottom sediments almost everywhere from depths of about 3,500-4,000 metres. This is also confirmed by microscopic study of foraminifers as well as by lithological data. Yet, the depth to which carbonate material penetrates in the solution zone varies rather considerably. The higher the rate of calcium carbonate supply and the lower its dilution, the greater is this depth.

Judging by the calcium carbonate distribution in suspension from oceanic depths, the maximum rate of calcium carbonate supply should take place in the carbonate suspension belts, i.e. in the three humid zones. However, in the temperate (northern and southern) zones the supply of terrigenous and siliceous (diatom) material to the sediments is high. This results in the fact that even in the solution zone the dilution of undissolved calcium carbonate is still taking place, and the concentration further decreases as does the critical depth.

In the arid zones dilution is scarcely observed but the rate of supply of carbonates is very low.

The maximum values of the critical depth are noted in the equatorial zones of the oceans, where the high rate of carbonate accumulation in suspension (the equatorial zone, the carbonate belt) coincides with insignificant dilution.

Table 11.2. Variations in the critical depth of carbonate accumulation (in metres) in the World Ocean by latitudinal zones.

Ocean		North latitude					South latitude		
Zone	>70	70-40	40-20	20-10	10-20	10-20	20-40	40-70	>70
Atlantic	3,650	5,000	5,900	6,000	5,600	5,500	5,400	5,000	–
Indian	–	–	3,400	5,000	5,500	5,200	5,100	5,000	–
Pacific	–	4,100	4,300	4,850	5,300	4,750	4,500	4,500	–

If one imagines the position of the critical depth, not just in a section but in space, separate critical depths will combine to form a surface. The surface for each of the oceans has the form of a bowl of irregular shape. Its brims rise steeply eastward and westward (to the continents) and more gently northward and southward. The deepest part of the bowl corresponds to the equatorial zone. Within this zone a sharp increase, of up to 500-700 metres of the critical depth is usually observed: on the smooth surface of the bowl a trench is formed. What is the cause of such fundamental changes in the critical depth? A number of papers deal with factors determining calcium carbonate solution with depth. Wiseman (1954) considers the main factors determining the solution to be:

1. carbon dioxide content of the water,
2. temperature,
3. pressure,
4. salinity.

In Pettersson's (1953) opinion, the supply of hydrochloric and carbonic acids with the products of the underwater volcanism may be of great importance. The material we have analysed does not confirm this point of view.

Through the whole water column of the Pacific Ocean, for instance, salinity only changes insignificantly in the vertical direction, pH from 7.9 to 8.2, the usual values at 4,000-5,000 metres depths being 8.0-8.1. To elucidate the possible effect of temperature of the water column the author has calculated the mean suspended temperatures along the meridional section through the ocean by ten latitudinal zones. The graph (Fig. 11.9) shows that the mean suspended temperatures change very smoothly from 0.5° to 1.7° C for the northern and southern periphery of the ocean to 4.55°-4.62° C for the Equatorial zone. With increasing temperatures the solubility of calcium carbonate decreases, which is likely to correspond to the increase of the critical depth at the Equator. However, the change of temperature is very smooth, and the critical depth changes sharply. Observations of the preservation of foraminiferal tests show that they have no traces of solution in suspension obtained even from the maximum depths exceeding the critical ones.

The solution proceeds in the surface layer of the bottom sediments. Whereas the settling time of the tests through the water column is measured by days, their 'exposure' in the upper layer of the sediments is measured by tens and sometimes by thousands of years. In this connection the physical-chemical properties of the near-bottom layer are of particular importance.

It may be assumed that at ocean depths close to the critical ones, some basins separated from each other and highly different in their physical-chemical environments exist at the bottom. This can be judged, first of all, by temperature. At 4,000-5,000 metre depths variation in the whole of the oceans is very insignificant – as little as 1.5° C, i.e. the conditions of a huge thermostat are found there. Thus the bottom temperature effect cannot account for so sharp a variation of the critical depth. Likewise other physical-chemical parameters (carbon dioxide content, pH, etc.) undergo only slight changes there.

The position of the solution zone is determined by a combination of a number of physical-chemical factors, by regular changes of the elements of the carbonate system with depth, of paramount importance being, apparently, hydrostatic pressure. Water aggressiveness with respect to calcium carbonate becomes common to the entire world's oceans with 3,500-4,000 metres depth; it is not associated with local anomalies. An almost horizontal surface of the beginning of calcium carbonate solution is formed.

Of exceptional interest are the experimental studies of calcite solution at oceanic depths made recently by Peterson (1966). He placed previously weighed small balls of optical calcite at different depths (from the surface down to five kilometres) for a period of three months. After the exposure was complete the balls were weighed again. The results of the experiment are shown in Fig. 11.10. It is clearly seen from the figure that in the upper part of the water column (from 0 to 3,500-3,700 metres) the solution of calcite is practically absent. At 3,700 metres depth solution increases sharply and grows with depth. The maximum values are recorded in the near-bottom layer. It is interesting to note that the maximum value of pH determined by the influence of hydrostatic pressure upon the dissociatiôn constant of the carbonic acid is 7.9 and was measured at a depth of about 3,500 metres (Park 1966).

Thus the depth with which the solution begins is determined by the physical-chemical factors common for

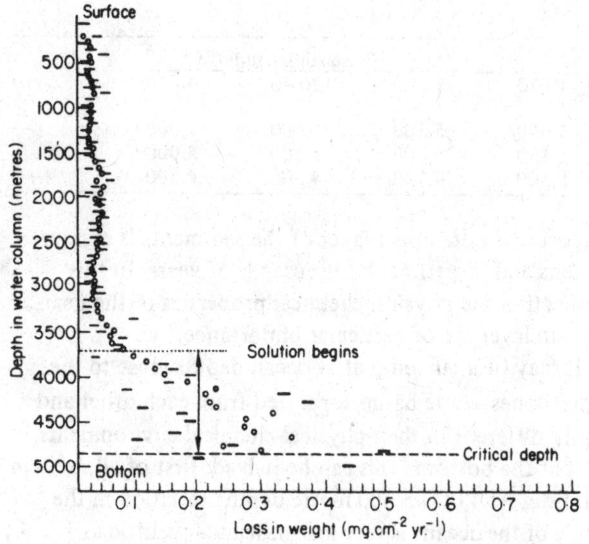

Surface
Solution begins
Critical depth
Bottom
Depth in water column (metres)
500
1000
1500
2000
2500
3000
3500
4000
4500
5000
0·1
0·2
0·3
0·4
0·5
0·6
0·7
0·8
Loss in weight (mg.cm.$^{-2}$ yr.$^{-1}$)

Fig. 11.10. **The solution of calcite balls relative to depth. The time exposure is three months. The horizontal axis per cent weight loss (mg/cm^2/year) (Peterson 1966).**

the vast space of the abyssal ocean. In different oceanic areas the carbonate material may be preserved in the bottom sediments, depending on the local conditions, at depths far greater than the depth of the beginning of solution. The 'slip' of carbonates below the solution depth may exceed 1,000 metres and sometimes reaches 2,000-2,300 metres. The critical depth of carbonate accumulation depends mainly not on the physical-chemical environment but on dynamic factors – the rate of calcium carbonate supply and its solution rate, the rate of supply of the diluting components and sedimentation as a whole. The higher the incoming part of the budget and the lower the expended one, the deeper is calcium carbonate penetration into the oceans and the greater the critical depth.

All the above-mentioned regularities have been revealed for the most widely-distributed materials in pelagic sediments – foraminiferal and coccolithophorid – which are composed of calcite. The solubility for carbonate material composed of aragonite (pteropods, coral material) is much higher than for calcite (at least twofold). This results in the critical depth of the aragonite material being much smaller than that of calcite material, i.e. 2,000-3,000 metres. Such a phenomenon causes a mineralogical separation of the carbonate microfossils, only the remains of organisms with calcite skeletons being distributed and preserved at great oceanic depths.

Under specific conditions in the oceans the zones of high calcium carbonate supply from suspension and of dilution are located somewhat differently due to local peculiarities, yet, the most general regularities hold.

The above-mentioned regularities in the vertical distribution of calcium carbonate make it possible to map the bottom sediments more accurately and with much assurance; in particular, to determine the areas of the distribution of red clay. Many authors explain the replacement of red clay in sediment cores by change of temperature of the bottom water due to the glacial fall of temperature or to depth changes. In the light of the facts discussed above it seems more correct, from our point of view, to relate this phenomenon to the changes in the rates of supply of carbonate and terrigenous (or siliceous) material.

Granulometric analysis of the bottom sediments and studies of calcium carbonate content of the separate fractions of the pelagic sediments show that the major part of calcium carbonate belongs to two main sediment fractions. In some cases to sandy-aleuritic and in the other cases to pelitic (Fig. 11.11). The first relation is attributable to the initial sizes of foraminiferal tests. If the conditions for preservation of the tests are unfavourable, they are dissolved, and break into separate calcite microcrystals of less than 0.01 millimetres in size. In this case the maximum of calcium carbonate is transferred from the sandy-aleuritic fraction to the pelitic one. As the critical depth is approached the 'pelitization' of foramin-

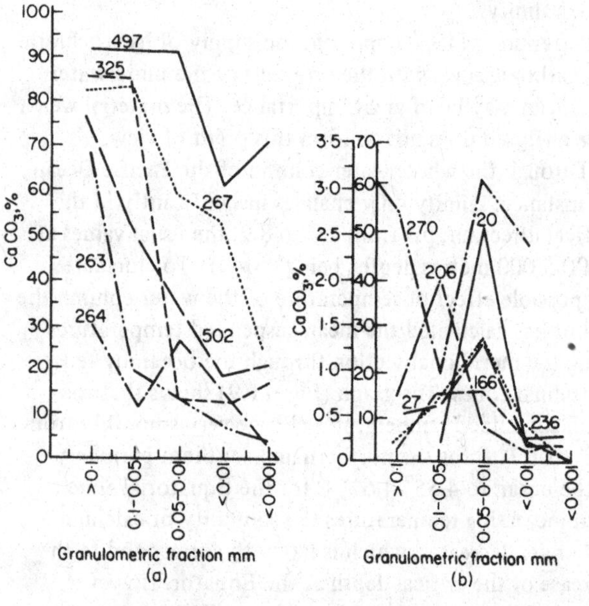

$CaCO_3$,%
100
90
80
70
60
50
40
30
20
10
0
497
325
267
263
264
502
>0·1
0·1–0·05
0·05–0·01
<0·01
<0·001
Granulometric fraction mm
(a)
3·5
3·0
2·5
2·0
1·5
1·0
0·5
270
20
206
166
27
236
Granulometric fraction mm
(b)

Fig. 11.11. **Types of calcium carbonate distribution by fractions of the bottom sediments of the Indian and Pacific oceans. (*a*) The enrichment of sandy-aleuritic fractions in shallow-sea sediments and at the tops of bottom elevations (pelagic sediments, depths less than 4,000 metres). (*b*) Compound distribution in deep-sea sediments (within the calcium carbonate solution zone) associated with the decay of foraminiferal tests. The figures denote station numbers of the ship *Ob*, Antarctic expedition.**

iferal tests becomes more and more complete. The sandy-aleuritic part of the histograms decreases sharply and in the subcolloidal fraction one can clearly see the beginning of solution.

The microscopic study of the sediment samples fully confirms such a 'step-wise' solution of foraminiferal tests.

The coccolith material is usually admixed with the foraminiferal. In accordance with the size of coccoliths practically the whole of it falls within the pelitic fractions of the sediment.

As indicated above, foraminifers are predominant in the pelagic sediments at depths from 500-1,000 metres down to the critical one (on an average, about 4,700 metres). Microscopic study of the foraminiferal sediments shows that the preservation of their tests is closely related to depth. In the depth range from the surface down to 3,700-4,000 metres the foraminiferal tests are usually very well preserved (except in high latitudes); that is why a close relationship between their amounts and calcium carbonate content of the sediments is observed.

Below 3,700 metres depth the solution of the tests and their breaking into separate crystals begin. Resistance to solution of different foraminiferal species varies. Thus at great oceanic depths not only the total amount of the sediment is decreased due to their solution but also their complexes are distorted because of the varying resistance of tests of different species. The most resistant are the species with thick tests: *Globorotalia tumida, Pulleniatina obliquiloculata, Hastigerina pelagica* and *Globigerinella aequilateralis* are broken down most rapidly (Ericson, Ewing, Wollin and Heezen 1961).

Proceeding from the foregoing, the curves of the vertical distribution of the amounts of planktonic foraminifers in the bottom sediments should run approximately parallel to those of calcium carbonate down to 3,700 metres below which depth they should run higher than the calcium carbonate curves since the breaking down of the tests begins much earlier (500-1,000 metres) than the complete disappearance of calcium carbonate from the sediment. The knowledge of the distribution of foraminiferal tests in suspension and of the conditions attending their burial in the sediments helps in correct understanding of the map of their quantitative distribution in the bottom sediments (Fig. 11.12). When inspecting this map one should keep in mind that far from always do the maximum amounts of the tests in the sediments correspond to their maximum development in the plankton. In some cases this is due to the solution of the tests or to the transfer of their carbonate material to the pelitic fractions, in other cases this is attributable to dilution by carbonate-free material. Of great importance for the latter process is the hydrodynamic environment. When settling from the ocean surface to the bottom the biogenic particles find themselves under different hydrodynamic conditions. Under conditions of high mobility of bottom waters (which is the case at bottom rises) sorting of incoming material occurs. The tops of sea mounts are found to be covered with the coarsest, usually foraminiferal, material. Under the calm conditions of the ocean basins the non-selective settling of both foraminiferal, diatom and other fine material occurs and dilution increases.

Thus the quantitative distribution of foraminiferal tests in the bottom sediments, like that of calcium carbonate depends on the combination of a number of factors. The most important of them are:

1. plankton productivity (controlled by the climatic zonation),
2. the dilution of tests by carbonate-free material,
3. the solution of the tests at depths greater than 3,700-4,000 metres,
4. mechanical differentiation

The map of the quantitative distribution of foraminifers (Fig. 11.12) enables one to trace all these processes in space.

The amount of tests in one gramme of sediment varies depending on a combination of conditions over very wide limits – from units and even from complete absence of tests to amounts of more than 10,000 specimens per gramme.

In the shelf sediments the tests of planktonic foraminifers are usually found in small amounts. This is due both to their small initial content of plankton and, mainly, to the powerful dilution processes.

At the continental slope the amount increases considerably, as it also does at the tops of bottom elevations (to 1,000-20,000 per gramme) in the areas of the carbonate suspension belts. The high initial foraminiferal content of suspension is combined here with the mechanical differentiation processes. The latter processes result in removal of all the fine diluting material and in a selective accumulation of the coarse material, mainly foraminifers. The tops of bottom elevations prove to be capped with carbonate sediments not only in the carbonate accumulation zone but also far beyond it (in the Arctic and Antarctic).

At depths greater than the depth of solution (3,700 metres in tropical water) the solution of foraminifers begins. Therefore the actual amount of the tests is determined here not only by the rate of their supply from suspension but also by the rate of solution and supply of diluting material. The enrichment of the sediments with foraminifers, as on bottom elevations, does not occur here.

The foraminiferal complexes in the bottom sediments

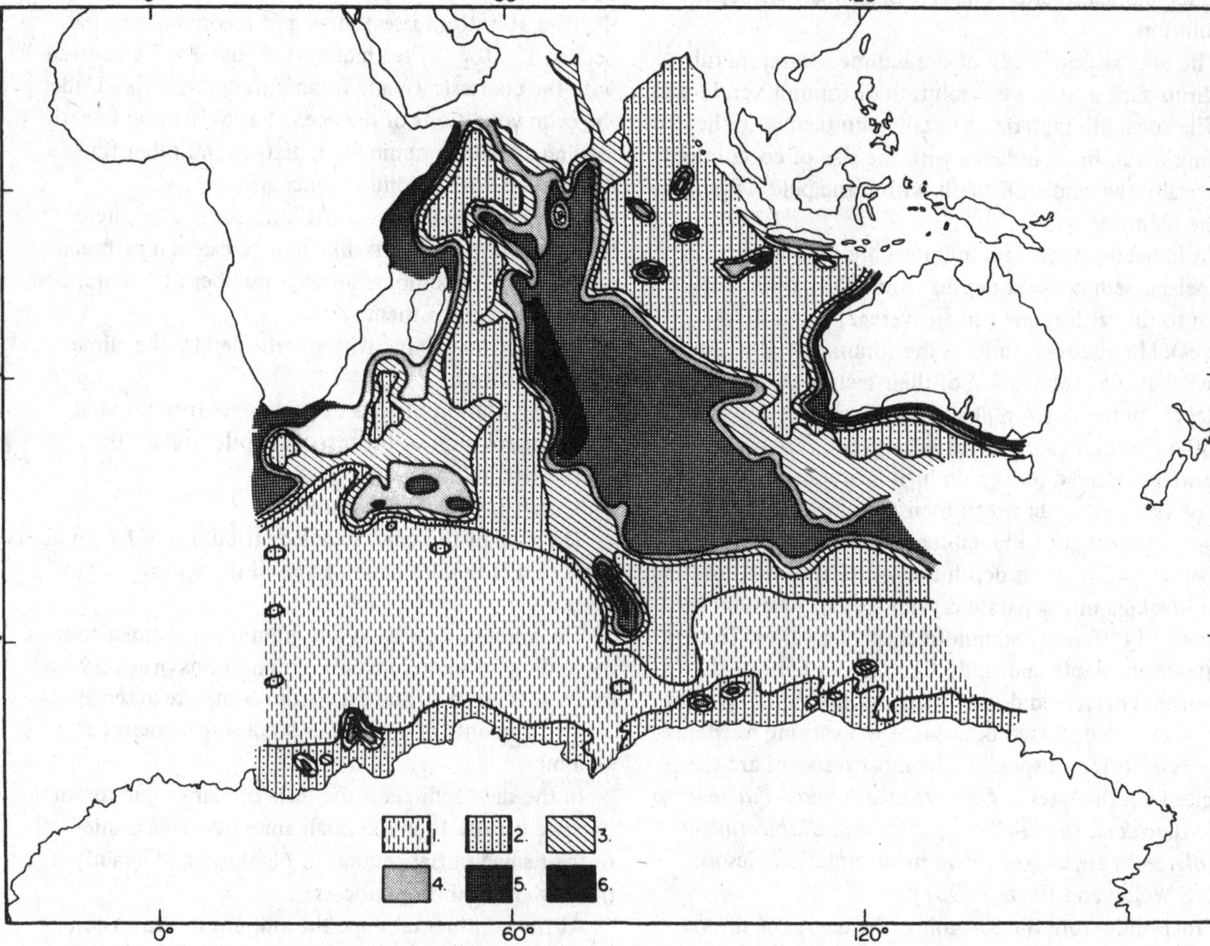
0°
60°
120°
0°
30°
60°
1.
2.
3.
4.
5.
6.

Fig. 11.12. Quantitative distribution of planktonic foraminifers in the surface layer of bottom sediments from the Indian Ocean (specimens per gramme of sediment). 1. foraminifers not found; 2. from 0 to 100; 3. from 100 to 500; 4. from 500 to 1,000; 5. from 1,000 to 10,000; 6. more than 10,000 specimens (Belyaeva 1963).

prove to be closely related to the climatic zonation (Belyaeva 1963, 1964, 1966). In the world's oceans the development of the *Globorotalia* complex is typical for 20° N-20° S latitudes, i.e. for the near-Equatorial humid zone, of the mixed complex for the arid zone (from 20° to 40°) and of the *Globigerina* complex for the Temperate humid zone.

Knowledge of the laws of calcium carbonate distribution in suspension and in bottom sediments is of great importance for micropalaeontological studies. It is evident that at great depths the quantity and specific composition of the carbonate micro-organisms are substantially distorted and, at depths greater than the critical ones, gaps exist where the remains of carbonate plankton are not found at all. Another important conclusion is that even above the critical depths the quantity of foraminiferal tests does not correspond to the distribution of their living forms in plankton. This is due chiefly to their dilution by carbonate free material. The influence of the diluting material can be eliminated by presenting the data in an open system (the absolute mass method).

CALCIUM CARBONATE ABSOLUTE MASSES IN THE BOTTOM SEDIMENTS AND ITS GENESIS

The present-day state of research makes it possible to compile reliable maps of absolute masses for the Atlantic

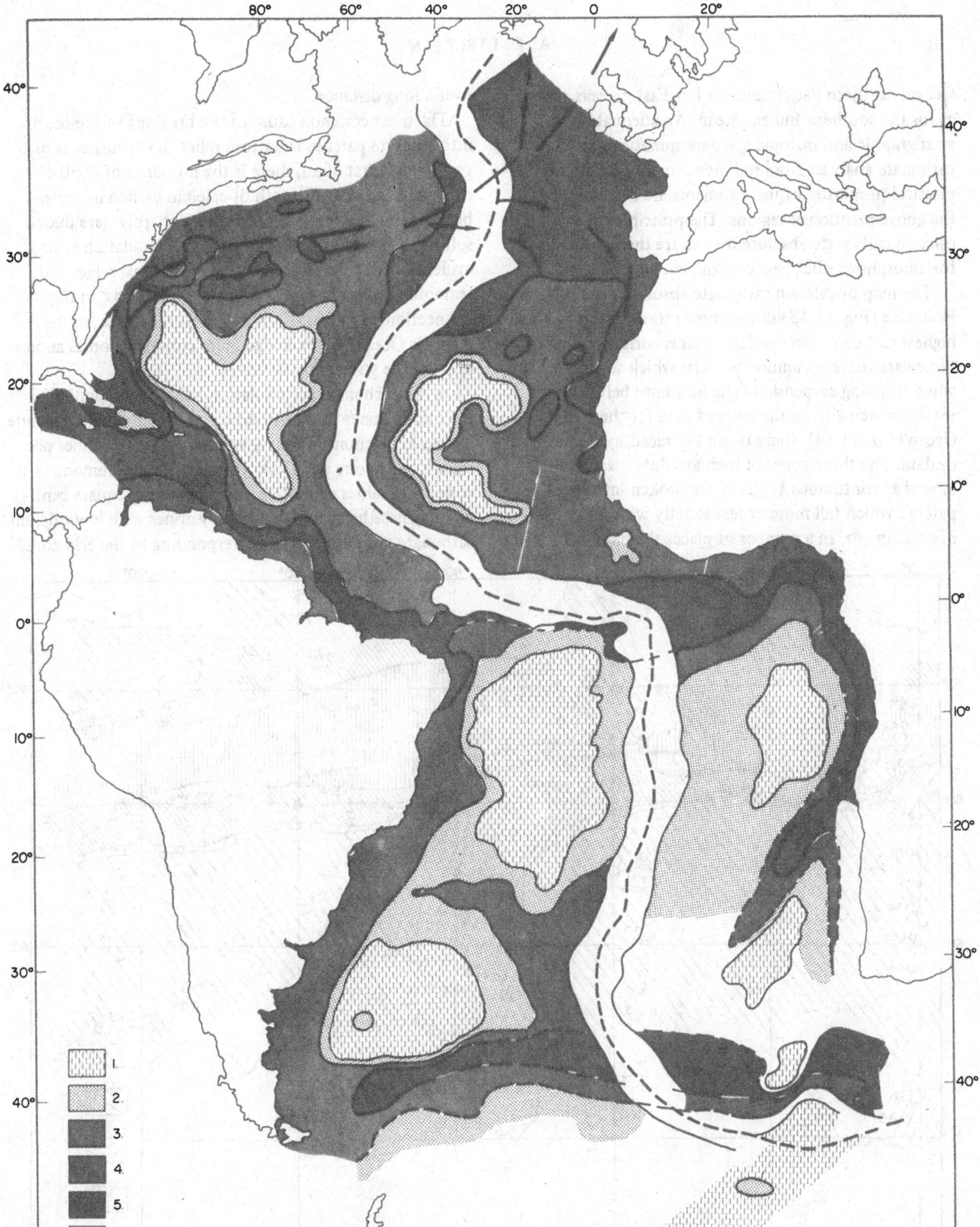
80°
60°
40°
20°
0
20°
40°
30°
20°
10°
0°
10°
20°
30°
40°
50°
1.
2.
3.
4.
5.
6.
7.
80°
60°
40°
20°
0°
20°

Fig. 11.13. The pattern of calcium carbonate absolute masses in sediments from the Atlantic Ocean (in g/cm^2/10,000 years); 1. less than 0.1; 2. from 0.1 to 1; 3. from 1 to 10; 4. from 10 to 30; 5. more than 30; 6. the crest of the Mid-Atlantic ridge; 7. the direction of the Gulf Stream.

Ocean, northern Pacific and for Far-Eastern seas as well as for the southern Indian Ocean. An adequate number of stratigraphic and isotopic age determinations, calcium carbonate analyses from sediment cores and data on the physical properties of the sediments are available for the above-mentioned regions. The principles of mapping calcium carbonate absolute masses are the same as those for amorphous silica (see Lisitzin, chapter 10).

The map of calcium carbonate absolute masses for the Holocene (Fig. 11.13) shows three rather distinct zones of highest carbonate accumulation rates corresponding to the three carbonate accumulation belts which were discussed when studying suspension. The Southern belt is represented principally on the basis of data for the Indian Ocean (Fig. 11.14) where it can be traced by a great body of data. The three zones of high absolute masses either appear as continuous bands or are broken into separate patches which fall more or less exactly within the band of enrichment. In a number of places the band is broken over a long distance.

The most common cause of the breaking of the band into separate patches is bottom relief. Its influence is of two kinds. First of all, there is the influence of depths exceeding the critical depth of calcium carbonate distribution. The absolute masses decrease sharply here due to solution. The decrease in carbonate accumulation is also evident at the tops of the bottom elevations, ridges and seamounts where the sedimentation rate drops. In this connection, the areas of the ridges (for instance, in the Atlantic Ocean) are excluded from consideration as anomalous for the given climatic zones.

In the carbonate accumulation belts corresponding to the humid zones of the Atlantic and Indian oceans absolute masses of calcium carbonate equal to 10-30 grammes per square centimetre per 1,000 years are most common, with some values exceeding 30 grammes per square centimetre. These belts are separated by zones with low calcium carbonate absolute masses corresponding to the arid zones.

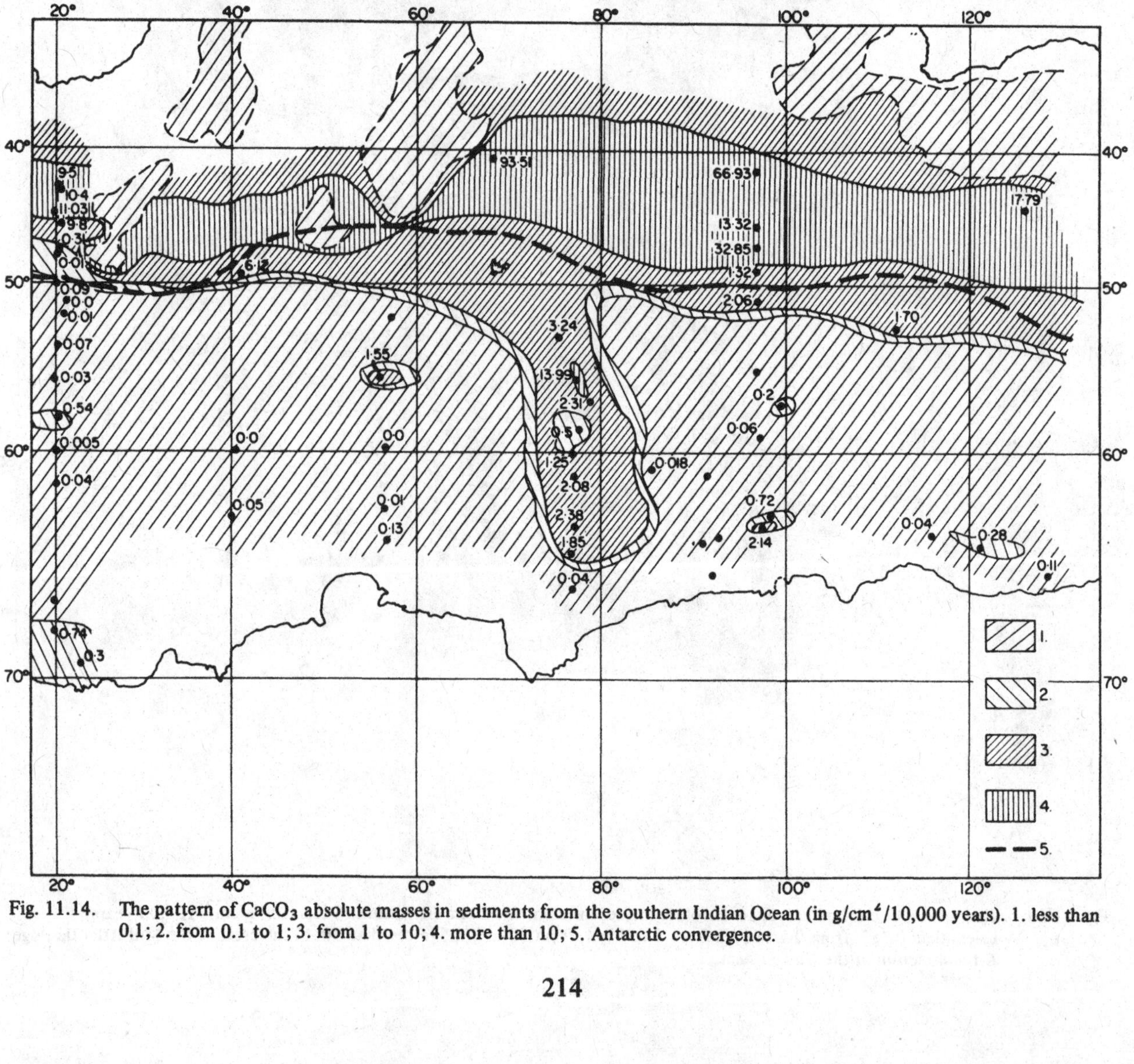

20°
40°
60°
80°
100°
120°
40°
50°
60°
70°
93·51
66·93
17·79
13·32
32·85
1·32
2·06
1·70
9·5
10·4
11·03
9·8
0·3
0·01
6·12
0·09
0·0
0·01
0·07
0·03
0·54
0·005
0·04
0·0
0·05
1·55
0·0
0·01
0·13
3·24
13·99
2·31
0·5
1·25
2·08
2·38
1·85
0·04
0·018
0·2
0·06
0·72
2·14
0·04
0·28
0·11
0·74
0·3
1.
2.
3.
4.
5.

Fig. 11.14. The pattern of $CaCO_3$ absolute masses in sediments from the southern Indian Ocean (in g/cm²/10,000 years). 1. less than 0.1; 2. from 0.1 to 1; 3. from 1 to 10; 4. more than 10; 5. Antarctic convergence.

Absolute masses here usually amount only to 0.1-1 gramme per square centimetre and sometimes even less.

Of great interest is the decrease of absolute masses in the cold parts of the temperate humid zones, as well as in the ice zones. Evidence of this decrease is also found in the southern portion of the Indian Ocean and in the Atlantic. This may be explained by the decrease in the rates of calcium carbonate supply. The suspension here contains few foraminifers, their development being limited by temperature.

Near the Antarctic Convergence zone and the northern Polar Front the high production of phytoplankton which is a food source and other conditions appropriate for foraminiferal development (in particular, temperature) are favourably combined. It is here that the carbonate belts of the Temperate humid zones are formed and here they are reflected in the sediments, if this is not hindered by depths exceeding the critical ones.

It is interesting to correlate our absolute mass determinations for the Atlantic Ocean with those made by Turekian (1963, 1964) and later by Turekian and Stuiver (1964). The values of absolute mass determinations given by these authors for the neighbouring stations differ from our results only slightly in the range 10 to 20 per cent. Thus the established zones are not the result of approximations inevitable in recalculations but reflect the natural phenomena. This is also indicated by the correlation between the zones of maximum carbonate accumulation (belts) in suspension and in bottom sediments.

Calcium carbonate absolute masses in the seas which lie within the carbonate accumulation zone are much higher than those in the oceans. For instance, the absolute masses for the Black Sea are 40-60 grammes per square centimetre and in a number of places more than 100 grammes per square centimetre (Strakhov 1963). The maximum values for the Caribbean Sea are as great as 120-150 grammes per square centimetre per 1,000 years and in the abyssal parts they drop to thirty and even lower, approaching the mean value for the Equatorial zone. The mean value of recent calcium carbonate deposition in the world's oceans based on the balance calculations of the discharge from land is 0.66 grammes per square centimetre per 1,000 years (Revelle and Fairbridge 1957). The rate of chemical-biological sedimentation over the Great Bahama Bank is about 800 grammes per square centimetre per 1,000 years, i.e. far in excess of the rate of the biogenous pelagic carbonate accumulation. Calcium carbonate absolute masses are very substantial in a number of places on the shelf. Thus after the influence of diluting material is eliminated the climatic zonation mentioned for suspension appears also in the distribution of calcium carbonate absolute masses in the bottom sediments; the carbonate accumulation zones in suspension and in sediments coincide. The climatic zonality is superimposed by the vertical zonality associated with calcium carbonate solution at great depths. Thus the seeming disordered pattern of calcium carbonate distribution in the recent sediments turns out to be regular, subjected to zonation. From the maps of calcium carbonate absolute masses one more important conclusion follows that the maximum rates of carbonate accumulation usually do not coincide with the maximum percentage of calcium carbonate in the sediments, neither with the maximum amount of foraminiferal tests per gramme of sediment. The methods of expression on a percentage basis, as well as of calculations of tests per gramme of sediment, sometimes do not reflect the existing carbonate accumulation; they should necessarily be supplemented by studies of absolute masses.

CONCLUSION

Despite the complexity and great variety of recent carbonate accumulation in the oceans, and the interaction of a number of factors, it has been possible to distinguish the main ones determining the specific character of carbonate sediments and the distribution of microfaunas. The relation to climate, depth and the influence of land (through diluting material) is invariably clearly defined. The climatic zonation studies show the existence of three global zones of highest rates of recent carbonate accumulation. These zones do not coincide with the regions of highest percentage of calcium carbonate in the sediments and (as in the case of silica) can be established only through the use of the method of absolute masses. The above-mentioned zones correspond to the humid climatic zones and are separated by latitudinal zones of low rates of carbonate accumulation (the arid zones). The peripheral parts of the carbonate accumulation belts and the silica accumulation belts are superimposed, which results in the formation of global zones of biogenous sedimentation. The latter zones are a reflection of the basic distribution of heat over the Earth, and of the atmospheric and hydrospheric circulation.

Received November 1967.

BIBLIOGRAPHY

Alekin, O. A. 1966. The discharge of dissolved matter into the World Ocean. *Second Internat. Oceanograph. Congress. Abstracts of papers.* Nauka: Moscow.

Alexandrov, A. N. 1964. Sediments of the Azov Sea. *Oceanology.* 5.

Arrhenius, G. 1952. Sediment cores from the East Pacific. *Rep. Swedish Deep-Sea Expedition.* 5, 1-227.

Atlas of the Antarctic, 1966, 1969. 1 (maps of suspended matter and of bottom sediments). 2 (explanatory notes): Leningrad.

Barash, M. S. 1964. The ecology of planktonic foraminifers in the northern Atlantic and their significance for stratigraphic constructions. *Trudy Inst. Oceanol. Acad. Sci. (U.S.S.R.)*, **65**.

Bé, A. W. H. 1960. Ecology of Recent planktonic foraminifera: Part 2 – Bathymetric and seasonal distributions in the Sargasso Sea off Bermuda. *Micropaleontology*, **6**, 373-92.

Bé, A. W. H. 1966. Distribution of planktonic foraminifera in the World Ocean. *Second Internat. Oceanograph. Congress. Abstracts of papers.* Nauka: Moscow.

Belov, N. A. and Lapina, N. N. 1961. *Bottom sediments of the Arctic basin.* Marine transport Publishers: Leningrad.

Belyaeva, N. V. 1962. The distribution of planktonic foraminifers through the water column of the Indian Ocean. *Bulletin de la Societe des naturalistes de Moscou. Sec. Geolog.* **3**.

Belyaeva, N. V. 1963. The distribution of planktonic foraminifers over the Indian Ocean bottom. *Problems of paleontology*, **7**.

Belyaeva, N. V. 1964. The distribution of planktonic foraminifera in the water and over the bottom of the Indian Ocean. *Trudy Inst. Oceanol. Acad. Sci. (U.S.S.R.)*, **68**.

Belyaeva, N. V. 1966. Climatic and vertical zonality in the distribution of planktonic foraminifera in sediments of the Pacific Ocean. *Second Internat. Oceanograph. Congress. Abstracts of papers.* Nauka: Moscow.

Belyaeva, N. V. and Saidova, Kh.M. 1965. The correlation between the benthic and planktonic foraminifers in the surface layer of the Pacific bottom sediments. *Oceanology*, **5** (6).

Bernard, F. 1966. Abondance du nannoplancton dans les couches aphotiques des océans-conséquences probables pour la productivité profonde des mers chaudes. *Second Internat. Oceanograph. Congress. Abstracts of papers.* Nauka: Moscow.

Bernikov, R. G., Vinogradova, L. A., Gruzov, L. N., Paliy, N. F., Sedykh, K. A., Sukhoruk, V. I., Chmyr, V. D., Fedoseev, A. F. and Yakovlev, V. N. 1966. Productive zones of the Equatorial Atlantic Ocean. *Second Internat. Oceanograph. Congress. Abstracts of papers.* Nauka: Moscow.

Bezrukov, P. L. 1959. Some problems of sedimentation zonality in the World Ocean. *First Internat. Oceanogr. Congress, Paper.*

Bezrukov, P. L. 1960. The bottom sediments of the Okhotsk Sea. *Trudy Inst. Oceanol. Acad. Sci. (U.S.S.R.)* 32.

Bezrukov, P. L. 1964. Sediments in the northern and central parts of the Indian Ocean. *Trudy Inst. Oceanol. Acad. Sci. (U.S.S.R.)*, **64**.

Bezrukov, P. L. and Lisitzin, A. P. 1966. Sedimentation in the Indian Ocean. *Second Internat. Oceanograph. Congress. Abstracts of papers.* Nauka: Moscow.

Bezrukov, P. L., Lisitzin, A. P., Petelin, V. P. and Skornyakova, N. S. 1961. Map of the recent sediments of the World Ocean. *Recent sediments of the seas and oceans.* Acad. Sci. (U.S.S.R.): Moscow.

Bezrukov, P. L., Lisitzin, A. P., Petelin, V. P. and Skornyakova, N. S. 1964. Sedimentation in the World Ocean. *The physical-geographical Atlas of the World,* Moscow.

Bezrukov, P. L., Lisitzin, A. P., Petelin, V. P. and Skornyakova, N. S. 1966. Sedimentation in the Pacific Ocean. *Second Internat. Oceanograph. Congress. Abstracts of papers.* Nauka: Moscow.

Bezrukov, P. L., Lisitzin, A. P., Romankevich, E. A. and Skornyakova, N. S. 1961. Recent sedimentation in the northern Pacific. *Recent sediments of the seas and oceans.* Acad. Sci. (U.S.S.R.): Moscow.

Bogorov, B. G., Bordovsky, O. K. and Vinogradov, M. E. 1966. The biogeochemistry of the oceanic plankton. The distribution of some chemical components of the plankton in the Indian Ocean. *Oceanology,* **2**.

Bradshaw, J. S. 1957. Laboratory studies on the rate of growth of the foraminifer *Streblus beccarii* (Linné). var. *tepida* (Cushman). *J. Palaeont.* **31**, 1138-47.

Bradshaw, J. S. 1959. Ecology of living planktonic Foraminifera in the North and Equatorial Pacific Ocean. *Contrib. Cushman Found. Foram. Res.* **10**, 25-64.

Burmistrova, I. I. 1967. Modern distribution of Foraminifera and stratigraphy of the Late Quaternary sediments in the Barents Sea. *Oceanology,* **2**.

Emelyanov, E. M. 1965. Carbon of the recent bottom sediments of the Mediterranean Sea. *Basic features of the geologic structure, hydrological regime and biology of the Mediterranean Sea.* Nauka: Moscow.

Emelyanov, E. M., Lucoshevichus, L. S., Svirenko, I. P., Soldatov, A. V., Koshelev, B. A., Lisitzin, A. P., Ilyin, A. V., Shurko, II., Litvin, V. M. and Senin, Yu,M. 1966. Sedimentation in the Atlantic Ocean. *Second Internat. Oceanograph. Congress. Abstracts of papers.* Nauka: Moscow.

Eriscson, D. B., Ewing, M., Wollin, G. and Heezen, B. C. 1961. Atlantic deep-sea sediment cores. *Bull. geol. Soc. Amer.* **72**, 193-286.

Fedosov, M. V. 1961. Some peculiarities of sedimentation in the sea of Azov. *Recent sediments of the seas and oceans.* Acad. Sci. (U.S.S.R.): Moscow.

Gershanovich, D. E. 1956. Silicic acid, calcium carbonate and organic carbon in the deep-sea sediments of the Japan Sea. *Trudy of the State Oceanographic Research Institute,* **31** (43).

Gershanovich, D. E. and Neiman, A. A. 1964. Bottom sediments and botton fauna in the East China Sea. *Oceanology,* **6**.

Gordeev, E. I. 1969. Chemical composition of suspension from the surface waters of the northern and central parts of the Indian Ocean. *Oceanological research,* **19**.

Korneva, F. R. 1966. Foraminifera distribution in the surface layer of sediments of the Eastern Mediterranean Sea. *Oceanology,* **5**.

Kozlov, V. F. 1964. On water rise in the area of the equator. *Oceanology,* **1**.

Lisitzin, A. P. 1959a. Recent data on the distribution and composition of substances in suspension in seas and oceans, related to geological problems. *Doklady Acad. Sci. (U.S.S.R.),* **126** (4).

Lisitzin, A. P. 1959b. Bottom sediments of the Bering Sea. *Trudy Inst. Oceanol. Acad. Sci. (U.S.S.R.),* **29**.

Lisitzin, A. P. 1960a. Distribution and composition of suspensions in the Indian Ocean. Communication 1. The quantitative distribution. *Oceanological research,*

2 (Xth Section of the IGY Programme). Acad. Sci. (U.S.S.R.): Moscow.

Lisitzin, A. P. 1960b. Sedimentation in the southern parts of the Pacific and Indian Oceans. *Internat. Geolog. Congr. Session XXI. Rep. Sov. Geologists.* Acad. Sci. (U.S.S.R.): Moscow.

Lisitzin, A. P. 1961a. The distribution and composition of suspended matter in the seas and oceans. *Recent sediments of the seas and oceans.* Acad. Sci. (U.S.S.R.): Moscow.

Lisitzin, A. P. 1961b. Distribution and composition of suspended matter in the Indian Ocean. Communication 2. Granulometric composition. *Oceanological research,* **3** (Xth Section of the IGY Programme). Acad. Sci. (U.S.S.R.): Moscow.

Lisitzin, A. P. 1961c. Processes of recent sedimentation in the southern and central parts of the Indian Ocean. *Recent sediments of the seas and oceans.* Acad. Sci. (U.S.S.R.): Moscow.

Lisitzin, A. P. 1961d. Processes of recent sedimentation in the Bering Sea. *Recent sediments of the seas and oceans.* Acad. Sci. (U.S.S.R.): Moscow.

Lisitzin, A. P. 1962a. Distribution and composition of suspended matter in the Indian Ocean. Communication 3. Correlation between the granulometric compositions of suspensions and of bottom sediments. *Oceanological research,* **5** (Xth Section of the IGY Programme). Acad. Sci. (U.S.S.R.): Moscow.

Lisitzin, A. P. 1962b. Suspended matter in the ocean. *Bulletin of the Oceanographic Commission,* **3**.

Lisitzin, A. P. 1963. Bottom sediments of the Antarctic shelf. *Delta and shallow-water marine bottom sediments.* Acad. Sci. (U.S.S.R.): Moscow.

Lisitzin, A. P. 1964. Distribution and chemical composition of suspended matter in the Indian Ocean. *Oceanological research,* **10** (Xth Section of the IGY Programme). Acad. Sci. (U.S.S.R.): Moscow.

Lisitzin, A. P. 1966. Processes of recent sedimentation in the Bering Sea. Nauka: Moscow.

Lisitzin, A. P. 1969. Bottom sediments of the Southern Ocean. *Atlas of the Antarctic,* **2**.

Lisitzin and Bogdanov (in press).

Lisitzin, A. P., Murdmaa, I. O. Petelin, V. P. and Skornyakova, N. S. 1966. Granulometric composition of the deep-sea sediments from the Pacific Ocean. *Lithology and mineral resources,* **2**.

Lisitzin, A. P. and Petelin, V. P. 1953. Recent carbonate sediments of the cold-water seas. *Bulletin de la Société der naturalistes de Moscou,* **28** (2).

Lisitzin, A. P. and Petelin, V. P. 1967. Calcium carbonate distribution and modification in the Pacific bottom sediments. *Lithology and mineral resources,* **5**.

Lohmann, H. 1920. Die Bevolkerung des Ozeans mit Plankton nach den Ergebnissen der Zentrifugenfänge während der Ausreise der Deutschland, 1911; zugleich ein Beitrag zur Biologie des Atlantischen Ozeans. *Archiv Biontologie Ges. Naturf. Freunde, Berlin,* **4** (3), 1-617.

Muhina, V. V. 1966. Siliceous organisms in suspension and in the surface layer of bottom sediments of the Indian Ocean. *Oceanology,* **5**.

Murray, J. and Renard, A. F. 1891. Deep-sea deposits. *Sci. Rep. Challenger Exped. 1873-76.*

Neeb, G. A. 1943. Composition and distribution of samples. *The 'Snellius' Exped. Geol. Res., sect.* **1**.

Park, K. 1966. Deep-sea pH. *Science,* **154** (3756).

Parker, F. L. 1960. Living planktonic foraminifera from the equatorial and southeast Pacific. *Sci. Rep. Tohoku Univ. (Geol.),* spec. **4**, 71-82.

Peterson, M. N. A. 1966. Calcite : rates of dissolution in a vertical profile in the Central Pacific. *Science,* **154**, 1542-4.

Petterson, H. 1953. The Swedish deep-sea expedition 1947-1948, *Deep-Sea Res.* **1**.

Phleger, F. B. 1945. Vertical distribution of pelagic Foraminifera. *Amer. Jour. Sci.* **243**, 377-83.

Pia, J. 1933. Die rezenten Kalksteine. *Zeits. f. Krist. Min. Pet., Abt. B., Erganzungsband, Leipzig.* 1-420.

Ponomarev, A. I. 1961. *Methods of chemical analysis of siliceous and carbonate rocks.* Acad. Sci. (U.S.S.R.): Moscow.

Revelle, R. R. 1944. Marine bottom samples collected in the Pacific Ocean by the *Carnegie* on its seventh cruise. *Carnegie Inst. Wash. Publ.* **556**, 1-182.

Revelle, R. and Fairbridge, R. 1957. Carbonates and carbon dioxide. *Mem. geol. Soc. Amer.* **67** (1), 244.

Saidova, Kh. M. 1961. *The ecology of foraminifers and palaeogeography of the Far-Eastern seas of the USSR and of the north-western Pacific.* Acad. Sci. (U.S.S.R.): Moscow.

Saidova, Kh. M. 1965. The distribution of bottom foraminifera in the Pacific Ocean. *Oceanology,* **1**.

Saidova, Kh. M. 1966a. The distribution of the species of benthonic agglutinating foraminifera in the Pacific Ocean. *Oceanology,* **1** (6).

Saidova, Kh. M. 1966b. Foraminiferal bottom fauna of the Pacific along 174° W meridian. Part II. The number of phytoplankton cells. *Trudy Inst. Oceanol. Acad. Sci. (U.S.S.R.),* **71**.

Strakhov, N. M. 1951a. The calcareo-dolomite facies of the recent and old basins. *Trudy Geol. Inst. Acad. Sci. (U.S.S.R.),* **124** (45).

Strakhov, N. M. 1951b. Descriptions of carbonate accumulation in the recent basins. *Collections of Papers 'In Memory of Academician A. D. Arkhangelsky': Moscow.*

Strakhov, N. M. (ed.). 1957. Methods for studying sedimentary rocks, 1, 2. *Gosgeoltehizdat.*

Strakhov, N. M. 1963. Types of lithogenesis and their evolution during the Earth's history. *Gosgeoltehizdat.*

Strakhov, N. M., Brodskaya, N. G., Knyazeva, L. M., Razzhivina, A. N., Rateev, M. A., Sapozhnikov, D. G. and Shishova, E. S. 1954. *Sedimentation in the recent basins.* Acad. Sci. (U.S.S.R.): Moscow.

Trask, P. D. 1932. *Origin and environment of source sediments of petroleum.* Amer. Petrol. Inst., Gulf Publ. Co.; Houston. 1-323.

Trask, P. D. 1937. Inferences about the origin of oil as indicated by the composition of the organic constituents of sediments. *Prof. Pap. U.S. geol. Surv.* **186**, 147-57.

Turekian, K. 1963. Rates of calcium carbonate deposition by deep-sea organisms, mollusc and coral-algae association. *Nature, Lond.* **197** (277).

Turekian, K. 1964. The geochemistry of the Atlantic

Ocean Basin. *Trans. N.Y. Acad. Sci.* **26.**

Turekian, K. and Stuiver, M. 1964. Clay and carbonate accumulation rates in three South Atlantic deep-sea cores. *Science,* **146** (3640).

Uschakova, M. G. 1966. The biostratigraphic significance of coccolithophorids in the bottom sediments of the Pacific Ocean. *Oceanology,* **1.**

Vikrenko, N. and Nikolaeva, V. 1962. Suspended matter of the North Atlantic from the data of the second and fourth cruises of the R/V 'Mikhail Lomonosov'. *Trudy Inst. Oceanol. Acad. Sci. (U.S.S.R.),* **56.**

Wiseman, J. D. H. 1954. Past temperatures of the upper equatorial Atlantic. In 'A discussion on the floor of the Atlantic Ocean. Part I, Sediments'. *Proc. Roy. Soc. A.* 222, 287-407.

12. ZONATION OF BIOGENOUS SEDIMENTATION IN THE OCEANS

P. L. BEZRUKOV

The progress achieved in the geological sciences and particularly in marine geology during the last few years has made possible a more comprehensive interpretation of the zonation of oceanic sediments. Some aspects of this problem were reported at the 1st International Oceanographic Congress in 1959 (Bezrukov 1959, 1962). During the last eight years geologists and micropalaeontologists of the Institute of Oceanology of the Academy of Sciences U.S.S.R. have continued to study various aspects of this problem, particularly the problem of the zonation of *biogenous* sedimentation. Some results of these investigations have been reported by A. P. Lisitzin and by A. P. Jousé (Lisitzin 1964, 1966; Jousé 1966; also Saidova 1966; Beljaeva 1964; Kozlova 1964, etc.)

Besides this, the Institute's geologists compiled a series of lithological and geochemical maps of all the oceans, which were demonstrated at the 2nd International Oceanographic Congress in Moscow in 1966. These maps illustrate the zonation of geological processes in the oceans. Some of them have been published (*Atlas of Antarctic,* 1966; *Physical-Geographical Atlas of the World,* 1964; Bezrukov and Lisitzin 1966; Bezrukov, Lisitzin, Petelin and Skornjakova 1966; Emelianov *et al.* 1966). Large contributions to the above problems were also made by marine geologists and micropalaeontologists of many other countries.

In the following I should like to refer briefly to some of the features of the zonation of biogenous sediments in the oceans. At least three types of zonation can be distinguished, namely: circum-continental, climatic and vertical zonations. At any point on the bottom, sediments are influenced more or less by these three types of zonation.

The *circum-continental zonation* of sedimentation can be characterized as a regular alteration in the composition and thickness of bottom sediments in relation to increasing distance from continents, reflected in a general replacement of terrigenous and benthonic biogenous sediments by pelagic biogenous oozes, or by the so-called red clays. The width of the fringing circum-continental zone of terrigenous sediments is variable, depending first of all upon the intensity of detritus transportation from the land, which depends in turn on topography and climatic conditions in the water-collecting basins. Outside the band of terrigenous sediments a hemi-pelagic type of sedimentation is taking place. The abyssal accumulative plains, where sediments consist partly or almost completely of terrigenous material, occur, as a rule, in the peripheral parts of the oceans, providing one of the vivid manifestations of circum-continental zonation of sedimentation. According to modern concepts, turbidity currents spreading down the continental slope take an active but not unique part in the formation of the accumulative plains (Heezen, Tharp and Ewing 1959). The accumulative plains are characterized by great thicknesses of sediments ; up to 2-3 kilometres and more. Outside the accumulative plains the thickness of unconsolidated sediments decreases to several hundreds of metres, and farther on it is almost independent of the distance from land.

Thus, circum-continental zonation, as its name indicates, is noticed mostly in the peripheral oceanic areas and its effect is very weak in the central parts. This feature makes it different from the climatic and vertical zonation of sedimentation. The distribution of some biogenic components of sediments is also controlled by circum-continental zonation, particularly organic matter, high concentrations of which are distributed mainly in peripheral parts of the oceans owing to the higher productivity of near-shore waters.

The *climatic zonation* of sedimentation can be characterized as a regular change from high to low latitudes of a number of genetic groups and types of sediments. This zonation has been known from the time of the *Challenger* expedition, and can be seen most obviously in the distribution of biogenic calcium carbonate, amorphous silica and organic matter. This zonation is reflected in the general distribution of calcareous and siliceous sediments in the oceans, as well as in the distribution of the main bio- and thanatocoenoses of sediment-building organisms such as planktonic foraminifera and diatoms.

In all oceans several highly productive, mostly latitudinal, zones can be distinguished, coinciding with upwelling of deep waters rich in biogenic elements. Such zones are also characterized by the high content of biogenous suspended matter in oceanic waters (Lisitzin 1964, 1966). These high productivity zones are, as a rule, projected onto the bottom as sediment belts where the content of biogenic amorphous silica (skeletal debris of diatoms and radiolarians) is relatively high. The largest latitudinal belts of silica accumulation cross the pelagic regions of all oceans and are interrupted only on submarine ridges.

The planktonic foraminifera are inhabitants of the upper water layers of tropical and temperate zones, and form foraminiferal oozes on the bottom if it is higher than the critical depth. In these zones skeletons of diatoms and radiolarians are dispersed amongst calcareous material, and do not produce high concentrations of amorphous silica in the sediments. The accumulation of siliceous sediments, with higher concentrations of amorphous silica (10 per cent and more) in tropical and subtropical regions of the Pacific and Indian oceans, takes place only deeper than the critical depth (beneath which carbonate material is being dissolved on the bottom in cold waters under high pressure conditions). This is one of the most eloquent evidences of vertical zonation in oceanic sedimentation.

The *vertical zonation* of sedimentation can be seen in the regular change of the granulometric and lithological composition of bottom sediments in relation to depth, irrespective of the distance from land. In the peripheral parts of the ocean depths usually increase with distance from shore. That is why the vertical zonation is more evident away from the continents where the circum-continental zonation practically disappears.

Vertical zonation in the ocean determines the distribution not only of calcareous sediments (coral, pteropodal, foraminiferal, etc.), but also of some siliceous ones (radiolarian and tropical diatom oozes), and of red clays. The critical or compensation depth of calcareous sediments varies in different latitudes. In the tropical and subtropical regions of the Pacific and Indian Oceans this depth is most often 4,600-5,000 metres, sometimes a little more. The granulometric composition of organogenous sediments also changes with depth (as for instance the replacement of foraminiferal sands by foraminiferal silts and muds), but a direct correlation between the granulometric composition and depth cannot be found. Vertical zonation of sedimentation is observed in all parts of the oceans from the coasts down to maximum depths. This zonation is mostly noticeable in bottom areas with rough topography.

Similar conditions are rather typical, not only for the continental slope, but also for the oceanic floor beyond the accumulative plains. Echo soundings, carried out during many oceanographic expeditions, clearly indicate that in the central parts of the oceans rough topography predominates. Sharp and frequent topography contrasts (with depths ranging over several kilometres in the regions of ridges, and over many dozens or hundreds of metres on the bottom of basins) determine not only the pronounced vertical zonation, but also the wide distribution over the bottom of hard rock and ancient sediment outcrops, or discontinuities of the sedimentary cover.

The vertical zonation, along with discontinuity in sedimentation under conditions of complex bottom topography, produces great *irregularity* in the distribution of deep-sea sediments, as well as sharp changes in their composition within short distances. Such changes are apparent at depths close to the critical depth, where calcareous sediments are replaced by non-calcareous ones, because at this depth the rate of sedimentation, and the thickness of sediments, decreases five to ten times (and even more) as a result of calcium carbonate solution.

It is necessary to note, however, that at the foot of submarine ridges in the Indian Ocean we have often observed the distribution of carbonate sand and ooze layers down to depths of 5,500-6,500 metres, i.e. considerably deeper than the critical depth. An explanation for the phenomenon should be sought in the rapid displacement of carbonate material by turbidity currents down the slopes of the ridges.

Thus, irregularities in deep-sea sedimentation under conditions of rough topography of the ocean bottom are displayed in various forms. It should be emphasized that such irregularities are not local phenomena, but extremely widespread ones additional to the zonation of sedimentation itself.

Received October 1967.

BIBLIOGRAPHY

Atlas of Antarctic, 1966. 1. Moscow (In Russian).

Bezrukov, P. L. 1959. Problems of zonation in sedimentation. *International Oceanographic Congress. Preprints. Amer. Assoc. for the Advanc. of Science.* 448-9: Washington, D.C.

Bezrukov, P. L. 1962. Some problems of sedimentation zonation in the World Ocean. *Trudy of the Oceanogr. Commission,* 3, 3-8 (In Russian).

Bezrukov, P. L. and Lisitzin, A. P. 1966. Sedimentation in the Indian Ocean. (New map of bottom sediments of the Indian Ocean, scale 1 : 20,000,000.) *Abstracts of Papers, 2nd Intern. Oceanogr. Congress, Moscow, 1966,* 42-3.

Bezrukov, P. L., Lisitzin, A. P., Petelin, V. P. and Skornjakova, N. S. 1966. Sedimentation in the Pacific Ocean. (New map of bottom sediments of the Pacific Ocean, scale 1:25,000,000.) *Abstracts of Papers, 2nd Intern. Oceanogr. Congress, Moscow, 1966,* 43-4.

Beljaeva, N. V. 1964. Distribution of planktonic foraminifera in the waters and on the bottom of the Indian Ocean. *Trudy Inst. Oceanol.* **68**, 12-83 (In Russian).

Emelijanov, E. M., Lucoshevichus, L. S., Svirenko, I. P., Soldatov, A. V., Koshelev, B. A., Lisitzin, A. P., Ilyin, A. V. Shurko, II., Litvin, V. M., and Senin, Yu, M. 1966. Sedimentation in the Atlantic Ocean. *Abstracts of Papers, 2nd Intern. Oceanogr. Congress, Moscow, 1966,* 108-9.

Heezen, B. C., Tharp, M. and Ewing, M. 1959. The floors of the oceans. 1. The North Atlantic. *Spec. Publ. Geol. Soc. Amer.* **65**, 1-122.

Jousé, A. P. 1966. Stratigraphic and palaeogeographic significance of planktonic algae. *Abstracts of Papers, 2nd Intern. Oceanogr. Congress, Moscow. 1966,* 183.

Kozlova, O. G. 1964. *Diatoms of the Indian and Pacific sectors of Antarctic.* 1964. 1-168. (In Russian). Nauka: Moscow.

Lisitzin, A. P. 1964. Distribution and chemical composition of suspended matter in the waters of the Indian Ocean. *Oceanology.* (Xth section of IGY Programme.) 1-136 (In Russian). Nauka: Moscow.

Lisitzin, A. P. 1966. Main regularities in the distribution of recent siliceous sediments and their relations with climatic zonality. *Geochemistry of Silica.* 90-191. (In Russian). Science: Moscow.

Physical-Geographical Atlas of the World. 1964. Sedimentation in the World Ocean. 16-17 (In Russian): Moscow.

Saidova, Kh. M. 1966. Foraminiferal bottom fauna of the Pacific. *Oceanology,* **6** (2), 276-84 (In Russian).

13. MICROPALAEONTOLOGY OF ANAEROBIC SEDIMENTS AND THE CALIFORNIA CURRENT

A. SOUTAR

Summary: A number of conveniently placed basins lie off the coast of North America along the southward path of the California Current. Within the sediments of these basins is a remarkably complete micropalaeontological record. The value of this record is further enhanced by the presence of annual layers or varves. From these sediments synoptic information may be retrieved, which relates to past biological, oceanographic and climatological conditions, and which provides a basis for projection of these conditions into the future. In particular, the problem of the Pacific sardine fishery is described.

It is intriguing to realize that everywhere in the ocean, year after year, the remains of many pelagic organisms find their way to the ocean bottom through the processes of sedimentation. The continued accumulation and preservation of these organisms and other sedimentary components constitutes a record of events in the overlying waters. The record is far from complete or proportional, in that many marine organisms are without skeletons or hard parts of sufficient resistance to survive the journey to the bottom. Even where hard parts are present, selective dissolution in the water, on the bottom, and in the sediments, can distort the record, as can the burrowing activities of the larger benthonic animals. The fact still remains that marine sediments contain a wealth of information about the past which has relevance to the present and the future.

One instance of this relationship is found beneath the California Current. The California Current system is described by Reid, Roden and Wyllie (1958) as 'part of the great clockwise circulation of the North Pacific Ocean. At high latitudes the waters move eastward under the influence of the strong westerly winds, and near the coast of North America divide into two branches. The smaller part turns northward into the Gulf of Alaska, and the larger part turns southeastward to become the California Current...The water which is brought south by the California Current system is cooler than the waters farther offshore. As it moves slowly south at speeds generally less than half a knot it becomes warmer under the influence of the sun and by mixing with the warmer waters to the west. As it nears latitude 25° N it begins to turn westward and its waters become part of the west flowing North Equatorial Current'.

The waters of the California Current form an important resource of North America; not only do they cool its western shores, but they contain an abundance of organic life potentially available to man. One relatively exploited resource has been that of the pelagic fishes, notably the Pacific sardine, *Sardinops caerulea.* The sardine fishery of the North American Pacific coast grew steadily from its inception in the early 1900s and by 1930 had become the largest single fishery on the American west coast. Development and utilization of the sardine resource continued until 1945 when a steady decline in available fish began. The decline continued until by 1960 Ahlstrom could state, 'We know that the sardine has not recently been a major component of the pelagic fish fauna of the California Current. It has often been a conspicuous component because of its habit of schooling relatively close to shore, but many other species of fish have been and are now greatly more abundant in the California Current waters. This includes several species of ecologically important deep-sea smelts, lantern fish, and the anchovy, hake, and jack-mackerel, which are also commercially important.' The virtual disappearance of the Pacific sardine caused much economic disruption in the United States as the sardine fishery failed progressively down the coast. In 1947 the fishery crisis led to an intense oceanographic and biologic study known as the California Cooperative Oceanic Fisheries Investigations, the objectives of which were in part 'to acquire knowledge and understanding of the factors governing the abundance, distribution, and variation of the pelagic marine fishes. The oceanographic and biologic factors affecting the sardine and its ecological

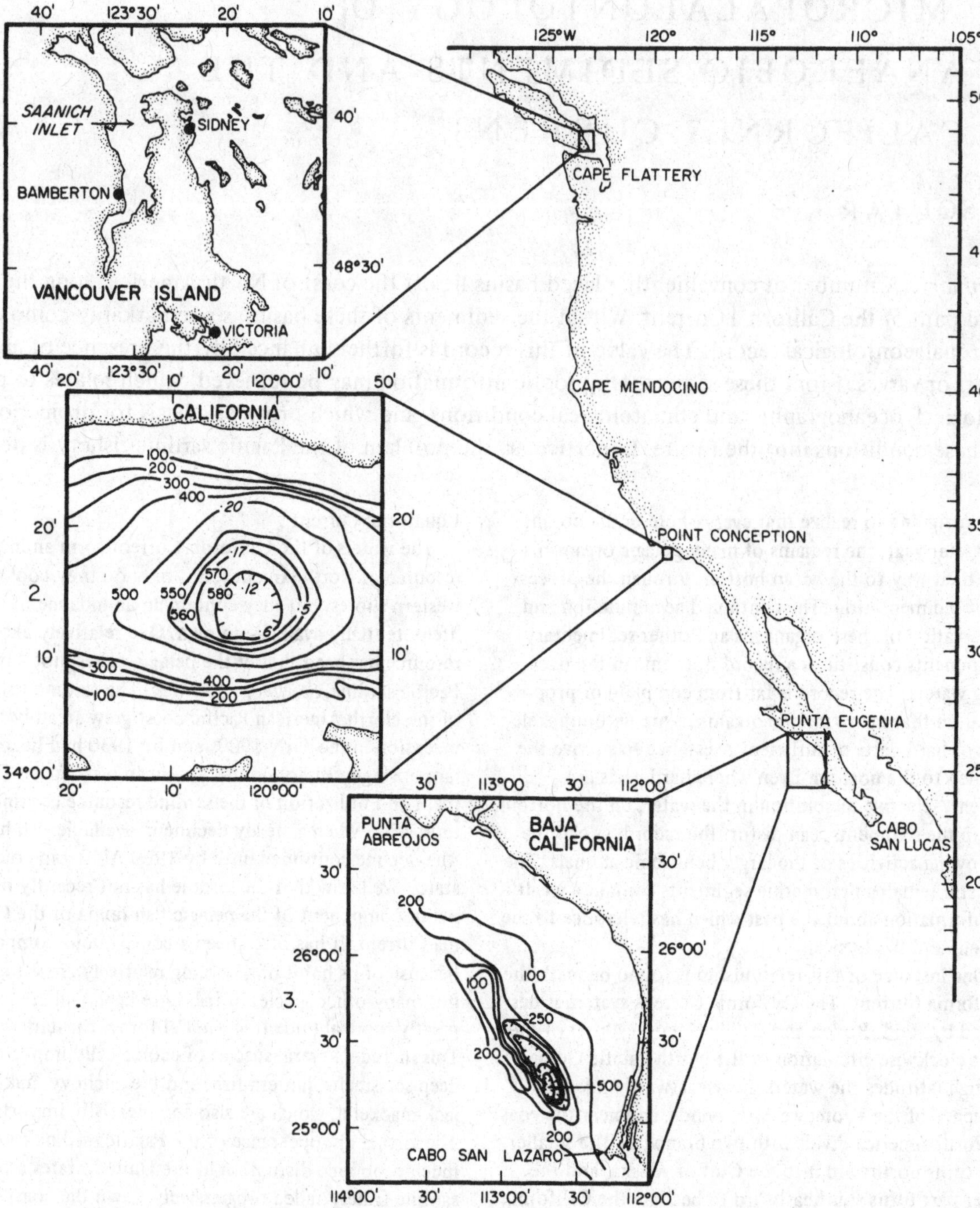
1.
SAANICH INLET
SIDNEY
BAMBERTON
VANCOUVER ISLAND
VICTORIA
2.
CALIFORNIA
3.
BAJA CALIFORNIA
PUNTA ABREOJOS
CABO SAN LAZARO
CAPE FLATTERY
CAPE MENDOCINO
POINT CONCEPTION
PUNTA EUGENIA
CABO SAN LUCAS
125°W
120°
115°
110°
105°
50°N
45°
40°
35°
30°
25°
20°

Fig. 13.1. Known regions of anaerobic sediment deposition along the coast of North America. 1.*Saanich Inlet* (after Gucleur and Gross 1964). An anaerobic fjord. Undisturbed and varved sediments are restricted to the central area of the fjord. 2.*Santa Barbara Basin* (after Hülsemann and Emery 1961). An anaerobic continental shelf basin. Undisturbed and varved sediments are present in the central region of the basin below the 550 metre contour. Locations of cores 5, 6, 12, 17, 20 are shown in italic. 3. *Basin north of Cabo San Lazaro* (after d'Angeljan-Castillon, 1965). An anaerobic continental shelf basin. Undisturbed sediments are found in the deeper part of the basin. Contours in metres.

associates will be given research emphasis...'. This study through the efforts of hundreds of researchers has resulted in this region being one of the best understood of ocean areas.

In 1957-8, there occurred within the California Current system, an unexpected warming associated with, among other unusual effects throughout the Pacific, a marked increase in the catch of sardines. The change was short lived but it helps to point out the limited span of time over which information is available. Extensive survey data have only been collected since 1949, general data on fish landings since 1915, and other pertinent information such as coastal temperatures since 1900. Questions on the frequency and duration of conditions similar to those of 1957-8 cannot be answered on the basis of available information. In fact it prompts the question what are the 'normal' biologic and oceanographic conditions over extended time spans?

It was suggested by John D. Isaacs of the Scripps Institution of Oceanography that a partial answer to these questions might be available from information in marine sediments. The particular sediments available were those of the Santa Barbara Basin (see Fig. 13.1). These sediments, described by Emery (1960), are deposited under a rare set of conditions in that there is little or no oxygen present in the bottom waters. The immediate effect of this condition is to exclude burrowing animals from the surface sediments and so allow the accumulation of undisturbed anaerobic sediments. From the standpoint of preservation of information in sediments this condition is most desirable. For one thing, organic matter is less likely to be degraded in an anaerobic environment. Complex organic structures are not physically disrupted and exposed to bacterial action; furthermore, the bacterial action in an anaerobic environment is less able to degrade complex molecules.

These factors account for the presence of a very interesting and useful component of the micropalaeontological record, fish scales. The following species of fish are represented by scales in the anaerobic sediments of the Santa Barbara Basin:

Colalabis saira – Pacific saury
Engraulis mordax – northern anchovy
Lampanyctus leucopsarsus – northern lampfish
Lampanyctus sp. – lampfish
Merluccius productus – Pacific hake
Physiculus rastrelliger
Sardinops caerulea – Pacific sardine
Sebastodes sp. – rock fish
Trachurus symmetricus – jack mackerel

Photomicrographs of some of the types of scales found in the sediments are shown in Plate 13.1.

Scales and other fish debris are found in the less intensely reducing sediments peripheral to the central basin sediments. David (1947) found scale material in the surface sediments of the Catalina Channel. However, counts in sediment cores suggest the preservation of scales is more strongly favoured in the intensely anaerobic environment. Table 13.1 presents the scale counts for five cores. The numbers represent the expected number of scales for cores with areas of 45 square centimetres and deposition times of about 1,500 years. Cores 5, 6, 12, and 17 were deposited on the anaerobic basin floor at water depths between 560 and 580 metres. Core 20 was deposited on the basin slope at a water depth of 515 metres. The locations of these cores are shown on Fig. 13.1. Examination of radiographs of these cores shows that only core 20 is extensively disturbed and homogenized. The average number of scales in highly anaerobic sediments of the basin floor is four times greater than the disturbed basin slope core indicating, in the case of scales, that there is at least four times more information preserved in the highly anaerobic sediments. The quality of this information is suggested by the data presented in Table 13.2. Here it is shown the relative abundance of the three major species of pelagic fish from sediment scales and larval fish populations compare favourably. That is to say the modern species composition of pelagic fishes of the California Current region is clearly reflected in the sediments of the Santa Barbara Basin. It is instructive to note that the bulk of the larval fish data was collected in very recent times; times when the sardine population has

Table 13.1. The total estimated number of fish scales in cores having areas of 45 square centimetres and deposition times of about 1,500 years.

Core	Water depth (metres)	Depositional environment	Number of scales
5	570	Highly anaerobic basin floor	1,290
6	578	Highly anaerobic basin floor	1,120
12	577	Highly anaerobic basin floor	1,440
17	562	Highly anaerobic basin floor	1,120
20	515	Slightly anaerobic basin slope (bottom water aerobic)	350

Table 13.2. Comparison of adjusted abundance percentages of fish scales found in cores 2, 3, 4 and 5 with fish larvae abundance ratios (from Ahlstrom 1959). The adjustment affects mainly the Pacific hake which, because of its greater number of scales per individual, is expected to contribute up to four times as many scales.

Species	Per cent scales in sediment Santa Barbara Basin	Per cent larvae in water Cal COFI region 1941, 1952-5
Northern anchovy	39	40
Pacific sardine	6	10
Pacific hake	29	18
Other species	26	32

Table 13.3. The estimated number of *Limacina helicina* (larger than 500μ) in Santa Barbara Basin cores having areas of 45 square centimetres and deposition times of about 1,500 years.

Core	Water depth (metres)	Depositional environment	Number of shells
6	578	Highly anaerobic basin floor	3,150
12	577	Highly anaerobic basin floor	3,400
17	562	Highly anaerobic basin floor	3,700
20	515	Slightly anaerobic basin slope (bottom water aerobic)	9

been conspicuously low.

The preservation of skeletal calcium carbonate is analogous to that of the organic material. The absence of physical attack and of solution combine to preserve a remarkable biogenic carbonate record. For example, the delicate foraminiferal species, *Hastigerina pelagica* and *H. digitata,* previously known only in the plankton in these latitudes, form a minor but consistent part of the record. Other planktonic species of Foraminifera are found in an excellent state of preservation (see Plate 13.2). These include the species *Globigerina quinqueloba, Globoquadrina eggeri, Globigerina bulloides, Globigerinoides ruber, Globigerinata glutinata, Globorotalia hirsuta, Globigerina calida, Hastigerina pelagica,* and *H. digitata,* and others (Berger and Soutar, 1967).

Also of great interest is the discovery that pelagic mollusc shells are preserved within these sediments. Table 13.3 presents the estimated numbers of *Limacina helicina* in cores 5, 6, 12, 17 and 20 and it can be seen that only in the intensely anaerobic sediments are they preserved. This is important in that, with the exception of the pelagic fish, this is the only group represented in the sediments for which extensive zoogeographical information is available in the California Current region (McGowan 1963, 1967).

Plate 13.3 shows a number of the pelagic mollusc, *Limacina helicina,* recovered from the sediments. This is the most common species in the waters and sediments of the Santa Barbara Basin. Among the other genera of pelagic molluscs present in the sediments of the Santa Barbara Basin are *Clio, Cavolinia, Atlanta,* and the species *Limacina inflata.*

The siliceous record is similarly well preserved. A photograph of a microfossil preparation is shown in Plate 13.4. The bulk of the diatom material is composed of *Coscinodiscus*-type diatoms of which *Coscinodiscus oculis iridis* is by far the most numerous. Radiolaria are well represented with numerous types of Spumellaria and Nassellaria. Fragments of Tripylea are occasionally encountered. Silicoflagellates are also present.

Although not yet investigated, a detailed coccolith and pollen record is anticipated.

Perhaps the most illuminating information preserved in the anaerobic sediments stems from the fact that any persistent variation in the amount and composition of the organic and inorganic sources of such sediments is likely to be preserved. Thus, in the Santa Barbara Basin sediments are regular fine layers which generally correspond to a biannual variation in the sediment source (Emery 1960). These layers are shown in Plate 13.5 which is a positive print of a core radiograph. During the winter part of the year, a denser layer (black on the photograph and in the sediment) is produced, reflecting the increased inorganic runoff from the winter rains. The summer produces a less dense layer (white on the photograph and olive green in the sediments). This layer contains a higher percentage of diatom frustules reflecting an increased organic production in the surface waters. The combination of the two layers of differing density constitutes a varve. The sedimentary record thus preserves the historical

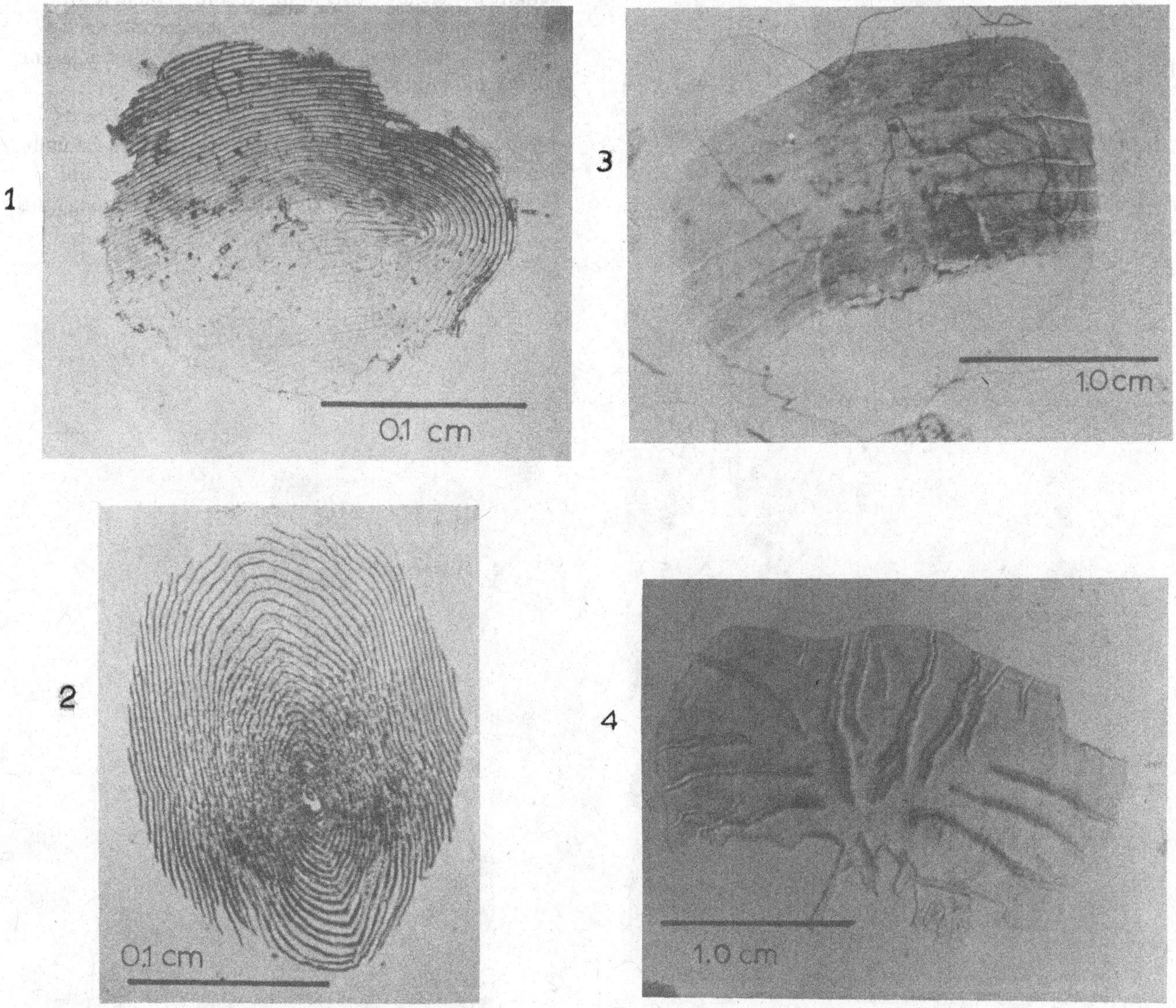
1
0.1 cm
3
1.0 cm
2
0.1 cm
4
1.0 cm

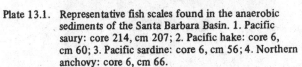

Plate 13.1. Representative fish scales found in the anaerobic sediments of the Santa Barbara Basin. 1. Pacific saury: core 214, cm 207; 2. Pacific hake: core 6, cm 60; 3. Pacific sardine: core 6, cm 56; 4. Northern anchovy: core 6, cm 66.

record with a high degree of resolution.

It was previously thought that this record was too discontinuous and incomplete to allow the extraction of meaningful climatological and palaeo-oceanographic information (Hulsemann and Emery 1961). However, with the development of refined sampling and observational techniques, it appears a rather complete time-biological record exists within these sediments. The varve series shown on Plate 13.5 is part of a continuous record extending over 37 centimetres and covering a period of about 140 years. This is a marked increase over the previously reported maximum record of 7.5 centimetres. In fact the section shown is a portion of the most valuable part of the record as it extends from the present back through modern times. Its discovery raises the very real possibility of matching the observed with the recent sediment record on a year to year basis. This, of course would allow a much better interpretation of deeper sections.

While it is of interest and importance that a record exists within the Santa Barbara sediments, with relevance to the California Current region, similar records from other portions of the region are needed to produce a coherent historical picture. Fortunately, two other anaerobic sediment basins are known to exist in this area. One of these lies near the southern end of the current off Cabo

1

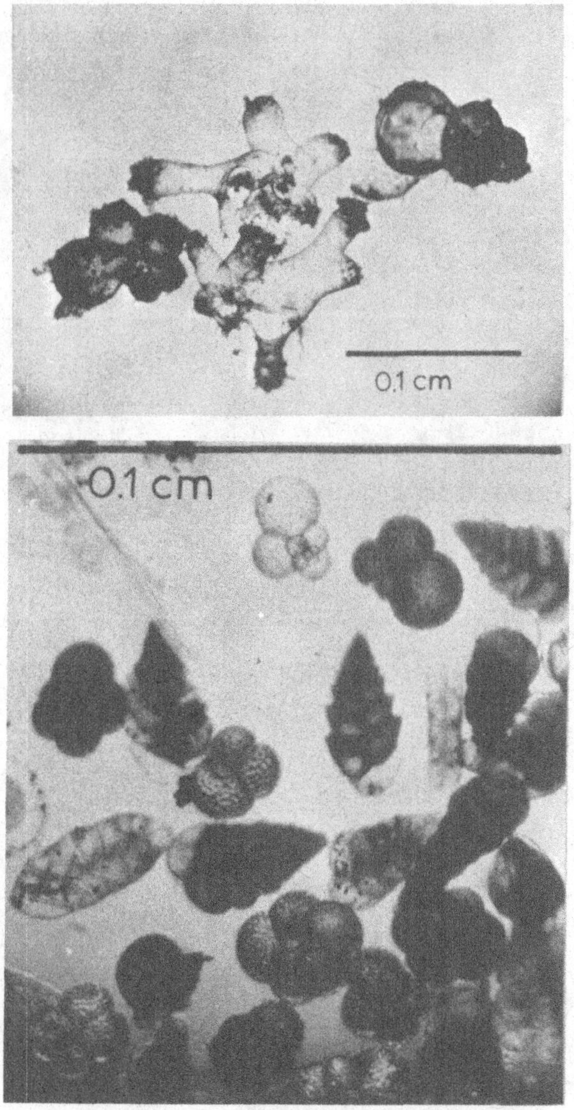
0.1 cm
0.1 cm

Plate 13.2. Representative planktonic Foraminifera present in the anaerobic sediments of the Santa Barbara Basin. 1. Core 214, cm 73: *Hastigerina pelagica* and *H. digitata.* 2. Core 214, cm 5: view of 124-177μ fraction. Planktonic species include: *Globigerina bulloides, Globigerina quinqueloba, Globoquadrina eggeri,* and *Globigerinoides ruber.*

San Lazaro, Baja California, Mexico (Fig. 13.1). The observation that the sediments in this shelf basin were accumulating under anaerobic conditions was first made by d' Anglejan-Castillon (1965). The micropalaeontological record is very similar in character to that of the Santa Barbara Basin. Fish scales and other fish debris are present. Pelagic molluscs and planktonic Foraminifera are well preserved, as are diatoms and radiolarians. The species composition reflects a more southerly fauna; for example, the dominant pelagic mollusc is *Limacina trochiformis.* A definite deficiency exists in this record as there is little development of annual varves. The absence of such layering most likely results from the extremely low adjacent rainfall and limited drainage area.

The other known anaerobic sediment record exists in the Saanich Inlet of Vancouver Island, British Columbia, Canada (see Fig. 13.1) – Gross, Gucleur, Creager and Dawson (1963), Gucleur and Gross (1964). Although this

2

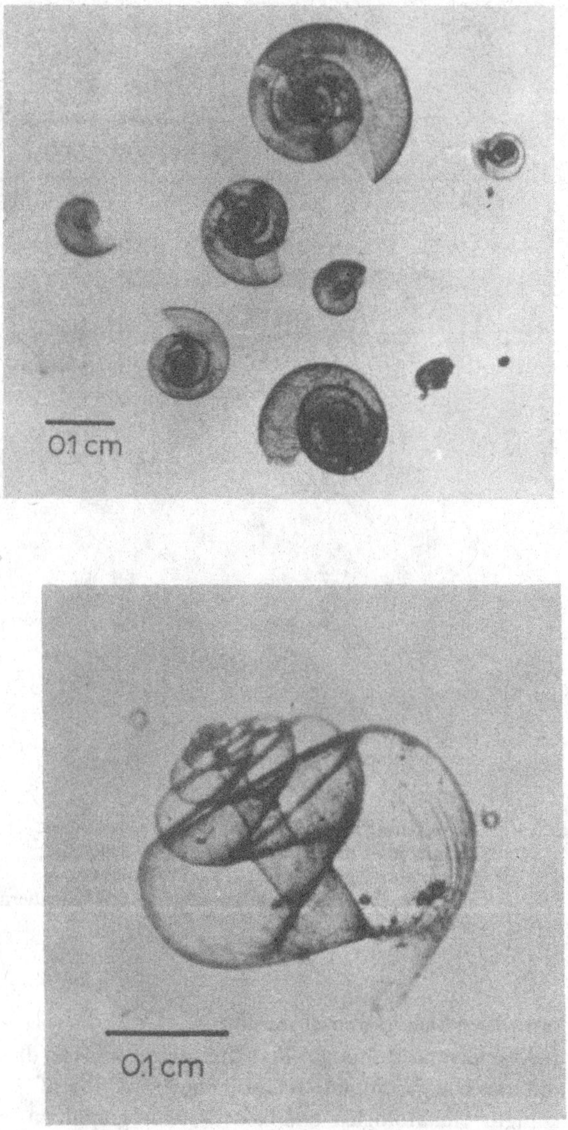
0.1 cm
0.1 cm

Plate 13.3. Representative pelagic molluscs found in the anaerobic sediments. These specimens are from core 214, cm 73. 1. *Limacina helicina* and one *L. inflata* (extreme left) The *L. helicina* shown are the variety AB of McGowan (1963). 2. *Limacina helicina* variety A of McGowan (1963).

1

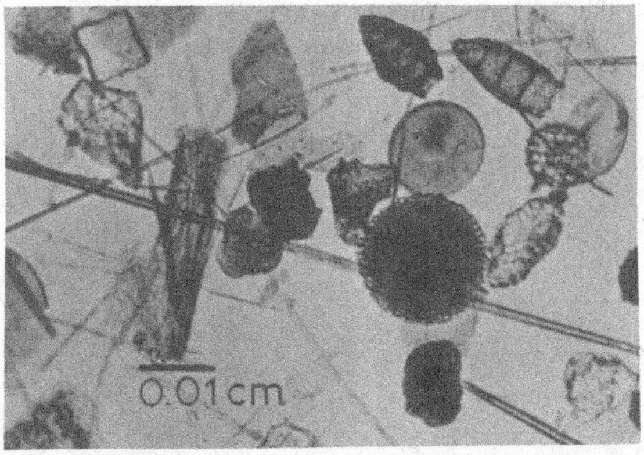
0.01cm

2

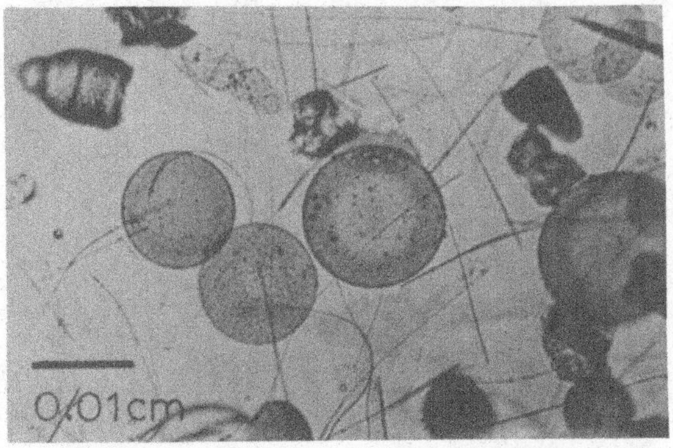
0.01cm

Plate 13.4. Representative photomicrographs of the siliceous microfossils in the Santa Barbara Basin. 1. Core 214, cm 141 Radiolaria. 2. Core 214, cm 81 Diatomacea.

inlet is well protected, the biological and sedimentary record seems to reflect the marine environment of the adjacent waters. The sediments are varved and should provide a valuable source of historical data in the more northerly regions of the current. In particular it is hoped that the occurrences of the Pacific sardine in these northerly waters is recorded in these sediments. This information would go far towards providing the statistical information necessary to help predict the frequency and intensity of this fishery.

Other areas which might provide comparable information are the Gulf of California and the coastal waters of Peru. The Gulf of California is a highly productive region of considerable fishery potential and is known to have an extensive varved sediment record (Calvert 1966). The vast populations of the anchovy off the coasts of Peru

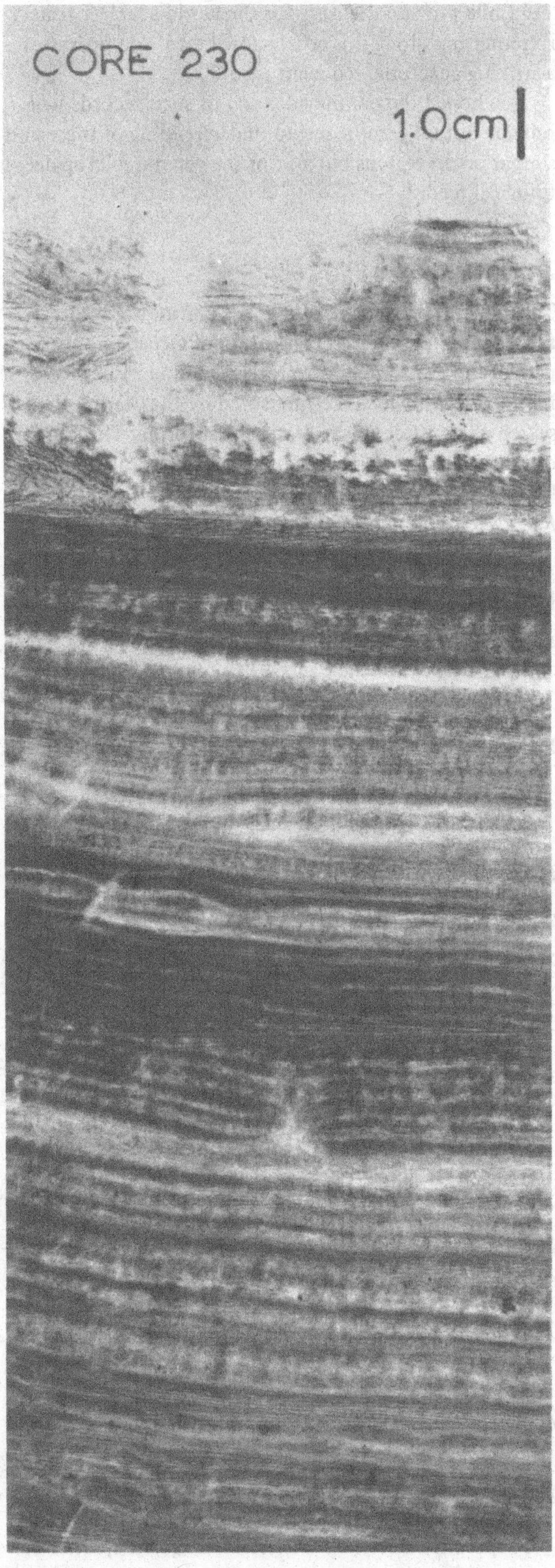
CORE 230
1.0cm

Plate 13.5. Core 230 (located near core 12), Santa Barbara Basin. Positive print of radiograph of core. Dark areas are the most dense; however, the relative density of the upper layers is less than the shading of the photograph suggests.

and Chile provide the basis for the world's largest fishery. Explorations along this coast show that it too possesses a striking anaerobic sediment record.

It is hoped the continued study of such records will provide not only an increased understanding of these and similar ocean regions but also of the general micropalaeontological record.

ACKNOWLEDGEMENT

This paper represents results of the Marine Life Research Programme, the Scripps Institution of Oceanography's part of the California Cooperative Oceanic Fisheries Investigations, which are sponsored by the Marine Research Committee of the State of California. The support of the National Science Foundation is gratefully acknowledged.

Received September 1967.

BIBLIOGRAPHY

Ahlstrom, E. H., Isaacs, J. D., Murphy, C. I. and Radovich, J. 1961. Report of the CalCOFI Committee. *Progress Report California Cooperative Oceanic Fisheries Investigations,* **8**, 5-8.

d'Anglejan-Castillon, Bruno, 1965. *Marine phosphorite deposits of Baja California, Mexico, present environment and recent history.* Unpublished Doctoral Thesis: Scripps Institution of Oceanography, University of California.

Berger, W. H., and Soutar, A. 1967. Planktonic Foraminifera: Field experiment on production rate. *Science,* **156** (3,781), 1495-7.

Calvert, S. E. 1966. Origin of diatom-rich, varved sediments from the Gulf of California. *Jour. Geol.* **64** (5) 546-64.

David, L. R. 1947. Significance of fish remains in recent deposits of Southern California. *Bull. Am. Assoc. Pet. Geol.* **31** (2), 367-70.

Emery, O. 1960. *The Sea off Southern California; A Modern Habitat of Petroleum.* 1-366. John Wiley and Sons, Inc.: New York.

Gross, M. G., Gucluer, S. M., Creager, J. S., and Dawson, W. A. 1963. Varved marine sediments in a stagnant fjord. *Science,* **141** (3584), 918-19.

Gucluer, S. M. and Gross, M. G. 1964. Recent marine sediments in Saanich Inlet, a stagnant marine basin. *Limnology and Oceanography,* **9** (3), 359-76.

Hülsemann, J. and Emery, K. O. 1961. Stratification in recent sediments of Santa Barbara Basin as controlled by organisms and water character. *Jour. Geol.* **69** (3), 279-90.

McGowan, J. A. 1963. Geographical variation in *Limacina helicina* in the North Pacific. *Systematics Assoc. Pub.* **5**, *Speciation in the Sea,* 109-28.

McGowan, J. A. 1967. Distributional Atlas of Pelagic Molluscs in the California Current Region. *California Cooperative Oceanic Fisheries Investigations, Atlas* **6**.

Reid, L., Jr., Roden, I., and Wyllie, G. 1958. Studies of the California Current System. *Progress Report California Cooperative Oceanic Fisheries Investigations* (January 1958), 27-55.

14. THE OCCURRENCE OF DINOFLAGELLATES IN MARINE SEDIMENTS

D. B. WILLIAMS

Summary: Using an example from the North Atlantic Ocean, the distribution of dinoflagellate cysts in marine sediments is discussed, and is shown to be a response to the surface environment.

INTRODUCTION

Chapter 4 in this book on the distribution of dinoflagellates in relation to physical and chemical characteristics in the water showed factors which might affect their distribution in the sediments. The purpose of this paper is to consider this distribution and to see how well it conforms with these factors.

MORPHOLOGY AND TAXONOMY

Dinoflagellates are found in sediments as the empty cysts formed by the motile stages. Some of these cysts are peridinioid in shape, closely resembling the motile form, but they possess a geometrically regular opening (the archeopyle) in the position of one or more of the thecal plates (Evitt 1967) through which the cell contents escaped on germination. Other cysts are spherical or egg-shaped and may be ornamented with spines, which may or may not reflect the tabulation pattern of the motile stage. However, the feature common to all the cysts is the presence of an archeopyle. The process of encystment has not been observed, but seems to be internal to the theca of the motile form, since Nordli (1951) and Evitt and Davidson (1964) record dinoflagellate cysts caught in the plankton, where the remains of the theca were still adhering to the spines of the cyst. Excystment of dinoflagellates from the cysts has been observed by Wall (1966) who gives a sequence of photographs of the process, taken over a span of a few minutes.

The wide range of variability in cyst morphology can lead to problems of nomenclature, for it is now known through Wall's studies that cysts which have been placed in three different genera (*Hystricohsphaera, Loptodinium, Lingulodinium*) all give rise to one genus of motile dinoflagellate (*Gonyaulax*). Fortunately for workers on these cysts, the Botanical Code of Nomenclature permits usage of organ taxa and form taxa, so a dual system of nomenclature may be employed.

METHODS OF STUDY

Dinoflagellate cysts are formed of chemically resistant organic compounds and can be relatively easily isolated from marine sediments by the normal methods of palynology. These include solution of the carbonate by hydrochloric acid, followed by solution of the silicates with hydrofluoric acid. Finely divided organic particles (kerogen) may then be removed by careful oxidation with Schulz reagent, followed by a dilute alkali to clear the residue of humic acids. Other workers replace the oxidation by acetylation with acetic anhydride and sulphuric acid, or the much simpler physical separation of fine particles by filtration (described by Neves and Dale 1963). The residue may be mounted on a microscope slide and examined, often using the highest possible magnifications to aid identification of specimens.

The area selected for study was the Atlantic Ocean north of the equator. Samples were supplied by the British Museum (Natural History) and the Department of Geology, Cambridge. The number of specimens on which the following conclusions are based is small (only 35), but the results show an overall pattern which is reasonable. So although many of the conclusions may be rather general it is possible that the same sort of relationships would hold for different ocean basins, and for the same ocean basin with a new set of data.

Most of the species found in the sediments have been described by Rossignol (1964) and Wall (1967) and the nomenclature of the latter will be used. Where the species are new or in doubt, open nomenclature (e.g. *Nematosphaeropsis* sp. C) will be used. Some of the forms observed were included in the counts, though their affinities are

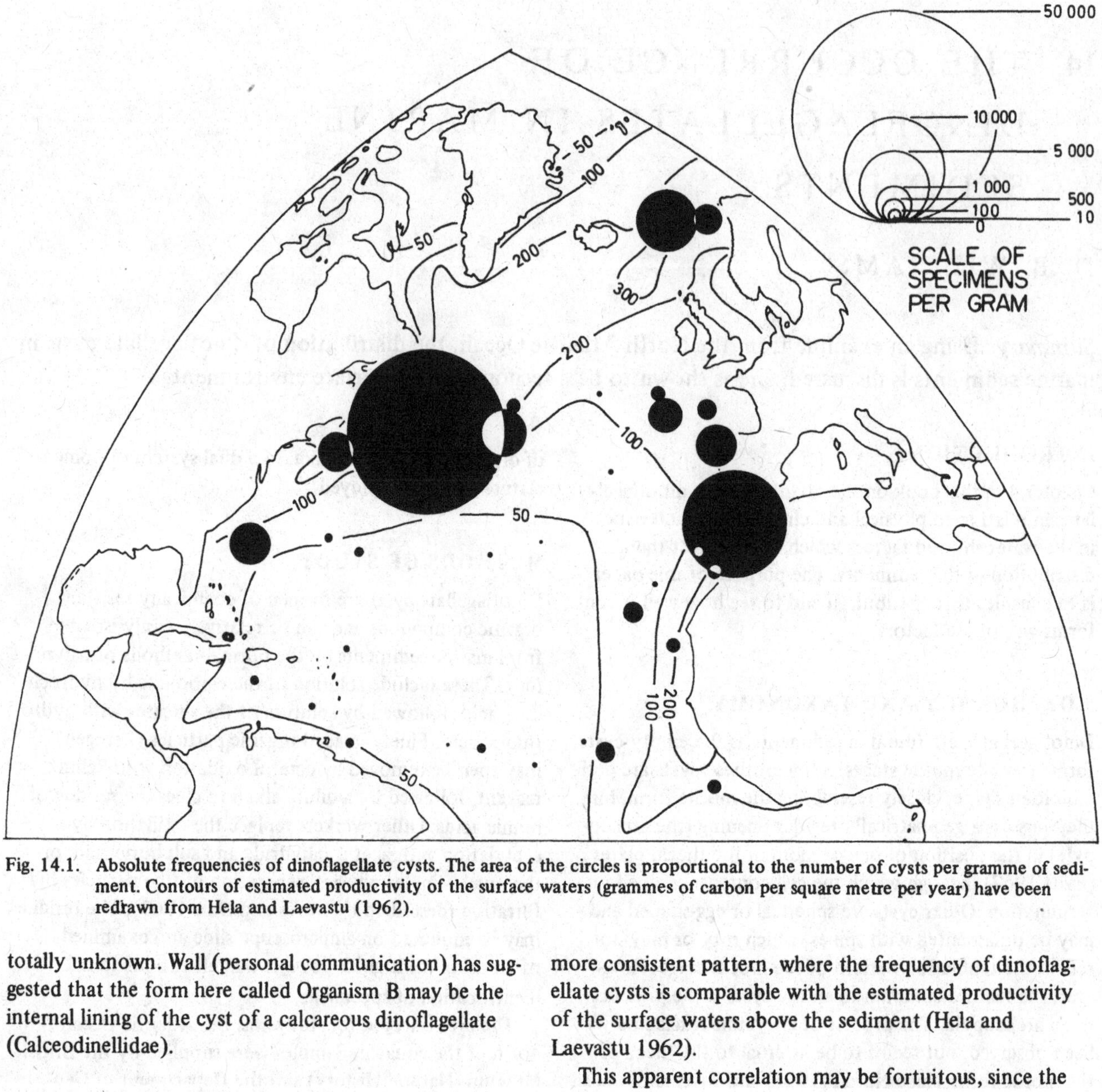
50 000
10 000
5 000
1 000
500
100
10
0
SCALE OF
SPECIMENS
PER GRAM
50
100
200
300
200
100
50
100
50
200
100
50

Fig. 14.1. Absolute frequencies of dinoflagellate cysts. The area of the circles is proportional to the number of cysts per gramme of sediment. Contours of estimated productivity of the surface waters (grammes of carbon per square metre per year) have been redrawn from Hela and Laevastu (1962).

totally unknown. Wall (personal communication) has suggested that the form here called Organism B may be the internal lining of the cyst of a calcareous dinoflagellate (Calceodinellidae).

QUANTITATIVE DISTRIBUTION OF DINOFLAGELLATE CYSTS

A modification of the preparation technique described above, permitted estimation of the frequencies of dinoflagellate cysts per unit weight of sediment. The results of this are shown on Fig. 14.1 as circles, whose area is proportional to the number of cysts per gramme, which gives a good representation of the range of variation of frequencies. However, frequencies represented this way are so extreme that it is difficult to determine any pattern in them. Use of a logarithmic scale of diameters makes the situation more comprehensible. Fig. 14.2 shows a more consistent pattern, where the frequency of dinoflagellate cysts is comparable with the estimated productivity of the surface waters above the sediment (Hela and Laevastu 1962).

This apparent correlation may be fortuitous, since the frequencies are probably further influenced by other factors, particularly the seasonal variation of the environment. The conditions which cause encystment are as yet unknown, but it is probable that it occurs in order to permit the organism to survive unsuitable conditions in the environment. In the subarctic regions, where there are high frequencies in the sediments, there would therefore be a seasonal encystment of the motile stages with the empty cysts sinking to the bottom after release of protoplasts at the beginning of the new growing season. The very low numbers of cysts in sediments under the central North Atlantic are probably a result of the stability of the environment over the year, further enhanced by a

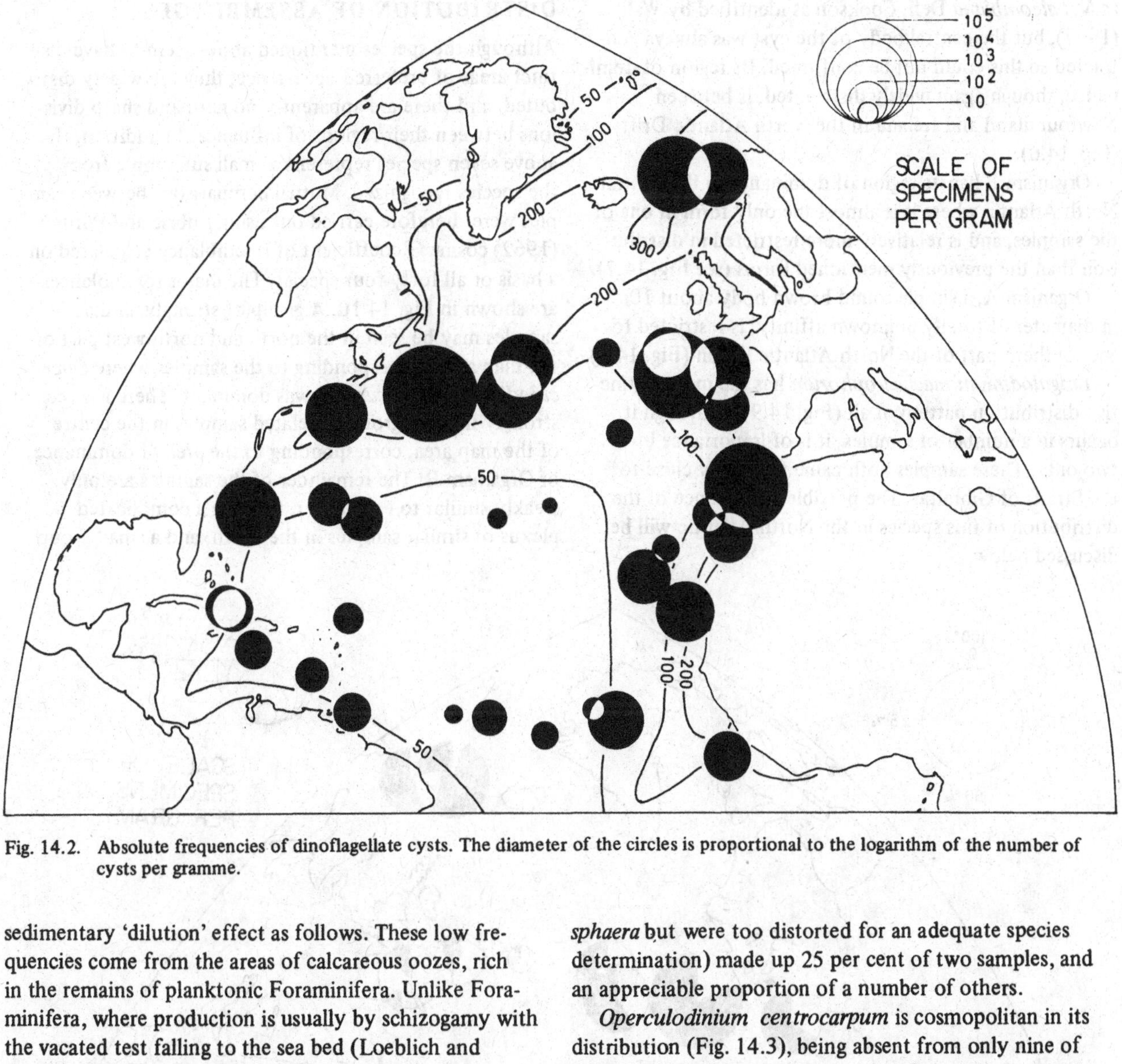
10^5
10^4
10^3
10^2
10
1
SCALE OF
SPECIMENS
PER GRAM
50
100
200
300
50
200
100
50
200
100
50

Fig. 14.2. Absolute frequencies of dinoflagellate cysts. The diameter of the circles is proportional to the logarithm of the number of cysts per gramme.

sedimentary 'dilution' effect as follows. These low frequencies come from the areas of calcareous oozes, rich in the remains of planktonic Foraminifera. Unlike Foraminifera, where production is usually by schizogamy with the vacated test falling to the sea bed (Loeblich and Tappan 1964), the dinoflagellates reproduce by binary fission, both halves survive, and no test is freed to sink to the bottom. Since the theca of the dinoflagellate motile stage is cellulose, it would in any case be destroyed by bacterial action before it reached the sediment.

THE DISTRIBUTION OF SPECIES

In all, forty-four different taxonomic groups ('species') were recognized, but the distribution patterns to be discussed here are restricted to those species which made up more than 25 per cent of the assemblage in at least one sample. There are eight of these out of the possible forty-four, but one category is excluded. This group (specimens which could only be attributed to the genus *Hystrichosphaera* but were too distorted for an adequate species determination) made up 25 per cent of two samples, and an appreciable proportion of a number of others.

Operculodinium centrocarpum is cosmopolitan in its distribution (Fig. 14.3), being absent from only nine of the samples examined. But in spite of its wide distribution, there are very definite areas where it is more important, namely along the coastal margin of North America and across the northernmost part of the ocean.

Leptodinium aculeatum is even more widely distributed than the previous species (31 samples, Fig. 14.4) but never reaches such a large proportion of the assemblage. Its region of major occurrence is off the north-west coast of Africa.

Hystrichosphaera mirabilis (Fig. 14.5), a very distinctive species, makes up a significant proportion of the assemblage in a very confined region off the western approaches of the English Channel, although it may also be found in samples from the coastal margin of the ocean.

Nematosphaeropsis sp. C is possibly the same species

as *N. balcombiana* Defl. Cookson as identified by Wall (1967), but the central body of the cyst was always contracted so this could not be confirmed. Its region of dominance, though again widely distributed, is between Newfoundland and Ireland in the North Atlantic Drift (Fig. 14.6).

Organism B has its region of dominance in the central North Atlantic where it is almost the only form in one of the samples, and is relatively more restricted in distribution than the previously mentioned forms (see Fig. 14.7).

Organism A, a simple round brown body about 10μ in diameter of totally unknown affinity, is restricted to the southern part of the North Atlantic Ocean (Fig. 14.8).

Lingulodinium machaerophorum has the most distinctive distribution pattern of all (Fig. 14.9). Although it occurs in a number of samples, it is of importance in two only. These samples both came from very close to the Straits of Gibraltar. The possible significance of the distribution of this species in the North Atlantic will be discussed below.

DISTRIBUTION OF ASSEMBLAGES

Although the species mentioned above seem to have distinct areas of preferred occurrence, they are widely distributed, and there are apparently no hard and sharp divisions between their regions of influence. In addition, the above seven species represent a small subsample from the species recognized. Mutual comparisons between samples were therefore carried out using Imbrie and Purdy's (1962) cosine Θ coefficient of resemblance calculated on a basis of all forty-four species. The major resemblances are shown in Fig. 14.10. A group of strongly similar samples may be seen in the north and north-west part of the map area, corresponding to the samples where *Operculodinium centrocarpum* was dominant. There is a second strongly marked group of related samples in the centre of the map area, corresponding to the area of dominance of Organism B. The remainder of the samples are only weakly similar to each other, forming a complicated plexus of similar samples in the south, and a small group

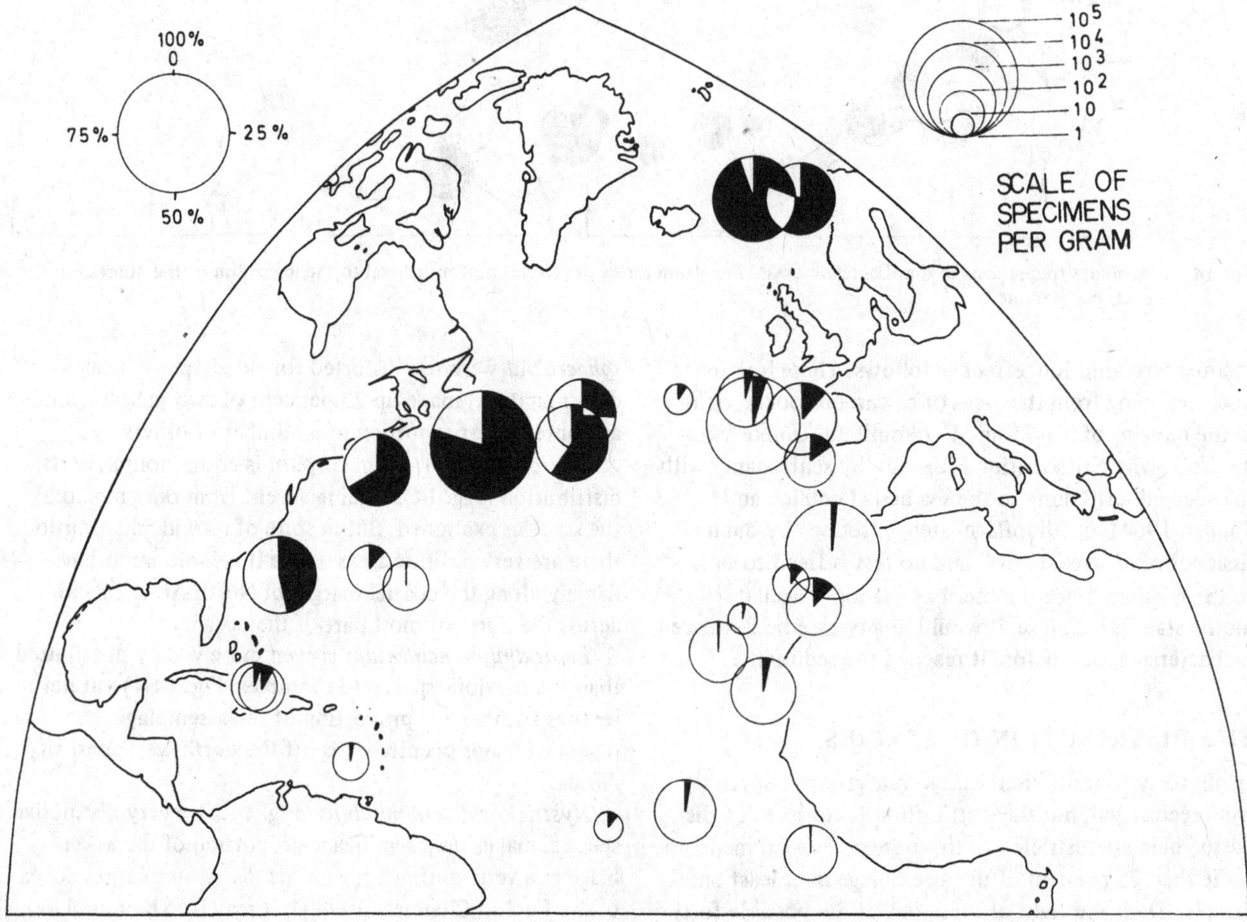
100%
0
25%
50%
75%
10^5
10^4
10^3
10^2
10
1
SCALE OF
SPECIMENS
PER GRAM

Fig. 14.3. Frequency of *Operculodinium centrocarpum*. The diameter of the circle is proportional to the logarithm of the number of cysts per gramme, area of black proportional to percentage of the species present in sample.

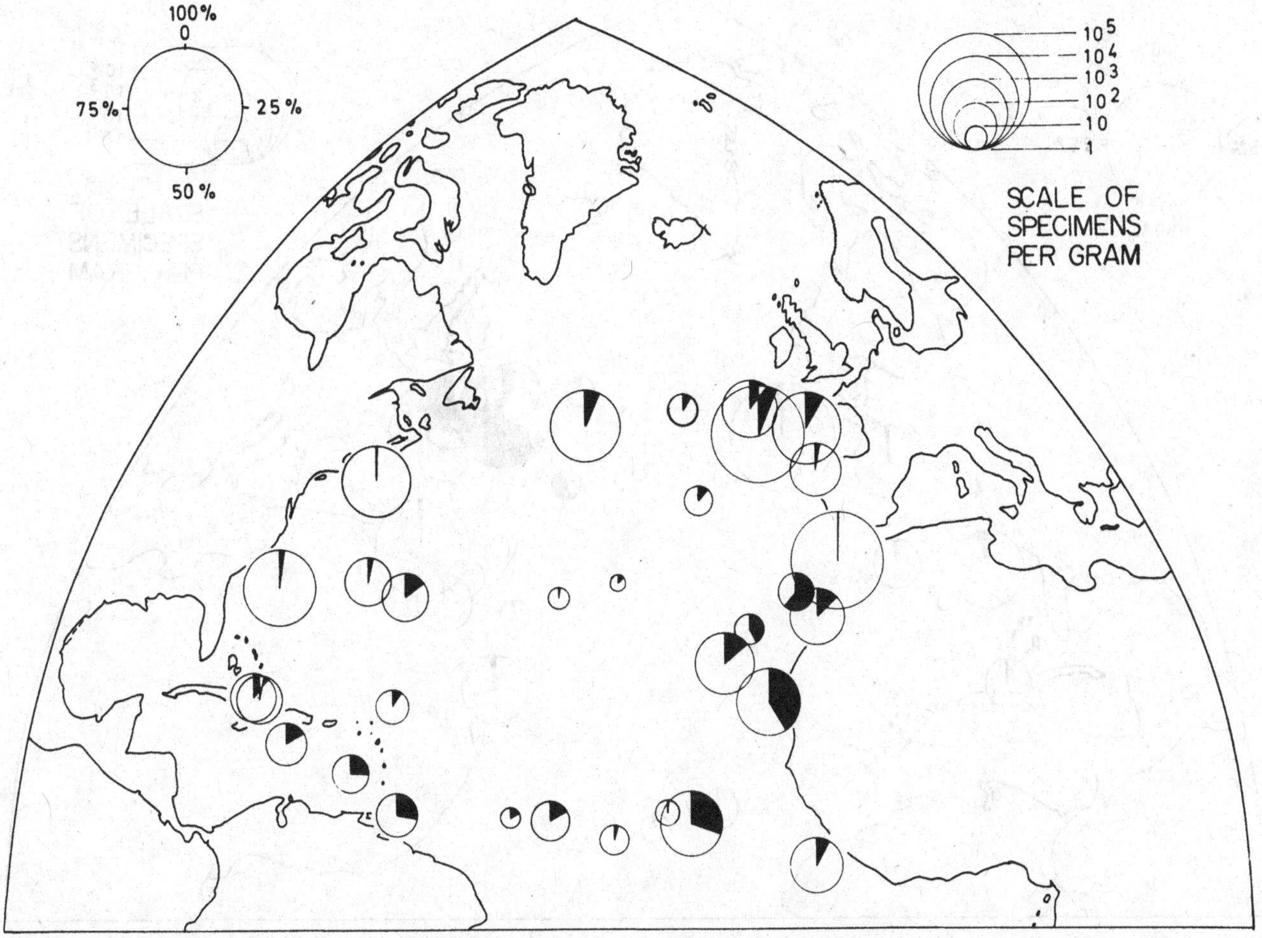
100%
0
75%
25%
50%
10⁵
10⁴
10³
10²
10
1
SCALE OF
SPECIMENS
PER GRAM

Fig. 14.4. Frequency of *Leptodinium aculeatum.* The diameter of the circles is proportional to the logarithm of the number of cysts per gramme, area of black proportional to percentage of the species present in sample.

of six weakly linked samples to the north of the central group. This network of similarities is sufficient to give a few ideas about the distribution of assemblages of species, but understanding would be made even more easy if we could identify an assemblage which is typical of a region, and which could be used for comparison with all other samples. By such comparison, the area of influence of the typical assemblage could be plotted out. It is, however, very difficult to decide on a method of formulating this ideal sample, and also difficult to decide how many of these ideal samples would be necessary. The mathematical technique of Principal Components Analysis (also known as Factor Analysis) can solve the first problem, and gives help towards solving the second.

Using the raw counts as data for Principal Components analysis followed by Varimax rotation (see Imbrie and van Andel 1964) ten ideal assemblages were found to be adequate to explain 98 per cent of the variation of the real assemblages. Contours were plotted of the similarities between the real samples and the principal components 'ideal' samples, and the resultant contour maps were superimposed to form the biofacies map illustrated in Fig. 14.11. There is remarkably little overlap of facies in the maps, suggesting that most of the samples belong to one facies only. Contours were drawn as objectively as possible, but there are doubts as to the limits of some of the facies, owing to the scatter of samples. This applies particularly to the area drawn for Facies II, between the latitudes of Newfoundland and Iceland, and the northern margin of Facies VII. Further samples from this region should help to clear up these difficulties. Facies IX and X are centred each round single samples which both had very low counts. These particular facies are therefore unreliable, and should probably be disregarded, pending better material from these localities.

Study of the distribution of the facies which were identified by principal components analysis, and comparison with the original data suggests that the areas of a maximum occurrence of a given facies coincides with the regions of maximum occurrence of one of the species groups

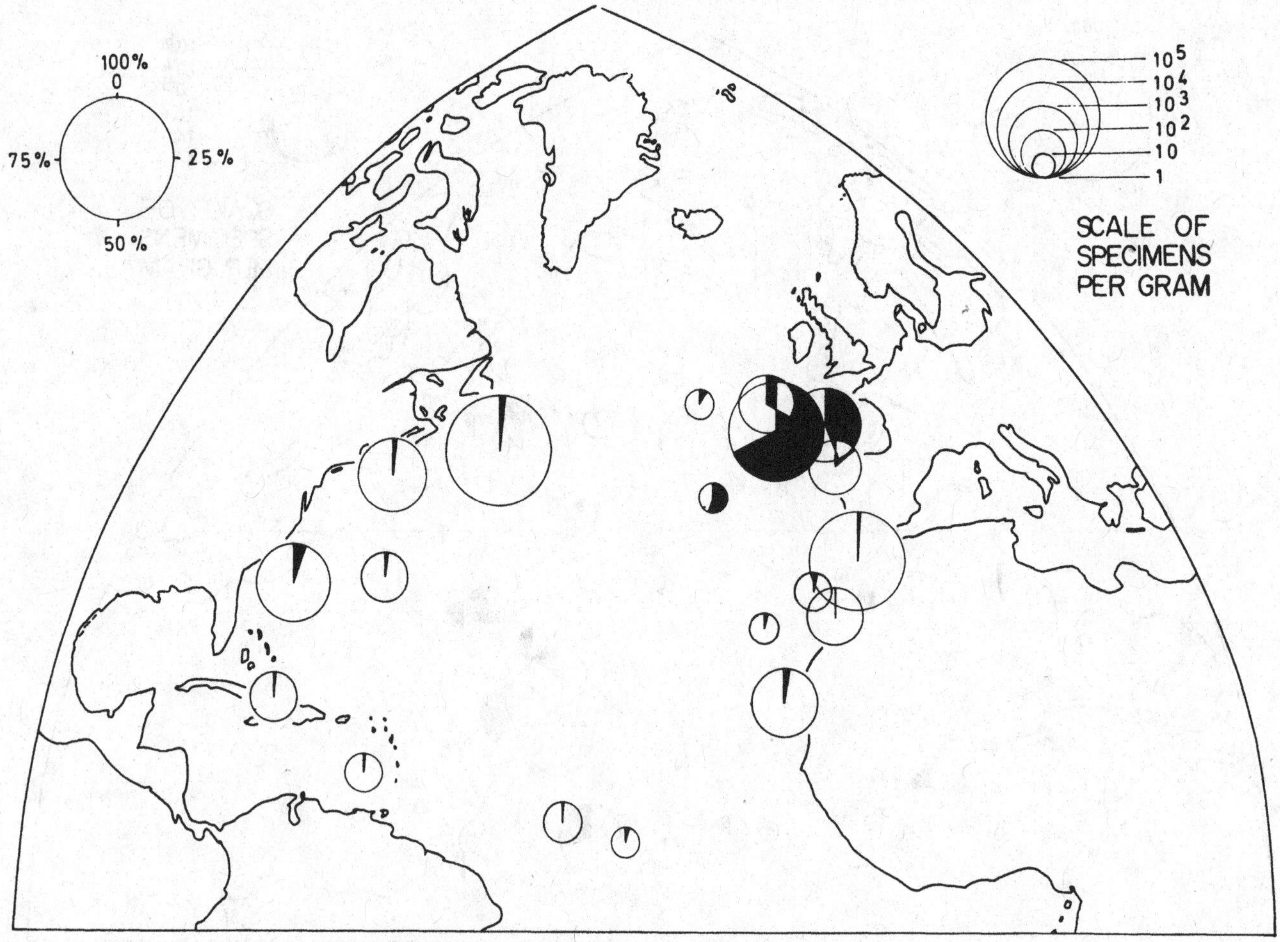
100%
0
75%
25%
50%
10^5
10^4
10^3
10^2
10
1
SCALE OF
SPECIMENS
PER GRAM

Fig. 14.5. Frequency of *Hystrichosphaera mirabilis.* The diameter of the circles is proportional to the logarithm of the number of cysts per gramme, area of black proportional to percentage of the species present in sample.

mentioned earlier in this section. Facies I coincides with samples having a high proportion of Organism A in their assemblage, Facies II with samples having a high proportion of *O. centrocarpum,* Facies III with a high proportion of Organism B, Facies IV with a high proportion of *Hystrichosphaera mirabilis,* Facies V with the undifferentiated *Hystrichosphaera* species. The dispersed distribution of this last facies also suggests that the species group on which it is based would merit taxonomic reinvestigation. Facies VI is represented by those samples which were rich in *Lingulodinium acuminatum.* Facies VII carries a high proportion of *Nematosphaeropsis* sp. C. Facies VIII is associated with dominant *Lingulodinium machaerophorum.* A further development of principal components analysis, which calculates the composition of the assemblages typical of the facies has now shown these identifications to be substantially correct.

ECOLOGICAL SIGNIFICANCE OF THE BIOFACIES

Comparison of the distribution of the various dinoflagellate biofacies with the distribution of different factors of the surface environment, has shown the following tentative correlations.

Facies I correlates quite well with the belt of low salinity which may be found in the southern part of the North Atlantic Ocean, which in its turn corresponds with part of the Equatorial Counter Current. The southern edge of Facies II closely follows the position of the boundary between the north-east flowing Gulf Stream and North Atlantic Drift and the south to south-west flowing colder currents. It would appear that this is an assemblage of cysts of cold-water dinoflagellates (notably *Operculodium centrocarpum* which tallies with the cysts of *Protoceratium reticulatum* obtained from culture by Braarud (1945)). Facies III is closely associated with the high salinity water in the centre of the North Atlantic current gyral (Sargasso Sea). There seems to be little correlation with other factors except the notable depletion of nutrients which exists in this region. Curiously, the inverse relationship between number of species and nutrients noted by Peters (1932) and Graham and Bronikovsky (1944) for species of *Ceratium* in the surface waters, does not seem to hold good for the cysts of dinoflagellates in the sediments. Facies III thus appears to be associated with the North Equatorial Current.

There seems to be nothing special in the recorded attributes of the surface environments where Facies IV and VII are found, except in the direction in which the currents are flowing. Facies IV underlies that part of the North Atlantic Drift which swings southwards, and Facies VII the part which continues its north-east trend.

Facies V has already been discussed from the point of view of subdivision of *Hystrichosphaera* species, since it is found to occur in three distinct regions, though its occurrence north of Cuba in three samples probably represents some sort of interaction between the Antilles Current and the Gulf Stream. Facies VI has its region of maximum importance below the region of seasonal upwelling off the coast of Africa. Its minor occurrence in the Caribbean may be fortuitous.

Facies VIII is particularly interesting in its distribution pattern. It is only well represented in two samples, one of which, situated very close to the Straits of Gibraltar, is almost exclusively composed of its typical species. *Lingulodinium machaerophorum.* The shape of distribution of Facies VIII follows very closely the lens of high salinity water which flows northwards after spilling over the sill from the Mediterranean Sea. It is possible therefore that this Facies represents a transported assemblage whose original source was the Mediterranean Sea. In her paper on the palynology of a Pleistocene bore in Israel, Rossignol (1962, pull-out chart) indicates that *Hystrichosphaeridium ashdodense* (= *Lingulodinium machaerophorum*) and *H. israelianum* (= *Operculodinium israelianum*) together make up between 70 and 80 per cent of

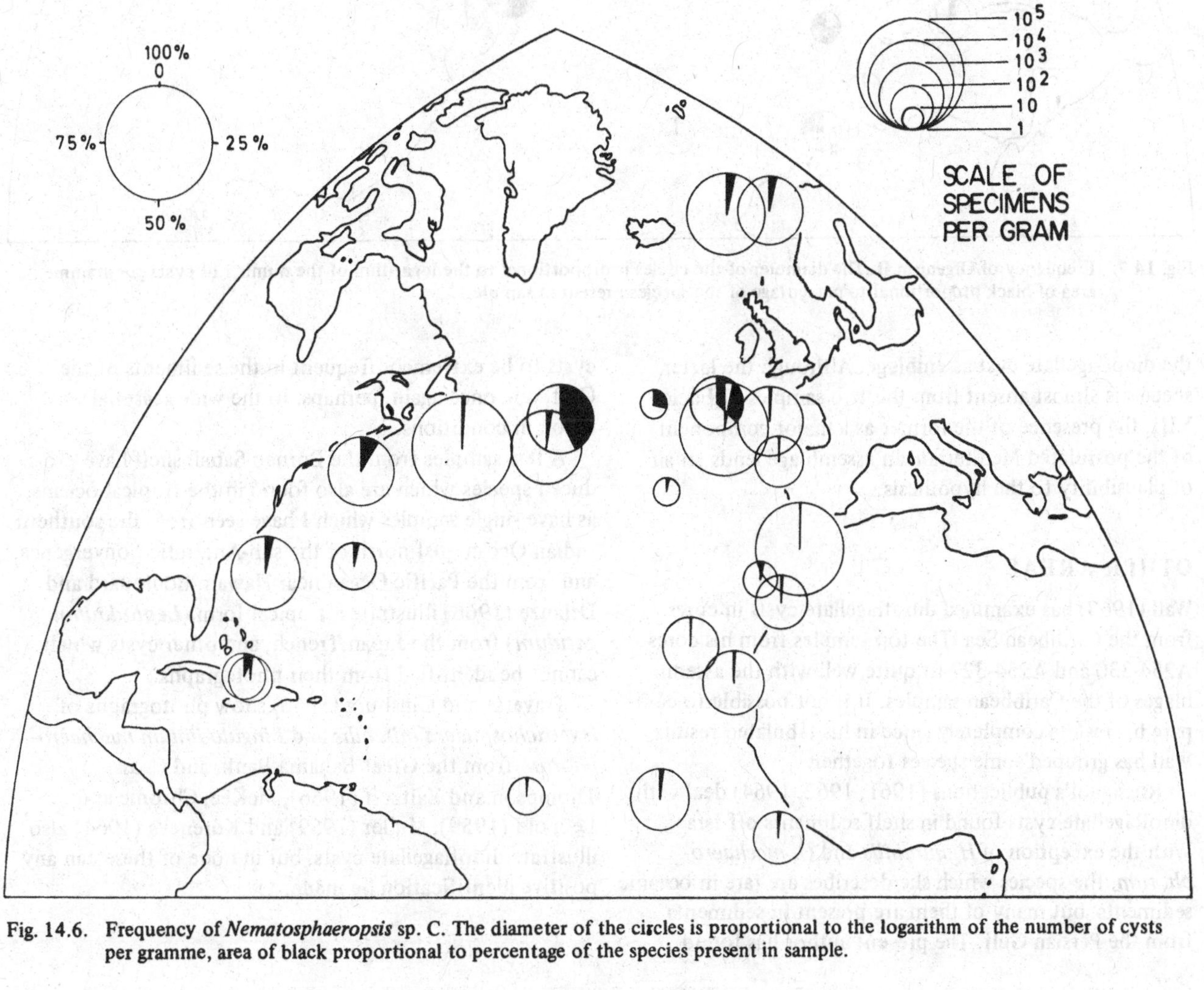
100%
0
75%
25%
50%
10^5
10^4
10^3
10^2
10
1
SCALE OF SPECIMENS PER GRAM

Fig. 14.6. Frequency of *Nematosphaeropsis* sp. C. The diameter of the circles is proportional to the logarithm of the number of cysts per gramme, area of black proportional to percentage of the species present in sample.

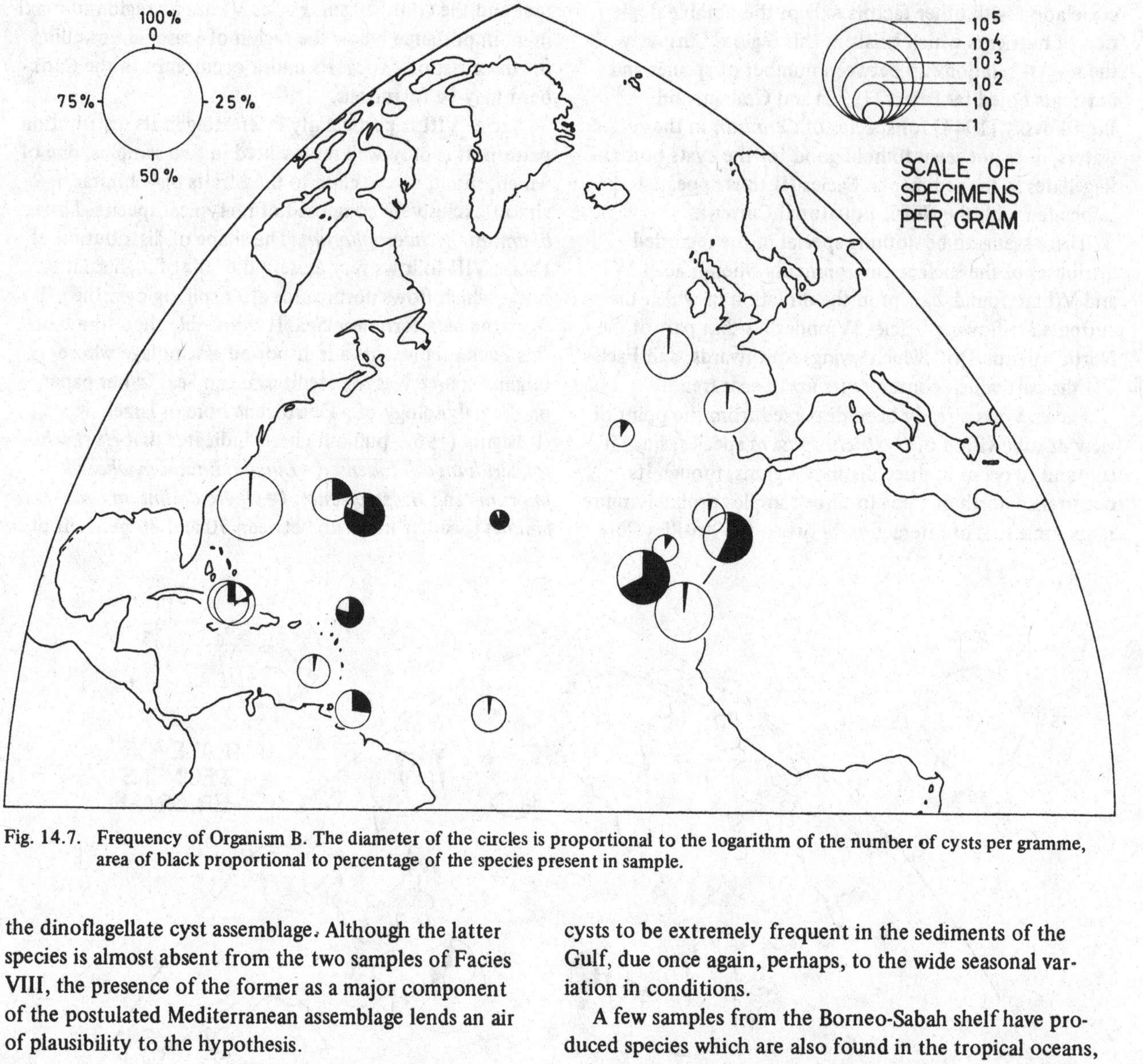
100%
0
75%
25%
50%
10^5
10^4
10^3
10^2
10
1
SCALE OF
SPECIMENS
PER GRAM

Fig. 14.7. Frequency of Organism B. The diameter of the circles is proportional to the logarithm of the number of cysts per gramme, area of black proportional to percentage of the species present in sample.

the dinoflagellate cyst assemblage. Although the latter species is almost absent from the two samples of Facies VIII, the presence of the former as a major component of the postulated Mediterranean assemblage lends an air of plausibility to the hypothesis.

OTHER AREAS

Wall (1967) has examined dinoflagellate cysts in cores from the Caribbean Sea. The top samples from his cores A254-330 and A254-327 fit quite well with the assemblages of the Caribbean samples. It is not possible to compare his results completely since in his tabulated results, Wall has grouped some species together.

Rossignol's publications (1961, 1962, 1964) deal with dinoflagellate cysts found in shelf sediments off Israel. With the exception of *H. mirabilis* and *O. machaerophorum,* the species which she describes are rare in oceanic sediments, but many of them are present in sediments from the Persian Gulf. The present author has found cysts to be extremely frequent in the sediments of the Gulf, due once again, perhaps, to the wide seasonal variation in conditions.

A few samples from the Borneo-Sabah shelf have produced species which are also found in the tropical oceans, as have single samples which I have seen from the southern Indian Ocean, just north of the sub-Antarctic Convergence, and from the Pacific Ocean near Hawaii. Boulouard and Delauze (1966) illustrate a tropical form (*Leptodinium patulum*) from the Japan Trench, and other cysts which cannot be identified from their photographs.

Traverse and Ginsburg (1966) show photographs of *Hystrichosphaera mirabilis* and *Lingulodinium machaerophorum* from the Great Bahama Bank, and Cross, Thompson and Zaitzeff (1966), McKee, Chronic and Leopold (1959), Muller (1959) and Koreneva (1964) also illustrate dinoflagellate cysts, but in none of these can any positive identification be made.

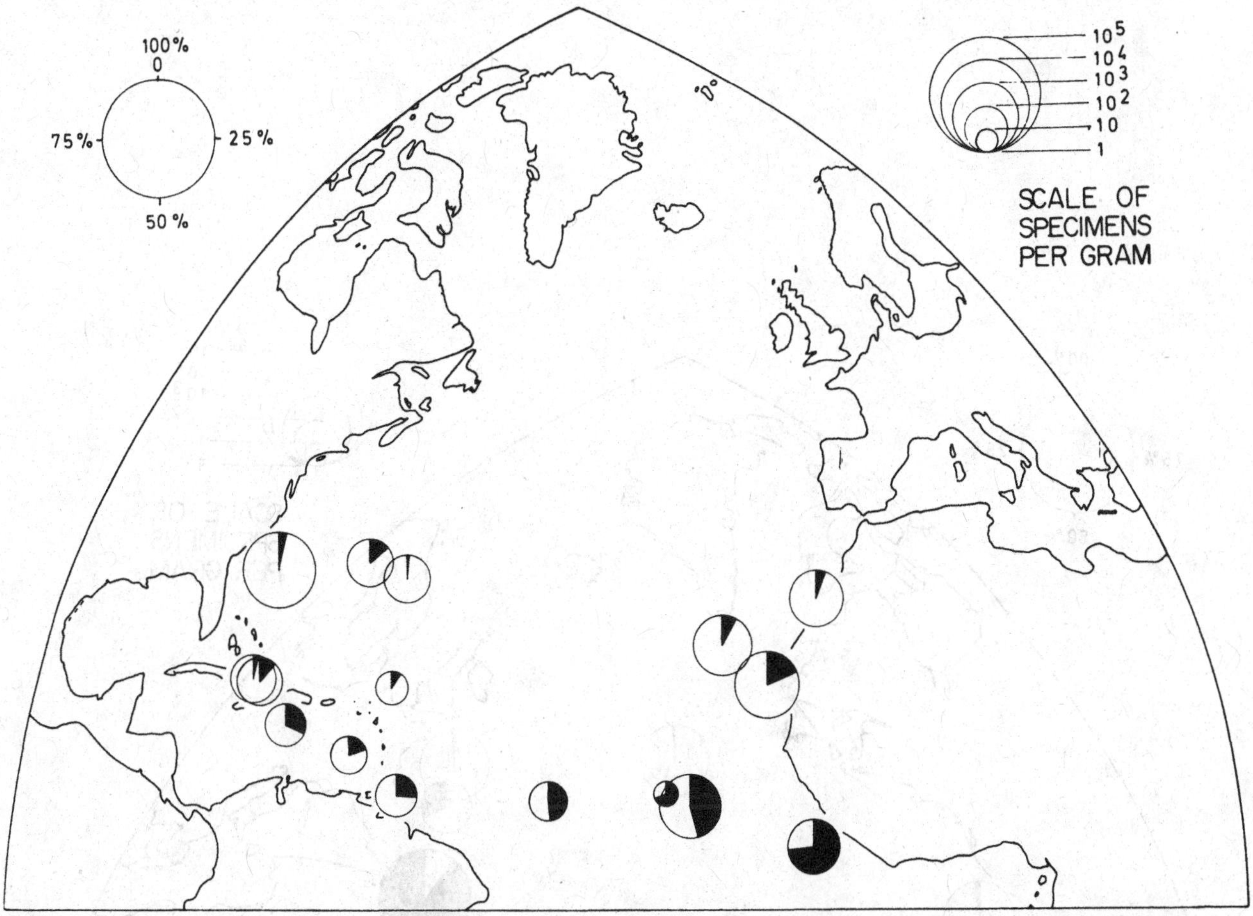
100%
0
25%
50%
75%
10^5
10^4
10^3
10^2
10
1
SCALE OF
SPECIMENS
PER GRAM

Fig. 14.8. Frequency of Organism A. The diameter of the circles is proportional to the logarithm of the number of cysts per gramme, area of black proportional to percentage of the species present in sample.

CONCLUSIONS

The above results show that the distribution of dinoflagellate cysts in marine sediments may be accounted for by reference to the conditions of the surface environment.

The absolute frequency of cysts per unit weight in the sediment is probably related to the conditions which control total productivity at the surface, the seasonal stability of the surface environment, and the relative rate of sediment accumulation.

The relative frequencies of the cysts is related more closely to the distribution of surface currents and water masses than to any other factors.

In spite of their low frequencies in sediments from tropical regions, dinoflagellate cysts have been found in all types of surface sediment, ranging from 'red clays' to a glacially rafted sand, made up dominantly of metamorphic minerals. Future work will show whether the same independence of sediment type will be true for cores. If so, there is a strong likelihood that dinoflagellate cysts may be useful in Quaternary correlations where Foraminifera and Radiolaria have failed as a result of solution.

ACKNOWLEDGEMENTS

I am grateful to the many people who supplied samples for use in this study. The major part of the work was carried out while the author held a D.S.I.R. Research Studentship at Reading University, and further work was done during the tenure of an N.E.R.C. Fellowship at Cambridge University.

Received March 1968.

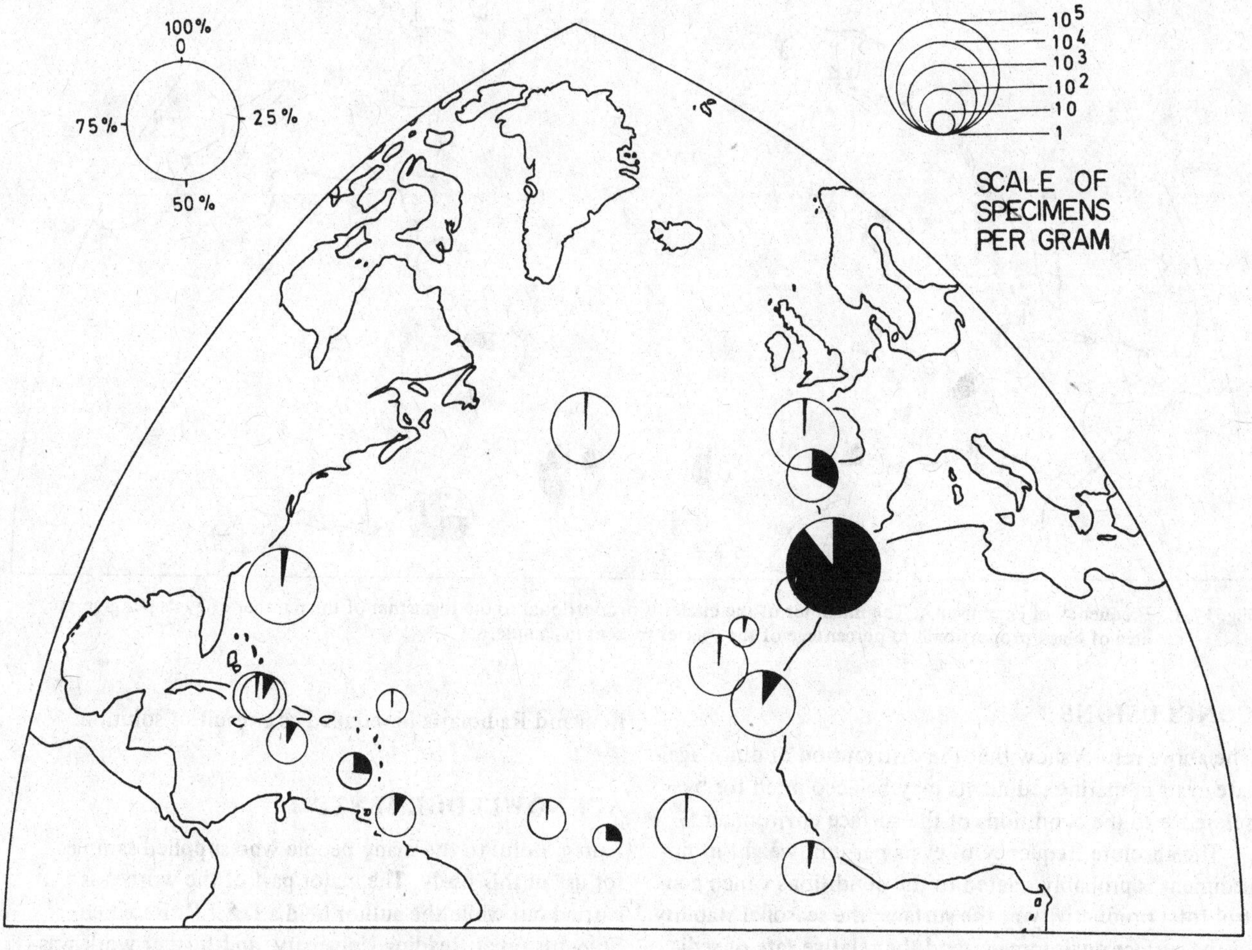
100%
0
25%
50%
75%
10^5
10^4
10^3
10^2
10
1
SCALE OF
SPECIMENS
PER GRAM

Fig. 14.9. Frequency of *Lingulodinium machaerophorum*. The diameter of the circles is proportional to the logarithm of the number of cysts per gramme, area of black proportional to percentage of the species present in sample.

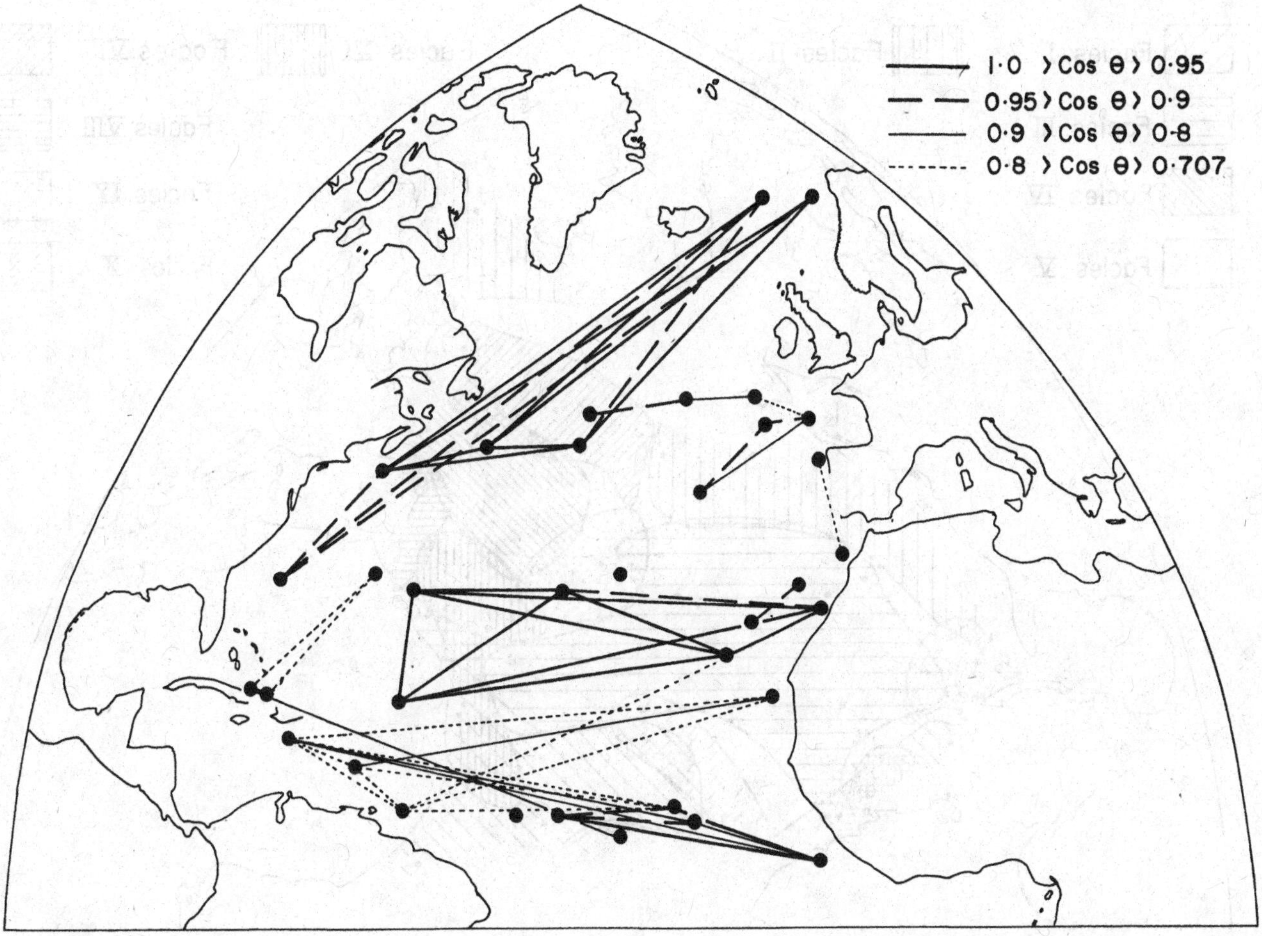
1·0 ⟩Cos Θ⟩ 0·95
0·95⟩Cos Θ⟩ 0·9
0·9 ⟩Cos Θ⟩ 0·8
0·8 ⟩Cos Θ⟩ 0·707

Fig. 14.10. Similarity web between samples. Imbrie and Purdy's cos Θ used as coefficient. Only values with Θ<45° shown.

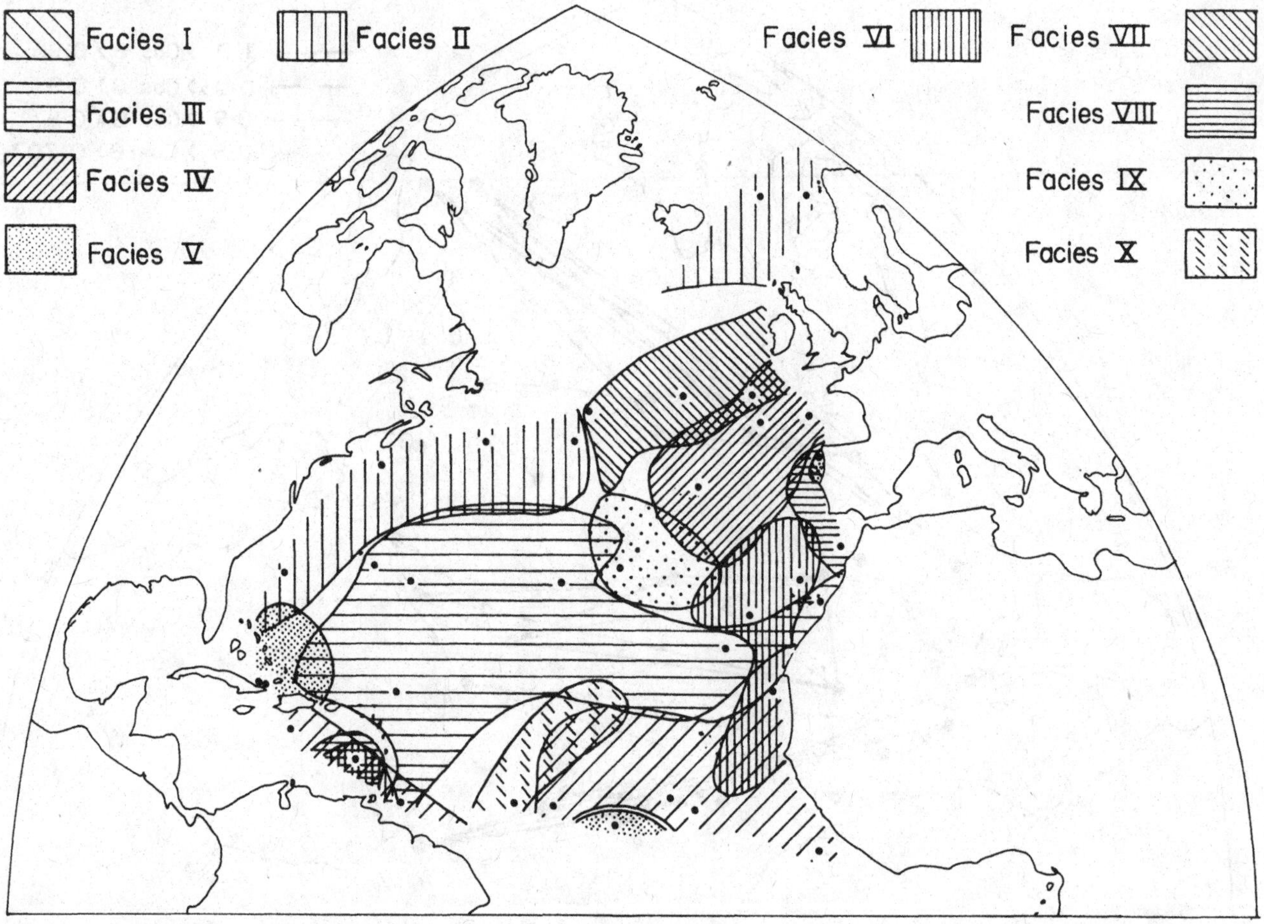
Facies I
Facies II
Facies III
Facies IV
Facies V
Facies VI
Facies VII
Facies VIII
Facies IX
Facies X

Fig. 14.11. Biofacies map obtained from superimposing plots of similarities between real assemblages and ideal assemblages I to X, resulting from Principal Components Analysis of assemblage data.

BIBLIOGRAPHY

Boulouard, C. and Delauze, H. 1966. Analyse palyno-planctologique de sédiments prélevés par le bathyscaphe *Archimède* dans la fosse du Japon. *Mar. Geol.* **4**, 461-6.

Braarud, T. 1945. Morphological observations on marine dinoflagellate cultures. *Avh. norske Videnskakad.* Oslo, 1944 (**11**), 18 pp.

Cross, A. T., Thompson, G. G. and Zaitzeff, J. B. 1966. Source and distribution of palynomorphs in bottom sediments, southern part of Gulf of California. *Mar. Geol.* **4**, 467-524.

Evitt, W. R. 1967. Dinoflagellate Studies. II. The archeopyle. *Stanford Univ. Publs Geol. Sci.* **10** (3), 83 pp.

Evitt, W. R. and Davidson, S. E. 1964. Dinoflagellate studies. I. Dinoflagellate cysts and thecae. *Stanford Univ. Publs Geol. Sci.* **10** (1), 12 pp.

Graham, H. W. and Bronikovsky, N. 1944. The genus *Ceratium* in the Pacific and North Atlantic Oceans. *Publs Carnegie Instn.* **542**, 209 pp.

Hela, I. and Laevastu, T. 1962. *Fisheries Hydrography.* Fishing News (Books) Ltd.: London, 137 pp.

Imbrie, J. and Purdy, E. G. 1962. Classification of modern Bahamian carbonate sediments. In Ham, W. E. (*ed.*), *Classification of carbonate rocks. Mem. Amer. Ass. Petrol. Geol.* **1**, 253-72.

Imbrie, J. and van Andel, T. H. 1964. Vector analysis of heavy mineral data. *Bull. geol. Soc. Amer.* **15**, 1131-56.

Koreneva, E. V. 1964. [Spores and pollen from bottom sediments in the western part of the Pacific Ocean.] (in Russian) *Trudy Geol. Inst. S.S.S.R.* **109**, 88 pp.

Loeblich, A. R. and Tappan, H. 1964. Protista 2. Sarcodina, chiefly 'Thecamoebians' and Foraminiferida. *Treatise on Invertebrate Paleontology.* **C**, 900 pp.

McKee, E. D., Chronic, J. and Leopold, E. B. 1959. Sedimentary belts in the lagoon of Kapingamarangi Atoll. *Bull. Amer. Ass. Petrol. Geol.* **43**, 501-62.

Muller, J. 1959. Palynology of Recent Orinoco Delta and shelf sediments: Reports of the Orinoco Shelf Expedition; volume 5. *Micropaleontology,* **5**, 1-32.

Neves, R. and Dale, B. 1963. A modified filtration system for palynological preparation. *Nature, Lond.* **198**, 775-6.

Nordli, G. 1951. Resting spores in *Goniaulax polyedra* Stein. *Nytt Mag. Naturvid.* **88**, 207-12.

Peters, N. 1932. Die Bevölkerung des Südatlantischen Ozeans mit Ceratien. *Wiss. Ergebn. dt. atlant. Exped. 'Meteor',* **11**, 1-69.

Rossignol, M. 1961. Analyse pollinique de sédiments marins Quaternaires en Israël. I: Sédiments Récents. *Pollen Spores,* **3**, 303-24.

Rossignol, M. 1962. Analyse pollinique de sédiments marine Quaternaires en Israël. II: Sédiments Pléistocènes. *Pollen Spores,* **4**, 121-48.

Rossignol, M. 1964. Hystrichosphères du Quaternaire en Mediterranée orientale, dans les sédiments Pléistocènes et les boues marines actuelles. *Rev. Micropaleont.* **7**, 83-99.

Traverse, A. and Ginsburg, R. N. 1966. Palynology of the surface sediments of Great Bahama Bank, as related to water movement and sedimentation. *Mar. Geol.* **4**, 417-59.

Wall, D. 1966. Modern hystrichospheres and dinoflagellate cysts from the Woods Hole region. *Grana palynol.* **6**, 297-314.

Wall, D. 1967. Fossil microplankton in deep-sea cores from the Caribbean Sea. *Palaeontology,* **10**, 95-123.

Wall, D. and Dale, B. 1966. Living fossils in western Atlantic plankton. *Nature, Lond.* **211**, 1025-6.

15. COCCOLITHS IN SUSPENSION AND IN THE SURFACE LAYER OF SEDIMENT IN THE PACIFIC OCEAN

M. G. USCHAKOVA

Summary: Coccoliths, skeletal remains of calcareous microscopic coccolithophorid algae, have been studied in water suspension samples, and in the surface layer of sediment from the ocean bottom. Taxonomic and quantitative data on coccoliths have been obtained with the aid of light and electron microscopes. The amount of coccoliths in the ocean waters agrees with the climatic and circum-continental zonation. Similar regularities in distribution are observed in the surface sedimentary layer. Electron microscope studies permitted the recognition of four complexes (associations): boreal, northern subtropical, equatorial and warm-water southern tropical zones.

INTRODUCTION

So far, available information concerning the distribution and composition of living coccolith forms in the waters of the world's oceans is very scarce. Quantitative estimates of the coccolith content of the plankton of the subtropical and equatorial zones have been carried out by Hasle (1959, 1960) and by Semina (1963).

During recent years coccoliths have been used, along with other micropalaeontological methods, for the purpose of examining the stratigraphy and palaeontology of sediments. Studies carried out at the Institute of Oceanology, U.S.S.R., include an investigation of the coccoliths' taxonomic composition and their distribution in water and sediments, and are based on the examination of water suspension, surface sedimentary layer and core samples. Both suspension and bottom sediment samples have been collected in the course of the numerous cruises of R.V. *Vitjaz* in the Pacific Ocean.

COCCOLITHS IN WATER SUSPENSION

For the study of coccoliths in water suspension samples the suspension was collected on to membrane filters, along vertical sections from the surface to the ocean bottom in the tropical area of the Pacific running along latitude 140° W, between 16° and 18° S (Figs. 15.1 and 15.2). The coccoliths were studied with the aid of a biological microscope equipped with a phase-contrast attachment.

It has been established that coccolithophorids are encountered in great numbers in the surface waters of warm and moderate latitudes. They are most frequent mainly in the 0-200 metre layer. The development of planktonic organisms in the oceans depends on the supply of essential nutrient materials to the surface waters. The enrichment of the surface layers with nutrient salts occurs by entry from the depths as a result of vertical mixing and, to a lesser extent, by the discharge of salts from dry land areas. The number of coccoliths increases in zones of upwelling (zones of divergence) and in zones with strongly pronounced vertical mixing.

According to the data available the waters of the Pacific Ocean allow one to distinguish a number of zones of diverse quantitative development of coccolithophorids. The suspensions reveal a much more frequent occurrence of separate coccoliths; intact cells of coccolithophorids are rare and not observed in all samples. The number of coccoliths in suspension samples, when recalculated in relation to a cubic metre, varies from 800 to 50,000. The quantitative maximum of coccolith development coincides with divergence zones and is ascribed to the 0-200 metre layer (Fig. 15.2). The most profuse area is situated next to the equatorial divergence (0°-8° S). The most impoverished areas are near the northern boundary of the tropical zone (5° W-18° N), and this seems to be connected with the influence of the northern tropical convergence.

The study of coccolith content along vertical sections (Fig. 15.2) allowed the establishment of a number of interesting facts relating to vertical distribution. The greatest number of coccoliths occurs, not in the surface

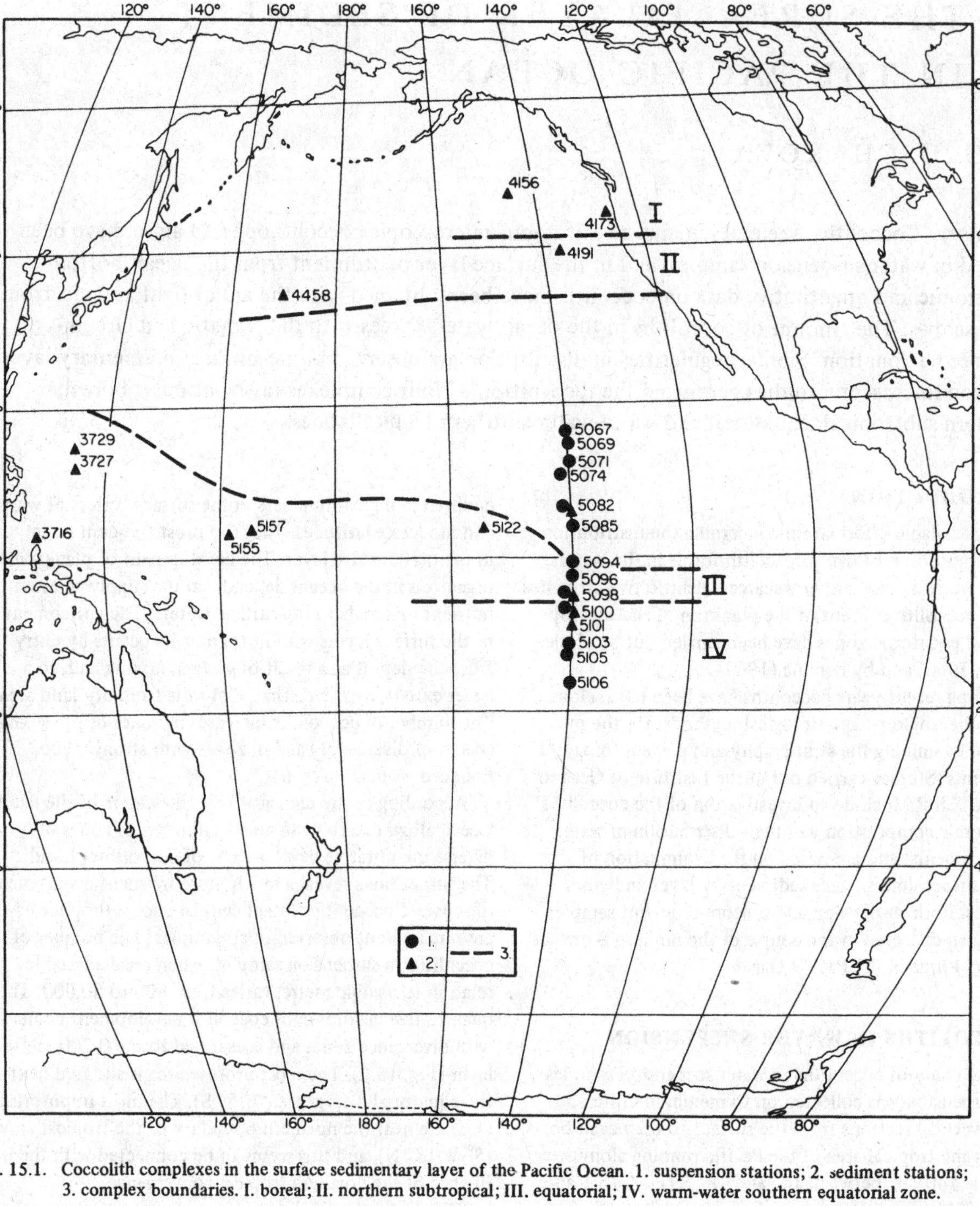
120°
140°
160°
180°
160°
140°
120°
100°
80°
60°
60°
40°
30°
0°
20°
40°
4156
4173
4191
4458
3729
3727
3716
5155
5157
5122
5067
5069
5071
5074
5082
5085
5094
5096
5098
5100
5101
5103
5105
5106
I
II
III
IV
1.
2.
3.

Fig. 15.1. Coccolith complexes in the surface sedimentary layer of the Pacific Ocean. 1. suspension stations; 2. sediment stations; 3. complex boundaries. I. boreal; II. northern subtropical; III. equatorial; IV. warm-water southern equatorial zone.

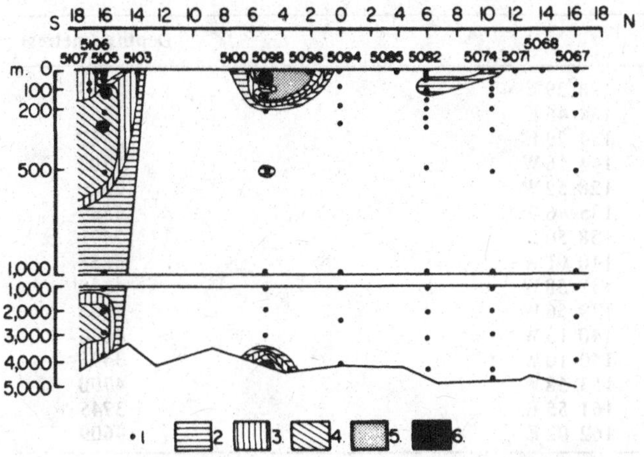

Fig. 15.2. Vertical distribution of coccoliths (in thousand-specimens/cubic metre) in the tropical waters of the Pacific Ocean along a section at 140° W. 1. sample levels; 2. coccoliths absent; 3. 1-10; 4. 10-100; 5. 100-1,000; 6. 1,000-10,000; 7. >10,000.

layer of water, but in the 50-200 metre zone. The greatest number of coccoliths in the near-bottom layers is marked in the equatorial zone (5°-6° S), which also reveals the maximum occurrence of coccolithophorids in the plankton. The area of the southern tropical zone (15°-18° S) is fifty times poorer than the equatorial zone. In this area coccoliths, and scarce intact coccospheres, are encountered throughout the entire water column.

The qualitative composition of coccoliths in suspension sampling from the Pacific and Indian Oceans shows a similarity of species composition (Uschakova 1967). About 75-95 per cent of the flora is made up of the following species: *Cyclococcolithus leptoporus* (Murr., Blackm.) Kampt., *Gephyrocapsa oceanica* Kampt., *Ceratolithus cristatus* Kampt. One of the distinguishing features of the coccolith composition of the Pacific Ocean lies in the absence of *Rhabdosphaera claviger* Murr. Blackm., a characteristic species of the Indian Ocean suspension. One of the peculiar traits of suspension of the Indian Ocean which is regarded as still not yet explained is the finding of the so-called *Discoasters* (Martini and Bramlette 1963; Riedel, Bramlette and Parker 1963). The entire water column reveals *Discoaster brouweri* Tan Sin Hok, *D. challengeri* Bramlette and Riedel, *D. crassus* Martini, *D. woodringi* Bramlette (Bogdanov and Uschakova 1966); this does not refer to the suspension collected from the Pacific Ocean. In the Pacific the suspension of near-bottom water layers contains all those species which are encountered in the near-surface waters. This fact testifies in favour of good preservation of coccoliths.

The results of coccolith studies within suspension show that the Pacific Ocean's equatorial zone is the richest area according to the number and variety of forms. The waters of the northern tropical zone proved to be the poorest.

COCCOLITHS IN THE SURFACE SEDIMENTARY LAYER

The distribution of coccoliths in the surface sedimentary layer reflects the general regularities which are characteristic of the distribution of coccoliths in the suspension. The area of the equatorial zone and that of the southern tropics appear to be the richest. Numerical variations in coccoliths depend on depth of sedimentation, dilution by terrigenous material and bottom relief.

Coccoliths form a characteristic part of calcareous sediments. In the surface layer of calcareous sediments one encounters all the coccolith species which were found in the waters of these areas. The distribution of species in the sediments is close to their geographical distribution in oceanic waters. The greatest number of all coccolith species are observed in the southern tropical zone. It is of interest to note the influence of depth on *Rhabdosphaera claviger*, a species which is encountered in the sediments of southern tropical and subtropical zones, but fails to occur below 4.3 kilometres. This fact seems to be connected with rapid dissolution of this species with depth.

The study of coccoliths in the surface sediments of the Pacific (Fig. 15.1) was carried out with the aid of an electron microscope. This instrument was applied for the first time to the coccoliths removed from sediments of station 5100, studied by Shumenko and Uschakova (1967) and Uschakova (1966). Coccoliths were studied in fifteen samples, eight of them obtained from a section at 140° W (50° N to 10° S), and seven from other parts of the tropical area (Table 15.1).

The systematic composition of coccoliths following electron microscopic studies permits the recognition of four complexes of coccoliths, which are characteristic for sediments of the ocean's four geographical zones. It is worth emphasizing the much worse preservation of coccoliths in the sediments of the Pacific Ocean compared with coccoliths in sediments from the same zones of the Indian Ocean.

The coccolith composition of the Pacific numbers about twenty five forms. The most characteristic of them are the following: *Ceratolithus cristatus* Kampt., *Coccolithus pelagicus* (Wall.) Schil., *C. huxleyi* (Lohm.) Kampt., *C.* aff. *obtusus* Kampt., *Cyclococcolithus leptoporus* (Murr., Black.) Kampt., *Cycloplacolithus laevigatus* Kampt., *Discoaster brouweri* Tan Sin Hok, *D. crassus* Mart., *Ellipsoplacolithus latifolius* Kampt., *Gephyrocapsa oceanica* Kampt., *Helicosphaera carteri* (Wall.) Kampt.,

Table 15.1

Stations	Coordinates		Depths (metres)
3716	2°38'N ×	128°29'E ×	563
3727	12 19 N	134 46 E	3451
3729	15 31 N	134 29 E	3482
4156	48 20 N	144 16 W	3703
4173	45 02 N	128 52 W	2620
4191	40 20 N	135 46 W	4569
4458	32 33 N	158 50 E	2491
5074	10 31 N	140 01 W	4858
5082	5 57 N	139 58 W	4842
5098	5 03 S	139 56 W	4268
5100	7 08 S	140 13 W	4076
5101	8 26 S	140 10 W	3491
5122	4 34 N	153 58 W	4790
5155	3 22 N	161 55 E	3745
5157	5 32 N	162 02 E	4609

Rhabdosphaera claviger Murr., Blackm., *Umbilicosphaera mirabilis* Lohm.

The first coccolith complex is represented in sediments of the boreal zone of the Pacific Ocean (49°-45° N). It is characterized by the presence of the dominant large form *Coccolithus pelagicus* (42 per cent), the cosmopolitan *Gephyrocapsa oceanica* (33.5 per cent) and *Helicosphaera carteri* (20 per cent). The boreal complex is made up of about nine forms. This complex is typical for the area north of the polar front, and it seems to correspond to the transitional diatom complex of the Pacific Ocean (Jousé, Kozlova and Muhina 1967).[1]

The following three complexes of coccoliths may be ascribed to the warmth-loving forms. They are encountered in the sediments of the Pacific in the areas in which suspension was examined (40° N-16° S). They form a uniform nucleus which is ordinarily made up of the following representatives: *Gephyrocapsa oceanica, Cyclococcolithus leptoporus, Coccolithus* aff. *obtusus*. However, these complexes may be distinguished from each other by the presence of some specific elements, which are not as a rule the dominant ones.

The second coccolith complex is represented in the sediments of the northern subtropical oceanic zone (40°-32° N). This complex is rather diverse quantitatively and qualitatively. *Cyclococcolithus leptoporus* (36 per cent), *Coccolithus* aff. *obtusus* (30 per cent),*Gephyrocapsa oceanica* (18 per cent) dominate the complex. The accompanying species in the complex are represented by: *Rhabdosphaera claviger* and *Umbilicosphaera mirabilis*. In all the complex is made up of about twenty forms. Here one encounters debris of *Coccolithus pelagicus* and it seems that the northern areas of the subtropics appear to be the southern boundary of its distribution in the Pacific. This complex may be named the northern subtropical complex.

The sediments, as well as the suspension, of the northern tropical divergence (6°-10° N) reveal single specimens of coccoliths: *Cyclococcolithus leptoporus, Coccolithus* aff. *obtusus, Rhabdosphaera* sp. (?).

The third complex of coccolith forms is represented in sediments of the equatorial zone of the Pacific Ocean (15° N-7° S). As regards richness and variety this complex is similar, to some extent, to the northern subtropical complex. The complex is dominated by *Cyclococcolithus leptoporus* (35 per cent), *Gephyrocapsa oceanica* (34 per cent) and *Coccolithus* aff. *obtusus* (12 per cent). The accompanying forms are: *Umbilocosphaera mirabilis* and *Coccolithus huxleyi*. The complex is made up of about eighteen forms. It is of interest to note the presence of rare specimens of *Discoaster brouweri* and *D. crassus* in the surface layer of sediments (stations 3727 and 3729). The disappearance of these forms facilitates the recognition of the Pliocene-Pleistocene boundary. The Pliocene form *Ellipsoplacolithus latifolius* is at times observed at these same stations. The presence of these forms in recent coccolith complexes may be abscribed, possibly, to the redeposition of the more ancient sediments. The third coccolith complex ought to be regarded as an equatorial one.

The fourth complex is represented in the sediments of the southern equatorial zone of the Pacific Ocean (7°-9° S). The complex is dominated by *Cyclococcolithus leptoporus* (50 per cent), *Ellipsoplacolithus productus* (10 per cent) and *Cycloplacolithus laevigatus* (10 per cent), accompanied by *Umbilicosphaera mirabilis* and *Ceratolithus cristatus*. The distinguishing feature of this complex is the presence of *Cycloplacolithus laevigatus* and *Ceratolithus cristatus* which is characteristic of these same áreas, situated south of the equator, in the Indian Ocean. The complex is made up of twelve forms. It may be regarded

[1] This volume, Chapter 17, pp. 263-69.

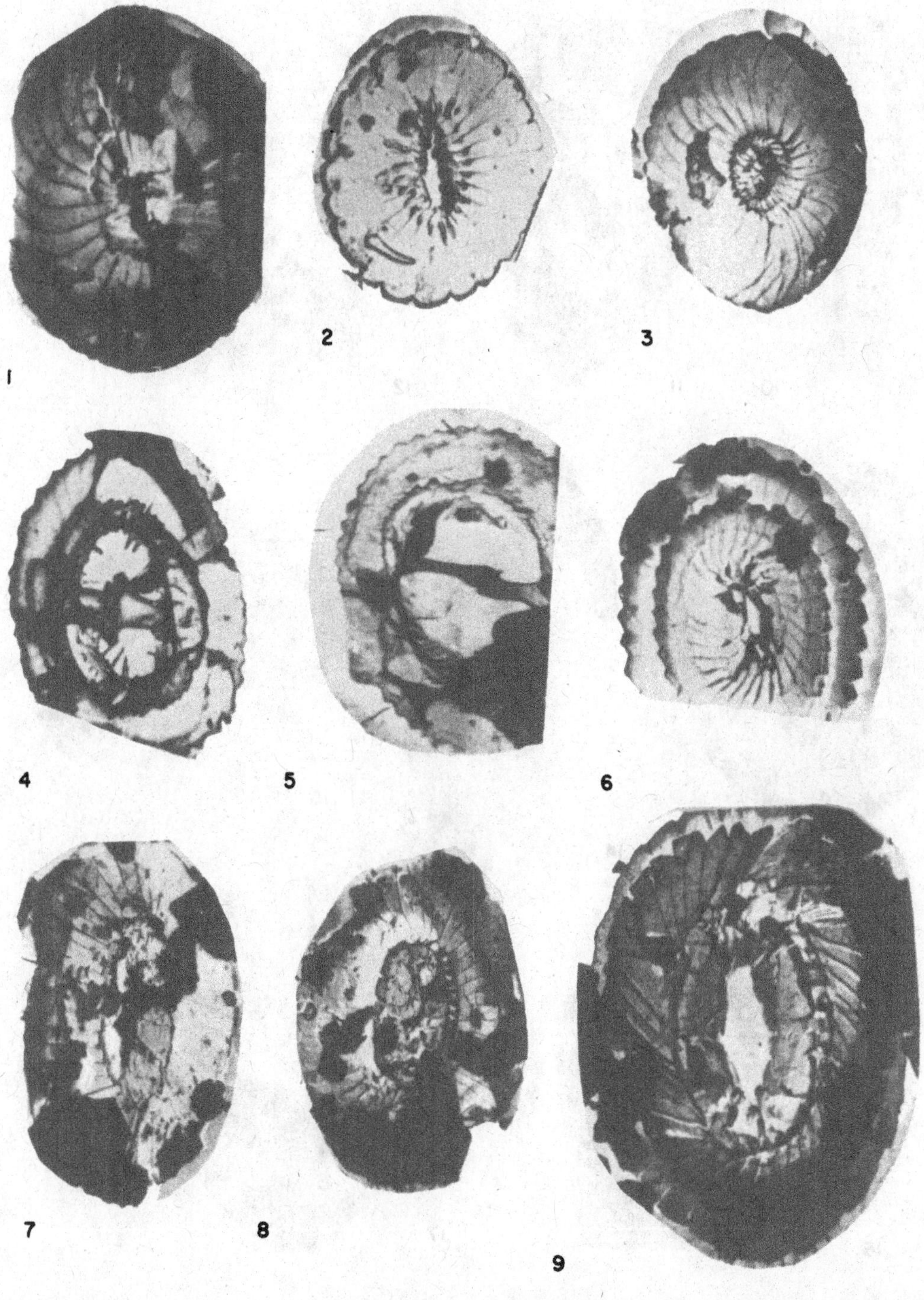
1
2
3
4
5
6
7
8
9

Plate 15.1. Coccoliths from the surface layer of bottom sediments from the Pacific Ocean (electron micrographs). 1, 3. *Cyclococcolithus leptoporus*: 1. st. 5098, x 6,600; 3. st. 5100, x 7,800. 2. *Coccolithus* aff. *obtusus*: st 3727, x 7,800. 4, 5. *Gephyrocapsa oceanica*: 4. st. 5100, x 13,200; 5. st. 3729, x 13,200. 6. *Ellipsoplacolithus productus*: st. 3729, x 19,200. 7-9. *Coccolithus pelagicus*: st. 4173; 7, 9. x 6,600; 8. x 5,400.

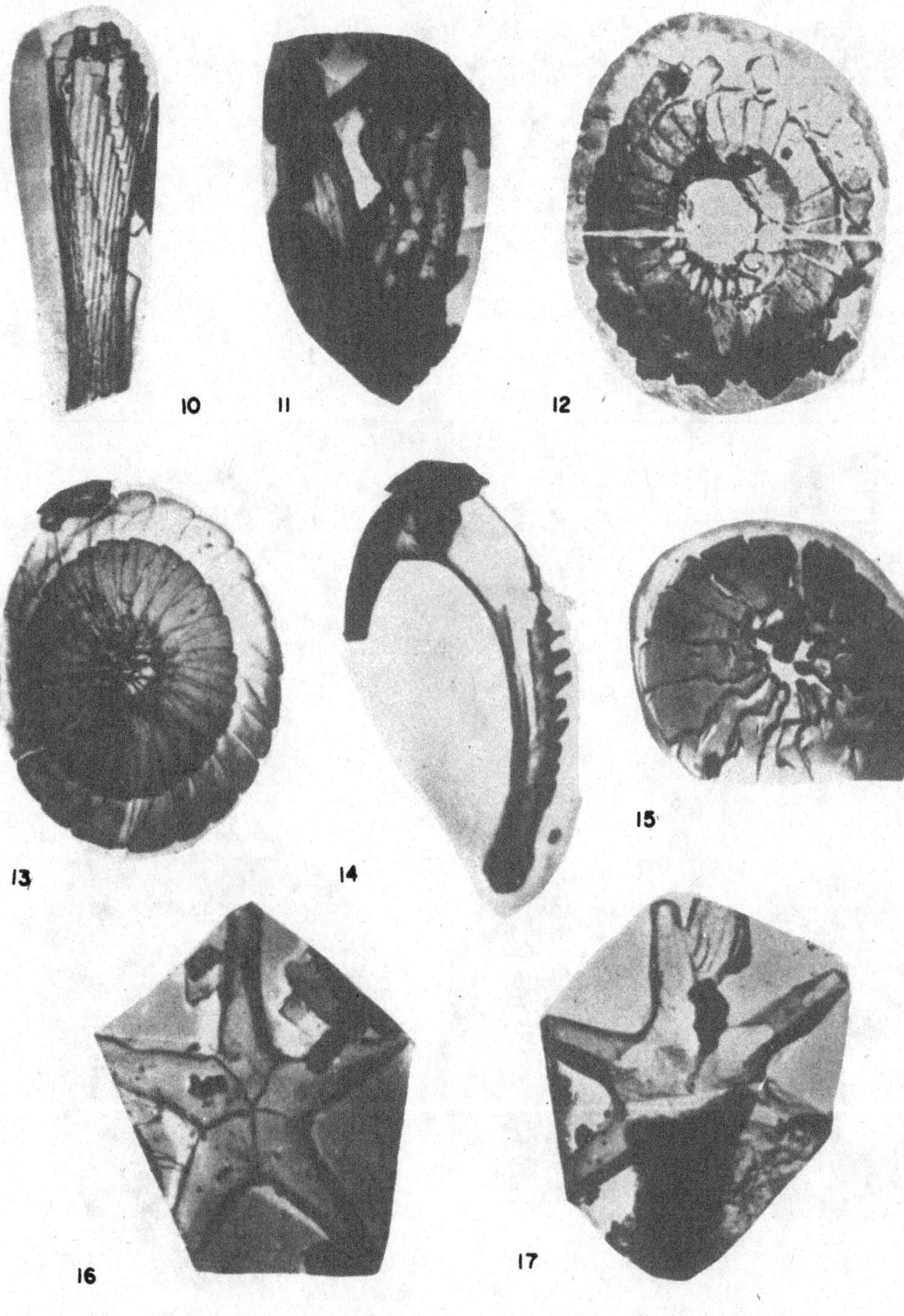
10
11
12
13
14
15
16
17

Plate 15.2. Coccoliths from the surface layer of bottom sediments of the Pacific Ocean (electron micrographs). 10. *Rhabdosphaera claviger*: st. 4191, x 12,600; 11. *Rhabdosphaera* sp. (?); st. 5082, x 5,400; 12. *Umbilicosphaera mirabilis*: st. 5157, x 1,400; 13. *Cycloplacolithus laevigatus*: st. 5100, x 7,800; 14. *Ceratolithus cristatus*: st. 5100, x 7,800; 15. *Ellipsoplacolithus latifolius*: st. 3727, x 19,200; 16. *Discoaster brouweri*: st. 3727, x 9,600; 17. *Discoaster crassus*: st. 3727, x 13,200.

as the warm-water complex of the south equatorial zone.

The study of coccoliths in suspension and in the surface sedimentary layer of the Pacific Ocean has shown that the distribution of these forms exhibits a well-developed zonation. Coccolith complexes are different in sediments of different climatic zones of the ocean. Each zone is characterized by specific forms. The boreal complex of coccoliths has its own distinctly defined characteristics.

One ought to pay special attention to the obvious necessity of a sequential study of coccolith forms: beginning with suspensions removed from different horizons of the water mass, continuing with the surface sedimentary layer, and concluding with the thicker bottom sediments. This latter stage leads to stratigraphical evaluation of the ocean's bottom sediments. Special attention ought to be drawn to the difficulty of comparing results of taxonomic studies of coccoliths in water suspensions and in sediments, obtained by optic and electron microscopic investigation respectively.

Received February 1968.

BIBLIOGRAPHY

Bogdanov, Y., Uschakova, M. 1966. Coccoliths of the *Discoaster* Tan Sin Hok group in the water suspension of the Pacific Ocean. *Doklady U.S.S.R. Acad. Sci.* **171** (2), 465-7.

Bramlette, M. 1961. Pelagic sediments. *Publ. Am. Ass. Sci.* **67**, 345-66.

Bramlette, M. and Riedel, W. 1954. Stratigraphic value of discoasters and some other microfossils related to Recent coccolithophores. *J. Paleont.* **28** (4), 385-403.

Hasle, G. 1959. A quantitative study of phytoplankton from the equatorial Pacific. *Deep-Sea Res.* **6**, 38-59.

Hasle, G. 1960. Plankton coccolithophorids from the Subantarctic and Equatorial Pacific. *Nytt Mag. Bot.* **8**, 77-88.

Jousé, A., Kozlova, O. and Muhina, V. 1967. Species composition and zonal distribution of diatoms in the surface sedimentary layers of the Pacific Ocean. *Doklady U.S.S.R. Acad. Sci.* **172** (5), 1183-6.

Kamptner, E. 1963. Coccolitheneen-Skelettreste aus Tiefseeblagerungen des Pazifischen Ozeans. *Ann. Naturh. Mus. Wien,* **66**, 139-204.

Martini, E. 1965. Mid-Tertiary calcareous nannoplankton from Pacific deep-sea cores. *Colston Papers,* **17**, 393-410.

Martini, E. and Bramlette, M. 1963. Calcareous nannoplankton from the experimental Mohole drilling. *J. Paleont.* **37** (4), 102-5.

Riedel, W., Bramlette, M. and Parker, F. 1963. Pliocene-Pleistocene boundary in deep-sea sediments. *Science,* **140** (3572), 1238-40.

Semina, G. 1963. Phytoplankton of the central part of the Pacific Ocean according to the section along 174° W1. *Trudy Institute of Oceanology, U.S.S.R. Acad. Sci.* **71**, 5-21.

Shumenko, S. and Uschakova, M. 1967. Coccoliths of bottom sediments of the Pacific Ocean. *Doklady U.S.S.R. Acad. Sci.* **176** (4) 200-2.

Uschakova, M. 1966. Biostratigraphic significance of coccolithophorids on the basis of bottom sediments of the Pacific Ocean. *Okeanologia,* **6**, 131-43.

Uschakova, M. G. 1967. Coccoliths of the suspension and of the surface sedimentary layer in the northern and central parts of the Indian Ocean. *Fossil algae of U.S.S.R. 'Nedra' Publ.* 84-90.

16. COCCOLITH CONCENTRATIONS AND DIFFERENTIAL SOLUTION IN OCEANIC SEDIMENTS[1]

A. AND R. McINTYRE

Summary: Coccoliths, the skeletal elements produced by the Coccolithophoridae, are present in the oceanic sediments above the maximum compensation depth between the Arctic and Antarctic Convergences and constitute from one to over 30 per cent of total Recent sediment weight. The highest values are found underlying high productivity areas of transitional and subtropical waters, averaging 26 per cent by weight. This figure drops to 18 per cent in equatorial sediments and often to less than 2 per cent in those of the subarctic and subantarctic. The correlation between productivity and sediment concentration may be useful for palaeoclimatic studies. A comparison of North Atlantic Recent and mid-glacial samples shows a poleward shift in coccolith carbonate concentrations from glacial to Recent.

Comparison of laboratory experiments and surface sediment samples indicates that of the three major structural types of heterococcoliths, the caneoliths are the least, and the placoliths the most resistant to destruction. Electron microscope examination of cores in the compensation zone shows that below 5,000 metres *Coccolithus huxleyi, Cyclococcolithus leptoporus, Gephyrocapsa oceanica,* and *Umbilicosphaera mirabilis* represent almost 100 per cent of the flora. With increasing solution of coccolith structure many species become almost indistinguishable. Below 5,000 metres *C. huxleyi* and *G. oceanica* are difficult to differentiate.

INTRODUCTION

The maximum biogeographical range of the Coccolithophoridae is from the Arctic to the Antarctic Convergences. The distribution of the palaeontologically important species in the Atlantic Ocean has been described in part in such publications as Gaarder (1954), Hulburt (1962, 1963, 1964), and McIntyre and Bé (1967). The species show definite water-mass preferences and may be grouped into 'tropical', 'subtropical', 'transitional', and 'subarctic-subantarctic' floristic zones. While there is some degree of species overlap between these zones, they are distinguishable by species type and concentration. The present oceanic surface sediment distribution and concentration of the calcite skeletal elements (coccoliths) produced by these golden-brown algae during a portion of their life-cycle is controlled by three factors. These are, floristic zonation (i.e. environmental preference of individual species), productivity in surface waters, and chemical solution. The effect of the latter two are the subjects of this paper.

[1] Lamont-Doherty Geological Observatory Contribution No. 1532.

PHYTOPLANKTON PRODUCTIVITY VERSUS COCCOLITH CONCENTRATION

Little information has been published with regard to the concentration of coccoliths in bottom sediments (Bramlette 1958). In the course of examining 123 Atlantic and thirty Indian Ocean cores, a number were processed for carbonate determination. The coccolith fraction was concentrated by short centrifugation (McIntyre, Bé and Preikstas 1967). This fraction was examined in the electron microscope and in all cases contained over 95 per cent coccoliths. Calcium carbonate gasometric determinations were made both on these fractions using Hulsemann's method (1966) with the values corrected for non-coccolith material in the centrifugate and on the fine fraction. Finally the weight of coccoliths per gramme of sediment was calculated.

Presumably the concentration of coccoliths in sediments above the compensation level is proportional to the productivity in the overlying surface waters as current transport in the open ocean is a minor factor (McIntyre and Bé 1967). Thus the highest coccolith concentrations should be in sediments beneath nutrient rich water bodies and boundaries. The plotted weights (Figs. 16.1 and 16.2) when compared to organic carbon productivity (Fleming and Laevastu 1956) are only in partial agreement.

In tropical, subtropical, and transitional areas a positive correlation exists. The highest concentrations of coccoliths in sediments of the North Atlantic underly the northern boundary of the subtropical gyre and adjacent transitional waters, and areas along the edge of the continental shelf. All are areas of high photosynthetic carbon production. The South Atlantic shows a similar pattern but modified as in the Indian Ocean by the solution effects of cold Antarctic bottom waters. The average weight for these high productivity areas is 0.26 gramme per gramme of sediment while the less productive Sargasso Sea and equatorial waters average 0.18 gramme per gramme.

The subantarctic and subarctic waters are areas of high organic productivity yet here the coccolith concen-

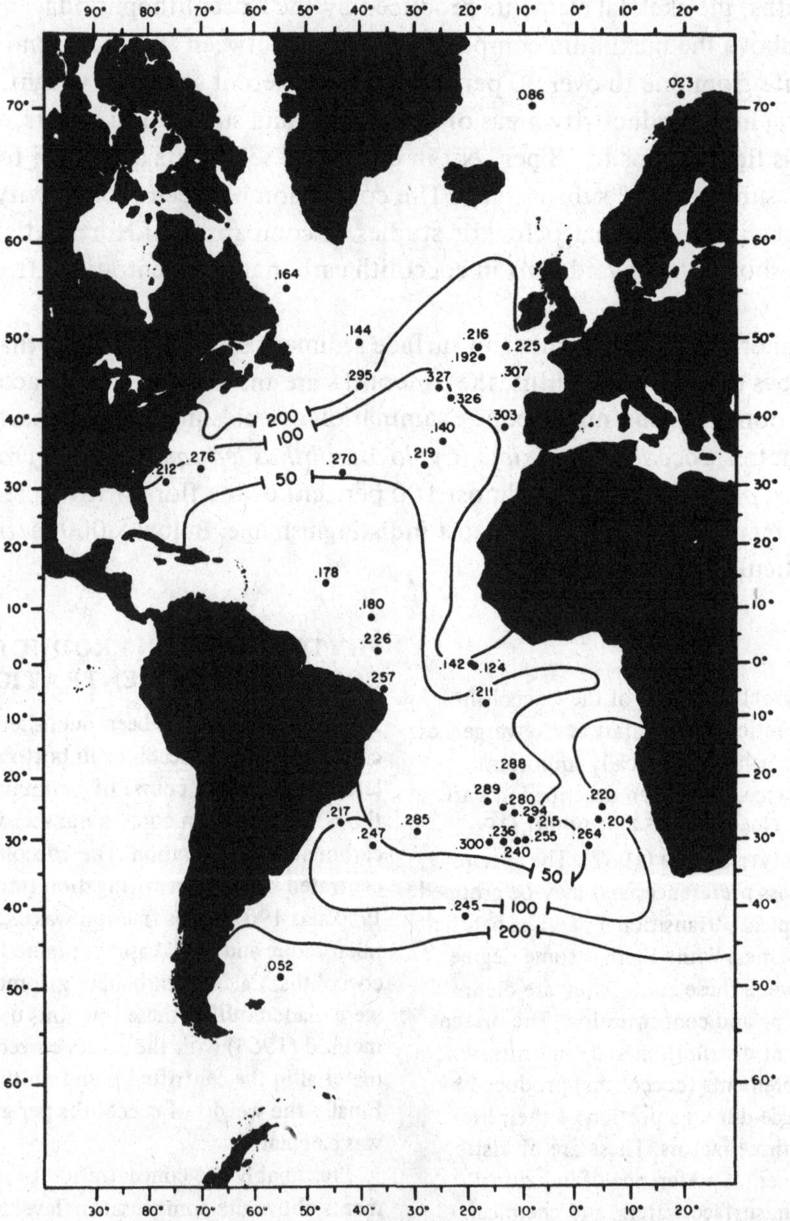
90° 80° 70° 60° 50° 40° 30° 20° 10° 0° 10° 20°
70°
60°
50°
40°
30°
20°
10°
0°
10°
20°
30°
40°
50°
60°
.023
.086
.164
.144
.216
.192
.225
.295
.327
.326
.307
200
100
.140
.303
.212
.276
.270
.219
50
.178
.180
.226
.142
.124
.257
.211
.288
.289
.280
.294
.220
.217
.247
.285
.236
.215
.255
.204
.300
.240
.264
50
.245
100
200
.052
30° 80° 70° 60° 50° 40° 30° 20° 10° 0° 10° 20°

Fig. 16.1. Atlantic Ocean. Coccolith concentration in surface sediments plotted in grammes per gramme of sediment. Organic carbon productivity contoured in grammes per square metre per year (after Fleming and Laevastu 1956).

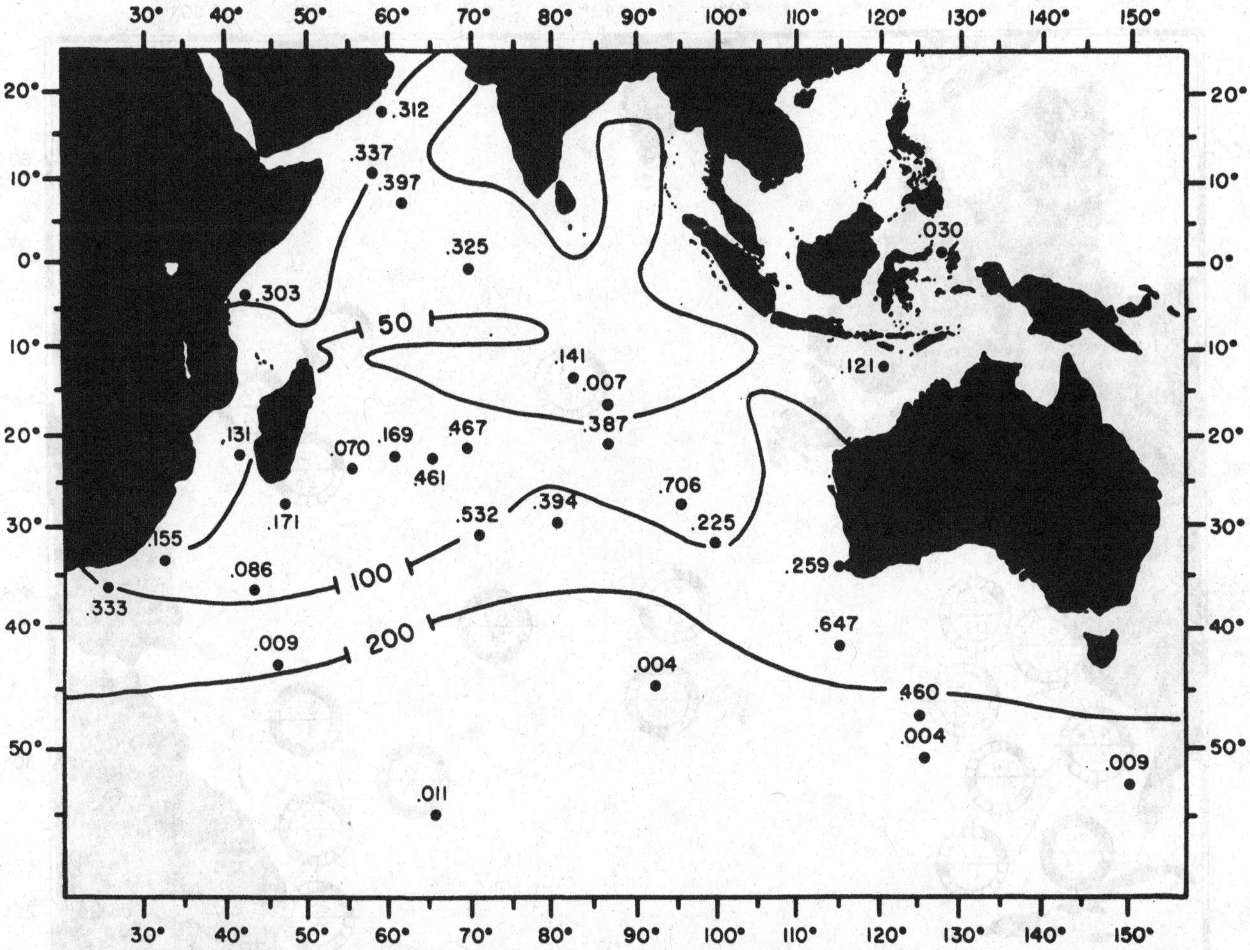
30° 40° 50° 60° 70° 80° 90° 100° 110° 120° 130° 140° 150°
20° 10° 0° 10° 20° 30° 40° 50°
.312
.337
.397
.325
.303
50
.141
.007
.387
.131
.070
.169
.467
.461
.171
.155
.532
.394
.706
.225
.086
100
.333
.259
.009
200
.647
.004
460
.004
.009
.011
.030
.121

Fig. 16.2. Indian Ocean. Coccolith concentration in surface sediments plotted in grammes per gramme of sediment. Organic carbon productivity contoured in grammes per square metre per year (after Fleming and Laevastu 1956).

trations are minimal, being as low as 0.01 gramme per gramme. This anomaly is due to the interaction of three factors. The majority of the Coccolithophoridae are limited to warm water. Only one, *Coccolithus pelagicus,* is indigenous to the Subarctic. The total common species are three in number in the North Atlantic and two in the South Atlantic. This limited flora is overshadowed by the diatoms, which being physiologically more efficient in cold conditions, limit coccolithophorid concentration by competition in the planktonic realm as well as by dilution of skeletal numbers in the sediment. Finally, the carbonate compensation zone is found at shallower depths resulting in a loss of carbonate by solution. This last effect can be found in the South Atlantic as far north as 35° S along the coast of Argentina.

Values obtained from the Indian Ocean while generally showing the trend previously mentioned, are not as clear because much of the area beyond the shelves is at or below the compensation depth. Many cores therefore have lower values than expected, while cores from shallow depths usually come from the slopes of oceanic islands which are typified by abnormally high productivity for their oceanographic position.

Even with the limitations outlined above it is possible, with sufficient core coverage, to contour coccolith concentrations in bottom sediments. The resultant map would indicate high productivity areas and by analogy water bodies and boundaries. This could and should be done with other planktonic organisms (e.g. Foraminifera) as well. While unnecessary in modern marine biology, it could be an important technique for palaeoceanographic studies. It is possible, using isotopic dating and palaeomagnetic reversals, to obtain good geographic distribution of sediment samples from relatively finite periods of time in the past. Coccolith carbonate values from these samples would show the migrations and variations in productivity with time.

To test this technique both glacial and Recent samples

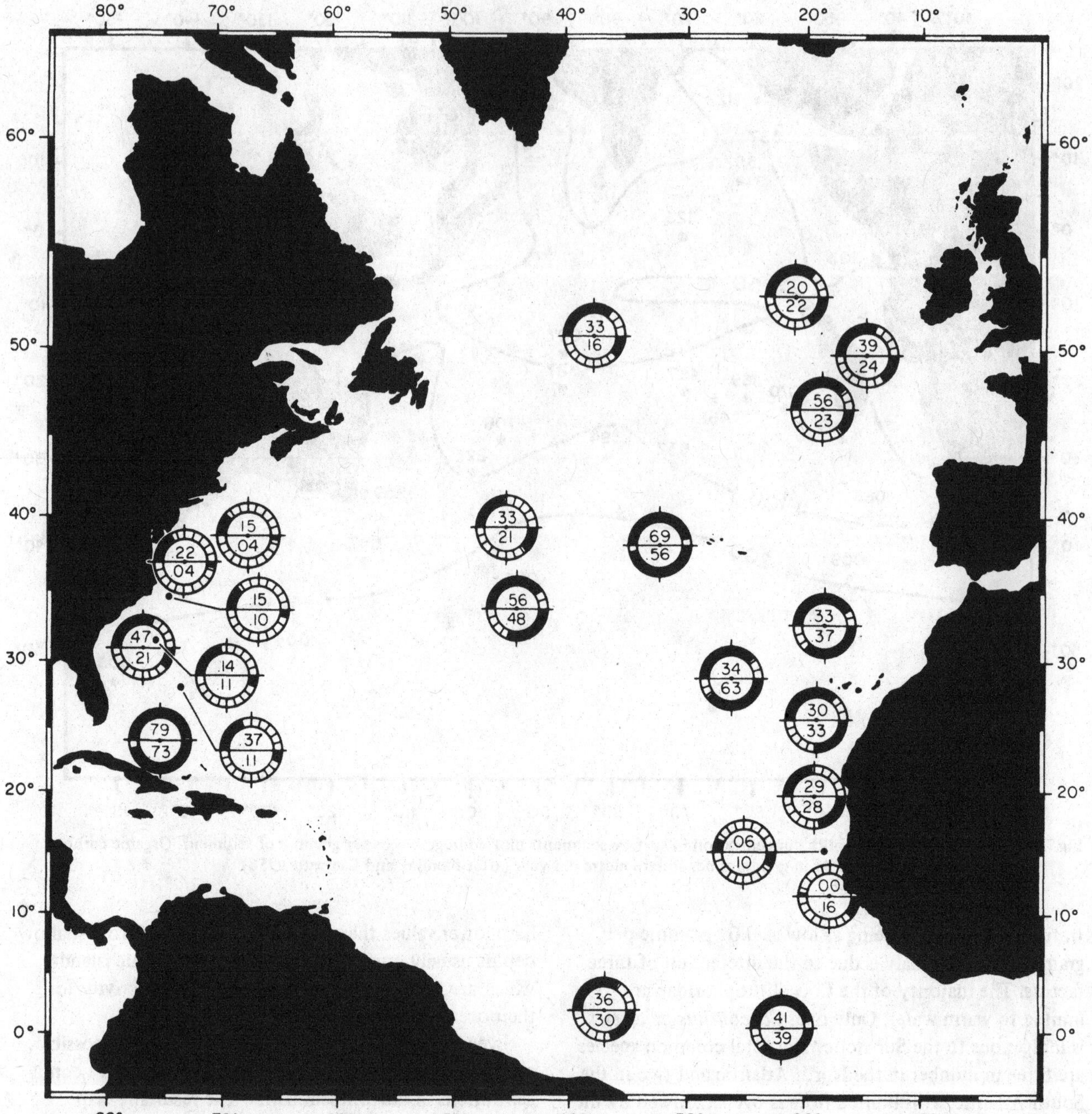
80°
70°
60°
50°
40°
30°
20°
10°
60°
50°
40°
30°
20°
10°
0°
.20
.22
.33
.16
.39
.24
.56
.23
.15
.04
.22
.04
.33
.21
.69
.56
.15
.10
.56
.48
.33
.37
.47
.21
.14
.11
.34
.63
.30
.33
.79
.73
.37
.11
.29
.28
.06
.10
.00
.16
.36
.30
.41
.39

Fig. 16.3. North Atlantic cores showing coccolith carbonate concentration in Recent and mid-Wisconsin glacial samples. The upper half of circles are Recent, lower half glacial. The outer half rings of the circle show the percent of coccolith carbonate for the less than 74μ fraction. The central numbers are the coccolith concentration in grammes per gramme of sediment.

from cores in the North Atlantic, previously studied for species migration in the last glacial period of the Pleistocene (McIntyre 1967), were treated as described above. Comparison, for each core, of the change in carbonate concentration from Recent to glacial sediments was then possible. While the number of cores is too low to allow contouring of values, the geographical distribution is more than adequate to show major geographical changes in coccolith carbonate concentration. None of these cores were near the present compensation depth so that a change in this level would not effect the carbonate values.

The floras from these samples (McIntyre 1967) show a migration poleward from glacial to Recent averaging 15° of latitude.[1] This presumably also indicates a shift in the

[1] The reader is referred to McIntyre (1967) for all data on floristic migration and core numbers, positions and depths.

1μ
1μ
1μ

Plate 16.1. Electron scanning micrographs of surface samples from cores in the compensation zone. A. *Cyclococcolithus leptoporus.* A well preserved placolith from 4,524 metres surrounded by clay particles. B, C. *Coccolithus pelagicus?* Dissaggregated placolith from 5,121 metres showing separation of shields and minor solution effects. D. *C. pelagicus.* Extreme solution of placolith from 5,466 metres.

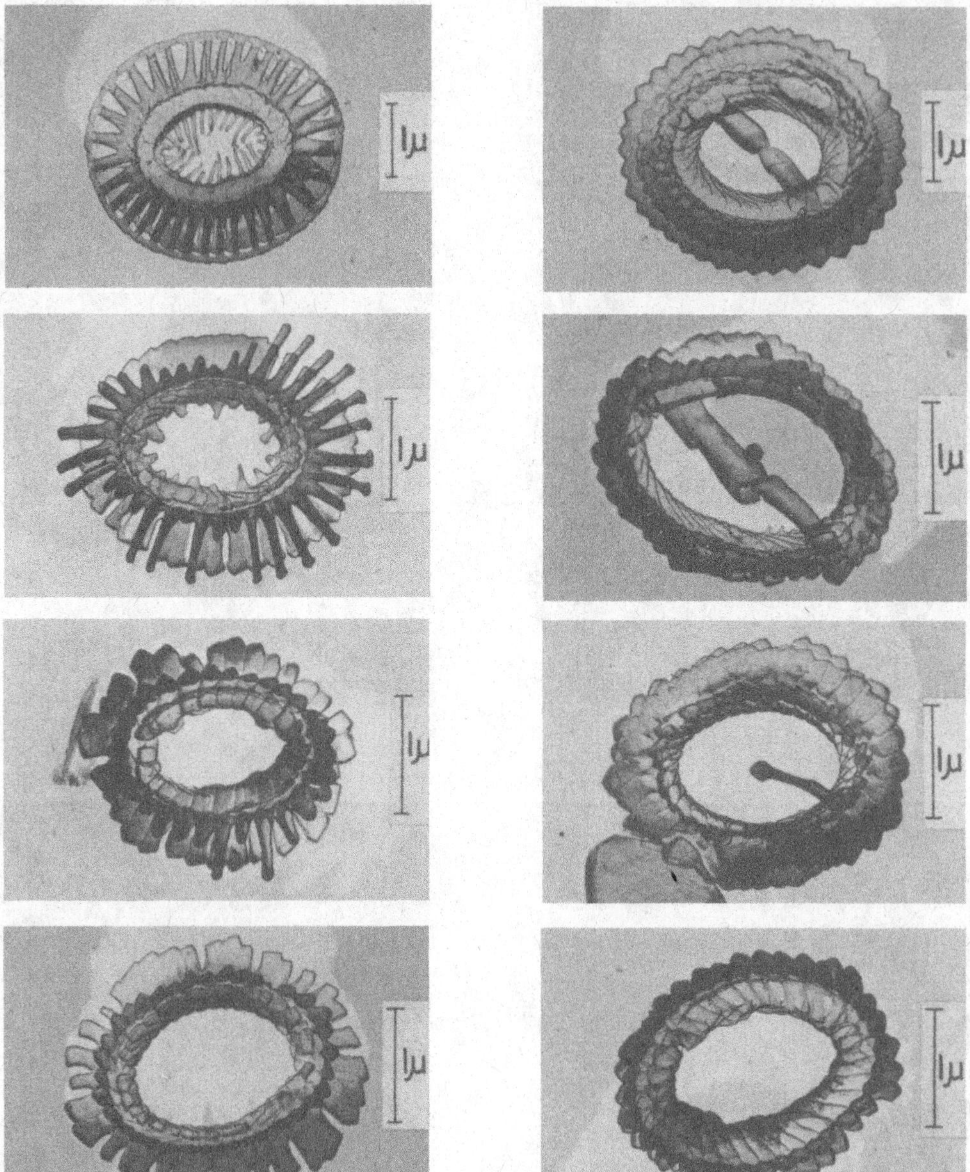
1μ
1μ
1μ
1μ
1μ
1μ
1μ
1μ
1μ
1μ

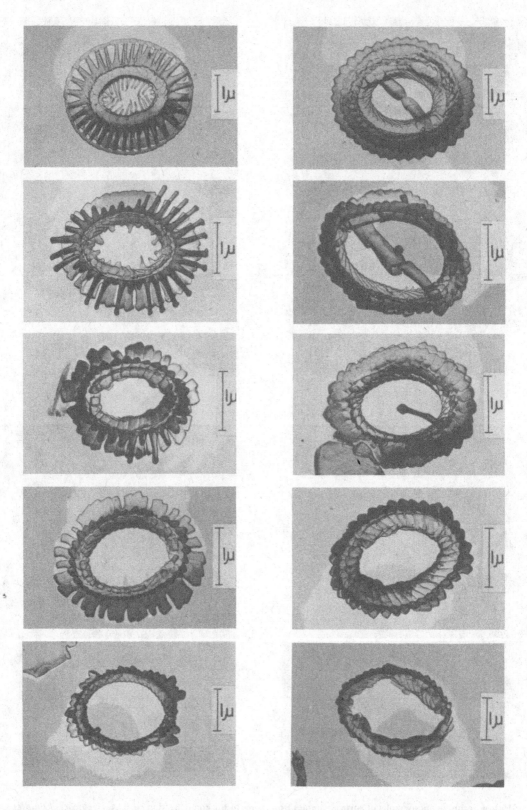

Plate 16.2. Electron transmission micrographs showing graded solution of *Coccolithus huxleyi* and *Gephyrocapsa oceanica* (left and right column respectively).

temperatures and water masses due to climatic changes associated with continental glaciation. The coccolith carbonate values when plotted as per cent of finer fraction (outer ring, Fig. 16.3) show analogous results. When plotted as grammes per gramme of sediment (inner circle, Fig. 16.3) the differences are not as sharp due to the effect of varying coarse fraction. In the north-eastern North Atlantic, today an area of high productivity, the values are markedly lower in the glacial samples. This change is found roughly north of 40° of latitude. South of this latitude both the Recent and glacial samples are more nearly equivalent. In near-shore cores the values are often erratic. Off North America the glacial samples are extremely low in carbonate. This could be due to a combination of cold bottom waters and clay dilution from heavy glacial front runoff. A few cores off the coast of Africa have the opposite effect with minimum values for Recent sediment. Since they are far above the compensation depth it appears to be the result of recent sediment dilution. The majority of the cores do not suffer from these discrepancies and it is clear that the line of carbonate equivalency for Recent and glacial correlates with the floral line separating subarctic from warmer species in the glacial samples. Thus coccolith concentrations as well as floristic parameters appear to have value as palaeoclimatic indicators.

DIFFERENTIAL SOLUTION EFFECTS

The Coccolithophoridae are classified largely on the basis of coccolith architecture. Halldal and Markali (1955) recognize two major divisions: holococcoliths and heterococcoliths. Within these groups, particularly the latter, a great structural range exists. It is these structural differences that are directly mirrored in coccolith resistivity to solution. All coccoliths are protected by a resistant proteinaceous membrane that covers the calcite elements and is one reason for their generally good preservation.

The holococcoliths are the most fragile, being built of discrete calcite rhombs or prisms weakly joined (Black 1963). Consequently, they are soon destroyed by the organic and inorganic processes operating in surface sediments and their fossil record is minimal. The heterococcoliths have a much better record but within the different structural types there is selective preservation in the depositional environment.

Laboratory experiments using ultrasonic vibration for varying periods of time to determine relative strength of modern heterococcoliths have shown that of the four major types (caneolith, scapholith, cyrtolith, and placolith), only the placoliths, and to a lesser extent the eyrtoliths, can really be considered strong.

The caneoliths are basket-shaped with a central area of thin weak lamellae surrounded by a girdle and rim of strong elements. All species are quickly destroyed by vibration, with the rims and girdles the last to disintegrate.

The scapholiths, similar in structure to the caneoliths, are also easily disintegrated. Only the narrow, heavy apical coccoliths show any degree of strength.

The complex intergrowth of crystal elements in the basal plate and appendix of many of the cyrtoliths, e.g. *Rhabdosphaera stylifera* and *R. clavigera*, produces a stronger structure. The attachment between appendix and basal plate is often the weakest point as in *Discosphaera tubifera*.

With few exceptions, the placoliths show the greatest resistance to disintegration. The heavy 'U'-shaped elements are strongly interlocked to form two oval to circular shields connected by a central tube. Some species, such as *Cyclococcolithus leptoporus*, show a tendency for the shields to separate by breakage of the central tube while others (e.g. *Coccolithus huxleyi, Gephyrocapsa oceanica)* show the opposite effect with the central tube the last to be destroyed.

In a parallel study, differential preservation of species in oceanic sediments was examined in two ways. Five closely spaced trigger weight cores from off the North American shelf, ranging from 4,500 to 5,400 metres, which covers the effective compensation zone, were sampled. The upper few millimetres of surface sediment were pipetted off immediately after core retrieval aboard ship. This is a much closer approximation to recent material than is usually obtainable in core work and should represent maximum limits of preservation since the long term effect of organic mixing is reduced. In the laboratory, the coarse fraction was removed by sieving and the flora in the fine fraction was counted using the electron microscope.

The data is in close agreement with predictions based on structure (Fig. 16.4). The caneoliths with the exception of *Syracosphaera pulchra* are the first to disappear; there is a diminution in cyrtoliths and the more fragile placoliths until the dominant forms at 5,400 metres are the placoliths, *Coccolithus huxleyi, Cyclococcolithus leptoporus, Gephyrocapsa oceanica,* and *Umbilicosphaera mirabilis.* However, at this depth all forms show significant solution, making identification difficult and the concentration is much lower per gramme of sediment (Plate 16.1).

With increasing solution of coccolith structure many species become almost unrecognizable even though, as with resistant placoliths, they may constitute a significant portion of the sediment. Even such distinctive though related forms as *Coccolithus huxleyi* and *Gephyrocapsa oceanica* can with progressive solution become difficult to differentiate (Plate 16.2).

Comparing this data with surface sediment floras from 153 trigger weight cores used in our study of the Atlantic Ocean (McIntyre and Bé 1967) and Pacific Ocean (McIntyre, Bé and Roche, in press) indicates that Fig. 16.4 does represent maximum preservation and that the more normal situation is for an upward displacement in most species' disappearance with depth. Nevertheless, among carbonate secreting invertebrates the coccoliths show the best resistance to solution with depth.

In summary then, the holococcoliths and caneoliths are the first to be lost, only *Syracosphaera pulchra* being resistant. This is borne out by the presence of *S. pulchra* at depth in cores, though often found with only the girdle and rim intact. Scapholiths, while more resistant, are commonly represented at depths, as well as in older sediments, by the robust apical coccoliths which cannot be used to distinguish species and in older sediments are called *Scapholithus fossilis.* Consequently, only the cyrtoliths and placoliths are important at depth and in strongly

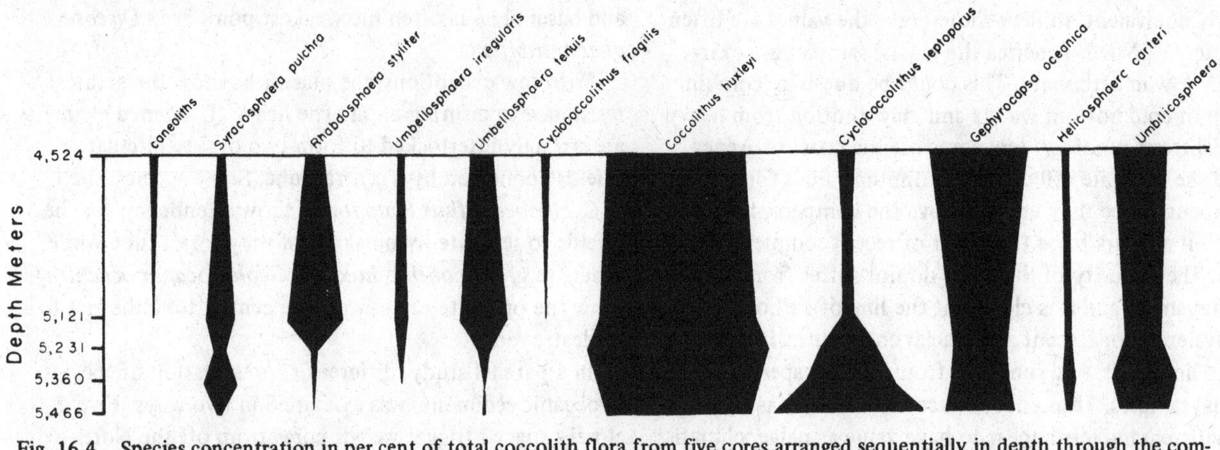
caneoliths
Syracosphaera pulchra
Rhabdosphaera stylifer
Umbellosphaera irregularis
Umbellosphaera tenuis
Cyclococcolithus fragilis
Coccolithus huxleyi
Cyclococcolithus leptoporus
Gephyrocapsa oceanica
Helicosphaerc carteri
Umbilicosphaera mirabilis
4,524
Depth Meters
5,121
5,231
5,360
5,466

Fig. 16.4. Species concentration in per cent of total coccolith flora from five cores arranged sequentially in depth through the compensation zone to show solution loss. With increasing depth the cores used were: RC11-2 30° 38′ N 64° 37′ W, RC11-5 21° 25′ N 57° 38′ W; RC10-2 34° 26′ N 66° 06′ W, RC10-3 29° 46′ N 62° 26′ W; RC10-9 26° 56′ N 55° 02′ W.

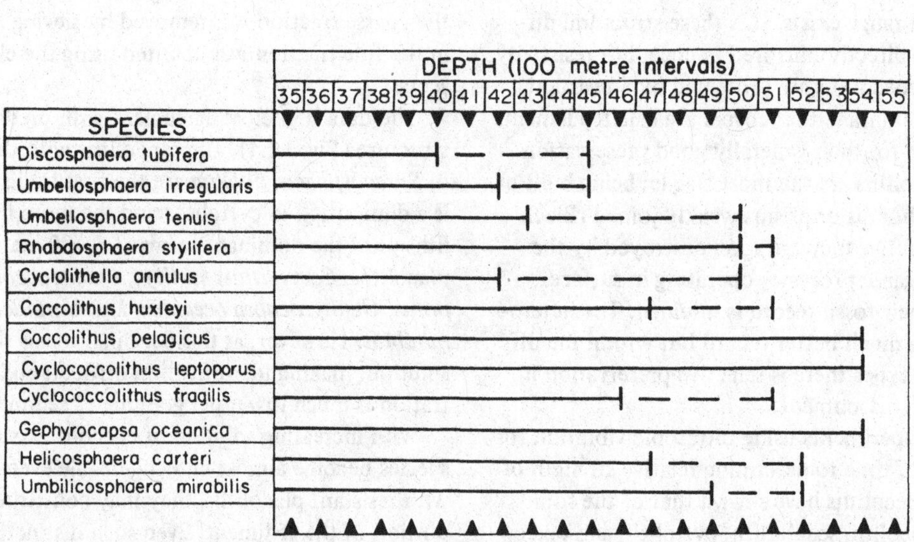
DEPTH (100 metre intervals)
35 36 37 38 39 40 41 42 43 44 45 46 47 48 49 50 51 52 53 54 55
SPECIES
Discosphaera tubifera
Umbellosphaera irregularis
Umbellosphaera tenuis
Rhabdosphaera stylifera
Cyclolithella annulus
Coccolithus huxleyi
Coccolithus pelagicus
Cyclococcolithus leptoporus
Cyclococcolithus fragilis
Gephyrocapsa oceanica
Helicosphaera carteri
Umbilicosphaera mirabilis

Fig. 16.5. The palaeontologically important modern coccoliths arranged to show effects of preferential preservation with water depth. The dotted lines indicate occasional preservation.

reworked sediments, and it is their variety and concentration that will give some indication of the severity of destructive forces active in the environment of deposition.

Fig. 16.5, a compilation of all data, represents the disappearance with depth of palaeontologically important modern species in normal sediments underlying tropical to transitional waters. Clearly, a sediment with long residence time at the compensation level would be typified by a limited flora of resistant placoliths showing solution effects. Floras from areas of low sedimentation rate and high organic reworking will consist of resistant placoliths and coccolith debris. Cores from the tropical Pacific where sedimentation rates are low are perfect examples of this.

ACKNOWLEDGEMENTS

Mr M. Roche assisted with carbonate determinations. This study was supported by National Science Foundation Grants GP 4768 and GA 1205. Indian Ocean data was secured under GA 668.

Received March 1968.

BIBLIOGRAPHY

Black, M. 1963. The fine structure of the mineral parts of Coccolithophoridae. *Proc. Linn. Soc. Lond.* **174**, 41-6.

Bramlette, M. N. 1958. Significance of coccolithophorids in calcium carbonate deposition. *Bull. Geol. Soc. Am.* **69** (1), 121-6.

Fleming, R. H. and Laevastu, T. 1956. The influence of hydrographic conditions on the behaviour of fish. *FAO Fisheries Bull.* **9** (4), 181-96.

Gaarder, K. R. 1954. Coccolithineae, Silicoflagellatae, Pterospermataceae and other forms from the Michael Sars North Atlantic Deep Sea Expedition 1910. Rep. Scient. *Results Michael Sars N. Atlantic Deep Sea Exped.* **2** (4), 20 pp.

Halldal, P. and Markali, J. 1955. Electron microscope studies on coccolithophorids from the Norwegian Sea, the Gulf Stream and the Mediterranean. *Skr. norske Vidensk-Akad.* 1955 (1), 5-29.

Hulburt, E. M. 1964. Succession and diversity in the phytoplankton flora of the western North Atlantic. *Bull. mar. Sci., Gulf Caribb.* **14** (1), 33-44.

Hulburt, E. M. 1963. The diversity of phytoplanktonic populations in oceanic, coastal, and estuarine regions. *J. mar. Res.* **21** (2), 81-93.

Hulburt, E. M. 1962. Phytoplankton in the southwestern Sargasso Sea and North Equatorial Current, February 1961. *Limnol. Oceanogr.* **7** (3), 307-15.

Hulsemann, J. 1966. On the routine analysis of carbonates in unconsolidated sediments. *Jour. Sed. Pet.* **36** (2), 622-5.

McIntyre, A. 1967. Coccoliths as palaeoclimatic indicators of Pleistocene glaciation. *Science,* **158** (3806), 1314-7.

McIntyre, A. and Bé, A. W. H. 1967. Modern Coccolithophoridae of the Atlantic Ocean-I, placoliths and cyrtoliths. *Deep-Sea Res.* **14**, 561-97.

McIntyre, A., Bé, A. W. H. and Roche, M. B. (in press). Modern Pacific Coccolithophoridae – a paleontologic thermometer. *Trans. N.Y. Acad. Sci.*

McIntyre, A., Bé, A. W. H. and Preikstas, R. 1967. Coccoliths and the Pliocene-Pleistocene boundary. *Progress in Oceanography,* **4**, 3-24.

17. DISTRIBUTION OF DIATOMS IN THE SURFACE LAYER OF SEDIMENT FROM THE PACIFIC OCEAN

A. P. JOUSÉ, O. G. KOZLOVA AND V. V. MUHINA

QUANTITATIVE DISTRIBUTION

The surface sediment layer is to a certain extent a reflection of recent diatom distribution, both in number and species composition. The multi-annual picture of diatom distribution within the plankton is projected on to the ocean floor. Diatom algae appear to be one of the basic sources for the formation of organic matter in ocean waters, and actively participate in sedimentation, not only during the Recent epoch but throughout the remote past (more precisely from the late Mesozoic onwards).

The number of diatoms which settle in to bottom sediments is naturally less than in the ocean's surface layer. This depends, first, on the incomplete preservation of many diatom species, owing to the solution of the valves towards the end of vegetation, or in the stomachs of microscopic animals. The inadequate silicification of valves in these species also facilitates solution. Secondly, the settled diatoms are diluted by terrigenous and other organic material on the ocean floor. Under the most favourable conditions, away from the source of terrigenous material, the organic dilution in the open ocean areas populated by highly silicified diatom species being rather low, a considerable proportion of the diatoms are buried in the sediments and diatomaceous oozes are formed.

Examination of the chart of quantitative diatom distribution in the surface sediment layer of the Pacific Ocean (Fig. 17.1) reveals an immense variation in diatom concentrations in different regions of the ocean – 200 surface sediment samples collected during the *Vityaz* and *Ob* expeditions have been investigated. All samples were studied according to a standard method. For the first time data were collected regarding the absolute valve content per gramme of sediment for the greater part of the Pacific Ocean floor. Quantitative data dealing with the Okhotsk and Bering Seas were taken from the papers of A. P. Jousé, P. L. Bezrukov and A. P. Lisitzin. The two last-named papers contain information on the distribution of amorphous silica in the surface sediment layer of the Okhotsk and Bering Seas. The variation of diatom number in the surface sediment layer of the Pacific Ocean is extremely high. In the most abundant areas their proportion in the sediments amounts to 50 per cent and even runs as high as 72 per cent, and in some areas one fails to encounter any trace of their remains.

It appears that Antarctic diatomaceous oozes are the richest in diatom content. In some stations the number of valves per gramme runs as high as from 227 to 442 million. The southern boundary of the Antarctic diatomaceous oozes extends approximately along 65° S. The northern boundary of diatom sediments in most cases coincides with the Antarctic Convergence; however, in a number of areas it is displaced somewhat to the north of it. The next vast area of intense recent diatom accumulation is that of the Okhotsk and Bering Seas. According to the amorphous silica content (up to 55 per cent) typical diatomaceous oozes are forming in their abyssal trenches. The calculation of diatom content in the aleurite fraction of the sediments also indicates exceptionally high concentrations of some species in the sediments of abyssal trenches in the southern parts of the Bering and Okhotsk Seas: e.g. *Coscinodiscus marginatus* Ehrenberg, *C. oculus-iridis* Ehrenberg. Other data from the Bering Sea reveal numbers amounting to 100-150 million per gramme. These numbers are close to the average for Antarctic diatomaceous oozes.

In the northern part of the ocean the diatom concentrations are two to three times lower than in the adjacent abyssal areas of the Okhotsk and Bering Seas. Areas of sedimentation with relatively high diatom content extend along the northwestern and northeastern coasts. At some stations the concentration rises to 50-55 million; however, in most cases it is below 35 million per gramme. In the area of the Emperor seamounts, where the sediments are calcareous, the diatom number decreases sharply. At station 3148, directly over these mounts, the number drops to 1.9 million per gramme, i.e. by about eight times compared with the level observed in the sediments of nearby stations. South of 35°-40° N, throughout an extensive area, the diatom content fails to exceed 25 million per gramme; however, it frequently equals 13-15 million.

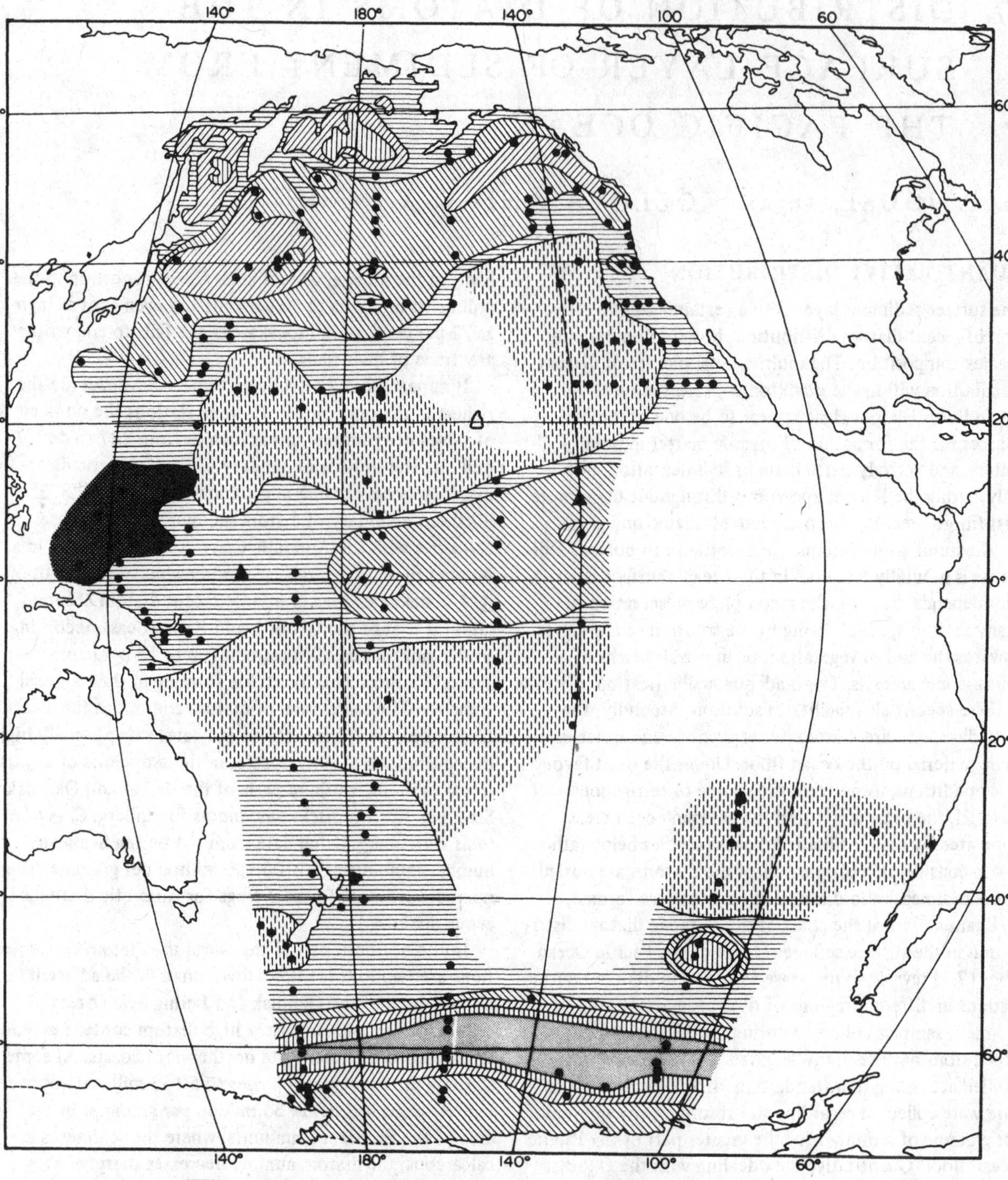
140°
180°
140°
100
60
60°
40°
20°
0°
20°
40°
60°
100°
60°
1.
2.
3.
4.
5.
6.
7.
8.
9.

Fig. 17.1. Diatom numbers in the surface bottom sediment layer (million valves per gramme of sediment). 1. devoid of diatoms; 2. 0.04-5; 3. 5-26; 4. 25-50; 5. 50-100; 6. >100; 7. *Ethmodiscus* oozes; 8. sediments of Tertiary age; 9. stations investigated.

Sediments on the western margins of the ocean's northern part are richer in their diatom content than the eastern ones. High diatom concentrations occur in a narrower and shorter zone along the coasts of North America. South of 20° N in the west and 40°-30° N in the east diatom-free sediments are observed. They extend to 15°-8° N where they are replaced by the organic silts of the equatorial zone. South of 5° S sediments are practically devoid of diatom remains. Diatom concentrations rise again to 30-35.5 million per gramme in a number of stations in the highly productive equatorial zone, and do not differ significantly from the slightly siliceous diatomaceous oozes in the ocean's northwestern part. The average valve number in the sediments of equatorial latitudes does not rise beyond 20 million per gramme owing to dilution by radiolarians, foraminifera and coccoliths. In the calcareous sediments of the equatorial zone the siliceous skeletons of the diatoms are well preserved and do not commonly show traces of corrosion.

Though the general picture of the numerical distribution of diatoms in the equatorial sediment belt is rather uniform, yet there are isolated abrupt deviations from this. For example, at station 5074 (10° N, 140° W) the diatom number drops to 1.86 million per gramme in comparison with 15-20 million per gramme at the nearby stations. (It appears that at the sediment surface at station 5074 there are Tertiary sediments assumed to be of Oligocene-Miocene age. Station 5074 is about 300 metres above the general level of the ocean floor in the area.) A similar picture is observed in some other stations in the tropical region, mainly in the transition zone extending from the organic silts of the equatorial zone towards the typical red clays. The so-called *Ethmodiscus* silts, which are formed mainly by the valves of the diatom *Ethmodiscus rex* Wallich (1.5-1.8 mm in diameter) occur in the tropical region. *Ethmodiscus* silts are encountered at the bottom of the Phillipine and Marianas trenches and their thickness is of the order of 4-5 metres.

In the Pacific Ocean highly siliceous (exceeding 50 per cent), siliceous (30-50 per cent) and weakly siliceous (10-30 per cent) oozes can be distinguished; this subdivision is based on the amorphous silica content of the surface layer of bottom sediments. Diatom oozes of the Antarctic region with a diatom concentration exceeding 100 million per gramme appear to be typical highly siliceous oozes, in which diatoms play the role of rock-forming organisms. In most cases the diatom oozes of the Okhotsk and Bering Seas, with a concentration of diatoms less than 100 million valves per gramme may be described as siliceous oozes. The argillaceous-diatom oozes of the north western and north eastern ocean areas, with a diatom content rarely amounting to 50 million per gramme, correspond to weakly siliceous oozes. Terrigenous sediments, including ice-rafted material, show that diatom content reaches a maximum of 10 million per gramme if the amorphous silica content is less than 10 per cent. Near-coastal terrigenous sediments are characterized by the greatest divergence between the diatom number observed in the plankton and that of the sediments.

As mentioned above, the factor responsible for this phenomenon, apart from dilution, lies in the poor preservation of diatom species inhabiting the near-coastal waters.

The red abyssal clays which form on the bottom in the ocean's tropical areas are practically devoid of diatom remains. However, this may be explained not merely by their dilution in sediments, but by the extremely high primary impoverishment of the plankton of tropical waters. Red clays commonly reveal strongly rounded semi-dissolved valves of Tertiary diatoms. Their secondary origin, in a number of cases, is certain. There is a possibility that single valves of Recent species which fall into sediments similar to the red clay type, do finally dissolve owing to unfavourable burial conditions.

We arrive therefore at the following conclusions. Diatoms play an active role in recent accumulation of organic siliceous sediments in the boreal, equatorial and Antarctic latitudes of the Pacific Ocean. In these areas there is a continuous and abundant supply from the ocean's productive layer. The dominance of diatoms in the plankton of boreal, equatorial and Antarctic areas is conditioned by the abundance of nutrient substances in the ocean's productive layer as a result of intense vertical mixing, and the emergence of essential salts, such as silica, phosphates, iron and nitrates, from the abyssal levels.

SPECIES DISTRIBUTION

Material collected in the course of the last ten years shows that the floristic composition of diatoms varies in relation to the geographical (climatic) zones of the ocean. Within the limits of each zone the species composition is influenced by the position of the station (its near-shore or open-ocean location).

The bottom sediments of the Pacific Ocean, from the northern boundaries of the Bering and Okhotsk seas to the shores of the Antarctic may be subdivided into seven ecologically different diatom complexes (thanatocoenoses) (Fig. 17.2). They correspond to diatom biocoenoses in the plankton, but owing to differences in the preservation of species, differ from them both in species composition and in the numerical ratios between them.

For drawing complex boundaries in the sediments of each station the percentage species content of a similar ecological group has been calculated. In all cases it exceeds

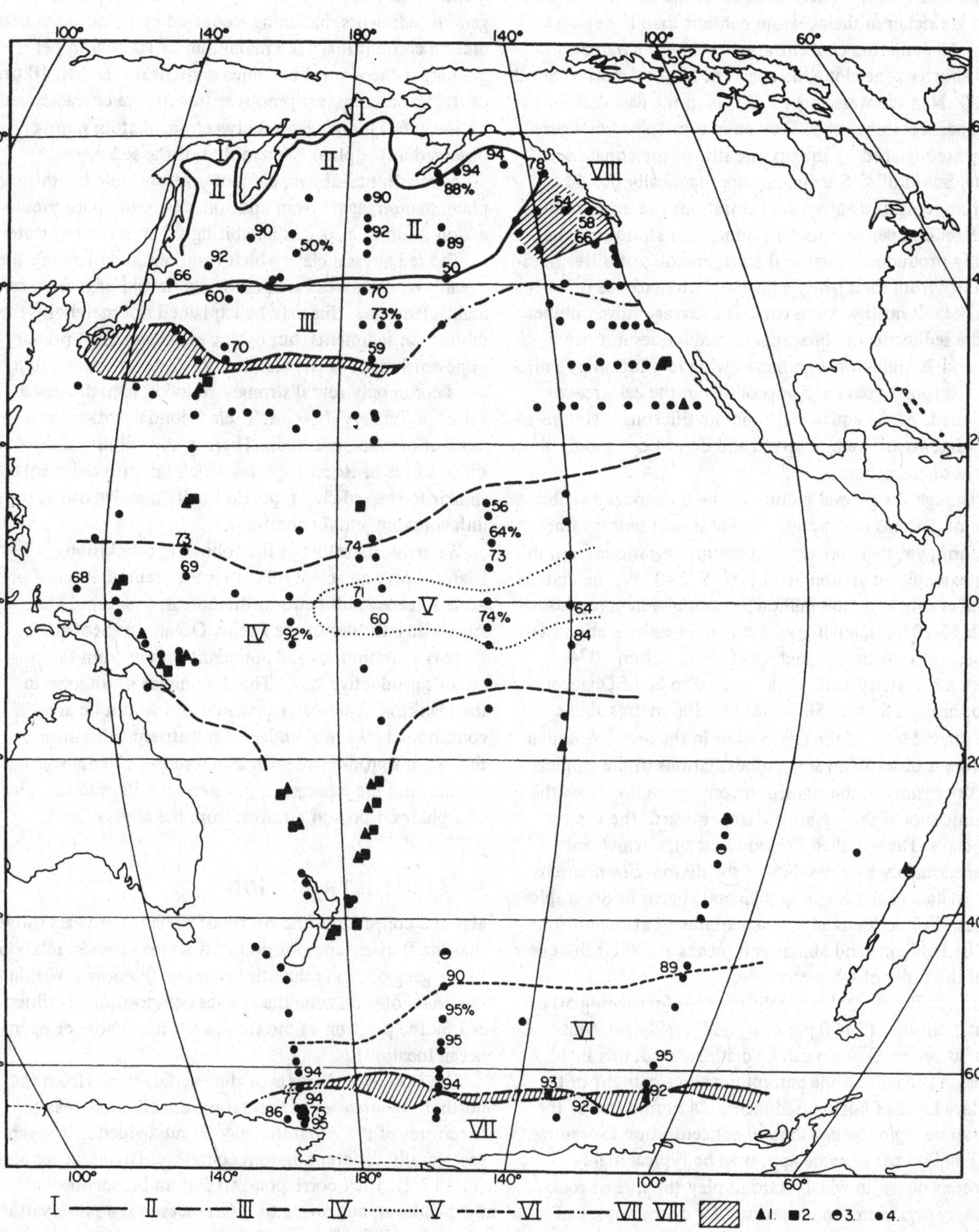
100°
140°
180°
140°
100°
60°
60°
40°
20°
0°
20°
40°
60°
I
II
III
IV
V
VI
VII
VIII
94
88%
90
92
89
50%
50
70
73%
59
70
66
60
92
90
78
54
58
66
56
64%
73
74
73
69
68
71
60
74%
92%
64
84
89
90
95%
95
94
94
77
94
86
75
95
93
92
92
95
I
II
III
IV
V
VI
VII
VIII
1.
2.
3.
4.

Fig. 17.2. Diatom complexes in the surface sediment layer of the Pacific Ocean. I. Arctoboreal; II. Boreal; III. Subtropical; IV. Tropical; V. Equatorial; VI. Subantarctic; VII. Antarctic; VIII. complexes of transitional type. 1.-3. Single findings of 1. subtropical; 2. tropical; and 3. subantarctic species; 4. stations investigated.

50 per cent of the total floral composition. In areas (between 30° and 15° N, at times up to 10° N and between 10° and 45° S) distinguished by extreme diatom impoverishment in sediments, isolated findings are marked by means of signs in the chart (Fig. 17.2). The percentage content of species was determined only in those cases when the preparation permitted counting of at least 200 diatom valves. Apart from the basic complexes it is possible to single out complexes of transient types which incorporate elements of different ecological groups.

I. The Arctoboreal diatom complex as observed in the sediments, is characterized by the presence of the following species: *Thalassiosira nordenskioldii* Cl., *T. gravida* Cl., *Bacterosira fragilis* Gran, *Chaetoceros furcellatus* Bail., *Biddulphia aurita* Breb. et Godey.

The Arctoboreal diatom complex is typical of sediments of the continental and insular shoals of the Okhotsk and Bering seas and also for the ocean's northern margins. Its southernmost occurrence is opposite to Hokkaido. This complex includes the neritic-glacial diatom flora which vegetates in spring during ice thawing and beneath ice following the melting of the ice cover. Diatoms of the complex are distinguished by their maximum cold-loving characteristics, superseding that of other plankton species in the ocean's northern areas.

II. The Northboreal diatom complex of sediments features the following species: *Thalassiosira excentrica* Cl., *Coscinodiscus curvatulus* Grun., *C. marginatus* Ehr., *Actinocyclus divisus* (Grun.) Hust, *Rhizosolenia hebetata* Gran (f. *hiemalis*), *Thalassiothrix longissima* Cl. and Grun., *Denticula seminae* Kanaya et Simonsen.

The area of mass occurrence of Northboreal diatoms in sediments coincides with the area of distribution of subarctic waters. In a number of stations the content of Northboreal species encountered in the complex composition reaches 90-94 per cent; they develop mainly in the plankton of the open ocean. In the near-Kamchatka and near-Kuril areas, as well as next to the coasts of North America, the Northboreal complex is admixed with neritic Arctoboreal species, *Thalassiosira gravida* Cl., *Biddulphia aurita* Breb. et Godey and *Melosira sulcata* (Ehr.) Kutz. The southern boundary of the area marked by profuse occurrence of Northboreal species coincides with the northern boundary of the Polar Front. In the northwestern margins of the ocean it extends to 38° N, but near the coast of North America it rises to 54° N. In the northwest of the ocean this boundary is shifted to the south under the influence of the cold Oyashio current. At the junction point of the cold Oyashio and the warm Kuroshio currents intense mixing of waters occurs, both in the vertical and horizontal directions, and this increases the diatom content of sediments. The mixed water forms a Polar Front, a transitory zone about 2° to 4° wide. The zone is precisely reflected in the sediments. The content of Northboreal species is reduced to 50 per cent. Near the coasts of North America, where all isotherms rise to the north compared with the western marginal areas, the participation of warm-loving species in the diatom complex is especially striking.

III. The Subtropical diatom complex of sediments is formed by the following species: *Thalassiosira decipiens* Jorg., *T. lineata* Jousé, *Coscinodiscus radiatus* Ehr., *Thalassionema nitzschioides* Grun. (a complexus of forms), *Pseudoeunotia doliolus* (Wall.) Grun., *Roperia tesselata* (Roper) Grun., *Nitzschia bicapitata* Cl., *N. interrupta* Heid., *N. sicula* (Castr.) Hust, etc. The greater proportion of these species occurs in the sediments of the open ocean.

The northern boundary of the zone marked by a high occurrence of subtropical species appears to be coincident with the southern boundary of the zone for the distribution of the Northboreal complex (see Fig. 17.2). Amongst the greater number of stations south of 40°-45° N a predominance of moderately warm-loving diatom subtropical species (55-60 per cent of the total number) has been noticed. The mass concentration of these species in sediments suggests that these latitudes are the areal centre for many species of the complex. This situation naturally leads to a question; what species are the remaining 30-45 per cent of sediments south of 40°-45° N made up of? Our investigations show that the northern boundaries of the diatom belt reveal Northboreal species, which infrequently comprise 30-40 per cent. In the south, next to the zone of subtropical convergence, the complex composition is built up of subtropical species 62 per cent, northboreal species 5.3 per cent, tropical species 26.3 per cent, and sublittoral benthic species 1.9 per cent.

There is an obvious relationship between the distribution of the subtropical complex in the surface sediment layer and the direction of the main Kuroshio current of the Pacific Ocean. The areal centres of some subtropical species is east of the Japanese Islands (see Fig. 17.2): *Pseudoeunotia doliolus, Nitzschia bicapitata, N. interrupta, N. sicula.* The maximum occurrence of forms of the *Thalassionema nitzschiodes* group is also related to the sediment zone lying between 40°-23° N. This is one of the most numerous diatom species in subtropical waters. *Coscinodiscus wailesii* Gran et Angst is a most characteristic representative of the moderate warm-loving complex. It is observed in the plankton near the coasts of North

America and also in the Sea of Japan. This species forms mass accumulations in the Middle Pleistocene sediments of the subtropical zone of the Pacific, and is relatively rarely detected in the surface sediment layer of the same latitudes. Many diatoms of the subtropical complex are inadequately studied. Some of them may be special ecological races of tropical species.

IV. The Tropical diatom complex is made up of the following species: *Coscinodiscus crenulatus* Grun., *C. nodulifer* A.S., *Hemidiscus cuneiformis* Wall., *Thalassiosira oestrupii* (Ostf.) Prosh.-Lavr., *Rhizosolenia bergonii* Perag., etc.

Coscinodiscus nodulifer and *Nitzschia marina* are the dominant forms in the complex. Owing to the good preservational features these species are accumulated in bulk in bottom sediments, and are relatively poorly developed in the plankton. The highest concentrations of *C. nodulifer* (up to 50 per cent of the total diatom number) are characteristic of the diatom-radiolarian oozes (5°-15° N). The tropical species *Ethmodiscus rex* Wallich is the most typical one for sediments east of 160° E. At some stations in the Marianas islands area the fragments of *Ethmodiscus* comprise almost 100 per cent. Such diatom species as *Planktoniella sol, Coscinodiscus crenulatus, Thalassiosira oestrupii* and *Thalassiothrix* spp. are similarly distributed in sediments to *Coscinodiscus nodulifer,* but are numerically inferior to it.

V. The Equatorial diatom complex is formed of the following species: *Asteromphalus imbricatus* Wall., *Coscinodiscus africanus* Janisch, *Triceratium cinnamoneum* Grev., *Asterolampra marylandica* Ehr., etc.

The sediment belt ranging between 5° N and 4° S in the western part of the ocean, and between 15° N and 7° S in the eastern areas, is characterized by the presence of this complex. The authors assume some species of this complex to be endemic forms in equatorial waters.

Diatom occurrences are very rare in sediments south of the equatorial zone. There are no special features relating to their composition. Tropical and subtropical species known in the northern hemisphere are sporadically encountered in the area between 5° and 45° S.

VI. The Subantarctic diatom complex is made up of: *Fragilariopsis antarctica* (Castr.) Hust., *Coscinodiscus lentiginosus* Janisch, *Thalassiothrix antarctica* Cl. and Grun., *Schimperiella antarctica* Karst, etc.

All the species indicated are found in the sediments of the Indian and Atlantic sectors of the Antarctic. The circumpolar distribution of most of the subantarctic diatoms may be attributed to the stability of the positioning, and the continuous movement of the basic water masses around the continent, abyssal waters included. The northern boundary of the zone, where the subantarctic species dominate in the sediments, extends along the New Zealand section at 58° S, rises to 54°-55° S in the central area, and to 50° S at 110° W. The species content of this complex varies between 85 and 97 per cent. The subantarctic complex is dominated by *Fragilariopsis antarctica* and *Coscinodiscus lentiginosus* – these are oceanic species well preserved in sediments.

It is necessary to emphasize the distinct isolation of the subantarctic diatom flora which is characterized by almost 100 per cent endemic features.

VII. The Antarctic diatom complex is composed of: *Fragilariopsis curta* Hust., *F. cylindrus* (Grun.) Helmcke et Krieger, *Eucampia balaustium* Castr., *Thalassiosira gracilis* (Karst.) Hust., *Charcotia actinochilus* (Ehr.) Hust., *C. oculoides* Karst., *C. symbolophorus* Grun., *Biddulphia weissflogii* Janisch, etc.

The greater part of the species occur in the shelf sediments and on the continental slope (*Eucampia balaustium, Fragilariopsis curta, Biddulphia weissflogii, Porosira pseudodenticulata, Coscinodiscus symbolophorus*); others are distributed more widely and are infrequently encountered in the open ocean areas (*Fragilariopsis rhombica, Coscinodiscus oculoides*). Most of the species of this complex are endemic forms of the neritic-glacial flora of the Antarctic zone. However, some species have adjusted themselves to more acute variations of temperature and salinity – the basic factors regulating geographical distributions of diatoms. The usual temperature of surface waters in the near-coastal waters of the Antarctic during spring vegetation of diatoms ranges from -1.7° to -1.9° C. The immediate influence of the ocean's ice regime on the biology and distribution of antarctic neritic diatoms allows them to be denoted as glacial species. The arctic and arctoboreal neritic diatoms, i.e. the most cold-loving plankton elements of the ocean's northern part may be regarded as analogues of these glacial species on account of their cold-loving features.

Thus, each geographical (climatic) zone of the ocean has a corresponding typical floristic diatom composition. The recent biogeographical zones of the Pacific Ocean can be traced in the surface sediment layer. Sediment cores contain direct evidence of boundary shift in these zones in the remote past. The diatom distribution within the cores allows the position of these boundaries to be interpreted for different epochs of the Quaternary and earlier

Tertiary periods.

Received September 1967.

18. THE MAIN FEATURES OF DIATOM AND SILICOFLAGELLATE DISTRIBUTION IN THE INDIAN OCEAN

O. G. KOZLOVA

Summary: The quantitative distribution of diatoms and silicoflagellates in the surface layers of the water and in the bottom sediments of the Indian Ocean is described. The coastal regions of the Antarctic abound in siliceous algae. Large quantities of diatoms are also found in the region of the Equatorial Divergence. In the Subtropical zone, and in the northern part of the Ocean, the number of diatoms is very low, which is the result of a low content of nutrient salts. Examination of suspended matter has established a much higher degree of preservation of oceanic, highly silicified species compared with neritic thinly silicified ones. Therefore planktonic oceanic diatoms are important in the sedimentation of Subantarctic diatom oozes, in spite of their quantitative development being lower than the neritic ones.

The quantitative distribution of diatoms in the bottom sediments is determined by such factors as their initial number in the surface water, their degree of preservation, and the quantity of terrigenous and organic dilution. The underwater relief is also important.

The following diatom biocoenoses were established: Antarctic, Subantarctic, Temperate, Subtropical and Tropical, each of them being typical of a certain biogeographical zone. Corresponding diatom complexes were found in the surface layer of bottom sediments.

INTRODUCTION

Samples of suspended matter and of surface layer sediments collected during expeditions of the R.V. *Vityaz* and *Ob* in the Indian Ocean have been studied. Suspended matter was collected from different horizons in the water mass. Samples from the surface layer of the ocean made it possible to study the biocoenoses of diatoms and silicoflagellates in the plankton, and to find out their quantitative distribution. In the surface layer of bottom sediments the thanatocoenoses of such algae were observed. By comparing the diatoms and silicoflagellates of the biocoenoses and thanatocoenoses with one another, conclusions can be drawn that are important for micropalaeontological analysis of oceanic deposits.

This article deals mainly with one aspect of the problem, namely the quantitative distribution of diatoms and silicoflagellates in the surface water and in the surface layer of the bottom sediments in the Indian Ocean. This problem is of interest in order to establish the role played by siliceous algae in sedimentation. Similar researches on material from bottom sediments from the southern part of the Indian Ocean were carried out earlier by Jousé, Koroleva and Nageava (1962), and later by Kozlova (1964) on material both from suspended matter and from bottom sediments.

Following the method evolved by Lisitzin (1956), the suspended matter was collected on membrane filters from the surface layer and throughout the water column. In addition, the suspended matter was obtained by separators from the surface layer, on sections from 50 to 400 miles long. On the one hand this material allows us to study the species composition of diatoms in the surface layer of the ocean, and on the other to determine which part of the plankton reaches the bottom and is buried in the sediments. From this follows some important conclusions concerning the part played by diatoms and silicoflagellates in the formation of siliceous sediments in different regions of the Indian Ocean.

QUANTITY OF DIATOMS AND SILICOFLAGELLATES IN THE SURFACE WATERS

It is easier to describe the quantitative distribution of siliceous algae by beginning with the near-Antarctic regions. Fig. 18.1 shows the data obtained by counting diatoms in the surface waters for the whole ocean. For the Antarctic zone (south of the Antarctic Divergence) the concentration of diatoms amounts to a maximum of 400-583 million cells per gramme of suspended matter. However, the zone as a whole is generally characterized by smaller quantities of diatoms in the range 200-400 million cells per gramme of suspended matter. The greatest amounts of diatoms are in the regions of the outer part of the divergence zone, and in the centre of the upwelling there is a considerable reduction – the quantity sometimes being five times smaller than in the outer part. Evidently this is the result of the purely mechanical drift of diatoms. Compared with the diatoms, the content of silicoflagellates is negligible, and does not exceed 5 million cells per gramme. On the average they

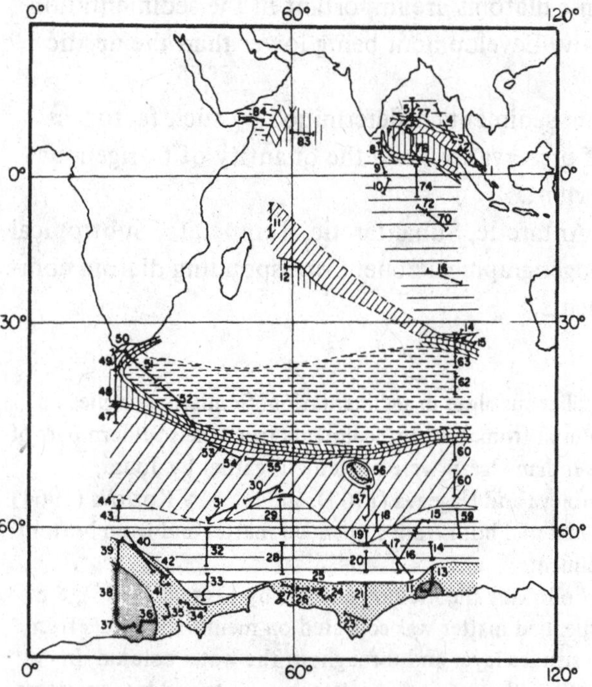
0°
60°
120°
30°
60°
1. 2. 3. 4. 5. 6. 7. 8. 9. 10.

Fig. 18.1. Quantity of diatoms in the surface waters of the Indian Ocean (million cells per gramme suspended matter). The second voyage of R. V. *Ob* summer-autumn 1956-7, thirty-sixth voyage of R.V. *Vityaz* winter 1964. 1. 1-5; 2. 5-10; 3. 10-20; 4. 20-40; 5. 40-100; 6. 100-200; 7. 200-400; 8. more than 400; 9. sections along which suspended matter was collected by separators by R.V. *Ob;* 10. sections along which suspended matter was collected by separators by R. V. *Vityaz*.

make up one million cells, but sometimes this falls to 0.1-0.5 million.

In the Subantarctic zone, the northern boundary of which is the zone of the Antarctic Convergence, the surface waters are less rich in diatoms than those adjacent to the continent of Antarctica. There are between 100 and 300 million cells per gramme near the southern boundary of the Subantarctic zone, but towards the northern limit there is a marked reduction in quantity and on average they do not exceed 50-100 million cells per gramme. Thus, the development of diatoms in the plankton of Subantarctic waters is 2.5 times less than in Antarctic regions. To a certain extent this is caused by strong turbulent mixing of the waters, which is not conducive to abundant plankton development. As far as silicoflagellates are concerned, their quantity greatly increases and attains 6-6.6 million cells per gramme. Such an amount is the highest for the whole of the Indian Ocean, and is shown by two samples of suspended matter from the open waters of this zone. The waters north of the zone of the Antarctic Convergence, that is from approximately 50° S to the zone of Subtropical Convergence, are distinguished by extreme scarcity of diatoms and silicoflagellates. However, expansions of their development take place in some regions caused by local upwelling of deep waters that supply the necessary nutrients.

In the Temperate belt of the Indian Ocean the quantity of diatoms per gramme of suspended matter varies between 1.9 and 50 million cells. This refers to the highest and lowest quantitative data for the whole temperate belt. The most typical quantity is 1.9-5 million cells, and this occurs mostly near the northern boundary. Such scarcity of life in surface waters is the result of the appropriate mineral salts being inadequate. It is common knowledge that in the waters of the Temperate belt the content of silica is five times less, of nitrate seven times less, and of phosphate three to five times less (data from st. 299), than in the Subantarctic regions of the Indian Ocean (st. 283). The same effect is observed in relation to the distribution of silicoflagellates. As a matter of fact, from the northern boundary of the Antarctic Convergence up to the northern limits of the Ocean they do not exceed 0.1-1 million cells per gramme, and on the whole concentrations remain constant. In the waters of Andaman Sea, however, the amounts of silicoflagellates may be much greater.

The content of diatoms and silicoflagellates in suspended matter from the surface of the ocean increases greatly in the Equatorial zone of the ocean. Moving north from the Antarctic continent this zone ranks third in abundance. The productive Equatorial zone in the Indian Ocean is situated approximately between 5° N and 10° S. The plentiful development of phytoplankton in this zone is nourished by the upwelling of salts from deep water horizons. Immediately overlying the centres of upwelling the content of diatoms in one gramme attains 100 million cells. But such high figures do not characterize the zone as a whole, for which a concentration of 20-40 million cells per gramme is more typical.

There is a wide range of quantitative data for the northern seas of the Indian Ocean: in the Bay of Bengal, in the Malacca Straits and in the Andaman Sea the content of diatoms is not homogeneous. For instance in the Andaman Sea, and in the east and southeast regions of the Bay of Bengal, the amount of diatoms is very high, and at some stations attains 100 million cells per gramme. However, in the waters of the southern regions of the Bay of Bengal there are only 14-15 million cells per gramme.

From this general picture of the distribution of diatoms and silicoflagellates in the surface waters of the Indian Ocean it can be concluded that their quantitative distribution is determined by hydrological conditions. Of great importance are the zones of divergence and convergence. In the first case the waters are rich in life, in the second they are characterized by an extreme scarcity of life as the amount of nutrient salts is insignificant.

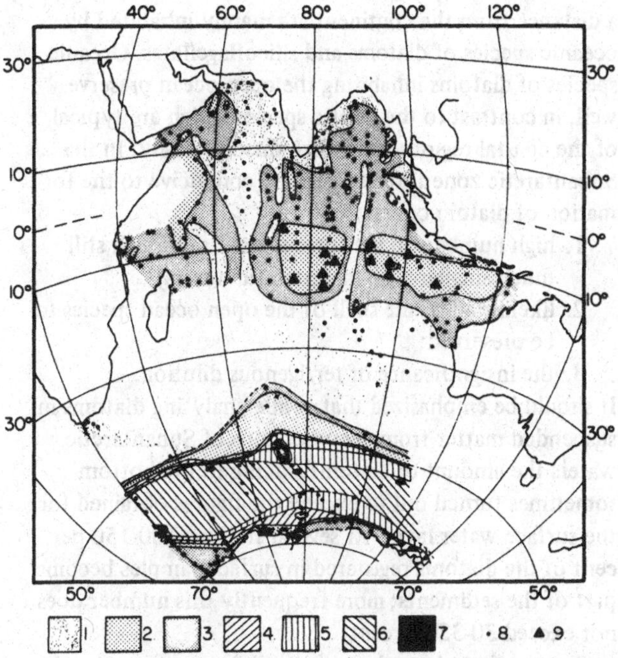
40°
60°
80°
100°
120°
30°
10°
0°
10°
30°
50°
70°
70°
50°
1
2
3
4
5
6
7
8
9

Fig. 18.2. Quantitative distribution of diatoms in the sediments of the Indian Ocean (million cells per gramme of sediment). 1. diatoms absent; 2. 0.1-1; 3. 1-5; 4. 5-15; 5. 15-45; 6. 45-100; 7. more than 100; 8. stations studied; 9. stations at which remains of *Ethmodiscus rex* Wall. are abundant in the sediments.

QUANTITY OF DIATOMS AND SILICOFLAGELLATES IN THE SURFACE LAYER OF SEDIMENTS

To what extent are inherent features of the distribution of diatoms in the surface waters preserved in contemporary sediments? Do the regions of plentiful development of diatoms in the plankton coincide with those where siliceous sediments are formed? To clarify these important problems the materials on which Fig. 18.2 is based have been studied. In Fig. 18.2 data on the amount of diatoms in the surface layer of sediments in the Indian Ocean are summarized. In describing the material from sediments the same order is adhered to that has been adopted for the description of materials from surface waters, i.e. moving from the Antarctic shore towards the northern limits of the ocean.

In the Antarctic zone, in the regions south of the Antarctic Divergence, in spite of the high development of diatoms in the plankton the sediments are poor in diatom remains. In sediments of the shelf bordering on the continent (iceberg and glacio-marine deposits), there are between 0.1 and 1 million cells of diatom per gramme; the quantity of diatoms increases to 1-5 million cells per gramme at the outer edge of the shelf. There is thus a sharp divergence between the content of diatoms in the plankton (maximum 583 million cells per gramme) and that in the surface layer of the sediments. The diatoms of the plankton are only partly preserved in the sediments. Neritic species of diatoms with only thin siliceous shells, such as are found in the waters of coastal regions, are mostly dissolved immediately after vegetation, and do not reach the bottom sediments. Only 1-5 per cent of the numbers observed in the active water layer reach the near-bottom horizons. In addition, those diatoms that do reach the bottom are afterwards diluted by a large amount of terrigenous material. In some deepwater spots on the shelf, however, there are diatomaceous and weakly diatomaceous oozes. They are usually found in a trench on the shelf with a maximum depth of 1,500 metres. According to Lisitzin's data, typical diatom oozes are accumulating in Olaf-Pruids Bay at a depth of 683 metres and in the Davis Sea at the bottom of the Lazarev trench. The concentration of diatoms in diatom oozes in Olaf-Pruids Bay amounts to 44-111 million cells per gramme. The content of diatoms in the sediments of the Antarctic continental slope varies between 5-15 million cells per gramme.

On the ocean bed of the Subantarctic zone, where typical diatom oozes are accumulating, the concentration of diatoms attains the highest figures of, on the average,

100 million cells per gramme. Subantarctic waters at a distance from the continent are mainly inhabited by oceanic species of diatoms and silicoflagellates. Oceanic species of diatoms inhabiting the open ocean preserve well, in contrast to the neritic species which are typical of the coastal regions and which dissolve easily. In the Subantarctic zone three factors are conducive to the formation of diatom oozes; they are:

1. high numbers in the surface waters (though still much smaller than in Antarctic waters);
2. likelihood of the shell of the open ocean species to be preserved;
3. the insignificance of terrigenous dilution.

It should be emphasized that, while analysing diatoms in suspended matter from deep horizons of Subantarctic waters the amount of diatoms sinking to the bottom sometimes turned out to exceed the figures obtained for the surface water layer. At several stations 100-150 per cent of the diatoms registered in surface samples become part of the sediments; more frequently this number does not exceed 30-33 per cent.

The northern boundary of the diatom oozes coincides with the Antarctic Convergence. North of it the quantity of diatoms in the surface waters decreases sharply, and siliceous sediments are replaced by weakly siliceous calcareous ones, and farther north by red clay. There is a sharp decrease in the amount of diatoms in the Temperate and Subtropical zones of the southern part of the Indian Ocean. In the weakly calcareous siliceous sediments of the southern part of the Subtropical zone the content of diatoms does not exceed 15-22 million cells per gramme, and in the calcareous oozes of the southern part of the Australian-Antarctic Rise it is reduced to 5-1 million cells. Diatoms are absent from the sediment off the southeast extremity of Africa, as well as from the larger part of the Australian-Antarctic Rise and the West Australian Submarine mountain range. Neither have diatoms been discovered in the red clays of the central and southern regions of the West Australian hollow. The low content, and sometimes complete absence, of diatoms typical of the sediments between 50°-18° S, is in keeping with their great scarcity in the plankton. The few diatom cells that sink to the bottom of this zone are diluted by foraminifers and coccoliths. Only 1-5 per cent of diatoms in the surface layer of the waters reach the near-bottom horizons.

In the northern regions of the ocean, between 20°-18° S to 20° N, the distribution of diatoms in sediments is uneven. There are no diatoms in the sediments on the Mascarene, Arabian-Indian, Maldive and East-Indian ridges. Nor are there any diatoms in the terrigenous sediments covering the northern part of the Bay of Bengal and in the larger part of the Arabian Sea (Fig. 18.2). The absence of diatoms in the sediments of the Subtropical and Tropical belts is a reflection of their low concentrations in the plankton – fewer than 100, or 100-1,000 cells per cubic metre (Zernova 1962). Also the neritic diatoms inhabiting the coastal waters of the Bay of Bengal and the Arabian Sea do not preserve well. Research has shown that only 0.2-2.2 per cent of their amount in the active layer of the ocean reaches near-bottom horizons. The dilution of diatoms at the bottom by terrigenous material and foraminifers greatly reduces their concentrations. For the sediments of the northern part of the Indian Ocean (20° N-20° S) a content of diatoms of 0.1-1 million cells per gramme is typical. A higher content of diatoms of 1-5 million cells per gramme is found in the sediments of the western part of the ocean, in the weakly siliceous oozes of the eastern part of Somali hollow, and in clayey calcareous oozes of the southern extremity of the Arabian peninsula.

In contrast to the general scarcity of diatoms in the sediments of the northern part of the Indian Ocean are the prominent central regions in the eastern part of the Equatorial zone (approximately 0°-15° S, 75°-115° E), whose sediments are rich in both radiolaria and diatoms. Here, at the bottom of the Central Coccos hollows, at depths of more than 4,600 metres, radiolarian-diatom ooze is accumulating.

The content of diatoms in the sediments of the Equatorial zone attains a maximum of 15 million cells per gramme. The average content of diatoms in the organic oozes of the Equatorial zone approximates 5-8 million per gramme; the content of amorphous silica averages 7.91 per cent. In the sediments of the Andaman Sea the content of diatoms also ranges between 3 and 8 million cells per gramme. The analysis of suspended matter collected at near-bottom horizons in the Equatorial zone has shown that there are three to five times more diatoms compared with the waters south and north of this productive region.

An important part in the accumulation of diatom oozes in the Equatorial zone of the Indian Ocean is played by *Ethmodiscus rex* Wall., a very large-sized tropical species of diatom (the diameter of the valves ranging up to 1,500-1,800μ), and *Thalassiothrix* sp. (the length of the valves of which range up to 300μ). *Ethmodiscus rex* was found in isolated samples of suspended matter on membrane filters at two stations: 4878 in a near-bottom horizon, and 318 at 25 metre depth. However, in samples of suspended matter collected with a separator *Ethmodiscus* is found more frequently. In the surface layer of sediments of the northern and central regions of the Indian Ocean *Ethmodiscus rex* was found in the sediments of 138 stations out of 200. *Ethmodiscus rex* is present in sediments in the form of separate fragments. The content of this species increases sharply in the radiolarian-diatom oozes of the Equatorial zone at a depth of more than 4,700 metres (Fig. 18.2). The sediments of a number of stations consist entirely of *Ethmodiscus rex* fragments. Kolbe described *Ethmodiscus* oozes from 4 to 6 metres thick in the Equatorial regions of the Indian Ocean.

Silicoflagellates are found sporadically in the sediments. Compared with diatoms the amounts in all Indian Ocean sediments are extremely small. On the average they amount to 0.1-1.5 million cells per gramme. In rare cases, at the stations near the northern fringe of the diatom oozes, there are up to 4 million cells. In the north of the Subtropical Convergence zone, and practically up to the northern boundary, their concentration reaches 0.03-0.05 million cells per gramme. The amount of silicoflagellates increases slightly in the sediments of the Andaman Sea.

The absolute content of silicoflagellates per gramme of sediments in the southern regions of the Indian Ocean is two orders, and in the northern one order, less than that of diatoms. Evidently some of the silicoflagellates that vegetate in the plankton are not preserved, and do not reach the sediments. The study of the components of suspended matter in a vertical cross-section at 97° E has shown that their vertical distribution in the water column is limited to the 0-300 metre layer, whereas the diatoms may be found along the whole cross-section down to depths of 5,000 metres. This fact seems to show that silicoflagellates cannot be expected to be as well preserved in the sediments as diatoms are. However, it may also be explained by the extreme primary scarcity of silicoflagellates in the surface layer of the ocean, and their subsequent gradual loss during sinking to the bottom, which is itself due to a number of causes. It is uncertain, being so scarce in the waters, that silicoflagellate cells should settle on the membrane filters at all.

A comparison of the maps of the distribution of diatoms in surface waters and in sediments (Figs. 18.1 and 18.2) shows a coincidence of data for most of the regions. This allows the conclusion to be drawn that the distribution of diatoms in sediments generally follows their live distribution in surface waters. Detailed comparisons have been made for the southern part of the ocean where the greatest amount of factual data is available, and where there are quite characteristic differences. Data on the quantity of diatoms in the plankton and in the surface layer of sediments in the Antarctic zone do not on the whole coincide. The abundance of diatoms in the plankton of the coastal waters of the Antarctic is not reflected in the sediments (with the exception of some localities, such as the Olaf-Pruids Bay). The causes of such differences were mentioned above. Regions where the data on a large quantity of diatoms in the plankton and in the sediments coincides, and are expressed by the relation 1.2:1, are situated far from terrigenous drift, in open regions of the ocean inhabited by oceanic species of diatoms. To such favourable regions for silica accumulation belong the Subantarctic and the Equatorial zones.

Contrary to the Pacific Ocean, in the Indian Ocean the quantity of diatom algae is smaller than that of other algae. In the waters of the central and northern regions Peridinae and blue-green algae prevail in the phytoplankton. According to the data of Suhanova (1962), in the waters between 25° and 30° S the phytoplankton contains 90 per cent of Peridinae, and in the waters between 30° and 10° S Zernova (1962) did not discover any diatoms at all. Thus, owing to the absence of diatoms and silicoflagellates over considerable areas of the Indian Ocean there are no typical diatom sediments in these regions.

Received September 1967.

BIBLIOGRAPHY

Bezrukov, P. L. 1964. Sedimentation in the northern and central parts of the Indian Ocean. *Intern. Geol. Congr. XXII sess.*

Jousé, A. P., Koroleva, G. S. and Nagaeva, G. A. 1962. Diatom algae in the surface layer of the bottom sediments from the Indian sector of Antarctic. *Trudy Inst. Oceanol. Acad. Sci. (U.S.S.R.)*, **61**.

Kozlova, O. G. 1964. *Diatom algae of the Indian and Pacific sectors of the Antarctic.* Nauka: Moscow.

Lisitzin, A. P. 1956. Methods of studying suspended matter for geological purposes. *Trudy Inst. Oceanol. Acad. Sci. (U.S.S.R.)*, **19**.

Suhanova, I. N. 1962. Species composition and distribution of phytoplankton in the northern part of the Indian Ocean. *Trudy Inst. Oceanol. Acad. Sci. (U.S.S.R.)*, **58**.

Zernova, V. V. 1962. Quantitative distribution of phytoplankton in the northern part of the Indian Ocean. *Trudy Inst. Oceanol. Acad. Sci. (U.S.S.R.)*, **58**.

19. PLANKTONIC FORAMINIFERAL ASSEMBLAGES OF THE EPIPELAGIC ZONE AND THEIR THANATOCOENOSES

E. BOLTOVSKOY

Summary: This article deals with the differences which exist between the foraminiferal epipelagic biocenoses and thanatocoenoses of the same area. Three main kinds of differences: (a) qualitative, (b) quantitative, and (c) morphological are discussed. Morphological differences are examined in detail as they are more diverse and complicated than the qualitative and quantitative ones. A given biocoenosis and its corresponding thanatocoenosis differ from each other because : (a) several species change morphologically during their life-cycle, (b) various species occupy specific water levels in the course of their ontogeny. The ontogenies of the most common and widely distributed Recent planktonic species are discussed and their shells at different life-stages are figured.

INTRODUCTION

The planktonic foraminiferal assemblages which are encountered in the epipelagic zone (200-0 metres) differ from the foraminiferal planktonic thanatocoenoses found on the sea floor of the same areas.

It is very important to be aware of this difference not only in order to understand the ontogenetic development of planktonic Foraminifera, but also in order to reconstruct the climatic conditions which existed in the oceans during the past epochs on the basis of those Foraminifera.

The following material was utilized in this study:

(1) The author's collections of planktonic Foraminifera collected mainly by the oceanographic ships of the Argentine Hydrographic Survey over many years in the waters, and on the sea floor, of the South Atlantic Ocean, and in the Argentine sector of the Antarctic seas. Some of these collections have been studied and described by the author previously.

(2) Plankton samples gathered in the Antarctic by the U.S. Oceanographic ship *Eltanin* during cruises 13, 14 and 15 (Boltovskoy 1966*c*), and collections of planktonic Foraminifera taken on the bottom of the same area and kindly sent to the writer by D. G. Blair and J. P. Kennett.

(3) Collections of planktonic Foraminifera generously sent by A. W. H. Bé, P. J. Bermúdez, M. D. Miro, F. L. Parker, P. J. Smith, I. Tinoco, R. Todd, J. D. H. Wiseman, and others from various parts of the world, and deposited in the foraminiferological laboratory of the Museo Argentino de Ciencias Naturales 'B. Rivadavia'.

(4) Data taken from the literature.

REASONS WHY BIOCOENOSES AND THANATOCOENOSES OF PLANKTONIC FORAMINIFERA OF THE SAME AREA DO NOT COMPLETELY CORRESPOND WITH ONE ANOTHER

The differences which exist between an epipelagic foraminiferal assemblage and its thanatocoenosis can be classified into three principal categories: qualitative, quantitative, and morphological.

Qualitative differences

These arise under the following conditions:

(*a*) There are some planktonic Foraminifera which do not live in the epipelagic zone, but occasionally a few specimens of these species are found there. In the tropical zone some cold-water Foraminifera as well as warm-water ones can be included in this group. The cold-water species cannot live in the epipelagic zone because the temperature of the water is too high. However, they may be encountered, usually as isolated individuals, at some depths where the water temperature is lower. This phenomenon is called 'tropical Submergence', and several times has been observed in *Globigerina bulloides* d'Orbigny, *Globorotalia inflata* (d'Orbigny) and some other cold or cold-temperate species (Schott 1935; Phleger 1954; Beljaeva 1964; Boltovskoy 1964; Jones 1967). To the warm-water species which avoid the uppermost water layer belong, e.g. *Globigerinoides trilobus* (Reuss) (f. *sacculifera*) (Beljaeva 1964; Boltovskoy 1964; Jones 1966*b*). *Hastigerina pelagica* (d'Orbigny) (Beljaeva 1964; Boltovskoy 1964; Jones 1966*b*), and *Globorotalia menardii* (d'Orbigny), f. *fimbriata*.

(*b*) Some planktonic species are extremely rare in the epipelagic zone during some seasons. *Globorotalia truncatulinoides* (d'Orbigny) can be cited as an example. Bé and Ericson (1963) state that between May and November the individuals of this foraminifer are present in the epipelagic zone in only very small numbers because almost the whole population descends to deeper water levels.

(*c*) After death the foraminiferal tests fall to the sea floor. As these tests are calcareous, if the depth is sufficiently great, they are dissolved. The critical depth at which this phenomenon takes place is variable, but taken on the average can be considered to be between 3,500 and 5,000 metres. The solution of different species undoubtedly depends very much on the character of the shell. The species most resistant to solution are *Globorotalia menardii* (d'Orbigny) (f. *tumida*), and particularly *Pulleniatina obliquiloculata* (Parker and Jones). This is quite understandable because the tests of these two species are very stout and thick-walled. Those which are the less resistant are: *Globigerinella aequilateralis* (Brady), *Globigerinoides trilobus* (Reuss) (f. *sacculifera*), *Orbulina universa* d'Orbigny and *Hastigerina pelagica* (d'Orbigny). The last-mentioned species possesses a particularly fragile test which is very easily dissolved.

(*d*) If the depth is relatively shallow, the foraminiferal shells are not attacked chemically, but they can be destroyed by mud-feeding bottom dwellers. In this case again *Hastigerina pelagica* (d'Orbigny) is most vulnerable.

All these factors can create a situation in which the foraminiferal assemblages of the epipelagic zone will differ qualitatively from the thanatocoenoses found on the bottom. If solution and destruction occurs in a given area, the thanatocoenosis will be dominated by heavy-shelled, stout forms, which may, in general, be rare in the upper-water layer.

Quantitative differences

These differences are conditioned by two factors: (*a*) the solution and destruction of foraminiferal shells on the bottom, and (*b*) the rate of sedimentation of non-foraminiferal particles. There are areas without foraminiferal tests on the bottom, where the water column above is very rich in living, floating foraminiferal specimens. There are also areas where, in spite of a poorly developed floating foraminiferal fauna, the thanatocoenosis on the bottom is quantitatively very rich as a result of many years of accumulation and the absence of destructive agencies.

Morphological differences

The differences in morphology of the specimens living in the epipelagic zone and those collected on the bottom are more diverse and more complicated than the differences in qualitative and quantitative character.

They originate under the influence of mechanical and/or chemical factors, and as a result of ontogenetic development. The former do not need much explanation. They are caused by the influence of the same factors (solution and other damage), which were discussed above, and are well understood. The most common morphological difference caused by mechanical and/or chemical factors is the lack of spines on the bottom specimens. Not all species and all specimens living in the epipelagic zone possess these spines. They partially disappear during the life-cycle when an individual reaches the mesopelagic zone. Furthermore, being extremely fragile, these spines are easily broken either at the time of collection, or afterwards. They also can be damaged when the specimens are lying on the bottom. According to Rhumbler (1911), these spines are composed of a large percentage of organic matter which facilitates their rapid and easy decomposition when the animal is dead.

However, it should be emphasized that the spines do not dissolve more easily than the test itself, as erroneously stated by many authors. The present writer has observed that, in many plankton samples with low pH, the shells were strongly attacked but the spines were still intact.

Morphological differences which occur during ontogenetic development are complicated and do need special explanation. The following discussion is dedicated to this problem.

BRIEF HISTORICAL SKETCH OF OBSERVATIONS ON THE MORPHOLOGICAL CHANGES OCCURRING DURING A LIFE-CYCLE

The first more or less detailed records concerning the differences between specimens of the same species taken from the surface water layer and on the bottom are those of Brady (1884). In his description, for example, of *Orbulina universa* d'Orbigny he noticed that bottom specimens differ from those taken at the surface in having a much thicker wall and in lacking the trochospiral spinose test inside, which is probably resorbed during the thickening of the outer wall.

Rhumbler (1911) paid much attention to the fact that among several planktonic Foraminifera the adult specimens possess much thicker walls than the young ones. He considered this phenomenon to be the result of the life activity of the organism.

The same fact of wall-thickening was also observed by other investigators but they attributed it to post-mortal wall recrystallization (e.g. Philippi 1910; Revelle 1944; Arrhenius 1952; Phleger, Parker and Peirson 1953).

Le Calvez (1936) observed differences between specimens of *Orbulina universa* d'Orbigny taken from the surface water and those taken at depth, but his observations were limited to the inside structure of the test and not to the wall thickness.

Since 1959 several authors have again started to investigate the problem of wall thickening in planktonic Foraminifera. In the interpretation of this phenomenon they accept Rhumbler's idea that it takes place when the animal is still alive. New observations and data have allowed them to describe the phenomenon in detail and to form a better understanding of its significance. Also publications have begun to appear in which the differences between adult and juvenile specimens are described not just with respect to the wall character, but also to other morphological features.

Here is a list of the most important papers which deal with differences between young and adult specimens of the same species: Ericson 1959; Wiseman and Todd 1959; Bé 1960; Bé and Lott 1964; Cifelli 1965; Bé 1965; Christiansen 1965.

MORPHOLOGICAL CHANGES IN DIFFERENT SPECIES DURING THEIR LIFE-CYCLE

The majority of present workers now believe that reproduction in many of the planktonic Foraminifera occurs at depth exceeding two hundred metres or probably, in some cases, even more. The small young specimens ascend to the epipelagic zone (200-0 metres) where food conditions, thanks to the phytoplankton, are much better. Besides, the photosynthetic activities of unicellular algae found in foraminiferal bodies can also take place there.

In the epipelagic zone the young individuals grow and a great portion of the whole population attains its normal size (or nearly so) in the surface waters. This is illustrated by a table showing the vertical distribution of specimens of different sizes (Boltovskoy 1964).

Afterwards, these young planktonic Foraminifera slowly begin to descend. Simultaneously some species begin to secrete calcium carbonate which covers the test with a coarsely crystalline calcite crust. This wall-thickening can be easily distinguished from the early bilamellar shell wall by its irregular thickness. It is thickest over the first chambers of the last coil and becomes gradually thinner over the following chambers. This calcite crust naturally increases the gross specific gravity of the organism. In addition, many specimens which possess spines during their stay in the surface layer begin to loose them. It is even possible that the organism uses part of the material of the spines for the thickening of the wall.

The increase in specific gravity of the foraminifers facilitates sinking, and the specimens then occupy greater depths. At the same time several other features develop during this last period in the life of the individual. These features are an aberrant last chamber (or chambers), which can be very small, radially elongated, or of quite another shape; unusual types of aperture; and the presence of *bullae*, which are tentatively interpreted by some investigators (Parker 1958; Hofker 1959) as reproductive features, but which probably should be considered only as supplementary weight which helps the specimen to sink.

In the ontogenetic development of planktonic Foraminifera the problem of their 'bottom stage' is especially interesting. Brady (1884) was one of the first to use the term 'bottom specimens'. He applied this term to various shells of planktonic Foraminifera figured in his work. It is not quite clear if he interpreted these specimens is really living on the bottom, or if he was referring to the place where they were found. Rhumbler (1911) also figured a 'bottom specimen' of *Orbulina*. Heton-Allen and Earland (1932) undoubtedly accepted the idea that bottom representatives of some planktonic species exist (e.g. *Orbulina universa*). Phleger (1960) recorded living planktonic Foraminifera on the sea floor without giving any details. But only Christiansen (1965), and only with respect to one species (*Globigerinoides ruber*), showed for the first time that a real 'bottom stage' exists. It is very probable, however, that the bottom stage, or bottom mode of life, exists also in other planktonic Foraminifera. Blanc-Vernet (1965), for instance, stated that she found in the sediment of the Mediterranean Sea, off Marseilles, some living specimens of *Globorotalia inflata* (d'Orbigny), *Globigerina bulloides* d'Orbigny, *Orbulina universa* d'Orbigny, and large numbers of *Globigerina pachyderma* (Ehrenberg). Undoubtedly, further investigation of the problem of the 'bottom stage' of planktonic Foraminifera will yield new and probably unexpected data.

Many other new problems arise in relation to the ontogenetic development of the planktonic Foraminifera. We know for example almost nothing about the physiological adaptation of the organism to life at such great depths. We do not know how a specimen can, after living on the surface in some specific temperature range, adjust itself to the much lower temperature, greater pressure, etc., of

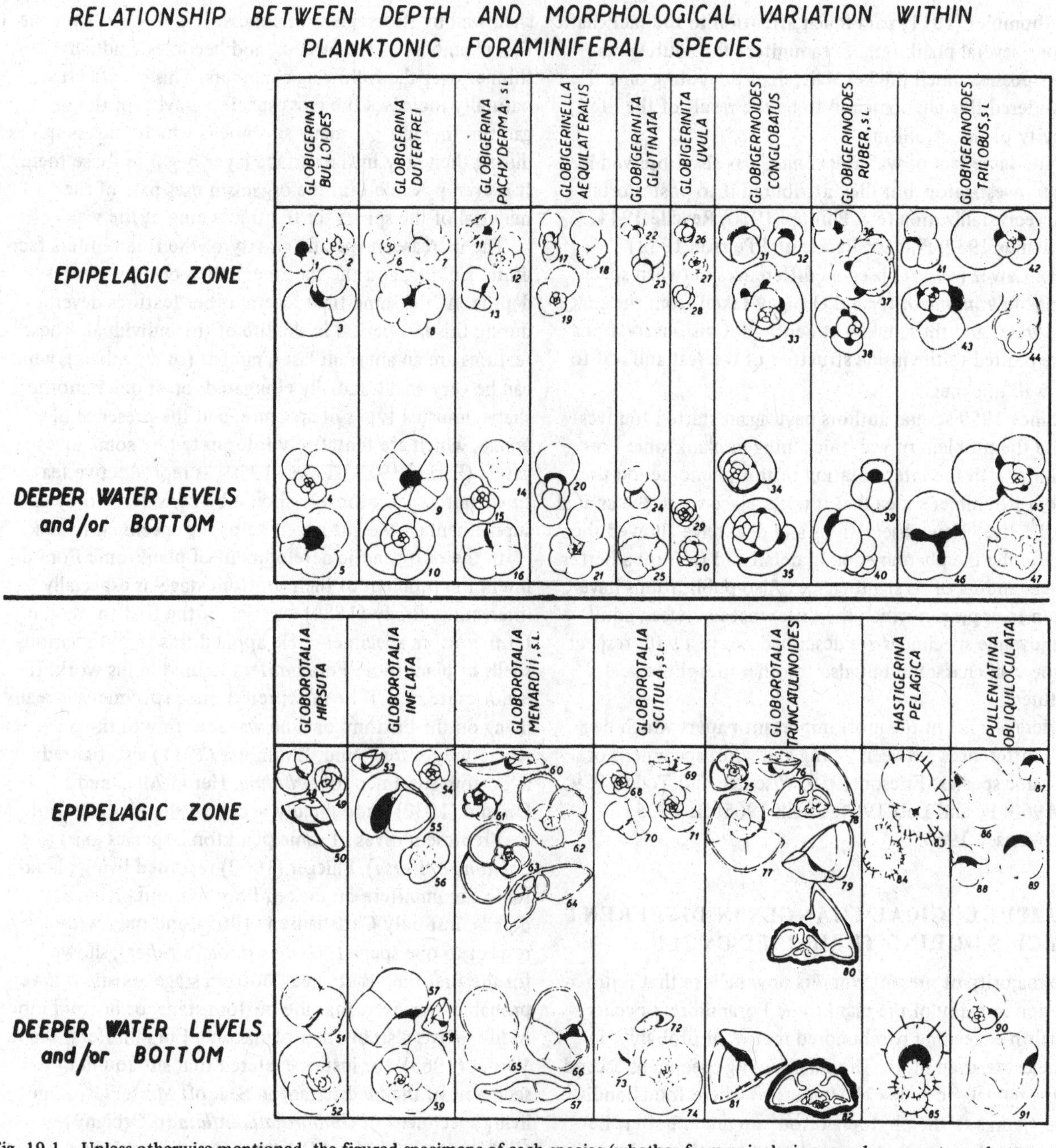
RELATIONSHIP BETWEEN DEPTH AND MORPHOLOGICAL VARIATION WITHIN PLANKTONIC FORAMINIFERAL SPECIES
GLOBIGERINA BULLOIDES
GLOBIGERINA DUTERTREI
GLOBIGERINA PACHYDERMA, s.l.
GLOBIGERINELLA AEQUILATERALIS
GLOBIGERINITA GLUTINATA
GLOBIGERINITA UVULA
GLOBIGERINOIDES CONGLOBATUS
GLOBIGERINOIDES RUBER, s.l.
GLOBIGERINOIDES TRILOBUS, s.l.
EPIPELAGIC ZONE
DEEPER WATER LEVELS and/or BOTTOM
GLOBOROTALIA HIRSUTA
GLOBOROTALIA INFLATA
GLOBOROTALIA MENARDII, s.l.
GLOBOROTALIA SCITULA, s.l.
GLOBOROTALIA TRUNCATULINOIDES
HASTIGERINA PELAGICA
PULLENIATINA OBLIQUILOCULATA
EPIPELAGIC ZONE
DEEPER WATER LEVELS and/or BOTTOM

Fig. 19.1. Unless otherwise mentioned, the figured specimens of each species (whether from epipelagic zone, deeper water or bottom sediments) were collected from the same geographical area.

The following specimens are from between 54°-55° S and 159°-160° W: nos. 1-5 (*Globigerina bulloides*); nos. 11-16 (*Globigerina pachyderma*, s.l.); nos. 22-25 (*Globigerinita glutinata*); nos. 26-30 (*Globigerinita uvula*).

The following specimens are from between 0°-4° N and 35°-48° W: nos. 6-10 (*Globigerina dutertrei*); nos. 17-21 (*Globigerinella aequilateralis*); nos. 31-35 (*Globigerinoides conglobatus*); nos. 36-40 (*Globigerinoides ruber*, s.l.); nos. 41-43, 45, 46 (*Globigerinoides trilobus*, s.l.); nos. 60-62, 64-66 (*Globorotalia menardii*, s.l.); nos. 84-85 (*Hastigerina pelagica*); nos. 86-91 (*Pulleniatina obliquiloculata*).

The following specimens are from between 35°-37° S and 43°-45° W: nos. 48-52 (*Globorotalia hirsuta*); nos. 53-59 (*Globorotalia inflata*); nos. 68-73 (*Globorotalia scitula*, s.l.).

The specimen illustrated in no. 74 (*Globorotalia scitula*, f. *gigantea*) was taken from the top 21 cm of a core obtained at 37° 52′ S, 53° 46′ W.

The following thin sections are reproduced from other studies: nos. 44 and 47 (*Globigerinoides trilobus*, f. *sacculifera* and *G. trilobus*, f. *dehiscens*, respectively) from Bé 1965; nos. 63 and 67 (*Globorotalia menardii*, s.l.) from Wiseman 1966; nos. 78, 80 and 82 (*Globorotalia truncatulinoides*) from Bé and Ericson 1963. In nos. 78, 80 and 82: white = original bilamellar test; black = calcite crust. In nos. 44, 47, 63 and 67: white = the total wall thickness.

the new environment which it encounters at a depth of 1,000 or more metres.

As the greatest percentage of young specimens is found in the epipelagic zone, and the greatest percentage of adult specimens is confined to the depths, it is quite logical to conclude that, taken on the average, the thanatocoenosis will differ from its living surface water assemblage in having a higher percentage of large adult specimens. From the ontogenetic development of many planktonic foraminiferal species sketched above we can also conclude that many of these specimens will be spineless, thick-walled, and will possess some aberrant features. Of course this is true only in very general terms, and does not apply equally well to each species. Several observations testify that the life-cycle of planktonic Foraminifera differs from species to species. It is known, for instance, that they have different seasonal and depth preferences. Their gerontic features are also dissimilar. Thus, to show the differences between the epipelagic and bottom foraminiferal fauna it is necessary to discuss each species separately.

The Foraminifera discussed here are arranged alphabetically. Not all the known species are included because, (*a*) for several species available data are still too incomplete, (*b*) some species are too rare, and (*c*) with respect to some species taxonomic confusion exists concerning their interpretation. Logically, in our discussion the main attention will be given to the age-stages of different species and to the morphological features of young and adult individuals which compose the bulk of the surface water assemblages and thanatocoenoses respectively.

Globigerina bulloides d'Orbigny (Fig. 19.1, nos. 1-5)

This species is found principally in the upper water layer and evidently decreases quantitatively with depth.

The most obvious differences between young and adult specimens are in size and sometimes in shape. Some adult specimens are less elongate, and possess a relatively shorter last chamber, and a large umbilical aperture opening into all chambers (no. 5). This less elongate shape is due to the tendency in the gerontic stage to diminish the relative size of the last chamber. In some areas another type of *G. bulloides* may also be encountered (no. 3). This type has a more compact shell which possesses a much smaller aperture and has a general shape resembling *Globigerinoides trilobus* (for details see Boltovskoy 1966*c*).

Globigerina dutertrei d'Orbigny (Fig. 19.1, nos. 6-10)

Because of an inadequately figured holotype, there was much confusion in the interpreation of this species until Banner and Blow (1960) established, figured, and described the lectotype.

Several morphological changes take place in *G. dutertrei* during its life-cycle. The juvenile individuals have a more regular and elongate configuration and a larger lip (nos. 6-7). Also they are flattened and their last chamber is usually obviously larger than the penultimate one. But the most important difference consists in the character of the aperture. The aperture of the adult specimens is umbilical (nos. 8-10), and that of the young specimens is umbilical-extra-umbilical, arched, interio-marginal (nos. 6-7).

Thanks to these morphological differences, the young specimens can be easily distinguished from the adult ones. In the surface water layer the majority of the specimens are young forms, whereas on the bottom these are less numerous, and the difference between the planktonic assemblage and the thanatocoenosis is therefore clear.

However, it should be mentioned that one important feature, the wall surface, is left unchanged. It is so roughly and characteristically pitted that even a piece of test can often be easily identified as belonging to *G. dutertrei.* The wall surface is left unchanged because a crystalline crust does not develop at all in this species.

Young specimens of *G. dutertrei* are undoubtedly morphologically very close to the right-coiled *G. pachyderma* (Ehrenberg). However, the exact relationship between these two species is still not quite clear.

Globigerina pachyderma (Ehrenberg), *s.l.* (Fig. 19.1, nos. 11-16)

Brief historical sketches of the development of our knowledge of the bathymetrical and geographical distribution of this species can be found in Uchio (1960) and Boltovskoy (1966*c*). The most important paper from the point of view of a better understanding of its life-cycle was published by Bé (1960). He was able to find a relationship between very distinct surface-water specimens of *G. pachyderma* and those taken from the sea bed. Taking into consideration all the data available concerning the life-cycle of *G. pachyderma, s.l.* it seems to be approximately as follows.

In the upper water column (200-0 metres) this species is thin-walled, often translucent, and possesses successively enlarged chambers, and a relatively large aperture situated between the umbilicus and the peripheral margin. There are usually five chambers in the last whorl (nos. 11-13, 15). These specimens were considered by several investigators as different species of *Globigerina* (*G. bulloides, G. dutertrei, G. groenlandica, G. incompta,* etc.). In previous papers (Boltovskoy 1966*a-c*), taking into account the fact that the ontogeny of the species discussed and its relationship with other species was still not quite clear, I preferred to call the specimens in question *Globigerina* ex gr. *pachyderma.* Now, as there is no doubt that they are an early and principally surface stage of *G.*

pachyderma, I propose to call them *Globigerina pachyderma* (f. *superficiaria*). The Latin adjective *superficiarius – a – um* means superficial in English. Although taxonomically it would be correct to leave these specimens under the name *G. pachyderma, s.l.*, it would not be as convenient since this name includes both surface- and deep-water individuals. Meanwhile, in many cases, for ecological or other reasons, it is very desirable to denote the type of *G. pachyderma* specimen which is being referred to. The taxonomic category *forma* is interpreted as being without status and thus not in conflict with the existing International Rules of Zoological Nomenclature. It is not included in the official name of an animal and therefore should be separated from it by a comma or put in parenthesis.

While *G. pachyderma* (f. *superficiaria*) is growing, it descends, and at a depth of approximately 300-500 metres the last chamber is formed. It is more or less the same size as before and possesses a small constricted aperture. At this depth the crystalline crust covers the whole test. Sometimes this crust becomes very thick. In a section of *G. pachyderma* taken from the bottom off the mouth of the Nile at a depth of about 300 metres (shown to me by Dr F. Werner), the wall thickness was equal to two-thirds of the greater diameter of the whole test.

After the calcite crust has covered the shell, the specimen possesses in the final whorl four thick-walled coalescing chambers. Thus, the specimen is transformed into *G. pachyderma* (f. *typica*). These shells compose the bulk of foraminiferal thanatocoenoses in high latitudes (nos. 14, 16).

However this is a rather schematic picture. There are some exceptions and additions which complicate the life-cycle sketched above. First of all, specimens of *G. pachyderma* (f. *typica*) have several times been encountered among the population of *G. pachyderma* (f. *superficiaria*) in the uppermost water level (Bradshaw 1959; Boltovskoy 1961, etc.). On the other hand, f. *superficiaria* is found everywhere on the sea bottom in high latitudes (Boltovskoy 1959*b*; Blair 1965, etc.). Why do these juvenile-stage specimens finish their life-cycle while still incomplete, and why don't they transform into *G. pachyderma* (f. *typica*)? Perhaps it is because of a high mortality rate in the young stage. It should be emphasized that although the ontogeny of *G. pachyderma* described above seems to be correct in general terms, several details are still not clear and need supplementary study.

Globigerinella aequilateralis (Brady) (Fig. 19.1, nos. 17-21)

Among the specimens of this species two *formae* can be recognized: f. *typica* and f. *involuta*. The latter was described as a variety by Cushman. It differs from f. *typica* in being more tightly coiled, more compact, and broader. At the present time is not possible to relate these *formae* to particular areas, conditions, or water levels.

G. aequilateralis apparently does not develop a crystalline crust during its life-cycle. However, another process takes place which allows the distinction of young (nos. 17-19) from the adult specimens (nos. 20-21). The former individuals are not only smaller in size, but they are trochospiral, and consequently they possess a laterally situated aperture. As these specimens acquire new chambers, the coiling becomes planispiral, the aperture moves to the peripheral position, and the whole test becomes more bilaterally symmetrical. The final chambers sometimes overlap and obscure the trochospiral portion of the test. One of the gerontic features of this species is a tendency of the final chambers to lose their tight coiling.

The tests of *G. aequilateralis* are rather fragile and not very resistant to chemical attack. Therefore in some cases they are absent from the sediments although they are present in the planktonic assemblage.

Globigerinita glutinata (Egger) (Fig. 19.1, nos. 22-25)

The principal differences between surface-water specimens (nos. 22-23) and bottom specimens (nos. 24-25) are in the test size and in the presence or absence of bullae. The surface-water specimens are somewhat smaller. In the thanatocoenoses of the Antarctic sector of the Pacific ocean the average size of the specimens is 249μ (Blair 1965). The average size of specimens of the same species found by this author in the upper water layer of the same area is about 170μ. Only the surface and bottom specimens from the same area can be compared, because this species also exhibits different average sizes in different latitudes.

As for the second difference, those who have studied bottom specimens have stated that *G. glutinata* exhibits tests both with and without bullae (Parker 1962; Kustanowich 1963; Blair 1965). At the same time in populations from epipelagic planktonic samples practically all the specimens lack bullae (Boltovskoy 1964, 1966*a-c*; Bé and Hamlin 1967), or those which possessed bullae comprise less than 1 per cent of the total population (Cifelli 1965).

The greater development of bullae in thanatocoenoses, compared with assemblages from the epipelagic zone indicates that this feature originates not in the epipelagic zone, but at greater depths, probably in the mesopelagic zone. Thus, the statement of Smith (1963) and Bé and Hamlin (1967) that *G. glutinata* is a surface-water dweller possibly needs reconsideration.

No crystalline crust is known in *G. glutinata*

Globigerinita uvula (Ehrenberg) (Fig. 19.1, nos. 26-30)

The ontogeny and vertical distribution of this species is almost unknown. Smith (1963) pointed out that *G. uvula* appears to be restricted to the upper 50 metres. Comparison of specimens from sediments collected in the Antarctic sector of the Pacific Ocean (Blair 1965) with those from the plankton samples in the same area (Boltovskoy 1966*c*) showed that the former are somewhat larger. Their maximum diameters are 210 and 150μ respectively.

Globigerinoides conglobatus (Brady) (Fig. 19.1, nos. 31-35)

In the upper-water layer the young individuals are chiefly encountered (nos. 31-32). They differ rather considerably from the adult specimens. The latter dwell at some depths and their shells form part of the thanatocoenoses (nos. 33-35). The differences consist not only in test size, but also in the outline of the chambers, and in the dimensions of the principal aperture. This aperture is relatively larger during the early stages of life of the individual. These differences were mentioned by Boltovskoy (1964) and Cifelli (1965). Adult specimens of *G. conglobatus* possess a crystalline crust (Ericson, Ewing, Wollin and Heezen 1961, Bé and Hamlin 1967). However, the life-cycle of this species and the water levels where the main development of the crystalline crust takes place are not at present well known.

Globigerinoides ruber (d'Orbigny), *s.l.* (Fig. 19.1, nos. 36-40)

This species evidently prefers the uppermost 100 metres of the water column (Beljaeva 1964; Boltovskoy 1964). The plankton hauls of this zone in low latitudes usually yield large numbers of specimens of quite different sizes and ages. Lee, Freudenthal, Kossoy and Bé (1965) observed that about 60-80 per cent of its protoplasm is composed of symbiotic algae (zooxanthellae). As Bé and Hamlin (1967) emphasized 'this indicates that *G. ruber* must be a near-surface dweller and restricts its distribution to the euphotic zone'.

However, Christiansen (1965) proved that this species also has a bottom dwelling stage. According to his study, conducted in the bay of Naples, these bottom dwelling specimens are high-spired, tightly coiled and represent a microspheric form (no. 40). Previously these specimens were considered to be an independent species *G. pyramidalis* (van den Broeck), *G. ruber* (f. *pyramidalis*) or, more often, as an unnamed variation of *G. ruber.* They have been recorded many times in sea bottom sediments (Parker 1958, 1962; Boltovskoy 1959*a*; Kustanowich 1963, etc.). It is rather strange that specimens of this bottom stage are also found in plankton hauls, for instance in the epipelagic zone of the South Pacific Ocean (Parker 1960). I have also found one very typical living specimen in the uppermost water layer (0-50 metres) in the Gulf of Guinea (unpublished data).

No crystalline crust has been observed on the walls of the adult *G. ruber.*

From the life-history outlined above it becomes evident that the main difference between the living population of *G. ruber* and its thanatocoenosis in a given area is the presence of a larger percentage of juvenile individuals and absence of f. *pyramidalis* in the former. However, this last difference – as mentioned above – is not always valid.

Globigerinoides trilobus (Reuss), *s.l.* (Fig. 19.1, nos. 41-47)

This species appears in three principal forms: f. *typica,* f. *sacculifera* and f. *dehiscens,* Bermudez (1961) was the first to state without any reserve that the first two undoubtedly belong to the same species. Bé (1965) showed that '*Sphaeroidinella dehiscens*' has no zoological value, but represents a growth stage of *G. sacculifera.* According to Bé, the following life-cycle characterizes the species discussed. Young specimens, which are encountered principally in the epipelagic zone, develop into four end-forms, listed here according to their relative abundance: (*a*) normal adult, (*b*) specimens with sac-like final chambers, (*c*) specimens in the '*dehiscens*' stage, which is achieved by the secretion of a thick translucent calcite crust (cortex), and *(d)* specimens in the '*dehiscens*' stage with sac-like final chamber. I prefer to call them: (*a*) *Globigerinoides trilobus* (f. *typica*) (nos. 41, 42, 45), (*b*) *G. trilobus* (f. *sacculifera*) (nos. 43, 44), and (*c*) *G. trilobus* (f. *dehiscens*) (nos. 46, 47), respectively. The (*c*) and (*d*) specimens as listed by Bé are included under f. *dehiscens.*

G. trilobus (f. *typica*) is more numerous in the uppermost water layer. Here all the age-stages – the large adult individuals, and those of median and quite small size – are encountered mixed with one another.

G. trilobus (f. *sacculifera*) also lives principally in the upper water layer, although somewhat deeper than f. *typica*, namely between 100 and 300 metres (Boltovskoy 1964; Jones 1966 *a* and *b*).

G. trilobus (f. *dehiscens*) was found almost exclusively in water deeper than 500 metres (Bé 1965), In the thanatocoenoses of the North Atlantic Ocean the following relationship was observed by Bé (1965) among these forms:

Globigerinoides trilobus	(f. *typica*)	73.3%
	(f. *sacculifera*)	24.8%
	(f. *dehiscens*)	1.9%

Summarizing, it is possible to conclude that the most striking difference among the surface population of *G. trilobus* (Reuss), *s.l.* and its corresponding thanatocoenosis is the following: the former is composed exclusively (or nearly so) of *G. trilobus* (f. *typica*). Many of them are spined, and all lack a crystalline crust. In addition the tests are on average smaller than those found on the bottom.

Globorotalia hirsuta (d'Orbigny) (Fig. 19.1, nos. 48-52).

This species varies considerably in the convexity of both sides, especially the ventral one. Its outline is also subject to some variation.

The adult specimens differ from the young ones in having their surface covered by pustules. The life-cycle of *G. hirsuta* is almost completely unknown.

Globorotalia inflata (d'Orbigny) (Fig. 19.1, nos. 53-59)

The variations which are typical of this species concern the character of the dorsal side (concave, plane, or convex), the general shell-outline, the number of chambers in the last coil, and the development of surface rugosity.

Compared with the adult ones the juvenile specimens are usually considerably flattened, their peripheral margin is somewhat more angular, and they are lacking a calcite crust and consequently the ventral rugosity over their walls, (nos. 53, 54, 58). This last feature is not always typical of the adult shells. Some are entirely smooth. The aperture of a juvenile specimen is smaller and – what is especially characteristic – rounder. They have four or even five chambers.

G. inflata is thought to spend the main part of its life in the upper water level because the plankton hauls from the 0-100 metre layer, and even from the surface, usually have a mixed size population. Nevertheless, the origin of the calcite crust takes place at a greater depth. It is interesting to note that several times I have found among bottom specimens some which lacked practically all the internal part of the test. It is not clear whether this was due to post-mortem solution or due to the utilization of shell material to construct shells for the new brood during reproduction.

Thus, we can summarize by saying that the adult individuals differ from the young ones in their size, shape, number of chambers, character of the aperture, and wall-thickness. It should be mentioned, however, that these features are often developed rather differently in different specimens.

Globorotalia menardii (d'Orbigny), *s.l.* (Fig. 19.1, nos. 60-67)

This species appears in the Recent oceans in three principal forms f. *typica* (nos. 60, 61, 63, 66, 67), f. *tumida* (no. 65) and f. *fimbriata* (no. 64).

Forma *tumida* was first described by Brady as a variety of *Pulvinulina menardii.* In his subsequent work Brady (1884) considered it as an independent species. Schmidt (1934) and Cosijn (1938) after a thorough statistical study of the variations of *G. menardii* and *G. tumida* in pre-Recent material, came to the conclusion that these names are synonyms.

Subsequent writers, on the basis of their interpretation of *G. menardii* and *G. tumida,* can be divided into the following categories:

(*a*) Those who considered them as belonging to the same species (Hamilton 1957; Asano 1957).

(*b*) Those who also considered them as belonging to the same species, but preferred to differentiate as *formae* or subspecies (Boltovskoy 1959*a*; Ericson *et al.* 1961).

(*c*) Those who considered them as two different species, but emphasized that biologically they are very closely related if not the same species (Kustanowich 1963; Todd 1965).

(*d*) Those who interpreted them as two different species (Hofker 1956; Bradshaw 1959; Parker 1962; Cheng and Cheng 1953; Beljaeva 1964; Jones 1966*b*).

Todd (1964) states 'In the later development of the complex *G. menardii* and *G. tumida* are clearly distinct and present no problem in distinction of the two forms. But in the early development of the complex, there is a tendency of the two forms to fuse into each other and become one, a form that is morphologically about halfway between *menardii* and *tumida*'.

As for f. *fimbriata* (first described by Brady as *Pulvinulina menardii,* var. *fimbriata,* nov., 1884) its biological similarity to *Globorotalia menardii* is still more evident. It should be mentioned, however, that some authors – for instance Hofker (1956, p. 194) – considered f. *fimbriata* to be an independent species.

Of the three forms mentioned above, *Globorotalia menardii* (f. *typica*) is undoubtedly the most numerous and widely distributed. There are some discrepancies with respect to the depth which it prefers. According to Beljaeva (1964) it occurs most abundantly between 50 and 200 metres. Boltovskoy (1964) and Jones (1966*b*) recorded it most commonly in the upper 100 metres. Parker (1960) found more specimens in deeper than in shallow water. Bé and Hamlin (1967) did not state its depth preference, but they found that the tests of *G. menardii* are generally larger in deeper water than at the surface.

Forma *tumida* is quantitatively considerably less common and insufficient observations have been made with respect to its vertical distribution. However, it seems to prefer a deeper water level than *G. menardii* (f. *typica*). Hofker (1962) stated that this foraminifer occurs at a depth greater than 1,000 metres. If such is the case, there are evidently several exceptions. Specimens have been obtained several times even in the uppermost water level.

Forma *fimbriata* apparently avoids the uppermost water layer. The present writer had the opportunity of studying a great number of surface planktonic samples taken during the *Equalant* Expeditions in the Tropical Atlantic and none of them contained any specimens of this form. Meanwhile *G. menardii* (f. *fimbriata*) is well known in the bottom sediments of the same area.

It is interesting to note that the general appearance of both quite young and adult specimens of *G. menardii* (f. *typica*) taken from the surface water level is practically identical, except for size. However, the crystalline crust is observed only on specimens collected at great depths and on the bottom. The walls of these specimens are as thick as 50μ because of encrusted calcite which, according to Wiseman (1966, p. 88) 'takes the form of overlapping rhombohedral crystals'. The surface water specimens are thin-walled (9-30μ) and in general are either lacking a calcite crust or are only partially covered with a very little secondary calcite. The calcite crust is especially coarse and thick on the walls of the specimens of *G. menardii* (f. *tumida*) found on the bottom.

Thus, the main differences among surface-water specimens of *Globorotalia menardii, s.l.* and the thanatocoenosis of the same area are the following: (*a*) bottom specimens on the average are larger, (*b*) their tests are covered by a calcite crust, (*c*) among bottom specimens the percentage of *G. menardii* (f. *tumida*) and *G. menardii* (f. *fimbriata*) is relatively greater than among the surface water ones.

Globorotalia scitula (Brady), *s.l.* (Fig. 19.1, nos. 68-74)

This species is morphologically variable principally with respect to its periphery (sharply angled or rounded) and the wall surface (smooth or roughened by pustules and short blunt spines). Usually there are only a relatively small number of representatives of this species in foraminiferal assemblages, and these live in cold-temperate waters.

According to Bé (in press), *G. scitula* is encountered very rarely in the epipelagic zone and appears mostly in depth ranges between 500 and 1,000 metres. Boltovskoy recorded it from many surface plankton samples, but always as an infrequent species, except at one station (39° 58′ S; 50° 21′ W; 9 July, 1966) where about ninety specimens were found representing more than a third of the whole assemblage. The surface plankton specimens which I have studied have been thin-walled and less than 0.32-0.33 millimetres in diameter.

Besides the above-mentioned small-sized specimens, which are widely distributed and rather well known, another type of *G. scitula* is sometimes encountered (no. 74). Specimens of this type are much larger (0.7-0.8 millimetres) and possess considerably thicker walls. The surface of all, except usually the last chamber, is coarsely pitted, and on the ventral side there are sometimes pustules. The test is more equally biconvex. I have found these specimens in some cores taken on the Argentine slope and also in two surface plankton samples. Blow (1959) has proposed the subspecific name *gigantea* for the very similar specimens he found in the Miocene sediments of Venezuela. Parker (1962) also found such shells in the bottom sediments of the South Pacific Ocean. But according to her, the subspecies *Globorotalia scitula gigantea* has no zoological value. I am inclined to share this point of view. I doubt if the differences mentioned above fulfil the requirements for the establishment of a new subspecies. Nevertheless, they are undoubtedly of importance in the study of this species as they are evidence of a development stage, or perhaps the result of some special environmental conditions. I believe it is worthwhile to separate the large individuals from the small-sized ones and thus I propose to designate the large specimens *G. scitula* (f. *gigantea*).

Thus, two formae can be distinguished in the population of *Globorotalia scitula*: f. *typica* and f. *gigantea*. The real significance of the latter is not yet quite clear. Forma *typica* is well known from plankton hauls as well as from the bottom sediments. Forma *gigantea*, as far as I know, has been found mainly in sediments, and very rarely in plankton samples.

Globorotalia truncatulinoides (d'Orbigny) (Fig. 19.1, nos. 75-82)

Thanks to studies conducted by Bé and Ericson (1963) and Bé and Lott (1964) the behaviour of this species at different depths is probably better known than that of many other planktonic foraminiferal species.

The specimens studied by them, caught in plankton tows from the upper 300 metres of the Sargasso Sea, attained a maximum test size of about 685μ. Their wall thickness was usually about 20μ (no. 78). The specimens collected from the layer deeper than 500 metres (no. 80), as well as those from the bottom sediment in the same area, had a maximum test dimension of 690μ, but their walls were as thick as 50μ. Sectioning these specimens showed that they have an obvious, well-developed calcite crust. According to Bé and Ericson (1963) this crust 'begins in the euphotic zone and probably continues to well below a depth of 1,000 metres' (no. 82). It is a very important distinguishing feature between surface water and bottom specimens of *G. truncatulinoides*.

Another difference consists in the following. Surface assemblages have a high percentage of small young specimens of different shape. They have no umbilicus and their ventral cone is much lower (nos. 75, 76). According to Cifelli (1965), at about the end of the second whorl in the ontogeny of each individual a sudden increase in size of the lateral dimension takes place. This creates the shape typical of adult specimens, characterized by a very large ventral cone (nos. 80, 81). The dorsal cone disappears. The outlines are very well illustrated by Cifelli (1965). The typical shape of a young individual is also figured by Boltovskoy (1966*c*).

According to Beljaeva (1964) and Jones (1966*b*), *G. truncatulinoides* shows a preference for a depth of 200-500 metres or 'greater than 100 metres', respectively. However, it is not quite clear from their papers whether these depths represent the real preference of the species, or whether they represent the depths at which the species is found in areas of tropical submergence.

Almost all the specimens of *G. truncatulinoides* found by Boltovskoy (1966*c*) in the epipelagic zone of the high latitudes of the South Pacific Ocean were quite small, without a crust and apparently young. The greater diameter of the majority of the specimens varied between 130 and 200μ but in one sample the diameter of one specimen was 380μ. The specimens of the same species encountered in the sediments are considerably larger; their average diameter was 447μ, and the maximum was 750μ (Blair 1965).

Hastigerina pelagica (d'Orbigny) (Fig. 19.1, nos. 83-85)

As mentioned above, the tests of this species are extremely thin and fragile and therefore very often they are found broken in the plankton samples, and are found in even worse condition when collected from bottom sediments. Parker (1960) and Jones (1966*b*) emphasize that although this species is encountered in many areas in comparatively large numbers in plankton tows, it is much less frequent in the sediments. They explained this phenomenon by the fragility of *H. pelagica.*

The main variation which this species exhibits occurs in the height:width relationship of the chambers.

Very little is known, in general, concerning the life-span of this species. According to Parker (1960) and Boltovskoy (1964) it seems that this species, in general, avoids the uppermost water layer. However, specimens have often been recorded in surface plankton samples.

Pulleniatina obliquiloculata (Parker & Jones) (Fig. 19.1, nos. 86-91)

In this species the coiling system changes during its life from trochospiral (no. 87) to streptospiral. This change is more or less evident starting from the termination of the second coil (nos. 88, 89). The last chambers of the adult specimen overlap the umbilical area (no. 90). This is well shown by Cifelli (1965). According to him, the stages which still conserve a pronounced trochospiral growth make up about 80 per cent of the total population of this species in the planktonic assemblages of the western part of the North Atlantic Ocean (Cifelli 1965). Thus, only about 20 per cent of the population looks like the typical specimens first described by Parker and Jones (nos. 86, 91).

Furthermore, Parker (1962), Todd (1965) and Cifelli (1965) have mentioned that the early chambers of this species have a hispid surface and the final ones are highly polished. According to Bradshaw (1959), immature specimens do not have the crescent-shaped aperture, and are more lobate and more coarsely perforate than the adult ones.

Received September 1967.

BIBLIOGRAPHY

Arrhenius, G. 1952. Sediment cores from the East Pacific. *Swedish Deep-Sea Exp. Rep.* **5** (1) 6-227.

Asano, K. 1957. The Foraminifera from the adjacent seas of Japan, collected by the S.S. *Soyo-maru*, 1922-1930. Pt.3 – Planktonic Foraminifera. *Tohoku Univ., Sci. Rep., Geol.* **28**, 1-26, pls. 1, 2.

Banner, F. T. and Blow, W. H. 1960. Some primary types of species belonging to the superfamily. Globigerinaceae. *Cushman Found. Foram. Res., Contr.* **11** (1), 1-41, pl. 1-8.

Bé, A. W. H. 1960. Some observations on Arctic planktonic Foraminifera. *Cushman Found. Foram. Res., Contr.* **11** (2), 64-8.

Bé, A. W. H. 1965. The influence of depth on shell growth in *Globigerinoides sacculifer* (Brady). *Micropaleontology*, **11** (1), 81-97, pls. 1,2.

Bé, A. W. H. (in press). Zoogeography of Antarctic and Subantarctic planktonic Foraminifera in the Atlantic and Pacific Ocean sectors. *Marin. Invert. Folio, Ant. Map Folio Ser., Amer. Geogr. Soc.;* MS, 1966.

Bé, A. W. H. and Ericson, D. B. 1963. Aspects of calcification in planktonic Foraminifers (Sarcodina). *New York Acad. Sci., An.* **109**, 65-81.

Bé, A. W. H. and Hamlin, W. H. 1967. Ecology of Recent planktonic Foraminifera. Pt. 3 – Distribution in the North Atlantic during the summer of 1962. *Micropaleontology*, **13** (1), 87-106.

Bé, A. W. H. and Lott, L. 1964. Shell growth and structure of planktonic Foraminifera. *Science* **145**, 823-4.

Beljaeva, N. V. 1964. Raspredelenie planktonnykh foraminifer v vodakh i na dne Indijskogo okeana. *Inst. Okeanologii, Trudy*, **68**, 12-83, pls.1-3.

Bermúdez, P. J. 1961. Los Foraminíferos planctónicos. *Soc. Cienc. Nat. La Salle, Mem.* **21** (59) 111-28, pls. 1-6.

Blair, D. G. 1965. The distribution of planktonic Foraminifera in deep sea cores from the Southern Ocean, Antarctica. *Florida State Univ., Dep. Geol., Contr.* **10**.

Blanc-Vernet, L. 1965. Note préliminaire sur quelques dragages effectués au large de Marseille (canyon de

Planier). *Rec. Trav. St. Mar. Endoume, Bull.* **36** (52), 185-90, pl.1.

Blow, W. H. 1959. Age, correlation, and biostratigraphy of the Upper Tocuyo (San Lorenzo) and Pozón formations, Eastern Falcón, Venezuela. *Bull. Amer. Pal.* **39**, 67-251, pls.6-19.

Boltovskoy, E. 1959*a*. Foraminiferos recientes del Sur de Brasil y sus relaciones con los de Argentina e India del Oeste. *Argentina, Serv. Hidr. Nav.*, **H.1005**, 1-124, pls.1-20.

Boltovskoy, E. 1959b. La corriente de Malvinas. *Ibid.* **H.1015**, 1-96, pls.1-3.

Boltovskoy, E. 1961. Línea de la convergencia subantártica en el Atlántico Sur y su determinación usando los indicadores biológicos – Foraminíferos. *Ibid.* **H.1018**, 1-35, pl.1.

Boltovskoy, E. 1964. Distribución de los Foraminíferos planctónicos vivos en el Atlántico Ecuatorial, parte oeste. *Ibid.* **H.639**, 1-54, pls.1 4

Boltovskoy, E. 1966a. La zona de convergencia subtropical/subantártica en el océano Atlántico (parte occidental). *Ibid.* **H.640**, 1-69, pl.1.

Boltovskoy, E. 1966b. Resultados oceanográficos sobre la base del estudio del plancton recogido durante la camaña *Cosetri II. Argentina, Serv. Hidr. Nav. Bol.* 3 (2), 105-14.

Boltovskoy, E. 1966c. Zonación en las latitudes altas del Pacífico Sur según los Foraminíferos planctónicos vivos. *Mus. Argent. Cienc. Nat., Hidrobiol.* 2 (1), 1-55, pls.1-4.

Bradshaw, J. S. 1959. Ecology of living planktonic Foraminifera in the north and equatorial Pacific Ocean. *Cushman Found. Foram. Res., Contr.* **10** (2), 25-64, pls.6-8.

Brady, H. B. 1884. Report on the Foraminifera dredged by H.M.S. *Challenger* during the years 1872-1876. *Rep. Sci. Res. Voyage H.M.S. Challenger, Zool.* **9** (1), 1-314, pls.1-115.

Cheng, T. and Cheng, S. 1964. The planktonic Foraminifera of the northern South China Sea. *Ocean. & Limnol. Sinica,* **6** (1), 38-78, pls.1-6.

Christiansen, B. O. 1965. A bottom form of the planktonic Foraminifer *Globigerinoides rubra* (d'Orbigny, 1839). *Publ. Staz. Zool. Napoli,* **34**, 197-202.

Cifelli, R. 1961. *Globigerina incompta,* a new species of pelagic Foraminifera from the North Atlantic. *Cushman Found. Foram. Res., Contr.* **12**, (3), 83-6, pl.4.

Cifelli, R. 1965. Planktonic Foraminifera from the western North Atlantic. *Smiths. Inst., Misc. Coll.* **148** (4), 1-35, pls.1-9.

Cosijn, A. J. 1938. Statistical studies on the phylogeny of some Foraminifera. *Cycloclypeus* and *Lepidocyclina* from Spain, *Globorotalia* from the East Indies. *Leidsche Geol. Meded.* **10** (1), 1-62, pls.1-5.

Ericson, D. B. 1959. The crystalline layer on the tests of planktonic Foraminifera. *First Intern. Oceanogr. Congress, Preprints,* pp.94-5.

Ericson, D. B., Ewing, M., Wollin, G. and Heezen, B. C. 1961. Atlantic Deep-Sea sediments cores. *Geol. Soc. Amer., Bull.* 72, 193-286.

Hamilton, E. 1957. Planktonic Foraminifera from an Equatorial Pacific core. *Micropaleontology,* **3** (1), 69-73.

Heron-Allen, E. and Earland, A. 1932. Foraminifera, Pt.1. The ice free area of the Falkland Islands and adjacent seas. *Discovery Rep.* **1**, 291-460, pls.6-17.

Hofker, J. 1956. Foraminifera dentata. Foraminifera of Santa Cruz and Thatch-Island, Virginia-Archipelago, West Indies. *Univ. Zool. Mus. København Skrift,* **15**, 9-237, pls.1-35.

Hofker, J. 1959. On the splitting of *Globigerina. Cushman Found. Foram. Res., Contr.* **10** (1), 1-9.

Hofker, J. 1962. Studien an planktonischen Foraminiferen. *N. Jb. Geol. Palaont., Abh.* **114**, 81-134. Figs. 1-85.

Jones, J. I. 1966a. The significance of the distribution of planktonic Foraminifera in the equatorial Atlantic undercurrent. *Inst. Mar. Sci., Univ. Miami, Florida, USA:* MS, 1966.

Jones, J. I. 1966b. Planktonic Foraminifera as indicator organisms in the Eastern Atlantic Equatorial Current system. *Ibid.:* MS, 1966.

Kustanowich, S. 1963. Distribution of planktonic Foraminifera in surface sediments of the South-West Pacific Ocean. *New Zealand Journ. Geol. Geophys.* **6** (4), 534-65, pls.1-3.

Le Calvez, J. 1936. Modifications du test des Foraminifères pélagiques en rapport avec la reproduction: *Orbulina universa* d'Orb. et *Tretomphalus bulloides* d'Orb. *Ann. Protistol.* 5 125-33.

Lee, J. J., Freudenthal, H. D., Kossoy, V. and Bé, A. W. H. 1965. Cytological observations on two planktonic Foraminifera, *Globigerina bulloides* d'Orbigny, 1826 and *Globigerinoides ruber* (d'Orbigny, 1839) Cushman, 1927. *J. Protozool.* **12** (4), 1-11, pls.1-5.

Parker, F. L. 1958. Eastern Mediterranean Foraminifera. *Swedish Deep-Sea Exp. Rep.* **8** (4), 219-83, pls.1-6.

Parker, F. L. 1960. Living planktonic Foraminifera from the Equatorial and South-East Pacific. *Tohoku Univ., Sci. Rep., Geol., Sp. Vol.* **4**, 71-82.

Parker, F. L. 1962. Planktonic foraminiferal species in Pacific sediments. *Micropaleontology,* **8** (2), 219-54, pls.1-10.

Philippi, E. 1910. *Die Grundproben der Deutschen Südpolar-Expedition, 1901-1903, vol. 2, no. 6, Geogr. Geol.* pp. 411-616. G. Reimer: Berlin.

Phleger, F. B. 1952. Foraminifera distribution in some sediment samples from the Canadian and Greenland Arctic. *Cushman Found. Foram. Res., Contr.* **3**, 80-9.

Phleger, F. B. 1954. Foraminifera and deep-sea research, *Deep-Sea Res.* 2, p.1-23, pl.1.

Phleger, F. B. 1960. *Ecology and distribution of Recent Foraminifera.* 297 pp. J. Hopkins Press: Baltimore.

Phleger, F. B., Parker, F. L., and Peirson, J. F. 1953. North Atlantic Foraminifera. *Swedish Deep-Sea Exp., Rep.* **7** (1), 3-122, pls.1-12.

Revelle, R. 1944. Marine bottom samples collected in the Pacific Ocean by the *Carnegie* on its seventh cruise. *Carnegie Inst. Washington, Publ.* **556**. Oceanogr., II, pt.1, pp.1-180, 12 pls.

Rhumbler, L. 1911. *Die Foraminiferen (Thalamophoren) der Plankton-Expedition. Pt. 1: Die allgemeinen Organizationsverhältnisse der Foraminifere.* pp. 1-331, pls.1-39, Kiel and Leipzig.

Schmidt, K. 1934. Biometrische Untersuchungen an Foraminiferen *Globorotalia menardii* (d'Orbigny) – *Globorotalia tumida* (Brady) und *Truncatulina margaritifera* (Brady) – *Truncatulina granulosa* Fischer aus dem Pliocaen von Nieder. Indien. *Eclog. Geol. Helvetiae*, **27**.

Schott, W. 1935. Die Foraminiferen in dem äquatorial Teil des Atlantischen Ozeans. *Wiss. Ergebn. Deutsch. Atlant. Exp. 'Meteor'*, 3 (3), 43-134, 1935.

Smith, A. B. 1963. Distribution of living planktonic Foraminifera in the northeastern Pacific. *Cushman Found. Foram Res., Contr.* **14** (1), 1-15, pls.1-2.

Stschedrina, Z. G. 1946. New species of Foraminifera from the Arctic Ocean. *Arkt. Nauchn.-Issled.Inst. Dreif. Eksp. Glavsevmorputi na Led. Par. 'G. Sedov'*, **3** *(Biology)*.

Todd, R. 1964. Planktonic Foraminifera from deep-sea cores off Eniwetok Atoll. *U.S. Geol. Surv., Prof. Pap.* **260-CC**, 1067-1100, pls. 289-95.

Todd, R. 1965. The Foraminifera of the Tropical Pacific collections of the *Albatross*, 1899-1900. *Smiths. Inst., Bull.* **161**, 1-127, pls.1-28.

Uchio, T. 1960. Planktonic Foraminifera of the Antarctic Ocean. *Japan. Antarct. Res. Exp., Biol. Res.* **11**, 3-9.

Wiseman, J. D. 1966. Evidence for recent climatic changes in cores from the ocean bed. *Roy. Meteor. Soc. Proc. Int. Symp. World Clim.* 84-98.

Wiseman, J. D. and Todd, I. 1959. Significance des variations du taux d'accumulation de *Globorotalia menardii* d'Orbigny dans une carotte de l'Atlantique Equatorial. *Colloq. Intern. Centr. Nat. Rech. Sci.* **78**, Topogr. Geol. Prof. Ocean. pp.193-206.

20. DISTRIBUTION OF PLANKTONIC FORAMINIFERA IN RECENT DEEP-SEA SEDIMENTS[1]

FRANCES L. PARKER

Summary: Distributions of thirty-eight planktonic foraminiferal species in Recent deep-sea sediments depend on distribution of calcareous sediments, selective solution of species, presence of pre-Recent deposits, sediment displacement, currents and water masses, and production rates. The species can be divided into groups according to latitudinal positions of greatest abundance. The total number of species decreases from low to high latitudes. Right- and left-coiling provinces occur for some species. Mixed faunas are found in current convergence areas. Anomalous distributions occur in small seas.

INTRODUCTION

A review of the literature on distribution of planktonic Foraminifera in sea-floor sediments reveals how little has been done since Murray and Renard (1891) and Murray (1897) published their papers on this subject. The list of species has grown longer since Murray's time and more precise distributional limits have been found for them. Nevertheless, Murray defined the general features of distribution, which subsequent work has not invalidated.

In making the generalized distribution charts for a few species (Figs. 20.3-20.6) it was found that enormous gaps in descriptive data occur. For example, very little is known of the South Atlantic. The same is true of the North Pacific but there the presence of vast non-calcareous areas is largely responsible.

This report contains a short résumé of factors controlling planktonic foraminiferal distribution and summarizes data on species distributions. This summary includes some unpublished information on South Pacific and Indian Ocean occurrences.

The bibliography includes papers dealing with various aspects of distribution in the bottom sediments, from the purely descriptive statements of the presence or absence of species to more general problems, such as the solution of calcium carbonate tests and attempts to classify the species into distributional groups. Papers on shallow-water deposits (continental shelf and slope) are omitted except where needed to establish distributional boundaries (Figs. 20.3-20.9). Papers on plankton are omitted except for those referred to in the text.

Most early papers lacking precise distributional data are also omitted. It is fitting, however, to point out that the work of the included authors depends to a great degree on that of such early workers as J. W. Bailey, W. B. Carpenter, C. G. Ehrenberg, T. R. Jones, A. D. d'Orbigny, W. K. Parker, and many others.

This work was supported by the Office of Naval Research under contract with the University of California, Nonr 2216 (23), and by the National Science Foundation.

PLANKTONIC SPECIES

For convenience, the nomenclature of an earlier paper (Parker 1962) will be used, except as noted; the figures in that publication illustrate these species. Planktonic foraminiferal species whose distributions are known, at least in part, in surface marine sediments are:

Candeina nitida d'Orbigny
Globigerina bulloides d'Orbigny
Globigerina calida Parker
Globigerina digitata Brady
Globigerina falconensis Blow
Globigerina pachyderma (Ehrenberg)
Globigerina quinqueloba Natland
Globigerina rubescens Hofker
Globigerinella adamsi (Banner and Blow) (Pacific and Indian Oceans only)
Globigerinella siphonifera (d'Orbigny)
Globigerinita glutinata (Egger)
Globigerinita iota Parker
Globigerinita uvula (Ehrenberg)

[1]Contribution from the Scripps Institution of Oceanography, University of California, San Diego, California.

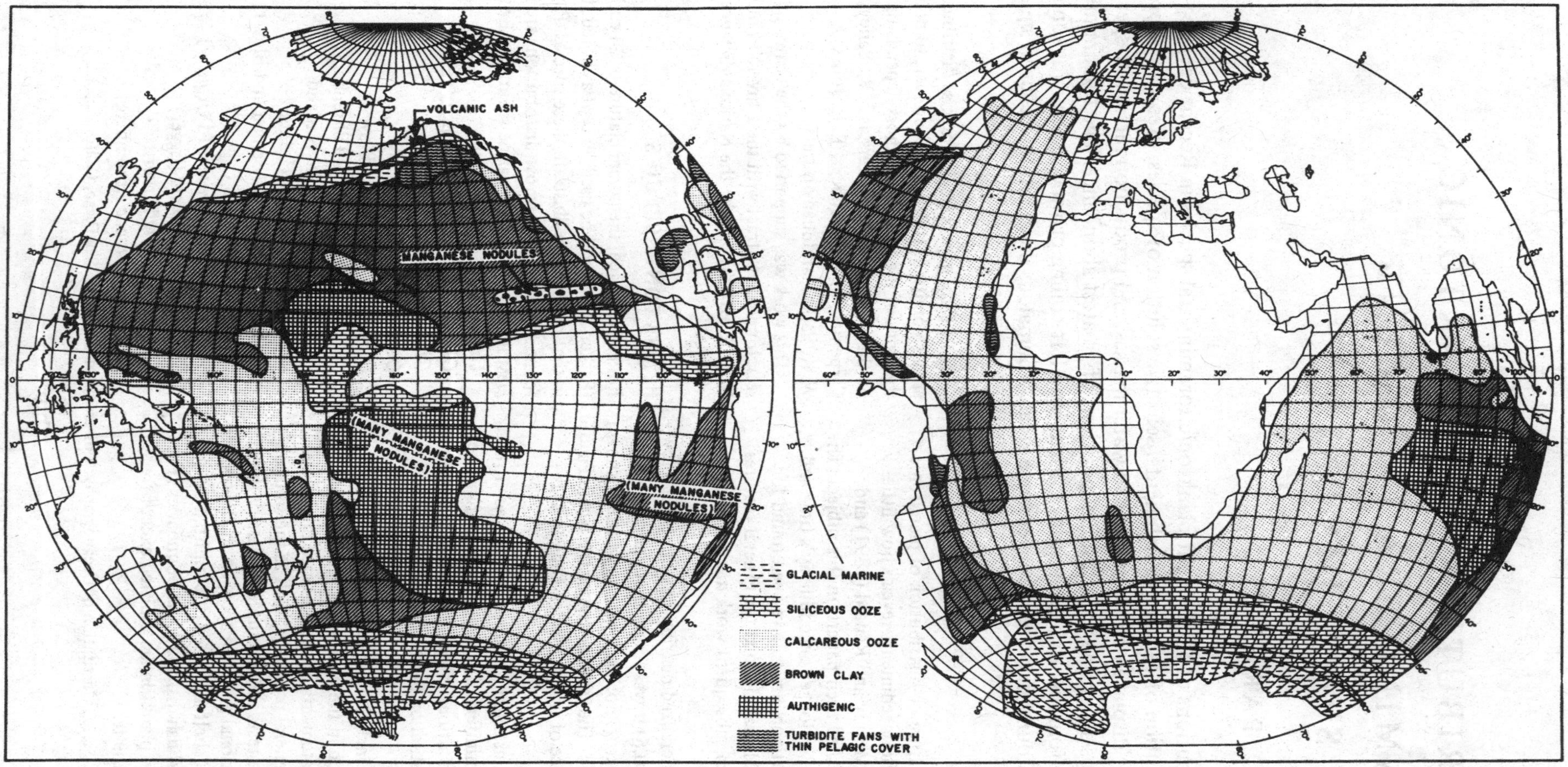
VOLCANIC ASH
MANGANESE NODULES
(MANY MANGANESE NODULES)
(MANY MANGANESE NODULES)
GLACIAL MARINE
SILICEOUS OOZE
CALCAREOUS OOZE
BROWN CLAY
AUTHIGENIC
TURBIDITE FANS WITH THIN PELAGIC COVER

Fig. 20.1 Distribution of deep-sea sediments. (From Shepard 1963, *Submarine Geology*, Harper and Row, Publishers.)

Globigerinoides conglobatus (Brady)
Globigerinoides ruber (d'Orbigny)
Globigerinoides sacculifer (Brady)
Globigerinoides tenellus Parker
Globoquadrina conglomerata (Schwager) (Pacific and Indian Oceans only)
Globoquadrina dutertrei (d'Orbigny)
Globoquadrina hexagona (Natland) (Pacific and Indian Oceans only)
Globorotalia crassaformis (Galloway and Wissler)
Globorotalia crassula (Cushman and R. E. Stewart) (=*G. hirsuta*, Group 3 of Parker, 1962)
Globorotalia cultrata (d'Orbigny)
Globorotalia fimbriata (Brady)
Globorotalia hirsuta (d'Orbigny)
Globorotalia inflata (d'Orbigny)
Globorotalia pumilio Parker
Globorotalia scitula (Brady)
Globorotalia truncatulinoides (d'Orbigny)
Globorotalia tumida (Brady)
'Orbulina universa' (d'Orbigny)
Pulleniatina obliquiloculata (Parker and Jones)
Sphaeroidinella dehiscens (Parker and Jones)
Turborotalita humilis (Brady) (=*Globigerinita humilis* of Parker, 1962)

Additional species have been described, but little is known of their distribution in the surface sediments; some of these are:

Globigerina bermudezi Seiglie: unreported except from off the coast of Venezuela.
Globorotalia anfracta Parker (recently described by Parker, 1967)
Hastigerina pelagica (d'Orbigny): usually removed by solution of calcium carbonate, especially at higher latitudes.
Hastigerinella digitata (Rhumbler): mostly seen in plankton.

FACTORS GOVERNING SPECIES DISTRIBUTIONS

Distribution of calcareous sediments

Fig. 20.1 shows the distribution of the calcareous sediments, which perhaps cover about half of the sea floor. Vast areas are covered by non-calcareous sediments, but small oases of sediments containing planktonic Foraminifera are found within these areas on topographic features raised above the sea floor. These occurrences enabled the construction of the generalized species distribution charts (Figs. 20.3-20.6) which include the non-calcareous areas within the distributional ranges shown. The most widespread non-calcareous areas are in the North Pacific, central South Pacific, western North Atlantic, and eastern Indian Oceans. Not all calcareous sediments are composed of foraminiferal tests, although the major part of the modern sediments is. Coccoliths, which are much more common in Tertiary oozes, are present in small amounts, as are pteropods.

Solution of calcium carbonate

Solution of calcium carbonate results in the elimination of foraminiferal tests from the sediment. The depth at which solution occurs is variable, depending on which ocean, latitude, or even locality is under consideration. In general, it takes place deeper than about 4,000 metres in the Pacific, and 4,500-5,000 metres in the Indian and Atlantic Oceans (Bramlette 1961; Beliaeva 1964*b*; Phleger, Parker and Peirson 1953). Beliaeva (1966) says that in the Pacific 'the critical depth' increases from 3,000 to 3,500 metres in the northern (north of Lat. 20° N) and southern (south of Lat. 40° S) latitudes to 4,800 metres in the tropics. In the Ross Sea the effects are seen much shallower, at about 500 metres (Kennett 1966). Special circumstances which are not entirely understood make these figures rather undependable; often solution is observed at shallower depths, and some well-preserved specimens are found much deeper than the depths cited above.

Recent experimental work on calcium carbonate solution by Berger (1967) has shown that for samples of foraminiferal ooze exposed at various depths (central Pacific) for four months, partial solution occurred at 1,000 metres with rapid increases below 3,000 metres and 5,000 metres. Selective solution of samples from different localities was also observed. These results show that there is still much to be learned about the factors affecting solution of planktonic Foraminifera.

Calcium carbonate solution is selective with respect to species. *Pulleniatina obliquiloculata, Globorotalia tumida* and *Sphaeroidinella dehiscens* are the most resistant, followed by other species of *Globorotalia* and members of the genus *Globoquadrina*. Unusual increase in frequencies of any of these species can be regarded with suspicion. In a normal *Globigerina* ooze the ratio of planktonic to benthonic specimens is more than 99 to 1. A decrease in this figure may mean that solution has occurred, because the benthonic forms are much more resistant than the planktonic ones.

Pre-Recent age of sediments

Pre-modern sediments are common on the ocean floor. It is often difficult to tell whether the planktonic Foraminifera are modern or not. The same species occur for the most part in the Pleistocene and many of them also are in

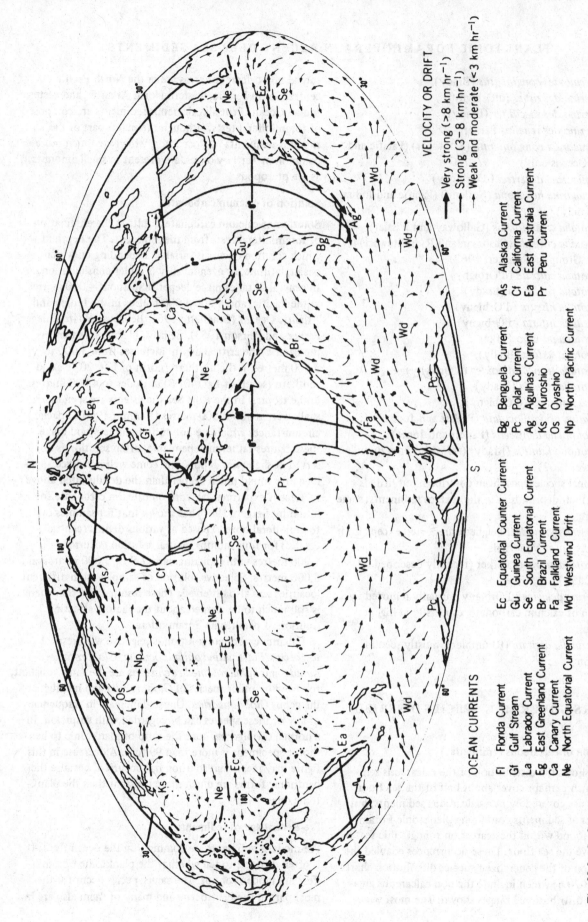

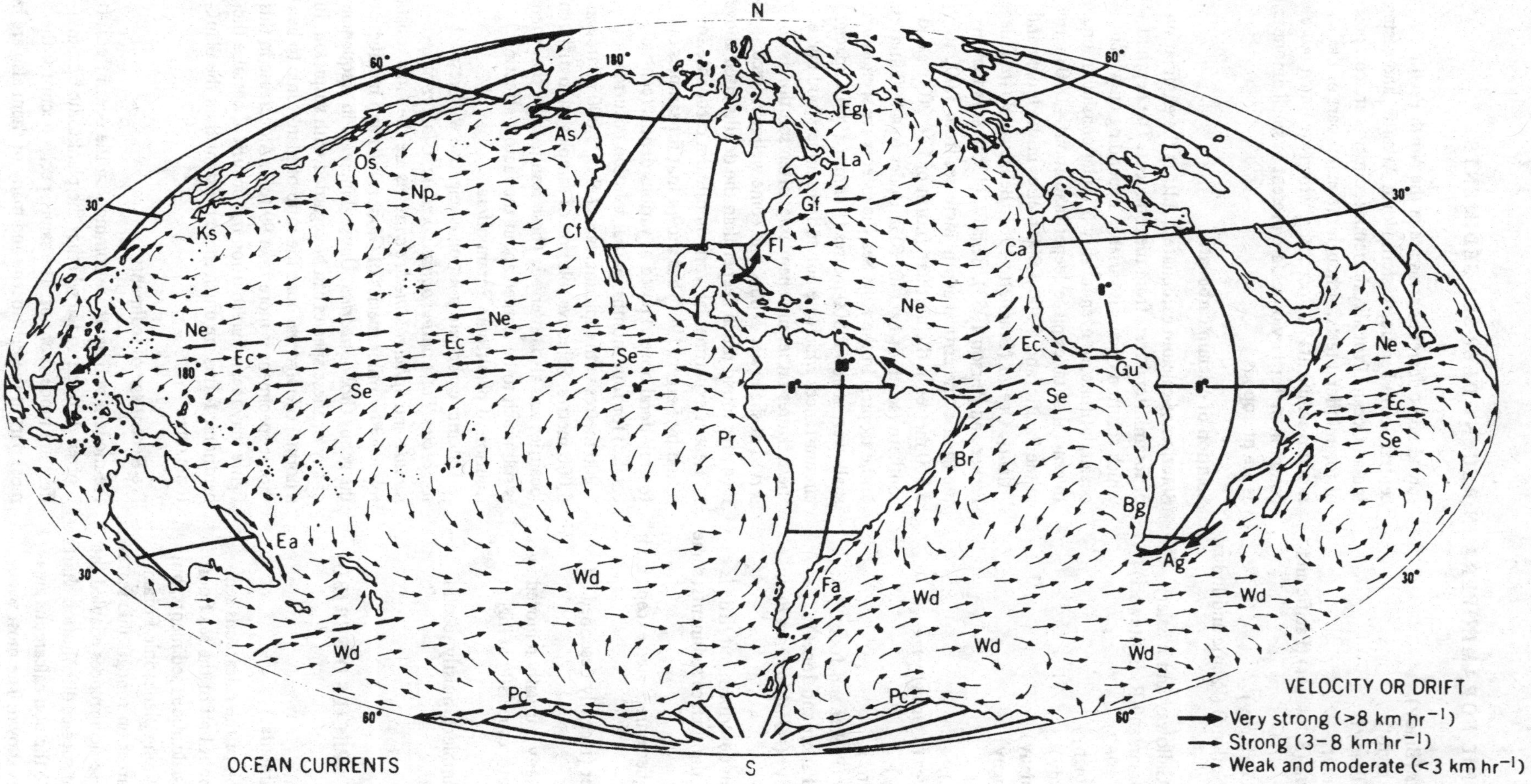
N
S
VELOCITY OR DRIFT
Very strong (>8 km hr^{-1})
Strong (3–8 km hr^{-1})
Weak and moderate (<3 km hr^{-1})

OCEAN CURRENTS

Fl	Florida Current	Ec	Equatorial Counter Current	Bg	Benguela Current	As	Alaska Current
Gf	Gulf Stream	Gu	Guinea Current	Pc	Polar Current	Cf	California Current
La	Labrador Current	Se	South Equatorial Current	Ag	Agulhas Current	Ea	East Australia Current
Eg	East Greenland Current	Br	Brazil Current	Ks	Kuroshio	Pr	Peru Current
Ca	Canary Current	Fa	Falkland Current	Os	Oyashio		
Ne	North Equatorial Current	Wd	Westwind Drift	Np	North Pacific Current		

Fig. 20.2 Surface currents during the northern winter. (After *Investigating the Earth*, 1965, American Geological Institute.)

the Pliocene. Older material is usually easy to recognize. Fortunately, there are a number of ways to detect the presence of Pliocene in the tropical faunas (Parker 1967). Almost nothing is known, however, about the deep-sea Pliocene faunas at higher latitudes.

Several things point to the possible presence of premodern specimens, including preservation, normality of the fauna for the given area, and unexpected coiling directions for some species.

In the Indo-Pacific tropical area many cores containing late Tertiary faunas have a thin layer of apparently fresh, modern material at the top, but others have mixed faunas of modern forms with some reworked older specimens. Such mixed faunas also may be due to the following factor.

Displaced sediments

The displacement of sediments by turbidity currents or other mechanisms, although especially prevalent near continents, cannot be discounted in areas far from them. Any topographic feature rising above the ocean floor may provide the source for displaced material by slumping. This fact probably accounts for much of the fossil material observed in the surface sediments. If the displacement consists of Quaternary material it may be hard to detect, and such material may be more common in surface sediments than has been supposed.

Currents and water-masses

The currents, divergences and convergences to a great extent control the distributional patterns of the species. One has only to compare the shape of the distributional patterns shown in Figs. 20.3-20.6 with those of the surface currents shown in Fig. 20.2 to observe this fact. These currents play an important part in shaping the water-masses, which have physical characteristics that are favourable or unfavourable to the species. Often currents with favourable water for the life of a species will carry it beyond its usual latitudinal limits. This is especially noticeable near continents. Where currents converge the planktonic faunas are mixed.

The planktonic species in the sediments follow these water patterns even as the living species do, but the boundaries are blurred because the sediment sample represents a smoothed out picture of many seasons and current shifts. Thus, the species on the sea bottom extend beyond their known boundaries in the plankton. Dead specimens may be carried beyond their known habitats by currents, but it is believed that they are not carried far. If they are, this displacement will further blur the boundaries of the reflected planktonic distributions.

Thus, we see that these water movements are among the most important factors in dictating species distributions, because it is across the current boundaries that the physical characteristics of the water-masses, which in turn dictate the living environment, show marked changes.

Production rates

Production rates do not have much bearing upon the present discussion except in so far as they insure a constant supply of material to the bottom and maintain a layer of modern specimens on the surface of the sea floor. Bé and Hamlin (1967), after a résumé of the literature dealing with the distributions of Foraminifera in the plankton, concluded that production rates are highest in the equatorial regions, in major current systems, and in some mixing areas between adjacent water-masses and current systems; they are lowest in the central water-masses. Berger and Soutar (1967) showed that the production rates for different species in a given area are not necessarily the same.

DISTRIBUTION OF PLANKTONIC SPECIES

Introduction

Much work remains to be done on planktonic species distribution on a worldwide basis. I intended to make distribution charts for all the species, like those in Figs. 20.3-20.6, but found it was difficult to make even an approximation for any species, and impossible for many, so that the number of species charted was limited to four. Some of the difficulties encountered are: non-uniformity of identification, enormous regional gaps in data, contradictions in the literature, and, to a lesser extent, different methods of presenting data.

Methods of data presentation

The early workers, and many later ones as well, gave distributional data in the form of lists, sometimes amplified by designations such as 'rare', 'few', 'abundant', etc. In more modern times, two types of presentation have been used. One method gives the data for each species in per cent of total planktonic population; the other gives the number of specimens present in one gramme dry weight of sediment. The disadvantage of the first of these methods is that, while showing the relative importance of a species at any given locality, it does not necessarily reveal where the species is most abundant. This is not a serious fault, because many factors of sedimentation mask this information, such as solution of calcium carbonate, the presence of sediments other than calcareous foraminiferal tests, etc. These factors also affect the reliability of the second method. One gramme dry weight of sediment may reflect deposition during a relatively long or short time,

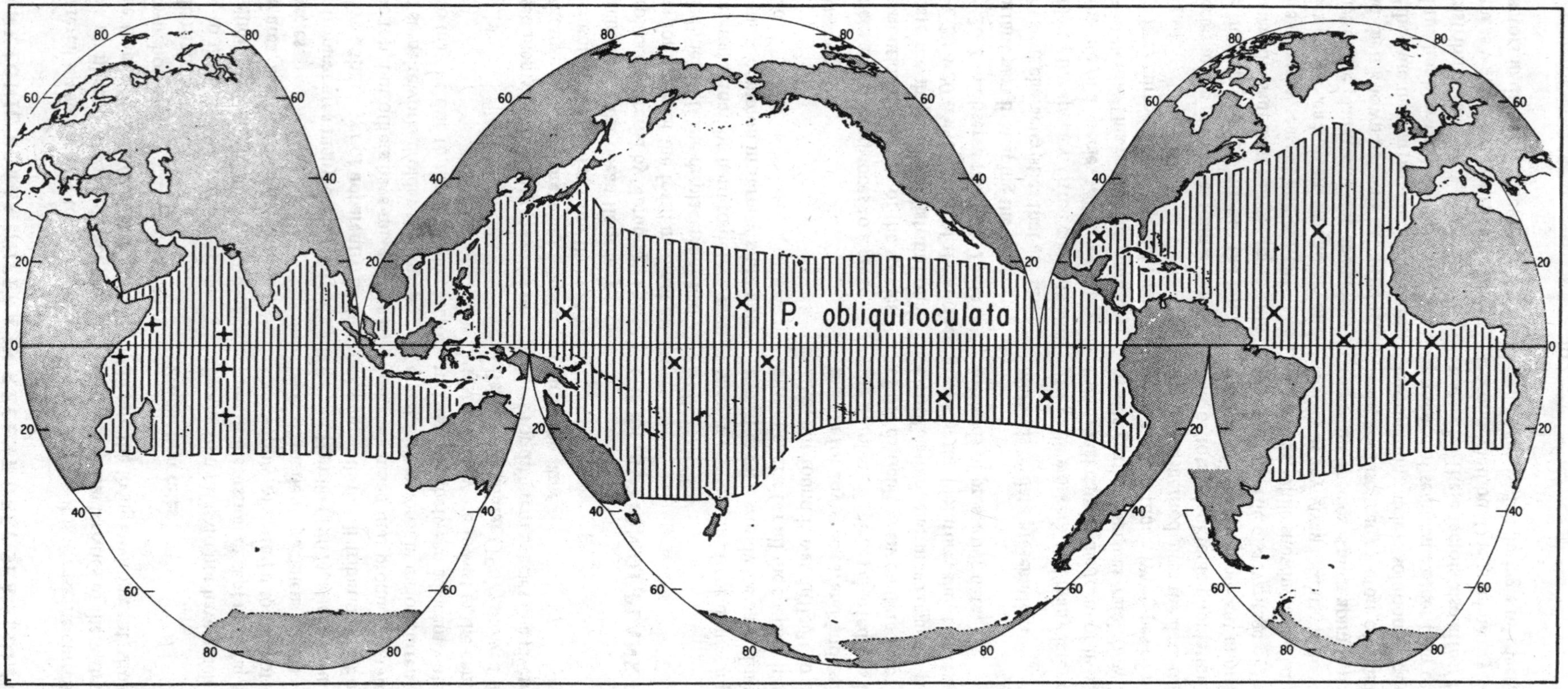
P. obliquiloculata
80
60
40
20
0
20
40
60
80

Fig. 20.3 Generalized distribution of *Pulleniatina obliquiloculata* (Parker and Jones) in deep-sea sediments. Dashed lines indicate hypothetical boundaries based on inadequate data; crosses indicate areas of greatest abundance. Sources indicated in the bibliography by*.

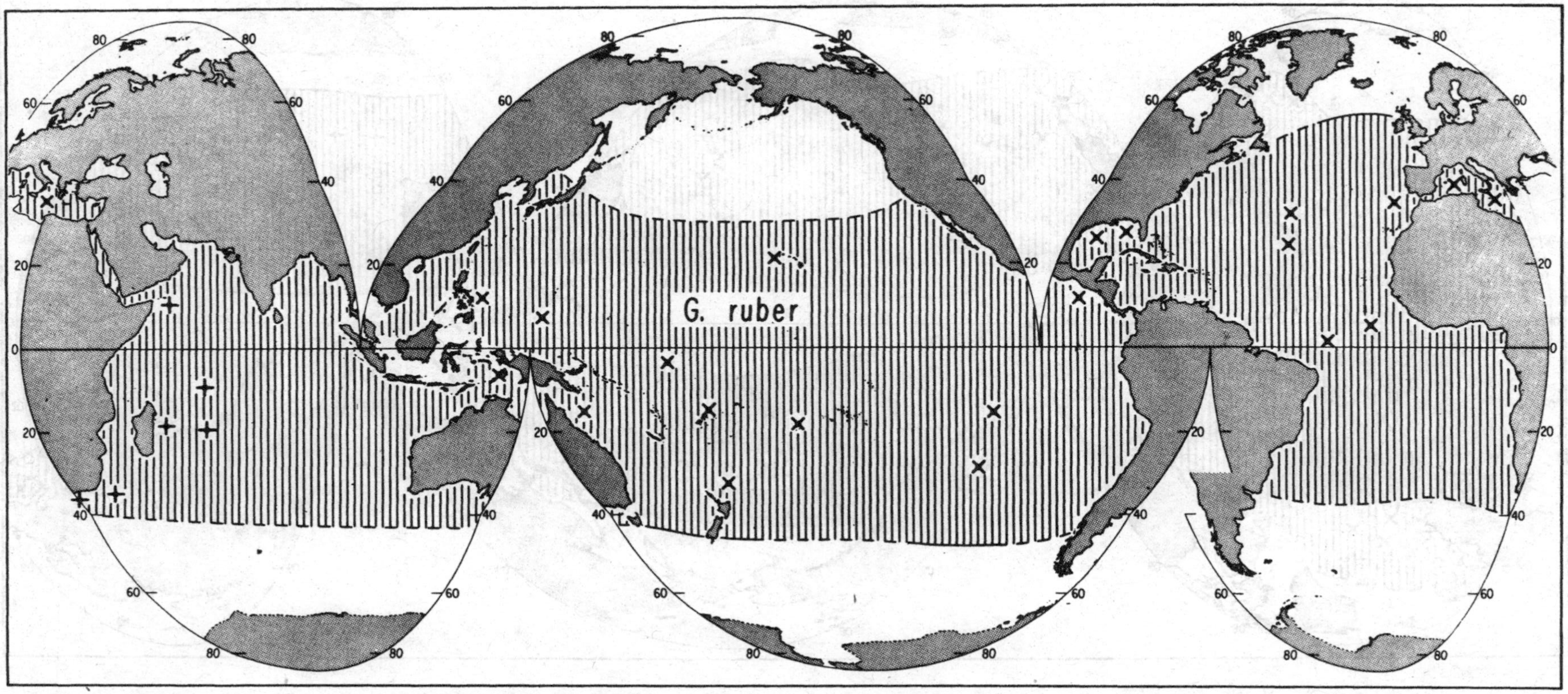
G. ruber

Fig. 20.4 Generalized distribution of *Globigerinoides ruber* (d'Orbigny) in deep-sea sediments. Dashed lines indicate hypothetical boundaries based on inadequate data; crosses indicate areas of greatest abundance. Sources indicated in the bibliography by *.

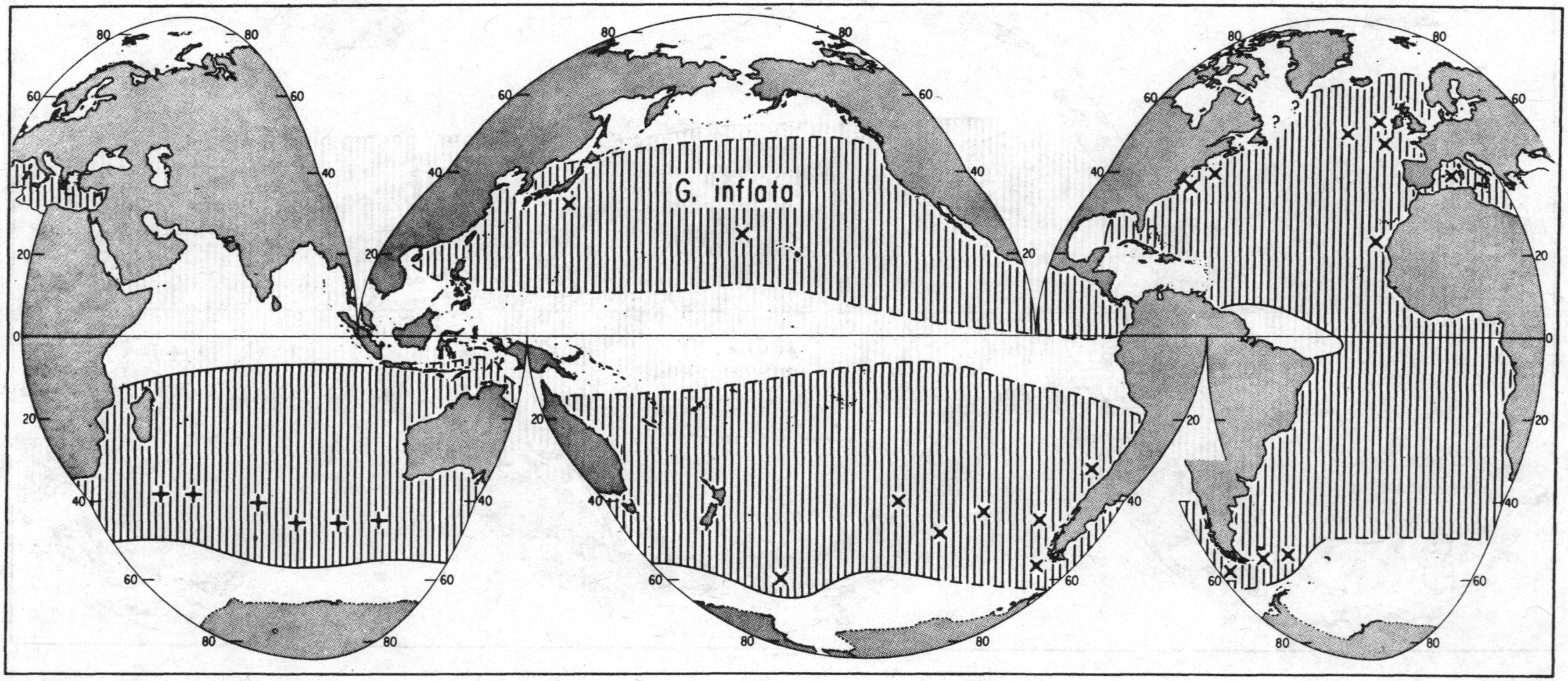
G. inflata

Fig. 20.5. Generalized distribution of *Globorotalia inflata* (d'Orbigny) in deep-sea sediments. Dashed lines indicate hypothetical boundaries based on inadequate data; crosses indicate areas of greatest abundance. Sources indicated in the bibliography by *.

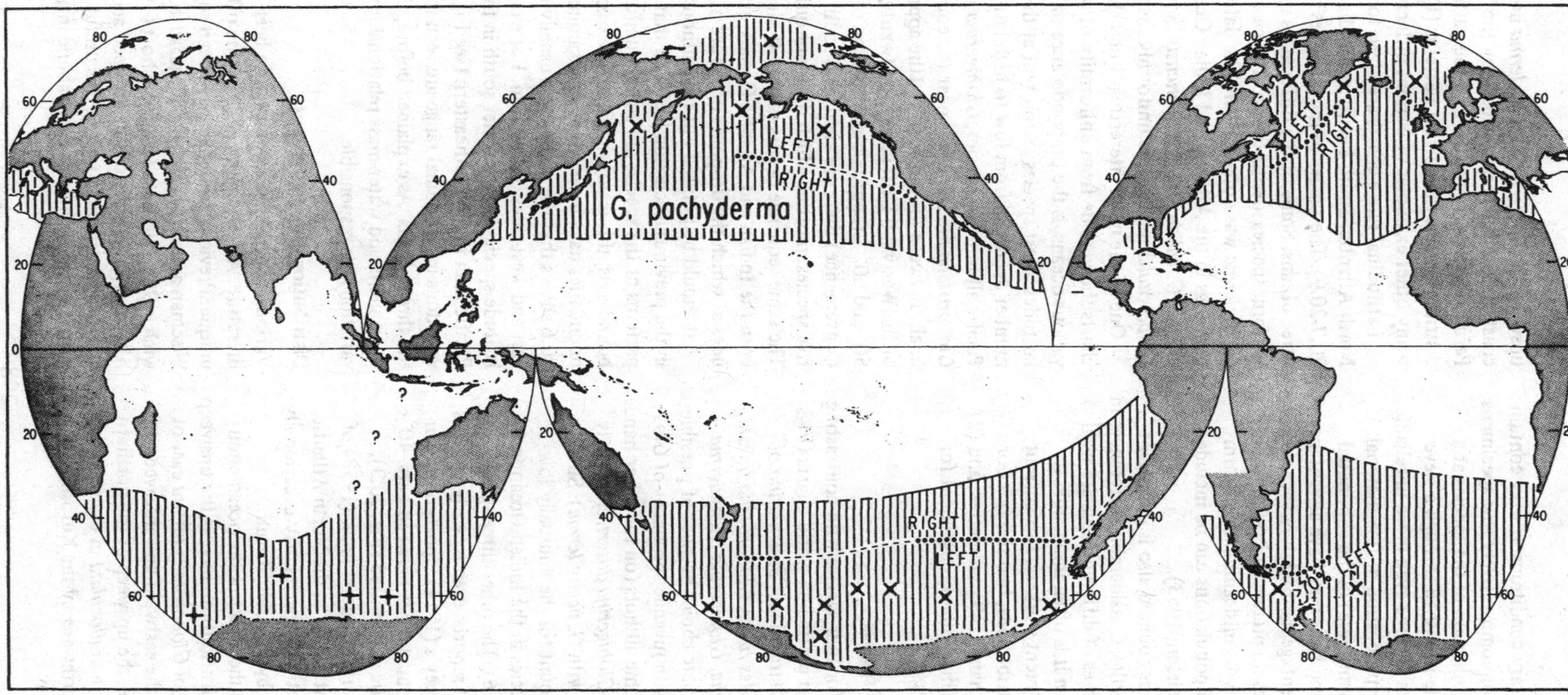
G. pachyderma
LEFT
RIGHT
LEFT
RIGHT
RIGHT
LEFT
30% LEFT
?
?
?

Fig. 20.6 Generalized distribution and coiling-direction provinces of *Globigerina pachyderma* (Ehrenberg) in deep-sea sediments. Dashed lines indicate hypothetical boundaries based on inadequate data; crosses indicate area of greatest abundance. Sources indicated in the bibliography by *.

depending on the sedimentary conditions. It may contain a pure foraminiferal ooze or comparatively few specimens, regardless of the rate of production of the species at a given locality. If the sample is washed through a sieve before weighing, the results become even more unreliable because of variation in the proportion of fine material (including planktonic Foraminifera) in the eliminated sediment. Either of these methods may have advantages in specific cases. The percentage method, however, seems preferable because it makes no pretence of establishing actual abundances that may be misleading. If uniform samples are used, the total populations can be included (see Phleger, Parker and Peirson, 1953).

The size of the specimens counted also is important in maintaining uniform results. Considerable differences in frequency may occur if sieves of different sizes are used for washing. For this reason it is often impossible to compare the results of different workers. In a current study of South Pacific distributions, two size fractions are being used: (1) larger than 0.149 millimetres, and (2) 0.062-0.149 millimetres. This method is excellent for evaluating the small species.

Latitudinal distribution of species

Generalized distributions for four species with contrasting latitudinal ranges have been plotted on world charts (Figs. 20.3-20.6). The species illustrated are: *Pulleniatina obliquiloculata, Globigerinoides ruber, Globorotalia inflata,* and *Globigerina pachyderma. Globigerina pachyderma* may have been an unfortunate choice because of possible confusion of this species with juvenile specimens of *Globoquadrina dutertrei* by some authors (on the other hand, however, many reports of *Globigerina dutertrei* by early authors are here included with *G. pachyderma*). Such misidentification may account for the unusually low-latitude reports of this species in the Indian Ocean (see question marks in Fig. 20.6). The other three species are easily identified, and most reports, even by early authors, are probably correct. Pearcey's (1914) reports of *Pulleniatina obliquiloculata* at four stations between Lats. 40°-51° S. in the South Atlantic are omitted (Fig. 20.3), perhaps wrongly. The report by Phleger *et al.* (1953) of this species from about Lat. 55° N in the North Atlantic is included, because low-latitude species may occasionally be carried into this region by the Gulf Stream.

It is interesting to note the effect of the ocean circulation on the patterns of distribution, such as the upsweep of the southern boundary of *Globorotalia inflata* east of South America following the eastward upward sweep of the Antarctic Convergence, the upsweep of the northern boundary of *Pulleniatina obliquiloculata* in the western Pacific, perhaps carried northward by the Kuroshio, the upsweep of *Globigerina pachyderma* northward along the coast of South America, caused by the Westwind Drift and Peru Current. A comparison of the patterns of species distribution and the current patterns (Fig. 20.2) will show many other examples of current influence.

Latitudinal range charts are given for species in the North Atlantic, South Pacific and Indian Oceans (Figs. 20.7-20.9). The ranges of most species are similar for the three oceans. Some of the differences may be due to non-uniform species identification. The case of *Globigerina pachyderma* was noted earlier. The data for *Globigerina bulloides* in the Atlantic and Indian Oceans probably include *G. calida* and *G. falconensis.* Some species of *Globorotalia* may not be uniformly identified.

One of the most interesting things that these range charts show, aside from uniformity of occurrence in the various oceans, is the preponderance of species in the low latitudes. Most species are present at the equator and the number decreases from low to high latitudes. In the South Pacific all but two species, *Globigerina pachyderma* and *Globorotalia inflata*, occur near the equator, thirty three in all. Between 40° and 50° S, (the approximate position of the Westwind Drift) there are twenty species; between 50° and 60° S (approximate position of the Subantarctic Convergence) there are eight; and south of 60° S, there are five species, most of which drop out quickly to leave one. The same sequence occurs in almost identical fashion in both the Indian and Atlantic Oceans, although the latitudes at which the dropouts occur are not exactly the same.

It should be pointed out that although the implication in the previous discussion has been that the distribution patterns of the species are strictly uniform along latitudinal boundaries, this is not the case. The maximal latitudinal boundaries have been given. An examination of Figs. 20.3-20.6 shows that the situation is usually different for the east and west sides of an ocean. For example, the low-latitude species range farther south in the southwestern Pacific than in the southeastern (see Fig. 20.3). Similar variations in latitudinal range are seen for most species in the three oceans. As explained before, the various water movements and water-mass relationships (sometimes depth variation) are responsible.

Distributional species groups

Various distributional groups have been suggested in the literature. All are similar and the one offered here is not unique. Previous groupings have been suggested by Wiseman and Ovey (1950), Phleger *et al.* (1953), Kustanowich (1963), Beliaeva (1966), Schott (1966), and others. Although the species can be grouped according to their maximum latitudinal ranges, it is useful to consider also the latitudes where they are most abundant. Table 20.1 gives a

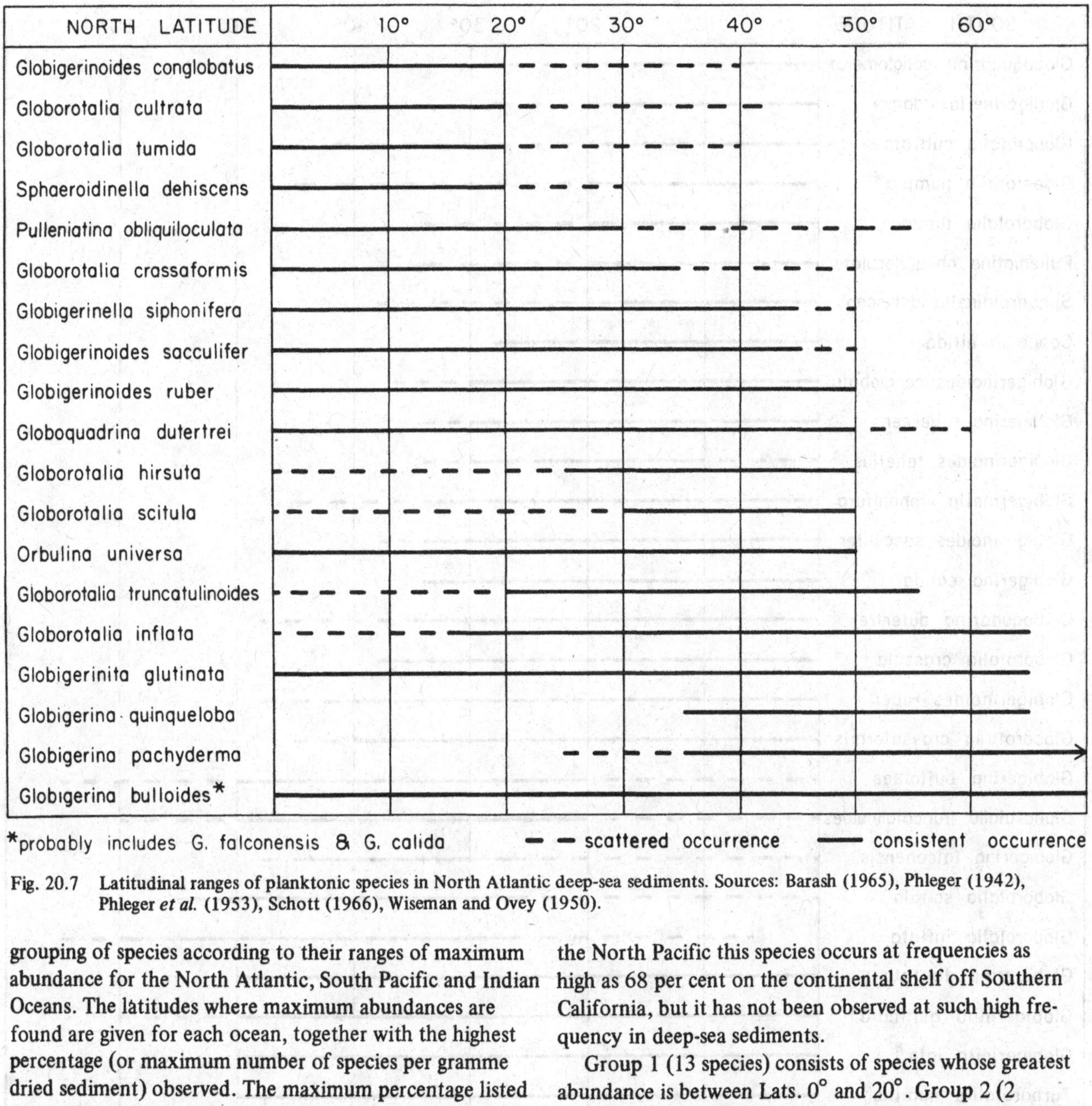
NORTH LATITUDE
10°
20°
30°
40°
50°
60°
Globigerinoides conglobatus
Globorotalia cultrata
Globorotalia tumida
Sphaeroidinella dehiscens
Pulleniatina obliquiloculata
Globorotalia crassaformis
Globigerinella siphonifera
Globigerinoides sacculifer
Globigerinoides ruber
Globoquadrina dutertrei
Globorotalia hirsuta
Globorotalia scitula
Orbulina universa
Globorotalia truncatulinoides
Globorotalia inflata
Globigerinita glutinata
Globigerina quinqueloba
Globigerina pachyderma
Globigerina bulloides*
*probably includes G. falconensis & G. calida
scattered occurrence
consistent occurrence

Fig. 20.7 Latitudinal ranges of planktonic species in North Atlantic deep-sea sediments. Sources: Barash (1965), Phleger (1942), Phleger *et al.* (1953), Schott (1966), Wiseman and Ovey (1950).

grouping of species according to their ranges of maximum abundance for the North Atlantic, South Pacific and Indian Oceans. The latitudes where maximum abundances are found are given for each ocean, together with the highest percentage (or maximum number of species per gramme dried sediment) observed. The maximum percentage listed does not necessarily persist throughout the maximum abundance range, but is given as a clue to relative abundances of the species. The high value (46 per cent) for *Pulleniatina obliquiloculata* in the South Pacific is undoubtedly from a sample that has undergone calcium carbonate solution, which is especially prevalent in Pacific deep-sea sediments. More normal figures would perhaps be from 2-6 per cent for this species in its maximum abundance area. The figure of 29 per cent in the Pacific for *Globoquadrina dutertrei* is also exceptionally high. This species occurs in the Pacific in greatest abundance at deep-sea stations fringing the continents. I cannot give any explanation for this. In the North Pacific this species occurs at frequencies as high as 68 per cent on the continental shelf off Southern California, but it has not been observed at such high frequency in deep-sea sediments.

Group 1 (13 species) consists of species whose greatest abundance is between Lats. 0° and 20°. Group 2 (2 species) is somewhat more widely ranging, between Lats. 0° and 35°. Group 3 (5 species) occurs most abundantly between Lats. 20° and 40°, Group 4 (5 species) between Lats. 20° and 50°, Group 5 (1 species) from Lats. 50° and 60° up. These latitudes vary to some extent between species and oceans and are generalized here; more accurate figures are shown in Table 20.1. Figs. 20.3-20.6 illustrate generalized distributions for species from four of these groups.

There are eight species listed at the end of Table 20.1 that do not fit any of these groups, because they are widely ranging and show no special areas of concentration.

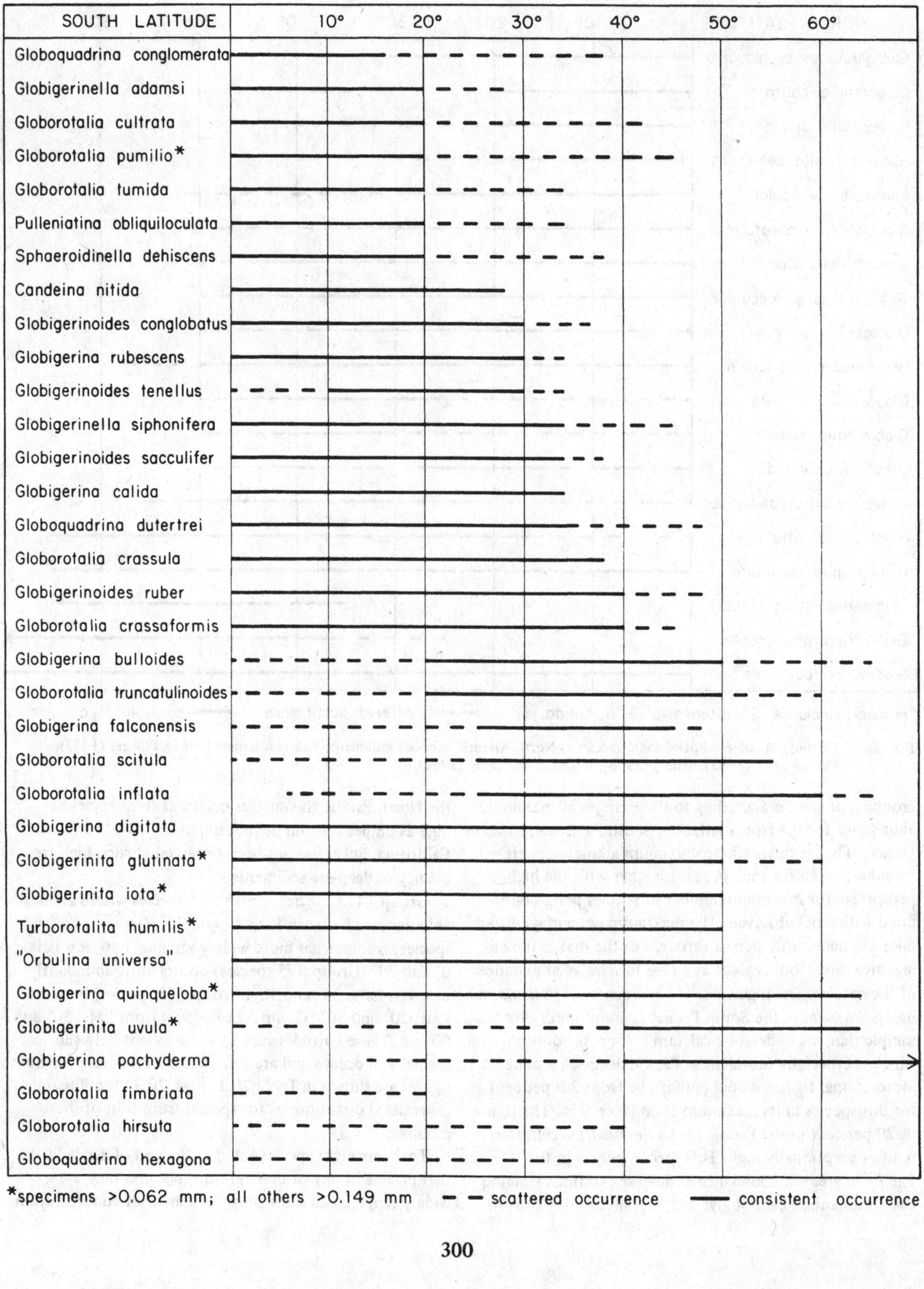
SOUTH LATITUDE
10°
20°
30°
40°
50°
60°
Globoquadrina conglomerata
Globigerinella adamsi
Globorotalia cultrata
Globorotalia pumilio*
Globorotalia tumida
Pulleniatina obliquiloculata
Sphaeroidinella dehiscens
Candeina nitida
Globigerinoides conglobatus
Globigerina rubescens
Globigerinoides tenellus
Globigerinella siphonifera
Globigerinoides sacculifer
Globigerina calida
Globoquadrina dutertrei
Globorotalia crassula
Globigerinoides ruber
Globorotalia crassaformis
Globigerina bulloides
Globorotalia truncatulinoides
Globigerina falconensis
Globorotalia scitula
Globorotalia inflata
Globigerina digitata
Globigerinita glutinata*
Globigerinita iota*
Turborotalita humilis*
"Orbulina universa"
Globigerina quinqueloba*
Globigerinita uvula*
Globigerina pachyderma
Globorotalia fimbriata
Globorotalia hirsuta
Globoquadrina hexagona

*specimens >0.062 mm; all others >0.149 mm — — scattered occurrence ——— consistent occurrence

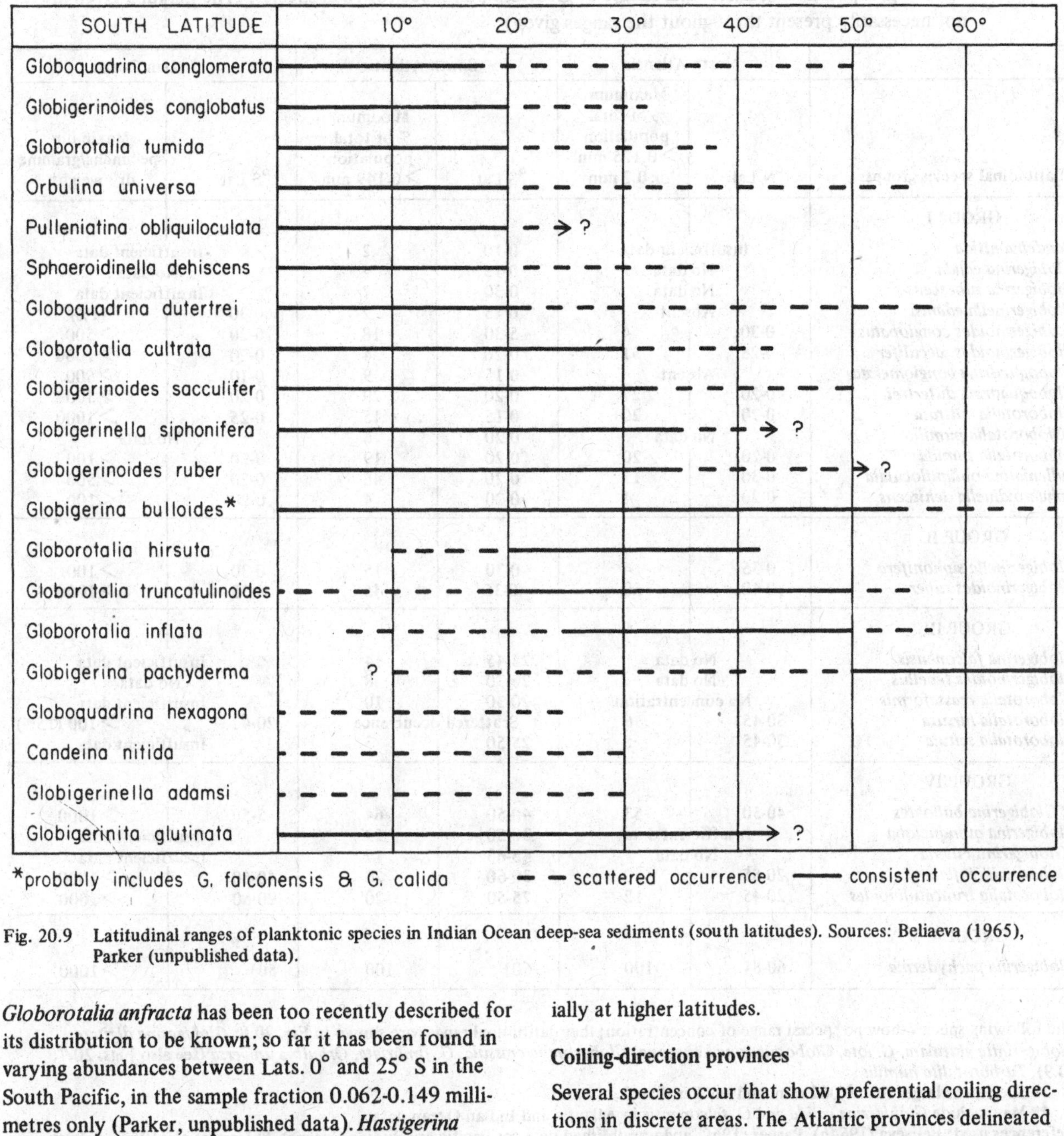
SOUTH LATITUDE
10°
20°
30°
40°
50°
60°
Globoquadrina conglomerata
Globigerinoides conglobatus
Globorotalia tumida
Orbulina universa
Pulleniatina obliquiloculata
?
Sphaeroidinella dehiscens
Globoquadrina dutertrei
Globorotalia cultrata
Globigerinoides sacculifer
Globigerinella siphonifera
?
Globigerinoides ruber
?
Globigerina bulloides*
Globorotalia hirsuta
Globorotalia truncatulinoides
Globorotalia inflata
Globigerina pachyderma
?
Globoquadrina hexagona
Candeina nitida
Globigerinella adamsi
Globigerinita glutinata
?
*probably includes G. falconensis & G. calida
scattered occurrence
consistent occurrence

Fig. 20.9 Latitudinal ranges of planktonic species in Indian Ocean deep-sea sediments (south latitudes). Sources: Beliaeva (1965), Parker (unpublished data).

Globorotalia anfracta has been too recently described for its distribution to be known; so far it has been found in varying abundances between Lats. 0° and 25° S in the South Pacific, in the sample fraction 0.062-0.149 millimetres only (Parker, unpublished data). *Hastigerina pelagica* occurs rarely in the sediments at low latitudes (Lats. 0°-30° S) in the South Pacific (Parker 1962) and in the tropical Indian Ocean (Beliaeva 1964). It is more widely ranging in the plankton, occurring at least as far south as Lat. 47° S in the Pacific (Parker 1960) and as far north as about Lat. 52° N in the Atlantic (Bé and Hamlin 1967). It is apparently fragile and easily dissolved, especially at higher latitudes.

Coiling-direction provinces

Several species occur that show preferential coiling directions in discrete areas. The Atlantic provinces delineated by Ericson, Wollin and Wollin (1954) for *Globorotalia truncatulinoides* are illustrated (Fig. 20.10), as well as those for the Pacific and Indian Oceans (Parker 1962, and unpublished data). Coiling ratios are based on counts of one hundred specimens, where possible. *G. truncatulinoides* occurs in left- and right-coiling provinces in the North Atlantic: a right-coiling one in the equatorial region

Fig. 20.8 Latitudinal ranges of planktonic species in South Pacific deep-sea sediments. Sources: Blair [1965], Kustanowich (1963), Parker (1962, and unpublished data). (Since going to press it has been established that *Globigerina bulloides* is not found north of *ca.* Latitude 20° S.)

Table 20.1 Species groups showing latitudinal ranges of greatest abundance for each species. (The maxima given are not necessarily present throughout the ranges given)

	North Atlantic		South Atlantic		Indian Ocean	
Latitudinal species groups	°N Lat.	Maximum % of total population >0.125 mm *or* 0.2 mm	°S Lat.	Maximum % of total population >0.149 mm	°S Lat.	Maximum specimens/gramme dry weight
GROUP I						
Candeina nitida	Insufficient data		0-10	2	Insufficient data	
Globigerina calida	No data		0-15	7	No data	
Globigerina rubescens	No data		0-30	7	Insufficient data	
Globigerinella adamsi	Absent		0.15	2	0-30	(1%)
Globigerinoides conglobatus	0-30	4	5-30	18	0-20	>500
Globigerinoides sacculifer	0-25	51	0-20	34	0-20	>1000
Globoquadrina conglomerata	Absent		0-15	9	0-10	>500
Globoquadrina dutertrei	0-20	27	0-20	29	0-20	>500
Globorotalia cultrata	0-20	20+	0-15	15	0-25	>1000
**Globorotalia pumilio*	No data		0-20	6	No data	
Globorotalia tumida	0-20	20	0-20	19	0-20	>100
Pulleniatina obliquiloculata	0-30	11	0-20	46	0-20	>500
Sphaeroidinella dehiscens	0-20	4	0-20	4	0-20	<100
GROUP II						
Globigerinella siphonifera	0-35	4	0-30	15	0-30	>100
Globigerinoides ruber	0-40	50	0-35	45	5-35	>1000
GROUP III						
Globigerina falconensis	No data		25-45	14	Insufficient data	
Globigerinoides tenellus	No data		20-30	8	No data	
Globorotalia crassaformis	No concentration		20-30	10	Insufficient data	
Globorotalia hirsuta	30-45	6	Scattered occurrence		20-42	>100 (15%)
Globorotalia scitula	30-45	5	25-50	3	Insufficient data	
GROUP IV						
***Globigerina bulloides*	40-50	53	40-50	64	5-50	>1000
Globigerina quinqueloba	Insufficient data		30-50	19	Insufficient data	
**Globigerinita uvula*	No data		45-63	17	Insufficient data	
Globorotalia inflata	20-55	30	30-60	20	40-50	>1000
Globorotalia truncatulinoides	20-45	12	25-50	20	20-50	>1000
GROUP V						
Globigerina pachyderma	60-84	100	60+	100	50+	>1000

The following species show no special range of concentration; their latitudinal ranges are shown in Fig. 20.8: *Globigerina digitata, Globigerinita glutinata, G. iota, Globoquadrine hexagona, Globorotalia crassula, G. fimbriata, Orbulina universa* (see also Figs. 20.7, 20.9), *Turborotalita humilis.*

* Specimens larger than 0.062 mm. and smaller than 0.149 mm.

** May include *Globigerina calida* and *G. falconensis* in Atlantic and Indian Ocean data.

References used: Beliaeva (1964*b*), Parker (1962, and unpublished data for Pacific and Indian Oceans), Phleger *et al.* (1953), Schott (1966).

and Gulf of Mexico, a left-coiling one in the mid-latitude region and Mediterranean Sea extending northward in the western Atlantic, and a right-coiling one northward from the entrance to the Mediterranean Sea in the eastern Atlantic. In the South Pacific the species appears to be left-coiling northward to about Lat. 40° S in the western region and 30° S in the eastern region; in the rest of the South Pacific and throughout the North Pacific it is apparently right coiling. In the Indian Ocean the species is right coiling in the northern part and left coiling south of about Lat. 30° S, as far as the rather limited data collected so far show.

Globorotalia crassaformis, G. crassula and *G. hirsuta* appear to have coiling provinces also, but the data are insufficient to delineate them accurately. For the first species the central and western South Pacific appear to form a left-

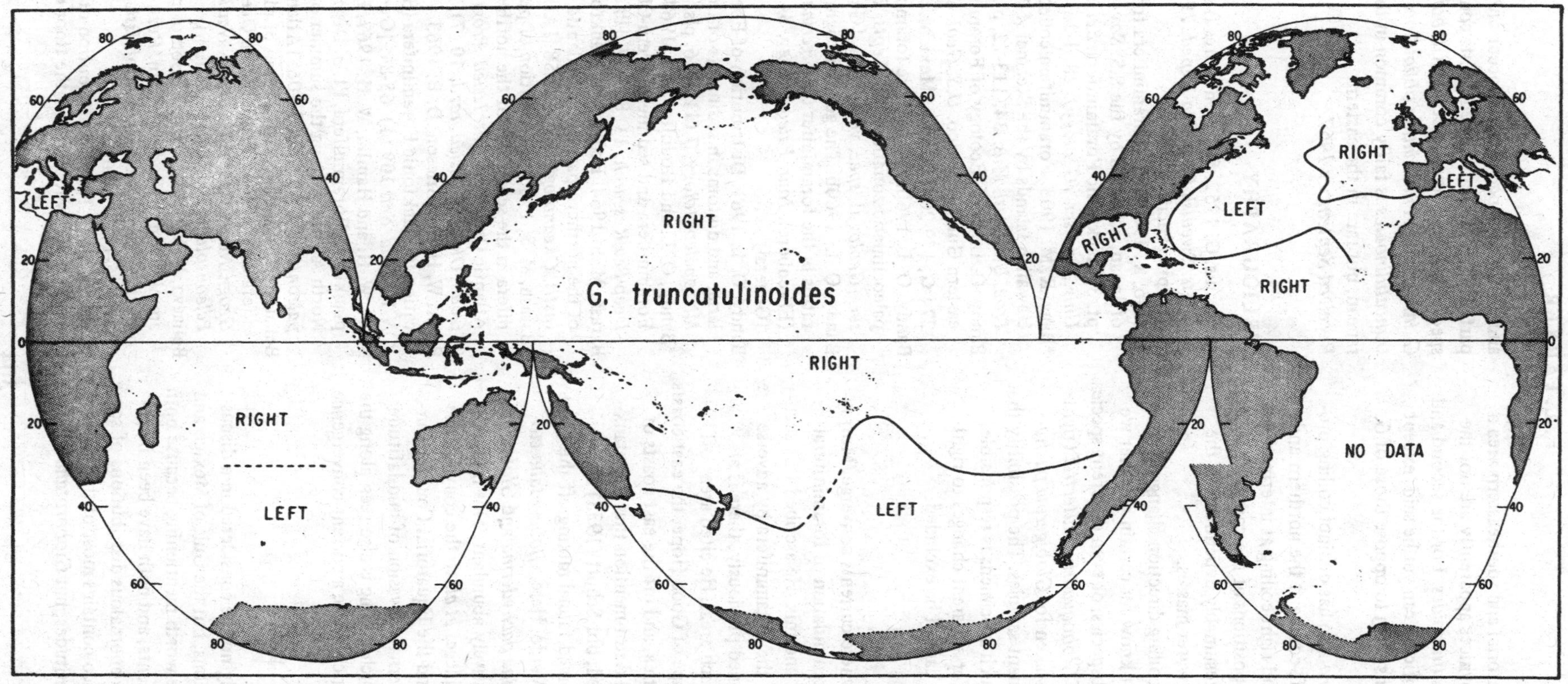
G. truncatulinoides
RIGHT
LEFT
RIGHT
LEFT
RIGHT
RIGHT
LEFT
RIGHT
LEFT
RIGHT
LEFT
RIGHT
NO DATA
80
60
40
20
0

Fig. 20.10. Coiling-direction provinces of *Globorotalia truncatulinoides* (d'Orbigny) in deep-sea sediments. Sources: *Atlantic Ocean,* Ericson, Wollin and Wollin (1954); *Pacific and Indian Oceans,* Parker (1962, and unpublished data).

coiling province, the equatorial and southeastern area a right-coiling one. The provinces apparently are not the same as those of *G. truncatulinoides.* For the second and third species nothing significant can yet be said, except that their coiling directions seem to oppose those of *G. crassaformis.*

Globigerina pachyderma also has distinct coiling provinces (Fig. 20.6). It is left-coiling at the northern and southern high latitudes and right-coiling at intermediate latitudes. The latitudinal positions of the coiling changes vary from east to west, presumably dependent on the positions of currents and water masses.

The reasons for these coiling-direction changes are not known. In fact, we do not know for certain whether we are dealing with races, subspecies, or even different species. For example, juveniles of *Globoquadrina dutertrei* (right-coiling) may be easily mistaken for *Globigerina pachyderma,* especially in sediment samples. The possibility that the misidentification of such specimens is responsible, either wholly or in part, for apparent changes to right-coiling of *G. pachyderma* cannot be excluded.

Mixed faunas

Mixed faunas are found where currents converge. Murray (1897) noted the rapid accumulation of foraminiferal tests where currents meet and mix. He ascribed this accumulation to the killing of the Foraminifera by adverse conditions but, as mentioned previously, there is also increased production in such areas. He cited areas off Agulhas Bank near the Cape of Good Hope, the east coasts of North and South America, and off the east coasts of Australia and Japan. He did not mention that the planktonic faunas are also mixed, but Schott (1935) and Phleger *et al.* (1953) observed faunal mixing off the coast of Africa, south of Cape Verde. Here, *Globorotalia cultrata, G. inflata, Globigerina pachyderma* and *Globoquadrina dutertrei* are all relatively abundant. This faunal mixing was ascribed by Phleger *et al.* to the convergence of the Canaries Current and the Equatorial Counter Current, which causes the alternate invasion of mid-latitude and low-latitude water. Such mixing is also seen along the west coast of South America where a current convergence also occurs.

Small seas

Small seas should probably not be considered in a discussion of deep-sea distributions, but the Gulf of Mexico and the Mediterranean Sea are worth mentioning because both contain deep-water sediments, and both have been intensively studied. They show anomalous distributions of some species. The Gulf of Mexico contains most of the species to be expected at those latitudes, but *Globorotalia inflata* and *Globigerina pachyderma* occur only in the eastern part. The Mediterranean sediments contain many of the species to be expected, but lack *Globorotalia crassaformis, G. hirsuta,* and *Pulleniatina obliquiloculata; Globorotalia truncatulinoides* is fairly common in the western Mediterranean, but rare in the eastern.

Received September 1967.

BIBLIOGRAPHY

Arrhenius, G. 1952. Sediment cores from the east Pacific. *Repts. Swedish Deep-Sea Exped. 1947-1948.* **5** (1-4), 228 pp. [Pacific]

*Asano, K. 1957. The Foraminifera from the adjacent seas of Japan, collected by the S.S. *Soyo-Maru,* 1922-1930, pt. 3, Planktonic Foraminifera. *Sci. Repts. Tohoku Univ., 2nd ser. (Geology).* **28**, 1-26, pls. 1, 2. [Pacific]

*Bagg, R. M. 1908. Foraminifera collected near the Hawaiian Islands by the Steamer *Albatross* in 1902. *Proc. U.S. Natl. Mus.* **34**, 113-72, pl. 5. [Pacific]

Bandy, O. L. 1956. Ecology of Foraminifera in northeastern Gulf of Mexico. *U.S. Geol. Surv., Prof. Paper* **274-G**, 179-204. [Gulf of Mexico]

Bandy, O. L. 1960a. Planktonic foraminiferal criteria for paleoclimatic zonation. *Sci. Repts. Tohoku Univ., 2nd ser. (Geology), spec. vol.* **4**, 1-8. [General]

Bandy, O. L. 1960b. The geologic significance of coiling ratios in the foraminifer *Globigerina pachyderma* (Ehrenberg). *Jour. Paleontology,* **34** (4), 671-81. [General]

Bandy, O. L. 1961. Distribution of Foraminifera, Radiolaria, and diatoms in sediments of the Gulf of California. *Micropaleontology,* **7** (1), 1-26, pls. 1-5. [Pacific]

*Bandy, O. L. and Rodolfo, K. S. 1964. Distribution of Foraminifera and sediments. Peru-Chile Trench area. *Deep-Sea Research,* **11**, 817-37. [Pacific]

Barash, M. S. 1964. Ecology of planktonic Foraminifera of the northern part of the Atlantic Ocean. *Trudy Instit. Okeanologii,* **65**, 229-58. [Atlantic]

*Barash, M. S. 1965. Distribution of planktonic Foraminifera in the sediments of the northern part of the Atlantic Ocean. *Rezult. Issled. Progr. Mezhd. Geofiz. Goda, Okeanol. Issled. razd,* **10** (7), 225-35. [Atlantic]

Bé, A. W. H. and Ericson, D. B. 1963. Aspects of calcification in planktonic Foraminifera (Sarcodina). *Ann. N.Y. Acad. Sci.* **109** (1), 65-81. [General]

Bé, A. W. H. and Hamlin, W. H. 1967. Ecology of Recent planktonic Foraminifera, Pt. 3. Distribution in the North Atlantic during the summer of 1962. *Micropaleontology,* **13** (1), 87-106. [Atlantic]

Beliaeva, N. V. 1963. Distribution of planktonic Foraminifera on the floor of the Indian Ocean. *Otdel. Geol.-Geogr. Nauk, Geol. Inst. Akad, Nauk SSSR, Voprosy Paleontologii,* **7**, 209-22. [Indian]

Beliaeva, N. V. 1964a. Distribution of planktonic Foraminifera in the Indian Ocean. *Rezult. Issled. Progr. Mezhd. Geofiz. Goda, Okeanol. Issled. Sbornik Stat.* **13**, 205-11. [Indian]

*Beliaeva, N. V. 1964b. Distribution of planktonic Foraminifera in the water and on the floor of the Indian Ocean.

Trudy Okeanologii, Akad. Nauk SSSR, **68**, 12-83, pls. 1-3. [Indian]

Beliaeva, N. W. 1966. Climatic and vertical zonality in the distribution of planktonic Foraminifera in the sediments of the Pacific Ocean. *2nd Internatl. Ocean. Cong. Abstracts of Papers*, 35, 36. [Pacific]

Beliaeva, N. V. and Saidova, Kh. M. 1965. Correlation of benthonic and planktonic Foraminifera in the surface layers of the sediments of the Pacific Ocean. *Akad. Nauk SSSR, Okeanologiya*, **5** (6), 1010-14. [Pacific]

Berger, W. H. 1967. Foraminiferal ooze: solution at depths. *Science*, **156** (3773), 383-5. [General]

Berger, W. H. and Soutar, A. 1967. Planktonic Foraminifera: field experiment on production rate. *Science*, **156** (3781), 1495-7. [Pacific]

*Bermúdez, P. J., 1961. Contribución al estudio de las Globigerinidea de la región Caribe-Antilla (Palaeoceno-Reciente). *Bol. Geología, Pub. Expec. no. 3; Mem. Terc. Cong. Geol. Venezolano*, **3**, 1119-393, pls. 1-20. [Caribbean Sea]

Blackman, A. [1966] *Pleistocene stratigraphy of cores from the southeast Pacific Ocean.* Ph.D. thesis, Univ. of California, San Diego, 200 pp. [Pacific]

*Blackman, A. and Somayajulu, B. L. K. 1966. Pacific Pleistocene: faunal analyses and geochronology. *Science*, **154** (3751), 886-9. [Pacific]

*Blair, D. G. [1965]. The distribution of planktonic Foraminifera in deep-sea cores from the Southern Ocean. Antarctica. *M.S. thesis, Fla. State Univ.* 141 pp. [Antarctic]

*Boltovskoy, E. 1961. Foraminíferos de la plataforma continental entre el Cabo Santo Tome y la desembocadura del Río de la Plata. *Rev. Mus. Argentino Cienc. Nat. 'Bernardino Rivadavia', Cienc, Zool.* **6** (6), 249-346, pls. 1-11. [Atlantic]

Bradshaw, J. S. 1959. Ecology of living planktonic Foraminifera in the North and equatorial Pacific Ocean. *Contr. Cushman Found. Foram. Research*, **10** (2), 25-64, pls. 6-8.

*Brady, H. B. 1884. Report on the Foraminifera dredged by H. M. S. *Challenger*, during the years 1873-1876. *Rept. Voy. 'Challenger', Zoology*, **9**, i-xxi, 814 pp, pls. 1-115. [Atlantic, Pacific, Indian]

Bramlette, M. N. 1961. Pelagic sediments. In *Oceanography*, (Mary Sears, *ed.*), Am. Assoc. Adv. Science, Washington, D.C., pp. 345-66.

Chapman, F. 1895. On some Foraminifera obtained by the Royal Indian Marine Survey's S. S. *Investigator* from the Arabian Sea, near the Laccadive Islands. *Proc. Zool. Soc. London, Jan. 15*, 55 pp., pl. 1. [Indian]

Chierici, M. A., Busi, M. T. and Cita, M. B. 1962. Contribution à une étude écologique des foraminifères dans la Mer Adriatique. *Rev. Micropaléontologie*, **5** (2), 123-42. [Mediterranean Sea]

Cita, M. B. and Chierici, M. A. 1962. Crociera talassografica Adriaica 1955. V. Ricerche sui foraminiferi contenuti in 18 carote prelevate sul fondo del Mare Adriaico. *Arch. Oceanografia e Limnologia*, **12** (3), 297-359, pls. 1-8. [Mediterranean Sea]

Colom, G. 1950. Estudio do los foraminíferos de muestras de fondo recogidas entre los Cabos Juby y Bojador. *Bol. Inst. Español de Oceanografia*, **28**, 45 pp. pls. 1-10. [Atlantic]

Colom, G. 1952. Foraminíferos de la costas de Galicia. *Bol. Inst. Español de Oceanografia*, **51**, 58 pp. pls. 1-8. [Atlantic]

*Cushman, J. A. 1914. A monograph of the Foraminifera of the North Pacific Ocean. Pt. 4. Chilostomellidae, Globigerinidae, Nummulitidae. *U.S. Natl. Mus., Bull.* **71** (4), 46 pp. pls. 1-19. [Pacific]

Cushman, J. A. 1915. A monograph of the Foraminifera of the North Pacific Ocean. Pt. 5. Rotaliidae. *U.S. Natl. Mus., Bull.* **71** (5), 87 pp. pls. 1-31. [Pacific]

Cushman, J. A. 1921. Foraminifera of the Philippine and adjacent seas. *U.S. Natl. Mus., Bull.* **100** (4), 608 pp. pls. 1-100. [Pacific]

*Cushman, J. A. 1924. The Foraminifera of the Atlantic Ocean. Pt. 5, Chilostomellidae and Globigerinidae. *U.S. Natl. Mus., Bull.* **104** (5), 55 pp. pls. 1-8. [Atlantic]

Cushman, J. A. 1931. The Foraminifera of the Atlantic Ocean. Pt. 8. Rotaliidae, Amphisteginidae, Calcarinidae, Cymbaloporettidae, Globorotaliidae, Anomalinidae, Planorbulinidae, Rupertiidae and Homotremidae. *U.S. Natl. Mus., Bull.* **104** (8), 179 pp. pls. 1-26. [Atlantic]

Cushman, J. A. 1941. A study of the Foraminifera contained in cores from the Bartlett Deep. *Am. Jour. Sci.* **239**, 128-47, pls. 1-6. [Caribbean Sea]

*Cushman, J. A. and Henbest, L. G. 1940. Geology and biology of North Atlantic deep-sea cores between Newfoundland and Ireland. Pt. 2., Foraminifera. *U.S. Geol. Surv., Prof. Paper* **196-A**, 35-50. [Atlantic]

Cushman, J. A., Todd, R. and Post, R. J. 1954. Recent Foraminifera of the Marshall Islands. *U.S. Geol. Surv., Prof. Paper* **260-H**, 319-84, pls. 82-93. [Pacific]

Di Napoli Alliata, E. 1952. Foraminiferi pelagici e facies in Italia. *'Atti' del VII Conv. Naz. Metano e Petrolio*, 1-34, pls. 1-5. [Mediterranean Sea]

Earland, A. 1933. Foraminifera, Part II. South Georgia. *Discovery Repts.* **7**, 27-138, pls. 1-7. [Antarctic]

*Earland, A. 1934. Foraminifera, Pt. III. The Falklands sector of the Antarctic (excluding South Georgie). *Discovery Repts.* **10**, 208 pp. pls. 1-10. [Antarctic]

Earland, A. 1936. Foraminifera, Pt. 4, Additional records from the Weddell Sea sector from material obtained by the S.Y. *Scotia. Discovery Repts.* **13**, 76 pp. pls. 1, 2, 2a. [Antarctic]

Egger, J. G. 1893. Foraminiferen aus Meeresgrundproben, gelothet von 1874 bis 1876 von. S. M. Sch. Gazelle. *Abhandl. k. bayer, Akad. Wiss. München, II, Cl. 18*, 2 (2), 195-458, pls. 1-21. [Atlantic, Pacific, Indian]

*Ericson, D. B. 1959. Coiling direction of *Globigerina pachyderma* as a climatic index. *Science*, **130** (3369), 219-20. [Atlantic, Arctic]

*Ericson, D. B., Ewing, M. and Wollin, G. 1964. Sediment cores from the arctic and subarctic seas. *Science*, **144** (3623), 1183-92. [Arctic]

Ericson, D. B., Ewing, M., Wollin, G. and Heezon, B. C. 1961. Atlantic deep-sea sediment cores. *Bull. Geol. Soc. America*, 72, 193-286. [Atlantic]

*Ericson, D. B. and Wollin, G. 1959. Micropaleontology and lithology of arctic sediment cores. *Geophys. Research Papers no. 63, 'Scientific Studies at Fletcher's Ice Island, T-3, 1952-1955'*, **1**, 51-8. [Arctic]

*Ericson, D. B., Wollin, G. and Wollin, J. 1954. Coiling

direction of *Globorotalia truncatulinoides* in deep-sea cores. *Deep-Sea Research,* 2, 152-8. [Atlantic]

Fierro, G. and Ceretti, P. 1965. I foraminiferi planctonici in alcuni campioni di fondo del mare Tirreno. *Atti Acad. Ligure Scienze e Lettere,* 22, 3-19, pls. 1, 2. [Mediterranean Sea]

Flint, J. M. 1899. Recent Foraminifera. *Ann. Rept. U.S. Natl. Mus., 1897,* 249-349, pls. 1-80. [Atlantic, Gulf of Mexico, Caribbean Sea, Pacific]

Grimsdale, T. F. and van Morkhoven, F. P. C. M. 1955. The ratio between pelagic and benthonic Foraminifera as a means of estimating depth of deposition of sedimentary rocks. *Proc. 4th World Petroleum Cong., Sec. I/D* (4), 473-91. [Gulf of Mexico]

Hamilton, E.L. 1953. Upper Cretaceous, Tertiary and Recent planktonic Foraminifera from Mid-Pacific flat-topped seamounts. *Jour. Paleontology,* **27** (2), 204-37, pls. 29-32. [Pacific]

Heron-Allen, Edward and Earland, Arthur 1922. Protozoa, Pt. 2. Foraminifera. *British Antarctic ('Terra Nova') Exped., 1910, Zoology,* **6** (2), 25-268, pls. 1-8. [Antarctic]

Hofker, Jan 1956. Foraminifera Dentata. Foraminifera of Santa Cruz and Thatch-Island, Virginia-Archipelago, West-Indies. *Spolia Zool. Mus. Haun.* **15**, 9-237, 35 pls. [Caribbean Sea, Pacific]

Kane, Julian 1953. Temperature correlations of planktonic foraminifera from the North Atlantic Ocean. *The Micropaleontologist,* 7 (3), 25-50. [Atlantic]

Kennett, J. P. 1966. Foraminiferal evidence of a shallow calcium carbonate solution boundary, Ross Sea, Antarctica. *Science,* **153** (3732), 191-3. [Antarctic]

Kustanowich, S. 1962. A foraminiferal fauna from Capricorn Seamount, south-west equatorial Pacific. *N.Z. Jour. Geology and Geophysics,* 5 (3), 423-34. [Pacific]

*Kustanowich, S. 1963. Distribution of planktonic Foraminifera in surface sediments of the south-west Pacific Ocean. *N.Z. Jour. Geology and Geophysics,* **6** (4), 534-65, pls. 1-3. [Pacific]

*Lipps, J. H. and Warme, J. E. 1966. Planktonic foraminiferal biofacies in the Okhotsk Sea. *Contr. Cushman Found. Foram. Research,* **17** (4), 125-34, pls. 9, 10. [Okhotsk Sea]

McKnight, W. M., Jr. 1962. The distribution of Foraminifera off parts of the Antarctic coast. *Bull. Am. Paleontology,* **44** (201), 65-158, pls. 9-23. [Antarctic]

Murray, J. 1885. No. 2. Reports on the results of dredging, under the supervision of Alexander Agassiz, in the Gulf of Mexico (1877-78), in the Caribbean (1878-79), and along the Atlantic Coast of the United States during the summer of 1880, by the U.S. Coast Survey Steamer *Blake,* Lieutenant-Commander C. D. Sigsbee, U.S.N., and Commander J. R. Bartlett, U.S.N., commanding-XXVII. Report on the specimens of bottom deposits. *Bull. Mus. Comp. Zool.* **12** (2), 37-61. [Atlantic, Gulf of Mexico, Caribbean Sea]

Murray, J. 1889. On marine deposits in the Indian, Southern and Antarctic Oceans. *Scottish Geogr. Mag.,* August, 1-32. [Indian, Antarctic]

Murray, J. 1897. On the distribution of the pelagic Foraminifera at the surface and on the floor of the ocean. *Nat. Science,* **11**, 17-27. [General]

Murray, J. 1902. Deep-sea deposits and their distribution in the Pacific Ocean. With notes on the samples collected by S.S. *Britannia,* 1901. *Geogr. Jour.,* June, 21 pp. [Pacific]

Murray, J. and Chumley, J. 1924. *The deep-sea deposits of the Atlantic Ocean.* Robert Grant and Son: Edinburgh, 252 pp. [Atlantic]

Murray, J. and Peake, R. E. 1904. On recent contributions to our knowledge of the floor of the North Atlantic Ocean. *Royal Geogr. Soc., Spec. Pub.* 35 pp. [Atlantic]

Murray, J. and Philippi, E. 1908. Die Grundproben der 'Deutschen Tiefsee-Expedition'. *Wiss, Ergebn, Deutschen Tiefsee-Exped. 'Valdivia' 1898-1899,* **10**, 77-206, pls. 16-22. [Atlantic]

Murray, J. and Renard, A. F. 1891. Deep-sea deposits based on the specimens collected during the voyage of H.M.S. *Challenger* in the years 1872 to 1876. *Repts. Voy. 'Challenger',* Longmans: London, 525 pp., 29 pls. [Atlantic, Pacific, Indian]

Norton, R. D. 1930. Ecologic relations of some Foraminifera. *Bull. Scripps Inst. Oceanography, Tech. Ser.* **2** (9), 331-88. [Atlantic]

*Parker, F. L. 1954. Distribution of the Foraminifera in the northeastern Gulf of Mexico. *Bull. Mus. Comp. Zoology,* **111** (1). 427-73, pls. 1-13. [Gulf of Mexico]

*Parker, F. L. 1955. Distribution of planktonic Foraminifera in some Mediterranean sediments. *Papers on Marine Biology & Oceanography: Deep-Sea Research, Supp. to vol. 3,* 204-11. [Mediterranean Sea]

Parker, F. L. 1958. Eastern Mediterranean Foraminifera. *Repts. Swedish Deep-Sea Exped.* **8** (4), 219-83, pls. 1-6. [Mediterranean Sea]

Parker, F. L. 1960. Living planktonic Foraminifera from the equatorial and southeast Pacific. *Sci. Repts. Tohoku Univ. 2nd ser. (Geology), spec. vol.* **4**, 71-82. [Pacific]

*Parker, F. L. 1962. Planktonic foraminiferal species in Pacific sediments. *Micropaleontology,* **8** (2), 219-54, pls. 1-10. [Pacific]

Parker, F. L. 1965. Irregular distributions of planktonic Foraminifera and stratigraphic correlation. *'Progress in Oceanography',* **3**, 267-72. [General]

Parker, F. L. 1967. Tertiary biostratigraphy (planktonic Foraminifera) of tropical Indo-Pacific deep-sea cores. *Bull. Am. Paleontology,* **52** (235), 115-208, pls. 17-32.

*Parr, W. J. 1950. Foraminifera. *B.A.N.Z. Antarct. Research Exped. 1929-1931, Repts., ser. B. (Zoology and Botany),* **5** (6), 233-392, pls. 3-15. [Antarctic]

Peake, R. E. 1901. *On the results of a deep-sea sounding expedition in the North Atlantic during the summer 1899. With notes on the temperature observations and depths and a description of the deep-sea deposits in this area, by Sir John Murray.* William Clowes and Sons: London, 44 pp. [Atlantic]

Pearcey, F. G. 1914. Foraminifera of the Scottish National Antarctic Expedition. *Trans. Roy. Soc. Edinburgh,* **49** (4), no. 19, 991-1044. pls. 1, 2. [Antarctic]

*Philippi, E. 1910. Die Grundproben der Deutschen Südpolar-Expedition 1901-1903. *Deutsche Südpolar-Exped. 1901-1903.* 2 (6), 415-616. [Atlantic, Indian]

Phleger, F. B., Jr. 1942. Foraminifera of submarine cores from the continental slope pt. 2. *Bull. Geol. Soc. America,* **53**, 1073-98, pls. 1-3. [Atlantic]

Phleger, F. B., Jr. 1947. Foraminifera of three submarine cores from the Tyrrhenian Sea. *Göteborgs Kungl. Vetensk.- och Vitterh.-Samhälles Handl, Sjätte Föl;* ser. B, **5** (5), 3-19. [Mediterranean Sea]

Phleger, F. B. 1951. Ecology of Foraminifera, northwest Gulf of Mexico, Pt. 1, Foraminifera distribution. *Geol. Soc. America, Mem.* **46**, 88 pp. [Gulf of Mexico]

Phleger, F. B. 1954. Foraminifera and deep-sea research. *Deep-sea Research,* 2, 1-23. [Atlantic, Gulf of Mexico]

Phleger, F. B., Jr. and Hamilton, W. A. 1946. Foraminifera of two submarine cores from the North Atlantic basin. *Bull. Geol. Soc. America,* **57**, 951-66. [Atlantic]

Phleger, F. B. and Parker, F. L. 1951. Ecology of Foraminifera, northwest Gulf of Mexico., Pt. 2. Foraminifera species. *Geol. Soc. America, Mem.* **46**, 64 pp. pls. 1-20. [Gulf of Mexico]

*Phleger, F. B., Parker, F. L. and Peirson, J. F. 1953. North Atlantic Foraminifera. *Repts. Swedish Deep-Sea Exped.* **7** (1), 122 pp., pls. 1-12. [Atlantic]

Riedel, W. R. and Funnell, B. M. 1964. Tertiary sediment cores and microfossils from the Pacific Ocean floor. *Quart. Jour. Geol. Soc. London,* **120**, 305-68, pls. 14-32. [Pacific]

*Rosenberg-Herman, Yvonne, 1965. Étude des sédiments Quaternaires de la Mer Rouge. *Ann. Inst. Océanographique,* **42** (3), 339-415, pls. 1-12. [Red Sea]

Ruscelli, Maria. 1949. Foraminiferi di due saggi di fondo del Mar Ligure, *Atti Accad. Ligure, Scienze e Lettere,* **6** (1), 31pp. pls. 1, 2. [Mediterranean Sea]

*Said, R. 1950. Additional Foraminifera from the northern Red Sea. *Contr. Cushman Found. Foram. Research,* **1** (1, 2), 4-9, pl. 1. [Red Sea]

Schott, W. 1934. Die jüngste Vergangenheit des äquatorialen Atlantischen Ozeans auf Grund von Untersuchungen an Bodenproben der *Meteor*-Expedition. *Sitz. Abh. Nat. Gesell. Rostock, Dritte Folge.* **4**, 11 pp. [Atlantic]

*Schott, W. 1935. Die Foraminiferen in dem äquatorialen Teil des Atlantischen Ozeans. *Wiss. Ergeb. Deutsch. Atlantisch. Exped. Forsch.-u Vermess. Meteor 1925-1927,* 43-134. [Atlantic]

Schott, W. 1952. On the sequence of deposits in the equatorial Atlantic. *Göteborgs Kungl. Vetensk.-och Vitterh.-Sämhalles Handl., Sjätte Földj., ser. B,* **6** (2), 3-15. [Atlantic]

Schott, W. 1954. Über stratigraphische Untersuchungsmethoden in rezenten Tiefseesedimenten. *Heidelberger Beitr. Mineralogie u. Petrographie,* **4**, 192-7. [General : Atlantic, Pacific]

*Schott, W. 1966. Foraminiiferenfauna und Stratigraphie der Tiefsee-Sedimente im Nord Atlantischen Ozean. *Repts. Swedish Deep-Sea Exped.* **7** (8), 357-469. [Atlantic]

Shepard, F. P. 1963. *Submarine Geology.* Second ed., Harper and Row: N.Y., 557 pp.

Silvestri, O. 1888. Le maggiori profondità del Mediterraneo recentemente esplorate ed analisi geologica dei relativi sedimenti marini. *Atti. Accad. Gioenia Sci. Nat., ser. 4,* **1**, 157-76, 1 pl. [Mediterranean Sea]

Stschedrina, Z. G. 1947. On the distribution of Foraminifera in the Greenland Sea. *Doklady Akad. Nauk. S.S.S.R.* **55** (9), 871-4. [Arctic]

Stubbings, H. G. 1939a. The marine deposits of the Arabian Sea. An investigation into their distribution and biology. *John Murray Exped. 1933-34, Sci. Repts.* **3** (2), 32-158, pls. 1-4. [Indian]

Stubbings, H. G. 1939b. Stratification of biological remains in marine deposits. *John Murray Exped. 1933-34, Sci. Repts.* **3** (3), 159-92. [Indian]

Tizard, T. H., Moseley, H. N., Buchanon, J. Y. and Murray, J. 1895. A summary of the scientific results. Pts. 1 and 2. *Repts. Voy. 'Challenger',* Longmans: London, 1,608 pp.

*Todd, Ruth 1958. Foraminifera from western Mediterranean deep-sea cores. *Repts. Swedish Deep-Sea Exped.* **8** (3), 169-215, pls. 1-3. [Mediterranean Sea]

Todd, Ruth 1964. Planktonic Foraminifera from deep-sea cores off Eniwetok Atoll. *U.S. Geol. Surv., Prof. Paper* **260-CC**, 1,067-1,100, pls. 289-95. [Pacific]

Todd, Ruth 1965. The Foraminifera of the tropical Pacific collections of the *Albatross,* 1899-1900, Pt. 4. Rotaliform families and planktonic families. *U.S. Natl. Mus., Bull.* **161** (4), 139 pp., pls. 1-28. [Pacific]

Uchio, T. 1960. Planktonic Foraminifera of the Antarctic Ocean. *Biol. results Japanese Antarct. Research Exped.,* **11**, *Spec. Pub., Marine Biol. Lab.* 3-9, pl. 1. [Antarctic]

*Waller, H. O. and Polski, W. 1959. Planktonic Foraminifera of the Asiatic shelf. *Contr. Cushman Found. Foram. Research.* **10** (4), 123-6, pl. 10. [Pacific]

Wallich, G. C. 1962. The North-Atlantic sea-bed: comprising a diary of the voyage on board H.M.S. *Bulldog* in 1860; and observations on the presence of animal life, and the formation and nature of organic depisits, at great depths in the ocean. Pt. 1. John Van Voorst: London, 160 pp., 6 pls. [Atlantic]

Warthin, A. S., Jr. 1934. Foraminifera from the Ross Sea. *Am. Mus. Novitates,* **721**, 4 pp. [Antarctic]

*Wiesner, H. 1931. Die Foraminiferen der Deutschen Südpolar-Expedition 1901-1903. *Deutsche Südpolar-Exped. 1901-1903,* **20**, Zoologie Bd. **XII**, 49-165, pls. 1-24. [Antarctic]

*Wiseman, J. D. H. and Ovey, C. D. 1950. Recent investigations on the deep-sea floor. *Proc. Geol. Assoc.* **61** (1), 28-84, pls. 2, 3.

(*Publications used in compiling Figs. 20-3-20.6. Unpublished data for the Pacific Ocean, and to a lesser extent for the Indian Ocean, were also used.)

21. SPUMELLARIAN AND NASSELLARIAN RADIOLARIA IN THE PLANKTON AND BOTTOM SEDIMENTS OF THE CENTRAL PACIFIC

M. G. PETRUSHEVSKAYA

Summary: In the plankton two radiolarian assemblages are found. One lives in the surface water (0-100 metres) at a temperature of 20-28° C (optimum 23-25° C), and at a salinity of 33.9-35.9‰ (optimum about 35‰). Another lives at a salinity of 34.2-36‰, at a temperature of 7-19° C (optimum 9-13° C) and at a depth of 75-100 metres and deeper. The second assemblage does not contain as many species and individuals as the first.

All Spumellarian and Nassellarian species found in the plankton can also be found in the bottom sediments of the same region. But the relative abundance of specimens of the different species in the sediment may differ to a great extent from those found in the plankton.

INTRODUCTION

Patterns of distribution of different radiolarian species in the oceanic plankton have not been as much investigated as those of many other planktonic organisms. The relation between the distribution in the plankton of spumelline and nasselline radiolarians, and the occurrence of their skeletons in bottom sediments, is not yet as clear as it is for diatoms.

This report is based on the investigation of materials obtained at ten stations occupied by R. V. *Vityaz* (Fig. 21.1). We have obtained radiolarians from the water column at different depths (0-5,000 metres) using bathometers of 2.5-40 litres capacity. In addition some plankton samples collected by a plankton net were also examined. Surface sediment samples were taken at the same stations.

1. The quantity of live Spumellaria and Nassellaria in the water column varied from 25-50 to 10,000-16,000 per cubic metre. The largest concentrations were found in the equatorial region. It may be seen (Fig. 21.2) that at 154° W the quantity of radiolarians southward from 10° S, and northward from 11° N, was less than that between these latitudes. The highest quantity was at station 5117 on the equator.

Living radiolarians occur in the whole water column (from the surface down to 5,000 metres), but most of them are in the upper horizons. In the vertical section (Fig. 21.2) a layer with abundant spumelline and nasselline radiolarians may be seen at depths of from 0-25 metres to 100-150 metres. At these depths there were 5,000-15,000 radiolarians per cubic metre; the temperature was 23°-28° C and salinity 34-36‰. At deeper horizons the number of living radiolarians is smaller than that in the upper horizons. At depths from 100-150

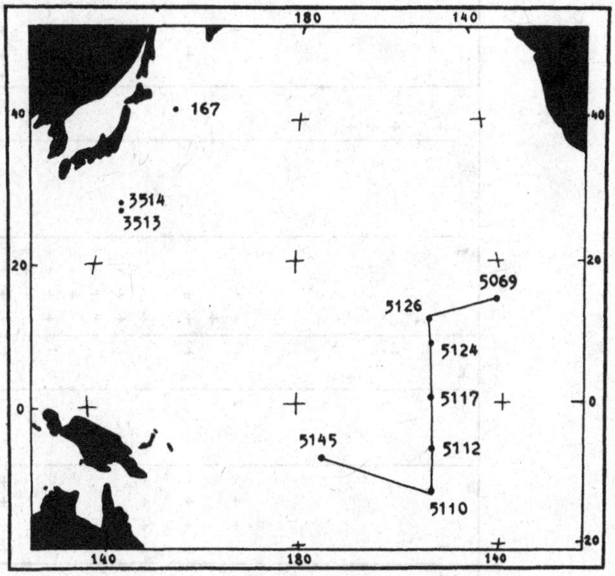
180
140
40
167
3514
3513
20
5069
5126
5124
5117
0
5145
5112
5110
140
180
140
20

Fig. 21.1. Position of *Vityaz* stations at which radiolarians were collected.

Table 21.1. Distribution of radiolarian species in the plankton (r = rare, + = present)

Station	5069										5110								5112									
depth (m) / Species	0	25	50	60	100	200	300	500	1,100	2,407	0	25	75	100	300	500	1,000	3,000	0	25	75	100-150	200	300	500	1,000	3,000	5,000
Warm-water surface species																												
Rhizosphaera serrata Hck.																			+									
Hexadoridium streptacanthium Hck.																												
Stylodictya tenuispina Joerg.(?)																			+									
Euchitonia mulleri Hck.																												
Panartus sp.																				+	+							
Pseudocubus obeliscus Hck.																			+		+							
Neosemantis distephanus Hck.																				+								
Zygocircus productus (Hertwig)																												
Zygocircus archicircus Popofsky											+	+	+	+					+	+	+	+						
Tetraspyris tetracorethra Hck.														+														
Psilomelissa calvata Hck.																			+	+	+							
Peromelissa phalacra Hck.														+					+									
Lithomelissa sp. X					+														+		+							
Trisulcus triacanthus Popofsky					+															+	+							
Pterocorys zancleus Müller	+				+																							
Pterocanium praetextum (Ehr.)																			+									
Spyrocyrtis scalaris Hck.																				+	+							
Eucyrtidium hexagonatum Hck.																												
Botryocyrtis scutum (Harting)																			+	+								
Cool-water																												
Stylatractus neptunus Hck.																								+				
Phormacantha hystrix Joergensen																												
Peridium longispinum Joergensen															+								+	+				
Litharachnium tentorium Hck.																								+				
Cornutella verrucosa Ehr.															+									+				
Artostrobus annulatus (Bailey)																												
Dictyophimus clevei Joergensen																								+				
Trisulcus sp.N					+	+																						

Station	5117									5124									
depth (m) / Species	0	25	50	100	150	300	1,000	2,000	4,500	0	25	50	75	150	300	500	1,000	3,000	4,500
Warm-water surface species																			
Rhizosphaera serrata Hck.																			
Hexadoridium streptacanthium Hck.		+									+								
Stylodictya tenuispina Joerg.(?)	+	+	+	+							+	+							
Euchitonia mulleri Hck.				+							+	+							
Panartus sp.		+	+		r					+	+		+						
Pseudocubus obeliscus Hck.	+	+	+	+	+						+	+							
Neosemantis distephanus Hck.		+	+	+															
Zygocircus productus (Hertwig)	+					r						+	+						
Zygocircus archicircus Popofsky		+	+	+	+					+	+	+	+						
Tetraspyris tetracorethra Hck.	+	+	+																
Psilomelissa calvata Hck.	+	+	+	+	r														
Peromelissa phalacra Hck.		+	+									+				r			
Lithomelissa sp. X	+	+	+	+															
Trisulcus triacanthus Popofsky		+	+	+															
Pterocorys zancleus Müller				+							+	+	+						
Pterocanium praetextum (Ehr.)																			
Spyrocyrtis scalaris Hck.			+	+	+					+	+	+	+						
Eucyrtidium hexagonatum Hck.				+															
Botryocyrtis scutum (Harting)		+	+	+	+		r			+	+	+				r			
Cool-water																			
Stylatractus neptunus Hck.																			
Phormacantha hystrix Joergensen						+									+				
Peridium longispinum Joergensen						+													
Litharachnium tentorium Hck.																+			
Cornutella verrucosa Ehr.					+										+				
Artostrobus annulatus (Bailey)				r															
Dictyophimus clevei Joergensen													r	+	+	+			
Trisulcus sp.N	r					+							+			+			

Station	5126								5145				
depth (m) / Species	0	30	50	100	200	250	1,142	1,914	0	1,000	2,000	3,000	5,000
Warm-water surface species													
Rhizosphaera serrata Hck.													
Hexadoridium streptacanthium Hck.													
Stylodictya tenuispina Joerg.(?)													
Euchitonia mulleri Hck.													
Panartus sp.													
Pseudocubus obeliscus Hck.		+											
Neosemantis distephanus Hck.													
Zygocircus productus (Hertwig)		+				r							
Zygocircus archicircus Popofsky		+	+	+									
Tetraspyris tetracorethra Hck.		+											
Psilomelissa calvata Hck.													
Peromelissa phalacra Hck.													
Lithomelissa sp. X													
Trisulcus triacanthus Popofsky		+											
Pterocorys zancleus Müller	+	+											
Pterocanium praetextum (Ehr.)													
Spyrocyrtis scalaris Hck.													
Eucyrtidium hexagonatum Hck.													
Botryocyrtis scutum (Harting)		+											
Cool-water													
Stylatractus neptunus Hck.													
Phormacantha hystrix Joergensen													
Peridium longispinum Joergensen						+					+		
Litharachnium tentorium Hck.													
Cornutella verrucosa Ehr.													
Artostrobus annulatus (Bailey)													
Dictyophimus clevei Joergensen					+						+		
Trisulcus sp.N													

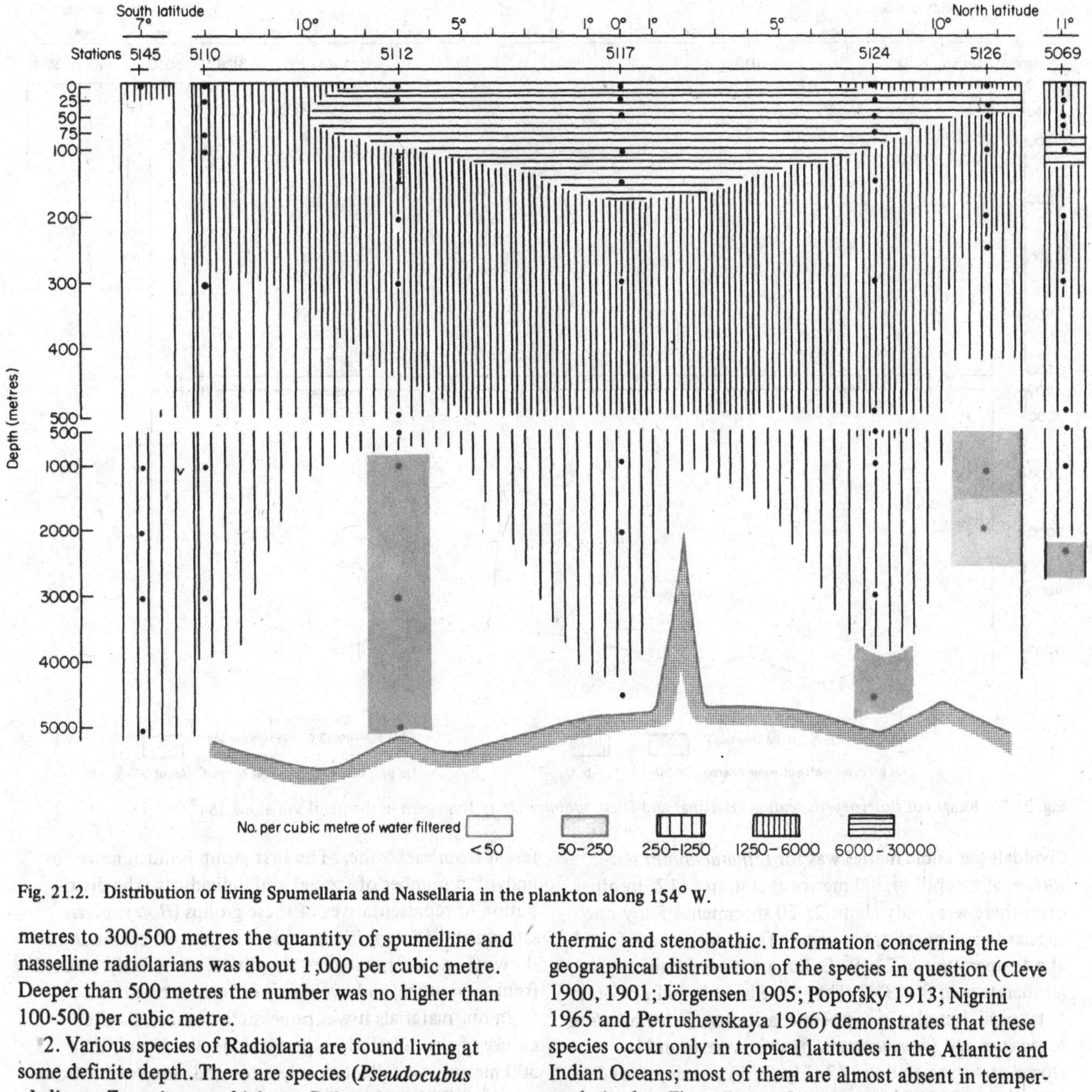
South latitude
North latitude
7°
10°
5°
1°
0°
1°
5°
10°
11°
Stations
5145
5110
5112
5117
5124
5126
5069
0
25
50
75
100
200
300
400
500
500
1000
2000
3000
4000
5000
Depth (metres)
No. per cubic metre of water filtered
<50
50–250
250–1250
1250–6000
6000–30000

Fig. 21.2. Distribution of living Spumellaria and Nassellaria in the plankton along 154° W.

metres to 300-500 metres the quantity of spumelline and nasselline radiolarians was about 1,000 per cubic metre. Deeper than 500 metres the number was no higher than 100-500 per cubic metre.

2. Various species of Radiolaria are found living at some definite depth. There are species (*Pseudocubus obeliscus, Zygocircus archicircus, Psilomelissa calvata, Lithomelissa* sp. X, *Botryocyrtis scutum* and others; see Table 21.1) which are abundant from the surface to a depth of about 100 metres. The number of individuals of some of these species reaches 200-600 per cubic metre of water. Their highest numbers were found at a temperature of 23°-25° C. At a temperature of 26°-28° C and also at 20°-22° C these species were less abundant. At depths of 150-300 metres and deeper there were no living radiolarians of these species, or at least they were represented only by single specimens (Table 21.1 and Fig. 21.3). One can conclude that these species are stenothermic and stenobathic. Information concerning the geographical distribution of the species in question (Cleve 1900, 1901; Jörgensen 1905; Popofsky 1913; Nigrini 1965 and Petrushevskaya 1966) demonstrates that these species occur only in tropical latitudes in the Atlantic and Indian Oceans; most of them are almost absent in temperate latitudes. The pattern of distribution of these species is connected with tropical surface waters. These species may be called warm-water surface species.

There is another group of species (*Cornutella verrucosa, Dictyophimus clevei, Phormacantha hystrix, Peridium longispinum, Litharachnium tentorium*), which was present at stations in the equatorial Pacific only at depths of 75-100 metres and deeper (Table 21.1). They were absent in the surface water. Individuals of these species are found most abundantly at a depth of 300 metres. But the number of individuals was much lower than that of many of the surface-water species. The greatest number (100 in-

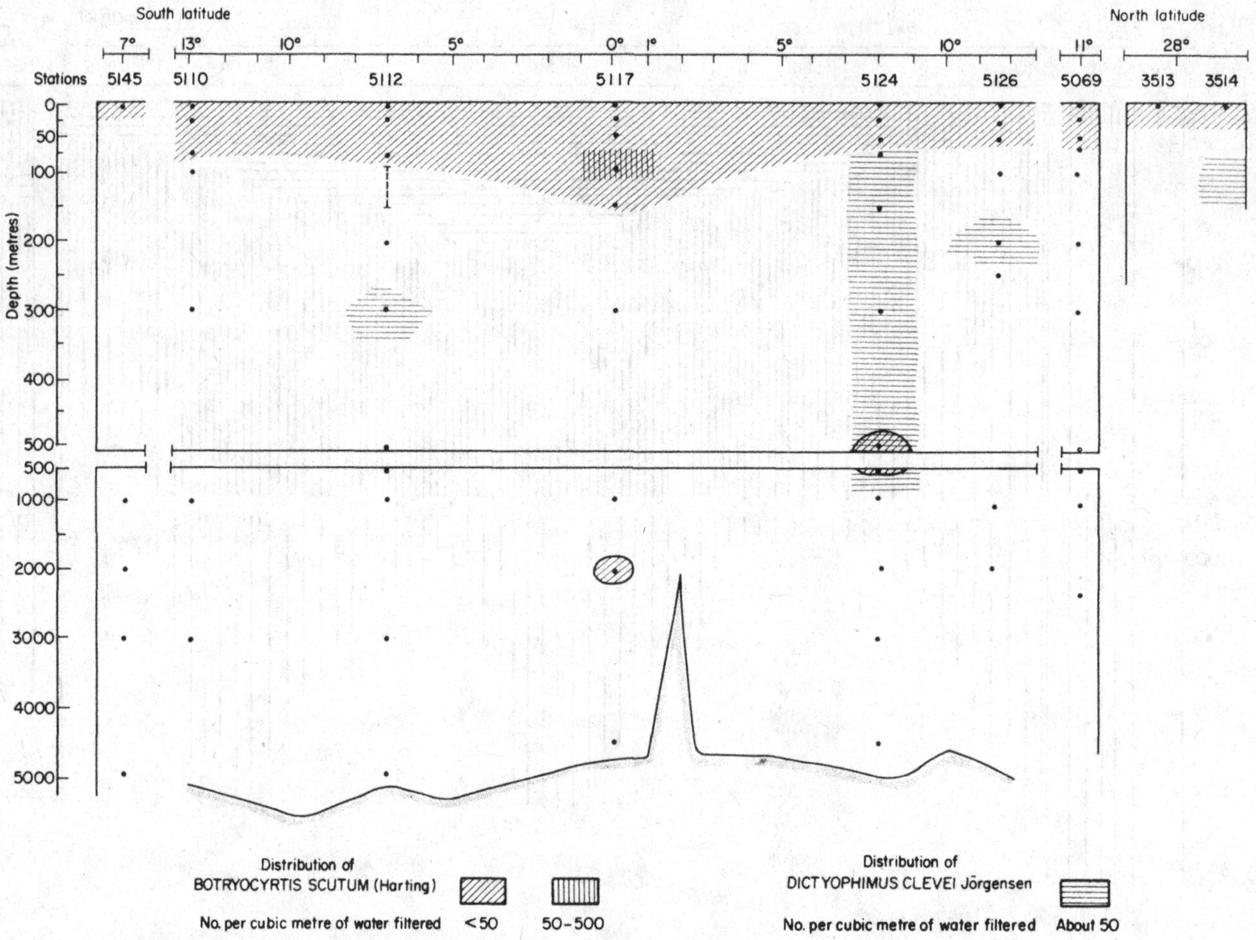
South latitude
North latitude
7°
13°
10°
5°
0°
1°
5°
10°
11°
28°
Stations
5145
5110
5112
5117
5124
5126
5069
3513
3514
Depth (metres)
0
50
100
200
300
400
500
500
1000
2000
3000
4000
5000
Distribution of
BOTRYOCYRTIS SCUTUM (Harting)
No. per cubic metre of water filtered
<50
50–500
Distribution of
DICTYOPHIMUS CLEVEI Jörgensen
No. per cubic metre of water filtered
About 50

Fig. 21.3. Ranges of *Botryocyrtis scutum* (Harting) and *Dictyophimus clevei* Jörgensen in the plankton along 154° W.

dividuals per cubic metre) was for *Litharachnium tentorium* at a depth of 300 metres at station 5112. In other cases there were only about 25-50 specimens of any one species in a cubic metre of water. These species are found at a temperature of 7°-19° C, but are more frequent at a temperature of 9°-13° C. They have been found not only in tropical, but also in temperate latitudes. *Litharachnium tentorium* and *Phormacantha hystrix* were found by Dogiel at *Vityaz* station 167 (Fig. 21.1) at a depth of 50-100 metres. *Litharachnium tentorium, Phormacantha hystrix* and *Dictyophimus clevei* were recorded (Cleve 1899, 1900; Jörgensen 1905) from the North Atlantic, and *Phormacantha hystrix, Artostrobus annulatus,* and perhaps *Dictyophimus clevei,* were also found in the Arctic Basin (Bernstein 1932, 1934, Hülsemann 1963). It is evident that these species can withstand various conditions, but as may be seen from Table 21.1 and Fig. 21.3, they avoid warm water (with a temperature of more than 19°-20° C). These species may be called cool-water species.

The two groups of species which have been established (warm-water surface and cool-water) are somewhat different from each other. The first group is much more abundant in number of species and individuals. The distribution of representatives of these groups (*Botryocyrtis scutum* and *Dictyophimus clevei*) in the vertical section shows (Fig. 21.3) that they are almost completely isolated from each other, and live under different conditions.

In our materials it was impossible to ascertain the nature of the radiolarian species at depths greater than 500 metres; we collected only a few individuals of various species at different stations. But we may suggest that there is a peculiar fauna of Spumellaria and Nassellaria, whose species live at depths of 1,000-5,000 metres. Specimens are very rare (less than 10-20 per cubic metre of water).

3. Besides living radiolarians there are also discarded skeletons of dead individuals of the same species. These shells are distributed throughout the whole water column (Fig. 21.4). In the upper horizons, where there are many living specimens, the number of discarded skeletons is less than that of living radiolarians. At depths of 1,000-5,000 metres, where there are few living individuals, discarded skeletons prevail. It is difficult to determine the reason

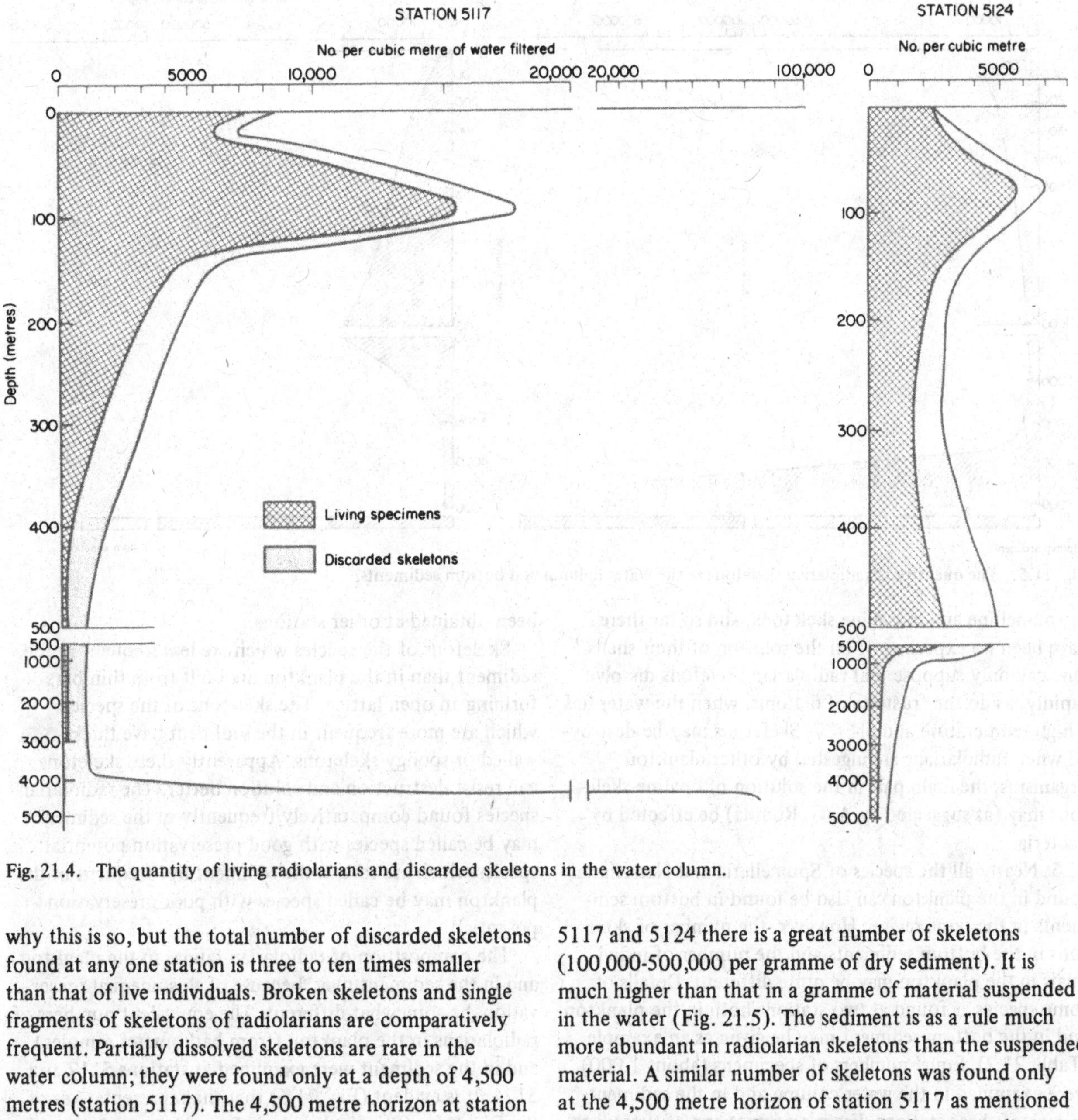
STATION 5117
No. per cubic metre of water filtered
0
5000
10,000
20,000
20,000
100,000
STATION 5124
No. per cubic metre
0
5000
Depth (metres)
0
100
200
300
400
500
500
1000
2000
3000
4000
5000
Living specimens
Discarded skeletons

Fig. 21.4. The quantity of living radiolarians and discarded skeletons in the water column.

why this is so, but the total number of discarded skeletons found at any one station is three to ten times smaller than that of live individuals. Broken skeletons and single fragments of skeletons of radiolarians are comparatively frequent. Partially dissolved skeletons are rare in the water column; they were found only at a depth of 4,500 metres (station 5117). The 4,500 metre horizon at station 5117 is of great interest because this is a layer where radiolarian skeletons are concentrated. The number of skeletons per cubic metre of water here is much higher than that at other horizons at the station (Fig. 21.4). The phenomenon of increasing concentration of skeletal remains in bottom-waters is also known for diatoms (Kozlova 1964). At the same depth (about 5,000 metres) at the other stations (5124, 5110, 5112) we have not found such a high concentration of radiolarian skeletons. At station 5117 there must have been some peculiar conditions.

In the surface layer of bottom sediment at stations 5117 and 5124 there is a great number of skeletons (100,000-500,000 per gramme of dry sediment). It is much higher than that in a gramme of material suspended in the water (Fig. 21.5). The sediment is as a rule much more abundant in radiolarian skeletons than the suspended material. A similar number of skeletons was found only at the 4,500 metre horizon of station 5117 as mentioned above. The number of skeletons in the bottom sediment at station 5117 (more than 500,000 per gramme of dry material) is itself much higher than that at the other stations, and the highest quantity of radiolarians in plankton was also found there. So there is some correspondence between the number of radiolarians living in the water column and the number of skeletons in the bottom sediment.

4. Many radiolarian skeletons reach the bottom sediment, but many of them do not reach the bottom and only pieces of them can be found in deep water. In addition many skeletons dissolve in the sediment. There is no exact information about the chemical composition

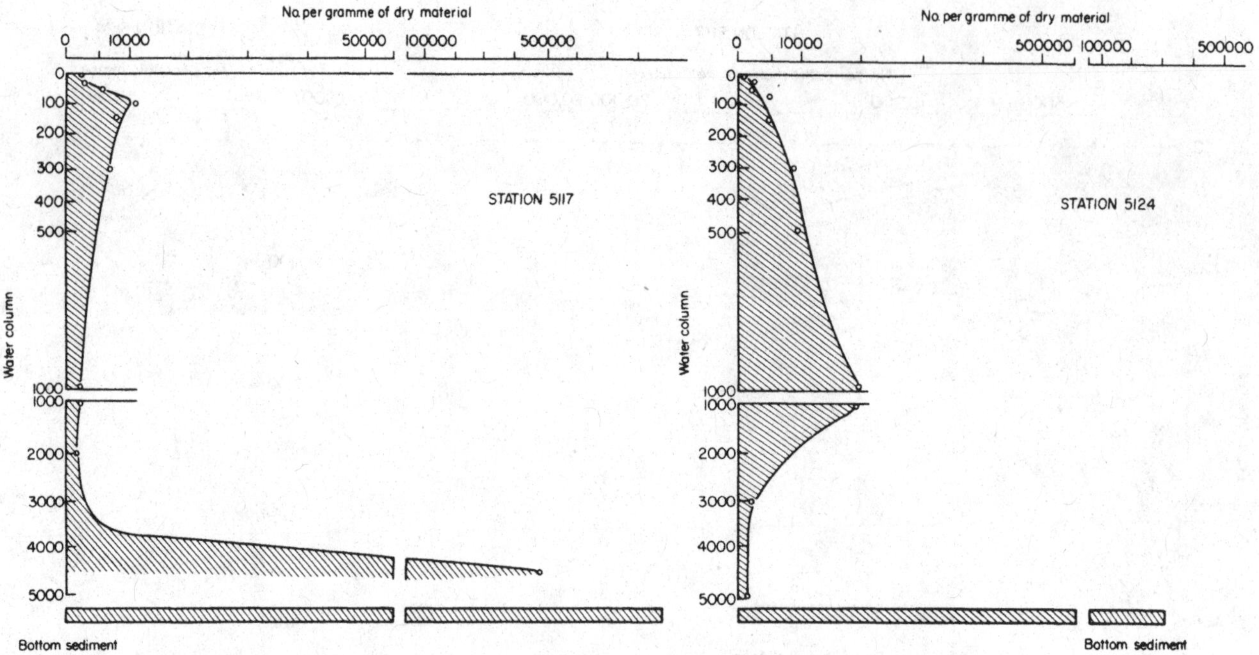
No. per gramme of dry material
0
10000
50000
100000
500000
STATION 5117
STATION 5124
Water column
100
200
300
400
500
1000
2000
3000
4000
5000
Bottom sediment

Fig. 21.5. The quantity of radiolarian skeletons in the water column and bottom sediments.

of spumelline and nasselline skeletons, and so far there have been no experiments on the solution of their shells.[1] One can only suppose that radiolarian skeletons dissolve rapidly, as do the frustules of diatoms, when the water has a high temperature and pH < 7. Skeletons may be destroyed when radiolarians are ingested by other plankton organisms; the main part in the solution of opaline skeletons may (as suggested by A. G. Rodina) be effected by bacteria.

5. Nearly all the species of Spumellaria and Nassellaria found in the plankton can also be found in bottom sediments in the same region. However, the number of skeletons in the bottom sediments and the number of individuals in the plankton may be quite different. Details of some species as found at two stations both in the plankton and in the bottom sediment may be given as an example (Table 21.2). Equal numbers of specimens (about 1,000) were examined in the water column and in the sediment at each of these stations. From a comparison of these lists it is evident that there are some species which are abundant in the plankton and rare in the sediment (*Arachnosphaera hexasphaera, Haliomma erinaceus, Phormacantha hystrix*, etc.). But others (*Artostrobus annulatus, Stylatractus neptunus, Spongodiscus resurgens*, etc.) are the opposite. There are also species (*Psilomelissa calvata, Peromelissa phalacra, Euchitonia mulleri, Hymeniastrum euclidis* and some others) which are almost equally frequent in plankton and in bottom sediment. This is not only the case at these two stations; similar results have been obtained at other stations.

Skeletons of the species which are less frequent in the sediment than in the plankton are built from thin bars forming an open lattice. The skeletons of the species which are more frequent in the sediment have thick-walled or spongy skeletons. Apparently these skeletons can resist destruction and solution better. The radiolarian species found comparatively frequently in the sediment may be called species with good preservation potential; species which are rare in the sediment but frequent in the plankton may be called species with poor preservation potential.

The composition of radiolarian faunas in the plankton and in the sediment may, because of this unequal preservation, be somewhat different. The equivalent numbers of radiolarians in the plankton (from bathometer samples) and in the sediment were examined at stations 5117 and 5124. It is evident (Fig. 21.6) that the representatives of the Discoidea (Spongodiscidae), Larcoidea and Botryoides are more frequent in the bottom sediment than in the water column. Radiolarians such as the Sphaeroidea, Plectoidae, Stephoidae and Crytoidae (Dicyrtida), whose skeletons consist as a rule of thin bars or thin-walled shells, are less frequent in the sediment than in the plankton.

6. A special problem is presented by cases (Riedel 1952, 1957) where Tertiary Radiolaria occur in the upper sediment layers. Skeletons of such radiolarians were found at stations 5069, 5124 and 5145. They were mixed (1 : 10 to 1 : 2) with skeletons of present-day species. Riedel

[1] But see Berger (1968).

(1957) stated that in such samples early and middle Tertiary forms have as a rule heavier skeletons than Quaternary Radiolaria. In the samples from stations 5069, 5124 and 5145 all skeletons of present-day forms were corroded, but the Tertiary species show no signs of solution. Skeletons of present-day radiolarians in such sediments may be dissolved more rapidly than the skeletons of Tertiary forms and therefore are not preserved there. The reworking of older radiolarians into younger sediments presents many difficulties in the investigation of Radiolaria in the surface layer of bottom sediments and in sediment cores. Therefore, although the occurrence of some radiolarian species in bottom sediments corresponds well with their occurrence in the plankton, the total distribu-

Table 21.2. Occurrence of radiolarian species in the plankton and in bottom sediments (r = rare, + = present)

Species	Station 3513		Station 5117	
	Plankton	Sediment	Plankton	Sediment
Species with poor preservation potential				
Arachnosphaera hexasphaera Popofsky	+	–	–	–
Haliomma erinaceus Hck.	+	–	–	–
Neosemantis distephanus (Hck.)	+	–	+	r
Phormacantha hystrix (Jörgensen)	–	–	+	–
Peridium longispinum Jörgensen	–	–	+	–
Pseudocubus obeliscus Haeckel	+	–	++	r
Spyrocyrtis scalaris Hck.	+	–	+	+
Litharachnium tentorium Hck.	–	–	+	–
Drymosphaera polygonalis Hck.	+	–	–	–
Centrocubus cladostylus Hck.	–	–	–	r
Euchitonia elegans (Ehrenberg)	+	–	r	r
Hexadoridium streptacanthum Hck.	+	–	+	+
Species with good preservation potential				
Species of average preservation potential				
Actinomma arcadoporum Hck.	+	+	–	–
Rhizosphaera serreta Hck.	+	+	–	–
Zygocircus productus (Hertwig)	+	+	+	r
Zygocircus archicircus Popofsky	+	+	++	+
Tetraspyris tetracorethra Hck.	+	+	–	–
Cornutella verrucosa Ehrenbrg	–	–	+	+
Trisulcus triacanthus Popofsky	–	–	+	+
Trisulcus sp. N.	–	–	+	+
Dictyophimus clevei Jörgensen	+	–	–	r
Psilomelissa calvata Hck.	+	–	++	++
Peromelissa phalacra Hck.	+	+	+	+
Lithomelissa thoracites Hck.	–	–	+	+
Lithomelissa nana Popofsky	–	–	+	+
Lithomelissa sp. X	–	–	++	++
Pterocorys zancleus Müller	+	+	+	+
Pterocanium praetextum (Ehrenberg)	+	+	r	r
Botryocyrtis scutum (Harting)	+	+	++	++
Euchitonia mulleri Hck.	+	+	+	+
Stylodictya tenuispina Jörgensen	+	+	+	+
Hymeniastrum angulatum (Ehrenberg)	+	+	+	+
Hymeniastrum profundum (Ehrenberg)	+	+	+	+
Spongaster tetras Ehrenberg	+	+	?	+
Spongotrochus longispinus Hck.	+	+	+	+
Heliodiscus asteriscus Hck.	+	+	–	–
Heliodiscus amphidiscus Hck.	+	+	–	–
Species with positive preservation potential				
Artostrobus annulatus (Bailey)	–	+	r	+
Lithomitra arachnea (Ehrenberg)	–	+	r	+
Lithomitra clevei Petrushevskaya	–	–	–	+
Theocalyptra davisiana (Ehrenberg)	–	+	–	+
Stylatractus neptunus Hck.	–	+	–	+
Cornutella bimarginata Hck.	–	+	–	+
Dictyocephalus papillosus (Ehrenberg)	–	+	–	+
Lithocampe sp.Nigrini	–	+	–	+
Lithocampe (?) *aquilonaris* (Bailey)	–	+	–	+
Spongodiscus resurgens Ehrenberg	–	+	–	+

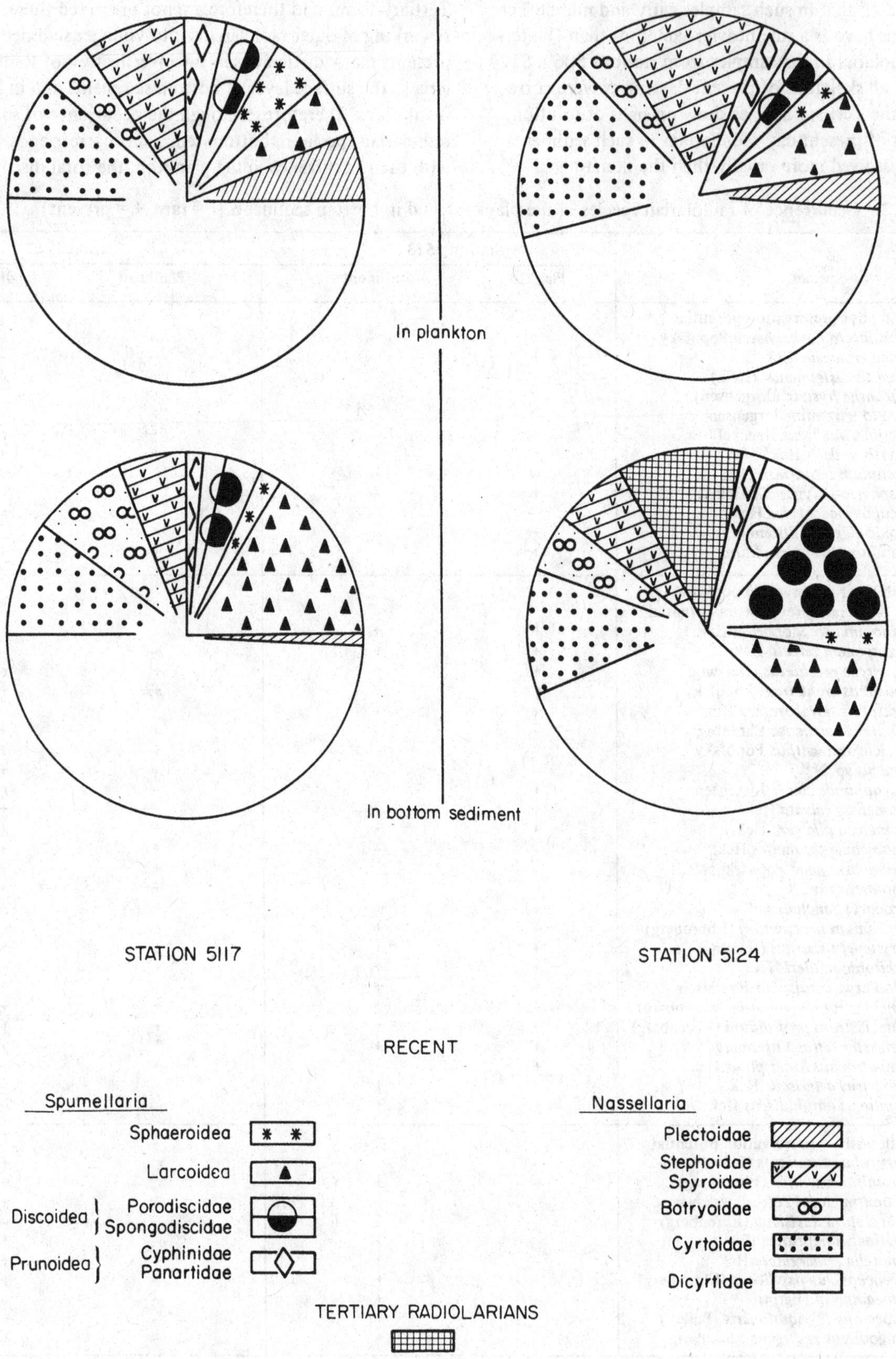
In plankton
In bottom sediment
STATION 5117
STATION 5124
RECENT
Spumellaria
Sphaeroidea
Larcoidea
Discoidea
Porodiscidae
Spongodiscidae
Prunoidea
Cyphinidae
Panartidae
Nassellaria
Plectoidae
Stephoidae
Spyroidae
Botryoidae
Cyrtoidae
Dicyrtidae
TERTIARY RADIOLARIANS

Fig. 21.6. Comparison of the composition of assemblages in plankton and in bottom sediments.

tion of radiolarian skeletons in sediments represents a very complicated problem.

Received September 1967.

BIBLIOGRAPHY

Berger, W. H. 1968. Radiolarian skeletons : solution at depths. *Science,* **159**, 1237-8.

Bernstein, T. 1932. Ueber einige arktische Radiolarien. *Archiv Protistenk.* **76**, 217-27.

Bernstein, T. 1934. Zooplankton Karskogo Morya po materialam ekspeditsii Arkticheskogo Instituta na *Sedove* 1930 goda i *Lomonosove* 1931 goda (Zooplankton des nördiichen Teiles des Karischen Meeres). *Trudy Arktichesk. Inst., Leningrad,* **9**, 3-58.

Cleve, P. T. 1899. Plankton collected by the Swedish expedition to Spitzbergen in 1898. *Handl. Kgl. Svenska Vetensk.-Akad.* **32**, 1-51.

Cleve, P. T. 1900. The seasonal distribution of Atlantic plankton organisms. *Handl. Göteborgs Kgl. Vettensk.-och Vitterh.-Samh., ser. 4,* **3**, 1-369.

Cleve, P. T. 1901. Plankton from the Indian Ocean and the Malay Archipelago. *Handl. Kgl. Svenska Vetensk.-Akad.* **35** (5), 1-58.

Hülsemann, K. 1963. Radiolaria in plankton from the Arctic Drifting station T-3, including the description of three new species. *Arctic Inst. Nth. Amer., Tech. Paper,* **13**, 1-52.

Jörgensen, E. 1905. The protist plankton and the diatoms in bottom samples. *Bergens Mus. Skr. 1905,* 114-51.

Kozlova, O. G. 1964. *Diatomovye vodorosly Indiyskogo i Ticho-okeanskogo sektorov Antarctiki.* 1-168. Nauka: Moscow.

Nigrini, C. 1965. Radiolaria in recent pelagic sediments from the Indian and the Atlantic oceans. Dissertation submitted for degree of D.Phil., University of Cambridge, 234 pp.

Petrushevskaya, M. G. 1966. Radiolarii v planktone i v donnykh osadkakh. in *Geokhimiya Kremnesema*, 219-45. Nauka: Moscow.

Popofsky, A. 1913. Die Nassellarien des Warmwassergebietes. *Deutsche Südpolar-Exped. 1901-1903,* **14** (Zool. vol. **6**), 217-416.

Riedel, W. R. 1952. Tertiary Radiolaria in western Pacific sediments. *Göteborgs Kgl. Vetensk.-och Vitterhets-Samhälles Handl., följ. 7, ser. B,* **6** (3), 1-18.

Riedel, W. R. 1957. Radiolaria : a preliminary stratigraphy. *Repts. Swedish Deep-Sea Exped.* **6** (3), 59-96.

22. RADIOLARIA IN THE PLANKTON AND RECENT SEDIMENTS FROM THE INDIAN OCEAN AND ANTARCTIC

M. G. PETRUSHEVSKAYA

Summary: Two groups of radiolarian species (Tropical and high-Antarctic) have been found to be confined to certain water-masses, the surface Tropical (0-100 metres) and the subsurface Antarctic (50-400 metres). These species may be used as indicators of hydrological conditions. Their skeletons in bottom sediments form characteristic tropical and antarctic radiolarian associations. The associations of radiolarian skeletons in recent sediments show variations within their total area of distribution.

INTRODUCTION

Radiolaria from bottom sediments of the Indian Ocean and Antarctic have been described by Riedel (1952, 1958), Hays (1965), Nigrini (1965) and Petrushevskaya (1966*a*, *b*, 1967). A striking contrast has been established between the radiolarian assemblages in equatorial and high-latitude sediments. Not only are different species found in antarctic and tropical sediments, but there is also a great difference in the preservation of skeletons and the quantity of skeletons per gramme of sediment. These investigations indicate that there must be zoned environmental control of radiolarian distribution.

ZOOGEOGRAPHICAL GROUPS

Several groups of radiolarian species are pointed out by the above-mentioned authors, namely:

(1) Antarctic species, which are confined to the cold water south of the Antarctic Convergence.
(2) Bipolar cold-water species which are found only in the Antarctic (and in the North Pacific or North Atlantic).
(3) Cosmopolitan species, which are apparently distributed continuously across the tropics submerged below the warmer surface water.
(4) Warm-water tropical species.
(5) Species indicative of cool-water conditions.

The concepts of these groups as described by different authors are somewhat variable; there are also cases where the same species are differently named in different papers.

The ecology of the Spumellaria and Nassellaria is still little known, and some of the groups described have only been established on the basis of their distribution in bottom sediments. Data on the ecology of some of these species are listed in Table 22.1. It can be seen that some species are listed which so far have not been encountered in the plankton at all (e.g.*Lithocampe* (?) *aquilonaris*, *Spongodiscus resurgens*, *Lithocampe* sp., etc.), but there are also species which have been adequately studied.

Certain species (*Euchitonia elegans*, *Botryocyrtis scutum*, *Pterocanium praetextum*, etc., Table 22.1) are found inhabiting the warm surface water of tropical regions in the Atlantic, Indian and Pacific Oceans (Cleve 1900, 1901, and the present author). They live at a temperature between 13° and 29° C and a salinity of 33-37‰, but they are more abundant at temperatures of 23°-25° C and a salinity of about 35‰. There is no indication of these species in antarctic waters. They may be called warm-water surface species.

As to the distribution of their skeletons in bottom sediments, it is known (Riedel 1952; Nigrini 1965, and some others) that they are widely distributed in low latitudes. The skeletons of such species as *Botryocyrtis scutum*, *Pterocanium praetextum* (Fig. 22.1) as well as *Euchitonia elegans*, *Eucyrtidium hexagonatum* and others (Table 22.1) are found only in the tropical zone of the Indian Ocean. As a rule their skeletons are absent south of the Subtropical Convergence. The distribution of their skeletons corresponds with their presence in tropical surface water. These species may be used as indicators of tropical waters.

The skeletons of *Heliodiscus asteriscus*, *Cornutella verrucosa* (Fig. 22.2) as well as of *Axoprunum staurax-*

onium, *Lamprocyclas maritalis*, *Lithocampe*(?) *aquilonaris* and *Lithocampe* sp. are more widely distributed in the bottom sediments of the Indian Ocean. They are found not only in tropical regions but also south of the Subtropical Convergence in the notal zone. Single specimens were found even south of the Antarctic Convergence in antarctic sediments, but they are absent at high latitudes south of the Antarctic Divergence. The distribution of the species in question allows them to be classified as subtropical species.

Some of these species were called warm-water species by Hays (1965); Nigrini (1965) called some of them indicators of cool-water. Some of them were found living in tropical subsurface waters (Haecker 1908, and the present author) at depths of about 100-400 metres and at temperatures of 9°-13° C (Table 22.1). Information

Table 22.1. Occurrence of radiolarian species in the water column and in sediment samples (r = rare, + = present, – = absent, ? = doubtful record.)

Occurrence of radiolarian species:	In the plankton (by Cleve 1900, 1901; Jörgensen 1905; Haecker 1908; Popofsky 1908; Bernstein 1934; Dogel and Reshetnyak 1952; Hülsemann 1963; Petrushevskaya 1966*a*, 1967 and present paper)								In sediment samples from different geographical zones				
	Depth (metres)					Salinity (‰)	Temperature °C		Tropical	Boreal	Notal	Antarctic	
	0-100	50-400	400-1,000	more than 1,000									
Euchitonia elegans (Ehr.)	+			–		32-35.2	24-29		+	–	–	–	
Eucyrtidium hexagonatum Hck.	+			–		33.5-34.3	23-28		+	–	–	–	
Botryocyrtis scutum (Harting)	+			r		34.3-35.2	20-29		+	–	–	–	
Panartus sp.	+			–		33.1-37.5	17-29		+	–	–	–	
Pterocanium praetextum (Ehr.)	+			–		33.9-37.2	16-29		+	–	–	–	tropical
Peromelissa phalacra Hck.	+			–	surface	34.3-35.2	23-28		+	–	–	–	
Theocorythium trachelium (Ehr.)	+			–		35.6	21-27	warm-water	+	–	r	–	
Siphonosphaera socialis Hck.	+			–		32.6-37.3	15-27		+	r	+	–	
Lithostrobus seriatus Hck.	+			?		-35.6	21-27		+	+	+	–	
Phorticium pylonium Hck.	+			–			25		+	–	+	–	
Eucyrtidium acuminatum (Ehr.)	+			–		34.4-37.2	12-28		+	r	+	–	
Hymeniastrum profundum (Ehr.)	+	+				34.2	13-28		+	+	+	r	
Cornutella verrucosa Ehr.	–	+				34.3-34.7	5-23		+	?	+	r	
Lamprocyclas maritalis Hck.	–	+			subsurface		7		+	+	+	–	
Heliodiscus asteriscus Hck.		+		?			20		+	?	+	r	subtropical
Axoprunum stauraxonium Hck.	–	+	+						+	+	+	r	
Cornutella bimarginata Hck.	–								+	?	+	–	
Lithocampe sp. Nigrini	–								+	?	+	–	
Lithocampe(?) *aquilonaris* (Bailey)	–								+	+	+	r	
Dictyocephalus papillosus (Ehr.)	–								+	+	+	+	
Stylatractus neptunus Hck.	–	+				35.2	15		+	+	+	+	
Spongodiscus resurgens Ehr.	–								+	+	+	+	
Lithomitra arachnea (Ehr.)	–	+	+	+		34.5	0.6-4.3		+	+	+	+	all-zones
Theocalyptra davisiana (Ehr.)	–	+		?					+	+	+	+	
Artostrobus annulatus (Bailey)	–	+		?		35.2	20		+	+	+	+	
Lithomitra clevei Petrushevsk.	–		+	+		34.5-34.7	2.2-4.5		+	+	+	r	
Androcyclas gamphonycha (Jörgensen)	–		+						–	+	+	–	
Theocalyptra craspedota Jörgensen	+	+				34.7-35.3	3-22		r	+	+	r	
Echinomma delicatulum (Dogel)	–	+		+			1.5		–	+	+	+	temperate
Dictyophimus gracilipes Bailey	+	+	?			35.0-35.6	0-12		?	+	+	+	
Lithomelissa borealis (Ehr.)	+	+				34	5		?	+	+	+	
Semantis micropora Popofsky	+	+							r	–	+	+	
Phorticium clevei (Jorgensen)		+				33.7	-1.5-5		–	+	+	+	
Theocalyptra bicornis (Popofsky)		+							r	–	+	+	
Antarctissa denticulata (Ehr.)		+		?		34.2	0.4-1.6		r	–	+	+	
Saccospyris antarctica Haecker		+		?	subsurface				–	–	+	+	
Lithamphora furcaspiculata Popof.		+							–	–	+	+	
Triceraspyris antarctica (Haecker)		+		?			1.2		–	–	r	+	
Spongotrochus glacialis Popofsky		+					-1.9-3	cold-water	–	?	r	+	antarctic
Spongodiscus favus Ehr. *maxima* Pop.		+					-1.7-2		–	–	r	+	
Antarctissa strelkovi Petrushevsk.		+				34.6	-1.9-2		–	–	r	+	
Lithelius nautiloides Popofsky		+				34.6	-1.8		–	–	r	+	

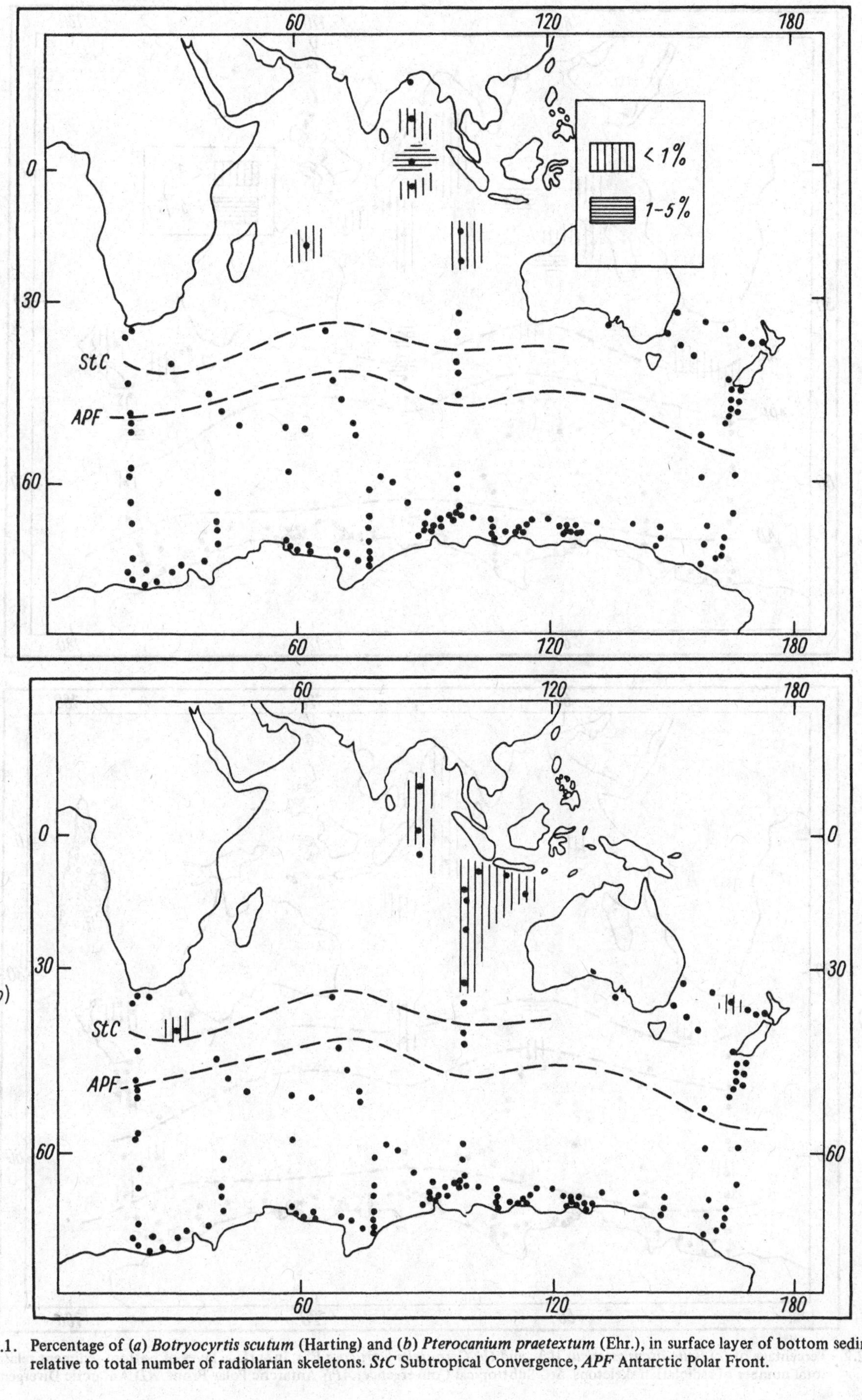

60
120
180
0
30
60
(a)
StC
APF
< 1%
1-5%
(b)

Fig. 22.1. Percentage of (*a*) *Botryocyrtis scutum* (Harting) and (*b*) *Pterocanium praetextum* (Ehr.), in surface layer of bottom sediment relative to total number of radiolarian skeletons. *StC* Subtropical Convergence, *APF* Antarctic Polar Front.

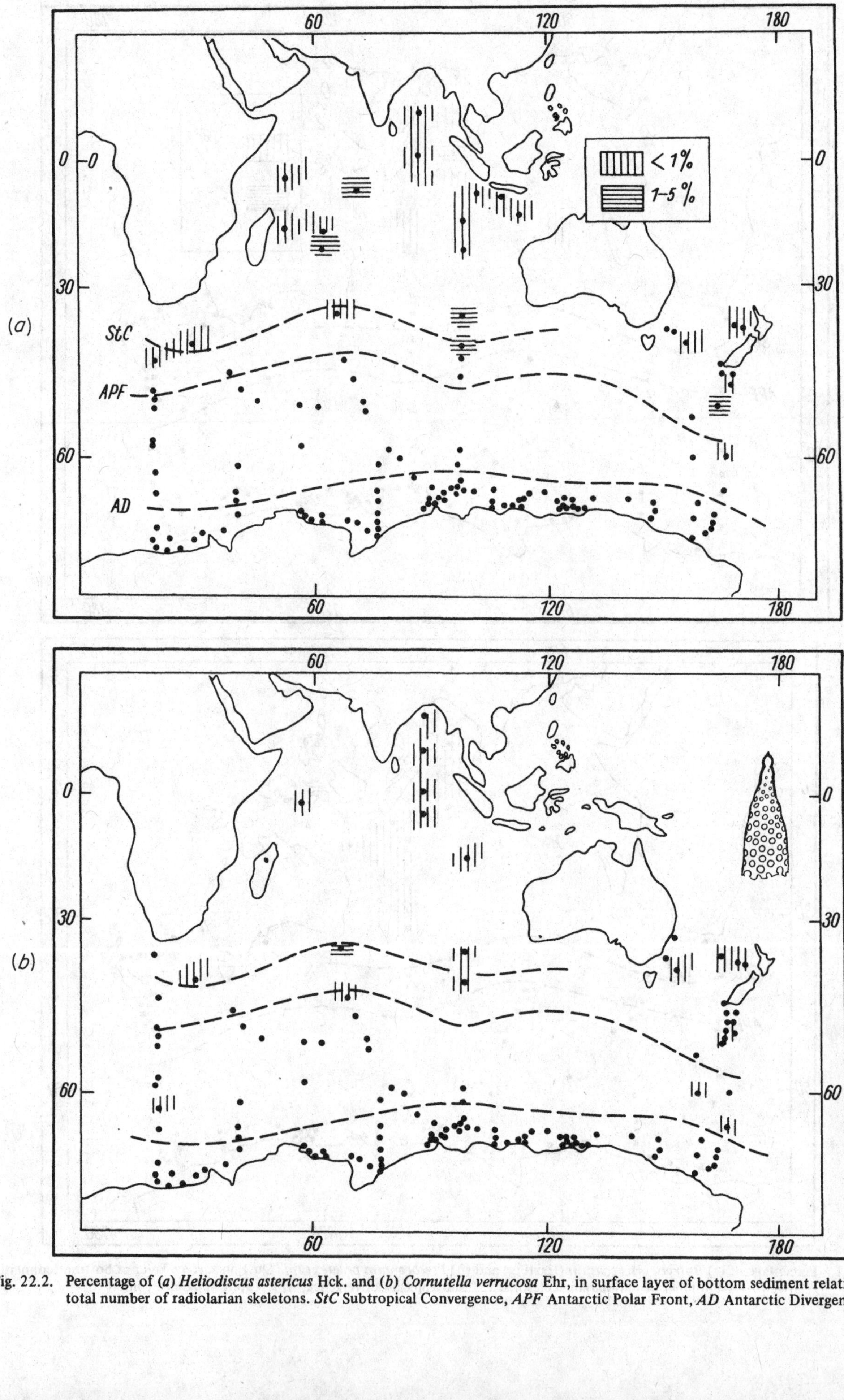

(a)
(b)
60
120
180
0
30
60
StC
APF
AD
< 1%
1–5 %

Fig. 22.2. Percentage of (*a*) *Heliodiscus astericus* Hck. and (*b*) *Cornutella verrucosa* Ehr, in surface layer of bottom sediment relative to total number of radiolarian skeletons. *StC* Subtropical Convergence, *APF* Antarctic Polar Front, *AD* Antarctic Divergence.

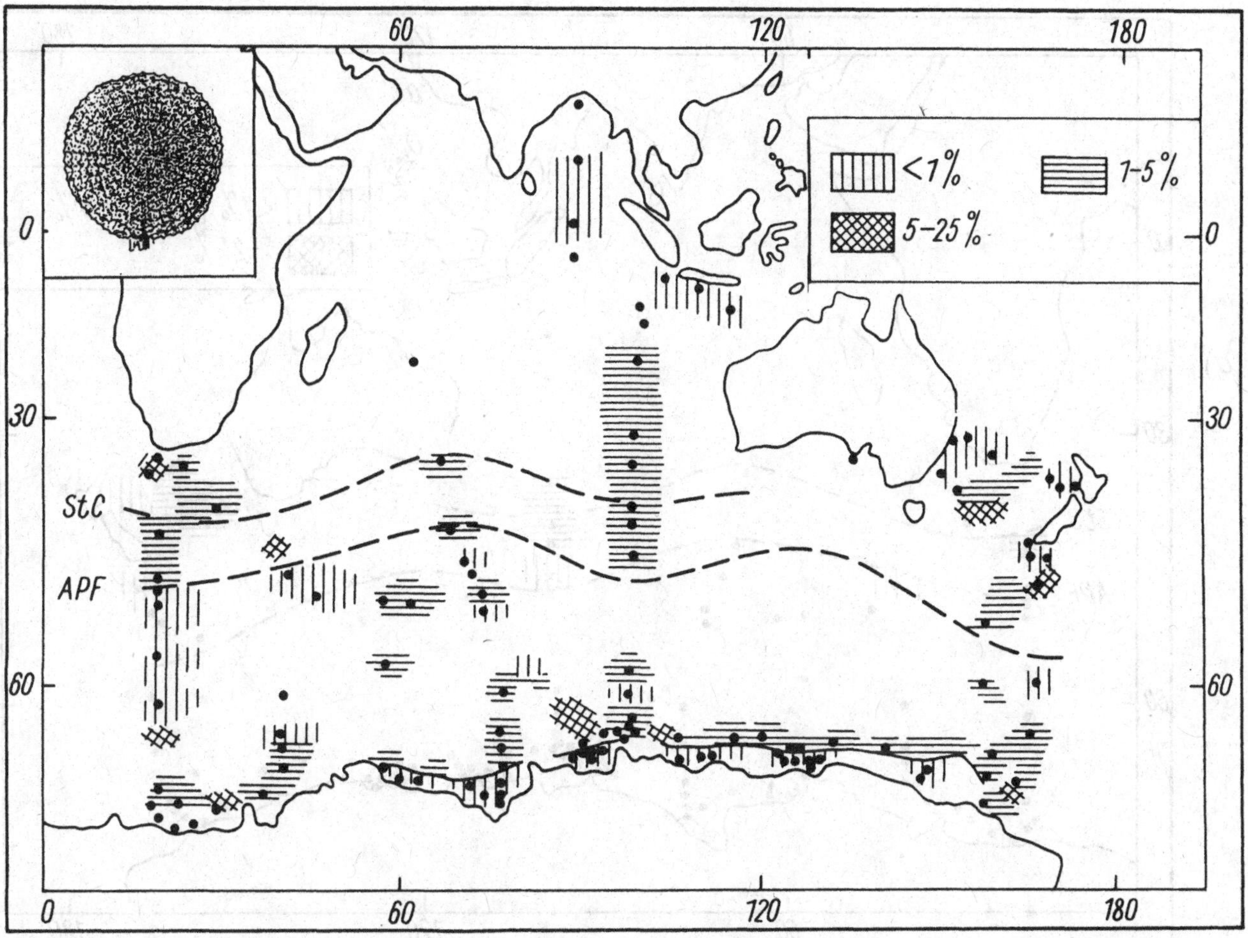

60
120
180
0
30
60
StC
APF
a)
<1%
1-5%
5-25%

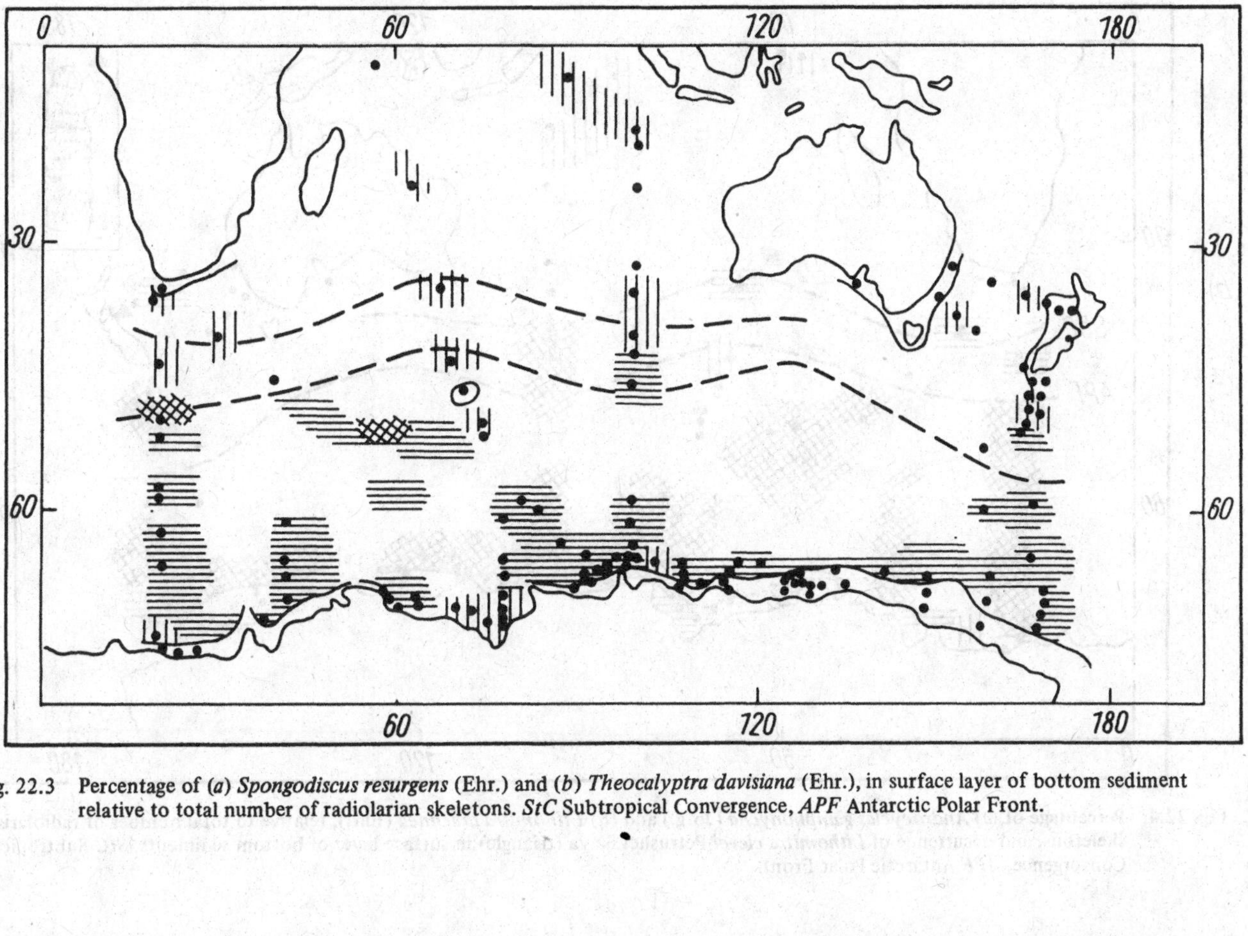

0
60
120
180
30
60
(b)

Fig. 22.3 Percentage of (*a*) *Spongodiscus resurgens* (Ehr.) and (*b*) *Theocalyptra davisiana* (Ehr.), in surface layer of bottom sediment relative to total number of radiolarian skeletons. *StC* Subtropical Convergence, *APF* Antarctic Polar Front.

(a)

(b)

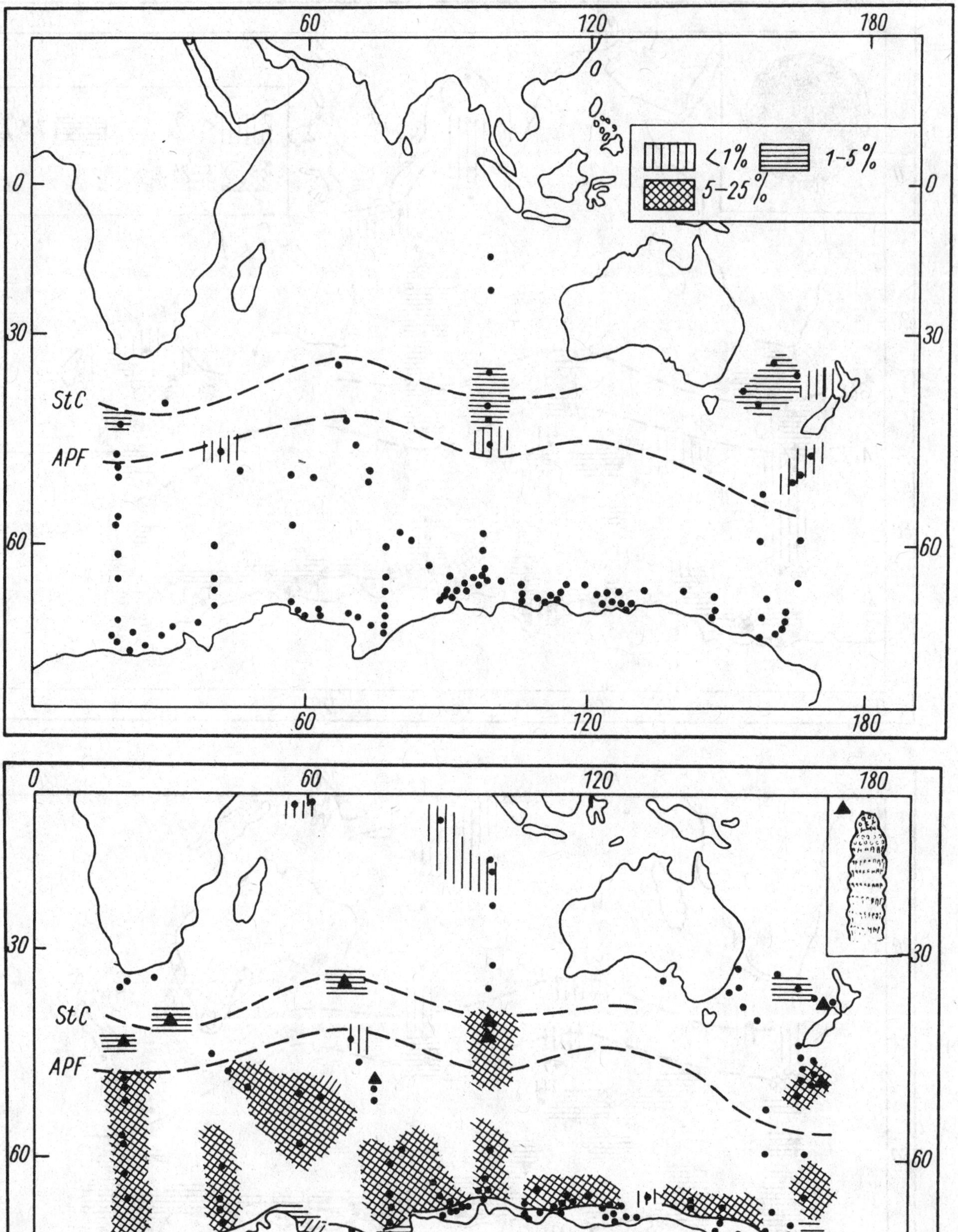
60
120
180
0
30
60
<1%
1-5%
5-25%
StC
APF
60
120
180
0
60
120
180
30
60
StC
APF
0
60
120
180

Fig. 22.4. Percentage of (*a*) *Androcyclas gamphonycha* (Jorg.) and (*b*) *Lithomitra arachnea* (Ehr.), relative to total number of radiolarian skeletons, and occurrence of *Lithomitra clevei* Petrushevskaya (triangle) in surface layer of bottom sediments. *StC* Subtropical Convergence, *APF* Antarctic Polar Front.

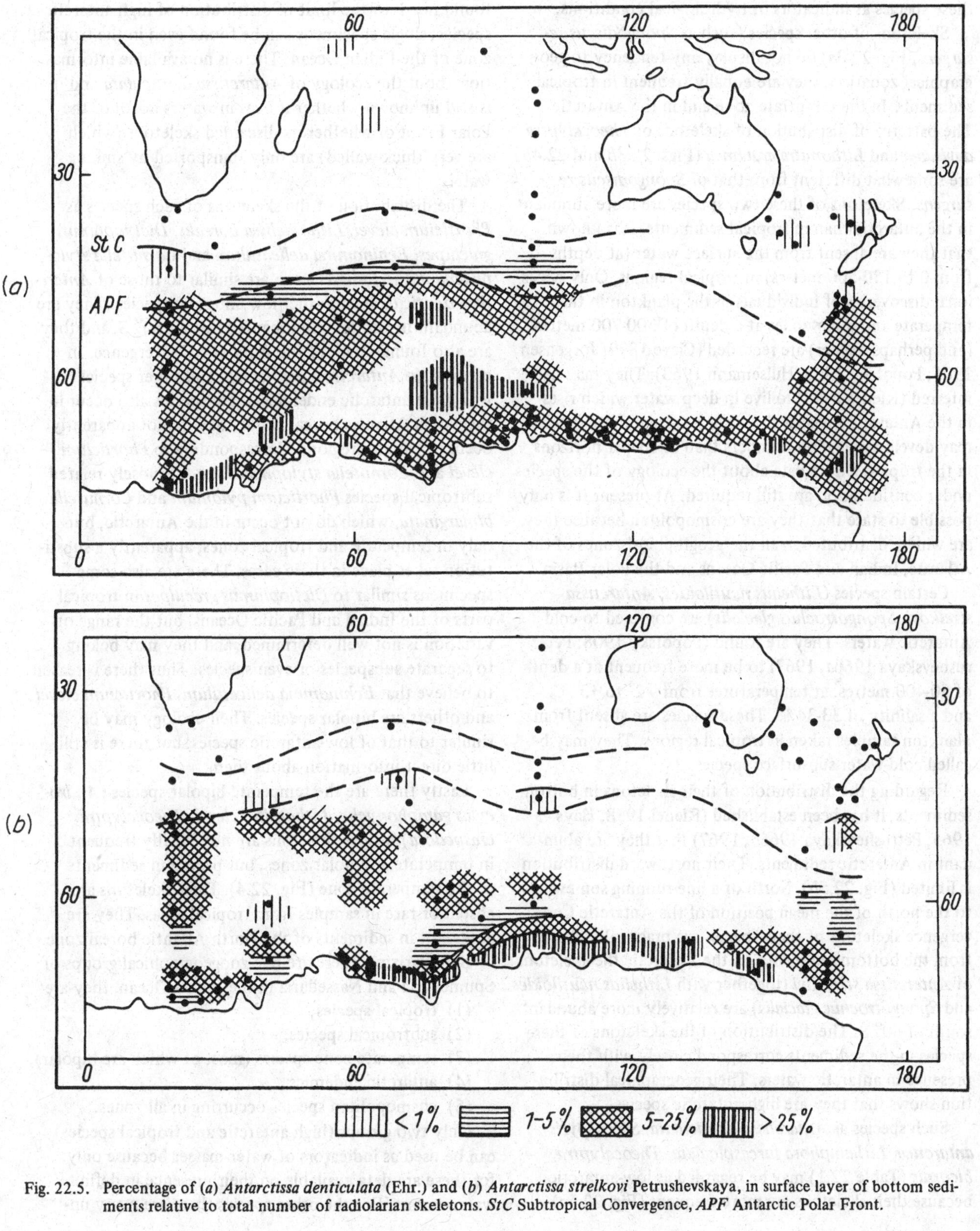
(a)
60
120
180
30
StC
APF
60
0
60
120
180
(b)
60
120
180
30
60
0
60
120
180
<1%
1-5%
5-25%
>25%

Fig. 22.5. Percentage of (*a*) *Antarctissa denticulata* (Ehr.) and (*b*) *Antarctissa strelkovi* Petrushevskaya, in surface layer of bottom sediments relative to total number of radiolarian skeletons. *StC* Subtropical Convergence, *APF* Antarctic Polar Front.

about them is scanty and special investigations of their ecology are required. At present it is difficult to consider these species as indicators of hydrological conditions.

Skeletons of other species (such as *Spongodiscus resurgens*, Fig. 22.3*a*) do not display any tendency to geographical zonation; they are equally frequent in tropical sediments, in the temperate zone and in the Antarctic. The patterns of distribution of skeletons of *Theocalyptra davisiana* and *Lithomitra arachnea* (Figs. 22.3*b* and 22.4*b*) are somewhat different from that of *Spongodiscus resurgens*. Skeletons of these two species are more abundant in the antarctic than in tropical sediments. It is known that they are absent from the surface water (at depths from 0 to 150-500 metres) in tropical regions. Only occasional discoveries of individuals in the plankton in the temperate and polar zones at a depth of 300-700 metres (and perhaps deeper) are recorded (Cleve 1899; Jorgensen 1905; Popofsky 1908; Hülsemann 1963). They may be inferred (Riedel 1958) to live in deep water which rises in the Antarctic towards the surface, where these species may develop in greater numbers than at deeper horizons in the tropics. Direct data about the ecology of the species under consideration are still required. At present it is only possible to state that they are cosmopolitan because they are widely distributed in all the geographical zones of the Atlantic, Indian and Pacific Oceans and the Polar Basin.

Certain species (*Lithelius nautiloides, Antarctissa strelkovi, Spongotrochus glacialis*) are confined to cold Antarctic waters. They are found (Popofsky 1908; Petrushevskaya 1966*a*, 1967) to be more frequent at a depth of 50-400 metres, at temperatures from –2° to +3° C, and a salinity of 33-34‰. These species are absent from plankton samples taken in tropical regions. They may be called cold-water subsurface species.

Regarding the distribution of their skeletons in bottom sediments, it has been established (Riedel 1958; Hays 1965; Petrushevskaya 1966*b*, 1967) that they are abundant in Antarctic sediments. Their northward distribution is limited (Fig. 22.5*b*). North of a line running somewhat to the north of the mean position of the Antarctic Convergence skeletons of these species are practically absent from the bottom sediments. In the Antarctic the skeletons of *Antarctissa strelkovi* (together with *Lithelius nautiloides* and *Spongotrochus glacialis*) are relatively more abundant south of 60° S. The distribution of the skeletons of these species in the sediments correspond closely with their presence in antarctic waters. Their geographical distribution shows that they are high-antarctic species.

Such species as *Antarctissa denticulata, Saccospyris antarctica, Lithamphora furcaspiculata, Theocalyptra bicornis* (Table 22.1) may be regarded as low-antarctic because their skeletons are more frequent (Fig. 22.5*a*) in sediments at about 60° S. They are relatively rare near the Antarctic Continent. Skeletons of *A. denticulata* are found north of the limit of distribution of high-antarctic species; single specimens can be found even in the tropical zone of the Indian Ocean. There is no available information about the ecology of *Antarctissa denticulata* and it is still unknown whether it lives in waters north of the Polar Front or whether its discarded skeletons (which are very thick-walled) are only transported by sinking water.

The distribution of the skeletons of such species as *Phorticium clevei, Lithomelissa borealis, Dictyophimus gracilipes, Echinomma delicatulum* and *Cornutella stylophaena* in the Indian Ocean are similar to those of *Antarctissa denticulata* and other low-antarctic species. They are abundant in antarctic sediments at about 60° S, and they are also found north of the Antarctic Convergence. In contrast to *Antarctissa denticulata* and other species which are antarctic endemics, these species also occur in the North Atlantic boreal zone. They do not apparently occur in tropical regions. Corresponding to *Phorticium clevei* and *Cornutella stylophaena* are the closely related subtropical species *Phorticium pylonium* and *Cornutella bimarginata*, which do not occur in the Antarctic, but only in temperate and tropical zones; apparently a substitution takes place in these cases. There are also some specimens similar to *Dictiophimus gracilipes* in tropical parts of the Indian and Pacific Oceans, but the range of variation is not well determined, and they may belong to separate subspecies or even species. Thus there is reason to believe that *Echinomma delicatulum, Phorticium clevei* and others are bipolar species. Their ecology may be similar to that of low-antarctic species, but there is still little direct information about them.

Lastly there are the temperate bipolar species (*Androcylas gamphonycha, Lithomitra clevei, Theocalyptra craspedota*), whose skeletons are not equally frequent in temperate and polar zones, but prevail in sediments of the temperate zone (Fig. 22.4). Their skeletons are absent or rare in samples from tropical areas. They are abundant in sediments of the North Atlantic boreal zone.

Summarizing, there are five zoogeographical groups of Spumellaria and Nassellaria in the Indian Ocean, they are:

(1) tropical species,
(2) subtropical species,
(3) temperate zone species (most of which are bipolar),
(4) antarctic endemics,
(5) cosmopolitan species occurring in all zones.

Only two groups (high-antarctic and tropical species) can be used as indicators of water-masses because only for these are data available on their presence in definite water-masses. For the other species there are many uncertainties about their distribution in the plankton, and at the present time it is difficult to use them as water-mass indicators.

ASSOCIATIONS

In the surface layer of bottom sediments in the Indian Ocean there are three main associations of radiolarian skeletons: antarctic, tropical and temperate.

The species which characterize the Antarctic association are the high-antarctic forms (*Lithelius nautiloides, Antarctissa strelkovi*). The northern limit of the Antarctic association coincides with the limit of distribution of these species (line I in Fig. 22.6). The Antarctic association has a circumpolar distribution (Hays 1965; Petrushevskaya 1966*b*). The preservation of radiolarian remains of this association is usually very good. It may be because of favourable physico-chemical conditions in the water column and in the sediments, or because of the morphology of the radiolarian skeletons themselves. Skeletons of many of the radiolarians occurring in the Antarctic are either spongy discs or heavy thick-walled shells of comparatively large size.

Within its area of distribution the Antarctic association is not homogeneous. Samples from the Antarctic shelf (especially east of Mirny) are poor in radiolarian skeletons. As a rule the total number of skeletons there is less than that in deep-sea sediments. Skeletons of low-antarctic species (*Botryopyle antarctica*) are comparatively rare, and those of the widely distributed cosmopolitan species (*Theocalyptra davisiana, Artostrobus annulatus*), and of *Dictyophimus gracilipes, Lithomelissa borealis, Cornutella stylophaena*, etc. are practically absent. The shelf samples contain abundant *Antarctissa strelkovi, Lithelius nautiloides*, and *Spongotrochus glacialis.* In terrigenous deep-sea sediments near the Antarctic continent there are more radiolarian shells (per gramme of dry sediment) than on the shelf. There is also a greater variety of radiolarian species, but *Antarctissa strelkovi, Lithelius nautiloides, Spongodiscus glacialis* remain in greatest relative abundance. In the diatom oozes away from the Antarctic continent skeletons of these species become comparatively less abundant, and those of *Antarctissa denticulata, Saccospyris antarctica, Phorticium clevei, Lithomitra arachnea, Theocalyptra davisiana, Cornutella stylophaena, Dictyophimus gracilipes* and of many other species, become relatively more frequent than they were at higher latitudes. The total number of skeletons per gramme in diatom oozes is always very high (scores of thousands).

At 58°-50° S and especially in the region of the Antarctic Convergence *Lithomelissa borealis, Theocalyptra craspedota* and even some subtropical species such as *Lithocampe* (?) *aquilonaris* and *Axoprunum stauraxonium* become comparatively frequent. North of the Antarctic Convergence skeletons of *Antarctissa denticulata, Lithomitra arachnea* and of some other species are found to be less abundant than south of it. Thus at the limits of its distribution the Antarctic association is somewhat different from in its central parts.

The Tropical association is characterized by the presence of tropical warm-water surface species (Table 22.1). The southern limit of distribution of these species forms the boundary of this association. This boundary lies in the region of the Subtropical Convergence.

In the siliceous-calcareous oozes of the Equatorial zone radiolarian skeletons are abundant (hundreds of thousands per gramme of dry sediment) and rather well preserved (Nigrini 1965; Petrushevskaya 1966*a*). Not only tropical species but also subtropical and widely distributed cosmopolitan species are present. In the calcareous oozes extending southwards (between 20° S and 32° S) there lies a belt where radiolarian skeletons are scanty (Nigrini 1965). In the sediment samples from that region (st. 302 and 307 *Ob*) only scores of skeletons were found per gramme.

The greater abundance of radiolarian remains in the equatorial sediments is evidently connected with their increased development in the plankton. Although there are practically no data about the amounts of Spumellaria and Nassellaria produced in these regions, in the suspended material of samples 69 and 74 *Ob* (3° N-12° S) there were more radiolarians than in sample 64 (26°-30° S).

In sediments from the region between 20° and 32° S the same species were present as in equatorial sediments. But such species as *Siphonosphaera socialis, Spongodiscus resurgens, Lithocampe* (?) *aquilonaris, Hymeniastrum asteriscus* and *Lamprocyclas maritalis* are present in greater relative abundance than at low latitudes. Conversely skeletons of such warm-water surface forms as *Euchitonia elegans* and *Botryocyrtis scutum* are more rare there than in equatorial sediments. All skeletons in these sediments are corroded. The increase in *Siphonosphaera socialis, Spongodiscus resurgens*, and other species may be explained not only by their higher production in that region, but also to some extent by their better preservation (compared with that of the delicate skeletons of *Euchitonia elegans*, etc.) in calcareous sediments. At the southern limits of distribution of the Tropical association some temperate zone species (*Theocalyptra craspedota*, etc.) and even some low-antarctic species become more frequent.

Between the Tropical and Antarctic associations there is a peculiar radiolarian association corresponding to the temperate zone, but its limits (Fig. 22.6) are displaced somewhat northwards compared with the mean positions of the Antarctic and Subtropical Convergences. In the Indian Ocean it represents a band constricted at about 100° E to 32°-47° S; it is somewhat broader at about 70° E. In the Pacific this association occupies nearly all the Tasman Sea. Skeletons of high-antarctic and tropical species are practically absent, and this association is characterized by the presence of skeletons of *Androcyclas gamphonycha, Lithomitra clevei, Lithomelissa borealis, Lithostrobus seriatus, Phorticium pylonium, Lithocampe* (?) *aquilonaris, Lithocampe*, sp., *Axoprunum stauraxonium, Eucyrtidium acuminatum, Siphonosphaera socialis* and some other colonial spumellarians; the cosmopolitan species *Spongodiscus resurgens* and *Stylatractus neptunus* are also abundant. Most of the species listed are common in the Tropical association. Such species as *Theocalyptra davisiana, Lithomitra arachnea, Artostrobus annulatus* and *Lithomelissa borealis* are also abundant in the Antarctic association. Only *Androcyclas gamphonycha* is confined to the zone.

The Notal association occurs in calcareous oozes. The number of skeletons of radiolarians in the association is measured in scores of thousands per gramme. The preservation of the shells is usually very poor. This association may be compared with the Boreal radiolarian association of the North Atlantic and Norwegian Sea sediments (Table

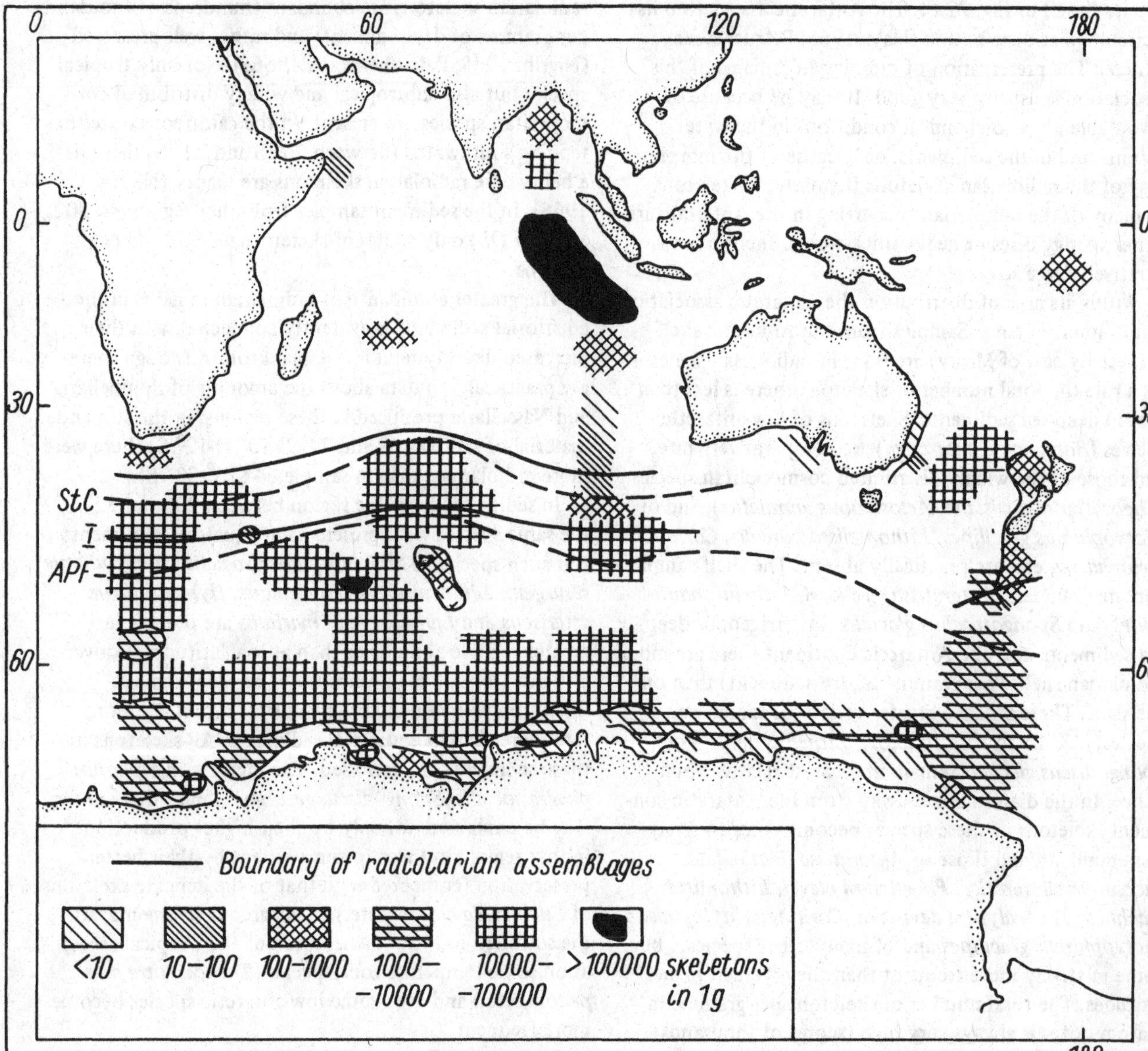
0
60
120
180
0
30
60
StC
I
APF
Boundary of radiolarian assemblages
<10
10–100
100–1000
1000–10000
10000–100000
>100000
skeletons in 1g

Fig. 22.6. Radiolarian abundance in surface layer of bottom sediments and boundaries of associations. I = northern limit of Antarctic association. *StC* Subtropical Convergence, *APF* Antarctic Polar Front.

22.1).

The decrease in radiolarian abundance south of 32° S, and the difference in the faunal content were noticed by Nigrini (1965). Hays (1965) described a zone which forms a transition from the antarctic fauna to the so-called warm-water fauna. Hays terms it a mixed zone. It seems to be the same as the association under consideration. The criteria used by Hays and the present author for the establishment of zones or radiolarian associations in bottom sediments are somewhat different. The mixed zone is limited (Hays 1965) by the 10 per cent and 90 per cent isopleths. These two isopleths were established on the total percentage of seven species: *Spongoplegma antarcticum, Lithelius nautiloides, Triceraspyris antarctica, Antarctissa strelkovi* [= *Helotholus histricosa*], *Antarctissa* [= *Peromelissa*] *denticulata, Theocalyptra davisiana* and *Saccospyris* [= *Botryopyle*?] *antarctica.* Amongst the species listed are high-antarctic, low-antarctic and even cosmopolitan forms. It would seem better to use species of one zoogeographical group for establishing isopleths.

In conclusion it is possible to say that there are some peculiarities in the distribution of radiolarians which may have great potential for their stratigraphical utilisation.

Received September 1967.

BIBLIOGRAPHY

Bernstein, T. 1932. Ueber einige arktische Radiolarien. *Archiv Protistenk.* **76**, 217-27.

Bernstein, T. 1934. Zooplankton Karskogo Morya po materialam ekspeditsii Arkticheskogo Instituta na *Sedove* 1930 goda i *Lomonosove* 1931 goda. *Trudy Arctichesk. Inst., Leningrad,* **9**, 3-58.

Cleve, P. T. 1899. Plankton collected by the Swedish expedition to Spitzbergen in 1898. *Handl. Kgl. Svenska Vetensk.-Acad.* **35** (5), 1-58.

Cleve, P. T. 1900. The seasonal distribution of Atlantic plankton organisms. *Handl. Göteborgs Kgl. Vettensk-och Vitterh.samh., ser. 4,* **3**, 1-369.

Cleve, P. T. 1901. Plankton from the Indian Ocean and the Malay Archipelago. *Handl. Kgl. Svenska Vetensk.-Akad.* **35** (5), 1-58.

Dogel, V. A. and Reshetnyak, V. V. 1952. Materialy po radiolariyam severozapadnoi chasti Tikhogo okeana. *Issledovanya Dalnevostochnykh Morei SSSR,* **3**, 5-36.

Haeckel, E. 1887. Report on the Radiolaria collected by H.M.S. *Challenger* during the years 1873-76. *Rept. Voyage 'Challenger', Zool.* **18**, clxxxviii + 1803 pp.

Haecker, V. 1908. Tiefsee-Radiolarien. *Wissenschaftl. Ergebn. Deutschen Tiefsee Expedition ('Valdivia'), Jena,* **14** (1), Spezieller Teil, 1-476; (2), Allgemeiner Teil, 417-706.

Hays, J. D. 1965. Radiolaria and late Tertiary and Quaternary history of Antarctic seas. *Biol. Antarctic Seas, Antarctic Research Ser. 5 (Amer. Geophys. Union),* **2**, 125-84.

Hülsemann, K. 1963. Radiolaria in plankton from the Arctic Drifting Station T-3, including the description of three new species. *Arctic Inst. Nth. Amer., Techn. Paper* **13**, 1-52.

Ivanov, Yu. A. 1961. O frontalnykh zonakh v antarcticheskikh vodakh. In : *Oceanological Research* (Xth section of the IGY programme oceanology), no. 3, pp. 30-53. Akademii Nauk SSSR: Moscow.

Jorgensen, E. 1905. The protist plankton and diatoms in bottom samples. *Bergen Mus. Skr., 1905,* 114-51.

Lisitzin, A. P. 1966. Osnovnye zakonomernosti raspredelenia sovremennykn kremnistykh osadkov i ich sviaz s klimatitcheskoy zonalnostyu. In: *Geokhimiva Kremnezema,* 90-192. Nauka: Moscow.

Nigrini, C. 1965. Radiolaria in recent pelagic sediments from the Indian and Atlantic Oceans. Dissertation submitted for degree of D.Phil.; University of Cambridge. 1-234.

Petrushevskaya, M. G. 1966a. Radiolarii v planktone i v donnykh osadkakh. In: *Geokhimiya Kremnezema,* 219-45. Nauka: Moscow.

Petrushevskaya, M. G. 1966b. Radiolaria in Antarctic bottom sediments. *SCAR, SCOR, IAPO, IUBS Symposium on the Antarctic Oceanography, Santiago,* 37-40.

Petrushevskaya, M. G. 1967. Radiolarii otryadov Spumellaria i Nassellaria Antarcticheskoy oblasti (po materialam Sovetskoi Antarcticheskoi Ekspeditsii). *Issledovanya fauny morei, t.IV (XII). Resultaty biologicheskikh issledovaniy Sovets-koi Antarctich. Eksped. 1955-1958,* **3**, 5-186.

Popofsky, A. 1908. Die Radiolarien der Antarktis (mit Ausnahme der Tripyleen). *Deutsche Südpolar-Exped. 1901-1903,* **10** (Zool. vol. 2), (3), 183-305.

Riedel, W. R. 1952. Tertiary Radiolaria in Western Pacific sediments *Göteborgs Kgl. Vetensk.-och Vittehets-Samhälles Handl, följ. 7, ser. B,* **6** (3), 1-18.

Riedel, W. R. 1958. Radiolaria in Antarctic sediments. Repts. B.A. N.Z. *Antarctic Research Exped., ser. B,* **6** (10), 217-55.

23. RADIOLARIANS AS INDICATORS OF PAST AND PRESENT WATER-MASSES

R. E. CASEY

Summary: Certain radiolarian distributions may be correlated with water-mass distributions. Not only do certain radiolarians seem to be endemic to specific water masses but they also, in some cases, tend to submerge with the diving water-masses , and therefore are indicators of the entire extent of those water-masses. From sediment and plankton studies, a scheme of biogeographical zones for polycystine radiolarians in the Pacific is proposed. A study in progress suggests that radiolarian water-mass indicators may be very useful in palaeo-oceanographic sediment studies.

INTRODUCTION

Radiolarian distribution has been correlated with water-masses or conditions that reflect water-mass relationships by many authors. These have been briefly reviewed in chapter 7 (Casey 1970).

The main thesis of the present paper is threefold: (1) to emphasize the importance of water masses to radiolarian distribution, (2) to give some examples of certain forms that may be indicative of specific water masses, (3) to note that the study of present day radiolarian water-mass indicators may be used in sediment studies.

INVESTIGATIONS OFF SOUTHERN CALIFORNIA

Sediment samples of Recent age from the Pacific Ocean (Fig. 21.1) were used to construct the table (Table 23.1) on radiolarian distribution in Pacific sediments (page 332) and to show some interesting natural zones of polycystine radiolarian distribution (Fig. 23.2). Since the main intentention of the sediment study is to aid in a seasonal study off Southern California mentioned above, most of the sediment samples are near or adjacent to the Southern California area. A few sediment samples are from the South Pacific in order to get a broader picture of radiolarian distribution in the entire Pacific.

The natural breaks in the sediments correspond very well with large scale hydrographic perimeters and to areas of water-mass formation and distribution. The subtropical and polar convergences seem to be natural breaks in the distribution of radiolarians in the North Pacific and this seems to be indicated also in the samples examined from the South Pacific. Other investigators such as Petrushevskaya (1966, 1967) and Kling (1966) have mentioned the same phenomena.

North of the polar (= subarctic) convergence *Cornutella profunda, Cyrtopera languncula, Pterocanium* sp., and *Spongotrochus glacialis* are dominant components of the sediments. However, *Cornutella profunda* and *Cyrtopera languncula* also occur (rarely) in the sediments of the equatorial and central regions and are abundant again in the sediments of the Antarctic region. These two forms *Cornutella profunda* and *Cyrtopera languncula* occur in the Southern California study (Casey 1966) at depths between 400 or 500-1,000 metres. The author believes these forms to be good indicators of the North Pacific Intermediate Water Mass which dives at the polar convergence. Since these two forms become dominant members of the radiolarian populations in the sediments of the Antarctic region (and are found in low numbers in equatorial and central waters), they are considered to be good tropical submergent forms and may be indicative of the intermediate waters of the Pacific in general. These two forms are here designated as Subarctic-Intermediate forms. *Pterocanium* sp. and *Spongotrochus glacialis* may dive to some degree; and, in fact, are also found to be dominant members in the sediments of the Transition region, along with *Pterocorys hirundo* and ? *Hexadoridium streptacantum* which seem to be abundant in the Transition region only. Therefore, *Pterocanium* sp. and *Spongotrocus glacialis* are here considered to be members of the Subarctic-Transition Fauna and *Pterocorys hirundo* and ? *Hexadoridium streptacantum* to be part of the Transition Fauna. *Sethophormis rotula* also occurs in the sediment of the Transition region but has been taken in the Arctic (Hulsemann 1963) and is here considered a component of the Subarctic-Transition Fauna.

The faunas considered Subarctic-Transition and Trans-

Table 23.1. Relative abundance of radiolarians in Pacific sediments (A = abundant; CA = common to abundant; C = common; RC = rare to common; R = rare).

Sediment zones	Subarctic			Transition				Central (north)				Equatorial						Calif. current								Central (south)		Subantarctic	
Biogeographic Faunal zones / Station no.	RUS 3324	RUS 3274	CK 8 G	CK 6	CK 13	CK 3	MUK B 22 G	RIS 125 G	RIS 118 G	MIDPAC C20-1	TET 72	TET 38	CHUB 4	CHUB 10	DWBG 12	DWBG 147 B	CHUB XI G	MUK B 31 G	FAN BG 27	FAN HMS 12 G	FAN BG 9	CUSP 24 G	AHF 8493	FAN HMS 5 G	ZAP 2 P	MSN 126 G	MSN 111 G	MSN 99 G	MSN 85 G
Subarctic Transition Fauna																													
Pterocanium sp.			C	R			A							R				C	C		R		R	R			R	R	
Sethophormis rotula																		A									C	C	R
Spongotrochus glacialis	A	A	A	A	A	RC	C			R	A	R		R	R			C	RC	A		R	C		R		R	C	A
Transition Fauna																													
Pterocorys hirundo	R	C	A	A	A	R	A				C	R	R	R				C	C		C		R	R	R		RC		R
? *Hexadoridium streptacanthum*					CA	A	C	R	R			R	R	R	R	R		C			C	RC		R	R		R		
Subarctic Intermediate Fauna																													
Cornutella profunda	R	RC	A	C	C	A		R	C	RC	C	R	R	RC	R	R		C	C	A	R	C	C	C	C	R	A	C	C
Cyrtopera langungula		R	A	A	CA	R	R		C	C	A		R	R	R				A			RC	R	R		R	RC	RC	
Transition Central Fauna																													
Peripyramis circumtexta	R	R		CA	C	C	A	R	R	R	C	R	RC	C	R	R		R	C	A	R	RC	C	C	R		RC	R	R
Siphocampe erucosa	A	R	A	C		A	A	R	C	RC	A	RC	C	RC	C	C		C	C	A	C	C	C	C	R	R	RC	C	R
Spongopyle osculosa	C		C	RC	CA	C	C				A	R	C	RC				RC		A	C	C	C	C	R	R	R	C	A
Central 'Shallow' Fauna																													
Calocyclas amicae						R		A	A	C	A	R		C	RC	R		RC				A	C	C	RC	R	CA		
Euchitonia furcata								A	A	C		CA	RC	C	C	C							C	C	C	A	R		
Eucyrtidium hertwigii									CA	R	A	RC			CA	C							C	C					
Equatorial Fauna																													
Acrosphaera murrayana									R	C		R	RC	A	C	A	A	C	C		R	C	C	C	CA				
Acrobotrissa cribrosa				R	R		R				A	R	R		A	C						R	C	C	C				
Amphispyris costata-thorax								C	C	RC		C		RC	C	CA			R			RC	C	C	C				
Anthocyrtidium cineraria								RC	C	C		R	C	C	A	CA						RC	C	C	C	R			
? *Clathrocorys murrayi*						R			C						C	R						R	C	R	RC				
Dictyoceras vichowii				R		R			RC			RC		R	R	C			C			R	C	C	C		R		
Eucecryphalus sp.																							R		R				
Lithomelissa monoceras										C	C		RC		CA	C			R				C		C				
Peridium spinipes		R		R	R	RC			R	RC	C	R	C	C	C	C		R	RC	R	R	R	A	RC	C				R
Tristylospyris scaphipes										R			RC	CA	C	C			R						C				
Equatorial Central 'Shallow' Fauna										CA	A	CA	C	C	A	C	C	R	C	C	R	RC	C	A					
Botrycyrtis sp.										CA	A	CA	C	C	A	C	C					R	C	C	C				
Carpocanium sp.						R		C	C	C	C	C	C	RC	RC	R	C	R	C	C	R	RC	C	A		A			
Clathrocanium ornatum									C	R	C	C			CA	R							C	R	RC				
Dictyocoryne profunda						C		R	R	RC	A	RC	R	RC	C	C		R	C	R	R		C	R	C		R	R	R
Eucyrtidium hexagonatum					R	R			C		A			C	C	C		C					C	C	C				
Lamprocyclas maritalis						C			R	R	C		RC	C	R	C	C		C	R	C		C	C	C		A	R	R
Larcospira quadrangula						RC		R		R			R		R	R	C	C	C			R	C		RC				
Lithamphora furcaspiculata		R	RC	C		C	C		C	RC	A	RC	C	C	C	C	C	C	C		C	C	C	C	C		A	C	C
Litharachnium tentorium				C	R		C		R	R		R	R		R			A	R			R		R	R		R	R	R
Pterocanium praetextum						R			R	R	A	C	C	C	A	C	R		R				C	R	C				
Pterocanium trilobum						C	C	A	A	R	A	R	R	R	C	C	R		C	R	R	R	C	C	R				R
Theoconus zancleus				R	A	A	A	R	A	R		RC	R	A	R	R	C	A	A	R	R	C	A	C	RC	R	R	RC	R
Theophormis callipilium															C										R				
Spongaster tetras	CA			R	C	C	A		R		A	RC	C	C	R	C			R					C	C	R		R	R
Spongocore puella					RC	R			R	R		R		C	R	C			R			R		C	RC	R	R		

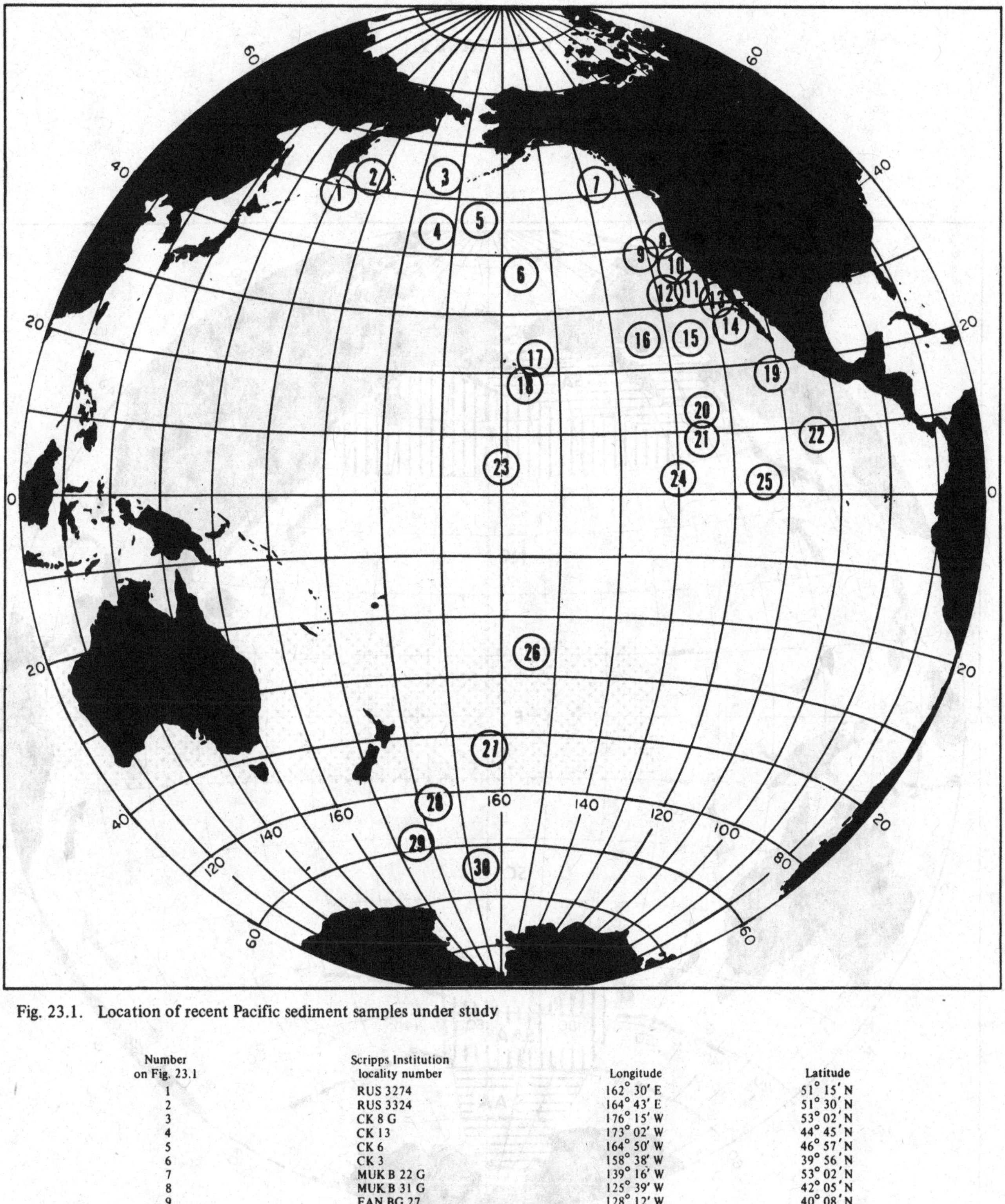

1
2
3
4
5
6
7
8
9
10
11
12
13
14
15
16
17
18
19
20
21
22
23
24
25
26
27
28
29
30
60
40
20
0
160
140
120
100
80

Fig. 23.1. Location of recent Pacific sediment samples under study

Number on Fig. 23.1	Scripps Institution locality number	Longitude	Latitude
1	RUS 3274	162° 30′ E	51° 15′ N
2	RUS 3324	164° 43′ E	51° 30′ N
3	CK 8 G	176° 15′ W	53° 02′ N
4	CK 13	173° 02′ W	44° 45′ N
5	CK 6	164° 50′ W	46° 57′ N
6	CK 3	158° 38′ W	39° 56′ N
7	MUK B 22 G	139° 16′ W	53° 02′ N
8	MUK B 31 G	125° 39′ W	42° 05′ N
9	FAN BG 27	128° 12′ W	40° 08′ N
10	FAN HMS 12 G	126° 24′ W	38° 03′ N
11	FAN BG 9	123° 19′ W	35° 08′ N
12	CUSP 24 G	126° 02′ W	34° 29′ N
13	AHF 8493	118° 30′ W	33° 10′ N
14	FAN HMS 5 G	118° 42′ W	28° 35′ N
15	RIS 125 G	126° 38′ W	28° 25′ N
16	RIS 118 G	135° 53′ W	28° 25′ N
17	MIDPAC C20-1	154° 55′ W	20° 27′ N
18	TET 72	157° 26′ W	19° 12′ N
19	ZAP 2 P	109° 31′ W	17° 52′ N
20	CHUB 4	125° 30′ W	14° 20′ N
21	CHUB 10	125° 26′ W	10° 21′ N
22	CHUB XI G	105° 09′ W	10° 53′ N
23	TET 38	160° 30′ W	5° 21′ N
24	DWBG 12	131° 31′ W	3° 12′ N
25	DWBG 147 B	116° 13′ W	1° 27′ N
26	MSN 126 G	154° 45′ W	24° 41′ S
27	MSN 111 G	164° 08′ W	40° 37′ S
28	MSN 99 G	178° 57′ W	52° 37′ S
29	MSN 85 G	169° 12′ E	57° 43′ S
30	MSN 91 G	165° 56′ W	64° 11′ S

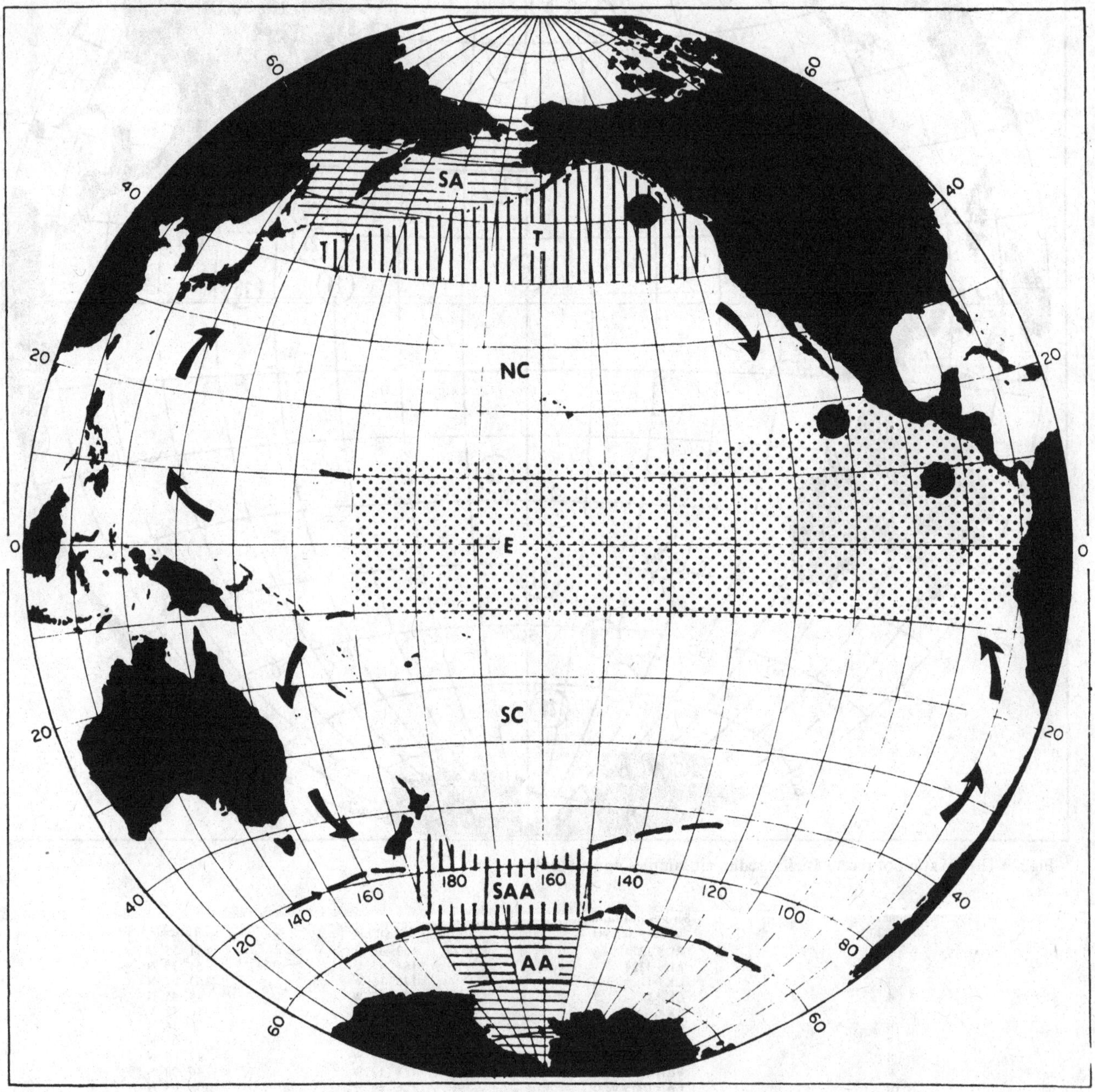
SA
T
NC
E
SC
SAA
AA
60
40
20
0
20
40
60
60
40
20
0
20
40
60
180
160
140
120
100
80
160
140
120

Fig. 23.2. Radiolarian zones from Pacific sediments. SA=Subarctic, T=Transition, NC=North Central, E=Equatorial, SC=South Central, SAA=Subantarctic, AA=Antarctic. (The large dots represent the three sub-zones mentioned in text.)

ition do not seem to be found to any great extent in the sediments of the central and equatorial Pacific and probably do not submerge substantially with the water-masses. However, since all except *Sethophormis rotula* are found to some extent in Equatorial, Central, and Antarctic sediments, they may be considered to be poor tropical submergent forms.

Another component found in the Transition sediments behaves differently; it is composed of *Peripyramis circumtexta, Siphocampe erucosa* and *Spongopyle osculosa.* These three are not only found in the Transition sediments but are rare in Central and Equatorial sediments and are dominant again in Subantarctic and Antarctic sediments. They are here considered to be tropical submergent and indicators of the North Pacific Central Water (and perhaps the South Pacific Central Water). Casey (1966) found them off Southern California at depths between 200 and 400 to 500 metres and considers them indicators of the North Pacific Central Water Mass.

Kling (1966), working with phaeodarian radiolarians in the North Pacific, found certain forms to be indicative of water-masses and, in fact, diving in the polar and subtropical convergent regions. The author's study off Southern California (Casey 1966) indicates the same trend, and the waters off Southern California between 200 to 400 or 500 metres are considered Central and those from 400 or 500 to 1,000 metres are considered Intermediate. The waters from the surface to 200 metres show a great seasonal fluctuation with invasions from Subarctic, Central 'surface' (not to be confused with the Central Water Mass of Svedrup, Johnson, and Fleming, but instead overlying this water in most regions), and Equatorial regions.

The three other zones found in the sediments may be termed Central 'surface', Equatorial, and Equatorial-Central 'surface' (living in both Equatorial and Central 'surface' areas) and are not diving forms (at least not to the degree the Subarctic-Intermediate and Transition-Central are). Off the coast of Southern California they are restricted, in the main, to the upper 200 metres.

These radiolarians that are here considered Central 'shallow' are *Calocyclas amicae, Euchitonia furcata* and *Eucrytidium hertwigii.* They occupy the Central 'shallow' faunal zone designated on Fig. 23.2. The Equatorial fauna consists of the following: *Acrosphaera murrayana, Acrobotrissa cribrosa, Amphispyris costata-thorax, Anthocyrtidium cineraria, ? Clathrocorys murrayi, Dictyoceras vichowii, Eucecryphalus* sp., *Lithomelissa monoceras, Peridium spinipes,* and *Tristylospyris scaphipes*, and they occupy the Equatorial faunal zone. Those occupying Equatorial and Central 'shallow' regions are termed the Equatorial-Central 'shallow' fauna and include: *Botrycyrtis* sp., *Carpocanium* sp., *Clathrocanium ornatum, Dictyocoryne profunda, Eucyrtidium hexagonatum, Lamprocyclas maritalis, Larcospira quadrangula, Lithampora furcaspiculata, Litharachnium tentorium, Pterocanium praetextum, Pterocanium trilobum, Theoconus zancleus, Theophormis callipilium, Spongaster tetras* and *Spongocore puella.*

Within the major areas mentioned, there seem to be subunits containing either species not generally found in the major area or species more abundant than in the major area. These subunits may be correlated with known oceanographically stable areas such as large eddy systems. Three of these areas were noted in the sediment study. One area was from the northeastern section of the Transition zone. In this zone, *? Hexadoridium streptacanthum*, and *Spongaster tetras* (the rectangular form) occurred in abundance and *Theoconus zancleus, Pterocanium trilobum,* and *Litharachnium tentorium* were present, although absent or rare in the rest of the Transition region. It is interesting to note here that the western transitional part of the zone did not have these individuals in abundance, or in some cases at all. However, the species *Peripyramis circumtexta, Siphocampe erucosa,* and *Spongopyle osculosa* were found in this western region and do dive with the central waters that form and submerge at this region. The individuals in the eastern region contribute to the fauna off Southern California in the oceanographic winter.

Another region that has subunits is the Equatorial region. There seems to be a zone south of Baja, California delimited by abundant *Eucecryphalus* sp., *Clathrocanium ornatum* and *Lithomelissa monoceras* and another near Central America delimited by abundant *Acrosphera murrayana, Botryocyrtis* sp., *Larcospira quadrangula* and *Lamprocyclas maritalis.* A more detailed study of this region has been completed by Nigrini (1968).

The only forms studied that seem to show a bipolar distribution (restricted to Subarctic and Transition water masses of the Northern Hemisphere and Antarctic and Subantarctic water masses of the Southern Hemisphere) are *Sethophormis rotula* (found in the North Pacific sediments in this study at station MUK B 31 G, which is located at the head of the California Current and considered a Transition Station) and *? Hexadoridium streptacanthum.* The latter may correspond to Petrushevskaya's *Rhizoflegma* (?) *boreale* which she suggests may be bipolar (Petrushevskaya 1967).

As has already been mentioned, there seem to be many forms that are tropical submergent and appear in common or abundant numbers in the Antarctic. However, it is worth noting that some of these faunas do not correspond directly with their counterparts in the opposing hemi-

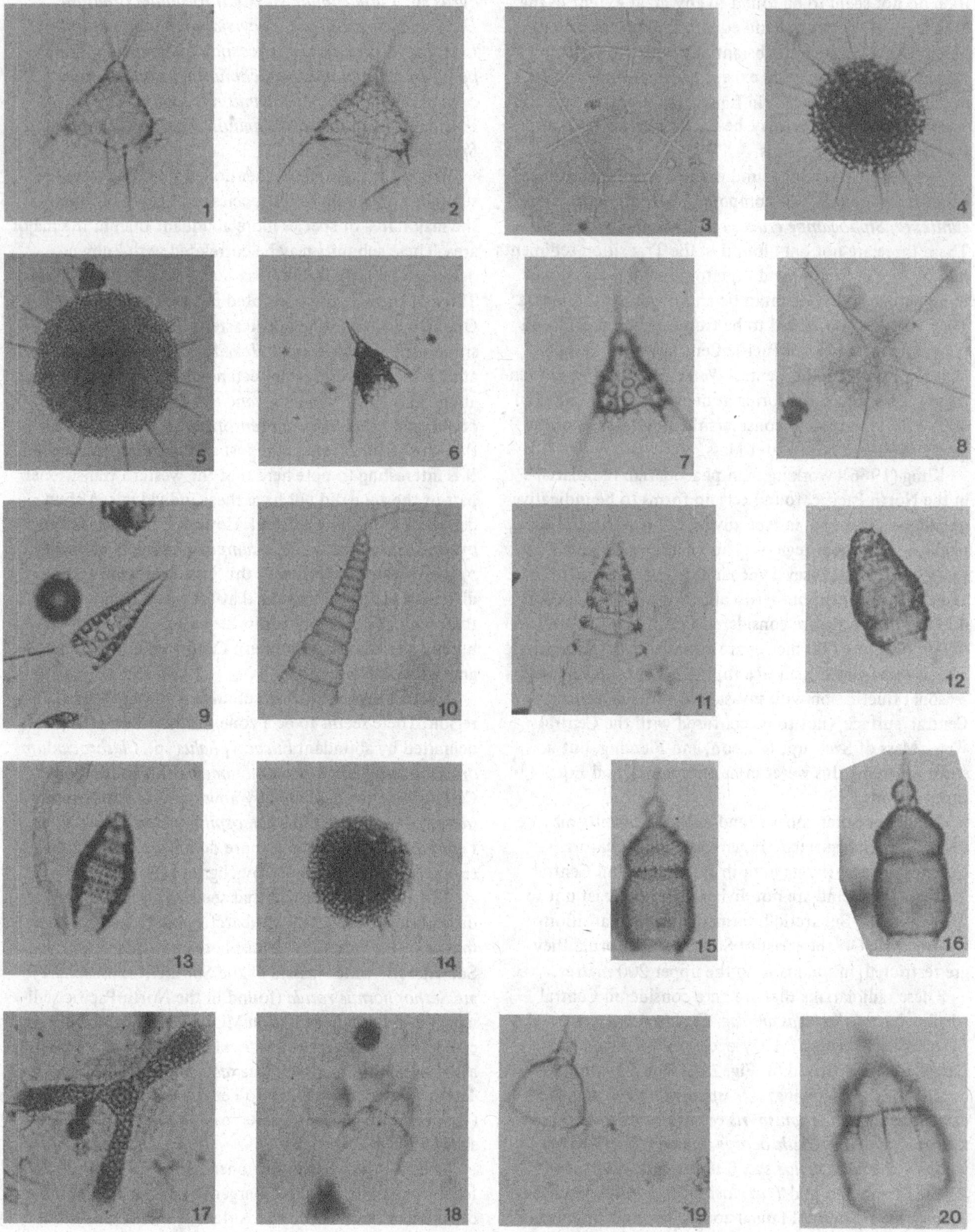
1
2
3
4
5
6
7
8
9
10
11
12
13
14
15
16
17
18
19
20

sphere. In other words, those radiolarians indicative of the Subarctic-Intermediate waters of the Northern Hemisphere do not necessarily correspond with those of the Subantarctic-Intermediate waters of the Southern Hemisphere. Rather they may occur in other diving water masses of the Southern Hemisphere, as seems to be the case with *Peripyramis circumtexta* which is here considered a Transition-Central form in the Northern Hemisphere but which may be an Antarctic-Intermediate form in the Southern Hemisphere. The author believes it is reasonable to assume that certain forms may dive with a water-mass in one hemisphere but not be found in the opposing hemisphere although no examples of this form of distribution were found.

In another study in progress, radiolarians from layered sediments within the anaerobic Santa Barbara Basin (off southern California) are being studied by the author. This work is being done in conjunction with investigations of these layered sediments by scientists of the Marine Life Research Programme at Scripps Institution of Oceanography under the direction of Professor J. D. Isaacs. Although the present author's work is still in its initial stages some aspects may be mentioned.

The layered sediments being studied vary in thickness, but average around 2 millimetres each. It is believed that two consecutive layers (light and dark in the core) probably represent summer and winter layers and may be designated as a 'varve'. These are usually distinguished by Equatorial forms being dominant in the lighter layers and Subarctic and Transition forms being dominant in the darker layers. However, the colour of the layer does not always correspond with the presence of the Equatorial or the Subarctic and Transition indicators. Usually the light layer correlates with abundant Equatorial radiolarians, and at least part of the light colour may be due to the abundant planktonic Foraminifera usually associated with that layer. In some light layers however, Subarctic and Transition indicators are dominant, and perhaps the light colour in these cases is due to the presence of abundant diatoms. Some invasions of Equatorial indicators seem to illustrate very strong invasions of Equatorial water; whereas, other Equatorial invasions are weaker. During what are considered strong Equatorial invasions, the collosphaerid component of the sediment is very apparent, contrasting with the weaker invasions where there are fewer Equatorial indicators in general and in some cases no collosphaerids.

On a larger scale the 'varves' are usually grouped into sets, and there appear to be about nine or ten of these sets or cycles in the core now being studied. If these are truly 'varves', the general cycle preserved seems to be as follows: warm 'years' with easily distinguishable layers of Equatorial indicators and layers of Subarctic and Transition indicators followed by cooler 'years' represented by poorer layering and poorer distinctions of what might represent summer and winter layers. Few Equatorial forms and good Subarctic and Transition forms are found in abundance in all layers. Subarctic-Intermediate and Transition-Central indicators are present in every layer and the only differences in assemblages come from the upper water layers. The author suggests that this basic pattern illustrated by the radiolarians is the same pattern that has been noticed for the local area with monotonous cool periods separated by warmer periods (Isaacs, paper given orally at the Symposium on the Marine Environment, Fullerton Junior College, 26 March, 1963).

Not only are the Equatorial and the Subarctic and Transition radiolarian invasions preserved in the sediments, but so far one example of an even shorter water mass and radiolarian invasion has been noted. This is an invasion of Central 'shallow' water indicators which lasted only a month in the plankton study of the author (Casey 1966). This invasion is not represented as an individual layer in the sediments but, rather, is preserved as Central 'shallow' water indicators mixed in a layer dominated by Equatorial species; indicating that this Central 'shallow' invasion was preserved but represented too small a thickness in the core to be observed visually or extracted intact. Therefore, what usually is found in the plankton as an Equatorial invasion replaced by a shortlived Central 'shallow' invasion and followed by a Subarctic and Transition in-

Plate 23.1.

Subarctic-Transition fauna
1-2 *Pterocanium* sp.
3 *Sethophormis rotula* Haeckel, 1887, p. 1246, pl. 57, fig. 9.
4-5 *Spongotrochus glacialis* Popofsky, 1908, p. 228, pl. 26, fig. 8; pl. 27, fig. 1; pl. 28, fig. 2.

Transitiona fauna
6-7 *Pterocorys hirundo* Haeckel, 1887, p. 1318, pl. 71, fig. 4.
8 *? Hexadoridium streptacanthum* Haeckel, 1887, p. 206, pl. 25, figs. 1 and 1a.

Subarctic-Intermediate fauna
9 *Cornutella profunda* Ehrenberg, 1859, p. 31.
10 *Cyrtopera languncula* Haeckel, 1887, p. 1451, pl. 75, fig. 10.

Transition-Central fauna
11 *Peripyramis circumtexta* Haeckel, 1887, p. 1162, pl. 54, fig. 5.
12-13 *Siphocampe erucosa* Haeckel, 1887, pp. 1500-1, pl. 79, fig. 11.
14 *Spongopyle osculosa* Dreyer, 1889, pp. 42-3, pl. 6, figs. 99 and 100.

Central 'shallow' fauna
15-16 *Calocyclas amicae* Haeckel, 1887, p. 1382, pl. 74, fig. 2.
17 *Euchitonia furcata* Ehrenberg, 1872, p. 308, pl. 6 (3), fig. 6.
18-20 *Eucyrtidium hertwigii* Haeckel, 1887, p. 1491, pl. 80, fig. 12.

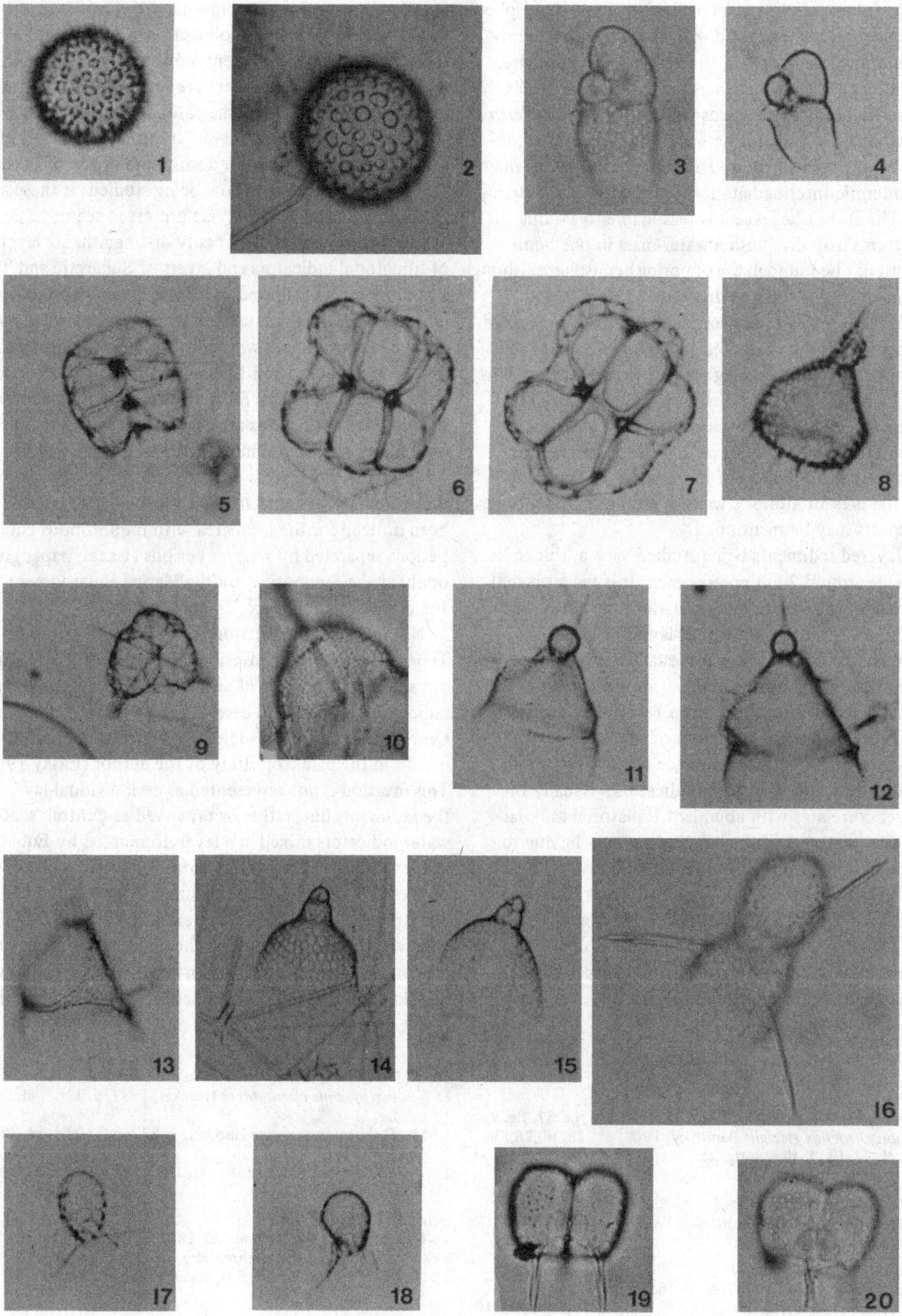
1
2
3
4
5
6
7
8
9
10
11
12
13
14
15
16
17
18
19
20

vasion, is recorded in the sediments as Central 'shallow' indicators mixed with a layer dominated by Equatorial indicators overlain by a layer dominated by Subarctic and Transition indicators.

In conclusion, it seems that certain polycystine radiolarians are endemic to water-masses to a greater or lesser degree (such forms as those of the Equatorial Water Mass to a high degree and those of the Equatorial-Central 'shallow' to a lesser degree) and may be used as indicators of both past and present water-mass and oceanographic conditions.

ACKNOWLEDGEMENT

My sincere thanks are extended to the Allan Hancock Foundation, University of Southern California, for providing the opportunity and facilities for the collection of plankton in the southern Californian areas and to Mr W. R. Riedel, Scripps Institution of Oceanography, for providing sediment samples from the Pacific for this study.
Received November 1967.

BIBLIOGRAPHY

Casey, R. E. 1966. *A seasonal study on the distribution of polycystine radiolarians from waters overlying the Catalina Basin, Southern California.* Ph.D. thesis: Univ. of So. Calif.

Casey, R. E. 1970. Distribution of polycystine radiolarians in the oceans. In: *The Micropalaeontology of Oceans* (B. M. Funnell and W. R. Riedel, *eds.*). Cambridge University Press. pp. 151-159.

Dreyer, F. 1889. Morphologische Radiolarienstudien. Heft 1; Die Pylobildundungen. *Jenaische Zeitschr. für Naturwiss.* **23.**

Ehrenberg, C. G. 1859. Kurze Characteristik der 9 neuen Genera und der 105 neuen Species des agaischen Meeres und des Tiefgrundes des Mittel-Meeres. *K. preuss. Akad. Wiss., Monatsber.*, Jahrg. 1858, 10-41.

Ehrenberg, C. G. 1872. Mikrogeologischen Studien als Zusammenfassung seiner Beobachtungen des kleinsten Lebens der Meeres-Tiefgrunde aller Zonen und dessen geologischen Einfluss. *Mber. preuss. Akad.Wiss.*, Jahrg. 1872, 265-322.

Haeckel, E. 1862. *Die Radiolarien. Eine Monographie.* Reimer: Berlin. xiv + 572 p.

Haeckel, E. 1887. Report on the radiolarian collected by H.M.S. *Challenger* during the years 1873-76. *Rept. Voyage 'Challenger', Zool.* **18**, clxxxviii + 1803 p.

Hülsemann, K. 1963. Radiolaria in plankton from the arctic drifting station T-3, including the description of three new species. *Arctic Inst. North Amer., Tech. Paper,* **13**, 1-52.

Kling, S. A. 1966. *Castanellid and circoporid radiolarians: systematics and zoogeography in the Eastern North Pacific.* Ph.D. thesis; Univ. of Calif., San Diego.

Nigrini, C. A. 1968. Radiolaria from eastern tropical Pacific sediments. *Micropaleont.* **14** (1), 51-63, pl.1.

Petrushevskaya, M. G. 1966. Radiolyarii v planktone i v donnykh osadkakh. *Geokhimiya Kremnezema.* 219-45. Nauka: Moscow.

Petrushevskaya, M. G. 1967. Nassellaria antarkticheskoe oblasti (Antarctic spumelline and nasseline radiolarians). *Issled. Fauny Morei,* **4** (12) (*Rez. biol. Issled. sov. antarkt. Eksped. 1955-58*), **3**, 5-186.

Popofsky, A. 1908. Die Radiolarien der Antarktis. *Dt. Sudpol. Exped. 1901-1903,* **10** (3), 183-305.

Popofsky, A. 1912. Die Sphaerellarien des Warmwassergebietes. *Dt. Sudpol. Exped. 1901-1903,* **14,** 217-416.

Plate 23.2.

Equatorial fauna
1-2 *Acrosphaera murrayana* (Haeckel) Hilmers, Popofsky, 1912, p. 259, figs. 22 and 23.
3-4 *Acrobotrissa cribrosa* Popofsky, 1913, p. 322, fig. 29.
5-7 *Amphispyris costata-thorax* group (could not distinguish between the forms *Amphispyris costata* Haeckel, 1887, p. 1097, pl. 88, fig. 3 and *Amphispyris thorax* Haeckel, 1887, p. 1096, pl. 88, fig. 4).
8 *Anthocyrtidium cineraria* Haeckel, 1887, p. 1278, pl. 62, fig. 16.
9-10 *? Clathrocorys murrayi* Haeckel, 1887, p. 1219, pl. 64, fig. 8.
11-13 *Dictyoceras vichowii* Haeckel, 1862, p. 333, pl. 8, figs. 1-5.
14-15 *Eucecryphalus* sp.
16 *Lithomelissa monoceras* Popofsky, 1913, p. 335, pl. 32, fig. 7.
17-18 *Peridium spinipes* Haeckel, 1887, p. 1154, pl. 53, fig. 9.
19-20 *Tristylospyris scaphipes* Haeckel, 1887, p. 1033, pl. 84, fig. 13.

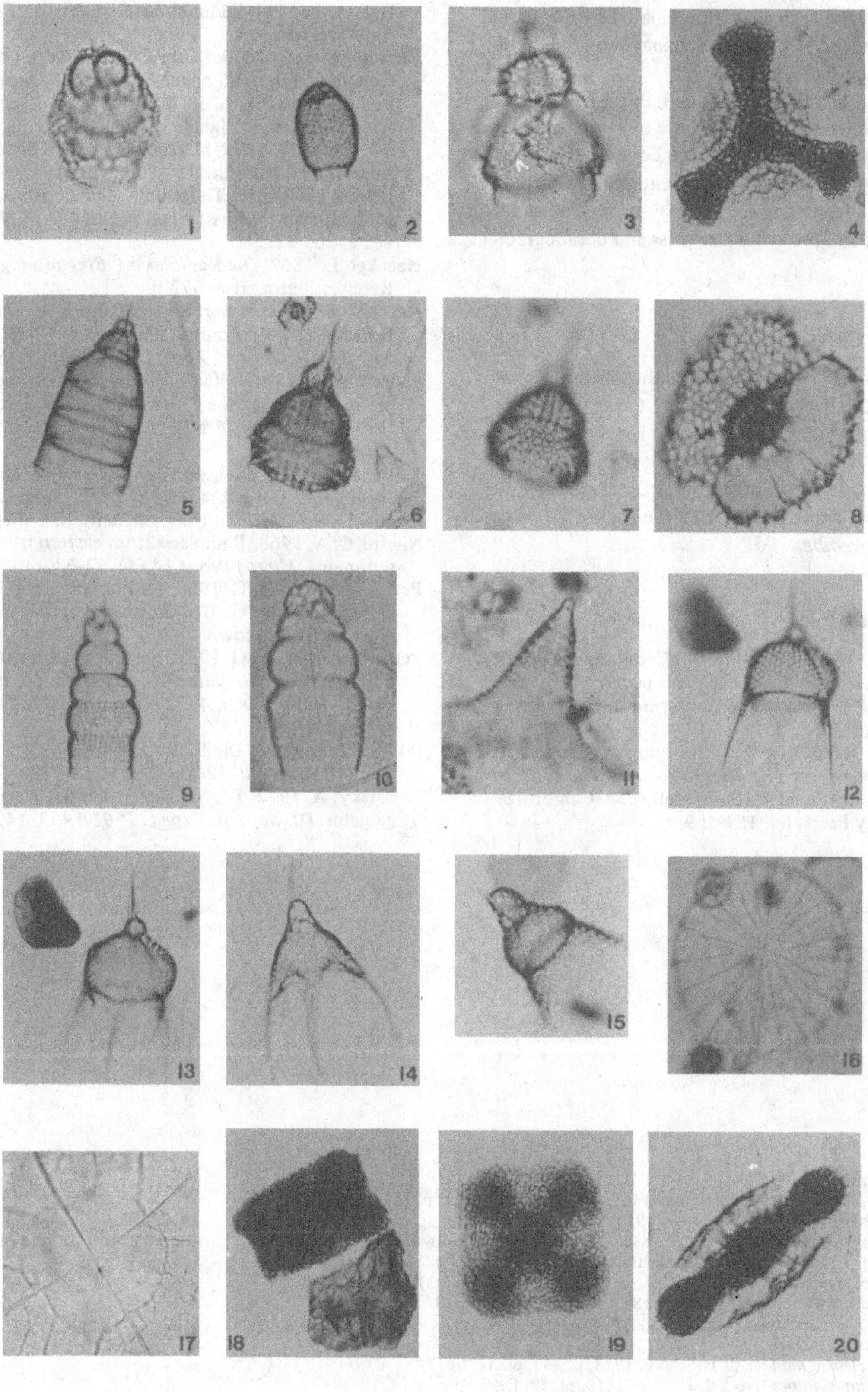
1
2
3
4
5
6
7
8
9
10
11
12
13
14
15
16
17
18
19
20

Plate 23.3.

Equatorial-Central 'shallow' fauna

1	*Botrycyrtis* sp.
2	*Carpocanium* sp.
3	*Clathrocanium ornatum* Popofsky, 1913, pp. 343-4, pl. 7, fig. 23.
4	*Dictyocoryne profunda* Ehrenberg, 1872, p. 307; 1873, pl. 7, fig. 23.
5	*Eucyrtidium hexagonatum* Haeckel, 1887, p. 1489, pl. 80, fig. 11.
6-7	*Lamprocyclas maritalis* Haeckel, 1887, p. 1390, pl. 74, figs. 13 and 14.
8	*Larcospira quadrangula* Haeckel, 1887, p. 696, pl. 49, fig. 3.
9-10	*Lithamphora furcaspiculata* Popofsky, 1908, p. 295, pl. 36, figs. 6-8.
11	*Litharachnium tentorium* Haeckel, 1862, Monogr. d.Radiol., p. 281, pl. 4, figs. 7-10.
12-13	*Pterocanium praetextum* (Ehrenberg) 1872; Haeckel, 1887, pp. 1330-1.
14	*Pterocanium trilobum* Haeckel, 1862, p. 340, pl. 8, figs. 6-10.
15	*Theoconus zancleus* (Muller) Haeckel, 1887, p. 1399.
16-17	*Theophormis callipilium* Haeckel, 1887, p. 1367, pl. 70, figs. 1-3.
18-19	*Spongaster tetras* Ehrenberg, 1872, p. 299, pl. 6 (3), fig. 8.
20	*Spongocore puella* Haeckel, 1887, p. 347, pl. 48, fig. 6.

24. OCCURRENCE OF PHAEODARIAN RADIOLARIA IN RECENT SEDIMENTS AND TERTIARY DEPOSITS

V. V. RESHETNJAK

Summary: So far eleven species of Phaeodaria have been found in Recent bottom deposits. Dumitrica (1964, 1965) records the presence of six more species in Tertiary deposits. The total number of Phaeodarian species found in either Recent sediments or Tertiary deposits is therefore very small, only seventeen species compared with a total of 606 species of Phaeodaria, otherwise known.

INTRODUCTION

Radiolaria of the order Phaeodaria are one of the most widely distributed groups of abyssal plankton. They have been investigated by numerous scientists, but their fragile shells are practically absent from bottom sediments, and there is therefore scant information on their occurrence there. The aim of the present report is to draw attention to those rare species of Phaeodaria whose shells may be preserved in marine deposits, and may be in some cases stratigraphically important.

OCCURRENCE IN BOTTOM SEDIMENTS

The first evidence of Phaeodarian shells in bottom sediments was found by Bailey (1856). After investigating sediments from the Kamchatka (Bering) sea, Bailey concluded that they were saturated with various siliceous remains. He gave a brief description of the genus *Cadium* and a picture of the species *C. marinum* Bailey, which he erroneously believed to be related to the Infusoria. Nevertheless, owing to the excellent illustrations of the shells of *C. marinum*, there is no doubt that Bailey was the first to study this order of radiolarians. Thirteen years later Wallich (1869), while studying recent abyssal sediments from the northern part of the Atlantic, also found phaeodarian shells, including *C. marinum* Bailey, in particular. Wallich established the new genus *Protocystis* with three new species: *P. aurita, P. lageniformis* and *P. spinifera* occurring in this area of investigation. But, judging from his pictures it is doubtful if the first two species belong to the Phaeodaria, because of the coarse, porous structure of the surface of the shells, and the third species *P. spinifera* greatly resembles *Euphysetta elegans*, particularly if it is turned over. Thus, only one of the three species of Phaeodaria mentioned above, namely *E. elegans* (=*P. spinifera*), can be regarded as an actual representative of the Phaeodaria. It is noteworthy, that *Cadium marinum* was originally found in the bottom sediments of the Bering sea and only later was it recorded from the plankton. Thus, Borgert (1901), dealing with Phaeodaria from the northern plankton, noted its occurrence in the plankton, as well as in sediments taken from arctic and tropical zones of the Atlantic ocean. Haecker (1908*a* and *b*) only referred to Bailey and Wallich's data in his description of *Cadium marinum.* Haecker himself reported this species only in plankton samples from the Atlantic and northern part of the Indian ocean. I have found this species in the plankton of the northwestern part of the Pacific Ocean and in Antarctica (Reshetnjak 1966, pp. 174-6).

Riedel (1963) has illustrated some unnamed Phaeodarian Radiolaria from a Quaternary sediment core from the south Pacific at 22° 03′ S, 178° 34′ E. These are the only records of the occurrence of the Phaeodaria in bottom sediments in the literature.

In order to study Phaeodaria in Recent bottom sediments materials from the surface sediment layer of the Antarctic (Pacific Ocean sector), cores and sediment grab samples obtained during the first and third cruises of R.V. *Ob* (1955-6 and 1957-8), were used. Also material from bottom sediments of the Norwegian Sea, obtained by R. V. *Sevastopol* (1958-9) was examined.

In the sediments from the Pacific sector of the Antarctic nine Phaeodarian species were distinguished, relating to three families: Aulacanthidae, Challengeridae and Medusettidae:

Euphysetta elegans Borgert (st. 33, 35) (Fig. 24.3)
Protocystis bicuspis Schroder (st. 34, 50) (Fig. 24.4)
P. micropelectus Haecker (st. 37) (Fig. 24.5)

P. swerei (Murray) (st. 44) (Fig. 24.6*a, b, c*)
Porospathis holostoma (Cleve) (st. 33, 413) (Fig. 24.7)
Aulospathis variabilis triodon Haecker (st. 370) (Fig. 24.8)
Aulographonium mediterraneum (Borgert) (st. 370) (Fig. 24.9)
Cadium marinum Bailey (st. 32, 35, 54, 123, 111) (Fig. 24.10)
Cadium melo (Cleve) (st. 32, 54, 123, 111) (Fig. 24.11)

In addition a small piece of the latticed net of a species of the Sagosphaeridae family (st. 37) (Fig. 24.12) was found.

In the Norwegian Sea only three Phaeodarian species were found, belonging to two families, Challengeridae and Porospathidae.

Cadium melo (st. 1125, 1378. 1706) (Fig. 24.11)

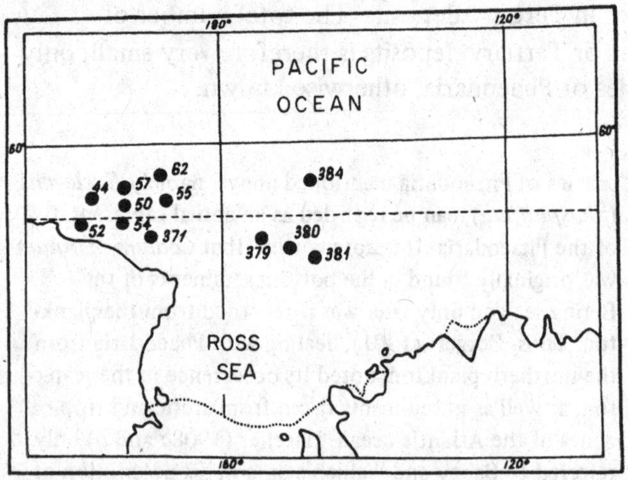
180°
120°
PACIFIC
OCEAN
60°
62
57
44
55
50
54
52
371
384
380
379
381
ROSS
SEA
180°
120°

Fig. 24.1. Map of Antarctic bottom sediment stations in which Phaeodaria were found.

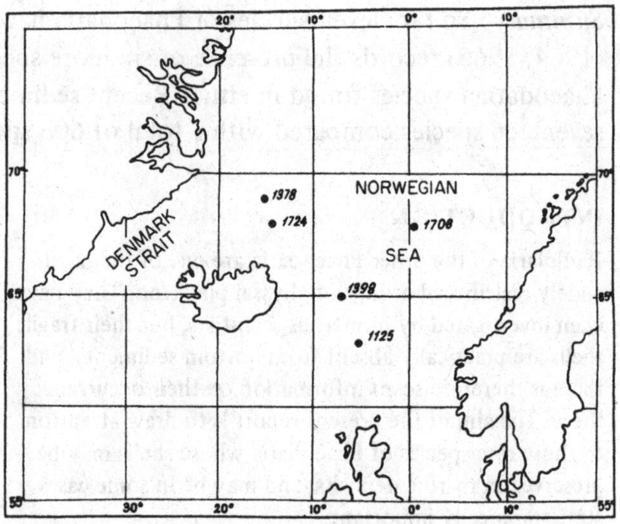
20°
10°
0°
10°
70°
NORWEGIAN
SEA
DENMARK
STRAIT
1378
1724
1706
1398
1125
65°
55°
30°
20°
10°
0°
10°

Fig. 24.2. Map of Norwegian Sea bottom sediment stations in which Phaeodaria were found.

Table 24.1. List of Phaeodarian species found in bottom sediments and their locations

Phaeodaria	Northern part of Atlantic (the Norwegian Sea)	Northern part of Pacific Ocean	Antarctica (Pacific Ocean sector)
I. Family Aulacanthidae			
Aulospathis variabilis triodon Haecker	–	–	+
Aulographonium mediterraneum (Borgert)	–	–	+
II. Family Porospathidae			
Porospathis holostoma (Cleve)	+		–
III. Family Challengeridae			
Challengeria naresi (J. Murray)	–	+	–
Protocystis bicuspis Schröder	–	–	+
P. swerei (J. Murray)	–	–	+
P. micropelectus Haecker	–	–	+
Cadium marinum Bailey	+	–	+
Cadium melo (Cleve)	+	–	–
IV. Family Medusettidae			
Euphysetta elegans Borgert	–	–	+
Euphysetta amphicodon Haeckel	–	+	–
V. Family Sagosphaeridae (piece of net)	–	–	+

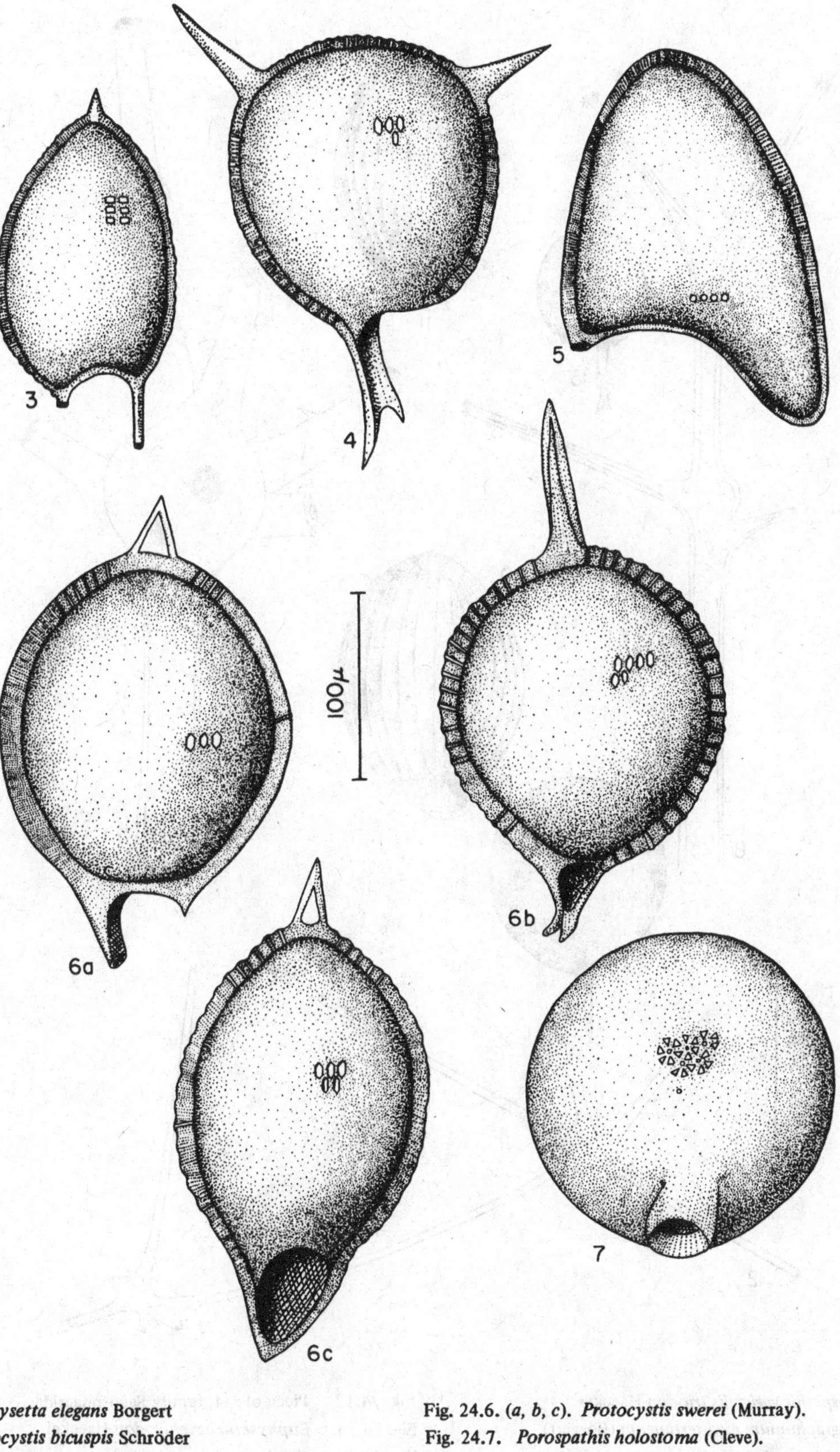
3
4
5
6a
100μ
6b
6c
7

Fig. 24.3. *Euphysetta elegans* Borgert

Fig. 24.4. *Protocystis bicuspis* Schröder

Fig. 24.5. *Protocystis micropelectus* Haecker.

Fig. 24.6. (*a, b, c*). *Protocystis swerei* (Murray).

Fig. 24.7. *Porospathis holostoma* (Cleve).

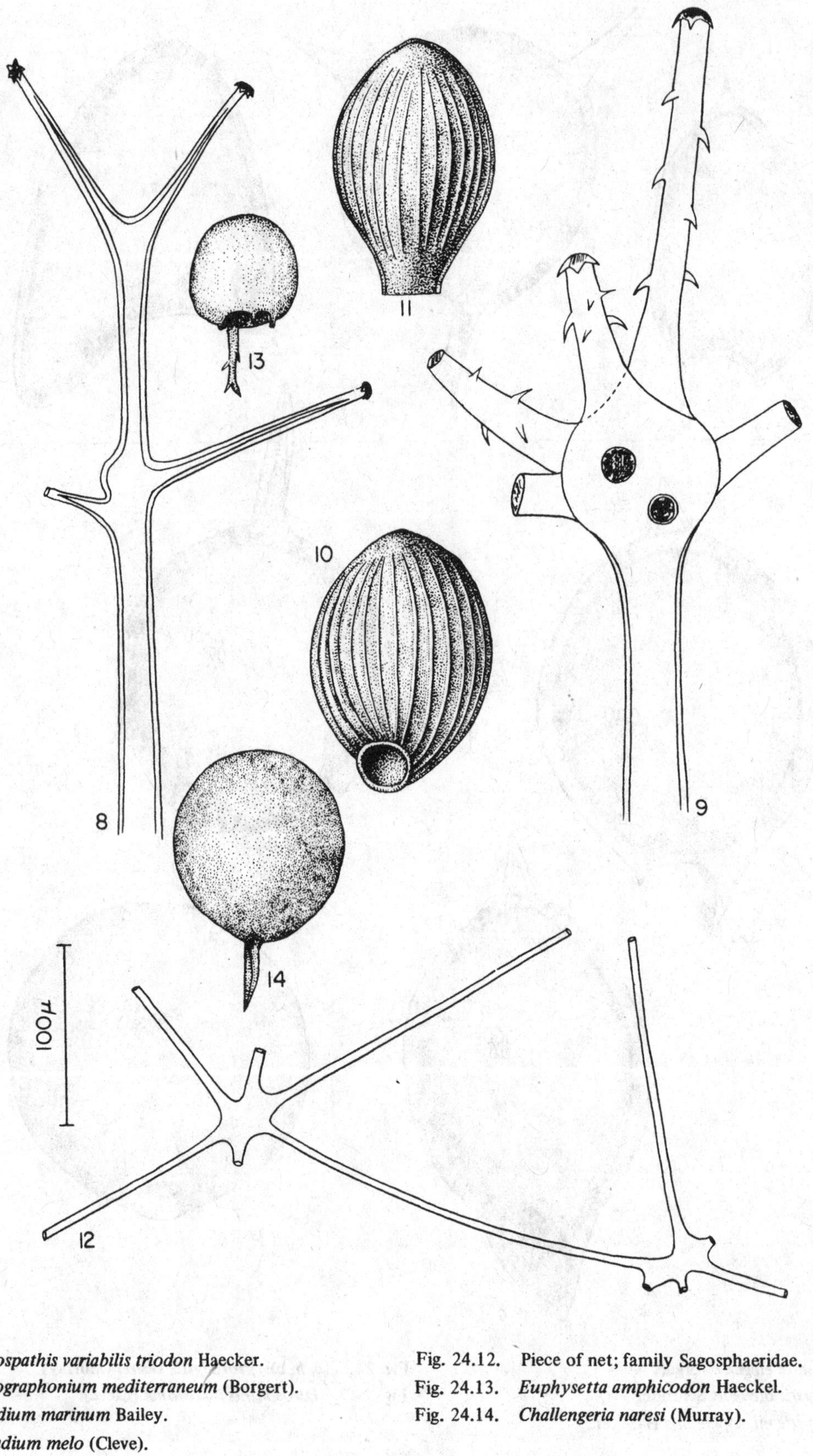
8
9
10
11
12
13
14
100μ

Fig. 24.8. *Aulospathis variabilis triodon* Haecker.
Fig. 24.9. *Aulographonium mediterraneum* (Borgert).
Fig. 24.10. *Cadium marinum* Bailey.
Fig. 24.11. *Cadium melo* (Cleve).
Fig. 24.12. Piece of net; family Sagosphaeridae.
Fig. 24.13. *Euphysetta amphicodon* Haeckel.
Fig. 24.14. *Challengeria naresi* (Murray).

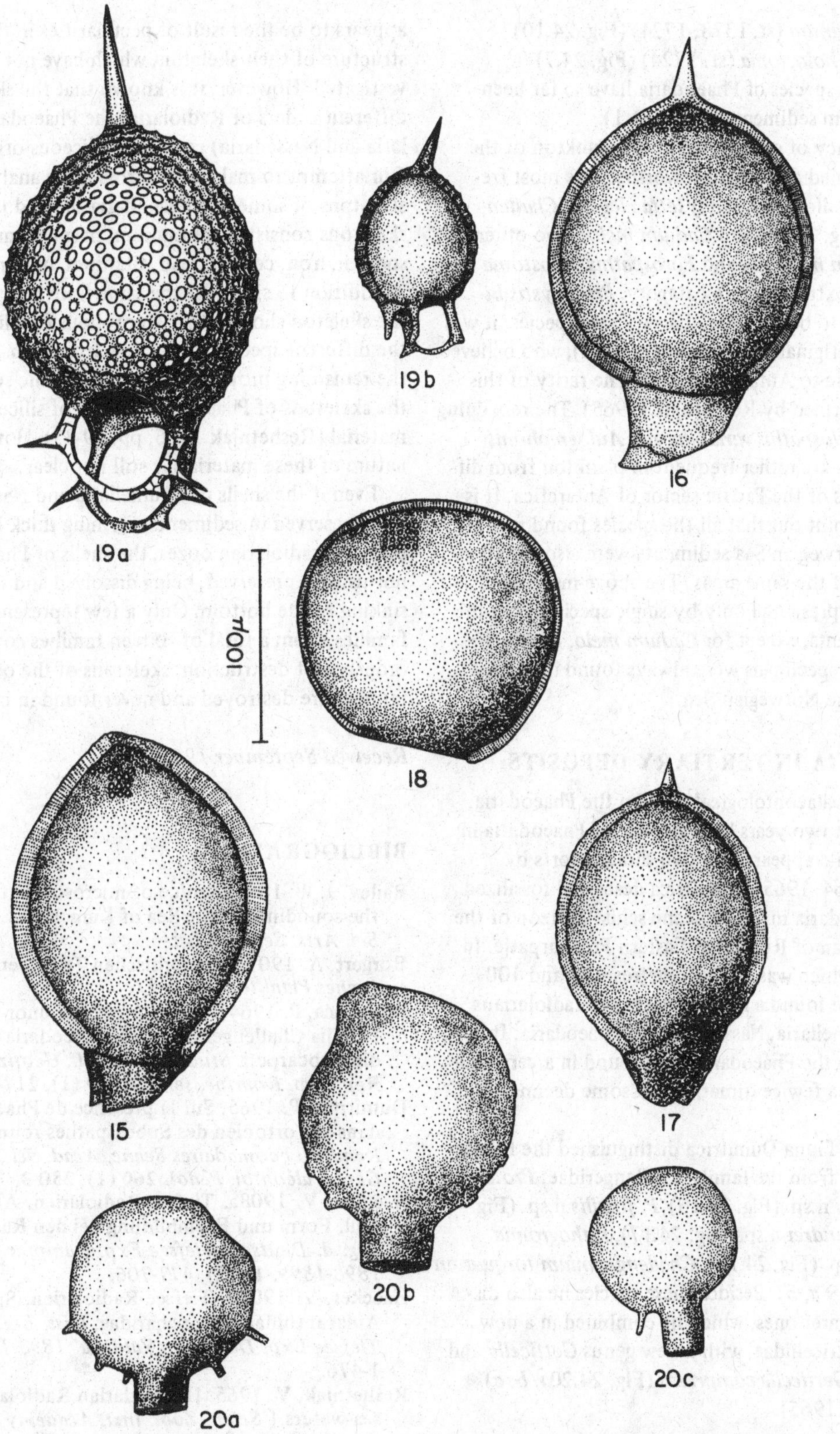
19a
19b
16
18
100μ
15
17
20a
20b
20c

Fig. 24.15. *Protocystis tortonica* Dumitrica.
Fig. 24.16. *Protocystis fossilis* Dumitrica.
Fig. 24.17. *Protocystis deflandrei* Dumitrica.
Fig. 24.18. *Lithogromia reschetnjak* Dumitrica.
Fig. 24.19 (*a*, *b*). *Challengerium torquatum* Dumitrica.
Fig. 24.20 (*a*, *b*, *c*). *Geticella compressa* Dumitrica.

Cadium marinum (st. 1378, 1724) (Fig. 24.10)
Porospathis holostoma (st. 1724) (Fig. 24.7)

Thus eleven species of Phaeodaria have so far been found in bottom sediments (Table 24.1).

The frequency of occurrence in the plankton of the species mentioned above is noteworthy. The most frequent are the following: *Euphysetta elegans, Challengeria naresi* (Fig. 24.14) and *Cadium melo.* Two other species, *Cadium marinum* and *Porospathis holostoma* belong to the extremely rare category. *Protocystis bicuspis* also has to be considered a very rare species; it was distinguished originally by Schröder (1913), who believed it to be endemic to Antarctic waters. The rarity of this species is confirmed by Reshetnjak (1965). The remaining two species *Aulospathis variabilis* and *Aulographonis mediterraneum* are rather frequent in plankton from different transects of the Pacific sector of Antarctica. It is necessary to point out that all the species found in Antarctic and Norwegian Sea sediments were also found in the plankton of the same areas. The above-mentioned species were represented only by single specimens in the bottom sediments, except for *Cadium melo,* of which thirty to forty specimens were always found in each sample from the Norwegian Sea.

PHAEODARIA IN TERTIARY DEPOSITS

There are few palaeontological data on the Phaeodaria. Only in the last two years has evidence of Phaeodaria in Tertiary deposits appeared, in two short reports by Dumitrica (1964, 1965). Dumitrica described fossilized Tertiary Phaeodaria in the Miocene schist horizon of the Upper Tortonian of Rumanian Eastern Predkarpatje. In this horizon, which was 100 kilometres long and 100 metres wide, he found a rich formation of radiolarians including Spumellaria, Nassellaria and Phaeodaria. It is interesting that the Phaeodaria were found in a very thin layer, between a few centimetres and some decimetres thick.

In the fossil fauna Dumitrica distinguished the following five species from the family Challengeridae: *Protocystis tortonica* n.sp. (Fig. 24.15), *P. fossilis* n.sp. (Fig. 24.16), *P. deflandrei* n.sp. (Fig. 24.17), *Lithogromia reschetnjak* n.sp. (Fig. 24.18), *Challengeranium torquatum* n.sp. (Fig. 24.19 *a, b*). Besides these species he also discovered some rarer ones, which he combined in a new family, the Getticellidae, with a new genus *Getticella* and a new species *Getticella compressa* (Fig. 24.20*a, b, c*) (see Dumitrica 1965).

CONCLUSIONS

The rare occurrence of Phaeodaria shells in sediments appears to be the result of peculiarities in the chemical structure of their skeleton, which have not yet been investigated. However, it is known that the skeletons of different orders of Radiolaria (the Phaeodaria, Spumellaria and Nassellaria) consist of siliceous-organic material. Our attempt to make a spectrographic analysis of the skeletons of some live Phaeodaria showed that their skeletons consist of aluminium, magnesium, calcium, sodium, iron, copper, lead, titanium, barium and silver, in addition to silica. At the same time X-ray analysis of the skeleton showed that amorphous (opaline) silica in the different species constitutes from 5 to 15 per cent; the remaining proportion is some organic substance, i.e. the skeletons of Phaeodaria consist of siliceous-organic material (Reshetnjak 1966, pp. 29-35). However, the nature of these materials is still not clear.

Even if the shells of Spumellaria and Nassellaria are well preserved in sediments, forming thick bottom deposits of radiolarian oozes, the shells of Phaeodaria are not usually preserved, being dissolved and destroyed while sinking to the bottom. Only a few representatives of four families, from a total of sixteen families constituting the order, resist destruction; skeletons of the other twelve families are destroyed and never found in bottom deposits.

Received September 1967.

BIBLIOGRAPHY

Bailey, J. W. 1856. Notice of microscopic forms found in the soundings of the Sea of Kamchatka. *Amer. Journ. Sci. Arts, Ser. 2,* 22, 1-6.

Borgert, A. 1901. Die nordischen Tripyleen. *Arten. Nordisches Plankton,* **15**, 1-52.

Dumitrica, P. 1964. Asupra prezentei unor Radilaria din Familia Challengeridae (ord. Phaeodaria) in Tortonianul din Subcarpeti. *Studi Cers. Geol., Geofiz., Geogr. Acad. Rep. Pop. Romfne, ser. Geol.* **9** (1), 217-22.

Dumitrica, P. 1965. Sur la présence de Phaeodaires fossiles dans le Tortonien des Subcarpathes roumaines. *Compt. Rend. Hebdomadaires Seanc. Acad. Sci., Gr.9, Miner Géol. Paléontol. Pédol.* **260** (1), 250-3.

Haecker, V. 1908a. Tiefsee-Radiolarien. Allgemeiner Teil. Form und Formbildung bei den Radiolarien. *Wiss. Erg. d. Deutsch. Tiefsee-Exp. Dampfer 'Valdivia' 1898-1899.* **14** (2), 477-706.

Haecker, V. 1908b. Tiefsee-Radiolarien. Speziller Teil. Aulacanthidae – Concharidae. *Wiss. Erg. d. Deutsch. Tiefsee-Exp. Dampfer 'Valdivia' 1898-1899,* **14** (1), 1-476.

Reshetnjak, V. 1965. [Phaeodarian Radiolaria of Antarctic sea-waters.] *Scien. Zool. Inst. Academy of Sciences,* **35**, 67-78. (In Russian)

Reshetnjak, V. 1966. [Deep-Sea Phaeodarian Radiolaria of the Northwest Pacific.] *Fauna SSSR, new ser.* **94**, 1-208. (In Russian)

Riedel, W. R. 1963. The preserved record: palaeontology of pelagic sediments. In *The Sea,* (M. N. Hill *ed.*), 3 Interscience: New York 866-87.

Wallich, G. 1869. On some undescribed testaceous rhizopods from North Atlantic deposits. *Monthly Microscop. Journ.* 1, 104-10.

25. OCCURRENCE OF PTEROPODS IN PELAGIC SEDIMENTS

C. CHEN

ABSTRACT

Pteropod shells are most common in Atlantic pelagic sediments with a high percentage of calcium carbonate, above depths of about 2,200 metres where the ocean water is supersaturated or saturated with respect to aragonite. Pteropods are rare or absent in North Pacific sediments because the water becomes undersaturated as shallow as 200 metres.

Occurrences of pteropods can be related to physiographic provinces. Pteropods are abundant in sediments of the following provinces in the North Atlantic: the Rift Mountains of the Mid-Atlantic Ridge, the Bermuda Pedestal, the Blake and Azores Plateaux, the Outer Ridge, the Bahama Banks, Gulf of Mexico, and the Caribbean and Mediterranean Seas. They are rare or absent in the sediments of the upper step of the Mid-Atlantic Ridge, Bermuda Rise, Bermuda Apron, Corner Rise, continental shelf, slope, and rise. No pteropods are found from the Abyssal Plain, Abyssal Hills, and Mid-Ocean Canyon. In the South Atlantic, they are present in the Walvis Ridge, Rio Grande Rise, and Mid-Atlantic Ridge sediments north of about 35° S latitude.

In the Pacific Ocean, pteropod shells are encountered more in the western than in the eastern side. They are reported from the Tasman Sea, East Indian Archipelago, and the South China Sea. In the Indian Ocean, pteropods are reported from the Arabian Sea, the Gulf of Aden, the Chagos Laccadive Plateau, and off the African and Australian coasts.

Pteropod shells are observed in the sediments in several modes of occurrence. They are mostly unaltered, a few have manganese-iron coating. Occasionally two types of altered shells are found; casts occur in the Pleistocene sediments of the Red Sea and Blake Plateau, and Pleistocene submarine limestones containing pteropods are reported in the Caribbean and Red Sea.

Orientation of pteropods is observed in cores from hot brine areas in the Red Sea. The C-axes (the longest dimension) of *Creseis acicula,* having a pencil-like conical shell, are parallel to the lamination of thin black layers, but random orientation appears in the thick grey layers. The black layers contain a greater amount of hydrogen sulphide indicating stagnant, reducing condition, during which pteropod shells were deposited on the ocean floor without disturbance. The grey layers were deposited under more oxygenated conditions, and the light pteropod shells were disturbed by very slight bottom currents.

Received September 1967.

BIBLIOGRAPHY

Chen, C. and Bé, A. 1964. Distribution of pteropods in the western North Atlantic sediments. *Bull. Amer. Assoc. Petrol. Geol.* **48**, 520-1.

Li, Y. H. 1967. *The degree of saturation of calcium carbonate in the oceans.* Ph.D. thesis; Columbia Univ. 1-145.

Murray, J. and Renard, A. 1891. Deep-sea deposits. *Rep. Voy. H.M.S. 'Challenger',* 525 pp.

Stubbings, H. G. 1939. The marine deposits of the Arabian Sea. *John Murray Exped. Science Rep.* **3**, 31-157.

26. OCCURRENCE OF OSTRACODES IN BOTTOM SEDIMENTS

H. S. PURI

Summary: The nature of bottom sediments, such as grain-size, sorting coefficient, bulk density, and presence or absence of bottom vegetation affect the distribution of ostracodes. Consequently, diverse sedimentary facies support distinct ostracode assemblages. Rate of compaction of littoral deposits affects the size of the interstices and thus the ability of ostracodes to crawl through them.

The faunal composition of marine and marginal marine environments varies from place to place. Sedimentary facies in the Mediterranean Basin, which exist as distinct units with chracteristic ostracode assemblages, are taken as an example.

INTRODUCTION

The nature of bottom sediments has a considerable effect on the distribution of ostracodes. The grain-size and shape of the sediment affects the distribution of interstitial ostracodes such as *Polycope, Leptocythere* and *Parvocythere, Cytheromorpha, Microcythere, Xestoleberis* and *Microloxoconcha.* The rate of compaction of littoral deposits affects the size of the interstices and thus the ability of the animal to crawl through the voids. Remane (1933), introduced the term 'phytal' and divided the fauna into species that showed preference for sand, mud and the phytal zone. This scheme was followed by Elofson (1941), who greatly refined Remane's classification and grouped the species as burrowers, clawers or creepers between sand grains. A direct correlation exists between the form and sculpture of the carapace, and the nature of the substrate. This correlation was shown for the first time by Elofson (1941), who observed that smooth shelled forms burrow either in the mud or were phytals, and coarsely sculptured animals were inhabitants of a sandy substrate.

It is interesting to note the association between the particle-size of the substrate and the ornamentation of the ostracode carapace. Several species, such as *Bosquetina* and *Cytheropteron,* have lateral spines or *alae* which are believed to aid in supporting the carapace on soft, fine-grained sediments. Coarsely ornamented carapaces are usually characteristic of forms living among large sediment grains, whereas carapaces which are smooth or with subdued ornamentation are generally found on ostracodes inhabiting fine sediment bottoms. Thick, heavy, highly ornamented carapaces of *Carinocythereis, Costa, Mutilus, Quadracythere,* and *Urocythereis* are characteristic of the near-shore environment with its coarser sediments.

Common forms of the offshore environment include the thin, smooth-shelled species of *Argilloecia, Cytheropteron, Erythrocypris, Krithe, Paracytherois, Pontocypris, Pseudocythere, Sclerochilus,* and *Xestoleberis.* Whereas the near-shore environment is largely restricted to large, highly ornamented species, the offshore environment has both the characteristic smooth forms and ornamented species such as *Henryhowella* and *Pterygocythereis.*

Deeper water assemblages are modified by the nature of the substrate, for instance in the Mediterranean, where a similar distinct fauna is found on *Globigerina,* and pteropod oozes.

Distribution of phytals also affects the distribution of ostracodes. At least fifteen species are found associated with marine grasses in the Eastern Gulf of Mexico. Marinov (1964) also shows the effect of vegetation on 'phytophile' assemblages in the Black Sea. One of the factors that controls distribution of phytals is transparency. In the transparent waters in the Gulf of Naples, sessile algae obtain sufficient light to live at depths greater than 100 metres (Harvey 1945). The zone of photosynthesis can be considered to extend below 100 metres. From our studies in the Gulf of Naples, marine plants are actually restricted to a depth of 100 metres. However, detritus of marine plants like *Posidonia* sometimes occurs at depths greater than 100 metres. The ostracode fauna supported by such detritus comprises a distinct assemblage.

SEDIMENTARY FACIES

The composition of the marine and marginal marine environments varies from one area to another, depending on local hydrographic and climatic conditions. The succession

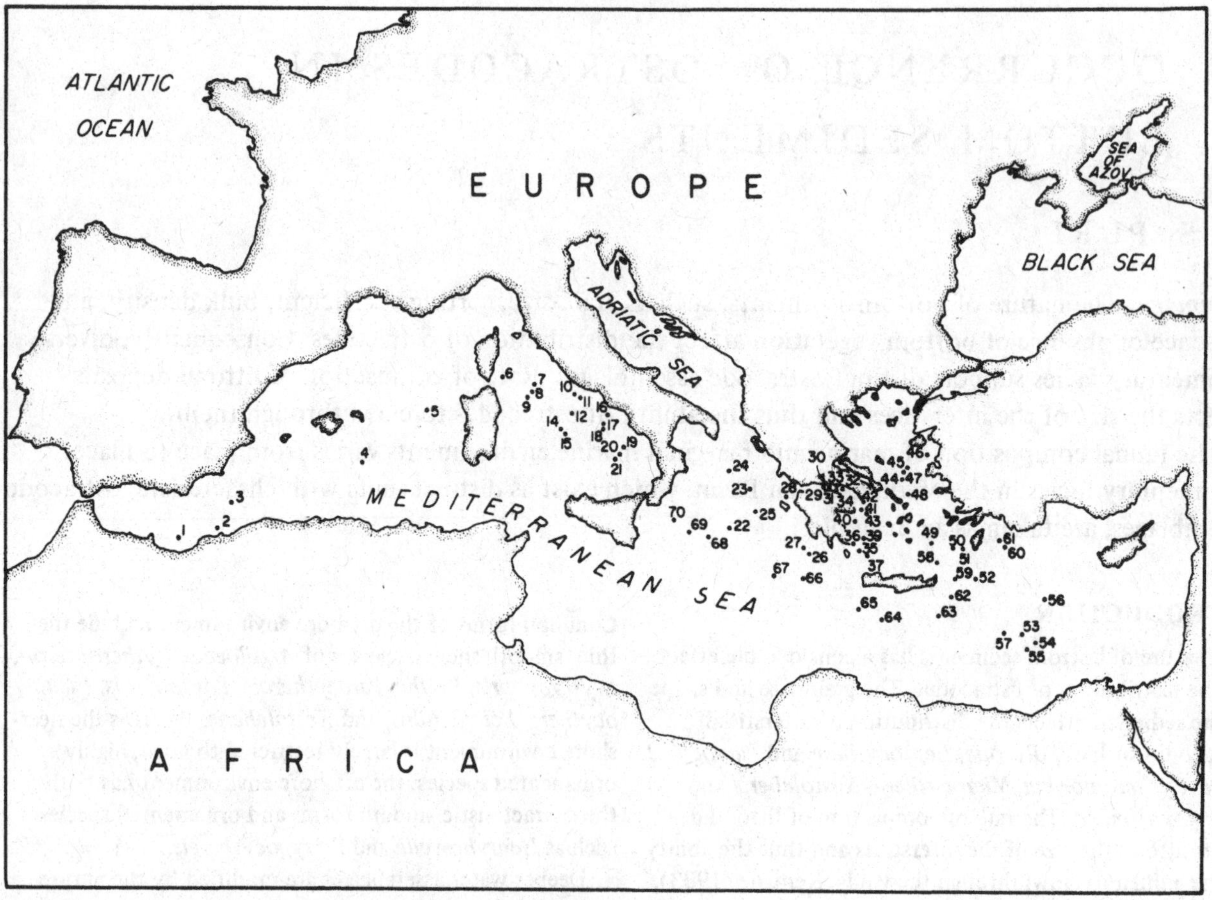
ATLANTIC
OCEAN
EUROPE
SEA OF AZOV
BLACK SEA
ADRIATIC SEA
MEDITERRANEAN SEA
AFRICA

Fig. 26.1. Map of the Mediterranean basin showing location of *Vema* cores.

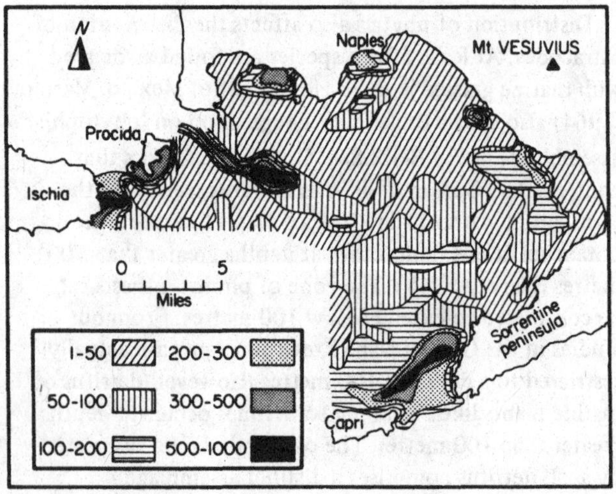
N
Naples
Mt. VESUVIUS
Procida
Ischia
0 5
Miles
Sorrentine peninsula
Capri
1-50
200-300
50-100
300-500
100-200
500-1000

Fig. 26.2. Distribution of total ostracode populations in the Gulf of Naples.

of sedimentary facies characterized by ostracode assemblages, from fluvial terrestrial to outer neritic, is well illustrated in the Mediterranean basin.

The Mediterranean Sea

The nature of the bottom in the Mediterranean varies from littoral sands, clays, muds and carbonate banks in the shallow shelf, to deep-water calcareous oozes. This region has provided classic works, such as G. W. Müller (1894), Rome (1939, 1942, 1964), Reys (1961) and Hartmann (1954*a-d*). Our present knowledge of this basin (Fig. 26.1) is summarized by Puri, Bonaduce and Gervasio (1969).

The surface temperature of the water in the western Mediterranean during summer is 20-25° C (subtropical), with a greater change in seasonal temperature than the eastern part, which has a summer temperature of 25-27° C (tropical). Salinity in the western part is 38‰ while in the eastern part it is 40‰. The Mediterranean Sea has a much lower fertility than the neighbouring Atlantic. The bottom sediments consist of sands, clays, muds and shell sands. Vegetation of the coastal areas in depths up to 100 metres consists of *Posidonia,* calcareous and fibrous algae. This uniformity in bottom conditions results in a remarkably similar fauna from the Balearic Sea in the West to the Levantine Coast in the East. Ecologically the ostracode fauna has been studied in much greater detail than anywhere else in the world.

The littoral areas of the western Mediterranean were studied by Hartmann (1954*a-d*) and a well-defined interstitial fauna was reported from the coasts of the western Mediterranean and Majorca. This interstitial fauna is typical of sandy substrates in this region.

The marine and marginal marine faunas show a remarkable uniformity in assemblages of ostracodes on similar substrates from shallow water lagoons to abyssal plains throughout this basin. The Gulf of Naples which has been studied in detail by Müller (1894), and Puri, Bonaduce and Malloy (1964) is typical.

Three distinct facies characterize the deeper water.

(*a*) *Sponge sand facies:* A sponge sand occurs in the Algean Province off the coast of Greece and Turkey, from which Brady (1866) described nineteen species and varieties.

(*b*) *Globigerina Ooze Facies.* The bottom sediments in *Vema* stations V10-1 to 9, 12, 27, 38, 40, 44 to 45, 48 to 52, 54 to 58, 61, and 63 to 65 consist primarily of *Globigerina* ooze (see Fig. 26.1 for location of stations).

The *Globigerina* ooze was encountered at depths between 496 and 4,286 metres; it supports a distinctive fauna consisting of *Bythocypris obtusata, B. bosquetina Polycopsis* n.sp. 1, *Paradoxostoma versicolor, Pterygocythereis* sp., *Cytherella* n.sp. A, and species of *Polycope.* This fauna is much more diversified than the fauna encountered on pteropod ooze bottoms.

(*c*) *Pteropod ooze facies.* The following *Vema* stations

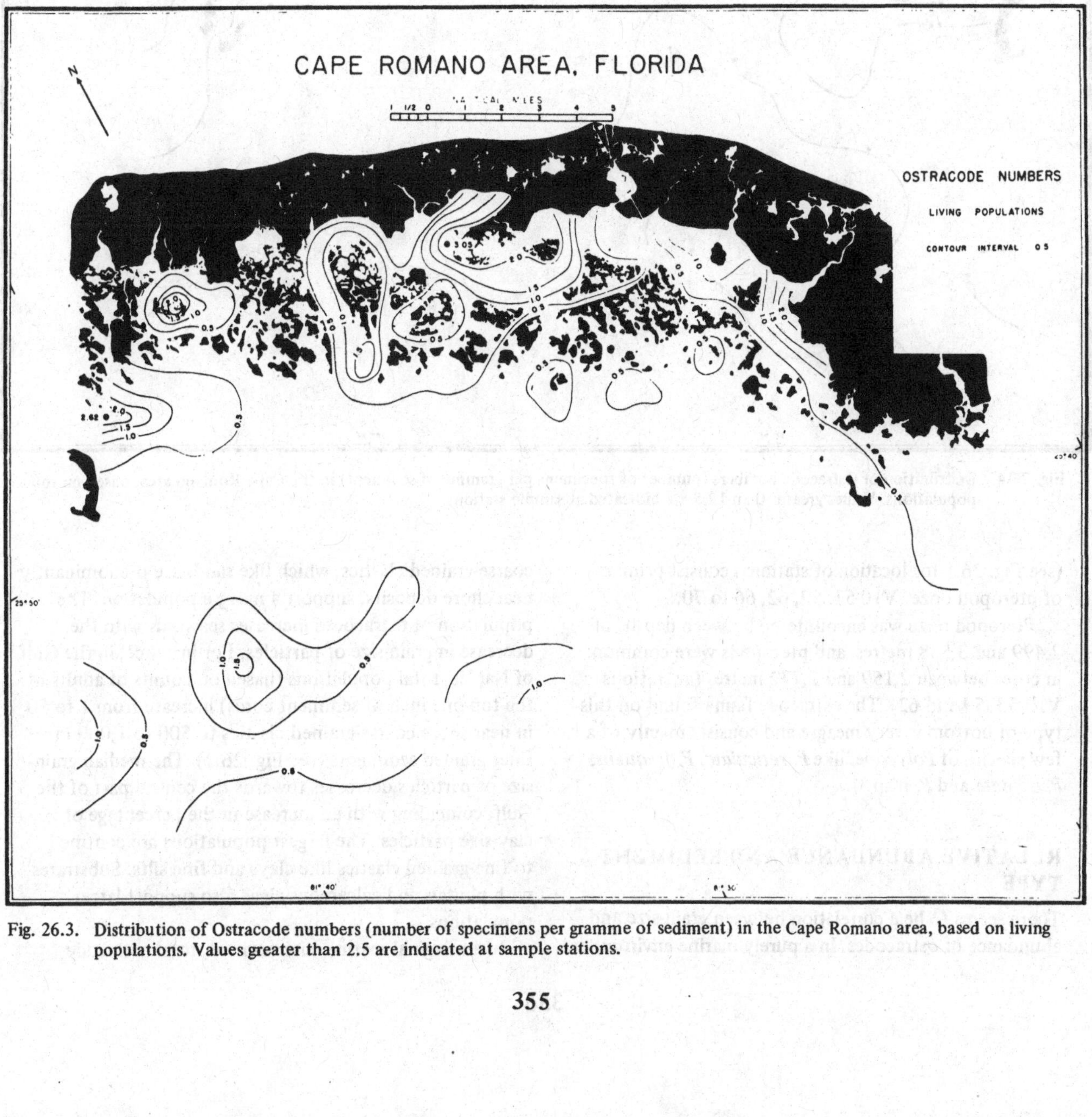
CAPE ROMANO AREA, FLORIDA
OSTRACODE NUMBERS
LIVING POPULATIONS
CONTOUR INTERVAL 0.5

Fig. 26.3. Distribution of Ostracode numbers (number of specimens per gramme of sediment) in the Cape Romano area, based on living populations. Values greater than 2.5 are indicated at sample stations.

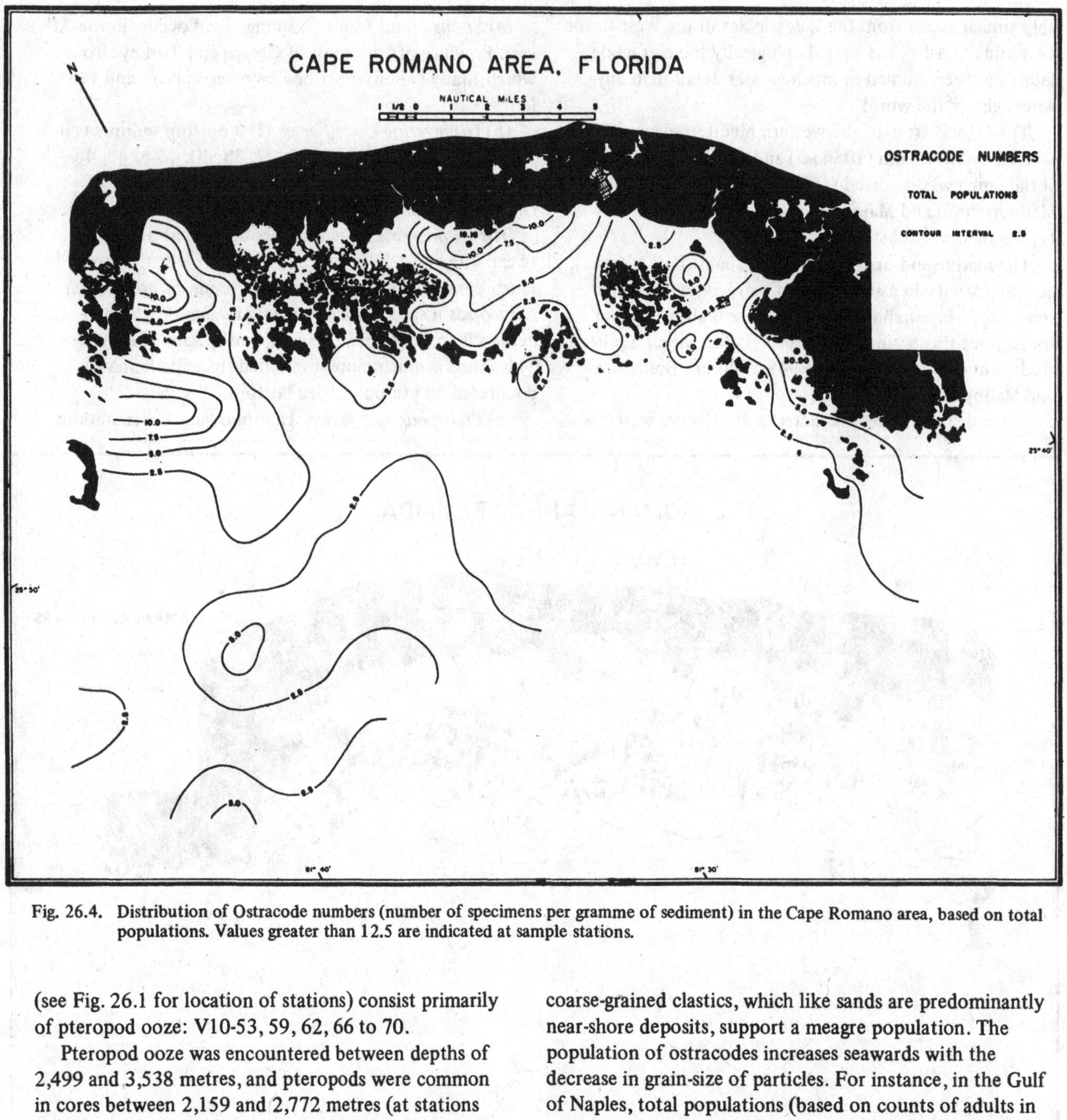
CAPE ROMANO AREA, FLORIDA
NAUTICAL MILES
OSTRACODE NUMBERS
TOTAL POPULATIONS
CONTOUR INTERVAL 2.5

Fig. 26.4. Distribution of Ostracode numbers (number of specimens per gramme of sediment) in the Cape Romano area, based on total populations. Values greater than 12.5 are indicated at sample stations.

(see Fig. 26.1 for location of stations) consist primarily of pteropod ooze: V10-53, 59, 62, 66 to 70.

Pteropod ooze was encountered between depths of 2,499 and 3,538 metres, and pteropods were common in cores between 2,159 and 2,772 metres (at stations V10-53, 59 and 62). The ostracode fauna found on this type of bottom is very meagre and consists mainly of a few species of *Polycope,* like *P. reticulata, P. frequens, P. dentata* and *P.* n.sp. 4.

RELATIVE ABUNDANCE AND SEDIMENT TYPE

There seems to be a correlation between grain-size and abundance of ostracodes. In a purely marine environment, coarse-grained clastics, which like sands are predominantly near-shore deposits, support a meagre population. The population of ostracodes increases seawards with the decrease in grain-size of particles. For instance, in the Gulf of Naples, total populations (based on counts of adults in the top one inch of sediment cores) increase from 1 to 50 in near-shore coarse-grained clastics to 500 to 1,000 in finer grained sediments (see Fig. 26.2). The median grain-size of particles decreases towards the central part of the Gulf, coinciding with an increase in the percentage of clay-size particles. The largest populations are confined to fine-grained clastics like clays and fine silts. Substrates with phytals and calcareous algae also support larger populations.

In a marginal marine environment, total ostracode

populations are confined to restricted areas with the greatest daily and temporal variation in salinity and temperature. In the Cape Romano area, Florida, the largest living (Fig. 26.3), dead, and total (Fig. 26.4) populations, based on the number of specimens per gramme of sediment, occur in areas with the greatest percentages of fine-grain sediments, such as very fine sand, silt and clay. Living, dead, and total populations are greatest in the lagoons, especially near the mouths of the rivers (Benda and Puri 1962). In the Bahamas, Kornicker (1964) also noted that ostracode abundance occurs in restricted areas and listed the following environments in order of increasing abundance: Outer Shelf, Restricted Shelf, Restricted Lagoon and Hypersaline Lagoon.

The bulk density of sediments has a direct bearing on ostracode population as burrowing ostracodes do not favour living in the denser sediments. In the Bimini area in the Bahamas, Kornicker (1964) thought that the scarcity of ostracodes in the Oolite facies, with a high bulk density is due to a low burrowing index. Compaction of sediments, and size and shape of the voids, also controls interstitial ostracode populations.

ACKNOWLEDGEMENTS

Studies in the Gulf of Naples were supported by National Science Foundation Grant Nos. G. 14562 and GB 2621; study of the *Challenger* Ostracoda was made possible by National Science Foundation Grant No. GB 6706, for which gratitude is expressed. I wish to thank Drs F. M. Swain and L. S. Kornicker, for their comments.
Received January 1968.

BIBLIOGRAPHY

Benda, W. K. and Puri, H. S. 1962. The distribution of Foraminifera and Ostracoda off the Gulf Coast of the Cape Romano area, Florida. *Trans. Gulf. Coast Assoc. Geol. Soc.* **12**, 303-41.

Brady, G. S. 1866. On new or imperfectly known species of Marine Ostracoda. *Zool. Soc. London. Trans.* **5**, 359-93, pls. 57-62.

Brady, G. S. 1880. Report on the Ostracoda dredged by H. M. S. *Challenger* during the years 1873-76. *Rept. Sci., Results Voyage H.M.S. 'Challenger', Zool.* **1** (3), 1-184, pls. 1-44.

Elofson, O. 1941. Zur Kenntnis der marinen Ostracoden Schwedens mit besonderer Berücksichtigung des Skageraks. *Zool. Bidr. Uppsala,* **19**, 215-534.

Funk, G. 1927. Die Algenvegetation des Golfes von Neapel. *Publ. Staz. Zool. Napoli,* **7** suppl.

Hartmann, G. 1954a. Ostracodes des eaux southerraines littorales de la Mediterranée et de Majorque. *Vie et Milieu,* **4**, 238-53.

Hartmann, G. 1954b. Les ostracodes de la zone d'algues de l'eulittoral de banyuls. *Vie et Milieu,* **4** (4), 608-12.

Hartmann, G. 1954c. Les ostracodes du sable à amphioxus de banyuls. *Vie et Milieu.* **4** (4), 648-58.

Hartmann, G. 1954d. Ostracodes des étangs mediterranéens. *Vie et Milieu,* **4** (4), 707-12.

Harvey, H. W. 1945. *Recent advances in the chemistry and biology of sea water.* Cambridge University Press. 164 pp.

Kornicker, L. S. 1964. Ecology of ostracods in the northwestern part of the Great Bahamas Bank. *Pubbl. Staz. Zool. Napoli,* **33** suppl., 345-60.

McKenzie, K. G. 1963. A brackish-water ostracod fauna from Lago di Patria, near Napoli. *Ann. Inst. Mus. Zool., Univ. Napoli,* **15** (1), 1-14, 2 pls.

Marinov, T. 1964. Untersuchungen über die Ostracodenfauna des Schwarzen Meeres. *Kieler Meeresfors.* **20** (1), 82-91.

Müller, G. W. 1894. Die Ostracoden des Golfes von Neapel und der angrenzenden Meeres-Abschnitte. *Fauna und Flora des Golfes von Neapel,* **21**, Monographie, i-viii, 1-404, pls. 1-40.

Müller, G. 1959. Die rezenten Sedimente im Golf von Neapel, Teil 1. *Pubbl. Staz. Zool. Napoli,* **31**, 1-27.

Müller, G. 1961. Die rezenten Sedimente im Golf von Neapel, Teil 2. *Beiträge zur Mineralogie und Petrographie,* **8**, 1-20.

Müller, G. 1963. *Die Korngrossenverteilung in den rezenten Sedimenten des Golfes von Neapel.* Unpublished MS.

Neale, J. W. 1964. Some factors influencing the distribution of Recent British Ostracoda. *Pubbl. Staz. Zool. Napoli,* **33** suppl., 247-307.

Puri, H. S. 1960. Recent Ostracoda from the West Coast of Florida. *Gulf Coast Assoc. Geol. Soc. Trans.* **10**, 107-49.

Puri, H. S. 1966. Ecologie distribution of Recent Ostracoda: In *Symposium on Crustacea, Part 1.* Marine Biological Association of India, pp. 557-95, 10 figs.

Puri, H. S., Bonaduce, G. and Gervasio, A. M., 1969. Distribution of Ostracoda in the Mediterranean. In: *The Taxonomy, Morphology and Ecology of Recent Ostracoda* (J.W. Neale, *ed.*). Oliver and Boyd: Edinburgh. pp. 356-411.

Puri, H. S., Bonaduce, G. and Malloy, J. 1964. Ecology of the Gulf of Naples. *Pubbl. Staz. Zool. Napoli,* **33** suppl., 87-199.

Puri, H. S. and Hulings, N. C. 1957. Recent ostracod facies from Panama City to Florida Bay Area. *Trans. Gulf. Coast Assoc. Geol. Soc.* **7**, 167-190.

Remane, A. 1933. Verteilung und organisation der benthonischen Mikrofauna der Kieler Bucht. *Wiss. Meeres. Kiel. N.F.* **21** (2), 161-221.

Reys, S. 1961. Note preliminare à l'étude des Ostracodes du Golfe de Marseille. *Rec. Trav. St. Mar. End. Bull.* **21** (34), 59-64.

Rome, D. R. 1939. Note sur des ostracodes marins des environs de Monaco. *Bull. Inst. Oceanogr.* **768.**

Rome, D. R. 1942. Ostracodes marins des environs de Monaco, 2me note. *Bull. Inst. Oceanogr.* **819**,

Rome, D. R. 1964. Ostracodes des environs de Monaco, Leur distribution en profondeur, nature des fonds marins explorés. *Pubbl. Staz. Zool. Napoli,* **33** suppl., 200-12.

Sacchi, C. F. 1961. L'évolution recente du milieu dans

l'étang saumatre dit *Lago Patria* (Naples) analysée par sa macrofauna invertebrae. *Vie et Milieu,* 12, 37-65.

27. DISTRIBUTION OF POLLEN AND SPORES IN THE OCEANS

J. J. GROOT

Up to now very little is known about the distribution of pollen and spores in the oceans, yet it must be recognized that this subject is of great importance to the interpretation of pollen diagrams from deep-sea sediments, and it is of special significance if we want to use palynological study as a means of determining the provenance of the lutites in which the pollen and spores are found.

If we are to identify possible source areas of water-transported lutites in ocean basins on the basis of the provenance of pollen and spores derived from a continental vegetation, we must be reasonably certain that the plant microfossils were also largely water transported. Available evidence, although rather scant, suggests that this is indeed the case.

It has been shown that by far the largest proportion of pollen produced by the vegetation of the continents is deposited from the atmosphere within a short distance from its source (Erdtman 1954). Therefore, the majority of the pollen grains which are not deposited on the land will fall into rivers, estuaries, and near-shore waters rather than at great distances from the continents. Once they are in suspension in the water, they can only be transported to the deep ocean by currents. Thus it appears that, *a priori*, the dispersion of pollen and spores must be mainly by water currents, or, in appropriate places, by turbidity currents.

Transportation of pollen grains and spores by water currents should be very effective considering the much higher density of the transporting medium compared with air, and the existence of surface currents having velocities of several knots. Although deep-sea currents generally have a smaller velocity, they must not be underestimated, as has been shown by the occurrence of deep-sea ripples (Heezen and Hollister 1964) and direct measurements (Wüst 1957). These indicate velocities capable of transporting mineral particles of fine sand size, and such currents are certainly competent to carry small, lightweight pollen grains. As a matter of fact, the current velocities reported in some areas are great enough to preclude the settling of these grains unless they are carried down attached to coagulated silt and clay particles.

Transportation of pollen grains by water currents was indirectly shown by Muller (1959) in his now classical study of the distribution of pollen and spores in the sediments of the Orinoco delta, an area subject to the onshore northeast trade wind. Cross and Shaefer (1965) pointed out that stream transportation plays an important role in palynomorph distribution in the sediments of the Gulf of California. Rossignol (1961), in her palynological study of sediments off the coast of Israel, found that many spores and pollen grains were transported first by the Nile, and subsequently by currents along the Israeli coast.

Some direct evidence of water transportation of pollen is also available. Federova (1952), in her study of the Volga River, reported the presence of 23,000-45,000 pollen grains per 100 litres of water, and pointed out that the river transported pollen and spores over very great distances, resulting in the mixing of the pollen from different vegetation zones. Groot (1966) studied some samples of water obtained from the estuary of the Delaware River in August 1964, finding 500-8,000 pollen grains per litre of water, and about 20,000 pollen grains per gramme of suspended sediment. It was shown that the pollen spectra from the suspended sediments were remarkably similar throughout the estuary, indicating thorough mixing of the pollen; they were also essentially the same as the spectra from the tops of bottom sediment cores. All the sediments investigated contained pollen and spores derived from the Delaware River Basin as a whole; the influence of the contribution of nearby, local vegetation was of very minor significance.

In a recent study of the palynology and sedimentology of the Great Bahama Bank, Traverse and Ginsburg (1966) found evidence of extensive water transportation of pollen, both on the basis of the distribution of pollen grains in bottom sediments and their concentration and geographical distribution in the water. Up to 2,000 pine pollen per 100 litres of water were obtained by centrifugation five months after flowering of the pine trees, proving that pollen remain suspended in the water for long periods of time. The concentration of pollen in the water at the time of collection had little if any relation to the source of the pollen had they been transported by the atmosphere; instead, there appeared to be a relationship with water movements, salinities and turbulence. Traverse and Ginsburg concluded that 'we must abandon aerodynamic explanations. Pollen of wind-pollinated plants is shed into the air, but most of it settles in a few miles unless the air is extremely turbulent... pollen behaves as any other particle with the same physical properties. The distribution of pollen is a result of sedimentation processes. As a sensitive indicator of sedimentary patterns palynology should be a useful tool in sedimentology.'

Available evidence, although still meagre, clearly indicates that pollen grains can stay in suspension in the water for long periods of time, and behave as other sedimentary particles which have the same hydrodynamic properties. Therefore, we may conclude that (1) pollen grains and continent-derived fine silts and clays are transported in the oceans in essentially the same manner and (2) pollen grains (and spores) may be indicative of the provenance of the lutites in which they are found. These conclusions should be tested by further investigations of suspended mineral matter and pollen and spores in the oceans, and their quantitative relationships.

Received December 1967.

BIBLIOGRAPHY

Cross, A. T. and Shaefer, B. L. 1965. Palynology of modern sediments, Gulf of California and environs. (abstract). *Bull. Am. Assoc. Petroleum Geol.* 49, 337.

Erdtman, G. 1954. *An introduction to pollen analysis.* The Ronald Press Co.: N.Y. 239 pp.

Federova, R. V. 1952. Dissémination des pollen et des spores par les eaux courantes. (Translation by S. Ketchian). *Travaux Inst. Géogr.* 52, *Données sur la géomorphologie et la paléogéographie d'U.R.S.S., Akad. Nauk S.S.S.R.*, 7, 46-72.

Groot, J. J. 1966. Some observations on pollen grains in suspension in the estuary of the Deleware River. *Marine Geol.* 4, 409-16.

Heezen, B. C. and Hollister, C. 1964. Deep-sea current evidence from abyssal sediments. *Marine Geol.* 1, 141-74.

Muller, J. 1959. Palynology of present Orinoco delta and shelf sediments. *Micropalaeontology*, 5, 1-32.

Rossignol, M. 1961. Analyse pollinique des sédiments marins quaternaires en Israël, I. Sédiments Récents. *Pollen et Spores,* 3, 303-24.

Traverse, A. and Ginsburg, R. N. 1966. Palynology of Great Bahama Bank, as related to water movement and sedimentation. *Marine Geol.* 4, 417-59.

Wüst, G. 1957. Stromgeschwindigkeiten und Strommengen in der Tiefen des Atlantischen Ozeans. *Wiss. Erg. Dtsch. Atl. Exp. 'Meteor' 1925-1927*, 6 (2), 261-420.

28. SPORES AND POLLEN IN MEDITERRANEAN BOTTOM SEDIMENTS

E. V. KORENEVA

Summary: Palynological studies of bottom sediments from the Mediterranean have been carried out on samples obtained by R. V. *Academician S.I. Vavilov*. The surface layer of bottom sediments and the underlying sequences in several cores have been investigated. During the time that the sediments in the cores were being deposited, considerable changes took place in the coastal vegetation of the Mediterranean sea. These changes were the result of large scale climatic fluctuations during the Upper Pleistocene.

DISTRIBUTION OF SPORES AND POLLEN IN SURFACE LAYER OF SEDIMENTS

Studies of the surface layer of sediments have shown a relatively low content of pollen and spores from terrestrial plants. The highest concentration – over 100 pollen grains per gramme – was found in two regions: in the western part (Alboran Sea) and in the northwestern part of the Adriatic Sea (Fig. 28.1). The lowest concentration was found in samples collected off the desert coast of Africa. In some samples from this region pollen and spores were not seen at all. This can be explained by the very low rate of deposition of terrigenous material, including pollen and spores, resulting from weak river flow from the African coast, and the wide distribution of strongly carbonaceous silty muds (Emelianov 1961).

The composition of the vegetation in the surrounding coastal areas is most clearly reflected by the spore-and-pollen spectra of samples collected close to the coast. With increasing distance from the coast, the number of species in the spectra decreases, owing to the disappearance of those pollen grains which are poorly transported over large distances. Among the arboreal pollen in the spectra, the pollen of *Pinus* predominates (Fig. 28.3). Its content is clearly higher than the contribution of *Pinus* trees to the vegetation of the Mediterranean region. Arboreal pollen dominates spectra of samples obtained in the Adriatic Sea, the Aegean Sea and the Tyrrhenian Sea. Non-arboreal pollen dominates spectra from the southern and southeastern Mediterranean (Fig. 28.2).

The richest spore-and-pollen spectra were obtained from the western Mediterranean (stations 478 and 480). In these, pollen grains were found from trees and shrubs belonging to various habitats that are characteristic of the adjacent parts of southern Spain and the western shore of North Africa. For example, pollen of *Cedrus atlantica, Abies pinsapo, Abies numidica, Quercus lusitanica, Castanea* sp., *Juglans* sp. and others were found. These are characteristic of the plant associations occurring in Numidia, the Northern Atlas Mountains, the Rif Mountains, and also in the southwestern part of the Iberian Peninsula. Pollen grains of evergreen shrubs that occur in the undergrowth of mountain forests are generally present. These shrubs also occur as major constituents of such special vegetation types as the 'maquis' or the 'garrigue'. To this group belong grains from the families Ericaceae, Leguminosae, Euphorbiaceae, Labiatae and from the genera *Pistacia, Cistus, Olea, Phyllirea* and others. Also abundantly represented in the spectra are pollen grains of various species of *Pinus*, such as *Pinus halepensis*, which are widespread in the lowermost mountain zone, below 1,300 metres elevation.

Spore-and-pollen spectra obtained from the northern and central parts of the Adriatic Sea also have a mixed composition. They consist of 70 per cent arboreal pollen, 20 per cent herbs and 10 per cent spores. Among the arboreal pollen, that of various species of *Pinus* predominates (75 per cent). The pollen of *Picea* and *Abies* together constitute 8 per cent, different types of pollen grains of *Quercus* 9 per cent. The pollen of broad-leaved genera is rather variable: *Tilia, Castanea, Juglans, Acer, Carpinus, Fagus, Ostrya, Fraxinus*, etc., total 6 per cent. The contribution of *Alnus* and *Betula* is small. It is possible to find only single pollen grains of *Olea europea, Cistus* and sclerophyllous *Quercus*.

When spectra from the upper layer of sediments of the Adriatic Sea and the vegetation of the surrounding land are compared it is possible to deduce the source of the pollen in the sediments. Thus it is possible that pollen of *Abies alba, Picea excelsa, Pinus silvestris* and *Fagus silvatica* have come to the basin from beech and mixed coniferous – beech forests in the mountains. The broad-leaved mountain forests with *Pinus* are represented in marine sediments by pollen of *Acer. Carpinus, Fagus, Fraxinus, Ostrya, Castanea, Quercus, Tilia, Juglans*, etc., and by different species of *Pinus* pollen. Pollen of sclerophyllous *Quercus (Quercus ilex* and *Q. subar*) and *Pinus pinea* characterize the true Mediterranean vegetation type of the coastal areas of the Apennine and Balkan Peninsulas.

The samples taken near the northern coast of Africa contain mainly the pollen of *Artemisia* and some other genera of the families Compositae, Chenopodiaceae, Gramineae, and *Ephedra*. Very rarely they contain the pollen of *Acacia* which characterizes the coastal desert area of Africa.

THE DISTRIBUTION OF POLLEN AND SPORES IN UPPER QUATERNARY SEDIMENTS

Only twelve cores of marine sediments from different parts of the Sea were investigated (Fig. 28.4).

One core was obtained in the Sea of Marmora, station 1069, in the northern depression, at a depth of 1,160 metres. The length of this core is 186 centimetres. Ten samples were investigated. The granular nature of the sediment is uniform. The core consists of grey mud. The colour changes at 152 centimetres depth to dark-grey. The quantity of pollen in the core is sufficient to make a full analysis of all specimens (Fig. 28.5).

In all samples the arboreal pollen predominates and among the arboreal pollen the pollen grains of *Pinus* are most numerous. The content of *Pinus* pollen decreases down the core. At the 152 centimetre level it falls very sharply to 47 per cent. The content of *Quercus* pollen on the other hand increases with depth. The maximum content of *Quercus* pollen (42 per cent) is seen in the lowest sample at a depth of 186 centimetres. A small quantity of the pollen of *Alnus* and *Betula* was found in the middle part of the core. Sometimes the pollen of *Picea* and *Abies* was found in very small quantities. In addition the pollen of broad-leaved and sclerophyllous trees was found: *Tilia, Castanea, Juglans, Ulmus, Olea, Fraxinus, Fagus, Acer, Ilex, Myrica*, etc. According to the spore-and-pollen spectra the core was not older than Holocene. The lower part of the core, below 152 centimetres, has a darker colour and contains the greatest quantity of pollen from *Quercus* and other broad-leaved trees, formed in a warmer and more humid climate than the pollen of the upper layers. This may correspond to the climatic optimum of the Holocene.

One core from the Aegean Sea was investigated. It was taken at station 1065, in one of the northern depressions,

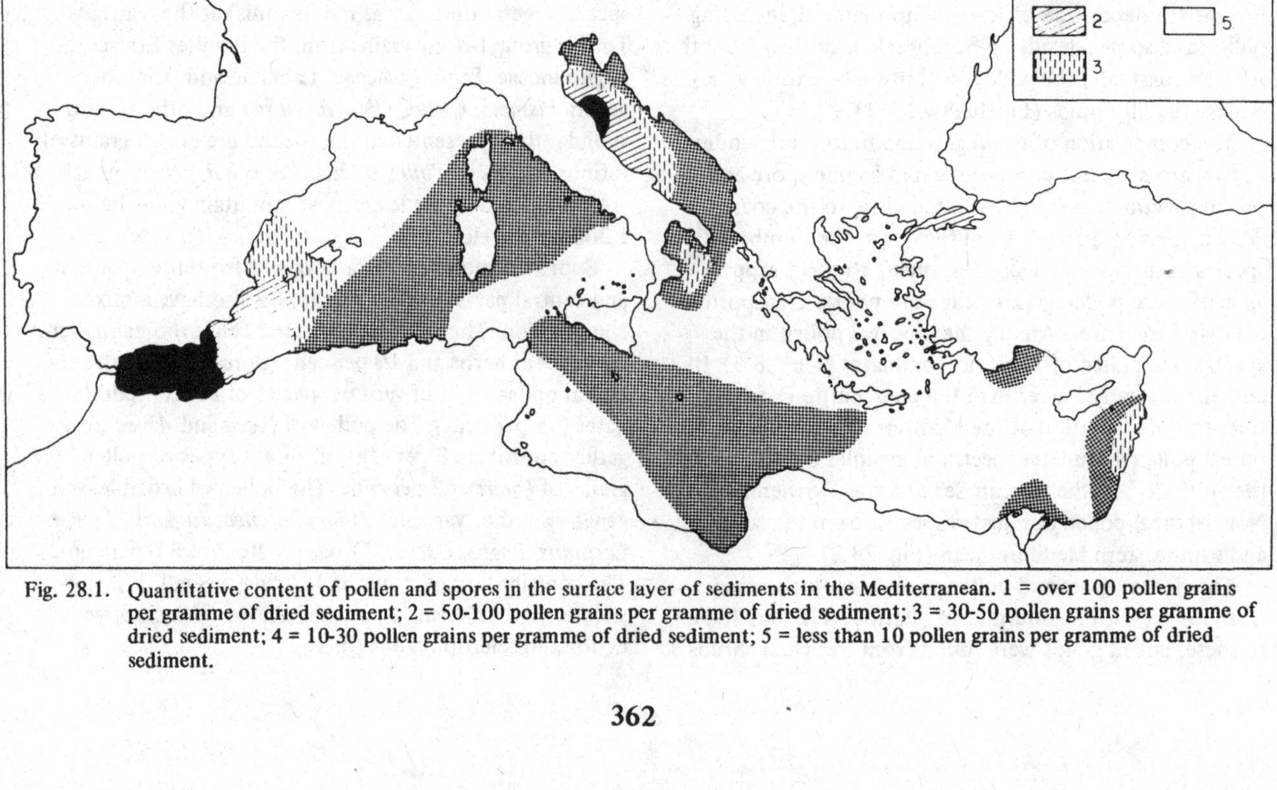
1
2
3
4
5

Fig. 28.1. Quantitative content of pollen and spores in the surface layer of sediments in the Mediterranean. 1 = over 100 pollen grains per gramme of dried sediment; 2 = 50-100 pollen grains per gramme of dried sediment; 3 = 30-50 pollen grains per gramme of dried sediment; 4 = 10-30 pollen grains per gramme of dried sediment; 5 = less than 10 pollen grains per gramme of dried sediment.

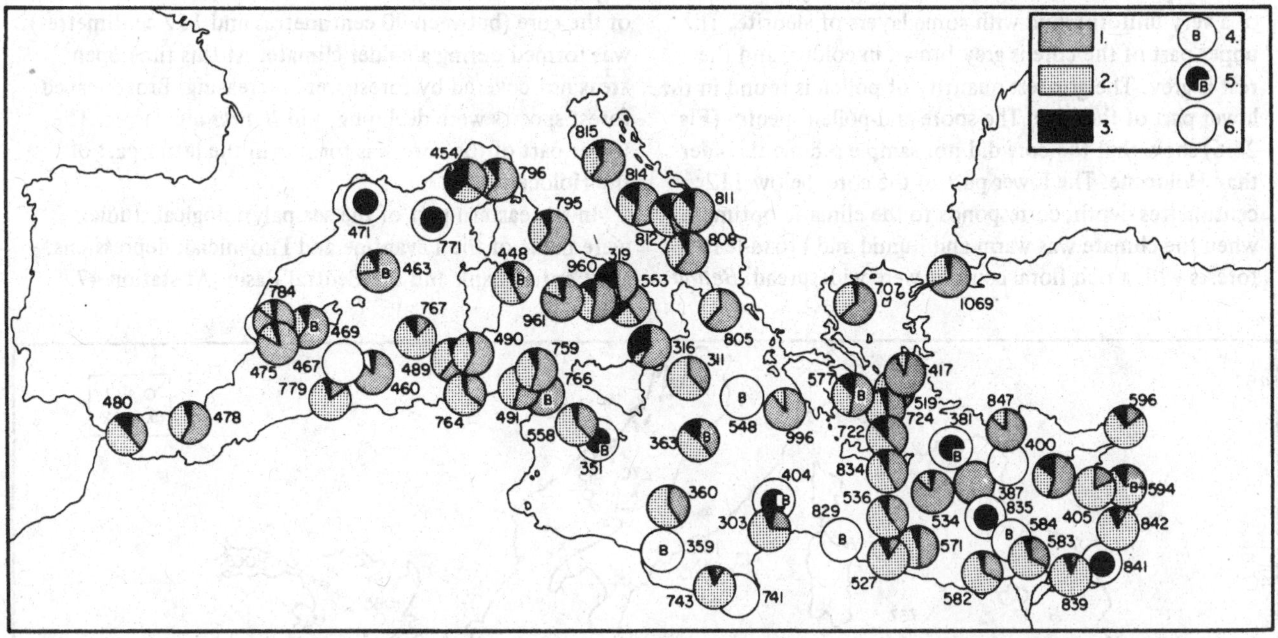
1.
2.
3.
B 4.
B 5.
6.
815
454
796
814
811
795
471
771
812
808
448
463
960
319
553
784
767
961
469
707
1069
490
759
316
805
475
467
489
311
779
460
766
577
417
480
478
764
491
558
548
519
381
847
596
363
996
722
724
400
351
834
404
536
387
835
594
360
829
534
405
584
842
303
571
583
359
841
527
743
741
582
839

Fig. 28.2. General composition of Mediterranean spore-and-pollen spectra. 1 = arboreal pollen; 2 = herbaceous pollen; 3 = spores; 4 = samples studied by V. A. Vronskiy; 5 = samples studied by V. A. Vronskiy, in which up to ten pollen grains of *Pinus* were found; 6 = samples in which no pollen grains or spores were found.

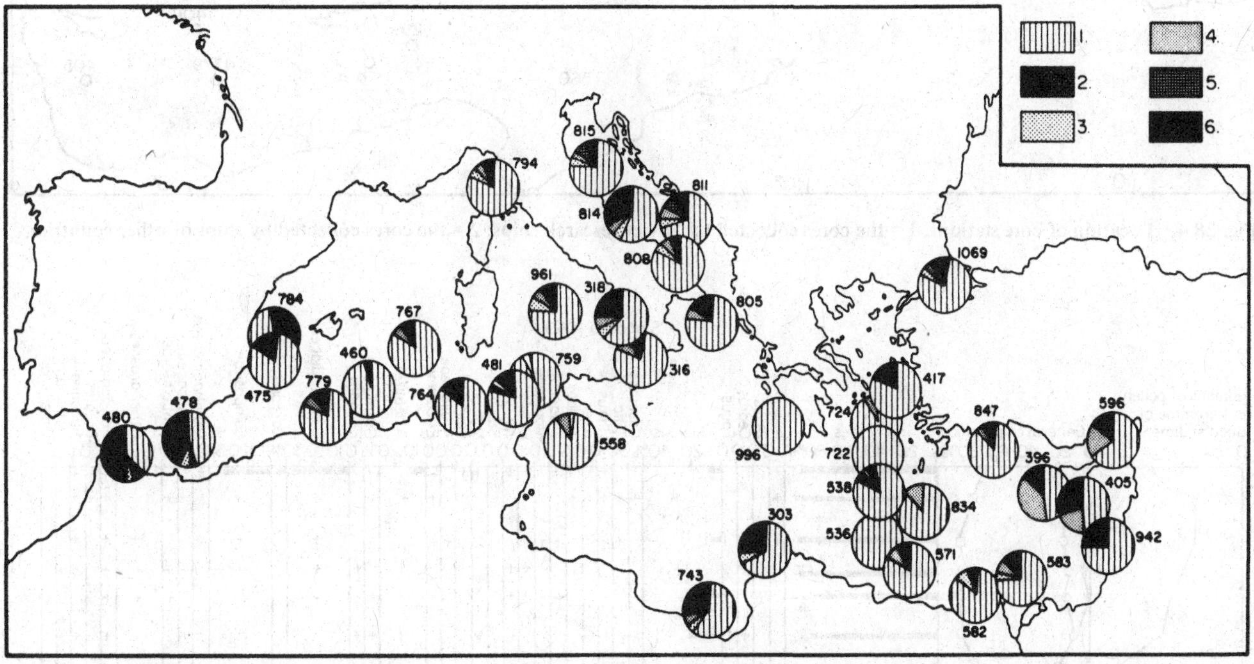
1.
2.
3.
4.
5.
6.
815
794
811
814
808
1069
961
318
784
767
805
460
759
475
779
481
316
417
480
478
764
558
724
847
596
996
722
396
538
405
834
303
536
571
942
583
743
582

Fig. 28.3. Composition of the arboreal pollen sum in samples collected in the Mediterranean. 1 = *Pinus*; 2 = *Cedrus*; 3 = *Abies* and *Picea*; 4 = *Betula* and *Alnus*; 5 = *Quercus*; 6 = total of pollen of broad-leaved and sclerophyllous species.

at a depth of 1520 metres, and was 180 centimetres long. Ten samples were investigated. Lithologically it consists of a very uniform mud with some layers of aleurite. The upper part of the core is grey-brown in colour, and the rest is grey. The greatest quantity of pollen is found in the lower part of the core. The spore-and-pollen spectra (Fig. 28.6) show that the core did not sample sediments older than Holocene. The lower part of the core, below 112 centimetres depth, corresponds to the climatic optimum, when the climate was warm and humid and broad-leaved forests with a rich floral content were widespread. *Betula* (boreal element) had a very limited distribution in the vegetation of the adjacent coastal area. The middle part of the core (between 90 centimetres and 112 centimetres) was formed during a colder climate. At this time open areas not covered by forest were increasing. Broad-leaved forest species were declining, and *Betula* increasing. The upper part of the core was formed in the latter part of the Holocene.

In the eastern part of the sea palynological studies were made of the Levantine and Phoenician depressions, the Central Bank and the Central Basin. At station 4779

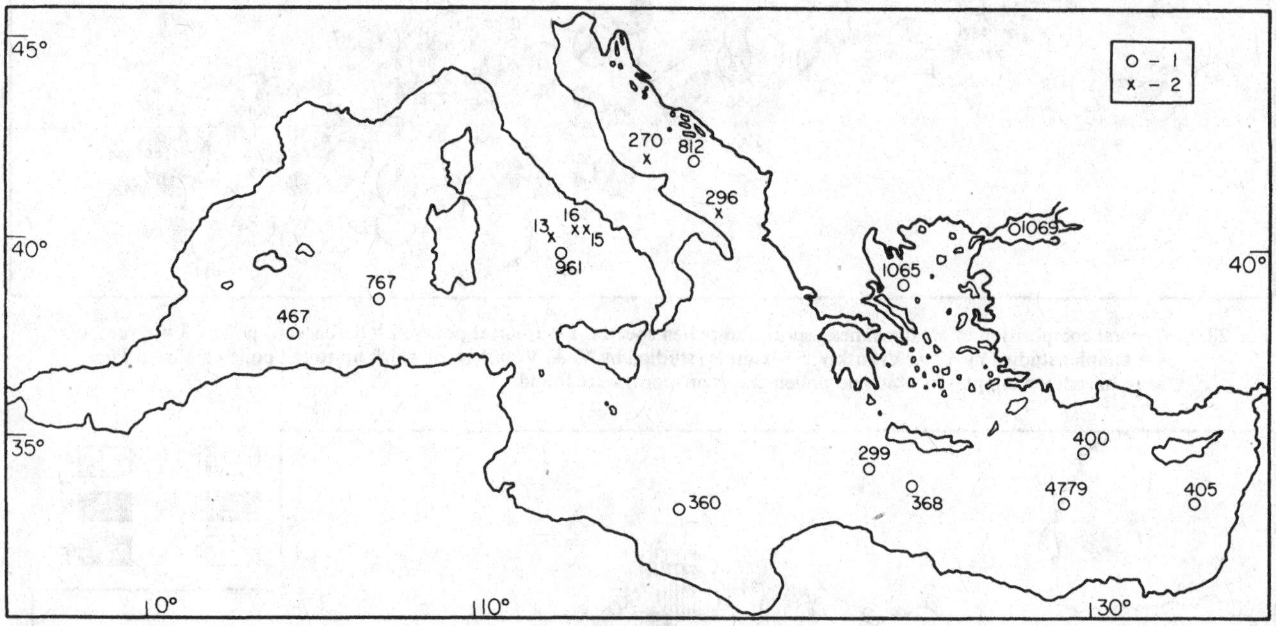
45°
40°
35°
40°
10°
110°
130°
O – 1
x – 2
270
812
296
16
13
15
961
767
467
1065
1069
299
400
360
368
4779
405

Fig. 28.4. Location of core stations. 1 = the cores collected by Soviet research ships; 2 = the cores collected by ships of other countries.

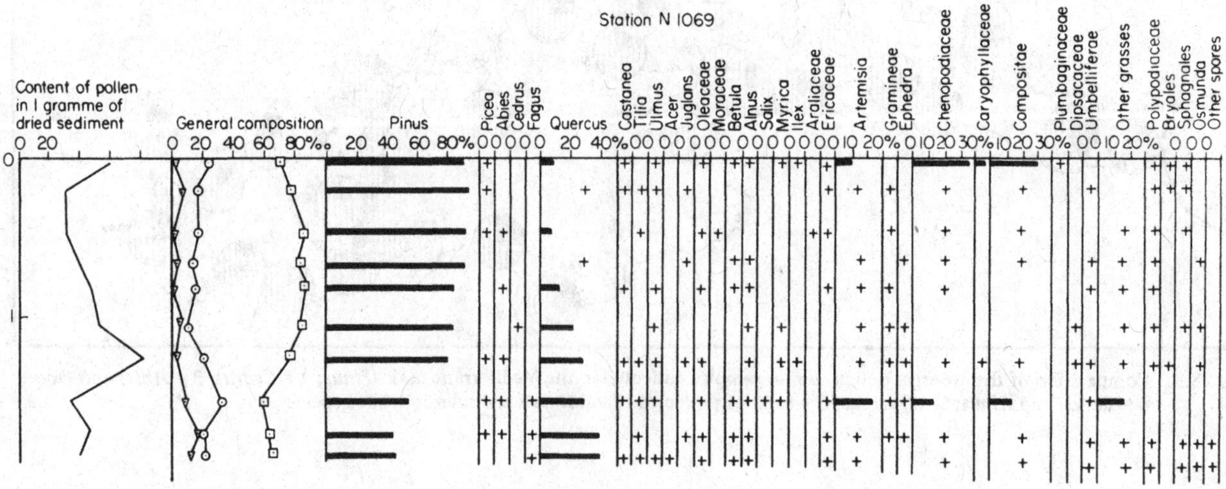
Station N 1069
Content of pollen in 1 gramme of dried sediment
General composition
Pinus
Picea
Abies
Cedrus
Fagus
Quercus
Castanea
Tilia
Ulmus
Acer
Juglans
Oleaceae
Moraceae
Betula
Alnus
Salix
Myrica
Ilex
Araliaceae
Ericaceae
Artemisia
Gramineae
Ephedra
Chenopodiaceae
Caryophyllaceae
Compositae
Plumbaginaceae
Dipsacaceae
Umbelliferae
Other grasses
Polypodiaceae
Bryales
Sphagnales
Osmunda
Other spores

Fig. 28.5. Spore-and-pollen diagram of the core from station N 1069.

a large diameter core was obtained by R. V. *Vityaz* in the Levantine depression from a depth of 3,090 metres. The length of the core was 650 centimetres. The lithology of the core is varied. The upper part (0-32.5 centimetres) is composed of highly carbonaceous dark-green mud enriched with organic material. In the interval 32.5-460 centimetres it is composed of a dark-grey mud with some layers of aleurite, and below 460 centimetres lime grey mud. The content of pollen is very low in the whole core. It is a little higher in the layers enriched with organic material (25-32 centimetres), and in the dark-grey mud between 645-650 centimetres. Herbaceous pollen predominates in nearly all samples. The only exception is for the interval 25-32 centimetres, where arboreal pollen predominates. When the sediments taken by this core were deposited, arid conditions prevailed in North Africa. A more humid climate prevailed during the deposition of the upper part of the core, between 25 and 32 centimetres. At this time Mediterranean vegetation was widespread on the North African coast, corresponding to the last pluvial period of the Holocene. The lower part of the core was formed in arid conditions and is dated as Würm.

The cores taken from the Phoenician depression (core 405), along the axis of the Central Bank (cores 400, 368, 299), and in the Central Basin (core 360) are very similar in both their lithology and their spore-and-pollen spectra (Figs. 28.8, 28.9, 28.10, 28.11 and 28.12). In all the cores, more or less, carbonate muds and brown aleurite with small quantities of organic material (content of organic carbon 0.2-0.5 per cent) alternate with dark-green or black layers rich in organic matter (content of organic carbon 1.6-3.5 per cent). These dark layers are only 20 centimetres thick, whereas the thickness of the carbonate muds and aleurite is 1 to 2 metres.

The pollen content of these cores is very low. Large quantities of pollen are found only in the layers rich in organic material. The upper horizon corresponds to the last pluvial period, coinciding with the Atlantic period of the Holocene. During the time of formation of this horizon the climate was warm and humid. The spore-and-pollen content shows that a very big area of mountains, not only in Asia Minor and the Near East, but also in North Africa, were covered by forests of Mediterranean type. The lower horizon corresponds to one of the pluvial periods of Riss-Würm age. A comparison of the spore-and-pollen spectra shows that the vegetation and climatic conditions of this pluvial period were very close to those of the last (Holocene) pluvial period. Carbonate muds separated by darker horizons were formed under completely different climatic conditions during the Würm glacial epoch. The climate was cold. The main arboreal species in the mountain forests was *Betula*. Now it has a very local distribution in the vegetation of the area.

One core, station 812, from the Adriatic Sea was investigated. It was taken at a depth of 543 metres on the edge of the trench. The core contains pelitic and aleuritic muds which are brown in the upper part and grey in the lower. According to the spore-and-pollen content (Fig. 28.13) the sediments of this core were formed during the Holocene. The lower horizons were formed during the pre-Boreal and Boreal periods when the climate was more or less cold. In the forests *Pinus* prevailed, but *Betula* formed a singificant part. Broad-leaved species were very rare. Optimum conditions for arboreal vegetation existed during the formation of the middle part of the core. Here *Quercus* and different kinds of broad-leaved species predominate. It is possible to date this horizon as the Atlantic period of the Holocene. The sediments of the upper part of the core were formed during the sub-Boreal and sub-Atlantic periods, when the climate was drier and cooler than in the preceding period. Bottema and van Straaten (1966) investigated two cores from the Adriatic Sea, and found the sediments corresponding to the Upper Würm were very deficient in arboreal pollen. In these cores the pollen of *Artemisia* and *Ephedra* predominate. The spore-and-pollen spectra corresponding to the Holocene are similar to those obtained for the core which we have investigated.

One core was taken in the Tyrrhenian Sea, station 961, at a depth of 3,328 metres. The length of this core is 361 centimetres. It consists mainly of pelitic and aleuritic-pelitic ooze, with some layers of aleurite enriched by volcanic material. The pollen content is very low, and it was impossible to make a reliable age determination by means of spore-and-pollen analysis (Fig. 28.14). Larsen (1948) investigated three cores from the Tyrrhenian Sea; the pollen content of these cores was also very low.

The core, station 767, was taken in the Alger-Provence region at a depth of 3130 metres. The length of this core is 333 centimetres, and it consists of alternating layers of dark-grey, grey and sandy pelitic oozes. The sediments in this core were formed during the Würm and Holocene. In the spore-and-pollen spectra (Fig. 28.15) for the Würm section (333-380 centimetres) arboreal pollen predominates, the pollen grains of *Pinus* accounting for more than 90 per cent. Single pollen grains of *Picea*, *Abies* and *Betula* are found consistently and among the herbs xerophytes are predominant. In the Holocene spectra the pollen of herbs predominates, and the arboreal pollen content varies. *Pinus* pollen predominates but does not exceed 60 per cent to 70 per cent. *Quercus*, *Cedrus* and *Oleaceae* pollen form a significant part. The herbaceous pollen is very different. The spore-and-pollen spectra may be interpreted as showing that in Würm times the mountains of

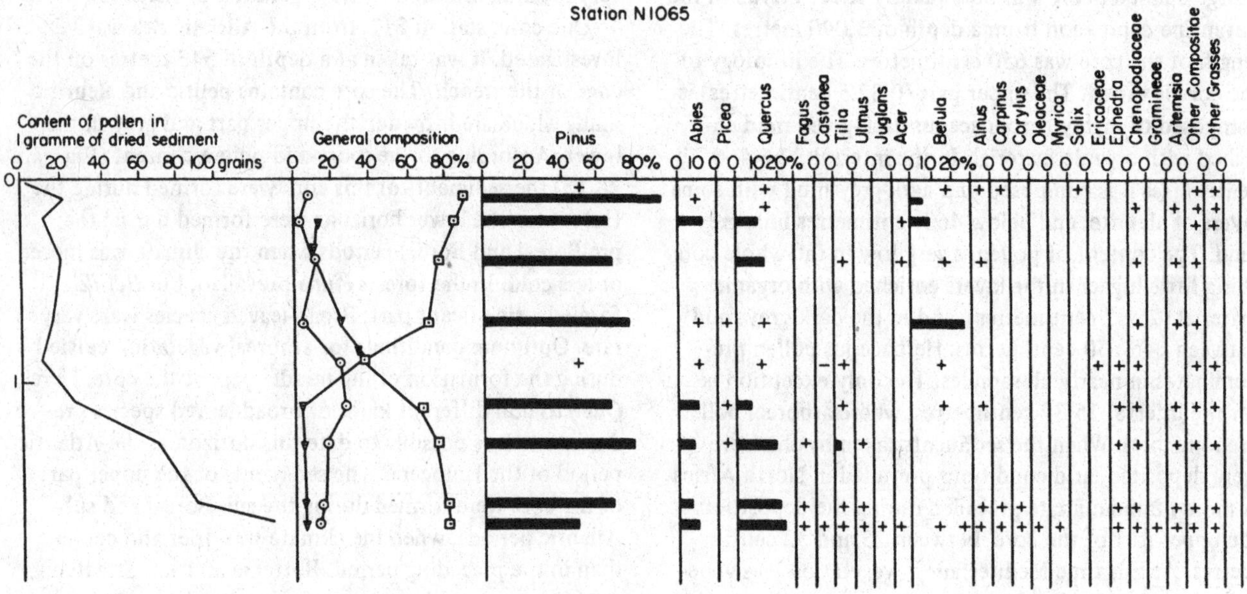
Station N1065
Content of pollen in 1 gramme of dried sediment
0 20 40 60 80 100 grains
General composition
0 20 40 60 80%
Pinus
0 20 40 60 80%
Abies
Picea
Quercus
Fagus
Castanea
Tilia
Ulmus
Juglans
Acer
Betula
Alnus
Carpinus
Corylus
Oleaceae
Myrica
Salix
Ericaceae
Ephedra
Chenopodiaceae
Gramineae
Artemisia
Other Compositae
Other Grasses

Fig. 28.6. Spore-and-pollen diagram of the core from station N 1065.

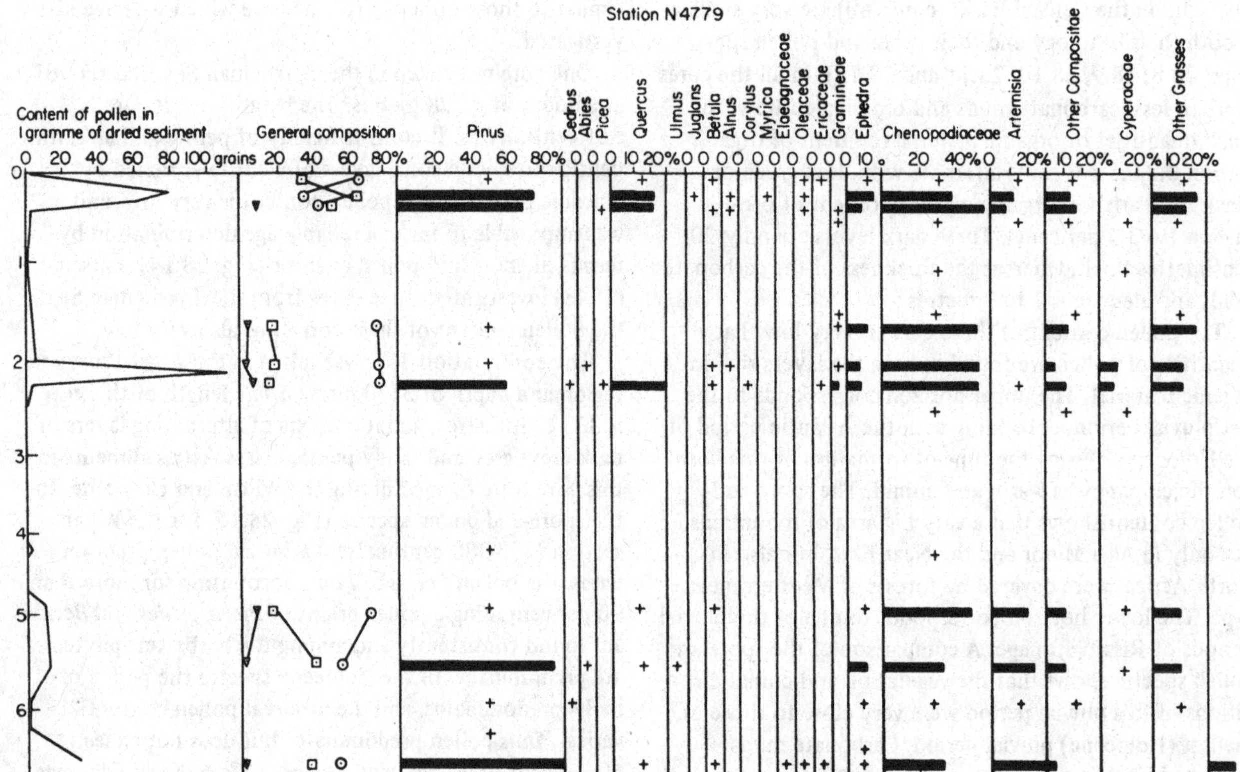
Station N4779
Content of pollen in 1 gramme of dried sediment
0 20 40 60 80 100 grains
General composition
20 40 60 80%
Pinus
20 40 60 80%
Cedrus
Abies
Picea
Quercus
Ulmus
Juglans
Betula
Alnus
Corylus
Myrica
Elaeagnaceae
Oleaceae
Ericaceae
Gramineae
Ephedra
Chenopodiaceae
Artemisia
Other Compositae
Cyperaceae
Other Grasses

Fig. 28.7. Spore-and-pollen diagram of the core from station N 4779.

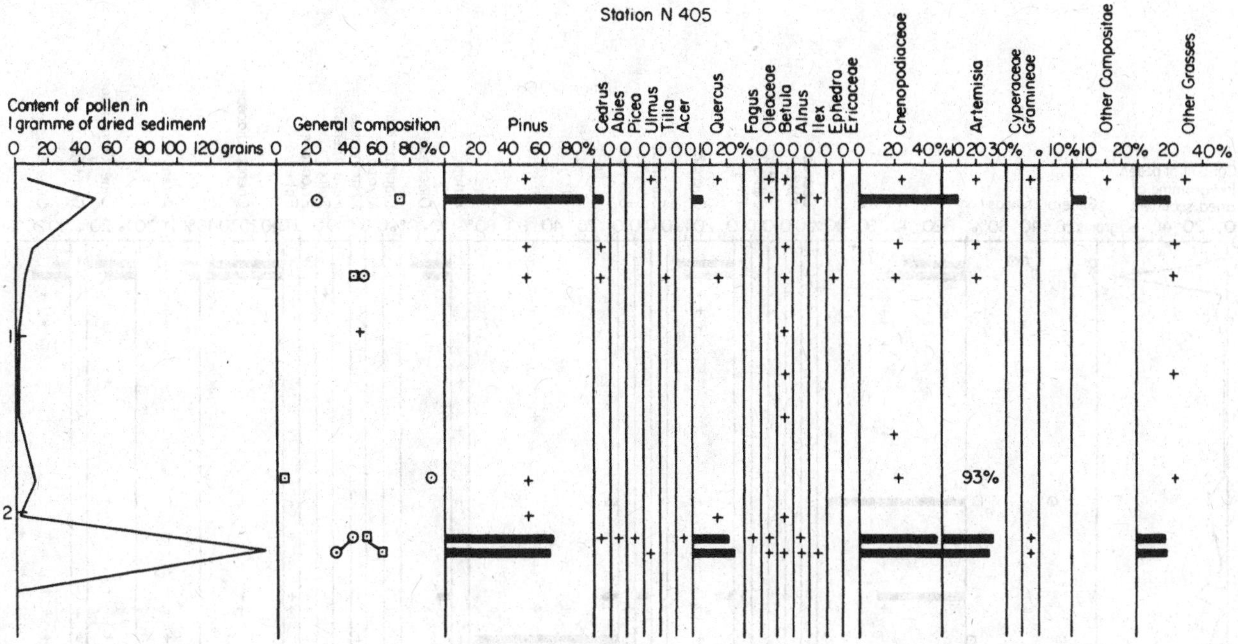

Station N 405
Content of pollen in 1 gramme of dried sediment
0 20 40 60 80 100 120 grains
General composition
0 20 40 60 80%
Pinus
0 20 40 60 80%
Cedrus
Abies
Picea
Ulmus
Tilia
Acer
Quercus
10 20%
Fagus
Oleaceae
Betula
Alnus
Ilex
Ephedra
Ericaceae
Chenopodiaceae
20 40%
Artemisia
10 20 30%
Cyperaceae
Gramineae
10%
Other Compositae
10 20%
Other Grasses
20 40%
1
2
93%

Fig. 28.8. Spore-and-pollen diagram of the core from station N 405.

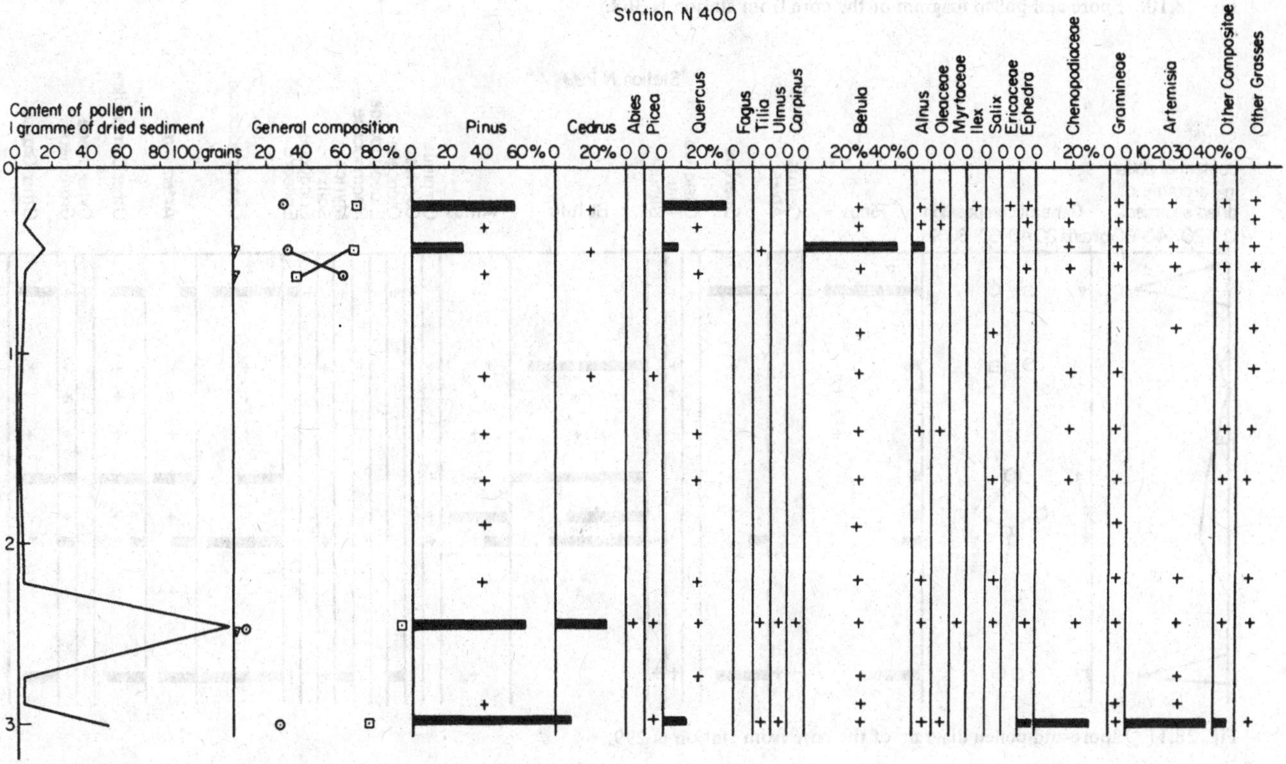

Station N 400
Content of pollen in 1 gramme of dried sediment
0 20 40 60 80 100 grains
General composition
20 40 60 80%
Pinus
0 20 40 60%
Cedrus
0 20%
Abies
Picea
Quercus
20%
Fagus
Tilia
Ulmus
Carpinus
Betula
20% 40%
Alnus
Oleaceae
Myrtaceae
Ilex
Salix
Ericaceae
Ephedra
Chenopodiaceae
20%
Gramineae
Artemisia
10 20 30 40%
Other Compositae
Other Grasses
0
1
2
3

Fig. 28.9. Spore-and-pollen diagram of the core from station N 400.

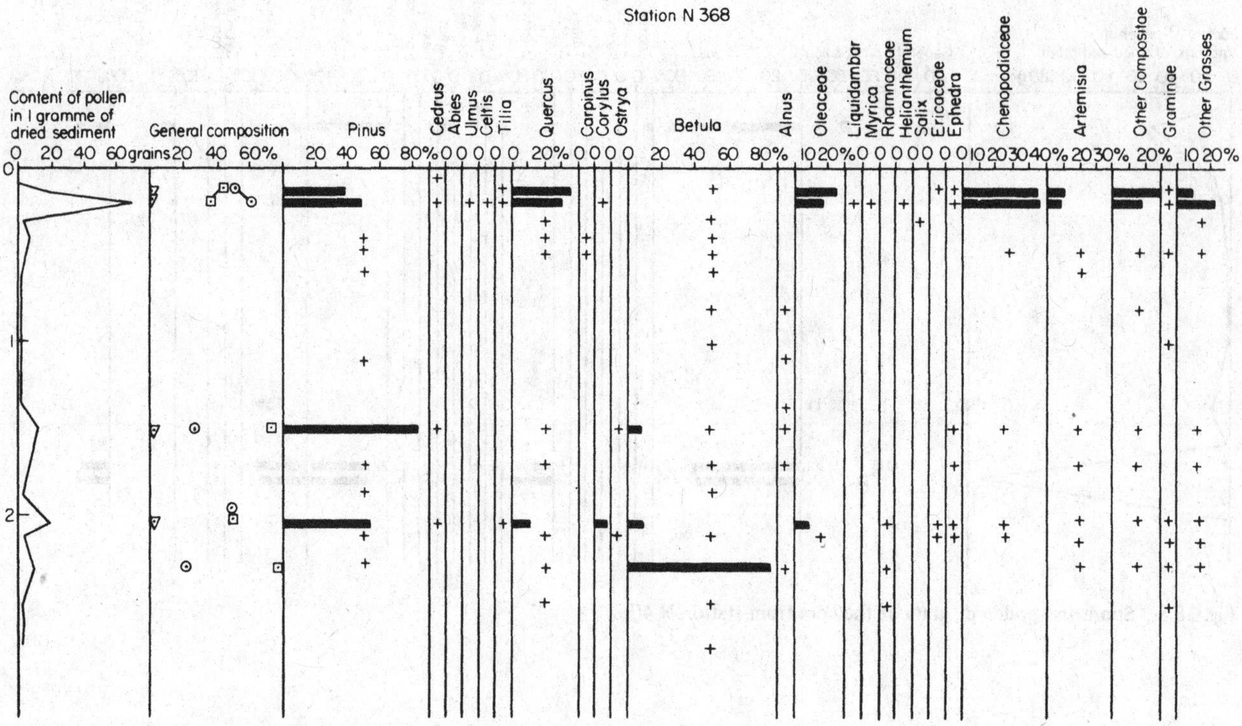
Station N 368
Content of pollen in I gramme of dried sediment
0 20 40 60 grains
General composition
20 40 60%
Pinus
20 40 60 80%
Cedrus
Abies
Ulmus
Celtis
Tilia
Quercus
20%
Carpinus
Corylus
Ostrya
Betula
20 40 60 80%
Alnus
Oleaceae
10 20%
Liquidambar
Myrica
Rhamnaceae
Helianthemum
Salix
Ericaceae
Ephedra
Chenopodiaceae
10 20 30 40%
Artemisia
20 30%
Other Compositae
20%
Graminae
Other Grasses
10 20%
0
1
2

Fig. 28.10. Spore-and-pollen diagram of the core from station N 368.

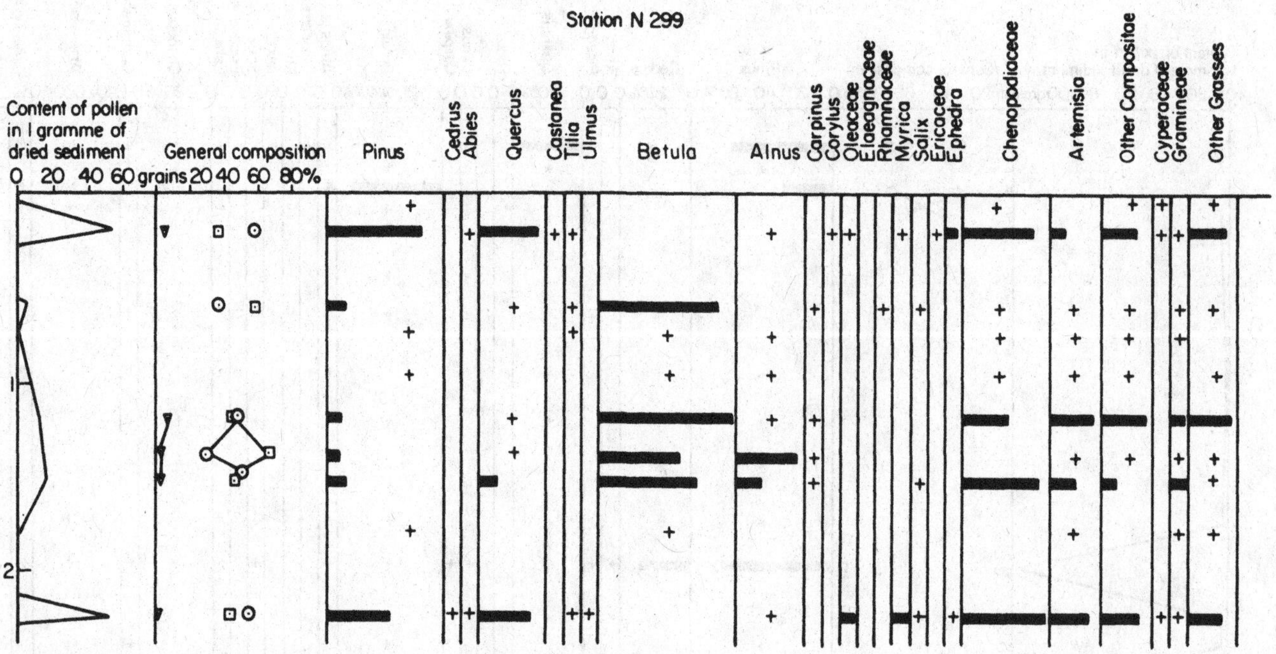
Station N 299
Content of pollen in I gramme of dried sediment
0 20 40 60 grains
General composition
20 40 60 80%
Pinus
Cedrus
Abies
Quercus
Castanea
Tilia
Ulmus
Betula
Alnus
Carpinus
Corylus
Oleaceae
Elaeagnaceae
Rhamnaceae
Myrica
Salix
Ericaceae
Ephedra
Chenopodiaceae
Artemisia
Other Compositae
Cyperaceae
Gramineae
Other Grasses
1
2

Fig. 28.11. Spore-and-pollen diagram of the core from station N 299.

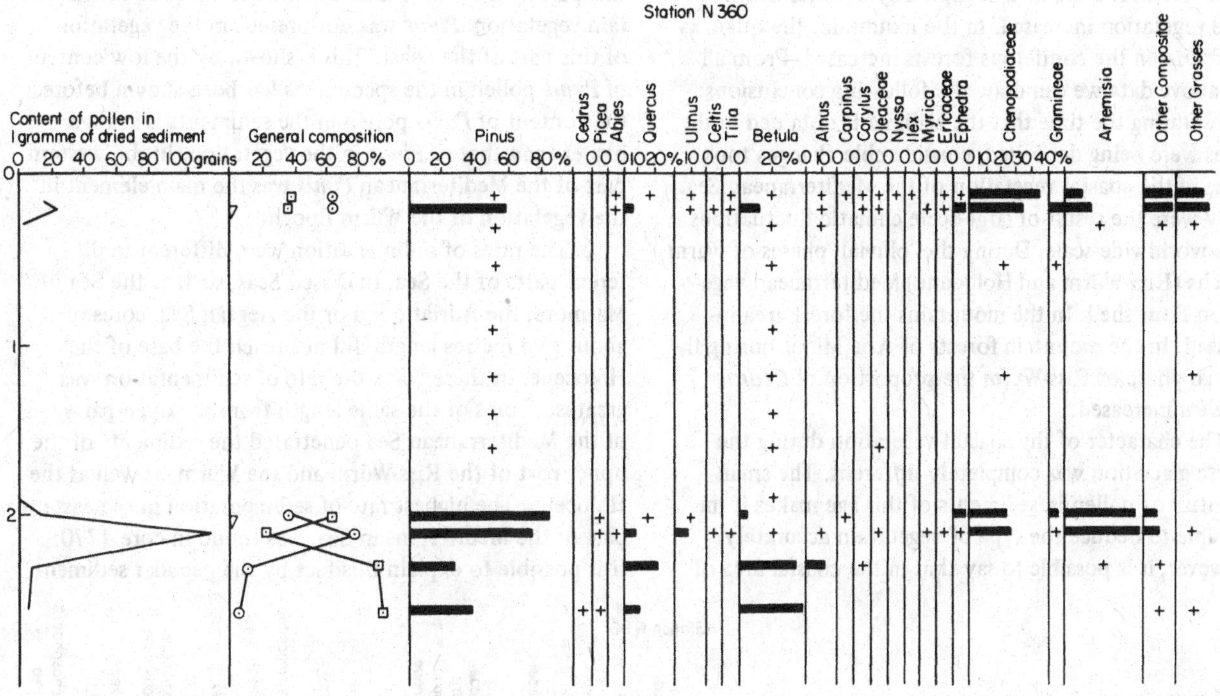
Station N 360
Content of pollen in 1 gramme of dried sediment
0 20 40 60 80 100 grains
General composition
20 40 60 80%
Pinus
0 20 40 60 80%
Cedrus
Picea
Abies
Quercus
Ulmus
Celtis
Tilia
Betula
20%
Alnus
Carpinus
Corylus
Oleaceae
Nyssa
Ilex
Myrica
Ericaceae
Ephedra
Chenopodiaceae
10 20 30 40%
Gramineae
Artemisia
Other Compositae
Other Grasses
0
1
2

Fig. 28.12. Spore-and-pollen diagram of the core from station N 360.

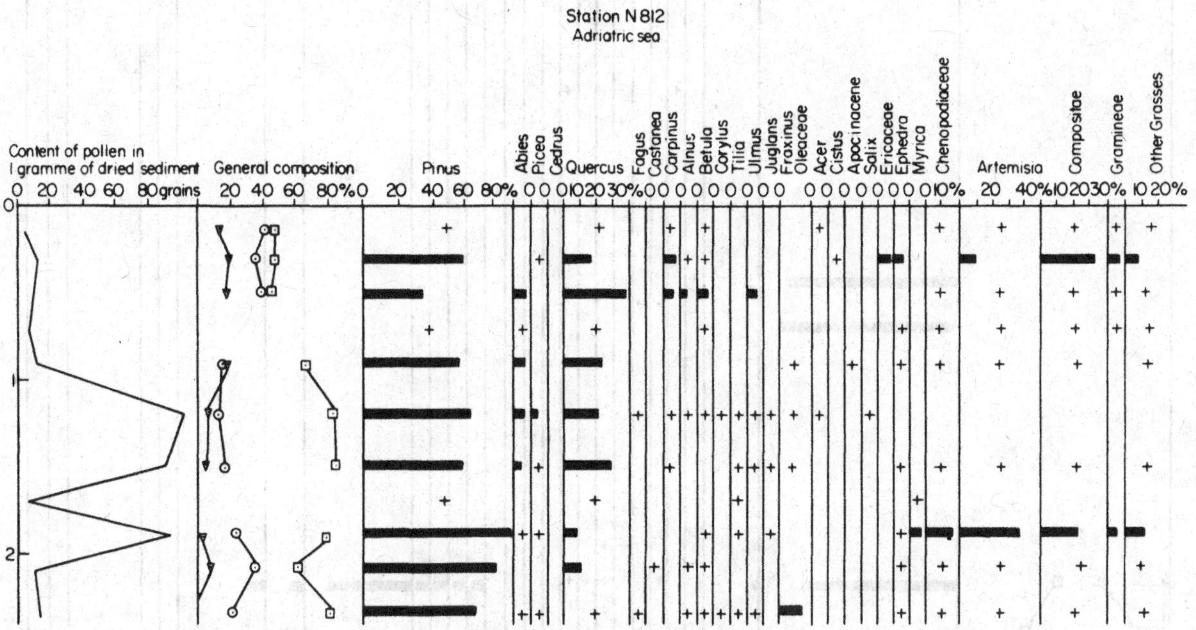
Station N 812
Adriatric sea
Content of pollen in 1 gramme of dried sediment
0 20 40 60 80 grains
General composition
20 40 60 80%
Pinus
0 20 40 60 80%
Abies
Picea
Cedrus
Quercus
10 20 30%
Fagus
Castanea
Carpinus
Alnus
Betula
Corylus
Tilia
Ulmus
Juglans
Fraxinus
Oleaceae
Acer
Cistus
Apocinacene
Salix
Ericaceae
Ephedra
Myrica
Chenopodiaceae
10%
Artemisia
20 40%
Compositae
10 20 30%
Gramineae
Other Grasses
10 20%
0
1
2

Fig. 28.13. Spore-and-pollen diagram of the core from station N 812.

Sardinia and the Atlas were covered by coniferous forests of *Pinus, Picea* and *Abies.* Low-lying areas were steppes. In the Holocene the area occupied by Mediterranean-type vegetation increased. In the mountains the quantity of *Cedrus* in the coniferous forests increased. From all the above data we can draw the following conclusions:

1. During the time that the sediments obtained in the cores were being deposited, considerable changes took place in the coastal vegetation of the Mediterranean Sea. They were the result of large-scale climatic fluctuations on a worldwide scale. During the 'pluvial' phases of warm epochs (Riss-Würm and Holocene) Mediterranean vegetation flourished. In the mountains the forest area increased. In the mountain forests of Asia Minor during the pluvial phase of Riss-Würm the proportion of *Cedrus libanica* increased.

The character of the coastal vegetation during the Würm glaciation was completely different. The small quantity of pollen in sediments of this age makes it impossible to deduce the type of vegetation accurately. However, it is possible to say that in the coastal area of the eastern part of the Mediterranean Sea *Betula* (whose pollen predominates in sediments in all the cores from this part of the sea) was an essential element of the mountain vegetation. *Pinus* was not present in the vegetation of this part of the coast. This is shown by the low content of *Pinus* pollen in the spectra. As has been shown before, the content of *Pinus* pollen in the sediments is always higher than that of *Pinus* in the vegetation. In the western part of the Mediterranean *Pinus* was the main element in the vegetation of the Würm Epoch.

2. The rates of sedimentation were different in different parts of the Sea. In closed Seas, such as the Sea of Marmora, the Adriatic Sea or the Aegean Sea, cores of about two metres length did not reach the base of the Holocene. In these cores the rate of sedimentation was greatest. Cores of the same length from the open parts of the Mediterranean Sea penetrated the sediments of the upper part of the Riss-Würm and the Würm, as well as the Holocene. The highest rate of sedimentation in the eastern part of the Mediterranean Sea, was found in core 4770. It is possible to explain this fact by the peculiar sedimen-

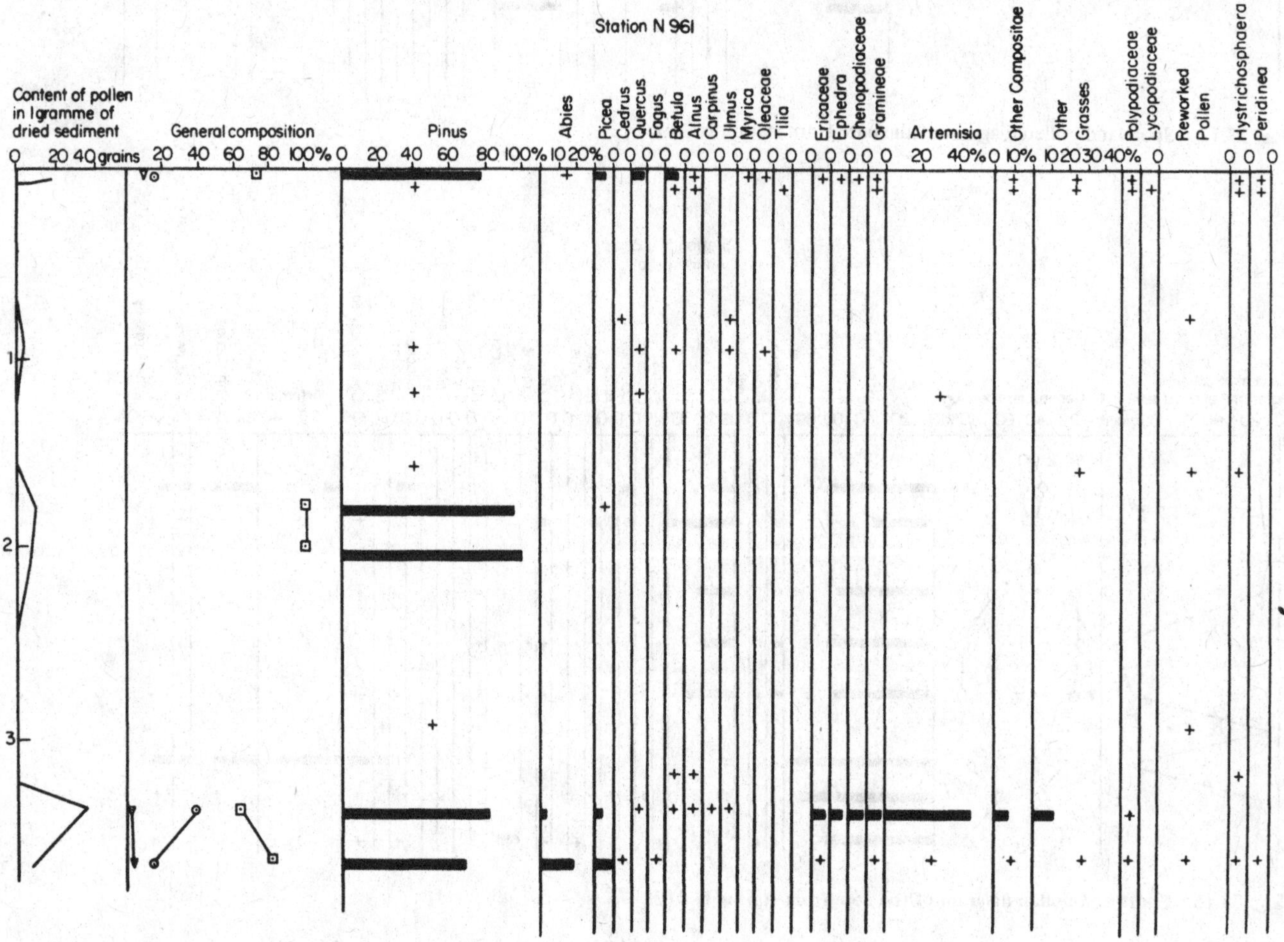
Station N 961
Content of pollen in 1gramme of dried sediment
grains
General composition
Pinus
Abies
Picea
Cedrus
Quercus
Fagus
Betula
Alnus
Carpinus
Ulmus
Myrica
Oleaceae
Tilia
Ericaceae
Ephedra
Chenopodiaceae
Gramineae
Artemisia
Other Compositae
Other Grasses
Polypodiaceae
Lycopodiaceae
Reworked Pollen
Hystrichosphaera
Peridinea

Fig. 28.14. Spore-and-pollen diagram of the core from station N 961.

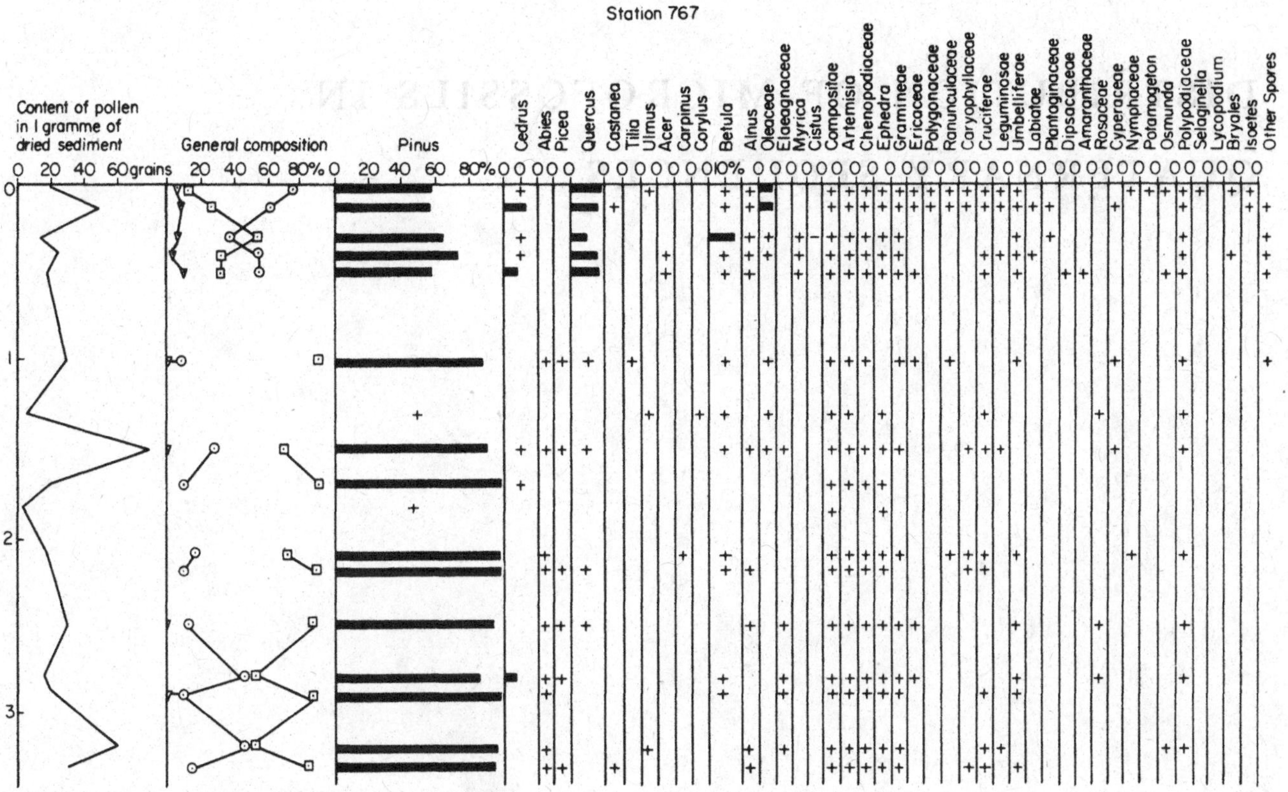
Station 767
Content of pollen in I gramme of dried sediment
General composition
Pinus
0 20 40 60grains
20 40 60 80%
0 20 40 60 80%
10%
Cedrus
Abies
Picea
Quercus
Castanea
Tilia
Ulmus
Acer
Carpinus
Corylus
Betula
Alnus
Oleaceae
Elaeagnaceae
Myrica
Cistus
Compositae
Artemisia
Chenopodiaceae
Ephedra
Gramineae
Ericaceae
Polygonaceae
Ranunculaceae
Caryophyllaceae
Cruciferae
Leguminosae
Umbelliferae
Labiatae
Plantaginaceae
Dipsacaceae
Amaranthaceae
Rosaceae
Cyperaceae
Nymphaceae
Potamogeton
Osmunda
Polypodiaceae
Selaginella
Lycopodium
Bryales
Isoetes
Other Spores
0
1
2
3

Fig. 28.15. Spore-and-pollen diagram of the core from station N 767.

tation in the Levantine depression.

ACKNOWLEDGEMENTS

I wish to acknowledge the scientists of the Black Sea Station of the Institute of Oceanology of the Academy of Sciences of the U.S.S.R. who helped in the collection of the material, and especially to thank K. Schimcus, who made a lithological investigation of these cores and helped me in their interpretation.

Received January 1968.

BIBLIOGRAPHY

Bottema, S. and van Straaten, L. M. J. U. 1966. Malacology and palynology of the two cores from the Adriatic Sea floor. *Marine Geology*, **6**, 553-64.

Emelianov, E. M. 1961. [New data about the sediments of the Mediterranean Sea.] *Dokl. Acad. Nauk, S.S.S.R.* **137** (6), 1435-40 (in Russian).

Larsen, G. 1948. Examination of pollen grains in three cores from the Tyrrhenian Sea. In: Three sediment cores from the Tyrrhenian Sea (ed. H. Petterson). *Göteborgs Kgl. Vetensk. Vitterhets-Samh., Handl. f. 6; ser. B,* **5** (13) 73-9.

SECTION C

DISTRIBUTION OF MICROFOSSILS IN QUATERNARY SEQUENCES

29. QUATERNARY CORRELATIONS AND THE GEOCHEMISTRY OF OOZES

E. OLAUSSON

Summary: Owing to the present drainage pattern the North Atlantic receives much more in the way of weathered products than the other oceans. Solution and oxidation at the bottom of the North Atlantic seem to change with time, being climatically controlled. During ice ages much more was dissolved, transported out to other oceans and laid down there than at present. These changes in the vertical circulation of the oceans cause variations in frequency of foraminiferal species in cores, owing to selective solution of Foraminifera, and changing sizes of *Coscinodiscus nodulifer* resulting from variations in the nutrient level.

The solution of carbonates in the North Atlantic is intensified during ice ages mainly in the zone of Polar Bottom Water, but is also evident in the zone of Upper North Atlantic Deep Water. The stagnation of the Eastern Mediterranean and its fauna during those phases are also discussed.

The Late Pleistocene history of the Red Sea is outlined. It is likely that the Red Sea during the period 18,000 to 20,000 C^{14} years ago was isolated from the rest of the world's oceans and that its water evaporated off.

The time-stratigraphy of some cores is discussed.

INTRODUCTION

The land area draining into the Atlantic is nearly twice as large as that draining into the Indian and the Pacific oceans taken together. Therefore, the Atlantic, particularly the North Atlantic, receives much more in the way of weathered products from rivers than the other oceans. This can be illustrated by the river discharge of calcium and silica to each ocean (Table 29.1). This is probably also the case for most other elements.

Now, if the chemical composition of sea water remains constant, the amount laid down on the ocean floor must be equal to the weathered products supplied. However, water exchange between the oceans more or less effectively smooths out the differences caused by the present drainage pattern, both with respect to the chemical composition of the sea water and the accumulation rates of each element. Elements with comparatively long passage times (e.g. zinc, nickel, cobalt) tend to be more concentrated in Pacific clays than shelf deposits, with Atlantic deposits at an intermediate level; for elements with short passage times (e.g. chromium) the reverse can be the case. Since the exchange of elements between the oceans seems to change with time, being climatically controlled, the present discussion concentrates on the distribution of some biogenically important elements in pelagic deposits and the interpretation of palaeontological data from a geochemical point of view.

In an earlier paper (Olausson 1967) it is stated that about half of the calcium discharged into the North Atlantic by rivers is transported out from it and deposited

Table 29.1. River discharge of dissolved calcium and silica (Olausson 1967)

	Ca		SiO_2	
	(x 10^{13} g/y)	%	(x 10^{13} g/y)	%
Atlantic Ocean	35.9	78	24.5	57
North Atlantic	32.2	70	17.2	40
South Atlantic	3.7	8	7.3	17
Indian Ocean	4.6	10	14.2	33
Pacific Ocean	5.5	12	4.3	10
	46.0		43.0	

Table. 29.2. The present accumulation of calcium carbonate in per cent of the total compared to the accumulation rate to be expected, if the inflow of calcium to an ocean were equal to the outflow. The figures are approximate.

	Accumulation of calcium carbonate if the inflow of calcium is equal to the outflow; in per cent of the total.	The present accumulation of calcium carbonate in per cent of the total deposition	Difference
North Atlantic	70	35	– 35
South Atlantic	8	20	+ 12
Indian Ocean	10	20	+ 10
Pacific Ocean	12	25	+ 13

in other oceans (Table 29.2). This process is essential in interpretation of the fossil record and must therefore be discussed in much more detail. First, we discuss the geo-economy and geochemistry of calcium and calcareous fossils and then the silica budget.

CALCIUM AND CALCAREOUS FOSSILS

General

There is no surface outflow from the North Atlantic. Therefore the net out-transport, 16×10^{13} grammes of calcium per year, occurs in the cool-water sphere. The calcium is removed from the surface water by organisms, chiefly Foraminifera and coccolithophorids. The settling time compared to the time during which the calcareous remains are in contact with the bottom water is negligible. Hence, the solution of the carbonates can be supposed to occur chiefly at the bottom (recycling due to predation in the warm-water sphere is here discounted). This solution can be described by the two following equations:

$$CaCO_3 + H_2O + CO_2 \rightarrow Ca^{2+} + 2HCO_3^- \quad (1)$$
$$CaCO_3 + CH_2O + O_2 \rightarrow Ca^{2+} + 2HCO_3^- \quad (2)$$

where CH_2O stands for organic matter (soft organic remains in various degrees of decomposition). Large, soft organic remains may reach the bottom due to their own weight, but smaller organic particles, colloidal organic matter, and organic matter in true solution seem to reach greater depths chiefly by descending water masses. In a litre of the water that sinks down to considerable depths or to the bottom in the northernmost Atlantic there may be some 0.5 milligrammes of carbon, 5-8 millilitres of oxygen and some 5 millilitres of free carbon dioxide. If 80×10^{13} grammes calcium carbonate settle down on the floor of the North Atlantic and adjacent seas every year (as an average) half of this quantity must be dissolved. In order to dissolve 40×10^{13} grammes of calcium carbonate all the free carbon dioxide in about 2×10^4 cubic kilometres sink water or all the carbon in 10^5 cubic kilometres is needed. The amount of this sink water is estimated by Sverdrup to be three times as large or 3×10^5 cubic kilometres per year (Sverdrup, Johnson and Fleming 1942). During its flux southward – in the direction of the Pacific – organic particles settle, adding more organic matter to the bottom. Therefore it seems possible that equation (2) is the most important one for carbonate solution, at least at lower latitudes, and it is also supposed to regulate the critical depth of calcium carbonate solution (Olausson 1967).

Thus we see that the carbonate content of sediments, the solution rate at the bottom, as well as the pH of the environment, depend on the supply (in moles per unit time) of calcium carbonate (chiefly the vertical supply), organic matter (both vertical and lateral supply), carbon dioxide and oxygen (both by descending water masses). In areas flushed by North Atlantic sink water oxygen is present in excess, and if we assume the proportion of calcium carbonate and organic matter settling down from the overlying surface water as rather constant, changes in the solution of calcium carbonate and thus also changes in the carbonate content of the final sediment are related to the intensity of the vertical circulation. This discussion also suggests that changes in the carbonate content of a sediment core might reflect changes in the intensity of the vertical circulation.

Changes in the accumulation rates of calcareous matter can be recognized by carbonate analyses of sediments, However, this method is limited by the assumption involved in this theory, namely that other components have much more constant rates of accumulation, which is not always so. Therefore, the percentage of calcium carbonate can be used only where solution and/or production have changed considerably. The first step in the solution of foraminiferal tests is fragmentation (cf. Fig. 29.5). I found that, in the Pacific, fragmentation of foraminiferal tests was a more useful method since it could also be used in areas where the carbonate content proved to be rather constant. The fragmentation depends, according to equations (1) and (2), on the molar supply of calcium carbonate, organic matter and carbon dioxide at the bottom. A low degree of solution and fragmentation is due to an excess of calcium carbonate but the solution and fragmentation increase if the molar ratio

$$\frac{CaCO_3}{CH_2O + CO_2} \quad (3)$$

at the bottom decreases. The number of entire tests per unit volume can be assumed to be rather constant for a certain locality, but changes occur from place to place, depending on the fertility of the overlying surface water and the intensity of the vertical circulation. An appreciable change in the number of Foraminifera indicates (in an undisturbed sequence) either a change in the supply of calcium carbonate (i.e. the fertility of the water) and/or in the lateral supply of carbon dioxide and organic matter (areas with oxygen deficiency are here disregarded). A climatic change could cause such variation, and therefore a curve of the number of Foraminifera per unit volume in a core sequence could be a climatic curve. Essentially the amount of matter $> 65\mu$ or $> 150\mu$ could show the same thing, but I have found that the number of entire tests is a more useful method.

Before going into further detail we must also consider the selective solution of Foraminifera. If the deposition of calcium carbonate in the North Atlantic is 3 milligrammes per square centimetre per year on average, the average weight of a test of an adult Foraminifera is 0.05 milligrammes (generally between 0.1-0.01 milligrammes), then about sixty tests settle to the bottom per square centimetre per year. Theoretically thirty of them are dissolved. In order to get a chemical expression of the selective solution of Foraminifera a series of experiments has been performed. From cores obtained from comparatively shallow waters well preserved tests of certain species of Foraminifera were isolated. The tests, 10 milligrammes of each, were then treated with 300 millilitres HAc – H_4NAc (pH~5.7) in a Millipore-funnel, and at certain intervals 5 millilitres of the solution were taken out and the calcium concentration measured by an atomic absorption spectrophotometer. The results, the average for three series, are given in Fig. 29.1. From these it appears that *Globigerina* and *Globigerinoides* are dissolved more easily than *Globorotalia* and *Pulleniatina*, which might be the last to be destroyed. This conclusion supports the observation of Phleger, Parker and Peirson (1953, p. 117). Another indication of the same thing is given in Fig. 29.2. Consequently, if the production of Foraminifera is constant, but the solution rate at the bottom is changing, changes in the relative frequency of the species will occur. If the pH at the bottom is constant, but temperature changes in the surface water favour some species more than others, the frequency of the species in the final sediment will reflect climatic events. If both the pH of the bottom water and temperature in the warm water sphere have varied the interpretation of the thanato-

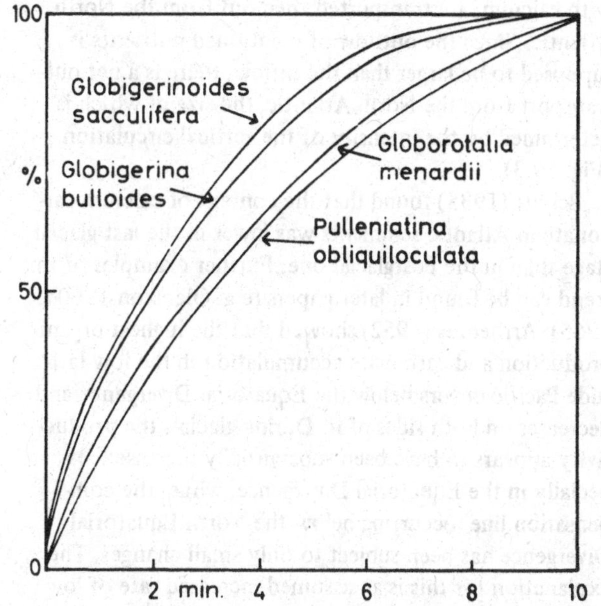
100
%
50
0
Globigerinoides sacculifera
Globorotalia menardii
Globigerina bulloides
Pulleniatina obliquiloculata
2 min. 4
6
8
10

Fig. 29.1. The dissolving rate of some foraminiferal tests. The tests, 10 milligrammes of each, having been dissolved by 300 millilitres HAc/H_4 NAc (pH ~5.7). The amount of calcium in the solution was measured at certain time intervals. The concentration is given in per cent of final one. Average of three series.

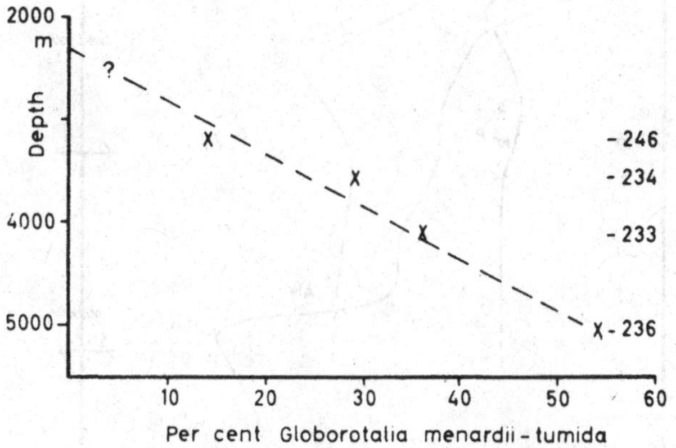
2000 m
Depth
4000
5000
?
-246
-234
-233
-236
10
20
30
40
50
60
Per cent Globorotalia menardii - tumida

Fig. 29.2. Average percentages for the *Globorotalia menardii-tumida* group during Eem-Brørup interval in mentioned Swedish cores.

coenosis will be difficult. This will become clear from the following discussion. Before that, however, we must discuss nutrient transport between the oceans.

Nutrient transport theory

Organic matter may be more exactly characterized as $(CH_2O)_{106} (NH_3)_{16} H_3PO_4$; siliceous organisms also contain silica. Oxidation of organic matter will thus not only release carbon dioxide but also compounds containing nitrogen, phosphorus, silicon, etc. These elements, together with calcium, are transported then out from the North Atlantic. Since the outflow of mentioned nutrients is supposed to be larger than the inflow, there is a net out-transport from the North Atlantic, the size of which is determined by the intensity of the vertical circulation (Fig. 29.3).

Schott (1935) found that the content of calcium carbonate in Atlantic sediments was lower in the last glacial stage than in the Postglacial one. Further examples of this trend can be found in later papers (e.g. Olausson 1960*d*, 1965). Arrhenius (1952) showed that the highest organic production and carbonate accumulation in the low-latitude Pacific occurs below the Equatorial Divergence and decreases on both sides of it. During glacials the productivity appears to have been substantially increased, especially in the Equatorial Divergence, while 'the compensation line' occurring below the North Equatorial Divergence has been subject to only small changes. The explanation for this is an assumed increased rate of low

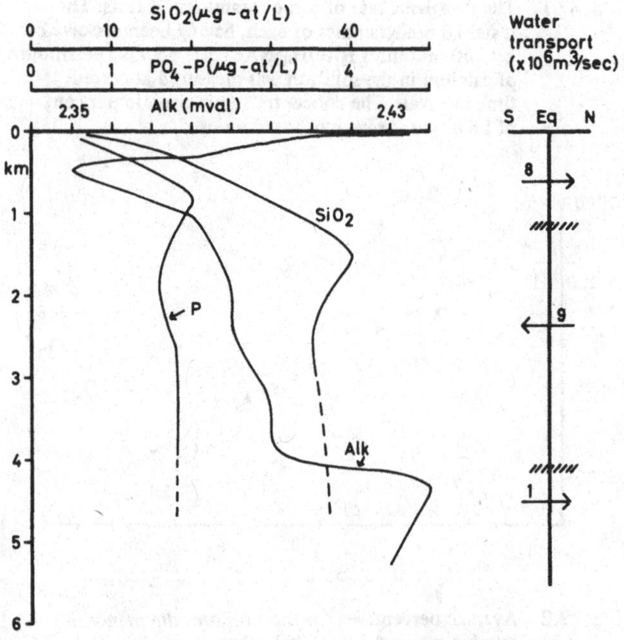

SiO2(μg-at/L)
0
10
40
PO4-P(μg-at/L)
0
1
4
Alk (mval)
2.35
2.43
km
0
1
2
3
4
5
6
SiO2
P
Alk
Water transport (x10^6 m^3/sec)
S Eq N
8
9
1

Fig. 29.3. The distribution of alkalinity at station 337 (0° 03′ S, 18° 13′ W, Bruneau *et al.* 1953), the distribution of PO_4-P and SiO_2 in the Atlantic and the water exchange across its equator (Sverdrup, Johnson and Fleming 1942, fig. 48.55 and table 76).

The alkalinity and the calcium content of sea-water may be related by the following equation:

[Ca] (mg-at./1 @ ½ Alk + 0.465 x Chlorosity. According to Lyman (1959) calcium must be the balancing cation both for the nitrate and the phosphate. Since both do not involve the alkalinity, the calcium concentration at large depths in the Atlantic will be slightly higher than that suggested by the equation given above.

latitude atmospheric circulation. As stated elsewhere (Olausson 1967) increased deposition is also a geoeconomic question and sea water itself cannot sustain increased deposition over a long period of time if other factors are constant. Further, wind velocity at low latitudes (5° S to 5° N) must be rather constant 'da hier der Luftdrucksgradient keine Rolle spielt und die verschwindende Coriolis-Kraft alle aufkommenden Sturme sehr rasch wieder abdampft' (Flohn 1952, p. 271). Therefore we would expect increased wind activity (if any) during glacials further out from the Equator and the increase in production would affect the area around the North Equatorial Divergence (10° N) more than the Equatorial Divergence region (0°). However, the largest change in accumulation occurred below the latter area. Therefore we have to look for another explanation which can also tell us something about the source of the increased amount of biogenically important elements available for production and deposition.

Olausson (1965, 1967) suggested that differences in carbonate solution and oxidation of organic matter in the North Atlantic create differences in production and deposition in the southern hemisphere of the other oceans. My curve from the Pacific (Olausson 1961*a*) is explained (Olausson 1967) as a result of changes in the calcium and net nutrient transport out from the North Atlantic. An ice age apparently starts the following chain reaction:

(i) intensified formation of deep and bottom water in the North Atlantic due to intensified cooling and ice sheet formations on adjacent continents, which

(ii) causes an increase in the supply of carbon dioxide, oxygen, and organic matter to the bottom, which

(iii) causes an increase in the oxidation of organic matter (or $C_{106}N_{16}P$), which

(iv) together with the supply of carbon dioxide causes an increase in the solution of calcite, apatite, etc, which

(v) results in an increase in the amount of calcium, phosphorus, silicon, nitrogen, etc, that are transported out from the North Atlantic (see Fig. 29.3), which

(vi) causes an increase in the rate of organic production in the other oceans due to an increased amount of nutrients there, which

(vii) causes an increase in the rate of accumulation of calcium carbonate and the numbers of entire foraminifera (Olausson 1961*a*, 1967), an increase in the deposition of phosphate, and probably also an increase in the size of *Coscinodiscus nodulifer* because of the increased supply of nutrients.

During interglacials oxidation and solution at the bottom of the North Atlantic are less intense; carbonate deposition increases there and decreases in other oceans.

An explanation of why the *South* Equatorial Diver-

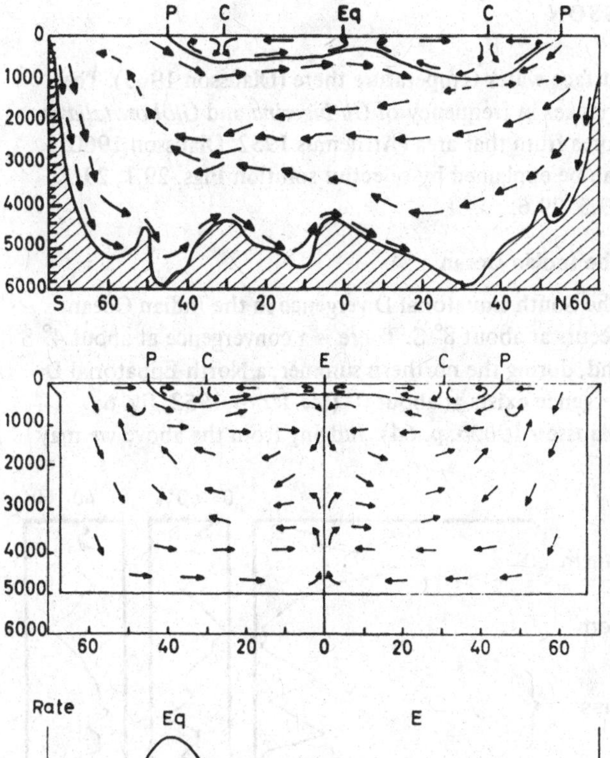

P C Eq C P
0
1000
2000
3000
4000
5000
6000
S 60 40 20 0 20 40 N60
P C E C P
0
1000
2000
3000
4000
5000
6000
60 40 20 0 20 40 60
Rate
Eq
E
a
e
Core 58
c
b
d
0° 5°N 10°

Fig. 29.4. Upper section: Longitudinal diagram of the circulation of the Atlantic Ocean (according to Wüst, 1936).

Middle section: Schematic representation of the meridional components of the oceanic circulation in a fully symmetrical ocean (according to Defant, 1961). The closest approach to this ideal case is found in the Pacific. Note that the boundary between the two closed circulations in each hemisphere occurs below the Equatorial Counter Current.

Lower section: The relative rate of accumulation of biogenic carbonates in the East Equatorial Pacific during a glacial age (a), an interglacial (e), and Lower Tertiary (c), and the solution of carbonates during a glacial (b) and Lower Tertiary (d; according to Arrhenius, 1952). P, Polar front; C, Subtropical Convergence; E, Equatorial Counter Current; Eq, Equator.

The large increase in the carbonate accumulation during an ice age is concentrated in the southern hemisphere, and the rapid decrease in carbonate content occurs below the Counter Current.

Fig. 29.5. An attempt to illustrate how changes in productivity influence the composition of the final sediment in the Pacific.

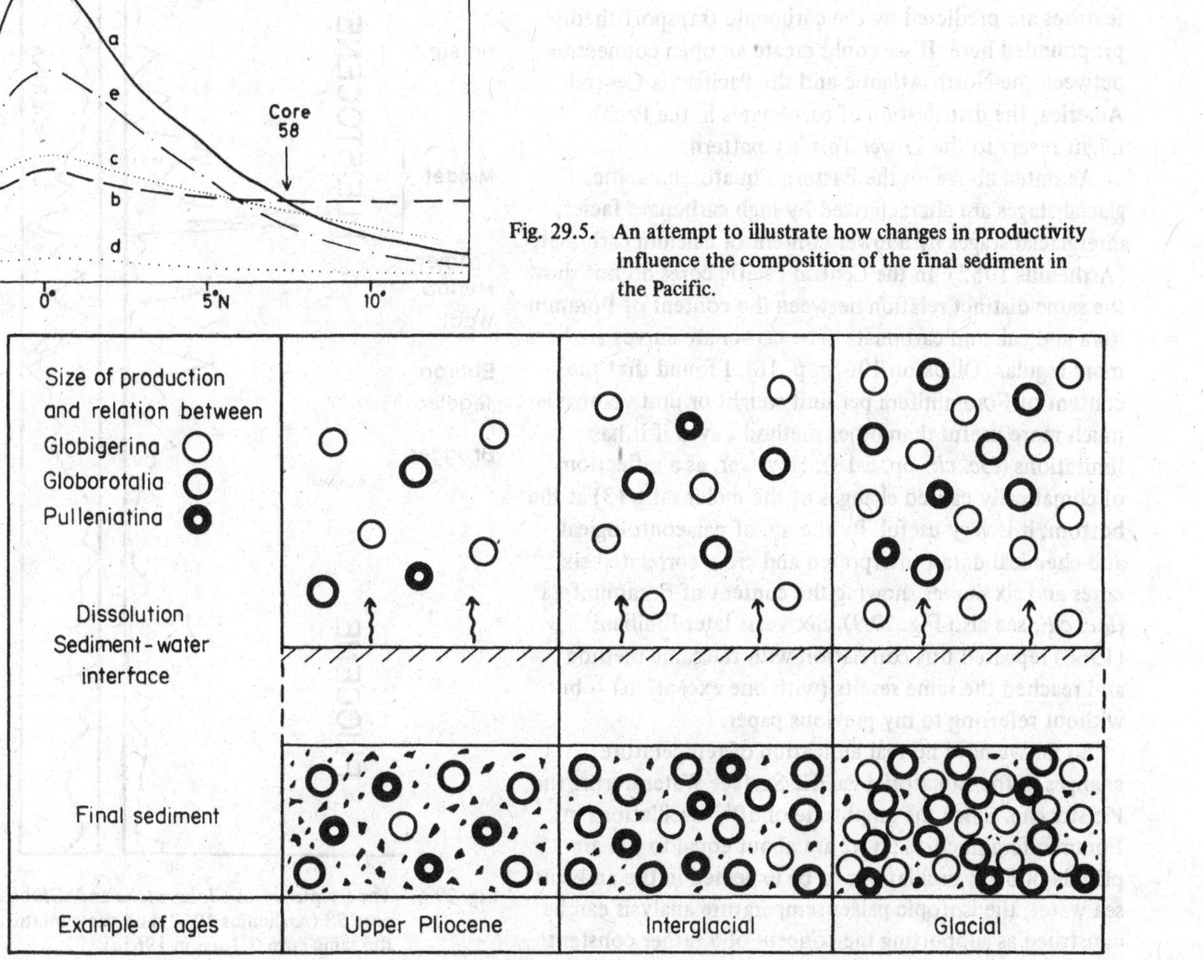

Size of production and relation between
Globigerina
Globorotalia
Pulleniatina
Dissolution
Sediment-water interface
Final sediment
Example of ages
Upper Pliocene
Interglacial
Glacial

gence region is much more affected than the North Equatorial Divergence zone is also provided by this theory. There is appreciable exchange of deep and bottom water across the Equator only in the Atlantic ocean. It is comparatively small in the Indian Ocean and practically absent in the Pacific (Fig. 29.4; Sverdrup, Johnson and Fleming 1942, p. 754; Defant 1961, fig. 323). The South Equatorial Divergence is the main cause of the formation of two independent deep-water circulations to the north and south of it (Ivanenkov and Gubin 1960). We can therefore assume that an increased outflow of nutrients chiefly or exclusively influences the area around and south of the South Equatorial Divergence in the Indian and Pacific Oceans.

The Pacific

As can be seen from Fig. 29.4 the area below the Counter Currents constitutes a transition zone between the high carbonate facies (below the Equatorial Divergence) and the low or non-carbonate facies to the north. Furthermore, the large increase in carbonate accumulation during an ice age is concentrated in the southern hemisphere. These features are predicted by the carbonate transport theory propounded here. If we could create an open connection between the North Atlantic and the Pacific via Central America, the distribution of carbonates in the Pacific might revert to the Lower Tertiary pattern.

As stated above, in the Eastern Equatorial Pacific, glacial stages are characterized by high carbonate facies, interglacial stages by a lower content of calcium carbonate (Arrhenius 1952). In the Central Pacific cores do not show the same distinct relation between the content of Foraminifera and calcium carbonate. The carbonate curves are here more regular (Olausson 1961*a*, p. 16). I found that the content of Foraminifera per unit weight or unit volume is much more useful than other methods, even if it has limitations (*loc. cit.* pp. 5-18). However, as a reflection of climatically caused changes of the molar ratio (3) at the bottom, it is very useful. By the aid of palaeontological and chemical data I interpreted and cross-correlated six cores and six curves showing the content of Foraminifera (*loc. cit.;* see also Fig. 29.9). Six years later Emiliani (1966) repeated this correlation with the same method and reached the same results (with one exception) – but without referring to my previous paper.

So far we have no real indication of temperature changes in the Equatorial Pacific Surface Water during the Pleistocene. Since the amplitude of δO^{18} oscillations in Foraminifera in cores 58-62 are about equal to the amplitude of δO^{18} oscillations to be expected in the ambient sea water, the isotopic palaeotemperature analysis can be construed as supporting the concept of a rather constant surface water temperature there (Olausson 1965). The changes in frequency of *Globigerina* and *Globorotalia* in cores from that area (Arrhenius 1952; Olausson 1960*a*) can be explained by selective solution Figs. 29.1, 29.2, 29.5, 29.6.

The Indian Ocean

The South Equatorial Divergence in the Indian Ocean occurs at about 8° S. There is a convergence at about 4° S and, during the northern summer, a North Equatorial Divergence exists at about 0° (see Jerlov 1953, fig. 6; Olausson 1960*b*, p. 64). Judging from the above we may

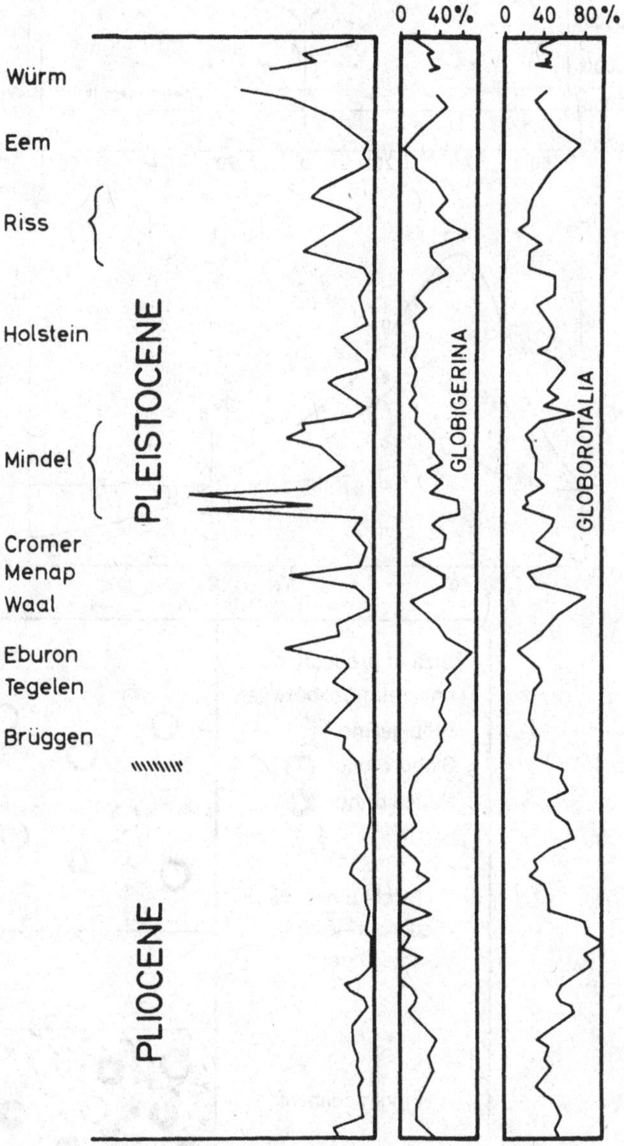
0 40%
0 40 80%
Würm
Eem
Riss
Holstein
Mindel
Cromer
Menap
Waal
Eburon
Tegelen
Brüggen
PLEISTOCENE
PLIOCENE
GLOBIGERINA
GLOBOROTALIA

Fig. 29.6. The frequency of *Globorotalia* and *Globigerina* in core 58 (Arrhenius 1952) and the climatic curve of the same core (Olausson 1961*a*).

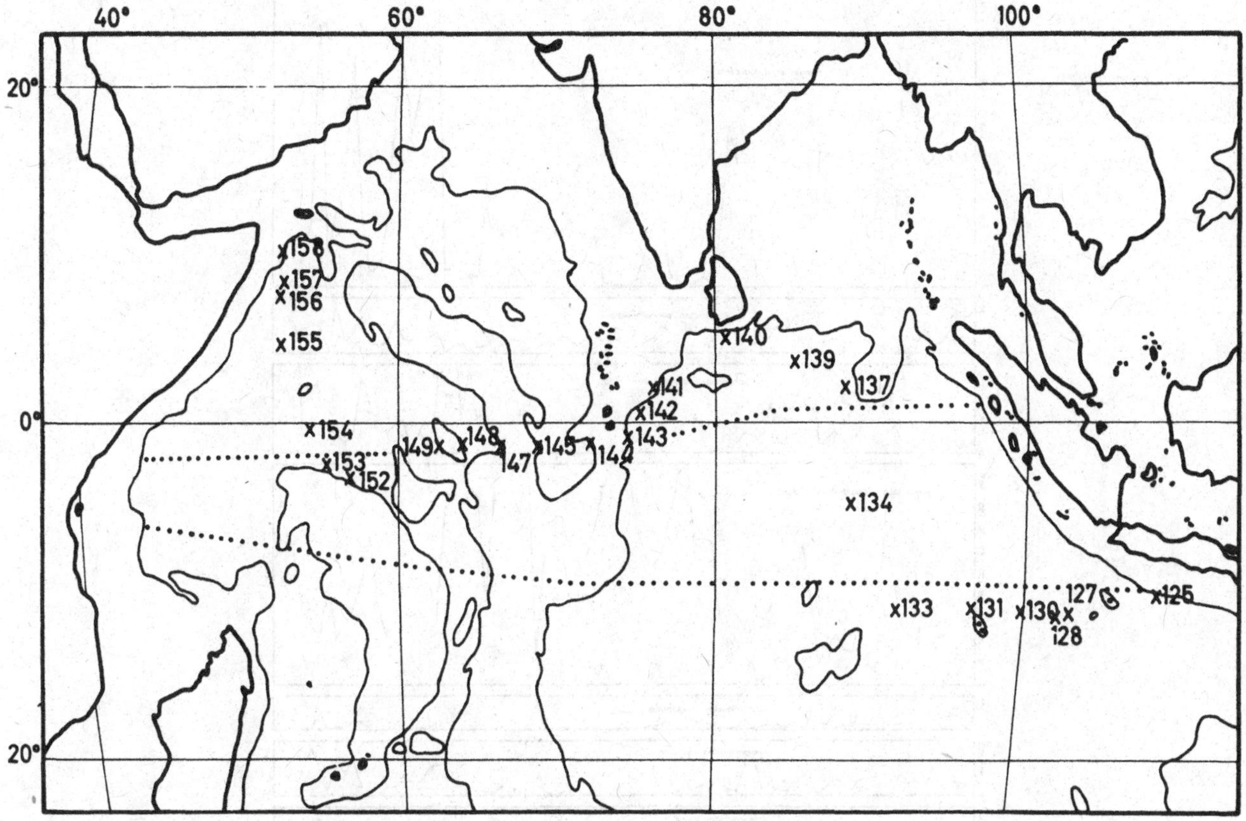
40°
60°
80°
100°
20°
0°
20°
x158
x157
x156
x155
x154
x153
x152
149x
x148
147
x145
x144
x143
x142
x141
x140
x139
x137
x134
x133
x131
x130
x127
x128
x125

Fig. 29.7. Location of cores from the Indian Ocean. Dotted lines indicate divergences.

assume that a glacial would create intensified production in the area below and south of the South Equatorial Divergence as was the case in the Pacific.

Glancing at the map (Fig. 29.7) and considering the coring depths, one might suppose that cores (128), 131, 152, 153, (154) would be likely to show a correlation between glacials and high carbonate facies. Cores 131 and 152-4 clearly show the same characteristics as cores 58-62, 74 and 75 from the Pacific, and core 128 has carbonate beds alternating with siliceous ones. Furthermore radiolarian ooze occurs in the deeper region below the eastern part of the South Equatorial Divergence (cores 127-33) while below the North Equatorial Divergence red clay ought to have been deposited if the carbonate had been dissolved (Olausson 1966, fig. 22). This suggests that the rate of production is higher below the South Equatorial Divergence than below the North Equatorial Divergence in both the Indian and Pacific Oceans (due to the present drainage system of the continents and the vertical circulation pattern of the oceans (Olausson 1967)).

The other Indian Ocean cores collected by the Swedish Deep-Sea Expedition can be divided into three groups, in which the correlations between glacials and high carbonate facies are less obvious. These are: 155-8 (under the influence of upwelling off the coast of Somaliland), 145-9, and 140-4. These three groups will be discussed in another paper in connection with O^{18}-analyses of Foraminifera from one core from the Maldive area and from core 152.

The curve obtained from analyses of the carbonate fraction of core 131 exhibits, as stated above, the same trend as the Pacific curves and is the reverse of carbonate curves in North Atlantic cores. It extends down to the Cromer stage. *Globigerina inflata* reaches a maximum percentage in the Upper Mindel in this as well as in the following cores (and core 128, 760-900 centimetres), constituting a good index horizon (Figs. 29.8, 29.9). Detailed foraminiferal analyses of the other cores will be published later (see also Olausson 1960*b*).

Core 152 extends approximately down to the boundary VI/VII (Cromer/Mindel). Core 153 penetrates down into the Holstein stage, approximately to its second cool substage. The trend of the last curve is close to that of core 152.

Core 154 is raised from a depth about 500 metres greater than that from which cores 131, 152 and 153 were obtained. Even if the solution of carbonates is less

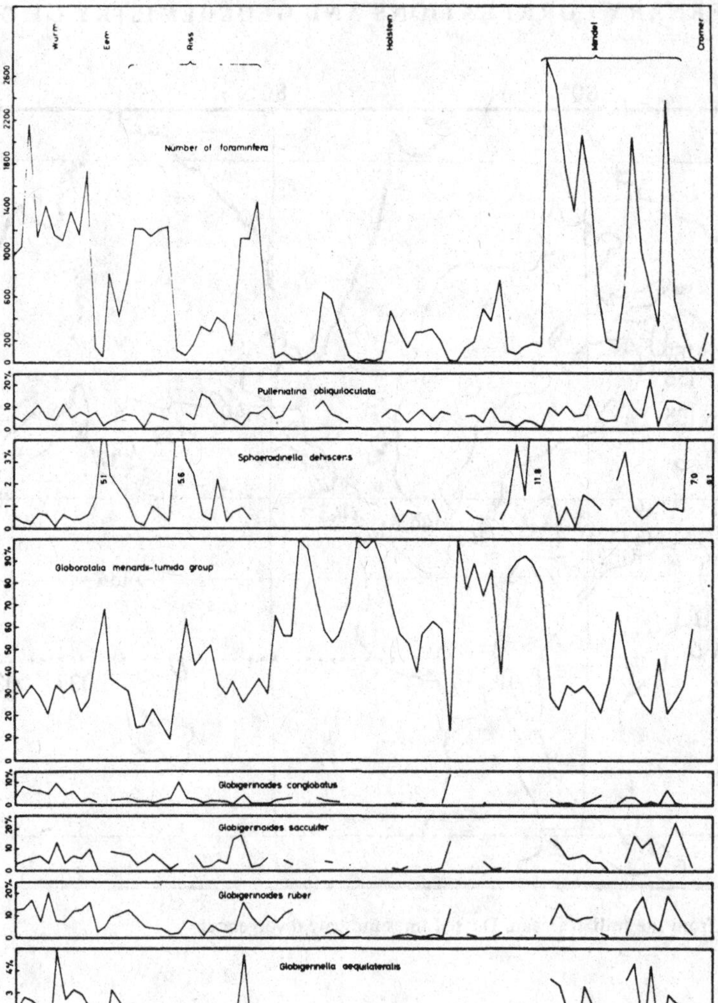
CORE 131
Würm
Eem
Riss
Holstein
Mindel
Cromer
Number of foraminifera
Pulleniatina obliquiloculata
Sphaeroidinella dehiscens
Globorotalia menardii-tumida group
Globigerinoides conglobatus
Globigerinoides sacculifer
Globigerinoides ruber
Globigerinella aequilateralis
Globoquadrina dutertrei
Orbulina universa
Globorotalia truncatulinoides
Globorotalia crassaformis
Globorotalia inflata
Globigerina bulloides

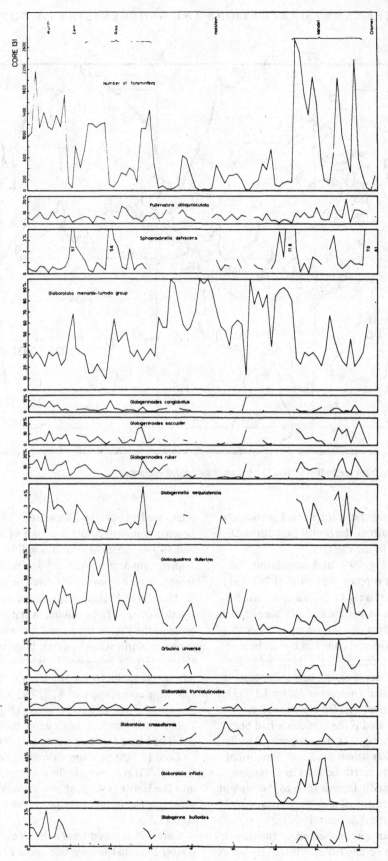

Fig. 29.8. Core 131 (11° 10′ S, 96° 42′ E. Depth 4,390 metres).

intense at 4,800-4,900 metres in the Indian Ocean than in the North Atlantic, core 154 exhibits large variations in the content of calcium carbonate and the number of entire Foraminifera per unit volume. It is in this respect similar to the North Atlantic cores 235-236 (Olausson 1965) which were collected from 5,000-5,200 metres and in which small climatic changes are recorded in an exaggerated way depending on nearness to the 'compensation depth' of calcium carbonate. This is so because the changes in the carbonate content generally decrease with increasing depth of the floor.

According to my interpretation core 154 consists of the whole of the Pleistocene, i.e. up from the base of the Brüggen (Praetiglian) glaciation.

The Atlantic Ocean

The uppermost stages of North Atlantic cores are characterized by changing carbonate content and certain trends in the frequency of planktonic Foraminifera (Olausson 1965). However, below stage 4^2 (Warthe) these trends are less conspicuous and in the Upper Holstein the 'warm' Foraminifera group *Globorotalia menardii-tumida* is absent in spite of the fact that other indicators suggest interglacial conditions. This is apparently a question of at least two main variables. One reflects the solution rate of carbonates at the bottom and this factor can in short be called 'pH'. The other concerns environmental conditions influencing the distribution and production of planktonic Foraminifera in the surface water during different ages and subages. The relative frequency of Foraminifera in the Pacific and Indian Ocean was only given as a function of 'pH' but for the North Atlantic we must also take changes in salinity and temperature into consideration. The carbonate content is supposedly a function of only one of these: the 'pH'. Apparently, studies of the carbonates are more useful for climatic studies than foraminiferal studies, at least on a large scale.

Starting from the present we realize that the 'pH' can change not only into lower values (into glacial age conditions) but also in the opposite direction (into very warm ages). This means that the solution of tests might have been less intense than at present and as a consequence

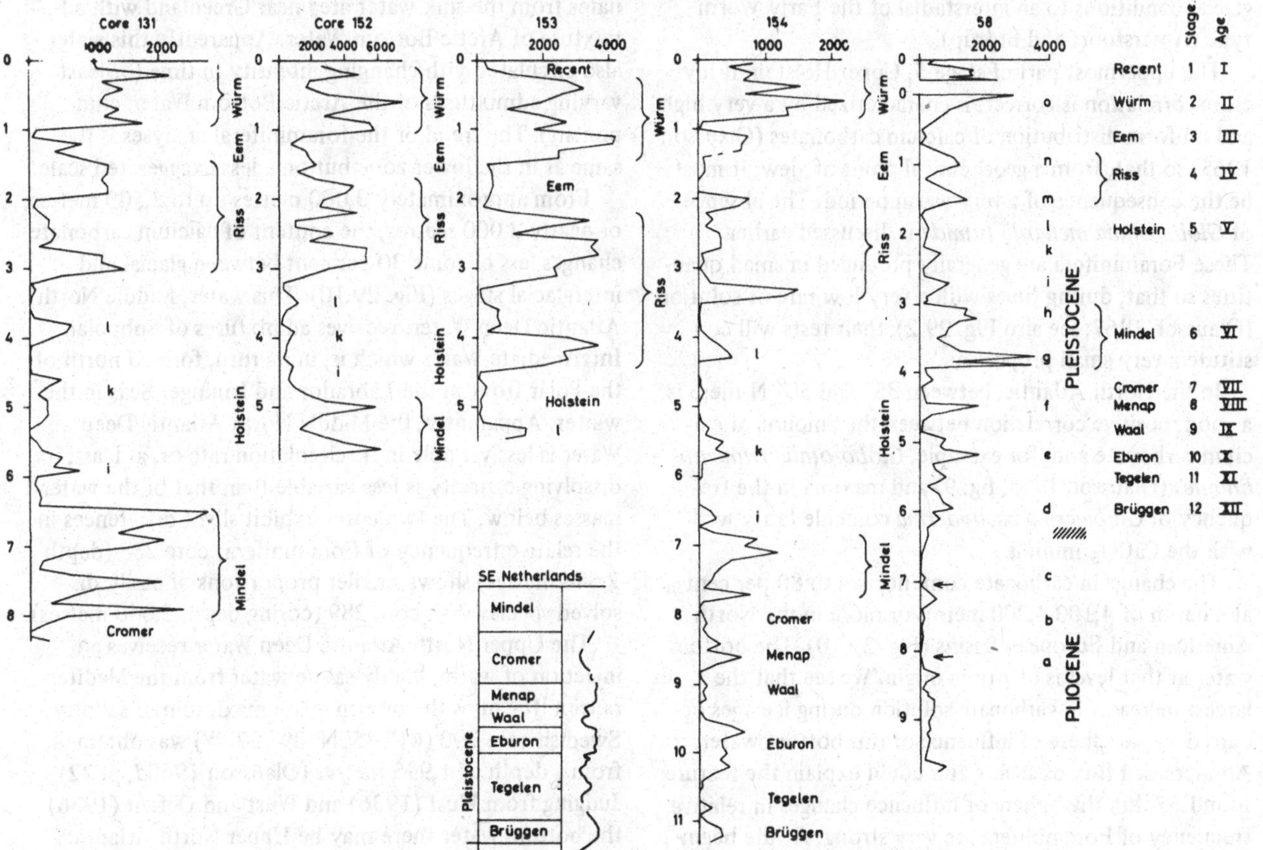
Core 131
Core 152
153
154
58
Recent
Würm
Eem
Riss
Holstein
Mindel
Cromer
SE Netherlands
Menap
Waal
Eburon
Tegelen
Brüggen
Reuver
Pleistocene
Pliocene
PLEISTOCENE
PLIOCENE
Stage
Age

Fig. 29.9. The content of entire Foraminifera in four Indian Ocean cores (131, 152-4) and in the Pacific cores 58 and 58B. The curves reflect the rate of carbonate accumulations. The climatic curve of the S.E. Netherlands (Zagwijn 1963) is redrawn in order to show complete agreement.

more calcium carbonate was at that time buried in the North Atlantic. The corresponding stages in the Pacific and Indian Oceans will be low-carbonate ones. Such conditions would, according to Fig. 29.1 and 29.2 lower the relative frequency of the *Globorotalia menardii-tumida* group in the Atlantic, but cause an increase in it in the Indian and Pacific core stages for that time. If a decrease in temperature gives the fauna a 'cool' impression, a temperature increase might increase the distribution of warm-water Foraminifera and diminish the area where 'cool' Foraminifera dominate. Changes in salinity with an amplitude of at least one per cent for average sea water are also evident, but the influence of this factor must be left open owing to lack of detailed data.

The distribution pattern of planktonic Foraminifera is illustrated for example by Schott (1935, 1966), Phleger, Parker and Peirson (1953), and Bé and Hamlin (1967). This pattern is useful for the southern North Atlantic for interpretation of cores up from the Warthe substage. The substage 4^1 has been discussed earlier (Olausson 1965, pp. 242-3). The climate was probably changing from glacial conditions to an interstadial of the Early Würm type (Amersfoort and Brørup).

The uppermost part of stage 5, Upper Holstein if my cross-correlation is correct, is characterized by a very high and uniform distribution of calcium carbonates (Olausson 1965) so that, from a geochemical point of view, it must be the consequence of a very warm period. The absence of *Globorotalia menardii-tumida* is discussed earlier. These Foraminifera are generally produced in small quantities so that, during times with a very low rate of solution (Olausson 1967; see also Fig. 29.2), their tests will constitute a very small proportion.

In the North Atlantic, between 35° and 50° N there is a good, positive correlation between the amount of calcium carbonate and, for example, *Globorotalia truncatulinoides* (Olausson 1965, fig. 9) and maxima in the frequency of *Globigerina pachyderma* coincide fairly well with the $CaCO_3$ minima.

The change in carbonate content rises to 80 per cent at a depth of 4,100-4,200 metres or more in the North American and European Basins (Fig. 29.10). The bottom water at that level is of Arctic origin. We see that the largest increases in carbonate solution during ice ages occurred in the sphere of influence of this bottom water. An increased flux of such water could explain the features found. Within this sphere of influence changes in relative frequency of Foraminifera are very strong. At the beginning of an interstadial for example, the surface temperature tends to increase and meltwater is discharged in large amounts into the North Atlantic, tending both to lower the density of the surface water and thus retard the vertical circulation (see also Olausson 1967). This will considerably decrease the solution rate at the bottom at the beginning of an interstadial, but this quasi-stagnation in the North Atlantic would disappear around the time of climatic optimum. Remembering the effect of selective solution (Fig. 29.1) we would expect to find for example a large increase of *Globigerinoides sacculifera* in an early interstadial or an increase of *Globorotalia menardii-tumida* later on, as was the case in the Postglacial in core 235 (Olausson 1965, fig. 3). This trend is noticeable further down in the same core (*loc. cit.* fig. 10). *G. sacculifera* therefore, is in that area more an indicator of the intensity of the vertical circulation than of surface water conditions. Further to the north, the thick-walled species are sparse and the frequency of Foraminifera is not as dependent on selective solution as in the Central Atlantic.

From about 3,000 to 4,100 (or 4,200) metres the variation in the carbonate content between glacial and interglacial stages is about 40 per cent. The water at this depth is Lower North Atlantic Deep Water, which originates from the sink water area near Greenland with admixture of Arctic Bottom Water. Apparently this water also circulates with changing intensity in time (or least varying admixtures of the Arctic Bottom Water component). The trend of the foraminiferal analyses is the same as in the lower zone but on a less exaggerated scale.

From approximately 3,000 metres up to 2,500 metres, or nearly 2,000 metres, the content of calcium carbonate changes less or some 30 per cent between glacial and interglacial stages (Fig. 29.10). This water, Middle North Atlantic Deep Water, receives admixtures of Subpolar Intermediate Water which is, in its turn, formed north of the Polar front in the Labrador and Irminger Seas in the winter. Apparently, the Middle North Atlantic Deep Water is less variable in its circulation rate or, at least, its dissolving capacity is less variable than that of the watermasses below. The two cores exhibit slight differences in the relative frequency of Foraminifera: core 288 (depth 2,684 metres) shows smaller proportions of easily dissolved species than core 289 (coring depth 2,640 metres).

The Upper North Atlantic Deep Water receives an injection of warm, highly-saline water from the Mediterranean. It causes the intermediate maximum of salinity. Swedish core 290 (41° 35′ N, 29° 37′ W) was obtained from a depth of 1,995 metres (Olausson 1960*d*, p. 72). Judging from Wüst (1936) and Wüst and Defant (1936), the bottom water there may be Upper North Atlantic Deep Water. As appears from my discussion of the Mediterranean, the 'pH' of its bottom water has been rather constant. We would therefore also expect that the area in the Atlantic in which the bottom water receives

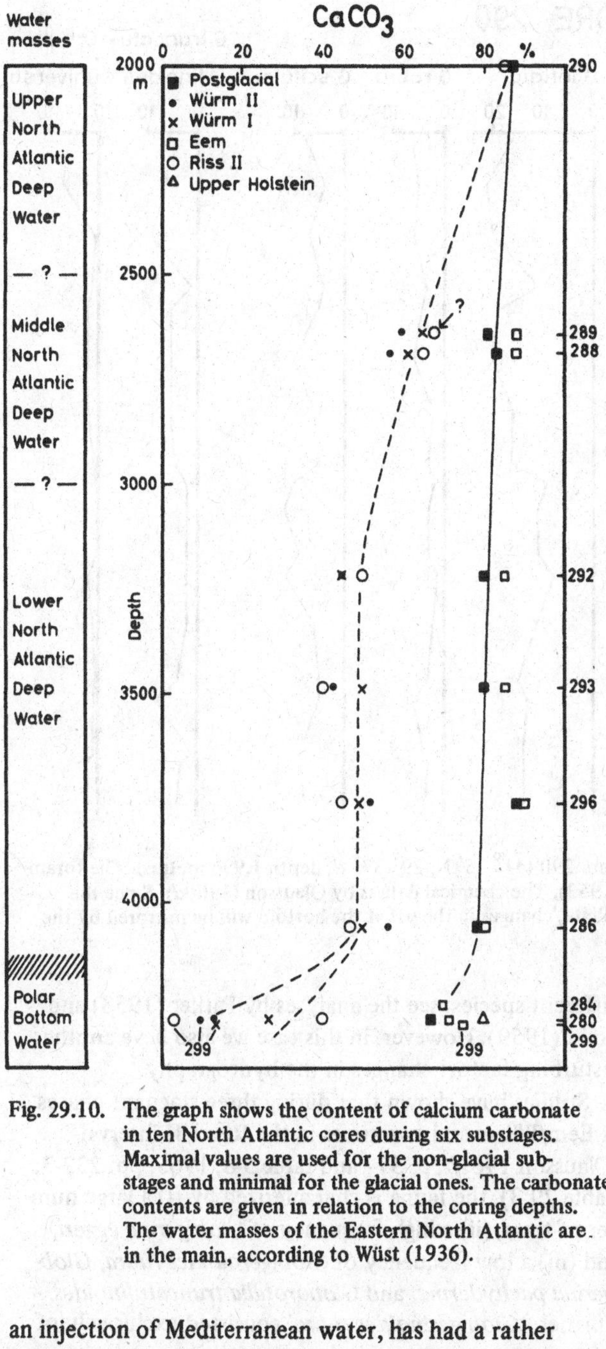
Water masses
Upper North Atlantic Deep Water
— ? —
Middle North Atlantic Deep Water
— ? —
Lower North Atlantic Deep Water
Polar Bottom Water
CaCO3
0
20
40
60
80
%
2000 m
2500
3000
3500
4000
Depth
■ Postglacial
• Würm II
x Würm I
□ Eem
O Riss II
Δ Upper Holstein
?
290
289
288
292
293
296
286
284
280
299
299
299
Cores

Fig. 29.10. The graph shows the content of calcium carbonate in ten North Atlantic cores during six substages. Maximal values are used for the non-glacial substages and minimal for the glacial ones. The carbonate contents are given in relation to the coring depths. The water masses of the Eastern North Atlantic are in the main, according to Wüst (1936).

an injection of Mediterranean water, has had a rather constant 'pH' at the bottom. During an ice age when the sea-level is lowered, so that the water depth in the Strait of Gibraltar is reduced by up to 125 metres or so, the Mediterranean component in the Upper North Atlantic Deep Water should be diminished. A slight increase in aggressiveness of the bottom water therefore seems likely to occur during ice ages. (I consider it unlikely that the outflowing Mediterranean Water during ice ages sank down to the bottom of the Eastern Atlantic, since mixing with the inflowing water at that time, as at present, would have reduced its density too much for that.) The solubility of apatite is slightly higher than that of calcite. A slight increase in the capacity to dissolve will thus lower the content of apatite (hydroxy apatite) more than that of calcite. This would be noticeable from the ratio $CaCO_3/P_2O_5$ in the sediments. Judging from the diagram of core 290 (Olausson 1960*d*; fig. 29.11) the trend of this curve is very close to that of calcium carbonate in cores from greater depths. Furthermore, we notice that in areas of rather constant rates of solution of carbonate – uniform distribution of $CaCO_3$ – the frequency of Foraminifera is rather constant. From these statements I draw the following conclusions: (*a*) the increase in solution of carbonates in the North Atlantic during ice ages occurred at a floor depth of more than 2,000 metres, (*b*) the changes in the surface distribution of planktonic Foraminifera during a climatic deterioration were much smaller than could be arrived at from analyses of cores from a depth of 2,500-4,500 metres, (*c*) the relative changes in the thanatocoenoses is mainly due to changes in the 'pH', and (*d*) the interpretation of thanatocoenoses is a complicated geochemical-palaeoceanographical problem.

The Mediterranean

The stratigraphy of the Mediterranean cores has been dealt with earlier (Kullenberg 1952; Parker 1958; Todd 1959, and Olausson 1961*b*, 1965, 1966). Because of slow rates of accumulation a 10 metre core in the Eastern Mediterranean penetrates into the Holstein stage. Sediment accumulation in the western basin was much more rapid and only one of the Swedish cores supposedly reaches further back in time than the last glacial (Olausson 1966). The Eastern Mediterranean has undergone a series of hydrographic changes indicated by alternating oxidized and reduced deposits. The Western Mediterranean has not, during post-Eemian times, undergone such a change.

The black horizons in Fig. 29.19 represent sapropelitic muds. The explanation of the stagnant or quasi-stagnant phases are given briefly in Fig. 29.12 (for further explanation see Olausson 1961*b*). Judging from the chemical analyses (Olausson 1960*c*, see for example core 192), both the carbonate and phosphate contents are rather constant, and apparently independent of the *Eh* changes. This suggests that oxygen deficiency has prevailed during the formation of the sapropelitic mud. Alternatively, possible oxidation of the high amount of carbonaceous matter, buried in the sediment (2-8 per cent), has dissolved the calcareous matter to an appreciable extent (cf. equation 2). The 'pH' at the bottom of the Eastern Mediterranean has been rather constant due to the large buffering capacity of so high a content of calcium carbonate (~40 per cent) and its uniform distribution. This

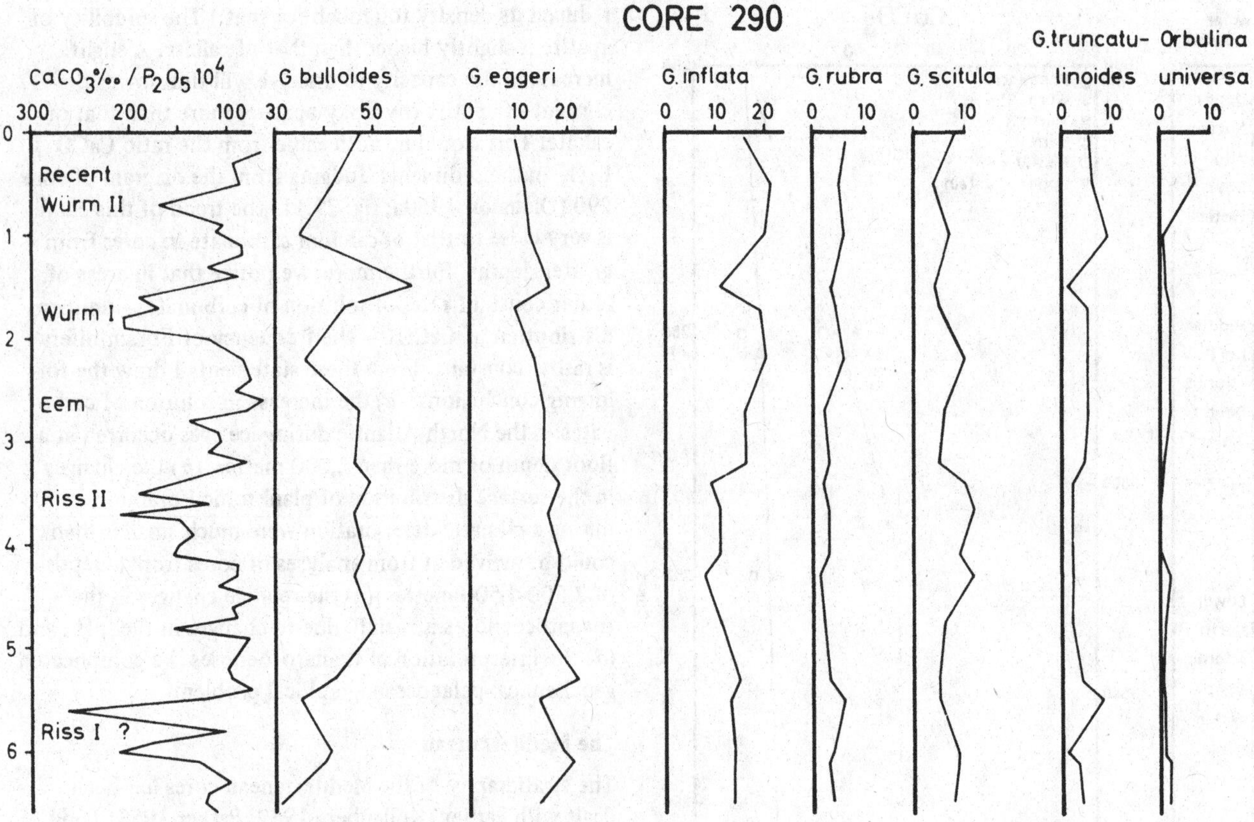
CORE 290
CaCO3‰ / P2O5 10⁴
G bulloides
G. eggeri
G. inflata
G. rubra
G. scitula
G. truncatu-linoides
Orbulina universa
300
200
100
30
50
0
10
20
0
10
20
0
10
0
10
0
10
0
10
Recent
Würm II
Würm I
Eem
Riss II
Riss I ?
0
1
2
3
4
5
6

Fig. 29.11. The frequency of Foraminifera and the climatic curve of core 290 (41° 35′ N, 29° 37′ W, depth 1,996 metres). The foraminiferal analyses are made by Phleger, Parker and Peirson (1953). The chemical data is by Olausson (1960*d*). Since the solubility of (hydroxy) apatite is slightly higher than for calcite, changes in the pH at the bottom will be mirrored by the ratio $CaCO_3/P_2O_5$ of the final sediment.

has an important bearing both on the chemistry of the outflowing water (see above) and the preservation of Foraminifera.

For the Equatorial Pacific I suggested that the temperature (T) had been rather constant but the 'pH' had varied. In the North Atlantic both have been changing, and in the Eastern Mediterranean T has apparently also fluctuated. The distribution of Foraminifera could therefore reflect changes in T, and since the amplitude of change might be of the order of 3-4° C, the percentage distribution indicates how such a change in T affects the growth of different species; see the analyses by Parker (1958) and Todd (1959). However, in this case we also have another disturbing factor: changes in the hydrography.[1]

Studies have shown that during three stagnant phases, at Eem/Würm and two times in the Riss I/II interval (Olausson 1961*b*, p. 377 and table 5-8; 1965, pp. 222-3; Table 29.3), the fauna is characterized by (i) a large number of tests, (ii) a high frequency of *Globigerina eggeri*, and (iii) a low frequency of *Globigerinoides rubra, Globigerina pachyderma*, and *Globorotalia truncatulinoides*. Further, '*Globigerinoides rubra*' consisted exclusively of

[1] Emiliani and Milliman (1966, p. 124) have summarized some of my comments on the Eastern Mediterranean cores in three points (*a*)-(*c*). Point (*a*) is incorrect since my statement about the disagreement between isotopic and foraminiferal analyses concerns only stage 3 of 189 and not the whole core. For the aforementioned section my opinion is correct. The percentages of 'warm' Foraminifera at 340 centimetres in core 189 is 3.5 per cent while at 320 centimetres it is 13 per cent (Parker 1958). The isotopic palaeotemperature of *G. rubra* is 32.8° C at 342 centimetres while at 312 centimetres 17.3° C is obtained. Point (*b*) concerns the correlation between δO^{18} of Foraminifera and sapropelitic muds. Emiliani and Milliman state that six of the eight sapropelitic layers do not correlate with any isotopic temperature maxima. The reason why no such a correlation exists for these six layers is that nobody has investigated the δO^{18} of Foraminifera from them! Point (*c*) is an incorrect summary. My opinion was and still is that both the foraminiferal fauna changes and changes in the δO^{18}_{aq} of the ancient waters of the basin are due to changes in temperature and hydrography.

Emiliani and Milliman (*loc. cit.* p. 123) also state that strict boundary conditions for glacial-interglacial variations in the average δO^{18}_{aq} of sea water can be established from Pacific benthonic Foraminifera. This opinion has been criticized by me (Olausson, in press) and I have concluded that Emiliani's own analyses of benthonic Foraminifera from two Pacific cores disprove that suggestion.

G. rubra rubra during the stagnant phases, while the subspecies *G. rubra gomitulus* disappeared during the stagnant phases. When it returned after a euxinic phase its size was at first smaller than before (Olausson 1965, p. 231). Since *Globigerinoides rubra* (d'Orbigny) and *Globigerina pachyderma* lives in near-surface waters during the juvenile stage (Bé and Hamlin 1967) while *Globigerina eggeri* is supposed to live at lower levels, the Foraminifera with moderate depth habitats seem to have been promoted by some factor or growth in a special current, while the others have been retarded by the abnormal milieu.

The Red Sea

The Red Sea is a subtropical adjacent sea. Owing to its small sill depth (about 125 metres according to Neumann and McGill 1962) and thus its limited water exchange, a high evaporation (2.1 metres per year, *loc. cit.*), and lack of fresh water supply, the salinity is very high. The surface inflow seems to be about 3 x 10^3 cubic kilometres per year. The outflowing, subsurface water is warm, high-saline, poor in oxygen, and rich in organic carbon in both particulate and dissolved states (during the middle or late summer a thin wind-driven outflow develops on top of the inflowing water). This outflowing water may be traced as far as 40° S (Clowes and Deacon 1935).

During a glacial age the sea-level may have dropped down to or below the sill depth. No surface inflow could then have compensated for the evaporation loss within the Red Sea. Supposing an evaporation rate during glacials of 2 metres per year and the Red Sea area at the sill depth of 175 x 10^9 square metre its water level falls below the sill depth when the inflow becomes lower than some 350

Table 29.3. Foraminiferal analyses made by Parker (1958) in relation to the hydrography of the Eastern Mediterranean (Olausson 1961*b*, Table 5) from a euxinic sequence in the Riss I/II interval. The stagnation is characterized by (i) a great number of tests, (ii) high frequency of *Globigerina eggeri*, and (iii) a low of *Globigerinoides rubra, Globigerina pachyderma* and *Globorotalia truncatulinoides.*

Core	Deposits (Analyses from just below 4a)	Depth in core (cm)	Σplanktonic forams	Σbenthonic forams	*Globigerina bulloides*	*Globigerina digitata*	*Globigerina eggeri*	*Globigerina inflata*	*Globigerina pachyderma*	*Globigerina quinqueloba*	*Globigerinella aequilateralis*	*Globigerinita glutinata*	*Globigerinoides rubra, etc.*	*Globigerinoides tenella*	*Globorotalia scitula*	*Orbulina universa*
187	Blue mud	788	4.000	8	20		33	14	5	7		3	12		4	2
	Sapropelitic mud	801	57.000	0	9		64		6	15		3	1			2
	Blue mud	814	87.000	0	13		63		4	16		3				1
	Gray mud	859	28.000	29	35	.5	18	3	27	2			5	.7	4	4
189	Blue mud	510	21.000	9	12		26	11	2	4	1	3	26	1	1	3
	Sapropelitic mud	520	46.000	10	16		52	1	3	22		1	1		.3	3
	Sapropelitic mud	528	46.000	0	3		67		2	26					2	
	Gray-blue mud	533	22.000	174	19	1	24	7	16	6		5	9	5	4	3
190	Blue-brown mud	600	12.000	18	20		33	17	3	4	2	5	7	1	2	5
	Sapropelitic mud	610	36.000	0	8		56	4	3	30		1				1
	Sapropelitic mud	620	37.000	0	.5		65			33						1
	Blue mud	630	5.000	5	28		30	7	9	8		4	7	5	1	1
195	Blue mud	770	29.000	9	23		16	18	1	1	1	.7	15	3		3
	Sapropelitic mud	780	25.000	14	14		63		4	16			1	1		.3
	Sapropelitic mud	800	34.000	14	5		60		.6	33						1
	Gray-blue mud	810	10.000	87	33		27		28	3		2			3	4
198	Greenish mud	1,010	7.100	0	16		9	19			1	.7	28	.7		3
	Sapropelitic mud	1,020	34.000	0	24		60			11		3	1			.6
	Blue mud	1,030	5.700	45	31	.3	21	8	13	2		3	10	3	5	4

cubic kilometres per year or about 10 per cent of the present inflow. Then, most of the present Red Sea area (about 90 per cent; see Fig. 29.13) would become dry after say 500 years, and after say 1,100 years of isolation only some deep holes would contain brine as long as

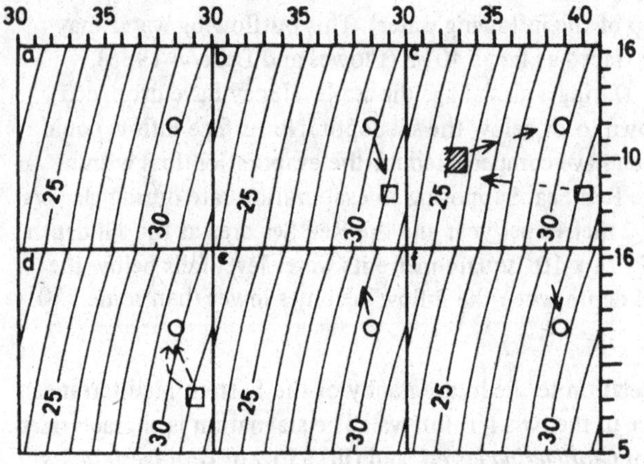
S, ‰
T
a
b
c
d
e
f
16
10
5

Fig. 29.12. Diagram showing the deduced changes in density of Eastern Mediterranean surface and bottom water (Olausson 1961). A renewal of the bottom water is possible, if the density of the surface water becomes larger than that of the bottom water.

(*a*) During severe winters the Eastern Mediterranean surface water becomes denser than the prevailing bottom water, sinks down and constitutes a new bottom water.

(*b*) During the time from a thermal maximum to the succeeding glacial maximum the density of the surface water is increased due to a temperature lowering (3-4° C) and a globally increasing salinity (up to 1‰ in the world oceans). This increase results in a good ventilation of the basin.

(*c-d*) During deglacial and early inter- or postglacial times (transgression phases) the increase in temperature and excess supply of fresh and less saline waters (meltwater from the Black Sea, precipitation, and Atlantic waters) have tended to increase the density difference between the bottom and surface waters in various degrees, causing stagnant conditions. (*c*) represents Early Eem (Olausson 1961; 1965, p. 230), (*d*) represents less intense stagnant phases or a period of time when the temperature factor dominates. A temperature rise from 8 to 12° C must be compensated by a salinity increase by 0.8‰ if density of the surface water is to remain constant; since the salinity instead decreases the ventilation is halted. The oxygen can have been used up within some 1,000 years and then black, sapropelitic mud accumulates (the black section in Fig. 29.18, right column).

(*e*) This graph represents a density change from normal interglacial into still warmer phases (Upper Holstein, Pliocene) which tends to slow down the renewal processes. Stagnant or quasistagnant conditions enter (cf. the Cariaco Trench from glacial→ postglacial age (Heezen *et al.* 1959)).

(*f*) This graph represents good ventilation when the temperature decreases from such a warm phase as was the case in (*b*).

connate water was added from below (with a net loss of 1.5 metres per year the aforementioned figures should be 700 and 1,450 years respectively). During such an evaporation phase calcium carbonate would have been precipitated initially. When the salinity had increased to 3.35 times the normal value (about 117 per mil) gypsum would have started to precipitate (see e.g. Degens 1965, p. 175). Theoretically this happened when the water level stood at about 600 metres. When the water volume was reduced to about one fifth compared to normal sea water (i.e. water level stood at about – 800 metres) anhydrite would have been the stable phase (*loc. cit.*). When the salinity of the residual water was about ten times the salinity of normal sea water (i.e. at a water level stand of about 1,000 metres) rock salts would also have been precipitated. Judging from Fig. 29.14 the pH of the basin water may have decreased a unit or more at the same time. If the salinity of the Red Sea at the isolation was 42 per mil and only one, rapid regression below the sill depth occurred, the amount of calcium carbonate precipitated during a total evaporation might have been only a few (6) milligrammes per square centimetre, the amount of gypsum and anhydrite 100-150 milligrammes per square centimetre, while the amount of rock salts could have been some 10 grammes per square centimetre if all minerals were distributed uniformly over the area where they could have been formed. During the evaporation trace

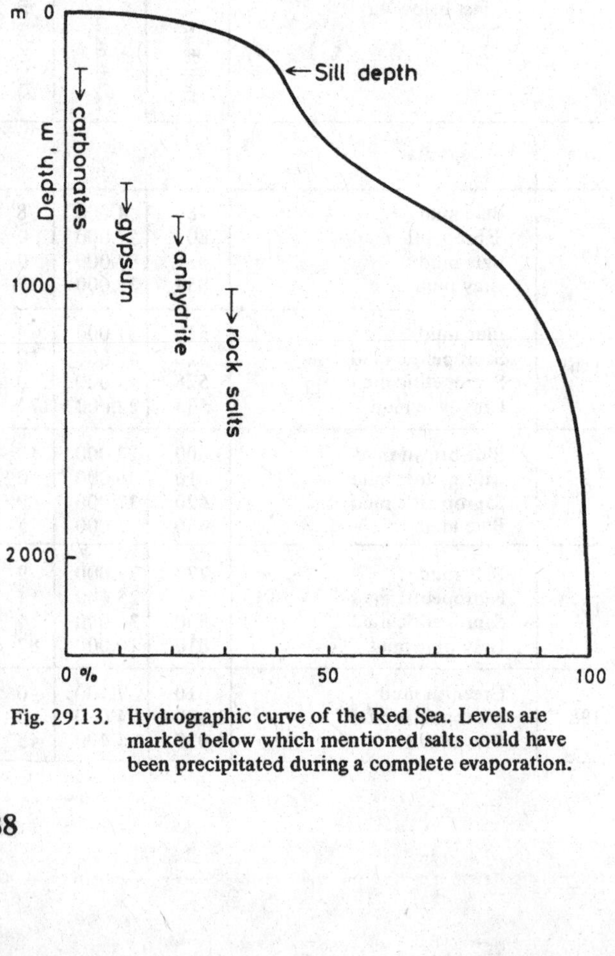
m 0
1000
2000
Depth, m
Sill depth
carbonates
gypsum
anhydrite
rock salts
0 %
50
100

Fig. 29.13. Hydrographic curve of the Red Sea. Levels are marked below which mentioned salts could have been precipitated during a complete evaporation.

elements would also have been concentrated in the brine and finally precipitated. The quantities can be estimated if we take the evaporated water volume at 13 x 10^{13} cubic metres and the concentrations, in milligrammes per cubic metre, would then be for iron and zinc 10, copper 3, manganese and cobalt 2, and cadmium 0.1 (Goldberg 1965; cf. Miller *et al.* 1965). Consequently, the sediment could be rich in siderite-rhodochrosite, sphalerite, etc. When the ocean sea-level reached higher than the sill depth, ocean water was transferred into the Red Sea basin. If we suppose an annual ice melting equivalent to a sea-level rise by 5 millimetres (some 7,000 cubic kilometres) then it took some twenty years to fill it up. During the transgression rock salts, precipitated theoretically below – 1,000 metres, and anhydrite and gypsum, precipitated below 600-800 metres, would have been dissolved when the salinity was lowered below the aforementioned concentrations. This solution would have increased the salinity of the basin water, particularly below 1,000 metres (the amount of calcium ions and sulphate ions added to the water between, say, 600-1,000 metres would be negligible compared to the amount of salts that went into solution below 1,000 metres). The aforementioned level coincides approximately with the edge of the axial trough. Theoretically the salinity of the water in the median valley could have reached about 30 per cent and because the inflow has probably occurred intermittently. Also evaporation would have contributed to making the bottom water very dense. Only the carbonates formed during the evaporation stage may not have been redissolved (that is if the pH did not go down too much). The limy crust may have been crushed by wave action constituting a crushed, hard layer sequence.

If the inflow during the evaporation phase was interrupted by periodical supplies from the Indian Ocean, if the regression took a much longer time than assumed above, and/or if the sill depth was left dry twice or more during a glacial (sub)age the carbonate precipitation might have been of a larger order. Also the figures of the amounts precipitated and the depths at which they were precipitated may have been different.

The stratigraphy of the Red Sea cores

Red Sea cores have been described earlier (Olausson 1961*b*). The basal part of cores 165 and 166, up to about 250 centimetres, consists of foraminiferal chalk and marl muds. Between 97 and 204 centimetres in core 165 and at 113-260 centimetres in core 166 (below 65 centimetres in core 169 and at least 199-221 centimetres in core 168) there are beds with large irregular fragments of calcareous layers interbedded with marl mud. The hard fragments sometimes have one surface smooth with infilled pteropods, the other is pitted. The matrix is a felt-like intergrowth of aragonite aggregates containing particularly pteropods and some Foraminifera, a few fragments of echinoids, etc. The orientation of the fragments of calcareous layers are more or less random, suggesting that it was formed by a crushed layer into which marl mud has been mixed. The lumps can be up to 5 millimetres thick and a length of up to 3 centimetres has been observed in a core diameter of 46 millimetres. The hard layers have darker bands, which, according to X-ray diffraction studies, consist chiefly of iron minerals. As stated above the bulk mineral of the hard layer is aragonite. Magnesium-calcite with approximately 15 mol-percent magnesium (the spacing distance between the two peaks is 0.4°; see Goldsmith *et al.* 1955) is the second most common mineral. Iron oxide minerals ($2\theta = 32.8$ and 33.2°) are recorded with peaks both much larger and smaller than the calcite peaks. Some quartz, feldspar, and small amounts of clay minerals are recorded.

The hard bed sequence is – apart from megascopic observations – clearly indicated by peaks in the $> 150\ \mu$-curve, particularly in cores from depths of 800 metres or more with a minimum number of Foraminifera. At higher levels the amount of hard fragmental material is small, which is in agreement with the theoretical model discussed above. However, the corresponding horizon above a depth of 800 metres is also characterized by a scarcity of Foraminifera (core 163: 125-275 cm; V14-118: 40-125 cm; V14-123: 40-130 cm). Above the hard layer sequence the palaeontological boundary Würm/Postglacial occurs, dated to 16,000 (± 500) years in core V14-123 and 15,400 (± 500) years in V14-115 (Herman 1965). This boundary is indicated by an increase in *Globigerinoides sacculifera* and *Limacina inflata.* The lower boundary of the hard layer sequence is observed not only visually, but also by a drop in the $> 150\ \mu$-curve, by a change in the number of Foraminifera, etc. A C^{14}-date on core V14-118 from *ca.* 25-55 centimetres below the layer poor in Foraminifera gave 22,690 years. These dates suggest that the hard layer sequence was formed somewhere between 16,000 and 22,000 C^{14}-years ago.

The indurated layer can be understood if we envisage that the sediment surface layer was cemented together by carbonates during an evaporite phase. Owing to the high salinity and comparatively high temperature the carbonate precipitated during the evaporation phase may have been aragonite, and the large amount of high magnesium-calcite can also be seen in the same light. It is likely that the hard layer was formed in near-shore environments (cf. Gevirtz and Friedman 1966, p. 147). Taking into account the porous nature and angular shapes of the lumps, and considering the wide-spread occurrence of the hard layer, it may constitute a sedentary deposit crushed by wave action and mixed with some other material. For the same reasons it is improbable that it could be a turbidite or that the horizon could be caused by interaction between hot, saline water and normal sea water from above.

A few gypsum crystals were found in this layer (Olausson 1961*b*; Gevirtz and Friedman 1966) and anhydrite was determined by X-ray analysis. The former may have been formed either during the original evaporation or during the storage of cores, while the anhydrite crystals are likely to be original (see Degens 1965; the cores have been stored up to eighteen years at about +4° C).

Pteropods, etc. were contributed to the interbedded marl mud. X-ray analyses of the silt and clay fractions generally show the following order of peak magnitudes: calcite > quartz>feldspar ~ aragonite>sphalerite>anhydrite (clay mineral not analysed). Towards the top of the hard layer sequence, e.g. in core 165, 101.5-103 centimetres, we have the following order: sphalerite>32.8° >calcite. The calcite is here too recorded by two peaks (0.4° spacing), both generally larger than the quartz peak. The fine fraction consists of small fragments of the dry crust, debris of fossils, authigenic minerals and terrestrial materials. All the samples studied by X-rays have been stored up to eighteen years and are now rather dry. More detailed mineralogical studies of fresh samples are being carried out.

In core 163 the boundary Würm/Postglacial is recognized by an increase in organic matter, phosphate, and calcium carbonate (Foraminifera). In cores 165 and 166 there are increases in carbon and nitrogen but a decrease in calcium carbonate. A change in the fauna is observed at the transition between the stages. Between 78 and 97 centimetres in core 165 there is a greenish, silty calcareous mud. The calcareous matter consists chiefly of high magnesium-calcite and calcite, but some aragonite and a small amount of dolomite also occur. The iron content is high (13 per cent), and in the dried samples peaks at (2θ) 31.4°, 32°, and 33.2° are found by X-ray diffraction studies. Feldspar and quartz dominate among the terrigenous matter. Montmorillonite and micas are recorded. The presence of anhydrite is of interest. The layer is rich in organic matter, particularly below 89 centimetres. In core 166 (101-113 centimetres) there is a sandy, silty calcareous mud resting above the indurated layer with magnesium-calcite, quartz, and feldspar. Between 70 and 101 centimetres there is a marl mud with pteropods, Foraminifera, magnesium-calcite, quartz, feldspar and some anhydrite. At the top of the core there is marl mud.

C^{14}-datings give some indication of sea-level stands during the Würm (Shepard 1963). Thus, a minor transgression up to 15 metres, apparent in some detail near Freeport, Texas, and on Padre Island, Texas, is dated to 23,400 B.P. (Fisk 1959) and 26,900 B.P. (Curray 1961). An intertidal shell obtained by Curray (1961) at a depth of 125 metres off the west coast of Mexico was dated to 19,100 B.P. The maximum extent of glaciers in Europe and North America occurred about 18,000 to 20,000 C^{14}-years ago. Calculations of ice volumes by Donn, Farrand and Ewing (1962) and Robin (1964) correspond to a water transfer from the oceans to the ice sheets equal to a layer of up to 160 metres thickness. Heezen *et al.* (1959) have summarized continental shelf terraces which range in depth from 142 to 154 metres. The sill depth of the Red Sea is given as 125 metres; a figure probably correct to within 10 per cent (after isostatic adjustment the sill would be about 80 metres below the present sea level if left dry). Consequently, it is likely that the Red Sea during, say, 18,000 to 20,000 C^{14}-years ago, was isolated from the world's oceans and that its water evaporated off. This is in agreement with the aforementioned C^{14}-datings. As stated above, only some 1,100 years of isolation would have been sufficient to make the Red Sea dry.

The scarcity of Foraminifera in the hard layer sequence and the general presence of pteropods, as well as the large quantities of calcium carbonate precipitated below *c.* 800 metres merit further discussion. The former may be due to unfavourable conditions, probably too high a salinity, and also to dilution by precipitated matter. The latter is understood if we realize that the evaporated volume might have been larger than the aforementioned theoretical value of 13×10^{13} cubic metres. If, during the time when the water level was below the sill depth, there was an inflow of half of the last-mentioned volume or so, the hard layer can have been formed by the calcium from the evaporated water. Such an inflow is possible since the ocean level may have dropped some 2.5 centimetres per year (Curray 1965, fig. 2) and it would have taken several years before the inflow had diminished from about 10,000 cubic metres per second to nearly nil. Concentrating into a smaller area a comparatively small amount of inflowing water would have meant an important supply per unit area when the water level stood several hundreds of metres below the sill depth (see Fig. 29.13). Another source of calcium carbonate is the formation of calcite by the degradation of gypsum through the activity of bacteria. This reaction might have contributed to carbonate precipitation below about 600 metres. Also a solution of calcium sulphate, in the presence of carbonic acid, can result in carbonate precipitation.

It is possible that a lower boundary to the vertical distribution of the calcareous hard layer in the Red Sea exists owing to a decrease in pH during the evaporation (Fig. 29.14). The hard, limy layer is recorded down to at least 1,830 metres (core 168). A non-calcareous hard layer might be presumed to exist at the bottom of the deep holes; this is also evident (Gevirtz and Friedman 1966; Miller *et al.* 1965).

I doubt whether the brine is the main source of the calcium because it is a question of too large a volume having been supplied from a well during so short a time, and the layer is spread over a very large area (Fig. 29.15). On the other hand, the pluvial phase in adjacent areas, together with the virtual absence of water in the Red Sea, could have increased the water supply from below. However, this problem is here left aside as it is being investigated by Woods Hole Oceanographic Institution.

In core 163 (and Lamont core V-14-118 (V-14-121)) no carbonate minimum exists in the Late Glacial substage (Fig. 29.16), the increase of carbon is much smaller than in 165 and 166 (Fig. 29.16), and the presence of fäkalier also suggests non-stagnant conditions at 163 (coring depth 300 metres; V-118 is from 516 metres). Core 169 (Fig. 29.5; see also V-125 (Herman 1965)) does not suggest stagnation at that depth, 860 metres.

Cores 165 (coring depth 1,160 metres) and 166 (1,280 metres) however, exhibit a carbonate minimum in the earliest Late Glacial seen in Fig. 29.4-5 (the Lamont cores V-14-115 (depth 1,194 metres), 116 (1,346 metres) belong to the same group and probably also 117 (depth 1,372 metres) and 120 (1,737 metres), while core 119 is truncated; data from Gevirtz and Friedman, 1966). If the water added to the axial trough was of normal salinity, then owing to solution of earlier precipitated rock salts, the salinity of the water may have been increased ten times or so, i.e. that of the present brines in the deep

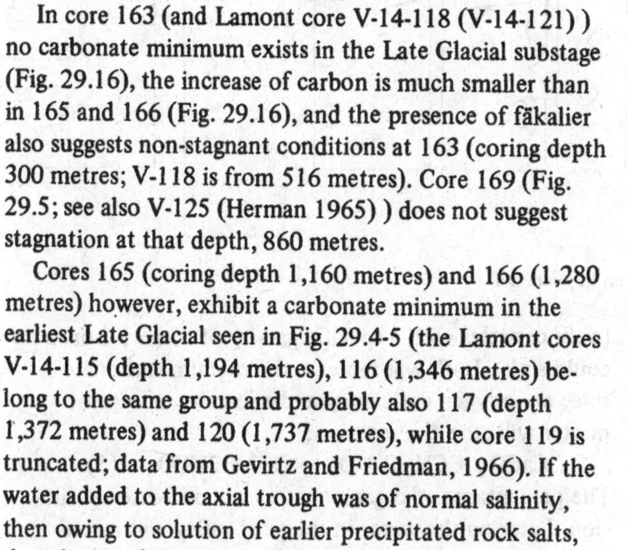
1000
800
600
400
200
Zero
-200
-400
-600
Eh
1
2
3
4
0 2 4 6 8 10 12
pH
1 bitterns (Mg-rich)
2 gypsum-halite
3 calcite-dolomite
4 sulphides

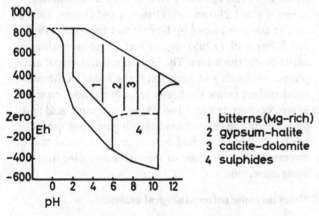

Fig. 29.14. **Suggested fields of formation of various evaporites (Baas Becking *et al.* 1960, fig. 17). The neutral point of a brine is not at pH = 7 but below it.**

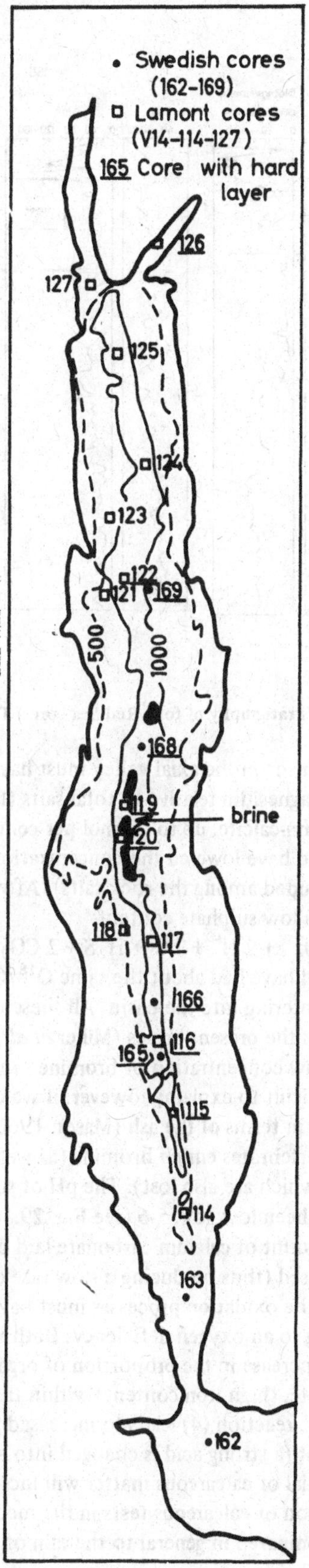
• Swedish cores
(162–169)
□ Lamont cores
(V14–114–127)
165 Core with hard
layer
126
127
125
124
123
122
121
169
500
1000
168
119
brine
120
118
117
166
116
165
115
114
163
162

Fig. 29.15. **Location of cores from the Red Sea.**

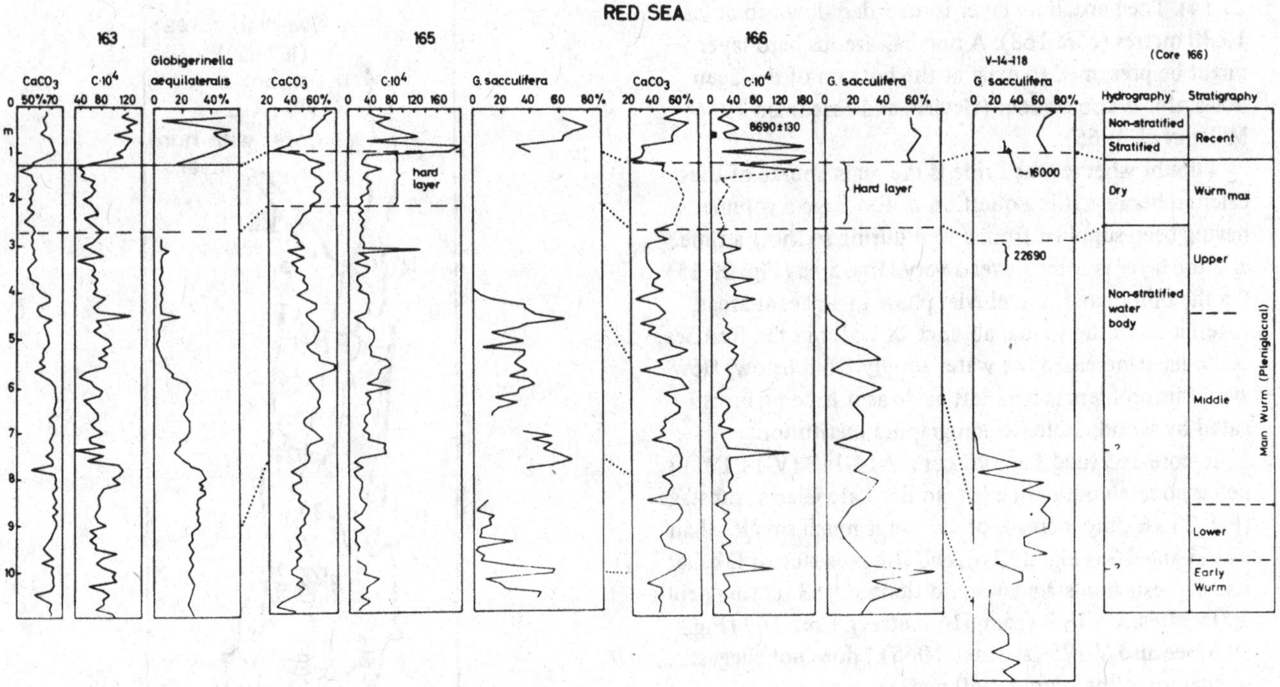
RED SEA
163
165
166
V-14-118
(Core 166)
CaCO3
C·10^4
Globigerinella aequilateralis
G. sacculifera
Hydrography
Stratigraphy
m
hard layer
Hard layer
8690±130
16000
22690
Non-stratified
Stratified
Recent
Dry
Wurm max
Upper
Non-stratified water body
Middle
Main Wurm (Pleniglacial)
Lower
Early Wurm

Fig. 29.16. Stratigraphy of four Red Sea cores. The C^{14} ages are given in years B.P.

holes. The brine in the axial valley must have been depleted in magnesium relative to total salts (the formation of magnesium-calcite, up to 20 mol-per cent is recorded, which would have lowered the concentration of magnesium included among the rock salts). After some time it attained a low sulphate content;

$$SO_4^{2-} + 2\,H^+ + 2\,C \rightarrow H_2S + 2\,CO_2 \quad (4)$$

but it would have had about the same O^{18}/O^{16} ratio as the water entering into the basin. All these characteristics are found in the present brines (Miller *et al.* 1965). The relatively low concentration of bromine in the present brine is difficult to explain; however, if we consider the enrichment in terms of the ash (Mason 1960, p. 222), marine invertebrates enrich bromine (as well as sodium and potassium which are also lost). The pH of such a brine might have been low, say, 5-6 (see Fig. 29.14). This means that the amount of calcium carbonate laid down must have decreased (thus producing a slow rate of accumulation) and the oxidation processes must have been diminished owing to an oxygen deficiency. Both these factors lead to an increase in the proportion of organic matter, sulphides, etc. (high iron content) within that horizon.

However, reaction (4) leads to increased pH in the environment (a strong acid is changed into a weak base), and the burial of calcareous matter will increase. The sedimentation of calcareous tests in the median valley has been compared in general to the rain of organic matter (only particles with a specific gravity of some 1.2 or more could sink). Both will increase the calcium carbonate since an oxygen deficiency and a scarcity of organic matter will stop the reaction;

$$CaCO_3 + CH_2O + O_2 \rightarrow Ca^{2+} + 2HCO_3^- \quad (5).$$

The percentages of organic matter will decrease by dilution. Later on the water in the axial valley was mixed with the overlying water, and, except for some deep holes, the Red Sea became oxidized and unrestricted some 8,700 C^{14}-years ago (core 166, 52.5-58.5 centimetres, coring depth 1,280 metres; Olausson and Olsson, 1969).

The cores described by Gevirtz and Friedman (1966) and Miller *et al.* (1965) suggest that a hard layer also exists below the brines. The high concentrations of amorphous iron oxide and some sphalerite found in the sediment surface below the brines are also recorded to some extent in cores 165 and 166. However, a more acid and longer enduring environment than higher up in the valley has resulted in intensified carbonate solution and thus in increased concentrations of minerals containing iron manganese, zinc, etc.

Notes on some palaeontological analyses

Foraminiferal analyses have been carried out on cores 163, 165 and 166, and pteropod analyses are being carried out by Dr Y. Herman.

Core 163 has a high frequency of *Globigerinella aequil-*

ateralis at the top (0-125 centimetres), moderate values below 7 metres and low ones from 5.5 metres up to the hard layer. *Globigerinoides sacculifera* is present in low percentages throughout the whole core, but especially below 10 metres and between the hard layer and the 8 metre level. *Globigerinoides rubra* cf. *pyramidalis* reaches its highest frequency in the Main Würm stage (from 5 metres up to the hard layer). Thus, 0-125 centimetres represents the Recent, the sequence down to 5 or 5.5 metres is Upper Main Würm in age, then an interstadial may be traced, and the sediment below 10 metres may have been deposited during another stadial.

The foraminiferal analyses of cores 165 and 166 reveal warm-water conditions in the uppermost 1-2 metres (Fig. 29.16). Below that they suggest a glacial stage. The boundary Würm/Postglacial may be tentatively fixed at about one metre in core 165, and at 130 cm in core 166, indicated by an increase in *Globigerinoides sacculifera.* This change occurs somewhat earlier than that of the pteropod *Limacina inflata* (Herman 1965). The increase in *G. sacculifera* is well marked and it reaches a much higher frequency in 166 than in 163. *Globigerinoides rubra* cf. *pyramidalis* and *Globigerinella aequllateralis* show the same trend as in 163. At 6 metres and at 10 metres there are indications of rather warm conditions, which may be interstadials in the Early and/or Middle Würm.

There is a tendency of Foraminifera like *G. sacculifera* and the pteropod *Limacina inflata* to increase in frequency from south to north, while other like *G. glutinata* and *G. rubra rubra* decrease in relative frequency in the core samples. It could be due to the decrease in temperature towards the north.

In the hard layer sequence *Globigerinoides sacculifera* predominates in the mud intermixed with the aragonite aggregates. This species grew at the water surface and its tests as well as those of the pteropods are rather brittle. These features suggest that Foraminifera at the beginning of the Post-Würmian phase had grown chiefly in the surface water zone, probably owing to the water stratification and that the hard layer is not a turbidite, but a sedentary sequence consisting of fragments of a dry crust with which tests of surface organisms have been mixed.

SILICA AND SILICEOUS ORGANIC REMAINS

The distribution of silica from weatherings on land appears in Table 29.1. Since the general distribution of siliceous deposits is discussed by Lizitsin only one aspect, namely the size distribution of *Coscinodiscus nodulifer* is considered here.

A diatom test such as *Coscinodiscus nodulifer* consists of a top valve (epitheca) and a bottom valve (hypotheca). In reproducing by cell division both valves become top valves in the next generation. One daughter cell is thus of reduced size. The reduction in size continues until a certain critical limit is attained, when auxospore formation sets in. The auxospore generally produces an individual of larger size than that which formed the auxospore. If this is not formed when the critical size is reached, the decrease in size appears to continue until an absolute minimum limit is attained when the cell dies (Arrhenius 1952, pp. 71-2).

The present studies indicate that auxospore formation in an early stage of the vegetative part of the reproduction cycle is favoured by an increased supply of nutrients. These biologically important elements may be phosphorus, nitrogen and/or silicon. The North Atlantic apparently receives 40 per cent of all silica from continental weathering, but a part of this must be lost in the same way as phosphorus, nitrogen and calcium are. It is therefore not necessary to distinguish between nitrogen and phosphorus, and silica since all will promote the production of siliceous organisms. We may expect to find the same cycles for accumulation of organic silica in the southern hemisphere of the Indian Ocean and the Pacific as for carbonates. Arrhenius (1952) has also demonstrated both variations in silica deposition in the Pacific and variations in the size of *Coscinodiscus nodulifer.*

Judging from the diagram of size distribution (Arrhenius 1952, p. 162, fig. 2.62.9, and diagrams; Arrhenius in Olausson 1960*a*, fig. 23) a pronounced maximum occurs in the Riss II substage of core: 58-62, and 74. Further, the maxima for Würm and Riss I are of about the same size. In the Holstein stage the maximum is found at index horizon *'l'*. The Pre-Holstein stages are not well marked, a feature which is not understood.

My transport theory suggests that similar cycles in the size of *Coscinodiscus nodulifer* should also be found in cores from the area around the South Equatorial Divergence in the Indian Ocean (although this was not found by Kolbe (1957)).

In cooperation with Miss Gunvor Karlsson analyses of this species of diatom have been performed for core 131 (Fig. 29.17; Olausson and Karlsson, in prep.). We found a fluctuating size distribution for this diatom. The difference between Riss II, Riss I and Würm is not as marked as in the Pacific, but the most pronounced size maximum is again found in the Holstein stage at horizon *'l'*. Further, above *'k'* the 1 per cent-line reaches much higher values during cooler phases than below *'k'*, both in the Indian and Pacific Oceans, a fact also supporting the cross-correlation suggested above, and indicating larger size amplitudes in the diatom above *'k'*. Below *'k'* the diagrams are

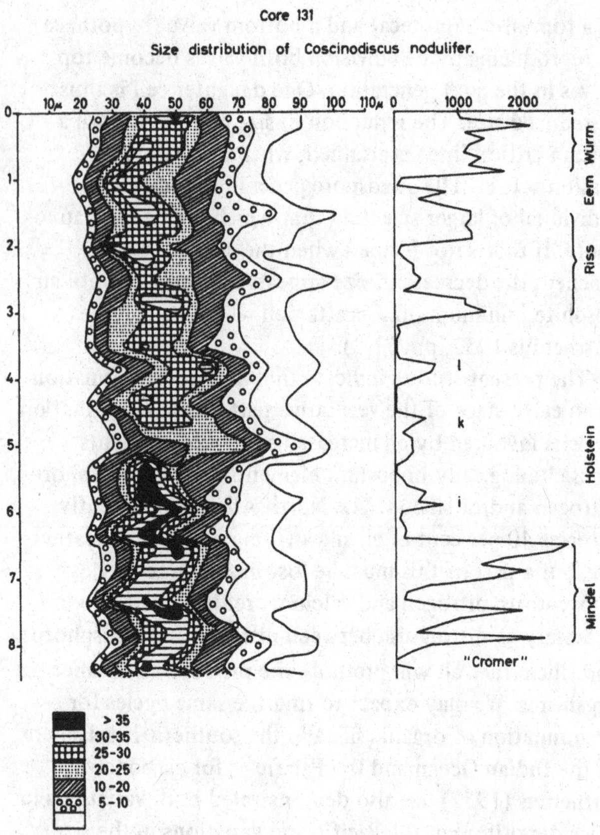
Core 131
Size distribution of Coscinodiscus nodulifer.
Würm
Eem
Riss
Holstein
Mindel
"Cromer"
> 35
30-35
25-30
20-25
10-20
5-10

Fig. 29.17. Size frequency distribution of *Coscinodiscus nodulifer* in core 131 from the Indian Ocean (11° 10′ S, 96° 42′ E). The abscissae correspond to the upper limit of the 10μ-size classes. The percentages found in these classes are given in the isopleth diagram.

frequency > 35 per cent in core 131. In the Mindel stage the diagram for core 131 shoes more marked size fluctuations than the Pacific cores do. Above '*k*' the productivity changes are better marked in the Pacific cores than in the Indian Ocean core; below '*k*' the reverse might possibly be the case.

Scarcity of diatoms in Central Atlantic cores (e.g. core 235) has made it difficult to carry out such an investigation there.

TIME-STRATIGRAPHIC PROBLEMS OF DEEP-SEA CORES

On the Plio-Pleistocene boundary and Pleistocene type localities

The Plio-Pleistocene boundary has been discussed by Flint (1965). For a type locality we must demand a clear definition of a floral and/or a faunal change, that it implies a climatic deterioration, and that the boundary can be easily transferred to other areas.

The validity of isotopic palaeotemperature analysis has been discussed elsewhere (Olausson 1965, 1967). I have shown that Emiliani's curves show from 70 to 100 per cent changes in the O^{18}/O^{16} ratio of ancient water. This is important in the discussion of the official Plio-Pleistocene boundary in the section at Le Castella, Calabria. Since the amplitude of temperature changes of the Mediterranean water in Early Pleistocene can only have been of the size of 3° C or less, and changes in the O^{18}/O^{16} ratio in Foraminifera are equivalent to changes in temperature of 15° C (Emiliani, Mayeda and Selli 1961) we realize that at least 12° C depend on changes in the isotopic composition of the water. We cannot at present correct the isotopic palaeotemperature curve and therefore I have concluded that no conclusion can be drawn from the aforementioned curve (Olausson 1965, p. 231). In my own work I have used the climatic curve from the S.E. Netherlands, published by Zagwijn (1957, 1963), as a standard of reference, and since that curve is tentatively correlated with the Leffe series in Italy (Zagwijn 1957) and with a Pleistocene sequence in Colombia (van der Hammen and Gonzales 1964), the Dutch sequence can be regarded as standard profile for Upper Pliocene-Lower Pleistocene. The extinction of some vascular plants in N.W. Europe (Zagwijn 1957, 1963) is a good indication of a (local) Pleistocene boundary, and in my view preferable to the first appearance of a single benthonic Foraminifera in an isolated sequence in Italy.

On Pleistocene stratigraphy

For correlation between continental and deep-sea stratigraphy we must – beyond the C^{14} scale – continue to base it on similarities between the shapes of the climatic curves, since other dating methods are still in the experimental stage or open to question (Damon 1965; Sacket 1966). A comparison between the climatic curves of cores 58 (62, 74) and that from the S.E. Netherlands (Olausson 1961*a*, fig. 5; fig. 29.6, 29.9) reveals that they are identical from the beginning of Pleistocene up to the Mindel glacial stage. The agreement is so complete that I cannot doubt the interpretation. Thus, for example, the climatic deterioration in the Tegelen and Waal interglacials, recognized by Zagwijn (1957), are also found in the deep-sea core sequences (Fig. 29.9, cores 154 and 58).

The Mindel-Eem sequences of the Netherlands are fragmentary. The pollen diagram from Sabana de Bogota (van der Hammen and Gonzales 1964) reveal a detailed climate for Mindel as do the deep-sea cores. It consists of two main substages, the first of which is divided into two phases (Olausson 1961*a*, 1965, 1967; Fig. 29.9). Elsewhere in the world the Mindel is also interpreted in the same way (Olausson 1961*a*).

The Holstein stage, as interpreted by me and recognized in cores from the Pacific, Indian and North Atlantic

Oceans (Olausson 1961*a*, 1965), consists of four warm subages and three cool intervals. The high carbonate content of stage 5 in North Atlantic cores, and the low rate of carbonate deposition in the Indian and Pacific Oceans (Fig. 29.9), prove that it is mainly an interglacial. The stage is in this case comparable to the uppermost Pliocene (Fig. 29.5). The (quasi) stagnant phases of Eastern Mediterranean in this stage (at or below 5c (Olausson 1961*b*, 1965; Fig. 29.16)) occurred during a warm period (cf. Fig. 29.12*e, f*). The cool intervals of the Holstein age may have been of the size of an 'Early Stadial' (*sensu* Zagwijn 1961) or more intense (particularly the index horizon '*i*'). The duration of the Holstein interglacial was estimated by Penck and Brückner (1909) sixty years ago to be about four times longer than the Eem interglacial which, in fact, accords with the present deep-sea floor data. As stated earlier and also apparent from Fig. 29.6, 29.9 the temperature maximum was in the Upper Holstein (Olausson 1967, in press). The lower Holstein was temperate-warm. In pollen diagrams from Northern Europe *Taxus, Ilex* and *Buxus* suggests a relatively high temperature, but the frequency of the mixed oak forest is much lower than in other interglacials. These features have been interpreted as the result of a poor soil from the beginning of the interglacial (Andersen 1965, p. 497). In other interglacials such a decrease in the fertility of the soils can only be traced at a later time. The question is now whether the traditional 'Holstein' pollen diagram represents only the upper part of my Holstein stage. If so, our knowledge of the subage from the index horizon *i* to the end of *k* ('Lower Holstein') from continental sequences is poor. That portion could be some half of the duration of the Holstein, during which time soils could have been weathered. In the Bogota-diagram the Holstein may be divided into three warm phases with cooler intervals between them (Fig. 29.18; see also the discussion of the Post-Mindel tills (?) in the Alps, the pollen profile of Zydowszczyzna, etc.; Woldstedt 1958, p. 126, 181; Olausson 1961*b*, pp. 349-50).

Stage 3 of North Atlantic cores exhibits three carbonate maxima and is generally divided into two phases in Indian Ocean and Pacific core sequences. The boundary between stage 4 (Riss) and 3 (Eem) is well defined, but whether the boundary 2/3 corresponds to the boundary Eem/Würm (*sensu* Zagwijn 1961) or is somewhat younger, has not been decided (Olausson 1965, p. 221). This important stratigraphical question will be dealt with below.

It has been considered by some authors that it is possible there was a cool phase within the last interglacial. However, more recent pollen analytical studies, e.g. by Zagwijn (1961) do not suggest any 'little ice age' within the Eem interglacial. This has an important bearing on the position of the boundary Würm/Eem in deep-sea cores. The Eem interglacial was succeeded by a time when the vegetation in the North Sea-Baltic region was alternating between the two types subarctic Park landscape and temperate forest (Zagwijn 1961; Andersen, De Vries and Zagwijn, 1960).

The Eastern Mediterranean cores exhibit several sapropelitic horizons (Olausson 1961*b*; Fig. 29.16). The last one is Postglacial in age. The penultimate stagnant phase is dated to $>$ 40,000 C^{14} years (Olsson in Olausson 1961*b*). Since stage 2 generally consists of two cool horizons and the lower one directly succeeds the penultimate sapropelitic horizon, it sounds reasonable to correlate that stagnant phase with the Brørup interstadial and the aforementioned cool subphase with the cool time after the Brørup (from 40,000 to 50,000 the climate was boreal, while 50,000-55,000 C^{14} years B.P. 'hochglazial' prevailed (Gross 1966)). Then, the third (antepenultimate) stagnant phase, occurring in the middle of Stage 3 (Olausson 1961*b*; 1965, fig. 1), may be tentatively correlated with the Amersfoort interstadial. That stagnant sequence also apparently ends a warm age (Parker 1958; Olausson 1961*b*).

In most North Atlantic cores there is a faunal change between the second and third carbonate peaks (Olausson 1960*d*; 1965, fig. 6-10), while no significant difference in the fauna exists between the first and second carbonate peaks. The isotopic palaeotemperature analysis reveals anomalies in the North Atlantic and reversed ones in the Pacific and it cannot solve our present problem (Olausson 1965; in press). The thickness of each of the carbonate beds of stage 3 suggests that they were all formed during a considerable length of time, the middle one may be somewhat shorter than the others. The traditional concept is that Eem began 100,000 Pa^{231}/Th^{230} years ago and that Brørup ended some 55,000 C^{14} years ago, leaving a time span of 10,000-20,000 astronomical years for each cycle.

A pollen diagram from Colombia (van der Hammen 1961, fig. 1) shows two peaks in the 'Eem' stage and a smaller one in Early Würm (Fig. 4). If we exclude the existence of a cool oscillation within the Eem interglacial, then my former stage 3 may be considered as consisting of the sequences laid down from the beginning of the Eem interglacial up to the end of the Brørup interstadial (Fig. 29.9; Gross 1966). Thus the stage boundary 2/3 must be put down, since every stage must represent a single glacial or interglacial sequence (the time units [ages] and the corresponding time-rock units [stages] appear in Fig. 29.9; the boundaries between the stages are given as fractions of adjoining stage numbers). In doing so there is an excellent correlation between the climatic curves from North

Atlantic cores, vegetational data from the Netherlands and Colombia, and hydrographic changes in the Eastern Mediterranean (Fig. 29.18).

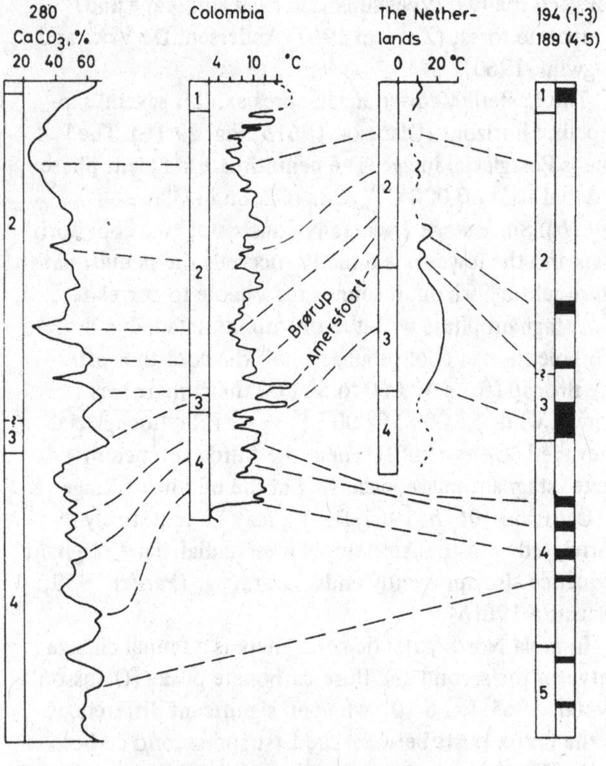
280
$CaCO_3$, %
20 40 60
Colombia
4 10 °C
The Nether-
lands
0 20 °C
194 (1-3)
189 (4-5)
Brørup
Amersfoort

Fig. 29.18. The figure consists of (*a*) the carbonate curve of core 280 (Olausson 1960) from 34° 57′ N, 44° 16′ W; (*b*) the curve of annual average temperature for the Upper Pleistocene in the Sabana de Bogota (van der Hammen 1961); (*c*) the estimate of mean temperature in July, the Netherlands (Zagwijn 1963); (*d*) the distribution of sapropelitic muds in the Eastern Mediterranean cores 194 (stages 1-3), and 189 (stages 4-5) (Olausson 1961, 1965). Stage 1 = Recent (Postglacial), 2 = Würm, 3 = Eem, 4 = Riss, and 5 = Holstein.

Since the carbonate minimum between the first and second post-Riss peaks in North Atlantic cores, and the corresponding productivity increase in Indian Ocean and Pacific cores, can be explained entirely by assuming an intensified convective sinking in the northernmost Atlantic (more organic matter and carbon dioxide are transported down to its floor by descending water masses), we must admit that the termperature of the subtropical and tropical Atlantic surface water did not decrease until the time when the second carbonate minimum was formed (the Amersfoort/Brørup stadial) and that the change in the oceans occurred after the Brørup interstadial. We still have no proof for a temperature decrease for the Equatorial Pacific during an ice age.

The first post-Eemian temperature drop (the Eem/Amersfoort stadial) is, if my correlation is correct, also noticeable in the Andes at 2,500 metres above sea level at 5° N (van der Hammen 1961). During the Amersfoort interstadial North America and Europe were marked by the coming ice age (Andersen *et al.* 1960; Flint 1963; Zagwijn 1961), and the Mediterranean cores reveal a lower water temperature during this subage than during the Eem age (Olausson 1961*b*; Parker 1958). In Colombia it represents a very humid time (pluvial) with a temperature at least as high as during the Eem interglacial. By the end of the Brørup the temperature dropped in Colombia as well as in the Atlantic.

The discussion so far suggests that the climatic changes in the earliest Würm (or a glacial age) are not the result of a simple temperature decrease either at higher latitudes or globally, but instead we must think in terms of a complex circulation pattern and latitudinal heat exchange for the onset of glaciation, as suggested by several authors.

The boundary Würm/Postglacial has been discussed earlier (Olausson 1965, pp. 232-7).

ACKNOWLEDGEMENTS

Laboratory work has been financed by the Swedish Natural Research Council, Magn. Bergvall's Foundation, and Karin and Herbert Jacobsson's Foundation.

Received September 1967.

BIBLIOGRAPHY

Andersen, Sv. Th. 1965. Interglacialer og Interstadialer i Danmarks kvartaer. *Medd. Dansk Geol. Forening,* **15** (4), 486-506.

Andersen, Sv. Th., De Vries, Hl. and Zagwijn, W. H. 1960. Climatic change and radiocarbon dating in the Weichselian glacial of Denmark and the Netherlands. *Geol. en Mijnbouw,* **39**, 38-42.

Arrhenius, G. 1952. Sediment cores from the East Pacific. *Rept. Swedish Deep-Sea Exp. 1947-1948,* **5** (1), 1-227.

Arrhenius, G. 1959. Sedimentation on the ocean floor. In *Research in geochemistry.* John Wiley: New York.

Baas Becking, L. G. M., Kaplan, I. R. and Moore, D. 1960. Limits of the natural environment in terms of pH and oxidation-reduction potentials. *J. Geol.* **68**, 243-84.

Bé, A. W. H. and Hamlin, W. H. 1967. Ecology of Recent planktonic Foraminifera. Part 3 – Distribution in the North Atlantic during the summer of 1962. *Micropaleontology,* **13** (1), 87-106.

Bruneau, L., Jerlov, N. G. and Koczy, F. F. 1953. Physical and chemical methods. Rept. *Swedish Deep-Sea Exped. 1947-48,* **3** (2).

Clowes, A. J. and Deacon, G. E. R. 1935. The deep water circulation of the Indian Ocean. *Nature, Lond.* **136**, 936-8.

Curray, J. R. 1961. Late Quaternary sea level; a discussion. *Geol. Soc. Amer. Bull.* **72**, 1707-12.

Curray, J.R. 1965. Late Quaternary history, continental shelves of the United States. In *The Quaternary of the*

United States (H. E. Wright and D. G. Frey, *eds.*). Princeton Univ. Press, Princeton. pp. 723-35.

Damon, P. E. 1965. Radioactive dating of Quaternary tephra. *Int. Ass. Quaternary Research, abstracts,* p. 92.

Defant, A. 1961. *Physical oceanography*. Pergamon Press: London.

Degens, E. T. 1965. *Geochemistry of sediments*. Prentice-Hall Inc.: New York.

Donn, W. L., Farrand, W. and Ewing, M. 1962. Pleistocene ice volumes and sea-level lowering. *J. Geol.* **70**, 206-14.

Emiliani, C. 1955. Pleistocene temperature variations in the Mediterranean, *Quaternaria,* **2**, 87-98.

Emilani, C. 1966. Isotopic paleotemperatures. *Science,* **154**, 851-7.

Emiliani, C., Mayeda, T. and Selli, R. 1961. Paleotemperature analysis of the Plio-Pleistocene section at Le Castella, Calabria, Southern Italy. *Bull. Geol. Soc. Am.* **72** (5), 679-88.

Emiliani, C. and Milliman, J. H. 1966. Deep-Sea sediments and their geological record. *Earth-Science Reviews,* **1**, 105-32.

Fisk, H. N. 1959. Padre Island and the Laguna Madre flats, coastal South Texas. *2nd Coastal Geography Conf., La. State Univ.* pp. 103-51.

Flint, R. F. 1965. The Pliocene-Pleistocene boundary. *Geol. Soc. Amer. Spec. Paper,* **84**, 497-533.

Flint, R. F. 1963. Status of the Pleistocene Wisconsin stage in Central North America. *Science,* **139**, 402-4.

Flohn, H. 1952. Studien über die atmosphärische Zirkulation in der letzten Eiszeit. *Erdkunde,* **7**, 266-75.

Gevirtz, J. L. and Friedman, G. M. 1966. Deep-sea carbonate sediments of the Red Sea and their implications on marine lithification. *J. Sed. Petrol.* **36**, 143-51.

Goldberg, E. D. 1965. Minor elements in sea water. In *Chemical oceanography, vol. 1.* (J.P. Riley and G. Skirrow *eds.*), ch. 5. Acad. Press: New York.

Gross, H. 1966. Der Streit um die Geochronologie des Spätpleistozäns und sein Ausgang. *Forschungen und Fortschritte,* **40**, 165-8.

Hammen, Th. van der, 1961. The Quaternary climatic changes of Northern South America. *An. New York Acad. Sciences,* **95**, 676-83.

Hammen, Th, van der and Gonzales, 1964. A pollen diagram from the Quaternary of the Sabana de Bogota (Colombia) and its significance for the geology of the Northern Andes. *Geol. en Mijnbouw,* **43**, 113-17.

Heezen, B. C., Menzies, R. J., Broocker, W. S. and Ewing W. M. 1959. Stagnation of the Cariaco Trench. *Int. Oceanogr. Congress, Proprints,* AAAS, Washington. pp. 99-102.

Herman, Y. 1965. Etudes des Sediments Quaternaires de la Mer Rouge. *Annales Inst. Ocean.* **42** (3), 343-415.

Ivanenkov, V. N. and Gubin, F. A. 1960. [Water masses and hydrochemistry of the western and southern parts of the Indian Ocean. (In Russian)] *Trudy Morsk. Gidrofis. Inst. Akad. Sci. SSSR,* 22, 33-115. (Translation (1963) Scripta Tecnica Inc., for Amer. Geophys. Union., 22, 27-99.)

Jerlov, N. 1953. Particle distribution in the Ocean. *Rept. Swedish Deep-Sea Exped. 1947-1948,* 3 (2), 73-97.

Kolbe, R. W. 1957. Diatoms from the equatorial Indian Ocean cores. *Rept. Swedish Deep-Sea Exped. 1947-48,* 9 (1).

Kullenberg, B. 1952. On the salinity of the water contained in marine sediments. *Medd. Occanografiska Inst. Goteborg,* **21**.

Lyman, J. 1959. Chemical considerations. *Nat. Acad. Sci. Nat. Res. Coun. Publ.* **600**, 87-97.

Mason, B. 1960. *Principles of geochemistry,* **2**. John Wiley: New York.

Mason, B. 1962. *Principles of geochemistry* (2nd. edn.) John Wiley: New York.

Menzel, D. W. 1964. The distribution of dissolved organic carbon in the Western Indian Ocean. *Deep-Sea Res.* **11**, 757-65.

Miller, A. R., Densmore, C. D., Degens, E. T., Hathaway, J. C., Manhein, F. T., McFarlin, P. F., Pocklington, R. and Jokela A. 1965. Hot brines and recent iron deposits in deeps of the Red Sea. *Woods Hole Oceanographic Institution,* ref. nr 65-38 (printed in Geoch. et Cosmochim, 1966).

Neumann, A. and McGill, D. A. 1962. Circulation of the Red Sea in early summer. *Deep-Sea Res.* 8, 223-35.

Olausson, E. 1960a. Description of sediment cores from Central and Western Pacific with the adjacent Indonesian region. *Rept. Swedish Deep-Sea Exped. 1947-1948,* **6** (5).

Olausson, E. 1960b. Description of sediment cores from the Indian Ocean. *Rept. Swedish Deep-Sea Exped. 1947-1948,* 9 (2), 53-88.

Olausson, E. 1960c. Description of sediment cores from the Mediterranean and the Red Sea. *Rept. Swedish Deep-Sea Exped. 1947-1948,* 8 (3), 287-334.

Olausson, E. 1960d. Description of sediment cores from the North Atlantic. *Rept. Swedish Deep-Sea Exped. 1947-1948,* 7 (5), 229-86.

Olausson, E. 1961a. Remarks on some Cenozoic core sequences from the Central Pacific, with a discussion of the role of coccolithophorids and Foraminifera in carbonate deposition. *Göteborgs Kungl. Vetenskaps- och Vitterhets-Sämhalles Handl., Sjätte Följden, (B),* 8 (10) [Medd. Oceanografiska Inst., Göteborg, 29,] 35 pp.

Olausson, E. 1961b. Studies of deep-sea cores. Sediment cores from the Mediterranean Sea and the Red Sea. *Rept. Swedish Deep-Sea Exped. 1947-1948,* 8 (4), 337-91.

Olausson, E. 1965. Evidence of climatic changes in North Atlantic deep-sea cores, with remarks on isotopic palaeotemperature analysis. *Progress in oceanography,* 3, 221-52.

Olausson, E. 1966. Sediments of the Atlantic Ocean. In *The Encyclopedia of oceanography* (ed. Rh. W. Fairbridge). Reinhold; New York. pp. 75-81.

Olausson, E. 1967. Climatological, geoeconomical, and palaeooceanographical aspects of carbonate deposition. *Progress in oceanography,* 4, 245-65.

Olausson, E. (in press). On the $0^{18}/0^{16}$ ratio of ancient sea water. Comments on C. Emiliani's paper 'Paleotemperature analysis of Caribbean cores P6304-8 and P6304-9 and a generalized temperature curve for the past 425,000 years'. *J. Geol.* (in press).

Olausson, E. and Karlsson, G. B. (in preparation).

Olausson, E. and Olsson, I. U. 1969. Varve Stratigraphy

in a core from the Gulf of Aden. *Palaeogeography, palaeoclimatology, palaeoecology*, **6**, 87-103.

Parker, F. L. 1958. Eastern Mediterranean Foraminifera. *Rept. Swedish Deep-Sea Exped. 1947-1948*, **5** (2).

Penck, A. and Brückner, E. 1909. *Die Alpen im Eiszeitalter*, 3 vols, Leipzig.

Phleger, F. B., Parker, F. L. and Peirson, J. F. 1953. North Atlantic Foraminifera. *Rept. Swedish Deep-Sea Exped. 1947-1948*. **7** (1), 1-122.

Robin, G. de Q. 1964. Glaciology. *Endeavour*, **23**, 102-7.

Sacket, W. M. 1966. Manganese nodules: Thorium-230: Protactinium-231 ratios. *Science*, **154**, 646-647.

Schott, W. 1935. Die Foraminiferen in dem aequatorialen Teil des Atlantischen Ozeans. *Wiss. Ergebn. Deutschen Atlant. Exped. 'Meteor' 1925-1927*, **3** (1) B.

Schott, W. 1966. Foraminiferenfauna und Stratigraphie der Tiefsee-sedimente im Nordatlantischen Ozean. *Rept. Swedish Deep-Sea Exped. 1947-1948*, **7** (8).

Shepard, F. 1963. Thirty-five thousand years of sea-level. In: *Essays in Marine Geology in honor of K. O. Emery*. Univ. of S. Calif. Press, Los Angeles, pp. 1-10.

Skopintsev, B. A. 1959. Organic matter in sea water. *Preprints Int. Oceanograph. Congr.*, AAAS, Washington, pp. 953-4.

Sverdrup, H. U., Johnson, M. W. and Fleming, R. H. 1942. *The Oceans, their physics, chemistry, and general biology*. Prentice-Hall Inc.: New York.

Todd, R. 1959. Foraminifera from Western Mediterranean deep-sea cores. *Rept. Swedish Deep-Sea Exped. 1947-1948*, **8** (3).

Van Andel, Tj. H. and Sachs, P. L. 1964. Sedimentation in the Gulf of Paris during the Holocene transgression; a subsurface acoustic reflection study. *J. Mar. Research*, **22**, 30-50.

Woldstedt, P. 1958. *Das Eiszeitalter, vol. 2*. Ferdinand Enke Verlag: Stuttgart.

Wüst, G. 1936. Die Stratosphare des Atlantischen Ozeans. *Wiss. Ergebn. Deutschen Atlant. Exped. 'Meteor' 1925-1927*, **6** (1), 105-288.

Wüst, G. and Defant A. 1936. Schichtung und Zirkulation des Atlantischen Ozeans. *Wiss. Ergebn. Deutsch. Atlantisch. Exped. 'Meteor', 1925-1927*, **6** (Atlas).

Zagwijn, W. H. 1957. Vegetation, climate and time-correlations in the Early Pleistocene of Europe. *Geol. en Mijnbouw*, **19**, 233-44.

Zagwijn, W. H. 1961. Vegetation, climate and radiocarbon datings in the Late Pleistocene of the Netherlands. *Memoirs of the Geol. Foundation in the Netherlands*, N.S. **14**.

Zagwijn, W. H. 1963. Pleistocene stratigraphy in the Netherlands, based on changes in vegetation and climate. *Ver. Kon. Ned. Geol. Mijnbouw. Gen., Geol. Ser.* **21**-2, 173-96.

30. THE LATERAL AND VERTICAL DISTRIBUTION OF DINOFLAGELLATES IN QUATERNARY SEDIMENTS

D. WALL

Summary: Dinoflagellates are numerous and widely distributed in Quaternary marine sediments in the form of resting spores (cysts), spinose types of which are in many instances identical with organisms formerly known as 'hystrichospheres'. These cysts occur on a world-wide basis in a variety of sedimentary facies: maximum concentrations (over 1,000 cysts per gramme) are usually found in relatively shallow water muddy sediments, but concentrations greater than 200 per gramme have been recorded in abyssal calcilutites. The lateral distribution of individual species is poorly known; it appears that many species are cosmopolitan, but are only prolific within more localized areas of preferred occurrence. This type of distribution has enabled facies-associations to be recognized, and these apparently reflect changing surface water environments. Similar facies-associations have been encountered in Lower Pleistocene sequences in East Anglia and southern Israel, and in a few piston cores from relatively young sediments. Some peculiarities of dinoflagellate distribution and major factors stabilizing it are discussed with particular reference to marine climates, ecological aspects of dinoflagellate life histories and sedimentation processes.

INTRODUCTION

Status of Quaternary dinoflagellate research

It is surprising that so little is known about the lateral and vertical distribution of Quaternary dinoflagellates because for decades planktonologists have studied the distribution of living dinoflagellates in the oceans, and for almost as long micropalaeontologists have studied fossil dinoflagellates as they are distributed in Mesozoic and Tertiary marine strata. Yet the study of Quaternary dinoflagellates has fallen between two disciplines, marine biology and micropalaeontology, to the detriment of both, and today a great deal of basic information concerning systematic, phylogenetic, ontogenetic, ecological and stratigraphical problems remains unknown.

Recent research has demonstrated that the resting spores (cysts) of many living dinoflagellates are homologues of fossil dinoflagellates as well as many types of 'hystrichospheres' (Evitt and Davidson 1964; Wall 1965; Wall and Dale 1966, 1967; Wall, Guillard and Dale 1967; Evitt 1967; Evitt and Wall 1968) and that only the resting spore phase in the life-history is fossilizable. The cell wall or theca of the planktonic phase as it is called, is cellulosic and cannot be fossilized. Other recent research has shown that fossil dinoflagellates are abundant in Quaternary sediments (West 1961; Rossignol 1962, 1964; Wall 1967; Wall and Dale 1968*a, b*). However, for many years these facts remained obscure as palaeontologists apparently were discouraged from working on Quaternary sediments by the widespread but not entirely universal belief that hystrichospheres were extinct. At that time, insufficient was known about the resting spores of living dinoflagellates to dispel this belief, and even today our knowledge of meroplankton (that is, resting spore-producing plankton) is sparse.

Fossilization of dinoflagellates

Extant species of *Gonyaulax, Protoceratium* and *Peridinium* produce the resting spores which are encountered most frequently in Quaternary sediments (Wall and Dale 1967). Other extant genera (*Diplopsalopsis, Diplopsalis, Diplopeltopsis, Scrippsiella*) produce similar resting spores but they have not been found abundantly as microfossils. However, future investigations will almost certainly reveal many new details of their occurrences and add more extant genera to the list of spore-producing dinoflagellates.

Palaeontologists have classified fossil dinoflagellates independently of modern ones but now that details of life-histories have been discovered it becomes apparent that taxonomic integration must be attempted. So far however, this has not been accomplished and so resting spores produced by different species of modern *Gonyaulax* receive the names *Hystrichosphaera, Nematosphaeropsis, Tectatodinium* and *Lingulodinium* in palaeontology. Similarly, fossilized spores of the extant genus *Protoceratium* (*P. reticulatum*) have been called *Operculodinium* in palaeontology (see Wall and Dale 1967, table 1). Resting spores of modern species of *Peridinium* also fossilize but generally they are delicate bodies and perhaps because of this they are less well known than *Gonyaulax* cysts in pre-Quaternary strata and have not been allocated palaeontological names, perhaps with the exception of *Lejeunia*, a relatively imprecisely known Neogene genus. A few extant species (*Peridinium trochoideum, Scrippsiella sweenyae*) produce calcareous resting spores which are homologous with those placed in the fossil family Calciodinellidae Defl. Systematic details of the resting spores found in Quaternary sediments are given in Rossignol (1964), Wall (1967) and Wall and Dale (1968*a*).

LATERAL DISTRIBUTION

Three aspects of the biogeography of dinoflagellates will be discussed here: (1) the lateral distribution of species in modern sediments; (2) the quantitative distribution of cysts in contrasting sedimentary environments; and (3) some of the factors affecting these distributions.

Species distribution

Williams[1] has shown that the overall distribution of several species (i.e. palaeontologically defined species) in North Atlantic sediments tends to be cosmopolitan but that individual species are most abundant in relatively localized areas. For example *Operculodinium centrocarpum* is found in sediments from close to 5° N to 70° N in the North Atlantic province, but is most abundant in an area stretching from the eastern coast of North America to the northwestern coast of Norway in a northeasterly direction. Other common species also have restricted areas of maximum abundance: *Leptodinium aculeatum* is most abundant off the north-west coast of Africa and in the Caribbean region, *Hystrichosphaera mirabilis* is most abundant in the westerly approaches to the English Channel, *Nematosphaeropsis* sp. cf. *balcombiana* is most abundant in the central and north-eastern North Atlantic region around 45° N. and *Lingulodinium machaerophorum* is very abundant near the Straits of Gibraltar.

Our observations complement those of Williams and appear to confirm his analysis. For example, *Operculodinium centrocarpum* is abundant in sediments we examined from an area between latitudes 35° to 40° N and longitudes 66° 30′ W to 70° 36′ W in the North Atlantic. We also observed an abundance of *Lingulodinium* in the eastern Mediterranean as Rossignol (1962) first reported. Concentrations of a *Peridinium* resting spore (*P.* sp. cf. *oblongum*) were found in the Gulf of Aqaba and concentrations of *Hemicystodinium zoharyi* around Bermuda. In Buzzards Bay, near Woods Hole, *Hystrichosphaera bentori, H. bulloidea* and *Operculodinium centrocarpum* are co-dominant in surface sediments. Observations by other authors confirm the general abundance and widespread occurrence of dinoflagellate cysts in modern sediments in the Pacific (McKee, Chronic and Leopold 1959; Evitt and Davidson 1964; Cross, Thompson and Zaitzeff 1966; Evitt 1967) and Atlantic (Nordli 1951; Wall 1965; Traverse and Ginsburg 1966) provinces, Caribbean region (Muller 1959; Wall 1967) and Sea of Japan (Boulouard and Delauze 1966) but do not generally amplify our knowledge of species distribution.

Quantitative distribution

The numerical abundance of dinoflagellate cysts on the sea floor relative to a gramme of sediment has been described locally in the Caribbean region (Muller 1959), the Bahamas (Traverse and Ginsburg 1966), along the coast of Israel (Rossignol 1961), in an equatorial Pacific atoll (McKee, Chronic and Leopold 1959) and in the Gulf of California (Cross, Thompson and Zaitzeff 1966).

Muller (1959, fig. 22) found that cysts were widespread on surface sediments around the region of the Orinoco Delta, Gulf of Paria and upon the continental shelf area north of Trinidad. Maximum concentrations with over 1,000 cysts per gramme were encountered along the western coast of Trinidad only a few miles offshore and in a zone approximately 100 miles square north of Dragons Mouth. Traverse and Ginsburg (1966, fig. 9) found a comparable pattern in the Bahamas. They noted maximum concentrations of cysts in a large area to the western, leeward side of Andros Island and several adjacent smaller islands such as Great Abaco Island, Eleuthera Island and the Exuma Islands; concentrations up to 1,800 cysts per gramme were recorded.

Rossignol (1961, fig. 15) found lower numbers reaching only 67 per gramme in a broad band running parallel to the coast of southern Israel and noted diminishing quantities to the north, east and west. McKee, Chronic and Leopold (1959, fig. 21) who were among the first to observe dinoflagellate cysts in modern sediments, found concentrations around 1,000 per gramme in two stations inside the reef of the Kapingamarangi Atoll which were

[1] This volume, chapter 4, pp. 91-5.

located in the deepest parts of the lagoon. Finally, Cross, Thompson and Zaitzeff (1966, figs. 12, 13) found that dinoflagellates were abundant in the Gulf of California: they recorded concentrations up to 17,000 per gramme in the southern part of the Gulf, southeast of the Baja Peninsula, and concentrations over 1,000 per gramme near Mazatlan and further north opposite the mouth of the River Mayo and east of Loreto.

Factors influencing the distribution of modern cysts

Factors influencing the distribution of modern cysts include first the environmental and biological controls which affect the primary biogeographic distribution of the living species in the planktonic phase and its encystment and secondly, the factors which determine the behaviour of both viable and non-viable cysts hydrodynamically as small organic particles in transportation and sedimentation.

Factors influencing the distribution of dinoflagellates species in the plankton are too numerous and complex for detailed discussion here: informative accounts have been presented by Graham (1941) and Wood (1964) who among many others, have contributed substantially to dinoflagellate ecology in the marine environment. A useful general review of the distribution of species in marine phytoplankton was given by Braarud (1962). He listed and discussed with selected examples from different groups (diatoms, dinoflagellates, coccolithophores), factors which theoretically may influence the distribution of phytoplanktonic species. These factors included autecological characteristics (those which vary interspecifically) such as temperature and salinity tolerance ranges and growth curves, nutrient requirements and tolerances, light growth curves, motility and flotation properties, life-cycle features, growth rate ranges and competitive characteristics. In addition, some environmental factors (those which may have a selective effect upon phytoplankton species composition in a given water mass) were listed: these included almost all of the autecological factors plus the effects of horizontal transport and migration, bathymetric conditions and physiographic barriers. The fact remains however, that the precise individual autecological preferences of the majority of modern dinoflagellate species are still poorly known or documented.

Factors controlling the encystment of dinoflagellates and even such important considerations as the genetic function of the resting spore are also for the most part unknown; despite their interest to biologists and palaeoecologists who study cysts in sediment. It seems fairly clear that there are two major functions of the cyst phase; first, to enable the dinoflagellate species to survive periods when water conditions are adverse towards the flagellated stage and secondly, to propagate the flagellated stage when favourable conditions for its growth are returned, that is, repopulation. However, how the requirements of individual species are related to environmental conditions is poorly known.

The existence of an empirical correlation between cyst biofacies (Williams, chapter 14) and marine climatic zones (Hall 1964) suggests that firstly, the major factors influencing the distribution of cyst biofacies are probably the same as those which determine the existence of marine climatic zones and secondly, a climatically determined or coincident distribution of modern cysts could provide a basis for palaeoecological interpretation of vertical fossiliferous sequences in cores or boreholes and outcrops of Quaternary or even older age.

The factors influencing the abundance of dinoflagellate cysts in sediment can be theorized along lines similar to those proposed by Cross (1964, tables 6, 7) to explain the abundance of the spores and pollen of higher plants in sediments. Thus there are factors controlling the supply of cysts: in the case of dinoflagellates they include the rate and density of cyst production by a parental population which itself is subject to numerous variables including seasonal ones, and influences introduced subsequently such as dissemination, sedimentation and preservation. In addition there are factors influencing the sedimentation of individual cysts which will relate to their sizes, shapes, specific gravity and condition (with or without protoplasts, living or dead) which together affect flotation and other environmental variables such as currents, turbulence and interception by biological agencies (predators).

Meroplanktonic dinoflagellates are neritic species by nature and dinoflagellate cysts, being very small organic sedimentary particles, possess very low settling velocities (as yet undetermined numerically) and thus will settle only in the absence of turbulence. In the natural environment we may expect to find maximum abundances of cysts where there is high cyst production and where cysts can sediment without dispersal or winnowing of the deposit. These conditions apparently have been met in areas such as the leeward side of Andros Island in the Bahamas (Traverse and Ginsburg 1966, 1967) and Trinidad in the West Indies (Muller 1959). The former authors note that the distribution of dinoflagellate cysts (hystrichosphaerids) in the Bahamas generally follows that of other small organic particles (pine pollen, microforaminifera shell linings) and that they accumulate most densely in areas of non-turbulence characterized by lime muds. Comparable abundances of cysts in oceanic provinces are found only in special cases where the sea floor is elevated locally as in the lagoon of the Kapingamarangi Atoll (McKee *et al.* 1959). High concentrations of cysts in the deepest part of this lagoon may be the result of selective sedimentation by cysts into the least turbulent area of the environment rather than a function of diurnal migrations or reproductive behaviour by their parental populations as suggested by those authors (p. 546).

A related point of interest is how closely accumulations of cysts on the sea floor coincide with areas of high productivity (of cysts in this instance) in surface waters because this is a matter of both ecological and palaeoecological concern. They probably are most highly coincidental in shallow water zones of little current action and little turbulence where resting spores can serve to repopulate the species. On the other hand, since meroplanktonic species are neritic organisms and apparently (but not certainly) do not form resting spores in oceanic regions, dinoflagellate cysts on abyssal plains must be allochthonous and their occurrences there unrelated to surface water primary productivity. The presence of cysts in abyssal sediments presumably is the result of transport either by surface currents or bottom currents which have traversed shelf areas or by turbidity flows. Certainly it seems most unlikely that the resting spores can function at great depths where there is no seasonal temperature fluctuation or light for photosynthesis.

A similar situation may prevail in certain basinal environments lying along continental margins which create deep water conditions locally, as for example, in the southern part of the Gulf of California. Here, very high concentrations of cysts were recorded by Cross, Thompson and Zaitzeff (1966) in water deeper than 1,000 fathoms and in a zone where temperatures never exceed 2° C at the bottom (compared with an annual temperature range of 14° C to 31° C at the surface, according to Roden 1964). Some information on local upwelling and plankton blooms in the Gulf of California, which is summarized in diagrams by van Andel (1964, figs. 37 A, B), although incomplete, indicates that they occur close to islands and in the lees of capes and bays on both sides of the Gulf. Thus areas which possibly represent the maximum productivity zones for cysts are not coincident with areas where maximum concentrations of cysts have been recorded in bottom sediment. Van Andel (1964, p. 267) remarks that locally diatoms are deposited in sheltered regions which also are bloom areas, but in general products of the blooms are dispersed towards deeper water. This relationship apparently is true for dinoflagellate cysts too.

VERTICAL DISTRIBUTION

The vertical distribution of dinoflagellates in Quaternary epicontinental marine successions has been described in boreholes through the Lower Pleistocene of Norfolk in England (West 1961; Wall and Dale 1968*a*) and southern Israel at Ashdod (Rossignol 1962). Their distribution in younger Quaternary sediments has been described from Late Pleistocene-Holocene (Flandrian) deposits in the coastal plain of Haifa (Rossignol 1963), Holocene (Atlantic-Sub Boreal) deposits in the Vilaine Valley of Brittany (Morzadec-Kerfourn 1966) and in a few piston cores from the Caribbean Sea (Wall 1967). There are also studies now in press or in progress, but the literature cited above represents almost our entire knowledge in this field which is clearly undeveloped.

West (1961) conducted a palynological investigation into the Lower Pleistocene marine sediments in the Royal Society borehole at Ludham, England. He divided this sequence into five local pollen stages (L1 to L5) and provisionally correlated them with the pollen stages of the Netherlands established earlier by Zagwijn (1957, 1960) and with local palaeontological horizons (LI to LVII) previously recognized at Ludham in an adjacent pilot borehole by Funnell (1961) who compared the latter sequence with the existing East Anglican Crag succession. West (1961) encountered numerous dinoflagellates and informally distinguished between several different species. Later, Wall and Dale (1968*a*) formally identified these species and elaborated upon their vertical distribution: they recognized five dinoflagellate species-associations in the Ludham sequence, one of which typified each of the local pollen stages. In the basal part of the sequence, the Ludhamian (L1), *Leptodinium multiplexum* was very abundant; it was accompanied in the basal Ludhamian (Lla) by *Hystrichosphaera bulloidea* and *Tectatodinium pellitum* as commonly associated forms and by *Operculodinium centrocarpum* and *Thalassiphora delicata* in the upper substage (Llb). *Lingulodinium machaerophorum* occurred in the Ludhamian but was virtually absent from overlying Icenian beds. In the Thurnian stage (L2), *Operculodinium israelianum* became extremely common and clearly dominant over all other species. In the overlying Antian stage (L3), *Hystrichosphaera bulloidea* became dominant and was associated with other species of this fossil genus (*H. furcata, H. mirabilis*) so that collectively this one group accounted for 75 per cent of the assemblage. Dinoflagellates were rare in the Baventian (L4) where only *Operculodinium israelianum* and *Tectatodinium pellitum* were common at all. The latter species became very abundant in the final stage in this local sequence (L5) with *Operculodinium centrocarpum* a commonly associated species.

Rossignol (1962) described the dinoflagellates in a Lower Pleistocene sequence from Ashdod on the coast of Israel. This included a series of sedimentary cycles initiated by a glacio-eustatic mechanism, each cycle comprising a transgressive and regressive phase. The ages of these cycles were determined as uppermost Calabrian, Sicilian, Ante-Tyrrhenian (=Milazzian) and Tyrrhenian in other stratigraphic and palaeontological studies (Reiss and Issar 1961; Itzhaki 1961). Rossignol (1962) recognized two alternating dinoflagellate species-associations: one was dominated by *Hemicystodinium zoharyi*, and the other by combinations of *Hystrichosphaera bentori, H. furcata, Lingulodinium machaerophorum* and *Operculodinium israelianum.* There were three horizons in the Sicilian characterized by the former type of association (Rossignol's stages IA2, IB2+3, IIB) and three alternating horizons of Calabrian and Sicilian age with the latter type of association (stages IA1, IB1, IIA). Finally, there was a Tyrrhenian stage (IIIB2) dominated by *H. zoharyi* var. *ktana* which was uncommon in the older horizons. Dinoflagellates were essentially absent locally in the Milazzian which was a limnic facies.

Wall (1967) identified over thirty species in two cores from the Caribbean Sea. Core A254-330 from the Yucatan Basin was dominated by species of *Hystrichosphaera* (*H. furcata, H. bulloidea, H. mirabilis*) with smaller numbers of *Lingulodinium, Leptodinium, Hemicystodinium* and *Operculodinium,* except for a thin horizon near 27 centimetres depth were *Leptodinium aculeatum* was dominant. This core appeared to comprise a Sangamon Interglacial lutite overlain by a thin early Wisconsin sequence. Core A254-327 contained similar assemblages dominated by *Hystrichosphaera* in a basal lutite sequence provisionally dated as a Wisconsin Interstadial deposit, and in Holocene lutites at the top of the core, but most of this core was a coarse ooze which was devoid of dinoflagellates. A core from the Cariaco Trench was spot sampled only; in it were assemblages including numerous *Peridinium* cysts and a relative abundance of *Operculodinium israelianum* and *O. psilatum. Leptodinium* was absent from these deposits.

In general Pleistocene assemblages from the Mediterranean and Caribbean areas are very similar to each other in species composition. They can differ rather strongly from assemblages found in more northerly provinces such as East Anglia, except that *Hystrichosphaera* or *Hystrichosphaera-Lingulodinium* dominated associations can be found in both regions (for example, the Calabrian in Israel, the Antian in East Anglia and the Holocene of Brittany). *Hemicystodinium*-dominated associations have been recorded primarily in the Middle East region, but do occur elsewhere (in the Miocene of Jamaica and modern sediments near Bermuda for instance).

Factors influencing the vertical distribution of fossil dinoflagellates

There is little doubt that the distribution of cysts in vertical Quaternary sequences is environmentally controlled and it may be palaeoclimatic to a large degree. The individual species which typify such sequences are extant, follow recurrent patterns of distribution and the associations they enter into are likewise recurrent and can be considered to be facies-associations. Such a recurrent pattern is well shown in the Israelian Sicilian (Rossignol 1962). Moreover, in the Ludham sequence in East Anglia, dinoflagellate species-associations can be correlated with comparable foraminiferal species-associations and pollen associations which have enabled palaeoclimatic intervals to be identified. Dinoflagellates, Foraminifera and pollen which are of different derivation (marine plankton and benthos, terrestrial plants) and whose distributions are influenced by some independent environmental parameters, show simultaneous fluctuations in composition at several horizons (Wall and Dale 1968*a*): thus it appears that the dinoflagellates also responded to changing palaeoclimatic conditions. On this comparative basis, dinoflagellates in Ludhamian (L1), Antian and L5 stages can be considered as temperate (interstadial or interglacial) associations and those in the Thurnian (L2) and Baventian (L4) as subarctic or colder water associations. A similar interpretation may be offered for the alternating species-associations recognized by Rossignol (1962) in the Israelian Pleistocene: the *Hemicystodinium zoharyi* association appears to be a tropical assemblage while the *Hystrichosphaera-Lingulodinium* association appears to be a warm temperate assemblage. These interpretations are consistent with our present knowledge of living dinoflagellate cyst-theca relationships and species biogeography but this is an area where much better data can be obtained in the future. However, no matter how good our understanding of contemporary ecological relationships becomes, it is unlikely we shall be able to isolate the individual factors which have controlled dinoflagellate distribution in ancient environments and interpretation will remain subjective as in other branches of palaeoecology.

The future use of dinoflagellates in Quaternary studies is one well worthy of exploration. Dinoflagellates doubtless are abundant in many basinal sediments where planktonic Foraminifera or calcareous nannoplankton which have been used to correlate pelagic deposits are absent and seem to form well defined species-associations with some palaeoclimatic significance. Thus they may be used to define stadia, interstadia or other Pleistocene events in marine sediments in a manner which parallels the use of pollen and spores in limnic deposits and so have limited use in local age determinations and stratal correlations in the absence of phylogenetic criteria. In addition, the possibility arises that it may be feasible to trace historical migration paths along continental margins throughout Pleistocene times and detect changes in current distribution in pelagic regions by studying fossilized dinoflagellates. These and many other aspects of dinoflagellate studies essential remain untouched and await future research.

ACKNOWLEDGEMENT

This work was supported by the National Science Foundation Grants GB-2881 and GB-5200 and Woods Hole Oceanographic Institution.

Received February 1968.

BIBLIOGRAPHY

Boulouard, C. and Delauze, H. 1966. Analyse palynoplanctologique de sédiments prélevés par le bathyscaphe *Archimède* dans la fosse du Japon. *Marine Geol.* **4**, 461-6.

Braarud, T. 1962. Species distribution in marine phytoplankton. *J. Oceanogr. Soc. Japan. 20th Anniversary Volume*, 628-49.

Cross, A. T. 1964. Plant microfossils and geology: an introduction. *In:* Palynology in Oil Exploration. *Soc. Econ. Paleontologists Mineralogists, Spec. Publ.* **11**, 3-13.

Cross, At. T., Thompson, G. G. and Zaitzeff, J. B. 1966. Source and distribution of palynomorphs in bottom sediments, southern part of Gulf of California. *Marine Geol.* **4**, 467-524.

Evitt, W. R. 1967. Dinoflagellate Studies: II. The Archeopyle. *Stanford Univ. Publs., Geol. Sci.* **10** (3), 88 pp.

Evitt, W. R. and Davidson, S. 1964. Dinoflagellate Studies: I. Dinoflagellate cysts and thecae. *Stanford Univ. Publs., Geol. Sci.,* **10** (1), 1-12.

Evitt, W. R. and Wall, D. 1968. Dinoflagellate Studies: IV. Thecae and cysts of Recent *Peridinium limbatum* (Stokes) Lemmermann. *Stanford Univ. Publs. Geol. Sci.,* **12** (2), 1-15.

Funnell, B. M. 1961. The Palaeogene and Early Pleistocene of Norfolk. *Trans. Norfolk Norw. Nat. Soc.* **19**, 340-64.

Graham, H. W. 1941. An oceanographic consideration of the dinoflagellate genus *Ceratium. Ecol. Monogr.* **11**, 99-116.

Hall, C. A. 1964. Shallow-water marine climates and molluscan provinces. *Ecology*, **45**, 226-34.

Itzhaki, Y. 1961. Contributions to the study of the Pleistocene in the coastal plain of Israel. Pleistocene shorelines in the coastal plain of Israel. *Bull. geol. Surv. Israel,* **32**, 1-9.

McKee, E. D., Chronic, J. and Leopold, E. B. 1959. Sedimentary belts in lagoon of Kapingamarangi Atoll. *Amer. Assoc. Petrol. Geol., Bull.* **43**, 510-62.

Morzadec-Kerfourn, M. T. 1966. Étude des Acritarches et Dinoflagellates de sédiments vaseux de la Vallée de la Vilaine aux environs de Redon (Ille-et-Vilaine). *Bull. Soc. géol. minér. Bretagne, 1964-5, N. Ser.* 137-146.

Muller, J. 1959. Palynology of Recent Orinoco delta and shelf sediments: reports of the Orinoco Shelf Expedition, 5, *Micropaleontology*, **5**, 1-32.

Reiss, Z. and Issar, A. 1961. Contributions to the study of the coastal plain of Israel. Subsurface Quaternary correlations in the Tel Aviv region. *Bull. geol. Surv. Israel*, **32**, 10-26.

Roden, G. I. 1964. Oceanographic aspects of the Gulf of California. In: *Marine Geology of the Gulf of California* (Tj. H. van Andel and Shor, G. G., *eds.*). Am. Assoc. Petrol. Geologists, Tulsa, Okla., 30-58.

Rossignol, M. 1961. Analyse pollinique de sédiments marins quaternaires en Israel. I. Sédiments récents. *Pollen et Spores*, **3**, 301-24.

Rossignol, M. 1962. Analyse pollinique de sédiments marins quaternaires en Israel. II. Sédiments pléistocènes. *Pollen et Spores*, **4**, 121-48.

Rossignol, M. 1963. Analyse pollinique de sédiments quaternaires dans la Plaine de Haifa-Israel. *Israel J. Earth-Sci.* **12**, 207-14.

Rossignol, M. 1964. Hystrichosphères du Quaternaire en Méditerranée orientale, dans les sédiments pléistocènes et les boues marines actuelles. *Rev. Micropaleont.* **7**, 83-99.

Traverse, A. and Ginsburg, R. N. 1966. Palynology of the surface sediments of Great Bahama Bank, as related to water movement and sedimentation. *Marine Geol.* **4**, 417-59.

Traverse, A. and Ginsburg R. N. 1967. Pollen and associated microfossils in the marine surface sediments of the Great Bahama Bank. *Rev. Palaeobotan. Palynol.* **3**, 243-54.

Van Andel, Tj. H. 1964. Recent marine sediments of Gulf of California. In: *Marine Geology of the Gulf of California.*(Tj. H. van Andel and Shor, G. G., *eds.*). Am. Assoc. Petrol Geologists, Tulsa, Okla., 216-310.

Wall, D. 1965. Modern hystrichospheres and dinoflagellate cysts from the Woods Hole region. *Grana Palynol.* **6**, 297-314.

Wall, D. 1967. Fossil microplankton in deep-sea cores from the Caribbean Sea. *Palaeontology*, **10**, 95-123.

Wall, D. and Dale, B. 1966. 'Living fossils' in western Atlantic plankton. *Nature, Lond.* **211**, 1025-6.

Wall, D. and Dale, B. 1967. The resting cysts of modern marine dinoflagellates and their palaeontological significance. *Rev. Palaeobotan. Palynol.* **2**, 349-54.

Wall, D. and Dale, B. 1968a. Early Pleistocene dinoflagellates from the Royal Society Borehole at Ludham, Norfolk, *New Phytol.* **67**, 315-26.

Wall, D. and Dale, B. 1968b. Quaternary calcareous dinoflagellates (Calciodinellidae) and their natural affinities. *J. Paleontology* **42** (6), 1395-408.

Wall, D. and Dale, B. 1968c. Modern dinoflagellate cysts and evolution of the Peridiniales. *Micropaleontology*, **14** (3), 265-304.

Wall, D., Guillard, R. R. L. and Dale, B. 1967. Marine dinoflagellate cultures from resting spores. *Phycologia*, **6**, 83-6.

West, R. G. 1961. Vegetational history of the Early Pleistocene of the Royal Society borehole at Ludham, Norfolk. *Proc. R. Soc. B,* **155**, 437-53.

Wood, Ferguson E. J. 1964. Studies in microbial ecology of the Australasian region. *Nova Hedwigia,* **8** (1/2), 1-20, 35-54, (3/4), 461-527, 548-68.

Zagwijn, W. H. 1957. Vegetation, climate and time-correlations in the Early Pleistocene of Europe. *Geologie Mijnb., N. Ser.,* **19**, 233-44.

Zagwijn, W. H. 1960. Aspects of the Pliocene and Early Pleistocene vegetation in the Netherlands. *Meded. geol. Sticht., Ser. C-III,* **1**, 3-78.

31. DIATOMS IN PLEISTOCENE SEDIMENTS FROM THE NORTHERN PACIFIC OCEAN

A. P. JOUSÉ

Summary: This paper presents the principal results obtained in the course of investigation of diatoms in some cores from north of 40° N. Three types of diatom variation are observed in the cores: quantitative, ecological and phylogenetic. These changes are quite clear-cut and constitute valid evidence, for establishing a stratigraphy for these cores, for the interpretation of palaeogeographical and palaeoecological conditions of sedimentation, and for the estimation of the age of separate horizons.

Quantitative variations are expressed in the alternation of layers which are either enriched or impoverished in diatoms. All cores collected north of 40° N, next to the northern boundary of the subtropical zone are characterized by identical quantitative curves. In all cores one can also observe the alteration of different ecological complexes of diatoms (and silicoflagellates): cold-loving, moderate cold-loving, moderate warm-loving and warm-loving. The change of temperature in surface waters during the sedimentation process led to intense development of different ecological groups.

Phylogenetic variations become quite obvious as soon as the lower boundary of the upper Pleistocene is crossed (in cores below horizon III). The brief Quaternary history of diatoms was marked both by the extinction of some species and the appearance of others. At present, owing to the long core V20-119, we have a definite picture of diatom evolution in the north Pacific throughout the whole Quaternary period.

INTRODUCTION

The geological history of the oceans has been greatly advanced by investigations carried out in the field of micropalaeontology. The analysis of diatom assemblages has proved to be a most promising component in the entire complex of micropalaeontological studies, especially in the stratigraphy and palaeogeography of siliceous and calcareous-siliceous deposits. In the remote past, as well as during the Recent epoch, the diatom algae of the Pacific dominated the phytoplankton, and played an important role in the process of sedimentation. The northern, pre-equatorial and antarctic areas of the ocean reveal the highest level of diatom remains. It is known that a number of areas form typical diatom oozes in which diatoms comprise over 50-70 per cent of the sediment by weight.

Planktonic diatoms, inhabiting the ocean's surface layer 0-100 metres, having fallen into sediments preserve to a considerable extent certain aspects which were characteristic of their life cycle. However, the picture reflected in the sediments has its own traits and it would be erroneous to equate the composition and quantity of diatoms in biocoenoses and thanatocoenoses. Under all conditions of preservation the fossil complex of diatoms is always poorer than complexes in the biocoenosis. These are variations both with regard to species composition and their quantitative distribution. Cases when a particular diatom species is accumulated in mass numbers, owing to favourable conditions of preservation, are frequent, and millions of valves are found in a gramme of sediment; whereas in the plankton diatoms are diluted by other species and no one species is dominant.

Hence, one ought to bear in mind that in a number of features: the species composition, number of separate components and total valve number – the fossil diatom complex differs from its life complex. The fossil complex reveals peculiar features which develop as a result of the valves' varied capacity for preservation. This conclusion has been reached in the course of a comparative study of diatoms from the plankton and those in the surface layer of ocean sediment. There is no reason to suppose that the situation differed in the remote past. The basic problem is to learn how close the fossil complexes resemble those particular biocoenoses from which they originated. Now there are grounds for solving this problem in an affirmative way. And thus diatom analysis enjoys equal standing with other methods in stratigraphical and palaeogeographical investigations of the oceans.

Material studied

All cores are plotted (Fig. 31.1). The number of Pacific cores investigated is about sixty. Most of the cores were collected during expeditions of R. V. *Vityaz* and *Ob*. One long core, V20-119, was donated by Lamont Geological Observatory.

In the following account attention is concentrated on results obtained during the investigation of diatom distribution in sediment cores collected north of 40° N. The tropical and equatorial belts are considered by Muhina. For the southern part of the Pacific only preliminary data are available, namely from cores from the New Zealand section, but these require further examination and hence will not be reviewed here.

BASIC FEATURES OF DIATOM DISTRIBUTION IN PLEISTOCENE CORES FROM THE NORTH PACIFIC

As mentioned above (see Summary) three types of diatom variations are observed in cores: quantitative, ecological and phylogenetic.

1. *Numerical variations*, can be compared by counting diatoms (per gramme of sediment) – see Fig. 31.2. Calculation of diatom numbers shows that sediments from the ocean's northern areas are neatly stratified according to numerical criteria, and may be subdivided into enriched and impoverished horizons. Here are some examples. Core 4112 (47°32′0″ N, 160° 03′0″ W), along the central section, in the region of subarctic waters, shows very sharp variations in diatom concentrations. Two quantitative maxima are observed in the surface sediment layer (0-4 centimetres) and at a depth of 246-272 centimetres. In the first case, there were about 48 million valves per gramme; in the second, the maximum was 36.2 million per gramme. The mid-part of the core shows noticeable maxima which are however inferior to the upper and lower peaks. The amplitude of diatom content ranges from 48 million valves per gramme in the surface sediment layer to 3.75 million per gramme at a depth of 170 centimetres. During the time required for the accumulation of 3 metres of sediment at station 4112 terrigenous sediments were repeatedly replaced by organogenic and other types of siliceous sediments. The same picture

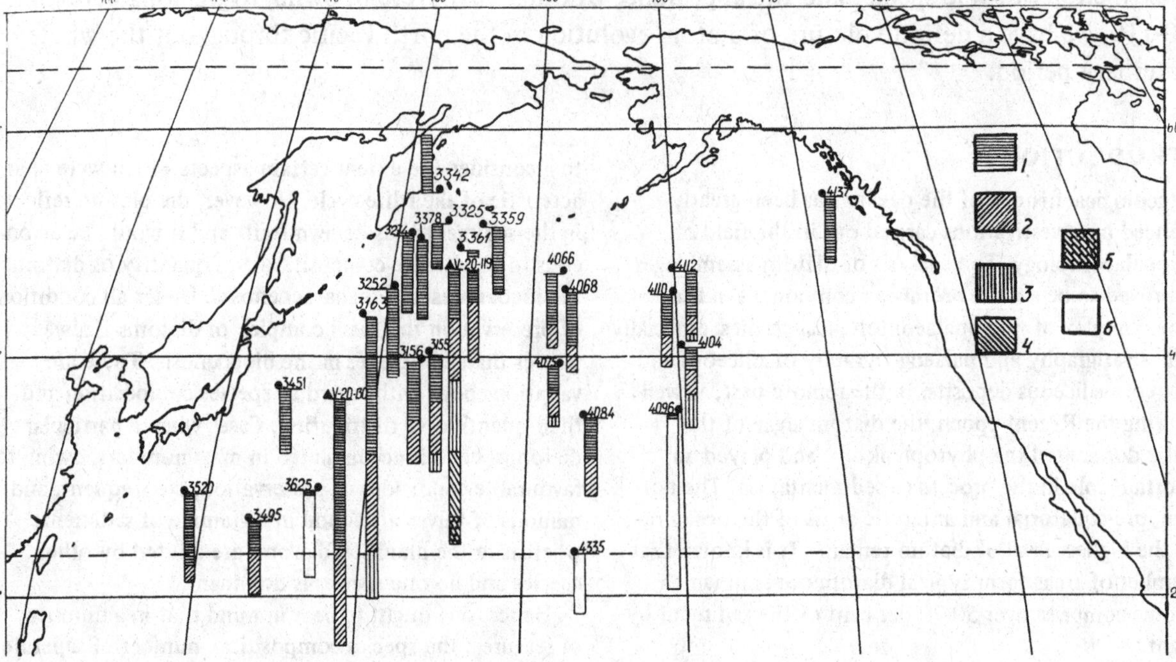

160°
180°
160°
140°
60°
40°
20°
3342
3325
3359
3378
3361
V-20-119
4066
4068
3252
3163
4112
4110
4104
3156
3155
4075
3451
V-20-60
4084
4096
3520
3625
3495
4335
1
2
3
4
5
6

Fig. 31.1. Distribution of cores and biostratigraphy of sediments. 1. Holocene and upper Pleistocene; 2. middle Pleistocene; 3. lower Pleistocene; 4-5. late Pliocene; 6. without stratigraphy.

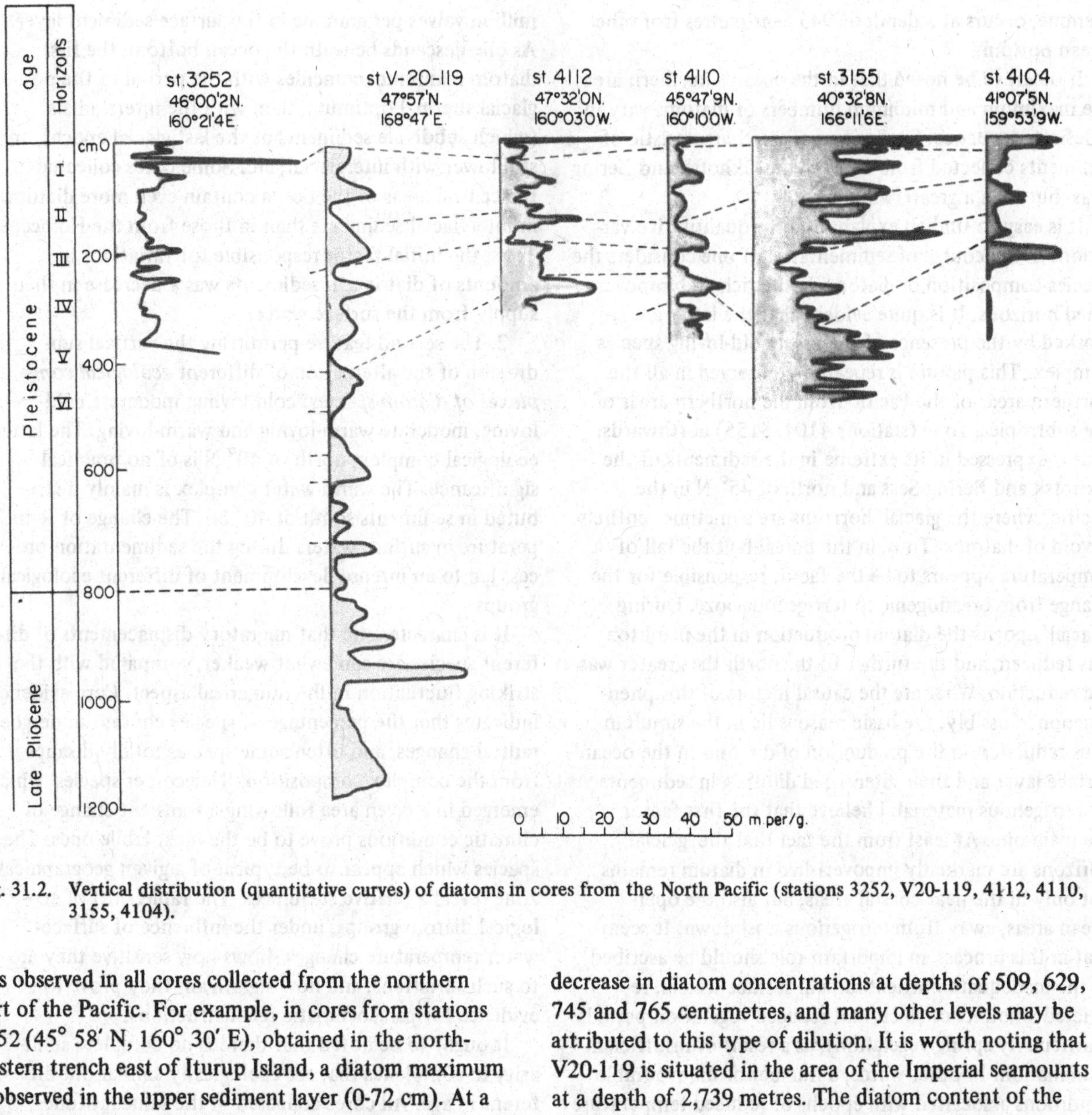
age
Horizons
st.3252
46°00'2N.
160°21'4E
st.V-20-119
47°57'N.
168°47'E.
st.4112
47°32'0N.
160°03'0W.
st.4110
45°47'9N.
160°10'0W.
st.3155
40°32'8N.
166°11'6E.
st.4104
41°07'5N.
159°53'9W.
cm0
I
II
III
IV
V
VI
200
400
600
800
1000
1200
Pleistocene
Late Pliocene
0 10 20 30 40 50 m.per/g.

Fig. 31.2. Vertical distribution (quantitative curves) of diatoms in cores from the North Pacific (stations 3252, V20-119, 4112, 4110, 3155, 4104).

was observed in all cores collected from the northern part of the Pacific. For example, in cores from stations 3252 (45° 58′ N, 160° 30′ E), obtained in the north-western trench east of Iturup Island, a diatom maximum is observed in the upper sediment layer (0-72 cm). At a depth of 40 centimetres the valve number reaches 47.5 million per gramme, at a depth of 70-72 centimetres, 21 million per gramme. In the mid-part of the core, at 210-270 centimetres, three not very high peaks (ranging from 14 to 18 million) are noted, separated by terrigenous sediments with very low diatom concentrations (up to 4 million per gramme).

Core V20-119 (47° 57′ N, 168° 47′ E, depth 2,379 metres) contains fewer diatoms than the cores from the bottom of deep trenches. This is commonly the case for sediments occurring on submarine elevations. In addition the sediments of this core are profusely packed with volcanic glass, which dilutes the diatom content. The decrease in diatom concentrations at depths of 509, 629, 745 and 765 centimetres, and many other levels may be attributed to this type of dilution. It is worth noting that V20-119 is situated in the area of the Imperial seamounts at a depth of 2,739 metres. The diatom content of the core's surface layer in the area of the station was not determined owing to lack of samples; however, it may be assumed the content is inferior to 10 million valves per gramme from the map of diatom distribution in the surface sediment layer of the Pacific. The amplitude of the variations in diatom number is not as great as at stations 4112 or 3252; it fluctuates from 1 million valves per gramme, at its minimum, to 13 million at its maximum in the core interval (0-805 cm), which according to our data corresponds to sediments of the Quaternary age. Repeated alternation of siliceous and terrigenous sediments is a characteristic feature of this core interval. Below 805 centimetres sediments are, on average, richer in diatom content. Their maximum, 33 million valves per gramme, occurs at a depth of 945 centimetres from the ocean bottom.

It ought to be noted that in the ocean's northern areas the maximum and minimum numbers of diatoms vary in a definite sequence. Similar curves are characteristic of sediments collected from cores in the Okhotsk and Bering Seas, but with a greater amplitude.

It is easy to find an explanation for quantitative variations in the course of sedimentation if one considers the species composition of diatoms in the rich and impoverished horizons. It is quite evident that the latter are marked by the presence of the most cold-loving species complex. This picture is repeatedly observed in all the northern areas of the Pacific from the northern areas of the subtropical zone (stations 4104, 3155) northwards. This is expressed in its extreme in the sediments of the Okhotsk and Bering Seas and north of 45° N in the Pacific, where the 'glacial' horizons are sometimes entirely devoid of diatoms. Thus, in the boreal belt the fall of temperature appears to be the factor responsible for the change from organogenic to terrigenous ooze. During 'glacial' epochs the diatom production in the plankton was reduced, and the further to the north the greater was the reduction. What are the causal factors of this phenomenon? Possibly, the basic reasons lie in the simultaneous reduction in the production of diatoms in the ocean's surface layer and their intensified dilution in sediments by terrigenous material. I believe that the first factor is the main one. At least from the fact that the 'glacial' horizons are markedly impoverished in diatom remains not only in the near-coastal areas, but also the open ocean areas, away from terrigenous washdown. It seems that in this process an important role should be ascribed to the inadequate nutrient salts in surface waters, restricted amounts of silica, and reduced vegetation periods (particularly spring vegetation), as a result of insufficient illumination. In other words, a number of unfavourable conditions associated with epochs of reduced temperature in the north of the Pacific. As a result, north of 40° N sediments were formed containing four to five times fewer diatoms than in the Recent epoch. One can find analogous conditions in the recent Polar basin (Bogorov 1967).

It would seem natural to expect that sediment horizons rich in diatoms will correspond to periods of warmer climate, and this is confirmed by the diatom complexes observed in the respective horizons. During these periods separate moderate warm-water elements developed, and the quantity of the most cold-loving species increased. Cores from the north-western areas of the ocean show that sediments of the last interglacial epoch contain up to 40 million valves per gramme. The same stations reveal 50-55 million valves per gramme in the surface sediment layer. As one descends beneath the ocean bottom, the first diatom maximum coincides with the period of the post-glacial thermal optimum, then with the interstadials (which subdivide sediments of the last glacial epoch), and still lower with interglacial, etc. Some cores collected in the central areas of the ocean contain even more diatoms in interglacial sediments than in those from the Holocene. Thus, the initial factor responsible for variations in amounts of diatoms in sediments was a decrease in their supply from the surface water.

2. The second feature permitting the vertical subdivision of the alternation of different *ecological complexes of diatom species*: cold-loving, moderate cold-loving, moderate warm-loving and warm-loving. The latter ecological complex, north of 40° N is of no practical significance. The warm-water complex is mainly distributed in sediments south of 40° N. The change of temperature in surface waters during the sedimentation process led to an intense development of different ecological groups.

It is characteristic that migratory displacements of different species are somewhat weaker, compared with the striking fluctuation in the numerical aspect. Core evidence indicates that the percentage of species content undergoes radical changes, and often some species totally disappear from the complex composition. 'Newcomer species' which emerged in a given area following a transient change of climatic conditions prove to be the most labile ones. The species which appear to be typical of a given geographical zone reveal a relative steadiness. The rapid shift of ecological diatom groups, under the influence of surface-water temperature changes shows how sensitive they are to such variations, and how significant they prove to be as evidence for palaeoclimatic reconstructions.

It ought to be noted that there is no complete similarity in complexes that are ecologically similar but different in age. All cores collected in the same latitudes, but for different horizons show that such diatom complexes are not identical. They differ in a series of features which can be ascribed to phylogenetic peculiarities and which will be dealt with later.

Core 4112 (Fig. 31.3) shows that four species occur throughout the entire core: *Coscinodiscus marginatus* Ehr., *Rhizosolenia hebetata* Gran, *Actinocyclus divisus* (Grun.) Hust. and *Denticula semina* Simonsen et Kanaya. Their representation in the diatom complex is expressed in per cent (Jousé 1963). A general review of diatom composition within the core shows that the highest percentage occurrence is shown by *Coscinodiscus marginatus* and *Rhizosolenia hebetata* and may be ascribed to sediments

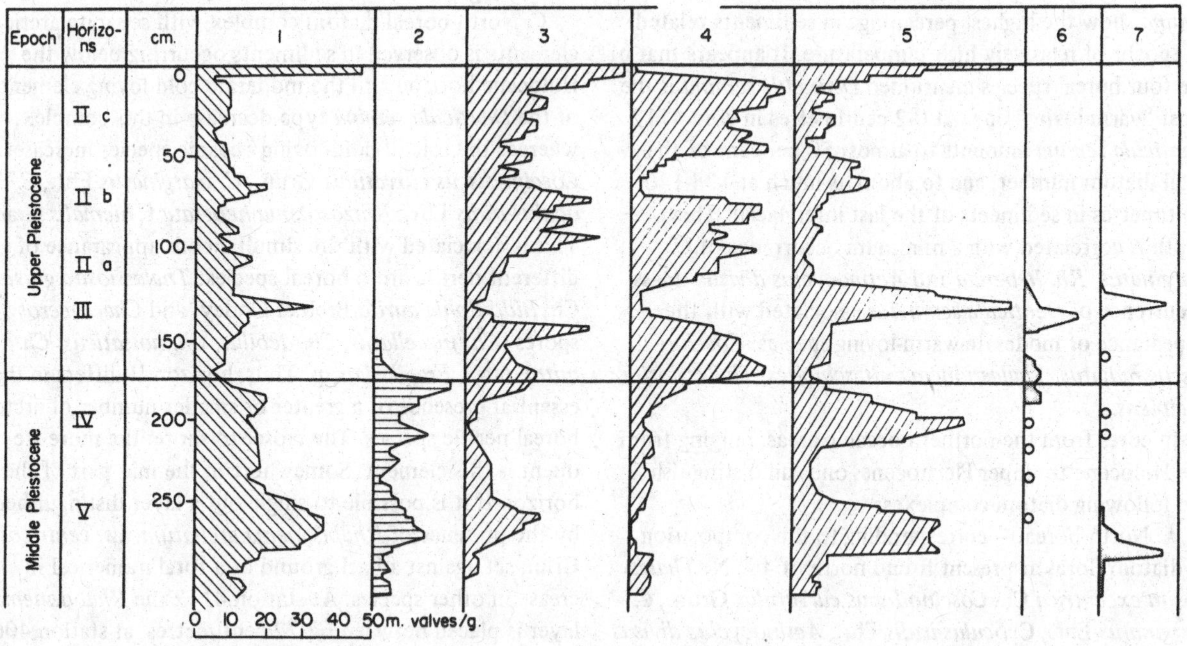
Epoch
Horizons
cm.
1
2
3
4
5
6
7
Upper Pleistocene
Middle Pleistocene
II c
II b
II a
III
IV
V
0
50
100
150
200
250
0 10 20 30 40 50m. valves/g.

Fig. 31.3. Station 4112. Vertical distribution of some characteristic forms. 1. Quantitative curve; 2. *Rhizosolenia curvirostris* Jousé; 3. *Coscinodiscus marginatus* Ehr., 4. *Rhizosolenia hebetata* Gran; 5. *Denticula semina* Simonsen et Kanaya; 6. *Coscinodiscus radiatus* Ehr.; 7. *Thalassionema nitzschioides* Grun.

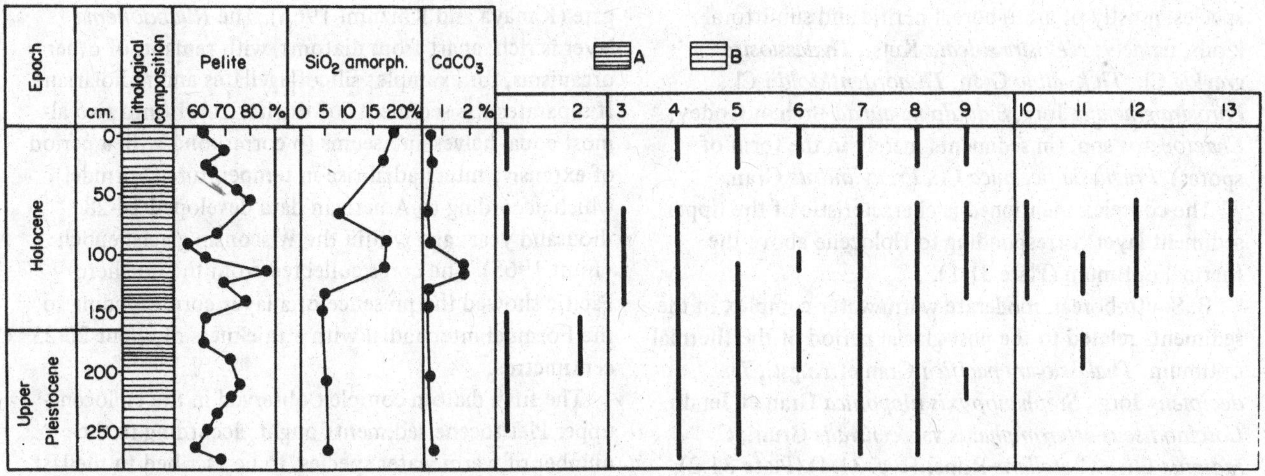
Epoch
Lithological composition
cm.
Pelite
60 70 80 %
SiO2 amorph.
0 5 10 15 20%
CaCO3
1 2 %
A
B
1
2
3
4
5
6
7
8
9
10
11
12
13
Holocene
Upper Pleistocene
0
50
100
150
200
250

Fig. 31.4. Station 3359 (51° 21′ 4″ N., 172° 14′ 1″ E, depth 4,880 metres). Diatom complex characteristic of the Holocene (postglacial, period of the thermal optimum). A. Clay; B. Aleurite clay.

Arctoboreal species: 1. *Thalassiosira gravida* Cl., 2. *Chaetoceros* spp. (*C. furcellatus*, *C. mitra*).

Boreal species: 3. *Stephanopyxis nipponica* Gran et Jendo; 4. *Thalassiosira excentrica* Cl.; 5. *Actinocyclus divisus* (Grun.) Hust.; 6. *Coscinodiscus oculus-iridis* Ehr.; 7. *C. marginatus* Ehr.; 8. *Rhizosolenia hebetata* Gran; 9. *Thalassiothrix longissima* Cl. and Grun.

Subtropical species; 10. *Thalassiosira pacifica* Gran et Angst; 11. *Coscinodiscus radiatus* Ehr.; 12. *C. stellaris* Roper. 13; *Distephanus octonarius* (Ehr.) Defl.

formed during relatively low temperature. On the other hand *Actinocyclus divisus* and especially *Denticula semina* show the highest percentage in sediments related to epochs of relatively high temperature. It appears that of the four boreal species mentioned *Denticula semina* is the most 'warm-loving' one. At 0-2 centimetres in core 4112 *Denticula semina* amounts to almost 50 per cent of the total diatom number, and to about as much at 138-140 centimetres in sediments of the last interglacial. This depth is correlated with a minimum occurrence of *C. marginatus, Rh. hebetata* and *Actinocyclus divisus.* Mass occurrence of *Denticula semina* is associated with the appearance of moderate-warm-loving species: *Coscinodiscus radiatus, Thalassionema nitzschiodes, Thalassiosira decipiens.*

In cores from the northern oceanic areas, ranging from the Holocene to upper Pleistocene, one can distinguish the following diatom complexes:

A. North-boreal – corresponding to the composition of diatom floras at present found north of 40° N: *Thalassiosira excentrica* Cl., *Coscinodiscus curvatulus* Grun., *C. marginatus* Ehr., C. *oculus-iridis* Ehr., *Actinocyclus divisus* (Grun.) Hust., *Rhizosolenia hebetata* Gran, *Thalassiothrix longissima C.* et Grun., *Denticula semina* Simonsen et Kanaya.

All species in this complex are of planktonic oceanic forms. In cores collected next to continental or insular shoals the complex may be supplemented by various species, mostly of arcto-boreal neritic and sublittoral kinds, namely: *Melosira sulcata* Kuts., *Thalassiosira gravida* Cl., *Th.hyalina* Gran, *Th.nordenskioldii* Cl., *Porosira gracialis* Jorg, *Biddulphia aurita* Breb.et Godey, *Chaetoceros* spp. (in sediments merely in the form of spores), *Fragilaria oceanica* Cl., *Fr. cylindrus* Gran.

The complex mentioned is characteristic of the upper sediment layer corresponding to Holocene above the thermal optimum (Plate 31.1).

B. South-boreal, moderate warm-water complex in the sediments related to the post-glacial period of the thermal optimum: *Thalassiosira pacifica* Gran et Angst., *Th. decipiens* Jorg., *Stephanopyxis nipponica* Gran et Jendo, *Coscinodiscus asteromphalus* var. *centralis* Grun., *C. radiatus* Ehr., *C. stellaris* Roper (Fig. 31.4) (Plate 31.2).

This sediment layer with a characteristic diatom complex indicating a temperature rise during the post-glacial epoch occurs directly beneath sediments containing north-boreal cold-loving diatoms. Sediments corresponding to the postglacial thermal optimum in the area near Kamtchatka reveal, in some cores, a thickness reaching 50 centimetres. In the central areas its thickness drops to 10-15 centimetres, and as a rule this layer is distinguished by an abrupt increase in diatom numbers (stations 4112, 4110, 4066, 4068 and 3252). E.A. Romankevich *et al.* (1966) regard this layer as an interstadial one.

C. North-boreal diatom complex with separate arctic elements is observed in sediments occurring below the Holocene horizon. All the moderate cold-loving elements of the *Denticula semina* type decrease in this complex, whereas the role of cold-loving oceanic species increases: *Coscinodiscus curvatulus* Grun., *C. marginatus* Ehr., *C. oculus-iridis* Ehr., *Rhizosolenia hebetata* f. *hiemalis* Gran. This is associated with the simultaneous appearance of different neritic arcto-boreal species: *Thalassiosira gravida* Cl., *Biddulphia aurita* Breb.et Godey, and *Chaetoceros* spores (*Ch. furcellatus, Ch. debillis, Ch. holsaticus, Ch. mitra,* etc.), *Fragilaria* spp. Thus, horizon II differs in the essential presence of a greater or smaller number of arcto-boreal neritic species. The closer to shore, the more frequent is this element. Somewhere in the mid-part of the horizon II it is possible to single out a layer distinguished by the presence of *Rhabdonema arcuatum* var. *ventricosa* Grun. set against a background of a total numerical increase in other species. At station 4112 the *Rhabdonema* layer is placed between 62-72 centimetres, at station 4066 it occurs at a depth of 70-90 centimetres, and at station V20-119 it is situated between 79-99 centimetres. The *Rhabdonema ventricosum* zone is rather consistent and is traced in most cores collected north of 40° N. However, in remote areas south of 40° N, e.g. V20-130 (36° 59′ N, 152° 36′ E) *Rhabdonema* seems to have failed to propagate (Kanaya and Koizumi 1966). The *Rhabdonema* layer is rich, apart from diatoms, with remains of other organisms, for example: silicoflagellates and radiolarians. It separates the sediments of the horizon II into two almost equal halves and seems to correspond with a period of extensive interstadial rise in temperature (Formdel), which according to American data developed 24-28 thousand years ago within the Wisconsin glacial epoch (Flint 1965). The cores collected from the northern Pacific showed the presence of a layer corresponding to the Formdel interstadial with a thickness of about 20-25 centimetres.

The fifth diatom complex observed in the Holocene-upper Pleistocene sediments ought, according to the number of warm water species, to be ascribed to the last interglacial epoch (horizon III). It is characterized by moderate warm-water species, *Coscinodiscus radiatus* Ehr., *Thalassiosira decipiens* Jorg., *Thalassionema nitzschioides* Grun., *Rhizosolenia styliformis* Bright. In the northern ocean areas *Denticula semina,* which is a moderate cold-loving oceanic species, predominates. Unlike the diatom complex in the sediments of the post-glacial thermal optimum, the interglacial complex is devoid of *Stephanopyxis nipponica, Coscinodiscus stellaris* and

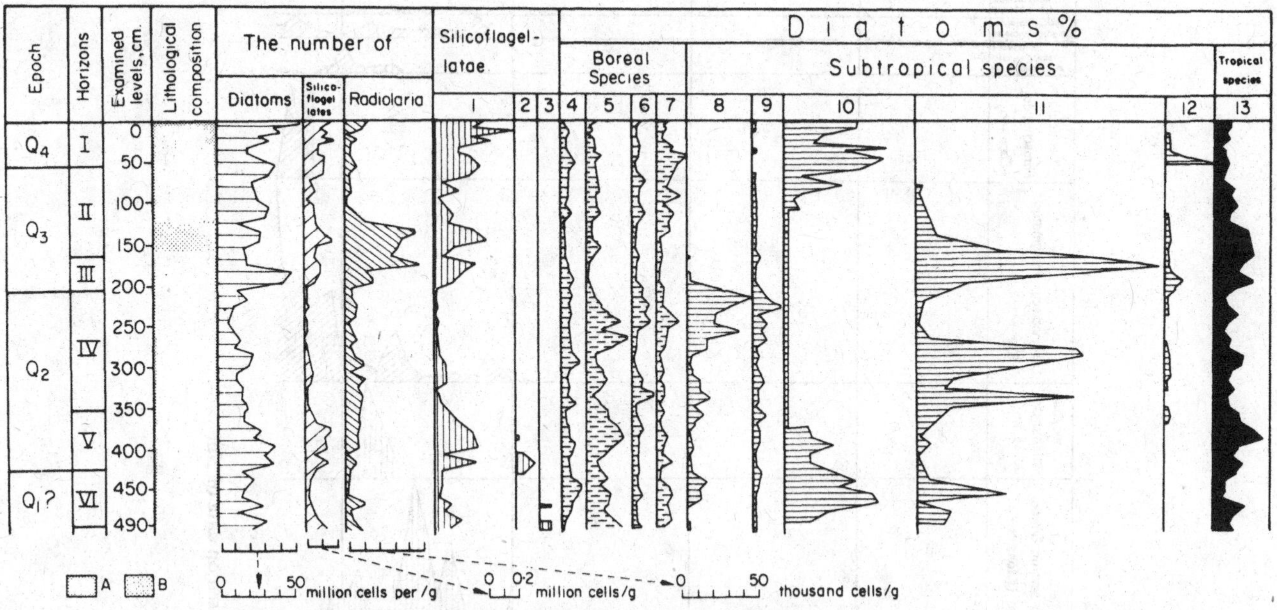
Epoch
Horizons
Examined levels, cm.
Lithological composition
The number of
Diatoms
Silicoflagellates
Radiolaria
Silicoflagel-latae
Diatoms %
Boreal Species
Subtropical species
Tropical species
Q4
Q3
Q2
Q1?
I
II
III
IV
V
VI
A
B
million cells per /g
million cells/g
thousand cells/g

Fig. 31.5. Station 3155. Vertical distribution of some characteristic forms. A. Clay; B. Aleurite clay. Silicoflagellates: 1. *Dictyocha fibula* Ehr. 2. *D. octonarius* (Ehr) Defl.; 3. *D. octonarius* var. *polyactis* (Jorg.) Gleser; Diatoms; 4. *Coscinodiscus marginatus* Ehr.; 5. *Denticula semina* Simonsen et Kanaya; 6. *Rhizosolenia hebetata* Gran; 7. *Thalassiosira excentrica* (Ehr.) Cl.; 8. *Coscinodiscus Wailesii* Gran A. Angst; 9. *Rhizosolenia styliformis* var. *latissima* Bright; 10. *Thalassiosira decipiens* Jorg; II. *Th. Oestrupii* (Ostenf.) Proch.-Lavr.; 12. *Pseudoeunotia doliolus* Grun.; 13. All tropical species.

some other species. It is worth noting that the interglacial complex is always distinguished by a mixed ecological composition, unlike the uniform cold water 'glacial' one.

The diatoms of these Holocene and upper Pleistocene sediments all fall within the limits of the Pacific Ocean boreal diatom flora. The observed differences in composition are attributable to migration and shifting of boundaries separating areas: towards the south during the colder epochs, and towards the north during the warmer epochs.

3. *Phylogenetic variations* in the diatom assemblages become quite obvious as soon as the lower boundary of the Upper Pleistocene is crossed (i.e. in cores below horizon III). Variations resulting from phylogenetic development of the entire group are expressed in the dying out of individual species and by the introduction of new forms. During the relatively short Quaternary history of the Pacific Ocean some species become extinct and disappear from the diatom assemblages: *Rhizosolenia curvirostris* Jousé, *Actinocyclus oculatus* Jousé, *Coscinodiscus Wailesii* Gran A. Angst (Fig. 31.5), others evolve, chiefly as a result of evolutionary morphological variations within the existing species (Plate 31.3). It has so far been impossible to follow the phylogenetic evolution of diatoms throughout the entire Pleistocene, on the evidence of existing core material (Jousé 1963). But, owing to the acquisition of core V20-119 from Lamont Geological Observatory this may soon be remedied. Core V20-119 (Fig. 31.6) allows the recognition of six floristic complexes. Each of them reveal dominant and accessory species. There are doubtless some new species which will be described shortly. In the lower parts of the core numerous ancestral forms of recent species are encountered, which differ from them morphologically. As we have done earlier, we intend to distinguish them as *fossilis,* although in a number of cases it would be more appropriate to rank them at a species level. The two lower floristic complexes refer to two successive Pliocene diatom complexes. The first of them is ascribed to the interval 1,165-1,075 centimetres. The guide species of this complex are: *Coscinodiscus marginatus* Ehr. (f. *fossilis*?), *Denticula kamtschatica* Zabelina, *Asteromphalus* aff. *robustus* Roper, *Stephanopyxis turris* (Arnott) Ralfs.

The second diatom complex is characteristic of the sediments between 1075 and 805 centimetres. It is populated by such mass species as: *Actinocyclus oculatus* Jousé, and *Denticula semina* Simonsen et Kanaya *fossilis* Jousé. The assemblage includes numerous Tertiary species, particularly *Thalassoisira: Th. Zabelinae* Jousé, *Th. nidulus* (Temp. et Brun) Jousé, *Th. gravida* f. *fossilis* Jousé, *Th. undulosum* (Mann.), *Chaetoceros incurvus* Bail., *Ch. cinctus* Gran (spores). The numerical maximum of diatoms occurs in this core interval. At a depth of 945 centimetres the valve count amounted to 33 million per gramme. Here one also finds a dominance of silicoflagellates (Fig. 31.7). Numerical abundance is intimately

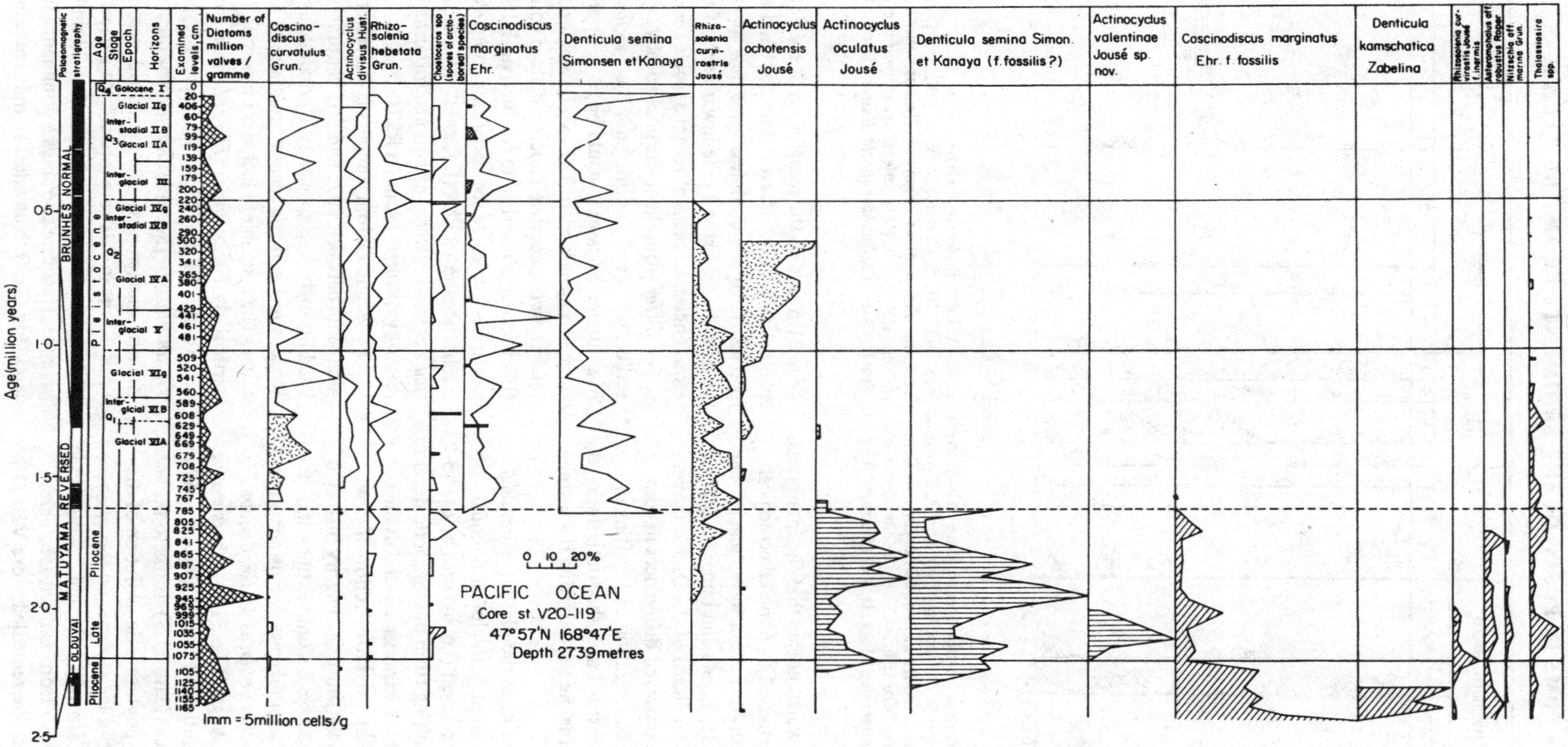
Palaeomagnetic stratigraphy
Age
Stage
Epoch
Horizons
Examined levels, cm
Number of Diatoms million valves / gramme
Coscino-discus curvatulus Grun.
Actinocyclus divisus Hust.
Rhizo-solenia hebetata Grun.
Chaetoceros spp (spores of arcto-boreal species)
Coscinodiscus marginatus Ehr.
Denticula semina Simonsen et Kanaya
Rhizo-solenia curvi-rostris Jousé
Actinocyclus ochotensis Jousé
Actinocyclus oculatus Jousé
Denticula semina Simon. et Kanaya (f. fossilis ?)
Actinocyclus valentinae Jousé sp. nov.
Coscinodiscus marginatus Ehr. f. fossilis
Denticula kamschatica Zabelina
Rhizosolenia cur-virostris Jousé f. inermis
Asteromphalus aff. robustus Roper
Nitzschia aff. marina Grun.
Thalassiosira spp.
BRUNHES NORMAL
MATUYAMA REVERSED
OLDUVAI
Pleistocene
Pliocene
Late
Golocene I
Glacial IIg
Inter-stadial IIB
Glacial IIA
Inter-glacial III
Glacial IVg
Inter-stadial IVB
Glacial IVA
Inter-glacial V
Glacial VIg
Inter-glacial VIB
Glacial VIA
Q4
Q3
Q2
Q1
Age (million years)
0·5
1·0
1·5
2·0
2·5
0 10 20%
PACIFIC OCEAN
Core st. V20-119
47°57'N 168°47'E
Depth 2739 metres
1mm = 5 million cells/g

Fig. 31.6. Station V20-119. Vertical distribution of some characteristic diatoms. (Palaeomagnetic stratigraphy based on data of Ninkovich, Opdyke, Heezen and Foster 1966.)

connected with a greater species variety.

The third complex (805-629 cm) is the least significant one. It apparently contains no dominant forms; all species occur in more or less equal numbers. It appears to reveal features of a transitional type and indicates the beginning of some new evolutionary stage. The complex is distinguished by some species that are ancestral to Recent ones; for example, *Coscinodiscus* aff. *curvatulus* Grun., *Thalassiosira* aff. *excentrica* Cl., etc. An entire *Actinocyclus* group lacking distinctive morphological characteristics is encountered. The lower boundary of this interval shows a mass occurrence of *Denticula semina* which is hardly distinguishable from the Recent one.

The fourth diatom complex (629-441 centimetres) is characterised by a maximum occurrence of *Rhizosolenia curvirostris* Jousé which became extinct by the end of the middle Pleistocene.

The fifth diatom complex (441-220 centimetres) is distinguished by the presence of *Actinocyclus ochotensis* Jousé and its numerous forms. It also contains *Rh. curvirostris,* albeit in an obviously regressive form; it is encountered at the upper boundary of occurrence in the form of a single specimen.

The sixth diatom complex (220-0 centimetres) is formed of Recent Pacific species of the boreal area.

The most ancient complex does not seem to contain any cold-loving elements. Only a single occurrence is recorded in the second complex. Above the second complex one notes an increase in cold-loving and a gradual extinction of warm-loving forms. One more aspect of the

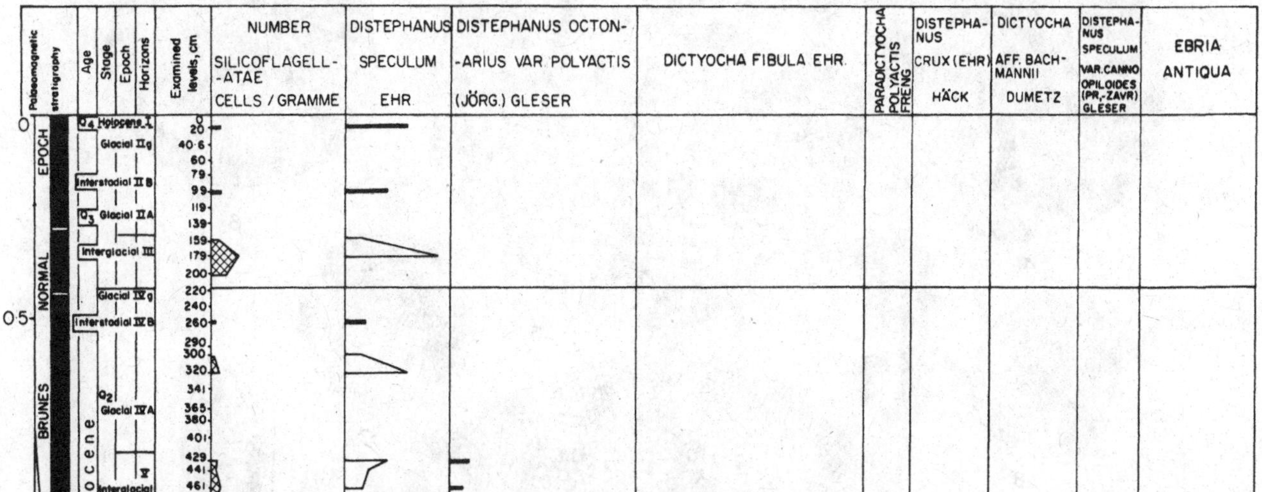
Palaeomagnetic stratigraphy
Age
Stage
Epoch
Horizons
Examined levels, cm
NUMBER SILICOFLAGELL-ATAE CELLS / GRAMME
DISTEPHANUS SPECULUM EHR
DISTEPHANUS OCTONARIUS VAR POLYACTIS (JÖRG.) GLESER
DICTYOCHA FIBULA EHR.
PARADICTYOCHA POLYACTIS FRENG
DISTEPHANUS CRUX (EHR) HÄCK
DICTYOCHA AFF. BACHMANNII DUMETZ
DISTEPHANUS SPECULUM VAR. CANNOPILOIDES (PR.-ZAVR) GLESER
EBRIA ANTIQUA
Age (million years)
NORMAL
BRUNES
EPOCH
REVERSED
MATUYAMA
OLDUVAI
Pleistocene
Late Pliocene
Holocene
Glacial
Interstadial
Interglacial
PACIFC OCEAN
CORE ST. V20-119
47°57'N 168°47'E
DEPTH 2739 metres
0
640
220
10³

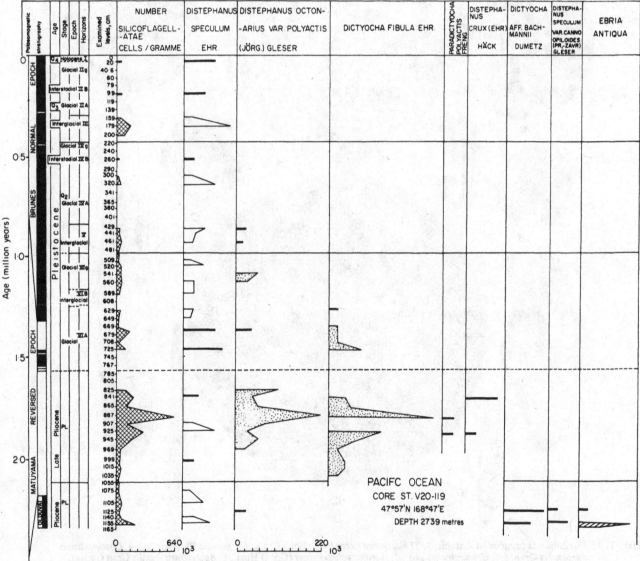

Fig. 31.7. Station V20-119. Vertical distribution of some characteristic Silicoflagellatae. (Palaeomagnetic stratigraphy based on data of Ninkovich, Opdyke, Heezen and Foster 1966.)

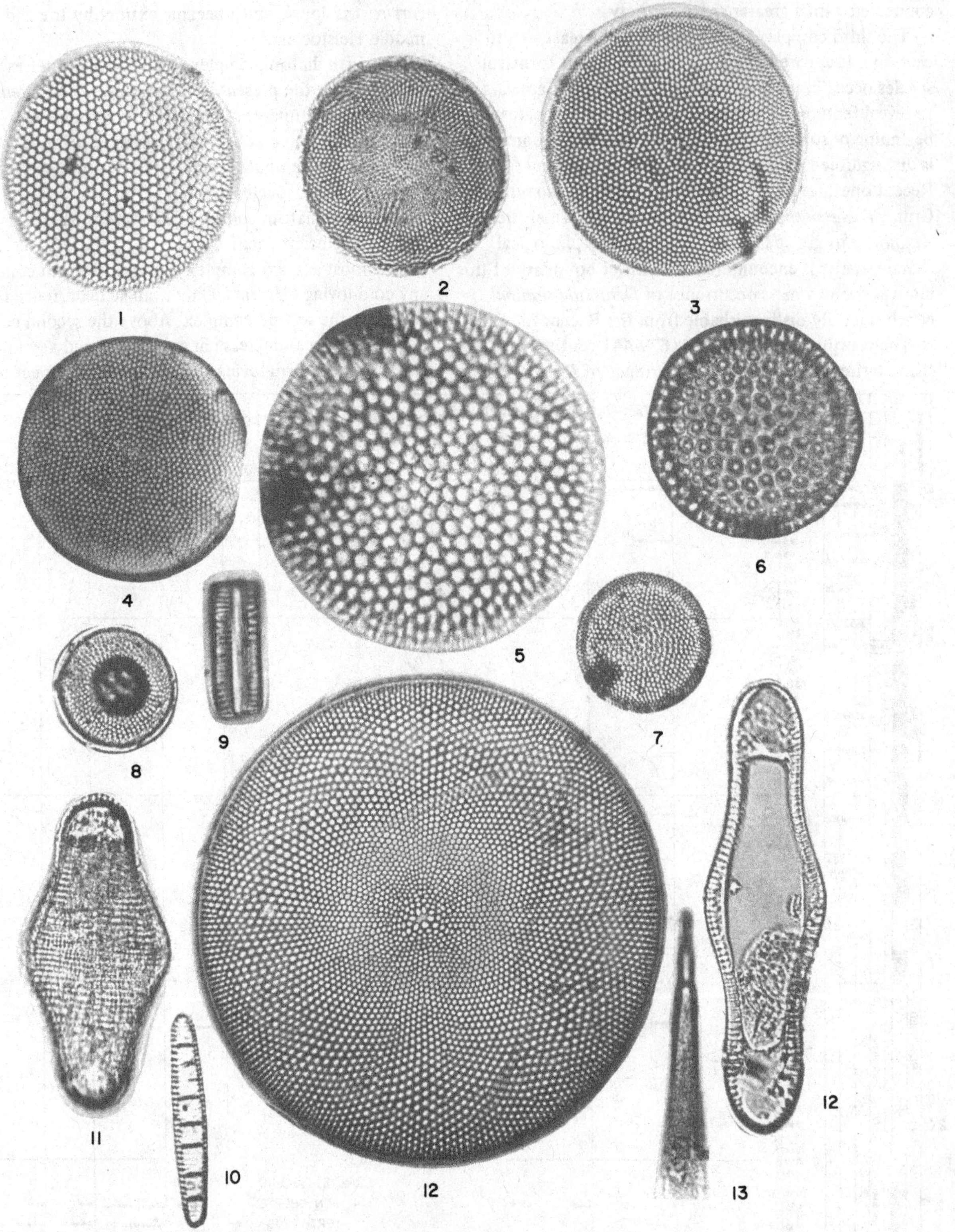
1
2
3
4
5
6
7
8
9
10
11
12
12
13

Plate 31.1. North-boreal complex of diatoms. 1. *Thalassiosira excentrica* (Ehr.) Cl.; 2. *Th. gravida* Cl. (spora); 3-4. *Coscinodiscus curvatulus* Grun; 5-6. *C. marginatus* Ehr.; 7. *Actinocyclus divisus* (Grun) Hust.; 8. *Bacterosira fragilis* Gran (spora); 9-10. *Denticula semina* Simonsen a. Kanaya; 11-12. *Rhabdonema arcuatum* var. *ventricosa* Grun.; 13. *Rhizosolenia hebetata* Grun; 14. *Coscinodiscus oculus-iridis* Ehr. magn. x 1,000.

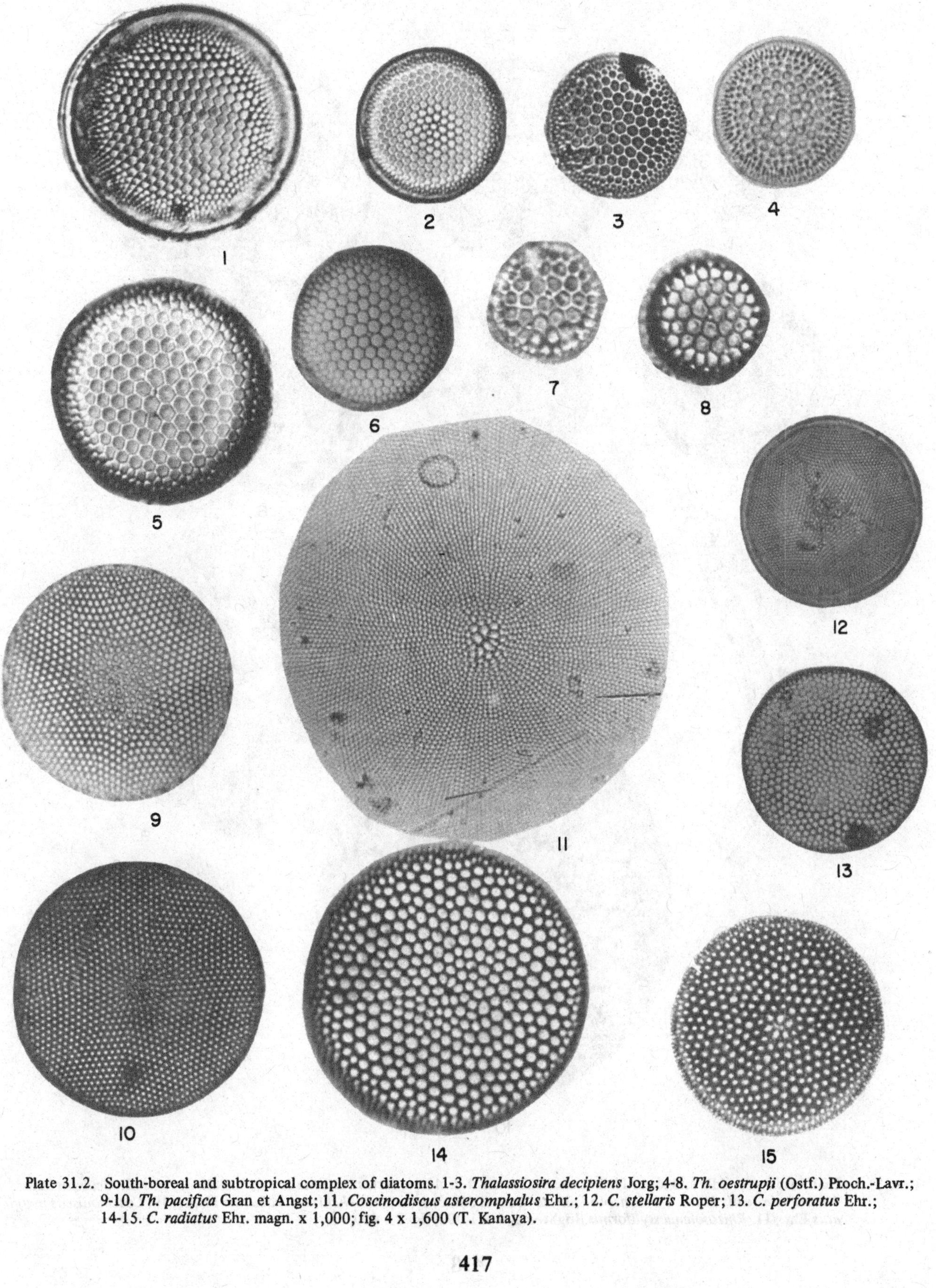

Plate 31.2. South-boreal and subtropical complex of diatoms. 1-3. *Thalassiosira decipiens* Jorg; 4-8. *Th. oestrupii* (Ostf.) Proch.-Lavr.; 9-10. *Th. pacifica* Gran et Angst; 11. *Coscinodiscus asteromphalus* Ehr.; 12. *C. stellaris* Roper; 13. *C. perforatus* Ehr.; 14-15. *C. radiatus* Ehr. magn. x 1,000; fig. 4 x 1,600 (T. Kanaya).

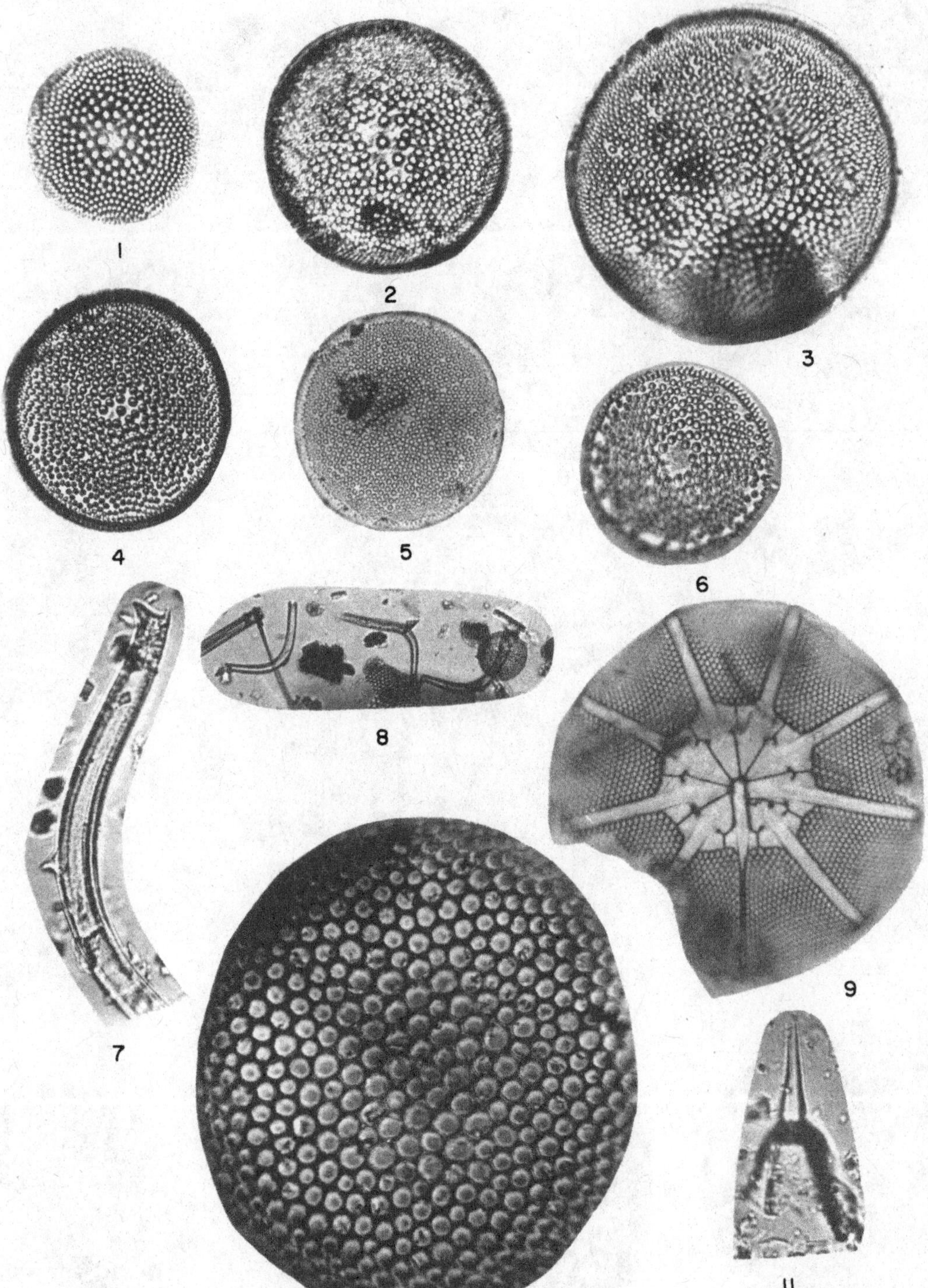

Plate 31.3. Diatoms from middle and lower Pleistocene sediments. 1-2. *Actinocyclus oculatus* Jousé; 3. *Actinocyclus* aff. *oculatus* Jousé; 4-6. *A. ochotensis* Jousé; 7-8. *Rhizosolenia curvirostris* Jousé; 9. *Asteromphalus* aff. *robustus* Roper; 10. *Coscinodiscus marginatus* Ehr.; 11. *Rhizosolenia styliformis* Bright. magn. x 1,000; fig. 8 x 400.

data obtained from the Late Pliocene and Pleistocene is worth noting: the pattern of diatom distribution preserved a zonation similar to the Recent one. During this time the diatom assemblages observed in different climatic zones proved to differ in floristic constituents. The available materials pertaining to the Miocene show a relative similarity of complexes encountered in the boreal, subtropical and equatorial zones of the ocean.

SEDIMENT STRATIGRAPHY

Most of the cores investigated correspond to the Holocene, upper Pleistocene and middle Pleistocene sedimentation cycles. Some (stations 3361, 3155 and 3156) closely approach the lower Pleistocene boundary. Fig. 31.6 of core V20-119 shows the sequence of stratigraphical horizons as well as the boundaries of sections. Within sections (stages) it is possible to distinguish horizons and subhorizons corresponding to glacials, interstadials and interglacials. Their extent is estimated on the basis of the displacement of ecological diatom complexes (also silicoflagellates) reflecting climatic variations (see Fig. 31.6 and Fig. 31.7).

Protracted epochs of low temperature shifted the boundaries of cold-loving diatom complexes and caused a marked expansion of some species south of the Recent boundary. There was a displacement of siliceous sediments by terrigenous ones (excluding station 3155) owing to the reduction of diatom numbers in the surface waters. The farther to the north, the more evident are these changes during periods of cooling. During periods of higher temperature the situation closely resembled the present one. The most characteristic features produced by the onset of colder epochs are observed in diatoms of horizons II and IV, i.e. in the sediments of the last and penultimate glacial periods (Wisconsin/Würm, Illinois/Riss).

The consequences of cooling are less evident in the sediments of horizon VIII, i.e. the one that may be assumed to synchronize with Nebraska or Gunz. The evidence of V20-119, as far as diatoms are concerned, leaves some doubts – some boundaries are indistinct and their features rather inadequate.

Horizons II and IV are subdivided by sediment layers which accumulated under conditions of transient temperature rise. Interstadials within horizons II and IV are distinguished by high numerical maxima, a change of composition caused by warming reminiscent of interglacial conditions. Thus, the tendency to regard some interstadials as independent interglacials is not altogether groundless.

Core V20-119 seems to correspond to the entire sequence of Pleistocene sediments. On the basis of the examination of about sixty diatom samples from along the length of core it seems that there is an unbroken sediment series. Changes in composition occur in a recognizable sequence. Pliocene and late Pliocene (Neo-Pliocene) sediments form the base of the core.

The most awkward problem arising at this point is connected with tracing the Plio-Pleistocene boundary. It is widely known that this problem has given rise to numerous investigations and views as to where the boundary ought to be traced vary greatly (Markov and Velichko 1967). For instance, the boundary line in the Plio-Pleistocene sediments of cores 58 and 62, initially established by Arrhenius (1952) has not yet been completely resolved (Flint 1965; Muhina, this volume).

CORRELATION OF THE RESULTS OF DIATOM STUDIES AND PALAEOMAGNETIC DATA

V20-119 and three other cores collected in the northern part of the Pacific Ocean have allowed the determination of the change of sign of their magnetic properties. Data relating to this problem will be found in a paper by Ninkovich, Opdyke, Heezen and Foster (1966). Palaeomagnetic epochs and events were established earlier for the continents where sedimentation has not been as continuous as might be expected in the oceans. According to Opdyke, Glass, Hays and Foster (1966) sediments obtained from some abyssal cores from the Southern Ocean contain an entire, or almost entire, record of the Earth's magnetic history over 3.5 million years.

The absolute age of magnetic reversals in sediments has been established by applying the K-A method to dry land rocks (Cox, Doell and Dalrymple 1963, 1965; Chramov and Sholpo 1967). The duration of the first epoch, with normal polarity (Brunhes epoch), amounts to 0.7 million years; the second epoch, (Matuyama epoch), with reversed polarity, is believed to be about 2.5 million years. The second epoch includes two relatively transient events of normal polarity, namely the Olduvai event, which is dated about 1.9 million years ago, and the Jaramillo event dated to about 1 million years ago.

According to Ninkovich *et al.* (1966, fig. 5) V20-119 covers a period which ends during the Olduvai event:

1. 0-638 cm (Brunhes epoch);
2. 638-1100 cm (Matuyama epoch) sediments ranging between 739 and 765 cm belong to the Jaramillo event.
3. 1,100-1,165 cm (core base) belongs to the Olduvai event.

It is important to see how far the data relating to diatom and silicoflagellate distribution in V20-119 compare with the palaeomagnetic stratigraphy. We shall begin by seeking comparisons from the bottom upwards.

The diatom and silicoflagellate complex between 1,165 and 1,075 centimetres is constituted exclusively of Tertiary species. Comparison with data relating to the Pliocene diatom and silicoflagellate floras of Sakhalin and Kamchatka (Jousé 1959, 1961; Glezer 1966; Sheshukova-Poretskaya 1967) allows the complex encountered in the interval 1,165-1,075 centimetres to be ascribed to the late Pliocene. Thus, the Olduvai event ends at the point where the typical Pliocene diatom complex disappears and no characteristic elements such as: *Coscinodiscus marginatus* f. *fossilis* Jousé and *Denticulata kamtshatica* Zabelina can be traced any longer.

The section from 1,075-638 centimetres corresponding with the Matuyama epoch coincides with the existence of two diatom complexes of diverse ages. A more ancient diatom complex is found in sediments between 1,075-800 centimetres, whereas a younger complex occurs in sediments above 800-638 centimetres. Diatoms and silicoflagellates between 1,075-800 centimetres are mainly of late Tertiary age. Many species are known in the late Pliocene deposits of Kamchatka, Iturup island, etc., but as a whole the diatom complex of this interval has so far been unknown from adjoining land areas and in deep-sea (abyssal) sediments. However, during the thirty-ninth cruise of R. V. *Vitjaz* in the Kurilo-Kamchatka trench, Y.A. Bogdanov discovered consolidated siliceous sediments at a depth of 2,500-6,000 metres on the inside slope of the trench (stations 5603, 5612, 5630, 5636). The diatoms in these sediments proved to be identical to those found between 1,075-800 centimetres in core V20-119. The similar age of both sediments is obvious. The difference in the diatom and silicoflagellate complexes are limited to customary discrepancies between the oceanic (station V20-119) and neritic (the Kurilo-Kamchatka trench) assemblages. In the second case there is a predominance of *Thalassiosira, Chaetoceros, Biddulphia,* and *Fragilaria* species and in the first *Actinocyclus, Coscinodiscus,* and *Asteromphalus* species prevail. The dominant species in the composition of both complexes is *Denticula semina* f. *fossilis* and *Actinocyclus oculatus.* The entire complex is characterized by the predominance of ancient forms of Recent species such as: *Thalassiosira gravida* f. *fossilis* Jousé (Jousé 1961), *Th. nidulus* (Temp. et Brun) Jousé, *Th. excentrica* Cl. (f. *fossilis*), *Porosira glacialis* Jorg. f. *fossilis,* etc., which existed during the late Pliocene. Simultaneously single species morphologically identical with Recent forms, for example – *Thalassiosira nordenskioldii* Cl., *Bacterosira fragilis* Gran, *Rhizosolenia hebetata* Gran and others, were encountered. Above 725 centimetres Tertiary elements are not found and the diatom complex contains mainly Quaternary species of Recent aspect.

It seems more appropriate to refer to the core interval 1075-800 centimetres as late Pliocene rather than early Pleistocene for the simple reason that the assemblages show a predominance of Pliocene diatom and silicoflagellate species. Thus, deposits between 1,075-638 centimetres ought to be ascribed to different periods, though there is no change in magnetic polarity. Core V20-119 reveals a short normal polarity event (Jaramillo) between 739-765 centimetres. The age of this is about 0.8-0.9 million years.

There are almost no typical Tertiary elements in the diatom assemblages above this layer and, hence, there are serious grounds for ascribing sediments which are younger than the Jaramillo event to the Pleistocene. Therefore the results of diatom studies of V20-119 core materials indicate that the transition from Pliocene to Pleistocene should be placed immediately under the Jaramillo event. Notwithstanding some obvious non-coincidence of polarity boundaries in core V20-119 with those based on diatom stratigraphy, it may be stated that on the whole these run close to each other. The entire coincidence of the Olduvai epoch with the predominance of totally Tertiary diatom assemblages beneath 1,075 centimetres is a matter of especial interest.

It is premature to draw any final conclusions based on the study of a single core and extend them to comparing palaeomagnetic events and variations in diatom and other assemblages. However, there is no doubt that palaeomagnetic stratigraphy combined with palaeontological data should be regarded as a new and most promising method for the study of oceanic sediments.

The change of polarity developed synchronously throughout the entire surface of the earth, and hence there is the possibility of correlating and synchronizing phenomena in different areas of the earth.

Received January 1968.

BIBLIOGRAPHY

Arrhenius, G. 1952. Sediment cores from East Pacific. *Rep. Swedish Deep-Sea Expedition 1947-1948,* 5, 1-227.

Bogorov, B. G. 1967. Biological transformation and exchange of energy and substances in the ocean. *J. Oceanology,* 7 (5), 839-58.

Chramov, A. N. and Sholpo, L. E. 1967. Palaemognetism. *Trudi Vsesoj, N. Issled. Geologorazved. Inst.* **256**, 251.

Cox A., Doell, R. and Dalrymple, G. 1963. Geomagnetic polarity epochs and Pleistocene Geochronometry. *Nature, Lond.* **198**, 1049-51.

Cox, A., Doell, R. and Dalrymple, B. 1965. Quaternary palaeomagnetic stratigraphy. In: *The Quaternary of the U.S.A.* VII Congress IAQR, Princeton Univ. Press. 817-30.

Flint, R. F. 1965. The Pliocene-Pleistocene Boundary. *Geolog. Soc. Amer., spec. paper,* **84**, 497-533.

Glezer, Z. I. 1966. Siliceous flagellate algae (silicoflagellates). In: *Spore plant flora of USSR,* vol. 8. Science Publ. p. 330.

Jousé, A. P. 1959. Basic stages of marine diatom algae flora (Diatoms) in the Far East during the Tertiary and Quaternary periods. *Botanical Jour.* **44** (1), 44-55.

Jousé, A. P. 1961. Marine diatom algae of Miocene and Pliocene age in the Far East. *Botanical materials of the Department of Spore Plants,* vol. **14**. Botanical Institute, Academy of Sciences. 59-70.

Jousé, A. P. 1963. Problems of stratigraphy and palaeogeography in the northern part of the Pacific (diatom analysis data). *J. Oceanology*, **3** (6), 1017-28.

Kanaya, T. and Koizumi, I. 1966. Interpretation of Diatom Thanatocoenoses from the North Pacific Applied to a study of core V20-130. *Sci. Rep. Tohoku Univ., 2nd ser. (Geol.),* **37** (2), 89-130.

Markov, K. K. and Velichko, A. A. 1967. *Quaternary period,* vol. 3. Nedra, Moscow. p. 435.

Ninkovich, D., Opdyke, N., Heezen, B. and Foster, J. 1966. Palaeomagnetic stratigraphy, rates of deposition and tephrachronology in North Pacific Deep-Sea sediments. *Earth and Planetary Science Letters,* **1**, 476-92.

Opdyke, N., Glass, B., Hays, J. and Foster, J. 1966. Palaeomagnetic study of Antarctic Deep-Sea Cores. *Science,* **154** (3748), 349-57.

Romankovich, E. A., Bezrukov, P. L., Baranov, V. I. and Khristianova, L. A. 1966. Stratigraphy and absolute age of deep-sea sediments in western Pacific. (Abstract of investigations along the lines of International Geophysics Projects.) *Oceanology Journ.* **14**, 165.

Sheshukova-Poretskaya, V. S. 1967. Diatoms of Neogene age in the Sakhalin and Kamchatka area. *Leningr. Univers.* p. 327.

32. PROBLEMS OF DIATOM AND SILICOFLAGELLATE QUATERNARY STRATIGRAPHY IN THE EQUATORIAL PACIFIC OCEAN

V. V. MUHINA

INTRODUCTION

With the aim of biostratigraphical elucidation of bottom sediments in the tropical region of the Pacific Ocean, the specific composition and quantitative distribution of diatoms and silicoflagellates was studied in 972 samples from twenty-eight deep-sea cores (Fig. 32.1, Table 32.1).

Sediment samples collected during the twenty-sixth, twenty-ninth, thirty-fourth and thirty-sixth cruises of R.V. *Vityaz* during 1957-65, as well as samples from two Swedish Deep-Sea Expedition cores and one Capricorn core from the collections of Scripps Institution of Oceanography, were used in the study.

Cores were collected from different ocean floor situations from weakly siliceous, calcareous-siliceous, calcareous and deep-sea red clay areas (Bezrukov, Lisitzin, Petelin and Skorniakova, 1966).

The core lengths ranged from 2 to 14.6 metres. The treatment of sediments and calculation of siliceous organisms has been carried out according to a standard method adopted at the Institute of Oceanology, USSR Academy of Sciences (Jousé, Kozlova and Muhina, 1969. By this method the calculation of siliceous organisms is always carried out for a definite weight with subsequent recalculation per gramme of dry sediment. Calcareous sediments have been subjected to additional treatment with 10% hydrochloric acid and, thus a sediment enriched in remains of siliceous organisms is obtained.

For the purpose of singling out biostratigraphical horizons in the cores the following characteristics were taken into account:

1. Variation in the substance and granulometric composition of the sediments.
2. Variation of diatom and silicoflagellate number (as calculated per gramme of dry sediment).
3. Phylogenetic variation of diatom and silicoflagellate species composition.
4. Variation of percentage correlation of diatom ecological complexes reflecting different temperature conditions in the surface waters during the course of sedimentation.

DISTRIBUTON OF BOTTOM SEDIMENTS

Ecological complexes were singled out on the basis of diatom distribution in trigger weight samples and in the surface layer of bottom sediments of the Pacific Ocean (Jousé, Kozlova and Muhina 1967).

Different geographical zones reveal the dominance of diatom complexes which differ according to their temperature requirements.

In the tropical area two ecologically distinguishable diatom complexes (thanatocoenoses) are found: the equatorial and the subtropical.

The equatorial diatom complex is characterized by the predominance of the following species: *Asteromphalus imbricatus* Wall., *Asterolampra marylandica* Ehr., *Actinocyclus ellipticus* var. *lanceolata* Grun., *A. ellipticus* v. *elongata* Grun., *Coscinodiscus africanus* Janisch, *C. nodulifer* A.S. *C. crenulatus* Grun., *Ethmodiscus rex* Wall., *Hemidiscus cuneiformis* Wall., *Nitzschia marina* Grun., *Planktoniella sol*., *Rhizosolenia bergonii* Per., *Triceratium cinnamomeum* Grev.

The subtropical complex shows a predominance of: *Pseudoeunotia doliolus* (Wall.) Grun., *Roperia tesselata* (Roper) Grun., *Thalassiosira lineata* Jouse, *Th. oestropii* (Ostf.), *Thalassionema nitzschioides* Grun., *Thalassiosira parva* Heiden, and *Nitzschia* (*N. sicula* (Castr.) Hust., *N. bicapitata* Cl., *N. braarudii, N. kolaschezkii).* Together with some south-boreal species: *Coscinodiscus asteromphalus* Ehr., *C. radiatus* Ehr., *Actinocyclus ehrenbergii* Ralfs, *Actinoptychus bipunctatus, Rhizosolenia styliformis* Bright, which are encountered rarely in this complex.

For the purpose of tracing the boundaries of complexes in the sediments of each station the percentage species content of similar ecological groups was calculated.

Representatives of the equatorial complex comprise

about 50-80 per cent of the total diatom composition throughout the entire series of stations investigated. The diatoms of the subtropical complex exhibit a subordinate position (in most of the stations they amount to less than 15-20 per cent). Only in the narrow equatorial zone, in the area of upwelling, where the temperature at the surface is reduced to 25-23.5° C, does the content of subtropical diatom species sometimes increase to 35-40 per cent.

It is possible to single out the following areas of distribution on the basis of diatom and silicoflagellate content: 1. Red deep-sea silts and calcareous sediments, and 2. calcareous-siliceous and weakly siliceous sediments.

1. Cores from red deep-sea silts and calcareous sediments were collected in the western part of the tropical area (13° N-14° S; 142° E-176° W) and in the eastern part, north of 12° N and south of 5° S.

Ten sediment cores from this area (5069, 5126, 5128, 5145, 5110, 3918, 3921, 5103, 5100, Cap. 8) contained an insubstantial number of diatoms and silicoflagellates (up to a maximum of six million valves, most frequently in surface layers), or were devoid of them.

This inadequacy of diatoms and silicoflagellates precludes, with rare exceptions, their subdivision. The paucity of remains of siliceous organisms in red deep-sea silts is a consequence of the low biological productivity of surface waters in these areas. In calcareous sediments siliceous material is strongly diluted by carbonate.

2. Cores collected in the area of weakly siliceous and calcareous-siliceous sediments range from 12° N to 5° S and between 180-120° W.

The near-equatorial belt is one of the three areas of the Pacific where there are at present favourable conditions for the accumulation of siliceous sediments. Lisitzin (1966) emphasizes that in this particular area an increased content of amorphous silica is noted in the surface layer of sediment (5-10 per cent, and in some sites 10-30 per cent).

STRATIGRAPHICAL DISTRIBUTION

Stratigraphical and palaeogeographical investigations were based on the study of thirteen cores. Cores from stations 5071, 5082, 5098, 5117, 62, 58, 3802 and 3797 collected in a narrow near-equatorial zone, contained a considerable content of diatoms and silicoflagellates throughout their entire length. The diatom number varied within limits ranging from 0.06 to 25.0 million valves per gramme in

Table 32.1. List of stations

Stations	Latitude and Longitude			Depth (m)	Length of cores (cm)	Samples examined	Type of sediment
5065	16° 23′ N	146° 36′ W	Pacific	5,363	340	33	red clay
5069	13° 55′ N	140° 36′ W		4,993	280	27	red clay
5128	12° 57′ N	176° 06′ W		5,090	545	12	red clay
5126	11° 17′ N	154° 07′ W		5,070	240	21	red clay
5145	8° 01′ S	175° 57′ W		5,520	800	82	red clay
5110	12° 59′ S	154° 06′ W		5,222	208	22	red clay
3918	0° 32′ N	142° 07′ E		3,164	252	16	calcareous
3921	0° 55′ S	142° 30′ E		3,075	288	18	calcareous
5100	7° 08′ S	140° 13′ W		4,076	240	25	calcareous
5103	11° 08′ S	140° 10′ W		4,136	260	27	calcareous
Cap 8	13° 39′ S	174° 58′ E		–	762	39	calcareous
SDSE 58	6° 44′ N	129° 28′ W		4,440	950	6	calcareous-siliceous
5082	5° 57′ N	139° 58′ W		4,830	805	75	calcareous-siliceous
5117	0° 03′ N	154° 14′ W		4,727	730	59	calcareous-siliceous
SDSE62	3° 00′ S	136° 26′ W		4,511	1460	36	calcareous-siliceous
5098	5° 03′ S	139° 56′ W		4,400	337	38	calcareous-siliceous
5071	12° 10′ N	140° 37′ W		4,900	373	41	siliceous
5074	10° 30′ N	140° 01′ W		4,850	300	36	siliceous
5080	8° 06′ N	139° 52′ W		5,090	199	23	siliceous
5124	7° 55′ N	153° 41′ W		5,050	894	90	siliceous
5133	5° 58′ N	176° 04′ W		5,370	960	70	siliceous
3797	2° 01′ N	172° 32′ W		5,328	300	26	siliceous
3802	3° 17′ S	172° 52′ W		5,329	266	21	siliceous
5113	5° 01′ S	154° 15′ W		5,046	290	39	siliceous
5315/1	8° 20′ S	80° 22′ E	Indian Ocean	5,162	345	44	siliceous
5315/6	8° 13′ S	80° 32′ E		5,218	382	21	siliceous
5315/7	8° 25′ S	80° 33′ E		5,024	113	16	siliceous
5315/8	8° 24′ S	80° 33′ E		5,150	35	9	siliceous

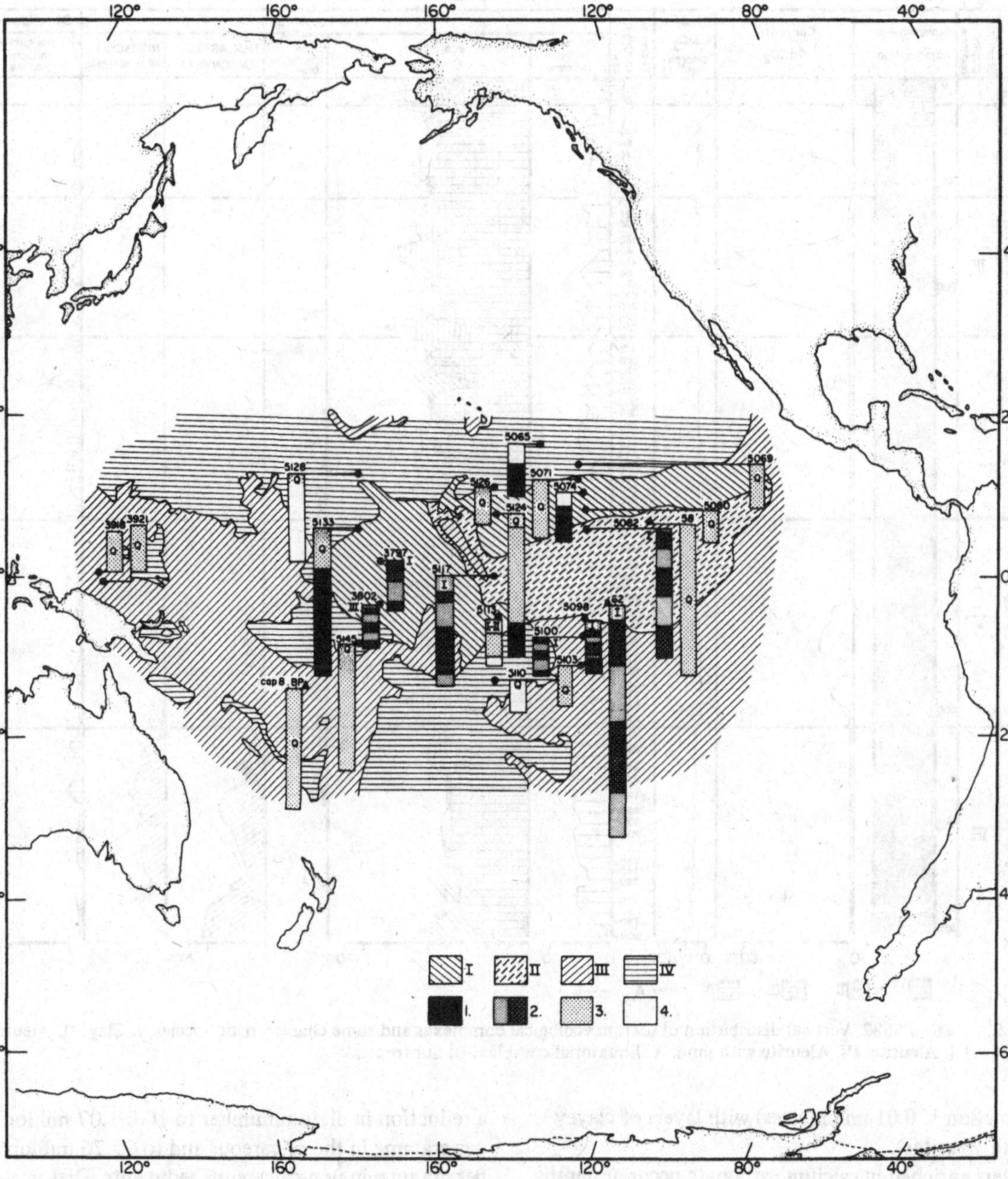
120°
160°
160°
120°
80°
40°
40°
20°
0°
20°
40°
60°
I
II
III
IV
1.
2.
3.
4.

Fig. 32.1. **Biostratigraphy of bottom sediments from the tropical Pacific. I. Siliceous sediments. II. Calcareous-siliceous. III. Calcareous. IV. Red-clay. 1. Pre-Quaternary. 2. I–VIII biostratigraphical horizons. 3. Without stratigraphy. 4. Diatoms absent.**

the radiolarian-silicoflagellate sediments and from 10.0 to 53.4 million valves per gramme in the diatomaceous-radiolarian silts. Silicoflagellates were encountered in similar amounts and ranged from 0.02 to 7.0 million specimens per gramme.

Data on the stratigraphy of sediments at station 5082 are given below.

Core 5082 (Fig. 32.2) was obtained by a corer of large diameter in the area of a northeastern trench. The bottom relief has the appearance of a hilly plain. Sediments are represented by radiolarian-foraminiferal and diatomaceous-radiolarian aleuritic-clayey silts (50.5-69.3 per

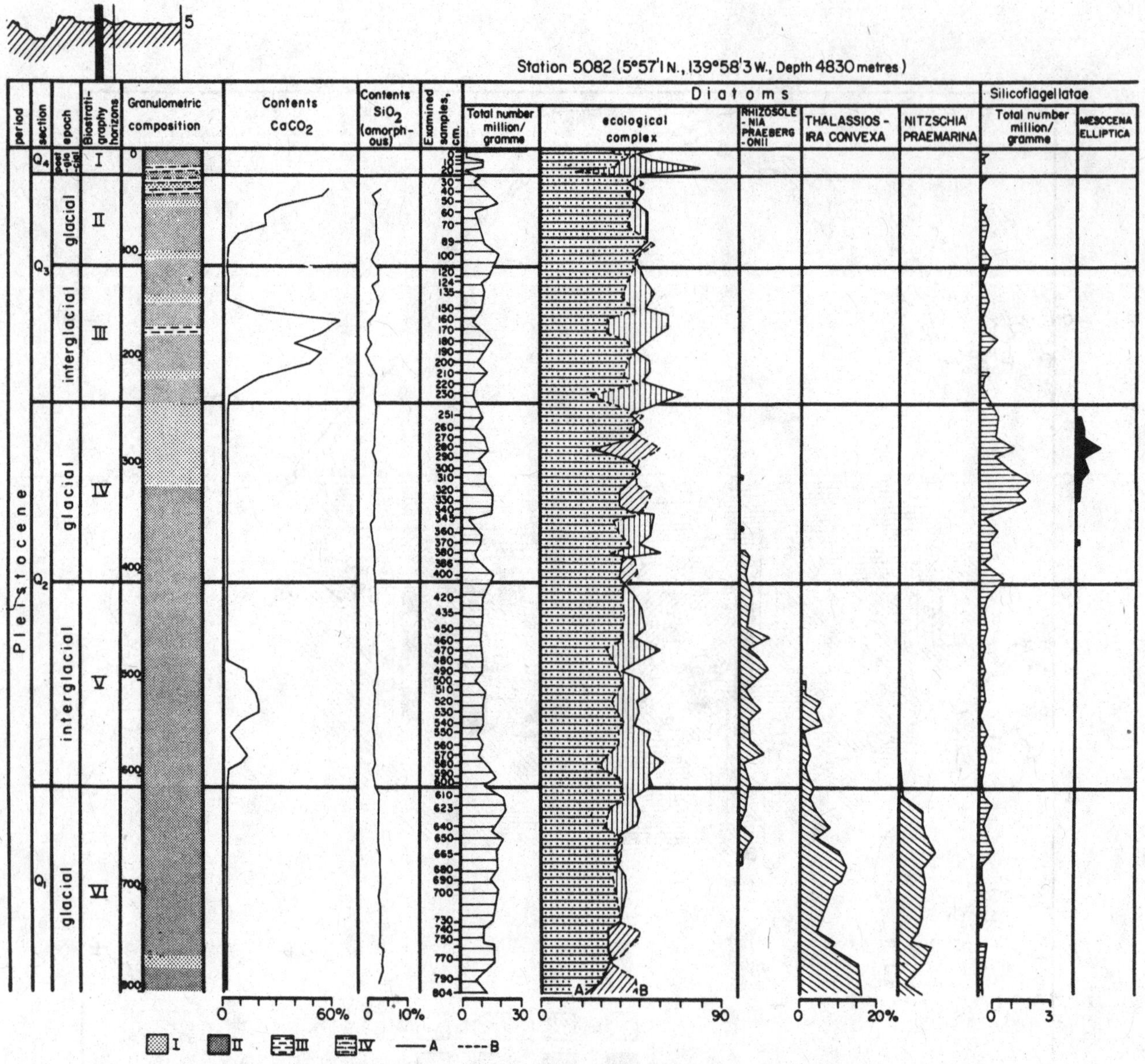
Station 5082 (5°57'I N., 139°58'3 W., Depth 4830 metres)
period
section
epoch
Biostratigraphy horizons
Granulometric composition
Contents CaCO2
Contents SiO2 (amorphous)
Examined samples, cm.
Diatoms
Total number million/gramme
ecological complex
RHIZOSOLENIA PRAEBERGONII
THALASSIOSIRA CONVEXA
NITZSCHIA PRAEMARINA
Silicoflagellatae
Total number million/gramme
MESOCENA ELLIPTICA
Pleistocene
glacial
interglacial
60%
10%
30
90
20%
3
I
II
III
IV
A
B

Fig. 32.2. Station 5082. Vertical distribution of diatom ecological complexes and some characteristic species. I. Clay. II. Aleuritic-clay. III. Aleurite. IV. Aleurite with sand. A. Equatorial complex. B. Subtropical.

cent fraction < 0.01 millimetres) with layers of clayey silts and aleurites).

Layers enriched in calcium carbonate occur at depths of 34-80 centimetres (up to 52.06 per cent), at 150-214 centimetres (up to 59.13 per cent) and at 493-583 centimetres (up to 19.1 per cent). The content of amorphous silica in the cores varies only slightly, between 5.47 and 11.13 per cent.

In almost all samples examined diatoms amount to over 50 per cent total siliceous organisms. The maximum number of diatoms was 35.4 million valves per gramme per 0-2 centimetre layer. The lower core portion showed a reduction in diatom number to 10.0-1.07 million valves per gramme in the calcareous and to 21.76 million valves per gramme in non-calcareous sediments. Diatoms are represented by equatorial and subtropical species; the proportions of these two groups varies along the length of the core. This is related to alternations of epochs of warm and cold climate. The 380-680 centimetre layer contains *Rhizosolenia praebergonii* Muhina up to 0.5-8.2 per cent. At a depth of 500-804 centimetres *Thalassiosira convexa* Muhina is found (0.7-16.7 per cent). At 590-804 centimetres *Nitzschia praemarina* (Jousé) occurs (0.6-9.4 per cent).

Ancient diatoms are encountered along almost the entire length of the core: *Aulacodiscus amoenus, Coscinodiscus paleaceus, C. lanceolatus, C. aeginensis, C. lewisianus, C. jabei, C. marginatus* f. *fossilis, C. exavatus, Asterolampra decora, A. grevilei, Denticula lauta.* These species are undoubtedly of a different age and are reworked into the sediment. Their total amount ranges from 0.2 to 6.0 per cent.

Silicoflagellates amount to 0.08-1.6 million valves per gramme. They are represented by three species. *Dictyocha fibula* is the prevailing species and is encountered along the entire length of the core. *Distephanus speculum* is present as single specimens from a depth of 251 centimetres to the bottom of the core. *Mesocena elliptica* was calculated to range from 0.16 to 1.28 million specimens per gramme in a layer at 251-330 centimetres. At depths of 450, 580, 650 and 755 centimetres single specimens of redeposited Tertiary silicoflagellates: *Naviculopsis biapiculata, Dictyocha crux* and *D. triacantha* were found.

According to the diatoms and silicoflagellates the deposits of core 5082 can be divided into six biostratigraphical horizons: I. 0-25 centimetres, II. 25-110 centimetres, III. 110-240 centimetres, IV. 240-410 centimetres, V. 410-605 centimetres, VI. 605-804 centimetres.

Horizon I (25 centimetres thick). Radiolarian-foraminiferal aleuritic-clayey silt.[1] Number of diatoms varies from 1.6 to 13.4 million valves per gramme. Equatorial diatom species prevail in this complex. Its maximum (83%) is observed in the 18-20 centimetre layer which corresponds to the postglacial climatic optimum. The number of subtropical diatoms in this layer is reduced to 17 per cent.

Horizon II (85 centimetres thick). Radiolarian-foraminiferal sediments of the upper part of this horizon pass into diatomaceous-radiolarian sediments in its lower part. Calcium carbonate content varies from 52 to 0.5 per cent. Amorphous silica varies from 5.9 to 10.5 per cent. The total number of diatoms increases to a maximum of 19.5 million valves per gramme, and the occurrence of subtropical species (up to 50%) increases in parallel. The 60-90 centimetre layer is marked by a more warm-water loving diatom composition than the higher and lower layers.

Horizon III (130 centimetres thick). Radiolarian-foraminiferal aleuritic-clayey silt. At a depth of 160 centimetres the maximum calcium carbonate content (59.1 per cent) in the entire core is observed. Amorphous silica varies from 5.5 to 10.6 per cent. The diatom number (6.6-16 million valves per gramme) is on average somewhat lower than in horizon II. Equatorial species amount to a maximum of 73.2 per cent and the complementary decrease in subtropical species is marked.

Horizon IV (170 centimetres thick). Non-calcareous sediment. Diatomaceous-radiolarian clayey and aleuritic-clayey silt. Calcium carbonate content varies from 0.2 to 0.8 per cent, and amorphous silica ranges from 8.3 to 10.2 per cent. There is a small increase in the total diatom number with a maximum up to 17.1 million valves per gramme. The flora is of a more cold-loving type than in horizon III. The subtropical diatoms in some layers comprise up to 60 per cent.

Horizon IV is characterized by the presence of *Mesocena elliptica.* This is characteristic of all sediments of this age in the Pacific Ocean, and is observed in all cores of the tropical area which accumulated during the penultimate glaciation. The lower part of the horizon is marked by the extinction of *Rhizosolenia praebergonii.*

Horizon V (195 centimetres thick). Diatomaceous-radiolarian aleuritic-clayey silts in the upper part of the horizon are replaced by radiolarian-foraminiferal sediments in its lower part. Calcium carbonate content increases to a maximum and reaches 19.1 per cent. Amorphous silica varies from 8.1 to 10.3 per cent. In diatom number sediments of the fifth horizon differ little from sediments of the fourth horizon. Equatorial diatoms (up to 62.5 per cent) prevail over the subtropical forms. The fifth horizon is characterized by the maximum occurrence (up to 8 per cent) of *Rhizosolenia praebergonii.* The upper limit of occurrence of *Thalassiosira convexa* runs through this horizon.

Horizon VI (199 centimetres thick). Diatomaceous-radiolarian aleuritic-clayey essentially non-calcareous silts (calcium carbonate ranging from 0.1 to 1.0 per cent). Amorphous silica varies from 9.5 to 11.1 per cent. The number of diatoms increases considerably and amounts to a maximum of 21.8 million valves per gramme with increased occurrence of subtropical species (up to 50.8 per cent). This horizon is characterized by the first occurrence of *Rhizosolenia praebergonii* in the core and a massive development of *Thalassiosira convexa* (up to 16 per cent). The upper boundary of horizon VI coincides with the extinction of *Nitzschia praemarina.*

Finally, it is possible to conclude that the core did not pass through the Pleistocene deposits – an inference based on the diatom and silicoflagellate assemblages. Phylogenetic variations in the diatoms and silicoflagellates indicate, according to our data, the lower and mid-Pleistocene period.

The Pleistocene series of deposits is also not passed through by other cores from the equatorial Pacific Ocean (stations 5071, 5117, 58, 3797). This testifies in favour of relatively high rates of sedimentation in the equatorial

[1] Information on the material and granulometric composition of sediments has been kindly provided by Dr. A. P. Lisitzin

area of the Pacific, in its eastern part especially, where relief forms are smoother and where calcareous-siliceous sediments are being accumulated. The contents of cores from stations 3802 and 5098 on the other hand, indicate that the Pleistocene has been completely traversed. These cores reveal sediments corresponding to the earliest cooling of the climate (in our case horizon VII).

In other cores from the equatorial belt it was possible to distinguish from four to eight biostratigraphical horizons which have been formed under different paleogeographical conditions during the Pleistocene and Holocene.

Horizons II, IV, VI and VII, the sediments of which accumulated during periods of ocean surface-water cooling are compared with the four glaciations established in the case of North America. A number of cores (stations 5098, 5082, 62) in sediments of horizon II permit the recognition of a stage of relative warming which corresponds to an interstadial. Sediments of horizons III, V and VII, as shown by diatom analysis, accumulated during periods of warming of surface waters, and are thought by the author to be synchronous with the interglacial periods. Sediments of horizon I were deposited during the Post-glacial or Holocene period.

Cores 5098, 5082 and 5117 show in horizon I a more warm-loving layer, according to its diatom composition, corresponding to maximum warming of the climate during the Post-glacial.

Cores from stations 5113, 5124, 5133, beyond the limits of the narrow equatorial zone, exhibit a change in the type of sediments; siliceous sediments are substituted by red deep-sea clays. The upper and lower layers of these cores are rich in the remains of siliceous organisms and their middle parts (red clays) contain practically no diatoms and silicoflagellates. Here one finds only sporadic occurrences of single and mostly ancient, poorly preserved, redeposited diatoms.

The farther away from the equatorial zone the station happens to be, the greater the intermediate layer of red deep-sea clays devoid of organic remains. For example, core 5113 shows that this stratum reaches a thickness of 125 centimetres. It seems most probable that sediments of this interval accumulated during the period of maximum cooling (the middle and greater part of the late Pleistocene). The lower part of this core (210-300 centimetres) according to its diatom and silicoflagellate content resembles deposits of horizon VI.

At station 5133 the barren series of red clays reaches a thickness of about 250 centimetres and at station 5124 the figure attains 610 centimetres. The intermediate stratum of red deep-sea clays in these two cores undoubtedly corresponds to a more lengthy intermediate period of time than at station 5113, and possibly corresponds to the greater part of the Pleistocene.

Core 5133 (Fig. 32.3) was obtained from the slope of an underwater elevation in an area of abyssal hills. It is composed of essentially non-calcareous, diatomaceous-radiolarian aleuritic-clayey silts with clayey silt layers between 268 and 925 centimetres. Calcium carbonate content amounts to 0.01-1 per cent. The amorphous silica varies in the 0-234 centimetre layer from 1.36 to 4.83 per cent; below 268 centimetres its content increases to 9.18-18.09 per cent.

It is possible to distinguish three distinct strata on the basis of the amounts of diatoms and silicoflagellates: the upper one (0-10 centimetres) with maximum diatom content 15.6 million valves per gramme; the middle one (10-260 centimetres) almost entirely devoid of siliceous algae (0-0.38 million valves per gramme); the lower (260-925 centimetres) with a diatom maximum of up to 69.4 million valves per gramme. As to the composition of diatom and silicoflagellate flora it is appropriate to single out two strata: the Pleistocene (0-260 centimetres) and the Pliocene (260-925 centimetres).

Sediments in the range 0-260 centimetres are characterized by Recent diatom species. The 0-170 centimetre layer exhibits a single occurrence of species of Tertiary age. At a depth of 170-260 centimetres the diversity and number of Tertiary species greatly increases. The Tertiary diatoms are mostly poorly preserved and are undoubtedly of a different age. The greater part of the latter are most likely secondarily deposited. The core interval between 260-925 centimetres is characterized by a most varied floristic composition by means of which the strata may be subdivided into three layers: an upper (260-380 centimetres), a middle (380-790 centimetres) and a lower (790-925 centimetres).

According to the diatoms and silicoflagellates the sediments of the 260-380 centimetre layer are not homogeneous. Recent species comprise 10-20 per cent of the diatoms. A number of Recent diatoms: *Coscinodiscus africanus, Thalassiosira lineata, Planktoniella sol, Roperia tesselata, Pseudoeunotia doliolus* are absent. Marked morphological deviations, compared with Recent forms, are observed in *Coscinodiscus nodulifer, C. crenulatus, Nitzschia marina, Rhizosolenia bergonii, Hemidiscus cunieformis.* This group of species, together with species known for certain back to the Miocene (i.e. *Thalassionema nitzschioides* and its varieties, *Asterolampra marylandica*) comprise a total of 40-60 per cent of the total diatom composition. A number of Tertiary species (7.9-20 per cent): *Nitzschia pliocena, Coscinodiscus lanceolatus, C. paleaceus, C. vigilens, Actinocyclus ellipticus* type and *A. ellipticus* var. *moronensis, Asterolampra moronensis, A. decora, A. Van Heurkii, Hemialus* sp., *Rouxia* sp. are

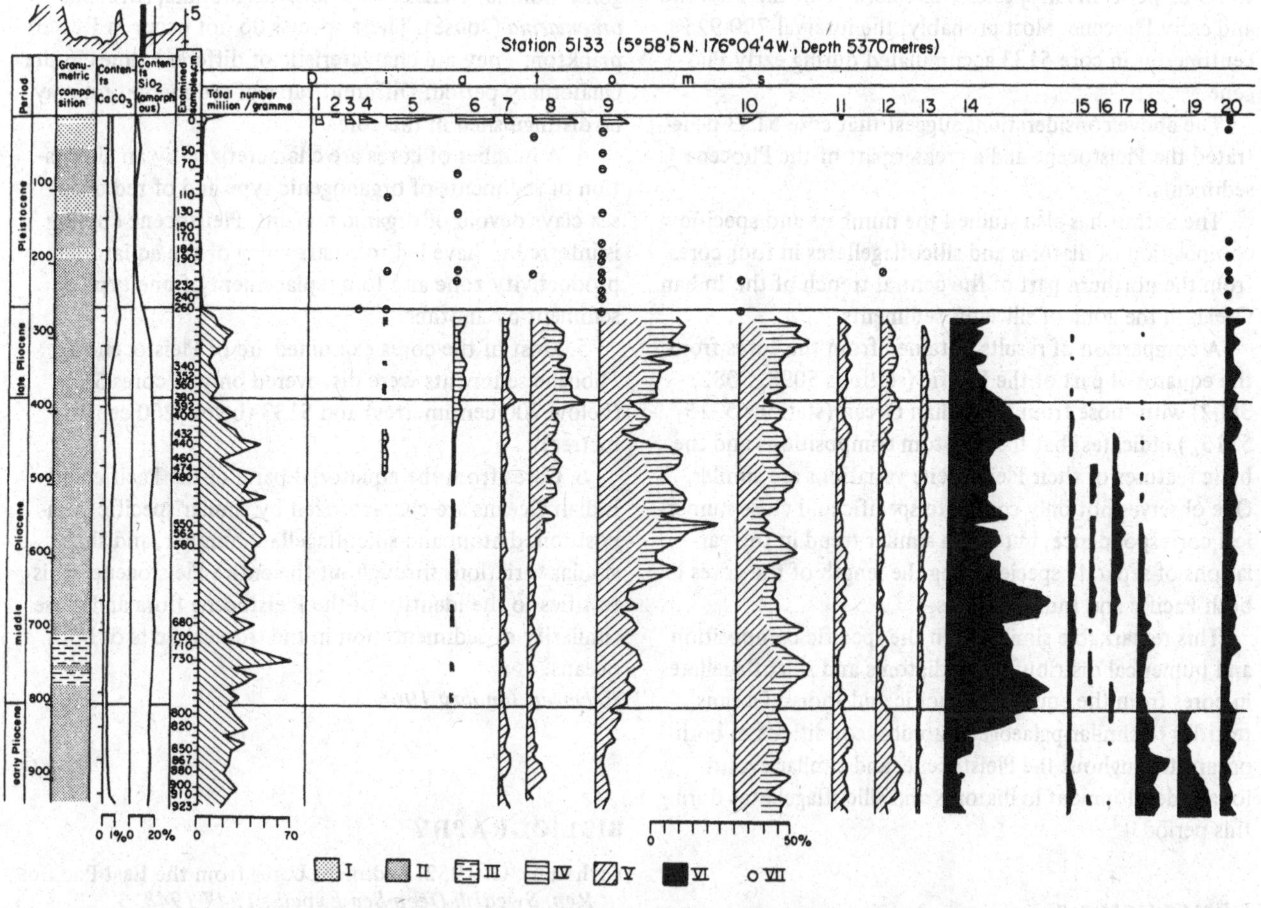

Fig. 32.3 Station 5133. Vertical distribution of some characteristic species of diatoms. I. Clay. II. Aleuritic-clay. III. Aleurite. IV. Pre-Quaternary diatoms. V. Recent species. VI. Tertiary species. VII. Single diatoms. 1. *Coscinodiscus africanus*, 2. *Roperia tesselata*, 3. *Thalassiosira lineata*, 4. *Pseudoeunotia doliolus*, 5. *Rhizosolenia bergonii*, 6. *Nitzschia marina*, 7. *Hemidiscus cuneiformis*, 8. *Coscinodiscus nodulifer*, 9. *Thalassionema nitzschioides*, 10. *Th. parva*, 11. *Triceratium*, 12. *Actinocyclus ellipticus*, 13. *Asterolampra marylandica*, 14. *Nitzschia pliocena*, 15. *Coscinodiscus paleaceus*, 16. *C. aeginensis*, 17. *Thalassiosira marujamica*, 18, *Coscinodiscus jabei*, 19. *Denticula hustedtii*, 20. Tertiary complex.

represented in the same layer.

The accumulated data allows us to maintain that sediments of the 260-380 centimetre layer of core 5133 correspond to the late Pliocene. On the basis of its floristic composition this layer is similar to the lower stratum (700-890 centimetres) of core 5124 which the author also regards as late Pliocene. Sediments between 380-790 centimetres are distinguished by their floristic composition from the layer above by the absence of Recent diatoms, increased occurrence of *Thalassionema* (up to 40-50 per cent), increased number and occurrence (up to 40-50%) of species typical of the Pliocene: *Nitzschia pliocena*, *Coscinodiscus paleaceus*, *C. aeginensis* (Kolbe 1957). Apart from this up to 7-10 per cent of species not encountered in sediments above 380 centimetres are present. *Actinocyclus tcugaruensis*, *Coscinodiscus jabei*, *C. marginatus fossilis*, *C. exavatus*, *Thalassiosira* aff. *marujamica* (Muhina). In literature (Sheshukova-Poretskaya 1967, Kanaya 1959) these species are known from the Miocene.

Among silicoflagellates, in the middle part of the core, one observes *Dictyocha fibula*, *D. fibula* f. *ausonia* and *D. fibula* f. *mutabilis*, all known from Tertiary sediments (Bachmann and Ishikava 1962).

According to its diatom and silicoflagellate complexes the author therefore relates the 380-790 centimetre layer to the middle Pliocene.

The number of Pliocene species is less in the lower layer (790-925 centimetres) including the *Thalassionema* forms (a total of up to 60-75 per cent). *Denticula hustedtii* Simonsen et Kanaya, characteristic of the Pliocene-Miocene complex (Simonsen and Kanaya 1961) occurs amongst the diatoms. There is a simultaneous increase (up to 20-25 per cent) in species characteristic of the Miocene and early Pliocene. Most probably, the interval 790-925 centimetres in core 5133 accumulated during early Pliocene.

The above considerations suggest that core 5133 penetrated the Pleistocene and a greater part of the Pliocene sediments.

The author has also studied the numbers and specific composition of diatoms and silicoflagellates in four cores from the northern part of the central trench of the Indian Ocean in the zone of siliceous sediments.

A comparison of results obtained from the cores from the equatorial part of the Pacific (stations 5098, 5082, 3802) with those from the Indian Ocean (stations 5315_1, 5315_6) indicates that their diatom compositions and the basic features of their Pleistocene variations are similar. One observes not only complete specific and close numerical correspondence, but also a similar trend in the variations of separate species along the length of the cores in both Pacific and Indian Oceans.

This remarkable similarity in the specific composition and numerical distribution of diatoms and silicoflagellates in cores from the equatorial Pacific and Indian Oceans testifies to similar palaeogeographical conditions in both oceans throughout the Pleistocene, and similar evolutionary development in diatoms and silicoflagellates during this period.

CONCLUSIONS

1. Diatoms and silicoflagellates are found in considerable amounts only in cores from the tropical Pacific Ocean which were obtained from areas of accumulation of weakly siliceous and calcareous-siliceous sediment. These types of sediment correspond to the productive zones of the ocean. Cores obtained from areas characterized by red deep-sea clays, are practically devoid of remains of siliceous organisms.

2. The Pleistocene strata reflect changes of complexes and total diatom and silicoflagellate numbers. Their sedimentary maxima coincide with an increase in the number of less warm-loving diatoms, particularly the subtropical forms and a simultaneous regression of equatorial species. This testifies to the fact that the equatorial zone, according to Arrhenius' concept (1952), has exhibited more favourable conditions for mass development of algae during cooler than during warmer periods.

3. The diatom and silicoflagellate composition shows well-featured phylogenetic variation during the Pleistocene. One may trace the appearance (and at times a complete cycle of development) of four species: *Mesocena elliptica* Lemm. (silicoflagellate), *Rhizosolenia praebergonii* Muhina, *Thalassiosira convexa* Muhina, *Nitzschia praemarina* (Jousé). These species do not occur in Recent plankton. They are characteristic of different times in the Quaternary period. Different rates of sedimentation may be distinguished in the cores.

4. A number of cores are characterized by an alternation of sediments of organogenic type and of red deep-sea clays devoid of organic remains. Pleistocene cooling is inferred to have led to a narrowing of the equatorial productivity zone and to a replacement of one type of sediment by another.

5. Most of the cores examined are of Pleistocene age. Pliocene sediments were discovered only in cores 5124 (below 700 centimetres) and 5133 (below 260 centimetres).

6. Cores from the equatorial parts of the Pacific and Indian Oceans are characterized by similar specific composition, diatom and silicoflagellate number, and show similar variations throughout the entire Pleistocene. This testifies to the identity of the Pleistocene flora and close similarity of sedimentation in the tropical belts of both oceans.

Received January 1968.

BIBLIOGRAPHY

Arrhenius, G. 1952. Sediment cores from the East Pacific. *Rep. Swedish Deep-Sea Exped., 1947-1948,* **5**.

Bachmann, A. and Ishikawa, W. 1962. The silicoflagellates in the Wakura beds, Nanao city, prefecture Ichikawa, Japan. *Sc. Rep. Kanazawa Univ.* **8** (1), 161-75.

Bezrukov, P. L., Lisitzin, A. P., Petelin, V. P. and Skorniakova, N. S. 1966. New chart of sediments of the Pacific Ocean, scale 1:25.000.000. *Physico-geographical Atlas of the World.*

Jousé, A. P., Kozlova, O. G. and Muhina, V. V. 1967. Species composition and zonal distribution of diatoms in the surface sedimentary layer of the Pacific Ocean. *Dokladi U.S.S.R. Acad. Sci.* **172** (5).

Jousé, A. P., Kozlova, O. G. and Muhina, V. V. 1969. Diatom algae in the surface sedimentary layer of the Pacific Ocean. *Monograph on the Pacific Ocean.*

Kanaya, T. 1959. Miocene diatom assemblages from the Onnagawa formation and their distribution in the correlative formations in north-east Japan. *Tohoku Univ., Sci. Rep., 2nd. ser. Geology,* **30**, 130.

Lisitzin, A. P. 1966. Basic regularities in the distribution of recent siliceous sediments and their relation with climatic zonation. In *Geochemistry of earth silicon.* Nauka: Moscow.

Sheshukova-Poretskaya, V. S. 1967. Diatoms of the Neogene in the Sakhalin and Kamchatka area. *Leningrad Univ. Publishers.* p. 360.

Simonsen, R. and Kanaya, T. 1961. Notes on the marine species of the diatom genus *Denticula. Kutz. Intern. Rev. Ges. Hydrobiol.* **46** (4).

33. THE VERTICAL AND HORIZONTAL DISTRIBUTION OF PLANKTONIC FORAMINIFERA IN QUATERNARY SEDIMENTS OF THE ATLANTIC OCEAN

M. S. BARASH

INTRODUCTION

The history of research on the distribution of planktonic Foraminifera in Atlantic ocean sediments, covers more than a hundred years. Many oceanologists have studied their distribution both in the surface layer of sediment and in cores. Quantitative methods have only been used in the past few years, commencing with the work of W. Schott (1935), who investigated the distribution of planktonic Foraminifera in the surface layer of sediments in the equatorial Atlantic, and who suggested stratigraphical divisions for cores in this region. Hitherto most investigations have been conducted in warm-water regions, in equatorial, tropical and subtropical zones. The sediments in the north and south Atlantic have been very little investigated from the point of view of the planktonic Foraminifera. But it is in these regions that the clearest reflection of the main stages of the Quaternary climatic changes could be expected.

As has been demonstrated by numerous investigators, the distribution of planktonic Foraminifera in the water, and the distribution of their tests on the bottom, depends chiefly on the temperature of the surface layer of ocean water, in which the overwhelming majority of these organisms live. We have tried to establish quantitative links between the ratio of species and the average annual temperature of the surface water. This provides the possibility of comparing fossil thanatocoenoses of planktonic Foraminifera with recent ones, and evaluating the temperature of the water during the accumulation of one or other horizon in the cores. The precision of palaeotemperature interpretations is determined by the temperature dispersion for a given thanatocoenosis type, being from 1 to 3. Such data have enabled us to plot palaeotemperature curves for several cores from the North Atlantic, which, along with lithological indications have been made the foundation of stratigraphical division. The cores from the central Atlantic areas, where such work had not been done before, were subjected to stratigraphical division on the basis of the change in the ratio between three groups of species; the warm-water group, the cold-water group, and the intermediate or neutral group. The last-named group included species whose quantitative content does not change in the warm-water and cold-water horizons of the cores. The change in the ratio between these three groups of species provides the possibility of plotting palaeoclimatic curves for the cores, which reflect the relative changes in climatic conditions (warming or cooling of the surface water) during the accumulation of the sediment.

In the North Atlantic, between the latitude of the Azores and Iceland, about 120 samples of surface sediment were collected and investigated. For the computation of numbers of tests, grades obtained by mechanical analysis were used, i.e. sand, more than 0.1 millimetres and aleurite, 0.1-0.05 millimetres. Nearly all the mature specimens of most of the species were in the sand fraction. Small juvenile shells, between 0.1 and 0.05 millimetres, do not possess distinguishing specific features, are carried far away by the water after the death of the organisms, and moved by bottom currents along the surface of the bottom concentrating in the small depressions, and are resistant to solution and mechanical destruction. For these reasons it is more expedient to use the grade of more than 0.1 millimetres for palaeontological analysis, as was done by W. Schott for the equatorial Atlantic and by V. G. Morozova and V. P. Androsova (1935) for sediments of northern seas.

The sand grade of each sample was divided until 300-600 specimens were obtained, which were then identified and counted. Then all the grade, usually containing thousands of shells, was examined to discover rare species. The species ratio in per cent, and also the total content 'per gramme of dry sediment' (the 'foraminiferal number') was then calculated.

DISTRIBUTION IN THE SURFACE SEDIMENT LAYER

In the topmost layer of North Atlantic sediments twenty-two species of planktonic Foraminifera were found. The 'foraminiferal number' in most samples ranges between 20,000 and 50,000 specimens. It increases on submarine rises. Near the southern end of the Reykjanes ridge the foraminiferal number reaches 149,000 specimens. The high numbers in this area can be explained as follows:

(1) the high productivity of zooplankton, and in particular of planktonic Foraminifera;

(2) the shallow depths which as a rule do not exceed the critical depth at which calcium carbonate is dissolved;

(3) the wide distribution of *Globigerina* ooze which consists mostly of their shells;

(4) the greatest concentration of species with small shells. The foraminiferal number drops sharply to a few hundred in areas of very intensive accumulation of terrigenous and volcanic material, and at great depths.

In order to establish connections between recent physicogeographical conditions and the distribution of species in the bottom sediment, and also for the stratigraphical analysis of the cores, it was most expedient to make use of species ratios given in percentages, as the number of shells per unit weight or volume depends upon factors which are not directly connected with climatic conditions.

Globigerina pachyderma (Fig. 33.1) is the most common species in North Atlantic sediments. It was found in all the samples in which planktonic Foraminifera were found. In the coldest water, north of the Subarctic Convergence, this species accounts for more than 60 per cent, and in many samples more than 90 per cent of the population. South of this line its relative numbers gradually decrease, and where the average annual temperature of the superficial water is between 15° and 18° C, it is no more than 20 per cent.

Globigerina quinqueloba (Fig. 33.2) is a warmer-water species. Its maximum concentration (55 per cent) was observed near the frontal zone of the Subarctic Convergence, where the water temperature is 8-9° C. In most samples from the central part of the North Atlantic we found 10-30 per cent of this species.

Globigerina bulloides commonly comprises 10-20 per cent of the planktonic Foraminifera. The maximum concentration (about 40 per cent) is also found where the temperature is 8-9° C. However, as a rule this species is more southerly in its distribution than *Globigerina quinqueloba.*

Globigerinita glutinata (Fig. 33.3) comprises 10-30 per cent generally reaching a maximum concentration of more than 40 per cent in the Iceland basin, where the water temperature is about 10° C.

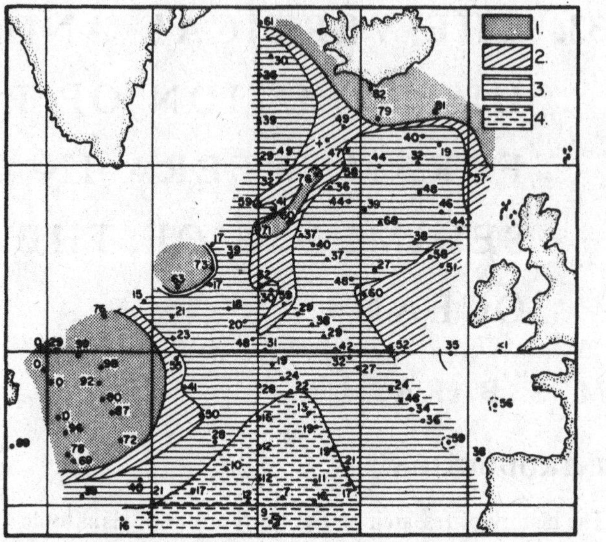

Fig. 33.1. Distribution of *Globigerina pachyderma* (Ehrenberg) in surface sediment samples. The number at each station is per cent of total population of planktonic Foraminifera. (1. more than 60 per cent; 2. 50-60 per cent; 3. 20-50 per cent; 4. less than 20 per cent.)

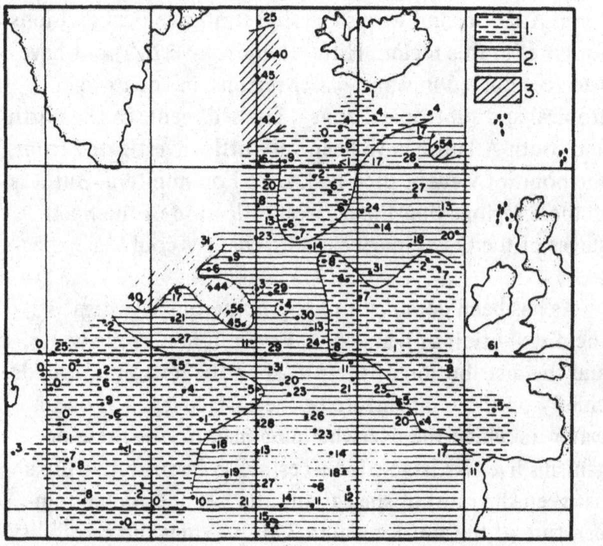

Fig. 33.2. Distribution of *Globigerina quinqueloba* Natland in surface sediment samples. (1. 0-10 per cent; 2. 10-30 per cent; 3. more than 30 per cent.)

Globigerina inflata (Fig. 33.4) accounts for 10-20 per cent where the water temperature ranges from 11° to 17° C, in some samples reaching 40 per cent.

Globorotalia scitula makes up 1-5 per cent in the

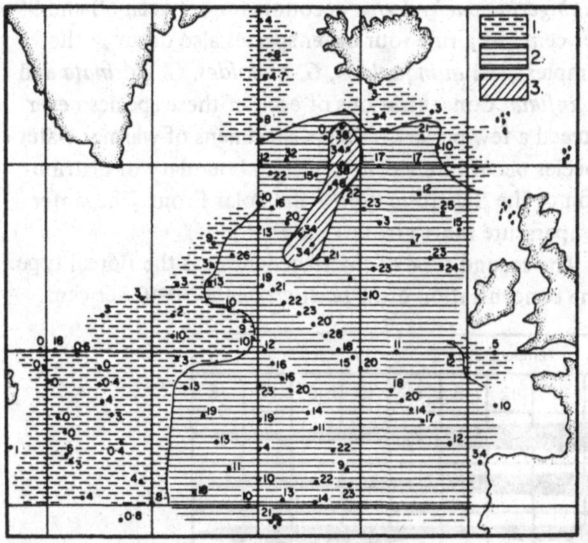
1.
2.
3.

Fig. 33.3. Distribution of *Globigerinita glutinata* Egger in surface sediment samples. (1. 0-10 per cent; 2. 10-30 per cent; 3. more than 30 per cent.)

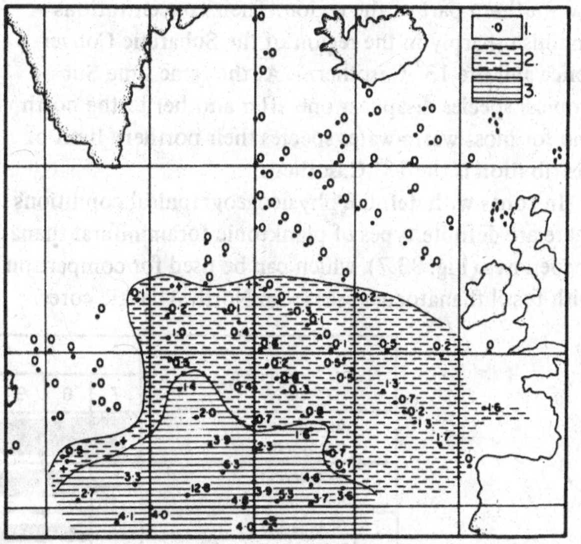
1.
2.
3.

Fig. 33.5. Distribution of *Globigerinoides ruber* d'Orbigny in surface sediment samples. (1. absent; 2. 0-1 per cent; 3. more than 1 per cent.)

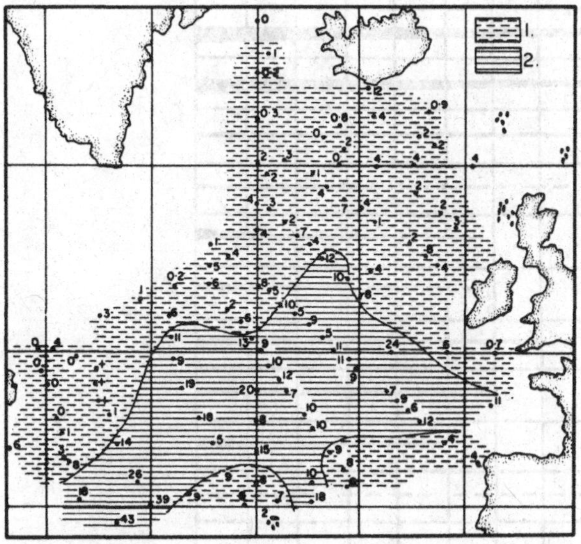
1.
2.

Fig. 33.4. Distribution of *Globigerina inflata* d'Orbigny in surface sediment samples. (1. 0-10 per cent; 2. more than 10 per cent.)

middle part of the North Atlantic, the concentration increasing to 14 per cent to the south.

Globigerinoides minuta is represented in the sand fraction by only a few specimens. However, in the fraction 0.1-0.05 millimetres this species occurs with frequencies of 20-40 per cent. Where the temperature is 8-9° C this species reaches 60-70 per cent. It declines quickly both north and south of this zone.

Globorotalia hirsuta and *G. truncatulinoides* occur with frequencies greater than 1 per cent only in the southern part of the North Atlantic, where the temperature exceeds 13° C. *Orbulina universa* was found in almost all samples, the concentration being usually no more than 1 per cent. Other species of planktonic Foraminifera do not occur in the northern part of the North Atlantic (Fig. 33.5). In the southern part, where the average annual temperature of the surface water is more than 13-14° C, warmer-water species occur rarely, and the more moderate do not exceed a few per cent.

From their distribution in the sediments, the species can be divided into three complexes:

1. The Subarctic complex. This consists of only one species, *Globigerina pachyderma,* which predominates over all other species and comprises the greater part of the planktonic foraminiferal fauna north of the Subarctic Convergence.

2. The Boreal complex: *G. quinqueloba, G. bulloides, G. glutinata, G. minuta. G. pachyderma*, accompanied by Boreal species, amounts to 90 per cent in the Boreal zone of the ocean where the temperature is 9-13° C. In this zone are the maximum concentrations of each of these species.

3. Subtropical complex: Includes all the remaining seventeen species whose maximum concentrations have been observed south of the 13° C isotherm.

The Subarctic and Boreal species extend broadly into the southern part of the region. Their concentrations diminish sharply in the region of the Subarctic Convergence and the 13° C isotherm. At the same time Subtropical species disappear one after another to the north and for most warm-water species their northern limit of distribution is the 13° C isotherm.

In zones with definite physicogeographical conditions there are definite types of planktonic foraminiferal thanatocoenoses (Fig. 33.7), which can be used for comparison with fossil thanatocoenoses from North Atlantic cores.

The first type of thanatocoenosis is the Subarctic type. *Globigerina pachyderma* accounts for between 60 and 99 per cent. As a rule four other species also occur in the samples – *G. quinqueloba, G. bulloides, G. glutinata* and *G. inflata.* Concentrations of each of these species never exceed a few per cent. Single specimens of warmer-water species occur rarely. The southern boundary of distribution of the Subarctic type is the Polar Front. The water temperature ranges from 1° to 9° C.

The second type of thanatocoenosis is the Boreal type. The concentration of *G. pachyderma* is 30-40 per cent.

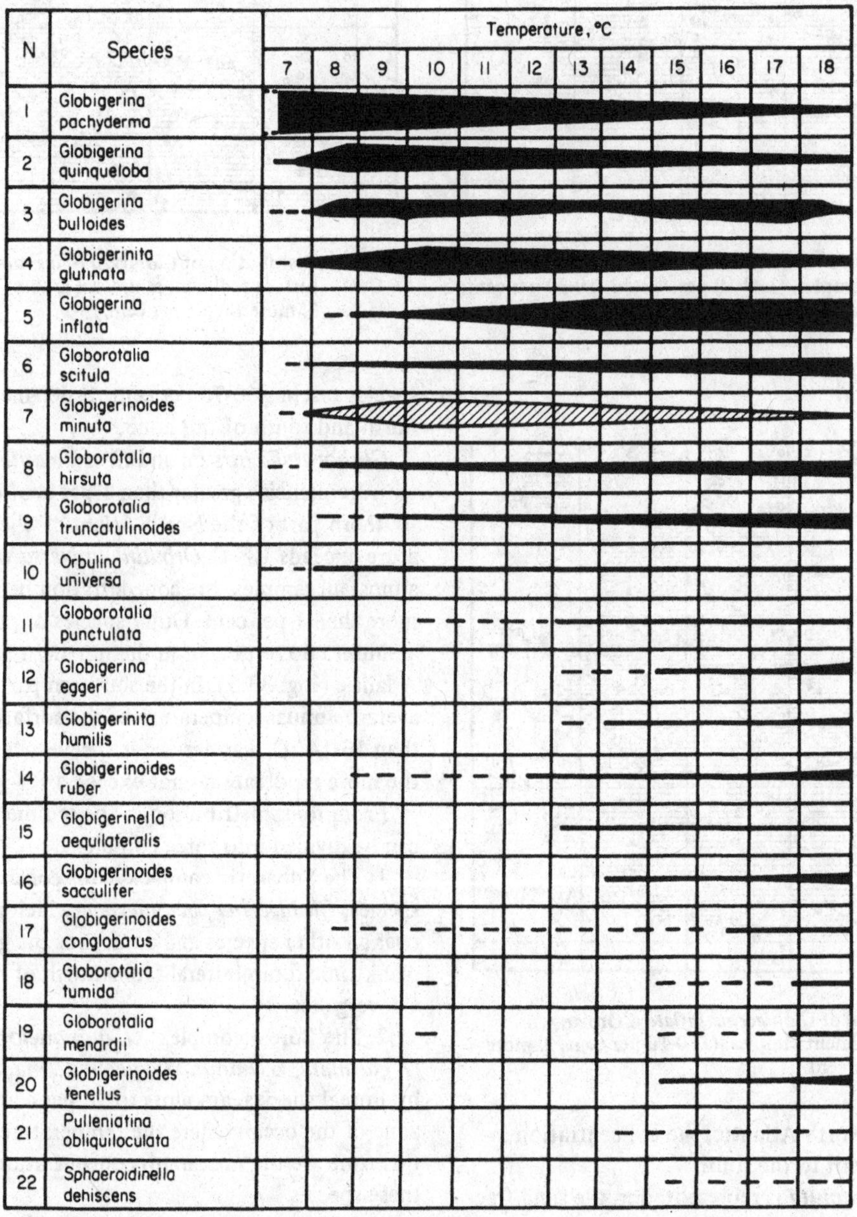
N
Species
Temperature, °C
7
8
9
10
11
12
13
14
15
16
17
18
1 Globigerina pachyderma
2 Globigerina quinqueloba
3 Globigerina bulloides
4 Globigerinita glutinata
5 Globigerina inflata
6 Globorotalia scitula
7 Globigerinoides minuta
8 Globorotalia hirsuta
9 Globorotalia truncatulinoides
10 Orbulina universa
11 Globorotalia punctulata
12 Globigerina eggeri
13 Globigerinita humilis
14 Globigerinoides ruber
15 Globigerinella aequilateralis
16 Globigerinoides sacculifer
17 Globigerinoides conglobatus
18 Globorotalia tumida
19 Globorotalia menardii
20 Globigerinoides tenellus
21 Pulleniatina obliquiloculata
22 Sphaeroidinella dehiscens

Fig. 33.6. Distribution of planktonic foraminiferal species in surface sediments in relation to average annual temperature of the surface water in the North Atlantic. The thickness of the lines corresponds to the abundance of the species. (The size of the shells is larger than 0.1 millimetres for all species, except *G. minuta* for which it is larger than 0.05 millimetres).

The concentration of each of the Boreal species: *G. quinqueloba, G. bulloides, G. glutinata,* ranges between 10 and 40 per cent. The following species also usually occur in small concentrations – *G. inflata, G. scitula, G. minuta, G. hirsuta, G. truncatulinoides, O. universa;* ten in all. In individual samples we occasionally meet several more Subtropical species. The total content of Subtropical species does not exceed 10 per cent. This type of thanatocoenosis is common in the Boreal zone with surface water temperatures ranging from 6° to 13° C, from the Polar Front to the 13° C isotherm (according to Schott 1944).

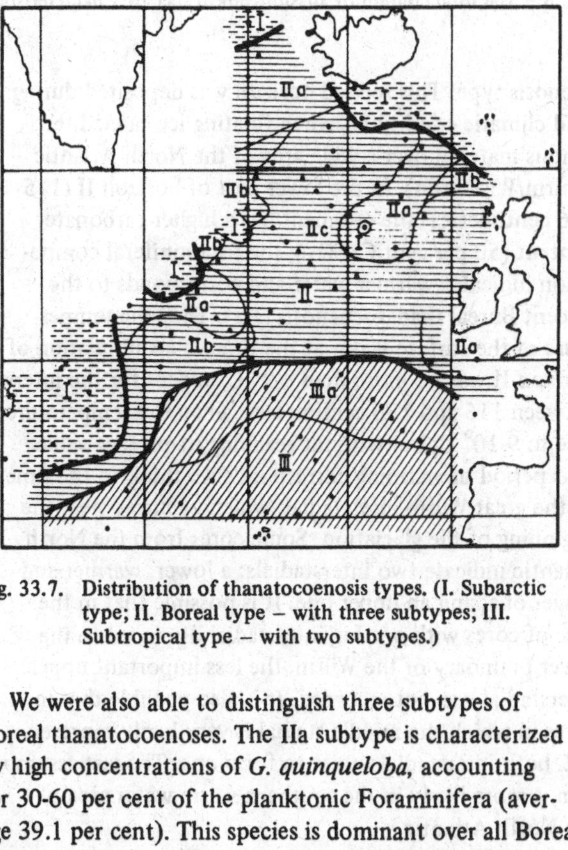
I
IIa
IIb
IIc
II
IIIa
III

Fig. 33.7. **Distribution of thanatocoenosis types. (I. Subarctic type; II. Boreal type – with three subtypes; III. Subtropical type – with two subtypes.)**

We were also able to distinguish three subtypes of Boreal thanatocoenoses. The IIa subtype is characterized by high concentrations of *G. quinqueloba,* accounting for 30-60 per cent of the planktonic Foraminifera (average 39.1 per cent). This species is dominant over all Boreal species, being second only to *G. pachyderma* in several samples. Samples of this sub-type are located south of the Subarctic zone, where the temperature is 9-10° C.

The IIb subtype is characterized by high concentrations of *G. bulloides,* which dominates all other species, yielding in several samples again to *G. pachyderma.* Samples with this subtype of thanatocoenosis are located south of subtypes IIa where typical water temperatures are 10-11° C.

The IIc subtype is characterized by high concentrations of *G. glutinata,* which prevails over all other species, yielding in some samples only to *G. pachyderma.* Samples of this subtype are common, and form an entire subzone IIc, in contrast to the IIa and IIb subzones, which are divided into separate portions. The IIc subzone includes the warmest-water portion of the Boreal zone with average annual temperatures of 9-13° C, typically 11-13° C.

The third thanatocoenosis type is a Subtropical one. The total content of Subtropical species accounts for 20-40 per cent, and samples usually contain fourteen to seventeen species. The Subtropical subzone is characterized by a temperature of 13-18° C. Two subtypes and two subzones conforming to them are distinguished.

The IIIa subtype is characterized by a total concentration of Subtropical species of about 20 per cent (average 21.0 per cent) mainly due to a sharp increase in the concentration of *G. inflata* and *G. scitula.* As a rule fourteen species are present (1-14 as given on Fig. 33.6). Samples of IIIa subtype are concentrated in the area with a temperature of 13-15° C.

The IIIb subtype is characterized by a further increase of the total content of warm-water species to 30-40 per cent, while the number of Subarctic and Boreal species decreases. As a rule seventeen species occur in samples of this subtype. The temperature of the surface water is 15-18° C.

DISTRIBUTION IN QUATERNARY SEDIMENT CORES

The above data along with published results of other investigators on the horizontal and vertical distribution of planktonic Foraminifera in Recent and Quaternary sediments from the Atlantic and other oceans, were used for the stratigraphical division of the cores collected on the R.V. *Lomonosov* and R.V. *Sedov* in the North and Central Atlantic by M.V. Klenova and colleagues.

The following criteria were used to identify stratigraphical horizons:

1. The visual lithological description of the cores.
2. Composition of the planktonic foraminiferal fauna reflecting the average annual temperature of the surface water during accumulation.
3. The calcium carbonate content of the sediment. (The ratio between calcareous and non-calcareous parts of the sediment approximately reflects the ratio between biogenic and terrigenous and volcanic constituents.)
4. The mechanical composition of the sediment (used in several cores).

For the correlation of the horizons between cores and the estimation of their relative age we used the following indications:

1. Succession of horizons ('warm-' and 'cold-water') from the surface of the sediment as datum.

Table 33.1 Types of planktonic foraminiferal thanatocoenosis and average annual temperature of the surface water.

Type	Sub-type	Number of species	Average percentage of main species					Temperature, °C
			1	2	3	4	5	
I. Subarctic	—	5	81.4	6.8	5.9	2.6	3.3	Less than 9
II.	IIa	10	30.5	39.1	16.0	9.5	4.9	9-10
Boreal	IIb	10	41.3	15.2	27.6	7.8	8.1	10-11
	IIc	10	41.8	12.7	10.6	26.0	8.9	11-13
III.	IIIa	14	31.8	15.7	16.5	15.0	21.0	13-15
Subtropical	IIIb	17	17.3	14.4	18.3	13.7	36.3	15-18

1 = *G. pachyderma*, 2 = *G. quinqueloba*, 3 = *G. bulloides*, 4 = *G. glutinata*, 5 = the total content of all Subtropical species. Characteristic species of Boreal subtypes are underlined.

2. Estimation of the degree of warming or cooling during accumulation of one or other horizon according to the ratio of the various species of planktonic Foraminifera by plotting of palaeotemperature and palaeoclimatic curves.
3. Appearance or disappearance of indicator species with a limited vertical distribution in cores from the central area of the ocean (south of 40° N).
4. Similarity of lithological indications of horizons (colour, calcium carbonate content, mechanical composition, presence or absence of terrigenous sand, gravel, pebbles, volcanic ash, etc.).
5. Comparison of results with those of other investigators of Atlantic sediments and with data about Pleistocene time in Europe and North America.

The following results were obtained from specific North Atlantic cores.

Core 198 was obtained near the south slope of the Rockall rise (57° 48′ N, 22° 51′ W), at a depth of 1,200 metres (length 578 centimetres; 558 centimetres dry); the average annual temperature of the surface water is 11-12° C. Five horizons were distinguished.

Horizon I (1-57 centimetres) is represented by white, highly calcareous ooze (60-67 per cent $CaCO_3$). In the surface sample a Boreal thanatocoenosis (subtype IIc) was found and in the middle and lower parts nearly Subtropical warmer-water thanatocoenoses. The sediment of this horizon accumulated in the last warm-water period and represents the Holocene including the Postglacial climatic optimum.

Horizon II (57-166 centimetres). This is a brown-grey mud with terrigenous sand, gravel and volcanic ash. The calcium carbonate content in most samples is 20-40 per cent. The foraminiferal composition indicates a cold climate: the quantity of Subarctic *G. pachyderma* increases to 71 per cent, compared with 26 per cent in horizon I; in other words it corresponds to the Subarctic thanatocoenosis type. This second horizon was deposited during cold climatic conditions, when floating ice carried terrigenous material over a wide area of the North Atlantic (Würm/Wisconsin). In the lower part of horizon II (115-136 centimetres) the sediment has a higher carbonate content (50 per cent $CaCO_3$), the foraminiferal composition indicates warmer water and corresponds to the Recent Boreal thanatocoenosis (IIa). Thus the temperature of the surface water at the time of accumulation of horizon II was no more than 8-9° C, except for the interval between 115 and 136 centimetres, when the temperature rose to 9-10° C, but without reaching its present value. This period of warming corresponds we think to the time of the great Würm interstadial, which occurred near the beginning of the glaciation. Some cores from the North Atlantic indicate two interstadials: a lower, warmer and longer one, and an upper one. It is possible that in the case of cores with one big interstadial displaced to the lower boundary of the Würm, the less important upper interstadial was not recorded. It is also possible that in cores in which the middle stadial is very badly represented, both interstadials unite to form one. The last assumption is more likely in the southern warm-water area of the North Atlantic.

Horizon III (167-309 centimetres). It is a high calcareous white ooze (up to 75 per cent $CaCO_3$). The foraminiferal composition is warm-water one, similar to the thanatocoenosis of horizon I: *G. pachyderma* is no more than 20 per cent. The thanatocoenosis of planktonic Foraminifera corresponds to the Boreal type (IIa). Thus horizon III was accumulated during a comparatively warm climatic period of long duration. The average annual temperature of the surface water was 9°-10° C (Riss-Würm Interglacial).

Horizon IV (310-423 centimetres). It is brown sandy mud with terrigenous sand, gravel and pebbles. The calcium carbonate content is 20-40 per cent. The upper part of horizon IV contains the Subarctic type of thanatocoenosis (the temperature of the water was no more than 8-9° C), the lower part contains mostly the cold-water subtype of the Boreal thanatocoenosis. Apparently horizon IV was accumulated during the penultimate (Riss) glaciation or during its upper stadial.

Horizon V (below 423 centimetres), is a highly calcareous ooze (up to 78.8 per cent $CaCO_3$). The Boreal type of thanatocoenosis indicates a warm climate. This horizon was only found in two cores and there is not sufficient data to define its age.

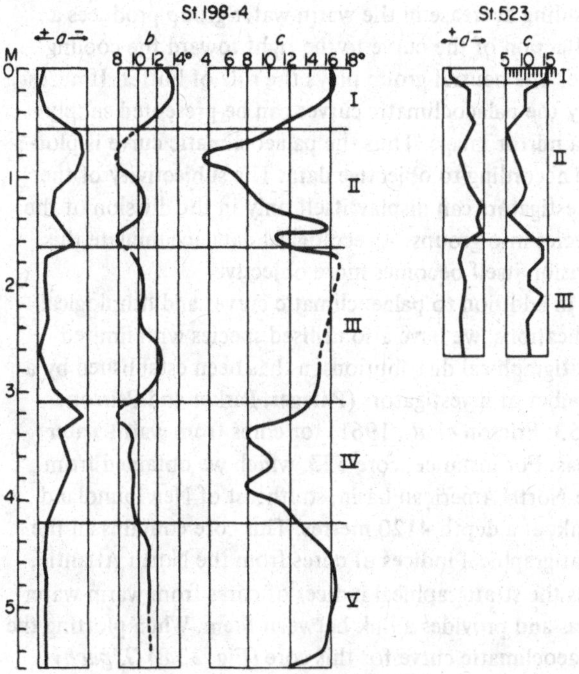
St.198-4
St.523
M
I
II
III
IV
V

Fig. 33.8. Palaeoclimatic and palaeotemperature curves for cores 198 and 523. a. palaeoclimatic curve based on ratio of warm- and cold-water species ('+' warmer, '–' colder); b. palaeotemperature curve based on fossil thanatocoenoses; c. palaeotemperature curve based on ratio of oxygen isotopes. (I-V are stratigraphical horizons.)

The comparison of fossil thanatocoenoses with Recent ones provides the possibility of plotting a palaeotemperature curve for core 198 (Fig. 33.8). Also T. S. Gromova and R. V. Teis, working under A. P. Vinogradov in the Isotopic Geochemistry laboratory of The Institute of Geochemistry have plotted palaeotemperature curves, based on the ratio of oxygen isotopes in the tests of planktonic Foraminifera for core 198. The curve is also given on Fig. 33.8. It shows that the stratigraphical horizons have been correctly defined. However it has some differences from the curve plotted on the basis of fossil thanatocoenoses. These differences are reduced to a minimum if the corrections suggested by Emiliani (1957) are taken into account. Thus two unconnected methods confirm one another.

One other core is also worth individual description. Core 66 was obtained in the European basin (40° 03.6′ N, 20° 18.4′W), at a depth of 3900 metres (length 318 centimetres; 298 centimetres dry). The average annual temperature of surface water in this area is 13-14° C. Four horizons were distinguished.

Horizon I (1-12 centimetres), is a layer of whitish yellow, highly calcareous ooze (more than 85 per cent $CaCO_3$). The foraminiferal composition indicates warm water, including 27 per cent of *G. pachyderma* and more than 20 per cent of *G. inflata* and *G. scitula,* and is interpreted as Holocene.

Horizon II (12-168 centimetres), is a layer of brown and grey mud with gravel and pebbles (up to 4 centimetres in size), with some volcanic glass. It is mainly terrigenous in composition and the calcium carbonate content is 10-30 per cent, with the exception of a layer of light-grey mud (134-149 centimetres) where it increases to 57.8 per cent. Most of the samples contained the Subarctic type of thanatocoenosis indicating a temperature below the present value in the area (with the exception of the layer at 120-149 centimetres, where the Boreal composition of the planktonic foraminiferal fauna indicated a warmer period). Thus the sediment of horizon II was accumulated during a stage of terrigenous sedimentation in conditions of cold, Subarctic climate during the last continental glaciation (Würm), with the period of slight warming (120-149 centimetres) corresponding to the interstadial.

Horizon III (169-225 centimetres), is a layer of white clayey mud, with a high carbonate content (74-82 per cent of $CaCO_3$). The planktonic Foraminifera indicate a relatively warm period, the climate resembling the present one: the subtypes of the Boreal and at 240-242 centimetres, the Subtropical thanatocoenosis were encountered. All samples contain Subtropical species, even including species which were absent from horizon I. Thus horizon III was accumulated during a long stage of warm climate (Riss-Würm Interglacial).

Horizon IV (below 255 centimetres) is a layer of terrigenous, low carbonate mud (about 18 per cent $CaCO_3$). Samples contain Subarctic thanatocoenoses and in the lower part the coldest subtype of the Boreal thanatocoenosis. Thus this horizon, like horizon II, was accumulated in cold Subarctic climatic conditions during the penultimate continental glaciation (Riss).

The palaeotemperature curve plotted for core 66 (Fig. 33.9) is like that plotted for core 198 and other

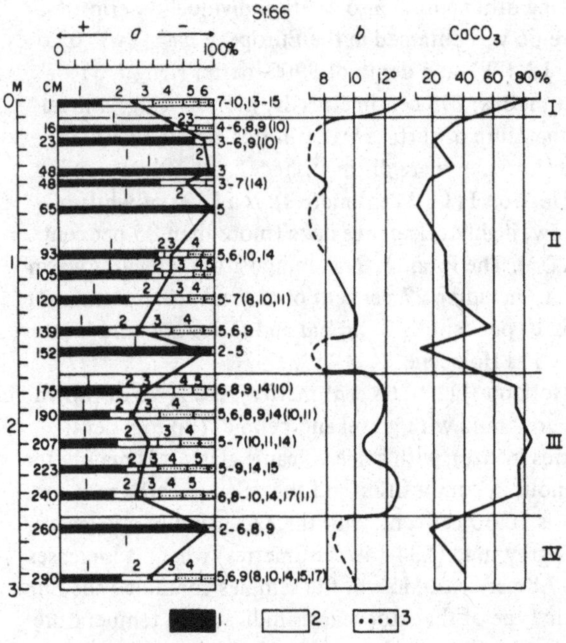
St.66
a
b
CaCO3
0
100%
CM
M
8 10 12°
20 40 60 80%
I
II
III
IV
1.
2.
3

Fig. 33.9. Palaeoclimatic and palaeotemperature curves for core 66. a. palaeoclimatic curve based on ration of cold-water (1), neutral (2), and warm-water (3) groups of species (species numbers correspond to numbers on Fig. 33.6); b. palaeotemperature curve based on fossil thanatocoenoses. (I-IV are stratigraphical horizons.)

cores from the North Atlantic.

The stratigraphical division of cores obtained south of 40° N was conducted on a different basis, as in this part of the ocean quantitative relationships between the types of thanatocoenosis and the temperature of the surface water are not yet established. As mentioned above after quantitative analysis the species were divided into three groups: cold-water, neutral and warm-water. On the basis of the ratio between these groups we plotted palaeoclimatic curves which make it possible to distinguish warm-water and cold-water horizons and estimate the degree of warming or cooling. In so doing we must take into consideration the fact that the groups of relatively cold- and warm-water species differ markedly in the different zones of the ocean. One species may be a member of a cold-water group in one region of the ocean, and of a warm-water group in another. For instance *G. inflata* in the Subarctic zone is a member of the warm-water group and in the Tropical zone a member of the cold-water group. In analysing the core samples we often met species which do not change their concentration, or fluctuate wihin the same limits, in cold-water horizons as in warm-water ones. The composition of this neutral group also changes in the different climatic zones.

We plotted our palaeoclimatic curves as follows. For each sample the percentage ratio between species was plotted, the species being arranged from the coldest-water forms (left) to the warmest-water forms (right). We have thus on the left side a group of cold-water species, on the right a group of warm-water species, and in the middle a neutral group. The palaeoclimatic curve links up the average points of the neutral group of each sample. An increase in the cold-water group, and corresponding decrease in the warm-water group produces a deflection of the curve to the right toward the cooling side. The neutral group plays the role of buffer. If necessary the palaeoclimatic curves can be presented simply as a mirror image. Thus the palaeoclimatic curve is plotted according to objective data. The subjectivity of the investigators can display itself only in the division of the species into groups. As ecological data accumulate this division itself becomes more objective.

In addition to palaeoclimatic curves and lithological indications, we have also utilised species with limited stratigraphical distributions, as has been established by a number of investigators (Phleger, Parker and Peirson, 1953; Ericson *et al.*, 1961) for cores from warm-water areas. For instance, core 523, which we obtained from the North American basin, southeast of Newfoundland bank, at a depth 4120 metres. This core contains all the stratigraphical indices of cores from the North Atlantic, plus the stratigraphical indices of cores from warm-water areas and provides a link between them. When plotting the palaeoclimatic curve for this core (Fig. 33.8) *G. pachyderma* and *G. quinqueloba* were included in the cold-water group, *G. bulloides, G. glutinata* and *G. inflata* were included in the neutral group and all other species in the relatively warm-water group. The length of the core is 267 centimetres, and it was divided into three horizons.

Horizon I (1-13 centimetres) is a light-brown, chalk-like ooze. The carbonate content is high, more than 80 per cent. The foraminiferal composition is a warm-water one and is interpreted as Holocene. The typical Holocene species, *G. tumida,* is present.

Horizon II (14-161 centimetres). Light-brown ooze with a carbonate content of 60-80 per cent. The palaeoclimatic curve for horizon II shows that this is a cold climate stage, with two cooling maxima (the stadials of the last continental glaciation). The horizon is characterized by the decrease in the carbonate content to 60-70 per cent, and by the appearance of terrigenous mineral grains of sand grade. The foraminiferal composition indicates that the lower cooling maximum was more significant in this part of the ocean than the upper. The intervening deposit accumulated during a warming (interstadial) phase is characterized by an increase in the carbonate content to 78 per cent. The foraminiferal composition indicates that the warming did not reach present values. All these indications, which are typical for deposits of the Würm interstadial in North Atlantic cores are confirmed by the presence of *Globorotalia menardii flexuosa* (Koch), which is found at this stratigraphical level in the majority of cores from the central Atlantic.

Horizon III. (161-221 centimetres) is a layer of brown ooze. (Below 221 centimetres we met a white chalk-like ooze, the carbonate content of which is 67-81 per cent.) The planktonic Foraminifera in the samples taken from this horizon indicated a cold climatic stage (Riss-Würm Interglacial).

As with core 198, we also plotted the palaeotemperature curve on the basis of the oxygen isotopes for core 523. This curve is parallel to the palaeoclimatic curve plotted on the basis of the ratio of species.

The analysis of the distribution of the shells of planktonic Foraminifera in cores from the Atlantic ocean permits the following general conclusions:

1. During the stages of the continental glaciations of Europe and North America Subarctic type thanatocoenoses, in which *Globigerina pachyderma* accounts for more than 60 per cent, were deposited over wide areas of the North Atlantic to approximately 40° N.
2. During the big interstadial of the last (Würm) glaciation the warming of the surface water did not reach present day values, and the thanatocoenoses were deposited a little further south than during the Holocene and at present.
3. During the Riss-Würm Interglacial the average annual temperature of the surface water in the North Atlantic was close to present values with some deviation one way or the other in different areas.

As was indicated by Phleger *et al.* (1953), the species *G. tumida* is present only in the upper warm-water sediment layer, and in the cold-water layer that preceded it. In most of the cores from the central Atlantic that we investigated this species is present only in the Holocene deposits; however, in some cores it was found in the upper part of horizon II (the upper stadial of the last glaciation) and in a smaller number of cores in the deposits of the upper Würm interstadial. Thus *G. tumida* is typical of the upper part of the Würm, and especially of the Holocene. As reported by Ericson *et al.* (1961) *G. menardii flexuosa* and *G. hexagona* do not extend above the last layer before the present, the age of which they suppose to be equivalent to the interstadial of the last glaciation (zone 'X'). Our data confirm that these species do not extend above the upper Würm interstadial. Both are warm-water species, and can be used for the correlation of the horizon of cores from the central, warm-water part of the ocean.

In many cores from the north and especially the northeast of the ocean *G. bulloides* has stratigraphical significance. In samples of the sediment accumulated during the Würm interstadial we found a large number of representatives of this species with large, massive shells. The number of *G. bulloides* at this stratigraphical level is more than at any other level, and in several cores accounts for 40-50 per cent of all the planktonic Foraminifera. By contrast with the recent sediment, the Boreal thanatocoenosis subtype IIb was widespread during the Würm interstadials.

In a number of cores from the northwestern part of the ocean in deposits of the last Riss-Würm Interglacial we encountered large quantities of *Biorbulina universa* shells. In some cores from the central part of the ocean at this horizon there are specimens of *G. inflata* with a sharp periphery resembling very convex shells of *G. punctulata.* It is possible that these shells should be separated as a distinct species. In sediment of the same horizon in cores from the Tropical part of the ocean we found a large quantity of *G. menardii* shells and their partially dissolved remains, up to 50-60 per cent.

In the horizons of cores from the central part of the ocean, which were accumulated in warm climate conditions (Mindel-Riss interglacial or Riss interstadial) we found a large quantity of shells of the species described by Phleger, Parker and Peirson (1953) as *Globigerina* sp. 2. This species occurs with frequencies up to 12-15 per cent.

The methods we have described for plotting palaeotemperature and palaeoclimatic curves, and the data on the horizontal and vertical distribution of planktonic Foraminifera in the Quaternary deposits of the Atlantic ocean, may find wide application in the analysis of cores from the Atlantic and other oceans.

Received January 1968.

BIBLIOGRAPHY

Androsova, V. P. 1935. Microfauna Severo-Dvinskogo postpliocena. *Trudy VNIRO,* **1**, 11-132.

Emiliani, C. 1955. Pleistocene temperature. *Jour. Geology.* **63**, 538-78.

Emiliani, C. 1957. Temperature and age analyses of deep-sea cores. *Science,* **125**, 383-7.

Ericson, D. B., Ewing, M., Wollin, G. and Heezen, B. C. 1961. Atlantic deep-sea sediment cores. *Bull. Geol. Soc. Amer.* 72, 193-286.

Phleger, F. B., Parker, F. L. and Peirson, J. F. 1953. North Atlantic foraminifera. *Rep. Swed. Deep-Sea Exped.* 7 (1), 1-112.

Schott, W. G. 1935. Die Foraminiferen in dem aequitorialen Teil des Atlantischen Ozeans. *Wiss. Ergebnisse der Deutschen Atl. Exp. auf dem Forschungs, und Vermessungschiff 'Meteor' 1925-1927,* 3 (3) teil I, Lieferung B, Berlin-Leipzig.

Schott, W. G. 1944. *Geographie des Atlantischen Ozeans.* Hamburg.

34. RADIOLARIAN ZONES IN THE QUATERNARY OF THE EQUATORIAL PACIFIC OCEAN

C. A. NIGRINI

Summary: A four-fold Quaternary stratigraphical sequence, based on the appearance or disappearance of two new and two established radiolarian species, is recognized in equatorial Pacific sediments. The faunal changes do not appear to be related to variations in climate. Nine of the seventeen cores examined extend into the 'Pliocene' as defined by the upper limit of occurrence of *Pterocanium prismatium* Riedel. Two additional radiolarian species are useful in determining the position of the 'Plio-Pleistocene' boundary. A similar stratigraphy occurs in four tropical Indian Ocean cores, but the additional forms used as criteria of the 'Plio-Pleistocene' boundary in the Pacific cannot be used in the Indian Ocean.

INTRODUCTION

Using Indian Ocean material, Nigrini (1967) found that it is possible to distinguish between low and middle latitude radiolarian assemblages, and her preliminary survey of Atlantic Ocean sediments showed that, in general, species restricted to low or middle latitudes in the Indian Ocean are similarly restricted in the Atlantic.

Recognition of low, middle and high latitude foraminiferal assemblages has provided a useful tool for the interpretation of Quaternary stratigraphy from deep-sea core material (Ericson, Ewing, Wollin and Heezen 1961). The present study was initiated with the idea of delineating similar Pleistocene sequences based on alternating low and middle latitude radiolarian assemblages. Such sequences, correlated with the established foraminiferal series, would allow the extension of this kind of Pleistocene stratigraphy into non-calcareous sediments. Even in calcareous sediments, the addition of a number of radiolarian species to those species of Foraminifera used stratigraphically would strengthen ecological interpretations.

To test this theory, samples were selected from two, well-documented cores known to contain relatively long and presumably continuous Pleistocene sequences with abundant Radiolaria. These were Swedish Deep-Sea cores 61 (0° 06′ S, 135° 58′ W, 4,437 metres) and 62 (3° 00′ S, 136° 26′ W, 4,510 metres). According to Arrhenius (1952), these cores contain cyclic variations in carbonate percentages which correspond to glacial maxima and minima; an increase in carbonate during glacial times and a decrease during interglacials. Correlation was made between the cores using the cumulative amounts of titanium oxide present in the sediments. Emiliani (1955) derived a rate of accumulation of titanium oxide almost double that calculated by Arrhenius. However, work by Sackett (1960) appears to confirm Arrhenius' original figure. There is some difference of opinion regarding the significance of changes in carbonate percentages. According to Emiliani and Flint (1963), cyclic variations in the calcium carbonate percentage probably reflect climatic fluctuations, but, depending upon the relative rates of accumulation of other components in the sediment, carbonate maxima may represent either glacial or interglacial conditions. In the Atlantic, increased carbonate is directly related to temperature (Emiliani, 1955), but the work of Arrhenius in the Pacific indicates an inverse relationship. Olausson (1965) suggested that near-surface water temperatures in the equatorial Pacific (in the region of Swedish Deep-Sea Cores 58-61) changed very little during the Pleistocene, and that carbonate fluctuations, in fact, reflect changes in the rate of influx of nutrients from the Atlantic which were, in turn, controlled by temperature changes in the North Atlantic. However, as a working hypothesis, the conclusions of Arrhenius were accepted by the author as correct.

EXAMINATION OF SDSE 61 AND SDSE 62

Sediment samples were taken at or near each of the stages and substages described by Arrhenius and several from approximately mid-way between stages (Tables 34.14, 34.15). The cores consist of calcareous ooze with a few to many siliceous microfossils. A few intercalated clayey strata occur in both cores.

Preliminary examination of strewn slides showed that most of those species characteristic of middle latitude assemblages in the Indian Ocean are absent from the two SDSE cores. Rather few *Actinomma medianum* were found at some levels in SDSE 61, but showed no consistent distribution pattern. Rare specimens of *Lithocampe* sp. and *Eucyrtidium acuminatum* were found at several levels in both cores, but this sparse occurrence is compatible with known low latitude assemblages.

To test the possibility that glacial and interglacial stages might be marked by variations in the abundance of species characteristic of the low latitude assemblage, counts of 350 individuals were made for thirteen low latitude species or species groups, namely:

Disolenia spp.
Polysolenia spp.
Panartus tetrathalamus
Euchitonia spp.
Spongaster tetras tetras
Centrobotrys thermophila
Botryocyrtis scutum
Lithopera bacca
Anthocyrtidium ophirense
Pterocanium praetextum praetextum
Theocorythium trachelium trachelium
Eucyrtidium hexagonatum
*Siphocampe corbula**

* All of the above species have been described by Nigrini (1967).

The resultant numerical data showed no convincing or consistent patterns which could be correlated with the stages of Arrhenius, or from one core to the other. It was apparent that in Swedish Deep-Sea Expedition cores 61 and 62, at least, the radiolarian fauna examined does not reflect cyclic fluctuations in ocean conditions during the Pleistocene. Characteristically middle latitude assemblages are absent and no consistent variation in the abundance of typically low latitude species could be found.

QUATERNARY ZONATION

During the above described study, it was found that a four-fold Quaternary zonation, based on the appearance or disappearance of certain species (*Collosphaera tuberosa* Haeckel, *Buccinosphaera invaginata* Haeckel, *Anthocyrtidium angulare* n.sp. and *Theocorythium vetulum* n.sp.) not included in the original survey, can be recognized. The zonation is further supported by changes in abundance of *Amphirhopalum ypsilon* Haeckel, and *Lithopera bacca* Ehrenberg.

In order to pursue this line of investigation, fifteen additional cores (both piston and gravity; see Table 34.1 and Fig. 34.1) from across the equatorial Pacific (between 10° N and 10° S) were sampled. Nine of these cores extend into the 'Pliocene', as defined by the presence of *P. prismatium* (Riedel 1957), and the absence, for all practical purposes, of *T. trachelium trachelium* and *A. angulare.* The relative abundances of those eight species which define a Quaternary zonation and the uppermost 'Pliocene', and the levels sampled in each core are recorded in Tables 34.3-34.19.

DESCRIPTIONS OF SPECIES USED IN ZONATION

The appearance or disappearance of the following species

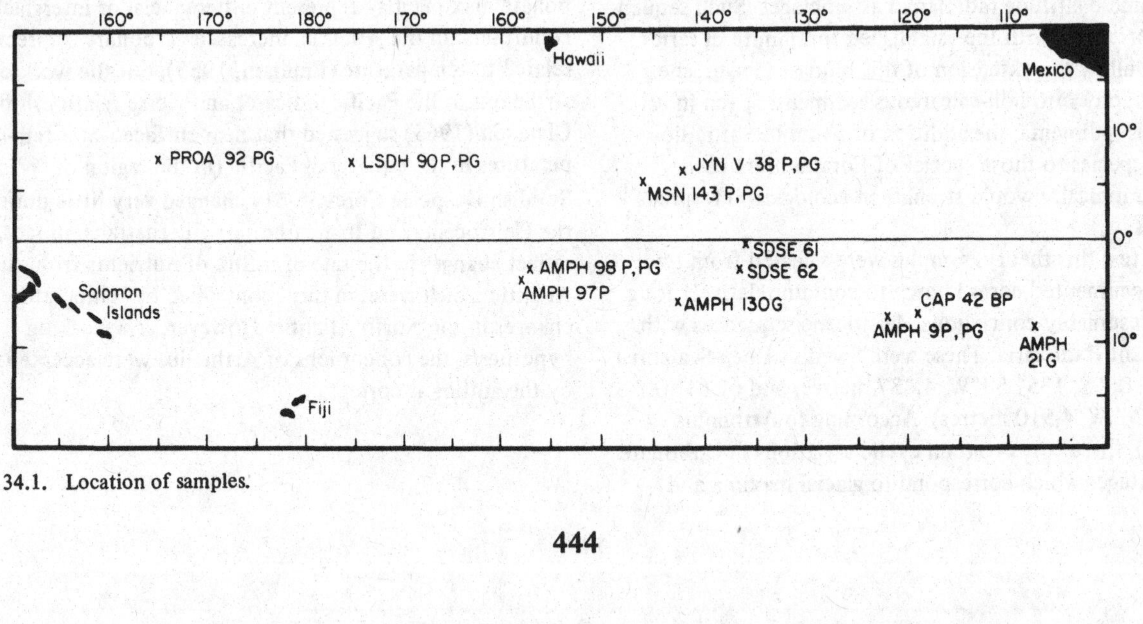
160°
170°
180°
170°
160°
150°
140°
130°
120°
110°
Hawaii
Mexico
10°
x PROA 92 PG
x LSDH 90 P, PG
x JYN V 38 P, PG
x MSN 143 P, PG
0°
x SDSE 61
x SDSE 62
x AMPH 98 P, PG
x AMPH 97 P
Solomon
Islands
x AMPH 130 G
CAP 42 BP
x AMPH 9 P, PG
x AMPH 21 G
10°
Fiji

Fig. 34.1. Location of samples.

Table 34.1. List of Pacific Ocean cores used in stratigraphical study

Expedition abbreviation and core number	Latitude	Longitude	Depth (m)	Core length (cm)
AMPH 21G*	8° 29′ S	107° 26′ W	3,120	79
CAP 42BP†	7° 19′ S	118° 40′ W	4,200	803
AMPH 9P†	7° 31′ S	121° 56′ W	4,410	594
AMPH 9PG+	7° 31′ S	121° 56′ W	4,410	140
SDSE 61	0° 06′ S	135° 58′ W	4,437	1029
SDSE 62	3° 00′ S	136° 26′ W	4,510	1479
JYN V 38P	6° 30′ N	141° 59′ W	5,018	475
JYN V 38PG	6° 30′ N	141° 59′ W	5,018	105
AMPH 130G	5° 58′ S	142° 43′ W	4,450	114
MSN 143P	5° 32′ N	146° 09′ W	5,100	691 (approx.)
MSN 143PG	5° 32′ N	146° 09′ W	5,100	156
AMPH 98P	2° 50′ S	157° 13′ W	5,225	569
AMPH 98PG	2° 50′ S	157° 13′ W	5,225	181
AMPH 97P	3° 42′ S	157° 40′ W	5,228	592
LSDH 90P	7° 19′ N	175° 28′ W	5,190	546
LSDH 90PG	7° 19′ N	175° 28′ W	5,190	145
PROA 92PG	7° 58′ N	164° 57′ E	5,219	169

Note: All cores housed at the Scripps Institution of Oceanography, La Jolla, California.

* G denotes a gravity core, gravity core only obtained.
† P denotes a piston core.
\+ PG denotes a gravity core, both piston and gravity cores obtained.

constitutes the basis for a division of the Quaternary into four zones, and definition of a 'Quaternary-Tertiary' boundary. Type and figured specimens will be deposited in the United States National Museum, Washington, D.C., and the USNM numbers used herein are from their Cenozoic Catalogue No. 132. Specimens are located according to sample number, slide letter (in the case of more than one slide from a particular location), and England Finder co-ordinates (Riedel and Foreman 1961).

Genus *Collosphaera* Muller 1855
Collosphaera tuberosa Haeckel 1887
(Pl. 34.1, fig. 1)

? 1851 *Thalassicolla punctata* Huxley (*partim.*), p. 434, pl. xvi, fig. 6.
1862 *Collosphaera huxleyi* Muller: Haeckel (*partim.*), pl. xxxiv, figs. 3,9.
1885 *Collosphaera huxleyi* Muller: Brandt, (*partim.*), p. 285, pl. 7, fig. 37.
1887 *non Collosphaera huxleyi* Muller: Haeckel, p. 96.
1887 *Collosphaera tuberosa* Haeckel, p. 97.
1905 *Collosphaera tuberosa* Haeckel: Brandt, p. 332, pl. 9, fig. 16.
? 1929 *Collosphaera huxleyi* Muller: Schroder, p.105, figs. 13, 14.
1962 *Collosphaera tuberosa* Haeckel: Strelkov and Reschetnjak, p. 128 and p. 136, fig. 10.
1964 *Collosphaera* ? *huxleyi* Muller: Nakaseko, p. 46, pl. 3, figs. 3,4.

Description: Shell is a smooth-surfaced, lumpy sphere, having numerous subcircular pores, irregular in size and distribution. Usually there is a rather larger pore where the shell indents. Figured specimen from AMPH 9P, 8-10 centimetres, USNM No. 650930.

Dimensions (based on twenty specimens): Maximum shell diameter 103-159μ.

Remarks: The author is indebted to Dr. Reschetnjak of the Academy of Sciences of the USSR for identification of this and the following collosphaerid species.

Genus *Buccinosphaera* Haeckel 1887
Buccinosphaera invaginata Haeckel 1887
(Pl. 34.1, fig. 2)

1887 *Buccinosphaera invaginata* Haeckel, p. 99, pl. 5, fig. 11.
1905 *Buccinosphaera invaginata* Haeckel: Brandt, p. 332, pl. 10, fig. 20.
1917 *Buccinosphaera invaginata* Haeckel: Popofsky, p. 248, text-figs. 7,8.
1962 *Buccinosphaera invaginata* Haeckel: Strelkov and Reschetnjak, p. 129 and p. 137, fig. 12.

Description: Similar to *C. tuberosa*, but with short (8-16μ), inwardly directed spines projecting from the larger pores found where the shell indents. Pores generally smaller and shell with a thinner wall than *C. tuberosa*. Figured specimen from SDSE 62, 22-4 centimetres, USNM No. 650931.

Dimensions (based on twenty specimens): Maximum shell diameter 88-119μ.

Genus *Anthocyrtidium* Haeckel 1881
Anthocyrtidium angulare n.sp.
(Pl. 34.1, figs. 3*a*, *b*)

Description: Cephalis trilobate, elongate, with subcircular pores, bearing a stout three-bladed apical spine of approximately the same length as the cephalis. Thorax shaped like a biretta. Three stout thoracic ribs, which may become external to form short, thorn-like wings, control the shape of the upper thorax. There is a sharp break in shell contour where the ribs terminate, and the lower thorax is approximately cylindrical. Pores circular to subcircular, usually arranged longitudinally. Eight to eleven, three-bladed subterminal teeth, directed outwards, are usually present, but may be absent or much reduced. Distally from the subterminal teeth, the thoracic wall curves sharply inwards and terminates at a narrow, poreless peristome which often bears numerous short, delicate, lamellar, terminal teeth, directed downwards and inwards. Figured specimens from SDSE 62, 1028-30 centimetres, USNM No. 650932 (holotype, fig. 3*a*), USNM No. 650933 (paratype, fig. 3*b*).

Dimensions (based on twenty specimens): Length of apical horn 18-27μ; of cephalis 18-36μ; of thorax 36-81μ; of subterminal

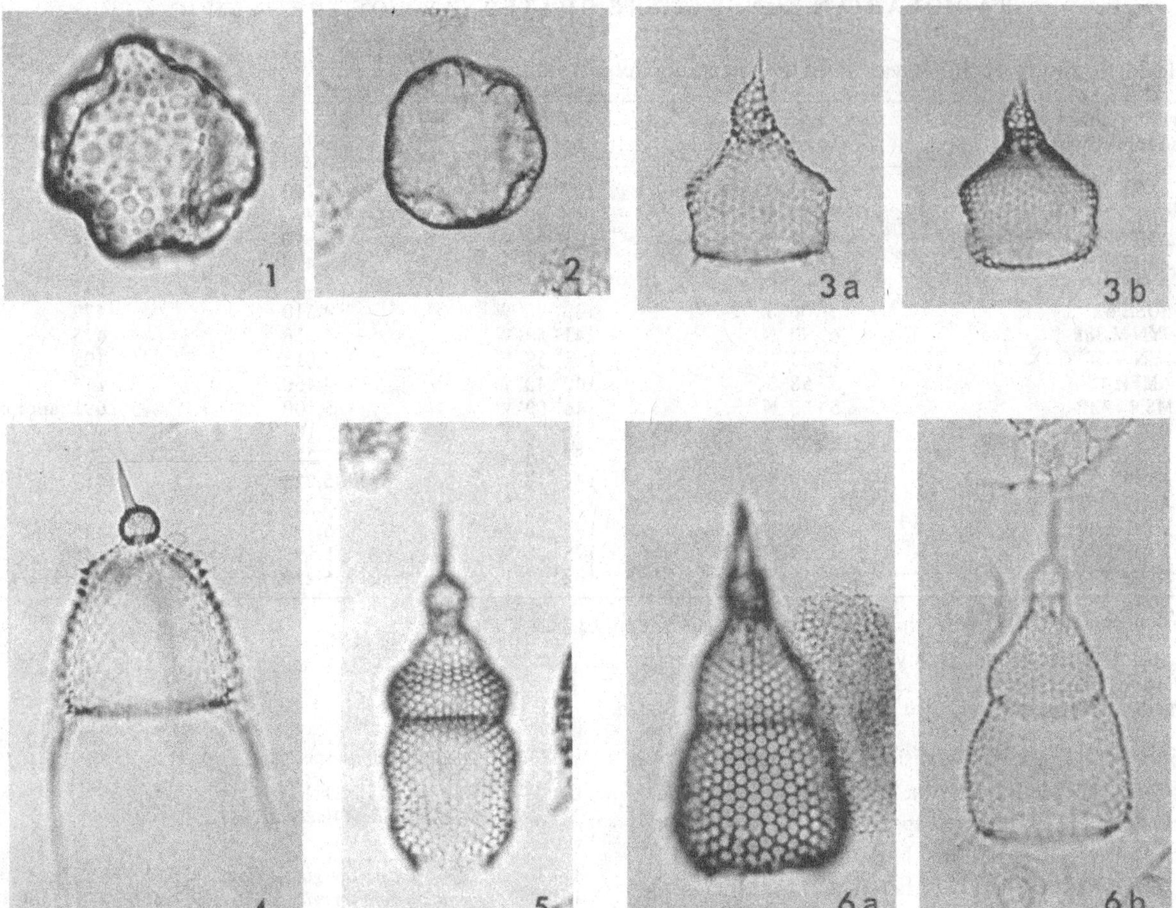
1
2
3a
3b
4
5
6a
6b
7a
7b
7c
8

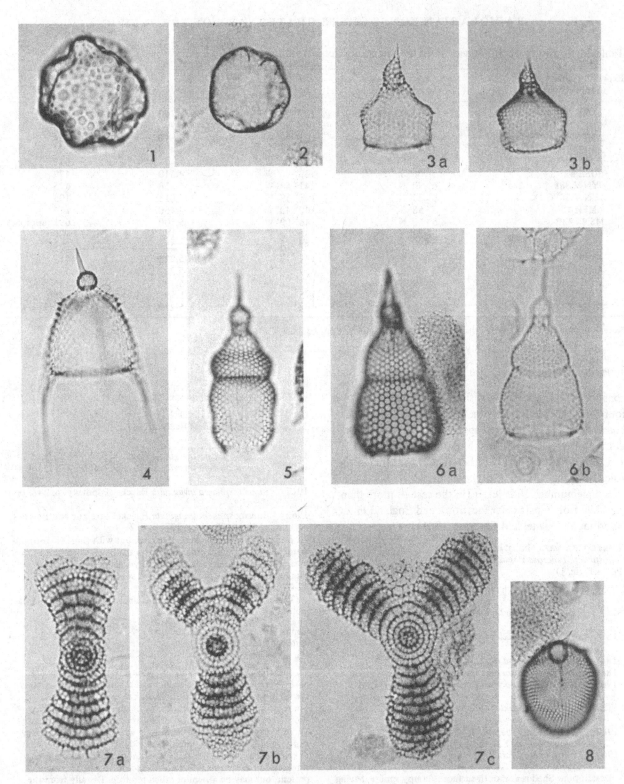

Plate 34.1. All figures x 215. Fig. 1. *Collosphaera tuberosa* Haeckel, AMPH 9P, 8-10 cm, T19/4; USNM No. 650930; fig. 2. *Buccinosphaera invaginata* Haeckel, SDSE 62, 22-4 cm, N29/4; USNM No. 650931; fig. 3*a*. *Anthocyrtidium angulare* n. sp., holotype, SDSE 62, 1028-30 cm, R53/3; USNM No. 650932; fig. 3*b*. *Anthocyrtidium angulare* n.sp., paratype, SDSE 62, 1028-30 cm, D57/4; USNM No. 650933; fig. 4. *Pterocanium prismatium* Riedel, AMPH 9P, 325-7 cm, U21/0; USNM No. 650934; fig. 5. *Theocorythium trachelium trachelium* (Ehrenberg), SDSE 62, 788-90 cm, B-T25/1; USNM No. 650935; fig. 6*a*. *Theocorythium vetulum* n.sp., holotype, SDSE 62, 1028-30 cm, S24/1; USNM No. 650936; fig. 6*b*. *Theocorythium vetulum* n.sp., paratype, AMPH 9P, 325-7 cm, E54/2; USNM No. 650937; fig. 7*a*. *Amphirhopalum ypsilon* Haeckel, showing 5 chambers on the forked arm before bifurcation; AMPH 9P, 8-10 cm, L54/2; USNM No. 650938; fig. 7*b*. *Amphirhopalum ypsilon* Haeckel, showing 2 chambers on the forked arm before bifurcation; SDSE 62, 1028-30 cm, M22/1; USNM No. 650939; fig. 7*c*. *Amphirhopalum ypsilon* Haeckel, showing 1 chamber on the forked arm before bifurcation; SDSE 62, 788-90 cm, A-H53/3; USNM No. 650940; fig. 8. *Lithopera bacca* Ehrenberg, SDSE 62, 788-90 cm, A-J37/3; USNM No. 650941.

teeth 9-18μ. Maximum breadth of cephalis 18-27μ; of thorax 72-100μ.

Remarks: *A. angulare* is similar to *A. ophirense* (cf. *A. cineraria* in Riedel, 1957, p. 84 and *A. ophirense* in Nigrini, 1967, p. 56). It may be distinguished from *A. ophirense* by its smaller size and by the distinctive shape of its thorax.

Genus *Pterocanium* Ehrenberg 1847
Pterocanium prismatium Riedel, 1957
(Pl. 34.1, fig. 4)

1957 *Pterocanium prismatium* Riedel, p. 87, pl. 3, figs. 4, 5.

Description: cf. Riedel, 1957, p. 87. Figured specimen from AMPH 9P, 325-7 centimetres, USNM No. 650934.

Genus *Theocorythium* Haeckel 1887
Theocorythium trachelium trachelium (Ehrenberg), 1872
(Pl. 34.1, fig. 5)

1967 *Theocorythium trachelium trachelium* (Ehrenberg): Nigrini, p. 79, pl. 8, fig. 2, pl. 9, fig. 2.

Description: cf. Nigrini, 1967, p. 79. Figured specimen from SDSE 62, 788-90 centimetres, USNM No. 650935.

Theocorythium vetulum n.sp.
(Pl. 34.1, figs. *6a, b*)

Description: Shell quite smooth and usually thin-walled. Cephalis trilocular, the paired lobes beneath and only slightly lateral to the larger unpaired lobe; pores small, subcircular. Stout, three-bladed apical horn, between equal to and twice the cephalic length. Primary lateral and dorsal spines continue as ribs in the thoracic wall for about half its length, but have not been observed to project externally.

Thorax cupola-shaped with circular to subcircular pores arranged longitudinally, 7-10 in a vertical series, 12-15 on a half-equator. Pronounced lumbar stricture.

Abdomen inflated conical. Pores similar in size, shape and arrangement to those of thorax, 5-12 in a vertical series, 12-17 on a half-equator. Distally, a row of three-bladed subterminal teeth is usually present. Slight terminal constriction to termination at a poreless peristome and up to eleven triangular terminal teeth which may or may not be well-developed. Figured specimens from SDSE 62, 1028-30 centimetres, USNM No. 650936 (holotype, fig. *6a*) and AMPH 9P, 325-7 centimetres, USNM No. 650937 (paratype, fig. *6b*).

Dimensions (based on twenty specimens): Total length (excluding apical horn and terminal teeth) 137-182μ. Length of cephalis 27-36μ; of thorax 45-63μ; of abdomen (excluding terminal teeth) 45-90μ. Maximum breadth of cephalis 27-36μ; of thorax 81-90μ; of abdomen 90-128μ.

Remarks: *T. vetulum* is similar to *Theocorythium trachelium dianae* (cf. Nigrini 1967, p. 77) except that the thorax is usually broader and the abdomen is inflated conical rather than cylindrical. Some specimens of *T. vetulum* are superficially similar to *Lamprocyclas maritalis polypora* (cf. Nigrini 1967, p. 76), but the paired cephalic lobes of members of the genus *Theocorythium* are directly beneath (or only slightly lateral to) the larger unpaired lobe, whereas they are decidedly lateral in *Lamprocyclas*. This distinction is difficult to recognize from all angles, and a more obvious difference is that *L. maritalis polypora* has a broader, shorter abdomen and generally larger abdominal pores.

The differences between *T. trachelium dianae*, *T. vetulum* and *L. maritalis polypora* can be shown most easily by comparing their average abdominal lengths and breadths and average thoracic breadths (based on twenty specimens of each species):

	Av abdominal length	Av abdominal breadth	Av thoracic breadth
T. trachelium dianae	95 μ	90 μ	72 μ
T. vetulum	80 μ	109 μ	82 μ
L. maritalis polypora	59 μ	123 μ	83 μ

Combining these values we get:

	Av abdominal length / Av abdominal breadth	Av abdominal breadth – av thoracic breadth
T. trachelium dianae	1.06	18 μ
T. vetulum	0.73	27 μ
L. maritalis polypora	0.48	40 μ

The following species support, by changes in their abundance, the zonation primarily defined by *C. tuberosa, B. invaginata, A. angulare, P. prismatium, T. trachelium trachelium* and *T. vetulum.*

Genus *Amphirhopalum* Haeckel 1887; *sens. emend.* Nigrini 1967
Amphirhopalum ypsilon Haeckel, 1887
(Pl. 34.1, figs. *7a, b, c*)

1967 *Amphirhopalum ypsilon* Haeckel: Nigrini, p. 35, pl. 3, figs. *3a-d*.

Description: cf. Nigrini 1967, p. 35. Specimens from the upper parts of the cores examined average four or five proximal chambers on the forked arm before it bifurcates. Lower down in the cores this number decreases, and forms with two or three (sometimes one) such chambers predominate. The decrease coincides approximately with an increase in abundance. Figured specimens from AMPH 9P, 8-10 centimetres, USNM No. 650938 (fig. *7a*), SDSE 62, 1028-30 centimetres, USNM No. 650939 (fig. *7b*), SDSE 62, 788-90 centimetres, USNM No. 650940 (fig. *7c*).

Genus *Lithopera* Ehrenberg 1847; *sens. emend.* Nigrini 1967
Lithopera bacca Ehrenberg, 1872
(Pl. 34.1, fig. 8)

1967 *Lithopera bacca* Ehrenberg: Nigrini, p. 54, pl. 6, fig. 2.

Description: cf. Nigrini 1967, p. 54. Figured specimen from SDSE 62, 788-90 centimetres, USNM No. 650941.

DETAILS OF ZONATION

Buccinosphaera invaginata Range Zone (Zone 1; Uppermost Quaternary)

This zone is defined by the range of *B. invaginata* in accordance with the American Stratigraphic Code (1961). *A. angulare, T. vetulum* and *P. prismatium* are absent. *L. bacca* is essentially absent from the zone in the eastern tropics; occurrences are sporadic and abundances relatively low ('rare'). West of about 157° W (AMPH 98) it is consistently present in small numbers ('rare' or 'few') throughout the zone. *A. ypsilon* occurs rarely and averages four or more proximal chambers on the forked arm before bifurcation. *C. tuberosa* and *T. trachelium trachelium* are present in varying abundances.

Zone 1 is found at the top of fourteen of the seventeen cores examined. It is missing from LSDH 90P and MSN 143P (but is present in the corresponding gravity cores) and from AMPH 97P for which there is no corresponding gravity core.

Collosphaera tuberosa Concurrent Range Zone (Zone 2)

The top of this zone is defined by the earliest appearance of *B. invaginata,* and the bottom by the earliest appearance of *C. tuberosa. B. invaginata, A. angulare, T. vetulum* and *P. prismatium* are absent. Within the zone *C.*

tuberosa, A. ypsilon and *T. trachelium trachelium* have concurrent ranges. *L. bacca* is essentially absent (as in zone 1) in the eastern tropics, but west of about 157° W (AMPH 98) it is present in small numbers ('rare' or 'few') throughout the zone. *A. ypsilon* is consistently present throughout the zone. It increases in abundance towards the bottom of the zone, and there is an approximately concomitant decrease in the number of proximal chambers on the forked arm before bifurcation. *T. trachelium trachelium* and *C. tuberosa* are present in varying abundances.

Zone 2 is found in all the cores examined. Five cores (possibly 6, since the bottom 80 centimetres of PROA 92PG contains an insufficient number of Radiolaria for a meaningful examination) bottom in zone 2.

Amphirhopalum ypsilon Assemblage Zone (Zone 3)

The top of this zone is defined by the earliest appearance of *C. tuberosa,* and the bottom by the latest appearance of *A. angulare. B. invaginata, C. tuberosa, A. angulare, T. vetulum* and *P. prismatium* are all absent. *A. ypsilon* is consistently present ('few' to 'common') and averages two to three proximal chambers on the forked arm before bifurcation. *L. bacca* occurs rarely throughout the zone, except in SDSE 62 where it is present ('rare') only towards the bottom of the zone. *T. trachelium trachelium* occurs throughout the zone in varying numbers. The zone has been called an assemblage zone in accordance with the American Stratigraphic Code (1961) although it might also be called a partial range zone in the nomenclature of stratigraphical code subcommittee of the Geological Society of London (1967).

Zone 3 is present in seven of the cores examined. MSN 143PG bottoms in the zone. In three additional cores (AMPH 98P, LSDH 90P and MSN 143P) the sampling interval (10 centimetres, 15 centimetres and 15 centimetres respectively) bridges zone 3 and the observable sequence jumps from zone 2 to zone 4. To miss a single zone over a 10-15 centimetre sampling interval is thought possible. In AMPH 130G the sampling interval (10 centimetres) bridges zones 3 and 4 and zonation appears to jump from zone 2 to 'Pliocene'. Since the sampling interval is small, and two entire zones are omitted, it is probable that the core contains an unconformity. At about 55 centimetres in the core (it is difficult to determine the level exactly because of the presence of a large manganese nodule) there is a change in colour from dark brown above to a higher brown below.

Anthocyrtidium angulare Concurrent Range Zone (Zone 4; Lowermost Quaternary)

The top of this zone is defined by the latest appearance of *A. angulare,* and the bottom by the latest appearance of *P. prismatium.* Within the zone *A. angulare, A. ypsilon* (with an average of two to three proximal chambers) *L. bacca* and *T. trachelium trachelium* have concurrent ranges.

In SDSE 62 (approximately 136° W) and CAP 42BP (approximately 118° W) the zone is further characterized by the presence of *T. vetulum.* In AMPH 9P (approximately 121° W) and LSDH 90P (approximately 175° W) *T. vetulum* occurs towards the bottom of the zone only. There is no readily apparent reason for the complete absence of this species between 136° W and 175° W. *C. tuberosa* and *B. invaginata* are absent.

Zone 4 is found in nine of the cores examined. JYN V 38PG bottoms in the zone.

'Pliocene'. The 'Quaternary-Tertiary' boundary is considered herein to be at the upper limit of occurrence of *P. prismatium* as has been recognized by Riedel, Bramlette and Parker (1963). In addition, it has been found by the present author that although *A. angulare* and *T. trachelium trachelium* are present in varying abundances in the lowermost Quaternary (zone 4), they are essentially absent from the uppermost 'Pliocene' of the tropical Pacific. *C. tuberosa* and *B. invaginata* are also absent. *A. ypsilon, L. bacca,* and *T. vetulum* (in the four cores cited in the discussion of zone 4) are present in varying abundances.

'Pliocene' sediments are present in nine of the cores examined. Once the presence of a 'Pliocene' fauna was established, no further examination was made of samples from lower levels in the cores.

INDIAN OCEAN CORES

Four cores (Table 34.2) from the Indian Ocean were examined to determine whether or not the criteria used for defining a fourfold Quaternary zonation in the equatorial Pacific could also be applied to Indian Ocean sediments. It was found (Tables 34.20-3) that zones 1 to 3 can be recognized in all of the cores.

Only one of the cores (V19-169) extends beyond zone 3. In this core zone 4 and 'Pliocene' material can also be recognized. As in the Pacific, the 'Plio-Pleistocene' boundary is defined by the uppermost level of occurrence of *P. prismatium.* In the tropical Pacific *A. angulare* and *T. trachelium trachelium* are essentially absent in the uppermost 'Pliocene', but in V19-169 both species occur, although in small numbers, in much of the upper 'Pliocene'.

In zones 1 and 2 in the tropical Pacific, east of about 157° W, *L. bacca* is practically absent; west of about 157° W it is consistently present in small numbers ('rare' or 'few'). In zone 3 it occurs rarely throughout zone 3 (except in SDSE 62 where it is present only towards the

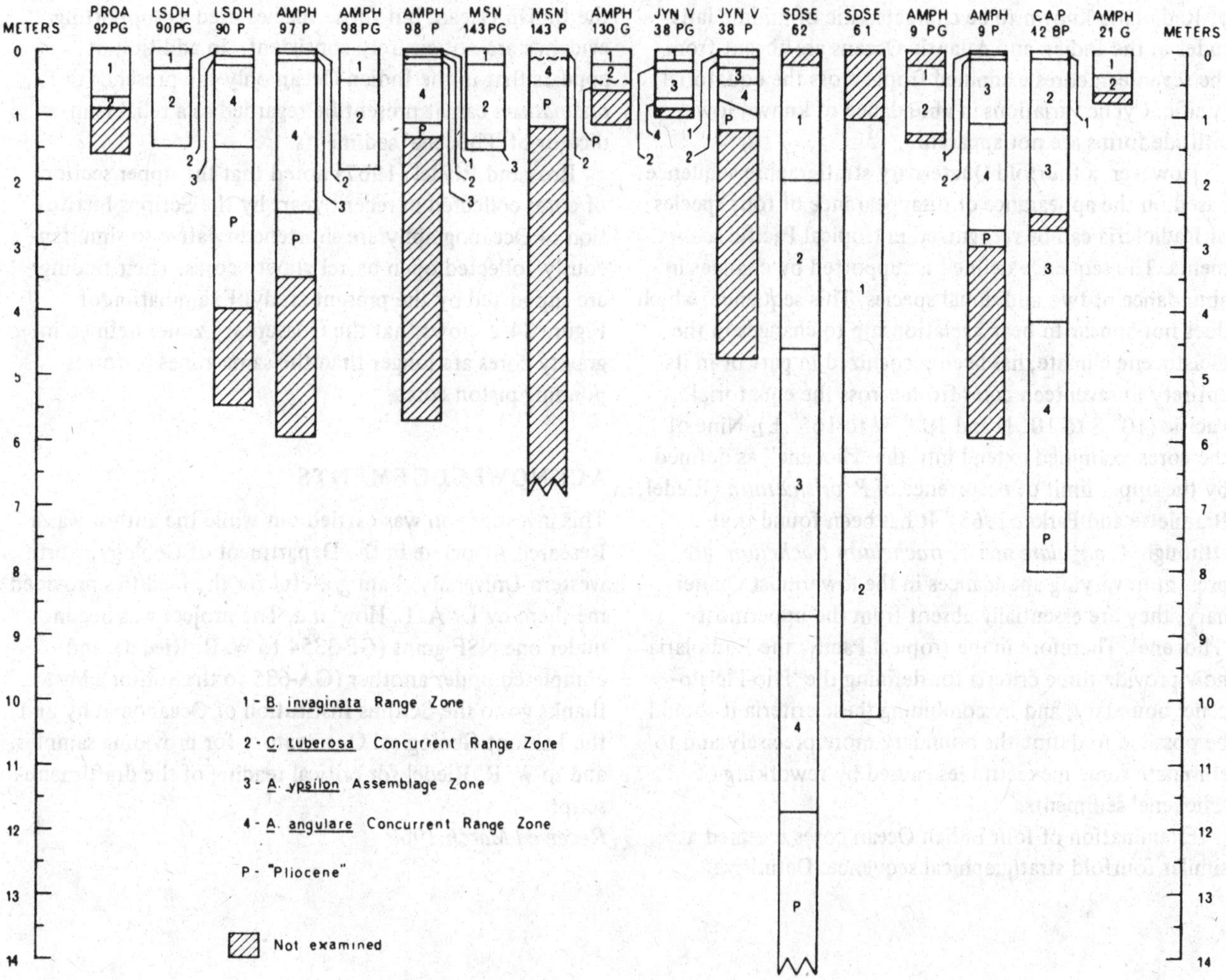
METERS
PROA 92PG
LSDH 90PG
LSDH 90 P
AMPH 97 P
AMPH 98PG
AMPH 98 P
MSN 143PG
MSN 143 P
AMPH 130 G
JYN V 38 PG
JYN V 38 P
SDSE 62
SDSE 61
AMPH 9 PG
AMPH 9 P
CAP 42 BP
AMPH 21 G
1 - B. invaginata Range Zone
2 - C. tuberosa Concurrent Range Zone
3 - A. ypsilon Assemblage Zone
4 - A. angulare Concurrent Range Zone
P - "Pliocene"
Not examined
Fig. 34.2. Quaternary zonation of deep-sea cores.

bottom of the zone). However, in the tropical Indian Ocean *L. bacca* is more abundant (usually 'few') in all three zones.

Throughout the examined Quaternary of the Indian Ocean *A. ypsilon* appears to be generally less abundant than in the Pacific. However, it does show the same downward decrease in the number of chambers before bifurcation.

SUMMARY AND CONCLUSIONS

The Quaternary radiolarian fauna of the equatorial Pacific does not show a cyclic fluctuation between warm and cold faunas, such as that described by Ericson *et al.* (1961) for

Table 34.2. Indian Ocean cores

Expedition abbreviation and core number	Latitude	Longitude	Depth (m)	Core length (cm)
V19-169	10° 13′ S	81° 37′ E	5,110	1,300 (approx)
MSN 48G	15°.15′ S	81° 08′ E	4,996	177
MSN 49G	14° 27′ S	78° 03′ E	5,214	170
MSN 50G	12° 58′ S	75° 01′ E	5,197	133

Note: V19-169 housed at the Lamont Geological Observatory, Palisades, New York; all other cores housed at the Scripps Institution of Oceanography, La Jolla, California.

foraminiferal assemblages in the Atlantic Ocean. Species of Radiolaria known to be characteristic of middle latitudes in the Indian and Atlantic Oceans are absent from the seventeen cores examined from across the equatorial Pacific. Cyclic variations in abundance of known low-latitude forms are not apparent.

However, a fourfold Quaternary stratigraphic sequence, based on the appearance or disappearance of four species of Radiolaria can be recognized in tropical Pacific sediments. The sequence defined is supported by changes in abundance of two additional species. This sequence, which does not appear to bear a relationship to changes in the Pleistocene climate, has been recognized in part or in its entirety in seventeen cores from across the equatorial Pacific (10° S to 10° N and 107° W to 165° E). Nine of the cores examined extend into the 'Pliocene', as defined by the upper limit of occurrence of *P. prismatium* (Riedel, Bramlette and Parker 1963). It has been found that although *A. angulare* and *T. trachelium trachelium* are present in varying abundances in the lowermost Quaternary, they are essentially absent from the uppermost 'Pliocene'. Therefore in the tropical Pacific the Radiolaria now provide three criteria for defining the 'Plio-Pleistocene' boundary, and by combining these criteria it should be possible to define the boundary more precisely and to eliminate some inexactitudes caused by reworking of 'Pliocene' sediments.

Examination of four Indian Ocean cores revealed a similar fourfold stratigraphical sequence. Definitive species used in the Pacific behave in a similar manner in the Indian Ocean, but those species used as supporting evidence are not entirely consistent. In addition, it appears that in the Indian Ocean only the presence of *P. prismatium* can at present be regarded as a reliable indicator of 'Pliocene' sediments.

Ross and Riedel (1967) noted that the upper sections of cores collected in recent years by the Scripps Institution of Oceanography are shortened relative to simultaneously collected open-barrel gravity cores. Their findings are supported by the present study. Examination of Figure 34.2 shows that the Quaternary zones defined in gravity cores are longer than the same zones in corresponding piston cores.

ACKNOWLEDGEMENTS

This investigation was carried out while the author was a Research Associate in the Department of Geology, Northwestern University. I am grateful for the facilities provided me there by Dr A. L. Howland. The project was begun under one NSF grant (GP-3354 to W. R. Riedel), and completed under another (GA-635 to the author). My thanks go to the Scripps Institution of Oceanography and the Lamont Geological Observatory for providing samples, and to W. R. Riedel for critical reading of the draft manuscript.

Received March 1968

Table 34.3.

PROA 92PG (cm)	*Buccinosphaera invaginata*	*Collosphaera tuberosa*	*Amphirhopalum ypsilon*	*Lithopera bacca*	*Theocorythium trachelium trachelium*	*Anthocyrtidium angulare*	*Theocorythium vetulum*	*Pterocanium prismatium*	– = absent R = rare F = few C = common
0-2	F	F	R	R	F	–	–	–	
10-2	R	F	R	R	F	–	–	–	
20-2	F	F	R	R	R	–	–	–	
30-2	R	R	R	F	F	–	–	–	
40-2	R	R	R	F	F	–	–	–	Zone 1
50-2	R	F	R	R	F	–	–	–	
62-4	R	F	R	R	F	–	–	–	
70-2	R	F	F	R	F	–	–	–	
80-2	–	F	F	F	R	–	–	–	Zone 2
90-2	–	F	F	R	F	–	–	–	

Table 34.4.

LSDH 90PG (cm)	*Buccinosphaera invaginata*	*Collosphaera tuberosa*	*Amphirhopalum ypsilon*	*Lithopera bacca*	*Theocorythium trachelium trachelium*	*Anthocyrtidium angulare*	*Theocorythium vetulum*	*Pterocanium prismatium*	– = absent R = rare F = few C = common
0-3	R	F	R	R	F	–	–	–	Zone 1
28-30	R	F	R	R	F	–	–	–	
60-2	–	R	F	R	F	–	–	–	Zone 2
78-80	–	R	F	R	F	–	–	–	
98-100	–	R	C	R	R	–	–	–	
135-7	–	R	C	–	R	–	–	–	

Table 34.5

LSDH 90P (cm)	*Buccinosphaera invaginata*	*Collosphaera tuberosa*	*Amphirhopalum ypsilon*	*Lithopera bacca*	*Theocorythium trachelium trachelium*	*Anthocyrtidium angulare*	*Theocorythium vetulum*	*Pterocanium prismatium*	– = absent R = rare F = few C = common
0-3	–	F	R	R	F	–	–	–	Zone 2
10-2	–	R	C	R	F	–	–	–	
25-7	–	–	C	R	F	R	–	–	Zone 3
34-6	–	–	C	R	R	R	–	–	
50-2	–	–	C	R	F	R	R	–	
74-6	–	–	F	R	F	R	R	–	Zone 4
100-2	–	–	F	R	F	R	R	–	
124-6	–	–	C	R	R	R	R	–	
150-2	–	–	C	R	R	–	–	R	'Pliocene'
174-6	–	–	F	R	–	R	R	F	
203-5	–	–	F	R	–	–	R	C	
251-3	–	–	F	R	R	–	R	C	
300-2	–	–	F	R	R	–	–	C	
354-6	–	–	C	R	–	–	R	F	
394-6	–	–	C	R	–	–	R	F	

Note: The upper limit of *P. prismatium* found in this core by the present author corresponds with that illustrated by Riedel, Bramlette, and Parker, (1963).

Table 34.6

AMPH 97P (cm)	*Buccinosphaera invaginata*	*Collosphaera tuberosa*	*Amphirhopalum ypsilon*	*Lithopera bacca*	*Theocorythium trachelium trachelium*	*Anthocyrtidium angulare*	*Theocorythium vetulum*	*Pterocanium prismatium*	– = absent R = rare F = few C = common
3-5	–	R	F	R	F	–	–	–	Zone 2
13-5	–	–	F	R	C	–	–	–	Zone 3
23-5	–	–	F	R	F	R	–	–	
33-5	–	–	F	R	F	F	–	–	
43-5	–	–	C	R	R	F	–	–	
53-5	–	–	F	R	F	F	–	–	
63-5	–	–	F	R	F	F	–	–	
73-5	–	–	F	R	F	F	–	–	
83-5	–	–	C	R	R	F	–	–	
93-5	–	–	F	R	F	R	–	–	
103-5	–	–	F	F	R	F	–	–	Zone 4
113-5	–	–	F	R	F	F	–	–	
123-5	–	–	F	R	R	F	–	–	
133-5	–	–	F	R	R	F	–	–	
143-5	–	–	C	R	F	R	–	–	
174-6	–	–	F	R	R	F	–	–	
199-201	–	–	F	F	R	F	–	–	
224-6	–	–	C	R	R	F	–	–	
249-51	–	–	F	–	–	–	–	C	
299-301	–	–	F	–	–	–	–	C	'Pliocene'
349-51	–	–	R	R	–	–	–	C	

Table 34.7.

AMPH 98PG (cm)	*Buccinosphaera invaginata*	*Collosphaera tuberosa*	*Amphirhopalum ypsilon*	*Lithopera bacca*	*Theocorythium trachelium trachelium*	*Anthocyrtidium angulare*	*Theocorythium vetulum*	*Pterocanium prismatium*	– = absent R = rare F = few C = common
9-11	R	F	–	F	F	–	–	–	
25-7	R	F	R	R	F	–	–	–	
35-7	R	F	R	R	F	–	–	–	Zone 1
45-7	R	F	R	R	F	–	–	–	
55-7	–	C	F	–	F	–	–	–	
65-7	–	F	F	R	R	–	–	–	
75-7	–	F	F	–	F	–	–	–	
85-7	–	R	F	R	C	–	–	–	
95-7	–	R	F	–	F	–	–	–	
105-7	–	F	F	R	R	–	–	–	
115-7	–	R	C	F	R	–	–	–	Zone 2
125-7	–	R	F	R	F	–	–	–	
135-7	–	R	F	F	R	–	–	–	
145-7	–	R	C	R	R	–	–	–	
155-7	–	R	C	–	R	–	–	–	
165-7	–	R	C	R	F	–	–	–	
175-7	–	R	F	R	C	–	–	–	

Table 34.8.

AMPH 98P (cm)	*Buccinosphaera invaginata*	*Collosphaera tuberosa*	*Amphirhopalum ypsilon*	*Lithopera bacca*	*Theocorythium trachelium trachelium*	*Anthocyrtidium angulare*	*Theocorythium vetulum*	*Pterocanium prismatium*	– = absent R = rare F = few C = common
0-2	R	F	R	–	F	–	–	–	Zone 1
10-2	–	R	F	–	C	–	–	–	Zone 2
20-2	–	–	F	R	R	F	–	–	
30-2	–	–	F	R	R	F	–	–	Zone 3
40-2	–	–	F	R	R	F	–	–	
50-2	–	–	F	R	R	F	–	–	
60-2	–	–	F	R	R	C	–	–	Zone 4
70-2	–	–	C	R	R	F	–	–	
82-4	–	–	F	R	R	F	–	–	
92-4	–	–	C	R	–	F	–	–	
102-4	–	–	C	R	–	F	–	–	
112-4	–	–	F	R	R	R	–	F	
122-4	–	–	F	–	–	–	–	C	
132-4	–	–	R	R	–	–	–	C	
142-4	–	–	F	R	–	–	–	C	

Table 34.9.

MSN 143 PG (cm)	*Buccinosphaera invaginata*	*Collosphaera tuberosa*	*Amphirhopalum ypsilon*	*Lithopera bacca*	*Theocorythium trachelium trachelium*	*Anthocyrtidium angulare*	*Theocorythium vetulum*	*Pterocanium prismatium*	– = absent R = rare F = few C = common
0-2	F	F	R	–	F	–	–	–	Zone 1
18-20	R	F	R	–	F	–	–	–	
37-9	–	F	F	–	F	–	–	–	
68-70	–	F	C	R	F	–	–	–	
90-2	–	F	F	–	F	–	–	–	Zone 2
118-20	–	F	C	R	R	–	–	–	
140-2	–	R	C	–	F	–	–	–	
149-51	–	–	C	R	F	–	–	–	Zone 3

Table 34.10.

MSN 143P (cm)	*Buccinosphaera invaginata*	*Collosphaera tuberosa*	*Amphirhopalum ypsilon*	*Lithopera bacca*	*Theocorythium trachelium trachelium*	*Anthocyrtidium angulare*	*Theocorythium vetulum*	*Pterocanium prismatium*	– = absent R = rare F = few C = common
0-2	–	R	F	R	R	R	–	C	Zone 2
4-6	–	R	F	R	R	R	–	F	
8-10	–	F	C	–	R	–	–	–	
25-7	–	–	F	R	F	F	–	R	Zone 4
35-7	–	–	F	R	R	F	–	–	
50-2	–	–	C	R	R	–	–	R	'Pliocene'
60-2	–	–	R	R	–	–	–	C	
68-70	–	–	F	R	–	–	–	C	
76-8	–	–	F	R	–	–	–	C	
116-8	–	–	R	R	–	–	–	C	

Note: The 'Plio-Pleistocene' boundary appears to be between 35 and 50 centimetres in this core. However, samples from 0-2 centimetres and 4-6 centimetres have anomalously high abundances of *P. prismatium*, and *A. angulare* appears rarely. Riedel and Funnell (1964) placed the 'Plio-Pleistocene' boundary in this core between 10 centimetres and 60 centimetres.

Table 34.11.

AMPH 130G (cm)	*Buccinosphaera invaginata*	*Collosphaera tuberosa*	*Amphirhopalum ypsilon*	*Lithopera bacca*	*Theocorythium trachelium trachelium*	*Anthocyrtidium angulare*	*Theocorythium vetulum*	*Pterocanium prismatium*	– = absent R = rare F = few C = common
9-12	F	F	R	–	F	–	–	–	Zone 1
18-20	R	F	R	–	C	–	–	–	
30-2	–	F	R	–	C	–	–	–	Zone 2
46-8	–	F	F	–	C	–	–	–	
58-60	–	–	F	R	R	–	–	C	'Pliocene'

Table 34.12.

JYN V 38PG (cm)	*Buccinosphaera invaginata*	*Collosphaera tuberosa*	*Amphirhopalum ypsilon*	*Lithopera bacca*	*Theocorythium trachelium trachelium*	*Anthocyrtidium angulare*	*Theocorythium vetulum*	*Pterocanium prismatium*	– = absent R = rare F = few C = common
4-6	R	C	R	–	R	–	–	–	Zone 1
14-6	R	F	F	R	F	–	–	–	
24-6	–	F	F	–	F	–	–	–	Zone 2
34-6	–	R	C	R	F	–	–	–	
44-6	–	–	C	R	F	–	–	–	Zone 3
54-6	–	–	C	R	R	–	–	–	
64-6	–	–	C	–	F	–	–	–	
74-6	–	–	C	R	F	R	–	–	Zone 4
84-6	–	–	F	R	F	R	–	–	

Table 34.13.

JYN V 38P (cm)	*Buccinosphaera invaginata*	*Collosphaera tuberosa*	*Amphirhopalum ypsilon*	*Lithopera bacca*	*Theocorythium trachelium trachelium*	*Anthocyrtidium angulare*	*Theocorythium vetulum*	*Pterocanium prismatium*	– = absent R = rare F = few C = common
8-10	F	F	R	R	R	–	–	–	Zone 1
18-20	–	C	F	R	R	–	–	–	Zone 2
28-30	–	R	C	–	F	–	–	–	
38-40	–	–	C	R	F	–	–	–	Zone 3
48-50	–	–	F	R	R	F	–	–	Zone 4
58-60	–	–	C	R	F	F	–	–	
68-70	–	–	F	R	R	F	–	–	
78-80	–	–	C	R	R	R	–	F	apparent mixing
88-90	–	–	F	R	–	C	–	R	
98-100	–	–	F	R	R	–	–	C	'Pliocene'
108-10	–	–	R	R	–	–	–	C	
118-20	–	–	R	R	–	–	–	C	

Table 34.14.

SDSE 62 (cm)	*Buccinosphaera invaginata*	*Collosphaera tuberosa*	*Amphirhopalum ypsilon*	*Lithopera bacca*	*Theocorythium trachelium trachelium*	*Anthocyrtidium angulare*	*Theocorythium vetulum*	*Pterocanium prismatium*	– = absent R = rare F = few C = common
22-4	R	C	R	–	F	–	–	–	
58-60	R	F	R	–	F	–	–	–	
77-8	R	F	R	–	F	–	–	–	
112-4	R	F	R	R	C	–	–	–	Zone 1
139-41	–	F	R	–	C	–	–	–	
181-3	R	F	R	R	C	–	–	–	
198-200	R	R	R	–	C	–	–	–	
318-20	–	R	F	–	C	–	–	–	
241-3	–	C	F	–	R	–	–	–	
261-3	–	C	F	–	R	–	–	–	
291-3	–	R	F	–	F	–	–	–	
311-3	–	R	C	R	F	–	–	–	Zone 2
348-50	–	R	C	R	F	–	–	–	
378-80	–	R	C	R	R	–	–	–	
408-10	–	R	F	R	F	–	–	–	
458-60	–	R	C	–	F	–	–	–	
508-10	–	–	C	–	R	–	–	–	
548-50	–	–	F	–	C	–	–	–	
608-10	–	–	C	–	F	–	–	–	
638-40	–	–	F	R	C	–	–	–	
668-70	–	–	F	R	C	–	–	–	Zone 3
688-90	–	–	F	R	C	–	–	–	
738-40	–	–	F	R	C	–	–	–	
788-90	–	–	F	R	F	–	–	–	
807-10	–	–	F	R	C	–	–	–	
868-70	–	–	C	R	R	R	R	–	
909-10	–	–	F	R	F	R	R	–	
953-5	–	–	C	R	F	R	R	–	
974-6	–	–	F	F	F	F	R	–	
1012-6	–	–	F	R	R	R	R	–	
1028-30	–	–	F	R	F	R	R	–	Zone 4
1066-70	–	–	F	R	R	F	R	–	
1092-5	–	–	F	R	R	F	R	–	
1124-8	–	–	F	R	F	F	R	–	
1144-7	–	–	F	R	R	F	R	–	
1157-60	–	–	F	F	R	F	R	–	
1167-70	–	–	C	R	R	R	R		
1217-20	–	–	F	R	R	R	R	F	
1232-6	–	–	F	R	–	–	R	C	
1268-70	–	–	F	R	R	–	R	C	
1321-3	–	–	F	–	–	–	R	C	'Pliocene'
1372-4	–	–	R	R	–	–	R	C	
1427-30	–	–	R	F	–	–	R	F	
1468-70	–	–	R	F	–	–	R	F	

Note: Riedel, Bramlette and Parker (1963).placed the upper limit of *P. prismatium* in this core between 1,170 centimetres and 1,140 centimetres. *P. prismatium* is absent from the author's 1,167-70 centimetres sample, but is present in Riedel's 1,170-71 centimetres sample. The upper limit of *P. prismatium* can, then, be precisely placed at 1,700 centimetres.

Table 34.15.

SDSE 61 (cm)	*Buccinosphaera invaginata*	*Collosphaera tuberosa*	*Amphirhopalum ypsilon*	*Lithopera bacca*	*Theocorythium trachelium trachelium*	*Anthocyrtidium angulare*	*Theocorythium vetulum*	*Pterocanium prismatium*	– = absent R = rare F = few C = common
110	F	C	R	–	R	–	–	–	
150	–	C	R	R	R	–	–	–	
202-4	R	C	R	–	–	–	–	–	
258-60	R	F	F	–	F	–	–	–	
290-2	R	R	F	R	C	–	–	–	
320-2	–	R	F	–	C	–	–	–	
368-70	F	F	R	–	F	–	–	–	Zone 1
408-10	–	R	R	–	R	–	–	–	
457-9	–	R	R	–	–	–	–	–	
512-4	F	R	F	–	F	–	–	–	
558-60	R	F	C	–	R	–	–	–	
609-10	–	R	R	–	C	–	–	–	
648-50	R	R	R	–	C	–	–	–	
724-6	–	F	F	–	F	–	–	–	
747-9	–	R	F	–	F	–	–	–	
774-6	–	F	F	–	C	–	–	–	
823-5	–	R	F	–	F	–	–	–	
841-3	–	F	R	–	F	–	–	–	Zone 2
902-4	–	F	F	–	R	–	–	–	
936-8	–	F	C	–	R	–	–	–	
979-81	–	R	C	–	R	–	–	–	
1,009-12	–	R	C	–	R	–	–	–	

Table 34.16.

AMPH 9PG (cm)	*Buccinosphaera invaginata*	*Collosphaera tuberosa*	*Amphirhopalum ypsilon*	*Lithopera bacca*	*Theocorythium trachelium trachelium*	*Anthocyrtidium angulare*	*Theocorythium vetulum*	*Pterocanium prismatium*	– = absent R = rare F = few C = common
22-4	R	F	R	–	F	–	–	–	
30-2	R	F	R	–	F	–	–	–	Zone 1
40-2	–	F	R	R	F	–	–	–	
49-51	R	R	R	R	C	–	–	–	
62-4	–	R	R	R	C	–	–	–	
70-2	–	F	R	R	F	–	–	–	
82-4	–	R	C	–	F	–	–	–	
90-2	–	R	F	–	F	–	–	–	Zone 2
104-6	–	R	F	–	C	–	–	–	
116-8	–	R	F	R	F	–	–	–	
126-8	–	R	F	R	F	–	–	–	

Table 34.17.

AMPH 9P (cm)	*Buccinosphaera invaginata*	*Collosphaera tuberosa*	*Amphirhopalum ypsilon*	*Lithopera bacca*	*Theocorythium trachelium trachelium*	*Anthocyrtidium angulare*	*Theocorythium vetulum*	*Pterocanium prismatium*	– = absent R = rare F = few C = common
0-2	R	F	R	–	C	–	–	–	Zone 1
8-10	–	F	R	–	F	–	–	–	Zone 2
25-7	–	–	F	R	C	–	–	–	
35-7	–	–	F	R	C	–	–	–	
50-2	–	–	F	R	C	–	–	–	
60-2	–	–	F	F	C	–	–	–	Zone 3
75-7	–	–	F	F	F	–	–	–	
85-7	–	–	F	F	F	–	–	–	
100-2	–	–	F	R	C	–	–	–	
110-2	–	–	F	R	F	R	–	–	
125-7	–	–	F	R	F	R	–	–	
132-4	–	–	F	R	F	R	–	–	
150-2	–	–	F	R	F	R	–	–	
160-2	–	–	F	R	F	R	–	–	
170-2	–	–	F	R	F	R	R	–	
190-2	–	–	R	R	C	R	R	–	Zone 4
200-2	–	–	F	R	F	R	R	–	
212-4	–	–	F	R	F	F	R	–	
225-7	–	–	F	R	F	R	R	–	
235-7	–	–	F	R	F	R	R	–	
250-2	–	–	F	R	F	R	R	–	
260-2	–	–	C	F	R	R	R	–	
275-7	–	–	F	R	R	–	R	F	
282-4	–	–	F	R	R	–	R	R	'Pliocene'
300-2	–	–	F	R	R	–	R	C	
325-7	–	–	R	R	R	–	R	C	

Table 34.18.

CAP 42BP (cm)	*Buccinosphaera invaginata*	*Collosphaera tuberosa*	*Amphirhopalum ypsilon*	*Lithopera bacca*	*Theocorythium trachelium trachelium*	*Anthocyrtidium angulare*	*Theocorythium vetulum*	*Pterocanium prismatium*	– = absent R = rare F = few C = common
0-8	R	F	R	–	F	–	–	–	Zone 1
49-51	–	R	R	–	C	–	–	–	
100-2	–	R	F	–	F	–	–	–	Zone 2
118-20	–	R	F	R	F	–	–	–	
137-206	–	–	F	R	F	–	–	–	
246-50	–	–	F	–	C	–	–	–	
275-345	–	–	F	R	F	–	–	–	Zone 3
345-414	–	–	F	R	C	–	–	–	
414-85	–	–	F	R	F	R	R	–	
485-555	–	–	F	R	F	R	R	–	
555-626	–	–	F	F	F	R	R	–	Zone 4
626-96	–	–	F	R	F	R	R	–	
696-757	–	–	F	F	R	–	R	R	
757-803	–	–	R	F	R	R	R	F	'Pliocene'

Table 34.19.

AMPH 21G (cm)	*Buccinosphaera invaginata*	*Collosphaera tuberosa*	*Amphirhopalum ypsilon*	*Lithopera bacca*	*Theocorythium trachelium trachelium*	*Anthocyrtidium angulare*	*Theocorythium vetulum*	*Pterocanium prismatium*	– = absent R = rare F = few C = common
2-5	F	F	R	–	F	–	–	–	Zone 1
12-5	R	F	R	–	F	–	–	–	
22-5	R	C	R	–	R	–	–	–	
32-5	R	C	R	–	F	–	–	–	
42-5	–	C	R	–	F	–	–	–	Zone 2
52-5	–	C	R	–	R	–	–	–	
62-5	–	F	R	–	F	–	–	–	

Table 34.20.

V 19-169 (cm)	*Buccinosphaera invaginata*	*Collosphaera tuberosa*	*Amphirhopalum ypsilon*	*Lithopera bacca*	*Theocorythium trachelium trachelium*	*Anthocyrtidium angulare*	*Theocorythium vetulum*	*Pterocanium prismatium*	– = absent R = rare F = few C = common
9-10	R	F	R	F	F	–	–	–	Zone 1
50-1	–	F	R	R	F	–	–	–	Zone 2
100-1	–	R	F	F	C	–	–	–	
150-1	–	R	F	R	C	–	–	–	
200-1	–	R	R	F	C	–	–	–	
251-2	–	–	R	F	C	–	–	–	Zone 3
300-1	–	–	R	F	C	–	–	–	
350-1	–	–	R	F	F	F	–	–	Zone 4
400-1	–	–	R	F	C	R	–	–	
450-1	–	–	R	F	C	R	–	–	
500-1	–	–	R	F	F	R	R	R	'Pliocene'
550-1	–	–	R	R	R	R	–	C	
600-1	–	–	F	F	–	R	–	C	

Table 34.21.

MSN 48G (cm)	*Buccinosphaera invaginata*	*Collosphaera tuberosa*	*Amphirhopalum ypsilon*	*Lithopera bacca*	*Theocorythium trachelium trachelium*	*Anthocyrtidium angulare*	*Theocorythium vetulum*	*Pterocanium prismatium*	– = absent R = rare F = few C = common
0-2	R	R	R	F	C	–	–	–	Zone 1
25-7	–	R	R	F	C	–	–	–	
45-7	–	R	R	F	C	–	–	–	
65-7	–	R	R	R	C	–	–	–	
85-7	–	R	R	F	F	–	–	–	Zone 2
105-7	–	R	R	F	C	–	–	–	
125-7	–	R	R	F	C	–	–	–	
145-7	–	–	R	R	C	–	–	–	Zone 3

Table. 34.22.

MSN 49G (cm)	*Buccinosphaera invaginata*	*Collosphaera tuberosa*	*Amphirhopalum ypsilon*	*Lithopera bacca*	*Theocorythium trachelium trachelium*	*Anthocyrtidium angulare*	*Theocorythium vetulum*	*Pterocanium prismatium*	– = absent R = rare F = few C = common
0-2	R	R	R	F	F	–	–	–	
19-21	R	R	R	R	C	–	–	–	Zone 1
38-40	R	R	R	F	F	–	–	–	
60-2	–	R	R	F	C	–	–	–	Zone 2
80-2	–	–	R	R	C	–	–	–	
100-2	–	–	F	R	C	–	–	–	
110-2	–	–	F	F	F	–	–	–	
120-2	–	–	R	F	C	–	–	–	Zone 3
140-2	–	–	R	F	C	–	–	–	
160-2	–	–	F	F	F	–	–	–	
168-70	–	–	R	R	C	–	–	–	

Table 34.23.

MSN 50G (cm)	*Buccinosphaera invaginata*	*Collosphaera tuberosa*	*Amphirhopalum ypsilon*	*Lithopera bacca*	*Theocorythium trachelium trachelium*	*Anthocyrtidium angulare*	*Theocorythium vetulum*	*Pterocanium prismatium*	– = absent R = rare F = few C = common
0-2	R	F	R	F	F	–	–	–	Zone 1
20-2	R	F	R	F	F	–	–	–	
40-2	–	R	R	R	C	–	–	–	Zone 2
60-2	–	R	R	R	C	–	–	–	
70-2	–	R	R	R	C	–	–	–	
83-5	–	R	R	R	C	–	–	–	
100-2	–	–	F	R	C	–	–	–	Zone 3
120-2	–	–	R	F	F	–	–	–	
140-2	–	–	F	R	C	–	–	–	

BIBLIOGRAPHY

American Commission on Stratigraphic Nomenclature. 1961. Code of stratigraphic nomenclature. *Bull. Amer. Ass. Petrol. Geol.* **45**, 645-65.

Arrhenius, G. 1952. Sediment cores from the east Pacific. *Rep. Swedish Deep-Sea Expedition,* **5** (3), 227 pp.

Brandt, K. 1885. Die koloniebildenden Radiolarien (Sphaerozoëen) des Golfes von Neapel. *Fauna u. Flora d. Golfes v. Neapel, monogr.* **13**, i-viii and 1-276, pls. 1-8.

Brandt, K. 1905. Zur systematik der koloniebildenden Radiolarien. *Zool. Jahrbuch, Jena, Suppl.* **8**, 311-52, pls. 9-10.

Emiliani, C. 1955. Pleistocene temperatures. *J. Geol.* **63** (6), 538-78.

Emiliani, C. and Flint, R. F. 1963. The Pleistocene record. In *The Sea* (M. N. Hill *ed.*), **3**, 888-927.

Ericson, D. B., Ewing, M., Wollin, G. and Heezen, B. C. 1961. Atlantic deep-sea sediment cores. *Bull. Geol. Soc. Am.* **72** (2), 193-285.

Geological Society of London. 1967. Report of the stratigraphical code sub-committee. *Proc. Geol. Soc. Lond.* no. **1,638**, 75-87.

Haeckel, E. 1862. *Die Radiolarien (Rhizopoda Radiolaria). Eine Monographie.* Reimer, Berlin. xiv + 572 pp., 35 pls.

Haeckel, E. 1887. Report on the Radiolaria collected by H.M.S. *Challenger* during the years 1873-76. *Rept. Voyage 'Challenger', Zool.* **18**, clxxxviii + 1,803 pp., 140 pls., 1 map.

Huxley, T. 1851. Zoological notes and observations made on board H.M.S. *Rattlesnake.* III. Upon *Thalassicolla,* a new zoophyte. *Ann. Mag. Nat. Hist., ser. 2,* **8** (48), 433-42, pl. 16.

Nakaseko, K. 1964. Liosphaeridae and Collosphaeridae (Radiolaria) from the sediment of the Japan Trench. (On Radiolaria from the sediment of the Japan Trench. 1.). *Osaka Univ. Sci. Repts.* **13** (1), 39-57, 3 pls.

Nigrini, C. 1967. Radiolaria in pelagic sediments from the Indian and Atlantic Oceans. *Bull. Scripps Inst. Oceanography,* **11**, 1-106, pls. 1-9.

Olausson, E. 1965. Evidence of climatic changes in North Atlantic deep-sea cores, with remarks on isotopic palaeotemperature analysis. *Progress in Oceanography* (ed. M. Sears), **3**, 221-52.

Popofsky, A. 1917. Die Collosphaeriden, mit Nachtrag zu den Spumellarien und Nassellarien. *Deutsche Südpolar-Exped. 1901-1903,* **16** (Zool. vol. 8), no. 3, 235-78, pls. 13-17.

Riedel, W. R. 1957. Radiolaria: a preliminary stratigraphy. *Rep. Swedish Deep-Sea Expedition,* **6** (3), 59-96, pls. 1-4.

Riedel, W. R., Bramlette, M. N. and Parker, F. L. 1963. 'Pliocene-Pleistocene' boundary in deep-sea sediments. *Science,* **140**, 1238-40.

Riedel, W. R. and Foreman, H. P. 1961. Type specimens of North American Paleozoic Radiolaria. *Journ. Paleont.* **35** (3), 628-32.

Riedel, W. R. and Funnell, B. M. 1964. Tertiary sediment cores and microfossils from the Pacific Ocean floor. *Quart. J. geol. Soc. Lond.* **120**, 305-68, pls. 14-32.

Ross, D. A. and Riedel, W. R. 1967. Comparison of upper parts of some piston cores with simultaneously collected open-barrel cores. *Deep Sea Res.* **14**, 285-94, 1 fig.

Sackett, W. M. 1960. Proactinium-231 content of ocean water and sediments. *Science,* **132**, 1761-2.

Schröder, O. 1929. Die nordischen Spumellarien: Sphaerocollida. In *Nordisches Plankton* (K. Brandt and C. Apstein *eds.*), pt. 16, pp. 91-120.

Strelkov, A. A. and Reschetnyak, V. V. 1962. Kolonialyne radiolyarii Spumellaria yuzhno-kitaiskogo morya (Raion yuzhnoi okonechnosti ostrova Hainan). *Studia Marina Sinica,* no. **1** (Aug. 1962), 121-39.

35. VERTICAL AND HORIZONTAL DISTRIBUTION OF PTEROPODS IN QUATERNARY SEQUENCES

Y. HERMAN

Summary: The spatial distribution of pteropods is correlatable with present-day climatic zones. This correlation is expressed by a decrease in number of species and genera from low to high latitudes, simultaneously with generic and specific changes. Vertical, temperate-dependent zonation is most obvious in the Mediterranean basins where, during glacials, high latitude faunas replaced low latitude elements.

Biostratigraphic correlations accompanied by radiometric datings indicate that climatic oscillations in the Mediterranean were synchronous with those in the Red Sea, but did not coincide with climatic fluctuations in the Arctic.

Knowledge of their depth habitat makes Pteropods with restricted depth ranges valuable indicators of past sea-level fluctuations in shallow continental shelf seas.

INTRODUCTION

Since the basic work of Meisenheimer (1905) on the distribution of pteropods in oceans and seas, a large number of publications have dealt with their horizontal and bathyal distributional patterns. Among the most comprehensive studies on this subject, the contribution of Tesch (1946, 1948) should be cited.

More recently, Hida (1957), Fager and McGowan (1963), and Rampal (1965) have attempted to correlate the occurrence of these holoplanktonic molluscs with various physical and chemical parameters of their environments. These studies have shown that pteropods, like other oceanic macroscopic zooplankton, can be used as indicators and tracers of specific water masses.

About thirty-five species and subspecies of living euthecosomatous pteropods have been described to date; the majority of these species are confined to the warm water provinces.

Although systematic large-scale sampling of the sea floor started less than two decades ago, the distributional patterns of many groups of planktonic shell-bearing organisms, in particular those of Foraminifera, are well known. Preservation of pteropod oozes, however, is limited to warm, shallow basins, and consequently information on their distribution and abundance is still limited.

The objectives of the present study were twofold:

1. To determine the areal distribution of pteropods in Postglacial sediments and to attempt to correlate the patterns obtained with known environmental factors.
2. To study the vertical distribution and abundance of euthecosomes in core sediments, and to interpret the changes in assemblages with time, in terms of their present-day distributional pattern.

MATERIAL AND METHODS

For faunal analysis dredge and piston core sediments were studied; the cores were sampled at 10-20 centimetre intervals, both the core and dredge sediments were washed through a 74μ sieve, and material covering a tray of 50 square centimetres was examined under a stereoscopic microscope. All pteropods and planktonic Foraminifera were identified and counted; from the population data obtained in these samples, percentages and ratios were calculated. Climatic oscillations were deduced from changes in population composition.

THE SPECIES OF EUTHECOSOMATOUS PTEROPODS

A detailed synonymy and taxonomic discussion has not been attempted here. The nomenclature used by Tesch (1946, 1948) will be followed in most cases.

The species whose distributions in marine sediments are known are listed in alphabetical order:

Limacina bulimoides (d'Orbigny)
Limacina helicina (Phipps)
Limacina inflata (d'Orbigny)
Limacina lesueuri (d'Orbigny)
Limacina retroversa (Fleming)
Limacina trochiformis (d'Orbigny)
Cavolinia gibbosa (Rang)
Cavolinia globulosa (Rang)
Cavolinia inflexa (Lesueur)
Cavolinia inflexa (Lesueur) f. *longa* (Boas)
Cavolinia longirostris (Lesueur)
Cavolinia tridentata Forskal
Cavolinia uncinata (Rang)
Clio cuspidata (Bosc)
Clio polita Pelseneer
Clio pyramidata Linné
Clio pyramidata Linné f. *pyramidata* Linné
Clio pyramidata Linné f. *convexa* Boas
Creseis acicula Rang
Creseis chierchiae (Boas)
Creseis conica Eschscholtz
Creseis virgula Rang
Creseis virgula Rang f. *constricta* Chen
Cuvierina columnella (Rang)
Diacria quadridentata (Lesueur)
Diacria trisponosa (Lesueur)
Hyalocylix striata (Rang)
Styliola subula (Quoy and Gaimard)

FACTORS CONTROLLING PTEROPOD DISTRIBUTION

Although pteropods are ubiquitous and abundant in oceans and seas, pteropod oozes are generally limited to shallow marine basins with high temperature and salinity (Sverdrup, Johnson and Fleming 1942), owing to their highly soluble and fragile tests.

In addition to the above-mentioned factor, preservation of planktonic microfaunal shells in bottom sediments depends upon their transport and removal by currents and rates of accumulation of detrital sediments (Arrhenius 1963). As a result, fewer tests are present in sediments than in the overlying water layers.

Solution of calcareous foraminiferal shells is selective; it varies with genera and species, the more delicate forms being eliminated first, thus changing species composition (Berger 1967; Boltovskoy in this volume). Solution is greater at high latitudes (Berger 1967) and differs in the various oceans (Sverdrup, Johnson and Fleming 1942).

The criteria which determine the selective solution of Foraminifera most probably also apply to pteropods. For instance, pteropods are present in a Caribbean core raised from 3,658 metres depth (V19-19) but are absent in equatorial and tropical Indian Ocean sediments, where water depth ranges from 1,150 to 4,200 metres as well as in eastern tropical and equatorial Pacific Ocean cores, raised from deep water (1,000-5,800 metres).

HORIZONTAL DISTRIBUTION IN CORES AND DREDGES

Pelseneer's study (1888) of H.M.S. *Challenger* material marks the beginning of investigation of pteropod distribution in marine sediments. More recently as new techniques have been developed, additional data on their distribution and abundance has become available. Their occurrence, plotted on a world map (Fig. 35.1) is based on piston core and dredge samples largely from the northern hemisphere, and on the following publications: Pelseneer (1888), Peck (1893), Stubbings (1937), Tesch, Helder and Holland (1912), Henbest (1942), Chen (1964), and Herman (1965, 1968). A synopsis of their distribution and abundance in several marine basins is given in Table 35.1.

Their latitudinal zonation, shown in Table 35.1 is apparently related to the temperature preference of various species and is expressed in the decrease in number of species and genera from low to high latitudes, simultaneously with changes in their genetic and specific composition. Table 35.1 summarizing their latitudinal range should be regarded as provisional and subject to modifications as additional quantitative and qualitative data become available.

Indian Ocean

Dredge samples collected off the west coast of India and from the Gulf of Oman (Tables 35.2, 35.3 and 35.4) are of particular interest because they were raised from shallow water and the salinity, temperature, and oxygen content for each station are available (Tables 35.3 and 35.4).

On the open shelf, off the west coast of India in the shallowest station (40 metres), the epipelagic *Creseis* sp. is dominant. The ratio of this species to *Limacina inflata,* the second most abundant form, is about 9 to 1; this ratio decreases with increasing water depth, *Limacina inflata* becoming more abundant than *Creseis.* At the deepest staion (310 metres) the ratio is 1 to 5 (Table 35.3).

The number of species and individuals increases from shallow to deep water (Table 35.3) and beyond 100 metres the assemblages are comparable to open ocean tropical faunas. Although the data are limited, this variation in distribution suggests that pteropods may be useful indicators of past sea-level fluctuations in shallow water sediments.

The uppermost 30 centimetres of fourteen cores raised in the equatorial and tropical region of the central basin were examined. Except for rare pteropod fragments, the sediment's coarse fraction ($> 74\ \mu$) is made up exclusively of planktonic foraminiferal shells. In core LSDA 114 G_b, pteropods were observed in two layers. Both layers, however, contain an *in situ* deep-sea fauna mixed with displaced shallow-water elements, such as *Halimeda* sp. and *Elphidium* spp.

With one exception (V14-111), pteropods are absent in the deep water cores of the Gulf of Aden. In the shallow water core V14-112, pteropods abound, the fauna of the Gulf closely resembles that of the neighbouring Red Sea. There are, however, a few quantitative differences; for example, the relative abundance of epipelagic *Creseis* sp. and *Limacina trochiformis* is greater here than in the Red Sea. The shallow depth of water (155 metres) may account for this difference.

Atlantic Ocean

Table 35.5 was compiled to illustrate the great variability in the relative abundance of different species in closely spaced dredges collected from the Blake Plateau (Table 35.6). These variations in the faunal composition of surface samples may be due to the effect of the Gulf Stream.

VERTICAL DISTRIBUTION IN PISTON CORES

Mediterranean and Red Seas

In inland seas, seasonal as well as long range climatic variations are relatively marked (Emiliani and Milliman 1966) because of their comparatively small size and proximity to land.

Preservation of pteropod tests in these warm, inland seas is excellent as a result of the relatively shallow water depth, high bottom water temperatures, and limited number of mud feeders restricted by the absence or meagre

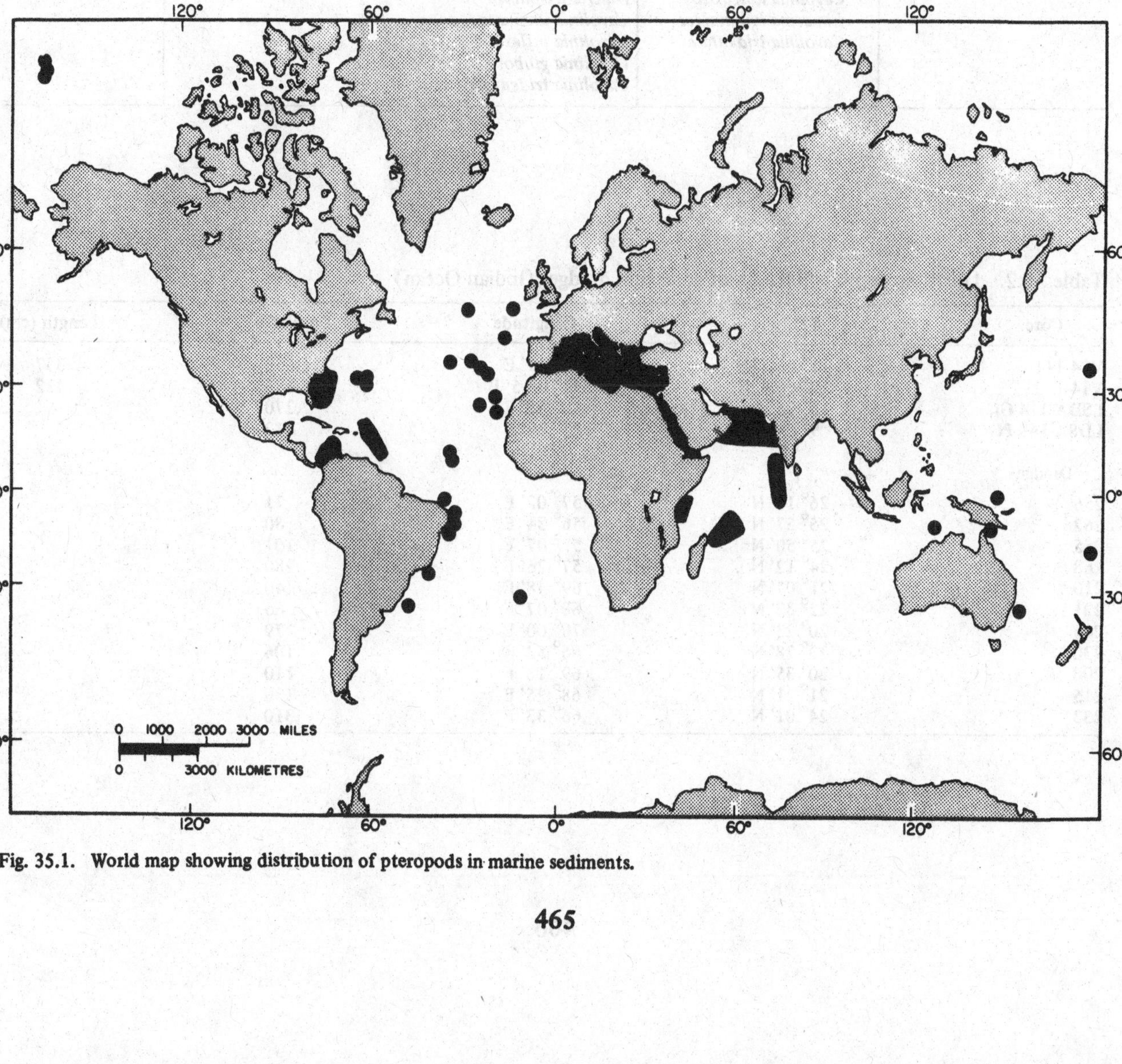
120°
60°
0°
60°
120°
60°
30°
0°
30°
60°
60°
30°
0°
30°
60°
0 1000 2000 3000 MILES
0 3000 KILOMETRES
120°
60°
0°
60°
120°

Fig. 35.1. World map showing distribution of pteropods in marine sediments.

Table 35.1. Latitudinal distribution of pteropod species in deep-sea sediments – species listed in order of decreasing abundance

Tropical Red Sea N.W. Indian Ocean (cores)	Tropical-Subtropical E. Mediterranean (cores)	Subtropical Atlantic (dredges)	Temperate North Atlantic (cores)	North Polar (cores)
		(Small shells or embryonic portion of large shells)		
Limacina inflata	*Limacina inflata*	*Styliola subula*	*Limacina retroversa*	*Limacina helicina*
Creseis sp.	*Creseis acicula*	*Creseis* sp.	*Clio pyramidata*	
Clio pyramidata f. *convexa*	*Creseis* sp.	*Limacina inflata*		
Diacria quadridentata	*Styliola subula*	*Clio pyramidata* f. *pyramidata*		
Hyalocylix striata	*Clio pyramidata*	*Limacina bulimoides*		
Limacina trochiformis	*Limacina trochiformis*	*Limacina lesueuri*		
Cavolinia longirostris	*Clio polita*	*Limacina trochiformis*		
Cavolinia uncinata	*Clio cuspidata*			
	Hyalocylix striata	(Large shells)		
	Cavolinia uncinata	*Diacria quadridentata*		
	Cavolinia longirostris	*Diacria trispinosa*		
	Limacina bulimoides	*Cavolinia uncinata*		
	Cavolinia tridentata	*Cavolinia inflexa*		
		Cavolinia gibbosa		
		Cavolinia tridentata		

Table 35.2. Location, depth and length of cores and dredges (Indian Ocean)

Core	Latitude	Longitude	Depth (m)	Length (cm)
V14-111	11° 56.5′ N	43° 41′ E	1,551	337
V14-112	12° 00.4′ N	44° 08.5′ E	155	412
LSDA-114 G_b	10° 36′ S	59° 52′ E	2,270	
LDSA-149 PG	9° 54′ S	127° 39′ E	432	
Dredge				
256	26° 16′ N	57° 02′ E	71	
262	25° 37′ N	56° 34′ E	80	
255	25° 50′ N	57° 07′ E	101	
268	24° 12′ N	57° 26′ E	280	
210	21° 07′ N	69° 48′ E	40	
221	22° 32′ N	68° 07′ E	66	
206	20° 23′ N	70° 00′ E	79	
230	23° 28′ N	66° 53′ E	106	
208	20° 35′ N	69° 18′ E	110	
215	21° 21′ N	68° 25′ E	130	
233	24° 01′ N	66° 33′ E	310	

Table 35.3. Distribution of pteropods on the Indian Ocean Shelf off the west coast of India

Water depth (m)	40	66	79	106	110	130	310
Cavolinia longirostris	1	2	2	20	–	4	1
Cavolinia sp. (tip)	–	–	–	–	3	2	2
Cavolinia uncinata	–	–	–	3	3		
Clio pyramidata f. *convexa*	–	–	–	–	3	12	1
Creseis acicula *Creseis* sp.	54	50	101	173	98	92	49
Diacria quadridentata	–	–	1	24	–	4	–
Diacria quadridentata (tip)	–	–	–	3	3	6	4
Hyalocylix striata	1	–	–	–	–	–	1
Limacina inflata	6	50	139	227	90	137	230
Surface water salinities Surface water temperatures	34.38‰ 27.8° C	36.17‰ 26.8° C	36.80‰ 28.0° C	36.68‰ 27.2° C	36.13‰ 28.3° C	35.05‰ 27.3° C	36.42‰ 25.7° C
Bottom water salinities Bottom water temperatures Bottom water oxygen (ml/l)	36.31‰ 25.7° C 2.31	36.25‰ 24.1° C 0.79	36.18‰ 22.9° C 0.99	36.24‰ 22.3° C 0.86	36.14‰ 22.8° C 1.21	36.18‰ 22.6° C 2.09	36.32‰ 22.6° C? 0.44

(Data obtained in mid-November 1963 by the Smithsonian Institution.)

Table 35.4. Distribution of pteropods in the Gulf of Oman

Water depth (m)	71	89*	101	280*
Cavolinia longirostris	19	2	252	62
Cavolinia uncinata	–	1	1	–
Creseis acicula	5	–	193	–
Creseis sp.	8	–	3	–
Diacria quadridentata	3	–	59	30
Hyalocylix striata	–	–	1	–
Limacina inflata	5	–	4	–
Limacina trochiformis	–	–	4	–
Surface water salinities Surface water temperatures	36.68‰ 26.3° C	36.82‰ 26.1° C	36.57‰ 26.0° C	36.55‰ 26.0° C
Bottom water salinities Bottom water temperatures Bottom water oxygen (ml/l)	36.55‰ 25.6° C 3.61	38.34‰ 26.1° C 2.05	36.35‰ 24.5° C 2.40	36.53‰ 16.2° C 0.28

(Data obtained in mid-November 1963.)
*Picked out at the Smithsonian Inst. Sorting Center

Table 35.5. Distribution and abundance of pteropods in Atlantic Ocean dredges

Water depth (m)	457	722	809	869	1015	1100	1127	1170
Pteropoda (small or embryonic portion of large shells)								
Styliola subula	145	130	83	160	126	26	140	108
Creseis sp.	185	22	21	50	62	7	250	137
Limacina inflata	21	32	42	90	47	4	105	186
Clio pyramidata f. *pyramidata*	12	11	11	24	22	1	18	–
Limacina bulimoides	7	1	9	12	10	2	18	56
Limacina lesueuri	3	1	–	4	1	–	7	8
Limacina trochiformis	2	–	2	4	3	1	2	11
(Large shells)								
Diacria quadridentata	8	7	19	16	26	2	16	3
Diacria trispinosa	4	4	–	4	10	1	5	6
Cavolinia uncinata	2	4	10	4	4	–	5	3
Cavolinia inflexa	2	7	3	4	3	3	4	1
Cavolinia gibbosa	–	1	2	2	1	–	3	–
Cavolinia tridentata	–	–	1	–	–	–	–	–
**Cavolinia longirostris*	8	80	20	10	8	–	3	1
**Cuvierina columnella*	–	3	4	8	7	–	2	1

*Random

Table 35.6. Location depth, and length of cores and dredges (Atlantic Ocean)

Core	Latitude	Longitude	Depth (m)	Length (cm)
V19-17	12° 40′ N	74° 02′ W	3,782	1,925
V19-18	13° 12′ N	76° 12′ W	3,802	1,037
V19-19	13° 14′ N	78° 22′ W	3,760	962
Dredge				
2451	27° 21′ N	79° 41′ W	457	
2445	26° 39′ N	79° 10′ W	700	
2460	28° 08′ N	79° 15′ W	722	
2473	30° 15′ N	79° 15′ W	809	
2456	27° 37′ N	78° 30′ W	869	
2350	29° 36′ N	77° 11′ W	965	
2351	29° 29′ N	77° 29′ W	1,015	
2334	33° 02′ N	76° 32′ W	1,100	
2353	29° 00′ N	77° 30′ W	1,127	
2354	28° 57′ N	77° 15′ W	1,170	

supply of oxygen (Kuenen 1950; Ekman 1953). Furthermore, as a result of rapid rates of sedimentation, seas offer higher resolution of Quaternary events than do ocean basins.

Succession of Climatic Events

Biostratigraphic correlation between cores supplemented by radiometric dating was used to determine the continuity in the sedimentary sequence and the synchroneity of climatic events in cores located thousands of miles apart. The following cyclic and non-cyclic criteria were used to demonstrate the equivalence of stratigraphical units and the continuity of the sedimentary record:

1. Duration and amplitude of successive warm and cool cycles based on planktonic population analysis.
2. Gross lithological character of cores.
3. Limited stratigraphical occurrence of several species (e.g. *Hyalocylix striata* and *Globigerinoides conglobata*).
4. Shifts in coiling ratio of *Globorotalia truncatulinoides.*
5. Occurrence of characteristic layers, including black sapropelitic layers, diatomaceous and ash layers, the latter two in the Mediterranean and the indurated calcareous layers in the Red Sea.
6. Radiomatric age determinations.

Results based on the aforementioned criteria indicate that the changes in the planktonic faunal composition with time, which took place as a response to climatic oscillations were contemporaneous in the two seas and were accompanied by comparable changes in hydrographic conditions. Alteration in the water circulatory system during low sea levels due to the sill barriers took place in both basins. This resulted in stagnation of the deeper water layers and consequent deposition of sapropelitic muds.

Red Sea

Of the seventeen cores studied, ten represent continuous or nearly continuous sequences of sedimentation. Their location, length, and water depth is indicated in Table 35.7 and Fig. 35.2. Inasmuch as a detailed discussion of most of the core sediments has been given in two previous publications (Herman 1965, 1968), the data presented herein is based principally on cores RC 9-169, RC 9-173 and SDSE 166.

Faunal analysis

Pteropods and planktonic Foraminifera generally constitute 90-95 per cent of the calcareous faunal remains, the remainder being made up of Heteropods. Table 35.8 summarizes the composition of faunal remains in southern Red Sea cores. Apart from a few exceptions, the fauna of the northern sector resembles the assemblages from the south. Progressively northward there is an increase in the relative abundance of *Limacina bulimoides* simultaneously with a decrease in *Diacria quadridentata* and *Hyalocylix striata. Cavolinia longirostris* and *Cavolinia uncinata* were not observed in the northern region.

The following discussion of the succession of climatic events is based on faunal analysis of core RC 9-169 (Fig. 35.3).

The Postglacial is contained in the upper 140 centimetres and has a fauna similar to that of the present (Table 35.8). Between 116 and 145 centimetres several dark grey sapropelitic layers are present. The fauna of the black layers differs from that of the over- and underlying beds; there is an increase in abundance of epipelagic *Creseis* sp. and *Limacina trochiformis* simultaneously with

Table 35.7. Location, depth, and length of cores (Red Sea)

Core	Latitude	Longitude	Depth (m)	Length (cm)
V14-114	15° 17′ N	42° 02′ E	768	611
*V14-115	16° 04.5′ N	41° 28′ E	1,194	590
*V14-116	17° 26′ N	40° 20.5′ E	1,346	490
*V14-117	18° 48′ N	39° 31′ E	1,372	190
*V14-118	18° 37′ N	39° 03′ E	516	635
V14-119	20° 50′ N	38° 17′ E	1,611	202
V14-120	20° 26′ N	38° 13′ E	1,737	90
V14-121	23° 36′ N	36° 14′ E	384	120
*V14-122	23° 55′ N	36° 28′ E	1,486	540
*V14-123	24° 02′ N	36° 02′ E	816	590
*V14-124	25° 20′ N	36° 14′ E	2,012	490
*V14-125	26° 57′ N	34° 38′ E	956	612
V14-126	28° 25.5′ N	34° 44.5′ E	1,183	295
V14-127	27° 32′ N	34° 11′ E	1,291	490
*RC9-169	16° 19.9′ N	41° 22.3′ E	940	1,199
RC9-173	25° 46.9′ N	35° 50.7′ E	1,269	1,146
*Alb 166	17° 56′ N	39° 57′ E	1,283	1,091

*Cores representing continuous or nearly continuous sedimentation.

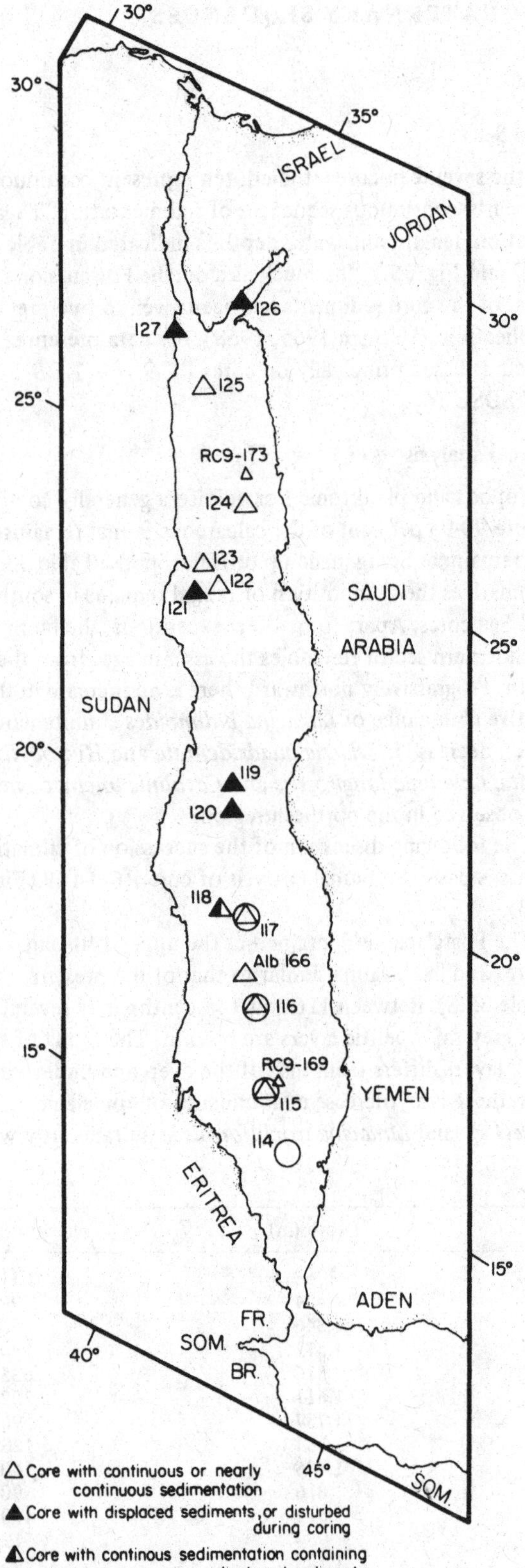
30°
30°
35°
ISRAEL
JORDAN
30°
126
127
125
25°
RC9-173
124
123
122
121
SAUDI
ARABIA
25°
SUDAN
20°
119
120
118
117
Alb 166
20°
116
15°
RC9-169
115
YEMEN
114
ERITREA
15°
ADEN
FR.
SOM.
BR.
40°
45°
SOM.
Core with continuous or nearly continuous sedimentation
Core with displaced sediments, or disturbed during coring
Core with continous sedimentation containing displaced sediments
Core containing volcanic ash
Core with continuous sedimentation containing volcanic ash

Fig. 35.2. Red Sea map showing location of cores.

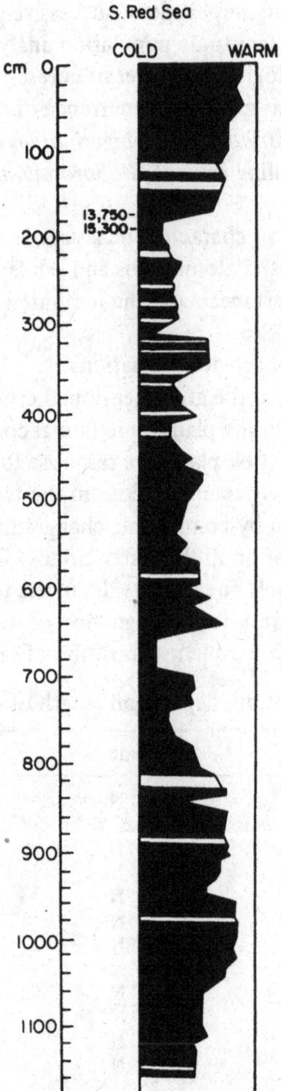
S. Red Sea
COLD
WARM
cm 0
100
13,750-
15,300-
200
300
400
500
600
700
800
900
1000
1100
1200
RC9-169
Black Sapropelitic Mud

Fig. 35.3. Climatic curve of Red Sea core RC9-169 based on variation in abundance of the planktonic fauna.

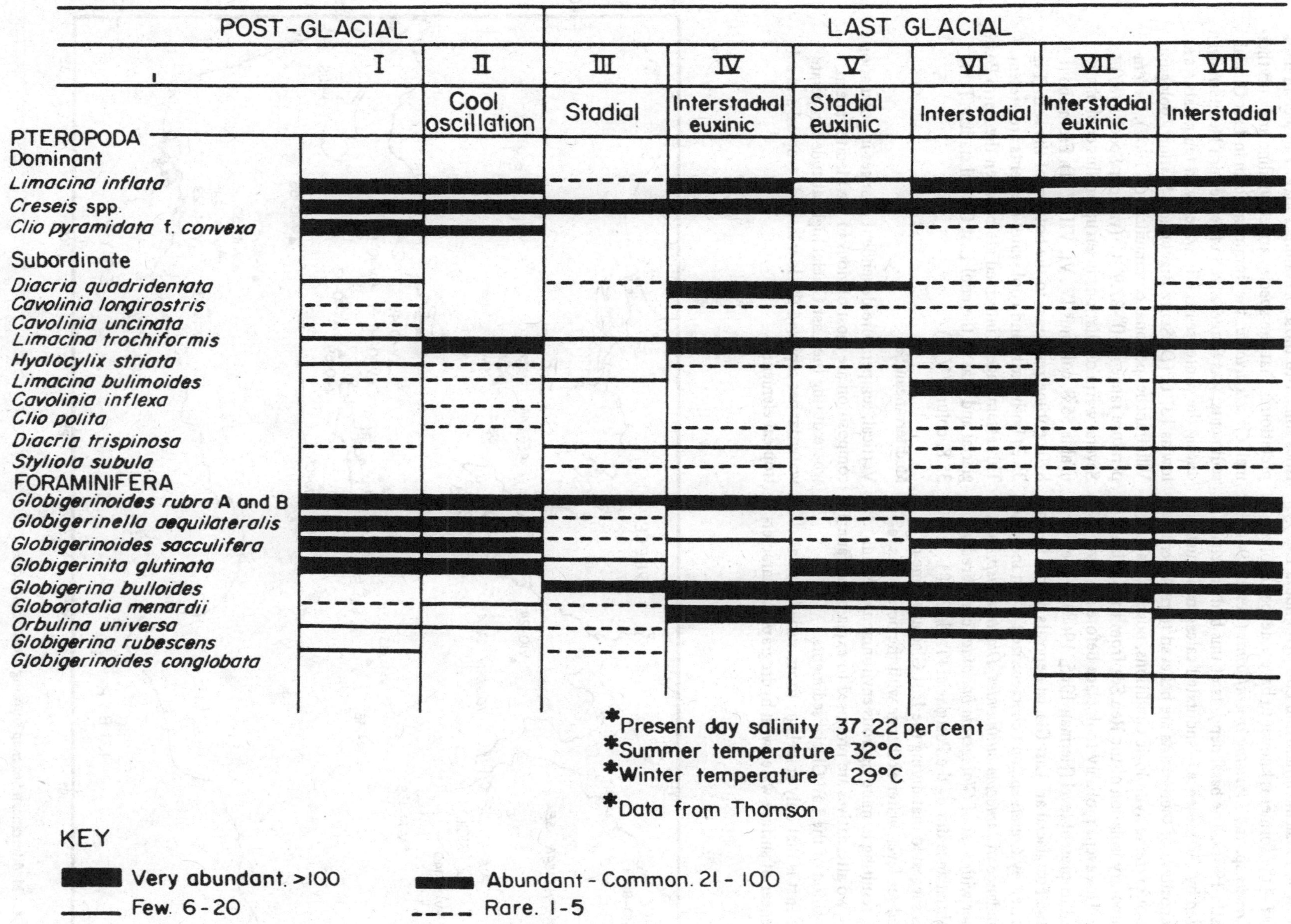
POST-GLACIAL
LAST GLACIAL
I
II
III
IV
V
VI
VII
VIII
Cool oscillation
Stadial
Interstadial euxinic
Stadial euxinic
Interstadial
Interstadial euxinic
Interstadial
PTEROPODA
Dominant
Limacina inflata
Creseis spp.
Clio pyramidata f. convexa
Subordinate
Diacria quadridentata
Cavolinia longirostris
Cavolinia uncinata
Limacina trochiformis
Hyalocylix striata
Limacina bulimoides
Cavolinia inflexa
Clio polita
Diacria trispinosa
Styliola subula
FORAMINIFERA
Globigerinoides rubra A and B
Globigerinella aequilateralis
Globigerinoides sacculifera
Globigerinita glutinata
Globigerina bulloides
Globorotalia menardii
Orbulina universa
Globigerina rubescens
Globigerinoides conglobata
*Present day salinity 37.22 per cent
*Summer temperature 32°C
*Winter temperature 29°C
*Data from Thomson
KEY
Very abundant. >100
Abundant - Common. 21 - 100
Few. 6-20
Rare. 1-5

Table 35.8. Distribution of planktonic fauna in Southern Red Sea cores.

31

a decrease in *Limacina inflata* and *Clio pyramidata* f. *convexa* – the former is known to have a deeper habitat than *Creseis* sp. and *Limacina trochiformis* (Tesch 1946; Furnestin 1961). The basal deposits of this bed are characterized by the presence of indurated calcareous crusts. The deposition of these crusts are believed to represent a period of extreme hydrologic conditions, possibly caused by temporary isolation of the Red Sea from the Indian Ocean. This event probably took place before sea level reached its present level (Herman 1965, 1968). Underlying the Postglacial are Last Glacial deposits with a fauna dominated by *Creseis* sp., and characterized by an increase in abundance of *Limacina bulimoides, Diactria trispinosa,* together with that of *Globigerina bulloides* compared with their abundance during the Postglacial (Table 35.8).

It is suggested that during the Last Glacial a time of lowered sea level, water exchange with the ocean was reduced, resulting in more rigorous environmental conditions. Two alternatives are proposed to explain the faunal composition of the Last Glacial sediments.

The first is that only epipelagic forms survived in a thin veneer of surface water with hydrographic characteristics similar to those prevailing today. The second interpretation is that the species are euryhaline and eurythermal. *Creseis acicula,* the dominant form in Last Glacial sediments, was found alive in the western Mediterranean during the winter season, in water with temperatures as low as 13° C. (Dr Sentz-Braconnot, Station Zoologique Villefranche, personal communication 1967). Its temperature range is 10°-27.9° C (van der Spoel 1967). Several warm oscillations interrupted this cool phase (Table 35.8, columns IV, VI, VII, VIII). From about 820 centimetres down to the bottom of the core (Fig. 35.3) both warm and cool water elements are present. This assemblage is intermediate between the warm Postglacial and that of the cool Last Glacial stadial (Table 35.8, column VIII).

Mediterranean Sea

Vertical, temperature-dependent changes in population composition are most obvious in the Mediterranean, where during the Last Glacial high latitude temperate water faunas replaced the low latitude tropical and subtropical elements.

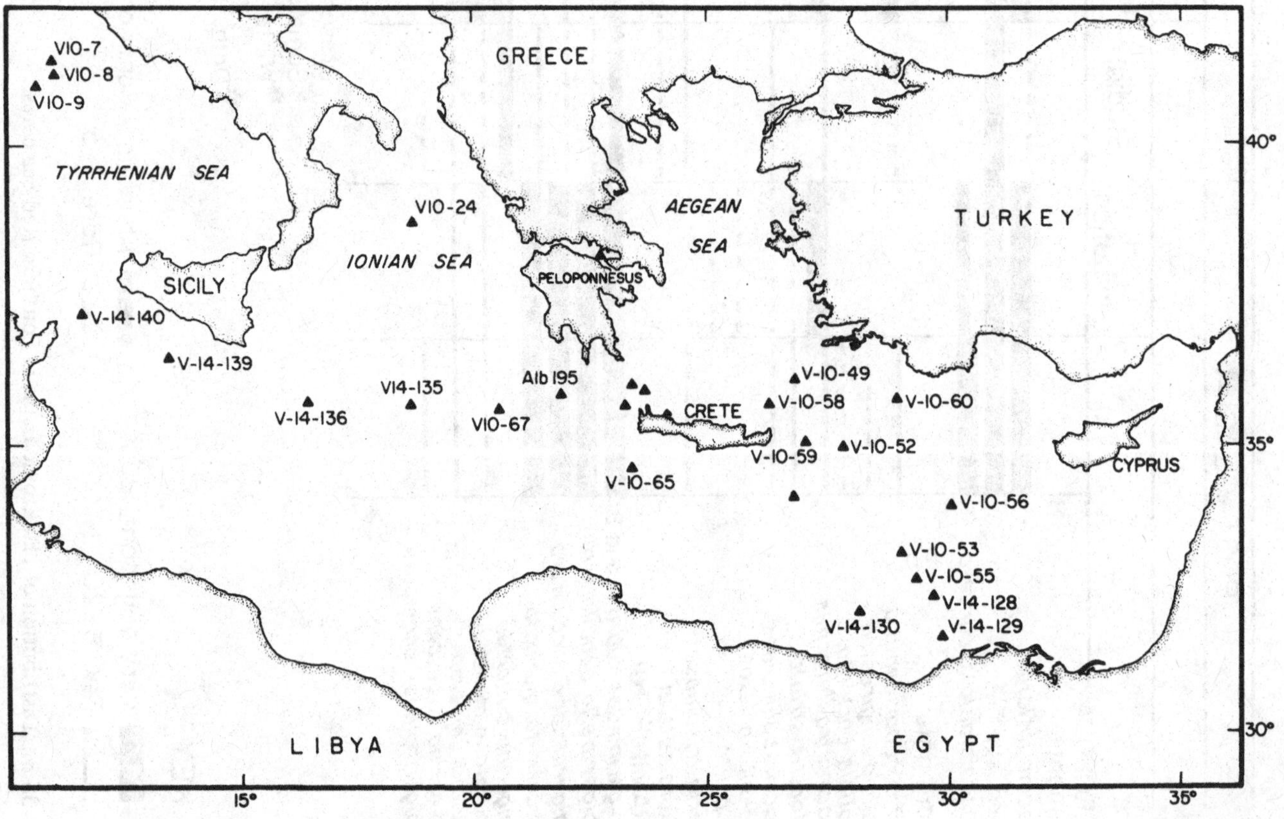
V10-7
V10-8
V10-9
GREECE
TYRRHENIAN SEA
V10-24
AEGEAN SEA
TURKEY
IONIAN SEA
PELOPONNESUS
SICILY
V-14-140
V-14-139
Alb 195
V14-135
V-14-136
V10-67
CRETE
V-10-49
V-10-58
V-10-60
V-10-59
V-10-52
CYPRUS
V-10-65
V-10-56
V-10-53
V-10-55
V-14-128
V-14-130
V-14-129
LIBYA
EGYPT
40°
35°
30°
15°
20°
25°
30°
35°

Fig. 35.4. Mediterranean Sea map showing location of cores.

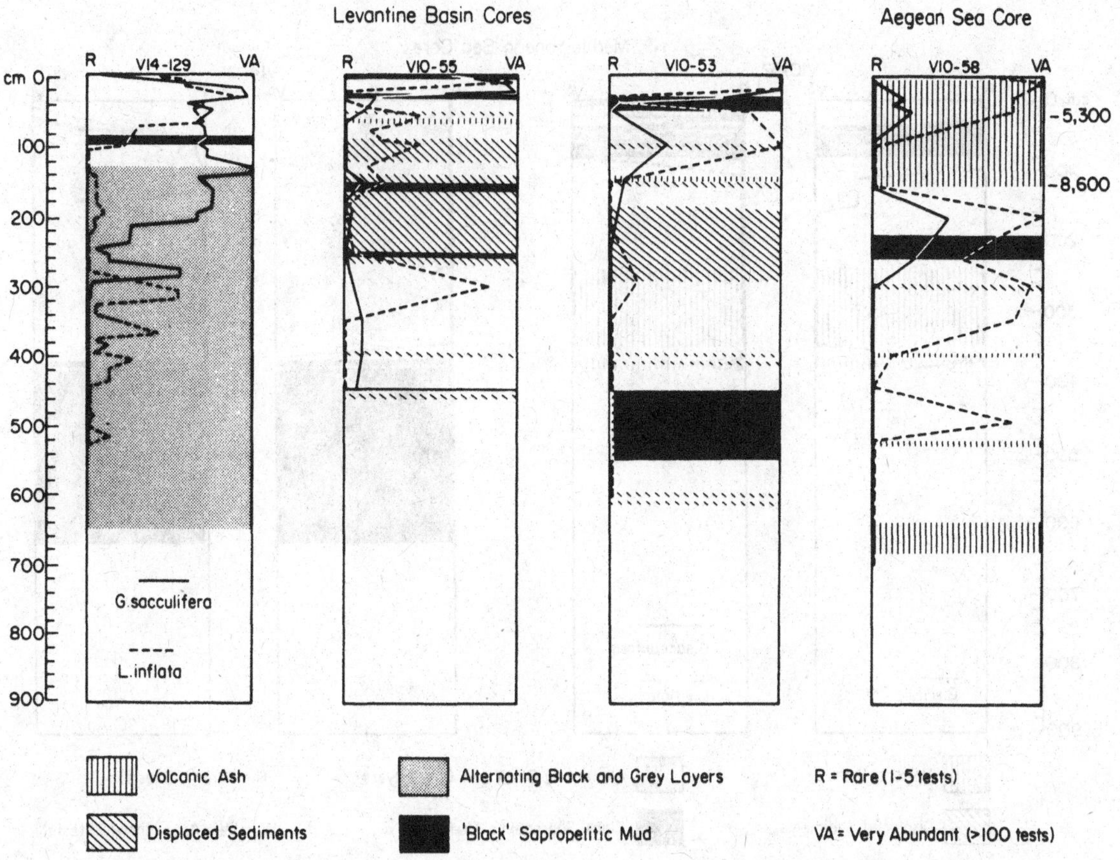

Levantine Basin Cores
Aegean Sea Core
cm
R
VA
V14-129
V10-55
V10-53
V10-58
-5,300
-8,600
G.sacculifera
L.inflata
Volcanic Ash
Displaced Sediments
Alternating Black and Grey Layers
'Black' Sapropelitic Mud
R = Rare (1-5 tests)
VA = Very Abundant (>100 tests)

Fig. 35.5. Levantine Basin and Aegean Sea cores showing sediment types and the variation in abundance of *Limacina inflata* and *Globigerinoides sacculifera*.

In Postglacial sediments, twenty species of pteropods were observed (Tables 35.9-16), whereas Last Glacial stadium deposits commonly contain only two forms; namely, the eurythermal *Clio pyramidata* f. *pyramidata* and temperate water *Limacina retroversa*.

Owing to its location in a region of tectonic activity, normal pelagic sedimentation was interrupted by the deposition of numerous, sporadically emplaced, terrigenous silts, sands, and ash layers. As a result, the majority of the cores raised from this sea encompass a much shorter time span than equally long Red Sea cores. Only six out of the fifty-four cores examined represent continuous sedimentation. Their location, length, and water depth are given in Table 35.17 and Fig. 35.4.

Sediments and microfaunal analysis

Two cores raised from the vicinity of the mouth of the Nile (V14-128, V14-129) contain high amounts of land-derived detritus and represent a short time interval. The sediment's dark grey colour is mainly due to finely divided organic debris.

Core V14-129 (Fig. 35.5) represents Postglacial deposits. The faunal composition remains essentially unchanged throughout the entire length of the core. The variation in abundance of *Limacina inflata* and *Globinerinoides sacculifera* is thought to represent mainly changes in rates of detrital sedimentation and/or adverse environmental conditions.

The same holds true for Core V14-128 (Figs. 35.6 and 35.7). In Fig. 35.7 the two curves represent the ratios of *Creseis* sp. to *Limacina inflata* and of *Globigerinoides sacculifera* to *Globigerinoides rubra.* The latter curve was constructed to verify the validity of the former.

As a result of high rates of detrital sedimentation, a detailed chronological sequence of events can be followed in V14-128 (Fig. 35.7). The Postglacial may be divided into three phases.

The first phase, contained in the upper 25 centimetres of the core, has a planktonic fauna dominated by *Limacina inflata, Limacina trochiformis, Creseis acicula, Styliola subula,* and *Clio pyramidata.*

The second phase, contained between 25-120 centi-

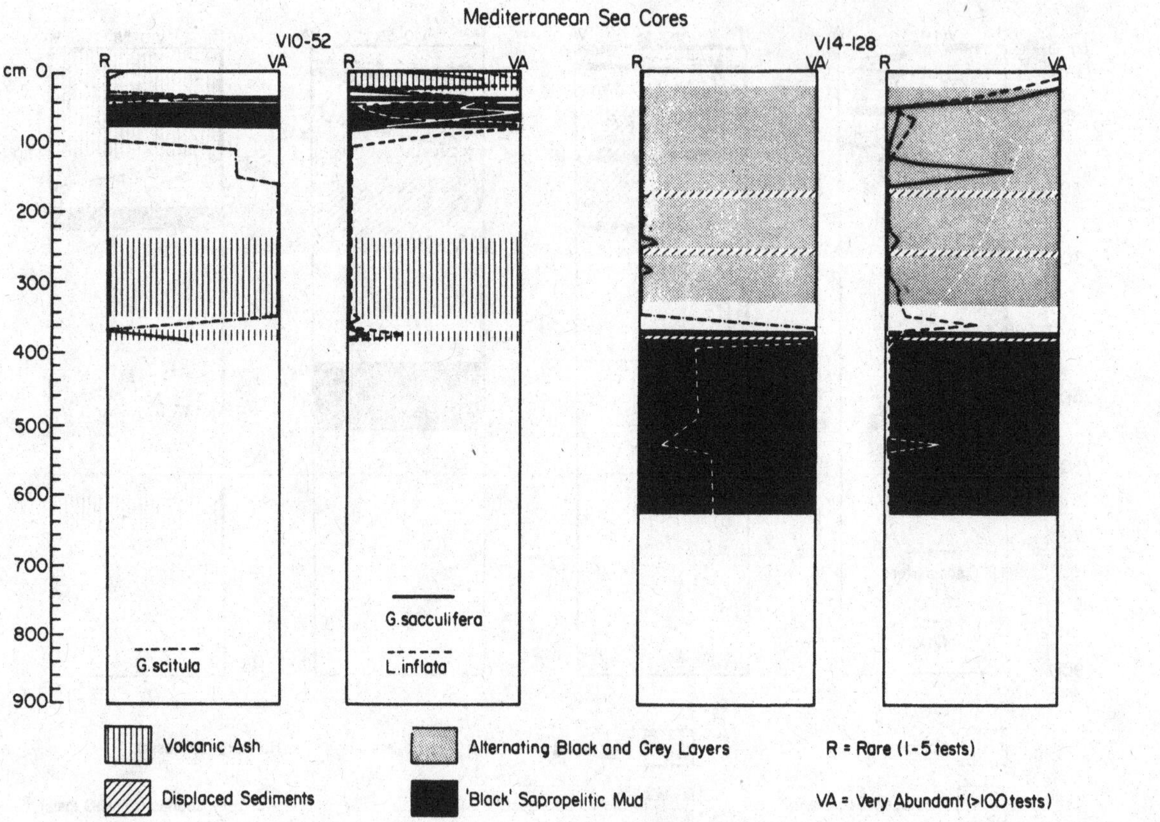
Mediterranean Sea Cores
V10-52
V14-128
cm 0
100
200
300
400
500
600
700
800
900
R
VA
G.scitula
G.sacculifera
L.inflata
Volcanic Ash
Displaced Sediments
Alternating Black and Grey Layers
'Black' Sapropelitic Mud
R = Rare (1-5 tests)
VA = Very Abundant (>100 tests)

Fig. 35.6. Eastern Mediterranean Sea cores showing sediment types and the variation in abundance of *Limacina inflata, Globigerinoides sacculifera,* and *Globorotalia scitula.*

metres is devoid of benthonic Foraminifera. The dominant pteropods are *Creseis* spp., *Clio pyramidata,* and *Diacria trispinosa.* The presence of black plant debris and iron sulphide minerals, in addition to the absence of benthonic Foraminifera, suggests an euxinic phase.

In the third phase, below 120 centimetres, there is a gradual increase in cold water Foraminifera. Pteropods are scarce or absent. The Postglacial/Last Glacial boundary is at about 360 centimetres and it is marked by a shift in the coiling direction of *Globoratalia truncatulinoides.* The Postglacial pteropods are replaced by *Limacina retroversa* and *Clio pyramidata.*

Fig. 35.8 represents the climatic variations in eastern Mediterranean cores, V10-65 and V10-67. The climatic curves were deduced from changes in the planktonic fauna. The right-hand curve (Core V10-67) is based on a continuous sequence of samples; the left-hand curve is drawn from core V10-65 analysed at 10 centimetre intervals.

The Postglacial/Last Glacial boundary is about 25 centimetres below the top, in both V10-65 and V10-67; at this level *Globoratalia truncatulinoides* shifts its coiling direction from dextral in the glacial, to sinistral in the Post-glacial sediments.

The Last Glacial deposits are characterized by the presence of northern temperate-water *Limacina retroversa* and the eurythermal *Clio pyramidata* f. *pyramidata,* and the absence of warm water planktonic faunas. This cold phase was interrupted by several mild oscillations of various lengths and amplitudes as indicated in the two curves of Fig. 35.8. Inasmuch as pteropods are scarce or absent in Last Glacial sediments, the climatic curves are based principally on the planktonic foraminiferal record. Core V10-67 penetrates sediments deposited during a long interstadial (760-880 centimetres), whereas core V10-65 located in the vicinity of Crete, where rates of sedimentation are thought to have been higher, did not reach the sediments laid down during the interstadial penetrated by core V10-67. Additional C^{14} determinations are awaited to confirm the biostratigraphic correlation between the two cores. A synopsis of the assemblages contained in Levantine, South Aegean, and Southern Ionian basins, cores is given in Table 35.14.

Table 35.9. Distribution of euthecosomatous pteropods in marine sediments

	Atlantic Ocean	Caribbean Sea	Indian Ocean	Pacific Ocean	Mediterranean Sea	Red Sea	North Polar Sea
Family Cavoliniidae							
Cavolinia gibbosa	+	–	+	+ *	–	–	–
Cavolinia globulosa	–	–	+ *	– *	–	+	–
Cavolinia inflexa	+	–	+ *	+ *	+	+	–
Cavolinia longirostris	+	–	+ *	+ *	+	+	–
Cavonilia tridentata	+	–	+ *	– *	+	+ *	–
Cavolinia uncinata	+	–	+	+ *	+	+	–
Clio cuspidata	+	+	+	– *	+	–	–
Clio polita	+	+	+	– *	+	+	–
Clio pyramidata	+	+	+	+ *	+	+ *	–
Clio pyramidata f. *convexa*	–	–	+	+	–	+	–
Creseis acicula	+	+	+	+ *	+	+	–
Creseis chierchiae	–	–	+	– *	–	+	–
Creseis conica	+	–	+	+ *	+	+	–
Creseis virgula	+	–	+	+ *	+	+	–
Creseis virgula constricta	+	+	+	– *	+	+	–
Cuvierina columnella	+	–	+ *	+ *	–	–	–
Diacria quadridentata	+	+	+	+ *	+	+	–
Diacria trispinosa	+	+	+	+ *	+	+	–
Hyalocylix striata	+	–	+	+ *	+	+	–
Styliola subula	+	+	+	+ *	+	+	–
Family Limacinidae							
Limacina bulimoides	+	+	+	+ *	+	+	–
Limacina helicina	–	–	+ *	– *	–	+ *	+
Limacina inflata	+	+	+	+ *	+	+	–
Limacina lesueuri	+	+	–	+ *	–	–	–
Limacina retroversa	+ *	–	–	+ *	+	–	–
Limacina trochiformis	+	+	+	+ *	+	+	–

*Sources indicated in the Bibliography.

Table 35.10. Pteropodal assemblages from Levantine Basin cores

Postglacial	Glacial
Dominant: >90 per cent	
Limacina inflata	
Creseis acicula	
Creseis sp.	
Styliola subula	
Clio pyramidata	*Clio pyramidata* f. *pyramidata*
Limacina trochiformis	*Limacina retroversa*
Subordinate: <10 per cent	
Clio polita	
Clio cuspidata	
Hyalocylix striata	
Cavolinia uncinata	
Cavolinia longirostris	*Present-day salinity: 38‰
Limacina bulimoides	*Summer temperature: 29° C
Cavolinia tridentata	*Winter temperature: 16° C
	*Data from Wüst.

Table 35.11. Pteropodal assemblages from southern Aegean Sea cores

Postglacial	Interstadial	Glacial
Dominant >90 per cent	Dominant >90 per cent	
Limacina inflata	*Limacina inflata*	
Creseis sp.	*Creseis* sp.	
Styliola subula	*Clio pyramidata*	
	Limacina trochiformis	
Subordinate <10 per cent	Subordinate <10 per cent	*Limacina retroversa*
Clio pyramidata	*Diacria quadridentata*	*Clio pyramidata* f. *pyramidata*
Limacina trochiformis	*Hyalocylix striata*	
Limacina bulimoides	*Cavolinia* sp.	
Diacria trispinosa		
Cavolinia sp.		*Present-day salinity: 39‰
Hyalocylix striata		*Summer temperature: 25° C
		*Winter temperature: 12° C
		*Data from Wüst.

Table 35.12. Pteropodal assemblages from Ionian Sea cores

Postglacial	Glacial
Dominant >90 per cent	
Limacina inflata	
Creseis acicula	
Creseis conica	
Creseis virgula constricta	
Clio pyramidata	*Limacina retroversa*
Styliola subula	*Clio pyramidata* f. *pyramidata*
Subordinate <10 per cent	
Clio polita	
Limacina trochiformis	
Limacina bulimoides	
Cavolinia inflexa f. *longa*	*Present-day salinity: 38.65‰ 39‰
Diacria trispinosa	*Summer temperature: 24° C
	*Winter temperature: 16° C
	*Data from Wüst.

Table 35.13. Faunal assemblages from eastern Mediterranean Sea cores – species listed in order of decreasing abundance

Postglacial	Interstadial	Glacial
Pteropoda		
Limacina inflata *Creseis acicula* *Creseis conica* *Styliola subula* *Clio pyramidata*	*Clio pyramidata* f. *pyramidata* *Limacina inflata* *Diacria trispinosa* *Creseis* sp. *Limacina trochiformis* *Clio cuspidata*	*Limacina retroversa* *Clio pyramidata* f. *pyramidata*
Foraminifera		
Globigerinoides rubra f. *B* *Globigerinoides sacculifera* *Globigerinella aequilateralis* *Globigerinoides rubra* (pink)	*Globigerinoides rubra* f. *B* *Globorotalia truncatulinoides* *Globigerinoides sacculifera* *Globigerinella aequilateralis* *Globigerina bulloides* *Globigerinoides conglobata*	*Globigerina bulloides* *Globorotalia scitula* *Globorotalia inflata* *Present-day salinity: 38->39‰ *Summer temperature 24° C *Winter temperature: 16° C *Data from Wüst.

Table 35.14. Pteropodal assemblages from Tyrrhenian Sea cores

Postglacial	Transition	Glacial
Dominant: $>$90 per cent *Limacina inflata* *Creseis acicula* *Creseis virgula constricta* *Styliola subula*	Dominant: $>$90 per cent *Creseis acicula* *Creseis virgula constricta* *Clio pyramidata*	*Limacina retroversa* *Clio pyramidata* f. *pyramidata*
Subordinate: $<$10 per cent *Clio pyramidata* *Clio cuspidata* *Clio polita* *Limacina trochiformis* *Limacina bulimoides* *Hyalocylix striata* *Cavolinia* sp.	Subordinate: $<$10 per cent *Limacina bulimoides* *Clio cuspidata* *Cavolinia* sp. *Limacina retroversa* *Diacria trispinosa*	Present-day salinity: 37‰

Table 35.15. Pteropodal assemblages from Balearic Sea cores

Postglacial	Glacial
Dominant: >70 per cent	
Limacina inflata	
Creseis acicula	
Creseis virgula constricta	
Subordinate: 20 per cent	*Limacina retroversa*
Styliola subula	*Clio pyramidata* f. *pyramidata*
Limacina trochiformis	
Clio pyramidata	
Subordinate: <10 per cent	
Cavolinia inflexa f. *longa*	
Clio cuspidata	

Table 35.16. Faunal assemblages from Tyrrhenian Sea cores – species listed in order of decreasing abundance

Postglacial	Transition	Cool Glacial
Pteropoda		
Limacina inflata	*Creseis acicula*	
Creseis acicula	*Creseis virgula constricta*	*Limacina retroversa*
Creseis conica	*Creseis conica*	*Clio pyramidata* f. *pyramidata*
Creseis virgula constricta	*Clio pyramidata*	
Styliola subula	*Limacina bulimoides*	
Foraminifera		
Globigerinoides rubra f. *B*	*Globorotalia truncatulinoides*	*Globigerina bulloides*
Globigerinoides sacculifera	*Globigerinoides rubra* f. *B*	*Globorotalia scitula*
Globigerinella aequilateralis	*Globorotalia inflata*	*Globorotalia inflata*
Globigerinoides rubra f. *A*	*Globigerina pachyderma*	
Globorotalia inflata	*Globigerina bulloides*	
Globigerina bulloides		

Table 35.17. Location, depth and length of cores (Mediterranean Sea)

Core	Latitude	Longitude	Depth (m)	Length (cm)
V10-8	41° 10′ N	11° 06′ E	1,384	945
V10-9	40° 58.30′ N	10° 46′ E	1,060	1,070
V10-52	35° 00.30′ N	27° 49.30′ E	2,595	360
V10-53	33° 11.30′ N	29° 06′ E	2,897	650
V10-55	32° 46.3′ N	29° 23′ E	2,520	422
V10-58	35° 40.30′ N	26° 18′ E	2,193	735
V10-65	34° 37′ N	23° 25′ E	2,586	960
V10-67	35° 42′ N	20° 43′ E	2,890	830
V14-128	32° 27′ N	29° 45′ E	1,931	626
V13-129	31° 47′ N	29° 55′ E	576	650

Cores V10-7, V10-8, and V10-9, raised in the northern Tyrrhenian Sea, represent continuous or nearly continuous sedimentation. As indicated in Fig. 35.9, cores V10-8 and V10-9 penetrate Last Glacial sediments. Several mild oscillations of short duration interrupted this cold phase. A synopsis of the faunal assemblages of this basin is given in Tables 35.14 and 35.16.

Several representative cores from the eastern basins are shown in Figs. 35.5 and 35.6.

Central Levantine basin cores are characterized by the intercalation between normal pelagic sediments of numerous sporadically emplaced terrigenous silts and sands, some of which are graded.

Typical cores from the southern Aegean and northern Levantine basins are shown in Figs. 35.5 and 35.6. In these regions, pelagic deposits are interbedded with volcanic ash layers. The close coincidence of the curves representing the variation in relative abundance of *Limacina inflata* and *Globigerinoides sacculifera,* and their inverse relationship to that of *Globorotalia scitula* in zones representing normal sedimentation is illustrated in Fig. 35.6.

Caribbean Sea

Three out of the five Caribbean deep-sea cores contain

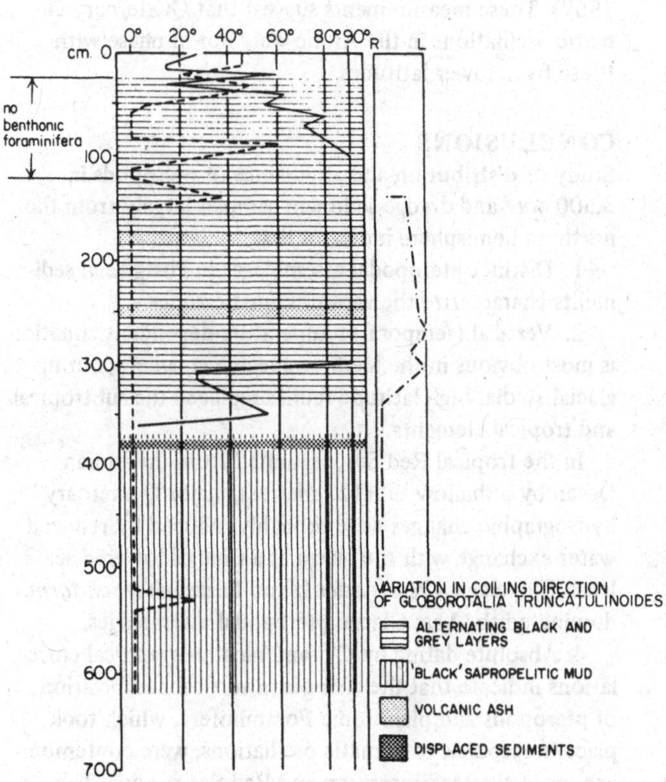

Levantine Basin
V14-128
cm. 0
0° 20° 40° 60° 80° 90°
R L
100
200
300
400
500
600
700
no benthonic foraminifera
VARIATION IN COILING DIRECTION OF GLOBOROTALIA TRUNCATULINOIDES
ALTERNATING BLACK AND GREY LAYERS
'BLACK' SAPROPELITIC MUD
VOLCANIC ASH
DISPLACED SEDIMENTS
PTEROPOD CURVE
FORAMINIFERA CURVE

Fig. 35.7. Levantine Basin core showing sediment types and climatic curves, based on percentages of warm and cool water microfauna. The solid line indicates the ratio of *Creseis* sp. to *Limacina inflata;* the broken line indicates the ratio of *Globigerinoides sacculifera* to *Globigerinoides rubra.* The ratios were converted into arc-tangents and plotted as degrees.

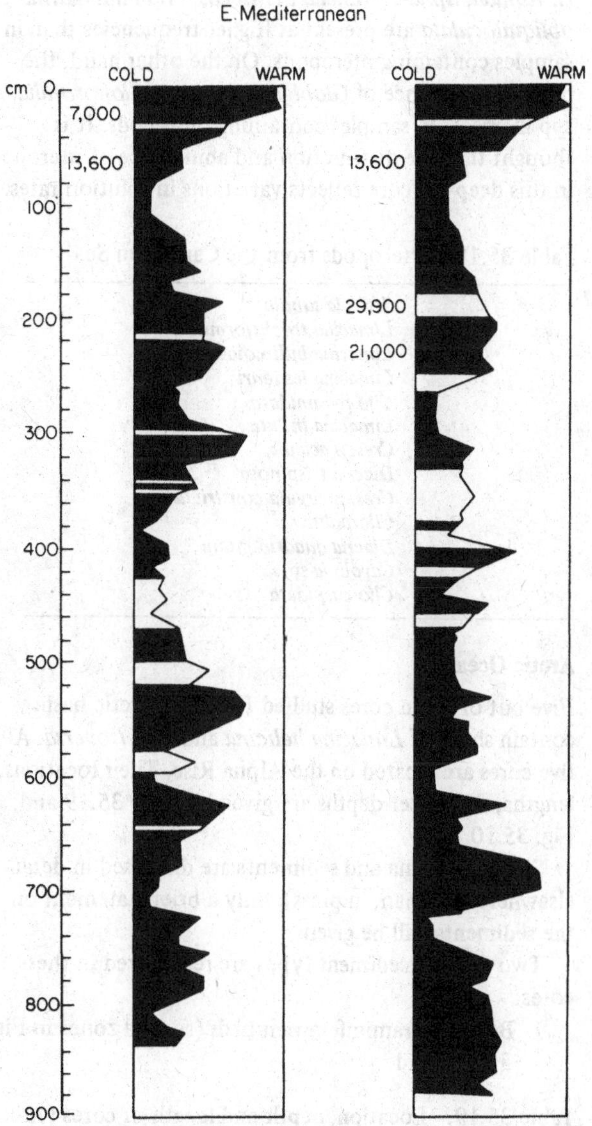

E. Mediterranean
cm 0
COLD WARM
COLD WARM
7,000
13,600
13,600
100
200
29,900
21,000
300
400
500
600
700
800
900
V10-65
V10-67
Black Sapropelitic Mud

Fig. 35.8. Climatic curve of Eastern Mediterranean Sea cores V10-65 and V10-67 based on variation in abundance of the planktonic fauna.

isolated pteropodal shells. They are V19-17, V19-18, and V19-19 (Table 35.6). Because the pteropod-bearing samples of cores V19-17 and V19-18 contain high percentages of detrital minerals and are believed to represent turbidity current deposits, they will not be considered here. In V19-19, pteropods were absent from the uppermost layers of the core but were seen at various levels throughout the core (Table 35.18). It should be mentioned that in the pteropod-barren zones, thick-shelled planktonic foraminifers, such as *Globorotalia menardii*, *G. tumida*, *Sphaeroidinella dehiscens*, and *Pulleniatina obliquiloculata* are present at higher frequencies than in samples containing pteropods. On the other hand, the relative abundance of *Globigerina* spp. to *Globorotalia* spp increases in samples containing pteropods. It is thought that the distribution and abundance of pteropods in this deep-sea core reflects variations in solution rates.

Table 35.18. Pteropods from the Caribbean Sea

Styliola subula
Limacina trochiformis
Limacina bulimoides
Limacina lesueuri
Clio pyramidata
Limacina inflata
Creseis acicula
Diacria trispinosa
Creseis virgula constricta
Clio polita
Diacria quadridentata
Cavolinia sp.
Clio cuspidata

Arctic Ocean

Five out of seven cores studied from the Arctic basin contain shells of *Limacina helicina* and *L. retroversa*. All five cores are located on the Alpha Rise. Their locations, lengths, and water depths are given in Table 35.19 and Fig. 35.10.

Since the fauna and sediments are discussed in detail elsewhere (Herman, in press), only a brief statement on the sediments will be given.

Two distinct sediment types are recognized in these cores:

1. Brown Foraminifera-rich beds (shaded zones in Fig. 35.11), and
2. Grey or tan Foraminifera-barren beds (with white intervals).

The diagram (Fig. 35.11) represents the variation in abundance of biogenic remains to clastic, the shaded area indicates the percentage of biogenic remains in the fraction $> 74\ \mu$. Circles on the right-hand side of each core show the presence of *Limacina helicina* and *L. retroversa*. Dark circles indicate that pteropods constitute 1-5 per cent of the planktonic population. Open circles indicate that they make up > 1 per cent of the fauna.

Micropalaeontological analysis and lithological examination of cores indicate that variations in fauna and sediment characteristics with time, were due to regional climatic and hydrographic oscillations. Available radiocarbon and uranium series isotope measurements indicate that rates of sedimentation in the Arctic are slow, about 0.2 centimetres/1,000 years, and they remained essentially unchanged over the past 150,000 years (Ku and Broecker 1967). These measurements suggest that Quaternary climatic oscillations in the Arctic were not in phase with those from lower latitudes.

CONCLUSIONS

Study of distribution and abundance of pteropods in 3,500 core and dredge sediment samples largely from the northern hemisphere indicates that:

1. Distinct pteropodal assemblages in Postglacial sediments characterize the various climatic zones.

2. Vertical (temporal) temperature-dependent zonation is most obvious in the Mediterranean Sea, where, during glacial stadia, high latitude faunas replaced the subtropical and tropical elements.

In the tropical Red Sea, separated from the Indian Ocean by a shallow sill (125 metres), major Quaternary hydrographic changes were probably induced by reduced water exchange with the ocean at a time of lowered sea level. Epipelagic *Creseis acicula* and *Limacina trochiformis* dominated the Last Glacial pteropodal assemblages.

4. Absolute dating by C^{14} and biostratigraphical correlations indicate that the changes in vertical composition of pteropods and planktonic Foraminifera, which took place in response to climatic oscillations, were contemporaneous in the Mediterranean and Red Sea regions. The Mediterranean Postglacial fauna has close affinities with

Table 35.19. Location, depth and length of cores (Arctic Ocean)

Core	Latitude	Longitude	Depth (m)	Length (cm)
D.st.A. 2	83° 52′ N	168° 12′ W	1,521	206
D.st.A. 3	84° 12′ N	168° 33′ W	2,409	95
D.st.A. 4	84° 21′ N	168° 49′ W	2,041	116
D.st.A. 5	84° 28′ N	169° 04′ W	1,934	125
D.st.A. 6	85° 15′ N	167° 54′ W	1,842	88

the Postglacial Red Sea fauna. In contrast with the Postglacial, during the Last Glacial stadia, the fauna of the Mediterranean and Red Seas were totally different, suggesting that the amplitude of climatic fluctuations were greater in the Mediterranean. Radiometric age determinations indicate that the climatic oscillations in these two seas did not coincide with those in the Arctic.

5. Knowledge of their depth habitat makes pteropods with restricted depth ranges valuable indicators of past sea-level fluctuations in shallow continental shelf cores.

6. Warm-water *Clio pyramidata* f. *convexa,* present in Red Sea, Indian, and Pacific Ocean sediments, has not been observed in Mediterranean nor Atlantic Ocean Quaternary deposits.

ACKNOWLEDGEMENTS

I thank Dr P. E. Rosenberg, Washington State University, for valuable suggestions during the preparation of the manuscript, Dr S. van der Spoel, Museum of Zoology, University of Amsterdam, Dr J. McGowan, Scripps Institution of Oceanography and Dr Chin Chen, Lamont Geological Observatory, for discussions of the pteropod species, and Drs W. S. Broecker, Lamont Geological Observatory and R. Chatters, Washington State University for the C^{14} determinations. This research was partially supported by Research Committee Grant No. 11C-2454-372 at Washington State University. Most of the cores discussed in this paper were studied during the author's residence at Lamont Geological Observatory.

I would like to thank Dr M. Ewing, director, for organizing the sea-going expeditions and the scientists who collected the cores, in particular, Drs C. L. Drake, K. Hunkins, H. Kutschale, and B. Heezen, as well as Mr Roy Capo of the core laboratory for providing samples. The coring operation was made possible by grants Nonr 266 (48) and Nonr 266 (82) and NSF (GA 588) to Lamont Geological Observatory.

I also wish to thank Dr W. Riedel, Scripps Institution of Oceanography, for core samples, Dr T. Pickett of the Oceanographic Sorting Center, Smithsonian Institution, for dredges collected by the Woods Hole Oceanographic Institution, and the Göteborgs Oceanographic Institute in Sweden, for cores collected during the Swedish Deep-Sea Expedition.

Received March 1968.

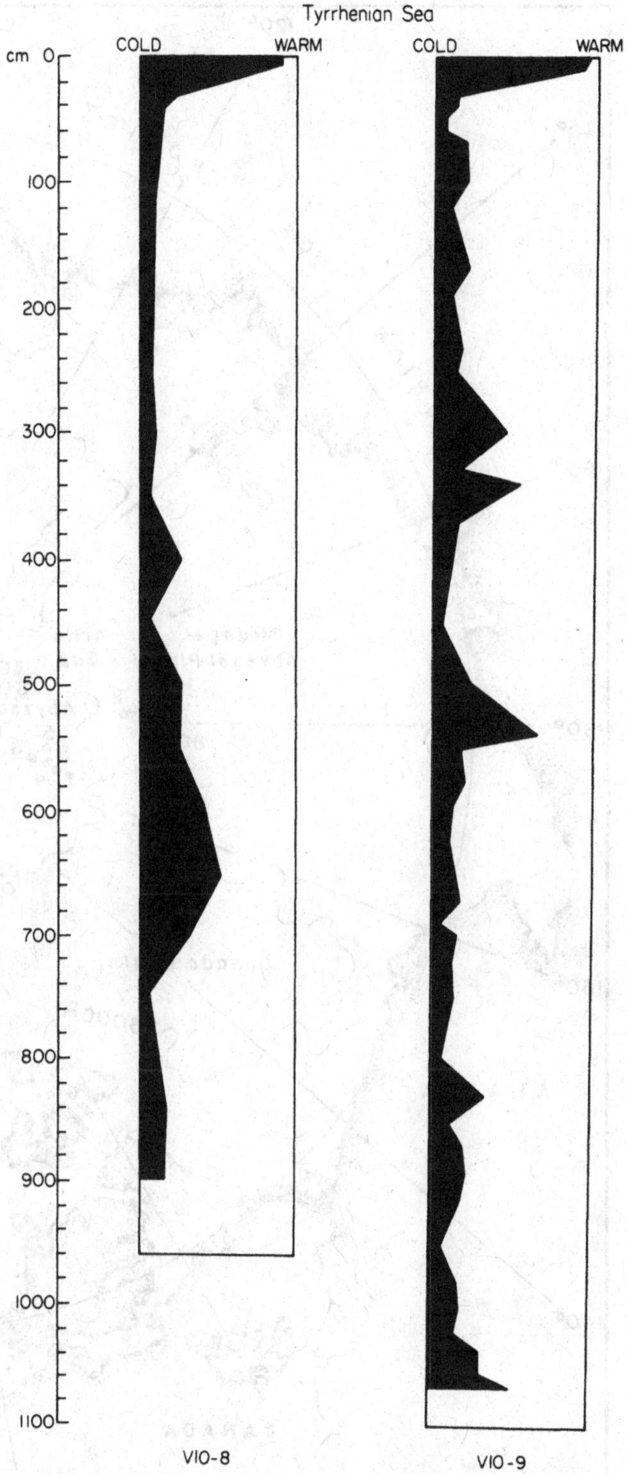
Tyrrhenian Sea
cm
COLD
WARM
COLD
WARM
0
100
200
300
400
500
600
700
800
900
1000
1100
V10-8
V10-9

Fig. 35.9. Climatic curve of Tyrrhenian Sea cores V10-8 and V10-9 based on variation in abundance of the planktonic fauna.

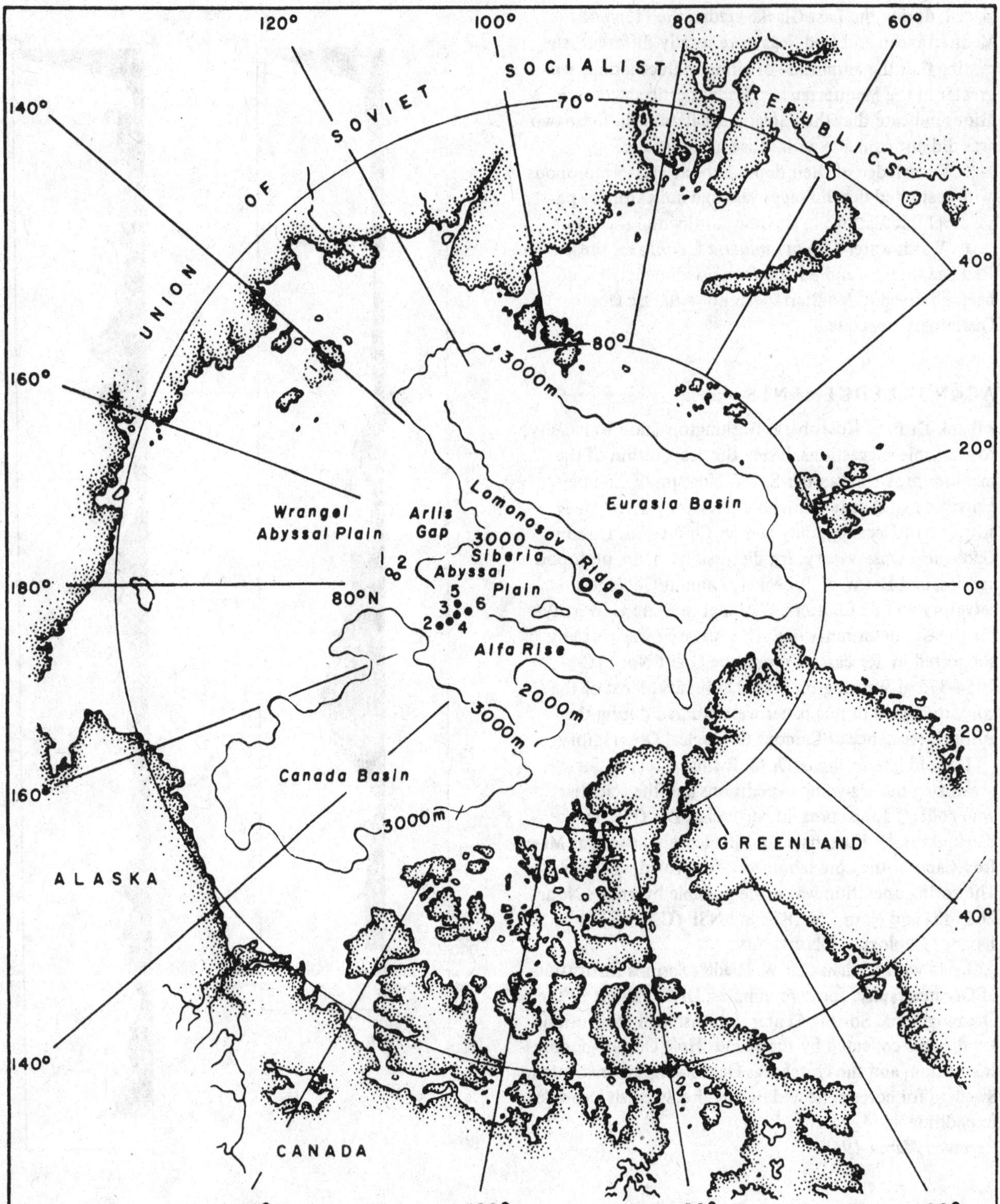

120°
100°
80°
60°
140°
70°
40°
SOCIALIST
REPUBLICS
SOVIET
OF
UNION
E
160°
80°
3000m
20°
Eurasia Basin
Lomonosov Ridge
Wrangel
Abyssal Plain
Arlis
Gap
3000
Siberia
Abyssal
Plain
1
2
180°
80°N
5
3
6
2
4
0°
Alfa Rise
2000m
3000m
20°
Canada Basin
W
160°
3000m
GREENLAND
ALASKA
40°
140°
CANADA
120°
100°
80°
60°

Fig. 35.10. Arctic Ocean map showing location of cores.

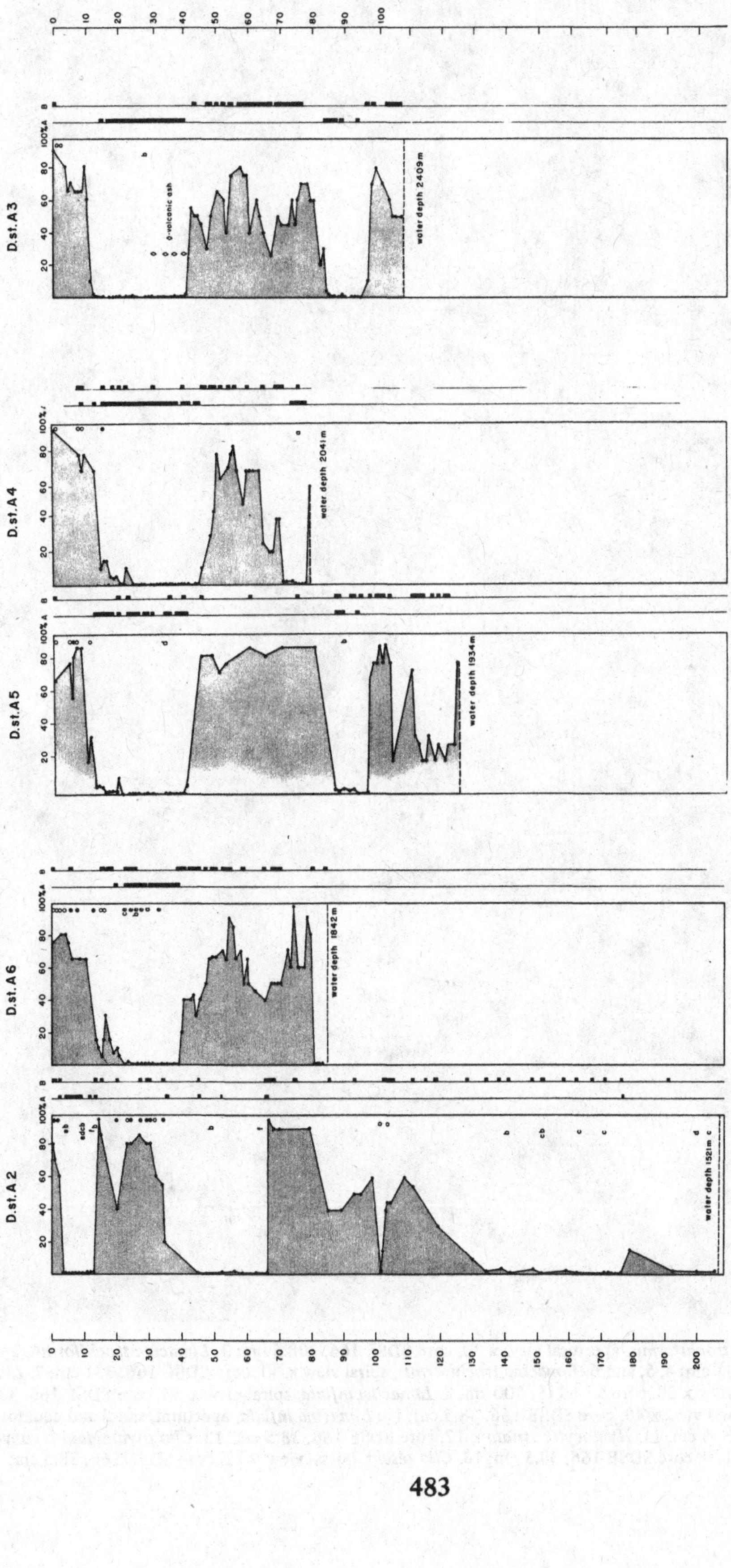
D.st.A3
D.st.A4
D.st.A5
D.st.A6
D.st.A2
volcanic ash
water depth 2409m
water depth 2041m
water depth 1934m
water depth 1842m
water depth 1521m

Fig. 35.11. Variation in abundance of clastic to biogenic remains in five Arctic cores. The shaded area indicates the percentage of biogenic remains in the fraction >74 μ. Circles on the right-hand side of each core show the presence of *Limacina helicina*. Dark circles indicate that pteropods constitute 1-5 per cent of the planktonic population; open circles indicate that pteropods constitute <1 per cent of the planktonic fauna. First column (A) to the right of each core indicates occurrence of plant debris, second column (B) to the right of each core indicates occurrence of granules and pebbles. Shaded area = microfauna; white area = mineral grains.

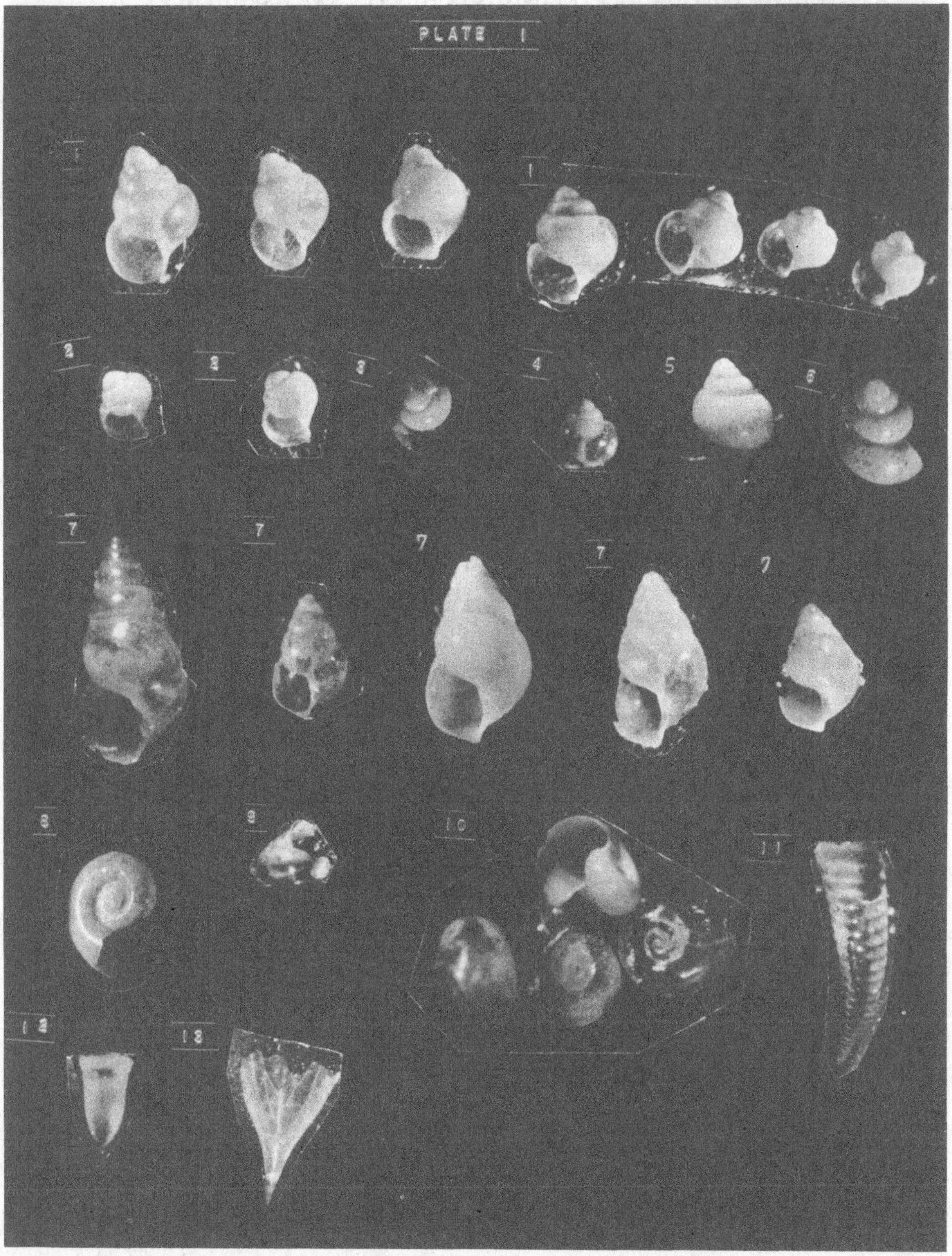
PLATE I

Plate 35.1. 1 and 2. *Limacina trochiformis,* apertural view x 30, core SDSE 166, 308.5 cm; 3. *Limacina trochiformis,* apical view x 30, core SDSE 166, 331 cm; 4, 5, and 6. *Limacina trochiformis,* spiral view x 30, core SDSE 166, 331 cm; 7. *Limacina bulimoides,* apertural view x 30, core V14-115, 500 cm; 8. *Limacina inflata,* spiral view x 30, core SDSE 166, 331 cm; 9. *Limacina inflata,* apertural view x 40, core SDSE 166, 38.5 cm; 10. *Limacina inflata,* apertural, spiral and equatorial side views, x 20, core V14-115, 5 cm; 11. *Hyalocylix striata* x 12, core SDSE 166, 38.5 cm; 12. *Clio pyramidata* f. *convexa,* embryonic portion of shell x 120, core SDSE 166, 38.5 cm; 13. *Clio teschi,* dorsal view x 12, core SDSE 166, 38.5 cm;

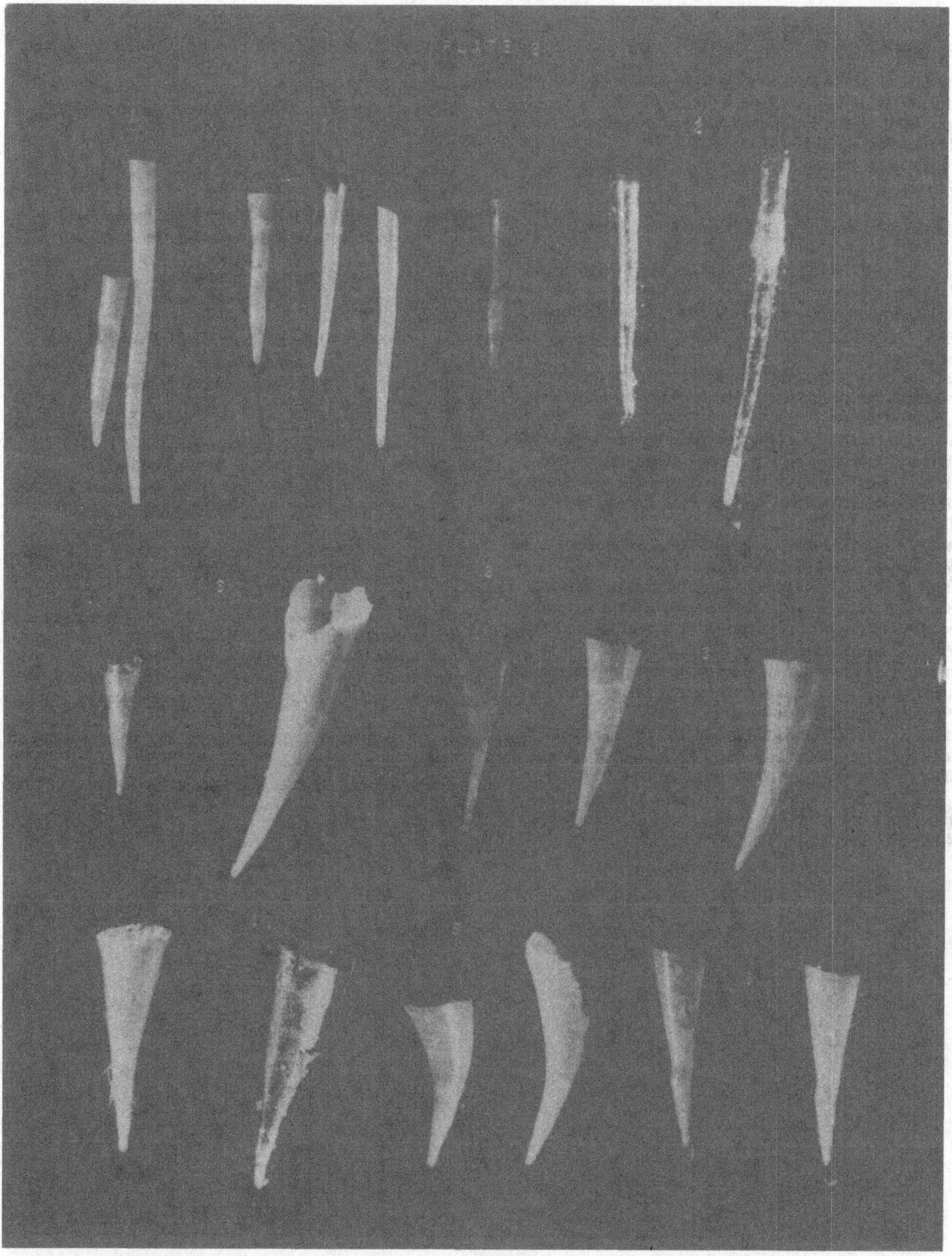

Plate 35.2. 1. *Creseis acicula* x 30, core SDSE 166, 331 cm; 2. *Creseis acicula* x 40, core SDSE 166, 18.5 cm; 3 and 4. *Creseis conica* x 30, core SDSE 166, 18.5 cm; 5. *Creseis virgula* x 30, core SDSE 166, 311 cm; 6. *Creseis virgula constricta* x 30, core SDSE 166, 331 cm.

BIBLIOGRAPHY

Arrhenius, G. 1963. Pelagic sediments. In *The Sea* (M. N. Hill *ed.*). Interscience: New York, **3**, 655-727.

Berger, W. H. 1967. Foraminiferal ooze: solution at depths. *Science,* **156** (3773), 383-5.

Chen, C. 1964. Pteropod ooze from Bermuda Pedestal. *Science,* **144** (3614), 60-2.

Emiliani, C. and Milliman, J. D. 1966. Deep-sea sediments and their geological record. *Earth-Science Reviews,* **1**, 105-32.

Ekman, S. 1953. *Zoogeography of the sea.* Sidgwick and Jackson: London. 417 pp.

Fager, E. W. and McGowan, J. A. 1963. Zooplankton species groups in the North Pacific. *Science,* **140** (3566), 453-60.

Furnestin, M. L. 1961. Pteropodes et heteropodes du plancton Marocain. *Rev. Trav. Inst. Peches marit.* **25** (3), 293-325.

Henbest, L. G. 1942. Geology and biology of the North Atlantic deep-sea cores between Newfoundland and Ireland. Part 7 – Miscellaneous fossils and significance of faunal distribution. *U.S.G.S. Prof. Paper* **196**, 119-33.

Herman, Y. 1965. Etude des sediments Quaternaires de la Mer Rouge. *Ann. Inst. Oceanog.* **42** (3), 339-415.

Herman, Y. 1968. Evidence of climatic changes in Red Sea cores. In *Means of correlation of Quaternary successions* (R. B. Morrison and H. E. Wright *eds.*). INQUA Proceedings, 8, University of Utah Press.

Herman, Y. In press. Arctic Ocean Quaternary microfauna and its relation to paleoclimatology. *Paleogeography, Paleoclimatology, Paleoecology.*

Hida, T. S. 1957. Chaetognaths and pteropods as biological indicators in the North Pacific. *U.S. Fish and Wildlife Service, Spec. Sci. Rept. Fisheries* **215**, 1-13.

Ku, T. L. and Broecker, W. S. 1967. Rates of sedimentation in the Arctic Ocean. *Progress in Oceanography,* **4**, 95-103. Pergamon Press: New York.

Kuenen, P. H. 1950. *Marine Geology.* John Wiley: New York, 568 pp.

Meisenheimer, J. 1905. Pteropoda. *Wissenschaftliche Ergebnisse der Deutschen Tiefsee-Expedition,* **9** (1), Jena, 314 pp.

Peck, J. I. 1893. Pteropods and heteropods 'of the *Albatross*' Norfolk to San Francisco. *Proc. U.S. Nat. Mus.,* **16** (943), 451-66.

Pelseneer, P. 1888. Report on the Pteropoda. *Challenger.* II. Thecosomata. *Sci. Rpts. 'Challenger', Zoology,* **23**, 132 pp.

Rampal, J. 1965. Pteropodes thecosomes indicateurs hydrologiques. *Rev. Trav. Inst. Peches marit,* **29** (4), 393-9.

Spoel, S. van der 1967. *Euthecosomata.* Noorduijn en Zoon N.V., Gorinchem. 375 pp.

Stubbings, H. G. 1937. Pteropoda. *Sci. Repts. John Murray Exped. 1933-34,* **5**, 15-33.

Sverdrup, H. U., Johnson, M. W. and Fleming, R. H. 1942. *The oceans, their physics, chemistry, and general biology.* Prentice-Hall: New York. 1059 pp.

Tesch, J. J. 1946. The thecosomatous pteropods. I. The Atlantic. *Dana Rept. No.* **28**. Carlsberg Found., Copenhagen, 1-82.

Tesch, J. J. 1948. The thecosomatous pteropods. II. The Indo-Pacific. *Dana Rept. No.* **30**. Carlsberg Found., Copenhagen, pp. 1-44.

Tesch, J. J., Helder and Holland 1912. Pteropoda and Heteropoda. In: *The Percy Sladen Trust Expedition to the Indian Ocean in 1905,* **3**, 165-88.

Thompson, E. F. 1939. The general hydrography of the Red Sea. *The John Murray Exped., Sci. Repts.* **2**, 83-102.

Wüst, G. 1960. Die Tiefenzirkulation des Mittelländischen Meeres in den Kernschichten des Zwischen- und des Tiefenwassers. *Deutschen Hydrogr. Zeitschrift,* **13** (3), 8-131.

36. PLEISTOCENE OSTRACODA FROM DEEP-SEA SEDIMENTS IN THE SOUTHEASTERN PACIFIC OCEAN

F. M. SWAIN

INTRODUCTION

On the basis of his studies of the planktonic Foraminifera Blackman (1966) recognized five zones in the Holocene and Late Pleistocene of the southeastern Pacific Ocean. The oldest of these zones began more than 200,000 years ago (Blackman and Somayajulu 1966). Dating of the cores in the various zones was by the ionium-thorium method, the protoactinium method, and the protoactinium-ionium method (Blackman and Somayajulu 1966; Arrhenius 1963). The principal features of the zones are as follows:

Zone 1. Holocene and latest Pleistocene extending back to between 8,000 and 12,000 years ago, probable average date of beginning 11,000 years ago; warming trend; *Globoquadrina dutertrei* nearly 100 per cent right-coiled in S.E. Pacific at 20° S Lat.; planktonic foraminiferal groups generally like those of Holocene surface samples.

Zone 2. Pleistocene; 11,000 to between 51,000 and 85,000 years ago; cooling trend; *Globoquadrina dutertrei* changes from 90 per cent right coiling to 44 per cent left coiling in equatorial region. *Globorotalia truncatulinoides* increases in percent right-coiling from south to north toward the equator; *G. inflata* moved into equatorial regions from high latitude regions.

Zone 3. Pleistocene; 51,000-85,000 to 110,000-140,000 years ago; warming trend; *Globorotalia crassaformis* changed to right-coiling in S.E. Pacific cores; other factors reverted to zone 1 conditions.

Zone 4. Pleistocene; 110,000-130,000 to 120,000-180,000 years ago; cooling trend; conditions similar to those of zone 2.

Zone 5. Pleistocene; 180,000-210,000 to ? years ago; warming trend; conditions like zones 1 and 2.

OSTRACODA FROM PLEISTOCENE CORES

Ostracoda were studied from the following of Blackman's core samples (see Appendix 36.1):

RIS 34g (2° 46′ S, 85° 28′ W; depth 3,210 metres). Blackman's zone 1 is at 0.45 centimetres, his zone 2 at 45-115 centimetres, and zone 3 at 115-60 centimetres.

AMPH 36 PG (18° 17′ S, 118° 17′ W; depth 3,550 metres). Blackman's zone 3 is at 45-95 centimetres, and his zone 4 at 95-120 centimetres.

As shown in Table 36.1, the shells of ten species of Ostracoda were found in these samples. Five of the species are believed to represent described forms: *Bradleya dictyon* (Brady), *Echinocythereis echinata* (Brady), *Krithe* cf. *K. tumida* Brady, *K.* cf. *K. glacialis* Brady, Crosskey and Robertson and *'Cythere' sulcatoperforata* Brady. *B. dictyon,* although a highly variable species and perhaps subject to subdivision, is widespread in the deep sea at nearly all latitudes (Brady 1880; Tressler 1941; Benson 1967). *E. echinata* and *K. glacialis* are more characteristic of the northern than the southern hemisphere at present, in both shallow cool waters and the deep sea. *K. tumida* and *'C.' sulcatoperforata* are at present found in deep waters in the southern hemisphere. The other five forms could not be assigned to described species. As far as is known all the ostracode species are benthonic in habitat.

All of the washed residues of two of Blackman's cores were studied for ostracodes; RIS 34g near the Equator and DWBG 70 near 30° S. No additional species were found. *Krithe* cf. *K. glacialis,* a northern hemisphere form, occurs in zone 2 (cool) in core DWBG 70 as well as in zone 4 (cool) in AMPH 36 PG (Table 36.1). Its presence in southeastern Pacific sediments may indicate intervals of glaciation.

Echinocythereis echinata seems to have been introduced into the southeast Pacific in zone 3 time from more northerly latitudes, and to have remained until the Holocene. *Bosquetina*? aff. *B.*? *fenestratum,* a southern type, appears to have migrated out of the southeastern Pacific in zone 3 time. *Echinocythereis* aff. *E. dasyderma, Krithe* cf. *K. glacialis* and *'Cythere' sulcatoperforata* possibly moved out of this part of the ocean in zone 4 time. Many additional cores will have to be examined, however, before

Table 36.1. Distribution of Plio-Pleistocene Ostracoda in South Pacific

Species	Living	Pleistocene				
		w Zone 1	c Zone 2	w Zone 3	c Zone 4	w Pliocene
Bradleya dictyon (Brady)	x	R	R	R	R	R
Echinocythereis echinata (Brady)	x[1]	R	C	C	–	–
Neomonoceratina sp.	–	–	R	–	–	–
Krithe cf. *K. tumida* Brady	x	–	R	?	–	R
Bosquetina? aff. *B.*? *fenestratum*[3]	–	–	–	R	R	R
Krithe aff. *K. bartonensis* (Jones)	–	R	–	A	R	R
Bradleya sp.	–	–	–	R	–	–
Echinocythereis aff. *E. dasyderma* (Brady)	–	–	–	–	R	R
Krithe cf. *K. glacialis* Brady, Crosskey, & Robertson	x[1]	–	R?	–	C	–
'Cythere' sulcatoperforata	x[2]	–	–	–	R	–
Echinocythereis? *ericea* (Brady)	x[2]	–	–	–	–	R
Trachyleberis aff. *T.* scabrocuneata (Brady)	–	–	–	–	–	R
Krithe bartonensis (Jones)	x[2]	–	–	–	–	R

Key: A = abundant (10 or more specimens); C = common (5-9 specimens); R = rare (1-4 specimens); W = warm; C = cold.
[1] Characteristic of Arctic or Boreal regions at present time, in shallow to deep cool waters.
[2] Characteristic of southern hemisphere at present time in deep cold waters.
[3] A similar living form *'Cytheropteron' mucronalatum* Brady is characteristic of deep cold water mainly in southern hemisphere, but ranges into North Atlantic; another similar living form *'Cytheropteron' fenestratum* Brady is now found in the South Atlantic and South Indian Oceans.

definitive statements can be made about the migrational history of ostracode species in this region.

ACKNOWLEDGEMENTS

Appreciation is expressed to W. R. Riedel and F. L. Parker of Scripps Institution of Oceanography for supplying samples and data for this study; also to the National Science Foundation for support provided by Grant GB 4110.

Received December 1967.

BIBLIOGRAPHY

Arrhenius, G. 1963. In *The Sea* (M. N. Hill *ed.*). Interscience: New York. pp. 710-715.

Benson, R. H. 1967. Oral presentation at Symposium on Recent Ostracoda, Hull, England, 10-14 July, 1967.

Blackman, A. 1966. Pleistocene stratigraphy of cores from the southeastern Pacific Ocean. Unpublished Ph.D. thesis, University of California, San Diego.

Blackman, A. and Somayajulu, B.L.K. 1966. Pacific Pleistocene cores: Faunal analyses and geochronology. *Science*, **154**, 886-9.

Brady, G. S. 1880. Report on the Ostracoda dredged by H.M.S. *Challenger*: *Rep. Sci. H.M.S. Challenger, Zool.* 1 (3).

Tressler, W. L. 1941. Geology and biology of North Atlantic deep-sea cores between Newfoundland and Ireland. Part 4. Ostracoda. *U.S. Geol. Survey Prof. Paper* **196-C**, 95-105.

Appendix 36.1. Pleistocene Ostracoda from the South Pacific and Indian Oceans

	No. of specimens
Holocene zone 1. Core RIS 34g, 6-40 cm (warm)	
Bradleya dictyon (Brady)	2
Pleistocene zone 2. Core RIS 34g, 50-110 cm (cool)	
Echinocythereis echinata (Brady)	6
Neomonoceratina sp.	1
Krithe cf. *K. tumida* Brady	2
Pleistocene zone 3. Core RIS 34g, 120-159 cm (warm)	
Bradleya dictyon (Brady)	3
Echinocythereis echinata (Brady)	6
Bosquetina? aff. *B.*? *fenestratum* (Brady)	2
Krithe aff. *K. bartonensis* (Jones)	7
AMPH 36 PG 50-90 cm	
Bradleya dictyon (Brady)	2
Bradleya sp.	1
Krithe aff. *K. bartonensis* (Jones) (large)	11
Krithe sp. (more like *tumida*)	1
Echinocythereis echinata (Brady)	1
Pleistocene zone 4. Core AMPH 36 PG 100-120 cm (cool)	
Bradleya dictyon (Brady)	2
Bosquetina? aff. *B.*? *fenestratum* (Brady)	1
Echincythereis? sp. aff. *E. dasyderma* (Brady)	1
Krithe cf. *K. glacialis* Brady, Crosskey and Robertson	7
Krithe aff. *K. bartonensis* (Jones)?	3
Cythere sulcatoperforata Brady	1

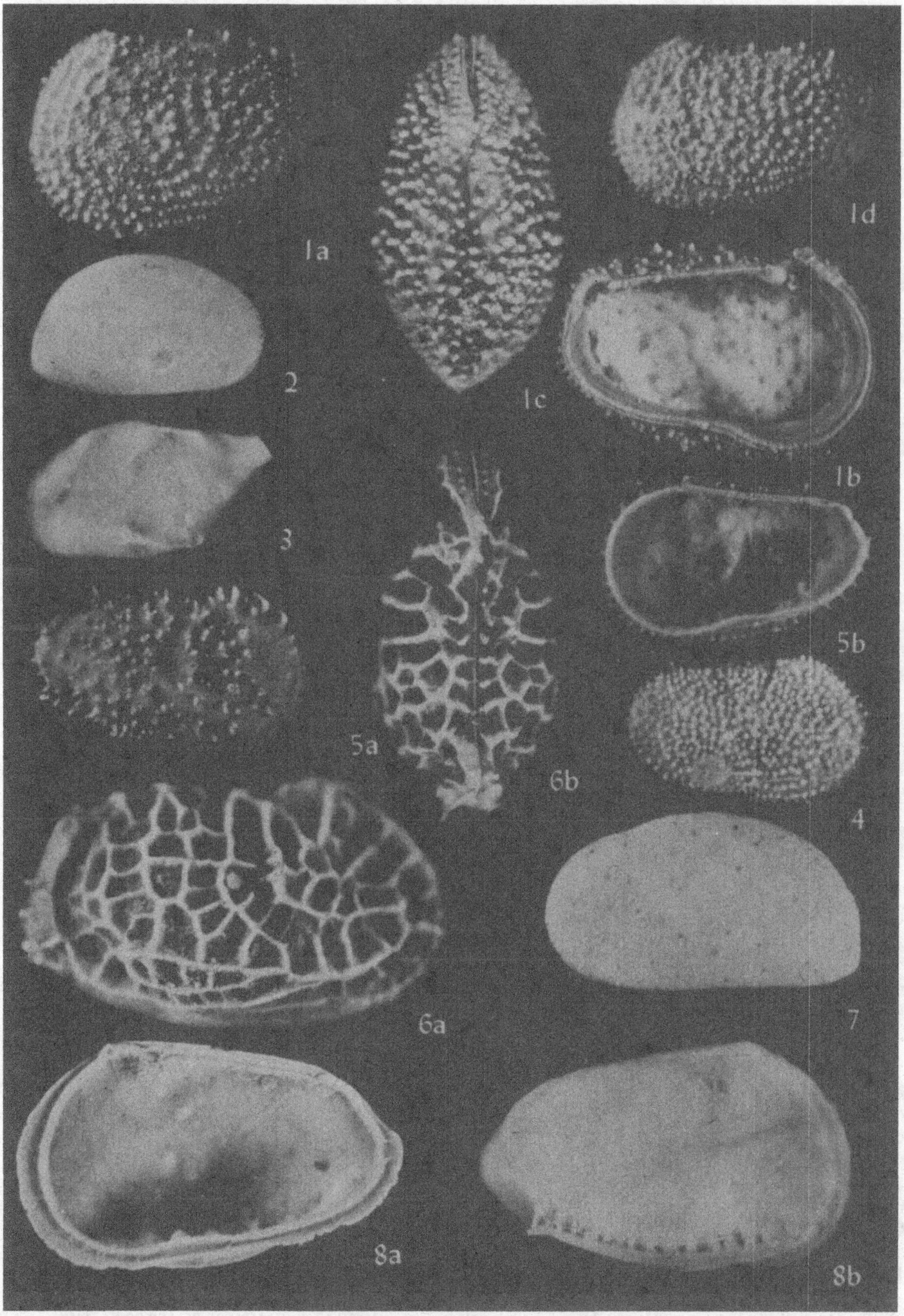
1a
1c
1d
1b
2
3
5a
6b
5b
4
6a
7
8a
8b

Plate 36.1. Figs. 1*a*, *d*. *Echinocythereis echinata* (Brady). *a*, *c*, right side and dorsal view of shell; *b*, interior of left valve; *d*, exterior of immature left valve. RIS 34 G, 50-110 cm, Pleistocene zone 2. x 50. Fig. 2. *Krithe* cf. *tumida* Brady. Exterior of right valve. RIS 34 G, 50-110 cm, Pleistocene zone 2. x 75. Fig. 3. *Neomonoceratina* sp. Exterior of left valve. RIS 34 G, 50-100 cm, Pleistocene zone 2, x 75. Fig. 4. *Echinocythereis echinata* (Brady). Exterior of immature right valve. RIS 34 G, 120-59 cm, Pleistocene zone 3. x 75. Fig. 5*a*, *b*. *Echinocythereis*? sp. Exterior and interior views of immature right valve. RIS 34 G, 120-59 cm, Pleistocene zone 3. x 75 and x 60. Fig. 6*a*, *b*. *Bradleya dictyon* Brady. Right side and dorsal views of shell. RIS 34 G, 120-59 cm, Pleistocene zone 3. x 50. Fig. 7. *Krithe* aff. *K. bartonensis* (Jones). Exterior of left valve showing scattered normal canal pits. RIS 34 G, 120-59 cm, Pleistocene zone 3. x 75. Figs. 8*a*, *b*. *Bosquetina*? aff. *B.*? *fenestratum* (Brady). Interior and exterior views of right valve showing weakly crenulate hinge teeth and small anterior rounded tooth. RIS 34 G, 120-59 cm, Pleistocene zone 3. x 60, 75.

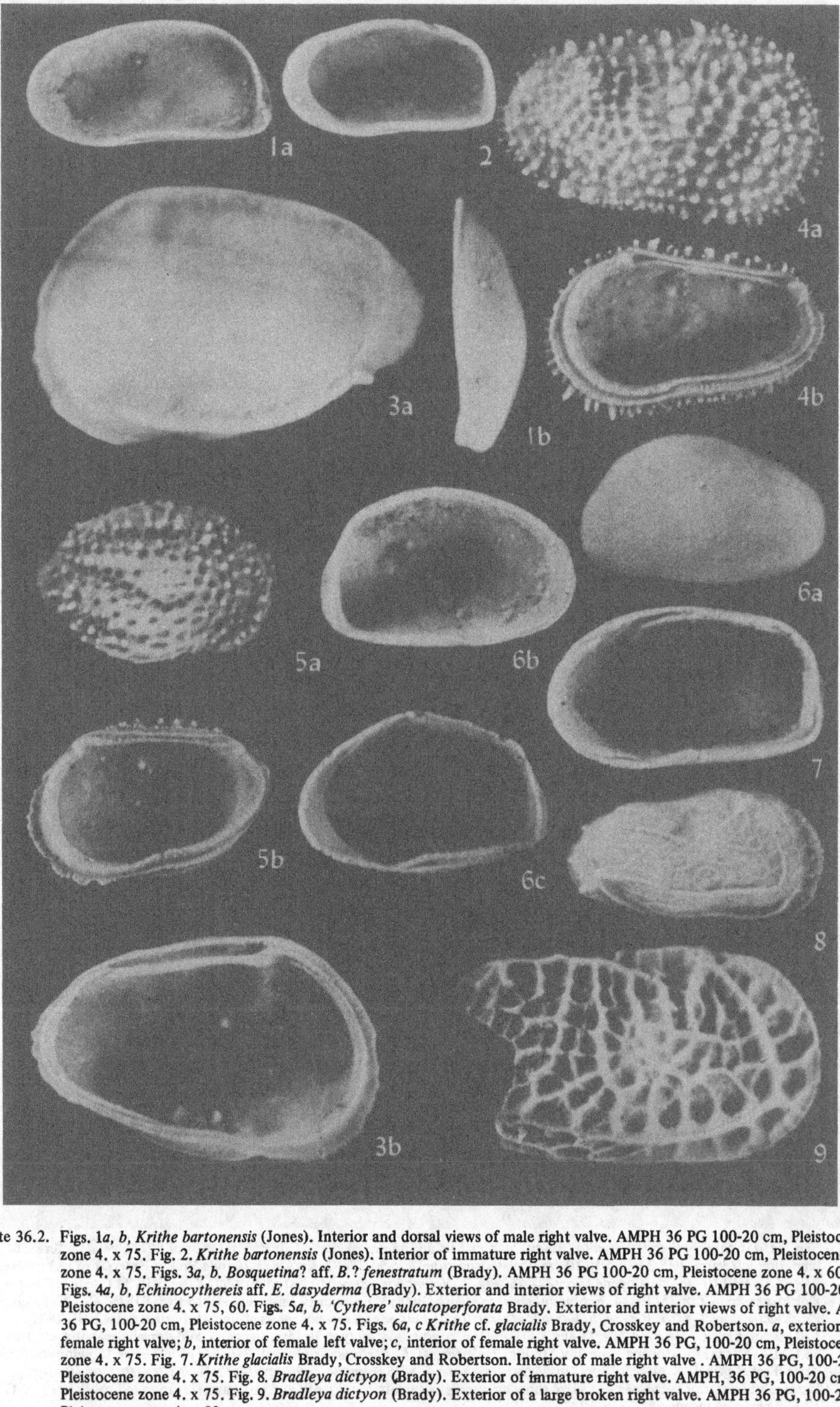
1a
2
4a
3a
1b
4b
6a
5a
6b
7
5b
6c
8
9
3b

Plate 36.2. Figs. 1*a*, *b*, ***Krithe bartonensis*** (Jones). Interior and dorsal views of male right valve. AMPH 36 PG 100-20 cm, Pleistocene zone 4. x 75. Fig. 2. ***Krithe bartonensis*** (Jones). Interior of immature right valve. AMPH 36 PG 100-20 cm, Pleistocene zone 4. x 75. Figs. 3*a*, *b*. ***Bosquetina***? aff. *B.*? ***fenestratum*** (Brady). AMPH 36 PG 100-20 cm, Pleistocene zone 4. x 60, 50. Figs. 4*a*, *b*, ***Echinocythereis*** aff. *E.* ***dasyderma*** (Brady). Exterior and interior views of right valve. AMPH 36 PG 100-20 cm, Pleistocene zone 4. x 75, 60. Figs. 5*a*, *b*. ***'Cythere' sulcatoperforata*** Brady. Exterior and interior views of right valve. AMPG 36 PG, 100-20 cm, Pleistocene zone 4. x 75. Figs. 6*a*, *c* ***Krithe*** cf. ***glacialis*** Brady, Crosskey and Robertson. *a*, exterior of female right valve; *b*, interior of female left valve; *c*, interior of female right valve. AMPH 36 PG, 100-20 cm, Pleistocene zone 4. x 75. Fig. 7. ***Krithe glacialis*** Brady, Crosskey and Robertson. Interior of male right valve . AMPH 36 PG, 100-20 cm, Pleistocene zone 4. x 75. Fig. 8. ***Bradleya dictyon*** (Brady). Exterior of immature right valve. AMPH, 36 PG, 100-20 cm, Pleistocene zone 4. x 75. Fig. 9. ***Bradleya dictyon*** (Brady). Exterior of a large broken right valve. AMPH 36 PG, 100-20 cm, Pleistocene zone 4. x 50.

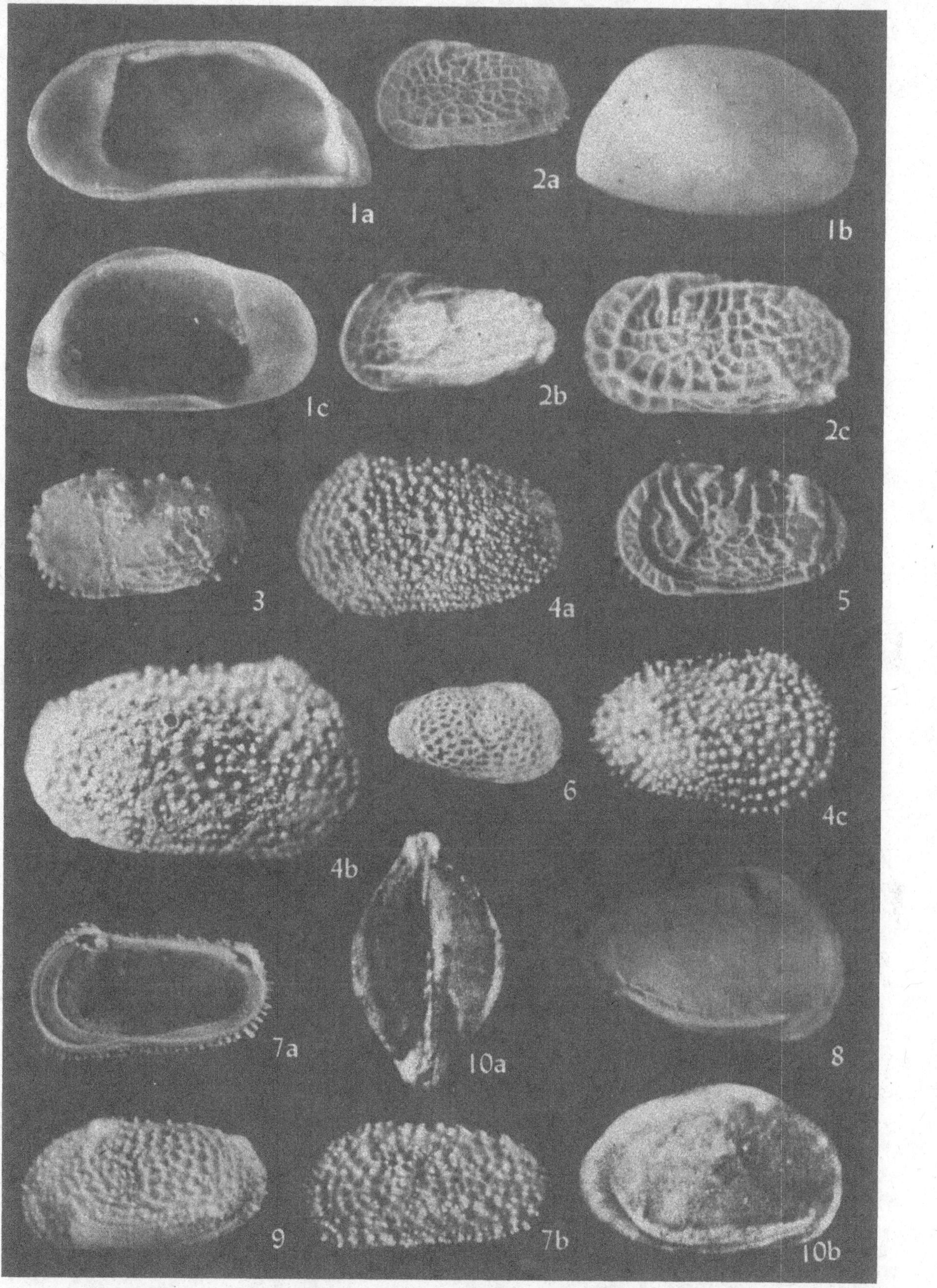
1a
2a
1b
1c
2b
2c
3
4a
5
4b
6
4c
7a
10a
8
9
7b
10b

Plate 36.3. Figs. 1*a-c*. *Krithe* aff. *K. bartonensis* (Jones). *a*, interior of male right valve; *b, c*, exterior and interior of female right and left valves. AMPH 36 PG, 50-90 cm, Pleistocene zone 3. x 40. Figs. 2*a-c*. *Bradleya* aff. *B. dictyon* (Brady). Exterior views of left valves of immature specimens. AMPH 36 PG, 50-90 cm, Pleistocene zone 3. x 40. Fig. 3. *Miracythere*? aff. *M. novaspecta* Hornibrook. Exterior of immature left valve. AMPH 36 PG, 50-90 cm, Pleistocene zone 3. x 40. Figs. 4*a-c*. *Echinocythereis echinata* (Brady). *a*, exterior of immature left valve; *b*, exterior of mature right valve; *c*, exterior of immature right valve. RIS 34 G, 50-110 cm. Fig. 5. *Bradleya*? aff. *B. dictyon* (Brady). Exterior of left valve. AMPH 36 PG 50-90 cm, Pleistocene zone 3. x 40. Fig. 6. *Rabilimus*? *clarkana* (Ulrich and Bassler). Right side of immature shell, St Marys Formation, Middle Miocene, Virginia, x 60; introduced for comparison with deep sea forms. Figs. 7*a-b*. *Henryhowella evax* (Ulrich and Bassler). Interior of right valve and exterior of left valve, St Marys Formation, Middle Miocene, Virginia, x 60; introduced for comparison with deep sea forms. Fig. 8. *Bosquetina*? aff. *B. fenestratum* (Brady). Exterior of right valve. AMPH 36 PG, 100-20 cm, Pleistocene zone 4. x 40. Fig. 9. *Echinocythereis planibasilis* (Ulrich and Bassler). Exterior of left valve, Calvert Formation, Middle Miocene, Maryland, x 60; introduced for comparison with deep-sea forms. Figs. 10*a, b*. *Brachycythere* cf. *B. hadleyi* Stephenson. Right side and dorsal views of shell, Paleocene, North Carolina. x 42; introduced for comparison with deep-sea forms.

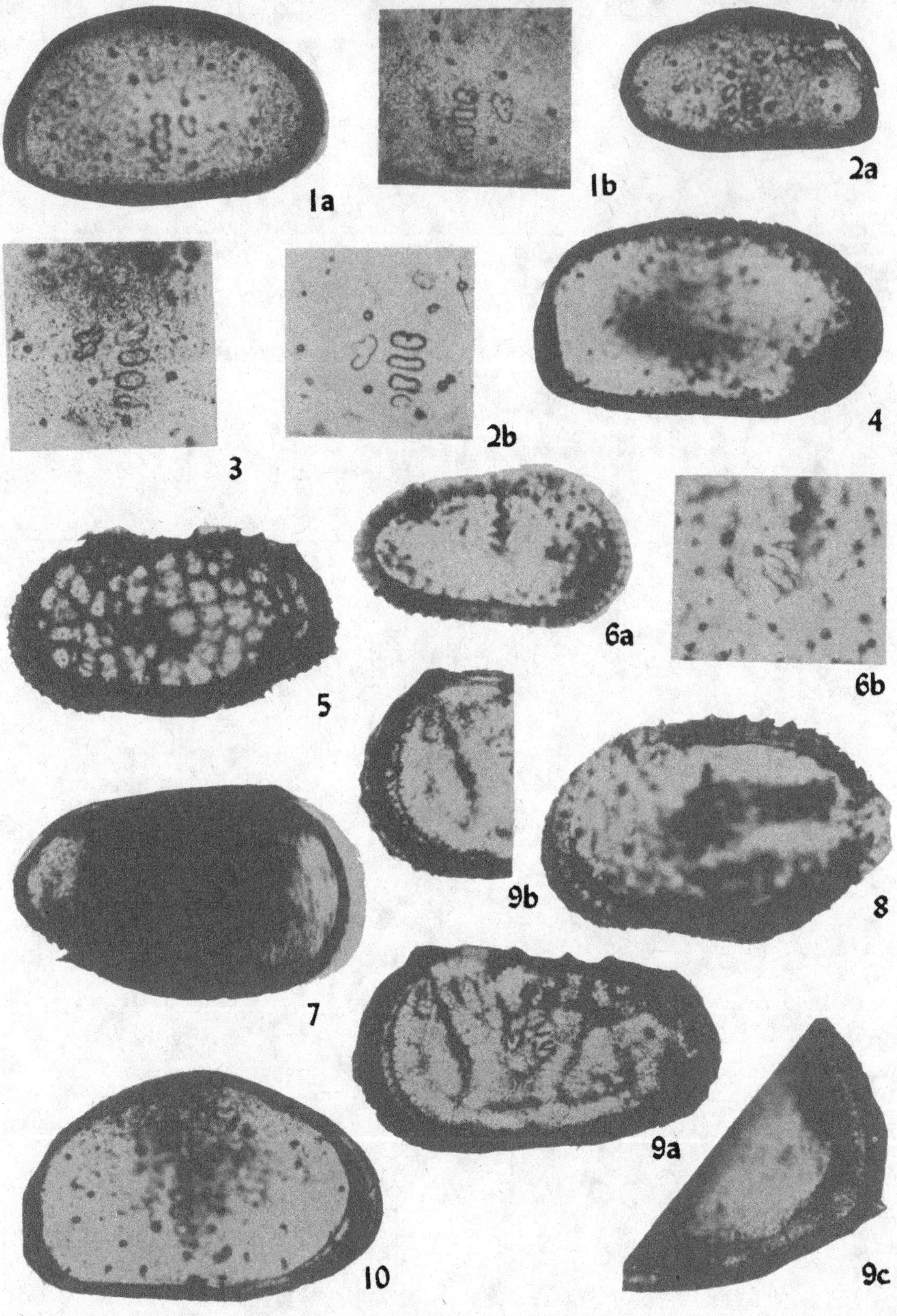
1a
1b
2a
3
2b
4
6a
5
6b
9b
8
7
9a
10
9c

Plate 36.4. Figs. 1*a, b. Krithe* cf. *K. glacialis* Brady, Crosskey and Robertson. Interior of left valve, x 76, and detail of muscle scar, x 103. AMPH 36 PG, 100-20 cm, Pleistocene zone 4. Figs. 2*a, b. Krithe bartonensis* (Jones). Interior of right valve, x 52, and detail of muscle scar, x 103. AMPH 36 PG, 100-20 cm. Pleistocene zone 4. Fig. 3 *Krithe* cf. *K. glacialis* Brady, Crosskey. and Robertson. Detail of muscle scar x 103. AMPH 36 PG, 100-20 cm, Pleistocene zone 4. Fig. 4 *Krithe glacialis* Brady, Crosskey, and Robertson. Interior of left valve, x 76. AMPH 36 PG, 100-20 cm, Pleistocene zone 4. Fig. 5. *Bradleya dictyon* Brady. Interior of right valve, x 52. RIS 34 G, 120-59 cm, Pleistocene zone 3. Figs. 6*a, b. Echinocythereis* aff. *E. dasyderma* (Brady). Interior of left valve, x 52, and detail of muscle scar, x 103. AMPH 36 PG 100-20 cm, Pleistocene zone 4. Fig. 7. *Bosquetina*? aff. *B*? *fenestratum* (Brady). Interior of left valve, x 52. RIS 34 G, 120-59 cm, Pleistocene zone 3. Fig. 8. *'Cythere' sulcatoperforata* Brady. Interior of right valve, x 103. AMPH 36 PG, 100-20 cm, Pleistocene zone 4. Figs. 9*a-c. Bradleya* aff. *B. dictyon* Brady. Interior of right valve, and parts of anterior and posterior end at different levels of focus to show radial canals. AMPH 36 PG, 50-90 cm, Pleistocene zone 3. x 52 (*a,b*), x 103 (*c*). Fig. 10. *Krithe* cf. *K. glacialis* Brady, Crosskey, and Robertson. Interior of right valve, AMPH 36 PG, 100-20 cm, Pleistocene zone 4. x 76.

37. HORIZONTAL AND VERTICAL DISTRIBUTION OF POLLEN AND SPORES IN QUATERNARY SEQUENCES

J. J. GROOT AND C. R. GROOT

Summary: The distribution of pollen and spores in deep-sea sediments appears to be affected mainly by oceanic circulation. The number of pollen grains per gramme of sediment is a function of distance from shore. Vertical distribution of pollen is chiefly determined by vegetational changes through time and offers an opportunity to study Quaternary climatic fluctuations. Pollen grains can be considered as indicators of the provenance of the fine-grained, continent-derived sediments in which they are found. The presence of reworked pollen is the greatest difficulty facing marine palynological research. Therefore, it should be carried out in conjunction with other investigations pertaining to the stratigraphy of deep-sea sediments.

INTRODUCTION

Palynological investigation of deep-sea sediments can make a unique contribution to the understanding of Quaternary events, because marine palynology, in conjunction with other fields of micropalaeontology, can provide a link between the Quaternary history of the continents and that of the ocean basins. In addition, the geographical distribution of pollen and spores from known continental sources can give information regarding sedimentary patterns and sediment provenance. It is therefore not surprising that there is a rapidly increasing interest in marine palynology (see *Marine Geology*, 4, no. 6).

Our studies have been mainly concerned with the palynology of deep-sea sediments which appear to offer the best opportunities for the investigation of Quaternary climatic events. Continental shelf sediments generally contain unconformities owing to Pleistocene sea-level fluctuations, and they are therefore as unsuitable for providing a continuous Quaternary pollen record as peats and other continental deposits. Sediments on the continental slope are probably largely unstable, and it does not seem to be a promising environment for palynological work. The sediments of abyssal plains and oceanic rises, however, offer the best chances of continuous accumulation and therefore of a continuous record of vegetational and climatic changes.

HORIZONTAL DISTRIBUTION

The horizontal or geographical distribution of pollen and spores in deep-sea sediments is related to many factors, including pollen production, dispersal by oceanic and/or atmospheric circulation, distance from continents, and sediment type.

Pollen Production

It is well known that pollen production varies greatly between different genera and even species of plants. For instance, one oak tree produces about 100 million pollen each year, whereas some species of pine produce 350 million pollen per tree (Faegri and Iversen 1950). In addition, the density of the vegetation in an area determines pollen production. Consequently, the number of pollen available for dispersion by wind or water currents varies widely. Desert areas produce few pollen, and it is therefore not surprising that deep-sea sediments adjacent to deserts usually contain very few pollen. This was shown by Vronskiy and Panov (1963) and Koreneva (1966) for sediments off the coast of North Africa. We investigated three cores off the West African Coast (at latitudes of about 23-25° N) and found nearly all samples barren. The same applies to samples off the Atacama Desert.

Oceanic Circulation

We have discussed elsewhere the matter of dispersal of pollen by water currents. It should be understood, however, that dispersal does not only occur by surface currents, but also, and perhaps even primarily, by currents at depth. If the idea that pollen are sedimentary particles with the same hydrodynamical properties as those of fine silts and clays is correct, transportation is probably determined mainly by bottom currents, since the greatest concentration of suspended matter in the oceans, apart from plankton, generally occurs in the lower few hundred metres of the water column (Ewing and Thorndike 1965; Ewing *et al.* in press). Therefore, an understanding of the dispersal of pollen grains in the oceans, and their horizontal distribution in the sediments requires knowledge of deep oceanic circulation.

The influence of oceanic circulation can be seen in the distribution of pollen and spores in the sediments of the Argentine Basin. On the basis of a study of samples from fourteen cores it was concluded that the vegetation of southern South America was the main source of the pollen in the Basin sediments, and that the vegetation of the subtropical areas (the Plata region) did not make a significant contribution, presumably because the Antarctic bottom water flows in a northeasterly direction (Wüst 1957). Groot *et al.* (1967) stated: 'Turbidity currents coming down the continental slope will transport material in an easterly direction and density currents occurring in the lower several hundred feet above the ocean bottom should move in a northeasterly direction under the influence of the Antarctic bottom current. The net result expected is transportation to the northeast.'

Recent studies by Ewing and co-workers at Lamont Geological Observatory indicate the presence of a large gyre of cold bottom water in the Argentine Basin, as evidenced by direct current measurements, and the presence and orientation of current lineations and ripples. Transportation of suspended matter by the bottom water is indicated by nephelometer measurements. Ewing *et al.* (in press) concluded that the deep current system in the Basin is responsible for the dispersion of lutites mainly derived from the southern part of the South American continent. This conclusion is in agreement with that based on the palynological study of the sediments by Groot *et al.* (1967), furnishing strong evidence that both pollen and small mineral particles are transported in the same fashion and have the same source area, and that pollen grains can serve as indicators of the provenance of the lutites.

Distance from Land

Since pollen and spores are land-derived particles, one would expect a general decrease in the number of pollen grains per gramme of sediment with distance from shore. D. B. Williams (personal communication) has found this to be true for the North Atlantic. Koreneva (1966) found in her study of Pacific Ocean sediments that at a distance of more than 500 kilometres from continental shores and major islands, the pollen and spore content of the sediments was small, and the composition of the pollen spectra did not reflect the character of the vegetation in the neighbouring lands.

Recently we have made a study of the number of pollen and spores per gramme of sediment in eighty-eight samples, nearly all from the upper 50 centimetres of a number of cores from the Atlantic, Pacific and Indian Oceans. Data pertaining to these cores are given in Table 37.1. Figure 37.1 shows the total number of pollen and spores per gramme of sediment versus the distance to the nearest shore.

A computer programme was set up to investigate the relationship between pollen numbers and distance from shore, and it was found that log-log expressions gave the best fit of the available data. (In order to plot all samples in Figs. 37.1-5, the 'barren' samples are assumed to contain 1 pollen grain.) The relationship of the number of pollen per gramme of dry sediment versus distance from shore can be expressed as follows:

$$\ln (N_1 + 1) = 12.2556 - 1.6639 \ln D \qquad (1)$$

The variance of $\ln (N_1 + 1) = 3.980$, and the standard deviation of $N_1 + 1$ is 5.8.

In some samples the number of pollen reworked from older deposits is quite large, as pointed out originally by Groot (1963), and since these reworked grains are not very useful in interpreting vegetational and climatic changes in the Quaternary, the number of non-reworked pollen and spores versus distance from shore is important. This relationship is shown in Fig. 37.2, and is expressed by

$$\ln (N_2 + 1) = 11.7182 - 1.6267 \ln D \qquad (2)$$

The variance of $\ln (N_2 + 1)$ is 3.494 and the standard deviation of $N_2 + 1$ is 7.37.

Figs. 37.1 and 37.2 are based on the *present* distance from the core locations to the *nearest* shore. We should recognize the fact, however, that distances from shore have varied considerably during the Quaternary, particularly where wide continental shelves occur which were occupied by vegetation during low sea-level stands. Many of the samples represented in Figs. 37.1 and 37.2 are pre-Recent, and they are certainly not all of the same age. Therefore, two or three samples from one core do not represent necessarily the same distance to the nearest land. Furthermore, other factors involved in pollen sedi-

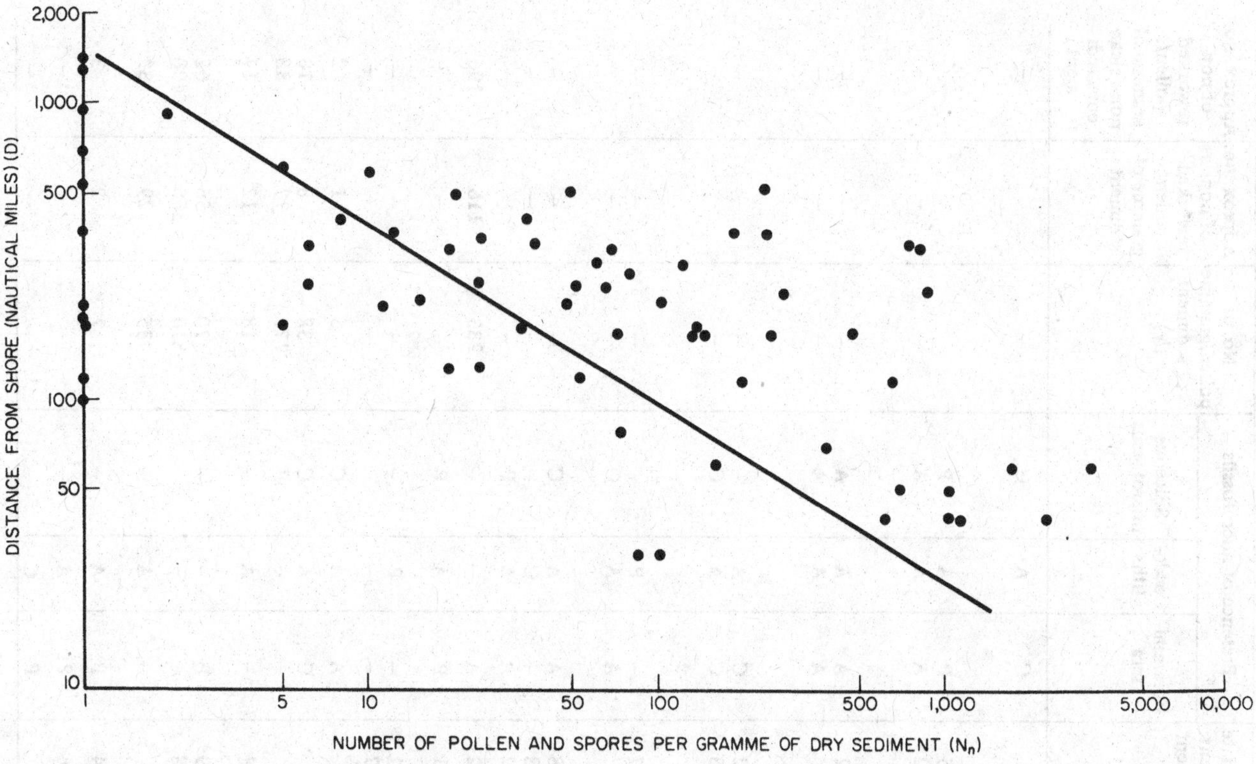

2,000
1,000
500
100
50
10
DISTANCE FROM SHORE (NAUTICAL MILES) (D)
1
5
10
50
100
500
1,000
5,000
10,000
NUMBER OF POLLEN AND SPORES PER GRAMME OF DRY SEDIMENT (N_n)

Fig. 37.1. The total number of pollen spores per gramme of sediment in cores from the Atlantic, Pacific and Indian Oceans, plotted against the distance from the nearest shore.

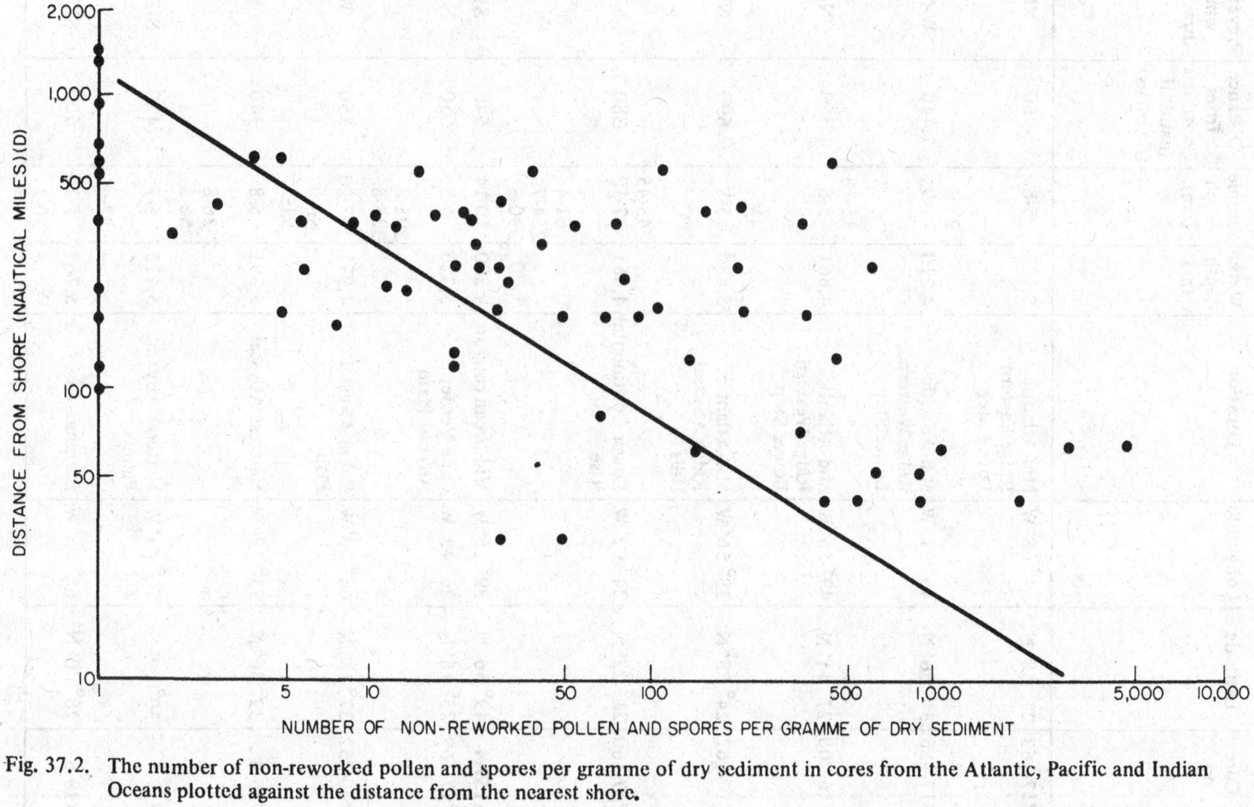

2,000
1,000
500
100
50
10
DISTANCE FROM SHORE (NAUTICAL MILES) (D)
1
5
10
50
100
500
1,000
5,000
10,000
NUMBER OF NON-REWORKED POLLEN AND SPORES PER GRAMME OF DRY SEDIMENT

Fig. 37.2. The number of non-reworked pollen and spores per gramme of dry sediment in cores from the Atlantic, Pacific and Indian Oceans plotted against the distance from the nearest shore.

Table 37.1.

Core no.	Latitude	Longitude	Location	Water depth (m)	Sample depth (cm)	Distance from Continent (nautical (*D*) miles)	Prevailing wind direction	Carbonate content per cent	Presence of microfossils			No. of pollen/gramme of sediment (*N*)	Approx. no. of non-reworked pollen/gramme of sediment	Approx. no. of non-reworked pollen/gramme of non-calcareous sediment
									Foraminifera	Coccoliths	Siliceous microfossils			
V17-163	27° 58′ N	34° 08′ W	Mid-Atlantic Ridge Eastern Lower Step	5,132	5-8	960	NE	56	C	A	C	–	–	–
					23-6			41.5	P	A	P			
V17-166	34° 56′ N	45° 21′ W	Mid-Atlantic Ridge Western Lower Step	4,210	7-9	1,440	SW-NW	75.2	A	A	A	–	–	–
					15-18			81.2	A	A	A	–	–	–
V16-208	27° 44′ N	49° 55′ W	Mid-Atlantic Ridge Western Lower Step	4,861	3-6	1,350	NE	68	A	A	A	–	–	–
					20-3			56.2	C	A	–	–	–	–
V17-162	24° 58′ N	28° 56′ W	Mid-Atlantic Ridge Abyssal Hills	5,480	5-8	690	NE	31.3	C	A	C	–	–	–
					43-45			9.7	–	P	–	–	–	–
V17-160	21° 22′ N	24° 02′ W	Lower Continental Rise	4,861	7-10	380	NE	84	A	A	C	–	–	–
					61-4			65.8	A	A	C	–	–	–
					254-7			29.8	–	C	–	–	–	–
					330-3			87.2	A	A	P	–	–	–
A157-4	49° 56′ N	39° 28′ W	Mid-ocean canyon	4,300	10-14	540	SW	41.2	A	–	–	233	116	196
V17-169	23° 16′ N	26° 43′ W	Cape Verde, Abyssal Plain	5,320	5-8	550	NE	35.1	P	A	P	–	–	–
					295-8			11.2	–	P	–	–	–	–
					335-8			10.7	–	–	–	–	–	–
V20-252	37° 14′ N	66° 32′ W	Sohm Abyssal Plain	4,991	18-20	350	W	51	A	A	C	5	1	2
					56-9			20	C	A	C	38	8	10
					91-5			13	–	–	P	775	77	88
V19-4	32° 36′ N	71° 19′ W	Hatteras Abyssal Plain	5,391	5-8	340	W	29.4	–	A	–	18	12	17
					36-8			9.9	P	–	–	70	56	62
					56-8			9.8	–	–	–	810	360	400
V19-5	30° 29′ N	70° 43′ W	Hatteras Abyssal Plain	5,415	5-7	420	S/SE	10	–	A	–	35	29	32
					22-4			24.4	P	A	–	7	2	3
V19-6	28° 20′ N	68° 06′ W	Bermuda Rise	5,262	7-10	600	S/SE	20.6	P	A	–	–	–	–
					30-2			14.	P	C	–	–	–	–

Table 37.1.

Core no.	Latitude	Longitude	Location	Water depth (m)	Sample depth (cm)	Distance from Continent (nautical (*D*) miles)	Prevailing wind direction	Carbonate content per cent	Presence of microfossils			No. of pollen/gramme of sediment (*N*)	Approx. no. of non-reworked pollen/gramme of sediment	Approx. no. of non-reworked pollen/gramme of non-calcareous sediment
									Foraminifera	Coccoliths	Siliceous microfossils			
V19-7	27° 28′ N	68° 27′ W	Bermuda Rise	5,362	5-8	600	S/SE	28.5	–	A	–	4	3	4
					16-19			18.3	–	C	–	9	4	5
A172-9	19° 48.5′ N	66° 11′ W	Puerto Rico	7,955	40-2	70	E/NE	44.4	P	P	P	390	350	630
A172-10	19° 45′ N	66° 37′ W	Puerto Rico	7,955	20-2	60	E/NE	9.3	C	A	A	162	150	166
V15-190	19° 49.5′ N	65° 53′ W	Puerto Rico	8,341	14-19	80	E/NE	15.3	–	–	P	75	68	80
RC8-106	14° 00′ N	74° 53′ W	North of Columbia	3,904	3-6	180	NE	51.4	A	A	A	–	–	–
					50-5			42.6	A	C	A	–	–	–
					100-3			37.5	A	A	A	33	30	48
					150-3			49.8	A	A	A	138	110	220
RC9-62	06° 49.6′ N	78° 43.6′ W	West of Columbia	3,250	12-15	40	S	18.8	P	P	A	1060	960	1190
					20-4			14.0	P	–	A	2310	2080	2420
					50-3			17.5	–	–	A	1120	560	675
					150-3			13.2	–	–	C	622	440	504
V19-18	13° 12′ N	76° 12′ W	North of Venezuela Abyssal Plain	3,914	30-3	250	NE	78	C	C	P	52	20	91
					70-3			–	P	–	–	23	23	23
					100-3			6.7	P	–	P	5	5	5
					140-3			22.4	P	–	P	63	30	39
RC9-73	10° 42.8′ S	81° 51.7′ W	West of Peru	4,352	12-14	210	SE/NE	8.7	P	–	A	14	13	14
					32-4			10.4	–	–	A	10	9	10
V20-78	47° 15′ N	131° 02′ W	Eastern North Pacific	2,983	5-10	240	W	25	P	C	A	274	206	275
					44-9			24.1	–	A	P	891	623	823
V20-79	46° 50′ N	133° 18′ W	Eastern North Pacific	3,711	5-9	300	W	10.2	–	–	A	62	43	48
					10-15			5.4	–	P	P	122	24	25
V20-80	46° 30′ N	135° 00′ W	Eastern North Pacific	3,801	12-15	360	W	62.4	P	A	P	55	5	13
					40-4			32.1	–	A	P	23	23	34
V14-102	10° 15′ N	57° 11′ E	Arabian Sea	3,915	16-20	380	NE and SW Monsoon	74.8	C	A	A	183	165	656
					26-30			73	A	A	A	239	215	795
V14-114	15° 17′ N	42° 02′ E	Red Sea	768	35-40	50	NW	62.2	A	A	P	708	650	1,670
					50-3			63.6	A	A	P	1,050	950	2,930
V20-125	43° 29′ N	154° 22′ E	East of Japan	5,545	10-15	220	SSE and NNW	12.2	–	–	A	103	83	94
					45-9			4.5	–	–	A	49	32	33

Table 37.1.

Core no.	Latitude	Longitude	Location	Water depth (m)	Sample depth (cm)	Distance from Continent (nautical (*D*) miles)	Prevailing wind direction	Carbonate content per cent	Presence of microfossils			No. of pollen/gramme of sediment (*N*)	Approx. no. of non-reworked pollen/gramme of sediment	Approx. no. of non-reworked pollen/gramme of non-calcareous sediment
									Foraminifera	Coccoliths	Siliceous microfossils			
V20-131	36° 20′ N	151° 00′ E	East of Japan	5,858	11-15	520	SSE and NNW	9	–	P	A	50	40	44
					34-8			0.8	–	–	A	19	15	15
V19-151	07° 43′ S	103° 15′ E	South-west of Java	6,419	15-18	170	W and E Moonsoon		P	–	C	136	91	212
									P	–	A	485	368	
					40-3 60-5			7.5	–	–	A	145	72	77
V19-152	07° 16′ S	102° 02′ E	South-west of Java	5,665	5-7	170	W and E Monsoon	13.2	P	–	A	72	50	56
					17-20			8.7	–	–	A	250	215	235
V19-153	08° 51′ S	102° 07′ E	South-west of Java	5,443	12-15	280	W and E Monsoon	13.5	–	P	A	80	50	58
V19-210	20° 39′ S	41° 43′ E	Mozambique Channel	3,241	14-19	100	SE	25.8	–	A	–	–	–	–
					29-32			28.8	–	A	–	–	–	–
V19-214	23° 22′ S	38° 51′ E	Mozambique Channel	3,092	6-9	220	SE	62.3	A	A	A	14	11	23
					23-6			55.8	A	A	A	–	–	–
V20-196	25° 11′ S	41° 37′ E	Mozambique Channel	4,036	10-14	130	SE	46	A	A	A	23	12	25
					38-42			45.2	A	A	A	18	15	27
V18-230	40° 43′ S	178° 42′ E	East of New Zealand	3,003	4-6	120	W	18.2	P	A	C	670	470	572
					45-8			16.8	P	A	C	200	140	168
V18-231	38° 49′ S	179° 10′ E	East of New Zealand	3,530	13-16	60	W	11.8	P	A	C	3,400	3,100	3,500
					31-4			13.8	–	A	C	1,800	1,100	1,280
V18-225	39° 09′ S	157° 29′ E	Between Australia and New Zealand	4,678	10-14	370	SW	73.5	P	A	P	10	10	38
					40-4			20.1	–	A	P	10	10	12
V18-222	38° 34′ S	140° 37′ E	South of Australia	1,904	6-9	30	W	68	A	A	C	101	30	93
					42-4			64.8	A	A	C	85	50	142
RC9-149	29° 56.4′ S	113° 30.8′E	West of Australia	4,334	7-10	120	S/SE and W	73.4	A	A	A	54	20	75
					37-41			86.8	A	A	A	–	–	–
RC9-152	30° 21.6′ S	112° 15.5′E	West of Australia	4,868	17-21	180	S/SE and W	11.5	–	–	–	–	–	–
					37-41			7.6	–	–	P	4	4	4

Notes: Presence of microfossils other than pollen and spores was determined by making three traverses at a magnification of 125 x and two traverses with a magnification of 1,250 x over a slide on which a small amount of unprocessed sediment was deposited in distilled water.
When I – 10 microfossils were encountered, the relative abundance was designated as P
When II – 50 microfossils were encountered, the relative abundance was designated as C
When > microfossils were encountered, the relative abundance was designated as A

mentation also varied during the Quaternary, and this is probably responsible for the wide scatter of points shown and the rather large variance and standard deviation.

An attempt was made to plot the number of non-reworked pollen and spores per gramme of sediment versus the leeward and windward distance from land (Fig. 37.3).

Difficulties are that pollen grains, if transported by wind, do not necessarily have to be transported by the *prevailing* wind only, and in cases of monsoons the prevailing wind in one season may make the sample location a leeward one, and in the other season a windward one. Therefore, only those samples were plotted whose location with respect to the prevailing wind is reasonably certain. Figure 37.3 does not show any clear influence of wind direction on pollen sedimentation, and suggests that oceanic circulation is the dominant factor in pollen dispersal.

Sediment Type

If pollen grains are mainly transported by currents, and can be considered as particles with the same hydrodynamical properties as those of fine-grained detrital grains derived from the continents, then we might expect that samples containing a high percentage of carbonate and/or abundant siliceous microfossils would have fewer pollen per gramme than samples consisting essentially of continent-derived lutite only. Figure 37.4 shows, however, that this is not the case, other factors than carbonate content apparently being of greater significance, particularly distance from shore again. This is suggested by the relationship between the number of non-reworked pollen grains per gramme of non-calcareous sediment and distance from shore (Fig. 37.5). This is expressed by

$$\ln (N_3 + 1) = 12.4667 - 1.7276 \ln D \qquad (3)$$

The variance of $\ln (N_3 + 1)$ is 4.037 and the standard deviation is 7.41.

Samples of *one core* generally do show an inverse relationship between pollen abundance and carbonate content, presumably because they were deposited under more or less similar conditions of oceanic circulation and distance from land.

POLLEN SPECTRA

Koreneva (1957, 1966) has shown that pollen spectra from superficial sediment samples in the Sea of Okhotsk reflect the vegetation of the neighbouring land, and the same was demonstrated by Vronskiy and Panov (1963) for the Mediterranean Sea, and Groot *et al.* (1967) for the Argentine Basin.

It should be understood that the pollen spectra only *reflect* the vegetation; they do not enable the palynologist to reconstruct it. Generally, only a limited number of plant genera or families are represented by pollen in deep-sea sediments; in addition, pollen percentages do not necessarily occur in proportion to the vegetation that produced them. However, this problem is not limited to the investigation of deep-sea sediments. It pertains also to palynological study of continental sediments.

Vertical Distribution

If pollen and spores reflect the vegetation of adjacent land masses, vertical changes in pollen spectra should indicate vegetational changes in time, provided that other factors, such as pollen source area, oceanic and/or atmospheric circulation, did not materially change during the time represented by the length of the core investigated. For the Quaternary this is probably true for most areas, because at the beginning of the Pleistocene the distribution of continents and oceans was essentially as it is at present, and atmospheric and oceanic circulation was probably similar to that of the Post-glacial. Fluctuations in circulation did undoubtedly occur in response to Quaternary temperature changes, but these were probably limited to latitudinal shifts, as described by Flohn (1952) for the atmosphere. Therefore, we may expect that the main variable responsible for vertical changes in pollen spectra during the Quaternary is the vegetation, which responds in turn to climatic change. Nevertheless, it would be unwise to assume that *all* changes in pollen spectra necessarily indicate climatic change, unless corroborative evidence can be obtained independent of the palynological data. Independent evidence may be based on the study of other microfossils (diatoms, radiolarians, foraminifers, or coccoliths), oxygen isotope ratios, or the carbonate content of the samples investigated.

An example of a study of the vertical distribution of pollen and spores is the investigation of the sediments of the Argentine Basin (Groot and Groot 1964, and Groot *et al.* 1967). The vertical distribution of the pollen is essentially the same in all fourteen cores investigated (except one which was apparently affected by slumping of the sediments). Representative pollen sequences are shown in Fig. 37.6 and 37.7. Core V17-121 is located at 43° 58′ S and 52° 09′ W, about 875 kilometres offshore, in the southwestern part of the Argentine Rise. The water depth is 5,786 metres. Core V15-141 is located at 45° 44′ S and 50° 45′ W, water depth 5,885 metres.

The most striking features of the pollen profiles are the general paucity of arboreal pollen and the alternation of zones with abundant pteridophyte spores on the one hand and zones with abundant Chenopodiaceae and *Ephedra*

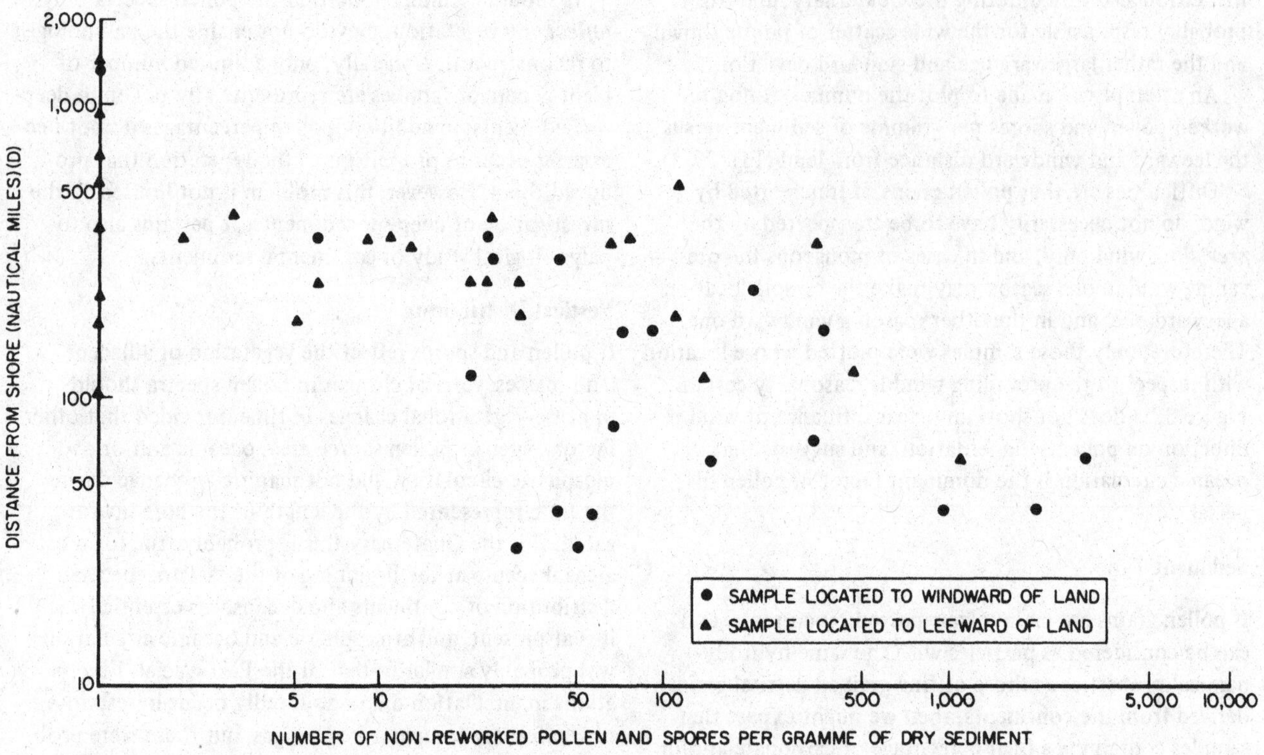
2,000
1,000
500
100
50
10
DISTANCE FROM SHORE (NAUTICAL MILES) (D)
1
5
10
50
100
500
1,000
5,000
10,000
NUMBER OF NON-REWORKED POLLEN AND SPORES PER GRAMME OF DRY SEDIMENT
SAMPLE LOCATED TO WINDWARD OF LAND
SAMPLE LOCATED TO LEEWARD OF LAND

Fig. 37.3. The number of non-reworked pollen and spores per gramme of dry sediment on both the leeward and windward sides of the land in the Atlantic, Pacific and Indian Oceans, plotted against the distance from the nearest shore.

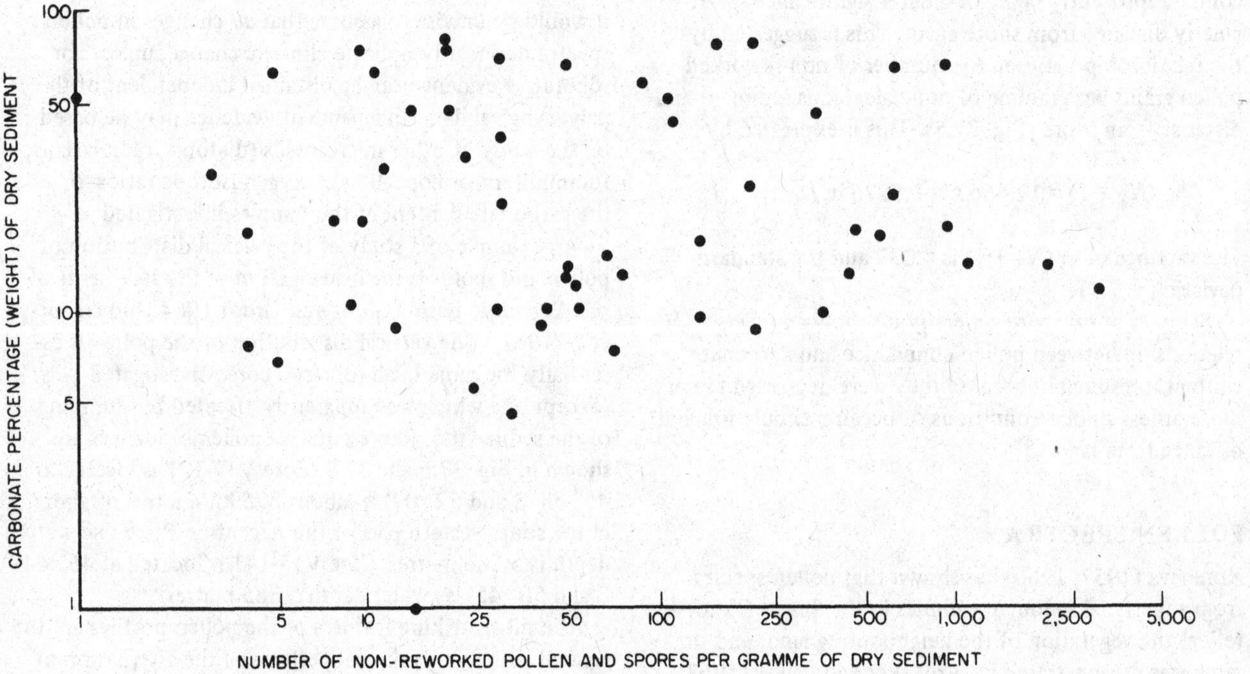
100
50
10
5
1
CARBONATE PERCENTAGE (WEIGHT) OF DRY SEDIMENT
1
5
10
25
50
100
250
500
1,000
2,500
5,000
NUMBER OF NON-REWORKED POLLEN AND SPORES PER GRAMME OF DRY SEDIMENT

Fig. 37.4. The number of non-reworked pollen and spores per gramme, plotted against the percentage carbonate content (by weight) of the dry sediment.

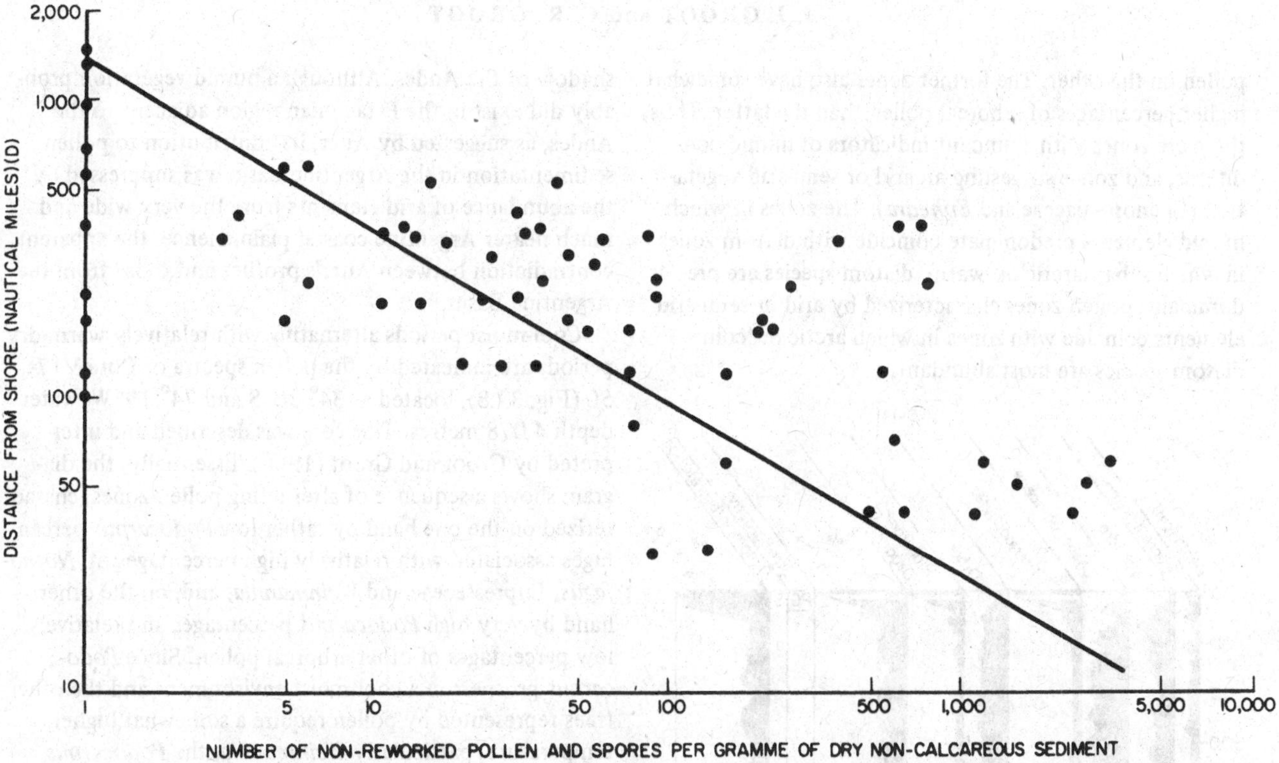

2,000
1,000
500
100
50
10
DISTANCE FROM SHORE (NAUTICAL MILES)(D)
1
5
10
50
100
500
1,000
5,000
10,000
NUMBER OF NON-REWORKED POLLEN AND SPORES PER GRAMME OF DRY NON-CALCAREOUS SEDIMENT

Fig. 37.5. **The number of non-reworked pollen and spores per gramme of dry non-calcareous sediment plotted against the distance from the shore.**

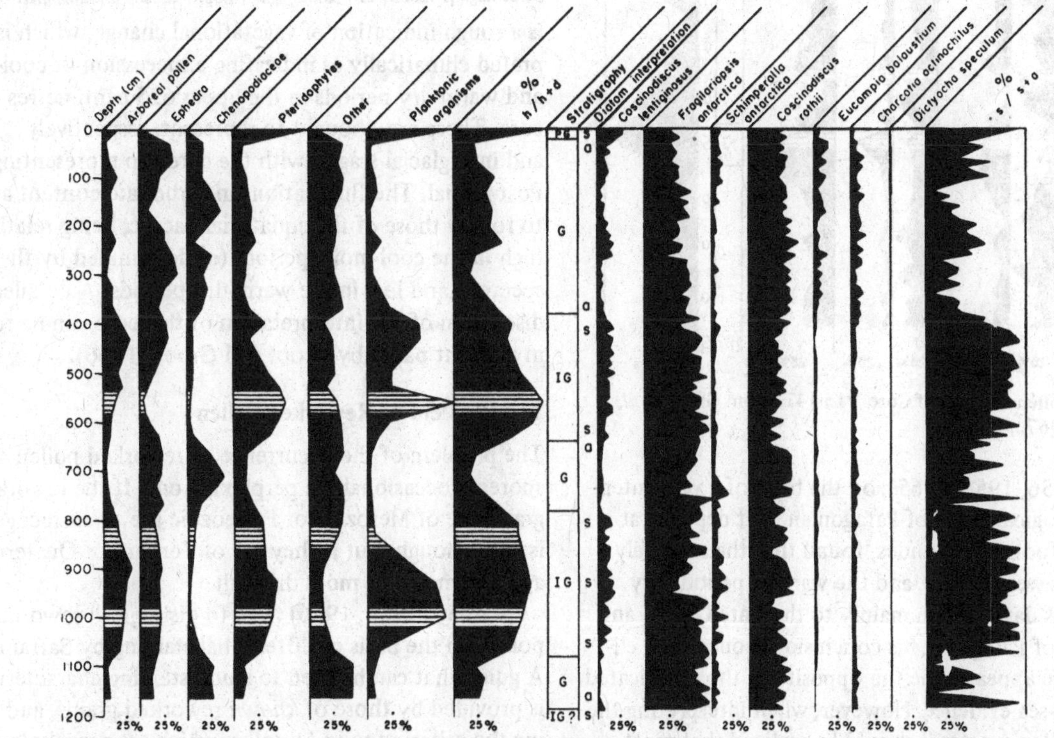

Depth (cm)
Arboreal pollen
Ephedra
Chenopodiaceae
Pteridophytes
Other
Planktonic organisms
% h / h+d
Stratigraphy
Diatom interpretation
Coscinodiscus lentiginosus
Fragilariopsis antarctica
Schimperiella antarctica
Coscinodiscus rothii
Eucampia balaustium
Charcotia actinochilus
Dictyocha speculum
% s / s+a
25%

Fig. 37.6. **Pollen diagram of Core V17-121. The column, %h/(h+d) represents the ratio of humid elements to the sum of humid elements plus arid elements, x 100. The abundance %s (s+a) indicates the percentage of the sub-antarctic diatom species** ***Coscinodiscus lentiginosus*** **to the number of diatoms of this species plus that of the Antarctic species** ***Eucampia balaustium*** **and** ***Charcotia actinochilus*****, x 100. The parts of the graphs of pollen abundance indicated by lines instead of by solid black are those for which the percentages are of lesser reliability, being based on slides in which less than 100 grains were available for counting (from Groot** ***et al.*** **1967, p. 191).**

pollen on the other. The former zones also have somewhat higher percentages of arboreal pollen than the latter. Thus, there are zones with abundant indicators of humid conditions, and zones suggesting an arid or semi-arid vegetation (Chenopodiaceae and *Ephedra*). The zones in which humid elements predominate coincide with diatom zones in which subantarctic or 'warm' diatom species are predominant; pollen zones characterized by arid or semi-arid elements coincide with zones in which arctic or 'cold' diatom species are most abundant.

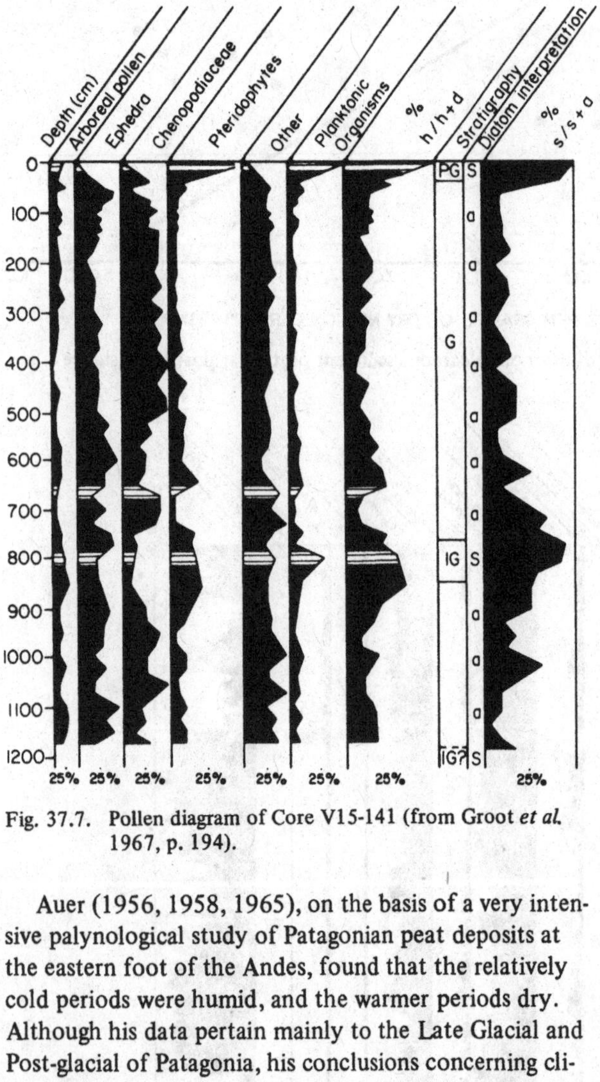
V 15-141
Depth (cm)
Arboreal pollen
Ephedra
Chenopodiaceae
Pteridophytes
Other
Planktonic Organisms
%
h/h+d
Stratigraphy
Diatom interpretation
%
s/s+a
0
100
200
300
400
500
600
700
800
900
1000
1100
1200
PG
G
IG
IG?
S
a
25%

Fig. 37.7. Pollen diagram of Core V15-141 (from Groot *et al.* 1967, p. 194).

Auer (1956, 1958, 1965), on the basis of a very intensive palynological study of Patagonian peat deposits at the eastern foot of the Andes, found that the relatively cold periods were humid, and the warmer periods dry. Although his data pertain mainly to the Late Glacial and Post-glacial of Patagonia, his conclusions concerning climatic change appear to be the opposite of those indicated by the deep-sea evidence. However, when interpreting the deep-sea pollen spectra it should be realized that most pollen and spores are derived from the vegetation of the coastal regions, and that during the glacial ages much of the wide continental shelf of Argentina was above sea level, supporting an arid or semi-arid vegetation in the rain shadow of the Andes. Although a humid vegetation probably did exist in the Patagonian region adjacent to the Andes, as suggested by Auer, its contribution to pollen sedimentation in the Argentine Basin was suppressed by the abundance of arid elements from the very wide and much nearer Argentine coastal plain. Hence, the apparent contradiction between Auer's profiles and those from the Argentine Basin.

Cool-moist periods alternating with relatively warm-dry periods are indicated by the pollen spectra of Core V17-50 (Fig. 37.8), located at 34° 30′ S and 74° 19′ W, water depth 4,078 metres. This core was described and interpreted by Groot and Groot (1966). Essentially, the diagram shows a sequence of alternating pollen zones, characterized on the one hand by rather low *Podocarpus* percentages associated with relatively high percentages of *Nothofagus*, Cupressaceae and *Weinmannia*, and, on the other hand by very high *Podocarpus* percentages and relatively low percentages of other arboreal pollen. Since *Podocarpus* prospers in a cool-moist environment and the other trees represented by pollen require a somewhat higher temperature, particularly *Weinmannia*, the *Podocarpus* ratio curve in Fig. 37.8

$$\frac{\text{percentage } \textit{Podocarpus}}{\text{percentage } [\textit{Podocarpus} + \textit{Nothofagus} + \text{Cupressaceae} + \textit{Weinmannia}]} \times 100\%$$

is a rough indication of vegetational change, which is interpreted climatically as indicating a succession of cool-moist and warm-dry periods in the upper 625 centimetres of the core. These are thought to represent, respectively, glacial and interglacial stages, with the core top representing the Post-glacial. The fluctuations in carbonate content appear to follow those of the equatorial Pacific, being relatively high in the cool-moist periods (as determined by the pollen spectra), and low in the warm-dry periods. A detailed discussion of the interpretation of this core can be found in a recent paper by Groot and Groot (1966).

The Problem of Reworked Pollen

The problem of the occurrence of reworked pollen and spores is occasionally a perplexing one. If the reworked grains are of Mesozoic or Palaeozoic age, their recognition is easy enough, but if they are of Tertiary or Quaternary age, the matter is more difficult.

Stanley (1965, 1966) tried to distinguish reworked pollen on the basis of differential staining by Safranin O. A guide that can be used to study staining characteristics is provided by those of *known* reworked grains, and applying the criteria found to pollen which are morphologically similar to those found in Pleistocene deposits. Unfortunately, the staining characteristics of known reworked pollen vary, as was shown by Stanley (1966), and some stain much like Recent pollen and spores. Consequently,

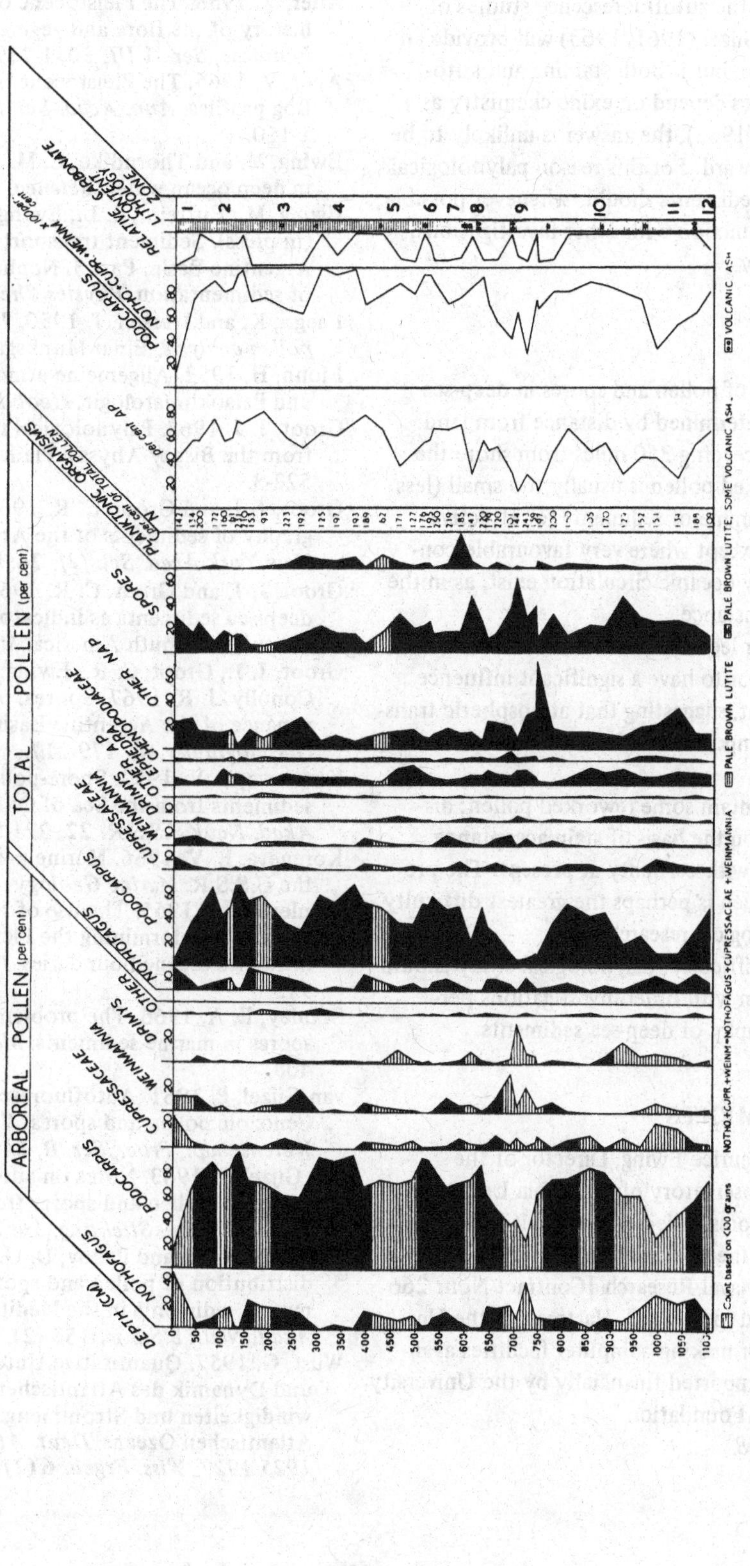
V17-50
ARBOREAL POLLEN (per cent)
TOTAL POLLEN (per cent)
DEPTH (CM)
NOTHOFAGUS
PODOCARPUS
CUPRESSACEAE
WEINMANNIA
DRIMYS
OTHER
NAP
SPORES
PLANKTONIC ORGANISMS
NOTHO + CUPR + WEINM + NOTHOFAGUS / CUPRESSACEAE + WEINMANNIA
PALE BROWN LUTITE
PALE BROWN LUTITE, SOME VOLCANIC ASH
VOLCANIC ASH

Fig. 37.8. Pollen diagram of Core V17-50 (from Groot and Groot 1966).

33

there is no way to be certain about the presence of reworked grains which are morphologically similar to Quaternary ones, and the determination of percentages of reworked material as shown in Table 37.1 can only be approximate. Perhaps the autofluorescence studies of pollen exines by van Gijzel (1961, 1963) will provide an answer to this problem, but if both staining and autofluorescence techniques depend on exine chemistry as suggested by Stanley (1966), the answer is unlikely to be simple and straightforward. For this reason palynological study of Quaternary sediments should, whenever possible, be carried out in conjunction with other investigations, as suggested previously.

CONCLUSIONS

1. The distribution of pollen and spores in deep-sea sediments is mainly determined by distance from land.

2. At a distance exceeding 350 miles from shore the number of non-reworked pollen is usually too small (less than ten grains per gramme of sediment) for effective palynological study, except where very favourable conditions for dispersal by oceanic circulation exist, as in the Argentine Basin, for instance.

3. The windward or leeward position of sediment samples does not appear to have a significant influence on their pollen content, suggesting that atmospheric transportation is of small importance relative to transportation by currents.

4. Most samples contain some reworked pollen; distinguishing the latter on the basis of stain acceptance cannot be carried out with certainty at present. The presence of reworked pollen is perhaps the greatest difficulty facing marine palynological research.

5. In view of this difficulty, palynological study should be done in conjunction with other investigations pertaining to the stratigraphy of deep-sea sediments.

ACKNOWLEDGEMENTS

We thank Professor Maurice Ewing, Director of the Lamont Geological Observatory of Columbia University, for making available cores collected under his direction aboard the *Vema* and the *Conrad* with the financial support of the Office of Naval Research [Contract Nonr 266 (48)]. We are indebted to Mr L. G. Hartman of the University of Delaware for making computer facilities available. The work was supported financially by the University of Delaware Research Foundation.

Received January 1968.

BIBLIOGRAPHY

Auer, V. 1956. The Pleistocene of Fuego-Patagonia, 1. The Ice and Interglacial ages. *Ann. Acad. Sci. Fennicae, Ser. A III*, **45**, 1-226.

Auer, V. 1958. The Pleistocene of Fuego-Patagonia, 2. The history of the flora and vegetation. *Ann. Acad. Sci. Fennicae, Ser. A III*, **50**, 1-239.

Auer, V. 1965. The Pleistocene of Fuego-Patagonia, 4. Bog profiles. *Ann. Acad. Sci. Fennicae, Ser. A III*, **80**, 1-160.

Ewing, M. and Thorndike, E. M. 1965. Suspended matter in deep ocean water. *Science*, **147** (3663), 1291-4.

Ewing, M., Eittreim, S. L., Ewing, J. I. and LePichon, X. (in press). Sediment transport and distribution in the Argentine Basin. Part 3. Nepheloid layer and processes of sedimentation. *Physics Chemistry Earth*, **8**.

Faegri, K. and Iversen, J. 1950. *Textbook of modern pollen analysis.* Ejnar Munksgaard: Copenhagen. 168 pp.

Flohn, H. 1952. Allgemeine atmosphärische Zirkulation und Palaoklimatologie. *Geol. Rundschau*, **40**, 153-78.

Groot, J. J. 1963. Palynological investigation of a core from the Biscay Abyssal Plain. *Science*, **141** (3580), 522-3.

Groot, J. J. and Groot, C. R. 1964. Quaternary stratigraphy of sediments of the Argentine Basin. *Trans. New York Acad. Sci. (2)*, **26**, 881-6.

Groot, J. J. and Groot, C. R. 1966. Pollen spectra from deep-sea sediments as indicators of climatic changes in southern South America. *Marine Geol.* **4**, 525-37.

Groot, J. J., Groot, C. R., Ewing, M., Burckle, L. and Conolly, J. R. 1967. Spores, pollen, diatoms and provenance of the Argentine Basin sediments. *Progress in Oceanography*, **4**, 179-217.

Koreneva, E. V. 1957. Spore-pollen analysis of bottom sediments from the Sea of Okhotsk. *Tr. Inst. Okeanol. Akad. Nauk S.S.S.R.* **22**, 221-51 (in Russian).

Koreneva, E. V. 1966. Marine palynological researches in the U.S.S.R. *Marine Geology*, **4** (6), 565-74.

Stanley, E. A. 1965. The use of reworked pollen and spores for determining the Pleistocene-Recent and the infra-Pleistocene boundaries. *Nature, Lond.* **206**, 289-91.

Stanley, E. A. 1966. The problem of reworked pollen and spores in marine sediments. *Marine Geology*, **4**, 397-408.

van Gijzel, P. 1961. Autofluorescence and age of some Cenozoic pollen and spores. *Koninkl. Ned. Akad. Wetenschap., Proc., Ser. B*, **64** (1), 56-63.

van Gijzel, P. 1963. Notes on autofluorescence of some Cenozoic pollen and spores from The Netherlands. *Mededel. Geol. Stichting*, **16**, 25-31.

Vronskiy, V. A. and Panov, D. G. 1963. Composition and distribution of pollen and spores in the surface layer of marine sediments in the Mediterranean Sea. *Dokl. Akad. Nauk S.S.S.R.* **153** (2), 447-9 (in Russian).

Wüst, G. 1957. Quantitative Untersuchungen zur Statik und Dynamik des Atlantischen Ozeans. Stromgeschwindigkeiten und Strommengen in den Tiefen des Atlantischen Ozeans. *Deut. Atlantische 'Meteor', 1925-1927. Wiss. Ergeb.* **6** (2), 420 pp.

SECTION D

DISTRIBUTION OF PRE-QUATERNARY MICROFOSSILS

38. THE OCCURRENCE OF PRE-QUATERNARY MICROFOSSILS IN THE OCEANS

B. M. FUNNELL

Summary: Well over 500 occurrences of pre-Quaternary microfossils have been recorded from the ocean basins. Their distribution is described, and briefly considered in relation to patterns of past oceanic circulation and the hypothesis of ocean-floor spreading. The relative usefulness of different groups of microfossils for ocean-floor stratigraphy is discussed, and some general conclusions on the relevance of studies of pre-Quaternary microfossils from the oceans for palaeontological theory and the understanding of the history of the oceans are suggested.

INTRODUCTION

This article reviews the state of knowledge of pre-Quaternary oceanic micropalaeontology up to August 1967.

Pre-Quaternary microfossils were first described and illustrated from the oceans by Haeckel (1887) and Murray and Renard (1891). These authors did not however recognize the pre-Quaternary nature of the microfossils they were describing, and assumed them to be part of the Recent assemblages usually encountered in surface sediment samples.

The Tertiary age of the discoasters illustrated by Murray and Renard (1891, pl. 11, fig. 4) from the South Atlantic, *Challenger* sample no. 338, was referred to by Tan Sin Hok in 1927, but it was not until 1954 that Bramlette and Riedel identified and named these forms as *Discoaster brouweri* and *D. challengeri* and recognized them as reworked Miocene forms. Later Black and Barnes (1961) also noted the occurrence of Tertiary species in this same sample, and recorded the presence of *D. perplexus* in addition to the species previously described.

Similarly, Tertiary radiolarians, described by Haeckel (1887) from a number of *Challenger* samples, were not recognized as such until much later. Riedel (1952) first recorded Tertiary radiolarians from western Pacific sediments, and later publications (1954, 1957*a*) shew that several species described by Haeckel (1887) and occurring in *Challenger* samples nos. 225, 244, 265-8, 272 and ?347 were Tertiary in age. The species concerned were: *Tympanidium binoctonum, Dipodospyris forcipata, Dorcadospyris dentata, Sethamphora mongolfieri, Phormocyrtis embolum, Calocyclas virginis, Stichocorys wolffii* and *Eusyringium fistuligerum,* of early and middle Tertiary age.

A few samples from relatively shallow submarine occurrences were recognized as Tertiary before the end of the nineteenth century (Verrill 1878), and preliminary work in the first half of the twentieth century brought to light a few occurrences on oceanic islands (e.g. Vaughan 1919), from canyons in the continental slope (Stetson 1949), and even from the deeper ocean (Wiseman 1936).

With the resurgence of interest in oceanography, including micropalaeontology, in the mid-twentieth century, and in particular with the development of better techniques in deep-sea coring, much more pre-Quaternary material was obtained and recognized as such, often leading as we have seen to re-evaluation of material collected earlier. Early results of these investigations were published by Arrhenius (1952), Ericson, Ewing and Heezen (1952), Phleger, Parker and Peirson (1953) and Riedel (1952). Considerable progress was made in the late fifties and early sixties, until now in the late sixties the subject seems to be set for further expansion under the influence of the United States Deep-Sea Drilling project and the general expansion of oceanographic activities throughout the world.

TYPES OF MICROFOSSILS

Most types of oceanic microfossil occur in pre-Quaternary sediments, except possibly the aragonitic pteropods. The principal representatives are, however, the calcareous nannoplankton, diatoms, planktonic and benthonic Foraminifera, and Radiolaria. Pollen, spores, dinoflagellates, silicoflagellates, ostrocods, fish otoliths, scales and teeth also occur. The list given represents the situation from the Jurassic onwards. Beginning with the Lower Cretaceous downwards such forms as nannoconids, *Calpionella*, tintinnids and *Halobia* may be expected. As the same time such important present-day groups as the planktonic Foraminifera and calcareous nannoplankton will presumably be found to disappear. No details of the above groups are given here, but a few general comments on their relative importance in current studies may be made.

The calcareous nannoplankton have stratigraphical ranges that are relatively well-known. Small quantities of material can be rapidly assessed under the light microscope, and detailed systematic study is possible using the electron microscope. They occur almost ubiquitously in calcareous sediments and recognizable remains may persist longer than those of planktonic (but not benthonic) Foraminifera near the limit of calcareous microfossil preservation. Their chief disadvantages are the extreme ease with which they can be re-worked (without sustaining any sign of mechanical damage), and their susceptibility to recrystallization in the course of submarine lithification or alteration by volcanic activity – whereas planktonic Foraminifera may still be identifiable after such vicissitudes and may even be identifiable in the form of moulds.

Planktonic Foraminifera have so far by far the best known stratigraphical ranges and a far more comprehensive descriptive literature exists. Compared with the calcareous nannoplankton rather more material has to be prepared for study but this is not usually critical even when only small amounts of core material are available. They are rather less easily reworked in small quantities than the calcareous nannoplankton, and in the grading of different-sized specimens in sediments may give a much more readily appreciated indication of winnowing or turbidity current accumulation.

Diatoms seem to be rather liable to solution of their more delicate representatives but they are potentially particularly useful for the elucidation of high latitude sequences – although they are also extremely abundant in calcareous-siliceous equatorial sediments.

Radiolaria again have relatively well-known stratigraphical ranges although generally speaking these are not yet available in published form, and adequate descriptions of all but a handful of the critical species are lacking. They are particularly useful in non-calcareous sediments deposited below the calcium carbonate compensation depth in the Pacific and Indian Oceans, but less so in the Atlantic. They should also prove to be invaluable in any pre-Mesozoic oceanic sediments where calcareous plankton will probably prove to be entirely absent. Radiolaria have the disadvantage that they are usually only preserved in sediments beneath zones of high productivity and they may also prove to be recrystallized to chert, at deeper levels beneath the ocean floor than those which have currently been sampled. They are also very persistant as reworked microfossils which tends to be confusing if the later Radiolaria are more susceptible to solution as they sometimes are.

Whilst, as has been pointed out, the principal groups of pre-Quaternary microfossils have various advantages and disadvantages in different circumstances, it is undoubtedly true that if all four groups are studied together the information gained will be more than the sum of the contributions by individual groups. This is true not only where questions of re-working are involved, when the different mechanical and chemical properties of the different groups may enable a more satisfactory distinction between reworked and *in situ* assemblages to be made, but it will also enable finer stratigraphical refinements to be made. Co-ordinated studies by different specialists on the same materials are therefore much to be preferred, although rarely so far achieved in practice.

The remaining groups of pre-Quaternary microfossils are each applicable in different, usually to some extent rather special circumstances, and their use is developed to various degrees.

TYPES OF OCCURRENCE

The occurrence of pre-Quaternary microfossils in the oceans is now known to be very widespread. Well over 500 *published* occurrences are known alone. In an Appendix to this article all published records (plus some unpublished ones) up to August 1967 are brought together in a geographically arranged index. The entries in this index are represented on Figs. 38.1-3 which depict the Atlantic, Indian and Pacific Oceans respectively. The manner of plotting has been to take the oldest occurrence (for this purpose *in situ* and reworked are not distinguished) in each 2° square. Records contained in three papers (Heirtzler and Hayes 1967; Burckle, *et al.* 1967; Riedel 1967), which contain insufficient detailed information for them to be incorporated in the index are represented by small symbols.

Pre-Quaternary microfossils occur at or near the surface of the ocean floor under a number of circumstances resulting from tectonic (including igneous) activity, non-deposition and probably actual erosion, some possible mechanisms for which were briefly discussed by Riedel and Funnell (1964, p. 364). These occurrences are usually sampled in three principal ways, by coring, dredging and collecting on oceanic islands.

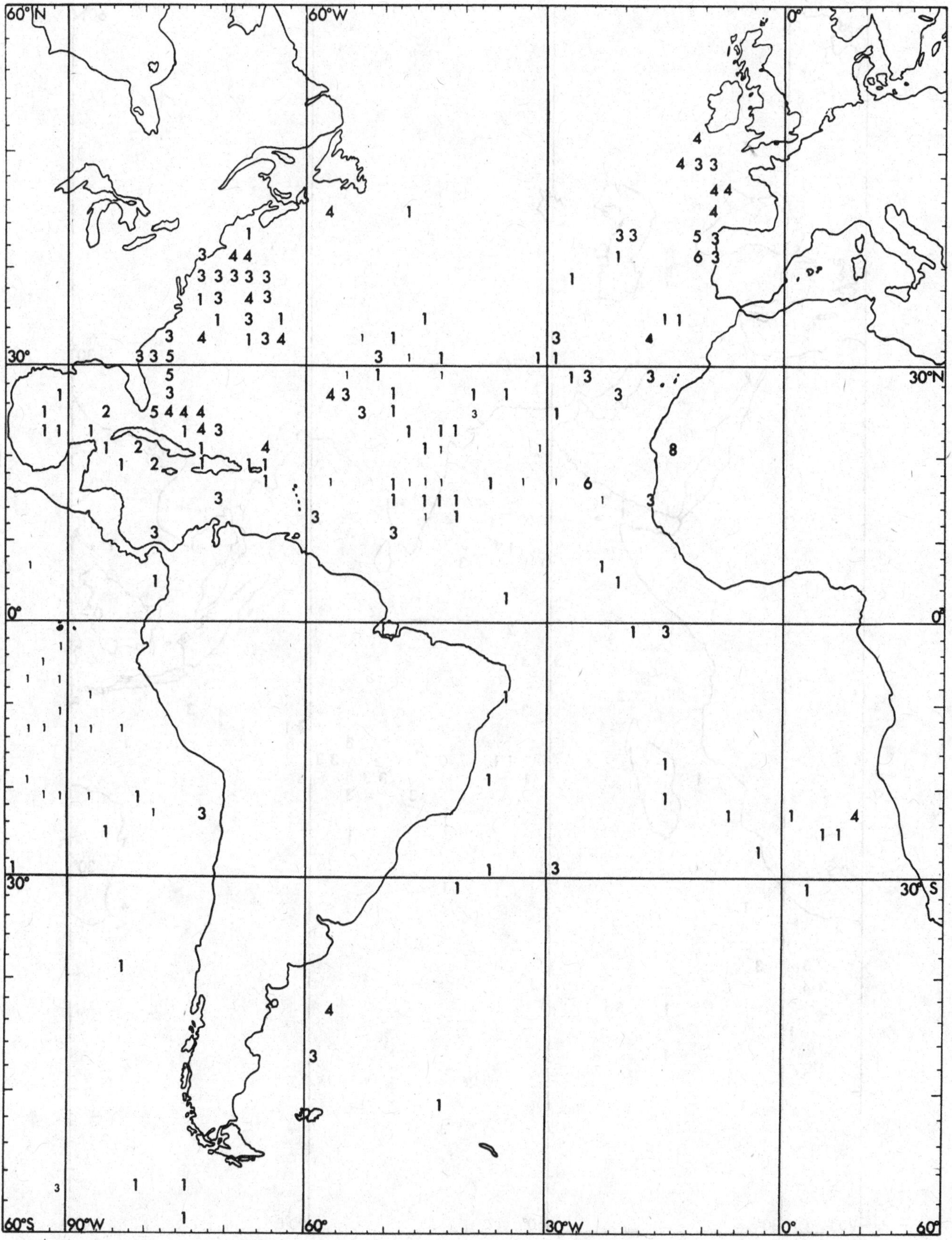
60°N
60°W
0°
30°
30°N
0°
30°
30°S
60°S
90°W
60°
30°W
0°
60°

Fig. 38.1. Pre-Quaternary microfossil occurrences in the Atlantic Ocean. 1 = upper Tertiary, 2 = mid Tertiary, 3 = lower Tertiary, 4 = Upper Cretaceous, 5 = Lower Cretaceous, 6 = Upper Jurassic.

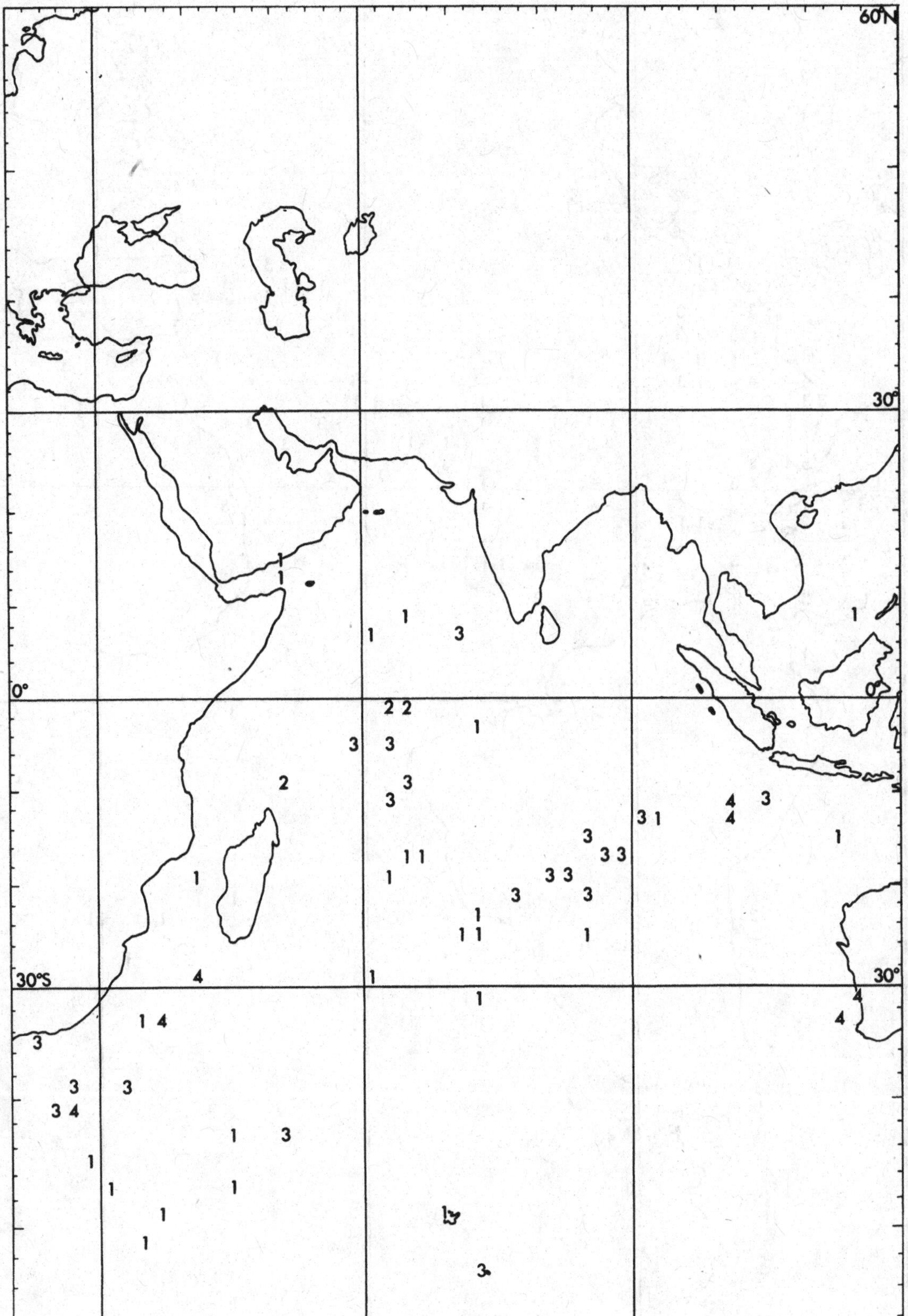
60°N
30°
0°
30°S
30°
20°E
30°
60°
90°E
60°S

Fig. 38.2. Pre-Quaternary microfossil occurrences in the Indian Ocean. 1 = upper Tertiary, 2 = mid Tertiary, 3 = lower Tertiary, 4 = Upper Cretaceous.

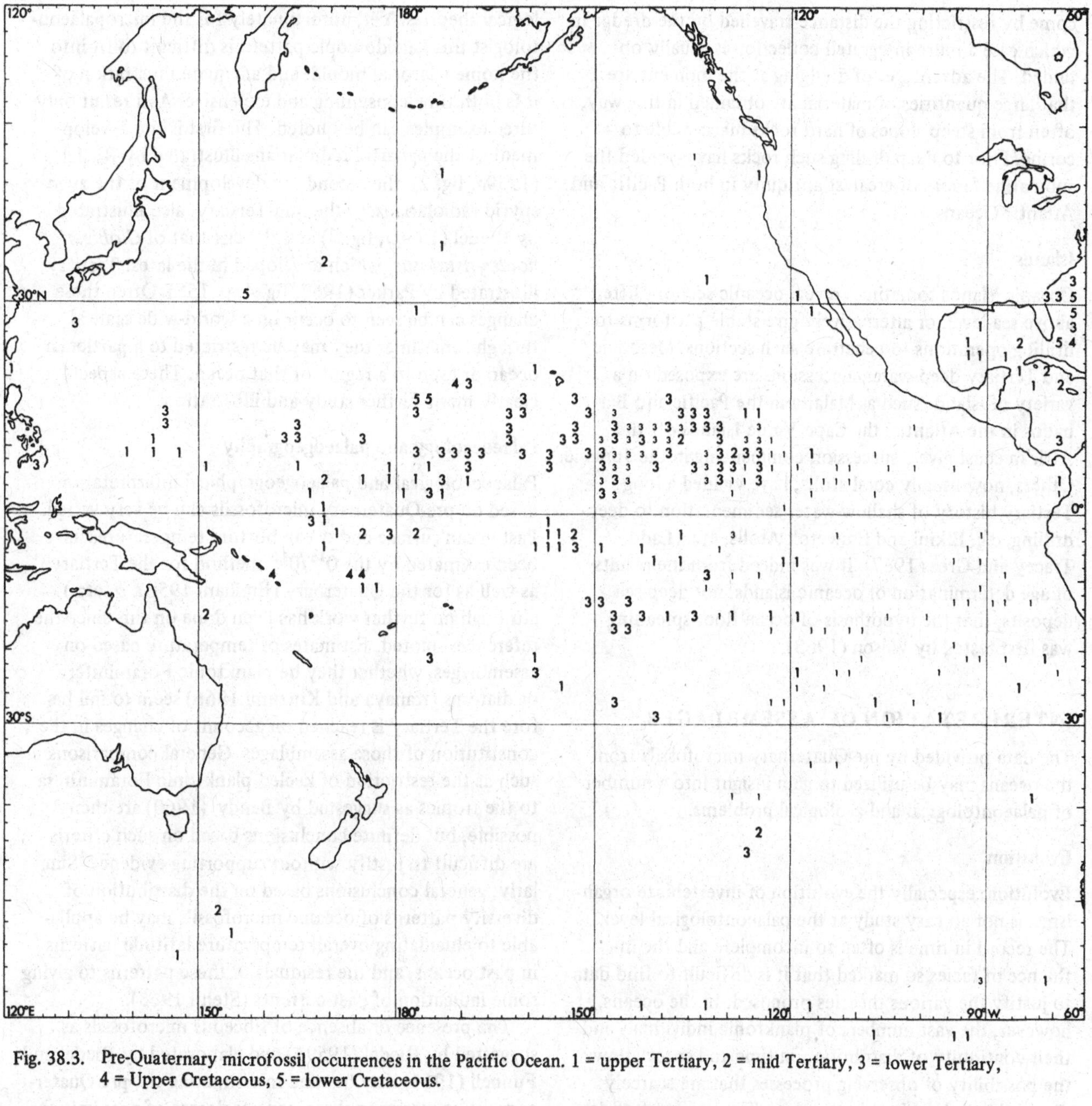
120°
180°
120°
60°
30°N
0°
30°S
30°
120°E
150°
180°
150°
120°
90°W
60°

Fig. 38.3. Pre-Quaternary microfossil occurrences in the Pacific Ocean. 1 = upper Tertiary, 2 = mid Tertiary, 3 = lower Tertiary, 4 = Upper Cretaceous, 5 = lower Cretaceous.

Cores

Samples obtained by coring are especially informative in giving information on disconformities, vertical changes in sedimentation and occasionally, e.g. in the Upper Tertiary, actual stratigraphical successions of faunas and floras. When used in conjunction with continuous seismic reflection profiling (Ewing *et al.* 1966; Saito, Ewing and Burckle 1966) older horizons can be sampled systematically. Otherwise coring relies on chance encounters with pre-Quaternary outcrops and depth of penetration (whereby Pliocene may be reached).

Dredge samples

Dredge samples are particularly effective in reconnaissance work and in obtaining hard rock samples. Steep slopes which can be detected and surveyed by routine P.D.R. apparatus are often productive of pre-Quaternary sediments (and/or igneous rocks) by dredging. In such terrain coring may be unsuitable unless the coring apparatus can be regarded as expendable. The disadvantages of such collecting is that a mixture of materials may be obtained, including the possibility of displaced (in many cases far-travelled, glacial) blocks. These disadvantages can be overcome by restricting the distance travelled by the dredge in which case a more integrated collection is usually obtained. The advantages of dredging at the moment are that large quantities of material are obtained in this way, often from steep slopes of hard rocks inaccessible to coring. Prior to deep drilling such rocks have yielded the submarine faunas of greatest antiquity in both Pacific and Atlantic Oceans.

Islands

Oceanic islands sometimes reveal oceanic sections lifted above sea-level, or alternatively give stable platforms for drilling operations to penetrate such sections. Mesozoic and Tertiary deep-water successions are exposed on a variety of islands such as Malaita in the Pacific and Barbados in the Atlantic; the Cape Verde Islands off the African coast give a succession commencing in the Jurassic. Others, now usually coral atolls, have yielded a long Tertiary history of shallow-water sedimentation to deep-drilling, e.g. Bikini and Eniwetok Atolls, etc. (Ladd, Tracey and Gross 1967). It was indeed from the results of age determination of oceanic islands, not deep-sea deposits, that the hypothesis of ocean floor spreading was first tested by Wilson (1963).

INTERPRETATION OF ASSEMBLAGES

The data provided by pre-Quaternary microfossils from the oceans may be utilized to gain insight into a number of palaeontological and geological problems.

Evolution

Evolution, especially the evolution of invertebrate organisms, is not an easy study at the palaeontological level. The record in time is often so incomplete and the influence of facies so marked that it is difficult to find data to justify the various theories proposed. In the oceans, however, the vast numbers of planktonic individuals and their continuity of distribution in time and space allow the possibility of observing processes that are scarcely discernible in land-based sequences. The low rates of deposition, and the constancy of the environment at any one place, allow the study of continuous sequences of morphological change in deep-sea cores. The world-encircling extent of populations allows these changes to be compared over very broad geographical fronts. In such cores as those recently studied by Parker (1967) it is possible to see evolution in action. Whole assemblages of planktonic Foraminifera can be seen changing in these cores, as the plexuses making up individual species are transmuted into other species. This is evolution recorded in detail that goes far towards satisfying genetic and biological theories. Yet, unfortunately for the micropalaeontologist this kaleidoscopic pattern is difficult to fit into the nomenclatorial mould, and adequate illustration of it is both time-consuming and expensive. As a result only three examples can be quoted. The first is the development of the panartid radiolarians illustrated by Riedel (1959*a*, fig. 2), the second the development of the zygospyrid radiolarians, in the mid-Tertiary, also illustrated by Riedel (1959*a*, fig. 3) and the last that of *Globigerinoides fistulosus,* which developed in the latest Tertiary, illustrated by Parker (1967, fig. 4, p. 155). Often these changes can be seen to occur on a world-wide scale although sometimes they may be restricted to a particular ocean or even to a region of that ocean. These aspects clearly merit further study and illustration.

Palaeoecology and palaeogeography

Palaeoecological and palaeogeographical interpretations based on pre-Quaternary microfossils can be very varied. Past ocean surface and ocean bottom temperatures have been estimated by the O^{16}/O^{18} method for the Tertiary as well as for the Quaternary (Emiliani 1956*a, b,* etc.) although no further work has been done on this since the references quoted. Estimates of temperature based on assemblages, whether they be planktonic Foraminifera or diatoms (Kanaya and Koizumi 1966) seem to fail before the Tertiary is reached on account of changes in the constitution of those assemblages. General comparisons such as the restriction of keeled planktonic Foraminifera to the tropics as suggested by Bandy (1960) are then possible, but definite conclusions based on such criteria are difficult to justify without supporting evidence. Similarly, general conclusions based on the distribution of diversity patterns of oceanic microfossils may be applicable to elucidating overall temperature-latitude patterns in past oceans, and the residuals of these patterns to giving some indication of past currents (Stehli 1965).

The presence or absence of siliceous microfossils as suggested by Riedel (1959*b*) and elaborated by Riedel and Funnell (1964) may be used in estimating the pre-Quaternary water masses in the oceans or degree of proximity to mid-oceanic ridges (Riedel 1967). Depth of deposition is another factor which may be indicated by microfossil assemblages. A large literature exists on the use of Foraminifera for interpreting depth of deposition, by many different methods. These range from inferences based on the known depth ranges of living benthonic forms, through the general association of larger Foraminifera (which like colonial corals often contain symbiotic algae in their protoplasm) with shallow depths, to general correlations of diversity of species and planktonic: benthonic ratios with depth. Spumellarian and Nassellarian Radiolaria may have distinctive depth ranges, although not so far applied to depth interpretation of deep-sea assemblages.

The solution of calcium carbonate which destroys calcareous microfossils at a remarkably uniform depth in some parts of the oceans (Bramlette 1961) has been used by some (Olausson 1961*a*; Riedel and Funnell 1964) to evaluate the possibility of changes in depth of the ocean floor during the Tertiary. Although, as it has been thought likely that surface productivity of calcareous plankton (Arrhenius 1952) and temperature of the bottom waters (Bramlette 1958) also effect the depth at which solution occurs, these speculations have been tentative. Recent work (Park 1966; Peterson 1966; Berger 1967; Kennett 1966) is leading to a better understanding of the process of solution which will enable a sounder view of its depth implications to prevail.

In the same way that the chemical effects of sea water acting on microfossil skeletons may possibly give an indication of depth of deposition, so to a limited extent may the physical effects of current movement in ocean waters. Coarser, presumably winnowed lag accumulations may be characteristic of past or present topographic highs, and winnowed out finer accumulations or graded (turbidity current) deposits of topographic lows. In this way (e.g. Saito, Ewing and Burckle 1966; Cann and Funnell 1967) some suggestions may be made regarding the past palaeogeography of the oceans.

Biostratigraphy

The principal groups employed for biostratigraphy in the oceans are the planktonic Foraminifera, calcareous nannoplankton, Radiolaria and diatoms. Other groups such as benthonic Foraminifera, orosphaerid Radiolaria and silicoflagellates, etc., can also be used to obtain general indications of age especially in particular facies. Thus benthonic Foraminifera are often particularly useful in continental slope and oceanic island occurrences and in partially dissolved deep-sea assemblages; orosphaerid Radiolaria in sparsely radiolarian red clays. The principal groups, however, are capable of separately producing a series of finely discriminated stratigraphical zones. These are best developed for the planktonic Foraminifera as summarised by Bolli (1966), with modifications for the Upper Tertiary as proposed by Banner and Blow (1965*b*; Parker 1967). Considerable refinement has also been obtained by means of the calcareous nannoplankton and Radiolaria although no coherent schemes based on these have yet been published. Diatoms are not so well known.

The application of calcareous nannoplankton to the pre-Quaternary stratigraphy of the oceans has been rather extensive, although much of the published data is in the form of expressions of opinion on ages of sediments with no explicit reference to the taxa on which these opinions are based. However, since the recognition of their importance in stratigraphy (Bramlette and Riedel 1954), a fair number of monographs have appeared in land-based sequences of coccoliths which do provide a basis for comparison. Important contributions on their occurrence in land-based sequences are those by Bramlette and Sullivan (1961), Bramlette and Martini (1964), Martini (1958, 1960), Stradner (1963, 1964) and Sullivan (1964, 1965). The principal papers which actually describe coccoliths (and discoasters) from pre-Quaternary deep-sea sediments are Bramlette and Riedel (1954), Black (1962, 1964), Martini (1965), Martini and Bramlette (1963) and Kamptner (1963). Certain distinctive associations and ranges emerge from these studies which allow a ready gross subdivision of the late Mesozoic and Caenozoic sediments encountered in the deep-sea. More subtle correlation, comparable with that established for the planktonic Foraminifera, depends on more detailed study for which, amongst other things, an appreciation of the ease of reworking of these forms is essential.

As yet no scheme of Mesozoic and Caenozoic zones has been established for the Radiolaria, and indeed no systematic account of their stratigraphical distribution has been published. A fair Caenozoic succession of radiolarian assemblages may however be inferred from the literature (Riedel 1957*a*, 1959*a*; Riedel and Funnell 1964; Riedel *et al.* 1963; Friend and Riedel 1967).

Both the calcareous nannoplankton and Radiolaria, and probably the diatoms also, are likely ultimately to yield a zonation scheme as discriminating as that already provided by the planktonic Foraminifera. In fact the possibly greater climatic tolerance of the calcareous nannoplankton, and the greater variety of form and larger numbers of taxa shown by the Radiolaria and diatoms may make them potentially better in some respects. The calcareous nannoplankton are particularly useful for very small quantities of material and detecting very small amounts of reworking; the Radiolaria are invaluable indicators in non-calcareous deposits, laid down, either originally or after reworking, beneath the calcium carbonate compensation depth; and diatoms will probably be found particularly useful in high latitudes.

Where all four groups occur together, such as in calcareous-siliceous deposits of the Equatorial Zone, not only will it be possible to establish the mutual ranges of the stratigraphical zones based on the four groups, but also probably to establish extreme stratigraphical refinement using the ranges of taxa from all four groups in conjunction.

Structural

A final application of pre-Quaternary microfossil studies is to the elucidation of structural problems concerning the ocean floors and ocean margins. Basically their application to these problems depends on their potential for determining (i) the *depth* of deposition, and (ii) the *age* of ancient sediments.

Depth assessments are important in interpreting the structural history of oceanic islands and ocean margins. Hamilton (1956) used the occurrence of Cretaceous (Aptian-Cenomanian) reef coral, rudistid, stromatoporoid assemblages near the tops of some of the Mid-Pacific (submarine) mountains to infer that these mountains were at or near sea-level during mid-Cretaceous times, whereas at the present their tops are 1,250 to 1,650 metres below sea-level. That much of the subsidence to which these mountains have been subjected had occurred by the end of the Cretaceous is indicated by the presence of reworked planktonic foraminiferal faunas of Upper Cretaceous (Campanian-Maastrichtian) age on the flanks, and planktonic foraminiferal oozes of Palaeocene and Eocene age on top of these sunken mountains. Similar evidence of progressive subsidence of oceanic islands, only this time accompanied by *pari passu* upgrowth of reef complexes, sustaining coral islands or atolls at or near the surface, has been proved by deep-drilling on such islands. The subject is summarized by Menard and Ladd (1963) and, more recently, by Ladd, Tracey and Gross (1967) in their account of the results of drilling on Midway Atoll. These give evidence of shallow-water limestones of various Tertiary ages resting on basalt platforms, the oldest being upper Eocene and ranging up to upper Miocene. At Eniwetok Atoll (Todd and Low 1960, pp. 812-15) there is evidence of shallow-water deposition extending out over earlier deeper-water deposits. The palaeontological and physiographical evidence of subsidence of oceanic islands over a large area of the Pacific led Menard (1958, 1964) to propose the previous existence of the Darwin Rise in that ocean. It should be noted in this context, that it is essentially the palaeontological data that has given evidence of the period at which this rise may have existed, and of the time-scale over which it has subsided.

An interesting corollary of the evidence that the summits of the Mid-Pacific mountains and the submerged basalt platforms of the Pacific Atolls were once at or near sea-level is that it is possible to gauge the depth of the deep ocean in their vicinity at that time. This calculation assumes that the volcanic piles forming these islands were in isostatic equilibrium with the surrounding sea-floor when their summits were at or near sea-level; for if this was so the present distance between the summits or basalt platforms and the surrounding deep-sea floor (plus something to allow for subsequent pelagic sedimentation on that deep-sea floor) is a measure of the depth of the ocean at the time they were at or near sea-level. For the Mid-Pacific mountains this depth amounts to about 3,100 metres.

In the Atlantic similar estimates can be made for the island of Bermuda, from which Vaughan (1919) recorded deposits containing shallow-water Foraminifera of probable upper Eocene age resting on basalt at −107 metres. In this case the amount of post-Eocene subsidence is small, but a depth of about 4,500 metres is indicated for the depth of the North American basin during the Eocene. Elsewhere in the Atlantic continuation of volcanic activity has brought Miocene marine deposits above sea-level in the Azores, and Jurassic *Aptychus* beds above sea-level in the Cape Verde Islands.

The behaviour of the submarine continental margins around the oceans can also be documented by fossil evidence. Funnell (1964) reviewed the evidence for the depth of deposition of Upper Cretaceous occurrences around the North Atlantic and concluded that although all these deposits are now at depths greater than 200 metres, in all cases where clear evidence of depth of deposition is available accumulation in water less than 200 metres deep is indicated. For the American side of the North Atlantic at least, this suggests a substantial subsidence of the continental margin during the Tertiary. Heezen and Sheridan (1966) have recently obtained Lower Cretaceous (Neocomian-Aptian) algal calcarenites from a depth of approximately 4,800 metres at the base of the Blake Escarpment which they conclude were accumulated at or near sea-level, indicating around 5,000 metres of post-Jurassic subsidence in this region. Shallow-water oolitic sediments of possible Aptian age occur up to a depth of 3,130 metres but are succeeded at a depth of 2,375 metres by Aptian-Albian assemblages of planktonic Foraminifera and nannoconids indicative of open-sea conditions. The JOIDES drill-holes 3 and 4 (Bunce *et al.* 1965) indicate a continuation of pelagic sedimentation on the outer part of the Blake Plateau from Palaeocene up until the Miocene. Here we have, therefore, a history of subsidence since the Jurassic of approximately 5,000 metres, for the first 2,000 metres of which, during the Lower Cretaceous, sedimentation kept pace with subsidence, thereafter failing to do so from about the mid-Cretaceous onwards so that the surface of the present, possibly intermittent, sediment accumulation is now some 1,000 metres below sea-level. In the Pacific Byrne, Fowler and Maloney (1966) have adduced micropalaeontological evidence for uplift of the continental margin off Oregon.

Changes of depth at the average level of the deep-sea, i.e. 4,000 to 5,000 metres, are more difficult to determine owing to the comparative uniformity of faunas below 2,000 metres in the oceans. Cann and Funnell (1967) speculated on the mid-Tertiary uplift of Palmer Ridge in the North Atlantic, from say 4,500 metres to its present level of 3,150 metres, on the grounds that the Upper Tertiary sediments on top of the ridge appeared to be winnowed compared with those accumulated during the Lower Tertiary. But the argument here is mainly sedimentological, the fact that the sedimentary particles are mostly microfossil skeletal remains being really only incidental.

The evidence of pre-Quaternary macro- and microfossil assemblages is also important in the evaluation of the concept of ocean-floor spreading. This idea was first introduced by Dietz (1961) and Hess (1962). Wilson (1963) tested this hypothesis against the ages (usually palaeontologically determined) of oceanic islands, and obtained an encouraging result – particularly for the Atlantic Ocean, where the older islands occur to the margins of the basin with younger ones nearer the midline. Evidence from deep-sea occurrences (Ericson *et al.* 1961; Riedel and Funnell 1964) was already sufficient to say that it was consistent with the hypothesis of ocean-floor spreading and set some limits on the rate at which it could have occurred, e.g. a North Atlantic of at least two-thirds its present size by the end of the Cretaceous was suggested (Funnell 1964). Further impetus was given to the idea of ocean-floor spreading by the Vine-Matthews (1963) hypothesis linking the linear magnetic patterns of the deep-sea floor with the record of reversals of the earth's magnetic field, an idea later elegantly elaborated by Vine in 1966. The contribution by Wilson (1965) on transform faulting was also helpful to acceptance of the idea of ocean-floor spreading. In 1965 van Andel *et al.*, on the basis of morphological studies and recovery of Miocene rocks adjacent to the crestal region of the Mid-Atlantic ridge, were still sceptical of the hypothesis; as were Saito, Burkle and Ewing (1966) when they described Tertiary sediments from the region of the Mid-Atlantic ridge. These authors, however, admitted possible average rates of Tertiary spreading of 0 to 1 centimetres per year or 1.6 centimetres per year respectively on the basis of their results. An alternative figure of 2.6 centimetres per year with no post-Lower Miocene spreading was also proposed by the latter authors. Cann and Funnell (1967) suggested a maximum average rate of spreading of the North Atlantic at approximately 43° N of 1.0 to 1.3 centimetres per year during the last 60 million years, based on a detailed study of the relations between igneous, meta-igneous and Lower Tertiary sediments outcropping on the flanks of Palmer ridge, north east of the Azores. Burckle *et al.* (1967) have found occurrences of Tertiary sediment across the East Pacific Rise consistent with a spreading rate of approximately 4.5 centimetres per year during the Upper Tertiary, and Riedel (1967) has suggested an average rate since the Eocene of 8 centimetres per year in the eastern Pacific based on microfossil occurrences between the equator and 20° N.

The use of age-determination by microfossils assemblages in evaluating the hypothesis of ocean-floor spreading is not particularly straightforward. The microfossils can, in the general case, only tell you that the volcanic ocean floor on which they rest is older than they are. How much older is uncertain; it may be up to at least 10 million years. Much of the Mid-Atlantic ridge system has little or no sediment cover although much of the exposed basaltic rock in this system must be of some antiquity and was probably extruded as long ago as the Miocene. On Palmer Ridge the youngest isotopic data obtained from amphibolite such as might have been produced at the active tectonic centre of the Mid-Atlantic ridge system at that time was 60 ± 6 million years, whereas the probable age of the oldest sediments recovered after intensive sampling approximated 50 million years. Thus microfossil assemblages can only be used to give the minimum age of the ocean floor, and therefore the minimum extent of the ocean at any one time. The actual age and extent may be considerably greater. Conversely the estimates of rates of any ocean-floor spreading will be maxima.

These under- and over-estimates will be further exaggerated if the oldest sediment in an area is not sampled. This may be due to a variety of factors, density of sampling, method of sampling, rate of sedimentation and age of sediment. The effects of density of sampling should be self evident. Because of the widespread extent of Quaternary deposits in the oceans the chances of sampling Quaternary sediments are high and it is only with repeated (or guided) sampling that the smaller areas of outcropping pre-Quaternary sediment will be encountered. Secondly different sampling methods have a different effectiveness in recovering older sediment; long piston cores will often encounter Pliocene or even late Miocene deposits in areas where short piston cores or gravity cores fail to penetrate the Quaternary cover. Deep-drilling programmes are of course virtually certain to encounter pre-Quaternary sediments wherever they are embarked upon. Dredging on steep slopes where older rocks are likely to be outcropping (and where use of coring devices would probably be inappropriate) often produces much older rocks than those obtained nearby by coring (viz. Heezen and Sheridan 1966). Thirdly, as Riedel (1967) has pointed out, other things being equal the rate of sedimentation affects availability of older sediments to normal ship-borne coring methods: they are for instance usually buried under Quaternary cover in the high productivity zone of the Pacific equatorial zone, but more frequently encountered beneath the slowly accumulating sediments further north. Lastly the chances of older sediments being covered by later sediments increases with age – the oceans still being overall a depositional environment. When this has been said it can be seen that the distribution of pre-Quaternary microfossils in the oceans, as at present known, is a reflection of many influences other than the age of the underlying ocean floor.

ACKNOWLEDGEMENTS

This account owes much to the work published, and sometimes unpublished, of many colleagues with whom I have had the privilege of collaborating from time to time, but in particular to M. N. Bramlette, F. L. Parker and W. R. Riedel of Scripps Institution of Oceanography, D. B. Ericson, L. H. Burckle, J. D. Hays and T. Saito of Lamont Geological Observatory, and M. Black, Michael Fisher, Jenny Friend and Tony Ramsay of the Department of Geology, Cambridge. Work in this field at Cambridge has been supported by a Natural Environment Research Council Grant.

Received September 1967, revised January 1968.

BIBLIOGRAPHY

Arkell, W. J. 1956. *Jurassic geology of the world.* Oliver and Boyd: London, 806 pp.

Arrhenius, G. 1952. Sediment cores from the East Pacific. *Rep. Swedish Deep-Sea Expedition.* **5**, 1-227.

Arrhenius, G. 1963. Pelagic sediments. In *The Sea* (M. N. Hill *ed.*), 3, 655-727. Interscience: New York and London.

Aubert de la Rue, E. 1956. *Chronique Mines d'Outre-Mer Paris,* **24**, 162.

Bandy, O. L. 1952. Miocene Foraminifera from Erben Bank. *Contrib. Cushman Found. Foram. Res.* **3**, 18-20.

Bandy, O. L. 1956. Palaeotemperatures of Pacific bottom waters and multiple hypotheses. *Science,* **123**, 459-60.

Bandy, O. L. 1960. Planktonic foraminiferal criteria for palaeoclimatic zonation. *Sci. Rep. Tohoku Univ., ser. 2 (Geol.),* spec. vol. **4**, 1-8.

Bandy, O. L. 1963a. Cenozoic planktonic foraminiferal zonation (Abstract). *J. Paleont.* **37**, 977.

Bandy, O. L. 1963b. Aquitanian planktonic Foraminifera from Erben Guyot. *Science,* **140**, 1402-3.

Bandy, O. L. 1963c. Miocene-Pliocene boundary in the Philippines as related to late Tertiary stratigraphy of deep-sea sediments. *Science,* **142**, 1290-2.

Bandy, O. L. 1964. Cenozoic planktonic foraminiferal zonation. *Micropaleont.* **10**, 1-17.

Bandy, O. L. 1966. Faunal evidence of Miocene-to-Recent palaeoclimatology in Antarctic (Abstract). *Bull. Amer. Assoc. Petrol. Geol.* **50**, 643-4.

Banner, F. T. and Blow, W. H. 1959. The classification and stratigraphical distribution of the Globigerinaceae. *Palaeontology,* **2**, 1-27.

Banner, F. T. and Blow, W. H. 1965a. Two new taxa of the Globorotaliinae (Globigerinacea, Foraminifera) assisting determination of the late Miocene/Middle Miocene boundary. *Nature, Lond.* **207**, 1351-4.

Banner, F. T. and Blow, W. H. 1965b. Progress in the planktonic foraminiferal biostratigraphy of the Neogene. *Nature, Lond.* **208**, 1164-6.

Bassler, R. S. 1936. Geology and palaeontology of the Georges Bank Canyons, Part III. Cretaceous Bryozoa from Georges Bank. *Bull. geol. Soc. Amer.* **47**, 411-12.

Beckmann, J. P. 1954. Die Foraminiferan der Oceanic Formation (Eocaen-Oligocaen) von Barbados, Kl. Antillen. *Ecl. geol. Helv.* **46**, 301-412.

Berger, W. H. 1967. Foraminiferal ooze: Solution at depths. *Science,* **156**, 383-5.

Black, M. 1956. The finer constituents of *Globigerina* ooze. *Resumenes de los trabajos presentados, XX Internat. Geol. Congr., Mexico City,* p. 173.

Black, M. 1962. Fossil coccospheres from a Tertiary outcrop on the continental slope. *Geol. Mag.* **99**, 123-7.

Black, M. 1964. Cretaceous and Tertiary coccoliths from Atlantic seamounts. *Palaeontology,* **7**, 306-16.

Black, M. and Barnes, B. 1961. Coccoliths and discoasters from the floor of the South Atlantic Ocean. *Jour. Royal Micros. Soc.* **80**, 137-47.

Black, M., Hill, M. N., Laughton, A. S. and Matthews, D. H. 1964. Three non-magnetic seamounts off the Iberian coast. *Quart. J. geol. Soc., Lond.* **120**, 477-517.

Bolli, H. M. 1957a. The genera *Globigerina* and *Globorotalia* in the Paleocene-Lower Eocene Lizard Springs Formation of Trinidad, B.W.I. *Bull. U.S. nat. Mus.* **215**, 61-81.

Bolli, H. M. 1957b. Planktonic Foraminifera from the Oligocene-Miocene Cipero and Lengua Formations of Trinidad, B.W.I. *Bull. U.S. nat. Mus.* **215**, 97-123.

Bolli, H. M. 1957c. Planktonic Foraminifera from the Eocene Navet and San Fernando Formations of Trinidad, B.W.I. *Bull. U.S. nat. Mus.* **215**, 155-72.

Bolli, H. M. 1959. Planktonic Foraminifera as index fossils in Trinidad, West Indies and their value for worldwide stratigraphic correlation. *Ecl. Geol. Helv.* **52**, 627-37.

Bolli, H. M. 1964. Observations on the stratigraphic distribution of some warm water planktonic Foraminifera in the Young Miocene to Recent. *Ecl. Geol. Helv.* **57**, 541-52.

Bolli, H. M. 1965. The planktonic Foraminifera in Well Bodjonegoro-1 of Java. *Ecl. geol. Helv.* **59**, 449-65.

Bolli, H. M. 1966. Zonation of Cretaceous to Pliocene marine sediments based on planktonic Foraminifera. *Bol. Inf. Assoc. Venez. Geol., Mineraria & Petrol.* **9**, 3-32.

Bolli, H. M. and Bermudez, P. J. 1965. Zonation based on planktonic Foraminifera of Middle Miocene to Pliocene warm-water sediments. *Bol. Inf. Assoc. Venez. Geol., Mineraria & Petrol.* **8**, 121-49.

Bourcart, J. and Marie, P. 1951. Sur la nature du 'Rebord Continental' a l'ouest de la Manche. *C.R. Acad. Sci., Paris,* **232**, 2346-8.

Bramlette, M. N. 1955. Changes in four groups of microfossils from the Eocene to Oligocene of the oceanic formation of Barbados (Abstract). *Bull. geol. Soc. Amer.* **66**, 1534.

Bramlette, M. N. 1957. Discoaster and some related microfossils. In: Geology of Saipan, Mariana Islands. *Prof. Pap. U.S. Geol. Survey*, **280-F**, 247-55.

Bramlette, M. N. 1958. Significance of coccolithophorids in calcium-carbonate deposition. *Bull. geol. Soc. Amer.* **69**, 121-6.

Bramlette, M. N. 1961. Pelagic sediments. In: Oceanography (ed. Sears, M.). *Publ. Amer. Ass. Advanc. Sci.* **67**, 345-66.

Bramlette, M. N. 1965. Massive Extinctions in Biota at the End of Mesozoic Time. *Science,* **148**, 1696-9.

Bramlette, M. N. and Bradley, W. H. 1940. Geology and biology of North Atlantic deep-sea cores between Newfoundland and Ireland. Pt. I. Lithology and geologic interpretations. *Prof. Pap. U.S. geol. Surv.* **196**, 1-34.

Bramlette, M. N., Faughn, J. L. and Hurley, R. J. 1959. Anomalous sediment deposition on the flank of Eniwetok Atoll. *Bull. geol. Soc. Amer.* **70**, 1549-52.

Bramlette, M. N. and Martini, E. 1964. The great change in calcareous nannoplankton fossils between the Maestrichtian and Danian. *Micropaleont.* **10**, 291-322.

Bramlette, M. N. and Riedel, W. R. 1954. Stratigraphic value of discoasters and some other microfossils related to Recent coccolithophores. *J. Paleont.* **28**, 385-403.

Bramlette, M. N. and Riedel, W. R. 1959. Stratigraphy of deep-sea sediments of the Pacific Ocean. *Preprints Int. Oceanogr. Congr. 1959*, AAAS: Washington, 86-7.

Bramlette, M. N. and Sullivan, F. R. 1961. Coccolithophorids and related nannoplankton of the early Tertiary in California. *Micropaleont.* **7**, 129-88.

Brotzen, F. and Dinesen, A. 1959. On the stratigraphy of some bottom sections from the central Pacific. *Rep. Swedish Deep-Sea Expedition,* **10**, 43-55.

Bryant, W. R. and Pyle, T. E. 1965. Tertiary sediments from Sigsbee knolls, Gulf of Mexico. *Bull. Amer. Assoc. Petrol. Geol.* **49**, 1517-18.

Bunce, E. T., Emergy, K. O., Gerard, R. D., Knott, S. T. Lidz, L., Saito, T. and Schlee, J. 1965. Ocean Drilling on the Continental margin. *Science,* **150**, 709-16.

Burckle, L. H. and Saito, T. 1966. An Eocene dredge haul from the Tuamotu ridge. *Deep-sea Res.* **13**, 1207-8.

Burckle, L. H., Ewing, J., Saito, T. and Leyden, R. 1967. Tertiary sediment from the East Pacific Rise. *Science,* **157**, 537-40.

Burckle, L. H. and Hays, J. D. 1966. Tertiary sediments on Falkland Platform and Argentine Continental Slope (Abstract). *Bull. Amer. Assoc. Petrol. Geol.* **50**, 607.

Byrne, J. V., Fowler, G. A. and Maloney, N. J. 1966. Uplift of the continental margin and possible continental accretion off Oregon. *Science,* **154**, 1654-6.

Cann, J. R. and Funnell, B. M. 1967. Palmer Ridge – a section through the upper part of the ocean crust? *Nature, Lond.* **213**, 661-4.

Carsola, A. J. and Dietz, R. S. 1952. Submarine geology of two flat-topped northeast Pacific seamounts. *Amer. J. Sci.* **250**, 481-7.

Cifelli, R. 1965. Late Tertiary planktonic Foraminifera associated with a basaltic boulder from the Mid-Atlantic Ridge. *J. Mar. Research,* **23**, 73-87.

Cloud, P. E. 1956. Provisional correlation of selected Cenozoic sequences in the western and central Pacific. *Proc. 8th Pacific Sci. Congr. Manila,* **2**, 555-74.

Cloud, P. E., Schmidt, R. G. and Burke, H. W. 1956. Geology of Saipan, Mariana Islands: Pt. 1. Geology. *Prof. Pap. U.S. geol. Surv.* **280-A**, i-xv, 1-265.

Cole, W. S. 1957. Bikini and nearby atolls. Larger Foraminifera from Eniwetok Atoll drill holes. *Prof. Pap. U.S. geol. Surv.* **260-V**, 743-84.

Cole, W. S. 1959. *Asterocyclina* from a Pacific seamount. *Contrib. Cushman Found. Foram. Res.* **10**, 10-4.

Cole, W. S. 1960. Upper Eocene and Oligocene larger Foraminifera from Viti Levu, Fiji. *Prof. Pap. U.S. geol. Surv.* **374-A**, 1-7.

Cole, W. S., Todd, R. and Johnson, C. G. 1960. Conflicting age determinations suggested by Foraminifera on Yap, Caroline Islands. *Bull. Amer. Paleont.* **41**, 73-112.

Colom, G. 1952. Foraminiferos de las costas de Galicia. *Boletin del Instituto Espanol de Oceanografia,* **51**.

Colom, G. 1955. Jurassic-Cretaceous pelagic sediments of the western Mediterranean zone and the Atlantic area. *Micropaleont.* **1**, 109-24.

Curry, D. 1965. The English Palaeogene pteropods. *Proc. Malacological Soc.* **36**, 357-71.

Curry, D., Martini, E., Smith, A. J. and Whittard, W. F. 1962. The geology of the western approaches of the English Channel. I. Chalky rocks from the upper reaches of the continental slope. *Phil. Trans. Royal Soc. Lond., ser. B*, **245**, 267-90.

Cushman, J. A. 1936. Geology and palaeontology of the Georges Bank Canyons, Part IV, Cretaceous and late Tertiary Foraminifera. *Bull. geol. Soc. Amer.* **47**, 413-40.

Cushman, J. A. 1939. Eocene Foraminifera from submarine cores off the eastern coast of North America. *Contrib. Cushman Lab. Foram. Res.* **15**, 49-76.

Cushman, J. A. and Henbest, L. G. 1940. Geology and biology of North Atlantic deep-sea cores between Newfoundland and Ireland, pt. 2. Foraminifera. *Prof. Pap. U.S. geol. Surv.* **196-A**, 35-56.

Cushman, J. A., Todd, R. and Post, R. J. 1954. Recent Foraminifera of the Marshall Islands. *Prof. Pap. U.S. geol. Surv.* **260-H**, i-iv, 319-84.

Dall, W. H. 1925. Tertiary fossils dredged off the northeastern coast of North America. *Amer. J. Sci.* **10**, 213-8.

Daly, R. A. 1927. The geology of Saint Helena Island. *Proc. Amer. Acad. arts Sci.* **62**, 31-92.

Day, A. A. 1959. The continental margin between Brittany and Ireland. *Deep-sea Res.* **5**, 249-65.

Dietz, R. S. 1961. Continent and Ocean Basin Evolution by spreading of the Sea Floor. *Nature, Lond.* **190**, 854-7.

Eames, F., Banner, F. T., Blow, W. H. and Clarke, W. J. 1962. *Fundamentals of Mid-Tertiary stratigraphical correlation.* Cambridge, i-vii, 1-163 pp.

Eardley, A. J. 1951. *Structural geology of North America.* Harper: New York. 624 pp.

Emiliani, C. 1954. Temperatures of Pacific bottom waters and Polar superficial waters during the Tertiary. *Science,* **119**, 853-5.

Emiliani, C. 1955. Pleistocene temperatures. *J. Geol.* **63**, 538-78.

Emiliani, C. 1956a. On palaeotemperatures of Pacific bottom waters. *Science,* **123**, 460-1.

Emiliani, C. 1956b, Oligocene and Miocene temperatures of the equatorial and subtropical Atlantic Ocean. *J. Geol.* **64**, 281-8.

Emiliani, C. 1961a. The temperature decrease of surface sea-water in high latitudes and of abyssal-hadal water in open ocean basins during the past 75 million years. *Deep-sea Res.* **8**, 144-7.

Emiliani, C. 1961b. Cenozoic climatic changes as indicated by the stratigraphy and chronology of deep-sea cores of globerigina-ooze facies. *Ann. N.Y. Acad. Sci.* **95**, 521-36.

Emiliani, C. and Edwards, G. 1953. Tertiary ocean bottom temperatures. *Nature, Lond.* **171**, 887.

Emiliani, C. and Milliman, J. D. 1966. Deep-sea sediments and their geological record. *Earth Science Reviews,* **1**, 105.

Ericson, D. B. 1953. Sediments of the Atlantic Ocean. *Columbia Univ. Lamont Geological Observatory, Techn. Rep., Submarine Geol.* **1**.

Ericson, D. B., Ewing, M. and Heezen, B. C. 1952. Turbidity currents and sediments in North Atlantic. *Bull. Amer. Assoc. Petrol. Geol.* **36**, 489-511.

Ericson, D. B., Ewing, M. and Wollin, G. 1963. Plio-Pleistocene boundary in deep-sea sediments. *Science,* **139**, 727-37.

Ericson, D. B., Ewing, M. and Wollin, G., and Heezen, B. C. 1961. Atlantic deep-sea sediment cores. *Bull. geol. Soc. Amer.* **72**, 193-286.

Ericson, D. B. and Wollin, G. 1964. *The deep and the past.* Knopf: New York. xii, 292 pp.

Ewing, M., Ericson, D. B. and Heezen, B. C. 1958. Sediments and topography of the Gulf of Mexico. In: Weeks, L. (ed.) Habitat of Oil. *Spec. Publ. Amer. Assoc. Petrol. Geol.* 995-1053.

Ewing, J. and Ewing, M. 1967. Sediment Distribution on the mid-ocean ridges with respect to spreading of the sea floor. *Science,* **156**, 1590-2.

Ewing, M. and Ewing, J. 1964. Distribution of oceanic sediments In *Studies on Oceanography: A volume dedicated to K. Hidaka* (Yoshida K. *ed.*), 525-37. Tokyo Univ. Press: Tokyo.

Ewing, M., Ewing, J. and Talwani, M. 1964. Sediment distribution in the oceans: The Mid-Atlantic Ridge. *Bull. geol. Soc. Amer.* **75**, 17-35.

Ewing, M., Saito, T., Ewing, J. I. and Burckle, L. H. 1966. Lower Cretaceous sediments from the Northwest Pacific. *Science,* **152**, 751-5.

Fisher, R. L. 1958. Preliminary report on expedition 'Downwind', University of California, Scripps Institution of Oceanography, IGY cruise to the southeast Pacific. *I.G.Y. World Data Center A General Rept. Ser. 2,* 1-58.

Fleming, C. A. 1957. A new species of fossil *Chlamys* from the Drygalski Agglomerate of Heard Island, Indian Ocean. *J. Geol. Soc. Austral.* **4**, 13-9.

Friend, J. K. and Riedel, W. R. 1967. Cenozoic orosphaerid radiolarians from tropical Pacific sediments. *Micropaleont.* **13**, 217-32.

Funnell, B. M. 1964. Studies in North Atlantic geology and palaeontology: 1. Upper Cretaceous. *Geol. Mag.* **101**, 421-34.

Furon, R. 1935. Notes sur la Paleogéographie de l'Océan Atlantique. 1. La géologie des îles du Cap Vert. *Bull. Mus. Nat. Hist. Paris,* **7**, 270-4.

Furon, R. 1963. *The Geology of Africa.* Oliver and Boyd: London. 377 pp.

Gagel, C. 1912. Studien uber den Aufbau und die Gesteine Madeiras. *Zeit. deut. Ges.* **64**, 344-491.

Gibson, T. G. 1965. Eocene and Miocene rocks off the north-eastern coast of the United States. *Deep-sea Res.* **12**, 975-81.

Goldberg, E. D. and Arrhenius, G. O. S. 1958. Chemistry of Pacific pelagic sediments. *Geochim. et Cosmochim. Acta,* **13**, 153-212.

Grimsdale, T. F. and van Morkhoven, F. P. C. M. 1955. The ratio between pelagic and benthonic Foraminifera as a means of estimating depth of deposition of sedimentary rocks. *Proc. 4th. World Petrol. Congr., section 1D,* 473-90.

Groot, J. J. 1963. Palynological investigation of a core from the Biscay Abyssal Plain. *Science,* **141**, 522-3.

Haeckel, E. 1887. Report on the Radiolaria collected by H.M.S. *Challenger* during the years 1873-1876. *Chall. Rep. Zool.* **18**, i-clxxxviii, 1-1803.

Hamilton, E. L. 1953. Upper Cretaceous, Tertiary and Recent planktonic Foraminifera from mid-Pacific flat-topped seamounts. *J. Paleont.* **27**, 204-37.

Hamilton, E. L. 1956. Sunken islands of the Mid-Pacific Mountains. *Mem. geol. Soc. Amer.* **64**.

Hamilton, E. L. 1959. Thickness and consolidation of deep-sea sediments. *Bull. geol. Soc. Amer.* **70**, 1399-1424.

Hamilton, E. L. and Rex, R. W. 1959. Lower Eocene phosphatized Globigerina ooze from Sylvania Guyot. *Prof. Pap. U.S. geol. Surv.* **260-W**, 785-98.

Hanzawa, S. 1933. Diatom (*Ethmodiscus*) Ooze obtained from the tropical southwestern north Pacific Ocean. *Record of Oceanographic Work in Japan,* **7**, 37-44.

Hanzawa, S. 1938. Studies on the foraminiferal fauna found in the bore cores from the deep well in Kita-Daito-Zima (North Borodino Island). *Proc. Imp. Acad. Tokyo,* **14**, 384-90.

Hanzawa, S. 1940. Micropalaeontological studies of drill cores from a deep well in Kita-Daito-Zima (North Borodino Island). *Jubilee Pub. in Commem. Prof. H. Yabe's 60th. Birthday, Tokyo,* **2**, 755-802.

Harrison, C. G. H. and Funnell, B. M. 1964. Relationship of palaeomagnetic reversals and micropalaeontology in two Late Cenozoic cores from the Pacific Ocean. *Nature, Lond.* **204**, 566.

Hausen, H. 1956. *Acta Geog.* (*Soc. Geog. Fennia*), **15**.

Hays, J. D. 1964. Antarctic Radiolaria and the late Tertiary and Quaternary history of the Southern Ocean. Ph.D. dissertation, Faculty of Geol., Columbia University, 1-215.

Hays, J. D. 1965. Radiolaria and Late Tertiary and Quaternary history of Antarctic seas. *Amer. geophys. Union Antarctic Res. Ser.* **5**, 125-84.

Heezen, B. C. and Menard, H. W. 1963. Topography of the deep-sea floor. In *The Sea* (M. N. Hill *ed.*), **3**,

233-80. Interscience: New York.

Heezen, B. C. and Sheridan, R. E. 1966. Lower Cretaceous rocks (Neocomian-Albian) dredged from Blake Escarpment. *Science,* **154**, 1644-7.

Heezen, B. C., Tharp, M. and Ewing, M. 1959. The floors of the oceans. 1. The North Atlantic. *Spec. Publ. geol. Soc. Amer.* **65**, 1-122.

Heirtzler, J. R. and Hayes, D. E. 1967. Magnetic boundaries in the North Atlantic Ocean. *Science,* **157**, 185-7.

Herman, Y. 1963. Cretaceous, Palaeocene and Pleistocene Sediments from the Indian Ocean. *Science,* **140**, 1316-17.

Hess, H. H. 1962. History of ocean basins. In *Petrologic studies : A volume to honor A. F. Buddington.* Geol. Soc. Amer., 599-620.

Jenkins, D. G. 1964. Location of the Pliocene-Pleistocene boundary. *Contrib. Cushman Found. Foram. Res.* **15**, 25-7.

Johnson, J. H. 1961. Bikini and nearby atolls. Fossil algae from Eniwetok, Funafuti and Kita-Daito-Jima. *Prof. pap. U.S. geol. Surv.* **260-Z**, 907.

Jones, E. J. W. and Funnell, B. M. 1968. Association of a seismic reflector and Upper Cretaceous sediment in the Bay of Biscay. *Deep-sea Res.* **15** (6), 701-9.

Kagami, H. *et al.* 1964. *J. Geol. Soc. Japan,* **70**, 414.

Kamptner, E. 1962. Tertiare und nach-tertiare Coccolithineen-Skelettreste aus Tiefseeablagerungen des östlichen Pazifischen Ozeans. *Palaont. Zeitschrift, Stuttgart,* **36**, 13.

Kamptner, E. 1963. Coccolithineen-Skelettreste aus Tiefseeablagerungen des Pacifischen Ozeans. *Ann. Naturhist. Mus. Wien,* **66**, 139-204.

Kanaya, T. and Koizumi, I. 1966. Interpretation of Diatom Thanatocoenoses from the North Pacific Applied to a Study of Core V20-130 (Studies of a Deep-Sea Core V20-130. Part IV). *Sci. Rep. Tohoku Univ., Sendai, ser. 2 (Geology),* **37**, 89-130.

Kennett, J. P. 1966. Foraminiferal evidence of a shallow calcium carbonate solution boundary, Ross Sea, Antarctica. *Science,* **151**, 191-3.

Kleinpell, R. M. 1954. Neogene smaller Foraminifera from Lau, Fiji. *Bull. B.P. Bishop Mus.* **211**, 1-96.

Kolbe, R. W. 1954. Diatoms from equatorial Pacific cores. *Rep. Swedish Deep-Sea Expedition,* **6**, 1-49.

Krejci-Graf, K. 1956. *Frankfurt Geol.* **30**, 1.

Krinsley, D. H. and Newman, W. S. 1965. Pleistocene glaciation: a criterion for recognition of its onset. *Science,* **149**, 442-3.

Ladd, H. S., Ingerson, E., Townsend, R. C., Russell, M. and Stephenson, H. K. 1953. Drilling on Eniwetok Atoll, Marshall Islands. *Bull. Amer. Assoc. Petrol. Geol.* **37**, 2257-80.

Ladd, H. S., Tracey, J. I. and Gross, M. G. 1967. Drilling on Midway Atoll, Hawaii. *Science,* **156**, 1088-94.

Ladd, H. S., Tracey, J. I. and Lill, G. G. 1948. *Science,* **107**, 51.

Lambeth, A. J. 1952. A geological account of Heard Island. *J. & Proc. Roy. Soc. New South Wales,* **86**, 14.

Langseth, M. G., Ewing, M. and Ewing, J. 1966. Seismic measurement of sediment thickness and sea floor structure in the Indian Ocean. *Abstracts of papers, Second International Oceanographic Congress, Moscow 1966,* 216.

Loeblich, A. R. and Tappan, H. 1957. Correlation of the Gulf and Atlantic Coastal Plain Palaeocene and Lower Eocene formations by means of planktonic Foraminifera. *J. Paleont.* **31**, 1109-37.

Loeblich, A. R. and Tappan, H. 1961. Cretaceous planktonic Foraminifera: Part 1 – Cenomanian. *Micropaleont.* **7**, 257-304.

Marshall, P. 1927. *Bull. B. P. Bishop Museum,* **36**.

Marshall, P. 1930. *Bull. B. P. Bishop Museum,* **72**, 1.

Martini, E. 1958. Discoasteriden und verwandte Formen im NW-deutschen Eozan (Coccolithophorida). *Senckberg. Leth.* **39**, 353-88.

Martini, E. 1960. Braarudosphaeriden, Discoasteriden und verwandte Formen aus dem Rupelton des Mainzer Beckens. *Natizbl. hess. L.-Amt Bodenforsch.* **88**, 67-87.

Martini, E. 1965. Mid-Tertiary calcareous nannoplankton from Pacific deep-sea cores. In *Submarine Geology and Geophysics* (Whittard, W. F. and Bradshaw, R. B. *eds.*). Butterworths: Lond. 393-410.

Martini, E. and Bramlette, M. N. 1963. Calcareous nannoplankton from the experimental Mohole drilling. *J. Paleont.* **37**, 845-56.

McIntyre, A., Bé, A. W. H. and Krinsley, D. H. 1964. Coccoliths and the Plio-Pleistocene Boundary in Deep-Sea Sediments. *Spec. Pap. Geol. Soc. Amer.* **76**, 113.

McIntyre, A., Bé, A. W. H. and Preikstas, R. 1967. Coccoliths and the Pliocene-Pleistocene boundary. *Progress in Oceanography,* **4**, 3-25.

McTavish, R. A. 1966. Planktonic Foraminifera from the Malaita Group, British Solomon Islands. *Micropaleont.* **12**, 1-36.

Mecarini, G., Shimaoka, G. and Krause, D. 1966. Submarine lithification of *Globigerina* ooze (Abstract). *Proc. geol. Soc. Amer.* **77**, 120.

Menard, H. W. 1958. Development of median elevations in ocean basins. *Bull. geol. Soc. Amer.* **69**, 1179-86.

Menard, H. W. 1964. *Marine geology of the Pacific.* McGraw Hill: New York. 271 pp.

Menard, H. W., Allison, A. C. and Durham, J. W. 1962. A drowned Miocene terrace in the Hawaiian Islands. *Science,* **138**, 896-7.

Menard, H. W. and Hamilton, E. L. 1964. Palaeogeography of the tropical Pacific. *Proc. 10th. Pacific Science Congress, Hawaii 1962.*

Menard, H. W. and Ladd, H. S. 1963. Oceanic islands, seamounts, guyots and atolls. In *The Sea* (M. N. Hill *ed.*), 3, 365-85. Interscience: New York.

Miller, E. T. and Ewing, M. 1956. Geomagnetic measurements in the Gulf of Mexico and in the vicinity of Caryn Peak. *Geophysics,* **21**, 406-32.

Milliman, J. D. 1966. Submarine lithification of carbonate sediments. *Science,* **153**, 994-6.

Murray, J. and Renard, A. F. 1891. Deep-sea deposits. *Rep. Sci. Res. Voy. H.M.S. 'Challenger'.* Eyre & Spottiswode: London.

Newell, N. D. 1955. Bahamian platforms. *Spec. Pap. geol. Soc. Amer.* **62**, 303-16.

Newell, N. D. and Rigby, J. K. 1957. Geological Studies on the Great Bahama Bank. *Spec. Publ. Soc. Econ. Paleont. and Mineral.* **5**, 15-72.

Northrop, J. and Heezen, B. C. 1951. An outcrop of

Eocene sediment on the continental slope. *J. Geol.* **59**, 396-9.

Northrop, J., Frosch, R. A., Frassetto, R. and Zeigler, J. M. 1959. The Bermuda-New England Seamount chain. *Preprints Int. Oceanographic Congr. 1959.* AAAS: Washington. 48.

Northrop, J. and Frosch, R. A. 1962. Bermuda-New England Seamount Arc. *Bull. geol. Soc. Amer.* **73**, 587-94.

Nuttall, W. L. F. 1926. A revision of the *Orbitoides* of Christmas Island (Indian Ocean). *Quart. J. geol. Soc. Lond.* **82**, 22-43.

Olausson, E. 1960. Description of sediment cores from the central and western Pacific with the adjacent Indonesian region. *Rep. Swedish Deep-Sea Expedition,* **6** (5), 163-214.

Olausson, E. 1961a. Remarks on some Cenozoic cores from the central Pacific, with a discussion of the roles of coccolithophorids and Foraminifera in carbonate deposition. *Göteborgs. Vetensk Samh. Handl. Följ. 7, ser. B,* **8** (10), 1-35.

Olausson, E. 1961b. Remarks on Tertiary sequences of two cores from the Pacific. *Bull. geol. Instn. Univ. Uppsala,* **40**, 299-303.

Opdyke, N. D., Glass, B., Hays, J. D. and Foster, J. 1966. Palaeomagnetic Study of Antarctic Deep-Sea Cores. *Science,* **154**, 349-57.

Owen, L. 1923. Notes on the phosphate deposit of Ocean Island; with remarks on the phosphates of the Equatorial Belt of the Pacific Ocean. *Quart. J. geol. Soc., Lond.* **79**, 1-15.

Park, K. 1966. Deep-Sea pH. *Science,* **154**, 1540-2.

Parker, F. L. 1964. Foraminifera from the experimental Mohole drilling near Guadelupe Island, Mexico. *J. Paleont.* **38**, 617-36.

Parker, F. L. 1965a. A new planktonic species (*Foraminiferida*) from the Pliocene of Pacific deep-sea cores. *Contrib. Cushman Found. Foram. Res.* **16**, 151-2.

Parker, F. L. 1965b. Irregular distributions of planktonic Foraminifera and stratigraphic correlation. *Progress in Oceanography,* **3**, 267-72.

Parker, F. L. 1967. Late Tertiary biostratigraphy (planktonic Foraminifera) of tropical Indo-Pacific deep-sea cores. *Bull. Amer. Paleont.* **52**, 111-208.

Part, G. M. 1950. Volcanic rocks from the Cape Verde Islands. *Bull. Brit. Mus. (Nat. Hist) Mineral.* **1**, 27-72.

Peach, B. N. 1922. Report on rock specimens dredged by the *Michael Sars* in 1910, by H.M.S. *Tritch* in 1882, and by H.M.S. *Knight Errant* in 1880. *Proc. Roy. Soc. Edinb.* **32**, 262-91.

Peterson, M. N. A. 1966. Calcite : rates of dissolution in a vertical profile in the Central Pacific. *Science,* **154**, 1542-4.

Phleger, F. B. 1939. Foraminifera of submarine cores from the continental slope. *Bull. geol. Soc. Amer.* **50**, 1395-419.

Phleger, F. B., Parker, F. L. and Peirson, J. F. 1953. North Atlantic Foraminifera. *Rep. Swedish Deep-sea Exped.* **7**, 3-122.

Pires Soares, J. M. 1948. Observations géologiques sur les îles du Cap Vert. *Bull. Soc. geol. France, ser. 5,* **18**, 383-9.

Pirsson, L. V. 1914a. Geology of Bermuda Island; the igneous platform. *Amer. J. Sci.* **38**, 195-206.

Pirsson, L. V. 1914b. Geology of Bermuda Island; petrology of the lavas. *Amer. J. Sci.* **38**, 331-4.

Pirsson, L. V. and Vaughan, T. W. 1913. A deep boring in Bermuda Island. *Amer. J. Sci.* **36**, 70-1.

Pratt, R. M. 1963a. Great Meteor Seamount. *Deep-sea Res.* **10**, 17-25.

Pratt, R. M. 1963b. Bottom currents on the Blake Plateau. *Deep-sea Res.* **10**, 245-9.

Pyle, T. E. 1966. Micropalaeontology and mineralogy of a Tertiary sediment core from the Sigsbee Knolls, Gulf of Mexico. *Texas A & M University, Department of Oceanography Technical Report* **66-13T**, vii, 106 pp.

Rehder, H. A. 1942. Geology and biology of North Atlantic deep-sea cores between Newfoundland and Ireland, part 5 Mollusca. *Prof. Pap. U.S. geol. Surv.* **196**, 107-9.

Reid, J. L. 1962. On circulation, phosphate-phosphorous content, and zooplankton volumes in the upper part of the Pacific Ocean. *Limnol. Oceanogr.* **7**, 287-306.

Repelin, J. 1919. *C. R. Acad. Sci. Paris,* **168**, 237.

Revelle, R. R., Bramlette, M. N., Arrhenius, G. and Goldberg, E. D. 1955. Pelagic sediments of the Pacific. *Spec. Pap. geol. Soc. Amer.* **62**, 221-36.

Reyment, R. A. 1961. A note on the Foraminifera in a sediment core from the Mindanao Trough. *Rep. Swedish Deep-Sea Expedition,* **6**, 139-56.

Ribeiro, O. 1958. *Bolm. Soc. Geol. Portugal,* **7**, 113.

Riedel, W. R. 1951. Sedimentation in the tropical Indian Ocean. *Nature, Lond.* **168**, 737.

Riedel, W. R. 1952. Tertiary Radiolaria in western Pacific sediments. *Goteborgs. Vetensk Samh. Handl. Folj 7, ser. B,* **6**, 1-18.

Riedel, W. R. 1953. Mesozoic and Late Tertiary Radiolaria of Rotti. *J. Paleont.* **27**, 805-13.

Riedel, W. R. 1954. The age of the sediment collected at *Challenger* (1875) Station 225 and the distribution of *Ethmodiscus rex* (Rattray). *Deep-sea Res.* **1**, 170-5.

Riedel, W. R. 1957a. Radiolaria: a preliminary stratigraphy. *Rep. Swedish Deep-sea Exped.* **6**, 59-96.

Riedel, W. R. 1957b. Eocene Radiolaria. In: Geology of Saipan, Marina Islands. *Prof. Pap. U.S. geol. Surv.* **280-G**, 257-63.

Riedel, W. R. 1959a. Oligocene and Lower Miocene Radiolaria in tropical Pacific sediments. *Micropaleont.* **5**, 285-302.

Riedel, W. R. 1959b. Siliceous organic remains in pelagic sediments. In: Silica in sediments. *Spec. Publ. Soc. Econ. Paleont. Mineral.* **7**, 80-91.

Riedel, W. R. 1963a. The preserved record: Palaeontology of Pelagic sediments. In *The Sea* (M. N. Hill *ed.*), vol. 3. Interscience: New York. 866-87.

Riedel, W. R. 1963b. *Radiolaria and deep-sea stratigraphy.* (unpublished report). 7 pp.

Riedel, W. R. 1964. *Microfossils in Indian Ocean sediments.* (unpublished report) 7 pp.

Riedel, W. R. 1967. Radiolarian evidence consistent with spreading of the Pacific floor. *Science,* **157**, 540-2.

Riedel, W. R. and Bramlette, M. N. 1959. Tertiary sediments in the Pacific Ocean basin. *International Oceanographical Congress 1959, Preprints,* 105-6. AAAS: Washington.

Riedel, W. R. and Bramlette, M. N. 1966. Some Tertiary sediments from the Indian Ocean floor. *Abstracts of Papers, Second International Congress, Moscow 1966.* 301.

Riedel, W. R., Bramlette, M. N. and Parker, F. L. 1963. 'Pliocene-Pleistocene' boundary in deep-sea sediments. *Science,* **140**, 1238-40.

Riedel, W. R. and Funnell, B. M. 1964. Tertiary sediment cores and microfossils from the Pacific Ocean floor. *Quart. J. Geol. Soc., Lond.* **120**, 305-68.

Riedel, W. R., Funnell, B. M., Bramlette, M. N. and Parker, F. L. 1967. Tertiary sediments from the Indian Ocean floor. (in preparation).

Riedel, W. R., Ladd, H. S., Tracey, J. I. and Bramlette, M. N. 1961. Preliminary drilling phase of Mohole Project: II. Summary of coring operations (Guadalupe Site). *Bull. Amer. Assoc. Petrol. Geol.* **45**, 1793-8.

Ross, D. A. and Riedel, W. R. 1967. Comparison of upper parts of some piston cores with simultaneously collected open-barrel cores. *Deep-sea Res.* **14**, 285-94.

Saito, T. and Be, A. W. H. 1964. Planktonic Foraminifera from the American Oligocene. *Science,* **145**, 702-5.

Saito, T., Burckle, L. H. and Ewing, M. 1966. Lithology and paleontology of the reflective layer Horizon A. *Science* **154**, 1173-6.

Saito, T., Ewing, M. and Burckle, L. H. 1966. Tertiary sediment from the Mid-Atlantic Ridge. *Science,* **151**, 1075-9.

Schlee, J. and Gerard, R. 1965. Cruise report and preliminary core log M.V. *Caldrill* I – 17 April to 17 May 1965. *J.O.I.D.E.S. Blake Panel Report* (unpublished), iv, 64 pp.

Shipek, C. J. 1960. Photographic study of some deep-sea environments in the eastern Pacific. *Bull. geol. Soc. Amer.* **71**, 1067-74.

Shor, G. G. 1964. Thickness of coral at Midway Atoll. *Nature, Lond.* **201**, 1207-8.

Smith, F. D. 1955. Planktonic Foraminifera as indicators of depositional environment. *Micropaleont.* **1**, 147-51.

Sousa Torres, A. and Pires Soares, J. M. 1946. Formacoes sedimentares do Arquipelago de Cabo Verde. *Mem. Minister. Colonias Port., ser. geol.* **3**, 397 pp.

Stahlecker, R. 1935. Neocom auf der Kapverden-Insel Maio. *Neues Jahrb. Min. Geol. Palaont. Abt. B,* **73**, 265-301.

Stainforth, R. M. 1948. Description, correlation, and palaeoecology of Tertiary Cipero Marl formation, Trinidad, B.W.I. *Bull. Amer. Assoc. Petrol. Geol.* **32**, 1292-330.

Stark, J. T. *et al.* 1958. *Military geology of Truk Islands.* U.S. Army Chief Eng., Intell. Div., H.Q.U.S. Army Pacific, Tokyo. 456 pp.

Stehli, F. G. 1965. Palaeontologic technique for defining ancient ocean currents. *Science,* **148**, 943-6.

Stehli, F. G. and Creath, W. B. 1964. Foraminiferal ratios and regional environments. *Bull. Amer. Assoc. Petrol. Geol.* **48**, 1810-27.

Stephenson, L. W. 1936. Geology and palaeontology of the Georges Bank Canyons, Part II. Upper Cretaceous fossils from Georges Bank (including species from Banquereau, Nova Scotia). *Bull. geol. Soc. Amer.* **47**, 367-410.

Stetson, H. C. 1936. Geology and palaeontology of the Georges Bank Canyons, Pt.I. *Bull. geol. Soc. Amer.* **47**, 339-66.

Stetson, H. C. 1949. The sediments and stratigraphy of the east coast continental margin; Georges Bank to Norfolk Canyon. *Mass. Inst. Techn. and Woods Hole Ocean. Inst., Papers in Physical Oceanography and Meteorology,* **11**, 1-60.

Stradner, H. 1963. New contributions to Mesozoic stratigraphy by means of nannofossils. *6th World Petrol. Congr. sect. 1, pap. 4 (preprint),* 1-16.

Stradner, H. 1964. Die Ergebnisse des Aufschlussarbeiten der ÖMV AG in der Molassezone Niederösterreichs in den Jahren 1957-1963. Ergebnisse der Nannofossil-Untersuchungen (Teil III). *Erdöl Z.* **80**, 133-9.

Sullivan, F. R. 1964. Lower Tertiary nannoplankton from the California Coast Ranges. I. Paleocene. *Publ. Univ. Calif. Geol. Sci.* **44**, 163-228.

Sullivan, F. R. 1965. Lower Tertiary nannoplankton from the California Coast Ranges. II. Eocene. *Publ. Univ. Calif. Geol. Sci.* **53**, 1-53.

Tan Sin Hok. 1927. Over de samenstelling en het ontstaan van Krijt- en mergel-gesteenten van de Molukken. *Jaarb. van het Mijnwezen in Ned. Oost-Indie, 1926,* 1-165, pls. 1-16.

Todd, R. 1957. Geology of Saipan, Mariana Islands. Smaller Foraminifera. *Prof. Pap. U.S. geol. Surv.* **280-H**, 265-320.

Todd, R. 1964. Planktonic Foraminifera from deep-sea cores off Eniwetok Atoll. *Prof. Pap. U.S. geol. Surv.* **260-CC**, 1067-100.

Todd, R., Cloud, P. E., Low, D. and Schmidt, R. G. 1954. Probable occurrence of Oligocene on Saipan. *Amer. J. Sci.* **252**, 673-82.

Todd, R. and Low, D. 1960. Bikini and nearby atolls, Marshall Islands. Small Foraminifera from Eniwetok Drill Holes. *Prof. Pap. U.S. geol. Surv.* **260-X**, i-iv, 799-861.

Todd, R. and Low, D. 1964. Cenomanian (Cretaceous) Foraminifera from the Puerto Rico Trench. *Deep-sea Res.* **11**, 395-414.

Todd, R. and Post, R. 1954. Bikini and nearby atolls. Marshall Islands. Smaller Foraminifera from Bikini Drill Holes. *Prof. Pap. U.S. geol. Surv.* **260-N**, 547-68.

Upham, W. 1894. The fishing banks between Cape Cod and Newfoundland. *Amer. J. Sci.* **47**, 123-9.

Uschakova, M. G. 1966. (Biostratigraphical significance of coccolithophorids in the bottom sediments of the Pacific Ocean.) *Okeanologia,* **6**, 136-43.

van Andel, Tj.H., Bowen, V. T., Sachs, P. L. and Siever, R. 1965. Morphology and sediments of a portion of the Mid-Atlantic Ridge. *Science,* **152**, 124-6.

Vaughan, T. W. 1919. Fossil corals from Central America, Cuba, and Porto Rico, with an account of the American Tertiary, Pleistocene, and Recent coral reefs. *Bull. U.S. Nat. Mus.* **103**, 189-524, pls. 68-152.

Verrill, A. E. 1878. Occurrence of fossiliferous Tertiary rocks on the Grand Banks and Georges Bank. *Amer. J. Sci.* **16**, 323-4.

Vine, F. J. 1966. Spreading of the ocean floor : new evidence. *Science,* **154**, 1405-15.

Watkins, N. D. and Goodell, H. G. 1967. Geomagnetic

polarity change and faunal extinction in the Southern Ocean. *Science,* **156**, 1083-1087.

Whipple, G. L. 1932. In: Hoffmeister, J. E. Geology of Eua, Tonga. *Bull. B.P. Bishop Mus.* **96**, 79-90, pls. 20-2.

Whittard, W. F. 1962. Geology of the western approaches of the English Channel : a progress report. *Proc. Roy. Soc., A,* **265**, 395-406.

Williams, D. B. 1965. The distribution and palaeocology of microplankton in Recent marine sediments. Ph.D. thesis, University of Reading, xii, 289 pp.

Wilson, J. T. 1965. A new class of faults and their bearing on continental drift. *Nature, Lond.* **207**, 343-7.

Wilson, J. T. 1963. Evidence from islands on the spreading of ocean floors. *Nature, Lond.* **197**, 536-8.

Wiseman, J. D. H. 1936. The petrography and significance of a rock dredged from a depth of 744 fathoms, near to Providence Reef, Indian Ocean. *Trans. Linn. Soc. London, ser. 2, Zoology.*

Wiseman, J. D. H. and Hendy N. I. 1953. The significance and diatom content of a deep-sea floor sample from the neighbourhood of the greatest oceanic depth. *Deep-sea Res.* **1**, 47-59.

Wiseman, J. D. H. and Ovey, C. D. 1950. Recent investigation on the deep-sea floor. *Proc. Geol. Assoc.* **61**, 28-84, pls. 2-3.

Wiseman, J. D. H. and Riedel, W. R. 1960. Tertiary sediments from the floor of the Indian Ocean. *Deep-sea Res.* **7**, 215-7.

Worzel, J. L. and Harrison, J. C. 1963. Gravity at sea. In: *The Sea* (M. N. Hill *ed.*), **3**, 134-74. Interscience: New York.

Wray, J. L. and Ellis C. H. 1965. Discoaster extinction in neritic sediments, northern Gulf of Mexico. *Bull. Amer. Assoc. Petrol. Geol.* **49**, 98-9.

Yabe, H. and Aoki, R. 1922. Reef conglomerate with small pellets of *Lepidocyclina* – Limestone found on the Atoll Jaluit. *Jap. J. Geol. Geog.* **1**, 40-3.

APPENDIX : Index of pre-Quaternary occurrences in the oceans (to August 1967)

The entries in the following list are arranged in serial order according to a numerical index similar to the Marsden square system. In the present system, however, the basic three digit (10° square) numbers run sequentially from the north to south poles, commencing with the 80-90° N, 0-10° E 'square' at the north pole (101) and continuing successively eastwards in 10° bands of latitude to the 80-90° S, 0-10° W 'square' (972) at the south pole. (Numbers above 172, 272, 372, are not used so that (i) the 90-70° N, 70-50° N, 50-30° N, etc. bands of latitude are represented by the initial digits 1, 2, 3, etc., with the initial digit 5 representing the band extending 10° N and S of the equator, and (ii) the last two digits repeat every other 10° in each 10° band of longitude.) The requisite index number can therefore be deduced from the latitude and longitude without recourse to a base map; although in practice reference to an appropriately numbered map may be preferred.

Each 10° square is subdivided into 1° squares represented in the index by two digits after a decimal point. The 10° square is always scanned from top to bottom and from left to right as shown in the key.

N	W 9	8	7	6	5	4	3	2	1	0	0
9	00	10	20	30	40	50	60	70	80	90	1
8	01	11	21	31	41	51	61	71	81	91	2
7	02	12	22	32	42	52	62	72	82	92	3
6	03	13	23	33	43	53	63	73	83	93	4
5	04	14	24	34	44	54	64	74	84	94	5
4	05	15	25	35	45	55	65	75	85	95	6
3	06	16	26	36	46	56	66	76	86	96	7
2	07	17	27	37	47	57	67	77	87	97	8
1	08	18	28	38	48	58	68	78	88	98	9
0	09	19	29	39	49	59	69	79	89	99	S
	0	1	2	3	4	5	6	7	8	9	E

Although here again this key is not essential and the approximate index number can be directly derived from the latitude and longitude.

Within a 1° square occurrences have been ordered by scanning from left to right and from top to bottom.

The purpose of devising these index numbers has been to provide an unequivocal numerical ordering where the sequence of numbers indicates a measure of geographical proximity, and where the index numbers are meaningful without the necessity of referring to a base map. At the same time the number of digits in the index has been kept to a minimum.

In the following catalogue Atlantic, Indian and Pacific Ocean occurrences are listed separately.

Atlantic Ocean

271.89 50° 22' N, 11° 44' W. 1,796 m. *Upper Cretaceous* (? *and Tertiary).* Peach 1912, p. 123; Curry *et al.* 1962, p. 283.

329.69 40° 15' N, 73° 59' W. 13 m. G13. *Palaeocene.* L.G.O. (unpublished).

329.69 40° 15' N, 73° 59' W. 9 m. G28. *Eocene.* L.G.O. (unpublished).

330.09 40° 09' N, 69° 04' W. 164-402 m. TOW 2-36 (west side of Hydrographer Canyon). *reworked Miocene.* Stetson 1949, pp. 11-12, 33.

330.09 40° 09' N, 69° 03' W. 140-319 m. TOW 1-36 (east side of Hydrographer Canyon). *reworked Miocene.* Stetson 1949, pp. 11-12, 33.

330.19 40° 24' N, 68° 07' W. 480-596 m. TOW 5-34 (east side of Oceanographer Canyon). *Upper Cretaceous and upper Pliocene or Pleistocene.* Stetson 1936, pp. 347-8; Stephenson 1936, pp. 367-410; Bassler 1936, pp. 411-12; Cushman 1936, pp. 413-40; Stetson 1949, pp. 8-9, 13, 33-4; Heezen *et al.* 1959, pp. 44-6, Funnell 1964, p. 425, table 1.

330.19 40° 23' N, 68° 08' W. 231-585 m. TOW 10-36 (east side of Oceanographer Canyon). *reworked Upper Cretaceous.* Stetson 1949, pp. 8-9, 33; Heezen *et al.* 1959, pp. 44-6; Funnell 1964, table 1.

330.19 40° 17' N, 68° 06' W. 950 m. TOW 2-39 (east side of Oceanographer Canyon). *Upper Cretaceous.* Stetson 1949, pp. 10-1, 33; Heezen, Tharp and Ewing 1959, pp. 44-6; Funnell 1964, p. 424, table 1.

330.27 42° 03' N, 67° 04' W. 54 m. G (north side of Georges Bank). *late Miocene.* Gibson 1965, p. 979-80.

330.28 41° 53' N, 67° 53' W. 32 m. B. (north side of Georges Bank). *late Miocene.* Gibson 1965, pp. 979-80.

330.28 41° 52' N, 67° 20' W. 37 m. J (north side of Georges Bank). *(?) late Miocene.* Gibson 1965, p. 979-80.

330.29 40° 30' N, 67° 42' W. 452-458 m. TOW 11-34 (east

side of Lydonia Canyon). *upper Pliocene or Pleistocene.* Cushman 1936, p. 413; Stetson 1949, pp. 13, 34.

330.29 40° 27′ N, 67° 39′ W. 512-640 m. TOW 9-34 (east side of Lydonia Canyon). *Pliocene, and upper Pliocene or Pleistocene.* Cushman 1936, pp. 413-40; Stetson 1949, pp. 12-13, 34.

330.29 40° 23′ N, 67° 38′ W. 283 m. TOW 20-36 (east side of Lydonia Canyon). *late Miocene.* Stetson 1949, pp. 11-12, 33; Gibson 1965, p. 979.

330-29 40° 21′ N, 67° 51′ W. 530-600 m. TOW 6-34 (east side of Gilbert Canyon). *Upper Cretaceous and upper Pliocene or Pleistocene.* Stetson 1936, p. 349; Cushman 1936, pp. 413-40; Stetson 1949, pp. 9, 13, 33-4; Heezen, Tharp and Ewing 1959, pp. 44-6; Funnell 1964, p. 424, table 1.

330.29 40° 19′ N, 67° 51′ W. 758 m. TOW 14-36 (east side of Gilbert Canyon). *Upper Cretaceous.* Stetson 1949, pp. 9-10, 33; Heezen, Tharp and Ewing. 1959, pp. 44-6; Funnell 1964, p. 424, table 1.

330.39 40° 21′ N, 66° 08′ W. 237-493 m. TOW 23-36 (east side of Corsair Canyon). *reworked Miocene.* Stetson 1949, pp. 11-12, 33.

331.25 approx. 44° 30′ N, 57° 30′ W. approx. 366 m. 'William Thomson'. Banquereau Bank. *Cenomanian.* Dall 1925, p. 215; Stephenson 1936, pp. 384-404, Heezen, Tharp and Ewing 1959, p. 47; Funnell 1964, p. 425, table 1.

332.25 44° 33′ N, 47° 33′ W. 3,658 m. R9-3. *late Tertiary or early Quaternary.* Ericson *et al.* 1961, pp. 213-235.

334.96 43° 01′ N, 20° 07′ W. 5,300 m. D5971. Palmer Ridge. *lower Tertiary (Eocene).* Funnell (unpublished).

334.97 42° 56′ N, 20° 07′ W. 3781 m. D5974. Palmer Ridge. *(?) upper Tertiary (possibly Pliocene).* Funnell (unpublished).

334.97 42° 55′ N, 20° 11′ W. approx. 3,380 to 3570 m. D5969. Palmer Ridge. *lower Tertiary (probably middle or upper Eocene) and upper Tertiary (probably Miocene or Pliocene.)* Cann and Funnell 1967, p. 663.

334.97 42° 55′ N, 20° 07′ W. 3,573 m. D5973. Palmer Ridge. *upper Tertiary (Miocene or Pliocene).* Funnell (unpublished).

334.97 42° 54′ N, 20° 16′ W. approx. 3,200 m. D5619. Palmer Ridge. *upper Tertiary (probably middle to upper Miocene or Pliocene).* Cann and Funnell 1967, p. 663.

334.97 42° 54′ N, 20° 16′ W. approx. 3,250 to 3,340 m. D5977. Palmer Ridge, *upper Tertiary (probably upper Miocene or Pliocene).* Cann and Funnell 1967, p. 663.

334.97 42° 54′ N, 20° 13′ W. approx. 3,200 m. D5975. Palmer Ridge. *upper Tertiary.* Cann and Funnell 1967, p. 663.

334.97 42° 54′ N, 20° 13′ W. approx. 3,150 to 3,340 m. D5983. Palmer Ridge. *upper Tertiary (probably middle to upper Miocene or Pliocene).* Cann and Funnell 1967, p. 663.

334.97 42° 52′ N, 20° 16′ W. approx. 3,930 to 4,300 m. D5608. Palmer Ridge. *lower Tertiary (lower or middle Eocene).* Cann and Funnell 1967, p. 663.

334.97 42° 52′ N, 20° 12′ W. approx. 4,210 to 4,750 m. D5985. Palmer Ridge. *middle Tertiary (Oligocene or lower Miocene).* Cann and Funnell 1967, p. 663.

334.97 42° 51′ N, 20° 16′ W. approx. 4,660 to 4,710 m. D5981. Palmer Ridge. *lower Tertiary (probably lower or middle Eocene).* Cann and Funnell 1967, p. 663.

334.97 42° 50′ N, 20° 12′ W. approx. 4,980 to 5,260 m. D5968 Palmer Ridge. *lower Tertiary (probably lower or middle Eocene).* Cann and Funnell 1967, p. 663.

335.06 43° 10′ N, 19° 39′ W. approx. 3,570 m. D5623. Palmer Ridge. *lower Tertiary (probably middle or upper Eocene).* Funnell (unpublished).

335.07 42° 51′ N, 19° 56′ W. approx. 2,740 m. D5627. Palmer Ridge. *lower Tertiary (Eocene) and (?) upper Tertiary.* Funnell (unpublished).

335.09 42° 51′ N, 20° 00′ W. approx. 2,560. D5626. Palmer Ridge. *upper Tertiary (probably middle to upper Miocene or Pliocene).* Funnell (unpublished).

335.60 49° 38′ N, 13° 28′ W. 1,955 m. LK13. Goban Spur. *reworked Upper Cretaceous.* Bramlette and Bradley. 1940, p. xiv; Day 1959, p. 260; Funnell 1964, p. 427, table 1.

335.60 49° 37′ N, 13° 34′ W. 3,230 m. LK12. Goban Spur. *reworked Upper Cretaceous.* Bramlette and Bradley. 1940 p. xiv; Day 1959, p. 260; Funnell 1964, p. 427, table 1.

335.87 42° 56′ N, 11° 47′ W. 1,189 m. D3809. Galicia Bank. *middle Eocene with Upper Cretaceous admixture.* Black 1964, pp. 306-7, 315; Black *et al.* 1964, p. 504.

335.87 42° 55′ N, 11° 47′ W. 1,097 m. D3808.8. *middle Eocene.* Black 1964, pp. 307, 315; Black *et al.* 1964, pp. 304-5.

335.87 42° 49′ N, 11° 37′ W. 969 to 1,244 m. D4272. Galicia Bank. *Lower Cretaceous, and Upper Cretaceous (Maastrichtian).* Black 1964, pp. 306, 315; Black *et al.* 1964, pp. 491, 503-4.

335.87 42° 36′ N, 11° 35′ W. approx. 695 m. (650 to 700 m). D3804-1. Galicia Bank. *upper Cretaceous (Maastrichtian) with middle Eocene admixture.* Funnell 1964, pp. 421-2, 428, table 1; Black 1964, pp. 306-7, 315; Black *et al.* 1964, p. 504.

335.87 approx. 42° 36′ N, 11° 36′ W. 695 m. D3806. Galicia Bank. *late Cretaceous.* Black *et al.* 1964, pp. 503-4.

335.91 48° 27′ N, 10° 07′ W. 1,061 to 1,244 m. AJS 15/1. *middle-upper Miocene.* Whittard 1962, pp. 400-1; Curry *et al.* 1962, p. 277.

335.91. 48° 25′ N, 10° 18′ W. 457 to 768 m. AJS 15/8. *upper Eocene.* Curry *et al.* 1962, p. 273.

335.98 approx. 41° 14′ N, 10° 42′ W. 2,067 to 2,140,m. D4279. *(?) middle Tertiary.* Black *et al.* 1964, p. 49.

335.98 approx. 41° 14′ N. 10° 42′ W. 2,743 m. D4280. *(?) Upper Jurassic.* Black *et al.* 1964, p. 491.

336.01 48° 38′ N, 09° 50′ W. 933 to 1061 m. AJS 15/12. *lower Miocene, with middle and upper Eocene admixture.* Curry *et al.* 1962, p. 278.

336.01 approx. 48° 25′ N, 09° 55′ W. 2,743 m. 'Alsace'. *Oligocene.* Bourcart and Marie 1951, p. 123; Day 1959, p. 260; Curry *et al.* 1962, p. 285.

336.07 42° 46′ N, 09° 36′ W. 648 m. Op. 597. [*? reworked*] *Eocene.* Colom 2952, pp. 11-12.

336.07 42° 07′ N, 09° 25′ W. 540 m. Op. 87. [*? reworked*] *Eocene.* Colom 1952, pp. 11-12.

336.08 41° 56′ N, 09° 20′ W. 215 m. Op. 381. [*? reworked*] *Eocene.* Colom 1952, pp. 11-12.

336.12 47° 48′ N, 08° 09′ W. 1,225 to 1,317 m. AJS 19/6. *middle Miocene.* Whittard 1962, pp. 400-1; Curry *et al.* 1962, p. 276.

336.12 approx. 47° 30′ N, 08° 45′ W. 2,743 m. H.M.T.S. 'Monarch' *Middle Eocene with Upper Cretaceous (Maastrichtian) admixture.* Wiseman and Ovey 1950, pp. 42-3; Day 1959, p. 260; Whittard, 1962, pp. 400-1; Curry *et al.* 1962, p. 285; Funnell 1964, p. 428, table 1.

336.22 47° 38′ N, 07° 30′ W. 658 to 823 m. BB Chalk Canyon. *Miocene.* Funnell (unpublished).

336.22 47° 37′ N, 07° 31′ W. 1,006 to 1,262 m. AZ Chalk Canyon. *Eocene.* Funnell (unpublished).

336.22 47° 36′-47° 37′ N, 07° 26′-07° 27′ W. 812 m. Sa (1959). *probably Pliocene.* Black 1962, pp. 123-7; Curry *et al.* 1962, p. 283.

336.22 47° 36′ N. 07° 33′ W. 896 to 1,737 m. AT Chalk Canyon. *Pliocene.* Funnell (unpublished).

336.22 47° 32′ N, 07° 34′ W. 1,646 to 1,682 m. AW-1 Chalk Canyon. *(?) Upper Cretaceous.* Funnell 1964, pp. 427-

8, table 1.
336.32 47° 31′ N, 06° 58′ W. 412 m. Sa 2. *Cretaceous or Tertiary [probably upper Tertiary].* Day 1959, pp. 259-60; Curry *et al.* 1962, p. 285.
336.42 47° 13′ N, 05° 45′ W. 914 to 1,280 m. AJS 17/1 *lower Miocene with Upper Cretaceous (Maastrichtian) admixture.* Curry *et al.* 1962, pp. 275-6.
336.94 45° 04′ N, 8° 00′ W. 4,526 m. D5946. (Cantabria seamount) *reworked Upper Cretaceous (Maastrichtian).* Jones and Funnell 1968.
336.94 45° 04′ N, 8° 00′ W. 4,389 m. D5947. (Cantabria seamount). *Quaternary, with Upper Cretaceous admixture.* Jones and Funnell 1968.
336.94 45° 05′ N, 8° 00′ W. 4,724-4,382 m. D5948. (Cantabria seamount). *Quaternary, with Upper Cretaceous admixture.* Jones and Funnell 1968.
364.89 30° 33′ N, 81° 00′ W. 25 m. JOIDES 1. *middle Eocene (at −208 m to −302 m below sea level), upper Eocene. Oligocene, Miocene and post Miocene.* Schlee and Gerard 1965, pp. 45-8; Bunce *et al.* 1965, pp. 709-16.
364-99 30° 23′ N, 80° 08′ W. 190 m. JOIDES 5 *upper Eocene (at −319 m to − 435 m), Oligocene and post-Miocene.* Schlee and Gerard 1965, pp. 59-62; Bunce *et al.* 1965, pp. 709-16.
364.99 30° 20′ N and 30° 21′ N, 80° 20′ W. 42 and 46 m. JOIDES 2 *middle Eocene (at 343 m to 364 m below sea-level), upper Eocene, Oligocene, Miocene and post-Miocene.* Schlee and Gerard 1965, pp. 49-52; Bunce *et al.* 1965, pp. 709-16.
365.09 30° 05′ N, 79° 15′ W. 805 m. JOIDES 6. Blake Plateau. *Palaeocene (− 922 m to −925 m below sea-level), Eocene, Oligocene and post-Miocene.* Schlee and Gerard 1965, pp. 63-4; Bunce *et al.* 1965, pp. 709-16.
365.27 32° 19′ N, 77° 39′ W. 543 m. RC9-5. *Eocene.* L.G.O. (unpublished).
365.28 31° 03′ N, 77° 45′ W. and 31° 02′ N, 77° 43′ W. 885 and 892 m. JOIDES 4. *Palaeocene (at −977 m to −1,067 m below sea level), lower Eocene, Oligocene, lower Miocene, and post-Miocene.* Schlee and Gerard 1965, pp. 56-8; Bunce *et al.* 1965, pp. 709-16.
365.29 30° 23′ N, 77° 04′ W. 1,865 m. A164-31. *Miocene.* Ericson *et al.* 1952, pp. 497, 504.
365.39 30° 04′ N, 76° 57′ W. 1,120 m. A164-30. *Miocene.* Ericson *et al.* 1952, pp. 497, 504; Emiliani 1956*b*, pp. 281-8; Ericson *et al.* 1961, pp. 208, 236; Heezen and Sheridan 1966, p. 1,646.
365.39 30° 15′ N, 76° 21′ W. 4,800 m. E-9-66-12. (Blake escarpment). *Lower Cretaceous (Neocomian-Aptian).* Heezen and Sheridan 1966, p. 1,645.
365-66 33° 13′ N, 73° 39′ W. 4,645 m. A167-8. *Quaternary, with Cretaceous admixture.* Ericson *et al.* 1952, pp. 492, 503; Funnell 1964, table 1.
365.71 38° 58′ N, 72° 28′ W. 1,565 m. Core 21-38. *Eocene.* Cushman 1939, p. 49; Phleger 1939, pp. 1,399, 1,409-10; Stetson 1949, pp. 11; 33; Riedel 1957*a*, pp. 81, 82, 89; Gibson 1965, pp. 976-8.
365.72 37° 17′ N, 72° 15′ W. 3,502 m. A185-61. *Pliocene.* L.G.O. (unpublished).
365.80 39° 25′ N, 71° 53′ W. 1,400 m. A156-10. *late Tertiary (or early Quaternary).* Ericson *et al.* 1961, pp. 207, 234.
365.80 39° 12′ N, 71° 48′ W. 2,167 m. A162-1. *upper Eocene or Oligocene.* Ericson *et al.* 1952, pp. 490, 494, 502.
365.90 39° 50′ N, 70° 57′ W. 880 m. Core 12-36. Marthas Vineyard. *Eocene.* Cushman 1939, p. 49; Stetson 1949, pp. 11, 33; Gibson 1965, pp. 976-7.
365.90 39° 43′ N, 70° 43′ W. 1,000 m. Dog-20. Marthas Vineyard. *Eocene.* Stetson 1949, pp. 11, 33; Northrop and Heezen 1951, pp. 396-9; Ericson *et al.* 1952, p. 502; Heezen *et al.* 1959, pp. 46-7; Gibson 1965, pp. 976-8.
365.91 38° 27′ N, 70° 59′ W. 3,475 m. A164-2. *late Tertiary (?Pliocene).* Ericson *et al.* 1961, pp. 207, 234.
365.91 38° 24′ N, 70° 55′ W. 3,109 m. A164-3. *late Tertiary (?Pliocene).* L.G.O. (unpublished).
365.91 38° 22′ N, 70° 56′ W. 3,570 m. C10-13. *upper Tertiary (upper Miocene or Pliocene).* Ericson *et al.* 1952, pp. 490, 494, 502; Ericson *et al.* 1961. pp. 213, 235.
365.91 38° 12′ N, 70° 51′ W. 3,329 m. A164-4. *upper Miocene.* Ericson *et al.* 1961, pp. 207, 234.
365.91 38° 12′ N, 70° 40′ W. 3,390 m. A158-1. *late Tertiary (or early Quaternary).* L.G.O. (unpublished).
365.92 37° 56′ N, 70° 44′ W. 3,823 m. A164-8 *upper Miocene with some Eocene admixture.* Ericson *et al.* 1952, pp. 490, 494, 502; Ericson *et al.* 1961, pp. 207, 235.
365.94 35° 57′ N, 70° 45′ W. 3,823 m. A164-7 *upper Miocene (or lower Pliocene).* Ericson *et al.* 1952, pp. 490, 494, 502.
366.00 39° 46′ N, 69° 44′ W. 1,695 m. Gosnold 49-2150. Cape Cod. *lower Eocene.* Gibson 1965, pp. 975-8, 981.
366.20 approx. 34° 55′ N, 67° 30′ W. 1,070 m. Bear seamount. *Eocene.* Northrop *et al.* 1959, p. 48; Northrop *et al.* 1962, p. 588.
366.20 approx. 39° 20′ N, 67° 20′ W. 2,368 m. Mytilus Seamount. *Eocene.* Northrop *et al.* 1959, p. 48; Northrop *et al.* 1962, p. 588.
366.23 36° 43′ N, 67° 56′ W. 3,548 m. A164-10. Caryn Peak. *reworked Cretaceous, and late Tertiary.* Ericson *et al.* 1952, p. 501; Miller and Ewing 1956, p. 424; Heezen Tharp and Ewing 1959, p. 78; Ericson *et al.* 1961, pp. 207, 240; Funnell 1964, table 1.
366.23 36° 42′ N, 67° 58′ W. 3,260 m. A158-2. Caryn Seamount *reworked Upper Cretaceous.* Ericson *et al.* 1952, p. 501; Miller and Ewing 1956, p. 424, Heezen, Tharp and Ewing 1959, p. 78; Worzel and Harrison 1963, p. 166; Funnell 1964, p. 427, table 1.
366.23 36° 41′ N, 67° 59′ W. 3,940 m. A158-4. Caryn Peak. *reworked Upper Cretaceous, and upper Tertiary.* Ericson *et al.* 1952, p. 501; Miller and Ewing 1956, p. 424; Heezen, Tharp and Ewing 1959, p. 78; Ericson *et al.* 1961, pp. 207, 239-40; Funnell 1964, table 1.
366.27 32° 17′ N, 67° 07′ W. 1,554 m. V5-31. Bermuda. *upper Miocene or Pliocene.* L.G.O. (unpublished).
366.36 33° 05′ N, 66° 06′ W. 4,580 m. A150-33. *pre-Quaternary* Ericson *et al.* 1952, pp. 499, 506.
366.40 39° 40′ N, 65° 55′ W. 2,166 to 2,496 m. V18-RD37. Picket Peak. *Eocene.* L.G.O. (unpublished).
366.47 32° 43′ N, 65° 16′ W. 4,517 m. A164-36. *Quaternary, with middle Tertiary admixture.* Ericson *et al.* 1952, pp. 499, 506.
366.52 37° 17′ N, 64° 38′ W. 1,097 m. C22-5. *Oligocene.* Ericson *et al.* 1952, pp. 499, 508; Ericson *et al.* 1961, pp. 213, 238.
366.57 32° 43′ N, 64° 34′ W. 1,710 m. C25-5. *Oligocene and Miocene.* Ericson *et al.* 1961, pp. 213, 239.
366.57 32° 35′ N, 64° 39′ W. 1,795 m. A150-30. *(?) pre-Quaternary (reworked limestone fragments).* Ericson *et al.* 1952, pp. 499, 506.
366.57 32° 33′ N, 64° 51′ W. 1,417 m. V5-24. Bermuda. *probably upper Miocene.* L.G.O. (unpublished).
366.57 32° 20′ N, 64° 34′ W. 1,150 m. C10-7. *Pliocene (or early Pleistocene).* Ericson *et al.* 1952, pp. 499, 507; Ericson *et al.* 1961, pp. 213, 239.
366.57 32° 20′ N, 64° 28′ W. 2,332 m. V5-15. *(?)Oligocene.* L.G.O. (unpublished).
366-57 32° 19′ N, 64° 36′ W. 1,005 m. C10-10. *Miocene.* Ericson *et al.* 1952, pp. 499, 507; Ericson *et al.* 1961, pp. 213, 239.
366-57 32° 19′ N, 64° 27′ W. 2,305 m. V5-14. *upper Oligocene*

or lower Miocene (upper *G. ciperoensis* or lower *C. dissimilis* zone). Ericson *et al.* 1961, pp. 214, 238.

366.57 approx. 32° 18′ N, 64° 45′ W. Bermuda Island. *Miocene and Pliocene (−32 to −46 m), Eocene or Oligocene (−63 to −79 m), ?Eocene (−70 to −107 m) resting on basalt* Pirsson and Vaughan 1913, pp. 70-1; Pirsson 1914*a*, pp. 195-206; Pirsson 1914*b*, pp. 331-44; Vaughan 1919, pp. 293-7; Wilson 1963, p. 536.

366.57 32° 16′ N, 64° 39′ W. 1,410 m. 1,376 m. V5-13. *(?)Oligocene.* L.G.O. (unpublished).

366.57 32° 16′ N, 64° 35′ W. 1,510 m. C22-6. *Oligocene* (upper *G. ciperoensis* or lower *C. dissimilis* zone). Ericson *et al.* 1961, pp. 213, 238-9.

366.57 32° 14′ N, 64° 32′ W. 2,230 m. C10-11. *upper Oligocene* (*G. ciperoensis* zone). Ericson *et al.* 1952, pp. 499, 507; Ericson *et al.* 1961, pp. 213, 239.

366.57 32° 13′ N, 64° 45′ W. 1,390 m. V5-36 Bermuda. *(?) upper Tertiary.* L.G.O. (unpublished).

366.57 32° 13′ N, 64° 31′ W. 2,953 m. A164-25. *Oligocene.* Ericson *et al.* 1952, pp. 499, 506; Ericson *et al.* 1961, pp. 208, 289.

366.57 32° 11′ N, 64° 45′ W. 1,940 m. C22-2. *Oligocene.* Ericson *et al.* 1952, pp. 499, 507-8; Ericson *et al.* 1961, pp. 213, 238.

366.57 32° 01′ N, 64° 58′ W. 2,780 m. A164-26. *Miocene.* Ericson *et al.* 1952, pp. 499, 506; Ericson *et al.* 1961, pp. 208, 239.

366.74 35° 54′ N, 62° 17′ W. 5,080 m. A152-134. *pre-Quaternary* L.G.O. (unpublished).

366.76 33° 47′ N, 62° 39′ W. 2,360 m. C25-6. Muir Seamount. *Oligocene or Miocene.* Ericson *et al.* 1961, pp. 213, 240.

366.76 33° 42′ N, 62° 30′ W. 1,555 m. A150-1. *middle Eocene* (*G. mexicana* zone or slightly older). Northrop *et al.* 1959, p. 48; (?) Bolli 1957*c*, pp. 164-5; Black 1964, pp. 307, 315.

366.76 33° 40′ N, 62° 30′ W. 1,480 m. C10-1. Muir Seamount *Eocene.* L.G.O. (unpublished); cf. Northrop *et al.* 1959, p. 48.

366.76 33° 40′ N, 62° 30′ W. 1,550 m. C10-4. Muir Seamount. *Miocene.* Ericson *et al.* 1961, pp. 213, 240.

366.76 33° 40′ N, 62° 27′ W. 2,370 m. C10-2. Muir Seamount. *Quaternary, with Cretaceous and upper Tertiary admixture.* Heezen, Tharp and Ewing 1959, p. 76; Funnell 1964, table 1.

366.76 33° 39′ N, 62° 27′ W. 2,840 m. C10-3. Muir Seamount. *(?)Pliocene.* L.G.O. (unpublished).

366.76 33° 37′ N, 62° 30′ W. 2,670 m. C10-5. Muir Seamount. *Upper Cretaceous.* Funnell 1964, p. 426-7, table 1.

368.16 33° 08′ 48° 08′ W. 4,850 m. A 153-144. *Pliocene.* Saito, Ewing and Burckle 1966, p. 1076.

368.45 34° 56′ N, 45° 21′ W. 4,210 m. V17-166. *Pliocene.* Saito, Ewing and Burckle 1966, p. 1,076.

368.78 31° 49′ N, 42° 25′ W. 3,700 m. A150 RD8. *lower and middle Miocene.* Saito, Ewing and Burkle 1966, p. 1,075.

368.79 30° 01′ N, 42° 04′ W. 4,280 m. A150 RD7. *lower and middle Miocene.* Saito, Ewing and Burckle 1966, p. 1,076.

369.99 30° 24′ N, 30° 47′ W. 4,444 m. SP8-3. *upper Tertiary.* Ericson *et al.* 1961, pp. 213, 241.

370.06 33° 05′ N, 29° 18′ W. 2,470 m. V4-53. Plato Seamount. *middle Miocene, with upper Eocene and lower Miocene admixture.* Saito, Ewing and Burckle 1966, p. 1078.

370.19 30° 15′ N, 28° 30′ W. 1,280 m. A180-25. *Miocene.* Saito, Ewing and Burckle 1966, p. 1,076.

370.21 approx. 38° 40′ N, 27° 15′ W above sea-level. Azores. *Miocene.* Krejci-Graf, 1956; Wilson 1963, p. 536.

371.27 approx. 32° 45′ N, 17° 00′ W above sea-level. Madeira Island. *(?) Cretaceous, and Miocene.* Gagel 1912, pp. 365-7, Daly 1927, p. 87; Ribeiro 1958, p. 123; Wilson 1963, p. 536.

371.44 35° 14′ N, 15° 12′ W. 2,926 m. V4-19. Ampere Bank. *upper Miocene or Pliocene.* L.G.O. (unpublished).

371.44 35° 11′ N, 15° 20′ W. 2,377 m. V4-15. Ampere Bank. *upper Miocene or Pliocene.* L.G.O. (unpublished).

371.64 35° 07′ N, 13° 04′ W. 1,719 m. V4-20. Ampere Bank. *upper Miocene or Pliocene.* L.G.O. (unpublished).

371.65 34° 58′ N, 13° 11′ W, 1,940 m. R5-50. Ampere Bank. *Miocene.* Ericson *et al.* 1961, pp. 213, 240.

371.74 35° 03′ N, 12° 57′ W. 1,280 m. V4-27. Ampere Bank. *Miocene.* L.G.O. (unpublished).

372.01 38° 12′ N, 09° 45′ W. V4-41. *Miocene.* L.G.O. (unpublished).

427.75 24° 34′ N, 92° 37′ W. 3,628 m. A185-35. *reworked upper Tertiary (Miocene).* Ewing *et al.* 1958, pp. 1,029, 1,036.

427.76 23° 50′ N, 92° 24′ W. 3,536 m. 64-A-9-5E. Sigsbee knolls, Gulf of Mexico. *Miocene and Pliocene.* Bryant and Pyle 1965, pp. 1,517-18; Pyle 1966, pp. iii, 19-53.

427.77 22° 49′ N, 92° 21′ W. 2,632 m. V3-74. *reworked upper Tertiary (Miocene).* Ewing *et al.* 1958, p. 1,037.

427.87 22° 42′ N, 91° 28′ W. 3,610 m. V3-129. *reworked upper Tertiary.* Ewing *et al.* 1958, p. 1037.

427.92 27° 25′ N, 90° 28′ W. 1,230 m. V3-48. *reworked upper Tertiary.* Ewing *et al.* 1958, p. 1044.

428.36 23° 29′ N, 86° 18′ W. 1,847 m. V3-137. Campeche Bank Canyon. *Miocene.* L.G.O. (unpublished).

428.36 23° 11′ N, 86° 15′ W. 1,691 m. A185-25. Campeche Bank. *upper Tertiary (upper Miocene or Pliocene).* L.G.O. (unpublished).

428.48 21° 32′ N, 85° 04′ W. 2,195 m. A185-23. *Miocene.* L.G.O. (unpublished).

428.53 26° 05′ N, 84° 52′ W. 1,582 m. V3-21. *middle Tertiary.* L.G.O. (unpublished).

428.89 20° 16′ N, 81° 29′ W. 2,802 m. RC10-33. *Oligocene-Pliocene.* L.G.O. (unpublished).

429.15 24° 53′ N, 78° 02′ W. Andros Island, Bahamas. *Lower Cretaceous to Quaternary (commencing at approx. −4450 m. below sea-level).* Eardley 1951, p. 573; Newell 1955, pp. 303, 304; Newell and Rigby 1957, pp. 63-5; Wilson 1963, p. 536; Funnell 1964, pp. 423-4, table 1.

429-21 28° 36′ N, 77° 10′ W. 1,005 m. A156-1. Blake Plateau. *upper Miocene and Pliocene.* Ericson *et al.* 1952, pp. 497, 504; Ericson *et al.* 1961; pp. 207, 235.

429.21 28° 30′ N, 77° 31′ W. 1,032 m. JOIDES 3. Blake Plateau *lower Eocene (at −1,193 m. to −1,210 m below sea-level), middle Eocene, Oligocene, Miocene and post-Miocene.* Schlee and Gerard 1965, pp. 53-5; Bunce *et al.* 1965, pp. 709-16.

429.21 28° 24′ N, 77° 56′ W. 969 m. V3-153. Blake Plateau. *upper Miocene or Pliocene overlain by Quaternary.* Ericson *et al.* 1963, pp. 728-34; Parker 1967, p. 123.

429.21 28° 10′ N, 77° 37′ W. 1,005 m. V3-152. Blake Plateau *upper Miocene or Pliocene and Quaternary.* Ericson *et al.* 1963, pp. 728-34.

429.22 27° 58′ N, 77° 23′ W. 1,130 m. V3-151. Blake Plateau *(?) Miocene and Pliocene, and Quaternary.* Ericson *et al.* 1963, pp. 728-32.

429.22 27° 08′ N, 77° 21′ W. 1,812 m. V15-195. *Eocene.* L.G.O. (unpublished).

429.24 25° 39′ N, 77° 21′ W. 2,560 m. A167-44. *upper Tertiary (upper Miocene or Pliocene).* Ericson *et al.* 1952, pp. 497, 505; Ericson *et al.* 1961, pp. 209, 237.

429.24 25° 27′ N, 77° 03′ W. 2,606 m. A167-43. *upper Tertiary (upper Miocene or Pliocene), and Quaternary.*

Ericson *et al.* 1952, pp. 497, 505; Ericson *et al.* 1961, pp. 209, 237; Ericson *et al.* 1963, pp. 728-31.

429.24 25° 23′ N, 77° 24′ W. 3,383 m. A167-51. mouth of Tongue of the Ocean. *Quaternary, with Upper Cretaceous* admixture. Ericson *et al.* 1952, pp. 497, 505; Heezen, Tharp and Ewing 1959, p. 47; Ericson *et al.* 1961, p. 209; Funnell 1964, table 1.

429.25 24° 36′ N, 77° 34′ W. 1,737 m. A167-49. *late Tertiary.* Ericson *et al.* 1952, pp. 497, 505.

429.30 29° 50′ N, 76° 28′ W. 2,213 m. A167-22. *lower Miocene (C. dissimilis* to *C. stainforthi* zone). Emiliani 1956*b*, pp. 281-8.

429.30 29° 49′ N, 76° 35′ W. 1,454 m. A167-21. *upper Eocene.* Ericson *et al.* 1952, pp. 497, 504; Ericson *et al.* 1961, pp. 209, 236; Saito and Be 1964, pp. 702-3; Heezen and Sheridan 1966, p. 1646.

429.30 29° 12′ N, 76° 49′ W. 2,140 m. A156-2. *upper Oligocene (G. ciperoensis* zone). ?Northrop and Heezen 1951, pp. 396-7; Ericson *et al.* 1952, pp. 497, 504; Ericson *et al.* 1961, pp. 207, 235-6; Heezen and Sheridan 1966, p. 1646.

429.31 28° 52′ N, 76° 47′ W. 1,747 m. A167-25. *Upper Cretaceous (Cenomanian).* Ericson *et al.* 1952, pp. 497, 500, 504; Heezen, Tharp and Ewing 1959, pp. 47, 49; Ericson *et al.* 1961, p. 236; Loeblich and Tappan 1961, pp. 263-301; Todd and Low 1964, pp. 397-8; Funnell 1964, p. 424, table 1; Heezen and Sheridan 1966, p. 1,646.

429.31 28° 42′ N, 76° 46′ W. 1,262 m. A167-28. Blake Plateau *upper Tertiary (Pliocene).* Ericson *et al.* 1961, pp. 209, 236; Heezen and Sheridan 1966, p. 1,646.

429.31 28° 26′ N, 76° 40′ W. 1,728 m. A167-29. *upper Tertiary (upper Miocene or Pliocene).* Ericson *et al.* 1961, pp. 209, 236-7.

429-31 28° 59′ N, 76° 45′ W. 3,200 m. E-9-66-3. (Blake escarpment) *Lower Cretaceous (Neocomian-Aptian).* Heezen and Sheridan 1966, p. 1,645.

429.31 28° 57′ N, 76° 45′ W. 2,375 m. E-9-66-7 (Blake escarpment) *Lower Cretaceous (Aptian-Albian).* Heezen and Sheridan 1966, p. 1,645.

429.31 28° 59′ N, 76° 43′ W. 3,130 m. E-9-66-4 (Blake escarpment). *Lower Cretaceous (?Aptian)* Heezen and Sheridan 1966, p. 1,645.

429.32 27° 10′ N, 76° 24′ W. 5,490 m. V15-203. *upper Tertiary.* L.G.O. (unpublished).

429.34 25° 39′ N, 76° 56′ W. 3,110 m. A167-41. *upper Tertiary (upper Miocene or Pliocene).* Ericson *et al.* 1952, pp. 497, 505; Ericson *et al.* 1961, pp. 209, 237.

429.45 24° 02′ N, 75° 22′ W. 1,719 m. A179-12. Cat Island, Exuma Sound. *upper Tertiary (probably upper Miocene).* Ericson *et al.* 1961, pp. 219, 237.

429.46 23° 56′ N, 75° 25′ W. 2,177m. A 179-11 *late Tertiary or early Pleistocene.* L.G.O. (unpublished).

429.55 24° 52′ N, 74° 01′ W. 5,239 m. V22-8. *Cretaceous (upper Cenomanian).* Saito, Burckle and Ewing 1966, p. 1,174, etc.

429.65 24° 43′ N, 73° 46′ W. 5,130 m. V22-10. *lower Miocene, with Cretaceous and Eocene admixture.* Saito, Burckle and ewing 1966, p. 1,174, etc.

429.65 24° 03′ N, 73° 33′ W. 5,158 m.V22-11. *middle to upper Miocene, with Cretaceous (Maastrichtian) and Eocene admixture.* Saito, Burckle and Ewing 1966, p. 1,174, etc.

429.65 24° 44′ N, 73° 23′ W. 5,187 m. V22-16. *Cretaceous (Maastrichtian).* Saito, Burckle and Ewing 1966, p. 1,174, etc.

429.65 24° 45′ N, 73° 10′ W. 5,244 m. V22-12. *Cretaceous (Maastrichtian).* Saito, Burckle and Ewing 1966, p. 1,174, etc.

429.65 24° 21′ N, 73° 01′ W. 2,564 m. V21-243. *lower Miocene, with Cretaceous and Eocene admixture.* Saito, Burckle and Ewing 1966, p. 1,174, etc.

429.66 23° 40′ N, 73° 51′ W. 4,782 m. V21-229. *middle to upper Miocene, with Cretaceous and Eocene admixture.* Saito, Burckle and Ewing 1966, p. 1,174, etc.

429.67 22° N, 73° W. 700-800 m. GG1 to GG7. Gerda Guyot. *upper Miocene and Plio-Pleistocene.* Milliman 1966, p. 995.

429.69 20° 31′ N, 73° 00′ W. 3,419 m. A185-6. Great Inagua Island. *Miocene.* Ericson *et al.* 1961, pp. 212, 237.

429.74 25° 12′ N, 72° 51′ W. 5,278 m. V21-239. *Cretaceous (Maastrichtian).* Saito, Burckle and Ewing 1966, p. 1,174, etc.

429.74 25° 15′ N, 72° 50′ W. 5,278 m. V21-241. *Cretaceous (Maastrichtian).* Saito, Burckle and Ewing 1966, p. 1,174, etc.

429.74 25° 14′ N, 72° 48′ W. 5,286 m. V21-238. *Cretaceous (Maastrichtian).* Saito, Burckle and Ewing 1966, p. 1,174, etc.

429.74 25° 15′ N, 72° 47′ W. 5,282 m. V21-236. *Cretaceous (Maastrichtian).* Saito, Burckle and Ewing 1966, p. 1,174, etc.

429.74 25° 16′ N, 72° 42′ W. 5,276 m. V21-237. *Cretaceous (Maastrichtian).* Saito, Burckle and Ewing 1966, p. 1,174, etc.

429.78 21° 07′ N, 72° 51′ W. 2,469 m. A185-7. Great Inagua Island. *Miocene.* Ericson *et al.* 1961, pp. 212, 238.

429.87 22° 27′ N, 71° 36′ W. 5,024 m. RC10-27. *Eocene.* L.G.O. (unpublished).

430.49 20° 17′ N, 65° 42′ W. 5,852 m. CH-34-D3-195, 224, 225, 226, 227. *Upper Cretaceous (Cenomanian).* Todd and Low 1964, pp. 395-400.

431.33 26° 35′ N, 56° 29′ W. 5,104 m. RC5-12. *Upper Cretaceous (Maastrichtian).* Saito, Ewing and Burckle 1966, pp. 1,076-7.

431.52 27° 52′ N, 54° 38′ W. 4,680 m. V10-96. *Eocene.* Saito, Ewing and Burckle 1966, p. 1,077.

431.65 24° 17′ N, 53° 04′ W. 5,009 m. V12-4. *Middle Eocene.* Saito, Ewing and Burckle 1966, p. 1,077.

431.80 30° 00′ N, 51° 52′ W. 4,673 m. V16-209. *lower Eocene.* Saito, Ewing and Burckle 1966, p. 1,077.

431.80 29° 02′ N, 51° 02′ W. 4,850. A150-24. *Pliocene.* Saito Ewing and Burckle 1966, p. 1,076.

432.02 27° 44′ N, 49° 55′ W. 4,861 m. V16-208. *Pliocene.* Saito, Ewing and Burckle 1966, p. 1,076.

432.15 24° 56′ N, 48° 59′ W. 4,260 m. V10-94. *Pliocene.* Saito, Ewing and Burckle 1966, p. 1,076.

432.36 23° 23′ N, 46° 24′ W. 3,540 m. V10-91. *Pliocene.* Saito, Ewing and Burckle 1966, p. 1,076.

432.36 23° 20′ N, 46° 29′ W. 3,733 m. V16-206. *upper Miocene.* Saito, Ewing and Burckle 1966, p. 1,076.

432.37 22° 56′ N, 46° 35′ W. approx. 3,125 m (2,893 to 3,586 m). Chain 17 (dredge). *Miocene.* Cifelli 1965, pp. 73-87; van Andel *et al.* 1965, p. 1,215.

432.37 approx. 22° 20′ N, 46° 20′ W. approx. 3,600 m. CH-44-DR-10. *late Tertiary.* van Andel *et al.* 1965, p. 1,215.

432.37 approx. 22° 20′ N, 46° 20′ W. 3,000 m. CH-44–PC-24. *late Tertiary.* van Andel *et al.* 1965, p, 1,215.

432.48 21° 12′ N, 45° 21′ W. 3,003 m. V12-5. *Pliocene and Quaternary.* Ericson *et al.* 1963, pp. 728-36; Saito, Ewing and Burckle 1966, p. 1,076.

432.66 23° 22′ N, 43° 39′ W. 4,565 m. V20-242. *Pliocene.* Saito, Ewing and Burckle 1966, p. 1,076.

432.66 23° 05′ N, 43° 40′ W. 3,525 m. V10-89. *Pliocene.* Saito, Ewing and Burckle 1966, p. 1,076.

432.87 22° 08′ N, 41° 30′ W. 4,372 m. V20-241. *Pliocene.* Saito, Ewing and Burckle 1966, p. 1,076.

432.87 22° 08′ N, 41° 30′ W. 4,372 m. V20-241. *Pliocene.*

Saito, Ewing and Burckle 1966, p. 1,076.

433.13 26° 22′ N, 38° 50′ W. 4,715 m. V19-397. *Pliocene.* Saito, Ewing and Burckle 1966, p. 1,076.

433.522 27° 58′ N, 34° 08′ W. 5,132 m. V17-163. *?Pliocene.* Saito, Ewing and Burckle 1966, p. 1,076.

434.10 approx. 30° 00′ N, 28° 30′ W. approx. 290 m. and 400 to 650 m. Great Meteor Seamount. *Miocene.* Pratt 1963*a*, pp. 17-19; Milliman 1966, p. 995.

434.15 24° 58′ N, 28° 56′ W. 5,480 m. V17-162. *?Pliocene.* Saito, Ewing and Burckle 1966, p. 1,076.

434.30 29° 07′ N, 26° 15′ W. 5,029 m. A180-32. *Pliocene.* Saito, Ewing and Burckle 1966, p. 1,076.

434.41 28° 56′ N, 25° 26′ W. 4,589 to 4,707 m. D4816. *Eocene.* Funnell (unpublished).

434.93 26° 01′ N, 20° 22′ W. 2,217 to 2,356 m. D4799. *middle Eocene.* Funnell (unpublished).

435.21 approx. 28° 40′ N, 17° 50′ W above sea-level. La Palma Island, Canary Islands. *Oligocene and Miocene.* Daly 1927, p. 87; Hausen 1956; Wilson 1963, p. 536.

464.70 19° 51′ N, 82° 00′ W. 2,158 m. A185-19. Grand Cayman Island. *Miocene.* Ericson *et al.* 1961, pp. 212, 238; Martini and Bramlette 1963, p. 848.

465.00 19° 25′ N, 79° 48′ W. 2,981 m. A185-16. Cayman Brac Island. *upper Oligocene and/or Miocene.* Ericson *et al.* 1961, pp. 212, 238.

465.08 11° 17′ N, 79° 14′ W. 3,395 m. V8-18. Grand Cayman. *lower Tertiary.* L.G.O. (unpublished).

465.60 19° 55′ N, 73° 50′ W. 2,926 m. A179-7. Windward Passage, between Cuba and Hispaniola. *upper Tertiary (upper Miocene or Pliocene).* L.G.O. (unpublished).

465.95 14° 42′ N, 70° 53′ W. 3,609 m. RC9-55. *Eocene.* L.G.O. (unpublished).

465.95 14° 38′ N, 70° 52′ W. 3,404 m. RC9-56. *Eocene.* L.G.O. (unpublished).

465.95 14° 38′ N, 70° 50′ W. 3,760 m. RC8-109. *Eocene.* L.G.O. (unpublished).

465.95 14° 37′ N, 70° 50′ W. 3,689 m. RC9-59. *Eocene.* L.G.O. (unpublished).

465.95 14° 33′ N, 70° 49′ W. 3,548 m. RC9-58. *Eocene.* L.G.O. (unpublished).

466.21 18° 51′ N, 67° 07′ W. 2,597 m. V3-3. *Pliocene.* Ericson *et al.* 1961, pp. 214, 238.

466.21 18° 49′ N, 67° 09′ W. 1,829 m. V3-2. *upper Miocene.* Ericson *et al.* 1961, pp. 214, 238.

466.40 19° 24′ N, 65° 07′ W. 6,401 m. A172-13. Virgin Islands. *Miocene.* Ericson *et al.* 1952, pp. 492, 509; Ericson *et al.* 1961, pp. 209, 237.

466.42 17° 47′ N, 65° 14′ W. 2,105 m. Swedish Deep-Sea Expedition 267. *reworked Miocene.* Phleger, Parker and Pierson 1953, pp. 98-9, 111.

467.06 13° 12′ N, 59° 30′ W above sea-level. Barbados. *Middle Eocene-Oligocene.* Beckman 1954, p. 301; Bramlette 1955, p. 1,534; Riedel 1959*a*, pp. 287-8, etc.; Eames *et al.* 1962, p. 44; Riedel and Funnell 1964, pp. 308-9.

468.05 14° 42′ N, 49° 33′ W. A153 CC163. *Pliocene.* L.G.O. (unpublished).

468.12 17° 16′ N, 48° 25′ W. 3,795 m. V16-21. *Pliocene and Quaternary.* Ericson *et al.* 1963, pp. 728-34; Saito, Ewing and Burckle 1966, p. 1,076.

468.18 11° 12′ N, 48° 05′ W. 4,614 m. RC8-2. *Middle Eocene.* Saito, Ewing and Burckle 1966, p. 1,077, table 1, fig. 4.

468.45 14° 10′ N, 45° 44′ W. 3,623 m. V9-32. *Pliocene.* Saito, Ewing and Burckle 1966, p. 1,076.

468.64 15° 24′ N, 43° 24′ W. 4,043 m. V16-205. *Pliocene.* Saito, Ewing and Burckle, 1966, p. 1,076.

468.86 13° 05′ N, 41° 02′ W. 4,875 m. B6.1. *Pliocene.* Funnell (unpublished).

468.94 15° 29′ N, 40° 31′ W. 4,473 m. V14-4. *Pliocene.* Saito, Ewing and Burckle 1966, p. 1,076.

468.96. 13° 15′ N, 40° 40′ W. 4,887 m. V16-23. *Pliocene.* Saito, Ewing and Burckle 1966, p. 1,076.

469.23 16° 00′ N, 37° 10′ W. 5,095 m. Cll. *Pliocene.* Funnell (unpublished).

469.33 16° 28′ N, 36° 19′ W. 5,233 m. V20-238. *Pliocene.* Saito, Ewing and Burckle 1966, p. 1,076.

470.53 approx. 16° 37′ N, 24° 20′ W. approx. + 250 m. Maio and Sao Nicolau Islands, Cape Verde Islands. *Jurassic, and Upper Cretaceous.* Stahlecker 1935, pp. 265-301; Furon 1935, pp. 270-4; Sousa Torres and Pires Soares 1946, p. 204; Pires Soares 1948, pp. 383-4; Part 1950, pp. 30-5; Colom 1955, p. 113; Arkell 1956, p. 602; Furon 1963, p. 1,823; Wilson 1963, p. 536; Funnell 1964, p. 429, table 1.

471.25 14° 57′ N, 17° 17′ W. 680 m. KM1-55. *Lower Tertiary* L.G.O. (unpublished).

533.56 03° 47′ N, 34° 47′ W. 4,675 m. V9-29. *lower Miocene* Saito, Ewing and Burckle 1966, p. 1,076.

534.72 07° 13′ N, 22° 38′ W. 4,125 m. Swedish Deep-Sea Expedition 233. *Miocene.* Phleger *et al.* 1953, pp. 81-3, 111.

534.84 05° 45′ N, 21° 43′ W. 3,577 m. Swedish Deep-Sea Expedition 234. *Miocene.* Phleger, Parker and Peirson 1953, p. 83-6, 110-11; Emiliani 1956*b*, pp. 281-8; Ericson *et al.* 1961, pp. 228-32.

569.59 09° 45′ S, 34° 24′ W. 3,380 m. V15-164. *upper Miocene or Pliocene and Quaternary.* Ericson, Ewing and Wollin 1963, pp. 728-33.

571.10 00° 07′ S, 18° 12′ W. 7,315 m. Swedish Deep-Sea Expedition 238. *Miocene.* Phleger, Parker and Peirson 1953, pp. 87-8, 110-11.

571.50 00° 15′ S, 14° 25′ W. 4,115 m. *'Challenger'* 347. *Eocene* Murray and Renard 1891, pp. 146-7; Riedel (unpublished see Riedel 1957*a*, p. 95).

633.39 19° 39′ S, 36° 04′ W. 4,023 m. A180-107. Abrolhos Bank. *(?)Pliocene.* L.G.O. (unpublished).

635.47 17° 39′ S, 15° 06′ W. 3,892 m. V16-35. *Pliocene.* Saito, Ewing and Burckle 1966, p. 1,076.

637.02 22° 06′ S, 00° 19′ E. 5,349 m. V20-207. *Pliocene.* Saito, Ewing and Burckle 1966, p. 1,076.

637.54 24° 07′ S, 05° 45′ E. 1,600 m. V19-250. *upper Miocene.* Saito, Ewing and Burckle 1966, p. 1,077.

637.65 25° 27′ S, 06° 28′ E. 1,626 m. V20-205. *upper Miocene.* Saito, Ewing and Burckle 1966, p. 1,077.

637.82 22° 59′ S, 08° 07′ E. 4,118 m. V12-65. *Upper Cretaceous (Maastrichtian), overlain by upper Tertiary.* Heezen and Menard 1963, p. 265; Ewing, Ewing and Talwani 1964, p. 21; Saito, Ewing and Burckle 1966, p. 1,078.

669.39 29° 52′ S, 36° 48′ W. 2,280 m. V12-19. *Miocene.* Saito, Ewing and Burckle 1966, p. 1,077.

670.08 28° 36′ S, 29° 01′ W. 3,601 m. V20-220. *upper Eocene.* Saito, Ewing and Burckle 1966, p. 1,078.

670.09 29° 28′ S, 29° 13′ W. 3,092 m. V20-219. *upper Miocene.* Saito, Ewing and Burckle 1966, p. 1,077.

671.52 21° 15′ S, 14° 02′ W. 3,639 m.*Challenger* 338. *Quaternary, with Miocene admixture.* Murray and Renard 1891, pp. 144-5, pl. 11 (fig. 4); Bramlette and Riedel 1954, pp. 389, 401-2; Black 1956, p. 126; Bramlette 1957, p. 248; Black and Barnes 1961, p. 138.

672.32 22° 59′ S, 06° 46′ W. 4,925 m. V16-38. *Pliocene.* Saito, Ewing and Burckle 1966, p. 1,076.

672.66 26° 16′ S, 03° 01′ W. 4,790 m. V16-40. *Pliocene.* Saito, Ewing and Burckle 1966, p. 1,076.

701.21 31° 27′ S, 02° 47′ E. 1,528 m. LSDA 162D. *Miocene.* Funnell (unpublished).

732.91 31° 28′ S, 40° 05′ W. 3,641 m. V16-187. *Pliocene.* Saito, Ewing and Burckle 1966, p. 1,077.

767.07 47° 29′ S, 59° 21′ W. 1,162 m. V12-46. Falkland Islands. *Eocene.* L.G.O. (unpublished).

767.22 42° 32′ S, 56° 29′ W. 3,731 m. *Challenger* 318. *Upper Cretaceous (?)-lower Tertiary.* Murray and Renard 1891, pp. 136-7; Williams (unpublished).
832.70 50° 48′ S, 42° 10′ W. V14-47. *Miocene.* L.G.O. (unpublished).
838.55 65° 43′ S, 16° 25′ E. 5,058 m. *'Quest'* D2.18. *Eocene* Williams (unpublished).

Indian Ocean

442.15 14° 22′ N, 51° 56′ E. 4,330-3,760 m. D6222. (Gulf of Aden). *(?)Pliocene.* Ramsay and Funnell (unpublished).
442.15 14° 20′ N, 51° 59′ E. 4,520-3,950 m. D6221. (Gulf of Aden). *Pliocene (N20).* Ramsay and Funnell (unpublished).
442.16 13° 56′ N, 51° 38′ E. 2,700-2,250 m. D6211. (Gulf of Aden). *Miocene (N17).* Ramsay and Funnell (unpublished).
442.16 13° 56′ N, 51° 39′ E. 4,100-3,000 m. D6210. (Gulf of Aden). *Pliocene (N19).* Ramsay and Funnell (unpublished).
442.16 13° 53′ N, 51° 45′ E. 4,800-4,140 m. D6216. (Gulf of Aden). *Miocene (N17).* Ransay and Funnell (unpublished).
442.16 13° 50′ N, 51° 48′ E. 3,950-3,000 m. D6213 (Gulf of Aden). *Miocene (N17).* Ramsay and Funnell (unpublished).
507.12 07° 26′ N, 61° 04′ E. 3,471 m. V19-184. *Miocene.* Lamont Geological Observatory (unpublished).
507.41 08° 16′ N, 64° 05′ E. 4,173 m. V19-182. *Miocene.* L.G.O. (unpublished).
508.02 07° 27′ N, 70° 39′ E. 4,110 m. LSDH 3G. *Miocene with Eocene admixture.* Riedel 1964, pp. 3, 5; Riedel *et al.* (in preparation).
542.09 09° 26′ S, 50° 57′ E. 1,362 m. *Sealark* (1905). *Eocene-Oligocene (?Tertiary 'e').* Cole 1959, p. 11; Wiseman and Riedel 1960, p. 215; Wilson 1963, p. 536.
542.95 05° 26′ S, 59° 15′ E. 4,010 m. LSDA 107 Ga. *Quaternary with some Eocene to Miocene admixture.* Riedel 1964, pp. 2-3, 5-6.
542.95 05° 26′ S, 59° 15′ E. 3,930 m. LSDA 107Gb. *Quaternary with Eocene to Miocene admixture.* Riedel 1964, pp. 3, 5-6.
543.21 01° 22′ S, 62° 38′ E. 4,750 m. Swedish Deep-Sea Expedition 149. *Tertiary.* Riedel 1951, p. 737; Olausson 1961*a*, p. 14.
543.35 05° 34′ S, 63° 43′ E. 4,090 m. LSDA 106G. *Quaternary with Eocene and (?) Oligocene admixture.* Riedel 1964, pp. 2, 5-6.
543.41 01° 20′ S, 64° 00′ E. 4,405 m. SDSE 148. *Tertiary.* Olausson 1961*a*, p. 14.
543.48 08° 45′ S, 64° 52′ E. 4,220 m. *Planet* station 127. *Eocene.* Wiseman and Riedel 1960, pp. 215-17; Riedel 1964, pp. 5-6.
544.32 02° 41′ S, 73° 12′ E. 2,960 m. LSDA 101G. *Pliocene.* Riedel 1964, pp. 2, 6; Parker 1967, p. 130, etc.; Riedel *et al.* (in preparation).
605.19 19° 55′ S, 41° 36′ E. 2,924 m. V19-209. *Middle Miocene.* L.G.O. (unpublished).
607.28 18° 21′ S, 62° 04′ E. 3,398 m. DODO 117 PG and P. *Pliocene.* Parker 1967, p. 130, etc.; Riedel *et al.* (in preparation).
607.30 10° 25′ S, 63° 15′ E. 3,115 m. DODO 123D. *Eocene.* Friend and Riedel 1967, p. 219; Riedel *et al.* (in preparation).
607.46 16° 58′ S, 64° 42′ E. 4,044 m. MSN 54V. *Quaternary with late Tertiary admixture.* Riedel 1964, pp. 4-5.
607.66 16° 25′ S, 66° 02′ E. 3,660 m. MSN 53G. *Quaternary with late Tertiary admixture.* Riedel 1964, pp. 4-5.
609.09 19° 29′ S, 80° 59′ E. 4,960 m. DODO 108 PG and P. *Upper Eocene.* Riedel *et al.* (in preparation).
609.19 19° 13′ S, 81° 25′ E. 4,815 m. DODO 107P. *Eocene.* Riedel *et al.* (in preparation).
609.39 19° 21′ S, 83° 25′ E. 5,060 m. DODO 105 PG and P. *Eocene.* Riedel *et al.* (in preparation).
609.55 15° 30′ S, 85° 19′ E. 4,850 m. DODO 83P. *Eocene.* Riedel *et al.* (in preparation).
609.55 15° 32′ S, 85° 04′ E. 4,755 m. DODO 86PG and P. *Lower Eocene.* Riedel *et al.* (in preparation).
609.66 16° 13′ S, 86° 06′ E. 5,127 m. DODO 87PG and P. *Eocene.* Riedel *et al.* (in preparation).
609.66 16° 17′ S. 86° 11′ E. 5,140 m. DODO 98PG and P. *Pliocene.* Riedel *et al.* (in preparation).
609.87 17° 59′ S, 88° 59′ E. 5,090 m. DODO 102P. *middle Eocene.* Riedel *et al.* (in preparation).
610.13 13° 52′ S, 91° 02′ E. 5,480 m DODO 78PG and P. *Eocene.* Riedel *et al.* (in preparation).
610.33 13° 09′ S, 93° 13′ E. 5,226 m. LSDA 144G. *Pliocene.* Riedel 1964, pp. 3, 6; Riedel *et al.* (in preparation).
611.11 11° 41′ S, 101° 40′ E. 4,964 m. V19-154. *Upper Cretaceous (Maastrichtian).* L.G.O. (unpublished).
611.12 12° 24′ S, 101° 32′ E. 4,731 m. V19-155. *Upper Cretaceous (Maastrichtian).* L.G.O. (unpublished).
611.50 10° 30′ S, 105° 35′ E above sea-level. Christmas Island. *Eocene and Miocene.* Nuttall 1926, pp. 23-5; Wilson 1963, p. 536.
612.25 15° 40′ S, 112° 44′ E. 3,660 m. DODO 57P. *Pliocene.* Parker 1967, p. 130, etc.; Riedel *et al.* (in preparation).
641.18 28° 34′ S, 41° 58′ E. 4,603 m. V14-80. *middle Cretaceous (Cenomanian).* L.G.O. (unpublished).
643.19 29° 54′ S, 61° 53′ E. 4,400 m. LSDA 122G. *late Tertiary (upper Miocene).* Riedel 1964, pp. 3, 6; Riedel *et al.* (in preparation).
644.04 24° 04′ S, 70° 07′-70° 12′ E. 3,700 m. DODO 115D. *Miocene.* Riedel *et al.* (in preparation).
644.33 23° 56′ S, 73° 53′ E. 3,700 m. MSN 56PG and P. *late Tertiary (Pliocene).* Riedel 1964, pp. 4-6; Parker 1967, p. 131, etc.; Riedel *et al.* (in preparation).
644.34 24° 42′ S, 73° 05′ E. 3,600 m. DODO 141G. *Pliocene.* Parker 1967, p. 130, etc.; Riedel *et al.* (in preparation).
644.71 21° 59′ S, 77° 22′ E. 4,700 m. DODO 111PG and P. *Oligocene.* Riedel *et al.* (in preparation).
645.50 20° 40′ S, 85° 29′ E. 4,722 m. *Egeria* sounding no. 10. *Eocene.* Wiseman and Riedel 1960, pp. 215-17; Riedel 1964, pp. 4-6.
645.55 25° 29′ S, 85° 09′ E. 4,546 m. V18-206. *lower Miocene.* L.G.O. (unpublished).
703.35 35° 45′ S, 23° 52′ E. 2,188 m. V19-226. *middle Eocene and upper Miocene.* L.G.O. (unpublished).
703.68 38° 30′ S, 26° 05′ E. 2,922 m. RC8-35. *middle Eocene.* L.G.O. (unpublished).
704.38 38° 20′ S, 33° 55′ E. 3,680 m. V18-187. *Palaeocene.* L.G.O. (unpublished).
704.43 33° 22′ S, 34° 24′ E. 2,005 m. V19-222. *Quaternary with Miocene admixture.* L.G.O. (unpublished).
704.72. 32° 32′ S, 37° 09′ E. 5,114 m. V19-221. *Mixture of Cretaceous, Eocene and Miocene.* L.G.O. (unpublished).
708.30 30° 50′ S, 73° 12′ E. 4,160 m. MSN 59G. *late Tertiary (Pliocene).* Riedel 1964, pp. 4-5.
712.23 33° 40′ S, 112° 40′ E. 3,063 m. RC8-56. *Upper Cretaceous (Turonian-Campanian).* L.G.O. (unpublished).
712.40 30° 48′ S, 114° 23′ E. 2,543 m. RC9-151. *Upper Cretaceous Miocene.* L.G.O. (unpublished).
739.50 40° 14′ S, 25° 15′ E. 2,772 m. V16-55. *Palaeocene.* Herman 1963, p. 1,316.
739.61 41° 21′ S, 26° 38′ E. 2,961 m. V16-56. *middle Cretaceous (Cenomanian).* Herman 1963, pp. 1,316-17.
739.95 45° 14′ S, 29° 29′ E. 5,289 m. V16-57. *Pliocene.* Hays 1965, pp. 154, 164; Opdyke *et al.* 1966, pp. 350, 354.

740.16 46° 30′ S, 31° 16′ E. 4,731 m. V16-58. *(?)Pliocene.* Hays 1965, pp. 160, 164.

740.69 50° 00′ S, 36° 46′ E. 4,581 m. V16-60. *Pliocene.* Hays 1965, pp. 151, 164; Opdyke *et al.* 1966, p. 350.

741.46 46° 01′ S, 44° 22′ E. 2,202 m. V16-64. *upper Miocene.* L.G.O. (unpublished).

741.52 42° 39′ S, 45° 40′ E. 2,996 m. V16-66. *Pliocene.* Ericson, Ewing and Wollin 1963, pp. 728-36; Krinsley and Newman 1965, pp. 442-3; Hays 1965, pp. 158, 164; Opdyke *et al.* 1966, p. 350.

742.13 43° 38′ S, 51° 16′ E. 2,897 m. RC8-41. *upper Eocene.* L.G.O. (unpublished).

743.99 approx. 49° 30′ S, 69° 30′ E above sea-level. Kerguelen Island. *Miocene.* Aubert de la Rue 1956, p. 162; Wilson 1963, p. 536.

804.50 50° 03′ S, 35° 11′ E. 4,878 m. V16-59. *Pliocene.* Hays 1965, pp. 151, 164.

808.33 approx. 53° 07′ S, 73° 20′ E above sea-level. Heard Island. *Palaeocene.* Lambeth 1952; Fleming 1957, p. 13, 18; Wilson 1963, p. 536.

815.75 55° 06′ S, 147° 29′ E. 3,296 m. V16-116. *Pliocene.* Hays 1965, pp. 156, 164.

816.21 51° 08′ S, 152° 56′ E. 4,356 m. RC8-67. *Tertiary.* L.G.O. (unpublished).

816.53 53° 29′ S, 155° 37′ E, 4,303 m. RC8-69. *Miocene.* L.G.O. (unpublished).

319.65 44° 45′ N, 173° 02′ W. 5,070 m. CK 13. *late Miocene or Pliocene.* Riedel and Bramlette 1959, p. 106; Riedel and Funnell 1964, p. 323.

324.55 approx. 44° 30′ N, 125° 00′ W. 300 to 2,700 m. Station nos. 14-137 (off Oregon, U.S.A.). *middle Miocene to Pliocene.* Byrne, Fowler and Maloney 1966, p. 1,655.

352.78 31° 51′ N, 157° 20′ E. 3,500 m. V21-143. *Albian.* Ewing *et al.* 1966, pp. 751-5.

353.94 35° 22′ N, 169° 53′ E. 5,304 m. *Challenger* 244. *(?) reworked middle Tertiary.* Murray and Renard 1891, pp. 114-15; Riedel 1957*a*, p. 93.

355.63 36° 30 N, 173° 16′ W. 4,290 m. CK16. *late Miocene or Pliocene.* Riedel and Bramlette 1959, p. 106; Riedel and Funnell 1964. pp. 323, 361.

359.77 32° 51′ N, 132° 32′ W. 686 m. NEL 667. (Erben Guyot). *early Miocene (C. stainforthi zone).* Carsola & Dietz 1952, pp. 485, 487, 490-1, pl. 3 (fig. 2); Bandy 1952, pp. 18-20; Menard and Hamilton 1964, table 1; Wilson 1963, p. 536; Bandy 1963*b*, pp. 1,402-3.

361.09 30° 43′ N. 119° 50′ W. 3,900 m. FAN BG7. *Quaternary, with Miocene admixture.* Kanaya and Koizumi 1966, p. 106.

413.19 20° 58′ N, 121° 33′ E. 2,575 m. HPGD 22. *upper Tertiary (Miocene or lower Pliocene).* Funnell (unpublished).

414.14 approx. 26° 00′ N, 131° 15′ E. 1 to 430 m. KDZ deep well. (Kita-Daito-Zima-North Borodino Island). *upper Oligocene (393-430 m), early Miocene (102-393 m), Plio-Pleistocene (1-102 m).* Hanzawa 1938, pp. 384-90; Hanzawa 1940, pp. 755-65; Johnson 1961, p. 943, Wilson 1963, p. 536; Ladd, Tracey and Gross 1967, p. 1,092.

419.21 approx. 28° 15′ N, 177° 30′ W. Reef borehole, Midway Atoll. *early Miocene (Tertiary 'e') resting on volcanic clay at* −338 m. Ladd, Tracey and Gross 1967, p. 1,091.

419.21 approx. 28° 15′N, 177° 30′ W. Sand Island borehole, Midway Atoll. *upper Miocene (Tertiary 'g')* resting on basalt at −148 m. Ladd, Tracey and Gross 1967, p. 1,091.

421.18 approx. 21° 10′ N, 158° 10′ W. 500 to 520 m. *Argo* (2 September 1961). (10 km southwest of Honolulu, Oahu, Hawaiian Islands). *Miocene.* Menard, Allison and Durham 1962, pp. 896-7; Wilson 1963, p. 538.

425.21 28° 59′ N, 117° 30′ W. 3,566 m. EM6, EM7, EM8, EM9, EM10. *middle Miocene to Holocene.* Riedel *et al.* 1961, pp. 1,793-8; Martini and Bramlette 1963, pp. 845-56; Parker 1964, pp. 617-36; Friend and Riedel 1967, p. 219 etc.; Parker 1967, p. 125.

451.38 11° 24′ N, 143° 16′ E. 8,184 m. *Challenger* 225. *lower Miocene (C. stainforthi/G. insueta zone).* Murray and Renard 1891, pp. 108-9, 204-5; pl. 15, fig. 3, pl. 27, fig. 5; Wiseman and Hendey 1953, pp. 54-5; Riedel 1954, pp. 170-5, pl. 1; Riedel 1957*a*, pp. 78, 91, 93; Riedel 1959*a*, pp. 286, 287, 288, etc.

451.46 approx. 13° 31′ N, 144° 40′ E above sea level. Guam, Mariana Islands. *upper Eocene (Tertiary ('b') to early Miocene (Tertiary 'f').* Cloud 1956, p. 555; Cole, Todd and Johnson 1960, p. 79, 87; Menard and Hamilton 1964, table 1.

451.54 approx. 15° 12′ N, 145° 45′ E above sea-level. Saipan, Mariana Islands. *upper Eocene (Tertiary 'b') to early Miocene (G. insueta zone).* Bramlette 1957, p. 247; Riedel 1957*b*, p. 257; Todd 1957, p. 265; Cole 1957, pp. 321-7; Cole 1960, pp. 1-3; Cole, Todd and Johnson 1960, p. 85; Cloud, Schmidt and Burke 1956, p. 123; Todd *et al.* 1954, pp. 678-82; Johnson 1961, p. 943; Menard and Hamilton 1964, table 1.

453.28 11° 40′ N, 162° 12′ E. Fl. Elugelab Island, Eniwetok Atoll. *upper Eocene (930-1,400 m), (?)Oligocene (820-930 m), Miocene (170-820 m), Plio-Pleistocene (0-170 m).* Cole 1957, pp. 743-50; Todd and Low 1960, pp. 799-815; Johnson 1961, p. 943; Wilson 1963, p. 536; Menard and Hamilton 1964, table 1; Ladd *et al.* 1967, p. 1,092.

453.28 11° 24′ N, 162° 22′ E. E1. Parry Islands, Eniwetok Atoll. *upper Eocene (845-1,280 m), (?)Oligocene (670-845 m), Miocene (170-670 m), Plio-Pleistocene (0-170 m), resting on basalt.* Cole 1957, pp. 743-50. Todd and Low 1960, pp. 799-815; Johnson 1961, p. 943; Wilson 1963, p. 536; Menard and Hamilton 1964, table 1; Ladd, Tracey and Gross 1967, p. 1,092.

453.29 10° 19′ N, 162° 06′ E. approx. 1,800 m. Eniwetok 27. *early Pliocene (N19).* Bramlette, Faughn and Hurley 1959, pp. 1,550-1; Todd 1964, pp. 1,067-71; Parker 1967, p. 124.

453.29 10° 19′ N, 162° 10′ E. approx. 1,550 m. Eniwetok 23. *early Pliocene (N19).* Bramlette, Faughn and Hurley 1959, pp. 1,550-1; Todd 1964, pp. 1,067-71; Parker 1967, p. 124.

453.29 10° 18′ N, 162° 06′ E. approx. 1,830 m. Eniwetok 4. *Pliocene (N21).* Bramlette, Faughn and Hurley 1959, pp. 1,550-1; Todd 1964, pp. 1,067-71; Parker 1967, p. 124.

453.29 10° 17′ N, 162° 10′ E. approx. 1,650 m. Eniwetok 18. *Pliocene (N20/N21).* Bramlette, Faughn and Hurley 1959, pp. 1,550-1; Todd 1964, pp. 1,067-71; Parker 1967, p. 124.

453.29 10° 16′ N, 162° 09′ E. approx. 1,650 m. Eniwetok 20. *Mio-Pliocene (N19).* Bramlette, Faughn and Hurley 1959, pp. 1,550-1; Todd 1964, pp. 1067-71; Parker 1967, p. 124.

453.47 12° 09′ N, 164° 51′ E. 1,481 to 1,884 m. MP43A. *lower Eocene.* Hamilton 1959, p. 1,419; Hamilton and Rex 1959, pp. 785-90; Wilson 1963, p. 536; Menard and Hamilton 1964, table 1.

453.58 11° 57′ N, 165° 08′ E. 1,500 to 2,103 m. MP43D *lower Miocene (G. insueta zone).* Hamilton and Rex 1959, pp. 785-9.

453.58 11° 49′ N, 165° 00′ E. 1,609 m. MP43DD. *upper Miocene or Pliocene.* Hamilton and Rex 1959, pp. 785-9.

453.58 11° 45′ N, 165° 10′ E. 1,317 m. Bikini 1176. *Tertiary.* Cushman, Todd and Post 1954, pp. 322, 327; Hamilton and Rex 1959, pp. 785 (fig. 255), 789-90.

453.58 approx. 11° 40′ N, 165° 15′ E. Bikini 2, 2A and 2B. Bikini Atoll. *Eocene or Oligocene (715-775m.) Miocene (185-715 m), Pliocene-Pleistocene (0-185 m)* Ladd, Tracey and Lill 1948, p. 51; Todd and Post 1954, pp. 547-53; Cole 1957, p. 743; Todd and Low 1960, pp. 799-802, pl. 264; Johnson 1961, p. 943; Wilson 1963, p. 536; Ladd, Tracey and Gross 1967, p. 1,092.

454.49 10° 54′ N, 174° 01′ E. 5,530 m. LSDH 83P. *Eocene.* Riedel 1963*b*, fig. 2.

455.22 17° 10′ N, 177° 10′ W. 1,929 to 2,048 m. MP37C just below break in slope on Cape Johnson Guyot. *middle Cretaceous and late Palaeocene.* Hamilton 1953, p. 213; Hamilton 1956, p. 17; Hamilton 1959, p. 1,409, pl. 1, fig. 2; Hamilton and Rex 1959, p. 787; Menard and Hamilton 1964, table 1; Wilson 1963, p. 536; Milliman 1966, p. 995.

455.22 17° 08′ N, 177° 16′ W. 1,796 m. MP37E. *Eocene to Quaternary.* Hamilton 1956, p. 17; Parker (unpublished).

455.22 17° 05′ N, 177° 13′ W. 1,812 m. MP37I. *Eocene-Quaternary Pliocene.* Parker (unpublished).

455.22 17° 04′ N, 177° 15′ W. 1,829-2.012 m. MP37A on slope of Cape Johnson Guyot. *middle Cretaceous.* Hamilton 1956, pp. 16, 20-2.

455.24 15° 32′ N, 177° 32′ W. 4,082 m. MP40-1. *Oligocene.* Bramlette 1958, p. 122; Olausson 1961*a*, p. 32; Kamptner 1963, p. 123; Martini 1965, p. 395.

455.52 17° 49′ N, 174° 17′ W. approx. 1,673 m. MP33C near top centre of Hess Guyot. *late Palaeocene.* Hamilton 1953, pp. 207, 209-14, 217; Bramlette and Riedel 1954, p. 396; Hamilton 1956, pp. 13, 18; Bolli 1957*c*, p. 170; Hamilton 1959, pp. 1,408-9; Milliman 1966, p. 995.

455.52 17° 48′ N, 174° 22′ W. 1,811 to 2,286 m. MP33K just below break in slope on southwest side of Hess Guyot. *Aptian to Cenomanian.* Hamilton 1956, pp. 16, 18; Wilson 1963, p. 536; Menard and Hamilton 1964, table 1; Milliman 1966, p. 995.

455.80 19° 34′ N, 171° 54′ W. 3,749 m. MP27 [MP27-2 and 2P]. *Cretaceous to Quaternary mixture.* Hamilton 1953, pp. 207-10, 214-18; Bramlette and Riedel 1954, p. 390; Hamilton 1956, pp. 6, 13-14.

455.80 19° 25′ N, 171° 00′ W. 1,317 to 1,408 m. MP26A-3. *lower-middle Eocene.* Hamilton 1953, pp. 207, 213-14; Hamilton 1956, p. 6; Hamilton 1959, pp. 1,408-9, pl. 1, fig. 1; Milliman 1966, p. 995.

456.00 19° 07′ N, 169° 44′ W. 1,710 to 1,756 m. MP25F-2. *mixture of early Tertiary to Quaternary.* Hamilton 1956, pp. 6, 9; Hamilton and Rex 1959, p. 787.

456.10 19° 40′ N, 168° 32′ W. 1,703 m. MP25E-1. *middle Eocene.* Hamilton 1953, pp. 207, 211-15, 217; Bramlette and Riedel 1954, p. 392; Hamilton 1956, p. 6; Bolli 1957*c*, pp. 169-70; Bramlette 1958, p. 122.

456.10 19° 40′ N, 168° 32′ W. 1,737 m. MP25E-2. *Eocene.* Hamilton 1953, pp. 207, 211-15, 217; Hamilton 1956, p. 6.

456.66 13° 05′ N, 163° 10′ W. 5,413 m. TET27A. *Eocene.* Riedel 1963, fig. 2; Riedel and Funnell 1964, pp. 349-50, 360.

456.66 13° 05′ N, 163° 10′ W. 5,430 m. TET27B. *Quaternary with Eocene admixture.* Riedel and Funnell 1964, p. 350.

456.67 12° 58′ N, 163° 09′ W. 5,430 m. TET28. *Quaternary with Eocene admixture.* Riedel and Funnell 1964, p. 350.

456.74 15° 11′ N, 162° 25′ W. 5,490 m. TET23. *Quaternary with Eocene admixture.* Riedel and Funnell 1964, p. 349.

456.78 11° 44′ N, 162° 41′ N, 162° 41′ W. 5,265 m. TET29. *Quaternary with Eocene admixture.* Riedel and Funnell 1964, pp. 350-1.

456.78 11° 32′ N, 162° 36′ W. 5,340 m. TET30. *Quaternary with Eocene admixture.* Riedel and Funnell 1964, p. 351.

456.79 10° 09′ N, 162° 05′ W. 5,063 m. TET32. *Quaternary with Eocene admixture.* Riedel and Funnell 1964, p. 351

456.85 14° 09′ N, 161° 08′ W. 5,652 m. MSN 7PG and P. *Quaternary with Eocene, (?)Oligocene and (?)Miocene admixture.* Riedel and Funnell 1964, p. 337.

456.92 17° 40′ N, 160° 39′ W. 5,393 m. TET18. *Quaternary with Eocene admixture.* Riedel and Funnell 1964, pp. 349, 364-5.

456.92 17° 36′ N, 160° 42′ W. 5,490 m. TET19. *Quaternary with Eocene admixture.* Riedel and Funnell 1964, pp. 349, 364-5.

457-458 17° 31′ N, 153° 36′ W-11° 07′ N, 152° 03′ W-13° 45′ N, 149° 15′ W. 4,440-5,660 m. *Eocene.* Riedel and Bramlette 1959, p. 106.

457.48 11° 27′ N, 155° 48′ W. 5,206 m. TET 53. *Eocene or Quaternary with Eocene admixture.* Riedel and Funnell 1964, p. 354.

457.49 10° 47′ N, 155° 54′ W. 5,234 m. TET 52. *Oligocene or early Miocene with Eocene admixture.* Riedel 1963, fig. 2; Riedel and Funnell 1964, pp. 353-4.

457.49 10° 39′ N, 155° 56′ W. 5,188 m. TET 51. *Oligocene or lower Miocene with Eocene admixture.* Riedel 1963, fig. 2; Riedel and Funnell 1964, pp. 353, 364.

457.77 12° 43′ N, 152° 06′ W. 5,549 m. JYN V 10P. *Eocene.* Riedel 1963, fig. 2.

457.77 12° 42′ N, 152° 01′ W. 5,304 m. *Challenger* 265. *reworked early Tertiary.* Murray and Renard 1891, pp. 120-1; Riedel 1957*a*, pp. 82, 89.

457.78 11° 07′ N, 152° 03′ W. 5,029 m. *Challenger* 266. *reworked early Tertiary.* Murray and Renard 1891, pp. 120-1; Riedel 1957*a*, pp. 82, 89, 94.

457.85 14° 41′ N, 151° 20′ W. 5,590 m. TET 11. *Eocene, or Quaternary with Eocene admixture.* Riedel and Funnell 1964, p. 349.

457.85 14° 38′ N, 151° 58′ W. 5,816 m. MP17-2. *Eocene, or Quaternary with Eocene admixture.* Riedel and Funnell 1964, p. 336.

458.26 13° 05′ N, 147° 19′ W. 5,455 m. JYN V 25P. *Eocene.* Riedel 1963, figs. 1, 2; Friend and Riedel 1967, p. 219.

458.29 10° 09′ N, 147° 33′ W. 5,241 m. JYN V 23G. *Eocene.* Riedel 1963, figs. 1, 2.

458.35 14° 13′ N, 146° 24′ W. 4,952 m. JYN V 28P. *Eocene.* Riedel 1963, figs. 1, 2; Friend and Riedel 1967, p. 219.

458.49 10° 44′ N, 145° 53′ W. 5,375 m. MP15-1. *Quaternary with Eocene admixture.* Riedel and Funnell 1964, pp. 335-66.

458.56 13° 15′ N, 144° 04′ W. 5,040 m. MSN 5G. *Eocene.* Riedel 1963, figs. 1, 2; Riedel and Funnell 1964, p. 336, 360, 364; Friend and Riedel 1967, p. 219.

458.58 11° 55′ N, 144° 54′ W. 5,539 m. JYN V 31P. *Eocene.* Riedel 1963, figs. 1, 2.

458.78 11° 03′ N, 142° 28′ W. 5,000 m. MSN 151P. *upper Oligocene; with some Eocene admixture.* Riedel 1963, figs. 1, 2; Riedel and Funnell 1964, pp. 346, 362-3.

458.78 11° 03′ N, 142° 28′ W. 5,000 m. MSN 151PG. *probably lower Miocene.* Riedel 1963, fig. 2; Riedel and Funnell 1964, pp. 345-6, 363-4; Martini 1965, p. 394; Friend and Riedel 1967, p. 219.

458.79 10° 59′ N, 142° 37′ W. 4,978 m. MSN 150G. *Eocene and Oligocene with Eocene admixture, or Quaternary with Eocene and Oligocene admixture.* Riedel and Funnell 1964, pp. 345, 360.

459.03 16° 55′ N, 139° 18′ W. 5,355 m. MSN 4G. *Quaternary with Eocene admixture.* Riedel and Funnell 1964, p. 336.

459.16 13° 07′ N, 138° 56′ W. 4,927 m. MSN 153PG and P. *upper Miocene with Eocene and Oligocene admixture.* Riedel and Funnell 1964, pp. 347-8, 361.

459.16 13° 04′ N, 138° 59′ W. 4,930 m. MSN 152G. *Eocene.* Riedel 1963, fig. 2; Riedel and Funnell 1964, pp. 346-7, 360; Friend and Riedel 1967, p. 219.

459.24 15° 01′ N, 137° 01′ W. 4,920 m. MSN 154PG and P. *late Oligocene.* Riedel 1963, fig. 2; Riedel and Funnell 1964, pp. 348, 361.

459.28 11° 07′ N, 137° 56′ W. 4,853 m. JYN V 44P. *middle Tertiary.* Riedel 1963, fig. 2.

459.64 15° 54′ N, 133° 57′ W. 4,606 m. JYN V 48PG and P. *Quaternary with Eocene and Oligocene admixture, overlying upper Eocene.* Riedel 1963, pp. 5, 7, fig. 2; Ross and Riedel 1967, p. 287: Riedel 1967, p. 542.

459.65 14° 55′ N, 133° 29′ W-134° 15′ W. 4,770 m. RIS 111P. *lower Oligocene.* Riedel 1963, fig. 2; Friend and Riedel 1967, p. 219.

459.65 14° 22′ N, 133° 07′ W. 4,800 m. MP5-1. *lower Oligocene.* Bramlette 1958, p. 122; Riedel 1959*a*, pp. 286, 287, 288, etc.; Olausson 1961*a*, p. 32; Riedel and Funnell 1964, p. 334; Friend and Riedel 1967, p. 219.

459.66 13° 38′ N, 133° 45′ W. 4,890 m. MP6-1. *Oligocene with Eocene admixture.* Riedel and Funnell 1964, pp. 334-5.

459.88 11° 42′ N, 131° 38′ W. 4,935 m. AMPH 144G. *lower lower Miocene.* Friend and Riedel 1967, p. 219.

459.96 13° 25′ N, 130° 18′ W. 4,843 m. DWBG 4. *Oligocene or Quaternary with Oligocene admixture.* Riedel and Funnell 1964, p. 325.

459.99 10° 26′ N, 130° 38′ W. 4,870 m. DWBG 5. *Oligocene, or Quaternary with Oligocene admixture.* Riedel and Funnell 1964, p. 325.

460-524 12° N, 125° W-4° N, 125° W. 4,315-4,730 m. *Oligocene -Miocene.* Riedel and Bramlette 1959, p. 106.

460.24 15° 34′ N, 127° 11′ W. 4,725 m. Swedish Deep-Sea Expedition 53. *Tertiary ? Miocene.* Arrhenius 1952, pp. 144-8, 195-7; Emiliani and Edwards 1953, p. 123; Bandy 1956, p. 123; Emiliani 1956*a*, p. 123; Bandy 1960, p. 6; Emiliani 1961*a*, p. 144; Olausson 1961*a*, p. 33; Riedel and Funnell 1964, p. 363; Riedel 1967, p. 542;

460.49 10° 21′ N, 125° 26′ W. 4,645 m. CHUB 10. *Tertiary probably Miocene.* Arrhenius 1963, p. 719.

460.49 10° 19′ N, 125° 27′ W. 4,545 m. CHUB 9. *middle Miocene (G. insueta zone).* Riedel 1959*a*, pp. 286, 287, 288, etc.; Arrhenius 1963, p. 719; Riedel 1963*b*, fig. 2.

460.55 14° 55′ N, 124° 12′ W. 4,260 m. CAP 50BP. *Tertiary.* Revelle *et al.* 1955, pp. 228-9; Goldberg and Arrhenius 1958, p. 203, table 20; Olausson 1961*a*, p. 27.

512.51 08° 36′ N, 115° 29′ E. 2,200 m. LSDA SCS 5G. *Pliocene (N21)* Parker 1967, p. 130, etc.

514.42 approx. 7° 30′ N, 134° 30′ E above sea-level. Palau Island, Caroline Islands. *late Eocene.* Cole 1950, p. 123; Menard and Hamilton 1964, table 1.

514.80 09° 35′ N, 138° 10′ E. +55 m and +45 m. YM-304 and 306. Map Island, Yap, Caroline Islands. *early Miocene (G. insueta zone).* Cole, Todd and Johnson 1960, pp. 80-91, 103-12; Menard and Hamilton 1964, table 1.

515.21 08° 04′ N, 142° 15′ E. 4,822 m. *Kosyu* 32a. *upper Tertiary.* Hanzawa 1933, p. 123; Wiseman and Hendey 1953, p. 55; Riedel 1954, pp. 173-4.

515.60 09° 21′ N, 146° 50′ E. 4,660 m. *Kosyu* 73. *upper Tertiary.* Hanzawa 1933, p. 123; Wiseman and Hendey 1953, p. 55; Riedel 1954, pp. 173-4.

516.12 approx. 07° 30′ N, 151° 45′ E above sea-level. Eo-102, Ud-170. Truk Islands. *Miocene.* Stark *et al.* 1958, p. 123; Cole, Todd and Johnson 1960, pp. 94-5; Wilson 1963, p. 536.

517.42 07° 58′ N, 164° 57′ E. 5,219 m. PROA 92P. *lower Pliocene.* Friend and Riedel 1967, p. 219.

517.77 02° 56′ N, 167° 14′ E. 4,428 m. PROA 88P. *upper Miocene.* Friend & Riedel 1967, p. 219, etc. Ross and Riedel 1967, p. 288; Parker 1967, p. 129, etc.

517.94 approx. 05° 55′ N, 169° 44′ E. approx. sea-level. Jaluit reef conglomerate. Jabor Island, Jaluit Atoll, Marshall Islands. *middle Tertiary.* Yabe and Aoki 1922, pp. 40-3; Wilson 1963, p. 536.

517.99 00° 47′ N, 169° 12′ E. 4,380 m. CAP 2BG 2. *Quaternary with some Miocene and Pliocene admixture.* Riedel and Funnell 1964, p. 315.

517.99 00° 47′ N, 169° 12′ E. 4,380 m. CAP 2BP 2. *middle Pliocene.* Riedel and Funnell 1964, pp. 315, 321; Friend and Riedel 1967, p. 219, etc.; Parker 1967, p. 129, etc.

518.54 05° 49′ N, 175° 11′ E. 5,225 m. PROA 95P. *late Tertiary.* Riedel 1963*b*, fig. 2.

518.93 06° 26′ N, 179° 36′ E. 5,836 m. PROA 100P. *late Tertiary.* Riedel 1963*b*, fig. 2.

518.95 04° 32′ N, 179° 45′ E. 5,712 m. PROA 97P. *middle Miocene.* Riedel 1963*b*, fig. 2; Friend and Riedel 1967, p. 219.

519.06 03° 37′ N, 179° 18′ W. 4,972 m. PROA 103P. *lower Pliocene.* Riedel 1963*b*, fig. 2; Friend and Riedel 1967, p. 219.

519.08 09° 01′ N, 171° 03′ W. 5,133 m. PROA 152G. *Oligocene or Miocene, with Eocene admixture.* Riedel 1963*b*, fig. 2; Friend and Riedel 1967, p. 230.

519.21 08° 33′ N, 177° 46′ W. 5,620 m. LSDH 88P. *Eocene.* Riedel 1963*b*, fig. 2; Friend and Riedel 1967, p. 219.

419.25 04° 58′ N, 177° 33′ W. 5,399 m. PROA 102P. *upper Miocene.* Riedel 1963, fig. 2; Friend and Riedel 1967, p. 219.

519.42 07° 19′ N, 175° 28′ W. 5,190 m. LSDH 90P. *upper Pliocene.* Riedel 1963, fig. 2; Riedel *et al.* 1963, p. 1,239; Friend and Riedel 1967, p. 219.

519.56 03° 01′ N, 174° 028′ W. 5,230 m. MSN 12G. *Pliocene with Eocene, (?)Oligocene and Miocene admixture.* Riedel 1963, fig. 2; Riedel and Funnell 1964, pp. 338-9; Harrison and Funnell 1964, p. 566.

519.67 02° 23′ N, 173° 50′ W. 5,560 m. Swedish Deep-Sea Expedition 87. *late Tertiary.* Riedel 1952, pp. 5-15; Riedel 1957*a*, pp. 63, 73, 76; Olausson 1960, p. 187; Riedel 1963, fig. 2; Riedel and Funnell 1964, p. 311.

519.70 09° 11′ N, 172° 03′ W. 5,510 m. LSDH 91P. *lower Oligocene.* Riedel 1963, fig. 2; Friend and Riedel 1967, p. 219.

519.74 05° 34′ N, 172° 12′ W. 5,590 m. Swedish Deep-Sea Expedition 85. *upper Tertiary, with early and middle Tertiary admixture.* Riedel 1952, pp. 6, 15; Riedel 1957*a*, pp. 63, 66, 72-3, 76; Olausson 1960, pp. 185-6; Riedel 1963, fig. 2; Riedel *et al.* 1963, p. 1,240; Riedel and Funnell 1964, p. 311.

520.01 08° 40′ N, 169° 28′ W. 5,440 m. Swedish Deep-Sea Expedition 83. *middle Tertiary, with early Tertiary admixture.* Riedel 1957*a*, pp. 63, 71, 76; Olausson 1960, pp. 184-5; Riedel 1963, fig. 2.

520.03 06° 04′ N, 169° 58′ W. 5,400 m. MSN 11G. *Quaternary with Eocene, Oligocene and Miocene admixture.* Riedel and Funnell 1964, p. 338.

520.05 04° 48′ N, 169° 50′ W. 5,747 m. PROA 133G. *(?) late Tertiary.* Riedel 1963, fig. 2.

520.12 07° 36′ N, 168° 06′ W. 4,994 m. MSN 10G. *Quaternary with Eocene, (?)Oligocene and Miocene admixture.* Riedel and Funnell 1964, p. 338.

520.20 09° 29′ N, 167° 51′ W. 5,106 m. PROA 142G. *(?)upper Oligocene, or lower Miocene.* Riedel 1963, fig. 2; Friend and Riedel 1967, p. 219.

520.21 08° 29′ N, 167° 12′ W. 5,142 m. MSN 9G. *Quaternary*

with Eocene, Miocene and Pliocene admixture. Riedel and Funnell 1964, p. 337.

520.81 08° 34′ N, 161° 39′ W. 5,070 m. TET 34. *Quaternary with Eocene, Oligocene and Miocene admixture.* Riedel and Funnell 1964, pp. 351-2, 360-2.

521.22 07° 53′ N, 157° 29′ W. 4,989 m. TET 44. *Quaternary with Eocene, Oligocene and Miocene admixture.* Riedel and Funnell 1964, p. 352.

521.30 09° 52′ N, 156° 09′ W. 5,277 m. TET 49. *Quaternary with Eocene admixture.* Riedel and Funnell 1964, p. 353.

521.31 08° 14′ N, 156° 38′ W. 5,266 m. TET 47. *early Miocene with Eocene and Oligocene admixture, or Quaternary with Eocene, Oligocene and Miocene admixture.* Riedel and Funnell 1964, pp. 352-3.

521.32 07° 56′ N, 156° 48′ W. 5,075 m. TET 46. *Miocene with Eocene and Oligocene admixture, or Quaternary with Eocene, Oligocene and Miocene admixture.* Riedel 1963, fig. 2; Riedel and Funnell 1964, p. 352.

521.90 09° 28′ N, 150° 49′ W. 4,838. *Challenger* 267. *reworked early Tertiary.* Murray and Renard 1891, pp. 120-1, Riedel 1957*a*, pp. 82, 89.

521.90 09° 20′ N, 150° 35′ W. 4,813 m. JYN V 14P. *Eocene.* Riedel 1963, fig. 2.

522.01 08° 02′ N, 149° 54′ W. 5,073 m. JYN V 15P. *middle and upper Tertiary.* Riedel 1963, figs. 1, 2.

522.02 07° 44′ N, 149° 44′ W. 5,168 m. JYN V 16P. *upper Oligocene.* Riedel 1963, figs. 1, 2; Friend and Riedel 1967, p. 219.

522.02 07° 35′ N, 149° 49′ W. 5,304 m. *Challenger* 268. *lower Miocene (G. insueta* zone) *and reworked early Tertiary.* Murray and Renard 1891, pp. 120-1 pl. 15, fig. 4; Riedel 1954, pp. 172, etc.; Riedel 1957*a*, pp. 79, 80, 82, 89, 91, 93; Riedel 1959*a*, pp. 286, 287, 288, etc. Olausson 1961*a*, p. 29.

522.06 03° 45′ N, 149° 44′ W. 5,155 m. Swedish Deep-Sea Expedition 76. *late Tertiary, with some early and middle Tertiary admixture.* Kolbe 1954, p. 18; Riedel 1957*a*, pp. 63, 66, 69-70, 76; Olausson 1960, pp. 179-83; Riedel 1952, p. 15; Olausson 1961*a*, pp. 25-9, 31-3; Riedel 1963, figs. 1, 2; Ross and Riedel 1967, p. 286.

522.10 09° 30′ N, 148° 40′ W. 5,358 m. Swedish Deep-Sea Expedition 77. *(?middle Tertiary, with some early Tertiary admixture.)* Riedel 1957*a*, pp. 63, 70-1; Olausson 1960, pp. 183-4.

522.12 07° 17′ N, 148° 12′ W. 4,925 m. JYN V 20P. *lower Miocene.* Riedel 1963, figs. 1, 2; Friend and Riedel 1967, p. 219.

522.34 05° 32′ N, 146° 09′ W. 5,100 m. MSN 143 PG and P. *middle Pliocene, with Miocene admixture.* Riedel 1963, figs. 1, 2; Riedel, Bramlette and Parker 1963, p. 1,239; Riedel and Funnell 1964, pp. 342-3; 368; Friend and Riedel 1967, p. 219; Ross and Riedel 1967, p. 286.

522.34 05° 20′ N, 146° 13′ W. 5,089 m. MSN 142G. *Pliocene with Miocene admixture.* Riedel 1963, figs. 1, 2; Riedel and Funnell 1964, p. 342; Harrison and Funnell 1964, p. 566.

522.40 09° 58′ N, 145° 13′ W. 5,130 m. MP14-1. *Quaternary with Eocene-Oligocene admixture.* Riedel and Funnell 1964, p. 335.

522.40 09° 23′ N, 145° 15′ W. 5,100 m. MSN 149G. *lower Miocene.* Riedel and Funnell 1964, pp. 344, 363-5.

522.40 09° 23′ N, 145° 15′ W. 5,100 m. MSN 149P. *upper Oligocene.* Riedel 1963, figs. 1, 2; Riedel and Funnell 1964, pp. 344-5, 362-5; Martini 1965, p. 394; Friend and Riedel 1967, p. 219.

522.42 07° 09′ N, 145° 35′ W. 5,100 m. MSN 146PG and P. *late Miocene and Pliocene with Eocene, Oligocene and Miocene admixture.* Riedel 1963, figs. 1, 2; Riedel and Funnell 1964, p. 344; Ross and Riedel 1967. p. 286.

522.50 09° 09′ N, 144° 26′ W. 5,236 m. MP13-1. *Quaternary with Eocene-Miocene admixture.* Riedel and Funnell 1964, p. 335.

522.50 09° 09′ N, 144° 26′ W. 5,236 m. MP13-2. *Quaternary with Eocene – (?)Oligocene admixture.* Riedel and Funnell 1964, p. 335.

522.83 96° 30′ N, 141° 59′ W. 5,018 m. JYN V 38P. *lower Pliocene.* Riedel 1963, figs. 1, 2.; Friend and Riedel 1967, p. 219.

523.01 09° 00′ N, 139° 51′ W. 5,000 m. JYN V 42P. *Oligocene or Miocene (probably middle or upper Miocene).* Riedel 1963, fig. 2; Friend and Riedel 1967, p. 230.

523.52 07° 39′ N, 134° 02′ W. 4,548 m. AMPH 140G. *upper Oligocene.* Friend and Riedel 1967, p. 219.

523.83 06° 54′ N, 131° 00′ W. 4,340 m. DWBG 10. *upper Oligocene (G. kugleri* zone). Riedel 1959*a*, pp. 286, 287, 288, etc.; Riedel 1963, fig. 2; Riedel and Funnell 1964, pp. 311, 326; Martini 1965, p. 394; Friend and Riedel 1967, p. 219.

523.84 05° 26′ N, 131° 19′ W. 4,155 m. DWBG 11. *Quaternary with late Oligocene and early Miocene admixture.* Riedel and Funnell 1964, p. 327.

523.91 08° 48′ N, 130° 48′ W. 4,917 m. DWBG 7. *late Miocene, with early Miocene and Oligocene admixture.* Riedel and Funnell 1964, pp. 325-6.

523.92 07° 39′ N, 130° 55′ W. 4,950 m. DWBG 9. *late Miocene, with early Miocene and Oligocene admixture.* Riedel 1963, fig. 2; Riedel and Funnell 1964, p. 326.

524.03 06° 44′ N, 129° 28′ W. 4,440 m. Swedish Deep-Sea Expedition 58. *Pliocene, Plio-Pleistocene boundary.* Arrhenius 1952, pp. 159-65, 194-5; Emiliani 1954; Emiliani 1955, pp. 560-4; Bandy 1956; Emiliani 1956*a*; Riedel 1957*a*, p. 65; Brotzen and Dinesen 1959, p. 52; Bandy 1960, p. 6; Emiliani 1961*a*, p. 144; Olausson 1961*a*, pp. 5-33; Riedel 1963, fig. 2; Riedel, Bramlette and Parker 1963, pp. 1,239-40; Uschakova 1966, p. 136; Parker 1967, p. 129, etc.

524.10 09° 50′ N, 128° 13′ W. 4,629 m. Swedish Deep-Sea Expedition 56. *middle Tertiary.* Arrhenius 1952, pp. 152-4, 195-7; Emiliani 1954; Olausson 1961*a*, pp. 31, 33; Riedel 1963, fig. 2; Riedel and Funnell 1964, p. 363.

524.11 08° 25′ N, 128° 48′ W. 4,607 m. Swedish Deep-Sea Expedition 57. *middle Tertiary.* Arrhenius 1952, pp. 154-9; 195-7; Emiliani 1954; Bandy 1956, p. 123; Emiliani 1956*a*; Bandy 1960, p. 6; Emiliani 1961*a*, p. 144; Olausson 1961*a*, pp. 29, 31, 33; Riedel 1963, fig. 2; Riedel and Funnell 1964, p. 363.

524.15 04° 52′ N, 128° 21′ W. 4,460 m. AMPH 6P. *middle Miocene.* Friend and Riedel 1967, p. 219.

524.31 08° 01′ N, 126° 58′ W. 4,440 m. CHUB 34. *upper Miocene or Pliocene.* Parker (unpublished).

524.32 07° 18′ N, 127° 24′ W. 3,640 m. CHUB 30. *Pliocene.* Riedel 1963, fig. 2; Riedel, Bramlette and Parker 1963, p. 1,240; Parker 1967, p. 129, etc.

524.41 08° 31′ N, 125° 25′ W. 4,462 m. CHUB 15. *lower Miocene (C. stainforthi* zone). Riedel 1959*a*, pp. 285, 286, 287, 288, etc.; Riedel 1963, fig. 2; Friend and Riedel 1967, p. 219.

524.41 08° 17′ N, 125° 19′ W. 4,334 m. CHUB 16. *middle Miocene.* Parker (unpublished).

524.41 08° 12′ N, 125° 19′ W. 4,315 m. CHUB 38. *upper Oligocene (G. kugleri* zone). Bramlette 1958, p. 122; Riedel 1959*a*, pp. 285, 286, 287, 288, etc.; Olausson 1961*a*, pp. 29, 32; Riedel 1963, fig. 2; Riedel and Funnell 1964, p. 311; Friend and Riedel 1967, p. 219.

524.41 08° 06′ N, 125° 26′ W. 4,415 m. CHUB 40. *middle*

Miocene (G. insueta-G. fohsi barisanensis zone). Bramlette 1961, pp. 349-50; Riedel 1959*a*, pp. 286, 287, 388, Riedel 1963, fig. 2; Friend and Riedel 1967, p. 219.

524.41 08° 05′ N, 125° 25′ W. 4,453 m. CHUB 17. *upper Oligocene (G. kugleri* zone). Riedel 1959*a*, pp. 285, 286, 287, 288, etc.; Bramlette 1961, pp. 349-51; Riedel 1963, fig. 2; Martini 1965, p. 399.

524.42 07° 22′ N, 125° 30′ W. 4,549 m. CHUB 20. *lower Miocene* (approx. *C. stainforthi* zone). Riedel 1959*a*, pp. 286, 287, 288, etc.; Riedel 1963, fig. 2; Friend and Riedel 1967, p. 219.

524.44 05° 29′ N, 125° 29′ W. 4,530 m. CHUB 24. *middle Miocene* (approx. *G.fohsi barisanensis* zone). Bramlette 1958, p. 122; Riedel 1959*a*, p. 285, 286, 287, 288, etc.; Olausson 1961*a*, p. 29, 32; Riedel 1963, fig. 2; Friend and Riedel 1967, p. 219.

524.50 09° 17′ N, 124° 09′ W. 4,410 m. CAP 49BP. *Tertiary.* Bramlette 1961, pp. 360-1.

525.21 08° 10′ N, 117° 53′ W. 3,530 m. RIS 12G. *upper Miocene.* Friend and Riedel 1967, p. 219, etc.; Parker 1967, p. 129, etc.

529.15 04° 25′ N, 78° 25′ W. 2,960-3,200 m. V15 SBT-39. *(?)Pliocene.* L.G.O. (unpublished).

553.19 09° 00′ S, 161° 00′ E above sea-level. BSIP 3535 and 3555. Malaita Island, Solomon Islands. *middle Tertiary.* Friend and Riedel 1967, p. 219.

553.19 09° 00′ S, 161° 00′ E above sea-level. Ulawa Island, Malaita Group, Solomon Islands. *upper Eocene.* McTavish 1966, p.1; Friend and Riedel 1967, p. 219.

553.71 01° 20′ S, 167° 23′ E. 3,965 m. Swedish Deep-Sea Expedition 93. *reworked early, middle and upper Tertiary.* Kolbe 1954, p. 14; Riedel 1957*a*, pp. 63, 75; Olausson 1960*a*, pp. 192-3.

553.84 04° 31′ S, 168° 02′ E. 3,208 m. LSDH 78P. *lower Pliocene.* Riedel *et al.* 1963, p. 1,240; Parker 1965*a*, pp. 151-2; Friend and Riedel 1967, p. 219, etc.; Ross and Riedel 1967, pp. 287-8; Parker 1967, p. 129, etc.

553.90 00° 52′ S, 169° 35′ E above sea-level. Ocean Island *probably Pliocene.* Owen 1923, p. 13; Wilson 1963, p. 53.

554.12 02° 50′ S, 171° 18′ E. 4,120 m. Swedish Deep-Sea Expedition 91. *lower Miocene* (approx. *G. insueta* zone). Riedel 1952, pp. 5, 12; Kolbe 1954, pp. 4, 14; Riedel 1957*a*, pp. 63, 74-6; Riedel 1959*a*, pp. 286, 287, 288, etc.; Olausson 1960, pp. 190-2; Olausson 1961*a*, p. 32; Olausson 1961*b*, pp. 299-393.

554.38 08° 44′ S, 173° 24′ E. 5,397 m. PROA 62PG and P. *Quaternary with Pliocene, Miocene and Cretaceous admixture, or Pliocene with Miocene and Cretaceous admixture.* Ross and Riedel 1967, p. 288.

554.43 03° 21′ S, 174° 12′ E. 4,830 m. Swedish Deep-Sea Expedition 90. *upper Oligocene-lower Miocene* (approx. *C. stainforthi* zone) *with Eocene admixture.* Olausson 1960, p. 190; Olausson 1961*a*, pp. 14, 30; Olausson 1961*b*, pp. 299-303.

554.49 09° 03′ S, 174° 52′ E. 4,960 m. CAP 5BG. *mixture of Cretaceous to late Tertiary.* Riedel and Funnell 1964, pp. 321, 361-2; Ross and Riedel 1967, p. 288.

554.49 09° 03′ S, 174° 52′ E. 4,960 m. CAP 5BP. *mixture of Cretaceous to Pliocene.* Riedel and Funnell 1964, p. 322, 361-2.

554.82 02° 04′ S, 178° 16′ E. 5,519 m. MSN 16G. *(?)Pliocene with early and middle Tertiary admixture.* Riedel and Funnell 1964, pp. 339, 361-3.

555.12 02° 48′ S, 178° 57′ W. 5,480 m. Swedish Deep-Sea Expedition 89. (*? middle Tertiary, with early Tertiary admixture*). Riedel 1957*a*, pp. 63, 74-5; Olausson 1960, pp. 188-9.

555.22 02° 37′ S, 177° 45′ W. 5,770 m. Swedish Deep-Sea Expedition 88. *middle Tertiary.* Riedel 1957*a*, pp. 63, 73-5; Olausson 1960, pp. 187-8.

557.14 04° 38′ S, 158° 10′ W. 5,140 m. AMPH 96P. *upper lower Miocene.* Friend and Riedel 1967, p. 219.

557.23 03° 42′ S, 157° 40′ W. 5,228 m. AMPH 97P. *upper Pliocene.* Friend and Riedel 1967, p. 219.

557.26 06° 02′ S, 157° 28′ W. 5,177 m. AMPH 91P. *middle Miocene* (approx. *G. fohsi barisanensis* zone). Friend and Riedel 1967, p. 219.

557.31 01° 52′ S, 156° 42′ W. 4,983 m. AMPH 99PG and P. *Quaternary, with Miocene and Pliocene admixture.* Ross and Riedel 1967, p. 287.

557.43 03° 51′ S, 155° 48′ W. 5,161 m. AMPH 101GV. *middle Miocene with Eocene admixture.* Friend and Riedel 1967, p. 219.

557.43 03° 52′ S, 155° 53′ W. 5,055 m. AMPH 105G. *middle Miocene* (approx. *G. fohsi barisanensis* zone) *with Eocene admixture.* Friend and Riedel 1967, p. 219.

557.43 03° 52′ S, 155° 43′ W. 5,172 m. AMPH 102P. *upper Miocene with Eocene admixture.* Friend and Riedel 1967, p. 219.

557.44 04° 49′ S, 155° 19′ W. 5,265 m. AMPH 109P. *upper lower Miocene with Eocene admixture.* Friend and Riedel 1967, p. 219.

557.73 03° 48′ S, 152° 56′ W. 4,755 m. *Challenger* 272. *(?) reworked middle Tertiary.* Murray and Renard 1891, pp. 120-1; Riedel 1957*a*, p. 80.

557.74 04° 04′ S, 152° 53′ W. 5,200 m. Swedish Deep-Sea Expedition 73. *upper Tertiary, with lower and middle Tertiary admixture.* Riedel 1952, p. 6; Riedel 1957*a*, pp. 63, 66-8, 76; Olausson 1960, pp. 174-5; Riedel, Bramlette and Parker 1963, p. 1,240; Riedel and Funnell 1964, p. 311; Ross and Riedel 1967, p. 286.

558.04 04° 26′ S, 149° 24′ W. 4,600 m. MSN 135PG and P. *lower Miocene (G. insueta zone) and Quaternary with lower Miocene admixture.* Riedel and Funnell 1964, pp. 341-2; Ross and Riedel 1967, p. 286.

558.05 05° 58′ S, 149° 33′ W. 5,115 m. MSN 132 PG and P. *(?) early, middle to late Miocene and Pliocene, with Miocene admixture.* Riedel and Funnell 1964, pp. 341, 364; Friend and Riedel 1967, p. 219.

558.75 05° 58′ S, 142° 43′ W. 4,450 m. AMPH 130G. *lower Pliocene.* Friend and Riedel 1967, p. 219.

558.97 07° 08′ S, 140° 13′ W. 4,076 m. USSR 5100. *Pliocene and Plio-Pleistocene boundary.* Uschakova 1966, pp. 136-43.

559.33 03° 00′ S, 136° 26′ W. 4,510 m. Swedish Deep-Sea Expedition 62. *Pliocene; Plio-Pleistocene boundary.* Arrhenius 1952, pp. 179-85, 194-5; Kolbe 1954, p. 14; Riedel 1957*a*, p. 65; Brotzen and Dinesen 1959, pp. 52-3; Emiliani 1961*a*, p. 144; Olausson 1961*a*, pp. 5-33; Riedel *et al.* 1963, pp. 1,239-40; Uschakova 1966; pp. 136, 141.

559.36 06° 23′ S, 136° 11′ W. 4,410 m. RIS 102G. *upper Miocene and lower Pliocene.* Friend and Riedel 1967, p. 220.

559.69 09° 59′ S, 133° 23′ W. 4,340 m. DWHH 13. *Quaternary with Eocene and later Tertiary admixture.* Riedel and Funnell 1964, p. 331.

560.87 07° 31′ S, 121° 56′ W. 4,410 m. AMPH 9P. *upper Pliocene.* Friend & Riedel 1967, p. 219.

561.17 07° 19′ S, 118° 40′ W. 4,200 m. CAP 42BP. *upper Pliocene.* Friend and Riedel 1967, p. 219.

616.14 14° 45′ S, 151° 14′ E. 4,400 m. MSN 21G. *Quaternary or Pliocene, with reworked middle Tertiary.* Riedel and Funnell 1964, pp. 339-40.

618.41 11° 33′/11° 51′ S, 174° 17′/174° 27′ E. 30 to 41 m. (Alexa Bank). *Miocene.* Bramlette (unpublished):

Riedel and Funnell 1964, p. 322.

618.51 11° 05′ S, 175° 10′ E. 2,560 m. CAP Dredge No. 2 (Alexa-Penguin Bank). *Miocene (N17) with Quaternary admixture.* Parker 1967, p. 130, etc.

618.77 approx. 17° 50′ S, 177° 25′ E above sea-level (0 to approx. +915 m). RB 174, 176, 183, and RB 220. Viti Levu, Fiji. *upper Eocene (Tertiary 'b') and Oligocene (Tertiary 'c').* Kleinpell 1954 (Lau); Cole 1960, p. 1-7; Johnson 1961, p. 943 (Lau, Fiji); Menard and Hamilton 1964, table 1.

618.88 approx. 18° S, 178° E. above sea-level. F1, F2a, F3b, F13b, F35 (Suva district Viti Levu, Fiji). *late Miocene and early Pliocene (N17-N19).* Parker 1967, p. 130, etc.

621.93 13° 53′ S, 150° 35′ W. 3,623 m. MSN 128G. *Pliocene with Eocene admixture.* Riedel and Funnell 1964, pp. 340-1.

621.94 14° 29′ S, 150° 01′ W. 1,628 m to 1,884 m. V18-RD29. (Tuamotu Ridge) *Middle and Upper Eocene.* Burckle and Saito 1966, pp. 1,207-8.

622.03 13° 25′ S, 149° 30′ W. 4,300 m. Swedish Deep-Sea Expedition 69. *Middle Eocene overlain by Upper Oligocene-Lower Miocene.* Olausson 1960, pp. 169-71; Olausson 1961*a*, pp. 29-32.

622.25 approx. 15° 50′ S, 147° 10′ W. above sea-level. Makatea, Tuamotu Islands. *Eocene.* Repelin 1919, p. 237; Wilson 1963, p. 536.

622.36 16° 23′ S, 146° 02′ W. 1,380 m. DWBG 25 (Tuamotu Ridge). *Middle Eocene.* Riedel and Funnell 1964, pp. 328, 362.

622.36 16° 47′ S, 146° 15′ W. 980 m. DWBD 4. from side of unnamed seamount on southwest flank of the Tuamotu Ridge. *upper Eocene (Tertiary 'b').* Cole 1959, pp. 10-11; Menard and Hamilton 1964, table 1.

622.46 16° 42′ S, 145° 48′ W. approx. 2,200 m. DWBG 23B. *Middle Eocene.* Riedel and Funnell 1964, pp. 327, 362; Black 1964, pp. 307, 315; Martini 1965, p. 395.

622.52 12° 00′ S, 144° 20′ W. 5,008 m. Swedish Deep-Sea Expedition 64. *Mio-Pliocene, with Eocene admixture.* Bramlette (unpublished).

622.52 12° 00′ S, 144° 21′ W. 5,000 m. DOLPHIN 2. *Mixture of early, middle and late Tertiary.* Wiseman and Riedel 1960, pp. 215-16; Riedel and Funnell 1964, pp. 324, 361-2.

622.52 12° 10′ S, 144° 25′ W. 4,332 m. DOLPHIN 1. *Pliocene or Quaternary with Pliocene admixture.* Riedel and Funnell 1964, p. 324.

622.75 15° 15′ S, 142° 27′ W. 3,675 m. RIS 84G. *Quaternary, with admixture of Pliocene and Eocene.* Parker 1967, p. 129, etc.

623.04. 14° 03′ S, 139° 35′ W. 3,900 m. RIS 82G. *Pliocene (lower N19).* Parker 1967, p. 129, etc.

623.44 14° 28′ S, 135° 29′ W. 4,400 m. DWHH 14. *Oligocene.* Riedel and Funnell 1964, pp. 331, 363; Martini 1965, p. 394.

624.14 14° 02′ S, 128° 29′ W. 3,985 m. RIS 77G. *late Miocene (?N18).* Parker 1967, p. 129, etc.

624.74 14° 01′ S, 122° 28′ W. 3,790 m. RIS 75G. *Pliocene (N21).* Parker 1967, p. 129, etc.

625.04 14° 16′ S, 119° 11′ W. 3,400 m. CAP 38BP. *Miocene, overlain by Pliocene (N17-19, N21).* Parker 1967, p. 129, etc.

655.51 approx. 21° 20′ S, 174° 55′ W. above sea-level. Eua, Tonga. *upper Eocene (Tertiary 'b').* Whipple 1932, pp. 79-90; Cole 1960, pp. 1-3; Menard and Hamilton 1964, table 1.

657.10 20° 00′ S, 158° 07′ W. above sea-level Atiu, Cook Islands. *Pliocene.* Marshall 1930, p. 123; Wilson 1963, p. 536.

657.12 approx. 22° 30′ S, 158° 30′ W. above sea-level. Mangaia, Cook Islands. *Oligocene-Miocene.* Marshall 1927; Wilson 1963, p. 536.

657.54 24° 41′ S, 154° 45′ W. 4,500 m. MSN 126G. *Pliocene.* Riedel and Funnell 1964, p. 340; Parker 1967, p. 129, etc.

658.26 26° 19′ S, 147° 07′ W. 3,680 m. DWBG 36. *Quaternary, with late Eocene or early Oligocene admixture.* Riedel and Funnell 1964, pp. 331-2.

661.13 23° 37′ S, 118° 14′ W. 3,440 m. DWHG 79. *Pliocene.* Riedel and Funnell 1964, pp. 333-4.

662.37 27° 54′ S, 106° 53′ W. 3,200 m. DWBP 119. *Pliocene.* Riedel and Funnell 1964, p. 330.

663.38 28° 02′ S, 96° 20′ W. 3,400 m. DWBG 118C. *Pliocene.* Riedel, Bramlette and Parker 1963, p. 1240; Riedel and Funnell 1964, p. 330.

664.45 25° 31′ S, 85° 14′ W. 920 m. DWHD 72. *Late Tertiary (?lower Pliocene).* Bramlette (unpublished).

665.00 approx. 20° S, 80° W. Nazca Seamount, *Miocene.* Fisher 1958, pp. 22-3; Wilson 1963, p. 536.

665.73 23° 29′ S, 72° 59′ W. 3,710 m. DWBG 94. *Quaternary with Oligocene and late Miocene to Pliocene admixture.* Riedel and Funnell 1964, p. 329.

728.68 38° 49′ S, 83° 21′ W. 4,080 m. DWHG 54. *middle Miocene.* Riedel and Funnell 1964, p. 333.

760.24 44° 13′ S, 127° 20′ W. 4,600 m. DWHG 34. *early Oligocene.* Riedel and Bramlette 1959, p. 106; Riedel and Funnell 1964, p. 332.

760.24 44° 21′ S, 127° 14′ W. 4,675 m. DWHH 35. *late Tertiary.* Riedel and Funnell 1964, pp. 332-3.

760.42 42° 50′ S, 125° 32′ W. 4,560 m. DWBG 57B. *Quaternary, with middle Tertiary admixture.* Shipek 1960, fig. 2; Riedel and Funnell 1964, pp. 328-9.

760.43 43° 07′ S, 125° 23′ W. 4,640 m. DWBG 58, *Quaternary with middle Tertiary admixture.* Riedel and Funnell 1964, p. 329.

823.79 59° 22′ S, 132° 46′ W. 3,910 m. V16-130. *Pliocene.* Hays 1965, pp. 157, 164.

828.86 56° 33′ S, 81° 45′ W. 5,000 m. V18-69. *Pliocene.* Hays 1965, pp. 155, 164.

829.57 57° 02′ S, 74° 29′ W. 4,064 m. V17-88. *Pliocene.* Hays 1965, pp. 157, 164.

862.20 60° 45′ S, 107° 29′ W. 4,898 m. V16-132. *Pliocene.* Hays 1965, pp. 153, 164; Opdyke *et al.* 1966, p. 350.

863.41 61° 57′ S, 95° 03′ W. 4,062 m. V16-133. *Pliocene.* Hays 1965, p. 164; Opdyke *et al.* 1966, p. 350.

863.81 61° 54′ S, 91° 15′ W. 5,145 m. V16-134. *Pliocene.* Hays 1965, pp. 150, 164; Opdyke *et al.* 1966, pp. 350, 355.

865.40 60° 29′ S, 75° 57′ W. 4,695 m. V18-72. *Pliocene.* Hays 1965, p. 164; Opdyke *et al.* 1966, p. 350.

39. THE OCCURRENCE OF PRE-QUATERNARY CALCAREOUS NANNOPLANKTON IN THE OCEANS

E. MARTINI

Summary: The oldest pre-Quaternary calcareous nannoplankton assemblage so far recovered from ocean sediments is of Lower Cretaceous age. Cretaceous and Palaeocene calcareous nannoplankton assemblages are rare, Eocene and Oligocene assemblages have been recovered at several localities, whereas Miocene and Pliocene assemblages are the most common pre-Quaternary occurrences to be encountered in the oceans. Distinctive changes in calcareous nannoplankton assemblages took place between the Maastrichtian and Danian, at the end of the Upper Eocene, and at the end of the Pliocene. The various calcareous nannoplankton assemblages from the Upper Cretaceous to the Pliocene are discussed, and attention is drawn to some phylogenetic sequences in the Tertiary.

INTRODUCTION

In recent years large numbers of deep-sea cores have been recovered by various deep-sea expeditions. Relatively few data on these cores and other samples from the deep-sea floor have been published. In the following paper a summary of the present knowledge of the occurrence of pre-Quaternary calcareous nannoplankton in deep-sea sediments is attempted.

The calcareous nannoplankton, a name commonly applied to coccoliths, discoasters and related forms, include many taxa which belong to the Protophyta, but others, especially extinct forms with no comparable Recent representatives, can only be classified at Protista. The systematic relationship within the calcareous nannoplankton is obscure in some parts, expressed in an unusual high rate of 'genera *incertae sedis*'. Zoological and botanical nomenclature have been used by various authors, but differences in nomenclature are avoided in many papers because authors have attempted no higher classification.

So far the oldest assemblage has been found in Lower Lias marine sediments. Reported occurrences in the Precambrian, Devonian and Pennsylvanian have not been confirmed. The oldest sample as yet recovered in the oceans is of Barremian age (L. H. Burckle, personal communication). The record of deep-sea calcareous nannoplankton assemblages is incomplete compared with the land-based sequences, but it seems to be only a matter of time before these gaps are closed by the recovery of more pre-Quaternary deep-sea samples.

In this paper a number of reference samples containing distinctive assemblages have been selected, from cores recovered by the Scripps Institution of Oceanography and the Lamont Geological Observatory. The author is indebted to M. N. Bramlette and W. R. Riedel of Scripps Institution of Oceanography, and to D. B. Ericson of Lamont Geological Observatory.

GENERAL REMARKS ON THE OCCURRENCE OF CALCAREOUS NANNOPLANKTON

Before discussing the various assemblages some general remarks are necessary. The distribution pattern of the Recent coccolithophorid-species is not yet fully understood, although efforts have been made to collate the scattered data available. A considerable number of species seem to prefer temperate or tropical waters, while others (e.g. *Coccolithus pelagicus*) are abundant in cooler waters. Several genera seem to prefer nearshore environments. The absence of representatives of the family Braarudosphaeridae, the rarity of the genera *Discolithina* and *Scyphosphaera* in Tertiary deep-sea cores, and the abundancy of these in the relatively nearshore deposits of Tertiary age in California, Mississippi, France, Germany, and the Mediterranean region has been already reported (Martini and Bramlette 1963; Martini 1965). Palaeocurrents in the oceans are of some importance, as shown in the middle Miocene to early Pliocene samples of the experimental Mohole cores, where a cold north-south directed 'California current' seems to have resulted in assemblages relatively poor in species, compared with corresponding deep-sea samples to the southwest or with samples from Trinidad, B.W.I. (Martini and Bramlette 1963). The latitude of the core locality is also of importance as indicated by Riedel and Funnell (1964, p. 332) and noted for some Oligocene samples below.

A serious problem, which has been discussed by various authors (e.g. Bramlette and Sullivan 1961, p. 133), is the reworking of calcareous nannoplankton into younger sediments. Coastal erosion of the continents and islands, especially, as well as underwater volcanic activity may lead to redistribution of older forms into younger sediments at any time. But in the deep-sea turbidity currents seem to be the main factor producing mixed calcareous nannoplankton assemblages containing forms of different age. Many cores with turbidity layers have been recovered from various parts of the oceans (Ericson, Ewing, Wollin and Heezen 1961, Riedel and Funnell 1964), including several with reworked Cretaceous nannoplankton in the Atlantic Ocean (Funnell 1964).

Secondary growth of calcite after deposition of the calcite skeletal elements, and excessive accretion of calcite during their lifetime, may result in unusual forms and inflated specimens, a problem which has been discussed by Bramlette and Sullivan (1961, p. 133) and Martini (1965, p. 397).

The calcareous nannoplankton assemblages found in deep-sea samples can be correlated with those of better known land-based sequences, despite the differences mentioned above, and in most cases an exact correlation is possible. Especially useful are sequences which have been chosen by others from extensive study of the Foraminifera or other groups, and where the foraminiferal zones can be used for stratigraphical comparisons (e.g. Trinidad, Cipero and Lengua Formations).

Within the evolution of the calcareous nannoplankton three major breaks are known. The most important one occurs at the end of the Cretaceous, where a large number of genera and species characteristic of Upper Cretaceous assemblages vanished and only a few taxa survived into Tertiary time (Danian). The complexly constructed coccoliths of the genera *Cretarhabdus, Deflandrius, Eiffellithus,* as well as the genera *'incertae sedis' Microrhabdulus* and *Micula* died out at the end of the Maastrichtian (Bramlette 1958; Bramlette and Martini 1964). The two other breaks occur at the end of the Upper Eocene, where amongst others the rosette-shaped discoasters disappear, and only few-rayed discoasters continue, and at the end of the Pliocene, where most discoasters disappear and some other taxa change their form (Ericson, Ewing and Wollin 1963; McIntyre, Bé and Krinsley 1964; Riedel, Bramlette and Parker 1963).

Among the calcareous nannoplankton in the Tertiary the discoasters are of special interest. These include a development from multi-radiate rosette-shaped forms in the Palaeocene (*Discoaster multiradiatus*) and Eocene (*D. barbadiensis*) to forms with fewer rays in the late Eocene and Oligocene (*D. tani; D. deflandrei),* and to slender rayed forms in the Miocene (*D. variabilis; D. challengeri*) and Pliocene (*D. brouweri*) (see Fig. 39.1). Some other genera show phylogenetic sequences which can be used for close age-determination of samples. Among these are the genera *Catinaster, Chiphragmalithus, Helicopontosphaera* (Fig. 39.2) and *Sphenolithus* (Bramlette and Sullivan 1961; Martini and Bramlette 1963). Together with short-ranged index fossils like *Heliolithus riedeli* in the late Palaeocene or *Isthmolithus recurvus* in the uppermost Eocene and lowest Oligocene, age determinations on the basis of the calcareous nannoplankton become quite reliable.

UPPER CRETACEOUS AND TERTIARY CALCAREOUS NANNOPLANKTON ASSEMBLAGES

The few samples of Upper Cretaceous age (Cenomaian, Turonian, Campanian and Maastrichtian) so far known from the deep-sea show only some of the calcareous nannoplankton assemblages found in equivalent land-based sequences. Most of the Cretaceous samples have been recovered along the continental margins or on seamounts, with a considerable number of reworked examples (Funnell 1964), and need study on the basis of additional material. In the following only the oldest Upper Cretaceous sample (Cenomanian) has been selected to show the differences from Tertiary assemblages.

The available Palaeocene sample contains an Upper Palaeocene calcareous nannoplankton assemblage. Lower Palaeocene (including the Danian) and Lower Eocene samples were not available at the time the present data were compiled. From the Middle Eocene onwards an increasing number of samples is available, and a more detailed description of the assemblages is possible. Reference samples and bibliography for each of the five subdivisions of the Tertiary period are listed, and the development of the calcareous nannoplankton within the subdivisions is discussed.

Upper Cretaceous (Plate 39.1, fig. 1)

Reference sample : A 167-25: 55 cm (Cenomanian).
28° 52′ N, 76° 47′ W, depth 1,745 m.

Reference : Black 1964 (Maastrichtian).

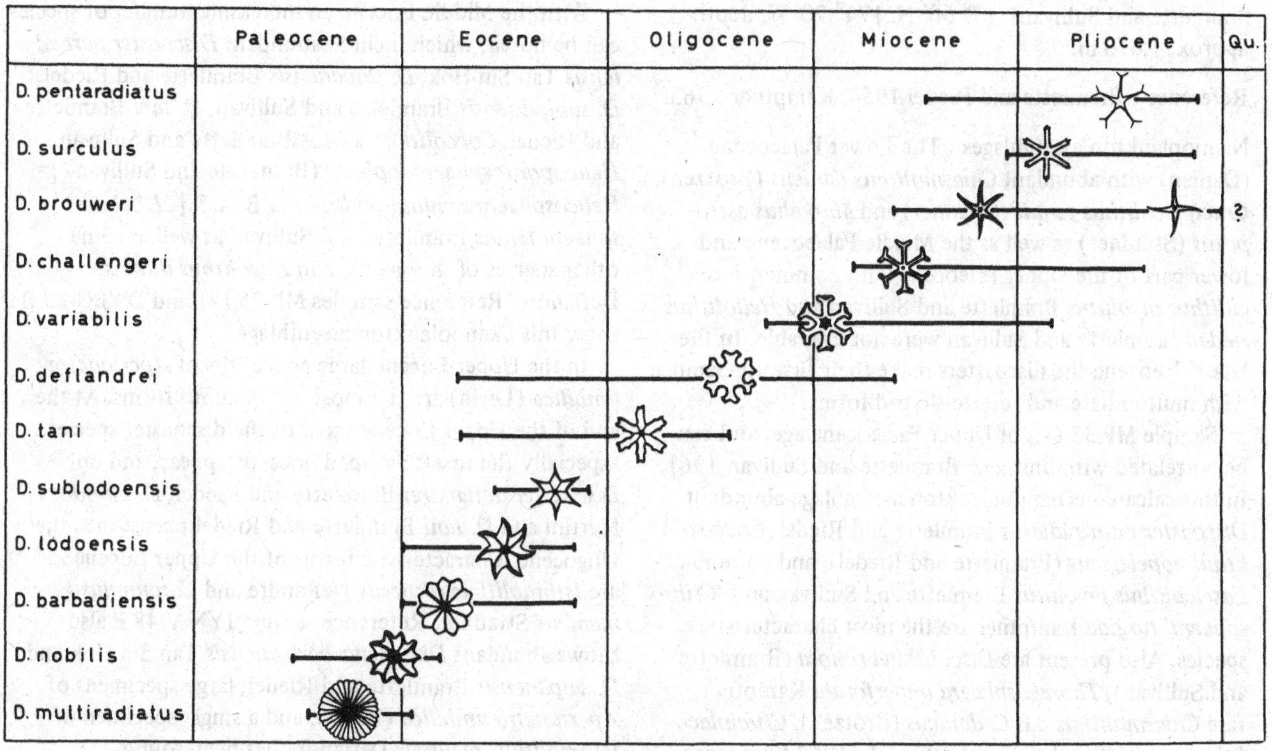
Paleocene
Eocene
Oligocene
Miocene
Pliocene
Qu.
D. pentaradiatus
D. surculus
D. brouweri
D. challengeri
D. variabilis
D. deflandrei
D. tani
D. sublodoensis
D. lodoensis
D. barbadiensis
D. nobilis
D. multiradiatus
?

Fig. 39.1. Selected discoasters, showing the general tendency to decrease the number and width of rays from the Palaeocene to the Pliocene.

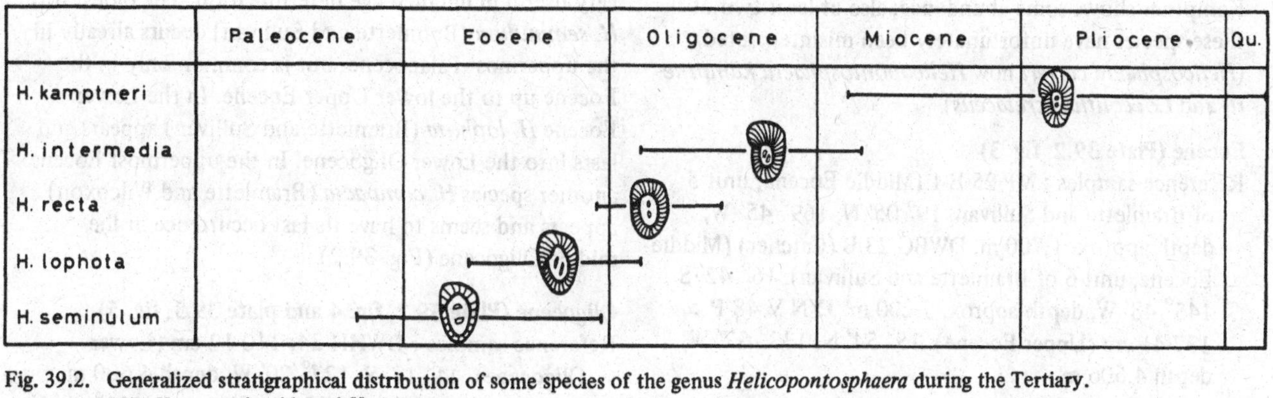
Paleocene
Eocene
Oligocene
Miocene
Pliocene
Qu.
H. kamptneri
H. intermedia
H. recta
H. lophota
H. seminulum

Fig. 39.2. Generalized stratigraphical distribution of some species of the genus *Helicopontosphaera* during the Tertiary. N.B. *H. recta* should read *H. compacta*

Nannoplankton assemblages : The Cenomanian assemblage of the reference sample shows abundant small *Zygolithus* sp., in outline similar to *Zygolithus fibula* (Lecal) in Stradner 1963, plate 4, fig. 5, and common *Glaukolithus diplogrammus* (Deflandre), *'Zygolithus' crux* (Deflandre and Fert), *Eiffellithus turriseiffeli* (Deflandre), *Eprolithus floralis* (Stradner), and an undescribed barrel-like form (see Plate 39.1, fig. 1, centre). In addition small *Coccolithus* spp., *Parhabdolithus embergeri* (Noel), and smaller *Nannoconus* spp. are present. Compared with the land-based assemblage described by Stradner (1963) no important difference can be seen. In the higher Upper Cretaceous representatives of the genera *Arkhangelskiella, Cretarhabdus, Deflandrius, Microrhabdulus, Micula* and others become common and are present up to the Maastrichtian. With the end of the Maastrichtian most of the genera and species disappear as mentioned above, and the succeeding Tertiary assemblages show an entirely different character.

Palaeocene (Plate 39.1, Fig. 2)

Reference sample: MP 33 C (Upper Palaeocene, unit 2 of

Bramlette and Sullivan). 17° 50′ N, 174° 20′ W, depth approx. 1,670 m.

Reference : Bramlette and Riedel 1954, Kamptner 1963.

Nannoplankton assemblages : The Lower Palaeocene (Danian) with abundant *Chiasmolithus danicus* (Brotzen), *Cruciplacolithus tenuis* (Stradner) and *Markalius astroporus* (Stradner) as well as the Middle Palaeocene and lower part of the Upper Paleocene with common *Fasciculithus involutus* Bramlette and Sullivan and *Heliolithus riedeli* Bramlette and Sullivan were not available. In the late Palaeocene the discoasters make their first appearance with multiradiate and rosette-shaped forms.

Sample MP 33 C is of Upper Palaeocene age, and can be correlated with unit 2 of Bramlette and Sullivan 1961. In this calcareous nannoplankton assemblage abundant *Discoaster multiradiatus* Bramlette and Riedel, *Coccolithus eopelagicus* (Bramlette and Riedel), and common *Fasciculithus involutus* Bramlette and Sullivan and *'Cribrosphaera' turgida* Kamptner are the most characteristic species. Also present are *Discolithina rimosa* (Bramlette and Sullivan), *Thoracosphaera imperforata* Kamptner, rare *Chiasmolithus* c.f. *C. danicus* (Brotzen), *Cruciplacolithus tenuis* (Stradner), and others. In 1963 Kamptner described sixteen coccolithophorid-species from the same sample of which only *'Cribrosphaera' turgida* Kamptner shows some abundance; also at least two of these species have unfortunately been misinterpreted (*Helicosphaera carteri* now *Helicopontosphaera kamptneri*, and *Coccolithus cretaceus*).

Eocene (Plate 39.2, fig. 3)

Reference samples : MP 25 E-1 (Middle Eocene, unit 5 of Bramlette and Sullivan. 19° 05′ N, 169° 45′ W, depth approx. 1,700 m. DWBG 23 B (Catcher) (Middle Eocene, unit 6 of Bramlette and Sullivan). 16° 42′ S, 145° 48′ W, depth approx. 2,200 m. JYN V 48 P : 137-41 cm (Upper Eocene). 15° 54′ N, 133° 57′ W, depth 4,606 m.

References : Black 1964; Bramlette and Riedel 1954; Kamptner 1963; Martini 1965.

Nannoplankton assemblages : Land-based Lower Eocene nannoplankton assemblages include three-rayed *Marthasterites tribrachiatus* (Bramlette and Riedel), *Discoaster lodoensis* Bramlette and Riedel, *Discoasteroides kuepperi* (Stradner), *Coccolithus crassus* Bramlette and Sullivan, *Coccolithites delus* Bramlette and Sullivan, *Helicopontosphaera seminulum* (Bramlette and Sullivan) [= *Helicosphaera seminulum seminulum* B. & S.], and *Rhabdosphaera perlonga* (Deflandre) as most common and characteristic species. No deep-sea sample showing this assemblage is as yet known.

With the Middle Eocene an increasing number of species can be noted, which include abundant *Discoaster barbadiensis* Tan Sin Hok, *D. saipanensis* Bramlette and Riedel, *D. sublodoensis* Bramlette and Sullivan, *D. tani* Bramlette and Riedel, *Coccolithus solitus* Bramlette and Sullivan, *Helicopontosphaera lophota* (Bramlette and Sullivan) [= *Helicosphaera seminulum lophota* B. & S.], *Rhabdosphaera tenuis* Bramlette and Sullivan as well as some other species of this genus, and *Zygolithus dubius* Deflandre. Reference samples MP 25 E-1 and DWBG 23 B show this nannoplankton assemblage.

In the Upper Eocene large coccoliths of *Apertapetra umbilica* (Levin) are the most conspicuous forms. At the end of the Upper Eocene most of the discoaster species, especially the rosette-shaped ones, disappear, and only *Discoaster deflandrei* Bramlette and Riedel, *D. binodosus* Martini and *D. tani* Bramlette and Riedel persist into the Oligocene. Characteristic forms of the Upper Eocene are *Isthmolithus recurvus* Deflandre and *Corannulus germanicus* Stradner. Reference sample JYN V 48 P also shows abundant *Discoaster barbadiensis* Tan Sin Hok and *D. saipanensis* Bramlette and Riedel, large specimens of *Apertapetra umbilica* (Levin), and a single specimen of *Isthmolithus recurvus* Deflandre has been found.

In the Eocene the widespread genus *Helicopontosphaera* shows a phylogenetic sequence which proved to be very useful in detailed age determination. The oldest form *H. seminulum* (Bramlette and Sullivan) occurs already in the uppermost Palaeocene, but is common only in the Eocene up to the lower Upper Eocene. In the Lower Eocene *H. lophota* (Bramlette and Sullivan) appears and lasts into the Lower Oligocene. In the uppermost Eocene another species *H. compacta* (Bramlette and Wilcoxon) appears and seems to have its last occurrence in the middle Oligocene (Fig. 39.2).

Oligocene (Plate 39.2, fig. 4 and plate 39.3, fig. 5)

Reference samples : DWHH 34: 110-12 cm (Lower Oligocene). 44° 13′ S, 127° 20′ W, depth 4,600 m. MSN 151 P: 127-30 cm (Upper Oligocene). 11° 03′ N, 142° 28′ W, depth 5,000 m.

References : Kamptner 1963; Martini 1965.

Nannoplankton assemblages : In the Lower Oligocene discoasters are relatively rare, and most of them belong to *Discoaster deflandrei* Bramlette and Riedel and *D. tani* Bramlette and Riedel. Coccoliths of the genus *Coccolithus*, including *C. pelagicus* (Wallich) are abundant. Among the coccolithophorids a number of undescribed species occur, which need study and description. In the higher Lower Oligocene the first specimens of another *Helicopontosphaera* species – *H. intermedia* (Martini) – occur, and

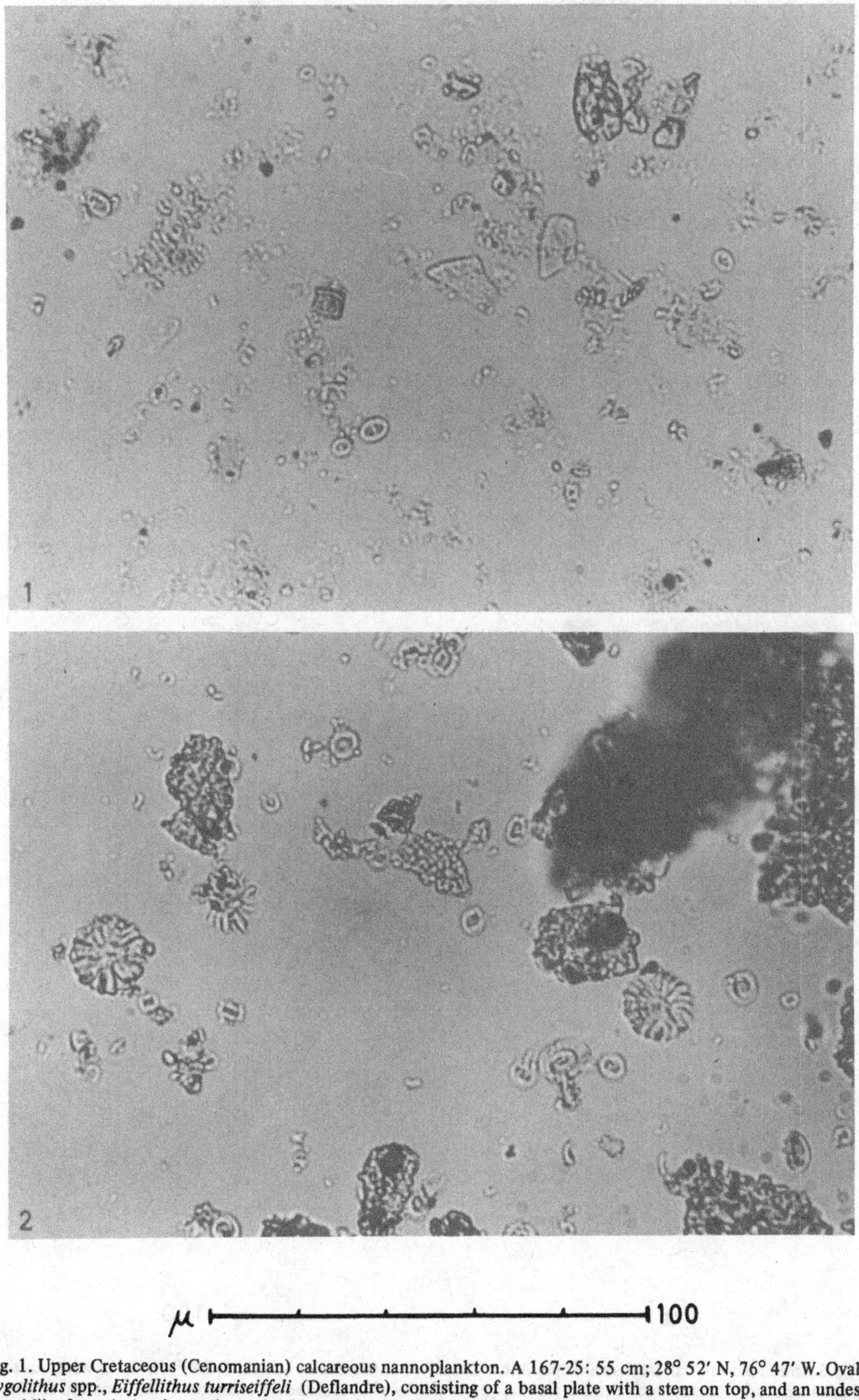
1
2
μ
100

Plate 39.1. Fig. 1. Upper Cretaceous (Cenomanian) calcareous nannoplankton. A 167-25: 55 cm; 28° 52′ N, 76° 47′ W. Oval-shaped *Zygolithus* spp., *Eiffellithus turriseiffeli* (Deflandre), consisting of a basal plate with a stem on top, and an undescribed barrel-like form (centre), are the main elements in this Cenomanian assemblage. Fig. 2. Palaeocene calcareous nannoplankton. MP 33 C; 17° 50′ N, 174° 20′ W. Rosette-shaped *Discoaster multiradiatus* Bramlette and Riedel and smaller *Coccolithus* spp. are (with *Fasciculithus involutus* Bramlette and Sullivan which is not shown in the present figure) the dominant forms in this late Palaeocene assemblage. (All specimens approx. x 690.)

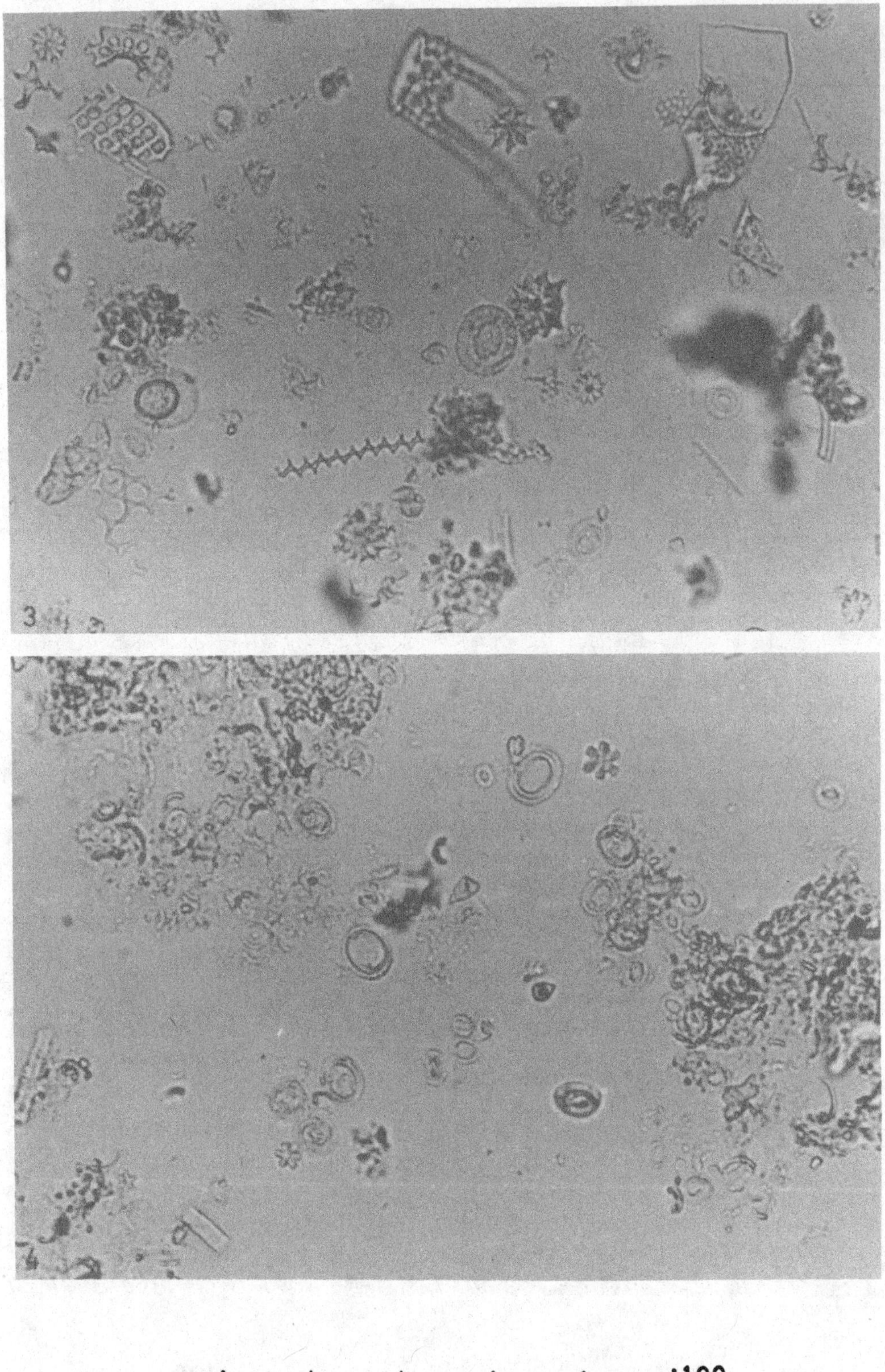
3
4
μ
100

Plate 39.2. Fig. 3. Middle Eocene calcareous nannoplankton. DWBG 23 B (Catcher); 16° 42′ S, 145° 48′ W. In addition to rosette-shaped *Discoaster barbadiensis* Tan Sin Hok, large coccoliths of *Coccolithus eopelagicus* (Bramlette and Riedel) (centre), and *Lophodolithus* (left side) are some of the more common forms in this late Middle Eocene assemblage. Fig. 4. Lower Oligocene calcareous nannoplankton. DWHH 34: 110-12 cm; 44° 13′ S, 127° 20′ W. Besides relatively large oval-shaped and ring-like *Chiasmolithus* sp., *Coccolithus eopelagicus* (Bramlette and Riedel), and star-like *Discoaster deflandrei* Bramlette and Riedel are the most common species in this high latitude assemblage. (All specimens approx. x 690.)

range up into the Lower Miocene. Such an assemblage is present in reference sample DWHH 34. Owing to its high southern latitude it shows a close correspondence with the New Zealand Whaingaroan calcareous nannoplankton assemblage.

In the Upper Oligocene discoasters again become more abundant, and belong to *Discoaster deflandrei* Bramlette and Riedel and undescribed related forms. In addition to these *D. variabilis* Martini and Bramlette appears in the higher part of the Upper Oligocene. Other forms of stratigraphical value in the uppermost Oligocene are the rod-like species of the genus *Triquetrorhabdulus.* Some difficulties discussed in an earlier paper (Martini 1965) arose from excessive accretion of calcite to the skeleton of certain calcareous nannoplankton in the equatorial region, which makes determination of species questionable. Reference sample MSN 151 P shows such an Upper Oligocene assemblage, and close correlation with the Trinidad Cipero Formation foraminiferal zones was possible. Other samples of Upper Oligocene age are those of the DWBG 10 core, in an earlier paper (Martini 1965) described as lowermost Miocene.

On the basis of the calcareous nannoplankton of several Oligocene samples, including those mentioned, it seems that during Oligocene times a more distinctive distribution pattern existed then than during the rest of Tertiary time, but this problem needs study on the basis of more samples.

Miocene (Plate 39.3, fig. 6 and plate 39.4, fig. 7)

Reference samples : DWHH 14 : 30-2 cm (Lower Miocene). 14° 28′ S 135° 29′ W, depth 4,400 m. EM 8-13 : 50-3 cm (Middle Miocene). 28° 59′ N 117° 30′ W, depth 3,566 m. EM 8-9 : 150-2 cm (Upper Miocene). 28° 59′ N 117° 30′ W, depth 3,566 m.

References : Bramlette and Riedel 1954; Kamptner 1963; Martini and Bramlette 1963; Martini 1965.

Nannoplankton assemblages : Many Miocene deep-sea samples, and the cores of the experimental Mohole drilling allow a detailed zonation corresponding with that of the higher Cipero and Lengua Formations of Trinidad, and later deposits elsewhere. In the Lower Miocene a large undescribed discoaster-species (see Martini 1965, pl. 37) has some significance, and is also present in the *Catapsydrax dissimilis* zone of the Cipero Formation, Trinidad. *Discoaster deflandrei* Bramlette and Riedel, *D. exilis* Martini and Bramlette, *D. variabilis* Martini and Bramlette, and *Coccolithus pelagicus* (Wallich) are the most common forms. Of special significance again is the phylogenetic sequence of the genus *Helicopontosphaera; H. intermedia* (Martini) seems to have its last occurrence in the *Globigerinatella insueta* zone, while *H. kamptneri* Hay and Mohler [= *Helicosphaera carteri* (Wallich) sensu Kamptner] first appears in the underlying *Catapsydrax stainforthi* zone. Reference sample DWHH 14, showing in addition to the mentioned species the rod-like form *Triquetrorhabdulus* sp. has been correlated with the *Catapsydrax dissimilis* zone of Trinidad.

The Middle Miocene calcareous nannoplankton assemblages, well represented in the experimental Mohole cores (reference sample EM 8-13), show abundant *Discoaster exilis* Martini and Bramlette, *D. variabilis* Martini and Bramlette, and in the lower part the last specimens of *D. deflandrei* Bramlette and Riedel. In the lower part *D. kugleri* Martini and Bramlette and *Cyclococcolithus rotula* Kamptner are distinctive forms. *Discoaster bollii* Martini and Bramlette, *D. hamatus* Martini and Bramlette and the first specimens of *D. brouweri* Tan Sin Hok, together with species of the genus *Catinaster,* are characteristic of the upper part. *Catinaster calyculus* Martini and Bramlette shows a tendency to increasing the length and developing a counter-clockwise curvature of the rays from the lower to the upper part of its occurrence.

In the Upper Miocene *Discoaster brouweri* Tan Sin Hok (mostly six-rayed specimens), *D. variabilis* Martini and Bramlette, and *D. pentaradiatus* Tan Sin Hok are the most common forms. Reference sample EM 8-9 shows such an assemblage. The stratigraphical value of several species of the genus *Sphenolithus* is noteworthy, but the study of their phylogenetic sequence is incomplete, and no detailed information is given here.

Pliocene (Plate 39.4, fig. 8)

Reference samples : EM 8-1 : 80-82 cm (Lower Pliocene). 28° 59′ N, 117° 30′ W, depth 3,566 m. CK 16 : 68-70 cm (Upper Pliocene). 36° 30′ N, 173° 16′ W, depth 4,290. m.

References : Ericson, Ewing and Wollin 1963; Kamptner 1963; Martini and Bramlette 1963; Riedel, Bramlette and Parker 1963.

Nannoplankton assemblages : In the Pliocene the most common forms are *Discoaster brouweri* Tan Sin Hok, *D. surculus* Martini and Bramlette, *D. pentaradiatus* Tan Sin Hok, *D. challengeri* Bramlette and Riedel, *Ceratolithus* aff. *cristatus* Kamptner, *Cyclococcolithus leptoporus* (Murray and Blackman), and one other *Cyclococcolithus* species. *Discoaster variabilis* Martini and Bramlette has its last occurrence in the Lower Pliocene, where *D. brouweri* Tan Sin Hok shows an increasing number of four-rayed specimens. Reference sample EM 8-1 contains these forms and is of Lower Pliocene age.

In the Upper Pliocene the portion of three- and four-

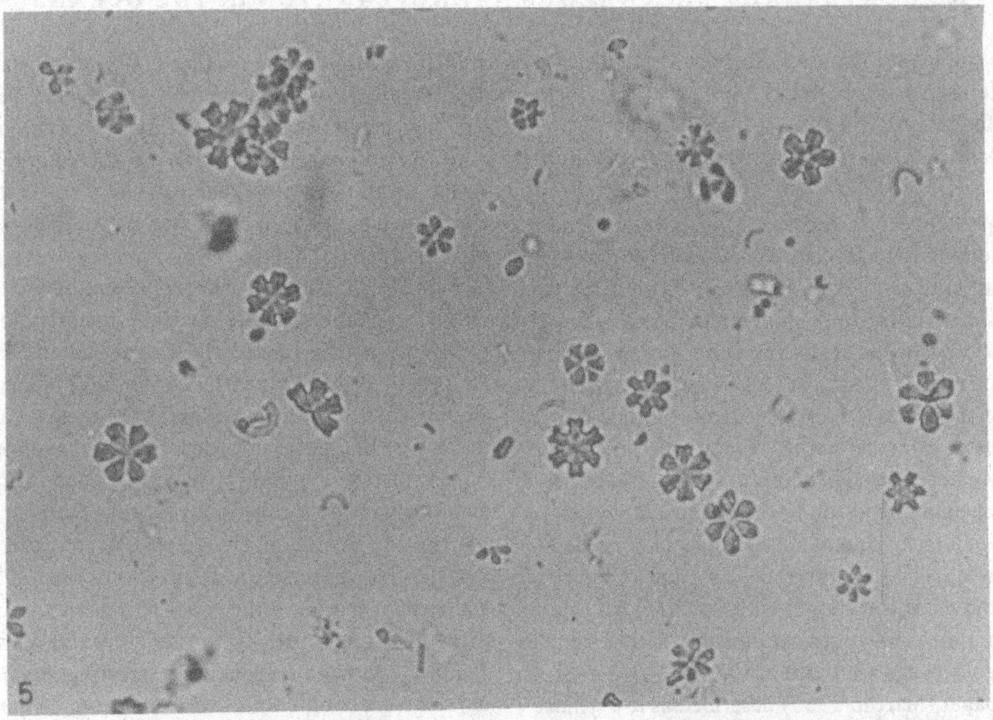
5

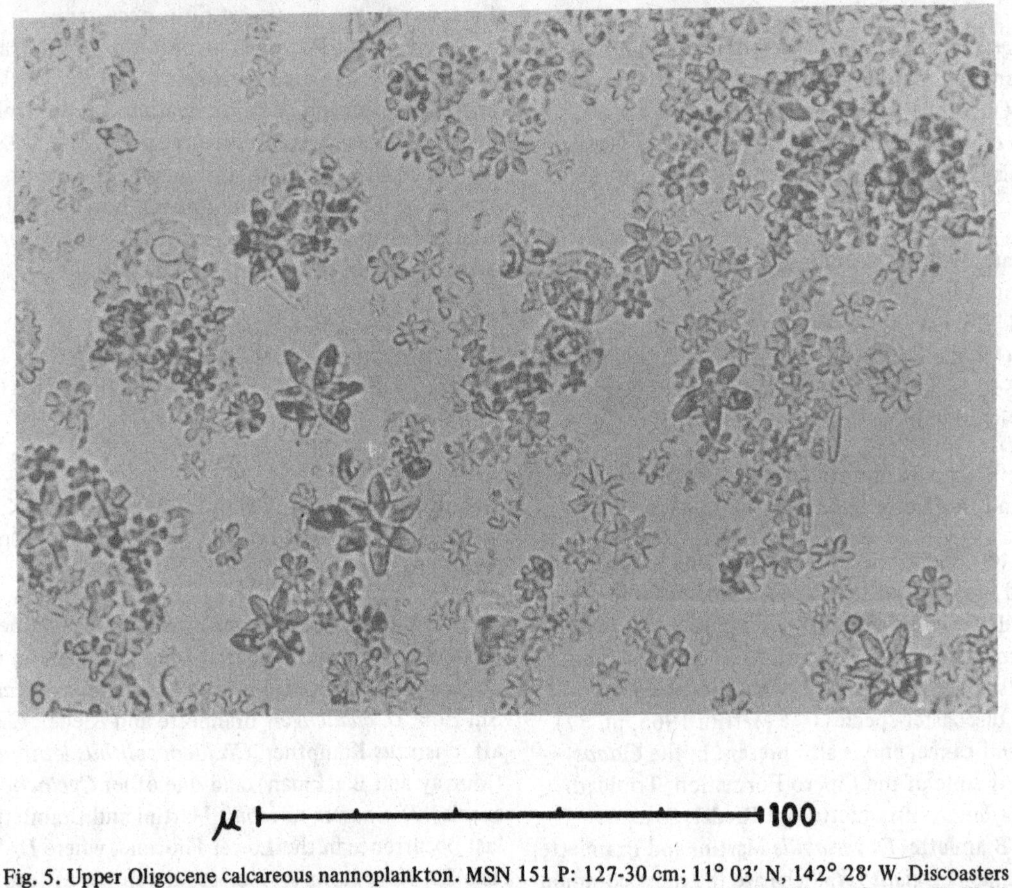
6
μ
100

Plate 39.3. Fig. 5. Upper Oligocene calcareous nannoplankton. MSN 151 P: 127-30 cm; 11° 03′ N, 142° 28′ W. Discoasters of the *deflandrei*-group are the dominant forms, but note broken pieces of oval-shaped *Coccolithus pelagicus* (Wallich). Fig. 6. Lower Miocene calcareous nannoplankton. DWHH 14: 30-2 cm; 14° 28′ S, 135° 29′ W. Oval-shaped coccoliths (*C. pelagicus*), discoasters of the *deflandrei-* and *ciperoensis*-group, and rod-like *Triquetrorhabdulus* sp. are abundant or common in this assemblage. (All specimens approx. x 690.)

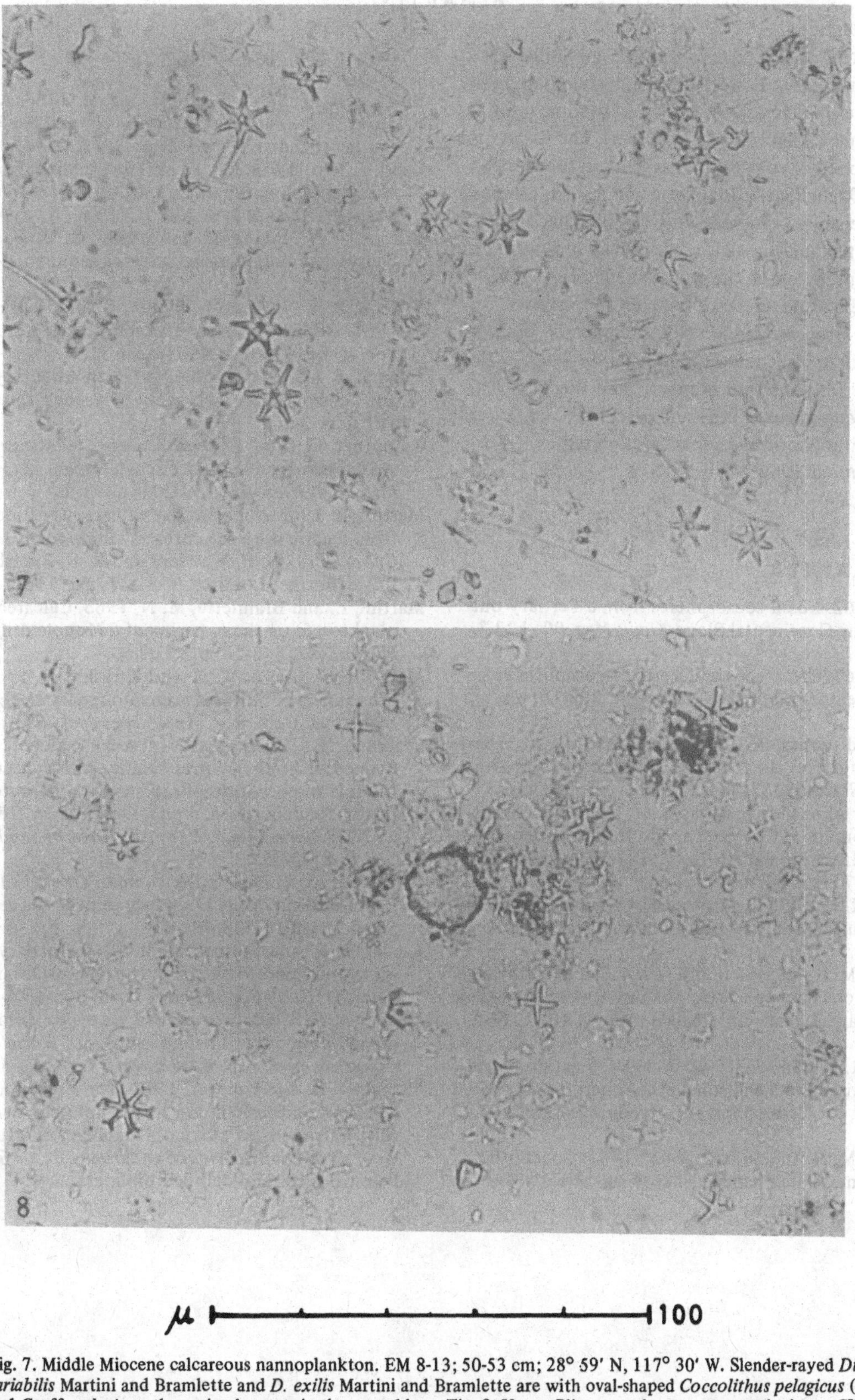
7
8
μ 100

Plate 39.4. Fig. 7. Middle Miocene calcareous nannoplankton. EM 8-13; 50-53 cm; 28° 59′ N, 117° 30′ W. Slender-rayed *Discoaster variabilis* Martini and Bramlette and *D. exilis* Martini and Bramlette are with oval-shaped *Coccolithus pelagicus* (Wallich and *C.* aff. *pelagicus,* the main elements in the assemblage. Fig. 8. Upper Pliocene calcareous nannoplankton. CK 16: 68-70 cm; 36° 30′ N, 173° 16′ W. Delicate four- and three-rayed specimens of *Discoaster brouweri* Tan Sin Hok are characteristic forms in this early Upper Pliocene assemblage. Note *D. challengeri* Bramlette and Riedel (lower left corner), *D. surculus* Martini and Bramlette (upper right corner) and circular *Cyclococcolithus leptoporus* (Murray and Blackman) (upper centre). (All specimens approx. x 690.)

rayed specimens of *Discoaster brouweri* Tan Sin Hok is still higher than in the Lower Pliocene, but in the uppermost Pliocene six-rayed specimens again become more abundant (core CAP 42 BP: 757-803 cm). The first of the above-mentioned discoasters to disappear in the Upper Pliocene is *D. challengeri* Bramlette and Riedel, followed at higher levels by *D. pentaradiatus* Tan Sin Hok and *D. surculus* Martini and Bramlette, whereas *D. brouweri* Tan Sin Hok is still present at the end of the Pliocene. The early Upper Pliocene reference sample CK 16 shows abundant *D. brouweri* Tan Sin Hok, frequent *D. surculus* Martini and Bramlette, rare *D. pentaradiatus* Tan Sin Hok, and very rare *D. challengeri* Bramlette and Riedel. *Cyclococcolithus leptoporus* Murray and Blackman and several other coccolithophorid-species are also present.

Received October 1967.

BIBLIOGRAPHY

Black, M. 1962. Fossil coccospheres from a Tertiary outcrop on the Continental Slope. *Geol. Mag.* **99**, 123-7, pls. 8-9.

Black, M. 1964. Cretaceous and Tertiary coccoliths from Atlantic seamounts. *Palaeontology*, **7**, 306-16, pls. 50-3.

Black, M. and Barnes, B. 1961. Coccoliths and discoasters from the floor of the South Atlantic Ocean. *J. Royal Microsc. Soc.* **80**, 137-47, pls. 19-26.

Bogdanov, J. A. and Uschakova, M. G. 1966. Coccoliths of the group of *Discoaster* Tan Sin Hok in aqueous suspensions from the Pacific. *Dokl. Akad. Nauk S.S.S.R.* **171**, 465-7 [in Russian].

Bramlette, M. N. 1958. Significance of coccolithophorids in calcium carbonate deposition. *Bull. Geol. Soc. Amer.* **69**, 121-6.

*Bramlette, M. N. and Martini, E. 1964. The great change in calcareous nannoplankton fossils between the Maastrichtian and Danian. *Micropaleontology*, **10**, 291-322, pls. 1-7.

Bramlette, M. N. and Riedel, W. R. 1954. Stratigraphical value of discoasters and some other microfossils related to Recent coccolithophores. *J. Paleont.* **28**, 385-403, pls. 38-9.

*Bramlette, M. N. and Sullivan, F. R. 1961. Coccolithophorids and related nannoplankton of the early Tertiary in California. *Micropaleontology*, **7**, 129-88, pls. 1-14.

Burckle, L. H., Saito, T. and Ewing, M. 1967. A Cretaceous (Turonian) core from the Naturaliste Plateau southeast Indian Ocean. *Deep-Sea Res.* **14**, 421-6.

Cann, J. R. and Funnell, B. M. 1967. Palmer Ridge : A section through the upper part of the ocean crust? *Nature, Lond.* **213**, 661-4.

Ericson, D. B., Ewing, M. and Wollin, G. 1963. Pliocene-Pleistocene boundary in deep-sea sediments. *Science*, **139**, 727-37.

Ericson, D. B., Ewing, M., Wollin, G. and Heezen, B. C. 1961. Atlantic deep-sea sediment cores. *Bull. Geol. Soc. Amer.* **72**, 193-286, pls. 1-3.

Funnell, B. M. 1964. Studies in North Atlantic geology and palaeontology 1. Upper Cretaceous. *Geol. Mag.* **101**, 421-34.

Kamptner, E. 1963. Coccolithineen-Skelettreste aus Tiefseeablagerungen des Pazifischen Ozeans. *Ann. Nat. hist. Mus. Wien*, **66**, 139-204, pls. 1-9.

Martini, E. 1965. Mid-Tertiary calcareous nannoplankton from Pacific deep-sea cores. In *Submarine Geology and Geophysics* (W. F. Whittard and R. B. Bradshaw). Butterworths: London. 393-411, pls. 33-7.

Martini, E. and Bramlette, M. N. 1963. Calcareous nannoplankton from the experimental Mohole drilling. *J. Paleont.* **37**, 845-56, pls. 102-5.

McIntyre, A., Bé, A. W. H. and Krinsley, D. 1964. Coccoliths and the Plio-Pleistocene boundary in deep-sea sediments. *Geol. Soc. Amer. spec. pap.* **76**, 113.

Olausson, E. 1961. Remarks on some Cenozoic core sequences from the central Pacific, with a discussion of the role of coccolithophorids and Foraminifera in carbonate deposition. *Medd. Ocean. Inst. Göteborg* **29** (*Göteborgs kungl. Vetensk. Vitterh.-Samh. Handl. Ser. B*, **8**), 3-35.

Riedel, W. R., Bramlette, M. N. and Parker, F. L. 1963. Pliocene-Pleistocene boundary in deep-sea sediments. *Science*, **140**, 1,238-40.

Riedel, W. R. and Funnell, B. M. 1964. Tertiary sediment cores and microfossils from the Pacific Ocean floor. *Quart. J. Geol. Soc. London*, **120**, 305-68, pls. 14-32.

*Stradner, H. 1963. New contributions to Mesozoic stratigraphy by means of nannofossils. *6th World Petrol. Congr., sect. 1, pap. 4 (preprint)*, 1-16, pls. 1-6.

*Stradner, H. and Papp, A. 1961. Tertiäre Discoasteriden aus Österreich und deren stratigraphische Bedeutung mit Hinweisen auf Mexico, Rumänien und Italien. *Jb. Geol. Bundesanst. Wien, Sonderband* **7**, 1-160, pls. 1-42.

*These titles contain detailed bibliographies.

40. SOME ASPECTS OF PRE-QUATERNARY DIATOMS IN THE OCEANS

T. KANAYA

Summary: The Miocene diatom assemblages of the experimental Mohole sequence, from a subtropical latitude, are intermediate in floral composition between those of California and equatorial latitudes of equivalent age. The correlation between the experimental Mohole sequence and an equatorial core (Swedish Deep-Sea Expedition 76) is presented as an example of the type of biostratigraphical correlation that is feasible by means of diatoms in the present state of knowledge. Correlation between core SDSE 76 and the Miocene sequence of California is possible through the intermediate experimental Mohole section. It appears that, when combined with the information of other groups of microfossils, diatoms are particularly useful in tying Tertiary biostratigraphical zones of lower latitudes in with their time-equivalents in higher latitudes.

INTRODUCTION

The first report on occurrence of Tertiary diatoms on the ocean floor was in a paper entitled 'Fossil diatoms dredged from Bering Sea' by Hanna in 1929, in which he called attention to the fact that a sediment sample from the Bering Sea at a depth of 913 fathoms (*Albatross* st. 4029 H: 54° 47′ 20″ N, 179° 80′ 00″ W, grey sand and clay) contained a considerable number of what are considered to be Tertiary species. Hanna noticed this during the scrutiny of a paper by A. Mann which, in 1907, reported the diatoms of the sediment collections made by the *Albatross* in the Pacific during 1888-1904.

The second report came almost twenty years later when more adequate material for micropalaeontological study was collected by the Swedish Deep-Sea Expedition 1947-48; the report by Kolbe (1954) on the diatoms of twenty-six piston cores from the Equatorial Pacific has a separate section devoted to the occurrence of Tertiary species in core 76. The vertical distribution of diatom species through the 13-metre sequence of this piston core was compared with the diatom sequences in the other cores. He showed that the species judged to be of Tertiary age were restricted to the lower levels of core 76, making the lower part of the core distinctly different in diatom composition from the upper part of the same core, as well as from the sequences in other piston cores studied for the report.

The next twenty years was a period of intensive deep-sea exploration. A great number of cores that are probably suitable for the study of pre-Quaternary diatoms have been collected from the Pacific (Riedel and Funnell 1964). This may be equally true for the Atlantic and Indian Oceans. The published records, however, are regrettably scanty. Confronted with the progress made during the last few years on pre-Quaternary Foraminifera, Radiolaria and calcareous nannoplankton from the deep-sea floor (*e.g.* Riedel and Funnell 1964; Parker, 1964, 1967; Riedel 1959; Friend and Riedel 1967; Martini and Bramlette 1963) it looks as if work on diatoms has fallen behind. However, in the nature of the diatoms must lie their unique value to the micropalaeontology of pre-Quaternary deep-sea sediments. Being autotrophic they should provide information regarding past events in the surface layers of the ocean; their silica cell walls give them a better chance than Foraminifera and calcareous nannoplankton for preservation in the bottom sediments at depths below the calcium carbonate compensation depth and their position as one of the main primary producers in the ecosystem of the oceans should not be forgotten either.

The purpose of the present paper is to illustrate the type of biostratigraphical correlations now feasible on the basis of diatoms, and their implications for deep-sea stratigraphy or palaeoceanography in the future. The examples are based on personal experience with limited material, both in numbers of samples and in the span of geological time covered.

STRATIGRAPHICAL INFORMATION ON FOSSIL DIATOMS

Basic information for the determination of the stratigraphical ranges of diatoms was summarized by Kanaya (1957, 1959). A number of important contributions have been added since, notably two textbooks on fossil diatoms by Proshkina-Lavrenko (1949) and Jousé and Sheshukova-Poretzkaya (1963). A brief summary of the present status of diatom stratigraphy is necessary before discussing the pre-Quaternary diatoms from the ocean floor.

The immense taxonomic value of classical monographs and atlases, mostly published in the latter half of the nineteenth century (*e.g.* Greville 1860, 1861-6; Grunow 1884; Van Heurck 1880-5; Schmidt 1875-; Grove and Sturt 1886-7; Rattray 1889; 1890; Brun 1891; Brun and Tempère 1889; Pantocsek 1886-92, etc.) must be taken into account in considering pre-Quaternary diatoms. It may be sufficient to say that more than 10,000 species, the large majority of which have fossil specimens as types, were already on record when De Toni (1891-4) compiled his synopsis of diatoms. The Latin descriptions were given for most of the species dealt with, and, under the heading 'habitat', the occurrences of the species were given. The inadequacy of the description of localities, drawn from original connotations, considerably lessens their stratigraphical value, and careful screening is necessary to eliminate doubtful records (Hanna 1936; Kanaya 1957, 1959; Sheshukova-Poretzkaya 1967).

The modern phase of work with the definite aim of laying the foundation for the stratigraphical use of fossil diatoms was started by G. D. Hanna when he described the Upper Cretaceous diatoms from the Moreno Shale of California (Hanna 1927*a, b*). Aided greatly by the refinements made in diatom taxonomy by Hustedt (1927-), descriptions of diatom floras from stratigraphically known horizons have been accumulated since, mainly by stratigrapher-palaeontologists. Knowledge is still fragmental; nevertheless the possibility of identifying by diatoms the upper Cretaceous and each Series of the Tertiary System can now be seriously considered. This progress may be summarized as follows:

1. On the basis of extensive research carried out in the U.S.S.R., the so-called phylogeny of diatoms was presented by Jousé and Sheshukova-Poretzkaya (1963). This paper gave the distribution and stratigraphical ranges of diatoms in terms of genera, and showed that the difference between the Mesozoic and Cenozoic diatoms are apparent even at a generic level. Further, it elucidated the fact that the Palaeogene flora lacks a number of superfamilies that exist during the Neogene.

2. To the records of Cretaceous diatoms recorded only from California (Hanna 1927*b*, 1934; Long, Fuge and Smith 1946), a number of records of upper Cretaceous diatoms have been added from the U.S.S.R., mainly from the Urals (Proshkina-Lavrenko 1949; Jousé 1951*a*; Strelnikova 1965*a*, 1965*b*, 1966*a*, 1966*b*).

3. Palaeocene and Oligocene floras which had not previously been adequately described were reported from the U.S.S.R. (Proshkina-Lavrenko 1949; Jousé 1951*b*, 1955; Strelnikova 1960; Gleser 1962; Jousé and Sheshukova-Poretzkaya 1963).

4. Subsequent to Hanna (1927*a*) knowledge of Eocene diatoms has been increased by reports from California (Kanaya 1957), Germany (Benda 1965) and the U.S.S.R. (Proshkina-Lavrenko 1949; Jousé 1955; Strelnikova 1960; Jousé and Sheshukova-Poretzkaya 1963; Gleser and Posnova 1964; Sheshukova-Poretzkaya and Gleser 1964).

5. The stratigraphical distribution of diatoms in the Miocene and Pliocene sequence of Java was reported by Reinhold (1937).

6. For the Soviet Far East a range chart has been compiled of the more important species from the late Miocene to Recent (Sakhalin, Kamtchatka, Kurile and Komandolski Islands: Jousé 1962), and a zonal subdivision of the Neogene of Sakhalin and Kamtchatka by means of diatoms has been presented by Sheshukova-Poretzkaya (1959, 1967).

7. In Japan, an assemblage zone of diatoms useful for correlation was recognized in the middle Miocene of Honshû (Kanaya 1959; Sawamura 1963*b*), and the adequacy of diatoms for the correlation of the Miocene strata of northern Honshû and Hokkaido was discussed (Sawamura and Yamaguchi 1961, 1963; Sawamura 1963*a*). The difference between the upper Miocene and Pliocene assemblages was defined for one section in northern Honshû, Japan Sea side (Koizumi 1966).

8. Type Delmontian (uppermost Miocene) and Pliocene diatoms from California were described (Wornardt 1967), adding previous knowledge of the Neogene diatoms of the West Coast of North America (Hanna and Grant 1926; Hanna 1928, 1930*a*, 1930*b*, 1930*c*, 1932; Lohman 1938, 1950 *in* Woodring and Bramlette). The stratigraphical distribution of marine species of the genus *Denticula* in Miocene and Pliocene samples from California suggest some value for correlation (Simonsen and Kanaya 1961).

Despite the considerable progress made during the past forty years, there still are many important questions remaining to be answered. For example, the geographical setting of certain fossil floras, the extent of which defines the value of the criteria for correlation; how far the standard set in the higher latitudes can be extended to the lower latitudes, or vice versa?; should effort be directed toward setting up sequences of concurrent range zones of diatoms, as has been done for planktonic Foraminifera?; or in case of diatoms should we rather rely upon the range zones of certain species? The availability of good sections either land-based or by deep-sea drilling in the oceans, and the availability of man power will decide these matters. Meanwhile, some observations can be made bearing these points in mind.

DIATOMS FROM THE EXPERIMENTAL MOHOLE DRILLINGS

Diatoms were studied from thirteen levels in the experimental Mohole drillings (Guadalupe site). The uppermost sample was barren of diatoms, and a total of 117 taxa have been distinguished from twelve samples containing diatoms. Of these 117 taxa, 22 are judged to be benthonic and tychopelagic species. No freshwater species were encountered.

The distributions of the more important species in the samples are shown in Fig. 40.1. These species were selected to illustrate the characteristics of the Tertiary diatom flora of the experimental Mohole site. Priority was given to the more frequent species; the stratigraphical value of the species, judged from occurrences in other localities, was also taken into consideration. The relative frequencies of the species, estimated from a count of approximately 1,000 specimens per sample, are also indicated in Fig. 40.1.

A distinctive feature of the flora listed is that it contains diatoms which are known in Tertiary records from higher as well as lower latitudes in the circum Pacific region. With a few exceptions, most of the twenty-six species listed occur in the California Miocene. To my knowledge, the exceptions are *Craspedodiscus coscinodiscus* Ehr., *Bruniopsis mirabilis* (Brun) Karst., *Coscinodiscus paleaceus* (Grun.) Ratt. and cf. *C. nodulifer* Schmidt. The story is more or less the same even if the comparison is extended to the Miocene records of the

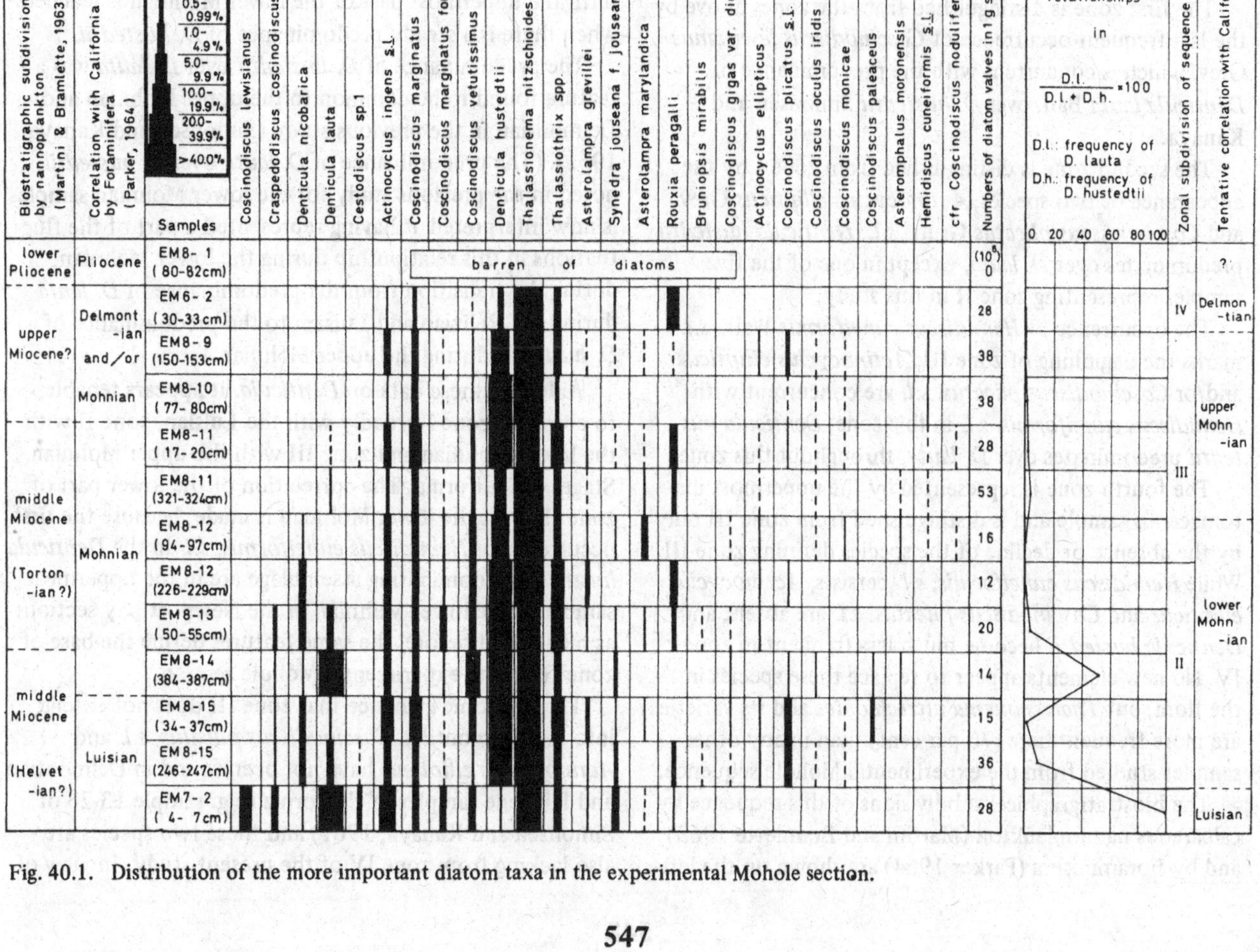
Biostratigraphic subdivision by nannoplankton (Martini & Bramlette, 1963)
Correlation with California Stages by Foraminifera (Parker, 1964)
<0.1%
0.1-0.49%
0.5-0.99%
1.0-4.9%
5.0-9.9%
10.0-19.9%
20.0-39.9%
>40.0%
Samples
Coscinodiscus lewisianus
Craspedodiscus coscinodiscus
Denticula nicobarica
Denticula lauta
Cestodiscus sp.1
Actinocyclus ingens s.l.
Coscinodiscus marginatus
Coscinodiscus marginatus forma
Coscinodiscus vetutissimus
Denticula hustedtii
Thalassionema nitzschioides and vars.
Thalassiothrix spp.
Asterolampra grevillei
Synedra jouseana f. jouseana
Asterolampra marylandica
Rouxia peragalli
Bruniopsis mirabilis
Coscinodiscus gigas var. diorama
Actinocyclus ellipticus
Coscinodiscus plicatus s.l.
Coscinodiscus oculus iridis var. borealis
Coscinodiscus monicae
Coscinodiscus paleaceus
Asteromphalus moronensis
Hemidiscus cuneiformis s.l.
cfr. Coscinodiscus nodulifer
Numbers of diatom valves in 1 g sediments
Denticula lauta and Denticula hustedtii relationships in
D.l. / D.l.+D.h. ×100
D.l.: frequency of D. lauta
D.h.: frequency of D. hustedtii
0 20 40 60 80 100
Zonal subdivision of sequence by diatoms
Tentative correlation with California Stages
lower Pliocene
Pliocene
EM8-1 (80-82cm)
barren of diatoms
(10⁶) 0
?
upper Miocene?
Delmontian and/or Mohnian
EM6-2 (30-33cm)
28
IV
Delmontian
EM8-9 (150-153cm)
38
EM8-10 (77-80cm)
38
upper Mohnian
EM8-11 (17-20cm)
28
middle Miocene (Tortonian?)
Mohnian
EM8-11 (321-324cm)
53
III
EM8-12 (94-97cm)
16
EM8-12 (226-229cm)
12
EM8-13 (50-55cm)
20
lower Mohnian
EM8-14 (384-387cm)
14
II
middle Miocene (Helvetian?)
Luisian
EM8-15 (34-37cm)
15
EM8-15 (246-247cm)
36
EM7-2 (4-7cm)
28
I
Luisian

Fig. 40.1. Distribution of the more important diatom taxa in the experimental Mohole section.

northern part of Japan. Similarly, when the list is compared with records from the Neogene of Java (Reinhold 1937) and from SDSE 76, it is again found that only five of the twenty-six are absent from the Miocene of these tropical latitudes. The five species that are missing are: *Denticula lauta* Bail., *Synedra jouseana jouseana* Sheshuk., *Rouxia peragalli* Brun and Herib., *Coscinodiscus oculus iridis borealis* (Bail.) Cleve, and *C. monicae* Grun. Many more species appear to be lacking, however, from the Tertiary records of subarctic latitudes. For example, a comprehensive study of Neogene marine diatoms from Sakhalin and Kamtchatka (Sheshukova-Poretzkaya 1967) does not mention thirteen of the twenty-six taxa under consideration.

As indicated in Fig. 40.1, the flora is largely dominated by *Thalassionema nitzschioides* and its varieties, diatoms which are almost ubiquitous in marine floras of the Miocene and younger ages. *Denticula hustedtii*, whose heyday appears to be in the Mohnian in California (Simonsen and Kanaya 1961), is the next frequent species, throughout the larger part of the experimental Mohole sequence.

The local sequence at the experimental Mohole site is subdivided into four zones, tentatively numbered I, II, III and IV in ascending order.

The first zone is distinguished from the zones above by the last frequent occurrence of *Coscinodiscus lewisianus* Grev. which is concurrent with the predominance of *Denticula lauta* Bail. over *D. hustedtii* Simonsen and Kanaya.

The second zone is distinguished from zone I by the appearance of two species, *Actinocyclus ellipticus* Grev. and *Coscinodiscus plicatus* Grun., *s.l.*; *Denticula hustedtii* predominates over *D. lauta*, except in one of the five samples representing zone II in this study.

The occurrence of *Hemidiscus cuneiformis* Wall., *s.l.* marks the beginning of zone III; *Actinocyclus ellipticus* and/or *Coscinodiscus plicatus, s.l.* are concurrent with *Hemidiscus cuneiformis, s.l.* in this zone; *Denticula hustedtii* predominates over *D. lauta* throughout this zone.

The fourth zone is represented by the uppermost diatomaceous sample and is distinguished from zone III only by the absence or decline of the species defining zone III. While *Hemidiscus cuneiformis, s.l.* persists, *Actinocyclus ellipticus* and *Coscinodiscus plicatus, s.l.* are absent and *Denticula hustedtii* become much less frequent in zone IV. No new elements appear to replace these species in the flora, but *Thalassionema nitzschioides* and its varieties are more frequent here (76 per cent) than in any other samples studied from the experimental Mohole sequence.

The biostratigraphical subdivisions of this sequence by calcareous nannoplankton (Martini and Bramlette 1963) and by Foraminifera (Parker 1964) are shown on the left-hand side of Fig. 40.1. On the right-hand side of the figure, the tentative correlation with California Stages made by the use of diatoms is indicated. The basis for this correlation is a comparison with available Californian records, relying particularly on the *Denticula lauta* and *Denticula hustedtii* relationships ascertained in California. The relationship between these two species in the experimental Mohole is also shown in Fig. 40.1.

Previous study of the stratigraphical distribution of marine species of the genus *Denticula* Kutz. in the Californian Tertiary by Simonsen and Kanaya (1961) was checked by further, more adequate samples. A series of thirteen samples collected in sequence from the upper Luisian and lower Mohnian parts of the Newport Bay section, California, represents the stratigraphical sequence ranging from sample no. 1 to no. 36 of Lipps (1964). Distribution of the species of *Denticula* in the Newport Bay samples has been studied by I. Koizumi for the present paper. It was found that the observations made by Simonsen and Kanaya (1961) generally hold true for the Newport Bay section. Namely, *Denticula lauta* predominates over *D. hustedtii* through seven samples from the upper Luisian; the lower Mohnian samples show fluctuations in the relationship between these two species until the uppermost part of the lower Mohnian is reached. when there is a steady predominance of *D. hustedtii.*

The predominance of *D. hustedtii* over *D. lauta* was a feature found to be common to the upper Mohnian and Delmontian in the previous study (Simonsen and Kanaya 1961). (The predominance of *D. lauta* over *D. hustedtii* noted in the previous study for the lower Mohnian samples is now interpreted as having represented a part of the fluctuations in this relationship during the lower Mohnian, during the transition from the predominance of *D. lauta* during the Relizian and Luisian to the predominance of *D. hustedtii* during the upper Mohnian.)

Aided by these data on *Denticula*, it appears tenable to correlate zone I broadly with the Luisian, zone II with the lower Mohnian and zone III with the upper Mohnian Stages of California. The correlation of the lower part of zone III with the lower Mohnian is made, because the first occurrence of *Hemidiscus cuneiformis, s.l.* in the *Denticula hustedtii*-predominating assemblage are in the uppermost sample of the lower Mohnian in the Newport Bay section; as mentioned before, the same features define the base of zone III in the experimental Mohole section.

There is some evidence that zone III does not extend into the Delmontian. *Coscinodiscus plicatus, s.l.* and *Actinocyclus ellipticus* have not been found in Delmontian and Pliocene samples in California (e.g. sample 13-23 of Simonsen and Kanaya, 1961) and these two species are also lacking from zone IV of the present study. In view of this homotaxial relationship, zone IV is correlated with the Delmontian Stage of California.

The possibility that zone I is Relizian can also very probably be ruled out by the diatom evidence, because of the lack of such good Relizian marker species as *Cymatogonia amblyoceras* (Ehr.), *Rhaphoidiscus marylandicus* Christ. and *Annelus californicus* Temp. These species, whose occurrence in the Californian Miocene was first reported by Hanna (1932) are consistently found in samples from the Relizian of California (e.g. sample nos. 1-6 of Simonsen and Kanaya 1961), but have not been found from samples younger than Luisian in age.

When the list in Fig. 40.1 is compared with the diatom assemblage from a late Oligocene sample from DWBG 10 (11-27 cm), the difference is distinct: only four of the species listed here. *Coscinodiscus lewisianus, Craspedodiscus coscinodiscus, Cestodiscus* sp. 1 and *Coscinodiscus vetutissimus,* are found in the late Oligocene sample (for age see Riedel and Funnell 1964). Of these, *Craspedodiscus coscinodiscus* is represented by very robust and coarsely areolated specimens (see Pl. 40.4, figs. *1a, b*). *Coscinodiscus lewisianus* is much more rare in the late Oligocene sample and is represented by the variety *C. lewisianus* Grun. var. *similis* Rattray (see Pl. 40.5, figs. 7, 8) and the obviously related species, *C. lanceolatus* Castr. (Pl. 40.5, fig. 9); both are common in the sample. Frequent occurrences in the late Oligocene sample of species that are lacking in the experimental Mohole material add to the differences between the two floras.

CORRELATION WITH CORE SDSE 76

The presence in the experimental Mohole sequence of species that have been recorded from core SDSE 76 (Kolbe 1954) gives hope for a correlation of the two sequences, from subtropical and equatorial latitudes respectively. For this purpose, core SDSE 76 was studied at fifteen levels to review the vertical distribution in the core of the species recorded by Kolbe. Kolbe's study was based on the examination of diatoms at sixty-eight levels, at almost every 20 centimetre throughout its 13 metre sequence. My revision (Fig. 40.2), therefore draws heavily

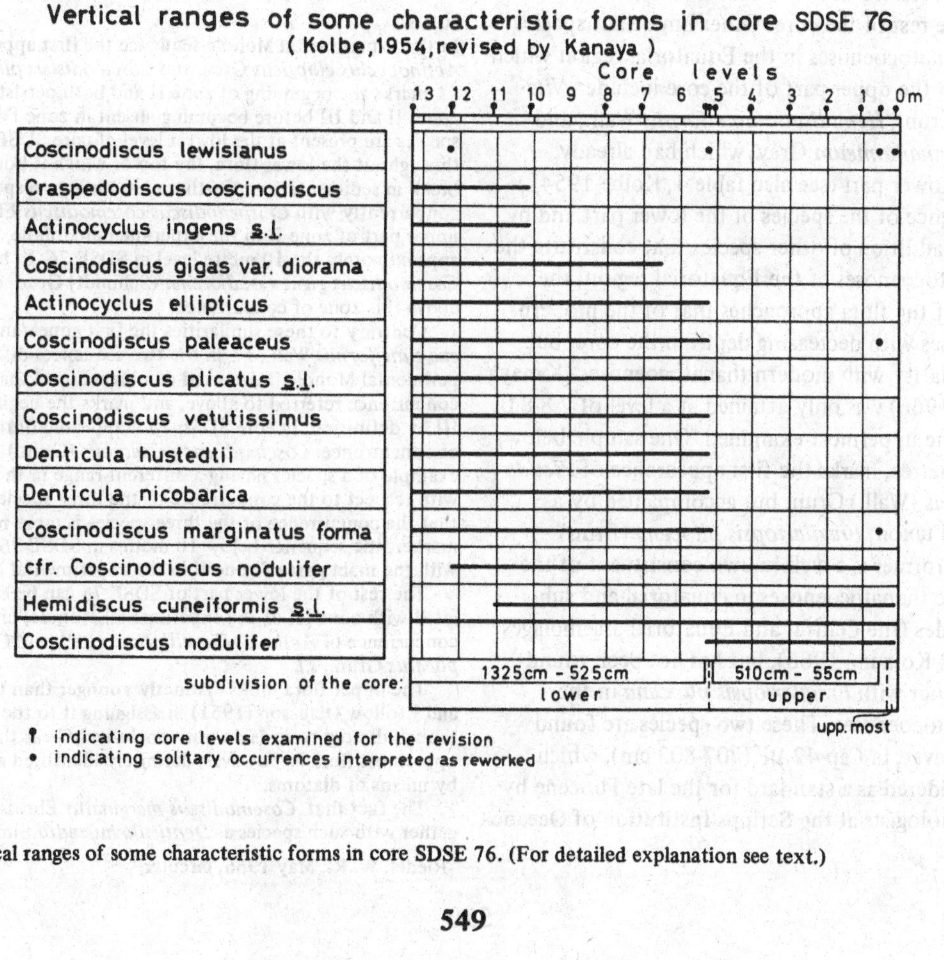
Vertical ranges of some characteristic forms in core SDSE 76
(Kolbe, 1954, revised by Kanaya)
Core levels
13 12 11 10 9 8 7 6 5 4 3 2 1 0m
Coscinodiscus lewisianus
Craspedodiscus coscinodiscus
Actinocyclus ingens s.l.
Coscinodiscus gigas var. diorama
Actinocyclus ellipticus
Coscinodiscus paleaceus
Coscinodiscus plicatus s.l.
Coscinodiscus vetutissimus
Denticula hustedtii
Denticula nicobarica
Coscinodiscus marginatus forma
cfr. Coscinodiscus nodulifer
Hemidiscus cuneiformis s.l.
Coscinodiscus nodulifer
subdivision of the core:
1325cm - 525cm
lower
510cm - 55cm
upper
upp. most
indicating core levels examined for the revision
indicating solitary occurrences interpreted as reworked

Fig. 40.2. Vertical ranges of some characteristic forms in core SDSE 76. (For detailed explanation see text.)

upon Kolbe's work. Since the purpose of the revision is to correlate two separate Tertiary sequences, the species which appear unimportant for this purpose were eliminated from Kolbe's original table (Kolbe 1954, table 4). Conversely, the distribution of some experimental Mohole species, chosen from those listed in Fig. 40.1, were added for comparison. Special attention was paid to three levels, 534-5 centimetres, 525-6 centimetres, and 509-509.5 centimetres, the levels between which important changes were likely to be detected, according to Olausson (1961).

The figure shows that several of the experimental Mohole species terminate at 525-6 centimetres, and do not extend into the 509-509.5 centimetre level. The boundary separating the lower and the upper part of the core is thus drawn between 525-6 centimetres and 509-509.5 centimetres. *Craspedodiscus coscinodiscus* Ehr. and *Coscinodiscus gigas* var. *diorama* (Schmidt) Grun. which terminate in the middle of the experimental Mohole sequence also terminate here at specific levels in the lower part of the core, and in the same order as they do at the experimental Mohole site.

As is shown on Fig. 40.2, *Coscinodiscus nodulifer* Schmidt, the most characteristic species of the uppermost levels in the Equatorial area (Kolbe 1954; Kanaya and Koizumi 1966) started at 534-5 centimetres, and persisted throughout the rest of the core. Other important species of modern thanatocoenoses in the Equatorial region which persist through the upper part of the core include: *Nitzschia marina* Grun., *Hemidiscus cuneiformis* Wall., and *Triceratium cinnamomeum* Grev. which had already started in the lower part (see also table 4, Kolbe 1954).

By the absence of the species of the lower part and by the successive addition of other species that constitute the modern thanatocoenoses of the Equatorial region, the composition of the flora approaches that of the modern thanatocoenoses with decreasing depth in the core, but complete similarity with modern thanatocoenoses (Kanaya and Koizumi 1966) was only attained at a level of 7.8-8.0 centimetres, the uppermost examined. One sample below, at 54-5 centimetres, marks the first appearance of *Pseudoeunotia doliolus* (Wall.) Grun. but accompanied by a closely related taxon, *Fragilariopsis pliocena* (Brun) Sheshuk. The former is a well-known constituent of the modern Pacific thanatocoenoses in equatorial and subtropical latitudes (the Central and Equatorial assemblages of Kanaya and Koizumi 1966), but has not been found to occur together with *Fragilariopsis pliocena* in the modern thanatocoenoses. These two species are found together, however, in Cap 42 BP (707-803 cm), which has been considered as a standard for the late Pliocene by micropalaeontologists at the Scripps Institution of Oceanography.[1] For this reason, the uppermost sample is distinguished from the part below comprising the sequence from 510 to 55 centimetres.

When Kolbe (1954) first studied SDSE 76, he found that the change in the diatom population takes place in a quite limited zone, between approximately 5 and 6 metres from the top. Because pre-Quaternary species occurred regularly below this zone, he assumed that the 5-6 metre zone represented the Plio-Pleistocene boundary (Kolbe 1954, p. 18). Later, aided by information provided by Radiolaria and coccolithophorids from the core, Olausson (1961) came to the conclusion that the boundary represented the Mio-Pliocene boundary. He suggested that the boundary could be located much more precisely, somewhere between 518 and 525 centimetres from the top; 'an undulating and somewhat inclined colour boundary' found at 519 centimetres was taken to indicate a break in sedimentation (Olausson 1961, p. 28). The present account of SDSE 76 accords with Olausson's view on the position and nature of the boundary; the disappearance of so many species between 525 centimetres and 509-509.5 centimetres can be attributed to the stratigraphical break at 519 centimetre postulated by Olausson.

Fig. 40. 3 illustrates a correlation of the two sequences, SDSE 76 and the experimental Mohole, by means of diatoms.

In the experimental Mohole sequence the first appearance of *Actinocyclus ellipticus* Grev. and *Coscinodiscus plicatus* Grun., *s.l.* marks the beginning of zone II and both persist throughout zones II and III before becoming absent in zone IV; these two species are present at the lowest level of core SDSE 76 and persist throughout the lower flora, the top of which is bounded by the break in sedimentation. Furthermore these two species are found concurrently with *Craspedodiscus coscinodiscus* Ehr. up to the upper part of zone II in the experimental Mohole, and up to approximately the 10 metre level in SDSE 76. In both sequences, *Coscinodiscus gigas* var. *diorama* (Schmidt) Grun. disappears above this zone of concurrence.

Contrary to these similarities the first appearance of *Hemidiscus cuneiformis* Wall., *s.l.* in the two sequences differs. In the experimental Mohole it occurred at a higher level than the zone of concurrence referred to above, and marks the beginning of zone III by definition; in core SDSE 76 it appeared during the period of concurrence. *Coscinodiscus paleaceous* (Grun.) Ratt. is another example of a species having a different range in the two sequences with respect to the concurrence of the three species. Assuming that the concurrence of the three species is more reliable as a time marker, the sequence below 10 metres in SDSE 76 is correlated with the main part of zone II in the experimental Mohole.

The rest of the lower part of SDSE 76 can be broadly correlated with zone III of the experimental Mohole, on the basis of the concurrence of *Actinocyclus ellipticus* Grev. and *Coscinodiscus plicatus* Grun., *s.l.*

The upper flora looks distinctly younger than the lower flora, and I follow Olausson (1961) in assigning it to the Pliocene, although diatom records do not conclusively bear this out. The uppermost flora is, however, definitely identified as Quaternary by means of diatoms.

The fact that *Coscinodiscus marginatus* Ehr. is present together with such species as *Denticula hustedtii* Simonsen and

[1] Riedel, W. R., May 1966, circular.

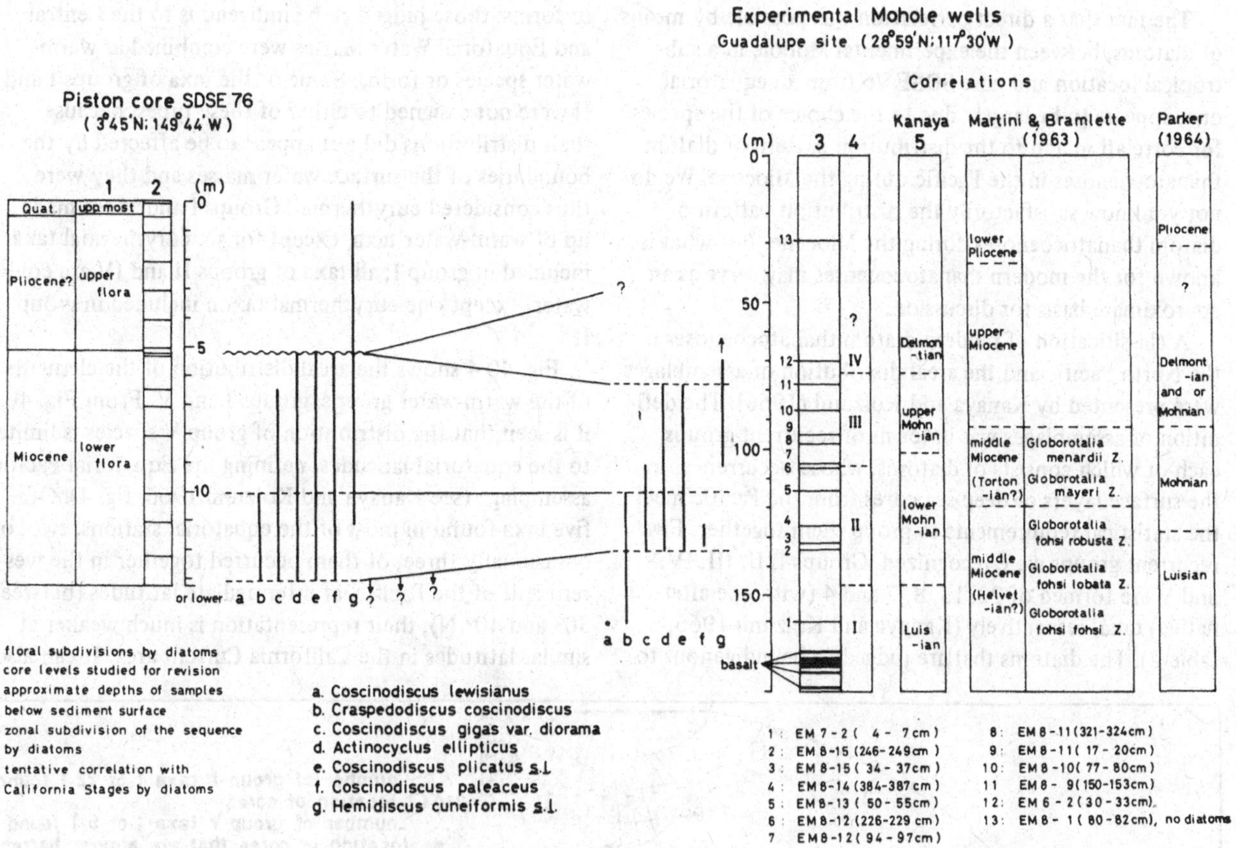
Experimental Mohole wells
Guadalupe site (28°59'N; 117°30'W)
Correlations
Kanaya
Martini & Bramlette (1963)
Parker (1964)
Piston core SDSE 76
(3°45'N; 149°44'W)
1 floral subdivisions by diatoms
2 core levels studied for revision
3 approximate depths of samples below sediment surface
4 zonal subdivision of the sequence by diatoms
5 tentative correlation with California Stages by diatoms
a. Coscinodiscus lewisianus
b. Craspedodiscus coscinodiscus
c. Coscinodiscus gigas var. diorama
d. Actinocyclus ellipticus
e. Coscinodiscus plicatus s.l.
f. Coscinodiscus paleaceus
g. Hemidiscus cuneiformis s.l.
1: EM 7-2 (4-7cm)
2: EM 8-15 (246-249cm)
3: EM 8-15 (34-37cm)
4: EM 8-14 (384-387cm)
5: EM 8-13 (50-55cm)
6: EM 8-12 (226-229 cm)
7: EM 8-12 (94-97cm)
8: EM 8-11 (321-324cm)
9: EM 8-11 (17-20cm)
10: EM 8-10 (77-80cm)
11: EM 8-9 (150-153cm)
12: EM 6-2 (30-33cm)
13: EM 8-1 (80-82cm), no diatoms

Fig. 40.3. Correlation between the experimental Mohole and SDSE 76.

Kanaya, *Rouxia peragalli* Brun et Herib. and *Actinocyclus ingens* Ratt., *s.l.* (Fig. 40.1) suggests that zone IV of the experimental Mohole sequence is still within the Miocene; whereas the upper flora of core SDSE 76 lacks Miocene aspects even in its basal horizon. The equivalent of zone IV appears to be missing from SDSE 76 in the hiatus between the upper and lower parts of the core. An upper limit to zone IV in the experimental Mohole is not definable because of the absence of diatoms in the higher deposits.

The equivalent of zone I in the experimental Mohole section does not appear to have been penetrated by the SDSE 76: *Coscinodiscus lewisianus,* whose occurrence is restricted to zone I is lacking from the lower part of SDSE 76, except for some broken and badly worn specimens occurring sporadically at various levels in the core (indicated by 'solitary occurrences' in Fig. 40.2). The frequency relationship between *Denticula lauta* Bail. and *D. hustedtii* Simonsen and Kanaya has not been used as a criterion for the correlation between the SDSE 76 and the experimental Mohole sequences, since *Denticula lauta* was totally lacking from SDSE 76.

The correlation of the experimental Mohole sequence in terms of planktonic Foraminifera zones has been mentioned by Martini and Bramlette (1963). They are given in the righthand side of Fig. 40.3. If their correlation is related to the diatom data, it is possible to suggest that the lower few metres of SDSE 76 lie within the zones defined by *Globorotalia fohsi* and its subspecies. This means that the lowermost part of the section, which Olausson (1961) assumed to be Oligocene, is in fact younger. The Miocene age of the lower flora as a whole seems to be established by the correlation with the experimental Mohole sequence presented here.

DISCUSSION AND GENERAL INTERPRETATION

The papers on the Foraminifera (Parker 1964) and calcareous nannoplankton (Martini and Bramlette 1963) from the experimental Mohole pointed out that tropical elements are mostly lacking from the Miocene part of the section, and suggested a cold current, comparable to that of the present-day, off the West Coast during the accumulation of that part of the sequence. According to Martini and Bramlette (1963) the experimental Mohole nannoplankton assemblages are rather meagre, like those of equivalent age in California, and in marked contrast to those in late Tertiary *Chubasco* cores from the eastern

Equatorial Pacific.

The fact that a direct correlation was possible by means of diatoms, between the experimental Mohole in a subtropical location and core SDSE 76 from an equatorial one, appears to be largely due to the choice of the species for correlation and to the distribution pattern of diatom thanatocoenoses in the Pacific during the Miocene. We do not yet know satisfactorily the distribution pattern of diatom thanatocoenoses during the Miocene. But what is known for the modern thanatocoenoses may serve as an approximate basis for discussion.

A classification of modern diatom thanatocoenoses in the North Pacific and the areal distribution of assemblages were presented by Kanaya and Koizumi (1966). The definition of assemblages was in terms of recurrent groups, each of which consists of diatoms, whose occurrences in the surface layers of deep-sea cores from the Pacific meet the statistical requirements to group them together. Five recurrent groups were recognized. Groups I, II, III, IV, and V are formed of 22, 13, 8, 7 and 4 (with one alternative) taxa, respectively (Kanaya and Koizumi 1966, table 2). The diatoms that are judged to be indigenous to the Subarctic Water Mass are called the cold-water species or forms; those judged to be indigenous to the Central and Equatorial Water Masses were combined as warm-water species or forms. Some of the taxa of groups I and II were not assigned to either of these types, because their distributions did not appear to be affected by the boundaries of the surface water masses and they were thus considered eurythermal. Groups I and V are made up of warm-water taxa, except for six eurythermal taxa included in group I; all taxa of groups II and IV are cold-water, except one eurythermal taxon included in group II.

Fig. 40.4 shows the areal distribution of the elements of the warm-water groups, groups I and V. From Fig. 40.4 it is seen that the distribution of group V species is limited to the equatorial latitudes, defining the Equatorial Pacific assemblage (see Kanaya and Koizumi 1966, fig. 4). Of five taxa found in most of the equatorial stations, two, or occasionally three, of them occurred together in the western half of the Pacific, at intermediate latitudes (between 30° and 40° N); their representation is much weaker at similar latitudes in the California Current area. It can also

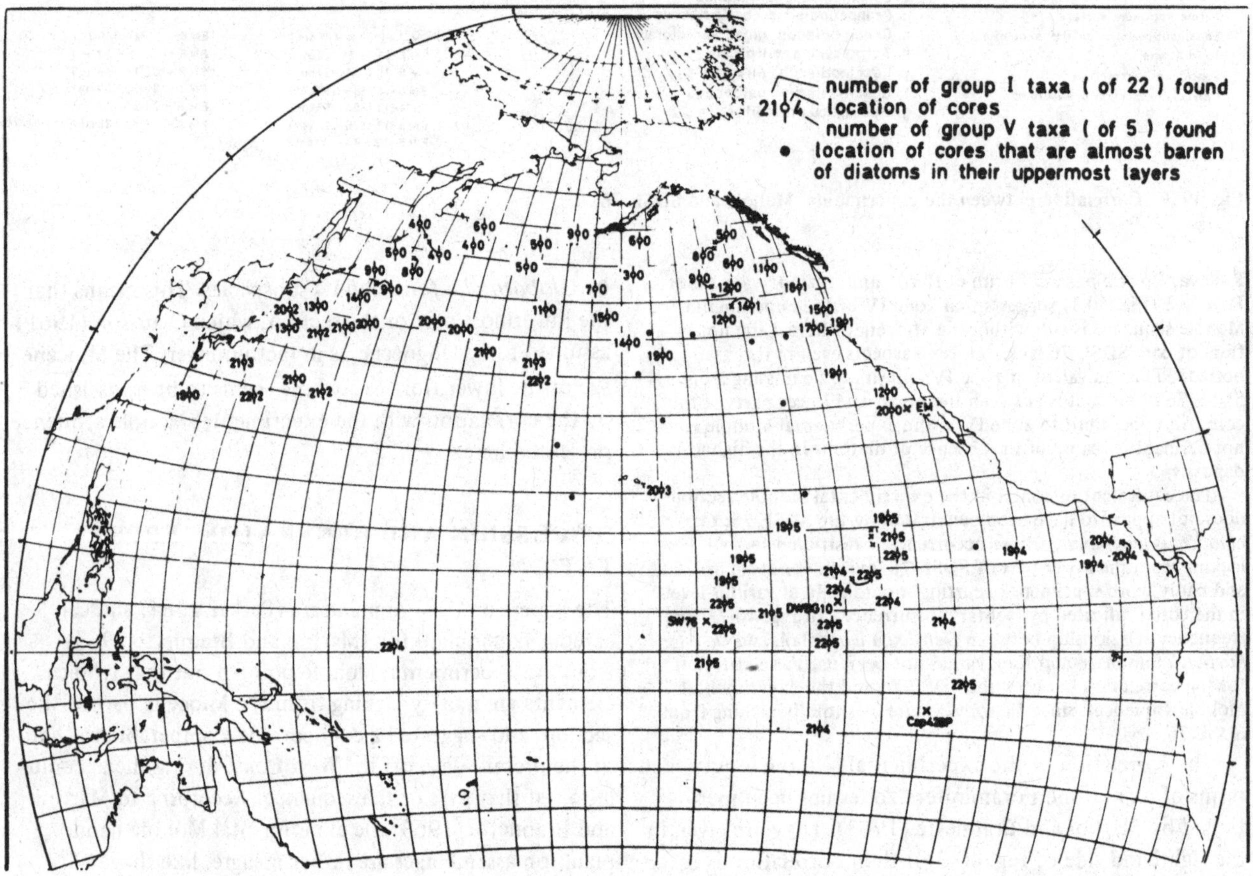
21
4
number of group I taxa (of 22) found
location of cores
number of group V taxa (of 5) found
location of cores that are almost barren of diatoms in their uppermost layers
EM
SW76
DWBG10
Cap43BP

Fig. 40.4. Distribution of group I and group V taxa (Kanaya and Koizumi, 1966) in the uppermost layers of deep-sea cores from the North Pacific.

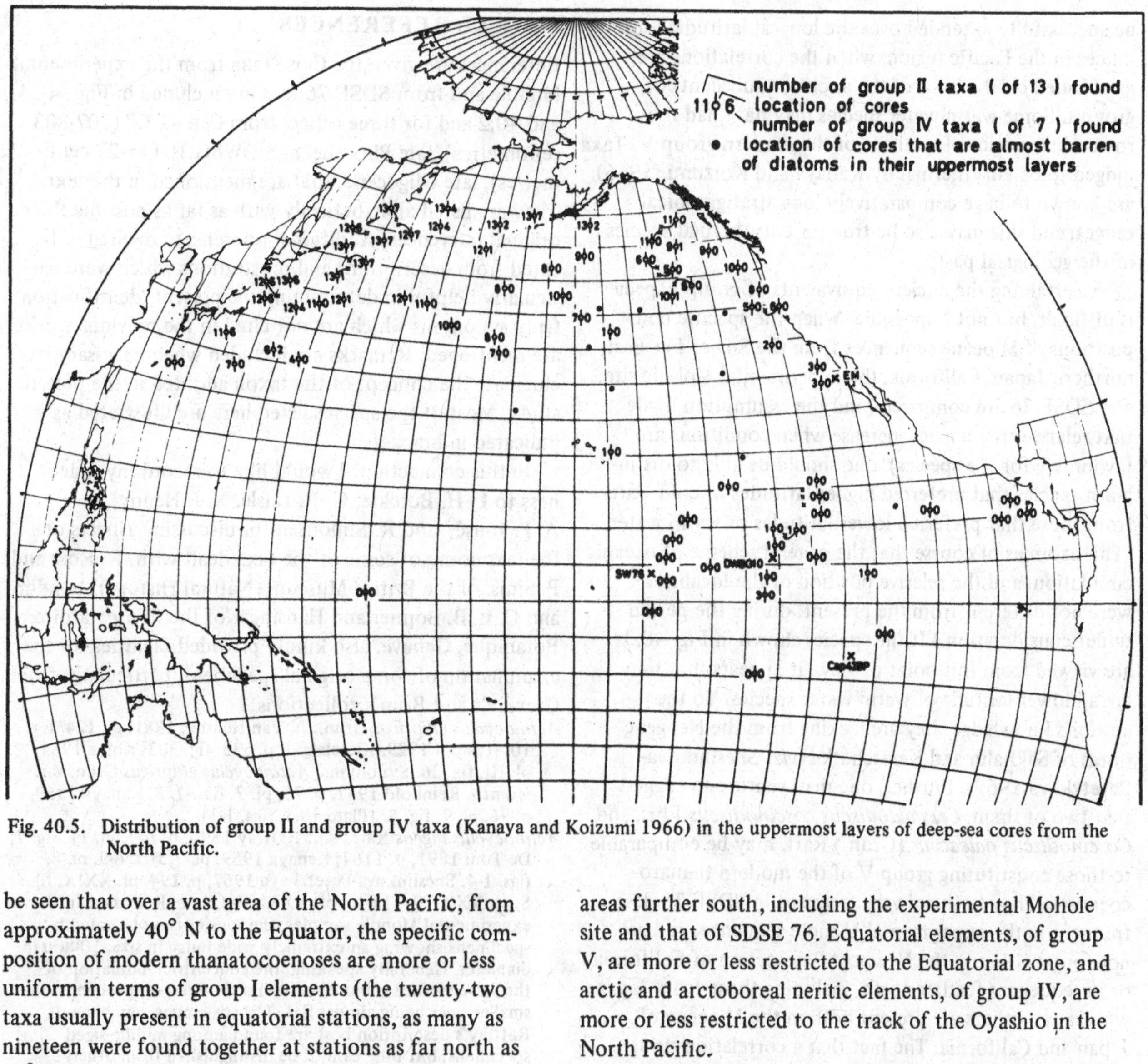
number of group II taxa (of 13) found
location of cores
number of group IV taxa (of 7) found
location of cores that are almost barren of diatoms in their uppermost layers

Fig. 40.5. Distribution of group II and group IV taxa (Kanaya and Koizumi 1966) in the uppermost layers of deep-sea cores from the North Pacific.

be seen that over a vast area of the North Pacific, from approximately 40° N to the Equator, the specific composition of modern thanatocoenoses are more or less uniform in terms of group I elements (the twenty-two taxa usually present in equatorial stations, as many as nineteen were found together at stations as far north as 40° N).

Figure 40.5 shows the areal distribution of the elements of the cold-water groups, groups II and IV. The group IV elements, mostly arctic and arcto-boreal neritic species do not quite reach the California Current area. More than half of the group II elements characterizing the thanatocoenoses below the Subarctic Water Mass are absent in the California Current area south of 40° N, including the site of the experimental Mohole.

From Figs. 40.4 and 40.5 it can be concluded that in the modern thanatocoenoses, the group II elements are poorly represented in the area off California and their representation is almost nil near the experimental Mohole site. Group I elements (mostly warm-water) maintain a similarly high degree of integrity in intermediate latitudes off both Japan and California, as well as throughout the areas further south, including the experimental Mohole site and that of SDSE 76. Equatorial elements, of group V, are more or less restricted to the Equatorial zone, and arctic and arctoboreal neritic elements, of group IV, are more or less restricted to the track of the Oyashio in the North Pacific.

The intermediate nature of the experimental Mohole diatom assemblages has been already pointed out. The comparison was crude, and the species taken into consideration were chosen rather subjectively. Nevertheless, it is the author's judgement that the Miocene floras of Sakhalin and Kamtchatka (Sheshukova-Poretzkaya 1967) and those of the experimental Mohole site have more species in common than comparable modern thanatocoenoses from the subarctic and subtropical areas of the North Pacific. The similarity in specific composition between the Miocene floras of SDSE 76, the experimental Mohole, California and northern Japan appears to be largely due to the presence of a group of warm-water species similar in their distribution to group I of the modern thanatocoenoses. If this is the case the correlation of Tertiary sequences by means of diatoms can be successfully extended over the longest latitudinal distances in the Pacific region, when the correlations are established on the basis of the ancient equivalents of group I. Some warm-water species may have had narrower distributions like those of the modern group V. Taxa judged to be eurythermal by Kanaya and Koizumi (1966) are known to have comparatively long stratigraphical ranges, and this may also be true for eurythermal species of the geological past.

Ascertaining the ancient equivalents of group I species is difficult, but not impossible. When the specific compositions of Miocene sequences from the Soviet Far East, northern Japan, California, the experimental Mohole site, and SDSE 76 are compared (and the assumption made that relative frequencies increase when conditions are favourable for the species), one should be able to distinguish species that preferred higher latitudes or cold water from those that preferred lower latitudes or warm water. (This assumes of course that the general scheme of oceanic circulation, and the relative position of the localities were not different from the present, during the period under consideration.) If the species shown in Fig. 40.3 are viewed from this point of view, it appears that they are all lower latitude or warm-water species. To the author's knowledge they are lacking from the Neogene floras of Sakhalin and Kamtchatka (*viz.* Sheshukova-Poretzkaya 1967). Further, one may venture to suggest that two of them, *Craspedodiscus coscinodiscus* Ehr., and *Coscinodiscus paleaceus* (Grun.) Ratt. may be comparable to those constituting group V of the modern thanatocoenoses. They both are very frequent in SDSE 76, less frequent in the experimental Mohole sequence, and are not found either in the Newport Bay section of California, or in sequences further north, including those from Japan; whereas all others are present both in the Miocene of Japan and California. The fact that a correlation using several warm-water species, including what appear to be equatorial ones, was possible between the station SDSE 76 and the experimental Mohole, and the fact that extremely high concentrations (see Fig. 40.1) of diatoms occurred at the experimental Mohole site during Middle and Upper Miocene times, strengthens the suggestion of Riedel and Funnell (1964) that the biologically productive Equatorial Current system was wider during part of the Miocene. On the other hand, the fact that the experimental Mohole sequence appears to have more elements in common with those of higher latitudes than might be expected from the distribution of diatom species in the modern thanatocoenoses of the North Pacific needs some explanation.

FLORAL REFERENCES

References are given for those taxa from the experimental Mohole and from SDSE 76 that are included in Figs. 40.1 and 40.2 and for three others from Cap 42 BP (707-803 centimetres), late Pliocene, and DWBG 10 (11-27 centimetres), late Oligocene, that are mentioned in the text. They are listed alphabetically with as far as possible their original references. Additional references are cited, selected from descriptions and illustrations which were particularly helpful in determining the present identifications. Only synonyms which are not cited in the previous works are mentioned. Remarks are included when necessary to elucidate the concept of the taxon adopted in the present study. Most of the species listed here are illustrated as indicated in brackets.

In this connection, I would like to record my indebtness to L. H. Burckle, G. R. Hasle. N. I. Hendy, A. P. Jousé, and R. Simonsen for discussing aspects of the taxonomy of some of the taxa dealt with. R. Ross and P. Sims, of the British Museum (Natural History), London, and C. E. B. Bonner and Hansinger of the Conservatoire Botanique, Genève, also kindly provided facilities for the examination of some original specimens, particularly from Greville's and Brun's collections.

Actinocyclus ellipticus Grun.: *in* Van Heurck, 1881, pl. 124, fig. 10; Hustedt 1929, Kieselalg. I, p. 533, fig. 303; Kolbe 1954, pl. III, fig. 26. Synonyms: *Actinocyclus ellipticus* Grun. var. *javanica* Reinhold 1937, p. 75, pl. 7, figs. 7, 8; Kanaya 1959, p. 96, pl. 7, fig. 5. [Plate 40.5, figs. 1-3]

Actinocyclus ingens Ratt., s.l.: Rattray 1890, p. 147, pl. 11, fig. 7; De Toni 1891, p. 1167; Kanaya 1959, pl. 7, figs. 6-9, pl. 8, figs. 1-4. Sheshukova-Poretzkaya 1967, p. 194, pl. XXIX, fig. 8, pl. XXX, figs. 1a-1e, pl. XXXI, figs. 1*a-e*. Remarks: In the experimental Mohole samples, the species is represented by specimens showing an extremely wide range in size, 10-80μ in diameter. Generally speaking, the concentric undulation of the valve surface is more pronounced on larger specimens, the smaller ones being almost flat. The specimens which meet Rattray's description best are found among middle-sized specimens, but they cannot be distinguished from others because of the presence of intermediate forms which connect individuals of various sizes in a continuous chain from one extreme to another. The presence of an ocelus, in a form like that described by Rattray, is always confirmed, when specimens are well preserved, even for the smaller and/or flatter forms. For the present paper a wide concept has been adopted to include specimens which are otherwise referable to *Coscinodiscus elegans* Greville (1866, v. 14, p. 3, pl. 1, fig. 6), if the ocelus is found to be absent. Also included are forms like those illustrated by Reinhold (1937, pl. 1, fig. 5) as *Actinocyclus neogenicus.* In the experimental Mohole sequence, the larger specimens are more common in lower horizons, i.e. in zones I and II, and they tend to give way to smaller ones in the upper horizons. Large and strongly undulated forms were not found from in SDSE 76. [Plate 40.6, figs. 1-8]

Asterolampra grevillei (Wall.) Greville: 1860, v. 8, p. 113, pl. IV, fig. 21; Hustedt 1929, Kieselalg. I, p. 489, fig. 274.

Asterolampra marylandica Ehr., 1845: Hustedt 1929, Kieselalg. I, p. 485, fig. 271.

Asteromphalus moronensis (Grev.) Rattray: 1889, p. 659; Kolbe 1954, p. 23, pl. I, fig. 3; Kanaya 1959, p. 92, pl. 6, fig. 7.

[Plate 40.2, fig. 7]

Bruniopsis mirabilis (Brun.) Karst.: Kolbe 1954, p. 24, pl. IV, figs. 44a, 44b. Remarks: Represented only by fragments in the lower part of SDSE 76 as well as in the experimental Mohole.

Cestodiscus sp. 1. Remarks: This form most closely resembles *Cestodiscus stokesianus* Greville (1866, v. 14, p. 123, pl. 11, fig. 4) reported from the 'Moron deposit'. [Plate 40.6, fig. 9]

Coscinodiscus gigas Ehr. var. *diorama* (Schmidt) Grunow: 1884, p. 76; Rattray 1889, p. 542; Warnardt 1967, p. 24, fig. 25. Remarks: Critical comparison is necessary with *Coscinodiscus fulgularis* Brun reported by Hanna (1932, p. 179, pl. 1, fig. 2). [Plate 40.1, fig. 1]

Coscinodiscus lanceolatus Castracane: 1886, p. 164, pl. 17, fig. 19; Kolbe 1954, p. 29, pl. II, figs. 23-5. Remarks: In the present study this species was found only from a late Oligocene sample, DWBG 10 (11-27 cm), although Kolbe reported its occurrence in SDSE 76. [Plate 40.5, fig. 9]

Coscinodiscus lewisianus Greville: 1886, v. 14, p. 78, pl. VIII, figs. 8-10; Rattray 1889, p. 598; Reinhold 1937, p. 96, pl. VIII, fig. 11; Lohman 1948, p. 161, pl. VI, fig. 1; Kolbe 1954, p. 31, pl. II, fig. 21. [Plate 40.5, figs. 4-6]

Coscinodiscus lewisianus Grev. var. *similis* Rattray 1889, p. 598, pl. III, fig. 10; Lohman 1948, p. 162, pl. VI, fig. 6: Kolbe 1954, p. 32, pl. II, fig. 22. Remarks: In the present study this variety was only found in DWBG 10 (11-27 cm), late Oligocene in age. Kolbe (1954) found it in SDSE 58, but not SDSE 76. According to Lohman (1948, p. 152, table 16), the variety was concurrent with *Coscinodiscus lewisianus* only in the lowermost horizon of the L. G. Hammond well no. 1. [Plate 40.5, figs. 7, 8]

Coscinodiscus marginatus Ehr. 1843: Hustedt 1928, Kieselalg. I, p. 416, fig. 223; Kanaya 1959, p. 80, pl. 4, figs. 4-6; Jousé 1962, pl. 3, fig. 2, 3; Wornardt 1967, p. 26, figs. 27, 28.

Coscinodiscus marginatus Ehr. forma. Remarks: Specimens with interstitial meshes on the valve, which otherwise are identifiable with *Coscinodiscus marginatus*, were counted separately. It is of interest that *C. marginatus* is represented only by the specimens with interstitial meshes in the lower part of the SDSE 76. [40.1, figs. 4, 5]

Coscinodiscus monicae Grun. *in* Rattray: 1889, p. 563; Hanna 1932, p. 182, pl. 9, fig. 2; Wornardt 1967, p. 27, fig. 29 [Plate 40.1, fig. 2]

Coscinodiscus nodulifer Schmidt: 1878, Atlas, pl. 59, figs. 20-3; Hustedt 1928, Kieselalg. I, p. 426, fig. 229; Kolbe 1954, p. 33, pl. III, figs. 35-7; Kolbe 1957, p. 30, pl. 1, fig. 101; Lohman 1941, pl. 14, figs. 3, 5. Synonym: *Coscinodiscus radiatus* Ehr. var. *nodulifer* Reinhold 1937, p. 100, fig. 6. Remarks: The specimens illustrated are all from Cap 42 BP (757-803 cm), late Pliocene in age. [Plate 40.3, figs. 1-4]

Cf. *Coscinodiscus nodulifer* Schmidt. Remarks: Specimens with a papilla or nodule near center but with margin more finely striated than *Coscinodiscus nodulifer* are separately identified. The margin of this form is similar to that of *Coscinodiscus vetutissimus* Pant. from which it has been distinguished by the lack of the fascicular arrangement of areolae and by the lack of marginal, 'process-like' markings (Randdornen by Hustedt, Kieselalg. I, p. 413). [Plate 40.2, fig. 6]

Coscinodiscus oculis iridis Ehr. var. *borealis* (Bail.) Cleve 1883: Hustedt 1928, Kieselalg. I, p. 456, fig. 253. [Plate 40.1, fig. 3]

Coscinodiscus paleaceus (Grun.) Rattray: 1889, p. 597; Van Heurck 1881, pl. 128, fig. 6 (as *Stoshia*); Kolbe 1954, p. 34, pl. III, figs. 32a, 32b. Not *Cymatosira andersoni* Hanna 1932, p. 187, pl. 10, fig. 6. [Plate 40.5, fig. 10]

Coscinodiscus plicatus Grun., *s.l.*: Grunow 1884, p. 73, pl. 3, fig. 10; Schmidt 1886, Atlas pl. 59, fig. 1; Rattray 1889, p. 530; Kolbe 1954, p. 34. Remarks: The examination of Brun's type of *Coscinodiscus flexuosus* Brun (1895, pl. XV, fig. 38, 39) in J. Brun's collection (slide no. 3551, labelled 'Abokiri') in the Conservatoire Botanique, Geneva, convinced me that *Coscinodiscus yabei* Kanaya (1959, p. 86, pl. 5, figs. 8, 9) is a synonym of *C. flexuosus;* they both have the distinctly established margin and coarse areolation of the valve. A strewn slide in Brun's collection (slide no. 618, labelled 'Polysystins mergel v. Nankoori'), most probably of the topotype material of Grunow's *Coscinodiscus plicatus,* contains specimens of what appear to be *C. plicatus;* they show somewhat finer areolation than on *C. yabei* or *C. flexuosus,* and I could not find specimens that quite qualify as *C. flexuosus* in marginal structure on a short examination of the slide. In the experimental Mohole, as well as in SDSE 76, both forms, one with relatively finer areolae and one with coarser areolae and a distinctly defined margin, are present, but they were identified with *C. plicatus.* A more thorough study is necessary to establish the relationship between *C. plicatus* and *C. flexuosus.* [Plate 40.4, figs. 4-6]

Coscinodiscus vetutissimus Pantocsek: 1886, Teil. I, p. 73, pl. 20, fig. 186; Rattray 1889, p. 477, pl. II, fig. 17; De Toni 1891, p. 1,220; Hustedt 1928, Kieselalg. I, p. 412, fig. 220; Kanaya 1959, p. 84, pl. 5, figs. 3-5. Not *C. inaqualis* Grove and Sturt (1887, p. 68). Remarks: The fascicular arrangement of areolae on the valve surface, the presence of one or two subcentral papillae, the presence of process-like marking at the outer end of the longest row of areolae in each fasciculus, and the presence of a single row of distinctly smaller areolae at the outer periphery of the valve mantle along the inner side of the finely striated margin are the features held in common by all the specimens here identified with *C. vetutissimus.* On these specimens the papillae are variable in size, ranging from hook-shaped ones, that probably end in two so-called nodules, to simple nodular ones, the latter being always more distinct but sometimes smaller than the areolae about them. Sizes of areolae are also variable, but range between 4.5 and 6 in 10μ half-way between the centre and margin. Subdivision of this group of specimens by the size and shape of papillae or by the size of subcircular and somewhat eccentric central area has failed in the present day. Specimens like those illustrated as *C. aeginensis* Schmidt by Kolbe (1954, pl. I, fig. 9, not fig. 10), *C. salisburyanus* Lohman by Lohman (1948, p. 164, pl. VII, fig. 5), *C. floridulus* Schmidt by Hanna and Brigger (1966, p. 228, fig. 8) and *C. curvatulus* Grun by Kolbe (1954, p. 28, pl. I, fig. 1) have been found among the experimental Mohole material. They are considered to fall within the range of morphological variation that is interpreted as *C. vetutissimus* on the basis of the experimental Mohole specimens. [Plate 40.2, figs. 1-5]

Craspedodiscus coscinodiscus Ehr., 1884 De Toni 1891, p. 1,199; Schmidt 1876, Atlas pl. 66, figs. 3, 4; Reinhold 1937, p. 110, pl. 12, figs. 1, 2; Boyer 1904, p. 500, pl. CXXXV, fig. 3; Lohman 1948, p. 166, pl. VIII, fig. 6; Kolbe 1954, p. 36, pl. I, fig. 4. Synonyms: *C. coscinodiscus* var. *nankoorensis* Grunow in Schmidt: 1876, Atlas, pl. 66, fig. 5; Reinhold 1937, p. 103, pl. 12, fig. 3. Remarks: In the material studied, those from the experimental Mohole and SDSE 76 agree better with var. *nankoorensis* Grun. in the size of areolation; whereas, the ones from a late Oligocene sample, DWBG 10 (11-27 cm), are much larger and more coarsely areolated like those from the Calvert Formation and its equivalent in the East Coast of the United States (Schmidt 1876, fig. 3, 4; Boyer 1904, p. 500, pl. CXXXV, fig. 3; Lohman 1948). Until more material is examined, particularly of Palaeogene samples from the Pacific region, I follow Kolbe (1954) in not recognizing the variety. [Plate 40.4, figs. 1-3]

Denticula hustedtii Simonsen and Kanaya: 1961, p. 501, pl. 1, figs. 19-25. [Plate 40.5, figs. 13, 14]

Denticula lauta Bailey: 1854, p. 9, figs. 1, 2; Simonsen and Kanaya, 1961, p. 500, pl. 1, figs. 1-10. [Plate 40.5, fig. 11]

Denticula nicobarica Grunow: 1868, p. 97, pl. 1a, fig. 5; Simonsen and Kanaya 1961, p. 503, pl. 1, figs. 11-13. [Plate 40.5, fig. 12]

Fragilariopsis pliocena (Brun) Sheshukova-Poretzkaya: 1959, pl. 3, fig. 6; Sheshukova-Poretzkaya 1967, p. 305, pl. XLVII, fig. 13, pl. XLVIII, fig. 7. Remarks: By comparison with Brun's type specimens (slide no. 3397 labelled 'Sendai Calcaire') of *Fragilaria pliocena* Brun (1891, p. 28, pl. XVII, fig. 7 and pl. XIV, fig. 7), the specimens here illustrated, from a supposed late Pliocene sample, Cap 42 BP (757-803 cm), must be identified with Brun's species, so far as light microscopic observation can tell. Several specimens in the slide range in outline from more or less lanceolate, as illustrated by Sheshukova-Poretzkaya (1967, pl. XLVII, fig. 13) to those with more or less bluntly rounded ends like that here illustrated. All specimens in Brun's slide are isopol in both apical and transapical axes; some are mounted in girdle view, showing a ribbon-shaped connection with neighbouring valves as illustrated in Brun's original figure (1891, pl. XVII, fig. 7). The canal raphe is very likely present on Brun's specimens, although it was not possible to establish its presence beyond any doubt; a feature suggesting a pseudonodule was vaguely defined at one side of a valve, however. The pseudonodule is not apparent on the specimens illustrated here.The structure of the intercostal membrane, of Brun's specimens and the present specimens alike, most resembles that of *Pseudoeunotia doliolus* (Wall.) Grun. More thorough study of this taxon by comparison with *P. doliolus* and *Fragilariopsis* (Hasle 1965) is now in progress by R. Simonsen and myself. Some *Nitzschia marina* Grun, from the SDE 76 and *Fragilariopsis australis* Muhina (1965, p. 25, pl. II, fig. 5) must be closely examined in this connection. For the present paper I have adopted the combination proposed by Sheshukova-Poretzkaya (1967) for this taxon. [Plate 40.3, figs. 7, 8]

Hemidiscus cuneiformis Wall., *s.l.*: Wallich 1860, p. 42, pl. 2, figs. 3, 4; Hustedt 1930, Kieselalg. I, p. 904, fig. 542; Hendy 1937, p. 264; Lohman 1941, p. 78, pl. 16, fig. 1, 2, 5; Kolbe 1954, p. 38; Hanna and Brigger 1965, p. 300, fig. 33; Wornardt 1967, p. 36, fig. 51. [Plate 40.3, figs. 5, 6]

Rouxia peragalli Brun and Herib: *in* Heribaud 1893, p. 156, pl. 1, fig. 12a: Hanna 1930*a*, p. 180, pl. 14, fig. 1, 5; Sheshukova-Poretzkaya 1967, p. 294, pl. XLIII, fig. 17, pl. XLVII, fig. 4. Remarks: Since the species is represented mostly by broken individuals, identifications of separate formae have not been attempted for this study.

Synedra jouseana Sheshuk. f. *jouseana* Sheshukova-Poretzkaya: 1962, p. 208, fig. 4; Sheshukova-Poretzkaya, 1967, p. 245, pl. XLII, 4a, 4b; pl. XLIII, figs. 12a, 12b. [Plate 40.5, fig. 15]

Thalassionema nitzschioides Grun. and vars. Remarks: The following taxa are represented by this category in this paper. *Thalassionema nitzschioides* Grun. *ex.* Hustedt (Kieselalg. II, 1932, p. 244, fig. 725); *Th. nitzschioides* var. *obtusa* Grun. (*in* Van Heurck, 1881, pl. 43, fig. 6); *Th. nitzschioides* var. *javanica* Grun. (*ibid.* pl. 43, fig. 11); *Th. nitzschioides* f. *parva* Heid. (*in* Heiden and Kolbe, 1928, p. 564, pl. 35, fig. 118); *Th. nitzschioides* f. *inflata* Heid. (*ibid.* fig. 116); *Th. nitzschioides* f. *incurvata* (*ibid.* fig. 117); *Th. nitzschioides* f. *gracilis* (*ibid.* fig. 115); *Spinigera capitata* Heid. (*ibid.* p. 565, pl. 35, fig. 119); *S. lanceolata* Heid. (*ibid.* fig. 120); *Fragilaria hirosakiensis* Kanaya (1959, p. 104, pl. 9, figs. 11-15).

Thalassiothrix. spp. Remarks: At least the following two species are represented. *T. longissima* Cleve and Grun. (see Hustedt, 1932, Kieselalg. II, p. 247, fig. 726) and *T. frauenfeldii* Grun. (*ibid.* fig. 727). They were found in fragments, and not all the specimens have been individually identified.

ACKNOWLEDGEMENTS

The deep-sea samples studied for the present paper are from four localities: experimental Mohole drilling at the Guadalupe site, Station 76 of the Swedish Deep-Sea Expedition; a late Pliocene sample from Cap 42 BP; a late Oligocene sample from DWBG 10. They were made available to me by M. N. Bramlette and W. R. Riedel of Scripps Institution of Oceanography. Also, in connection with this paper, a preliminary survey was made of a series of samples from the Miocene section in Newport Bay, California. These samples were made available by J. C. Ingle Jr. of the Department of Geology, University of Southern California, and Y. Takayanagi of the Institute of Geology and Palaeontology, Tohoku University, who collected samples from the upper Luisian and lower Mohnian of the section from which Lipps (1964) has described the planktonic foraminifera. The distribution of the species of the genus *Denticula* in the Newport Bay samples was studied by I. Koizumi of the Institute of Geology and Palaeontology, Tohoku University. F. L. Parker of Scripps Institution of Oceanography and K. Hatai of the Institute of Geology and Palaeontology read the manuscript critically.

Received January 1968.

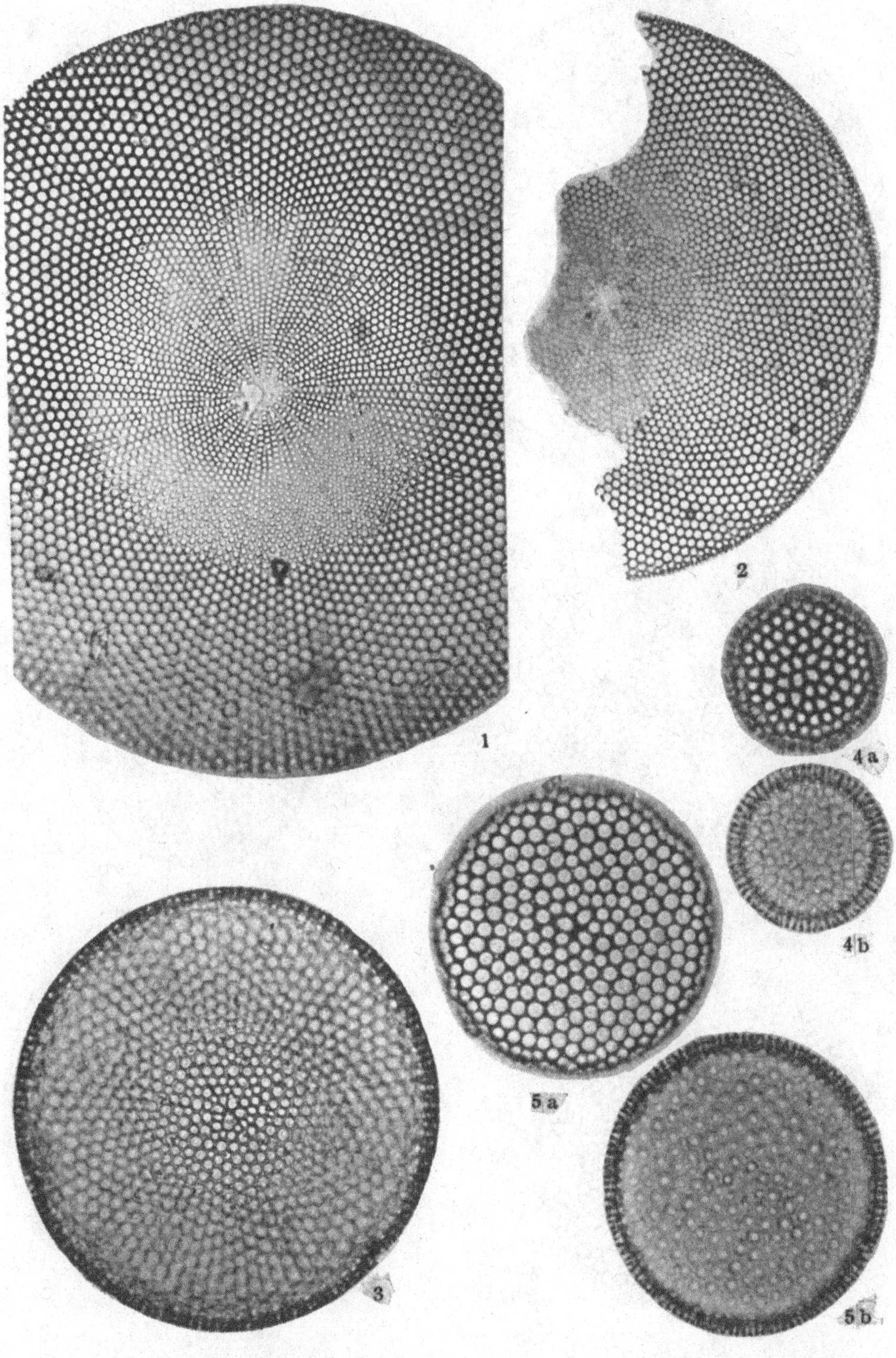
1
2
3
4a
4b
5a
5b

Plate 40.1. Fig. 1. *Coscinodiscus gigas* var. *diorama* (Schmidt) Grun. IGPS coll. cat. no. 90364, 260μ in diameter, from EM 8-15 (34-7 cm). x 390. Fig. 2. *Coscinodiscus monicae* Grun. IGPS coll. cat. no. 90372, 100μ in radius, from EM 8-9 (150-3 cm). x 390. Fig. 3. *Coscinodiscus oculus iridis* var. borealis (Bail.) Cleve. IGPS coll. cat. no. 90378, 150μ in diameter, from EM 8-9 (150-3 cm). x 390. Fig. 4*a*, *b*, *Coscinodiscus marginatus* Ehr. forma. IGPS coll. cat. no. 90371, 37μ in diameter, from EM 7-2 (4-7 cm). x 625. *b*, with the focus lower than *a*. Fig. 5*a*, *b*. *Coscinodiscus marginatus* Ehr. forma. IGPS coll. cat. no. 90370, 66μ in diameter, from EM 7-2 (4-7 cm). x 625. *b*, with the focus lower than *a*.

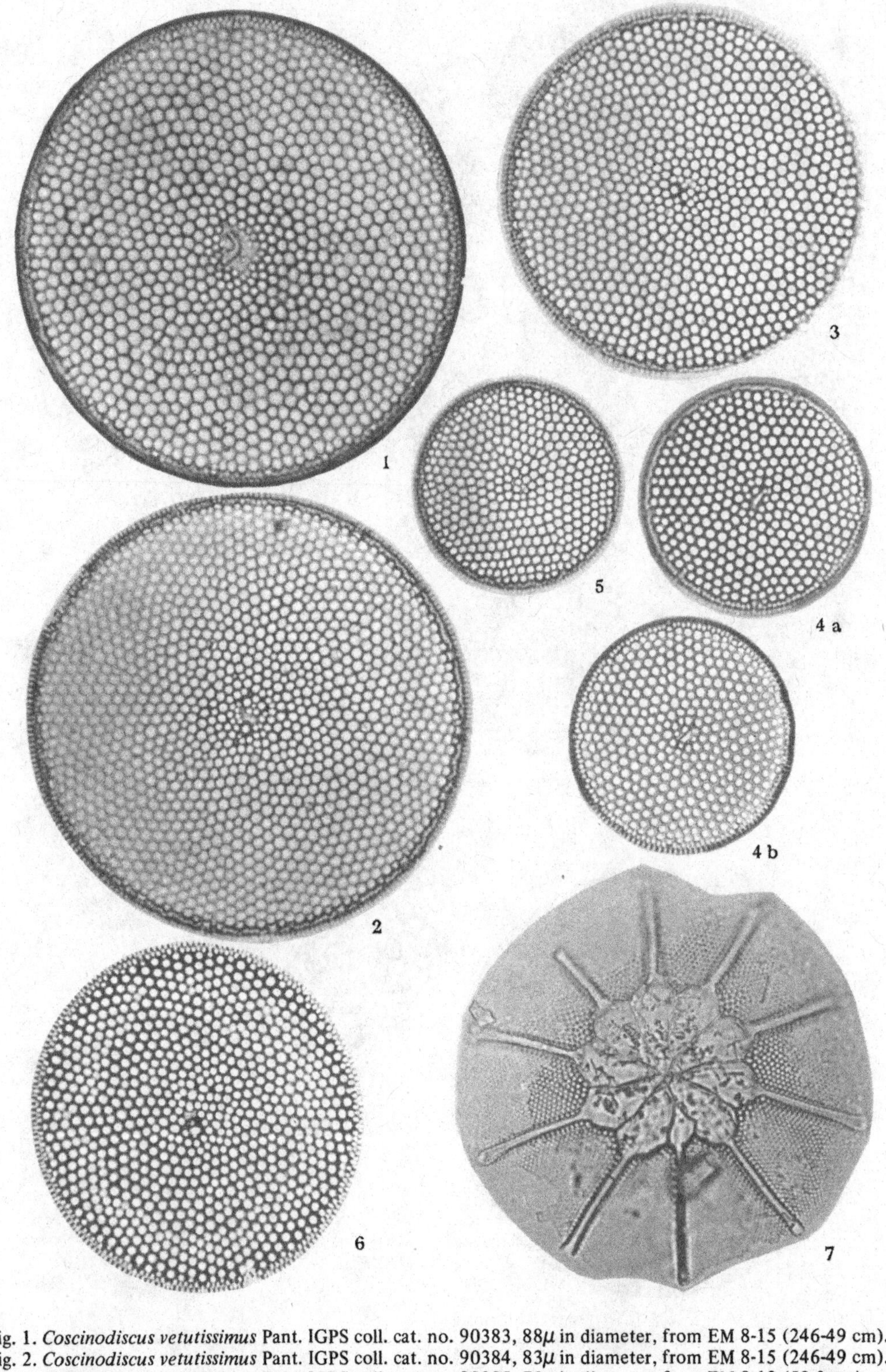
1
2
3
4 a
4 b
5
6
7

Plate 40.2. Fig. 1. *Coscinodiscus vetutissimus* Pant. IGPS coll. cat. no. 90383, 88μ in diameter, from EM 8-15 (246-49 cm). x 800. Fig. 2. *Coscinodiscus vetutissimus* Pant. IGPS coll. cat. no. 90384, 83μ in diameter, from EM 8-15 (246-49 cm). x 800. Fig. 3. *Coscinodiscus vetutissimus* Pant. IGPS coll. cat. no. 90385, 70μ in diameter, from EM 8-13 (50-3 cm). x 800. Fig. 4*a*, *b*. *Coscinodiscus vetutissimus* Pant. IGPS coll. cat. no. 90386, 42μ in diameter, from EM 7-2 (4-7 cm) x 800. *b*, with the focus lower than *a*. Fig. 5. *Coscinodiscus vetutissimus* Pant. IGPS coll. cat. no. 90387, 39μ in diameter, from EM 7-2 (4-7 cm). x 800. Fig. 6. Cfr. *Coscinodiscus nodulifer* Schmidt. IGPS coll. cat. no. 90377, 65μ in diameter, from EM 8-12 (94-7 cm). x 800. Fig. 7. *Asteromphalus moronensis* (Grev.) Ratt. IGPS coll. cat. no. 90362, approx. 45μ in the diameter of central field, margin entirely broken, from EM 8-15 (34-7 cm). x 625.

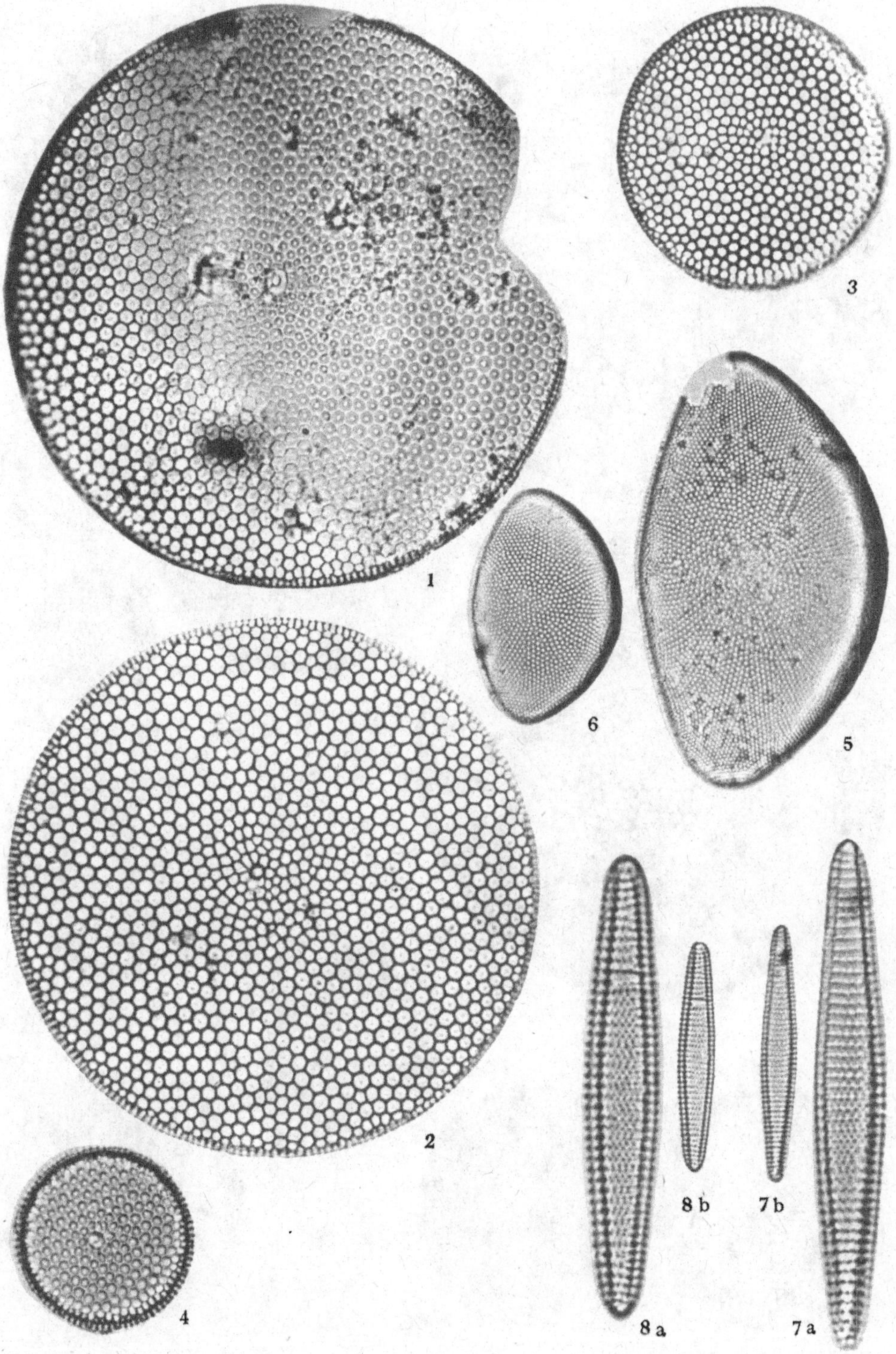
1
2
3
4
5
6
7a
7b
8a
8b

Plate 40.3 Fig. 1. *Coscinodiscus nodulifer* Schmidt. IGPS coll. cat. no. 90374, 125μ in diameter, margin partly broken, from Cap 42 BP (757-803 cm). x 625. Fig. 3. *Coscinodiscus nodulifer* Schmidt. IGPS coll. cat. no. 90375, 65μ in diameter, from Cap 42 BP (757-803 cm). x 625. Fig. 4. *Coscinodiscus nodulifer* Schmidt. IGPS coll. cat. no. 90376, 40μ in diameter, from Cap 42 BP (757-803 cm). x 625. Fig. 5. *Hemidiscus cuneiformis* Wall., *s.l.* IGPS coll. cat. no. 90396, 97μ in length, from EM 8-9 (150-3 cm). x 625. Fig. 6. *Hemidiscus cuneiformis* Wall., *s.l.* IGPS coll. cat. no. 90397, 52μ in length, from EM 8-9 (150-3 cm). x 625. Fig. 7*a*, *b*. *Fragilariopsis pliocena* (Brun.) Sheshuk. IGPS coll. cat. no. 90394, 46μ in length, from Cap 42 BP (757-803 cm). *a*, x 1600; *b*, x 800. Fig. 8*a*. *b*. *Fragilariopsis pliocena* (Brun.) Sheshuk. IGPS coll. cat. no. 90395, 40μ in length, from Cap 42 BP (757-803 cm). *a*, x 1600; *b*, x 800.

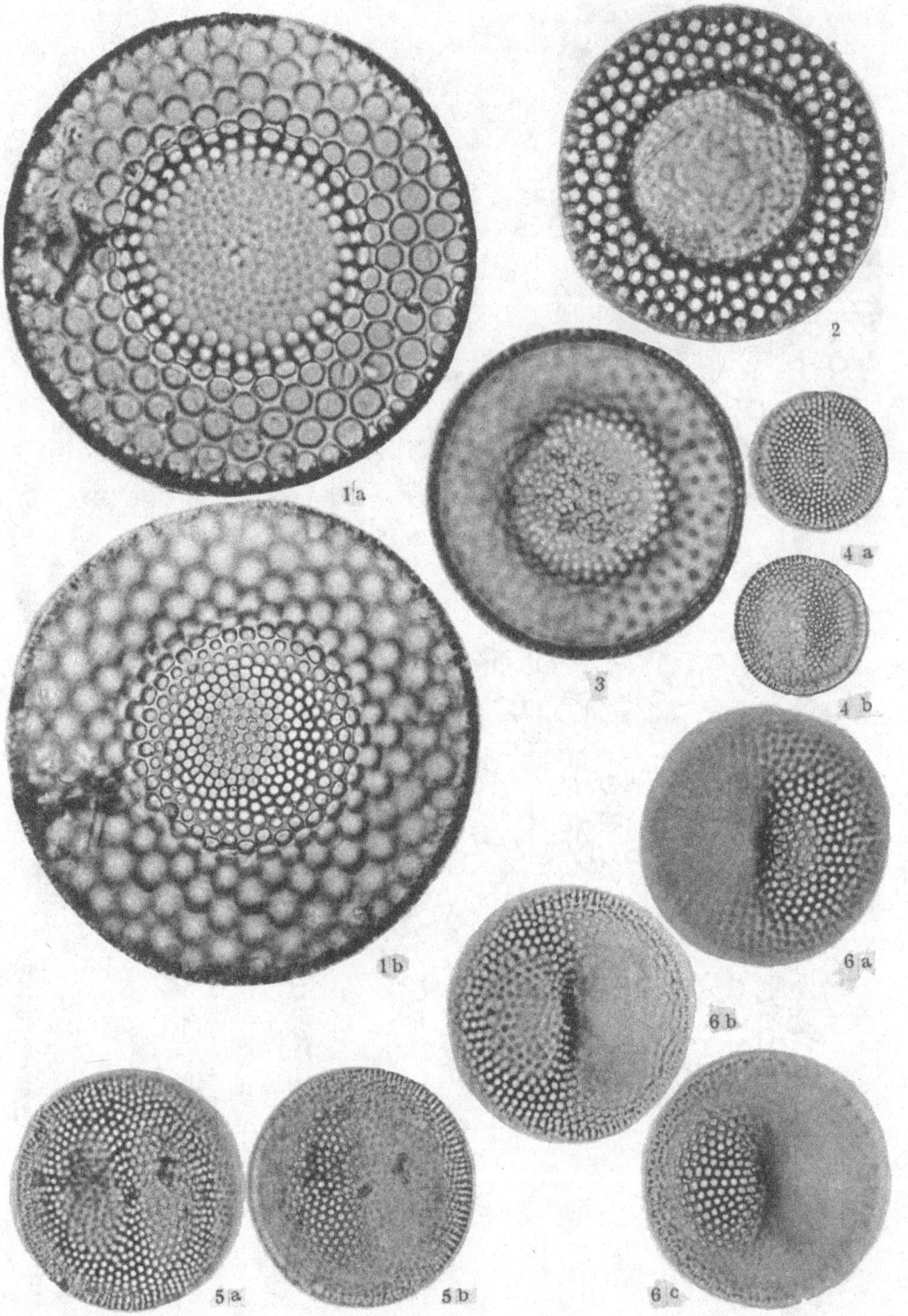
2
1a
3
4 a
4 b
1 b
6 a
6 b
5 a
5 b
6 c

Plate 40.4. Fig. 1*a*, *b*. *Craspedodiscus coscinodiscus* Ehr. IGPS coll. cat. no. 90389, 110μ in diameter, from DWBG 10 (11-27 cm). x 625. *b*, with the focus lower than *a*. Fig. 2. *Craspedodiscus coscinodiscus* Ehr. IGPS coll. cat. no. 90388, 74μ in diameter, from EM 7-2 (4-7 cm). x 625. Fig. 3. *Craspedodiscus coscinodiscus* Ehr. IGPS coll. cat. no. 90399, 73μ in diameter, from EM 7-2 (4-7 cm). x 625. Fig. 4*a*, *b*. *Coscinodiscus plicatus* Grun., *s.l.* IGPS coll. cat. no. 90382, 30μ in diameter, from EM 8-9 (150-3 cm). x 625. *b*, with the focus lower than *a*. Fig. 5*a*, *b*, *Coscinodiscus plicatus* Grun., *s.l.* IGPS coll. cat. no. 90381, 44μ in diameter, from EM 8-9 (150-3 cm). x 800. *b*, with the focus lower than *a*. Fig. 6*a*, *b*, *c*. *Coscinodiscus plicatus* Grun., *s.l.* IGPS coll. cat. no. 90380, 45μ in diameter, from EM 8-15 (34-7 cm). x 800. *b*, with the focus lower than *a*; *c* with the focus lower than *b*.

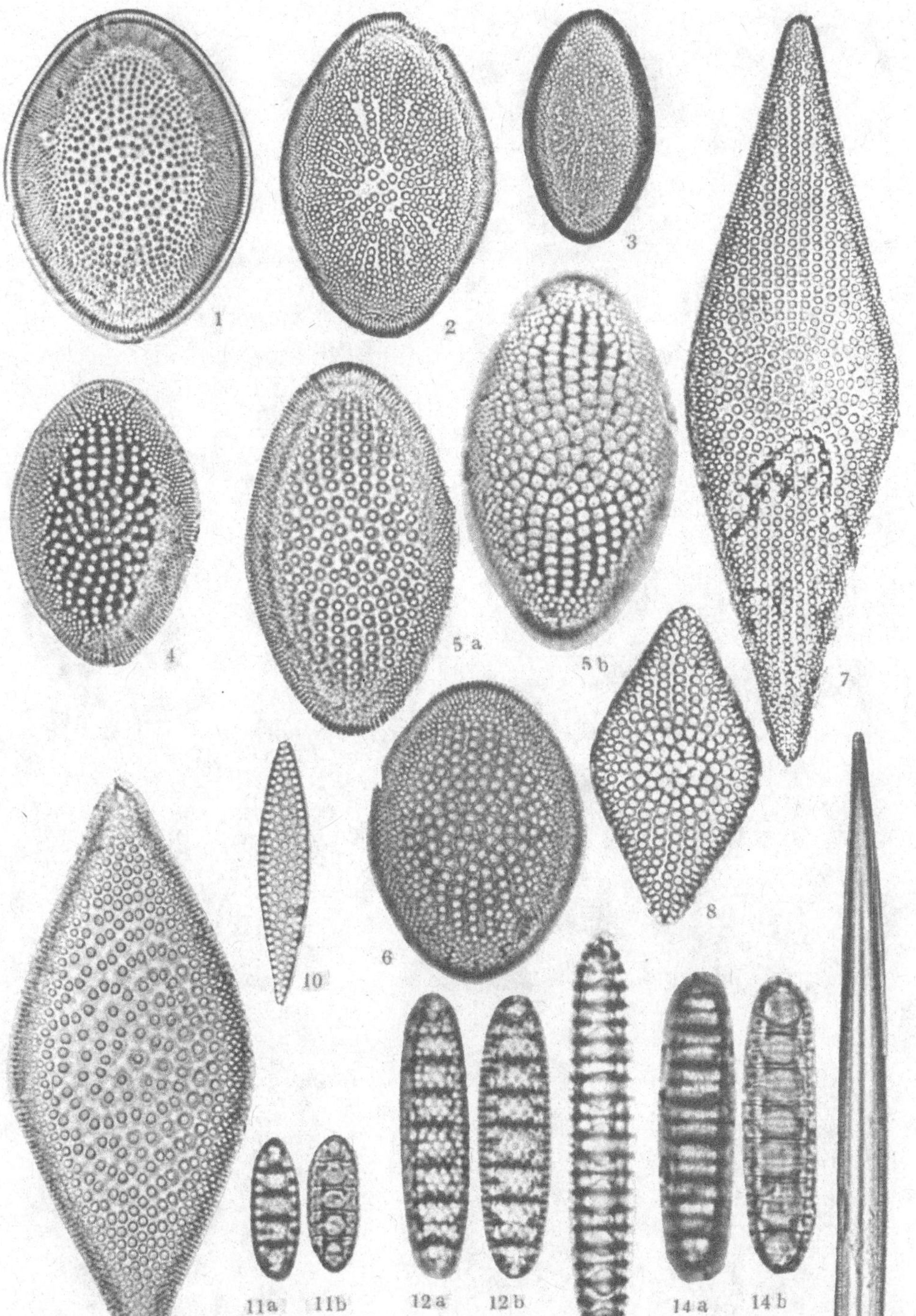
1
2
3
4
5 a
5 b
7
6
8
10
9
11a
11b
12 a
12 b
13
14 a
14 b
15

Plate 40.5. Fig. 1. *Actinocyclus ellipticus* Grun. IGPS coll. cat. no. 90351, 60μ in length, from SDSE 76 (1,324-5 cm). x 800. Fig. 2. *Actinocyclus ellipticus* Grun. IGPS coll. cat. no. 90352, 58μ in length, from SDSE 76 (1,324-5 cm). x 800. Fig. 3. *Actinocyclus ellipticus* Grun. IGPS coll. cat. no. 90353, 40μ in length, from SDSE 76 (1,324-5 cm). x 800. Fig. 4. *Coscinodiscus lewisianus* Grev. IGPS coll. cat. no. 90367, 50μ in length, from EM 7-2 (4-7 cm). x 800. Fig. 5*a*, *b*. *Coscinodiscus lewisianus* Grev. IGPS coll. cat. no. 90366, 63μ in length, from EM 7-2 (4-7 cm). x 800. *b*, with the focus lower than *a*. Fig. 6. *Coscinodiscus lewisianus* Grev. IGPS coll. cat. no. 90400, 50μ in length, from EM 7-2 (4-7 cm). x 800. Fig. 7. *Coscinodiscus lewisianus* var. *similis* Ratt. IGPS coll. cat. no. 90369, 172μ in length, from DWBG 10 (11-27 cm). x 625. Fig. 8. *Coscinodiscus lewisianus* var. *similis* Ratt. IGPS coll. cat. no. 90368, 43μ in length, from DWBG 10 (11-27 cm). x 800. Fig. 9. *Coscinodiscus lanceolatus* Castr. IGPS coll. cat. no. 90365, 95μ in length, from DWBG 10 (11-27 cm). x 800. Fig. 10. *Coscinodiscus paleaceus* (Grun.) Ratt. IGPS coll. cat. no. 90379, 48μ in length, from EM 8-15 (34-7 cm). x 800. Fig. 11*a*, *b*. *Denticula lauta* Bail. IGPS coll. cat. no. 90392, 12μ in length, from EM 7-2 (4-7 cm). x 1600. *b*, with the focus lower than *a*.
Fig. 12*a*, *b*. *Denticula nicobarica* Grun. IGPS coll. cat. no. 90393, 26μ in length, from EM 8-14 (384-7 cm). x 1,600. *b*, with the focus lower than *a*. Fig. 13. *Denticula hustedtii* Sim. and Kan. IGPS coll. cat. no. 90391, 53μ in length, from EM 8-9 (150-3 cm). x 1,600. Fig. 14*a*, *b*. *Denticula hustedtii* Sim. and Kan. IGPS coll. cat. no. 90390, 25μ in length, from EM 8-9 (150-3 cm). x 1,600. *b*, with the focus lower than *a*. Fig. 15. *Synedra jouseana* f. *jouseana* Sheshuk. IGPS coll. cat. no. 90398, approx. one half of a valve, 108μ in length, from EM 7-2 (4-7 cm). x 800.

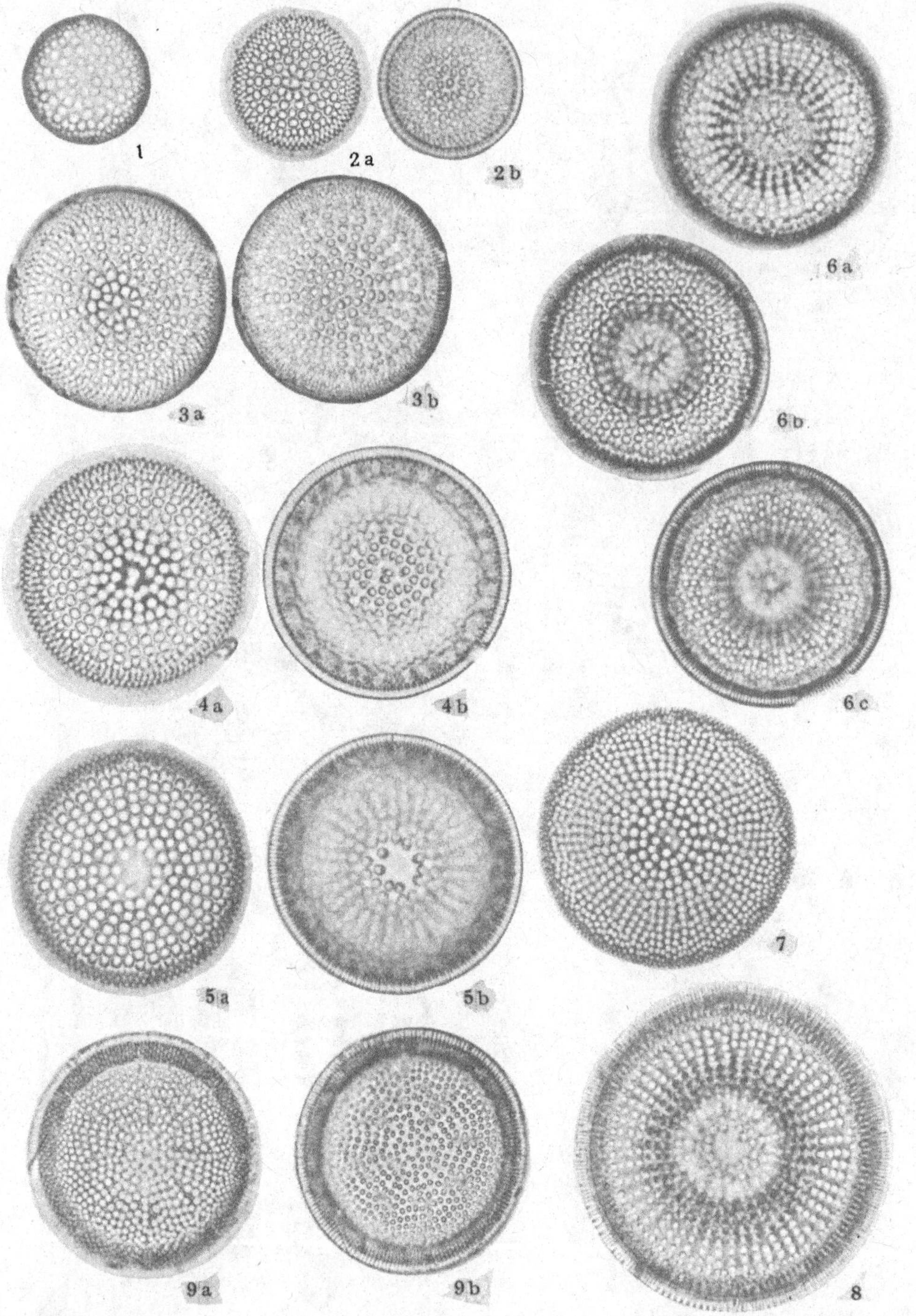
1
2a
2b
6a
3a
3b
6b
4a
4b
6c
7
5a
5b
9a
9b
8

Plate 40.6. Fig. 1. *Actinocyclus ingens* Ratt., *s.l.* IGPS coll. cat. no. 90354, 24μ in diameter, from EM 8-9 (150-3 cm). x 800. focused to show ocellus. Fig. 2*a, b. Actinocyclus ingens* Ratt., *s.l.* IGPS coll. cat. no. 90355, 27μ in diameter, from EM 8-9 (150-3 cm). x 800. *a*, focused to show ocellus; *b*, focused lower, on margin. Fig. 3*a, b. Actinocyclus ingens* Ratt., *s.l.* IGPS coll. cat. no. 90357, 41μ in diameter, from EM 8-9 (150-3 cm). x 800. *a*, focused to show ocellus; *b*, focused lower, on margin. Fig. 4*a, b. Actinocyclus ingens* Ratt., *s.l.* IGPS coll. cat. no. 90356, 46μ in diameter, from EM 8-9 (150-3 cm). x 800. *a*, focused to show ocellus; *b*, focused lower, on margin. Fig. 5*a, b. Actinocyclus ingens* Ratt., *s.l.* IGPS coll. cat. no. 90361, 47μ in diameter, from EM 8-13 (50-3 cm). x 800. *a*, focused to show ocellus; *b*, focused lower, on margin. Fig. 6*a, b, c. Actinocyclus ingens* Ratt., *s.l.* IGPS coll. cat. no. 90360, 56μ in diameter, from EM 7-2 (4-7 cm). x 625. *a*, focused on centre; *b*, focused lower to show ocellus; *c*, focused still lower, on margin. Fig. 7. *Actinocyclus ingens* Ratt., *s.l.* IGPS coll. cat. no. 90359, 58μ in diameter, from EM 7-2 (4-7 cm). x 625. Focused to show ocellus. Fig. 8. *Actinocyclus ingens* Ratt., *s.l.* IGPS coll. cat. no. 90358, 73μ in diameter, from EM 7-2 (4-7 cm). x 625. Focused to show ocellus. Fig. 9*a, b. Cestodiscus* sp. 1. IGPS coll. cat. no. 90363, 54μ in diameter, from EM 7-2 (4-7 cm). x 625. *b*, with the focus lower than *a*.

BIBLIOGRAPHY

Bailey, J. W. 1854. Notes on new species and localities of microscopical organisms. *Smithsonian Contrib. Knowledge*, 7, 1-16.

Benda, L. 1965. Diatomeen aus dem Eozän Norddeutschlands. *Paläont. Zeistschr.* **39** (3/4), 165-87.

Boyer, C. S. 1904. Thallophyta-Diatomacea. *In* Miocene. *Maryland Geological Survey*, pp. 487-507. Baltimore.

Brun, J. 1891. Diatomées espèces nouvelles marines, fossiles ou pélagiques. *Mém. Soc. Phys. et Hist. Nat. Genève*, **31**, 1-47.

Brun, J. 1895. Diatomeés lacustres, marines ou fossiles, espèces nouvelles ou insuffisamment connues. *Le Diatomiste*, 2, pl. 14-17.

Brun, J. and Tempère, J. 1889. Diatomées fossiles du Japon. Espèces marines et nouvelles des calcaires argileux de Sendai et de Yedo. *Mém. Soc. Phys. et Hist. Nat. Genève*, **30**, 1-75.

Castracane, F. 1886. Report on the Diatomaceae collected by H.M.S. *Challenger* during the years 1873-76. *Sci. Rep. Challenger, Botany*, **2** (4), i-iii, 1-178.

De Toni, G. S. 1891-4. *Sylloge algarum omnium hucusque cognitarium*, 2, Bacillariaceae. 1,556 pp. Padova.

Friend, J. K. and Riedel, W. R. 1967. Cenozoic orospherid Radiolaria from tropical Pacific sediments. *Micropaleontology*, **13** (2), 217-232.

Gleser, Z. I. 1962. [New diatom genus *Poretzkia* Jousé from Paleocene deposit of the eastern slope of Ural.] *Bot. Mater., Bot. Inst., Akad. Nauk S.S.S.R.* **15**, 32-4 (in Russian).

Gleser, Z. I. and Posnova, A. N. 1964. Diatomeae novae marinae ex Eoceno Kazachstaniae occidentalis. *Nov. Systemat. Plant. Niz. Rast. 1964, Bot. Inst., Acad. Nauk, S.S.S.R.* pp. 59-66 (in Russian).

Greville, R. K. 1860. A monograph of the genus *Asterolampra* including *Asteromphalus* and *Spatangidiam. Trans. Micr. Soc. London, new ser.* **8**, 102-24.

Greville, R. K. 1861-6. Description of new and rare diatoms. Ser. I-IV: *Trans. Roy. Micr. Soc. London, new ser.* **9**, 39-45, 67-87, 1861–Ser. V, VI: *ibid.* **10**, 18-29, 89-96, 1862–Ser. VII: *Quart. Journ. Micr. Sci., new ser.* **2**, 231-6, 1862–Ser. VIII, IX: *Trans. Roy. Micr. Soc. London, new ser.* **11**, 13-21, 63-76, 1863–Ser. X: *Quart. Journ. Micr. Sci., new ser.* **3**, 227-37, 1863–Ser. XI-XIII: *Trans. Micr. Soc. London, new ser.* **12**, 8-14, 81-94, 1864–Ser. XIV-XVII: *ibid.* **13**, 1-10, 24-34, 43-75, 97-105, 1865–Ser. XVII-XX: *ibid.* 1-9, 77-86, 121-30, 1886.

Grove, E. and Sturt, G. 1886-7. On a fossil marine diatomaceous deposit from Oamaru, Otago, New Zealand. *Jour. Quekett Micr. Club.* **2**, 321-30; **3**, 7-12, 63-78, 131-48.

Grunow, A. 1868. Reise seiner Majestät Fregatte 'Novara' um die Erde. *Bot. Teil.* vol. 1, *Algen.*

Grunow, A. 1884. Die Diatomeen von Franz Josefs Land. *Denkschr. Math.-Naturw. K. Akad. Wiss. (Wien)*, **48**, 1-60.

Hanna, G. D. 1927a. The lowest known Tertiary diatoms in California. *Jour. Paleont.* **1**, 103-27.

Hanna, G. D. 1927b. Cretaceous diatoms from California. *California Acad. Sci., Occasional papers*, **8**, 5-39.

Hanna, G. D. 1928. The Monterey shale of California at its type locality with a summary of its fauna and flora. *Amer. Assoc. Petrol. Geol., Bull.* **12** (10), 969-83.

Hanna, G. D. 1929. Fossil diatoms dredged from Bering Sea. *Trans. San Diego Soc. Nat. Hist.* **5** (20), 287-96.

Hanna, G. D. 1930a. Review of genus *Rouxia. Jour. Palaeont.* **4** (2), 179-88.

Hanna G. D. 1930b. Observations on *Lithodesmium cornigerum* Brun., *ibid.* 189-91.

Hanna, G. D. 1930c. The growth of *Omphalotheca. ibid.* 192.

Hanna, G. D. 1931. Diatoms and Silicoflagellata of the Kryenhagen Shale. *Mining in Calif.* **27** (2), 187-201.

Hanna, G. D. 1932. The diatoms of Sharktooth Hill, Kern County, California. *Calif. Acad. Sci., Proc., 4th ser.* **20** (6), 161-262.

Hanna, G. D. 1934. Additional notes on diatoms from the Cretaceous of California. *Jour. Paleont.* **8** (3), 352-5.

Hanna, G. D. 1936. Notes on localities of fossil diatoms in California. *Soc. Francaise Microscopie, Bull.* **5** (3), 109-11.

Hanna, G. D. and Grant, W. M. 1926. Miocene marine diatoms from Maria Madre Island, Mexico (Expedition to the Revillagigedo Island, Mexico in 1925). *Calif. Acad. Sci., Proc. 4th ser.* **15** (2), 115-93.

Hanna, G. D. and Brigger, A. L. 1966. Fossil diatoms from southern Baja California. *ibid.* **30** (15), 285-308.

Hasle, G. R. 1965. *Nitzschia* and *Fragilariopsis* species studied in the light and electron Microscopes, III, The genus *Fragilariopsis. Skrif. utgitt av Det Norske Videnskap-Akad. Oslo, Mat.-Naturv. Kl. new ser.* **21**, 1-49.

Heribaud, J. 1893. *Les diatomées d'Auvergne.* 255 pp. Paul Klincksieck: Paris.

Heiden, H. and Kolbe, R. W. 1928. Die Marinen Diatomeen der Deutschen Südpolar-Expedition, 1901-1903. In *Deutsche Südpolar-Expedition, 1901-1903*, 8, *Botanik*. pp. 447-715. Berlin.

Hendy, N. I. 1937. The plankton diatoms of the southern sea. *Discovery Reports,* **16**, 151-364.

Hustedt, Fr. 1927 onwards. Die Kieselalgen Deutschland, Oesterreichs und der Schweiz mit Berücksichtigung der ubrigen Länder Europas sowie der angrenzenden Meeresgebiete. In Rabenhorst's *Kryptogamenflora von Deutschland, Oesterreichs und der Schweiz.* **7**, (I), 920 pp., (II), 1105 pp.; (III), pp. 1-816, not yet completed. Leipzig.

Jousé, A. P. 1951a. [Diatoms and Silicoflagellata of the upper Cretaceous of Northern Ural.] *Bot. Mater. Spor. Rast., Bot. Inst., Akad. Nauk, S.S.S.R.* **7**, 42-65 (in Russian).

Jousé, A. P. 1951b. [Palaeocene diatoms from northern Ural.] *ibid.* 24-42 (in Russian).

Jousé, A. P. 1955. Species novae diatomacearum astatis Paleogenae. *ibid.* **10**, 81-103 (in Russian with descriptions in Latin).

Jousé, A. P. 1962. [Stratigraphic and palaeogeographic studies in the northwestern part of the Pacific Ocean.] *Izd. Akad. Nauk S.S.S.R.* 258 pp. Moscow (in Russian with English abstract).

Jousé, A. P. and Sheshukova-Poretzkaya, V. S. 1963. Class Bacillariophyta. In *Osnovi Paleontologii.* Akad. Nauk S.S.S.R.: Moscow. pp. 55-151 (in Russian).

Kanaya, T. 1957. Eocene diatom assemblages from the

Kellogg and 'Sidney' Shales, Mt. Diablo area, California. *Sci. Rept. Tohoku Univ., Sendai, 2nd ser. (Geol.)* **28**, 27-124.

Kanaya, T. 1959. Miocene diatom assemblages from the Onnagawa formation and their distribution in the correlative formations in Northeast Japan. *ibid.* **30**, 1-130.

Kanaya, T. and Koizumi, I. 1966. Interpretation of diatom thanatocoenoses from the North Pacific applied to a study of core V20-130 (Studies of a deep-sea core V20-130. Part IV). *ibid.* **37** (2), 89-130.

Koizumi, I. 1966. Tertiary stratigraphy and diatom flora of the Ajigasawa district, Aomori Prefecture, Northeast Japan. *Contr. from Inst. Geology and Paleontology, Tohoku Univ.* **62**, 1-34 (in Japanese, English abstract).

Kolbe, R. W. 1954. Diatoms from Equatorial Pacific cores. *Rep. Swedish deep-sea Exped.* **6** (1), 1-48.

Kolbe, R. W. 1957. Diatoms from Equatorial Indian Ocean cores. *ibid.* **9** (1), 1-50.

Lipps, J. H. 1964. Miocene planktonic Foraminifera from Newport Bay, California. *Tulane Studies in Geology,* 2 (4), 109-33.

Lohman, K. E. 1938. Pliocene diatoms from the Kettleman Hills, California. *U.S. Geol. Survey. Prof. Papers,* **189-C**, 81-94.

Lohman, K. E. 1941. Diatoms. In: Geology and Biology of North Atlantic deep-sea cores, between Newfoundland and Ireland. *ibid.* **196-B**, v-xiv, 55-85.

Lohman, K. E. 1948. Middle Miocene diatoms from the Hammond Well. In *Cretaceous and Tertiary subsurface geology.* pp. 151-87. Maryland Board of Natur. Resources.

Long, J. A., Fuge, D. P. and Smith, J. 1946. Diatoms of the Moreno shale. *Jour. Paleont.* **20** (2), 89-118.

Mann, A. 1907. Report on the diatoms of the 'Albatross' voyage in the Pacific Ocean, 1888-1904. *U.S. Nat. Herbarium Contr.* **10** (5), 220-419.

Martini, E. and Bramlette, M. N. 1963. Calcareous nannoplankton from the experimental Mohole drilling. *Jour. Paleont.* **37** (4), 845-56.

Mertz, D. 1966. Micropaläontologische und Sedimentologishe Untersuchung der Pisco-Formation Südperus. *Paleontographica,* **118**, Abt. B., 1-51.

Muhina, V. V. 1965. [New species of diatoms from the bottom sediments of the equatorial region of the Pacific.] *Nov. System. Plant. non Vascularium, 1965, Bot. Inst. Akad. Nauk, S.S.S.R.* pp. 22-5 (in Russian).

Olausson, E. 1961. Remarks on some Cenozoic core sequences from the Central Pacific with a discussion of the role of coccolithophorids and Foraminifera in carbonate deposition. *Göteborgs Vetensk. Samh. Handl. Följ. 7, ser. B.* **8** (10), 1-35.

Pantocsek, J. 1886-1892. *Beiträge zur Kenntnis der fossilen Bacillarien Ungarns.* **I**, 77 pp., 30 pls.; II, 123 pp., 30 pls. III, 42 pls. Berlin.

Parker, F. L. 1964. Foraminifera from the experimental Mohole drilling near Guadalupe Island, Mexico. *Jour. Paleont.* **38**, 617-36.

Parker, F. L. 1967. Late Tertiary biostratigraphy (planktonic Foraminifera) of tropical Indo-Pacific deep-sea cores. *Bull. Amer. Paleont.* **52**, 111-208.

Proshkina-Lavrenko, A. I. (ed.) 1949. *'Diatomovyi analis', Gosgeolisdat.,* **1, 2 and 3**, Akad. Nauk. S.S.S.R.: Moscow.

Rattray, J. 1889. A revision of the genus *Coscinodiscus* and some allied genera. *Roy. Soc. Edinb. Proc.* **16**, 449-692.

Rattray, J. 1890. A revision of the genus *Actinocyclus* Ehr. *Jour. Quekett Micr. Club., 2d ser.* **4**, 137-212.

Reinhold, Th. 1937. Fossil diatoms of the Neogene of Java and their zonal distribution. *Nederland en Kolonien Geol. Mijnb. Genoot., Verh., Geol. Ser.* **12**, 43-132.

Riedel, W. R. 1959. Oligocene and lower Miocene Radiolaria in tropical Pacific sediments. *Micropaleontology,* **5** (3), 285-302.

Riedel, W. R. and Funnell, B. M. 1964. Tertiary sediment cores and microfossils from the Pacific Ocean floor. *Quart. Journ. Geol. Soc. London,* **120**, 305-68.

Sawamura, K. 1963a. On the correlation of the Miocene of the Joban, Tomamae coal fields and that of East Hokkaido. *Geol. Surv. Japan, Bull.* **14** (1), 91-4 (in Japanese with English abstract).

Sawamura, K. 1963b. Fossil diatoms in the Oidawara formation of the Mizunami Group. *ibid.* **14** (5), 387-90. (in Japanese with English abstract).

Sawamura, K. and Yamaguchi, S. 1961. Correlation of the Hard Shales by diatoms in the Abashiri- Urahoro area, East Hokkaido. *ibid.* **12** (11), 885-90 (in Japanese with English abstract).

Sawamura, K. and Yamaguchi, S. 1963. Subdivision of Miocene by fossil diatoms in the Tsubetsu area, East Hokkaido. *ibid.* **14** (10), 777-82 (in Japanese with English abstract).

Schmidt, A. 1875-1959. *Atlas der Diatomaceenkunde.* pls. 1-480. Leipzig.

Sheshukova-Poretzkaya, V. S. 1959. [About fossil diatom floras of South Saghalin (Marine Neogene).] *Leningr. Univ., Vest. no. 15, ser. Biol.* **3**, 36-55 (in Russian with summary in English).

Sheshukova-Poretzkaya, V. S. 1962. [New and rare diatoms from formations of Saghalin.] *Leningr. Gos. Univ., Uchen. zap. no. 313, Biol. Inst. ser. Biol. Nauk. Vup.* **49**, 203-11 (in Russian with descriptions in Latin).

Sheshukova-Poretzkaya, V. S. 1967. [Neogene marine diatoms of Saghalin and Kamtchatka.] *Izd. Leningr. Univ.* 327 pp. Leningrad.

Sheshukova-Poretzkaya, V. S. and Gleser, S. I. 1964. Diatomeae marinae novae e paleogeno Ucrainea. *Nov. Systemat. Plant. non Vascular.* **1964**, 78-91 (in Russian).

Simonsen, R. and Kanaya, T. 1961. Notes on the marine species of the diatom Genus *Denticula* Kütz. *Int. Revue ges. Hydrobiol.* **46** (4), 498-513.

Strelnikova, N. I. 1960. [Diatoms and Sillicoflagellata from the Palaeogene formations of Obi-Pir interfleuve.] *Trud. VNIGRI,* **158**, 33-45 (in Russian).

Strelnikova, N. I. 1965a. [Upper Cretaceous diatoms from the northwestern part of the western Sibursk lowland.] *Bot. Jour., Akad. Nauk S.S.S.R.* **50**, 986-90 (in Russian).

Strelnikova, N. I. 1965b. De Diatomeis Cretae Superioris paris et novis declivis orientalis montium Uralesium Polarium. *Nov. Systemat. Plant. non Vasicular.* **1965**, 29-37 (in Russian).

Strelnikova, N. I. 1966a. Revisio specierum generum

Gladius Schulz et *Pyxilla* Grev. (Bacillariophyta) e sedimentis Cretae Superioris. *ibid.* **1966**, 23-36 (in Russian).

Strelnikova, N. I. 1966b. Species generis *Aulacodiscus* Ehr. (Bacillaroophyta) novae e sedimentis Cretae Superior declivis orientalis partis borealis Jugi Uralensis. *ibid.* **1966**, 36-7 (in Russian).

Van Heurck, H. 1880-5. *Synopsis des diatomées de Belgique.* Atlas, 3 pp., 132 pls., 1880-81; text, 235 pp., 3 suppl. pls. and alphabetical table, 1885. Anvers.

Wallich, G. C. 1860. On the siliceous organisms found in the digestive cavities of the Salpae and their relation to the flint nodules of the Chalk formation. *Trans. Micr. Soc., London, new ser.* **8**, 36-54.

Woodring, W. P. and Bramlette, M. N. 1950. Geology and Palaeontology of the Santa Maria District, California. *U.S. Geol. Survey, Prof. Paper,* 222, 1-142.

Wornardt, W. W. 1967. Miocene and Pliocene marine diatoms from California. *Calif. Acad. Sci., Occasional Papers,* **63**, 1-108.

41. THE OCCURRENCE OF PRE-QUATERNARY RADIOLARIA IN DEEP-SEA SEDIMENTS[1]

W. R. RIEDEL

Summary: Virtually all known occurrences of pre-Quaternary radiolarians in ocean sediments are tabulated and charted. Through most of the Tertiary, the pattern of distribution of radiolarian sediments appears to have been generally similar to the Quaternary pattern. Differences are apparently due to displacements of the boundaries of persistent water masses and large-scale current systems, with some distortion probably introduced by lateral movements of portions of the sea floor.

INTRODUCTION

Pre-Quaternary radiolarians in sea-floor sediments could be investigated from many different points of view. A few aspects of their investigation that seem to be the most fundamental, and to offer the greatest promise for results of broad significance, are listed below.

1. Their occurrence in open-ocean sediment cores provides a large number of widely distributed localities to supplement the relatively few land-based fossil radiolarian occurrences. Thus these submarine localities now provide an important part of our knowledge of the world-wide radiolarian fauna throughout the Cenozoic, and deep-sea drilling will almost certainly add significantly to our knowledge of Mesozoic assemblages. Investigation of the geological history of radiolarians on the world-wide scale thus made possible will provide the most satisfactory basis for following their evolutionary history, for formulating a natural system of classification, and for making palaeoecological interpretations.

2. Radiolarians occurring in non-calcareous sediments perform the practical function of providing a basis for age assignments. As indicated by Friend and Riedel (1967), they can be used to subdivide the Cenozoic into smaller units than the epochs indicated in the present paper.

3. Quaternary radiolarian sediments evidently reflect areas of high biological productivity in the overlying water, and it seems reasonable to assume that the same relationship obtained in earlier geological time. Thus the distribution of pre-Quaternary radiolarian sediments, as such, provides information regarding the past distribution of large-scale current systems and water masses characterized by high biological productivity.

However, a word of caution seems necessary in this connection. It does not seem meaningful to compare the distribution of Recent radiolarians in the *uppermost few centimetres* of sea-floor sediments with their distribution in older sediments, since there is evidently a tendency for the siliceous skeletons to dissolve while they are at the sediment surface or only shallowly buried. Over large areas of the sea floor, the Quaternary sediments contain radiolarians in the uppermost few centimetres, but none below (as mentioned by Riedel and Funnell 1964, p. 361). In such areas, the sediments now accumulating will apparently not contain preserved radiolarians several million years from now. Chart 41.1 shows the distribution of sediments in which radiolarians are common and well preserved to a depth of at least several decimetres in the Quaternary sediment column in the Pacific and Indian Oceans, and it is this pattern which will be used as a basis for comparison with Tertiary radiolarian sediments.

4. As our coverage of pre-Quaternary radiolarian sediments increases, it will become possible to make more detailed palaeoecological interpretations of the successive Tertiary epochs. It is now apparent that various zoogeographical provinces are recognizable on the basis of Quaternary radiolarian assemblages, and that similar differentiations will be recognizable throughout the Tertiary. Their elucidation will require much painstaking investi-

[1] Contribution from the Scripps Institution of Oceanography, University of California, San Diego, California.

gation at the species and subspecies level.

5. The frequent occurrence of older radiolarians reworked into younger sediments can cause a great deal of trouble for some types of investigations, but brings considerable advantages to others. A single Quaternary sample often contains sufficient reworked microfossils to indicate the composition of the radiolarian assemblages during several epochs of the Tertiary in its general vicinity. And a detailed study of reworked forms provides insight into the general effects of reworking processes which move inorganic sediment components along with the microfossils. The relatively robust siliceous radiolarian skeletons are presumably more durable tracers than are calcareous microfossils. Charts 41.2-6 indicate the extent to which occurrences of reworked radiolarians supplement information from *in situ* occurrences.

TABLE AND CHARTS

Table 41.1 and Charts 41.2-6 show sea-floor occurrences of Pliocene, Miocene, Oligocene, Eocene and Cretaceous radiolarians. These include all localities at which I have seen pre-Quaternary radiolarians, with only few exceptions. A few cores that exactly duplicate others from the same locality are omitted – thus CAP 5 BG is tabulated, but not CAP 5 BP (Riedel and Funnell 1964). Some apparently late Tertiary occurrences in high southern latitudes recorded by Hays (1965) are not included. Some Eocene radiolarians recorded in the western tropical Atlantic, at 11° 12′ N, 48° 05′ W, depth 4,614 metres, by Saito, Ewing and Burckle (1966) are also omitted.

Most of the samples recorded here have been collected by expeditions of Scripps Institution of Oceanography, and the inclusion of some samples collected by other institutions will permit the system of age assignments used here to be calibrated with those used by other investigators. The indicated ages are based on the criteria used by Friend and Riedel (1967) and Riedel (1967, footnote 9).

Table 41.2 lists occurrences of *in situ* Tertiary calcareous sediments lacking siliceous microfossils, since these provide the only positive evidence on areas in which the sediments deposited during the various Tertiary epochs do not contain radiolarians. This table is not complete, since emphasis has been placed on those regions near or in which pre-Quaternary radiolarians occur.

PATTERNS OF DISTRIBUTION

Comparison of Charts 41.1-5 permit interpretations along the lines indicated in paragraph 3 of the introduction to this paper, in somewhat greater detail and over a larger area than the interpretation by Riedel and Funnell (1964). During the various epochs, the distribution of radiolarian sediments shows a pattern broadly similar to that in the Quaternary. Generally, in the regions occupied by the North Pacific, South Pacific and southern Indian Ocean Central Water Masses, Tertiary sediments with sufficient microfossils to permit determination of their ages are calcareous oozes lacking siliceous microfossils. Apparent displacements, with time, of the boundaries separating radiolarian-bearing from radiolarian-free sediments have been interpreted by Riedel and Funnell as reflecting changes in the areal extent of some major current systems and water masses with high and low biological productivity, respectively, but such an interpretation must now be approached with more caution because of the strong possibility that parts of the sea floor may have been displaced by movements along fracture zones or by crustal spreading originating at oceanic rises (Riedel 1967).

Although the boundaries separating radiolarian from non-radiolarian sediments are generally distinct, there are a few occurrences violating these boundaries. These anomalies require further investigation – they may be due to differences of age within the Tertiary epochs concerned, or, if this is not the case, they may provide clues to displacements of one part of the sea-floor crust relative to another.

Charts. 41.2-5 show also the extent to which deep-sea occurrences of Tertiary radiolarian sediments are now available to supplement our geographical coverage in the large areas separating land-based occurrences (paragraphs 1. and 4. above). Twenty years of intensive coring in the tropical Pacific by Scripps Institution has resulted in a good representation of radiolarian sediments throughout the Tertiary in that region, and there are also scattered occurrences resulting from less intensive sampling in the Indian Ocean, and in higher latitudes in the Pacific.

In some parts of the tropical Pacific, the data are sufficient to allow the mapping of types and (by estimating rates of accumulation) thicknesses of sediments constituting the uninterrupted Caenozoic column. The ages indicated for many of the calcareous sediments are based on evaluations of calcareous nannoplankton and Foraminifera by M. N. Bramlette, F. L. Parker and B. M. Funnell, while ages of non-calcareous sediments are based on the radiolarian criteria used by Friend and Riedel (1967) and Riedel (1967, footnote 9) which depend originally on correlation with calcareous microfossil sequences.

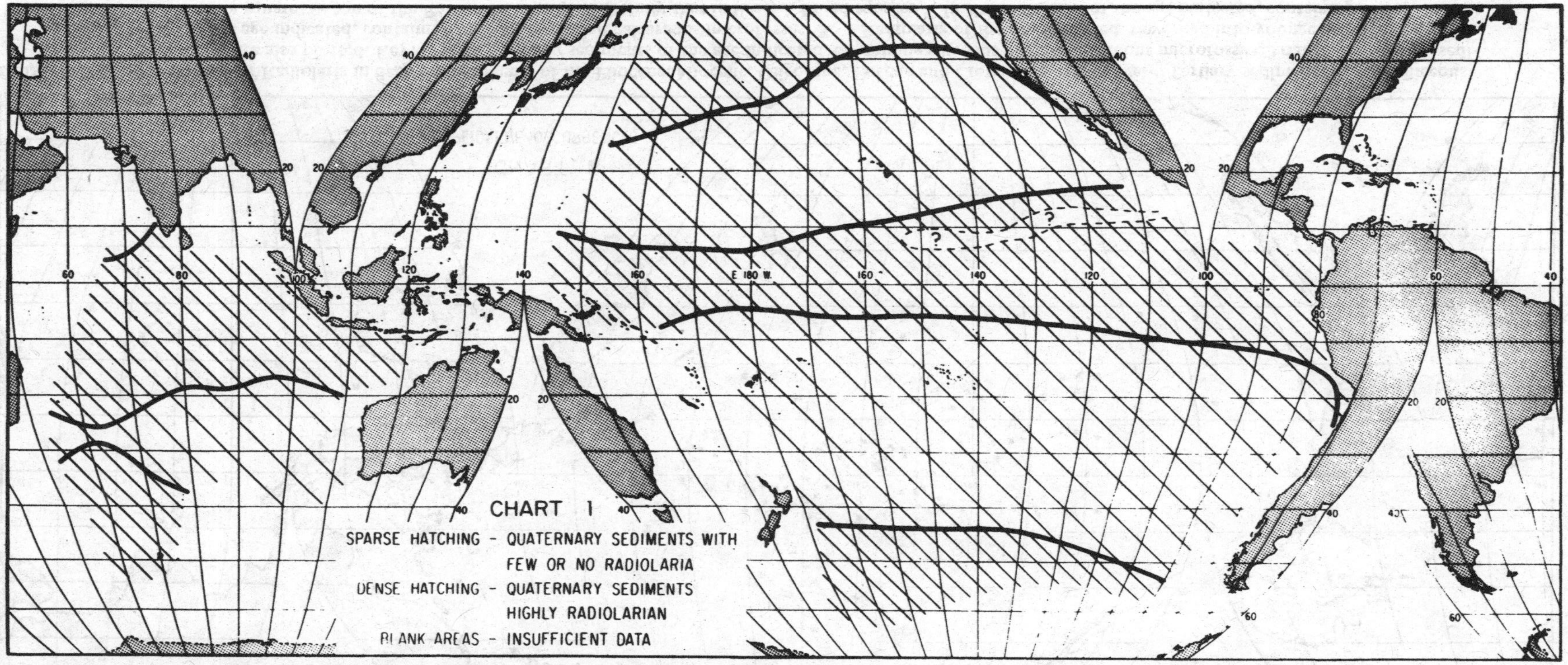
CHART 1
SPARSE HATCHING - QUATERNARY SEDIMENTS WITH FEW OR NO RADIOLARIA
DENSE HATCHING - QUATERNARY SEDIMENTS HIGHLY RADIOLARIAN
BLANK AREAS - INSUFFICIENT DATA
E 180 W.

Chart 41.1. **Distribution of Radiolaria in Quaternary sediments, at a depth of at least several decimetres below the sediment surface.**

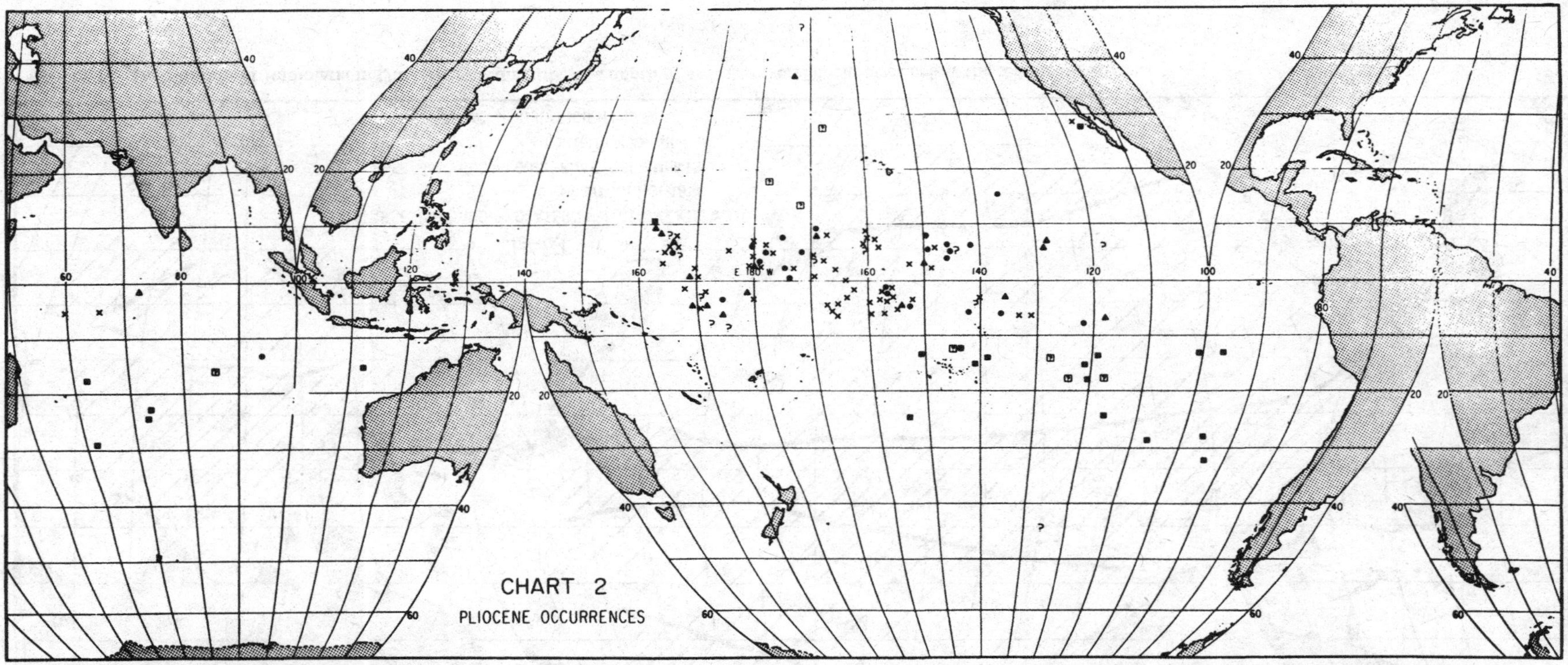
CHART 2
PLIOCENE OCCURRENCES

Charts 41.2-6. Distribution of Radiolaria in deep-sea sediments of the Pliocene, Miocene, Oligocene, Eocene and Cretaceous, respectively. Tertiary sediments lacking siliceous microfossils are also plotted. Key: Circles. – *In situ* sediments of the age indicated, containing siliceous but no calcareous microfossils. Triangles. – *In situ* sediments of the age indicated, containing both siliceous and calcareous microfossils. X. – Radiolaria of the age indicated, reworked into younger sediments. ? not enclosed in a square. – Age of the Radiolaria is only doubtfully that indicated. Filled squares – *In situ* sediments of the age indicated, containing calcareous but no siliceous microfossils. ? enclosed in a square. – Calcareous sediment, lacking siliceous microfossils, is only doubtfully *in situ* or doubtfully of the age indicated. Some localities listed in Table 41.1 are omitted from the charts, where plotting them would have caused overcrowding.

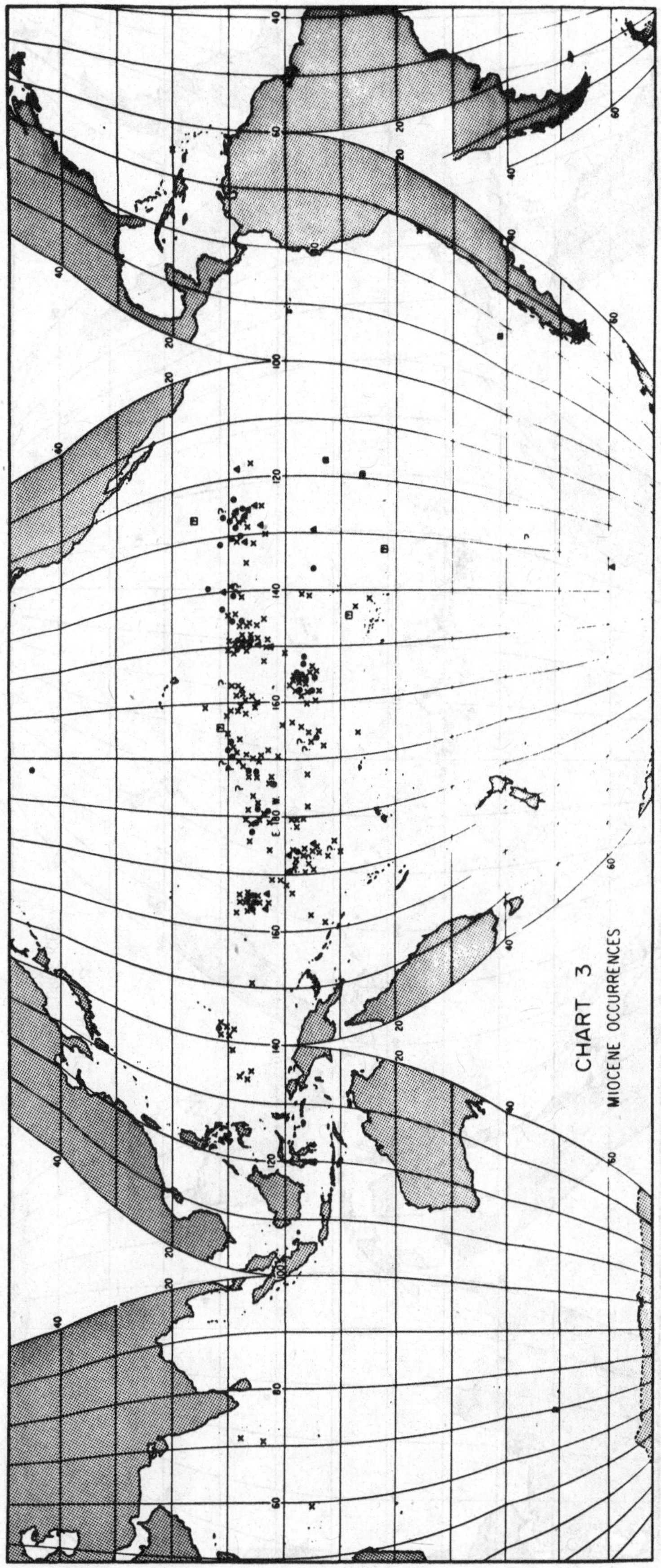

CHART 3
MIOCENE OCCURRENCES

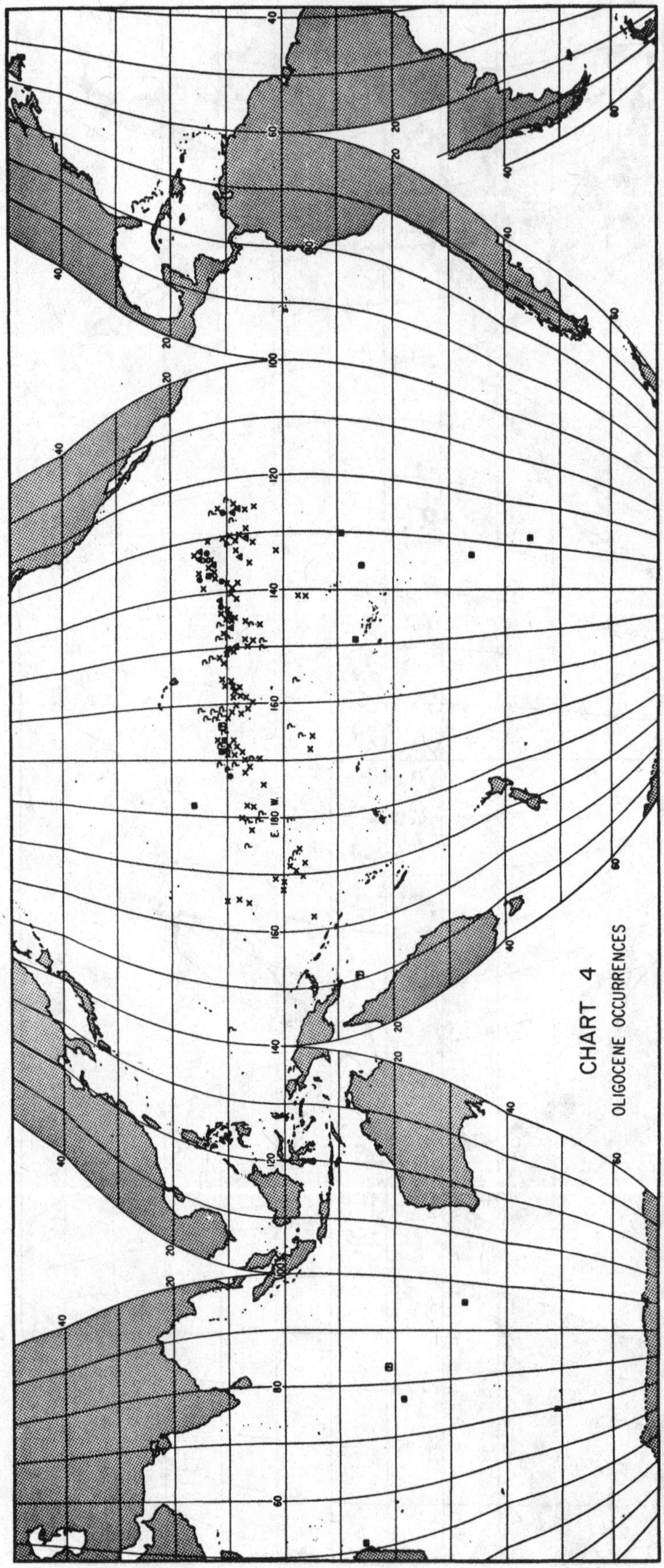

CHART 4
OLIGOCENE OCCURRENCES

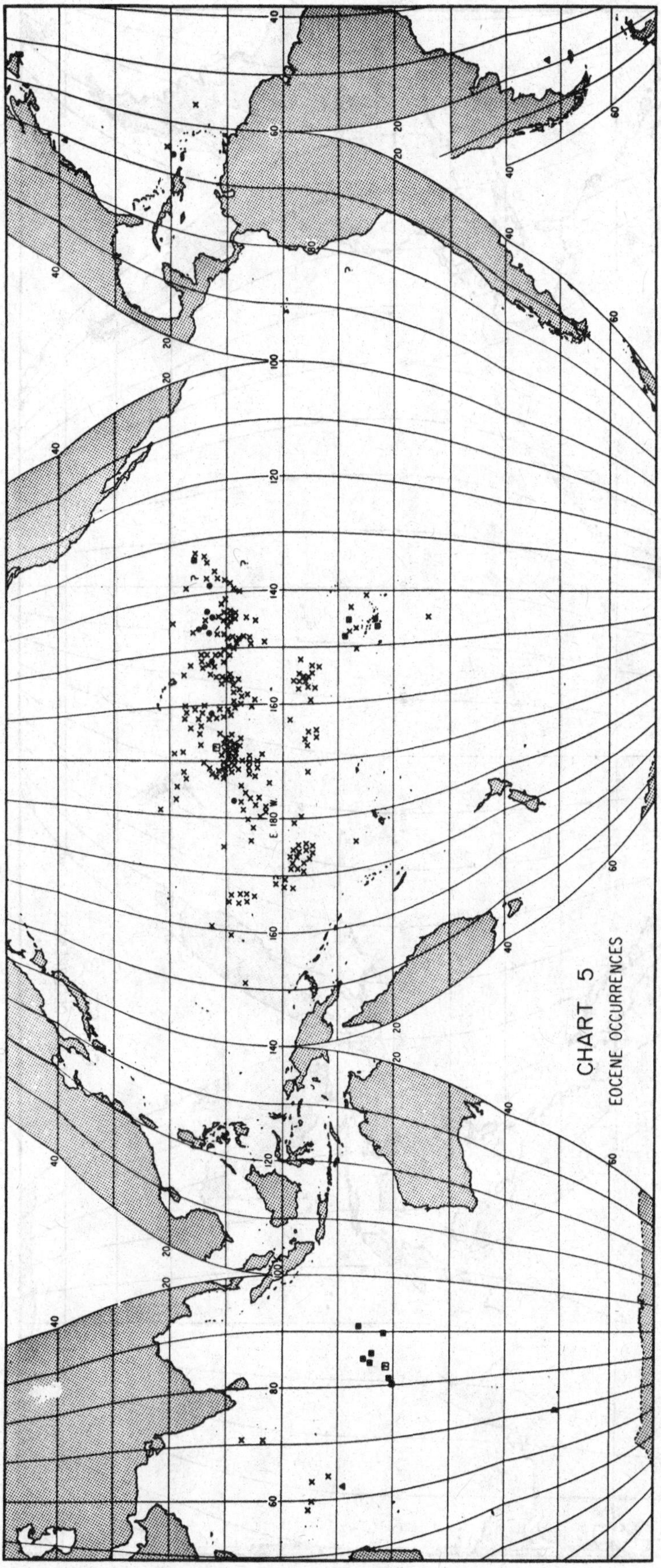

CHART 5
EOCENE OCCURRENCES

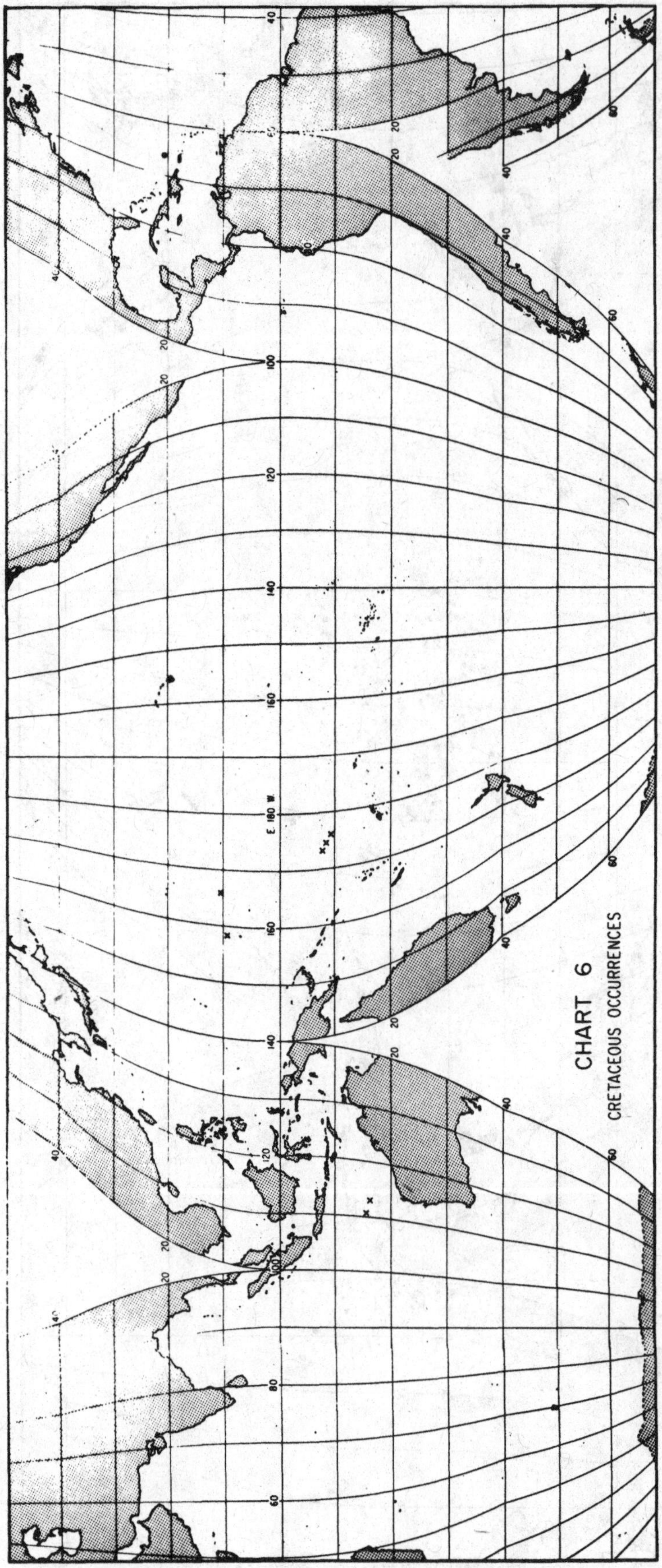
E 180 W

CHART 6
CRETACEOUS OCCURRENCES

ACKNOWLEDGEMENTS

Most of the samples forming the basis for this paper were collected by expeditions supported by the National Science Foundation and the Office of Naval Research. Lore Dobler, Heidi Häusermann, Phyllis Helms, Kathleen McCauley and Steve Moran assisted in organizing and plotting the results. I am grateful also to M. N. Bramlette, B. M. Funnell and F. L. Parker for determinations of the ages of some calcareous oozes, and to colleagues at other institutions (Lamont Geological Observatory, Oceanografiska Institutet in Göteborg, Institute of Oceanology in Moscow, British Museum Natural History, Tohoku University and Woods Hole Oceanographic Institution) for providing some of the samples. Part of the research reported in this paper was supported by the National Science Foundation grant no. GA-658.

Received September 1967.

BIBLIOGRAPHY

Bowin, C. O., Nalwalk, A. J. and Hersey, J. B. 1966. Serpentinized peridotite from the north wall of the Puerto Rico Trench. *Bull. geol. Soc. Am.* **77**, 257-70, 4 pls. *(1)*

Bramlette, M. N., Faughn, J. L. and Hurley, R. J. 1959. Anomalous sediment deposition on the flank of Eniwetok Atoll. *Bull. geol. Soc. Am.* **70**, 1,549-52. *(2)*

Dymond, J. R. 1966. Potassium-argon geochronology of deep-sea sediments. *Science, N.Y.* **152**, 1,239-41. *(3)*

Friend, J. K. and Riedel, W. R. 1967. Cenozoic orosphaerid radiolarians from tropical Pacific sediments. *Micropaleontology,* **13**, 217-32. *(4)*

Hays, J. D. 1965. Radiolaria and late Tertiary and Quaternary history of Antarctic Seas. *Biology of Antarctic Seas* II, *Antarctic Research Series* **5** *(American Geophysical Union),* 124-84. *(5)*

Martini, E. 1965. Mid-Tertiary calcareous nannoplankton from Pacific deep-sea cores. In *Submarine Geology and Geophysics* (W. F. Whittard and R. B. Bradshaw *eds.*). **17**, 393-411, pls. 33-7. Butterworths: Lond. *(6)*

Martini, E. and Bramlette, M. N. 1963. Calcareous nannoplankton from the experimental Mohole drilling. *J. Paleont.* **37**, 845-56, pls. 102-5. *(7)*

Parker, F. L. 1964. Foraminifera from the experimental Mohole drilling near Guadalupe Island, Mexico. *J. Paleont.* **38**, 617-36, pls. 97-102. *(8)*

Parker, F. L. 1967. Late Tertiary biostratigraphy (planktonic Foraminifera) of tropical Indo-Pacific deep-sea cores. *Bull. Am. Paleont.* **52** (235), 111-208, pls. 17-32. *(9)*

Rex, R. W. and Goldberg, E. D. 1958. Quartz contents of pelagic sediments of the Pacific Ocean. *Tellus,* **10** (1), 153-9. *(10)*

Riedel, W. R. 1954. The age of the sediment collected at *Challenger* (1875) Station 225 and the distribution of *Ethmodiscus rex* (Rattray). *Deep Sea Res.* **1**, 170-5. *(11)*

Riedel, W. R. 1957. Radiolaria: a preliminary stratigraphy. *Rept. Swed. deep-sea Exped.* **6** (3), 59-96, pls. 1-4. *(12)*

Riedel, W. R. 1959. Oligocene and Lower Miocene Radiolaria in tropical Pacific sediments. *Micropaleontology,* **5**, 285-302. *(13)*

Riedel, W. R. 1967. Radiolarian evidence consistent with spreading of eastern Pacific floor. *Science,* **157**, 540-42. *(14)*

Riedel, W. R., Bramlette, M. N. and Parker, F. L. 1963. 'Pliocene-Pleistocene' boundary in deep-sea sediments. *Science,* **140**, 1,238-40. *(15)*

Riedel, W. R. and Funnell, B. M. 1964. Tertiary sediment cores and microfossils from the Pacific Ocean floor. *Q. Jl. geol. Soc. Lond.* **120**, 305-68, pls. 14-32. *(16)*

Riedel, W. R., Ladd, H. S., Tracey, J. I. jr. and Bramlette M. N. 1961. Preliminary drilling phase of Mohole Project: summary of coring operations. *Bull. Am. Ass. Petrol. Geol.* **45**, 1,793-8. *(17)*

Ross, D. A. and Riedel, W. R. 1967. Comparison of upper parts of some piston cores with simultaneously collected open-barrel cores. *Deep Sea Res.* **14**, 285-94, 1 fig. *(18)*

Saito, T., Ewing, M. and Burckle, L. H. 1966. Tertiary sediment from the Mid-Atlantic Ridge. *Science,* **151**, 1,075-9. *(19)*

Wiseman, J. D. H. and Riedel, W. R. 1960. Tertiary sediments from the floor of the Indian Ocean. *Deep Sea Res.* **7**, 215-17. *(20)*

Note: numbers in brackets in the Bibliography are used for reference in the following Tables 41.1 and 41.2.

TABLES 41.1 and 41.2

Table 41.1. Samples containing pre-Quaternary Radiolaria.
Table 41.2. Pre-Quaternary samples lacking Radiolaria.

Abbreviations used are: B. M. F., age determined by B. M. Funnell; calc., calcareous; Cret., Cretaceous; Eoc., Eocene; exped., expedition; F. L. P., age determined by F. L. Parker; foram., foraminiferal; inst., institution; IOAN, Institute of Oceanology, Akad. Nauk; lat., latitude; LGO, Lamont Geological Observatory; long., longitude; Mioc., Miocene; M. N. B., age determined by M. N. Bramlette; Oligoc., Oligocene; oz., ooze; Plioc., Pliocene; Quat., Quaternary; radiol., radiolarian; silic., siliceous; SIO, Scripps Institution of Oceanography; spl., sample; stn., station; Swed. D-S Exped., Swedish Deep-Sea Expedition 1947-48; and WHOI, Woods Hole Oceanographic Institution.

In the first column, the indicated depths in the cores are those specifically examined for this tabulation. The total core interval containing the pre-Quaternary microfossils is commonly longer.

In the column 'Previous references' are included references only to the ages of the samples, and not to other aspects of their investigation. Papers are referred to by numbers indicated in the bibliography.

Table 41.1.

Core or sample number and depth in core	Collecting Institute	Latitude and Longitude	Water depth (m)	Age of pre-Quaternary Radiolaria	Age of sediment	Nature of sediment	Previous references
ALB Dredge, Stn. 4666	Albatross	11° 56′ S, 84° 20′ W	4,755	Eoc?	Eoc?	Pebble of silic. silty clay	
AMPH 4G, 0-27 cm	SIO	8° 31′ N, 127° 25′ W	4,740	Oligoc. ?, Mioc.	Quat.	Silic. oz.	
AMPH 6P, 15-60 cm	SIO	4° 52′ N, 128° 21′ W	4,460	Mioc.	Mioc.	Silic. calc. oz.	(*4*), p. 219
AMPH 9P, 210-578 cm	SIO	7° 31′ S, 121° 56′ W	4,410	Plioc.	Plioc.	Silic. oz.	(*4*), p. 219
AMPH 77G, 15-17 cm	SIO	12° 59′ S, 165° 34′ W	5,646	Mioc.	Quat. ?	Zeolitic clay	
AMPH 91P, 118-573 cm	SIO	6° 02′ S, 157° 28′ W	5,177	Mioc.	Mioc.	Radiol. clay	(*4*), p. 219
AMPH 92P, 0-13 cm	SIO	5° 56′ S, 157° 39′ W	5,283	Eoc., Mioc., Plioc.	Quat	Zeolitic radiol. clay	
AMPH 93G, 0-96 cm	SIO	5° 52′ S, 159° 55′ W	5,110	Mioc., Plioc.	Quat. ?	Zeolitic radiol. clay	
AMPH 94G, 0-144 cm	SIO	4° 44′ S, 159° 57′ W	5,204	Eoc., Mioc., Plioc.	Quat.	Silic. oz.	
AMPH 94G, 169-71 cm	SIO	4° 44′ S, 159° 57′ W	5,204	Mioc., Plioc.	Plioc. ?	Silic. oz.	
AMPH 95 PG, 0-2 cm	SIO	4° 44′ S, 158° 05′ W	5,380	Eoc., Mioc., Plioc.	Quat	Zeolitic silic. oz.	
AMPH 95P, 28-44 cm	SIO	4° 44′ S, 158° 05′ W	5,380	Eoc.	?	Zeolotic radiol. clay	
AMPH 96 PG, 0-2 cm	SIO	4° 38′ S, 158° 10′ W	5,140	Mioc., Plioc.	Quat.	Zeolitic silic. oz.	
AMPH 96P, 0-569 cm	SIO	4° 38′ S, 158° 10′ W	5,140	Mioc.	Mioc.	Zeolitic silic. oz.	(*4*), p. 219
AMPH 97P, 208-592 cm	SIO	3° 42′ S, 157° 40′ W	5,228	Plioc.	Plioc.	Zeolitic silic. oz.	(*4*), p. 219
AMPH 97P, 0-87 cm	SIO	3° 42′ S, 157° 40′ W	5,228	Eoc., Mioc., Plioc.	Quat.	Silic. oz.	
AMPH 98P, 158-569 cm	SIO	2° 50′ S, 157° 13′ W	5,225	Mioc., Plioc.	Plioc.	Silic. oz.	
AMPH 99P, 0-482 cm	SIO	1° 52′ S, 156° 42′ W	4,983	Eoc., Mioc., Plioc.	Quat.	Silic. oz. to silic. calc. oz.	(*18*), p. 287
AMPH 100G, 0-154 cm	SIO	3° 40′ S, 156° 37′ W	5,115	Eoc., Mioc., Plioc.	Quat.	Silic. oz. to silic. calc. oz.	
AMPH 101GV, 14-58 cm	SIO	3° 51′ S, 155° 48′ W	5,161	Eoc, Mioc.	Mioc.	Silic. oz.	(*4*), p. 219
AMPH 102P, 20-2 cm	SIO	3° 52′ S, 155° 43′ W	5,172	Eoc., Mioc., Plioc.?	Plioc. ?	Zeolitic silic. oz.	
AMPH 102P, 115-487 cm	SIO	3° 52′ S, 155° 43′ W	5,172	Eoc., Mioc.	Mioc.	Silic. oz.	(*4*), p. 219
AMPH 105G, 0-2 cm	SIO	3° 52′ S, 155° 53′ W	5,050	Eoc., Oligoc. ?, Mioc.	Quat.	Silic. oz.	
AMPH 105G, 8-145 cm	SIO	3° 52′ S, 155° 53′ W	5,050	Mioc.	Mioc.	Zeolitic silic. oz.	(*4*), p. 219
AMPH 107G, 0-18 cm	SIO	3° 52′ S, 155° 53′ W	4,915	Eoc., Oligoc. ? Mioc.	Quat.	Silic. calc. oz. to silic. oz.	
AMPH 107G, 30-122 cm	SIO	3° 52′ S, 155° 53′ W	4,915	Eoc., Oligoc. ?	Oligoc. ?	Silic. oz.	

Table 41.1. (continued)

Core or sample number and depth in core	Collecting Institute	Latitude and Longitude	Water depth (m)	Age of pre-Quaternary Radiolaria	Age of sediment	Nature of sediment	Previous references
AMPH 108GV, 0-86 cm	SIO	4° 32′ S, 155° 30′ W	5,100	Eoc., Mioc.	Quat.	Silic. oz. to zeolitic radiol. clay	
AMPH 109P, 0-2 cm	SIO	4° 49′ S, 155° 19′ W	5,265	Eoc., Mioc., Plioc.	Quat.	Zeolitic silic. oz.	
AMPH 109P, 14-223 cm	SIO	4° 49′ S, 155° 19′ W	5,265	Eoc., Mioc., Plioc.	Plioc.	Zeolitic calc. clay to silic. oz.	
AMPH 109P, 234-524 cm	SIO	4° 49′ S, 155° 19′ W	5,265	Eoc., Mioc.	Mioc.	Silic. oz.	(*4*), p. 219
AMPH 110PG, 0-2 cm	SIO	5° 49′ S, 154° 39′ W	4,768	Eoc., Mioc.	Quat.	Zeolitic silic. calc. oz.	
AMPH 117PG, 0-2 cm	SIO	12° 24′ S, 148° 24′ W	5,151	Eoc.	Quat.	Zeolitic clay	
AMPH 119C2, 0-18 cm	SIO	13° 27′ S, 147° 24′ W	4,525	Eoc.	Quat.	Zeolitic calc. clay	
AMPH 130G, 58-107 cm	SIO	5° 58′ S, 142° 43′ W	4,450	Plioc.	Plioc.	Silic. oz.	(*4*), p. 219
AMPH 132G, 0-142 cm	SIO	4° 23′ S, 141° 54′ W	4,430	Oligoc., Mioc., Plioc. ?	Quat.	Silic. calc. oz.	
AMPH 133G, 0-156 cm	SIO	3° 32′ S, 141° 26′ W	4,149	Oligoc., Mioc., Plioc.	Quat.	Silic. calc. oz.	
AMPH 139G, 0-114 cm	SIO	6° 05′ N, 134° 58′ W	4,550	Eoc. ?, Oligoc., Mioc.	Quat.	Silic. calc. oz. to calc. silic. oz.	
AMPH 140G	SIO	7° 39′ N, 134° 02′ W	4,548	Oligoc.	Oligoc.	Silic. calc. oz.	(*4*), p. 219
AMPH 142C1, 0-24 cm	SIO	8° 23′ N 133° 39′ W	4,950	Oligoc.	Quat.	Silic. oz.	
AMPH 144G, 0-9 cm	SIO	11° 42′ N, 131° 38′ W	4,935	Mioc.	Mioc.	Radiol. clay	(*4*), p. 219
CAP 1BG, 54-86 cm	SIO	1° 57′ N, 169° 29′ E	4,280	Mioc., Plioc.	Quat.	Silic. calc. oz.	
CAP 2BG2, 0-70 cm	SIO	0° 47′ N, 169° 12′ E	4,270	Eoc?, Mioc., Plioc.	Quat.	Silic. calc. oz.	(*16*),p.315
CAP 2BP2, 122-549 cm	SIO	0° 47′ N, 169° 12′ E	4,270	Mioc., Plioc.	Plioc.	Silic. calc. oz.	(*16*),p.315; (*9*), p. 132
CAP 3BG, 0-70 cm	SIO	2° 41′ S, 170° 46′ E	3,940	Eoc., Oligoc., Mioc. ?	Quat.	Silic. calc. oz.	
CAP 4BP, 0-11 cm	SIO	4° 08′ S, 171° 46′ E	3,670	Mioc., Plioc.	Quat.	Silic. calc. oz.	
CAP 4BP, 26-31 cm	SIO	4° 08′ S, 171° 46′ E	3,670	Mioc., Plioc.	Plioc.	Silic. calc. oz.	
CAP 5BG, 32-59 cm	SIO	9° 03′ S, 174° 52′ E	4,960	Mioc., Cret.	Mioc. or Plioc.	Graded calc. oz.	(*16*),p.321
CAP 42BP, 696-803 cm	SIO	7° 19′ S, 118° 40′ W	4,200	Plioc.	Plioc.	Silic. calc. oz.	(*4*), p. 219
CAP 49BP, 508-832 cm	SIO	9° 17′ N, 124° 09′ W	4,410	Oligoc. ?, Mioc.	Mioc.	Radiol. clay	(*10*),p.155
CAP 1HG, 98-100 cm	SIO	2° 06′ N, 169° 01′ E	4,320	Eoc., Oligoc., Mioc., Plioc.	Quat.	Calc. oz.	
CAP 3HG, 0-72 cm	SIO	0° 00′, 168° 35′ E	4,364	Eoc., Oligoc., Mioc.	Quat.	Calc. oz.	
CAP 6HG, 77-79 cm	SIO	4° 15.5′ S, 171° 55.5′ E	3,363	Eoc., Oligoc., Mioc.	Quat.	Silic. calc. oz.	

Table 41.1. (continued)

Core or sample number and depth in core	Collecting Institute	Latitude and Longitude	Water depth (m)	Age of pre-Quaternary Radiolaria	Age of sediment	Nature of sediment	Previous references
CAP 7HG, 28-30 cm	SIO	5° 13′ S, 172° 28′ E	4,600	Eoc., Mioc.?	Quat.	Silty calc. clay	
CAP 27HG, 0-27 cm	SIO	14° 50′ S, 146° 08′ W	2,763	Eoc.	Quat.	Calc. oz.	
CARN VII, Spl. 79	*Carnegie*	12° 40′ N, 137° 32′ W	4,918	Eoc., Oligoc.	Quat.	Silic. oz.	
CARN VII, Spl. 80	*Carnegie*	7° 45′ N, 141° 24′ W	5,003	Oligoc., Mioc.	Quat.	Silic. oz.	
CARN VII, Spl. 81	*Carnegie*	3° 01′ N, 149° 46′ W	4,953	Oligoc. ?, Mioc., Plioc.	Quat.	Silic. calc. oz.	
Chain 19: D3	WHOI	20° 00′ N, 66° 25′ W	approx. 6,400	Eoc.	Eoc.	Radiol. clay	(*12*), p. 266
Chain 34: D3	WHOI	20° 17′ N, 65° 42′ W	approx. 6,000	Cret.	Cret.	Radiol. clay.	(*12*), p. 266
CHALL., Stn. 225	Challenger Exped.	11° 24′ N, 143° 16′ E	8,184	Mioc.	Mioc.	Silic. oz.	(*11*); (*13*), p. 287
CHALL., Stn. 264	Challenger Exped.	14° 19′ N, 152° 37′ W	5,459	Eoc.	Quat. ?	Silty clay	
CHALL., Stn. 265	Challenger Exped.	12° 42′ N, 152° 01′ W	5,304	Eoc., Oligoc. ?	Quat.	Radiol. clay	
CHALL., Stn. 266	Challenger Exped.	11° 07′ N, 152° 03′ W	5,019	Eoc., Oligoc.	Quat.	Silic. oz.	
CHALL., Stn. 267	Challenger Exped.	9° 28′ N, 150° 49′ W	4,946	Eoc.	Quat.	Silic. oz.	
CHALL., Stn. 268	Challenger Exped.	7° 35′ N, 149° 49′ W	5,304	Mioc.	Mioc.	Silic. oz.	(*13*), p. 287
CHALL., Stn. 272	Challenger Exped.	3° 48′ S, 152° 56′ W	4,755	Eoc., Mioc., Plioc.	Quat.	Silic. oz.	
CHUB 6, 140-8 cm	SIO	12° 03′ N, 125° 31′ W	4,636	Oligoc. ?	Oligoc. ?	Zeolitic radiol. clay	
CHUB 9, 80-100 cm	SIO	10° 19′ N, 125° 27′ W	4,545	Oligoc.	Oligoc.	Radiol. clay	(*13*), p. 287
CHUB 10, 20-4 cm	SIO	10° 21′ N, 125° 26′ W	4,645	Oligoc., Mioc. ?	Quat.	Radiol. clay	
CHUB 15, 48-52 cm and 70-86 cm	SIO	8° 31′ N, 125° 25′ W	4,462	Mioc.	Mioc.	Silic. calc. oz.	(*13*), p. 287
CHUB 17, 11-20 cm	SIO	8° 05′ N, 125° 25′ W	4,453	Oligoc.	Oligoc.	Silic. calc. oz.	(*13*), p. 287
CHUB 18, 80-164 cm	SIO	7° 55′ N, 125° 29′ W	4,609	Oligoc., Mioc.	Mioc.	Radiol. clay	
CHUB 20, 36-46 cm	SIO	7° 22′ N, 125° 30′ W	4,549	Mioc.	Mioc.	Silic. calc. oz.	(*13*), p. 287
CHUB 24, 16-28 cm	SIO	5° 29′ N, 125° 29′ W	4,530	Oligoc., Mioc.	Mioc.	Silic. calc. oz.	(*13*), p. 287
CHUB 25, 0-152 cm	SIO	4° 37′ N, 125° 26′ W	4,480	Mioc.	Quat.	Calc. silic. oz.	
CHUB 30, 43-50 cm	SIO	7° 18′ N, 127° 25′ W	3,640	Mioc., Plioc.	Plioc.	Silic. calc. oz.	(*9*), p. 132; (*15*), p. 1240
CHUB 35, 0-164 cm	SIO	8° 11′ N, 126° 58′ W	4,517	Mioc.	Quat.	Calc. silic. oz. to silic. oz.	

Table 41.1. (continued)

Core or sample number and depth in core	Collecting Institute	Latitude and Longitude	Water depth (m)	Age of pre-Quaternary Radiolaria	Age of sediment	Nature of sediment	Previous references
CHUB 38, 0-18 cm	SIO	8° 12′ N, 125° 19′ W	4,315	Oligoc.	Oligoc.	Silic. calc. oz.	(*13*), p. 287
CHUB 39, 0-144 cm	SIO	8° 09′ N, 125° 20′ W	4,360	Oligoc., Mioc.	Quat.	Silic. calc. oz. to silic. oz.	
CHUB 40, 12-64 cm	SIO	8° 06′ N, 125° 26′ W	4,415	Mioc.	Mioc.	Radiol. clay to silic. calc. oz.	(*13*), p. 287
CK 13, 100-109 cm	SIO	44° 45′ N, 173° 02′ W	5,070	Mioc. or Plioc.	Mioc. or Plioc.	Silic. clay	(*16*), p. 323
CK 16, 60-73 cm	SIO	36° 30′ N, 173° 16′ W	4,290	Plioc.	Plioc.	Calc. oz.	(*16*), p. 323
DODO 17PG, 25-27 cm	SIO	approx. 10° 00′ N, 168° 10′ W	5,310	Eoc.	?	Zeolitic radiol. clay	
DODO 20PG, 0-181 cm	SIO	10° 02′ N, 167° 50′ W	5,280	Eoc., Oligoc?	Quat.	Zeolitic radiol. clay	
DODO 21P, 185-517 cm	SIO	10° 00′ N, 168° 00′ W	5,165	Eoc., Mioc.	Mioc.	Zeolitic radiol. clay	
DODO 22P, 186-572 cm	SIO	approx. 9° 50′ N, 168° 00′ W	5,245	Mioc.	Mioc. ?	Zeolitic clay	
DODO 23P, 185-573 cm	SIO	9° 45′ N, 168° 05′ W	5,100	Eoc., Oligoc. ?, Mioc.	Mioc. ?	Silic. oz.	
DODO 26PG, 0-17 cm	SIO	approx. 9° 20′ N, 169° 00′ W	5,140	Eoc.	Quat.	Silic. oz.	
DODO 30P, 187-292 cm	SIO	9° 10′ N, 168° 50′ W	5,175	Eoc., Oligoc.	Quat.	Zeolitic radiol. clay	
DODO 30P, 401-574 cm	SIO	approx. 9° 10′ N, 168° 50′ W	5,175	Eoc., Oligoc.	Oligoc.?	Zeolitic radiol. clay	
DODO 31P, 297-527 cm	SIO	9° 11′ N, 168° 48′ W	5,220	Eoc., Oligoc.	Oligoc. ?	Zeolitic radiol. clay	
DODO 33P, 477-9 cm	SIO	approx. 9° 30′ N, 169° 00′ W	5,064	Eoc., Plioc.	Plioc.	Zeolitic radiol. clay	
DODO 33P, 583-887 cm	SIO	9° 30′ N, 169° 00′ W	5,064	Eoc., Oligoc.	Oligoc. ?	Zeolitic radiol. clay	
DODO 36P, 0-53 cm	SIO	6° 00′ N, 165° 55′ E	5,010	Eoc., Oligoc., Mioc., Plioc.	Quat.	Silic. calc. oz. to silic. oz.	
DODO 36P, 202-442 cm	SIO	6° 00′ N, 165° 55′ E	5,010	Mioc.	Mioc.	Radiol. clay to calc. silic. oz.	
DODO 37P, 0-117 cm	SIO	6° 00′ N, 165° 55′ E	5,010	Eoc., Oligoc. Mioc., Plioc.	Quat.	Radiol. clay to silic. calc. oz.	
DODO 37P, 142-516 cm	SIO	6° 00′ N, 165° 55′ E	5,010	Mioc.	Mioc.	Silic. oz.	
DODO 38P, 7-22 cm	SIO	6° 00′ N, 166° 00′ E	5,010	Eoc., Mioc., Plioc.	Quat.	Radiol. clay to silic. calc. oz.	
DODO 38P, 130-457 cm	SIO	6° 00′ N, 166° 00′ E	5,010	Mioc.	Mioc.	Silic. oz.	
DODO 39P, 18-20 cm	SIO	6° 00′ N, 166° 00′ E	5,000	Eoc., Mioc., Plioc.	Quat.	Silic. calc. oz.	

Table 41.1. (continued)

Core or sample number and depth in core	Collecting Institute	Latitude and Longitude	Water depth (m)	Age of pre-Quaternary Radiolaria	Age of sediment	Nature of sediment	Previous references
DODO 39P, 103-475 cm	SIO	6° 00′ N, 166° 00′ E	5,000	Mioc.	Mioc.	Silic. oz.	
DODO 40G, 0-175 cm	SIO	6° 00′ N, 165° 00′ E	5,010	Mioc.	Quat.	Silic. calc. oz. to radiol. clay	
DODO 41PG, 117-19 cm	SIO	3° 20′ N, 165° 10′ E	4,225	Mioc., Plioc.	Quat. ?	Silic. calc. oz.	
DODO 41P, 26-444 cm	SIO	3° 20′ N, 165° 10′ E	4,225	Mioc.	Mioc.	Silic. calc. oz.	
DODO 56P, 17-216 cm	SIO	15° 31′ S, 113° 45′ E	5,377	Cret.	Quat.	Radiol. clay to silic. calc. oz.	
DODO 58P, 117-19	SIO	15° 42′ S, 110° 37′ E	5,710	Cret.	Quat.	Silic. calc. oz.	
DODO 123 D	SIO	10° 25′ S, 63° 15′ E	3,115	Eoc.	Eoc.	Silic. calc. oz.	(*4*), p. 219
DWBG 4, 2-140 cm	SIO	13° 25′ N, 130° 18′ W	4,712	Oligoc.	Oligoc. ?	Radiol. clay	(*16*), p.325
DWBG 5, 0-54 cm	SIO	10° 26′ N, 130° 38′ W	4,733	Oligoc.	Quat.	Radiol. clay	(*16*). p.325
DWBG 6, 0-22 cm	SIO	9° 36′ N, 130° 41′ W	5,003	Oligoc.	Quat.	Silic. oz.	
DWBG 7, 73-100 cm	SIO	8° 48′ N, 130° 48′ W	4,917	Oligoc. ?, Mioc.	Mioc.	Radiol. clay	(*16*), p.325
DWBG 8, 0-147 cm	SIO	7° 51′ N, 130° 55′ W	5,069	Oligoc., Mioc.	Quat.	Silic. oz.	
DWBG 9, 62-118 cm	SIO	7° 39′ N, 130° 55′ W	4,950	Oligoc., Mioc.	Mioc.	Zeolitic radiol. clay to silic. oz.	(*16*), p.326
DWBG 10, 2-27 cm	SIO	6° 54′ N, 131° 00′ W	4,340	Oligoc.	Oligoc.	Silic. calc. oz.	(*13*), p. 287; (*16*), p. 326; (*6*), p. 398
DWBG 11, 0-70 cm	SIO	5° 26′ N, 131° 19′ W	4,155	Oligoc., Mioc.	Quat.	Silic. calc. oz.	(*16*), p.327
DWBG 16, 0-102 cm	SIO	6° 05′ S, 132° 53′ W	4,855	Plioc.	Quat.	Silic. calc. oz. to silic. oz.	
DWBG 23B, 4.5-13.5 cm, 14-19 cm, 22-22.5 cm	SIO	16° 42′ S, 145° 48′ W	approx. 2,200	Eoc.	Eoc.	Silic. calc. oz.	(*16*), p.327
DWBG 36 31-3 cm	SIO	26° 19′ S, 147° 07′ W	3,680	Eoc.	Plioc. ?	Calc. oz.	(*16*), p. 328
DWBG 89, entire sample	SIO	24° 14.5′ S, 70° 59′ W	2,765	Mioc.	Mioc. ?	Sandy silt	
DWHG 64, hard mud sample	SIO	24° 30′ S, 70° 49′ W	1,520	Mioc. ?	Mioc. ?	Clayey silt with silic. and calc. microfossils	
DWHH 12, 0-64 cm	SIO	6° 02′ S, 131° 36′ W	4,700	Plioc.	Quat.	Silic. calc. oz. to radiol. clay	
DWHH 35, 12-52 cm	SIO	44° 21′ S, 127° 14′ W	4,675	Plioc. or Mioc.	Plioc. or Mioc.	Silic. oz. to zeolitic radiol. clay	(*16*), p. 332
EM, 42-170 metres	Mohole Project.	28° 59′ N, 117° 30′ W	3,566	Mioc.	Mioc.	Silic. calc. to silic. oz.	(*17*), p. 1793; (*7*), p. 847; (*8*), p. 619; (*3*), p. 1239
FAN HMS 2, 16-86 cm	SIO	30° 17′ N, 118° 14′ W	2,970	Mioc. ?	Mioc. ?	Calc. and silic. clay	

Table 41.1. (continued)

Core or sample number and depth in core	Collecting Institute	Latitude and Longitude	Water depth (m)	Age of pre-Quaternary Radiolaria	Age of sediment	Nature of sediment	Previous references
FAN BP 2, 380-402 cm	SIO	28° 32.5′ N, 117° 20′ W	3,580	Mioc.	Mioc. ?	Slightly radiol. clay	
FAN BG4, 80-2 cm	SIO	28° 52′ N, 117° 20′ W	3,560	Mioc.	Mioc. ?	Slightly radiol. clay	
FAN BG5, 42-117 cm	SIO	29° 00′ N, 117° 30′ W	3,590	Mioc.	Mioc. ?	Slightly radiol. clay	
FAN BG6-1, 90-117 cm	SIO	28° 59′ N, 117° 28′ W	3,535	Mioc. or Plioc.	Mioc. ? or Plioc. ?	Slightly radiol. clay	
Hz. 12 (Manshu Stn. 1463-39)	Tohoku Univ.	6° 20′ N, 135° 26′ E	4,825	Mioc.	Quat.	Radiol. clay	
Hz. 79 (Manshu Stn. 1561-28-247)	Tohoku Univ.	9° 37′ N, 158° 40′ E	5,300	Cret., Eoc.	Quat. ?	Zeolitic radiol. clay	
Hz. 80 (Manshu Stn. 1558-25-242)	Tohoku Univ.	12° 55′ N, 159° 55′ E	5,576	Eoc.	Quat. ?	Zeolitic radiol. clay	
Hz 95 (Manshu Stn. 1624-91-329)	Tohoku Univ.	9° 26′ N, 126° 41′ E	5,844	Eoc.	Quat.	Radiol. clay	
Hz 98 (Manshu Stn. 1674-142-393)	Tohoku Univ.	6° 40′ N, 150° 25′ E	5,648	Eoc. and Mioc.	Quat.	Zeolitic radiol. clay	
Hz 101 (Manshu Stn. 1681-140-405)	Tohoku Univ.	11° 50′ N, 145° 36′ E	6,388	Mioc.	Quat.	Zeolitic clay with silic. microfossils	
Hz 121 (Manshu Stn. 682)	Tohoku Univ.	9° 38′ N, 166° 10′ E	4,702	Eoc.	Quat.	Silic. calc. oz.	
Hz 122 (Manshu Stn. 684)	Tohoku Univ.	9° 13′ N, 163° 12′ E	5,214	Plioc.	Plioc.	Silic. calc. oz.	
JYN IV 14G, 0-3 cm	SIO	22° 01′ N, 179° 51′ W	5,172	Eoc.	Quat.	Zeolitic radiol. clay	
JYN V 2G, 32-4 cm	SIO	17° 32′ N, 154° 32′ W	5,023	Eoc.	Quat.	Pyroclastic silt	
JYN V 3G, 0-8 cm	SIO	16° 40′ N, 153° 31′ W	5,170	Eoc.	Quat.	Pyroclastic silt	
JYN V 5G, 0-72 cm	SIO	14° 29′ N, 152° 41′ W	5,514	Eoc.	Quat.	Zeolitic radiol. clay	
JYN V 8G, 0-2 cm	SIO	13° 26′ N, 152° 17′ W	5,746	Eoc.	Quat.	Zeolitic radiol. clay	
JYN V 9G, 0-8 cm	SIO	12° 44′ N, 152° 03′ W	5,510	Eoc., Oligoc. ?	Quat.	Zeolitic radiol. clay	
JYN V 10P, 0-2 cm	SIO	12° 43′ N, 152° 06′ W	5,549	Eoc., Oligoc. ?	Quat.	Zeolitic radiol. clay	
JYN V 10P, 279-836 cm	SIO	12° 43′ N, 152° 06′ W	5,549	Eoc.	Eoc.	Zeolitic radiol. clay	
JYN V 11G, 0-42 cm	SIO	11° 08′ N, 152° 01′ W	5,408	Eoc. and Oligoc.	Quat.	Zeolitic silic. oz.	
JYN V 12G, 0-2 cm	SIO	10° 46′ N, 151° 44′ W	5,011	Eoc.	Quat.	Silic. oz.	
JYN V 13G, 0-6 cm	SIO	9° 27′ N, 150° 42′ W	5,100	Eoc., Oligoc., Mioc.	Quat.	Radiol. clay	
JYN V 14G, 0-33 cm	SIO	9° 20′ N, 150° 35′ W	5,073	Eoc., Oligoc., Mioc.	Quat.	Calc. silic. oz. to zeolitic radiol. clay	

Table 41.1. (continued)

Core or sample number and depth in core	Collecting Institute	Latitude and Longitude	Water depth (m)	Age of pre-Quaternary Radiolaria	Age of sediment	Nature of sediment	Previous references
JYN V 15P, 0-100 cm	SIO	8° 02′ N, 149° 54′ W	5,073	Eoc., Oligoc., Mioc.	Quat.	Silic. oz.	
JYN V 15P, 515-667 cm	SIO	8° 02′ N, 149° 54′ W	5,073	Eoc., Oligoc., Mioc., Plioc.	Plioc.	Silic. oz.	
JYN V 15P, 876-989 cm	SIO	8° 02′ N, 149° 54′ W	5,073	Eoc., Oligoc., Mioc.	Mioc.	Silic. oz.	
JYN V 16P, 0-250 cm	SIO	7° 44′ N, 149° 44′ W	5,168	Oligoc.	Oligoc.	Silic. oz. to silic. calc. oz.	(*4*), p. 219
JYN V 17G, 32-72 cm	SIO	6° 05′ N, 148° 52′ W	5,036	Eoc., Oligoc., Mioc.	Mioc.	Silic. oz.	
JYN V 18G, 150-2 cm	SIO	5° 01′ N, 148° 33′ W	4,952	Oligoc., Mioc.	Quat.	Silic. oz.	
JYN V 19G, 94-6 cm	SIO	6° 28′ N, 148° 23′ W	4,690	Eoc., Oligoc., Mioc., Plioc.	Plioc. ?	Silic. oz.	
JYN V 20PG, 0-2 cm	SIO	7° 17′ N, 148° 12′ W	4,925	Eoc., Oligoc., Mioc.	Quat.	Calc. silic. oz.	
JYN V 20P, 7-459 cm	SIO	7° 17′ N, 148° 12′ W	4,925	Mioc.	Mioc.	Silic. oz. to silic. calc. oz.	(*4*), p. 219
JYN V 22G, 0-32 cm	SIO	8° 34′ N, 147° 47′ W	5,184	Eoc., Mioc.	Mioc.	Silic. oz.	
JYN V 23P, 55-89 cm	SIO	10° 09′ N, 147° 33′ W	5,241	Eoc., Oligoc.	Oligoc. ?	Zeolitic radiol. clay	
JYN V 24G, 56-155 cm	SIO	11° 49′ N, 147° 26′ W	5,417	Eoc., Oligoc. ?	Oligoc. ?	Silic. oz. to radiol. clay	
JYN V 25P, 0-502 cm	SIO	13° 05′ N, 147° 19′ W	5,455	Eoc.	Eoc.	Zeolitic radiol. clay	(*4*), p. 219
JYN V 27P, 45-7 cm	SIO	15° 29′ N, 147° 04′ W	5,424	Eoc.	Eoc. ?	Zeolitic radiol. clay	
JYN V 28P, 0-473 cm	SIO	14° 13′ N, 146° 24′ W	4,952	Eoc.	Eoc.	Silic. oz.	(*4*), p. 219
JYN V 29G, 75-151 cm	SIO	13° 32′ N, 146° 02′ W	5,285	Eoc.	Eoc.	Zeolitic radiol. clay	
JYN V 30G, 47-9 cm	SIO	12° 52′ N, 145° 35′ W	5,104	Eoc.	Eoc. ?	Zeolitic radiol. clay	
JYN V 31P, 126-281 cm	SIO	11° 55′ N, 144° 54′ W	5,539	Eoc.	Eoc.	Zeolitic radiol. clay	
JYN V 32G, 0-3 cm	SIO	11° 11′ N, 144° 26′ W	5,501	Eoc.	Quat.	Zeolitic silic. oz.	
JYN V 33G, 0-145 cm	SIO	10° 34′ N, 144° 00′ W	5,184	Eoc., Oligoc.	Quat.	Silic. oz.	
JYN V 38P, 119-453 cm	SIO	6° 30′ N, 141° 59′ W	5,018	Plioc.	Plioc.	Silic. oz.	(*4*), p. 219
JYN V 39G, 0-69 cm	SIO	7° 59′ N, 140° 50′ W	5,188	Mioc.	Mioc.	Radiol. clay	
JYN V 41G, 0-102 cm	SIO	8° 35′ N, 140° 21′ W	5,087	Eoc., Oligoc., Mioc. ?	Quat.	Silic. oz. to zeolitic radiol. clay	
JYN V 42P, 0-33 cm	SIO	9° 00′ N, 139° 51′ W	5,000	Eoc., Oligoc., Mioc. ?	Quat.	Silic. oz. to radiol. clay	
JYN V 43G, 0-11 cm	SIO	9° 53′ N, 138° 56′ W	4,893	Eoc., Oligoc., Mioc.	Quat.	Zeolitic radiol. clay	
JYN V 44P, 457-91 cm	SIO	11° 07′ N, 137° 56′ W	4,853	Eoc., Oligoc.	Oligoc.	Silic. oz. to radiol. clay	

Table 41.1. (continued)

Core or sample number and depth in core	Collecting Institute	Latitude and Longitude	Water depth (m)	Age of pre-Quaternary Radiolaria	Age of sediment	Nature of sediment	Previous references
JYN V 45G, 0-52 cm	SIO	12° 18′ N, 137° 04′ W	5,043	Eoc., Oligoc.	Quat.	Silic. oz. to zeolitic radiol. clay	
JYN V 46P, 12-14 cm	SIO	13° 24′ N, 136° 11′ W	4,822	Eoc., Oligoc.	Quat.	Zeolitic radiol. clay	
JYN V 47P, 0-3 cm	SIO	14° 39′ N, 135° 04′ W	4,813	Eoc., Oligoc.	Quat.	Zeolitic radiol. clay	
JYN V 48P, 3-5 cm	SIO	15° 54′ N, 133° 57′ W	4,606	Eoc., Oligoc.	Quat.	Zeolitic calc. clay	
LGO A172-15, 0-12 cm	LGO	20° 13′ N, 64° 56′ W	5,960	Eoc., Mioc.	Quat.	Radiol. clay	
LGO V12-46, 50-174 cm	LGO	47° 29′ S, 59° 21′ W	1,185	Eoc.	Eoc.	Calc. silic. oz.	
LGO V18-307, at approx. 1,100 cm	LGO	60° 33′ S, 131° 57′ W	4,581	Mioc.	Mioc.	Silic. calc. oz.	
LGO V18-310, at approx. 269 cm	LGO	04° 26′ S, 129° 02′ W	4,651	Mioc.	Mioc.	Silic. calc. oz.	
LGO V18-316, at approx. 400 cm	LGO	1° 03′ N, 120° 46′ W	4,316	Mioc.	Mioc.	Silic. calc. oz.	
LGO V20-39, at approx. 268 cm	LGO	12° 17′ N, 135° 27′ W	4,715	Oligoc.	Oligoc.	Silic. calc. oz.	
LGO, 20 D	LGO	39° 50′ N, 70° 50′ W	?	Eoc.	Eoc.	Silic. calc. oz.	
LSDA 99G, 31-3 cm	SIO	3° 58′ N, 70° 48′ E	4,130	Eoc., Mioc.	Quat.	Calc. oz.	
LSDA 101G, 48-50 cm	SIO	2° 41′ S, 73° 12′ E	2,960	Plioc.	Plioc.	Silic. calc. oz.	(*9*), p. 132
LSDA 105G, 0-2 cm	SIO	5° 40′ S, 66° 36′ E	4,365	Plioc.	Quat.	Silic. calc. oz.	
LSDA 106G, 55-7 cm	SIO	5° 34′ S, 63° 43′ E	4,090	Eoc.	Quat.	Silic. calc. oz.	
LSDA 107Ga, 0-56 cm	SIO	5° 26′ S, 59° 15′ E	4,010	Eoc., Mioc., Plioc.	Quat.	Silic. calc. oz.	
LSDA 144G, 48-50 cm	SIO	13° 09′ S, 93° 13′ E	5,226	Plioc.	Plioc.	Silic. oz.	
LSDH 3G, 130-2 cm	SIO	7° 27′ N, 70° 39′ E	4,110	Eoc., Mioc.	Quat. ?	Calc. oz.	
LSDH 13G, 0-44 cm	SIO	5° 26′ S, 58° 29′ E	3,950	Eoc.	Quat.	Silic. calc. oz.	
LSDH 73G, 0-48 cm	SIO	8° 04′ S, 160° 23′ E	2,640	Mioc.	Quat.	Silic. calc. oz.	
LSDH 76P, 12-110 cm	SIO	6° 40′ S, 163° 13′ E	3,560	Oligoc., Mioc.	Quat.	Silic. calc. oz.	
LSDH 78P, 58-522 cm	SIO	4° 31′ S, 168° 02′ E	3,208	Plioc.	Plioc.	Silic. calc. oz.	(*15*), p.1,240; (*9*), p. 132; (*4*), p. 219; (*18*), p. 288
LSDH 83P, 210-522 cm	SIO	10° 54′ N, 174° 01′ E	5,530	Eoc. Mioc. ?	Mioc. ?	Silic. oz.	
LSDH 88P, 0-422 cm	SIO	8° 33′ N, 177° 46′ W	5,620	Eoc.	Eoc.	Silic. oz.	(*4*), p. 219
LSDH 89P, 0-2 cm	SIO	8° 08′ N, 177° 10′ W	5,435	Eoc.	Quat.	Silic. oz.	

Table 41.1. (continued)

Core or sample number and depth in core	Collecting Institute	Latitude and Longitude	Water depth (m)	Age of pre-Quaternary Radiolaria	Age of sediment	Nature of sediment	Previous references
LSDH 90P, 0-102 cm	SIO	7° 19′ N, 175° 28′ W	5,190	Eoc., Mioc. ?	Quat.	Calc. silic. to silic. oz.	
LSDH 90P, 150-542 cm	SIO	7° 19′ N, 175° 28′ W	5,190	Eoc., Plioc.	Plioc.	Silic. oz.	(*4*), p. 219; (*15*), p. 1239
LSDH 91PG, 0-2 cm	SIO	9° 11′ N, 172° 03′ W	5,510	Eoc., Oligoc.	Quat.	Zeolitic silic. oz.	
LSDH 91P, 20-518 cm	SIO	9° 11′ N, 172° 03′ W	5,510	Eoc., Oligoc.	Oligoc.	Silic. oz.	(*4*), p. 219
LSDH 95G, 6-8 cm	SIO	11° 28′ N, 168° 51′ W	5,285	Eoc.	Quat.	Zeolitic radiol. clay	
LSDH 96PG, 0-2 cm	SIO	15° 10′ N, 164° 39′ W	5,523	Eoc.	Quat. ?	Zeolitic clay	
LSDH 99G, 0-2 cm	SIO	20° 06′ N, 145° 18′ W	5,370	Eoc.	Quat.	Zeolitic calc. clay	
MP 5-1, 0-43 cm	SIO	14° 22′ N, 133° 07′ W	approx. 4,800	Oligoc.	Oligoc.	Silic. calc. oz.	(*13*), p. 287; (*16*), p. 334
MP 6-1, 77-9 cm	SIO	13° 38′ N, 133° 45′ W	4,890	Oligoc.	Oligoc.	Zeolitic radiol. clay	(*16*), p. 334
MP 7-2, 0-107 cm	SIO	12° 47′ N, 134° 26′ W		Oligoc.	Quat.	Radiol. clay	
MP 13-1, 0-68 cm	SIO	9° 09′ N, 144° 26′ W	5,236	Eoc., Oligoc., Mioc.	Quat.	Zeolitic radiol. clay	(*16*), p. 335
MP 14-1, 0-10 cm	SIO	9° 58′ N, 145° 13′ W	5,130	Eoc., Oligoc.	Quat.	Zeolitic radiol. clay	(*16*), p. 335
MP 15-1, 0-20 cm	SIO	10° 44′ N, 145° 53′ W	5,375	Eoc., Oligoc. ?	Quat.	Zeolitic radiol. clay	(*16*), p. 335
MP 17-2, 0-96 cm	SIO	14° 40′ N, 151° 56′ W	5,816	Eoc.	Quat.	Zeolitic radiol. clay	(*16*), p. 336
MP 30, 0-3 cm	SIO	18° 20′ N, 173° 20′ W	approx. 3,910	Eoc.	Quat.	Calc. clay	
MP 32, 0-13 cm	SIO	18° 20′ N, 173° 20′ W	approx. 3,859	Eoc.	Quat.	Zeolitic radiol. clay	
MP 35-1, 0-5 cm	SIO	19° 21′ N, 174° 58′ W	approx. 4,938	Eoc.	Quat.	Zeolitic radiol. clay	
MP 39, 0-5 cm	SIO	19° 02′ N, 177° 18′ W	approx. 4,838	Eoc.	Quat.	Zeolitic clay	
MP 45, 0-9 cm	SIO	10° N, 166° E (very approx.)	4,298	Cret., Eoc., Oligoc.	Quat.	Zeolitic calc. radiol. clay	
MSN 4G, 0-2 cm	SIO	16° 55′ N, 139° 18′ W	5,355	Eoc.	Quat.	Zeolitic radiol. clay	(*16*), p. 336
MSN 5G, 0-102 cm	SIO	13° 15′ N, 144° 04′ W	5,040	Eoc.	Eoc.	Silic. oz.	(*16*), p. 336
MSN 7PG, 0-2 cm	SIO	14° 09′ N, 161° 08′ W	5,652	Eoc., Oligoc. ?, Mioc.	Quat.	Zeolitic radiol. clay	(*16*), p. 337
MSN 9G, 0-154 cm	SIO	8° 29′ N, 167° 12′ W	5,142	Eoc., Oligoc., Mioc., Plioc.	Quat.	Zeolitic radiol. clay	(*16*), p. 337
MSN 10G, 0-164 cm	SIO	7° 36′ N, 168° 06′ W	4,994	Eoc., Oligoc., Mioc.	Quat.	Silic. calc. oz. to silic. oz.	(*16*), p. 338
MSN 11G, 0-52 cm	SIO	6° 04′ N, 169° 58′ W	5,400	Eoc., Oligoc., Mioc.	Quat.	Silic. oz. to zeolitic radiol. clay	(*16*), p. 338
MSN 12G, 127-52 cm	SIO	3° 01′ N, 174° 02′ W	5,230	Eoc., Oligoc., Mioc., Plioc.	Plioc.	Silic. oz.	(*16*), p. 338

Table 41.1. (continued)

Core or sample number and depth in core	Collecting Institute	Latitude and Longitude	Water depth (m)	Age of pre-Quaternary Radiolaria	Age of sediment	Nature of sediment	Previous references
MSN 13V	SIO	3° 01′ N, 174° 02′ W	5,250	Eoc., Oligoc., Mioc., Plioc.	Quat.	Silic. oz.	
MSN 16G, 19-154 cm	SIO	2° 04′ S, 178° 16′ E	5,519	Eoc., Mioc., Plioc.	Plioc.	Silic. calc. oz.	(*16*), p. 339
MSN 128G, 0-2 cm	SIO	13° 53′ S, 150° 35′ W	3,623	Eoc.	Quat.	Foram. oz.	(*16*), p. 340
MSN 132PG, 38-72 cm	SIO	5° 58′ S, 149° 33′ W	5,115	Mioc, Plioc.	Plioc.	Silic. oz.	(*16*), p. 341
MSN 132P, 13-360 cm	SIO	5° 58′ S, 149° 33′ W	5,115	Mioc.	Mioc.	Silic. oz.	(*16*), p. 341
MSN 135P, 0-813 cm	SIO	4° 26′ S, 149° 24′ W	4,600	Mioc.	Mioc.	Silic. oz. to silic. calc. oz.	(*16*), p. 342
MSN 141G, 140-2 cm	SIO	3° 32′ N, 146° 47′ W	4,577	Oligoc., Mioc.	Quat.	Silic. calc. oz.	
MSN 142G, 0-76 cm	SIO	5° 20′ N, 146° 13′ W	5,089	Eoc., Oligoc. ?, Mioc., Plioc. ?	Quat.	Silic. oz.	(*16*), p. 342
MSN 142G, 94-142 cm	SIO	5° 20′ N, 146° 13′ W	5,089	Mioc., Plioc.	Plioc.	Silic. oz.	(*16*), p. 342
MSN 143P, 60-904 cm	SIO	5° 32′ N, 146° 09′ W	5,100	Mioc., Plioc.	Plioc.	Silic. oz.	(*16*), p. 343; (*15*), p. 1239
MSN 144G, 0-92 cm	SIO	6° 01′ N, 145° 58′ W	5,200	Mioc.	Quat.	Silic. oz.	
MSN 146P, 28-284 cm	SIO	7° 09′ N, 145° 35′ W	5,100	Eoc., Oligoc., Mioc., Plioc.	Plioc.	Silic. oz.	(*16*), p. 344
MSN 146P, 349-585 cm	SIO	7° 09′ N, 145° 35′ W	5,100	Mioc.	Mioc.	Silic. oz.	(*16*), p. 344
MSN 148G, 0-17 cm	SIO	9° 06′ N, 145° 18′ W	5,400	Eoc., Oligoc.	Quat.	Zeolitic radiol. clay to silic. oz.	
MSN 149G, 1-46 cm	SIO	9° 23′ N, 145° 15′ W	5,100	Mioc.	Mioc.	Zeolitic radiol. clay	(*16*), p. 344
MSN 149P, 0-318 cm	SIO	9° 23′ N, 145° 15′ W	5,100	Oligoc.	Oligoc.	Silic. oz. to silic. calc. oz.	(*16*), p. 344; (*6*), p. 398
MSN 150G, 40-2 cm	SIO	10° 59′ N, 142° 37′ W	4,978	Eoc., Oligoc.	Oligoc.	Zeolitic radiol. clay	(*16*), p. 345
MSN 151P, 0-148 cm	SIO	11° 03′ N, 142° 28′ W	5,000	Eoc., Oligoc.	Oligoc.	Silic. oz. to calc. silic. oz.	(*16*), p. 346; (*6*), p. 398
MSN 152G, 64-6 cm	SIO	13° 04′ N, 138° 59′ W	4,930	Eoc.	Eoc.	Silic. oz.	(*16*), p. 346
MSN 153P, 46-200 cm	SIO	13° 07′ N, 138° 56′ W	4,927	Eoc., Oligoc., Mioc.	Mioc.	Silic. oz.	(*16*), p. 347
MSN 154P, 110-46 cm	SIO	15° 01′ N, 137° 01′ W	4,920	Eoc. ?, Oligoc., Plioc.	Plioc.	Zeolitic radiol. clay	(*16*), p. 348
MSN 154P, 561-641 cm	SIO	15° 01′ N, 137° 01′ W	4,920	Oligoc.	Oligoc.	Zeolitic radiol. clay	(*16*), p. 348
Planet Stn. 127	Planet Exped.	8° 45′ S, 64° 52′ E	4,220	Eoc.	Quat.	Silic. calc. oz.	(*20*), p. 216
PROA 4GA	SIO	11° 22′ N, 142° 31′ E	10,900	Mioc.	Quat.	Zeolitic radiol. clay	
PROA 7G, 0-2 cm	SIO	10° 29′ N, 143° 06′ E	5,080	Mioc.	Quat.	Radiol. clay	
PROA 12G, 0-80 cm	SIO	7° 00′ N, 135° 00′ E	5,055	Mioc.	Quat.	Silic. oz. to radiol. clay	

Table 41.1. (continued)

Core or sample number and depth in core	Collecting Institute	Latitude and Longitude	Water depth (m)	Age of pre-Quaternary Radiolaria	Age of sediment	Nature of sediment	Previous references
PROA 13P, 0-7 cm	SIO	7° 44′ N, 134° 59′ E	8,040	Mioc.	Quat.	Silic. calc. oz.	
PROA 19G, 0-102 cm	SIO	9° 24′ N, 139° 00′ E	4,660	Mioc.	Quat.	Radiol. clay to silic. calc. oz.	
PROA 21G, 0-20 cm	SIO	9° 05′ N, 143° 07′ E	4,100	Oligoc. ?, Mioc.	Quat.	Silic. calc. oz.	
PROA 58G, catcher	SIO	12° 40′ S 176° 19′ E	2,996	Eoc.	Quat.	Silic. calc. oz.	
PROA 61P, 0-2 cm	SIO	9° 58′ S, 173° 52′ E	5,151	Mioc.	Quat.	Silty clay with silic. and calc. microfossils.	
PROA 62P, 29-31 cm	SIO	8° 44′ S, 173° 24′ E	5,397	Cret., Mioc., Plioc. ?	Quat.	Silty clay with silic. and calc. microfossils.	(*18*), p. 288
PROA 65P, 0-2 cm	SIO	9° 13′ S, 176° 11′ E	5,186	Cret., Mioc., Plioc. ?	Quat.	Calc. oz.	
PROA 78PG, 0-2 cm	SIO	5° 40′ S, 174° 06′ E	5,550	Eoc., Mioc.	Quat.	Radiol. clay	
PROA 78P, 30-2 cm	SIO	5° 40′ S, 174° 06′ E	5,550	Eoc., Mioc., Plioc.	Plioc.	Silic. calc. oz.	
PROA 78P, 170-442 cm	SIO	5° 40′ S, 174° 06′ E	5,550	Eoc., Mioc.	Mioc.	Zeolitic clay to calc. silic. oz.	
PROA 79PG, 0-168 cm	SIO	4° 51′ S, 174° 01′ E	4,895	Eoc., Mioc.	Quat.	Silic. calc. oz. to zeolitic radiol. clay	
PROA 80P, 0-2 cm	SIO	4° 04′ S, 174° 08′ E	4,862	Eoc., Mioc.	Quat.	Silty clay	
PROA 81P, 0-2 cm	SIO	3° 23′ S, 174° 13′ E	4,723	Eoc., Mioc.	Mioc.	Silic. oz.	
PROA 81PG, 0-87 cm	SIO	3° 23′ S, 174° 13′ E	4,723	Eoc., Oligoc. ?, Mioc.	Quat.	Silic. calc. oz.	
PROA 82P, 364-472 cm	SIO	3° 05′ S, 174° 27′ E	4,951	Eoc., Oligoc., Mioc., Plioc.	Plioc.	Silic. oz.	
PROA 83P, 0-132 cm	SIO	2° 04′ S, 172° 29′ E	4,290	Eoc., Oligoc. ?, Mioc.	Quat.	Silic. calc. oz.	
PROA 83P, 210-293 cm	SIO	2° 04′ S, 172° 29′ E	4,290	Eoc., Oligoc., Mioc., Plioc.	Plioc. ?	Silic. calc. oz.	
PROA 85PGI, 0-116 cm	SIO	2° 16′ S, 170° 18′ E	4,205	Eoc., Oligoc., Mioc., Plioc. ?	Quat.	Silic. calc. oz.	
PROA 86P, 0-352 cm	SIO	0° 02′ N, 168° 17′ E	4,238	Eoc., Oligoc., Mioc.	Quat.	Silic. calc. oz.	
PROA 88P, 108-454 cm	SIO	2° 56′ N, 167° 14′ E	4,428	Mioc.	Mioc.	Calc. silic. oz.	(*4*), p. 219; (*18*), p. 288; (*9*), p. 132
PROA 89PG, 0-156 cm	SIO	5° 01′ N, 166° 24′ E	4,800	Eoc., Mioc., Plioc.	Quat.	Calc. silic. oz.	
PROA 89P, 480-548 cm	SIO	5° 01′ N, 166° 24′ E	4,800	Mioc, Plioc.	Plioc.	Silic. oz.	
PROA 90P, 60-2 cm	SIO	5° 59′ N, 165° 53′ E	5,021	Mioc., Plioc.	Plioc. ?	Radiol. clay	
PROA 90P, 232-4 cm	SIO	5° 59′ N, 165° 53′ E	5,021	Mioc.	Mioc.	Radiol. clay	
PROA 91PG, 0-9 cm	SIO	7° 02′ N, 165° 21′ E	5,208	Eoc., Mioc., Plioc. ?	Quat.	Calc. silic. to silic. oz.	

Table 41.1. (continued)

Core or sample number and depth in core	Collecting Institute	Latitude and Longitude	Water depth (m)	Age of pre-Quaternary Radiolaria	Age of sediment	Nature of sediment	Previous references
PROA 91P, 38-40 cm	SIO	7° 02′ N, 165° 21′ E	5,208	Plioc.	Plioc.	Silic. calc. oz.	
PROA 92P, 80-474 cm	SIO	7° 58′ N, 164° 57′ E	5,219	Mioc., Plioc.	Plioc.	Silic. calc. oz.	(*4*), p. 219
PROA 93PG2, 0-2 cm	SIO	6° 59′ N, 166° 26′ E	5,117	Eoc., Oligoc., Mioc., Plioc.	Quat.	Silic. oz.	
PROA 95P, 160-305 cm	SIO	5° 49′ N, 175° 11′ E	5,225	Eoc., Oligoc. ?, Mioc., Plioc.	Plioc. ?	Silic. oz. to radiol. clay	
PROA 96P, 0-544 cm	SIO	5° 12′ N, 177° 43′ E	5,828	Eoc., Mioc.	Mioc.	Silic. oz.	
PROA 97P, 0-2 cm	SIO	4° 32′ N, 179° 45′ E	5,712	Eoc., Oligoc. ?, Mioc., Plioc.	Quat.	Silic. oz.	
PROA 97P, 137-478 cm	SIO	4° 32′ N, 179° 45′ E	5,712	Mioc.	Mioc.	Silic. oz.	(*4*), p. 219
PROA 98G, 0-121 cm	SIO	2° 45′ N, 179° 42′ E	5,490	Mioc., Plioc.	Quat.	Silic. oz.	
PROA 100PG, 0-21 cm	SIO	6° 26′ N, 179° 36′ E	5,836	Eoc., Oligoc., Mioc., Plioc.	Quat.	Silic. oz. to radiol. clay	
PROA 100P, 340-571 cm	SIO	6° 26′ N, 179° 36′ E	5,836	Eoc., Oligoc. ?, Mioc., Plioc.	Plioc.	Silic. oz.	
PROA 101P, 0-2 cm	SIO	6° 02′ N, 178° 35′ W	5,097	Eoc., Oligoc., Mioc., Plioc.	Quat.	Calc. radiol. clay	
PROA 102P, 0-2 cm	SIO	4° 58′ N, 177° 33′ W	5,399	Eoc., Oligoc., Mioc., Plioc.	Quat.	Radiol. clay	
PROA 102P, 13-15 cm	SIO	4° 58′ N, 177° 33′ W	5,399	Eoc., Mioc., Plioc.	Plioc.,	Silic. oz.	
PROA 102P, 67-499 cm	SIO	4° 58′ N, 177° 33′ W	5,399	Eoc., Mioc.	Mioc.	Silic. oz.	(*4*), p. 219
PROA 103PG, 0-2 cm	SIO	3° 37′ N, 179° 18′ W	4,972	Eoc., Oligoc. ?, Mioc., Plioc.	Quat.	Silic. calc. oz.	
PROA 103P, 123-515 cm	SIO	3° 37′ N, 179° 18′ W	4,972	Mioc., Plioc.	Plioc.	Silic. oz.	(*4*), p. 219
PROA 107PG, 0-2 cm	SIO	2° 27′ S, 179° 59′ E	5,490	Eoc., Mioc., Plioc.	Quat.	Silic. oz.	
PROA 107PG, 53-5 cm	SIO	2° 27′ S, 179° 59′ E	5,490	Eoc., Oligoc. ?, Mioc., Plioc.	Plioc.	Silic. calc. oz.	
PROA 109P, 440-544 cm	SIO	4° 30′ S, 177° 06′ W	5,710	Mioc.	Mioc. ?	Zeolitic radiol. clay	
PROA 117P, 0-2 cm	SIO	6° 44′ S, 167° 21′ W	5,067	Eoc., Mioc.	Quat.	Zeolitic radiol. clay	
PROA 117P, 160-514 cm	SIO	6° 44′ S, 167° 21′ W	5,067	Mioc.	Mioc.	Silic. oz.	
PROA 120G, 0-2 cm	SIO	4° 32′ S, 167° 32′ W	5,534	Eoc., Oligoc., Mioc. ?, Plioc.	Quat.	Silty radiol. clay	
PROA 121G, 0-2 cm	SIO	4° 54′ S, 167° 00′ W	5,214	Mioc., Plioc.	Quat.	Silic. oz.	
PROA 121G, 70-145 cm	SIO	4° 54′ S, 167° 00′ W	5,214	Eoc., Mioc.	Mioc. ?	Zeolitic radiol. clay	
PROA 122G, 124-6 cm	SIO	5° 29′ S, 166° 06′ W	4,127	Eoc., Oligoc., Mioc., Plioc.	Plioc. ?	Calc. oz.	
PROA 124G-1, 0-120 cm	SIO	4° 45′ S, 165° 23′ W	4,165	Eoc., Oligoc. ?, Mioc., Plioc.	Quat.	Silic. calc. oz.	

Table 41.1. (continued)

Core or sample number and depth in core	Collecting Institute	Latitude and Longitude	Water depth (m)	Age of pre-Quaternary Radiolaria	Age of sediment	Nature of sediment	Previous references
PROA 126G, 0-127 cm	SIO	2° 56′ S, 164° 33′ W	5,312	Mioc., Plioc.	Quat.	Silic. oz.	
PROA 127G, 0-165 cm	SIO	1° 35′ S, 163° 55′ W	5,512	Eoc., Oligoc. ?, Mioc., Plioc.	Quat.	Silic. oz.	
PROA 129G, 0-59 cm	SIO	0° 45′ N, 164° 10′ W	5,298	Mioc., Plioc.	Quat.	Silic. calc. oz.	
PROA 131G2, 37-9 cm	SIO	2° 37′ N, 166° 39′ W	5,523	Plioc.	Quat.	Silic. calc. oz.	
PROA 132G, 0-137 cm	SIO	3° 40′ N, 168° 22′ W	5,441	Eoc., Mioc., Plioc.	Quat.	Silic. oz.	
PROA 133G, 0-100 cm	SIO	4° 48′ N, 169° 50′ W	5,747	Eoc., Oligoc., Mioc., Plioc. ?	Quat.	Silic. oz. to radiol. clay	
PROA 134G, 0-166 cm	SIO	5° 00′ N, 170° 16′ W	5,810	Eoc., Oligoc. ?, Mioc., Plioc.	Quat.	Silic. oz.	
PROA 135G, 35-94 cm	SIO	5° 37′ N, 171° 24′ W	5,717	Eoc., Oligoc., Mioc.	Mioc.	Silic. oz.	
PROA 136G, 0-114 cm	SIO	6° 23′ N, 172° 33′ W	5,832	Eoc., Oligoc., Mioc.	Quat.	Silic. oz. to radiol. clay	
PROA 138G, 0-90 cm	SIO	7° 54′ N, 170° 39′ W	5,613	Eoc., Oligoc. ?, Mioc.	Quat.	Silic. oz.	
PROA 139G, 0-55 cm	SIO	8° 06′ N, 170° 25′ W	5,444	Eoc., Mioc.	Quat.	Silic. oz.	
PROA 141G, 0-2 cm	SIO	9° 08′ N, 168° 41′ W	5,205	Eoc., Oligoc. ?	Quat.	Calc. silic. oz.	
PROA 142G, 101-8 cm	SIO	9° 29′ N, 167° 51′ W	5,106	Eoc., Oligoc., Mioc.	Mioc.	Silic. oz.	(*4*), p. 219
PROA 143G, 0-2 cm	SIO	9° 33′ N, 167° 48′ W	5,140	Eoc.	Quat.	Zeolitic radiol. clay	
PROA 144G, 0-2 cm	SIO	9° 49′ N, 167° 21′ W	4,946	Eoc.	Quat.	Zeolitic silic. oz.	
PROA 145G, 0-66 cm	SIO	10° 00′ N, 167° 04′ W	5,265	Eoc., Oligoc. ?	Quat.	Silic. oz. to zeolitic radiol. clay	
PROA 146G, 0-99 cm	SIO	10° 22′ N, 166° 00′ W	4,773	Eoc.	Quat.	Silic. calc. oz.	
PROA 149G, 0-2 cm	SIO	10° 01′ N, 164° 59′ W	3,170	Oligoc. ?, Mioc.	Quat.	Calc. oz.	
PROA 150G, 0-2 cm	SIO	9° 35′ N, 166° 54′ W	5,139	Eoc., Oligoc.	Quat.	Calc. radiol. clay	
PROA 151G, 0-69 cm	SIO	8° 34′ N, 168° 52′ W	4,397	Eoc., Oligoc. ?	Quat.	Silic. calc. oz.	
PROA 151G, 126-35 cm	SIO	8° 34′ N, 168° 52′ W	4,397	Eoc., Plioc.	Plioc.	Silic. calc. oz.	
PROA 152G, 30-2 cm	SIO	9° 01′ N, 171° 03′ W	5,133	Eoc., Oligoc. ?, Mioc. ?	Mioc. or Oligoc.	Radiol. clay	(*4*), p. 230
PROA 153G, 0-172 cm	SIO	9° 06′ N, 171° 16′ W	4,900	Eoc.	Quat.	Silic. to calc. silic. oz.	
PROA 156G, 0-62 cm	SIO	10° 23′ N, 170° 57′ W	4,469	Eoc., Oligoc. ?	Quat.	Silic. calc. oz.	
PROA 156G, 123-5 cm	SIO	10° 23′ N, 170° 57′ W	4,469	Eoc., Mioc.	Quat. ?	Silic. calc. oz.	
PROA 157G, 0-2 cm	SIO	10° 20′ N, 172° 06′ W	5,106	Eoc.	Quat.	Zeolitic silic. calc. oz.	

Table 41.1. (continued)

Core or sample number and depth in core	Collecting Institute	Latitude and Longitude	Water depth (m)	Age of pre-Quaternary Radiolaria	Age of sediment	Nature of sediment	Previous references
PROA 158G, 0-2 cm	SIO	10° 20′ N, 172° 37′ W	5,865	Eoc.	Quat.	Silic. oz.	
PROA 159G, 0-2 cm	SIO	11° 23′ N, 172° 47′ W	5,380	Eoc.	Quat.	Zeolitic radiol. clay	
PROA 160G, 0-2 cm	SIO	11° 31′ N, 172° 49′ W	4,837	Eoc.	Quat.	Silic. calc. oz.	
PROA 162G, 0-2 cm	SIO	12° 31′ N, 172° 51′ W	5,464	Eoc.	Quat.	Zeolitic radiol. clay	
PROA 163G, 0-2 cm	SIO	13° 17′ N, 172° 58′ W	5,728	Eoc.	Quat.	Zeolitic clay	
PROA 164G, 0-2 cm	SIO	13° 33′ N, 171° 56′ W	5,863	Eoc.	Quat.	Zeolitic clay	
PROA 165G, 0-2 cm	SIO	13° 45′ N, 171° 12′ W	5,865	Eoc.	Quat.	Silic. oz.	
PROA 168G, 0-2 cm	SIO	14° 56′ N, 167° 34′ W	4,315	Eoc.	Quat.	Zeolitic radiol. clay	
PROA 170G, 0-2 cm	SIO	16° 04′ N, 166° 04′ W	5,435	Eoc.	Quat.	Zeolitic radiol. clay	
PROA 171G, 0-2 cm	SIO	16° 43′ N, 165° 10′ W	5,556	Eoc.	Quat.	Zeolitic radiol. clay	
PROA 173G, 0-2 cm	SIO	17° 55′ N, 163° 25′ W	5,386	Eoc.	Quat.	Zeolitic radiol. clay	
RIS 12G, 13-20 cm	SIO	8° 10′ N, 117° 53′ W	3,930	Mioc.	Mioc.	Silic. calc. oz.	(*4*), p. 219; (*9*), p. 132
RIS 14G, 0-92 cm	SIO	5° 20′ N, 117° 55′ W	4,330	Mioc., Plioc. ?	Quat.	Silic. calc. oz.	
RIS 84G, 0-34 cm	SIO	15° 15′ S, 142° 27′ W	3,675	Eoc., Mioc.	Quat.	Calc. oz.	
RIS 95G, 14-121 cm	SIO	13° 03′ S, 144° 03′ W	4,980	Eoc., Mioc.	Plioc. ?	Calc. oz.	
RIS 102G, 45-52 cm	SIO	6° 23′ S, 136° 11′ W	4,410	Plioc.	Plioc.	Silic. oz.	(*4*), p. 220
RIS 102G, 81-87 cm	SIO	6° 23′ S, 136° 11′ W	4,410	Mioc.	Mioc.	Silic. oz.	(*4*), p. 220
RIS 111P, 280-698 cm	SIO	14° 55′ N, 133° 29′ - 134° 15′ W	4,770	Oligoc.	Oligoc.	Silic. calc. oz.	(*4*), p. 219
SDSE 55, 102-4 cm	Swed. D-S Exped.	11° 30′ N, 127° 37′ W	4,196	Oligoc.	Oligoc.	Zeolitic radiol. clay	
SDSE 56, 85-102 cm	Swed. D-S Exped.	9° 50′ N, 128° 13′ W	4,629	Mioc.	Mioc.	Radiol. clay	
SDSE 57, 10-484 cm	Swed. D-S Exped.	8° 25′ N, 128° 48′ W	4,607	Mioc.	Mioc.	Silic. oz.	
SDSE 58, 765-985 cm	Swed. D-S Exped.	6° 44′ N, 129° 28′ W	4,440	Oligoc., Mioc., Plioc.	Plioc.	Calc. silic. oz.	(*15*), p.1239; (*9*), p. 132
SDSE 62, 1,170-471 cm	Swed. D-S Exped.	3° 00′ S, 136° 26′ W	4,511	Plioc.	Plioc.	Calc. silic. oz.	(*15*), p.1239
SDSE 73, 11-848 cm	Swed. D-S Exped.	4° 04′ S, 152° 53′ W	5,200	Eoc., Mioc., Plioc.	Plioc.	Silic. oz.	(*12*), p. 67
SDSE 73, 949-1,471 cm	Swed. D-S Exped.	4° 04′ S, 152° 53′ W	5,200	Eoc., Mioc.	Mioc.	Silic. oz.	(*12*), p. 67
SDSE 76, 108-509 cm	Swed. D-S Exped.	3° 45′ N, 149° 44′ W	5,155	Eoc., Mioc., Plioc.	Plioc.	Calc. silic. oz.	(*12*), p. 70

Table 41.1. (continued)

Core or sample number and depth in core	Collecting Institute	Latitude and Longitude	Water depth (m)	Age of pre-Quaternary Radiolaria	Age of sediment	Nature of sediment	Previous references
SDSE 76, 609-1,330 cm	Swed. D-S Exped.	3° 45′ N, 149° 44′ W	5,155	Oligoc. ?, Mioc.	Mioc.	Calc. silic. oz.	(*12*), p. 70
SDSE 77, 9-10 cm	Swed. D-S Exped.	9° 30′ N, 148° 40′ W	5,384	Eoc., Oligoc. ?	Quat.	Radiol. clay	(*12*), p. 71
SDSE 77, 109-1,310 cm	Swed. D-S Exped.	9° 30′ N, 148° 40′ W	5,384	Eoc., Mioc.	Mioc. ?	Zeolitic radiol. clay	(*12*), p. 71
SDSE 83, 99-290 cm	Swed. D-S Exped.	8° 40′ N, 169° 28′ W	5,456	Eoc., Mioc.	Mioc.	Silic. oz.	(*12*), p. 71
SDSE 85, 299-630 cm	Swed. D-S Exped.	5° 34′ N, 172° 12′ W	5,590	Eoc., Mioc., Plioc.	Plioc.	Silic. oz.	(*12*), p. 72
SDSE 85, 699-1,050 cm	Swed. D-S Exped.	5° 34′ N, 172° 12′ W	5,590	Eoc., Mioc.	Mioc.	Silic. oz.	(*12*), p. 72
SDSE 87, 0-30 cm	Swed. D-S Exped.	2° 23′ N, 173° 50′ W	5,560	Mioc., Plioc.	Plioc.	Silic. oz.	(*12*), p. 73
SDSE 87, 39-900 cm	Swed. D-S Exped.	2° 23′ N, 173° 50′ W	5,560	Mioc.	Mioc.	Silic. oz.	(*12*), p. 73
SDSE 90, 56-405 cm	Swed. D-S Exped.	3° 21′ S, 174° 12′ E	4,830	Eoc., Mioc.	Mioc.	Silic. calc. oz.	
SDSE 91, 309-1,410 cm	Swed. D-S Exped.	2° 50′ S, 171° 18′ E	4,096	Eoc., Mioc.	Mioc.	Silic. calc. oz.	(*12*), p. 74; (*13*), p. 287
SDSE 93, 0-1,509 cm	Swed. D-S Exped.	1° 20′ S, 167° 23′ E	3,965	Eoc., Mioc., Plioc.	Quat.	Silic. calc. oz.	(*12*), p. 75
SDSE 263A, 10-12 cm	Swed. D-S Exped.	15° 44′ N, 56° 50′ W	5,320	Eoc.	Quat.	Calc. clay	
TET 4, 0-113 cm	SIO	4° 34′ N, 130° 37′ W	4,478	Mioc.	Quat.	Silic. calc. oz.	
TET 11, 151-3 cm	SIO	14° 41′ N, 151° 20′ W	5,590	Eoc.	Quat. ?	Zeolitic radiol. clay	(*16*), p. 348
TET 18, 22-4 cm	SIO	17° 40′ N, 160° 39′ W	5,393	Eoc.	Quat.	Pyroclastic silt	(*16*), p. 349
TET 19, 37-9 cm	SIO	17° 36′ N, 160° 42′ W	5,490	Eoc.	Quat.	Pyroclastic silt	(*16*), p. 349
TET 21, 0-2 cm	SIO	16° 13′ N, 161° 40′ W	5,285	Eoc.	Quat.	Pyroclastic silt	
TET 23, 0-3 cm	SIO	15° 11′ N, 162° 25′ W	5,490	Eoc.	Quat.	Zeolitic radiol. clay	(*16*), p. 349
TET 24, 0-2 cm	SIO	15° 02′ N, 162° 31′ W	5,666	Eoc.	Quat.	Zeolitic radiol. clay	
TET 25, 0-3 cm	SIO	13° 50′ N, 163° 23′ W	5,589	Eoc., Oligoc. ?	Quat.	Zeolitic radiol. clay	
TET 27A, entire spl.	SIO	13° 05′ N, 163° 10′ W	5,413	Eoc.	Eoc.	Silic. oz.	(*16*), p. 349
TET 28, 0-2 cm	SIO	12° 58′ N, 163° 09′ W	5,430	Eoc., Oligoc. ?	Quat.	Zeolitic radiol. clay	(*16*), p. 350
TET 29, 0-32 cm	SIO	11° 44′ N, 162° 41′ W	5,265	Eoc., Oligoc. ?	Quat.	Zeolitic radiol. clay	(*16*), p. 350
TET 30, 0-3 cm	SIO	11° 32′ N, 162° 36′ W	5,340	Eoc., Oligoc. ?	Quat.	Zeolitic radiol. clay	(*16*), p. 351
TET 31, 0-2 cm	SIO	10° 19′ N, 162° 07′ W	5,151	Eoc., Oligoc. ?, Mioc.	Quat.	Silty radiol. clay	
TET 32, 0-3 cm	SIO	10° 09′ N, 162° 05′ W	5,063	Eoc., Oligoc., Mioc.	Quat.	Zeolitic radiol. clay	(*16*), p. 351

Table 41.1. (continued)

Core or sample number and depth in core	Collecting Institute	Latitude and Longitude	Water depth (m)	Age of pre-Quaternary Radiolaria	Age of sediment	Nature of sediment	Previous references
TET 34, 0-102 cm	SIO	8° 34′ N, 161° 39′ W	5,070	Eoc., Oligoc., Mioc., Plioc.	Quat.	Silic. calc. oz.	(*16*),p. 351
TET 35, 0-130 cm	SIO	7° 14′ N, 161° 03′ W	3,935	Eoc., Oligoc., Mioc., Plioc.	Quat.	Silic. calc. oz.	
TET 36, 0-138 cm	SIO	7° 04′ N, 160° 58′ W	4,143	Eoc., Oligoc., Mioc., Plioc.	Quat.	Silic. calc. oz.	
TET 37, 0-65 cm	SIO	5° 33′ N, 160° 33′ W	3,600	Eoc., Oligoc., Mioc., Plioc. ?	Quat.	Silic. calc. oz.	
TET 40, 0-161 cm	SIO	5° 18′ N, 160° 05′ W	3,816	Eoc., Oligoc., Mioc.	Quat.	Silic. calc. oz.	
TET 41, 0-122 cm	SIO	6° 52′ N, 158° 44′ W	4,407	Eoc., Oligoc., Mioc., Plioc.	Quat.	Silic. calc. oz.	
TET 42, 0-182 cm	SIO	7° 03′ N, 158° 35′ W	4,407	Eoc., Oligoc., Mioc., Plioc.	Quat.	Silic. calc. oz.	
TET 43, 0-48 cm	SIO	8° 14′ N, 157° 51′ W	4,675	Eoc., Oligoc., Mioc.	Quat.	Calc. oz. to radiol. clay	
TET 44, 0-162 cm	SIO	7° 53′ N, 157° 29′ W	4,989	Eoc., Oligoc., Mioc.	Quat.	Silic. calc. oz. to radiol. clay	(*16*),p. 352
TET 46, 0-152 cm	SIO	7° 56′ N, 156° 48′ W	5,075	Eoc., Oligoc., Mioc.	Quat. ?	Silic. oz. to zeolitic radiol. clay	(*16*),p. 352
TET 47, 70-164 cm	SIO	8° 14′ N, 156° 38′ W	5,266	Eoc., Oligoc., Mioc.	Mioc. ?	Zeolitic radiol. clay	(*16*),p. 352
TET 48, 0-2 cm	SIO	9° 10′ N, 156° 25′ W	5,285	Eoc., Oligoc., Mioc.	Quat.	Radiol. clay	
TET 49, 0-96 cm	SIO	9° 52′ N, 156° 09′ W	5,277	Eoc.	Quat.	Radiol. clay	(*16*),p. 353
TET 51, 51-149 cm	SIO	10° 39′ N, 155° 56′ W	5,188	Eoc., Oligoc.	Oligoc. ?	Zeolitic radiol. clay	(*16*),p. 353
TET 52, 50-106 cm	SIO	10° 47′ N, 155° 54′ W	5,234	Eoc., Oligoc., Mioc. ?	Mioc. ?	Zeolitic radiol. clay	(*16*), p. 353
TET 53, 0-2 cm	SIO	11° 27′ N, 155° 48′ W	5,206	Eoc.	Quat.	Radiol. clay	(*16*), p. 354
TET 54, 0-2 cm	SIO	12° 05′ N, 155° 29′ W	4,940	Eoc.	Quat.	Calc. radiol. clay	
TET 55, 0-2 cm	SIO	13° 07′ N, 155° 01′ W	5,518	Eoc.	Quat.	Radiol. clay	
USSR Stn. 3680 grab, 0-5 cm	IOAN	5° 35′ N, 147° 33′ E	3,944	Mioc.	Quat.	Silic. calc. oz.	
USSR Stn. 3802 grab, lower part	IOAN	3° 17′ S, 172° 52′ W	5,329	Eoc., Mioc.	Quat.	Silic. oz.	
USSR Stn. 3903, 8-36 cm	IOAN	10° 00′ N, 141° 52′ E	3,825	Mioc.	Quat.	Silic. calc. oz. to silty clay	
USSR Stn. 5074, 169-300 cm	IOAN	10° 31′ N, 140° 01′ W	?	Mioc.	Mioc.	Silic. calc. oz.	
W. Scoresby, 474 cm	Discovery Exped.	61° 03′ S, 56° 42′ W	2,813	Cret.	Quat.	?	

Table 41.2.

Core or sample number and depth in core	Collecting Institute	Latitude and Longitude	Water depth (m)	Age of sediment and how determined	Nature of sediment	Previous references
AMPH 10P, 138-389 cm	SIO	7° 31′ S, 117° 43′ W	4,286	Mioc. Calc. nannoplankton, M.N.B.	Calc. oz.	
AMPH 36P, 519 cm	SIO	18° 17′ S, 118° 17′ W	3,550	Plioc. ? Calc. nannoplankton, M.N.B.	Calc. oz.	
AMPH 37P, 125-582 cm	SIO	18° 16′ S, 121° 05′ W	3,720	Plioc. Calc. nannoplankton	Zeolitic calc. oz.	
AMPH 38P, 155-472 cm	SIO	18° 30′ S, 124° 30′ W	3,860	Plioc. Calc. nannoplankton, M.N.B.	Zeolitic calc. oz.	
AMPH 39P, 252-474 cm	SIO	18° 35′ S, 126° 25′ W	4,030	Plioc. ? Calc. nannoplankton, M.N.B.	Calc. oz.	
AMPH 42P, 145-437 cm	SIO	18° 37′ S, 133° 03′ W	3,930	Mioc. ? Calc. nannoplankton, M.N.B.	Zeolitic calc. oz.	
AMPH 116P, 495-569 cm	SIO	11° 26′ S, 149° 17′ W	5,106	Eoc. Calc. nannoplankton, M.N.B.	Calc. oz.	
AMPH 120P, 251-475 cm	SIO	12° 52′ S, 146° 23′ W	4,826	Eoc. Calc. nannoplankton, M.N.B.	Calc. oz.	
CAP 38 BP, 413-700 cm	SIO	14° 16′ S, 119° 11′ W	3,400	Plioc. Forams., F.L.P.	Calc. oz.	(*9*), p. 132
CAP 36 HG, 100-9 cm	SIO	11° 00′ S, 130° 06′ W	4,320	Oligoc. Calc. nannoplankton, M.N.B.	Calc. oz.	
CAP 38 BP, 800-62 cm	SIO	14° 16′ S, 119° 11′W	3,400	Mioc. Forams., F.L.P.	Calc. oz.	(*9*), p. 132
CARO II 2 G, 75-7 cm	SIO	32° 10′ S, 94° 49′ W	3,450	Plioc. Calc. nannoplankton, M.N.B.	Zeolitic calc. oz.	
CHALL. Stn. 286	Challenger Exped.	33° 29′ S, 133° 22′ W	4,270	Oligoc. Calc. nannoplankton, M.N.B.	Zeolitic calc. oz.	
CK 22, 156-60 cm	SIO	26° 22′ N, 168° 53′ W	4,450	Plioc. ? Calc. nannoplankton, M.N.B.	Calc. clay	
DODO 17P, 285-7 cm	SIO	10° 00′ N, 168° 10′ W	5,310	Oligoc. Calc. nannoplankton, M.N.B.	Calc. oz.	
DODO 17P, 551-3 cm	SIO	10° 00′ N, 168° 10′ W	5,310	Eoc. ? Calc. nannoplankton, M.N.B.	Calc. oz.	
DODO 20P, 184-287 cm	SIO	10° 02′ N, 167° 50′ W	5,280	Oligoc. Calc. nannoplankton, M.N.B.	Calc. oz.	
DODO 57P, 21-115 cm	SIO	15° 40′ S, 112° 44′ E	3,660	Plioc. Forams. and Calc. nannoplankton, B.M.F. and M.N.B.	Calc. oz.	(*9*), p. 132
DODO 78P, catcher, light	SIO	13° 52′ S, 91° 02′ E	5,460	Eoc. Calc. nannoplankton	Calc. oz.	
DODO 83P, approx. 20-40 cm	SIO	15° 30′ S, 85° 19′ E	4,850	Eoc. Calc. nannoplankton and forams., M.N.B. and B.M.F.	Calc. oz.	
DODO 86P, 20-153 cm	SIO	15° 32′ S, 85° 04′ E	4,755	Eoc. Calc. nannoplankton and forams., M.N.B. and B.M.F.	Calc. oz.	
DODO 87P, approx. 10-41 cm	SIO	16° 13′ S, 86° 06′ E	5,127	Eoc. Calc. nannoplankton, M.N.B.	Calc. oz.	
DODO 98P, 15-197 cm	SIO	16° 17′ S, 86° 11′ E	5,140	Plioc. ? Calc. nannoplankton and forams., M.N.B. and B.M.F.	Calc. oz.	
DODO 102P, 1-16 cm	SIO	17° 59′ S, 88° 59′ E	5,090	Eoc. Calc. nannoplankton and forams, M.N.B. and B.M.F.	Calc. oz.	

Table 41.2. (continued)

Core or sample number and depth in core	Collecting Institute	Latitude and Longitude	Water depth (m)	Age of sediment and how determined	Nature of sediment	Previous references
DODO 105P, 23-252 cm	SIO	19° 21′ S, 83° 25′ E	5,060	Up. Eoc. or low. Oligoc. Calc. nannoplankton, M.N.B.	Calc. oz.	
DODO 107P, 21-94 cm	SIO	19° 13′ S, 81° 25′ E	4,815	Eoc. Calc. nannoplankton and forams., M.N.B. and B.M.F.	Calc. oz.	
DODO 108P, 15-278 cm	SIO	19° 29′ S, 80° 59′ E	4,960	Eoc. Calc. nannoplankton and forams., M.N.B. and B.M.F.	Calc. oz.	
DODO 111P, 83-253 cm	SIO	21° 59′ S, 77° 22′ E	4,700	Oligoc. Calc. nannoplankton and forams., M.N.B. and B.M.F.	Calc. oz.	
DODO 117P, 17-285 cm	SIO	18° 21′ S, 62° 04′ E	3,398	Plioc. Calc. nannoplankton and forams., M.N.B. and B.M.F.	Calc. oz.	(*9*), p. 132
DODO 141G, 97-120 cm	SIO	24° 42′ S, 73° 05′ E	3,600	Plioc. Calc. nannoplankton and forams., M.N.B. and B.M.F.	Calc. oz.	(*9*), p. 132
DOLPH 1, 40-88 cm	SIO	12° 10′ S, 144° 25′ W	4,332	Plioc. ? Calc. nannoplankton, M.N.B.	Calc. oz. to calc. clay	(*16*),p.323
DOLPH 2, 1-104 cm	SIO	12° 00′ S, 144° 21′ W	5,000	Mioc. or Plioc., Calc. nannoplankton and forams., M.N.B. and B.M.F.	Calc. oz.	(*20*),p.215; (*16*),p.324
DWBG 25, 0-9 cm	SIO	16° 23′ S, 146° 02′ W	1,380	Eoc. Calc. nannoplankton and forams. M.N.B. and B.M.F.	Calc. oz.	(*16*),p.328
DWBG 118C, 64-68 cm	SIO	28° 02′ S, 96° 20′ W	3,400	Plioc. Calc. nannoplankton and forams, M.N.B., B.M.F., F.L.P.	Zeolitic calc. oz.	(*15*),p.1240; (*16*), p. 330
DWBP 119, 38-768 cm	SIO	27° 54′ S, 106° 53′ W	3,200	Plioc. Calc. nannoplankton, M.N.B.	Calc. oz.	(*16*),p.330
DWHG 34, 103-12 cm	SIO	44° 13′ S, 127° 20′ W	4,600	Oligoc. Calc. nannoplankton and forams, M.N.B. and B.M.F.	Calc. oz.	(*16*),p.332
DWHG 54, 78-84 cm	SIO	38° 49′ S, 83° 21′ W	4,080	Mioc. Calc. nannoplankton and forams, M.N.B. and B.M.F.	Calc. oz.	(*16*), p. 333
DWHG 79, 130-3 cm	SIO	23° 37′S, 118° 14′ W	3,440	Plioc. Calc. nannoplankton, M.N.B.	Calc. oz.	(*16*),p.333
DWHH 14, 30-2 cm	SIO	14° 28′ S, 135° 29′ W	4,400	Oligoc. Calc. nannoplankton, M.N.B.	Calc. oz.	(*16*), p. 331
EM, 28-34 metres	Mohole Project	28° 59′ N, 117° 30′ W	3,566	Plioc. Calc. nannoplankton and forams., M.N.B. and F.L.P.	Calc. clay	(*17*),p.1793; (*7*), p. 847; (*8*), p. 622
Eniwetok, several cores	SIO	approx. 10° 18′ N, 162° 08′ E	approx. 1,650	Plioc. Calc. nannoplankton	Calc. oz.	(*2*), p. 1551
JYN V 46P, 472-4 cm	SIO	13° 24′ N, 136° 11′ W	4,822	Oligoc. Calc. nannoplankton, M.N.B.	Calc. oz.	
JYN V 48P, 40-506 cm	SIO	15° 54′ N, 133° 57′ W	4,606	Eoc. Calc. nannoplankton, M.N.B.	Calc. oz.	(*18*), p. 287
LSDA 122G, 117-19 cm	SIO	29° 54′ S, 61° 53′ E	4,400	Plioc. Calc. nannoplankton, M.N.B.	Calc. oz.	
LSDA 132P, 140-302 cm	SIO	33° 47′ S, 96° 00′ E	4,328	Oligoc. Calc. nannoplankton, M.N.B.	Calc. oz.	
LSDA 178G, 96-8 cm	SIO	24° 03′ S, 15° 33′ W	4,045	Plioc. ? Calc. nannoplankton, M.N.B.	Calc. oz.	
LSDH 59P, 20-2 cm	SIO	14° 53′ S, 151° 12′ E	4,140	Oligoc. Calc. nannoplankton, M.N.B.	Calc. oz.	

Table 41.2. (continued)

Core or sample number and depth in core	Collecting Institute	Latitude and Longitude	Water depth (m)	Age of sediment and how determined	Nature of sediment	Previous references
MP 25 E-1, 2-38 cm	SIO	19° 05′ N, 169° 45′ W	approx. 1,703	Eoc. Calc. nannoplankton	Calc. oz.	
MP 37 I 250-64 cm	SIO	17° 05′ N, 177° 15′ W	approx. 1,812	Plioc. ? Calc. nannoplankton, M.N.B.	Calc. oz.	
MP 40, 89-380 cm	SIO	15° 35′ N, 177° 30′ W	approx. 4,080	Oligoc. Calc. nannoplankton, M.N.B.	Calc. oz.	(*6*), p. 398
MSN 21G, entire spl.	SIO	14° 45′ S, 151° 14′ E	4,400	Oligoc. ? Calc. nannoplankton, M.N.B.	Calc. oz.	(*16*), p. 339
MSN 56P, 80-250 cm	SIO	23° 56′ S, 73° 53′ E	3,700	Plioc. Calc. nannoplankton and forams., M.N.B. and B.M.F.	Calc. oz.	(*9*), p. 132
MSN 126G, 60-105 cm	SIO	24° 41′ S, 154° 45′ W	4,542	Plioc. Calc. nannoplankton and forams, M.N.B. and F.L.P.	Zeolitic calc. oz.	(*9*), p. 132; (*16*), p. 340
MSN 128G, 54-61 cm	SIO	13° 53′ S, 150° 35′ W	3,623	Plioc. Calc. nannoplankton M.N.B.	Zeolitic calc. oz.	(*16*), p. 340
PROA 149G, 16-18 cm	SIO	10° 01′ N, 164° 59′ W	3,170	Oligoc. or Mioc. Calc. nannoplankton, M.N.B.	Calc. oz.	
PROA 164G, 68-70 cm	SIO	13° 33′ N, 171° 56′ W	5,863	Plioc ? Calc. nannoplankton, M.N.B.	Calc. oz.	
RIS 50G, 113-16 cm	SIO	13° 36′ S, 96° 42′ W	4,120	Plioc. Calc. nannoplankton, M.N.B.	Calc. oz.	
RIS 52G, 95-8 cm	SIO	13° 24′ S, 100° 29′ W	4,210	Plioc. Calc. nannoplankton, M.N.B.	Zeolitic calc. oz.	
RIS 75G, 67-94 cm	SIO	14° 01′ S, 122° 28′ W	3,790	Plioc. forams., F.L.P.	Zeolitic calc. oz.	(*9*), p. 132
RIS 77G, 35-66 cm	SIO	14° 02′ S, 128° 29′ W	3,985	Plioc. ? forams, F.L.P.	Zeolitic calc. clay to oz.	(*9*), p. 132
RIS 82G, 92-126 cm	SIO	14° 03′ S, 139° 35′ W	3,900	Plioc. forams, F.L.P.	Zeolitic calc. oz.	(*9*), p. 132
RIS 84G, 45-8 cm	SIO	15° 15′ S, 142° 27′ W	3,675	Plioc. forams, F.L.P.	Calc. oz.	(*9*), p. 132
SDSE 53, 2-328 cm	Swed. D-S Exped.	15° 34′ N, 127° 11′ W	4,730	Mioc. ? Calc. nannoplankton, M.N.B.	Calc. oz.	
SDSE 69, 392-562 cm	Swed. D-S Exped.	13° 25′ S, 149° 30′ W	4,635	Oligoc. Calc. nannoplankton, M.N.B.	Zeolitic calc. oz.	

42. THE OCCURRENCE OF PRE-QUATERNARY PTEROPODS

D. CURRY

The pteropods, a small group of pelagic molluscs, swim with a flapping motion by means of extensions of the foot and may possess an external aragonitic shell. Their anatomy suggests links with the opisthobranchs and they are generally placed as an Order, the Pteropoda, of the Sub-Class Opisthobranchia of the gastropod molluscs.

Two Families are known fossil, both with a range from Eocene to Recent (though there are doubtful records from Jurassic and Cretaceous beds). These Families are the Spiratellidae, which are minute forms provided with a shell coiled in a sinistral spiral, and the Cavolinidae, which are relatively large (up to 15 millimetres long) and possess a shell which may be conical, pyramidal or nearly spherical and which is most commonly bilaterally symmetrical.

Shells of pteropods are extremely thin and fragile and, being made of aragonite, they are readily dissolved. For these reasons pteropod shells are very rarely preserved fossil in their original state and even moulds or casts are uncommon. Derivation of pteropod remains is likely therefore to be a rare event and indeed the writer knows of no record of the occurrence of pteropods, either in the form of solid fossils or of moulds, as derivatives in younger deposits.

There are few references to pteropods in the palaeontological literature; such records as do occur are mostly to be found in general descriptions of complete mollusc faunas. Works dealing solely with fossil pteropods are very few.

BIBLIOGRAPHY

Avnimelech, M. 1945. Revision of fossil Pteropoda from Southern Anatolia, Syria and Palestine. *J. Paleont.* **19**, 637-47.

Bellardi, L. 1872. *I molluschi dei terreni terziari del Piemonte e della Liguria,* **1**, 25-37.

Blanckenhorn, M. 1889. Pteropodenreste aus der Oberen Kreide Nord-Syriens und aus dem hessischen Oligocän. *Z. dtsch. geol. Ges.* **41**, 593-602.

Collins, R. L. 1934. A monograph of the American Tertiary pteropod mollusks. *Johns Hopk. Univ. Stud. Geol.* **11**, 137-234.

Curry, D. 1965. The English Palaeogene pteropods. *Proc. malac. Soc. Lond.* **36**, 357-71.

Dollfus, G. and Ramond, G. 1886. Liste des ptéropodes du terrain tertiaire parisien. *Mém. Soc. malac. Belg.* **20**, 36-44.

Kittl, E. 1886. Ueber die miocenen Pteropoden von Oesterreich-Hungarn. *Ann. k.k. naturh. Hofmus.* **1**, 47-74.

Korobkov, I. A. 1966. Krylonogie (Mollusca Pteropoda) paleogenovykh otlozheniy yuga SSSR. *Vopr. Paleont.* **5**, 71-92.

Korobkov, I. A. and Makarova, R. K. 1962. Novyy krylonogiy Mollyusk iz verkhneeotsenovikh otlozheniy SSSR. *Paleont. Zh.* **4**, 83-7.

Pelseneer, P. 1888. Report on the Pteropoda. *Report on the scientific results of the voyage of H.M.S. 'Challenger',* 1873-1876. *Zoology,* part 65.

Seguenza, G. 1867. Paleontologia malacologica dei terreni terziari del distretto di Messina (Pteropodi ed Eteropodi). *Mem. Soc. ital. Sci. nat.* **2**, no. 9.

Tesch, J. J. 1946, 1948. The Thecosomatous pteropods, I. The Atlantic, II. The Pacific. *Dana-Reports,* **28**, **30**.

Troelsen, J. 1937. Pteropoden-Reste aus dem oberen Senon Dänemarks. *Medd. dansk geol. Foren.* **9**, 183-85.

Watelet, A. and Lefèvre, T. 1885. Note sur des ptéropodes du genre *Spirialis* découverts dans le Bassin de Paris. *Ann. Soc. malac. Belg.* **15**, 100-3.

43. PLIOCENE OSTRACODES FROM DEEP-SEA SEDIMENTS IN THE SOUTHWEST PACIFIC AND INDIAN OCEAN

F. M. SWAIN

INTRODUCTION

Using the Neogene planktonic foraminiferal zonation of Banner and Blow (1962, 1965*a*, 1965*b*, 1967), Parker (1967), recognized Pliocene planktonic assemblages in several Pacific and Indian Ocean cores. The zones in question range from N.19 to N.21.

Of the Pliocene cores studied by Parker (1967) and by Riedel, Bramlette and Parker (1963) the following were examined for Ostracoda:

DODO 57P (15° 40′ S, 112° 44′ E, Indian Ocean, water depth 3,660 m) – Pliocene at 20-111 cm.
DODO 117P (18° 21′ S, 62° 04′ E, Indian Ocean, water depth 3,398 m) – Pliocene at 20-286 cm.
LSDH 78P (4° 31′ S, 168° 02′ E, Pacific Ocean, water depth 3,208 m) – Pliocene at 100-470 cm.

Most of the planktonic Foraminifera obtained by Parker (1967) from the Upper Miocene and Pliocene cores she studied (Parker 1967, p. 140) 'appear to be those typical of the tropics, although not necessarily confined to this area'. Near the base of Zone N. 21 (upper Pliocene), Parker recorded a few planktonic Foraminifera more common in temperate waters of today. This may indicate a late Pliocene glaciation. See Akers (1965) for similar evidence in the Gulf of Mexico area.

Recent papers by Burckle, Ewing, Saito and Leyden (1967) and Riedel (1967) provide palaeontological documentation of the proposition that the Pacific floor has undergone spreading in Neogene time, as a result of which Pliocene and older sediments now occur at or near the ocean floor on the flanks of the East Pacific Rise and elsewhere. Whether this broadly diastrophic process or more localized tectonic events are responsible for the Pliocene occurrences in the cores discussed here cannot be determined at present.

PLIOCENE OSTRACODA

Eight benthonic ostracode species were obtained from the Pliocene core samples. Five of the species also occur in Pleistocene cores of the South-east Pacific (Swain 1970) and three of the five are still living. The other three species were not found in the Pleistocene cores but two are still living. In addition to the ubiquitous *Bradleya dictyon,* the other living species are represented by southern hemisphere forms.

Both Lower and Upper Pliocene are represented in the cores according to Parker (1967). The Upper Pliocene (N21) species (all from DODO 57P:30-70 cm) include the following:

Echinocythereis? sp. aff. *E. dasyderma* (Brady)	2 specimens
Bradleya cf. *dictyon* (Brady)	1 specimen
Krithe sp. aff. *K. bartonensis* (Jones)	1 specimen
Trachyleberis sp. aff. *T. scabrocuneata* Brady	1 specimen

The Lower Pliocene (N19) species are:

DODO 57P:70-110 cm	
Echinocythereis? sp. aff. *E. dasyderma* (Brady)	1 specimen
Bosquetina? aff. *B.*? *fenestratum* (Brady)	1 specimen
Bradleya dictyon (Brady) mature	1 specimen
Bradleya dictyon? (Brady) immature	1 specimen
Krithe bartonensis (Jones)	1 specimen
Krithe tumida (Brady)	1 specimen
LSDH 78P:160-200 cm	
Echinocythereis ericea (Brady)	
Bradleya dictyon (Brady)	2 specimens

There is no evidence of northern hemisphere elements in either the Upper or Lower Pliocene assemblages.

If a glaciation occurred during the late Pliocene it would plausibly have occurred in the southern hemisphere as was suggested by Rutford, Craddock and Bastien (1966).

The ostracode species will be described formally in a later publication.

ACKNOWLEDGEMENTS

Appreciation is expressed to W. R. Riedel and F. L. Parker of Scripps Institution of Oceanography who supplied the samples and data for this study. The writer is also grateful for support provided by Grant GB 4110 of the National Science Foundation.

Received December 1968.

BIBLIOGRAPHY

Akers, W. H. 1965. Pliocene-Pleistocene boundary, northern Gulf of Mexico. *Science,* **149**, 741-52.

Banner, F. T. and Blow, W. H. 1962. In *Fundamental mid-Tertiary stratigraphic correlations.* 61-151, pls. 8-17, Cambridge Univ. Press.

Banner, F. T. and Blow, W. H. 1965a. Two new taxa of the Globorotaliinae (Globigerinacea, Foraminifera) assisting determination of the late Miocene-Middle Eocene boundary. *Nature, Lond.* **207**, 1,351-4.

Banner, F. T. and Blow, W. H. 1965b. Progress in the planktonic foraminiferal biostratigraphy of the Neogene. *Nature, Lond.* **208**, 1,164-6.

Banner, F. T. and Blow, W. H. 1967. The origin evolution and taxonomy of the Foraminiferal genus *Pulleniatina* Cushman 1927. *Micropaleontology,* **13**, 133-62, pls. 1-4.

Burckle, L. H., Ewing, J., Saito, T. and Leyden, R. 1967. Tertiary sediment from the East Pacific Rise. *Science,* **157**, 537-40.

Parker, F. L. 1967. Late Tertiary biostratigraphy (planktonic Foraminifera) of tropical Indo-Pacific deep-sea cores. *Bull. Am. Paleont.* 52, 115-208, pls. 17-32.

Riedel, W. R., Bramlette, M. N. and Parker, F. L. 1963. 'Pliocene-Pleistocene' boundary in deep-sea sediments. *Science,* **140**, 1,238-40.

Riedel, W. R. 1967. Radiolarian evidence consistent with spreading of the Pacific floor. *Science,* **157**, 540-2.

Rutford, R. H., Craddock, C. and Bastien, T. W. 1966. Possible late Tertiary glaciation, Jones Mountains, Antarctica. *Geol. Soc. Amer. Spec. Paper,* **87**, 144-5.

Swain, F. M. 1970. Pleistocene Ostracoda from deep-sea sediments in the southeastern Pacific Ocean. *This volume,* 487-92.

Plate 43.1. Ostracoda from South-west Pacific and Indian Ocean Pliocene Sediments. Fig. 1*a, b. Bradleya dictyon* (Brady), Exterior and dorsal views of immature left valve. LSDH 78 P, 160-200 cm, Pliocene. x 50. Fig. 2*a, b. Bradleya dictyon* (Brady). Exterior and interior views of right valve. LSDH 78 P, 160-200 cm, Pliocene. x 50. Fig. 3. *Krithe* aff. *K. bartonensis* (Jones). Exterior of imperfect left valve showing scattered normal canal pits. DODO 57 P, 30-70 cm, Pliocene x 75. Fig. 4. *Krithe tumida* Brady. (*a*) Exterior and (*b*) dorsal views of right valve; (*c*) interior of left valve. DODO, 57 P 70-110 cm. Pliocene. x 75. Fig. 5. *Bradleya dictyon*? (Brady). Exterior of immature right valve of one of the forms assigned by Brady (1,880, pl. 34, fig. 1 q) to this species, DODO 117 P, 70-110 cm, Pliocene. x 75. Fig. 6. *Krithe bartonensis* (Jones). Exterior of right valve. DODO 117 P, 70-110 cm, Pliocene. x 75. Figs. 7*a, b. Trachyleberis* sp. aff. *T. scabrocuneata* (Brady) Exterior and interior views of left valve. DODO 57 P, 30-70 cm. Pliocene. x 50. Figs. 8*a, b. Echinocythereis ericea* (Brady). Exterior and interior views of left valve. LSDH 78 P, 160-200 cm, Pliocene. x 50.

1a
1b
2a
3
2b
4a
4b
6
4c
5
7a
7b
8a
8b

SECTION E

GENERAL METHODS AND SYSTEMATICS

44. PROBLEMS IN ISOTOPE PALAEOCLIMATOLOGY AND MICROPALAEONTOLOGY

A. LONGINELLI

Summary: A short review is given of oxygen isotopic work carried out for palaeotemperature measurements since 1951. The problem of possible variation in the oxygen isotopic composition of oceanic water through geological time is considered. Some remarks are also made on the possibility of establishing isotopic temperature scales for sulphate and silica from living marine organisms. The possibility of using isotopic techniques for micropalaeontological problems is also briefly discussed.

INTRODUCTION

The origin and development of the isotopic method of palaeotemperature analysis was described in papers by various authors (Urey *et al.* 1951; Epstein *et al.* 1951; Epstein *et al.* 1953; Emiliani 1958*a*). A complete bibliography up to 1965 is given by Bowen (1966). Since the abundance of O^{18} in calcium carbonate varies with the temperature of deposition, variations in O^{18} abundance can be used as a thermometer when the deposition process takes place under isotopic equilibrium conditions. A relationship between temperature of precipitation, oxygen isotopic composition of the carbonate, and oxygen isotopic composition of the water in which the carbonate is deposited was provided by Epstein *et al.* in 1953. Such a relationship was established empirically measuring present-day marine organisms grown under controlled temperature conditions. The relevant equation is the following:

$$t = 16.5 - 4.3\,(\delta_C - \delta_W) + 0.14\,(\delta_C - \delta_W)^2,$$

where t is the temperature (° C) at which the calcium carbonate was precipitated from water, i.e. the growth temperature of the marine organism being considered;

– δ_C is the oxygen isotopic composition of the carbonate

$$\left(\text{where } \delta = \frac{O^{18}/O^{16}\ \text{sample} - O^{18}/O^{16}\ \text{standard}}{O^{18}/O^{16}\ \text{standard}} \times 1000\right)$$

– δ_W is the oxygen isotopic composition of CO_2 isotopically equilibrated at 25.2° C with the environmental water and measured against the same standard gas as used to measure δ_C.

Different standard gases are now being used in different laboratories but oxygen isotope measurements on carbonates are generally referred to PDB-1 Chicago standard. PDB-1 standard is a *Belemnitella americana* from the Peedee formation of South Carolina (Upper Cretaceous-Maastrichtian), whose oxygen isotopic composition corresponds to that of a carbonate precipitated under isotopic equilibrium conditions at 16.5° C from a water whose δ is equal to zero.

VARIATION IN OXYGEN ISOTOPIC COMPOSITION OF OCEAN WATER

During the last fifteen years a number of papers have been published dealing with measurements of the oxygen isotopic composition of carbonate fossils, ranging in age from Palaeozoic to Quaternary. Almost all the authors gave their results in terms of temperature values assuming that there were no major variations of the oxygen isotopic composition of ocean water during the last 200-250 million years. This is an arbitrary hypothesis which was generally accepted mainly because, using $\delta_W = 0$ in the Epstein equation the O^{18}/O^{16} ratios measured on fossil shells yielded 'reasonable' temperatures. The isotopic method was applied mainly to Mesozoic belemnites and Tertiary and Quaternary molluscs and Foraminifera. The 'reasonable' temperatures obtained were accepted in full, in spite of some evident discrepancies, such as:

1. temperature ranges of 10° C and even more in fossils from the same fauna, or from faunas of similar age from the same area (Urey *et al.* 1951; Bowen 1961; Fritz 1964);
2. temperatures as low as about 13° C obtained from belemnites that lived in tropical (or subtropical) Jurassic waters (Fritz 1964);

3. differences of up to 10-15° C between average palaeotemperature values obtained, e.g. from belemnite faunas of different ages from the Jurassic (Bowen and Fritz, 1963; Fritz 1964). Such an enormous variation of temperature is absurd when attributed to climatic variations alone, considering that each δ value from a fossil belemnite should represent an average value for ocean water over a period of a few years;
4. temperatures as high as 30-35° C (*Globigerinoides sacculifera,* Upper Pliocene) and 32° C (*G. rubra,* Lower Pleistocene) from Mediterranean pelagic Foraminifera (Emiliani, Mayeda and Selli 1961). The authors state in this case that 'the temperatures above 30° C may be due *in part* to isotopic effects of the sea water' but even in such a case the difference from modern temperatures seems to be a very large one.

The fact that temperature values were obtained within the range of 'possible' temperatures was considered proof of the hypothesis of constancy of oxygen isotopic composition of ocean water through geological time, and proof of a very low rate of isotopic exchange between carbonate fossils and ground water during diagenetic processes. Because of this the major portion of research work in this field was made in the direction of providing further figures for palaeoclimatic reconstructions. It now seems rather strange that almost everybody admitted that isotopic exchange processes had taken place between massive limestone formations and ground waters but not between single fossils and ground water. The possibility of such an exchange was generally accepted only when 'strange' or 'impossible' temperatures (35-40° C and even more) were obtained measuring badly preserved specimens. Very little work was carried out towards a better understanding of the general problems still remaining unsolved, and towards providing further methods for development of research in this field. In my judgement two major problems had to be considered – both very important from a general point of view, and not only for palaeoclimatological interpretation of the data obtained. These problems are: (1) the possibility of evolution of the oxygen isotopic composition of ocean water through geological time: (2) our lack of detailed knowledge of the biological behaviour of both molluscs and Foraminifera. These organisms have been those mainly used for isotopic measurements on Tertiary and Quaternary material.

NON-CARBONATE TEMPERATURE SCALES

The uncertainty inherent in palaeotemperature measurements due to the possibility of variation in the oxygen isotopic composition of oceanic water could be removed, Urey suggested, by establishing a second relationship between growth temperature and isotopic abundance of oxygen for a compound co-precipitated in equilibrium with the water and the carbonate. Thus, by measuring the O^{18}/O^{16} ratios of carbonate and of a second compound co-existing in the same shell, and inserting these two values into the Epstein equation, and in a second one of the same type with the same unknowns, both the average growth temperature and the oxygen isotopic composition of the environmental water could be calculated. A paper published by Tudge (1960) was the only contribution in this direction. However, a complete technique for the purification of the phosphate of shells was given in this paper. Through a long purification process all the oxygen-bearing contaminant compounds are removed and the phosphate is transformed into a form of definite composition containing no non-phosphate oxygen.

A few years ago I started a palaeotemperature project with the purpose of establishing a phosphate-water isotopic temperature scale measuring specimens of living molluscs. I followed Tudge's technique for the purification of the samples and the fluorination of the final product, $BiPO_4$, with BrF_3 for the removal of the oxygen. The conversion of the oxygen to carbon dioxide for spectrometric measurements was made following the technique described by Clayton and Mayeda (1963) cycling the oxygen over a hollow graphite cylinder heated inductively (Longinelli 1965). A tentative equation was calculated for the straight-line relationship between δO^{18} (PO_4^{3-}) and t° C (Longinelli 1966). From the similarity in the slopes of the phosphate and carbonate equations and from the precision one can obtain in measuring ($\delta_P - \delta_C$) the uncertainty with which temperature could be calculated is quite large (±4° C) for reasonably accurate palaeotemperature measurements. Considering that the uncertainty in the calculation of δ_W (±0.8‰) is small enough to permit the evaluation of whether or not there have been major variations of the oxygen isotopic composition of ocean water through geological time, a set of fossils from lower Jurassic to Quaternary was measured for the oxygen isotopic composition of both carbonate and phosphate. The results obtained (Longinelli 1966, in preparation) lead to the conclusion that it is possible that a quite large isotopic variation of ocean water very likely took place through the Mesozoic and Tertiary.

Trying to calculate both temperature and δ_W from the Epstein equation and my own equation for phosphates, it can be seen that the system is solved with the values obtained from living specimens, but cannot be solved with the values obtained from the measurements on belemnites or other fossil organisms. Moreover, carbonate δs are constantly more negative than phosphate δs. My interpretation of this fact is that substantial isotopic exchange took place preferentially between carbonate and ground water, thus giving a variation with time of the initial isotopic composition. This is better preserved by phosphate owing to the characteristics of the oxygen-phosphorus bond. The isotopic variation of ocean water could be determined by sedimentation processes. In fact, assuming an average value of 1.03 for the fractionation factor of oxygen (a) for the precipitation of marine carbonate and other chemical compounds, a variation of several per mil in the O^{18}/O^{16} ratio of ocean water could be obtained over a period of about 200 million years solely by the sedimentation of *Globigerina* ooze. Further measurements are needed to confirm this hypothesis and they are now in progress. At any rate, it will be possible, within a reasonable time, to make a substantial contribution towards better knowledge of the history of the oceans.

It would be very desirable to be able to determine experimentally more isotopic temperature scales based on the deposition, under isotopic equilibrium conditions, of other chemical compounds than carbonate and phosphate. It might be possible, for example, to set up a thermometer for present-day organisms establishing an empirical temperature scale for sulphate and silica. The possibility of measuring different chemical compounds and of using similar equations with different slopes should enable one to obtain more reliable data both in terms of temperature and oxygen isotopic composition of the water in the case of fossil material. In the case of sulphate the possibility of measuring δO^{18} ($\bar{S}\bar{O}_4$) with reasonable accuracy has been recently demonstrated (Longinelli and Craig 1967; Lloyd 1967; Rafter 1967) but, so far no substantial progress has been made with the problem of the chemical purification of sulphate, generally present in very low percentages in the shells of marine organisms. In the case of silica it is theoretically possible to establish a temperature scale using very common organisms like radiolarians or diatoms. However, there is a serious technical difficulty which has so far prevented the establishment of a temperature scale for these marine organisms. In fact, their skeletons and shells are made of hydrous silica and it is almost impossible to separate silica from water without oxygen isotope exchange. There is still a possibility of overcoming this difficulty using the following procedure. The sample should be well homogenized and then divided into three portions. Each portion could be used to measure respectively: (1) the isotopic composition of hydrogen (D/H ratio) of the water present in the hydrous silica; (2) the water percentage by weight; (3) the oxygen isotopic composition of the mixture of water and silica. From Craig's data (1961) it should be possible to evaluate the oxygen isotopic composition of the water of hydration starting from the D/H value. On the basis of the calculated isotopic composition of the oxygen, and of the water percentage by weight, it should be possible to correct the δO^{18} obtained from the total sample (H_2O+SiO_2), thus calculating δO^{18} for silica alone. Obviously the proposed technique would not yield very precise results, however it is important to start measurements in this direction to evaluate the practicability of such a procedure. A substantial improvement of our knowledge in this field should be obtained by further work on phosphate, and by careful experimentation on the sulphate and silica contained in the shells and skeletons of living marine organisms.

BIOLOGICAL FACTORS

As regards our lack of detailed knowledge of the biological behaviour of both molluscs and foraminifers we must admit that in this field research has not made substantial progress. On the one hand isotopic work for palaeoclimatic purposes needs much better knowledge on this point while, on the other hand, knowledge of the biological behaviour of marine organisms can be improved by oxygen-isotope methods. In his palaeotemperature work Emiliani extensively applied the isotopic technique to Tertiary and Quaternary Foraminifera and the results of his detailed work are published in a number of papers (Emiliani 1954*a*, 1954*b*, 1955, 1956, 1957, 1958*b*, 1961*a*, 1961*b*, 1964, 1966*a*; Emiliani, Mayeda and Selli 1961: Emiliani and Mayeda 1961).

Unfortunately this intensive research work for palaeotemperature purposes was not supported by similar effort in other directions by other authors. Even now micropalaeontologists do not fully admit the possibility of openings in classical palaeontological research in the direction of using isotopic techniques for solving biological and palaeontological problems. Moreover, aside from the application of isotopic techniques, biological research within the field of micropalaeontology has not been very active. To give an example of the present situation let us focus attention for a moment on the planktonic Foraminifera. Shells of planktonic foraminiferal species have been used extensively in isotopic studies of Plio-Pleistocene temperatures but little is known, e.g. about their life-cycles and productivity. A recent paper (Berger and Soutar 1967) is illuminating from this point of view. Previously evidence was presented in favour of yearly life cycles, with reproduction of some species taking place at great depth. Berger and Soutar found that, at least in the case of the four species studied (*Globigerina bulloides, Globoquadrina eggeri, Globigerinoides ruber* and *Globigerina quinqueloba*) their life spans are of the order of a few weeks while reproduction seems to take place generally in the upper water layers. No doubt, even when considering the possible variations with time of the oxygen isotopic composition of the ocean water, the interpretation of isotopic data obtained from these species would change drastically using Berger and Soutar's conclusions rather than previous hypotheses. It must be pointed out here that *Globigerinoides ruber* is one of the species used for palaeotemperature measurements.

OXYGEN ISOTOPIC ANALYSES OF FORAMINIFERA

To summarize very briefly the isotopic work carried out on Foraminifera I wish first to recall that each measurement generally represents the average isotopic composition of some 100 to 400 Foraminifera (owing to the necessity of obtaining a minimum amount of about 1 c.c. carbon dioxide for a correct mass spectrometric measurement). Only recently has a different technique of measurement been introduced (Shackleton 1965) through which an oxygen isotopic determination is possible on between 0.3 and 0.4 milligrammes calcium carbonate with an analytical precision not much lower than that obtained using much larger samples. The technique generally used for the chemical treatment of the samples is the same as that used for all carbonates and was described by Epstein *et al.* (1953). It involves the grinding of the carbonate to a fine powder and roasting it in a stream of helium at 475° C. Roasting is carried out in order to destroy organic matter and the helium flow has the purpose of removing volatiles to prevent oxygen exchange between the volatiles and calcium carbonate. Other techniques were developed in subsequent years. Naydin, Teys and Chupakhin (1956) roasted the powdered samples at 475° C in vacuum. A third method, developed by Lowenstam and Epstein (1957), involved digesting the powdered carbonate in commercial Clorox. Emiliani (1966) tested these different procedures and compared results with those obtained from similar material which was not roasted at all. He claims that no roasting and vacuum roasting appear to yield results closer to real temperatures than those obtained with helium roasting.

The carbon dioxide gas for mass spectrometric measurements is always obtained by reacting the calcium carbonate with 100 per cent phosphoric acid at 25° C in a thermostat. The carbon dioxide is then transferred to a sample tube and analysed on the mass spectrometer.

In my opinion the results obtained from the measurement of Quaternary Foraminifera from deep-sea cores are by far the most interesting and stimulating from several points of view. There have been, and there still are discussions about the validity of Emiliani's work. Personally I do not accept in full Emiliani's interpretation of the data obtained, partially summarized in Fig. 44.1 and Fig. 44.2 (Emiliani, 1966*a*, 1966*b*). Owing to the difficulty of precise evaluation of changes in the isotopic composition of ocean water during glacial and interglacial periods and to some uncertainty about the biological behaviour of different species, I assume that a rigorous interpretation of the data in Fig. 44.1 and Fig. 44.2 in terms of temperature is almost impossible. Anyhow, whatever the interpretation of the variations observed can be – temperature changes, isotopic composition of the water (Olausson 1965; Shackleton 1967) or a composite effect of these and other factors – there is no doubt that there is a direct connection between isotopic variations and climatic changes and that the amplitude of the possible variations determined by the alternation of glacial and interglacial periods was relatively small.

Another critical point in the interpretation of palaeotemperature data is the correlation between the age determinations made on Quaternary marine sediments and those made on continental deposits significantly related to the established Quaternary stratigraphy (Evernden, Curtis and Kistler 1957; Evernden *et al.* 1964; Evernden and Curtis 1965; Rósholt *et al.* 1961; Rosholt *et al.* 1962; Emiliani 1966*b*). Large discrepancies exist between the two groups of results. In spite of all these difficulties, I take Emiliani's research to be one of the best pieces of work ever carried out in this field. I hope that more work will be done in the future to improve our knowledge of both the climatic variations which took place during the Quaternary and the behaviour of Foraminifera facing such oscillations in environmental conditions.

To conclude I would like to recall briefly some cases in which isotopic studies contributed to the knowledge of events affecting foraminiferal species.

During 1954 a paper by Emiliani was published on the depth habitats of some species of pelagic Foraminifera. He showed that several species grow at essentially the same temperature even in different areas. For example *Globorotalia tumida* lives at depths of 140 metres in the Gulf of Mexico and at 55 metres in the equatorial Atlantic. Isotopic temperatures and density of the sea water are almost the same in both areas. From this one may deduce that this species appears to be adapted to waters of the same densities in spite of large pressure differences. Emiliani also considered the possibility of variation with time of the relative distances between zones of maximum density of two specific populations. He reported this to have occurred in his reference species *Globorotalia tumida* and *Pulleniatina obliquiloculata.*

During 1964 Longinelli and Tongiorgi published a short

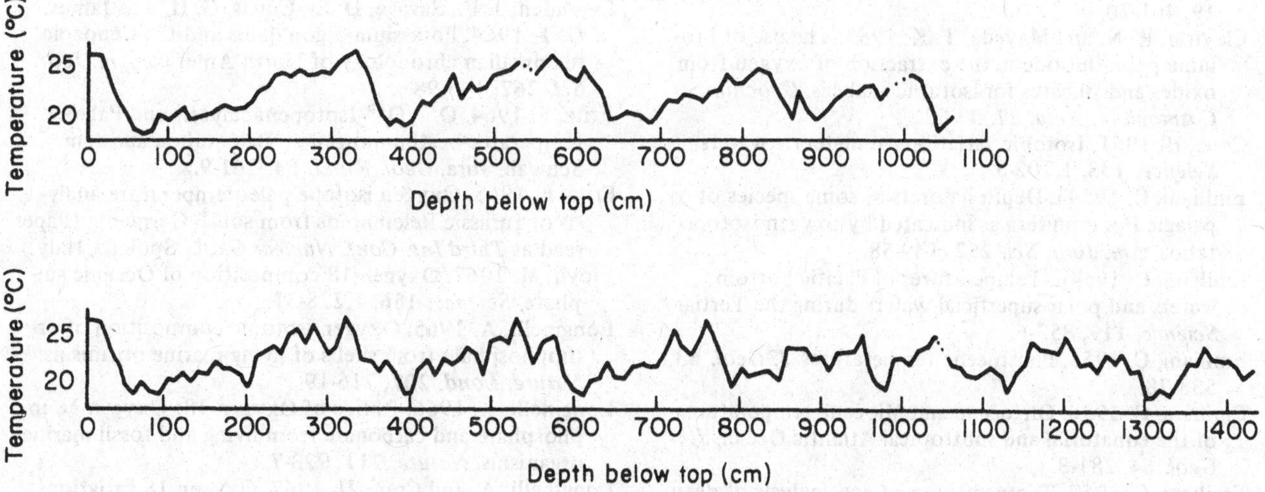
Temperature (°C)
25
20
0 100 200 300 400 500 600 700 800 900 1000 1100
Depth below top (cm)
Temperature (°C)
25
20
0 100 200 300 400 500 600 700 800 900 1000 1100 1200 1300 1400
Depth below top (cm)

Fig. 44.1. Temperatures calculated from the isotopic analysis of two Caribbean cores (from Emiliani, 1966*a*).

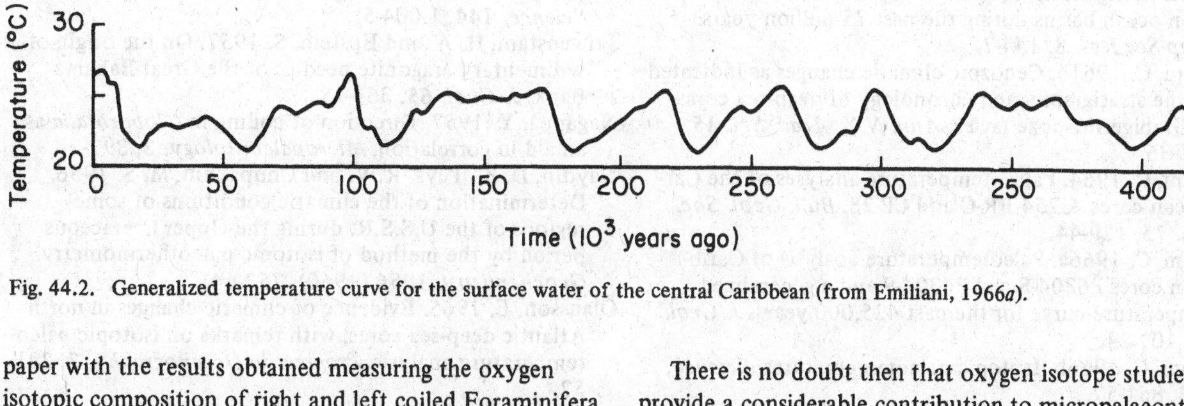
Temperature (°C)
30
20
0 50 100 150 200 250 300 250 400
Time (10^3 years ago)

Fig. 44.2. Generalized temperature curve for the surface water of the central Caribbean (from Emiliani, 1966*a*).

paper with the results obtained measuring the oxygen isotopic composition of right and left coiled Foraminifera. This research followed previous papers by various authors (Nagappa 1957; Ericson 1959; Longinelli and Tongiorgi 1960) demonstrating a direct relationship between the percentages of these two groups and climatic conditions. A difference was found in the O^{18}/O^{16} ratios in right and left coiled specimens of the same species and from the same area. Such a difference is significant because its order of magnitude is much larger than the experimental error. Moreover, this difference was found both in the case of pelagic and benthonic Foraminifera. It is quite difficult to interprete these results in terms of difference in the biological behaviour of the two groups, however there is little doubt about the existence of a significant variation in one or more of the factors affecting the oxygen isotopic composition of the shells of Foraminifera.

There is no doubt then that oxygen isotope studies can provide a considerable contribution to micropalaeontology and related problems. It is a new and powerful method which is waiting to be used by palaeontologists to improve knowledge of palaeontology, ecology and palaeoclimatology.

Received October 1967.

BIBLIOGRAPHY

Berger, W. H. and Soutar, A. 1967. Planktonic Foraminifera: field experiment on production rate. *Science*, **156**, 1,495-7.

Bowen, R. 1961. Paleotemperature analyses of Belemnoidea and Jurassic paleoclimatology. *J. Geol.* **69** (3), 309-20.

Bowen, R. 1966. *Paleotemperature analysis.* Elsevier: The Netherlands.

Bowen R. and Fritz, P. 1963. Oxygen isotope paleotemperature analyses of Lower and Middle Jurassic fossils from Pliensbach, Wurttemberg (Germany). *Experientia,* **19,** 461-70.

Clayton, R. N. and Mayeda, T. K. 1963. The use of bromine pentafluoride in the extraction of oxygen from oxides and silicates for isotopic analysis. *Geochim. Cosmochim. Acta,* **27,** 43-52.

Craig, H. 1961. Isotopic variations in meteoric waters. *Science,* **133,** 1,702-3.

Emiliani, C. 1954a. Depth habitats of some species of pelagic Foraminifera as indicated by oxygen isotope ratios. *Am. Jour. Sci.* **252,** 149-58.

Emiliani, C. 1954b. Temperatures of Pacific bottom waters and polar superficial waters during the Tertiary. *Science,* **119,** 853-6.

Emiliani, C. 1955. Pleistocene temperatures. *J. Geol.* **63,** 538-78.

Emiliani, C. 1956. Oligocene and Miocene temperatures of the equatorial and subtropical Atlantic Ocean. *J. Geol.* **64,** 281-8.

Emiliani, C. 1957. Temperature and age analysis of deep sea cores. *Science,* **125,** 383-7.

Emiliani, C. 1958a. Ancient temperatures, *Sci. Am.* **198** (2), 54-63.

Emiliani, C. 1958b. Paleotemperature analysis of core 280 and Pleistocene correlations. *J. Geol.* **66,** 264-75.

Emiliani, C. 1961a. The temperature decrease of surface water in high latitudes and of abyssal-hadal water in open ocean basins during the past 75 million years. *Deep-Sea Res.* 8, 144-7.

Emiliani, C. 1961b. Cenozoic climatic changes as indicated by the stratigraphy and chronology of deep-sea cores of Globigerina-ooze facies. *Ann. N.Y. Acad. Sci.* **15,** 521-36.

Emiliani, C. 1964. Palaeotemperature analyses of the Caribbean cores A 254-BR-C and CP-28. *Bull. Geol. Soc. Am.* **75,** 129-44.

Emiliani, C. 1966a. Paleotemperature analysis of Caribbean cores P6304-8 and P6304-9 and a generalized temperature curve for the past 425,000 years. *J. Geol.* **74,** 109-24.

Emiliani, C. 1966b. Isotopic paleotemperatures. *Science,* **154,** 851-57.

Emiliani, C., Mayeda, T. and Selli, R. 1961. Paleotemperature analysis of the Plio-Pleistocene section at le Castella, Calabria, Southern Italy. *Bull. Geol. Soc. Am.* **72,** 679-88.

Emiliani, C. and Mayeda, T. 1961. Carbonate and oxygen isotopic analysis of core 241 A. *J. Geol.* **69,** 729-32.

Epstein, S., Buchsbaum, R., Lowenstam, H. A. and Urey, H. C. 1951. Carbonate-water isotopic temperature scale. *Bull. Geol. Soc. Am.* **62,** 417-26.

Epstein, S., Buchsbaum, R., Lowenstam, H. A. and Urey, H. C. 1953. Revised-carbonate-water isotopic temperature scale. *Bull. Geol. Soc. Am.* **64,** 1,315-26.

Ericson, D. B. 1959. Coiling direction of *Globigerina pachyderma* as a climatic index. *Science,* **130,** 219-22.

Evernden, J. F. and Curtis, G. H. 1965. The potassium-argon dating of late Cenozoic rocks in East Africa and Italy. *Current Anthropology,* **6,** 343-64.

Evernden, J. F., Curtis, G. H. and Kistler, R. 1957. Potassium-argon dating of Pleistocene volcanics. *Quaternaria,* **4,** 13-17.

Evernden, J. F., Savage, D. E., Curtis, G. H. and James, G. T. 1964. Potassium-argon dates and the Cenozoic mammalian chronology of North America. *Am. Jour. Sci.* **262,** 145-98.

Fritz, P. 1964. O^{18}/O^{16} Isotopenanalysen und Paläotemperatur-bestimmungen an Belemniten aus dem Schwab. Jura. *Geol. Rund.* **54,** 261-9.

Fritz, P. 1965. Oxygen isotope paleotemperature analysis of Jurassic Belemnoids from south Germany. (Paper read at *Third Int. Conf. Nuclear Geol.,* Spoleto, Italy.)

Lloyd, M. 1967. Oxygen-18 composition of Oceanic sulphate. *Science,* **156,** 1,228-31.

Longinelli, A. 1965. Oxygen isotopic composition of orthophosphate from shells of living marine organisms. *Nature, Lond.* **207,** 716-19.

Longinelli, A. 1966. Ratios of Oxygen 18: Oxygen 16 in phosphate and carbonate from living and fossil marine organisms. *Nature,* **211,** 923-7.

Longinelli, A. and Craig, H. 1967. Oxygen 18 variations in sulphate ions in sea water and saline lakes. *Science,* **156,** 56-9.

Longinelli, A. and Tongiorgi, E. 1960. Frequenza degli individui destrogiri in diverse popolazioni di *Rotalia beccarii* Linneo. *Boll. Soc. Paleont. Ital.* **1,** 5-16.

Longinelli, A. and Tongiorgi, E. 1964. Oxygen isotopic composition of some right and left coiled Foraminifera. *Science,* **144,** 1,004-5.

Lowenstam, H. A. and Epstein, S. 1957. On the origin of sedimentary aragonite needles of the Great Bahama Bank. *J. Geol.* **65,** 364-75.

Nagappa, Y. 1957. Direction of coiling in *Globorotalia* as an aid in correlation. *Micropaleontology,* **3,** 393-7.

Naydin, D. P., Teys, R. V. and Chupakhin, M. S. 1956. Determination of the climatic conditions of some regions of the U.S.S.R. during the Upper Cretaceous period by the method of isotopic paleothermometry. *Geochemistry,* 1956 (1960), 752-64.

Olausson, E. 1965. Evidence of climatic changes in north Atlantic deep-sea cores, with remarks on isotopic paleotemperature analysis. *Progress in Oceanography,* **3,** 221-52.

Rafter, A. 1967. Oxygen isotopic composition of sulphates. Part I. *N. Zeal. Journ. of Science,* **10,** 493-510.

Rosholt, J. N., Emiliani, C., Geiss, J., Koczy, F. F. and Wangersky, P. J. 1961. Absolute dating of deep sea cores by the Pa^{231}/Th^{230} method. *J. Geol.* **69,** 162-85.

Rosholt, J. N., Emiliani, C., Geiss, J., Koczy, F. F. and Wangersky, P. J. 1962. Pa^{231}/Th^{230} dating and O^{18}/O^{16} temperature analysis of core A 254-BR-C. *J. Geoph. Res.* **67,** 2,907-11.

Shackleton, N. J. 1965. Some variations in the technique for measuring carbon and oxygen isotope ratios in small quantities of calcium carbonate. (Paper read at *Third Int. Conf. Nuclear Geol.,* Spoleto, Italy.)

Shackleton, N. J. 1967. Oxygen isotope analyses and Pleistocene temperatures re-assessed. *Nature, Lond.* **215,** 15-17.

Tudge, A. P. 1960. A method of analysis of oxygen isotopes in orthophosphate, its use in the measurement of palaeotemperatures. *Geochim. Cosmochim. Acta,* **18**, 81-93.

Urey, H. C., Lowenstam, H. A., Epstein, S. and McKinney, C. R. 1951. Measurements of paleotemperatures and temperatures of the Upper Cretaceous of England, Denmark and the southeastern United States. *Bull. Geol. Soc. Am.* **62**, 399-416.

45. THE SYSTEMATICS OF COCCOLITHS IN RELATION TO THE PALAEONTOLOGICAL RECORD

M. BLACK

Summary: The coccolith-bearing algae, and certain related forms with uncalcified scales, have recently been shown to differ from the rest of the Chrysophyceae sufficiently to justify their removal to a separate class, known as the Haptophyceae. Within the Haptophyceae, coccolith-morphology provides the most practicable means of classification for the taxa with calcified scales, and can be applied to living and fossil forms alike. Several living families can be traced back to the Mesozoic, and the systematic status of these and of a few extinct families is discussed in the light of their geological history. The Coccolithophoraceae are shown to include several phylogenetic stocks, which are probably best treated as subfamilies.

SYSTEMATIC STATUS OF THE COCCOLITHOPHORALES

The first comprehensive classification of coccolith-bearing organisms was published by Lohmann (1902). Adopting a zoological nomenclature, he placed them in a single family, the Coccolithophoridae. Since Lohmann's time, this name has come to be used in an informal way, not necessarily having any strict taxonomic implication. Many workers followed Lohmann in treating the Coccolithophoridae as Protozoa, but on the basis of their physiology and life-histories, they find a natural place among the scale-bearing Chrysophyceae, and have come to be accepted as unicellular algae; several families are now recognized and have been united into the order Coccolithophorales (Lemmermann 1908; Schiller 1926).

The traditional method of subdividing the Chrysophyceae into orders depends largely upon the number and character of the flagella in the motile phase, but recent work has shown that if the nature of the scales were to be made the basis of classification instead of the flagella, the ultimate result would be much the same (Parke 1961*a*). All those Coccolithophorales whose motile phases have been examined possess a similar flagellar apparatus, with two acronematic flagella and one haptonema, an arrangement which is also found in *Chrysochromulina* and a few other general with uncalcified scales (Parke and Adams 1960; Parke 1961*b*).

The possession of a haptonema is regarded by Christensen (1962) and Parke (in Parke and Dixon 1964) as a character of fundamental systematic importance which justifies the removal of these taxa from the remainder of the Chrysophyceae and the creation of a new class, the Haptophyceae, for their reception. The present situation thus appears to be that, in regard to flagellar apparatus, the Coccolithophorales form a homogeneous order, distinguished from other Haptophyceae by the presence of calcified scales at some stage in their life-history.

BASIS OF CLASSIFICATION

The detailed classification of the Coccolithophorales is based almost entirely upon the structure and arrangement of their coccoliths, with only minor emphasis upon other features such as cell-shape and the presence or absence of an apical depression. The well-known schemes drawn up by Lohmann (1902), Lemmermann (1908), Schiller (1930) and Kamptner (1958) are all based upon this principle, and the classifications used by most workers today are natural developments from one or other of these.

This basis of classification has great advantages for the marine palaeontologist, because it enables him to identify and name isolated coccoliths discovered in samples of bottom sediments. Further, it can be expanded to accommodate extinct taxa within the same framework that is used for living forms. These are advantages not to be lightly sacrificed for the sake of other classifications based upon more sophisticated principles.

Lohmann (1902, p. 127) divided the Coccolithophoridae into two subfamilies, the Syracosphaerinae with unpierced, and the Coccolithophorinae with perforated coccoliths. Lemmermann (1908) recognized the same two major subdivisions, but elevated them to the rank of suborders to make room for several newly-created families. Schiller (1930, p. 171) made the same fundamental distinction, and recognized three families with unpierced coccoliths, the Syracosphaeraceae, Deutschlandiaceae and Halopappaceae, and two families with perforated coccoliths, the Thoracosphaeraceae and Coccolithaceae.

In his classification of the living Coccolithineae of the northern Adriatic, Kamptner (1941, p. 71) made a slight departure from the strict separation of imperforate from perforate forms by including the newly-discovered *Tergestiella* in the otherwise perforate Coccolithaceae. His grounds for doing this were that although *Tergestiella* has no perforation, it can theoretically be regarded as a derivative of perforated ancestors by shrinkage and obliteration of the pore. Recent work with the electron microscope has tended to alter our views on the taxonomic importance of perforations more radically than this, for many Syracosphaeraceae are now known to be riddled with pores, and too many Coccolithaceae have solid centres for this anomaly to be explained away as an exception to a general rule.

In his most recent scheme of classification, Kamptner (1958, p. 68) has strongly emphasized the phylogenetic principle. On theoretical grounds he sought out what he considered to be the most primitive design in coccolith-structure, and with this as a starting-point, grouped the known genera into a series of sub-tribes and higher taxa according to their position in a hypothetical phylogenetic scheme, depending upon the departure of their structure from that of *Tergestiella,* as representing the assumed ancestral form. Although fossil coccoliths were included in the scheme, the underlying phylogenetic framework was constructed without reference to the geological record, which at that time was insufficiently well-known to be of much help.

Since 1958 an enormous amount of new information about the geological history of the coccolith-bearing algae has become available, and the time-element can no longer be ignored; this has inevitably modified our ideas about the interrelationships of many taxa and their phylogeny. It is too early to predict what the ultimate effect upon the systematics of coccoliths is likely to be; the main purpose of this paper is to make an interim report on a few well-known taxa that have already been examined in this way, and to suggest one or two taxonomic changes that seem to be desirable in the light of these studies.

THE FOSSIL RECORD

We can trace the history of the coccoliths back to about the middle of the Cretaceous Period in marine bottom sediments under the present oceans, and on land we can go a little further, to the beginning of the Jurassic. This history is by no means a record of steady progress; like other groups of fossils, the coccoliths have had their phases of exuberant proliferation and their times of mass-extinction. The pre-Jurassic record is shrouded in obscurity. At the beginning of the Jurassic a modest number of species already existed; there is a spectacular influx of new arrivals in the early Oxfordian, and once more in the Albian. Bramlette and Martini (1964) have shown that there is a drastic extinction of Mesozoic species at the end of the Maastrichtian, followed by a new diversification that started in the Palaeocene and led up to the remarkably rich assemblages of the Middle and Upper Eocene. The majority of the Tertiary species died out before the Pleistocene, and the richly varied population of the present-day oceans is very largely a post-Tertiary development.

The main events in this history have an important bearing upon the classification of coccoliths, and we may now look at some of the more general implications. In the first place, the modern calcareous phytoplankton includes two distinct groups of species from this point of view; one of these consists of survivals from the Tertiary, and the other consists of purely modern forms for which we can find no obvious ancestors in pre-Quaternary bottom deposits. Some of the survivals belong to very ancient lineages which became specialized at an early stage in coccolith history, and have maintained their individuality ever since.

The Coccolithophoraceae provide an outstanding example of an ancient stock which was already in existence early in the Jurassic, and has flourished from that time to the present day. The Mesozoic members of this stock being mostly too small for comfortable study under the ordinary microscope, were difficult to classify until electron microscopes became available. Denise Nöel's work (1965) on the Jurassic species has cleared up a great deal of misunderstanding, and has shown that these early forms are in several respects different from the Tertiary and living members of the family. A general picture seems to be emerging of two main evolving complexes in the Mesozoic, and a new set of complexes in the Tertiary, one of which leads up to the living *Coccolithus pelagicus* (Fig. 3),[1] and the other to *Gephyrocapsa oceanica* (Fig. 15) and *Cocco-*

[1] Figures 1 to 42 illustrating this article will be found on Plates 45.1 to 45.4 following the Bibliography.

lithus huxleyi. In the detailed discussion which follows on a later page, these complexes will be treated as separate families.

The geological history of several living genera such as *Helicosphaera* and *Rhabdosphaera,* which in some classifications have been included in the same family as *Coccolithus,* can also be traced back into the Tertiary. The ancestral forms retain most of the peculiarities of the living species, and show no tendency to become more like *Coccolithus* as they are traced back in geological time. If there is any phylogenetic connection with the Coccolithophoraceae, it must be very remote, and far beyond the limit of the known stratigraphical record. The stratigraphical evidence is thus in favour of regarding the living *Helicosphaera* and *Rhabdosphaera* as representatives of two separate families, each with a long record of independence.

Braarudosphaera bigelowi is the sole surviving species of a vigorous family which flourished greatly during the Tertiary, and is now otherwise extinct. *B. bigelowi* itself probably has a longer range than any other species of coccolithophore, since coccoliths that cannot be distinguished from those of the living alga are well-known in Cretaceous rocks.

The remarkably-shaped Calciosoleniaceae (Figs. 1, 2) have an equally long geological record. Never abundant, members of this family have been found in rocks of various ages back to the middle of the Cretaceous, and there has been astonishingly little change in their appearance during this long interval. Their palaeontological record emphasizes the homogeneity of the family, but gives no help towards solving the problem of their relationship to other families of coccolithophores.

The purely Holocene members of the living plankton raise an entirely different set of problems, for which no solution is forthcoming from a study of the fossil record. Nevertheless, the mystery of their sudden appearance in the post-Tertiary oceans is not unique, for it can be paralleled to some extent by the sudden bursts of new forms in the Oxfordian, Albian and Eocene.

At various levels in the geological record, new types of coccoliths appear in this way, without any recognizable ancestors, and they become extinct equally abruptly without leaving any known descendants. Some of these, for example the Podorhabdaceae, include a great variety of morphological variants, and flourished for a substantial length of time. At the other extreme are single cryptogenic species such as *Goniolithus fluckigeri* which existed for a short time only, and cannot be fitted into any previously known family. Each of these appears to be the sole representative of an otherwise unknown family. Other 'small families', as Deflandre (1966) has called them, consist of just a few closely similar species; the Deflandriaceae, for example, include no more than half-a-dozen species, yet these are so abundant in the Upper Cretaceous and have such a distinctive appearance that rocks of this age may be dated by their presence (Figs. 26-8).

LIVING FAMILIES

Of the living families whose ancestry can be traced back into the Mesozoic, the Coccolithophoraceae are much the most prolific both in species and in individuals, and they are discussed below in some detail; a few other families with fossil records are dealt with more briefly. The Zygosphaeraceae, which are generally not found in bottom sediments, are mentioned only to emphasize their independence from the extinct Zygolithaceae.

Coccolithophoraceae Lemmermann 1908; (Figs. 3-7)

Sediments of all ages from the Jurassic to the most recent contain coccoliths which are constructed on the same general plan as those of *Coccolithus pelagicus* and *Cyclococcolithus leptoporus.* They are either elliptical or circular in outline and consist of two shields, each made of radial elements arranged like the petals of a flower round a central pillar. In some species this pillar is tubular with an obvious central pore; in others it is solid, or may take the form of a wide elliptical opening, either empty or ornamented in various ways. There are innumerable reports of *C. pelagicus* in the literature from rocks going as far back as the Jurassic, but even a cursory examination of the Mesozoic forms in polarized light is sufficient to show that they are optically quite different from the living species; under the electron microscope even the Tertiary fossils can be distinguished from *C. pelagicus.* Indeed, it seems likely that *C. pelagicus,* as understood by a marine biologist, is a very recent addition to the oceanic plankton. It is clearly desirable to bring the nomenclature of these *pelagicus*-like forms into closer correspondence with our present knowledge of their biology and fine-structure. From a biological point of view, it is tolerably certain that all the specimens found in bottom sediments are remains of non-motile resting stages, analogous with the cyst-phase of *C. pelagicus*, and we can employ a classification based upon the assumption that we are all the time comparing encysted coccospheres, or their component coccoliths. The International Code of Botanical Nomenclature (Lanjouw 1966) has provisions for accommodating any motile phases that may be discovered, such as the *Crystallolithus* phase of *C. pelagicus.*

The first necessity is to restrict the genus *Coccolithus* to species that do not differ too widely from the type species, *C. pelagicus.* There are at present between 80 and 100 valid species referred to this genus, and these include forms with a very wide range in the details of their structure. Noel's critical revision of the Jurassic species has shown that none of these can reasonably be retained in the genus *Coccolithus*, and a similar rigorous study of Cretaceous species would inevitably lead to the same result. Revision of the Lower Tertiary forms has already led to the removal of further species into new genera such as *Apertapetra* and *Chiasmolithus*, so that *Coccolithus* is in danger of being left with an ill-defined residue of species awaiting revision.

It is proposed to restrict the genus *Coccolithus* to those species in which the coccoliths of the cyst-phase agree with those of *C. pelagicus* in the following respects: two elliptical shields united by a connecting pillar or a tube which may be open or concealed; distal shield consisting of strongly imbricated petaloid elements, without a corona; proximal shield and connecting tube strongly birefringent, distal shield having little or no effect on polarized light passing through it at right angles to the elliptical plane of the coccolith.

Most of these characters are shared by living species of *Cyclococcolithus* (Fig. 7); this genus differs from *Coccolithus* mainly in the circular outline of the placoliths, the different structure of the central pillar, and the curved radial margins of the petaloid elements. Both genera differ from comparable Mesozoic forms in the special optical orientation of the crystals in the distal shield, and it is proposed to restrict the family Coccolithophoraceae Lemmermann to taxa which share this peculiarity. As thus delimited, the Coccolithophoraceae appear to be confined to the Caenozoic, and are most abundant in the Upper Tertiary and Quaternary.

Gephyrocapsaceae fam. nov. (Figs. 15-18)

In the living genus *Gephyrocapsa* the rays of the distal shield do not overlap each other as they do in the genera so far considered, but lie side by side, with the sutures between adjacent rays nearly at right angles to the surface of the shield. The same arrangement is seen in *Coccolithus huxleyi* and in the species-group of *C. doronicoides* in the late Tertiary and Quaternary. *Ellipsoplacolithus exsectus* Kamptner and *E. lacunosus* Kamptner from Pleistocene cores in the Pacific are again very similar in the structure of their shields (Kamptner 1963, figs. 50-2); indeed, almost identical forms in Pleistocene cores from the North Atlantic have recently been figured under the name of *Coccolithus doronicoides* (McIntyre and Bé 1967, pl. 2B and pl. 3A).

In the Pleistocene, McIntyre (1967) has found coccoliths intermediate in form between *Gephyrocapsa oceanica* and *C. huxleyi*, and he has given reasons for believing that *C. huxleyi* has evolved from the *Gephyrocapsa* stock. Recent work suggests that all these forms are closely related, and that they belong to a family that was actively evolving during the Pleistocene, independently of the Coccolithophoraceae. The name Gephyrocapsaceae is proposed for this family, with the following diagnosis: Coccolithophorales bearing placoliths with a large central opening and with the shields constructed of non-imbricate radial elements. Typical genus *Gephyrocapsa* Kamptner. (Gephyrocapsaceae: Coccolithophorales placolithos ellipticos ferentes; in medio coccolitho foramen amplum; pali geminorum scutorum commissuris orthogoniis. Genus typicum: *Gephyrocapsa* Kamptner.)

Coccoliths with a similar structure are abundant in the Tertiary; *C. marismontium* in the Middle Eocene has much in common with some variants of *C. doronicoides*, and there are forms in the Miocene and Pliocene which appear to link the two together. An independent generic name is needed for the *C. doronicoides* and *C. marismontium* complex of species, in which the placoliths have a large open centre and in which both shields consist of numerous non-imbricate rays. The earliest generic name other than *Coccolithus* which has been proposed for a coccolith of this kind appears to be *Ellipsoplacolithus* Kamptner. Loeblich and Tappan (1966, p. 139), invoking Article 34 of the International Code of Botanical Nomenclature, have regarded this name as not having been validly published. Nevertheless, when Kamptner (1963, p. 171) proposed *Ellipsoplacolithus* as a new genus, he gave a diagnosis in which it is treated as an organ-genus, and designated *E. lacunosus* Kamptner as type species. The name is thus provisional only in the same sense that all organ-genera are provisional, but not within the meaning of Article 34. If we therefore accept it as a legitimate name, *Coccolithus doronicoides* becomes *Ellipsoplacolithus doronicoides* (Black and Barnes) n. comb., and *C. marismontium* becomes *E. marismontium* (Black) n. comb.

The central opening of *Gephyrocapsa* is spanned distally by an oblique bridge, and proximally by a grille. A similar grille, but without a bridge, is present in *C. huxleyi* and in a large number of Tertiary species of *Tremalithus* as defined by the genotype, *T. placomorphus. Reticulofenestra* Hay, Mohler and Wade, and *Dictyococcites* Black differ very little from *Tremalithus* Kamptner, and probably ought to be merged into this genus.

All the forms discussed above fall quite naturally into the Gephyrocapsaceae. In the Cretaceous, there is an interesting complex of species whose relationships are not so clear-cut. They reach their greatest diversity in the Albian and Cenomanian, where there are forms with a bewildering combination of characters, some suggestive of the Gephyrocapsaceae, others of the Ellipsagelosphaeraceae. Much work will be needed before these can be satisfactorily sorted out; it may well be that we have here a record of an evolutionary divergence amongst whose products were the ancestors of the gephyrocapsoid stock that did not become clearly differentiated until the Tertiary.

Rhabdosphaeraceae Lemmermann 1908; (Figs. 19-20)

In early classifications, when great emphasis was laid upon the presence or absence of a central pore, the rhabdoliths and placoliths were regarded as homologous structures. Because *Rhabdosphaera claviger* and similar species were believed to have a narrow canal running through the centre of the spine, it was argued that the placoliths and rhabdoliths could be regarded as divergent modifications of a single structural plan, the one with a short connecting-tube and well-developed distal shield, the other with a much exaggerated connecting-tube and no distal shield. For this reason, *Rhabdosphaera* and *Cyclococcolithus* were regarded as being closely enough related to be put in the same family or even the same subtribe.

Recent observations have removed the force of these arguments. Broken specimens of *R. claviger* show that the spine consists of a solid bundle of calcite fibres, and the basal plate has no resemblance to the proximal shield of *Cyclococcolithus*, the method of construction being quite different in the two genera (compare Figs. 7 and 20).

Rhabdoliths with a similar structure are known throughout most of the Tertiary, but have not yet been discovered in Mesozoic rocks. The spine-bearing forms in the Cretaceous and Jurassic are of a different kind, with basal disks quite unlike the disk of a modern rhabdolith.

The living rhabdospheres and their Tertiary relatives clearly belong to a separate family, whose earliest representatives show no resemblance to placoliths in the generally accepted sense, and there is no longer any justification for sinking Lemmermann's Rhabdosphaeraceae into the Coccolithophoraceae.

Helicosphaeraceae fam. nov. (Figs. 21-3)

Diagnosis: Coccolithophorales whose coccoliths have a large elliptical central shield surrounded by a spiral or otherwise asymmetrical flange constructed of radial elements. Typical genus *Helicosphaera* Kamptner. (Coccolithophorales coccolithis irregulariter ellipticis; in medio coccolitho scutum magnum ellipticum, ora inaequali circumdatum; ora palis angustis structa, vel in cochleam serpens, vel aliter inaequalis. Genus typicum: *Helicosphaera* Kamptner.)

The peculiar coccoliths for which Kamptner created the genus *Helicosphaera* were for many years regarded as asymmetrically developed placoliths, and many authors have included them in *Coccosphaera* or *Coccolithophora*. The internal structure of the massive central area is still not fully understood, but is clearly quite different from the pillar or tube of a placolith, and the petaloid elements are arranged in a single spiral band and not in two separate annular shields.

Looking back into the Tertiary, we find a succession of species that share these peculiarities, and the genus *Helicosphaera* can confidently be traced back to the Eocene. The early species are associated with other asymmetrical coccoliths included in the genus *Lophodolithus*; in these the petaloid flange is apparently not spiral, but is more strongly developed at one sector of the periphery than elsewhere. *Lophodolithus* is unknown from pre-Tertiary rocks, but *Kamptnerius* in the Upper Cretaceous shares some of its peculiarities.

With the information available at present, it is difficult to assess the phylogenetic significance of an asymmetrical flange. Bramlette and Sullivan (1961) have given evidence of a gradation between the two genera in the less extreme forms of *Lophodolithus* and *Zygodiscus*, but there is as yet no connecting link between either of these and *Kamptnerius*. At no time in the known geological history of these genera is there any sign of derivation from a placolith-like structure.

Braarudosphaeraceae Deflandre 1947; (Figs. 24, 25)

The geological history of this family is in several respects remarkable. The single living species, *Braarudosphaera bigelowi*, has peculiar coccoliths which are very easy to recognize in oceanic sediments. Similar coccoliths, indistinguishable from those of *B. bigelowi*, have been found in deposits of many ages going back to the Cretaceous, and there is a strong probability that this species has indeed survived virtually unaltered from the Mesozoic. The unusual pentalith-construction of the coccoliths is unique among living forms, but is found in great diversity in numerous species in the Lower Tertiary; some of these can be referred to *Braarudosphaera*, and a larger number have been placed in the related genera *Micrantholithus* and *Pemma*.

In the condition in which it is collected in plankton-hauls, the cell of *B. bigelowi* is completely enclosed in a shell constructed of twelve coccoliths which fit together to make a regular pentagonal dodecahedron. No flagella are present, and the cells are presumably in an encysted condition. Two other genera are known which produce dodecahedral shells. In *Goniolithus fluckigeri* from the Eocene, the coccoliths are pentagonal, but are otherwise quite unlike those of *B. bigelowi*. *Tergestiella adriatica*, now living in the Adriatic Sea, has circular coccoliths, different in structure from both the others. These genera appear to belong to three distinct families, each of which has independently stabilized the number of coccoliths at twelve, and has adopted the dodecahedral method of shell-construction for the cyst-phase.

Zygosphaeraceae fam. nov.

Diagnosis: Coccolithophorales bearing zygoform or calyptroform holococcoliths. Typical genus *Zygosphaera* Kamptner. (Coccolithophorales holococcolithos vel zygoformes vel calyptroformes ferentes. Genus typicum: *Zygosphaera* Kamptner.)

Zygosphaera was first described by Kamptner (1936, 1937) together with several other genera for which he created the subfamily Zygosphaeroideae. The coccoliths were described and figured as elliptical rings spanned on the distal side by a bridge which was more or less strongly arched and often carried a knob or spine. Kamptner used the name zygolith for coccoliths built on this plan; they are easily recognized under the light-microscope, and fossils with a similar appearance have long been known to be common in the Chalk, both in Europe and North America. For these, Deflandre (in Deflandre and Fert 1954) used the generic name *Zygolithus,* originally proposed by Kamptner for zygoliths found in the Upper Tertiary of the East Indies.

The fine-structure of zygoliths remained unknown so long as their study was confined to the light-microscope. Examination under the electron microscope has shown that two totally different types of fine-structure are involved. The fossil zygoliths are typical heterococcoliths, and will be discussed later (p. 617); examples are shown in Figs. 29-33. All the Zygosphaeraceae that have been examined at high magnifications, on the other hand, have holococcoliths; these are constructed of very small calcite crystals, either rhombohedra or short hexagonal prisms, which have a remarkably uniform size and shape in any particular species (Halldal and Markali 1955; Gaarder 1962). These authors have shown that the coccoliths of *Calyptrosphaera* and *Sphaerocalyptra* also are constructed in the same way.

Noël (1965) has created the family Zygolithaceae for the fossil species (mostly Mesozoic) which have zygoliths constructed on the heterococcolith plan. It is here proposed that the living forms discussed above, which have zygoform or calyptroform holoccoliths, should be brought together into a separate family, the Zygosphaeraceae.

EXTINCT FAMILIES

The extinct coccoliths are dealt with here according to the International Code of Botanical Nomenclature (Lanjouw 1966), which allows the creation of organ-genera for fragmentary specimens of fossil plants such as isolated coccoliths. Since the Coccolithophorales are classified on the basis of coccolith-structure, there are no problems about using such organ-genera, which can be assigned to a family that may be either living or extinct. It is thus unnecessary to use a separate classification into parataxa for fossil coccoliths.

Work on the systematics of fossil coccoliths is still in the early stages, and a large number of the genera which have been described still remain unassigned to any family. Four families which are represented in marine bottom sediments are discussed below.

Ellipsagelosphaeraceae Noël 1965; (Figs. 8-14)

The Mesozoic coccoliths that have sometimes been recorded as *C. pelagicus* are readily distinguishable from this species in polarized light. Both shields are strongly birefringent, and between crossed nicols there is a complicated interference pattern formed by the superposition of two black crosses, one produced by each shield. In the Coccolithophoraceae, only the proximal shield and the central structure contribute to the interference pattern. A second distinction is that on the distal surface of most Mesozoic forms there is a ring of quadrate or keystone-shaped granules separating the outer zone of petaloid elements from the central area (Fig. 8, 9). This corona is very characteristic of Mesozoic species, but is not present in the Coccolithophoraceae.

Ellipsagelosphaeroideae Noël 1965; (Figs. 8-11)

The Jurassic forms have recently been monographed by Noël (1965). In the most abundant genus, *Ellipsagelosphaera,* the petaloid elements are strongly imbricated, although not in exactly the same way as in *C. pelagicus,* and their optical orientation is different, as can readily be seen in smear-slides viewed in polarized light. Circular coccoliths with a similar ultrastructure and similar optical properties are placed in a separate genus, *Cyclagelosphaera,* but are retained within the same sub-family as the elliptical forms. Complete coccospheres are not uncommon; the component coccoliths are interlocked, and the spheres are so robust that it seems likely that the shell-membrane was calcified. In these respects they much resemble the spheres of living species of *Coccolithus,* and like them are probably cysts.

Actinosphaeroideae Noël 1965; (Figs. 12-14)

The two genera *Ellipsagelosphaera* and *Cyclagelosphaera* are placed in a subfamily, the Ellipsagelosphaeroideae Noël, in which the two shields are united by a short tubular column. In a second subfamily, the Actinosphaeroideae, the two shields are united at the inner border of the distal shield, without the development of a special tubular structure. The distal shield of *Actinosphaera deflandrei* Noël, as commonly preserved, is without a corona. In the other genus, *Calolithus* Noël, there is a zone in which the proximal rays are contracted so as to form a ring of slots or pits, concentric with the large central opening. The distal shield of the only species named by Noël, *C. martelae*, is described as having no corona, but this may possibly be only a question of preservation, since other species which probably belong to this genus have a well-developed but rather loosely attached corona. The same is probably true of certain unnamed species of *Actinosphaera* in the Upper Cretaceous.

The Ellipsagelosphaeroideae and Actinosphaeroideae are characteristic of Jurassic and Lower Cretaceous rocks, and often occur in such overwhelming abundance that coccoliths of other kinds are reduced to obscurity. A few species of *Actinosphaera* persist into the Upper Cretaceous, but neither sub-family is at present known to survive into the Tertiary.

Deflandriaceae fam. nov. (Figs. 26-8)

Diagnosis: Coccoliths with an open ring of sixteen primary granules spanned by a cross from the centre of which arises a stalk constructed of two bundles of elongated crystals placed end-to-end. Typical genus *Deflandrius* Bramlette and Martini 1964, p. 300.

The genus *Deflandrius* was proposed by Bramlette and Martini for coccoliths of a type which had long been known in the European white chalks, where they are important rock-builders. They were recognized by Sorby (1861) as parts of an organic skeleton, and he published drawings showing the essential features of the genus, but unfortunately did not name it. *Deflandrius* first appears in the Albian, where it is usually not very common; from the Cenomanian to the Maestrichtian it is extraordinarily abundant, and turns up in most core-samples that penetrate Upper Cretaceous sediments. It became extinct quite suddenly at the end of the Maestrichtian.

Throughout the short span of its existence, *Deflandrius* preserved a remarkable uniformity of structure, which is quite distinct from that of any of its contemporaries. The family Deflandriaceae contains but the single genus with about half-a-dozen species; it has no recognizable ancestors, and has apparently left no descendants. The architectural pattern upon which its coccoliths are built is unusual amongst calcified Haptophyceae, but an extraordinarily similar pattern has recently been discovered in the uncalcified scales of *Chrysochromulina pringsheimii*, which was taken living in plankton-hauls from the English Channel off Plymouth (Parke and Manton 1962). This species bears organic scales of several kinds; the largest consist of an elliptical base from the margin of which arise four flying buttresses which unite to support a slender spine. If the margin of the basal scale were calcified by growing a ring of sixteen calcite crystals, and the spine with its supporting buttresses were also mineralized, the result would be a coccolith very much like a generalized *Deflandrius*.

The unmineralized scales of *Chrysochromulina* are much too delicate to be preserved as fossils, and if a genus with scales of this kind had existed in Mesozoic times, it is unlikely that it could have left any recognizable traces. As a hypothesis to explain the cryptogenic appearance of the Deflandriaceae, it is suggested that the ancestors of this family may have borne uncalcified scales built on the same general pattern, and that the stock to which they belonged may have persisted after the calcified members had become extinct at the end of the Cretaceous period.

Zygolithaceae Noël 1965; (Figs. 29-35)

A very large number of fossil coccoliths are built after a pattern that incorporates one fundamental component which for convenience may be called a loxolith-ring (Fig. 29). This consists of a short tube, elliptical in cross-section, with a wall composed of numerous steeply inclined staves. It is seen without any modifications or additions in *Loxolithus*. In *Zygodiscus* there is a bridge on the proximal side, usually following the shorter diameter of the ellipse, and thus dividing the central opening into two D-shaped pores (Fig. 30).

In other genera the space within the ring may be spanned by cross-bars, or closed by a floor which may be solid or perforated in various patterns (Figs. 32-4); in some genera there is a spine either rising from the floor or supported by a bridge (Fig. 35). Several of these features may be combined in different ways to give a very wide structural diversity. Nevertheless, the unifying presence of a loxolith-ring in all these coccoliths suggests a reasonably close relationship, and for our present purpose, there is much in favour of including them all in a single family. A separate family, the Discolithaceae, has been proposed by Noël for genera in which the loxolith-ring is closed by a floor. This distinction is theoretically sound, but in practice it is difficult to maintain. For example, forms with partially developed floors are known, and bridged species such as '*Discolithus theta*' are indistinguishable from *Zygodiscus* when the delicate floor has been destroyed, as usually has happened under normal conditions of preservation. For this reason it is expedient to include both these architectural types within the single genus *Zygodiscus*.

The Zygolithaceae, interpreted in this broad sense, are present in Jurassic rocks; they are extremely common in the Cretaceous, and continue into the Lower Tertiary. Records from the Upper Tertiary are very few, and need confirmation; some of them are known to be based upon specimens of Mesozoic species that have been reworked into Tertiary sediments. Some species of *Zygosphaera* and related genera living in modern oceans have bridged coccoliths which have been called zygoliths, and under the light-microscope superficially resemble coccoliths of *Zygodiscus.* This resemblance, however, is deceptive, for the fine-structure is entirely different, and it is most unlikely that the *Zygosphaeraceae* have any close relationship with the *Zygolithaceae.*

Podorhabdaceae Noël 1965; (Figs. 36-41)

This characteristically Mesozoic family makes its first appearance in the Oxfordian; it is abundantly represented in the Lower Cretaceous, and if we include *Cretarhabdus,* it continues in strength to the top of the Maestrichtian. The unifying feature of the family is the possession of an elliptical shield with a narrow rim consisting of two layers of nearly quadrate, non-imbricate crystals surrounding a very large central area. Many species have a tall spine or stalk arising from the centre of the shield, giving them a rhabdolith-like appearance.

In *Podorhabdus, Hexapodorhabdus* and *Octopodorhabdus* the area within the rim is occupied by a finely granular vaulted carapace which is pierced by a number of large windows, and is surmounted by a hollow spine. These three genera are very much alike, and for convenience may be referred to as typical Podorhabdaceae. Other genera which have been assigned to this family with varying degrees of confidence show a much greater range of structure. In *Ethmorhabdus* the granular carapace is pierced by numerous small perforations, but has no large windows. The central structure of *Polypodorhabdus* departs still further from the type. There is no granular carapace, and the spine is supported by four heavy buttresses made of calcite fibres. Very similar buttresses are conspicuous in many species of *Cretarhabdus.* As Noël (1965, p. 115) has pointed out, this genus has strong affinities with the Podorhabdaceae.

Reinhardt (1967, p. 164) has proposed that the Podorhabdaceae be treated as a subfamily of the Ahmuellerellaceae, distinguished from the rest of the family by having a double as opposed to a single 'Randscheibe'. The distinction, however, is more fundamental than this, for the 'Randscheiben' are of entirely different kinds in the two families. In *Ahmuellerella* the marginal wall is apparently in the form of a loxolith-ring, in no way resembling the radially constructed rim of the Podorhabdaceae; it is unlikely that the two structures are homologous. The affinities of *Ahmuellerella* are at present rather obscure, but they are clearly not with the Podorhabdaceae.

DIAGNOSES

To conform with the Rules of the International Code of Botanical Nomenclature, details are given below of those taxa for which formal diagnoses could not conveniently be inserted in the text.

Actinosphaera sera sp. nov. (Fig. 14)

Diagnosis: A species of *Actinosphaera* with small coccoliths usually between 4 and 5μ in greatest length, and with 22 to 24 radial elements in each shield.

Holotype: No. 10512 from the Santonian Chalk of Shudy Camps, Cambridgeshire.

Dimensions of Holotype: Distal shield 4.7 x 3.7μ; 24 rays.

Remarks: In Britain this species is found in the Upper Chalk, ranging from the Santonian of Cambridgeshire to the Lower Maestrichtian of the Norfolk coast. It is also known from the Upper Maestrichtian of Stevns Klint, Denmark.

Calolithus speetonensis sp. nov. (Fig. 12)

Diagnosis: A species of *Calolithus* with relatively large radial slits, and a central opening not more than one fifth of the breadth of the distal shield.

Holotype: No. 12034 from the Speeton Clay, Division A3 (Albian), of Speeton, Yorkshire.

Dimensions of Holotype: Distal shield 5.3 x 4.2μ; proximal shield 4.5 x 3.3μ; central opening 1.8 x 0.9μ; 31 rays in each shield.

Remarks: This species is constructed on exactly the same plan as the Jurassic *C. martelae* Noël (Fig. 13), but with different proportions. The coccolith of *C. speetonensis* is larger, but the central opening is proportionally smaller, and the inter-radial slits are larger than in *C. martelae.* There are commonly 30 or more rays in each shield of *C. speetonensis,* and 30 or fewer in *C. martelae.*

Helicosphaera burkei sp. nov. (Fig. 23)

Diagnosis: A species of *Helicosphaera* with large coccoliths whose radial elements are roughly and irregularly shaped, and whose central shield has an uneven outline with no special ornamentation at the outer margin.

Holotype: No. 16171 from the Cobre Member of the Upper white Limestone, Jamaica (Miocene, N.11).

Dimensions of Holotype: Outer flange 10.0 x 6.5μ; central shield 6.3 x 3.9μ.

Remarks: This species resembles *H. carteri* in shape,

but is of larger size,and differs in its more rugged and untidy construction, and further in the complete lack of any specialized structure at the margin of the central shield.

Helicosphaera orientalis sp. nov. (Fig. 22)

Diagnosis: A species of *Helicosphaera* with small coccoliths having a narrow marginal flange and a peculiar, almost rectangular, outline.

Holotype: No. 15353 from the Miocene (N.17) of Bebalain, Indonesia.

Dimensions of Holotype: Outer flange 4.2 x 2.9μ; central shield 2.9 x 1.8μ.

Remarks: The marginal flange increases very gradually in width, and does not terminate in an expanded wing. The direction of coiling is right-handed, as in other species of the genus; the illustration, Fig. 22, shows the coccolith in mirror-image.

Polypodorhabdus madingleyensis Black (Fig. 37)

Diagnosis: A species of *Polypodorhabdus* with a solid spine and four strong fibrous buttresses following the axes of the ellipse, 24 to 30 rays in the distal shield, and 14 to 20 grid bars, always fewer in number than the rays of the distal shield.

Holotype: No. 17413 from the Upper Oxford Clay (Lower Oxfordian), Cambridge Experimental Borehole, 220 ft 3 in. to 221 ft 3 in. Black 1968, p. 806, pl. 150, fig. 2.

Dimensions of Holotype: Distal shield 6.8 x 5.0μ, 30 rays; central area 4.2 x 3.0μ, 20 grid bars.

Remarks: This species is conspicuous in samples from the Upper Oxford Clay and the Kimeridge Clay; a very similar, and possibly identical species is also present in the Speeton Clay. *P. madingleyensis* differs from *P. escaigi* of the Oxford Clay in having a solid spine, and in the much stronger development of the striated buttresses.

Genus *Pontolithina* Black

Diagnosis: Elliptical coccoliths with a two-layered floor from which a thin one-layered wall of steeply imbricate laths arises on the distal side; proximal surface of the floor with a pattern of segments arranged pinnately about the long axis, distal surface with a concentric or spiral pattern of very narrow fibres.

Type species: *Pontolithina moorevillensis* Black from the Santonian of Alabama.

Remarks: In the structure of its wall, this genus closely resembles the Cretaceous Zygolithaceae. The floor, on the other hand, has a structural pattern unknown in other genera of the Zygolithaceae, and is much more like the floor of some Tertiary species of *Pontosphaera*, such as *P. scutellum* Kamptner; a similar floor-pattern is also characteristic of many species of *Helicosphaera*.

Pontolithina moorevillensis Black

Diagnosis: A species of *Pontolithina* with two small perforations in the floor, and 80-90 imbricate elements in the wall.

Holotype: No. 22362 from the Mooreville Chalk (Santonian) near Eutaw, Alabama. Black 1968, p. 806, pl. 149, fig. 4.

Dimensions of Holotype: 11.0 x 8.4μ; floor 8.0 x 5.8μ; pores 0.5μ; 85 imbricate elements in the wall.

ACKNOWLEDGEMENTS

The illustrations (plates 45.1 to 45.4) are of carbon-replicas shaded with chromium and photographed with an EM 6 (AEI) electron microscope, provided by a grant from the Department of Scientific and Industrial Research (now NERC). Numbers up to 16172 photographed by Mr Peter Hyde, others by Mr Derek Stubbings.

I am indebted to Dr M. N. Bramlette, Mr A. G. Brighton, Prof. O. M. B. Bulman, Dr Kevin Burke, Dr. D. B. Ericson, Dr C. L. Forbes, Dr Helge Gry, Dr W. R. Riedel, and the late Dr M. N. Hill for generously supplying samples which have been used in preparing some of the illustrations.

Received March 1968.

BIBLIOGRAPHY

Black, M. 1967. New names for some coccolith taxa. *Proc. geol. Soc. Lond.* no. 1640, 139-45.

Black, M. 1968. Taxonomic problems in the study of coccoliths. *Palaeontology*, **11**, 793-813.

Bramlette, M. N. and Martini, E. 1964. The great change in calcareous nannoplankton fossils between the Maestrichtian and the Danian. *Micropaleontology,* **10**, 291-322.

Bramlette, M. N. and Sullivan, F. R. 1961. Coccolithophorids and related nannoplankton of the early Tertiary in California. *Micropaleontology,* **7**, 129-88.

Christensen, T. 1962. Alger. In *Botanik* 2 (2), 178 pp. Copenhagen.

Deflandre, G. 1959. Sur les nannofossiles calcaires et leur systématique. *Rev. Micropaléont.* **2**, 127-52.

Deflandre, G. 1966. Commentaire sur la systématique et la nomenclature des nannofossiles calcaires. 1. Généralités. *Cahiers Micropaléont.*, ser. 1, no. **3**. Arch. Orig. Centre Docum., C.N.R.S. 433. 9 pp. Paris.

Deflandre, G. and Fert, C. 1954. Observations sur les coccolithophoridés actuels et fossiles en microscopie ordinaire et électronique. *Ann. Paléont.* **40**, 115-76.

Fritsch, F. E. 1935. *The structure and reproduction of the algae.* Vol. 1. 791 pp. Cambridge.

Gaarder, Karen R. 1962. Electron microscope studies on holococcolithophorids, *Nytt Mag. Bot.* **10**, 35-51.

Halldal, P. and Markali, J. 1955. Electron microscope studies on coccolithophorids from the Norwegian Sea, the Gulf Stream and the Mediterranean. *Avh. norske VidenskAkad.*, Mat.-Naturv. Kl. 1955 (1), 1-30.

Hay, W. W., Mohler, H. and Wade, Mary E. 1966. Calcareous nannofossils from Nal'chik (Northwest Caucasus). *Ecl. Geol. Helv.* **59**, 379-400.

Kamptner, E. 1928. Uber das System und die Phylogenie des Kalkflagellaten. *Arch. Protistenk.* **64**, 19-43.

Kamptner, E. 1936. Uber die Coccolithineen der Südwestküste von Istrien. *Anz. Akad. Wiss. Wien*, Math.-Naturw. Kl. **73**, 243-7.

Kamptner, E. 1937. Neue und bemerkenswerte Coccolithineen aus dem Mittelmeer. *Arch. Protistenk.* **89**, 279-316.

Kamptner, E. 1941. Die Coccolithineen der Südwestküste von Istrien. *Ann. Naturh. Mus. Wien* **51**, 54-149.

Kamptner, E. 1958. Betrachtungen zur Systematik der Kalkflagellaten, nebst Versuch einer neuen Gruppierung der Chrysomonadales. *Arch. Protistenk*, **103**, 54-116.

Kamptner, E. 1963. Coccolitheen-Skelettreste aus Tiefseeablagerungen des Pazifischen Ozeans. *Annln naturh. Mus. Wien*, **66**, 139-204.

Lanjouw, J. 1966. International Code of Botanical Nomenclature. *Regnum Vegetabile*, **46**, 372 pp. Utrecht.

Lemmermann, E. 1908. Flagellatae, Chlorophyceae, Coccosphaerales und Silicoflagellatae. XXI in *Nordisches Plankton. Botanischer Teil.* ed. K. Brandt and C. Apstein. Kiel and Leipzig. 40 pp.

Loeblich, A. R. and Tappan, Helen. 1966. Annotated bibliography of the calcareous nannoplankton. *Phycologia*, **5**, 81-216.

Lohmann, H. 1902. Die Coccolithophoridae, eine Monographie der Coccolithen bildenden Flagellaten, zugleich ein Beitrag zur Kenntnis der Mittelmeer-auftriebs. *Arch. Protistenk.* **1**, 89-165.

McIntyre, A. and Bé, A. W. H. 1967. Modern Coccolithophoridae of the Atlantic Ocean – I. Placoliths and Cyrtoliths. *Deep Sea Res.* **14**, 561-97.

McIntyre, A. 1967. Coccoliths as Paleoclimatic Indicators of Pleistocene Glaciation. *Science*, **158**, ,1314-17.

Noël, D. 1965. *Coccolithes Jurassiques.* Essai de classification des coccolithes fossiles. C.N.R.S., Paris, 209 pp.

Papenfuss, 1955. Classification of the Algae. In *A century of progress in the natural sciences*, 115-224. Calif. Acad. Sci. San Francisco.

Parke, Mary 1961a. Some remarks concerning the class Chrysophyceae. *Brit. phycol. Bull.* **2**, 47-55.

Parke, M. 1961b. Electron microscope observations on scale-bearing Chrysophyceae. *Recent Advances in Botany*, Sect. 3, 226-9. Toronto.

Parke, M. and Adams, Irene, 1960. The motile (*Crystallolithus hyalinus* Gaarder and Markali) and non-motile phases in the life history of *Coccolithus pelagicus* (Wallich) Schiller, *J. mar. biol. Ass. U.K.* **39**, 263-74.

Parke, M. and Dixon, P. S. 1964. A revised check-list of British marine algae. *J. mar. biol. Ass. U.K.* **44**, 499-542.

Parke, M. and Manton, Irene 1962. Studies on marine flagellates. VI. *Chrysochromulina pringsheimii* sp. nov. *J. mar. Ass. U.K.* **42**, 391-404.

Reinhardt, P. 1967. Fossile Coccolithen mit rhagoidem Zentralfeld. *N. Jb. Geol. Palaont. Mh.* **3**, 163-78.

Schiller, J. 1926. Uber Fortpflanzung, geissellose Gattungen und die Nomenklatur der Coccolithophoraceen nebst Mitteilung über Copulation bei *Dinobryon. Arch. Protistenk.* **53**, 326-42.

Schiller, J. 1930. Coccolithineae. In Rabenhorst's *Kryptogamen-Flora.* **10** (2). Leipzig. 79 pp.

Sorby, H. C. 1861. On the organic nature of the so-called 'crystalloids' of the Chalk. *Ann. Mag. nat. Hist.* **3** (8), 193-200.

Specimen numbers refer to the collection of electron micrographs at the Sedgwick Museum, Cambridge.

Numbers in brackets refer to entries in Funnell's *Index of pre-Quaternary occurrences in the oceans*, pp. 522-34 in this volume, where details of locality and references to literature may be found.

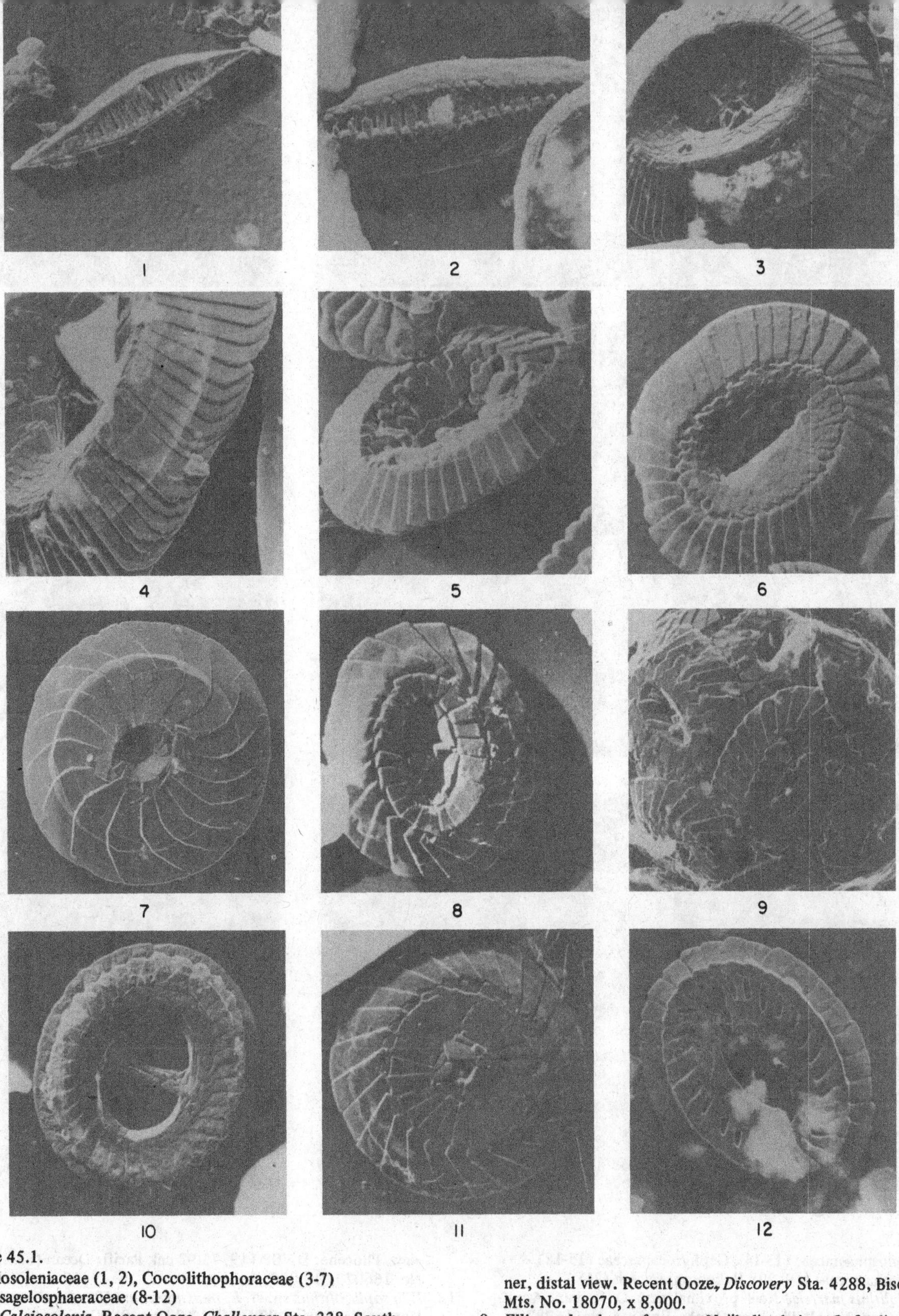

Plate 45.1.

Calciosoleniaceae (1, 2), Coccolithophoraceae (3-7)
Ellipsagelosphaeraceae (8-12)

1. *Calciosolenia.* Recent Ooze, *Challenger* Sta. 338, South Atlantic (671.52). No. 13135, x 8,000.
2. *Scapholithus.* Cenomanian: Chloritic Marl. Folkestone, England. No. 20678, x 13,000.
3. *Coccolithus pelagicus* (Wallich) Schiller, oblique distal view. Recent Ooze, *Discovery* Sta. 4269, Biscay Abyssal Plain. No. 21084, x 4,000.
4. *Coccolithus pelagicus,* lateral view of edge of distal shield. *Discovery* Sta. 4269. No. 21074, x 5,300.
5. *Coccolithus* sp., oblique distal view. Upper Oligocene: white calcareous ooze, DWBG 10, 11-27 cm. Pacific Ocean (523.83). No. 22292, x 8,000.
6. *Coccolithus* sp. cf. *C. sarsiae* Black, distal view, Oligocene. Odder, Denmark. No. 11157, x 6,000.
7. *Cyclococcolithus leptoporus* (Murray and Blackman) Kamptner, distal view. Recent Ooze, *Discovery* Sta. 4288, Biscay Mts. No. 18070, x 8,000.
8. *Ellipsagelosphaera frequens* Noël, distal view. Oxfordian: Upper Oxford Clay. Cambridge Experimental Borehole, Cambridge, England. No. 17383, x 8,000.
9. *Ellipsagelosphaera frequens* Noël, surface of coccosphere. Kimeridgian: Upper Kimeridge Clay. Ringstead Bay, Dorset, England. No. 20472, x 5,400.
10. *Ellipsagelosphaera lucasi* Noël, proximal view, Kimeridgian: Lower Kimeridge Clay. Osmington Mills, Dorset, England. No. 15178, x 6,000.
11. *Cyclagelosphaera margereli* Noël, distal view. Kimeridgian: Lower Kimeridge Clay. Osmington Mills, Dorset, England. No. 15163, x 10,000.
12. *Calolithus speetonensis* sp. nov., proximal view. Albian: Speeton Clay, base of A 3, Speeton, Yorks, England. No. 12034, x 9,000.

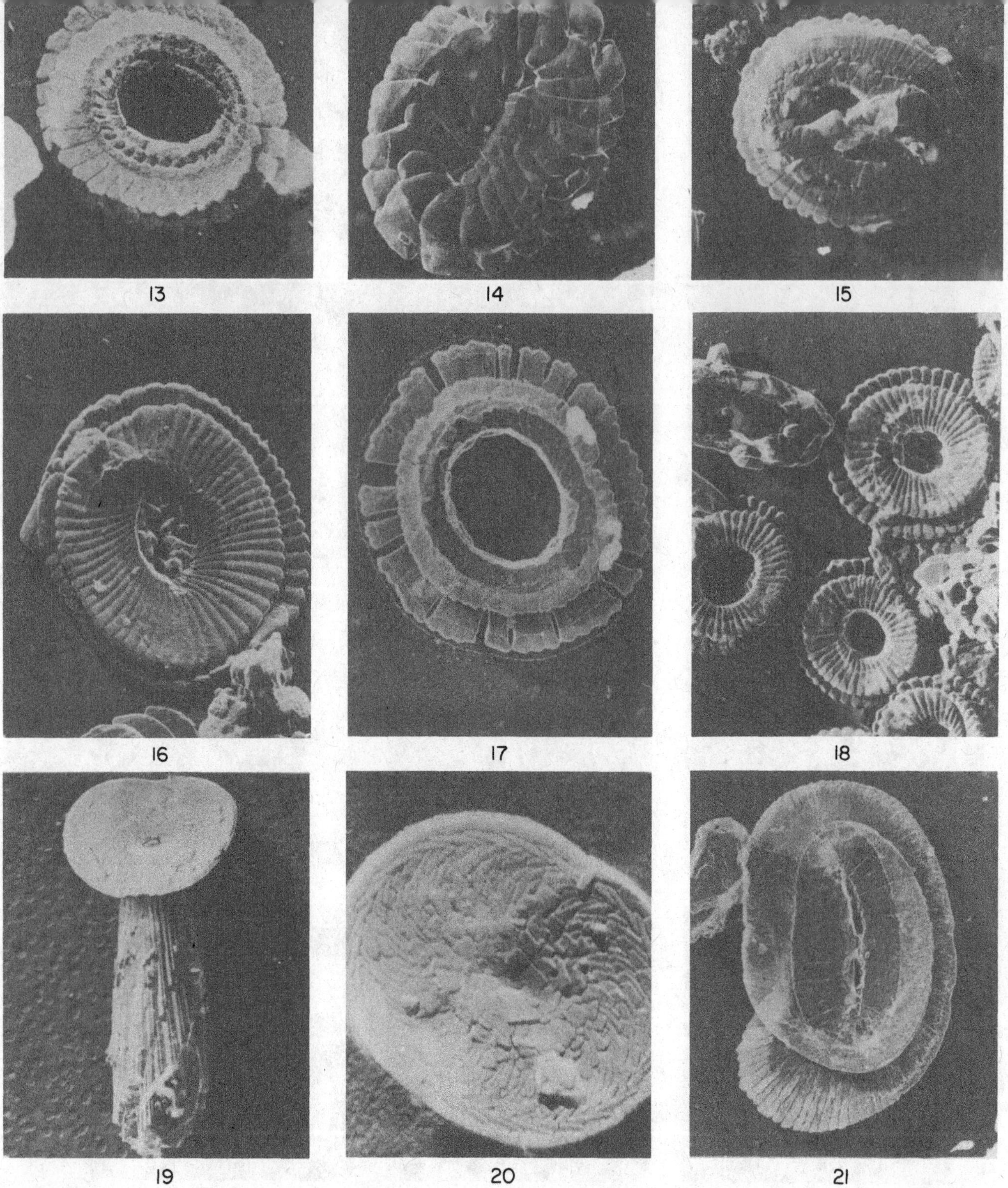
13
14
15
16
17
18
19
20
21

Plate 45.2.

Ellipsagelosphaeraceae (13-14), Gephyrocapsaceae (15-18), Rhabdosphaeraceae (19, 20), Helicosphaeraceae (21)

13. *Calolithus martelae* Noël, proximal view. Oxfordian: Ampthill Clay, Near Willingham, Cambs, England. No. 16282, x 10,000.
14. *Actinosphaera sera* sp. nov., proximal view. Santonian: Upper Chalk, Shudy Camps, Cambs, England. No. 10512, x 10,000.
15. *Gephyrocapsa oceanica* Kamptner, distal view. Recent Ooze, *Challenger* Sta. 338, South Atlantic (671.52). No. 13121, x 10,000.
16. *Ellipsoplacolithus doronicoides* (Black) n. comb., proximal view. Lower Pliocene: LSDH 78 P, 250-70 cm, Pacific Ocean (553.84). No. 22215, x 6,000.
17. *Ellipsoplacolithus exsectus* (Kamptner) n. comb., proximal view. Pliocene: DWBP 119, 750-2 cm, Pacific Ocean (662.37). No. 16607, x 10,000.
18. *Ellipsoplacolithus* sp. cf. *E. marismontium* (Black) n. comb. Upper Oligocene: DWBG 10, 11-27 cm, Pacific Ocean (523.83). No. 22284, x 3,600.
19. *Rhabdosphaera claviger* Murray and Blackman, lateral view. Recent Ooze, *Challenger* Sta. 338, South Atlantic (671.52). No. 11272. x 6,000.
20. *Rhabdosphaera claviger*, proximal view of basal disk, Recent Ooze, *Challenger* Sta. 338, South Atlantic (671.52). No. 11391, x 16,000.
21. *Helicosphaera* sp. cf. *H. carteri* (Wallich) Kamptner, proximal view. Upper Pliocene. Cisano, nr. Albenga, N. Italy. No. 16859, x 5,300.

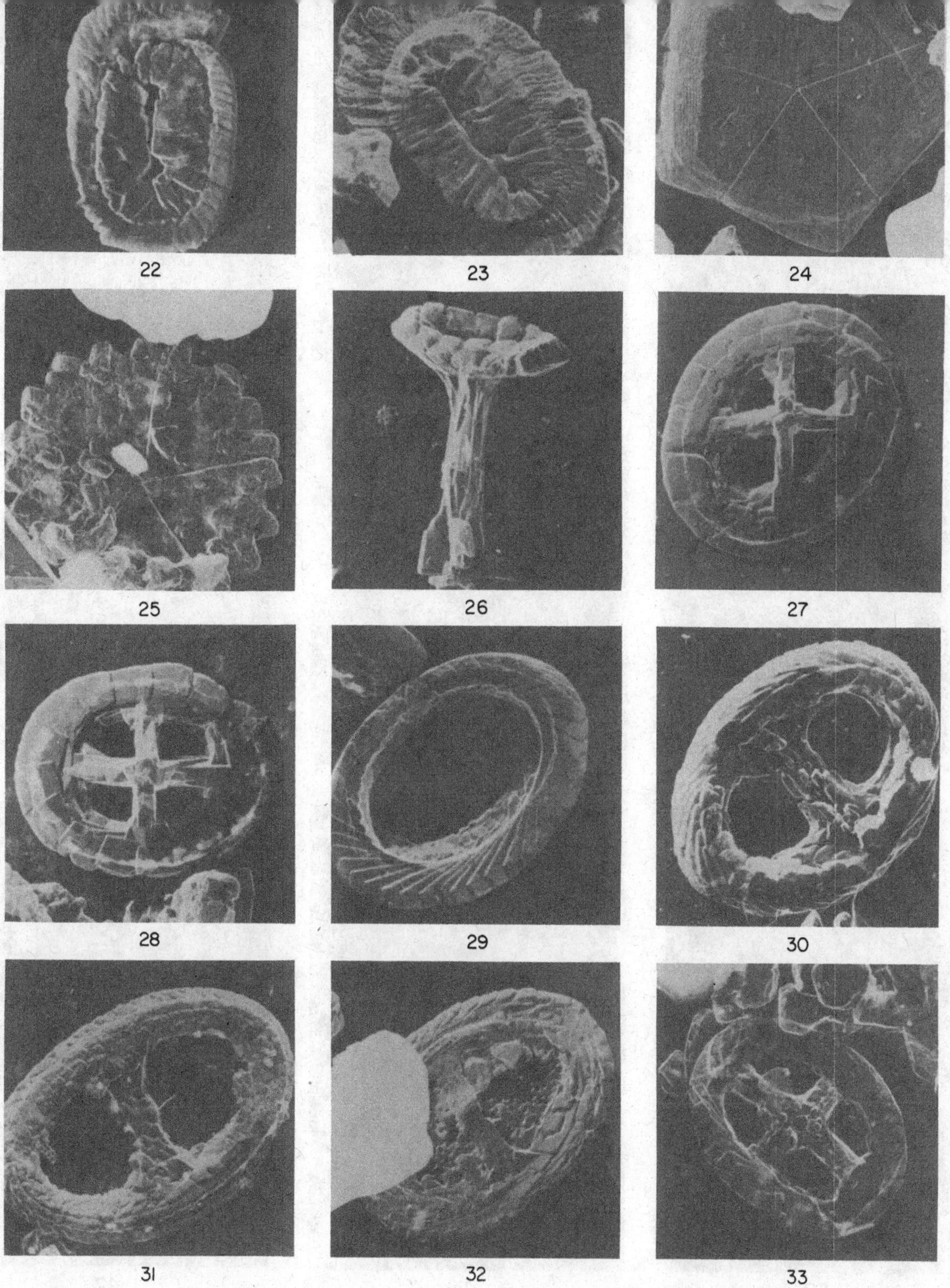
22
23
24
25
26
27
28
29
30
31
32
33

Plate 45.3.
Helicosphaeraceae (22, 23), Braarudosphaeraceae (24, 25), Deflandriaceae (26-28) Zygolithaceae (29-33)

22. *Helicosphaera orientalis* sp. nov., proximal view. Upper Miocene. Bebalain, Rotti, Indonesia. No. 15353, x 10,000.
23. *Helicosphaera burkei* sp. nov., proximal view. Middle Miocene: Upper White Limestone. Jamaica. No. 16171, x 5,400.
24. *Braarudosphaera bigelowi* (Gran and Braarud) Deflandre, distal view, with the outline of the proximal surface showing through the replica. Upper Eocene: Upper Bracklesham Beds. Bracklesham Bay, Sussex, England. No. 19865, x 5,400.
25. *Pemma papillatum* Martini, Upper Eocene: Upper Bracklesham Beds. Bracklesham Bay, Sussex, England. No. 19920, x 6,000.
26. *Deflandrius cretaceus* (Arkhangelski) Bram. and Martini, lateral view. Turonian: Belemnite Marl. Cherry Hinton, Cambridge, England. No. 17159, x 6,000.
27. *Deflandrius cretaceus*, distal view of basal shield. Maestrichtian: Prairie Bluff Formation. Wilcox County, Alabama. No. 22118, x 8,000.
28. *Deflandrius spinosus* Bram. and Martini, distal view of basal shield. Maestrichtian: Prairie Bluff Formation. Wilcox County, Alabama. No. 22124, x 6,000.
29. Loxolith - ring, distal view, isolated from other structural elements. Cenomanian: Cambridge Greensand, Hauxton, Cambs, England. No. 22422, x 6,000.
30. *Zygodiscus ponticulus* (Deflandre) Reinhardt, showing loxolith - ring in proximal view. Turonian: Belemnite Marl. Cherry Hinton, Cambridge, England. No. 17150, x 8,000.
31. *Zygodiscus theta* (Black) n. comb., proximal view, normal state of preservation with floor missing. Cenomanian: Burwell Rock. Burwell, Cambs, England. No. 18827, x 5,400.
32. *Zygodiscus theta,* distal view, with remains of the delicately perforated floor preserved. Cenomanian: Cambridge Greensand. Cherry Hinton Fields, near Cambridge, England. No. 21761, x 6,000.
33. *Staurolithites* sp. indet., distal view. Cenomanian V22-8, Western Atlantic (429.55). No. 21925, x 8,000.

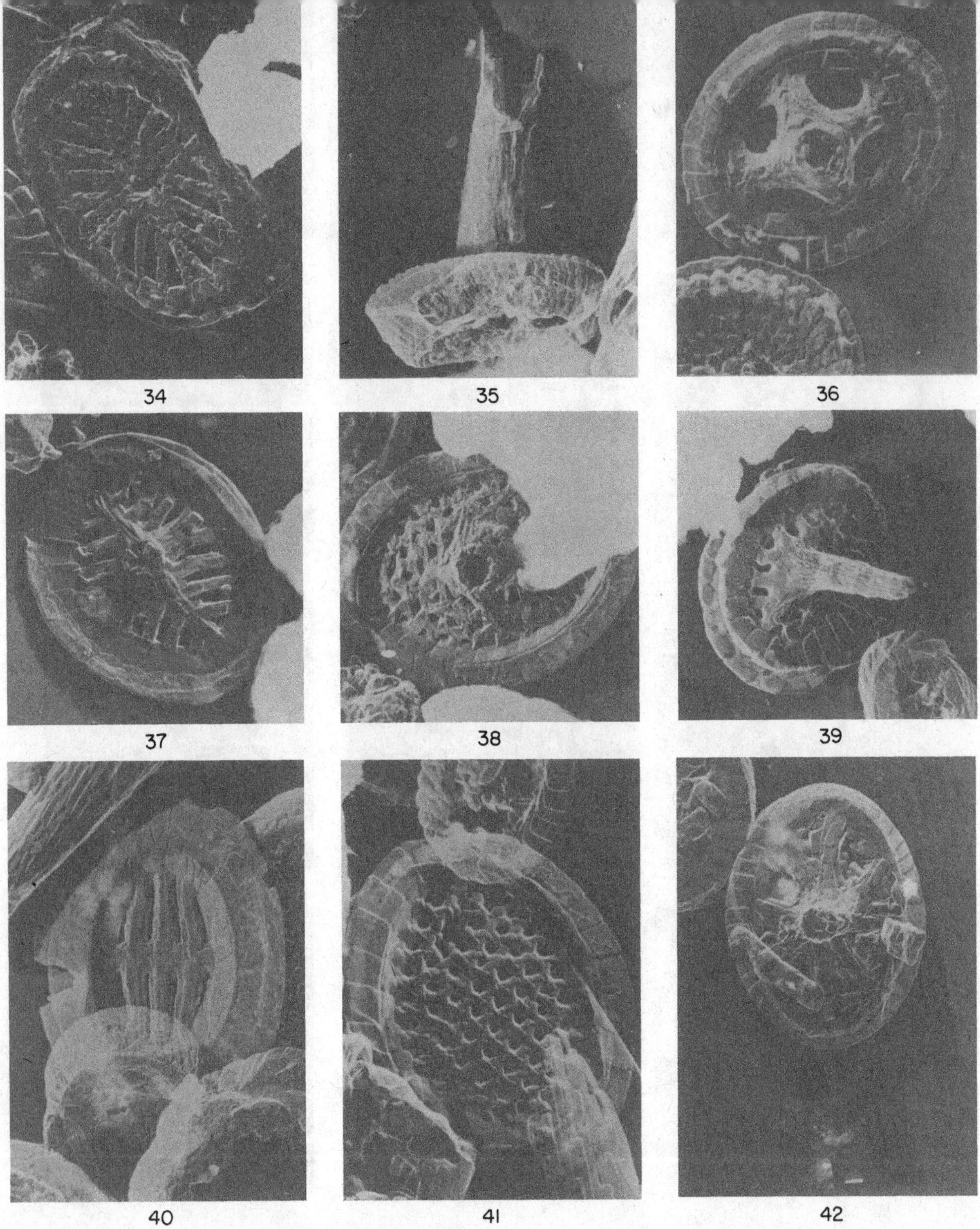
34
35
36
37
38
39
40
41
42

Plate 45.4.

Zygolithaceae (34-35), Podorhabdaceae (36-41), Ahmuellerellaceae (42)

34. *Pontilithus flabellosus* (Stradner) Black, oblique proximal view. Cenomanian: Cambridge Greensand. Barrington, Cambs. No. 14680, x 8,000.
35. *Rhagodiscus* sp. cf. *R. asper* (Stradner) Reinhardt, lateral view. Cenomanian: V22-8, Western Atlantic (429.55). No. 22629, x 6,000.
36. *Podorhabdus* sp. indet., distal view, showing broken remnant of hollow spine. Oxfordian: Upper Oxford Clay. Cambridge Experimental Borehole, Cambridge, England. No. 22782, x 4,000.
37. *Polypodorhabdus madingleyensis* Black, distal view. Oxfordian: Upper Oxford Clay. Cambridge Experimental Borehole, Cambridge, England. No. 22800, x 8,000.
38. *Ethmorhabdus gallicus* Noël, distal view. Oxfordian: Upper Oxford Clay. Cambridge Experimental Borehole, Cambridge, England. No. 22805, x 6,000.
39. *Cretarhabdus* sp. indet., oblique distal view. Cenomanian: Cambridge Greensand, Cherry Hinton Fields, near Cambridge, England. No. 21766, x 5,300.
40. *Sollasites horticus* (Stradner) Black, distal view. Cenomanian: Cambridge Greensand, Hauxton, Cambs, England. No. 22445, x 8,000.
41. *Cribrosphaera ehrenbergii* Arkh., distal view, Campanian: Upper Demopolis Formation. Near Dayton, Alabama. No. 22585, x 9,000.
42. *Ahmuellerella* sp. cf. *A. octoradiata* (Gorka) Reinhardt, distal view. Campanian: Upper Demopolis Formation. Near Dayton, Alabama. No. 22534, x 5,300.

46. ELECTRON MICROSCOPE STUDIES AND THE CLASSIFICATION OF DIATOMS

N. I. HENDEY

Summary: The classification of diatoms in the past has been largely influenced by the work of Schütt, who divided them into two groups based upon the radial or bilateral symmetry of their valves or shells. Recently, however, this system has been abandoned for a less rigid interpretation of diatom structure.

Electron microscope studies of diatoms show that four basic types of ultimate structure exist, and that any classificatory system based upon these would separate taxa that are recognized to have strong structural, ecological and phylogenetic relationships. Therefore, it is argued that data from electron microscope studies alone are unlikely to provide the basis for a new classificatory system.

INTRODUCTION

With the advent of the electron microscope and electron microscope studies of the siliceous elements of the diatom cell, it was thought by some workers that this new and powerful research tool would reveal hitherto unknown structural characters that might provide a basis for a new classificatory system for this group of plants. Twenty years or more have elapsed since those early days and several hundred species representing almost every Family of diatoms have been electron micrographed. The purpose of this communication is to examine whether those hopes have been sustained.

For a new classification of diatoms to be established and so replace any existing one, it would have to: (*a*) be more rational, (*b*) take cognizance of all diagnostic characters so that minor as well as major groups of differences would receive due weighting, (*c*) perceive and express phylogenetic relationships more closely than those expressed heretofore, and (*d*) provide a system that is more purposeful and workable, as well as one that must be seen to be so, and (*e*) be clearly based on characters that are readily discernible. In the short space permitted it will be possible to consider only the position of the higher taxonomic categories and even these but briefly.

CURRENT CLASSIFICATION

Before turning to any possible new system, we must first consider the classification system now in current use. If we exclude from our consideration classification systems used by Agardh (1830-2), de Brébisson (1838), Ehrenberg (1840), Kützing (1844) and Wm. Smith (1853) as being of historical interest only, most will agree that all modern systems are based on, or influenced by, that described by Schütt (1896) in Engler and Prantl, *Die natürlichen Pflanzenfamilien.* Schütt's classification was based entirely upon the structure of the siliceous elements, that is, upon the valves or shells of the frustule, their shape and the symmetry of their markings. Schütt's first division was a clear dichotomy based upon valves whose structure was arranged either with reference to a central point, that is, the structure was either concentric with or radial about that point (Centricae), or the structure was arranged with reference to an axial line (real or imaginary); that is, the structure was arranged symmetrically in mirror image, upon either side of the major axis (Pennatae). Such a system of classification has a good deal to commend it, particularly as it may be applied to fossil and extant forms alike, but it has serious limitations when applied to many plankton species whose structure is not easily referable to either group or at best can only be fitted in by ignoring some of their salient structural characters.

Although minor modifications to Schütt's scheme have been suggested from time to time, such as a change of rank to Centrales and Pennales, or a change of names to Radiales and Bilaterales (Heiden and Kolbe 1928) it has dominated thought until relatively recently. Such a classification is purely morphological and is little more than a recognition of two 'shape-groups' where the outline of the valves is of major significance and where the structure on the valve surface is conditioned very largely by the outline. Such a system was soon found to be inadequate and too rigid to accommodate species with irregular structures that were neither radial nor bilaterally symmetrical.

Peragallo (1897-1908, 1921) adopted in part an earlier

system proposed by H. L. Smith (1872) which embraced Schütt's groups, but recognized four more divisions on the symmetry of the whole cell rather than upon that of the individual valve or shell. He separated the Pennatae into two groups, those possessing a true raphe or slit in the apical axis and those which do not, and he separated from the Centricae those tubular species that have little that is either centric or radial about them (e.g. the Rhizosoliniaceae).

The dominance of Schütt's scheme was not seriously challenged until Mereschkowsky (1903) proposed a system of classification based on motility, and divided the diatoms into those that possess the power of movement, Mobiles, and those that do not, Immobiles. All diatoms that have the power of movement possess a raphe or true cleft in the apical axis of the valve while those without motile power are devoid of this structural character.

The Mobiles are thus included in Schütt's Pennatae while the Immobiles included the elongated or linear forms of the Pennatae that do not possess a raphe, together with those forms included in Schütt's Centricae.

Forti (1912) pointed out that the Mobiles formed auxospores sexually whereas the Immobiles formed asexual auxospores. Thus it seems that the classification that was originally based entirely on 'shape-groups', with a little modification, has received some support from physiological and 'reproduction method' aspects. This system placed together the Fragilariaceae, Coscinodiscaceae and the Biddulphiaceae, Families that bear no morphological resemblance to each other whatsoever.

Hendey (1937, 1964) sought to abandon the rigid divisions proposed by Schütt as they tended to limit the full appreciation of the structure of the diatom cell, particularly the structure of many marine plankton genera, and considered diatoms as belonging to one Class, the Bacillariophyceae, containing one Order, the Bacillariales, but adopted previously established Suborders and Families, with some additions and modifications, much in the same sense as they had been used by previous workers. Patrick and Reimer (1966) adopted Hendey's classification in the main by abandoning Schütt's Pennatae and Centricae, but they raised Hendey's Suborders to the rank of Orders.

Without pursuing the problem further the question now is, have electron microscope studies revealed any deeper or more fundamental characters that could form the basis of a more meaningful classificatory system?

EFFECTS OF ELECTRON MICROSCOPE STUDIES

It is quite obvious from the nature of the problem that the major characters upon which existing systems have been based are well within the capabilities of the modern microscope of classical design using light, and that the electron microscope could add little to the better understanding of them. Electron studies have added very considerably however, to the more precise knowledge of the ultimate structure of the diatom valve, though even here, its use has been largely confirmatory. While the structure of the diatom valve was reasonably well known in the larger or more coarsely marked species, e.g. *Triceratium* of the *favus* group (Müller 1871) and the larger species of *Coscinodiscus* and *Pinnularia*, to name but a few, finer structure of small species of diatoms was less well known and though in the latter case the structure was thought to be more simple or at least, less complicated, the precise details were largely beyond the range of resolution obtainable with the light microscope; despite this the perforate nature of many small species was assumed.

During the last twenty years or so several hundred species of diatoms have been electron micrographed and their ultimate structure revealed by Helmcke and Krieger (1951, 1953, 1954), Kolbe (1948, 1951), Hustedt (1945), Desikachary (1957), Toman and Rozsival (1948, 1950, 1953), Hendey, Cushing and Ripley (1954), Hendey (1959), Hasle (1964, 1965), Okuno (1948-63) and others (see Bibliography).

Okuno (1953) was one of the first to classify diatom structure as seen in the electron microscope and distinguished twenty-three types of cell wall. Later, Okuno (1954), he simplified the system slightly. Desikachary (1954, 1957) illustrated thirteen types of structure and suggested that these could be further divided. Hendey (1959) reviewed all the previous work in this field and described seventeen types of valve structure which corresponded very closely to the groups recognized by Okuno.

Hendey divided all structures into two main groups: (1) Cells with a laminar wall consisting of one layer of siliceous substance only, which in some instances may be strengthened by thickening or ribs. Such a wall is usually perforated by a series of holes which may or may not be partially occluded by marginal outgrowths. (2) Cells with a locular wall consisting of a double layer of siliceous substance separated by vertical walls or spacers. These vertical walls are usually arranged in a hexagon or in something approximating a hexagon, thus giving the cell wall the appearance of a single layer of honeycomb cells. This loculate structure produces a series of chambers and in some species pass-pores are sometimes present in the vertical side walls connecting one hexagonal loculus with another. The outstanding feature of this type of structure is that the outer layer of silica usually differs from the inner one. In some species the outer layer is a multi-perforate membrane, with rows of fine sieve-pores, while the inner (lower) layer has one large, more or less circular opening or cover pore. In other species the positions of these structures are reversed with the cover pore being on the outside and the sieve pore membrane being the inner layer.

Despite the fact that numerous variations and modifications of ultimate microstructure exist, it is possible to simplify all existing classifications of electron microscope

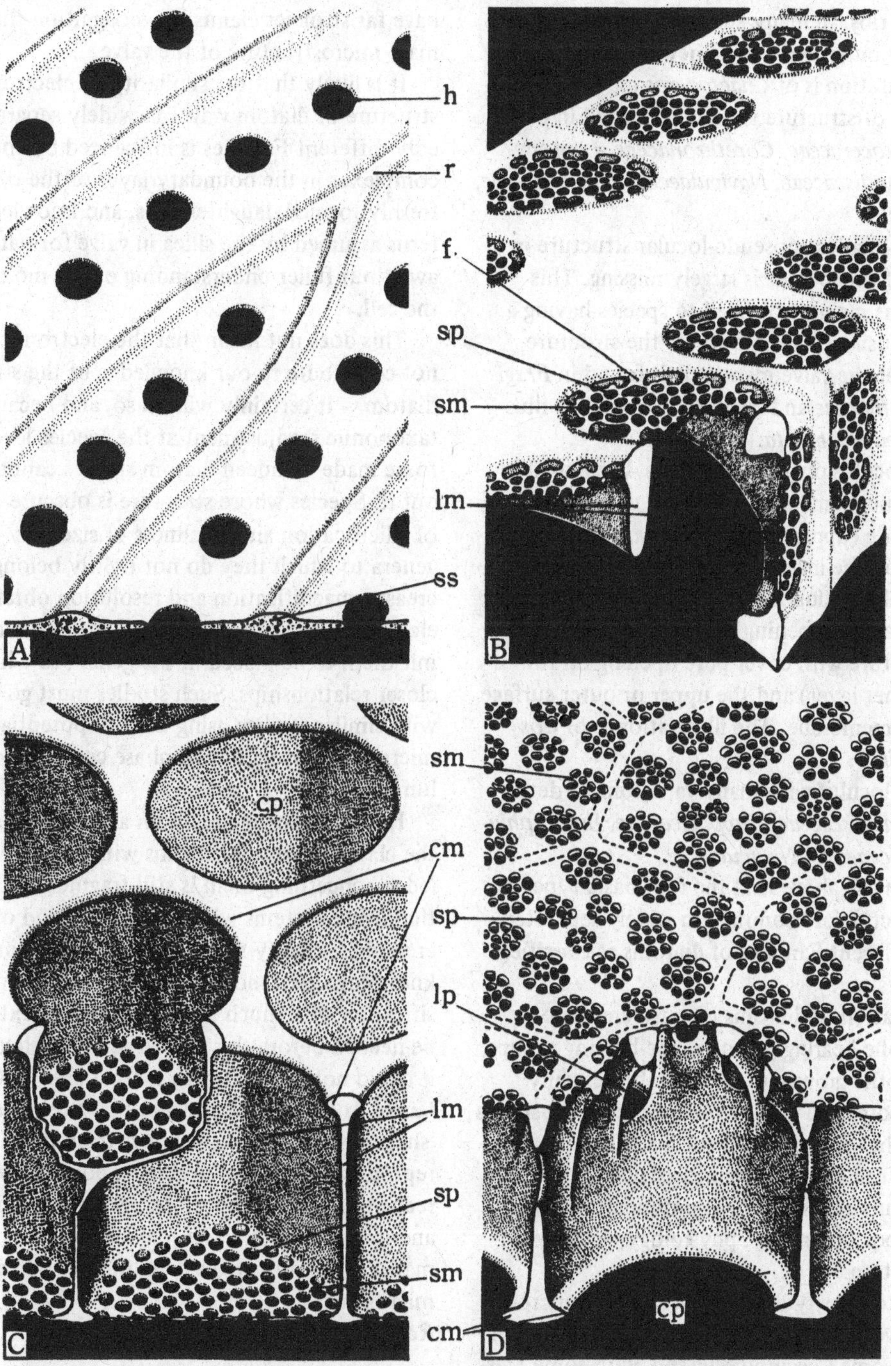
h
r
f.
sp
sm
lm
ss
A
B
cp
cm
sm
sp
lp
lm
sp
sm
cm
C
D

Fig. 46.1. Types of frustule pores. A, Hole (*Chaetoceros didymus* var. *anglica*). B, Incomplete loculus (*Synedra tabulata*). C, Complete loculus opened outwards (*Coscinodiscus lineatus*). D, Complete loculus opened inwards (*C. wailesii*). (cm, cover membrane. op. cover pore. F, fulcrum. h, hole. lm, lateral membrane. lp, lateral pore. r, rib. sm, sieve membrane. sp, sieve pore. ss, spongy porous structure, which seems common to diatom frustules.)

structures by reducing them to four basic types. These are shown in the excellent drawings kindly prepared for the author by Dr Okuno of Kyoto University, Japan, and shown in Fig. 46.1.

A. Represents the true laminar type of valve substance showing lines of circular holes of uniform size separated by thickenings or ribs. In many species the holes are partially occluded by outgrowths of somewhat complicated pattern. The illustration is of *Chaetoceros didymus* var. *anglica.* This type of structure is however found in other members of *Chaetoceraceae, Corethronaceae, Leptocylindraceae, Coscinodiscaceae, Naviculaceae, Fragilariaceae,* and *Bacillariaceae.*

B. Incomplete loculus, a pseudo-locular structure in which the floor of the loculus is largely missing. This type of structure is mainly confined to species having a narrow linear or fusiform outline where the structure extends down over the valve mantle. It is found in *Fragilariaceae, Achnanthaceae* and *Biddulphiaceae.* The illustration is of *Synedra tabulata.*

C. Complete loculus opening outwards, hexagonal or sub-hexagonal honeycomb structure with vertical walls, open on the outside (upper surface) with a large cover-pore, but closed on the inside (lower surface) by a sieve membrane. The illustration is of *Coscinodiscus lineatus.*

D. Complete loculus opening inwards. Hexagonal honeycomb structure with cover pore opening on the lower surface (inner layer) and the upper or outer surface closed by a sieve membrane. The illustration is of *Coscinodiscus wailesii.*

The complete loculus with many variations of detail is found in *Coscinodiscaceae, Eupodiscaceae, Biddulphiaceae, Actinodiscaceae* and *Naviculaceae.*

It is clear from the above that the four basic types of diatom valve structure are common in widely separated and strikingly different Families of diatoms as classified by any current system.

If the classification of diatoms is considered upon shape-groups, raphe-bearing or non-raphe-bearing groups, motile or non-motile groups, or groups with sexually produced auxospores and groups with asexual auxospores it will be found that any arrangement made by virtue of similarity in electron microscope microstructure of the valve substance cuts across all other groupings whether they are based upon symmetry, physiology, ecology, or reproduction pattern.

Any attempt to classify diatoms by electron microscope structure would lead to the ludricrous situation of placing together some *Coscinodiscus* spp. with some *Mastogloia* spp. because they both possess a loculate microstructure and some *Eucampia* spp. with some *Achnanthes* spp. because they share the same type of microstructure, while other species of *Achnanthes* would have to be grouped with some *Corethron* spp. for the same reason. Such a grouping would be entirely without reason, and would disregard all ideas of structural, physiological and ecological affinities that are well recognized in the systems of classification in use at the present time, and which have far stronger claims for recognition than does the ultimate microstructure of the valve.

It is likely that the similarity of electron microscope structure of diatom valves in widely separated and markedly different Families is influenced by the amino-acid complexes in the boundary layer of the cytoplasm of the freshly formed daughter cells, and knowledge of the patterns assumed by the silica in valve formation will have to await our fuller understanding of the molecular biology of the cell.

This does not mean that the electron microscope will not contribute to our knowledge of the structure of the diatom – it certainly will do so, and because of it much taxonomic readjustment at the species level will continue to be made – indeed diatom studies cannot progress without it. Species whose structure is obscure due to weakness of silicification and smallness of size may be placed in genera to which they do not rightly belong and the increased magnification and resolution obtainable with the electron microscope will reveal their true nature and permit them to be placed in the genera to which they have closer relationships. Such studies must go hand in hand with similar studies using the full potential of the light microscope, using direct, phase contrast and dark field illumination.

It is difficult to state with any degree of certainty how the classification of diatoms will develop while our knowledge concerning them is still fragmentary but it is likely that other systems will be proposed and one based on increasing sexuality might be put forward in the light of new knowledge, as a more rational and more meaningful classification. But much more research on extant forms will be needed before this can be accomplished and even then it could not be applied to extinct species with any certainty. It is likely, therefore, that the system based upon 'shape-groups', having support from physiological and reproduction pattern evidence, now in use, and easily recognized by means of the light microscope will continue, and it is not likely to be replaced by one based on information gained from electron microscope studies of valve microstructure.

Received September 1967.

BIBLIOGRAPHY

Ardenne, M. von 1940. Ergebnisse einen Electronenmikroskop Anlage. *Naturwissenschaften*, **28**, 113.

Agardh, C. A. 1830-2. *Conspectus Criticus Diatomacearum*. Lund.

Brébisson, A. de 1838. *Considérations sur les diatomées et essai d'une classification des genres et des espèces apartment à cette famille*. Falaise 1838.

Desikachary, T. V. 1952. Electron microscope study of diatom wall structure. *J. sci. industr. Res.* **11B**, 491.

Desikachary, T. V. 1954a. Electron microscope study of diatom wall structure III. *Amer. J. Bot.* **41**, 616.

Desikachary, T. V. 1954b. Electron microscope study of diatom wall structure. VI. *Microscopie*, **9**, 168-78.

Desikachary, T. V. 1954c. The structure of the aerolae in diatoms. *Rapp. Comm. VIIIe Congr. int. Bot., sect.* **17**, 125.

Desikachary, T. V. 1957. Electron microscope studies on diatoms. *J. R. micr. Soc.* **76**, 9-36.

Desikachary, T. V. and Aleem, A. A. 1955. Electron microscope study of diatom wall structure. VII. *J. Sci. industr. Res.* **14C**, 42.

Desikachary, T. V. and Kanwar Bahadur 1954a. Electron microscope study of diatom wall structure. II. *J. sci. industr. Res.* **13B**, 92.

Desikachary, T. V. and Kanwar Bahadur 1954b. Electron microscope study of diatom wall structure. IV. *J. sci. industr. Res.* **13B**, 240.

Desikachary, T. V. and Kanwar Bahadur 1954c. Electron microscope study of diatom wall structure. V. *Trans. Amer. micr. Soc.* **73**, 274.

Ehrenberg, C. G. 1840. *Über noch jetzt zahlreich lebende Thierarten der Kreidebildung und den Organismus der Polythalamien*. Berlin 1840, and in *Abhandl. Akad. Wiss. Berlin*. **1839**. 1841. Berlin.

Fott, B. and Rozsival, M. 1950. Frustules of *Attheya zachariasii* in electron microscope. *Studia bot. cech.* **11**, 262.

Forti, A. 1912. Contributione Diatomalogiche. *Atti del R. Instit. Veneto di Sci. Lett. ed Arti*. **71** (2), 677-731.

Gerloff, J. and Golz., E. 1944. Über den Feinbau der Kieselschalen bei einigen zentrischen Diatomeen. *Hedwigia*, **81**, 283.

Hamly, D. H. and Watson, J. H. L. 1942. Electron and optical microscope interpretation of the wall of *Pleurosigma angulatum*. *J. opt. Soc. Amer.* **32**, 433.

Hasle, G. R. 1964. *Nitzschia* and *Fragilariopsis* species studied in the light and electron microscopes. I. Some marine species of the groups Nitzschiella and Lanceolatae. *Skr. Norske Vidensk.-Acad I. Mat.-Nat. Kl. N.S.* **16**, 1-48.

Hasle, G. R. 1965. *Nitzschia* and *Fragilariopsis* species studied in the light and electron microscopes. II. The group Pseudonitzchia. *Skr. Norske Vidensk.-Akad. I. Mat.-Nat. Kl. N.S.* **18**, 1-45.

Helmcke, J. G. 1954. Die Feinstruktur de Kieselsaure und ihre physiologische Bedeutung in Diatomeenschalen. *Naturwissenschaften*, **41**, 254.

Helmcke, J. G. and Krieger, W. 1951a. Feinbau von Diatomeenschalen in Einzeldarstellung. 1. *Cocconeis placentula. Z. wiss. Mikr.* **60**, 85.

Helmcke, J. G. and Krieger, W. 1951b. Feinbau von Diatomeenschalen in Einzeldarstellung. 2. Die Gattung *Achnanthes* Bory. *Z. wiss. Mikr.* **60**, 197.

Helmcke, J. G. and Krieger, W. 1951c. Demonstration einiger raumrichtiger Rekonstruktionszeichnungen von Diatomeenschalen. *Berh. dtsch. zool. Ges.* 438.

Helmcke, J. G. and Krieger, W. 1951d. Elektronenmikroskopische Untersuchungen über den Feinbau der Diatomeenmembran. *Ber. dtsch. bot. Ges.* **64**, 29.

Helmcke, J. G. and Krieger, W. 1951e. Elektronenmikroscopische Untersuchungen über der Kammerbau der Diatomeenmembran. *Ber. dtsch. bot. Ges.* **64**, 29-30.

Helmcke, J. G. and Krieger, W. 1952a. Feinbau de Kieselschalen der Diatomeen.*Cyclotella comta* (Ehrenb.) *Kütz. Ber. dtsch. bot. Ges.* **65**, 70.

Helmcke, J. G. and Krieger, W. 1952b. Neue Erkenntnisse über dem Schalenbau von Diatomeen. *Naturwissenschaften*, **39**, 146.

Helmcke, J. G. and Krieger, W. 1952c. Feinbau von Diatomeenschalen in Einzeldarstellungen. 3. Die Gattung *Melosira Ag. Z. wiss. Mikr.* **61**,

Helmcke, J. G. and Krieger, W. 1952d. Kieselalgen in Elektronenmikroskops. *Kosmos. Stuttgart*, **48**, 405.

Helmcke, J. G. and Krieger, W. 1953-64. Diatomeenschalen im eleketronenmikroscopischen. *Bild.* I–V. Wienheim.

Helmcke, J. G. and Krieger, W. 1954. Elektronenmikroskopische Untersuchungen über Feinstrukturen der Diatomeenschalen. *Rapp. Comm. VIIIe. Congr. int. bot., Sect.* **17**, 126.

Helmcke, J. G. and Richter, H. 1951a. Photogrammetrische Ausmessung elektronenmikroskopischer Stereobilder. *Z. wiss. Mikr.* **60**, 189.

Helmcke, J. G. and Richter, H. 1951b. Elektronenmikroskopie in der Photogrammetrie. *Allg. Vermess Nach.* **6**, 142.

Hendey, N. I. 1937. The plankton diatoms of the southern seas. *'Discovery' Rep.* **16**, 151-364. Cambridge.

Hendey, N. I. 1954. Note on the Plymouth *Nitzschia* culture. *J. mar. Biol. Ass. U.K.* **33**, 335-9.

Hendey, N. I. 1959. The structure of the diatom cell wall as revealed by electron microscope. *J. Quek. Micr. Club*, **5**, 147-75.

Hendey, N. I. 1964. An introductory account of the smaller algae of British Coastal Waters. *Fishery Invest. Series IV.* Pt. **5**, *Bacillariophycea* (Diatoms). H.M.S.O.

Hendey, N. I., Cushing, D. H. and Ripley 1954. Electron microscope studies of diatoms. *J. R. micr. Soc.* **74**, 22.

Hendey, N. I., and Wiseman, J. D. H. 1953. The significance and diatom content of a deep-sea floor sample from the neighbourhood of the greatest oceanic depth. *Deep-sea Res.* **1**, 47-59.

Hustedt, F. 1945. Die Struktur der Diatomeen und die Bedetung des Elektronenmikroskops für ihre Analyse. *Arch. Hydrobiol.* **41**, 31[illegible].

Hustedt, F. 1952. Die Struktur der Diatomeen und die Bedeutung des Elektronen mikroskops für ihre Analyse. II. *Arch. Hydrobiol.* **47**, 295.

Kolbe, R. W. 1948. Elektronenmikroskopische Untersuchungen von Diatomeenmembran. I. *Ark. Bot.* **33A** (17).

Kolbe, R. W. 1951. Elektronenmikroskopische Untersuchungen von Diatomeenmembran. II. *Sevnsk bot. Tidskr.*

45, 635-47.

Kolbe, R. W. 1954. Einige Bemerkungen zu drei Aufsätzen von Fr. Hustedt. *Bot. Notiser,* **43**, 217.

Kolbe, R. W. and Golz, E. 1943. Elektronenoptische Diatomeenstudien. *Ber. dtsch. bot. Ges.* **61**, 93.

Kützing, F. T. 1844. *Die Kieselschaligen Bacillarien oder Diatomeen.* Nordhausen, 1844.

Krause, F. 1936. Elektronenoptische Aufnahmen von Diatomeen mit magnetischen Elektronenmikroskop. *Z. Phys.* **102**, 417.

Lewin, J. C. 1957. Studies with the diatom *Phaeodactylum. Phyc. Soc. America, L.M.S. Bull.* **10**, No. 32, 73-4.

Locquin, M. 1953. Structure fine des frustules de la Diatomée. *Proc. 1st Congr. Electron Microscopy,* 751.

Mahl, H. 1949. Diatomeen in Elektronenmikroskop. *Orion,* **4**, 416.

Mereschkowsky, C. 1903. *Zur Morphologie der Diatomeen Kasan.* pp. 1-427.

Muhlethaler, K. and Braun, R. 1946. Elektronenoptische Diatomeen-Untersuchungen. *Ber. schweiz. bot. Ges.* **56**, 360.

Müller, H. O. and Pasewaldt, C. W. A. 1942. Der Feinbau der Test-diatomee *Pleurosigma angulatum* S. Sm. nach Beobachtungen und stereiskopischen Aufnahmen in Übermikroskop. *Naturwissenschaften,* **30**, 55.

Müller, Otto 1871. Über den feineren Bau der Zellwand der Bacillariaceen, inbesondere des *Triceratum favus,* Ehrg. und der *Pleurosigma. Arch. Anat. Physiol* (du Bois-Reymond).

Nemoto, T. 1956. On the diatoms of the skin film of whales in the Northern Pacific. *Sci. Rep. Whales Res. Inst.* **11**, 99-132.

Nemoto, T. 1958. *Cocconeis* diatoms infected on whales in the Antarctic. *Sci. Rep. Whales Res. Inst.* **13**, 185-91.

Neilson, J. E. 1947. Electron microscope reveals a possible valve structure of *Amphipleura pellucida. Trans. Amer. micr. Soc.* **66**, 140.

Okuno, H. 1948a. Electron microscopic studies on diatom frustules. *Shimadzu Rev.* **5**, 45.

Okuno, H. 1948b. Electron microscopic studies on diatom frustules. II. *Shimadzu Rev.* **5**, 100.

Okuno, H. 1949a. Electron microscopical study on fine structures of diatom frustules. VI. *Bot. Mag. Tokyo,* **62**, 97.

Okuno, H. 1949b. Electron microscopical study on fine structures of diatom frustules. VII. *Bot. Mag., Tokyo,* **62**, 136.

Okuno, H. 1950a. On the electron microscopical fine structure of *Pinnularia* cell wall. *Bot. Mag., Tokyo,* **63**, 34.

Okuno, H. 1950b. Electron microscopical study on fine structures of diatom frustules. VIII. *Bot. Mag., Tokyo,* **63**, 97.

Okuno, H. 1950c. On the electron microscopical fine structures of fossil *Coscinodiscus oculus-iridis* Ehrenberg. *Bot. Mag., Tokyo,* **63**, 232.

Okuno, H. 1951. Electron microscopical study of antarctic diatoms. I. *J. Jap. Bot.* **26**, 305.

Okuno, H. 1952a. Electron microscopical study of antarctic diatoms. II. *J. Jap. Bot.* **27**, 46.

Okuno, H. 1952b. Electron microscopical study of antarctic diatoms. III. *J. Jap. Bot.,* **27**, 347.

Okuno, H. 1952c. Electron microscopical study on fine structures of diatom frustules. IX. *Bot. Mag., Tokyo,* **65**, 158.

Okuno, H. 1952d. *Atlas of fossil diatoms from Japanese diatomite deposits.* Tokyo.

Okuno, H. 1953a. Electron microscopical study of antarctic diatoms. IV. *J. Jap. Bot.* **28**, 171.

Okuno, H. 1953b. Electron microscopical study on fine structures of diatom frustules. X. *Bot. Mag., Tokyo,* **66**, 5.

Okuno, H. 1953c. Electron microscopical study on fine structures of diatom frustules. XI. *Bot. Mag., Tokyo,* **66**, 121-4.

Okuno, H. 1953d. Study on super-fine structure of fossil diatoms. *Proc. 1st int. Congr. Electron Microscopy,* 758.

Okuno, H. 1954a. Electron microscopical study of antarctic diatoms. V. *J. Jap. Bot.* **29**, 18.

Okuno, H. 1954b. Electron microscopical study on fine structures of diatom frustules. XII. *Bot. Mag., Tokyo,* **67**, 172.

Okuno, H. 1954c. Electron microscopic fine structure of some marine diatoms. *Rev. Cytol, Paris,* **15**, 237-44.

Okuno, H. 1954d. Electron microscopic fine structure of fossil diatoms. I. *Trans. Proc. palaeont. Soc. Japan, N.S.* **13**, 125.

Okuno, H. 1954e. Electron microscopic fine structure of fossil diatoms. I. *Trans. Proc. palaeont. Soc. Japan, N.S. N.S.* **14**, 143.

Okuno, H. 1954f. Four marine diatoms under electron microscope. *Rapp. Comm. VIIIe Congr. int. Bot., Sect.* **17**, 124.

Okuno, H. 1955a. Electron microscopical study on fine structures of diatom frustules. XII. *Bot. Mag., Tokyo,* **68**, 125.

Okuno, H. 1955b. Fine structure of diatom frustules. *Sobi.,* **21**, 1.

Okuno, H. 1955c. Electron microscopic fine structure of fossil diatoms. III. *Trans. Proc. palaeont. Soc. Japan, N.S.,* **19**, 53-8.

Okuno, H. 1956. Electron microscopical study on fine structures of diatom frustules. XIV. *Bot. Mag., Tokyo,* **69**, 814, 193-8.

Okuno, H. 1956. Electron microscopic fine structures of fossil diatoms. IV. *Trans. Proc. palaeont. Soc. Japan, N.S.* **21**, 133-9.

Okuno, H. 1957a. Electron microscopical study on fine structures of diatom frustules. XV. *Ibid.* **70**, no. 826.

Okuno, H. 1957b. Electron microscopical study on fine structures of diatom frustules. XVI. *Ibid.* **70**, no. 829-30.

Patrick, R. and Reimer, C. W. 1966. The diatoms of the United States. *Acad. Nat. Sci. Philad.* **13**, 1-688.

Peragallo, H. and Peragallo, M. 1897-1908. *Diatomées marines de France, et des Districts maritimes voisins.* Grez-sur-Loing.

Peragallo, M. 1921. Diatomées d'eau douce et diatomées d'eau salée. *Deuxième Exped. Antarct. Française (1908-1910). Botanique,* pp. 1-98.

Reiman, B. E. F., Lewin, J. C. and Volcani, B. E. 1965. Studies on the biochemistry of silica shell formation in diatoms. *J. Cell Biol.* **24**, 39-55.

Schutt, F. 1896. In: Engler and Prantl, *Die Natürlichen*

Pflanzenfamilien, 1, 1b, Bacillariales, 31-150. Leipzig.

Smith, Wm. 1853-56. *A synopsis of the British Diatomaceae.* Vol. 1, 1853. Vol. 2, 1856. London.

Toman, M. 1948. *Cyclotella Pratti* n.sp. *Studia bot. cech.* **9**, 49.

Toman, M. 1950. *Navicula seminuloides* Hust. u nas. *Cesk. bot. Listy,* **2**, 128.

Toman and Rozsival, M. 1948. The structure of the raphe of *Nitzschia. Studia bot. cech.* **9**, 26.

Toman and Rozsival, M. 1950. The structure of the cell-wall in *Melosira varians* Ag. *Studia bot. cech.* **11**, 65.

Toman and Rozsival, M. 1953. Pvspevky k poznani bunecne blany ras. *Preslia,* **25**, 43.

Werzner, V. N. 1944. Some structure details of diatom *Pleurosigma elongatum* as revealed with the aid of electron microscope. *C. R. Acad. Sci. U.R.S.S.* **44**, 118-20.

47. PROGRESS AND PROBLEMS OF FORAMINIFERAL SYSTEMATICS

Z. REISS

Summary: A brief review of some recent findings on the life-cycle, cytology, ecology, and on the structure of the test, tends to show that knowledge on Foraminifera has greatly advanced, although many problems still await solution, and new and complex ones have arisen. Further detailed study by conventional and new methods is required. All observable features, not single or selected ones, must be used in classification, which remains for the time being largely empirical. New nomenclatural procedures are necessary to express properly the natural relationships recognized.

During the last few decades the usefulness of Foraminifera has been increasingly recognized in stratigraphy and structural geology; in the interpretation of palaeoenvironments; in the reconstruction of tectonic events and palaeogeographical patterns; as well as in the determination of temperature changes and current-paths in ancient seas. Extension of foraminiferal research to all fossiliferous strata on all continents and on many rises below the sea, as well as to sediments and water bodies of Recent oceans has resulted in the accumulation of considerable documentation (Loeblich and Tappan 1964*a*).

More than ever the necessity for accurate taxonomy is apparent in every phase of this field of research, but it is precisely here that some of the greatest difficulties arise. This has happened despite – and, one may add, partly because of – the considerable progress that has been made in the study of Foraminifera during the last few years.

As in most other branches of science, modern research on Foraminifera has clarified many baffling problems, but has drawn attention to many unsolved ones, old and new, and has emphasized the great difficulties encountered by any attempt to systematize natural phenomena in an all-embracing and all-applicable manner. With every addition of data on living and fossil Foraminifera the lack of sufficient knowledge, the inadequacy of present classifications, and the difficulties in reconciling much seemingly conflicting evidence become more and more apparent. Many so-called 'laws' seem invalid and many 'facts' are far from what they were believed to be only a short time ago.

Use of characters of known biological significance only in classification does not suffice at present to build a satisfactory or natural system, while various structural features of the test, the function of which is unknown and the formation of which can at best be guessed at, are apparently taxonomically important. Thus, the dilemma of choosing between proposed hierarchies of criteria in classification assumes both theoretical and practical importance (Drooger 1956, 1963; Glaessner and Wade 1959).

A few examples may illustrate the problems mentioned.

The 'genus' *Orbulina* is now known to represent the normal end-stage of different species of *Globigerina* and *Globigerinoides*, while *Sphaeroidinella* has been shown to be nothing but a depth-conditioned, ecophenotypic variant of adult *Globigerinoides* (Adshead 1966*a*, Bandy 1966, Bé 1965). *Eponides, Poroeponides,* and *Sestranophora* were shown to be ontogenetic stages in the development of a single species (Resig 1962). Various 'planorbulinid', 'acervulinid', uncoiling, and annular forms were shown to be nothing but phases in the life-cycle of *Cibicides* (Nyholm 1961). Analysis of Recent populations of *Brizalina* indicates that width/length ratio, as well as strength and distribution of ornamentation are strongly dependent upon environmental factors (probably oxygen content) in relation to amount of calcium carbonate secreted and to the rate of growth (Lutze 1964). Cytological studies on *Globigerina* and *Globigerinoides* have shown that these genera possess heterokaryotic schizonts similar to such genera as *Spirillina, Rotaliella,* and particularly *Glabratella* (Lee *et al.* 1965). Axopodia were shown to be present in planktonic Foraminifera in addition to rhizopodia (Adshead 1966*b*).

There is no simple explanation which would relate the indisputable facts concerning living *Orbulina* to the equally indisputable fact of the *first* derivation of *Orbulina* from *Globigerinoides* through a well-documented bioseries during a time-interval to be measured probably in millions of years.[1] The relatively considerable time-lag between the

[1] The often repeated assumption that the first sequence leading to *Orbulina* was initiated at different times in different places (Bandy 1966) is due to stratigraphical confusion and is unwarranted.

first *Globigerinoides* and the first *Sphaeroidinella* is also not easily explained in the light of findings on their Recent representatives.

Polymorphism in the life-cycle of *Cibicides* is certainly much more pronounced than suspected hitherto, but must be carefully considered with reference to the stratigraphical distribution of the various forms and to the assumed adaptive significance of the tendency observed in the phylomorphogeny of many groups to develop cyclical or annular tests during geological time (Drooger 1956, 1963; Reiss 1963*b*; Smout 1954).

The ontogeny of *Eponides* may recapitulate phylogeny, as the stratigraphical distribution of its various forms suggests, but the *'Sestranophora'*-form seems to be geographically restricted. The curves obtained from a statistical evaluation of the phenotypic variation of *Brizalina* populations from basin- to slope-biotopes strongly resemble those obtained from analysis of successive Upper Cretaceous populations of *Bolivinoides.* It is almost impossible at the present time to provide a well-founded explanation of the similarity between Recent infrapopulational, ecophenotypic variation on the one hand, and fossil morphogenetic lineages interpreted as a result of orthogenesis through orthoselection (Bettenstaedt 1962) on the other. The suggestion that the fossil lineages are in fact simulated, and represent successions of phenotypic variation caused by long-range progressive changes in environmental conditions (Lutze 1964) must be examined in the light of the fact that similar morphogenetic lineages were traced for *Bolivinoides* in many parts of the world and in various facies.

The interesting similarity in the cytology of schizonts in planktonic Foraminifera with that of other groups hardly indicates any close relationship when all other features are considered. The presence in planktonic Foraminifera of axopodia in addition to rhizopodia emphasizes the individuality of this group, but lessens considerably the value of pseudopodia in classifying protozoans (Adshead 1966*b*).

There is hardly any doubt about the taxonomic importance of wall structure of the test, the latter having to be regarded as a metabolic product of the living organism, determined by the stereochemistry of organic matrices (Degens and Schmidt 1966; Towe and Cifelli 1967; Wilbur 1960). The chemistry, mineralogy, and microstructure of foraminiferal tests has become known in greater detail only during the last few years, particularly by the use of new methods and techniques, like X-ray diffraction, electronmicroprobe analysis, and ultramicroscopy.

It has been shown that magnesium (like strontium) occurs in combined form in the secreted calcareous test and is quantitatively independent of macro-environmental factors (Blackmon and Todd 1959; Lipps and Ribbe 1967) the content being apparently genetically controlled (Blackmon and Todd 1959; Reiss 1963*b*, in press). The genetic stability of aragonite in foraminiferal tests has been repeatedly emphasized and considerable taxonomic value is attributed to it (Bandy 1954; Loeblich and Tappan 1964, 1964*b*, 1964*c*; McGowran 1966; Reiss 1963*b*, in press). Advances have been made lately in the analysis of the ultrastructure of secreted tests and important results obtained (Bé 1965; Bé and Ericson 1963; Bé, McIntyre and Breger 1966; Hay, Towe and Wright 1963, Towe and Cifelli 1967). The discovery that the 'radial' and 'granular' structures of so-called hyaline Foraminifera as observed under the polarizing microscope are quite similar from a crystallographic point of view has reconciled conflicting statements with regard to the test microstructure of such genera as *Ammonia, Elphidium, Nonium, Globigerina,* and *Globorotalia* (Bolli, Loeblich and Tappan 1957; Cifelli 1962; Hay, Towe and Wright 1963; Loeblich and Tappan 1964*b*, 1964*c*; Pessagno 1964; Reiss 1958, 1963*b*; Towe and Cifelli 1967; Wood, 1949; Wood, Haynes and Adams 1963). The clarification of the ultrastructure of so-called 'porcelaneous' Foraminifera has solved various observational problems and has emphasized the individuality of the porcelaneous group.

However, many problems of an analytical nature arise and much more information is needed before generalizations can be made or evolutionary patterns postulated. Numerous difficulties arise in connection with fossil and particularly extinct Foraminifera. The taxonomic significance of magnesium content is inferred by correlation with other features in Recent Foraminifera, and application to fossil tests is still unexplored. Original magnesium content may be of interest from a palaeobiochemical point of view (Degens and Schmidt 1966), as well as in connection with diagenesis of the tests and with dolomitisation of sediments.

Although the diagenetic stability of aragonite in foraminiferal tests is at present well-known, replacement of aragonite by calcite does occur in highly calcareous, particularly hard, rocks and at least some extinct (now calcitic) genera are still suspected to have been aragonitic. The clarification of the true meaning of 'radial' and 'granular' structure has lessened the classificatory significance of this feature to a considerable degree and, e.g. classification of planktonic foraminifera on the basis of size of individual 'crystals' (in fact arrays of mineral units) as recently proposed (Lipps 1964; Lipps and Ribbe 1967), must also be viewed in the light of ultrastructural studies (Bé, McIntyre and Breger 1966). There are differences in ultrastructure between various members of the 'porcelaneous' group and further study is necessary to assess correctly the significance of these differences. Nearly nothing is known on the details of composition and structure of so-called 'spicular', 'siliceous' or 'single crystal' tests; the ultra-structure of aragonitic tests is unknown to date; while many so-called 'microgranular' Foraminifera are repeatedly shifted from group to group, ranging from agglutinated to porcelaneous, on the basis of similarity in appearance of the diagenetically changed test.

Of particular interest, especially if considered under the aspects of macromolecular chemistry, are layering and lamination of foraminiferal tests, and great taxonomic value is currently attributed to these features which seem to be genetically stable. The more detailed study of these features has shown that they are intimately connected with various objective features used for a long time in classifying Foraminifera. Thus, various projections and corrugations called double septa, septula, secondary septa, subepidermal partitions, alveoli, etc., are directly related to layering, while inflational ornamentation of various types, formation of grooves (marginal cord), gradual constriction and deflection of pores, lateral chamberlets, umbilical and intramural cavities, etc., are consequential to lamination irrespective of the biological causes underlying their formation (Reiss 1963*b*; Smout 1954; Towe and Cifelli 1967). Lamellar chamber formation also indicates that the so-called 'non-septate' or 'biloculinid' tests are not produced by continuous growth, but as successive instars, and thus there is less fundamental difference between non-septate and septate Foraminifera than it would appear from the literature.

Many problems concerning lamination and layering still remain to be investigated, many of them being complicated by diagenetic features. This the question whether or not each lamella covers the whole or only part of the previously formed test in certain genera may arise in part from observations on diagenetically changed tests; statements with regard to double walls present in various extinct genera, may be based in fact on the observable differences between primary lamella and laminated secondary skeleton. The diagenetic destruction of layers and difficulties in observing particularly thin layers are the cause of unwarranted dispute concerning the 'mono'- or 'bilamellar' nature of certain genera (Hofker 1962, 1964; Lipps 1966; McGowran 1966; Reiss 1958, 1960, 1963*a*, 1963*b*). It is still not clear why the assumedly postontogenetic 'calcite crust' present in many planktonic Foraminifera follows a thickness pattern identical with that produced by lamination. The relationship between lamination and true pores, as well as between the latter and interlamellar sheets of ectoplasm which interrupt the mineral phase between instars, are not yet clarified. The presence of a sheet of organic material between the primary layers in 'bilamellar' Foraminifera distinguishes them from other multilayered (and non-lamellar) groups in which layering is expressed by differences in arrangement or size of the grains (secreted and agglutinated) in the individual layers, no evidence of separation by organic material being apparent between the layers in this latter group. Attention is also drawn to the ultramicroscopic layering on both hyaline and porcelaneous Foraminifera (Towe and Cifelli 1967) which must be investigated in its relationship to processes of calcification (Arnold 1964; Reiss 1963*b*) and postulated models for them. The formation of thin and long spines in some planktonic Foraminifera as a result of the calcification of cores of axopodia (Adshead 1966*b*, and cf. Lipps 1966; Parker 1962; Reiss 1963*b*; Smout 1954) may be noteworthy in this connection.

Features which are apparently completely independent of ecological factors and seem to be genetically stable, but on the formation and function of which nothing is known, are various plate-like, folded structures referred to as 'toothplates', 'median septa', 'siphons', 'buttresses', 'septal flaps', etc. These structures are extremely valuable in distinguishing otherwise similar forms, they exhibit evolutionary patterns, and they are closely connected with chamber-forming apertures and foramina. Many taxonomically important features, like canals and passages, various 'stellar chamberlets', 'secondary chamberlets' of certain genera, 'retral processes', etc., are formed by these plates.

Nevertheless, the homology of toothplates and septal flaps is much disputed and attempts to classify Foraminifera on the basis of these as single characters proved failures. It is noteworthy that both toothplate and septal flap of the bilamellar, aragonitic Robertinacea have been shown recently to be formed by the inner lining (McGowran 1966) which again draws attention to the significance of primary layering.

The foregoing examples may suffice to show that wider application of conventional techniques and the use of new ones in the study of the detailed structure of the foraminiferal test; the application of statistical methods in population analysis; observations on the cytology, life-cycle, and ecology of living Foraminifera, all have lately greatly contributed to our knowledge of this group. In the light of modern research many older ideas and earlier observations, which came nearly to be forgotten in the course of time, have assumed new significance and much apparently conflicting evidence could be reconciled. The true interrelationship between taxonomy and ecology — as discussed in Glaessner's penetrating analysis twelve years ago (Glaessner 1955) — has also become more apparent. Modern research has proven that in living Foraminifera many characters of the test on which classification has been traditionally based are merely ecophenotypic modifications or stages in ontogeny, while various 'genera' are merely phases in the life-cycle of individual species. Such long-accepted laws as that of 'recapitulation', which served among others to interpret relationships, have been shown to be invalid, juvenile similarities being apparently the result of similar selective pressures on the young stages (Lipps 1966; Mayr 1963; Wade 1964). In the search for a more natural classification greater taxonomic value is currently attributed to chemistry, mineralogy, micro- and ultra-structure of the test as well as to various so-called internal characters. As a result of recent findings a wider species concept is being advocated. Emphasis varies in the selection of characters of proven or assumed taxonomic significance for classification, although the importance of some of these characters both in classification and in the interpretation of evolutionary patterns seems generally accepted.

On the other hand, recent studies indicate that certain features regarded as of great taxonomic importance are far less clearly defined, and of lesser classificatory significance, than it was generally believed hitherto, while other characters – the formation, function, and biological significance of which are admittedly unknown or can only be guessed at – have proven to be highly valuable in distinguishing clearly and objectively between similar forms and also appear to be genetically stable. It seems furthermore that findings on the test of living Foraminifera and its relationship to ecology are not easily reconciled with evidence provided by numerous investigations on large assemblages of fossil Foraminifera from closely sampled stratigraphical sequences in various facies, while present knowledge on the cytology of Foraminifera is far from warranting any general conclusions, and supports present classifications to a lesser degree than it would appear from certain recent statements (Loeblich and Tappan 1964*a*, 1964*b*).

It becomes more and more clear that taxa cannot be defined in terms of a rigid statement of criteria, placed in a certain hierarchical order (Drooger 1956, 1963; Reiss 1963*b*) and there is little doubt that no single character and no combination of a few *selected* features should be used in classifying Foraminifera at the present stage.

It is necessary to use *all* observable characters and features with particular reference to their correlation. The whole range of character combinations must be analyzed against the background of geographical and stratigraphical distribution, as well as that of knowledge on biology and ecology of living species (Glaessner and Wade 1959; Gordon 1966; Reiss 1963*b*). Natural classifications 'associate forms that resemble one another in a maximum of features' (Michener 1963) and the greater the number of features used in empirical classification the more closely related and 'natural' the resulting groups will be. The use of all observable characters makes it also possible to separate taxa by means of objective character-combinations and character-discontinuities. The artificiality of many 'traditional' taxa is due to their being based on certain selected characters with disregard of others which are not less important.

Many features of the foraminiferal test seem to be iterative and even those features known or regarded at present to be primitive may have arisen repeatedly in various stocks at different times. Poly-phyletism, convergence, iso- and homoeomorphism, archallaxis, deviation, and anaboly, geographical polytypy, as well as lack of sufficient knowledge on various groups, and confusing stratigraphical correlation and nomenclature introduce additional difficulties in tracing evolutionary patterns.

Little is known on living Foraminifera, and much has yet to be learned on fossil ones; although the 'present is the key to the past' there are – as pointed out by Glaessner (1955) – 'hidden in the past record further keys which may be quite different from those supplied by the present fauna'. Much more analytical (and experimental) work is still to be done, and a more integrated approach to the foraminiferal test in terms of modern biology is needed.

Ways and means must be found to reconcile natural classification with the necessity of typological fixation (which by no means implies a typological or legalistic approach to systematics) and to express natural relationships by an appropriate nomenclature. There is little doubt that new approaches in zoological nomenclature will have to be finally adopted for Foraminifera (and certainly not only for them) and such recent proposals as that of Sigal (1966) deserve full consideration.

It is also probable that with the recognition of true relationships between foraminiferal taxa and with the adoption of new nomenclatural procedures for them, systematics of Foraminifera will lose in years to come much of the 'familiar' aspect which even the most up-to-date classifications still preserve.

Received September 1967.

BIBLIOGRAPHY

Adshead, P. C. 1966a. Observations on living Foraminifera in cultures. *Abstr., Geol. Soc. Am., Annual Meeting,* (Pacific Section). pp. 603-4.

Adshead, P. C. 1966b. Taxonomic significance of pseudopodial development in living planktonic Foraminifera. *Abstr., Geol. Soc. Am. Annual Meeting* (Pacific Session). pp. 642-3.

Arnold, Zach M. 1964. Biological observations on the foraminifer *Spiroloculina Hyalina* Schulze. *Univ. Calif. Publ. Zool.* 72.

Bandy, O. L. 1954. Aragonite tests among the Foraminifera. *J. Sed. Petrology*, **24** (1), 60-1.

Bandy, O. L. 1966. Restrictions of the *'Orbulina'* datum. *Micropaleontology*, 12 (1), 79-86.

Bé, A. W. H. 1965. The influence of depth on shell growth in *Globigerinoides sacculifer* (Brady). *Micropaleontology*, **11** (1), 81-97.

Bé, A. W. H. and Ericson, D. B. 1963. Aspects of calcification in planktonic Foraminifera (Sarcodina). *Ann. New York Acad. Sci.* **109** (1), 65-81.

Bé, A. W. H., McIntyre, A. and Breger, D. L. 1966. Shell microstructure of a planktonic foraminifer, *Globorotalia menardii* (d'Orbigny). *Eclogae geol. Helv.*, **59** (2), 885-96.

Bettenstaedt, F. 1962. Evolutionsvorgaenge bei fossilen foraminiferen. *Mitt. Geol. Staatsinstit. Hamburg*, **31**, 385-460.

Blackmon, P. D. and Todd, R. 1959. Mineralogy of some Foraminifera as related to their classification and ecology. *J. Paleontology*, **33** (1), 1-15.

Bolli, H. M., Loeblich, A. R., Jr. and Tappan, H. 1957. Planktonic foraminiferal families Hantkeninidae, Orbulinidae, Globorotaliidae and Globotruncanidae. *U.S. Nat. Mus. Bull.* **215**, 3-50.

Cifelli, R. 1962. The morphology and structure of *Ammonia beccarii* (Linné). *Contrib. Cushman Found. Foram. Res.* **13** (4), 119-26.

Degens, E. T. and Schmidt, H. 1966. Die Paläobiochimie, ein neues Arbeitsgebiet der Evolutionsforschung. *Palaeont. Z.* **40** (3-4), 218-29.

Drooger, C. W. 1956. Transatlantic correlation of the Oligo - Miocene by means of Foraminifera. *Micropaleontology*, 2 (2), 183-92.

Drooger, C. W. 1963. Evolutionary trends in the Miogypsinidae. In *Evolutionary Trends in Foraminifera.* Elsevier: The Netherlands. pp. 315-49.

Glaessner, M. F. 1955. Taxonomic, stratigraphic and ecologic studies of Foraminifera, and their interrelations. *Micropaleontology*, **1** (1), 3-8.

Glaessner, M. F. and Wade, M. 1959. Revision of the foraminiferal family Victoriellidae. *Micropaleontology*, 5 (2), 193-212.

Gordon, W. A. 1966. Variation and its significance in classification of some English Middle and Upper Jurassic nodosariid Foraminifera. *Micropaleontology*, **12** (3), 325-33.

Hay, W. W., Towe, K. M. and Wright, R. C. 1963. Ultramicrostructure of some selected foraminiferal tests. *Micropaleontology*, **9** (2), 171-95.

Hofker, J. 1962. Studien an planktonischem Foraminiferen. *Neue Geol. Jahrb. Pal. Abh.* **114** (1), 81-134.

Hofker, J. 1964. Wall structure of Globotruncanidae, Globorotalia and Gavelinella. *Micropaleontology*, **10** (4), 453-6.

Lee, J. J., Freudenthal, H. D., Kossoy, V. and Bé, A. 1965. Cytological Observations on two planktonic Foraminifera, *Globigerina bulloides* d'Orbigny, 1826, and *Globigerinoides ruber* (d'Orbigny, 1839) Cushman, 1927. *J. Protozool.* **12** (4), 531-2.

Lipps, J. H. 1964. Miocene planktonic Foraminifera from Newport Bay, California. *Tulane Studies in Geology*, 2 (4), 109-33.

Lipps, J. H. 1966. Wall structure, systematics, and phylogeny of Cenozoic planktonic Foraminifera. *J. Paleontology*, **40** (6), 1,257-74.

Lipps, J. E. and Ribbe, P. H. 1967. Electron-probe microanalysis of planktonic Foraminifera. *J. Paleontology*, **41** (2), 492-6.

Loeblich, A. R., Jr. and Tappan, H. 1964a. Foraminiferal facts, fallacies, and frontiers. *Geol. Soc. Am. Bull.* **75**, 367-92.

Loeblich, A. R., Jr. and Tappan, H. 1964b. Sarcodina chiefly 'Thecamoebians' and Foraminiferida. In: *Treatise on Invertebrate Palaeontology* (R. C. Moore, *ed.*), Part C, Protista 2. Geol. Soc. Amer. and Univ. Kansas Press.

Loeblich, A. R., Jr. and Tappan, H. 1964c. Foraminiferal classification and evolution. *Journ. Geol. Soc. India*, **5**, 5-40.

Lutze, G. F. 1964. Statistical investigations on the variability of *Bolivina argentea* Cushman. *Contrib. Cushman Found. Foram. Res.* **15** (3), 105-16.

Mayr, E. 1963. *Animal species and evolution.* Belknap Press of Harvard Univ.

Mayr, E., Linsley, E. G. and Usinger, R. L. 1953. *Methods and principles of systematic zoology.* McGraw Hill.

McGowran, B. 1966. Bilamellar walls and septal flaps in the Robertinacea. *Micropaleontology*, **12** (4), 477-88.

Michener, C. D. 1963. Some future development in taxonomy. *System Zool.* **12** (4), 151-72.

Nyholm, K.-G. 1961. Morphogenesis and biology of *Cibicides lobatulus. Zoologiska Bidrag fran Uppsala.* **33**, 157-96.

Parker, F. L. 1962. Planktonic foraminiferal species from Pacific sediments. *Micropaleontology*, 8 (2), 219-54.

Pessagno, E. A. 1964. Form analysis of sectioned specimens of *Globorotalia* s.s. *Micropaleontology*, **10** (2), 217-30.

Reiss, Z. 1958. Classification of lamellar Foraminifera. *Micropaleontology*, **4** (1), 51-70.

Reiss, Z. 1960. Structure of so-called Eponides and some other rotaliiform Foraminifera. *Geol. Survey of Israel Bull.* **29**.

Reiss, Z. 1963a. Note sur la structure des foraminifères planctoniques. *Rev. de Micropaleont.* **6** (3), 127-9.

Reiss, Z. 1963b. Reclassification of perforate Foraminifera. *Geol. Survey of Israel Bull.* **35**.

Reiss, Z., in press. 'Biomineralogy' – presidential address for 1966. *Israel Geol. Soc.*

Resig, J. M. 1962. The morphological development of *Eponides repandus* (Fichtel and Moll), 1798. *Contrib. Cushman Found. Foram. Res.* **13** (2), 55-7.

Sigal, J. 1966. The concept taxinomique du spectre. *Soc. géol. France, Mém. hors-série*, **3**.

Smout, A. H. 1954. Lower Tertiary Foraminifera of the Qatar Peninsula. *Brit. Mus. (Nat. Hist.) Mem.*

Tappan, H. N. and Lipps, J. H. 1966. Wall structures, classification and evolution in planktonic Foraminifera *Abstr., Geol. Soc. Am. Annual Meeting*, p. 637.

Towe, K. M. and Cifelli, R. 1967. Wall ultrastructure in

the calcareous Foraminifera: crystallographic aspects and a model for calcification. *J. Paleontology*, **41** (3), 742-63.

Wade, M. 1964. Application of the lineage concept to the biostratigraphic zoning based on planktonic Foraminifera. *Micropaleontology*, **10** (3), 273-90.

Wilbur, K. M. 1960. Shell structure and mineralization in molluscs. In: Calcification in Biological Systems (*ed.* A. F. Sognnaes). *Amer. Ass. Adv. Sci.* pp. 15-40.

Wood, A. 1949. The structure of the wall of the test in the Foraminifera; its value in classification. *Quart. J. Geol. Soc. London*, **104**, 229-55.

Wood, A., Haynes, J. and Adams, T. D. 1963. The structure of *Ammonia beccarii* (Linné). *Contrib. Cushman Found. Foram. Res.* **14** (4), 157-7.

48. THE DIRECTION OF COILING IN PLANKTONIC FORAMINIFERA

H. M. BOLLI

Summary: Four different trends of preferential coiling directions are distinguished in trochospiral planktonic Foraminifera. Some may be applied to stratigraphical correlation and/or habitat interpretation. Similar coiling trends are repeated in the Cretaceous, the Palaeocene to Eocene, and the Oligocene to Recent. Each interval is characterized by an early period of random coiling that is followed by a period in which many genera and species develop a distinct preference for either dextral or sinistral coiling.

INTRODUCTION

This paper is a review of coiling trends in planktonic Foraminifera and is based largely on data already published by the author. The first observations on the direction of coiling in planktonic Foraminifera were made in the Upper Cretaceous and Palaeocene, and date back to Tschachtli (1941), Gandolfi (1942) and Cita (1948). Bolli (1950, 1951) dealt with the subject in two papers. He showed that for some genera and species certain trends exist in the preferential direction of coiling. These trends are apparently governed by either evolutionary rules and/or ecological conditions. As a consequence, numerous papers appeared, dealing with the coiling of planktonic Foraminifera in the Cretaceous, in the Tertiary and particularly in the Quaternary.

The conclusions to be drawn from the coiling trends recognized, and the rules deduced from them, include the determination of evolutionary stages within genera and species, as well as phylogenetic relationships between them. It is possible to arrive at certain stratigraphical correlations, of world-wide application in the Cretaceous, Palaeocene and Eocene. A more restricted regional, or local, scale can be applied to the Oligocene to Recent. Preferred coiling directions, or sudden changes from one preferred direction to the opposite, in particular in younger Tertiary to Recent species, appear to be valuable aids in the determination of the number and duration of climatic changes, caused in the youngest geological history by glacial and interglacial periods.

COILING TRENDS AND RULES

Not all trochospiral planktonic Foraminifera attain a preference for one or the other direction of coiling during their stratigraphical range. Such preference does not develop significantly in most primitive species belonging to the genus *Globigerina.* Most species of higher evolved genera, such as *Rotalipora, Globotruncana* and *Globorotalia,* however, do develop a distinct preference during their evolution.

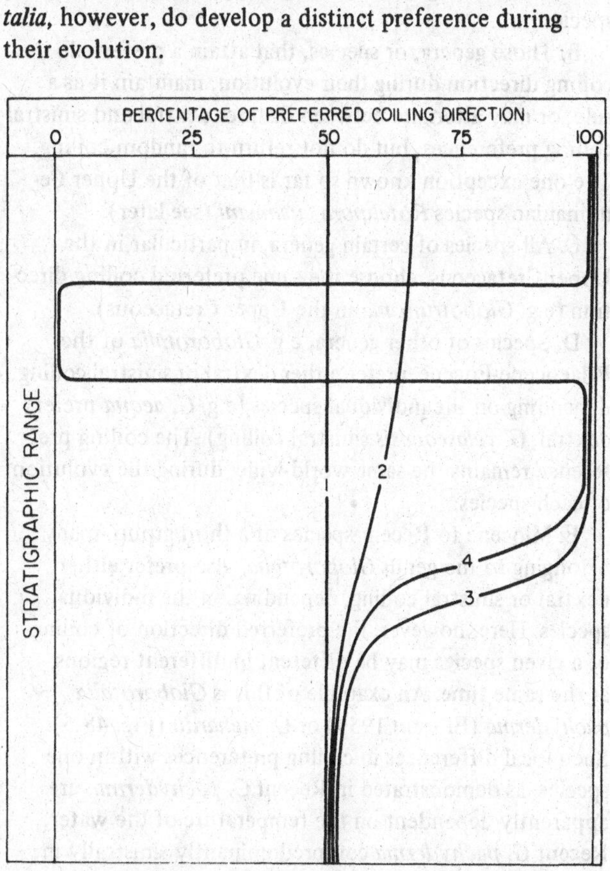
PERCENTAGE OF PREFERRED COILING DIRECTION
0
25
50
75
100
STRATIGRAPHIC RANGE
1
2
3
4

Fig. 48.1. The four principal coiling trends in planktonic Foraminifera. (For explanation of curves 1–4 see text.)

The following four principal coiling trends can be distinguished (see Fig. 48.1):

1. Genera or species that do not develop a noticeable preference throughout their evolution.

2. Genera or species that coil at random during their early evolutionary stages, changing later to a moderate preference for one direction, which may be just over 50 per cent, or higher, depending on the species and its stage of evolution (Bolli 1951).

3. Genera or species that change rapidly after an initial random stage to an almost exclusively preferred direction, to be maintained throughout the remaining evolution.

4. Genera or species that, after an early random stage, attain a distinctly preferred direction like those mentioned under 3. The chosen preferred direction is, however, not maintained in this group throughout the remaining evolution. It can alternate once or more often with the opposite direction.

The following general rules can be deduced from these four coiling trends:

A. Random coiling persists during the early evolutionary stage of a genus, related groups of species, or species.

B. Those genera, or species, that attain a preferred coiling direction during their evolution, maintain it as a rule, or may alternate between distinct dextral and sinistral coiling preferences, but do not return to random coiling. The one exception known so far is that of the Upper Cenomanian species *Rotalipora cushmani* (see later).

C. All species of certain genera, in particular in the Upper Cretaceous, choose the same preferred coiling direction (e.g. *Globotruncana* in the Upper Cretaceous).

D. Species of other genera, e.g. *Globorotalia* of the Palaeocene-Eocene, prefer either dextral or sinistral coiling, depending on the individual species (e.g. *G. aequa* prefers dextral, *G. velascoensis* sinistral coiling). The coiling preference remains the same world-wide, during the evolution of each species.

E. Miocene to Recent species of a third group, many belonging to the genus *Globorotalia,* also prefer either dextral or sinistral coiling, depending on the individual species. Here, however, the preferred direction of coiling of a given species may be different in different regions at the same time. An example of this is *Globorotalia pachyderma* (Ericson 1959) or *G. menardii* (Fig. 48.5). Such local differences in coiling preferences within one species, as demonstrated in Recent *G. pachyderma,* are apparently dependent on the temperature of the water. Recent *G. pachyderma* coil predominantly sinistrally in colder waters, dextrally in warmer (Ericson 1959).

CRETACEOUS

Our knowledge of coiling-trends of pre-Cenomanian planktonic Foraminifera is still conjectural. Better results are available from the higher Cretaceous (Bolli 1951, 1957*a*). After an initial random stage in the Albian and early Cenomanian, species of the genera *Hedbergella, Praeglobotruncana, Rotalipora,* and later of *Globotruncana* and *Rugoglobigerina,* all follow the same trend, that is, they coil almost exclusively dextrally. This is clearly seen in Fig. 48.2, where species of *Rotalipora* in the Albian and early Cenomanian coil at random. A relatively rapid change to strongly preferred dextral coiling takes place with the appearance of *R. appenninica,* s.l. It is maintained by all subsequent species, the only exception being *R. cushmani* (*R. turonica* of authors), which appears late in the Cenomanian. The species has fewer and more inflated chambers than the other contemporary *Rotalipora* species. It may be regarded as a gerontic stage, or as a branch at the end of the *Rotalipora* lineage, that, for one reason or another, reverted to the random coiling of the early *Rotalipora* species.

All *Globotruncana* species coil almost exclusively dextrally, from their first appearance in the Turonian to their extinction at the end of the Cretaceous. From this, it may be assumed that the genus evolved from *Praeglobotruncana,* a genus whose pre-Upper Cenomanian species coil at random, while those in the Upper Cenomanian, and prior to the appearance of the first *Globotruncana* species, had already developed a strong preference for dextral coiling. Similarly, all *Rugoglobigerina* species from the Turonian upwards coil almost exclusively dextrally.

PALAEOCENE to EOCENE

Coiling patterns of some Palaeocene-Eocene planktonic Foraminifera were discussed by Bolli, first in 1950, and in more detail in 1957*b* and 1957*c*. Figure 48.3 of this paper, of which the lower part was included in the 1957*b* paper, shows that the coiling trend of several of the selected species is different from that of Cretaceous species. Some forms prefer the same dextral coiling but others the opposite sinistral direction. The coiling pattern in the early Palaeocene is the same as in the Lower Cretaceous, i.e. random for all species. The coiling of one group of apparently highly developed Middle Eocene *Globorotalia* (*G. lehneri, G. spinulosa*) remains at random, contrary to the general trend. Species of some Eocene genera other than *Globorotalia,* and possibly closely interrelated, clearly prefer either dextral (*Globigerapsis, Globigerinatheka, Porticulasphaera*), or sinistral coiling (*Truncorotaloides*).

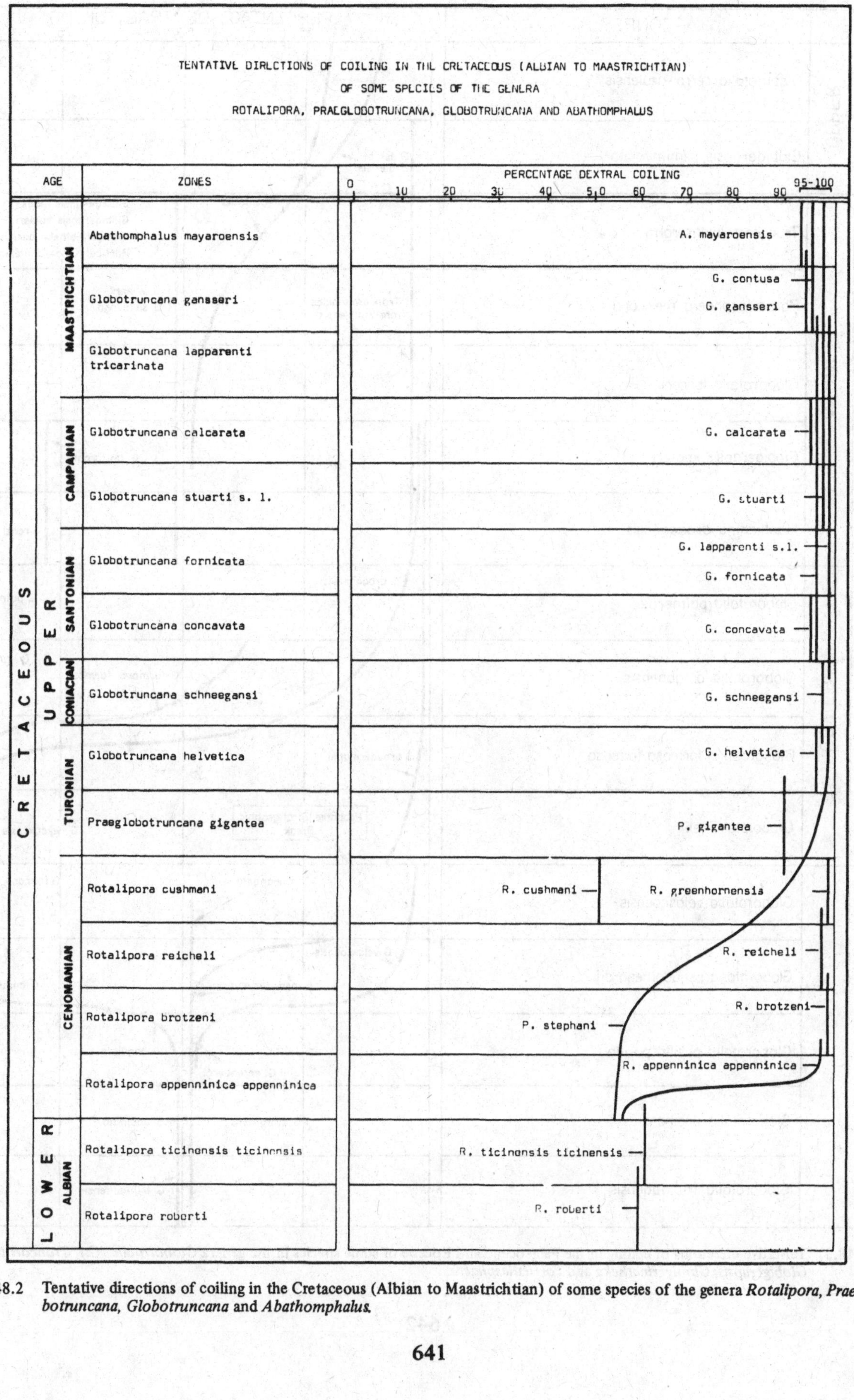
TENTATIVE DIRECTIONS OF COILING IN THE CRETACEOUS (ALBIAN TO MAASTRICHTIAN)
OF SOME SPECIES OF THE GENERA
ROTALIPORA, PRAEGLOBOTRUNCANA, GLOBOTRUNCANA AND ABATHOMPHALUS
AGE
ZONES
PERCENTAGE DEXTRAL COILING
0
10
20
30
40
50
60
70
80
90
95-100
CRETACEOUS
UPPER
LOWER
MAASTRICHTIAN
CAMPANIAN
SANTONIAN
CONIACIAN
TURONIAN
CENOMANIAN
ALBIAN
Abathomphalus mayaroensis
Globotruncana gansseri
Globotruncana lapparenti tricarinata
Globotruncana calcarata
Globotruncana stuarti s. l.
Globotruncana fornicata
Globotruncana concavata
Globotruncana schneegansi
Globotruncana helvetica
Praeglobotruncana gigantea
Rotalipora cushmani
Rotalipora reicheli
Rotalipora brotzeni
Rotalipora appenninica appenninica
Rotalipora ticinensis ticinensis
Rotalipora roberti
A. mayaroensis
G. contusa
G. gansseri
G. calcarata
G. stuarti
G. lapparenti s.l.
G. fornicata
G. concavata
G. schneegansi
G. helvetica
P. gigantea
R. cushmani
R. greenhornensis
R. reicheli
R. brotzeni
P. stephani
R. appenninica appenninica
R. ticinensis ticinensis
R. roberti

Fig. 48.2 Tentative directions of coiling in the Cretaceous (Albian to Maastrichtian) of some species of the genera *Rotalipora, Praeglobotruncana, Globotruncana* and *Abathomphalus*.

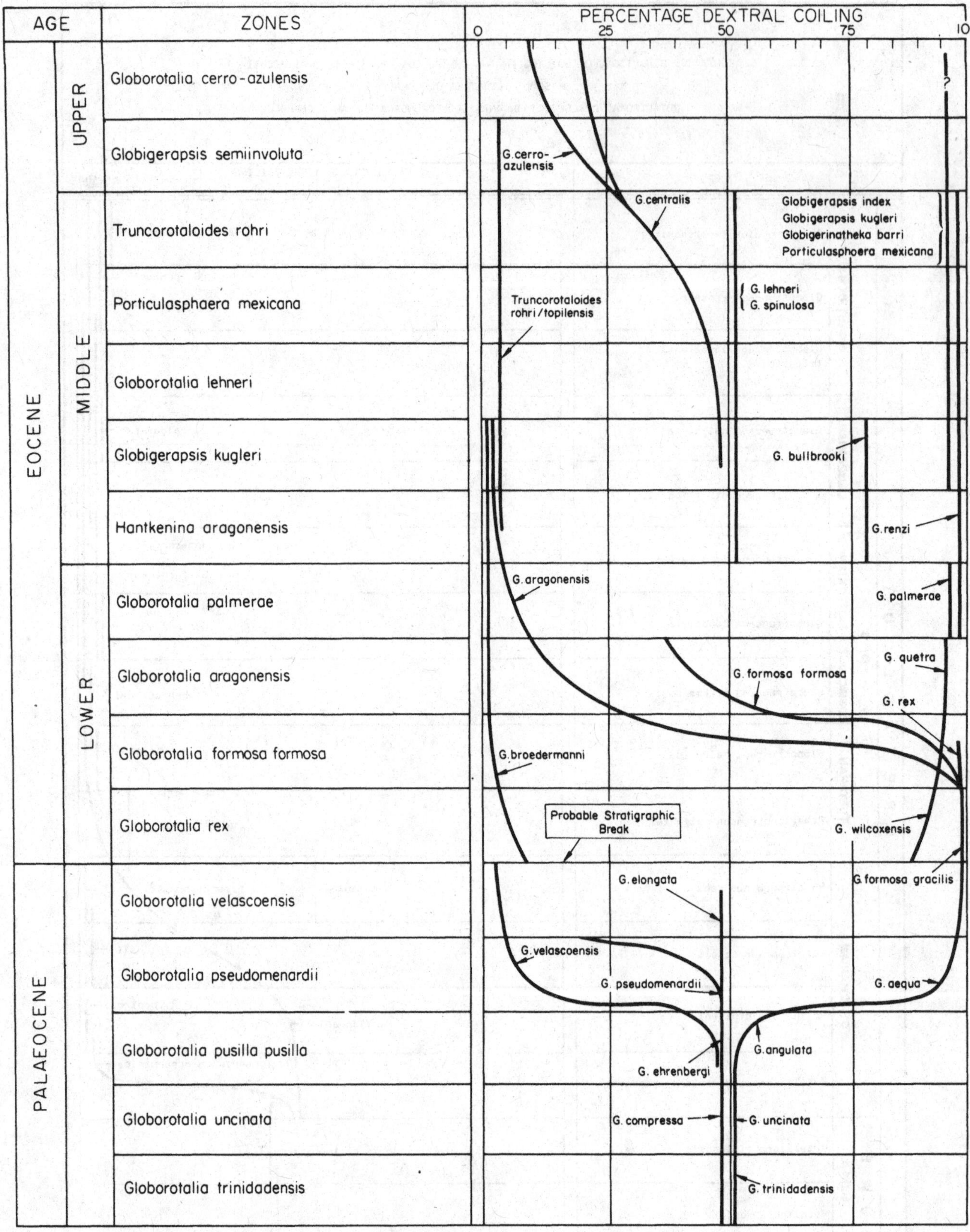
AGE
ZONES
PERCENTAGE DEXTRAL COILING
0
25
50
75
100
EOCENE
UPPER
MIDDLE
LOWER
PALAEOCENE
Globorotalia cerro-azulensis
Globigerapsis semiinvoluta
Truncorotaloides rohri
Porticulasphaera mexicana
Globorotalia lehneri
Globigerapsis kugleri
Hantkenina aragonensis
Globorotalia palmerae
Globorotalia aragonensis
Globorotalia formosa formosa
Globorotalia rex
Globorotalia velascoensis
Globorotalia pseudomenardii
Globorotalia pusilla pusilla
Globorotalia uncinata
Globorotalia trinidadensis
?
G. cerro-azulensis
G. centralis
Globigerapsis index
Globigerapsis kugleri
Globigerinatheka barri
Porticulasphoera mexicana
G. lehneri
G. spinulosa
Truncorotaloides rohri/topilensis
G. bullbrooki
G. renzi
G. aragonensis
G. palmerae
G. quetra
G. formosa formosa
G. rex
G. broedermanni
Probable Stratigraphic Break
G. wilcoxensis
G. formosa gracilis
G. elongata
G. velascoensis
G. pseudomenardii
G. aequa
G. angulata
G. ehrenbergi
G. compressa
G. uncinata
G. trinidadensis

Fig. 48.3. Tentative directions of coiling in the Palaeocene and Eocene of some species in the genera *Globorotalia (G.)*, *Truncorotaloides*, *Globigerapsis*, *Globigerinatheka* and *Porticulasphaera*.

Some subsequent observations on coiling preferences in Palaeocene-Eocene planktonic Foraminifera were reported from sections in the Alpine/Mediterranean region, such as Paderno d'Adda (Bolli and Cita 1960*a*, 1960*b*), Pessagno (Cita and Bolli 1966), and Molinetto di Pederobba (Proto Decima and Zorzi 1965) from the Southern Alps. From the Central Apennines, the Gubbio and other sections were also studied in detail (Luterbacher and Premoli Silva 1962; Luterbacher 1964). These observations largely confirm the trends described from the Caribbean region. Of particular interest is the fact that the change in *Globorotalia aragonensis* from almost exclusively dextral to sinistral coiling, as observed in Trinidad, was also found in the Lower Eocene of the Central Apennines (Luterbacher and Premoli Silva 1962, p. 265; Luterbacher 1964, p. 698). This identical coiling change in *G. aragonensis* in the Lower Eocene of the Caribbean and the Mediterranean regions indicates that it is a world-wide feature and may as such be used as a datum line for long range correlation.

OLIGOCENE to PLIOCENE

In his 1950 and 1951 papers the present author described from Trinidad the coiling trends of the *Globorotalia fohsi* subspecies, and also of a number of planktonic species whose coiling preferences are only moderately developed. Later, in 1964, he compared coiling patterns of some Miocene to Recent planktonic species between the tropical Pacific and the Caribbean regions. Coiling trends in some planktonic species were followed in detail in the Miocene-Pliocene section of well Bodjonegoro-1, Java (Bolli 1966), and in well Cubagua-1, Cubagua Island, Venezuela (Bolli in Bermúdez and Fuenmayor 1966). Some of the results contained in the above mentioned publications are given in Figs 48.4 and 48.5 of this paper.

Rounded, unkeeled *Globorotalia* species coil at random during the Oligocene, and are hence comparable to the coiling of *Rotalipora* species in the Lower Cretaceous and to *Globorotalia* species in the Lower Palaeocene. *Çatapsydrax dissimilis,* which originates in the Eocene, is the only planktonic foraminiferal species with a distinct (dextral) coiling preference in the Oligocene and Lower Miocene.

The renewed appearance of more highly evolved *Globorotalia,* and other planktonic genera of the late Lower Miocene, leads again to preferential coiling in many species. Of particular interest is the fact that all species so far studied in Trinidad and Venezuela, which develop a coiling preference in the late Lower and early Middle Miocene, prefer sinistral coiling, some to a lesser (Bolli 1951), others to a higher degree (Bolli 1950, 1951). A similar, but still more pronounced preference for one direction of coiling is only found in the Upper Cretaceous, where practically all planktonic Foraminifera coil exclusively dextrally.

Preferences for dextral or sinistral coiling in certain Miocene and younger species are not necessarily identical on a world-wide scale. For instance, the subspecies *Globorotalia fohsi fohsi, G. fohsi lobata* and *G. fohsi robusta* coil sinistrally throughout the Lower Miocene in Trinidad and Venezuala (Bolli 1950). The same sequence of subspecies was found to follow different coiling trends in well Bodjonegoro-1, Java (Bolli 1966). There, several changes from a strong preference for sinistral to dextral coiling exist during the corresponding vertical range.

Coiling patterns of the species studied can become rather complex from the higher Miocene onward, as can readily be seen from Figs. 48.4 and 48.5. Certain species continue with a sinistral preference, while new ones appear to follow the same trend. Other species develop a strong preference for dextral coiling, while a third group, to which *G. menardii* belongs, may switch several times from one preferred direction to the opposite one. Such sudden changes were pointed out by Bolli (1950) and were followed closely in well Bodjonegoro-1 of Java (Bolli 1966) and well Cubagua-1 of Venezuela (Bolli in Bermúdez and Fuenmayor 1966) and are here compared on Fig. 48.5. This figure shows that preferred coiling directions and changes from sinistral to dextral and back to sinistral in *Globorotalia menardii* do not always coincide between the Bodjonegoro-1 and the Cubagua-1 section. Especially in the *Globorotalia acostaensis* Zone and the *Globorotalia dutertrei* Zone, there are several coiling changes in the Cubagua section, while at the same time the species continues to coil sinistrally in Bodjonegoro. If water temperature influenced the coiling direction of *Globorotalia menardii,* at that time, as it does some Recent species (but not including *G. menardii*), then the southern part of the Caribbean Sea would seem to have had several marked temperature changes during the later Miocene. At the same time, the temperature would seem to have remained more stable in the tropical Pacific.

Such intermittent colder temperatures for the southern Caribbean are also indicated by other planktonic foraminiferal species accompanying *Globorotalia menardii* in Cubagua-1. *G. tumida* s.l., a typical living tropical species, appears in Cubagua-1 only sporadically during the interval under discussion, while the species is continuously present in Bodjonegoro-1.

PLEISTOCENE to RECENT

Much information on the distribution of planktonic Foraminifera in the present oceans, as well as on coiling

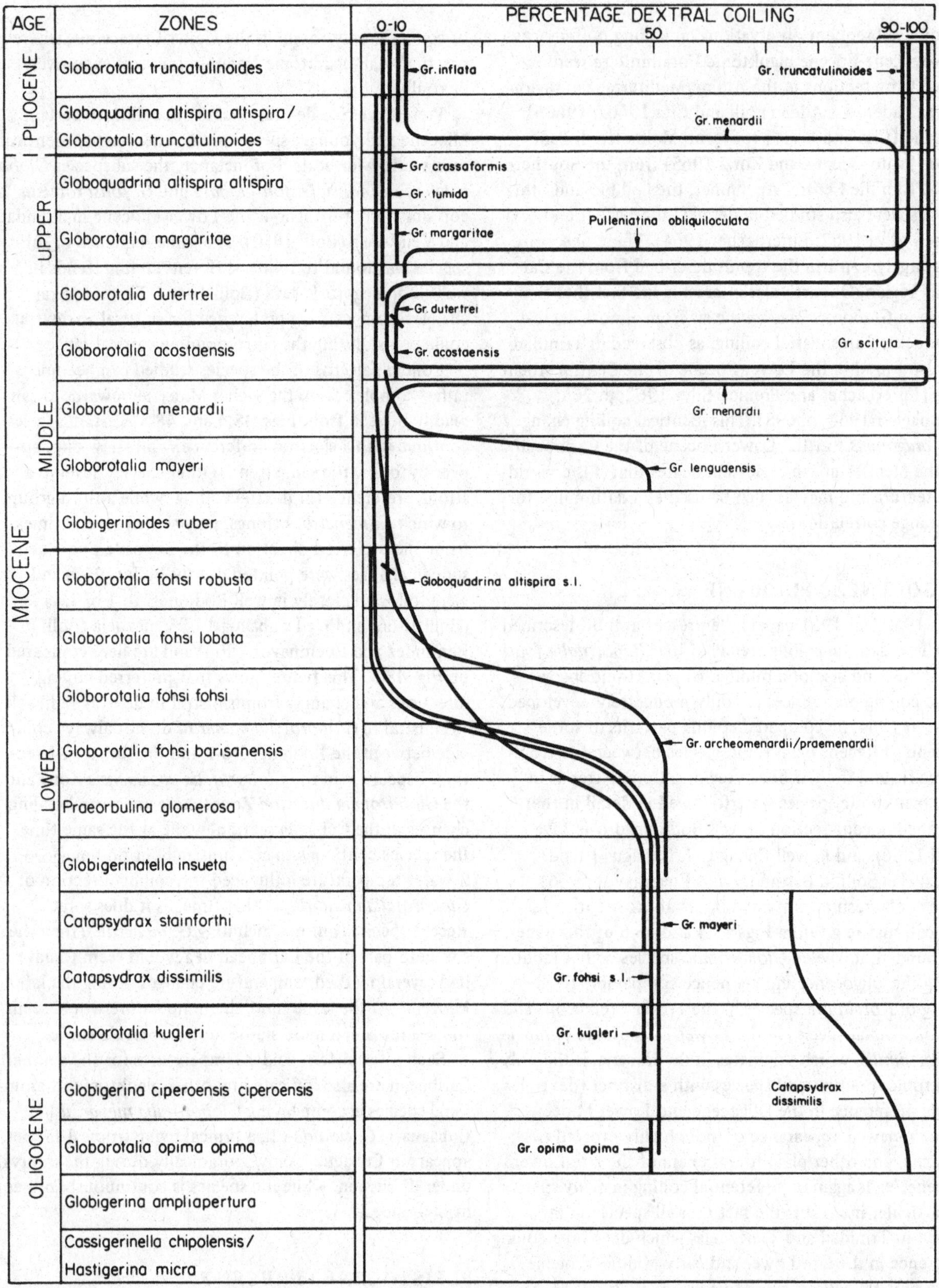
AGE
ZONES
PERCENTAGE DEXTRAL COILING
0-10
50
90-100
PLIOCENE
UPPER
MIDDLE
MIOCENE
LOWER
OLIGOCENE
Globorotalia truncatulinoides
Globoquadrina altispira altispira/
Globorotalia truncatulinoides
Globoquadrina altispira altispira
Globorotalia margaritae
Globorotalia dutertrei
Globorotalia acostaensis
Globorotalia menardii
Globorotalia mayeri
Globigerinoides ruber
Globorotalia fohsi robusta
Globorotalia fohsi lobata
Globorotalia fohsi fohsi
Globorotalia fohsi barisanensis
Praeorbulina glomerosa
Globigerinatella insueta
Catapsydrax stainforthi
Catapsydrax dissimilis
Globorotalia kugleri
Globigerina ciperoensis ciperoensis
Globorotalia opima opima
Globigerina ampliapertura
Cassigerinella chipolensis/
Hastigerina micra
Gr. inflata
Gr. truncatulinoides
Gr. menardii
Gr. crassaformis
Gr. tumida
Pulleniatina obliquiloculata
Gr. margaritae
Gr. dutertrei
Gr. scitula
Gr. acostaensis
Gr. menardii
Gr. lenguaensis
Globoquadrina altispira s.l.
Gr. archeomenardii/praemenardii
Gr. mayeri
Gr. fohsi s.l.
Gr. kugleri
Catapsydrax dissimilis
Gr. opima opima

Fig. 48.4. Tentative directions of coiling in the Oligocene to Pliocene of some species of the genera *Globorotalia, Catapsydrax, Globoquadrina* and *Pulleniatina*.

50. OBSERVATIONS ON THE BIOSTRATIGRAPHY OF PELAGIC SEDIMENTS[1]

M. N. BRAMLETTE AND W. R. RIEDEL

Summary: The basic principles of biostratigraphy, developed during more than a century from the study of strata on land, are obviously the same as those to be applied to the deep-sea floor, even though pelagic deposits are rare, indeed, on land. These principles are briefly discussed, and some consideration given to aspects peculiar to pelagic deposits which result from: low rates of sediment accumulation, a high proportion of planktonic microfossils, uniformity over wide areas, extensive effects of reworking processes, and the small number of continuous sequences representing long periods of time so far available.

INTRODUCTION

It is impossible to present here an adequate discussion of such a large subject as the biostratigraphy of pelagic sediments, and this brief paper therefore includes some rather dogmatic statements. These may, it is hoped, serve as a basis for the discussion of different points of view.

Arkell (1933) has aptly referred to 'the monstrous and sterile problems of stratigraphical nomenclature'. However, the excellent discussion in the first 37 pages of his *Jurassic System in Great Britain* clarifies the concepts expressed by various stratigraphical terms that are basic to some principles of classification and to an understanding of the subject. Whether one uses the nomenclature suggested by Arkell, based largely on priority and clearly defined concepts, or that of the American, Russian, or some other Code, this need only be stated – or, if no established code is strictly followed, it is necessary to make clear the degree of departure from such a code. No system is likely to be immediately acceptable to all workers, and continued discussion of principles, at least, seems important and profitable.

Most will probably agree that stratigraphy involves three kinds of stratal units, commonly termed lithostratigraphical, biostratigraphical and chronostratigraphical, besides the purely time units corresponding to the chronostratigraphical stratal units. As lithostratigraphical units are apparently of less importance to studies of deep-sea sediments than to studies of shallow-water deposits, they are not discussed here. Although chronostratigraphical units (stages, etc.) are basically different from biostratigraphical ones (various types of zones), they remain at present largely dependent on the biostratigraphical units for their delimitation away from their stratotype locality. This distinction seems subtle or even unnecessary to some biostratigraphers, but non-palaeontological methods for determining time relations between stratal units are developing rapidly. When (or if) such methods produce results demonstrably more precise and refined than is possible by biostratigraphy, the biostratigrapher may concentrate more of his attention on the many fascinating problems of distribution of the organic remains at precise times, in the oceans as well as on land, and on the many palaeoecological implications. At present, biostratigraphers interested primarily in ecology and facies tend to overlook the fact that their interpretations of facies are usually of little significance to earth history unless these can be placed in a time framework indicating relations to conditions elsewhere.

BIOSTRATIGRAPHICAL UNITS

Biostratigraphical units may be grouped into three principal kinds of zones. Those strata characterized by a certain assemblage of fossils which, regardless of their varying time ranges, distinguish the stratal unit as a particular kind of facies deposit (littoral, lagunal, etc.), have been termed 'assemblage zones' in the American Code – these, again, are of less importance to investigators con-

[1] Contribution from the Scripps Institution of Oceanography, University of California, San Diego, California.

trends of many species in the Pleistocene and Recent, has been published in numerous papers. They include Wiseman and Ovey (1950), Bradshaw (1959), Parker (1962), Ericson (1959), Ericson, Ewing and Wollin (1963), Ericson, Wollin and Wollin (1954), Bandy (1960), and Jenkins (1967). It has been found that the preferred coiling direction of species such as *G. pachyderma* is controlled, at least to some degree, by water temperature.

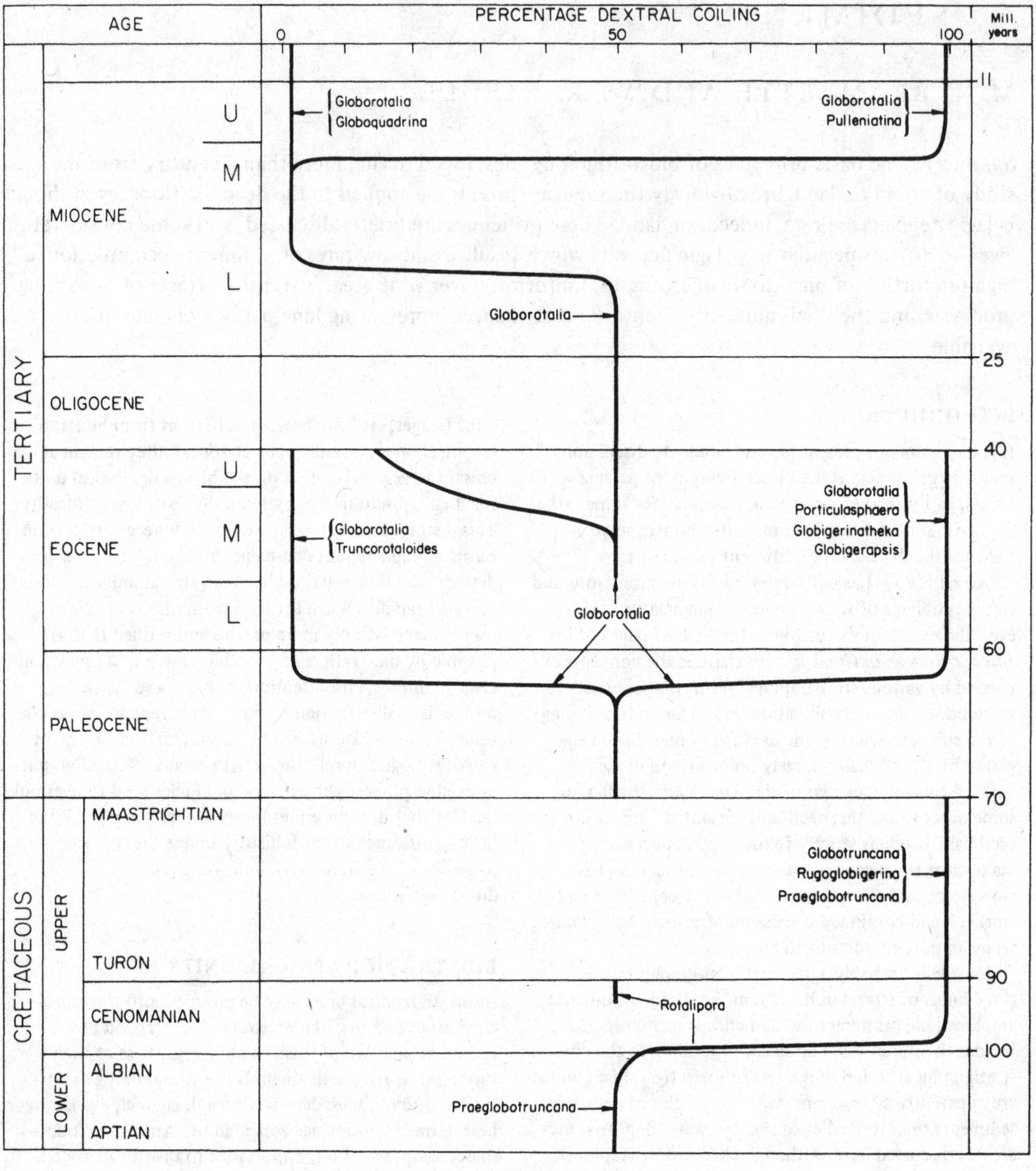

AGE
PERCENTAGE DEXTRAL COILING
0
50
100
Mill. years
TERTIARY
MIOCENE
U
M
L
OLIGOCENE
EOCENE
PALEOCENE
CRETACEOUS
UPPER
LOWER
MAASTRICHTIAN
TURON
CENOMANIAN
ALBIAN
APTIAN
Globorotalia
Globoquadrina
Globorotalia
Pulleniatina
Globorotalia
Globorotalia
Truncorotaloides
Globorotalia
Porticulasphaera
Globigerinatheka
Globigerapsis
Globorotalia
Globotruncana
Rugoglobigerina
Praeglobotruncana
Rotalipora
Praeglobotruncana
11
25
40
60
70
90
100

Fig. 48.6. Trends in the direction of coiling of some planktonic foraminiferal genera.

This species today coils predominantly sinistrally below a certain water temperature, dextrally above it. Such an interdependence of preferred coiling direction and water temperature makes the coiling changes of such species a useful tool for the relative dating and correlation of warmer and colder climates, i.e. glacial and interglacial periods during the Upper Caenozoic (see e.g. Bandy 1960; Jenkins 1967).

CRETACEOUS to RECENT

Figure 48.6 is a combination of Figs. 48.2, 48.2 and 48.4, giving a simplified picture of prevailing coiling directions of planktonic Foraminifera from the Cretaceous to the Pliocene. Three principal intervals can clearly be distinguished, each beginning with random coiling (during the early Cretaceous, the Lower Palaeocene and the Oligocene to early Lower Miocene), followed by a period of preferred coiling that ends abruptly at the Cretaceous/Palaeocene and at the Eocene/Oligocene boundary. The Oligocene-Recent interval (since the upper Lower Miocene) shows distinctly preferred coiling directions for species of several genera.

The reason for the sudden change from preferred back to random coiling at the end of the Cretaceous, and at the end of the Eocene, lies in the sudden extinction at these levels of the planktonic genera and species that distinctly preferred a coiling direction. Surviving or newly appearing, more primitive, species at the base of the Palaeocene and the Oligocene respectively, coil at random.

Some absolute dates in millions of years are included on Fig. 48.6, to show the duration of these cycles of random and preferred coiling. The first change from random to a distinctly preferred direction took place just above the base of the Cenomanian, or about 100 million years ago. It lasted about 20-25 million years, i.e. until the end of the Cretaceous. During the early Palaeocene, for nearly 10 million years, we have a second period with no preferred coiling, followed again by 20-25 million years, from the younger Palaeocene to the end of the Eocene, when many genera and species strongly preferred one or the other direction of coiling. A third period of practically ubiquitous random coiling, with the exception of *Catapsydrax dissimilis,* followed during the Oligocene and into the early Miocene. The duration of this interval was 15 to 20 million years, somewhat longer than the corresponding random period during the early Palaeocene. This third interval of random coiling was again followed by one with distinct coiling preferences in many genera and species, beginning in the younger Lower Miocene and continuing until today, i.e. for about 20 million years.

Received December 1967.

BIBLIOGRAPHY

Bandy, O. L. 1960. The geologic significance of coiling ratios in the foraminifer *Globigerina pachyderma* (Ehrenberg). *J. Pal.* **34** (4), 671-81.

Bermúdez, P. J. and Fuenmayor, A. N. 1966. Consideraciones sobre los sedimentos del Mioceno Medio al Reciente de las costas central y oriental de Venezuela. Segunda parte. Los foraminiferos bentonicos. *Bol. de Geol.* **7** (14), 333-611.

Bolli, H. M. 1950. The direction of coiling in the evolution of some Globorotaliidae. *Cushman Found. Foram. Res., Contr.* **1** (3-4), 82-9.

Bolli, H. M. 1951. Notes on the direction of coiling of rotalid foraminifera. *Cushman Found. Foram. Res., Contr.* **2** (4), 139-43.

Bolli, H. M. 1957a. The genera *Praeglobotruncana, Rotalipora, Globotruncana,* and *Abathomphalus* in the Upper Cretaceous of Trinidad, B.W.I. *U.S. Nat. Mus. Bull.* **215**, 51-60.

Bolli, H. M. 1957b. The genera *Globigerina* and *Globorotalia* in the Palaeocene-Lower Eocene Lizard Springs Formation of Trinidad. B.W.I. *U.S. Nat. Mus. Bull.* **215**, 61-81.

Bolli, H. M. 1957c. Planktonic Foraminifera from the Eocene Navet and San Fernando formations of Trinidad B.W.I. *U.S. Nat. Mus. Bull.* **215**, 155-98.

Bolli, H. M. 1957d. Planktonic Foraminifera from the Oligocene-Miocene Cipero and Lengua formations of Trinidad. B.W.I. *U.S. Nat. Mus. Bull.* **215**, 97-123.

Bolli, H. M. 1964. Observations on the stratigraphic distribution of some warm water planktonic Foraminifera in the young Miocene to Recent. *Eclogae Geol. Helv.* **57** (2), 541-52.

Bolli, H. M. 1966. The planktonic Foraminifera in well Bodjonegoro-1 of Java. *Eclogae Geol. Helv.* **59** (1), 449-65.

Bolli, H. M. and Cita, Maria B. 1960a. Globigerine e Globorotalie de Palaeocene di Páderno d'Adda (Italia). *Riv. Ital. di Pal.* **66** (3), 2-38.

Bolli, H. M. and Cita, Maria B. 1960b. Upper Cretaceous and Lower Tertiary planktonic Foraminifera from the Paderno d'Adda section, northern Italy. *Int. Geol. Congr. Copenhagen. 21st Session,* Pt. **5**, 150-61.

Bolli, H. M., Loeblich, A. R. and Tappan, Helen 1957. Planktonic foraminiferal families Hantkeninidae, Orbulinidae, Globorotaliidae and Globotruncanidae. *U.S. Nat. Mus. Bull.* **215**, 3-50.

Bradshaw, J. S. 1959. Ecology of living planktonic Foraminifera in the north and equatorial Pacific. *Cushman Found. Foram. Res., Contr.* **10** (2), 25-64.

Cita, Maria B. 1948. Ricerche stratigrafiche e micropaleontologiche sul Cretacico e sull'Eocene di Tignale (Lago di Garda). *Riv. Ital. di Pal.* **54** (3 & 4), 117-34, 143-68.

Cita, Maria B. and Bolli, H. M. 1966. Biostratigrafia della serie paleocenico-eocenica di Possagno, Treviso. *Boll. Soc. Geol. Ital.* **85**, 231-9.

Ericson, D. B. 1959. Coiling direction of *Globigerina pachyderma* as a climatic index. *Science,* **130**, 219-20.

Ericson, D. B., Ewing, M. and Wollin, G. 1963. Pliocene-Pleistocene boundary in deep-sea sediments. *Science.* **139**, 727-37.

Ericson, D. B., Wollin, G. and Wollin, J. 1954. Coiling direction of *Globorotalia truncatulinoides* in deep-sea cores. *Deep-sea Res.* 2, 152-8.

Gandolfi, R. 1942. Ricerche micropaleontologiche e stratigrafiche sulla Scaglia e sul Flysch cretacici dei dintorni de Balerna (Canton Ticino). *Riv. Ital. di Pal.* 48 (4), 1-160.

Jenkins, D. G. 1967. Recent distribution, origin, and coiling ratio changes in *Globorotalia pachyderma* (Ehrenberg). *Micropaleontology,* 13 (2), 195-203.

Luterbacher, H. P. 1964. Studies in some *Globorotalia* from the Paleocene and Lower Eocene of the Central Apennines. *Eclogae Geol. Helv.* 57 (2), 631-730.

Luterbacher, H. P. and Premoli Silva, Isabella 1962. Note préliminaire sur une révision de profil de Gubbio, Italie. *Riv. Ital. di Pal.* 68 (2), 253-88.

Parker, Frances L. 1962. Planktonic foraminiferal species in Pacific sediments. *Micropaleontology,* 8 (2), 219-54.

Proto Decima, Franca and Zorzi, P. 1965. Studio micro-paleontologico-stratigrafico della serei Cretaceo-Terziaria del Molinetto de Pederobba. *Mem. Ist. Geol. Min. Univ. Padova.* 25, 1-44.

Tschachtli, B. S. 1941. *Ueber Flysch und Couches Rouges in den Decken der östlichen Prealpes romandes (Simmental-Saanen).* Mettler & Salz A.G.: Bern. pp. 1-78.

Wiseman, J. D. H. and Ovey, C. D. 1950. Recent investigations on the deep-sea floor. *Geol. Assoc. Proc.* 61 (1), 28-84.

49. SYSTEMATIC CLASSIFICATION OF POLYCYSTINE RADIOLARIA[1]

W. R. RIEDEL

Summary: It now seems possible to abandon the unsatisfactory, artificial classification of Haeckel. In its place is proposed a suprageneric classification which is believed to reflect natural relationships, but which cannot yet be comprehensive.

INTRODUCTION

The artificiality of Haeckel's radiolarian classification, and the difficulty of assigning species to generic and even suprageneric taxa, has been pointed out by many authors, including Bütschli (1889), Deflandre (1953), Guppy (1909), Haeckel (1887), Khabakov (1937) and Tan Sin Hok (1927). In spite of this, the Haeckelian system has remained in use, with only minor modifications, to the present day – the reason apparently being that no one researcher has been able to familiarize himself with all of the very numerous species described, and to recast them into a fundamentally different classification. It now seems possible, however, to proceed with this major task of reclassification, if we are prepared to endure the inconvenience that results from discarding the comprehensiveness and geometrical regularity of Haeckel's system, and to use for a while a developing system which cannot cover all known species until a great deal of fundamental research is done.

Because the needed revision is so far-reaching, it is necessary to keep clearly in mind the differences between the taxonomic work and the nomenclatural work, that will together produce a better classification. Taxonomy has fundamentally two stages – (1) the grouping together of related forms (preferably by using fossil evidence to deduce phylogenetic series), and (2) the dividing of these groups into families, genera and species. When the limits of these taxa have been defined, the literature and the International Code of Zoological Nomenclature must be used to determine whether a name already exists that can be applied to each taxon, and which one must be used if there are several possibilities.

The *definitions* of taxa in the Haeckelian system will have to be largely ignored for the purposes of this reclassification, though of course many of the *names* will remain in use. The type species of genus-group taxa, and the type genera of family-group taxa are of great practical importance in determining the nomenclature to be used. The species of many Haeckelian genera will be distributed among other genera in the new classification, and the generic names must follow the type species in this redistribution – no matter how 'atypical' the type species may be of the forms that have come to be included under those generic names.

There are several ways in which the revision has begun, and all are fruitful –

(1) Detailed investigations of intraspecific variation are revealing that the type species of some genus-group taxa are conspecific or very closely related. As a result, some generic and subgeneric names can be placed in synonymy and the taxonomic system simplified to that extent. Examples of this are the synonymizing of *Sethocyrtis* with *Anthocyrtidium* by Riedel (1957), of *Spongopyramis* with *Peripyramis* by Riedel (1958), of *Anthocyrtissa* with *Anthocyrtidium* by Nigrini (1967), etc.

(2) Evolutionary lineages are being worked out on the basis of evidence from the fossil record, to reveal phylogenetic relationships. This results in a very significant and useful taxonomy, if the deductions regarding the phylogeny are sound. Examples are the revisions of the Trissocyclidae (= Acanthodesmiidae) by Goll (1967) and Artiscinae (Riedel 1959, and this paper).

Reconsiderations of family-group taxa without evidence of evolutionary lineages can also result in useful revisions, as in the treatment of the botryoids by Petrushevskaya (1965), and of the orosphaerids by Friend and Riedel (1967).

(3) The entire polycystine radiolarian literature, and all available living and fossil assemblages, can be reviewed, and an attempt made to recast completely the broad lines of the classification, as is being done by Riedel (1967*b*) and in this paper. The resulting new classification is thus

[1] Contribution from the Scripps Institution of Oceanography, University of California, San Diego, California.

not very firmly based, and will undoubtedly require modification in the near future, but it provides a conceptual framework different from that which has hampered progress during the past eighty years, and enables us to think of radiolarians in new groupings which may reflect natural relationships.

A proposed classification to the family level is outlined below, with some examples of redefined subfamilies and genera. This represents a somewhat extended and more detailed treatment of the draft classification presented previously (Riedel 1967*b*). As in that paper, a few of the smaller and lesser-known families are also omitted here. Some of the families included here are used as 'waste-baskets', to contain numerous forms of which the relationships are still unclear, while others contain only forms which are believed to be related. There are also many described genera which cannot yet be placed in any of the listed families. A measure of the incompleteness of the treatment that is now possible is the fact that in this paper only about one-fifth of the 1,200 described genus-group taxa of polycystine radiolarians are mentioned. The situation can be compared with a saturated solution, in which crystallization is commencing at isolated points – it is possible to accommodate a small proportion of the known species in newly defined 'natural' family-group taxa, but the true relationships of many others are not yet understood. Hopefully, this process can be continued until all radiolarians can be placed in a satisfactory system of classification.

One practical result of this revision is that the genus and family-group taxa have time-ranges that are more restricted and therefore more useful stratigraphically than those in the Haeckelian system – this becomes evident from a comparison of the time-ranges indicated herein with those listed by Campbell (1954).

SYSTEMATIC SECTION

Subclass **Radiolaria** Müller 1858
Order **Polycystina** Ehrenberg 1838, emend. Riedel 1967*b*.

Definition: Radiolaria with skeleton of opaline silica without admixed organic compounds.

Suborder **Spumellaria** Ehrenberg 1875

Family **Entactiniidae** Riedel 1967*a*

Definition: Palaeozoic spumellarians with single or multiple, spherical or ellipsoidal lattice-shell, with internal spicule of 4-6 (rarely more) rays which extend to (and often beyond) the shell wall.

Genera include:

? *Druppalonche* Hinde 1899*b*, p. 217. Type species: *D. clavigera* Hinde 1899*b*.

Ellipsostigma Hinde 1899*a*, p. 51. Type species: *E. australe* Hinde 1899*a*.

Entactinia Foreman 1963, p. 271. Type species: *E. herculea* Foreman 1963.

Entactinosphaera Foreman 1963, p. 274. Type species: *E. esostrongyla* Foreman 1963.

Haplentactinia Foreman 1963, p. 270. Type species: *H. rhinophyusa* Foreman 1963.

Polyentactinia Foreman 1963, p. 281. Type species: *P. craticulata* Foreman 1963.

? *Spongocoelia* Hinde 1899*a*, p. 52. Type species: *S. citreum* Hinde 1899*a*.

Stigmosphaerostylus Rüst 1892, p. 142. Type species: *S. notabilis* Rüst 1892.

Tetrentactinia Foreman 1963, p. 282. Type species: *T. barysphaera* Foreman 1963.

Stratigraphical range: Devonian to Carboniferous. Extending back to Ordovician if the assemblage described by Hinde (1899*b*) is of that age, as he presumed.

Remarks: No forms with internal spicule are known from the Mesozoic or Tertiary, and therefore it seems unlikely that the Entactiniidae are related to the extant genera *Centrolonche* and *Centracontium* of Popofsky (1912) and the spumellarian genera with internal spicules described by Hollande and Enjumet (1960).

Family **Orosphaeridae** Haeckel 1887
Stratigraphical range: Eocent to Recent.

Remarks: The genera of this family have been revised by Friend and Riedel (1967).

Family **Collosphaeridae** Müller 1858
Definition: Colonial spumellarians with lattice-shells (and one genus with no skeletal elements).

Genera: Hilmers (1906) proposed modifications of Haeckel's system, on the basis of his examination of specimens from plankton samples, and Popofsky (1917) generally confirmed Hilmers' interpretations. The modified classification depends partly on the soft parts of the organisms, but only the skeletal characteristics are summarized here. Because the investigations of these authors were confined to plankton samples, they were unable to evaluate some of Haeckel's forms based on occurrences in sediments. The summary below represents the conclusions of Hilmers and Popofsky, regarding the relationships of type species of genus-group taxa. Generic names are modified in accordance with the ICZN, where necessary.

Genus **Collosphaera** Müller 1855*a*
Collosphaera Müller 1855*a*, p. 238. Type species: *C. huxleyi* Müller 1855*a*.

Definition: Inner and outer surfaces of shells smooth.

Genus **Myxosphaera** Brandt 1885

Myxosphaera Brandt 1885, p. 254. Type species: *Sphaerozoum coeruleum* Haeckel 1860 (? *Thalassicolla caerulea* Schneider 1858).

Definition: No skeletal elements.

Genus **Polysolenia** Ehrenberg 1860

Polysolenia Ehrenberg 1860, p. 832. Type species: *P. setosa* Ehrenberg 1872*a*.

Acrosphaera Haeckel 1881, p. 471. Type species: *Polysolenia setosa* Ehrenberg 1872*a*.

Choenicosphaera Haeckel 1887, p. 102. Type species: *C. murrayana* Haeckel 1887.

Choenicosphaerium Haeckel 1887, p. 103. Type species: *Choenicosphaera flammabunda* Haeckel 1887.

Clathrosphaera Haeckel 1881, p. 472. Type species: *C. circumtexta* Haeckel 1887.

? *Coronosphaera* Haeckel 1887, p. 117. Type species: *C. diadema* Haeckel 1887.

Trypanosphaera Haeckel 1887, p. 109. Type species: *T. trepanata* Haeckel 1887.

? *Trypanosphaerium* Haeckel 1887, p. 110. Type species: *Trypanosphaera coronata* Haeckel 1887.

Definition: Shells with external spines.

Genus **Siphonosphaera** Müller 1858

Siphonosphaera Müller 1858, p. 59. Type species: *S. tubulosa* Müller 1858.

Merosiphonia Haeckel 1887, p. 106. Type species: *Siphonosphaera socialis* Haeckel 1887.

? *Odontosphaera* Haeckel 1887, p. 101. Type species: *O. monodon* Haeckel 1887.

? *Otosphaera* Haeckel 1887, p. 116. Type species: *O. polymorpha* Haeckel 1887.

Definition: Shells generally regularly spherical, with some or all of the circular pores (which are unequal in size) extended as solid tubes.

Genus **Solenosphaera** Haeckel 1887

Solenosphaera Haeckel 1887, p. 112. Campbell (1954) indicated *Disolenia follis* as the type species of both *Solenosphaera* and *Disolenia,* but the only information about that species is the name (Ehrenberg 1872*b*, p. 289) and generic diagnosis (Ehrenberg 1860). *Disolenia* may be the correct name for this genus.

? *Solenosphenia* Haeckel 1887, p. 114. Type species: *Solenosphaera ascensionis* Haeckel 1887.

Tetrasolenia Ehrenberg 1860, p. 833. Type species: *T. venosa* Ehrenberg 1872*b*.

Definition: Shells generally with more or less pronounced projections, each of which terminates in one large opening which is either smooth or provided with one or more spines.

Genus **Tribonosphaera** Haeckel 1881

Tribonosphaera Haeckel 1881, p. 471. Type species: *T. centripetalis* Haeckel 1887.

Buccinosphaera Haeckel 1887, p. 99. Type species: *B. invaginata* Haeckel 1887.

Definition: Shells with outer surface smooth, and with truncated cones or conical spines on the inner surface.

Stratigraphical range of family: Apart from the record of *Acrosphaera hirsuta* in the Bohemian Cretaceous (Perner 1891), this family is known only from Lower Miocene to Recent.

Family **Actinommidae** Haeckel 1862, emend. Riedel 1967*b*

Definition: Solitary spumellarians with shells spherical or ellipsoidal (or modifications of those shapes), not discoidal, generally without internal spicule, generally much smaller than orosphaerids.

Remarks: This is a polyphyletic group containing many genera, the relationships of most of which have not yet been worked out. A few of the genera can be grouped into the two subfamilies below.

Subfamily **Saturnalinae** Deflandre 1953

Definition: Spherical latticed or spongy shell, with two opposite spines joined by a ring. In some species, there are no spines and the ring (or two incomplete half-rings) joins the shell directly.

Genera include:

Acanthocircus Squinabol 1903, p. 124. Type species: *A. irregularis* Squinabol 1903.

Saturnalis Haeckel 1881, p. 450. Type species: *S. circularis* Haeckel 1887.

Saturnalium Haeckel 1881, p. 450. Type species: *Saturnalis rotula* Haeckel 1887.

Saturninus Haeckel 1887, p. 146. Type species: *S. triplex* Haeckel 1887.

Saturnulus Haeckel 1879, p. 705. Type species: *S. planeta* Haeckel 1879.

Spongosaturnalis Campbell and Clark 1944*b*, p. 7. Type species: *S. spiniferus* Campbell and Clark 1944*b*.

Spongosaturninus Campbell and Clark 1944*b*, p. 7. Type species: *S. ellipticus* Campbell and Clark 1944*b*.

Stratigraphical range: Jurassic to Recent.

Remarks: Reported Cenozoic occurrences of saturnalins with spiny rings (*Saturnalis rotula* Haeckel 1887 from the surface at *Challenger* Station 244, and 'a *Haliomma*' of Bury 1867 from Barbados) are puzzling, since all other information indicates that such forms are restricted to the Mesozoic. I have searched for saturnalins with spiny rings in assemblages from the sediment sample from *Challenger* Station 244, and from the Oceanic Formation of Barbados, without success.

Subfamily **Artiscinae** Haeckel 1881, emend. Riedel 1967*b*

Definition: Actinommids with ellipsoidal cortical shell, almost always equatorially constricted, and generally enclosing a single or double medullary shell; opposite poles of the cortical shell generally bear spongy columns and/or single or multiple latticed caps.

Stratigraphical range: Oligocene to Recent.

Genus **Cannartus** Haeckel 1881, emend. herein

Cannartus Haeckel 1881, p. 462. Type species: *C. violina* Haeckel 1887.

Cannartidissa Haeckel 1887, p. 375. Type species: *Cannartidium mammiferum* Haeckel 1887.

Cannartidium Haeckel 1887, p. 373. Type species: *C. amphiconicum* Haeckel 1887.

Cannartiscus Haeckel 1887, p. 372. Type species: *C. amphiconiscus* Haeckel 1887.

Pipetta Haeckel 1887, p. 337. Type species: *P. fusus* Haeckel 1887, p. 337.

Pipettaria Haeckel 1887, p. 338. Type species: *P. tubaria* Haeckel 1887.

Pipettella Haeckel 1887, p. 304. Type species: *P. prismatica* Haeckel 1887.

Emended definition: Artiscinae with spongy polar columns, and no polar caps.

Stratigraphical range: Lower Oligocene (approx. *Globorotalia kugleri* zone of Bolli 1957) to upper lower Miocene (*Globorotalia fohsi barisanensis* zone of Bolli).

Remarks: This genus may have arisen from an ancestor related to *Trigonactura angusta* Riedel 1959 by reduction in the number of spongy protuberances from three to two, and relocation of the bars joining the medullary structures to the cortical shell. The lineage appears to have ended as polar caps were developed between the cortical shell and spongy columns.

Genus **Ommatartus** Haeckel 1881, emend. herein

Ommatartus Haeckel 1881, p. 463. Type species: *O. amphicanna* Haeckel 1887.

Cyphocolpus Haeckel 1887, p. 368. Type species: *C. virginis* Haeckel 1887.

Desmartus Haeckel 1887, p. 398. Type species: *D. larvalis* Haeckel 1887.

Ommatacantha Haeckel 1887, p. 395. Type species: *Ommatocampe amphilonche* Haeckel 1887.

Ommatocampula Haeckel 1887, p. 394. Type species: *Ommatocampe nereis* Haeckel 1887.

Ommatocorona Haeckel 1887, p. 394. Type species: *Ommatocampe chaetopodum* Haeckel 1887.

Panarium Haeckel 1881, p. 463. Type species: *P. facettarium* Haeckel 1887.

Panaromium Haeckel 1887, p. 389. Type species: *Panarium tubularium* Haeckel 1887.

Panartidium Haeckel 1887, p. 385. Type species: *Panicium coronatum* Haeckel 1887.

Panartissa Haeckel 1887, p. 379. Type species: *Panartus diploconus* Haeckel 1887.

Panartoma Haeckel 1887, p. 380. Type species: *Panartus quadriceps* Haeckel 1887.

Panartura Haeckel 1887, p. 381. Type species: *Panartus pluteus* Haeckel 1887.

Panartus Haeckel 1887, p. 376. Type species: *P. tetrathalamus* Haeckel 1887.

Panicium Haeckel 1887, p. 384. Type species: *P. amphacanthum* Haeckel 1887.

Peripanarium Haeckel 1887, p. 390. Type species: *P. cenoconicum* Haeckel 1887.

Peripanartium Haeckel 1887, p. 384. Type species: *Peripanartus atractus* Haeckel 1887.

Peripanartus Haeckel 1887, p. 382. Type species: *P. amphiconus* Haeckel 1887.

Peripanicium Haeckel 1887, p. 386. Type species: *P. amphixiphus* Haeckel 1887.

Peripanicula Haeckel 1887, p. 387. Type species: *Peripanicium amphicorona* Haeckel 1887.

Zygartus Haeckel 1881, p. 463. Type species: *Z. doliolum* Haeckel 1887.

Zygocampe Haeckel 1887, p. 399. Type species: *Z. chrysalidium* Haeckel 1887.

Emended definition: Artiscinae with polar caps, and sometimes spongy columns as well.

Stratigraphical range: Middle Miocene (*Globorotalia fohsi fohsi* zone of Bolli 1957) to Recent.

Remarks: Most species of this genus seem to belong to a lineage beginning with *O. antepenultimus* Riedel, in press (the undescribed *Panarium antepenultimum* of Riedel and Funnell 1964, p. 311) and proceeding by increase in size of the polar caps and reduction of the spongy columns to the present-day *O. tetrathalamus* (originally described in *Panartus*). The form described as *Ommatocampe hughesi* Campbell and Clark may not belong to this lineage, but until its relationships are elucidated it seems desirable to retain it in this genus.

Artiscinae incertae sedis

The following genera apparently belong in this subfamily, but information available on their type species is insufficient to determine whether they belong to either of the above genera, or to other genera.

Artidium Haeckel 1881, p. 462. Type species: *Artiscus nodosus* Haeckel 1887.

Artiscus Haeckel 1881, p. 462. Type species: *A. paniscus* Haeckel 1887.

Astromma Ehrenberg 1847*b*, p. 54. Type species: *A. entomocora* Ehrenberg 1854*b*.

Cypassis Haeckel 1887, p. 366 (= *Astromma*).

Cyphanta Haeckel 1887, p. 360. Type species: *C. colpodes* Haeckel 1887.

Cyphantissa Haeckel 1887, p. 362. Type species: *Cyphanta hispida* Haeckel 1887.

Cyphinidium Haeckel 1887, p. 371. Type species: *C. amphistylium* Haeckel 1887.

Cyphinidura Haeckel 1887, p. 372. Type species: *Cyphinidium coronatum* Haeckel 1887.

Cyphinura Haeckel 1887, p. 370. Type species: *Cyphinus amphilophus* Haeckel 1887.

Cyphinus Haeckel 1881, p. 463. Type species: *C. amphacanthus* Haeckel 1887.

Didymospyris Haeckel 1887, p. 366. Type species: *Cypassis palliata* Haeckel 1887.

Ommatocyrtis Haeckel 1887, p. 364. Type species: *Cyphonium hexagonium* Haeckel 1887.

Stylartura Haeckel 1887, p. 357. Type species: *Stylartus palatus* Haeckel 1887.

Stylartus Haeckel 1881, p. 462. Type species: *S. bipolaris* Haeckel 1887.

Remarks: Some of the forms described as having several spines (or one small one) on each pole may be rather simply accounted for, since it is not uncommon for the terminal structures to be spines similar to (or a little larger than) the spines on the cortical shell surface, or those supporting the polar caps. However, the forms described as having a single large spine at each pole pose a special problem. I have examined approximately 10,000-20,000 radiolarians each from the sediment samples collected at *Challenger* Stations 241, 244 and 224 (type localities of the type species of *Cyphinidium, Cyphinus* and *Stylartus,* respectively), but have found no specimens with a single large spine at each pole. If such forms exist, they must be excessively rare.

Forms that have been regarded as related to artiscins, but are excluded here:

Cyphonium Haeckel 1887, p. 362. No type species has yet been designated. Campbell (1954, D75) indicated *Ommatospyris apicata* Ehrenberg 1872*a* as the type species, but this is not among the species included by Haeckel at the time of his original description of the genus.

Desmocampe Haeckel 1887, p. 396. Type species: *D. catenula* Haeckel 1887. Apparently related to *Ommatocampe* (see below).

Diplellipsis Popofsky 1908, p. 221. Type species: *D. lapidosa* Popofsky 1908. The relationships of this species are quite uncertain.

Monaxonium Popofsky 1912, p. 125. Type species: *M. perforatum* Popofsky 1912. Apparently a spongodiscid.

Ommatocampe Ehrenberg 1860, p. 832. Type species: *O. polyarthra* Ehrenberg 1872*a*. Although this species superficially resembles *O. hughesi* Campbell and Clark 1944*a* (an artiscin), it seems unlikely that it has a true cortical twin-shell, and is therefore probably a spongodiscid related to *Monaxonium* (see above).

Ommatospyris Ehrenberg 1860, p. 832. Type species: *O. apicata* Ehrenberg 1872*a*. The relationships of this unfigured species are quite uncertain.

Family **Phacodiscidae** Haeckel 1881

Remarks: It is uncertain whether the Cenodiscidae should be included with this family. No satisfactory division into subfamilies is yet possible.

Family **Coccodiscidae** Haeckel 1862

Genus-group taxa include:

Amphiactura Haeckel 1881, p. 458. Type species: *A. amphibrachia* Haeckel 1887.

Amphicyclia Haeckel 1881, p. 458. Type species: *A. chronometra* Haeckel 1887.

Astractinium Haeckel 1887, p. 476. Type species: *Astromma aristotelis* Ehrenberg 1847*b*.

Astractura Haeckel 1881, p. 459. Type species: *A. ordinata* Haeckel 1887.

Astrococcura Sutton 1896, p. 138. Type species: *A. concinna* Sutton 1896.

Astrocyclia Haeckel 1881, p. 458. Type species: *S. solaster* Haeckel 1887.

Coccocyclia Haeckel 1881, p. 458. Type species: *C. liriantha* Haeckel 1887.

Coccodiscus Haeckel 1862, p. 485. Type species: *C. darwinii* Haeckel 1862.

Dicoccura Carter 1896*b*, p. 163. Type species: *D. brevibrachia* Carter 1896*b*.

Diplactinium Haeckel 1887, p. 470. Type species: *Diplactura diploconus* Haeckel 1887.

Echinactura Haeckel 1887, p. 480. Type species: *E. culcita* Haeckel 1887.

Hymenactinium Haeckel 1887, p. 475. Type species: *Hymenactura copernici* Haeckel 1887.

Hymenactura Haeckel 1881, p. 459. Type species: *H. archimedis* Haeckel 1887.

Hymeniastrum Ehrenberg 1847*a*, p. 54. Type species: *H. pythagorae* Ehrenberg 1854*b*.

Lithocyclia Ehrenberg 1847*a*, p. 54. Type species: *L. ocellus* Ehrenberg 1854*b*.

Pentactura Haeckel 1881, p. 459. Type species: *Astromma pentactis* Ehrenberg 1873.

Stauractinium Haeckel 1887, p. 478. Type species: *Stauractura medusina* Haeckel 1887.

Stauractura Haeckel 1881, p. 459. Type species: *S. octogona* Haeckel 1887.

? *Staurococcura* Carter 1896*a*, p. 96. Type species: *S. quaternaria* Carter 1896*a*.

Staurocyclia Haeckel 1881, p. 458. Type species: *S. cruciata* Haeckel 1887.

Stylocyclia Ehrenberg 1847b, p. 54. Type species: *S. dimidiata* Ehrenberg 1873.

Trigonactinium Haeckel 1887, p. 472. Type species: *Trigonactura triacantha* Haeckel 1887.

Trigonocyclia Haeckel 1887, p. 464 (= *Tripocyclia* Haeckel 1881, p. 458). Type species: *Trigonocyclia triangularis* Haeckel 1887.

Genera sometimes included in this family, but here excluded because of uncertainty regarding structure:

Diplactura Haeckel 1881, p. 458. Type species: *D. longa* Rüst 1885.

Trigonactura Haeckel 1881, p. 459. Type species: *T. weissmannii* Rüst 1885. (May be a hagiastrin.)

Stratigraphical range: Believed to be Cretaceous to early Oligocene (Bath Beds of Barbados, and equivalents in tropical Pacific). Many forms described by Haeckel 1887 are apparently reworked from early Tertiary into younger sediments. The few forms that have been recorded as living (including the type species of the type genus) are puzzling, since no coccodiscids are known from uncontaminated sediments younger than Oligocene.

Family **Spongodiscidae** Haeckel 1862, emend. Riedel 1967*b*

Emended definition: Discoidal, spongy or finely-chambered skeleton, with or without surficial pore-plate, often with radiating arms or marginal spines, and without a large central phacoid shell.

Remarks: This polyphyletic family contains a large number of genera, most of which cannot yet be satisfactorily grouped into subfamilies. There are, however, two subfamilies which it appears justifiable to separate out at this stage.

Subfamily **Myelastrinae** Riedel in press

Definition of subfamily: Spongodiscidae with arms much more delicately constructed than the small central area, which is the only part of the skeleton sufficiently robust to be preserved in sediments.

Genus-group taxa include:

Dicranastrum Haeckel 1881, p. 460. Type species: *D. furcatum* Haeckel 1887.

Myelastromma Haeckel 1887, p. 553. Type species: *Myelastrum octocorne* Haeckel 1887.

Myelastrum Haeckel 1887, p. 460. Type species: *M. medullare* Haeckel 1887.

Pentophiastromma Haeckel 1887, p. 558. Type species: *Pentophiastrum caudatum* Haeckel 1887.

Pentophiastrum Haeckel 1887, p. 558. Type species: *P. dicranastrum* Haeckel 1887.

Tetracranastrum Haeckel 1887, p. 552. Type species: *Dicranastrum bifurcatum* Haeckel 1887.

Triastrum Cleve 1901, p. 53. Type species: *T. aurivillii* Cleve 1901.

Tricranastrum Haeckel 1879, p. 705. Type species: *T. wyvillei* Haeckel 1879.

The following three genus-group taxa have a much larger central area, and therefore, their relationships are less certain:

Dictyocorynium Haeckel 1887, p. 593. Type species: *Spongodiscus charybdaeus* Haeckel 1860.

Spongasteriscus Haeckel 1862, p. 474. Type species: *Spongodiscus quadricornis* Haeckel 1860.

Spongastromma Haeckel 1887, p. 598. Type species: *Spongodiscus orthogonus* Haeckel 1860.

Stratigraphical range: Known only from Recent.

Subfamily **Hagiastrinae** Riedel in press

Definition of subfamily: Spongodiscidae in which the principal structural elements of the arms are several straight, strong, longitudinal elements from which arise (approximately at right angles) branches which form a regular spongy meshwork.

Remarks: Although there are in the literature numerous descriptions of species with this characteristic structure of the arms, there is only one genus-group taxon with a type species that appears to belong in this subfamily, namely, *Hagiastrum* Haeckel 1881, p. 542. Type species: *H. plenum* Rüst 1885.

Another genus that may belong to this subfamily is *Trigonactura* Haeckel 1881 (p. 459), the type species of which is *T. weismannii* Rüst 1885.

Stratigraphical range: Members of this subfamily have been found only in Mesozoic rocks.

Family **Pseudoaulophacidae** Riedel 1967*a*

Definition: Spongy discoidal spumellarians, with all or part of the surface covered by a regular meshwork of equilateral triangular frames.

Taxa include:

Pseudoaulophacus Pessagno 1963, p. 200. Type species: *P. florensis* Pessagno 1963.

Spongotrochus ehrenbergi Bütschli 1882*a*.

Theodiscus superbus Squinabol 1914.

Stratigraphical range: Cretaceous and probably Eocene.

Family **Pyloniidae** Haeckel 1881
Remarks: No modification is here proposed, to the established concept of this family. G. E. Kozlova has pointed out (personal communication) an asymmetry in the internal skeletal structure of at least some members of this family, which was noticed by early researchers (e.g. Muller, 1858, pl. 3, fig. 6; Hertwig, 1879, pl. 6, fig. 5) but has apparently been overlooked by almost all recent workers.

Stratigraphical range: Eocene to Recent, but well developed and common only in Miocene and later assemblages.

Family **Tholoniidae** Haeckel 1887
Remarks: No modification is here proposed, to Haeckel's concept of this family. As he states, representatives are rarely found.

Stratigraphical range: Pliocene to Recent.

Family **Litheliidae** Haeckel 1862
Stratigraphical range: Carboniferous to Recent.

Suborder **Nassellaria** Ehrenberg 1875

Family **Plagoniidae** Haeckel 1881, emend. Riedel 1967*b*
Definition: Skeleton consisting entirely of spicule with median bar, apical and dorsal spines, vertical spine, primary lateral spines and sometimes other spines; or having a lattice skeleton including a large cephalis within which this spicule is well developed.

Genus-group taxa include:

?*Acanthocoronium* Haeckel 1887, p. 1263. Type species: *Arachnocorys umbellifera* Haeckel 1860.

?*Arachnocorallium* Haeckel 1887, p. 1265. Type species: *Arachnocorys hexaptera* Haeckel 1887 (? = *Arachnocorys circumtexta* Hertwig 1879).

?*Arachnopilium* Haeckel 1881, p. 435. Type species: *Pteropilium clathrocanium* Haeckel 1887.

?*Archibursa* Haeckel 1881, p. 429. Type species: *A. tripodiscus* Haeckel 1887.

?*Archipera* Haeckel 1881, p. 429. Type species: *A. cortiniscus* Haeckel 1887.

Archiperidium Haeckel 1881, p. 429. Type species: *Peridium spinipes* Haeckel 1887.

Archiscenium Haeckel 1881, p. 429. Type species: *A. quadrispinum* Haeckel 1887.

Astrophormis Haeckel 1887, p. 1248. Type species: *Sethophormis aurelia* Haeckel 1887.

Callimitra Haeckel 1881, p. 431. Type species: *C. carolotae* Haeckel 1887.

Clathrocanium Ehrenberg 1860, p. 829. Type species: *C. squarrosum* Ehrenberg 1872*a*.

?*Clathrocorona* Haeckel 1881, p. 431. Type species: *Clathrocanium diadema* Haeckel 1887.

Clathrocorys Haeckel 1881, p. 432. Type species: *C. murrayi* Haeckel 1887.

Clathrolychnus Haeckel 1881, p. 431. Type species: *C. araneosus* Haeckel 1887.

Clathromitra Haeckel 1881, p. 432. Type species: *C. pterophormis* Haeckel 1887.

Cyrtostephanus Popofsky 1913, p. 289. Type species: *C. globosus* Popofsky 1913.

Deflandrella Loeblich and Tappan 1961 (for *Camplyacantha* Joergensen 1905). Type species: *Campylacantha cladophora* Joergensen 1905.

Dicorys Popofsky 1913, p. 369. Type species: *D. architypus* Popofsky 1913.

Dimelissa Campbell 1951, p. 529 (for *Micromelissa* Haeckel 1881 sens. emend., in Haeckel 1887, p. 1205). Type species: *Lithomelissa thoracites* Haeckel 1862.

Dumetum Popofsky 1908, p. 264. Type species: *D. rectum* Popofsky 1908.

Enneaphormis Haeckel 1881, p. 432. Type species: *E. rotula* Haeckel 1887.

Euscenium Haeckel 1887, p. 1146. Type species: *E. plectaniscus* Haeckel 1887.

Helotholus Joergensen 1905, p. 137. Type species: *H. histricosa* Joergensen 1905.

? *Hexaphormis* Haeckel 1881, p. 432. Type species: *Sethophormis hexalactis* Haeckel 1887.

Obeliscus Popofsky 1913, p. 279. Type species: *O. pseudocuboides* Popofsky 1913.

?*Octophormis* Haeckel 1887, p. 1245. Type species: *Sethophormis octalactis* Haeckel 1887.

?*Pentaphormis* Haeckel 1881, p. 432. Type species: *Sethophormis pentalactis* Haeckel 1887.

?*Peridium* Haeckel 1887, p. 1153. Type species: *P. lasanum* Haeckel 1887.

Periplecta Haeckel 1881, p. 424. Type species: *P. cortina* Haeckel 1887.

?*Phaenoscenium* Haeckel 1887, p. 1174. Type species: *P. hexapodium* Haeckel 1887.

Phormacantha Joergensen 1905, p. 132. Type species: *Peridium hystrix* Joergensen 1899.

?*Plagiocarpa* Haeckel 1881, p. 424. Type species: *P. procortina* Haeckel 1887.

Plagonidium Haeckel 1881, p. 424. Type species: *P. bigeminum* Haeckel 1887.

Plagonium Haeckel 1881, p. 423. Type species: *P. sphaerozoum* Haeckel 1887.

Plectacantha Joergensen 1905, p. 131. Type species: *P. oikiskos* Joergensen 1905.

Plectanium Haeckel 1881, p. 424. Type species: *P. trigeminum* Haeckel 1887.

Polyplagia Haeckel 1887, p. 917. Type species: *P. septenaria* Haeckel 1887.

Protoscenium Joergensen 1905, p. 133. Type species: *Plectanium simplex* Cleve 1899.

? *Sethopera* Haeckel 1881, p. 433. Type species: *S. tricostata* Haeckel 1887.

Spongomelissa Haeckel 1887, p. 1209. Type species: *Lithomelissa spongiosa* Bütschli 1882b.

Tetraphormis Haeckel 1881, p. 432 (= *Sethophormis* Haeckel 1887, p. 1243). Type species: *Sethophormis cruciata* Haeckel 1887.

Theophormis Haeckel 1881, p. 436. Type species: *T. callipilium* Haeckel 1887.

Tripocyrtis Haeckel 1887, p. 1201. Type species: *T. plagoniscus* Haeckel 1887.

Verticillata Popofsky 1913, p. 281. Type species: *V. hexacantha* Popofsky 1913.

Stratigraphical range: Cretaceous to Recent.

Remarks: This is a large, probably polyphyletic taxon, but it is judged to be a less unsatisfactory grouping than occurs in the Haeckelian system.

The spicule in this family is apparently not homologous with that of the Devonian *Cyrtentactinia* Foreman (1963, p. 285) and the Carboniferous *Pylentonema* Deflandre (1963, which appears to be a synonym of *Crytentactinia).*

Family **Acanthodesmiidae** Haeckel 1862

Definition: Nassellaria possessing a sagittal ring.

Stratigraphical range: Palaeocene to Recent.

Remarks: This family (under the name 'Trissocyclidae') has been extensively revised by Robert M. Goll, in a Ph.D. thesis at Ohio State University. The distinguishing characters of the revised genera cannot be summarized briefly, and the thesis is being prepared for early publication – for these reasons I only list the genera which Goll proposes to retain, and the taxa he places in synonymy with them.

Genus **Dendrospyris** Haeckel 1881, emend. Goll

Dendrospyris Haeckel 1881, p. 441. Type species: *Ceratospyris stylophora* Ehrenberg 1873.

Corythospyris Haeckel 1881, p. 443. Type species: *Elaphospyris damaecornis* Haeckel 1887.

Genus **Liriospyris** Haeckel 1881, emend. Goll.

Liriospyris Haeckel 1881, p. 443. Type species: *L. hexapoda* Haeckel 1887.

Amphispyridium Haeckel 1887, p. 1096. Type species: *Amphispyris sternalis* Haeckel 1887.

Trissocyclus Haeckel 1881, p. 446. Type species: *T. stauroporus* Haeckel 1887.

Genus **Tholospyris** Haeckel 1881, emend. Goll.

Tholospyris Haeckel 1881, p. 441. Type species: *T. tripodiscus* Haeckel 1887.

Tricolospyris Haeckel 1881, p. 443. Type species: *T. kantiana* Haeckel 1887.

Genus **Giraffospyris** Haeckel 1881, emend. Goll.

(Note: It appears more correct to regard Haeckel 1887 as the first reviser of the names *Elaphospyris* and *Giraffospyris* which he proposed in 1881, and therefore to follow his use of *Elaphospyris* as the name of this genus.)

Giraffospyris Haeckel 1881, p. 442. Type species: *Ceratospyris heptaceros* Ehrenberg 1873.

Aegospyris Haeckel 1881, p. 442. Type species: *A. aequispina* Haeckel 1887.

Dipocubus Haeckel 1887, p. 993. Type species: *Acrocubus arcuatus* Haeckel 1887.

Semantidium Haeckel 1887, p. 960. Type species: *S. hexastoma* Haeckel 1887.

Genus **Dorcadospyris** Haeckel 1881, emend. Goll

Dorcadospyris Haeckel 1881, p. 441. Type species: *D. dentata* Haeckel 1887.

Lophospyris Haeckel 1881, p. 443. Type species: *Ceratospyris pentagona* Ehrenberg 1872*a*.

Patagospyris Haeckel 1881, p. 443. Type species: *P. confluens* Ehrenberg 1873.

Family **Theoperidae** Haeckel 1881, emend. Riedel 1967*b*

Definition: Cephalis relatively small, approximately spherical, often poreless or sparsely perforate. The internal spicule, homologous with that of plagoniids, reduced to a less conspicuous structural element than in the latter group.

Stratigraphical range: Triassic to Recent.

Remarks: This is a large, probably polyphyletic taxon, which includes the majority of cyrtoids. No satisfactory division into subfamilies is yet possible.

Family **Carpocaniidae** Haeckel 1881, emend. Riedel 1967*b*

Definition: Cephalis small, not sharply distinguished in contour from thorax, and tending to be reduced to a few bars within top of thorax.

Genus-group taxa include:

Asecta Popofsky 1913, p. 373. Type species: *A. prunoides* Popofsky 1913.

?*Carpocanistrum* Haeckel 1887, p. 1170. The type species

indicated by Campbell 1954 is *Lithocarpium pyriforme* Stöhr 1880, but this contravenes Art. 67 (h) of the ICZN since Haeckel recognized its generic position as doubtful. The other species described in this genus probably belong in this family.

Carpocanium Ehrenberg 1847*a*, table facing p. 385. Type species: *Lithocampe solitaria* Ehrenberg 1838.

Carpocanobium Haeckel 1887, p. 1282. Type species: *Carpocanium trepanium* Haeckel 1887.

Cyrtocalpis Haeckel 1860, p. 835. Type species: *C. amphora* Haeckel 1860.

Sethamphora Haeckel 1887, p. 1249. Type species: *S. favosa* Haeckel 1887.

Stratigraphical range: Eocene to Recent.

Family **Pterocoryidae** Haeckel 1881, emend. Riedel 1967*b*

Definition: Cephalis subdivided into three lobes by two obliquely downwardly directed lateral furrows arising from the apical spine, in the manner described for *Anthocyrtidium cineraria* Haeckel and *Calocyclas virginis* Haeckel by Riedel (1957).

Genus-group taxa include:

Androcyclas Joergensen 1905, p. 139. Type species: *Pterocorys gamphonyxos* Joergensen 1899.

Anthocyrtidium Haeckel 1881, p. 431. Type species: *A. cineraria* Haeckel 1887.

Anthocyrtissa Haeckel 1887, p. 1270. Type species: *Anthocyrtis ophirensis* Ehrenberg 1872*a*.

Anthocyrtonium Haeckel 1887, p. 1274. Type species: *Anthocyrtium campanula* Haeckel 1887.

Anthocyrtura Haeckel 1887, p. 1271. Type species: *Anthocyrtis ovata* Haeckel 1887.

Calocycletta Haeckel 1887, p. 1381. Type species: *Calocyclas veneris* Haeckel 1887.

? *Conarachnium* Haeckel 1881, p. 430 (= *Sethoconus* Haeckel 1887, p. 1290). Type species: *Eucyrtidium trochus* Ehrenberg 1872*a*.

Craterocyclas Haecker 1908, p. 454. Type species: *C. robustissima* Haecker 1908.

Lamprocyclas Haeckel 1881, p. 434. Type species: *L. nuptialis* Haeckel 1887.

Lamprocycloma Haeckel 1887, p. 1392. Type species: *Lamprocyclas bajaderae* Haeckel 1887.

Podocyrtis Ehrenberg 1847*a*, p. 54. Type species: *P. papalis* Ehrenberg 1854*b*.

Pterocorys Haeckel 1881, p. 435. Type species: *P. campanula* Haeckel 1887.

Sethocorys Haeckel 1881, p. 430. Type species: *S. achillis* Haeckel 1887.

Sethocyrtis Haeckel 1887, p. 1298. Type species: *S. oxycephalis* Haeckel 1887.

Theoconus Haeckel 1887, p. 1399. Type species: *Eucyrtidium zancleum* Müller 1855*b*.

Theocorbis Haeckel 1887, p. 1401. Type species: *Theoconus jovis* Haeckel 1887.

Theocorythium Haeckel 1887, p. 1416. Type species: *Theocorys dianae* Haeckel 1887.

Theocyrtis Haeckel 1887, p. 1405. Type species: *Eucyrtidium barbadense* Ehrenberg 1873.

Stratigraphical range: Eocene to Recent.

Family **Amphipyndacidae** Riedel 1967*a*

Definition: Cephalis, generally with poreless wall, divided into two chambers by a transverse internal ledge.

Genus-group taxa include:

Amphipyndax Foreman 1966, p. 355. Type species: *A. enesseffi* Foreman 1966.

Stratigraphical range: Restricted to Cretaceous.

Family **Artostrobiidae** Riedel 1967*a*

Definition: Cephalis bears a lateral tubule, its relation to the internal spicular structure being generally similar to that described by Riedel (1958) for *Siphocampe* sp. and *Dictyocephalus papillosus* (Ehrenberg).

Genus-group taxa include:

Artostrobium Haeckel 1887, p. 1482. Type species: *Lithocampe aurita* Ehrenberg 1844.

Botryostrobus Haeckel 1887, p. 1475. Type species: *Lithostrobus botryocyrtis* Haeckel 1887.

? *Dictyoprora* Haeckel 1881, p. 430. Type species: *Dictyocephalus amphora* Haeckel 1887.

? *Lithomitra* Bütschli 1882*b*, p. 529. Type species: *Eucyrtidium pachyderma* Ehrenberg 1873.

? *Siphocampula* Haeckel 1887, p. 1499. Type species: *Siphocampe tubulosa* Haeckel 1887.

? *Theocamptra* Haeckel 1887, p. 1424. Type species: *Theocampe collaris* Haeckel 1887.

Theocorusca Haeckel 1887, p. 1407. Type species: *Theocyrtis macroceros* Haeckel 1887.

Tricolocapsa Haeckel 1881, p. 436. Type species: *T. theophrasti* Haeckel 1887.

? *Tricolocapsium* Haeckel 1887, p. 1433. Type species: *Tricolocapsa schleidenii* Haeckel 1887.

Stratigraphical range: Cretaceous to Recent.

Family **Cannobotryidae** Haeckel 1881, emend. Riedel 1967*b*

Definition: Cephalis consisting of two or more unpaired lobes, only one of which is homologous with the cephalis of theoperids.

Genus-group taxa include:

Acrobotrissa Popofsky 1913, p. 321. Type species: *A. cribrosa* Popofsky 1913.

Amphimelissa Joergensen 1905, p. 136. Type species: *Botryopyle setosa* Cleve 1899.

Bisphaerocephalus Popofsky 1908, p. 284. Type species: *B. minutus* Popofsky 1908.

Botryocella Haeckel 1887, p. 1116. Type species: *Lithobotrys nucula* Ehrenberg 1873.

Botryocyrtis Ehrenberg 1860, p. 829. Type species: *B. caput-serpentis* Ehrenberg 1872*a*.

Glycobotrys Campbell 1951, p. 530. Type species: *Lithobotrys geminata* Ehrenberg 1873.

Monotubus Popofsky 1913, p. 322. Type species: *M. microporus* Popofsky 1913.

Neobotrys Popofsky 1913, p. 320. Type species: *N. quadritubulosa* Popofsky 1913.

Phormobotrys Haeckel 1881, p. 440. Type species: *P. trithalamia* Haeckel 1887.

Pylobotrys Haeckel 1881, p. 440. Type species: *P. putealis* Haeckel 1887.

Saccospyris Haecker 1907, p. 124. Type species: *S. calva* Haecker 1907.

Stratigraphical range: Eocene to Recent.

Radiolaria incertae sedis

Family **Albaillellidae** Deflandre 1952

Genus-group taxa include:

Albaillella Deflandre 1952. Type species. *A. paradoxa* Deflandre 1952.

? *Ceratoikiscum* Deflandre 1953, p. 409. Type species: *C. avimexpectans* Deflandre 1953.

? *Holoeciscus* Foreman 1963, p. 294. Type species: *H. auceps* Foreman 1963.

Stratigraphical range: May be restricted to Carboniferous, but the two genera doubtfully assigned here occur in the Devonian.

Family **Palaeoscenidiidae** Riedel 1967*a*

Definition: Four divergent basal spines connected proximally by flat lamellae, and 2-4 shorter apical spines; no additional regular structures at junction of these spines.

Only one genus-group taxon is included, namely: *Palaeoscenidium* Deflandre 1953, p. 408. Type species: *P. cladophorum* Deflandre 1953.

Stratigraphical range: Devonian to Carboniferous.

ACKNOWLEDGEMENT

This paper includes results of research supported by the National Science Foundation (Grants GP-3354 and GA-658).

Received September 1967.

BIBLIOGRAPHY

Bolli, H. M. 1957. Planktonic Foraminifera from the Oligocene-Miocene Cipero and Lengua Formations of Trinidad, B.W.I. *Bull. U.S. natn. Mus.* **215**, 97-123, pls. 22-9.

Brandt, K. 1885. Die Koloniebildenden Radiolarien (Sphaerozoëen) des Golfes von Neapel. *Fauna Flora Golf. Neapel, Monogr.* **13**, i-viii, 1-276, pls. 1-8.

Bury, Mrs. 1867. *Polycystins, figures of remarkable forms, etc., in the Barbados chalk deposit* (2nd ed.). Weldon: London. 4 pp., 25 pls.

Bütschli, O. 1882a. Radiolaria. In: *Klassen und Ordnungen des Thier-Reichs* (ed. H. G. Bronn), **1**, pt. 1, 332-478, pls. 17-32.

Bütschli, O. 1882b. Beiträge zur Kenntnis der Radiolarienskelette, insbesondere der der Cyrtida. *Z. wiss. Zool.* **36**, 485-540, pls. 31-3.

Bütschli, O. 1889. Kurze Uebersicht des Systems der Radiolaria. In: *Klassen und Ordnungen des Thier-Reichs* (ed. H. G. Bronn), **1**, pt. 3, 1946-2004.

Campbell, A. S. 1951. New genera and subgenera of Radiolaria. *J. Paleont.* **25** (4), 527-30.

Campbell, A. S. 1954. Radiolaria. In: *Treatise on Invertebrate Paleontology* (R. C. Moore *ed.*), Part. D, Protista 3. Geol. Soc. Amer. and Univ. Kansas Press. 11-163.

Campbell, A. S. and Clark, B. L. 1944a. Miocene radiolarian faunas from Southern California. *Spec. Pap. geol. Soc. Amer.* **51**, i-vii, 1-76, pls. 1-7.

Campbell, A. S. and Clark, B. L. 1944b. Radiolaria from Upper Cretaceous of Middle California. *Spec. Pap. geol. Soc. Amer.* **57**, i-viii, 1-61, pls. 1-8.

Carter, F. B. 1896a. Radiolaria: a new genus from Barbados. *Amer. mon. microsc. J.* **17**, 96-7.

Carter, F. B. 1896b. Radiolaria: a new genus and new species. *Amer. mon. microsc. J.* **17**, 163-4.

Cleve, P. T. 1899. Plankton collected by the Swedish expedition to Spitzbergen in 1898. *K. svenska Vetensk - Akad. Handl.* **32** (3), 1-51, pls. 1-4.

Cleve, P. T. 1901. Plankton from the Indian Ocean and the Malay Archipelago. *K. svenska Vetensk - Akad. Handl.* **35** (5), 1-58, 8 pls.

Deflandre, G. 1952. *Albaillella* nov. gen., Radiolaire fossile du Carbonifère inferieur, type d'une lignée aberrante éteinte. *C. r. hebd. Séanc. Acad. Sci., Paris,* **234**, 872-4.

Deflandre, G. 1953. Radiolaries fossiles. In: *Traité de Zoologie* (ed. P.-P. Grassé, **1**, pt. 2. Masson: Paris. 389-436.

Deflandre, G. 1963. *Pylentonema,* nouveau genre de Radiolaire du viséen: Sphaerellaire ou Nassellaire? *C. r. hebd. Séanc. Acad. Sci., Paris,* **257**, 3,981-4.

Ehrenberg, C. G. 1838. Über die Bildung der Kreidefelsen und des Kreidemergels durch unsichtbare Organismen. *Abh. preuss. Akad. Wiss.,* Jahrg. **1838**, 59-147, pls. 1-4.

Ehrenberg, C. G. 1844. Über 2 neue Lager von Gebirgsmassen aus Infusorien als Meeres-Absatz in Nord-Amerika und eine Vergleichung derselben mit den organischen Kreide-Gebilden in Europa und Afrika. *Mber. preuss. Akad. Wiss.*, Jahrg. **1844**, 57-97.

Ehrenberg, C. G. 1847a. Über eine halibiolithische, von Herrn R. Schomburgk entdeckte, vorherrschend aus mikroskopischen Polycystinen gebildete, Gebirgsmasse von Barbados. *Mber. preuss. Akad. Wiss.*, Jahrg. **1846**, 382-5.

Ehrenberg, C. G. 1847b. Über die mikroskopischen kieselschaligen Polycystinen als mächtige Gebirgsmasse von Barbados und über das Verhältnis der aus mehr als 300 neuen Arten bestehenden ganz eigenthümlichen Formengruppe jener Felsmasse zu den lebenden Thieren und zur Kreidebildung. Eine neue Anregung zur Erforschung des Erdlebens. *Mber. preuss. Akad. Wiss.*, Jahrg. **1847**, 40-60, 1 pl.

Ehrenberg, C. G. 1854a. Über das organische Leben des Meeresgrundes in bis 10,800 und 12,000 Fuss Tiefe. *Mber. preuss. Akad. Wiss.*, Jahrg. **1854**, 54-75.

Ehrenberg, C. G. 1854b. *Mikrogeologie.* xxviii + 374 pp. Atlas, 31 pp.+41 pls. Fortsetzung (1856), 89 pp. Leipzig.

Ehrenberg, C. G. 1860. Über den Tiefgrund des stillen Oceans zwischen Californien und den Sandwich-Inseln aus bis 15,600' Tiefe nach Lieut. Brooke. *Mber. preuss. Akad. Wiss.*, Jahrg. **1860**, 819-33.

Ehrenberg, C. G. 1861. Beitrag zur Übersicht der Elemente des Tiefen Meeresgrundes im Mexikanischen Golfstrome bei Florida. *Mber. preuss. Akad. Wiss.*, Jahrg. **1861**, 222-40, 1 table.

Ehrenberg, C. G. 1872a. Mikrogeologischen Studien als Zusammenfassung seiner Beobachtungen des kleinsten Lebens der Meeres-Tiefgrunde aller Zonen und dessen geologischen Einfluss. *Mber. preuss. Akad. Wiss.*, Jahrg. **1872**, 265-322.

Ehrenberg, C. G. 1872b. Mikrogeologischen Studien über das kleinste Leben der Meeres-Tiefgrunde aller Zonen und dessen geologischen Einfluss. *Abh. preuss. Akad. Wiss.*, Jahrg. **1872**, 131-399, pls. 1-12, 1 chart.

Ehrenberg, C. G. 1873. Grössere Felsproben des Polycystinen-Mergels von Barbados mit weiteren Erläuterungen. *Mber. preuss. Akad. Wiss.*, Jahrg. **1873**, 213-63.

Ehrenberg, C. G. 1875. Fortsetzung der mikrogeologischen Studien als Gesammt-Uebersicht der mikroskopischen Paläontologie gleichartig analysirter Gebirgsarten der Erde, mit specieller Rücksicht auf den Polycystinen-Mergel von Barbados. *Abh. preuss. Akad. Wiss.*, Jahrg. **1875**, 1-226, pls. 1-30.

Foreman, H. P. 1963. Upper Devonian Radiolaria from the Huron member of the Ohio shale. *Micropaleontology*, **9** (3), 267-304, pls. 1-9.

Foreman, H. P. 1966. Two Cretaceous radiolarian genera. *Micropalaeontology*, **12** (3), 355-9.

Friend, J. K. and Riedel, W. R. 1967. Cenozoic orosphaerid radiolarians from tropical Pacific sediments. *Micropalaeontology*, **13**, 217-32.

Goll, R. M. 1967. The classification and phylogeny of the Trissocyclidae (Radiolaria) in the Cenozoic of the Pacific and Caribbean Basins. Ph.D. Thesis, Ohio State University.

Guppy, R. J. L. 1909. The geological connections of the Caribbean region. *Trans. R. Can. Inst.* **8** (18), 373-91, 1 chart.

Haeckel, E. 1860. Abbildungen und Diagnosen neuer Gattungen und Arten von lebenden Radiolarien des Mittelmeeres. *Mber. preuss. Akad. Wiss.*, Jahr. **1860**, 835-45.

Haeckel, E. 1982. *Die Radiolarien (Rhizopoda Radiolaria).* Eine Monographie. Reimer: Berlin. xiv + 572 pp., 35 pls.

Haeckel, E. 1879. *Natürliche Schöpfungsgeschichte* (7th ed.). Reimer: Berlin.

Haeckel, E. 1881. Entwurf eines Radiolarien-Systems auf Grund von Studien der Challenger – Radiolarien. *Jena. Z. Naturw.* **15** (new ser. 8), no. 3, 418-72.

Haeckel, E. 1887. Report on the Radiolaria collected by H.M.S. *Challenger* during the years 1873-76. *Rept. Voyage 'Challenger', Zool.* **18**, clxxxviii+1803 pp., 140 pls., 1 map.

Haecker, V. 1907. Altertümliche Sphärellarien und Cyrtellarien aus grossen Meerestiefen.*Arch. Protistenk.* **10**, 114-26.

Haecker, V. 1908. Tiefsee- Radiolarien. Spezieller Teil, Lfg. 2. Die Tripyleen, Collodarien und Mikroradiolarien der Tiefsee. *Wiss. Ergebn. dt. Tiefsee-Exped. 'Valdivia'*, **14**, 337-476, pls. 63-85.

Hertwig, R. 1879. *Der Organismus der Radiolarien.* G. Fischer: Jena. iv + 149 pp., 10 pls.

Hilmers, C. 1906. Zur Kenntnis der Collosphaeriden. Doctoral Dissertation, Kgl. Christian-Albrecht-Univ. Kiel, 95 pp., 1 pl.

Hinde, G. J. 1899a. On the Radiolaria in the Devonian rocks of New South Wales. *Q. Jl. geol. Soc. Lond.* **55**, 38-64, pls. 8-9.

Hinde, G. J. 1899b. On Radiolaria in chert from Chypons Farm, Mullion Parish (Cornwall). *Q J. geol. Soc. Lond.* **55**, 214-19, pl. 16.

Hollande, A. and Enjumet, M. 1960. Cytologie, évolution et systématique des Sphaeroïdés (Radiolaries). *Archs Mus. natn. Hist. nat., Paris, ser. 7*, **7**, 1-134, pls. 1-64.

Joergensen, E. 1899. Protophyten und Protozoen im Plankton aus der norwegischen Westkuste. *Bergens Mus. Årb.* for **1899**, no. 6, 1-112, pls. 1-5.

Joergensen, E. 1905. The protist plankton and the diatoms in bottom samples. *Bergens Mus. Skr.* **1905**, 49-151, 195-225, pls. 6-18.

Khabakov, A. V. 1937. Fauna radiolyarii nizhemelovykh i verkhneyurskikh fosforitov basseina verkhnei vyatki i kamy. (Die Radiolarien-Fauna aus mesozoischen Phosphoriten des Kama- und Wiatka-Gebietes.) *Ezheg. vses. paleont. Obshch.* **11**, 90-120, pls. 11-14.

Loeblich, A. R., Jr. and Tappan, H. 1961. Remarks on the systematics of the Sarkodina (Protozoa), renamed homonyms and new and validated genera. *Proc. biol. Soc. Wash.* **74**, 213-14.

Müller, J. 1855a. Über *Sphaerozoum* und *Thalassicolla.* *Mber. preuss. Akad. Wiss.* Jahrg. **1855**, 229-53.

Müller, J. 1855b. Über die im Hafen von Messina beobachteten Polycystinen. *Mber. preuss. Akad. Wiss.* Jahrg. **1855**, 671-74.

Müller, J. 1858. Über die Thalassicollen, Polycystinen und Acanthometren des Mittelmeeres. *Abh. preuss. Akad. Wiss.* Jahrg. **1858**, 1-62, pls. 1-11.

Nigrini, C. 1967. Radiolaria in pelagic sediments from the Indian and Atlantic Oceans. *Bull. Scripps Instn Oceanogr.* **11**, 1-106, pls. 1-9.

Perner, J. 1891. O radiolariích z českého útvaru křídového. *Sber. k. böhm. Ges. Wiss.* **1891**, 255-69, pl. 10.

Pessagno, E. A., Jr. 1963. Upper Cretaceous Radiolaria from Puerto Rico. *Micropaleontology*, **9** (2), 197-214, pls. 1-7.

Petrushevskaya, M. G. 1965. Osobennosti konstruktsii skeleta radiolyarii Botryoidae (otr. Nassellaria). *Trudy zool. Inst. Leningr.* **35**, 79-118.

Popofsky, A. 1908. Die Radiolarien der Antarktis (mit Ausnahme der Tripyleen). *Dt. Südpol.-Exped.* 1901-1903, **10** (*Zool.* vol. 2), no. 3, 183-305, 1 table, pls. 20-36.

Popofsky, A. 1912. Die Sphaerellarien des Warmwassergebietes. *Dt. Südpol.-Exped.* 1901-1903, **13** (*Zool.* vol. 5), no. 2, 73-159, pls. 1-8.

Popofsky, A. 1913. Die Nassellarien des Warmwassergebietes. *Dt. Südpol.-Exped.* 1901-1903, **14**, (*Zool.* vol. 6), pp. 217-416, pls. 28-38.

Popofsky, A. 1917. Die Collosphaeriden, mit Nachtrag zu den Spumellarien und Nassellarien. *Dt. Südpol.-Exped.* 1901-1903, **16** (*Zool.* vol. 8) no. 3, 235-78, pls. 13-17.

Riedel, W. R. 1957. Radiolaria: a preliminary stratigraphy. *Rep. Swed. deep Sea Exped.* **6** (3), 59-96, pls. 1-4.

Riedel, W. R. 1958. Radiolaria in Antarctic sediments. *Rep. B.A.N.Z. antarct. Res. Exped. ser. B*, **6** (10), 217-55.

Riedel, W. R. 1959. Oligocene and Lower Miocene Radiolaria in tropical Pacific sediments. *Micropaleontology*, **5** (3), 285-302.

Riedel, W. R. 1967a. Some new families of Radiolaria. *Proc. geol. Soc. Lond.* **1640**, 148-9.

Riedel, W. R. 1967b. Subclass Radiolaria. In: *The Fossil Record* (ed. W. B. Harland *et al.*), 291-298. Geological Society: London.

Riedel, W. R. in press.

Riedel, W. R. and Funnell, B. M. 1964. Tertiary sediment cores and microfossils from the Pacific Ocean floor. *Q. J. geol. Soc. Lond.* **120**, pp. 305-68, pls. 14-32.

Rüst, D. 1885. Beiträge zur Kenntniss der fossilen Radiolarien aus Gesteinen des Jura. *Paleontographica*, **31** (*ser. 3*, vol. **7**), 269-321, pls. 26-45.

Rüst, D. 1892. Beiträge zur Kenntniss der fossilen Radiolarien aus Gesteinen der Trias und der palaeozoischen Schichten. *Palaeontographica*, **38**, 107-92, pls. 6-30.

Schneider, A. 1858. Ueber 2 neue Thalassicollen von Messina. *Arch. Anat. Physiol., Leipzig*, **1858**, 38-42, pl. 3B.

Squinabol, S. 1903. Le Radiolarie dei Noduli selciosi nella Scaglia degli Euganei. *Riv. ital. Paleont.* **9**, 105-50, pls. 8-10.

Squinabol, S. 1914. Contributo alla conoscenza dei Radiolarii fossili del Veneto. Appendice- Di un genere di Radiolari caratteristico del Secondario. *Memorie Ist. geol. Univ. Padova*, **2**, 249-306 (and Corrigenda), pls. 20-4.

Stöhr, E. 1880. Die Radiolarienfauna der Tripoli von Grotte, Provinz Girgenti in Sicilien. *Paleontographica*, **26**, (*ser. 3*, vol. 2), 69-124 (and Corrigenda), pls. 17-23 (1-7).

Sutton, H. J. 1896. Radiolaria; a new genius [*sic*, W. R. R.] from Barbados. *Am. mon. microsc. J.* **17**, 138-9.

Tan Sin Hok. 1927. Over de samenstelling en het onstaan van krijt- en mergelgesteenten van de Molukken. *Jaarb. Mijnw. Ned.-Oost-Indië*, jaarg. **1926**, *Verhand.* pt. 3, 5-165, pls. 1-16.

INDEX

This is an index of the genus-group (and a few species-group) taxa classified in this paper, with an indication of the family or subfamily to which they are assigned. Abbreviations are as follows: Acanth., Acanthodesmiidae; Alb., Albaillellidae; Amphi, Amphipyndacidae; Artisc., Artiscinae; Artost., Artostrobiidae; Canno., Cannobotryidae; Carpo., Carpocaniidae; Cocco., Coccodiscidae; Collosph., Collosphaeridae; Entact., Entactiniidae; Hag., Hagiastrinae; Myel., Myelastrinae; *non* Artisc., forms previously regarded as related to artiscins; *non* Cocco., forms previously regarded as coccodiscids but excluded here; Palaeo., Palaeoscenidiidae; Plag., Plagoniidae; Pseudo., Pseudoaulophacidae; Ptero., Pterocoryidae; Saturn., Saturnalinae.

SYSTEMATIC CLASSIFICATION OF POLYCYSTINE RADIOLARIA

Enneaphormis, Plag.
Entactinia, Entact.
Entactinosphaera, Entact.
Euscenium, Plag.
Giraffospyris, Acanth.
Glycobotrys, Canno.
Hagiastrum, Hag.
Haplentactinia, Enctact.
Helotholus, Plag.
Hexaphormis, ?Plag.
Holoeciscus, ?Alb.
Hymenactinium, Cocco.
Hymenactura, Cocco.
Hymeniastrum, Cocco.
Lamprocyclas, Ptero.
Lamprocycloma, Ptero.
Liriospyris, Acanth.
Lithocyclia, Cocco.
Lithomitra, ?Artost.
Lophospyris, Acanth.
Merosiphonia, Collosph.
Micromelissa, Plag.
Monaxonium, *non* Artisc.
Monotubus, Canno.
Myelastromma, Myel.
Myelastrum, Myel.
Myxosphaera, Collosph.
Neobotrys, Canno.
Obeliscus, Plag.
Octophormis, ?Plag.
Odontosphaera, Collosph.
Ommatacantha, Artisc.
Ommatartus, Artisc.
Ommatocampe, *non* Artisc.
Ommatocampula, Artisc.
Ommatocorona, Artisc.
Ommatocyrtis, Artisc.
Ommatospyris, *non* Artisc.
Otosphaera, Collosph.
Palaeoscenidium, Palaeo.
Panarium, Artisc.
Panaromium, Artisc.
Panartidium, Artisc.
Panartissa, Artisc.
Panartoma, Artisc.
Panartura, Artisc.
Panartus, Artisc.
Panicium, Artisc.
Patagospyris, Acanth.
Pentactura, Cocco.
Pentaphormis, ?Plag.
Pentophiastromma, Myel.
Pentophiastrum, Myel.
Peridium, ? Plag.
Peripanarium, Artisc.
Peripanartium, Artisc.
Peripanartus, Artisc.
Peripanicium, Artisc.
Peripanicula, Artisc.
Periplecta, Plag.
Phaenoscenium, ?Plag.
Phormacantha, Plag.
Phormobotrys, Canno.
Pipetta, Artisc.
Pipettaria, Artisc.
Pipettella, Artisc.
Plagiocarpa, ?Plag.
Plagonidium, Plag.
Plagonium, Plag.
Plectacantha, Plag.
Plectanium, Plag.
Podocyrtis, Ptero.
Polyentactinia, Entact.
Polyplagia, Plag.
Polysolenia, Collosph.
Protoscenium, Plag.
Pseudoaulophacus, Pseudo.
Pterocorys, Ptero.
Pylobotrys, Canno.
Saccospyris, Canno.
Saturnalis, Saturn.
Saturnalium, Saturn.
Saturninus, Saturn.
Saturnulus, Saturn.
Semantidium, Acanth.
Sethamphora, Carpo.
Sethoconus, ?Ptero.
Sethocorys, Ptero.
Sethocyrtis, Ptero.
Sethopera, ?Plag.
Sethophormis, Plag.
Siphocampula, ?Artost.
Siphonosphaera, Collosph.
Solenosphaera, Collosph.
Solenosphenia, Collosph.
Spongasteriscus, ?Myel.
Spongastromma, ?Myel.
Spongocoelia, ?Entact.
Spongomelissa, Plag.
Spongosaturnalis, Saturn.
Spongosaturninus, Saturn.
Spongotrochus ehrenbergi, Pseudo.
Stauractinium, Cocco.
Stauractura, Cocco.
Staurococcura, ?Cocco.
Staurocyclia, Cocco.
Stigmosphaerostylus, Entact.
Stylartura, Artisc.
Stylartus, Artisc.
Stylocyclia, Cocco.
Tetracranastrum, Myel.
Tetraphormis, Plag.
Tetrasolenia, Collosph.
Tetrentactinia, Entact.
Theocamptra, ?Artost.
Theoconus, Ptero.
Theocorbis, Ptero.
Theocorusca, ?Artost.
Theocorythium, Ptero.
Theocyrtis, Ptero.
Theodiscus superbus, Pseudo.
Theophormis, Plag.
Tholospyris, Acanth.
Triastrum, Myel.
Tribonosphaera, Collosph.
Tricolocapsa, Artost.
Tricolocapsium, ?Artost.
Tricolospyris, Acanth.
Tricranastrum, Myel.
Trigonactinium, Cocco.
Trigonactura, *non* Cocco.; ?Hag.
Trigonactura angusta, ?Artisc.
Trigonocyclia, Cocco.
Tripocyclia, Cocco.
Tripocyrtis, Plag.
Trissocyclus, Acanth.
Trypanosphaera, Collosph.
Trypanosphaerium, Collosph.
Verticillata, Plag.
Zygartus, Artisc.
Zygocampe, Artisc.

SECTION F

STRATIGRAPHICAL BASES FOR OCEANIC MICROPALAEONTOLOGY

50. OBSERVATIONS ON THE BIOSTRATIGRAPHY OF PELAGIC SEDIMENTS[1]

M. N. BRAMLETTE AND W. R. RIEDEL

Summary: The basic principles of biostratigraphy, developed during more than a century from the study of strata on land, are obviously the same as those to be applied to the deep-sea floor, even though pelagic deposits are rare, indeed, on land. These principles are briefly discussed, and some consideration given to aspects peculiar to pelagic deposits which result from: low rates of sediment accumulation, a high proportion of planktonic microfossils, uniformity over wide areas, extensive effects of reworking processes, and the small number of continuous sequences representing long periods of time so far available.

INTRODUCTION

It is impossible to present here an adequate discussion of such a large subject as the biostratigraphy of pelagic sediments, and this brief paper therefore includes some rather dogmatic statements. These may, it is hoped, serve as a basis for the discussion of different points of view.

Arkell (1933) has aptly referred to 'the monstrous and sterile problems of stratigraphical nomenclature'. However, the excellent discussion in the first 37 pages of his *Jurassic System in Great Britain* clarifies the concepts expressed by various stratigraphical terms that are basic to some principles of classification and to an understanding of the subject. Whether one uses the nomenclature suggested by Arkell, based largely on priority and clearly defined concepts, or that of the American, Russian, or some other Code, this need only be stated – or, if no established code is strictly followed, it is necessary to make clear the degree of departure from such a code. No system is likely to be immediately acceptable to all workers, and continued discussion of principles, at least, seems important and profitable.

Most will probably agree that stratigraphy involves three kinds of stratal units, commonly termed lithostratigraphical, biostratigraphical and chronostratigraphical, besides the purely time units corresponding to the chronostratigraphical stratal units. As lithostratigraphical units are apparently of less importance to studies of deep-sea sediments than to studies of shallow-water deposits, they are not discussed here. Although chronostratigraphical units (stages, etc.) are basically different from biostratigraphical ones (various types of zones), they remain at present largely dependent on the biostratigraphical units for their delimitation away from their stratotype locality. This distinction seems subtle or even unnecessary to some biostratigraphers, but non-palaeontological methods for determining time relations between stratal units are developing rapidly. When (or if) such methods produce results demonstrably more precise and refined than is possible by biostratigraphy, the biostratigrapher may concentrate more of his attention on the many fascinating problems of distribution of the organic remains at precise times, in the oceans as well as on land, and on the many palaeoecological implications. At present, biostratigraphers interested primarily in ecology and facies tend to overlook the fact that their interpretations of facies are usually of little significance to earth history unless these can be placed in a time framework indicating relations to conditions elsewhere.

BIOSTRATIGRAPHICAL UNITS

Biostratigraphical units may be grouped into three principal kinds of zones. Those strata characterized by a certain assemblage of fossils which, regardless of their varying time ranges, distinguish the stratal unit as a particular kind of facies deposit (littoral, lagunal, etc.), have been termed 'assemblage zones' in the American Code – these, again, are of less importance to investigators con-

[1] Contribution from the Scripps Institution of Oceanography, University of California, San Diego, California.

centrating on the deep sea. Secondly, zones delimited by the known range of a single fossil taxon ('biozone' of Arkell and many others; 'range zone' of the American Code) obviously receive the name of that one taxon. A third kind of biostratigraphical zone is delimited by the concurrent and/or overlap relations of the ranges of more than one taxon ('faunizone' of Arkell; 'concurrent range zone' of the American Code) – these are the most reliable in helping to delimit any chronostratigraphical stratal units (Eocene Series, Lutetian Stage, etc.). Unfortunately, much confusion and misunderstanding has resulted from the accepted custom of also naming these zones by the name of only one of the several or many taxa that define such a zone. Therefore it seems most important to make clear that this kind of zone is based not on the range of the one taxon which was selected to name it, but on the ranges of associated taxa as well. It can thus remain a stable and valuable zone for correlation even though the range of the single named taxon later needs to be modified in relation to the associated taxa that help define the zone. Although the earliest occurrence of any taxon marks only a horizon for possible correlation, rather than a stratal zone, this, combined with the earliest occurrence of a later taxon, defines a kind of concurrent range zone – it is defined, however, on the basis of only these two events, and without the relations to ranges of other taxa it offers little more assurance of synchroneity than does a single taxon range zone.

A datum defined by the earliest occurrence of a taxon may seem more dependable than one based on a latest occurrence, since the latter may be difficult to recognize where reworking has occurred. However, unless lineages are clearly determined, first appearances may represent only the earliest invasion of a locality by a species that evolved earlier elsewhere. And, when a group of microfossils is sufficiently well known to permit recognition of reworking, a latest occurrence can be at least as useful as an earliest occurrence, particularly if sufficient data indicate an extinction of a widespread taxon.

Many lineages are being recognized among planktonic microfossils, although among unicellular forms without special skeletal parts of functional value the morphological similarities are perhaps not as significant as we would like to believe when postulating phylogenetic series. Arkell has remarked 'Unfortunately opportunities for making use of lineages in zonal work are extremely rare' – and, regarding the ammonites of the Jurassic seas of Europe, he says 'Conditions of life have been normally too unstable for any stocks to evolve *in situ.* They continually move to pastures new.' However, we are fortunate in dealing with large populations, rather commonly in continuous sequences, which offer considerable assurance of demonstrating the lineages that are so important in determining the total range zones of taxa.

DISTINCTIVE FEATURES OF DEEP-SEA BIOSTRATIGRAPHY

Although the biostratigraphical principles appropriate to the investigation of submarine sediments do not differ from those applied to land-based sequences of sedimentary rocks, there are marked differences in the kinds of fossils involved and the factors affecting their preservation. Other characteristics of pelagic stratigraphy are the relative inaccessibility of the materials, the small number of extended sequences available so far, and the smaller variety of facies represented.

One of the outstanding characteristics of deep-sea sediments is their low rate of accumulation. Of the widely distributed sediment types, that accumulating most rapidly is siliceous calcareous ooze in areas of high biological productivity, with an accumulation rate of approximately 15 metres per million years, while the lowest rates of accumulation are about 1 metre or less per million years for unfossiliferous 'red' clays. Therefore strata representing entire epochs of the Tertiary may be remarkably thin, and the effects of physical and biological disturbance are profound. The removal of small sediment thicknesses by erosional processes causes large time-gaps in sequences, and commonly results in the admixture of older material into younger deposits. Unless microfossils or isotopic tracers are present, the presence of the resulting hiatuses are difficult to detect, and the reworked admixtures cannot be evaluated.

Difficulties in interpreting pelagic microfossil assemblages are particularly serious in the case of non-calcareous sediments, since the evolutionary sequences of radiolarian and diatom assemblages are not yet sufficiently well known to permit precise age-assignments. Sediments containing calcareous nannoplankton, and especially planktonic Foraminifera (less easily reworked), can at present be correlated more readily. The greater usefulness of the Foraminifera derives in large measure from the fact that they can more commonly be correlated with well known land-based sequences where stratotypes have been designated. Even if a comparable amount of work and knowledge on the other groups, such as the coccolithophorids, should permit a more refined zonation, their occurrence in time and in space is more limited. Graptolites, or other groups, may be of importance only in the few cases when penetration to such old strata on the ocean floor may prove possible. Hystrichosphaerids, pollen, etc., seem to be largely destroyed by chemical or biological oxidation in the slowly accumulated pelagic deposits, but they do survive in some hemipelagic sediments and deep-sea turbidity deposits.

It is essential that our planktonic zonations be related as thoroughly as possible to those in stratotypes on land, if we are to use the well known series names of Lyell, or the recognized stage names in common use. In the early phases of investigation of an unfamiliar microfossil group, it may be necessary to correlate the sea-floor occurrences with well known sequences in areas such as, for example, the Caribbean or Indonesia. When, however, such correlatable zones are assigned to the series and stages, the degree of reliability of correlation with the European stratotypes should be stated. When a microfossil group is not even well enough known to permit correlation with secondary, extra-European reference sections, it may nevertheless be possible to make useful local correlations by the use of faunal or floral zones defined and named in accordance with one or other of the national stratigraphical codes.

The concept of thanatocoenosis seems much more important in deep-sea stratigraphy than in land-based investigations. Accumulation may be in quite a different area from that of the surface population. Surprisingly, however, data on diatoms (Kanaya and Koizumi 1966) and coccoliths (McIntyre 1967) do not indicate the transport over many thousands of kilometres which might be theoretically deduced from rates of settling and current velocities. The most reasonable explanation now known is that most of the surface plankton reaching the bottom was incorporated in the settling dead bodies or the faecal pellets of various predators – and thus in aggregates with high settling rates.

Even more important in ecological interpretations of slowly accumulated skeletal remains are the profound effects of varying preservation, which depend on the size or robustness and composition of the skeletal material. Non-calcareous red clay at depths greater than about 5,000 metres is evidently no indication of absence of calcareous plankton in the upper water layers, nor the number and variety of diatom frustules a complete representation of their surface occurrence, and so on. In some areas of low biological productivity, highly calcareous Quaternary sediments contain no Quaternary siliceous microfossils but may contain Tertiary radiolarians. Thus examination of the siliceous microfossils alone would lead to an erroneous age assignment. The situation arises from the fact that most Quaternary radiolarian skeletons are constructed of thinner elements than most of the Tertiary forms; and therefore the siliceous microfossils deposited contemporaneously with the Quaternary calcareous microfossils tend to dissolve, while any Tertiary radiolarians reworked into such a locality are more resistant to solution. An analogous situation obtains in the case of some non-calcareous clays with sparse siliceous microfossils – the sediment may have accumulated during the Quaternary and the only microfossils present may be older, reworked radiolarians.

Despite various special problems presented to the stratigrapher investigating deep-sea deposits, there are important advantages for him. He is commonly working with an abundance of well preserved microfossils of four major and widespread planktonic groups. With fewer and less complex facies variations to contend with, one of his major contributions can be in developing, extending and relating the zonation of these planktonic groups, to offer a more secure and precise time-correlation of world events. In recent years, advances in the stratigraphical usefulness of all four major microfossil groups have been based to a considerable extent on investigations of open-ocean sediments.

Widespread as many planktonic groups are, however, do we tend to assume too much synchroneity in their distribution in time and space? Certainly some forms are limited to one ocean, or arrive in one at a distinctly different time than elsewhere. Parker (1965) has reviewed some evidence for this among the Foraminifera, and it is becoming more evident for the coccolithophorids. (Loeblich and Tappan 1964) and others have emphasized the wide and rapid transport of the floating bottle, but such an object is little affected by the temperature or fertility of the water through which it moves, nor by predators. The subject of ecological effects, and their bearing on rates of migration, etc., cannot be pursued at length here, interesting and important as it is to biostratigraphers.

Another matter of concern to the deep-sea stratigrapher is the fact that sediment sequences may be distorted during the coring operation. The most common distortion is a shortening of the sequence in open-barrel gravity cores, and an even greater shortening of the upper portions of cores collected by some types of piston cores. The degree of such shortening remains unknown in many cases, but can sometimes be evaluated by comparison of the cores obtained by several different types of sampling equipment in a limited area. Also, the piston coring process sometimes causes water-filled gaps in the core, and material from other levels may fill such gaps and thus form artificial layers – this effect is relatively easily recognized if the core is examined immediately after collection, but may be difficult to detect after the core has been stored for a number of years.

In attempting to foresee the future of stratigraphical studies of pelagic sediments it seems evident that a major part of the value of such research will derive from the fact that palaeontological information will become available from the extensive water-covered areas of the globe separating the better known land-based sequences. Thus we shall be able to fill in large geographical gaps in the record of earth history. And drilling at many localities in the deep sea opens up exciting prospects of extending the record further back in time.

Received September 1967.

BIBLIOGRAPHY

Arkell, W. J. 1933. *The Jurassic system of Great Britain.* Clarendon Press: Oxford. xii+681 pp., 41 pls.

Kanaya, T. and Koizumi, 1966. Interpretation of diatom thanatocoenoses from the North Pacific applied to a study of core V20-130 (Studies of a deep-sea core V20-130. Part IV.) *Sci. Rep. Tohoku Univ. ser. 2 (Geology)*, 37 (2), 89-130.

Loeblich, A. R., Jr. and Tappan, Helen, 1964. Foraminiferal facts, fallacies, and frontiers. *Bull. geol. Soc. Amer.* 75, 367-92.

McIntyre, A. 1967. *The Coccolithophoridae of the Atlantic Ocean.* Ph.D. Thesis: Faculty of Pure Science, Columbia University, New York.

Parker, Frances L. 1965. Irregular distributions of planktonic Foraminifera and stratigraphic correlation. *Prog. Oceanography*, 3, 267-72.

51. QUATERNARY BOUNDARIES AND CORRELATIONS[1,2]

J. D. HAYS AND W. A. BERGGREN

Summary: Recent investigations have yielded data which allow recognition of the Pliocene/Pleistocene boundary in deep-sea cores and correlation with the type section of the Calabrian in southern Italy. Palaeomagnetic data have aided in the formulation of a time-scale for the past 4-5 million years to which palaeontological-biostratigraphical events can be related. Criteria can now be established for recognizing the Pliocene/Pleistocene boundary in all parts of the world's oceans.

The initiation of glaciation has no direct bearing on the definition of the lower limit of the Quaternary period; however, the criteria used by vertebrate palaeontologists may closely approximate the boundary based on marine faunas.

INTRODUCTION

As the most recent period of geological history, the Quaternary has been one of the most extensively studied. Within this time unit the fields of anthropology, glaciology invertebrate palaeontology and vertebrate palaeontology have met; however, each has chosen to define its limits in a way most suitable to its respective discipline. The conflict between specialists has been particularly strong with respect to the definition of the base of the Quaternary, with each discipline feeling strongly that it should be marked by some auspicious event in its own field. Consequently, since the Quaternary has been considered the age of man, the anthropologists thought its base should be marked by the first evidence of man. The glaciologists, believing the Quaternary to be synonymous with ice ages, advanced the first evidence of glaciation as a suitable criterion. Vertebrate palaeontologists recommended the first appearance of elephants and horses, while invertebrate palaeontologists, influenced by glaciologists, advocated the first appearance in Mediterranean sediments of certain cool water species. Much of this special pleading was silenced by the 18th International Geological Congress whose Temporary Committee for the Study of the Pliocene/Pleistocene Boundary recommended (*Int. Geol. Congr., Repts. 18th Sess., 1948, Pt. 10,* 1950) that the 'Pliocene/Pleistocene boundary should be based upon changes in marine faunas, since this is the classic method of grouping fossiliferous strata'. It also recommended that the 'Lower Pleistocene should include as its lower member in the type-area the Calabrian formation (marine) together with its terrestrial (continental) equivalent the Villafranchian'. This boundary was believed to coincide with the first indication of climatic deterioration in the Italian Neogene succession.

Gignoux (1910) introduced the Calabrian as the youngest stage of his *Pliocène supérieur.* At the base of the Calabrian a change in the molluscan and foraminiferal fauna occurred with the appearance of species from cooler water regions in northern Europe (the so-called *nordische Gaste* of Suess). *Arctica islandica* and *Hyalinea baltica* are the most commonly cited guide-fossils for recognizing this boundary in Italy. In littoral and epineritic sections *A. islandica* appears somewhat earlier than *H. baltica* (a deeper water species) but the difference is best explained as a difference in facies distribution (Selli 1967).

In a paper in a subsequent report of the commission nominated by the Geological Society of Italy to study the Pliocene/Pleistocene boundary in Italy, Gignoux (1954) expressed his agreement with the placement of the Calabrian in the Quaternary. The stratotype section of the Calabrian Stage is at Santa Maria di Catanzaro (Gignoux 1913), on the southeast coast of Italy (Fig. 51.1). The proposal made at the 7th INQUA Congress in Denver, Colorado, in 1965, that the type section of the Pliocene/Pleistocene boundary should be established at Le Castella (Calabria), Italy, not far from Santa Maria di Catanzaro, would appear to have been approved (see reply of Emiliani to Richmond in *Science*, 1967, vol. **156**, p. 410).

Until recently the various criteria for recognizing the base of the Quaternary (palaeoclimatic, anthropological, palaeontological, glaciological) have developed relatively independently of each other. Lack of overlapping data rendered impossible attempts to determine whether any

[1] Lamont – Doherty Geological Observatory Contribution No. 1550.
[2] Woods Hole Oceanographic Institution Contribution No. 2066.

or all of the boundaries thus recognized were indeed isochronous. Various lines of evidence, which will be discussed in the sections below, now make it possible to estimate the age of the Pliocene/Pleistocene boundary and to formulate a relatively accurate chronometric scale of Quaternary events.

THE CONCEPT OF THE QUATERNARY – HISTORICAL REVIEW

Subdivision of geological time as recorded in the rock record began with the investigations of Arduino, who in 1759 recognized a Primary and Secondary subdivision (which later became known as the Palaeozoic and Mesozoic), a third subdivision (low mountains and hills of sand and gravel, clays and marls with abundant marine fossils) and a fourth subdivision (earthy and rocky materials and alluvial debris).

In 1810 Brongniart applied the term *tertiaire* to the strata which lay above the chalks of the Cretaceous of the Paris Basin. The term Tertiary, although a relic of early geological science, has proved durable and is used at present for all pre-Quaternary - post-Mesozoic rocks. The fourth subdivision recognized by Arduino later became known as the Quaternary.

In 1833 Sir Charles Lyell introduced the term Pliocene Period as the youngest division of the Tertiary and subdivided it, on the basis of the percentages of molluscan species still living, into two parts:

Newer Pliocene: 90-95 per cent still living
Older Pliocene: more than 50 per cent still living

He subsequently introduced the term Pleistocene as a substitute for Newer Pliocene, although later recommending that the term Pleistocene be abandoned.

In 1824 Marcel de Serres (*in* Creuze de Lesser, p. 174) discussed the presence in France of relatively recent calcareous sediments which he attributed to his *quatrième formation d'eau douce.* He later (1855) claimed, somewhat unjustly it would appear, priority for the creation of the term Quaternary. He had used the term *Quaternaire* in 1830 (a year after its introduction by Desnoyers) in considering it a synonym of Diluvium.

Desnoyers (1829) suggested that the seas had remained longer in the region of Touraine and Languedoc than in the Seine Basin. He proposed the term *Quaternaire* or *Tertiaire récent* for marine deposits younger than those in the Seine Basin and divided it into three parts (from the top):

3. *Récent*
2. *Diluvium*
1. *Faluns de Touraine, la molasse suisse, le Pliocène marin de Languedoc*

We see, thus, that the Quaternary originally included a heterogeneous assemblage of rocks (even within a given subdivision such as the lower one) and corresponded, essentially, to the Neogene of Hoernes and later authors in its general application.

Morlot (1854) introduced the word *Quaternaren* into the German language and literature and subsequently (1858) modified it to *Quartären,* a translation of the term *Quartaire* which he had proposed that same year.

Marcel de Serres (1830) had affirmed that early man was contemporaneous with these deposits of the Quaternary. Thus within a year of its creation by Desnoyers, de Serres had considerably modified the concept of Quaternary to that form in which it was subsequently used by most stratigraphers. The classic treatises of Reboul (1833) and d'Archiac (1849), in which the *Terrain Quaternaire ou diluvien* is discussed (in 439 pages) gave added weight to the general acceptance of the Quaternary in geological literature. The term *Pliostène* was introduced by Buteux (1843) as a brief epoch of approximately post-Pliocene time to denote the Quaternary, but it is a superfluous term.

In general, it appears that the term Quaternary has been used primarily by geologists working on continental and non-marine sections, whereas Pleistocene has generally been applied to corresponding marine deposits. More accurately the Quaternary designates a period in the chronostratigraphical scale of the Cenozoic, whereas Pleistocene is an epoch/series term the limits of which happen to coincide with the Quaternary.

Modifications of the term Quaternary were suggested from the beginning. Anthropological investigations yielded recurring evidence of the relationship of early hominids and the Quaternary and among the various terms proposed we can recall the following: *Période anthropeienne* (Reboul 1833), *Période Lamozooique* (Vezian 1865), *Terrain humain* (Mercey 1874-7), *temps anthropiques primitifs* (Piette 1880), *Psychozoique* (LeConte 1887, *in* Munier-Chalmas 1908). At the first International Geological Congress in London (1888) Gaudry, with the approval of Prestwich and de Lapparent, made the proposal that Man (represented by his artifacts in particular) was the characteristic element of the Quaternary, which justified the creation of a geological period distinct from the Tertiary period. Thus, some geologists (particularly those working with anthropological sources) would define the base of the Quaternary at the level of first human industry. In Europe, however, these earliest levels (Abbevillian) are considerably younger than the earliest levels discovered in

East Africa (Tanzania) by Dr L. Leakey (pre-Olduvai).

Among vertebrate palaeontologists it has been common to consider the base of the Quaternary at the base of the Villafranchian Stage on the basis of the occurrence in this stage of *Equus, Bos (Leptobos)* and *Elephas.* The inclusion of the Villafranchian within the Quaternary was first made by Julien in 1869, adopted by Haug (1900) in his treatise, and more recently by the Temporary Commission on the Pliocene/Pleistocene Boundary at the 18th International Geological Congress in London in 1948. However, recent data have shown that the lower (type) Villafranchian is older than most deposits ascribed to it elsewhere and that the genera listed above have not been identified with certainty from the *type* Villafranchian.

The concept of a glacial epoch came from Switzerland, where Agassiz had boldly suggested in 1838 that the European continent had been recently covered by ice.

The concept that the Pleistocene epoch is synonymous with glaciation was first expressed by Forbes, who in 1846 (p. 402) redefined the Pleistocene of Lyell (1839) as equivalent to the 'glacial epoch', i.e. 'the time distinguished by severe climatal conditions through a great part of the northern hemisphere'. Much of the evidence for this supposed equivalence is physical in nature (glacial drift, etc.) and some is faunal (appearance of temperate marine and terrestrial faunas in the Lower Pleistocene). However, definitive correlations between these various occurrences have not been possible to date and the evidence cited elsewhere in this text should help to dispel this prevalent misconception.

Most early glaciologists refused to believe in more than a single glaciation, but the pioneering work of Penck (1895-1903) in the Swiss Alps led to the recognition in the Danube Basin of the four classic glaciations: Gunz, Mindel, Riss, Würm. It has been generally accepted among glaciologists that the base of the Quaternary coincides approximately with the earliest glacial deposits (pre-Gunz glacial deposits have been recognized in Europe).

Determination of the base of the Quaternary (Pleistocene) in marine sedimentary sections using palaeontological criteria has been a goal which has eluded stratigraphers for many years. The criteria suggested are numerous, but correlation with the stratotype section in southern Italy has proved difficult. The base of the Quaternary (Pleistocene) is interpreted here as the base of the Calabrian Stage of Gignoux, 1910. The generally accepted type section is at Santa Maria di Catanzaro in southern Italy where the base of the Pleistocene (*Pliocène supérieur* of Gignoux) was placed at the base of a calcarenite unit, G-G, in which *Cyprina islandica* was recorded. The section at Le Castella, subsequently described by Emiliani, Mayeda and Selli (1961) was later proposed to serve as the type section for the Pliocene/Pleistocene boundary. As thus defined the Pleistocene represents the time which has elapsed since the first appearance of *Hyalinea baltica* (Schroeter) in the section at Le Castella (above sample 50). Correlation between the two areas is generally good. The most apparent weakness of this type of criterion is that it utilizes an event which may be strongly controlled by facies influences to define a chronostratigraphical boundary. Chronostratigraphical boundaries are better founded upon clearly defined phylogenetic fossil populations.

Thus, four independent criteria for recognizing the base of the Quaternary have developed over the years: earliest evidence of man and his culture (anthropological), earliest horse-elephant faunas (vertebrate palaeontologists), earliest glaciation (glaciologists), and earliest appearance of cool-water molluscan and foraminiferal species in southern Europe (invertebrate palaeontologists).

The 18th International Geological Congress in Great Britain, Section H, (1950) recommended the following fundamental criteria for determining the Pliocene/Pleistocene boundary:

a. the appearance of northern immigrants in the marine fauna;

b. the appearance of new mammalian genera in terrestrial faunas such as *Equus* s.s., *Elephas* s.l., and *Bos;*

c. traces of the earliest considerable climatic deterioration.

Since that time a considerable body of information has accumulated which has added to our knowledge of late Cenozoic earth history and, in some cases, drastically modified earlier conceptions. For example:

1. it is now generally accepted that the lower part of the Villafranchian *s.s.* is of Pliocene age, equivalent, at least in part, to the Astian Stage;
2. oxygen isotope (palaeotemperature) studies by Emiliani have revealed a general cooling trend across the Pliocene/Pleistocene boundary at Le Castella in Calabria, southern Italy from an average summer surface water temperature of 23-25° C in the Late Pliocene to about 15° C in the Calabrian;
3. palaeobotanical investigations of lacustrine lignites in the Upper Val d'Arno area, near Bastardo and Liffe, have revealed a warm flora in the Lower Villafranchian (cited in Venzo 1965, p. 313), whereas Lona (1963) has discovered a pre-Calabrian cold phase designated the 'Arquatian' within the Astian (in strata below the *Amphistegina* limestone) which may herald the onset of regional cooling in the Mediterranean region.

Nine criteria were listed for determining the Pliocene/Pleistocene boundary in a report of the Subcommission on the Pliocene/Pleistocene boundary at the VI International Congress on Quaternary in Warsaw, 1961 (see Grichuk, Hey and Venzo 1965, p. 324). In addition to those cited previously, they included regional changes in palaeogeography, appearance of man as a social being, geomorphological criteria, and others.

Strong emphasis was placed upon using the initial thermal decrease which led to an irreversible change in the flora as a distinctive criterion in determining the base of the Pleistocene. However, as pointed out elsewhere in this article, such a criterion must be used with caution for a pre-Pleistocene 'Arquatian' cool-phase has been reported within the late Pliocene of Italy and ice rafting has been observed in sediments dated as early as four million years which may well pre-date this early cool phase (Hays and Opdyke 1967).

Grichuk, Hey and Venzo (1965, p. 327) observed that, 'Italian investigators came to the conclusion that the Lower Villafranchian deposits containing *Elephas meridionalis* and *Bos etruscus* should be referred to the Pliocene. Hence, the criterion introduced over fifty years ago by Haug distinguishing the Pleistocene from the Pliocene (i.e. Quaternary from the Tertiary) (the first known representatives of *Elephas* s.l., *Equus* and *Bos*) should be revised as obsolete'. However, Savage and Curtis (1967) point out that 'Villafranchian = early Pleistocene' is the age designation that has been given to strata throughout Eurasia-Africa which supposedly contain these faunal elements. They point out that the *Equus – Bos (Leptobos) – Elephas* fauna does not in fact occur in the lower (type) Villafranchian (= late Pliocene).

With regard to the upper limits of the Quaternary, two ideas have generally developed. One considers that post-

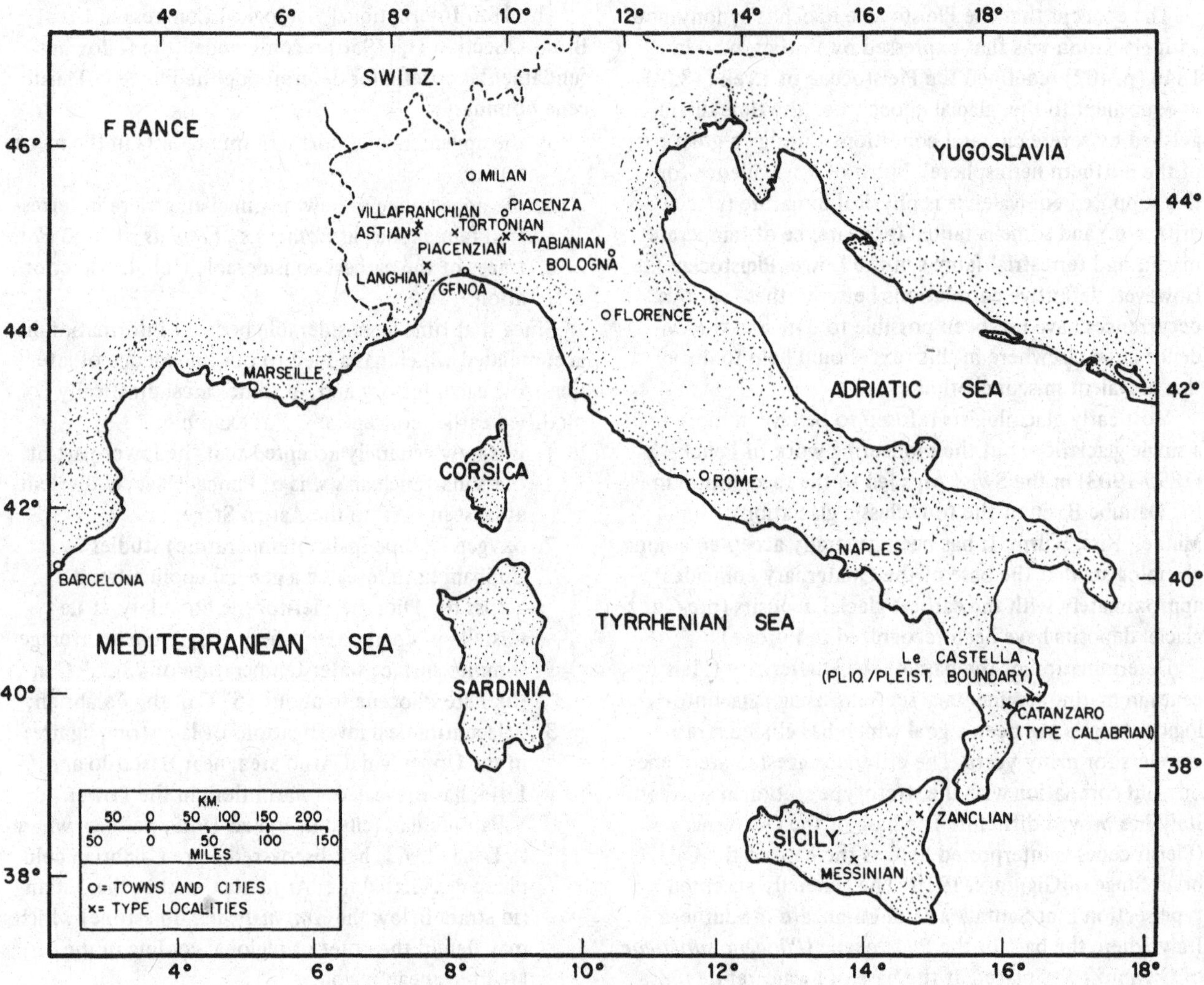

4° 6° 8° 10° 12° 14° 16° 18°
46° 44° 42° 40° 38°
SWITZ.
FRANCE
YUGOSLAVIA
MILAN
PIACENZA
VILLAFRANCHIAN
ASTIAN
TORTONIAN
TABIANIAN
PIACENZIAN
BOLOGNA
LANGHIAN
GENOA
FLORENCE
MARSEILLE
ADRIATIC SEA
CORSICA
ROME
NAPLES
BARCELONA
TYRRHENIAN SEA
MEDITERRANEAN SEA
Le CASTELLA
(PLIO./PLEIST. BOUNDARY)
SARDINIA
CATANZARO
(TYPE CALABRIAN)
KM.
50 0 50 100 150 200
50 0 50 100 150
MILES
ZANCLIAN
SICILY
MESSINIAN
O= TOWNS AND CITIES
X= TYPE LOCALITIES
LATE CENOZOIC TYPE LOCALITIES IN ITALY

Fig. 51.1. Location of late Cenozoic type localities in Italy and Sicily.

glacial, or Recent, or Holocene (Gervais 1867-9) constitutes a fifth era the *Quinquennaire* (= Quinquenary) of Parandier (1891), which would appear completely unnecessary. A second school of thought considers the Holocene as synonymous with postglacial and as a subdivision of the Quaternary. Indeed, in terms of the glacial and chronostratigraphical scale of the Quaternary developed in recent years, we are living in an interglacial period within the Quaternary.

Table 51.1. The commonly accepted subdivision of the marine Pleistocene in Italy

Tyrrhenian	(Issel 1914)
Milazzian	(Deperet 1918)
Silician	(Doderlein 1872)
Emilian	(Ruggieri and Selli 1950)
Calabrian	(Gignoux 1910)

THE PLIOCENE/PLEISTOCENE BOUNDARY IN DEEP-SEA CORES (1952-66) – A REVIEW

The 'Pliocene/Pleistocene' boundary was first defined in deep-sea cores by Arrhenius (1952) primarily on the basis of a sharp upward increase in calcium carbonate content in some east equatorial Pacific cores. Riedel (1957) subsequently suggested that the level of extinction of two radiolarian species (*Pterocanium prismatium* and *Eucyrtidium elongatum peregrinum*) in tropical Pacific cores may serve to mark the 'top of Pliocene'. Riedel (*in* Riedel *et al.* 1963) subsequently used only the extinction of *P. prismatium* in defining the Pliocene/Pleistocene boundary in Equatorial Pacific cores. This boundary was correlated by Riedel, Parker and Bramlette (1963) approximately with the Pliocene/Pleistocene boundary of Ericson, Ewing and Wollin (1963, 1964).

The Pliocene/Pleistocene boundary determined by Ericson, Ewing and Wollin (1963) was the result of detailed and laborious investigations on numerous deep-sea cores. It was based primarily on the following criteria:

1. extinction of discoasters;
2. first appearance of abundant *Globorotalia truncatulinoides;*
3. change in *G. menardii* complex, from diverse (below) to a more uniform lineage (above), increase in average test size and reduction in number with respect to total population above the boundary and change from 95 per cent dextrally coiled (below) to 95 per cent sinistrally coiled (above);
4. disappearance of *Globigerinoides sacculifera fistulosa* above the boundary in cores from the Atlantic.

Ericson, Ewing and Wollin (1964, p. 727) suggested that 'this faunal boundary, which records the beginning of the first glaciation, be chosen to denote the beginning of the Pleistocene epoch'. Ericson and his colleagues also presented a Pleistocene time-scale and suggested that the base of the Pleistocene, defined by the onset of the first glaciation, occurred about 1.5 million years ago. Dates for the various glacial and interglacial stages were obtained from a combination of radiocarbon dating (to 35,000 years), the protactinium-ionium and the protactinium methods (to 175,000 years) and calculated sedimentation rates. More recently, Ericson, Ewing and Wollin's stratigraphy has been dated back to 320,000 years by Th^{230} measurements (Ku and Broecker 1966).

Bandy (1963) compared changes in planktonic foraminiferal faunas across what he determined to be the Miocene/Pliocene boundary in the Philippines with those in the material studied by Ericson, Ewing and Wollin (1963). He suggested that the Pliocene was missing and that the forms present below the boundary determined by Ericson, Ewing and Wollin (1963) were Miocene. He subsequently modified his opinions to allow the presence of some Pliocene but still was of the opinion that a major hiatus (representing about ten million years) was represented in the cores studied by Ericson, Ewing and Wollin.

Wray and Ellis (1965) and Akers (1965) discussed the pattern of discoaster extinction in the northern Gulf of Mexico and suggested that the time of extinction of discoasters approximates the Pliocene/Pleistocene boundary in that area. The latter author suggested that three species of discoasters persisted into the early Quaternary and that they became extinct between the Nebraskan glacial period and the Aftonian interglacial period.

McIntyre, Bé and Preikstas (1967) have recently examined calcareous nannofossil assemblages in the deep-sea cores upon which Ericson, Ewing and Wollin (1963, 1964) had based their determination of the Pliocene/Pleistocene boundary. A major floral and faunal boundary was confirmed but the modern aspect of the species composition and the transition from a cold-water assemblage below the boundary to a warm-water assemblage above it suggested to the authors that the boundary in these cores lies somewhere within the lower Pleistocene (probably Nebraskan-Aftonian) rather than between the Pliocene/Pleistocene. Here again, we see an illustration of attempts to recognize the Pliocene/Pleistocene on the assumption of the base Pleistocene = initiation of glaciation.

Harrison and Funnell (1964) related the last occurrence of *Pterocanium prismatium* to the most recent (Matuyama/Brunhes) palaeomagnetic reversal, then thought to occur at about 0.99 million years, since redetermined and shown to occur at 0.7 million years. Discoasters were not found in the two cores examined by Harrison and Funnell (1964) (letter from Dr Funnell, dated 22 May 1967). The likelihood that the last occurrence of discoasters would coincide with the Matuyama/Brunhes reversal was based on the approximate coincidence of the disappearance (i.e. extinction) of *Pterocanium prismatium* and *Discoaster* spp. in Pacific cores (Riedel, Parker and Bramlette 1963).

In a study of radiolarians from deep-sea cores in Antarctic seas, Hays (1965) recognized four faunal zones, the boundary between the lower two (Φ/X) in cores south of the Polar Front being marked by a striking faunal change, and by a change from red clay (below) to diatom ooze (above). This lithological change was believed to have been ultimately caused by climatic deterioration which eventually produced the total glaciation of Antarctica. Hays (1965) suggested that the Φ/X boundary coincided approximately with the Pliocene/Pleistocene boundary of Ericson *et al.* (1963), primarily on the basis of determining a Φ/X boundary at 722 centimetres in core V16-66. Ericson, Wollin and Ewing (1963) had determined a 'Pliocene/Pleistocene' boundary in this core at 270 centimetres on the basis of the extinction of discoasters below and the appearance of *Globorotalia truncatulinoides* above it. The disappearance of discoasters was said to be gradual between 800 centimetres and 270 centimetres. Hays (1965) assumed that his Φ/X boundary in Antarctic cores corresponded approximately with that of Ericson, Wollin and Ewing (1963), the difficulty in relating the two boundaries in core V16-66 being due to difficulties in drawing a boundary in a core where discoasters decrease gradually. Palaeomagnetic measurements on Antarctic cores (Opdyke, Glass, Hays and Foster 1966) have shown that the Φ/X boundary occurs just below the Olduvai event (*ca.* two million years).

Emiliani (1961, 1964, 1966*a*, *b*) has studied temperature variations in the Pleistocene as reflected in the oxygen isotope ratios in fossil planktonic Foraminifera. His results have led him to estimate the base of the glacial Pleistocene variously at 300,000 years (1961) to 425,000 years (1966*a*, *b*) and the base of the Pleistocene epoch at 600,000-800,000 years (1961, 1964, 1966*a*, *b*). Emiliani (1955) placed the base of the Pleistocene in Swedish Deep Sea Core 58 (eastern Equatorial Pacific) at 610 centimetres on the basis of the palaeotemperature record alone. Riedel, Parker and Bramlette (1963) observed that in this core the change in nannoplankton occurs 'near or within the zone of low carbonate between 7.9 and 8.1 metres'. They cite the occurrence of *Globorotalia multicamerata* at 9 metres, of *G. truncatulinoides* at 7.6 metres. The last occurrences of *Discoaster* spp. are shown to be in the interval within which the only specimen of *G. truncatulinoides* was found. Thus the Pliocene/Pleistocene boundary as determined by Emiliani (1955) at 610 centimetres lies above that determined by Riedel, Parker and Bramlette (1963) at about 8 metres, and by extrapolation above that determined by Ericson, Wollin and Ewing (1963).

Riedel, Parker and Bramlette (1963) suggested that the faunal changes recognized by Ericson and his colleagues were poorly defined and that it was doubtful whether it (the boundary) represented a change as drastic as that of the onset of a glacial period. However, they did agree that 'there was considerable agreement' between the boundary recognized by Ericson and his colleagues in the Atlantic Ocean and that provisionally drawn by them in the Pacific Ocean.

The failure by specialists to relate the criteria discussed above to the type sections of the Pliocene/Pleistocene in Italy has made it difficult to evaluate the relationship of these deep-sea boundaries to the Pliocene/Pleistocene boundary.

BASIC CRITERIA FOR RECOGNIZING THE PLIOCENE/PLEISTOCENE BOUNDARY: PLANKTONIC FORAMINIFERA

Recent work by Banner and Blow (1965) has shown that the evolutionary appearance of the planktonic foraminiferal species *Globorotalia truncatulinoides* from its immediate ancestor, *G. tosaensis,* occurs near the base of the holostratotype Calabrian at Santa Maria di Catanzaro, in southern Italy. These authors have used the first evolutionary appearance of *G. truncatulinoides* as the criterion for recognizing the base of their zone N 22 and have suggested this as a suitable criterion for recognizing the Pliocene/Pleistocene boundary.

Dr D. D. Bayliss[1] has recently recollected the stratigraphical section exposed at Santa Maria di Catanzaro. Good specimens of *G. tosaensis* (advanced morphological development) were found (together with *Globigerina pachyderma*) in the lower, pre-Calabrian part of the section (Gignoux 1913, p. 35). The calcarenite unit, designated G-G on the NE face of Pta dei Briganti by Gignoux (1913, p. 35), in his section at Santa Maria di Catanzaro, is laterally equivalent to the sand stringers in the middle of the section measured by Bayliss. Gignoux (1913) recorded *Cyprina islandica* from this *'grès calcaire très fossilifère'*; this was the base of his *'Pliocène supérieur'*. This cold-water interval contains neither *G. tosaensis* nor *G. truncatulinoides.* The first appearance of *G. truncatu-*

[1] Documentation of the faunal criteria upon which we have based our Pliocene/Pleistocene boundary has been provided through the kindness of Dr W. H. Blow, British Petroleum Co., Ltd, Sunbury-on-Thames and Dr D. D. Bayliss, British Museum (Natural History), London. We would like to express our sincere gratitude to Drs Blow and Bayliss for permission to use this information.

linoides is recorded in a sample (No. 42) about 75 feet stratigraphically above the laterally equivalent thin-bedded clays, silts and sands and about 15 feet above the base of Gignoux's type Calabrian.

The value of Banner and Blow's (1965) criterion is that the evolutionary transition from one species to another happens only once and denotes isochroneity. This is not true of extinctions and non-evolutionary appearances, the criteria previously applied in deep-sea stratigraphy. Extinctions and (non-evolutionary) appearances can and commonly do transgress time lines. It is conceivable for a species to disappear in one ocean and continue living in another. The (non-evolutionary) appearance of a species may occur at any level within its total range dependent upon ecological and/or climatic conditions. Consequently, we believe that at this time, the evolutionary transition from *G. tosaensis* to *G. truncatulinoides* is the most satisfactory palaeontological criterion for recognizing the Pliocene/Pleistocene boundary.

RADIOMETRIC DATING, PALAEOMAGNETISM AND THE PLIOCENE/PLEISTOCENE CHRONOMETRIC SCALE

Since most microfossil species suitable for deep-sea stratigraphy have limited geographical distribution, no single criterion for recognizing the Plio/Pleistocene boundary can be applied in all deep-sea sediments. The situation is further complicated by the absence of calcareous microfossils in the deeper parts of the ocean and their rare occurrence in impoverished assemblages in high latitudes. In these latter areas Radiolaria are frequently abundant but cores containing both well-preserved calcareous fossils and Radiolaria are rare. The problem of correlating microfossil assemblages between high and low latitudes and correlating siliceous with calcareous assemblages has recently been partially solved through simultaneous palaeomagnetic and faunal studies.

Cox, Doell and Dalrymple (1963) and MacDougall and Tarling (1963) have shown that Pliocene and Pleistocene lava flows are magnetized either in accordance with the present direction of the earth's magnetic field (normal) or in the opposite direction (reversed). The importance of this bimodal distribution of direction of magnetization is that it has its origin in reversals of the earth's magnetic field (Cox, Doell and Dalrymple 1965). Because of this, normally and reversely magnetized rocks can be expected to occur in alternating sequences traceable as precise time-stratigraphical units around the world. The longer intervals of constant polarity are termed normal and reversed polarity epochs (Cox, Doell and Dalrymple 1963) while intervals of short duration are called events. Epochs have a duration of the order of a million years while events are about a tenth as long. By carrying out simultaneous palaeomagnetic and radiometric studies of Pliocene and Pleistocene lava flows it has been possible to date the changes in polarity of the earth's magnetic field through the past 3.5 million years. These have been named in order of increasing age (in millions of years): the Brunhes normal epoch (0 to 0.7), the Matuyama reversed epoch (0.7-2.4 ± 0.1), the Gauss normal epoch (2.4 ± 0.1 to 3.35). Within the Matuyama reversed epoch, two short periods of normal polarity occurred at about 0.9 and 1.9 million years and these have been termed the Jaramillo and Olduvai events respectively. The Olduvai event was identified by Grommé and Hay (1963) in a basalt from Olduvai Gorge, Tanzania and has a duration of about 0.1 million years. Within the Gauss normal epoch there was a short period of reversed polarity at 3.0 million years, called the Mammoth event, (Cox, Doell and Dalrymple 1965; Doell, Dalrymple and Cox 1966).

With this basic dated sequence it has been possible to correlate between continents and relate rocks of similar polarity. Although palaeomagnetic studies have not been made on the type Calabrian, considerable data is available from continental sections. Using both radiometric and palaeomagnetic data Cox, Doell and Dalrymple (1965) have shown that the Blancan age began within the Gilbert reversed epoch and terminated within the Matuyama reversed epoch. At least 0.4 million years elapsed between the end of the Blancan age and the end of the Matuyama polarity epoch.

The palaeomagnetic studies of Roche (1951, 1956) are very important in dating the upper boundary of the Villafranchian. In the Auvergne, lava flows of late Villafranchian age are reversely magnetized, as are the younger flows of Saint Prestian age. Thus, the upper boundary of the Villafranchian like that of the Blancan, lies within the Matuyama reversed epoch and is greater than one million years old. The Saint Prestian (= Cromerian), in turn, is at least in part equivalent with the late Matuyama reversed epoch and hence may be correlative with the Bruneau Formation in Idaho, which is Irvingtonian or younger. A radiometric age of greater than 3 million years for the base of the Villafranchian *s.l.* (Curtis 1965) indicates that the base of the Villafranchian cannot be much younger than the base of the Blancan and therefore also probably began in the Gilbert reversed polarity epoch.

Until very recently most palaeomagnetic studies have been of continental rocks. Within the past few years a major effort has been directed toward the study of the palaeomagnetic stratigraphy of deep-sea cores and the use of the absolute ages obtained for dating microfossil assemblages and correlation between deep-sea cores. A few early palaeomagnetic works on deep-sea sediments were made by McNish and Johnson (1938), Johnson, Murphy and Torreson (1948), Keen (1960, 1963), Harrison and Funnell (1964) and Fuller, Harrison and Nayudu (1966). Of these only Harrison and Funnell (1964) and Fuller, Harrison and Nayudu (1966) found reversals of polarity. Opdyke, Glass, Hays and Foster (1966) studying Antarctic deep-sea cores were able to reproduce Cox, Doell and Dalrymple's (1965) palaeomagnetic stratigraphy back to the Gilbert reversed polarity epoch. Using the known ages of the reversals it was possible to calculate rates of sedimentation and approximately date radiolarian faunal zones previously established by Hays (1965). Hays and Opdyke (1967), studying long Antarctic cores which contain a record extending well below the Gauss normal series, identified and estimated the age of three events within the Gilbert reversed series and a new epoch lying below the Gilbert. This work extended the continuous record of the Earth's magnetic field back to more than five million years.

Since the reversals of the Earth's magnetic field are global phenomena, independent of latitude or water depth, their record in the sediments offers an independent means of correlation between disparate fossil assemblages. With this new tool it is now possible to set up multiple criteria based on different fossil groups that can be used to recognize the Pliocene/Pleistocene boundary in various parts of the world's oceans.

THE PLIOCENE/PLEISTOCENE BOUNDARY

Central North Atlantic

In a recent study (Berggren, Phillips, Bertels and Wall 1967; Phillips, Berggren, Bertels and Wall 1968) the Pliocene/Pleistocene boundary was identified in a deep-sea core from the central North Atlantic (latitude 26° N, longitude 39° W), about 470 kilometres east of the Mid-Atlantic Ridge. The evolutionary transition from *Globorotalia tosaensis* to *G. truncatulinoides* was found to occur within the Olduvai normal event and the boundary, as determined by the first evolutionary appearance of *G. truncatulinoides,* was estimated to be about 1.85 million years old. The major extinction of discoasters was found to occur about 90 centimetres above the Plio/Pleistocene boundary (as palaeontologically determined) and about 70 centimetres above the top of the Olduvai event; an age of 1.60 million years was estimated for this level. *G. truncatulinoides* first became abundant at this level as well. A brief reappearance of discoasters was found within the Jaramillo event at about 0.9 million years. The extinction of *G. fimbriata, G. multicamerata* and *G. miocenica* (all right coiling) was observed within the interval of zone N21 (*G. tosaensis* zone). Typical *G. menardii* (left coiling) were found only above the N 21/N 22 boundary.

The first appearance of *G. inflata,* a cool-temperate water indicator, was found to coincide with the first evolutionary occurrence of *G. truncatulinoides.* A general cooling trend is observed in the lower half of the Pleistocene of several cores near 26° N based on relative ratios of *Globigerinoides sacculifera* and *G. rubra, Pulleniatina obliquiloculata* and *Sphaeroidinella dehiscens* (warm water indicators) and *Globorotalia inflata* and *G. hirsuta* (temperate water indicators). A significant (stronger) cooling was suggested above a level within the Jaramillo event (0.9 million years) on the basis of a sharp reduction of the former group and a sharp increase in the latter two species. The suggestion was made that this stronger cooling may have been related to the onset of Pleistocene glaciation in continental Europe and that the 'pre-glacial' and 'glacial' Pleistocene may thus have an approximate ratio of 1:1 (see discussion page 685).

Glass *et al.* (1967) made palaeomagnetic measurements on a number of cores studied by Ericson, Wollin and Ewing (1963). Although most of the cores studied by Ericson *et al.* contain hiatuses, Glass *et al.* present data which lead them to conclude that when Ericson's criteria are applied to cores that do not contain hiatuses his boundary falls near to or within the Olduvai event. Ericson's boundary therefore is very close to the age of the boundary in the type section in Italy.

THE ANTARCTIC

Five faunal zones have been established for Antarctic deep-sea sediments (Hays 1965; Hays and Opdyke 1967). These zones designated, from oldest to youngest, Υ, Φ, X, Ψ and Ω are based primarily on the upward sequential disappearance of radiolarian species (Fig. 51.2). It has been considered that the upper three zones (X, Ψ and Ω) represent Quaternary sedimentation while the Φ zone represents late Tertiary (Hays 1965). The Ω zone contains only those species found in recent Antarctic sediments and presumably all are still living today. The upper limit of the Ψ zone is based on the last occurrence of *Stylatractus* sp. The upper limit of the X zone is based on the last occurrence in Antarctic sediments south of the Polar Front of the species *Saturnulus planetes* and *Pterocanium trilobum.* Both of these species are still living north of the Polar Front. The disappearance of these species at the X-Ψ boundary probably indicates a change from warmer conditions to cooler conditions. The Φ-X boundary is based on the last common occurrence of *Eucyrtidium calvertense* in Antarctic sediments. Near or at the Φ-X boundary there is a change from clay sediments below containing few or no diatoms or Radiolaria to highly diatomaceous sediments above. This change has been interpreted to indicate the initiation of large-scale upwelling around the Antarctic continent in response to cooling engendered by the full glaciation of Antarctica (Hays 1965).

Opdyke, Glass, Hays and Foster (1966) measured the remanent magnetism on a number of Antarctic cores and compared the magnetic stratigraphy with previously determined radiolarian zones. The Ψ-Ω boundary falls at about the middle to slightly below the middle of the Brunhes normal polarity epoch and was therefore assigned an age of 0.4-0.5 million years old. The X- Ψ boundary falls near to, but below, the Matuyama-Brunhes boundary and has an age of slightly more than 0.7 million years B.P. The Φ-X boundary falls just below the base of the Olduvai event and therefore has an age of about 2.0 million years B.P.

The Φ-X boundary or the last common occurrence in Antarctic sediments of *Eucyrtidium calvertense* represents a close approximation to the Pliocene/Pleistocene boundary shown, in North Atlantic cores, to fall within the Olduvai event. The X- Ψ boundary with its change from warmer conditions to cooler conditions generally corresponds with a similar change in North Atlantic cores described above.

North Pacific

The Quaternary sediments of the North Pacific, as well as

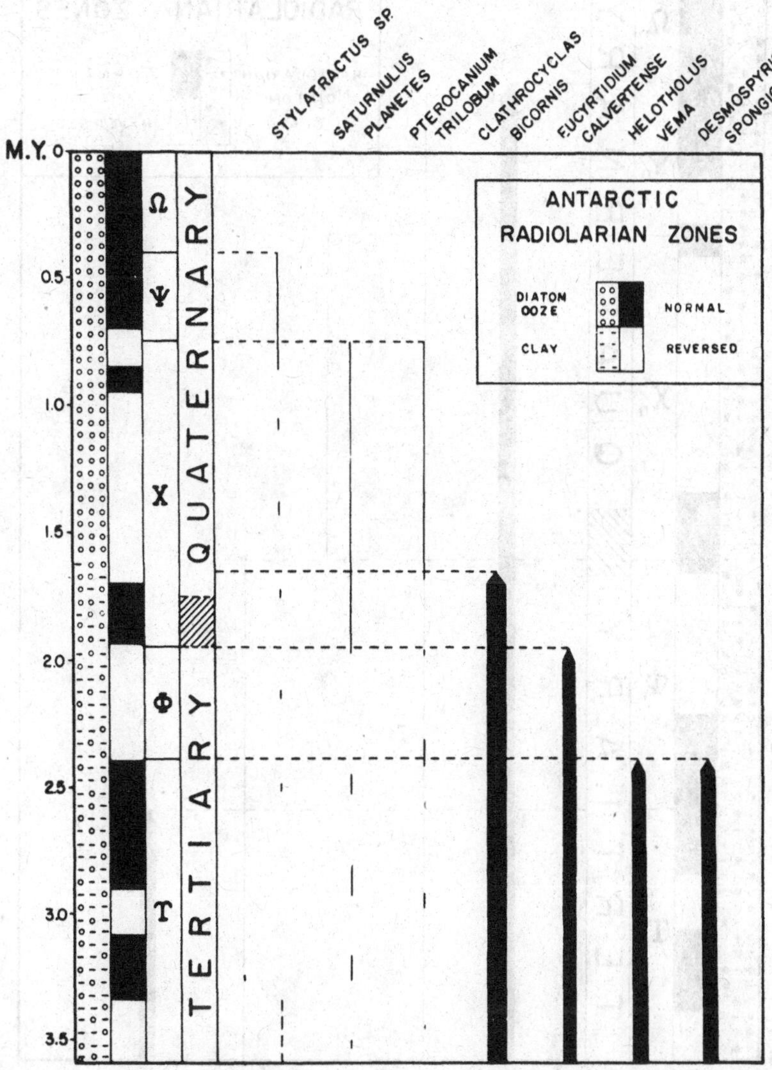

STYLATRACTUS SP.
SATURNULUS PLANETES
PTEROCANIUM TRILOBUM
CLATHROCYCLAS BICORNIS
EUCYRTIDIUM CALVERTENSE
HELOTHOLUS VEMA
DESMOSPYRIS SPONGIOSA
M.Y. 0
0.5
1.0
1.5
2.0
2.5
3.0
3.5
Ω
Ψ
X
Φ
Υ
QUATERNARY
TERTIARY
ANTARCTIC RADIOLARIAN ZONES
DIATOM OOZE
CLAY
NORMAL
REVERSED

Fig. 51.2. Stratigraphical ranges of certain radiolarian species present in Antarctic sediments; compared with generalized lithology of Antarctic sediments and magnetic stratigraphy (columns far left). Greek letters represent faunal zones.

those of the Antarctic, can be divided into three zones based on their radiolarian content. Two late Tertiary zones can also be recognized. These zones from oldest to youngest are designated Υ_n, Φ_n, χ_n, Ψ_n, and Ω_n. The subscript 'n' indicates that these zones are restricted to the northern hemisphere, are based on different species to the Antarctic zones, and do not necessarily represent the same period of time as the Antarctic zone with the same Greek letter designation. A paper describing the taxonomy of the North Pacific Radiolaria is in preparation.

The Ω_n zone contains only species that are common in the surface sediments and are presumably still living throughout most of the North Pacific north of 40° north latitude.

The upper limit of the Ψ_n zone is defined by the upper limit of *Stylatractus* sp., a species present throughout the earlier North Pacific zones. This species is morphologically very similar to *Stylatractus* sp. found in Antarctic sedi-

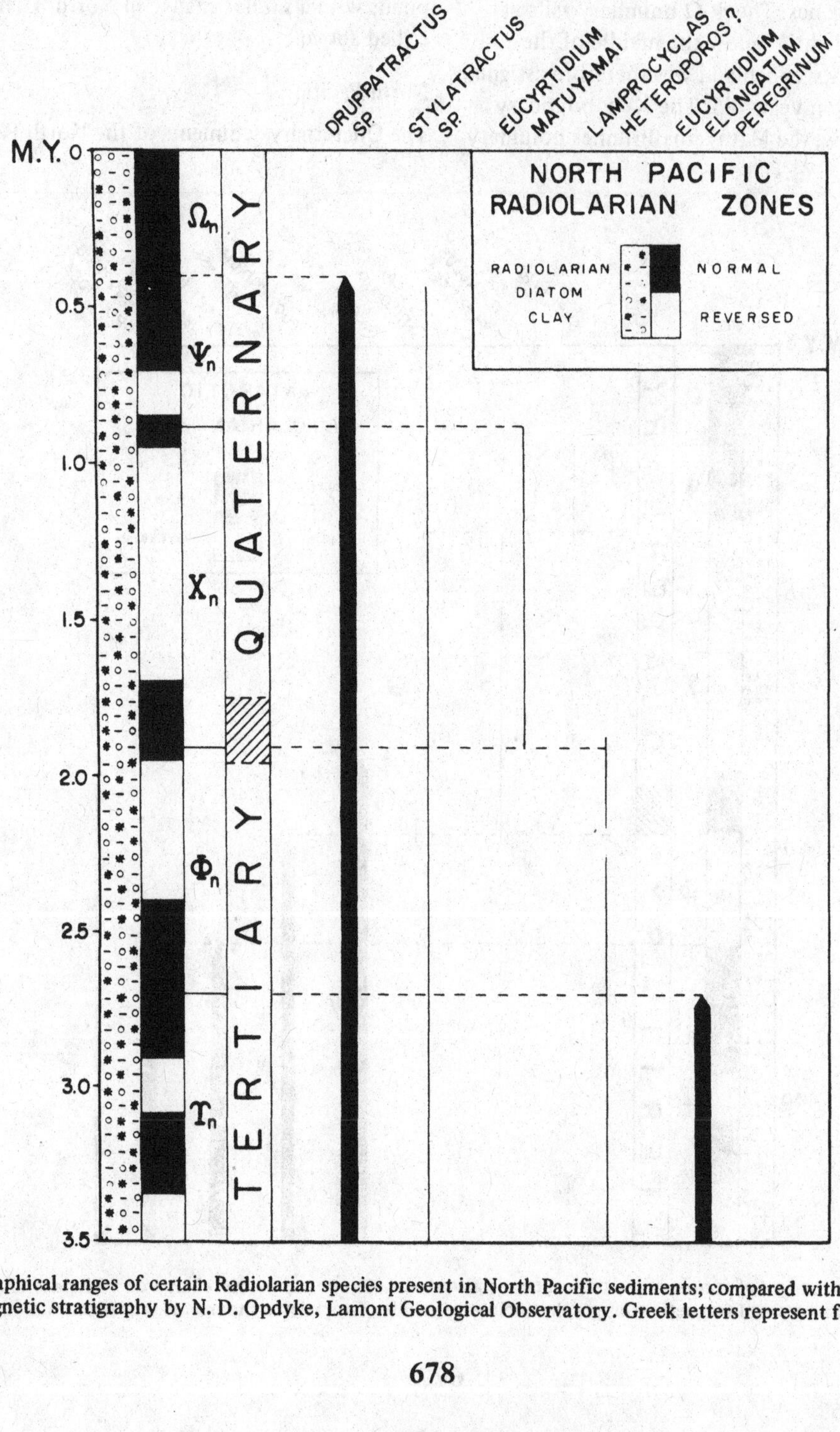
DRUPPATRACTUS SP.
STYLATRACTUS SP.
EUCYRTIDIUM MATUYAMAI
LAMPROCYCLAS HETEROPOROS ?
EUCYRTIDIUM ELONGATUM PEREGRINUM
M.Y.
0
0.5
1.0
1.5
2.0
2.5
3.0
3.5
Ωn
Ψn
Χn
Φn
ϒn
QUATERNARY
TERTIARY
NORTH PACIFIC RADIOLARIAN ZONES
RADIOLARIAN
DIATOM
CLAY
NORMAL
REVERSED

Fig. 51.3. Stratigraphical ranges of certain Radiolarian species present in North Pacific sediments; compared with generalized lithology and magnetic stratigraphy by N. D. Opdyke, Lamont Geological Observatory. Greek letters represent faunal zones.

ments and may well be the same species. In both the Antarctic and the North Pacific the upper limit of *Stylatractus* sp. falls near the middle of the Brunhes normal polarity epoch and so the Ω_n zone in the North Pacific is roughly correlative with the Ω zone in the Antarctic and has an extrapolated duration of about 400,000 years.

The X_n zone is defined by the range in North Pacific sediments of the new species *Eucyrtidium matuyamai.* This species first appears near the base of the Olduvai event and disappears in the Jaramillo event. This range has been confirmed by an examination of ten cores that contain its full range. The appearance of this species within the Olduvai event roughly corresponds in time with the transition from *Globorotalia tosaensis* to *G. truncatulinoides* observed by Berggren in a North Atlantic core, and therefore provides a criterion in North Pacific sediments which can be used to identify the Pliocene/Pleistocene boundary.

The X_n zone in several cores, such as V20-105, contains a warmer water fauna than the overlying two zones. Although the upper limit of X_n is not isochronous with the Antarctic X zone the upper boundaries of these zones are not far removed from one another. Therefore the warmer period of time they both represent is roughly equivalent and also equivalent to the warm zones noted in North Atlantic cores (Berggren 1967).

Equatorial Pacific

In a detailed investigation of deep-sea cores taken by the *Albatross* during the Swedish Deep-Sea Expedition (1947-8) in the eastern equatorial region of the Pacific Ocean, Arrhenius (1952) attempted a correlation between cyclical changes in abundance in microfossils, abundance and rate of solution of calcium carbonate and glacial cycles. He suggested a correlation, of stratigraphical intervals rich in calcium carbonate relative to clay with glacial periods, and of those poorer in calcium carbonate with non-glacial periods. He placed the Pliocene/Pleistocene boundary at 395 centimetres in Swedish deep-sea core 58 on the basis of a large and sudden increase in both microfossils and total calcium carbonate. Emiliani (1955) suggested a Pliocene/Pleistocene boundary at 610 centimetres in this core based upon isotopic studies. A palaeontological boundary was placed at about 8 metres by Riedel, Parker and Bramlette (1963) based on the extinction of *Pterocanium prismatium* (cf. Riedel 1957). This latter boundary is judged here to be very nearly correct on the basis of the criteria discussed here and in Berggren, Phillips, Bertels and Wall (1967). The boundary recognized by Arrhenius (1952) may be related to some phase of Pleistocene glaciation (i.e. late Pleistocene in the sense used here).

Olausson (1961) subsequently modified Arrhenius's carbonate curve as a plot of abundance of Foraminifera against core depth in cores 58 and 58B. He found minor fluctuations in abundance below 400 centimetres and more pronounced fluctuations above this depth. He agreed with the correlation between increased carbonate, Foraminifera and glaciation suggested by Arrhenius (1952), while accepting the Pliocene/Pleistocene boundary determined by Emiliani (1955) at 610 centimetres.

In a series of investigations on isotopic palaeotemperatures in deep-sea cores, Emiliani (1955-66) has presented a generalized temperature curve (with eight cycles for the Caribbean and equatorial Atlantic) which he suggested extends from the present to about 425,000 years ago. He suggested a correspondence between the carbonate cycles recognized by Arrhenius in the eastern equatorial Pacific cores and the temperature cycles in the equatorial Atlantic and Caribbean cores. On the basis of correlating the 400 centimetre level in Pacific core 58 with the bottom layer of core 59 (based, in turn, on the size distribution of the diatom, *Coscinodiscus nodulifer*), and with cycle 9 in his temperature curve at about 400,000 years, Emiliani (1966, p. 853) suggested an age of about 1.1 million years for the bottom of core 58 (991 centimetres). But, as we have seen above (pp. 674 and 679) the Pliocene/Pleistocene boundary as determined palaeontologically by Riedel, Parker and Bramlette (1963) in this core is at about 8 metres, giving this level an age of about 2 million years. On this basis, and on the assumption that the rate of sedimentation did not drastically change within the Pleistocene, it is suggested that Emiliani's Pleistocene chronology may be low by a factor of 2. Emiliani's investigations were done prior to the advent of refined palaeomagnetic stratigraphy, so that he did not have the independent control of this method as a guide. This is perhaps an illustration of the extremely tenuous nature of age estimates in deep-sea cores based upon estimated rates of sedimentation and correlation between cores based on abundance of species.

Riedel (1957), investigating Radiolaria in tropical Pacific cores, found that two species (*Pterocanium prismatium* and *Eucyrtidium elongatum peregrinum*) become extinct at an horizon approximately correlated with that believed to represent the top of the Pliocene in cores described by Arrhenius (1952). Later Riedel, Parker and Bramlette (1963) approximately correlated this boundary with the Pliocene/Pleistocene boundary drawn by Ericson, Wollin and Ewing (1963) in low-latitude Atlantic cores. Harrison and Funnell (1964) compared the palaeomagnetic stratigraphy with the radiolarian stratigraphy of two equatorial Pacific cores and concluded that *Pterocanium prismatium* disappeared coincidentally with the last reversal of the Earth's magnetic field. Recent work at Lamont Geological Observatory has shown that *P. prismatium* in four long oriented cores from the equatorial Pacific consistently disappears just above the top of the Olduvai event (Fig. 51.4). The discrepancy between our results and Harrison and Funnell's is probably due to hiatuses in their considerably shorter cores. The proximity of the upper limit of the range of *P. prismatium* to the top of the Olduvai event makes it, as originally suggested by Riedel, Parker and Bramlette (1963), a good approximate marker for the Pliocene/Pleistocene boundary in equatorial Pacific sediments. The somewhat earlier disappearance of discoasters (Riedel, Parker and Bramlette 1963) also corresponds approximately with the Pliocene/Pleistocene boundary.

MULTIPLE CRITERIA FOR RECOGNITION OF THE PLIOCENE/PLEISTOCENE BOUNDARY IN DEEP-SEA CORES

It is now possible with the aid of palaeomagnetic stratigraphy to relate the ranges of stratigraphically important microfossil species to an independent time scale. The Pliocene/Pleistocene boundary has been shown to fall within the Olduvai and has an age of approximately 1.85 million years. The palaeomagnetic correlations have in large part verified the estimates of earlier workers as to the stratigraphical position of the Pliocene/Pleistocene boundary in deep-sea cores (Fig. 51.5). This is fortunate, although surprising when the diversity of criteria used and the tentative nature of the earlier correlations are considered.

The most reliable criterion, where it can be found, is the evolutionary transition from the foraminiferal species *Globorotalia tosaensis* to *G. truncatulinoides*. Since *G. truncatulinoides* is a temperate water form this criterion will probably be most useful in mid-latitudes (Fig. 51.6).

In high latitudes the richness of the radiolarian assemblages compared with Foraminifera suggest that these will be most useful. The last occurrence of *Eucyrtidium calvertense* falls near the base of the Olduvai event and can be used to mark approximately the Pliocene/Pleistocene boundary in Antarctic sediments.

The first evolutionary appearance of *E. matuyamai* comes within the Olduvai event and therefore can also be used to designate the Pliocene/Pleistocene boundary in North Pacific sediments above 40° N latitude.

The last appearance of *Pterocanium prismatium* and the extinction of discoasters serve as indicators of the Pliocene/Pleistocene boundary in equatorial Pacific cores (Figs. 51.7 and 51.8).

Planktonic Foraminifera criteria which have been found useful in recognizing the approximate location of the Pliocene/Pleistocene boundary in low latitudes are the following:

1. extinction of *Globigerinoides sacculifera fistulosa* near the top of Zone N 21;

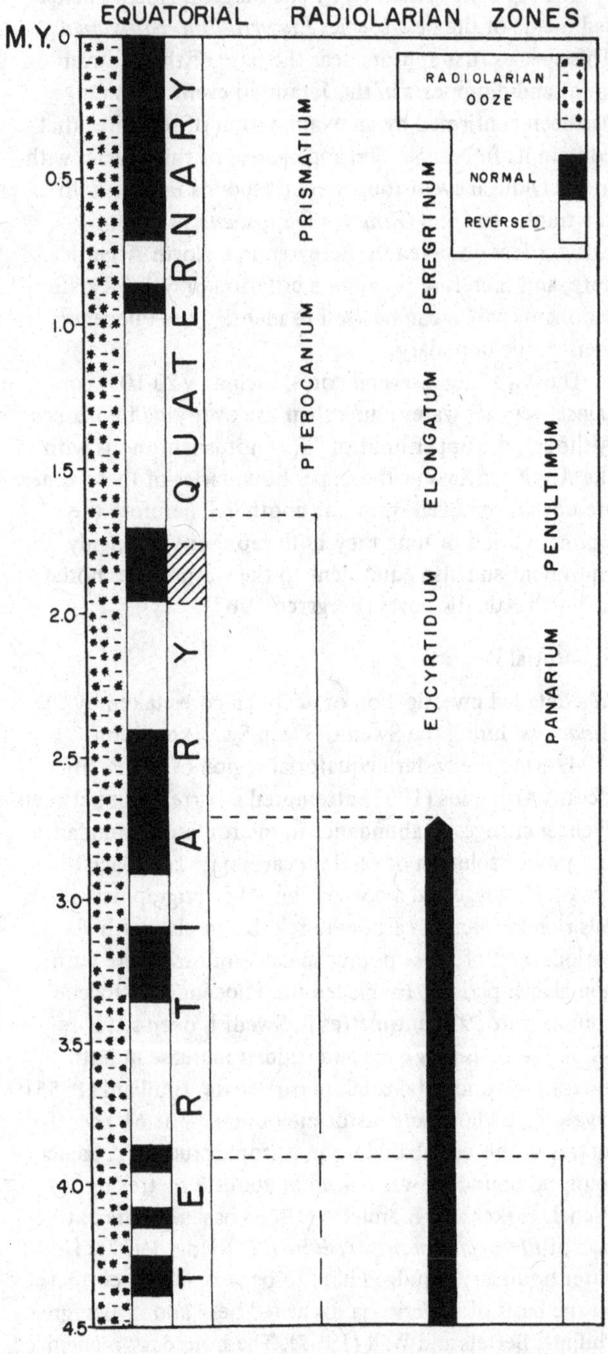

EQUATORIAL RADIOLARIAN ZONES
M.Y.
0
0.5
1.0
1.5
2.0
2.5
3.0
3.5
4.0
4.5
QUATERNARY
TERTIARY
PTEROCANIUM PRISMATIUM
EUCYRTIDIUM ELONGATUM PEREGRINUM
PANARIUM PENULTIMUM
RADIOLARIAN OOZE
NORMAL
REVERSED

Fig. 51.4. **Stratigraphical ranges of certain radiolarian species present in tropical sediments. Columns at far left show generalized lithology and magnetic stratigraphy by N. D. Opdyke, Lamont Geological Observatory.**

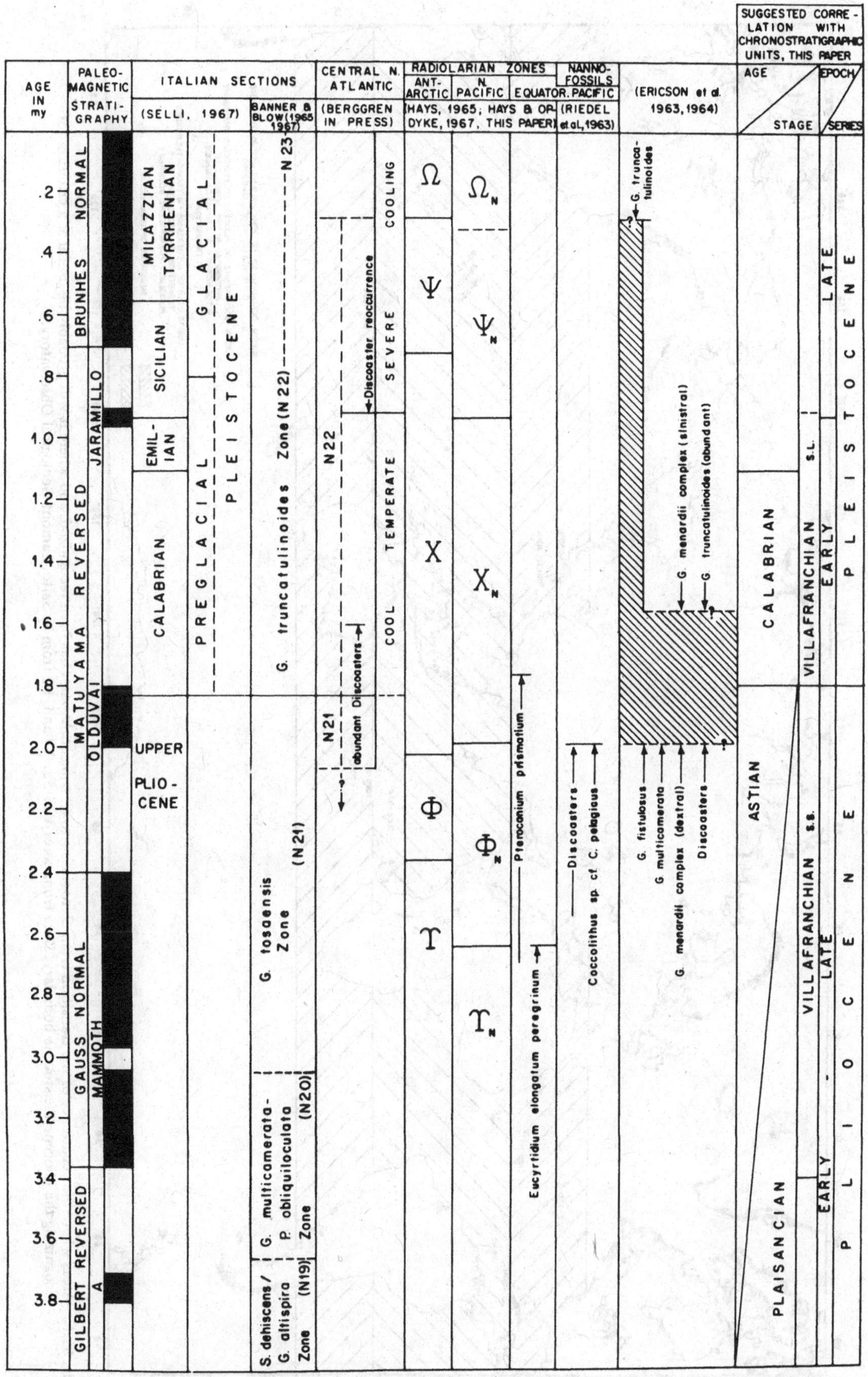
AGE IN my
PALEO-MAGNETIC STRATI-GRAPHY
ITALIAN SECTIONS
(SELLI, 1967)
BANNER & BLOW (1965 1967)
CENTRAL N. ATLANTIC
(BERGGREN IN PRESS)
RADIOLARIAN ZONES
ANT-ARCTIC
N. PACIFIC
EQUATOR. PACIFIC
(HAYS, 1965; HAYS & OPDYKE, 1967, THIS PAPER)
NANNO-FOSSILS
(RIEDEL et al, 1963)
(ERICSON et al. 1963, 1964)
SUGGESTED CORRELATION WITH CHRONOSTRATIGRAPHIC UNITS, THIS PAPER
AGE
STAGE
EPOCH
SERIES
.2
.4
.6
.8
1.0
1.2
1.4
1.6
1.8
2.0
2.2
2.4
2.6
2.8
3.0
3.2
3.4
3.6
3.8
BRUNHES NORMAL
MATUYAMA REVERSED
JARAMILLO
OLDUVAI
GAUSS NORMAL
MAMMOTH
GILBERT REVERSED
A
MILAZZIAN
TYRRHENIAN
SICILIAN
EMILIAN
CALABRIAN
UPPER PLIOCENE
GLACIAL
PREGLACIAL
PLEISTOCENE
N23
G. truncatulinoides Zone (N22)
G. tosaensis Zone (N21)
G. multicamerata-P. obliquiloculata Zone (N20)
S. dehiscens / G. altispira Zone (N19)
N22
N21
Discoaster reoccurrence
abundant Discoasters
COOLING
SEVERE
COOL TEMPERATE
Ω
Ψ
X
Φ
Υ
Ω_N
Ψ_N
X_N
Φ_N
Υ_N
Pterocanium prismatium
Eucyrtidium elongatum peregrinum
Discoasters
Coccolithus sp. cf. C. pelagicus
G. truncatulinoides
G. menardii complex (sinistral)
G. truncatulinoides (abundant)
G. fistulosus
G. multicamerata
G. menardii complex (dextral)
Discoasters
CALABRIAN
ASTIAN
PLAISANCIAN
VILLAFRANCHIAN s.l.
VILLAFRANCHIAN s.s.
LATE
EARLY
LATE
EARLY - LATE
EARLY
PLEISTOCENE
PLIOCENE

Fig. 51.5. Late Pliocene-Pleistocene biostratigraphy in deep-sea cores.

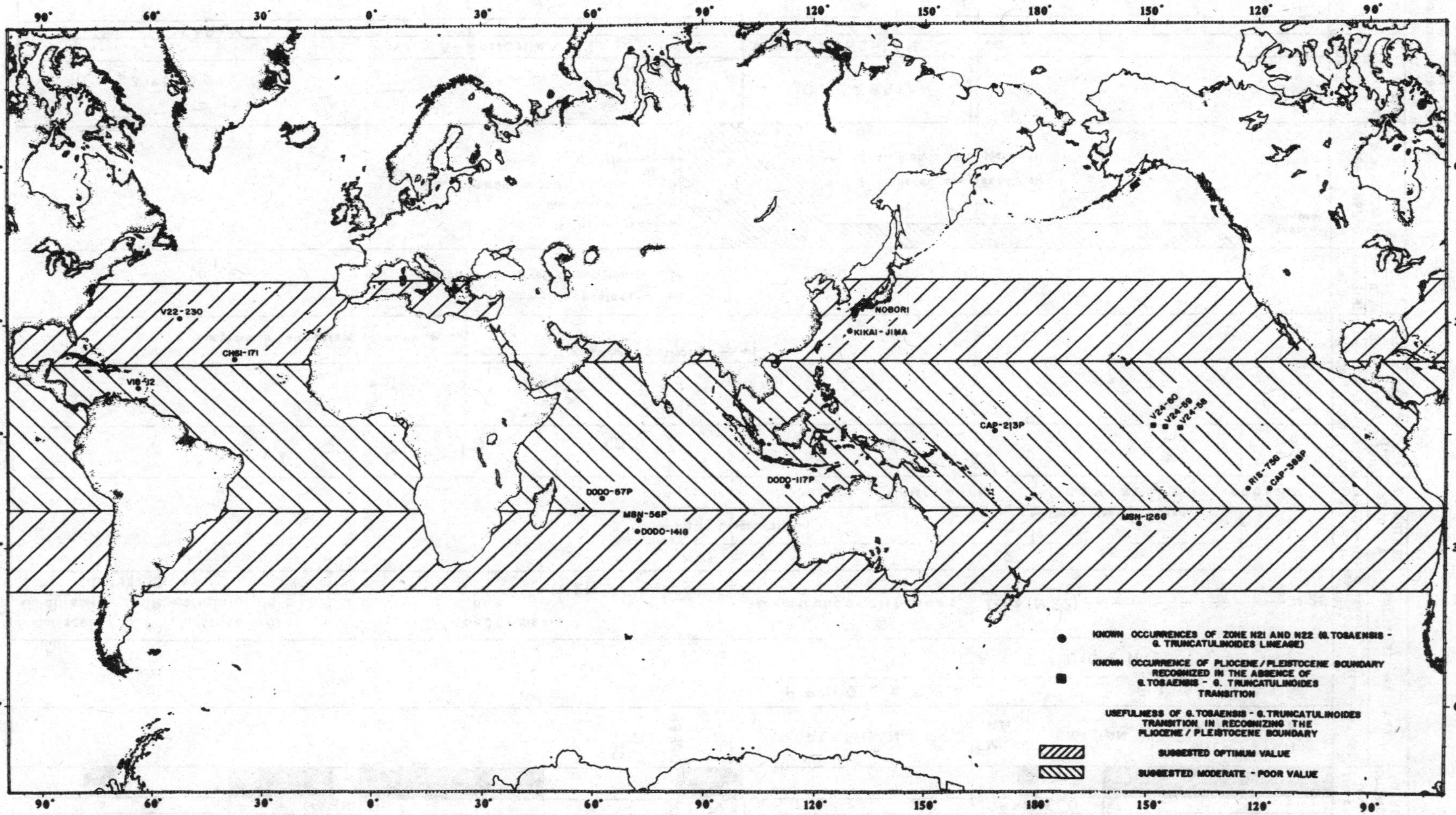
V22-230
CHBI-171
V16-12
NOBORI
KIKAI-JIMA
CAP-213P
V24-60
V24-59
V24-58
RIS-75G
CAP-38BP
DODO-117P
DODO-57P
MSN-56P
DODO-141G
MSN-126G
KNOWN OCCURRENCES OF ZONE N21 AND N22 (G. TOSAENSIS - G. TRUNCATULINOIDES LINEAGE)
KNOWN OCCURRENCE OF PLIOCENE/PLEISTOCENE BOUNDARY RECOGNIZED IN THE ABSENCE OF G. TOSAENSIS - G. TRUNCATULINOIDES TRANSITION
USEFULNESS OF G. TOSAENSIS - G. TRUNCATULINOIDES TRANSITION IN RECOGNIZING THE PLIOCENE / PLEISTOCENE BOUNDARY
SUGGESTED OPTIMUM VALUE
SUGGESTED MODERATE - POOR VALUE

Fig. 51.6. Areas where the occurrence of the transition from *G. tosaensis* to *G. truncatulinoides* has been recognized, and where it will probably be useful as a criterion for recognizing the Pliocene/Pleistocene boundary. Data from cores V24-58, -59 and -60 from T. Saito, Lamont Geological Observatory.

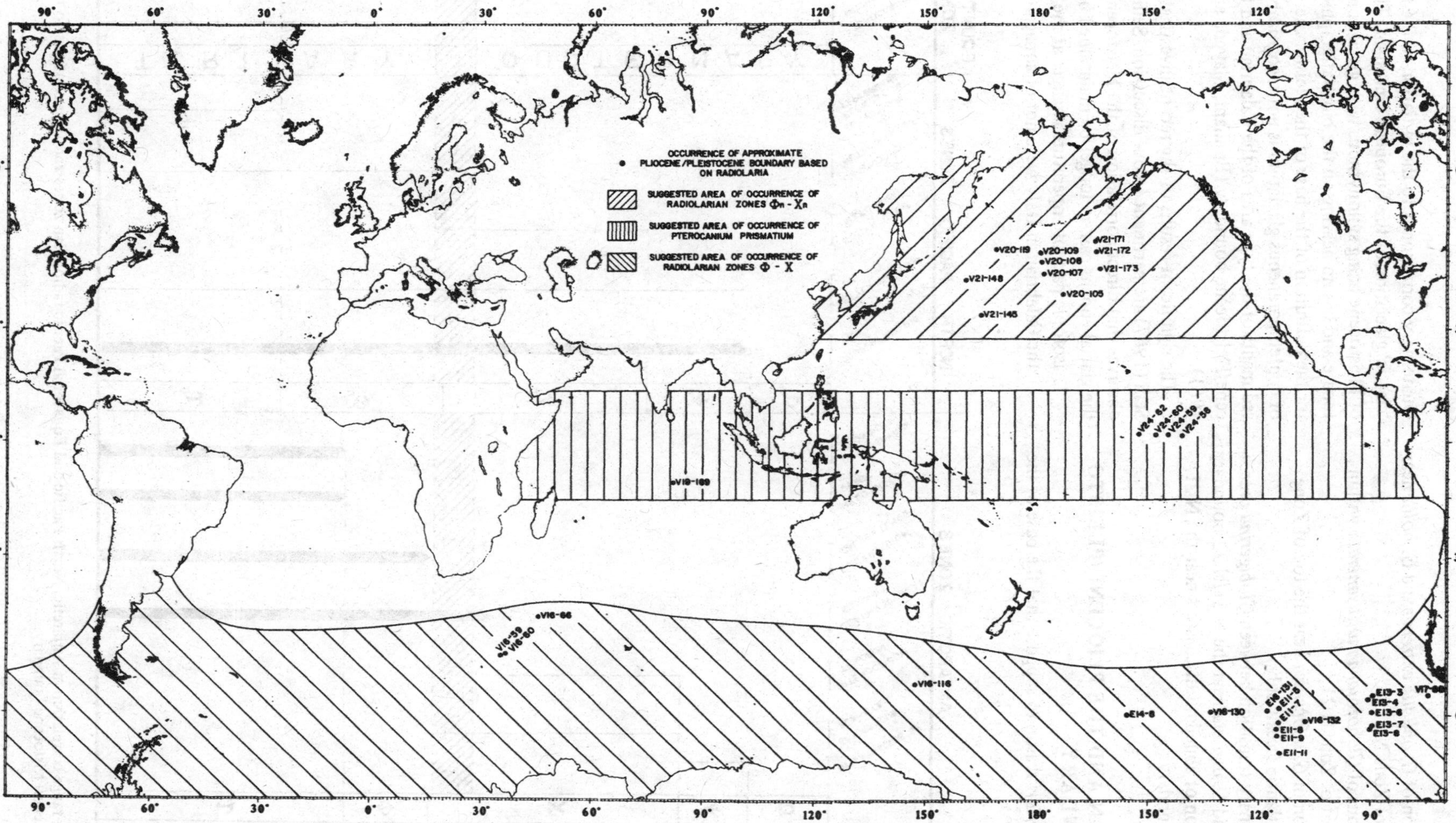
OCCURRENCE OF APPROXIMATE PLIOCENE/PLEISTOCENE BOUNDARY BASED ON RADIOLARIA
SUGGESTED AREA OF OCCURRENCE OF RADIOLARIAN ZONES Φn - Χn
SUGGESTED AREA OF OCCURRENCE OF PTEROCANIUM PRISMATIUM
SUGGESTED AREA OF OCCURRENCE OF RADIOLARIAN ZONES Φ - Χ
V20-119
V20-109
V20-108
V20-107
V21-171
V21-172
V21-173
V21-148
V20-105
V21-145
V24-62
V24-60
V24-59
V24-58
V19-169
V16-66
V16-59
V16-60
V16-116
E14-8
V16-130
E11-7
E11-8
E11-9
E11-11
V16-132
E13-3
E13-4
E13-6
E13-7
E13-8
V17-66
90°
60°
30°
0°
120°
150°
180°

Fig. 51.7. Areas where various species of Radiolaria are useful in recognizing the Pliocene/Pleistocene boundary.

2. extinction of *G. obliqua extrema* and *G. bollii* near the top of Zone N 21;
3. extinctions of *Globorotalia multicamerata* within lower half of Zone N 21;
4. extinction of *G. miocenica* near the top of Zone N 21 (Atlantic Ocean only).
5. evolutionary transition between *Globigerina praedigitata* and *G. digitata* near the N 21/N 22 boundary;
6. extinction of abundant discoasters near the N 21/N 22 boundary.

GLACIATION AND THE PLIOCENE/PLEISTOCENE BOUNDARY

There is no *a priori* reason to suppose that the base of the Quaternary corresponds to the initiation of the first major 'classic' glaciation of the European continent. The concept of Pleistocene being synonymous with glaciation is unfortunate and has no bearing on the biostratigraphical basis for the definition of the base of the Pleistocene. Evidence of contemporaneous glaciation is not recognizable in the sediments of the type area of the Calabrian and the Plio-cene/Pleistocene boundary (Emiliani, Mayeda and Selli 1961).

The significant results of the recent investigations of Selli (1967) are pertinent to our discussion. Selli (1967) observes that direct correlation of the Pleistocene with the glacial epochs can no longer be maintained because:

1. isotopic palaeotemperature changes at the base of the Calabrian in Italy were not of a magnitude to

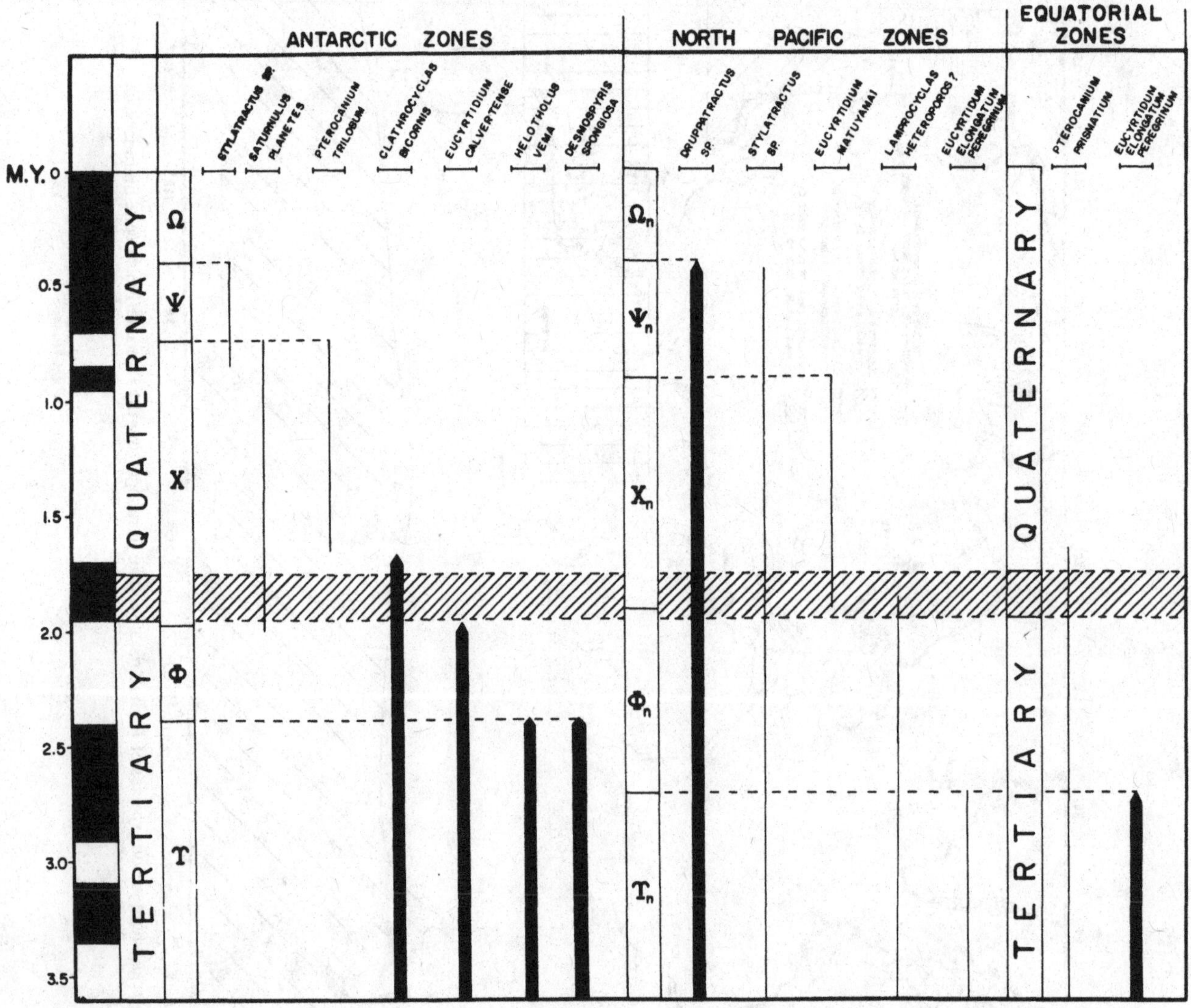
ANTARCTIC ZONES
NORTH PACIFIC ZONES
EQUATORIAL ZONES
STYLATRACTUS SP.
SATURNULUS PLANETES
PTEROCANIUM TRILOBUM
CLATHROCYCLAS BICORNIS
EUCYRTIDIUM CALVERTENSE
HELOTHOLUS VEMA
DESMOSPYRIS SPONGIOSA
DRUPPATRACTUS SP.
STYLATRACTUS SP.
EUCYRTIDIUM MATUYAMAI
LAMPROCYCLAS HETEROPOROS ?
EUCYRTIDIUM ELONGATUM PEREGRINUM
PTEROCANIUM PRISMATIUM
EUCYRTIDIUM ELONGATUM PEREGRINUM
M.Y. 0
0.5
1.0
1.5
2.0
2.5
3.0
3.5
QUATERNARY
TERTIARY
Ω
Ψ
X
Φ
Υ
Ω_n
Ψ_n
X_n
Φ_n
Υ_n

Fig. 51.8. Long-range correlation of Antarctic, North Pacific and Equatorial Pacific radiolarian zones and their relationship to the Pliocene/Pleistocene boundary.

have caused great glacial expansions;

2. climatic oscillations in Italian marine sections (Le Castella), and in the oceans, were more numerous than the generally accepted glacial expansions. It is thus extremely difficult to correlate the climatic marine temperature fluctuations with glacial areas;
3. eustatic marine terraces, caused by glacial expansions, are not known in Italy earlier than the Milazzian;
4. in the Po Plain, in the foothills of the Alps, and near Brescia, Calabrian and Emilian deposits have been found unconformably overlain by Mindel glacial deposits (Castenedolo) and in the case of the Brescia occurrence, perhaps Donau and Gunz also;
5. if one accepts the Milankovitch chronology of the glacial epochs (beginning of glaciation 0.8 million years ago) and the correlation of Calabrian with Gunz or Donau-Gunz, an excessively high sedimentation rate for this part of the Pleistocene results;
6. in front of the advancing glaciers (in the terminal moraines of the Po Plain) considerable transport of coarse detrital material would be expected. In deep wells in the Po Plain region the Calabrian and Emilian are represented only by clays.

Selli calculates that the thickness ratio of Milazzian-post-Milazzian: Quaternary has an almost constant average of 0.3. Thus Milazzian and post-Milazzian strata represent only approximately 30 per cent of the whole Quaternary. The Milazzian is generally correlated with either the interglacial Mindel/Riss or the interglacial Gunz/Mindel (i.e. with either the end of the Mindel or end of the Gunz glaciation).

The time from the end of the Mindel to the present represents an average of 62 per cent of the whole 'glacial Pleistocene' and from the end of the Gunz to the present an average 80 per cent of the 'glacial Pleistocene'. Thus an average of 62 per cent or 80 per cent of the 'glacial Pleistocene' should have elapsed since the beginning of the Milazzian, depending on whether the Milazzian is correlated with the Mindel/Riss or Gunz/Mindel. On the assumption of a relatively constant sedimentation and subsidence rate, and calculating the ratio time: thickness from the beginning of Milazzian (i.e. 62:30 and 80:30) the Quaternary is shown to have lasted 2 to 2.6 times the whole 'glacial Pleistocene' (depending on our basis of correlation); in other words, the pre-glacial Pleistocene appears to have lasted 1 to 1.5 times longer than the 'glacial Pleistocene'.

Selli accepts the Milankovitch curve as the chronological basis for the Pleistocene glaciations and assumes that the Donau glaciation is the first (earliest) of the major Pleistocene glaciations. Selli accepts the correlation of the Milazzian with the Gunz/Mindel because it agrees better with rates of mammalian evolution and recent radiometric dates (Fig. 51.9). On this basis, Selli (1967) suggests a date of 1.8 million years for the base of the Pleistocene, a date which is in remarkable close agreement with that suggested here. The evidence which Selli (1967) has presented is of considerable significance in illustrating the approximate relationships between glacial and preglacial parts of the Pleistocene and has clarified some of the problems arising out of attempts to recognize the Pliocene/Pleistocene boundary in deep-sea cores.

Movius (1949) inferred first glacial and first interglacial conditions for certain mammalian fossil assemblages of 'Villafranchian' age in the Auvergne of France. These deposits are intercalated between reversely magnetized lava flows and are hence within the Matuyama (Cox, Doell and Dalrymple 1965). This indicates that glaciation in Europe must have begun during the latter part of the Matuyama if the inferences of Movius are correct.

The evidence we have presented from deep-sea cores showing a cooling between 0.7 and 1.0 million years ago in the Atlantic, Antarctic and North Pacific may reflect a worldwide cooling associated with the first glaciation in Europe.

The evidence now available indicates that the onset of glaciation in high latitudes occurred well before the beginning of the Pleistocene. The earliest glacial till dated in Iceland occurs in the middle of the Gauss normal polarity series (Rutten and Wensink 1960) giving it an age of about 3 million years B.P. This palaeomagnetic age has recently been confirmed by potassium-argon dating (MacDougall and Wensink 1966). In all probability the glaciation of Antarctica preceded that of Iceland. Craddock, Bastieu and Rutford (1964) have reported evidence for glacial action probably older than 10 million years in the Jones mountain area. Ice rafted debris in Antarctic deep-sea cores has been dated in excess of 4 million years old (Hays and Opdyke 1967).

On the North Island of New Zealand lava flows have been dated at approximately 2.5 million years (Stipp, Chappell and MacDougall 1967) which have been correlated as approximately equivalent in age to the earliest glacial tills on South Island.

The climatic criteria for the definition of the base of the Pleistocene is seen to be useful in a local sense only and is not applicable on a worldwide scale.

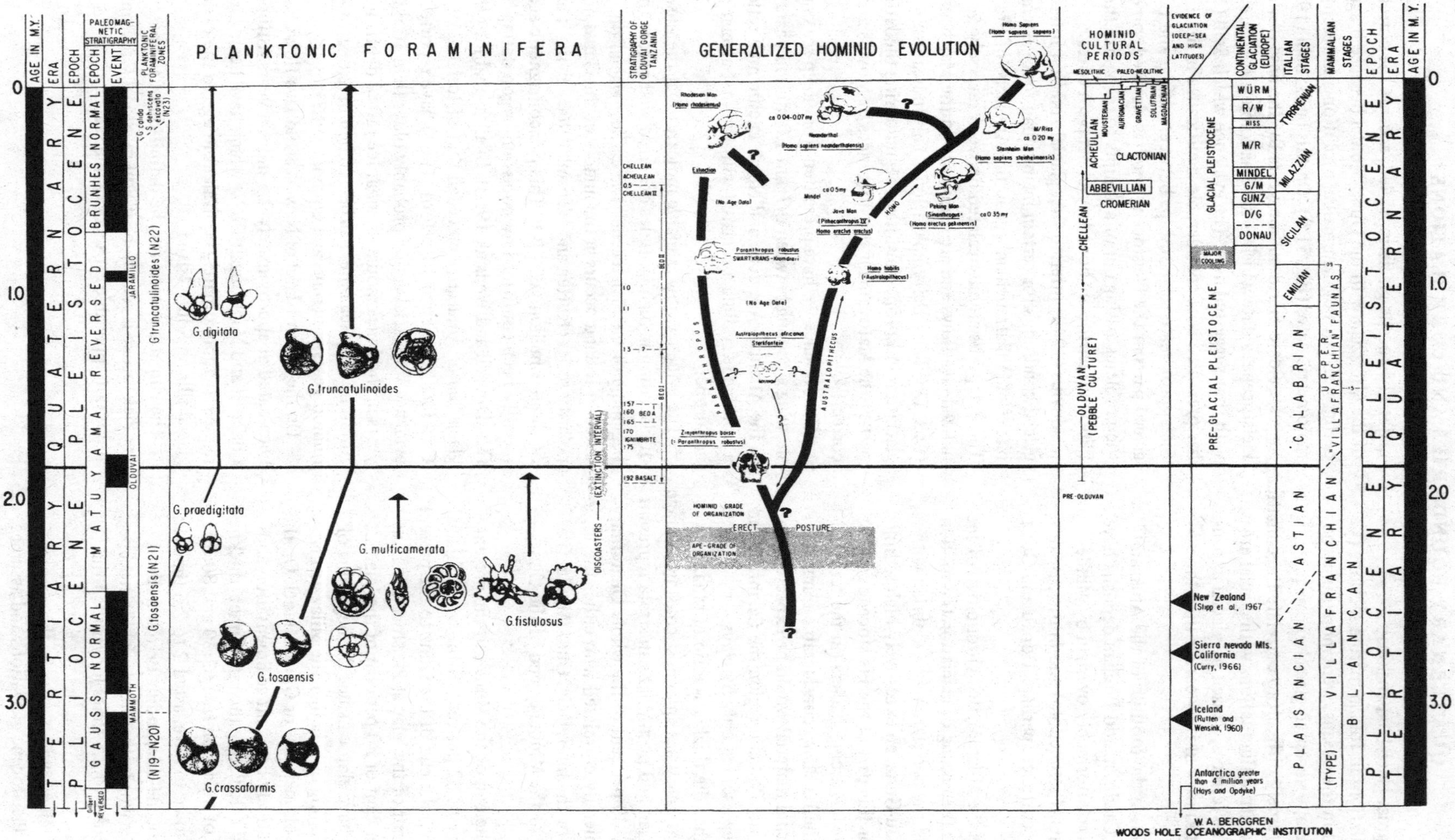

PLANKTONIC FORAMINIFERA
GENERALIZED HOMINID EVOLUTION
AGE IN M.Y.
ERA
EPOCH
PALEOMAGNETIC STRATIGRAPHY
EVENT
PLANKTONIC FORAMINIFERAL ZONES
STRATIGRAPHY OF OLDUVAI GORGE TANZANIA
HOMINID CULTURAL PERIODS
EVIDENCE OF GLACIATION (DEEP-SEA AND HIGH LATITUDES)
CONTINENTAL GLACIATION (EUROPE)
ITALIAN STAGES
MAMMALIAN STAGES
TERTIARY
QUATERNARY
PLIOCENE
PLEISTOCENE
GAUSS NORMAL
MATUYAMA REVERSED
BRUNHES NORMAL
MAMMOTH
OLDUVAI
JARAMILLO
G. tosaensis (N21)
G. truncatulinoides (N22)
(N19–N20)
G. digitata
G. truncatulinoides
G. praedigitata
G. multicamerata
G. tosaensis
G. crassaformis
G. fistulosus
DISCOASTERS (EXTINCTION INTERVAL)
192 BASALT
BED I
BED II
PARANTHROPUS
AUSTRALOPITHECUS
HOMO
Homo habilis
Zinjanthropus boisei
Paranthropus robustus SWARTKRANS–Kromdraai
Australopithecus africanus Sterkfontein
(No Age Data)
Java Man
Peking Man
Neanderthal
Steinheim Man
Rhodesian Man
Homo Sapiens
HOMINID GRADE OF ORGANIZATION
ERECT POSTURE
APE-GRADE OF ORGANIZATION
OLDUVAN (PEBBLE CULTURE)
PRE-OLDUVAN
CHELLEAN
ACHEULIAN
ABBEVILLIAN
CROMERIAN
CLACTONIAN
MOUSTERIAN
AURIGNACIAN
GRAVETTIAN
SOLUTRIAN
MAGDALENIAN
MESOLITHIC
PALEO-NEOLITHIC
GLACIAL PLEISTOCENE
PRE-GLACIAL PLEISTOCENE
MAJOR COOLING
New Zealand (Stipp et al., 1967
Sierra Nevada Mts. California (Curry, 1966)
Iceland (Rutten and Wensink, 1960)
Antarctica greater than 4 million years (Hays and Opdyke)
WÜRM
R/W
RISS
M/R
MINDEL
G/M
GUNZ
D/G
DONAU
TYRRHENIAN
MILAZZIAN
SICILIAN
EMILIAN
CALABRIAN
ASTIAN
PLAISANCIAN
UPPER "VILLAFRANCHIAN" FAUNAS
VILLAFRANCHIAN (TYPE)
BLANCAN
W. A. BERGGREN
WOODS HOLE OCEANOGRAPHIC INSTITUTION

Fig. 51.9. Comparison of late Cenozoic magnetic stratigraphy with foraminiferal evolution, hominid evolution, earliest evidences of glaciation, and the four classic glaciations of Europe. (Hominid chart drawn from *'Time and stratigraphy in the evolution of man'. National Academy of Sciences, National Research Council, Washington, D.C., 1967.*)

CONTINENTAL AND MARINE LATE PLIOCENE AND PLEISTOCENE

Europe

The Pliocene of Italy is classically divided (from the bottom) into the Tabianian (or Zanclian), Piacenzian and Astian Stages. Stratigraphical work has shown that the Piacenzian and Astian are, essentially, lateral facies of each other and that they overlap the limits of the Tabianian in some cases. Recent biostratigraphical evidence of this has been provided by Banner and Blow (1965, 1967) who recognized three planktonic foraminiferal zones in the Pliocene (N 19-N 21), drew the Miocene/Pliocene boundary within the upper part of Zone N 18 and related their zones to the stratotype sections. Zones N 18 and N 19 were recognized in the type Trubi Marl of Sicily, the lowest strata of the Pliocene (Zanclian Stage). Zones N 18, N 19 and probably N 20 were recognized in the stratotype Piacenzian at Castell'Arquato. Zone N 21 was determined as 'probably latest Pliocene' since it was found in pre-Calabrian horizons in Italy. Thus we see ample justification for the suggestion by Ruggieri and Selli (1950) that the Pliocene be divided into a lower, middle and upper part (originally correlated with the Tabianian, Piacenzian and Astian, but subsequently said to bear little relationship to these 'stages'). Palaeontological criteria for the definition of this three-fold subdivision of the Pliocene have been given by various Italian investigators, so that it now corresponds more closely to a true chronostratigraphical subdivision of the Pliocene epoch than did the classic stage subdivision. However, the stage terms Piacenzian, Astian and Villafranchian will be used in this discussion because of their familiarity to stratigraphers.

The Piacenzian-Astian sequence is well developed in northern Italy in the vicinity of Castell'Arquato. This is the type area of the Piacenzian and the succession contains a continuous sequence across the Pliocene/Pleistocene boundary, as classically defined and recommended by the Commission of the Eighteenth International Geological Congress in London in 1948. The appearance of *Cyprina islandica* is followed abruptly by the appearance of *Globigerina pachyderma* (a cold-water species), and the disappearance of several warm-water molluscan and foraminiferal species (Napoli-Alliata 1954).

The recognition of the Pleistocene in marine sections in Italy is generally made on the first appearance of *Hyalinea baltica* and/or *Cyprina islandica*. In littoral and epineritic marine sections *C. islandica* appears prior to *H. baltica* (Ruggieri 1961, 1962).

The recognition of the Pliocene/Pleistocene boundary in marine and continental sections is intimately linked with the problem of the Villafranchian Stage (Pareto 1865). The type section of this stage is at Villafranca d'Asti, near Turino, not, as often supposed, the more familiar Villefranche-Sur-Mer, near Nice, on the French Riviera. The Villafranchian was originally included within the Pliocene (Haug 1911) on the basis of its mammalian fauna. The type Villafranchian has been generally characterized as containing early elephants (*Elephas meridionalis* and *E. antiquus*), but these occurrences have since been discredited. Faunal similarity indicates that Villafranchian time represents essentially the same time interval as the Blancan mammalian stage of North America, but that earliest Blancan faunas may be slightly earlier than any known fauna referable to the Villafranchian. Savage and Curtis (1967) have recently presented a radiometric time-scale for late Pliocene/Pleistocene in which they suggest a date of approximately 3.5 million years for base Blancan, and 3.4 million years for base Villafranchian at Etouaires (Auvergne, France). They point out that the supposed occurrence of *Equus* – *Bos* (*Leptobos*) – *Elephas* has not been verified in the type Villafranchian (i.e. the lower part of the Villafranchian). These genera are to be considered essentially Pleistocene in terms of their first appearance and development. Thus we have rather conclusive evidence that the type Villafranchian (based on mammalian faunas) is older than the base of the Calabrian (1.85 million years) and that the Villafranchian *s.l.* is essentially a continental facies of the late Pliocene-early Pleistocene marine sections in Italy.

An excellent summary of the Pliocene/Pleistocene boundary in marine sections in Italy and its relationship to continental (non-marine) sections has recently been presented by Selli (1967) to which the reader is referred for more details (see also Flint 1965).

In northwestern Europe (Netherlands, Belgium, Germany) the Pliocene/Pleistocene boundary is generally recognized on the basis of the appearance of boreal molluscan or foraminiferal faunas. In the Netherlands, for instance, the boundary is determined on the basis of the first appearance of the foraminifers *Elphidiella* sp.cf. *E. arctica* and *Eponides frigidus* (Van Voorthuysen 1957) and on the basis of bryozoans (Lagaaij 1952). Pollen studies in the Pliocene/Pleistocene succession in this region have demonstrated a gradual cooling with time followed by a change to more temperate climate in later Pleistocene time. The earliest recognized cold-period (pre-Tiglian) has been regarded as equivalent to the first glacial age in Europe. Implicit in the definition of the Pliocene/Pleistocene boundary as determined by most investigators in northwestern Europe is the assumption that base Pleistocene is equivalent to first continental glaciation, an assumption which, as we have seen above, has led to unfortunate blind alleys in attempts to recognize equivalents of the base of the Pleistocene (as stratotypified at Santa Maria di Catanzaro, Calabria, Italy). The base of the Pleistocene is generally placed at the base of the Scaldisian Stage in Belgium, the Amstelian Stage in the Netherlands and the Red Crag in Great Britain (East Anglia). In the Red Crag, teeth of both zebrine and caballine horses and of *Mammuthus (Archidiskodon) meridionalis* – unknown in the underlying Coralline Crag – have been found, suggesting that this boundary may, indeed, lie close to the true Pliocene/Pleistocene boundary.

North America

The Pleistocene record of mammals and reptiles in North America indicates a progressive cooling from beginning to end, with warm and cold fluctuations superimposed upon this trend (Hibbard *et al.* 1965, Auffenberg and Milstead 1965). North American Pleistocene molluscs show a similar trend, at least to the extent that the Wisconsin glaciation was more severe than previous glaciations (Taylor 1965). Contrary to this interpretation is the fact that in the mid-continent region the earlier glaciations extended farther south than the later ones, suggesting the earlier glaciations were more severe.

Successive phases in the evolution of the North American mammalian fauna are the basis of the land-mammal ages. The most recent three of these are the Blancan, Irvingtonian and Rancholabrean (Hibbard 1958; Savage 1951; Wood *et al.* 1941). Recent potassium-argon radiometric dates (Evernden *et al.*, 1964) can be correlated roughly with the mammalian ages, and thus indirectly with glacial-interglacial events. At present only the earliest Blancan faunas (about 2.5 - 4 x 10^6 years old) seem to be older than the earliest recognized continental glaciation that reached mid-latitudes. These faunas also are pre-Villafranchian, according to the occurrence in North America of more primitive members of phyletic series whose later members occur in Villafranchian faunas of Europe (Hibbard *et al.* 1965). Hibbard *et al.* (1965) place the lower limit of the Pleistocene at the earliest palaeontological evidence of a markedly cooler climate in mid-latitudes. The few radiometric dates now available indicate that the base of a Pleistocene so defined would probably lie between two and three million years old and would therefore lie within the Blancan.

The North America mammalian fossil record is so discontinuous that virtually no well-documented evolutionary sequence of American Pleistocene mammals has been described (Hibbard *et al.* 1965). The most detailed studies of Pleistocene faunal sequences come from the Great Plains, mainly southwestern Kansas and northwestern Oklahoma. A general hypothesis of major climatic shifts in this region has been developed by Hibbard and Taylor (Hibbard 1944, 1949, 1960, 1963; Taylor 1960, 1965; Hibbard and Taylor 1960).

The first major southward faunal shift of ecological significance is shown by some of the mammals of the Cudahy fauna (Hibbard 1949) of late Kansan age (basal Irvingtonian, Hibbard *et al.*, 1965). Some living species of mammals of the Boreal subregion extended their range southward during Illinoian and/or Wisconsin time. The late Blancan, early Irvingtonian was also a time of stress for reptiles. At this time the highest degree of speciation and/or extinction per unit time occurred (Auffenberg 1963 and Auffenberg and Milstead 1965). The Irvingtonian and Rancholabrean are characterized by successive expansions and contractions of ranges. Among reptiles the group most affected by Pleistocene climatic fluctuations were the tortoises; these show two periods of extinction, first at the end of the late Blancan land mammal age, and secondly from the Wisconsin maximum to the waning phases of the Wisconsin, approximately 11,000 years ago. There is no strong evidence of climatic extremes in pre-Irvingtonian mammalian or reptilian sequences.

Although the exact age of the end of the Blancan is not known, palaeomagnetic and radiometric dating would place it in the Matuyama reversed polarity epoch. Evernden *et al.* (1964) have dated the Bruneau basalt of Elmore County, Idaho at 1.36 million years. The flow dated is the youngest of a sequence of flows interbedded with *Mammuthus*-bearing sediments which are placed in the Irvingtonian mammalian age. The Irvingtonian then is at least as old as 1.36 million years. The cooling represented by the fauna of the Cudahy formation would then be older than the cooling noted in deep-sea cores at about 1 million years.

CONCLUSIONS

The various faunal criteria previously advanced, by Ericson, Wollin and Ewing (1963), Riedel, Parker and Bramlette (1963), Hays (1965) and Berggren *et al.* (1967), to mark a Pliocene/Pleistocene boundary in deep-sea sediments all define a boundary, approximately equivalent in time, which can be correlated with the Pliocene/Pleistocene boundary in the type section in Calabria, Italy.

The lower boundary of the type Villafranchian is not equivalent to the Pliocene/Pleistocene boundary as defined in Calabria. It is older by at least 1.5 million years (Savage and Curtis 1967). However, the widespread deposits of Eurasia and Africa assigned to the Villafranchian on the basis of the occurrence in these strata of *Equus, Elephas* and *Bos (Leptobos)* may well be younger than the type Villafranchian where these species are absent, and therefore may be roughly equivalent to the Calabrian of Italy. It has become increasingly evident that the transition from Pliocene to Pleistocene was not marked by a major environmental change of world-wide significance.

Like Flint (1965) we would agree that no significant stratigraphical or faunal discontinuity occurs between the Pliocene and Pleistocene as at present defined. Indeed, the conspicuous climatic fluctuations long used as a criterion for distinguishing between the two epochs may characterize only the upper half of the Pleistocene. It may well be that as absolute dating becomes more refined, and palaeomagnetic stratigraphy is applied more extensively, other commonly accepted boundaries will be shown to have significance for only one group of organisms and therefore not represent major events in earth history. There are boundaries, which, because of the simultaneous changes in a number of groups of organisms, represent some major change in the environment. Such a boundary as the Cretaceous/Tertiary boundary certainly must represent this kind of change. We strongly feel that boundaries between periods should represent real changes in the environment, not just a minor change in the lineage of a few organisms.

The onset of glaciation which has long been thought to be a major environmental change of sufficient magnitude to separate the Tertiary from the Quaternary period has no value for world-wide correlation. The world climate has apparently deteriorated gradually, with glaciation beginning in high latitudes several million years before it reached temperate regions. We would tend to agree then with Flint (1965) that distinction between a Tertiary and a Quaternary period is invalid and rather artificial. It might be more suitable to recognize simply a Cenozoic period co-extensive with the Cenozoic era. As for the Pleistocene epoch, work is at present advancing so rapidly in the late Tertiary that it will soon be possible to make a better evaluation of the boundaries between Miocene/Pliocene and Pliocene/Pleistocene.

ACKNOWLEDGEMENTS

The writers are grateful to W. S. Broecker and D. B. Ericson, who read the manuscript and made helpful suggestions. The work was made possible through support to Lamont Geological Observatory, from the National Science Foundation (Grants NSF GA-558, GA-861) and to Woods Hole Oceanographic Institution from the National Science Foundation (Grant GA-676), and Office of Naval Research (Grant Nonr 4029).
Received January 1968.

BIBLIOGRAPHY

Akers, W. H. 1965. Pliocene-Pleistocene boundary, northern Gulf of Mexico. *Science,* **149**, 741-2.

Archiac, A. d' 1849. *Histoire des progrès de la Géologie de 1834 à 1845,* 2, 441-1,100. Soc. Geol. France: Paris.

Arrhenius, G. 1952. Sediment cores from the East Pacific. *Repts. Swedish Deep-Sea Exped.* **5**, 89.

Auffenberg, W. 1963. The fossil snakes of Florida. *Tulane Stud. Zool.* **10**, 131-216.

Auffenberg, W. and Milstead, W. W. 1965. Reptiles in the Quaternary of North America. In *Quaternary of the United States,* (ed. Wright, H. E. Jr. and Frey, D. G.), 557-68. Princeton Univ. Press.

Bandy, O. L. 1963. Miocene-Pliocene boundary in the Philippines as related to Late Tertiary stratigraphy of deep-sea sediments. *Science,* **142**, 1,290-2.

Banner, F. T. and Blow, W. H. 1965. Progress in the planktonic foraminiferal biostratigraphy of the Neogene. *Nature, Lond.* **208**, 1,164-6.

Banner, F. T. and Blow, W. H. 1967. The origin, evolution and taxonomy of the foraminiferal genus *Pulleniatina* Cushman, 1927. *Micropaleont.* **13**, 133-62.

Berggren, W. A. Phillips, J. D., Bertels, A. and Wall, D. 1967. Late Pliocene-Pleistocene stratigraphy in deep-sea cores from the south-central North Atlantic. *Nature, Lond.* **216**, 253-5.

Buteux, C. 1843. Esquisse géologique du départment de la Somme. *Mém. Acad. Sc. Agric., Dept. de la Somme,* 187-322.

Cox, A., Doell, R. R. and Dalrymple, G. B. 1963. Geomagnetic polarity epochs and Pleistocene geochronometry. *Nature, Lond.* **198**, 1,049-51.

Cox, A., Doell, R. R. and Dalrymple, G. B. 1965. Quaternary paleomagnetic stratigraphy. In *Quaternary of the United States,* (ed. Wright, H. E. Jr. and Frey, D. G.), 817-30. Princeton Univ. Press.

Craddock, C., Bastien, T. W. and Rutford, R. H. 1964. Geology of the Jones Mountains area. In *Antarctic Geology,* 171-87. Interscience Publishers: New York.

Creuze de Lesser, H. 1824. *Statistique du départment de l'Hérault.* 606 pp. Montpellier.

Curtis, G. H. 1965. Potassium-argon date for Early Villafranchian of France. *Trans. Amer. Geophys. Union,* **46**, 178.

Deperet, M. C. 1918. Essai de coordination chronologique des temps quaternaires. *C.R. Acad. Sc., Paris,* **166** (2), 480-6.

Desnoyers, J. 1829. Observations sur un ensemble de dépôts marins plus récents que les terrains tertiaires du bassin de la Seine, et constituant une formation géologique distincte; précedées d'un aperçu de la non-simultanéité des bassins tertiaires. *Annales des Sciences Naturelles,* **16**, 171-214, 402-19.

Doderlein, P. 1872. *Note illustrative della carta geologica del Modenese e del Reggiano, Memoria terza.* 74 pp. Modena.

Doell, R. R., Dalrymple, G. B. and Cox, A. 1966. Geomagnetic polarity epochs: Sierra Nevada Data, 3. *J. Geophys. Res.* **71**, 531-41.

Eberl, B. 1930. *Die Eiszeintenfalge im nordlichen Alpenvorlande.* vii-427 pp. Augsburg.

Emiliani, C. 1955. Pleistocene temperatures. *J. Geol.* **63**, 538-78.

Emiliani, C. 1961. Cenozoic climatic changes as indicated by the stratigraphy and chronology of deep-sea cores of *Globigerina*-ooze facies. *New York Acad. of Sci., Ann.* **95**, 521-36.

Emiliani, C. 1964. Paleotemperature analysis of the Caribbean cores A254-BR-C and CP-28. *Geol. Soc. Amer. Bull.* **75**, 129-44.

Emiliani, C. 1966a. Paleotemperature analysis of the Caribbean cores P6304-8 and P6304-9 and a generalized temperature curve for the last 425,000 years. *J. Geol.*, **74**, 109-26.

Emiliani, C. 1966b. Isotopic paleotemperatures. *Science*, **154**, 851-7.

Emiliani, C., Mayeda, T. and Selli, R. 1961. Paleotemperature analysis of the Plio-Pleistocene section at Le Castella, Calabria, southern Italy. *Geol. Soc. Amer. Bull.* **72**, 679-88.

Ericson, D. B., Ewing, M. and Wollin, G. 1963. Pliocene-Pleistocene boundary in deep-sea sediments. *Science*, **139**, 727-37.

Ericson, D. B., Ewing, M. and Wollin, G. 1964. The Pleistocene epoch in deep-sea sediments. *Science*, **146**, 723-32.

Evernden, J. F., Savage, D. E., Curtis, G. H. and James, G. T. 1964. Potassium-argon dates and the Cenozoic mammalian chronology of North America. *Amerc. J. Sci.* **262**, 145-98.

Flint, R. F. 1947. *Glacial geology and the Pleistocene epoch.* 589 pp. John Wiley: New York.

Flint, R. F. 1965. The Pliocene-Pleistocene boundary. In: International Studies on the Quaternary. *Geol. Soc. Amer. Sp. Paper*, **84**, 497-533.

Forbes, E. 1846. On the connexion between the distribution of the existing fauna and flora of the British Isles and the geographical changes which have affected their area, especially during the epoch of the Northern Drift. *Great Britain Geol. Survey Memoir*, **1**, 336-432.

Fuller, M. D., Harrison, C. G. A. and Nayudu, Y. R. 1966. Magnetic and petrologic studies of sediment found above basalt in experimental Mohole core EM 7. *Amer. Assoc. Petrol. Geologists Bull.* **50**, 566-73.

Gervais, P. 1867-1869. *Zoologie et paléontologie générales nouvelles recherches sur les animaux vertébrés vivants et fossiles.* 263 pp. Paris.

Gignoux, M. 1910. Sur la classification du Pliocène et du Quaternaire de l'Italie du Sud. *C.R. Acad. Sc., Paris*, **150**, 841-44.

Gignoux, M. 1913. Les formations marines pliocènes et quaternaires de l'Italie du Sud et de la Sicile. *Ann. Un. Lyon, N.S.* **36**, 693 pp.

Gignoux, M. 1954. Pliocene et Quaternaire marines de la Méditerranée occidentale. *XIX Int. Geol. Congress (Algeria, 1952), Sect. 13, Questions diverses de Géologie générale*, **3** (15), 249-58.

Glass, B., Ericson, D. B., Heezen, B. C., Opdyke, N. D. and Glass, J.A. 1967. Geomagnetic reversals and Pleistocene chronology. *Nature, Lond.* **216**, 437-41.

Grichuk, V. P., Hey, R. W. and Venzo, S. 1965. Report of the Subcommission on the Plio-Pleistocene Boundary. *Rept. of the VI Intern. Congrs. on the Quaternary*, 311-29. Warsaw: Poland.

Grommé, C. S. and Hay, R. L. 1963. Magnetization of basalt of Bed I, Olduvai Gorge, Tanganyika, *Nature, Lond.* **200**, 560-1.

Harrison, C. G. A. and Funnell, B. M. 1964. Relationship of palaeomagnetic reversals and micropalaeontology in two late Caenozoic cores from the Pacific Ocean. *Nature, Lond.* **204**, 566.

Haug, E. 1900. *Traité de Géologie.* Armand Colin: Paris.

Haug, E. 1911. *Traité de Géologie.* 2,024 pp. Armand Colin: Paris.

Hays, J. D. 1965. Radiolaria and Late Tertiary and Quaternary history of Antarctic Seas. *Ant. Arctic Res. Ser.* **5**, *Am. Geophys. Union*, 125-84.

Hays, J. D. and Opdyke, N. D. 1967. Antarctic Radiolaria magnetic reversals and climatic change. *Science*, **158**, 1,001-11.

Hibbard, C. W. 1944. Stratigraphy and vertebrate paleontology of Pleistocene deposits of southwestern Kansas. *Geol. Soc. Amer. Bull.* **55**, 707-54.

Hibbard, C. W. 1949. Pleistocene vertebrate paleontology in North America. *Geol. Soc. Amer. Bull.* **60**, 1,417-28.

Hibbard, C. W. 1958. Summary of North American Pleistocene mammalian local faunas. *Michigan Acad. Sci. Pap.* **43** (1957), 3-32.

Hibbard, C. W. 1960. An interpretation of Pliocene and Pleistocene climates in North America. *Michigan Acad. Sci. Ann. Rep.* **62**, 5-30.

Hibbard, C. W. 1963. A late Illinoian fauna from Kansas and its climatic significance. *Michigan Acad. Sci. Pap.*, **49**, 115-27.

Hibbard, C. W., Ray, D. E., Savage, D. E., Taylor, D. W. and Guilday, J. E. 1965. Quaternary mammals of North America: In: *Quaternary of the United States*, (ed, Wright, H. E. Jr. and Frey, D. G.), 509-26. Princeton Univ. Press.

Hibbard, C. W. and Taylor, D. W. 1960. Two late Pleistocene faunas from southwestern Kansas. *Univ. Mich. Mus. Paleont. Contr.* **16**, 1-223.

Issel, A. 1914. Lembi fossiliferi quaternari e recenti osservati nella Sardegna meridionale dal prof. D. Lovisato. *Atti R. Acc. Lincei, Rendiconti, Roma, (5)*, **23**, 759-70.

Johnson, E. A., Murphy, T. and Torreson, O. W. 1948. Prehistory of the Earth's magnetic field. *Terr. Magn. Atmos. Elec.* **53**, 349-72.

Keen, M. J. 1960. Magnetization of sediment cores from the eastern Atlantic Ocean. *Nature, Lond.* **187**, 220-2.

Keen, M. J. 1963. The magnetization of sediment cores from the eastern basin of the North Atlantic Ocean. *Deep-sea Res.* **10**, 607-22.

Ku, T. L. and Broecker, W. S. 1966. Atlantic deep-sea stratigraphy: Extension of absolute chronology to 320,000 years. *Science*, **151**, 448-50.

Lagaaij, R. 1952. The Pliocene bryozoa of the low countries and their bearing on the marine stratigraphy of the North Sea region. *Geologische Strichting, Med. Ser. C*, **5**, 233 pp.

Lona, F. 1963. Prime analisa pollinologiche sui depositi terziari-quaternari di Castell'Arquato; reperti di vegitazione da clima freddo sotto le formazioni calcaree ad *Amphistegina. Soc. Geol. Italiana Boll.*, **81**.

Lyell, C. 1833. *Principles of Geology*, London.

Lyell, C. 1839. *Nouveaux éléments de géologie.* 648 pp. Pitois-Levrault: Paris.

MacDougall, I. and Tarling, D. H. 1963. Dating of polarity zones in the Hawaiian Islands. *Nature, Lond.* **200**, 54-6.

MacDougall, I. and Wensink, H. 1966. Paleomagnetism and geochronology of the Pliocene-Pleistocene lavas in Iceland. *Earth and Planet. Sci. Letters,* **1**, 232.

McIntyre, A., Bé, A. W. H. and Preikstas, R. 1967. Coccoliths and the Pliocene-Pleistocene boundary. *Progr. in Oceanography,* **4**, 3-25.

McNish, A. G. and Johnson, E. A. 1938. Magnetization of sediments from the bottom of the Atlantic Ocean (Abstract). *Am. Geophys. Union Trans. 19th Ann. Meeting, Pt. 1,* 204-5.

Mercey, A. de 1874-7. Sur la classification de la Période Quaternaire en Picardie. *Mem. Soc. Linn. Nord de la France,* **4**, 18-29.

Morlot, A. 1854. Ueber die quaternaren Gebilde des Rhonegebiets. *Verhandl. Schweiz. Gesel. ges. Naturw.* **39**, 161-4.

Morlot, A. 1858. Sur le terrain quaternaire du bassin du Léman. *Bull. Soc. vaudoise Sc. nat.* **6**, 101-8.

Morlot, A. 1858. Ueber die quaternaren Gebilde des Rhonegebietes. *Verhandl. Schweiz. Gesel. ges. Naturw.* **43**, 144-50.

Movius, H. L., Jr. 1949. Villafranchian stratigraphy in southern and southwestern Europe. *J. Geol.* **57**, 380-412.

Napoli-Alliata, E. 1954. Foraminiferi pelagici e facies in Italia. *Convegno Nuz. Metano e Petrolio, 7th Atti.,* **1**, 221-54.

Olausson, E. 1961. Remarks on Tertiary sequences of two cores from the Pacific. *Bull. Geol. Inst. Uppsala,* **40**, 299-303.

Opdyke, N. D., Glass, B., Hays, J. D. and Foster, J. 1966. Paleomagnetic study of Antarctic deep-sea cores. *Science,* **154**, 349-57.

Parandier, 1891. Notice géologique et paléontologique sur la nature des terrains traversés par le chemin de fer entre Dijon et Châlon-sur-Saône. *Bull. Soc. Géol. France, Sér. 3,* **19**, 794-818.

Pareto, L. 1865. Note sur les subdivision que l'on pourrait établir dans les terrains tertiaires de l'Apennin Septentrional. *Bull. Soc. Géol. France, Paris (2),* **22**, 210-77.

Penck, A. and Brückner, E. 1909. *Die Alpen im Eiszeitalter,* 3 vols. Leipzig: Tauchnitz.

Phillips, J. D., Berggren, W. A., Bertels, A. and Wall, D. 1968. Paleomagnetic stratigraphy and micropaleontology of three deep sea cores from the central North Atlantic Ocean. *Earth and Planet. Sci. Letters,* **4**, 118-30.

Piette, E. 1880. *Nomenclature des temps anthropiques primitifs.* 78 pp. Laon.

Reboul, H. 1833. *Géologie de la période Quaternaire.* 222 pp. Paris.

Riedel, W. R. 1957. Radiolaria: A preliminary stratigraphy. *Swedish Deep-Sea Exped. Rept.* **6**, 61-96.

Riedel W. R., Parker, F. L. and Bramlette, M. N. 1963. 'Pliocene-Pleistocene' boundary in deep-sea sediments. *Science,* **140**, 1,238-40.

Roche, Alexandre 1951. Sur les inversions de l'aimanation rémante des roches volcaniques dans les monts d'Auvergne. *Acad. Sci. (Paris) Compt. Rend.* **233**, 1,132-4.

Roche, Alexandre 1956. Sur la date de la dernière inversion du champ magnétique terrestre. *Acad. Sci. (Paris) Compt. Rend.* **243**, 812-14.

Ruggieri, G. 1961. Alcune zone biostratigrafiche del Pliocene e del Pleistocene italiano. *Riv. It. Pal. e Strat., Milano,* **67**, 405-17.

Ruggieri, G. 1962. La serie Marina Pliocenica e Quaternaria della Romagna. *Pubbl. Camera Comm. Ind. e Agricultura, Forli.* 79 pp.

Ruggieri, G. and Selli, R. 1950. Il Pliocene e il Post-pliocene dell'Emilia. *XVIII Intern. Geol. Congr. Great Britain, 1948, Rept., Lond.* **9**, 85-93.

Rutten, M. G. and Wensink, H. 1960. Palaeomagnetic dating, glaciations and the chronology of the Plio-Pleistocene in Iceland. *Int. Geol. Congr. Sess. 21, Pt. IV,* 62.

Savage, D. E. 1951. Late Cenozoic vertebrates of the San Francisco Bay region. *Univ. Calif. Dept. Geol. Bull.* **28**, 215-314.

Savage, D. E. and Curtis, G. H. 1967. The Villafranchian age and its radiometric dating. *Amer. Assoc. Petrol. Geologists* 41st Annual Meeting, April 1967 (preprint). *Amer. Assoc. Petrol. Geologists Bull.* **51**, 479-80. (abstract).

Selli, R. 1967. The Pliocene-Pleistocene boundary in Italian marine sections and its relationship to continental stratigraphies. *Progr. in Oceanography,* **4**, 67-86.

Serres, M. de 1830. De la simultanéité des terrains de sédiment supérieurs. In *La Géographie Physique de l'Encyclopédie Méthodique,* **5**, 125 pp., 1 pl.

Serres, M. de 1855. Des caractères et de l'importance de la période Quaternaire. *Bull. Soc. géol. France, sér. 2,* **12**, 1,257-63.

Stipp, J. J., Chappell, J. A. and MacDougall, I. 1967. Potassium-argon age estimate of the Pliocene/Pleistocene boundary in New Zealand. *Amer. J. Sci.* **265**, 462-74.

Taylor, D. W. 1960. Late Cenozoic molluscan faunas from the High Plains. *U.S. Geol. Surv. Prof. Pap.* **337**, 94 pp.

Taylor, D. W. 1965. The study of Pleistocene nonmarine molluscs in North America. In *Quaternary of the United States,* (ed.Wright, H. E. Jr. and Frey, D. G.), 597-611. Princeton Univ. Press.

Venzo, S. 1965a. The Plio-Pleistocene boundary in Italy. *VI Congr. INQUA Warsaw 1961,* 367-92.

Vezian, A. 1865. *Prodrome de Géologie,* **3**. Paris.

Voorthuysen, J. H. van 1957. The Plio-Pleistocene boundary in the Netherlands based on the ecology of Foraminifera. *Geologie en Mijnbouw (N.S.),* **12**, 26-30.

Wood, H. E., Chaney, R. W., Clark, J., Colbert, E. H., Jepsen, G. L., Reeside, J. B. Jr. and Spock, C. 1941. Nomenclature and correlation of the North American continental Tertiary. *Geol. Soc. Amer. Bull.* **52**, 1-48.

Wray, J. L. and Ellis, C. H. 1965. Discoaster extinction in neritic sediments northern Gulf of Mexico. *Bull. Amer. Assoc. Petr. Geol.* **49**, 98-9.

Zeuner, F. E. 1959. *The Pleistocene period; its climate, chronology and faunal succession.* 447 pp. Hutch. Scient. Tech.: London.

52. TERTIARY BOUNDARIES AND CORRELATIONS[1]

W. A. BERGGREN

Summary: An historical summary of the original definition and subsequent usage of stage and epoch boundaries of the Tertiary Period is presented. The distinction between lithostratigraphy and chronostratigraphy was seldom made by geologists of the last century and, indeed, is not realized by some stratigraphers to-day. Thus, many of the Tertiary Stages, as originally defined, are lithostratigraphical units (defined on the basis of their relationship to cycles of sedimentation) rather than true chronostratigraphical units. Recent biostratigraphical studies on various stratotype sections of the Tertiary have allowed correlation of these units with empirically determined planktonic foraminiferal zones in tropical and sub-tropical regions. Definition of the stratotypes in terms of significant palaeontological datum-levels will allow the stages to assume a chronostratigraphical basis.

The most satisfactory method of measuring geological time itself, is through the erection of zones based on phylogenetic successions in rapidly evolving forms which have been, in turn, related to an 'absolute' chronometric (i.e. radiometric) scale. Within this framework geological history can be related to a continuous secular time-scale.

A radiometric scale for the Tertiary is constructed on the basis of available data from marine and non-marine sections. The approximate relationship between stage and epoch boundaries and planktonic foraminiferal zones to this time-scale is shown below (Tables 52.31, 52.39 and 52.40).

OUTLINE

INTRODUCTION

Geologists and natural scientists in the eighteenth and nineteenth centuries were engaged in the description of

[1] Woods Hole Oceanographic Institution Contribution No. 2016

Author	Terminology (youngest to oldest)
BRONGNIART (1807)	"TERRAINS TERTIAIRES" / TERTIARY
MANTELL (1822) BUCKLAND (1823)	ALLUVIUM; DILUVIUM
DESNOYERS (1829)	QUATERNAIRE (=QUATERNARY) (=approximately equivalent to NEOGENE of HOERNES and later authors)
LYELL (1833, 1839)	RECENT; PLIOCENE (NEWER [PLEISTOCENE (1839)], OLDER); MIOCENE; EOCENE (Base defined as oldest TERTIARY in London and Paris Basins)
MARCEL DE SERRES (1832) MORLOT (1854)	QUATERNAIRE (=QUATERNARY)
SCHIMPER (1837)	PÉRIODE GLACIARE
PHILLIPS (1841)	CÉNOZOIQUE = CENOZOIC
D'ORBIGNY (1852)	SUBAPENNIN; FALUNIAN S.L. (FALUNIAN S.S.; "FALUNIAN" (=TONGRIAN & STAMPIAN)); PARISIAN; SUESSONIAN; PARISIAN S.L. (DENIZOT, 1957, LEX. STRAT. INTERNAT.) = MONTIAN - STAMPIAN & AQUITANIAN
HOERNES (1853)	NÉOGÈNE (NEOGENE) (AQUITANIAN - QUATERNARY - Recent; Interpretation of DENIZOT, 1957, LEX. STRAT. INTERNAT.); PALEOGENE - (DENIZOT, 1957, LEX. STRAT. INTERNAT.)
NAUMANN (1866)	PALÉOGÈNE (PALEOGENE) (EOCENE & OLIGOCENE, excluding MONTIAN, including CHATTIAN & AQUITANIAN)
RENEVIER (1873)	SYSTÈME NUMMULITIQUE (THANETIAN-STAMPIAN, 1873 and MONTIAN, 1896); (1873); (1896)
HAUG (1911)	PÉRIODE QUATERNAIRE; PÉRIODE NÉOGÈNE (NÉOMÉDITERRANEAN; MESOMÉDITERRANEAN; EOMÉDITERRANEAN); PÉRIODE NUMMULITIQUE (NEONUMMULITIQUE; MESONUMMULITIQUE; EONUMMULITIQUE)
HAUG (1911) SUGGESTED CORRELATIONS	ASTIAN; PLAISANCIAN; SAHELIAN-PONTIAN; VINDOBONIAN; BURDIGALIAN =AQUITANIAN; CHATTIAN; RUPELIAN; SANNOISIAN; LUDIAN (UPPER, MIDDLE, LOWER); BARTONIAN; AUVERSIAN; LUTETIAN (UPPER, LOWER); CUISIAN; SPARNACIAN; THANETIAN; MONTIAN
GIGNOUX (1950-1955)	QUATERNARY; NEOGENE; NUMMULITIC OR PALEOGENE (MONTIAN to and including AQUITANIAN)
SUGGESTED USAGE IN THIS PAPER	CENOZOIC: QUATERNARY (PLEISTOCENE: CALABRIAN); TERTIARY: NEOGENE (PLIOCENE: ASTIAN / PIACENZIAN, ZANCLIAN; MIOCENE: MESSINIAN, TORTONIAN, LANGHIAN, BURDIGALIAN, AQUITANIAN, BORMIDIAN); PALEOGENE (OLIGOCENE: RUPELIAN AND CHATTIAN / VICKSBURGIAN; EOCENE: BARTONIAN, LUTETIAN, YPRESIAN; PALEOCENE: THANETIAN, MONTIAN s.s., DANIAN s.s.)

Table 52.1. Historical perspective of the development of Cenozoic supra-stage terminology.

outcrop sections and in establishing the horizontal and vertical relationships between them. In this manner stratigraphy may be said to have had its birth and early development. The terms Primary and Secondary were introduced by Arduino in 1759 in private correspondence to Vallisnieri and encompassed essentially the crystalline, sedimentary and metamorphic rocks then known. He further recognized a third and fourth subdivision; the third was represented by low mountains and hills composed of gravel and sand, clay and marl, etc. with abundant marine fossils. He also included volcanic rocks in this category. In his fourth category he recognized earthy and rocky materials and alluvial debris. The term Primary was abandoned when it was realized that the various crystalline rocks assigned to it were of different ages; the term Palaeozoic was substituted partially for some of these rocks. The term Secondary gradually evolved into Mesozoic.

In 1810 Brongniart applied the term *tertiaires* to the strata which followed the chalks of the Cretaceous in the Paris Basin. The term Tertiary, although a relic of early geological science, has proved durable and is used at present for all pre-Quaternary–post-Mesozoic rocks. The fourth division recognized by Arduino later became known as Quaternary. The Cenozoic embraces the Tertiary and Quaternary, although the term Neozoic was used for a brief time as a synonym of Tertiary, and even as a substitute for the entire Cenozoic; it had originally been proposed by Edward Forbes to comprise both Mesozoic and Cenozoic.

In passing it is interesting to observe that the Cenozoic has been broadly characterized in the following manner in general textbooks on historical geology: Tertiary – Age of Mammals; Quaternary – Age of Man. We shall have occasion to return to this point at the conclusion of this article.

A threefold subdivision of the Tertiary was introduced by Lyell in the first edition of his *Principles of Geology* (1833, vol. 3). It was based primarily on his own investigations in England and Europe (and in particular Italy)and on those of the French conchologist, Deshayes, who was at the time in the process of monographing the Tertiary molluscs of the Paris Basin. From an examination of nearly 40,000 specimens, the proportion of living to extinct species was accepted as the distinctive character of the subdivisions. These subdivisions and the properties adopted for the approximate limits are as follows:

1. Older Pliocene: more than half the species living
2. Newer Pliocene: 90-95 per cent still living
3. Miocene period: 20-40 per cent living
4. Eocene period: no species, or less than 5 per cent living

The Miocene and Pliocene were sometimes united under the name Neocene by Lyell's contemporaries, especially in instances where the divisions were not well differentiated. The Oligocene was proposed by the German geologist Beyrich, in 1854, and was created from the upper part of the Eocene and the lower part of what had been referred to the Miocene prior to that time.

In 1864 the Austrian geologist Hörnes used the term Palaeogene (see also Naumann 1866) for the combined Eocene and Oligocene and Neogene for the Miocene and Pliocene; Eocene was briefly used as a synonym of Palaeogene. The lower Eocene was distinguished as Palaeocene by Schimper in 1874. In 1889 Dawson adopted the terms Orthrocene for the Lower Eocene, Nummulitic for the Middle Eocene, and Proicene for the Upper Eocene (or Vicksburg Epoch to use his term).

That Lyell realized the approximate and imperfect nature of his subdivision of the Tertiary is brought out in the following statement:

> In regard to distinct zoological periods, the reader will understand, from our observations in the third chapter, that we consider the wide lines of demarcation that sometimes separate different Tertiary epochs, as quite unconnected with extraordinary revolutions of the surface of the globe, and as arising, partly, like chasms in the history of nations, out of the present imperfect state of an information, and partly from the irregular manner in which geological memorials are preserved, as already explained. We have little doubt that it will be necessary hereafter to intercalate other periods, and that many of the deposits, now referred to a single era, will be found to have been formed at very distinct periods of time, so that, notwithstanding our separation of tertiary strata into four groups, we shall continue to use the term *contemporaneous* with a great deal of latitude.
>
> We throw out these hints, because we are apprehensive lest zoological periods in Geology, like artificial divisions in other branches of Natural History, should acquire too much importance, from being supposed to be founded on some great interruptions in the regular series of events in the organic world, whereas, like the genera and orders in zoology and botany, we ought to regard them as invented for the convenience of systematic arrangement, always expecting to discover intermediate graduations between the boundary lines that we have first drawn (Lyell 1833, pp. 56-57).

And a little further on:

> If intermediate formations should hereafter be found between the Eocene and Miocene, and between these of the last period and the Pliocene, we may still find an appropriate place for all, by forming subdivisions on the same principle as that which has determined us to separate the lower from the upper Pliocene groups. Thus, for example, we might have three divisions of the Eocene epoch – older, middle, and lower; and three similar divisions, both of Miocene and Pliocene epochs. For that case, the formations of the middle period must be considered as the types from which the assemblage of organic remains in the groups immediately antecedent or subsequent will diverge (Lyell 1833, pp. 57-58).

Lyell's original subdivision of the Tertiary is presented below (Tables 52.2-52.4).

It should be made clear to the reader at the beginning that one of the main problems in presenting a summary discussion of the various stage units of the European Tertiary and their correlatives elsewhere is the inconsistent usage on the part of investigators in referring to these

Synoptical Table of Recent and Tertiary Formations.

Periods.		Character of Formations.	Localities of the different Formations.
I. Recent.		Marine.	Coral formations of Pacific. Delta of Po, Ganges, &c.
		Freshwater.	Modern deposits in Lake Superior—Lake of Geneva—Marl lakes of Scotland—Italian travertin, &c.
		Volcanic.	Jorullo — Monte Nuovo — Modern lavas of Iceland, Etna, Vesuvius, &c.
II. Tertiary.	1. Newer Pliocene.	Marine.	Strata of the Val di Noto in Sicily. Ischia, Morea? Uddevalla.
		Freshwater.	Valley of the Elsa around Colle in Tuscany.
		Volcanic.	Older parts of Vesuvius, Etna, and Ischia—Volcanic rocks of the Val di Noto in Sicily.
	2. Older Pliocene.	Marine.	Northern Subapennine formations, as at Parma, Asti, Sienna, Perpignan, Nice—English Crag.
		Freshwater.	Alternating with marine beds near the town of Sienna.
		Volcanic.	Volcanos of Tuscany and Campagna di Roma.
	3. Miocene.	Marine.	Strata of Touraine, Bordeaux, Valley of the Bormida, and the Superga near Turin—Basin of Vienna.
		Freshwater.	Alternating with marine at Saucats, twelve miles south of Bordeaux.
		Volcanic.	Hungarian and Transylvanian volcanic rocks. Part of the volcanos of Auvergne, Cantal, and Velay?
	4. Eocene.	Marine.	Paris and London Basins.
		Freshwater.	Alternating with marine in Paris basin—Isle of Wight—purely lacustrine in Auvergne, Cantal, and Velay.
		Volcanic.	Oldest part of volcanic rocks of Auvergne.

Table 52.2. (from Lyell, 1833, p. 61)

stages. The stage, the fundamental unit in the chronostratigraphical scale, is a 'stratigraphical synthesis' to some (Gignoux 1950, 1955; see Gignoux's discussion of the nature of the stage as early as 1913). The stage, for Gignoux, and with him many European geologists, is a sequence of rock-strata representing a certain unspecified interval of geological time, intimately linked with and, indeed, reflecting cycles of sedimentation. It would be fair to state that most of the European Tertiary stage names in common use today were originally defined as lithostratigraphical (facies) units within cycles of sedimentation. On the other hand, as defined in article 26 of the 'Code' of the American Commission of Stratigraphic Nomenclature (1961, p. 657) a chronostratigraphical unit is 'a subdivision of rocks considered solely as the record of a specific interval of geologic time'. The boundaries of chronostratigraphical units are to be based upon physical criteria, palaeontological criteria, radiometric and isotopic methods (Article 28). As the 'Code' points out (1961, p. 658), 'ideally, these boundaries are independent of lithology, fossil content, or any other material bases of stratigraphic division. In actual practice, the geographic extent of a time-stratigraphic unit is influenced and generally controlled by stratigraphic features.'

Thus in deep marine sequences representing (relatively) continuous deposition and containing rich microfaunal assemblages it is relatively easy to follow evolutionary changes in various fossil groups and to formulate a biostratigraphical subdivision of the sedimentary rock succession. In this instance biostratigraphy approximates chronostratigraphy and it is possible to recognize relatively unequivocal time-significant *Datum-points* between which are represented finite and, within the limits of accuracy obtainable by subjective palaeontological methods, equal intervals of time. Thus the interval of time between the first (evolutionary) appearance of *Pseudohastigerina* and *Hantkenina* is the same in marine sections and represented by a finite accumulation of sediment. This interval of time (represented by sediment accumulation) would correspond well to a chronostratigraphical unit, say a stage, as defined by the 'Code'.

On the other hand, in sedimentary sequences deposited

Showing the Relations of the Alluvial, Aqueous, Volcanic, and Hypogene Formations of different ages.

	Periods.	Formations.		Some of the Localities where the Formations occur.
I. Recent.	A.	Alluvial.	. . . . .	Beds of existing rivers, &c., vol. ii. ch. xiv.
		Aqueous.	*a.* Marine.	Coral reefs of the Pacific, vol. ii. ch. xviii.
			b. Freshwater.	Bed of Lake Superior, &c., vol. i. ch. xiii.
		Volcanic.	. . . . .	Etna, Vesuvius, vol. i. ch. xix. xx. xxi.
		Hypogene.	*a.* Plutonic.	*Concealed;* foci of active volcanos, vol. iii. ch. xxv.
			b. Metamorphic.	*Concealed;* around the foci of active volcanos, vol. iii. ch. xxvi.
II. Tertiary.	1. Newer Pliocene. B.	Alluvial.	. . . . .	Loess of the Rhine—gravel covering the Newer Pliocene strata of Sicily.
		Aqueous.	*a.* Marine.	Val di Noto, Sicily.
			b. Freshwater.	Colle, in Tuscany.
		Volcanic.	. . . . .	Val di Noto, Sicily.
		Hypogene.	*a.* Plutonic.	*Concealed;* foci of Newer Pliocene volcanos—underneath the Val di Noto, vol. iii. p. 107, and ch. xxv.
			b. Metamorphic.	*Concealed;* near the foci of Newer Pliocene volcanos—underneath the Val di Noto, vol. iii. p. 109, and ch. xxvi.
	2. Older Pliocene. C.	Alluvial.	. . . . .	Norfolk? vol. iii. p. 173.
		Aqueous.	*a.* Marine.	Subapennine formations.
			b. Freshwater.	Near Sienna, vol. iii. p. 160.
		Volcanic.	. . . . .	Tuscany, vol. iii. p. 159.
		Hypogene.	*a.* Plutonic.	*Concealed;* foci of Older Pliocene volcanos—beneath Tuscany.
			b. Metamorphic.	*Concealed;* probably near the same foci.
	3. Miocene. D.	Alluvial.	. . . . .	Mont Perrier, Auvergne—Orleanais, vol. iii. p. 217.
		Aqueous.	*a.* Marine.	Bordeaux. Dax.
			b. Freshwater.	Saucats, near Bordeaux, vol. iii. p. 207.
		Volcanic.	. . . . .	Hungary, vol. iii. ch. xvi.
		Hypogene.	*a.* Plutonic.	*Concealed;* foci of Miocene volcanos—beneath Hungary.
			b. Metamorphic.	*Concealed;* probably around the same foci.
	4. Eocene. E.	Alluvial.	. . . . .	Summit of North and South Downs? vol. iii. p. 311.
		Aqueous.	*a.* Marine.	Paris and London basins.
			b. Freshwater.	Isle of Wight—Auvergne.
		Volcanic.	. . . . .	Oldest volcanic rocks of the Limagne d'Auvergne, vol. iii. ch. xix.
		Hypogene.	*a.* Plutonic.	*Concealed;* foci of Eocene volcanos—beneath the Limagne d'Auvergne.
			b. Metamorphic.	*Concealed;* probably near the same foci.

Table 52.3. (from Lyell, 1833, Appendix)

Showing the Order of Superposition, or Chronological Succession, of the principal Sedimentary Deposits or Groups of Strata in Europe.

Periods and Groups.		Names of the principal Members and general Mineral nature of the Formation.		Some of the Localities where the Formation occurs.
I. Recent Period.		The deposits of this period are for the most part concealed under existing lakes and seas.		
I. Recent Period.	A	Consolidated sandy and gravelly beds (*a*), travertin limestones (*b*), calcareous sandstones with broken shells (*c*), coral limestone, consisting of corals, shells, &c. (*d*)		*a.* Delta of the Rhone. *b.* Tivoli, and other parts of Italy. *c.* Shore of island of Guadaloupe. *d.* Coral reefs in Pacific, &c.
II. Tertiary Period.	B Newer Pliocene.	Marine. *Limestone*, sands, clays, sandstones, conglomerates, marls with gypsum; containing *marine* fossils (*a*).	Freshwater. Sands, clays, sandstones, lignites, &c.; containing *land* and *freshwater* fossils (*b*).	*a.* Sicily, Ischia, Morea? *b.* Colle in Tuscany.
II. Tertiary Period.	C Older Pliocene.	*Subapennine marl*, *Subapennine yellow sand*, *English 'crag,'* and other deposits, as in B, containing *marine* fossils (*a*).	Similar deposits to B; containing *land* and *freshwater* fossils (*b*).	*a.* Subapennine formations, Perpignan, Nice, Norfolk and Suffolk. *b.* Near Sienna, &c.
II. Tertiary Period.	D Miocene.	*Faluns of the Loire*, and other deposits of similar mineral composition with B and C, containing *marine* fossils (*a*).	Similar deposits to B and C; containing *land* and *freshwater* fossils (*b*).	*a.* Touraine, Bordeaux, Valley of Bormida, Superga near Turin, Basin of Vienna. *b.* Saucats, twelve miles south of Bordeaux.

Periods and Groups.		Names of the principal Members and general Mineral nature of the Formation.		Some of the Localities where the Formation occurs.
II. Tertiary Period, *continued.*	E Eocene.	*Calcaire Grossier* (*a*), *plastic clay*, sands, sandstones, &c., with *marine* fossils (*b*).	*Calcaire siliceux*—sandstones and conglomerates, red marl, green and white marls, limestone, gypseous marls, —with land and freshwater fossils (*c*).	*a.* Paris basin. *b.* Paris, London, and Hampshire basins, Isle of Wight. *c.* Paris Basin, Isle of Wight, Auvergne, Velay, Cantal.
III. Secondary Period.	F Cretaceous Group.	1. *Maestricht Beds.*—Earthy white limestone with siliceous masses, resembling chalk (marine).		St. Peter's Mount, Maestricht.
III. Secondary Period.	Cretaceous Group.	2. *Chalk with flints* (marine).		North and South Downs, and parts of the intervening Weald of Kent, Surrey, and Sussex. Isle of Wight, coasts of Hampshire and Dorsetshire, Yorkshire, North of Ireland.
III. Secondary Period.	Cretaceous Group.	3. Chalk without flints (marine).		
III. Secondary Period.	Cretaceous Group.	4. *Upper green sand* (marine).—Marly stone, and sand with green particles; layers of calcareous sandstone.		
III. Secondary Period.	Cretaceous Group.	5. *Gault* (marine).—Blue clay, with numerous fossils, passing into calcareous marl in the lower parts.		
III. Secondary Period.	Cretaceous Group.	6. *Lower green sand* (marine).—Grey, yellowish, and greenish sands, ferruginous sands and sandstones, clays, cherts, and siliceous limestones.		
III. Secondary Period.	G Wealden Group.	1. *Weald clay* (freshwater).—Clay, for the most part without intermixture of calcareous matter, sometimes including thin beds of sand and shelly limestone.		1, 2. Extensively developed in the central parts of Kent, Surrey, and Sussex. 3. Isle of Purbeck, in Dorsetshire.
III. Secondary Period.	Wealden Group.	2. *Hastings sands* (freshwater).—Grey, yellow, and reddish-brown sands, sandstones, clays, calcareous grits passing into limestone.		
III. Secondary Period.	Wealden Group.	3. *Purbeck beds* (freshwater).—Various kinds of limestones and marls.		

Table 52.4. (from Lyell, 1833, Appendix)

on the margins of the continents (continental shelves) the alternating transgressions and regressions of the seas have resulted in shifting changes in sediments (facies) and, concomitantly, faunal composition. Continuous sequences are rare for any prolonged period of time in this environment and, as a result, the subdivisions of geological time recognized in the rock record preserved in this environment have been generally linked to physical events which are not time-equivalent (e.g. transgressive-regressive cycles).

The formation of a given lithostratigraphical sequence occupied a finite period of time. But, whereas secular time is the continuous link which connects all rock sequences, these same rock sequences represent discontinuous time intervals. Gaps of various magnitudes are represented in all rock sequences and it is unlikely that a complete time-stratigraphical sequence covering all of geological time will ever be assembled. Nevertheless, considerable emphasis is placed by many geologists upon the stratotype locality and its character. They are considered definitive in the abstract formulation of the concept of a given time-stratigraphical entity (e.g. stage). But it is not difficult to understand that the rocks (and their contained fossils) at a given locality cannot be wholly typical in terms of mondial correlation of the time span involved in their formation.

It would appear that a more satisfactory method of subdividing geological time would be through zones based on phylogenetic successions in rapidly evolving forms, such as planktonic Foraminifera, which can in turn be related to radiometrically determined dates. In this manner significant palaeontological events, or datum-points, can be related to a continuous secular time-scale. This combination will provide a proper frame of reference within which geological history can be delineated. The fundamental unit for measuring the span of geological time remains, as it has since Oppel first defined it, the zone.

Hornibrook (1966) has recently presented a cogent discussion on the subject of the use of planktonic foraminiferal zones in the subdivision of the Tertiary. He observes rightly that a single 'world system of Tertiary planktonic zones is too simple an object'. Two sets of zones, one for tropical regions, another for temperate zones may be more satisfactory. In the face of the continuing difficulty in achieving regional interlatitudinal correlation with planktonic Foraminifera, he suggests that greater effort should be directed towards establishing a number of 'correlation levels or reference horizons or datum planes'. This writer is in full agreement with Hornibrook (1966) in his suggestion that these datum planes should be based upon evolutionary sequences, i.e. bioseries, which can be considered to be isochronous in their development. The most familiar example in the Cenozoic is the *Orbulina* datum. Another datum-point of regional significance is the *Pseudohastigerina* datum (Berggren 1964*a*; Berggren, Olsson and Reyment 1967) at the base of the Eocene. Hornibrook (1966) appears to have mislocated his *Globanomalina micra* datum; in Fig. 1 he has placed the Palaeocene/Eocene boundary in New Zealand at the first appearance of *Globanomalina wilcoxensis* (vel *Pseudohastigerina wilcoxenis*) and within the upper part of the Waipawan Stage; this is the customary placement of the Palaeocene/Eocene boundary in New Zealand. In Fig. 3 he has drawn the Palaeocene/Eocene boundary at the first (evolutionary) appearance of *G. micra* (which is shown in Fig. 1 to correspond to Waipawan/Mangaorapan boundary and to lie within the Lower Eocene). Hornibrook (1966, Fig. 3) suggests several 'datum-levels' within the Tertiary which he believes can be used in intercontinental stratigraphical correlation. Jenkins (1966*c*) has recently correlated some twenty-nine homotaxial datum-points within the Tertiary in various successions around the world. However, there are several points at which the present author would disagree with Jenkins' scheme, but discussion of this is beyond the scope of this paper. In connection with the preparation of this paper I have presented several datum-points in the Tertiary which have been adopted by the J.O.I.D.E.S. micropalaeontology panel for use in shipboard and preliminary studies. These datum planes are shown in Tables 52.30 and 52.38. The suggested relationship of these datum-planes to the Cenozoic radiometric time scale is also shown.

LITHOSTRATIGRAPHICAL, BIOSTRATIGRAPHICAL and CHRONOSTRATIGRAPHICAL TERMINOLOGY AND USAGE

In connection with current attempts to formulate a biostratigraphical and chronostratigraphical subdivision of the Tertiary a general discussion is presented of the methods used by this author in achieving these aims, and his interpretation of the concepts involved. Biostratigraphical classification is a subdivision of strata into units based on significant aspects of the fauna or flora. Two major types of units have been generally used in palaeontological literature: *faunizone* (or *florizone*), a stratigraphical unit characterized by an assemblage of organisms, one of which is chosen as an index and lends its name to the unit, although its stratigraphical range need not be restricted to the interval; subdivision of the faunizone into subfaunal zone or even further into zonules may be practicable. The zonule implies a biostratigraphical unit of dubious (that is, unproven) time-stratigraphical signifi-

Proposed Modification of the Table of Fossiliferous Strata,

Periods and Groups.	British Examples.	Foreign Equivalents and Synonyms.
POST-TERTIARY.		TERRAINS CONTEMPORAINS.
1. POST-PLIOCENE.	1. *Recent.* — Peat of British Isles, with human remains. (Principles of Geology, ch. 45.) Alluvial plains of the Thames, Mersey, and Rother, with buried ships, p. 120., and Principles, ch. 48. 2. *Post-Pliocene.* — Deposits, with fossil shells of living species, in which no human remains have yet been found. Shell-marl of Scotch and Irish Lakes.	1. Marine Strata inclosing Temple of Serapis at Puzzuoli. Principles, ch. 29. 1. Freshwater Strata inclosing Temple in Cashmere. *Ibid.* 9th ed. p. 762. 2. Volcanic tuff of Ischia and Naples, with living species of marine shells, and as yet without human remains, p. 118. 2. Newer part of boulder-formation of Sweden, with brackish-water shells of species now living in the Baltic, p. 130.
TERTIARY. PLIOCENE		TERRAINS TERTIAIRES.
2. NEWER PLIOCENE, or Pleistocene.	1. *Glacial.* — Drift or boulder-formation, with remains of *Elephas primigenius* and shells, nearly all of living species. Ochreous gravel of valley of Thames, p. 154. *Supp.* p. 7. *Glacial* deposits of the Clyde, p. 131.; of North Wales, p. 155. 2. *Preglacial* deposits of Grays, Thurroch, and Ilford (valley of Thames), with *Elephas antiquus*, Falc., and shells, nearly all of recent species, p. 154. *Supp.* p. 4. 3. Norwich Crag, with marine shells (85 per cent. of recent species), p. 155., and *Supp.* p. 4. Cave deposits of Britain, chiefly Newer Pliocene, p. 161.	Terrain quaternaire, diluvium. Terrains tertiaires supérieurs, p. 139. 1. Glacial drift of Northern Europe, p. 129.; and of Northern United States, p. 140.; and Alpine erratics, p. 149. 3. Limestone of Girgenti, p. 159. Australian cave-breccias, p. 162.
3. OLDER PLIOCENE.	1. Red Crag of Suffolk, pp. 169—171., and *Supp.* p. 2. 2. Coralline Crag of Suffolk, pp. 169—172., and *Supp.* p. 2.	Subapennine strata, p. 174. Hills of Rome, Monte Mario, &c., p. 176. and p. 535. Antwerp and Normandy crag, p. 174. Aralo-Caspian deposits, p. 176.
MIOCENE.		TERRAINS TERTIAIRES MOYENS.
4. UPPER MIOCENE.	Wanting in the British Isles.	Faluns of Touraine, p. 176. Bolderberg Strata in Belgium, p. 179. Sansans, near Pyrenees, South of France. Basin of Vienna, p. 180.
5. LOWER MIOCENE.	Hempstead Beds, Isle of Wight, p. 193.	Grès de Fontainebleau, p. 195. Calcaire de la Beauce. *Ibid.* Mayence basin, p. 191. Limburg beds, Belgium, p. 189. "Oligocene" strata of North Germany. Nebraska beds in United States, p. 207.
EOCENE.		TERRAINS TERTIAIRES INFÉRIEURS.
6. UPPER EOCENE.	1. Bembridge Beds, Isle of Wight, p. 209. 2. Osborne Series, p. 211. 3. Headon Series. *Ibid.* 4. Barton Clay, p. 213.	1. Gypseous Series of Montmartre, p. 224. 2 & 3. Calcaire Siliceux, p. 226; or Travertin inférieur. 4. Grès de Beauchamp, or Sables Moyens, p. 227. 4. Lacken beds, Belgium.
7. MIDDLE EOCENE.	1. Bagshot and Bracklesham Beds, p. 214. 2. Wanting.	1. Calcaire Grossier of Paris basin, p. 227. 2. Upper Soissonnais, Sands of Cuisse-Lamotte, p. 229. 1 & 2. Nummulitic formation of Europe, Asia, &c., p. 230.
8. LOWER EOCENE.	As in the table, p. 106.	

Table 52.5. (from Lyell, 1857, Suppl. to 5th ed., p. 13)

cance and may, more often than not, be found to represent merely facies-controlled (ecological) aspects of a fauna. A *biozone,* the second commonly used unit of biostratigraphy, implies a body of strata characterized by the stratigraphical range of a selected species. A common practice involves a certain amount of overlap among the ranges of the selected guide species, with the zonal boundaries falling within the interval of overlap (range-zone). The American Commission on Stratigraphic Nomenclature has clarified usage of biostratigraphical terminology over the past few years and in the Report of the International Subcommission of Stratigraphical Terminology at the Twenty-first International Geological Congress in Copenhagen (1960) the statement of principles of stratigraphical classification and terminology further defined the suggested usage of biostratigraphical units. The term biozone is now generally replaced by the Range-zone (Acrozone); faunizone is replaced by Assemblage-zone (Cenozone).

It has been common practice in the past for many palaeontologists to use the term *zone* in a chronostratigraphical sense, that is, as a subdivision of a stage. The Subcommission, meeting in Copenhagen in August 1960, suggested revision of the adoption by the Eighth International Geological Congress, Paris, 1900, of the term zone as a subdivision of stage, and the substitution of the term substage as the fifth order chronostratigraphical term. In the words of the Subcommission (Part XXV, ed. Hedberg, pp. 13-14):

> There is a fundamental difference between chronostratigraphic units and biostratigraphic units and the terms used for these units should be distinct from one another. A chronostratigraphic unit may frequently coincide in its type section with the scope of a biostratigraphic unit, but they differ in that the geographic extent of the biostratigraphic unit is limited to, and defined by, the extent of the physical occurrences of the particular fossils on which it was based in its type section or by the range of which it is defined, whereas the chronostratigraphic unit extends geographically to include all strata anywhere having the same age as the unit in its type section, regardless of what their fossil content may be.

It is in this sense which this writer conceives the term *zone* and in which he has applied it in his work. A zone may help to recognize and define the limits of chronostratigraphical units but is not, in itself, a chronostratigraphical unit.

Time-stratigraphical (or chronostratigraphical) units are lithological (material) units whose boundaries are theoretically independent of physical characters. Their boundaries are time-surfaces and are based on the duration of geological time which the particular material unit represents. Type sections serve as the criteria of reference within which the limits of the time-stratigraphical units are understood. Ideally the boundaries of time-stratigraphical units are isochronous. In order to facilitate formulation of a uniform regional (and mondial) time-stratigraphical sequence, chronostratigraphical units are usually made to coincide with biostratigraphical or lithostratigraphical units. However this is not always the case, as the recently (and unfortunately) defined Ilerdian Stage (which spans a part of upper Palaeocene-lower Eocene time) amply demonstrated (Hottinger and Schaub 1960).

Geological time units are wholly conceptual in that they represent merely the time during which their material counterparts – rock units – were deposited; they are not stratigraphical units in themselves.

Period/System represent the highest members of the chronostratigraphic hierarchy below Era/Erathem. The term Erathem was recently proposed by the International Subcommission on Stratigraphical Terminology (Part XXV, pp. 13, 27, 1960). Epoch/Series forms the third order. The adjectival ending, *-ian* or *-an,* is generally (but not exclusively) added to the root of the geographical place upon which the units are based. Lower, Middle or Upper are customarily applied to recognizable parts of a series, whereas Early, Middle and Late are applied to the parts of an epoch.

Fourth in order are Age/Stage. These are the basic units in local time-stratigraphical correlation. Of paramount importance in the concept of a stage is a well chosen and defined reference section, or type-locality, containing tangible features characteristic of the unit as a whole (such as distinct, characteristic fauna and/or lithological features).

The term substage has often been applied to subdivisions of a stage. Although the significance of the term substage as a formal time-rock unit may be open to question in some cases, its usage as the proper time-stratigraphical term would appear correct. The term stage has often been applied to rock units (and facies of rock units) after which these same 'stage' units are 'demoted' to the rank of 'substage'; thus they sometimes lose their significance in the time-stratigraphical sense. (The current question of the nature and limits of the Thanetian 'Stage', Sparnacian 'Stage', Ilerdian 'Stage', and Biarritzian 'Stage' illustrate this point. The stratigraphical equivalence of at least a part of the type Danian and Montian Stages, but the apparently slightly younger age of the upper Montian *s.s.* raises the question of the validity of the Montian Stage as an independent stratigraphical entity of stage rank and further illustrates this point.)

The use of the term biozone as a time-stratigraphical unit was introduced into American literature by Schenck and Muller (1941), but this usage is not followed here. Biozone is a biostratigraphical term and implies correlation of rock strata by means of organic criteria. It is through the biostratigraphical zonation which a palaeontologist is able to construct that the various rock-stratigraphical units are fitted into a time-stratigraphical framework of reference. Biostratigraphy is an aid in achieving a unified time-stratigraphical sequence as mentioned above, not a substitute. However, it is often common practice to speak of biostratigraphal zones as if they were chronostratigraphical units. This is particularly common among palaeontologists and stratigraphers in speaking of regional correlations based upon planktonic microfossils. The distinction between the use of *zone* as a biostratigraphical unit and as a chronostratigraphical unit can best be exemplified by asking a question such as: 'Where is the *Globorotalia velascoensis* Zone in the fluviatile sand', or 'where is the boundary between zones N21/N22 in those glacial tills?', etc. It is quite clear that biostratigraphical units are valid and applicable only within the limits of their development. Beyond the area of development one can scarcely speak (with any degree of certainty) of the time-equivalent limits of a given biostratigraphical unit. A biostratigraphical zone may approximate a chronostratigraphical unit within the limits of development, but is limited by the parameters of its occurrence.

In order to provide a term to encompass the time-stratigraphical limits of units based on the time span of fossil ranges, the term *Chronozone* was recently proposed by Henningsmoen and accepted by the International Subcommission on Stratigraphical Terminology (Hedberg 1966, p. 561). The distinction in usage between biozone and chronozone should be understood by stratigraphical palaeontologists for it is of fundamental importance in discussions of regional stratigraphical correlations.

As Hedberg has so succinctly observed (1958, p. 1896) the ultimate criterion of a stratigraphical term is its *utility*. In the succeeding pages we shall observe that some zones, stages and other terms have different degrees of utility, indeed.

THE PALAEOGENE

Palaeocene

Danian and Montian Stages The question of the affinities of the Danian stage has been the subject of animated discussion for over 100 years, for already at the meeting of the Geological Society of France (16 November, 1846) at which Desor proposed the establishment of the *terrain danien* (from its typical development on the islands of Denmark), there were dissenters. That Desor (1847, p. 181) considered his Danian stage to belong in the Upper Cretaceous is clear from the following quotation from his communication:

Il est évident que le terrain dont il s'agit n'est point une simple forme locale de la craie blanche, puisqu'il se trouve superposé à cette dernière, en Danemarck aussi bien qu'à Laversine et à Vigny, et qu'il contient des espèces qu'on n'a pas trouvées jusqu'à present dans la craie blanche. D'un autre côté, la presence de genres tels que les *Ananchytes,* les *Holaster* et les *Micraster,* ne permet pas de rapporter ce terrain à l'étage tertiaire.

He equated the *calcaire de Faxoe* and limestone at Stevns Klint with the *calcaire pisolithique* of the Paris basin on the basis of lithological similarity, stratigraphical position and echinoid fauna. It is interesting to note in this connection that Forchhammer (1825, p. 263), who had made the first systematic studies of the chalk and limestone sequence at Stevns Klint and Faxe, compared the *Cerithium* Limestone and Faxe Limestone with the *calcaire grossier* of the Paris Basin on the basis of similarity in molluscan fauna and considered them to be of Tertiary age. With the exception of Gardner (1884, p. 829), who stated that with few exceptions the coral limestone at Faxe contained no distinctively Cretaceous molluscs, he found little support for his contention for nearly seventy-five years.

An idea of the emotional intensity which the so called 'Danian problem' has engendered may be obtained from the following significant quotations from J. Gardner (1926, pp. 453-55) in a criticism of Scott's (1926*a, b*) correlation of the Midway of Texas and the Danian Stage of Europe. She stated (p. 453) that 'such a correlation indicates courage and imagination on the part of the author rather than field or laboratory experience'. Scott (1926*a, b*) had based his correlations on the alleged identity between *Enclimatoceras ulrichi* White in the Midway of Texas, and *Hercoglossa danica* (Schlotheim) in the Danian of Denmark. She continues (p. 454), 'The hiatus in Texas between the Midway and the underlying beds is not bridged by a single recorded molluscan species. So sharply separated are the Cretaceous and Midway coral and foraminiferal faunas that with the crudest sort of knowledge of these groups I have been able to make correct age determinations in the field on the evidence of the corals and foraminifera alone. According to Stephenson, the break at the close of the Cretaceous is equally great in the other groups.' She points out that *E. ulrichi* probably does not occur in Texas, so that only the Midway of Arkansas and Alabama can be included in his (Scott's) classification and notes that the Texas species was separated in 1923 as *Enclimatoceras vaughani* by herself. She concludes (pp. 454, 456), 'The author apparently assumes that '*Enclimatoceras ulrichi*' is restricted to the basal Midway of Texas and that the remainder of the section may be allowed to rest in the Tertiary. In Solomon's Branch, six miles south of Elgin in Bastrop County, I have found young forms which I have been unable to separate from *E. vaughani* at the very top of the Midway section while the adults abound in the lower but not the basal Midway of the same section. If the mere presence of this species is sufficient reason for shifting the formation to the Cretaceous, apparently the entire Midway section must be involved. This, of course, takes us back to the pre-Roemer days of Texas geology.

'The Midway is filled with quicksands of disputable problems but the base at least now rests on solid ground and there is no possibility of reviving the Danian controversy in Texas.'

Scott's correlation of the Midway and the Danian stage was a piece of remarkable insight. However, like Gardner he was misled by the acceptance of the Danian as being the youngest stage of the Cretaceous. The Danian controversy in Texas was revived by Loeblich and Tappan (1957*a*) and the various stratigraphical units in the Gulf Coast were discussed within the proper frame of reference.

The type localities of the Danian Stage 'are Stevns Klint and

Faxse (=Faxe,=Faxoe), both of which are located some forty miles south-southwest of Copenhagen in eastern Denmark' (Troelsen 1957). The section exposed at Stevns Klint extends from the base of the Danian (in contact with the underlying Maestrichtian chalk) up into the Middle Danian (basal part of *Tylocidaris bruennichi* Zone; the section at Faxe spans the middle Danian (*T. abildgaardi* and *T. bruennichi* zones) (see Berggren 1962, fig. 2).

The stratigraphical and palaeontological relationships of the Maestrichtian, Danian and Montian stages and the boundary problems associated with them have recently been reviewed by the present author (Berggren 1962, 1964) and Pozaryska (1965). Inasmuch as we are concerned here with Tertiary boundaries, discussion of the Maestrichtian Stage is deemed unnecessary in the context of the present discussion, except as it relates to the problem of the Cretaceous/Tertiary boundary discussed below.

The planktonic Foraminifera exhibit an abrupt change in generic composition at the Maestrichtian-Danian boundary in southern Scandinavia (area of the type Danian). *Globotruncana* s.s., *Globotruncana (Rugotruncana), Heterohelix, Planomalina (Globigerinelloides), Praeglobotruncana* s.s., *Praeglobotruncana (Hedbergella), Pseudotextularia* and *Rugoglobigerina* from the Maestrichtian are replaced in the Danian by *Globigerina, Globoconusa, Globorotalia* and *Chiloguembelina.* This change is considered by many palaeontologists to have worldwide significance (see Glaessner, 1934, 1937*a, b*; Morozova 1946*b*, 1959; Cita 1955; Reiss 1955; Loeblich and Tappan 1957*a, b*; Bolli 1957*a, b*; Troelsen 1957; Bolli and Cita 1960; Hay 1960; Olsson 1960; Said and Kerdany 1961, and papers in pt. 5, The Cretaceous/Tertiary boundary, XXI Int. Geol. Congr., Copenhagen, 1960). The boundary between these two stages is here considered to represent the demarcation between the Mesozoic and Cenozoic eras.

The Danian Stage of Denmark has been equated with the *Globoconusa daubjergensis* Zone by the present author (Berggren 1962). The upper Danian is characterized by the evolutionary appearance of *Globorotalia compressa* from *G. pseudobulloides.* In lower latitudes this level has been found to correspond to the base of Bolli's (1957) *G. trinidadensis* Zone. Recent investigations by the present author on the Midway Group of the Gulf Coastal Plain has revealed that the first evolutionary occurrences of *G. compressa* and *G. inconstans* coincides with the base of the Wills Point Formation, which serves as a further point of reference in regional correlations in the Upper Danian. (Previously I had suggested that the *G. inconstans* Zone was wholely correlative with the *G. uncinata* Zone of Bolli; see Berggren 1965*a, b.*)

As has been the case with the Maestrichtian and Danian Stages, interpretation of the concept of the Montian Stage has varied widely with different authors. Although known to earlier workers, the designation *Calcaire grossier de Mons* was first applied by Cornet and Briart (1865) to the coarse-grained, buff-coloured, detrital chalk found in wells drilled by Goffint and Coppée (city of Mons, etc.) and recognized as a 'New System' of the Eocene of Belgium. Formalization of the Montian sediments into the *'système montien'* (Montian Stage) was not made until the hesitant designation by Dewalque (1868, p. 185).

In a subsequent work Cornet and Briart (1880) expanded the original limits of the Montian by adding to it a fresh water facies above (*Calcaire lacustre à* Physa) discovered by them in 1877, and (below) the *Calcaire grossier de Cuesmes à grands Cérithes* (Cornet and Briart 1877; Briart and Cornet 1880). These units were designated (respectively) M_1, M_2 and M_3 from the base up.

At this time (and earlier) the *Tuffeau de Ciply* was considered to belong to the Maestrichtian Stage (D'Archiac 1850; van den Binckhorst 1859; Cornet and Briant 1866*a, b*, 1877; Briart and Cornet 1870, 1873, 1880) on the basis of reworked fossils found abundantly in the *Poudingue de la Malogne* at the base of the *Tuffeau de Ciply* and an unfortunate series of errors in the use of stratigraphical terminology.

The acute observations of Rutot and Van den Broeck (1885, 1886) have formed the basis of present day concepts of the Montian Stage. These authors were able to demonstrate that the lower boundary of the Tertiary lay at the base of the *Tuffeau de Ciply* (i.e. at the base of the *Poudingue de la Malogne*). The subjacent chalk was said to be characterized by *Thecidia papillata* and *Belemnitella mucronata* and to be Cretaceous in all aspects. The two sequences were said to be separated from each other by a disconformity and marked by channels and rolled nodules. They proposed (1855*d*, p. 93) the name *Tuffeau de Saint Symphorien* for this lower part of the *Tuffeau de Ciply* (*sensu* Briart and Cornet 1865, 1866, 1880; Cornet and Briart 1873, 1874). They further demonstrated that the *Tuffeau de Ciply* passes gradually into the *Calcaire de Mons* through the transitional bed, the *Calcaire de Cuesmes.* It is in this form that the Montian Stage is recognized today (see Marlière; 1955, 1957, 1958).

Rutot and Van den Broeck (1885*d*, pp. 94, 95, 1885*e*, pp. 109, 110) were also able to show that there had been a misapplication of stratigraphical terminology in regard to so called *poudingues* in the type Montian area. The *Poudingue de la Malogne* (previously considered to contain an abundant and characteristic Cretaceous fauna) was said to contain a Tertiary (Montian) fauna (pelecypods, gastropods and foraminifers) with minor reworked Cretaceous elements (*Terebratulina, Argiope, Thecidea* and bryozoans, *inter alia.*). Cornet and Briart 1865, 1866, 1880 had distinguished two *poudingues* (rubble beds): *Poudingue de Cuesmes* at the base of *la craie brune phosphatée* (in the absence of the *Craie de Spiennes*). A third *poudingue* was shown by Rutot and Van den Broeck to occur at the base of the *Tuffeau de St. Symphorien.* Thus there were three *poudingues,* two of Cretaceous age (*Poudingue de Cuesmes* and *Poudingue de St. Symphorien*) and one of Tertiary age (*Poudingue de la Malogne*). The *poudingue* which had been previously considered the type of the *Poudingue de la Malogne* was not the one with the Tertiary fauna at the base of the *Tuffeau de Ciply* (*sensu* Rutot and Van den Broeck) but the one at the base of the *craie phosphatée* (i.e. the *Poudingue de Cuesmes;* Briart and Cornet 1880). Rutot and Van den Broeck (1885*e*, p. 110) suggested that the denotation of the name *Poudingue de la Malogne* should be applied only to the rubble bed at the base of the *Tuffeau de Ciply* (Montian) and that of *Cuesmes* to the rubble bed at the base of the *craie phosphatée,* although the type of the *Poudingue de la Malogne* should be that same *Poudingue de Cuesmes.* Indeed, they suggested that distinction be made between them by referring to them as *Poudingue base du tuffeau de Ciply* and *Poudingue base de la craie brune phosphatée.* A comparative history of the change in terminology of these stratigraphical units is given by Marlière (1955, p. 298) and reproduced in Berggren (1964*b*, as table 4).

Rutot (1902, p. 605, 1908) remained adamant about the lateral facies relationships between the *Tuffeau de Ciply, Calcaire de Cuesmes* and the *Calcaire de Mons,* the former two being interpreted as exclusively marine and the latter deposited under littoral or lagoonal conditions. The various stratigraphical units of the type Montian have not been observed in direct succession (*cf.* Vincent 1928, 1930*a*; Marlière 1955, 1957, 1958). The type *Calcaire de Mons* is found in the northeast part of the basin of Mons (wells of Goffint); the type *Tuffeau de Ciply* is found in the southwest part of the basin where it is exposed near Ciply. In the vicinity of the town of Mons, which lies over the central axis of the basin, the maximum thickness of the Montian Stage is about 90 metres (Marlière 1957, p. 153). There, in artesian walls drilled in 1903 by l'École des Mines de Mons (now Faculté Polytechnique de Mons), Marlière (1957) aided by an investigation of the ostracod fauna was able to demonstrate conclusively the essential vertical sequence and facies relationships of the type Montian.

top .... an upper sequence with *Triginglymus* which corresponds to the beds in the wells at Goffint (Briart and Cornet 1865, 1866) (i.e. *Calcaire de Mons*).
.... a middle sequence with *Cytheretta,* up to the present unidentified in outcrop.
base .... a lower series with *Cytherelloidea,* which represents the *Tuffeau de Ciply* (Rutot and Van den Broeck 1885*d*).

Of particular interest here is Marlière's (1957, p. 164) record of a distinct horizon – a '*Terebratula*' bank – which occurs from 17 to 23 metres above the base of the *Tuffeau de Ciply* in the wells at Mons (in the centre of the basin). Owing to the effects of transgression it occurs near the base of the *Tuffeau de Ciply* at Ciply. A similar *'Terebratula'* bank composed primarily of the species *Chatwinothyris incisa* (Von Buch) can be seen in the middle Danian in the now abandoned quarry Hvideland, adjacent to the present Faxe quarry in Denmark. It is only of local significance but a comparison of the faunas might be of stratigraphical value.

Marlière (1957, 1958) has thus been the first to present a palaeontological zonation of the type Montian, to define the *Tuffeau de Ciply,* other than by gross lithology, and to demonstrate conclusively the superposition of the *Calcaire de Mons* upon the *Tuffeau de Ciply.* A marine environment varying between 150 metres to less than 20 metres was postulated for the type Montian, transgression at the base being followed by a gradual regression towards the top (1957, p. 161; 1958, pp. 46-8, fig. 4). Marlière (1955, p. 302) believed that the stratigraphical succession does not follow directly in vertical order but is rather the result of a gradual migration of faunal environments, which were displaced under the influence of the mobility of subsiding zones. Thus the Montian Stage represents a gradually regressing sea (after the initial transgression) with the gradual transition from a neritic environment, to a littoral, lagoonal and, finally, continental (lacustrine) one. The Montian is separated from the Landenian (above) and the Maestrichtian (below) by distinct disconformities (cf. Nakkady 1957, p. 431) and thus occurs in a condition quite similar to that in Denmark (cf. Brotzen 1940, 1945, 1948, 1960; Berggren 1960*b*, 1962*a*). A comparison of the ostracod fauna of the type Danian and Montian would be of great significance in further attempts to determine the time-stratigraphical relationships between these two stages.

The Montian is of Tertiary age according to Marlière (1958, p. 49) based on the occurrence in the lower part of the *Tuffeau de Ciply* of such Tertiary ostracods as *Trachyleberis aculeata* (Bosquet) and *T. ciplyensis* Marlière, *Monsmirabilia* and the *Cytheretta* beds being immediately below. These conclusions would appear to reinforce those of Loeblich and Tappan (1957*a*) who reported (from the same material from which Marlière recorded his ostracods) finding *Globigerina daubjergensis* and *G. trilocuLinoides* in the *Tuffeau de Ciply* (see Marlière 1958, p. 49, footnote 1; cf. Hofker 1959*a*). Berggren (1962*a*, 1963) and Gohrbandt (1963) have reported finding similar Danian planktonic faunal elements in the *Poudingue de la Malogne* and lower part of the *Tuffeau de Ciply.*

It will be recalled that Schimper (1874) placed the Lignites de Soissons and the Sables de Bracheux in the *Période paléocène.* It was Von Koenen (1885*b*), however, who first gave the Palaeocene its faunistic-stratigraphical basis with his investigations of the foraminiferal fauna of the glauconites at Copenhagen. He placed the Montian, Thanetian and Sparnacian in the Palaeocene (i.e. all beds older than the London Clay, Ypresian, Lower Eocene). He also showed (1897) that the *Glaukonitformation,* which lay above the Danian chalks, was equivalent to the Copenhagen Palaeocene and not the Cretaceous (i.e. upper Danian). He correlated (1885) the Palaeocene of Copenhagen with the Bracheux sands of northern France (lower part of his middle Palaeocene) – a correlation which has stood essentially to the present day. Haynes (1958, p. 90), for instance, regards the Swedish Palaeocene (=glauconites at Copenhagen) as essentially equivalent to the Thanetian Substage of England (cf. Barr and Berggren 1964).

The question of the relationship of the Danian and Montian Stages depends to a great extent upon the nature of the contact between the Danian and Selandian in Denmark. Grönwall (1899*a*) demonstrated the petrographic differences in the upper Danian beds of the *Crania tuberculata* Zone. Subsequently (1904) he placed the boundary above the *Crania*-kalk. It was not until 1920 that Rosenkrantz demonstrated that the *Crania*-kalk, as previously recognized, was not a stratigraphically homogeneous unit, but rather separated in its midst by a disconformity. The upper part was shown to be a glauconitic conglomerate which contained reworked elements from the lower *Crania*-kalk. The Danian/Palaeocene boundary was placed between the lower and upper *Crania*-kalk and the hiatus was said to have been of short duration. The main reason for this conclusion was the fact that all of the molluscs with aragonitic shells known from the upper Danian Kalksand (Limesand) are identical with or closely related to those species found in the basal Selandian (Rosenkrantz 1924*b*, p. 30). Only a few molluscs with shells of calcite are known from the Selandian, apart from derived Danian elements. Ravn considered the duration of the break to have been considerably longer and forming the boundary between the Cretaceous and Tertiary. The glauconites above the boundary belong to the Selandian Stage (Rosenkrantz 1924*b*) which, it would now seem, can be correlated with the lower Landenian (Thanetian) Stage of England (Haynes 1958).

In France the Danian is known only on the south side of the foothills of the Pyrenees; the most complete sections are developed in marine facies towards the west on the Tercis anticline and nearby areas. According to Seunes, Lambert, Abrard, and others, the marly limestones of the Maestrichtian, with *Pachydiscus, Scaphites, Seunaster, Stegaster* and other forms, are conformably overlain by a sequence of white and, in places, reddish, glauconitic limestones of the Garumnian Stage (which Lambert correlated with the Danian Stage). Seunes also included part of the underlying Maestrichtian in his Danian Stage, as well as the Garumnian, Lambert (1907) subdivided this sequence as follows:

(a) lower and middle Garumnian with *Hercoglossa danica, Echinocorys pyrenaicus, E. cotteaui* Lambert (=*E. sulcatus* Goldfuss), *Cyclaster pyriformis* Cotteau (= *C. gindrei* Seunes) and without *Micraster tercensis* (= *Protobrissus depressus* Kongiel + *P. tercensis* Cotteau).
(b) upper Garumnian divided into a lower part with *Hercoglossa danica, Micraster tercensis, Isaster aquitanicus, Echinocorys cotteaui, Cyphosoma pseudomagnificus* Cotteau (= ? *Rachiosoma krimica* Weber) and an upper part with *Micraster tercensis, Isaster aquitanicus* but without *H. danica.*

Above this lie glauconitic sandstones and limestones with *Operculina heberti* Munier-Chalmas and containing in the lower part conglomerates with Garumnian echinoids: *Isaster aquitanicus, Ceraster, Echinocorys* and other forms. According to Seunes (1890) these remnant forms appear to have been derived from different levels. Seunes at first dated this level as Eocene. Lambert (1907) included it in the upper Garumnian and thus correlative with the Danian Stage of northern Europe. More recently Cuvillier (1945) found *Operculina heberti* and *Discocyclina seunesi* at this level and dated it as Thanetian or Ypresian.

Eastwards a rather complete sequence of an Upper Cretaceous-

Lower Tertiary section is developed along the Garonne river on the Little Pyrenées anticline. This area is the type of the Garumnian Stage of Leymerie (1877). The lower part of the Garumnian consists of variegated clays, sands and marly limestones, with remains of a marine fauna and interbedded limestones with rudistids of Maestrichtian age. In the middle Garumnian lacustrine limestones occur which are equivalent to the Rognacian of Provence and the Danian Stage, according to Naidin and Moskvin (1960, p. 35; but cf. Jeletzky 1962, p. 1,009, fig. 2, in which the dinosaur-bearing Rognacian Substage of the Begudo-Rognacian Stage is shown to be of Upper Maestrichtian age. In the upper Garumnian – considered by most authors to be of Montian age – a threefold division has been recognized: lower marly limestone with *Natica brevispira, Venus striatissima;* limestone with *Echinanthus carinatus, E. pouechi, Cerithium inopinatum, Cerithium montense,* and others; glauconitic limestone with *Echinanthus, Ostrea uncifer* and 'Cretaceous' echinoids (considered an isolated colony by Leymerie): *Micraster tercensis* (=*Protobrissus*), *Echinocorys semiglobus* Cotteau (=*E. sulcatus* Goldfuss), *Hemiaster nasutulus, H. canaliculatus* (=*Protobrissus*), *Cyclaster coloniae* (=*Isopneustes*), and others. Naidin and Moskvin (1960) point out that all these species are found in the Danian strata of Crimea and many of them in the northern Caucasus and in the Transcaspian oblast (Moskvin and Naidin 1960, table between pp. 36-7).

These authors note the interesting, if not somewhat inconsequent, usage of *Micraster tercensis* as a guide form of the 'Montian' in the Little Pyrenees and for the Danian on the Tercis anticline. The faunistic complex in the region of southern France extends eastwards towards the longitude of Ashkbad in southwestern Russia. The vertical development of the echinoids in sections in the Crimea, Caucasus and the Transcaspian oblast allow correlation with similar sequences in western Europe. The authors suggest that it is possible to correlate with sufficient confidence beds with *Cyclaster danicus, C. gindrei, Hercoglossa danica* (the two lower zones of the Danian of southern Dagestan) with the Danian Stage of Denmark and a large part of the Garumnian Stage on the Tercis anticline. The beds with *Protobrissus tercensis* can then be correlated with the uppermost part of the Garumnian, and equivalent beds would then apparently be lacking in Denmark.

The problem of the definition of the upper stratigraphical boundary of the Danian Stage remains. Incomplete sections in western Europe and the difference in interpretation by various workers have hampered attempts to resolve the Danian and Montian or Dano-Montian controversy. Moskvin and Naidin (1960, p. 36) suggest that the change in species complexes in the fuller sections in the Crimea and Caucasus and the Transcaspian oblast offer two alternatives for drawing the top of the Danian Stage: (*a*) at the top of beds with *Protobrissus tercensis, Coraster ansaltensis, Globigerina inconstans* and below beds with the (first) abundant appearance of *Globorotalia angulata.* Placement of the boundary here would, they suggest, appear to correspond to the concept of Seunes and with present day usage in the Tercis anticline and nearby regions of the Pyrenees; (*b*) below beds with *Protobrissus tercensis* and above beds with *Cyclaster gindrei* and *Protobrissus depressus.* In this interpretation the Danian Stage would correspond to the vertical development of *Hercoglossa danica;* this would correspond to its development in the type area of Denmark.

The authors note that in southwestern Russia it has been possible to distinguish strata of lower Palaeocene age between Danian and Thanetian which correspond to the Montian Stage of Europe. Thus according to them correlation of the Danian and Montian Stages *in toto,* as claimed by some (cf. Loeblich and Tappan 1957*a*), is impossible. But, on the other hand, the similarity in fauna under uniform facies conditions raises the question of the validity of the Montian Stage as an independent stratigraphical entity with the rank of stage. The authors suggest that the Montian be retained only as the upper substage of the Danian, as suggested earlier by Munier-Chalmas and De Lapparent (1893), Munier-Chalmas (1897) and Kongiel (1935). This writer would caution only that from evidence presented elsewhere (Voigt 1960; Wienberg-Rasmussen 1962) and his own investigations it is apparent that the lower Montian *Tuffeau de Ciply* is time-stratigraphically equivalent with a part of the Danian (probably the lower and middle Danian) of Denmark; the relationship between the upper part of the Montian (*Calcaire de Mons*) with the upper Danian remains uncertain. By extrapolation it is probable that the upper part of the Montian is correlative in part with the upper Danian, and, in part, younger than any known Danian exposed in Denmark. The recent demonstration that the *Globorotalia uncinata* Zone of Bolli is equivalent in part to the *Globigerina inconstans* Zone of Subbotina (Berggren 1965*a, b*) has resolved some of these problems. It is also possible that subsurface younger Danian in Denmark may fill the missing void in our information and allow confirmation of one of the two alternatives suggested by Moskvin and Naidin (1960).

Cretaceous/Tertiary boundary. Until the end of the nineteenth century the Danian Stage remained, by almost universal consent, at the top of the Cretaceous. It was Grossouvre (1897) who made the suggestion that the Mesozoic-Tertiary boundary be placed at the upper limit (i.e. disappearance) of ammonites, rudistids, belemnites, inocerami, dinosaurs, mosasaurs and other characteristic Mesozoic animals. However, the fact that these organisms became extinct at different stratigraphical levels in various geographical regions led to considerable inaccuracy in the delimitation of the boundary in some places (see the discussions by Reiss 1955, pp. 115, 116; Jeletzky 1960, 1962; cf. Reyment 1956, pp. 42-6). Indeed, neglecting to study the type Danian section led micropalaeontologists to place the break in the planktonic microfauna between the Danian-Palaeocene and to include the *Pseudoguembelina-Globotruncana* assemblage in the Danian (Henson 1938) (cf. Renz 1936; Glaessner 1936).
Grossouvre's suggestion, as we have seen above, found support in Scandinavia from Brunnich-Nielsen (1919, 1920), Rosenkrantz (1920, 1924*a, b*, 1937, 1960) and Harder (1922). However, Ravn remained opposed to this and placed the Danian in the Upper Cretaceous. The quest for determining the age and affinities of the Danian Stage was pursued elsewhere. Scott (1926*a, b*, 1934), although recognizing the strong affinities between the molluscan faunas of the type Danian and the lower Midway of Texas, correlated the two but placed them in the Upper Cretaceous.

In the Soviet Union Bezrukov (1934, 1936) was the first to suggest that the Danian should be placed in the Tertiary. Basing his opinion on field and palaeontological observations of the extensive Danian deposits on the left bank of the Ural River and in the western part of Obshchiyy Syrt, he recorded (1934) a distinctly Danian microfauna in coccolithic-bryozoan limestones above the belemnite-bearing Maestrichtian chalks. The contact with the overlying Syzranian Stage (lower Landenian) was discussed and the suggestion made that the Danian sea was connected with the Polish basin by way of the Dneiper-Donetz and Black Sea depressions (cf Jeletzky 1958; Naidin 1959, 1960). Soviet micropalaeontologists have generally considered the Danian to belong in the Upper Cretaceous (Subbotina 1936*a*, 1939, 1947, 1953*b*; Sjutskaja, 1956; cf. Keller 1936; see discussion in Berggren 1960*b*). Morozova (1946*a*), on the basis of studies of the microfauna in the Caucasus, Crimea, Russian Platform, Mangyshlak Peninsula, Kopet Dag, and elsewhere, observed that the change in the foraminiferal fauna occurs at the boundary between the Danian and Maestrichtian Stages as an abrupt break in the plank-

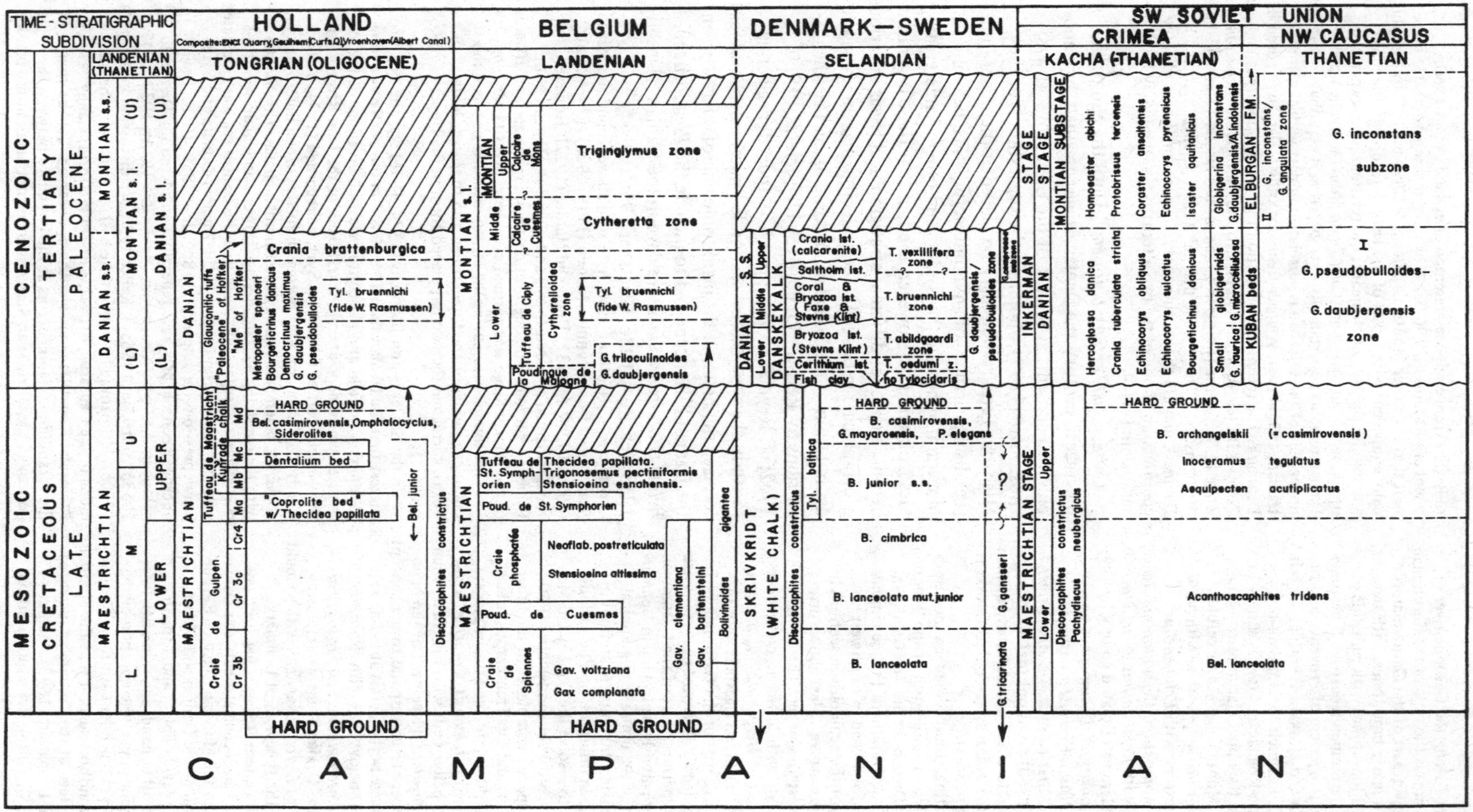
TIME - STRATIGRAPHIC SUBDIVISION
HOLLAND
Composite: ENCI Quarry, Geulhem (Curfs Q.), Vroenhoven (Albert Canal)
BELGIUM
DENMARK—SWEDEN
SW SOVIET UNION
CRIMEA
NW CAUCASUS
CENOZOIC
TERTIARY
PALEOCENE
LANDENIAN (THANETIAN)
TONGRIAN (OLIGOCENE)
LANDENIAN
SELANDIAN
KACHA (THANETIAN)
THANETIAN
MESOZOIC
CRETACEOUS
LATE
MAESTRICHTIAN
HARD GROUND
HARD GROUND
C A M P A N I A N

DATA SOURCES:

Holland: Hofker (1955b; 1956f; 1959f; 1961f) Schmid (1959); Kruytzer and Meijer (1958); Meijer (1959); Wienberg Rasmussen (1962)

Belgium: Marlière (1957); Schmid (1959); Hofker (see bibl.); Wienberg Rasmussen (1962)

Denmark-Sweden: Troelsen (1937); Birkelund (1957); Rosenkrantz (1937); Berggren (1960a; 1962a)

SW Soviet Union

Crimea: Morozova (for references see Berggren 1960c; 1965b); Moskvin and Naidin (1962)

NW Caucasus: Subbotina (1953); Leonov and Alimarina (1961)

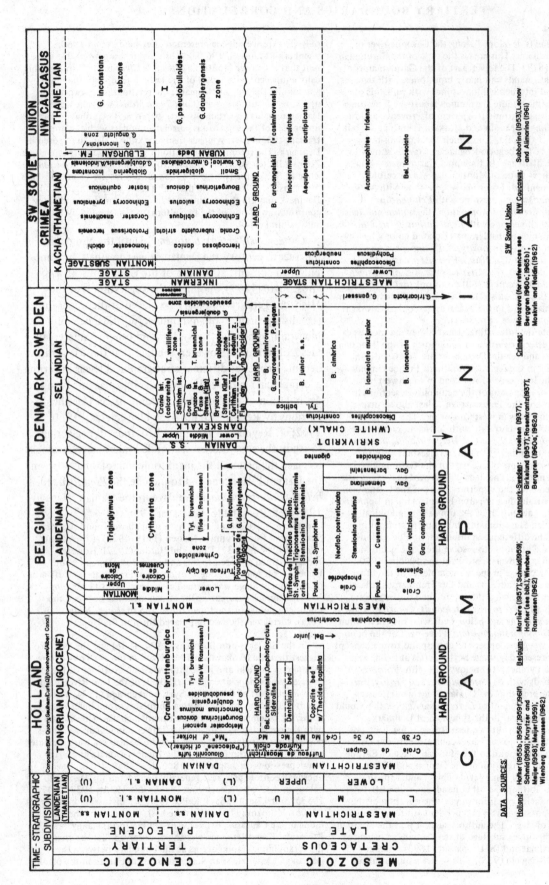

Table 52.6. Suggested correlation of Maestrichtian, Danian and Montian Stages in their type areas and S.W. Soviet Union (Crimea and N.W. Caucasus). Conventional and recently applied terminology are shown together, viz. Crimea, as well as alternatively accepted interpretation of subdivisions of stages, viz. Maestrichtian, Danian, Montian, left column.

tonic fauna and a gradual alteration in the benthonic fauna. She noted the change from a *Globotruncana-Pseudotextularia* assemblage in the Cretaceous to three and four-chambered globigerinids and the Tertiary genus *Globorotalia*. She further noted the marked change in the macrofauna, the gradual alteration in sedimentation (increase in terrigenous material), its association with orogenic disturbances and the quantitative increase of agglutinated benthonic foraminifers (suggesting a gradual lowering of the temperature). She concluded perceptively: 'From all the aforesaid it follows that confined to the boundary between the Maestrichtian and Danian were great changes in the palaeogeographical setting. No such sharp demarcation line can be observed either between the stages of the Senonian or between the stages of the Palaeocene. This demarcation line should be considered as a boundary between the Cretaceous and Tertiary systems. Having no right arbitrarily to change the boundaries between the systems, we are nevertheless justified in advancing the question as to the necessity of a revision of the position of the Danian stage on the geochronological scale, since it is not merely a question of nomenclature, but one of great importance for the study of geological history.'

Deroo (1966) has recently presented the results of his investigations on the ostracods of the type Maestrichtian for the Maestrichtian Stratigraphical Committee. Some 202 species were discussed (56 new or renamed) belonging to 59 genera (13 new) and 16 subfamilies (1 new). The Cretaceous/Tertiary boundary is drawn at the Maestrichtian/Danian boundary by Deroo (1966) in accordance with the suggestion of Grossouvre (1897, 1901). Deroo (1966) concludes, among other things, that:

(1) The ostracods exhibit a pronounced diversification in the *Craie Grossière* and the *Tuffeau de Maastricht* whose equivalence is not known in other parts of the Cretaceous.

(2) Some of the genera present in the upper Maestrichtian indicate transitional trends to the Tertiary. Thus an increase in the Cytherettinae heralds the appearance of the Tertiary genus *Cytheretta;* certain genera occur which are transitional between the typically Cretaceous *Cythereis* and the Tertiary genera *Trachyleberis, Hermanites* and *Trachyleberidea;* presence and, probably appearance, of *Clithrocytheridea, Kingmaina,* presence of Leguminocythereidinae such as *Anticythereis.*

(3) On the specific level a complete change in the ostracod fauna occurs between the Maestrichtian and the overlying *Tuffeau Glauconieux* (which is equivalent to at least a part of the *Tuffeau de Ciply* and the Danian).

(4) Boreal European type sedimentation is suggested to have occurred from the Campanian to lower part of upper Maestrichtian; meridional (subtropical) type sedimentation from the upper part of Maestrichtian to the overlying Danian and Montian.

The Maestrichtian/Danian boundary in Denmark and southern Sweden is represented by a surface of discontinuity (i.e. disconformity). This discontinuity in the stratigraphical record has been recognized in many parts of the world. In some areas it is marked by a sharp lithological change, in others by subtle lithological change and in some areas there are reports of no apparent lithological change, but there is general agreement that a distinct change occurred among the marine planktonic and nektonic faunal elements. The duration of the hiatus in the area of the type Danian (where palaeontological data indicates the presence of youngest Maestrichtian below basal Danian) would appear to be relatively short, geologically speaking, probably less than the duration of a single planktonic foraminiferal zone, i.e. probably a million years or less.

Recently Luterbacher and Premoli Silva (1964) have introduced the *Globigerina eugubina* Zone for a thin stratigraphical interval (less than 1 metre thick) characterized by minute globigerinids (averaging 0.1 millimetre in diameter) which occurs between Maestrichtian limestones with globotruncanids, heterohelicids and rugoglobigerinids and Danian limestones of the *Globigerina pseudobulloides/Globigerina daubjergensis* Zone in the Central Appenines of Italy. They have subsequently recorded this in the Southern Alps of Italy (Premoli Silva and Luterbacher 1966). Luterbacher and Premoli Silva (1964) have suggested that the *G. eugubina* Zone is the oldest planktonic foraminiferal zone of the Tertiary, older than the basal Danian in the type region as well as older than the oldest part of Zone I (*Eoglobigerina*) of Morozova (1960) in the Crimea (which the present author has suggested is correlative with the basal Danian in the type area). Although this zone may ultimately prove to represent the oldest part of the Cenozoic, an alternative possibility should also be considered. I have suggested elsewhere (Berggren 1965c) that the stratigraphical interval represented by the *G. eugubina* Zone may represent the latest Maestrichtian, that the minute forms present in this interval may be juvenile forms of Maestrichtian species, and represent a stratigraphical condensate through the action of winnowing currents. This is little more than speculation at present. In the light of the uncertainty which surrounds the affinities of this zone the J.O.I.D.E.S. deep-sea drilling programme, scheduled to begin in late 1968, assumes greater significance. Perhaps in continuous marine sections it will be possible to solve some of the problems which have perplexed palaeontologists for so many years with regard to the faunal break at the Cretaceous/Tertiary boundary.

The molluscan fauna of the *Tuffeau de Ciply* (lower Montian) is closely related to that in the *Calcaire de Mons* (upper, type Montian) which has led Rosenkrantz (in Rasmussen 1964) to suggest that the type Montian may be regarded as representing the upper Danian. This similarity in fauna is a direct reflection of the continuity of sedimentation (and concomitantly homogeneity of sediments) in the Mons Basin during Dano-Montian time. In the same manner several groups of invertebrates in the type Danian show a marked affinity with Maestrichtian forms below (bryozoans, molluscs, echinoids). The chalk-on-chalk facies in southern Scandinavia, although probably representing different ecological conditions such as depth, etc., was evidently similar enough to provide suitable conditions for the more facies-bound benthonic organisms to maintain an essential continuity across the boundary in this area. No such faunal continuity, on the other hand, exists in the Gulf Coast Region where there is an almost complete change in the benthonic invertebrate fauna between the Navarro (Maestrichtian) and Midway (Danian-Palaeocene).

Thus, if we return to Grossouvre's (1897) suggestion to place the boundary between the Mesozoic and the Cenozoic at the upper limits of the extinction of the nektonic organisms of the Cretaceous, and if we accept the recently documented changes in the planktonic faunas at the end of the Maestrichtian as definitive, it would seem appropriate to place the Cretaceous/Tertiary and, by extension, Mesozoic/Cenozoic boundary at the Maestrichtian/Danian boundary. This boundary, as thus defined, is essentially independent of the facies conditions which have resulted in the anomalous occurrences of Cretaceous and Tertiary faunal elements together in strata of Danian age.

In this connection it is interesting that a current trend noticeable in palaeontology is leading to a reconciliation and unification of once widely opposed points of view with regard to the nature of the macrofauna of the Danian Stage. The recent discussion of type Danian molluscs by Rosenkrantz (1960) was mentioned above. The recent investigations by Poslavskaya and Moskvin (1960) on the spatangoid echinoids of the Crimea, Caucasus and the Transcaspian region is significant in this connection. They note that a sharp change occurs in the echinoid fauna at the Maestrichtian-Danian boundary, that the Danian-Palaeocene (and possibly Eocene) spatangoids form a unified complex distinctly different from the underlying Upper Cretaceous forms and from the true Neogene-Recent forms, and that the peculiar character of the Danian-Palaeocene spatangoids are of considerable importance with regard to the emplacement of the Cretaceous-Tertiary boundary. As they so rightly mention, one of the main reasons that early investigators considered the Danian as belonging to the Cretaceous was the 'Cretaceous' character of its echinoid fauna. This position is no longer tenable, they assert, as Danian echinoids are shown to have their closest affinities with Palaeocene forms. The decision to transfer the Danian Stage formally, from its original position at the top of the Cretaceous to the base of the Tertiary, is a juridical matter for the international subcommittee on stratigraphical nomenclature. It is hoped that the discussion presented here will serve as supporting evidence for such a recommendation.

In retrospect it is an interesting fact that the strata of the Danian Stage, although placed in the Upper Cretaceous by Desor in 1846 (1847), and correlated with the *Calcaire pisolithique* of the Paris Basin (now regarded as Dano-Montian in age), were earlier considered to be of Tertiary age by Forchhammer (1825) who made the first systematic study of them. Recent palaeontological and stratigraphical studies have vindicated both Forchhammer and Desor.

Landenian, Thanetian stages. The Landenian Stage was created by Dumont (1839, pp. 467-70) to encompass, in Belgium, the *tuffeau de Lincent à Pholadomya oblitterata*, the *argiles schistoides*, and the overlying *sables fins* (sands of Landen). Dumont subsequently (1849, pp. 368, 369) transferred the latter to his Ypresian Stage. The term was applied to clays in the north of France by Meugy (1852) and later to the *Sables de Bracheux* by Hébert (1853). Lyell (1852, p. 279) separated the lower Landenian from the upper Landenian which he equated with the Lower London Tertiaries and the *Lignite du Soissonnais* in the Paris Basin, and placed the upper Landenian in the Lower Eocene below the lower Ypresian.

The Thanet Sands which outcrop in Kent and Surrey, England were described by Prestwich (1852), and were formally elevated to stage status by Renevier (1873). Renevier (1897) thought that the Thanetian represented only the lower part of the Landenian. Dollfus (1880) created the Sparnacian Stage in the Paris Basin, for the *argile plastique* near Paris and the lignites in the regions of Soissons and Epernay. Subsequently he added (1905) at the top the *fausses glaises*, the level just beneath the *sables d'Auteuil* being considered equivalent to the *sables de Sinceny* at the base of the Cuisian. Leriche (1905, *et seq.*) equated the Sparnacian in the Paris Basin with the upper Landenian of Belgium and northern France, despite the fact that Dollfus (1880) had clearly insisted on the stratigraphical independence of the Sparnacian. The concept of Leriche was followed by Farchad (1936), and more recently by Barr and Berggren (1965). This opinion was contrary to earlier ones expressed by Gosselet (1874), and was contested by Dollfus (1905), and more

		Bassin de Paris	Belgique	Angleterre
Oligocène		? Aquitanien ? Stampien (Sannoisien)	Sables de Voort Rupélien Tongrien sup.	Hamstead beds
Éocène	Bartonien	Ludien	Tongrien inf.	Bembridge beds Headon beds
		Marinésien Auversien	Asschien	Barton beds
	Lutétien	(« Biarritzien ») supérieur	Wemmelien Lédien	Upper Bracklesham beds
		inférieur	Bruxellien	Middle Bracklesham beds
	Yprésien	Cuisien	Yprésien	Lower Bracklesham beds
		Sparnacien		Argiles de Londres
		Thanétien	Landénien	Thanet sands
		Montien	Calcaire de Mons Tuffeau de Ciply	Lacune

Table. 52.7. General stratigraphy and principal correlations in Paris Basin, Belgium and England (Blondeau *et al.* 1965, p. 216, tab. 8)

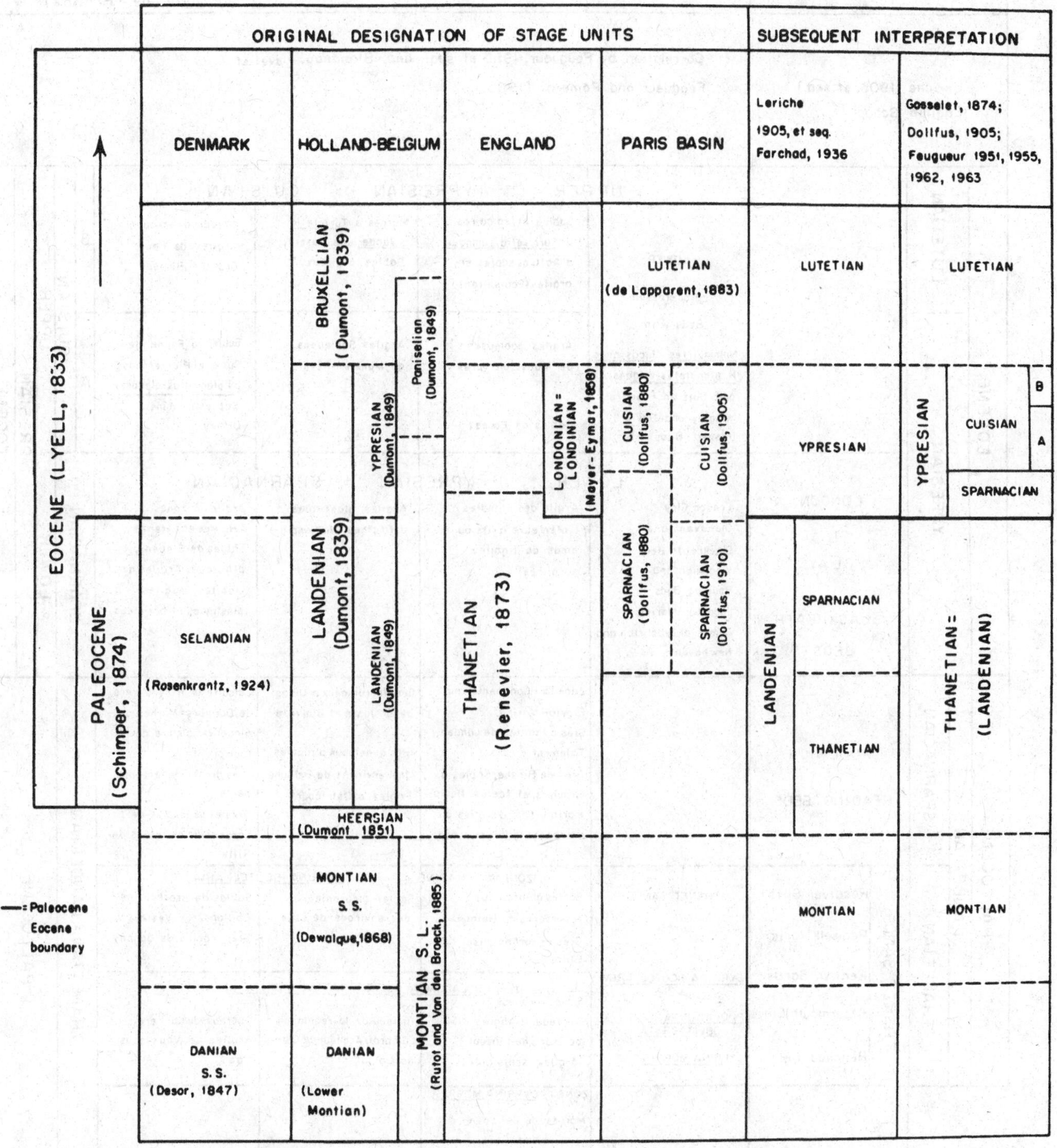
ORIGINAL DESIGNATION OF STAGE UNITS
SUBSEQUENT INTERPRETATION
DENMARK
HOLLAND-BELGIUM
ENGLAND
PARIS BASIN
Leriche 1905, et seq. Farchad, 1936
Gosselet, 1874; Dollfus, 1905; Feugueur 1951, 1955, 1962, 1963
EOCENE (LYELL, 1833)
PALEOCENE (Schimper, 1874)
SELANDIAN
(Rosenkrantz, 1924)
DANIAN S. S. (Desor, 1847)
BRUXELLIAN (Dumont, 1839)
Paniselian (Dumont, 1849)
YPRESIAN (Dumont, 1849)
LANDENIAN (Dumont, 1839)
LANDENIAN (Dumont, 1849)
HEERSIAN (Dumont 1851)
MONTIAN S. S. (Dewalque, 1868)
MONTIAN S. L. (Rutot and Van den Broeck, 1885)
DANIAN (Lower Montian)
LONDONIAN = LONDINIAN (Mayer-Eymar, 1858)
THANETIAN (Renevier, 1873)
LUTETIAN (de Lapparent, 1883)
CUISIAN (Dollfus, 1880)
CUISIAN (Dollfus, 1905)
SPARNACIAN (Dollfus, 1880)
SPARNACIAN (Dollfus, 1910)
LUTETIAN
YPRESIAN
LANDENIAN
SPARNACIAN
THANETIAN
MONTIAN
LUTETIAN
YPRESIAN
CUISIAN
B
A
SPARNACIAN
THANETIAN = (LANDENIAN)
MONTIAN
— = Paleocene Eocene boundary

Table 52.8. Chronostratigraphical concepts of the Palaeocene and lower-middle Eocene

Epoch	Stage	Substage	ENGLAND Leriche (1905, et seq.) English School	ENGLAND Correlation by Feugueur, 1955 et seq., and Blondeau, Cavelier Feugueur and Pomerol (1965)	BELGIUM	NORTH FRANCE	PARIS BASIN (Ile - de - France)	Paris Basin divisions
				UPPER YPRESIAN = CUISIAN				
EOCENE	LUTETIAN			LOWER Bracklesham Beds with *Nummulites lucasianus*, *N. planulatus-elegans* and Cuisian Fauna of Wrigley and Davis (Fisher Beds 1-5)	Sables et calcaires à *Turritella solanderi* d'Aeltre, sables et argiles (Paniselian)	Marnes à *Turritella solanderi* de Cassel Sables argileux	Niveau d'Herouval = Argiles de Laon Grés de Belleu	B — CUISIAN — UPPER — YPRESIAN — EOCENE
EOCENE	YPRESIAN				Argiles sableuses de Mouscron-Ypres Sables de Forest	Argiles Sableuses de Mons-en-Pévêle	Sables à Faune de Cuise et Pierrefonds (*N planulatus-elegans*) Sables à Faune D'Aizy	A — CUISIAN — UPPER — YPRESIAN — EOCENE
				LOWER YPRESIAN = SPARNACIAN				
EOCENE	YPRESIAN		LONDON CLAY BLACKHEATH BEDS	London Clay Oldhaven and Blackheath Beds Woolwich and Reading Beds Lagoonal-Marine Faunas of Woolwich and Newhaven	Argile des Flandres (inférieure a lits ou amas de lignites localisés)	Argiles plastiques d'Orchies-Phalempin	Argiles à bancs d'Huitres et Cyrènes Sables de Sinceny, d'Auteuil, Argiles à Lignites, Argiles plastiques et bariolées	SPARNACIAN — LOWER — YPRESIAN — EOCENE
PALEOCENE	LANDENIAN	SPARNACIAN	WOOLWICH READING BEDS		Zone III - Continental and Fluvio-Marine Grès à Plantes de Landen, Tirlemont Grès de Binche, Sables à graviers et facies fluvio-marin à lits d'argiles ou brèches argilo-sableuses	Grès ferrugineux à Unios de St. Josse et Blaiville Grès quartzeux à plantes des environs de Béthune Sables d'Ostricourt	Calcaire de Rilly = Marnes de Dormans = Marnes de Sinceny = Calcaire de Clairoix Grès de Molinchart (pro parte) Grès et poudinque de Laniscourt et Sables de Rilly	THANETIAN = LANDENIAN — PALEOCENE
PALEOCENE	LANDENIAN	THANETIAN	THANET SANDS: Reculver Silts Pegwell Marls Kentish Sands	THANET SANDS WITH CYPRINA SCUTELLARIA	ZONE III - MARINE *CYPRINA SCUTELLARIA* Sables glauconieux fossilifères du Hainaut Grès de Grandglise de Blaton	Sables glauconieux des environs de Lille	Sables de Bracheux de Châlons-sur-Vesle Grès marins de Gonnes	THANETIAN = LANDENIAN — PALEOCENE
PALEOCENE	LANDENIAN	THANETIAN	THANET SANDS: Stourmouth Clays Bullhead Cgl.	PHOLADOMYA OBLITTERATA CYPRINA MORRISI	ZONE II *PHOLADOMYA OBLITTERATTA* (= *P. KONINCKI*) Sables argileux Tuffeau d'Angres de Cherq, de Lincent Argiles schistoides	Tuffeau de la région de Cambrai, Argiles de Clary, de Louvil	Tuffeau de La Fère, Argiles de Vaux-sous Laon	THANETIAN = LANDENIAN — PALEOCENE
					ZONE I - *CYPRINA MORRISI* Marnes de Gelinden and Sables d'Orp-le-Grand			

Table 52.9. Palaeocene-lower Eocene stratigraphical correlations in Northwest Europe

recently by Feugueur (1955, 1962, 1963), LeCalvez and Feugueur (1956) and Blondeau *et al.* (1965). The essential argument of these authors is that the Thanetian and Landenian Stages are equivalent, and that the Sparnacian is of Lower Eocene age and equivalent to the lower part of the Ypresian. The discussion presented below is based, in large part, on the recent investigations of Feugueur and his colleagues in the Paris Basin.

In the opinion of Feugueur (1955, *et seq.*) the Thanetian in Belgium begins with the Heersian deposits (*Marnes de Gelinden* and *Sables d'Orp-le-Grand*) (cf. Gulinck 1965, p. 223, who has opposed this, and suggests that the Heersian is an independent unit comprising the *Marnes de Gelinden* and the *Infra-Heersien* brackish water sediments whose stratigraphical affinities are still uncertain). The Thanetian of the Paris Basin begins with strata containing *Pholadomya oblitterata* (Zone II of Leriche 1905*b*); *Tuffeau de la Fère* and *Argile de Vaux-sous-Laon.* This series corresponds to the Belgium sequence: *Tuffeau de Lincent* and *Argiles de Campine* which lie above the Heersian sediments.

Above, in the Paris Basin, follow the *Sables de Bracheux* with a shallow, neritic fauna (Rouvillois 1960), which is represented towards the east by the *Sables de Chalons-sur-Vesle* and the *Sables de Rilly* (the *Sables à Cyrènes*) with a lagoonal-marine fauna, containing brackish-fresh water molluscs, which oocur in the continental beds above that terminate the Thanetian (*Calcaires de Mortemer, de Clairoix, de Rilly, Marnes de Sinceny*). The *Sables de Bracheux* correspond to the marine glauconitic sands of Hainaut, Grand glise and Lille (Zone III, marine, of Leriche 1905*b*).

The *Sables à Cyrènes* are important in determining the spatial-temporal relationships of the Upper Palaeocene of the Paris Basin. They occur near the mouth of the Canche River at St Josse and St Aubin where they contain *Cyrena cordata,* 'type anglais de *Cyrena veneriformis* de sables de Châlons-sur-Vesle' (Blondeau *et al.* 1965), which disappears at the end of the Thanetian with the renewed transgression of the Sparnacian (*la transgression lagunaire du Sparnacian* of Blondeau *et al.* 1965). At Ostende (Belgium) an equivalent sandy series, which lies below the *Argile de Flandres,* overlies beds of Zone II and correlates stratigraphically with the *sables de Bracheux-Châlons-sur-Vesle.* This series contains a marine molluscan fauna (*Arca, Pitar, Turritella*) associated with a brackish fauna with abundant *Cyrena cuneiformis* and *Melania inquinata.*

Blondeau *et al.* (1965) point out that it is because of these molluscs that the sands at Ostende are generally considered Sparnacian against all stratigraphical logic, and without taking into consideration the marine faunas. The presence of the brackish molluscs (*Cyrena* and *Melania*) indicates merely that these forms have appeared earlier (Thanetian) in the north of Belgium, and that they have been displaced towards the south by the advancing transgression (*Argile de Flandres*) of the Sparnacian. There they have continued to develop in the lagoonal deposits of the Sparnacian.

The *Sables de Bracheux* (marine) and *Châlons-sur-Vesle* (brackish) are terminated at the top by the *grés blancs mamelonnes* which are particularly well developed in the region of Laon-Soissons-Noyon. These beds contain plants (*Dryophyllum* and *Laurus*) in particular. They always occur in the same stratigraphical position, i.e. at the top of the Thanet sands, from the northern part of the Paris Basin to the vicinity of Landen (Belgium), east of Bruxelles (Zone III, continental and fluviatile, of Leriche 1905*b*). At some localities in the Paris Basin these beds contain a lacustrine fauna with *Physa gigantea* and other freshwater molluscs (*Calcaire de Mortemer, Calcaire de Clairoix, Calcaire de Rilly, Marnes de Sinceny*). In the northern part of France these beds pass laterally into the *grés de Blainville à Unios* at Saint-Josse and Saint Aubin which contain *Cyrena cordata* from the underlying sands. This latter bed lies immediately below the lagoonal transgression of the Sparnacian.

Recent vertebrate palaeontological studies by Gurr (1962), on fishes of the Woolwich Bottom Beds of Herne Bay, and Russell (1964), on mammals of Cernay-les-Reims, has confirmed the earlier conclusions of Teilhard de Chardin (1922) concerning the importance of the Thanetian-Sparnacian boundary. The Thanetian mammals found at Cernay differ markedly from those found at Meudon. *Hyracotherium* has been found in the London Clay (=*Argile de Flandres*) whence it has probably been transported from mearby continental deposits. In France this fossil occurs in Sparnacian beds.

In the light of the foregoing summary we see that the detailed investigations in the Paris Basin by Feugueur and his colleagues have led to the following results and conclusions:

(1) In the Paris Basin the Belgian Landenian Stage is equivalent to the Thanetian.

(2) The lagoonal sequence of the Landenian (upper part) in Belgium corresponds to the lagoonal sequence of the same age (upper part of the Thanetian) which extends in France along the eastern margin of the Paris Basin. The marine transgression of the *Argile de Flandres* in Belgium and northern France caused the migration of certain brackish water molluscs towards the south, where they continued to develop in the lagoonal deposits of the basal part of the Sparnacian. These lagoonal deposits are then somewhat younger than the Landenian deposits at Ostende (Belgium) which correspond to the lower part of the *Argile de Flandres.*

EOCENE

Lower Eocene

The Ypresian Stage was created by Dumont (1849) as an upper subdivision of his Landenian (1839) to encompass the deposits (the *Argile de Flandres s.l.*) between the Landenian and Bruxellian. He subsequently (1851) intercalated the Paniselian between the Ypresian and Bruxellian. Considered to contain an ensemble of facies, the Paniselian has been subsequently subdivided into two parts: lower Paniselian (Mt. Panisel) belonging to the Ypresian (or Cuisian), and an upper Paniselian (*Sables d'Aeltre*) to the Bruxellian.

Dollfus (1880) erected the Cuisian Stage (Stratotype: Cuise-Lamotte near Compiègne) which he considered equivalent to the Belgian Ypresian. In 1905 he augmented the Cuisian by adding to its lower part the *Sables de Sinceny.* Dollfus insisted on the stratigraphical independence of the Sparnacian in the Paris Basin, whereas Leriche (1905*b*) considered the Sparnacian merely a brackish water formation which he placed in the upper part of the Landenian.

The use of the term Cuisian has been uneven. Although it has been more thoroughly defined and contains a better fauna in its type area, it has generally been neglected in favour of its (supposed) synonym, the Ypresian. The type level of the Cuisian is apparently higher than that of the Ypresian and this should perhaps be taken into consideration in further attempts to reach agreement on definition and usage of Lower Eocene chronostratigraphical terminology.

The Londinian Stage was created by Mayer-Eymar (1857)[1]

[1] (*Londonische stufe* or Londonian, 1857; = Londinian, 1874).

and included the Sparniacian and Ypresian, although in its original definition it did not include the Woolwich and Reading Beds (which were added at the base of the Sparnacian = Lower Eocene by Blondeau *et al.* 1965). It is, in reality, more or less equivalent to the Ypresian *s.l.* (=Cuisian, as emended by Dollfus 1909).

Above the Thanet sands (Landenian) there are two distinct sequences in northwestern Europe: lagoonal-lacustrine beds (Sparnacian) in the Paris Basin, and lagoonal-marine beds in northern France and Belgium (lower Ypresian).

The lowest member is the *Conglomerat de Meudon* with *Coryphodon* and other Sparnacian mammals. The lignites and sands which extend from Paris to the region of Provins-Sézanne correspond to the conglomerate. This basal detrital and lignitic series lies beneath the *argiles plastiques* which are developed in the valley of the Seine up to Mantes (west) and Sézanne (east).

The *argiles plastiques* extend northward in the Paris Basin, primarily in the form of a lignitic facies. Marine deposits are also intercalated in the upper part of the Sparnacian: *Sables de Sinceny, d'Auteuil* and *Argiles de Sarron.* The Sparnacian is terminated in general by clays with oyster and *Cyrena* banks, from Sinceny (in the extreme north of the Paris Basin) to the vicinity of the Paris (Mantes and Paris).

In northern France and Belgium the Sparnacian facies grades into the marine facies of the *Argile de Flandres* (lower part) and is equivalent to *Argiles d'Orchies* in northern France. The *Argile d'Orchies* lies beneath the *Sables de Mons-en-Pévèle* with *Nummulites planulatus-elegans,* which corresponds precisely with the Ypresian. These clays grade upwards into sandy clays containing Cuisian faunas, known in northern France as the *Argiles sableuses de Roubaix* and *Sables de Mons en Pévèle,* and in Belgium as the *Sables de Forest* and *Argiles sableuses de Mouscron-Ypres.*

The Woolwich and Reading Beds with a Sparnacian fauna at the base of the London Clay (lower Ypresian) are a lagoonal facies of this unit. However, both units have a thin marine band at the base, with a distinctive planktonic fauna (*fide* D. Curry pers. comm.). Blondeau *et al.* (1965) attach them to the lower Ypresian *s.l.*, as well as to the Sparnacian of the Paris Basin, and beds with *Cyrena cuneiformis* near Dieppe.

The Cuisian in France has recently been subdivided into two parts by Feugueur: a lower Cuisian (fine grained sands, in places argillaceous and locally fossiliferous; and an upper Cuisian, itself subdivided into two parts: a lower part (*niveau de Pierrefonds - falun à Nummulites planulatus-elegans, Turritella solanderi* and *Velates schmiedeli*), called *Cuisien supérieur* A by Feugueur (1963), and an upper part (*niveau de Hérouval, argiles de Laon, sables à Gisortia* near Laon, *sables du Paniselien supérieur* of Belgium = *sables d'Aeltre*) called *Cuisien supérieur* B by Feugueur.

The Ypresian in Flanders is divided into a lower Ypresian (*Argile de Flandre*) and an upper Ypresian (sands and clays with *Nummulites planulatus-elegans* and *Turritella solanderi*). The foraminiferal fauna of the Franco-Belgian Ypresian have been studied by LeCalvez (in LeCalvez and Feugueur 1956). The fauna of the upper Ypresian in a well at Mouscron was said to be characterized by miliolids, rotaliids, textulariids and anomalinids; the middle Ypresian by abundant lagenids, anomalinids, buliminids, and rotaliids, as well as by planktonic species (the distinctive character of this interval was emphasized); the lowest part of the Ypresian (*argile de Flandres*) was said to contain solely arenaceous benthonic forms: *Haplophragmoides, Trochammina* and *Ammodiscus.* The authors point out that the microfauna of the upper part of the Ypresian is Cuisian, and that it contains a strong affinity with those of the beds at Aizy and Pierrefonds in the Paris Basin.

The present author has observed in samples collected from the Ypresian at various localities in Belgium (near the cemetery of Forest, Mont Saint Aubert) an association of *Nummulites planulatus-elegans* and planktonic foraminifers. The planktonic forms belong to the species *Globigerina yeguaensis* Berggren (= *G. patagonica* Todd and Kniker), *Acarinina esnaensis, A. pentacamerata* (=*gravelli*) and *A. soldadoensis.* This fauna is similar to that which occurs in the Lower Eocene 3 of northwestern Germany and the Rosnaes Clay of Denmark (Berggren 1960), both of which are approximately equivalent to at least a part of the *Globorotalia formosa formosa* Zone of Bolli.

Palaeocene/Eocene boundary problems and the use of stage names. The investigations by Feugueur and his colleagues in France have an important bearing on the placement of the Palaeocene-Eocene boundary and the use of chronostratigraphical terms. Virtually all vertebrate palaeontologists, most continental geologists, and some invertebrate palaeontologists and micropalaeontologists have favoured placing the Palaeocene/Eocene boundary at the top of the Thanetian-base Sparnacian. On the other hand, most invertebrate palaeontologists and micropalaeontologists, and some English and continental geologists (following Leriche 1905, *et seq.*) have preferred to place the boundary at the top of the Sparnacian (which was considered an upper part of the Landenian). When it is realized that the Thanetian and Landenian are equivalent to each other in a chronostratigraphical sense, and that the Sparnacian is but the lowermost part of the Ypresian (Lower Eocene), it will then be apparent that both protagonists in this long polemic have been using the same boundary all along. The boundary at the base of the Sparnacian based on the *Hyracotherium-Coryphodon* fauna is the same as that determined, and generally agreed upon, by invertebrate palaeontologists and micropalaeontologists – the junction *Globorotalia velascoensis/G. rex* Zone, which occurs at the base of the Ypresian.

The Palaeocene-Eocene boundary is generally placed between the *G. velascoensis and G. rex* Zones by micropalaeontologists (in the Soviet Union this level is approximately equivalent to the base of the *G. subbotinae* Zone). Luterbacher (1964) has recently recognized a *G. aequa* Zone between these two zones characterized by the overlap in occurrence of *G.* sp. aff. *velascoensis, G. marginodentata, G. formosa gracilis* and *G. subbotinae.* The present author has observed this overlap in several localities (Lodo Formation, California; upper part of Esna Shale, Luxor, Egypt). Associated with this overlap is the first appearance of *Pseudohastigerina wilcoxensis* (Cushman and Ponton), the first planispiral planktonic foraminifer in the Tertiary. As such it is a distinct form, whose first appearance can be recognized in many parts of the world. I have proposed the term *Pseudohastigerina*-datum (=*Globanomalina*-datum) for this level (Berggren 1964, 1965; Berggren, Olsson and Reyment 1967). This level corresponds to the boundary of Bramlette and Sullivan (1961) between the *Discoaster multiradiatus* Zone (Unit 2) and *Marthasterites tribrachiatus* Zone (Unit 3) which they also considered to be the Palaeocene/Eocene boundary. Hay (1962) placed this level in his *D. contortus* Subzone (*D. multiradiatus* Zone).

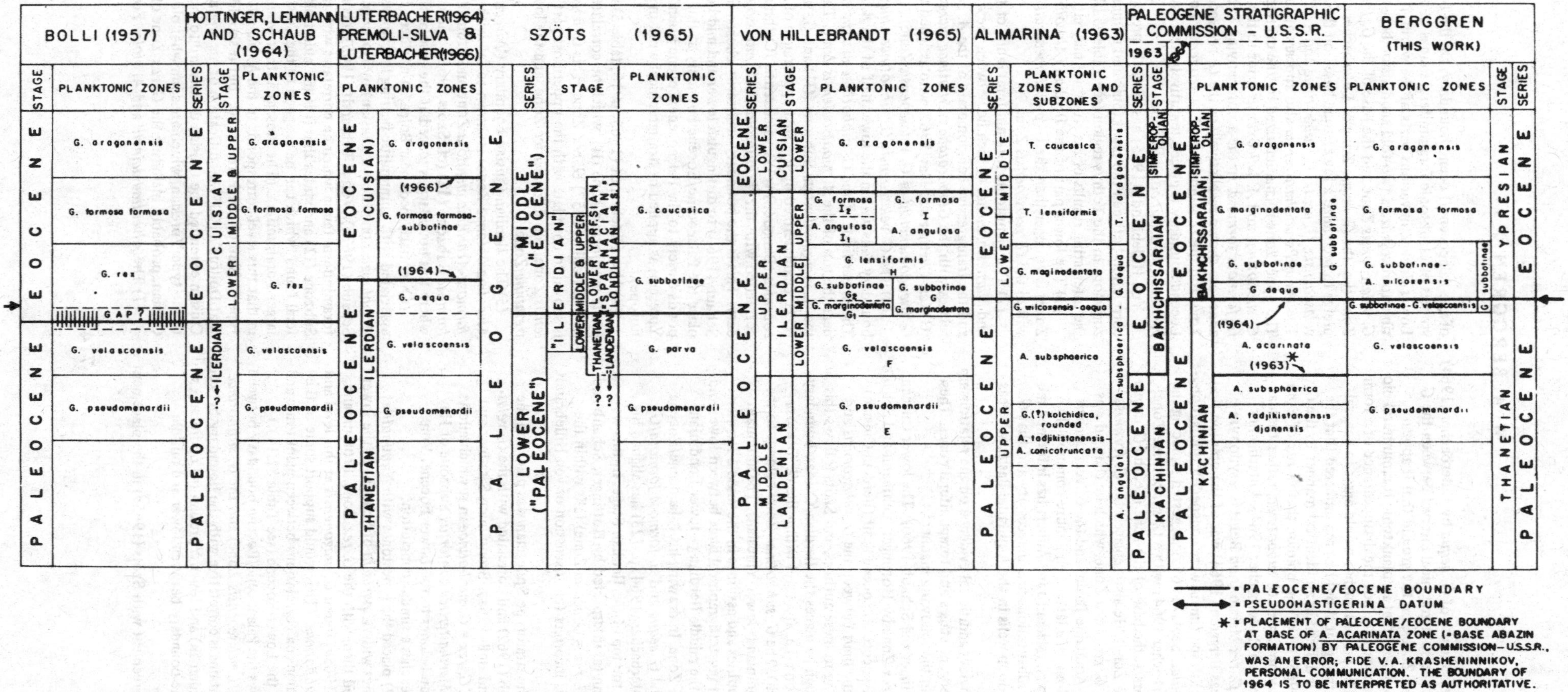

BOLLI (1957)
STAGE
PLANKTONIC ZONES
EOCENE
PALEOCENE
G. aragonensis
G. formosa formosa
G. rex
GAP ?
G. velascoensis
G. pseudomenardii
HOTTINGER, LEHMANN AND SCHAUB (1964)
SERIES
STAGE
PLANKTONIC ZONES
EOCENE
PALEOCENE
LOWER CUISIAN MIDDLE & UPPER
? ← ILERDIAN
G. aragonensis
G. formosa formosa
G. rex
G. velascoensis
G. pseudomenardii
LUTERBACHER (1964) PREMOLI-SILVA & LUTERBACHER (1966)
PLANKTONIC ZONES
EOCENE (CUISIAN)
PALEOCENE
ILERDIAN
THANETIAN
G. aragonensis
(1966)
G. formosa formosa-subbotinae
(1964)
G. aequa
G. velascoensis
G. pseudomenardii
SZÖTS (1965)
SERIES
STAGE
PLANKTONIC ZONES
PALEOGENE
MIDDLE ("EOCENE")
LOWER ("PALEOCENE")
"ILERDIAN"
LOWER MIDDLE & UPPER
THANETIAN LOWER YPRESIAN
"LANDENIAN" "SPARNACIAN" LONDINIAN S.S.
? ?
G. caucasica
G. subbotinae
G. parva
G. pseudomenardii
VON HILLEBRANDT (1965)
SERIES
STAGE
PLANKTONIC ZONES
EOCENE LOWER
PALEOCENE UPPER MIDDLE
CUISIAN LOWER
ILERDIAN UPPER MIDDLE LOWER
LANDENIAN
G. aragonensis
G. formosa I2
A. angulosa I1
G. formosa I
A. angulosa
G. lensiformis H
G. subbotinae G2
G. marginodentata G1
G. subbotinae G
G. marginodentata
G. velascoensis F
G. pseudomenardii E
ALIMARINA (1963)
SERIES
PLANKTONIC ZONES AND SUBZONES
EOCENE MIDDLE LOWER
PALEOCENE UPPER
T. caucasica
T. lensiformis
G. marginodentata
G. wilcoxensis - aequa
A. subsphaerica
G.(?) kolchidica, rounded
A. tadjikistanensis - A. conicotruncata
T. aragonensis
G. aequa
A. subsphaerica
A. angulata
PALEOGENE STRATIGRAPHIC COMMISSION - U.S.S.R.
1963
1964
SERIES
STAGE
PLANKTONIC ZONES
EOCENE
PALEOCENE
SIMFEROPOLIAN
BAKHCHISSARAIAN
KACHINIAN
G. aragonensis
G. marginodentata
G. subbotinae
G. subbotinae
G. aequa
(1964)
A. acarinata
(1963)
A. subsphaerica
A. tadjikistanensis djanensis
BERGGREN (THIS WORK)
PLANKTONIC ZONES
STAGE
SERIES
G. aragonensis
G. formosa formosa
G. subbotinae - A. wilcoxensis
G. subbotinae
G. subbotinae - G. velascoensis
G. velascoensis
G. pseudomenardii
YPRESIAN
THANETIAN
EOCENE
PALEOCENE
= PALEOCENE/EOCENE BOUNDARY
= PSEUDOHASTIGERINA DATUM
* = PLACEMENT OF PALEOCENE/EOCENE BOUNDARY AT BASE OF A. ACARINATA ZONE (= BASE ABAZIN FORMATION) BY PALEOGENE COMMISSION—U.S.S.R., WAS AN ERROR; FIDE V.A. KRASHENINNIKOV, PERSONAL COMMUNICATION. THE BOUNDARY OF 1964 IS TO BE INTERPRETED AS AUTHORITATIVE.

Table 52.10. Different interpretations of planktonic foraminiferal zones and the position of the Palaeocene-Eocene boundary

Szöts (1965*a*) criticizes the use of *G. aequa* by Luterbacher (1964) as the guide fossil for the stratigraphical interval between the *G. velascoensis* and *G. rex* Zones, on the ground that it appears earlier than in this zone, and that in planktonic foraminiferal biostratigraphy it is the appearance not the disappearance of a form which is important. The present author would agree in general with this criticism, although it should be remembered that the *G. velascoensis* Zone is defined in an identical manner to the *G. aequa* Zone. Szöts (1965*a*) is of the opinion that *G. rex* of Bolli (*non* Martin) is synonymous with *G. subbotinae*, rather than a keeled variety of *G. aequa* (Luterbacher 1964). Luterbacher (1964) believed that *G. rex* Martin (*non* Bolli) is a synonym of *G. subbotinae*. whereas Berggren (1964) was of the opinion that both Martin's and Bolli's specimens were conspecific and synonymous with *G. subbotinae*. Szöts (1965*a*) suggested replacing the *G. aequa* Zone (Luterbacher 1964) with a *G. subbotinae* Zone, since the latter appears at the base of the zone (i.e. the *G. aequa* Zone = *G. subbotinae* Zone). He also replaced the *G. formosa formosa* Zone with a *G. caucasica* Zone, which he claimed was more frequent in the Aquitaine Basin, and the *G. velascoensis* Zone by a *G. parva* Zone (the latter being more common than the rare *G. velascoensis*). (Madam Jane Aubert has informed me that these two latter substitutions are unnecessary and that *G. velascoensis* is common in wells in the Aquitaine Basin, whereas *G. parva* is rare.)

Szöts (1965*a*) provides additional evidence on the relationship between Lower Palaeogene stages and foraminiferal zones. These may be summarized in the following manner:

(1) Ilerdian (Hottinger and Schaub 1960): The lower Cuisian was placed in the *G. rex* Zone by Hottinger, Lehmann and Schaub (1964, p. 647). The Palaeocene/Eocene limit (Ilerdian-Cuisian) thus corresponds to the limit between the *G. velascoensis* and *G. rex* Zones according to these authors, and Szöts follows this interpretation, although he places the Ilerdian/Cuisian boundary in the *G. caucasica* Zone (i.e. at a level within the *G. formosa formosa* Zone of Bolli). The *G. subbotinae* Zone belongs to the basal part of the Londinian (=lower Ypresian). The abundant appearance of nummulites and assilines in the middle Ilerdian denotes an important biostratigraphical limit between lower and middle Ilerdian, and the middle Ilerdian (= lower Londinian) belongs to the *G. rex* Zone. In drawing his Palaeocene/Eocene boundary between the *G. aequa* and *G. formosa formosa*/*G. subbotinae* Zone, Luterbacher (1964, p. 723) has aligned his top Palaeocene with the top of the Ilerdian Stage. In this way even the lower Cuisian (as interpreted by Hottinger, Schaub and Lehmann (1964) (=*G. rex* = *G. aequa* Zone) falls within the 'Palaeocene', which is incorrect (cf. discussion of von Hillebrandt 1965, *below*).

(2) Spileccian: The stratotype Spileccian was placed in the Palaeocene by Schaub (1962) and correlated with the *G. velascoensis* Zone (Cita and Bolli 1961). Bronniman, Stradner and Szöts (1965*a*) found *G. rex* = *G. subbotinae* in it which allows correlation with the *Marthasterites contortus* Subzone of Hay (1964). The Spileccian belongs to the Lower Eocene. Von Hillebrandt (1965) reaches a similar conclusion.

(3) Schaub (1962) placed the limestones with *Nummulites irregularis* (=limestones with *N. partschi* of Schaub) in the lower Cuisian and correlated them with the *G. rex* Zone. Bronnimann, Stradner and Szöts (1965*a*) found *G. aragonensis* at this level and no sign of *G. rex* = *subbotinae*. (This would support some of the present author's opinions on correlations between planktonic and nummulitic zones in the Lower Eocene (see Table 52.11).)

Szöts (1965) draws the Palaeocene/Eocene boundary between the *G. parva* Zone (=*G. velascoensis* Zone) and the *G. subbotinae* Zone, which he maintains is correlative with the boundary between the Thanetian (=Landenian) and lower Ypresian (=Londinian *s.s* = Sparnacian) and the boundary between lower and middle Ilerdian.

I am in general agreement with Szöts (1965*a*) in the placement of the Palaeocene/Eocene boundary. On the basis of current studies of planktonic foraminiferal faunas of the Palaeocene/Lower Eocene in various parts of the world, it would seem possible to recognize a two-fold subdivision between the top of the *G. velascoensis* Zone and the base of the *G. formosa formosa* Zone (i.e. the *G. aequa* Zone of Luterbacher 1964). Thus I would prefer to recognize a lower zone based on the overlap of *G. velascoensis, G. acuta, G. subbotinae* and *Acarinina wilcoxensis*. The genus *Pseudohastigerina* also appears at the top of this zone. The upper zone is characterized by the overlap in range between *G. subbotinae* and *A. wilcoxensis* prior to the first occurrence of *G. formosa formosa*. In the accompanying Table (52.10) these two zones are shown respectively as the *G. velascoensis*-*G. subbotinae* Zone and the *G. subbotinae*-*A. wilcoxensis* Zone. The Palaeocene/Eocene boundary is placed tentatively near the top of the *G. velascoensis*-*G. subbotinae* Zone.

Von Hillebrandt (1965) has recently presented evidence for correlation between the planktonic and nummulitic foraminiferal zonations in the Tethys region (see Tables 52.10, 52.11). Among the interesting results of his study we may cite the following:

(1) The lower part of the Ilerdian (*Alveolina cucumiformis* Zone) is younger than the *G. pseudomenardii* Zone and is placed within the *G. velascoensis* Zone.

(2) The planktonic foraminiferal fauna of the *Marnes bleues* and *Marnes blanches* (which lie between the *A. ellipsoidalis* Zone and the transgressive conglomerate of the *A. corbarica* Zone) correlates with the *Globorotalia subbotinae*.*marginodentata* Zone (G) (cf. Gartner and Hay 1962, who assigned this interval to the *G. pseudomenardii* or *G. velascoensis* Zone).

(3) The *G. velascoensis*/*G. subbotinae* - *marginodentata* Zone boundary falls within the lower half of the *A. ellipsoidalis* Zone. The upper limit of the *G. subbotinae*-*marginodentata* Zone lies near the top of the *A. moussoulensis* Zone. The boundary between the *G. marginodentata* Subzone (G1) and *G. subbotinae* Subzone (G2), characterized by the first appearance of *Pseudohastigerina eocenica* (Berggren) vel *P. wilcoxensis* (Cushman and Ponton), is correlated with the *A. ellipsoidalis*-*A. moussoulensis* Zone boundary. This is the boundary arbitrarily chosen by Hottinger and Schaub (1960) to distinguish lower and middle Ilerdian. I would place the Palaeocene/Eocene boundary at this level. It corresponds closely to the Palaeocene/Eocene boundary, i.e. base Ypresian, as applied in stratigraphical work in northwestern Europe).

(4) Von Hillebrandt (1965, p. 18, table 5) recognizes a *Globorotalia lensiformis* Zone (H), which he correlates with the upper half of Bolli's (1957) *G. rex* Zone. This *G. lensiformis* Zone (H) is shown to correlate with the uppermost part of the *A. moussoulensis* Zone, the *A. corbarica* Zone, and the lower half of the *A. trempina* Zone.

(5) The Ilerdian-Cuisian boundary ('*Niveau de Coudures*' – *Alveolina oblonga* Zone) is shown to fall within the *Globorotalia formosa*-*Acarinina angulosa* Zone and probably within the *G. formosa* Subzone (I2) (cf. Szöts 1965*a*, and discussion above, where this boundary is placed at the base of the *G. rex* Zone).

(6) In marly beds near Bolca, in northern Italy, which are associated with nummulitic-bearing limestones attributed to the *A. oblonga* Zone (Lower Cuisian) by Hottinger (1960*b*) and Schaub (1962*a*), Von Hillebrandt (1965) records a planktonic foraminiferal fauna which he correlates with the *G. formosa* Subzone (I2) and states that it is the oldest planktonic foraminiferal fauna which can be assigned to the Cuisian on the basis of larger Foraminifera. (The present author would agree with this, in that this level corresponds to the *N. planulatus* level in the type Cuisian of northern France, but it should be remembered that Dollfus' (1905) modification of his original concept of Cuisian extended the Cuisian downward to include beds equivalent to the type Ypresian, which are somewhat older than the type Cuisian, probably including the *G. rex* Zone of Bolli.)

(7) The *Alveolina dainelli* and *A. violae* Zones are correlated

with the '*Globorotalia*' *palmerae* Zone of Bolli, and both are placed within the Cuisian.

It is interesting to observe that Von Hillebrandt (1965) recognizes a two-fold subdivision of the *G. subbotinae-marginodentata* Zone (G), into a lower *G. marginodentata* Subzone (G1) and an upper *G. subbotinae* Subzone (G2). In the Soviet Union those zones are recognized in reverse order (Palaeogene Stratigraphical Committee, U.S.S.R. 1962, 1963, 1964; Krasheninnikov 1964). An illustration of the various interpretations of the Palaeocene/Eocene boundary and planktonic foraminiferal zones used in defining this boundary is presented in Table 52.10. It will be seen that the studies of von Hillebrandt (1965) have allowed a more precise correlation of the top Ilerdian with the planktonic foraminiferal zonation. Thus its location has been raised from its assumed position at the base of the *G. rex* Zone (Schaub *et al.*) to the upper part of the *G. rex* Zone (Luterbacher) and then to the upper part of the *G. formosa formosa* Zone (Szöts, von Hillebrandt).

In the light of the discussion above it would seem that the term Thanetian or Landenian (in the sense of Leriche 1905) could be used satisfactorily in a chronostratigraphical sense for the interval between top Montian *s.s.* (Danian *s.l.*) and base Ypresian. The Landenian has priority in terms of original definition. However, the term Thanetian has been far more widely used in stratigraphical literature, and the faunas, particularly the microfaunas, are better known than those of the Landenian.

As it has been originally defined and subsequently used over a period of nearly a century, the Sparnacian consists of lagoonal-lacustrine beds which grade gradually southwards into fluvio-lacustrine beds. Brackish-water facies with marine intercalations are developed only in the northeastern part of the Paris Basin. The faunas are rather poor, as might be expected, with the exception of the upper part where greater diversity is seen, a presage, albeit transitory, of the developing Ypresian transgression from the north and west. The Sparnacian does not have the characteristics which warrant its retention as a chronostratigraphical unit. It has been defined on the basis of its nonmarine characters, among other things, which makes direct correlation with marine sequences difficult. It is, in reality, little more than a lower facies of the lithostratigraphical units which belong to the Ypresian. Exclusion from the Palaeocene of the Sparnacian, which was after all what Schimper (1874) originally based his concept of Palaeocene upon, in describing the fossil floral assemblages from the *Travertin de Sézanne* of the Paris Basin, raises questions concerning the use of Palaeocene in the chronostratigraphical hierarchy. Perhaps a redefinition of Palaeocene to include strata of pre-Sparnacian – pre-Ypresian age in the manner suggested by Blondeau *et al.* (1965, p. 203) would be sufficient.

The conclusions drawn here represent a departure from those I have previously expressed (Berggren 1965*c*). At that time the possibility of eliminating the Sparnacian as a chronostratigraphical unit was not considered, since it was interpreted as an upper subdivision of the Landenian Stage. It will be seen at the same time that in placing the Palaeocene/Eocene boundary at the base of the 'Sparnacian facies' of the Ypresian Stage, a modification of Schimper's original definition of the Palaeocene is made. This change is warranted by consideration of the regional geology of the type area which has allowed a reconciliation of apparently anomalous palaeontological data. The Palaeocene epoch, thus modified, is a unified subdivision of the Cenozoic.

Middle Eocene

The Middle Eocene is generally considered to be represented by a single stage, the Lutetian. However, interpretations have varied: Munier-Chalmas and de Lapparent (1893, pp. 474, 475), for example, interpreted the Middle Eocene as consisting of the Lutetian and Bartonian and considered the latter to be little more than a subdivision of the former.

As a result of field observations in the Paris Basin by Cuvier, Brongniart, d'Archiac and Graves in the first half of the nineteenth century, d'Orbigny (1852) created one of the first stages (in a time-stratigraphical sense) of the Tertiary: The Parisian. In his original definition of the term d'Orbigny subdivided his Parisian into two parts: a lower unit A and an upper unit B. In his Parisian Stage he included all rock units in the vicinity of Paris between the *Argile plastique* (bottom) and the so-called *Sables inférieurs* (top), including thus at the top of his stage unit the *Marne blanche* above the *Gypse de Montmartre.* As it was originally defined d'Orbigny's Parisian includes rocks presently grouped in the Lutetian, Bartonian, and a part of the Tongrian by French stratigraphers.

Mayer-Eymar (1858), in using the term Parisian, limited it to d'Orbigny's unit A, and erected a Bartonian Stage which he included in d'Orbigny's Parisian B. In order to avoid the confusion in usage occasioned by Mayer-Eymar's (1858) modification of an original definition without a formal change in nomenclature, de Lapparent (1883, p. 989) defined the Lutetian Stage (Lutetia = Paris) which is the exact equivalent of d'Orbigny's Parisian A = Parisian Mayer-Eymar, 1858. The Parisian A = Lutetian was originally defined and subsequently used to denote the *Calcaire grossier* and its various member units in the Paris Basin. The macro- and microfossils of this unit have been studied in greater detail than any other corresponding unit in the Paris Basin (molluscs: Deshayes, Cossmann and Pisarro; Foraminifera: Terquem, LeCalvez).

The investigations of Leriche (1912) and Abrard (1925) have led to an understanding of the Lutetian of the Paris Basin. Abrard (1925) presented a biostratigraphical subdivision of the Lutetian. The *calcaire grossier* was divided into a lower part (Zones I-III), a middle part (Zone IVa) and an upper part (Zone IVb). Zones I–III were grouped into the lower Lutetian; Zones IVa and IVb in the upper Lutetian. The primary distinction made by Abrard (1925) between his lower and upper Lutetian was the presence of *Nummulites laevigatus* and *N. lamarcki* in Zone I, *N. laevigatus* in Zone II, and *Orbitolites complanatus* in Zone IVa. Investigations by Blondeau and Cavelier (1962*a*), Blondeau and Curry (1963), Blondeau, Cavelier and Pomerol (1964), Blondeau (1964, 1965*b*), Cavelier and LeCalvez (1965) and Blondeau *et al.* (1965) have shown that (*a*) the distinction of Zone III is superficial; (*b*) Zone IVb can be defined palaeontologically by *Arenagula kernfornei;* (*c*) distinction between Zones I and II can be made by the coexistence in Zone II of *N. variolarius* and *N.*

Table 52.11. Suggested correlation of some Palaeogene planktonic foraminiferal, nummulitic and nannofossil biozones

Column	Zones (top to bottom, as printed)
Epoch / Series	EOCENE; PALEOCENE
Age / Stage	BARTONIAN; LUTETIAN; YPRESIAN; THANETIAN; DANIAN S.L. (MONTIAN S.S.; DANIAN S.S.)
BERGGREN (1967) General-Composite — Planktonic zones	G. gortanii/G. centralis; C. danvillensis; G. mexicana; T. pseudodubia; Orbulinoides beckmanni; G. lehneri; G. kugleri; H. aragonensis zone; A. bullbrooki zone; G. aragonensis zone; G. formosa-G. subbotinae zone; G. subbotinae-A. wilcoxensis; G. velascoensis-G. subbotinae; G. velascoensis zone; G. pseudomenardii zone; G. pusilla pusilla-G. angulata zone; G. uncinata-G. spiralis zone; G. compressa / G. praecursoria; G. daubjergensis zone; G. pseudobulloides subzone
LUTERBACHER (1964) Central Apennines	?; G. bullbrooki; ?; G. aragonensis zone; ?; G. formosa formosa/G. subbotinae; G. aequa; G. velascoensis zone; G. pseudomenardii zone; G. pusilla pusilla zone; G. uncinata; G. trinidadensis zone; G. pseudobulloides/G. daubjergensis; G. eugubina
BOLLI (1957), BOLLI AND BERMÚDEZ (1966) Trinidad — Planktonic zones	? ?; G. cocoaensis (G. cerroazulensis); G. semi-involuta; T. rohri; P. mexicana; G. lehneri; G. kugleri; H. aragonensis; G. palmerae; G. aragonensis; G. formosa formosa; G. rex; [hatched interval]; G. velascoensis; G. pseudomenardii; G. pusilla pusilla; G. angulata; G. uncinata; G. trinidadensis; G. pseudobulloides; G. eugubina
GOHRBANDT (1963) Austria	F; ?; not zoned (non-planktonic facies); E; D; C; B; A; ?
HILLEBRANDT (1965) NW Spain	G. caucasica M; G. palmerae-G. aspensis L; G. aragonensis K; G. formosa I_2 / A. angulosa I_1 (= G. formosa-A. angulosa I); G. lensiformis H; G. subbotinae G_2 / G. marginodentata G_1 (= G. marginodentata-G. subbotinae G); G. velascoensis F; G. pseudomenardii E; G. pusilla D; G. angulata C; G. inconstans-uncinata B; G. trinidadensis A_2 / G. edita A_1 (= G. edita-trinidadensis A)
BRAMLETTE & SULLIVAN (1961) California	?; UNIT 5 D. sublodoensis; ?; UNIT 4 D. lodoensis; UNIT 3 M. tribrachiatus zone; UNIT 2 D. multiradiatus zone; (?) absent; UNIT 1 H. riedeli zone; ?
HAY (1962, 1964) Switzerland	?; D. sublodoensis; ?; D. lodoensis zone; M. tribrachiatus zone (2 D. lodoensis subzone; 1 D. diastypus); D. contortus subzone; D. multiradiatus zone; D. gemmeus z. ?; D. delicatus (D. delicatus; D. megastypus); ? ?
SCHAUB (1964) — Stage	PRIABONIAN; BIARRITZIAN; LUTETIAN; CUISIAN; ILERDIAN; MONTIAN-THANETIAN
SCHAUB (1964) — Biozones based on Nummulites	fabianii; brongniarti, perforatus, praefabianii; aturicus, carpenteri; crassus, millecaput; boussaci, sismondai; laevigatus (uranensis; gallensis); polygyratus, distans; planulatus, burdigalensis; "Niveau de Coudures"; involutus; atacicus-exilis; praecursor; n. sp. minervensis; deserti-fraasi; Operculina heberti
SCHAUB (1964) — Biozones based on Assilina	exponens (gigantea; planospira; spira); tenuimarginata; major; placentula; "Niveau de Coudures"; n. sp. tremp; leymeriei; pyrenaica; n. sp. cayta; n. sp. harmana
SCHAUB (1964) — Biozones based on Alveolina	(neoalveolinids); elongata; n. sp. gargano; prorrecta; munieri; stipes, tenuis; violae; oblonga; "Niveau de Coudures"; trempina; corbarica; moussoulensis; ellipsoidalis; cucumiformis; levis; primaeva

laevigatus.

The Lutetian constitutes a well-defined sedimentary cycle in the Paris-Belgian Basin which is clearly delimited at its base and summit by erosional unconformities. The great shortcoming of the Lutetian is in the fact that its upper part is developed as a regressive sequence containing continental and brackish-water deposits.

The lower part of the Lutetian is characterized by *Nummulites laevigatus* (beds 1 and 2 of Leriche 1912; Zones I and II of Abrard 1925). The recent investigations by Blondeau and Cavelier (1962 *a, b*), Blondeau and Curry (1964) and Blondeau (1965) have shown that Abrard's Zone I and II can be distinguished by the coexistence in Zone II of *N. laevigatus* and *N. variolarius,* and that *N. variolarius* characterizes the upper Lutetian as well. The original stratotype of the Lutetian, represented by Zones I and II, is no longer in existence. Blondeau (1962) has presented a sedimentological analysis of the *Calcaire grossier* in the vicinity of Cramoisy Saint-Vaast-les-Malo (N.W. of the Oise River), and suggests it may serve as a neostratotype of the Lutetian Stage.

Bouché (1962) has studied the calcareous nannofossils in the Lutetian of the Paris Basin. An informal three-fold subdivision of the Lutetian into a lower, middle and upper part was formulated on the basis of the distribution of various species in samples from different localities. *Discoaster lodoensis,* generally agreed to be characteristic of the Middle Eocene, was recorded from the 'lower' Lutetian only. *D. barbadiensis* was recorded throughout the Lutetian, but was found to be most common in the 'middle' Lutetian. Several forms whose stratigraphical range is known to extend elsewhere from Lower Eocene into Middle as well as Upper Eocene, were recorded from the Lutetian. The following forms, restricted elsewhere to the Middle Eocene, were recorded from the Lutetian of the Paris Basin: *Micrantholithus basquensis cressus, Discoaster sublodoensis, D. gemmifer* and *Trochoaster hohnensis.*

The Bruxellian Stage of Belgium (Dumont 1839) is generally accepted as the equivalent of the lower part of the Lutetian (Blondeau *et al.* 1965). Although Kaasschieter (1961, pp. 7, 132, 133) equated it with most of the Lutetian, including in the lower part of the Lutetian the upper part of the Panisel beds. The Ledian Stage (Mourlon 1887) is characterized by the occurrence of *Nummulites variolarius* with a basal conglomerate of rolled, worn *N. laevigatus.* Belgian geologists generally place the Ledian in the Upper Eocene and correlate it with the Auversian Stage (Dollfus 1880, 1907), which is considered to correspond to the lower part of the Bartonian. Kaasschieter (1961, pp. 7, 10) correlated the *Sables d'Auvers* (the Auversian was originally created by Dollfus (1880) for the lower part of these sands, later (1905) modified and expanded to include the entire sandy sequence at Beauchamp and Ezanville) in addition to the *Calcaire de Ducy* with the lower part of the Ledian. The *Sables de Marines* above were correlated with the upper Ledian, although they are generally correlated with the Bartonian. He united the Ledian and Bartonian into a single stage, the upper part including the Sands of Wemmel, and Clays and Sands of Asse. Pomerol (1961) placed the Ledian in the Lutetian on the basis of sedimentological, palaeontological and palaeogeographical considerations. Pomerol (1961) concluded that: (1) the detrital sands of the Auversian (Paris Basin) have no sedimentological relationship with those of the Ledian (Belgium); (2) the Ledian sands contain typical Lutetian fossils not found in the Auversian (*Ditrupa strangulata, Echinolampas,* inter alia; the Ledian microfauna contains *Orbitolites complanatus* and *Fabularia discolithes* which characterize the upper Lutetian but are not known in the Auversian. Curry (1961) has reported the occurrence of *Nummulites variolarius* in the upper Lutetian of the Paris Basin and it occurs in the type Ledian as well (Kaasschieter 1961, p. 29). Blondeau *et al.* (1965) placed the Ledian in the lower part of the upper Lutetian and included the Sands of Wemmel in the upper part of the upper Lutetian, but Banner and Eames (1966, pp. 163, 165) consider the Ledian a synonym of the Auversian and place it in the lower part of the Upper

Sous-étages	Bassin de Paris	Belgique	Angleterre Whitecliff Bay, île de Wight
Lutétien supérieur	Faluns de Foulangues à *Arenagula kerfornei* Calc. à Cérithes et Potamides	Sables de Wemmel	(Fisher bed XVI et XVII)
	Calc. zoogène à *Miliolidae* *Orbitolites complanatus* *Fabularia discolithes* *N. variolarius* rares	Lédien *N. variolarius* *Orbitolites complanatus*	Upper Bracklesham beds et équivalents Fisher bed IX
Lutétien inférieur	*N. laevigatus* et *N. variolarius* rares	*N. laevigatus* et *N. variolarius* rares	Fisher bed VIII
	N. laevigatus	Bruxellien	Partie supérieure des Lower Bracklesham beds (Fisher bed VII-VI)
Yprésien supérieur	Argile de Laon, niveau d'Hérouval	Sables d'Aeltre	Fisher bed V

Tableau 3. — Les formations du Lutétien dans le bassin de Paris. Essai de corrélation.

Table 52.12. Suggested correlations of Lutetian units in the Paris Basin and Hampshire Basin (Blondeau *et al.* 1965, p. 207, tab. 3)

		Bassin de Paris				Bassin de Bruxelles	Hampshire
		(L. et J. Morellet 1948)	(Gignoux 1945)	(Abrard 1950 & Söyer 1953)	(Pomerol 1961)		
Éocène supérieur		Ludien	Ludien	Bartonien: Sup. Ludien ou Wemmelien	Bartonien: Zone III ou Ludien	Asschien, Wemmelien	Lower Headon Beds; Barton Beds
Éocène supérieur		Bartonien	Lédien	Bartonien: Inf. ou Lédien	Bartonien: Zone II ou Marinesien	Lédien	Upper Bracklesham Beds
Éocène supérieur		Bartonien	Lédien	Bartonien: Inf. ou Lédien	Bartonien: Zone I ou Auversien	Lédien	Upper Bracklesham Beds
Éocène moyen = Lutétien	(R. Abrard 1925)	Zone IV		(A. Blondeau & C. Cavelier 1962) Supérieur		Lacune	Middle Bracklesham Beds
Éocène moyen = Lutétien		Zone III		Supérieur		Lacune	Middle Bracklesham Beds
Éocène moyen = Lutétien		Zone II		Inférieur		Bruxellien	Middle Bracklesham Beds
Éocène moyen = Lutétien		Zone I		Inférieur		Bruxellien	Middle Bracklesham Beds

Table 52.13. Generally accepted subdivisions and stratigraphical correlations in the Middle Eocene and Paris Basin, Brussels Basin and Hampshire Basin (Pomerol 1964, p. 161, tab. 2).

	Bassin de Paris	Dome de l'Artois	Bassin de Bruxelles	Hampshire
Oligocène Stampien inf.	Marnes à huîtres; Calcaire de Sannois; Marnes vertes; Marnes à Cyrènes		Tongrien sup. niveau d'Hoogbutzel	Hamstead Beds
Éocène supérieur = Bartonien	Zone III: Ludien = Lattorfien?		Tongrien inf. sable de Neerepen	Bembridge Beds; Headon Beds; Upper Barton Beds
Éocène supérieur = Bartonien	Zone II: Marinésien		Asschien	Lower Barton Beds
Éocène supérieur = Bartonien	Zone I: Auversien		Wemmellien	Upper Bracklesham Beds (Fisher Beds IX à XIX)
Éocène moyen = Lutétien	Faciès lagunaires (caillasses); Supérieur (Zones III & IV)		Lédien	Upper Bracklesham Beds (Fisher Beds IX à XIX)
Éocène moyen = Lutétien	Inférieur (Zones I & II)		Bruxellien	Middle Bracklesham Beds (Fischer Beds à VI à VIII)
Éocène inférieur				

Table 52.14. Proposed stratigraphical correlations between different stages of Middle Eocene in Paris, Brussels and Hampshire Basins. Lutetian/Bartonian boundary does not necessarily correspond to an isochronous surface in the different basins. The same is true, with stronger reason, for correlations between the subdivisions of the stages (Pomerol 1964, p. 163, tab. 3).

Eocene equivalent to the early Priabonian.

The occurrence of transitional Middle Eocene-early Upper Eocene nummulites and alveolinids in southern Europe (*N. praefabianii, N. perforatus, N. brongniarti* Zones) led Hottinger and Schaub (1960) to intercalate a new Biarritzian Stage between the Lutetian and Priabonian. Mangin (1961) and Szöts (1961) have criticized the validity of the Biarritzian as a chronostratigraphical unit. Szöts (1961) considered that the faunas at Biarritz have a Lutetian character and Blondeau *et al.* (1965, p. 206) have concluded that the Biarritzian can be correlated with the uppermost Lutetian. Szöts (1964) has recently obtained some interesting results from a study of the Middle Eocene of the southwest part of the Aquitaine Basin. A general subdivision of the Lutetian of the Adour Basin prepared by Burger, Cuvillier and Schoeffler (1945) was presented by Szöts (1964) and is reproduced here.

Upper Lutetian	Couches de la grotte de Brassempouy
	Couches à grandes *Nummulites*
Middle Lutetian	Couches de Nousse
Lower Lutetian	Marne de Donzacq
	Marne à *Xanthopsis dufouri*

The lower two units were shown to belong in the Lower Eocene, the former belonging to the *G. rex* Zone, the latter to the *G. formosa formosa* and *G. aragonensis* Zones.

The appearance of *Assilina exponens*, discocyclinids, small *Nummulites* and *Ganella neumannae* along with *G. gravelli, G. soldadoensis* and *G. soldadoensis angulosa* and *G. aragonensis-G. crater* transition was cited in the lower part of the *Couches de Nousse. G. bullbrooki, G. spinuloinflata, G. soldadoensis, G. soldadoensis angulosa,* and *G. higginsi* were reported from the upper part of the *Couches de Nousse,* alternating with massive occurrences of *A. exponens,* and indicate basal Lutetian according to Szöts (1964, p. 402). Elsewhere *G. crater, G. aragonensis* (very rare), *G. bullbrooki, G. soldadoensis angulosa* (very rare), and *G. linaperta* were reported from the upper part of the *Couches de Nousse.* An association of *G. bullbrooki, G. spinuloinflata, G. aragonensis-crater* transition, *G. broedermanni, G. turgida* and *G. trilocularis* was reported from the *Couches à Nummulites millecaput.*

Szöts (1964) considered that the lower part of the *Couches de Nousse* corresponds to the *Hantkenina aragonensis* Zone and that the planktonic fauna indicates the proximity of the Londinian-Lutetian boundary, although he does suggest (Szöts 1964, p. 405) that the *H. aragonensis* Zone may belong to the Lower Eocene. He observes that he was unable to discern the *Globorotalia palmerae* Zone, but this is not surprising. This species belongs to the genus *Pararotalia* and has a very limited distribution. It is more than likely that this stratigraphical interval is represented by the lower part of the *Couches de Nousse* and that it is represented by a part of the *H. aragonensis* Zone of Szöts.

The occurrence of *G. spinulosa, G. spinuloinflata, Truncorotaloides topilensis* and the appearance of *T. rohri* and *Catapsydrax echinatus* in marly-sandy beds intercalated with the *Couches à grandes Nummulites* is taken to signify the *G. lehneri* Zone and the top of the lower Lutetian.

The *P. mexicana* (vel *Orbulinoides beckmanni*) and *T. rohri* recte *T. pseudodubia,* teste Bandy 1964) Zones are recognized in the type Biarritzian (Szöts, 1964). The microfauna from the strata at Villa Marbella at Biarritz also indicates the presence of these two zones (Szöts 1964, p. 405). According to Szöts (1964, p. 405) the Lutetian/Bartonian boundary is denoted by the extinction of *Truncorotaloides.* He observes that in beds immediately above these at Villa Marbella (base of the *Marnes à Pentacrines*) the lower Bartonian fauna with *Globigerapsis semiinvoluta* and *Globorotalia* cf. *cocoaensis* appears, which excludes the possibility of intercalating an Auversian or Biarritzian Stage between the Lutetian and Bartonian. The 'Auversian' at Biarritz can be correlated with the upper Lutetian-Bartonian according to the author. While agreeing in general with Szöts' conclusions, the present author would point out that the Lutetian/Bartonian is not marked by the disappearance of *Truncorotaloides* (Szöts has in mind no doubt the disappearance of *T. rohri* (recte *T. pseudodubia,* teste Bandy 1964). However, *Globigerina collactea* Finlay has recently been shown to belong to *Truncorotaloides* by Jenkins (1965*d*); I believe it is a junior synonym of *Acarinina rotundimarginata* Subbotina and the same as *Globorotalia spinuloinflata* Bolli (*non* Bandy). This species ranges from upper Lower Eocene-Upper Eocene in New Zealand (Jenkins 1965*d*) and I have observed it in association with planktonic foraminifers of latest Eocene age in Denmark.

Schaub, and Veillon and Vigneaux (in Szöts 1964), questioned some of Szöts' conclusions, in particular the correlation of various units in the Adour Basin with the Lutetian in the Paris Basin. The latter two question in particular, the assignation of the *Couches de la Grotte de Brassempouy* (which Szöts had assigned to the *P. mexicana* Zone) to the Lutetian. This unit was said to correspond to the *Calcaire de Blaye* in the Bordeaux Basin and to belong to the Upper Eocene. The marine forms characteristic of the Lutetian of the Paris Basin were said to be restricted to levels *below* the transgression of the *Calcaires de Blaye.* If we may be allowed to extrapolate from this, it is possible that the Lutetian/Upper Eocene boundary may fall within the *Truncorotaloides pseudodubia* Zone. The assignation of the *Couches de la Grotte de Brassempouy* to the *P. mexicana* (=*O. beckmanni*) Zone, made by Szöts, could just as easily have been made to the *T. rohri* (=*T. pseudodubia*) Zone. The discovery of Cavelier and LeCalvez (1965) of faunas with Biarritzian affinities in upper Lutetian strata at Foulongues (Oise) has obviated the necessity for the Biarritzian Stage.

The relationship between the limits of the Middle Eocene (Lutetian) and planktonic foraminiferal zones. In the Paris Basin and surrounding areas the Lutetian is transgressive upon Ypresian and older beds. It is regressive towards the top and marked both at its base and top by significant disconformities. Thus a clear and unequivocal correlation of the boundaries of the Middle Eocene with the standard planktonic foraminiferal zones is virtually impossible.

Nummulites laevigatus is generally accepted as the characteristic larger foraminiferal guide fossil of at least the lower part of the Lutetian of the Paris Basin. But a certain stratigraphical interval is certainly missing in the Paris Basin, and is represented by the stratigraphical hiatus which separates the oldest Lutetian strata from earlier Eocene strata.

In the Sirte Basin of Libya faunas assignable to the *Hantkenina aragonensis* and *Globigerapsis kugleri* Zone were found to overlie limestones assigned to the *Nummulites burdigalensis-partschi* Zone. The upper part of this zone was denominated the *N. planulatus* Subzone and the entire zone was placed in the Lower Eocene (Ypresian). The *H. aragonensis-G. kugleri* and lower part of the *G. lehneri* Zone were correlated towards the margin of the basin with the *Nummulites beaumonti-discorbinus* Zone. Above this the upper part of the *G. lehneri* Zone, the *Porticulasphaera mexicana* (vel *Orbulinoides beckmanni* *) Zone and the *Truncorotaloides rohri* (vel *T. pseudodubia*) Zone were correlated with the *Nummulites gizehensis* Zone, which was in turn subdivided into a

*See footnote on p. 720.

(lower) *N. lorioli* Subzone and an (upper) *N. praefabianii* Subzone.

Direct correlation between Libya and the Paris Basin is impossible because of the difference in the species of larger Foraminifera, but if a relationship could be demonstrated between the *N. laevigatus* faunas of the Paris Basin and the *N. gizehensis* faunas of Libya, it would suggest that the time-stratigraphical interval represented by the *N. beaumonti-discorbinus* Zone (=*H. aragonensis-G. kugleri* Zone) is missing in the Paris Basin. By any criteria it is unlikely that the five planktonic foraminiferal zones commonly assigned to the Middle Eocene (Bolli 1957) correspond to the limits of the Lutetian Stage in the Paris Basin.

In stratigraphical sections containing planktonic Foraminifera the appearance of *Hantkenina* is tentatively taken as a convenient horizon for the base of the Middle Eocene, and the extinction of keeled globorotaliids (*G. spinuloinflata*) and appearance of *Globigerapsis semi-involuta* (vel *G. mexicana* *) is used to denote the top of the Middle Eocene.

Upper Eocene

Two stage names are commonly applied to the Upper Eocene by stratigraphers – The Bartonian (Mayer-Eymar 1857) and the Priabonian (Munier-Chalmas and de Lapparent 1893). These stages have found widespread acceptance in the areas in which they were originally defined – the former in northern Europe, the latter in the Mediterranean region. The superiority of one over the other in terms of worldwide chronostratigraphical applicability has not been demonstrated, inasmuch as each is developed in a restricted, somewhat provincial region.

A thorough analysis of the vicissitudes to which the term Bartonian has been subjected has recently been given by Feuillée (1964) and summarized by Blondeau *et al.* (1965).

Feuillée (1964, p. 37) claims there are fifty different interpretations of Upper Eocene and about eight or nine of 'Bartonian'. Mayer-Eymar (1858) erected the Bartonian Stage for the strata in the vicinity of Paris which he believed showed a close affinity to those in the Hampshire Basin of England. Although the stage was based and defined primarily on studies by Mayer-Eymar of strata cropping out in the Paris Basin, he took the name of his stage and its type section from the vicinity of Barton, England.

In 1857 Mayer-Eymar (1858) subdivided the Middle and Upper Eocene into several stages. It will be recalled from the discussion above that he restricted d'Orbigny's (1852) Parisian to represent only the lower part, the *Calcaire grossier.* The name Bartonian was given to the so-called *Sables moyens* above that sequence of beds in the Paris Basin between the *Calcaire grossier* and the *Gypse de Montmartre* (which Mayer-Eymar included in his Ligurian Stage). As it was originally defined the Bartonian Stage of Mayer-Eymar included (from the base):

1. *Sables d'Auvers, Beauchamp* and *Ezanville* (subsequently used to define the Auversian Stage by Dollfus, 1880), the upper Bracklesham Beds of Hampshire, and the Sands of Lede in Belgium.
2. *Sables de Mortefontaine, Calcaire de Saint-Ouen, Sables de Monceau-Argenteuil;* Barton Clay of England with *Nummulites prestwichianus;* Sands of Wemmel in Belgium with *N. wemmelensis* (considered to be a junior synonym of *N. prestwichianus*).
3. *Gypse inférieur parisien* (3rd and 4th massive beds) with intercalations of the *Marne à Pholadomya ludensis; Marne du Vouast* in Vexin (*Cerithium concavum* level); Sands of Long Mead End, near Barton; and Headon Hill on the Isle of Wight.

This latter unit comprises the Ludian *s.s.* of Munier-Chalmas and de Lapparent (1893, p. 477). In its original definition the Bartonian begins with strata containing *Potamides perditus* above the Lutetian in the lower Loire region; in the region of Gironde it commences with beds containing *Nummulites variolarius* and *Ostrea cucullaris;* at Biarritz its lower part is equivalent to beds with *Nummulites striatus* (at the southern end of the Biarritz section), and then beds with *N. fabianii* from the Côte des Basques (Villa Marbella up to Cachaou). The equivalency of the lower part of the Bartonian with the Auversian is shown by the occurrence of *N. striatus* (Denizot 1957, *Lex. Strat. Internat.*, vol. *1*, *fasc. 4a vii*, p. 24).

Mayer-Eymar (1869) subsequently radically altered his original concept of the Bartonian by excluding from it (without explanatory discussion) the upper beds listed under 3 (above), i.e. the *Marnes à Pholadomya ludensis.* The Bartonian thus reduced included: (1) *Sables d'Auvers* (Seine et Oise); (2) *Couches de Mortefontaine* (Oise). It was for these marls containing *Pholadomya ludensis* (which represents a brief transgression in the Paris Basin, associated with the deposition of gypsiferous beds) that Munier-Chalmas and de Lapparent (1893, p. 477) subsequently created the stage name Ludian which they considered equivalent to the Priabonian in the Mediterranean area, while stating that the Bartonian was little more than a subdivision of the Lutetian, so close were the faunal affinities between the two stages (Munier-Chalmas and de Lapparent 1893, p. 475).

In 1880 Dollfus erected a number of terms (Auversian, Andoenian, Mortefontian, Ermenonvillian, Argentian, Montimartrian) for each of the rock-stratigraphical units of the *Sables moyens,* without specifying their connotation or hierarchical rank. Dollfus, unable to discern a stratigraphical subdivision of the Barton Beds in the Hampshire Basin in England, advocated the usage of his own terms. However, it is clear that these terms, as defined, are little more than facies terms applicable on a local, not regional, level.

Haug (1905, 1911, p. 1436) equated the term Auversian (Dollfus 1880) with the *Sables moyens* of the Paris Basin, believing it to be equivalent to and a senior synonym of the Ledian (Mourlon 1887) which Leriche (1903) had also used for the *Sables moyens.* In 1901 Dollfus expanded his concept of Auversian and created the term Marinesian for the beds above: *Sables de Mortefontaine, Sables* of *Marines* and *Cresnes,* and the *Marnes à Pholadomya ludensis.*

The Ludian was subsequently suppressed by de Lapparent himself (1906) and the original definition of the Bartonian (Mayer-Eymar 1857, 1858) adopted.

Boussac (1907) however, interpreted the Bartonian in a more restrained sense (*Calcaire de St. Ouen* and *Sables de Cresnes*) and added the Auversian (below) and Ludian (above). The latter was based on the individuality of the fauna in the *Marnes à Pholadomya ludensis,* despite the fact that the author of the stage (de Lapparent) had himself suppressed the term the year before.

Although most continental geologists subdivided the *Sables moyens* into a lower and upper part, Leriche (1903) had not insisted on the retention of the Auversian against the opposition

*The holotype of *P. mexicana* is conspecific with *Globigerapsis mexicana.* Bolli's (1957) *P. mexicana* has been renamed *P. beckmanni* by Saito (1963). Saito and Blow are currently assigning *P. beckmanni* to *Orbulinoides beckmanni. Globigerapsis semi-involuta* thus becomes a junior synonym of *Globigerapsis mexicana* (pers. comm. Saito and Blow 1967).

of Haug and Dollfus. However Abrard (1925) demonstrated the lithological homogeneity of the *Sables moyens,* and that they contain *Nummulites variolarius* and no evidence of *N. prestwichianus-wemmelensis* (typical Bartonian species). Leriche (1925) thus proposed the abandonment of the term Auversian in favour of the Ledian and restricted the Bartonian to the beds above.

Therefore:

Marnes à Pholadomya ludensis = *Sables de Wemmel* = Bartonian (=*N. wemmelensis*)

Marnes de Beauchamp and *Cresnes* = *Sables de Lède* = Ledian (with *N. variolarius*)

This usage of Bartonian is identical to that which Mayer-Eymar (1869) rejected in altering the original concept of his stage.

Abrard (1925, 1933) accepted the Ledian for the lower and middle parts of *Sables moyens,* but applied the term Wemmelian (Rutot 1878) to the *Marnes à Pholadomya ludensis* and the associated gypsiferous beds. The term Bartonian *s.l.* was used for the whole sequence, a return essentially to the original definition of Mayer-Eymar (1857, 1858).

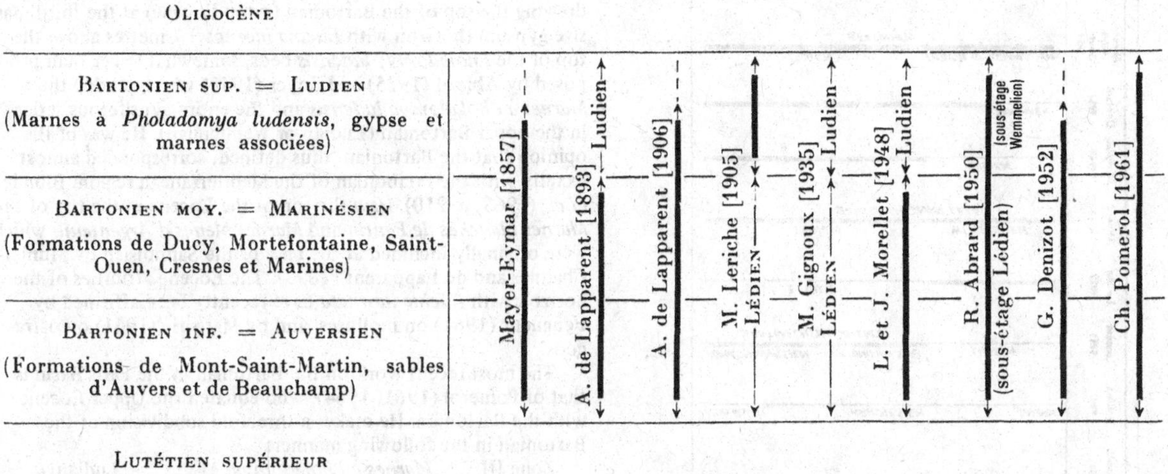
OLIGOCÈNE
BARTONIEN SUP. = LUDIEN
(Marnes à *Pholadomya ludensis*, gypse et marnes associées)
BARTONIEN MOY. = MARINÉSIEN
(Formations de Ducy, Mortefontaine, Saint-Ouen, Cresnes et Marines)
BARTONIEN INF. = AUVERSIEN
(Formations de Mont-Saint-Martin, sables d'Auvers et de Beauchamp)
LUTÉTIEN SUPÉRIEUR
Mayer-Eymar [1857]
de Lapparent [1893]
Ludien
A. de Lapparent [1906]
M. Leriche [1905]
LÉDIEN
M. Gignoux [1935]
LÉDIEN
Ludien
L. et J. Morellet [1948]
Ludien
R. Abrard [1950]
(sous-étage Lédien)
(sous-étage Wemmelien)
G. Denizot [1952]
Ch. Pomerol [1961]

Table 52.15. Principal determinations of the term Bartonian (straight bars) compared with subdivision of Bartonian proposed by Pomerol (1961) (Blondeau *et al.* 1965, p. 208, tab. 4).

	Bassin de Paris	Belgique		Angleterre
Ludien	Marnes blanches de Pantin Marnes bleues d'Argenteuil Haute masse du gypse Marnes et gypse Marnes à *Pholadomya ludensis*	Sables d'Assche	Niveau de Neerrepen Niveau de Grimmertingen	Bembridge marls Bembridge limestone Osborne beds Middle et Upper Headon beds Brockenhurst beds
Marinésien	Sables de Marines à *Corbula costata* Sables de Cresnes / Calcaire de Saint-Ouen Sables de Mortefontaine Calcaire de Ducy Sables d'Ezanville	Argile d'Assche		Lower Headon beds Barton beds
Auversien	Sables d'Auvers-Beauchamp-Le Guépelle Formation de Mont-Saint-Martin			
Lutétien supérieur	Falun de Foulangues	Sables de Wemmel		Upper Brackelsham beds

Table 52.16. Correlation of the formations of the Bartonian in the Paris Basin, Belgium and England (Blondeau *et al.* 1965, p. 211, tab. 5).

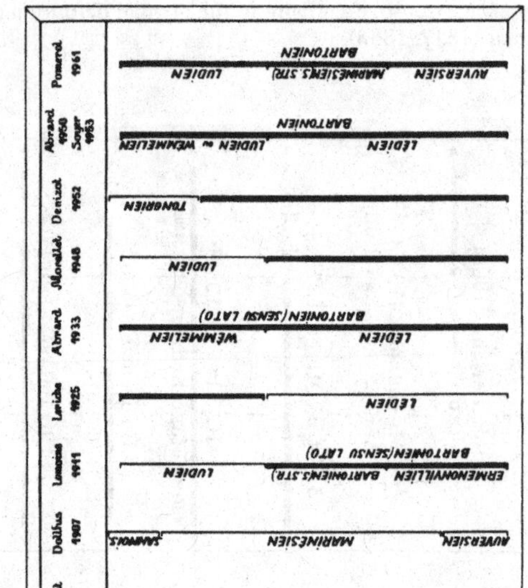
BASSIN DE PARIS
GYPSE
MARNES A PHOLADOMYA LUDENSIS
SABLES DE CRESNES, MARINES
CALCAIRE DE ST OUEN
SABLES DE MORTEFONTAINE
SABLES DE BEAUCHAMP

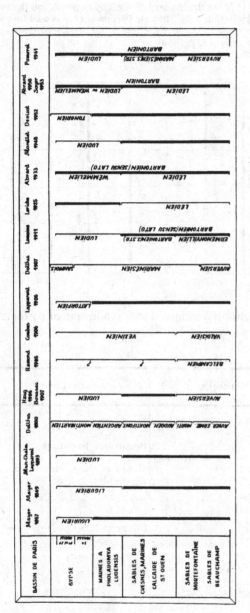

Table 52.17. Various interpretations of the Bartonian (modified from Feuillée 1955; Feuillée 1964, p. 42).

As a result of detailed studies of the Upper Eocene of the Paris Basin L. and T. Morellet (1934, 1940, 1948) have adopted the opinion of Boussac (1907) of the equivalence of the Barton Clay and the *Sables de Cresnes* and *Marines* and the exclusion of the *Marnes à Pholadomya ludensis* from the Bartonian. These authors have thus returned to the modified definition of the Bartonian (Mayer-Eymar 1869).

Denizot (1952 and 1957, *Lex. Strat. Internat.* pp. 23-6), unable to recognize any significant faunal break within the Upper Eocene adopted the original definition of the Bartonian (Mayer-Eymar 1857, 1858 = de Lapparent 1906), and suggested drawing the top of the Bartonian (= top Eocene) at the third massive gypsum (horizon with *Lucina inornata*) 7 metres above the top of the *Pholadomya ludensis* beds, somewhat lower than proposed by Abrard (1925) and Soyer (1943) who included the *Marnes à Pholadomya ludensis* and the entire gypsiferous series in the upper Bartonian (Ludian or Wemmelian). He was of the opinion that the Bartonian, thus defined, corresponded almost exactly with the Priabonian of the Mediterranean region. Blondeau *et al.* (1965, p. 210) draw the top of the Eocene at the top of the *Marnes blanches de Pantin* and *Marnes bleues d'Argenteuil,* which were originally included at the base of the Sannoisian by Munier-Chalmas and de Lapparent (1893). The Eocene affinities of the horizon with *Lucina inornata* have recently been affirmed by Feugueur (1961) on molluscs, and by Margerie (1961) on ostracods.

The most recent work on the Bartonian in the Paris Basin is that of Pomerol (1961, 1964), who equated the Upper Eocene with the Bartonian. He makes a threefold subdivision of the Bartonian in the following manner:

Zone III	*Marnes à Pholadomya*	-	Ludian
Zone II	*Sables de Cresnes* et de *Marines* et *Calcaire de Saint-Ouen*	-	Marinesian (Dollfus, *s.s.*)
Zone I	(*b*) *Sables d'Auvers, et de Mortefontaine*	]	Auversian (Dollfus, *s.l.*)
	(*a*) *Sables de Mont Saint-Martin*		

Pomerol (1961, 1964) maintained that the term Ledian was unsatisfactory in the Paris Basin because correlations between the *Sables moyens* and the Sands of Lede were inexact.

A diagrammatic illustration of the vicissitudes in the interpretations of the Bartonian in the Paris Basin was presented by Yanshin (1953, table 6, p. 221) and Feuillée (1955). A modified version presented by Feuillée (1964, p. 42) and Yanshin's table 6 are given in Tables 52.17 and 52.18 respectively.

Abrard (1925) has pointed out the strong faunal affinites between the *Sables moyens* and the Lutetian (over 50 per cent of molluscan species is common). Blondeau *et al.* (1965, p. 209) give the following figures for other faunal groups; Foraminifera, 9/10 of Auversian species occur in the Lutetian; ostracods, 3/4; charophyta, about half. Curry (1966, p. 466) has presented additional data on the relationship of some Eocene mollusc faunas of the Anglo-Paris-Belgian Basin. The following figures are significant: Lutetian molluscan species 1780; Auversian 1029; in common 462. The mammalian fauna at Guepelle bears a closer relationship to the Lutetian than to the middle and upper Bartonian. Blondeau *et al.* admit that various new faunal elements appear in the Auversian, but that it is not possible to recognize a fine stratigraphical subdivision within the Auversian. Indeed these authors recognize three substages of the Bartonian: the Auversian, Marinesian, Ludian.

Pomerol (1964) has given a summary of the palaeontological sedimentological and palaeogeographical characteristics of the Bartonian of the Paris Basin. His findings may be summarized in the following manner:

(1) The Belgian and Paris Basins were separated in the Upper Eocene by the Dôme de l' Artois (cf. Feugueur, 1961, p. 168, who points out that this separation existed as early as in the Ypresian).

(2) The Ledian has its strongest faunal affinities with the Lutetian (*Orbitolites complanatus-Ditrupa strangulata*). *Nummulites variolarius* occurs commonly in the Ledian and the Auversian. It

Таблица сопоставления различных стратиграфических расчленений верхнего эоцена Парижского бассейна

Разрез верхнеэоценовых отложений в районе непосредственно севернее Парижа	Гипсы Монмартра, переслаивающие их мергели и замещающий их пресноводный известняк Шампиньи	Мергели с Pholadomya ludensis Desh. и замещающие их песчано-глинистые отложения Вексена	Пресноводные известняки Буа-дю-Мюло и Нуази-ле-Сек	Пески Марине, Кресне, Шара и Рюэля с морской фауной	Пресноводный известняк Сент-Оуэн	Морские пески зоны Мортфонтен	Пресноводный известняк Дюци	Морские пески Бошан: Зона Эзанвиль, Зона Бошан, Зона Эрменонвиль, Зона Гепель, Зона Овер, Зона Мон-Сен-Мартен
Майер-Эймар, 1857	Лигурийский	Бартонский						
Майер-Эймар, 1868	Лигурийский		Бартонский					
Дольфюс, 1880	Монтимартский (олигоцен)		Одоенский			Морфонтский		Эрменонвильский; Оверский
Мюнье-Шальма и Лаппаран, 1893—1900	Людский		Бартонский (средний эоцен)					
Лериш, начало 1905	Людский (олигоцен)		Асшийский и веммельский			Ледский		
Ог, 1905 и 1911; Буссак, 1906 и 1912 а	Людский		Бартонский			Оверский		
Лериш, конец 1905	Людский (олигоцен)		Бартонский			Ледский		
Рамон и Комбе, 1906	Людский		Вексенский			Валуазский		
Дольфюс, 1907—1909	Маринезский					Оверский		
Буссак, 1907	Приабонский: Людские слои		Приабонский: Бартонские слои			Оверский (средний эоцен)		
Лемуан, 1911	Людский		Бартонский: Бартонский подъярус			Бартонский: Оверский подъярус		
Английские геологи, 1906—1920	Бартонский					Оверский		
Абрар, 1925—1942	Веммельский		Ледский					
Лериш, 1925—1939	Бартонский		Ледский (средний эоцен)					
Братья Морелле, до 1939	Людский		Бартонский					
Братья Морелле, после 1939 и Абрар, 1948	Бартонский							

Table 52.18. Various stratigraphical subdivisions of the Upper Eocene of the Paris Basin (Yanshin 1953, tab. 6).

was not thought to occur in the Lutetian, but Curry (1961), Le-Calvez (1961) and observations of Blondeau and Curry (1963) have recently recorded *N. variolarius* from about a dozen localities in the Paris Basin, not only in the upper Lutetian, but also in the lower Lutetian (Zone II) where it occurs with *N. laevigatus*, as it does in Bed VIII of the Bracklesham beds of Whitecliffe Bay. As a result Pomerol (1961) had earlier suggested a correlation between the Ledian and the upper Lutetian.

(3) Following Curry (1962), who concluded that the fauna of the Sands of Wemmel was similar to that of the Upper Bracklesham Beds of England, and indicated an age somewhat older than lower Bartonian, if not upper Lutetian, Pomerol concluded that both the Ledian and Wemmelian would have to be lowered somewhat in the stratigraphical scale, leaving room, perhaps, for the lower Tongrian in the Upper Eocene = Ludian (see Table 52.15).

(4) The weight of evidence favours correlating the Upper Bracklesham Beds with the Lutetian (Beds XVI and XVII are the only units which contains *Orbitolites complanatus* and *Rotalia trochidiformis*, the former of which is a diagnostic Lutetian form) rather than the Auversian.

(5) The Bruxellian/Ledian boundary in Belgium is difficult to trace in some places, but where observable the uppermost Bruxellian displays Ledian characteristics in term of morphoscopy, granulometry, heavy minerals, and foraminiferal and ostracod faunas. The basal conglomerate does not always represent the beginning of the Ledian but represents a detrital continental deposit determined by climatic factors.

(6) The anomaly caused by the occurrence of *N. variolarius* and *N. laevigatus* in Bed VIII in the upper part of the Middle Bracklesham Beds (which was the top of the Lutetian according to White), and *N. variolarius* and a macrofauna which Wrigley (1940) considered definitely Lutetian in Bed IX in the basal part of the Upper Bracklesham Beds (considered Auversian by most English geologists today) is only apparent, because in France the two nummulites occur together in the Lutetian (Zone II); thus the assignation of Bed IX to the upper Lutetian is quite normal.

In Belgium the Wemmelian Stage (Vincent and Rutot 1878) was created for the *Sables de Wemmel*, the *Argile d'Assche* and the *Sables chamois*, which were separated from the Laekenian of Dumont, placed in the Upper Eocene and correlated with the Bartonian. The Upper Eocene of Belgium was subsequently subdivided into a lower Wemmelian and upper Asschian Stage by Rutot and Van den Broeck (1883). (The Asschian had been created by Rutot in 1882.) The introduction of the Ledian by Mourlon (1881) between the Laekenian and Wemmelian led to problems of identification of stratigraphical units and correlation (see discussion by de Heinzelin and Glibert, in *Lex. Strat. Internat.* 1957, vol. *1*, fasc. *4a-vii*, p. 204). It would appear that the terms Wemmelian and Asschian are but facies units of the Upper Eocene; the lower part of the Asschian is inseparable from the Wemmelian, and the main part of the Assche Formation is considered equivalent to the Lower Tongrian by Batjes (1958). The Wemmelian is characterized by the occurrence of *Nummulites wemmelensis*, which replaced *N. variolarius* in the Ledian below. The general aspect of the microfauna shows a strong affinity with the Ledian (Kaasschieter 1961; Keij 1957). On the basis of the marked similarity between the molluscan fauna of the Wemmelian, the Lutetian and the Auversian, Blondeau *et al.* (1965) place the Wemmelian in the upper part of the Lutetian (= 'Biarritzian').

It would seem to the present author that on the basis of the discussion above a threefold chronostratigraphical subdivision of the Upper Eocene as proposed by some workers is presumptive, optimistic, and unjustified.

The Priabonian Stage was erected by Munier-Chalmas and de Lapparent (1893, p. 479) as a Mediterranean equivalent of the Ludian (Upper Eocene) of the Paris Basin. The recent study by Roveda (1962) has allowed a subdivision of the Priabonian into a lower-middle part characterized by abundant *Nummulites fabianii* and a middle-upper part characterized by a rich association of *N. chavennesi, N. stellatus, Operculina alpina, Pellatispira madaraszi, Spiroclypeus granulosus* and abundant discocyclinids. As it is generally used today, the Priabonian is interpreted as approximately time-correlative with the Bartonian Stage.

The review of Banner and Eames (1966, pp. 163-5) has shown that:

(1) The nummulitic zonation of the Upper Eocene of the Soviet Union (Yanshin *et al.* 1960) can be correlated quite satisfactorily with the planktonic zones recognized by themselves.

(2) Although the exact equivalence of the *Cribrohantkenina danvillensis* and *Globigerina turritilina turritilina* (= vel *G. gortanii*) Zones to the type Bartonian (*s.s.*) has not been demonstrated, the total equivalence of the three Upper Eocene planktonic foraminiferal zones to the type Priabonian seems very probable (see Fig. 52.1).

The authors concluded that the best usage seems to be to regard Ledian and Bartonian as comprising the early and late Priabonian (but see Pomerol 1964, who considers the Ledian to be equivalent to the upper part of the Lutetian).

Blow and Banner, in Eames *et al.* (1962) have modified Bolli's (1957) *G. cocoaensis* (vel *G. cerroazulensis*) Zone and proposed two new Upper Eocene planktonic foraminiferal zones. The three Upper Eocene planktonic zones and the characteristics which distinguish them are shown below:

	Planktonic Zones	Characteristics	Larger Foraminifera Associations
Lower Oligocene	*Globigerina oligocaenica* (=*G. sellii* = *G. sakitoensis*) Zone		*Nummulites vascus-intermedius*
Upper Eocene	*Globigerina turritilina turritilina* (=*G. gortanii*) Zone	1. No *G. centralis, cerro-azulensis, Hantkenina, Cribrohantkenina* 2. *G. tripartita* common *G. pseudoampliapertura*	*Nummulites fabianii* *Nummulites lormoensis* *Pellatispira madaraszi* *Discocyclina* spp. *Aktinocyclina* spp. *Chapmanina gassinensis*
	Cribrohantkenina danvillensis Zone	1. presence of *Hantkenina* and *Cribrohantkenina* 2. absence of *Globigerapsis, Globigerinatheka*	
	Globigerapsis semiinvoluta Zone	1. *G. semiinvoluta* limited to this interval 2. *G. lindiensis* limited to this interval	

The boundary of the Middle Eocene/Upper Eocene remains problematical. By placing the Ledian (with *N. variolarius* and *Orbitolites complanatus*) and the Wemmelian in the upper Lutetian, Blondeau *et al.* (1965, table 3) achieve a unified scheme for the Middle Eocene which allows a distinct separation from the Bartonian. The Middle Eocene/Upper Eocene boundary is thus equated with the upper limit of *Nummulites perforatus* and *Orbitolites* and *Alveolina.* On the other hand, Banner and Eames (1966) consider the Ledian to be correlative with the Auversian, and place it in the Upper Eocene, and believe that the Upper Eocene/Middle Eocene boundary of the Far East may lie within the Ledian of Europe. The questions of chronostratigraphical boundaries are a matter of convention. Since they are most useful when based on significant changes in phylogenetic series of fossils which can be recognized on a regional-mondial basis, the interpretation of Blondeau *et al.* (1965) is tentatively accepted in this paper.

Definitive proof has not been presented that these 'substages' have the characteristic of chronostratigraphical units. We are dealing here with rapid facies changes which occurred in a regressive marginal sea. The attempt to recognize isochronous surfaces in these lithological units is rendered all the more difficult when we realize that it is based upon overlapping incursions of nummulitic species, which usually occupy but small parts of the total stratigraphical record. If, following over a century of investigations on the subdivision and correlation of the various lithostratigraphical units assigned to the Bartonian, satisfactory correlation has yet to be achieved between the Paris, Belgian and Hampshire Basins, then it seems rather presumptuous to recognize a threefold chronostratigraphical subdivision in the Upper Eocene. At best the threefold subdivision thus recognized has only local application and cannot be applied on a regional or world wide scale.

It would seem more advisable to the present author to recognize a single, broad, well-defined chronostratigraphical unit for the Upper Eocene, and to subdivide the sequence into recognizable biostratigraphical zones which may be expanded into chronozones if and when their regional correspondence to the limits of the stage can be demonstrated (Hedberg 1965, p. 561).

Eocene/Oligocene boundary. Blondeau *et al.* (1965) have correlated the *Argile d'Asse* with the Barton Beds and lower Headon Beds in England, and with the Auversian-Marinesian sequence in the Paris Basin (see Table 52.17). The *Sables d'Asse* were correlated with the Lower Tongrian (Sands of Grimmertingen and Sands of Neerepen) (as Drooger 1964, based on the works of Batjes 1958 and Kaasschieter 1961, had also done) with the Brockenhurst Beds, and the beds above up to the base of the Hamstead Beds, the Lattorfian of Germany and the Ludian of the Paris Basin. The Ludian was shown to include (from the base): *Marnes à Pholadomya, marnes et gypse, haute masse du gypse, marnes bleues d'Argenteuil, marnes blanches de Pantin.* Whereas the Eocene/Oligocene boundary is placed by the majority of French workers within the gypsum mass, the inclusion of the two stratigraphical units above the main gypsum mass heralds a return to the original definition of the Stampian by d'Orbigny (1852), in which the base of the Stampian was defined by the *Argile verte de Romainville.* The two units above the main gypsum mass are, along with the strata *Argile verte de Romainville* and the *Calcaire de Sannois,* generally placed in the Sannoisian, but this is placed in the lower Stampian by Blondeau *et al.* (1965), the Sannoisian being considered but a lower facies of the Stampian. It is of interest to note that the *Marnes bleues d'Argenteuil* in the Paris Basin, and the Bembridge Marls in England lie above the main gypsum mass and the Bembridge Limestone respectively, both of which contain the so-called mammalian fauna of Montmartre.

The stratigraphical correlation by Blondeau *et al.* (1965) of the lower Stampian with the Upper Tongrian *Niveau de Hoogbutsel* and the lower Rupelian (*Sables de Berg*), and of the Upper Stampian with the Upper Rupelian (*Argile de Boom*), is based upon several lines of reasoning which are summarized by the authors (Blondeau *et al.* (1965) pp. 214, 215). The variation in molluscan and foraminiferal and ostracod faunas in the various stratigraphical units of the uppermost Eocene/Oligocene is strongly connected with rapid facies changes. In the Belgian Basin, as well as the Paris Basin and the Isle of Wight, the Stampian = Upper Tongrian-Rupelian begins with a brackish water facies, which passes gradually upwards into more marine facies. Thus it is quite apparent that we are dealing with a single sedimentary cycle in the Oligocene, representing a relatively short time interval. The use of several stage names in this area where definite stratigraphical correlations are difficult to make would appear unnecessary, and the extension of the Rupelian = Stampian (=Upper Tongrian of Dumont 1849) to include Lower Oligocene, would seem entirely justified to the present author on the basis of regional palaeogeography and stratigraphy in northern Europe (see also Batjes 1958). It would obviate the necessity of creating a new stage for the relatively short time-stratigraphical interval which exists between top Eocene and base Middle Oligocene (i.e. 'Lattorfian' of Mayer-Eymar 1893, recently shown by Krutzsch and Lotsch 1957, to be of Upper Eocene age).

In terms of planktonic foraminiferal faunas the extinction of the genus *Hantkenina* has long been considered one of the definitive criteria for determining the Eocene/Oligocene boundary. Blow and Banner in Eames *et al.* (1962) described the *Globigerina turritilina turritilina* (vel *G. gortanii s.s.)* Zone from East Africa, which was shown to contain a Late Eocene fauna above the extinction of *Hantkenina* and *Cribrohantkenina.* These authors now suggest drawing the Eocene/Oligocene boundary at the base of the *Globigerina tapuriensis* Zone (P18) which lies directly above the *G. gortanii* Zone (Blow, pers. comm.).

Palaeogene stages in the S.W. Soviet Union

The resolutions of the 5th Conference of the Permanent Stratigraphical Commission of the Palaeogene of the U.S.S.R. which were held 15-19 May 1962, were presented and published in *Sovetskaya Geologiya,* 1963, no. 4, and by Korobkov and Mironova at the Palaeogene Colloquium in Bordeaux in 1962.

The Palaeogene Commission of the Interdepartmental Stratigraphical Committee of the Soviet Union has recently recommended a new stage subdivision of the Palaeocene/Eocene of the Soviet Union (*Sov. Geologiya,* no. 4, 1963). The stratotype section was chosen in the region of Bakhchisaray, in the Crimea; a parastratotype section was chosen along the Kuban River, northern Caucasus. The committee has proposed that this section serve as a new type section for the Palaeogene of Europe. The various stage names accepted by the Commission are shown in Tables 52.10, and 52.12 along with suggested correlations with other chronostratigraphical and biostratigraphical units.

Criticism of this scheme has been offered by Leonov (1963) and by Leonov, Alimarina and Naidin (1965). Leonov 1963 makes several important points concerning the decisions of the Committee.

(1) According to data presented by the Commission only the Lower Eocene (Bakhchisarian Stage) is documented by both planktonic and nummulitic zones. In the Middle Eocene the zonation presented is based on nummulites alone, and in the Upper Eocene by smaller foraminifers (inlcuding planktonics). No zonal subdivision was given for the Palaeocene (Inkermanian and Kachaian Stages).

(2) The Bakhchisaray section is not complete. Two important hiatuses occur in the course of this section: one between lower (Inkermanian = Montian) and middle (Kachaian = Thanetian) Palaeocene and one between the Palaeocene and Lower Eocene.

(3) Strict correlation between the stratotype section (Crimea) and the parastratotype section (W. Caucasus) is not completely satisfactory.

(*a*) The Commission correlates the upper part of the Dano-Montian limestones (Crimea) with the upper part of the Elburgan Formation (N. Caucasus). This correlation is questionable. It is possible that the Inkermanian Stage of the southwest Crimea corresponds to the lower, rather than to the upper part of the Elburgan Formation. The upper part of the Elburgan Formation may correspond to a hiatus in the Bakhchisaray region.

(*b*) The Bakhchisarian Stage in the Crimea corresponds to two zones: *Globorotalia subbotinae* and *G. aequa* according to the Committee. On the basis of their micropalaeontological characters these two zones correspond fully to one zone: *G. subbotinae* Zone in the Kuban River region (N. Caucasus) (*G. marginodentata–G. subbotinae* Subzone). These beds in the Crimea can thus be correlated with the Novogeorgi beds in the Kuban River section. At the same time the Thanetian beds (Kachaian Stage) in the Crimea correspond solely to the *Acarinina tadjikistanensis* Zone in the scheme presented by the Palaeogene commission. Equivalents of the *A. subsphaerica, A. acarinata* and *G. aequa* Zones cannot be found in the Crimean region. It is quite possible and more than likely true, that a part of the upper Palaeocene of the Kuban River section (Labin Group of Leonov and Alimarina) corresponds to a hiatus in the Bakhchisaray section.

(c) In the Bakhchisaray section the Bodrakian Stage (Upper Eocene) in the Palaeogene Commission's scheme, begins with a Zone of small *Nummulites;* above follows the *Acarinina rotund–imarginata* Zone. In the Kuban River section this same stage begins with the *Acarinina rotundimarginata* Zone. The Zone of small *Nummulites* here corresponds to the *A. crassaformis* Zone, which belongs to the Simferopolian Stage (Middle Eocene).

Leonov (1963, p. 35) concludes by suggesting that the scheme proposed by the Palaeogene Commission is in need of some revision, and that further work on the regional stratigraphy of the Palaeogene strata of the Soviet Union is needed.

Some further difficulties in the Soviet zonation are seen in a comparison of different versions of the standard section of the Palaeogene Commission. The Abazin Formation is generally included in the Upper Palaeocene (*Acarinina subsphaerica* Zone) (Yanshin *et al.* 1960; Leonov and Alimarina 1962). In the original presentation of the Palaeogene Commission of the U.S.S.R. (*Sovetskaya Geologiya,* vol. 4, 1963, and English translation in *Internat. Geol. Rev.* vol. 7, no. 4, 1965) the Abazin Formation is correlated with a lower *A. acarinata* Zone and an upper *G. aequa* Zone. The Palaeocene/Eocene boundary was drawn at the base of the *A. acarinata* Zone, thus including all the Abazin in the Lower Eocene (Bakhchisarian Stage) (cf. Leonov and Alimarina 1962). In the version presented at the Palaeogene Colloquium at Bordeaux (Korobkov and Mironova 1964) the Abazin Formation was correlated with the *A. acarinata* Zone and included in the Palaeocene (Kachinian Stage).

In a communication which I received in early April 1967, from Dr V. A. Krasheninnikov of the Geological Institute of the Academy of Sciences of the Soviet Union in Moscow, the following information pertinent to the anomalous placement of the Palaeocene/Eocene boundary by the Soviet Palaeogene Commission is presented (present author's free translation).

'In Palaeogene sections of the Central Pre-Caucasus it is quite difficult to determine the Palaeocene/Lower Eocene boundary owing to the rare occurrences of [suitable] lithological section... The upper part of the Palaeocene (Goryachi Kliuch) and, apparently, the lower part of the Abazin suite consist of sandy 'aleurolites' (siltsones) and non-carbonaceous shales with agglutinated Foraminifera and radiolarians. Pelagic foraminifers are practically absent. The Lower Eocene (Novogeorgi suite, *Globorotalia subbotinae* Zone) is [composed of] carbonaceous siltsones and marls with abundant planktonic foraminifers.

'As a result two opinions have arisen regarding the Palaeocene/Eocene boundary. The first considers the top of the Palaeocene to be best placed at the top of the Abazin suite, where lithological sections are rarely developed. This concept is popular among geologists because it is practical for mapping purposes. It must not be forgotten that the benthonic foraminifers of the Goryachi Kliuch are almost identical to those of the Abazin.

'The second opinion, popular among many micropalaeontologists, is that the Palaeocene/Eocene boundary is [to be drawn] within the Abazin suite. In the upper beds of this (Abazin) suite individual specimens of *G. aequa, G. wilcoxensis, A. pseudotopilensis* occur, which indicates the development of Lower Eocene... The general poverty of the microfauna prevents a unanimous decision [on the determination of the boundary]. *The main boundary (Palaeocene/Eocene) at the base of the Abazin Suite in Sov. Geol., no. 4, 1963 was a mistake, committed either at the printers or in the preparation of the chart.'* The boundary as drawn in the charts presented by Korobhov and Mironova in 1962 at the Palaeogene Colloquium in Bordeaux is thus to be considered as definitive. (Dr Krasheninnikov observes that he does not know a geologist or a micropalaeontologist in the Soviet Union who can resolve this dilemma.)

Dr Krasheninnikov further points out in his communication that *Pseudohastigerina wilcoxensis (eocenica, voluta)* appears at the base of the *G. subbotinae* Zone. It may also occur in the upper part of the Abazin Suite but the pelagics are rare here. Examination of samples which I collected in 1962 from the Abazin and Novogeorgi Formations in the N. Caucasus fully corroborate Dr Krasheninnikov's discussion.

From the data presented as characteristic of the *G. aequa* Zone in the Soviet zonation, and from an examination of material from the upper part of the Abazin Formation in my collections, it would appear that the Abazin Formation is near the Palaeocene/Eocene boundary. The occurrence of *G. velascoensis* (Shutskaya, Schwemberger and Khasina 1965), *G. acuta, G. aequa (Internat. Geol. Rev.* vol. 7, no. 4) and *G. marginodentata* (my own collection) in the *G. aequa* Zone would seem to indicate that this zone is approximately equivalent (in part at least) to the *G. aequa* Zone of Luterbacher (1964), *G. marginodentata* Subzone (G_1) of Hillebrandt (1965), and the *G. velascoensis-subbotinae* Zone of the present author.

It might be added here that Wade, Mohler and Hay (1964, p. 215) have recently recorded some Palaeogene nannofossils from the Crimea and Caucasus. Their results are presented below, taken from their abstract:

'Study of nannofossils from Palaeogene sections in the Crimea and northern Caucasus permits correlation of Soviet stages with those of western Europe and North America. The Bakhchisaray section in the Crimea, standard for the Palaeogene of the southern Soviet Union, contains: (1) *Heliolithus riedeli* but no discoasters in strata of the Kachinskian stage, suggesting an age slightly older than the type Thanetian and Unit 1 of the Lodo Formation of California; (2) *Marthasterites tribrachiatus* in the *Operculina semiinvoluta* zone of the Bakhchisarian stage. indicating correlation with the lower part of the type Ypresian and lower part of Unit 3 of the Lodo; (3) *Marthasterites tribrachiatus* and *Discoaster lodoensis* in the *Nummulites crimensis* zone and lower part of the *Assilina placentula* zone of the Bakhchisarian stage, indicating correlation with the upper part of the type Ypresian and the upper part of Unit 3 of the Lodo; (4) *Discoaster lodoensis* present but *Marthasterites tribrachiatus* absent at the top of the *Assilina placentula* zone and in the *Acarinina rotundimarginata* zone of the Bodrakskian stage, indicating Cuisian age for the top of the Bakhchisarian, the Simferopolskian, and Bodrakskian stages. At Cherkessk, *Discoaster lodoensis* occurs in the *Acarinina crassaformis* zone of the Simferopolskian stage, confirming its Cuisian age. At Cherkessk and at Nal'chik in the northern Caucasus, the *Globigerinoides conglobatus* and *Bolivina* zones of the Al'minskian stage contains *Isthmolithus recurvus,* indicating a late Eocene age for these strata. A Lutetian nannofossil assemblage was not encountered.'

It will be seen that the dating of the *Assilina placentula* Subzone of the *Nummulites planulatus* Zone (=*Globorotalia subbotinae* Zone approximately) which is at the top of the Bakhchisarian Stage as Cuisian (upper Ypresian) is consistent with this usage. This level is approximately equivalent to the *Globorotalia formosa formosa* Zone of Bolli.

The correlation of the Simferopolian stage (upper part = *A. crassaformis* Zone) with the Cuisian is in sharp disagreement with its correlation by the Soviet Palaeogene Commission with the Lutetian (Middle Eocene). The *A. crassaformis* Zone – as applied by Soviet micropalaeontologists – is roughly equivalent to a part of Bolli's '*Globorotalia*' *palmerae* Zone (=*Acarinina bullbrooki* Zone) the *Hantkenina aragonensis* Zone, and perhaps a part of the *G. kugleri* Zone.

Even greater disparity is seen in comparing the assignation of the *A. rotundimarginata* Zone of the Bodrakian Stage to the Cuisian. The Soviet Committee places the Bodrakian in the Upper Eocene. The *A. rotundimarginata* Zone corresponds approximately to the upper part of Bolli's *Globigerapsis kugleri* and the *G. lehneri* Zones. Assignment of these zones to the Cuisian is somewhat surprising, since it has been generally regarded that these zones are equivalent to a part of the Lutetian. The absence of *Discoaster sublodoensis* – a guide form of the Lutetian – in the samples from the *Acarinina rotundimarginata* Zone, is considered definitive by Hay (pers. comm.) in assigning these strata to the Cuisian. Upon the presentation of the resolution of the 5th Conference of the Permanent Commission of the Palaeogene of the U.S.S.R. Palaeogene Colloquium in Bordeaux by Korobkov and Mironova (1964); Schaub has also questioned the correlation of the Simferopolian Stage with the Lutetian, pointing out that all the nummulitic zones in the Soviet scheme (supposedly including the *Nummulites incrassatus* Zone which Soviet palaeontologists correlate with the lower part of the *A. rotundimarginata* Zone) are older than the *N. laevigatus* Zone of the Lutetian of the Paris Basin.

The correlations by the Committee on Palaeogene Stratigraphy of the Soviet Union (1962, 1964) of the nummulitic and planktonic foraminiferal zones are, indeed, difficult to resolve. The *N. polygyratus* Zone was placed at the top of the Simferopolian Stage (Middle Eocene) equivalent (by inference) to the *A. crassaformis* Zone. Earlier *N. perforatus* had been chosen as the guide form for the upper part of the 'Middle' Eocene and the zone correlated with the *A. crassaformis* Zone (Yanshin *et al.* 1960). Thus *N. perforatus* and *N. polygyratus* are inferred to be contemporaneous forms, but Schaub (1964) has shown that the *N. polygyratus* Zone is upper Cuisian, the *N. perforatus* Zone 'Biarritzian'. The *N. distans* Zone (below) in the Crimea has been correlated with the *N. laevigatus* and *G. aragonensis* Zones in the Crimea-Caucasus area, Armenia, and elsewhere (Yanshin *et al.* 1960). The *N. incrassatus* Zone was correlated with the lower part of the *A. rotundimarginata* Zone by the Soviet Committee (1962, 1964) and placed in the lower part of the Bodrakian Stage (Upper Eocene of Soviet geologists). The *Nummulites polygyratus* Zone can, indeed, be correlated with the *A. crassaformis* Zone, but most probably only with the lower part.

The *Acarinina rotundimarginata* Zone would appear, on the basis of the discussion presented above, to be of Middle Eocene age, and older than levels containing *N. perforatus (sensu* Schaub 1964, i.e. Biarritzian, or late Lutetian of French stratigraphers). However, Schaub (in Korobkov and Mironova 1964, p. 939, 1966, p. 300) has suggested that on the evidence of the nummulites the Middle Eocene is missing in the Crimea. The *N. distans-planulatus-polygyratus* Zones (which Soviet palaeontologists generally correlate with Lower Eocene to upper part of Middle Eocene) are all placed by Schaub in the Lower Eocene. The beds with *N. incrassatus* above are placed in the Upper Eocene.

I have re-examined a series of samples collected in 1962 in the Crimea and North Caucasus. A summary of the results is presented here. In the Crimea a biostratigraphical zonation of Palaeocene/Eocene is difficult to make on the basis of planktonic Foraminifera. The acarininids are common in the Palaeocene-Lower Eocene; keeled globorotaliids are scarce except at certain levels near the Palaeocene/Eocene boundary (*G. velascoensis-G. subbotinae* Zones). Lower Eocene was recognized on the basis of the acarininid assemblages. In the region of Bakhchisaray some samples of the *Lyrolepis* marls contain *A. bullbrooki, A. rotundimarginata, A. aspensis,* and various small globigerinids, *Globigerapsis kugleri* and *G. index.* This fauna correlates approximately with the *G. kugleri lehneri* Zones of Bolli.

In the North Caucasus (Kuban River section) the Novogeorgi Group is overlain by the Cherkessk Group. The boundary lies approximately at the base of the *Globorotalia aragonensis* Zone. Samples from the upper part of the Cherkessk Group have yielded a fauna containing *inter alia, A. coalingensis* (=*A. primitiva* = *A. triplex*), *A. pentacamerata-aspensis* group, *A. bullbrooki-Truncorotaloides topilensis* transition, *Globorotalia aragonensis, G. pseudoscitula, Pseudohastigerina micra,* and *Globigerina frontosa* (= *G. boweri*). Significant here is the transition from *A. bullbrooki* to *T. topilensis* for this occurs within the upper part of the *G. kugleri* Zone in the Caribbean. The upper part of the Cherkessk Group is generally equated with the *A. crassaformis* Zone of Subbotina (which is approximately equivalent to the *H. aragonensis* Zone and a part of the *G. kugleri* Zone (at least). The Kumsk Group above contains the *Lyrolepis* Marl with a relatively poor planktonic fauna among which are, *inter alia.; A. rotundimarginata, Truncorotaloides topilensis* and *Pseudohastigerina micra.* The association suggests that this level is not younger than the *P. mexicana* (vel *Orbulinoides beckmanni*) Zone of Bolli. The fauna is still Middle Eocene in its affinities. The Beloglin clays above contain a rich and diversified globigerinid and globigerapsid microfauna which correlates well with the Upper Eocene.

Thus we see that although a sharp, sequential, biostratigraphical zonation of the Crimean and Caucasus sections is difficult, the presence of Middle Eocene is quite apparent. What is more likely is that the correlation between nummulitic and planktonic foraminiferal zones is still unsatisfactory.

The base of the Bodrakian Stage and its relationship to the underlying Simferopolian Stage is still open to discussion (Leonov, Alimarina and Naidin 1965). A part of the fauna of this Bodrakian fauna may be reworked in the Bakhchisaray section. The co-occurrence of *Nummulites globulus, N. atacicus, N. rotularius, N. distans* (typical Lower Eocene, Cuisian forms) in these beds (*Nummulites incrassatus* Zone) (see Nemkov and

Barkhatova 1961) are, if determined correctly, an indication of such reworking. The identity of *N. incrassatus* of Nemkov and Barkhatova (1959) should also be verified (cf. Bolli and Cita 1960, pp. 158-60, where an association of nummulites identified by Vialli as Middle-Upper Eocene was found to occur with a planktonic foraminiferal fauna of the *Hantkenina aragonensis* Zone).

Nemkov (1964) has presented a summary of the standard nummulitic zonation in the Soviet Union which has been adopted by the Committee on Palaeogene Stratigraphy. His summary includes a discussion of the zonation used in areas outside those chosen as the stratotype (Crimea) and parastratotype (North Caucasus) sections, where nummulites continue to characterize the Middle and Upper Eocene. A graphic summary of Nemkov's discussion is presented on Table 52.19 accompanying this paper for the sake of comparison with the nummulitic zonations discussed in Yanshin *et al.* (1960), the Palaeogene Stratigraphical Commission of the Soviet Union (1962, 1964), and various planktonic foraminiferal zonations developed in the Soviet Union.

On the basis of available data it would seem that the anomalous results and discrepancies to which strict correlation of planktonic and nummulitic zones in the Soviet Union leads can be attributed to a combination of factors: (*a*) different interpretation of species by specialists in different countries; (*b*) inaccurate correlation between nummulitic and planktonic foraminiferal zones (the fossiliferous horizons on which some nummulitic zones are based may represent but a small part of supposedly equivalent planktonic foraminiferal zones); (*c*) zonation in shallow water facies by larger Foraminifera may, in some instances, represent only partial ranges of the forms as reflected by local distribution; (*d*) reworking.

Banner and Eames (1966, p. 163) point out the anomalous Eocene dating and correlations of the nummulitic sequences in the Soviet Union (Yanshin *et al.* 1960, table 1). *Nummulites laevigatus* was accepted as the index fossil of the lower part of the Middle Eocene and the zone was correlated with the *T. aragonensis* Zone (Caucasus and other areas in the southwestern Soviet Union) and the *Nummulites distans* Zone in the Crimea, and the *N. atacicus-N. murchisoni* Zones in the Georgian Republic. *Nummulites perforatus* was chosen as the index fossil for the upper part of the Middle Eocene, and the zone was correlated with the *Acarinina crassaformis* Zone of Subbotina (1953). In the stratigraphical tables presented by the Committee on Palaeogene Stratigraphical Nomenclature of the U.S.S.R. (Korobkov 1964, tables I, II) the Lower/Middle Eocene boundary is placed at the base of the Simferopolian Stage (i.e. *G. aragonensis* Zone) following the procedure adopted earlier in Yanshin *et al.* (1960). The Middle/Upper Eocene boundary is drawn at the top of the *A. crassaformis* Zone. As Banner and Eames (1966, pp. 162, 163) indicate, some anomalous results are apparent if all records are to be credited: (1) the Middle Eocene/Lower Eocene boundary should be placed at the base of the *G. aragonensis* Zone; (2) the Upper Eocene/Middle Eocene boundary should be placed between the *Globorotalia lehneri* and *Globigerapsis kugleri* Zones of Bolli; (3) or the *Nummulites perforatus* Zone lies within the Middle Eocene rather than at its top.

A few words of caution would seem appropriate at this point with regard to the subdivision of Eocene by microfossils in the Soviet Union. As I have shown elsewhere (Berggren 1965) the *G. aragonensis* Zone as applied in the Soviet Union is roughly the same as the *G. aragonensis* Zone of Bolli (1957) and is a modified version of Subbotina's (1953) zone of conical globorotaliids. This zone was subsequently subdivided into two parts by Subbotina (1960, p. 28, fig. 2 on p. 33): a lower zone 'd' (*T. lensiformis*, although the illustration used was a prior illustration of *G. aragonensis*), and an upper zone 'e' (*T. caucasica*). Subbotina (*loc. cit.*) drew the Lower/Middle Eocene boundary between these two zones. Her *A. crassaformis* zone above (zone 'f') was interpreted as Middle Eocene (*op. cit.* p. 29). Banner and Blow (1966, p. 163) note that Subbotina (1960, fig. 3, p. 34) still considered her *A. crassaformis* Zone to range 'no higher than the early Middle Eocene', but this is not quite accurate. In placing the Lower/Middle Eocene boundary at the base of her zone 'd' (*T. caucasica*) and the Middle/Upper Eocene boundary at the top of her zone 'f' (*A. crassaformis*), Subbotina is left little choice other than considering zone 'f' to be Upper/Middle Eocene in age (at least in part). The use of the term *Acarinina crassaformis* Zone in the Soviet Union must be taken into consideration in attempts to evaluate stratigraphical results. As it was originally defined (Subbotina 1953), and subsequently used by that author and most other Soviet micropalaeontologists, the *A. crassaformis* Zone corresponds approximately to a part of the '*Globorotalia*' *palmerae* Zone and the *Hantkenina aragonensis* Zone of Bolli (1957) (see Berggren 1965). Its upper limits may correspond also to the lower part of the *Globigerapsis kugleri* Zone of Bolli (see Krasheninnikov and Ponikarov 1963, p. 66). The base of the overlying *Acarinina rotundimarginata* Zone (Subbotina 1953; = zone 'g', 1960) is considered by most Soviet palaeontologists to denote the base of the Upper Eocene (Bodrakian Stage). Characteristic forms occurring in this zone include *A. rotundimarginata, G. lehneri, G. spinulosa, Hantkenina lehneri, H. liebusi, Globigerapsis kugleri, G. index, Globigerina boweri*, and *Truncorotaloides topilensis* (Krasheninnikov and Ponikarov 1964, p. 66). This zone can be approximately equated with the *G. kugleri* and *G. lehneri* Zones of Bolli, which are placed well within the limits of the Middle Eocene by most stratigraphers in Europe and North and South America. That *Nummulites laevigatus* characterizes at least the lower part of the Lutetian, would seem to be well documented. The correlation of the *N. laevigatus* Zone with the *G. aragonensis* Zone would appear to be incorrect. The *G. aragonensis* Zone can be correlated with the upper Ypresian which is equivalent to Cuisian, and occupies a position approximately equivalent to the *N. planulatus-burdigalensis* Zone of Schaub, i.e. well below the level of *Nummulites laevigatus* in the Middle Eocene.

An attempt to correlate Palaeogene planktonic foraminiferal zones in the Soviet Union with those of the Caribbean is presented in Table 52.19., Table 52.11 shows a suggested correlation of various Palaeogene zonations based on nummulites, planktonic Foraminifera and nannofossils.

Oligocene

It will be recalled that Lyell (1833) subdivided the Tertiary strata recognized by Cuvier and Brongniart (1809) in the Paris Basin, as well as younger Tertiary strata recognized elsewhere in Europe, into four parts: Eocene, Miocene, Older Pliocene, Newer Pliocene. At the time he expressed the belief that missing intervals in his scheme might well be expected and that adjustments should be made accordingly as data warranted them. Subsequent studies in the middle part of the last century, in regions where Eocene and Miocene sediments were recognized, and particularly in northern Europe, Holland and N. Germany, demonstrated that there were extensive marine, brackish, fresh-water and continental sediments intermediate between the Eocene and Miocene of Lyell. In the belief that these intermediate sediments were the result of a large marine transgression which covered a large part of northern Europe, Beyrich (1854) created the name Oligocene for this period of earth history.

A year earlier Beyrich had correlated the Magdeburg Sands with the marine sediments in the lower part of the Tongrian (Dumont 1839) of Belgium. These Magdeburg Sands lie beneath the Septaria Clay (which belong to the Rupelian of Dumont 1849) and form the basal part of the marine Oligocene (and its transgression) in north-central Germany as defined by Beyrich. In a lecture before the Royal Prussian Academy of Science on 27 November, 1854, Beyrich discussed the difficulties encountered in attempts to relate the Tertiary strata of the Mainz Basin, northern Germany, and Belgium with the Tertiary subdivision proposed by Lyell in 1833. Beyrich subsequently devoted a considerable number of papers to further studies of his Oligocene, subdividing the Oligocene into three parts in a lecture read before the Royal Prussian Academy of Science on 19 July 1855 (published 1856). The Oligocene series of north-central Germany was said to consist of five marine and two fresh-water strata. His reasons for formulating the Oligocene were stated in the following words (Beyrich 1856, p. 11), 'Die mittlere Tertiärzeit, welcher die norddeutschen Lager angehoren, theile ich in eine ältere oligocäne und eine jungere miocäne Zeit. Die verschiedenen oligocänen Glieder, welche auf der Karte unterschieden wurden, sind dieselben, deren Altersfolge in der Abhandlung uber die Stellung der hessischen Tertiärbildungen festgestellt wurde. Ihre zusammenhängende und selbstständige, von dem Vorhandensein älterer eocäner Tertiärbildungen unabhängige geognistische Verbreitung ist der vornehmliche Grund, der mich abhalt, sie mit Lyell nur als einen oberen Abschnitt der eocänen Tertiärreihe zuzuzahlen; ihre mannigfältige Gliederung und ihr grosser erst in Deutschland in seinem ganzen Umfange bekannt werdender Reichthum an ihnen eigenthümlich zukommenden organischen Resten bestimmen mich, sie lieber als einen besonderen Abschnitt der tertiären Periode getrennt zu betrachten, als sie dem miocänen Tertiärgebirge anzureihen, wie ich es früher mit d'Orbigny und andern Autoren gethan habe.'

One of the German marine beds, and one of the fresh water beds in Belgium were said to be absent, while marine deposits are present in Belgium which were not known with certainty in Germany. The Middle Oligocene was said to begin with the marine *Lager von Alzey* (Mainz Basin) which Beyrich believed was absent in both Belgium and northern Germany. The second Middle Oligocene marine layer was said to be that at Kleyn-Spauwen (the *Nucula*-bearing sand, R1G of de Heinzelin and Glibert 1957, which lies between the Sand of Berg and the Boom Clay), and its equivalent was thought to be absent in Germany. A table of the units assigned to the Oligocene by Beyrich (1856) and his suggested correlations has been compiled from his paper and is presented below (Table 52.21). Beyrich was not the most concise of scientific writers and as a result it is sometimes difficult to extrapolate his meanings and express them graphically or in tabular form.

Although the concept of the Oligocene was presented by Beyrich in 1854, it was not until 1856 that he indicated more precisely the limits of his system. The Upper Oligocene was denoted by the *'Lager vom Alter des Sternberger Gesteins'*. He considered as lower Miocene the *'Lager vom Alter des Holsteiner Gesteins'*. Beyrich drew the Oligocene/Miocene boundary between these two units and this should serve as a point of departure in further interpretations of this boundary. However, it does not follow necessarily that this level *must* be chosen as *the* Oligocene/Miocene boundary on the basis of the priority of Beyrich's (1854) determination, any more than the reference by Mayer-Eymar in Gressly (1853) to deposits in Switzerland should serve as the stratotype of his Aquitanian as Durham (1944) and Dehm (1949) insisted. We are in full accord with Rutsch (1952, pp. 354-5) who says, 'Beide Autoren (Durham and Rutsch) wenden das in der zoologischen Nomenklatur geltende Prioritäts-Prinzip strikte auf stratigraphische Begriffe an. Es muss nun aber mit allem Nachdruck betont werden: Ein Prioritäts-Gesetz im Sinne der zoologischen Nomenklatur existiert in der Stratigraphie nicht. Einer solchen Bindung müsste man sich sogar auf das entschiedenste entgegensetzen, da sie unfehlbar zu so unhaltbaren Interpretationen stratigraphischer Begriffe fuhren müsste, wie dies das Beispiel des Aquitanien im Sinne von Durham and Dehm zeigt.

Und selbst wenn für stratigraphische Begriffe das Prioritätsprinzip strikte angewendet würde, dann müssten offenbar auch die Ausnahmen gestattet sein, wie sie der Codex für die zoologische Nomenklatur vorsieht.'

The best determination of a geological boundary is one which is consonant with known facts on the regional geology and with palaeontological information. As such it may be subject to change as additional information is made available. It is not a fixed, static entity, allowing of no modification once formulated.

The subsequent detailed studies by von Koenen on the molluscs of the glauconitic sands of the Lower Oligocene (to which he gave the name Latdorf=Lattorf Beds) has played an important role in the problems associated with the lower boundary of the Oligocene. The beds at Latdorf have generally been considered as typical for the Lower Oligocene by most workers, and were chosen as the stratotype of the Lower Oligocene Lattorfon=Lattorfian Stage by Mayer-Eymar (1893). In doing so Mayer-Eymar replaced the locality at Montmartre in the Paris Basin by that at Latdorf in Saxony, considered the Lattorfian Stage as equivalent to the lower part of his Ligurian Stage, and implied correlation between the deposits at the two localities. De Lapparent was of a similar opinion, and the deposits at Latdorf are generally considered to be correlative with the Sands of Vliermaal (=Sands of Grimmertingen) and the Sands of Neerrepen in the Leuven-Tongeren region (Belgium-Holland) which correspond to the lower

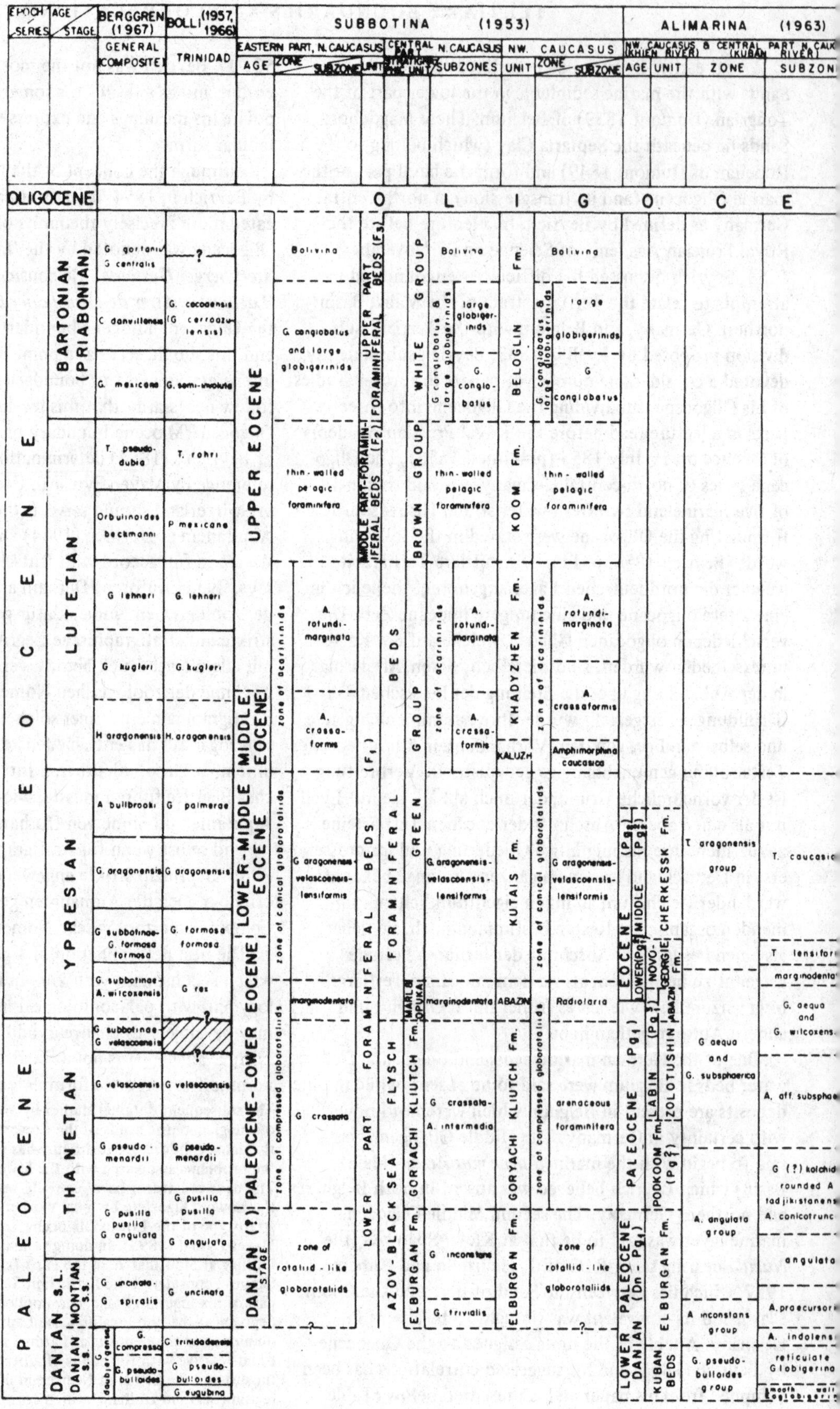
EPOCH SERIES
AGE STAGE
BERGGREN (1967)
GENERAL COMPOSITE
BOLLI (1957, 1966)
TRINIDAD
SUBBOTINA (1953)
EASTERN PART, N. CAUCASUS
AGE
ZONE
SUBZONE
UNIT
CENTRAL PART N. CAUCASUS
STRATIGRAPHIC UNIT
SUBZONES
NW. CAUCASUS
UNIT
ZONE
SUBZONE
ALIMARINA (1963)
NW. CAUCASUS (KHIEN RIVER)
CENTRAL PART N. CAUCASUS (KUBAN RIVER)
AGE
UNIT
ZONE
SUBZONE
OLIGOCENE
O L I G O C E N E
E O C E N E
P A L E O C E N E
BARTONIAN (PRIABONIAN)
LUTETIAN
YPRESIAN
THANETIAN
DANIAN S.L.
MONTIAN S.S.
DANIAN S.S.
G. gortanii
G. centralis
C. danvillensis
G. mexicana
T. pseudo-dubia
Orbulinoides beckmanni
G. lehneri
G. kugleri
H. aragonensis
A. bullbrooki
G. aragonensis
G. subbotinae-G. formosa formosa
G. subbotinae-A. wilcoxensis
G. subbotinae-G. velascoensis
G. velascoensis
G. pseudo-menardii
G. pusilla pusilla
G. angulata
G. uncinata-G. spiralis
G. compressa
G. daubjergensis
G. pseudo-bulloides
G. cocoaensis (G. cerroazulensis)
G. semi-involuta
T. rohri
P. mexicana
G. lehneri
G. kugleri
H. aragonensis
G. palmerae
G. aragonensis
G. formosa formosa
G. rex
G. velascoensis
G. pseudo-menardii
G. pusilla pusilla
G. angulata
G. uncinata
G. trinidadensis
G. pseudo-bulloides
G. eugubina
UPPER EOCENE
LOWER-MIDDLE EOCENE
MIDDLE EOCENE
PALEOCENE-LOWER EOCENE
DANIAN (?) STAGE
Bolivina
G. conglobatus and large globigerinids
thin-walled pelagic foraminifera
zone of acarininids
A. rotundimarginata
A. crassaformis
zone of conical globorotaliids
G. aragonensis-velascoensis-lensiformis
marginodentata
zone of compressed globorotaliids
G. crassata
zone of rotaliid-like globorotaliids
UPPER PART FORAMINIFERAL BEDS (F3)
MIDDLE PART FORAMINIFERAL BEDS (F2)
LOWER PART FORAMINIFERAL BEDS (F1)
AZOV-BLACK SEA FLYSCH
WHITE GROUP
BROWN GROUP
GROUP BEDS
GREEN FORAMINIFERAL
MARL & OPOKA
GORYACHI KLIUTCH Fm.
ELBURGAN Fm.
Bolivina
G. conglobatus & large globigerinids
large globigerinids
G. conglobatus
thin-walled pelagic foraminifera
A. rotundimarginata
A. crassaformis
G. aragonensis-velascoensis-lensiformis
marginodentata
G. crassata-A. intermedia
G. inconstans
G. trivialis
BELOGLIN Fm.
KOOM Fm.
KHADYZHEN Fm.
KALUZH
KUTAIS Fm.
ABAZIN
GORYACHI KLIUCH Fm.
ELBURGAN Fm.
zone of acarininids
zone of conical globorotaliids
zone of compressed globorotaliids
Bolivina
large globigerinids
G. conglobatus
thin-walled pelagic foraminifera
A. rotundimarginata
A. crassaformis
Amphimorphina caucasica
G. aragonensis-velascoensis-lensiformis
Radiolaria
arenaceous foraminifera
zone of rotaliid-like globorotaliids
EOCENE
LOWER (Pg2)
MIDDLE (Pg2)
PALEOCENE (Pg1)
LOWER PALEOCENE DANIAN (Dn Pg1)
NOVO-GEORGIE
ABAZIN Fm.
CHERKESSK Fm.
ZOLOTUSHIN
LABIN (Pg1?)
PODKOOM Fm. (Pg1?)
ELBURGAN Fm.
KUBAN BEDS
T. aragonensis group
A. angulata group
A. inconstans group
G. pseudo-bulloides group
T. caucasica
T. lensiformis
marginodentata
G. aequa and G. wilcoxensis
G. aequa and G. subsphaerica
A. aff. subsphaerica
G. (?) kolchidica
rounded A. tadjikistanensis
A. conicotruncata
angulata
A. praecursoria
A. indolensis
reticulate Globigerina

Table 52.19. Suggested correlation between Palaeogene stages, planktonic foraminiferal and nummulitic zones in the S.W. Soviet Union, and their relationships to European stages.

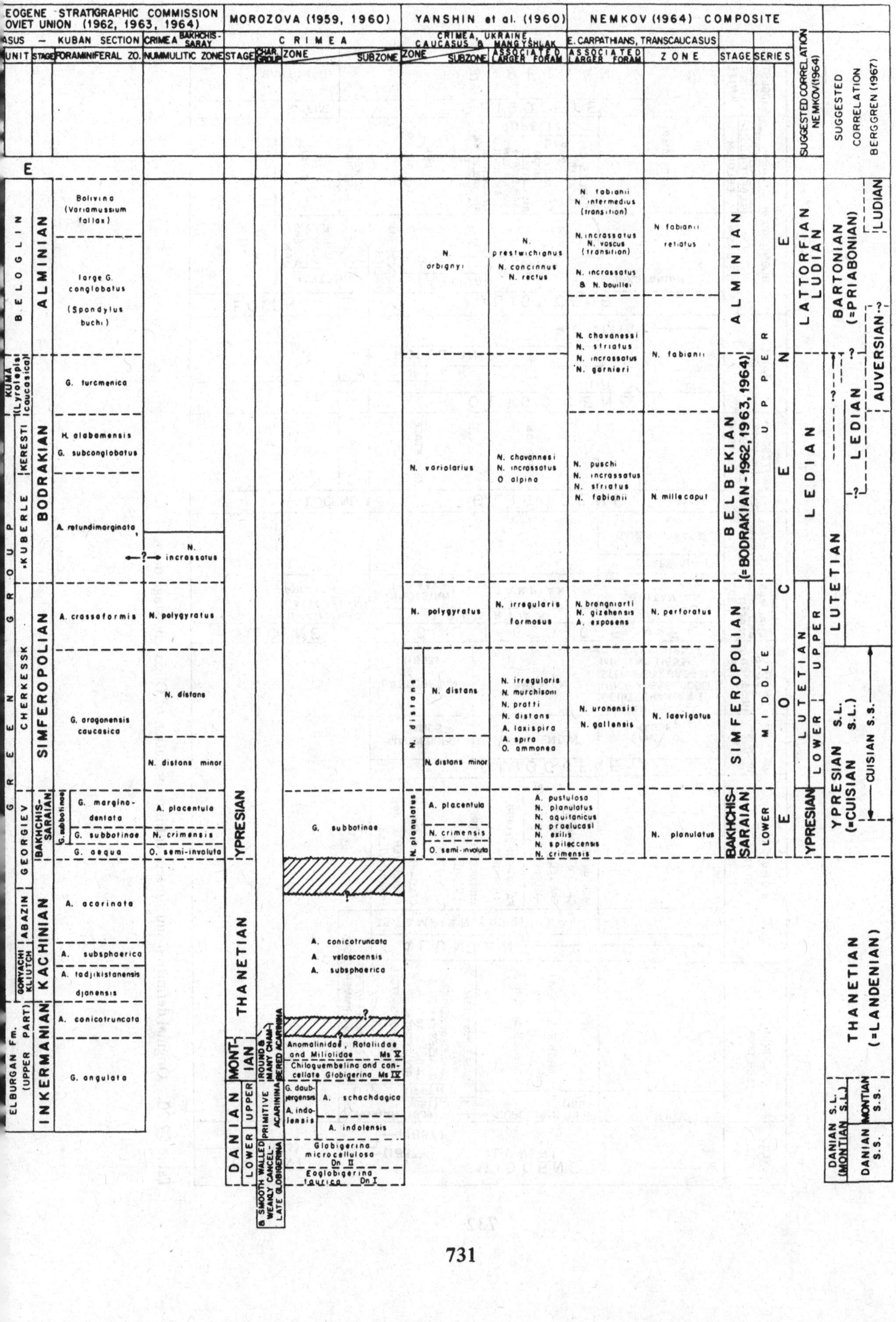
EOGENE STRATIGRAPHIC COMMISSION OVIET UNION (1962, 1963, 1964)
ASUS — KUBAN SECTION
CRIMEA
BAKHCHIS-SARAY
UNIT
STAGE
FORAMINIFERAL ZO.
NUMMULITIC ZONE
MOROZOVA (1959, 1960)
C R I M E A
STAGE
CHAR. GROUP
ZONE
SUBZONE
YANSHIN et al. (1960)
CRIMEA, UKRAINE, CAUCASUS & MANGYSHLAK
ZONE
SUBZONE
ASSOCIATED LARGER FORAM
NEMKOV (1964) COMPOSITE
E. CARPATHIANS, TRANSCAUCASUS
ASSOCIATED LARGER FORAM
ZONE
STAGE
SERIES
SUGGESTED CORRELATION NEMKOV (1964)
SUGGESTED CORRELATION BERGGREN (1967)
E
G R E E N   G R O U P
B E L O G L I N
KUMA (Lyrolepis caucasica)
KERESTI
KUBERLE
CHERKESSK
GEORGIEV
ABAZIN
GORYACHI KLIUTCH
ELBURGAN Fm. (UPPER PART)
ALMINIAN
BODRAKIAN
SIMFEROPOLIAN
BAKHCHIS-SARAIAN
KACHINIAN
INKERMANIAN
Bolivina (Variamussium fallax)
large G. conglobatus (Spondylus buchi)
G. turcmenica
H. alabamensis
G. subconglobatus
A. rotundimarginata
A. crassaformis
G. aragonensis caucasica
G. subbotinae
G. marginodentata
G. subbotinae
G. aequa
A. acarinata
A. subsphaerica
A. tadjikistanensis djanensis
A. conicotruncata
G. angulata
N. incrassatus
?
N. polygyratus
N. distans
N. distans minor
A. placentula
N. crimensis
O. semi-involuta
YPRESIAN
THANETIAN
MONTIAN
DANIAN
LOWER
UPPER
SMOOTH WALLED, WEAKLY CANCELLATE GLOBIGERINA
PRIMITIVE ACARININA
ROUND & MANY CHAMBERED ACARININA
G. subbotinae
A. conicotruncata
A. velascoensis
A. subsphaerica
Anomalinidae, Rotaliidae and Miliolidae Ms V
Chiloguembelina and cancellate Globigerina Ms IV
G. daubjergensis
A. schachdagica
A. indolensis
A. indolensis
Globigerina microcellulosa Dn II
Eoglobigerina taurica Dn I
N. orbignyi
N. prestwichianus
N. concinnus
N. rectus
N. variolarius
N. chavannesi
N. incrassatus
O. alpina
N. polygyratus
N. irregularis formosus
N. distans
N. distans
N. distans minor
N. irregularis
N. murchisoni
N. pratti
N. distans
A. laxispira
A. spira
O. ammonea
N. planulatus
A. placentula
N. crimensis
O. semi-involuta
A. pustulosa
N. planulatus
N. aquitanicus
N. praelucasi
N. exilis
N. spileccensis
N. crimensis
N. fabianii
N. intermedius (transition)
N. incrassatus
N. vascus (transition)
N. incrassatus & N. bouillei
N. chavanessi
N. striatus
N. incrassatus
N. garnieri
N. puschi
N. incrassatus
N. striatus
N. fabianii
N. brongniarti
N. gizehensis
A. exponens
N. uronensis
N. gallensis
N. fabianii retiatus
N. fabianii
N. millecaput
N. perforatus
N. laevigatus
N. planulatus
ALMINIAN
BELBEKIAN (=BODRAKIAN-1962, 1963, 1964)
SIMFEROPOLIAN
BAKHCHIS-SARAIAN
UPPER
MIDDLE
LOWER
EOCENE
LATTORFIAN LUDIAN
LEDIAN
LUTETIAN
LUTETIAN UPPER
LUTETIAN LOWER
YPRESIAN
BARTONIAN (=PRIABONIAN)
LUDIAN
AUVERSIAN
LEDIAN
LUTETIAN
YPRESIAN S.L. (=CUISIAN S.L.)
CUISIAN S.S.
THANETIAN (=LANDENIAN)
DANIAN S.L. (MONTIAN S.L.)
DANIAN S.S.
MONTIAN S.S.

LYELL (1833, 1857)	DUMONT (1839, 1849, 1851)	d'ORBIGNY (1852)	BEYRICH (1854 et seq.)	MAYER-EYMAR (1853, 1857, 1893)	MUNIER-CHALMAS AND de LAPPARENT (1893)	FUCHS (1894)	KRUTZSCH & LOTSCHE (1957) KOROBKOV (1951–1964)	BATJES (1958)	BLONDEAU et al (1965) PARIS BASIN	de HEINZELIN & GILBERT (1957) BELGIUM	SUGGESTED USAGE (THIS WRITER)
MIOCENE (LOWER) ↑ (1833) ↓ (1857) EOCENE (1833) ↓	RUPELIAN (1849) (1851): ARGILE de BOOM; SABLES de BERG TONGRIAN (1839) (1851): UPPER: SABLES de VIEUX JONCS, SABLES de KERKOM, GLAISES de HENIS, SABLES de BOUTERSEM, HOR. de HOOGBUTSEL; L: SABLES de NEEREPEN, SABLES de GRIMMERTINGEN	FALUNIAN ↑ STAMPIAN (1852): MOL. du GÂTINAIS, CALC. d'ÉTAMPES, SABLES d'ORMOY, FALUNS de PIERREFITTE; FALUNS de JEURRE-MORIGNY; MARNES à OSTREA LONGIROSTRIS; CALCAIRE de SANNOIS, ARGILE VERTE de ROMAINVILLE	OLIGOCENE: SANDS OF KASSEL; "SEPTARIENTON"; MAGDEBURG SANDS	AQUITANIAN = MEDITTERANIAN I (INCL. KASSEL, DOBERG, STERNBERG, BAZAS) MIOCENE (1853) OLIGOCENE (1857) TONGRIAN LATTORFIAN (1893)	OLIGOCENE: AQUITANIAN; TONGRIAN: STAMPIAN; SANNOISIAN (1893) EOCENE: LUDIAN (1893) (LATTORFIAN) PRIABONIAN (1893)	CHATTIAN; SANDS OF KASSEL	OLIGOCENE: CHATTIAN; RUPELIAN; (NEW STAGE ?) EOCENE: LATTORFIAN	OLIGOCENE: ← CHATTIAN →; TONGRIAN: RUPELIAN EOCENE	OLIGOCENE: AQUITANIAN; STAMPIAN: UPPER, LOWER; SANNOISIAN FACIES EOCENE: BARTONIAN: LUDIAN (? LATTORFIAN), MARINESIAN, AUVERSIAN	(CHATTIAN ?) ? – ? RUPELIAN: SABLES de VOORT; ? – ? ; R2 = ARGILE de BOOM; R1b = ARGILE SABLEUSE à NUCULA COMTA ou à NUCUNELLA TAXANDRICA; R1a = SABLES de BERG UPPER TONGRIAN LOWER TONGRIAN ASSIAN	OLIGOCENE: RUPELIAN EOCENE: BARTONIAN

Table 52.20. Original definition and subsequent interpretation of Oligocene stage units.

Tongrian of Dumont (1839, 1849) and to the Lower Oligocene in age (but see Batjes 1958, and Kaasschieter 1961, where evidence is presented for considering the lower part of the Tongrian to be equivalent to the Clays and Sands of Asse and of Upper Eocene age; cf. also Krutzsch and Lotsch 1957, who present strong evidence that the type Latdorf beds are the sandy, near-shore facies of the more clayey Upper Eocene 5 in the central part of the North German Basin).

Attempts to recognize and define the lower and upper limits of the Oligocene have been the subject of discussion and debate for about a century. It will be recalled that Mayer-Eymar (1858) defined the Bartonian Stage essentially for strata in the Paris Basin, but took as its stratotype the beds at Barton (Hampshire), England. Although originally including at the top of his Bartonian the *Marnes à Pholadomya ludensis*, he subsequently (1869) excluded these beds from the concept of his stage, which led Munier-Chalmas and de Lapparent (1893) to create the Ludian Stage for these beds plus the gypsum beds above (de Lapparent, himself, later dropped the term Ludian in the fifth edition of his Treatise, 1906).

According to Munier-Chalmas and de Lapparent (1893) the Oligocene consisted of two stages: Tongrian and Aquitanian. They included in their Tongrian (from the base upwards)Sands of Wemmel, Sands of Asse, Sands of Vliermaal (which demonstrates that they were including in their concept of Oligocene, beds which are now considered correlative with Bartonian=Upper Eocene, while at the same time recognizing strata of Upper Eocene age (=Ludian and Priabonian Stages) which are today generally correlated with the lower Tongrian, at least in part). The Tongrian was divided into two substages (the term substage was not used, but as the authors stated that the Oligocene consisted of two stages: Tongrian and Aquitanian, the term substage is inescapable): Sannoisian and Stampian. The Sannoisian was created for the supragypsiferous marls and the *Calcaire de Sannois.*

Epoch	Series / Sub-division	Suggested correlation by Beyrich	North-Central Germany: Area of development	Suggested correlation by Beyrich (1854, 1856): Belgium	Mainz Basin	Rheinisch-Hessische Tertiary Basin
OLIGOCENE	UPPER		Die Lager von Alter des Sternberger Gesteins; Mecklenburg Bezirk (loose glacially transported material); Mitteldeutsche Bezirk (Mergellager vonOsnabrück, Bünde Lemgo, Dickholzen, Bodenwerder, Luithorst, Sand von Göttingen und Cassel); Niederung des Rheinthales (Crefeld)	Absent	Absent	*
	MIDDLE	RUPELIEN SUPÉRIEUR (BEYRICH)	SEPTARIENTON (Thon von Boom und Rupelmonde) Stettiner Tertiärgestein (Sande) und Magdeburg Sande		Brackwasser und Süsswasserbildung (Cyrenen mergel, Cerithium Kalkstein)	
		TONGRIEN SUPÉRIEUR (BEYRICH); RUPELIEN INFÉRIEUR (DUMONT)	Absent	Kleyn - Spauwen		Rheinisch-hessische Braunkohlenbildung
			Absent	Absent	Lager von Alzey	
	LOWER	TONGRIEN INFÉRIEUR (BEYRICH)	Braunkohlenfuhrenden Tertiärbildung (Nordostdeutschen Braunkohlenbildung) / Lager von Egeln	marine Tongrien inférieur von Dumont	Absent	*

* UNITS FROM THIS AREA NOT DISCUSSED IN 1854 OR 1856.

Table 52.21. The three-fold subdivision of the Oligocene according to Beyrich (1854-6).

The Stampian Stage (*sensu* Munier-Chalmas and de Lapparent 1893) begins with the *Marnes à Ostrea longirostris* and *O. cyathula* and is terminated by the beds of Ormoy above which follow the first Aquitanian strata (*Calcaire de Beauce*). The Stampian thus corresponds to the strata called *Sables de Fontainebleau* in the Paris Basin (Munier-Chalmas and de Lapparent 1893, pp. 480, 481). Now the problems begin. The Sannoisian beds in the Paris Basin contain a few marine fossils, the large majority of those which have been recorded to date being brackish or fresh-water forms. The faunas in the basal part are primarily continental, whereas those which might serve to distinguish the stage are localized in the upper stratigraphical levels (*Argile verte de Romainville* and *Calcaire de Sannois*) (Cavelier 1965; Blondeau *et al.* 1965). Cavelier (1965) has suggested that the *Marnes blanches de Pantin* (near the base of the Sannoisian) be placed in the Lattorfian (or Ludian) and that the original definition of Stampian (d'Orgibny 1852) be followed: base Stampian=*Argile verte de Romainville*. The Sannoisian thus loses its value as a chronostratigraphical entity, but Cavelier suggests that the section at Sannois (upper part of Sannoisian) can be retained as a facio-stratotype of the *Argile verte de Romainville-Calcaire de Sannois*, while retaining the term Stampian for pre-Aquitanian Oligocene strata. The lower limit of the 'Sannoisian' in the Paris Basin is quite elastic with regard to its interpretation by various authors. The majority of investigators in the Paris Basin (such as de Lapparent, Dollfus, Denizot, *inter alia*) have placed the Eocene/Oligocene boundary - base Sannoisian at the base of the second mass of gypsum; others have placed it at the base of the *Glaises à Cyrènes* at the base of the *Argile verte de Romainville* (Stehlin, Vasseur, Soyer, Feugueur, *inter alia*).

The benthonic foraminiferal fauna of the Sannoisian of the Paris Basin has been recently studied by Le Calvez (1966). In agreement with Cavelier (1965) she places the *Argile verte de Romainville* – the first marine transgressive unit of the Oligocene – at the base of the Oligocene. The Lower Oligocene in the Paris Basin thus corresponds to those beds which lie between the first post-Eocene transgression and the *Marnes à Huitres* (Stampian of all authors); i.e. *Argile verte de Romainville*, the *Caillasses d'Orgemont* and the *Calcaire de Sannois* = Sannoisian of Munier-Chalmas and de Lapparent 1893 = Lower Stampian of d'Orbigny, 1852. Most of the species described or recorded have been found to occur in the Oligocene of northern Germany (Ruess 1866) and Alsace (Andreae 1884). Although most of the forms are of use only for local stratigraphical correlation, such forms as *Bolivina beyrichi, B. melettica, B. fastigia, B. oligocaenica, Elphidium hiltermanni, Caucasina coprolithoides*, all typical of the Oligocene of northern Europe, were recorded.

A distinct separation between Eocene and Oligocene is found in the foraminiferal faunas according to LeCalvez (1966). The majority of the species appear in the Lower Oligocene and continue their development in the Oligocene. Thus the essential homogeneity of the Stampian as a stage unit would appear to be demonstrated, the relegation of the Sannoisian to a lower, brackish-water facies of Stampian suggested by Cavelier (1962) is confirmed, and the need for a multiple chronostratigraphical subdivision of the Oligocene (at least lower and middle part) is obviated.

In conclusion an interesting comparison was made with the foraminiferal faunas of the Upper Tongrian beds (Vieux-Joncs) in Belgium; these were shown to have been deposited in a more marine environment than those of the 'Sannoisian' of the Paris Basin.

The Stampian above, although containing a rich microfauna in some parts, is regressive at the top, and an upper boundary with the Chattian is difficult to determine (the Chattian is considered a northern facies of the Aquitanian by most French stratigraphers). Demarcq (1964) has suggested a possible solution for the Paris Basin Oligocene: suppression of the Sannoisian as a chronostratigraphical unit; use of Tongrian *sensu stricto* (Dumont 1849, 1851) and Stampian d'Orbigny (1852) (with the addition of a supplementary marine, fossiliferous stratotype in its upper part).

The extent of the Stampian as a chronostratigraphical unit is thus a matter of considerable variation in interpretation. Whereas the French (Demarcq 1964; Blondeau *et al.* 1965; Cavelier 1965) suggest that the Stampian corresponds to the Lower and Middle Oligocene (and by implication the Upper Oligocene also, at least in part; inasmuch as they regard the Chattian as a boreal facies of the Mediterranean Aquitanian, while including the Stampian strata generally placed in the Chattian), studies in the Mainz Basin (where Beyrich 1854, p. 665, had originally thought Oligocene was missing) have shown that the Chattian is equivalent to the Middle and Upper Stampian *s.l.* (Zöbelein 1960, p. 249). If we accept the conclusions of both these sources, then we see strong evidence for considering the Stampian = Rupelian = Oligocene.

A large part of the problem in determining the Eocene/Oligocene boundary can be traced to the following causes:

(1) varying opinions on the significance and extent of the Oligocene transgression;

(2) varying opinions on the affinities of the Lattorfian microfauna (even von Koenen, in the last of his seven volumes on the molluscan fauna at Latdorf in 1894, recognized a large number of species endemic to Latdorf and having Eocene affinities);

(3) complicated nomenclatorial problems associated with the stratigraphical units near the Eocene/Oligocene boundary.

Mayer-Eymar himself failed to distinguish clearly between the limits of the Lattorfian (1893) and his Ligurian Stage (1858). Mayer-Eymar believed that his Ligurian Stage and the Tongrian Stage of his classification corresponded to the Lower and Middle Oligocene, respectively, of Beyrich. The Lattorfian, in turn, was considered to be the Lower Ligurian, the name Lattorfian replacing the ill-chosen Montimartrian of the Paris Basin. It will be well to recall here that Dumont's (1839) original definition of the Tongrian included strata placed today between Tongrian *s.s.* and Rupelian. He modified the concept of Tongrian somewhat in 1846, and again in 1849, but retained the *Sables de Berg* at the top, then subsequently (1851) placed them in the Rupelian.

From 1852 the Tongrian may be said to have been accurately defined; it included in its lower part the *Sables de Grimmertingen* and *Sables de Neerrepen*, which are today considered Upper Eocene in age by most continental stratigraphers and broadly equivalent to the *Sables d'Asse* and the Ludian (Blondeau *et al.* 1965; see Table 52.17, this paper). Thus Mayer-Eymar's (1893) concept of Tongrian is decidedly different from that of Dumont (1839, 1849). His (Mayer-Eymar's) 'Tongrian' did, indeed, correspond to the Middle Oligocene of Beyrich and not to the Tongrian of Dumont which included rocks of Upper Eocene/Lower Oligocene age. Munier-Chalmas and de Lapparent (1893, pp. 478-80) recognized the transitional nature of the faunas in beds variously attributed to the Ludian, Bartonian or Tongrian Stages, and noted that it may be necessary to recognize a Lattorfian Stage (above the Ludian) as a transitional unit between the Eocene and Oligocene.

Korobkov (1964) has discussed the history of the definition of the Oligocene. As early as 1951-2 he reached the conclusion that the Eocene/Oligocene boundary was to be drawn between the Lattorfian and Rupelian. He discussed evidence in the Mandrikovka Beds in the region of Dnepropetrovsk in the S.W. Soviet Union where the molluscan fauna shows a striking similarity to that of the Lattorfian. Associated with this fauna is *Discocyclina*, a reliable Eocene marker.

The Oligocene was originally described in northern Germany, however, not the Paris Basin, and it is there we should seek criteria for the recognition of its limits.

The evidence discussed by Krutszch and Lotsch (1957) on the Lattorfian Beds of Germany is now familiar to most specialists and was reviewed by me in 1963. The Lattorfian was said to represent a shallow water facies of the more normal marine Upper Eocene '5' of the North German Eocene stratigraphical subdivision.

With the removal of the Lattorfian as the stratotype of the Lower Oligocene we are faced with a dilemma; either to trace the Eocene/Oligocene boundary at the base of the Rupelian *s.l.* or to locate a new stratotype for the Lower Oligocene. Indeed Krutszch and Lotsch (1957) have suggested that the Conow Beds in eastern Mecklenburg (which are the lateral equivalents of the *Neuengammer Gassande,* which are in turn generally considered by micropalaeontologists in the German Democratic Republic as a lower, sandy facies of the Rupelian) be chosen as a stratotype for Lower Oligocene. Korobkov (1964) has maintained, on the contrary, that a stratigraphical interval containing geological, and palaeontological criteria significant for the recognition of a new chronostratigraphical subdivision (i.e. a stage) does not exist. In the S.W. Soviet Union (N. Caucasus) there is certainly a stratigraphical section (which represents geological time) missing between the white marls of the Beloglin Formation (Upper Eocene) (with a rich pelagic foraminiferal fauna) and the brown sandy marl and clay of the Khadum Formation (Maykop Group) (with a molluscan and ostracod fauna) of Repulian Age. However, the interval is probably short and corroboration of this is seen in examining the results of Batjes's (1958) study on the Oligocene of Belgium. His examination of microfaunas and facies of the Upper Eocene-Oligocene of Belgium revealed that it is 'unlikely that the sequence Tongrian-Rupelian-Chattian represents a clear-cut subdivision of Oligocene time' (Batjes 1958, p. 97); the Boom Clay is synchronous with the Rupelian marginal deposits as well as part of the Tongrian and the Chattian. Batjes concluded that the Tongrian is partially synchronous with late Bartonian and early Rupelian, whereas the Chattian may have a similar relationship to the late Rupelian. The lagoonal-brackish Sands of Berg are best interpreted as lateral equivalents of 'some part of the Rupel Formation in western Belgium' (Batjes 1958, p. 95). Batjes (1958, p. 95) considers it likely that the time-stratigraphical equivalents of the Upper Tongrian (which is absent in western Belgium) are to be found in the Berg Sand which underlies the Boom clay. Here about 20 metres of 'uncharacteristic sand' that is to say Berg Sand, overlie the Late Eocene Asse Sand with a vague, indistinct boundary. This represents a return to the ideas expressed by Beyrich (1854, 1856) and Ortlieb and Dollfus (1873, cited in Forir 1901), who believed that the Upper Tongrian Beds, the Berg Sand and *Nucula* clay in Limburg were facies equivalents of the Boom clay in western Belgium. Beyrich, of course, believed in the existence of a Lower Oligocene (=Magdeburg Sands) below the Septaria Clay horizon of the Rupelian, but these strata are probably correlative with the Lattorfian of Mayer-Eymar (1893) and equivalent to the Lower Tongrian. Thus we see that there is evidence for extending the concept of Rupelian down to include strata which have normally been called Lattorfian, but which represent the time interval between late Eocene and the deposition of the Rupel Clay in its type area. In this way the palaeogeographical homogeneity of the stage in its type area is emphasized, with the lower part representing the early, shallow-marginal phase of the Oligocene transgression, and the middle part that is to say Rupel Clay, representing the maximum transgression of the Oligocene in northern Europe.

Merklin (1964) has reached somewhat different conclusions. In a discussion of the marine Aquitanian in the southern part of the Soviet Union and the Palaeogene/Neogene boundary, he places the sands and clays in Cisarabia and Ustyurt with *inter alia Glycymeris obovata, Nucula comta, Dosiniopsis sublaevigata* in the Rupelian, and correlates them with the Lower Maykopian in the North Caucasus (as has Korobkov). The Solenovsky horizon above is correlated with the upper part of Rupelian. The Batgubeck horizon above is correlated with the Chattian-Aquitanian which are both included in Upper Oligocene. Merklin (1964) noted the transitional nature of the molluscan fauna in the Baygubeck horizon and includes it in the Oligocene since, according to him the Burdigalian equivalents contain faunas of definite Miocene affinities.

If we turn now to the upper limit of the Oligocene we are faced with an equally frustrating situation. In a summary of the Neogene stratigraphy of Austria Janoschek (1960, p. 151) observes that, 'Unter den vielfach mit Lokalnamen bezeichneten Schichtgliedern seien besonders hervorzuheben die Chattischen Linzer Sande, der untermiozane Haller Schlier und der durch R. Hoernes bekannt gewordene helvetische Ottnanger Schlier. Leider is es jedoch bischer noch nicht gelungen, das exact Alter der einzelnen Schichtglieder und deren Grenzen durch Leitfossilien genau festzulegen. *Ungeklart ist vor allem die Oligo/Miozangrenze*'. Janoschek is here referring to the Oligocene/Miocene boundary in a specific region, but his despair has been echoed in the works of numerous investigators faced with the problem of this boundary in many parts of the world. Mayer-Eymar (1853) had created his Aquitanian Stage in order to justify a distinction between a lower and an upper Miocene (of Lyell, 1833), corresponding respectively to Mediterranean I (=Aquitanian) for which Depéret (1892) created the Burdigalian Stage, and Mediterranean II (which Depéret later believed was equivalent to the Helvetian-Tortonian) of Suess. He placed in his new stage the stratigraphical units of the following localities: Bazas (subsequently fixed as the stratotype of the Aquitanian by Dollfus, Munier-Chalmas and de Lapparent; see also Drooger, Kaasschieter and Keij 1955), Merignac, Kassel, Doberg and Sternberg (the last three in north Germany). Mayer-Eymar (1858) subsequently transferred his Aquitanian Stage to Beyrich's Oligocene which had been created in 1854. Fuchs (1894), in overlooking the fact that Mayer-Eymar (1858) had transferred his Aquitanian Stage to the Oligocene, created the Chattian Stage with the section at Kassel as the stratotype. Further confusion has resulted from the fact that Mayer-Eymar included in his lower Aquitanian the lacustrine limestones of Étampes and the so-called '*Meulières*' in the vicinity of Paris, which are generally considered older and a part of the Stampian.

This situation, in addition to several other factors, has led to a considerable amount of disagreement and confusion as to the nature of the Oligocene/Miocene boundary and the relationship between the Chattian and Aquitanian Stages. A detailed historical summary of this problem has been presented by Csepreghy-Meznerics (1962) and myself (1963).

In northern Germany the Rupelian deposits grade upwards into a sandy facies characterized by *Asterigerina guerichi,* the so-called '*Asterigerina* horizon' of German palaeontologists. This corresponds to the lowermost part of the Chattian Stage, and represents a shallow, more neritic environment than that in which the Rupelian deposits were formed. Although the Chattian deposits of northern Germany contain both macro- and microfossils it has not been possible to present a biostratigraphical subdivision or zonation of these deposits (even in the most complete sections at Doberg and Astrup), or to recognize significant elements in common with age-correlative deposits in the Mediterranean region (see Grossheide and Trunko 1965; Kiesel 1962; Batjes 1958). However, the Chattian (in the type area) lies conformably and transitionally upon Rupelian clays (see footnote 2 in Zobelein 1960, pp. 247, 248) and is the regressive phase of a single sedimentary cycle, equivalent to the Oligocene.

The Chattian Stage was created by Fuchs (1894) for the upper Oligocene of Beyrich (1854). Although he chose the '*Kasseler Meeressand*' at Gelben Berg near Niederkaufingen, Germany as the stratotype, he was actually motivated to erect the stage for the *Pectunculus* sands of Hungary, which he believed to be older than Aquitanian and younger than Rupelian. In this manner we have the dual problem of the definition of a stage unit and its stratigraphical correlation in areas outside the type region.

Although the type locality of the Chattian is the marine sands at Gelben Berg, there has been a tendency in recent times to include the outcrop at Doberg near Bunde, where a more complete marine section is exposed, in determining the characteristics of the Chattian Stage.

Hubach (1922, 1957) studied the stratigraphical section at Doberg, and subdivided the Chattian there into two parts:

Eochattikum (=Eochattian) and Neochattikum (=Neochattian). These have been called the '*Unterer Dobergerschichten*' and '*Oberer Dobergerschichten*', respectively, by Schmidt in Görges (1957) and by Hubach (1957) himself. Anderson (1961) has discouraged the use of the terms Eochattian and Neochattian and proposed a threefold biostratigraphical subdivision of the Doberg section on the basis of the pectinids and other molluscs. The foraminiferal fauna of the *Kasseler Meeressand* has been described by Reuss, Roemer, Speyer and von Koenen (see further historical background in Anderson 1961). The Foraminifera of the Doberg section were studied by Hosius (1895). Recently Grossheide and Trunko (1965) have studied the Foraminifera of the sections at Doberg and Astrup, but were unable to recognize a biostratigraphical subdivision of the Upper Oligocene in these localities. Based on a comparative evaluation of data presented by Hubach (1922, 1957), Görges (1941, 1951) and Anderson (1958, 1961) an attempt has been made to relate approximately the stratotype Chattian (as well as the Eochattian-Neochattian of Hubach) to the foraminiferal zonations currently in use (Table 52.23). Hubach (1922, 1957) suggested that the stratotype *Kasseler Meeressand* corresponds to his beds 7-16 at Doberg; Anderson (1961) said only that it was within his molluscan Zone A. Bed 7 has recently been shown to contain the *Asterigerina guerichi* assemblage (Indans 1965), which had been placed in 'Zone E' = basal Chattian by Indans (1958).

Recent investigations by myself (Berggren, in press) on Oligocene planktonic foraminiferal faunas of northern Europe suggests that the Chattian Stage is partly, if not wholly, correlative with Zone P20/N1. *Globorotalia opima s.s.* and forms which can be assigned to *Globigerina ex interc.angulisuturalis* (but not possessing the full morphological development of that species) have been found in Eochattian samples from Astrup. This suggests a high level in Zone P20/N1.

Drooger (1960a) has recorded the earliest miogypsinids from the upper part of the Eochattian (Zone B of Anderson 1958*b*, 1961) at Doberg, and from the Neochattian (Zone C of Anderson 1958*b*, 1961) at Astrup. No miogypsinids were found in the Kassel Sands. Thus it would appear that the earliest record of miogypsinids in the Chattian of northern Germany is within Zone P20/N1, and somewhat lower than the base of the *Globorotalia opima opima* Zone (=N2 of Banner and Blow 1965*b*).

It is not possible to relate the uppermost part of the Neochattian at Doberg to a planktonic foraminiferal zone, but it is more than likely that it is either in the uppermost part of Zone P20/N1 or low in Zone N2. Thus the top of the Chattian Stage is estimated here to lie approximately at the N1/N2 boundary. The base of the stratotype section of the Aquitanian lies approximately at the base of the N4 Zone. Between the uppermost Chattian and the base of the Aquitanian Stage *s.s.* there is still the N2 and N3 Zone between these two limits. It would seem feasible to consider extending the concept of Aquitanian downwards to include strata in the vicinity of Bordeaux-Dax, which are known to be older than those exposed in the Saint Jean d'Etampes type section so as to allow a solution to the Oligocene/Miocene boundary in terms of planktonic foraminiferal zones. Alternately there is the possibility of an additional stage between the Chattian *s.s.* and Aquitanian *s.s.* such as the Bormidian of Pareto (Banner and Blow 1965*b*) for which further documentation will soon be presented.

If the Oligocene/Miocene boundary is placed at the base of the N2 Zone, it probably corresponds to the upper chronostratigraphical boundary of the Chattian Stage. Thus the boundary would coincide closely with the lowest occurrence of miogypsinids, and antedate by a short interval of time the last occurrence of *Nummulites (petites Nummulites* of French authors). In this sense the upper limit of the Palaeogene coincides approximately with that of the Nummulitique, and serves to define a significant and easily recognizable part of Tertiary earth history.

In northern Europe (Germany) the Chattian is uncomformably overlain by the '*Hemmoor-Stufe*' (Gripp 1919) which is generally correlated with the Burdigalian Stage. In northwestern Germany (Hannover-Hamburg region) and in Denmark-Schleswig-Holstein the marine sands of the Upper Oligocene (=*Sternberger Gestein*) are overlain by the marine sands of Vierlande, and the so-called *Glimmerton* (*Vierland Stufe*, Gripp 1919) which are generally correlated with the Aquitanian.

Anderson (1963, p. 114) presents evidence (continuity of some elements of the Chattian molluscan fauna into the *Vierland Stufe*) for considering the Vierlandian Stage (Kowing 1956) to lie directly upon the Chattian without stratigraphical hiatus, and to define the Oligocene/Miocene boundary in northern Germany. By extension the Vierlandian is probably equivalent (in this interpretation) to the Aquitanian *s.l.* (of Mayer-Eymar) and to the Bormidian as used by Banner and Blow (1965*b*).

However, the '*Vierland Stufe*' has not so far been accurately stratotypified; palaeontological and stratigraphical subdivision and the limits of the unit have not been set, and the term seems useful, at best, in a rock-stratigraphical, not a chronostratigraphical sense. In short we have no positive evidence of the Chattian-Aquitanian transition (i.e. nature of the Oligocene/Miocene boundary) in northern Europe.

In a study of the Bavarian Molasse Hagn (1955) found that it was possible to utilize the uvigerinids in determining the Oligocene/Miocene boundary. He recognized a late Chattian in the Rainer Muhle in Priental, as well as in the Ortenburg CF1002 boring, on the basis of the distinctive ribbed *Uvigerina hantkeni*. The Aquitanian was said to be characterized by its smooth uvigerinids. *Asterigerina guerichii* was also shown to be characteristic of the 'Chattian' in Bavaria, which allows a general correlation with the type Chattian of northern Germany (see *above*).

In the Sirte Basin of Libya, I have utilized the uvigerinids in preparing a biostratigraphical subdivision of the post-Eocene sequence. An *Uvigerina rustica* Zone was recognized, which corresponded approximately to the *G. ciperoensis* and *G. kugleri* Zones. An *U. hantkeni* Subzone, corresponding approximately to the *G. ciperoensis* Zone, was also distinguished. Thus the Oligocene/Miocene boundary, as determined by Hagn (1955), would appear to lie within or near the top of *G. ciperoensis* Zone.

In northern France (Paris Basin) the distinction between Chattian and Aquitanian has not been proven to the satisfaction of all specialists. However, most French stratigraphers consider the Chattian a synonym of the Aquitanian (Denizot 1957). The *Calcaire de Beauce, Calcaire de l'Orléanais* and associated deposits are generally attributed to the Aquitanian by French stratigraphers. The *Sables de Voort* (in Belgium), generally placed in the Chattian by Belgian geologists, were placed in the upper part of the Stampian which is equivalent to Rupelian by Blondeau *et al.* (1965), which essentially implies the absence of Chattian in Belgium (but see Batjes 1958, p. 97).

In the Bordelais region of the Aquitaine Basin the Oligocene/Miocene boundary (i.e. base Aquitanian) is drawn at an erosional disconformity separating the shallow-water limestones below (*Calcaire à Astéries*) from the *Marnes à Cèrites = Couches de la Brède* and the *Couches de Mérignac* and *Lariey* (above). Elsewhere the passage appears to exist in a more normal marine environment. Separation of the two is difficult on the basis of the microfauna but a general scheme has been presented by Alvinerie, Caralp, Moyes and Vigneax (1964). These authors suggest that it may be advisable to unite the Chattian Stage in the Upper Rupelian (see also Doebl 1964, pp. 597, 598, where the Chattian is considered an equivalent of the upper Stampian *Marnes à Cyrènidés* and freshwater marls and limestones with *Helix ramondi*, overlain by the Aquitanian *Cerithium* beds in the Mayence Basin and the Lower Rhine Valley and Alsace).

Lorenz (1963, 1965) and Szöts (1965) have discussed the relationship of Oligocene 'stages' and the Oligocene as a chronostratigraphical unit and have both reached the conclusion that the Oligocene corresponds to a single chronostratigraphical stage:

BEYRICH (1854, 1856)	FUCHS (1894)	HUBACH (1927, 1957) STAGE	HUBACH ROCK UNIT		HUBACH BEDS		HUBACH BIOSTRATIGRAPHIC UNITS	HUBACH SOME CHARACTERISTIC MOLLUSCS	GÖRGES (1940, 1951) STAGE	GÖRGES ROCK UNIT	GÖRGES THICK-NESS	GÖRGES SOME CHARACTERISTIC MOLLUSCS			ANDERSON (1958, 1961, 1962) BIOSTRATI-GRAPHIC UNITS	ANDERSON SOME CHARACTERISTIC MOLLUSCS
OLIGOCENE (UPPER)	CHATTIAN	NEOCHATTIKUM = NEOCHATTIAN	DOBERGERSHICHTEN	OBERER DOBERGERSHICHTEN	32 - 53	42-53	F	P. semistriatus	NEOCHATTICUM	OBERER DOBERGERSHICHTEN	33 m	UPPER	Chlamys semistriata s.s.		C	Pecten (Hilberia) hoffmanni P. (H.) semicostatus Chlamys (Camptonectes) semistriata
						35-41	E	Maretia hoffmanni (to bed 38)				MIDDLE				
						32-34	D	last C. neovenosa P. hoffmanni				LOWER		C. hoffmanni		
		EOCHATTIKUM = EOCHATTIAN		UNTERER DOBERGERSHICHTEN	1 - 31	21-31	C	Echinolampas. kleinii (to bed 29) C. decussata (to bed 25) P. semistriatus	EOCHATTICUM	UNTERER DOBERGERSHICHTEN	37 m	UPPER	Amussium corneum Chlamys hausmanni s.s. C. hausmanni exlaevigata C. crinita C. striatocostata C. decussata (lower part)		B	P.(H.) bifidus praehoffmanni C.(C.) hausmanni s.s. & ex-laevigata, C.(C.) ambigna
	* CASSEL SANDS = STRATOTYPE					7-20	B	Pecten hauchecornei Amussium pygmaeum C. decussata				MIDDLE	Chlamys decussata (upper part) C. janus C. bifida s.s. C. pictae aquaetranquillae C. pygmae C. hauchecornei		A	P. (H.) bifidus s.s. P. (H.) bifidus lucidus P. (H.) bifidus ocuticostatus C. (C.) aquaetranquillae C. (C.) cancellata C. (C.) decussata Palliolum (Similipecten) hauchecornei
						1-6	A	Isocardia subtransversa plana Pecten bifidus Chlamys pictae aquaetranquillae				LOWER	Chlamys bifidus acuticostata			

* THE POSITION OF THE CASSEL SANDS-STRATOTYPE OF THE CHATTIAN STAGE OF FUCHS(1894)-IS AFTER HUBACH(1957) WHO SUGGESTED THAT THE CASSEL SANDS CORRESPOND APPROXIMATELY TO HIS BEDS 7-16, UNIT B OF THE EOCHATTIKUM = EOCHATTIAN.

Table 52.22. Biostratigraphical subdivision and correlation of the Chattian in the type area – North Central Germany

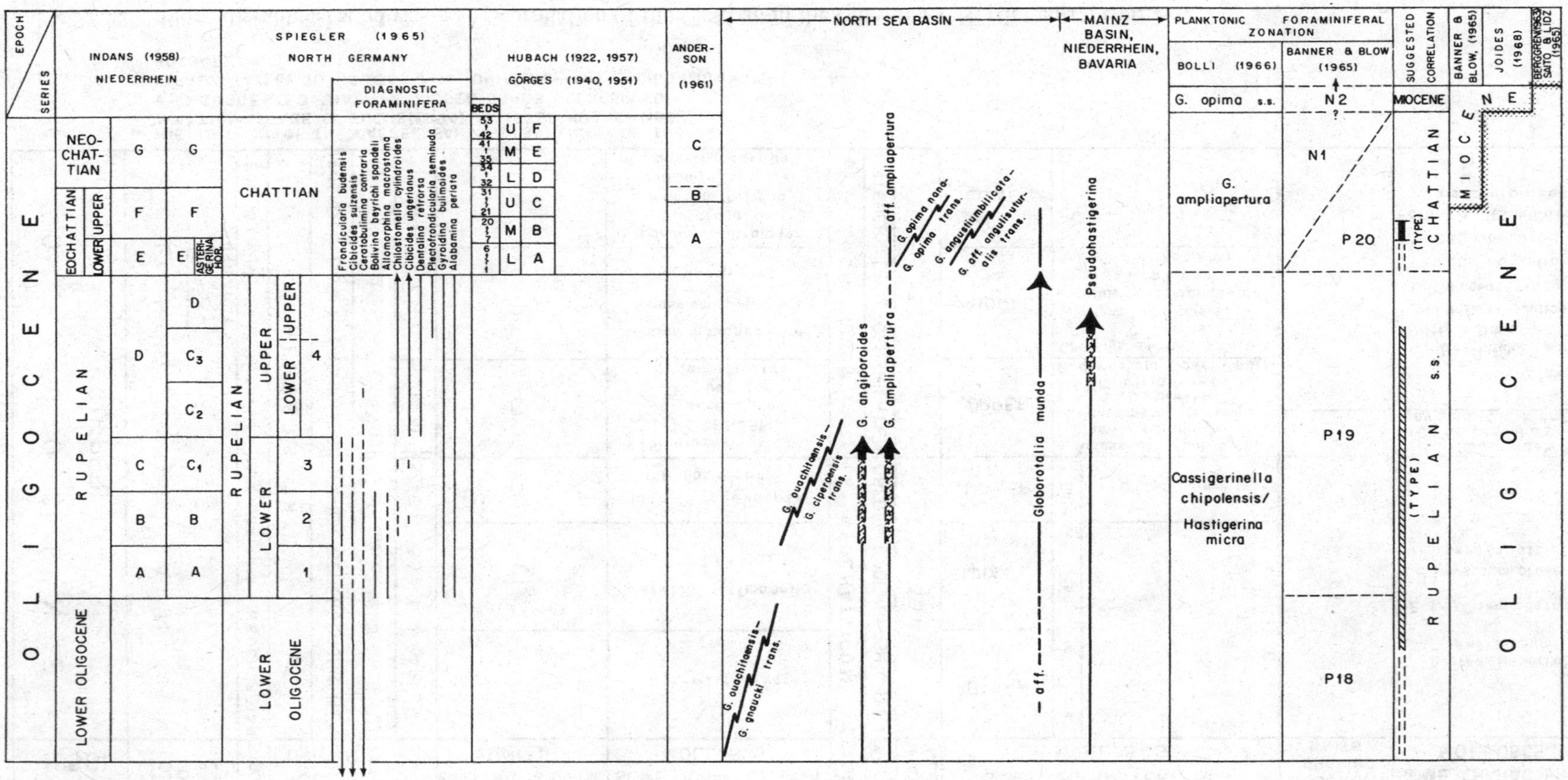
EPOCH
SERIES
INDANS (1958)
NIEDERRHEIN
SPIEGLER (1965)
NORTH GERMANY
DIAGNOSTIC FORAMINIFERA
HUBACH (1922, 1957)
GÖRGES (1940, 1951)
BEDS
ANDER-SON (1961)
NORTH SEA BASIN
MAINZ BASIN, NIEDERRHEIN, BAVARIA
PLANKTONIC FORAMINIFERAL ZONATION
BOLLI (1966)
BANNER & BLOW (1965)
SUGGESTED CORRELATION
BANNER & BLOW, (1965)
JOIDES (1968)
BERGGREN(1963) SAITO & LIDZ (1965)
OLIGOCENE
NEO-CHAT-TIAN
EOCHATTIAN
LOWER
UPPER
RUPELIAN
LOWER OLIGOCENE
ASTERI-GERINA HOR.
CHATTIAN
Frondicularia budensis
Cibicides sulzensis
Ceratobulimina contraria
Bolivina beyrichi spandeli
Allomorphina macrostoma
Chilostomella cylindroides
Cibicides ungerianus
Dentalina retrorsa
Plectofrondicularia seminuda
Gyroidina bulimoides
Alabamina perlata
G. ouachitaensis–G. gnaucki trans.
G. ouachitaensis–G. ciperoensis trans.
G. angiporoides
G. ampliapertura
aff. ampliapertura
G. opima nana–G. opima trans.
G. angustiumbilicata–G. aff. angulisuturalis trans.
aff. Globorotalia munda
Pseudohastigerina
G. opima s.s.
G. ampliapertura
Cassigerinella chipolensis/ Hastigerina micra
N2
N1
P20
P19
P18
MIOCENE
CHATTIAN
(TYPE)
RUPELIAN s.s.
MIOCENE

Table 52.23. Biostratigraphy of the Oligocene of Northern Europe

Stampian.

Recent studies by Drooger (1960) and Butt (1966) have led to the suggestion that the section at Escornebéou (Aquitaine Basin), considered by many to be Aquitanian in age, is actually Upper Oligocene and correlative with the Chattian in northern Europe. This was based upon the discovery of *Miogypsina complanata* and *Lepidocyclina (Nephrolepidina) morgani* at Escornebéou. Butt (1966) has recorded an association of *Nummulites bouillei, Spiroclypeus blanckenhorni* var., *Gryzbowskia assilinoides, Lepidocyclina (Nephrolepidina) dilitata, L. (N.) morgani, Miogypsina (Miogypsinoides) complanata* and a planktonic foraminiferal fauna containing *inter alia,* the first *Globigerinoides, G. primordius* Blow and Banner. This foraminiferal association is certainly anomalous. An examination of the planktonic foraminiferal collections deposited by Butt at the Department of Geology, University of Utrecht by me, in 1966, has led to the following interpretation:

Butt's determination	Berggren's remarks
1. *Globigerinoides trilobus primordius*	*G. primordius* is present, but specimens which show the more advanced morphological development of *G. trilobus trilobus* are also included.
2. *Globigerina bulloides* d'Orbigny	*G. praebulloides* and *G. occlusa* are included here.
3. *Globigerina opima nana* Bolli	The majority of specimens deposited by Butt, included the specimen figured as fig. 16, pl. 7, are *Globorotaloides suteri* Bolli
4. *Globigerina globularis* Roemer	*G. praebulloides* and *G. occlusa,* predominantly the latter, are included here
5. *Globigerina obesa* (Bolli)	Some specimens appear to be referable to *G. obesa;* some may be transitional forms.
6. *Globigerina concinna* Reuss	Different forms observed on different slides; specimens appear to be related to *G. ciperoensis* group; typical and atypical forms are present.
7. *Globigerina ciperoensis ciperoensis* Bolli	*G. ciperoensis* predominantly. *G. angustiumbilicata* and *G. angulisuturalis* also associated.
8. *Globigerina* cf. *ampliapertura* Bolli	*Globigerina woodi* Jenkins.

The evidence of the planktonic Foraminifera at Escornebéou suggests to me that faunas as young as upper *G. kugleri* Zone equivalent are present, and that faunas as old as *G. ciperoensis* Zone age may be present, or reworked.

The association is interesting in that it is similar to that observed by Jenkins (1964, 1965) from the type Aquitanian and correlated with the *Globorotalia kugleri* and *G. ciperoensis* Zone of Bolli (1957). Blow and Banner (1962) have claimed also to have seen evidence of the *Globigerina ouachitaensis ciperoensis* Zone in a sample from Escornebéou. Jenkins (1965) has recently suggested, on the basis of a study of type Aquitanian-Burdigalian planktonic foraminifers, that the Oligocene/Miocene boundary be drawn at the base of the *Globigerina ciperoensis* Zone.

Drooger (1960) has recorded the earliest *Miogypsina* and *Miogypsinoides* in the Chattian and has recognized related forms at Escornebéou in the Aquitaine Basin, suggesting that these deposits are older than Aquitanian. Eames *et al.* (1962, p. 42) note that in the Caribbean region the first *Miogypsina (s.s.)* occur in the *Globorotalia opima opima* Zone (see also Hottinger 1964; cf. Stainforth 1960).

The coexistence of *Globorotalia opima opima* and *Miogypsina septentrionalis* and *Miogypsinoides complanata s.s.* (characteristic of the Chattian) has been reported in the Prerif region of Morocco (Choubert, Hottinger, Marcais and Suter 1964, pp. 254, 255). This is important evidence for narrowing the Oligocene/Miocene boundary in terms of the planktonic zonation which is widely used today. It suggests that the Oligo-Miocene boundary could be placed within or at the base of the *Globorotalia opima* Zone (approximately Zone N2 of Banner and Blow 1965). Berggren (1963), Lagaaij (1963) and Saito and Lidz (1965) have placed the Oligocene/Miocene boundary within the *Globorotalia opima* Zone; Jenkins (1965) places it at the top of the Zone, Banner and Blow (1965) place it somewhat below the first evolutionary appearance of *G. angulisuturalis* (Zone N2), i.e. in their *Globigerina ampliapertura* Zone after the last *Pseudohastigerina* and before the first evolutionary appearance of *G. angulisuturalis* (=?Chattian). I have observed *G. ampliapertura* and *Pseudohastigerina* together in core samples within the lower half of the *G. ampliapertura* Zone in Libya (North Africa) (Berggren 1963); approximately 300 feet of section characterized by the presence of *G. ampliapertura* and the absence of *Pseudohastigerina,* and followed above by the *G. opima opima* Zone would probably correspond to this interval (N1) which is equivalent to *G. ampliapertura* Partial-Range-Zone of Banner and Blow (1965).

The isolated occurrences of species of larger Foraminifera are not in themselves sufficient for more than approximate correlation over long distances, and Drooger himself has recognized these limitations. For this purpose the distinct assemblages which characterize the planktonic foraminiferal zones are distinctly superior. These zones represent a distinct sequence in time (the distinction between planktonic foraminiferal zones as biostratigraphical units, i.e. biozones, and chronostratigraphical units, i.e. chronozones, is rather subtle), and it will be admitted by most palaeontologists that the sequential superposition of planktonic foraminiferal zones affords the most accurate method for subdividing geological time in marine facies.

Eames *et al.* (1962) have presented the results of an exhaustive comparative study of the Oligocene/Miocene in East Africa, the Mediterranean Region, northern Europe, the Near East, the Far East and the Caribbean Region. A thorough critique of this work is beyond the scope of this paper (a general review and discussion of their work has been presented by Berggren 1963; Szöts 1963; Drooger 1964, *inter alia*; and the authors of that work are themselves now in the midst of a detailed revision and expansion of much of their original work). The Lattorfian Stage was recognized by the presence of *Nummulites vascus-intermedius, Palaeonummulites incrassatus* and no *Eulepidina.* The Rupelian was recognized by the co-occurrence of the *N. vascus-intermedius* and *Eulepidina.* The Chattian was said to be characterized by *Palaeonummulites incrassatus* and *Eulepidina* after the extinction of *Nummulites s.s.; Miogypsinella* was also noted as occurring at this level in some areas. The Aquitanian was said to be characterized by *Eulepidina, Miogypsina, Miogypsinella, Miogypsinoides.*

The most controversial aspect of the results presented by Eames *et al.* (1962) is their suggestion that a distinct hiatus exists within the San Fernando Formation in Central Trinidad, and that the Oligocene is generally absent in the Caribbean region (with the exception of the Alazan Formation in Mexico, and an unnamed formation in the Dominican Republic), the Gulf Coast, the Mediterranean Region, the Middle East and the Far East. The absence of true *Nummulites,* from the sequence generally attributed to the Oligocene (Vicksburgian) in the Gulf Coastal Plain of the United States, was used as an argument for including the Vicksburgian in the Lower Miocene. Since the authors are in the process of preparing for publication a comprehensive study on world-wide post-Eocene biostratigraphy, it would not be appropriate for me to comment in any detail on this work in advance. Suffice it to say that the Vicksburgian is looked upon as a useful Oligocene stage, if not superior to those commonly used in Europe (Blow, pers. comm.). Indeed, in the micropalaeontological manual currently being prepared for the J.O.I.D.E.S. deep-sea drilling programme, the term Vicksburgian is being suggested for standard use (equivalent to the Lattorfian, Rupelian and Chattian as used by various authors in Europe).

Stainforth (1965) has recently presented a summary discussion on mid-Tertiary diastrophism in northern South America and its relationship to planktonic foraminiferal zones.

Three major cycles of deposition are recognized (from the bottom up): *Hantkenina – G. ciperoensis* interval (*ciperoensis* Zone here includes *G. ampliapertura – G. kugleri* Zones of Bolli); *G. dissimilis* Zone into *G. fohsi* Zone; *G. menardii* interval. Sedimentation was said to be continuous on a regional scale from the *Hantkenina* Zone into the *G. ciperoensis* Zone (but Blow criticized this opinion verbally at the 4th Caribbean Conference). The zonal boundary was also said to correspond to a phase of maximum marine transgression near the midpoint of a sedimentary cycle. Stainforth (1965) rejects the emphasis placed on *Pliolepidina tobleri* as a Miocene (Aquitanian) index fossil. In the entire Caribbean area this form is generally acknowledged to be a valuable Eocene marker, where it is normally found in association with discocyclinids. The absence in the Caribbean of two planktonic zones which in East Africa separate the *Hantkenina* Zone (*G. cocoaensis* Zone of Bolli) and the *G. ciperoensis* Zone (*G. ampliapertura* Zone of Bolli) is also used by Eames *et al.* (1962) as a strong argument for their postulation that the Oligocene is missing in the Caribbean. However, I have questioned the validity of this argument (Berggren 1963). In the Sirte Basin, Libya, continuous deposition at bathyal depths occurred across the Eocene/Oligocene boundary. Normal, transitional fauna development can be observed in the benthonic and planktonic fauna. In three years of work on these faunas I was unable to recognize and separate the *G. sellii* Zone from the *G. ampliapertura* Zone. Therefore, while partially agreeing with Stainforth (1965, p. 19) in his suggestion that the 'distinctive East African species were geographically confined by the climatic provincialism which affects distribution of all planktonic Foraminifera'. I would point out that these forms are now known to have a fairly wide distribution within tropical-subtropical regions. The occasional presence in the Caribbean of *G. sellii* (Eames *et al.* 1962, pp. 35, 49, 71), in the Gulf Coast (*fide* Olsson, pers. comm.) and in the J.O.I.D.E.S. drill holes, indicates that it did occur as an element in Oligocene faunas in these areas. The planktonic faunas indicative of Zones P18 and P19 (=Oligocene of Blow and Banner) have been observed by me in the J.O.I.D.E.S. cores from the Blake Plateau, in deep-sea cores from the South Pacific, and in the Caribbean (Barbados). The presence of Oligocene is, indeed, very real in the Caribbean-Gulf Coast, but the actual limits of the Oligocene remain a topic of vigorous debate.

In a recent paper Butterlin (1965) has presented a discussion of the larger Foraminifera of the Caribbean region and their bearing on the limits of the Oligocene. The author takes as a point of departure the fourfold zonation of the Oligo-Miocene based on larger Foraminifera presented by Gravell and Hanna (1938). These are (from the bottom up):

(1) *Lepidocyclina mantelli*
(2) *Lepidocyclina supera*
(3) *Lepidocyclina (Eulepidina)*
(4) *Miogypsina-Heterostegina*

Since *L. mantelli* was only found in the Gulf Coast, it was thought at the time that the basal marine Oligocene was represented only in the Gulf Coast, and absent in the remainder of the Caribbean area. The evolution of thought on the geological history of the Oligocene is seen to have turned full circle with the recent suggestion by Eames *et al.* (1962) that the Oligocene System is missing in the Caribbean-Gulf Coast Region. In fairness to the authors it should be observed here that they (Eames *et al.*) have since modified their position on the development of Oligocene in the Gulf Coast-Caribbean region. Full documentation of their views will soon appear in print.

However, Butterlin draws attention to the fact that on Haiti he has observed the *Lepidocyclina (Eulepidina)* Zone immediately above levels with Upper Eocene larger Foraminifera. Butterlin (1965) suggests that in the Antilles the three lower zones of Gravell and Hanna (1938) are actually a single zone involving an identical stratigraphical extension. Cole (1957, p. 34) has suggested that *L. supera* should be included in the synonymy of *L. mantelli,* which strengthens the conclusion of Butterlin. Indeed Cole (in Cole and Applin 1961, p. 131) included the lower three zones of Gravell and Hanna (1938) into a single zone, which he equated with the Oligocene. Thus:

Lower Miocene	-	*Lepidocyclina-Miogypsina* Zone
Oligocene-*Eulepidina* Zone	-	upper-*Miogypsina* Subzone lower-*Lepidocyclina* Subzone

Sachs (1959) had earlier subdivided the *Lepidocyclina-Miogypsina* Zone into a lower *Miogypsina-Lepidocyclina s.s.* Subzone (Aquitanian) and an upper *Miogypsina* Zone (without *Lepidocyclina s.s.*) (Burdigalian).

On the basis of his work in the Caribbean, and particularly in Haiti, Butterlin has proposed a post-Eocene zonation based on larger Foraminifera which is shown below for the sake of comparison with zonations developed by other workers.

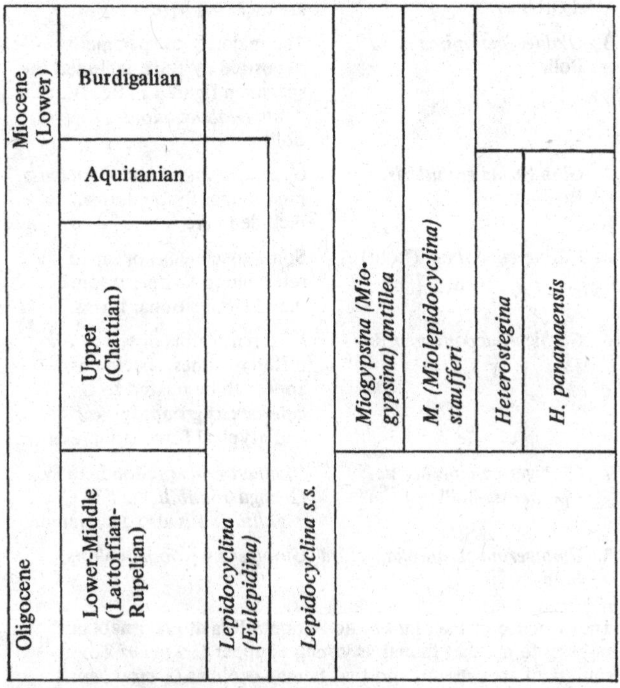
Miocene (Lower)
Burdigalian
Aquitanian
Oligocene
Upper (Chattian)
Lower-Middle (Lattorfian-Rupelian)
Lepidocyclina (Eulepidina)
Lepidocyclina s.s.
Miogypsina (Miogypsina) antillea
M. (Miolepidocyclina) staufferi
Heterostegina
H. panamaensis

Zonation of Oligocene-Lower Miocene in the Caribbean based on larger Foraminifera (Bürgl 1965, p. 15).

Bürgl (1965) has stated that the *Cibicides perlucidus* Zone of Petters and Sarmiento (1956) in Colombia contains a planktonic foraminiferal assemblage characteristic of the *G. oligocaenica* (vel *G. sellii* Zone. It is overlain by a zone with a fauna transitional between the *G. oligocaenica* and *G. ampliapertura* Zones, which is interpreted as Chattian in age by the author. The next zone above contains *Globigerinoides, G. kugleri* and *G. mayeri (*vel *G. siakensis)* and corresponds to the upper part of the *G. kugleri* Zone of Bolli. The author states that the *G. ampliapertura, G. opima, G. ciperoensis* and lower part of the *G. kugleri* Zone have never been observed in Colombia and suggests that sediments of this interval were never deposited there (cf. Stainforth 1965).

Bürgl (1965) observed that in Colombia the *G. kugleri* Zone contains 'numerous planktonic and benthonic species known from other regions to be of Oligocene age and, in lesser number, of Eocene age', and he explains their presence by redeposition at the beginning of the Miocene transgression (*G. kugleri-G. dissimilis* Zones).

The results of work carried out by myself in the Sirte Basin of Libya (1963-5) are pertinent to the remarks by Bürgl (1965). *Cibicides perlucidus* was found to be a distinctive guide fossil in this area as well. It was found that the *C. perlucidus* Zone extended from the upper part of the *G. semi-involuta* Zone (Upper Eocene) to the basal part of the *G. ouachitaensis ciperoensis* Zone (my Lower Miocene). The Oligocene occurs within this interval, although its upper limit is open to discussion. The *G. sellii* Zone of Blow and Banner (1962) was not recognized in this sequence of continuous deposition. I have observed a remarkable parallelism in the development of Oligocene-Lower Miocene benthonic and planktonic faunas between North Africa (Mediterranean) and the Caribbean Region.

In concluding our discussion on the Oligocene it may be well to recall that Lyell (1833) had created his Miocene primarily for the rocks he had observed in the provinces of Touraine and Gironde in Southwestern France. Lyell (1857, p. 9) subsequently adopted the prevailing opinion of many of his colleagues that a significant break existed between the *Gypse de Montmartre* and the '*Sables de Fontainebleau*' Thus the gypsum beds were arbitrarily separated from marine beds which in England and Belgium are actually correlative. In the later editions of his 'Elements of Geology' the Miocene of western Europe was shown to begin with the first marine beds *above* the gypsum beds of Paris and its equivalents.

Lyell (1833, p. 389; Table 52.2) had also divided his Miocene into two parts, a lower and an upper. The upper (*a*) included the *Faluns de Touraine, Bordeaux* (which include rocks which range in age according to current opinion and usage from 'Helvetian'-Burdigalian), Valley of Bormida, Superga near Turin (an extensive section, which on the basis of subsequent work by Pareto, Sacco and Rovereto indicates the presence of Oligocene-Aquitanian), Basin of Vienna (Middle Miocene, post Langhian = ? Serravallian-Tortonian) and (*b*) Saucats, 'twelve miles south of Bordeaux' (Burdigalian) (Table 52.4).

There is no mention of the strata at Merignac and Bazas and the outcrops along the river Saint-Jean-d'Etampes which have subsequently been chosen as the Aquitanian stratotype. These, as well as the *Calcaire à Astéries*, (Aquitaine) *Calcaire de Beauce, Calcaire d'Etampes* (Paris Basin) and the various units in Belgium and Germany (see *above*) were included by Beyrich (1854) in his Oligocene. With due regard to the lack of knowledge at the time (Lyell would no doubt have included all the outcrops in the Bazadais-Bordelais region of the Aquitaine Basin now included in Chattian (or Stampian), Aquitanian and Burdigalian in his Miocene had he been more familiar with their regional development), we see that Lyell's lower subdivision (*b*) of the Miocene is essentially the Oligocene of Beyrich (1854) (as he himself observed, Lyell 1857, pp. 9, 13). The Aquitanian of Mayer-Eymar (in Gressly 1853) was included in the concept of Oligocene by Beyrich (1854) and Mayer-Eymar (1858) himself was quick to transfer his stage to the Oligocene in accord with Beyrich's opinions. This brings us back full circle to the extensive discussions on the Chattian-Aquitanian by Csepreghy-Meznerics (1962) and myself (1963) referred to above.

The transfer of the Aquitanian Stage to the Miocene (now accepted by most invertebrate palaeontologists) reduces the Oligocene to a relatively short interval of time in comparison to the other epoch-series of the Tertiary. If, as seems possible, the Oligocene can be demonstrated to correspond approximately to a single stage, the Rupelian (cf. *inter alia*, Douvillé and Rutten who reduced the Oligocene to Stampian), the question may well arise as to whether the Oligocene merits the rank of epoch-series. Indeed Haug, in his Treatise, included it in his 'Neonummulitique' (which he equated somewhat erroneously with Tongrian), the third and youngest group of his 'Période Nummulitique'.

The Oligocene Epoch probably represents an interval of time of approximately five million years (see section on Cenozoic radiometric time scale below), and is thus equivalent in its duration to some of the stages of the Palaeogene and Neogene. Indeed it is considerably shorter than some (e.g. Ypresian and Langhian). The microfauna and macrofauna of the Oligocene are, in general, strongly related to those in the Late Eocene. Indeed, the faunal picture is one of gradual, but persistent decrease and extinction of various Palaeogene elements towards the later part of the Oligocene. The marked change in both micro- and macrofauna occurs within or near the top of the Chattian and base of the Aquitanian. This is where the typical Neogene faunal elements first appear. In the light of this, in view of the considerable problems which the Oligocene has caused stratigraphers, and in view of the disproportionate extent of the Oligocene when compared with other chronostratigraphical units of equivalent rank in the Cenozoic hierarchy, it might be well to pause and consider whether it would be wiser to abandon the use of the term Oligocene altogether, and return to the original subdivision of the middle part of the Cenozoic by Lyell into Eocene and Miocene. The Oligocene is, in terms of its fauna, essentially 'Late Eocene', but more significantly it shows transitional features between the Palaeogene and the Neogene.

It would appear from the discussion presented in Eames *et al.* (1962), by Drooger (1960 *et seq.*), by myself (1963, and this paper) and various authors in the current polemics concerning the determination of an unambiguous Oligocene/Miocene boundary, that the main difficulties in the way of arriving at a clear-cut, unambiguous determination are the failures on the part of earlier workers, (*a*) to describe clearly the time-stratigraphical limits of their stage units, (*b*) to specify in an unambiguous manner a stratotype locality and, (*c*) poor choice of type areas for stages. Perhaps the main difficulty is to be found in the fact that most 'chronostratigraphical' units were originally defined as biostratigraphical units. Thus we see an explanation to the apparent dilemma in the different approaches towards solving stratigraphical problems in European stratigraphy so clearly discussed by MacNeill (1966, p. 2357). Whereas some authors (Csepreghy-Meznerics 1962), in discussing the age relationships between different stages, base their opinion on the distribution of (molluscan) faunas, which are strongly related to facies changes due to transgressive and regressive cycles, others (Drooger 1960 *et seq.*) record microfaunas from selected stratotypes of the stages. But the

TABLE 52.24 HISTORICAL PERSPECTIVE OF THE VICISSITUDES OF THE OLIGOCENE/MIOCENE BOUNDARY FROM 1945–1968

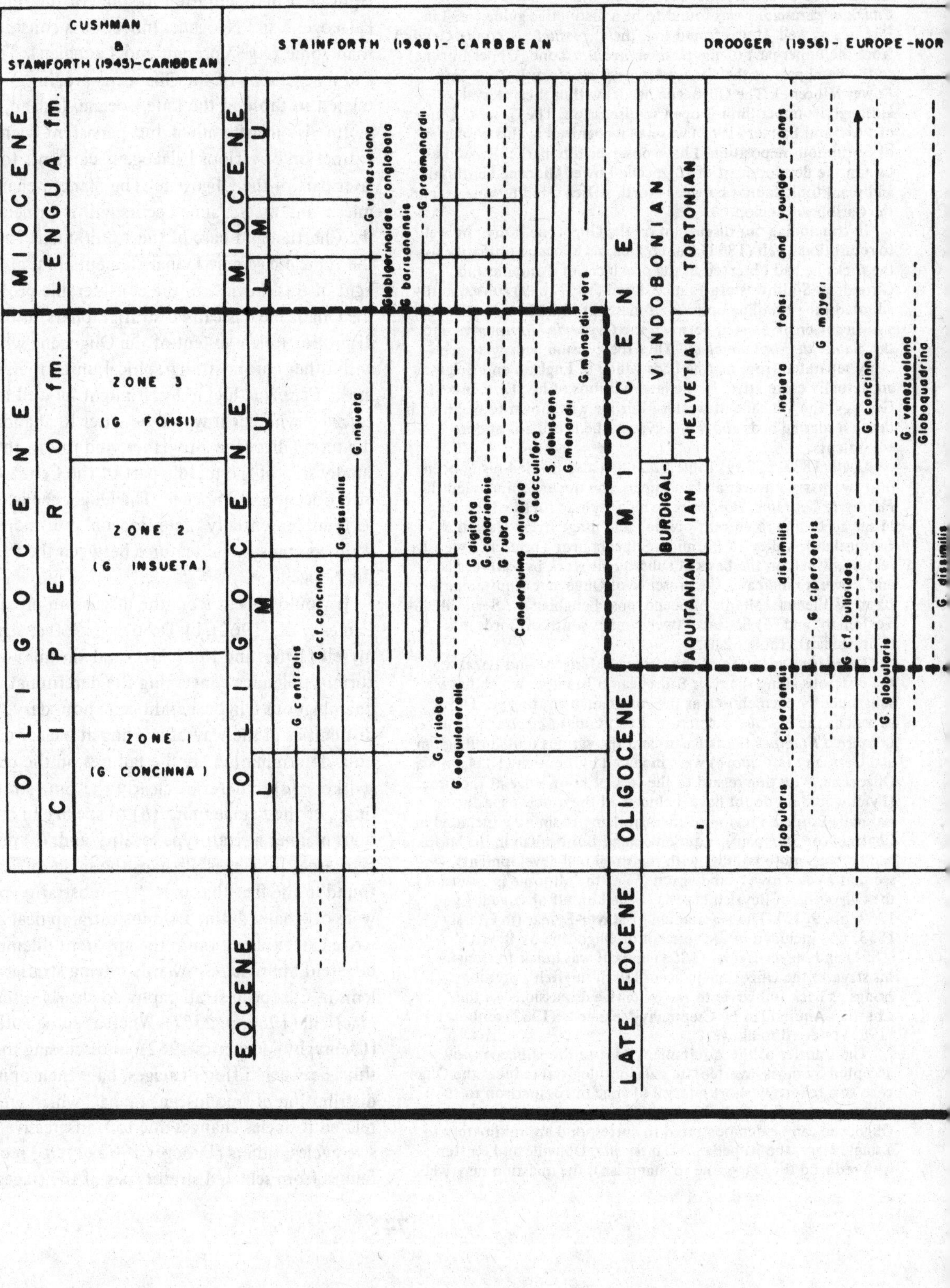

CUSHMAN & STAINFORTH (1945)-CARIBBEAN
STAINFORTH (1948)-CARRIBEAN
DROOGER (1956)-EUROPE-NOR
MIOCENE
OLIGOCENE
LENGUA fm.
CIPERO fm.
ZONE 3
(G. FOHSI)
ZONE 2
(G. INSUETA)
ZONE 1
(G. CONCINNA)
L
U
M
EOCENE
G. centralis
G. cf. concinna
G. dissimilis
G. insueta
G. venezuelana
Globigerinoides conglobata
S. barisanensis
G. praemenardii
G. triloba
G. aequilateralis
G. digitata
G. canariensis
G. rubra
Candorbulina universa
G. sacculifera
S. dehiscens
G. menardii
G. menardii vars
LATE EOCENE
OLIGOCENE
MIOCENE
AQUITANIAN
BURDIGALIAN
HELVETIAN
VINDOBONIAN
TORTONIAN
globularis
ciperoensis
dissimilis
insueta
fohsi
and younger zones
G. mayeri
G. ciperoensis
G. cf. bulloides
G. concinna
G. globularis
G. venezuelana
Globoquadrina
G. dissimilis

TABLE 52.24 continued

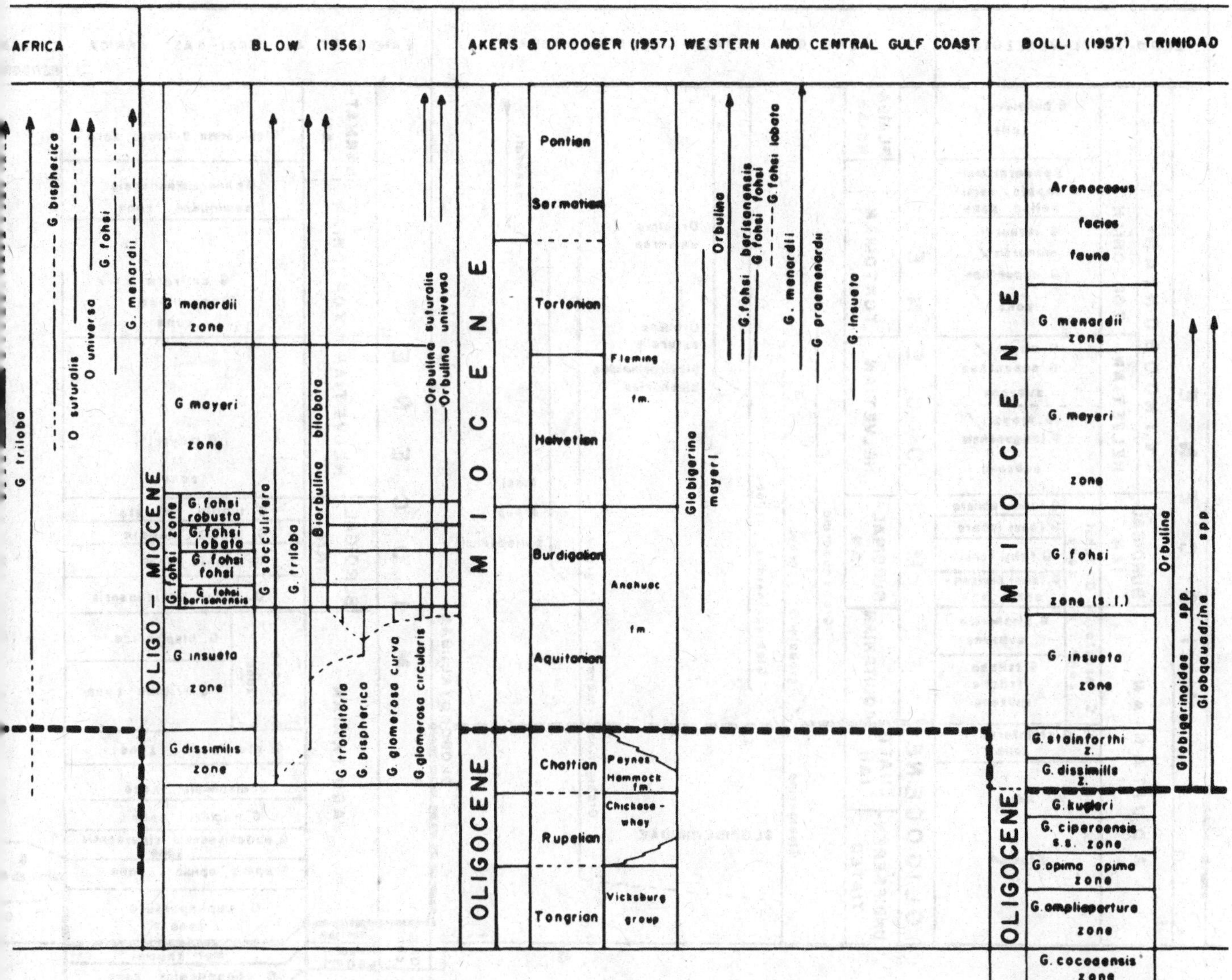
AFRICA
BLOW (1956)
AKERS & DROOGER (1957) WESTERN AND CENTRAL GULF COAST
BOLLI (1957) TRINIDAD
G. trilobo
O. suturalis
O. universa
G. bispherica
G. fohsi
G. menardii
OLIGO - MIOCENE
G. menardii zone
G. mayeri zone
G. fohsi zone
G. fohsi robusta
G. fohsi lobata
G. fohsi fohsi
G. fohsi barisanensis
G. insueta zone
G. dissimilis zone
G. sacculifera
G. trilobo
Biorbulina bilobata
G. transitoria
G. bispherica
G. glomerosa curva
G. glomerosa circularis
Orbulina suturalis
Orbulina universa
MIOCENE
OLIGOCENE
Pontian
Sarmatian
Tortonian
Helvetian
Burdigalian
Aquitanian
Chattian
Rupelian
Tongrian
Fleming fm.
Anahuac fm.
Paynes Hammock fm.
Chickasawhay
Vicksburg group
Globigerina mayeri
Orbulina
G. fohsi barisanensis
G. fohsi fohsi
G. fohsi lobata
G. menardii
G. praemenardii
G. insueta
Arenaceous facies fauna
G. menardii zone
G. mayeri zone
G. fohsi zone (s.l.)
G. insueta zone
G. stainforthi Z.
G. dissimilis Z.
G. kugleri
G. ciperoensis s.s. zone
G. opima opima zone
G. ampliapertura zone
G. cocoaensis zone
G. semi-involuta zone
EOCENE
Orbulina
Globigerinoides spp.
Globoquadrina spp.

TABLE 52.24 continued

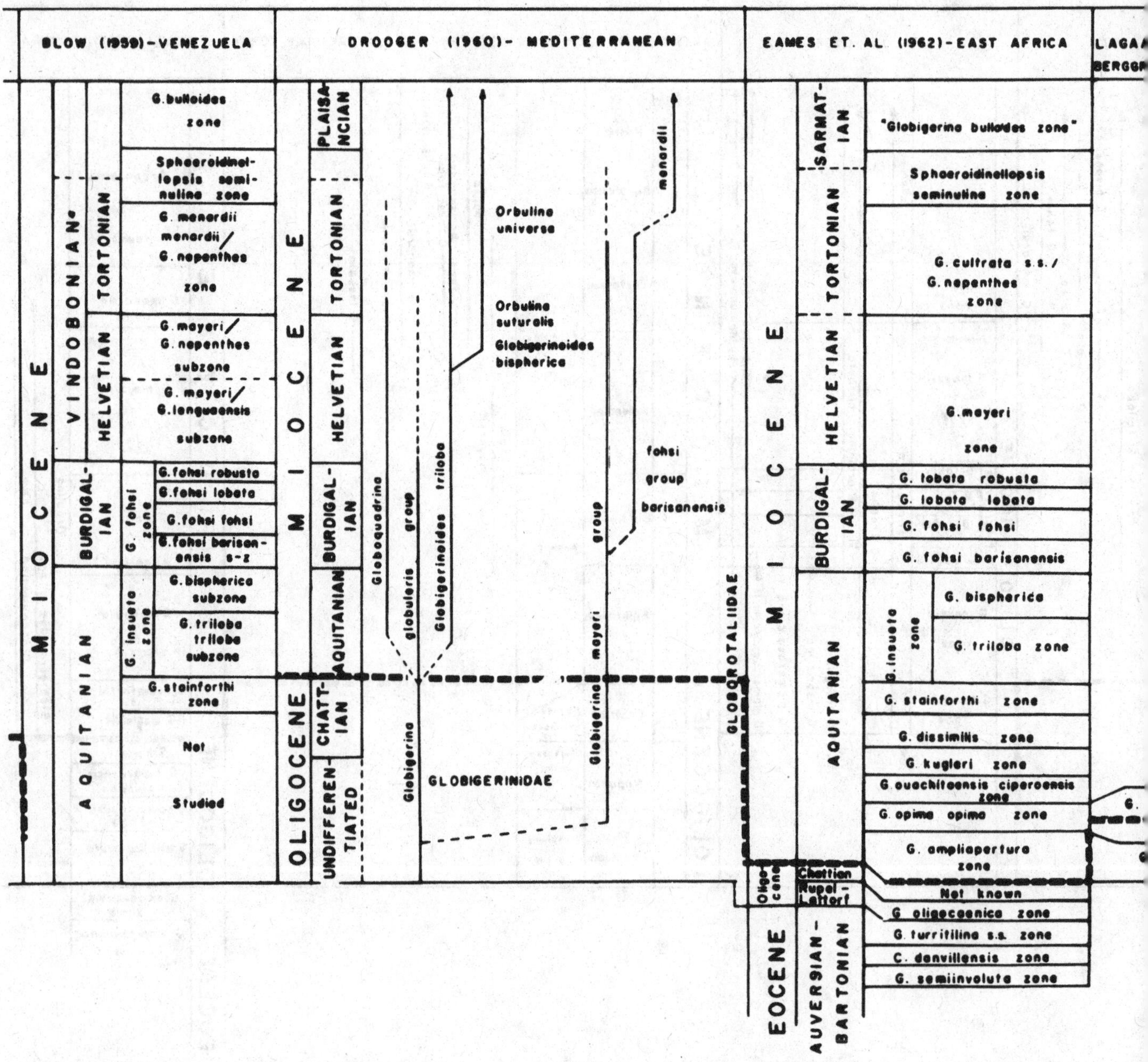
BLOW (1959)-VENEZUELA
DROOGER (1960)- MEDITERRANEAN
EAMES ET. AL. (1962)-EAST AFRICA
LAGAA
BERGGR
MIOCENE
VINDOBONIAN*
TORTONIAN
HELVETIAN
BURDIGAL-IAN
AQUITANIAN
G. bulloides zone
Sphaeroidinellopsis seminulina zone
G. menardii menardii / G. nepenthes zone
G. mayeri / G. nepenthes subzone
G. mayeri / G. lenguaensis subzone
G. fohsi zone
G. fohsi robusta
G. fohsi lobata
G. fohsi fohsi
G. fohsi barisanensis s-z
G. insueta zone
G. bispherica subzone
G. trilobа trilobа subzone
G. stainforthi zone
Not Studied
PLAISANCIAN
TORTONIAN
HELVETIAN
BURDIGAL-IAN
AQUITANIAN
OLIGOCENE
CHATT-IAN
UNDIFFERENTIATED
Globoquadrina
globularis group
Globigerinoides trilоba
Orbulina universa
Orbulina suturalis
Globigerinoides bispherica
menardii
fohsi group
barisanensis
Globigerina mayeri group
Globigerina
GLOBIGERINIDAE
GLOBOROTALIIDAE
SARMATIAN
TORTONIAN
HELVETIAN
BURDIGAL-IAN
AQUITANIAN
"Globigerina bulloides zone"
Sphaeroidinellopsis seminulina zone
G. cultrata s.s. / G. nepenthes zone
G. mayeri zone
G. lobata robusta
G. lobata lobata
G. fohsi fohsi
G. fohsi barisanensis
G. insueta zone
G. bispherica
G. trilоba zone
G. stainforthi zone
G. dissimilis zone
G. kugleri zone
G. ouachitaensis ciperoensis zone
G. opima opima zone
G. ampliapertura zone
Oligocene
Chattian
Rupel - Lattorf
Not known
G. oligocaenica zone
G. turritilina s.s. zone
C. danvillensis zone
G. semiinvoluta zone
EOCENE
AUVERSIAN - BARTONIAN
G.

TABLE 52.24 continued

(1963) CARIBBEAN-VENEZUELA-GULF COAST (1963) N. AFRICA; SAITO AND LIDZ (1965)-JOIDES			BECKMANN (1962)-GENERAL		JENKINS (1964, 1966) AQUITAINE BASIN HOFKER (1963) N. GERMANY (JENKINS, 1966) PLANKTONIC ZONES	STAGES	VAN DER BOL (1966) TRINIDAD
			G. fohsi fohsi to G. fohsi robusta	Helvetian to (?) lower (?) Tortonian			
			G. fohsi barisanensis	lower Helvetian			
			G. insueta zone	Burdigalian - Helvetian		BURDIGALIAN	
			G. stainforthi zone	Burdigalian -	G. stainforthi zone	BURDIGALIAN	
			G. dissimilis zone	Aquitanian - Burdigalian	G. dissimilis zone	BURDIGALIAN	
			G. kugleri zone	Aquitanian	G. kugleri zone	AQUIT-ANIAN	G. kugleri zone
			G. ouachitaensis ciperoensis zone	Aquitanian	G. ciperoensis zone	AQUIT-ANIAN	G. ciperoensis z
ma opima zone	MIOCENE	Chickasaway, Frio & Post-Vicksburg f.ms.	G. opima s.s. z.	Aquitanian - Chattian	G. opima zone		G. opima zo
mpliapertura zone	OLIGO-CENE	Vicksburg Group	G. ampliapertura zone	Chattian	G. ampliapertura Zone		
			G. oligocaenica zone	Lattorfian - Rupelian			

JACKSON GROUP & EQUIVALENTS

TABLE 52.24 continued

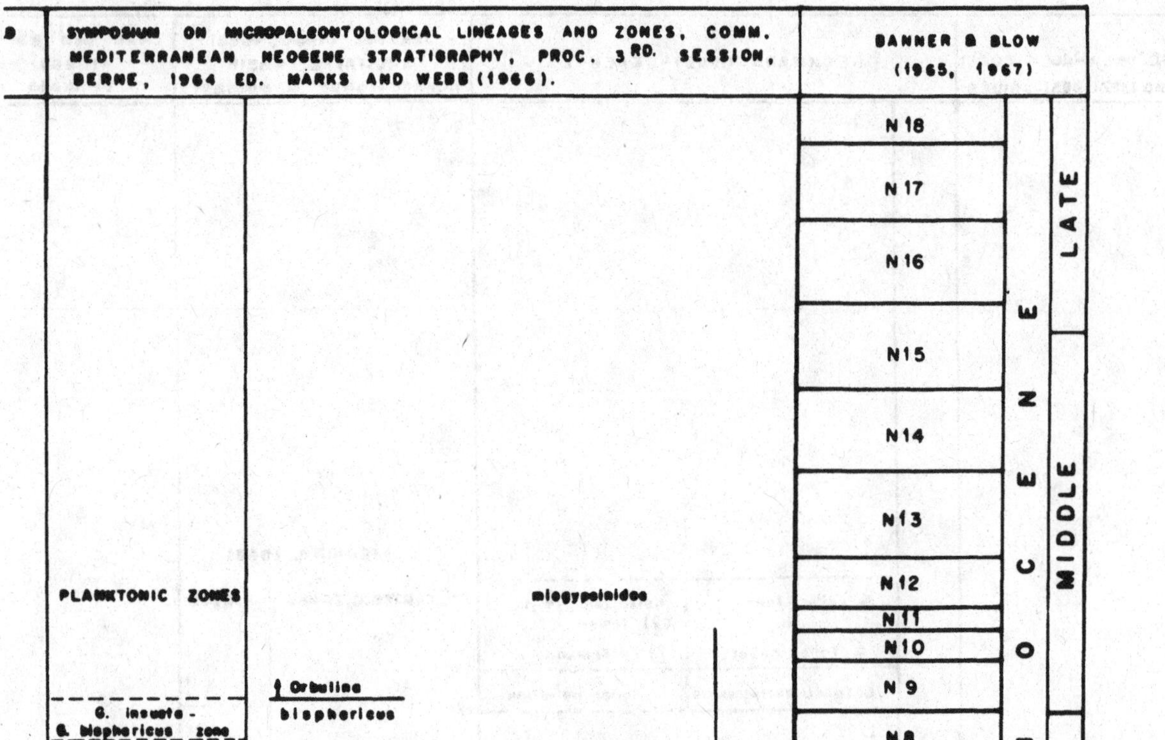
SYMPOSIUM ON MICROPALEONTOLOGICAL LINEAGES AND ZONES, COMM. MEDITERRANEAN NEOGENE STRATIGRAPHY, PROC. 3RD. SESSION, BERNE, 1964 ED. MARKS AND WEBB (1966).
BANNER & BLOW (1965, 1967)
PLANKTONIC ZONES
G. insueta - G. bisphericus zone
G. insueta - G. trilobus zone
G. stainforthi zone
G. dissimilis zone
G. kugleri zone
G. ciperoensis zone
G. opima zone
Orbulina
bisphericus
Globigerinoides
negrii
burdigalensis
socini
bantamensis
formosensis
complanata
miogypsinidae
mediterranea
cushmani
intermedia
globulina
tani
gunteri
basraensis
septentrionalis
CHATTIAN
AQ.
BURDIG.
N 18
N 17
N 16
N 15
N 14
N 13
N 12
N 11
N 10
N 9
N 8
N 7
N 6
N 5
N 4
N 3
N 2
P20/N 1
P 19
P 18
G. gortanii zone
C. danvillensis zone
G. semi-involuta zone
MIOCENE
LATE
MIDDLE
EARLY
OLIGOCENE
EOCENE
LATE

relationships of these faunas and type sections to the limits of the transgressive-regressive phases are virtually impossible to determine. Although we may be able to demonstrate the superposition of stratotypes (and thus their real difference in age), the critical factor is the determination of the limits of the stages (and this is directly related to the temporal interval represented by the rock units attributed and attributable to the stage). In the case of the Oligocene there is no *a priori* justification for a threefold subdivision of the unit. It is generally agreed that the Oligocene in northern Europe represents a single sedimentary cycle and recent evidence suggests that there is overlap in time between parts of the three subdivisions of the Oligocene (lower, middle and upper, Batjes 1958). It would seem sensible to this writer to recognize but a single stage – Rupelian – corresponding to the time represented by this cycle with the three subdivisions (as represented by their stratotypes) representing facies of this stage. The stratotype Rupelian – developed in a marine facies, would serve as type of the stage; the stratotype Chattian (and a stratotype of the Lower Oligocene if such should prove necessary) could be used as types of the upper and lower facies, respectively, of the Rupelian. The fundamental unit of stratigraphy remains, as it always has been, the zone. Subdivision of the Oligocene into several zones based on benthonic and planktonic forms appears possible and their recognition over large areas would avoid the polemics which continue over the currently ill-defined stages.

THE NEOGENE

Miocene

Lyell (1833) named the Miocene as the second (from the bottom) of his four subdivisions of the Tertiary and considered the outcrops in the Touraine (S.W. Paris Basin) as the type of the Miocene. The upper limits of the Miocene in Europe are generally agreed upon, as it corresponds to a widespread regression, but direct unequivocal correlation of the Miocene/Pliocene boundary with recognized micropalaeontological zones has not been generally achieved (cf. however, Banner and Blow 1965, 1967, who have recognized the Miocene/Pliocene boundary in Sicily near the top of their Zone N18).

In 1853 Mayer-Eymar prepared a list of stage names to encompass the then known Tertiary strata of Europe. While he adopted the Suessonian and Parisian of d'Orbigny for the (essentially) Eocene deposits of the Paris Basin and the region of Switzerland in which he was working, he formulated his own terminology for other stage names, some of which are still in common use today and to which we shall have recourse during the course of this paper. His original list is presented below.

1. Pedemontien (=Piedmontian) – upper Pliocene of Mayer-Eymar
2. Placentien (=Plaisancien *Fr.;* Piacenziano *It.;* Piacenzian) – Lower Pliocene of Mayer-Eymar
3. Dertonien (=Dertonian) - Upper Miocene of Mayer-Eymar
4. Helvetien (=Helvetian) – Middle Miocene of Mayer-Eymar
5. Aquitanien (=Aquitanian) – Lower Miocene of Mayer-Eymar.
6. Moguntien (=Moguntian) – Tongrian-Rupelian of Dumont
7. Parisien (=Parisian — Upper Eocene or d'Orbigny – Mayer-Eymar
8. Suessonien (=Suessonian) — Lower Eocene of Mayer-Eymar

Although these terms were published in 1853 by Mayer-Eymar (in Gressly), formal definition and elucidation of these stage terms was not presented until the Congress at Trogen in 1857, the results of which were published in 1858. The latter date is the correct date of citation when ascribing various stage names to Mayer-Eymar. In this work Mayer-Eymar subdivided the Tertiary into twelve stages which he equated with d'Orbigny's stages. It is of historical interest to note that he rejected some of Dumont's stage names in Belgium (Landenian, Laekenian, Bolderian, Diestian) and de Rouville's (1853) Ligerian and Sestian in southern France. He did not use the five-part division of the Cenozoic (Eocene-Oligocene-Miocene-Pliocene-Pleistocene) which Lyell and Beyrich had by this time effected. Rather he subdivided the Tertiary into a lower and upper part and referred six stages to each subdivision as follows:

Tertiary	Upper	Astian Piacenzian Tortonian Helvetian Mainzian (Mayencian) Aquitanian
	Lower	Tongrian Ligurian Bartonian Parisian Londonian Soissonian

In discussing the stratal sequence in northern Europe Mayer-Eymar (1853) included in his Aquitanian Stage, *inter alia,* the *Steinberger Gestein, Lager von Osnabruck, Lager von Crefeld*-strata which has been explicitly included by Beyrich (1854, 1856) in his (Upper) Oligocene. This has led, in no small way, to confusion regarding the affinities of the Aquitanian and assignment of Aquitanian to the Oligocene by some, to the Miocene by others.

Mayer-Eymar (1853, 1858) created the Aquitanian Stage for the Lower Miocene deposits in the Aquitaine Basin; although including in it some Oligocene-Stampian of d'Orbigny (1852)-strata in the Paris Basin and beds in northern Germany later placed in the Chattian by Fuchs. He subsequently (1858) transferred his Aquitanian to the Oligocene, which had been described in the interim by Beyrich (1854). Mayer-Eymar (1853, 1858) thus considered that the Aquitanian and the 'Upper Oligocene' of Beyrich, for which the stage name Chattian was created by Fuchs in 1894, were equivalent in age (at least in part). Actually Mayer-Eymar originally included in his Aquitanian beds which are now agreed to range in age from Rupelian-Burdigalian.

Mayer-Eymar believed that the fifth, sixth and seventh 'Formations' of the Tertiary were equivalent to the Oligocene of Beyrich. The latter was of the opinion (1858) that the Oligocene/Miocene boundary fell within the Aquitanian of Mayer-Eymar. Mayer-Eymar (1888, 1889) himself referred to his Aquitanian as 'Lower Miocene-Upper Oligocene' and still later (1900) as Upper Oligocene. The confusion which subsequent authors have had in attempting to determine the affinities of the Aquitanian Stage

MAYER-EYMAR (1853)	MAYER-EYMAR (1857, 1858)	MAYER-EYMAR (1868)	MAYER-EYMAR (1881)	MAYER-EYMAR (1884, 1888)	DEPÉRET (1892)	DOUVILLÉ (1909)
LOWER MIOCENE — AQUITANIAN (↑ ? — ?)	OLIGOCENE — MAYENCIAN: 10 III Falun de Saucats; 9 II Falun de Léognan; 8 I Pectines-Schicht	MAYENCIAN (1857) (= LANGHIAN OF MAYER-EYMAR, NON PARETO)		LÉOGNANIEN (1888)	LOWER MIOCENE BURDIGALIAN MEDITERRANEAN I: Molasse marno-calcaire à Echinolampas hemisphoericus de Martignas; Faluns de Saucats et de Léognan	BURDIGALIAN (LOWER): Sables à Pecten burdigalensis et P. beudanti, du Moulin de l'Eglise, La Cassagne, etc. (Molasse ossifère de Léognan)
	AQUITANIAN: 7 Weisser Thon u. Ober S-W Kalk	MÉRIGNAC: 2b Argile blanche et calcaire lacustre supérieur	MÉRIGNACIEN (1857, 1865): Falun de Lariey	MÉRIGNACIEN	UPPER OLIGOCENE: couches à M. lainei de Lariey, Mérignac	AQUITANIAN: VII Marne et calcaire lacustre de la route de Son, couches à Cerites du Moulin de l'Eglise. Calcaire gris de l'Agenais
	6 Falun de Lariey, etc.	Saucats, Mauras, et Martillac				
	5 Susswasserkalk von Lariey	2a Falun de Lariey - Saucats, Mauras, Mérignac, Marnes à Ostrea hyotis de Saint Avit				VI Falun de Lariey à Mytilus aquitanicus
	4 Unt. Cerithien-Schicht von Moulin de l'Eglise					V Calcaire lacustre irrégulier
	3 Molasse ossifère	BAZAS: 1e calcaire lacustre de Lariey et de Merignac	BAZAS: Falun de la Brède	BAZASIN (1884) (BAZASON 1888)		II-IV Sable et molasse du Moulin de Bernachon
	2 Konkretionen Schicht	1d couches à Cerithes et à Congéries de Lariey				I Marne grise à Neritina ferussaci
		1c Molasse sableuse jaune ossifère de Lariey				
	1 Blauer Mergel	1b Roches concretionnées de Bazas St. Morillon, Gradignan				
		1a Marnes et Sables blanc du Moulin de la Brède				
OLIGOCENE (TONGRIAN-RUPELIAN)	TONGRIAN	Calcaire blanc de l'Agenais	Calcaire blanc de l'Agenais	Calcaire blanc de l'Agenais	Calcaire blanc de l'Agenais	STAMPIAN (UPPER): Calcaire blanc de l'Agenais à Helix ramondi

Table 52.25. Early stratigraphical terminology and interpretation in the Aquitaine Basin

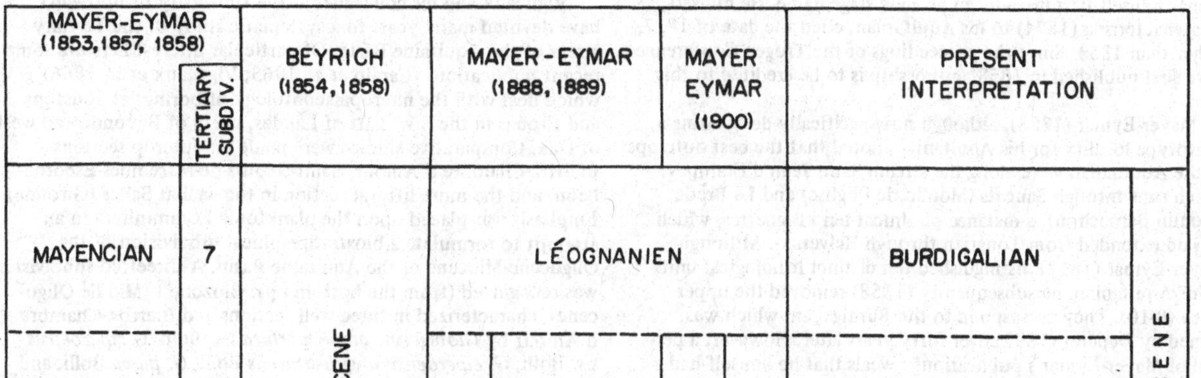
MAYER-EYMAR (1853,1857, 1858)
TERTIARY SUBDIV.
BEYRICH (1854,1858)
MAYER-EYMAR (1888, 1889)
MAYER EYMAR (1900)
PRESENT INTERPRETATION
MAYENCIAN
7
LÉOGNANIEN
BURDIGALIAN
MIOCENE
LOWER MIOCENE
MIOCENE
AQUITANIAN
6
AQUITANIAN
MIOCENE (1853)
OLIGOCENE (1857, 1858)
5
OLIGOCENE
AQUITANIAN
UPPER OLIGOCENE
AQUITANIAN
OLIGOCENE
RUPELIAN & CHATTIAN
OLIGOCENE

Table 52.26. Variation in age designation by Mayer-Eymar of the Aquitanian stage

may be seen to stem in no small way from the uncertainty evinced by Mayer-Eymar himself. A graphic representation of these different interpretations is presented below (Table 52.26). In more recent times this position may be said to be represented by the 'French School'. On the other hand, Dollfus (1909) considered the Chattian and Aquitanian to represent two independent sequential stages. This position has been represented in more recent times by *inter alia,* Abrard. Drooger, Hagn and Eames *et al.*; though these authors, in general agreement that the Aquitanian belongs in the Miocene, are far from agreement on the location of the Oligocene/Miocene boundary in terms of fossil faunas. It may be said, in general, that vertebrate palaeontologists interpret the Aquitanian as Oligocene, and recognize a marked change in the mammalian fauna within the Burdigalian, whereas invertebrate palaeontologists (particularly malacologists) place strong emphasis on the decidedly Neogene character of the Aquitanian molluscan faunas.

As a result of the different interpretations of the Aquitanian there has been a tendency to use the Aquitanian Stage in two senses – Aquitanian (*s.s.*) and Aquitanian Stage (*s.l.*), the latter including the Chattian as well within a single stage. If we are able to relate our position to the respective stage stratotypes this makes for less ambiguity, but it does not necessarily enable us to correlate precisely with phases of the Chattian-Aquitanian sedimentary cycles which may indeed overlap in time. The situation is not unlike that which exists between the Danian and Montian Stages (see *above*).

The stage name Aquitanian was introduced into the literature by Thurmann in Gressly (1853, p. 259) from a manuscript name of Mayer-Eymar in which he was referring to supposed lower Miocene rocks in Switzerland. There can be no doubt, however, that the type area of his Aquitanian was the molluscan-rich beds of the Aquitaine Basin (*Faluns des Départements de la Gironde*). As Rutsch (1952) has observed, Mayer-Eymar was already familiar with the outcrops in the Aquitaine Basin (having made a seven month trip through southwestern France in 1852 at the conclusion of his studies in Paris), and it is hardly likely that he would have chosen the name Aquitanian to typify rocks other than those in the Aquitaine Basin. The correct date for the Aquitanian Stage is 1858, for it was at the Trogen Conference (1857) that Mayer-Eymar himself first introduced his new stage name. He himself, in later referring (1874) to his Aquitanian, cited the date of 1857, rather than 1853. Since the proceedings of the Trogen Conference were first published in 1858, authorship is to be credited to this date.

Mayer-Eymar (1858), although not specifically designating a stratotype locality for his Aquitanian, noted that the best outcrops of the Aquitanian were along the stream Saint Jean d'Etampes which runs through Saucats (Moulin de l'Eglise) and La Brède (Moulin Bernachon), a distance of almost ten kilometres, which he said extended from Tongrian through Helvetian. Although Mayer-Eymar (1857) distinguished ten distinct lithological units in his Aquitanian, he subsequently (1858) removed the upper three (8-10). They correspond to the Burdigalian, which was named by Depéret (1892) over forty years later. However, a perusal of Mayer-Eymar's publications reveals that he himself had a broader concept of the Aquitanian than that generally accepted today, since he included strata in the area of Dax, S. W. France, and also beds in northern Europe, which Beyrich (1854) had included in his Oligocene and are now placed in the Chattian.

Mayer-Eymar subjected his ideas to revision as a result of continuing work by himself and others in southwestern France, and these are reflected in the various tables of stratigraphical classification which he published. He attempted a twofold subdivision of his Aquitanian into a lower *Bazasin* Substage (1884), subsequently changed to *Bazason* (1888), and an upper *Mérignacin* (1865, 1888). He had also recognized a Mayencian Stage (1858, 1865) for the *Mainzer Stufe* which is actually Chattian-Stampian (*s.l.*) in age. The beds in the Aquitaine Basin (*Faluns de Léognan*) with which Mayer-Eymar correlated his Mayencian are actually younger (Burdigalian) than Aquitanian and he, himself, altered this inexplicable error in 1868. He used the term Langhian for these beds, and eventually (1888) erected the substage *Léognanon* (=Leognanian). However, the beds at Léognan have been shown by Drooger *et al.* (1955) to contain a mixed Aquitanian-Burdigalian microfauna, and these terms have long since been relegated to the list of superfluous chronostratigraphical terms and the obscurity they deserve.

The type locality and affinities of the Aquitanian Stage have been discussed recently by several workers (Drooger, Kaasschieter and Key 1955; Drooger 1960*d*; Zöbelein 1960; Drooger, in Marks and Webb 1966). The Saint Jean d'Etampes section in the Aquitaine Basin has been accepted as the stratotype section of the Aquitanian (Lower Miocene) by the Committee on Mediterranean Neogene Stratigraphy (1958 meeting at Aix-en-Provence, 1959 meeting in Vienna, see also 1964 meeting in Berne). Szöts (1965) has recently presented evidence supporting this choice in the form of an unnumbered plate illustration by Mayer-Eymar (1857-8) of the section along the stream between Saucats and La Brède, which he found in the library of the Geological Society of France.

Blow and Banner in Eames *et al.* (1962) record a planktonic fauna from Escornebéou, S.W. France (near Dax), which they believe is correlative with their *G. ouachitaensis ciperoensis* Zone (Zone N3 in Banner and Blow 1965). However, Butt (1966) has recorded *Globigerinoides primordius* (and other planktonic forms, see above) from the sections at Escornebéou, S.W. France. Several alternative conclusions can be drawn: (1) *G. primordius* evolves prior to the *G. kugleri* Zone, and appears at different levels for ecological reasons; and the Escornebéou section is pre-*G. kugleri* Zone in age; (2) the Escornebéou section is of *G. kugleri* Zone age or possibly younger; (3) a mixed fauna occurs at different levels within the Escornebéou section. From an examination of Butt's material (see above) and my familiarity with the development of *Globigerinoides* in normal marine sediments in the Mediterranean area, I am led to the latter conclusion. It is quite possible that we are dealing here with a marginal deposit containing a rather condensed (and thus composite) fauna, so that elements of more than a single planktonic zone may be included in a given sample or in closely spaced samples. Reworking may also be expected.

Vigneaux and his colleagues at the University of Bordeaux have devoted many years to a systematic study of the Tertiary strata of the Aquitaine Basin. Of particular interest here are some recent publications (Caralp *et al.* 1965; Vigneaux *et al.* 1966) which deal with the micropalaeontology of borings at Soustons and Frouas in the S.W. part of Landes, north of Bayonne and west of Dax. Comparative studies were made on outcrop sections at Biarritz-Chambre d'Amour, Saint-Géours-de-Maremnes-Escornebéou, and the more littoral section in the well at Salles (Gironde). Emphasis was placed upon the planktonic Foraminifera in an attempt to formulate a biostratigraphical subdivision of the Oligocene-Miocene of the Aquitaine Basin. A threefold subdivision was recognized (from the bottom up): Biozone I (Middle Oligocene), characterized in three well sections and Biarritz-Chambre d'Amour by *Globigerina ampliapertura s.s.* Bolli, *G. ciperoensis s.s.* Bolli, *G. ciperoensis angulisuturalis* Bolli, *G. parva* Bolli, and *Globorotalia opima opima* Bolli – all of which are restricted to this interval. Biozone I was correlated with the *G. ampliapertura* and *G. opima s.s.* zones of Bolli.

Biozone II (Upper Oligocene) was found in the three well sections and the Saint-Géours-de-Maremnes outcrops. It was characterized by *G. opima nana* Bolli, *G. ciperoensis angustiumbilicata* Bolli, *G. euapertura* Jenkins, *G. praebulloides occlusa* Blow and Banner, *G. rohri* Bolli, *G. tripartita s.s.* Koch, *G. venezuelana* Hedberg, and *G. dissimilis* (Cushman and Bermúdez). Three species limited to the top of Biozone II include *Globigerinoides* sp. 1, *G.* sp. 2, *Globorotalia kugleri* Bolli. The authors indicate (Caralp *et al.* 1965, p. 62) 'the presence of these three forms is one of the factors which contributed to assigning this population to the Oligocene fauna'. Biozone II was correlated with the *G. ciperoensis s.s.* and *G. kugleri* zones of Bolli.

Biozone III (Lower Miocene) is characterized by *Globorotalia menardii* (d'Orbigny), *Globigerinoides obliqua* Bolli, *Orbulina bilobata* (d'Orbigny), *O. suturalis* Bronnimann, *O. universa* d'Orbigny. In the Frouas well the presence of forms with short stratigraphical ranges permitted the distinction of three subzones characterized by presence of either *G. fohsi barisanensis* Le Roy, *G. fohsi fohsi* Cushman and Ellisor, or carinate species such as *G. fohsi lobata* Bermúdez, *G. fohsi robusta,* or *G. menardii* (d'Orbigny). In the Soustons well two subzones can be recognized: *G. fohsi barisanensis* and *G. fohsi* and its subspecies. Biozone III was correlated with the *G. dissimilis* to top of *G. fohsi robusta* zones of Bolli.

The Upper Miocene Biozone IV is marked by the extinction of the subspecies of *G. fohsi* and the expansion of *Orbulina.* Zone V was correlated with the *G. mayeri* and *G. menardii* zones of Bolli. Biozone V corresponds to the Pliocene and is only represented at Frouas and Soustons. *Globorotalia pliocenica* Mistretta and *O. universa* d'Orbigny characterize this unit.

An anomalous finding in these investigations has been the recording of *Globigerinoides triloba s.s.*, *G. triloba sacculifera*, *G. triloba immatura* in Zone I – down to the bottom in Soustons well (Vigneaux *et al.* 1966, pl. 5). Some of the anomalous stratigraphical ranges presented by Vigneaux *et al.* (1966, pl. 6) compared with those of Bolli (1957) probably have their logical explanation in different taxonomic concepts. Unfortunately no direct attempt at correlation with stratotype units in the Aquitaine Basin was made, and this must await further, more detailed investigations.

The strata included in the stratotype Aquitanian as currently agreed upon by the Neogene Committee include the sections at Moulin Bernachon and Moulin de l'Eglise, which would appear to correspond to the *G. kugleri* Zone. The base of the stratotype Aquitanian does not necessarily correspond to the base of the Miocene, however. It is quite possible that outcrops in other areas of the Aquitaine Basin which were included by Mayer-Eymar in his concept of Aquitanian (and which would probably have been included by Lyell in his Miocene and by d'Orbigny within his Falunian) can be included on a stratigraphical and faunal basis within the Aquitanian Stage-Lower Miocene. Indeed the base of the Aquitanian has been a subject of considerable discussion, although it seems best to interpret it as lying above the *Calcaire à Astéries* (Denizot 1958; Zöbelein 1960). The marly layer just above the *Calcaire à Astéries* (=Mayer-Eymar's *'grüne tongrische Tone und Mergel'*) is best interpreted as an equivalent of the *Calcaire blanc de l'Agenais,* which contains the characteristic fresh-water snail assemblage, *Helix ramondi,* which is also found in the *Calcaire d'Etampes* (Paris Basin) and the *Mainzer Stufe* (Mainz Basin).

It will be recalled that Lyell (1833), in proposing his fourfold subdivision of the Tertiary, included in his Miocene the marine strata in the Valley of the Bormida (table on p. 61; pp. 202, 211; see Tables 52.2, 52.3 this paper). Lyell (1833) recognized the separation of the older Eocene limestone at Pauillac and Blaye, which crop out along the right bank of the Gironde (the *Calcaire de Blaye,* considered Middle Eocene by French geologists today), and the Miocene formations in the neighbourhood of Bordeaux, and believed that they were separated by the 'fresh-water lime-stone at Saucats' (Lyell 1833, pp. 207-9).

Lyell (1833, p. 211) included in his Miocene the exposures in the hills of Mont Ferrat and the Superga, and observed that the exposures were particularly well-developed in the basin of the Bormida. He recognized that the conglomerates, 'interstratified with green sands, containing rounded blocks of serpentine and chlorite schist', constitute the basal part of this sequence, lying transgressively above the basement complex (Lyell 1833, pp. 211, 212). In a footnote of historical interest (Lyell 1833, p. 212) Lyell entertained the hope that Pareto, Parsini and LaMarmora 'will devote their attention to the relative position of the several groups of Tertiary strata in Piedmont, by instituting a comparison be-tween their respective organic remains'. His hopes were not in

	ZÖBELEIN (1960)		DENIZOT (1957) FRENCH SCHOOL		BATJES (1958) LORENZ (1963), SZÖTS (1965) THIS WORK
MIOCENE	BURDIGALIAN	MIOCENE	BURDIGALIAN	MIOCENE	BURDIGALIAN
MIOCENE	AQUITANIAN	MIOCENE	AQUITANIAN S.L.	MIOCENE	AQUITANIAN
OLIGOCENE	CHATTIAN	MIOCENE	AQUITANIAN S.L.	OLIGOCENE	CHATTIAN
OLIGOCENE	RUPELIAN S.S.	OLIGOCENE	RUPELIAN S.L. (STAMPIAN S.L.)	OLIGOCENE	RUPELIAN S.L. (= STAMPIAN S.L.)
OLIGOCENE	LATTORFIAN	OLIGOCENE	RUPELIAN S.L. (STAMPIAN S.L.)	OLIGOCENE	RUPELIAN S.L. (= STAMPIAN S.L.)

Table 52.27. Recent interpretations of the connotation of Oligocene – Lower Miocene stages

vain for it was Pareto who, some thirty-two years later (1865), erected several stage names for the Tertiary stratigraphical sequence in the Piedmont area of northern Italy.

In 1865 Pareto presented a subdivision of the 'Miocene' of the Piedmont region in northern Italy. He introduced three stages:

Serravallian (p. 232)
Langhian (p. 229)
Bormidian (p. 220)

Sacco (1887) subsequently subdivided the Bormidian into Tongrian and Stampian and placed both in the Oligocene. The Bormidian has not found many adherents. In a recent summary of the Bormidian Stage Cita and Piccoli (1964) have pointed out that various authors (Trabucco, Rovereto and Haug) have correlated the Bormidian with various parts of the Oligocene. On the other hand Lorenz (1963, 1965) has recently shown that the limits of the Bormidian actually extend from Oligocene (Stampian) to Miocene (Aquitanian). In discussing the placement of the Oligocene/Miocene boundary Lorenz (1963, 1965) suggests the abandonment of the term Chattian, uses the term Stampian *s.l.* for the Oligocene, and places his Oligocene/Miocene boundary between his Zone b (*Nummulites* and large lepidocyclinids) = Oligocene, and Zone c (large lepidocyclinids and rare *Miosypsina* sp.)= Aquitanian. Rovereto (1910) had earlier reached similar conclusions in attributing the Bormidian to the Sannoisian, Stampian and Aquitanian. The recent study by Vervloet (1966) has confirmed the opinion of Rovereto and Lorenz. The Bormidian appears to extend from a level within the Rupelian to a level within the Aquitanian (*Globigerinoides-Globoquadrina* Zone of Vervloet).

Banner and Blow (1965) mention that their Zones N2-N3 and the upper part of N1 correspond to the Bormidian Stage, and are older than the stratotype Aquitanian. Two possibilities suggest themselves: (*a*) to equate the Oligocene with but a single stage – Rupelian (or Stampian) *s.l.* or Rupelian-Chattian, and to include in the Aquitanian strata somewhat older than those exclusively in the stratotype section; (*b*) to utilize the Bormidian for the time-stratigraphical interval between Rupelian *s.l.* or Rupelian-Chattian and Aquitanian *s.s.*

If the former procedure is accepted it would seem possible to arrive at a satisfactory placement of the Oligocene/Miocene boundary between the Rupelian *s.l.* and the Aquitanian. Approximate correlation with the planktonic foraminiferal zonations, although not agreed upon by all specialists, is nevertheless a reasonable prospect (see above). The intercalation of the Bormidian Stage, with its attendant modifications and redefinition (lectostratotypification), awaits further documentation. It has been accepted for use in the micropalaeontological manual prepared for use in the J.O.I.D.E.S. drilling programme, following discussions with Dr W.H. Blow.

The affinities of the Aquitanian invertebrate fauna have been recently reviewed by Zöbelein (1960) and Eames *et al.* (1962). The foraminiferal and ostracod faunas of the Aquitanian have been discussed by Drooger, Kaasschieter and Key (1955). Without exception all groups show pronounced Neogene affinities in contrast to the stronger Palaeogene affinities in strata of Rupelian – Stampian and Chattian age below. In open-sea facies the planktonic Foraminifera exhibit a gradual and somewhat monotonous evolutionary development in the Oligocene-Lower Miocene (Blow and Banner in Eames *et.al.* 1962; Berggren 1966). Many vertebrate palaeontologists agree that the Aquitanian contains a Palaeogene mammalian fauna, that the Lower Burdigalian contains a fauna transitional between Oligocene and Miocene, and that in the Upper Burdigalian the typical Miocene fauna first appears. On the other hand numerous other vertebrate palaeontologists include the Aquitanian in the Miocene (see discussion in Zöbelein 1960). In making a decision as to where to include the Aquitanian we should allow the facts to speak for themselves, and then draw conclusions from a synthesis of information. We should not allow ourselves to be influenced by rigid, static rules of priority or original designations. Mayer-Eymar, himself, originally included the Aquitanian in the Miocene, subsequently in the Oligocene. Beyrich had determined the upper limit of the Oligocene however, and referred to Mayer-Eymar's Aquitanian as Miocene as well. Thought processes, leading to advancement in information and knowledge, are self-facilitating, and we should be guided in our interpretations by the most logical, coherent picture which arises as a result of an evaluation of available information. On the basis of the information available at the present time the Aquitanian Stage is most satisfactorily placed in the Miocene.

The Burdigalian (Depéret 1892 [1893]) and the Chattian (Fuchs 1894 [1892-4]) were created almost simultaneously. The Chattian was created for the upper Oligocene of Beyrich (1854), which Mayer-Eymar (1858 *et seq.*) continued to include in his Aquitanian. (For the sake of stratigraphical sanity it is best to merely exclude from the Aquitanian the beds of Chattian age which Mayer-Eymar erroneously included in his Aquitanian.)

The Burdigalian was created for strata above the Aquitanian in the Aquitaine Basin, and received its name from the rich fauna in the so-called *Faluns de Bordeaux.* Depéret (1893) correlated his Burdigalian Stage with the Mediterranean II of Suess, excluded the Aquitanian from the Mediterranean I, and drew the Oligocene/Miocene boundary between the Aquitanian and Burdigalian. He was of the opinion that his Burdigalian was correlative, at least in part, with the Langhian of the Piedmont area in northern Italy and the *schlier* of Austria. Although Depéret did not accurately define a stratotype locality for his Burdigalian Stage (he mentioned that the horizon of the *faluns de Saucats et de Léognan* constituted the basal part, and the *molasse calcaire du bassin du Rhone à Pecten praecasbriusculus* the upper part), it would seem satisfactory to accept the suggestion of Szöts (1965, p. 745) to designate as 'neostratotype' the section along the Saint-Jean d'Etampes stream between Moulin de l'Eglise and Pont-Pourquey (near Saucats). There are two good reasons for this: (1) The section is a continuation (upwards) of the Aquitanian stratotype; (2) The beds correspond to those which Mayer-Eymar (1857-8) placed in his Mayencian Stage (later named Leognanian).

The outcrop at Le Coquillat (a small ditch in the forest), about 2 kilometres S.W. of Léognan (Bordelais), S.W. France is generally acknowledged to be the type locality of the Burdigalian. A poorer type locality cannot be imagined, since no significant stratigraphical succession is exposed, it being necessary to dig beneath the forest cover to uncover the shell beds.

The Burdigalian is limited upwards by the *Faluns de Salles* (*Cardita jouanneti* beds) in the Aquitaine Basin, which are generally attributed to the Helvetian by most French geologists, but which also served as the type of the Sallomacian Stage (Fallot 1893).

Attempts to determine the limits of the Burdigalian Stage in terms of foraminiferal biostratigraphical zonation have proved frustrating. Drooger (1960) has shown that the Burdigalian is characterized by *Miogypsina globulina-intermedia-burdigalensis.* Blow and Banner in Eames *et al.* (1962), having placed the *Orbulina* datum in the uppermost Aquitanian, correlated the Burdigalian with the *Globorotalia fohsi s.s.-G.f. lobata-G.f. robusta* zones of Bolli. However, they believed that Burdigalian equivalents in East Africa and in the Middle and Far East (Tf_{1-2}) are characterized by the occurrence of *Austrotrillina howchini, Taberina malabarica, Miogypsina* and the first *Borelis melo.* However, this dating would appear to be incorrect, and the authors appear to have modified these correlations more recently (Banner and Blow 1965; Blow, pers. comm. 1966, 1967). The stratigraphical range of *Borelis melo* appears to be restricted to beds of post-Helvetian pre-Messinian age, and its first occurrence to coincide closely with the *Orbulina*-datum, according to Reiss and Gvirtzman (1966). I am in agreement with this conclusion on the basis of observations in the Miocene of the Sirte Basin of Libya, North Africa. Recent work by Sourdillon (1960) and Jenkins (1964, 1966) has demonstrated that the lineage taxa leading from *Globigerinoides bis-*

phericus (vel *G. sicanus*) to *Orbulina universa* are not present in rocks of Burdigalian Age in the Aquitaine Basin. This has been confirmed by recent examination of material which I collected from the type region of the Aquitanian-Burdigalian in S.W. France. The Burdigalian appears to be approximately equivalent to the *G. dissimilis* and *G. stainforthi* Zones of Bolli (N5 and 6 of Banner and Blow) and possibly the lower part of the *G. insueta* Zone (N7).

The investigations by Cita and Elter (1960) (see also Cita and Premoli Silva 1960; Cita 1964; Vervloet 1966) reveals that the Burdigalian and Langhian are not time-stratigraphically equivalent. The Langhian Stage is shown by Cita to straddle the *Orbulina*-datum with indefinite lower and upper boundaries. The Langhian was defined by Pareto (1865) for outcrops in the valley of the Langhe, north of Ceva, northern Italy. Mayer-Eymar (1868) emended the Langhian, relegating the basal part to his Aquitanian and the upper part to the Langhian *s.s.*. Sacco (1887) subsequently modified the concept of the stage by including transitional beds at the top in the area near Cissole (Spigno Monferrato area). The base of the Langhian starts approximately at the base of the *Orbulina*-datum according to Vervloet (1966, p. 48), whereas Cita and Premoli-Silva (1960*a*) place the base of the Langhian at the base of their '*Globoquadrina dehiscens* Zone' (= *G. dehiscens* Partial-range-zone) which they correlate with the *Globigerinatella insueta* Zone of Cushman and Stainforth (1945), Bolli (1957) and others. Cita and Elter (1960, p. 15) have suggested that only the lower part of the Langhian (*G. dehiscens* Zone) can be correlated with the Burdigalian. The younger zones may be referred to the 'Helvetian (?)' in their opinion. In view of the work by Jenkins (1964, 1966) it would appear that the Burdigalian is pre-*G. sicanus* Zone in age and from the work of Cita and Elter (1960) it appears that the base of the Langhian may be satisfactorily drawn at the base of the *G. sicanus* Zone (N8 of Banner and Blow 1965), which is where, in fact, the latter have placed the base Burdigalian, although subsequent modifications are planned (Blow 1966, 1967, pers. comm.).

There appears to be a general consensus of opinion that in Europe *Miogypsina* is restricted to the Aquitanian-Burdigalian (although the lower part of its range might be included in the Oligocene by some, Vervloet 1966; Marks and Webb, Neogene Colloquium, Bern, 1966, p. 142). Its first appearance is in the *G. opima opima* Zone (N2), or perhaps as low as the upper part of Zone (P20/N1). A somewhat more detailed discussion of the relationship of Neogene larger Foraminifera to stage boundaries is presented below.

Between the Burdigalian and the Tortonian a considerable period of time (represented by stratal accumulation) is represented. Considerable confusion has resulted in attempts to recognize and correlate chronostratigraphical units within this interval.

The Helvetian, defined by Mayer-Eymar (1858) in Switzerland for the beds at Randen and Saint-Gall, was subsequently modified (1868) to include beds at present included in Tortonian. It corresponds in its original definition to the *Faluns de la Touraine* and to the *Couches de Salles* and *Martignas* in S.W. France. The Helvetian is generally developed in a sandy, poorly fossiliferous facies. Rutsch (1958) has selected a neostratotype, Imihubel, but here also the meagre fauna has rendered inter-regional correlation difficult. However, on the basis of megafossils it has been possible to demonstrate that the Helvetian is pre-*Orbulina*-datum in age and can probably be correlated with the *G. sicanus* Zone (Cita and Elter 1960; Cita and Premoli-Silva 1960*a*; Buday, Cicha, Ctyroky and Fejfar 1964; Papp 1960, 1963; Turnovsky 1963). Cicha, Tejkal and Senes (1960) have suggested that the Helvetian in the type-section, as well as in sections correlatable with the stratotype in the entire Paratethys region, is so closely related to the Burdigalian (on megafossil evidence primarily) that its independence as a chronostratigraphical unit is open to question. They suggest uniting the Burdigalian and Helvetian into a single stage. Rutsch, in Marks and Webb (1966, p. 144), states that the molluscan fauna of the type Helvetian is younger than that of the type Burdigalian and older than that of the Sallomacian. Papp (1960, 1963) maintains that the Helvetian lies beneath the *G. bisphericus* Zone. A synthesis of these views reveals that the Helvetian probably overlaps in time the upper part of the Burdigalian and the lower part of the Langhian (approximately Zones N7 and N8 of Banner and Blow). However its continued use as a chronostratigraphical unit appears unnecessary, in view of the more satisfactory correlation through the use of the Burdigalian and Langhian stages, and the approximate coincidence of their upper and lower boundaries respectively.

In the Vienna Basin a marine transgression with *Orbulina* and small globorotaliids (*G. scitula* group) occurs, and has signified to some (Papp and Drooger) the Helvetian/Tortonian boundary. However, a considerable time interval remains from the top Helvetian (which is still pre-*Orbulina*-datum in age) and the base of the Tortonian (which is within Zone N15 of Banner and Blow 1965), and somewhat below the first evolutionary appearance of *Globorotalia acostaensis*. The so-called 'Upper Helvetian' of the Vienna Basin, the Badenian, the 'Elveziano' and Serravallian of Italian geologists, and the Carpathian Stage of Czechoslovakian geologists, are among the names most often used within this interval.

The genus *Orbulina* appears (with its evolutionary praecursors) near the upper boundary of the 'Upper Helvetian' in the Vienna Basin. Cicha and Senes (1964) have created the Carpathian Stage for this rock-stratigraphical sequence (Laaer Serie, Karpatische Formation). It underlies the Badener Series which is considered Lower Tortonian by Austrian geologists, but which is definitely older than the Tortonian in its type area (Rio Mazzapiedi-Castellania-Santa Agata Fossili). The *P. glomerosa*-*O. suturalis*-*O. universa* development occurs within this interval (Papp 1966), and Reiss and Gvirtzman (1966) used the term Badenian for this interval (between *Orbulina*-datum and *G. nepenthes* Zone).

As Reiss and Gvirtzman (1966) point out, discussions about the placement of the Helvetian-Tortonian boundary are superfluous since they are nowhere near contiguous. From the above it follows that the Carpathian is equivalent to the lower part of the Langhian, and is probably best considered as superfluous in terms of chronostratigraphy. The 'Elveziano' of Italian geologists is seen to be decidedly younger than the Helvetian of Switzerland.

The term Badenian is useful to characterize the interval between the *Orbulina*-datum and the *G. mayeri-lenguaensis* Zone, or approximately first occurrence of *G. nepenthes*. However, an alternative solution may be suggested in the work of Vervloet (1966). The Langhian was shown to correspond to a greater part of the '*Orbulina* Zone'. The Serravallian Stage was shown to correspond to the upper part of the *Orbulina* Zone and the *G. menardii s.l.* Zone. The Serravallian Stage was created by Pareto (1865, pp. 232, 233) for outcrops on the Scrivia River, near Serravalle in the Piedmont area, northern Italy. The type locality, east of the village of Serravalle Scrivia, lies approximately fifty kilometres east of the type locality of the Langhian.

The evidence presented by Vervloet (1966) is not unequivocal. It is difficult to determine whether the Serravallian follows sequentially upon the Langhian, or is, in part, laterally correlatable with it. Since *G. menardii* and *G. nepenthes* occur simultaneously within the Serravallian it would seem that the Serravallian corresponds to the *G. fohsi s.l.* Zone (above the *G. barisanensis* Zone = *G. peripheroronda*). This is approximately the Badenian of Reiss and Gvirtzman (1966). It is also equivalent to the *O. suturalis* Zone of the upper part of the Langhian according to Cita and Premoli-Silva (1960, pp. 9-11, fig. 4). They correlated this zone with the *G. fohsi s.l.* Zone (above the *G. barisanensis* Zone) and showed that, as in Trinidad, it lay beneath the *G. mayeri* Zone.

The Langhian thus appears to extend from the *G. sicanus* Zone to approximately the top of the *G. fohsi s.l.* Zone of Bolli (approximately top of Zone N12 of Banner and Blow 1965).

Vervloet (1960, pp. 48, 49) suggested that the use of two stages between the Helvetian and Tortonian is unwarranted, and recommended using the Serravallian because of the common perpetuation of the incorrect correlation of the Langhian and the Burdigalian. However, in view of the fact that the foraminiferal fauna of the Langhian appears to be better known than that of the Serravallian, and since the two stages would appear to be laterally equivalent (at least in part), it would seem more suitable to use the name Langhian. There may be a short interval between the top of the Langhian and the base of the Tortonian, but extension of the chronostratigraphical limits of the Langhian is clearly preferable to the creation of an additional time-stratigraphical unit which only represents a short time-interval. Thus extended the Langhian stage corresponds approximately to the upper part of the 'Te' stage and most of the 'Tf' stage of the Far East (cf Van der Vlerk 1966; Banner and Blow 1965).

Haug (1911, p. 1607) observed that a great disproportion existed between Miocene and Pliocene, the Miocene representing a considerably larger interval than the Pliocene. He thus distinguished a lower, middle and upper Neogene, the latter corresponding to the Pliocene. He was of the opinion that these three units were more or less equivalent to the first (lower) three Mediterranean Stages of Suess. The relationship which Haug (1911) suggested between the Neogene Stages and Suess's stages is shown in Table 52.28 below.

As we shall see in the section on radiometric dating and biostratigraphy the Pliocene was indeed a relatively short interval of time, comparable in its extent to the Oligocene.

Late Miocene and Pliocene chronostratigraphical terminology has been the subject of vigorous debate for many years. It is doubtful whether any comparable time-interval in geological history has as many stage names in current use vying for superiority. A detailed analysis of each of these stage names and their affinities is beyond the scope of this paper. However, a general discussion is presented below which is aimed, at acquainting the reader with the complexities of the problem, and at offering a possible workable and usable scheme for stratigraphical studies in the late Tertiary.

The seemingly impossibly complex problem of Late Miocene/Pliocene age-stage terminology can be traced to a single main factor: towards the end of the Miocene a widespread regression occurred throughout the Mediterranean region. Its effects were felt from the Caspian Sea area in the east to the western limits of southern Europe (Spain) and northwestern Africa (Morocco), and from the Vienna Basin in the north to the southern shores of the Mediterranean in north Africa (Libya, Algeria). At its maximum the Tethys Sea (palaeo-Mediterranean) was cut off from the Atlantic Ocean to the west, and a series of gypsum-anhydrite beds, alternating with clays with marine fossils, was deposited in some areas.

This major regression resulted in a widespread alteration in the existing geographical relationships between marine, non-marine and intermediate areas. The resulting changes were reflected in highly complex stratigraphical sections, which,when studied in various areas which once formed a part of the great Tethys Sea, proved difficult to correlate with other (supposedly equivalent) sections in the Tethyan region. In the western Mediterranean marine sections extend up to the Quaternary, whereas in the Pannonian Basin of southeastern Europe dominantly brackish-limnic sequences characterize the late Miocene/Pliocene. Thus various local stage names have been formulated by different workers in attempts to delineate the local stratigraphical succession in the late Tertiary of the Mediterranean region.

The name Tortonian is accepted by nearly all stratigraphers as a useful stage in the Miocene, although it is doubtful whether many could delineate its exact chronostratigraphical limits in regards to palaeontological criteria. The name of Mayer-Eymar is intimately

GROUP	MEDITERRANEAN STAGES OF SUESS (1869)		STAGES
QUATERNARY	MEDITERRANEAN V		
	MEDITERRANEAN IV		CALABRIAN
→ (UPPER NEOGENE) (NEOMEDITERRANEAN)	MEDITERRANEAN III		← ASTIAN PLAISANCIAN
MIDDLE NEOGENE (MESOMEDITERRANEAN)	MEDITERRANEAN II	VINDOBONIAN	SARMATIAN (=SAHELIAN=PONTIAN) TORTONIAN, HELVETIAN
LOWER NEOGENE (EOMEDITERRANEAN)	MEDITERRANEAN I		BURDIGALIAN AQUITANIAN

HAUG (1911) DID NOT INCLUDE UNITS ABOVE THE ARROWS →← ; SUGGESTED CORRELATION IS MADE HERE BY THE WRITER BASED ON HAUG'S WRITINGS.

Table 52.28. Relationship between Neogene units and stages of Haug and Suess (After Haug, 1911).

associated with the Tortonian, for it was at the Trogen Congress in 1857 (1858) that he first defined and used the term. His original definition of Tortonian was quite clear. Included originally in the Tortonian were:

(1) marine strata of Tortona (Italy), Baden (Vienna Basin) and Cabrières d'Aigues (Vaucluse, France);
(2) continental beds above, with *Hipparion* at Eppelsheim; beds at Cucurou (Mont-Léberon, Vaucluse) and Orignac (Hautes-Pyrénées) with the same fauna. It should be pointed out that *Hipparion* appears earlier in marine Tortonian at a horizon containing *Nassa michaudi* at Dauphiné (Isère).

We have seen above that Mayer-Eymar was in the habit of modifying his concepts of the limits of his Tertiary stage names over the years. The Tortonian was no exception. In 1868 he essentially decapitated his Tortonian Stage, creating the term Messinian for the regressive, increasingly brackish-water strata above the marine Tortonian. (Actually three stage names were created for the Late Miocene/Pliocene within two to three years of each other: Messinian (Mayer-Eymar 1867, 1858); Zanclian (Seguenza 1868); Sarmatian (Suess 1866).) However, in dividing his Messinian into three parts into (from the bottom up): 1. Billovitz strata; Inzersdorf strata; and Eppelsheim strata (in Austria), he considerably weakened the usefulness of this term in regional stratigraphical correlation. However, he did refer a series of marine units in southern Italy (Messina-Calabria) to his Messinian, which allows a more objective comparison and evaluation. Mayer-Eymar was of the opinion that the strata attributed by Seguenza (1868) to his Zanclian were of the same age as his own Messinian; indeed, finding that the name Zanclian (from *Zancla,* the name of the Greek colony of Messina before the Roman conquest) was too classic for his taste, he essentially replaced it with his own Messinian. As a result the use of the term has led to confusion as it has been applied variously to the gypsum-anhydrite series ('*gessoso-solfifera*') and to the deposits of Pontian age.

Because of tectonic and sedimentological complications in the vicinity of Messina, near Gesso, southern Italy, Selli (1960) has recently selected and described a neostratotype section for the Messinian Stage of Mayer-Eymar, the Pasquasia-Capodarso series in the province of Enna, between Enna and Caltanissetta, in central Sicily. The sequence is bounded below by marls of Tortonian age and above by the '*trubi*', generally accepted as representing the basal Pliocene (=Zanclian) in this area.

A discussion of the lithology and preliminary foraminiferal faunal list of the units exposed in the Pasquasia-Capodarso sections was given by Selli (1960). In its original concept the Messinian included the diatomaceous marls ('*tripoli*'), evaporites, and the '*trubi*' (deep-sea lutites with rich planktonic faunas). Seguenza (1868, 1879), restricted the Messinian to the evaporitic series and the '*tripoli*'. He considered the overlying unit, the '*trubi*', to be Pliocene, and introduced the term Zanclian for the calcareous-marly sediments with uniform fossil content and lithology up to, but excluding, the *Amphistegina*-bearing sands. According to Seguenza the Zanclian represents a distinct interval between the Tortonian and the Pliocene (Astian-Piacenzian), and he placed it at the base of the Pliocene, which he believed could be subdivided into three parts; Zanclian, Piacenzian, Astian. (It should be remembered that at the time Seguenza introduced the term Zanclian, the Tortonian of Mayer-Eymar extended up to the base of the Piacenzian, which Mayer-Eymar considered Pliocene.)

The main shortcoming of the Messinian Stage is its general lack of fossils, these being restricted primarily to the marine marly intercalations between the gypsum-anhydrite beds. Thus a distinct palaeontological definition of the stage appears difficult. However, as Selli (1960) points out, the chronostratigraphical limits of the stage can probably best be set by the palaeontological character of the adjacent Tortonian and Piacenzian *s.l.* (i.e. Tabianian or Zanclian).

The biostratigraphy and planktonic foraminiferal fauna of the type Tortonian have been described by Gianotti (1953), and more recently by Cita, Silva and Ross (1965). The foraminiferal marls of the Tortonian are justly famous for their variety and abundance of forms.

Banner and Blow (1965, 1967) have recognized the upper part of Zone N15 and N16 in the type Tortonian. The Tortonian and Messinian were said to overlap within Zone N17. Zones N17, and at least part of Zone N18, were said to occur in the stratotype Messinian, whereas N18 and N19 were recognized in the type Trubi Marl of Sicily.

Depéret (1895), just three years after having introduced the term Burdigalian, proposed that the combined Helvetian and Tortonian stages should be collectively referred to as Vindobonian, which he equated with Suess's 2nd Mediterranean Stage, below the Sarmatian. He was of the opinion that in N. Italy, southern France and the Vienna Basin it was possible to distinguish two subdivisions, the Helvetian and Tortonian, but that they were actually facies, partially superimposed upon each other, rather than distinct time-sequential units.

Pareto (1865) had considered the Tortonian as Pliocene and introduced the Serravallian for Late Miocene. Depéret excluded the Sarmatian from his Vindobonian. Mayer-Eymar consistently interpreted his Messinian Stage as equivalent to the Sarmatian-Pontian, and equated it with the Late Miocene. Thus the Helvetian-Tortonian have generally been regarded as synonymous with Middle Miocene.

The terms Lower, Middle, Upper, or Early, Middle, Late, are however merely conventions, based on the propensity of scientists to categorize things and events into three-fold subdivisions on a linear scale. This accession to convention has actually little to do with the geological and faunal evidence upon which subdivisions of geological time should be more properly based. The Vindobonian has, to this day, not been stratotypified, so that its use as a stage term is to be discouraged. However, the Tortonian and Messinian are intimately linked both in terms of sedimentology and fauna (indeed, they were originally included by Mayer-Eymar under the name Tortonian), so that inclusion of these two stages in a Late Miocene would be more appropriate than the distinction between Middle and Late Miocene being made *between* the two stages. A threefold subdivision is suggested here in the following manner:

Miocene	Late	Messinian		Development of planktonic foraminiferal fauna of late Cenozoic character
		Tortonian		
	Middle	Langhian		First appearance of *Orbulina* (near base), and continued radiation of Neogene faunas including keeled globorotaliids
	Early	Burdigalian		First appearance and radiation of Neogene faunas (molluscs, larger Foraminifera etc.)
		Aquitanian *s.l.*	Aquitanian *s.s.*	
			Bormidian	

The names commonly used to designate Late Miocene are, in addition to Messinian, the Pontian, Sahelian, and Sarmatian.

The Sahelian (Pomel 1858) was defined in Algeria where it was believed to be the only area in the eastern Mediterranean where the Late Miocene was represented by marine strata. Haug (1911) championed its use in his Treatise, but the term has been subsequently discredited and shown to be composite in nature, with faunal affinities with the Piacenzian (Anglada 1966).

The term Sarmatian was introduced by Suess (1866, p. 232) for the *Cerithien-Schichten* and *Hernalser-Tegel* of the Vienna Basin, although the name was derived from southeastern Russia and the molluscan faunas of that area were cited as typical. Thus the Sarmatian is younger than the Vindobonian = Suess's Mediterranean Stage II (cf. Haug 1911, p. 1608, who included the Sarmatian at the top of his threefold subdivision of the Vindobonian), which it overlies, and older than the succeeding beds assigned to the Pliocene. The Late Miocene sections in eastern Europe cannot in all cases be satisfactorily correlated with western European (Vienna Basin) sequences. Thus various interpretations have developed regarding the identity and affinities of the Sarmatian Stage. Renevier (1897) interpreted the Sarmatian as a brackish water facies of the Tortonian, and in this he appears to have been followed by Gignoux (1950). Munier-Chalmas and de Lapparent (1893), Depéret (1893, 1895), Janoschek (1951), and Glaessner (1953) consider the Sarmatian as definitely younger than the marine Tortonian. The stratotypes of the two stages are indeed of different ages; on the other hand the Sarmatian represents the regressive phase of a marine sedimentary cycle, of which the Tortonian represents the early, transgressive phase. It is this basic dichotomy which has caused the difference in opinions, in much the same manner as the various interpretations of the relationships between the Rupelian and Chattian discussed above.

The Pontian Stage was introduced by LePlay (1842) for limestone strata near Odessa and Novocherkassk and Taganrog. Barbot de Marny, in Suess (1866) restricted the term Pontian to include only the former two, the outcrops at the latter locality being shown to be of Sarmatian age. The Pontian was thought to contain beds transitional between the Miocene and Pliocene in southern Russia. In Eastern Europe the term is generally applied to strata which occupy at least part of this time-interval, but accurate correlation has not been possible. In the last century it was common practice to correlate the Pontian with the marine Plaisancian, which lies above the Messinian gypsums and shales. Variation in its age assignation can be seen to range from its almost unanimous inclusion in the Pliocene in eastern Europe (Soviet Union), and Upper Miocene in western Europe. Whereas Gignoux (1950) placed the Pontian above the Tortonian, Munier-Chalmas and de Lapparent (1893) earlier placed it above the Sarmatian; Laffitte (1948) even correlated it with the Tortonian.

Andrusov (1899) divided the Sarmatian into a lower, middle and upper part; these were in turn, subsequently named Volhynian, Bessarabian and Chersonian by Simionescu (1903).

As early as 1913 Winkler-Hermaden expressed the opinion that the Lower Sarmatian (Volhynian) of Russia corresponds to the Lower-Middle Sarmatian of Steiermark, and that the Middle-Upper Sarmatian of Russia (*Nubecularia* horizon of Bessarabia) corresponds to the Upper Sarmatian of Steiermark. On the basis of comparative studies by various investigators (cited in Winkler-Hermaden 1960, p. 227) the Sarmatian of the Vienna Basin area is shown to correspond to the Lower and Middle Sarmatian of the euxinic basins in the Paratethys region, and the Upper Sarmatian and Maeotian of the Paratethys corresponds to the Lower Pannonian in the Pannonian Basin. Winkler-Hermaden (1924) suggested that the marine Piacenzian of the Mediterranean region corresponds to the thinner and less widely developed Dacian Stage (=*Dazische stufe*) of the Vienna Basin, and that the Pontian Stage (*s.l.*) and the Piacenzian are broadly correlatable (see also Winkler-Hermaden 1943, 1952, 1960). Papp (1959) has correlated the Pontian Stage with the upper Pannonian and upper Messinian and placed the Miocene/Pliocene boundary at the top of the Pontian *s.s.*

The widespread regression which characterizes the Late Miocene in Europe and the Mediterranean area is intimately connected with the process of expansion of continental margins, uplift and erosion. The presence of *Hipparion* in beds of late Tortonian-early Sarmatian age rules out this criterion in defining the Miocene/Pliocene boundary. Continental sediments have been shown to extend from the upper Tortonian into the Lower Pliocene (Gouvernet 1958) so that *Hipparion* cannot be used to define the base of the Pliocene. At the Miocene Colloquium (1958) in France it was suggested that the base of the Pliocene be drawn at the first appearance of the Pliocene marine transgression. The Miocene would thus include all continental beds which characterize the end of the Late Miocene sedimentary cycle. Winkler-Hermaden (1960) has pointed out that if the Plaisancian is considered equivalent to the Pontian Stage of Russia (=Upper Pannonian of the Pannonian Basin), which begins with a regional transgression, then correlation can be carried across the Mediterranean area into the Paratethys. A table showing suggested correlations between stage units of the Late Miocene/Pliocene in various regions is presented below (Table 52.29). It is more than likely that most of the various terms which have been used to designate time-stratigraphical units represent little more than rock-stratigraphical entities in local areas. From the point of view of chronostratigraphy most of the terms suggested as stage names in the late Miocene-early Pliocene are unsatisfactory. They are represented by brackish or fresh-water sequences with restricted faunas, which do not allow satisfactory correlations outside the immediate area of their development.

Thus Denizot (1957) suggested a return to the original concept of Tortonian of Mayer-Eymar (1858), i.e. extending Tortonian up to the base of the Piacenzian *s.l.* (or Tabianian-Zanclian). Selli (1960, footnote p. 313) objected to this on the grounds that the Tortonian would then represent the time interval from Helvetian to Piacenzian *s.l.*, and would include perhaps the major part of the Miocene, particularly since Drooger, Papp and others tend to place the base of the Tortonian approximately at the first appearance of *Orbulina*. This argument has been rendered superfluous to some extent by the recent realization that Helvetian and Tortonian are not closely related from the point of view of time. The Helvetian would appear to be pre-*Orbulina* in age (pre-Zone N9) and broadly correlatable with (at least a part of) the Burdigalian, whereas the base of the Tortonian is within Zone N15. Thus the late Miocene (including the Tortonian) would include parts of Zones N15, N16, N17, and a part of Zone N18 – a relatively short interval of time with respect to the remainder of the Miocene, and approximately equivalent to the Pliocene. Banner and Blow (1965, p. 1165) report that the upper Tortonian and at least a part of the Messinian belong to Zone N17, and thus overlap in time. The Messinian, or latest Miocene, is then a very short interval of time, equivalent probably to little more than a single planktonic foraminiferal zone. The stage name Messinian is accepted here with reservation, but the suggestion by Denizot (1957) that we return to Mayer-Eymar's original concept of Tortonian is considered appropriate and worthy of consideration.

Miocene/Pliocene boundary. Anyone who has had the good fortune of visiting Sicily and examining it in the field cannot help being struck by the dazzling beauty of the massive '*trubi*' sequence as the sun is reflected from its white surface along the shores of the azure-green Mediterranean. It is a unit that is widespread in Sicily, where it lies above the Messinian marls and gypsum beds. It corresponds well with the Tabianian of central and northern Italy. According to Selli (1960) the faunas of the Tabianian-Zanclian have a Miocene rather than Pliocene aspect, and thus a Miocene/Pliocene boundary might better be placed between Tabianian and Piacenzian *s.s.* than at the base of the Tabianian. However, the necessity of emending the definition of the Pliocene is here outweighed by the desirability of maintaining stability in circum-Mediterranean stratigraphical concepts.

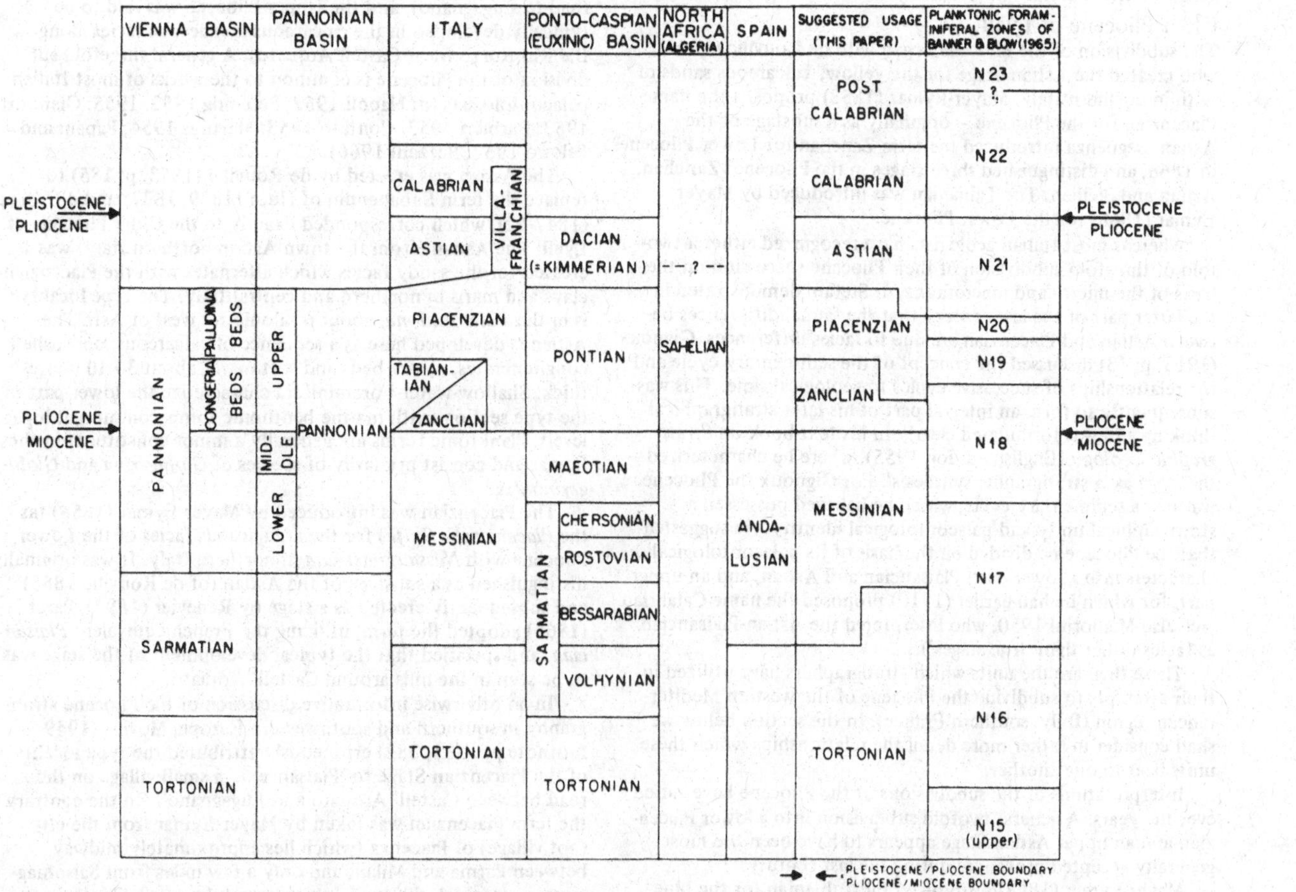
VIENNA BASIN
PANNONIAN BASIN
ITALY
PONTO-CASPIAN (EUXINIC) BASIN
NORTH AFRICA (ALGERIA)
SPAIN
SUGGESTED USAGE (THIS PAPER)
PLANKTONIC FORAMINIFERAL ZONES OF BANNER & BLOW (1965)
PLEISTOCENE
PLIOCENE
PLIOCENE
MIOCENE
PANNONIAN
CONGERIA BEDS
PALUDINA BEDS
UPPER
MIDDLE
LOWER
PANNONIAN
SARMATIAN
TORTONIAN
CALABRIAN
VILLAFRANCHIAN
ASTIAN
PIACENZIAN
TABIANIAN
ZANCLIAN
MESSINIAN
TORTONIAN
DACIAN (=KIMMERIAN)
— — ? — —
PONTIAN
MAEOTIAN
SARMATIAN
CHERSONIAN
ROSTOVIAN
BESSARABIAN
VOLHYNIAN
TORTONIAN
SAHELIAN
ANDALUSIAN
POST CALABRIAN
CALABRIAN
ASTIAN
PIACENZIAN
ZANCLIAN
MESSINIAN
TORTONIAN
N 23
?
N22
N 21
N20
N19
N 18
N 17
N16
N 15 (upper)
PLEISTOCENE/PLIOCENE BOUNDARY
PLIOCENE/MIOCENE BOUNDARY

Table 52.29. Suggested correlation of late Miocene-Pliocene stages in Mediterranean and Paratethys regions

Zone N17 has been found in the neostratotype Messinian, Zones N18 and N19 in the Zanclian and Piacenzian, and Zone N18 apparently straddles the Miocene/Pliocene boundary (Banner and Blow 1965, p. 1,165; 1967, p. 151). The general paucity of planktonic Foraminifera in the Tabianian (of northern Italy) renders this stage less useful in comparative stratigraphical work than the more richly endowed Zanclian of southern Italy. The actual Miocene/Pliocene boundary does not appear to be marked by any significant criteria among the planktonic Foraminifera. However, the appearance of *Sphaeroidinella dehiscens* (at the base of N19), occurs in the lower Pliocene and the overlap in ranges of *Globigerina nepenthes, Globorotalia tumida, G. multicamerata, G. miocenica* and *Sphaeroidinellopsis subdehiscens, inter alia,* can be used in distinguishing latest Miocene-earliest Pliocene (Banner and Blow 1967; Blow, personal communication; Parker 1967).

Pliocene

Lyell (1833) proposed the term Pliocene for the youngest Tertiary deposits which he recognized at the time. He divided his Pliocene into an older Pliocene (based on more than 50 per cent species still living), which would correspond to the Astian-Piacenzian as subsequently recognized in Italy, and a Newer Pliocene (with 90-95 per cent of species still living), for which he subsequently (Lyell 1839) introduced the term Pleistocene. Although he later suggested abandonment of the term Pleistocene in favour of post-Tertiary, or post-Pliocene, the Pleistocene has become firmly rooted in stratigraphical literature and is used as a virtual synonym of Quaternary.

The term *'deposito subappenino'* was used by Huot (1837) to characterize the late Tertiary clays and marls in the hilly regions in the vicinity of Castell' Arquato. Adoption of this term was based on earlier work in this area by Brocchi, who in 1814 described the calcareous sands and marls and their molluscan faunas in this region.

d'Orbigny adopted the term *Subapennin* as the youngest stage of the Tertiary, essentially substituting it for the Older Pliocene of Lyell.

The subdivision of the Pliocene began with de Rouville (1853), who created the Astian Stage for the yellow, calcareous sands of Asti, in northern Italy. Mayer-Eymar (1858) proposed the name Piacenzian for the Pliocene – originally as a substage of the Astian. Seguenza introduced the term Zanclian for Lower Pliocene in 1868, and distinguished three stages in the Pliocene: Zanclian, Astian and Sicilian. The Tabianian was introduced by Mayer-Eymar (1868) for the Lower Pliocene.

Whereas most Italian geologists have recognized either a twofold or threefold subdivision of their Pliocene succession on the basis of the micro- and macrofauna, di Stefani demonstrated in the latter part of the last century that the faunal differences between Astian and Piacenzian are due to facies differences. Gignoux (1913, p. 13) discussed the concept of the sedimentary cycle and the relationships of successive cycles to geological time. This was subsequently to form an integral part of his later stratigraphical thinking and was formulated clearly in his text-book on *Stratigraphic Geology* (English edition 1955), where he characterized the stage as 'a stratigraphic synthesis'. For Gignoux the Pliocene was just a sedimentary cycle, which he believed possessed a stratigraphical unity and palaeontological identity. He suggested that the Pliocene be divided on the basis of its palaeontological characters into a lower part, Plaisancian and Astian, and an upper part, for which he had earlier (1910) proposed the name Calabrian (see also Migliorini 1950, who interpreted the Astian-Plaisancian as facies rather than 'true stages').

These then are the units which stratigraphers have utilized in their attempts to subdivide the Pliocene of the western Mediterranean region (Italy, southern France). In the section below we shall consider in rather more detail the relationships which these units bear to one another.

Interpretations of the subdivisions of the Pliocene have varied over the years. A general twofold subdivision into a lower Piacenzian and an upper Astian Stage appears to have been the most generally accepted procedure during the last century.

Mayer-Eymar (1868) introduced the Tabianian for the blue marls which outcrop at Tabiano Bagni, near Salsomaggiore, in northern Italy as the lower stage of the Pliocene. The Tabianian was designated as *Tabbianer Schichten* in 1874, as *Tabbianin* in 1884 and as *Tabbianon* in 1888.

The Zanclian, introduced by Seguenza (1868), has its type locality in the region of Monti Peloritani on Sicily. The microfauna of the typelevel of the Zanclian – the *trubi* marl and its equivalents – is in Sicily rich and diversified, and consists to a large extent of planktonic Foraminifera and other pelagic organisms (nannofossils). Correlation with other areas can be achieved by comparison with the Zanclian, whereas the cooler water faunas of the Tabianian do not contain the rich faunal elements essential for intercontinental stratigraphical correlation. Indeed it was on the basis of Zanclian faunas that Banner and Blow (1965) were able to recognize their Zone N18 in the Mediterranean area, and were able to demonstrate a partial overlap between the Piacenzian and Zanclian Stages. The development of *Sphaeroidinella dehiscens* from *Sphaeroidinellopsis subdehiscens*, as well as the overlap of *Globigerina nepenthes* and *S. dehiscens* in the lower part of the '*trubi*' of Sicily, has been considered characteristic of the Lower Pliocene (Blow, personal communication; Parker 1967; cf Blow 1959, p. 196). For this reason the use of Zanclian for the Lower Pliocene is preferred to Tabianian.

Seguenza (1868) proposed a threefold subdivision of the Pliocene: Zanclian (Lower): Piacenzian (Middle) and Astian (Upper). Ruggieri and Selli (1950) have suggested a threefold subdivision of the Pliocene into Lower, Middle and Upper. The Lower was said to correspond with the Tabianian of Doderlein (1872) and Cocconi (1873). The Middle Pliocene was said to be typically developed in the argillaceous facies at Santa Maria Maddalena (several kilometres west of Castell' Arquato on the road to Lugagnano), and the Upper Pliocene was said to be typically developed in the calcareous-arenaceous facies along the Rio Rorzo (near Castell'Arquato). A general threefold subdivision of the Pliocene is common to the works of most Italian palaeontologists (di Napoli 1952; Perconig 1952, 1955; Giannoti 1953; Barbieri 1953; Contato 1953; Martinis 1954; Papani and Pelosio 1962; Bezzani 1966).

The Astian was erected by de Rouville (1853, p. 185) to replace the term Subapennin of Huot (1829, 1837) and d'Orbigny (1847-52), which corresponded exactly to the Older Pliocene of Lyell. The Astian (from the town Asti in northern Italy) was created for the sandy facies which alternates with the Piacenzian clays and marls in northern and central Italy. The type locality is in the *Valle Andona*, about 6 kilometres west of Asti. The Astian is developed here as a sequence of calcareous sands, shell conglomerates, oyster beds and soft marls, about 30-40 metres thick. Shallow-water Foraminifera characterize the lower part of the type section, with marine benthonic forms common at higher levels. Planktonic forms are generally a minor constituent in these faunas and consist primarily of species of *Globigerina* and *Globigerinoides*.

The Piacenzian was introduced by Mayer-Eymar (1858) (as the *Piacenzische Stufe*) for the argillaceous facies of the Lower Pliocene with *Nassa semistriata* in northern Italy. It was originally distinguished as a substage of the Astian (of de Rouville 1853) and subsequently erected as a stage by Renevier (1897). Pareto (1865) adopted the term, utilizing the French equivalent *Plaisancian*, and specified that the typical development of the stage was to be seen in the hills around Castell' Arquato.

In an otherwise informative discussion of the Pliocene stratigraphy in southern and southwestern Europe, Movius (1949 footnote pp. 381, 382) erroneously attributed the type locality of the Piacenzian Stage to 'Plaisance . . .a small village on the road between Castell' Arquato and Lugagnano'. On the contrary the term Piacenzian was taken by Mayer-Eymar from the city (not village) of Piacenza (which lies approximately midway between Parma and Milan, and only a few miles from Salsomaggiore, near which the type Tabianian can be seen). Contrary to Movius's statement that the Piacenzian, erroneously attributed to Gignoux (1913), 'never came into general use', Piacenzian is widely used in Italy. It is only correct that this name be used, since the type locality and most typical development of the sediments of this age are to be seen in north-central Italy. As pointed out above, the term Plaisancian is merely the French translation of the word Piacenza, and Pareto (1865) appears to have been the first to use this term.

The Foraminifera of the Castell' Arquato region were described by d'Orbigny in 1826, and Malagoli (1892) and more recently by di Napoli (1952) and Barbieri and Medioli (1964).

The Piacenzian consists of sands, with marly intercalations at the base, followed upwards by marl and claystones (the typical blue clays of the Piacenzian), which grade upward into sandy and silty claystones, heralding the presence of the Astian facies. Marine benthos and benthonic foraminiferal faunas are abundant in these blue marls; faunal lists from localities in the lower, middle and upper Pliocene in the Castell' Arquato-Lugagnano-Vernasca area were given by Barbieri in the guidebook to the 4th International Micropalaeontological Colloquium in Italy (1958). Jenkins (1964) presented a list of planktonic foraminiferal species observed in samples from the 'Upper Pliocene' of Castell' Arquato. *Globorotalia* cf. *G. hirsuta, G. scitula* and *G. inflata* were among the forms cited but *G. menardii* was not found, nor were *G. fistulosus* or *Discoaster* spp.

Both Depéret and Gignoux have suggested that the Astian-Piacenzian be grouped into a single 'Older Pliocene' – a sandy and an argillaceous facies – the term Astian being suitable for this combination. On the other hand de Lapparent and Haug accepted the two stages as being distinguished by Renevier: Astian, de Rouville (1853); Piacenzian, Mayer-Eymar (1858).

Ruggieri (1961) proposed the following biostratigraphical zonation of the Pliocene and Pleistocene in Italy: lower zone (without *Cyprina islandica*) and two subzones, A. *Globorotalia hirsuta;* B. absence of both forms, which corresponds to Gignoux's (1913) 'Pliocene antico', and an upper zone (with *Cyprina islandica*) with two subzones, C. (without *Anomalina baltica*); D. (with *A. baltica*). The base of D was considered to be the base of the Pleistocene (see also Ruggieri and Selli 1950). This type of zonation, though of interest in terms of local stratigraphical studies, can hardly be applied outside the immediate area of investigation as already pointed out by Banner and Eames (1966).

Banner and Blow (1965) have shown that the Pliocene can be subdivided into approximately three planktonic foraminiferal zones: the upper part of N18, and N19-N21. The Pliocene, as we shall see below, is actually a relatively short period of time, comparable to the Oligocene. Zone N20, if it exists, represents a very short interval of time, so that the Pliocene consists really of the equivalent of about two normal zones.

Parker (1967) has demonstrated the applicability of the Neogene planktonic foraminiferal zonation of Banner and Blow (1965) to the late Tertiary (Late Miocene-Pliocene) in some deep-sea cores from the Pacific Ocean. A summary of her results is presented here.

(1) The base of the Pliocene is placed between Zones N18/N19 (i.e. at the first evolutionary appearance of *Sphaeroidinella dehiscens*).

(2) The Pliocene/Quaternary boundary is drawn between Zones N21/N22 (i.e. base of first evolutionary appearance of *Globorotalia truncatulinoides*. The base of the Quaternary was shown to be characterized by the first evolutionary appearances of the following forms: *Globigerinoides tenellus* (? from *G. rubescens*), *Globorotalia truncatulinoides* (from *G. tosaensis*), *Globigerina digitata* (from *G. praedigitata*), *Globanomalina* (?) *pumilio* (? from *G. praepumilio*); *Globoquadrina dutertrei* was shown to evolve from *G. humerosa* in the upper part of Zone N21.

(3) Zone N20 (*G. multicamerata-Pulleniatina obliquiloculata* partial-range-zone) was found to be of questionable value in the material studied. The anomalous occurrence together of *G. tosaensis* and *G. altispira* in two cores, and the general difficulty of recognizing this zone in other cores was discussed.

A review and discussion of the various late Cenozoic planktonic foraminiferal zonations which have been proposed during the past few years is also presented, and the reader is referred to these for further information (Parker 1967).

Pliocene/Pleistocene boundary. Gignoux (1910, p. 841) introduced his Calabrian Stage for the post-Pliocene formations in Italy (characterized by the appearance of *Cyprina islandica*), which precede the Sicilian, overlie the Astian, and are equivalent to the Newer Pliocene of Lyell. As localities where the stratigraphical succession was said to be typical of the Calabrian he cited the following (1913, p. 23): 'collines toscanes à Vallebiaja, celle du Monte Mario, près de Rome, et surtout celles de Calabre (Gravina di Puglia, détroit de Catanzaro, environs de Messine et de Reggio)' He included his Calabrian in the Pliocene, believing that it represented the upper, concluding part of the Pliocene sedimentary cycle in Italy.

At the 19th International Geological Congress in Algeria (1952) the Calabrian was designated the oldest stage of the marine Pleistocene (Quaternary). This was based upon a recommendation of a special temporary Commission of the 18th International Geological Congress in Great Britain that the 'Lower Pleistocene' should include as its basal member in the type-area the Calabrian formation (marine) together with its terrestrial (continental) equivalent the Villafranchian (*Internat. Geol. Congress, 19 Sess., 1952, sect.* H, p. 6).

Migliorini (1950) presented a review of studies on the Pliocene/Pleistocene boundary in Italy. He reached several important conclusions, several of which are enumerated below:

(1) In Italy the Pliocene/Pleistocene boundary may be placed either between Plaisancian-Astian and the Calabrian, or between the Calabrian and the Sicilian.

(2) The Calabrian can be distinguished satisfactorily from the Plaisancian-Astian by means of both molluscan and foraminiferal faunas.

(3) The Calabrian rests uncomformably upon the Astian-Plaisancian in some areas, and in areas where deposition is continuous there is evidence that the Calabrian is transgressive. Significant orogeny occurred in the Apennine Region between the late Pliocene and Early Pleistocene.

(4) Fossil faunas and floras indicate that pronounced climatic cooling occurred between the Astian-Plaisancian and the Calabrian. Migliorini was of the opinion that this cooling coincided with the onset of the 'Ice Age', but this seems unlikely (see Zeuner 1950, who reviewed the evidence which suggests that the Calabrian is wholly pre-Günzian).

In a report prepared by the commission nominated by the Italian Geological Society to study the Pliocene/Pleistocene boundary in Italy, Gignoux (1954) expressed his general agreement with the placement of the boundary at the base of his Calabrian Stage and with the discussion presented by Migliorini (1950).

Selli (1962) has admirably summarized our general knowledge of the marine Quaternary in the Adriatic-Ionian region of the Italian peninsula. However, it is not within the scope of this paper to present a comprehensive discussion of the Quaternary.

Studies by Selli on the palaeontology and stratigraphy, and by Emiliani and Mayeda on isotopic palaeotemperatures at Le Castella (Calabria, southern Italy) have shown that, although in stratigraphical continuity, the Pliocene and Pleistocene can be differentiated on various lines of evidence. Average surface temperature of the late Pliocene seas showed values of about 23-25° C, whereas Pleistocene temperatures have a value of about 15-16° C. No sharp temperature change was observed across the Pliocene/Pleistocene boundary, but the boundary is characterized by the appearance here, as elsewhere in Italian sections, by the appearance of *Hyalinea baltica* and *Cyprina islandica.*

Papani and Pelosio (1962) have studied a 970 metres thick section of Pliocene-Pleistocene strata along the Stirone River, near Parma, central Italy. They drew the Pliocene/Pleistocene boundary at the first appearance of *Cyprina islandica, Pholadidea vibonensis* and *Hyalinea baltica.* A regressive, brackish phase in the Calabrian was also described.

Dondi (1962) has summarized the faunas and stratigraphy of the late Cenozoic of the Pedeappenino Padano. He points out that the 18th International Geological Congress in London fixed the base of the Pleistocene at the first appearance of *Cyprina islandica,* that Ruggieri prefers to place it at the first appearance of *Hyalinea baltica,* and that Selli, *in* Emiliani *et al.* (1961) points out that the two forms occur almost simultaneously, but that their association together is rare because of their ecological preferences. *C. islandica* prefers a shallow, sandy bottom; *H. baltica* prefers a somewhat deeper, muddy bottom.

Dondi and Papetti (1966) described the Pliocene-Pleistocene section at Valle de Santerno (Emilia), Italy. The Pliocene/Quaternary boundary was fixed on the basis of a statistical approach, i.e. at the point of increase in percentage of some cold-water species. *Globigerina pachyderma* in sample 58 was considered indicative of Calabrian. The presence of *Hyalinea baltica* and *Bolivina quadrilatera* in samples above was said to be characteristic of the Quaternary. The occurrence of *Globorotalia scitula, G. inflata, Orbulina universa, G. naparimaensis* and *Globigerinoides* spp. in samples below was said to characterize the Upper Pliocene and indicate high ambient temperatures.

Banner and Blow (1965; confirmed in discussions with Blow 1966) have recognized the important phylogenetic trend from *Globorotalia tosaensis* to *G. truncatulinoides* near the base of the type Calabrian at Santa Maria di Catanzaro, southern Italy. The first evolutionary appearance of *G. truncatulinoides* was used to define the base of the Pleistocene (base Zone N22). This significant occurrence has made it possible to recognize and distinguish Pliocene and Pleistocene in marine sequences over large areas, and has furnished palaeontologists with the first relatively unequivocal (easily recognizable) criterion for recognizing the base of the Pleistocene.

The present author has recently recognized the transition from *G. tosaensis* to *G. truncatulinoides* in deep sea cores from the Atlantic Ocean, and with the aid of colleagues at the Woods Hole Oceanographic Institution has been able to relate the Pliocene/Pleistocene boundary thus defined to a level within the Olduvai Normal Event, at approximately 1.85 million years ago (see following discussion).

BIOSTRATIGRAPHY OF POST-EOCENE LARGER FORAMINIFERA

In the Far East the Tertiary has been subdivided into a series of letter stages, first introduced by van der Vlerk and Umbgrove (1927), subsequently revised by van der Vlerk (1955, 1966). Precise correlation between the letter stages of the Far East and the classic stages of European stratigraphy has been hampered by several factors:

(1) Lack of adequately described fossiliferous sequences in different provinces, particularly across stage boundaries.

(2) The generally accepted stratigraphical ranges of many Tertiary larger Foraminifera are still inadequately known.

(3) Disagreement on identification of various species of larger Foraminifera amongst specialists lessens their importance.

(4) Although sophisticated techniques have been developed for dating strata on the evolutionary stages reached by large Foraminifera (Tan Sin Hok, Drooger, van der Vlerk), the primary stratigraphical evidence upon which this evidence is dependent is weak. It is, in many cases, difficult to ascertain the stratigraphical level from which a sample has been taken relative to the stage stratotype, or its location within a transgressive-regressive cycle (which may define the limits of a particular stage unit). Studies in areas where relatively continuous successions are available are an obvious necessity.

The following brief discussion is taken from the summary presented in Eames *et al.* (1962), Adams (1965), and from information which Dr Adams has kindly conveyed in private correspondence.

According to Dr Adams 'over the whole Indo-Pacific (and possibly over the whole Tethyan region outside Europe) it is possible to divide the Oligocene into three parts – Upper, Middle and Lower. These divisions almost certainly do not correspond to the European stages in general use' (letter dated 15 February 1967). The main criteria which Dr Adams uses are as follows:

Upper Oligocene, *Lepidocyclina* ± *Miogypsinoides s.l.*

Middle Oligocene, reticulate *Nummulites* ± striate *Nummulites* with *Eulepidina* and *Nephrolepidina*

Lower Oligocene, reticulate and striate *Nummulites.* Absence of lepidocyclinids.

It is generally agreed that the letter stage 'Tc' corresponds to the 'Lattorfian' as it is used in the Middle East (as used by Eames *et al.* (1962) and to the Lower Oligocene as recognized in Europe.

In the Far East *Miogypsinella complanata* and *Lepidocyclina (Eulepidina)* range throughout the Td and Lower Te (Eames *et al.* 1962, although as Adams points out in a footnote; 1965, p. 320, there is no published evidence confirming this in the case of *M. complanata*). The Td is generally correlated with the Rupelian Stage of Europe; its upper limit was defined by the extinction of the nummulites.

Recognition of the Chattian in the Far East is virtually impossible owing to the absence of positive criteria. In southern Europe *M. complanata* appears for the first time in beds of Late Oligocene age; in northern Europe *M. septentrionalis* (which is interpreted as related to *M. complanata*) appears in the Chattian.

The Te stage in the Far East can be divided into two parts by the first appearance of *Flosculinella, Miogypsina* and *Miogypsinoides dehartii,* and by a change in the composition of the *Lepidocyclina* fauna. The lower part of the Lower Te is correlated with the Chattian by Adams, since 'by definition, Te began when the last nummulites became extinct' (Adams 1965, p. 321).

Adams (1965, pp. 321, 322) has presented evidence which suggests that *Austrotrillina howchini* and *A. striata,* although they appear near the Td/Te boundary, are not reliable markers for the base of the Aquitanian and lower Te stages. He records the earliest *Austrotrillina,* together with *Nummulites,* in the Melinau Gorge,

Sarawak (Borneo). *Miogypsinoides* and *Spiroclypeus*, on the other hand, appear to be useful in defining the base of Te.

The Aquitanian-Burdigalian boundary, as determined by Eames *et al.* (1962), was placed just *above* the *Orbulina*-datum. The Aquitanian was said to include all planktonic zones from the base of the *G. ampliapertura* Zone to the top of the *G. fohsi barisanensis* Zone. Subsequent work on the type Aquitanian-Burdigalian section by Jenkins (1964, 1966), by various other workers in Europe, and by the authors themselves (Banner and Blow 1965, and pers. comm.) has resulted in considerable modifications to this scheme. It would appear that the Burdigalian Stage is older than the *Orbulina*-datum and that the Aquitanian/Burdigalian boundary falls approximately at the top of the *G. kugleri* Zone (Jenkins 1966). *Orbulina* appears in the lower part of the Langhian Stage (Zone N9 of Banner and Blow).

In the Melinau Gorge, Sarawak (Borneo) the Melinau Limestone includes beds of lower Te age at its top; the upper part of the Lower Te and the Upper Te are not found there. Planktonic foraminiferal faunas from the strata just above this limestone suggests that the top of the limestone cannot be younger in age than the *G. dissimilis* Zone of Bolli (Adams, 1965, p. 323).

Adams (1965, p. 322, and pers. comm.) would equate the Te/Tf boundary with the Aquitanian/Burdigalian boundary as defined by Eames *et al.* (1962). Dr Adams has informed me that according to his work the so-called 'Burdigalian' of the Middle East is the same as the Lower Tf (= Tf 1-2) in the Far East. Tf 1-2 are characterized by the appearance of *Flosculinella bontangensis*, *Taberina malabarica* and *Borelis melo*.

According to Adams (pers. comm.) the following tabular presentation is possibly relevant to the appearance of *Orbulina* in the Far East succession, based on his examination of faunas from the Bikini and Enitwetok drill holes, limestones of Saipan, limestones of Australia and elsewhere:

Orbulina

3. Frequently seen in strata datable as Tf_3 on larger Foraminifera.
2. Occasionally seen in strata datable as Tf 1-2 on larger Foraminifera.
1. Never seen in strata datable as Te on larger Foraminifera.

Adams is of the opinion that *Orbulina* appears at or near the Te/Tf boundary. This was also the opinion of Eames *et al.* (1962), and still is (Banner and Blow 1965). However, it is the relationship of this level to the stage units in Europe which has been drastically revised. The *Orbulina*-datum and the Te/Tf-datum is one of the most distinct biostratigraphical markers in the Tertiary, and it now appears that we can state with relative certainty that it is post-Burdigalian, pre-Tortonian in age, and occurs in the lower part of the Langhian Stage of Italy.

The relationship between the stratigraphical ranges of *Miogypsina, Miogypsinoides* and *Orbulina* remains controversial. Eames *et al.* (1962) have argued that these forms occur together, whereas Drooger (1964, 1966) has maintained that the miogypsinids became extinct *prior* to the first evolutionary appearance of *Orbulina*. Dr Adams has informed me that 'it seems probable that in some areas *Miogypsina* was extinct before *Orbulina* appeared, but I don't think that this is generally true. There is good evidence of overlap from Australia and New Guinea' (letter of 28 February 1967). It would appear from a survey of the literature that the miogypsinids became extinct in central and southern Europe prior to, but near, the *Orbulina*-datum, whereas in the Far East they persisted into somewhat younger strata. The last *Miogypsinoides* appear in f_2, the last *Miogypsina* and lepidocyclinids at the top of the f stage (f_3). A suggested correlation of the Far East letter 'stages' with European stages and planktonic foraminiferal zones is shown in Table 52.31 accompanying this paper.

CENOZOIC CHRONOSTRATIGRAPHY, PLANKTONIC FORAMINIFERAL ZONATION AND THE RADIOMETRIC TIME-SCALE

The desire for precise correlation is one of the main goals of stratigraphical geology. Subdivision of geological time, as represented in the sedimentary rock record, has occupied the talents and tested the ingenuity of geologists for over two centuries (if we take the work of William Smith as the first significant contribution in this respect). The ability to recognize and distinguish smaller and smaller units of time by biostratigraphical methods is a tribute to an increased understanding of evolutionary patterns, improved instrumentation, and a growing body of informative, descriptive and interpretative literature.

Within the space of the last thirty years we have witnessed the development of a more refined biostratigraphical zonation of the Cenozoic using the planktonic Foraminifera. This work, originally begun in the Soviet Union in the 1930's, was later undertaken in the Caribbean region, primarily by oil company palaeontologists. The information obtained by specialists in these various countries lay fallow for many years owing to lack of communication, but with the recent increase in communication between specialists in all parts of the world, we have seen in the past ten years greater advances in intercontinental biostratigraphical zonation and correlation of the Cenozoic using planktonic Foraminifera than in the preceding thirty.

Several radiometric time-scales have also been proposed for the Cenozoic. A comparative scheme of these scales is shown in Table 52.37. The various Cenozoic radiometric time-scales which have been proposed so far are based primarily on dating of continental ash beds, etc., which are associated with terrestrial mammals. Dates on marine beds, though available, are in the minority. Recently some dates have been made on submarine ash layers intercalated in deep sea cores.

In the following discussion terms of reference should be established for the sake of clarity and uniformity. When speaking of Cenozoic planktonic foraminiferal zones I have adopted the following scheme:

Neogene	Zones N2 - N23 (Banner and Blow 1965)	
Palaeogene	Oligocene	P18-P19-N1 (Banner and Blow 1965)
	Eocene	Upper, 3 planktonic zones (Blow and Banner 1962)
		Middle, 5 (Bolli 1957)
		Lower, 5 (Bolli 1957, plus a lower *G. velascoensis-G. subbotinae* zone; see charts this paper)
	Palaeocene	5 (Bolli 1957; modified by the present author

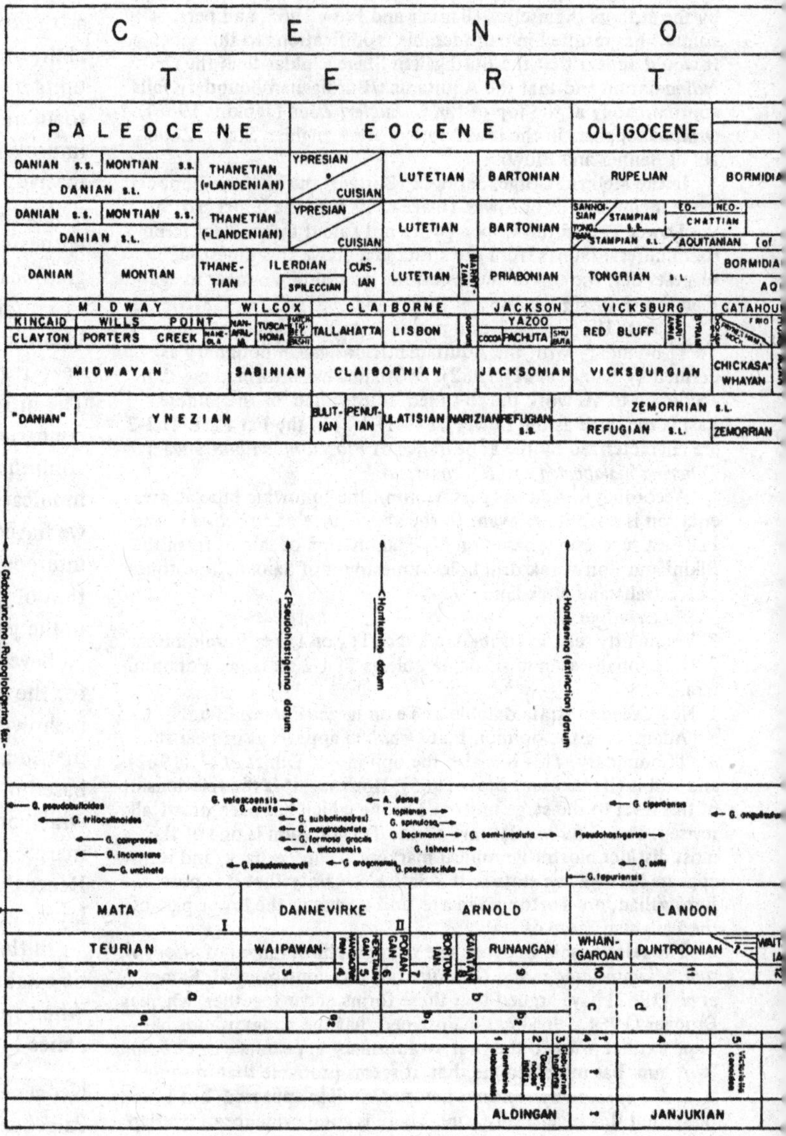
C E N O
T E R T
PALEOCENE
EOCENE
OLIGOCENE
DANIAN s.s.
MONTIAN s.s.
DANIAN s.l.
THANETIAN (=LANDENIAN)
YPRESIAN
CUISIAN
LUTETIAN
BARTONIAN
RUPELIAN
BORMIDIA
SANNOI-SIAN
STAMPIAN s.s.
TONG-RIAN
STAMPIAN s.l.
EO-
NEO-
CHATTIAN
AQUITANIAN (of
DANIAN
MONTIAN
THANE-TIAN
ILERDIAN
SPILECCIAN
CUIS-IAN
BIARRITZIAN
PRIABONIAN
TONGRIAN s.l.
MIDWAY
WILCOX
CLAIBORNE
JACKSON
VICKSBURG
CATAHOU
KINCAID
WILLS POINT
CLAYTON
PORTERS CREEK
TUSCA-HOMA
TALLAHATTA
LISBON
YAZOO
COCOA
PACHUTA
SHU-BUTA
RED BLUFF
FRIO
MIDWAYAN
SABINIAN
CLAIBORNIAN
JACKSONIAN
VICKSBURGIAN
CHICKASA-WHAYAN
"DANIAN"
YNEZIAN
BULIT-IAN
PENUT-IAN
ULATISIAN
NARIZIAN
REFUGIAN s.s.
ZEMORRIAN s.l.
REFUGIAN s.l.
ZEMORRIAN
Globotruncana–Rugoglobigerina (extinction) datum
Pseudohastigerina datum
Hantkenina datum
Hantkenina (extinction) datum
G. pseudobulloides
G. triloculinoides
G. compressa
G. inconstans
G. uncinata
G. velascoensis
G. acuta
G. subbotinae (rex)
G. marginodentata
G. formosa gracilis
A. wilcoxensis
G. pseudomenardii
G. aragonensis
A. densa
T. topilensis
G. spinulosa
G. beckmanni
G. semi-involuta
G. lehneri
G. pseudoscitula
Pseudohastigerina
G. ciperoensis
G. angulisuturalis
G. selli
G. tapuriensis
MATA
DANNEVIRKE
ARNOLD
LANDON
I
II
TEURIAN
WAIPAWAN
PORAN-GAN
BORTON-IAN
RUNANGAN
WHAIN-GAROAN
DUNTROONIAN
2
3
4
5
6
7
8
9
10
11
12
a
b
c
d
a_1
a_2
b_1
b_2
1
2
3
4
4
5
Hantkenina alabamensis
Globigerina linaperta
Victoriella conoidea
ALDINGAN
JANJUKIAN

Table 52.30. Suggested correlation between Cenozoic chronostratigraphical units in various regions and relationship to planktonic foraminiferal events

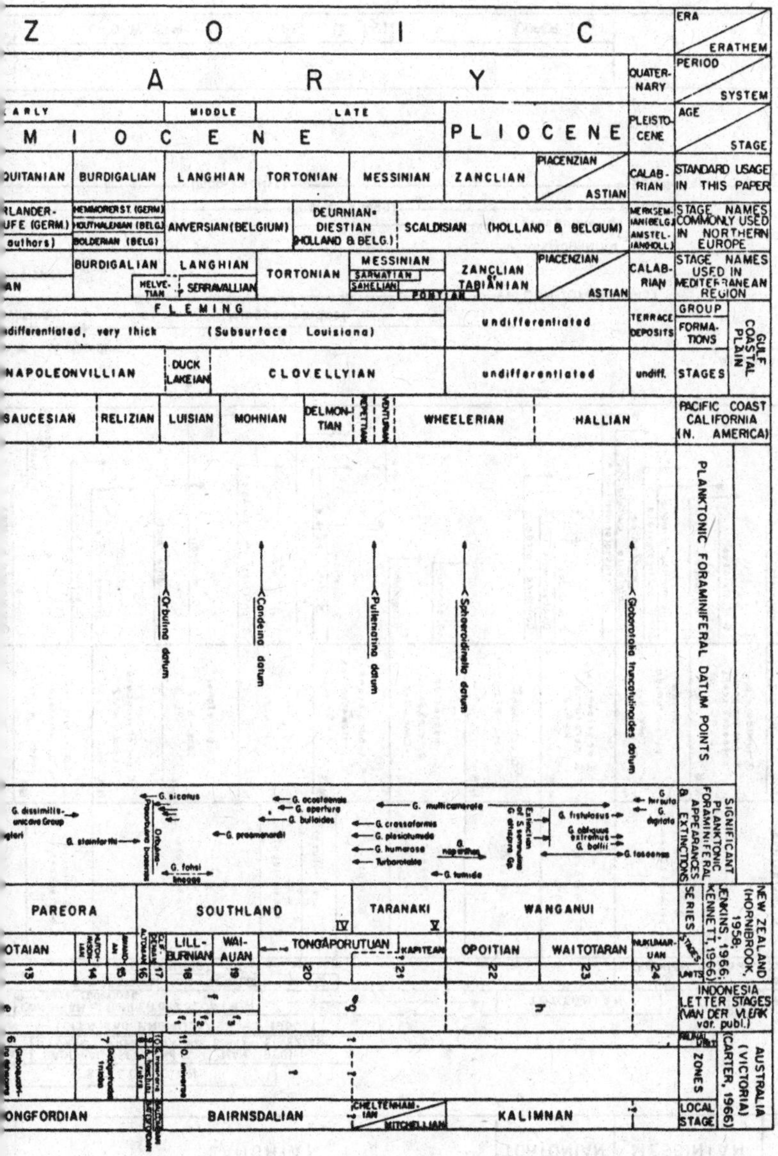

Z O I C
ERA
ERATHEM
A R Y
QUATER-NARY
PERIOD
SYSTEM
EARLY
MIDDLE
LATE
MIOCENE
PLIOCENE
PLEISTO-CENE
AGE
STAGE
QUITANIAN
BURDIGALIAN
LANGHIAN
TORTONIAN
MESSINIAN
ZANCLIAN
PIACENZIAN
ASTIAN
CALAB-RIAN
STANDARD USAGE IN THIS PAPER
RLANDER-UFE (GERM.) authors)
HEMMOORERST. (GERM.)
HOUTHALENIAN (BELG.)
BOLDERIAN (BELG.)
ANVERSIAN (BELGIUM)
DEURNIAN-DIESTIAN (HOLLAND & BELG.)
SCALDISIAN (HOLLAND & BELGIUM)
MERKSEM-IAN (BELG.)
AMSTEL-IAN (HOLL.)
STAGE NAMES COMMONLY USED IN NORTHERN EUROPE
BURDIGALIAN
LANGHIAN
HELVE-TIAN
SERRAVALLIAN
TORTONIAN
MESSINIAN
SARMATIAN
SAHELIAN
PONTIAN
ZANCLIAN TABIANIAN
PIACENZIAN
ASTIAN
CALAB-RIAN
STAGE NAMES USED IN MEDITERRANEAN REGION
FLEMING
undifferentiated, very thick (Subsurface Louisiana)
undifferentiated
TERRACE DEPOSITS
GROUP
FORMA-TIONS
GULF COASTAL PLAIN
NAPOLEONVILLIAN
DUCK LAKEIAN
CLOVELLYIAN
undifferentiated
undiff.
STAGES
SAUCESIAN
RELIZIAN
LUISIAN
MOHNIAN
DELMON-TIAN
REPETTIAN
VENTURIAN
WHEELERIAN
HALLIAN
PACIFIC COAST CALIFORNIA (N. AMERICA)
Orbulina datum
Candeina datum
Pulleniatina datum
Sphaeroidinella datum
Globorotalia truncatulinoides datum
PLANKTONIC FORAMINIFERAL DATUM POINTS
G. dissimilis-unicava Group
G. siakensis
G. stainforthi
Orbulina
Praeorbulina bisaerata
G. praemenardii
G. fohsi lineage
G. acostaensis
G. apertura
G. bulloides
G. multicamerata
G. crassaformis
G. plesiotumida
G. humerosa
Turborotalia
G. nepenthes
G. tumida
G. fistulosus
G. obliquus extremus
G. bolli
G. fossensis
G. hirsuta
G. digitata
SIGNIFICANT PLANKTONIC FORAMINIFERAL APPEARANCES & EXTINCTIONS
PAREORA
SOUTHLAND
TARANAKI
WANGANUI
SERIES
IV
V
OTAIAN
ALTONIAN
LILL-BURNIAN
WAI-AUAN
TONGAPORUTUAN
KAPITEAN
OPOITIAN
WAITOTARAN
NUKUMAR-UAN
STAGES
13
14
15
16
17
18
19
20
21
22
23
24
UNITS
NEW ZEALAND (HORNIBROOK, 1958; JENKINS, 1966; KENNETT, 1966)
INDONESIA LETTER STAGES (VAN DER VLERK var. publ.)
ZONES
ONGFORDIAN
BAIRNSDALIAN
CHELTENHAM-IAN
MITCHELLIAN
KALIMNAN
LOCAL STAGE
AUSTRALIA (VICTORIA) (CARTER, 1966)

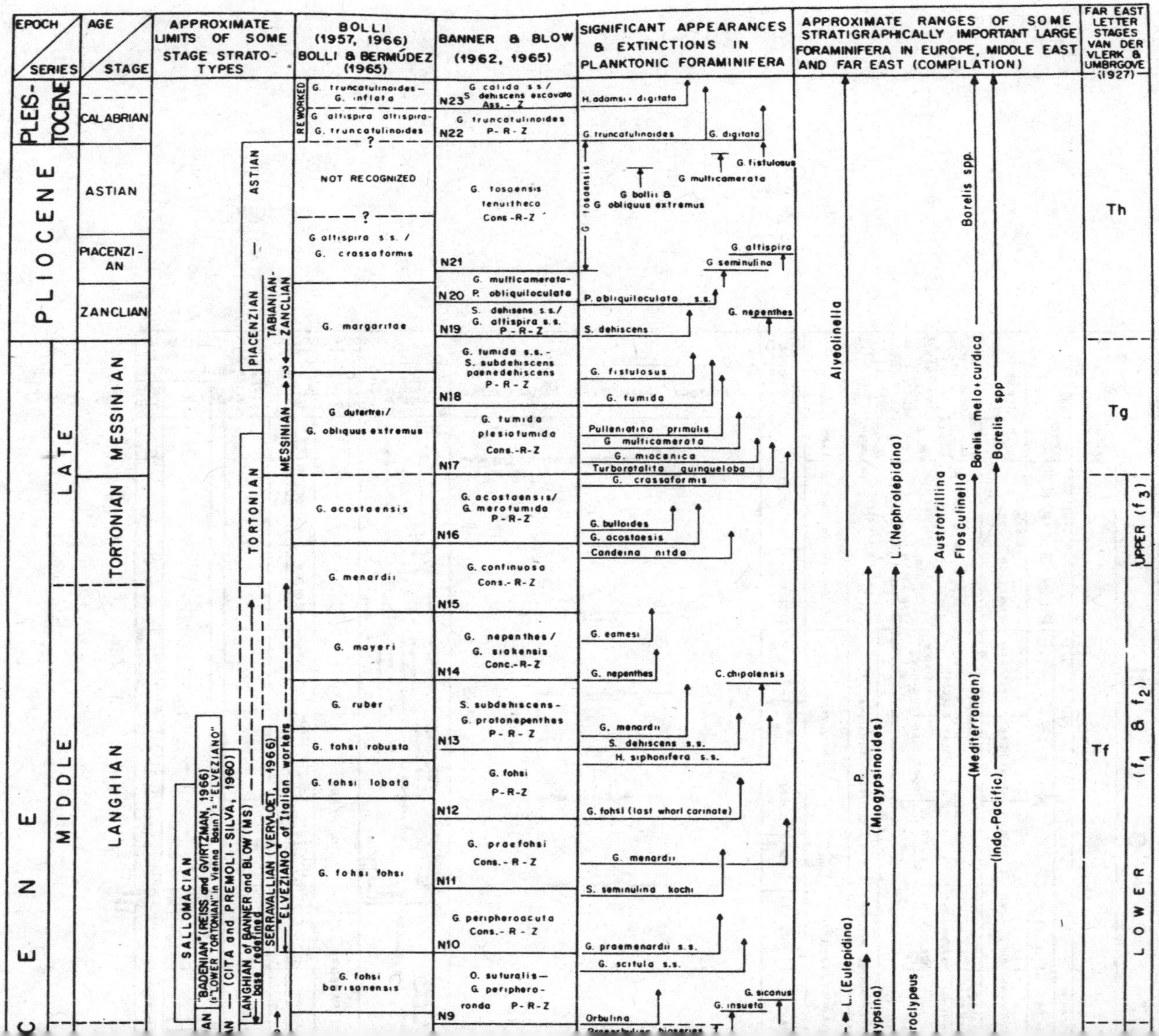
EPOCH
SERIES
AGE
STAGE
APPROXIMATE LIMITS OF SOME STAGE STRATO-TYPES
BOLLI (1957, 1966) BOLLI & BERMÚDEZ (1965)
BANNER & BLOW (1962, 1965)
SIGNIFICANT APPEARANCES & EXTINCTIONS IN PLANKTONIC FORAMINIFERA
APPROXIMATE RANGES OF SOME STRATIGRAPHICALLY IMPORTANT LARGE FORAMINIFERA IN EUROPE, MIDDLE EAST AND FAR EAST (COMPILATION)
FAR EAST LETTER STAGES VAN DER VLERK & UMBGROVE (1927)
PLEIS-TOCENE
PLIOCENE
LATE
MIDDLE
CALABRIAN
ASTIAN
PIACENZI-AN
ZANCLIAN
MESSINIAN
TORTONIAN
LANGHIAN
SALLOMACIAN
"BADENIAN" (REISS and GVIRTZMAN, 1966) (="LOWER TORTONIAN" in Vienna Basin) = "ELVEZIANO"
(CITA and PREMOLI-SILVA, 1960)
LANGHIAN of BANNER and BLOW (MS) base redefined
SERRAVALLIAN (VERVLOET, 1966)
"ELVEZIANO" of Italian workers
ASTIAN
PIACENZIAN
TABIANIAN-ZANCLIAN
MESSINIAN
TORTONIAN
REWORKED
G. truncatulinoides – G. inflata
G. altispira altispira – G. truncatulinoides
NOT RECOGNIZED
G. altispira s.s. / G. crassaformis
G. margaritae
G. dutertrei / G. obliquus extremus
G. acostaensis
G. menardii
G. mayeri
G. ruber
G. fohsi robusta
G. fohsi lobata
G. fohsi fohsi
G. fohsi barisanensis
N23 G. calida s.s. / S. dehiscens excavata Ass.-Z
N22 G. truncatulinoides P-R-Z
G. tosaensis tenuitheca Cons.-R-Z
N21
N20 G. multicamerata – P. obliquiloculata
N19 S. dehisens s.s. / G. altispira s.s. P-R-Z
N18 G. tumida s.s. – S. subdehiscens paenedehiscens P-R-Z
N17 G. tumida plesiotumida Cons.-R-Z
N16 G. acostaensis / G. merotumida P-R-Z
N15 G. continuosa Cons.-R-Z
N14 G. nepenthes / G. siakensis Conc.-R-Z
N13 S. subdehiscens – G. protonepenthes P-R-Z
N12 G. fohsi P-R-Z
N11 G. praefohsi Cons.-R-Z
N10 G. peripheroacuta Cons.-R-Z
N9 O. suturalis – G. peripheroronda P-R-Z
H. adamsi + digitata
G. truncatulinoides
G. digitata
G. fistulosus
G. multicamerata
G. bollii & G. obliquus extremus
G. tosaensis
G. altispira
G. seminulina
P. obliquiloculata s.s.
G. nepenthes
S. dehiscens
G. fistulosus
G. tumida
Pulleniatina primalis
G. multicamerata
G. miocenica
Turborotalita quinqueloba
G. crassaformis
G. bulloides
G. acostaesis
Candeina nitda
G. eamesi
G. nepenthes
C. chipolensis
G. menardii
S. dehiscens s.s.
H. siphonifera s.s.
G. fohsi (last whorl carinate)
G. menardii
S. seminulina kochi
G. praemenardii s.s.
G. scitula s.s.
G. siconus
G. insueta
Orbulina
Alveolinella
Borelis spp.
Borelis melo + curdica
Borelis spp
L. (Nephrolepidina)
Austrotrillina
Flosculinella
(Mediterranean)
(Indo-Pacific)
P. (Miogypsinoides)
L. (Eulepidina)
ypsina)
rocly peus
Th
Tg
Tf
UPPER (f3)
LOWER (f1 & f2)

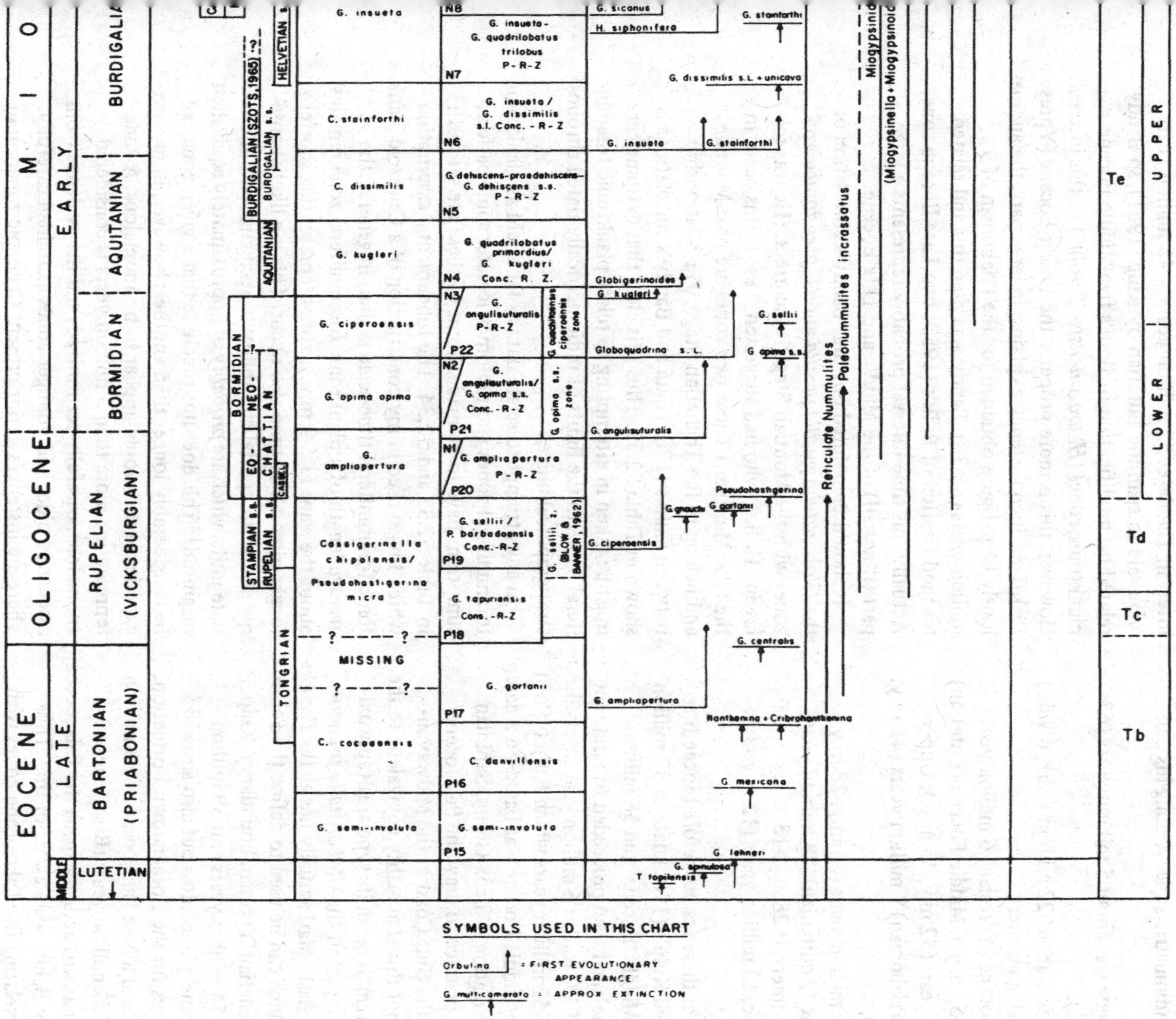
EOCENE
OLIGOCENE
M I O
LATE
EARLY
MIDDLE
LUTETIAN
BARTONIAN (PRIABONIAN)
RUPELIAN (VICKSBURGIAN)
BORMIDIAN
AQUITANIAN
BURDIGALIA
TONGRIAN
STAMPIAN s.s.
RUPELIAN s.s.
CASSEL
EO-
NEO-
CHATTIAN
BORMIDIAN
AQUITANIAN
BURDIGALIAN s.s.
BURDIGALIAN (SZOTS, 1965) ?
HELVETIAN s.
G. semi-involuta
C. cocoaensis
MISSING
Pseudohastigerina micra
Cassigerinella chipolensis/
G. ampliapertura
G. opima opima
G. ciperoensis
G. kugleri
C. dissimilis
C. stainforthi
G. insueta
P15
G. semi-involuta
P16
C. danvillensis
P17
G. gortanii
P18
G. tapuriensis Conc.-R-Z
P19
G. sellii / P. barbadoensis Conc.-R-Z
P20
G. ampliapertura P-R-Z
N1
P21
G. angulisuturalis/ G. opima s.s. Conc.-R-Z
N2
P22
G. angulisuturalis P-R-Z
N3
N4
G. quadrilobatus primordius/ G. kugleri Conc. R. Z.
N5
G. dehiscens-praedehiscens- G. dehiscens s.s. P-R-Z
N6
G. insueta / G. dissimilis s.l. Conc. - R - Z
N7
G. insueta - G. quadrilobatus trilobus P-R-Z
N8
G. sellii z. (BLOW & BANNER, 1962)
G. opima s.s. zone
G. quadrilobata ciperoensis zone
T. topilensis
G. spinulosa
G. lehneri
G. mexicana
Hantkenina + Cribrohantkenina
G. ampliapertura
G. centralis
G. ciperoensis
G. gortanii
G. gravelli
Pseudohastigerina
G. angulisuturalis
Globoquadrina s. l.
G. opima s.s.
G. sellii
G. kugleri
Globigerinoides
G. insueta
G. stainforthi
G. dissimilis s.l. + unicava
H. siphonifera
G. siconus
G. stainforthi
Reticulate Nummulites
Paleonummulites incrassatus
(Miogypsinella + Miogypsinoides)
Miogypsina
Tb
Tc
Td
Te
LOWER
UPPER
SYMBOLS USED IN THIS CHART
Orbulina = FIRST EVOLUTIONARY APPEARANCE
G. multicamerata = APPROX. EXTINCTION

Table 52.31. Suggested relationship between late Eocene-Recent planktonic foraminiferal zones, significant appearances and extinctions of planktonic Foraminifera and ranges of some larger Foraminifera

Thus 43 Cenozoic planktonic foraminiferal zones are recognized.

An attempt to relate Cenozoic planktonic foraminiferal zones to the radiometric scale was made by Bandy (1964). Using Kulp's (1961) data Bandy related the Cenozoic planktonic zones to various stages, and suggested some relationships to the radiometric scale. Among his conclusions were:

(1) The Palaeocene was about 5 million years (63 m.y.-58 m.y.).

(2) The Eocene was about 22 million years, divided in the following manner:

(*a*) Lower Eocene (Ypresian) 6 million years (58 m.y.-52 m.y.); Middle Eocene (Lutetian) 7 million years (52 m.y.-45 m.y.); Upper Eocene (Priabonian) 9 million years (45 m.y. -36 m.y.).

(3) Oligocene, 8 million years (36 m.y.-28 m.y.).

(4) Base Miocene (Aquitanian), 28 million years (Miocene interval: 28 m.y.-13 m.y.).

(5) Base Pliocene, 13 million years (13 m.y.-3.5 m.y.).

Let us look closer at these results before proceeding. Bandy (1964) accepted Kulp's (1961) date of 52 million years for the Lower/Middle 'Eocene' and 58 million years for base 'Eocene', and then proceeded to relate the five zones of Bolli (1957) to this 6 million year interval. However, the date of 52 million years which Kulp (1961) used to define Lower/Middle Eocene was based on a date on the Bashi Marl Formation (Wilcox Group, Sabinian Stage). It is common practice to draw the Palaeocene/Eocene boundary in the Gulf Coast at the Midwayan-Sabinian boundary, but this is in reality equivalent to the middle of the Thanetian Stage in its type area (approximately equivalent to a part of the *Globorotalia pseudomenardii* Zone). The Bashi Marl actually lies at the *Pseudohastigerina* Datum which can be used to define the Palaeocene/Eocene (i.e. Thanetian/Ypresian boundary). Kulp (1961) gave dates of 54 million years and 59 million years for 'Lower Eocene' glauconites but also showed a date of 62 million years for the Hornerstown Formation, New Jersey; he suggested a base Palaeocene of 63 million years. Kulp's data of 52 million years (Bashi) and 63 million years (Hornerstown) actually spans the Palaeocene, though neither he nor Bandy were aware of this. Thus Bandy (1964), in correlating five Palaeocene zones with what was essentially a time interval of Middle Thanetian-Montian (58-63) arrived at an anomalously low average length for his Palaeocene zones (6.5 or 1.2 m.y. each). Bandy's (1964) correlations between foraminiferal zonations are thus seen to be superfluous in the lower Middle Eocene. The date of 52 million years is from Bashi Formation (base Eocene = *G. rex* zone of Bolli). The date of 58 million years is from within the Palaeocene, although Kulp (1961) presented it as Palaeocene/Eocene boundary. Thus these dates span the interval between *G. pseudomenardii*-base *G. rex* zone approximately. Yet Bandy related his *G. rex* and *G. palmerae* zones to this interval. The dates 52 million years – 45 million years were said to span the Lutetian by Kulp (1961), so Bandy related his modification of Bolli's five Middle Eocene planktonic zones (*H. aragonensis*- *T. rohri*) to this interval. However, this is more properly the Lower Eocene (Ypresian). This inconsistency explains, in large part, the anomalously large figure obtained for the Priabonian of 9 million years (45-36). Bandy accepted this and related his modification of Bolli's (1957) two Late Eocene zones. Actually this time-interval probably represents a large part (if not all) of the Middle and Late Eocene.

It should be pointed out that Bandy's correlation of the *Globorotalia pusilla pusilla* and *G. pseudomenardii* zones with the Montian Stage is incorrect. (He was following Hofker who had suggested these forms occur in the type Montian.) These two zones are post-Montian, and belong to the Thanetian Stage. The relationship between Kulp's (1961) data and Bandy's correlations is shown in Table 52.32. Thus it is clear that extreme care must be taken in attempting to relate planktonic foraminiferal zones to the limits of radiometrically dated chronostratigraphical units.

In any attempt to relate data as far afield as planktonic foraminiferal zones and radiometric dates some basic assumptions have to be made. Some of these are set forth in Table 52.33 and 52.34. The fundamental assumption which I have used in my construction of a Cenozoic planktonic foraminiferal time-scale is that, in general, the *average* length of a planktonic foraminiferal zone remains about the same (1.5-2 m.y.): zones representing shorter and longer intervals of time also occur. This is a likely consequence of the specialist's ability to recognize and discriminate units which are *probably* of approximately equivalent magnitude. This does not preclude that a given zone can be considerably longer than another; however, in my own experience this would appear to be anomalous. A zone representing more than 3 million years is considered anomalous, carefully scrutinized, and the evidence from associated older and younger zones examined carefully. There are instances in our present Cenozoic zonation in which some zones are of considerably shorter length than the average (N21-N23), but most major planktonic zones should represent approximately equivalent intervals of time. When related to the most reliable estimates on the radiometric scale, at intervening points in the Cenozoic, this has been found to be a valid premise.

RAD. DATES	STAGE	DATED FORMATIONS	STAGE CORRELATION BY BANDY (1964)	PLANKTONIC FORAMINIFERAL CORRELATIONS BY BANDY (1964)	ACTUAL RELATIONSHIPS BETWEEN PLANKTONIC ZONES AND CHRONOSTRATIGRAPHIC BOUNDARIES AND RADIOMETRIC DATES	RAD. DATES
	OLIGOCENE					
36	(UPPER) PRIABONIAN			G. cerroazulensis-G. increbescens	G. cerroazulensis G. semi-involuta	
						39
				G. cerroazulensis-G. semi-involuta	T. rohri O. beckmanni (=P. mexicana) G. lehneri G. kugleri H. aragonensis	
45	(MIDDLE) LUTETIAN			G. pseudodubia P. mexicana	A. densa G. aragonensis	45
				T. topilensis G. frontosa	G. formosa G. rex	
52	(LOWER) YPRESIAN	BASHI		Globorotalia palmerae	G. velascoensis	52
				Globorotalia rex	G. pseudomenardii	
58	PALEOCENE		LANDENIAN (THANETIAN)	G. velascoensis	G. pusilla s.s.	58
			MONTIAN	G. pseudomenardii G. pusilla s.s.	G. uncinata	
		HORNERS-TOWN	DANIAN	G. uncinata G. daubjergensis		
63						

Table 52.32. Suggested relationship of Palaeogene biostratigraphical zones, chronostratigraphical units and radiometric dates of Bandy (1964)

In Table 52.34 the upper two rows (1 and 1a) show the approximate average duration of a Cenozoic planktonic zone according to the time-scale adopted. For the purposes of this work a date of 65 million years suggested by Funnell (1964) has been used. An average length of 1.51 million years is seen for the 43 Cenozoic planktonic foraminiferal zones. Table 52.34, column 2 reveals that the suggestion of Evernden and Curtis (1961) that the Palaeocene was 12 million years long yields an anomalously high average for the Palaeocene zones (3 million years). This does not in itself necessarily mean much but, when considered together with their dates for the Eocene (22 million years; average zone length 1.69) it does suggest that either a high degree of discrimination has been made in Eocene zonation, less in the Palaeocene, or that some of the dates are incorrect. The high degree of discrimination is probably true in the Middle Eocene only. Following Funnell's (1964) comment on the uncertainty of the radiometric date on the Bashi glauconite (it may be considered a minimum), a Palaeocene/Eocene boundary is drawn here at 54 million years. This gives an average Palaeocene zone duration of 2.20 million years (see Table 52.33, column DII). The approximate length of the various Palaeocene zones have been drawn on the basis of my personal experience. The Danian *G. daubjergensis* zone for instance, is accorded a somewhat longer duration; the *G. uncinata* zone occupies a somewhat shorter time span than the other Palaeocene zones. This is a highly subjective approach, but within the limits of accuracy of the estimated boundaries based upon radiometric data, it is believed that it yields a valid estimate for the duration of a sequence of planktonic zones.

	REMARKS	A BASIC ASSUMPTION	B CONCLUSION	REMARKS	C SUGGESTED BASIC ASSUMPTION	D MODIFICATIONS CONCLUSION
I	(1)	1. LENGTH OF CENOZOIC = 65 my 2. NUMBER CENOZOIC PLANKTONIC FORAMINIFERAL ZONES = 43	AVERAGE LENGTH OF CENOZIC PLANKT. FORAMINIFERAL ZONE = 1.51 my	→	ACCEPTED	
II	(2)	1. BASE PALEOCENE = 65 my 2. TOP PALEOCENE = 52 my 3. NUMBER OF PALEOCENE PLANKT. FORAM. ZONES = 5	AVERAGE LENGTH PALEOCENE PLANKT. FORAM. ZONE = 2.6 my	1. GLAUCONITE DATE ON BASHI MARL (52 my) CONSIDERED A MINIMUM. 2. DOWNWARD REVISION OF PALEOCENE/EOCENE RESULTS IN MORE PROPORTIONATE RELATIONSHIP BETWEEN LENGTH OF PALEOCENE AND EOCENE PLANKTONIC ZONES. →	1. BASE PALEOCENE = 65 my 2. TOP PALEOCENE = 54 my 3. NUMBER PALEOCENE PLANKT. FORAM. ZONES = 5	AVERAGE LENGTH OF PALEOCENE PLANKT. FORAM. ZONES = 2.20 my
III		1. BASE EOCENE = 52 my 2. TOP EOCENE = 37 my 3. NUMBER EOCENE PLANKT. FORAM. ZONES = 13	AVERAGE LENGTH EOCENE PLANKT. FORAM. ZONE = 1.15 my	ANOMALOUSLY SHORT AVERAGE LENGTH OF EOCENE PLANKT. FORAM. ZONE SUGGESTS THAT ONE OF THE 2 BOUNDARIES MAY BE INCORRECT. →	1. BASE EOCENE = 54 my 2. TOP EOCENE = 35 my 3. NUMBER EOCENE PLANKT. FORAM. ZONES = 13	AVERAGE LENGTH EOCENE PLANKT. FORAM. ZONE = 1.46 my
IV		1. BASE EOCENE = 52 my. 2. EARLY/MIDDLE EOCENE BOUNDARY = 44 my 3. NUMBER LOWER EOCENE PLANKT. FORAM. ZONES = 5	AVERAGE LENGTH LOWER EOCENE PLANKT. FORAM. ZONE = 1.6 my	→	1. BASE EOCENE = 54 my 2. EARLY/MIDDLE EOCENE BOUNDARY = 45 my 3. NUMBER LOWER EOCENE PLANKT. FORAM. ZONES = 5	AVERAGE LENGTH LOWER EOCENE PLANKT. FORAM. ZONE = 1.8 my
V	(3)	1. E/M EOCENE BOUNDARY = 44 my 2. M/L EOCENE BOUNDARY = 39 my 3. MIDDLE EOC. ZONES = 5	AVERAGE LENGTH OF MIDDLE EOCENE PLANKT. FORAM. ZONE = 1 my	→	1. E/M EOCENE BOUNDARY = 45 my 2. M/L EOCENE BOUNDARY = 39 my 3. NUMBER MIDDLE EOCENE PLANKT. ZONES = 5	AVERAGE LENGTH MIDDLE EOCENE PLANKT. FORAM. ZONE = 1.2 my
VI		1. M/L EOCENE BOUNDARY = 39 my 2. EOC./OLIG. BOUNDARY = 37 my 3. U. EOC. PLANKT. ZONES = 3	AVERAGE LENGTH LATE EOCENE PLANKT. FORAM. ZONE = 0.66 my	ANOMALOUSLY LOW VALUE FOR AVERAGE VALUE OF LATE EOCENE PLANKT. FORAM. ZONES SUGGESTS THAT ONE OR BOTH DATES MAY BE INCORRECT →	1. M/L EOCENE BOUNDARY = 39 my 2. EOC./OLIG. BOUNDARY = 35 my 3. NUMBER LATE EOCENE PLANKT. ZONES = 3	AVERAGE LENGTH LATE EOCENE PLANKT. FORAM. ZONE = 1.33 my
VII		1. BASE PALEOCENE = 65 my 2. TOP EOCENE = 37 my 3. NUMBER PALEOCENE–EOCENE ZONES = 18	AVERAGE LENGTH PALEOCENE & EOCENE PLANKT. FORAM. ZONE = 1.55 my	→	1. BASE PALEOCENE = 65 my 2. TOP EOCENE = 35 my 3. NUMBER OF ZONES (SAME) = 18	AVERAGE LENGTH PALEOCENE-EOCENE PLANKT. FORAM. ZONE = 1.66 my.
VIII		1. EOC./OLIG. BOUNDARY = 37 my 2. BASE AQUITANIAN = 26 my 3. DATED BASE APPROX. EQUIVALENT TO BASE N4 4. NUMBER PLANKT. ZONES = 5	AVERAGE LENGTH OF "OLIGOCENE" PLANKT. ZONES = 2.20 my	→	1. EOC./OLIG. BOUNDARY = 35 my 2. BASE AQUITANIAN = 27 my 3. SAME ASSUMPTION 4. SAME ZONES = 5	AVERAGE LENGTH "OLIGOCENE" ZONES 1.60 my
IX		1. EOC./OLIG. BOUNDARY = 37 my 2. NUMBER POST-EOCENE PLANKT. FORAM. ZONES = 25	AVERAGE LENGTH POST-EOCENE PLANKT. FORAM. ZONES = 1.48 my	→	1. EOC./OLIG. BOUNDARY = 35 my 2. NUMBER ZONES = 25	AVERAGE LENGTH POST-EOCENE PLANKT. FORAM. ZONES = 1.40 my
X	(4)	1. BASE AQUITANIAN = 27 my 2. DATED BASE APPROX. EQUIV. TO BASE N4 3. ORBULINA DATUM = 20 my 4. NUMBER PLANKT. ZONES = 5	AVERAGE LENGTH PLANKT. FORAM. ZONES N4–N8 (incl.) = 1.40	→	ACCEPTED	
XI	(5)	1. ORBULINA DATUM = 20 my 2. HELV./TORTONIAN BOUNDARY IN MOHOLE = N12/N13 & 12 my 3. NUMBER PLANKT. FORAM. ZONES = 4	AVERAGE LENGTH OF ZONES N9–N12 (INCL.) = 2.0 my	1. IF ORBULINA DATE CORRECT, N9–N12 ZONE LENGTHS SOMEWHAT LONG. 2. IF N12/N13 CORRELATION IS CORRECT, AND RADIOMETRIC DATE IS CORRECT, THEN N13–N18 ZONE LENGTHS ANOMALOUSLY SHORT BOTH IF BASE PLIOCENE IS TAKEN AS 10 my OR 6 my (SEE XIII BELOW)	1. ORBULINA DATUM = 20 my 2. "HELV-TORTONIAN" DATE OF 12 my IS A MINIMUM. 3. N12/N13 DATE SUGGESTED = 14 my	AVERAGE LENGTH OF N9–N12 PLANKT. FORAM. ZONES = 1.5 my
XII		1. ORBULINA DATUM = 20 my 2. NUMBER POST-ORBULINA PLANKT. FORAM. ZONES = 15	AVERAGE LENGTH OF POST-ORBULINA DATUM PLANKT. FORAM. ZONES = 1.33 my		ACCEPTED	
XIII	(6)	1. ZONE 12/13 = 12 my 2. BASE PLIOCENE = 10 my 3. NUMBER PLANKT. FORAM. ZONES = N13–N18 = 6	AVERAGE LENGTH OF PLANKT. FORAM. ZONES N13–N18 = 0.33 my	EXCEEDINGLY LOW AVERAGE LENGTH OF PLANKT. FORAM. ZONES SUGGESTS: a) ASSUMPTION 1 IS INCORRECT b) ASSUMPTION 2 IS INCORRECT c) BOTH →	1. ZONE 12/13 = 14 my 2. BASE PLIOCENE = 6 my 3. NUMBER ZONES = 6	AVERAGE LENGTH OF ZONES N13–N18 = 1.33 my
XIV					1. BASE PLIOCENE = 6 my 2. NUMBER POST-MIOCENE PLANKT. FORAM. ZONES = (N19–23) = 5	AVERAGE LENGTH OF POST-MIOCENE PLANKT. FORAM. ZONES = 1.1 my

REMARKS

(1) TOP PALEOCENE DATE IS TAKEN FROM 52 my DATE FOR BASHI MARL (WILCOX), GULF COAST. (EVERNDEN et al., 1961)

(2) TOP EOCENE DATE OF 37 my IS FROM FUNNELL (1964).

(3) M/L EOCENE DATE OF 39 my IS FROM 39 my DATE FOR BASE JACKSON, GULF COAST. (EVERNDEN et al, 1961)

(4) ORBULINA DATUM DATE OF 20 my IS FROM SAITO (1967, PERS. COMMUNIC.) ESTIMATED ON BASIS OF SEA-FLOOR SPREADING HYPOTHESIS IN EAST PACIFIC.

(5) DATE OF 12 my FOR "HELVETIAN-TORTONIAN" BOUNDARY IN EXPERIMENTAL MOHOLE BY DYMOND (1966, Unpubl. PhD THESIS). LEVEL CORRELATED BY THIS WRITER WITH N12/N13 ZONES.

(6) BASE PLIOCENE DATE OF 10 my IS FROM BANDY (1963, 1964) AND GENERALLY ACCEPTED DATE FOR BASE PLIOCENE IN CALIFORNIA.

Table 52.33. Some basic assumptions used in suggesting some approximate relationships between radiometrically dated Cenozoic chronostratigraphical units and the average length of planktonic foraminiferal zones.

	TIME IN my	HOLMES (1959)	KULP (1961)	EVERNDEN & CURTIS (1961)	FUNNELL (1964)	REMARKS
1	LENGTH OF CENOZOIC	70	63±2	67	65	
1a	AVERAGE LENGTH OF CENOZOIC PLANKTONIC FORAMINIFERAL ZONE	1.62	1.46	1.55	1.51	
2	LENGTH OF PALEOCENE	10	5±2	12	12	* KULP USED DATE BY EVERNDEN ET AL. (1961) OF 52 MY FOR BASHI FM. (WILCOX GP.) WHICH HE SAID WAS TOP OF LOWER EOCENE. THIS FOLLOWS COMMON PRACTICE IN GULF COAST TO EQUATE PALEOCENE/EOCENE BOUNDARY WITH MIDWAYAN/SABINIAN BOUNDARY. THIS BOUNDARY IS ACTUALLY MID-THANETIAN IN AGE. THE BASHI FM. IS AT THE *Pseudohastigerina* DATUM AND IS BASAL EOCENE. KULP'S DATES OF 58-52 MY FOR LOWER EOCENE REFER TO MID-THANETIAN TO BASE LOWER EOCENE. HIS DATES OF 52-45 (LUTETIAN) PROBABLY INCLUDES LOWER EOCENE; HIS DATES OF 45-36 (UPPER EOCENE) MAY INCLUDE MIDDLE AND UPPER EOCENE. SEE OTHER DISCUSSION IN TEXT CONCERNING BANDY'S (1964) USE OF THESE DATES.
2a	AVERAGE LENGTH OF PALEOCENE PLANKTONIC FORAMINIFERAL ZONE	2	* 2.0	3	2.4	
3	LENGTH OF EOCENE	20	22±4	22	16-17	
3a	AVERAGE LENGTH OF EOCENE PLANKTONIC FORAMINIFERAL ZONE	1.66	1.40	1.69	1.23	FUNNELL'S DATA GIVES ANOMALOUSLY SHORT AVERAGE SPAN TO ZONES. IN CALCULATING RELATIONSHIP OF ZONES TO KULP'S DATA, THE ASSUMPTION IS MADE THAT HIS LOWER DATE (58 MY) CORRESPONDS TO MID-THANETIAN; HIS UPPER DATE TO TOP EOCENE. THUS THE 13 EOCENE PLUS 2 PALEOCENE ZONES HAVE BEEN USED IN CALCULATING THE AVERAGE LENGTH OF ZONES FROM KULP'S FIGURES.
4	LENGTH OF PALEOCENE & EOCENE	30	27±4	34	28	
4a	AVERAGE LENGTH OF PALEOCENE-EOCENE PLANKTONIC FORAMINIFERAL ZONE	1.66	1.50	1.86	1.55	
5	LENGTH OF OLIGOCENE	15	11±3	8	11-12	
5a	AVERAGE LENGTH OF "OLIGOCENE" PLANKTONIC FORAMINIFERAL ZONE	3	2.2	1.6	2.2-2.5	BASIC ASSUMPTION HERE IS THAT BASE AQUITANIAN DATE (26 MY) IS RELIABLE AND CORRESPONDS WITH BASE N4. USING ZONES P18-P19 AND N1-N3 FOR INTERVAL TOP EOCENE-BASE AQUITANIAN GIVES THE FIGURES IN COLUMN TO LEFT. IT CAN BE SEEN THAT HOLMES' DATA RESULT IN AN ANOMALOUSLY LONG DURATION FOR EACH OF THE PLANKTONIC FORAMINIFERAL ZONES. IN FACT, THE ONLY DATA CONSISTENT WITH AVERAGE ZONE LENGTHS ARE THOSE OF EVERNDEN ET AL. (1961).
6	LENGTH OF MIOCENE RECENT	25	25±1	25	26	
6a	AVERAGE LENGTH OF MIOCENE-RECENT PLANKTONIC FORAMINIFERAL ZONES N4-N23	1.25	1.25	1.25	1.3	

NUMBER OF PLANKTONIC FORAMINIFERAL ZONES		
NEOGENE		23 - (N2-23)
OLIGOCENE		3-(P18, 19, N1)
EOCENE U	13	3
EOCENE M		5
EOCENE L		5
PALEOCENE		5
TOTAL CENOZOIC		43

Table 52.34. Calculations of average length of Cenozoic planktonic foraminiferal zones and chronostratigraphical units using various radiometric scales.

If we examine Table 52.34, columns 3 and 3a, we see that using Funnell's limits for the Eocene (53/54-37/38 m.y., or 16/17 m.y.) yields an anomalously low average duration for the Eocene zones of 1.23 million years. The crux of the problem lies in the Eocene/Oligocene boundary. Radiometric dates for the vicinity of the Eocene/Oligocene boundary vary from 33 million years to 38 million years and stratigraphical control is generally poor, or at least equivocal, in connection with these samples (see discussion in Evernden and Curtis 1961; Funnell 1964). Thus if Funnell's date of 37/38 million years is accepted we get an average Eocene zone length of 1.15 million years (Table 52.33 column A, BIII). Evernden and Curtis (1961) suggest 33 million years which, using their data gives an average of 1.69 (Table I, column 3, 3a). This writer has adopted a figure of 35 million years for the Eocene/Oligocene boundary (Table 52.33, column C, DIII)), with which vertebrate palaeontologists would agree (according to Dr M. McKenna of the American Museum of Natural History, pers. comm.), and which gives an average duration of 1.46 million years for the Eocene zones. On the basis of radiometric data, which suggests that the Eocene was somewhat longer than the Palaeocene (an estimated 9 million years longer), and the fact that thirteen zones have been differentiated in the Eocene the average of 1.46 is considered reasonable. It also suggests, that if one of the three subdivisions of the Eocene is demonstrably shorter than the others, and the number of zones is approximately the same, the relative length of the zones will be shorter in that interval. Such appears to have been the case in the Middle Eocene. On the basis of available data the Early/Middle Eocene boundary is drawn at 45 million years (see Table 52.33, columns A, B, C, DIV). A Middle/Late Eocene boundary is drawn at 39 million years (see Table 52.33, columns A, B, C. DV). The relationship between the average duration of the Early Eocene planktonic foraminiferal zones (1.8 m.y.) and the Middle Eocene ones (1.2 m.y.) is interesting, and believed to be valid on the basis of my own experience.

There is an interesting biostratigraphical-chronostratigraphical problem connected with this 'Middle Eocene'. Bolli (1957) equated his five planktonic zones (*H. aragonensis-T. rohri*) with the Middle Eocene (Lutetian). However no direct correlation has been possible between all these zones and the type Lutetian. Szöts has been able to demonstrate that the upper two zones probably correspond to the Biarritzian Stage, but then the Biarritzian is probably equivalent to the upper part of the Lutetian. The Lutetian rests unconformably upon the equivalents of the Ypresian in the Paris Basin. It is quite possible, and in fact more than likely, that the *H. aragonensis* and *G. kugleri* zones at least, represent time which is represented by a stratigraphical hiatus in the Paris-Belgian Basins. The date of 47 million years on Lutetian from Fosse, Paris Basin, is considered anomalous here. A date of 46.1 million years for an ash bed level within the *G. aragonensis* Zone in JOIDES J-3-(561 ft) is of interest. It is here considered a minimum age.

In the scheme presented here the Late Eocene corresponds to about 4 million years. It has been divided into two zones by Bolli (1957). Blow and Banner (1962) have distinguished three zones in the Late Eocene and on the basis of data used here an average length of 1.33 million years is suggested (Table 52.33, column C, DVI).

Various assumptions were made in an attempt to arrive at a satisfactory scheme for the Oligocene and Miocene. Let us start with the assumption that the Eocene/Oligocene boundary, as determined, is approximately correct. Then (Table 52.33, column C, DIX) the post-Eocene planktonic foraminiferal zones should average 1.40 million years in length.

Most authors appear to agree that the Lower Miocene (Aquitanian) boundary should be placed near 25-26 million years. Because of the possibility that the dated materials of the Whitneyan stage may be correlatable with the 'Lower Aquitanian' in the Aquitaine Basin (i.e. below the stratotype Aquitanian of Mayer-Eymar 1858), and because Durham, Jahns and Savage (1954) correlated the Whitneyan with the Chattian and Thenius (1959) correlated the Whitneyan with the Stampian-Aquitanian, a somewhat older age for the base of the Aquitanian would appear to be reasonable. Bandy (1964) suggested 28 million years for the base Aquitanian; 27 million years is suggested here (Table 52.33, column C, DVIII), and an average length is obtained of 1.6 million years for zones P18, P19, N1-N3.

Table 52.38 shows an attempt to relate Cenozoic planktonic foraminiferal zones to radiometric dates. On the right it will be seen that a top Eocene of 36 million years has been used (close to Funnell's figure of 37/38 million years). A radiometric date of 30 million years was obtained on an ash bed in JOIDES J6 (107 ft. 7 ins) (Lidz 1967, *Amer. Assoc. Petrol. Geol., Ann. Conv., Los Angeles*). An examination of this sample, and several others above and below, has shown that it lies approximately at the P18/P19 boundary. *Pseudohastigerina* becomes extinct between 32 ft. 7 ins. and 52 ft. 6 ins. in this core. Sample 32 ft. 7 ins. is probably in zone P20/N1. Thus if we assume that the top of the Eocene is at 36 million years and the date of 30.1 million years for P18/P19 is correct, we have an anomalously high figure of about 6 million years for zone P18. There is no direct evidence known to this writer to suggest that either of these zones should be so anomalously long. In fact P18 probably occupied a shorter time than P19, so that P19 would be that much longer. P18 and P19 together essentially correspond to the Rupelian-Stampian in northern Europe. The Oligocene of this region constitutes a single, sharply defined cycle of sedimentation which was concluded by the regressive deposits of the Chattian to which zone N1 is, at least in part, equivalent. This interval of time is followed by zones N2, N3 (the top of which was estimated above the approximate 27 million years). It would seem then that either the date for top Eocene (36/37 m.y.) is incorrect, the date of the ash bed is incorrect, or both. A date of 35 million years has been taken for the Eocene/Oligocene boundary in this paper on the basis of estimating the base Aquitanian at about 27 million years. The intervening zones have been drawn in on Table 52.39. The top of the Oligocene, for the purposes of this paper, is drawn between zone N1/N2, i.e. base *G. angulisuturalis* Zone. The Oligocene/Miocene boundary (as defined here) is close to, but not identical with Banner and Blow's Oligocene/Bormidian boundary = Lower Aquitanian *s.l.* of authors), and is placed at 30 million years on the basis of the estimates presented here. The date of 30.1 million years for P18/P19 is here considered a minimum date. Strict acceptance of this date would yield a distorted elongation of the Oligocene zones which appears unwarranted. The Oligocene, as defined and estimated here, is a relatively short interval of time of 5 million years. If we were to include zones N2, N3 (to the base holostratotype Aquitanian) it would still represent only about 8 million years. In retrospect it is interesting to note that von Linstow (1922, pp. 34, 93), in his large work on the Tertiary of Germany, believed that since the beginning of the Oligocene and Miocene, 8 and 6 million years respectively have elapsed, giving a duration of 2 million years for the Oligocene.

An approximate date of 20 million years has been accepted for the *Orbulina*-datum. This is based on recent evidence (Burckle, Ewing, Saito and Leyden, 1967, p. 539) from deep sea core V20-80, in the vicinity of the East Pacific Rise, in which the *Orbulina* bioseries was observed in sediment which lay above basement which was estimated to be 22 million years old on the basis of the magnetic anomalies sea-floor spreading hypothesis).

On the basis of a base Aquitanian boundary at 27 million years and an *Orbulina*-datum of 20 million years, an average duration is obtained for zones N4-N8 of 1.4 million years which is somewhat lower than for the preceding five zones (1.6 m.y.). If the date of 20 million years is accepted as valid we obtain an average duration for post-*Orbulina* Cenozoic zones of 1.33 million years. (Table 52.33, column A, BXII). This would seem to signify that a somewhat higher degree of differentiation has been possible in Late Cenozoic planktonic foraminiferal zonation during a given time span. This appears to have been the case when we consider the late Miocene-Pliocene zonation, and the fact that Late Pliocene-Pleistocene zones are probably of relatively short duration (see zones N21-N23 below). Similar agreement is seen in estimating the approximate duration of post-N3 Cenozoic zones (N3-N23), 1.39 million years each, using the various dates for base Miocene suggested by authors (see Table 52.34, columns 6a, b). Actually 27 million years was chosen as probably more reliable for this level and this merely brings the average up somewhat higher.

Several problems have presented themselves which have made determination of the duration of Middle-Late Miocene planktonic foraminiferal zones difficult. The Miocene/Pliocene boundary has been variously estimated at 6/7 million years (Tongiorgi and Tongiorgi 1964), 11 million years (Holmes 1959), 12 million years (Evernden and Curtis 1961; Evernden *et al* 1965); and 13 million years (Kulp 1961; Bandy 1964). If we assume an age of 13 million years for base Pliocene, and our assumption of 20 million years for *Orbulina*-datum is valid, then we are faced with the interesting result that the ten planktonic foraminiferal zones in this interval (7 m.y.) average 0.7 million years each, which appears quite irregular. If we assume a base Pliocene age of 10 million years we obtain an average figure of 1.43 million years, which is more reasonable. The date of 12 million years by Evernden *et al.* (1960, 1961) is based upon a dating of the Lower Clarendonian with the first *Hipparion* fauna in California, which was believed by Savage to represent earliest Pliocene. Data in Europe has shown that *Hipparion* occurs in beds as old as Tortonian,and that the first appearance of *Hipparion* (a function of migration patterns) cannot be used in determining a chronostratigraphical boundary.

An additional problem has arisen in the terms of the date of 11.3-12.3 million years for samples straddling a boundary in the experimental Mohole core, which Martini and Bramlette (1963) designated the 'Helvetian-Tortonian' boundary. On the basis of the fauna in this sample, I would correlate this level with the N12/N13 zone boundary. If the date of 12 million years determined for this boundary is correct (Dymond 1966, unpublished Ph.D. Thesis), then we obtain the following figures: zones N9-N12 average 2 million years each (20-12 = 8; 8/4 = 2 m.y.). However, if this is correct, and if we assume an age of 10 million years for base Pliocene, we obtain the following results: zones N13-N18 (6) average out to .033 million years each (12-10 = 2; 2/6 = 0.33 m.y.) This implication is clear. If the date for the *Orbulina*-datum is correct (20 m.y.), then the dates for base Pliocene of 10-13 million years, based on terrestrial sequences in which biostratigraphical control in the stratigraphical correlation with the Miocene/Pliocene boundary in Italy has not been accurate, are not correct. The date by Tongiorgi and Tongiorgi (1964) of 6-7 million years for the base Pliocene would appear to fit our assumptions much more satisfactorily on the premise that there is no major variation in the length of the individual planktonic foraminiferal zones between *Orbulina*-datum and base Pliocene (N9-N18). Using the estimate of Tongiorgi and Tongiorgi (1964) of 6-7 million years for base Pliocene (6 million years is chosen here), we arrive at the following figures: 20 - 6 = 14 million years; 14/10 zones = 1.4 million years each zone. Using these figures we arrive at an estimated N12/N13 age of about 14 million years. This estimate is used here (Table 52.33, column C, DXIII). An age of 12 million years for N12/N13 is thus interpreted here as a minimum and thought to be actually somewhat older (14 m.y.). The remaining zones (N13-N18) have been inserted at calculated intervals of 1.33 million years.

On the assumption that the base of the Pliocene is 6 million years we arrive at the following figures: 6 million years 5 (zones N19-23) = 1.1 million years each (Table 52.33 column C, DXIV). This can be shown to be consistent with recent observations in deep sea cores. In a study of deep sea cores I and my colleagues at the Woods Hole Oceanographic Institution have recognized a relationship between the Pliocene/Pleistocene boundary (defined by the first evolutionary appearance of *Globorotalia truncatulinoides*, zone N22) and the Olduvai Normal Event within the Matuyama Reversed Epoch, which has been dated at about 1.85 million years. Zone N23 probably corresponds to only a short part of the Late Pleistocene-Recent. Thus the Pliocene-Pleistocene zonation (as suggested by Banner and Blow 1965) has been drawn on our chart. Zone N19 probably represents the greater part of the Pliocene, N21 is somewhat shorter, and N20 very short.

The date of approximately 1.85 million years, which we have determined for the base of the Pleistocene, is of particular interest in the light of recent results presented by Savage and Curtis (1967). The conclusions of these authors are summarized below:

(1) The genera *Equus, Elephas, Bos (Leptobos)*, generally assumed to characterize the Villafranchian Stage, have not been reported with certainty from the *type* Villafranchian. The type Villafranchian is notably earlier than most of the well-known sites and faunas which are usually cited as typical of the Villafranchian.

(2) The Villafranchian began before the Calabrian and the earliest Villafranchian is pre-Pleistocene, pre-Quaternary. Fauna-taxonomic similarity and potassium-argon dates indicate that Villafranchian represents about the same time interval as the Blancan mammalian age in North America.

(3) A date of approximately 3.4 million years was shown for the base of the Villafranchian, based on radiometric dating of rocks in southern France (Auvergne) equivalent to the Lower Villafranchian of Italy. Dates of 2.5 million years (Roca Neyra) and 1.9 million years (Coupet) were given within the Villafranchian. The youngest date for Villafranchian was said to be that at Ceyssac, 0.9 million years. The date of 1.9 million years is seen to correspond to the Olduvai Bed I date and to the base of the Pleistocene as we recognize it in this paper.

	ESTIMATED LENGTH OF CHRONOSTRATIGRAPHIC UNITS	my.	ESTIMATED AVERAGE LENGTH OF PLANKTONIC FORAMINIFERAL ZONES WITHIN THESE LIMITS IN my.
1	CENOZOIC	65	1.51
2	PALEOCENE	11	2.20
3	EOCENE	19	1.46
4	EARLY EOCENE	9	1.80
5	MIDDLE EOCENE	6	1.20
6	LATE EOCENE	4	1.33
7	OLIGOCENE (P18-P19-N1)	5	1.66
8	MIOCENE (N2-N18)	24	1.47
9	PLIOCENE	4.2	1.40
10	PALEOGENE (DANIAN-N1)	35	1.77
11	NEOGENE (N2-N23)	30	1.42

Table 52.35. Summary of estimated average duration of Cenozoic planktonic zones in relation to chronostratigraphical units. The anomalously high average values for Palaeocene and Lower Eocene may indicate: (1) that the estimated dates for the boundaries of the chronostratigraphical units within which these zones are believed to lie are incorrect, or (2) that a lower level of biostratigraphical discrimination has been made in these intervals. (3) Both. Until it can be demonstrated otherwise the present author is of the opinion that the correct explanation lies in (3).

This, then is the reasoning behind the assumptions upon which an attempted correlation between the radiometric time-scale and planktonic foraminiferal zonation has been made. It is fully realized that this system is highly subjective, and not unlike a house of cards. It is my belief however, that the basic assumptions are sound, that they are supported by a reasonable interpretation of available radiometric dates, so that *major* deviations between the relationships expressed here and future findings will not be found. The Cenozoic planktonic foraminiferal zonation used here and its suggested relationship to chronostratigraphical boundaries and the radiometric time-scale is shown in Table 52.39.

In the early part of this paper it was mentioned that the

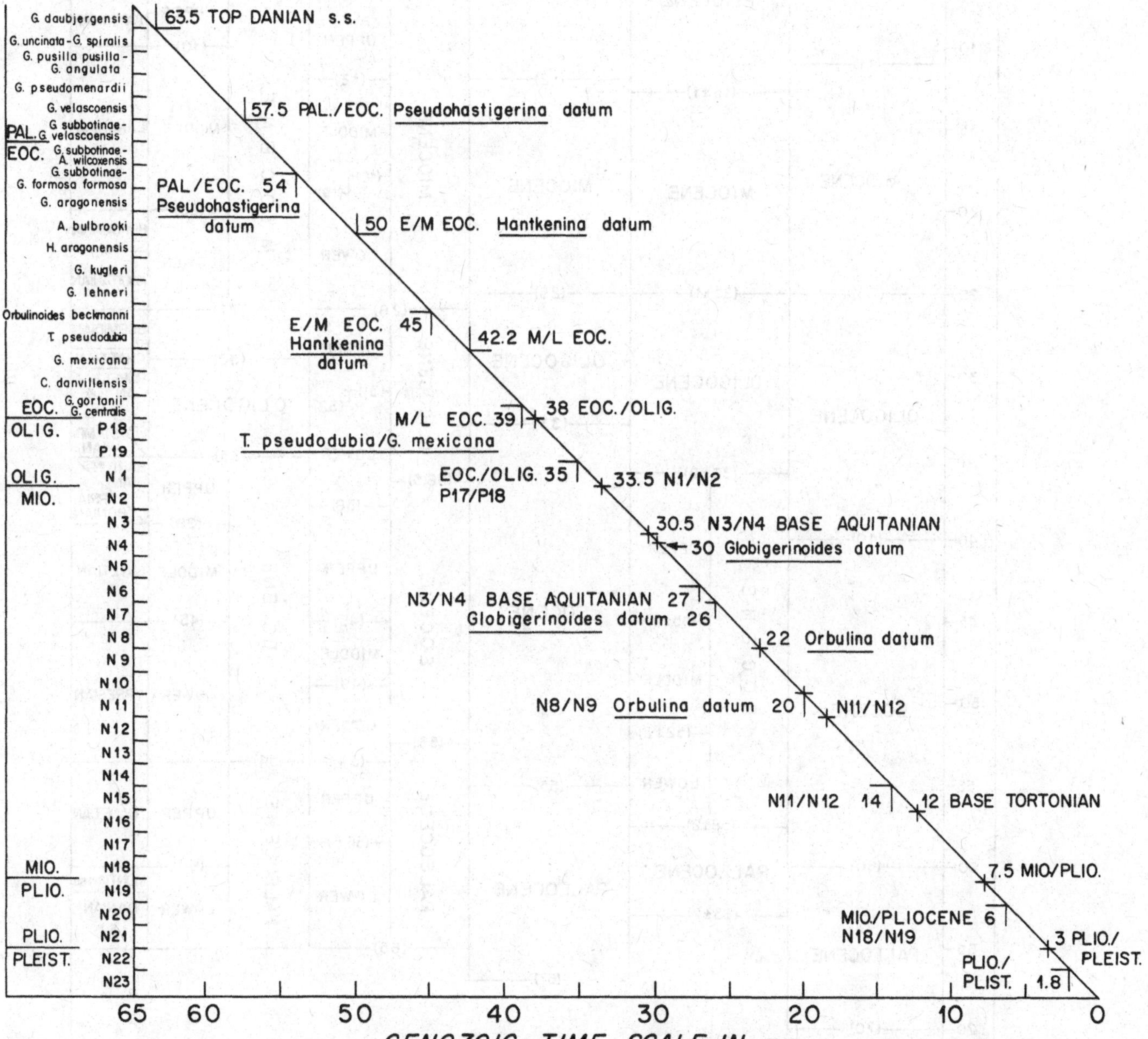
G. daubjergensis
G. uncinata - G. spiralis
G. pusilla pusilla - G. angulata
G. pseudomenardii
G. velascoensis
PAL. EOC.
G. subbotinae - G. velascoensis
G. subbotinae - A. wilcoxensis
G. subbotinae - G. formosa formosa
G. aragonensis
A. bulbrooki
H. aragonensis
G. kugleri
G. lehneri
Orbulinoides beckmanni
T. pseudodubia
G. mexicana
C. danvillensis
EOC. OLIG.
G. gortanii - G. centralis
P18
P19
OLIG. MIO.
N1
N2
N3
N4
N5
N6
N7
N8
N9
N10
N11
N12
N13
N14
N15
N16
N17
MIO. PLIO.
N18
N19
N20
PLIO. PLEIST.
N21
N22
N23
63.5 TOP DANIAN s.s.
57.5 PAL./EOC. Pseudohastigerina datum
PAL/EOC. 54
Pseudohastigerina datum
50 E/M EOC. Hantkenina datum
E/M EOC. 45
Hantkenina datum
42.2 M/L EOC.
M/L EOC. 39
38 EOC./OLIG.
T. pseudodubia/G. mexicana
EOC./OLIG. 35
P17/P18
33.5 N1/N2
30.5 N3/N4 BASE AQUITANIAN
30 Globigerinoides datum
N3/N4 BASE AQUITANIAN 27
Globigerinoides datum 26
22 Orbulina datum
N8/N9 Orbulina datum 20
N11/N12
N11/N12 14
12 BASE TORTONIAN
7.5 MIO/PLIO.
MIO./PLIOCENE 6
N18/N19
3 PLIO./PLEIST.
PLIO./PLIST. 1.8
65
60
50
40
30
20
10
0
CENOZOIC TIME SCALE IN my

Table 52.36. Occurrence in time of Cenozoic planktonic foraminiferal zones on the assumption that each zone is of equal duration (1.5 m.y.) and Cenozoic = 65 m.y., 1. Numbers above hypotenuse are plotted values on the straight line. 2. Numbers below the hypotenuse (inside the triangle, are those estimated by the present author on the basis of various assumptions.

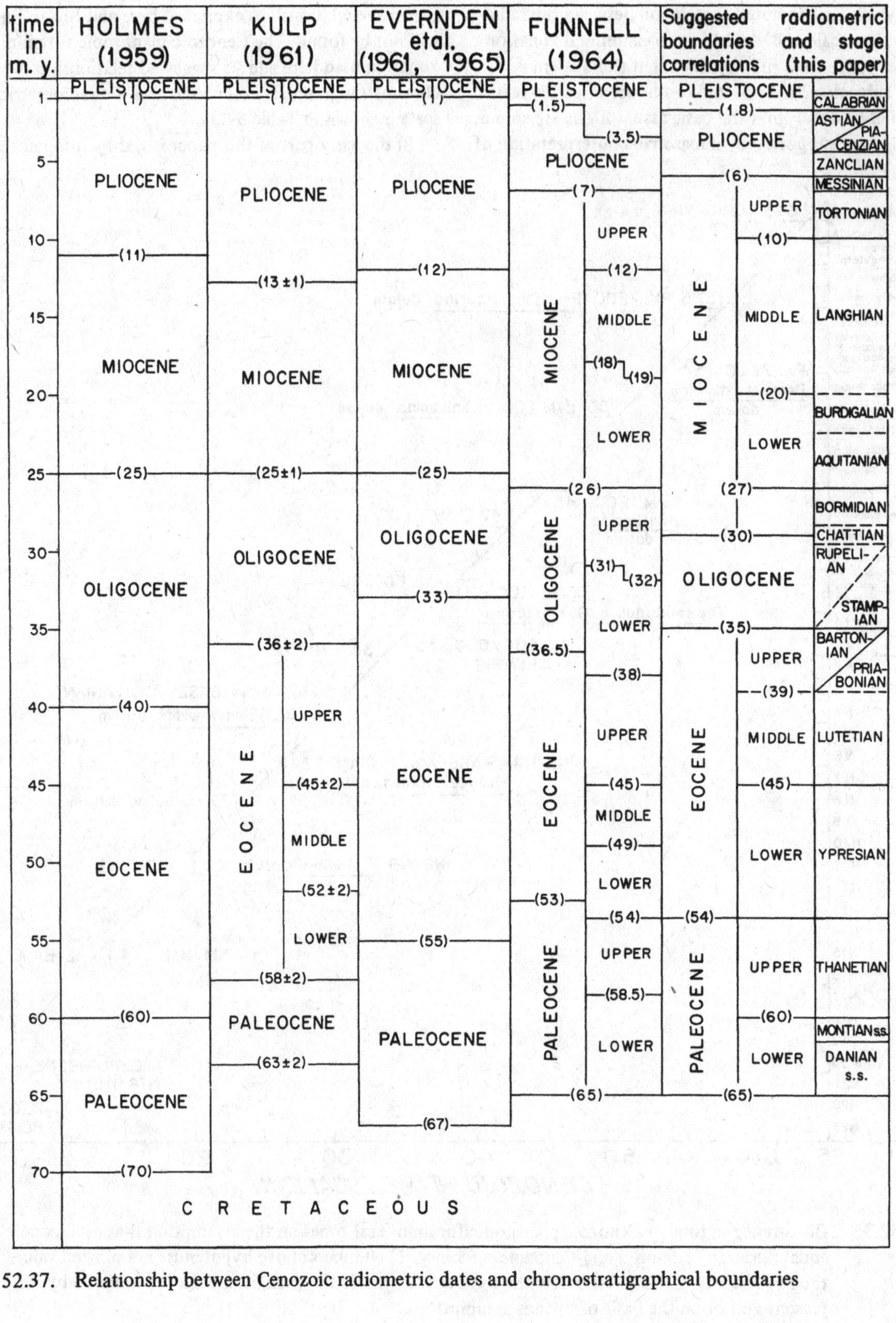
time in m. y.
1
5
10
15
20
25
30
35
40
45
50
55
60
65
70
HOLMES (1959)
PLEISTOCENE
(1)
PLIOCENE
(11)
MIOCENE
(25)
OLIGOCENE
(40)
EOCENE
(60)
PALEOCENE
(70)
KULP (1961)
PLEISTOCENE
(1)
PLIOCENE
(13 ±1)
MIOCENE
(25±1)
OLIGOCENE
(36±2)
EOCENE
UPPER
(45±2)
MIDDLE
(52±2)
LOWER
(58±2)
PALEOCENE
(63±2)
EVERNDEN etal. (1961, 1965)
PLEISTOCENE
(1)
PLIOCENE
(12)
MIOCENE
(25)
OLIGOCENE
(33)
EOCENE
(55)
PALEOCENE
(67)
FUNNELL (1964)
PLEISTOCENE
(1.5)
(3.5)
PLIOCENE
(7)
MIOCENE
UPPER
(12)
MIDDLE
(18)
(19)
LOWER
(26)
OLIGOCENE
UPPER
(31)
(32)
LOWER
(36.5)
(38)
EOCENE
UPPER
(45)
MIDDLE
(49)
LOWER
(53)
(54)
PALEOCENE
UPPER
(58.5)
LOWER
(65)
Suggested radiometric boundaries and stage correlations (this paper)
PLEISTOCENE
(1.8)
PLIOCENE
(6)
MIOCENE
UPPER
(10)
MIDDLE
(20)
LOWER
(27)
(30)
OLIGOCENE
(35)
EOCENE
UPPER
(39)
MIDDLE
(45)
LOWER
(54)
PALEOCENE
UPPER
(60)
LOWER
(65)
CALABRIAN
ASTIAN
PIA-CENZIAN
ZANCLIAN
MESSINIAN
TORTONIAN
LANGHIAN
BURDIGALIAN
AQUITANIAN
BORMIDIAN
CHATTIAN
RUPELI-AN
STAMP-IAN
BARTON-IAN
PRIA-BONIAN
LUTETIAN
YPRESIAN
THANETIAN
MONTIAN s.s.
DANIAN s.s.
C R E T A C E O U S

Table 52.37. Relationship between Cenozoic radiometric dates and chronostratigraphical boundaries

Quaternary has been characterized as the Period of Man. Although relationships between the development of Man and the limits of the Pliocene/Pleistocene have fascinated and occupied anthropologists for many years, it would appear that we are now in a position to provide some information of a definite character on this subject. The relationship of '*Zinjanthropus*' to the basalts dated as Olduvai Gorge at 1.8-1.9 million years has been made clear in various publications by Leakey, Evernden, Curtis, Hay and others. With the recent recognition at the Woods Hole Oceanographic Institution that the base of the Pleistocene (defined on micropalaeontological criteria) can be related to the Olduvai Normal Event (dated at *ca.* 1.85 million years), we can say with some assurance that *Homo habilis* and *Zinjanthropus* lived at a time equivalent with the Lower Calabrian (= Upper Villafranchian of Leakey). The fossil evidence at Olduvai has shown that these creatures had evolved an erect bipedal posture, a criterion distinguishing the hominid from the ape degree of organization.

POSTCRIPT 1

Since the completion of this manuscript data concerning the relationship of Pliocene planktonic foraminiferal zones to palaeomagnetic stratigraphy have come to hand through the kindness of Dr T. Saito and his colleagues at the Lamont Geological Observatory. Perhaps the most significant is the observation, which I have been able to verify, that the stratigraphical ranges of *Globoquadrina altispira altispira* and *Globorotalia tosaensis* appear to overlap within the Mammoth Reversed event (*ca.* 3 million years B.P.). Blow (pers. comm.) has also observed the overlap of *G. altispira (s.l.)* and *G. tosaensis* in some deep-sea core samples, but Blow reports he has not seen the overlap in land-based sections in either the Indo-Pacific or Atlantic/Caribbean Provinces. However, Blow (in press) has now redefined the base of Zone N20 in terms of the first appearance of *Globorotalia acostaensis pseudopima.* Zone N20 as now redefined includes the uppermost part of Zone N19. The last occurrence of *G. altispira* is at about 2.9 million years B.P. Originally, Zone N20 was defined on the basis of the concurrent range of *G. multicamerata* and *Pulleniatina obliquiloculata,* subsequent to the extinction of *Globoquadrina altispira* and prior to the advent of *G. tosaensis,* but the new redefinition of Zone N20 is independent of the final range of *G. altispira altispira.* The consistent occurrence together of *Globorotalia tosaensis, G. multicamerata, Globoquadrina altispira, Pulleniatina obliquiloculata, Sphaeroidinella seminulina* and *S. dehiscens* in deep-sea cores from both the Indo-Pacific region and the North Atlantic suggests that Zones N19 and N21 follow each other sequentially in terms of the original definition. Zone 20 seems to represent a very short interval of time. Parker (1967, pp. 119, 138) discussed problems in recognizing the N20 zone in tropical Indo-Pacific Pliocene cores, and the data presented here appear to substantiate her observations as far as the original definition was concerned.

Further data conveyed to me by Dr T. Saito indicates that sediments which have been dated at approximately 4.5 million years B.P. (within the Gilbert Reversed Epoch) are low in Zone N19, which suggests that the Miocene/Pliocene boundary may lie somewhere between 5-6 million years B.P.

In the calculations presented in this paper, Zone N20 has been included as one of the Neogene zones and the date of 6 million years has been accepted for the Miocene/Pliocene boundary. Zone N21 is now considered to be about 1.15 million years long (3.0 m.y.-1.85 m.y.). Zone N19 is thus seen to represent a greater part of the Pliocene and probably has a duration of between 2-2.5 million years. Eventual modifications in the Pliocene zonation of Banner and Blow (1965) will not alter to any significant degree the estimates presented here.

POSTSCRIPT 2

In a letter received after the completion of this manuscript Dr W. H. Blow made several points which have a bearing on this paper. They are briefly:

(Item 1) Zone N3 has been emended and redefined: The top of this zone is placed at the base of Zone N4, emended (see below, Item 2).

(Item 2) Zone N4 has been emended and redefined as follows: *Globigerinoides quadrilobatus primordius/Globorotalia (Turborotalia) kugleri* Concurrent-range Zone. The base of this zone is placed at the first evolutionary stratigraphical occurrence of *G. quadrilobatus primordius,* and is thus somewhat higher than the base of the *G. kugleri* Zone of Bolli 1957, and Zone N4 of Banner and Blow 1965.

(Item 3) Zone N20 has been redefined (see above) as follows: *Globorotalia (G.) multicamerata-Pulleniatina obliquiloculata* Partial-range Zone. Base placed at first evolutionary appearance of *G. (T.) acostaensis pseudopima,* which occurs within the lower part of the range of *P. obliquiloculata s.s.*

(Item 4) Dr Blow has decided, upon a review of available evidence, to accept the recommendations of the Mediterranean Neogene Committee in 1959 to define the base of the Miocene at the base of the Aquitanian Stage, or virtually at the base of Zone N4 as he has now redefined it (see above under Item 2).

These modifications do not appear in the text of this paper, but they have been incorporated as far as possible in the pertinent text figures.

ADDENDUM

Since the completion of the major part of the manuscript for this paper a considerable body of data on radiometric dates and their relationship to biostratigraphic and time-stratigraphic units within the Cenozoic has come to hand. As a result it has become necessary to modify to some extent the relationships shown in Table 52.39. An initial modification of this table was made in Berggren (1969). More recently, in a paper in *Nature, Lond.,* **224** (5224), late-1969, a further modification was made incorporating information from the following sources:

1. D. L. Turner (1969).
2. Data presented at a symposium at Johannes Gutenberg University, Mainz, at the conclusion of an excursion to the type localities of the Oligocene.

The new basic information pertinent to the Cenozoic timescale includes the following:

(*a*) the base of the Lattorfian Stage has been dated at about 37.5 million years. This date is in good agreement with other dates estimated for the boundary between the Eocene and Oligocene in non-marine sections,

(*b*) the base of Miocene (i.e. the base of the Aquitanian, as defined in its type section) is about 22.5 million years. This is the level of the first evolutionary appearance of the genus *Globigerinoides,* otherwise known as the *Globigerinoides*-datum,

(*c*) the age of the Early/Middle Miocene boundary, denoted by the first evolutionary appearance of the genus *Orbulina,* is about 14 million years.

It will be noticed that the dates of 22.5 million years for the Oligocene/Miocene boundary and of 14 million years for the Early/Middle Miocene boundary are considerably younger than

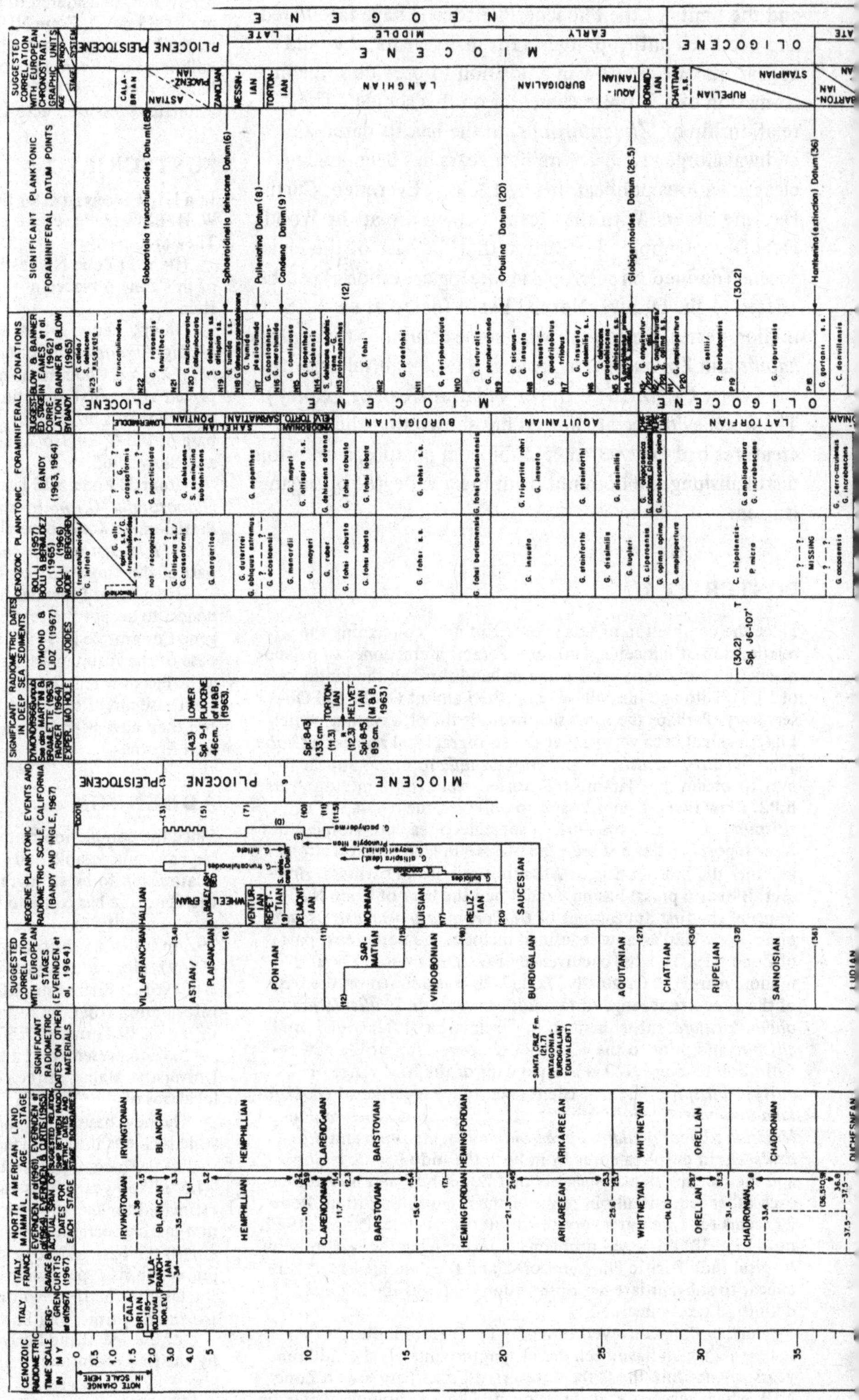
NEOGENE
PLEISTOCENE
PLIOCENE
MIOCENE
OLIGOCENE

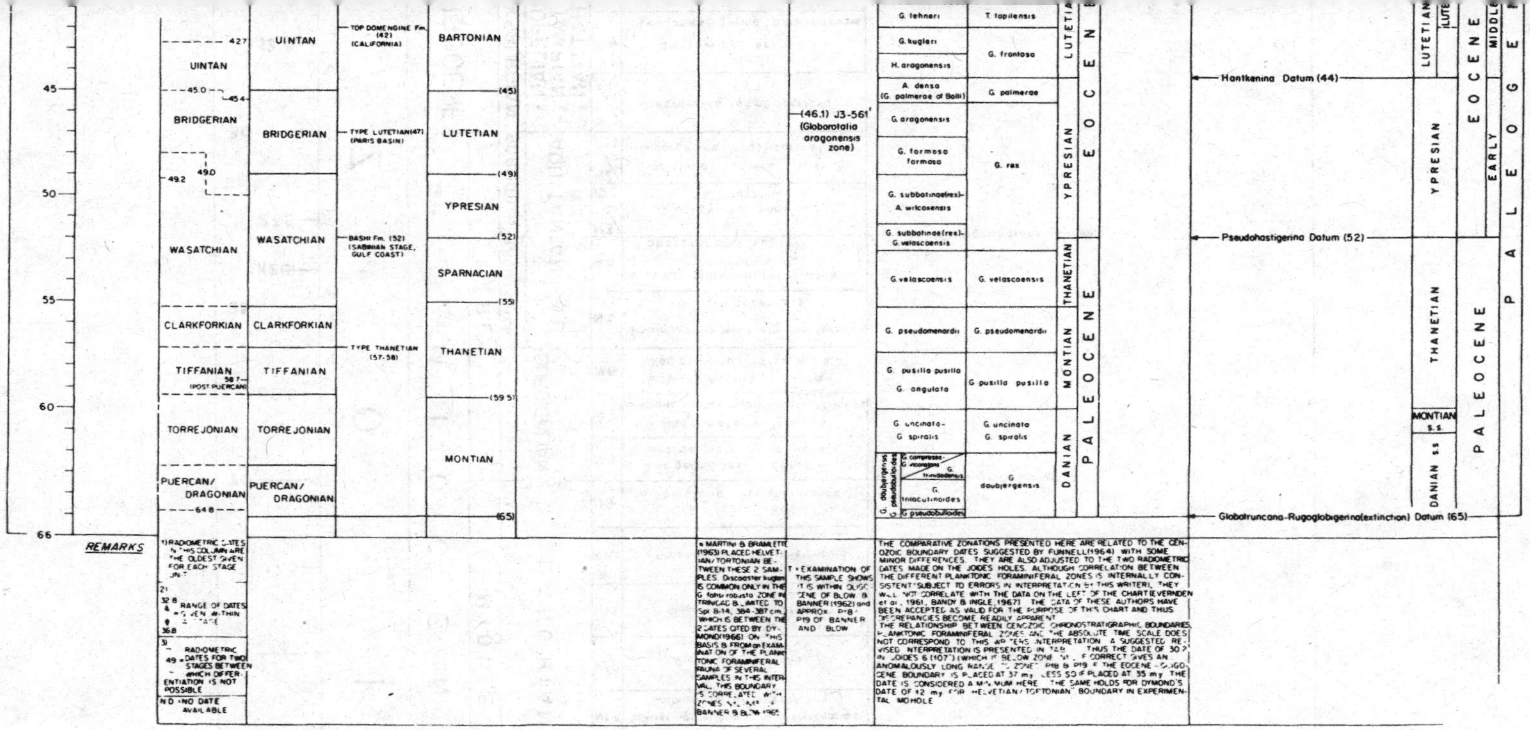

45
50
55
60
66
UINTAN
427
45.0
45.4
BRIDGERIAN
49.0
49.2
WASATCHIAN
CLARKFORKIAN
TIFFANIAN
58.7
(POST PUERCAN)
TORREJONIAN
PUERCAN/ DRAGONIAN
64.8
UINTAN
BRIDGERIAN
WASATCHIAN
CLARKFORKIAN
TIFFANIAN
TORREJONIAN
PUERCAN/ DRAGONIAN
TOP DOMENGINE Fm. (42) (CALIFORNIA)
TYPE LUTETIAN(47) (PARIS BASIN)
BASHI Fm. (52) (SABINIAN STAGE, GULF COAST)
TYPE THANETIAN (57-58)
BARTONIAN
(45)
LUTETIAN
(49)
YPRESIAN
(52)
SPARNACIAN
(55)
THANETIAN
(59.5)
MONTIAN
(65)
(46.1) J3-561' (Globorotalia aragonensis zone)
G. lehneri
G. kugleri
H. aragonensis
A. densa (G. palmerae of Bolli)
G. aragonensis
G. formosa formosa
G. subbotinae(rex)-A. wilcoxensis
G. subbotinae(rex)-G. velascoensis
G. velascoensis
G. pseudomenardii
G. pusilla pusilla
G. angulata
G. uncinata-G. spiralis
G. compressa-inconstans
G. trinidadensis
G. triloculinoides
G. pseudobulloides
G. daubjergensis
G. pseudobulloides
T. topilensis
G. frontosa
G. palmerae
G. rex
G. velascoensis
G. pseudomenardii
G. pusilla pusilla
G. uncinata
G. spiralis
G. daubjergensis
LUTETIAN
YPRESIAN
THANETIAN
MONTIAN
DANIAN
EOCENE
PALEOCENE
Hantkenina Datum (44)
Pseudohastigerina Datum (52)
Globotruncana-Rugoglobigerina(extinction) Datum (65)
LUTETIAN
LUTE
YPRESIAN
THANETIAN
MONTIAN s.s.
DANIAN s.s.
EOCENE
PALEOCENE
MIDDLE
EARLY
PALEOGENE
REMARKS
1) RADIOMETRIC DATES IN THIS COLUMN ARE THE OLDEST GIVEN FOR EACH STAGE UNIT
2) 32.8 ↕ 36.8 = RANGE OF DATES GIVEN WITHIN A STAGE
3) 49 = RADIOMETRIC DATES FOR TWO STAGES BETWEEN WHICH DIFFERENTIATION IS NOT POSSIBLE
N.D. = NO DATE AVAILABLE
* MARTINI & BRAMLETTE (1963) PLACED HELVETIAN/TORTONIAN BETWEEN THESE 2 SAMPLES. Discoaster kugleri IS COMMON ONLY IN THE G. fohsi robusta ZONE IN TRINIDAD & LIMITED TO Spl 8-14, 384-387 cm, WHICH IS BETWEEN THE 2 DATES CITED BY DYMOND (1966). ON THIS BASIS & FROM an EXAMINATION OF THE PLANKTONIC FORAMINIFERAL FAUNA OF SEVERAL SAMPLES IN THIS INTERVAL, THIS BOUNDARY IS CORRELATED WITH ZONES N 11 AND ... OF BANNER & BLOW 1965
T = EXAMINATION OF THIS SAMPLE SHOWS IT IS WITHIN OLIGOCENE OF BLOW & BANNER (1962) and APPROX. P 18/P19 OF BANNER AND BLOW
THE COMPARATIVE ZONATIONS PRESENTED HERE ARE RELATED TO THE CENOZOIC BOUNDARY DATES SUGGESTED BY FUNNELL (1964) WITH SOME MINOR DIFFERENCES. THEY ARE ALSO ADJUSTED TO THE TWO RADIOMETRIC DATES MADE ON THE JOIDES HOLES. ALTHOUGH CORRELATION BETWEEN THE DIFFERENT PLANKTONIC FORAMINIFERAL ZONES IS INTERNALLY CONSISTENT (SUBJECT TO ERRORS IN INTERPRETATION BY THIS WRITER), THEY WILL NOT CORRELATE WITH THE DATA ON THE LEFT OF THE CHART (EVERNDEN et al., 1961, BANDY & INGLE, 1967). THE DATA OF THESE AUTHORS HAVE BEEN ACCEPTED AS VALID FOR THE PURPOSE OF THIS CHART AND THUS DISCREPANCIES BECOME READILY APPARENT.
THE RELATIONSHIP BETWEEN CENOZOIC CHRONOSTRATIGRAPHIC BOUNDARIES, PLANKTONIC FORAMINIFERAL ZONES AND THE ABSOLUTE TIME SCALE DOES NOT CORRESPOND TO THIS WRITER'S INTERPRETATION. A SUGGESTED REVISED INTERPRETATION IS PRESENTED IN TAB. ... THUS THE DATE OF 30.2 IN JOIDES 6 (107') (WHICH IS BELOW ZONE ...), IF CORRECT GIVES AN ANOMALOUSLY LONG RANGE TO ZONES P18 & P19 IF THE EOCENE-OLIGOCENE BOUNDARY IS PLACED AT 37 my, LESS SO IF PLACED AT 35 my. THE DATE IS CONSIDERED A MINIMUM HERE. THE SAME HOLDS FOR DYMOND'S DATE OF 12 my FOR HELVETIAN/TORTONIAN BOUNDARY IN EXPERIMENTAL MOHOLE

Table 52.38. Relationships between Cenozoic radiometric scale for Continental (Terrestrial) and Marine Strata

CENOZOIC RADIOMETRIC TIME SCALE in my	ERA	EPOCH / SERIES	AGE / STAGE	CENOZOIC PLANKTONIC FORAMINIFERAL ZONES		SIGNIFICANT CENOZOIC PLANKTONIC FORAMINIFERAL DATUM POINTS
	QUAT.	PLEISTOCENE		N23	Globigerina calida s.s./Sphaeroidinella dehiscens excavata	
				N22	Globorotalia truncatulinoides	
(1.8)		(1.8)				Globorotalia truncatulinoides Datum
2.5	C	PLIOCENE	ASTIAN / PIACENZIAN	N21	Globorotalia tosaensis tenuitheca	
				N20	Globorotalia multicamerata-Pulleniatina obliquiloculata	
5			ZANCLIAN	N19	Sphaeroidinella dehiscens dehiscens/Globoquadrina altispira altispira	
		(6)				Sphaeroidinella dehiscens s.s. Datum
		MIOCENE — LATE	MESSINIAN	N18	Globorotalia tumida s.s.-Sphaeroidinellopsis dehiscens paenedehiscens	
7.5				N17	Globorotalia tumida plesiotumida	Pulleniatina Datum
			(C.8) TORTONIAN	N16	Globorotalia acostaensis s.s.-G. merotumida	
10						Candeina nitida Datum
			(C.11)	N15	Globorotalia continuosa	
12.5	I	MIDDLE	LANGHIAN	N14	Globigerina nepenthes/Globorotalia siakensis	
				N13	Sphaeroidinellopsis subdehiscens ss-Globigerina protonepenthes	
15				N12	Globorotalia fohsi	
				N11	Globorotalia praefohsi	
17.5				N10	Globorotalia peripheroacuta	
				N9	Orbulina suturalis-Globorotalia peripheroronda	
20	O	(20) EARLY	(20) BURDIGALIAN			Orbulina Datum
				N8	Globigerinoides sicanus-Globigerinatella insueta	
				N7	Globigerinatella insueta-Globigerinoides quadrilobatus trilobus	
22.5				N6	Globigerinatella insueta/Globigerinita dissimilis	
			(24) AQUITANIAN S.L. / AQUITANIAN s.s.	N5	Globoquadrina dehiscens praedehiscens-G. dehiscens s.s.	
25						
(26)				N4	Globigerinoides quadrilobatus primordius/Globorotalia kugleri	
			(27) BORMIDIAN			Globigerinoides Datum
27.5	Z			N3 / P22	Globigerina angulisuturalis	
				N2 / P21	Globorotalia opima s.s./Globigerina angulisuturalis	
30		(30) OLIGOCENE	KASBURGIAN; RUPELIAN s.l.; STAMPIAN s.s.; CHATTIAN s.s.	N1 / P20	Globigerina ampliapertura	
32.5				P19	Globigerina sellii/Pseudohastigerina barbadoensis	

Age (m.y.)	Era	Epoch	Stage		Planktonic foraminiferal zone	Datum
35.6	C E N O Z O I C	E O C E N E	BARTONIAN / PRIABONIAN		Globorotalia centralis	Hantkenina (extinction) Datum
37.5			BARTONIAN / PRIABONIAN		Cribrohantkenina danvillensis	
			BARTONIAN / PRIABONIAN		Globigerapsis mexicana	
39				(39)		
40			LUTETIAN		Truncorotaloides pseudodubia	
			LUTETIAN		Orbulinoides beckmanni	
42.5			LUTETIAN		Globorotalia lehneri	
			LUTETIAN		Globigerapsis kugleri	
			LUTETIAN		Hantkenina aragonensis	
45				(45)		Hantkenina Datum
			YPRESIAN		Acarinina bulbrooki	
47.5			YPRESIAN		Globorotalia aragonensis	
50			YPRESIAN		Globorotalia formosa formosa	
52.5			YPRESIAN		Globorotalia subbotinae (rex)- Acarinina wilcoxensis	
				(54)		Pseudohastigerina Datum
55		P A L E O C E N E	THANETIAN		Globorotalia subbotinae (rex)- Globorotalia velascoensis	
			THANETIAN		Globorotalia velascoensis	
57.5			THANETIAN		Globorotalia pseudomenardii	
			THANETIAN		Globorotalia pusilla pusilla- Globorotalia angulata	
60				(60)		
			DANIAN s.l. / MONTIAN s.s.		Globorotalia uncinata- Globigerina spiralis	
				(62)		
62.5			DANIAN s.l. / DANIAN s.s.		Globoconusa daubjergensis / Globorotalia pseudobulloides	G. compressa- G. inconstans / G. trinidadensis
			DANIAN s.l. / DANIAN s.s.		Globoconusa daubjergensis / Globorotalia pseudobulloides	G. triloculinoides
			DANIAN s.l. / DANIAN s.s.		Globoconusa daubjergensis / Globorotalia pseudobulloides	G. pseudobulloides
65						Globotruncana–Rugoglobigerina (extinction) Datum

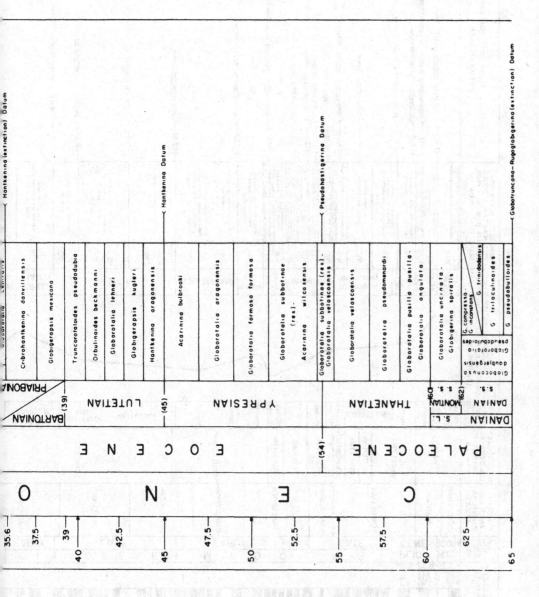

Table 52.39 Suggested relationship between Cenozoic radiometric time-scale, Chronostratigraphical boundaries and planktonic foraminiferal zonation

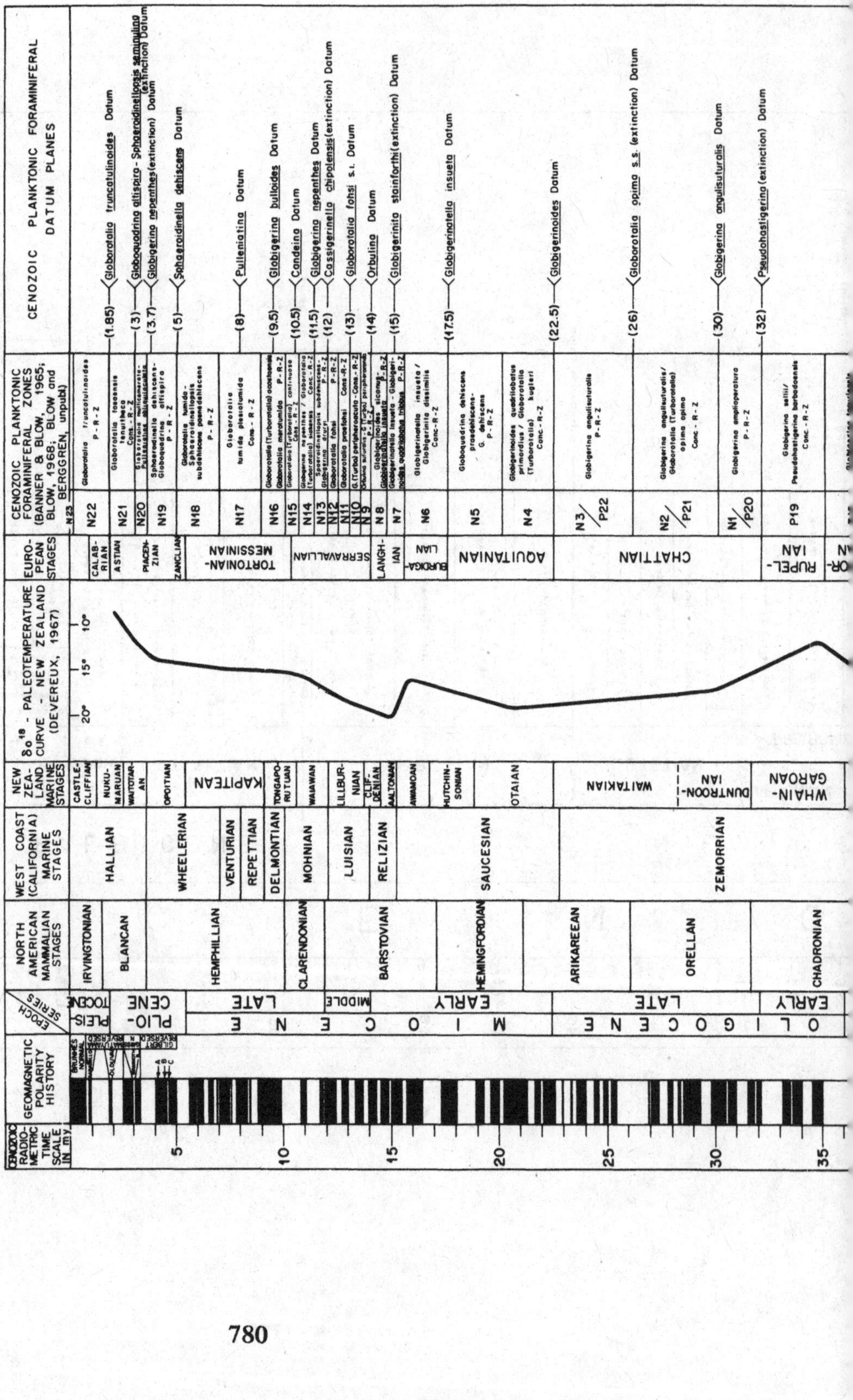
CENOZOIC RADIOMETRIC TIME SCALE IN m.y.
GEOMAGNETIC POLARITY HISTORY
EPOCH SERIES
NORTH AMERICAN MAMMALIAN STAGES
WEST COAST (CALIFORNIA) MARINE STAGES
NEW ZEALAND MARINE STAGES
δO[18] - PALEOTEMPERATURE CURVE - NEW ZEALAND (DEVEREUX, 1967)
EUROPEAN STAGES
CENOZOIC PLANKTONIC FORAMINIFERAL ZONES (BANNER & BLOW, 1965; BLOW, 1968; BLOW and BERGGREN, unpubl.)
CENOZOIC PLANKTONIC FORAMINIFERAL DATUM PLANES
5
10
15
20
25
30
35
PLEISTOCENE
PLIOCENE
MIOCENE
OLIGOCENE
LATE
MIDDLE
EARLY
IRVINGTONIAN
BLANCAN
HEMPHILLIAN
CLARENDONIAN
BARSTOVIAN
HEMINGFORDIAN
ARIKAREEAN
ORELLAN
CHADRONIAN
HALLIAN
WHEELERIAN
VENTURIAN
REPETTIAN
DELMONTIAN
MOHNIAN
LUISIAN
RELIZIAN
SAUCESIAN
ZEMORRIAN
CASTLECLIFFIAN
NUKUMARUAN
WAITOTARAN
OPOITIAN
KAPITEAN
TONGAPORUTUAN
WAIAUAN
LILLBURNIAN
CLIFDENIAN
ALTONIAN
AWAMOAN
HUTCHINSONIAN
OTAIAN
WAITAKIAN
DUNTROONIAN
WHAINGAROAN
20°
15°
10°
CALABRIAN
ASTIAN
PIACENZIAN
ZANCLIAN
TORTONIAN-MESSINIAN
SERRAVALLIAN
LANGHIAN
BURDIGALIAN
AQUITANIAN
CHATTIAN
RUPELIAN
N23
N22
N21
N20
N19
N18
N17
N16
N15
N14
N13
N12
N11
N10
N9
N8
N7
N6
N5
N4
N3/P22
N2/P21
N1/P20
P19
(1.85) Globorotalia truncatulinoides Datum
(3) Globoquadrina altispira - Sphaeroidinellopsis seminulina (extinction) Datum
(3.7) Globigerina nepenthes (extinction) Datum
(5) Sphaeroidinella dehiscens Datum
(8) Pulleniatina Datum
(9.5) Globigerina bulloides Datum
(10.5) Candeina Datum
(11.5) Globigerina nepenthes Datum
(12) Cassigerinella chipolensis (extinction) Datum
(13) Globorotalia fohsi s.l. Datum
(14) Orbulina Datum
(15) Globigerinita stainforthi (extinction) Datum
(17.5) Globigerinatella insueta Datum
(22.5) Globigerinoides Datum
(26) Globorotalia opima s.s. (extinction) Datum
(30) Globigerina angulisuturalis Datum
(32) Pseudohastigerina (extinction) Datum

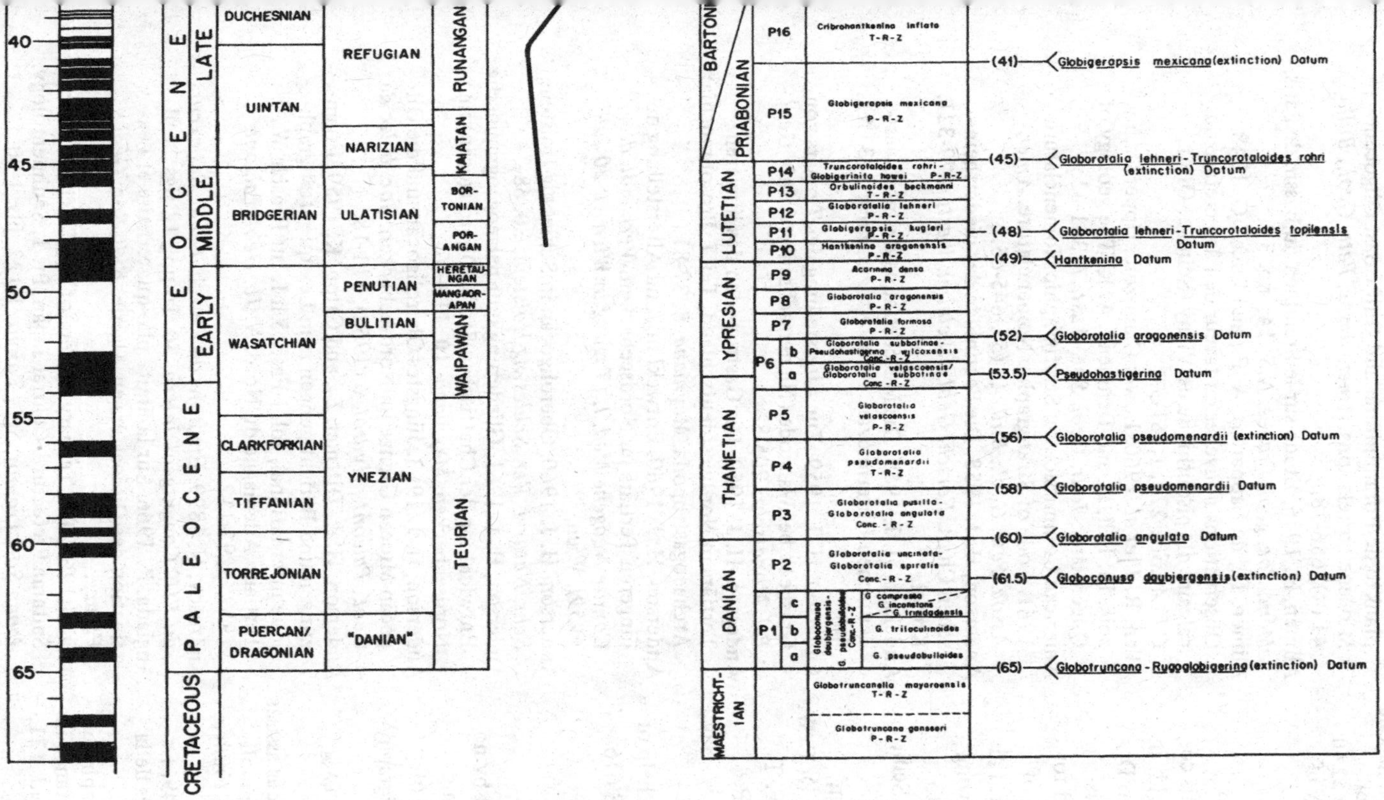
40
45
50
55
60
65
E O C E N E
LATE
MIDDLE
EARLY
P A L E O C E N E
CRETACEOUS
DUCHESNIAN
UINTAN
BRIDGERIAN
WASATCHIAN
CLARKFORKIAN
TIFFANIAN
TORREJONIAN
PUERCAN/ DRAGONIAN
REFUGIAN
NARIZIAN
ULATISIAN
PENUTIAN
BULITIAN
YNEZIAN
"DANIAN"
RUNANGAN
KAIATAN
BOR-TONIAN
POR-ANGAN
HERETAUNGAN
MANGAOR-APAN
WAIPAWAN
TEURIAN
BARTONIAN
PRIABONIAN
LUTETIAN
YPRESIAN
THANETIAN
DANIAN
MAESTRICHTIAN
P16 Cribrohantkenina inflata T-R-Z
P15 Globigerapsis mexicana P-R-Z
P14 Truncorotaloides rohri - Globigerinita howei P-R-Z
P13 Orbulinoides beckmanni T-R-Z
P12 Globorotalia lehneri P-R-Z
P11 Globigerapsis kugleri P-R-Z
P10 Hantkenina aragonensis P-R-Z
P9 Acarinina densa P-R-Z
P8 Globorotalia aragonensis P-R-Z
P7 Globorotalia formosa P-R-Z
P6 b Globorotalia subbotinae - Pseudohastigerina wilcoxensis Conc-R-Z
P6 a Globorotalia velascoensis/ Globorotalia subbotinae Conc-R-Z
P5 Globorotalia velascoensis P-R-Z
P4 Globorotalia pseudomenardii T-R-Z
P3 Globorotalia pusilla - Globorotalia angulata Conc. - R - Z
P2 Globorotalia uncinata - Globorotalia spiralis Conc - R - Z
P1 c G. compressa G. inconstans G. trinidadensis
P1 b G. triloculinoides
P1 a G. pseudobulloides
Globoconusa daubjergensis - G. pseudobulloides Conc-R-Z
Globotruncanella mayaroensis T-R-Z
Globotruncana gansseri P-R-Z
(41) Globigerapsis mexicana (extinction) Datum
(45) Globorotalia lehneri - Truncorotaloides rohri (extinction) Datum
(48) Globorotalia lehneri - Truncorotaloides topilensis Datum
(49) Hantkenina Datum
(52) Globorotalia aragonensis Datum
(53.5) Pseudohastigerina Datum
(56) Globorotalia pseudomenardii (extinction) Datum
(58) Globorotalia pseudomenardii Datum
(60) Globorotalia angulata Datum
(61.5) Globoconusa daubjergensis (extinction) Datum
(65) Globotruncana - Rugoglobigerina (extinction) Datum

Table 52.40 Suggested relationship between Cenozoic radiometric time-scale, Chronostratigraphical boundaries and planktonic foraminiferal zonation (revised)

the dates originally suggested in this paper. Confirmation of the accuracy of the relationship of the time-stratigraphic boundaries, the biostratigraphic datum points and the absolute ages suggested in this Addendum have recently been obtained from the results of the United States Deep-Sea Drilling Program.

The most recent revision of the Cenozoic radiometric time-scale prepared by the present author, which is a modification of the one which appeared in *Nature* 1969, is given in Table 52.40; it should be compared with Table 52.39 in the main body of the paper.

ACKNOWLEDGEMENTS

This work was only accomplished with the generous co-operation of several colleagues throughout the world. In particular a debt of sincere gratitude is expressed to Dr. W. H. Blow, Sunbury-on-Thames for stimulating discussions on the subject of Tertiary boundaries, and for access to unpublished information which has been of considerable value in the compilation of this paper. I also wish to acknowledge the co-operation of the following colleagues who transmitted valuable data relating to various phases of this study: Dr F. Barbieri, Dr R. Selli, Dr G. Ruggieri, Italy; Dr H. P. Luterbacher, Dr M. Vigneaux, France; Dr W. Hinsch, Dr H. Hiltermann, Germany; Dr A. Papp, Austria; Dr Z. Reiss, Israel; Dr C. W. Drooger. Mr. J. E. Meulenkamp, Netherlands; Dr F. T. Banner, Dr C. G. Adams, United Kingdom; Dr D. P. Naidin, Dr V. A. Krasheninnikov, U.S.S.R.; Dr W. W. Hay, Dr M. N. Bramlette, Mr J. Lamb, U.S.A.

This investigation has been completed with the help of funds from Nonr contract 4029 and NSF grant GS 676.

Received January 1968.

BIBLIOGRAPHY

The following bibliography provides a general, but by no means comprehensive, summary of literature dealing primarily with marine Tertiary stratigraphy. A list of symposia treating various aspects of Tertiary stratigraphy is presented at the end of the bibliography.

Abrard, R. 1925. *Le Lutétien du bassin de Paris.* Thèse, Angers.

Abrard, R. 1933. Nomenclature et synchronisme des assises de l'Eocène moyen et supérieur des bassins nummulifigues de l'Europe Occidentale. *Bull. Soc. Géol, France,* sér 5, **3**, 227-37.

Abrard, R., Furon R., Marie, P., and Soyer, R. 1949. Le calcaire pisolithique de Vigny n'est pas un faciès de la craie. *C.R. Ac. Sc.,* **228**, 189-90.

Adams, C. G. 1965. The Foraminifera and stratigraphy of the Melinau Limestone, Sarawak, and its importance in Tertiary correlation, *Quar. Jour. Geol. Soc. Lond.* **121**, 283-338, pls. 21-30.

Adams, C. G. 1968. Tertiary Foraminifera in the Tethyan, American and Indo-Pacific Provinces. *System. Assoc. Publ.* **7**, *Aspects of Tethyan Biogeography.*

Agip, Mineraria 1957. *Foraminifera padani (Terziario e Quaternario); Atlante iconografico e distribuzione stratigrafica,* 52 pls. Milan.

Akers, W. H. 1955. Some planktonic Foraminifera of the American Gulf Coast and suggested correlations with the Caribbean Tertiary, *Jour. Paleont.* **29** (4), 647-64.

Akers, W. H. and Drooger, C. W. 1957. Miogypsinids, planktonic Foraminifera, and Gulf Coast Oligocene-Miocene correlations, *Amer. Assoc. Petr. Geol., Bull.* **41** (4), 656-78.

Alimen, H. 1936. Étude sur le Stampien du bassin de Paris. *Mém. Soc. géol. France, N.S.* **14**, no. 31.

Alimen, H., de Lapparent, A. F., and Lucas, G. 1948. Observations nouvelles en faveur de l'âge crétacé du calcaire dit pisolithique de Vigny (Seine-et-Oise). *C.R.Ac.Sc.,* **227**, 1,161-3.

Allen, R. J. *et al.* 1960. Lower Cainozoic, Upper Cainozoic. In: Hill, D. and Denmead, A. K. The geology of Queensland. *Jour. Geol. Soc. Aust.* **7**, 341-423.

American Commission on Stratigraphic Nomenclature 1961. Code of Stratigraphic Nomenclature. *Amer. Assoc. Petr. Geol., Bull.* **45** (5), 645-65.

Anderson, H. J. 1958. Die Pectiniden des niederrheinischen Chatt. *Forsch. Geol. Rheinl. u. Westf.* 279-321. Krefel.

Anderson, H. J. 1959. Die Muschelfauna des nordwestdeutschen Untermiozan. *Palaeontographica,* **113** (A), 4-6.

Anderson, H. J. 1959. Zur Alterstellung des Tertiärs von Grasbeck bei Walsrode (Niedersachsen). *N. Jhb. Geol.-Pal. Monatschr.* 205-8.

Anderson, H. J. 1959. Die Gastropoden des jüngeren Tertiärs in Nordwestdeutschland, Teil I. Prosobranchia: Archaeogastropoda. *Meyniana,* **8**, 37-81.

Anderson, H. J. 1960. Entwcklung und Alterstellung des jüngeren Tertiärs im Nordseebecken. *Verhandl. d. Comite Neogene Med. I. (Tagung in Wien 10-20 Juli 1959).* Wien.

Anderson, H. J. 1960. Oberoligocän in Schleswig-Holstein. *Schr. Naturw. Ver. Schleswig-Holstein,* **30**, 68-74.

Anderson, H. J. 1961. Gliederung und palaeogeographische Entwicklung der Chattischen Stufe (Oberoligocän) im Nordseebecken. *Meyniana,* **10**, 118-46.

Anderson, H. J. 1963. Jüngstes Oberoligocän und die Oligocän-Miocan Grenze im Nordseebecken. *Soc. belge de Géol., Paléont., Hydrol,* **6** (1962), 111-18.

Anderson, H. J., Dittmer, E., and Gripp, K. 1959. Ergänzungen und Berichtigungen zum Lexique stratigraphique international, vol. I Fasc. 5 h I. In: Hinsch, W. Tertiaire Allemagne du Nord. *N. Jb. Geol. Palaont. 1959,* **3**, 97-112.

Andrusov, N. 1899. Environs de Kertch. *Guide des excurs. du VIIe Congr. géol. intern.* **30**, 16 pp., 12 figs., 1 pl.

Anglada, R. 1966. Sur la limite plio-quaternaire et l'extinction des Discoasters en Algérie. *Bull. Soc. Géol., France.*

Aria, J. 1960. The Tertiary System of the Chichibu Basin, Saitama Prefecture, Central Japan, Pt. 1, Sedimentology. *Japan Soc. Promot. Sci., Tokyo,* 1-122, pls. 1-30.

Archiac, A. d' 1850. *Histoire des progrès de la Géologie de 1834 à 1845,* **II**, 2 part, 441-1100, Terrain tertiaire III, 624 pp. (1849-50) Paris.

Asano, K. 1962a. Faunal change of planktonic Foraminifera through the Neogene of Japan. *Koninkl. Nederl.*

Akade. Wetenschappen-Amsterdam, Proc. ser. B, **65** (1), 1-16.

Asano, K. 1962b. Tertiary globigerinids from Kyushu, Japan. *Tohoku Univ., Sci. Rep., Ser. 2 (Geol.), Spec. Vol.* **5**, 49-65, pls. 19-23.

Asano, K. 1962c. Japanese Paleogene from the view-point of Foraminifera with descriptions of several new species. *Tohoku Univ., Inst. Geol. Pal., Contr.* **57**, 1-32, 1 pl.

Asano, K. and Takayanagi, Y. 1965. Stratigraphic significance of the planktonic Foraminifera from Japan *Tohoku Univ. Sci. Rep., Ser. 2, (Geol.)*, **37** (1), 1-14.

Asano , K. and Hatai, K. 1967. Micro- and macropalaeontological Tertiary correlations within Japanese Islands and with planktonic foraminiferal sequences of foreign countries. In: Tertiary correlations and climatic changes in the Pacific. *11th Pacific Sci. Congr., Tokyo (1966)*, 77-87.

Ascoli, P. 1956. Microfauna della serie eocenica di Rio Repregoso e della serie oligocenica superiore di Mombisaggio-Mongariolo (Tortona-Alessandria). *Riv. Ital. Pal. Strat.* **62** (3), 153-96.

Ascoli, P. 1957. Microfauna del Tortoniano di Mombisaggio e della serie pliocenica di Volpeglino. *Riv. Ital. Pal. Strat.* **63** (1), 3-30.

Avias, M. J. 1958. La limite inférieure du Miocène. *Compt. rend. du Congr. des Soc. Savantes, Colloque sur le Miocène.* Paris.

Bandy, O. L. 1949. Eocene and Oligocene Foraminifera from Little Stave Creek, Clarke County, Alabama. *Bull. Amer. Pal.* **32** (131), 210 pp., 27 pls.

Bandy, O. L. 1959. The geologic significance of coiling ratios in the foraminifer *Globigerina pachyderma* (Ehrenberg), *Geol. Soc. Amer. Bull.* **70** (2), 1,708.

Bandy, O. L. 1960a. Planktonic foraminiferal criteria for paleoclimatic zonation. *Tohoku Univ., Sci. Rep., Ser. 2 (Geol.), Spec. Vol.* **4**, 1-8.

Bandy, O. L. 1960b. The geologic significance of coiling ratios in the foraminifer *Globigerina pachyderma* (Ehrenberg). *Jour. Paleont.* **34** (4), 671-681.

Bandy, O. L. 1961. Distribution of foraminifera, radiolaria, and diatoms in sediments from the Gulf of California. *Micropaleontology*, **7** (1), 1-26.

Bandy, O. L. 1963. Miocene-Pliocene boundary in the Philippines as related to Late Tertiary stratigraphy of deep-sea sediments. *Science*, **142** (3,597), 1,290-2.

Bandy, O. L. 1964. Cenozoic planktonic foraminiferal zonation. *Micropaleontology*, **10** (1), 1-17.

Bandy, O. L. 1966a. Restrictions of the *'Orbulina'* datum. *Micropaleontology*, **12** (1), 79-86.

Bandy, O. L. 1966b. Faunal evidence of Miocene-to-Recent paleoclimatology in Antarctic. *Bull. Am. Assoc. Petrol. Geologists*, **50** (3), 643-4.

Bandy, O. L. 1966c. Base of the Pleistocene in Los Angeles Basin, California. *Bull. Am. Assoc. Petrol. Geologists*, **50** (3), 604-5.

Bandy, O. L. 1967. Problems of Tertiary foraminiferal and radiolarian zonation, circum-Pacific area. In: Tertiary correlations and climatic changes in the Pacific. *11th Pacific Sci. Congr., Tokyo*, (1966), 95-102.

Banks, M. R. 1962. Cainozoic. Marine succession, in Spry and Banks, The geology of Tasmania. *Jour. Geol. Soc. Aust.* **9** (2), 233-6.

Banner, F. T. and Blow, W. H. 1959. Classification and stratigraphical distribution of the Globigerinaceae. *Palaeontology*, **2** (1), 1-27, pls. 1-3.

Banner, F. T. and Blow, W. H. 1960a. The taxonomy, morphology and affinities of the genera included in the subfamily Hastigerininae. *Micropaleontology* **6** (1), 19-31.

Banner, F. T. and Blow, W. H. 1960b. Some primary types of species belonging to the superfamily Globigerinaceae. *Cushman Found. Foram. Res., Contr.* **11** (1), 1-41, 8 pls.

Banner, F. T. and Blow, W. H. 1965a. Progress in the planktonic foraminiferal biostratigraphy of the Neogene. *Nature, Lond.* **208** (5,016), 1,164-6.

Banner, F. T. and Blow, W. H. 1965b. Two new taxa of the Globorotaliinae (Globigerinacea, Foraminifera) assisting determination of the Late Miocene/Middle Miocene boundary. *Nature, Lond.* **207** (5004), 1,351-4.

Banner, F. T. and Blow, W. H. 1966. The origin, evolution and taxonomy of the foraminiferal genus *Pulleniatina* Cushman, 1927. *Micropaleontology*, **13** (2), 133-62, pls. 1-4.

Banner, F. T. and Blow, W. H. 1967. The origin evolution and taxonomy of the foraminiferal genus *Pulleniatina* Cushman, 1927. *Micropaleontology*, **13** (2), 133-62, pls. 1-4.

Banner, F. T. and Eames, F. E. 1966. Recent progress in world-wide Tertiary stratigraphical correlation. *Earth-Sci. Rev.* 2 (1966), 157-9.

Barbieri, F. 1953. Il Pliocene di val Recchio (Parma). *Boll. Soc. Geol. Ital.* **74** (1), 171-9.

Barbieri, F. 1958. La serie pliocenica di Castell 'Arquato. *Guida Coll. Internaz. Micropal. in Italia*, 23-33, Milan.

Barbieri, F. 1967. The Foraminifera in the Pliocene section Vernasca-Castell 'Arquato including the "Piacenzian Stratotype" (Piacenza Province). *Mem. Soc. Ital. Science natur. e del Mus. Civ. Stor. natur. Milano*, **15** (3), 145-63.

Barbieri, F. and Medioli, F. 1964a. Nota preliminare sullo studio micropaleontologico della serie pliocenica Vernasca-Castell' Arquato (Piacenza). *Boll. Soc. Geol. Italiana*, **83** (1), 3-8.

Barbieri, F. and Medioli, F. 1964b. Significato paleoecologico di alcuni generi di foraminifera nella serie pliocenica Vernasca-Castell' Arquato. *Ist. Geol. dell Univers. di Parma (Lab. di Micropal.), L'Ateneo Parmeuse*, **35**, *Suppl. 1*, 1-27.

Barbieri, F. and Petrucci, F. 1967. La série stratigraphique du Messinien au Calabrien dans la vallée du T. Crostolo (Reggio Emilia-Italia Sept.). *Mem Soc. Ital. Sc. Nat. Milano*, **15** (3), 181-8, 1 pl.

Barr, F. T. and Berggren, W. A. 1964. Planktonic Foraminifera from the Thanet formation (Paleocene) of Kent, England. *Stokholm Contrib. Geol.* **13** (2), 9-26.

Barr, F. T. and Berggren, W. A. 1965. Lower Tertiary planktonic Foraminifera from the Thanet Formation of Kent, England. *XXII Int. Geol. Congress (New Delhi) Stockholm Contr. Geology*, **13** (2), 9-26.

Batjes, D. A. J. 1958. Foraminifera of the Oligocene of Belgium. *Inst. Roy. Sci. Nat. Belg., Mem.* **143**, 188 pp.

Beckmann, J. P. 1957. *Chiloguembelina* Loeblich and

Tappan and related Foraminifera from the Lower Tertiary of Trinidad, B.W.I. *U.S. Nat. Mus. Bull.* **215**, 83-95, pl. 21.
Belford, D. J. 1962. Miocene and Pliocene planktonic Foraminifera, Papua-New Guinea. *Bur. Mine. Res., Geol. Geophy., Austral., Bull.* **62-1**, 1-50.
Belford, D. J. 1966. Miocene and Pliocene smaller Foraminifera from Papua and New Guinea. *Dept. Nat. Devel., Bur. Mineral. Res., Geol. Geophy., Bull.* **79**, 1-305, 38 pls.
Berggren, W. A. 1960a. Biostratigraphy, planktonic Foraminifera and the Cretaceous-Tertiary boundary in Denmark and southern Sweden. *XXI Int. Geol. Congress (Copenhagen), proc. sect.* **5**, 181-92.
Berggren, W. A. 1960b. Paleogene biostratigraphy and planktonic Foraminifera of Nigeria (West Africa). *XXI Int. Geol. Congress (Copenhagen), proc. sect.* **6**, 41-55.
Berggren, W. A. 1960c. Paleogene biostratigraphy and planktonic Foraminifera of the SW Soviet Union: An analysis of recent Soviet investigations. *Stockholm Contr. Geol.* **6** (5), 64-125.
Berggren, W. A. 1962a. Some planktonic Foraminifera from the Maestrichtian and type Danian stages of Denmark and southern Sweden. *Stockholm Contr. Geol.* **9** (1), 1-102, pls. 1-14.
Berggren, W. A. 1962b. Stratigraphic and taxonomic-phylogenetic studies of Upper Cretaceous and Lower Tertiary planktonic Foraminifera. *Stockholm Contr. Geol.* **9** (2), 103-29.
Berggren, W. A. 1963. Some problems of Paleocene stratigraphic correlation. *Rev. Inst. Franc, Pétrole,* **18** (10), 134-43.
Berggren, W. A. 1964a. Paleocene-Lower Eocene biostratigraphy of Luxor and nearby Western Desert. *Petr. Expl. Soc. Libya, 6th Ann. Field Conf.* 149-76, 2 pls.
Berggren, W. A. 1964b. The Maestrichtian, Danian and Montian Stages and the Cretaceous-Tertiary boundary. *Stockholm Contr. Geol.* **11** (5), 103-76.
Berggren, W. A. 1965a. The recognition of the *Globorotalia uncinata* Zone (Lower Paleocene) in the Gulf Coast. *Micropaleontology,* **11** (1), 111-13.
Berggren, W. A. 1965b. Some problems of Paleocene-Lower Eocene planktonic foraminiferal correlations. *Micropaleontology,* **11** (3), 278-300.
Berggren, W. A. 1965c. Paleocene – A micropaleontologist's point of view. *Bull. Amer. Assoc. Pet. Geol.* **49** (9), 1,473-84.
Berggren, W. A. 1966. [Phylogenetic and taxonomic problems of some Tertiary planktonic foraminiferal lineages.] *Voprosy Mikropal.,* **10**, 309-32, (in Russian).
Berggren, W. A. 1968a. Paleogene biostratigraphy and planktonic Foraminifera of northern Europe. *1st Planktonic Conf., Geneva, Sept. 1967* (in press), 8 pls.
Berggren, W. A. 1968b. Biostratigraphy and planktonic foraminiferal zonation of the Tertiary System of the Sirte Basin, Libya, North Africa. *1st Planktonic Conf., Geneva, Sept. 1967.* (in press).
Berggren, W. A. 1968c. Micropaleontology and the Pliocene/Pleistocene boundary in a deep-sea core from the South-Central North Atlantic. *4th Meeting of Comm. Mediterr. Neog. Strat., Bologna (Sept. 1967), Proc.* (in press).
Berggren, W. A. 1969. Rates of evolution in some Cenozoic planktonic Foraminifera. *Micropaleont.* **15** (3)
Berggren, W. A., Olsson, R. K., and Reyment, R. A. 1967. Origin and development of the foraminiferal genus *Pseudohastigerina* Banner and Blow, 1959. *Micropaleontology,* **13** (3), no. 3, 265-88, 1 pl.
Bermúdez, P. J. 1950. Contribución al estudio del Cenozóico cubano. *Soc. Cubana Hist. Nat., Mem.* **19** (3), 205-375.
Bermúdez, P. J. 1952. Estudio Sistematico de les Foraminiferos Rotaliformes. *Boln. Geol. Minist. Minas Venez,* **2** (4), 1-230, pls. 1-35.
Bermúdez, P. J. 1961. Contribucion al estudio de las Globigerinidea de la region Caribe-Antillana (Paleocene-Reciente). *Memoria del III Congreso Geologico Venezolano, Tomo III Boletin de Geologia, Publicacion especial* 3, *1960,* 1,119-393. 20 pls.
Berthelsen,O. 1962. Cheilostome Bryozoa in the Danian deposits of East Denmark. *Danm. Geol. Unders., II Rk,* **83**, 290 pp., 28 pls.
Bettenstaedt, F. and Wicher, C. A. 1955. Stratigraphic correlation of Upper Cretaceous and Lower Cretaceous in the Tethys and Boreal by the aid of microfossils. *IV World Petrol. Congr. Proc., sect. I/D, pap. 5,* 493-516, 5 text-pls.
Beyrich, E. 1848. Zur Kenntnis der tertiären Bodens der Mark Brandenburg, Karstens u. Dechens. *Arch. Bergb. usw.* **22** (1), 3-102.
Beyrich, E. 1853. Sternberger Kuchen im Oderbett bei Cunitz. *Z. dt. Geol. Ges.* **5**, 7.
Beyrich, E. 1853, 1854, 1856. Die Conchylien des norddeutschen Tertiargebirges. *Z. dt. Geol. Ges.* **5**, 273-358, pls. 4-8; **6**, 409-500, pls. 9-14, 726-81, pls. 15-18; **8**, 21-88, pls. 1-10, 553-88, pls. 15-17.
Beyrich, E. 1853, 1854, 1856, 1857. *Die Conchylien des norddeutschen Tertiargebirges.* **I.** 1-82, pls. 1-5, 1853; **2./3.**, 83-176, pls. 6-15, 1854; **4./5.**, 177-296, pls. 16-25, 1856; **6.**, 297-336, pls. 26-30, 1857. Hertz: Berlin.
Beyrich, E. 1854. Über die Stellung der Hessischen Tertiärbildungen. *Ber. Verh. kgl. preuss. Akad. Wiss. Berlin 1854,* 640-66.
Beyrich, E. 1855. Über die Verbreitung tertiärer Ablagerungen in der Gegend von Düsseldorf. *Z. dt. Geol. Ges.* **7**, 451-2.
Beyrich, E. 1856a. Tertiäre Conchylien aus Bohrungen bei Neuss. *Z. dt. Geol. Ges.* **8**, 10.
Beyrich, E. 1856b. Über das Alter tertiärer Eisensteine, welche bei Rothenburg an der Saale vorkommen. *Z. dt. Geol. Ges.* **8**, 309.
Beyrich, E. 1856c. Über den Zusammenhang der norddeutschen Tertiärbildungen. *Abh. Akad. Wiss. Berlin 1856 (Physik. Abh.),* 1-20.
Beyrich, E. 1858. Über die Abgrenzung der oligocanen Tertiärzeit. *Monatsber, kgl. preuss. Akad. Wiss. Berlin 1858,* 54-69.
Beyrich, E. 1886. 'Graues Sternberger Gestein' aus der Umgebung von Mittenwalde. *Z. dt. Geol. Ges.* **38**, 245.
Bezrukov, P. L. 1934. Verkhnernelooge i paleogenooye okolzhenia basseina Verkhovev r. tobola. Byul. *Mosk. obshch. ispty. privody, otd. geol.* **12** (2), (Summary in English).

Bezrukov, P. L. 1936. Datskii yarus vostoch-nevropeiskoi plakforraiy. *Izd. Akad. Nauk U.S.S.R.* ser. geol., no. 5, 657-88 (English summary), 685-8).

Bhatia, S. B. 1955. The foraminiferal fauna of the late Paleogene sediments of the Isle of Wight, England. *Journ. Paleont.* **29** (4), 665-93.

Bignot, G. 1962. Etude micropaléontologique de la formation de Varengeville du gisement éocène du Cap-d'Ailly (Seine-Mme). *Rev. Micropal.* **5** (3), 161.

Binckhorst, J. T. Van den 1859. *Esquisse géologique et paléontologique des couches crétacées du Limbourg.* 1 vol. in 8°, 270 p., 5 pls., 1 map.

Blondeau, A. 1964a. Rapport sur le Lutétien du bassin de Paris. In: Colloque sur le Paléogène, Bordeaux, 1962. *Mém. Bur. Rech. géol. et min. no.* **28** (1), 15-19.

Blondeau, A. 1964b. Le Lutétien du Nord de l'Ile-de-France, entre le Soissonnais et le Vexin. *Mém. Bur. Rech. géol. et min., no.* **28** (1), 27-35.

Blondeau, A. 1965a. Etude biométrique et statistique de *Nummulites laevigatus* dans les bassins de Paris et du Hampshire. Implications stratigraphiques. *Bull. Soc. géol. France, Sér. 7*, **7**, 268-72.

Blondeau, A. 1965b. *Le Lutétien des bassins de Paris, de Belgique et du Hampshire, Etude sédimentologique et paléontologique.* Thèse, Paris.

Blondeau, A. and Cavelier, C. 1962a. Etude du Lutétien inférieur de Saint-Leu-d'Esserent (Oise). *Bull. Soc. géol. France, Sér. 7*, **4**, 55-63.

Blondeau, A. and Cavelier, C. 1962b. Etude du Lutétien inférieur à Boissy-Sant-Leu-d'Esserent. *Bull. Soc. géol. France, Sér. 7*, **4**, 222-5.

Blondeau, A., Cavelier, C. and Pomerol, Ch. 1964. Influence de la tectonique du Pays de Bray sur les formations paléogènes au voisinage de sa terminaison orientale. *Bull. Soc. géol. France, Sér. 7*, **6**, 357-67.

Blondeau, A., Cavelier, C., Feugueur, L., and Pomerol, C. 1965. Stratigraphie du Paléogène du bassin de Paris en relation avec les bassins avoisinants. *Bull. Soc. géol. France, Sér. 7*, **7** (2), 200-21.

Blondeau, A. and Curry, D. 1963. Sur la présence de *Nummulites variolarius* (Lmk) dans les diverses zones du Lutétien des bassins de Paris, de Bruxelles et du Hampshire. *Bull. Soc. géol. France, Sér. 7*, **5**, 275-7.

Blow, W. H. 1956. Origin and evolution of the foraminiferal genus *Orbulina* d'Orbigny. *Micropaleontology*, **2**, 57-70.

Blow, W. H. 1957. Transatlantic correlation of Miocene sediments. *Micropaleontology*, **3** (1), 77-9.

Blow, W. H. 1959. Age, correlation and biostratigraphy of the Upper Tocuyo (San Lorenzo) and Poson formations, Eastern Falcon, Venezuela. *Amer. Paleont., Bull.* **39** (178), 59-251, pls. 6-19.

Blow, W. H. and Banner, F. T. 1962. The mid-Tertiary (Upper Eocene to Aquitanian) Globigerinaceae. In: Eames, F. E. *et al., Fundamentals of mid-Tertiary stratigraphical correlation*, 61-151, pls. 8-17. C.U.P.

Blow, W. H. and Banner, F. T. 1966. The morphology taxonomy and biostratigraphy of *Globorotalia barisanensis* LeRoy, *Globorotalia fohsi* Cushman and Ellisor, and related taxa. *Micropaleontology*, **12** (3), 286-302, pls. 1-2.

Blow, W. H. and Banner, F. T. 1966. Zonation of Cretaceous to Pliocene marine sediments based on planktonic Foraminifera – a comment. *Bol. Inform. Assoc. Venez. Geol. Miner., Petrol.* **9** (2), 55.

Bolli, H. M. 1950. The direction of coiling in the evolution of some Globorotaliidae. *Cushman Found. Foram. Res., Contr.* **1** (3 and 4), 82-9.

Bolli, H. M. 1951. Notes on the direction of coiling of rotalid Foraminifera. *Cushman Found. Foram. Res., Contr.* **2** (4), 139-43.

Bolli, H. M. 1957a. The genera *Globigerina* and *Globorotalia* in the Paleocene-Lower Eocene Lizard Springs Formation of Trinidad B.W.I. *U.S. Nat. Mus. Bull.* **215**, 61-81.

Bolli, H. M. 1957b. Planktonic Foraminifera from the Oligocene-Miocene Cipero and Lengua Formations of Trinidad, B.W.I. *U.S. Nat. Mus., Bull.* **215**, 97-123.

Bolli, H. M. 1957c. Planktonic Foraminifera from the Eocene Navet and San Fernando Formations of Trinidad, B.W.I. *U.S. Nat. Mus., Bull.* **215**, 155-72.

Bolli, H. M. 1959. Planktonic Foraminifera as index fossils in Trinidad, West Indies, and their value for worldwide stratigraphic correlation. *Eclogae Geol. Helv.* **52** (2), 627-37.

Bolli, H. M. 1962. *Globigerinopsis*, a new genus of the foraminiferal family Globigerinidae. *Eclogae Geol. Helv.* **55** (1), 281-4, 1 pl.

Bolli, H. M. 1964. Observations on the stratigraphic distribution of some warm water planktonic Foraminifera in the young Miocene to Recent. *Eclogae Geol. Helv.* **57** (2), pp. 541-52.

Bolli, H. M. and Bermúdez, P. J. 1965. Zonation based on planktonic Foraminifera of Middle Miocene to Pliocene warm-water sediments. *Asoc. Venezolana Geol. Min. Petr., Bol. Informativo*, **8** (5), 121-49, 1 pl.

Bolli, H. M. and Cita, M. B. 1960a. Globigerine e Globorotalie del paleocene di Paderno d'Addo (Italia). *Riv. Ital. Pal.* **66** (3), 361-402, 3 pls.

Bolli, H. M. and Cita, M. B. 1960b. Upper Cretaceous and Lower Tertiary planktonic Foraminifera from the Paderno d'Adda section, northern Italy. *Internat. Geol. Congr., 21st, pt. 5, sect. 5*, 150-61, 3 figs.

Bolli, H., Loeblich, Jr., A. R., and Tappan, H. 1957. Planktonic foraminiferal families Hantkeninidae, Orbulinidae, Globorotaliidae and Globotruncanidae. *U.S. Nat. Mus., Bull.* **215**, 3-50, pls. 1-11.

Borsetti, A. M. 1959. Tre nuovi foraminiferi plactonici dell'Oligocene Piacentino. *Giorn. Geol., Ann. Museo. Geol. Bologna*, **27**, 205-11.

Bouché, P. M. 1962. Nannofossiles calcaires de Lutetien du Bassin de Paris. *Rev. de Micropaléontologie*, **5** (2), 75-103, 4 pls.

Boulanger, D. and Poignant, A. 1964. Le passage Eocène supérieur-Oligocène en Aquitaine occidentale. *Compt. Rend. Soc. Géol. France*, 1964, **2**, 85-6.

Bourdon, M. and Lys, M. 1954. Microfaune du Calcaire à Astéries (Stampien) de la carrière de la Souys-Floirac (Gironde). *C.R. somm. Soc. Géol. France*, 337-8.

Bourdon, M. and Lys, M. 1955. Foraminifères du Stampien de la carrière de la Souys-Floirac (Gironde). *C.R. somm. Soc. Géol. France*, 336-8.

Bramlette, M. N. 1965. Massive extinctions in biota at the end of Mesozoic time. *Science*, **148**, 1,696-9.

Bramlette, M. N. and Sullivan, F. R. 1961. Coccolithophorids and related nannoplankton of the early Tertiary in California. *Micropaleontology*, **7** (2), 129-80, 14 pls.

Bramlette, M. N. and Wilcoxon, J. A. 1967. Middle Tertiary calcareous nannoplankton of the Cipero section, Trinidad, W. I. *Tulane Studies Geol.* **5** (3), 93-131, 10 pls.

Brebion, P. 1961. Etude du Miocène supérieur (Redonien) de l'Ouest de la France. *Meyniana*, **10**, 104-12.

Briart, A. and Cornet, F. L. 1870. Description des fossiles du Calcaire grossier de Mons. Première partie: Gastéropodes (pro parte). Mém. présenté le 11 mai 1869. *Mém. courronnés et Mém. des savants étrangers de l'Acad. roy. de Belgi.* **36**.

Briart, A. and Cornet, F. L. 1873. Description des fossiles du Calcaire grossier de Mons. Deuxième partie: Gastéropodes (pro parte). Mém. presente le 1er fevrier 1873. *Ibid.* **37**.

Briart, A. and Cornet, F. L. 1880. Note sur la carte géologique de la partie centrale de la Province de Hainaut exposée à Bruxelle en 1880. *Ann. Soc. Géol. de Belgi.* **7**, 139-48.

Brocchi, G. 1814. *Conchiologia fossile subappenina, 1–III*, Milano.

Brönnimann, P. 1950. The genus *Hantkenina* Cushman in Trinidad and Barbados, B.W.I. *Jour. Paleont.* **24** (4), 397-420.

Brönnimann, P. 1952a. Globigerinidae from the Upper Cretaceous (Cenomanian-Maestrichtian) of Trinidad, B.W.I. *Bull. Am. Paleontology*, **34** (140), 5-61, 4 pls.

Brönnimann, P. 1952b. Trinidad Paleocene and Lower Eocene Globigerinidae. *Bull. Am. Paleontology*, **34** (143), 5-34, 3 pls.

Brönnimann, P. 1953. Note on planktonic Foraminifera from Danian localities of Jutland, Denmark. *Eclog. Geol. Helv.* **45** (1952), 339-41.

Brönnimann, P. and Brown, N. 1953. Observations on some planktonic Heterohelicidae from the Upper Cretaceous of Cuba. *Contr. Cush. Found. Foram. Res.* **4** (4), 150-6.

Brönnimann, P. and Brown, N. 1956. Taxonomy of the Globotruncanidae. *Eclog. Geol. Helv.* **48** (2), 503-61, pls. 20-4.

Brönnimann, P. and Brown, N. 1958. *Hedbergella*, a new name for a Cretaceous planktonic foraminiferal genus. *Jour. Wash. Acad. Sci.* **48** (1), 15-17.

Brönnimann, P. and Rigassi, D. 2963. Contribution to the Geology and Paleontology of the area of the city of La Habana, Cuba and its surroundings. *Ecl. geol. Helv.* **56**, 13-480, pls. 1-26.

Brönnimann, P., Stradner, H. and Szöts, E. 1965. Sur les microfossiles planctoniques du stratotype du Spilecciano de Purgia di Bolca. *Arch. Sciences, Genève*, **18**, 93-102.

Brönnimann, P. and Stradner, H. 1960. Die Foraminiferen und Discoasteridenzonen von Kuba und ihre interkontinentale Korrelation. *Erdoel-Zeitschr.* **10**, pp. 3-8.

Brotzen, F. 1936a. Einige Bemerkungen zur Stratigraphie Schonens: 1) Zur Mächtigkeit des Daniens. *Geol. Foren. Forhandl.* **58** (1), 16-21.

Brotzen, F. 1936b. Foraminiferen aus dem schwedischen untersten Senon von Eriksdal in Schonen. *Sver. Geol. Unders., ser. C, Arsb.* **30** (3), 204 pp., 14 pls.

Brotzen, F. 1938. Der postkimmerische Bau des südlichsten Schweden. *Ibid.* **60**, 73-87.

Brotzen, F. 1940. Flintrännans och Trindelrännans geologi (Oresund). *Sver. Geol. Unders., ser. C. Arsb.* **34**, (5), 1-33, 1 pl.

Brotzen, F. 1945. De geologiska resultaten fran borrningarna vid Hollviken. *Sver. Geol. Unders., ser. C, Arsb.* **38** (7), 1-64, 2 pls.

Brotzen, F. 1948. The Swedish Paleocene and its foraminiferal fauna. *Sver. Geol. Unders., ser. C, Arsb.* **42** (2), 140 pp., 19 pls.

Brotzen, F. 1949a. Zur Gliederung der oberen Kreide in Skandinavien. *Geol. Foren. Forhandl.* **71** (1), 181-3.

Brotzen, F. 1949b. Zur Diskussion mit J. A. Jeletzky. *Geol. Foren. Forhandl.* **71** (1), 501, 502.

Brotzen, F. 1959. On *Tylocidaris* species (Echinoidea) and the stratigraphy of the Danian of Sweden: with a bibliography of the Danian and Paleocene. *Sver. Geol. Unders., ser. C, Arsb.* **54** (2), 81 pp., 3 pls.

Brotzen, F. and Pozaryska, K. 1957. The Paleocene in central Poland. *Acta. Geol. Polonica, 7.*

Brünnich-Nielsen, K. 1909. Brachiopoderne i Danmarks Kridtaflejringer. *Kgl. Dansk Vid. Selsk. Skr., 7, Rk., Nat. -math. Afd.* **6** (4).

Brünnich-Nielsen, K. 1910. Om det i København Havn ved Knipelsbro fundne Yngste Danien. *Medd. Dansk. Geol. Foren.* **3**.

Brünnich-Nielsen, K. 1911. Brachiopoderne i Faxe. *Medd. Dansk Geol. Foren.* **3**.

Brünnich-Nielsen, K. 1913a. Crinoiderne i Danmarks Kridtaflejringer. *Danmarks Geol. Unders., II Rk.* **26**.

Brünnich-Nielsen, K. 1913b. Moltkia isis, Steenstrup og andre Octocorallia fra Danmarks Kridtaflejringer. *Mindeskr. for Japetus Steenstrup,* **18**.

Brünnich-Nielsen, K. 1914. Some remarks on the brachiopods in the Chalk of Denmark. *Medd. Dansk Geol. Foren.* **4**.

Brünnich-Nielsen, K. 1915. *Rhizocrinus maximus* n. sp. og nogle Bemaerkninger om *Bourgueticrinus maximus. Br. N., Medd. Dansk Geol. Foren.* **4**.

Brünnich-Nielsen, K. 1917a. Cerithiumkalk i Stevns Klint. *Danmarks Geol. Unders., 4, Rk.* **1** (7), (see also *Medd. Dansk Geol. Foren.* **5** (7).

Brünnich-Nielsen, K. 1917b. *Heliopora incrustans* nov. sp.: with a survey of the Octocorallia in the deposits of the Danian in Denmark. *Med. Dansk Geol. Foren.*, **5** (8).

Brünnich-Nielsen, K. 1918. Slaegten '*Moltkia*' og andre Octocoraller i Sveriges Krittidsaflejringer. *Geol. Foren Forhandl.* **40**.

Brünnich-Nielsen, K. 1919. En Hydrocoralfauna fra Faxe og Bemaerkninger om Danien'ets geologiske Stilling. *Medd. Dansk Geol. Foren.* **5** (16), 66 pp., 2 pls.

Brünnich-Nielsen, K. 1920. Inddelingen af Danien'et i Danmark og Skaane. *Ibid.* **5** (19), 3-16.

Brünnich-Nielsen, K. 1921. Nogle Bemaerkninger om de store Terebratler i Danmarks Kridt- og Danienaflejringer. *Ibid.* **6** (3), 3-18.

Brünnich-Nielsen, K. 1943. The asteroids of the Senonian and Danian deposits of Denmark. *Biol. Skr. Dan. Vid. Selsk,* **2** (5).

Buday, T. and Cicha, I. 1956. Neue Ansichten über die

Stratigraphie unteren und mittleren Miozäns des Inneralpinen Wiener Beckens und des Waagtaales. *Geolog. prace*, **43**.

Buday, T., Cicha, I. and Senes, J. 1958. Les relations du Miocène inférieur de la Molasse de l'Autriche et de la Bavière, des Carpathes occidentales et du Bassin Intracarpathique. *Congrès des Sociétés savantes, Paris.*

Burckle, L. H., Ewing, M., Saito, T. and Leyden, R. 1967. Tertiary sediment from the East Pacific Rise. *Science*, **157**, 537-40.

Bürgl, H. 1965. The Oligo-Miocene boundary in the marine Tertiary of Colombia. *4th Caribbean Geol. Conf., Trinidad, 1965*, 1 p.

Burollet, P. F. and Magnier, Ph. 1960. Remarques sur la limite Crétacé-Tertiaire en Tunisie et en Libye. *21st Int. Geol. Congr., Copenhagen*, **5**, 136-44.

Burrows, H. and Holland, R. 1897. The Foraminifera of the Thanet Beds of Pegwell Bay. *Proc. Geol. Assoc.* **15**, 19-52, pls. 1-5.

Butt, A. A. 1966. *Late Oligocene Foraminifera from Escornébeou, SW France*, 123 pp., 8 pls. Thesis, Utrecht Univ. (Shotanus & Jens, Utrecht.)

Butterlin, J. 1965. Données fournies par les Macroforaminifères pour l'établissement des limites de l'Oligocène dans la région des Caraibes. *4th Caribbean Geol. Conf., Trinidad, 1965.* 23 pp.

Bykova, N. K. 1960. Danian and Palaeogene deposits of north Mangyshlak and south Emba area. In: Yanshin, A. L. and Menner, V. V. (Editors), The Cretaceous-Tertiary Boundary. *Intern. Geol. Cong., 21st, Copenhagen, 1960, Rept. Sov. Geologists*, **5**, 159-68.

Bykova, N. K. *et al.* 1959. Otryad Rotaliida. In: Rauzer-Chernousova, D. M., and Fursenko, A. V., *Osnovy paleontologii. Obhchaya chast', Prosteishie*, 265-307. Akad. Nauk U.S.S.R: Moscow.

Caralp, M., Valeton, M. and Vigneaux, M. 1965. Les foraminifères pélagiques du Tertiaire terminal sud-aquitain. *C.R. Acad. Sci. Paris*, **261**, 3,431-4.

Caralp, M., Julius, Ch., and Vigneaux, M. 1960. Considérations sur le Miocène inférieur aquitain. *Verhandl der Comité du Néogène Méditerranéen*, **I.** *Tagung in Wien 10-20 Juli 1959.*

Caralp, M. and Vigneaux, M. 1961. Nouvelle interprétation stratigraphique des 'étages' du Miocène inférieur en Aquitaine. *Compt. Rend. Soc. Géol. France*, **5**, 140-2.

Carter, A. N. 1958a. Tertiary Foraminifera from the Aire district, Victoria. *Bull. Geol. Surv. Vict.* **55**, 1-76, pls. 1-10.

Carter, A. N. 1958b. Pelagic Foraminifera in the Tertiary of Victoria. *Geol. Mag.* **95** (4), 297-304.

Carter, A. N. 1959. Guide Foraminifera of the Tertiary stages in Victoria. *Min. Geol. J. (Victoria)*, **6** (3), 48-54.

Carter, A. N. 1964. Tertiary Foraminifera from Gippsland, Victoria, and their stratigraphical significance. *Mem. Geol. Surv. Vict.* **23**, 1-154, pls. 1-16.

Cavelier, C. 1963a. L'Oligocène inférieur au Sud de la Butte Pincon (Seine), Présence du Calcaire de Sannois. *Bull. Soc. géol. France, Sér. 7*, **5**, 917-25.

Cavelier, C. 1963b. L'Eocène supérieur et l'Oligocène de la Butte de Villers-Cotterets (Aisne). *Ann. Soc. géol. Nord.* **83**, 203-13.

Cavelier, C. 1964a. L'Oligocène inférieur du bassin de Paris. In: Colloque sur le Paléogène, Bordeaux, 1962, *Mém. Bur. Rech. géol. et min. no.* **28** (1), 65-73.

Cavelier, C. 1964b. L'Oligocène marin des buttes de Cormeilles, Sannois, Argenteuil (Seine-et-Oise). *Ibid. no.* **28** (1), 75-123.

Cavelier, C. 1964c. Sur le classement des 'Upper Hamstead Beds' de l'île de Wight (Angleterre) dans le Stampien inférieur et leur parallélisme avec le bassin de Paris. *Ibid. no.* **28** (1), 585-90.

Cavelier, C. 1965. Le Sannoisien de Sannois (Seine-et-Oise) dans le cadre du bassin de Paris et sa signification stratigraphique. *Bull. Soc. géol. France, Sér. 7*, **7**, 228-38.

Cavelier, C. and Le Calvez, Y. 1964. Les marnes à Huîtres et le calcaire de Sannois du Mont-Valérien (Seine). *Mém. Bur. Rech. géol. et Min. no.* **28** (1), 125-43.

Cavelier, C. and Le Calvez, Y. 1965. Présence d'*Arenagula kerfornei* Allix, espèce 'biarritzienne' à la partie terminale du Lutétien supérieur de Foulangues (Oise). *Bull. Soc. géol. France, Sér. 7*, **7**, 284-6.

Cavelier, C. and Pomerol, Ch. 1962. Le Bartonien de Ronquerolles (Seine-et-Oise). *Ibid.* **4**, 170-81.

Cechovic, V. 1959. Quelques remarques sur la valeur stratigraphique de l'Aquitanien. *C.R. Sommaire des Séances de la Soc. Géol. de France*, 143.

Chang, L. S. 1953. Tertiary Cyclammina from Taiwan and their stratigraphic significance. *Bull. Geol. Surv. Taiwan*, **4**, 27-37.

Chang, L. S. 1959. A biostratigraphic study of the Miocene in western Taiwan based on smaller Foraminifera (Part I: Planktonics). *Proc. Geol. Soc. China*, **2**, 47-72.

Chang, L. S. 1960a. Tertiary biostratigraphy of Taiwan with special reference to smaller Foraminifera and its bearing on the Tertiary geohistory of Taiwan. *Proc. Geol. Soc. China*, **3**, 7-30.

Chang, L. S. 1960b. A biostratigraphic study of the Miocene in western Taiwan based on smaller Foraminifera (Part II: Benthonics). *Bull. Geol. Surv. Taiwan*, **12**, 67-91.

Chang, L. S. 1962a. A biostratigraphic study of the Oligocene in northern Taiwan based on smaller Foraminifera. *Proc. Geol. Soc. China*, **5**, 47-64.

Chang, L. S. 1962b. Some planktonic Foraminifera from the Suo and Urai Groups of Taiwan and their stratigraphic significance. *Proc. Geol. Soc. China*, **5**, 127-33.

Chang, L. S. 1962c. Tertiary planktonic foraminiferal zones of Taiwan and overseas correlation. *Mem. Geol. Soc. China*, **1**, 107-12.

Chang, L. S. 1963a. A biostratigraphic study of the so-called Hori Slate in central Taiwan based on smaller Foraminifera. *Proc. Geol. Soc. China*, **6**, 3-17.

Chang, L. S. 1963b. Mid-Tertiary planktonic foraminiferal zones of the Hengchun Peninsula, Taiwan. *Proc. Geol. Soc. China*, **6**, 61-6.

Chang, L. S. 1964. A biostratigraphic study of the Tertiary in the Hengchun Peninsula, Taiwan, based on smaller Foraminifera (I: Northern Part). *Proc. Geol. Soc. China*, **7**, 48-62.

Chang, L. S. 1965. A biostratigraphic study of the Tertiary in the Hengchun Peninsula, Taiwan, based on smaller Foraminifera (II: Middle Part). *Proc. Geol. Soc. China*, **8**, 9-18.

Chang, L. S. 1966. A biostratigraphic study of the Tertiary in the Hengchun Peninsula, Taiwan, based on smaller Foraminifera (III: Southern Part). *Proc. Geol. Soc. China,* **9**, 55-63.

Chang, L. S. 1967a. Tertiary biostratigraphy of Taiwan and its correlation. In: Tertiary correlations and climatic changes in the Pacific. *11th Pacific Sci. Congress, Tokyo,* 57-65.

Chang, L. S. 1967b. A biostratigraphic study of the Tertiary in the Coastal Range, Eastern Taiwan, based on smaller Foraminifera (I: Southern Part). *Proc. Geol. Soc. China,* **10**, 64-76.

Chang, L. S. and Yen, T. P. 1958. The Oligocene Tawuan foraminiferal faunule from southern Taiwan. *Proc. Geol. Soc. China,* **1**, 41-53.

Chang, Y. M. 1963. Biostratigraphic study of smaller Foraminifera from the Wu-chi section, Kuohsing, Nantou, Taiwan. *Petrol. Geol. Taiwan,* **2**, 183-207.

Choubert, G., Hottinger, L., Marcais, J. and Suter, G. 1964. Stratigraphie et micropaléontologie du Néogène au Maroc septentrional. *Inst. 'Lucas Mallada', C.S.I.C. (Espana), Cursillos y Conferencias,* **9**, 229-57.

Cicha, I. 1960a. Bericht über das Vorkommen des marinen Untermiozäns bei Lubietova, westlich vom Banska-Bystrica. *Vestnik UUG,* **35**, 361-3.

Cicha, I. 1960b. Neue stratigraphische Auswertung der Mikrofauna aus den sog. kattischen Schichten der Südslovakei in Beziehung zu den Ablagerungen der Paratethys. *Geol. prace,* **57**.

Cicha, I., Paulik, J. and Tejkal, J. 1957. Bemerkungen zur Stratigraphie des Miozäns des südwestlichen Teiles des ausserkarpatischen Becken in Mähren. *Sbornik UUG.* 23, *odd. paleont.*

Cicha, I. and Serreš, J. 1964. Zur Definition eines neuen Zeitabschnites – des sogenannten Karpatien – im Mittelmiozän. *Comm. Mediter. Neogene Strat. Proc. 3rd session,* 191-3.

Cicha, I. and Tejkal, J. 1959. Zum Problem des sog. Oberhelvets in den karpatischen Bekken. *Vestnik UUG,* **34** (2).

Cicha, I. and Zapletalova, I. 1960. Stratigraphische Verbreitung der planktonischen Foraminiferen im Miözan der karpatischen Becken. *Vestnik UGG,* **35** (5).

Cita, M. B. 1948. Ricerche stratigrafiche e micropaleontologiche sul Cratacico e sull'Eocene di Tignale (Lago di Garda). *Riv. Ital. Pal. Strat.* **54**, 49-74, 117-33, 143-68, pls. 2-4.

Cita, M. B. 1959. Stratigrafia micropaleontologica del Miocene siracusano. *Soc. Geol. Ital., Boll.* **77**, 1-97.

Cita, M. B. 1964. Considérations sur le Langhien des Langhe et sur la stratigraphie miocene du bassin tertiaire du Piemont, Instituto 'Lucas Mallada'. *C.S.I.C. (Espana), Cursillos y Conferencias, IX (1964),* 203-10.

Cita, M. B. and Bolli, H. 1961. Nuovi dati sull'eta paleocenica della Spilecciano di Spilecco. *Riv. Ital. Paleontol.* **67**, 369-92.

Cita, M. B. and Elter, 1960. La posizione stratigrafica delle marne a Pteropodi della Langhe della Collina di Torino ed il significato cronologico del Langhiano. *Accad. Naz. dei Lincei, ser. 8,* **29** (5), 360-9.

Cita, M. B. and Gelati, R. 1960. *Globoquadrina langhiana* n. sp. del Langhiano-tipo. *Riv. It. Pal. Strat.* **66**, 241-6.

Cita, M. B. and Piccoli, G. 1964. Les stratotypes du Paléogène d'Italie. *Colloque sur le Paléogène, Bur. Rech. géol. Min.* **28**, 653-84.

Cita, M. B. and Premoli Silva, I. 1960a. Pelagic Foraminifera from the type Langhian. *Int. Geol. Cong. Repts. XXI, Sess., Pt. XXII, Internat. Paleont. Union, Proc.* 39-50.

Cita, M. B. and Premoli Silva, I. 1960b. *Globigerina bolii* nov. sp. della Langhe. *Ist. Geol. Pal. Geograf. fis. Univ. Milano, ser. P,* **104**, 119-26.

Cita, M. B., Silva, I. P. and Rossi, P. 1965. Foraminifera planctonici del Tortoniano-tipo. *Riv. Ital. Paleontol.* **71**, 3-10.

Cocconi, G. 1873. Enumera zione sisternatica dei ncolluschi miocenici e plioceniu della provincie di Parma e Piacenza. *Mem. acc. Sci. Bodogna,* **3** (3).

Colalonga, M. L. and Sartoni, S. 1967. *Globorotalia hirsuta aemiliana* nuova sottospecie chronologica del Pliocene in Italia. *Giorn. Geol., ser. 2a,* **34** (1), 1-15, 2 pls.

Cole, W. S. 1957. Variations in American Oligocene species of *Lepidocyclina. Bull. Amer. Paleontology,* **38** (166), 31-51, pls. 1-5.

Cole, W. S. and Applin, E. R. 1961. Stratigraphic and geographic distribution of larger Foraminifera occurring in a well in Coffee County, Georgie. *Contr. Cushman Found. Foram. Res.* **12** (4), 127-35, pls. 6, 7.

Colom, G. 1946. Los sedimentos burdigalienses de las Baleares. *Estudios Geologicos,* **3**, 21-112.

Colom, G. 1954. Estudio de la biozonas con foraminiferos del Terciario de Alicante. *Inst. Geol. Min. Espana, Bol.* **66**, 1-279.

Colom, G. 1956. Los foraminiferos del Burdigaliense de Mallorca. *R. Acad. Cien. Artes Barcelona, Mem.* **32** (5), 92-230.

Colom, G. and Gamundi, J. 1951. Sobre la extension e importancia de las "Moronitas" a lo largo de las formaciones aquitano-burdigalienses del estrecho nortbetico. *Inst. Invest. Geol. Lucas Mallada, Publ., Rev. Estud. Geol.* **14**, 331-85.

Conato, V. 1953. Una microfauna pliocenica del subappenino romagnolo. *Boll. Serv. Geol. Ital.* **74**, 155-68.

Cornet, C. 1963. Sur l'existence d'une phase tectonique oligocène d'origine profonde en Provence. *Compt. Rend. Soc. Géol. France,* **2**, 59.

Cornet, J. 1900. Documents sur l'extension souterraine du Maestrichtien et du Montien dans la vallée de la Haine. *Bull. Soc. Belg. de Géol., de Paléontologie et d'Hydrologie,* **14**, 249-57.

Cornet, J. 1902. Note sur la présence du Calcaire de Mons, du Tuffeau de Saint-Symphorien et de la Craie phosphatée de Ciply au Sondage des Herbières (Commune de Tertre). *Ibid.* **16**, 39-42.

Cornet, J. 1903. Documents sur l'extension souterraine du Maestrichtien et du Montien dans la vallée de la Haine (deuxième note). *Ibid.* **17**, 184-8.

Cornet, J. 1906. Documents sur l'extension souterraine du Maestrichtien et du Montien dans la vallée de la Haine (troisième note). *Ibid.* **20**, 81-6.

Cornet, J. 1928. Les mouvements saxoniens dans le Hainaut. *Bull. Cl. Sci. Acad. Roy. de Belg., sér. 5,* **14**, 109-26.

Cornet, F. L. and Briart, A. 1866a. Notice sur l'extension

du Calcaire grossier de Mons dans la vallée de la Haine. *Bull. Acad. Roy. de Belg.* 22 (12), 523-38, 1 pl.
Cornet, F. L. and Briart, A. 1866b. Descriptions minéralogique, paléontologique et géologique du terrain Cretacé de la Province de Hainaut. *Mém. Soc. des Sciences du Hainaut,* 1 (1865-6).
Cornet, F. L. and Briart, A. 1873. Compte rendu de l'excursion faite aux environs de Ciply par la Société malacologique de Belgique le 20 avril 1873. *Ann. Soc. Malac. Belg.* 20, 21-35.
Cornet, F. L. and Briart, A. 1874. Aperçu sur la géologie des environs de Mons. *Bull. Soc. Géol. France, sér. 3,* 2, 534-53.
Cornet, F. L. and Briart, A. 1877. Note sur l'existence d'un Calcaire d'eau douce dans le terrain tertiaire du Hainaut. *Bull. Acad. Roy. de Belgique,* 1 (9).
Cornet, F. L. and Briart, A. 1885. Sur l'âge du Tuffeau de Ciply (Réponse a MM. Rutot et van den Broeck). *Bull. des Séances de la Soc. Roy. Malac. de Belgique,* 20 (séance du 7 novembre 1885), 100-8.
Cossmann, M. 1908. Pélécypodes du Montien de Belgique. *Mém. Mus. Roy. d'Hist. Natur. de Belgique.* 5 (19).
Cossmann, M. 1913, 1914. Scaphopodes, Gastropodes et Céphalopdes du Montien de Belgique, Pt. 1. *Ibid.* 6 (24); Pt. 2. *Ibid.* 8 (34).
Csepreghy-Meznerics, I. 1953. La faune et l'âge des couches du mur des gisements de charbon à Salgotarjan. *Földtani Közlöny,* 83, 35-36. Budapest.
Csepreghy-Meznerics. I. 1959. Die Burdigalfauna in den Liegendschichten des Braunkohlenflözes von Egercsehi. *Ozd. Ann. hist. nat. Mus. Nat. hungarici,* 51.
Csepreghy-Meznerics, I. 1956. Stratigraphische Gliederung des Ungarischen Miozäns im Lichte der neuen Faunauntersuchungen. *Acta Geologica,* 4 (2). Budapest.
Csepreghy-Meznerics, I. 1960. Das marine Neogen Ungarns in seiner Beziehung zum Wiener Becken. *Verhandl. d. Comité du Néogène Mediterranéen, I. Tagung in Wien, 10-20 Juli.*
Csepreghy-Meznerics, I. 1961. Pectinidés néogènes de la Hongrie et leur importance biostratigraphique. *Mém. Soc. géol. de France (Nouvelle Série), Mém.* 92.
Csepreghy-Meznerics, I. 1962. Das problem des 'chatt' – Aquitans in Missenschaftgeschichtlicher Beleuchtung. *Ann. Mist-Natur. Musei nation. Hungarici,* 54, 57-71.
Csepreghy-Meznerics, I. 1964a. L'analyse de la faune de Peyrère (basin de l'Adour) et de l'Aquitanien du Bordelais et du Bazadais. *Mém. Bur. Rech. Géol. Min. no.* 28, 455-66.
Csepreghy-Meznerics, I. 1964b. Le problème du 'Chattien' – Aquitanien du point de vue de l'histoire de la subdivision du Miocène. *Ibid.* 893-907.
Csepreghy-Meznerics, I. and Senĕs, J. 1957. Neue Ergebnisse der stratigraphischen Untersuchungen miozäner Schichten in der Sudslovakei und Nordungarn. *Neues Jhb. Geol. Palaont. Mh.* 1-13.
Curry, D. 1961. Sur la découverte de *Nummulites variolarius* dans le Lutétien des bassins de Paris et du Hampshire. *C.R. somm. Soc. géol. France,* 247-8.
Curry, D. 1962. A lower Tertiary outlier in the central English Channel, with notes on the beds surrounding it. *Quart. J. Geol. Soc. London,* 118, 177-205.
Curry, D. 1966. Problems of correlation in the Anglo-Paris-Belgian Basin. *Proc. Geol. Assoc.* 77 (4), 437-67.
Curry, D., Hersey, J. B., Martini, E. and Whittard, W. F. 1965. The geology of the western approaches of the English Channel, Geological interpretation aided by boomer and sparker records. *Phil. Trans. roy. Soc. London, ser. B,* 248 (749).
Cushman, J. A. and Bermúdez, P. 1949. Some Cuban species of *Globorotalia. Contr. Cush. Lab. Foram. Res.* 25 (2), 26-45, pls. 5-8.
Cushman, J. A. and Jarvis, P. W. 1928. Cretaceous Foraminifera from Trinidad. *Ibid.* 4, 84-103, pls. 12-14.
Cushman, J. A. and Jarvis, P. W. 1931. Some new Eocene Foraminifera from Jamaica. *Ibid.* 7, 75-8, pl. 10.
Cushman, J. A. and Renz, H. H. 1942. Eocene Midway, Foraminifera from Soldado Rock, Trinidad. *Contr. Cushman Lab. Foramin. Res.* 18, 1-20, pls. 1-3.
Cushman, J. A. and Renz, H. H. 1946. The foraminiferal fauna of the Lizard Springs Formation of Trinidad B.W.I. *Cushman Lab. Foram. Res. Spec. Publ.,* 18, 1-48, pls. 1-8.
Cuvillier, J., Dalbiez, F., Glintzboeckel, C., Lys, M., Magné, J., Perebaskine, V. and Rey, M. 1955. Études micropaléontologiques de la limite Crétacé-Tertiaire dans les mers mesogéennes. *Proc. IV World Petrol. Congr., sect. I/D,* 517-44, 2 text-pls.
Dam, A. Ten 1944. Die stratigraphische Gliederung der niederländischen Palaozäns und Eozäns nach Foraminiferen. *Med. Geol. Sticht. C,* 5 (3), 1-142.
Dam, A. Ten, 1947. (Rec.) R. C. van Bellen: Foraminifera from the Middle Eocene in the southern part of the Netherlands Province of Limburg. *J. Paleont.* 21 (2), 187a.
Dam, A. Ten and Reinhold, Th. 1941. Die stratigraphische Gliederung des niederländischen Oligo-Miozäns nach Foraminiferen (mit Ausnahme Von S. Limburg). *Med. Geol. Sticht.* 5 (1), 1-106.
Dam, A. Ten and Sigal, J. 1950. Some new species of Foraminifera from the Dano-Montian of Algeria. *Contr. Cush. Found. Foram. Res.* 1, 31-7.
Damotte, R. 1964. Contribution à l'étude des 'calcaires montiens' du bassin de Paris: la faune d'Ostracodes. *C.R. somm. Soc. géol. France,* 208.
Damotte, R. and Feugueur, L. 1963. L'âge du calcaire de Vigny (S.-et-O.) à partir de données paléontologiques nouvelles. *C.R. Ac. Sc.* 256, 3,864-6.
Dehm, R. 1949. Zur Oligozän-Miozän-Grenze. *N. Jb. Mineral., Geol. u. Palaont., Monatsh., Jg. 1949, Abt. B, H. 4-6,* 141-6.
Demarcq, G. 1958. Observation sur le Burdigalien du Bassin de Valréas (Drôme-Vaucluse). *Bull. Serv. Cart. Géol. France, C.R. Coll. Camp.* 56 (258).
Demarcq, G. 1959a. Contributions à l'étude des faciés du Miocène de la Vallée du Rhône. *Conférences Internat. sur le Néogène Méditerranéen, Vienne (Autriche) 10-18 Juillet.*
Demarcq, G. 1959b. Le Miocène du bassin de Crest (Drôme). *Bull. Soc. Géol. France, Sér. 7,* 1.
Demarcq, G. 1960. Contribution à l'étude des facies du Miocène de la Vallée du Rhône. *Verhandl. d. Comité Néogène Med., Wien.*
Demarcq, G. 1964. Etude critique générale sur la valeur des étages du Paléogène, In: Colloque sur le Paléogène,

Bordeaux, 1962. *Mém. Bur. Rech. géol. et min. no.* 28 (1), 7-13.
Denizot, M. G. 1929. Les horizons continentaux du Stampien et de l'Aquitanien. *Bull. Soc. Géol. France, Sér. 4,* **29**.
Denizot, M. G. 1952. Le classement des terrains tertiaires en Europe occidentale. (*Trav. Lab. géol. Montpellier,* 3)
Denizot, G. 1957. *Lexique stratigraphique International* 1 fasc. VII, Tertiaire: France, Belgique, Pays-Bas, Luxembourg. *Centre nat. rech. sci. Paris.*
Denizot, M. G. 1958a. L'étage Aquitanien et la limite Oligocène-Miocène. *Compt. rend. Congr. d. Soc. Sav. Colloque sur le Miocène.* Paris.
Denizot, G. 1958b. Sur la convenance de rétablir le Tortonien dans son acception première comportant équivalence de tout le Miocène supérieur. *Ibid.* 139-41.
Denizot, G. 1964. Commission du Calcaire pisolithique, *Bull. Inf. Géologues Bass. Paris.* 2, 249-50.
Depéret, Ch. 1892. Note sur la classification et le parallélisme du système Miocène. *C.R.Somm. Séance Soc. Géol. France,* **13**, 165-8.
Depéret, Ch. 1893. Note sur la classification et le parallélisme du système Miocène. *Bull. Soc. Géol. France, Sér. 3,* **21**, 170-266, 1 pl.
Depéret, M. 1895. Observations à propos de la note sur la nomenclature des terrains sédimentaires, par MM. Munier-Chalmas et de Lapparent. *C.R. Soc. Géol. France, Sér. 3,* **23**, 33-6.
De Porta, J. 1962. Consideraciones sobre el estado actual de la estratigrafia del Terciario en Colombia. *Bol. Geol. Univ. Ind. Santander,* **9**, 5-43.
Deroo, G. 1959. Réparition stratigraphique de quelques Ostracodes des 'craies tuffeaux' des tranchées du canal Albert (Belgique). *Ann. Soc. Géol. de Belg.* 5 (82), 283-92.
Deroo, G. 1966. Cytheracea (Ostracodes) du Maastrichtien de Maastricht (Pays-Bas) et des régions voisines; résultats stratigraphiques et paléontologiques de leur étude. *Meded. Geol. Sticht., Ser. C,* 2 (2), 197 p., pls. 1-27.
de Rouville, P. G. 1853. *Description géologique des environs de Montpellier.* Boehm (Montpellier), 185.
Desor, E. 1847. Sur le terrain Danien, nouvel étage de la craie. *Bull. Soc. Géol. France, sér. 2,* **4**, 179-81.
Dewalque, G. 1868. *Prodrome d'une déscription géologique de la Belgique, Brussels and Liege.*
Dieci, G. 1959. I Foraminiferi tortoniani di Montegibbio e Castelvetro (App. Modenese). *Pal. Ital.* **54**,. 1-113.
di Napoli, Alliata, E. 1951. Considerazione sulla microfaune del Miocene superiore italiano. *Riv. Ital. Pal. Strat.* **57**, 91-119.
di Napoli, Alliata, E. 1952. Foraminifera pelagici e facies in Italia. *Atti 7° Conv. Naz. Met. Petrolio Taormina,* **1**, 231-54.
di Napoli, Alliata, E. 1953. Microfauna della parte superiore della serie oligocenica del Monte San Vito e del Rio Mazzapiedi-Castellania (Tortona-Alessandria). In: Gino *et al.* (1953), 25-98.
Doderlein, P. 1872. *Note illustrative della carta geologica del Modenese e del Reggiano.* Memoria terza, Madena, 74 p.
Dollfus, G. F. 1880. Essai sur l'étendue des terrains tertiaires dans le bassin anglo-parisien. *Bull. Soc. géol. Normandie,* **6**, 584-605.
Dollfus, G. F. 1905. Critique de la classification de l'Eocene inférieur (lettre à M. Leriche). *Ann. Soc. géol. Nord,* **34**, 378-82.
Dollfus, G. F. 1909. Essai sur l'étage Aquitanien. *Bull. Serv. Carte Géol. de la France,* **19**.
Dollfus, G. F. 1918-1919. L'Oligocène supérieur marin dans le Bassin de l'Adour. *C.R. Sommaire et Bull. de la Soc. Géol. de France, Sér. 4,* **17**.
Dollfus, G. F. 1930. *Essai d'une histoire géologique de la Seine et de la Loire, Livre jubilaire 1830-1930* (Centaire de la Soc. Géol. de France). **1**. Paris.
Dondi, L. 1962. Nota paleontologico-stratigrafica sul pede-apennino padano. *Boll. Soc. Geol. Ital.* **81** (4), 113-229.
Dondi, L. and Papetti, I. 1966. Studio paleoecologico e stratigrafico sul passagio Pliocene-Quaternario nella bassa valle del Santorno (Emilia). *Riv. Ital. Paleont. e Strat.* 72, 231-44.
Douvillé, H. 1917. Sur l'âge des couches a Lépidocyclines. *Mém. Soc. Géol. France, n.s. 2,* **1**, 1-115.
Douvillé, H. 1924-1925. Revision des Lépidocyclines. *Mém. Soc. Géol. France, n.s. 2,* **1**, 1-49; **2**, 51-115.
Douvillé, M. H. 1899. Sur les couches à Orbitoides du Bassin de l'Adour. *Bull. de la Soc. Géol. de France,* **27**.
Douvillé, R. 1970a. Sur la variation chez les Foraminifères du genre *Lepidocyclina. Bull. Soc. Géol. France, Sér. 4,* **7**, 51-7.
Douvillé, R. 1907b. Sur les Lépidocyclines nouvelles. *Bull. Soc. Géol. France, Sér. 4,* **7**, 307-13.
Douvillé, R. 1908. Observations sur les faunes à Foraminifères du sommet du Nummulitique Italien. *Bull. Soc. Géol. France, Sér. 4,* **8**, 88-95.
Drooger, C. W. 1954a. *Miogypsina* in northern Italy. *K. Nederl. Akad. Wetensch., Proc., ser. B,* **57** (2), 227-49, pls. 1-2.
Drooger, C. W. 1954b. The Oligocene-Miocene boundary on both sides of the Atlantic. *Geological Magazine,* **111** (6).
Drooger, C. W. 1956. Transatlantic correlation of the Oligo-Miocene by means of Foraminifera. *Micropaleontology,* 2 (2), 183-92, 1 pl.
Drooger, C. W. 1960a. *Miogypsina* in northwestern Germany. *K. Nederl. Akad. Wetensch., Proc., ser. B,* **63** (1), 38-50, 2 pls.
Drooger, C. W. 1960b. Some early rotaliid Foraminifera I. *Ibid.* **63** (3), 287-334, 5 pls.
Drooger, C. W. 1960c. Microfauna and age of the Basses Plaines Formation of French Guyana, I. *Ibid.* **63** (4), 449-68, 4 pls.
Drooger, C. W. 1960d. Die biostratigraphischen Grundlagen der Gliederung des marinen Neogens an den Typlokalitaten. *Mitt. geol. Ges. Wien,* **52**, 105-14.
Drooger, C.W. 1961. *Miogypsina* in Hungary. *Ibid.* **64** (3), 417-27.
Drooger, C. W. 1963. Evolutionary trends in the Miogypsinidae. In: *Evolutionary trends in Foraminifera,* 317-49. Elsevier Publ. Co.: Amsterdam.
Drooger, C. W. 1964a. Les microfaunes éocènes-oligocènes du bassin nordique. In: Colloque sur le Paléogène,

Bordeaux, 1962. *Mém. Bur. Rech. géol. et min. no.* **28** (2), 547-52.

Drooger, C. W. 1964b. Problems of mid-Tertiary stratigraphic interpretation. *Micropaleontology*, **10** (3), 369-74.

Drooger, C. W. 1964c. Les biozones des foraminifères du Miocène. *Estud. Geol., Inst. Invest. Geol. 'Lucas Mallada' (Madrid)*, **9**, 209-10.

Drooger, C. W. 1966. Miogypsinidae of Europe and North Africa. *Comm. Med. Neogene. Strat. Proc. 3rd session (Berne, 1964)*, 51-4.

Drooger, C. W. and Batjes, D. A. J. 1959. Planktonic Foraminifera in the Oligocene and Miocene of the North Sea Basin. *K. Nederl. Akad. Wetensch., Proc. ser. B*, **62** (3), 172-86.

Drooger, C. W. and Felix, R. 1961. Some variations in foraminiferal assemblages from the Miocene of the North Sea Basin. *K. Nederl. Wetensch. Akad., Proc. ser. B*, **64** (2), 316-24.

Drooger, C. W., Kaasschieter, J. P. H. and Key, A. J. 1955. The microfauna of the Aquitanian-Burdigalian of southwestern France. *K. Nederl. Akad. Wetensch., Afd. Natvurk., Verh. ser. 1*, **21** (2), 1-136, pls. 1-20.

Drooger, C. W. and Magné, J. 1959. Miogypsinids and planktonics of the Algerian Oligocene-Miocene. *Micropaleontology*, **5** (3), 273-84, pls. 1, 2.

Drooger, C. W., Papp, A. and Socin, C. 1957. Über die Grenze zwischen den stufen Helvet und Torton. *Österr. Akad. Wiss., Anz., Math.-Naturw. Kl.* **94**, 1-10.

Dumont, A. 1832. Mémoire sur la constitution géologique de la province de Liège. *Mém. couronnés en 1829 et 1830 par l'Acad. Roy. des Sciences et Belles-Lettres de Bruxelles.*

Dumont. A. 1839. Rapport sur les travaux de la carte géologique en 1839, avec une carte géologique des environs de Bruxelles. *Bull. Acad. R. Belg.* **6** (2), 464-85.

Dumont, A. 1849. Rapport sur la carte géologique du Royaume. *Bull. Acad. R. de Belge.* **16** (2), 351-73.

Dumont, A. 1851. Note sur la position géologique de l'argile rupélienne et sur le synchronisme des formations tertiaires de la Belgique, de l'Angleterre et du nord de la France. *Bull. Acad. R. Sci. Belgique*, **18** (2), 179-95.

Durham, J. W. 1944. The type section of the Aquitanian. *Amer. Journ. Sci.* **242** (5), 246-250.

Durham, J.W., Jahns, R. H., and Savage, D. E. 1954. Marine – non-marine relationships in the Cenozoic section of California. *California Dept. Nat. Resources Div. Minec. Bull.* **170**, 59-71.

Dymond, J. R. 1966. *Potassium-argon geochronology of deep-sea sedimentary materials.* Univ. California (San Diego), Unpubl. Ph.D. dissert., 58 pp.

Dymond, J. R., Lidz, L. and Bonatti, E. 1967. Absolute age, stratigraphic correlation and mineralogy of ash layers in Tertiary sediments from Atlantic off Florida. *Bull. Amer. Assoc. Petrol. Geol.* **51** (3), 462.

Eames, F. E. 1953. The Miocene/Oligocene boundary and the use of the term Aquitanian. *Geol. Mag.* **110** (6), 388-92.

Eames, F. E., Banner, F. T., Blow, W. H. and Clarke W. J. 1962. *Fundamentals of Mid-Tertiary Stratigraphical Correlation.* 163 pp. Cambridge Univ. Press.

Eckert, H. R. 1963. Die obereozänen Globigerinien-Schiefer (Stad und Schimbergschiefer) zwischen Pilatus und Schrattenfluh. *Eclogae Geol. Helv.* **56**, 1,001-72.

Edwards, A. R. 1966. Calcareous nannoplankton from the uppermost Cretaceous and lowermost Tertiary of the mid-Waipara section, South Island, New Zealand, N. 2, *Jour. Geol. Geophys.* **9** (4), 481-90, 3 pls.

El-Naggar, Z. R. 1966a. Stratigraphy and planktonic Foraminifera of the Upper Cretaceous-Lower Tertiary succession in the Esna-Idfu region, Nile Valley,Egypt, U.A.R. *Brit. Mus. Nat. History Bull. Suppl.* **2**, 1-279, pls. 1-23.

El-Naggar, Z. R. 1966b. Stratigraphy and classification of type Esna Group of Egypt. *Amer. Assoc. Petrol. Geol.* **50** (7), 1,455-77.

Emiliani, C., Mayeda, T. and Selli, R. 1961. Paleotemperature analysis of the Plio-Pleistocene section of the Castella, Calabria, Southern Italy. *Geol. Soc. Amer. Bull.* **72**, 679-88.

Evernden, J. F. and Curtis, G. H. 1961. Present state of potassium-argon dating of Tertiary and Quaternary rocks. INQUA Congr. Warsaw, *Trev. Serv. geol. Pologue.*

Evernden, J. F. and Curtis, G. H. 1965. The potassium-argon dating of Late Cenozoic rocks in East Africa and Italy. **6** (4), 343-85.

Evernden, J. F., Savage, D. E., Curtis, G. H. and James, G. T. 1964. Potassium-argon dates and the Cenozoic Mammalian chronology of North America. *Amer. Journ. Sci.* **262**, 145-98.

Faber, J. 1961. Paléogéographie et sédimentologie du Danien et du Paléocène de la région de Pau. *Rev. Inst. Franc. Pétrole Ann. Combust. Liquides*, **16**, 907-22.

Fabiani, R. 1912. Nuove osservazioni sul Terziario fra il Brenta e l'Astico. *Atti. Acad. Sci. Veneto-Trentino-Istriana*, **1**, 1-36.

Fallot, E. 1893. Sur la classification de Néogène inférieur. *Bull. Soc. Géol. France, Sér. 3*, **21**.

Farchad, H. 1936. Etude du Thanetien (Landenien marin) du bassin de Paris. *Mém. Soc. géol. France, nouv. sér.* **13** (30), 103 pp.

Feugueur, L. 1951. Sur l'Yprésien des bassins français et belge et l'âge des Sables d'Aeltre. *Bull. Soc. belge, Géol. Hydr., Pal.* **60** (2), 216-42, 3 pls.

Feugueur, L. 1955. Essai de synchronisation entre les assises saumâtres du Thanétien-Landénien (Gand-Ostende) et du Sparnacien (Ile-de-France). *Ibid.* **64** (1), 67-92.

Feugueur, L. 1961. Présence d'une faune marine dans les assises infragypseuses (Ludien) de la butte de l'Hautil (S.-et-O.). *Bull. Soc. géol. France, Sér. 7*, **3**, 514-17.

Feugueur, L. 1962. Définition et valeur stratigraphique des termes Yprésien et Landénien. *C.R. Ac. Sc.* **254**, 3,317-19.

Feugueur, L. 1963. L'Yprésien du bassin de Paris. Essai de monographie stratigraphique, Thèse, 1958. *Mém. Expl. Carte géol. dét.France.*

Feugueur, L. and Pomerol, Ch. 1962. L'Éocène du bassin de Paris. In: Compte rendu de la Session extraordinaire de la Société belge de Géologie, de Paléontologie et d'Hydrologie et de la Société géologique de Belgique dans l'Eocène du bassin de Paris, du 21 au 24 septembre

1962. *Bull. Soc. belge Géol.* **71** (3), 385-446.
Feuillée, P. 1955. *Les étages de l'Eocène parisien: problèmes de nomenclature stratigraphique.* Dipl. Et. sup., Paris.
Feuillée, P. 1964. Historique de l'utilisation du terme 'Bartonien' dans le bassin de Paris. In: Colloque sur le Paléogène, Bordeaux, 1962. *Mém. Bur. Rech. géol. et min. no.* **28** (1), 37-46.
Finlay, H. J. 1939a. New Zealand Foraminifera; Key species in stratigraphy No. I. *Trans. Roy. Soc. N.Z.* **68**, 504-43.
Finlay, H. J. 1939b. New Zealand Foraminifera: Key species in Stratigraphy No. 2. *Trans. Roy. Soc. N.Z.* **69** (I), 89-128.
Finlay, H. J. 1939c. New Zealand Foraminifera: Key species in Stratigraphy, No. 3. *Trans. Roy. Soc. N.Z.* **69** (3), 309-29.
Finlay, H. J. 1940. New Zealand Foraminifera: Key species in Stratigraphy, No. 4. *Trans. Roy. Soc. N.Z.* **69** (4), 448-72, pls. 62-7.
Fisher, W. L. 1961. Stratigraphic names in the Midway and Wilcox Groups of the Gulf Coastal Plain. *Coast Assoc. Geol. Soc. Trans.* **11**, 263-95.
Forchhammer, G. 1825. Om de geognostiske Forhold i en Deel af Sjelland og Naboeoerne. *Kong. Dan. Vid. Selsk. Skr.* **2**.
Forchhammer, G. 1835. *Denmarks geognostiske Forhold, Universitets-program.* Copenhagen.
Forchhammer, G. 1843. Über Geschiebebildungen in Dänemark und einem Theile von Schweden. *Poggendorf's Annalen,* **58**.
Forchhammer, G. 1847. Det nyere Kridt i Danmark. *Forh. Skand. Naturforsk, 5:te Mode.*
Forchhammer, G. 1858a. Den vestlige Deel af Liimfjordens Omgivelser. *Danmarks illustrered Almanak.*
Forchhammer, G. 1858b. Bidrag til Skildringen af Danmarks geographiske Forhold i deres Aghaengighed af Landets indre geognostiske Bygning, Indbydelsesskrift till Kjøbenhavns Universitets Fest i Anledning af H. M. Kongens Fødelsedag, Copenhagen.
Forchhammer, G. 1860. Om Lejringforholdene og Sammensaetningen af det nyere Kridt i Danmark, Beretning om det 8. skandinaviske Naturforskermøde, 1860, Copenhagen.
Forchhammer, G. 1863. Danmarks geognostiske Forhold, Naturforskerforsamlingen i Stockholm, 1863, Stockholm.
Francken, C. 1947. Bijdrage tot de Kennis van het Boven-Senoon in Zuid Limburg. *Meded. Geol. Sticht., ser. C,* **6** (5), 148 pp.
Franke, A. 1911. Die Foraminiferen des Unter-Eocantones der Ziegelei Schwarzenbeck. *Jb. preuss. geol. Landesanst, B.* **32**, 106-11.
Franke, A., 1927. Die Foraminiferen und Ostracoden des Paläocäns von Rugaard in Jutland und Sundkrogen bei Kopenhagen. *Danm. Geol. Unders.* **2** (46), 1-49.
Fuchs, Th. 1878. Studien über die Gliederung der jüngeren Tertiär-bildungen Oberitaliens. *Sitzungsber, d. Kais. Akad. d. Wiss., I Abth.* **77**.
Fuchs, Th. 1885. *Der Versuch einer Gliederung des unteren Neogen im Gebiete des Mittelmeeres.*
Fuchs, Th. 1893. Harmadkori kövületek Krapina és Radoboj környékének széntartalmu miocén-képzodményeibol és au ugynevezett 'akvitani emelet' geologiai helyzéteröl. *Foldt. Int. Evk.* **10** (5).
Fuchs, Th. 1892-94. Tertiärfossilien aus den Kohlenführenden Miozänablagerungen der Umgebung von Krapina und Radoboj und über die Stellung der sogenannten 'Aquitanischen' Stufe. *A.M. k. Földt. Int. Evkonyve,* **10** (5).
Funnell, B. M. 1964. The Tertiary Period. In: The Phanerozoic Time Scale: a Symposium. *Quart. J. Geol. Soc. London,* **120** suppl., 179-191.
Furon, P. and Soyer, R. 1947. Catalogue des fossiles tertiares du bassin de Paris. *Guides techniq. natur.* **6**.
Furst, M. 1964. Die Oberkreide – Paleozän Transgression in Östlichen Fezzan. *Geol. Rundschau,* **54**, 1,060-88.
Gage, M. 1966. Geological divisions of time, *N.Z. Jour. Geol. Geophys.* **9**, 399-407.
Gardner, J. 1926. On Scott's new correlation of the Texas Midway, *Amer. Jour. Sci.* **5** (12), 453-5.
Gardner, J. 1931. Relation of certain foreign faunas to Midway fauna of Texas. *Amer. Assoc. Petr. Geol., Bull.* **15**, 149-60.
Gardner, J. S. 1884. On the relative ages of the American and English Cretaceous and Eocene series. *Geol. Mag. N.S., Dec. 3,* **1**.
Gartner, S. Jr. and Hay, W. W. 1962. Planktonic foraminifera from the type Ilerdian, *Eclogae Geol. Helv.,* vol. 55, no. 2, pp. 553-572, 2 pls., Basel.
Gianotti, A. 1953. Microfaune della serie tortoniana del Rio Mazzapiedi-Castellània (Tortona-Alessandria). *Rev. Ital. Pal. Strat., Mem.* **6**, 167-302.
Gignoux, M. 1910. Sur la classification du Pliocene et du Quaternaire de l'Italie du Sud. *C.R. Acad. Sci. Paris,* **150**, 841-4.
Gignoux, M. 1913. Les formations marines pliocènes et quaternaires de l'Italie du Sud et de la Sicile. *Lyon. Univ., Ann., n.sér.* **36**, 1-690.
Gignoux, M. 1914. L'étage Calabrien (Pliocène supérieur marin) sur le versant nord-est de l'Apennin, entre le Monte Gargano et Plaisance. *Bull. Soc. géol. France, sér. 4,* **4** (6), 324-48.
Gignoux, M. 1936. *Géologie stratigraphique.* 2 ed., 709 pp. Paris.
Gignoux, M. 1950. *Géologie stratigraphique,* 4 ed., 735 pp. Paris.
Gignoux, M. 1954. Pliocène et Quaternaire marines de la.Méditerranée occidentale. *Congr. Géol. Intern. d'Alger (1952), section 13, Questions diverses de Geologie générale, 3 partie, n. 15,* 249-58.
Gignoux, M. 1955. *Stratigraphic Geology.* 682 pp., Freeman and Co.: San Francisco.
Gino, G. F. *et al.* 1953. Studi stratigrafice e micropaleontologiche sull'Apennino Tortonese. In: Osservazioni geologiche sui Dintorni di Sant' Agata Fossili (Tortona-Alessandria). *Riv. Ital. Pal. Strat., Memoria VI: Milano, 1953.* 7-24.
Ginsburg, L., Montenat, C. and Pomerol, Ch. 1965. Découverte d'une faune de Mammifères dans l'Auversien (Bartonien inférieur) au Guépelle (Seine-et-Oise). *C.R. Ac. Sc.* **260**, 3,445-6.
Girard d'Albissin, M. 1955. Étude du Sannoisien de l'Ile-de-France. *Ann. Centre et Doc. paléont. fr.,* **11**.

Glaessner, M. F. 1934. Stratigraphie des unteren Paleogens des nordlichen und ostlichen Kaukasus im Lichte des Studiums der Mikrofauna. *Inf. Sb. N.G.R.I.* **4**, 110-29.

Glaessner, M. F. 1936. Die Foraminiferengattungen *Pseudotextularia* und *Amphimorphina. Probl. Paleontology, Moscow Univ. Lab. Paleontology,* **1**, 116-35, pls. 1, 2.

Glaessner, M. F. 1937a. Planktonforaminiferen aus der Kreide und dem Eozän und ihre stratigraphische Bedeutung, Studies in Micropaleontology. *Moscow Univ., Lab. Paleontology,* **1** (1), 27-46. (German), 46-52 (Russian), pls. 1, 2.

Glaessner, M. F. 1937b. Studien uber Foraminiferen aus der Kreide und dem Tertiar des Kaukasus, I-Die Foraminiferen der altesten Tertiarschichten des Nordwestkaukasus. *Probl. Paleontology, Moscow Univ. Lab. Paleontology,* **2-3**, 349-410, pls. 1-5.

Glaessner, M. F. 1948. Problems of stratigraphic correlation in the Indo-Pacific region. *Proc. Roy. Soc. Vict.* **55**, 41-80.

Glaessner, M. F. 1953. Time-stratigraphy and the Miocene Epoch. *Bull. Geol. Soc. Amer.* **64**, 647-58.

Glaessner, M. F. 1959a. Die Indo-Pazifische region, In: Papp, A., *Handb. Stratigraph. Geol.* **3** (1), 288-310.

Glaessner, M. F. 1959b. Tertiary stratigraphic correlation in the Indo-Pacific region and Australia. *J. Geol. Soc. India,* **1**, 53-67.

Glaessner, M. F. 1960. West-Pacific stratigraphic correlation. *Nature, Lond.* **186** (4,730–, 1,039-40.

Glaessner, M. F. 1967. Time scales and Tertiary correlations. In: Tertiary correlations and climatic changes in the Pacific. *11th Pacific Science Congress, Tokyo, 1966, Sympos. no. 25,* 1-5.

Glibert, M. 1933. Faune malacologique du Bruxellian du bassin de Bruxelles. *Mém. Mus. roy. Hist. nat. Belgique,* **53**, 214 pp., 11 pls.

Glibert, M. 1938. Faune malacologique du Wemmelien. *Ibid.* **85**.

Glibert, M. 1957. Pélécypodes et Gastropodes du Rupélien supérieur et du Chattien de la Belgique. *Mém. Inst. r. Sc. Natur. Belg.* **137**, 97 pp. 6 pts.

Glibert, M. and De Heinzelin, J. 1954. L'Oligocène inférieur belge. In: Volume jubilaire Van Straelen, **1**. *Bruxelles, Inst. r. Sc. nat.*

Gohrbandt, K. 1963. Zur Gliederung des Paläogen im Helvetikum nordlich Salzburg nach planktonischen Foraminiferen; Erste Teil-Paleozän und tiefstes Untereozän mit Beiträgen von Adolf Papp (Grossforaminiferen) und Herbert Stradner (Nannofloren). *Geol. Ges. Wien, Mitt.,* **56** (1), 116 pp., 11 pls.

Gohrbandt, K. 1966. Upper Cretaceous and Lower Tertiary stratigraphy along the western and southwestern edge of the Sirte Basin, Libya. *Petrol. Expl. Soc. Libya, 8th Ann. Field Conf.* 33-41.

Gohrbandt, K. H. A. 1967. Some new planktonic foraminiferal species from the Austrian Eocene. *Micropaleontology,* **13** (3), 319-26, 1 pl.

Gordon, W. A. 1961. Planktonic Foraminifera and correlation of the middle Tertiary rocks of Puerto Rico. *Micropaleontology,* **7**, 451-60.

Gordon, W. A. 1963. Middle Tertiary echinoids of Puerto Rico. *J. Paleont.* **37**, 628-42.

Görges, J. 1941. Die Oberoligocänfauna von Rumeln am Niederrhein. *Decheniana* **100**, 115-86, 3 pls.

Görges, J. 1951. Die oberoligozänen Pectiniden des Doberges bei Bunde und ihre stratigraphische Bedeutung. *Palaeont. Z.* **24**, 9-22, 3 pls.

Görges, J. 1952a. Die Lamellibranchiaten und Gastropoden des Oberoligozänen Meeressandes von Kassel. *Abh. Hess. L.-Amt. Bodenforsch.* **4**, 1-134, 3 pls.

Görges, J. 1952b. Neue Invertebraten aus dem norddeutschen Oberoligozän. *Palaeont. Z.* **26**, 1-9, 2 pls.

Görges, J. 1957. Die Mollusken der oberoligozänen Schichten des Doberges bei Bunde in Westfalen. *Palaeont. Z. 31,* 116-34. 2 pls.

Görges, J. and Penndorf, H. 1952. Das niederhessische Tertiär und seine marinen Ablagerungen. *Notizbl. Hess. L.-Amt Bodenforsch,* **6** (3), 138-46.

Gosselet, J. 1874. L'étage éocène inférieur dans le Nord de la France et en Belgique. *Bull. Soc. géol. France,* **3**, 598-617.

Grambast, L. 1962. Sur l'intérêt stratigraphique des Charophytes fossiles; exemples d'application au Tertiaire parisien. *C.R. somm. Soc. géol. France,* 207-9.

Grambast, L. 1964. Indications fournies par les Charophytes pour la stratigraphie du Paléogène. In: Colloque sur le Paléogène, Bordeaux 1962. *Mém. Bur. Rech. géol. et min., no.* **28** (2), 1,009-11.

Gravell, D. W. and Hanna, M. A. 1938. Subsurface Tertiary zones of correlation through Mississippi, Alabama and Florida. *Bull. Amer. Assoc. Petrol. Geol.* **22**, 984-1,013, 7 pls.

Gressly, A. 1853. Nouvelles données sur les faunes tertiaires d'Ajoie, avec les déterminations de M. Mayer, *Act. Soc. helv. Sci. nat.,* **38**. *Sess.*

Grichuk, V. P., Hey, R. W and Venzo, S. 1965. Report of the subcommission on the Plio-Pleistocene boundary, Rep. 6° *Int. Congr. Quat. Warsaw (1961),* **1**, 311-29.

Grimsdale, T. F. 1951. Correlation, age determination, and the Tertiary pelagic Foraminifera. *World Petrol. Congr., 3rd, Proc., Sec. 1,* 463-75.

Gromov, V. I., Vangengeym, E. A. and Nikiforova, K. V. 1963. Etapyrazvitiya Antropogenovoi fauni miskopitayuschchikh kak ostrazhenue etapov razvitiya zemlii. *Izv. Akad. Nauk S.S.S.R. Ser. Geol.* **1**, 46-64.

Grönwall, K. A. 1899a. Danmarks yngsta krit- och aldsta tertiaraflagringar. *For handl. vid 15de Skandinaviska Naturforskaremotet i Stockholm, 1898,* p. 223.

Grönwall, K. A. 1899b. Några anmärkningar om lagserien i Stevns Klint. *Geol. Fören. Förhandl.* **21** (4), 365-73.

Grönwall, K. A. 1899c. Smånotiser om Jyllands krita. *Medd. Dansk. Geol. Foren.* **1** (5), 65-72.

Grönwall, K. A. and Harder, P. 1907. Paleocaen ved Rugaard i Jydland. *Danmarks Geol. Unders.* **2** (18), 1-102, 2 pls.

Grossheide, K. and Trunko, L. 1965. Die Foraminiferen des Doberges bei Bunde und von Astrup mit Beitropen zur Geologie dieser Profile (Oligozän, N.W. Deutschland), *Beih. Géol. Jb.* **60**, 213 pp.

Grossouvre, A. De 1897. Sur la limite du crétacé et du tertiaire. *Bull. Soc. Géol. France, sér. 3,* **25**, 57-81.

Grossouvre, A. De, 1901/1903. Recherches sur la craie supérieure, Stratigraphie générale, Fasc. 1 (1901). *Mém.*

Carte Géol. France, 1,013 pp., 3 pls., Fasc. II (1903).
Grossouvre, A. De, 1908. Description des Ammonitidés du Crétacé supérieur du Limbourg Belge et Hollandais et du Hainaut. *Mém. du Musée Roy. d'Hist. Nat. de Belgi.* 7, 39 pp., 11 pls.
Guembel, K. W. 1888. *Grundzüge der Geologie.* 504 pp. Kassel.
Gulinck, M. 1964. Aperçu général sur les dépôts éocènes de la Belgique. *Bull. Soc. géol. France, Sér. 7,* **7** (2), 222-7.
Gulinck, M. 1965. Aperçu général sur les dépôts éocènes de la Belgique. *Bull. Soc. géol. France,* **7** (2), 221-7.
Gulinck, M. and Hacquaert, A. 1954. L'Eocène. In: Prodrôme d'une description géologique de la Belgique, 451-93. *Liège, Soc. géol. Belgique.*
Gurr, P. R. 1962. A new fish fauna from the Woolwich Bottom Bed (Sparnacian) of Herne Bay, Kent. *Proc. Geol. Assoc.* **73** (4), 419-47.
Hagn, H. 1955. Paläontologische Untersuchungen am Bohrgut der Bohrungen Ortenburg CF 1001, 1002 und 1003 in Niederbayern. *Z. deutsch. geol. Ges.* **105** (2), 1953, 324-59, pl. 10.
Hagn, H. 1960. Die Gliederung der bayrischen Miozän-Molasse mit Hilfe von Kleinforaminiferen. *Verhandl. d. Comité Néogène Med., Wien.*
Hagn, H. and Hölzl, O. 1952. Geologisch paläontologische Untersuchungen in der subalpinen Molasse der östl. Oberbayerns zwischen Prien und Sur mit Berücksichtigung des im S. anschliessenden Helvetikums. *Geologica Bavarica,* **10**, 208 pp., 8 pls.
Hagn, H. and Hölzl, O. 1954. Zur Grenzziehung Katt (Akvitan in der bayerischen Molasse). *Jahrb. f. Geol. u. Palaont.* **I**.
Hagn, H. and Lindberg, H. G. 1966. Reviziya *Globigerina (Subbotina) eocaena* Gümbel iz Eocena predgorii Bavarskikh Alp. *Vopr. Mikropal.* **10**, 342-58, 1 pl.
Harder, P. 1922. Om Graensen mellem Saltholmskalk og Lellinge Gronsand og nogle Bemaerkninger om Inddelningen of Denmarks aeldre Tertiaer Danm. *Geol. Unders. II Rk,* **38**, 79-108.
Hartono, H. M. S. 1964. Coiling direction of *Pulleniatina obliquiloculata trochospira* n. var. and *Globorotalia menardii. Indonesia, Geol. Survey, Bull.* **1** (1), 5-12.
Hatai, K. 1960. Japanese Miocene reconsidered. *Tohoku Univ., Sci. Rep., Ser. 2, (Geol.), Spec. Vol.* **4**, 127-53.
Hatai, K. 1962. Mizuho To. *Tohoku Univ., Sci. Rep., Ser. 2, (Geol.), Spec. Vol.* **5**, 329-48.
Haug, E. 1905. Observations diverses sur la classification de l'Eocène supérieur et moyen. *Bull Soc. géol, France, Sér.* 4, **5**, 659-84.
Haug, E. 1908/1911. *Traité de Géologie II, fasc. 2, Les Périodes géologiques,* 540-1,396. Paris.
Haug, E. 1911. *Traité de Géologie II, fasc. 3, Les Périodes géologiques,* 1397-2024. Paris.
Hay, W. W. 1960. The Cretaceous-Tertiary boundary in the Tampico Embayment. Mexico. *Internat. Geol. Congr., 21st, pt. 5,* 70-7.
Hay, W. W. 1962. Zonation of the Paleocene and Lower Eocene utilizing discoasterids. In: Colloque Paléogène. *Bordeaux, Mém. B.R.G.M., no.* 28, 885-9.
Hayasaka, S. 1962. Summary of the geology and paleontology of the Atsumi peninsula, Aichi Prefecture, Japan. *Tohoku Univ., Sci. Rep., Ser. 2 (Geol.), Spec. Vol.* **5**, 195-217.
Haynes, J. 1955. Pelagic Foraminifera in the Thanet beds and the use of Thanetian as a stage name. *Micropaleontology,* **1** (2), 189.
Haynes, J. 1956. Certain smaller British Paleocene Foraminifera, Part I., Nonionidae, Chilostomellidae, Epistominidae, Discorbidae, Amphistegenidae, Globigerinidae, Globorotaliidae, and Gumbelinidae. *Contr. Cushman Found. Foram. Research,* **7** (3), 79-101, pls. 16-18.
Haynes, J. 1958. Certain smaller British Paleocene Foraminifera, Part V., Distribution. *Contr. Cushman Found. Foram. Research,* **9** (4), 83-92.
Hébert, E. 1853. Note sur l'âge des sables blancs et des marnes à *Physa gigantea* de Rilly, avec coupes. *Bull. Soc. géol. France, Sér. 2,* **10**, 436.
Hedberg, H. D. 1948. Time-Stratigraphic classification of sedimentary rocks. *Bull. Geol. Soc. Amer.* **59**, 447-62.
Hedberg, H. D. 1958. Stratigraphic classification and terminology. *Bull. Amer. Assoc. Petrol. Geol.* **42** (8), 1,881-96.
Hedberg, H. D. (ed.) 1961. Stratigraphic classification and terminology. *Rept. 21st Session Int. Geol. Congr., Norden, pt. 25,* 7-38.
Hedberg, H. D. 1961. The stratigraphic panorama. *Bull. Geol. Soc. Amer.* **72**, 499-518.
Hedberg, H. D. 1962. Les zones stratigraphiques: remarques sur un article de P. Hupé (1960). *Bull. Trimestr. Serv. d'Inform. géol. B.R.G.M.* **14** (54), 6-11.
Hedberg, H. D. (ed.) 1964. Definition of geologic systems. *Rept. 22nd Sess. Int. Geol. Congr., India, pt. 18,* 5-26.
Hedberg, H. D. 1965a. Earth history and the record in the rocks. *Proc. Amer. Philos. Soc.* **109** (2), 99-104.
Hedberg, H. D. 1965b. Chronostratigraphy and biostratigraphy. *Geol. Mag.* **102** (5), 451-61.
Hedberg, H. D. 1965c. American Commission on Stratigraphic Nomenclature, Definition of geologic systems, by International Subcommission on stratigraphic terminology. *Bull. Amer. Assoc. Petrol. Geol.* **49** (10), 1,694-703.
Hedberg, H. D. 1966, Note 33. Application to American Commission on Stratigraphic Nomenclature for amendments to articles 29, 31, and 37 to provide for recognition of Erathem, Substage, and Chronozone as time-stratigraphic terms in the Code of Stratigraphic Nomenclature. *Bull. Amer. Assoc. Petrol. Geol.* **50** (3), 560-1.
Henson, F. R. S. 1938. Stratigraphical correlation by small Foraminifera in Palestine and adjoining countries. *Geol. Mag.* **75**.
Herb. R. und Schaub, H. 1963. Zur Nummulitenfauna des Mitteleozäns von Sorde-l'Abbaye (Landes, Frankreich). *Eclogae Geol. Helv.* **56**, 973-1,000.
Hillebrandt, A. von 1962a. Das Alttertiär im Becken von Reichenhall und Salzburg (Nordliche Kalkalpen). *Z.D. Geol. Ges.* **113**, p. 339-58, 7 pls.
Hillebrandt, A. von 1962b. Das Paleozän und seine Foraminiferenfauna im Becken von Reichenhall und Salzburg. *Bayer. Akad. Wiss., Math.-Nat. Kl., N.F.* **108**, 182 pp. 15 pls.
Hillebrandt, A. von 1962c. Das Alttertiar im Mont-

Perdu-Gebiet (Spanische Zentralpyrenaen). *Eclogae geol. Helv.* **55** (2), 295-315, 6 Taf.

Hillebrandt, A. von 1964. Zur Entwicklung der planktonischen Foraminiferen im Alttertiär und ihre stratigraphische Bedeutung. *Paläont. Z.* **38** (3/4), 189-206.

Hillebrandt, A. von 1965. Foraminiferen – Stratigraphie im Alttertiär von Zumaya (Provinz Guipuzcoa, N. W. – Spanien) und ein Vergleich mit anderen Tethys – Gebieten. *Bayer. Akad. Wiss., Math.-Nat. Kl., Abh., N.F.* **123**, 62pp.

Hinte, J. E. van 1963. Zur Stratigraphie und Mikropaläontologie der Oberkreide und des Eozäns des Krappfeldes (Karnten). *Geol. Bundesanstalt, Jahrb.* 8, 1-147, pls. 1-22.

Ho, C. S. 1961. Geological relationships and comparison between Taiwan and the Philippines. *Geol. Soc. China, Proc.* **4**, 3-31.

Hoffmann, F. 1825. Über die geognostischen Verhältnisse des linken Weserufers. *Poggendorf Ann. Phys. usw.* **1**, 1-12.

Hoffmann, F. 1830. *Übersicht der orographischen und geognostischen Verhältnisse vom nordwestlichen Deutschland.* 676 pp., 3 pls. Barth: Leipzig.

Hofker, J. 1955-1962. (A list of references can be found in the reference lists in Berggren (1962a), Deroo (1966) and El Naggar (1966).)

Hofmann, K. 1871. Die geologischen Verhältnisse des Ofen-Kovacsier Gebirges. *M. K. Földt. Int. Evkönyve,* **1** (2), Budapest.

Holmes, A. 1959. A revised geologic time-scale. *Trans. Edinb. Geol. Soc.* **17**, 183-216.

Hornibrook, N. de B. 1958a. New Zealand Upper Cretaceous and Tertiary foraminiferal zones and some overseas correlations. *Micropaleontology,* 4 (1), 25-38, 1 pl.

Hornibrook, N. de B. 1958b. New Zealand Foraminifera: Key species in stratigraphy – No. 6. *N.Z. Jour. Geol. Geophys.* **1** (4), 653-76.

Hornibrook, N. de B. 1961. Tertiary Foraminifera from Oamaru district (N.Z.), Part 1, Systematics and description. *N.Z. Geol. Surv. Pal. Bull.* **34** (1), 1-192, pls. 1-28.

Hornibrook, N. de B. 1962. The Cretaceous-Tertiary boundary in New Zealand. *N.Z. Jour. Geol. Geophys.* **5**, 295-303.

Hornibrook, N. de B. 1964. A record of *Globigerinatella insueta* Cushman and Stainforth, from New Zealand. *N.Z. Jour. Geol. Geophys.* 7 (4), 891-2.

Hornibrook, N. de B. 1965a. A preliminary statement on the types of the New Zealand Tertiary Foraminifera described in the reports of the Novara Expedition in 1865. *N.Z. Jour. Geol. Geophys.* 8 (3), 530-1.

Hornibrook, N. de B. 1965b. A viewpoint on stages and zones. *N.Z. Jour. Geol. Geophys.* 8 (4), 1,195-212.

Hornibrook, N. de B. 1966. The stratigraphy of Landers (or Boundary) Creek, Oamaru. *N.Z. Jour. Geol. Geophys.* **9** (4), 458-70.

Hornibrook, N. de B. 1967. New Zealand Tertiary microfossil zonation, correlation and climate. In: Tertiary correlations and climatic changes in the Pacific. *11th Pacific Science Congr., Tokyo (1966),* 29-39.

Hornibrook, N. de B. and Harrington, H. J. 1957. The status of the Wangaloan Stage. *N.Z. Jour. Sci. Techn. sect. B,* **38** (6), 655-70.

Hornibrook, N. de B. and Schofield, J. C. 1963. Stratigraphic relations in the Waitemata group of the Lower Waikato District. *N.Z. Jour. Geol. Geophys.* **6** (1), 38-51.

Hornibrook, N. de B. and Jenkins, D. G. 1965. *Candeina zeocenica* Hornibrook and Jenkins, a new species of Foraminifera from the New Zealand Eocene and Oligocene. *N.Z. Jour. Geol. Geophys.* 8, 839-842.

Hornstein, F. F. 1906. Neues vom Kasseler Tertiär. *Z. dt. Geol. Ges.* **58** (B), 114-18.

Horusitzky, F. 1926. Neue Daten zur Miozän-Stratigraphie der Umgebung von Budapest. *Földtani Közlöny,* **56**. Budapest.

Horusitzky, F. 1934. Remarques sur la question du Burdigalien des environs de Budapest. *Földtäni Közlöny,* **64**.

Horustizky, F. 1941. Erdgeschichtliche Gliederung und paleo-geographische Verbindungen des unteren Miozäns im Karpathenbecken. *Beszamolo a Földt. Int. Vitaülésëinek munkalatairol, 1940.* Függeléke: Budapest.

Hosius, A. 1895. Beitrag zur Kenntnis der Foraminiferenfauna des Oberoligocäns vom Doberg bei Bunde. *Jb. naturwiss. Ver. Osnabruck* **10**, 75-124, 159-84.

Hottinger, L. 1958. Geologie du Mont Cayla (Aude, Aquitaine orientale) *Eclogae Geol. Helv.* **51** (2), 437-51, 1 pl.

Hottinger, L. 1960a. Über paleocaene und eocaene Alveolinen. *Eclogae Geol..Helv.* **53** (1), 265-83, 21 pls.

Hottinger, L. 1960b. Recherches sur les Alvéolines paléocènes et éocènes. *Mém. Suisses de Pal.* **75/76**, 1-243, pls. 1-18.

Hottinger, L. Lehmann, R. and Schaub, H. 1964. Données actuelles sur la biostratigraphie du Nummulitique méditérranean. *Colloque sur le Paléogène, Mém. Bur. Rech. Géol. Min.* 28, 611-52.

Hottinger, L. and Schaub, H. 1960. Zur Stufenteilung des Paleocaens und des Eocaens. Einführung des Ilerdien und des Biarritzien. *Eclogae Geol. Helv.* **53**, 454-79.

Huang, T. 1960. The Foraminifera from the Liuchiuhsu off the southwestern coast of Taiwan. *Proc. Geol. Soc. China,* **3**, 59-66.

Huang, T. 1963. Planktonic Foraminifera from the Peikang PK-3 well in the Peikang shelf area, Yunlin. *Taiwan, Petrol, Geol. Taiwan,* 2, 153-81.

Huang, T. 1964. Smaller Foraminifera from the Sanhsienchi, Taitung, eastern Taiwan. *Proc. Geol. Soc. China,* 7, 63-72.

Huang, T. 1966. Planktonic Foraminifera from the Somachi Formation, Kikai-Jima, Kagoshima Prefecture, Japan. *Trans. Proc. Palaeont. Soc. Japan, n.s.* **62**, 217-233, pls. 27, 28.

Hubach, H. 1922. *Das Oberoligozän des Doberges bei Bunde in Westf.,* Diss. Berlin 1922 und 1957: *Ber. Naturhist. Ges. Hannover* **103**, 1-69, 4 pls.

Iaccarino, S. 1963. Il Pliocene inferiore del Rio Lombasino (S. Andrea Bagni-Parma). *Tiv. Ital. Paleont. Strat.* **69** (2), 261-84, 1 pl.

Iaccarino, S. 1967. Les Foraminifères du stratotype du Tabianien (Pliocène inférieur) de Tabiano Bagni (Parme). *Mem. Soc. Ital. Sc. Nat. Milano,* **15** (3), 165-80, 1 pl.

Iaccarino, S. and Papani, G. 1967. La transgressione del Pliocene inferiore ('Tabianiano') sul Tortoniano del Colle di Vigoleno (Piacenza). *Riv. Ital. Paleont.* **73** (2), 679-700, pl. 59.

Ikebe, N. 1954. Cenozoic biochronology of Japan. *Osaka City Univ., Polytech. Jour. Inst. Ser. G*, **1**, 73-86.

Indans, J. 1956. Zur mikropaläontologischen Gliederung des Oligozäns in der Bohrung Kuhlerhof bei Erkelenz. *N. Jb. Geol. Paläont. Mh.* 173-84.

Indans, J. 1958. Mikrofaunistische Korrelationen im marinen Tertiär der Niederrheinischen Bucht. *Fortschr. Geol. Rheinld. Westf. I*, 223-38, 8 pls.

INQUA (International Geological Congress Commission on the Quaternary), 1950. The Pliocene-Pleistocene boundary. *Intern. Geol. Congr. 18th, London, 1948, Rept. Session*, **9**, 6.

Janoschek, R. 1951. Das inneralpine Wiener Becken. In *Geologie von Österreich*, Verl. F. Dueticke, Vienna.

Janoschek, R. 1960. Überblick über den Afbau der Neogengebiete Österreichs. *Verh. Com. Néogène Méditerr., 1 tag. Wien (1959)*, 149-58.

Janoschek, R. 1964. Das Tertiär in Österreich. *Mitt. Geol. Ges. Wien*, **56**, 319-60.

Jekelius, E. 1943. Das Pliozän und die sarmatische Stufe im mittleren Donaubecken. *Annuarul Inst. Geol. al Romaniei*, **22**.

Jeletzky, J. 1958. Die jüngere Oberkreide (Oberconiac bis Maastricht) Südwestrusslands und ihr Vergleich mit der Nordwest- und Westeuropas. *Geol. Jb.* **1**, 157 pp.

Jeletzky, J. 1960. Youngest marine rocks in western interior of North America and the age of the *Triceratops*-beds; with remarks on comparable dinosaur-bearing beds outside North America. *21st Int. Geol. Congr. (Copenhagen)*, **5**, 25-40.

Jeletzky, J. 1962. The allegedly Danian dinosaur-bearing rocks of the globe and the problem of the Mesozoic-Cenozoic boundary. *J. Paleont.* **36**, 1,005-18, pl. 141.

Jenkins, D. G. 1958. Pelagic Foraminifera in the Tertiary of Victoria. *Geol. Mag.* **95**, 438-9.

Jenkins, D. G. 1960. Planktonic Foraminifera from the Lakes Entrance Oil Shaft, Victoria, Australia. *Micropaleontology*, **6** (4), 345-71, pls. 1-5.

Jenkins, D. G. 1963. The Eocene-Oligocene boundary in New Zealand. *N.Z. Jour. Geol. Geophys.* **6** (5), 707.

Jenkins, D. G. 1964a. Location of the Pliocene-Pleistocene boundary. *Contrib. Cushman Found. Foram. Res.* **15** (1), 25-7.

Jenkins, D. G. 1964b. Preliminary account of the type Aquitanian-Burdigalian planktonic Foraminifera. *Ibid.* **15** (1), 28-9.

Jenkins, D G. 1964c. A new planktonic foraminiferal subspecies from the Australasian lower Miocene. *Micropaleontology*, **10** (1), 72.

Jenkins, D. G. 1964d. Panama and Trinidad Oligocene rocks. *Jour. Paleont.* **38** (3), 606.

Jenkins, D. G. 1964e. New Zealand Mid-Tertiary stratigraphical correlation. *Nature, Lond.* **203** (4941). 180-2.

Jenkins, D. G. 1964f. A history of the holotype, ontogeny and dimorphism of *Globorotaloides turgida* (Finlay). *Contrib. Cushman Found. Foram. Res.* **15** (3), 117-21, pls. 7-8.

Jenkins, D. G. 1965a. The genus *Hantkenina* in New Zealand. *N.Z. Jour. Geol. Geophys.* **8** (3), p. 518-26.

Jenkins, D. G. 1965b. The origin of the species *Globigerinoides trilobus* (Reuss) in New Zealand. *Contrib. Cushman Found. Foram. Res.*, **16** (3), 116-19, pl. 17.

Jenkins, D. G. 1965c. Planktonic Foraminifera and Tertiary inter-continental correlations. *Micropaleontology*, **11** (3), 265-77.

Jenkins, D. G. 1965. A re-examination of *Globorotalia collactea* Finlay, 1939. *N.Z. Jour. Geol. Geophys.* **8**, 843-8.

Jenkins, D. G. 1965e. Planktonic foraminiferal zones and new taxa from the Danian to Lower Miocene of New Zealand. *N.Z. Jour. Geol., Geophys.* **8** (6), 1,088-126.

Jenkins, D. G. 1966a. Planktonic Foraminifera from the type Aquitanian-Burdigalian of France. *Contrib. Cushman Found. Foram. Res.* **17** (1), 1-15, 3 pls.

Jenkins, D. G. 1966b. Standard Cenozoic stratigraphic zonal scheme. *Nature*, **211** (5045), 178.

Jenkins, D. G. 1966c. Planktonic foraminiferal datum planes in the Pacific and Trinidad Tertiary. *N.Z. Jour. Geol. Geophys.*, **9** (4), 424-7.

Jodot, P. 1945. Les Mollusques du Ludien continental de la montagne de Reims (Marne). *C.R. somm. Soc. géol. France*, 99-101.

Jodot, P. 1949. Role d'*Helix ramondi* dans la classification du Stampien du bassin de Paris et de la Limagne *Ibid.* 65-7.

Jodot, P. and Nicolesco, C. P. 1933. Découverte d'un bloc de calcaire lacustre ludien dans le port du Havre (S.-I.). *Ibid.* 152-4.

Jordan, R. R. 1962. Planktonic Foraminifera and the Cretaceous/Tertiary boundary in central Delaware. *Delaware Geol. Surv., Rept. Invest.* **5**, 1-13.

Kaasschieter, J. P. H. 1961. Foraminifera of the Eocene of Belgium. *Inst. Roy. Sci. Nat. de Belgique, mém.* **147**, 271 pp. 16 pls.

Kacharava, I. and Kacharava, M. 1960. Danian stage of Georgia and relations with analogous sediments of the Mediterranean province. In: Yanshin, A. L. and Menner V. V. (Editors). The Cretaceous-Tertiary Boundary. *Intern. Geol. Congr., 21st, Copenhagen, 1960, Rept. Sov. Geologists*, **5**, 136-45.

Kapounek, J., Papp, A. and Turnovsky, K. 1960. Grundzüge der Gliederung von Oligozän und älteren Miozän in Nieder-Österreich nördlich der Donau. *Verhandl. Geol. Bundesanstalt*, **1960** (2), 217-26.

Keij, A. J. 1957. Eocene and Oligocene Ostracoda of Belgium. *Mem. Inst. Roy. Sci. Nat. Belg.* no. 136.

Keller, B. M. 1935. Microfauna of the upper chalk of the Dniepr-Donets depression and several other adjacent districts. *Byul. Mosk. O-va. ispyt. prirody otd. geol.* **13** (4).

Keller, B. M. 1936. Stratigraphy of the upper chalk of the Western Caucasus. *Izv. Acad. Nauk. U.S.S.R. ser. geol.* **5**, 619-56.

Keller, B. M. 1946. The Foraminifera of the Upper Cretaceous deposits in the Sotchi region. *Bull. Mosk. obsch. ispyt. prir., Moscow, 51 otd. geol.* **21**, 83-108, pls. 1-3.

Kellough, G. R. 1959. Stratigraphic and paleontologic study of Midway Foraminifera along Tehuacana

Creek, Limestone County, Texas. *Gulf Coast Assoc. Geol. Soc. Trans.* 9, 147-60, 1 pl.

Kennett, J. P. 1962. The Kapitean Stage (Upper Miocene) at Cape Foulwind, West Coast. *N.Z. Jour. Geol. Geophys.* 5 (4), 620-5.

Kennett, J. P. 1966a. Faunal succession in two Upper Miocene-Lower Pliocene sections, Marlborough, New Zealand. *Trans. Roy. Soc. N.Z.* 3 (15), 197-213.

Kennett, J. P. 1966b. The *Globorotalia crassaformis* bioseries in north Westland and Marlborough, New Zealand *Micropaleontology.* 12 (2), 235-45, pls. 1-2.

Kennett, J. P. 1966c. Stratigraphy and fauna of the type section and neighbouring sections of the Kapitean Stage, Greymouth. *N.Z. Trans. roy. Soc. N.Z.* 4 (1), 1-77, 16 pls.

Kennett, J. P. 1966d. Biostratigraphy and paleoecology in Upper Miocene-Lower Pliocene sections in Wairarapa and Southern Hawke's Bay. *Ibid.* 4 (3), 83-102.

Khalilov, D. M. 1956. On the pelagic foraminiferal fauna of the Paleogene of Azerbaidzhan. *Akad. Nauk. Azerb. S.S.R., Inst. Geol. Baku,* 17, 234-361.

Khalikov, D. M. 1960. Danian stage of Azerbaidzhan, In: A. L. Yanshin and V. V. Menner (Editors), The Cretaceous-Tertiary Boundary. *Intern. Geol. Congr., 21st, Copenhagen, 1960, Rept. Sov. Geologists,* 5, 146-58.

Kiesel, Y. 1962. Die oligozänen Foraminiferen der Tiefbohrung Dobbertin (Mecklenberg). *Freib. forsch-H.* 122, 1-123.

Kihara, T. 1960. Relation between the Paleogene coal-bearing and marine formations along the eastern coast of the Ariake Sea, Kyushu. *Tohoku Univ., Sci. Rep., Ser., 2, (Geol.), Spec. Vol.* 4, 515-22.

Kikuchi, Y. 1954. Biostratigraphy of the Neogene and Quaternary deposits based upon the smaller foraminifera in the Southern Kanto Region. *Tohoku Univ., Contr., Inst., Geol. Pal.* 59, 1-36.

Koenen, A. von 1864. On the correlation of the Oligocene deposits of Belgium, Northern Germany and the South of England. *Quart. Jour. Geol. Soc. London,* p. 98.

Koenen, A. von 1866a. Über das Alter der Tertiärschichten bei Bünde in Westfalen. *Z. dt. Geol. Ges.* 18, 287-91.

Koenen, A. von 1866b. Über das Alter der Tertiärschichten bei Bunde in Westfalen. *Verh. Naturhist. Ver. Rheinld. Westf,* 58.

Koenen, A. von 1867a. Beitrag zur Kenntnis der Mollusken des norddeutschen Tertiärgebirges. *Palaeontogr.* 16, 145-58, 12-14.

Koenen, A. von 1867b. Das marine Mittel-Oligocän Norddeutschlands und seine Molluskenfauna. 1. Teil. *Palaeontogr.* 16, 53-128, pls. 6-7; 2. Teil. *Palaeontogr.* 16, 223-96, pls. 26-30.

Koenen, A. von. 1869. Über das Oberoligocän von Wiepke. *Arch. Ver. Fr. Naturgesch. Mecklenb.* 22, 106-13.

Koenen, A. von. 1872. Das Miocaen Nord-Deutschlands und seien Molluskenfauna, I. Einleitung und siphonostome Gastropoden. *Schr. Ges. Beford. Naturw. Marburg* 10 (3), 262 pp., 3 pls.

Koenen, A. von 1879. Über das Alter und die Gliederung der Tertiärbildungen zwischen Guntershausen und Marburg. *Rektoratsprogramm Marburg.*

Koenen, A. von 1880. Tertiär zwischen Guntershausen und Marburg. *N. Jb. Mineral. usw,* 1, 95-6.

Koenen, A. von 1882. Die Gastropoda Holostoma und Tectibranchiata, Cephalopoda und Pteropoda des Norddeutschen Miocän, (2. Teil von 'Das norddeutsche Miocän und seine Molluskenfauna'). *N. Jb. Mineral. usw,* 2, 223-363, pls. 5-7.

Koenen, A. von 1883. Über das Alter der Eisensteine bei Hohenkirchen. *Nachr. kgl. Ges. Wiss. Göttingen,* 346-9.

Koenen, A. von 1885a. Comparaison des couches de l' Oligocène supérieure et du Miocène de l'Allemagne septentrionale avec celles de la Belgique. *Ann. Soc. Géol. Belg.* 12, 194-206.

Koenen, A. von 1885b. Über eine palaocäne Fauna von Kopenhagen. *Abh. Kgl. Ges. f. Wiss. in Gottingen,* 32.

Koenen, A. von 1886a. Über fas Mitteloligocän von Aarhus in Jutland. *Z. d. d. Geol. Ges.* 38.

Koenen, A. von 1886b. Über das norddeutsche und belgische Ober-Oligocän und Miocän. *N. Jb. Mineral. usw.* 1, 81-4.

Koenen, A. von 1887. Über die ältesten und jüngsten Tertiärbildungen bei Cassel. *Nachr. Kgl. Ges. Wiss. Gottingen,* 7, 123-128.

Koenen, A. von. 1892. Über die Casseler Tertiärbildungen. *N. Jb. Mineral. usw.* 161-2.

Koenen, A. von. 1895. Blatt Moringen, Gradabt. 55 Nr. 16. *Erl. Geol. Spec.-Kart. Preuss. Lfrg.* 71, 16 pp.

Koenen, A. von 1900. Blatt Dransfeld, Gradabt. 55 Nr. 27. *Erl. Geol. Spec.-Kart. Preuss. Lfrg.* 91, 15 pp.

Koenen, A. von 1906. Blatt Dassel, Gradabt. 55 Nr. 9. *Erl. Geol. Kart. Preuss. Lfrg.* 127, 24 pp.

Koenen, A. von 1909. Das Tertiärgebirge des nordwestlichen Deutschland. *Jber. Nieders. Geol. Ver.* 2, 80-96.

Koenen, A.von 1012. Über die geologischen Verhältnisse des südlichen Reinhardswaldes und Bramwaldes, besonders auf Blatt Munden. *Nachr. kgl. Ges. Wiss. Göttingen,* 1-4.

Koenen, A. von and Grupe, O. 1906. Blatt Hardegsen, Gradabt. 55 Nr. 21. *Erl. Geol. Kart. Preuss. Lfrg.* 127, 16 pp.

Koenen, A. von and Grupe, O. 1910. Blatt Eschershausen, Gradabt. 55 Nr. 2. *Erl. Geol. Kart. Preuss. Lfrg.* 152, 36 pp, 1 pl.

Koenen, A. von, Grupe, O. and Seidl, E. 1915. Blatt Sibbesse, Gradabt. 41 Nr. 58. *Erl. Geol. Kart. Preuss. Lfrg.* 182, 53.

Koenen, A. von and Müller, G. 1900. Blatt Gross-Freden, Gradabt. 55 Nr. 4 *Erl. Geol. Kart. Preuss. Lfrg.* 91, 28 pp.

Kongiel, R. 1935. W. sprawie wiekv 'siwaka' w okolicah Pulaw. *Prace Tow. przyj. nauk w Wilnie,* 9.

Korobkov, I. A. 1947. O vozraste Elburganskogo horizonta paleogenovikh otlozhenii Severnogo Kavkaza. *Dokl. Akad. Nauk U.S.S.R.,* 58 (3), 439-41.

Korobkov, I. A. 1964. Historique de la définition de l'Oligocène, Colloque sur le Paléogène, Bordeaux (1962). *Mém. Bur. Rech. Géol. Min.* 28 (2), 747-60.

Korobkov, I. A. and Mironova, L. V. 1964. Résolution de la cinquième conférence de la commission permanente pour la stratigraphie du Paléogène de l'U.R.S.S. 9 mai, 1962, consacrée à la subdivision du Paléogène en étages. Colloque sur le Paléogène. *Bur. Rech. géol. Min.* 28, 937-9.

Kowing, K. 1956. Ausbildung und Gliederung des Miozins im Raum vom Bremen. *Abh. naturwiss. Ver. Bremen,* **34** (2), 69-171.

Krasheninnikov, V. A. 1960. Elfidiidy Miotsenovykh otlozhenii podolii. *Akad. Nauk S.S.S.R., Tr. Geol. Inst.* **21, 141** pp., 11 pls.

Krasheninnikov, V. A. 1964. Znachenie foraminifer otrytykh tropicheskikh basseinov datskogo i paleogenovogo vremeni dlya razrabotki mezhdunarodnoi stratigraficheskoi shkaly. *Vopr. Mikropal.* **8**, 190-213.

Krasheninnikov, V. A. 1965. Zonalnaya stratigrafiya Paleogena Vostochnogo Sredizemnomorya. *Akad. Nauk S.S.S.R., Geol. Inst.* 1-75.

Krasheninnikov, V. A. and Ponikarov, V. P. 1964a. Stratigrafiya Mezozoiskikh i Paleogenovykh otlozhenii Egipta. *Sovet. Geol.* **2**, 42-71.

Krutzsch, W. and Lotsch, D. 1964. Proposition à l'appui d'une tentative en vue de subdiviser les dépôts de l'Éocène supérieur et ceux de l'Oligocène inférieur et moyen et de mettre en parallèle des dépôts d'Europe occidentale entre eux et avec ceux d'Europe centrale et l'étude de la position à assigner à la limite entre l'Éocène et l'Oligocène dans ce regions. In: Colloque sur le Paléogène, Bordeaux 1962. *Mém. Bur. Rech. géol. et min., no.* **28** (2), 949-63.

Kulp, J. L. 1961. Geologic time scale. *Science,* **133** (3459), 1,105-14.

Laffitte, R. 1948. Sur l'étage Sahélien Pomel. *Bull. Soc. Hist. Nat. Afrique du N.,* **39**, 31-56.

LeCalvez, Y. 1947. Révision des Foraminifères lutétiens du bassin de Paris. *Mém. Expl. Carte géol. det. France.*

LeCalvez, Y. and Feugueur, L. 1956. L'Ypresien franco-belge; essai de corrélation stratigraphique et micropaléontologique. *Bull. Soc. Géol. France, sér. 6,* **6**, 735-51.

LeCalvez, Y. and Pomerol, Ch. 1961. Sur la séparation des bassins de Bruxelles et de Paris par l'anticlinal de l'Artois à l'Eocène. *C.R. Ac. Sc.* **252**, 2,268-70.

Leonov, G. P. 1963. K probleme yarusnogo deleniya paleogenovykh otlozhenii S.S.S.R. *Vest. Mock. Univ.* **4**, 34-5.

Leonov, G. P. and Alimarina, V. P. 1961. Problems on the boundary between the Cretaceous and Paleogene Systems. *Moscow, Univ., Geol. Fac., Sbornik Trudov,* 29-60, 7 pls.

Leonov, G. P. and Alimarina, V. P. 1962. Problema V. Granitsa melovoi i paleogenovoi sistem. *Sb. Trudov. geol. fak. MGU,* 29-53.

Leonov, G. P., Alimarina, V. P. and Naidin, D. P. 1965. O printsipe i metodakh vydeleniya yarusnykh podrazdelenii etalonni shkaly. *Vestn. Mosk. Univ.* **4**, 15-28.

Leriche, M. 1905a. Sur la signification des termes Landénien et Thanétien. *Ann. Soc. géol. Nord.* **34**, 201-5.

Leriche, M. 1905b. Observations sur le synchronisme des assises éocènes dans le bassin anglo-franco-belge. *Bull. Soc. géol. France, Sér. 4,* **5**, 683-4.

Leriche, M. 1912. L'Éocène des bassins parisien et belge. *Ibid. Sér. 4,* **7**, 692-724.

Leriche, M. 1922. Les terrains tertiaires de la Belgique, Livret-guide. *Congr. géol. intern., 23 sess., Belgique.*

Leriche, M. 1936. Sur l'importance des Squales fossiles dans l'établissement des synchronismes de formations à grandes distances et sur la répartition stratigraphique et géographique de quelques espèces tertiaires. *Mém. Mus. r. Hist. natur. Belg. (2e sér.), 3 (Mélanges-Paul Pelseneer),* 739-75.

Leriche, M. 1937. Les sables d'Aeltre. Leur place dans la classification des assises éocènes du bassin anglo-franco-belge. *Ann. Soc. géol. Nord,* **112**, 96.

Leriche, M. 1942. Contribution à l'étude des faunes ichthyologiques marines des terrains tertiaires de la plaine côtière atlantique et du centre des Etats-Unis. *Mém. Soc. géol. France, N.S. 20,* **45** (2-4), 1-111, pls. 5-12.

LeRoy, L. W. 1941a. Small Foraminifera from the Late Tertiary of Siberoet Island, off the west coast of Sumatra, Nederlands East Indies. *Colorado School Mines, Quart.* **36** (1), no. 1, 63-105, pls. 1-7.

LeRoy, L. W. 1941b. Some small Foraminifera from the type locality of the Bantimein Substage, Bodjong beds, Bantam Residency. West Java, Nederlands East Indies. *Colorado School Mines, Quart.* **36** (1), 107-27, pls. 1-3.

LeRoy, L. W. 1948. The foraminifer *Orbulina universa* d'Orbigny, a suggested Middle Tertiary time indicator. *Jour. Paleont.* **22** (4), 500-8.

LeRoy, L. W. 1952. *Orbulina universa* d'Orbigny in central Sumatra. *Jour. Paleont.* **26** (4), 576-84.

LeRoy, L. W. 1953. Biostratigraphy of the Maqfi section, Egypt. *Geol. Soc. America, Mem.* **54**, 73 pp., 13 pls.

LeRoy, L. W. 1964. Smaller Foraminifera from the Late Tertiary of Southern Okinawa. *U.S. Geol. Survey Prof. Paper* **454-F**, 1-58, 16 pls.

Leupold, W. and Vlerk, van der, I. M. 1931. The Tertiary. In: De stratigraphie van Nederlandsch Oost-Indie, In: Feestbundel uitgegeven ter eere van Prof. Dr. K. Martin. *Leidsche Geol. Meded.* **5** (1), 611-48.

Leymerie, M. 1877. Mémoire sur le type Garumnien. *Ann. Soc. Géol. France,* **9**.

Lienenklaus, E. 1891. Die Ober-Oligocän-Fauna des Doberges. *8. Jber. Naturw. Ver. Osnabrück (1889/90),* 43-174, pls. 1-2.

Lienenklaus, E. 1894. Monographie der Ostracoden des norddeutschen Tertiärs. *Z. dt. Geol. Ges.* **46**, 158-268, pls. 13-18.

Lienenklaus, E. 1900a. Über das Tertiär des Doberges bei Bunde. *Verh. Naturhist. Ver. Rheinld. Westf.* **57**, 55-8.

Lienenklaus, E. 1900b. Die Tertiär-Ostrakoden des mittleren Norddeutschland. *Z. dt. Geol. Ges.* **52**, 497-550, pls. 19-22.

Linstow, O. von 1899. Die Tertiärablagerungen im Reinhardswalde bei Cassel. *Jb. preuss. geol. L.-Anst. (1898),* 1-23, 1 pl.

Linstow, O. von 1907. Beiträge zur Geologie von Anhalt. – *Festschr.* **70**. *Geb. A. v. Koenen,* 19-64, 2 pls.

Linstow, O. von 1912. Die Tertiärbildungen auf dem Grafenhainichen-Schmiedeberger Plateau (Dubener Heide z. T.). *Jb. preuss. geol. L.-Anst.* **29** (2), 254-300, pls. 20-1.

Linstow, O. von 1914. Die geologische Stellung der sog. oberoligocänen Meeressande. *Jb. preuss. geol. L.-Anst. (1911)* **32** (2), 198-200.

Linstow, O. von 1922. Die Verbreitung der tertiären und diluvialen Meere in Deutschland. *Abh. preuss. geol. L. Anst. (N.F.)* **87**, 242 pp, 14 pls.

Lipps, J. H. 1965. Oligocene in California? *Nature, Lond.* **208**, 885-6.

Lipps, J. H. 1967. Planktonic Foraminifera, intercontinental correlation and age of California mid-Cenozoic microfaunal stages. *Jour. Paleont.* **41** (4), 994-1005, pls. 131-2.

Loeblich, A. R. Jr. and Tappan, H. 1957a. Correlation of the Gulf and Atlantic Coastal Plain Paleocene and lower Eocene formations by means of planktonic Foraminifera. *Jour. Paleont.* **31** (6), 1109-1137.

Loeblich, A. R. Jr. and Tappan, H. 1957b. Planktonic Foraminifera of Paleocene and early Eocene age from the Gulf and Atlantic Coastal Plain, *U.S. Nat. Mus. Bull.* **215**, 173-97, pls. 40-64.

Lona, F. 1962. Prime analisi pollinologiche sui depositi Terziari-Quaternari di Castell 'Arquato. *Bol. Soc. Geol. Italia*, **81**, 89-91.

Lorenz, C. 1959. Les couches à Lépidocyclines de Mollere (Près de Ceva, Piémont, Italie). *Rev. Micropal.* **2**, 181-91.

Lorenz, C. 1960. Sur la présence des Miogypsines dans les grès de Montezémolo (Aquit.) entre Ceva et Millesimo. *Com. Rend. As. Fr. Sci.* **1225**, 3001.

Lorenz, C. 1963. Le Stampien et l'Aquitanien ligures. *Bull. Soc. Géol. France, sér. 7*, **4**, 657-65.

Lorenz, C. 1965. La serie Aquitanienne de Millesimo (Italie, prov. de Savone). *Bull. Soc. géol. France, sér. 7*, **6**, 192-204.

Lorenz, J. and Pomerol, Ch. 1965. Caractères sédimentologiques et micropaléontologiques des formations de Ducy, Mortefontaine, Saint-Ouen et des marnes à *Pholadomya ludensis* dans le Bartonien du Bassin de Paris. *Bull. géol. Soc. France, Sér. 7*, **7**, 292-5.

Ludbrook, N. H. 1961. Stratigraphy of the Murray Basin in South Australia. *Bull. Geol. Surv. S. Aust.* **36**, 1-96, pls. 1-8.

Ludbrook, N. H. 1963. Correlation of the Tertiary rocks of South Australia. *Trans. R. Soc. S. Aust.* **87**, 5-15.

Ludbrook, N. H. 1967. Correlation of Tertiary rocks of the Australasian region. In: Tertiary correlations and climatic changes in the Pacific. *11th Pacific Science Congress, Tokyo, 1966, Sympos. no. 25*, 7-19.

Ludbrook, N. and Lindsay, J. M. 1966. The Aldingan Stage. *Quart. Geol. Notes, Geol. Surv. S. Aust.* **19**.

Luterbacher, H. P. 1964. Studies on some *Globorotalia* from the Paleocene and Lower Eocene of the Central Apennines. *Ecl. Geol. Helv.* **57** (2), 631-730.

Luterbacher, H. 1966. K razvitiyo nekotorykh globorotalii v paleotsene tsentralnykh apennina. *Vopr. Mikropal.* **10**, 333-41, 3 pls.

Luterbacher, H. P. and Premoli Silva, I. 1964. Biostratigrafia del limite Cretaceo-Terziario nell 'appennino centrale. *Riv. Ital. Paleont.* **20** (1), 67-128, pls. 2-7.

Luterbacher, H. P. and Premoli Silva, I. 1966. The Cretaceous-Tertiary boundary in the southern Alps (Italy). *Riv. Ital. Paleont.* **72** (4), 1183-1266, pls. 91-9.

Lyell, C. 1833. *Principles of Geology (1st Edn.), vol. 3*, 398 pp. John Murray: London.

Lyell, C. 1839. *Nouveaux éléments de géologie*, Pitois-Levranet, Paris. 648 pp.

Lyell, C. 1852. On the Tertiary strata of Belgium and French Flanders, Part II: The Lower Tertiary of Belgium. *Quart. Jour. Geol. Soc. London*, **8**, 277-370, pls. 17-20.

Lyell, C. 1857. *Supplement to the Fifth Edition*, 40 pp. John Murray: London.

Majzon, L. 1942. Ujabb adatok az egri oligocénrétegek faunajahoz és a paleogén-neogén hatarkérdes. *Földt. Közl.* **72**, 29-39.

Majzon, L. 1960. Magyarorszagi paleogén foraminiferaszintek, *Földt. Közl.* **90**, 355-62.

Malagoli, A. 1892. Foraminifera riscentrati nel Pliocene di Castell 'Arquato. *Boll. Soc. Geol. Ital.*

Mallory, V. S. 1959. Lower Tertiary biostratigraphy of the California Coast ranges. *Am. Assoc. Petrol. Geol.. Spec. Publ.*, 1-416, pls. 1-42.

Mangin, J. P. 1957a. La limite Crétacé-Tertiaire sur le versant sud des Pyrénées occidentales. *C.R. Acad. Sciences*, **244**, 1,227-9.

Mangin, J. P. 1957b. Remarques sur le terme Paléocène et sur la limite Crétacé-Tertiare. *C.R. Soc. Géol. France*, **14**, 319.

Margerie, P. 1961. Ostracodes de la carrière Lambert à Cormeilles-en-Parisis. *Bull. Soc. amic. Géol. amateurs*, **2021**, 1-24, 4 pls.

Marie, P. 1964. Les faciès du Montien (France, Belgique, Hollande). In: Colloque sur le Paléogène, Bordeaux 1962. *Mém. Bur. Rech. géol. et min. no.* **28** (2), 1,077-97.

Marks, P. 1951. A revision of the smaller Foraminifera from the Miocene of the Vienna Basin. *Cushman Found. Foram. Res., Contr.* **2** (2), 33-73, pls. 5-8.

Marks, P. and Webb, P. N. (Editors) 1966. Discussion to the 'Symposium on Micropaleontological lineages and zones'. *Comm. Medit. Néogène Strat. Proc. 3rd session, Berne*, 140-5.

Marlière, R. 1955. Definition actuelle et gisement du Montien dans le bassin de Mons. *Ann. Soc. Géol. de Belge*, **78**, 297-316.

Marlière, R. 1957. Sur le 'Montien' de Mons et de Ciply. *Bull. Soc. Belge de Géol., Pal. et Hydr.* **66** (1), 153-66.

Marlière, R. 1958. Ostracodes du Montien de Mons et resultats de leur étude. *Mém. Soc. Belge de Géol., Pal. et Hydr.* **5**, 3-53.

Marlière, R. 1964. Le Montien de Mons; état de la question. In: Colloque sur le Paléogène, Bordeaux 1962. *Mém. Bur. Rech. géol. et min. no.* **28** (2), 875-83.

Martini, E. and Bramlette, M. N. 1963. Calcareous nannoplankton from experimental Mohole drilling. *Jour. Paleontology*, **37** (4), 845-56, pls. 102-5.

Martin, L. T. 1943. Eocene Foraminifera from the type Lodo Formation, Fresno County, California. *Stanford Univ. Pub., Geol. Sci.* **3**, 93-125, pls. 5-9.

Martinis, B. 1954. Richerche stratigrafiche e micropaleontologiche sue Pliocene piemontese. *Riv. Ital. Paleont. Strat.* **60** (2), 45-114, and **60** (3), 125-94.

Mayer-Eymar, K. 1858. Versuch einer synchronistichen Tabelle der Tertiär-Gebilde Europas. *Verh. Schweiz. natur. Gessell., 1857*, 32 pp.

Mayer-Eymar, K. 1867. *Catalogue systématique et descriptif des fossiles des terrains tertiaires qui se trouvent au Musée fédéral de Zürich.* Zürich.

Mayer-Eymar, K. 1868. *Tableau des terrains tertiaires supérieurs, (4th edn.).* Zürich.

Mayer-Eymar, K. 1878. Zur Geologie der mittleren Ligurien, etc. *Vierteljahrl zürch. nat.-forsch. Gesell.* **23**, 21 pp.

Mayer-Eymar, K. 1889. Tableau des terrains de sédiment. *Soc. Hist. Nat. Croatica,* 35 pp.

McGowran, B. 1964. Foraminiferal evidence for the Paleocene age of the King's Park Shale (Perth Basin, Western Australia). *J. R. Soc. W. Aust.* **47** (3), 81-6.

McGowran, B. 1965. Two Paleocene foraminiferal faunas from the Wangerrip Group, Pebble Point Coastal section Western Victoria. *Proc. R. Soc. Vict.* **79** (1), 9-74, pls. 1-5.

McTavish, R.A. 1966. Planktonic Foraminifera from the Malaita Group, British Solomon Islands. *Micropaleontology,* **12** (1), 1-36, pls. 1-7.

McWhae, J. R. H., Playford, P. E., Lindner, A. W., Glenister, B. F. and Balme, B. E. 1958. The stratigraphy of Western Australia. *J. Geol. Soc. Aust.* **4** (2), 1-161.

Meijer, M. 1959. Sur la limite supérieure de l'étage Maastrichtien dans la région type. *Bull. Acad. Roy. de Belgi. (Class. des Sciences), sér. 5,* **45**, 316-38.

Merklin, R. L. 1964. Sur la stratigraphie de l'Oligocène moyen et supérieur dans le sud de l'U.R.S.S. *Colloque sur le Paléogène Bur. Rech. géol. Min.* **28**, 771-6.

Migliorini C. 1950. The Pliocene-Pleistocene boundary in Italy. *Atti 18th Congr. Int. Londra (1948), pt.* **9**, 66-72 London.

Mistretta, F. 1962. Foraminiferi planctonici del Pliocene inferiore di Altavilla Milicia (Palermo, Sicilia). *Riv. Ital. Pal. Strat.* **68** (1), 97-114, pls. 8-11.

Mohsenul Haque, A. F. M. 1956. The Foraminifera of the Ranikot and the Laki of the Nammal Gorge, Salt Range. *Palaeontol. Pakist.* **1**, 1-300, pls. 1-35.

Morozova, V. G. 1939. K stratigrafii verchnego mela i paleogena Embenskoj oblasti po faune foraminifer. *Bjul. Mosk. Obsc. Isp. Prir., Otd. geol.* **17** (4/5), 59-86,

Morozova, V. G. 1946a. On the age of the lower foraminiferous beds of the North Caucasus. *Dokl. Acad. Nauk U.S.S.R.* **54** (1), 53-5.

Morozova, V. G. 1946b. The boundary between Cretaceous and Tertiary deposits in the light of the study of foraminifers. *Dokl. Acad. Sci. U.S.S.R.* **54** (2), 153-5.

Morozova, V. G. 1957. Foraminiferal Superfamily Globigerinidea, superfam. nov., and some of its representatives. *Akad. Nauk S.S.S.R., Doklady,* **114** (5), 1,109-12.

Morozova, V. G. 1958. K sistematike i morfologii paleogenovych predstavitelej nadsemejstva Globigerinidea. *Vopr. Mikropaleont.* 2, 22-52.

Morozova, V. G. 1959. Stratigraphy of the Danian-Monian deposits of the Crimea based on Foraminifera. *Akad. Nauk S.S.S.R., Doklady,* **124** (5), 1,113-16.

Morozova, V. G. 1960. [Stratigraphic zonation of the Danian-Montian deposits of the U.S.S.R. and the Cretaceous-Paleogene boundary.] *Internat. Geol. Congr., 21st, Rept. Sov. Geol., Prob.* 5, 83-100. (in Russian)

Morozova, V. G. 1961. Danian-Montian planktonic Foraminifera of the southern part of the U.S.S.R. *Pal. Zhurnal, no.* 2, 8-19, pls. 1-2.

Morellet L., Morellet J. (1940). – Diverses interprétations du terme de Bartonien. *Bull. Soc. géol. Fr.,* sér 5, **10**.

Morellet L., Morellet, J. (1948). – *Le Bartonien du Bassin de Paris.* (Mém. à l'expl. Carte Géol. detaillée de la France), 430 pp.

Moskvin, M. M. and Naidin, D. P. 1960. Danian and adjoining deposits of Crimea, Caucasus and Transcaspian region and the southeastern part of the Russian Platform. *Internat. Geol. Congr., 21st, Rept. Sov. Geol., Prob.* 5, 15-40.

Mouratov, M. V. and Nemkov, G. I. 1960. Paleogenovye otlozheniya okrestnostei Bakhchisaraya i ikh znachenie dlya stratigrafii paleogena yuga S.S.S.R. In: Paleogenovye otlozheniya yuga evrop. chasti S.S.S.R. (ed. Yanshin, A. L., Vyalov, O. S., Dolgopolov, N. N. and Menner, V. V.), 15-23, *Izd. Akad. Nauk S.S.S.R.*

Mourlon, M. 1887. Sur une nouvelle interprétation de quelques dépôts tertiaires. *Bull. Ac. roy. belge,* **3** (14), 15-19; and *Bull. Assoc. Sc. belge,* **4** (3), 16.

Movius, H. L., Jr. 1949. Villafranchian stratigraphy in southern and southwestern Europe. *Jour. Geol.* **57**, 380-412.

Muller, S. W. and Schenck, H. H. 1943. Standard of the Cretaceous system. *Bull. Am. Assoc. Petrol. Geol.* **27**, 262-78.

Muller, T. 1937. *Das marine Palaozän und Eozän in Norddeutschland und Sudskandinavian.* 1-120. Gebr. Borntraeger: Berlin.

Munier-Chalmas, E. 1897. Note préliminaire sur les assises montienne du bassin de Paris. *Bull. Soc. Géol. France, sér. 3,* **21**.

Munier-Chalmas, M. and de Lapparent, A. 1893. Note sur la nomenclature des terrains sédimentaires. *Bull. Soc. Géol. France, sér. 3,* **21**, 438-88.

Murata, S. 1961. Paleogene microbiostratigraphy of North Kyushu, Japan. *Kyushu Inst. Tech., Bull.* **8**, 1-90, 1 pl.

Murray, G. E. 1953. History and development of Paleocene-Lower Eocene nomenclature, central Gulf Coastal Plain. *Geol. Soc. Mississippi, Guidebook, Tenth Field Trip,* 48-60.

Murray, G. E. 1955. Midway Stage, Sabine Stage and Wilcox Group. *Amer. Assoc. Petrol. Geol. Bull.* **39** (5), 671-96.

Murray, G.E. 1961. *Geology of the Atlantic and Gulf Coastal Province of North America.* Harper and Bros., New York.

Nagao, T. 1928a. A summary of the Paleogene stratigraphy of Kyushu, Japan, with some accounts on the fossiliferous zones. *Tohoku Imp. Univ., Sci. Rep., Ser., 2 (Geol.),* **12** (1), 1-10, 1 pl.

Nagao, T. 1928b. Paleogene fossils of the island of Kyushu, Japan. Pt. II. *Ibid.* **12** (1), 11-140, 17 pls.

Nagappa, Y. 1959. Foraminiferal biostratigraphy of the Cretaceous-Eocene succession of the India-Pakistan-Burma region, *Micropaleontology,* **5** (2), 145-92, pls. 1-11.

Nagappa, Y. 1960. The Cretaceous-Tertiary boundary in the India-Pakistan subcontinent. *21st Int. Geol. Congr. (Copenhagen),* 5, 41-9.

Naidin, D. P. 1960. Concerning the boundary between the Maastrichtian and the Danian stages. In: A. L. Yanshin and Menner, V. V. (editors), The Cretaceous-Tertiary Boundary. *Intern. Geol. Congr., 21st, Copenhagen, 1960, Rept. Sov. Geologists,* **5**, 45-78.

Nakkady, S. E. 1949. The foraminiferal fauna of the Esna shales of Egypt, part I. *Bull. Inst. Egypt, LeCaire,* **31**, 209-47.

Nakkady, S. E. 1950. A new foraminiferal fauna from the

Esna shales and Upper Cretaceous chalk of Egypt. *J. Paleont.* **24**, 675-92, pls. 89-90.
Nakkady, S. E. 1951a. Zoning the Mesozoic-Cenozoic transition of Egypt by the Globorotaliidae. *Bull. Fac. Sci., Alexandria,* **1**, 45-58, pls. 1-2.
Nakkady, S. E. 1951b. Stratigraphical study of the Mahamid district. *Bull. Fac. Sci., Alexandria,* **1**, 17-43.
Nakkady, S. E. 1952. The foraminiferal fauna of the Esna shales of Egypt, part II. *Bull. Inst. Egypt, LeCaire,* **33**, 397-430, pls. 1-8.
Nakkady, S. E. 1955. The stratigraphical implication of the accelerated tempo of evolution in the Mesozoic-Cenozoic transition of Egypt. *J. Paleont.* **29**, 702-6.
Nakkady, S. E. 1957. Biostratigraphy and inter-regional correlation of the Upper Senonian and Lower Paleocene of Egypt. *J. Paleont.* **31**, 428-47.
Nakkady, S. E. 1959. Biostratigraphy of the Um Elghanayem section, Egypt. *Micropaleontology,* **5**, 453-72, pls. 1-7.
Nemkov, G. I. 1964. Distribution zonale des assisses éocènes de l'U.R.S.S. d'après les nummulitidés. In: Colloque sur le Paléogène, Bordeaux (1962). *Bur. Rech. géol. Min., Mém. no.* **28** (2), 761-5.
Nemkov, G. I. and Barkhatova, N. N. 1959. Zony krupnykh foraminifer eotsenovykh otlozhenii Kryma. *Vestnik Leningr. Gos. Univ.* **2**.
Olsson, R. 1958. *Late Cretaceous-Early Tertiary stratigraphy of New Jersey.* 280 pp. Doct. Dissert.: Princeton Univ.
Olsson, R. 1960. Foraminifera of latest Cretaceous and earliest Tertiary age in the New Jersey Coastal Plain. *Jour. Paleont.* **34** (1), 1-58, 12 pls.
Onofrio, S. d' 1964. I foraminiferi del neostratotipo del Messiniano. *Giorn. Geool., ser. 2,* **32** (2), no. 2, 409-54.
Orbigny, A. d' 1846. *Foraminifères fossiles du Bassin Tertiaire de Vienne (Autriche).* Gide et Comp.: Paris.
Papani, G. and Pelosio, G. 1962. La serie Plio-Pleistocenica del T. Stirone (Parmense occidentale). *Boll. Soc. Geol. Ital.* **81** (4), 283-335.
Papulov, G. N. and Kupryanova, F. V. 1960. Danian stage of the eastern slope of the Urals and of the Transurals. In: A. L. Yanshin and Menner, V. V. (editors), The Cretaceous-Tertiary Boundary. *Intern. Geol. Congr., 21st, Copenhagen, 1960, Rept. Sov. Geologists,* **5**, 179-87.
Papp, A. 1951. Das Pannon des Wiener Beckens. *Mitteilungen Geol. Ges. Wien,* **39-41**.
Papp, A. 1956. Fazies und Gliederung des Sarmats im Wiener Becken. *Ibid.* **47**.
Papp, A. 1959. *Grundzuge regionaler Stratigraphie. Tertiär 1 Teil. Handb. Stratigr. Geologie, III,* 411 pp. Stuttgart.
Papp, A. 1963. Die biostratigraphische Gliederung des Neogens im Wiener Becken. *Mitt. Geol. Ges. Wien,* **56** (1), 225-317.
Pareto, M. F. 1852. Sur quelques alternances de couches marines et fluviatiles dans les dépots supérieurs des Collines subapennines. *Bull. Soc. Géol. France, sér. 2,* **9**, 257.
Pareto, M. F. 1855. Sur le terrain nummulitique du pied des Apennins, *Bull. Soc. Géol. France, sér. 2,* **12**, 370-95.
Pareto, M. F. 1865. Note sur la subdivision, que l'on pourrait établir dans les terrains tertiaires de l'Apennin septentrional. *Bull. Soc. Géol. France, sér. 2,* **22**, 210-77.
Parker, F. 1967. Late Tertiary biostratigraphy (planktonic Foraminifera) of tropical Indo-Pacific deep-sea cores. *Bull. Amer. Paleontology,* **52** (235), 115-208, pls. 17-31.
Perconig, E. 1952. Stratigrafia del sondaggio profondo n.29 di Catemaggiore. *Acti. del VII Conv. Naz. Mekano Petidio, Taormina.* **1**, 135-48.
Perconig, E. 1955. Richerche stratigrafiche ę micropaleontologiche nella regione marchigiora (foglio fermo). *Boll. Serv. geol. Ital.* **77** (2-3), 199-269.
Perconig, E. 1966. Sull 'esistenza del Miocene superiore in facies narina nella Spogna meridionale. *Com. Med. Neog. Strat., Proc. 3rd Sess., Berne, 8-13 June 1964,* 288-302.
Perconig, E. 1967a. Biostratigrafia della sezione di Carmona (Andalusia, Spagna) in base ai foraminiferi plactonici. *4th Int. Congr. Comm. Medit. Neog. Strat., Bologna (Sept., 1967).*
Perconig, E. 1967b. Nuove specie di foraminifera planctonici della Sezione di Carmona (Andalusia, Spagna). *Ibid.*
Perier, S. 1941. *Contribution à l'étude du Ludien du bassin de Paris: la faune des marnes à Pholadomya ludensis.* Dipl. Et. sup.: Paris.
Pessagno, E. A. Jr. 1960. Stratigraphy and micropaleontology of the Cretaceous and Lower Tertiary of Puerto Rico. *Micropaleontology,* **6** (1), 87-110, pls. 1-5.
Pessagno, E. A. 1963. Planktonic Foraminifera from the Juana Diaz formation, Puerto Rico. *Micropaleontology,* **9**, 53-60.
Petters, V. and Sarmiento, R. 1956. Oligocene and Lower Miocene biostratigraphy of the Carmen–Zambrano area, Colombia, *Micropaleontology-,* **2** (1), 7-35.
Petri, S. 1957. Foraminiferos Miocenicos da Formacao Pirabas. *Univ. São Paulo, Fac. Filosof., Ciênc. Letràs, Bol.* **216** (16), 1-79.
Pezzani, F. 1963. Studio micropaleontologico di un campione della serie Messiniani di Tabiano Bagni (Parma). *Riv. Ital. Paleontol.* **69**, 559-662.
Pishvanova, L. S. 1965a. K voprosy sopostovleniya miotsena zapadnykh oblastei U.S.S.R., i Italii po planktonnym foraminiferam. *Pal. Sborn.* **2**, 8-15. *Izd. Lvov. Univer., Lvov.*
Pishvanova, L. S. 1965b. Zony planktonnykh foraminifer i ikh znachenie dlya Raslchlenehiya molassovykh otlozhenii. *Karpato-balkanskaya geol. Assots., VII Kongr., Sofia, Santyabr, 1965, Dokl., chast 11,* **2**, 109-13.
Pishvanova, L. S. 1966a. Novaya mikrofaunisticheskaya xona v predkarpate zona Globigerina bollii. *Byul. Mosk. O-va. isp. prirody. otd. geol.*.**41** (2), 94-7.
Pishvanova, L. S. 1966b. Sistematicheskoe polozhenie roda *Candorbulina* i ego znachenie dlya biostratigrafii. *Voprosy Mikropal.* **10**, 393-8.
Plummer, F. B. 1932. Pt. 3: The Cenozoic Systems in Texas. In: Sellards, E. H., Adkins, W. S. and Plummer, F. B. The geology of Texas; vol. 1 – Stratigraphy. *Texas, Univ., Bull., no.* **3232**, 519-818, 10 pls.
Plummer, H. J. 1926. Foraminifera of the Midway For-

mation in Texas. *Texas, Univ., Bull.* **2644**, 9-198, pls. 2-15.

Poag, C. W. and Akers, W. H. 1967. *Globigerina nepenthes* Todd of Pliocene age from the Gulf Coast. *Contr. Cushman Found. Foram. Res.*, **18** (4), 168-75, pls. 16-17.

Poignant, A. 1964. Position stratigraphique du niveau d'Escornébeou (Landes) et quelques gisements analogues. *Mém. Bur. Rech. Géol. Min. no.* **28**, 425-31.

Poignant, A. 1965. Deux nouvelles espèces de foraminifères d'Aquitaine méridionale. *Rev. Micropal.* **8** (2), 103-5.

Pokorný, V. 1956. The zone with *Globigerinoides mexicanus* (Cushman) in the Eocene of Moravia, Czechoslovakia. *Universitas Carolina (Prague), Geologica,* 2 (3), 287-93.

Pokorný, V. 1960. Microstratigraphie et biofacies du Flysch carpatique de la Moravie méridional (Tchecoslovaquie). *Rev. Inst. Franc. Pétrole Ann. Combust. Liquides,* **15** (7-8), 1099-1141.

Pomel, A. 1858. Sur le système de montagnes du Mermaucha et sur le terrain sahélien. *C.R. Ac. Sci.* **67**, 852-5.

Pomerol, Ch. 1961. Correlation entre le Lédien du bassin de Bruxelles et le Lutétien supérieur du bassin de Paris. *C.R. Ac. Sc.* **252**, 3,389-91.

Pomerol, Ch. 1964a. Découverte de paléosols de type podzol au sommet de l'Auversien (Bartonien inférieur) de Moisselles (Seine-et-Oise), *Ibid.* **258**, 974-6.

Pomerol, Ch. 1964b. Sur la paléographie des 'sables d'Ezanville' et la présence de paléopodzols à la limite Auversien-Marinésien dans le Bartonien du bassin de Paris. *C.R. somm. Soc. geol. France,* 48-51.

Pomerol, Ch. 1964c. Le Bartonien du bassin de Paris, Interpretation stratigraphique et essai de correlation avec les bassins de Belgique et du Hampshire. In: Colloque sur le Paléogène, Bordeaux 1962, *Mém. Bur. Rech. géol. et min. no.* **28** (1), 153-68.

Pomerol, Ch. 1964d. Contribution sédimentologique à la stratigraphie du Bartonien dans le bassin de Paris, *Ibid. no.* 28 (1), 169-73.

Pomerol, Ch. 1965. Les sables de l'Éocène supérieur (étages Lédien et Bartonien) des bassins de Paris et de Bruxelles. (Thèse, 1961.) *Mém. Expl. Carte géol. de France.*

Poslavskaya, H. A. and Moskvin, M. M. 1960. Echinoids of the order Spatangoida in Danian and adjacent deposits of Crimea, Caucasus and the Transcaspian region. *Internat. Geol. Congr., 21st, Rept. Sov. Geol.* 47-82, 8 pls.

Pozaryska, K. 1952. Zagadnienia sedymentologiczne gornego Mastrychtu i Danu okolic Pulaw (The sedimentological problems of Upper Maestrichtian and Danian of the Pulawy environment, Middle Vistula). *Biul. P. Inst. Geol.* **81**, 1-104.

Pozaryska, K. 1954. O przewodnich otwornicach z kredy gornej Polski srodkowej (The Upper Cretaceous Index foraminifers from central Poland). *Acta Geol. Pol.*, 4 (2), 249-76.

Pozaryska, K. 1957. Lagenidae du Crétacé supérieur de Pologne (Lagenidae z kredy gornej Polski). *Palaeont. Pol.* 8, 1-190.

Pozaryska, K. 1964. On some Foraminifera from the Boryszew boring, central Poland (O pewnych otwornicach z wiercenia w Boryszewie). *Acta Palaeont. Pol.* **9** (4), 539-48.

Pozaryski, W. 1938. Stratygrafia senonu w przelomie Wisly miedzy Rachowem i Pulawami (Senonstratigraphie im Durchbruch der Weichsel zwischen Rachow und Pulawy in Mittelpolen). *Biul. P. Inst. Geol.* **6**, 1-94.

Pozaryski, W. 1957. Poludniowo-zachodnia krawedz Fennosarmacji. The southwestern margin of Fenno-Sarmatia. *Kwart. Geol.* **1** (3/4), 383-424.

Pozaryski, W. and Pozaryska, K. 1959. Comparaison entre le Crétacé de la Belgique et de la Pologne. *Am. Soc. Géol. Belg.* **82**, 1-14.

Pozaryski, W. and Pozaryska, K. 1960. On the Danian and Lower Paleocene sediments in Poland. *Int. Geol. Congr. (Copenhagen), 21 Sess., Norden,* **5**, 170-80.

Premoli-Silva, I. 1964. Le Microfaune del Pliocene di Balerna (Canton Ticino). *Ecl. geol. Helv.* **57** (2), 731-41, 1 pl.

Premoli-Silva, I. and Palmieri, V. 1962. Osservazioni stratigrafiche sul Paleogene della val di Non (Trento). *Mem. Soc. Geol. Ital., Pavia,* **3**, 191-212.

Prestwich, J. 1852. On the structure of the strata between the London clay and the chalk. *Quart. Jour. geol. Soc. Lond.* **8**, 235-268.

Prestwich, J. 1855. On the correlation of the Eocene Tertiaries of England, France and Belgium. *Quart. Jour. geol. Soc. Lond.* **11**, 206-46.

Prestwich, J. 1888. Further observations on the correlation of the Eocene strata in England, Belgium and the North of France. *Ibid.* **44**, 88-111.

Rainwater, E. H. 1960. Paleocene of the Gulf Coastal Plain in the United States of America. *Internat. Geol. Congr., 21st, pt. 5, sect. 5,* 97-116.

Rao, L. Rama 1964. The problem of the Cretaceous-Tertiary Boundary with special reference to India and adjacent countries. *Mysore Geol. Assoc.* 66 pp. Bangalore, India.

Rasmussen, H. W. 1964. Les affinités du Tuffeau de Ciply en Belgique et du post-Maestrichtien 'Me' des Pays-Bas avec le Danien. *Colloque sur le Paléogène, Mém. Bur. Rech. Géol. Min.* No 28, 865-73.

Ravn, J. P. J. 1897. Nogle Bemaerkninger om danske Tertiaeraflejringers Alder. *Medd. Dansk Geol. Foren.* **1** (4), 1-16.

Ravn, J. P. J. 1899. Et Par Bemaerkninger i Anledning af A. Henning: Studier ofver den baltiska yngre kritans bildningshistoria. *Geol. Foren. Forhandl.* **21** (3), 265-7.

Ravn, J. P. J. 1902a. Molluskerne i Danmarks Kridtaflejringer. I. Lamellibranchiater. *Kgl. Danske Vid. Selsk. Skr. Rk. 6,* **11** (2).

Ravn, J. P. J. 1902b. Same. II. Scaphopoder, Gastropoder og Cephalopoder, *Ibid. Rk. 6,* **11** (4).

Ravn, J. P. J. 1903. Same. III. Stratigrafiske Undersøgelser, *Ibid. Rk. 6,* **11** (6).

Ravn, J. P. J. 1907. Molluskfaunen i Jyllands Tertiaeraflejringer. En palaeontologisk-stratigrafisk Undersøgelser. *Acad. Roy. Danmark, Mem. 7, ser.* **3** (2).

Ravn, J. P. J. 1922. Études sur les Pelecypodes et Gastropodes Danien du calcaire du Faxe. *Mém. Acad. Roy.*

Sci. Lett. Denmark, 9.

Ravn, J. P. J. 1925. Sur le placement géologique du Danien. *Danmarks Geol. Unders. Rk. 2,* **43**, 5-48.

Ravn, J. P. J. 1933. Etudes sur les Pélécypodes et Gastropodes daniens du calcaire de Faxe. *Kgl. Danske Vid. Selsk. Skr. Rk. 9,* **5** (2).

Ravn, J. P. J. 1939. Etudes sur les Mollusques du Paléocène de Copenhagen. *Biol. Skr. Danske Vid. Selsk.* **1** (1).

Ravn, J. P. J., Brünnich-Nielsen, K., Grönwall, K. A. and Ødum, H. 1926. Diskussion om Daniets geologiske Stilling. *Medd. Dansk. Geol. Foren.* **7**, 55-82.

Reed, K. J. 1965. Mid-Tertiary smaller Foraminifera from a bore at Heywood, Victoria, Australia. *Bull. Amer. Paleontol.* **49** (220), 40-104.

Riechel, M. 1952. Remarques sur les Globigerines du Danien de Faxe (Danemark) et sur celles des couches de passage du Crétacé au Tertiaire dans la Scaglia de l'Apennin. *Ecl. geol. Helv.* **45** (2), 341-9.

Reiss, Z. 1952. On the Upper Cretaceous and Lower Tertiary microfaunas of Israel. *Bull. Res. Counc. Israel,* **2** (1), 37-50.

Reiss, Z. 1954. Upper Cretaceous and Lower Tertiary Bolivinoides from Israel. *Contr. Cush. Found. Foram. Res.* **5** (4), 154-64.

Reiss, Z. 1955. Micropaleontology and the Cretaceous-Tertiary boundary in Israel. *Bull. Res. Counc. Israel, B, 8, 5 B,* **1**, 105-20.

Reiss, Z. and Gvirtzman, G. 1962. Subsurface Neogene stratigraphy of the Israel coastal plain. *State Israel. Min. Develop., Geol. Surv., Rept., Pa 1/4/62,* 30 pp.

Reiss, Z. and Gvirtzman, G. 1964. Subsurface Neogene stratigraphy of Israel. *Comm. Stratigraphie Congr. Geol. Intern., Comité Néogène Mediterranéen, 3e Assemblée,* 19 pp.

Reizz, Z. and Gvirtzman, G. 1966. *Borelis* from Israel. *Eclog. geol. Helv.* **59** (1), 437-47.

Renevier, E. 1873. *Tableau des terrains sédimentaires (in-4°) plus un texte explicatif,* **Tab. III**, 1897, Chrono. géol. G. Bridel: Lausanne.

Renevier, E. 1897. Chronologie géologique. *Congrès géologique international, IV, Sess., Zürich, 1894,* 523-695.

Renz, R. A. 1936. Stratigraphische und mikropaläontologische Untersuchungen der Scaglia (Oberkreide-Tertian) im Zentral-Apennin. *Ecl. Geol. Helv.* **29**.

Renz, H. H. 1948. Stratigraphy and fauna of the Agua Salada group, State of Falcon, Venezuela. *Geol. Soc. Amer., Mem.* **32**, 1-129, pls. 1-12.

Reuss, A. E. 1855. Beiträge zur Charakteristik der Tertiärschichten des nördlichen und mittleren Deutschlands. *Sizt.-Ber. math.-naturw. Kl. k. k. Akad. Wiss. Wien* **18**, 197-273, 12 pls.

Reuss, A. E. 1865. Zur Fauna des deutschen Oberoligocäns. *Sitz-Ber. math.-naturw. Kl. k. k. Akad. Wiss. Wien* **50**, 435-82, 5 pls.; 614-90, 10 pls.

Rey, M. 1958. Le genre *Almaena* en Aquitaine occidentale. *Rev. Micropal.* **1** (2), 59-63.

Rey, R. 1964. Corrélations entre différents bassins de l'Oligocène. In: Colloque sur le Paléogène, Bordeaux 1962. *Mém. Bur. Rech. géol..et min., no.* **28** (1), 917-20.

Reyment, R. A. 1956. On the stratigraphy and paleontology of Nigeria and the Cameroons, British West Africa. *Geol. Foren. Forhandl.* **78** (1), 17-96.

Reyment, R. A. 1963. Studies on Nigerian Upper Cretaceous and Lower Tertiary Ostracoda, Part 2: Danian, Paleocene, and Eocene Ostracoda. *Stockholm Contr. Geol.* **10**, 286 pp. 23 pls.

Reyment, R. A. 1964. Biostratigraphie et micropaléontologie du Paléogène de la Nigeria occidentale, Paléogène Colloque Bordeaux (1962). *Mém. B.R.G.M. no.* **28** (2), 829-47.

Riedel, W. R. and Funnell, B. M. 1964. Tertiary sediment cores and microfossils from the Pacific Ocean floor. *Quart. Jour. Geol. Soc. Lond.* **120**, 305-368.

Rosenkrantz, A. 1920a. Craniakalk fra København Sydhavn, Denmarks. *Geol. Unders. Rk. 2,* **36**, 5-79, 2 pls.

Rosenkrantz, A. 1920b. En ny københavnsk Lokalitet for forstenings-førende Paleocaen. *Medd. Dansk Geol. Foren.* **5** (20), 3-10.

Rosenkrantz, A. 1924a. De københavnske Grønsandslag og deres Placering i den danske Lagraekke, *Ibid.* **6** (23), 3-39.

Rosenkrantz, A. 1924b. Nye iagttagelser over Cerithiumkalken i Stevns Klint med Bemaerkninger om Graensen mellem Kridt og Tertiaer, *Ibid.* **6**, 28-31.

Rosenkrantz, A. 1925. Undergrundens tektoniske Forhold i København og deres naermeste Omegn, *Ibid.* **6** (26), 3-26, 1 pl.

Rosenkrantz, A. 1930. Den paleocaene Lagserie ved Vestre Gasvaerk. *Medd. Dansk Geol. Foren.,* **7** (5), 371-90.

Rosenkrantz, A. 1937. Bemaerkninger om det Østjaellandske Daniens Stratigrafi og Tektonik. *Ibid.* **9** (2), 199-212.

Rosenkrantz, A. 1939. Faunaen i Cerithiumkalken og det haerdnede Skrivekridt i Stevns Klint. *Ibid.* **9** (4), 509-14.

Rosenkrantz, A. 1940. Exkursion til Møns Klint. *Ibid.* **9** (5), 664, 665.

Rosenkrantz, A. 1944. Smaabidrag til Danmarks Geologie, *Ibid.* **10** (4), 436-59.

Rosenkrantz, A. 1960. Danian Mollusca from Denmark. *XXI Int. Geol. Congr. (Copenhagen), pt. 5, proc. sect. 5,* 193-8.

Rørdam, K. 1897. Kridtformationen i Sjaelland i Terranet mellem København og Køge, og paa Saltholm. *Danmarks Geol. Unders. Rk. 2.* **6**, 152 pp.

Rouvillois, A. 1960. *Le Thanétien du bassin de Paris (étude hydrogéologique et micropaléontologique).* Thèse, Paris.

Roveda, V. 1959. *Nummulites retiatus,* nouvelle espèce de nummulite réticulée des Abruzzes (Italie). *Rev. Micropaléontol.* **1**, 201-7.

Roveda, V. 1962. Contributo allo studio di alcune macroforaminiferi di Priabona. *Riv. Ital. Paleontol.* **67**, 153-224.

Rovereto, G. 1910. Conclusions d'une étude sur l'Oligocène des Apennins de la Ligurie. *Bull. Soc. Géol. France,* **10**, 62-72.

Ruggieri, G. 1940. Il Calabriano dell'Appennino romagnolo. *Rend. R. Acc. Ital., Cl. Sc. Fis. Mat. Nat., s. 7,*

1 (1-5), 60-62.
Ruggieri, G. 1944. Il Calabriano e il Sicillano della valle del Santerno (Imola). *Giorn. Geol., s. 2,* **17**, 95-113.
Ruggieri, G. 1949. Il Pliocene superiore di Capocolle (Forlì). *Giorn. Geol., s. 3,* **20**, 19-38.
Ruggieri, G, 1950. Gli ultimi capitoli della storia geologica della Romagna. *Studi romagnoli,* **1**, 303-11.
Ruggieri, G. 1954. La limite entre Pliocène et Quaternaire dans la série Plio-Pleistocène du Santerno. *Congr. Géol. Intern. d'Alger (1952), section 13, Questions diverses de géologie générale, 3 partie, n. 15,* 235-40.
Ruggieri, G. 1957a. L'interesse paleontologico della Romagna. *Studi romagnoli,* **8**, 627-37. Faenza.
Ruggieri, G. 1957b. Nuovi dati sul contatto Pliocene-Calabriano nella sezione del Santerno (Imola). *Giorn. Geol., s. 2,* **26**, 8pp.
Ruggieri, G. 1960. Segnalazione di *Globoquadrina altispira* nei trubi di Buonfornello (Palermo). *Riv. Mineraria Siciliana,* **11**, 3-7.
Ruggieri, G. 1961. Alcune zone biostratigrafiche del Pliocene e del Pleistocene italiano. *Riv. Ital. Paleontol.* **67**, 405-17.
Ruggieri, G. 1962. La serie marina Pliocenica e Quaternaria della Romagna. *Publ. Camera Com., Ind. Agr., Forli,* **79** pp.
Ruggieri, G. and Greco, A. 1965. Studi geologici e paleontologici su Capo Milazzo con particolare riguardo al Milazziano. *Geol. Romana,* **4**, 41-88, 11 pls.
Ruggieri, G. and Selli, R. 1950. Il Pliocene e il Postpliocene dell'Emilia. *Int. Geol. Congr., XVIII Sess., London, 1948, part 9,* 85-93.
Ruscelli, M. 1952. I foraminiferi del deposito tortoniano di Marentino (Torino). *Riv. Ital. Pal. Strat.* **58**, 39-58.
Ruscelli, M. 1953. Microfaune della serie elveziano del Rio Mazzapiedi-Castellania (Tortona-Alessandria). *Riv. Ital. Pal. Strat., Mem.* **6**, 99-166.
Ruscelli, M. 1956. Il serie aquitaniano-elveziano del Rio Maina (Asti). *Riv. Ital. Pal. Strat.,* **62** (2), 11-108.
Russell, D. E. 1964. *Les Mammifères paléocènes d'Europe.* Thèse, Paris, *Mém. Mus. nat. Hist. nat., sér. C, Sc. Terre,* **23**, 324 pp., 16 pls.
Rutot, A. 1894. Montien et Maestrichtien. *Bull. Soc. Belge Géol. Pal. et Hydrol.* **8**, 126-32.
Rutot, A. 1908. Stratigraphie. In: Cossmann, Pélécypodes du Montien de Belgique. *Mém. Mus. Roy. d'Hist. Nat. de Belgique.* **19** (5).
Rutot, A. and Broeck, E. Van den 1885a. Note sur la division du Tuffeau de Ciply en deux termes stratigraphiques distincts. *Ann. Soc. Géol. de Belgique,* 12, 201-7.
Rutot, A. and Broeck, E. Van den 1885b. Résumé de nouvelles recherches dans le Crétacé supérieur des environs de Mons. *Ibid.* **12**, 207-11.
Rutot, A. and Broeck, E. Van den 1885c. Sur l'âge tertiare de la masse principale du Tuffeau de Ciply. *Mém. Soc. Géol. Belgique,* 3-13.
Rutot, A. and Broeck, E. Van den 1885d. Note préliminaire sur l'âge des diverses couches confondues sous le nom de Tuffeau de Ciply. *Ann. Soc. Roy. Malacol. de Belgique,* **20**, 93-96.
Rutot, A. and Broeck, E. Van den 1885e. Sur l'âge tertiaire du Tuffeau de Ciply. *Ibid.* **20**, 108-10.
Rutot, A. and Broeck, E. Van den 1885f. Nouveaux documents relatifs à la détermination de la masse principale du Tuffeau de Ciply. *Ibid.* 112-17.
Rutot, A. and Broeck, E. Van den, 1885g. Résultats de nouvelles recherches relatives à la fixation de l'âge de la masse principale du Tuffeau de Ciply. *Ann. Soc. Géol. de Belgique.* **13**, 94-8.
Rutot, A. and Broeck, E. Van den 1886a. Sur les relations stratigraphiques du Tuffeau de Ciply avec le Calcaire de Cuesmes à grands Cérithes. *Ibid.* **13**, 99-124.
Rutot, A. and Broeck, E. Van den 1886b. La géologie de Mesvin-Ciply. *Ibid.* **13**, 197-260.
Rutot, A. and Broeck, E. Van den 1886c. La géologie des territoires de Spiennes, Saint-Symphorien et Havre. *Ibid.* **13**, 306-35.
Rutot, A. and Broeck, E. Van den 1887. Documents nouveaux sur la base du terrain tertiare en Belgique et sur l'âge du Tuffeau de Ciply. *Bull. Soc. Géol. France, sér. 3,* **15**, 157-62.
Rutsch, R. F. 1952. Das Typusprofil des Aquitanien. *Eclog. geol. Helv.* **44** (2), 352-5.
Rutsch, R. F. 1958. Das Typusprofil des Helvetien. *Eclog. geol. Helv.* **51**, 107-18.
Rutsch, R. F. 1965. Comité du Néogène méditerranéen. *Geol. Mijnbouw,* **44**, 96-7.
Sacal, V. and Debourle, A. 1957. Foraminifères d'Aquitaine. pt. 2: Peneroplidae à Victoriellidae. *Mém. Soc. Géol. France,* **36** (78).
Sacco, F. 1887. Classification des terrains tertiaires conforme à leurs faciès. *Bull. Soc. Belge Géol. Pal. Hydr. I,* **11**, 276-94.
Sacco, F. 1905. Les étages et les faunes du Bassin tertiaire du Piemont. *Bull. Soc. Géol. France, Sér. 4,* **1**, 893-916.
Sachs, K. N. jr. 1959. Puerto Rican Upper Oligocene layer Foraminifera. *Bull. Am. Paleontol.* **39** (183), 399-416, pls. 34-6.
Sachs, K. N. jr. and Gordon, W. A. 1962. Stratigraphic distribution of Middle Tertiary larger Foraminifera from southern Puerto Rico. *Bull. Am. Paleontol.* **44**, 5-24.
Said, R. 1960. Planktonic Foraminifera from the Thebes Formation, Luxor Egypt. *Micropaleontology,* **6**. 277-86.
Said, R. and Kenawy, A. 1956. Upper Cretaceous and lower Tertiary Foraminifera from northern Sinai, Egypt. *Micropaleontology,* 2, 105-73.
Said, R. and Kerdany, M. T. 1961. The geology and micropaleontology of Farafra Oasis. *Micropaleontology,* **7**, 317-36.
Said, R. and Sabry, H. 1964. Planktonic Foraminifera from the type locality of the Esna Shale, Egypt. *Micropaleontology,* **10** (3), 375-95, pls. 1-3.
Saito, M. and Bando, Y. 1960. Plio-Pleistocene strata of the Inner Zone of Shikoku, Japan. *Tohoku Univ., Sci. Rep., Ser. 2 (Geol.), Spec. Vol.* **4**, 576-82.
Saito, R. 1957. Pre-Kishima (Pre-Aquitanian) crustal deformation of Japan and the adjacent circum Pacific regions. *Kumamoto Univ., Jour. Sci. Ser. B, Sec. 1, Geol.* 2 (2), 33-50.
Saito, R. 1958. Poronai Group, the Paleogene formation of Hokkaido. *Kumamoto Univ., Jour. Sci., Ser. B, Sec. 1, Geol.,* 3 (1), 1-15.

Saito, T. 1962. Eocene planktonic Foraminifera from Hahajima (Hillsborough Island). *Trans. Proc. Paleont. Soc. Japan, N.S.* **45**, 209-25.

Saito, T. 1963. Miocene planktonic Foraminifera from Honshu, Japan. *Sci. Rept. Tohoku Univ., 2nd ser. (Geol.),* **35** (2), 123-209, 4 pls.

Saito, T. and Bé, A. W. H. 1964. Planktonic Foraminifera from the American Oligocene. *Science,* **145**, 702-5.

Sastri, V. V. and Bedi, T. S. 1962. On the occurrence of *Miogypsina, Cycloclypeus* and *Orbulina* in the Miocene of the Andaman Islands. *Current Sci. (India),* **31**, 20-1.

Savage, D. E. and Curtis, G. H. 1967. The Villafranchian Age and its radiometric dating. *Bull. Amer. Assoc. Petrol. Geol.* **51** (3), 479 (Abstract).

Schaub, H. 1951. Stratigraphie und Paläontologie des Schlierenflysches mit besonderer Berücksichtigung der paleocaenen und untereocaenen Nummuliten und Assilinen. *Schweiz. Pal. Ab.* **68**, 222 pp., 9 pls.

Schaub, H. 1955. Zur Nomenklatur und Stratigraphie der europaischen Assilinen, (Vorläufige Mitteilung), *Eclogae Geol. Helv.* **48** (2), 409-13.

Schaub, H. 1960. Uber einige Nummuliten und Assilinen der Monographie und der Sammlung d'Archiac. *Eclogae Geol. Helv.* **53** (1), 443-51, 4 pls.

Schaub, H. 1962a. Contribution à la stratigraphie du Nummulitique du Veronais et du Vicentin. *Mém. Soc. Géol. Italiana,* **3**, 59-66.

Schaub, H. 1962b. Uber einige stratigraphisch wichtige Nummuliten-Arten. *Eclogae Geol. Helv.* **55** (2), 527-51, 8 pls.

Schaub, H. 1963. Uber einige Entwicklungsreihen von *Nummulites* und *Assilina* und ihre stratigraphische Bedeutung. In: Koenigswald, G. H. R. v., Emeis, J. D., Buning, W. L. and Wagner, C. W. *Evolutionary trends in Foraminifera,* 282-97. Elsevier: Amsterdam.

Schaub, H. and Schweighauser, J. 1951. Nummuliten und Discocyclinen aus dem tiefsten Untereocaen von Gan. *Eclogae Geol. Helv.* **43** (2), 236-42.

Schenk, H. G. and Muller, S. W. 1941. Stratigraphic terminology. *Geol. Soc. Amer. Bull.* **52**, 1,419-26.

Schimper, W. P. 1874. *Traité de Paléontologie végétable.* **3**. Paris.

Schwager, C. 1883. Die Foraminiferen aus den Eocaenablagerungen der libyschen Wüste und Aegyptens. *Palaeontographica,* **30**, 81-153. pls. 24-29.

Scott, G. 1926a. On a new correlation of the Texas Cretaceous. *Amer. Jour. Sci., ser. 5,* **12**, 157-61.

Scott, G. 1926b. Etudes stratigraphiques et paléontologiques sur les terraines Crétacés du Texas. *Ann. Univ. de Grenoble, n. s.* **3** (1-2), 93-307, 3 pls.

Scott, G. 1934. Age of the Midway group. *Bull. Geol. Soc. Amer.* **45**, 1,111-58, 3 pls.

Scott, G. H. 1960. The type locality concept in time-stratigraphy. *N.Z. Jour. Geol. Geophys.* **3** (4), 580-4.

Scott, G. H. 1965. Homotaxial stratigraphy. *N.Z. Jour. Geol. Geophys.* **8**, 859-62.

Scott, G. H. 1966. Description of an experimental class within the Globigerinidae (Foraminifera) - 1. *N.Z. Jour. Geol. Geophys.* **9**, 513-40.

Seguenza, G. 1862. Prime ricerche intorno ai rizopodi fossili delle argille Pleistoceniche dei dintorni di Catania. *Accad. Gioenia Sci. Nat. Catania, Atti, ser. 2,* **8**, 85-125, pls. 1-2.

Seguenza, G. 1868. La formation zancléanne, ou recherches sur une nouvelle formation tertiaire. *Bull. Soc. Géol. France.* Sér. 2, **25**, 465-86.

Seguenza, G. 1873a. Studii stratigrafici sulla Formazione pliocenica dell'Italia meridionale. Il Pliocene del Messinese. *Boll. R. Com. Geol. Italia,* **4**, 84-103.

Seguenza, G. 1873b. Brevissimi cenni intorno la serie terziaria della provincia di Messina. *Boll. R. Com. Geol. Italia,* **4**, 231-8 and 259-70.

Seguenza, G. 1879. Le formazioni terziarie nella provincia di Reggio Calabria. *Mem. R. Acc. Lincei, Cl. Sc. Fis. Mat. Nat.* **3** (6), 1-446.

Seguenza, G. 1902. I Vertebrati fossili della provincia di Messina. Parte seconda, Mammiferi e Geologia del piano Pontico. *Boll. Soc. Geol. It.* **21**, 115-75.

Seguenza, G. 1908. Il Miocene della provincia di Messina. *Atti. R. Acc. Lincei* **6** (17), 379-85.

Selli, R. 1954. Il Bacino del Metauro. *Giornale Geologia* **2** (24), 1-268.

Selli, R. 1957. Sulla trasgressione del Miocene nell'Italia meridionale. *Giornale Geologia* **2** (26), 1-54.

Selli, R. 1960. Il Messiniano Mayer-Eymar, 1867. Proposta di un neostratotipo. *Giorn. Geol., ser. 2,* **28**, 1-33.

Selli, R. 1964. The Mayer-Eymar Messinian, 1867: Proposal for a neostratotype. *Intern. Geol. Congr., 21st, Copenhagen, 1960, Rept. Session, Norden, 28,* 311-33.

Selli, R. 1967. The Pliocene-Pleistocene boundary in Italian marine sections and its relationship to continental stratigraphies. *Progress in Oceanography,* **4**, 67-82.

Senes, J. 1951. Géologie de la région située entre Rapovce et Cakanovce en Slovaquie du Sud. *Geolog. Sbornik,* **2**.

Senes, J. 1956. Bemerkungen zur Stratigraphie des Untermiozäns der Sudslovakei auf Grund neuer Forschungen in Mitteleuropa. *Geolog. Sbornik,* **7**.

Senes, J. 1958a. *Pectunculus*-Sande und egerer Faunentypus im Tertiär bei Kovacev in Karpatenbecken. *Geol. Prace, Monogr., Ser. 1,* 1-232.

Senes, J. 1958b. Kritische Bemerkungen zu den Stratotypen des Oligozäns und Pliozäns und zur Frage der Neostratotypen. *Geolog. Sbornik,* **9** (1).

Senes, J. 1958c. Considération sur la nécessité de créer des stratotypes nouveaux du Tertiaire de l'Europe (raison et critères). *Extrait du C. R. Sommaire des Séances de la Société Géologique de France,* **9**.

Senes, J. 1959. Sucasne znalosti o paleogeografii contelnej paratetydy. *Geologicke Prace,* **55**.

Serova, M. Ya. 1967. Zonal scale of Paleogene deposits from the north-west part of the Pacific province and their correlation with Tethys deposits. In: Tertiary Correlations and climatic changes in the Pacific. *11th Pacific Science Congress, Tokyo, 1966, Sympos. no. 25,* 21-7..

Seunes, J. 1890. Recherches géologiques sur les terrains secondaires et l'Eocène inférieur de la région Sous-Pyrénéenne du Sud-Ouest de la France (Basses Pyrénées et Landes). *Ann. Mines, sér. 8,* **18**.

Sigal, J. 1949. Sur quelques foraminifères de l'Aquitanien des environs de Dax; leur place dans l'arbre phylétique des rotaliiformes. *Rev. Inst. Franc. Pétr.* **4** (5), 155-65.

Sigal, J. 1950. Les genres *Queraltina* et *Almaena* (Foraminiferes); leur importance stratigraphique et paleontologique. *Bull. Soc. Géol. France, sér. 5,* **20**, 63-71.
Sigal, J. 1958. Réflexions à propos des termes Paléocène et Danien. *Compt. Rend. Soc. Géol. France,* **5**, 94-7.
Silvestri, A. 1904. Ricerche strutturali su alcune forme dei Trubi di Bonfornello. *Mem. Acc. Pont. N. Lincei,* 22, 235-76.
Sjutskaja, E. K. 1956. Stratigrafiya nizhnikh gorizontov paleogena Tsentrolnovo Predkavkazya po foraminiferam. *Tr. Inst. Geol. Nauk, U.S.S.R.* **164** (70), *geol. ser.,* 3-119, 4 pls.
Sorgenfrei, T. 1940. Marint Nedre-Miocaen i Klintinghoved paa Als, *Danm. Geol. Unders.* 2 (65), 143 pp. 8 pls.
Sorgenfrei. T. 1957. *Lexique stratigraphique international I, 2 d, Danemark.* 44 pp. Paris (CNRS).
Soyer, R. 1943. Recherche sur l'extension du Montien dans la bassin de Paris. *Bull. Serv. Carte géol. France,* **64** (213), 106 pp.
Speyer, O. 1863. Die fossilen Ostracoden aus den Casseler Tertiärbildungen. *Jber. Ver. Naturkd.* **13**, 1-62.
Speyer, O. 1862-1864, 1867, 1869, 1870. Die Conchylien der Casseler Tertiärbildungen. *Palaeontogr.* **9**, 91-141, pls. 18-22; 153-198, pls. 30-4; **16**, 175-218, pls. 16-24; **18**, 297-339, pls. 31-35; **19**, 47-101, pls. 10-15, 159-202, pls. 18-21.
Speyer, O. 1866. Die ober-oligocaenen Tertiärgebilde und deren Fauna im Fürstenthum Lippe-Detmold. *Palaeontogr.* **16**, 1-52, pls. 1-5.
Speyer, O. 1878. Uber die durch das fiskalische Bohrloch Priorfliess bei Cottbus erschlossenen Tertiärschichten. *Z. dt. Geol. Ges.* **30**, 534-5.
Speyer, O. 1879. Die aus dem Bohrloche Nr. VII bei Gross-Strobitz geforderten Tertiärversteinerungen. *Z. dt. Geol. Ges.* **31**, 213-5.
Speyer, O. and Koenen, A. von 1884. Die Bivalven der Casseler Tertiärbildungen, 31 Taf. mit Vorwort u. Taf. Erkl. v. A. v. Koenen. *Abh. Geol. Spec.-Kart. Preuss.* **4** (4), 4.
Spraul, G. L. 1962a. Current status of the Upper Eocene foraminiferal guide fossil, *Cribrohantkenina. Amer. Assoc. Petrol. Geol., Bull.* **46** (10), 1969.
Spraul, G. L. 1962b. Current status of the Upper Eocene foraminiferal guide fossil, *Cribohantkenina. Gulf Coast Assoc. Geol. Soc., Trans.* **12**, 343-47.
Stainforth, R. M. 1948. Applied micropaleontology in coastal Equador. *Jour. Paleont.* 22 (2), 113-51, pls. 24-6.
Stainforth, R. M. 1960. Current status of transatlantic Oligo-Miocene correlation by means of planktonic Foraminifera. *Rev. Micropal.* 2 (4), 219-30.
Stefani, C. De. 1893. *Les terrains tertiaires supérieurs du bassin de la Méditerranée.* Cermanne: Liege.
Stefani, C. De. 1895. Sulla posizione del Langhiano nelle Langhe. *Proc. verb. Test. Sac. Nat. Adun.*
Stehlin, H. G. 1909. Remarques sur les faunules de Mammifères des couches éocènes et oligocènes du Bassin de Paris. *Bull. géol. Soc. France, Sér. 4,* **9**, 488-520.
Subbotina, N. N. 1934. Distribution of microfauna in the foraminiferous strata in the region of Nalchik and the Black Mountains (North Caucasus). *Inf. Sbornik Nepht. Geol.-Razved. Inst.* 1933-4.
Subbotina, N. N. 1936a. Stratigraphy of the lower Paleogene and the upper chalk of the Northern Caucasus from the species of Foraminifera. *Tr. Nepht. Geol.-Razved. Inst.* series A, **96**, 3-32. (French summary).
Subbotina, N. N. 1936b. The discovery of Foraminifera in the upper Maikop in the Northern Caucasus. *Tr. Nepht. Geol.-Razved. Inst.* series B, **60**, 1-15.
Subbotina, N. N. 1938. Distribution of microfauna in the Maikop and Khadum sediments of the Northern Caucasus. *Tr. Nepht. Geol.-Razved. Inst.* series A, 194, 43-62.
Subbotina, N. N. 1939. Foraminifera of the Lower Tertiary sediments of the U.S.S.R. Collection of papers on microfauna. *Tr. Nepht. Geol.-Razved. Inst.* series A, 31-69.
Subbotina, N. N. 1947.Foraminifera of the Danian and Paleogene deposits of the northern Caucasus. Microfauna of the oil fields of the Caucasus, Emba, and Central Asia. *Vses. Neft. Nauchno-Issled. Geol.-Razved. Inst. (VNIGRI),* 39-160, pls. 1-9 (In Russian).
Subbotina, N. N. 1949. Microfauna of the chalk on the southern slope of the Caucasus. *Vses. Neft. Naucho-Issled. Geol-Razved. Inst.* (new series), **34**, 5-36.
Subbotina, N. N. 1950. [Microfauna and stratigraphy of the Elburgan and Goryachy Klyuch horizons. Microfauna of the U.S.S.R.] *Vses. Neft. Nauchno-Issled. Geol.-Razved. Inst. (VNIGRI), Trudy, sbornik, 4,* **51**, 5-112, pls. 1-5 (in Russian).
Subbotina, N. N. 1952. Pelagic Foraminifera and their significance in the stratigraphy and paleogeography of the Palaeogene of the Northern Caucasus (doctoral thesis). Vses. Neft. Naucho-Issled. Geol-Razved. Inst. 31 pp.
Subbotina, N. N. 1953a. Upper Eocene lagenids and buliminids of southern U.S.S.R. *Vses. Neft. Nauch-Issled. Geol.-Razved. Inst.* (new series), **69** (6), 5-112.
Subbotina, N. N. 1953b. [Globigerinidae, Hantkeninidae, and globorotaliidae. Fossil Foraminifera of the U.S.S.R.] *Vses. Neft. Nauchno-Issled. Geol.-Razved. Inst. (VNIGRI), Trudy, n.s.* **6**, **9**, 1-296, 41 pls.
Subbotina, N. N. 1960. [Pelagic Foraminifera of the Paleogene deposits of the southern part of the U.S.S.R., Paleogene deposits of the south european part of the U.S.S.R.] *Akad. Nauk U.S.S.R.,* 24-36 (in Russian).
Subbotina, N. N., Pischvanova, L. S. and Ivanova, L. V. 1960. Mikrofauna Oligothenovykhi Miothenovykh Otlozhenij R. Vorotysche (Predkarpat'e). *Vses. neft. nauchno-issl. geol. instit.* Mikrofauna S.S.S.R, *Trudy, Sbornik 11,* 157-263, pls. 1-10.
Suess, E. 1866. Untersuchungen über den Charakter der Österreichischen Tertiärablagerungen. *Sitzungsber. d. k. Akad. d. Wissensch. I. Abt.* **54**, 66 + 40 pp., 2 pls.
Szöts, E. 1956. Les problèmes de la limite entre le Paléogène et le Néogène et les étages Chattien et Aquitanien. *Acta Geologica* **4**.
Szöts, E. 1958. Notes sur la stratigraphie du Miocène. *Compt. Rend. Congr. d. Soc. Sav. Colloque sur le Miocène.*
Szöts, E. 1961a. Remarques critiques sur l' 'Ilerdien' et sur le 'Biarritzien', nouveaux étages introduits par L. Hottinger et H. Schaub (1960). *C.R. Soc. Géol. France,*

1961(2), 24-5.

Szöts, E. 1961b. Note sur la position stratigraphique des couches d'Escornebéou (Saubusse près de Saint-Geours-de-Maremne, Landes). *C.R. Somm. Soc. Géol. France*, 215-16.

Szöts. E. 1962. Remarques sur le problème de l'Oligocène americain et sur les zones planctoniques de l'Oligocène et du Miocène inférieure. *Ibid.* 236-7.

Szöts, E. 1964. *Nouvelles remarques critiques sur les zones planctoniques de l'Oligocène et du Miocène inférieur.* Annemasse.

Szöts, E. 1965. Note critique sur la position stratigraphique des Calcaires de Blaye. *Compt. Rend. Soc. Géol. France, 1965 (1)*, 19-20.

Szöts, E. 1965a. Sur la limite entre la partie inférieure (Paléocène) et la partie moyenne (Eocène) du Paléogène. *Bull. Soc. géol. France*, **7**, 773-6.

Szöts, E., Malmoustier, G., and Magné, J. 1964. Observations sur le passage Oligocène-Miocène en Aquitaine et sur les zones de foraminifères planctoniques de l'Oligocène. *Mém. Bur. Rech. Géol. Min. no.* **28**, 433-54.

Takayanagi, Y. and Saito, T. 1962. Planktonic Foraminifera from the Nobori Formation, Shikoku, Japan. *Tohoku Univ., Sci. Repts., ser. 2 (Geol.), spec. vol.* **5**, 67-106.

Tavernier, R. and Heinzelin, J. De 1961. La Neogène de la Belgique. *Meyniana*, **10**, 94-101.

Teilhard de Chardin, P. 1922. Mammifères de l'Eocène inférieur français. *Ann. Paléont.* **11**, 9-116.

Telegdi-Roth, K. 1914. Eine oberoligozäne Fauna aus Ungarn. *Geologica Hungarica.*

Terquem, M. 1882. Les Foraminifères de l'Eocène des environs de Paris. *Mém. Soc. géol. France, Géol., 3e sér.*, 2 (3), 1-193, pls. 9-27.

Thalmann, H. E. 1942. *Hantkenina* in the Eocene of East Borneo. *Stanford Univ. Publ., Geol. Sci.* **3** (1), 5-24.

Thenius, E. 1959. *Tertiari: Wirheltier Faunen.* F. Enike Verlag, Stuttgart.

Tipsword, H. L. 1962. Tertiary Foraminifera in Gulf Coast petroleum exploration and development. In: Rainwater E. H. and Zingula, R. P. (eds), Geology of the Gulf Coast and central Texas. *Geol. Soc. Amer. and Houston Geol. Soc., Guidebook*, 16-57.

Tobien, H. 1958. Relations stratigraphiques entre la faune mammalogique pontienne et les faciès marins en Europe et Afrique du Nord. *C.R. Congr. Soc. Savantes Paris, Colloque sur le Miocene*, 299-303.

Todd, R. 1957. Smaller Foraminifera. In: Geology of Saipan, Mariana Islands, pt. 3, Paleontology. *U.S. Geol. Surv. Prof. Paper*, **280-H**, 265-320.

Todd, R. 1964. Planktonic Foraminifera from deep-sea cores off Eniwetok Atoll (in Bikini and nearby atolls, Marshall Islands). *U.S. Geol. Surv. Prof. Paper*, **260-CC**, 1067-1100.

Todd, R. 1966. Smaller Foraminifera from Guam. *U.S. Geol. Surv. Prof. Paper*, **403-I**, 11-141, 19 pls.

Todd, R. Cloud Jr., P. E., Low, D. and Schimidit, R. G. 1954. Probable occurrence of Oligocene on Saipan. *Amer. Jour. Sci.* **252** (11), 673-82, 1 pl.

Tongiorgi, E. and Tongiorgi, M. 1964. Age of the Miocene-Pliocene limit in Italy. *Nature, Lond.* **201**, 365-7.

Trabucco, G. 1891. *Sulla vera posizione del Calcare di Aqui, Firenze.*

Trabucco, G. 1895. Se si debba sostituire il termine di Burdigaliano a quello de Langhiano della serie miocenica. *Prob. verb. Soc. Tosc. Sc. Nat. Adun.*

Trabucco, G. 1908. Fossili, stratigrafia ed eta del calcare di Aqui (Alto Montferrato). *Boll. Sec. Geol. Ital.* **27**.

Trevisan, L. 1951. Sul complesso sedimentarie del Miocene superiore e Pliocene della Val Cecina e sui movimenti tettonici tardivi in rapporto ai giacimenti di lignite e di salgemma. *Boll. Soc. Geol. It.* **70**, 65-78.

Troelsen, J. C. 1937. Om den stratigraphiske Indeling af Skrivekridtet i Danmark. *Medd. Dansk Geol. Foren.* **6** (2), 260-3.

Troelsen, J. C. 1955. *Globotruncana contusa* in the White Chalk of Denmark. *Micropaleontology*, **1** (1), 76-82.

Troelsen, J. C. 1957. Some planktonic Foraminifera of the type Danian and their stratigraphic importance. *U.S. Nat. Mus. Bull.* **215**, 125-31, pl. 30.

Tromp, S. W. 1949. The determination of the Cretaceous-Eocene boundary by means of quantitative generic microfaunal determinations and the conception 'Danian' in the near East. *J. Paleont.* **23**, 673-6.

Tromp, S. W. 1952. Tentative compilation of the micropaleontology of Egypt. *J. Paleont.* **26**, 661-7.

Turner, D. L. 1969. Potassium-argon dates concerning the Tertiary foraminiferal time scale and San Andreas fault displacement. *Geol. Soc. Amer., Spec. Pap.*

Turnovsky, K. 1963. Zonengliederung mit Foraminiferenfaunen und Okologie im Neogen des Wiener Beckens. *Mitt. Geol. Ges. Wien*, **56** (1), 211-24.

Van der Vlerk, I. M. 1959. Problems and principles of Tertiary and Quaternary stratigraphy. *Quart. Journ. Geol. Soc. Lond.* **115** (1), 49-63.

Van der Vlerk, I. M. 1966. Stratigraphie du Tertiaire des domaines Indo-Pacifique et Mesogéen. *Koninkl. Nederl. Akad. Wetensh., Proc., Ser. B*, **69** (3), 336-44.

Van der Vlerk, I. M. and Postma, J. A. 1967. Oligo-Miocene lepidocyclinas and planktonic Foraminifera from East Java and Madura, Indonesia. *Koninkl. Nederl. Akad. Wetensch., Proc., Ser. B*, **70** (4), 391-8. 1 pl.

Van der Vlerk, I. M. and Umbgrove, J. H. F. 1927. Tertiare gidsforaminiferen van Nederlandsch Oost-Indie. *Wetensch. Meded. Dienst. Mijnb. Ned.-Indie*, **9**, 44 pp.

Venables, E. M. 1962. The London Clay of Bognor Regis. *Proc. Geol. Assoc.* **73**, 245-71.

Venzo, S. 1965. The Plio-Pleistocene boundary in Italy. *Rept. VI Int. Congr. Quaternary, Warsaw, 1961*, 367-92.

Vervloet, C. C. 1966. *Stratigraphical and micropaleontological data on the Tertiary of Southern Piemont (Northern Italy).* Schotanus & Jens: Utrecht (Univ. thesis).

Vezzani, L. 1966. La sezione tortoniana di Perosa sue fiurme sinni presso Episcopia (Potenza). *Geol. Romana*, **5**, 263-90.

Vigneaux, M., Alvinerie, M., Caralp, M., Julius, C., La Touche, C., Moyes, J., Rechiniac, A., and Valeton, S. 1966. Une succession stratigraphique en milieu marin épicontinental: principes et méthodes d'interprétation. *Bull. Inst. Géol. Bassin d'Aquitaine*, **1**, 1-61.

Vincent, E. 1928. Observations sur les couches montiennes traversées au puits no. 2 du charbonnage d'Eysden, près de Maaseyk (Limbourg). *Bull. Acad. Roy. de Belge (Cl. des Sciences), ser. 5,* **14**, 554-68.

Vincent, E. 1930a. Etudes sur les Mollusques montiens du Poudingue et du Tuffeau de Ciply. *Mem. Mus. Roy. Hist. Nat. Belg.* **46**.

Vincent, E. 1930b. Mollusques des couches à Cyrènes (Paléocène du Limburg), *Ibid.* **43**.

Vincent, G. and Rutot, A. 1878. Coup d'oeil sur l'état d'avancement des connaissances géologiques relatives aux terrains tertiaires de la Belgique. *Ann. Soc. géol. Belgique. 7 (Mém.)*,69-167.

Vinogradov, C. 1960. Limita cretacic-paleogen in bazinul vaii Prahova. *Acad. epubl. pop. Romina, sect. Geol., Geogr. si Inst. de Geol., Geofiz.ši Geogr. Bucharest,* 5, 299-324, pls. 1-6.

Voigt, S. 1925. Gehört das Danien zum Tertiär? *Zeitschr. Geschiebef.* **1**, 172-86.

Voigt, S. 1929. Die Lithogenese der Flach- und Tiefwassersedimente des jungeren Oberkreidemeeres. *J. B. Halle'schen Verb. Erforschung der mitteldt. Bodenschatze und ihre Verwertung,* **8**, 1-136, 1-13.

Voigt, S. 1956. Zur Frage der Abgrenzung der Maastricht-Stufe. *Palaont. Z.* **30**, Sonderh. 11-17.

Voigt, S. 1959. Die ökologische Bedeutung der Hartgrunde ('Hardgrounds') in der oberen Kreide. *Ibid.* **33** (3), 129-47, 4 pls.

Voigt, S. 1960. Zur Frage der stratigraphischen Selbständigkeit der Danienstufe. *XXI Int. Geol. Congr. (Copenhagen), pt. 5, proc. sect. 5,* 199-209.

Vokes, E. H. 1963. Cenozoic Muricidae of the Western Atlantic region, 1. *Tulane Studies Geol.* **1**, 93-123.

Wade, M. 1964. Application of the lineage concept to biostratigraphic zoning based on planktonic Foraminifera. *Micropaleontology,* **10**, 273-90.

Walters, R. 1965. The *Globorotalia zealandica* and *G. miozea* lineages. *N.Z. Jour. Geol. Geophys.* **8** (1), 109-27.

Wezel, F. C. 1966. La sezione tipo del Flysch Numidico: stratigrafia preliminare della parte sosstante al Complesso Panormide (Membro di Portello Calla). *Atti della Accad. Gioenia di Scienze Naturali in Cataria, Serie Sesta,* **18**, 71-92, 1 pl.

Wicher, C. A. 1949. On the age of the higher Upper Cretaceous of the Tampico embayment area in Mexico, as an example of the world wide existence of microfossils and the practical consequences arising from this. *Bull. Mus. d'Hist. nat., ser. A-2,* 76-105, pls. 2-8.

Wicher, C. A. 1953. Mikropaläontologische Beobachtungen in der höheren borealen Oberkreide, besonders im Maastricht. *Geol. Jb.* **68**, 1-2.

Wicher, C. A. 1956. Die Gosau-Schichten im Becken von Gams (Österreich) und die Foraminiferengliederung der höheren Oberkreide in der Tethys. *Paläont. Z.* **30**, Sonderh., 87-136, 2 pls.

Wienberg-Rasmussen, H. 1962. The Danian affinities of the Tuffeau de Ciply in Belgium and the post-Maestrichtian 'Me' in the Netherlands. In: Colloq, Paléogène, Bordeaux. *Mém. B.R.G.M., no.* **28**.

Winkler-Hermaden, A. 1913. Das eruptivgebiet von Gleichenberg. *Jahrb. geol. Reichanst., Wien,* **63**.

Winkler-Hermaden, A. 1924. Bezieh zw sediment u. Tekt. u. Morph. jungtert. Entwicklungs-geschichte ostalp. *Sitzungsber., Akad. Wiss. Wien, m.-naturw. Kl.* **132**.

Winkler-Hermaden, A. 1960. Der Vergleich der obermiozan-pliozanen Schichtfolgen im Mediterränbereich mit jenen in den pannonisch-pontischen Gebieten. *Verh. Com. Neogene Mediterr., 1. Tag. Wien (1959),* 225-43.

Wrigley, A. G. 1940. The English Eocene Campanile. *Proc. malac. Soc. Lond.* **24**, 97-112.

Wrigley, A. and Davis, A. G. 1937. The occurrence of *Nummulites planulatus* in England, with a revised correlation of the strata containing it. *Proc. Geol. Assoc.* **48**, 203-27.

Yabe, H. and Hanzawa, S. 1930. Tertiary foraminiferous rocks of Taiwan (Formosa). *Sci. Rept. Tohoku Imp. Univ., 2nd ser. (Geol.),* **14** (1), 1-46.

Yanshin, A. L. 1953. Nekotorye voprosy stratigraficheskogo raschleneniya i stratigraficheskoi nomenklatury paleogenovykh otlozhenii [Some problems of stratigraphical subdivision and stratigraphical nomenclature of Paleogene deposits], in Geologiya severnogo Priaraliya [Geology of northern Aral-Land]. *Press of Moscow Society of Naturalists,* 180-243.

Yanshin, A. L. 1960. Stratigraphic position of the Danian stage and the problem of the Cretaceous-Paleocene boundary. *XXI Int. Geol. Congr. (Copenhagen) proc. sect. 5,* 210-16.

Yanshin, A. L. and Menner, V. V. (eds) 1960. The Cretaceous-Tertiary Boundary. *Intern. Geol. Congr., 21st, Copenhagen, 1960, Rept. Sov. Geologists,* **5**, 218 pp.

Yanshin, A. L., Vyalov, O., Dolgopolov, H. N. and Menner, V. V. (eds.), 1960. [Paleogene deposits of the south european part of the U.S.S.R.] *Akad. Nauk U.S.S.R.,* 3-12 (in Russian).

Zaklinskaya, E. D. 1960. Subdivision of the Maastrichtian-Danian-Palaeocene sediment of western Siberia on the basis of data on spore and pollen analysis. In: A. L. Yanshin and Menner, V. V. (eds.). The Cretaceous-Tertiary Boundary. *Intern. Geol. Congr., 21st, Copenhagen, 1960, Rept. Sov. Geologists,* **5**, 188-200.

Zenner, F. E. 1958. *Dating the past.* Methuen, London. 4th edition.

Zenner, F. E. 1959. *The Pleistocene Period, its climate, chronology and faunal successions.* Hutchinson and Co., London.

Zobelein, H. K. 1952. Die Bunte Molasse bei Rottenbuch und ihre Stellung in der Subalpinen Molasse. *Geol. Bavarica,* **12**, 86 pp.

Zobelein, H. K. 1957. Kritische Bemerkungen zur Stratigraphie der Subalpinen Molasse Oberbayerns. *Abh. hess. L.-Amt Bodenforsch,* **23**, 1-76.

Zobelein, H. K. 1958. Land- und Susswasserschnecken aus dem Chattien und Aquitanien der Subalpinen Molasse des westlichen Allgaus. In: Vollmayr, Th.; *Erl. Geol. Karte Bayern 1:25000, Bl. Nr. 8426 Oberstaufen, (41-55), Munchen (Bayer. Geol. L.-Amt).*

Zobelein, H. K. 1960. Uber chattische und aquitanische Stufe und die Grenze Oligozän/Miozän (Palaeogen/Neogen) in Westeuropa. *Verh. Com. Neog. Mediterr. (Mitt. Geol. Ges. Wien,* 52), 245-65.

Symposia on Tertiary Stratigraphy

1. Seminar on Cretaceous-Tertiary Formations of

South India. *Proceedings, Geol. Soc. India, Bull. Spec. Number.* Bangalore, 1966 (4 July).

2. Colloque sur le Paléogène (Bordeaux, Sept. 1962). *Mém. Bur. Rech. Géol. Min. no.* **28**, *pts. 1 and 2, 1107 pp.* Paris, 1964.
3. Colloque Internat. de Micropaleontologie (Dakar, 6-11 May 1963). *Mém. Bur. Rech. Géol. Min. no.* **32**, 369 pp. Paris, 1965.
4. Second West African Micropaleontological Colloquium, Ibadan, 1965. *Proceedings (ed. J. van. Hinte), 294 pp.* E. J. Brill, Leiden, 1966.
5. La géologie du bassin de Paris. *Bull. Soc. Géol. France,* **7**, *no. 2, pp. 197-340, pls. 1-6.* Paris, 1965.
6. Paleogene in Italia (Roma, 1960). *Mem. Soc. Geol. Italiana,* **3**, 389 pp. Pavia, 1962.
7. Verhandlungen des Comité du Neogène Méditerranéen (I Tag., 1959), *265 pp.* Vienna, 1960.
8. Actas de la Sequnda Reunion del Comite del Neogene Mediterraneo y del Symposium de la Union Paleontologico Internacional, (1961). *Cursillos y Conferencias del Inst. 'Lucas Mallada', fasc.* **9**, *318 pp.* Madrid, 1964.
9. Committee on Mediterranean Neogene Stratigraphy. Third Session (Berne, 1964). *346 pp. Proceedings,* E. J. Brill, Leiden, 1966.
10. Symposium sur la stratigraphie du Neogène nordique (Gand, 1961). *Mém. Soc. Belge Géol. Paléontol., Hydrol, no.* **6**, *248 pp.* Brussels, 1962.
11. Ber. über das Internat. Symposium zur Stratigraphie des Miocän im Nordseebecken. *Meyniana,* **10**, *pp. 1-188.* Kiel, 1961.
12. Tertiary correlations and Climatic changes in the Pacific (ed. K. Hatai) *11th Pacific Science Congress, Tokyo, 1966, Symposium No. 25, 102 pp.* Sendai, 1966.

GENERAL INDEX to contents of chapters

All references in this General Index are to *chapter* numbers.

SYSTEMATIC INDEX

This index is divided into sections for ease of reference. These sections are: *General* (pp. 813-14), *Dinoflagellates* (p. 814), *Calcareous nannoplankton* (pp. 814-16), *Diatons and Silicoflagellates* (pp. 816-17), *Foraminifera* (pp. 817-18), *Radiolaria* (pp. 818-19), *Pteropoda* (p. 820), *Ostracoda* (pp. 819-20), and *Pollen and spores* (pp. 820-1). This represents the same order as the arrangement within each of the main Sections A-E of this work.

Page numbers in normal type indicate the first reference only in any one chapter.

Page numbers of illustrations are given in *italics*.

Principal references to a taxon within a chapter are in **bold face**, and where a group forms the subject matter of an entire chapter the chapter number is given in **(bold face within brackets)**.

(N.B. No attempt has been made to include Chapter 52 in this Systematic Index, and a leading entry only is provided for Chapter 49 – see under *Radiolaria*.)

GENERAL

DINOFLAGELLATES

CALCAREOUS NANNO-PLANKTON

DIATOMS AND SILICOFLAGELLATES

FORAMINIFERA

RADIOLARIA

(No entries of detail are provided in this Index for Chapter 49, which itself presents a systematic treatment of polycystine Radiolaria; a list and index of genera of polycystine Radiolaria will be found at the end of that chapter, pp. 660-61.)

OSTRACODA

PTEROPODA

POLLEN AND SPORES

AUTHOR-REFERENCE INDEX

This index gives the pages on which authors are referred to in the main text. Only the first occurrence in any one chapter is indexed; publications made in the same year are not distinguished. Authors entered in Bibliographies, but not referred to in the main text are not included here. By main text is meant all text except Appendices or any other text set in small type.

Contributions to this work are set in *italics*.

Printed in the United States
By Bookmasters